Disha's **Errorless**

183 New Syllabus Chapter-wise + Topic-wise

24 Years JEE Main CHEMISTRY

(2025 - 2002) Previous Year Solved Papers

DISHA Publication Inc.

A-23 FIEE Complex, Okhla Phase II

New Delhi-110020

Tel: 49842349/ 49842350

Edited By

Kalpana Bhargav

Mohammad Salman

Typeset By

DISHA DTP Team

Write To Us At

feedback_disha@aiets.co.in

www.dishapublication.com

CONTENTS

CLASS XI — A1 - A146

Hints & Solutions (Class XI) — A147 - A286

For future updates on JEE Main & Advanced Scan the QR Code.

https://bit.ly/38eW4Dg

You will also get Latest Syllabus, Past Papers, Mock Tests and much more.

CRACK JEE
Deep JEE Analysis
ANALYZING THE EXAM

No. of students who scored in **100 Percentile**

2022	2023	2024	2025
24	42	56	24

State-Wise Distribution of Toppers (2025)

State	Session 1	Session 2	Total
Rajasthan	5	2	7
Uttar Pradesh	2	1	3
Delhi (NCT)	1	0	1
Maharashtra	2	2	4
Gujarat	1	0	1
Telangana	2	2	4
Karnataka	1	0	1
Andhra Pradesh	1	0	1
West Bengal	1	1	2
Total	16	8	24

Total qualifying candidates for JEE (Advanced) 2,50,000

Category-wise no. of Qualified Candidates for JEE Adv.

Category	Percentile		Candidate
	From	To	
UR-ALL	100	93.1023262	97,321
UR-PwBD	93.0950208	0.0079349	3,950
EWS-ALL	93.0950208	80.3830119	25,009
OBC-ALL	93.0950208	79.4313582	67,614
SC-ALL	93.0950208	61.1526933	37,519
ST-ALL	93.0950208	47.9026465	18,823

Percentile Vs Marks JEE Main 2025

Marks Range	Percentile Range
300-281	100 - 99.99989145
271 - 280	99.994681 - 99.997394
263 - 270	99.990990 - 99.994029
250 - 262	99.977205 - 99.988819
241 - 250	99.960163 - 99.975034
231 - 240	99.934980 - 99.956364
221 - 230	99.901113 - 99.928901
211 - 220	99.851616 - 99.893732
201 - 210	99.795063 - 99.845212
191 - 200	99.710831 - 99.782472
181 - 190	99.597399 - 99.688579

Marks Range	Percentile Range
171 - 180	99.456939 - 99.573193
161 - 170	99.272084 - 99.431214
151 - 160	99.028614 - 99.239737
141 - 150	98.732389 - 98.990296
131 - 140	98.317414 - 98.666935
121 - 130	97.811260 - 98.254132
111 - 120	97.142937 - 97.685672
101 - 110	96.204550 - 96.978272
91 - 100	94.998594 - 96.064850
81 - 90	93.471231 - 94.749479
71 - 80	91.072128 - 93.152971

Jan Attempt:
Total Registered- 13,11,544
Total Appeared- 12,58,136

April Attempt:
Total Registered- 10,61,840
Total Appeared- 9,92,350

Total Registered (Unique): 15.39 lakhs
Total Appeared (Unique): 14.75 lakhs

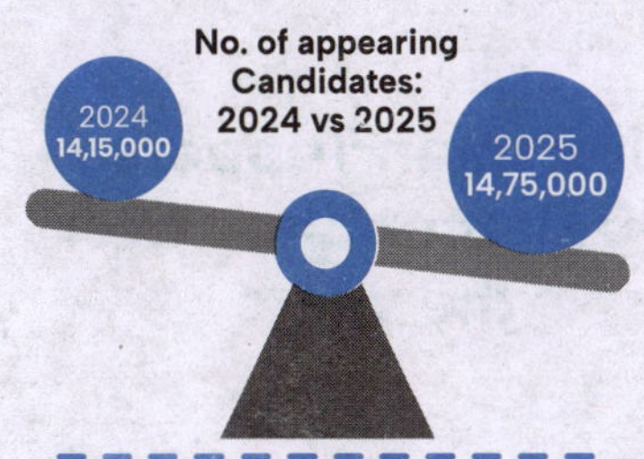

Slight increase from 2024

Cut off for writing JEE (Adv) (in percentile)

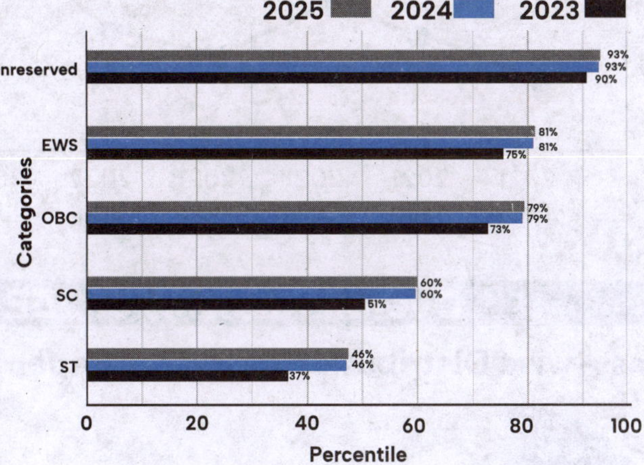

JEE Main cut-off from the past 5 Yrs

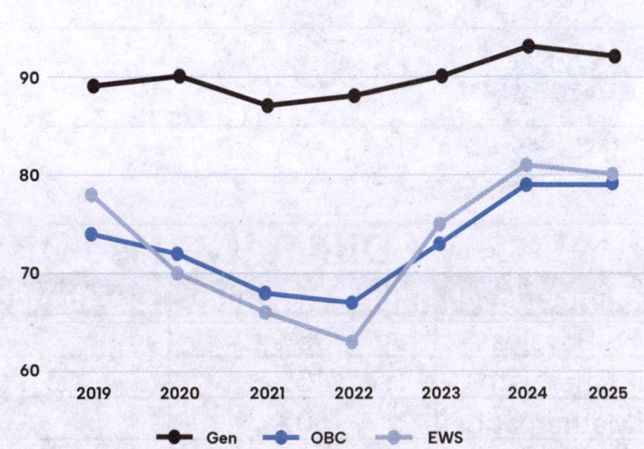

Percentile-Marks-Rank JEE 2025

Percentile Range	Rank Range
100 - 99.99989145	1 - 20
99.994681 - 99.997394	80 - 24
99.990990 - 99.994029	83 - 55
99.977205 - 99.988819	210 - 85
99.960163 - 99.975034	367 - 215
99.934980 - 99.956364	599 - 375
99.901113 - 99.928901	911 - 610
99.851616 - 99.893732	1367 - 920
99.795063 - 99.845212	1888 - 1375

Percentile Range	Rank Range
99.710831 - 99.782472	2664 - 1900
99.597399 - 99.688579	3710 - 2700
99.456939 - 99.573193	5003- 3800
99.272084 - 99.431214	6706 - 5100
99.028614 - 99.239737	8949 - 6800
98.732389 - 98.990296	11678 - 9000
98.317414 - 98.666935	15501 - 11800
97.811260 - 98.254132	20164 - 15700
97.142937 - 97.685672	26321 - 20500

Percentile Range	Rank Range
96.204550 - 96.978272	34966 - 26500
94.998594 - 96.064850	46076 - 35000
93.471231 - 94.749479	60147 - 46500
91.072128 - 93.152971	82249 - 61000
87.512225 - 90.702200	115045 - 83000
82.016062 - 86.907944	165679 - 117000
73.287808 - 80.982153	246089 - 166000
58.151490 - 71.302052	385534 - 264383

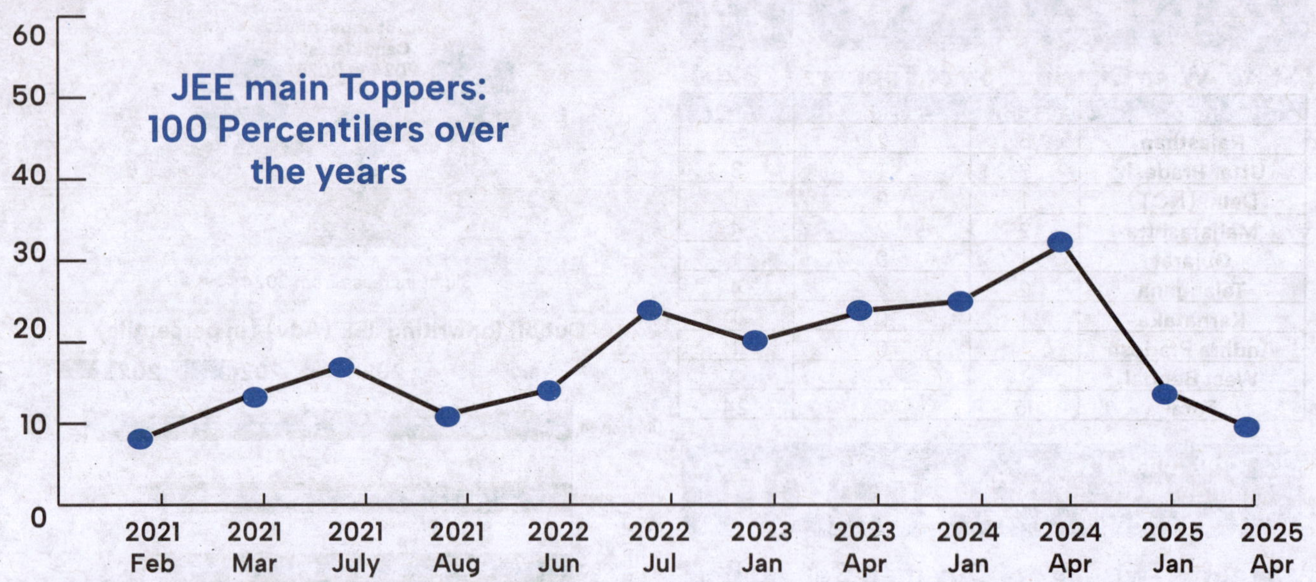

Overall Analysis of JEE Main 2025 Shift Papers

Class-wise Distribution of Overall Paper

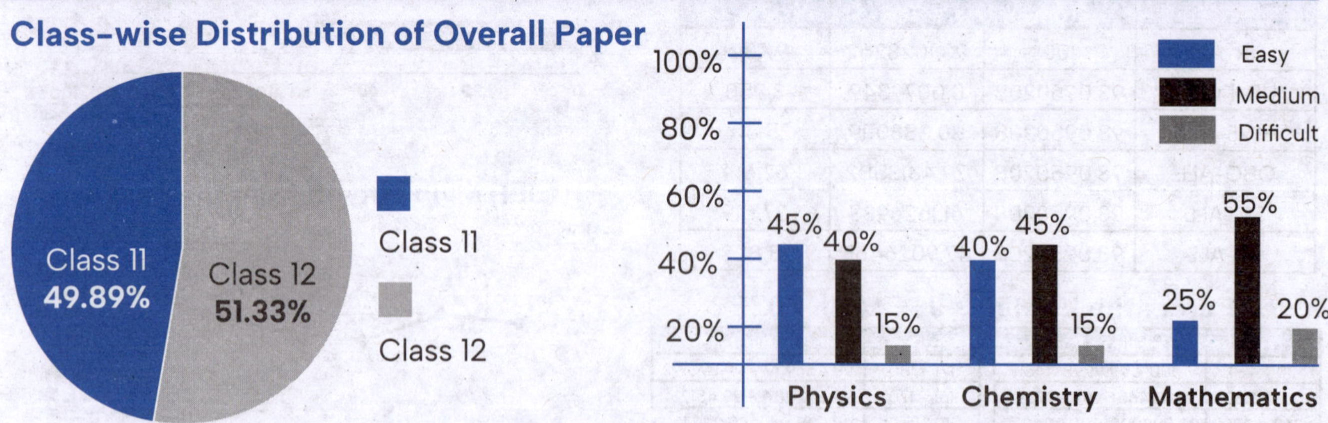

Difficulty-wise Distribution of Overall Paper

Subject level	Moderate, tricky	Easy to moderate, NCERT based Qns	Moderate to difficult, lengthy
Physics	35%	50%	15%
Chemistry	45%	50%	5%
Mathematics	60%	25%	15%

CHEMISTRY

Class-wise Analysis of JEE Main 2025 Shift Papers

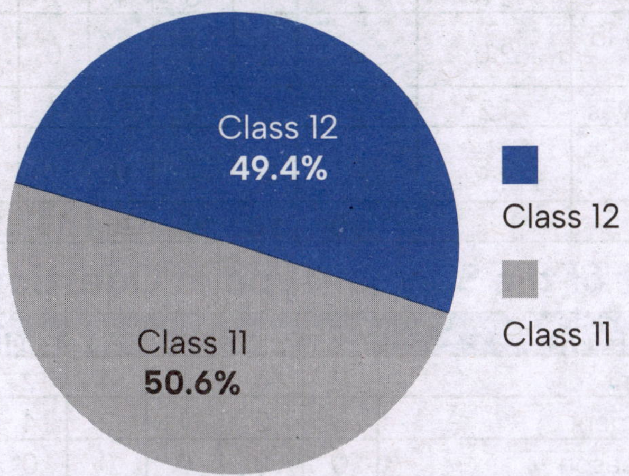

Chapter-wise Analysis of JEE Main 2025 Shift Papers

Class XI

Chapter Name	Chapter Weightage	Total no. of Qns	22s1	22s2	23s1	23s2	24s1	24s2	28s1	28s2	29s1	29s2
General Organic Chemistry	9.2	23	4	2	4	0	2	3	1	2	3	2
Hydrocarbons	5.2	13	1	1	1	0	2	0	0	3	1	4
p Block Elements (Group 13 & 14)	0.8	2	0	0	0	1	0	0	1	0	0	0
Thermodynamics (C)	6.8	17	1	2	2	2	2	2	2	2	1	1
Some Basic Concepts of Chemistry	5.2	13	0	2	1	2	2	2	2	1	1	0
Structure of Atom	5.2	13	1	1	1	2	0	1	1	1	2	3
Chemical Bonding and Molecular Structure	5.2	13	1	1	1	0	2	3	1	1	2	1
Classification of Elements and Periodicity	4.4	11	2	1	1	0	1	1	1	1	1	2
Ionic Equilibrium	3.2	8	0	1	2	1	1	0	1	1	0	1
Chemical Equilibrium	2.4	6	1	0	0	1	1	1	0	0	1	1
Redox Reactions	2.0	5	1	1	0	0	0	0	2	0	0	1

Class XII

Chapter Name	Chapter Weightage	Total no. of Qns	22s1	22s2	23s1	23s2	24s1	24s2	28s1	28s2	29s1	29s2
Coordination Compounds	7.6	19	2	3	3	1	1	3	1	1	2	2
d and f Block Elements	5.6	14	1	1	1	2	2	1	2	3	1	0
Electrochemistry	5.2	13	1	1	1	1	1	1	1	1	3	2
Chemical Kinetics	4.8	12	2	1	1	1	0	2	1	2	1	1
Biomolecules	4.4	11	1	1	1	1	1	1	1	2	1	1
Solutions	4.0	10	1	0	0	2	1	1	1	1	2	1

Aldehydes and Ketones	4.0	10	0	1	1	2	2	1	2	0	1	0
Haloalkanes and Haloarenes	4.0	10	1	3	0	1	1	1	3	0	0	0
Amines	3.6	9	2	1	2	1	0	1	0	1	1	0
Alcohols Phenols and Ethers	2.8	7	1	0	1	2	0	0	0	0	1	2
p Block Elements (Group 15, 16, 17 & 18)	2.0	5	1	1	1	0	1	0	0	1	0	0
Carboxylic Acid Derivatives	1.2	3	0	0	0	1	0	0	1	1	0	0
Practical Chemistry	0.8	2	0	0	0	1	1	0	0	0	0	0
Total	**100**	**250**	**25**	**25**	**25**	**25**	**25**	**25**	**25**	**25**	**25**	**25**

Number of the Special Type of Questions per Shift

Type of Question	22s1	22s2	23s1	23s2	24s1	24s2	28s1	28s2	29s1	29s2
Statement	2	4	2	4	2	3	3	2	1	4
Multi Statement	3	2	1	2	4	2	6	4	2	5
Assertion Reason	0	0	0	0	0	0	0	0	0	0
Graphical	0	1	0	1	1	1	1	2	0	1
Incorrect Type	2	1	2	0	0	0	1	0	0	0
Matrix Match	1	2	2	2	0	3	1	2	3	1
How many correct/incorrect – Numerical	1	0	0	0	1	0	0	2	0	1
Total Special Questions	9	10	7	9	8	9	12	12	6	12

Difficulty-wise Distribution of JEE Main 2025 Shift Papers

Chapter Name	Easy	Medium	Difficult	Total
General Organic Chemistry	10	8	6	24
Coordination Compounds	8	8	3	19
Thermodynamics (C)	8	7	2	17
d and f Block Elements	6	6	2	14
Hydrocarbons	6	5	2	13
Chemical Bonding and Molecular Structure	6	5	2	13
Structure of Atom	6	5	2	13
Electrochemistry	5	5	3	13
Some Basic Concepts of Chemistry	6	5	2	13
Chemical Kinetics	5	5	2	12
Biomolecules	5	5	1	11
Classification of Elements and Periodicity	5	5	1	11
Solutions	5	4	1	10
Aldehydes and Ketones	4	4	2	10
Haloalkanes and Haloarenes	4	4	2	10
Amines	4	3	2	9
Ionic Equilibrium	3	3	2	8
Alcohols Phenols and Ethers	3	3	1	7
Chemical Equilibrium	3	2	1	6
p Block Elements (Group 15, 16, 17 & 18)	2	2	1	5
Redox Reactions	2	2	1	5
Carboxylic Acid Derivatives	1	1	1	3
p Block Elements (Group 13 & 14)	1	1	0	2
Practical Chemistry	1	1	0	2
Total	109	99	42	250

CLASS XI

Some Basic Concepts of Chemistry

TREND ANALYSIS
JEE MAIN (2025-2021)

	2025				2024				2023				2022				2021			
	Session II		Session I		Session II		Session I		Session II		Session I		Session II		Session I		Session II		Session I	
	MCQs	NVQs	MCQs	NVQs	MCQs	NVQs	MCQs	NVQs	MCQs	NVQs	MCQs	NVQs	MCQs	NVQs	MCQs	NVQs	MCQs	NVQs	MCQs	NVQs
Topic 1	1	1	2	–	3	–	–	1	2	–	–	–	1	3	1	2	–	1	–	–
Topic 2	2	2	1	1	1	–	1	–	1	3	–	2	–	1	3	2	–	–	1	1
Topic 3	2	4	5	7	1	–	–	7	–	4	1	8	5	—	2	6	–	8	–	10
Topic 4																				
Total	**5**	**7**	**8**	**8**	**5**	**–**	**1**	**8**	**3**	**7**	**1**	**10**	**6**	**4**	**6**	**10**	**–**	**9**	**1**	**11**

How to Code Qns for Better & Faster Revision →

Type 1: Able to solve (No code Req)

Type 2: Solved with struggle or by seeing solution **(Use code NR means Needs Revision or orange stickers)**

Type 3: Tricky/ Could not solve on my own. Need help from Teacher/ Friend. **(Use code TQ means Tricky Question or red stickers)**

You can either hand write the code in boxes or else paste coloured stickers given at end of book.

Topic 1 - Laws of Chemical Combinations and Mole Concept

1. Among 10^{-9} g (each) of the following elements, which one will have the highest number of atoms? Element : Pb, Po, Pr and Pt **[NCERT : P L-18 | April 3, 2025 (I)]**

(a) Po (b) Pr

(c) Pb (d) Pt

2. 0.1 mol of the following given antiviral compound (P) will weigh ______ $\times 10^{-1}$ g **[NCERT : P L-18 | NV, April 2, 2025 (I)]**

(Structure of compound (P): a pyrimidine ring bearing I, NH, two C=O groups, attached through N to a sugar ring bearing HO–CH₂, HO and F.)

(P)

(Given : molar mass in g mol^{-1} H: 1, C : 12, N : 14, O : 16, F : 19, I : 127)

3. Choose the correct statements. **[NCERT : P L-5 | Jan. 29, 2025 (I)]**

(A) Weight of a substance is the amount of matter present in it.

(B) Mass is the force exerted by gravity on an object.

(C) Volume is the amount of space occupied by a substance.

(D) Temperatures below 0ºC are possible in Celsius scale, but in Kelvin scale negative temperature is not possible.

(E) Precision refers to the closeness of various measurements for the same quantity.

(a) (B), (C) and (D) Only (b) (A), (B) and (C) Only

(c) (A), (D) and (E) Only (d) (C), (D) and (E) Only

4. 2.8×10^{-3} mol of CO_2 is left after removing 10^{21} molecules from its 'x' mg sample. The mass of CO_2 taken initially is Given : $N_A = 6.02 \times 10^{23}$ mol^{-1}. **[NCERT : P L-18 | Jan. 23, 2025 (I)]**

(a) 196.2 mg (b) 98.3 mg

(c) 150.4 mg (d) 48.2 mg

5. The candela is the luminous intensity, in a given direction, of a source that emits monochromatic radiation of frequency 'A' × 10^{12} hertz and that has a radiant intensity in that direction of $\frac{1}{\text{'B'}}$ watt per steradian. 'A' and 'B' are respectively **[NCERT : P L-8 | April 9, 2024 (II)]**
(a) 540 and $\frac{1}{683}$
(b) 540 and 683
(c) 450 and $\frac{1}{683}$
(d) 450 and 683

6. The **incorrect** postulates of the Dalton's atomic theory are : **[NCERT : P L-16 | April 5, 2024 (I)]**
(A) Atoms of different elements differ in mass.
(B) Matter consists of divisible atoms.
(C) Compounds are formed when atoms of different element combine in a fixed ratio.
(D) All the atoms of given element have different properties including mass.
(E) Chemical reactions involve reorganisation of atoms.
Choose the **correct** answer from the options given below:
(a) (B), (D), (E) only (b) (A), (B), (D) only
(c) (C), (D), (E) only (d) (B), (D) only

7. Choose the **Incorrect** Statement about Dalton's Atomic Theory **[NCERT : P L-16 | April 4, 2024 (II)]**
(a) Compounds are formed when atoms of different elements combine in any ratio
(b) All the atoms of a given element have identical properties including identical mass
(c) Matter consists of indivisible atoms
(d) Chemical reactions involve recorganization of atoms

8. 0.05 cm thick coating of silver is deposited on a plate of 0.05 m^2 area. The number of silver atoms deposited on plate are ____ × 10^{23}. (At mass Ag = 108, d = 7.9g cm^{-3}) **[NCERT : P L-18 | NV, Jan. 30, 2024 (I)]**

9. The number of molecules and moles in 2.8375 litres of O_2 at STP are respectively **[NCERT : P L-18 | April 10, 2023 (I)]**
(a) 7.527×10^{22} and 0.250 mol
(b) 1.505×10^{23} and 0.250 mol
(c) 7.527×10^{23} and 0.125 mol
(d) 7.527×10^{22} and 0.125 mol

10. Match List I with List II **[NCERT : P L-18 | April 10, 2023 (II)]**

List I	List II
A. 16g of CH_4(g)	I. Weighs 28g
B. 1g of H_2(g)	II. 60.2×10^{23} electrons
C. 1 mole of N_2(g)	III. Weighs 32g
D. 0.5 mol of SO_2(g)	IV. Occupies 11.4 L volume at STP

Choose the correct answer from the options given below:
(a) A-I, B-III, C-II, D-IV
(b) A-II, B-III, C-IV, D-I
(c) A-II, B-IV, C-III, D-I
(d) A-II, B-IV, C-I, D-III

11. In the given reaction, $X + Y + 3Z \rightleftharpoons XYZ_3$
if one mole of each of X and Y with 0.05 mol of Z gives compound XYZ_3. (Given : Atomic masses of X, Y and Z are 10, 20 and 30 amu, respectively). The yield of XYZ_3 is ________ g. (Nearest integer) **[NCERT : P L-19 | NV, July 28, 2022 (I)]**

12. Chlorophyll extracted from the crushed green leaves was dissolved in water to make 2L solution of Mg of concentration 48 ppm. The number of atoms of Mg in this solution is $x \times 10^{20}$ atoms. The value of x is ____. (Nearest Integer) (Given: Atomic mass of Mg is 24 g mol^{-1}; $N_A = 6.02 \times 10^{23}$ mol^{-1}) **[NCERT : P L-18 | NV, July 26, 2022 (I)]**

13. Hemoglobin contains 0.34% of iron by mass. The number of Fe atoms in 3.3 g of hemoglobin is : (Given : Atomic mass of Fe is 56 u, N_A in 6.022×10^{23} mol^{-1}) **[NCERT : P L-18 | July 26, 2022 (II)]**
(a) 1.21×10^5 (b) 12.0×10^{16}
(c) 1.21×10^{20} (d) 3.4×10^{22}

14. 56.0 L of nitrogen gas is mixed with excess of hydrogen gas and it is found that 20 L of ammonia gas is produced. The volume of unused nitrogen gas is found to be____ L. **[NCERT : P L-20 | NV, July 25, 2022 (II)]**

15. Two elements A and B which form 0.15 moles of A_2B and AB_3 type compounds. If both A_2B and AB_3 weigh equally, then the atomic weight of A is _____ times of atomic weight of B. **[NCERT : P L-23 | NV, June 27, 2022 (I)]**

16. A commercially sold conc. HCl is 35% HCl by mass. If the density of this commercial acid is 1.46 g/mL, the molarity of this solution is : **[NCERT : P L-23 | June 26, 2022 (I)]** (Atomic mass : Cl = 35.5 amu, H = 1 amu)
(a) 10.2 M (b) 12.5 M
(c) 14.0 M (d) 18.2 M

17. The number of N atoms in 681 g of $C_7H_5N_3O_6$ is $x \times 10^{21}$. The value of x is ________ ($N_A = 6.02 \times 10^{23}$ mol^{-1}) (Nearest Integer) **[NCERT : P L-18 | NV, June 25, 2022 (I)]**

18. The number of atoms in 8 g of sodium is $x \times 10^{23}$. The value of x is________. (Nearest integer)
[Given : $N_A = 6.02 \times 10^{23}$ mol^{-1}
Atomic mass of Na = 23.0 u] **[NCERT : P L-18 | NV, Sep. 1, 2021 (II)]**

19. A solution of two components containing n_1 moles of the 1st component and n_2 moles of the 2nd component is prepared. M_1 and M_2 are the molecular weights of component 1 and 2 respectively. If d is the density of the solution in g mL^{-1}, C_2 is the molarity and x_2 is the mole fraction of the 2nd component, then C_2 can be expressed as: **[Sep. 06, 2020 (I)]**

(a) $C_2 = \frac{1000\, x_2}{M_1 + x_2(M_2 - M_1)}$

(b) $C_2 = \frac{d\, x_2}{M_2 + x_2(M_2 - M_1)}$

(c) $C_2 = \frac{1000\, d\, x_2}{M_1 + x_2(M_2 - M_1)}$

(d) $C_2 = \frac{d\, x_1}{M_2 + x_2(M_2 - M_1)}$

20. The molarity of HNO_3 in a sample which has density 1.4 g/mL and mass percentage of 63% is ______. (Molecular Weight of $HNO_3 = 63$)
[NCERT : P L-23 | NV, Jan. 09, 2020 (I)]

21. Amongst the following statements, that which was not proposed by Dalton was:
[NCERT : P L-16 | Jan. 07, 2020 (I)]
(a) Chemical reactions involve reorganization of atoms. These are neither created nor destroyed in a chemical reaction.
(b) All the atoms of a given element have identical properties including identical mass. Atoms of different elements differ in mass.
(c) When gases combine or reproduced in a chemical reaction they do so in a simple ratio by volume, provided all gases are at the same *T* & *P*.
(d) Matter consists of indivisible atoms.

22. 5 moles of AB_2 weigh 125×10^{-3} kg and 10 moles of A_2B_2 weigh 300×10^{-3} kg. The molar mass of A (M_A) and molar mass of B (M_B) in kg mol^{-1} are:
[NCERT : P L-18 | April 12, 2019 (I)]
(a) $M_A = 10 \times 10^{-3}$ and $M_B = 5 \times 10^{-3}$
(b) $M_A = 50 \times 10^{-3}$ and $M_B = 25 \times 10^{-3}$
(c) $M_A = 25 \times 10^{-3}$ and $M_B = 50 \times 10^{-3}$
(d) $M_A = 5 \times 10^{-3}$ and $M_B = 10 \times 10^{-3}$

23. The percentage composition of carbon by mole in methane is : **[NCERT : P L-19 | April 8, 2019 (II)]**
(a) 75% (b) 80% (c) 25% (d) 20%

24. 8 g of NaOH is dissolved in 18 g of H_2O. Mole fraction of NaOH in solution and molality (in mol kg^{-1}) of the solution respectively are : **[NCERT : P L-23 | Jan. 12, 2019 (II)]**
(a) 0.2, 22.20 (b) 0.2, 11.11
(c) 0.167, 11.11 (d) 0.167, 22.20

25. The amount of sugar ($C_{12}H_{22}O_{11}$) required to prepare 2 L of its 0.1 M aqueous solution is:
[NCERT : P L-23 | Jan. 10, 2019 (II)]
(a) 136.8 g (b) 17.1 g (c) 68.4 g (d) 34.2 g

26. An unknown chlorohydrocarbon has 3.55% of chlorine. If each molecule of the hydrocarbon has one chlorine atom only, chlorine atoms present in 1g of chlorohydrocarbon are: (Atomic wt. of Cl = 35.5 u; Avogadro constant = 6.023×10^{23} mol^{-1}) **[NCERT : P L-19 | Online April 16, 2018]**
(a) 6.023×10^{9} (b) 6.023×10^{23}
(c) 6.023×10^{21} (d) 6.023×10^{20}

27. Excess of NaOH (aq) was added to 100 mL of $FeCl_3$ (aq) resulting into 2.14 g of $Fe(OH)_3$. The molarity of $FeCl_3$ (aq) is : **[NCERT : P L-23 | Online April 8, 2017]**
(Given molar mass of Fe=56 g mol^{-1} and molar mass of Cl = 35.5 g mol^{-1})
(a) 0.2 M (b) 0.3 M (c) 0.6 M (d) 1.8 M

28. 5 L of an alkane requires 25 L of oxygen for its complete combustion. If all volumes are measured at constant temperature and pressure, the alkane is :
[Online April 9, 2016]
(a) Isobutane (b) Ethane
(c) Butane (d) Propane

29. The molecular formula of a commercial resin used for exchanging ions in water softening is $C_8H_7SO_3^-Na^+$ (Mol. wt. 206. What would be the maximum uptake of Ca^{2+} ions by the resin when expressed in mole per gram resin ?
[NCERT : P L-18 | 2015]
(a) $\frac{2}{309}$ (b) $\frac{1}{412}$ (c) $\frac{1}{103}$ (d) $\frac{1}{206}$

30. Dissolving 120 g of a compound of mol. wt. 60 in 1000 g of water gave a solution of density 1.12 g/mL. The molarity of the solution is:
[NCERT : P L-23 | Online April 9, 2014]
(a) 1.00 M (b) 2.00 M (c) 2.50 M (d) 4.00 M

31. The amount of oxygen in 3.6 moles of water is:
[NCERT : P L-18 | Online April 9, 2014]
(a) 115.2 g (b) 57.6 g (c) 28.8 g (d) 18.4 g

32. The density of 3M solution of sodium chloride is 1.252 g mL^{-1}. The molality of the solution will be :
(molar mass, NaCl = 58.5 g mol^{-1})
[NCERT : P L-24 | Online April 22, 2013]
(a) 260 m (b) 2.18 m (c) 2.79 m (d) 3.00 m

33. Number of atoms in the following samples of substances is the largest in :
[NCERT : P L-18 | Online April 23, 2013]
(a) 4.0 g of hydrogen
(b) 71.0 g of chlorine
(c) 127.0 g of iodine
(d) 48.0 g of magnesium

34. 10 mL of 2M NaOH solution is added to 200 mL of 0.5 M of NaOH solution. What is the final concentration ?
[NCERT : P L-23 | Online April 25, 2013]
(a) 0.57 M (b) 5.7 M
(c) 11.4 M (d) 1.14 M

35. The concentrated sulphuric acid that is peddled commercial is 95% H_2SO_4 by weight. If the density of this commercial acid is 1.834 g cm^{-3}, the molarity of this solution is **[NCERT : P L-23 | Online May 7, 2012]**
(a) 17.8 M (b) 12.0 M
(c) 10.5 M (d) 15.7 M

36. The ratio of number of oxygen atoms (O) in 16.0 g ozone (O_3), 28.0 g carbon monoxide (CO) and 16.0 oxygen (O_2) is (Atomic mass : C = 12, O = 16 and Avogadro's constant $N_A = 6.0 \times 10^{23}\ mol^{-1}$) **[NCERT : P L-19 | Online May 7, 2012]**

(a) 3 : 1 : 2 (b) 1 : 1 : 2 (c) 3 : 1 : 1 (d) 1 : 1 : 1

37. The ppm level of F^- in a 500 g sample of a tooth paste containing 0.2 g F^- is **[Online May 12, 2012]**

(a) 400 (b) 1000 (c) 250 (d) 200

38. 5 g of benzene on nitration gave 6.6 g of nitrobenzene. The theoretical yield of the nitrobenzene will be **[NCERT : P L-19 | Online May 12, 2012]**

(a) 4.5 g (b) 5.6 g (c) 8.09 g (d) 6.6 g

39. How many moles of magnesium phosphate, $Mg_3(PO_4)_2$ will contain 0.25 mole of oxygen atoms? **[NCERT : P L-18 | 2006]**

(a) 1.25×10^{-2} (b) 2.5×10^{-2}
(c) 0.02 (d) 3.125×10^{-2}

40. If we consider that 1/6, in place of 1/12, mass of carbon atom as the relative atomic mass unit, the mass of one mole of the substance will **[NCERT : P L-18 | 2005]**

(a) be a function of the molecular mass of the substance
(b) remain unchanged
(c) increase two fold
(d) decrease twice

41. 6.02×10^{20} molecules of urea are present in 100 mL of its solution. The concentration of urea solution is **[NCERT : P L-18 | 2004]**

(a) 0.02 M (b) 0.01 M (c) 0.001 M (d) 0.1 M
(Avogadro constant, $N_A = 6.02 \times 10^{23}\ mol^{-1}$)

42. 25 mL of a solution of barium hydroxide on titration with a 0.1 molar solution of hydrochloric acid gave a titre value of 35 mL. The molarity of barium hydroxide solution was **[NCERT : P L-23 | 2003]**

(a) 0.14 (b) 0.28 (c) 0.35 (d) 0.07

43. With increase of temperature, which of these changes? **[NCERT : P L-23 | 2002]**

(a) molality (b) weight fraction of solute
(c) molarity (d) mole fraction

44. Number of atoms in 558.5 gram Fe (at. wt. of Fe = 55.85 g mol^{-1}) is **[NCERT : P L-18 | 2002]**

(a) twice that in 60 g carbon
(b) 6.023×10^{22}
(c) half that in 8 g He
(d) $558.5 \times 6.023 \times 10^{23}$

Topic 2 - Percent Composition and Empirical Formula

45. On combustion 0.210 g of an orgainc compound containing C, H and O gave 0.127 g H_2O and 0.307 g CO_2. The percentages of hydrogen and oxygen in the given organic compound respectively are: **[NCERT : P L-23 | April 8, 2025 (II)]**

(a) 53.41, 39.6 (b) 6.72, 53.41
(c) 7.55, 43.85 (d) 6.72, 39.87

46. Thyroxine, the hormone has given below structure

I, I, HO, O, I, I, COOH, NH_2

The percentage of iodine in thyroxine is .... %. (nearest integer)
(Given molar mass in g mol–1 C:12. H:1, O:16, N:14, I:127) **[NCERT : P L-23 | NV, April 7, 2025 (I)]**

47. An organic compound weighing 500 mg, produced 220 mg of CO_2. on complete combustion. The percentage composition of carbon in the compound is .... %. (nearest integer) **[NCERT : P L-23 | NV, April 7, 2025 (I)]**
(Given molar mass in g mol^{-1} of C : 12, O : 16)

48. On complete combustion 1.0 g of an organic compound (X) gave 1.46 g of CO_2 and 0.567 g of H_2O. The empirical formula mass of compound (X) is _________ g.
(Given molar mass in g mol^{-1} C : 12, H : 1, O : 16) **[NCERT : P L-19 | NV, April 2, 2025 (I)]**

(a) 30 (b) 45 (c) 60 (d) 15

49. The hydrocarbon (X) with molar mass 80 g mol^{-1} and 90% carbon has _________ degree of unsaturation. **[NCERT : P L-19 | NV, Jan. 24, 2025 (II)]**

50. The elemental composition of a compound is 54.2%, C, 9.2% H and 36.6% O. If the molar mass of the compound is 132 g mol^{-1}, the molecular formula of the compound is :
[Given : The relative atomic mass of C : H : O = 12 : 1 : 16] **[NCERT : P L-19 | Jan. 24, 2025 (II)]**

(a) $C_4H_9O_3$ (b) $C_6H_{12}O_6$
(c) $C_6H_{12}O_3$ (d) $C_4H_8O_2$

51. An organic compound has 42.1% carbon, 6.4% hydrogen and remainder is oxygen. If its molecular weight is 342, then its molecular formula is : **[NCERT : P L-19 | April 5, 2024 (I)]**

(a) $C_{11}H_{18}O_{12}$ (b) $C_{12}H_{20}O_{12}$
(c) $C_{14}H_{20}O_{10}$ (d) $C_{12}H_{22}O_{11}$

52. A sample of $CaCO_3$ and $MgCO_3$ weighed 2.21 g is ignited to constant weight of 1.152 g. The composition of mixture is :
(Given molar mass in g mol^{-1} $CaCO_3$:100, $MgCO_3$: 84) **[Jan. 31, 2024 (II)]**

(a) 1.187 g $CaCO_3$ + 1.023 g $MgCO_3$
(b) 1.023 g $CaCO_3$ + 1.023 g $MgCO_3$
(c) 1.187 g $CaCO_3$ + 1.187 g $MgCO_3$
(d) 1.023 g $CaCO_3$ + 1.187 g $MgCO_3$

53. An organic compound gives 0.220 g of CO_2 and 0.126 g of H_2O on complete combustion If the % of carbon is 24 then the % hydrogen is ______ $\times 10^{-1}$. (Nearest integer) **[NCERT : P L-19 | NV, April 13, 2023 (I)]**

54. A metal chloride contains 55.0% of chlorine by weight. 100 mL vapours of the metal chloride at STP weight 0.57 g. The molecular formula of the metal chloride is

(Given : Atomic mass of chlorine is 35.5u)
[NCERT : P L-19 | April 12, 2023 (I)]

(a) MCl_2 (b) MCl_4 (c) MCl_3 (d) MCl

55. A solution of sugar is obtained by mixing 200 g of its 25% solution and 500 g of its 40% solution (both by mass). The mass percentage of the resulting sugar solution is ______. (Nearest integer)
[NCERT : P L-19 | NV, April 11, 2023 (I)]

56. A sample of a metal oxide has formula $M_{0.83}O_{1.00}$. The metal M can exist in two oxidation states +2 and +3. In the sample of $M_{0.83}O_{1.00}$, the percentage of metal ions existing in +2 oxidation state is _____ % (nearest integer)
[NV, Jan. 31, 2023 (II)]

57. Uracil is base present in RNA with the following structure. % of N in uracil is _______. **[NV, Jan. 24, 2023 (I)]**

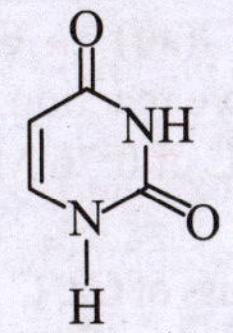

Given :
Molar mass N = 14 g mol^{-1}; O = 16 g mol^{-1};
C = 12 g mol^{-1}; H = 1 g mol^{-1}

58. Number of hydrogen atoms per molecule of a hydrocarbon A having 85.8% carbon is ________ (Given : Molar mass of A = 84 g mol^{-1})
[NCERT : P L-19 | NV, Jan. 25, 2023 (II)]

59. The complete combustion of 0.492 g of an organic compound containing 'C', 'H' and 'O' gives 0.793g of CO_2 and 0.442 g of H_2O. The percentage of oxygen composition in the organic compound is __________. (nearest integer)
[NCERT : P L-19 | NV, June 28, 2022 (II); Same NV, July 28, 2022 (I)]

60. Compound A contains 8.7% Hydrogen, 74% Carbon and 17.3% Nitrogen. The molecular formula of the compound is,
Given : Atomic masses of C, H and N are 12, 1 and 14 amu respectively. **[NCERT : P L-19 | [June 28, 2022 (II)]**
The molar mass of the compound A is 162 g mol^{-1}.
(a) $C_4H_6N_2$ (b) C_2H_3N (c) C_5H_7N (d) $C_{10}H_{14}N_2$

61. 116 g of a substance upon dissociation reaction, yields 7.5 g of hydrogen, 60g of oxygen and 48.5 g of carbon. Given that the atomic masses of H, O and C are 1, 16 and 12 respectively. The data agrees with how many formulae of the following______ **[NCERT : P L-19 | NV, June 27, 2022 (II)]**
(A) CH_3COOH (B) HCHO
(C) CH_3OOCH_3 (D) CH_3CHO

62. On complete combustion 0.30 g of an organic compound gave 0.20 g of carbon dioxide and 0.10 g of water. The percentage of carbon in the given organic compound is _______ (Nearest Integer)
[NCERT : P L-18, 19 | NV, June 26, 2022 (I)]

63. 120 of an organic compound that contains only carbon and hydrogen gives 330 g of CO_2 and 270 g of water on complete combustion. The percentage of carbon and hydrogen, respectively are.
[NCERT : P L-18, 19 | June 24, 2022 (II)]
(a) 25 and 75 (b) 40 and 60
(c) 60 and 40 (d) 75 and 25

64. Complete combustion of 750 g of an organic compound provides 420 g of CO_2 and 210 g of H_2O. The percentage composition of carbon and hydrogen in organic compound is 15.3 and ____ respectively. (Round off to the Nearest Integer). **[NCERT : P L-20 | NV, March 16, 2021 (I)]**

65. Complete combustion of 1.80 g of an oxygen containing compound ($C_xH_yO_z$) gave 2.64 g of CO_2 and 1.08 g of H_2O. The percentage of oxygen in the organic compound is:
[NCERT : P L-19 | Feb. 25, 2021 (I)]
(a) 50.33 (b) 53.33 (c) 51.63 (d) 63.53

66. The average molar mass of chlorine is 35.5 g mol^{-1}. The ratio of ^{35}Cl to ^{37}Cl in naturally occurring chlorine is close to: **[Sep. 06, 2020 (II)]**
(a) 4 : 1 (b) 3 : 1 (c) 2 : 1 (d) 1 : 1

67. The ratio of the mass percentages of 'C & H' and 'C & O' of a saturated acyclic organic compound 'X' are 4 : 1 and 3 : 4 respectively. Then, the moles of oxygen gas required for complete combustion of two moles of organic compound 'X' is ________.
[NCERT : P L-19 | NV, Sep. 02, 2020 (II)]

68. The ratio of mass percent of C and H of an organic compound ($C_XH_YO_Z$) is 6 : 1. If one molecule of the above compound ($C_XH_YO_Z$) contains half as much oxygen as required to burn one molecule of compound C_XH_Y completely to CO_2 and H_2O. The empirical formula of compound $C_XH_YO_Z$ is : **[2018]**
(a) $C_3H_6O_3$ (b) C_2H_4O
(c) $C_3H_4O_2$ (d) $C_2H_4O_3$

69. The most abundant elements by mass in the body of a healthy human adult are : **[NCERT : P L-19 | 2017]**
Oxygen (61.4%) ; Carbon (22.9%), Hydrogen (10.0%) ; and Nitrogen (2.6%). The weight which a 75 kg person would gain if all ^1H atoms are replaced by ^2H atoms is
(a) 15 kg (b) 37.5 kg (c) 7.5 kg (d) 10 kg

70. What quantity (in mL) of a 45% acid solution of a mono-protic strong acid must be mixed with a 20% solution of the same acid to produce 800 mL of a 29.875% acid solution? **[Online April 9, 2017]**
(a) 320 (b) 325 (c) 316 (d) 330

71. At 300 K and 1 atm, 15 mL of a gaseous hydrocarbon requires 375 mL air containing 20% O_2 by volume for complete combustion. After combustion the gases occupy 330 mL. Assuming that the water formed is in liquid form and the volumes were measured at the same temperature and pressure, the formula of the hydrocarbon is:
[NCERT : P L-20 | 2016]
(a) C_4H_8 (b) C_4H_{10} (c) C_3H_6 (d) C_3H_8

72. In Carius method of estimation of halogens, 250 mg of an organic compound gave 141 mg of AgBr. The percentage of bromine in the compound is : **[NCERT : P L-19 | 2015]**
(at. mass Ag = 108; Br = 80)
(a) 48 (b) 60 (c) 24 (d) 36

73. Choose the incorrect formula out of the four compounds for an element X below :
[NCERT : P L-19 | Online April 11, 2015]
(a) X_2O_3 (b) X_2Cl_3 (c) $X_2(SO_4)_3$ (d) XPO_4

74. A gaseous compound of nitrogen and hydrogen contains 12.5% (by mass) of hydrogen. The density of the compound relative to hydrogen is 16. The molecular formula of the compound is:
[NCERT : P L-19 | Online April 11, 2014]
(a) NH_2 (b) N_3H (c) NH_3 (d) N_2H_4

75. The amount of $BaSO_4$ formed upon mixing 100 mL of 20.8% $BaCl_2$ solution with 50 mL of 9.8% H_2SO_4 solution will be:
[NCERT : P L-19 | Online April 12, 2014]
(Ba = 137, Cl = 35.5, S = 32, H = 1 and O = 16)
(a) 23.3 g (b) 11.65 g (c) 30.6 g (d) 33.2 g

76. A gaseous hydrocarbon gives upon combustion 0.72 g of water and 3.08 g of CO_2. The empirical formula of the hydrocarbon is : **[NCERT : P L-19 | 2013]**
(a) C_2H_4 (b) C_3H_4 (c) C_6H_5 (d) C_7H_8

77. A transition metal M forms a volatile chloride which has a vapour density of 94.8. If it contains 74.75% of chlorine, the formula of the metal chloride will be
[NCERT : P L-19 | Online May 26, 2012]
(a) MCl_3 (b) MCl_2 (c) MCl_4 (d) MCl_5

78. In a compound C, H and N atoms are present in 9 : 1 : 3.5 by weight. Molecular weight of compound is 108. Molecular formula of compound is **[NCERT : P L-19 | 2002]**
(a) $C_2H_6N_2$ (b) C_3H_4N (c) $C_6H_8N_2$ (d) $C_9H_{12}N_3$.

Topic 3 – Stoichiometric Calculations

79. 20 mL of sodium iodide solution gave 4.74 g silver iodide when treated with excess of silver nitrate solution. The molarity of the sodium iodide solution is ______ M. (Nearest Integer value) (Given : Na = 23, I = 127, Ag = 108, N = 14, O = 16 g mol^{-1})
[NCERT : P L-20 | NV, April 8, 2025 (II)]

80. Butane reacts with oxygen to produce carbon dioxide and water following the equation given below
[NCERT : P L-20 | NV, April 7, 2025 (II)]

$$C_4H_{10}(g) + \frac{13}{2}O_2(g) \rightarrow 4CO_2(g) + 5H_2O(l)$$

If 174.0 kg of butane is mixed with 320.0 kg of O_2, the volume of water formed in litres is ______. (Nearest integer)
[Given : (a) Molar mass of C, H, O are 12, 1, 16 g mol^{-1} respectively, (b) Density of water = 1 g mL^{-1}]

81. The amount of calcium oxide produced on heating 150 kg limestone (75% pure) is ______kg. (Nearest integer)
Given : Molar mass (in g mol^{-1}) of Ca-40, O-16, C-12.
[NCERT : P L-20 | NV, April 4, 2025 (II)]

82. Mass of magnesium required to produce 220 mL of hydrogen gas at STP on reaction with excess of dil. HCl is
Given : Molar mass of Mg is 24 g mol^{-1}.
[NCERT : P L-20 | April 3, 2025 (II)]
(a) 235.7 g (b) 0.24 mg
(c) 236 mg (d) 2.444 g

83. 0.5 g of an organic compound on combustion gave 1.46 g of CO_2 and 0.9 g of H_2O. The percentage of carbon in the compound is ____. (Nearest integer)
[Given : Molar mass (in g mol^{-1}) C : 12, H : 1, O : 16]
[NCERT : P L-20 | April 3, 2025 (I)]

84. $CaCO_3(s) + 2HCl(aq) \rightarrow CaCl_2(aq) + CO_2(g)\ H_2O(l)$
Consider the above reaction, what mass of $CaCl_2$ will be formed if 250 mL of 0.76 M HCl reacts with 1000 g of $CaCO_3$?
(Given : Molar mass of Ca, C, O, H and Cl are 40, 12, 16, 1 and 35.5 g mol^{-1}, respectively)
[NCERT : P L-20 | April 2, 2025 (I)]
(a) 3.908 g (b) 2.636 g
(c) 10.545 g (d) 5.272 g

85. 0.1 M solution of KI reacts with excess of H_2SO_4 and KIO_3 solution. According to equation

$$5I^- + IO_3^- + 6H^+ \rightarrow 3I_2 + 3H_2O$$

Identify the correct statements :
[NCERT : P L-19 | Jan. 29, 2025 (II)]
(A) 200 mL of KI solution reacts with 0.004 mol of KIO_3
(B) 200 mL of KI solution reacts with 0.006 mol of H_2SO_4
(C) 0.5 L of KI solution produced 0.005 mol of I_2
(D) Equivalent weight of KIO_3 is equal to

$$\left(\frac{\text{Molecular weight}}{5}\right)$$

Choose the correct answer from the options given below :
(a) (A) and (D) only (b) (B) and (C) only
(c) (A) and (B) only (d) (C) and (D) only

86. Concentrated nitric acid is labelled as 75% by mass. The volume in mL of the solution which contains 30 g of nitric acid is ________. **[NCERT : P L-23 | Jan. 28, 2025 (II)]**
Given : Density of nitric acid solution is 1.25 g/mL
(a) 45 (b) 55 (c) 32 (d) 40

87. The molarity of a 70% (mass/mass) aqueous solution of a monobasic acid (X) is ______ $\times 10^{-1}$ M(Nearest integer)
[Given : Density of aqueous solution of (X) is 1.25 g mL^{-1}
Molar mass of the acid is 70 g mol^{-1}]
[NCERT : P L-20 | NV, Jan. 28, 2025 (I)]

88. Consider the following reaction occurring in the blast furnace. **[NCERT : P L-20 | NV, Jan. 24, 2025 (I)]**

$$Fe_3O_{4(s)} + 4CO_{(g)} \rightarrow 3Fe_{(l)} + 4CO_{2(g)}$$

'x' kg of iron is produced when 2.32×10^3 kg Fe_3O_4 and 2.8 $\times 10^2$ kg CO are brought together in the furnace. The value of 'x' is_____. (nearest integer)

{Given:

Molar mass of $Fe_3O_4 = 232$ g mol^{-1}

Molar mass of CO = 28 g mol^{-1}

Molar mass of Fe = 56 g mol^{-1}}

89. Xg of benzoic acid on reaction with aq. $NaHCO_3$ release CO_2 that occupied 11.2 L volume at STP.

X is_____g. **[NCERT : P L-20 | NV, Jan. 24, 2025 (I)]**

90. When 81.0 g of aluminium is allowed to react with 128.0 g of oxygen gas, the mass of aluminium oxide produced in grams is_____. (Nearest integer)

Given:

Molar mass of Al is 27.0 g mol^{-1}

Molar mass of O is 16.0 g mol^{-1}

[NCERT : P L-20 | NV, Jan. 23, 2025 (II)]

91. 0.01 mole of an organic compound (X) containing 10% hydrogen, on complete combustion produced 0.9 g H_2O. Molar mass of (X) is_____ g mol^{-1}.

[NCERT : P L-20 | NV, Jan. 23, 2025 (II)]

92. What amount of bromine will be required to convert 2 g of phenol into 2, 4, 6-tribromophenol?

(Given molar mass in g mol^{-1} of C, H, O, Br are 12, 1, 16, 80 respectively) **[NCERT : P L-20 | Jan. 23, 2025 (I)]**

(a) 10.22 g (b) 6.0 g

(c) 4.0 g (d) 20.44 g

93. 20 mL of 2M NaOH solution is added to 400 mL of 0.5 M NaOH solution. The final concentration of the solution is _____ $\times 10^{-2}$ M. (Nearest integer).

[NCERT : P L-23 | Jan. 22, 2025 (II)]

94. Density of 3 M NaCl solution is 1.25 g/mL. The molality of the solution is : **[NCERT : P L-23 | Jan. 22, 2025 (II)]**

(a) 1.79m (b) 2m (c) 3m (d) 2.79m

95. Some CO_2 gas was kept in a sealed container at a pressure of 1 atm and at 273 K. This entire amount of CO_2 gas was later passed through an aqueous solution of $Ca(OH)_2$. The excess unreacted $Ca(OH)_2$ was later neutralized with 0.1M of 40 mL HCl. If the volume of the sealed container of CO_2 was x, then x is_____ cm^3 (nearest integer).

[Given: The entire amount of CO_2(g) reacted with exactly half the initial amount of $Ca(OH)_2$ present in this aqueous solution.] **[NCERT : P L-23 | NV, Jan. 22, 2025 (I)]**

96. Given below are two statements:

[NCERT : P L-20 | Jan. 22, 2025 (I)]

Statement I: One mole of propyne reacts with excess of sodium to liberate half a mole of H_2 gas.

Statement II : Four g of propyne reacts with $NaNH_2$ to liberate NH_3 gas which occupies 224 mL at STP.

In the light of the above statements, choose the most appropriate answer from the options given below:

(a) Statement I is correct but Statement II is incorrect.

(b) Both Statement I and Statement II are incorrect

(c) Statement I is incorrect but Statement II is correct

(d) Both Statement I and Statement II are correct.

97. The number of moles of methane required to produce 11 g CO_2(g) after complete combustion is :

(Given molar mass of methane in g mol^{-1} : 16)

[NCERT : P L-18 | April 5, 2024 (II)]

(a) 0.75 (b) 0.25 (c) 0.35 (d) 0.5

98. Consider the following reaction :

$3PbCl_2 + 2(NH_4)_3PO_4 \rightarrow Pb_3(PO_4)_2 + 6NH_4Cl$

If 72 mmol of $PbCl_2$ is mixed with 50 mmol of $(NH_4)_3PO_4$, then the amount of $Pb_3(PO_4)_2$ formed is_______ mmol (nearest integer). **[NCERT : P L-20 | NV, Feb. 1, 2024 (I)]**

99. 10 mL of gaseous hydrocarbon on combustion gives 40 mL of CO_2(g) and 50 mL of water vapour. Total number of carbon and hydrogen atoms in the hydrocarbon is _______.

[NV, Feb. 1, 2024 (II)]

100. Number of moles of methane required to produce 22g CO_2(g) after combustion is x $\times 10^{-2}$ moles. The value of x is_____. **[NCERT : P L-18 | NV, Jan. 31, 2024 (I)]**

101. If 50 mL of 0.5M oxalic acid is required to neutralise 25 mL of NaOH solution, the amount of NaOH in 50 mL of given NaOH solution is ____ g.

[NCERT : P L-23 | NV, Jan. 29, 2024 (II)]

102. Mass of methane required to produce 22 g of CO_2 after complete combustion is_____g.

(Given Molar mass in g mol^{-1} C = 12.0, H = 1.0, O = 16.0)

[NCERT : P L-22 | NV, Jan. 27, 2024 (I)]

103. Volume of 3 M NaOH (formula weight 40 g mol^{-1}) which can be prepared from 84 g of NaOH is________ $\times 10^{-1}$ dm^3. **[NCERT : P L-23 | NV, Jan. 27, 2024 (II)]**

104. 9.3 g of aniline is subjected to reaction with excess of acetic anhydride to prepare acetanilide. The mass of acetanilide produced if the reaction is 100% completed is _______ $\times 10^{-1}$ g.

(Given molar mass in g mol^{-1} N : 14, O : 16, C : 12, H : 1)

[NCERT : P L-18 | NV, Jan. 27, 2024 (II)]

105. 1 g of a carbonate (M_2CO_3) on treatment with excess HCl produces 0.01 mol of CO_2. The molar mass of M_2CO_3 is _______ g mol^{-1}. (Nearest integer)

[NCERT : P L-18 | NV, April 13, 2023 (II)]

106. The volume of hydrogen liberated at STP by treating 2.4 g magnesium with excess of hydrochloric acid is_____ $\times 10^{-2}$ L.

Given: Molar volume of gas is 22.4 L at STP. Molar mass of magnesium is 24 g mol^{-1}.

[NCERT : P L-18 | NV, April 11, 2023 (II)]

107. 0.5 g of an organic compound (X) with 60% carbon will produce_____ $\times 10^{-1}$ g of CO_2 on complete combustion.

[NCERT : P L-19 | NV, April 08, 2023 (I)]

108. If 5 moles of $BaCl_2$ is mixed with 2 moles of Na_3PO_4, the maximum number of moles of $Ba_3(PO_4)_2$ formed is.......... (Nearest integer) **[NCERT : P L-20 | NV, April 06, 2023 (I)]**

109. On complete combustion, 0.492 g of an organic compound gave 0.792 g of CO_2. The % of carbon in the organic compound is _____ (Nearest integer) **[NCERT : P L-19 | NV, Jan. 31, 2023 (I)]**

110. Zinc reacts with hydrochloric acid to give hydrogen and zinc chloride. The volume of hydrogen gas produced at STP from the reaction of 11.5 g of zinc with excess HCl is ____ L (Nearest integer)
(Given : Molar mass of Zn is $65.4 g mol^{-1}$ and Molar volume of H_2 at STP = 22.7L)
[NCERT : P L-18 | NV, Jan. 31, 2023 (I)]

111. Assume carbon burns according to following equation : $2C(s) + O_2(g) \rightarrow 2CO(g)$
When 12 g carbon is burnt in 48 g of oxygen, the volume of carbon monoxide produced is _____ $\times 10^{-1}$ L at STP [nearest integer]
[Given : Assume CO as ideal gas, Mass of C is $12 g mol^{-1}$, Mass of O is $16 g mol^{-1}$ and molar volume of an ideal gas at STP is $22.7 L mol^{-1}$]
[NCERT : P L-20 | NV, Jan. 31, 2023 (II)]

112. A 300 mL bottle of soft drink has 0.2 M CO_2 dissolved in it. Assuming CO_2 behaves as an ideal gas, the volume of the dissolved CO_2 at STP is _______ mL. (Nearest integer) Given: At STP, molar volume of an ideal gas is $22.7 L mol^{-1}$.
[NCERT : P L-18 | NV, Jan. 30, 2023 (I)]

113. The strength of 50 volume solution of hydrogen peroxide is ______ g/L (Nearest integer).
Given: Molar mass of H_2O_2 is $34 g mol^{-1}$
Molar volume of gas at STP = 22.7 L.
[NV, Jan. 30, 2023 (II)]

114. When 0.01 mol of an organic compound containing 60% carbon was burnt completely, 4.4 g of CO_2 was produced. The molar mass of compound is ________ $g mol^{-1}$ (Nearest integer) **[NCERT : P L-18 | NV, Jan. 29, 2023 (II)]**

115. The volume of HCl, containing $73 g L^{-1}$, required to completely neutralise NaOH obtained by reacting 0.69 g of metallic sodium with water, is _______ mL. (Nearest Integer)
(Given : Molar masses of Na, Cl, O, H are 23, 35.5, 16 and $1 g mol^{-1}$ respectively)
[NCERT : P L-20 | NV, Jan. 29, 2023 (II)]

116. What is the mass ratio of ethylene glycol ($C_2H_6O_2$, molar mass = 62 g/mol) required for making 500 g of 0.25 molal aqueous solution and 250 mL of 0.25 molar aqueous solution ? **[NCERT : P L-23 | Jan. 25, 2023 (II)]**
(a) 1 : 1 (b) 3 : 1 (c) 2 : 1 (d) 1 : 2

117. When $Fe_{0.93}O$ is heated in presence of oxygen, it converts to Fe_2O_3. The number of correct statement/s from the following is _______. **[NV, Jan. 24, 2023 (I)]**
A. The equivalent weight of $Fe_{0.93}O$ is $\frac{\text{Molecular weight}}{0.79}$
B. The number of moles of Fe^{2+} and Fe^{3+} in 1 mole of $Fe_{0.93}O$ is 0.79 and 0.14 respectively.
C. $Fe_{0.93}O$ is metal deficient with lattice comprising of cubic closed packed arrangement of O^{2-} ions.
D. The % composition of Fe^{2+} and Fe^{3+} in $Fe_{0.93}O$ is 85% and 15% respectively.

118. $N_2(g) + 3H_2(g) \rightleftharpoons 2NH_3(g)$
20 g 5 g
Consider the above reaction, the limiting reagent of the reaction and number of moles of NH_3 formed respectively are: **[NCERT : P L-18, 20 | July 29, 2022 (I)]**
(a) H_2, 1.42 moles (b) H_2, 0.71 moles
(c) N_2, 1.42 moles (d) N_2, 0.71 moles

119. Consider the reaction
$4HNO_3(l) + 3KCl(s) \longrightarrow Cl_2(g) + NOCl(g) + 2H_2O(g) + 3KNO_3(s)$
The amount of HNO_3 required to produce 110.0 g of KNO_3 is:
(Given : Atomic masses of H, O, N and K are 1, 16, 14 and 39, respectively.) **[NCERT : P L-20 | July 29, 2022 (II)]**
(a) 32.2 g (b) 69.4 g
(c) 91.5 g (d) 162.5 g

120. $C(s) + O_2(g) \rightarrow CO_2(g) + 400 kJ$
$C(s) + \frac{1}{2}O_2(g) \rightarrow CO(g) + 100 kJ$
[NCERT : P L-20 | July 29, 2022 (II)]
When coal of purity 60% is allowed to burn in presence of insufficient oxygen, 60% of carbon is converted into 'CO' and the remaining is converted into 'CO_2'. The heat generated when 0.6 kg of coal is burnt is _____.
(a) 1600 kJ (b) 3200 kJ
(c) 4400 kJ (d) 6600 kJ

121. 250 g solution of D-glucose in water contains 10.8% of carbon by weight. The molality of the solution is nearest to (Given: Atomic Weights are H, 1u ; C, 12u ; O, 16u)
[NCERT : P L-24 | July 27, 2022 (I)]
(a) 1.03 (b) 2.06
(c) 3.09 (d) 5.40

122. SO_2Cl_2 on reaction with excess of water results into acidic mixture
$SO_2Cl_2 + 2H_2O \rightarrow H_2SO_4 + 2HCl$
16 moles of NaOH is required for the complete neutralisation of the resultant acidic mixture.
The number of moles of SO_2Cl_2 used is:
[NCERT : P L-18 | July 25, 2022 (I)]
(a) 16 (b) 8 (c) 4 (d) 2

123. Blister copper is produced by reaction of copper oxide with copper sulphide.
$2Cu_2O + Cu_2S \rightarrow 6Cu + SO_2$
When 2.86×10^3 g of Cu_2O and 4.77×10^3 g of Cu_2S are used for reaction, the mass of copper produced is ____ g. (nearest integer) **[NCERT : P L-20 | NV, June 30, 2022 (I)]**
(Atomic mass of Cu = 63.5 a.m. u)
S = 32.0 a.m. u
O = 16.0 a.m. u)

124. Production of iron in blast furnace follows the following equation

$Fe_3O_4(s) + 4CO(g) \rightarrow 3Fe(l) + 4CO_2(g)$

when 4.640 kg of Fe_3O_4 and 2.520 kg of CO are allowed to react then the amount of iron (in g) produced is :

[Given : Molar Atomic mass (g mol^{-1}): Fe = 56

Molar Atomic mass (g mol^{-1}) : O = 16

Molar Atomic mass (g mol^{-1}): C = 12

[NCERT : P L-18 | June 29, 2022 (I)]

(a) 1400 (b) 2200

(c) 3360 (d) 4200

125. CNG is an important transportation fuel. When 100 g CNG is mixed with 208g oxygen in vehicles, it leads to the formation of CO_2 and H_2O and produces large quantity of heat during this combustion, then the amount of carbon dioxide, produced in grams is______. [nearest integer] [Assume CNG to be methane]

[NCERT : P L-20 | NV, June 26, 2022 (II)]

126. The moles of methane required to produce 81 g of water after complete combustion is ______ × 10^{-2} mol. [nearest integer] **[NCERT : P L-22 | NV, June 26, 2022 (II)]**

127. Number of grams of bromine that will completely react with 5.0g of pent-1-ene is ________ × 10^{-2}g. (Atomic mass of Br = 80 g/mol) (Nearest Integer)

[NCERT : P L-20 | NV, June 25, 2022 (I)]

128. 1 L aqueous solution of H_2SO_4 contains 0.02 m mol H_2SO_4. 50% of this solution is diluted with deionized water to give 1 L solution (A). In solution (A), 0.01 m mol of H_2SO_4 are added. Total m mols of H_2SO_4 in the final solution is______ × 10^3 m mols.

[NCERT : P L-18 | NV, June 25, 2022 (I)]

129. A protein 'A' contains 0.30% of glycine (molecular weight 75). The minimum molar mass of the protein 'A' is ______ × 10^3 g mol^{-1} [nearest integer]

[NCERT : P L-18 | NV, June 25, 2022 (II)]

130. The neutralization occurs when 10 mL of 0.1 M acid 'A' is allowed to react with 30 mL of 0.05 M base $M(OH)_2$. The basicity of the acid 'A' is_______. [M is a metal]

[NCERT : P L-23 | NV, June 25, 2022 (II)]

131. If a rocket runs on a fuel ($C_{15}H_{30}$) and liquid oxygen, the weight of oxygen required and CO_2 released for every litre of fuel respectively are:

(Given: density of the fuel is 0.756 g/mL)

[NCERT : P L-20 | June 24, 2022 (I)]

(a) 1188 g and 1296 g (b) 2376 g and 2592 g

(c) 2592g and 2376 g (d) 3429 g and 3142 g

132. When 35 mL of 0.15 M lead nitrate solution is mixed with 20 mL of 0.12 M chromic sulphate solution, ________ × 10^{-5} moles of lead sulphate precipitate out. (Round off to the Nearest Integer).

[NCERT : P L-20 | NV, March 16, 2021 (II)]

133. Sodium oxide reacts with water to produce sodium hydroxide. 20.0 g of sodium oxide is dissolved in 500 mL of water. Neglecting the change in volume, the concentration of the resulting NaOH solution is _____ × 10^{-1} M. (Nearest integer)

[Atomic mass : Na = 23.0, O = 16.0, H = 1.0]

[NCERT : P L-18 | NV, Aug. 31, 2021 (II)]

134. 100 g of propane is completely reacted with 1000 g of oxygen. The mole fraction of carbon dioxide in the resulting mixture is $x \times 10^{-2}$. The value of x is _______. (Nearest integer)

[Atomic weight : H = 1.008; C = 12.00; O = 16.00]

[NCERT : P L-23 | NV, Aug. 27, 2021 (II)]

135. The density of NaOH solution is 1.2 g cm^{-3}. The molality of this solution is _______ m.

(Round off to the Nearest Integer)

[Use : Atomic masses : Na : 23.0 u O : 16.0 u H : 1.0 u

Density of H_2O : 1.0 g cm^{-3}]

[NCERT : P L-24 | NV, July 27, 2021 (I)]

136. When 10 mL of an aqueous solution of Fe^{2+} ions was titrated in the presence of dil H_2SO_4 using diphenylamine indicator, 15 mL of 0.02 M solution of $K_2Cr_2O_7$ was required to get the end point. The molarity of the solution containing Fe^{2+} ions is $x \times 10^{-2}$ M. The value of x is_____. (Nearest integer) **[NCERT : P L-23 | NV, July 25, 2021 (I)]**

137. Consider the complete combustion of butane, the amount of butane utilized to produce 72.0 g of water is _______ × 10^{-1} g. (in nearest integer)

[NV, July 25, 2021 (I); Same March 18, 2021 (I)]

138. If the concentration of glucose ($C_6H_{12}O_6$) in blood is 0.72 g L^{-1}, the molarity of glucose in blood is _______ × 10^{-3} M. (Nearest integer)

[Given : Atomic mass of C = 12, H = 1, O = 16 u]

[NCERT : P L-23 | NV, July 22, 2021 (II)]

139. 250 mL of 0.5 M NaOH was added to 500 mL of 1 M HCl. The number of unreacted HCl molecules in the solution after complete reaction is ______ ×10^{21}. (Nearest integer) ($N_A = 6.022 \times 10^{23}$)

[NCERT : P L-23 | NV, July 20, 2021 (I)]

140. 4 g equimolar mixture of NaOH and Na_2CO_3 contains x g of NaOH and y g of Na_2CO_3. The value of x is _______ g. (Nearest integer)

[NCERT : P L-18 | NV, July 20, 2021 (II)]

141. _______ grams of 3-Hydroxypropanal (MW = 74) must be dehydrated to produce 7.8 g of acrolein (MW = 56) (C_3H_4O) if the percentage yield is 64. (Rounded off to the Nearest Integer).

[NCERT : P L-19 | NV, March 18, 2021 (I)]

142. A reaction of 0.1 mole of Benzylamine with bromomethane gave 23 g of Benzyl trimethyl ammonium bromide. The number of moles of bromomethane consumed in this reaction are $n \times 10^{-1}$, when n = _____ (Rounded off to the Nearest Integer).

[Given : Atomic masses are C : 12.0 u, H : 1.0 u, N : 14.0 u, Br : 80.0 u] **[NCERT : P L-18 | NV, March 18, 2021 (I)]**

143. Complete combustion of 3 g of ethane gives x × 10^{22} molecules of water. The value of x is ________. (Rounded off to the Nearest Integer).
[Use : $N_A = 6.023 \times 10^{23}$;
Atomic masses in u : C : 12.0 ; O : 16.0 ; H : 1.0]
[NCERT : P L-20 | NV, March 18, 2021 (I)]

144. The number of chlorine atoms in 20 mL of chlorine gas at STP is ________ × 10^{21}. (Round off to the Nearest integer)
[Assume chlorine is an ideal gas at STP]
$R = 0.083$ L bar mol^{-1} K^{-1}, $N_A = 6.023 \times 10^{23}$
[NCERT : P L-18 | NV, March 17, 2021 (II)]

145. 15 mL of aqueous solution of Fe^{2+} in acidic medium completely reacted with 20 mL of 0.03 M aqueous $Cr_2O_7^{2-}$. The molarity of the Fe^{2+} solution is ________ × 10^{-2} M.
(Round off to the Nearest Integer). **[NV, March 17, 2021 (I)]**

146. The $NaNO_3$ weighed out to make 50 mL of an aqueous solution containing 70.0 mg Na^+ per mL is ________ g. (Rounded off to the nearest integer)
[Given : Atomic weight in g mol^{-1} - Na : 23; N : 14; O : 16]
[NV, Feb. 26, 2021 (II)]

147. 4.5 g of compound A (MW = 90) was used to make 250 mL of its aqueous solution. The molarity of the solution is $x \times 10^{-1}$. The value of x is ________ (Rounded off to the nearest integer) **[NCERT : P L-23 | NV, Feb. 24, 2021 (I)]**

148. The formula of a gaseous hydrocarbon which requires 6 times of its own volume of O_2 for complete oxidation and produces 4 times its own volume of CO_2 is C_xH_y. The value of y is ________. **[NCERT : P L-20 | NV, Feb. 24, 2021 (II)]**

149. The minimum number of moles of O_2 required for complete combustion of 1 mole of propane and 2 moles of butane is ________. **[NCERT : P L-20 | NV, Sep. 05, 2020 (I)]**

150. The mass of ammonia in grams produced when 2.8 kg of dinitrogen quantitatively reacts with 1 kg of dihydrogen is ________. **[NCERT : P L-18 | NV, Sep. 04, 2020 (I)]**

151. A 100 mL solution was made by adding 1.43 g of $Na_2CO_3 \cdot xH_2O$. The normality of the solution is 0.1 N. The value of x is ________. **[NV, Sep. 04, 2020 (II)]**
(The atomic mass of Na is 23 g/ mol)

152. The mole fraction of glucose ($C_6H_{12}O_6$) in an aqueous binary solution is 0.1. The mass percentage of water in it, to the nearest integer, is ________.
[NCERT : P L-23 | NV, Sep. 03, 2020 (I)]

153. 6.023 × 10^{22} molecules are present in 10 g of a substance 'x'. The molarity of a solution containing 5 g of substance 'x' in 2 L solution is ________ × 10^{-3}.
[NCERT : P L-23 | NV, Sep. 03, 2020 (II)]

154. The volume (in mL) of 0.1 N NaOH required to neutralise 10 mL of 0.1 N phosphinic acid is ________.
[NV, Sep. 03, 2020 (II)]

155. 10.30 mg of O_2 is dissolved into a liter of sea water of density 1.03 g/mL. The concentration of O_2 in ppm is ________. **[NCERT : P L-23 | NV, Jan. 09, 2020 (II)]**

156. Ferrous sulphate heptahydrate is used to fortify foods with iron. The amount (in grams) of the salt required to achieve 10 ppm of iron in 100 kg of wheat is ________. Atomic weight: Fe = 55.85; S = 32.00; O = 16.00
[NV, Jan. 08, 2020 (I)]

157. $NaClO_3$ is used, even in spacecrafts, to produce O_2. The daily consumption of pure O_2 by a person is 492 L at 1 atm, 300 K. How much amount of $NaClO_3$, in grams, is required to produce O_2 for the daily consumption of a person at 1 atm, 300 K? ________.
$NaClO_3(s) + Fe(s) \rightarrow O_2(g) + NaCl(s) + FeO(s)$
$R = 0.082$ L atm mol^{-1} K^{-1} **[NV, Jan. 08, 2020 (II)]**

158. The ammonia (NH_3) released on quantitative reaction of 0.6 g urea (NH_2CONH_2) with sodium hydroxide (NaOH) can be neutralized by: **[Jan. 07, 2020 (II)]**
(a) 200 mL of 0.4 N HCl (b) 200 mL of 0.2 N HCl
(c) 100 mL of 0.2 N HCl (d) 100 mL of 0.1 N HCl

159. The minimum amount of $O_2(g)$ consumed per gram of reactant is for the reaction : (Given atomic mass : Fe = 56, O = 16, Mg = 24, P = 31, C = 12, H = 1)
[NCERT : P L-20 | April 10, 2019 (II)]
(a) $4\,Fe(s) + 3\,O_2(g) \rightarrow 2\,Fe_2O_3(s)$
(b) $P_4(s) + 5\,O_2(g) \rightarrow P_4O_{10}(s)$
(c) $C_3H_8(g) + 5\,O_2(g) \rightarrow 3\,CO_2(g) + 4\,H_2O\,(l)$
(d) $2\,Mg(s) + O_2(g) \rightarrow 2\,MgO(s)$

160. For a reaction, $N_2(g) + 3\,H_2(g) \rightarrow 2\,NH_3(g)$; identify dihydrogen ($H_2$) as a limiting reagent in the following reaction mixtures. **[NCERT : P L-20 | April 9, 2019 (I)]**
(a) 56 g of N_2 + 10 g of H_2 (b) 35 g of N_2 + 8 g of H_2
(c) 28 g of N_2 + 6 g of H_2 (d) 14 g of N_2 + 4 g of H_2

161. 0.27 g of a long chain fatty acid was dissolved in 100 cm^3 of hexane. 10 mL of this solution was added dropwise to the surface of water in a round watch glass. Hexane evaporates and a monolayer is formed. The distance from edge to centre of the watch glass is 10 cm. What is the height of the monolayer?
[Density of fatty acid = 0.9 g cm^{-3} ; π = 3]
[April 8, 2019 (II)]
(a) 10^{-6} m (b) 10^{-8} m
(c) 10^{-2} m (d) 10^{-4} m

162. 50 mL of 0.5 M oxalic acid is needed to neutralize 25 mL of sodium hydroxide solution. The amount of NaOH in 50 mL of the given sodium hydroxide solution is:
[Jan. 12, 2019 (I)]
(a) 40 g (b) 10 g (c) 20 g (d) 80 g

163. A 10 mg effervescent tablet containing sodium bicarbonate and oxalic acid releases 0.25 mL of CO_2 at T = 298.15 K and P = 1 bar. If molar volume of CO_2 is 25.0 L under such condition, what is the percentage of sodium bicarbonate in each tablet ?

[Molar mass of $NaHCO_3$ = 84 g mol^{-1}]
[NCERT : P L-18 | Jan. 11, 2019 (I)]
(a) 0.84 (b) 33.6
(c) 16.8 (d) 8.4

164. 25 mL of the given HCl solution requires 30 mL of 0.1 M sodium carbonate solution. What is the volume of this HCl solution required to titrate 30 mL of 0.2 M aqueous NaOH solution? **[NCERT : P L-23 | Jan. 11, 2019 (II)]**
(a) 25 mL (b) 75 mL
(c) 50 mL (d) 12.5 mL

165. For the following reaction, the mass of water produced from 445 g of $C_{57}H_{110}O_6$ is: **[NCERT : P L-18 | Jan. 9, 2019 (II)]**
$2\,C_{57}H_{110}O_6(s) + 163\,O_2(g) \rightarrow 114\,CO_2(g) + 110\,H_2O(1)$
(a) 490 g (b) 445 g
(c) 495 g (d) 890 g

166. A sample of $NaClO_3$ is converted by heat to NaCl with a loss of 0.16 g of oxygen. The residue is dissolved in water and precipitated as AgCl. The mass of AgCl (in g) obtained will be: (Given: Molar mass of AgCl = 143.5 g mol^{-1}) **[NCERT : P L-18 | Online April 15, 2018 (I)]**
(a) 0.35 (b) 0.54
(c) 0.41 (d) 0.48

167. 1 gram of a carbonate (M_2CO_3) on treatment with excess HCl produces 0.01186 mole of CO_2. The molar mass of M_2CO_3 in g mol^{-1} is : **[NCERT : P L-18 | 2017]**
(a) 1186 (b) 84.3
(c) 118.6 (d) 11.86

168. The amount of arsenic pentasulphide that can be obtained when 35.5 g arsenic acid is treated with excess H_2S in the presence of conc. HCl (assuming 100% conversion) is : **[NCERT : P L-18 | Online April 9, 2016]**
(a) 0.25 mol (b) 0.50 mol
(c) 0.333 mol (d) 0.125 mol

169. The volume of 0.1 N dibasic acid sufficient to neutralize 1 g of a base that furnishes 0.04 mole of OH^- in aqueous solution is : **[Online April 10, 2016]**
(a) 400 mL (b) 600 mL (c) 200 mL (d) 800 mL

170. A sample of a hydrate of barium chloride weighing 61 g was heated until all the water of hydration is removed. The dried sample weighed 52 g. The formula of the hydrated salt is : (atomic mass, Ba = 137 amu, Cl = 35.5 amu) **[NCERT : P L-19 | Online April 10, 2015]**
(a) $BaCl_2.4H_2O$ (b) $BaCl_2.3H_2O$
(c) $BaCl_2.H_2O$ (d) $BaCl_2.2H_2O$

171. $A + 2B + 3C \rightleftharpoons AB_2C_3$
Reaction of 6.0 g of A, 6.0×10^{23} atoms of B, and 0.036 mol of C yields 4.8 g of compound AB_2C_3. If the atomic mass of A and C are 60 and 80 amu, respectively, the atomic mass of B is (Avogadro no. = 6×10^{23}): **[NCERT : P L-19 | Online April 11, 2015]**
(a) 50 amu (b) 60 amu
(c) 70 amu (d) 40 amu

172. 6 litres of an alkene require 27 litres of oxygen at constant temperature and pressure for complete combustion. The alkene is : **[NCERT : P L-20 | Online April 25, 2013]**
(a) Ethene (b) Propene
(c) 1-Butene (d) 2 -Butene

173. The molality of a urea solution in which 0.0100 g of urea, $[(NH_2)_2CO]$ is added to 0.3000 dm^3 of water at STP is : **[NCERT : P L-24 | 2011RS]**
(a) 5.55×10^{-4}m (b) 33.3 m
(c) 3.33×10^{-2} m (d) 0.555 m

174. The density (in g mL^{-1}) of a 3.60 M sulphuric acid solution that is 29% H_2SO_4 (molar mass = 98 g mol^{-1}) by mass will be **[NCERT : P L-23 | 2007]**
(a) 1.45 (b) 1.64
(c) 1.88 (d) 1.22

175. In the reaction, **[NCERT : P L-20 | 2007]**
$2Al(s) + 6HCl(aq) \rightarrow 2Al^{3+}(aq) + 6Cl^-(aq) + 3H_2(g)$
(a) 11.2 L H_2(g) at STP is produced for every mole of HCl(aq) consumed
(b) 6 L HCl(aq) is consumed for every 3 L of H_2(g) produced
(c) 33.6 L H_2(g) is produced regardless of temperature and pressure for every mole of Al that reacts
(d) 67.2 H_2(g) at STP is produced for every mole of Al that reacts.

176. Density of a 2.05M solution of acetic acid in water is 1.02 g/mL. The molality of the solution is **[NCERT : P L-23 | 2006]**
(a) 2.28 mol kg^{-1} (b) 0.44 mol kg^{-1}
(c) 1.14 mol kg^{-1} (d) 3.28 mol kg^{-1}

177. Two solutions of a substance (non electrolyte) are mixed in the following manner: 480 mL of 1.5 M first solution + 520 mL of 1.2 M second solution. What is the molarity of the final mixture ? **[NCERT : P L-23 | 2005]**
(a) 2.70 M (b) 1.344 M
(c) 1.50 M (d) 1.20 M

178. To neutralise completely 20 mL of 0.1 M aqueous solution of phosphorous acid (H_3PO_3), the valume of 0.1 M aqueous KOH solution required is **[2004]**
(a) 40 mL (b) 20 mL
(c) 10 mL (d) 60 mL

179. What volume of hydrogen gas, at 273 K and 1 atm. pressure will be consumed in obtaining 21.6 g of elemental boron (atomic mass = 10.8) from the reduction of boron trichloride by hydrogen ? **[NCERT : P L-20 | 2003]**
(a) 67.2 L (b) 44.8 L
(c) 22.4 L (d) 89.6 L

ANSWER KEYS

1	(b)	19	(c)	37	(a)	55	(36)	73	(b)	91	(100)	109	(44)	127	(1143)	145	(24)	163	(d)
2	(372)	20	(14.00)	38	(c)	56	(59)	74	(d)	92	(a)	110	(4)	128	(0)	146	(13)	164	(a)
3	(d)	21	(c)	39	(d)	57	(25)	75	(b)	93	(57)	111	(227)	129	(25)	147	(2)	165	(c)
4	(a)	22	(d)	40	(b)	58	(12)	76	(d)	94	(d)	112	(1362)	130	(3)	148	(8)	166	(d)
5	(b)	23	(d)	41	(b)	59	(46)	77	(c)	95	(45)	113	(150)	131	(c)	149	(18)	167	(b)
6	(d)	24	(c)	42	(d)	60	(d)	78	(c)	96	(a)	114	(200)	132	(525)	150	(3400)	168	(d)
7	(a)	25	(c)	43	(c)	61	(2)	79	(a)	97	(b)	115	(15)	133	(13)	151	(10)	169	(a)
8	(11.00)	26	(d)	44	(a)	62	(18)	80	(138	98	(24)	116	(c)	134	(19)	152	(47)	170	(d)
9	(d)	27	(a)	45	(b)	63	(d)	81	(63)	99	(14)	117	(4)	135	(5)	153	(25)	171	(a)
10	(d)	28	(d)	46	(65)	64	(3)	82	(c)	100	(50)	118	(c)	136	(18)	154	(10)	172	(b)
11	(2)	29	(b)	47	(12)	65	(b)	83	(80)	101	(4)	119	(c)	137	(464)	155	(10.00)	173	(a)
12	(24)	30	(b)	48	(a)	66	(b)	84	(c)	102	(8)	120	(d)	138	(4)	156	(4.96)	174	(d)
13	(c)	31	(b)	49	(3)	67	(5)	85	(a)	103	(7)	121	(b)	139	(226)	157	(2130)	175	(a)
14	(46)	32	(c)	50	(c)	68	(d)	86	(c)	104	(135)	122	(c)	140	(1)	158	(c)	176	(a)
15	(2)	33	(a)	51	(d)	69	(c)	87	(125)	105	(100)	123	(3810)	141	(16)	159	(a)	177	(b)
16	(c)	34	(a)	52	(a)	70	(c)	88	(420)	106	(224)	124	(c)	142	(3)	160	(a)	178	(a)
17	(5418)	35	(a)	53	(56)	71	(d)	89	(61)	107	(11)	125	(143)	143	(18)	161	(a)	179	(a)
18	(2)	36	(d)	54	(a)	72	(c)	90	(153)	108	(1)	126	(225)	144	(1)	162	(N)		

Structure of Atom

2

TREND ANALYSIS
JEE MAIN (2025-2021)

	2025				2024				2023				2022				2021			
	Session II		Session I		Session II		Session I		Session II		Session I		Session II		Session I		Session II		Session I	
	MCQs	NVQs	MCQs	NVQs	MCQs	NVQs	MCQs	NVQs	MCQs	NVQs	MCQs	NVQs	MCQs	NVQs	MCQs	NVQs	MCQs	NVQs	MCQs	NVQs
Topic 1	1	–	–	–	–	–	–	1	2	2	–	2	–	1	2	1	–	4	–	1
Topic 2	2	1	3	–	–	5	1	1	2	2	3	2	–	1	1	1	1	1	1	–
Topic 3	–	–	2	–	–	2	1	–	1	1	–	1	–	2	–	1	–	2	–	3
Topic 4	6	–	5	–	1	2	4	1	1	3	3	1	5	–	6	1	–	2	2	2
Total	**9**	**1**	**10**	**–**	**1**	**9**	**6**	**3**	**6**	**8**	**6**	**6**	**5**	**4**	**9**	**4**	**1**	**9**	**3**	**6**

How to Code Qns for Better & Faster Revision →

Type 1: Able to solve (No code Req)
Type 2: Solved with struggle or by seeing solution
(Use code NR means Needs Revision or orange stickers)

Type 3: Tricky/ Could not solve on my own. Need help from Teacher/ Friend.
(Use code TQ means Tricky Question or red stickers)

You can either hand write the code in boxes or else paste coloured stickers given at end of book.

Topic 1 – Dual Nature of Electromagnetic Radiation

1. Visible spectrum

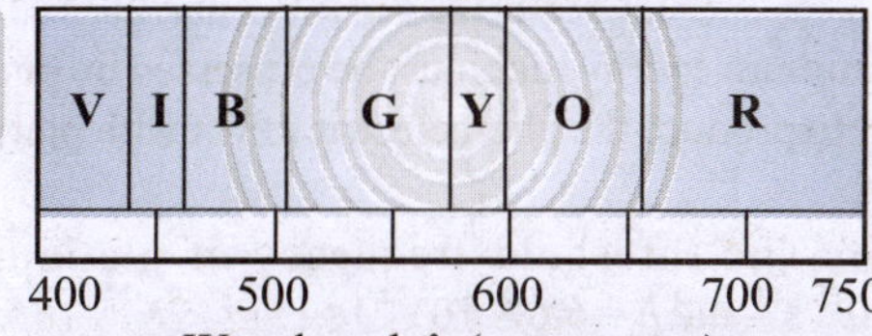

Wavelength λ (nanometers)

Which of the following statements are correct, if the threshold frequency of caesium is 5.16×10^{14} Hz ?

[NCERT : P L-44 | April 7, 2025 (I)]

A. When *Cs* is placed inside a vacuum chamber with an ammeter connected to it and yellow light is focused on *Cs* the ammeter shows the presence of current.

B. When the brightness of the yellow light is dimmed, the value of the current in the ammeter is reduced.

C. When a red light is used instead to the yellow light, the current produced is higher with respect to the yellow light.

D. When a blue light is used, the ammeter shows the formation of current.

E. When a white light is used, the ammeter shows formation of current.

Choose the **correct** answer from the options given below :

(a) A, D and E Only
(b) B, C and D Only
(c) A, C, D and E Only
(d) A, B, D and E Only

2. The ionization energy of sodium in kJ mol^{-1}, if electromagnetic radiation of wavelength 242 nm is just sufficient to ionize sodium atom is _____.

[NCERT : P L-39 | NV, Jan. 31, 2024 (I)]

3. The number of incorrect statement/s about the black body from the following is _______

[NCERT : P L-40 | NV, April 10, 2023 (I)]

(A) Emit or absorb energy in the form of electromagnetic radiation

(B) Frequency distribution of the emitted radiation depends on temperature

(C) At a given temperature, intensity vs frequency curve passes through a maximum value

(D) The maximum of the intensity vs frequency curve is at a higher frequency at higher temperature compared to that at lower temperature activation energy.

4. Following figure shows spectrum of an ideal black body at four different temperatures. The number of correct statement/s from the following is ______.

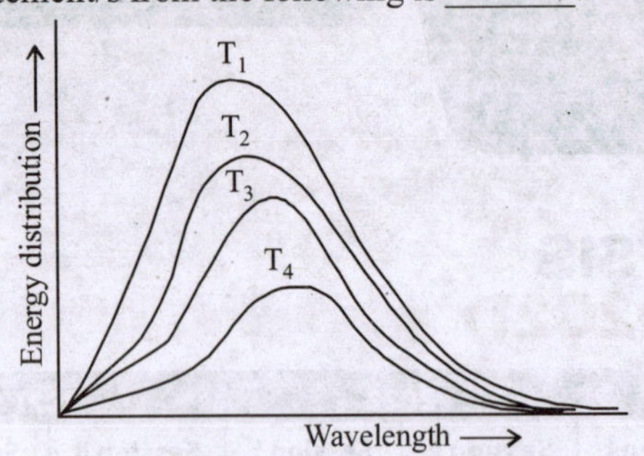

[NCERT : P L-40 | NV, Jan. 24, 2023 (II)]

(A) $T_4 > T_3 > T_2 > T_1$

(B) The black body consists of particles performing simple harmonic motion.

(C) The peak of the spectrum shifts to shorter wavelength as temperature increases.

(D) $\frac{T_1}{\nu_1} = \frac{T_2}{\nu_2} = \frac{T_3}{\nu_3} \neq$ constant

(E) The given spectrum could be explained using quantisation of energy.

5. Values of work function (W_0) for a few metals are given below **[NCERT : P L-41 | NV, April 12, 2023 (I)]**

Metal	Li	Na	K	Mg	Cu	Ag
W_0 **/ eV**	2.42	2.3	2.25	3.7	4.8	4.3

The number of metals which will show photoelectric effect when light of wavelength 400 nm falls on it is ______

Given : $h = 6.6 \times 10^{-34}$ J s

$c = 3 \times 10^8$ m s^{-1}

$e = 1.6 \times 10^{-19}$ C

6. Given below are two statements : one is labelled as Assertion A and the other is labelled as Reason R :

Assertion A : In the photoelectric effect, the electrons are ejected from the metal surface as soon as the beam of light of frequency greater than threshold frequency strikes the surface.

Reason R : When the photon of any energy strikes an electron in the atom, transfer of energy from the photon to the electron takes place.

In the light of the above statements, choose the most appropriate answer from the options given below :

[NCERT : P L-41 | April 11, 2023 (I)]

(a) Both A and R are correct but R is NOT the correct explanation of A.

(b) A is correct but R is not correct.

(c) Both A and R are correct and R is the correct explanation of A.

(d) A is not correct but R is correct.

7. Given below are two statements: one is labelled as assertion and the other is labelled as reason.

[NCERT : P L-42 | April 06, 2023 (I)]

Assertion: Loss of electron from hydrogen atom results in nucleus of ~1.5×10^{-3} pm size.

Reason: Proton (H^+) always exists in combined form

In the light of the above statements, choose the most appropriate answer from the options given below:

(a) Both A and R are correct R is the correct explanation of A.

(b) A is correct but R is not correct.

(c) A is not correct but R is correct.

(d) Both A and R are correct but R is NOT the correct explanation of A.

8. The energy of one mole of photons of radiation of frequency 2×10^{12} Hz in Jmol^{-1} is ______. (Nearest integer) (Given: $h = 6.626 \times 10^{-34}$ Js $N_A = 6.022 \times 10^{23}$ mol^{-1})

[NCERT : P L-39 | NV, Jan. 30, 2023 (I)]

9. The energy of one mole of photons of radiation of wavelength 300 nm is

(Given : $h = 6.63 \times 10^{-34}$ Js, $N_A = 6.02 \times 10^{23}$mol^{-1}, $c = 3 \times 10^8$ ms^{-1}) **[NCERT : P L-39 | June 24, 2022 (II)]**

(a) 235 kJ mol^{-1} (b) 325 kJ mol^{-1}

(c) 399 kJ mol^{-1} (d) 435 kJ mol^{-1}

10. If the wavelength for an electron emitted from H-atom is 3.3×10^{-10} m, then energy absorbed by the electron in its ground state compared to minimum energy required for its escape from the atom, is ______ times. (Nearest integer)

[Given : $h = 6.626 \times 10^{-34}$ J s]

Mass of electron = 9.1×10^{-31} kg

[NCERT : P L-39 | NV, July 28, 2022 (II)]

11. If the work function of a metal is 6.63×10^{-19} J, the maximum wavelength of the photon required to remove a photoelectron from the metal is ______ nm. (Nearest integer)

[Given : $h = 6.63 \times 10^{-34}$ J s, and $c = 3 \times 10^8$ m s^{-1}]

[NCERT : P L-41 | NV, June 28, 2022 (I)]

12. The minimum energy that must be possessed by photons in order to produce the photoelectric effect with platinum metal is:

[**Given:** The threshold frequency of platinum is 1.3×10^{15} s^{-1} and $h = 6.6 \times 10^{-34}$ J s.]

[NCERT : P L-41 | June 25, 2022 (II)]

(a) 3.21×10^{-14} J (b) 6.24×10^{-16} J

(c) 8.58×10^{-19} J (d) 9.76×10^{-20} J

13. A 50 watt bulb emits monochromatic red light of wavelength of 795 nm. The number of photons emitted per second by the bulb is $x \times 10^{20}$. The value of x is ______.

[Given : $h = 6.63 \times 10^{-34}$ Js and $c = 3.0 \times 10^8$ ms^{-1}]

[NCERT : P L-39 | NV, Sep. 1, 2021 (II)]

14. The number of photons emitted by a monochromatic (single frequency) infrared range finder of power 1 mW and wavelength of 1000 nm, in 0.1 second is $x \times 10^{13}$. The value of x is _____. (Nearest integer)
($h = 6.63 \times 10^{-34}$ Js, $c = 3.00 \times 10^8$ ms^{-1})
[NCERT : P L-39 | NV, Aug. 27, 2021 (II)]

15. A metal surface is exposed to 500 nm radiation. The threshold frequency of the metal for photoelectric current is 4.3×10^{14} Hz. The velocity of ejected electron is ________ $\times 10^5$ ms^{-1} (Nearest integer)
[Use : $h = 6.63 \times 10^{-34}$ Js, $m_e = 9.0 \times 10^{-31}$ kg]
[NCERT : P L-41 | NV, Aug. 26, 2021 (II)]

16. A source of monochromatic radiation of wavelength 400 nm provides 1000 J of energy in 10 seconds. When this radiation falls on the surface of sodium, $x \times 10^{20}$ electrons are ejected per second. Assume that wavelength 400 nm is sufficient for ejection of electron from the surface of sodium metal. The value of x is _______. (Nearest integer)
($h = 6.626 \times 10^{-34}$ Js)
[NCERT : P L-39 | NV, July 25, 2021 (I)]

17. Electromagnetic radiation of wavelength 663 nm is just sufficient to ionise the atom of metal A. The ionization energy of metal A in kJ mol^{-1} is _____. (Rounded-off to the nearest integer)
[$h = 6.63 \times 10^{-34}$ Js, $c = 3.00 \times 10^8$ ms–1, $N_A = 6.02 \times 10^{23}$ mol^{-1}] **[NCERT : P L-39 | NV, Feb. 25, 2021 (II)]**

18. The work function of sodium metal is 4.41×10^{-19} J. If photons of wavelength 300 nm are incident on the metal, the kinetic energy of the ejected electrons will be
($h = 6.63 \times 10^{-34}$ J s; $c = 3 \times 10^8$ m/s) _______ $\times 10^{-21}$ J.
[NCERT : P L-41 | NV, Sep. 02, 2020 (II)]

19. If p is the momentum of the fastest electron ejected from a metal surface after the irradiation of light having wavelength λ, then for $1.5p$ momentum of the photoelectron, the wavelength of the light should be:
(Assume kinetic energy of ejected photoelectron to be very high in comparison to work function):
[NCERT : P L-41 | April 8, 2019 (II)]

(a) $\frac{3}{4}\lambda$ (b) $\frac{1}{2}\lambda$ (c) $\frac{2}{3}\lambda$ (d) $\frac{4}{9}\lambda$

20. What is the work function of the metal if the light of wavelength 4000 Å generates photoelectrons of velocity 6×10^5 ms^{-1} from it ? **[NCERT : P L-41 | Jan. 12, 2019 (I)]**
(Mass of electron $= 9\times10^{-31}$ kg
Velocity of light $= 3\times10^8$ ms^{-1}
Planck's constant $= 6.626\times10^{-34}$ Js
Charge of electron $= 1.6\times10^{-19}$ JeV^{-1})
(a) 0.9 eV (b) 3.1 eV
(c) 2.1 eV (d) 4.0 eV

21. Which of the graphs shown below does not represent the relationship between incident light and the electron ejected from metal surface?
[NCERT : P L-41 | Jan. 10, 2019 (I)]

(a)
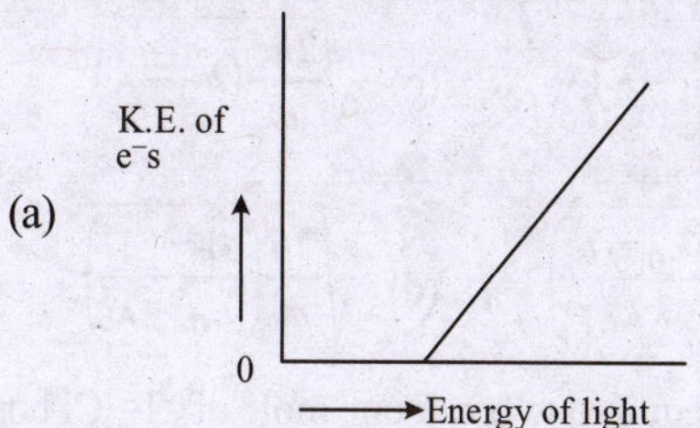

(b)
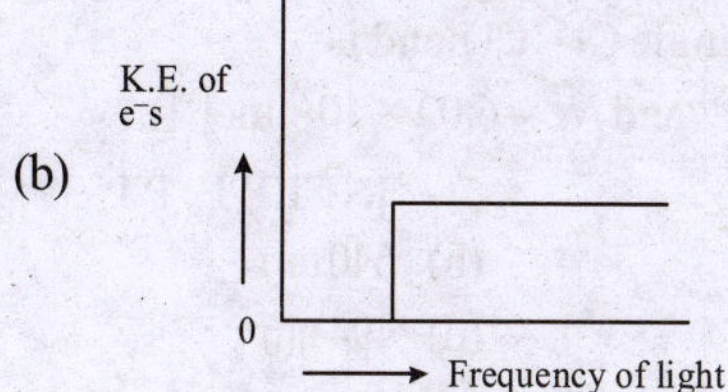

(c)
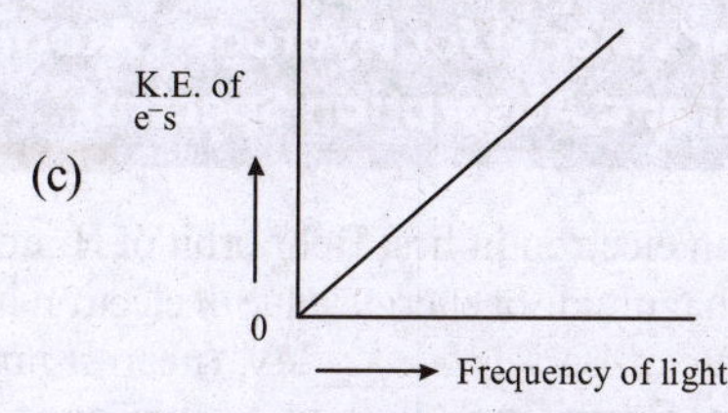

(d)
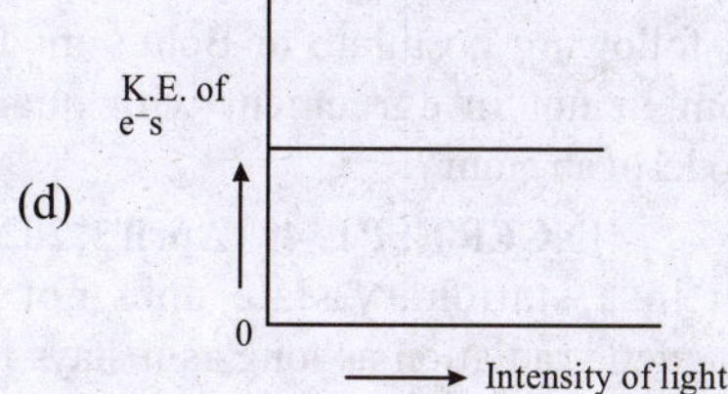

22. Ejection of the photoelectron from metal in the photoelectric effect experiment can be stopped by applying 0.5V when the radiation of 250 nm is used. The work function of the metal is :
[NCERT : P L-40 | Online April 15, 2018 (I)]
(a) 4 eV (b) 5.5 eV
(c) 4.5 eV (d) 5 eV

23. Excited hydrogen atom emits light in the ultraviolet region at 2.47×10^{15} Hz. With this frequency, the energy of a single photon is: **[NCERT : P L-41 | Online April 12, 2014]**
($h = 6.63 \times 10^{-34}$ Js)
(a) 8.041×10^{-40} J (b) 2.680×10^{-19} J
(c) 1.640×10^{-18} J (d) 6.111×10^{-17} J

24. Ionization energy of gaseous Na atoms is 495.5 kJ mol^{-1}. The lowest possible frequency of light that ionizes a sodium atom is
($h = 6.626 \times 10^{-34}$ Js, $N_A = 6.022 \times 10^{23}$ mol^{-1})
[NCERT : P L-41 | Online April 19, 2014]
(a) 7.50×10^4 s^{-1} (b) 4.76×10^{14} s^{-1}
(c) 3.15×10^{15} s^{-1} (d) 1.24×10^{15} s^{-1}

25. If λ_0 and λ be threshold wavelength and wavelength of incident light, the velocity of photoelectron ejected from the metal surface is:

[NCERT : P L-42 | Online April 11, 2014]

(a) $\sqrt{\frac{2h}{m}(\lambda_0 - \lambda)}$ (b) $\sqrt{\frac{2hc}{m}(\lambda_0 - \lambda)}$

(c) $\sqrt{\frac{2hc}{m}\left(\frac{\lambda_0 - \lambda}{\lambda\lambda_0}\right)}$ (d) $\sqrt{\frac{2h}{m}\left(\frac{1}{\lambda_0} - \frac{1}{\lambda}\right)}$

26. The energy required to break one mole of Cl – Cl bonds in Cl_2 is 242 kJ mol^{-1}. The longest wavelength of light capable of breaking a single Cl – Cl bond is
($c = 3 \times 10^8$ ms^{-1} and $N_A = 6.02 \times 10^{23}$ mol^{-1}).

[NCERT : P L-39 | 2010]

(a) 594 nm (b) 640 nm
(c) 700 nm (d) 494 nm

Topic 2 - Bohr's Model for Hydrogen Atom (Emission and Absorption Spectra)

27. The energy of an electron in first Bohr orbit of H-atom is –13.6 eV. The magnitude of energy value of electron in the first excited state of Be^{3+} is ______ eV. (nearest integer value) **[NCERT : P L-48 | NV, April 8, 2025 (II)]**

28. Which of the following postulate of Bohr's model of hydrogen atom in not in agreement with quantum mechanical model of an atom?

[NCERT : P L-49 | April 3, 2025 (I)]

(a) An atom in a stationary state does not emit electromagnetic radiation as long as it stays in the same state
(b) An atom can take only certain distinct energies E_1, E_2, E_3, etc. These allowed states of constant energy are called the stationary states of atom
(c) When an electron makes a transition from a higher energy stationary state to a lower energy stationary state, then it emits a photon of light
(d) The electron in a H atom's stationary state moves in a circle around the nucleus

29. According to Bohr's model of hydrogen atom, which of the following statement is **incorrect**?

[NCERT : P L-48 | April 2, 2025 (I)]

(a) Radius of 3rd orbit is nine times larger than that of 1st orbit.
(b) Radius of 8th orbit is four times larger than that of 4th orbit.
(c) Radius of 6th orbit is three time larger than that of 4th orbit.
(d) Radius of 4th orbit is four times larger than that of 2nd orbit.

30. For hydrogen like species, which of the following graphs provides the most appropriate representation of E vs Z plot for a constant n? **[NCERT : P L-48 Jan. 29, 2025 (II)]**
[E : energy of the stationary state, Z : atomic number, n = principal quantum number]

(a) 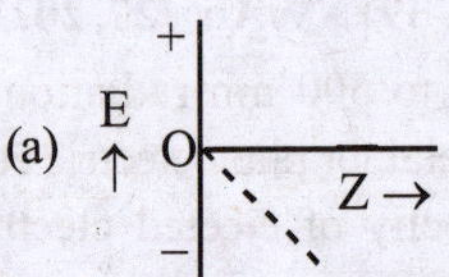

(b) 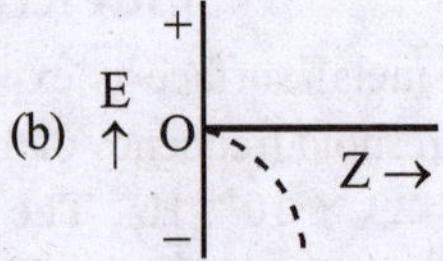

(c) 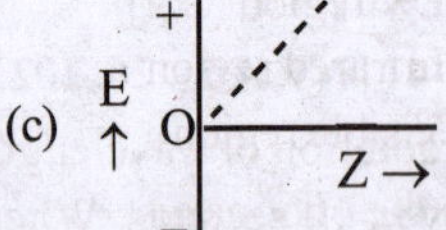

(d) 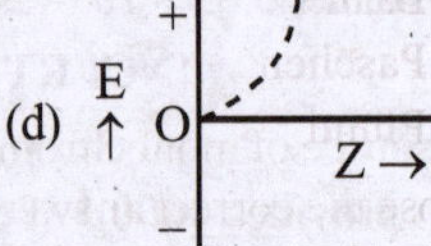

31. Heat treatment of muscular pain involves radiation of wavelength of about 900 nm. Which spectral line of H atom is suitable for this?
Given: Rydberg constant
$R_H = 10^5$ cm^{-1}, $h = 6.6 \times 10^{-34}$ J s, $c = 3 \times 10^8$ m/s)

[NCERT : P L-45 | Jan. 23, 2025 (I)]

(a) Paschen series, $\infty \rightarrow 3$ (b) Lyman series, $\infty \rightarrow 1$
(c) Balmer series, $\infty \rightarrow 2$ (d) Paschen series, $5 \rightarrow 3$

32. Radius of the first excited state of Helium ion is given as:
$a_0 \rightarrow$ radius of first stationary state of hydrogen atom.

[NCERT : P L-48 | Jan. 22, 2025 (I)]

(a) $r = \frac{a_0}{2}$ (b) $r = \frac{a_0}{4}$
(c) $r = 4a_0$ (d) $r = 2a_0$

33. A hypothetical electromagnetic wave is show below.

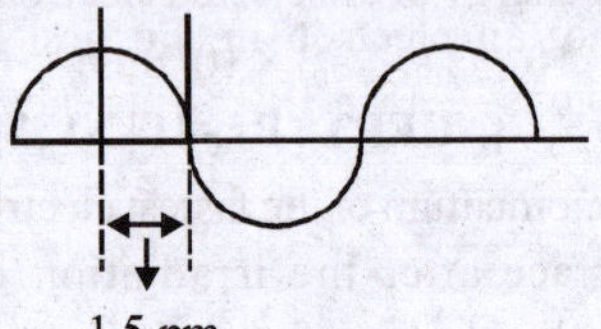

The frequency of the wave is $x \times 10^{19}$ Hz.
x = ______ (nearest integer)

[NCERT : P L-38 | NV, April 8, 2024 (I)]

34. Wavenumber for a radiation having 5800 Å wavelength is $x \times 10$ cm^{-1}. The value of x is ______.

[NCERT : P L-39 | NV, April 8, 2024 (II)]

35. For hydrogen atom, energy of an electron in first excited state is – 3.4 eV, K.E. of the same electron of hydrogen atom is x eV. Value of x is ______ $\times 10^{-1}$ eV. (Nearest integer)

[NCERT : P L-46 | NV, April 6, 2024 (II)]

36. The value of Rydberg constant (R_H) is 2.18×10^{-18} J. The velocity of electron having mass 9.1×10^{-31} kg in Bohr's first orbit of hydrogen atom = ... $\times 10^5$ ms^{-1} (nearest integer) **[NCERT : P L-47 | NV, April 5, 2024 (I)]**

37. The de-Broglie's wavelength of an electron in the 4th orbit is ______ πa_0. (a_0 = Bohr's radius)

[NCERT : P L-48 | NV, April 4, 2024 (I)]

38. Number of spectral lines obtained in He^+ spectra, when an electron makes transition from fifth excited state to first excited state will be
[NCERT : P L-44 | NV, Jan. 30, 2024 (II)]

39. Match List I with List II
[NCERT : P L-45 | Jan. 29, 2024 (II)]

LIST I (Spectral Series for Hydrogen)		LIST II (Spectral Region / Higher Energy State)	
A.	Lyman	I.	Infrared region
B.	Balmer	II.	UV region
C.	Paschen	III.	Infrared region
D.	Pfund	IV.	Visible region

Choose the correct answer from the options given below:
(a) A-I, B-III, C-II, D-IV (b) A-II, B-III C-I, D-IV
(c) A-II, B-IV, C-III, D-I (d) A-I, B-II, C-III, D-IV

40. Given below are two statements:
[NCERT : P L-46 | April 15, 2023 (I)]
Statement – I : According to Bohr's model of hydrogen atom, the angular momentum of an electron in a given stationary state is quantised.
Statement – II : The concept of electron in Bohr's orbit, violates the Heisenberg uncertainty principle.
In the light of the above statements, choose the most appropriate answer from the options given below:
(a) Both Statement I and Statement II are correct.
(b) Statement I is correct but Statement II is incorrect.
(c) Statement I is incorrect but Statement II is correct.
(d) Both Statement I and Statement II are incorrect.

41. The energy of an electron in the first Bohr orbit of hydrogen atom is -2.18×10^{-18} J. Its energy in the third Bohr orbit is ______.
[NCERT : P L-46 | April 13, 2023 (I)]
(a) $\frac{1}{27}$ of this value
(b) One third of this value
(c) Three times of this value
(d) $\frac{1}{9}$th of this value

42.
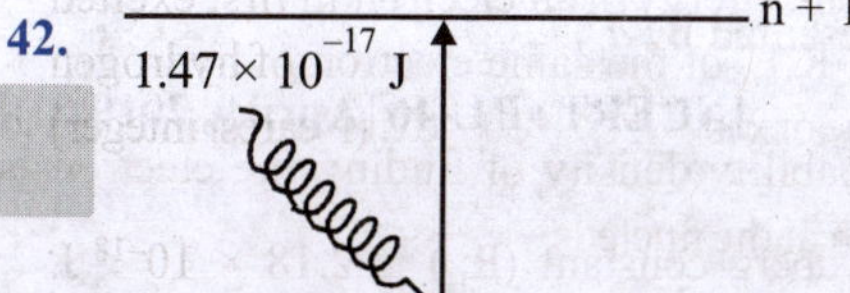

The electron in the nth orbit of Li^{2+} is excited to (n + 1) orbit using the radiation of energy 1.47×10^{-17} J (as shown in the diagram). The value of n is ______.
Given $R_H = 2.18 \times 10^{-18}$ J
[NCERT : P L-48 | NV, April 10, 2023 (II)]

43. The number of following statement/s which is/are incorrect is ______
[NCERT : P L-48 | NV, April 08, 2023 (I)]
(A) Line emission spectra are used to study the electronic structure
(B) The emission spectra of atoms in the gas phase show a continuous spread of wavelength from red to violet
(C) An absorption spectrum is like the photographic negative of an emission spectrum
(D) The element helium was discovered in the sun by spectroscopic method

44. Which transition in the hydrogen spectrum would have the same wavelength as the Balmer type transition from n = 4 to n = 2 of He^+ spectrum
[NCERT : P L-45 | Jan. 31, 2023 (I)]
(a) n = 2 to n = 1 (b) n = 1 to n = 3
(c) n = 1 to n = 2 (d) n = 3 to n = 4

45. The shortest wavelength of hydrogen atom in Lyman series is λ. The longest wavelength in Balmer series of He^+ is **[NCERT : P L-45 | Jan. 29, 2023 (I)]**
(a) $\frac{5}{9\lambda}$ (b) $\frac{9\lambda}{5}$ (c) $\frac{36\lambda}{5}$ (d) $\frac{5\lambda}{9}$

46. Assume that the radius of the first Bohr orbit of hydrogen atom is 0.6 Å. The radius of the third Bohr orbit of He^+ is ______ picometer. (Nearest Integer)
[NCERT : P L-46 | NV, Jan. 29, 2023 (II)]

47. The radius of the 2nd orbit of Li^{2+} is x. The expected radius of the 3rd orbit of Be^{3+} is
[NCERT : P L-46 | Jan. 25, 2023 (I)]
(a) $\frac{9}{4}x$ (b) $\frac{4}{9}x$ (c) $\frac{27}{16}x$ (d) $\frac{16}{27}x$

48. If wavelength of the first line of the Paschen series of hydrogen atom is 720 nm, then the wavelength of the second line of this series is ______ nm.
(Nearest integer) **[NCERT : P L-45 | NV, Jan. 24, 2023 (I)]**

49. When the excited electron of a H atom from n = 5 drops to the ground state, the maximum number of emission lines observed are ______.
[NCERT : P L-48 | NV, July 25, 2022 (II)]

50. If the radius of the 3rd Bohr's orbit of hydrogen atom is r_3 and the radius of 4th Bohr's orbit is r_4.
Then : **[NCERT : P L-46 | June 26, 2022 (I)]**
(a) $r_4 = \frac{9}{16}r_3$ (b) $r_4 = \frac{16}{9}r_3$
(c) $r_4 = \frac{3}{4}r_3$ (d) $r_4 = \frac{4}{3}r_3$

51. The longest wavelength of light that can be used for the ionisation of lithium atom (Li) in its ground state is $x \times 10^{-8}$ m. The value of x is ______. (Nearest Integer)
(Given : Energy of the electron in the first shell of the hydrogen atom is -2.2×10^{-18} J; $h = 6.63 \times 10^{-34}$ Js and $c = 3 \times 10^8$ ms^{-1}) **[NCERT : P L-48 | NV, June 25, 2022 (I)]**

52. The kinetic energy of an electron in the second Bohr orbit of a hydrogen atom is equal to $\frac{h^2}{xma_0^2}$. The value of $10x$ is ________. (a_0 is radius of Bohr's orbit) (Nearest integer) [Given : $\pi = 3.14$]

[NCERT : P L-46 | NV, Aug. 27, 2021 (I)]

53. Given below are two statements.

Statement I : According to Bohr's model of an atom, qualitatively the magnitude of velocity of electron increases with decrease in positive charges on the nucleus as there is no strong hold on the electron by the nucleus.

Statement II : According to Bohr's model of an atom, qualitatively the magnitude of velocity of electron increases with decrease in principal quantum number.

In the light of the above statements, choose the most appropriate answer from the options given below

[NCERT : P L-46 | Aug. 26, 2021 (I)]

(a) Both Statement I and Statement II are false.
(b) Both Statement I and Statement II are true.
(c) Statement I is false but Statement II is true.
(d) Statement I is true but Statement II is false.

54. According to Bohr's atomic theory:

[NCERT : P L-46 | Feb. 24, 2021 (II)]

(A) Kinetic energy of electron is $\propto \frac{Z^2}{n^2}$.

(B) The product of velocity (v) of electron and principal quantum number (n), 'vn' $\propto Z^2$.

(C) Frequency of revolution of electron in an orbit is $\propto \frac{Z^3}{n^3}$.

(D) Coulombic force of attraction on the electron is $\propto \frac{Z^3}{n^4}$.

Choose the most appropriate answer from the options given below :

(a) (C) only (b) (A) only
(c) (A), (C) and (D) only (d) (A) and (D) only

55. The difference between the radii of 3rd and 4th orbits of Li^{2+} is ΔR_1. The difference between the radii of 3rd and 4th orbits of He^+ is ΔR_2. Ratio $\Delta R_1 : \Delta R_2$ is :

[NCERT : P L-46 | Sep. 05, 2020 (I)]

(a) 8 : 3 (b) 3 : 8
(c) 2 : 3 (d) 3 : 2

56. The region in the electromagnetic spectrum where the Balmer series lines appear is :

[NCERT : P L-45 | Sep. 04, 2020 (I)]

(a) Visible (b) Microwave
(c) Infrared (d) Ultraviolet

57. The shortest wavelength of H atom in the Lyman series is λ_1. The longest wavelength in the Balmer series of He^+ is:

[NCERT : P L-45 | Sep. 04, 2020 (II); Same Online, April 8, 2017]

(a) $\frac{36\lambda_1}{5}$ (b) $\frac{5\lambda_1}{9}$ (c) $\frac{9\lambda_1}{5}$ (d) $\frac{27\lambda_1}{5}$

58. The radius of the second Bohr orbit, in terms of the Bohr radius, a_0, in Li^{2+} is: **[Jan. 08, 2020 (II)]**

(a) $\frac{2a_0}{3}$ (b) $\frac{4a_0}{9}$ (c) $\frac{4a_0}{3}$ (d) $\frac{2a_0}{9}$

59. For the Balmer series in the spectrum of H atom, $\bar{v} = R_H\left\{\frac{1}{n_1^2} - \frac{1}{n_2^2}\right\}$, the correct statements among (I) to (IV) are: **[NCERT : P L-45 | Jan. 08, 2020 (I)]**

(I) As wavelength decreases, the lines in the series converge
(II) The integer n_1 is equal to 2
(III) The lines of longest wavelength corresponds to $n_2 = 3$
(IV) The ionization energy of hydrogen can be calculated from wave number of these lines

(a) (I), (III), (IV) (b) (I), (II), (III)
(c) (I), (II), (IV) (d) (II), (III), (IV)

60. The radius of the second Bohr orbit in terms of the Bohr radius, a_0, in Li^{2+} is: **[NCERT : P L-46 | Jan. 08, 2020 (II)]**

(a) $\frac{2a_0}{3}$ (b) $\frac{4a_0}{9}$ (c) $\frac{4a_0}{3}$ (d) $\frac{2a_0}{9}$

61. The ratio of the shortest wavelength of two spectral series of hydrogen spectrum is found to be about 9. The spectral series are : **[NCERT : P L-46 | April 10, 2019 (II)]**

(a) Lyman and Paschen (b) Balmer and Brackett
(c) Brackett and Pfund (d) Paschen and Pfund

62. For any given series of spectral lines of atomic hydrogen, let $\Delta\bar{v} = \bar{v}_{max} - \bar{v}_{min}$ be the difference in maximum and minimum frequencies in cm^{-1}. The ratio $\Delta\bar{v}_{Lyman}/\Delta\bar{v}_{Balmer}$ is :

[NCERT : P L-46 | April 9, 2019 (I)]

(a) 4 : 1 (b) 9 : 4 (c) 5 : 4 (d) 27 : 5

63. Which one of the following about an electron occupying the 1s orbital in a hydrogen atom is incorrect? (The Bohr radius is represented by a_0).

[NCERT : P L-46 | April 9, 2019 (II)]

(a) The probability density of finding the electron is maximum at the nucleus.
(b) The electron can be found at a distance $2a_0$ from the nucleus.
(c) The magnitude of the potential energy is double that of its kinetic energy on an average.
(d) The total energy of the electron is maximum when it is at a distance a_0 from the nucleus.

64. Heat treatment of muscular pain involves radiation of wavelength of about 900 nm. Which spectral line of H-atom is suitable for this purpose ?
[NCERT : P L-45 | Jan. 11, 2019 (I)]

$[R_H = 1 \times 10^5\ cm^{-1}, h = 6.6 \times 10^{-34}$ Js, $c = 3 \times 10^8 ms^{-1}]$

(a) Paschen, $\infty \to 3$ (b) Paschen, $5 \to 3$
(c) Balmer, $\infty \to 2$ (d) Lyman, $\infty \to 1$

65. The ground state energy of hydrogen atom is –13.6 eV. The energy of second excited state of He^+ ion in eV is:
[NCERT : P L-48 | Jan. 10, 2019 (II), Same 2002]

(a) –54.4 (b) –3.4
(c) –6.04 (d) –27.2

66. For emission line of atomic hydrogen from $n_i = 8$ to $n_f = n$, the plot of wave number ($\bar{\nu}$) against $\left(\frac{1}{n^2}\right)$ will be (The Rydberg constant, R_H is in wave number unit)
[NCERT : P L-48 | Jan. 9, 2019 (I)]

(a) Linear with intercept $-R_H$
(b) Non linear
(c) Linear with slope R_H
(d) Linear with slope $-R_H$

67. Which of the following statements is false?
[NCERT : P L-44 | Online April 16, 2018]

(a) Splitting of spectral lines in electrical field is called Stark effect
(b) Frequency of emitted radiation from a black body goes from a lower wavelength to higher wavelength as the temperature increases
(c) Photon has momentum as well as wavelength
(d) Rydberg constant has unit of energy

68. The electron in the hydrogen atom undergoes transition from higher orbitals to orbital of radius 211.6 pm. This transition is associated with :
[NCERT : P L-48 | Online April 9, 2017]

(a) Lyman series (b) Balmer series
(c) Paschen series (d) Brackett series

69. If the shortest wavelength in Lyman series of hydrogen atom is A, then the longest wavelength in Paschen series of He^+ is :
[Online April 8, 2017]

(a) $\frac{5A}{9}$ (b) $\frac{9A}{5}$ (c) $\frac{36A}{5}$ (d) $\frac{36A}{7}$

70. The radius of the second Bohr orbit for hydrogen atom is : (Plank's const. $h = 6.6262 \times 10^{-34}$ Js ; mass of electron $= 9.1091 \times 10^{-31}$ kg ; charge of electron $e = 1.60210 \times 10^{-19}$ C ; permittivity of vaccum $\epsilon_0 = 8.854185 \times 10^{-12}\ kg^{-1} m^{-3} A^2$)
[NCERT : P L-48 | 2017]

(a) 1.65Å (b) 4.76Å
(c) 0.529Å (d) 2.12Å

71. If the shortest wavelength in Lyman series of hydrogen atom is A, then the longest wavelength in Paschen series of He^+ is :
[NCERT : P L-45 | Online April 8, 2017]

(a) $\frac{5A}{9}$ (b) $\frac{9A}{5}$ (c) $\frac{36A}{5}$ (d) $\frac{36A}{7}$

72. Which of the following is the energy of a possible excited state of hydrogen ?
[NCERT : P L-48 | 2015]

(a) –3.4 eV (b) +6.8 eV
(c) +13.6 eV (d) –6.8 eV

73. The energy of an electron in first Bohr orbit of H-atom is – 13.6 eV. The energy value of electron in the excited state of Li^{2+} is:
[NCERT : P L-48 | Online April 9, 2014]

(a) –27.2 eV (b) 30.6 eV
(c) –30.6 eV (d) 27.2 eV

74. Based on the equation: $\Delta E = -2.0 \times 10^{-18} J\left(\frac{1}{n_2^2} - \frac{1}{n_1^2}\right)$ the wavelength of the light that must be absorbed to excite hydrogen electron from level $n = 1$ to level $n = 2$ will be: ($h = 6.625 \times 10^{-34}$ Js, $c = 3 \times 10^8\ ms^{-1}$)
[NCERT : P L-43 | Online April 11, 2014]

(a) 1.325×10^{-7} m (b) 1.325×10^{-10} m
(c) 2.650×10^{-7} m (d) 5.300×10^{-10} m

75. If m and e are the mass and charge of the revolving electron in the orbit of radius r for hydrogen atom, the total energy of the revolving electron will be:
[NCERT : P L-46 | Online April 12, 2014]

(a) $\frac{1}{2}\frac{e^2}{r}$ (b) $-\frac{e^2}{r}$ (c) $\frac{me^2}{r}$ (d) $-\frac{1}{2}\frac{e^2}{r}$

76. Energy of an electron is given by $E = -2.178 \times 10^{-18} J\left(\frac{Z^2}{n^2}\right)$.

Wavelength of light required to excite an electron in an hydrogen atom from level $n = 1$ to $n = 2$ will be :
[NCERT : P L-48 | 2013]

($h = 6.62 \times 10^{-34}$ Js and $c = 3.0 \times 10^8\ ms^{-1}$)

(a) 1.214×10^{-7} m (b) 2.816×10^{-7} m
(c) 6.500×10^{-7} m (d) 8.500×10^{-7} m

77. The wave number of the first emission line in the Balmer series of H-spectrum is : (R = Rydberg constant) :
[NCERT : P L-48 | Online April 22, 2013]

(a) $\frac{5}{36}R$ (b) $\frac{9}{400}R$ (c) $\frac{7}{6}R$ (d) $\frac{3}{4}R$

78. Energy of an electron is given by $E = -2.178 \times 10^{-18} J\left(\frac{Z^2}{n^2}\right)$.

Wavelength of light required to excite an electron in an hydrogen atom from level $n = 1$ to $n = 2$ will be :
[NCERT : P L-48 |2013]

($h = 6.62 \times 10^{-34}$ Js and $c = 3.0 \times 10^8\ ms^{-1}$)

(a) 1.214×10^{-7} m (b) 2.816×10^{-7} m
(c) 6.500×10^{-7} m (d) 8.500×10^{-7} m

79. The limiting line in Balmer series will have a frequency of (Rydberg constant, $R_\infty = 3.29 \times 10^{15}$ cycles/s)
[NCERT : P L-45 | Online May 7, 2012]

(a) $8.22 \times 10^{14}\ s^{-1}$ (b) $3.29 \times 10^{15}\ s^{-1}$
(c) $3.65 \times 10^{14}\ s^{-1}$ (d) $5.26 \times 10^{13}\ s^{-1}$

80. The frequency of light emitted for the transition $n = 4$ to $n = 2$ of the He^+ is equal to the transition in H atom corresponding to which of the following ?
[NCERT : P L-48 | 2011RS]
(a) $n = 2$ to $n = 1$ (b) $n = 3$ to $n = 2$
(c) $n = 4$ to $n = 3$ (d) $n = 3$ to $n = 1$

81. Ionisation energy of He^+ is 19.6×10^{-18} J atom^{-1}. The energy of the first stationary state ($n = 1$) of Li^{2+} is
[NCERT : P L-48 | 2010]
(a) 4.41×10^{-16} J atom^{-1} (b) -4.41×10^{-17} J atom^{-1}
(c) -2.2×10^{-15} J atom^{-1} (d) 8.82×10^{-17} J atom^{-1}

82. The ionization enthalpy of hydrogen atom is 1.312×10^6 J mol^{-1}. The energy required to excite the electron in the atom from $n = 1$ to $n = 2$ is **[NCERT : P L-48 | 2008]**
(a) 8.51×10^5 J mol^{-1} (b) 6.56×10^5 J mol^{-1}
(c) 7.56×10^5 J mol^{-1} (d) 9.84×10^5 J mol^{-1}

83. According to Bohr's theory, the angular momentum of an electron in 5th orbit is **[NCERT : P L-46 | 2006]**
(a) $10\,h/\pi$ (b) $2.5\,h/\pi$
(c) $25\,h/\pi$ (d) $1.0\,h/\pi$

84. The wavelength of the radiation emitted, when in a hydrogen atom electron falls from infinity to stationary state 1, would be (Rydberg constant = 1.097×10^7 m^{-1})
[NCERT : P L-45 | 2004]
(a) 406 nm (b) 192 nm
(c) 91 nm (d) 9.1×10^{-8} nm

85. In Bohr series of lines of hydrogen spectrum, the third line from the red end corresponds to which one of the following inter-orbit jumps of the electron for Bohr orbits in an atom of hydrogen **[NCERT : P L-45 | 2003]**
(a) $5 \rightarrow 2$ (b) $4 \rightarrow 1$
(c) $2 \rightarrow 5$ (d) $3 \rightarrow 2$

86. In a hydrogen atom, if energy of an electron in ground state is 13.6. eV, then that in the 2nd excited state is
[NCERT : P L-48 | 2002]
(a) 1.51 eV (b) 3.4 eV (c) 6.04 eV (d) 13.6 eV.

87. In a hydrogen atom, if energy of an electron in ground state is 13.6. eV, then that in the 2nd excited state is
[NCERT : P L-48 | 2002]
(a) 1.51 eV (b) 3.4 eV (c) 6.04 eV (d) 13.6 eV.

Topic 3 - Dual Behaviour of Matter and Heisenberg Uncertainty Principle

88. Given below are two statements :
[NCERT : P L-50 | Jan. 29, 2025 (II)]
Statement (I): It is impossible to specify simultaneously with arbitrary precision, both the linear momentum and the position of a particle.
Statement (II) : If the uncertainty in the measurement of position and uncertainty in measurement of momentum are equal for an electron, then the uncertainty in the measurement of velocity is $\geq \sqrt{\frac{h}{\pi}} \times \frac{1}{2m}$.
In the light of the above statements, choose the correct answer from the options given below :
(a) Statement I is true but Statement II is false.
(b) Both Statement I and Statement II are true.
(c) Statement I is false but Statement II is true.
(d) Both Statement I and Statement II are false.

89. If a_0 is denoted as the Bohr radius of hydrogen atom, then what is the de-Broglie wavelength (λ) of the electron present in the second orbit of hydrogen atom ? [n : any integer]
[NCERT : P L-50 | Jan. 29, 2025 (I)]
(a) $\frac{2a_0}{n\pi}$ (b) $\frac{8\pi a_0}{n}$ (c) $\frac{4\pi a_0}{n}$ (d) $\frac{4n}{\pi a_0}$

90. Based on Heisenberg's uncertainty principle, the uncertainty in the velocity of the electron to be found within an atomic nucleus of diameter 10^{-15} m is ______ $\times 10^9$ ms^{-1} (nearest integer)
[Given : mass of electron = 9.1×10^{-31} kg, Plank's constant (h) = 6.626×10^{-34} Js]
(Value of π = 3.14) **[NCERT : P L-18 | NV, April 9, 2024 (II)]**

91. Frequency of the de-Broglie wave of electron in Bohr's first orbit of hydrogen atom is ______ $\times 10^{13}$ Hz (nearest integer).
[Given : R_H (Rydberg constant) = 2.18×10^{-18} J. h (Plank's constant) = 6.6×10^{-34} J.s.]
[NCERT : P L-18 | NV, April 6, 2024 (I)]

92. According to the wave particle duality of matter by de - Brogile, which of the following graph plot presents ,most appropriate relationship between wavelength of electron (λ) and momentum of electron (p)?
[NCERT : P L-50 | Feb. 1, 2024 (I)]

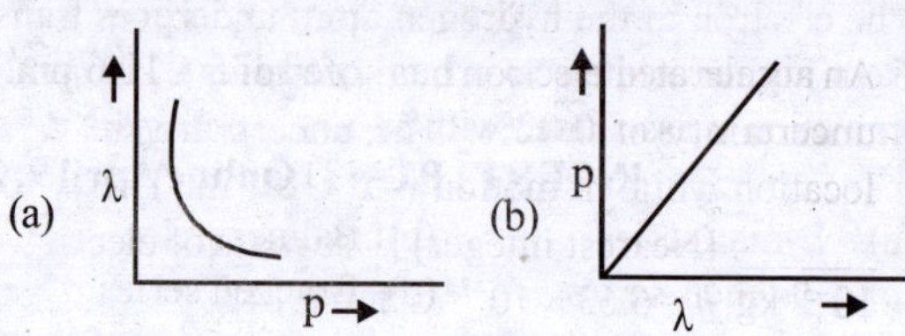

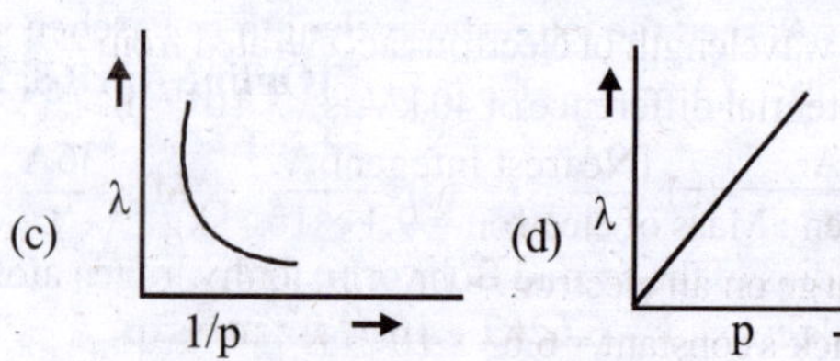

93. The wavelength of an electron of kinetic energy 4.50×10^{-29} J is .... $\times 10^{-5}$ m. (Nearest integer)
Given : mass of electron is 9×10^{-31} kg, h = 6.6×10^{-34} J s
[NCERT : P L-50 | NV, April 06, 2023 (I)]

94. If the radius of the first orbit of hydrogen atom a_0, then de-Broglie's wavelength of electron in 3rd orbit is
[NCERT : P L-48 | April 06, 2023 (II)]
(a) $\frac{\pi a_0}{6}$ (b) $\frac{\pi a_0}{3}$ (c) $6\pi a_0$ (d) $3\pi a_0$

95. Electrons in a cathode ray tube have been emitted with a velocity of 1000 m s^{-1}. The number of following statements which is/are true about the emitted radiation is ______.
Given : $h = 6 \times 10^{-34}$ J s, $m_e = 9 \times 10^{-31}$ kg.
[NCERT : P L-50 | NV, Feb. 01, 2023 (I)]
(A) The deBroglie wavelength of the electron emitted is 666.67 nm.
(B) The characteristic of electrons emitted depend upon the material of the electrodes of the cathode ray tube.
(C) The cathode rays start from cathode and move towards anode.
(D) The nature of the emitted electrons depends on the nature of the gas present in cathode ray tube.

96. The minimum uncertainty in the speed of an electron in an one-dimensional region of length $2a_0$ (Where a_0 = Bohr radius 52.9 pm) is_____km s^{-1}.
(Given : Mass of electron $= 9.1 \times 10^{-31}$ kg, Planck's constant $h = 6.63 \times 10^{-34}$ Js)
[NCERT : P L-50 | NV, July 29, 2022 (I)]

97. The wavelength of an electron and a neutron will become equal when the velocity of the electron is x times the velocity of neutron. The value of x is ______. (Nearest Integer) (Mass of electron is 9.1×10^{-31} kg and mass of neutron is 1.6×10^{-27} kg)
[NCERT : P L-50 | NV, July 26, 2022 (I)]

98. If the uncertainty in velocity and position of a minute particle in space are, 2.4×10^{-26} (ms^{-1}) and 10^{-7} (m) respectively. The mass of the particle in g is ________ (Nearest integer)
(Given : $h = 6.626 \times 10^{-34}$ Js)
[NCERT : P L-50 | NV, June 27, 2022 (I)]

99. An accelerated electron has speed of 5×10^6 ms^{-1} with an uncertainty of 0.02%. The uncertainty in finding its location while in motion is $x \times 10^{-9}$m. The value of x is ____. (Nearest integer) [Use mass of electron = 9.1×10^{-31} kg, $h = 6.63 \times 10^{-34}$Js, $\pi = 3.14$]
[NCERT : P L-50 | NV, July 25, 2021 (II)]

100. The wavelength of electrons accelerated from rest through a potential difference of 40 kV is $x \times 10^{-12}$ m. The value of x is ______. (Nearest integer)
Given : Mass of electron $= 9.1 \times 10^{-31}$ kg
Charge on an electron $= 1.6 \times 10^{-19}$ C
Planck's constant $= 6.63 \times 10^{-34}$ Js
[NCERT : P L-50 | NV, July 20, 2021 (II)]

101. When light of wavelength 248 nm falls on a metal of threshold energy 3.0 eV, the de-Broglie wavelength of emitted electrons is ____ Å. (Round off to the Nearest Integer).
[Use : $\sqrt{3} = 1.73, h = 6.63 \times 10^{-34}$ Js,
$m_e = 9.1 \times 10^{-31}$kg; $c = 3.0 \times 10^8$ ms^{-1}; 1 eV $= 1.6 \times 10^{-19}$ J]
[NCERT : P L-50 | NV, March 16, 2021 (I)]

102. A ball weighing 10 g is moving with a velocity of 90 ms^{-1}. If the uncertainty in its velocity is 5%, then the uncertainty in its position is __________ $\times 10^{-33}$ m. (Rounded off to the nearest integer)
[Given : $h = 6.63 \times 10^{-34}$ Js]
[NCERT : P L-50 | NV, Feb. 26, 2021 (II)]

103. A proton and a Li^{3+} nucleus are accelerated by the same potential. If λ_{Li} and λ_P denote the de Broglie wavelengths of Li^{3+} and proton respectively, then the value of $\frac{\lambda_{Li}}{\lambda_P}$ is $x \times 10^{-1}$. The value of x is ______
(Rounded off to the nearest integer)
(Mass of Li^{3+} = 8.3 times mass of proton)
[NCERT : P L-50 | NV, Feb. 24, 2021 (I)]

104. The figure that is **not** a direct manifestation of the quantum nature of atoms is : **[Sep. 02, 2020 (I)]**

(a)
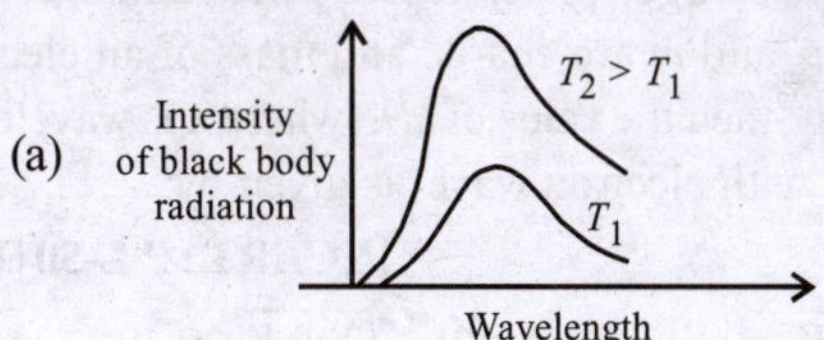

(b)
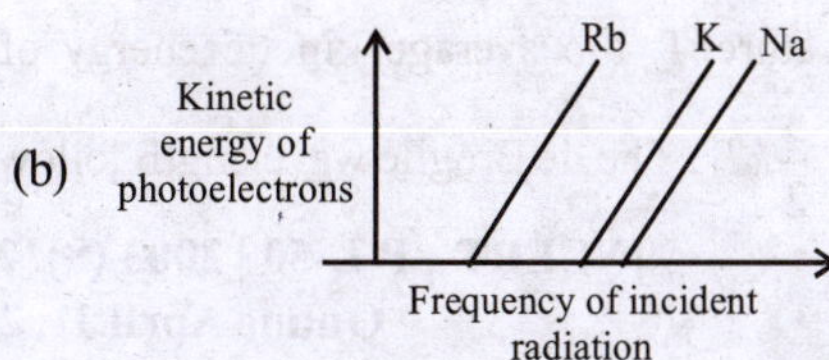

(c)
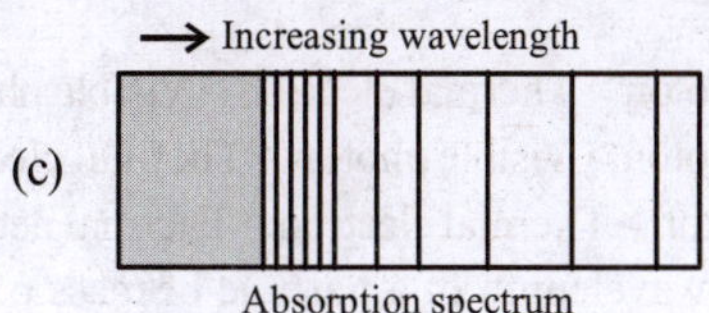

(d)
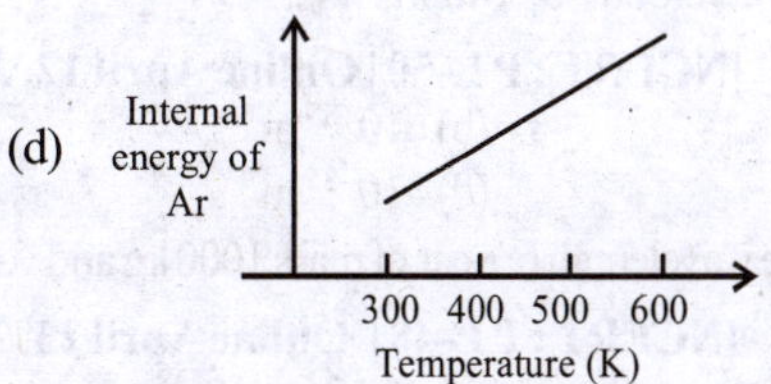

105. The de Broglie wavelength of an electron in the 4th Bohr orbit is: **[NCERT : P L-50 | Jan. 09, 2020 (I)]**
(a) $2\pi a_0$ (b) $4\pi a_0$
(c) $6\pi a_0$ (d) $8\pi a_0$

106. If the de Broglie wavelength of the electron in n^{th} Bohr orbit in a hydrogenic atom is equal to $1.5\ \pi a_0$ (a_0 is Bohr radius), then the value of n/Z is :
[NCERT : P L-50 | Jan. 12, 2019 (II)]
(a) 0.40 (b) 1.50 (c) 1.0 (d) 0.75

107. The de Broglie wavelength (λ) associated with a photoelectron varies with the frequency (v) of the incident radiation as, [v_0 is threshold frequency]:

[NCERT : P L-50 | Jan. 11, 2019 (II)]

(a) $\lambda \propto \frac{1}{(v - v_0)}$ (b) $\lambda \propto \frac{1}{(v - v_0)^{\frac{1}{4}}}$

(c) $\lambda \propto \frac{1}{(v - v_0)^{\frac{3}{2}}}$ (d) $\lambda \propto \frac{1}{(v - v_0)^{\frac{1}{2}}}$

108. The de-Broglie's wavelength of electron present in first Bohr orbit of 'H' atom is:

[NCERT : P L-48 | Online April 15, 2018 (II)]

(a) 4×0.529Å (b) $2\pi \times 0.529$Å

(c) $\frac{0.529}{2\pi}$Å (d) 0.529Å

109. A stream of electrons from a heated filament was passed between two charged plates kept at a potential difference V volt. If 'e' and m are charge and mass of an electron, respectively, then the value of h/λ (where λ is wavelength associated with electron wave) is given by:

[NCERT : P L-50 |2016]

(a) $\sqrt{meV}$ (b) $\sqrt{2meV}$

(c) meV (d) $2meV$

110. At temperature T, the average kinetic energy of any particle is $\frac{3}{2}kT$. The de Broglie wavelength follows the order :

[NCERT : P L-50 | 2003 (S), 2013, Online April 11, 2015]

(a) Visible photon > Thermal neutron > Thermal electron

(b) Thermal proton > Thermal electron > Visible photon

(c) Thermal proton > Visible photon > Thermal electron

(d) Visible photon > Thermal electron > Thermal neutron

111. The de-Broglie wavelength of a particle of mass 6.63 g moving with a velocity of 100 ms^{-1} is:

[NCERT : P L-50 | Online April 12, 2014]

(a) 10^{-33} m (b) 10^{-35} m

(c) 10^{-31} m (d) 10^{-25} m

112. The de Broglie wavelength of a car of mass 1000 kg and velocity 36 km/hr is: **[NCERT : P L-48 | Online April 23, 2013]**

(a) 6.626×10^{-34} m (b) 6.626×10^{-38} m

(c) 6.626×10^{-31} m (d) 6.626×10^{-30} m

113. If the kinetic energy of an electron is increased four times, the wavelength of the de-Broglie wave associated with it would become **[NCERT : P L-50 | Online May 19, 2012]**

(a) one fourth (b) half

(c) four times (d) two times

114. If the radius of first orbit of H atom is a_0, the de-Broglie wavelength of an electron in the third orbit is

[NCERT : P L-48 | Online May 12, 2012]

(a) $4\pi a_0$ (b) $8\pi a_0$ (c) $6\pi a_0$ (d) $2\pi a_0$

115. Calculate the wavelength (in nanometer) associated with a proton moving at 1.0×10^3 m-s^{-1}. (Mass of proton = 1.67×10^{-27} kg and $h = 6.63 \times 10^{-34}$ Js) **[NCERT : P L-50 | 2009]**

(a) 0.40 nm (b) 2.5 nm

(c) 14.0 nm (d) 0.32 nm

116. In an atom, an electron is moving with a speed of 600 m/s with an accuracy of 0.005%. Certainity with which the position of the electron can be located is ($h = 6.6 \times 10^{-34}$ kg m^{2}s^{-1}, mass of electron, $e_m = 9.1 \times 10^{-31}$ kg) :

[NCERT : P L-50 | 2002 (S), 2006(S), 2009]

(a) 5.10×10^{-3} m (b) 1.92×10^{-3} m

(c) 3.84×10^{-3} m (d) 1.52×10^{-4} m

117. Uncertainty in the position of an electron (mass = 9.1×10^{-31} kg) moving with a velocity 300 ms^{-1}, accurate upto 0.001% will be **[NCERT : P L-50 | 2006]**

(a) 1.92×10^{-2} m (b) 3.84×10^{-2} m

(c) 19.2×10^{-2} m (d) 5.76×10^{-2} m

($h = 6.63 \times 10^{-34}$ Js)

118. The de Broglie wavelength of a tennis ball of mass 60 g moving with a velocity of 10 metres per second is approximately **[NCERT : P L-48 | 2003]**

(a) 10^{-31} metres (b) 10^{-16} metres

(c) 10^{-25} metres (d) 10^{-33} metres

Planck's constant, $h = 6.63 \times 10^{-34}$ Js

119. Uncertainty in position of a minute particle of mass 25 g in space is 10^{-5} m. What is the uncertainty in its velocity (in ms^{-1})? ($h = 6.6 \times 10^{-34}$ Js) **[NCERT : P L-50 | 2002]**

(a) 2.1×10^{-34} (b) 0.5×10^{-34}

(c) 2.1×10^{-28} (d) 0.5×10^{-23}

Topic 4 - Quantum Mechanical Model of Atom

120. Correct statements for an element with atomic number 9 are **[NCERT : P L-55 | April 8, 2025 (II)]**

A. There can be 5 electrons for which $m_s = +\frac{1}{2}$ and 4 electrons for which $m_s = -\frac{1}{2}$

B. There is only one electron in p_z orbital

C. The last electron goes to orbital with n = 2 and l = 1

D. The sum of angular nodes of all the atomic orbitals is 1.

Choose the **correct** answer from the options given below:

(a) C and D only (b) A and C only

(c) A, C and D only (d) A and B only

121. The extra stability of half-filled subshell is due to

[NCERT : P L-64 | April 7, 2025 (II)]

(A) Symmetrical distribution of electrons

(B) Smaller coulombic repulsion energy

(C) The presence of electrons with the same spin in non-degenerate orbitals

(D) Larger exchange energy
(E) Relatively smaller shielding of electrons by one another
Identify the **correct** statements
(a) (B), (D) and (E) only (b) (A), (B), (D) and (E) only
(c) (B), (C) and (D) only (d) (A), (B) and (D) only

122. Consider the ground state of chromium atom (Z = 24). How many electrons are with Azimuthal quantum number $l = 1$ and $l = 2$ respectively ?
[NCERT : P L-55 | April 4, 2025 (II)]
(a) 12 and 4 (b) 16 and 4
(c) 12 and 5 (d) 16 and 5

123. Which one of the following about an electron occupying the 1*s* orbital in a hydrogen atom is **incorrect** ? (Bohr's radius is represented by a_0)
[NCERT : P L-55 | April 4, 2025 (I)]
(a) The probability density of finding the electron is maximum at the nucleus
(b) The electron can be found at a distance $2a_0$ from the nucleus
(c) The 1*s* orbital is spherically symmetrical
(d) The total energy of the electron is maximum when it is at a distance a_0 from the nucleus

124. For electron in '2s' and '2p' orbitals, the orbital angular momentum values, respectively are :
[NCERT : P L-55 | April 3, 2025 (II)]
(a) $\sqrt{2}\frac{h}{2\pi}$ and 0 (b) $\frac{h}{2\pi}$ and $\sqrt{2}\frac{h}{2\pi}$
(c) 0 and $\sqrt{6}\frac{h}{2\pi}$ (d) 0 and $\sqrt{2}\frac{h}{2\pi}$

125. Which of the following statements are **true** ?
[NCERT : P L-55 | April 2, 2025 (II)]
(A) The subsidiary quantum number l describes the shape of the orbital occupied by the electron.
(B) 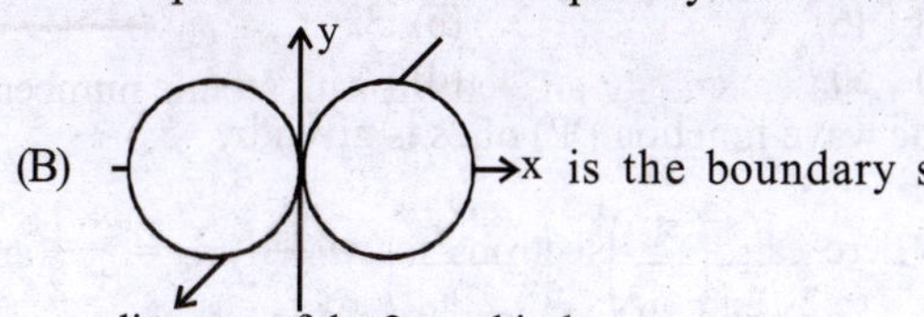

is the boundary surface diagram of the $2p_x$ orbital.
(C) The + and – signs in the wave function of the $2p_x$ orbital refer to charge.
(D) The wave function of $2p_x$ orbital is zero everywhere in the xy plane.
(a) (B) and (D) only (b) (A), (B) and (C) only
(c) (C) and (D) only (d) (A) and (B) only

126. Which of the following is/are not correct with respect to energy of atomic orbitals of hydrogen atom?
[NCERT : P L-61 | Jan. 28, 2025 (II)]
(A) 1s < 2p < 3d < 4s (B) 1s < 2s = 2p < 3s = 3p
(C) 1s < 2s < 2p < 3s < 3p (D) 1s < 2s < 4s < 3d
Choose the correct answer from the options given below :
(a) (B) and (D) only (b) (A) and (C) only
(c) (C) and (D) only (d) (A) and (B) only

127. In a multielectron atom, which of the following orbitals described by three quantum numbers will have same energy in absence of electric and magnetic fields?
[NCERT : P L-55 | Jan. 28, 2025 (I)]
(A) $n = 1, l = 0, m_l = 0$ (B) $n = 2, l = 0, m_l = 0$
(C) $n = 2, l = 1, m_l = 1$ (D) $n = 3, l = 2, m_l = 1$
(E) $n = 3, l = 2, m_l = 0$
Choose the correct answer from the options given below :
(a) A and B only (b) B and C only
(c) C and D only (d) D and E only

128. For hydrogen atom, the orbital/s with lowest energy is/are:
[NCERT : P L-61 | Jan. 24, 2025 (II)]
(A) 4*s* (B) $3p_x$
(C) $3d_{x^2-y^2}$ (D) $3d_{z^2}$
(E) $4p_z$
Choose the correct answer from the options given below :
(a) (A) and (E) only (b) (B) only
(c) (A) only (d) (B), (C) and (D) only

129. Given below are two statements:
[NCERT : P L-55 | Jan. 23, 2025 (II)]
Statement (I): For a given shell, the total number of allowed orbitals is given by n^2.
Statement (II): For any subshell, the spatial orientation of the orbitals is given by $-l$ to $+l$ values including zero.
In the light of the above statements, choose the correct answer from the options given below
(a) Statement I is true but Statement II is false
(b) Statement I is false but Statement II is true
(c) Both Statement I and Statement II are true
(d) Both Statement I and Statement II are false

130. Given below are two statements:
[NCERT : P L-45, 61 | Jan. 22, 2025 (II)]
Statement (I): A spectral line will be observed for a $2p_x \rightarrow 2p_y$ transition.
Statement (II): $2p_x$ and $2p_y$ are degenerate orbitals.
In the light of the above statements, choose the correct answer from the options given below:
(a) Both Statement I and Statement II are true.
(b) Both Statement I and Statement II are false.
(c) Statement I is true but Statement II is false.
(d) Statement I is false but Statement II is true.

131. Compare the energies of following sets of quantum numbers for multielectron system.
[NCERT : P L-61 | April 9, 2024 (I)]
(A) n = 4, l = 1 (B) n = 4, l = 2
(C) n = 3, l = 1 (D) n = 3, l = 2
(E) n = 4, l = 0
Choose the correct answer from the options given below:
(a) (B) > (A) > (C) > (E) > (D)
(b) (E) > (C) < (D) < (A) < (B)
(c) (E) > (C) > (A) > (D) > (B)
(d) (C) < (E) < (D) < (A) < (B)

132. In an atom, total number of electrons having quantum numbers $n = 4$, $|m_l| = 1$ and $m_s = -\frac{1}{2}$ is

[NCERT : P L-55 | NV, April 5, 2024 (II)]

133. The maximum number of orbitals which can be identified with $n = 4$ and $m_l = 0$ is_____

[NCERT : P L-55 | NV, April 4, 2024 (II)]

134. The number of radial nodes for 3p orbital is:

[NCERT : P L- 55 | Feb. 1, 2024 (II)]

(a) 3 (b) 2 (c) 1 (d) 4

135. The four quantum numbers for the electron in the outer most orbital of potassium (atomic no. 19) are

[NCERT : P L-55 | Jan. 31, 2024 (II)]

(a) $n = 4, l = 2, m = -1, s = +\frac{1}{2}$

(b) $n = 4, l = 0, m = 0, s = +\frac{1}{2}$

(c) $n = 3, l = 0, m = 1, s = +\frac{1}{2}$

(d) $n = 2, l = 0, m = 0, s = +\frac{1}{2}$

136. Given below are two statements:

[NCERT : P L-60 | Jan. 30, 2024 (I)]

Statement I : The orbitals having same energy are called as degenerate orbitals.

Statement II : In hydrogen atom, 3p and 3d orbitals are not degenerate orbitals.

In the light of the above statements, choose the **most appropriate** answer from the options given below:

(a) **Statement I** is true but **Statement II** is false.
(b) Both **Statement I** and **Statement II** are true.
(c) Both **Statement I** and **Statement II** are false.
(d) **Statement I** is false but **Statement II** is true.

137. The correct set of four quantum numbers for the valence electron of rubidium atom (Z = 37) is :

[NCERT : P L-55 | Jan. 29, 2024 (I)]

(a) $5, 1, 0, +\frac{1}{2}$ (b) $5, 1, 1, +\frac{1}{2}$

(c) $5, 0, 1, +\frac{1}{2}$ (d) $5, 0, 0, +\frac{1}{2}$

138. The number of electrons present in all the completely filled sub shells having n=4 and $s = +\frac{1}{2}$ is _____.

(Where n = principal quantum number and s = spin quantum number) **[NCERT : P L-55 | NV, Jan. 27, 2024 (I)]**

139. The orbital angular momentum of an electron in 3s orbital is $\frac{xh}{2\pi}$. The value of x is _____.

[NCERT : P L-55 | NV, April 13, 2023 (II)]

140. Given below are two statement : one is labelled as **Assertion A** and the other is labelled as **Reason R**

Assertion A : 5f electrons can participate in bonding to a far greater extent than 4f electrons

Reason R : 5f orbitals are not as buried as 4f orbitals

In the light of the above statements, choose the *correct* answer from the options given below.

[NCERT : P L-61 | April 12, 2023 (I)]

(a) Both A and R are true and R is NOT the correct explanation of A.
(b) Both A and R are true but R is the correct explanation of A.
(c) A is false but R is true.
(d) A is true but R is false.

141. The number of correct statements from the following _____.

[NCERT : P L-61 | NV, April 11, 2023 (II)]

A. For 1s orbital, the probability density is maximum at the nucleus.
B. For 2s orbital, the probability density first increases to maximum and then decreases sharply to zero.
C. Boundary surface diagrams of the orbitals encloses a region of 100% probability of finding the electron.
D. p and d-orbitals have 1 and 2 angular nodes respectively.
E. Probability density of p-orbital is zero at the nucleus.

142. The number of atomic orbitals from the following having 5 radial nodes is _____.

7s, 7p, 6s, 8p, 8d

[NCERT : P L-61 | NV, April 08, 2023 (II)]

143. Arrange the following orbitals in decreasing order of energy? **[NCERT : P L-60 | Jan. 31, 2023 (II)]**

(A) $n = 3, l = 0, m = 0$ (B) $n = 4, l = 0, m = 0$
(C) $n = 3, l = 1, m = 0$ (D) $n = 3, l = 2, m = 1$

The correct option for the order is :

(a) B > D > C > A (b) D > B > C > A
(c) A > C > B > D (d) D > B > A > C

144. Maximum number of electrons that can be accommodated in shell with $n = 4$ are:

[NCERT : P L-55 | Jan. 30, 2023 (II)]

(a) 16 (b) 32
(c) 50 (d) 72

145. The wave function (Ψ) of 2s is given by

$$\Psi_{2s} = \frac{1}{2\sqrt{2\pi}}\left(\frac{1}{a_0}\right)^{1/2}\left(2 - \frac{r}{a_0}\right)e^{-r/2a_0}$$

At $r = r_0$, radial node is formed. Thus, r_0 in terms of a_0

[NCERT : P L-59 | Jan. 30, 2023 (II)]

(a) $r_0 = a_0$ (b) $r_0 = 4a_0$

(c) $r_0 = \frac{a_0}{2}$ (d) $r_0 = 2a_0$.

146. The number of given orbitals which have electron density along the axis is _____

$p_x, p_y, p_z, d_{xy}, d_{yz}, d_{xz}, d_{z^2}, d_{x^2-y^2}$

[NCERT : P L-57 | NV, Jan. 25, 2023 (II)]

147. Given below are the quantum numbers for 4 electrons.

A. $n = 3, l = 2, m_1 = 1, m_s = +1/2$
B. $n = 4, l = 1, m_1 = 0, m_s = +1/2$
C. $n = 4, l = 2, m_1 = -2, m_s = -1/2$
D. $n = 3, l = 1, m_1 = -1, m_s = +1/2$

The correct order of increasing energy is :

[NCERT : P L-61 | July 29, 2022 (II)]

(a) D < B < A < C (b) D < A < B < C
(c) B < D < A < C (d) B < D < C < A

148. Identify the incorrect statement from the following.

[NCERT : P L-46 | July 28, 2022 (I)]

(a) A circular path around the nucleus in which an electron moves is proposed as Bohr's orbit.
(b) An orbital is the one electron wave function (ψ) in an atom.
(c) The existence of Bohr's orbits is supported by hydrogen spectrum.
(d) Atomic orbital is characterised by the quantum numbers n and *l* only

149. Given below are two statements. One is labelled as **Assertion A** and the other is labelled as **Reason R**.

Assertion A : Energy of 2s orbital of hydrogen atom is greater than that of 2s orbital of lithium.

Reason R : Energies of the orbitals in the same subshell decrease with increase in the atomic number.

In the light of the above statements, choose the **correct** answer from the options given below.

[NCERT : P L-60 | July 27, 2022 (I)]

(a) Both A and R are true and R is the correct explanation of A.
(b) Both A and R are true but R is NOT the correct explanation of A.
(c) A is true but R is false.
(d) A is false but R is true.

150. The correct decreasing order of energy, for the orbitals having, following set of quantum numbers:

[NCERT : P L-61 |July 27, 2022 (II)]

(A) $n = 3, l = 0, m = 0$ (B) $n = 4, l = 0, m = 0$
(C) $n = 3, l = 1, m = 0$ (D) $n = 3, l = 2, m = 1$

(a) (D) > (B) > (C) > (A) (b) (B) > (D) > (C) > (A)
(c) (C) > (B) > (D) > (A) (d) (B) > (C) > (D) > (A)

151. Which of the following sets of quantum numbers is not allowed? **[NCERT : P L-55 |July 25, 2022 (I)]**

(a) $n = 3, l = 2, m_l = 0, s = +\frac{1}{2}$

(b) $n = 3, l = 2, m_l = -2, s = +\frac{1}{2}$

(c) $n = 3, l = 3, m_l = -3, s = -\frac{1}{2}$

(d) $n = 3, l = 0, m_l = 0, s = -\frac{1}{2}$

152. The number of radial nodes and total number of nodes in 4p orbital respectively are: **[NCERT : P L-59 | June 30, 2022 (I)]**

(a) 2 and 3 (b) 2 and 2
(c) 3 and 4 (d) 4 and 4

153. Which of the following statements are **correct** ?

(A) The electronic configuration of Cr is $[Ar]\, 3d^5\, 4s^1$.
(B) The magnetic quantum number may have a negative value.
(C) In the ground state of an atom, the orbitals are filled in order of their increasing energies.
(D) The total number of nodes are given by n – 2 .

Choose the **most appropriate** answer from the options given below :

[NCERT : P L-55, 59, 60 | June 29, 2022 (I)]

(a) (A), (C) and (D) only
(b) (A) and (B) only
(c) (A) and (C) only
(d) (A), (B) and (C) only

154. Which of the following is the correct plot for the probability density $\psi^2(r)$ as a function of distance '*r*' of the electron form the nucleus for 2*s* orbital?

[NCERT : P L-59 | June 29, 2022 (II)]

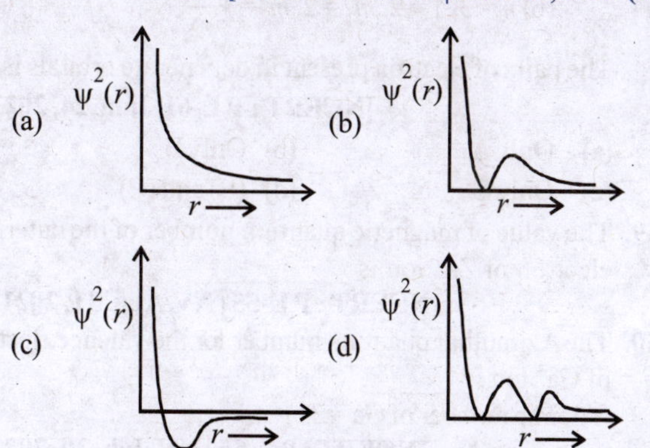

155. Consider the following statements :

[NCERT : P L-55 | June 28, 2022 (II)]

(A) The principal quantum number '*n*' is a positive integer with values of '*n*' = 1, 2, 3, ...
(B) The azimuthal quantum number '*l*' for a given '*n*' (principal quantum number) can have values as '*l*' = 0, 1, 2, ... *n*
(C) Magnetic orbital quantum number 'm_l' for a particular '*l*' (azimuthal quantum number) has (2*l* + 1) values.
(D) ±1/2 are the two possible orientations of electron spin.
(E) For *l* = 5, there will be a total of 9 orbital.

Which of the above statements correct?

(a) (A), (B) and (C) (b) (A), (C), (D) and (E)
(c) (A), (C) and (D) (d) (A), (B), (C) and (D)

156. Consider the following set of quantum numbers:

	n	*l*	m_l
A.	3	3	–3
B.	3	2	–2
C.	2	1	+1
D.	2	2	+2

The number of correct sets of quantum numbers is ______. **[NCERT : P L-55 | NV, June 27, 2022 (II)]**

157. The number of radial and angular nodes in 4d orbital are respectively **[NCERT : P L-59 | June 26, 2022 (II)]**

(a) 1 and 2 (b) 3 and 2
(c) 1 and 0 (d) 2 and 1

158. Consider the following pairs of electrons

(A) (a) $n = 3, 1 = 1, m_1 = 1, m_s = +\frac{1}{2}$

(b) $n = 3, 1 = 2, m_1 = 1, m_s = +\frac{1}{2}$

(B) (a) $n = 3, 1 = 2, m_1 = -2, m_s = -\frac{1}{2}$

(b) $n = 3, 1 = 2, m_1 = -1, m_s = -\frac{1}{2}$

(C) (a) $n = 4, 1 = 2, m_1 = 2, m_s = +\frac{1}{2}$

(b) $n = 3, 1 = 2, m_1 = 2, m_s = +\frac{1}{2}$

The pairs of electron present in degenerate orbitals is/are: **[NCERT : P L-61| June 24, 2022 (I)]**

(a) Only A (b) Only B
(c) Only C (d) (B) and (C)

159. The value of magnetic quantum number of the outermost electron of Zn^+ ion is ________. **[NCERT : P L-55 | NV, Aug. 31, 2021 (II)]**

160. The Azimuthal quantum number for the valence electrons of Ga^+ ion is ______.
(Atomic number of Ga = 31) **[NCERT : P L-55 | NV, July 20, 2021 (I)]**

161. A certain orbital has no angular nodes and two radial nodes. The orbital is **[NCERT : P L-59 | March 18, 2021 (I)]**

(a) $2p$ (b) $3p$
(c) $3s$ (d) $2s$

162. A certain orbital has $n = 4$ and $m_l = -3$. The number of radial nodes in this orbital is ________.
(Round off to the Nearest Integer). **[NCERT : P L-59 | NV, March 17, 2021 (I)]**

163. The number of orbitals with $n = 5, m_l = +2$ is ________.
(Round off to the Nearest Integer). **[NCERT : P L-55 | NV, March 16, 2021 (II)]**

164. The orbital having two radial as well as two angular nodes is: **[NCERT : P L-59 | Feb. 26, 2021 (I)]**

(a) $3p$ (b) $4d$
(c) $5d$ (d) $4f$

165. In the sixth period, the orbitals that are filled are : **[NCERT : P L-61 | Sep. 05, 2020 (I)]**

(a) $6s, 4f, 5d, 6p$ (b) $6s, 5d, 5f, 6p$
(c) $6s, 5f, 6d, 6p$ (d) $6s, 6p, 6d, 6f$

166. The correct statement about probability density (except at infinite distance from nucleus) is : **[NCERT : P L-59 | Sep. 05, 2020 (II)]**

(a) It can be zero for 1s orbital.
(b) It can be negative for 2p orbital.
(c) It can be zero for 3p orbital.
(d) It can never be zero for 2s orbital.

167. Consider the hypothetical situation where the azimuthal quantum number, l, takes values 0, 1, 2, ... $n + 1$, where n is the principal quantum number. Then, the element with atomic number : **[NCERT : P L-55 | Sep. 03, 2020 (II)]**

(a) 9 is the first alkali metal
(b) 13 has a half-filled valence subshell
(c) 8 is the first noble gas
(d) 6 has a 2p-valence subshell

168. The number of subshells associated with $n = 4$ and $m = -2$ quantum numbers is : **[NCERT : P L-55 | Sep. 02, 2020 (II)]**

(a) 8 (b) 2 (c) 16 (d) 4

169. The number of orbitals associated with quantum numbers $n = 5, m_s = +\frac{1}{2}$ is: **[NCERT : P L-55 | Jan. 07, 2020 (I)]**

(a) 11 (b) 25 (c) 50 (d) 15

170. Among the following, the energy of 2s orbital is lowest in: **[NCERT : P L-61 | 2017 (S); April 12, 2019 (II)]**

(a) K (b) H (c) Li (d) Na

171. The electrons are more likely to be found : **[April 12, 2019 (I)]**

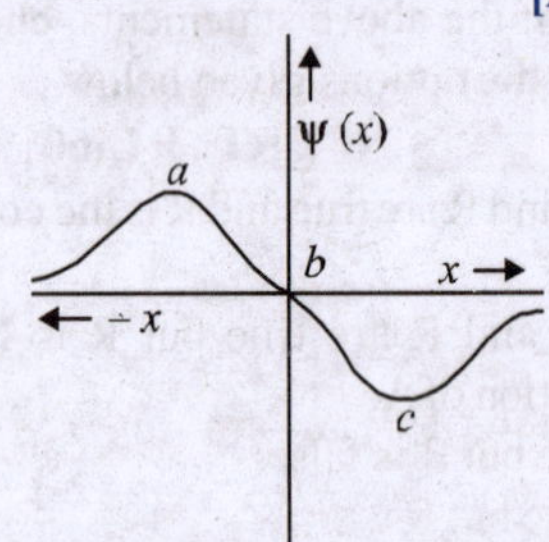

(a) in the region a and c (b) in the region a and b
(c) only in the region a (d) only in the region c

172. The graph between $|\psi|^2$ and r (radial distance) is shown below. This represents : **[NCERT : P L-57 | April 10, 2019 (I)]**

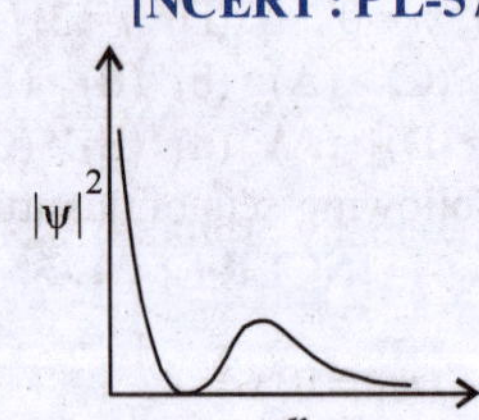

(a) 3s orbital (b) 2s orbital
(c) 1s orbital (d) 2p orbital

173. The size of the iso-electronic species Cl^-, Ar and Ca^{2+} is affected by : **[NCERT : P L-55 | April 8, 2019 (I)]**

(a) azimuthal quantum number of valence shell
(b) electron-electron interaction in the outer orbitals
(c) principal quantum number of valence shell
(d) nuclear charge

174. The quantum number of four electrons are given below :

I. $n = 4, l = 2, m_l = -2, m_s = -1/2$
II. $n = 3, l = 2, m_l = 1, m_s = +1/2$
III. $n = 4, l = 1, m_l = 0, m_s = +1/2$
IV. $n = 3, l = 1, m_l = 1, m_s = -1/2$

The correct order of their increasing energies will be : **[NCERT : P L-55 | 2007(S), April 8, 2019 (I)]**

(a) IV < III < II < I (b) I < II < III < IV
(c) IV < II < III < I (d) I < III < II < IV

175. Which of the following combination of statements is true regarding the interpretation of the atomic orbitals?

[NCERT : P L-54 | Jan. 9, 2019 (II)]

(1) An electron in an orbital of high angular momentum stays away from the nucleus than an electron in the orbital of lower angular momentum.

(2) For a given value of the principal quantum number, the size of the orbit is inversely proportional to the azimuthal quantum number.

(3) According to wave mechanics, the ground state angular momentum is equal to $\frac{h}{2\pi}$.

(4) The plot of ψ vs r for various azimuthal quantum numbers, shows peak shifting towards higher r value.

(a) (1), (4) (b) (1), (2)
(c) (1), (3) (d) (2), (3)

176. The total number of orbitals associated with the principal quantum number 5 is :

[NCERT : P L-55 | Online April 9, 2016]

(a) 20 (b) 25
(c) 10 (d) 5

177. If the principal quantum number $n = 6$, the correct sequence of filling of electrons will be :

[NCERT : P L-55 | Online April 10, 2015]

(a) $ns \rightarrow (n-2)f \rightarrow np \rightarrow (n-1)d$
(b) $ns \rightarrow (n-2)f \rightarrow (n-1)d \rightarrow np$
(c) $ns \rightarrow np \rightarrow (n-1)d \rightarrow (n-2)f$
(d) $ns \rightarrow (n-1)d \rightarrow (n-2)f \rightarrow np$

178. The correct set of four quantum numbers for the valence electrons of rubidium atom ($Z = 37$) is:

[NCERT : P L-55 | 2014]

(a) $5, 0, 0, +\frac{1}{2}$ (b) $5, 1, 0, +\frac{1}{2}$
(c) $5, 1, 1, +\frac{1}{2}$ (d) $5, 0, 1, +\frac{1}{2}$

179. In an atom how many orbital(s) will have the quantum numbers; $n = 3$, $l = 2$ and $m_l = +2$?

[NCERT : P L-61 | Online April 9, 2013]

(a) 5 (b) 3
(c) 1 (d) 7

180. Given

(A) $n = 5, m_\ell = +1$
(B) $n = 2, \ell = 1, m_\ell = -1, m_s = -1/2$

The maximum number of electron(s) in an atom that can have the quantum numbers as given in (A) and (B) are respectively:

[NCERT : P L-61 | Online April 25, 2013]

(a) 25 and 1 (b) 8 and 1
(c) 2 and 4 (d) 4 and 1

181. The increasing order of the ionic radii of the given isoelectronic species is : **[NCERT : P L-35 | 2012]**

(a) $Cl^-, Ca^{2+}, K^+, S^{2-}$ (b) $S^{2-}, Cl^-, Ca^{2+}, K^+$
(c) $Ca^{2+}, K^+, Cl^-, S^{2-}$ (d) $K^+, S^{2-}, Ca^{2+}, Cl^-$

182. The following sets of quantum numbers represent four electrons in an atom.

[NCERT : P L-55 | Online May 26, 2012]

(i) $n = 4, l = 1$ (ii) $n = 4, l = 0$
(iii) $n = 3, l = 2$ (iv) $n = 3, l = 1$

The sequence representing increasing order of energy, is

(a) (iii) < (i) < (iv) < (ii) (b) (iv) < (ii) < (iii) < (i)
(c) (i) < (iii) < (ii) < (iv) (d) (ii) < (iv) < (i) < (iii)

183. Which one of the following constitutes a group of the isoelectronic species? **[NCERT : P L-35 | 2008]**

(a) C_2^{2-}, O_2^-, CO, NO (b) $NO^+, C_2^{2-}, CN^-, N_2$
(c) $CN^-, N_2, O_2^{2-}, C_2^{2-}$ (d) N_2, O_2^-, NO^+, CO

184. Which of the following sets of quantum numbers represents the highest energy of an atom?

[NCERT : P L-55 | 2007]

(a) $n = 3, l = 0, m = 0, s = +1/2$
(b) $n = 3, l = 1, m = 1, s = +1/2$
(c) $n = 3, l = 2, m = 1, s = +1/2$
(d) $n = 4, l = 0, m = 0, s = +1/2.$

185. Which one of the following sets of ions represents a collection of isoelectronic species?

[NCERT : P L-35 | 2006]

(a) $N^{3-}, O^{2-}, F^-, S^{2-}$ (b) $Li^+, Na^+, Mg^{2+}, Ca^{2+}$
(c) $K^+, Cl^-, Ca^{2+}, Sc^{3+}$ (d) $Ba^{2+}, Sr^{2+}, K^+, Ca^{2+}$

186. Of the following sets which one does NOT contain isoelectronic species? **[NCERT : P L-35 | 2005]**

(a) $BO_3^{3-}, CO_3^{2-}, NO_3^-$ (b) $SO_3^{2-}, CO_3^{2-}, NO_3^-$
(c) CN^-, N_2, C_2^{2-} (d) $PO_4^{3-}, SO_4^{2-}, ClO_4^-$

187. The electrons identified by quantum numbers n and ℓ :

[NCERT : P L-61 | 2005]

(A) $n = 4, l = 1$ (B) $n = 4, l = 0$
(C) $n = 3, l = 2$ (D) $n = 3, l = 1$

can be placed in order of increasing energy as :

(a) (C) < (D) < (B) < (A)
(b) (D) < (B) < (C) < (A)
(c) (B) < (D) < (A) < (C)
(d) (A) < (C) < (B) < (D)

188. In a multi-electron atom, which of the following orbitals described by the three quantum members will have the same energy in the absence of magnetic and electric fields?

[NCERT : P L-55 | 2005]

(A) $n = 1, l = 0, m = 0$ (B) $n = 2, l = 0, m = 0$
(C) $n = 2, l = 1, m = 1$ (D) $n = 3, l = 2, m = 1$
(E) $n = 3, l = 2, m = 0$

(a) (D) and (E) (b) (C) and (D)
(c) (B) and (C) (d) (A) and (B)

189. Which of the following sets of quantum numbers is correct for an electron in 4f orbital ? **[NCERT : P L-55 | 2004]**

(a) $n = 4, l = 3, m = +1, s = +½$
(b) $n = 4, l = 4, m = -4, s = -½$
(c) $n = 4, l = 3, m = +4, s = +½$
(d) $n = 3, l = 2, m = -2, s = +½$

190. Which one of the following sets of ions represents the collection of isoelectronic species? **[NCERT : P L-35 | 2004]**

(a) K^+, Cl^-, Mg^{2+}, Sc^{3+}
(b) Na^+, Ca^{2+}, Sc^{3+}, F^-
(c) K^+, Ca^{2+}, Sc^{3+}, Cl^-
(d) Na^+, Mg^{2+}, Al^{3+}, Cl^-

(Atomic nos. : F = 9, Cl = 17, Na = 11, Mg = 12, Al = 13, K = 19, Ca = 20, Sc = 21)

191. Consider the ground state of Cr atom (X = 24). The number of electrons with the azimuthal quantum numbers, $l = 1$ and 2 are, respectively **[NCERT : P L-61 | 2004]**

(a) 16 and 4 (b) 12 and 5
(c) 12 and 4 (d) 16 and 5

192. The number of d-electrons retained in Fe^{2+} (At. no. of Fe = 26) ion is **[NCERT : P L-61 | 2003]**

(a) 4 (b) 5 (c) 6 (d) 3

193. The orbital angular momentum for an electron revolving in an orbit is given by $\sqrt{l(l+1)}.\dfrac{h}{2\pi}$. This momentum for an s-electron will be given by **[2003]**

(a) zero (b) $\dfrac{h}{2\pi}$ (c) $\sqrt{2}.\dfrac{h}{2\pi}$ (d) $+\dfrac{1}{2}.\dfrac{h}{2\pi}$

194. Which one of the following groupings represents a collection of isoelectronic species ?(At. nos. : Cs : 55, Br : 35) **[NCERT : P L-35 | 2003]**

(a) N^{3-}, F^-, Na^+ (b) Be, Al^{3+}, Cl^-
(c) Ca^{2+}, Cs^+, Br (d) Na^+, Ca^{2+}, Mg^{2+}

ANSWER KEYS

1	(d)	21	(b, c)	41	(d)	61	(a)	81	(b)	101	(9)	121	(b)	141	(3)	161	(c)	181	(c)
2	(494)	22	(c)	42	(1)	62	(b)	82	(d)	102	(1)	122	(c)	142	(3)	162	(0)	182	(b)
3	(0)	23	(c)	43	(1)	63	(d)	83	(b)	103	(2)	123	(d)	143	(b)	163	(3)	183	(b)
4	(2)	24	(d)	44	(a)	64	(a)	84	(c)	104	(d)	124	(d)	144	(b)	164	(c)	184	(c)
5	(3)	25	(c)	45	(b)	65	(c)	85	(a)	105	(d)	125	(d)	145	(d)	165	(a)	185	(c)
6	(a)	26	(d)	46	(270)	66	(c)	86	(a)	106	(d)	126	(c)	146	(5)	166	(c)	186	(b)
7	(d)	27	(54)	47	(c)	67	(b)	87	(a)	107	(d)	127	(d)	147	(b)	167	(c)	187	(b)
8	(798)	28	(d)	48	(492)	68	(b)	88	(b)	108	(b)	128	(d)	148	(d)	168	(b)	188	(a)
9	(c)	29	(c)	49	(10)	69	(d)	89	(b)	109	(b)	129	(c)	149	(a)	169	(b)	189	(a)
10	(2)	30	(b)	50	(b)	70	(d)	90	(58)	110	(d)	130	(d)	150	(a)	170	(a)	190	(c)
11	(300)	31	(a)	51	(N)	71	(d)	91	(661)	111	(a)	131	(d)	151	(c)	171	(a)	191	(b)
12	(c)	32	(d)	52	(3155)	72	(a)	92	(a)	112	(b)	132	(6)	152	(a)	172	(b)	192	(c)
13	(2)	33	(5)	53	(c)	73	(c)	93	(7)	113	(b)	133	(4)	153	(d)	173	(d)	193	(a)
14	(50)	34	(1724)	54	(d)	74	(a)	94	(c)	114	(c)	134	(c)	154	(b)	174	(c)	194	(a)
15	(5)	35	(34)	55	(c)	75	(d)	95	(2)	115	(a)	135	(b)	155	(c)	175	(a)		
16	(2)	36	(22)	56	(a)	76	(a)	96	(548)	116	(b)	136	(a)	156	(2)	176	(b)		
17	(181)	37	(8)	57	(c)	77	(a)	97	(1758)	117	(a)	137	(d)	157	(a)	177	(b)		
18	(222)	38	(10)	58	(c)	78	(a)	98	(22)	118	(d)	138	(16)	158	(b)	178	(a)		
19	(d)	39	(c)	59	(b)	79	(a)	99	(58)	119	(c)	139	0	159	(0)	179	(c)		
20	(c)	40	(a)	60	(c)	80	(a)	100	(6)	120	(b)	140	(b)	160	(0)	180	(b)		

Classification of Elements and Periodicity in Properties

3

TREND ANALYSIS
JEE MAIN (2025-2021)

	2025				2024				2023				2022				2021			
	Session II		Session I		Session II		Session I		Session II		Session I		Session II		Session I		Session II		Session I	
	MCQs	NVQs	MCQs	NVQs	MCQs	NVQs	MCQs	NVQs	MCQs	NVQs	MCQs	NVQs	MCQs	NVQs	MCQs	NVQs	MCQs	NVQs	MCQs	NVQs
Topic 1	1	–	4	–	1	–	–	1	1	–	2	–	–	–	–	–	–	–	1	–
Topic 2	5	–	7	–	9	–	10	1	9	1	4	1	8	–	11	–	2	–	5	–
Topic 3																				
Topic 4																				
Total	6	-	11	-	10	-	10	2	10	1	6	1	8	-	11	-	2	-	6	-

How to Code Qns for Better & Faster Revision

Type 1: Able to solve (No code Req)
Type 2: Solved with struggle or by seeing solution
(Use code NR means Needs Revision or orange stickers)
Type 3: Tricky/ Could not solve on my own. Need help from Teacher/ Friend.
(Use code TQ means Tricky Question or red stickers)

You can either hand write the code in boxes or else paste coloured stickers given at end of book.

Topic 1 - Modern Periodic Table

1. Match the **LIST-I** with **LIST-II**.
[NCERT : P L-84 | April 3, 2025 (II)]

LIST-I (Family)	LIST-II (Symbol of Element)
A. Pnictogen (group 15)	I. Ts
B. Chalcogen	II. Og
C. Halogen	III. Lv
D. Noble gas	IV. Mc

Choose the **correct** answer from the options given below:
(a) A-IV, B-I, C-II, D-III (b) A-IV, B-III, C-I, D-II
(c) A-III, B-I, C-IV, D-II (d) A-II, B-III, C-IV, D-I

2. Given below are two statements :
Statement (I) : According to the Law of Octaves, the elements were arranged in the increasing order of their atomic number. **[NCERT : P L-78 | Jan. 28, 2025 (II)]**
Statement (II) : Meyer observed a periodically repeated pattern upon plotting physical properties of certain elements against their respective atomic numbers.
In the light of the above statements, choose the correct answer from the options given below :
(a) Statement I is false but Statement II is true
(b) Both Statement I and Statement II are true
(c) Statement I is true but Statement II is false
(d) Both Statement I and Statement II are false

3. Which of the following Statements are NOT true about the periodic table? **[NCERT : P L-78 | Jan. 24, 2025 (I)]**
A. The properties of elements are function of atomic weights.
B. The properties of elements are function of atomic numbers.
C. Elements having similar outer electronic configuration are arranged in same period.
D. An element's location reflects the quantum numbers of the last filled orbital.

E. The number of elements in a period is same as the number of atomic orbitals available in energy level that is being filled.

Choose the correct answer from the options given below:

(a) A, C and E Only (b) D and E Only

(c) A and E Only (d) B, C and E Only

4. Given below are two statements about X-ray spectra of elements : **[NCERT : P L-78 | Jan. 23, 2025 (II)]**

Statement (I) : A plot of $\sqrt{\nu}$ (ν = frequency of X-rays emitted) vs atomic mass is a straight line.

Statement (II) : A plot of ν (ν = frequency of X-rays emitted) vs atomic number is a straight line.

In the light of the above statements choose the correct answer from the options given below:

(a) Statement I is true but Statement II is false

(b) Both Statement I and Statement II are true

(c) Both Statement I and Statement II are false

(d) Statement I is false but Statement II is true

5. The element that does not belong to the same period of the remaining elements (modern periodic table) is: **[NCERT : P L-79 | Jan. 23, 2025 (I)]**

(a) Palladium (b) Iridium

(c) Osmium (d) Platinum

6. Match **List-I** with **List-II**. **[NCERT : P L-81 | April 9, 2024 (II)]**

List-I (Element)	**List-II (Electronic Configuration)**
(A) N	(I) [Ar] $3d^{10}4s^2 4p^5$
(B) S	(II) [Ne] $3s^2 3p^4$
(C) Br	(III) [He] $2s^2 2p^3$
(D) Kr	(IV) [Ar] $3d^{10} 4s^2 4p^6$

Choose the correct answer from the options given below:

(a) A-(IV), B-(III), C-(II), D-(I)

(b) A-(III), B-(II), C-(I), D-(IV)

(c) A-(I), B-(IV), C-(III), D-(II)

(d) A-(II), B-(I), C-(IV), D-(III)

7. If IUPAC name of an element is "Unununnium" then the element belongs to n^{th} group of Periodic table. The value of n is ____. **[NCERT : P L-82 | NV, Jan. 30, 2024 (I)]**

8. Henry Moseley studied characteristic X-ray spectra of elements. The graph which represents his observation correctly is : **[NCERT : P L-78 | April 08, 2023 (II)]**

(Given v = frequency of X-ray emitted; Z = atomic number)

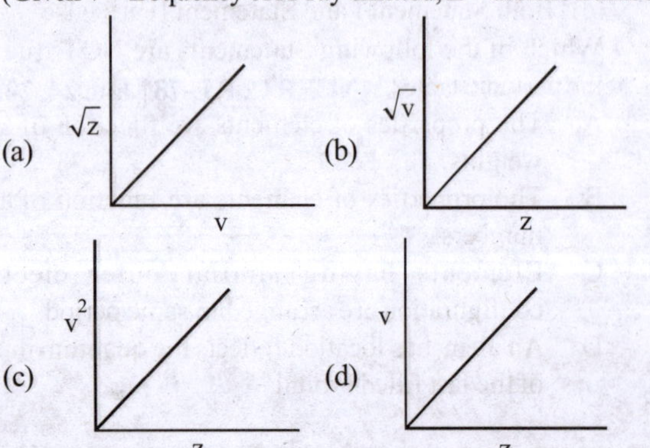

9. It is observed that characteristic X-ray spectra of elements show regularity. When frequency to the power '*n*' i.e. v^n of X-rays emitted is plotted against atomic number 'Z', following graph is obtained. **[NCERT : P L-78 | Jan. 24, 2023 (I)]**

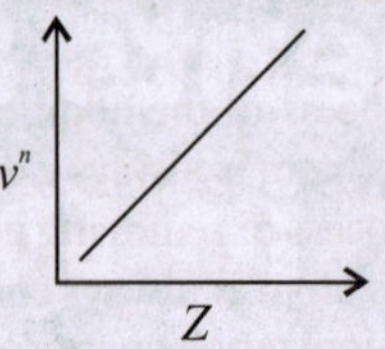

The value of '*n*' is

(a) 1 (b) 2 (c) $\frac{1}{2}$ (d) 3

10. Match **List I** with **List II** **[NCERT : P L-82 | Jan. 30, 2023 (I)]**

List-I (Atomic number)	**List-II (Block of periodic table)**
(A) 37	I. p-block
(B) 78	II. d-block
(C) 52	III. f-block
(D) 65	IV. s-block

Choose the correct answer from the options given below:

(a) A – II, B – IV, C – I, D – III

(b) A – I, B – III, C – IV, D – II

(c) A – IV, B – III, C – II, D – I

(d) A – IV, B – II, C – I, D – III

11. The characteristics of elements X, Y and Z with atomic numbers, respectively, 33, 53 and 83 are **[NCERT : P L-83 | March 16, 2021 (II)]**

(a) X and Y are metalloids and Z is a metal.

(b) X is a metalloid, Y is a non-metal and Z is a metal.

(c) X and Z are non-metals and Y is a metalloid.

(d) X, Y and Z are metals.

12. Similarity in chemical properties of the atoms of elements in a group of the periodic table is most closely related to: **[NCERT : P L-78 | Online April 12, 2014; Same 2003]**

(a) atomic numbers

(b) atomic masses

(c) number of principal energy levels

(d) number of valence electrons

13. According to the periodic law of elements, the variation in properties of elements is related to their

(a) nuclear masses **[NCERT : P L-78 | 2003]**

(b) atomic numbers

(c) nuclear neutron-proton number ratios

(d) atomic masses

Topic 2 - Periodic Trends in Properties of Elements

14. The atomic number of the element from the following with lowest 1st ionisation enthalpy is : **[NCERT : P L-87 | April 8, 2025 (II)]**

(a) 32 (b) 35 (c) 87 (d) 19

15. Choose the incorrect trend in the atomic radii (r) of the elements : **[NCERT : P L-87 | April 7, 2025 (II)]**
(a) $r_{Br} < r_K$ (b) $r_{Mg} < r_{Al}$
(c) $r_{Rb} < r_{Cs}$ (d) $r_{At} < r_{Cs}$

16. Which of the following statements are correct? **[NCERT : P L-87 | April 3, 2025 (I)]**
A. The process of the addition an electron to a neutral gaseous atom is always exothermic
B. The process of removing an electron from an isolated gaseous atom is always endothermic
C. The 1st ionization energy of the boron is less than that of the beryllium
D. The electronegativity of C is 2.5 in CH_4 and CCl_4
E Li is the most electropositive among elements of group I
Choose the **correct** answer from the options gives below
(a) B and C only (b) A, C and D only
(c) B and D only (d) B, C and E only

17. The nature of oxide (TeO_2) and hydride (TeH_2) formed by Te, respectively are : **[NCERT : P L-92 | April 2, 2025 (II)]**
(a) Oxidising and acidic (b) Reducing and basic
(c) Reducing and acidic (d) Oxidising and basic

18. Electronic configuration of four elements A, B, C and D are given below : **[NCERT : P L-84 | April 2, 2025 (II)]**
(A) $1s^22s^22p^3$ (B) $1s^22s^22p^4$
(C) $1s^22s^22p^5$ (D) $1s^22s^22p^2$
Which of the following is the correct order of increasing electronegativity (Pauling's scale) ?
(a) A < D < B < C (b) A < C < B < D
(c) A < B < C < D (d) D < A < B < C

19. The type of oxide formed by the element among Li, Na, Be, Mg, B and Al that has the least atomic radius is : **[NCERT : P L-92 | Jan. 29, 2025 (II)]**
(a) A_2O_3 (b) AO_2 (c) AO (d) A_2O

20. Given below are two statements : **[NCERT : P L-87 | Jan. 29, 2025 (I)]**
Statement (I) : The radii of isoelectronic species increases in the order.
$Mg^{2+} < Na^+ < F^- < O^{2-}$
Statement (II) : The magnitude of electron gain enthalpy of halogen decreases in the order.
Cl > F > Br > I
In the light of the above statements, choose the most appropriate answer from the options given below :
(a) Statement I is incorrect but Statement II is correct
(b) Both Statement I and Statement II are incorrect
(c) Statement I is correct but Statement II is incorrect
(d) Both Statement I and Statement II are correct

21. The incorrect decreasing order of atomic radii is : **[NCERT : P L-87 | Jan. 28, 2025 (I)]**
(a) Mg > Al > C > O (b) Al > B > N > F
(c) Be > Mg > Al > Si (d) Si > P > Cl > F

22. The successive 5 ionisation energies of an element are 800, 2427, 3658, 25024 and 32824 kJ/mol, respectively. By using the above values predict the group in which the above element is present : **[NCERT : P L-87 | Jan. 24, 2025 (II)]**
(a) Group 2 (b) Group 13
(c) Group 4 (d) Group 14

23. Given below are two statements : **[NCERT : P L-94 | Jan. 22, 2025 (II)]**
Statement (I) : An element in the extreme left of the periodic table forms acidic oxides.
Statement (II) : Acid is formed during the reaction between water and oxide of a reactive element present in the extreme right of the periodic table.
In the light of the above statements, choose the correct answer from the options given below :
(a) Statement I is false but Statement II is true.
(b) Both Statement I and Statement II are false.
(c) Statement I is true but Statement II is false.
(d) Both Statement I and Statement II are true.

24. Match List-I with List-II **[NCERT : P L-87 | Jan. 22, 2025 (I)]**

List-I	List-II
(A) $Al^{3+} < Mg^{2+} < Na^+ < F^-$	(I) Ionisation Enthalpy
(B) B < C < O < N	(II) Metallic character
(C) B < Al < Mg < K	(III) Electronegativity
(D) Si < P < S < Cl	(IV) Ionic radii

Choose the correct answer from the options given below:
(a) A-IV, B-I, C-III, D-II
(b) A-II, B-III, C-IV, D-I
(c) A-IV, B-I, C-II, D-III
(d) A-III, B-IV, C-II, D-I

25. Which of the following electronegativity order is incorrect? **[NCERT : P L-90 | Jan. 22, 2025 (I)]**
(a) Al < Mg < B < N (b) Al < Si < C < N
(c) Mg < Be < B < N (d) S < Cl < O < F

26. Match **List-I** with **List-II** : **[NCERT : P L-85, 89 | April 8, 2024 (I)]**

List-I (Elements)	List-II (Properties in their respective groups)
(A) Cl,S	(I) Elements with highest electronegativity
(B) Ge, As	(II) Elements with largest atomic size
(C) Fr, Ra	(III) Elements which show properties of both metals and non metal
(D) F, O	(IV) Elements with highest negative electron gain enthalpy

Choose the **correct** answer from the options given below:
(a) A-(II), B-(III), C-(IV), D-(I)
(b) A-(III), B-(II), C-(I), D-(IV)
(c) A-(IV), B-(III), C-(II), D-(I)
(d) A-(II), B-(I), C-(IV), D-(III)

27. Identify the correct statements about p-block elements and their compounds. **[NCERT : P L-85 | April 8, 2024 (II)]**
(A) Non metals have higher electronegativity than metals.

(B) Non metals have lower ionisation enthalpy than metals.
(C) Compounds formed between highly reactive nonmetals and highly reactive metals are generally ionic.
(D) The non-metal oxides are generally basic in nature.
(E) The metal oxides are generally acidic or neutral in nature.
(a) (D) and (E) only (b) (A) and (C) only
(c) (B) and (E) only (d) (B) and (D) only

28. The electron affinity value are negative for :
A. $Be \rightarrow Be^-$
B. $N \rightarrow N^-$
C. $O \rightarrow O^{2-}$
D. $Na \rightarrow Na^-$
E. $Al \rightarrow Al^-$
Choose the most appropriate answer from the options given below : **[NCERT : P L-90 | April 6, 2024 (I)]**
(a) D and E only (b) A, B, D and E only
(c) A and D only (d) A, B and C only

29. Match **List-I** with **List-II**. **[NCERT : P L-87 |April 6, 2024 (II)]**

List-I (Alkali Metal)	**List-II (Emission Wavelength in nm)**
(A) Li	(I) 589.2
(B) Na	(II) 455.5
(C) Rb	(III) 670.8
(D) Cs	(IV) 780.0

Choose the **correct** answer from the options given below:
(a) (A)-(I), (B)-(IV), (C)-(III), (D)-(II)
(b) (A)-(III), (B)-(I), (C)-(IV), (D)-(II)
(c) (A)-(IV), (B)-(II), (C)-(I), (D)-(III)
(d) (A)-(II), (B)-(IV), (C)-(III), (D)-(I)

30. Evaluate the following statements related to group 14 elements for their correctness. **[NCERT : P L-93 | April 6, 2024 (II)]**
(A) Covalent radius decreases down the group from C to Pb in a regular manner.
(B) Electronegativity decreases from C to Pb down the group gradually.
(C) Maximum covalence of C is 4 whereas other elements can expand their covalence due to presence of d orbitals.
(D) Heavier elements do not form $p\pi$-$p\pi$ bonds.
(E) Carbon can exhibit negative oxidation states.
Choose the **correct** answer from the options given below:
(a) (C), (D) and (E) Only
(b) (A) and (B) Only
(c) (A), (B) and (C) Only
(d) (C) and (D) Only

31. The statement(s) that are **correct** about the species O^{2-}, F^-, Na^+ and Mg^{2+}. **[NCERT : P L-87 | April 5, 2024 (I)]**
(A) All are isoelectronic
(B) All have the same nuclear charge
(C) O^{2-} has the largest ionic radii
(D) Mg^{2+} has the smallest ionic radii
Choose the **most appropriate** answer from the options given below :
(a) (B), (C) and (D) only (b) (A), (B), (C) and (D)
(c) (C) and (D) only (d) (A), (C) and (D) only

32. Given below are two statements: **[NCERT : P L-87 | April 5, 2024 (II)]**
Statement I : The metallic radius of Na is 1.86 Å and the ionic radius of Na^+ is lesser than 1.86 Å.
Statement II : Ions are always smaller in size than the corresponding elements.
In the light of the above statements, choose the **correct** answer from the options given below :
(a) **Statement I** is correct but **Statement II** is false
(b) Both **Statement I** and **Statement II** are true
(c) Both **Statement I** and **Statement II** are false
(d) **Statement I** is incorrect but **Statement II** is true

33. The correct order of the first ionization enthalpy is **[NCERT : P L-87 |April 4, 2024 (II)]**
(a) Al > Ga > Tl (b) Ga > Al > B
(c) B > Al > Ga (d) Tl > Ga > Al

34. Given below are two statements: **[NCERT : P L-88 | April 4, 2024 (II)]**
Statement I : The correct order of first ionization enthalpy values of Li, Na, F and Cl is Na < Li < Cl < F.
Statement II : The correct order of negative electron gain enthalpy values of Li, Na, F and Cl is Na < Li < F < Cl
In the light of the above statements, choose the **correct** answer from the options given below :
(a) Both Statement I and Statement II are true.
(b) Both Statement I and Statement II are false.
(c) Statement I is false but Statement II is true.
(d) Statement I is true but Statement II is false.

35. In case of isoelectronic species the size of F^- Ne and Na^+ is affected by: **[NCERT : P L-87 | Feb. 1, 2024 (I)]**
(a) Principal quantum number (n)
(b) None of the factors because their size is the same
(c) Electron - electron interaction in the outer orbitals
(d) Nuclear charge (z)

36. The lowest oxidation number of an atom in a compound A_2B is –2. The number of electrons in its valence shell is ______. **[NCERT : P L-92 | NV, Feb. 1, 2024 (I)]**

37. Given below are two statements: **[NCERT : P L-90, 91 | Feb. 1, 2024 (II)]**
Statement (I) : Both metals and non - metals exist in *p* and *d* - block elements.
Statement (II) : Non - metals have higher ionisation enthalpy and higher electronegativity than the metals.
In the light of the above statements, choose the most appropriate answer from the options given below:
(a) Both Statement I and Statement II are false.
(b) Both Statement I and Statement II are true.
(c) Statement I is false but Statement II is true.
(d) Statement I is true but Statement II is false.

38. The correct sequence of electron gain enthalpy of the elements listed below is
[NCERT : P L-89 | Jan. 31, 2024 (I)]
A. Ar B. Br
C. F D. S
Choose the **most appropriate** from the options given below:
(a) A > D > B > C (b) D > C > B > A
(c) A > D > C > B (d) C > B > D > A

39. Consider the following elements.

Group ↓ A' B' ⟶ Period
C' D'

Which of the following is/are true about A', B', C' and D' ?
[NCERT : P L-85 | Jan. 31, 2024 (II)]
A. Order of atomic radii: B' < A' < D' < C'
B. Order of metallic character : B' < A' < D' < C'
C. Size of the element : D' < C' < B' < A'
D. Order of ionic radii : $B'^{+} < A'^{+} < D'^{+} < C'^{+}$
Choose the correct answer from the options given below:
(a) A only (b) A, B and D only
(c) A and B only (d) B, C and D only

40. Given below are two statements : One is labelled as Assertion **(A)** and the other is labelled as Reason **(R)**.
[NCERT : P L-86, 87 | Jan. 30, 2024 (I)]
Assertion (A) : There is a considerable increase in covalent radius from N to P. However from As to Bi only a small increase in covalent radius is observed.
Reason (R) : Covalent and ionic radii in a particular oxidation state increases down the group.
In the light of the above statements, choose the **most appropriate** answer from the options given below:
(a) **(A)** is false but **(R)** is true.
(b) Both **(A)** and **(R)** are true but **(R)** is not the correct explanation of **(A)**.
(c) **(A)** is true but **(R)** is false.
(d) Both **(A)** and **(R)** are true and **(R)** is the correct explanation of **(A)**.

41. Given below are two statements:
[NCERT : P L-94 | Jan. 30, 2024 (II)]
Statements – I: Along the period, the chemical reactivity of the elements gradually increases from group 1 to group 18.
Statement – II: The nature of oxides formed by group 1 elements is basic while that of group 17 elements is acidic.
In the light of the above statements, choose the most appropriate from the options given below:
(a) Both Statement I and Statement II are true.
(b) Statement I is true but Statement II is false.
(c) Statement I is false but Statement II is true.
(d) Both Statement I and Statement II are false.

42. Given below are two statements : one is labelled as **Assertion A** and the other is labelled as **Reason R :**
[NCERT : P L-87 | Jan. 29, 2024 (I)]
Assertion A : The first ionisation enthalpy decreases across a period.
Reason R : The increasing nuclear charge outweighs the shielding across the period.
In the light of the above statements, choose the **most appropriate** from the options given below:
(a) **A** is false but **R** is true.
(b) **A** is true but **R** is false.
(c) Both **A** and **R** are true and **R** is the correct explanation of **A**.
(d) Both **A** and **R** are true but **R** is NOT the correct explanation of **A**.

43. Given below are two statements
[NCERT : P L-83, 90 | Jan. 29, 2024 (I)]
Statement I : The electronegativity of group 14 elements from Si to Pb, gradually decreases.
Statement II : Group 14 contains non-metallic, metallic, as well as metalloid elements.
In the light of the above statements, choose the **most appropriate** from the options given below:
(a) Statement I is true but Statement II is false.
(b) Statement I is false but Statement II is true.
(c) Both Statement I and Statement II are false.
(d) Both Statement I and Statement II are true.

44. The element having the highest first ionization enthalpy is
[NCERT : P L-87 | Jan. 29, 2024 (II)]
(a) C (b) Al (c) Si (d) N

45. Given below are two statements :
[NCERT : P L-84 | Jan. 27, 2024 (I)]
Statement I : The 4f and 5f - series of elements are placed separately in the Periodic table to reserve the principle of classification.
Statement II : s-block elements can be found in pure form in nature.
In the light of the above statements, choose the **most appropriate** answer from the options given below:
(a) Statement I is false but Statement II is true.
(b) Both Statement I and Statement II are true.
(c) Statement I is true but Statement II is false.
(d) Both Statement I and Statement II are false.

46. Which one of the following sets of ions represents a collection of isoelectronic species ?
(Given : Atomic number : F : 9, Cl : 17, Na = 11, Mg = 12, Al = 13, K = 19, Ca = 20, Sc = 21)
[NCERT : P L- 87 | Feb. 01, 2023 (II)]
(a) $N^{3-}, O^{2-}, F^{-}, S^{2-}$
(b) $Ba^{2+}, Sr^{2+}, K^{+}, Ca^{2+}$
(c) $K^{+}, Cl^{-}, Ca^{2+}, Sc^{3+}$
(d) $Li^{+}, Na^{+}, Mg^{2+}, Ca^{2+}$

47. The total number of isoelectronic species from the given set is ______.
[NCERT : P L-81 | NV, April 15, 2023 (I)]
$O^{2-}, F^{-}, Al, Mg^{2+}, Na^{+}, O^{+}, Mg, Al^{3+}, F$

48. Which of the following statements are **not** correct?
[NCERT : P L-89, 90 | April 13, 2023 (I)]
A. The electron gain enthalpy of F is more negative than that of Cl

B. Ionization enthalpy decreases in a group of periodic table

C. The electronegativity of an atom depends upon the atoms bonded to it.

D. Al_2O_3 and NO are examples of amphoteric oxides.

Choose the most appropriate answer from the options given below:

(a) A, B, C and D (b) A, C and D Only

(c) B and D Only (d) A, B and D Only

49. Given below are two statements : **[NCERT : P L-94 | April 12, 2023 (I)]**

Statement – I : $SbCl_5$ is more covalent than $SbCl_3$.

Statement – II : The higher oxides of halogens also tend to be more stable than the lower ones.

In the light of the above statements, choose the most appropriate answer from the options given below

(a) Both Statement I and Statement II are correct

(b) Both Statement I and Statement II are incorrect

(c) Statement I is correct but Statement II is incorrect

(d) Statement I is incorrect but Statement II is correct

50. For elements B, C, N, Li, Be, O and F the correct order of first ionization enthalpy is **[NCERT : P L-88 | April 11, 2023 (I)]**

(a) Li < Be < B < C < N < O < F

(b) B > Li > Be > C > N > O > F

(c) Li < B < Be < C < O < N < F

(d) Li < Be < B < C < O < N < F

51. In the following reactions, the total number of oxygen atoms in X and Y is ______.

[NCERT : P L-94 | NV, April 10, 2023 (I)]

$Na_2O + H_2O \rightarrow 2X$

$Cl_2O_7 + H_2O \rightarrow 2Y$

52. Given below are two statements: one is labelled as **Assertion A** and the other is labelled as **Reason R.**

Assertion A: The energy required to form Mg^{2+} from Mg is much higher than that required to produce Mg^+.

Reason R: Mg^{2+} is small ion and carry more charge than Mg^+.

In the light of the above statements, choose the **correct** answer from the options given below:

[NCERT : P L-89 | April 10, 2023 (II)]

(a) Both A and R are true but R is NOT the correct explanation of A.

(b) A is true but R is false.

(c) A is false but R is true.

(d) Both A and R are true and R is the correct explanation of A.

53. The correct order of metallic character is:

[NCERT : P L-85 | April 10, 2023 (II)]

(a) Be > Ca > K (b) Ca > K > Be

(c) K > Ca > Be (d) K > Be > Ca

54. The correct order of electronegativity for given elements is: **[NCERT : P L-91 | April 08, 2023 (I)]**

(a) C > P > At > Br (b) Br > P > At > C

(c) P > Br > C > At (d) Br > C > At > P

55. The difference between electron gain enthalpies will be maximum between: **[NCERT : P L-90 | April 06, 2023 (I)]**

(a) Ne and F (b) Ne and Cl

(c) Ar and Cl (d) Ar and F

56. Group-13 elements react with O_2 in amorphous form to form oxides of type M_2O_3 (M = element). Which among the following is the most basic oxide?

[NCERT : P L-94 | April 06, 2023 (II)]

(a) Al_2O_3 (b) Ga_2O_3 (c) Tl_2O_3 (d) B_2O_3

57. The correct increasing order of the ionic radii is

[NCERT : P L-87 | Jan. 31, 2023 (I)]

(a) $Cl^- < Ca^{2+} < K^+ < S^{2-}$ (b) $K^+ < S^{2-} < Ca^{2+} < Cl^-$

(c) $S^{2-} < Cl^- < Ca^{2+} < K^+$ (d) $Ca^{2+} < K^+ < Cl^- < S^{2-}$

58. Given below are two statements: one is labelled as Assertion (A) and the other is labelled as Reason (R)

Assertion (A) : The first ionization enthalpy of 3d series elements is more than that of group 2 metals

Reason (R) : In 3d series of elements successive filling of d-orbitals takes place.

In the light of the above statements, choose the correct answer from the options given below :

[NCERT : P L-87 | Jan. 31, 2023 (II)]

(a) Both (A) and (R) are true and (R) is the correct explanation of (A).

(b) Both (A) and (R) are true but (R) is not the correct explanation of (A).

(c) (A) is false but (R) is true.

(d) (A) is true but (R) is false.

59. Total number of acidic oxides among N_2O_3, NO_2, N_2O, Cl_2O_7, SO_2, CO, CaO, Na_2O and NO is ______. **[NCERT : P L-94 | NV, Jan. 29, 2023 (II)]**

60. Which of the following represents the correct order of metallic character of the given elements ?

[NCERT : P L-85 | Jan. 25, 2023 (II)]

(a) Si < Be < Mg < K (b) Be < Si < Mg < K

(c) K < Mg < Be < Si (d) Be < Si < K < Mg

61. The first ionization enthalpy of Na, Mg and Si, respectively, are: 496, 737 and 786 kJ mol^{-1}. The first ionization enthalpy (kJ mol^{-1}) of Al is:

[NCERT : P L-87 | July 29, 2022 (I)]

(a) 487 (b) 768 (c) 577 (d) 856

62. In which of the following pairs, electron gain enthalpies of constituent elements are nearly the same or identical ?

[NCERT : P L-90 | July 28, 2022 (I)]

(A) Rb and Cs (B) Na and K

(C) Ar and Kr (D) I and At

Choose the correct answer from the options given below :

(a) (A) and (B) only (b) (B) and (C) only

(c) (A) and (C) only (d) (C) and (D) only

63. The correct decreasing order for metallic character is

[NCERT : P L-85 | July 28, 2022 (II)]

(a) Na > Mg > Be > Si > P

(b) P > Si > Be > Mg > Na

(c) Si > P > Be > Na > Mg

(d) Be > Na > Mg > Si > P

64. The **incorrect** statement is **[NCERT : P L-87 | July 27, 2022 (I)]**
(a) The first ionization enthalpy of K is less than that of Na and Li
(b) Xe does not have the lowest first ionization enthalpy in its group
(c) The first ionization enthalpy of element with atomic number 37 is lower than that of the element with atomic number 38.
(d) The first ionization enthalpy of Ga is higher than that of the d-block element with atomic number 30.

65. Outermost electronic configurations of four elements A, B, C, D are given below: **[NCERT : P L-87 | July 27, 2022 (II)]**
(A) $3s^2$ (B) $3s^2 3p^1$
(C) $3s^2 3p^3$ (D) $3s^2 3p^4$
The **correct** order of first ionization enthalpy for them is:
(a) (A) < (B) < (C) < (D) (b) (B) < (A) < (D) < (C)
(c) (B) < (D) < (A) < (C) (d) (B) < (A) < (C) < (D)

66. An element A of group 1 shows similarity to an element B belonging to group 2. If A has maximum hydration enthalpy in group 1 then B is: **[July 27, 2022 (II)]**
(a) Mg (b) Be
(c) Ca (d) Sr

67. Given two statements below: **[NCERT : P L-86, 87 | July 26, 2022 (I)]**
Statement I: In Cl_2 molecule the covalent radius is double of the atomic radius of chlorine.
Statement II: Radius of anionic species is always greater than their parent atomic radius.
Choose the **most appropriate** answer from options given below:
(a) Both statement I and statement II are correct.
(b) Both statement I and statement II are incorrect.
(c) Statement I is correct but statement II is incorrect.
(d) Statement I is incorrect but statement II is correct.

68. The first ionization enthalpies of Be, B, N and O follow the order **[NCERT : P L-88 | July 25, 2022 (II)]**
(a) O < N < B < Be (b) Be < B < N < O
(c) B < Be < N < O (d) B < Be < O < N

69. The correct order of electron gain enthalpy (– ve value) is: **[NCERT : P L-90 | June 30, 2022 (I)]**
(a) O > S > Se > Te (b) O < S < Se < Te
(c) O < S > Se > Te (d) O < S > Se < Te

70. Given below are two statements. One is labelled as **Assertion A** and the other is labelled as **Reason R**.
Assertion A : The first ionization enthalpy for oxygen is lower than that of nitrogen.
Reason R : The four electrons in 2*p* orbitals of oxygen experience more electron-electron repulsion.
In the light of the above statements, choose the correct answer from the options given below.
[NCERT : P L-87 | June 29, 2022 (II)]
(a) Both A and R are correct and R is the correct explanation of A.
(b) Both A and R are correct but R is NOT the correct explanation of A.
(c) A is correct but R is not correct.
(d) A is not correct but R is correct.

71. Among the following basic oxide is : **[NCERT : P L-94 | June 28, 2022 (II)]**
(a) SO_3 (b) SiO_2 (c) CaO (d) Al_2O_3

72. Match List-I with List-II. **[NCERT : P L-94 | June 28, 2022 (II)]**

List-I (Oxide)	**List-II (Nature)**
(A) Cl_2O_7	(I) Amphoteric
(B) Na_2O	(II) Basic
(C) Al_2O_3	(III) Neutral
(D) N_2O	(IV) Acidic

Choose the correct answer from the options given below:
(a) (A) – (IV), (B) – (III), (C) – (I), (D) – (II)
(b) (A) – (IV), (B) – (II), (C) – (I), (D) – (III)
(c) (A) – (II), (B) – (IV), (C) – (III), (D) – (I)
(d) (A) – (I), (B) – (II), (C) – (III), (D) – (IV)

73. Given below are two statements: one is labelled as **Assertion (A)** and the other is labelled as **Reason(R)**
Assertion (A) : The ionic radii of O^{2-} and Mg^{2+} are same.
Reason (R) : Both O^{2-} and Mg^{2+} are isoelectronic species
In the light of the above statements, choose the correct answer from the options given below : **[NCERT : P L-87 | June 27, 2022 (I)]**
(a) Both (A) and (R) are true and (R) is the correct explanation of (A).
(b) Both (A) and (R) are true but (R) is not the correct explanation of (A).
(c) (A) is true but (R) is false.
(d) (A) is false but (R) is true.

74. The correct order of increasing ionic radii is **[NCERT : P L-87 | June 27, 2022 (II)]**
(a) $Mg^{2+} < Na^+ < F^- < O^{2-} < N^{3-}$
(b) $N^{3-} < O^{2-} < F^- < Na^+ < Mg^{2+}$
(c) $F^- < Na^+ < O^{2-} < Mg^{2+} < N^{3-}$
(d) $Na^+ < F^- < Mg^{2+} < O^{2-} < N^{3-}$

75. The correct order of melting point is : **[June 26, 2022 (I)]**
(a) Be > Mg > Ca > Sr (b) Sr > Ca > Mg > Be
(c) Be > Ca > Mg > Sr (d) Be > Ca > Sr > Mg

76. Which of the following elements is considered as a metalloid? **[NCERT : P L-85 | June 26, 2022 (II)]**
(a) Sc (b) Pb (c) Bi (d) Te

77. The pair, in which ions are isoelectronic with Al^{3+} is : **[NCERT : P L-87 | June 25, 2022 (I)]**
(a) Br^- and Be^{2+} (b) Cl^- and Li^+
(c) S^{2-} and K^+ (d) O^{2-} and Mg^{2+}

78. The correct order of electron gain enthalpies of Cl, F, Te and Po is **[NCERT : P L-90 | June 25, 2022 (II)]**
(a) F < Cl < Te < Po (b) Po < Te < F < Cl
(c) Te < Po < Cl < F (d) Cl < F < Te < Po

79. Metals generally melt at very high temperature. Amongst the following, the metal with the highest melting point will be **[June 24, 2022 (II)]**
(a) Hg (b) Ag (c) Ga (d) Cs

80. The ionic radii of K^+, Na^+, Al^{3+} and Mg^{2+} are in the order: **[NCERT : P L-87 | July 25, 2021 (I)]**
(a) $Na^+ < K^+ < Mg^{2+} < Al^{3+}$ (b) $Al^{3+} < Mg^{2+} < K^+ < Na^+$
(c) $Al^{3+} < Mg^{2+} < Na^+ < K^+$ (d) $K^+ < Al^{3+} < Mg^{2+} < Na^+$

81. The set in which compounds have different nature is : **[July 20, 2021 (I)]**
(a) $B(OH)_3$ and H_3PO_3 (b) $B(OH)_3$ and $Al(OH)_3$
(c) NaOH and $Ca(OH)_2$ (d) $Be(OH)_2$ and $Al(OH)_3$

82. The set of elements that differ in mutual relationship from those of the other sets is **[NCERT : P L-93 | March 17, 2021 (II)]**
(a) B – Si (b) Li – Na
(c) Be – Al (d) Li – Mg

83. The first ionization energy of magnesium is smaller as compared to that of elements X and Y, but higher than that of Z. The elements X, Y and Z, respectively, are **[NCERT : P L-87 | March 18, 2021 (II)]**
(a) Chlorine, lithium and sodium
(b) argon, lithium and sodium
(c) argon, chlorine and sodium
(d) neon, sodium and chlorine

84. Which pair of oxides is acidic in nature? **[NCERT : P L-94 | Feb. 26, 2021 (II)]**
(a) CaO, SiO_2 (b) B_2O_3, CaO
(c) B_2O_3, SiO_2 (d) N_2O, BaO

85. The correct order of electron gain enthalpy is: **[NCERT : P L-90 | Feb. 26, 2021 (II)]**
(a) O > S > Se > Te (b) Te > Se > S > O
(c) S > O > Se > Te (d) S > Se > Te > O

86. Consider the elements Mg, Al, S, P and Si, the correct increasing order of their first ionization enthalpy is : **[NCERT : P L-87 | February 24, 2021 (I)]**
(a) Mg < Al < Si < S < P (b) Al < Mg < Si < S < P
(c) Mg < Al < Si < P < S (d) Al < Mg < S < Si < P

87. The set that contains atomic numbers of only transition elements, is : **[NCERT : P L-84 | Sep. 06, 2020 (I)]**
(a) 37, 42, 50, 64 (b) 21, 25, 42, 72
(c) 9, 17, 34, 38 (d) 21, 32, 53, 64

88. The correct order of the ionic radii of O^{2-}, N^{3-}, F^-, Mg^{2+}, Na^+ and Al^{3+} is : **[NCERT : P L-87 | Sep. 04, 2020(I) (S)]**
(a) $N^{3-} < O^{2-} < F^- < Na^+ < Mg^{2+} < Al^{3+}$
(b) $Al^{3+} < Na^+ < Mg^{2+} < O^{2-} < F^- < N^{3-}$
(c) $Al^{3+} < Mg^{2+} < Na^+ < F^- < O^{2-} < N^{3-}$
(d) $N^{3-} < F^- < O^{2-} < Mg^{2+} < Na^+ < Al^{3+}$

89. The process that is NOT endothermic in nature is : **[NCERT : P L-89 | Sep. 04, 2020 (II)]**
(a) $Ar(g) + e^- \rightarrow Ar^-(g)$
(b) $H(g) + e^- \rightarrow H^-(g)$
(c) $O^-(g) + e^- \rightarrow O^{2-}(g)$
(d) $Na(g) \rightarrow Na^+(g) + e^-$

90. The ionic radii of O^{2-}, F^-, Na^+ and Mg^{2+} are in the order: **[NCERT : P L-87 | Sep. 04, 2020 (I)]**
(a) $F^- > O^{2-} > Na^+ > Mg^{2+}$ (b) $O^{2-} > F^- > Na^+ > Mg^{2+}$
(c) $Mg^{2+} > Na^+ > F^- > O^{2-}$ (d) $O^{2-} > F^- > Mg^{2+} > Na^+$

91. Among the statements (I - IV), the **correct** ones are : **[NCERT : P L-85 | Sep. 03, 2020 (II)]**
(I) Be has smaller atomic radius compared to Mg.
(II) Be has higher ionization enthalpy than Al.
(III) Charge/radius ratio of Be is greater than that of Al.
(IV) Both Be and Al form mainly covalent compounds.
(a) (I), (II) and (IV) (b) (I), (III) and (IV)
(c) (II), (III) and (IV) (d) (I), (II) and (III)

92. The five successive ionization enthalpies of an element are 800, 2427, 3658, 25024 and 32824 kJ mol^{-1}. The number of valence electrons in the element is : **[NCERT : P L-87 | Sep. 03, 2020 (II)]**
(a) 5 (b) 4 (c) 3 (d) 2

93. In general, the property (magnitudes only) that shows an opposite trend in comparison to other properties across a period is : **[NCERT : P L-85 | Sep. 02, 2020 (I); Same Jan. 09, 2019 (I)]**
(a) Ionization enthalpy (b) Electronegativity
(c) Electron gain enthalpy (d) Atomic radius

94. Three elements X, Y and Z are in the 3rd period of the periodic table. The oxides of X, Y and Z, respectively, are basic, amphoteric and acidic. The correct order of the atomic numbers of X, Y and Z is : **[NCERT : P L-94 | Sep. 02, 2020 (II)]**
(a) Z < Y < X (b) X < Y < Z
(c) X < Z < Y (d) Y < X < Z

95. B has a smaller first ionization enthalpy than Be. Consider the following statements:
(I) it is easier to remove 2p electron than 2s electron
(II) 2p electron of B is more shielded from the nucleus by the inner core of electrons than the 2s electrons of Be
(III) 2s electron has more penetration power than 2p electron
(IV) atomic radius of B is more than Be
(atomic number B = 5, Be = 4)
The correct statements are: **[NCERT : P L-88, 89 | April 12, 2019 (II) (S), Jan. 09, 2020 (I)]**
(a) (I), (II) and (IV) (b) (II), (III) and (IV)
(c) (I), (II) and (III) (d) (I), (III) and (IV)

96. The acidic, basic and amphoteric oxides, respectively, are: **[NCERT : P L-94 | Jan. 09, 2020 (I)]**
(a) Na_2O, SO_3, Al_2O_3 (b) Cl_2O, CaO, P_4O_{10}
(c) N_2O_3, Li_2O, Al_2O_3 (d) MgO, Cl_2O, Al_2O_3

97. The first and second ionisation enthalpies of a metal are 496 and 4560 kJ mol^{-1}, respectively. How many moles of HCl and H_2SO_4, respectively, will be needed to react completely with 1 mole of the metal hydroxide? **[NCERT : P L-87 | Jan. 09, 2020 (II)]**
(a) 1 and 1 (b) 2 and 0.5 (c) 1 and 2 (d) 1 and 0.5

98. In general, the properties that decrease and increase down a group in the periodic table, respectively, are: **[NCERT : P L-94 | Jan. 9, 2019 (I)]**

(a) atomic radius and electronegativity.
(b) electron gain enthalpy and electronegativity.
(c) electronegativity and atomic radius.
(d) electronegativity and electron gain enthalpy.

99. The first ionization energy (in kJ/mol) of Na, Mg, Al and Si respectively, are: **[NCERT : P L-87 | Jan. 08, 2020 (I)]**
(a) 496, 737, 577, 786 (b) 496, 577, 737, 786
(c) 786, 737, 577, 496 (d) 496, 577, 786, 737

100. The increasing order of the atomic radii of the following elements is: **[NCERT : P L-86 | Jan. 08, 2020 (II)]**
(i) C (ii) O (iii) F (iv) Cl (v) Br
(a) (ii) < (iii) < (iv) < (i) < (v)
(b) (iv) < (iii) < (ii) < (i) < (v)
(c) (iii) < (ii) < (i) < (iv) < (v)
(d) (i) < (ii) < (iii) < (iv) < (v)

101. The electron gain enthalpy (in kJ/mol) of fluorine, chlorine, bromine and iodine, respectively, are:
[NCERT : P L-89 | Jan. 07, 2020 (I); Same 2011]
(a) –296, –325, –333 and –349
(b) –349, –333, –325 and –296
(c) –333, –349, –325 and –296
(d) –333, –325, –349 and –296

102. Within each pair of elements F & Cl, S & Se, and Li & Na, respectively, the elements that release more energy upon an electron gain are: **[NCERT : P L-90 | Jan. 07, 2020 (II)]**
(a) Cl, Se and Na (b) Cl, S and Li
(c) F, S and Li (d) F, Se and Na

103. In comparison to boron, berylium has:
[NCERT : P L-87 | April 12, 2019 (II)]
(a) lesser nuclear charge and lesser first ionisation enthalpy.
(b) greater nuclear charge and lesser first ionisation enthalpy.
(c) greater nulear charge and greater first ionisation enthalpy.
(d) lesser nuclear charge and greater first ionisation enthalpy.

104. The isoelectronic set of ions is
[NCERT : P L-81 | [2003 (S), 2004 (S), 2005 (S), 2006(S), 2008 (S), April 10, 2019 (I)]
(a) N^{3-}, O^{2-}, F^- and Na^+ (b) N^{3-}, Li^+, Mg^{2+} and O^{2-}
(c) F^-, Li^+, Na^+ and Mg^{2+} (d) Li^+, Na^+, O^{2-} and F^-

105. In general, the properties that decrease and increase down a group in the periodic table, respectively, are:
[NCERT : P L-87, 90 | Jan. 9, 2019 (I)]
(a) atomic radius and electronegativity.
(b) electron gain enthalpy and electronegativity.
(c) electronegativity and atomic radius.
(d) electronegativity and electron gain enthalpy.

106. The element having greatest difference between its first and second ionization energies, is:
[NCERT : P L-87 | April 9, 2019 (I)]
(a) Ca (b) Sc (c) Ba (d) K

107. The correct order of the atomic radii of C, Cs, Al, and S is:
[NCERT : P L-86 | Jan. 11, 2019 (I)]
(a) C < S < Al < Cs (b) S < C < Cs < Al
(c) S < C < Al < Cs (d) C < S < Cs < Al

108. The correct option with respect to the Pauling electronegativity values of the elements is:
[NCERT : P L-90 | Jan. 11, 2019 (II)]
(a) Te > Se (b) Ga < Ge
(c) Si < Al (d) P > S

109. When the first electron gain enthalpy($\Delta_{eg}H$) of oxygen is – 141 kJ/mol, its second electron gain enthalpy is:
[NCERT : P L-89 | Jan. 09, 2019 (II)]
(a) a more negative value than the first
(b) almost the same as that of the first
(c) negative, but less negative than the first
(d) a positive value

110. For Na^+, Mg^{2+}, F^- and O^{2-}; the correct order of increasing Ionic radii is: **[NCERT : P L-87 | Online April 15, 2018 (I)]**
(a) $O^{2-} < F^- < Na^+ < Mg^{2+}$
(b) $Na^+ < Mg^{2+} < F^- < O^{2-}$
(c) $Mg^{2+} < Na^+ < F^- < O^{2-}$
(d) $Mg^{2+} < O^{2-} < Na^+ < F^-$

111. The correct order of electron affinity is:
[NCERT : P L-94 | Online April 15, 2018 (II)]
(a) O > F > Cl (b) F > O > Cl
(c) F > Cl > O (d) Cl > F > O

112. Both lithium and magnesium display several similar properties due to the diagonal relationship; however, the one which is incorrect is : **[NCERT : P L-93 | 2017]**
(a) Both form basic carbonates
(b) Both form soluble bicarbonates
(c) Both form nitrides
(d) Nitrates of both Li and Mg yield NO_2 and O_2 on heating

113. The group having isoelectronic species is :
[NCERT : P L-87 | 2017]
(a) O^{2-}, F^-, Na^+, Mg^{2+}
(b) O^-, F^-, Na, Mg^+
(c) O^{2-}, F^-, Na, Mg^{2+}
(d) O^-, F^-, Na^+, Mg^{2+}

114. The electronic configuration with the highest ionization enthalpy is : **[NCERT : P L-88 | Online April 9, 2017]**
(a) [Ne] $3s^2 3p^1$ (b) [Ne] $3s^2 3p^2$
(c) [Ne] $3s^2 3p^3$ (d) [Ar] $3d^{10} 4s^2 4p^3$

115. Consider the following ionization enthalpies of two elements 'A' and 'B'.
[NCERT : P L-98 | Online April 8, 2017]

Element	Ionization 1st	enthalpy 2nd	(kJ/mol) 3rd
A	899	1757	14847
B	737	1450	7731

Which of the following statements is correct ?
(a) Both 'A' and 'B' belong to group–1 where 'B' comes below 'A'.
(b) Both 'A' and 'B' belong to group–1 where 'A' comes below 'B'.
(c) Both 'A' and 'B' belong to group–2 where 'B' comes below 'A'.
(d) Both 'A' and 'B' belong to group–2 where 'A' comes below 'B'.

116. Which of the following atoms has the highest first ionization energy? **[NCERT : P L-87 | 2016]**
(a) K (b) Sc (c) Rb (d) Na

117. The following statements concern elements in the periodic table. Which of the following is true ?

[NCERT : P L-89 | Online April 10, 2016]

(a) For Group 15 elements, the stability of +5 oxidation state increases down the group

(b) Elements of Group 16 have lower ionization enthalpy values compared to those of Group 15 in the corresponding periods.

(c) The Group 13 elements are all metals

(d) All the elements in Group 17 are gases.

118. The ionic radii (in Å) of N^{3-}, O^{2-} and F^- are respectively :

[NCERT : P L-87 | 2015]

(a) 1.71, 1.40 and 1.36 (b) 1.71, 1.36 and 1.40

(c) 1.36, 1.40 and 1.71 (d) 1.36, 1.71 and 1.40

119. Which of the following arrangements represents the increasing order (smallest to largest) of ionic radii of the given species O^{2-}, S^{2-}, N^{3-}, P^{3-}?

[NCERT : P L-87 | Online April 12, 2014(s), 2020]

(a) $O^{2-} < N^{3-} < S^{2-} < P^{3-}$ (b) $O^{2-} < P^{3-} < N^{3-} < S^{2-}$

(c) $N^{3-} < O^{2-} < P^{3-} < S^{2-}$ (d) $N^{3-} < S^{2-} < O^{2-} < P^{3-}$

120. Which of the following series correctly represents relations between the elements from X to Y?

$X \rightarrow Y$ **[NCERT : P L-88 | Online April 11, 2014]**

(a) $_3Li \rightarrow {}_{19}K$ Ionization enthalpy increases

(b) $_9F \rightarrow {}_{35}Br$ Electron gain enthalpy (negative sign) increases

(c) $_6C \rightarrow {}_{32}Ge$ Atomic radii increases

(d) $_{18}Ar \rightarrow {}_{54}Xe$ Noble character increases

121. Which one of the following has largest ionic radius?

[NCERT : P L-87 | 2004 (S), Online April 19, 2014]

(a) Li^+ (b) O_2^{2-} (c) B^{3+} (d) F^-

122. Which of the following represents the correct order of increasing first ionization enthalpy for Ca, Ba, S, Se and Ar ?

[NCERT : P L-87 | 2013]

(a) Ca < S < Ba < Se < Ar

(b) S < Se < Ca < Ba < Ar

(c) Ba < Ca < Se < S < Ar

(d) Ca < Ba < S < Se < Ar

123. The order of increasing sizes of atomic radii among the elements O, S, Se and As is :

[NCERT : P L-94 | Online April 22, 2013]

(a) As < S < O < Se (b) Se < S < As < O

(c) O < S < As < Se (d) O < S < Se < As

124. Which is the correct order of second ionization potential of C, N, O and F in the following ? **[NCERT : P L-87 | 2006 (S), Online May 26, 2012; April 23, 2013]**

(a) O > N > F > C (b) O > F > N > C

(c) F > O > N > C (d) C > N > O > F

125. Which among the following elements has the highest first ionization enthalpy?

[NCERT : P L-87 | Online May 12, 2012]

(a) Nitrogen (b) Boron

(c) Carbon (d) Oxygen

126. The increasing order of the ionic radii of the given isoelectronic species is : **[NCERT : P L-87 | 2012]**

(a) Cl^-, Ca^{2+}, K^+, S^{2-} (b) S^{2-}, Cl^-, Ca^{2+}, K^+

(c) Ca^{2+}, K^+, Cl^-, S^{2-} (d) K^+, S^{2-}, Ca^{2+}, Cl^-

127. Which among the following elements has the highest first ionization enthalpy?

[NCERT : P L-87 | Online May 12, 2012]

(a) Nitrogen (b) Boron

(c) Carbon (d) Oxygen

128. Which among the following elements has the highest first ionization enthalpy?

[NCERT : P L-87 | Online May 12, 2012]

(a) Nitrogen (b) Boron

(c) Carbon (d) Oxygen

129. The electron affinity of chlorine is 3.7 eV. 1 gram of chlorine is completely converted to Cl^- ion in a gaseous state. (1 eV = 23.06 kcal mol^{-1}).

Energy released in the process is

[NCERT : P L-89 | Online May 7, 2012]

(a) 4.8 kcal (b) 7.2 kcal (c) 8.2 kcal (d) 2.4 kcal

130. In which of the following arrangements, the sequence is not strictly according to the property written against it?

[2005 (S), 2008, Online May 7, 2012]

(a) $CO_2 < SiO_2 < SnO_2 < PbO_2$: increasing oxidising power

(b) $NH_3 < PH_3 < AsH_3 < SbH_3$: increasing basic strength

(c) HF < HCl < HBr < HI: increasing acid strength

(d) B < C < O < N: increasing first ionisation enthalpy.

131. The correct order of electron gain enthalpy with negative sign of F, Cl, Br and I, having atomic numbers 9, 17, 35 and 53 respectively, is : **[NCERT : P L-90 | 2011RS]**

(a) F > Cl > Br > I (b) Cl > F > Br > I

(c) Br > Cl > I > F (d) I > Br > Cl > F

132. The correct sequence which shows decreasing order of the ionic radii of the elements is

[NCERT : P L-87 | 2010]

(a) $Al^{3+} > Mg^{2+} > Na^+ > F^- > O^{2-}$

(b) $Na^+ > Mg^{2+} > Al^{3+} > O^{2-} > F^-$

(c) $Na^+ > F^- > Mg^{2+} > O^{2-} > Al^{3+}$

(d) $O^{2-} > F^- > Na^+ > Mg^{2+} > Al^{3+}$

133. The increasing order of the first ionization enthalpies of the elements B, P, S and F (Lowest first) is

[NCERT : P L-87 | 2006]

(a) B < P < S < F (b) B < S < P < F

(c) F < S < P < B (d) P < S < B < F

134. Following statements regarding the periodic trends of chemical reactivity of alkali metals and halogens are given. Which of these statements gives the correct picture?

[NCERT : P L-94 | 2006]

(a) Chemical reactivity increases with increase in atomic number down the group in both the alkali metals and halogens

(b) In alkali metals, the reactivity increases but in the halogens, it decreases with increase in atomic number down the group

(c) The reactivity decreases in the alkali metals but increases in the halogens with increase in atomic number down the group

(d) In both, alkali metals and halogens, chemical reactivity decreases with increase in atomic number down the group

135. In which of the following arrangements, the order is NOT according to the property indicated against it? **[NCERT : P L-87 | 2005]**

(a) $Li < Na < K < Rb$: Increasing metallic radius

(b) $I < Br < F < Cl$: Increasing electron gain enthalpy (with negative sign)

(c) $B < C < N < O$ Increasing first ionization enthalpy

(d) $Al^{3+} < Mg^{2+} < Na^{+} < F^{-}$ Increasing ionic size

136. Which of the following oxides is amphoteric in character? **[NCERT : P L-94 | 2003 (S), 2005]**

(a) SnO_2 (b) SiO_2

(c) CO_2 (d) CaO

137. Among Al_2O_3, SiO_2, P_2O_3 and SO_2, the correct order of acid strength is **[NCERT : P L-94 | 2004]**

(a) $Al_2O_3 < SiO_2 < SO_2 < P_2O_3$

(b) $SiO_2 < SO_2 < Al_2O_3 < P_2O_3$

(c) $SO_2 < P_2O_3 < SiO_2 < Al_2O_3$

(d) $Al_2O_3 < SiO_2 < P_2O_3 < SO_2$

138. The formation of the oxide ion O^{2-}(g) requires first an exothermic and then an endothermic step as shown below

$$O(g) + e^{-} = O^{-}(g) \ \Delta H° = -142 \ \text{kJmol}^{-1}$$

$$O^{-}(g) + e^{-} = O^{2-}(g) \ \Delta H° = +844 \ \text{kJmol}^{-1}$$

This is because **[NCERT : P L-89 | 2004]**

(a) O^- ion will tend to resist the addition of another electron

(b) Oxygen has high electron affinity

(c) Oxygen is more elecronegative

(d) O^- ion has comparatively larger size than oxygen atom

139. Which one of the following ions has the highest value of ionic radius ? **[NCERT : P L-87 | 2004]**

(a) O^{2-} (b) B^{3+}

(c) Li^+ (d) F^-

140. Which one of the following is an amphoteric oxide ? **[NCERT : P L-94 | 2003]**

(a) Na_2O (b) SO_2

(c) B_2O_3 (d) ZnO

ANSWER KEYS

1	(b)	15	(b)	29	(b)	43	(b)	57	(d)	71	(c)	85	(d)	99	(a)	113	(a)	127	(a)
2	(d)	16	(a)	30	(a)	44	(d)	58	(b)	72	(b)	86	(b)	100	(c)	114	(c)	128	(a)
3	(a)	17	(a)	31	(d)	45	(c)	59	(4)	73	(d)	87	(b)	101	(c)	115	(c)	129	(d)
4	(c)	18	(d)	32	(a)	46	(c)	60	(a)	74	(a)	88	(c)	102	(b)	116	(b)	130	(b)
5	(a)	19	(a)	33	(d)	47	(5)	61	(c)	75	(d)	89	(b)	103	(d)	117	(b)	131	(b)
6	(b)	20	(d)	34	(a)	48	(b)	62	(c)	76	(d)	90	(b)	104	(a)	118	(a)	132	(d)
7	(11)	21	(c)	35	(d)	49	(a)	63	(a)	77	(d)	91	(a)	105	(c)	119	(a)	133	(b)
8	(b)	22	(b)	36	(6)	50	(c)	64	(d)	78	(b)	92	(c)	106	(d)	120	(c)	134	(b)
9	(c)	23	(a)	37	(c)	51	(5)	65	(b)	79	(b)	93	(d)	107	(a)	121	(b)	135	(c)
10	(d)	24	(c)	38	(a)	52	(d)	66	(a)	80	(c)	94	(b)	108	(b)	122	(c)	136	(a)
11	(b)	25	(a)	39	(b)	53	(c)	67	(d)	81	(b)	95	(c)	109	(d)	123	(d)	137	(d)
12	(d)	26	(c)	40	(b)	54	(d)	68	(d)	82	(b)	96	(c)	110	(c)	124	(b)	138	(a)
13	(b)	27	(b)	41	(c)	55	(b)	69	(c)	83	(c)	97	(d)	111	(d)	125	(a)	139	(a)
14	(c)	28	(d)	42	(a)	56	(c)	70	(a)	84	(c)	98	(c)	112	(N)	126	(c)	140	(d)

Chemical Bonding and Molecular Structure

4

TREND ANALYSIS
JEE MAIN (2025-2021)

	2025				2024				2023				2022				2021			
	Session II		Session I		Session II		Session I		Session II		Session I		Session II		Session I		Session II		Session I	
	MCQs	NVQs	MCQs	NVQs	MCQs	NVQs	MCQs	NVQs	MCQs	NVQs	MCQs	NVQs	MCQs	NVQs	MCQs	NVQs	MCQs	NVQs	MCQs	NVQs
Topic 1	–	–	–	–	–	–	–	–	–	–	–	–	1	–	–	–	–	–	–	–
Topic 2	–	1	3	1	1	4	2	–	–	–	4	2	1	–	3	–	1	–	–	–
Topic 3	1	–	2	–	3	2	4	3	3	1	2	–	–	–	1	2	–	–	–	–
Topic 4	8	–	3	1	8	2	2	3	2	5	3	1	2	2	6	2	1	3	2	1
Topic 5	1	–	1	1	3	3	3	5	5	–	–	–	3	3	3	–	1	4	2	1
Total	**10**	**1**	**9**	**3**	**15**	**11**	**11**	**11**	**10**	**6**	**9**	**3**	**7**	**5**	**13**	**4**	**3**	**7**	**4**	**2**

How to Code Qns for Better & Faster Revision

Type 1: Able to solve (No code Req)
Type 2: Solved with struggle or by seeing solution **(Use code NR means Needs Revision or orange stickers)**
Type 3: Tricky/ Could not solve on my own. Need help from Teacher/ Friend. **(Use code TQ means Tricky Question or red stickers)**

You can either hand write the code in boxes or else paste coloured stickers given at end of book.

Topic 1 – Electrovalent, Covalent and Co-ordinate Bonding

1. Arrange the following in increasing order of their covalent character. **[NCERT : P L-112 | July 26, 2022 (II)]**
(A) CaF_2 (B) $CaCl_2$ (C) $CaBr_2$ (D) CaI_2
Choose the correct answer from the options given below.
(a) B < A < C < D (b) A < B < C < D
(c) A < B < D < C (d) A < C < B < D

2. Which of the following compounds contain(s) no covalent bond(s)? KCl, PH_3, O_2, B_2H_6, H_2SO_4 **[2018]**
(a) KCl, B_2H_6, PH_3 (b) KCl, H_2SO_4
(c) KCl (d) KCl, B_2H_6

3. Which of these statements is not true? **[NCERT : P L-109 | Online April 19, 2014]**
(a) NO^+ is not isoelectronic with O_2
(b) B is always covalent in its compounds
(c) In aqueous solution, the Tl^+ ion is much more stable than Tl (III)
(d) $LiAlH_4$ is a versatile reducing agent in organic synthesis.

4. Lattice energy of an ionic compound depends upon **[NCERT : P L-107 | [2005]**
(a) Charge on the ion and size of the ion
(b) Packing of ions only
(c) Size of the ion only
(d) Charge on the ion only

Topic 2 – Octet Rule, Resonance and Hydrogen Bonding

5. Resonance in X_2Y can be represented as

$$\overset{\ominus}{\ddot{\underset{\cdot\cdot}{X}}} = \overset{\oplus}{X} = \ddot{\underset{\cdot\cdot}{Y}} \leftrightarrow :X \equiv \overset{\oplus}{X} - \overset{\ominus}{\ddot{\underset{\cdot\cdot}{Y}}}:$$

The enthalpy of formation of

$$X_2Y\left(X \equiv X(g) + \frac{1}{2}Y = Y(g) \rightarrow X_2Y(g)\right)$$ is 80 kJ mol^{-1}.

The magnitude of resonance energy of X_2Y is ______ kJ mol^{-1} (nearest integer value)

Given : Bond energies of $X \equiv X, X = X, Y = Y$ and $X = Y$ are 940, 410, 500 and 602 kJ mol^{-1} respectively.
valence X : 3, Y : 2 **[NCERT : PL-109 | NV, April 8, 2025 (II)]**

6. Total number of non-bonded electrons present in NO_2^- ion based on Lewis theory is _____.
[NCERT : P L-102 | NV, Jan. 29, 2025 (II)]

7. The steam volatile compounds among the following are :
[NCERT : P L-106 | Jan. 29, 2025 (I)]

(A) OH, NO_2 (B) NH_2, NO_2

(C) OH, H_2N (D) NH_2, HO

Choose the correct answer from the options given below :
(a) (B) and (D) only (b) (A) and (C) only
(c) (A) and (B) only (d) (A),(B) and (C) only

8. Given below are two statements :
[NCERT : P L-109 | Jan. 24, 2025 (II)]

Statement(I) : Experimentally determined oxygen-oxygen bond lengths in the O_3 are found to be same and the bond length is greater than that of a O = O (double bond) but less than that of a single (O – O) bond.

Statement (II) : The strong lone pair-lone pair repulsion between oxygen atoms is solely responsible for the fact that the bond length in ozone is smaller than that of a double bond (O=O) but more than that of a single bond (O – O).

In the light of the above statements, choose the correct answer from the options given below:
(a) Statement I is true but Statement II is false
(b) Both Statement I and Statement II are true
(c) Both Statement I and Statement II are false
(d) Statement I is false but Statement II is true

9. Match the List-I with List-II
[NCERT : P L-105 | Jan. 23, 2025 (I)]

List-I (Classification of molecules based on octet rule)		List-II (Example)	
A.	Molecules obeying octet rule	I.	NO, NO_2
B.	Molecules with incomplete octet	II.	BCl_3, $AlCl_3$
C.	Molecules with incomplete octet with odd electron	III.	H_2SO_4, PCl_5
D.	Molecules with expanded octet	IV.	CCl_4, CO_2

Choose the correct answer from the options given below:
(a) A-IV, B-II, C-I, D-III (b) A-III, B-II, C-I, D-IV
(c) A-IV, B-I, C-III, D-II (d) A-II, B-IV, C-III, D-I

10. Number of molecules from the following which are exceptions to octet rule is _____.
CO_2, NO_2, H_2SO_4, BF_3, CH_4, SiF_4, ClO_2, PCl_5, BeF_2, C_2H_6, $CHCl_3$, CBr_4. **[NCERT : P L-101 | NV, April 8, 2024 (I)]**

11. Number of molecules from the following which can exhibit hydrogen bonding is _____. (nearest integer)
[NCERT : P L-131 | NV, April 6, 2024 (I)]

CH_3OH, H_2O, C_2H_6, C_6H_6, (NO_2, OH), HF, NH_3

12. Consider the following reactions
$NiS + HNO_3 + HCl \rightarrow A + NO + S + H_2O$

$$A + NH_4OH + \begin{matrix} H_3C-C=N-OH \\ | \\ H_3C-C=N-OH \end{matrix} \rightarrow B + NH_4Cl + H_2O$$

The number of protons that do not involve in hydrogen bonding in the product B is_____.
[NCERT : P L-131 | NV, April 6, 2024 (II)]

13. In the lewis dot structure for NO_2^-, total number of valence electrons around nitrogen is ....
[NCERT : P L-104 | NV, April 5, 2024 (I)]

14. The correct statement(s) about Hydrogen bonding is/are :
[NCERT : P L-131 | April 4, 2024 (II)]
(A) Hydrogen bonding exists when H is covalently bonded to the highly electronegative atom.
(B) Intermolecular H bonding is present in *o*-nitro phenol
(C) Intramolecular H bonding is present in HF.
(D) The magnitude of H bonding depends on the physical state of the compound.
(E) H-bonding has powerful effect on the structure and properties of compounds.

Choose the correct answer from the options given below:
(a) A only (b) A, D, E only
(c) A, B, D only (d) A, B, C only

15. Select the compound from the following that will show intramolecular hydrogen bonding.
[NCERT : P L-132 | Feb. 1, 2024 (II)]
(a) (NO_2, OH) (b) H_2O
(c) C_2H_5OH (d) NH_3

16. The difference in energy between the actual structure and the lowest energy resonance structure for the given compound is **[NCERT : P L-109 | Jan. 29, 2024 (I)]**
(a) electromeric energy
(b) ionization energy
(c) hyperconjugation energy
(d) resonance energy

17. Resonance in carbonate ion (CO_3^{2-}) is

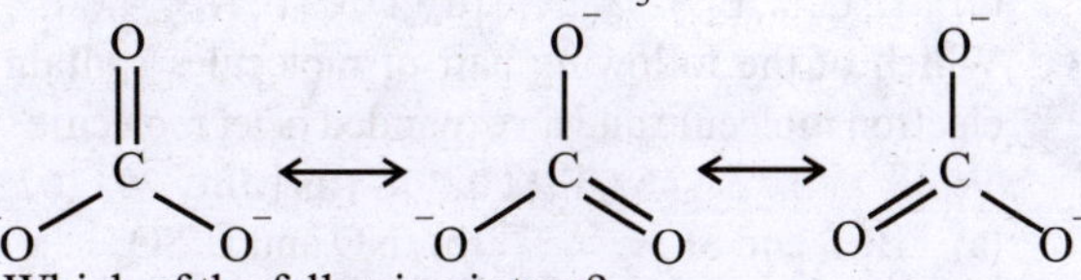

Which of the following is true?
[NCERT : P L-110 | Feb. 01, 2023 (I)]
(a) It is possible to identify each structure individually by some physical or chemical method.
(b) All these structures are in dynamic equilibrium with each other.

(c) CO_3^{2-} has a single structure i.e., resonance hybrid of the above three structures.

(d) Each structure exists for equal amount of time.

18. All structures given below are of vitamin C. Most stable of them is: **[NCERT : P L-131 | Feb. 01, 2023 (II)]**

(a) 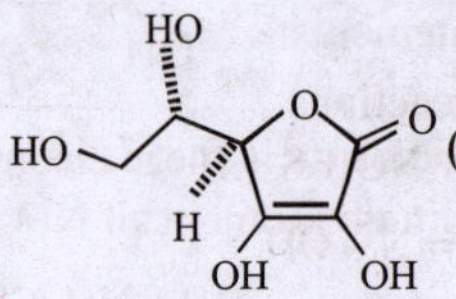 (b)

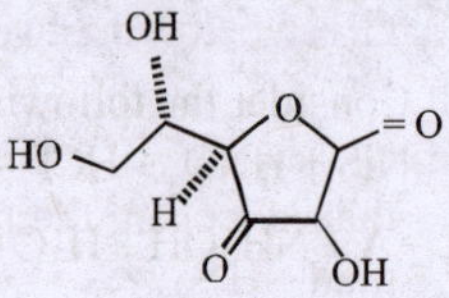

(c) 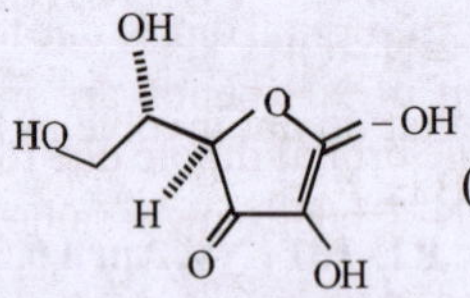 (d)

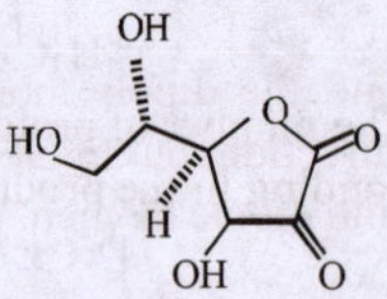

19. The number of molecules or ions from the following, which do not have odd number of electrons are ______. **[NCERT : P L-102 | NV, Jan. 29, 2023 (I)]**

(A) NO_2 (B) ICl_4^- (C) BrF_3 (D) ClO_2

(E) NO_2^+ (F) NO

20. The total number of lone pairs of electrons on oxygen atoms of ozone is ______. **[NCERT : P L-103 | NV, Jan. 25, 2023 (I)]**

21. Decreasing order of the hydrogen bonding in following forms of water is correctly represented by **[NCERT : P L-131 | Jan. 24, 2023 (I)]**

A. Liquid water B. Ice

C. Impure water

(a) A = B > C (b) B > A > C

(c) C > B > A (d) A > B > C

22. Increasing order of stability of the resonance structure is:

A. 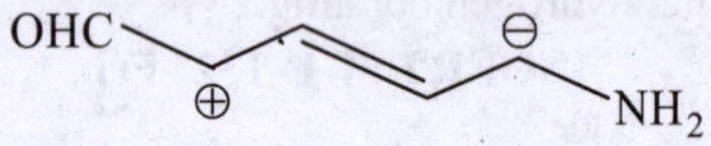 OHC ... NH_2

B. 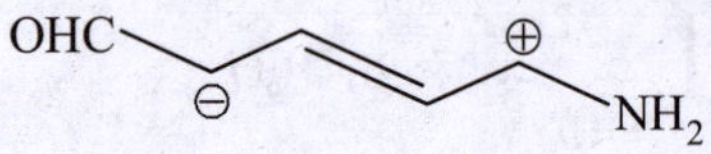 OHC ... NH_2

C. 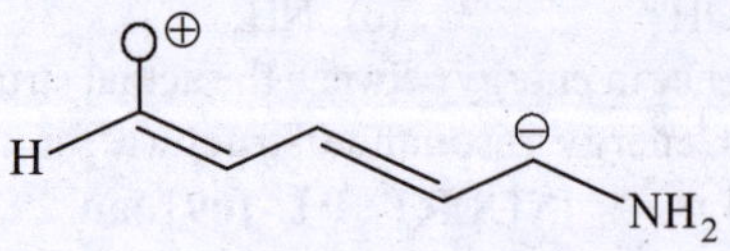 H ... NH_2

D. 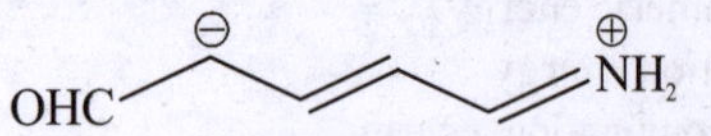 OHC ... NH_2

[NCERT : P L-109 | Jan. 24, 2023 (I)]

(a) C, D, B, A (b) C, D, A, B

(c) D, C, A, B (d) D, C, B, A

23. Which of the following pair of molecules contain odd electron molecule and an expanded octet molecule? **[NCERT : P L-105 | July 29, 2022 (I)]**

(a) BCl_3 and SF_6 (b) NO and H_2SO_4

(c) SF_6 and H_2SO_4 (d) BCl_3 and NO

24. The correct order of increasing intermolecular hydrogen bond strength is **[NCERT : P L-131 | June 27, 2022 (II)]**

(a) $HCN < H_2O < NH_3$ (b) $HCN < CH_4 < NH_3$

(c) $CH_4 < HCN < NH_3$ (d) $CH_4 < NH_3 < HCN$

25. Number of electron deficient molecules among the following: PH_3, B_2H_6, CCl_4, NH_3, LiH and BCl_3 is **[NCERT : P L-108 | June 25, 2022 (I)]**

(a) 0 (b) 1 (c) 2 (d) 3

26. PCl_5 is well known but NCl_5 is not. Because. **[NCERT : P L-105 | June 24, 2022 (II)]**

(a) nitrogen is less reactive than phosphorous.

(b) nitrogen doesn't have d-orbitals in its valence shell.

(c) catenation tendency is weaker in nitrogen than phosphorous.

(d) size of phosphorous is larger than nitrogen.

27. Identify the species having one π-bond and maximum number of canonical forms from the following : **[NCERT : P L-109 | July 25, 2021 (II)]**

(a) SO_3 (b) O_2 (c) SO_2 (d) CO_3^{2-}

28. *ortho*-Nitrophenol is less soluble in water than *p*- and *m*- nitrophenols because : **[NCERT : P L-132 | 2012]**

(a) *o*-nitrophenol is more volatile steam than those of *m*- and *p*-isomers.

(b) *o*-nitrophenol shows intramolecular H-bonding

(c) *o*-nitrophenol shows intermolecular H-bonding

(d) Melting point of *o*-nitrophenol is lower than those of *m*-and *p*-isomers.

29. Which of the following hydrogen bonds is strongest? **[NCERT : P L-131 | 2007]**

(a) O – H - - - F (b) O – H - - - H

(c) F – H - - - F (d) O – H - - - O

30. An ether is more volatile than an alcohol having the same molecular formula. This is due to **[NCERT : P L-131 | 2003]**

(a) alcohols having resonance structures

(b) intermolecular hydrogen bonding in ethers

(c) intermolecular hydrogen bonding in alcohols

(d) dipolar character of ethers

Topic 3 -Dipole Moment and Bond Polarity

31. Given below are two statements.

Statement I : The dipole moment of

$\overset{4}{C}H_3-\overset{3}{C}H=\overset{2}{C}H-\overset{1}{C}H=O$ is greater than

$\overset{4}{C}H_3-\overset{3}{C}H_2-\overset{2}{C}H_2-\overset{1}{C}H=O$

Statement II : C1–C2 bond length of

$\underset{4}{C}H_3-\underset{3}{C}H=\underset{2}{C}H-\underset{1}{C}H=O$ is greater than C_1–C_2 bond length of $\underset{4}{C}H_3-\underset{3}{C}H_2=\underset{2}{C}H_2-\underset{1}{C}H=O$

In the light of the above statements, choose the **correct** answer from the options given below: **[NCERT : P L-111 | April 4, 2025 (I)]**

(a) Statement I is false but Statement II is true

(b) Both Statement I and Statement II are false

(c) Statement I is true but Statement II is false

(d) Both Statement I and Statement II are true

32. An element 'E' has the ionisation enthalpy value of 374 kJ mol^{-1}. 'E' reacts with elements A, B, C and D with electron gain enthalpy values of –328, –349, –325 and –295 kJ mol^{-1}, respectively. **[NCERT : P L-111 | Jan. 29, 2025 (I)]**
The correct order of the products EA, EB, EC and ED in terms of ionic character is
(a) EB > EA > EC > ED (b) ED > EC > EA > EB
(c) EA > EB > EC > ED (d) ED > EC > EB > EA

33. Arrange the following compounds in increasing order of their dipole moment : **[NCERT : P L-111 | Jan. 22, 2025 (II)]**
HBr, H_2S, NF_3 and $CHCl_3$
(a) $NF_3 < HBr < H_2S < CHCl_3$
(b) $HBr < H_2S < NF_3 < CHCl_3$
(c) $H_2S < HBr < NF_3 < CHCl_3$
(d) $CHCl_3 < NF_3 < HBr < H_2S$

34. The correct increasing order for bond angles among BF_3, PF_3 and ClF_3 is : **[April 9, 2024 (II)]**
(a) $PF_3 < BF_3 < ClF_3$ (b) $BF_3 < PF_3 < ClF_3$
(c) $ClF_3 < PF_3 < BF_3$ (d) $BF_3 = PF_3 < ClF_3$

35. Given below are two statements : one is labelled as **Assertion (A)** and the other is labelled as **Reason (R)**. **[NCERT : P L-111 | April 5, 2024 (II)]**
Assertion (A) : NH_3 and NF_3 molecule have pyramidal shape with a lone pair of electrons on nitrogen atom. The resultant dipole moment of NH_3 is greater than that of NF_3.
Reason (R) : In NH_3, the orbital dipole due to lone pair is in the same direction as the resultant dipole moment of the N–H bonds. F is the most electronegative element.
In the light of the above statements, choose the **correct** answer from the options given below:
(a) Both **(A)** and **(R)** are true and **(R)** is the correct explanation of **(A)**
(b) **(A)** is false but **(R)** is true
(c) **(A)** is true but **(R)** is false
(d) Both **(A)** and **(R)** are true but **(R)** is NOT the correct explanation of **(A)**

36. Number of compounds from the following with zero dipole moment is ____. **[NCERT : P L-112 | NV, April 5, 2024 (II)]**
HF, H_2, H_2S, CO_2, NH_3, BF_3, CH_4, $CHCl_3$, SiF_4, H_2O, BeF_2

37. Which one of the following molecules has maximum dipole moment ? **[NCERT : P L-112 | April 4, 2024 (I)]**
(a) NF_3 (b) CH_4 (c) NH_3 (d) PF_5

38. Number of compounds/species from the following with non-zero dipole moment is ____
$BeCl_2$, BCl_3, NF_3, XeF_4, CCl_4, H_2O H_2S, HBr, CO_2, H_2, HCl
[NCERT : P L-113 | NV, April 4, 2024 (II)]

39. Arrange the bonds in order of increasing ionic character in the molecules. LiF, K_2O, N_2, SO_2 and ClF_3:
[NCERT : P L-112 | Feb. 1, 2024 (I)]
(a) $ClF_3 < N_2 < SO_2 < K_2O < LiF$
(b) $LiF < K_2O < ClF_3 < SO_2 < N_2$
(c) $N_2 < SO_2 < ClF_3 < K_2O < LiF$
(d) $N_2 < ClF_3 < SO_2 < K_2O < LiF$

40. Which of the following is least ionic ?
[NCERT : P L-106 | Jan. 31, 2024 (II)]
(a) $BaCl_2$ (b) AgCl (c) KCl (d) $CoCl_2$

41. A diatomic molecule has a dipole moment of 1.2 D. If the bond distance is 1Å, then fractional charge on each atom is ______ $\times 10^{-1}$ esu .
(Given 1D = 10^{-18} esu cm)
[NCERT : P L-111 | NV, Jan. 31, 2024 (II)]

42. Given below are two statements:
[NCERT : P L-112 | Jan. 30, 2024 (II)]
Statements – I: Since Fluorine is more electronegative than nitrogen, the net dipole moment of NF_3 is greater than NH_3.
Statement – II: In NH_3, the orbital dipole due to lone pair and the dipole moment of NH bonds are in opposite direction, but in NF_3, the orbital dipole due to lone pair and dipole moments of N- F bonds are in same direction.
In the light of the above statements, choose the most appropriate from the options given below:
(a) Statement I is true but Statement II is false.
(b) Both Statement I and Statement II are false.
(c) Both Statement I and Statement II are true.
(d) Statement I is false but Statement II is true.

43. The total number of molecules with zero dipole moment among CH_4, BF_3, H_2O, HF, NH_3, CO_2 and SO_2 is _____.
[NCERT : P L-111 | NV, Jan. 29, 2024 (II)]

44. Choose the polar molecule from the following :
[NCERT : P L-111 | Jan. 27, 2024 (I)]
(a) CCl_4 (b) CO_2
(c) $CH_2=CH_2$ (d) $CHCl_3$

45. The number of non - polar molecules from the following is______ **[NCERT : P L-111 | NV, Jan. 27, 2024 (II)]**
HF, H_2O, SO_2, H_2, CO_2, CH_4, NH_3, HCl, $CHCl_3$, BF_3

46. Among the following compounds, the one which shows highest dipole moment is
[NCERT : P L-112 | April 13, 2023 (I)]

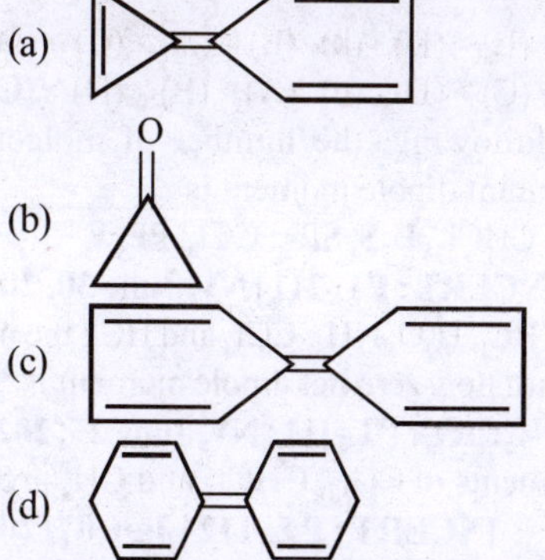

47. Which one of the following pairs is an example of polar molecular solids? **[NCERT : P L-111 | April 11, 2023 (II)]**
(a) $SO_2(s)$, $NH_3(s)$ (b) $SO_2(s)$, $CO_2(s)$
(c) HCl(s), AlN(s) (d) MgO(s), $SO_2(s)$

48. The pair from the following pairs having both compounds with net non-zero dipole moment is
(a) Benzene, anisidine **[NCERT : PL-111 | April 10, 2023 (I)]**
(b) 1,4-Dichlorobenzene, 1,3-Dichlorobenzene
(c) CH_2Cl_2, $CHCl_3$
(d) cis-butene, trans-butene

49. The number of following factors which affect the percent covalent character of the ionic bond is____
[NCERT : P L-112 | NV, April 08, 2023 (I)]
(A) Polarising power of cation
(B) Extent of distortion of anion
(C) Polarisability of the anion
(D) Polarising power of anion

50. **Statement I :-** Dipole moment is a vector quantity and by convention it is depicted by a small arrow with tail on the negative centre and head pointing towards the positive centre. **[NCERT : P L-111 | Jan. 25, 2023 (II)]**
Statement II :- The crossed arrow of the dipole moment symbolizes the direction of the shift of charges in the molecules.
In the light of the above statements, choose the most appropriate answer from the options given below :-
(a) Both Statement I and Statement II are correct.
(b) Statement I is incorrect but Statement II is correct.
(c) Both Statement I and Statement II are incorrect.
(d) Statement I is correct but Statement II is incorrect.

51. Order of Covalent bond;
[NCERT : P L-112 | Jan. 24, 2023 (I)]
A. KF > KI; LiF > KF
B. KF < KI; LiF > KF
C. $SnCl_4 > SnCl_2$; CuCl > NaCl
D. LiF > KF; CuCl < NaCl
E. KF < KI; CuCl > NaCl
(a) C, E only (b) B, C only
(c) B, C, E only (d) A, B only

52. Arrange the following in the decreasing order of their covalent character : **[NCERT : P L-112 | June 29, 2022 (I)]**
(A) LiCl (B) NaCl (C) KCl (D) CsCl
Choose the **most appropriate** answer from the options given below :
(a) (A) > (C) > (B) > (D) (b) (B) > (A) > (C) > (D)
(c) (A) > (B) > (C) > (D) (d) (A) > (B) > (D) > (C)

53. Amongst the following, the number of molecule/(s) having net resultant dipole moment is ______.
$NF_3, BF_3, BeF_2, CHCl_3, H_2S, SiF_4, CCl_4, PF_5$
[NCERT : P L-111 | NV, June 30, 2022 (I)]

54. Amongst BeF_2, BF_3, H_2O, NH_3, CCl_4 and HCl, the number of molecules with non-zero net dipole moment is ____.
[NCERT : P L-111 | NV, June 25, 2022 (II)]

55. The dipole moments of CCl_4, $CHCl_3$ and CH_4 are in the order: **[NCERT : P L-112 | Jan. 07, 2020 (I)]**
(a) $CHCl_3 < CH_4 = CCl_4$ (b) $CCl_4 < CH_4 < CHCl_3$
(c) $CH_4 < CCl_4 < CHCl_3$ (d) $CH_4 = CCl_4 < CHCl_3$

56. Which compound exhibits maximum dipole moment among the following ?
[NCERT : P L-111 | Online April 11, 2015]

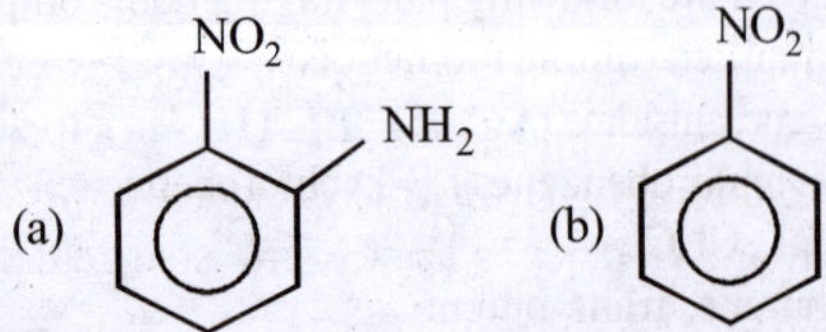

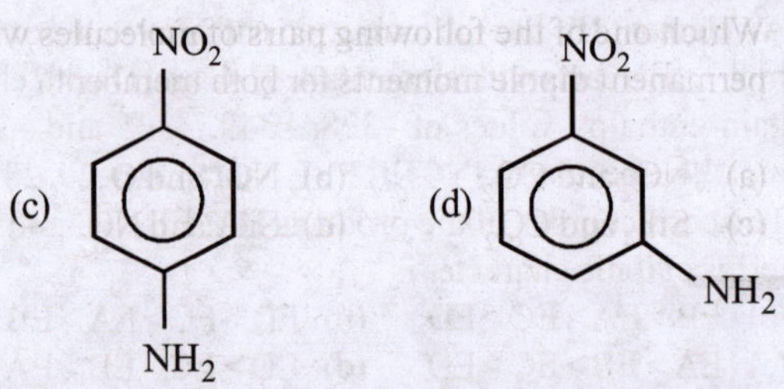

57. Molecule AB has a bond length of 1.617Å and a dipole moment of 0.38 D. The fractional charge on each atom (absolute magnitude) is : ($e_0 = 4.802 \times 10^{-10}$ esu)
[NCERT : P L-111 | Online April 11, 2015]
(a) 0.5 (b) 0.05 (c) 0 (d) 1.0

58. Amongst LiCl, RbCl, $BeCl_2$ and $MgCl_2$ the compounds with the greatest and the least ionic character, respectively are: **[NCERT : P L-112 | Online April 19, 2014]**
(a) LiCl and RbCl (b) RbCl and $BeCl_2$
(c) $MgCl_2$ and $BeCl_2$ (d) RbCl and $MgCl_2$

59. Which one of the following molecules is polar ?
[NCERT : P L-112 | Online April 9, 2013]
(a) XeF_4 (b) IF_5 (c) SbF_5 (d) CF_4

60. Bond distance in HF is 9.17×10^{-11} m. Dipole moment of HF is 6.104×10^{-30} Cm. The percentage ionic character in HF will be : (electron charge = 1.60×10^{-19} C)
[NCERT : P L-111 | Online April 23, 2013]
(a) 61.0% (b) 38.0% (c) 35.5% (d) 41.5%

61. Dipole moment is shown by
[NCERT : P L-111 | Online May 26, 2012]
(a) 1, 2-dichlorobenzene (b) *trans*-2, 3-dichloro-2-butene
(c) 1, 4-dichlorobenzene (d) *trans*-1, 2-dinitroethene

62. Among the following, the molecule with the lowest dipole moment is **[NCERT : P L-111 | Online May 19, 2012]**
(a) $CHCl_3$ (b) CH_3Cl (c) CH_2Cl_2 (d) CCl_4

63. Among the following chloro-compound having the lowest dipole moment is **[NCERT : P L-111 | Online May 12, 2012]**
(a) CH_3Cl (b) Cl(H₃C)C=C(H)Cl
(c) CH_2Cl_2 (d) Cl(H₃C)C=C(Cl)H

64. Although CN^- ion and N_2 molecule are isoelectronic, yet N_2 molecule is chemically inert because of
[NCERT : P L-111 | Online May 12, 2012]
(a) presence of more number of electrons in bonding orbitals
(b) low bond energy
(c) absence of bond polarity
(d) uneven electron distribution.

65. The charge/size ratio of a cation determines its polarizing power. Which one of the following sequences represents the increasing order of the polarizing power of the cationic species, $K^+, Ca^{2+}, Mg^{2+}, Be^{2+}$?
[NCERT : P L-112 | 2007]
(a) $Ca^{2+} < Mg^{2+} < Be^+ < K^+$ (b) $Mg^{2+} < Be^{2+} < K^+ < Ca^{2+}$
(c) $Be^{2+} < K^+ < Ca^{2+} < Mg^{2+}$ (d) $K^+ < Ca^{2+} < Mg^{2+} < Be^{2+}$.

66. Which one of the following pairs of molecules will have permanent dipole moments for both members ? **[NCERT : P L-111 | 2003]**

(a) NO_2 and CO_2 (b) NO_2 and O_3
(c) SiF_4 and CO_2 (d) SiF_4 and NO_2

Topic 4 - VSEPR Theory and Hybridisation

67. In SO_2, NO_2^- and N_3^- the hybridizations at the central atom are respectively : **[NCERT : P L-120 | April 7, 2025 (II)]**

(a) sp^2, sp^2 and sp (b) sp^2, sp and sp
(c) sp^2, sp^2 and sp^2 (d) sp, sp^2 and sp

68. Match the **LIST-I** with **LIST-II.** **[NCERT : P L-117 | April 7, 2025 (I)]**

LIST-I Molecule/ion		LIST-II Bond pair : lone pair (on the central atom)	
A.	ICl_2^-	I.	4 : 2
B.	H_2O	II.	4 : 1
C.	SO_2	III.	2 : 3
D.	XeF_4	IV.	2 L 2

Choose the **correct** answer from the options given below :

(a) A-IV, B-III, C-II, D-I (b) A-III, B-IV, C-II, D-I
(c) A-III, B-IV, C-I, D-II (d) A-II, B-I, C-IV, D-III.

69. Given below are two statements: **[NCERT : P L-117 | April 4, 2025 (II)]**

Statement (I) : for $\ddot{C}l F_3$, all three possible structures may be drawn as follows.

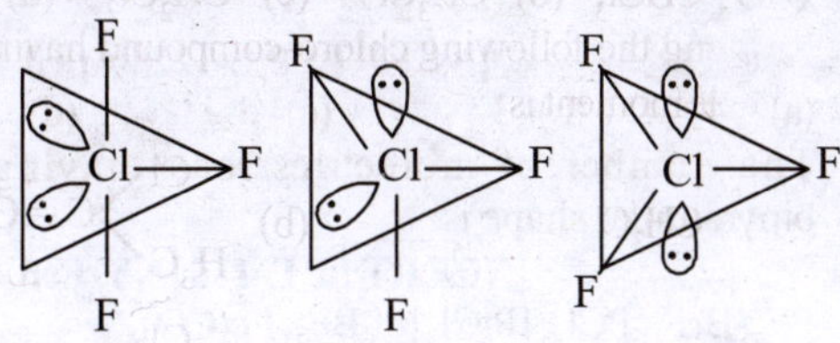

Statement (II) : Structure III is most stable, as the orbitals having the lone pairs are axial, where the lp – bp repulsion is minimum.

In the light of the above statements, choose the **most appropriate answer** from the options given below:

(a) Statement I is incorrect but statement II is correct.
(b) Statement I is correct but statement II is incorrect.
(c) Both Statement I and statement II are correct.
(d) Both Statement I and statement II are incorrect.

70. Match the **LIST-I** with **LIST-II** **[NCERT : P L-120 | April 3, 2025 (I)]**

LIST-I (Molecules/ion)		LIST-II (Hybridisation of central atom)	
A.	PF_5	I	dsp^2
B.	SF_6	II.	sp^3d
C.	$Ni(CO)_4$	III.	sp^3d^2
D.	$[PtCl_4]^{2-}$	IV.	sp^3

Choose the **correct** answer from the options given below:

(a) A-II, B-III, C-IV, D-I (b) A-IV, B-I, C-II, D-III
(c) A-I, B-II, C-III, D-IV (d) A-III, B-I, C-IV, D-II

71. Which among the following molecules is (a) involved in sp^3d hybridization, (b) has different bond lengths and (c) has lone pair of electrons on the central atom ? **[NCERT : P L-120 | April 2, 2025 (II)]**

(a) PF_5 (b) XeF_4 (c) SF_4 (d) XeF_2

72. In 3,3-dimethylhex-1-ene-4-yne, there are _____ sp^3, _____ sp^2 and _____ sp hybridised carbon atoms respectively : **[NCERT : P L-120 | April 2, 2025 (II)]**

(a) 4, 2, 2 (b) 3, 3, 2 (c) 2, 4, 2 (d) 2, 2, 4

73. A molecule with the formula AX_4Y has all it's elements from p-block. Element A is rarest, monoatomic, non-radioactive from its group and has the lowest ionization enthalpy value among A, X and Y. Elements X and Y have first and second highest electronegativity values respectively among all the known elements. The shape of the molecule is : **[NCERT : P L-120 | April 2, 2025 (I)]**

(a) Square pyramidal (b) Octahedral
(c) Pentagonal planar (d) Trigonal bipyramidal

74. Among SO_2, NF_3, NH_3, XeF_2, ClF_3 and SF_4, the hybridization of the molecule with non-zero dipole moment and highest number of lone-pairs of electrons on the central atom is **[NCERT : P L-120 | April 2, 2025 (I)]**

(a) sp^3 (b) dsp^2 (c) sp^3 d^2 (d) sp^3 d

75. The molecules having square pyramidal geometry are **[NCERT : P L-114 | Jan. 28, 2025 (I)]**

(a) BrF_5 & $XeOF_4$ (b) SbF_5 & $XeOF_4$
(c) SbF_5 & PCl_5 (d) BrF_5 & PCl_5

76. In the given structure, number of sp and sp^2 hybridized carbon atoms present respectively are : **[NCERT : P L-120 | Jan. 24, 2025 (II)]**

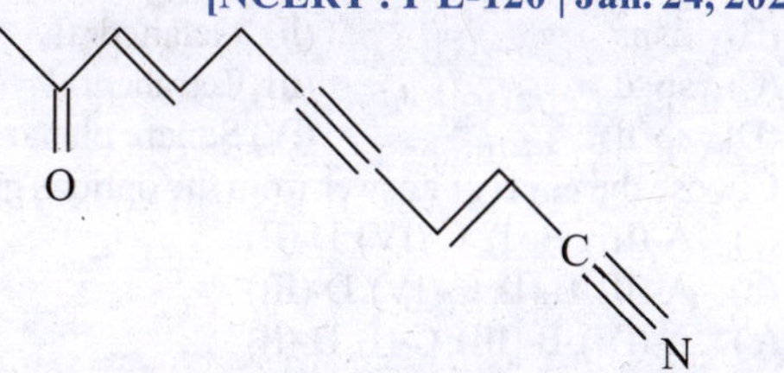

(a) 3 and 6 (b) 3 and 5 (c) 4 and 6 (d) 4 and 5

77. Which of the following statement is true with respect to H_2O, NH_3 and CH_4? **[NCERT : P L-114 | Jan. 24, 2025 (I)]**

A. The central atoms of all the molecules are sp^3 hybridized.
B. The H-O-H, H-N-H and H-C-H angles in the above molecules are 104.5°, 107.5° and 109.5° respectively.
C. The increasing order of dipole moment is $CH_4 < NH_3 < H_2O$.
D. Both H_2O and NH_3 are Lewis acids and CH_4 is a Lewis base
E. A solution of NH_3 in H_2O is basic. In this solution NH_3 and H_2O act as Lowry-Bronsted acid and base respectively.

Choose the correct answer from the options given below:
(a) A, B and C only (b) C, D and E only
(c) A, D and E only (d) A, B, C and E only

78. The number of molecules/ions that show linear geometry among the following is____. **[NCERT : P L-114 | Jan. 22, 2025 (I)]**
SO_2, $BeCl_2$, CO_2, N_3^-, NO_2, F_2O, XeF_2, NO_2^+, I_3^-, O_3

79. In which one of the following pairs the central atoms exhibit sp^2 hybridization ? **[NCERT : P L-121 | April 9, 2024 (I)]**
(a) BF_3 and NO_2^- (b) NH_2^- and H_2O
(c) H_2O and NO_2 (d) NH_2^- and BF_3

80. The total number of species from the following in which one unpaired electron is present, is _______.
$N_2, O_2, C_2^-, O_2^-, O_2^{2-}, H_2^+, CN^-, He_2^+$
[NCERT : P L-129 | NV, April 9, 2024 (I)]

81. Match List-I with List-II : **[NCERT : P L-114, 115 | April 8, 2024 (I)]**

List-I (Molecule)	List-II (Shape)
(A) NH_3	(I) Square pyramid
(B) BrF_5	(II) Tetrahedral
(C) PCl_5	(III) Trigonal pyramidal
(D) CH_4	(IV) Trigonal bipyramidal

Choose the **correct** answer from the options below:
(a) A-(IV), B-(III), C-(I), D-(II)
(b) A-(II), B-(IV), C-(I), D-(III)
(c) A-(III), B-(I), C-(IV), D-(II)
(d) A-(III), B-(IV), C-(I), D-(II)

82. Match List-I with List-II : **[NCERT : P L-124 | April 6, 2024 (I)]**

List-I (Hybridization)	List-II (Orientation in Space)
(A) sp^3	(I) Trigonal bipyramidal
(B) dsp^2	(II) Octahedral
(C) sp^3d	(III) Tetrahedral
(D) sp^3d^2	(IV) Square planar

Choose the **correct** answer from the options given below:
(a) A-(III), B-(I), C-(IV), D-(II)
(b) A-(II), B-(I), C-(IV), D-(III)
(c) A-(IV), B-(III), C-(I), D-(II)
(d) A-(III), B-(IV), C-(I), D-(II)

83. Match List-I with List-II : **[NCERT : P L-114 | April 6, 2024 (I)]**

List-I (Compound/ Species)	List - II (Shape/ Geometry)
(A) SF_4	(I) Tetrahedral
(B) BrF_3	(II) Pyramidal
(C) BrO_3^-	(III) See saw
(D) NH_4^+	(IV) Bent T-shape

Choose the **correct** answer from the options given below:
(a) A-(II), B-(III), C-(I), D-(IV)
(b) A-(III), B-(IV), C-(II), D-(I)
(c) A-(II), B-(IV), C-(III), D-(I)
(d) A-(III), B-(II), C-(IV), D-(I)

84. Given below are two statements:
Statement I: PF_5 and BrF_5 both exhibit sp^3d hybridisation.
Statement II: Both SF_6 and $[Co(NH_3)_6]^{3+}$ exhibit sp^3d^2 hybridisation. **[NCERT : P L-124 | April 6, 2024 (II)]**
In the light of the above statements, choose the **correct** answer from the options given below:
(a) **Statement I** is true but **Statement II** is false
(b) Both **Statement I** and **Statement II** are true
(c) Both **Statement I** and **Statement II** are false
(d) **Statement I** is false but **Statement II** is true

85. Total number of species from the following with central atom utilising sp^2 hybrid orbitals for bonding is..........
NH_3, SO_2, SiO_2, $BeCl_2$, C_2H_2, C_2H_4, BCl_3, HCHO, C_6H_6, BF_3, $C_2H_4Cl_2$ **[NCERT : P L-121 | April 6, 2024 (II)]**

86. Match List-I with List-II **[NCERT : P L-124 | April 5, 2024 (II)]**

List-I	List-II
(A) ICI	(I) T-Shape
(B) ICl_3	(II) Square pyramidal
(C) ClF_5	(III) Pentagonal bipyramidal
(D) IF_7	(IV) Linear

Choose the **correct** answer from the options given below:
(a) A–(I), B–(IV), C–III, D–(II)
(b) A–(I), B–(III), C–(II), D–(IV)
(c) A–(IV), B–(I), C–(II), D–(III)
(d) A–(IV), B–(III), C–(II), D–(I)

87. Number of molecules/ions from the following in which the central atom is involved in sp^3 hybridization is______.
NO_3^-, BCl_3, ClO_2^-, ClO_3^-
[NCERT : P L-122 | April 4, 2024 (I)]
(a) 2 (b) 4 (c) 3 (d) 1

88. The number of species from the following that have pyramidal geometry around the central atom is_______
$S_2O_3^{2-}, SO_4^{2-}, SO_3^{2-}, S_2O_7^{2-}$
[NCERT : P L-122 | April 4, 2024 (II)]
(a) 4 (b) 3 (c) 1 (d) 2

89. The number of molecules/ions/ having trigonal bipyramidal shape is ____. **[NCERT : P L-115 | NV, Feb. 1, 2024 (I)]**
PF_5, BrF_5, PCl_5, $[PtCl_4]^{2-}$, BF_3, $Fe(CO)_5$

90. The number of species from the following in which the central atom uses sp^3 hybrid orbitals in its bonding is____.
NH_3, SO_2, SiO_2, $BeCl_2$, CO_2, H_2O, CH_4, BF_3
[NCERT : P L-122 | NV, Jan. 31, 2024 (I)]

91. Match List - I with List - II **[NCERT : P L-115 | Jan. 30, 2024 (I)]**

List - I (Molecule)	List - II (Shape)
(A) BrF_5	(I) T - shape
(B) H_2O	(II) See saw
(C) ClF_3	(III) Bent
(D) SF_4	(IV) Square pyramidal

Choose the correct answer from the options given below:
(a) (A) - (I), (B) - (II), (C) - (IV), (D) - (III)
(b) (A) - (II), (B) - (I), (C) - (III), (D) - (IV)
(c) (A) - (III), (B) - (IV), (C) - (I), (D) - (II)
(d) (A) - (IV), (B) - (III), (C) - (I), (D) - (II)

92. Number of compounds with one lone pair of electrons on central atom amongst following is ______
O_3, H_2O, SF_4, ClF_3, NH_3, BrF_5, XeF_4
[NCERT : P L-113 | NV, Jan. 29, 2024 (I)]

93. Identify from the following species in which d^2sp^3 hybridization is shown by central atom:
[NCERT : P L-124 | Jan. 27, 2024 (II)]
(a) $[Co(NH_3)_6]^{3+}$ (b) BrF_5
(c) $[Pt(Cl)_4]^{2-}$ (d) SF_6

94. Given below are two statements:
[NCERT : P L-122 | April 13, 2023 (II)]
Statement-I: SO_2 and H_2O both possess V-shaped structure.
Statement-II: The bond angle of SO_2 is less than that of H_2O.
In the light of the above statements, choose the **most appropriate** answer from the options given below:
(a) Both Statement I and Statement II are correct.
(b) Statement I is correct but Statement II is incorrect.
(c) Both Statement I and Statement II are incorrect.
(d) Statement I is incorrect but Statement II is correct.

95. Match List-I with List-II :
[NCERT : P L-115 | April 11, 2023 (I)]

List - I Species	List - II Geometry / Shape
A. H_3O^+	I. Tetrahedral
B. Acetylide anion	II. Linear
C. NH_4^+	III. Pyramidal
D. ClO_2^-	IV. Bent

Choose the correct answer from the options given below:
(a) A-III, B-II, C-I, D-IV (b) A-III, B-I, C-II, D-IV
(c) A-III, B-IV, C-I, D-II (d) A-III, B-IV, C-II, D-I

96. The maximum number of lone pairs of electrons on the central atom from the following species is ______
ClO_3^-, XeF_4, SF_4, and I_3^-
[NCERT : P L-113 | NV, April 11, 2023 (II)]

97. The number of bent-shaped molecule/s from the following is ______ **[NCERT : P L-121 | NV, April 10, 2023 (I)]**
N_3^-, NO_2^-, I_3^-, O_3, SO_2

98. The number of molecules from the following which contain only two lone pair of electrons is ____
H_2O, N_2, CO, XeF_4, NH_3, NO, CO_2, F_2
[NCERT : P L-122 | NV, April 10, 2023 (II)]

99. The number of species from the following which have square pyramidal structure is
PF_5, BrF_4^-, IF_5; BrF_5, $XeOF_4$, ICl_4^-
[NCERT : P L-124 | NV, April 06, 2023 (I)]

100. The number of species having a square planar shape from the following is ______
[NCERT : P L-124 | NV, April 06, 2023 (II)]
XeF_4, SF_4, SiF_4, BF_4^-, BrF_4^-, $[Cu(NH_3)_4]^{2+}$,
$[FeCl_4]^{2-}$, $[PtCl_4]^{2-}$

101. Match List I with List II.
[NCERT : P L-125 | Jan. 31, 2023 (I)]

List I	List II
A. XeF_4	I. See-saw
B. SF_4	II. Square planar
C. NH_4^+	III. Bent T-shaped
D. BrF_3	IV. Tetrahedral

Choose the correct answer from the options given below:
(a) A-IV, B-III, C-II, D-I
(b) A-II, B-I, C-III, D-IV
(c) A-IV, B-I, C-II, D-III
(d) A-II, B-I, C-IV, D-III

102. Amongst the following, the number of species having the linear shape is __. **[NCERT : P L-121 | NV, Jan. 31, 2023 (II)]**
XeF_2, I_3^+, C_3O_2, I_3^-, CO_2, SO_2, $BeCl_2$ and BCl_2^-

103. Match **List I** with **List II.** **[Jan. 30, 2023 (I)]**

List-I (molecules/ions)	List-II (No. of lone pairs of e^- on central atom)
(A) IF_7	I. Three
(B) ICl_4^-	II. One
(C) XeF_6	III. Two
(D) XeF_2	IV. Zero

Choose the **correct** answer from the options given below:
(a) A – II, B – III, C – IV, D – I
(b) A – IV, B – III, C – II, D – I
(c) A – II, B – I, C – IV, D – III
(d) A – IV, B – I, C – II, D – III

104. For OF_2 molecule consider the following:
[NCERT : P L-117 | Jan. 30, 2023 (I)]
(A) Number of lone pairs on oxygen is 2.
(B) FOF angle is less than 104.5°.
(C) Oxidation state of O is –2.
(D) Molecule is bent 'V' shaped.
(E) Molecular geometry is linear.
Correct options are:
(a) C, D, E only (b) B, E, A only
(c) A, C, D only (d) A, B, D only

105. Number of lone pairs of electrons in the central atom of SCl_2, O_3, ClF_3 and SF_6, respectively, are:
[NCERT : P L-117 | July 29, 2022 (I)]
(a) 0, 1, 2 and 2 (b) 2, 1, 2 and 0
(c) 1, 2, 2 and 0 (d) 2, 1, 0 and 2

106. Consider, PF_5, BrF_5, PCl_3, SF_6, $[ICl_4]^-$, ClF_3 and IF_5. Amongst the above molecule(s)/ion(s), the number of molecule(s)/ion(s) having sp^3d^2 hybridisation is____.
[NCERT : P L-125 | NV, July 29, 2022 (II)]

107. The number of molecule(s) or ion(s) from the following having non-planar structure is______.
NO_3^-, H_2O_2, BF_3, PCl_3, XeF_4,
SF_4, XeO_3, PH_4^+, SO_3, $[Al(OH)_4]^-$
[NCERT : P L-117 | NV, July 27, 2022 (II)]

108. Match **List - I** with **List - II**

[NCERT : P L-124 | July 26, 2022 (I)]

List - I (Compound)	**List - II (Shape)**
(A) BrF_5	(I) bent
(B) $[CrF_6]^{3-}$	(II) square pyramidal
(C) O_3	(III) trigonal bipyramidal
(D) PCl_5	(IV) octahedral

Choose the **correct** answer from the options given below:

(a) (A) – (I), (B) – (II), (C) – (III), (D) – (IV)
(b) (A) – (IV), (B) – (III), (C) – (II), (D) – (I)
(c) (A) – (II), (B) – (IV), (C) – (I), (D) – (III)
(d) (A) – (III), (B) – (IV), (C) – (II), (D) – (I)

109. Consider the species CH_4, NH_4^+ and BH_4^-. Choose the correct option with respect to the three species:

(a) They are isoelectronic and only two have tetrahedral structures **[June 29, 2022 (II)]**
(b) They are isoelectronic and all have tetrahedral structures
(c) Only two are isoelectronic and all have tetrahedral structures
(d) Only two are isoelectronic and only two have tetrahedral structures

110. Number of lone pair(s) of electrons on central atom and the shape of BrF_3 molecule respectively, are :

[NCERT : P L-117 | June 29, 2022 (II)]

(a) 0, triangular planar. (b) 1, pyramidal.
(c) 2, bent T-shape. (d) 1, bent T-shape.

111. The hybridization of P exhibited in PF_5 is sp^xd^y. The value of y is ________.

[NCERT : P L-124 | NV, June 28, 2022 (I)]

112. In the structure of SF_4, the lone pair of electrons on S is in

(a) equatorial position and there are two lone pair-bond pair repulsions at 90° **[NCERT : P L-115 | June 28, 2022 (II)]**
(b) equatorial position and there are three lone pair-bond pair repulsions at 90°
(c) axial position and there are three lone pair-bond pair repulsion at 90°
(d) axial position and there are two lone pair-bond pair repulsion at 90°

113. Based upon VSEPR theory, match the shape (geometry) of the molecules in **List-I** with the molecules in **List-II** and select the most appropriate option.

[NCERT : P L-115 | June 27, 2022 (I)]

List-I (Shape)	**List-II (Molecules)**
(A) T-shaped	(I) XeF_4
(B) Trigonal planar	(II) SF_4
(C) Square planar	(III) ClF_3
(D) See-saw	(IV) BF_3

(a) (A) – (I), (B) – (II), (C) – (III), (D) – (IV)
(b) (A) – (III), (B) – (IV), (C) – (I), (D) – (II)
(c) (A) – (III), (B) – (IV), (C) – (II), (D) – (I)
(d) (A) – (IV), (B) – (III), (C) –(I), (D) – (II)

114. Identify the **incorrect** statement for PCl_5 from the following. **[NCERT : P L-125 | June 27, 2022 (II)]**

(a) In this molecule, orbitals of phosphorous are assumed to undergo sp^3d hybridization.
(b) The geometry of PCl_5 is trigonal bipyramidal.
(c) PCl_5 has two axial bonds stronger than three equatorial bonds.
(d) The three equatorial bonds of PCl_5 lie in a plane.

115. Amongst SF_4, XeF_4, CF_4 and H_2O, the number of species with two lone pairs of electrons ______.

[NCERT : P L-115 | NV, June 26, 2022 (II)]

116. Match List-I with List-II

[NCERT : P L-124 | June 24, 2022 (I)]

List-I	**List-II**
(A) $[PtCl_4]^{2-}$	(I) sp^3d
(B) BrF_5	(II) d^2sp^3
(C) PCl_5	(III) dsp^2
(D) $[Co(NH_3)_6]^{3+}$	(IV) sp^3d^2

(a) (A) → (II), (B) → (IV), (C) → (I), (D) → (III)
(b) (A) → (III), (B) → (IV), (C) → (I), (D) → (II)
(c) (A) → (III), (B) → (I), (C) → (IV), (D) → (II)
(d) (A) → (II), (B) → (I), (C) → (IV), (D) → (III)

117. The number of species having non–pyramidal shape among the following is______.

[NCERT : P L-117 | NV, Aug. 27, 2021 (II)]

(A) SO_3 (B) NO_3^- (C) PCl_3 (D) CO_3^{2-}

118. A central atom in a molecule has two lone pairs of electrons and forms three single bonds. The shape of this molecule is **[NCERT : P L-117 | March 17, 2021 (I)]**

(a) trigonal pyramidal (b) see-saw
(c) T-shaped (d) planar triangular

119. AB_3 is an interhalogen T-shaped molecule. The number of lone pairs of electrons on A is ______.

[NCERT : P L-117 | NV, Aug. 26, 2021 (I)]

120. The hybridisations of the atomic orbitals of nitrogen in NO_2^-, NO_2^+ and NH_4^+ respectively are:

[NCERT : P L-117 | July 20, 2021 (II)]

(a) sp^3, sp^2 and sp (b) sp, sp^2 and sp^3
(c) sp^3, sp and sp^2 (d) sp^2, sp and sp^3

121. The number of lone pairs of electrons on the central I atom in I_3^- is ______.

[NCERT : P L-117 | NV, July 20, 2021 (I)]

122. The number of species below that have two lone pairs of electrons in their central atom is ______.

SF_4, BF_4^-, ClF_3, AsF_3, PCl_5, BrF_5, XeF_4, SF_6

[NCERT : P L-117 | NV, March 18, 2021 (II)]

123. Which among the following species has unequal bond lengths?

[NCERT : P L-117 | Feb. 25, 2021 (II); Same 2006]

(a) XeF_4 (b) BF_4^- (c) SF_4 (d) SiF_4

124. The compound that has the largest H – M – H bond angle (M = N, S, C), is : **[NCERT : P L-117 | Sep. 05, 2020 (II)]**

(a) H_2O (b) NH_3 (c) H_2S (d) CH_4

125. If AB_4 molecule is a polar molecule, a possible geometry of AB_4 is : **[NCERT : P L-117 | Sep. 02, 2020 (I)]**

(a) Square pyramidal (b) Tetrahedral
(c) Rectangular planar (d) Square planar

126. The shape/ structure of $[XeF_5]^-$ and XeO_3F_2, respectively, are : **[NCERT : P L-117 | Sep. 02, 2020 (II)]**
(a) pentagonal planar and trigonal bipyramidal
(b) octahedral and square pyramidal
(c) trigonal bipyramidal and pentagonal planar
(d) trigonal bipyramidal and trigonal bipyramidal

127. The molecular geometry of SF_6 is octahedral. What is the geometry of SF_4 (including lone pair(s) of electrons, if any)? **[NCERT : P L-115 | Sep. 02, 2020 (II)]**
(a) Tetrahedral (b) Trigonal bipyramidal
(c) Pyramidal (d) Square planar

128. The correct statement about ICl_5 and ICl_4^- is : **[NCERT : P L-121 | April 8, 2019 (II)]**
(a) both are isostructural.
(b) ICl_5 is trigonal bipyramidal and ICl_4^- is tetrahedral.
(c) ICl_5 is square pyramidal and ICl_4^- is tetrahedral.
(d) ICl_5 is square pyramidal and ICl_4^- is square planar.

129. The ion that has sp^3d^2 hydridisation for the central atom, is: **[NCERT : P L-125 | April 8, 2019 (II)]**
(a) $[ICl_4]^-$ (b) $[ICl_2]^-$ (c) $[IF_6]^-$ (d) $[BrF_2]^-$

130. In XeO_3F_2, the number of bond pair(s), π-bond(s) and lone pair(s) on Xe atom respectively are: **[NCERT : P L-125 | Online April 15, 2018 (II)]**
(a) 5, 3, 0 (b) 5, 2, 0 (c) 4, 2, 2 (d) 4, 4, 0

131. The decreasing order of bond angles in BF_3, NH_3, PF_3 and I_3^- is : **[NCERT : P L-117 | Online April 15, 2018 (I)]**
(a) $I_3^- > BF_3 > NH_3 > PF_3$ (b) $BF_3 > I_3^- > PF_3 > NH_3$
(c) $BF_3 > NH_3 > PF_3 > I_3^-$ (d) $I_3^- > NH_3 > PF_3 > BF_3$

132. Total number of lone pair of electrons in I_3^- ion is : **[2018]**
(a) 3 (a) 6
(c) 9 (d) 12

133. Which of the following conversions involves change in both shape and hybridisation? **[NCERT : P L-120 | Online April 16, 2018]**
(a) $H_2O \rightarrow H_3O^+$ (b) $BF_3 \rightarrow BF_4^-$
(c) $CH_4 \rightarrow C_2H_6$ (d) $NH_3 \rightarrow NH_4^+$

134. The incorrect geometry is represented by __________. **[NCERT : P L-122 | Online April 16, 2018]**
(a) NF_3- trigonal planar (b) BF_3- trigonal planar
(c) AsF_5-trigonal bipyramidal (d) H_2O- bent

135. Identify the pair in which the geometry of the species is T-shape and square pyramidal, respectively **[NCERT : P L-125 | Online April 15, 2018 (I)]**
(a) ICl_2^- and ICl_5 (b) IO_3^- and $IO_2F_2^-$
(c) ClF_3 and IO_4^- (d) $XeOF_2$ and $XeOF_4$

136. sp^3d^2 hybridisation is not displayed by : **[NCERT : P L-125 | Online April 8, 2017]**
(a) BrF_5 (b) SF_6 (c) $[CrF_6]^{3-}$ (d) PF_5

137. The group having triangular planar structures is : **[NCERT : P L-114 | Online April 9, 2017]**
(a) BF_3, NF_3, CO_3^{2-} (b) CO_3^{2-}, NO_3^-, SO_3
(c) NH_3, SO_3, CO_3^{2-} (d) NCl_3, BCl_3, SO_3

138. The bond angle H–X–H is the greatest in the compound : **[NCERT : P L-117 | Online April 10, 2016]**
(a) PH_3 (b) CH_4 (c) NH_3 (d) H_2O

139. The species in which the N atom is in a state of *sp* hybridisation is : **[2016]**
(a) NO_3^- (b) NO_2
(c) NO_2^+ (d) NO_2^-

140. The group of molecules having identical shape is : **[NCERT : P L-125 | Online April 9, 2016]**
(a) PCl_5, IF_5, XeO_2F_2 (b) BF_3, PCl_3, XeO_3
(c) SF_4, XeF_4, CCl_4 (d) $ClF_3, XeOF_2, XeF_3^+$

141. The type of hybridisation and number of lone pair(s) of electrons of Xe in $XeOF_4$, respectively, are: **[NCERT : P L-125 | Online April 10, 2015 (S)]**
(a) sp^3d^2 and 1 (b) sp^3d and 2
(c) sp^3d^2 and 2 (d) sp^3d and 1

142. The number and type of bonds in C_2^{2-} ion in CaC_2 are: **[NCERT : P L-129 | Online April 9, 2014; Same 2011 RS]**
(a) One σ bond and one π-bond
(b) One σ bond and two π-bond
(c) Two σ bond and two π-bond
(d) Two σ bond and one π-bond

143. Which of the following molecules has two sigma (σ) and two pi (π) bonds? **[NCERT : P L-128 | Online April 12, 2014]**
(a) C_2H_4 (b) N_2F_2
(c) $C_2H_2Cl_2$ (d) HCN

144. The shape of IF_6^- is : **[NCERT : P L-125 | Online April 23, 2013]**
(a) Trigonally distorted octahedron
(b) Pyramidal
(c) Octahedral
(d) Square antiprism

145. In which of the following sets, all the given species are isostructural ? **[NCERT : P L-114 | 2012 Online April 25, 2013]**
(a) CO_2, NO_2, ClO_2, SiO_2 (b) $PCl_3, AlCl_3, BCl_3, SbCl_3$
(c) BF_3, NF_3, PF_3, AlF_3 (d) $BF_4^-, CCl_4, NH_4^+, PCl_4^+$

146. The formation of molecular complex $BF_3 \leftarrow NH_3$ results in a change in hybridisation of boron **[NCERT : P L-114, 116 | Online May 12, 2012]**
(a) from sp^2 to dsp^2 (b) from sp^2 to sp^3
(c) from sp^3 to sp^2 (d) from sp^3 to sp^3d

147. Which of the following has the square planar structure? **[Online May 19, 2012]**
(a) XeF_4 (b) NH_4^+ (c) BF_4^- (d) CCl_4

148. Among the following species which two have trigonal pyramidal shape? **[NCERT : P L-117 | Online May 26, 2012]**
(I) NI_3 (II) I_3^- (III) SO_3^{2-} (IV) NO_3^-
(a) I and III (b) III and IV
(c) I and IV (d) II and III

149. In which of the following pairs, the two species are not isostructural ? **[NCERT : P L-113 | 2012]**
(a) CO_3^{2-} and NO_3^- (b) PCl_4^+ and $SiCl_4$
(c) PF_5 and BrF_5 (d) AlF_6^{3-} and SF_6

150. The bond dissociation energy of B – F in BF_3 is 646 kJ mol^{-1}, whereas that of C – F in CF_4 is 515 kJ mol^{-1}. The correct reason for higher B – F bond dissociation energy as compared to that of C – F bond is **[2008]**
(a) stronger σ bond between B and F in BF_3 as compared to that between C and F in CF_4.
(b) significant pπ – pπ interaction between B and F in BF_3, whereas there is no possibility of such interaction between C and F in CF_4.
(c) lower degree of pπ – pπ interaction between B and F in BF_3 than that between C and F in CF_4.
(d) smaller size of B– atom as compared to that of C– atom.

151. The decreasing values of bond angles from NH_3 (106°) to SbH_3 (101°) down group-15 of the periodic table is due to **[NCERT : P L-117 | 2006]**
(a) decreasing *lp-bp* repulsion
(b) decreasing electronegativity
(c) increasing *bp-bp* repulsion
(d) increasing *p*-orbital character in *sp*^{3}

152. The states of hybridization of boron and oxygen atoms in boric acid (H_3BO_3) are respectively **[NCERT : P L-120 | 2004]**
(a) sp^3 and sp^2 (b) sp^2 and sp^3
(c) sp^2 and sp^2 (d) sp^3 and sp^3

153. The maximum number of 90° angles between bond pair-bond pair of electrons is observed in **[NCERT : P L-120 | 2004]**
(a) dsp^2 hybridization (b) sp^3d hybridization
(c) dsp^3 hybridization (d) sp^3d^2 hybridization

154. Which one of the following has the regular tetrahedral structure ? **[NCERT : P L-114 | 2004]**
(a) BF_4^- (b) SF_4
(c) XeF_4 (d) $[Ni(CN)_4]^{2-}$
(Atomic nos. : B = 5, S = 16, Ni = 28, Xe = 54)

155. The pair of species having identical shapes for molecules of both species is **[NCERT : P L-117 | 2003]**
(a) XeF_2, CO_2 (b) BF_3, PCl_3
(c) PF_5, IF_5 (d) CF_4, SF_4

156. Hybridisation of the underline atom changes in: **[NCERT : P L-117 | 2002]**
(a) $\underline{Al}H_3$ changes to AlH_4^-
(b) $H_2\underline{O}$ changes to H_3O^+
(c) $\underline{N}H_3$ changes to NH_4^+
(d) in all cases

157. In which of the following species the interatomic bond angle is 109° 28'? **[NCERT : P L-117 | 2002]**
(a) NH_3, BF_4^- (b) NH_4^+, BF_3
(c) NH_3, BF_4 (d) NH_2^-, BF_3.

Topic 5 - Valence Bond Theory and Molecular Orbital Theory

158. Which of the following molecules(s) show/s paramagnetic behavior ? **[NCERT : P L-125 | April 4, 2025 (I)]**
(A) O_2 (B) N_2
(C) F_2 (D) S_2 (E) Cl_2
Choose the **correct** answer from the options given below :
(a) B only (b) A & C only
(c) A & E only (d) A & D only

159. Total number of molecules/species from following which will be paramagnetic is ______.
O_2, O_2^+, O_2^-, NO, NO_2, CO, K_2, $[NiCl_4]$, $[Co(NH_3)_6]Cl_3$, $K_2[Ni(CN)_4]$
[NCERT : P L-129 | NV, Jan. 28, 2025 (II)]

160. Which of the following linear combination of atomic orbitals will lead to formation of molecular orbitals in homonuclear diatomic molecules [internuclear axis in z-direction]? **[NCERT : P L-125 | Jan. 24, 2025 (I)]**
A. $2p_z$ and $2p_x$ B. 2s and $2p_x$
C. $3d_{xy}$ and $3d_{x^2-y^2}$ D. 2s and $2p_z$
E. $2p_z$ and $3d_{x^2-y^2}$
(a) E Only (b) A and B Only
(c) D Only (d) C and D Only

161. Given below are two statements : **[NCERT : P L-129 | April 9, 2024 (I)]**
Statement (I) : The oxidation state of an element in a particular compound is the charge acquired by its atom on the basis of electron gain enthalpy consideration from other atoms in the molecule.
Statement (II) : pπ-pπ bond formation is more prevalent in second period elements over other periods.
In the light of the above statements, choose the **most appropriate** answer from the options given below :
(a) Both **Statement I** and **Statement II** are incorrect
(b) **Statement I** is correct but **Statement II** is incorrect
(c) Both **Statement I** and **Statement II** are correct
(d) **Statement I** is incorrect but **Statement II** is correct

162. Total number of electron present in (π^*) molecular orbitals of O_2, O_2^+ and O_2^- is ______.
[NCERT : P L-131 | NV, April 9, 2024 (II)]

163. When Ψ_A and Ψ_B are the wave functions of atomic orbitals, then σ* is represented by : **[April 8, 2024 (II)]**
(a) $\Psi_A - 2\Psi_B$ (b) $\Psi_A - \Psi_B$
(c) $\Psi_A + 2\Psi_B$ (d) $\Psi_A + \Psi_B$

164. Number of molecules having bond order 2 from the following molecule is ______.
$C_2, O_2, Be_2, Li_2, Ne_2, N_2, He_2$
[NCERT : P L-130 | NV, April 8, 2024 (II)]

165. Match **List-I** with **List-II**:

[NCERT : P L-121 | April 6, 2024 (I)]

List-I (Molecule / Species)	**List-II (Property / Shape)**
(A) SO_2Cl_2	(I) Paramagnetic
(B) NO	(II) Diamagnetic
(C) NO_2^-	(III) Tetrahedral
(D) I_3^-	(IV) Linear

Choose the **correct** answer from the options given below:

(a) A-(IV), B-(I), C-(III), D-(II)
(b) A-(III), B-(I), C-(II), D-(IV)
(c) A-(II), B-(III), C-(I), D-(IV)
(d) A-(III), B-(IV), C-(II), D-(I)

166. Number of molecules/species from the following having one unpaired electron is ________.

$O_2, O_2^-, NO, CN^-, O_2^{2-}$

[NCERT : P L-129 | NV, April 4, 2024 (I)]

167. Given below are two statements:

[NCERT : P L-127 | Feb. 1, 2024 (II)]

Statement (I) : A π bonding MO has lower electron density above and below the inter - nuclear axis.

Statement (II) : The π^* antibonding MO has a node between the nuclei.

In the light of the above statements, choose the most appropriate answer from the options given below:

(a) Both Statement I and Statement II are true
(b) Both Statement I and Statement II are false
(c) Statement I is true but Statement II is false
(d) Statement I is false but Statement II is true

168. The linear combination of atomic orbitals to form molecular orbitals takes place only when the combining atomic orbitals **[NCERT : P L-127 | Jan. 31, 2024 (I)]**

A. have the same energy
B. have the minimum overlap
C. have same symmetry about the molecular axis
D. have different symmetry about the molecular axis

Choose the **most appropriate** from the options given below:

(a) A, B, C only (b) A and C only
(c) B and D only (d) B, C, D only

169. The total number of molecular orbitals formed from 2s and 2p atomic orbitals of a diatomic molecule is ______. **[NCERT : P L-127 | NV, Jan. 30, 2024 (I)]**

170. The molecule /ion with square pyramidal shape is

[NCERT : P L-124 | Jan. 30, 2024 (II)]

(a) $[Ni(CN)_4]^{2-}$ (b) PCl_5
(c) BrF_5 (d) PF_5

171. The number of species from the following which are paramagnetic and with bond order equal to one is _____.

$H_2, He_2^+, O_2^+, N_2^{2-}, O_2^{2-}, F_2, Ne_2^+, B_2$

[NCERT : P L-131 | NV, Jan. 29, 2024 (I)]

172. The total number of anti bonding molecular orbitals, formed from 2s and 2p atomic orbitals in a diatomic molecule is _____.

[NCERT : P L-127 | NV, Jan. 29, 2024 (II)]

173. The total number of 'Sigma' and 'Pi' bonds in 2-formylhex-4-enoic acid is ____.

[NCERT : P L-119 | NV, Jan. 29, 2024 (II)]

174. Sum of bond order of CO and NO^+ is _______.

[NCERT : P L-125 | NV, Jan. 27, 2024 (I)]

175. Consider the following statements:

[NCERT : P L-129 | April 15, 2023 (I)]

(A) NF_3 molecule has a trigonal planar structure.
(B) Bond length of N_2 is shorter than O_2.
(C) Isoelectronic molecules or ions have identical bond order.
(D) Dipole moment of H_2S is higher than that of water molecule.

Choose the correct answer from the option below:

(a) (A) and (D) are correct (b) (C) and (D) are correct
(c) (A) and (B) are correct (d) (B) and (C) are correct

176. In which of the following processes, the bond order increases and paramagnetic character changes to diamagnetic one?

[NCERT : P L-130 | April 13, 2023 (I); Same 2019 (II)]

(a) $O_2 \rightarrow O_2^{2-}$ (b) $NO \rightarrow NO^+$
(c) $N_2 \rightarrow N_2^+$ (d) $O_2 \rightarrow O_2^+$

177. The bond order and magnetic property of acetylide ion are same as that of **[NCERT : P L-131 | April 12, 2023 (I)]**

(a) NO^+ (b) O_2^+ (c) O_2^- (d) N_2^+

178. According to MO theory the bond orders for O_2^{2-}, CO and NO^+ respectively, are

[NCERT : P L-125 | Jan. 29, 2023 (II)]

(a) 1, 3 and 3 (b) 1, 3 and 2
(c) 1, 2 and 3 (d) 2, 3 and 3

179. What is the number of unpaired electron(s) in the highest occupied molecular orbital of the following species :

$N_2; N_2^+; O_2; O_2^+$? **[NCERT : P L-125 | Jan. 24, 2023 (II)]**

(a) 0, 1, 2, 1 (b) 2, 1, 2, 1
(c) 0, 1, 0, 1 (d) 2, 1, 0, 1

180. The number of paramagnetic species among the following is _______. **[NCERT : P L-125 | NV, July 28, 2022 (I)]**

$B_2, Li_2, C_2, C_2^-, O_2^{2-}, O_2^+$ and He_2^+

181. Given below are two statements: One is labelled as **Assertion A** and the other is labelled as **Reason R**.

Assertion A: Zero orbital overlap is an out of phase overlap.

Reason R: It results due to different orientation/direction of approach of orbitals.

In the light of the above statements, choose the correct answer from the options given below: **[July 28, 2022 (II)]**

(a) Both A and R are true and R is the correct explanation of A.
(b) Both A and R are true but R is NOT the correct explanation of A.
(c) A is true but R is false.
(d) A is false but R is true.

182. According to MO theory, number of species/ions from the following having identical bond order is_____:

$CN^-, NO^+, O_2, O_2^+, O_2^{2+}$ **[NV, July 27, 2022 (I)]**

183. Match **List-I** with **List-II**

[NCERT : P L-125 | July 27, 2022 (II)]

List-I	**List-II**
(A) $\Psi_{MO} = \Psi_A - \Psi_B$	(I) Dipole moment
(B) $\mu = Q \times r$	(II) Bonding molecular orbital
(C) $\frac{N_b - N_a}{2}$	(III) Anti-bonding molecualr orbital
(D) $\Psi_{MO} = \Psi_A + \Psi_B$	(IV) Bond order

(a) (A)-(II), (B)-(I), (C)-(IV), (D)-(III)
(b) (A)-(III), (B)-(IV), (C)-(I), (D)-(II)
(c) (A)-(III), (B)-(I), (C)-(IV), (D)-(II)
(d) (A)-(III), (B)-(IV), (C)-(II), (D)-(I)

184. Among the following species

[NCERT : P L-129 | NV, July 25, 2022 (I)]

N_2, N_2^+, N_2^-, N_2^{2-}, O_2, O_2^+, O_2^-, O_2^{2-} the number of species showing diamagnetism is _______.

185. Consider the ions/molecule O_2^+, O_2, O_2^-, O_2^{2-}

For increasing bond order the correct option is :

[NCERT : P L-129 | June 26, 2022 (I); Same July 25, 2021 (II)]

(a) $O_2^{2-} < O_2^- < O_2 < O_2^+$ (b) $O_2^- < O_2^{2-} < O_2 < O_2^+$
(c) $O_2^- < O_2^{2-} < O_2^+ < O_2$ (d) $O_2^- < O_2^+ < O_2^{2-} < O_2$

186. Bonding in which of the following diatomic molecule(s) become(s) stronger, on the basis of MO Theory, by removal of an electron ? **[NCERT : P L-125 | June 25, 2022 (I)]**

(A) NO (B) N_2 (C) O_2 (D) C_2
(E) B_2

Choose the most appropriate answer from the options given below :

(a) (A), (B), (C) only (b) (B), (C), (E) only
(c) (A), (C) only (d) (D) only

187. The correct order of bond orders of C_2^{2-}, N_2^{2-} and O_2^{2-} is, respectively. **[NCERT : P L-129 | June 24, 2022 (II)]**

(a) $C_2^{2-} < N_2^{2-} < O_2^{2-}$ (b) $O_2^{2-} < N_2^{2-} < C_2^{2-}$
(c) $C_2^{2-} < O_2^{2-} < N_2^{2-}$ (d) $N_2^{2-} < C_2^{2-} < O_2^{2-}$

188. According to molecular orbital theory, the number of unpaired electron(s) in O_2^{2-} is :

[NCERT : P L-127 | NV, Aug. 31, 2021 (II); Same July 27, 2021 (II)]

189. The bond order and magnetic behaviour of O_2^- ion are, respectively : **[NCERT : P L-129 | Aug. 26, 2021 (II)]**

(a) 1.5 and paramagnetic (b) 1.5 and diamagnetic
(c) 2 and diamagnetic (d) 1 and paramagnetic

190. The difference between bond orders of CO and $NO^{\oplus}$ is $\frac{x}{2}$ where $x =$ _______.

(Round off to the Nearest Integer)

[NCERT : P L-127 | NV, July 27, 2021 (I)]

191. The total number of electrons in all bonding molecular orbitals of O_2^{2-} is _______.

(Round of to the nearest integer)

[NCERT : P L-129 | NV, July 27, 2021 (II)]

192. AX is a covalent diatomic molecule where A and X are second row elements of periodic table. Based on Molecular orbital theory, the bond order of AX is 2.5.

The total number of electrons in AX is __________.

[NCERT : P L-131 | NV, March 18, 2021 (I)]

193. Match **List-I** with **List-II**

List-I (Molecule)	**List-II (Bond order)**
(A) Ne_2	(I) 1
(B) N_2	(II) 2
(C) F_2	(III) 0
(D) O_2	(IV) 3

Choose the correct answer from the options given below

[NCERT : P L-127 | Feb. 26, 2021 (II)]

(a) (A)-(IV); (B)-(III); (C)-(II); (D)-(I)
(b) (A)-(II); (B)-(I); (C)-(IV); (D)-(III)
(c) (A)-(I); (B)-(II); (C)-(III); (D)-(IV)
(d) (A)-(III); (B)-(IV); (C)-(I); (D)-(II)

194. According to molecular orbital theory, the species among the following that does not exist is

[NCERT : P L-127 | Feb. 25, 2021 (I)]

(a) He_2^- (b) Be_2 (c) He_2^+ (d) O_2^{2-}

195. The molecule in which hybrid MOs involve only one *d*-orbital of the central atom is :

[NCERT : P L-124 | Sep. 04, 2020 (II)]

(a) $[Ni(CN)_4]^{2-}$ (b) BrF_5
(c) XeF_4 (d) $[CrF_6]^{3-}$

196. The potential energy curve for the H_2 molecule as a function of internuclear distance is : **[Sep. 05, 2020 (I)]**

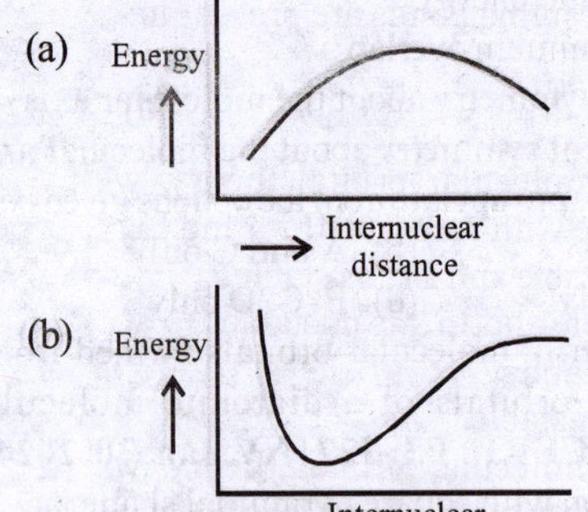

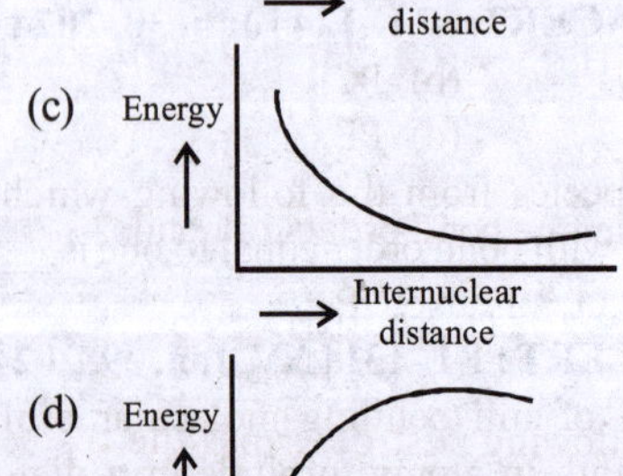

197. The structure of PCl_5 in the solid state is:
[NCERT : P L-125 | Sep. 05, 2020 (I)]
(a) tetrahedral $[PCl_4]^+$ and octahedral $[PCl_6]^-$
(b) square planar $[PCl_4]^+$ and octahedral $[PCl_6]^-$
(c) square pyramidal
(d) trigonal bipyramidal

198. Of the species, NO, NO^+, NO^{2+} and NO^-, the one with minimum bond strength is :
[NCERT : P L-129 | Sep. 03, 2020 (I)]
(a) NO^+ (b) NO (c) NO^{2+} (d) NO^-

199. If the magnetic moment of a dioxygen species is 1.73 B.M, it may be: [NCERT : P L-129 | Jan. 09, 2020 (I)]
(a) O_2^- or O_2^+ (b) O_2 or O_2^+
(c) O_2 or O_2^- (d) O_2, O_2^- or O_2^+

200. The bond order and the magnetic characteristics of CN^- are: [NCERT : P L-129 | Jan. 07, 2020 (II)]
(a) $2\frac{1}{2}$, diamagnetic (b) 3, diamagnetic
(c) 3, paramagnetic (d) $2\frac{1}{2}$, paramagnetic

201. During the change of O_2 to O_2^-, the incoming electron goes to the orbital :
[NCERT : P L-129 | April 10, 2019 (I)]
(a) $\pi 2p_y$ (b) $\sigma^* 2p_z$ (c) $\pi^* 2p_x$ (d) $\pi 2p_x$

202. Among the following, the molecule expected to be stabilised by anion formation is:
C_2, O_2, NO, F_2 [NCERT : P L-129 | April 9, 2019 (I)]
(a) C_2 (b) F_2 (c) NO (d) O_2

203. Among the following species, the diamagnetic molecule is: [NCERT : P L-131 | April 9, 2019 (II)]
(a) NO (b) CO (c) B_2 (d) O_2

204. Among the following molecules/ions, C_2^{2-}, N_2^{2-}, O_2^{2-}, O_2 Which one is diamagnetic and has the shortest bond length? [NCERT : P L-131 | April 8, 2019 (II)]
(a) O_2 (b) N_2^{2-} (c) O_2^{2-} (d) C_2^{2-}

205. Two pi and half sigma bonds are present in:
[Jan. 10, 2019 (I)]
(a) O_2^+ (b) N_2 (c) O_2 (d) N_2^+

206. According to molecular orbital theory, which of the following is true with respect to Li_2^+ and Li_2^-?
(a) Li_2^+ is unstable and Li_2^- is stable [Jan. 9, 2019 (I)]
(b) Li_2^+ is stable and Li_2^- is unstable
(c) Both are stable
(d) Both are unstable

207. According to molecular orbital theory, which of the following will not be a viable molecule? [2018]
(a) He_2^{2+} (b) He_2^+ (c) H_2^- (d) H_2^{2-}

208. (I) (II)
H – N - - - N - - - N
In hydrogen azide, the bond orders of bonds (I) and (II) are ________. [Online April 15, 2018 (I)]
(a) I < 2, II > 2 (b) I > 2, II > 2
(c) I > 2, II < 2 (d) I < 2, II < 2

209. Which of the following best describes the diagram of molecular orbital?
[NCERT : P L-128 | Online April 15, 2018 (II)]

(a) A bonding π orbital
(b) A non-bonding orbital
(c) An antibonding σ orbital
(d) An antibonding π orbital

210. In the molecular orbital diagram for the molecular ion, N_2^+, the number of electrons in the σ_{2p} molecular orbital is: [NCERT : P L-131 | Online April 15, 2018 (I)]
(a) 0 (b) 2 (c) 3 (d) 1

211. Which of the following is paramagnetic ?
[NCERT : P L-131 | Online April 9, 2014(s), Online April 8, 2017(s)]
(a) NO^+ (b) CO (c) O_2^{2-} (d) B_2

212. Which of the following species is not paramagnetic ?
[NCERT : P L-129 | 2017]
(a) NO (b) CO
(c) O_2 (d) B_2

213. After understanding the assertion and reason, choose the correct option.
[NCERT : P L-126 | Online April 10, 2015]
Assertion : In the bonding molecular orbital (MO) of H_2, electron density is increased between the nuclei.
Reason : The bonding MO is $\Psi_A + \Psi_B$, which shows destructive interference of the combining electron waves.
(a) Assertion is incorrect, reason is correct.
(b) Assertion is correct, reason is incorrect.
(c) Assertion and reason are correct and reason is the correct explanation for the assertion.
(d) Assertion and reason are correct, but reason is not the correct explanation for the assertion.

214. The correct order of bond dissociation energy among N_2, O_2, O_2^- is shown in which of the following arrangements?
[NCERT : P L-125 | Online April 11, 2014]
(a) $N_2 > O_2^- > O_2$ (b) $O_2^- > O_2 > N_2$
(c) $N_2 > O_2 > O_2^-$ (d) $O_2 > O_2^- > N_2$

215. Which one of the following properties is **not** shown by NO? [2014]
(a) It is diamagnetic in gaseous state
(b) It is neutral oxide
(c) It combines with oxygen to form nitrogen dioxide
(d) Its bond order is 2.5

216. Which one of the following molecules is paramagnetic?
[NCERT : P L-129 | Online April 19, 2014]
(a) N_2 (b) NO (c) CO (d) O_3

217. Which of the following has unpaired electron(s)?
[NCERT : P L-129 | Online April 9, 2014]
(a) N_2 (b) O_2^- (c) N_2^{2+} (d) O_2^{2-}

218. In which of the following ionization processes the bond energy has increased and also the magnetic behaviour has changed from paramagnetic to diamagnetic ?
[NCERT : PL-129 | Online April 9, 2013; Same 2007]
(a) $NO \rightarrow NO^+$ (b) $N_2 \rightarrow N_2^+$
(c) $C_2 \rightarrow C_2^+$ (d) $O_2 \rightarrow O_2^+$

219. Which one of the following molecules is expected to exhibit diamagnetic behaviour ? [NCERT : P L-129 | 2013]
(a) C_2 (b) N_2 (c) O_2 (d) S_2

220. Which of the following is the wrong statement **[2013]**
(a) ONCl and ONO^- are isoelectronic.
(b) O_3 molecule is bent
(c) Ozone is violet-black in solid state
(d) Ozone is diamagnetic gas.

221. In which of the following pairs of molecules/ions, both the species are not likely to exist ? **[NCERT : P L-129 | 2013]**
(a) H_2^+, He_2^{2-} (b) H_2^-, He_2^{2-}
(c) H_2^{2+}, He_2 (d) H_2^-, He_2^{2+}

222. Stability of the species Li_2, Li_2^- and Li_2^+ increases in the order of : **[NCERT : P L-127 | 2013]**
(a) $Li_2 < Li_2^+ < Li_2^-$ (b) $Li_2^- < Li_2^+ < Li_2$
(c) $Li_2 < Li_2^- < Li_2^+$ (d) $Li_2^- < Li_2 < Li_2^+$

223. Bond order normally gives idea of stability of a molecular species. All the molecules viz. H_2, Li_2 and B_2 have the same bond order yet they are not equally stable. Their stability order is **[NCERT : PL-129, 130, 131 | Online April 22, 2013]**
(a) $H_2 > B_2 > Li_2$ (b) $Li_2 > H_2 > B_2$
(c) $Li_2 > B_2 > H_2$ (d) $B_2 > H_2 > Li_2$

224. The internuclear distances in O – O bonds for O_2^+, O_2, O_2^- and O_2^{2-} respectively are : **[NCERT : P L-129 | Online April 25, 2013]**
(a) 1.30 Å, 1.49 Å, 1.12 Å, 1.21 Å
(b) 1.49 Å, 1.21 Å, 1.12 Å, 1.30 Å
(c) 1.21 Å, 1.12 Å, 1.49 Å, 1.30 Å
(d) 1.12 Å, 1.21 Å, 1.30 Å, 1.49 Å

225. Among the following, the species having the smallest bond length is **[NCERT : P L-129 | Online May 7, 2012]**
(a) NO^- (b) NO^+ (c) O_2 (d) NO

226. Using MO theory, predict which of the following species has the shortest bond length? **[NCERT : P L-129 | 2008]**
(a) O_2^+ (b) O_2^- (c) O_2^{2-} (d) O_2^{2+}

227. Which one of the following pairs of species have the same bond order? **[NCERT : P L-129 | 2008]**
(a) CN^- and NO^+ (b) CN^- and CN^+
(c) O^-_2 and CN^- (d) NO^+ and CN^+

228. Which of the following species exhibits the diamagnetic behaviour ? **[NCERT : P L-129 |2007]**
(a) NO (b) O_2^{2-} (c) O_2^+ (d) O_2

229. Which of the following molecules/ions does not contain unpaired electrons? **[NCERT : P L-127 | 2006]**
(a) N_2^+ (b) O_2 (c) O_2^{2-} (d) B_2

230. Which of the following species is diamagnetic in nature? **[NCERT : P L-129 | 2005]**
(a) H_2^- (b) H_2^+ (c) H_2 (d) He_2^+

231. The bond order in NO is 2.5 while that in NO^+ is 3. Which of the following statements is true for these two species ? **[2004]**
(a) Bond length in NO^+ is equal to that in NO
(b) Bond length in NO is greater than in NO^+
(c) Bond length in NO^+ is greater than in NO
(d) Bond length is unpredictable

232. Which of the following are arranged in an increasing order of their bond strength? **[NCERT : P L-129 | 2002]**
(a) $O_2^- < O_2 < O_2^+ < O_2^{2-}$ (b) $O_2^{2-} < O_2^- < O_2 < O_2^+$
(c) $O_2^- < O_2^{2-} < O_2 < O_2^+$ (d) $O_2^+ < O_2 < O_2^- < O_2^{2-}$

ANSWER KEYS

1	(b)	**25**	(c)	**49**	(3)	**73**	(a)	**97**	(3)	**121**	(3)	**145**	(d)	**169**	(8)	**193**	(d)	**217**	(b)
2	(c)	**26**	(b)	**50**	(b)	**74**	(d)	**98**	(4)	**122**	(2)	**146**	(b)	**170**	(c)	**194**	(b)	**218**	(a)
3	(N)	**27**	(d)	**51**	(c)	**75**	(a)	**99**	(3)	**123**	(c)	**147**	(a)	**171**	(1)	**195**	(a)	**219**	(a, b)
4	(a)	**28**	(b)	**52**	(c)	**76**	(b)	**100**	(4)	**124**	(d)	**148**	(a)	**172**	(4)	**196**	(b)	**220**	(a)
5	(98)	**29**	(c)	**53**	(3)	**77**	(a)	**101**	(d)	**125**	(a)	**149**	(c)	**173**	(22)	**197**	(a)	**221**	(c)
6	(12)	**30**	(c)	**54**	(3)	**78**	(6)	**102**	(5)	**126**	(a)	**150**	(b)	**174**	(6)	**198**	(d)	**222**	(b)
7	(c)	**31**	(c)	**55**	(d)	**79**	(a)	**103**	(b)	**127**	(b)	**151**	(b)	**175**	(d)	**199**	(a)	**223**	(N)
8	(a)	**32**	(a)	**56**	(c)	**80**	(4)	**104**	(d)	**128**	(d)	**152**	(b)	**176**	(b)	**200**	(b)	**224**	(d)
9	(a)	**33**	(a)	**57**	(b)	**81**	(c)	**105**	(b)	**129**	(a)	**153**	(d)	**177**	(a)	**201**	(c)	**225**	(b)
10	(6)	**34**	(c)	**58**	(b)	**82**	(d)	**106**	(4)	**130**	(a)	**154**	(a)	**178**	(a)	**202**	(a)	**226**	(d)
11	(5)	**35**	(a)	**59**	(b)	**83**	(b)	**107**	(6)	**131**	(a)	**155**	(a)	**179**	(a)	**203**	(b)	**227**	(a)
12	(12)	**36**	(6)	**60**	(d)	**84**	(c)	**108**	(c)	**132**	(c)	**156**	(a)	**180**	(4)	**204**	(d)	**228**	(b)
13	(8)	**37**	(c)	**61**	(a)	**85**	(6)	**109**	(b)	**133**	(b)	**157**	(a)	**181**	(a)	**205**	(d)	**229**	(c)
14	(b)	**38**	(5)	**62**	(d)	**86**	(c)	**110**	(c)	**134**	(a)	**158**	(d)	**182**	(3)	**206**	(c)	**230**	(c)
15	(a)	**39**	(c)	**63**	(b)	**87**	(a)	**111**	(1)	**135**	(d)	**159**	(6)	**183**	(c)	**207**	(d)	**231**	(b)
16	(d)	**40**	(b)	**64**	(c)	**88**	(c)	**112**	(a)	**136**	(d)	**160**	(N)	**184**	(2)	**208**	(a)	**232**	(b)
17	(c)	**41**	(2.5)	**65**	(d)	**89**	(3)	**113**	(b)	**137**	(b)	**161**	(d)	**185**	(a)	**209**	(d)		
18	(a)	**42**	(b)	**66**	(b)	**90**	(4)	**114**	(c)	**138**	(b)	**162**	(6)	**186**	(c)	**210**	(d)		
19	(3)	**43**	(3)	**67**	(a)	**91**	(d)	**115**	(2)	**139**	(c)	**163**	(b)	**187**	(b)	**211**	(d)		
20	(6)	**44**	(d)	**68**	(b)	**92**	(4)	**116**	(b)	**140**	(d)	**164**	(2)	**188**	(0)	**212**	(b)		
21	(b)	**45**	(4)	**69**	(b)	**93**	(a)	**117**	(3)	**141**	(a)	**165**	(b)	**189**	(a)	**213**	(b)		
22	(N)	**46**	(a)	**70**	(a)	**94**	(b)	**118**	(c)	**142**	(b)	**166**	(2)	**190**	(0)	**214**	(c)		
23	(b)	**47**	(a)	**71**	(c)	**95**	(a)	**119**	(2)	**143**	(d)	**167**	(d)	**191**	(10)	**215**	(a)		
24	(c)	**48**	(c)	**72**	(a)	**96**	(3)	**120**	(d)	**144**	(a)	**168**	(b)	**192**	(15)	**216**	(b)		

5 Thermodynamics

TREND ANALYSIS
JEE MAIN (2025-2021)

	2025				2024				2023				2022				2021			
	Session II		Session I		Session II		Session I		Session II		Session I		Session II		Session I		Session II		Session I	
	MCQs	NVQs	MCQs	NVQs	MCQs	NVQs	MCQs	NVQs	MCQs	NVQs	MCQs	NVQs	MCQs	NVQs	MCQs	NVQs	MCQs	NVQs	MCQs	NVQs
Topic 1	5	1	3	–	–	4	1	3	1	3	–	3	1	2	–	4	–	3	–	1
Topic 2	4	1	2	4	1	3	–	2	1	3	–	5	–	4	2	3	–	6	–	3
Topic 3	1	–	2	–	–	2	–	–	–	2	–	1	–	–	–	–	–	1	–	3
Topic 4	2	–	2	1	–	–	1	–	–	1	1	–	1	–	2	1	–	3	–	–
Total	12	2	9	5	1	9	2	5	2	9	1	9	2	6	4	8	–	13	–	7

How to Code Qns for Better & Faster Revision →

Type 1: Able to solve (No code Req)

Type 2: Solved with struggle or by seeing solution **(Use code NR means Needs Revision or orange stickers)**

Type 3: Tricky/ Could not solve on my own. Need help from Teacher/ Friend. **(Use code TQ means Tricky Question or red stickers)**

You can either hand write the code in boxes or else paste coloured stickers given at end of book.

Topic 1 - First Law and Fundamentals of Thermodynamics

1. One mole of an ideal gas expands isothermally and reversibly from 10 dm^3 to 20 dm^3 at 300 K. ΔU, q and work done in the process respectively are :
Given : $R = 8.3\ JK^{-1}$ and mol^{-1}
In $10 = 2.3$
$\log 2 = 0.30$
$\log 3 = 0.48$ **[NCERT : PL-142 | April 4, 2025 (I)]**
(a) 0, 21.84 kJ, –1.26 kJ (b) 0, –17.18 kJ, 1.718 J
(c) 0, 21.84 kJ, 21,84 kJ (d) 0,1.78 kJ, –1.718 kJ

2. 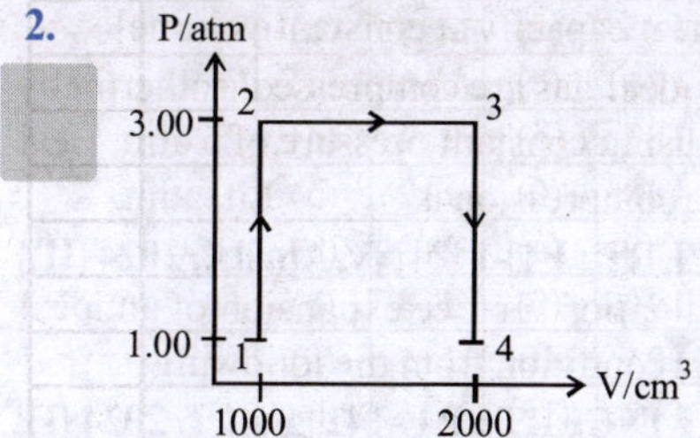

A perfect gas (0.1 mol) having $\bar{C}_V = 1.50\ R$ (independent of temperature) undergoes the above transformation from point 1 to point 4. If each step is reversible, the total work done (w) while going from point 1 to point 4 is (–) ___ J (nearest integer)
[Given : $R = 0.082\ L\ atm\ K^{-1}\ mol^{-1}$]
[NCERT : PL-142 | NV, April 3, 2025 (II)]

3. Which of the following properties will change when system containing solution 1 will become solution 2 ?
[NCERT : PL-138 | April 3, 2025 (I)]

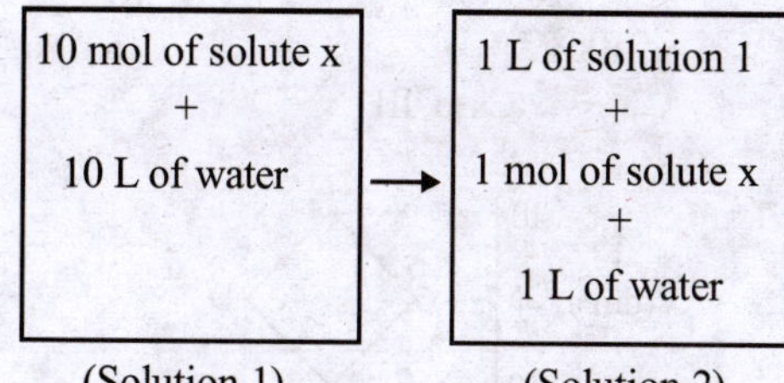

(Solution 1) (Solution 2)

(a) Molar heat capacity (b) Density
(c) Concentration (d) Gibbs free energy

4. Arrange the following in order of magnitude of work done by the system / on the system at constant temperature :

[NCERT : P L-142 | April 2, 2025 (II)]

(a) $|w_{reversible}|$ for expansion in infinite stage.
(b) $|w_{irreversible}|$ for expansion in single stage.
(c) $|w_{reversible}|$ for compression in infinite stage.
(d) $|w_{irreversible}|$ for compression in single stage.

Choose the **correct** answer from the options given below :

(a) $a > b > c > d$ (b) $d > c = a > b$
(c) $c = a > d > b$ (d) $a > c > b > d$

5. 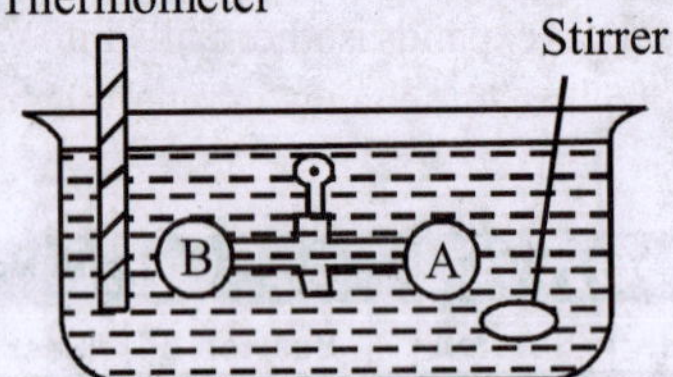

Two vessels A and B are connected via stopcock. The vessel A is filled with a gas at a certain pressure. The entire assembly is immersed in water and is allowed to come to thermal equilibrium with water. After opening the stopcock the gas from vessel A expands into vessel B and no change in temperature is observed in the thermometer. Which of the following statement is **true**?

[NCERT : P L-138 | April 2, 2025 (I)]

(a) $dw \neq 0$
(b) $dq \neq 0$
(c) $dU \neq 0$
(d) The pressure in the vessel B before opening the stopcock is zero.

6. 500 J of energy is transferred as heat to 0.5 mol of Argon gas at 298 K and 1.00 atm. The final temperature and the change in internal energy respectively are :
Given : $R = 8.3\ JK^{-1}\ mol^{-1}$ **[NCERT : P L-143 | Jan. 29, 2025 (I)]**

(a) 348 K and 300 J (b) 378 K and 300 J
(c) 368 K and 500 J (d) 378 K and 500 J

7. **[NCERT : P L-142 | Jan. 28, 2025 (II)]**

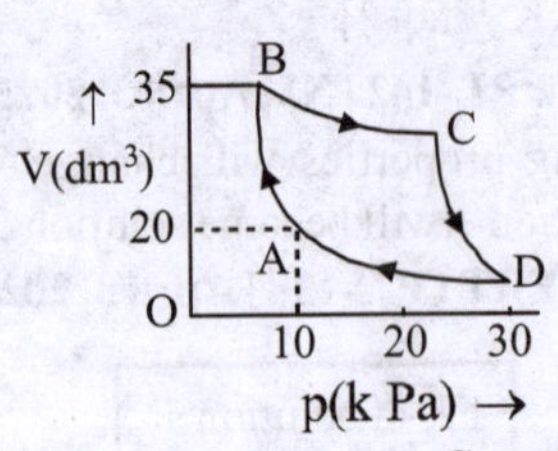

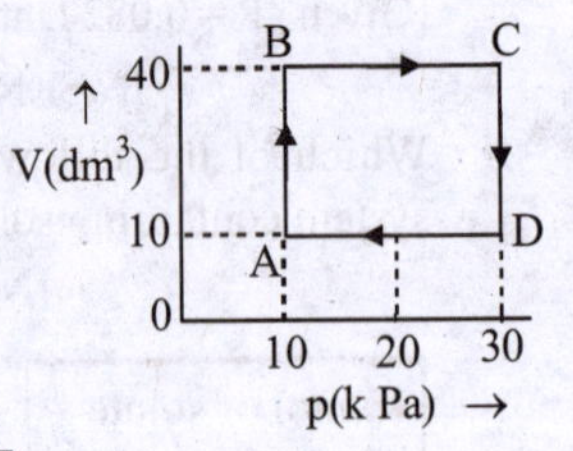

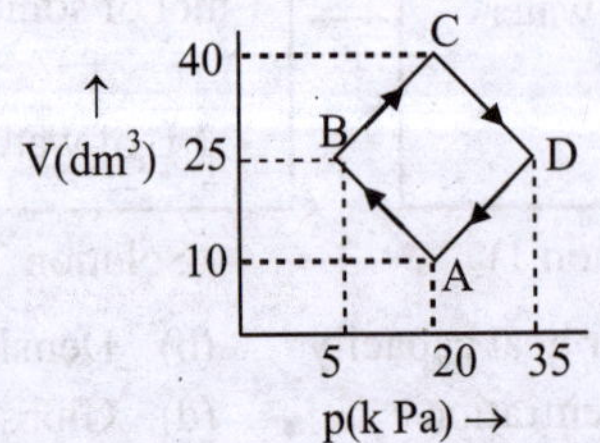

An ideal gas undergoes a cyclic transformation starting from the point A and coming back to the same point by tracing the path $A \to B \to C \to D \to A$ as shown in the three cases above.

Choose the correct option regarding ΔU.

(a) ΔU (Case-III) > ΔU (Case-II) > ΔU (Case-I)
(b) ΔU (Case-I) > ΔU (Case-II) > ΔU (Case-III)
(c) ΔU (Case-I) > ΔU (Case-III) > ΔU (Case-II)
(d) ΔU (Case-I) = ΔU (Case-II) = ΔU (Case-III)

8. A liquid when kept inside a thermally insulated closed vessel at 25°C was mechanically stirred from outside. What will be the correct option for the following thermodynamic parameters? **[NCERT : P L-143 | Jan. 22, 2025 (I)]**

(a) $\Delta U > 0, q = 0, w > 0$ (b) $\Delta U = 0, q = 0, w = 0$
(c) $\Delta U < 0, q = 0, w > 0$ (d) $\Delta U = 0, q < 0, w > 0$

9. 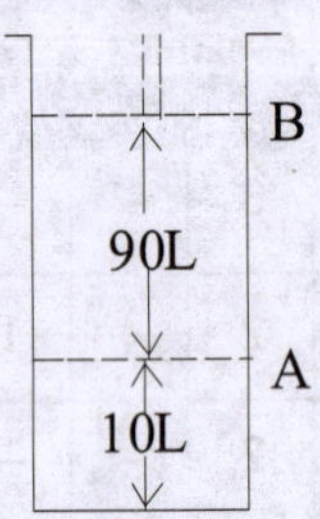

Consider the figure provided.

[NCERT : P L-142 | NV, April 8, 2024 (I)]

1 mol of an ideal gas is kept in a cylinder, fitted with a piston, at the position A, at 18°C. If the piston is moved to position B, keeping the temperature unchanged, then 'x' L atm work is done in this reversible process.

x = ______ L atm. (nearest integer)

[Given : Absolute temperature = °C + 273.15, R = 0.08206 L atm $mol^{-1}\ K^{-1}$]

10. $\Delta_{vap}H^{\ominus}$ for water is $+40.49\ kJ\ mol^{-1}$ at 1 bar and 100°C. Change in internal energy for this vapourisation under same condition is ______ $kJ\ mol^{-1}$. (Integer answer)
(Given $R = 8.3\ JK^{-1}\ mol^{-1}$)

[NCERT : P L-143 | NV, April 8, 2024 (II)]

11. An ideal gas, $\bar{C}v = \frac{5}{2}R$, is expanded adiabatically against a constant pressure of 1 atm untill it doubles in volume. If the initial temperature and pressure is 298 K and 5 atm, respectively then the final temperature is ______ K (nearest integer).

[NCERT : P L-145 | NV, April 6, 2024 (I)]

[$\bar{C}v$ is the molar heat capacity at constant volume]

12. Three moles of an ideal gas are compressed isothermally from 60 L to 20 L using constant pressure of 5 atm. Heat exchange Q for the compression is ______ Lit. atm.

[NCERT : P L-142 | NV, April 4, 2024 (II)]

13. Choose the correct option for free expansion of an ideal gas under adiabatic condition from the following:

[NCERT : P L-142 | Feb. 1, 2024 (I)]

(a) $q = 0, \Delta T \neq 0, w = 0$
(b) $q = 0, \Delta T < 0, w \neq 0$
(c) $q = \neq 0, \Delta T = 0, w = 0$
(d) $q = 0, \Delta T = 0, w = 0$

14. If 5 moles of an ideal gas expands from 10 L to a volume of 100 L at 300 K under isothermal and reversible condition then work, w, is –x J. The value of x is ______.
(Given $R = 8.314\ J K^{-1} mol^{-1}$)
[NCERT : P L-142 | NV, Jan. 31, 2024 (II)]

15. An ideal gas undergoes a cyclic transformation starting from the point A and coming back to the same point by tracing the path A → B → C → A as shown in the diagram above. The total work done in the process is ______ J.
[NCERT : P L-140 | NV, Jan. 30, 2024 (I)]

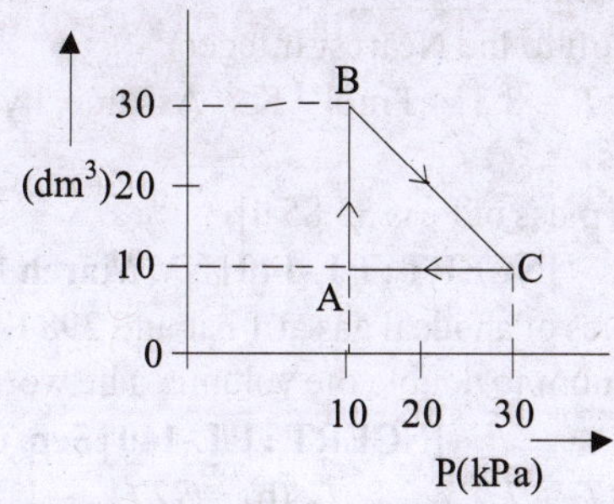

16. If three moles of an ideal gas at 300 K expand isothermally from 30 dm^3 to 45 dm^3 against a constant opposing pressure of 80 kPa, then the amount of heat transferred is ______ J.
[NCERT : P L-142 | NV, Jan. 27, 2024 (I)]

17. What happens when methane undergoes combustion in systems A and B respectively?
[NCERT : P L-138 | April 13, 2023 (II)]

Adiabatic system	Diathermic container
System A	System B

(a)

System A	System B
Temperature rises	Temperature remains same

(b)

System A	System B
Temperature falls	Temperature rises

(c)

System A	System B
Temperature falls	Temperature remains same

(d)

System A	System B
Temperature remains same	Temperature rises

18. The total number of intensive properties from the following is ______
Volume, Molar heat capacity, Molarity, E° cell, Gibbs free energy change, Molar mass, Mole.
[NCERT : P L-144 | NV, April 11, 2023 (II)]

19. The number of endothermic process/es from the following is ______ **[NCERT : P L-150 | NV, April 10, 2023 (II)]**
A. $I_2(g) \rightarrow 2I(g)$
B. $HCl(g) \rightarrow H(g) + Cl(g)$
C. $H_2O(l) \rightarrow H_2O(g)$
D. $C(s) + O_2(g) \rightarrow CO_2(g)$
E. Dissolution of ammonium chloride in water

20. When a 60 W electric heater is immersed in a gas for 100s in a constant volume container with adiabatic walls, the temperature of the gas rises by 5°C. The heat capacity of the given gas is ______ $J K^{-1}$ (Nearest integer)
[NCERT : P L-140 | NV, April 08, 2023 (I)]

21. When 2 litre of ideal gas expands isothermally into vacuum to a total volume of 6 litre, the change in internal energy is ______ J. (Nearest integer)
[NCERT : P L-140 | NV, Jan. 30, 2023 (I)]

22. 1 mole of ideal gas is allowed to expand reversibly and adiabatically from a temperature of 27°C. The work done is 3 kJ mol^{-1}. The final temperature of the gas is ______ K (Nearest integer). Given $C_v = 20\ J mol^{-1} K^{-1}$.
[NCERT : P L-140 | NV, Jan. 30, 2023 (II)]

23. One mole of an ideal monoatomic gas is subjected to changes as shown in the graph. The magnitude of the work done (by the system or on the system) is ______ J (nearest integer). **[NCERT : P L-140 | NV, Jan. 24, 2023 (II)]**

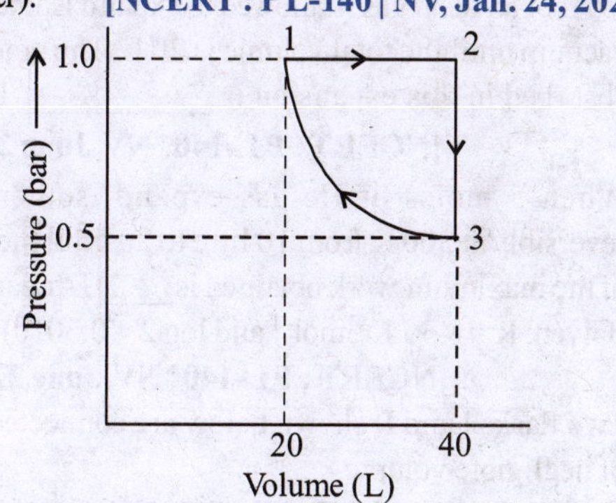

Given : log 2 = 0.3, ln 10 = 2.3

24. Which of the following relation is not correct?
[NCERT : P L-143 | NV, July 28, 2022 (I)]
(a) $\Delta H = \Delta U - P \Delta V$
(b) $\Delta U = q + W$
(c) $\Delta S_{sys} + \Delta S_{surr} \geq 0$
(d) $\Delta G = \Delta H - T \Delta S$

25. Among the following the number of state variables is ______. **[NCERT : P L-138 | NV, July 28, 2022 (II)]**
Internal energy (U)
Volume (V)
Heat (q)
Enthalpy (H)

26. The molar heat capacity for an ideal gas at constant pressure is 20.785 $J K^{-1} mol^{-1}$. The change in internal energy is 5000 J upon heating it from 300K to 500K. The number of moles of the gas at constant volume is ___ [Nearest integer]
(Given: $R = 8.314\ J K^{-1} mol^{-1}$)
[NCERT : P L-138 | NV, July 27, 2022 (I)]

27. 1.0 mol of monoatomic ideal gas is expanded from state 1 to state 2 as shown in the figure. The magnitude of the work done for the expansion of gas from state 1 to state 2 at 300 K is _______ J. (Nearest integer)
(Given : $R = 8.3\ J K^{-1} mol^{-1}$, ln 10 = 2.3, log 2 = 0.30)
[NCERT : PL-140 | NV, June 30, 2022 (I)]

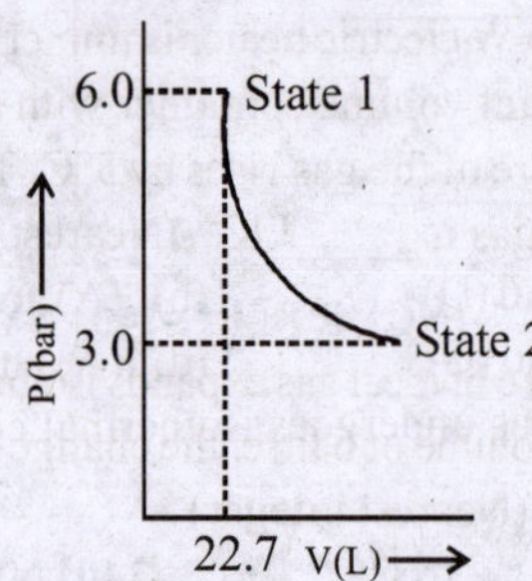

28. 2.2 g of nitrous oxide (N_2O) gas is cooled at a constant pressure of 1 atm from 310 K to 270 K causing the compression of the gas from 217.1 mL to 167.75 mL. The change in internal energy of the process, ΔU is '–x' J. The value of 'x' is __. [nearest integer]
(Given: atomic mass of N = 14 g mol^{-1} and of O = 16 g mol^{-1}. Molar heat capacity of N_2O is 100 $JK^{-1}\ mol^{-1}$)
[NCERT : PL-140 | NV, June 29, 2022 (II)]

29. 4.0 L of an ideal gas is allowed to expand isothermally into vacuum until the total volume is 20 L. The amount of heat absorbed in this expansion is __________ L atm.
[NCERT : PL-140 | NV, June 28, 2022 (I)]

30. When 5 moles of He gas expand isothermally and reversibly at 300 K from 10 litre to 20 litre, the magnitude of the maximum work obtained is ____ J. [nearest integer]
(Given: R = 8.3 $J K^{-1} mol^{-1}$ and log 2 = 0.3010)
[NCERT : PL-140 | NV, June 27, 2022 (II)]

31. Two flasks I and II shown below are connected by a valve of negligible volume.
[NCERT : PL-140 | NV, Aug. 27, 2021 (II)]

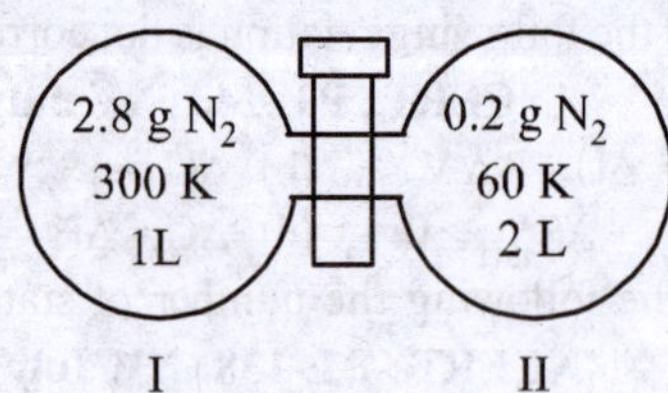

When the valve is opened, the final pressure of the system in bar is $x \times 10^{-2}$. The value of x is ______. (Integer answer)
[Assume–Ideal gas; 1 bar = 10^5 Pa; Molar mass of $N_2 = 28.0$ g mol^{-1} ; $R = 8.31$ J $mol^{-1}K^{-1}$]

32. For water $\Delta_{vap}H = 41$ kJ mol^{-1} at 373 K and 1 bar pressure. Assuming that water vapour is an ideal gas that occupies a much larger volume than liquid water, the internal energy change during evaporation of water is _____ kJ mol^{-1}
[Use : $R = 8.3$ J $mol^{-1} K^{-1}$]
[NCERT : PL-140 | NV, Aug. 26, 2021 (II)]

33. For water at 100°C and 1 bar,
$\Delta_{vap}H - \Delta_{vap}U$ = ________ $\times 10^2$ J mol^{-1}.
(Round off to the Nearest Integer)
[Use : R = 8.31 J $mol^{-1} K^{-1}$]
[Assume volume of $H_2O(l)$ is much smaller than volume of $H_2O(g)$. Assume $H_2O(g)$ treated as an ideal gas]
[NCERT : PL-143 | NV, July 27, 2021 (I)]

34. A system does 200 J of work and at the same time absorbs 150 J of heat. The magnitude of the change in internal energy is _____ J. (Nearest integer)
[NCERT : PL-140 | NV, July 25, 2021 (II)]

35. At 25 °C, 50 g of iron reacts with HCl to form $FeCl_2$. The evolved hydrogen gas expands against a constant pressure of 1 bar. The work done by the gas during this expansion is _______ J.
(Round off to the Nearest Integer).
[Given : $R = 8.314$ J $mol^{-1} K^{-1}$. Assume, hydrogen is an ideal gas]
[Atomic mass of Fe is 55.85 u]
[NCERT : PL-140 | NV, March 16, 2021 (II)]

36. Five moles of an ideal gas at 1 bar and 298 K is expanded into vacuum to double the volume. The work done is :
[NCERT : PL-140 | Sep. 04, 2020 (II)]
(a) $C_V(T_2 - T_1)$ (b) $-RT(V_2 - V_1)$
(c) $-RT \ln V_2/V_1$ (d) zero

37. For one mole of an ideal gas, which of these statements must be **true**? **[NCERT : PL-140 | Sep. 04, 2020 (I)]**
(1) U and H each depends only on temperature
(2) Compressibility factor Z is not equal to 1
(3) $C_{P,m} - C_{V,m} = R$
(4) $dU = C_V dT$ for any process
(a) (1) and (3) (b) (2), (3) and (4)
(c) (3) and (4) (d) (1), (3) and (4)

38. The magnitude of work done by a gas that undergoes a reversible expansion along the path ABC shown in the figure is _____
[NCERT : PL-140 | NV, Jan. 08, 2020 (I)]

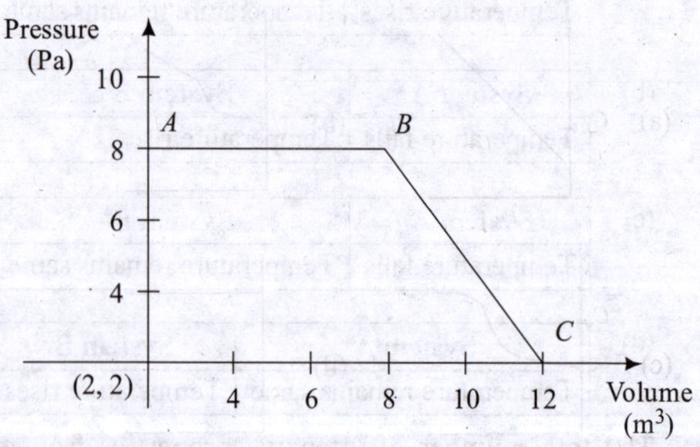

39. At constant volume, 4 mol of an ideal gas when heated from 300 K to 500 K changes its internal energy by 5000 J. The molar heat capacity at constant volume is ______.
[NCERT : PL-144 | NV, Jan. 08, 2020 (II)]

40. An ideal gas is allowed to expand from 1 L to 10 L against a constant external pressure of 1 bar. The work done in kJ is : **[NCERT : P L-140 | April 12, 2019 (I)]**

(a) −9.0 (b) +10.0
(c) −0.9 (d) −2.0

41. Among the following, the set of parameters that represents path functions, is: **[NCERT : P L-138 | April 9, 2019 (I)]**

(A) $q + w$ (B) q (C) w (D) $H - TS$

(a) (B) and (C) (b) (B), (C) and (D)
(c) (A) and (D) (d) (A), (B) and (C)

42. During compression of a spring the work done is 10 kJ and 2 kJ escaped to the surroundings as heat. The change in internal energy, U (in kJ) is: **[NCERT : P L-140 | April 9, 2019 (II)]**

(a) −12 (b) −8
(c) 8 (d) 12

43. Which one of the following equations does not correctly represent the first law of thermodynamics for the given processes involving an ideal gas ? (Assume non-expansion work is zero) **[NCERT : P L-138 | April 8, 2019 (I)]**

(a) Cyclic process : $q = -w$
(b) Adiabatic process : $\Delta U = -w$
(c) Isochoric process: $\Delta U = q$
(d) Isothermal process: $q = -w$

44. 5 moles of an ideal gas at 100 K are allowed to undergo reversible compression till its temperature becomes 200 K. If $C_V = 28\ J\ K^{-1}\ mol^{-1}$, calculate ΔU and ΔPV for this process. ($R = 8.0\ J\ K^{-1} mol^{-1}$) **[NCERT : P L-144 | April 8, 2019 (II)]**

(a) $\Delta U = 14$ kJ; $\Delta(PV) = 18$ kJ
(b) $\Delta U = 14$ kJ; $\Delta(PV) = 0.8$ kJ
(c) $\Delta U = 14$ kJ; $\Delta(PV) = 4$ kJ
(d) $\Delta U = 14$ kJ; $\Delta(PV) = 8.0$ kJ

45. For a diatomic ideal gas in a closed system, which of the following plots does not correctly describe the relation between various thermodynamic quantities? **[NCERT : P L-144 | Jan. 12, 2019 (I)]**

(a) 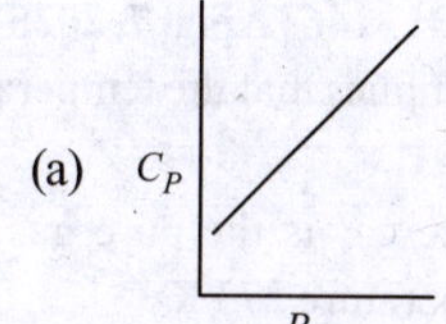

(b) 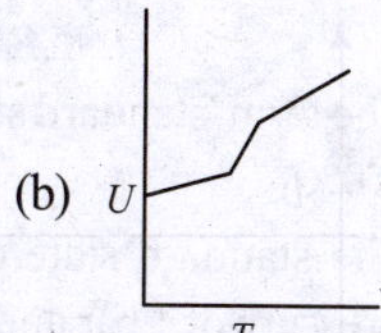

(c) 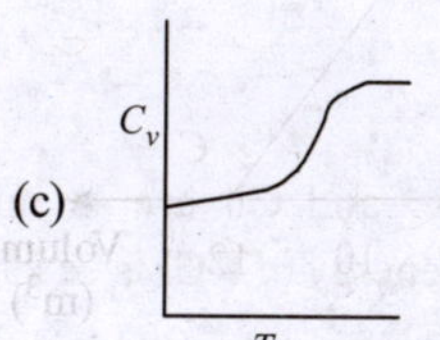

(d) 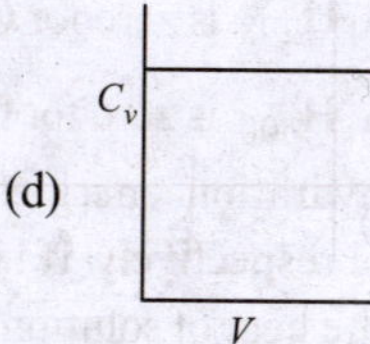

46. The combination of plots which does not represent isothermal expansion of an ideal gas is: **[NCERT : P L-140 | Jan. 12, 2019 (II)]**

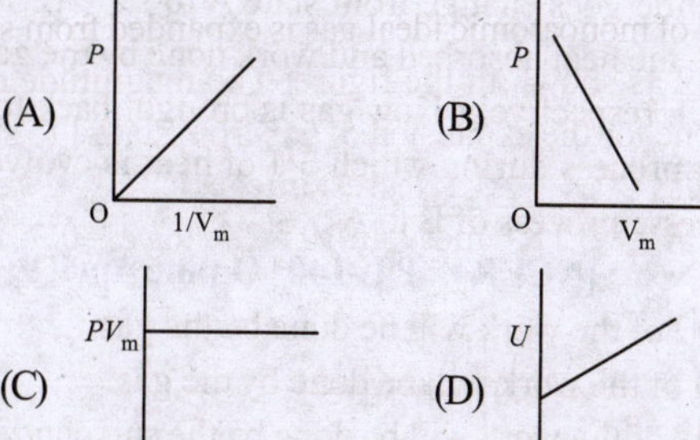

(a) (B) and (D) (b) (A) and (C)
(c) (B) and (C) (d) (A) and (D)

47. An ideal gas undergoes isothermal compression from 5 m^3 to 1 m^3 against a constant external pressure of 4 Nm^{-2}. Heat released in this process is used to increase the temperature of 1 mole of Al. If molar heat capacity of Al is 24 $J\ mol^{-1}\ K^{-1}$, the temperature of Al increases by: **[NCERT : P L-144 | Jan. 10, 2019 (II)]**

(a) $\frac{3}{2}$ K (b) 2 K (c) $\frac{2}{3}$ K (d) 1 K

48. Consider the reversible isothermal expansion of an ideal gas in a closed system at two different temperatures, T_1 and T_2 ($T_1 < T_2$). The correct graphical depiction of the dependence of work done (w) on the final volume (V) is: **[NCERT : P L-140 | Jan. 9, 2019 (I)]**

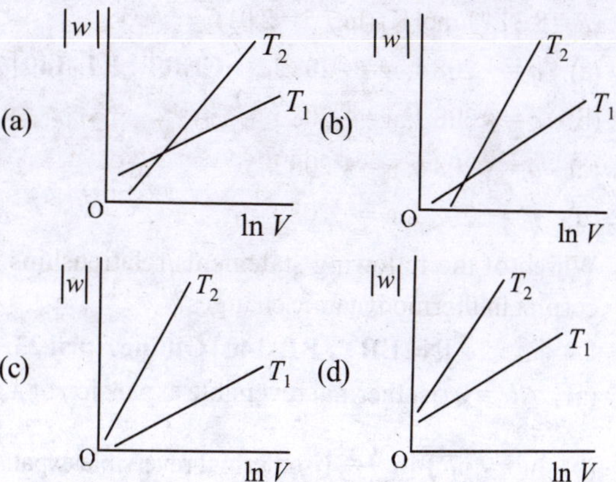

49. An ideal gas undergoes a cyclic process as shown in figure.

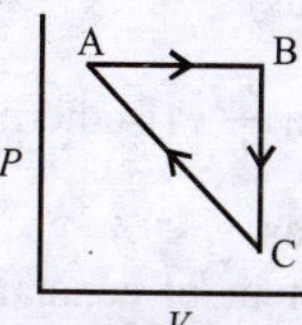

$\Delta U_{BC} = -5\ kJ\ mol^{-1}$, $q_{AB} = 2\ kJ\ mol^{-1}$
$W_{AB} = -5\ kJ\ mol^{-1}$, $W_{CA} = 3\ kJ\ mol^{-1}$

Heat absorbed by the system during process CA is: **[NCERT : P L-140 | Online April 15, 2018 (I)]**

(a) $-5\ kJ\ mol^{-1}$ (b) $+5\ kJ\ mol^{-1}$
(c) $18\ kJ\ mol^{-1}$ (d) $-18\ kJ\ mol^{-1}$

50. ΔU is equal to **[NCERT : P L-140 | 2017]**

(a) Isochoric work (b) Isobaric work
(c) Adiabatic work (d) Isothermal work

51. A gas undergoes change from state A to state B. In this process, the heat absorbed and work done by the gas is 5 J and 8 J, respectively. Now gas is brought back to A by another process during which 3 J of heat is evolved. In this reverse process of B to A :

[NCERT : P L-140 | Online April 9, 2017]

(a) 10 J of the work will be done by the gas.
(b) 6 J of the work will be done by the gas.
(c) 10 J of the work will be done by the surrounding on gas.
(d) 6 J of the work will be done by the surrounding on gas.

52. If 100 mole of H_2O_2 decomposes at 1 bar and 300 K, the work done (kJ) by one mole of $O_2(g)$ as it expands against 1 bar pressure is :

[NCERT : P L-140 | Online April 10, 2016]

$$2H_2O_2(l) \rightleftharpoons 2H_2O(l) + O_2(g)$$

$(R = 8.3\ JK^{-1}\ mol^{-1})$

(a) 124.50 (b) 249.00
(c) 498.00 (d) 62.25

53. A piston filled with 0.04 mol of an ideal gas expands reversibly from 50.0 mL to 375 mL at a constant temperature of 37.0 °C. As it does so, it absorbs 208 J of heat. The values of q and w for the process will be:

$(R = 8.314\ J/mol\ K)$ $(\ln 7.5 = 2.01)$

(a) $q = +208\ J, w = -208\ J$ **[NCERT : P L-140 | 2013]**
(b) $q = -208\ J, w = -208\ J$
(c) $q = -208\ J, w = +208\ J$
(d) $q = +208\ J, w = +208\ J$

54. Which of the following statements/relationships is **not** correct in thermodynamic changes?

[NCERT : P L-140 | Online April 23, 2013]

(a) $\Delta U = 0$ (isothermal reversible expansion of a gas)
(b) $w = -nRT \ln \frac{V_2}{V_1}$ (isothermal reversible expansion of an ideal gas)
(c) $w = nRT \ln \frac{V_2}{V_1}$ (isothermal reversible expansion of an ideal gas)
(d) For a system of constant volume, heat involved directly changes to internal energy.

55. Assuming that water vapour is an ideal gas, the internal energy change (ΔU) when 1 mol of water is vapourised at 1 bar pressure and 100 °C, (given : molar enthalpy of vapourisation of water at 1 bar and 373 K = 41 kJ mol^{-1} and R = 8.3 J mol^{-1} K^{-1}) will be **[NCERT : P L-151 | 2007]**

(a) 41.00 kJ mol^{-1} (b) 4.100 kJ mol^{-1}
(c) 3.7904 kJ mol^{-1} (d) 37.904 kJ mol^{-1}

56. An ideal gas is allowed to expand both reversibly and irreversibly in an isolated system. If T_i is the initial temperature and T_f is the final temperature, which of the following statements is correct?

(a) $(T_f)_{rev} = (T_f)_{irrev}$ **[NCERT : P L-140 | 2006]**
(b) $T_f = T_i$ for both reversible and irreversible processes
(c) $(T_f)_{irrev} > (T_f)_{rev}$
(d) $T_f > T_i$ for reversible process but $T_f = T_i$ for irreversible process

57. An ideal gas expands in volume from 1×10^{-3} to $1 \times 10^{-2}\ m^3$ at 300 K against a constant pressure of $1 \times 10^5\ Nm^{-2}$. The work done is **[NCERT : P L-140 | 2004]**

(a) 270 kJ (b) –900 kJ
(c) –900 J (d) 900 kJ

58. The internal energy change when a system goes from state A to B is 40 kJ/mole. If the system goes from A to B by a reversible path and returns to state A by an irreversible path what would be the net change in internal energy ? **[NCERT : P L-140 | 2003]**

(a) > 40 kJ (b) < 40 kJ
(c) Zero (d) 40 kJ

59. A heat engine abosrbs heat Q_1 at temperature T_1 and heat Q_2 at temperature T_2. Work done by the engine is $(Q_1 + Q_2)$. This data **[NCERT : P L-140 | 2002]**

(a) violates 1st law of thermodynamics
(b) violates 1st law of themodynamics if Q_1 is –ve
(c) violates 1st law of thermodynamics of Q_2 is –ve
(d) does not violate 1st law of themodynamics.

60. The heat required to raise the temperature of body by 1 K is called **[NCERT : P L-144 | 2002]**

(a) specific heat (b) thermal capacity
(c) water equivalent (d) none of these.

Topic 2 - Laws of Thermochemistry

61. The correct statement amongst the following is :

[NCERT : P L-146 | April 7, 2025 (II)]

(a) The term 'standard state' implies that the temperature is 0°C
(b) The standard state of pure gas is the pure gas at a pressure of 1 bar and temperature 273 K
(c) $\Delta_f H^0_{298}$ is zero for O(g)
(d) $\Delta_f H^0_{500}$ is zero for $O_2(g)$

62. The hydration energies of K^+ and Cl^- are $-x$ and $-y$ kJ/mol respectively. If lattice energy of KCl is $-z$ kJ/mol, then the heat of solution of KCl is :

[NCERT : P L-146 | April 7, 2025 (II)]

(a) $+x - y - z$ (b) $x + y + z$
(c) $z - (x + y)$ (d) $-z - (x + y)$

63. Total enthalpy change for freezing of 1 mol of water at 10°C to ice at – 10°C is ______

[NCERT : P L-151 | April 7, 2025 (I)]

(Given : $\Delta_{fus}H = x$ kJ/mol
$C_p [H_2O(l)] = y\ J\ mol^{-1}\ K^{-1}$
$C_p [H_2O(s)] = z\ J\ mol^{-1}\ K^{-1}$

(a) $-x - 10y - 10z$ (b) $-10(100x + y + z)$
(c) $10(100x + y + z)$ (d) $x - 10y - 10z$

64. Consider the given data :

[NCERT : P L-156 | April 4, 2025 (II)]

(A) $HCl(g) + 10H_2O(l) \rightarrow HCl \cdot 10H_2O$
$\Delta H = -69.01\ kJ\ mol^{-1}$

(B) $HCl(g) + 40H_2O(l) \rightarrow HCl \cdot 40H_2O$
$\Delta H = -72.79\ kJ\ mol^{-1}$

Choose the **correct** statement :

(a) Dissolution of gas in water is an endothermic process
(b) The heat of solution depends on the amount of solvent.
(c) The heat of dilution for the HCl ($HCl \cdot 10H_2O$ to $HCl \cdot 40H_2O$) is 3.78 kJ mol^{-1}.
(d) The heat of formation of HCl solution is represented by both (A) and (B)

65. Given : **[NCERT : P L-153 | NV, April 3, 2025 (I)]**

$\Delta H^{\ominus}_{sub}$ [C(graphite)] = 710 kJ mol^{-1}
$\Delta_{C-H}H^{\ominus} = 414\ kJ\ mol^{-1}$
$\Delta_{H-H}H^{\ominus} = 436\ kJ\ mol^{-1}$
$\Delta_{C=H}H^{\ominus} = 611\ kJ\ mol^{-1}$

The $\Delta H^{\ominus}_f$ for $CH_2 = CH_2$ is ______ kJ mol^{-1} (nearest integer value)

66. If C(diamond) → C(graphite) + X kJ mol^{-1}
C(diamond) + O_2(g) → CO_2(g) + Y kJ mol^{-1}
C(graphite) + O_2(g) → CO_2(g) + Z kJ mol^{-1}
At constant temperature. Then

[NCERT : P L-151 | Jan. 29, 2025 (II)]

(a) X = Y + Z (b) –X = Y + Z
(c) X = –Y + Z (d) X = Y – Z

67. Consider the following data :

Heat of formation of CO_2(g) = –393.5 kJ mol^{-1}
Heat of formation of H_2O(l) = – 286.0 kJ mol^{-1}
Heat of combustion of benzene = –3267.0 kJ mol^{-1}
The heat of formation of benzene is ______ kJ mol^{-1}. (Nearest integer)

[NCERT : P L-151 | NV, Jan. 28, 2025 (II)]

68. The formation enthalpies, $\Delta H^{\ominus}_f$ for $H_{(g)}$ and $O_{(g)}$ are 220.0 and 250.0 kJ mol^{-1}, respectively, at 298.15 K, and ΔH_f^{-} for $H_2O_{(g)}$ is –242.0 kJ mol^{-1} at the same temperature. The average bond enthalpy of the O–H bond in water at 298.15 K is ______ kJ mol^{-1} (nearest integer).

[NCERT : P L-151 | NV, Jan. 28, 2025 (I)]

69. $S(g) + \frac{3}{2}O_2(g) \rightarrow SO_3(g) + 2x\ kcal$

$SO_2(g) + \frac{1}{2}O_2(g) \rightarrow SO_3(g) + y\ kcal$

The heat of formation of SO_2(g) is given by :

[NCERT : P L-151 | Jan. 24, 2025 (II)]

(a) $\frac{2x}{y}$ kcal (b) $y - 2x$ kcal
(c) $2x + y$ kcal (d) $x + y$ kcal

70. The bond dissociation enthalpy of X_2 ΔH^o_{bond} calculated from the given data is ______ kJ mol^{-1}. (Nearest integer)

$M^+X^-(s) \rightarrow M^+(g) + X^-(g)\ \Delta H^o_{lattice} = 800\ kJ\ mol^{-1}$
$M(s) \rightarrow M(g)\ \Delta H^o_{sub} = 100\ kJ\ mol^{-1}$
$M(g) \rightarrow M^+(g)^- + e^-(g)\ \Delta H^o_i = 500\ kJ\ mol^{-1}$
$X(g) + e^-(g) \rightarrow X^-(g)\ \Delta H^o_{eg} = -300\ kJ\ mol^{-1}$

$M(s) + \frac{1}{2}X_2(g) \rightarrow M^+X^-(s)\ \Delta H^o_f = -400\ kJ\ mol^{-1}$

[Given : M^+X^- is a pure ionic compound and X forms a diatomic molecule X_2 is gaseous state]

[NCERT : P L-155 | NV, Jan. 23, 2025 (II)]

71. Consider the following cases of standard enthalpy of reaction [ΔH^0_r in kJ mol^{-1}]

$C_2H_6(g) + \frac{7}{2}O_2(g) \rightarrow 2CO_2(g) + 3H_2O(\ell)\ \Delta H^0_1 = -1550$
$C(graphite) + O_2(g) \rightarrow CO_2(g)\ \Delta H^0_2 = -393.5$

$H_2(g) + \frac{1}{2}O_2(g) \rightarrow H_2O(\ell)\ \Delta H^0_3 = -286$

The magnitude of $\Delta H^0_{fC_2H_6(g)}$ is ______ kJ mol^{-1} (Nearest integer).

[NCERT : P L-151 | NV, Jan. 22, 2025 (II)]

72. Combustion of glucose ($C_6H_{12}O_6$) produces CO_2 and water. The amount of oxygen (in g) required for the complete combustion of 900 g of glucose is: [Molar mass of glucose in g mol^{-1} = 180] **[NCERT : P L-152 | April 8, 2024 (I)]**

(a) 480 (b) 960 (c) 800 (d) 32

73. The heat of combustion of solid benzoic acid at constant volume is –321.30 kJ at 27°C. The heat of combustion at constant pressure is (–321.30 – xR) kJ, the value of x is …………… **[NCERT : P L-152 | NV, April 5, 2024 (I)]**

74. Combustion of 1 mole of benzene is expressed at

$C_6H_6(1) + \frac{15}{2}O_2(g) \rightarrow CO_2(g) + 3H_2O(1)$.

The standard enthalpy of combustion of 2 mol of benzene is – 'x' kJ.

x = ______. **[NCERT : P L-152 | NV, April 5, 2024 (II)]**

(1) standard Enthalpy of formation of 1 mol of $C_6H_6(1)$, for the reaction
6C(graphite) + $3H_2$(g) → C_6H_6(1) is 48.5 kJ mol^{-1}.
(2) Standard Enthalpy of formation of 1 mol of CO_2(g), for the reaction
C(graphite) + $O_{2(g)}$ → CO_2(g) is –393.5 kJ mol^{-1}.
(3) Standard and Enthalpy of formation of 1 mol of H_2O(1), for the reaction

$H_2(g) + \frac{1}{2}O_2(g) \rightarrow H_2O(1)$ is –286 kJ mol^{-1}.

75. The enthalpy of formation of ethane (C_2H_6) from ethylene by addition of hydrogen where the bond-energies of C–H, C–C, C=C, H–H are 414 kJ, 347 kJ, 615 kJ and 435 kJ respectively is – ______ kJ.

[NCERT : P L-149, 154 | NV, April 4, 2024 (I)]

76. Two reactions are given below:

[NCERT : P L-150 | NV, Jan. 30, 2024 (II)]

$2Fe(s)+\frac{3}{2}O_2(g)\rightarrow Fe_2O_3(s), \Delta H^o = -822 kJ/mol$

$C(s)+\frac{1}{2}O_2(g)\rightarrow CO(g), \Delta H^o = -110 kJ/mol$

Then enthalpy change for following reaction $3C(s) + Fe_2O_3(s)\rightarrow 2Fe(s)+3CO(g)$ is ______ kJ/mol.

77. Standard enthalpy of vapourisation for CCl_4 is 30.5 kJ mol^{-1}. Heat required for vapourisation of 284g of CCl_4 at constant temperature is ______ kJ.
(Given molar mass in g mol^{-1} ; C = 12, Cl = 35.5)

[NCERT : P L-150 | NV, Jan. 29, 2024 (II)]

78. $A_2 + B_2 \rightarrow 2AB, \Delta H_f^\circ = -200 kJmol^{-1}$

AB, A_2 and B_2 are diatomic molecule. If the bond enthalpies of A_2, B_2 and AB are in the ratio 1 : 0.5 : 1, then the bond enthalpy of A_2 is ______ $kJmol^{-1}$ (Nearest integer)

[NCERT : P L-150 | NV, April 13, 2023 (I)]

79. Solid fuel used in rocket is a mixture of Fe_2O_3 and Al (in ratio 1 : 2). The heat evolved (kJ) per gram of the mixture is ______ (Nearest integer)

Given : $\Delta H_f^\circ(Al_2O_3) = -1700$ kJ mol^{-1}

$\Delta H_f^\circ(Fe_2O_3) = -840$ kJ mol^{-1}

Molar mass of Fe, Al and O are 56, 27 and 16 g mol^{-1} respectively. **[NCERT : P L-150 | NV, April 11, 2023 (I)]**

80. Given **[NCERT : P L-150 | April 10, 2023 (I)]**

(A) $2CO(g)+O_2(g)\rightarrow 2CO_2(g)\ \Delta H_1^\circ = -x\ kJ\ mol^{-1}$

(B) $C(graphite)+O_2(g)\rightarrow CO_2(g)\ \Delta H_2^\circ = -y\ kJ\ mol^{-1}$

The ΔH° for the reaction

$C(graphite)+\frac{1}{2}O_2(g)\rightarrow CO(g)$ is

(a) $\frac{x-2y}{2}$ (b) $\frac{x+2y}{2}$ (c) $\frac{2x-y}{2}$ (d) $2y-x$

81. Consider the following data

[NCERT : P L-150 | NV, April 06, 2023 (II)]

Heat of combustion of $H_2(g) = -241.8$ kJ mol^{-1}

Heat of combustion of $C(s) = -393.5$ kJ mol^{-1}

Heat of combustion of $C_2H_5OH(l) = -1234.7$ kJ mol^{-1}

______ kJ mol^{-1} (Nearest integer)

The heat of formation of $C_2H_5OH(l)$ is (–)

82. At 25ºC, the enthalpy of the following processes are given:

$H_2(g)+O_2(g)\rightarrow 2OH(g)\quad \Delta H^o = 78\ kJ\ mol^{-1}$

$H_2(g)+1/2\,O_2(g)\rightarrow H_2O(g)\quad \Delta H^o = -242\ kJ\ mol^{-1}$

$H_2(g)\rightarrow 2H(g)\ \Delta H^o = 436\ kJ\ mol^{-1}$

$1/2\,O_2(g)\ \rightarrow\ O(g)\ \Delta H^o = 249\ kJ\ mol^{-1}$

What would be the value of X for the following reaction? ______ (Nearest integer)

$H_2O(g)\rightarrow H(g)+OH(g)\ \Delta H^o = X\ kJ\ mol^{-1}$

[NCERT : P L-150 | NV, Feb. 01, 2023 (I)]

83. The enthalpy change for the conversion of $\frac{1}{2}Cl_2(g)$ to $Cl^-(aq)$ is (-) ______ kJ mol^{-1} (Nearest integer)

Given : $\Delta_{dis}H^\circ_{Cl_{2(g)}} = 240 kJmol^{-1}$.

$\Delta_{eg}H^\circ_{Cl_{(g)}} = -350 kJmol^{-1}$, $\Delta_{hyd}H^\circ_{Cl^-_{(g)}} = -380 kJmol^{-1}$

[NCERT : P L-150 | NV, Jan. 31, 2023 (I)]

84. Enthalpies of formation of $CCl_4(g)$, $H_2O(g)$, $CO_2(g)$ and HCl(g) are –105, –242, –394 and –92 KJ mol^{-1} respectively. The magnitude of enthalpy of the reaction given below is ______ kJ mol^{-1} (nearest integer)

$CCl_4(g)+2H_2O(g)\rightarrow CO_2(g)+4HCl(g)$

[NCERT : P L-150 | NV, Jan. 31, 2023 (II)]

85. 28.0 L of CO_2 is produced on complete combustion of 16.8 L gaseous mixture of ethene and methane at 25°C and 1 atm. Heat evolved during the combustion process is ______ kJ. Given : $\Delta H_C(CH_4) = -900$ kJ mol^{-1}
$\Delta H_C(C_2H_4) = -1400$ kJ mol^{-1}

[NCERT : P L-150 | NV, Jan. 25, 2023 (II)]

86. An athlete is given 100 g of glucose ($C_6H_{12}O_6$) for energy. This is equivalent to 1800 kJ of energy. The 50% of this energy gained is utilized by the athlete for sports activities at the event. In order to avoid storage of energy, the weight of extra water he would need to perspire is ______ g (Nearest integer)

Assume that there is no other way of consuming stored energy.

Given : The enthalpy of evaporation of water is 45 kJ mol^{-1}

Molar mass of C, H & O are 12. 1 and 16 g mol^{-1}.

[NCERT : P L-150 | NV, Jan. 25, 2023 (I)]

87. When 600 mL of 0.2 M HNO_3 is mixed with 400 mL of 0.1M NaOH solution in a flask, the rise in temperature of the flask is ______ $\times 10^{-2}$ °C.

(Enthalpy of neutralisation = 57 kJ mol^{-1} and Specific heat of water = 4.2 JK^{-1} g^{-1}) (Neglect heat capacity of flask)

Density of water = 1.00 g cm^{-3}]

(Assume no volume change on mixing)

[NCERT : P L-150 | NV, July 29, 2022 (I)]

88. For the reaction

$H_2F_2(g)\rightarrow H_2(g)+F_2(g)$

$\Delta U = -59.6$ kJ mol^{-1} at 27 °C.

The enthalpy change for the above reaction is (–) ______ kJ mol^{-1} [nearest integer]

Given : R = 8.314 JK^{-1} mol^{-1}.

[NCERT : P L-150 | NV, July 26, 2022 (II)]

89. 17.0 g of NH_3 completely vapourises at – 33.42 °C and 1 bar pressure and the enthalpy change in the process is 23.4 kJ mol^{-1}. The enthalpy change for the vapourisation of 85 g of NH_3 under the same conditions is ________ kJ.

[NCERT : P L-150 | NV, June 29, 2022 (I)]

90. The enthalpy of combustion of propane, graphite and dihydrogen at 298 K are –2220.0 kJ mol^{-1}, –393.5 kJ mol^{-1} and – 285.8 kJ mol^{-1} respectively. The magnitude of enthalpy of formation of propane (C_3H_8) is ____ kJ mol^{-1}. (Nearest integer) **[NCERT : P L-150 | NV, July 25, 2022 (I)]**

91. While performing a thermodynamics experiment, a student made the following observations,

$HCl + NaOH \rightarrow NaCl + H_2O, \Delta H = -57.3$ kJ mol^{-1}

$CH_3COOH + NaOH \rightarrow CH_3COONa + H_2O$

$\Delta H = -55.3$ kJ mol^{-1}

The enthalpy of ionization of CH_3COOH as calculated by the student is _____ kJ mol^{-1}, (nearest integer)

[NCERT : P L-150 | NV, July 25, 2022 (II)]

92. For combustion of one mole of magnesium in an open container at 300 K and 1 bar pressure, $\Delta_C H° = -601.70$ kJ mol^{-1}, the magnitude of change in internal energy for the reaction is _____ kJ. (Nearest integer)

(Given : $R = 8.3$ J K^{-1} mol^{-1})

[NCERT : P L-150 | NV, June 28, 2022 (II)]

93. A fish swimming in water body when taken out from the water body is covered with a film of water of weight 36 g. When it is subjected to cooking at 100 °C, then the internal energy for vaporization in kJ mol^{-1} is _______. [nearest integer]

[Assume steam to be an ideal gas. Given $\Delta_{vap} H^{\ominus}$ for water at 373 K and 1 bar is 41.1 kJ mol^{-1} ; R = 8.31 $JK^{-1}mol^{-1}$]

[NCERT : P L-150 | NV, June 26, 2022 (II)]

94. At 25°C and 1 atm pressure, the enthalpy of combustion of benzene (*l*) and acetylene (g) are –3268 kJ mol^{-1} and –1300 kJ mol^{-1}, respectively. The change in enthalpy for the reaction $3\ C_2H_2(g) \rightarrow C_6H_6(l)$, is

[NCERT : P L-150 | June 25, 2022 (II)]

(a) + 324 kJ mol^{-1} (b) + 632 kJ mol^{-1}
(c) – 632 kJ mol^{-1} (d) – 732 kJ mol^{-1}

95. At 25°C and 1 atm pressure, the enthalpies of combustion are as given below:

Substance	H_2	C (graphite)	$C_2H_6(g)$
$\frac{\Delta_c H°}{kJmol^{-1}}$	–286.0	–394.0	–1560.0

The enthalpy of formation of ethane is

[NCERT : P L-150 | June 24, 2022 (II)]

(a) + 54.0 kJ mol^{-1} (b) – 68.0 kJ mol^{-1}
(c) – 86.0 kJ mol^{-1} (d) + 97.0 kJ mol^{-1}

96. 200 mL of 0.2 M HCl is mixed with 300 mL of 0.1 M NaOH. The molar heat of neutralization of this reaction is –57.1 kJ. The increase in temperature in °C of the system on mixing is $x \times 10^{-2}$.

The value of x is ________. (Nearest integer)

[Given : Specific heat of water = 4.18 J g^{-1} K^{-1}

Density of water = 1.00 g cm^{-3}]

(Assume no volume change on mixing)

[NCERT : P L-156 | NV, Aug. 27, 2021 (I)]

97. These are physical properties of an element

(A) Sublimation enthalpy
(B) Ionisation enthalpy
(C) Hydration enthalpy
(D) Electron gain enthalpy

The total number of above properties that affect the reduction potential is ________ (Integer answer)

[NCERT : P L-146 | NV, Aug. 26, 2021 (I)]

98. The Born-Haber cycle for KCl is evaluated with the following data :

$\Delta_f H°$ for KCl = –436.7 kJ mol^{-1}; $\Delta_{sub} H°$ for K = 89.2 kJ mol^{-1};

$\Delta_{ionization} H°$ for K = 419.0 kJ mol^{-1}; $\Delta_{electron\ gain} H°$ for $Cl_{(g)}$ = –348.6 kJ mol^{-1}; $\Delta_{bond} H°$ for Cl_2 = 243.0 kJ mol^{-1}

The magnitude of lattice enthalpy of KCl in J mol^{-1} is _______ (Nearest integer)

[NCERT : P L-151 | NV, Aug. 26, 2021 (I)]

99. When 400 mL of 0.2 M H_2SO_4 solution is mixed with 600 mL of 0.1 M NaOH solution, the increase in temperature of the final solution is ______ $\times 10^{-2}$ K. (Round off to the nearest integer).

[Use : $H^+ (aq) + OH^- (aq) \rightarrow H_2O : \Delta_r H = -57.1$ kJ mol^{-1}]

Specific heat of H_2O = 4.18 J K^{-1} g^{-1}.

Density of H_2O = 1.0 g cm^{-3}

Assume no change in volume of solution on mixing.

[NCERT : P L-156 | NV, July 27, 2021 (II)]

100. At 298 K, the enthalpy of fusion of a solid (X) is 2.8 kJ mol^{-1} and the enthalpy of vaporisation of the liquid (X) is 98.2 kJ mol^{-1}. The enthalpy of sublimation of the substance (X) in kJ mol^{-1} is ________. (in nearest integer)

[NCERT : P L-147 | NV, July 25, 2021 (I)]

101. If the standard molar enthalpy change for combustion of graphite powder is -2.48×10^2 kJ mol^{-1}, the amount of heat generated on combustion of 1 g of graphite powder is ________ kJ. (Nearest integer)

[NCERT : P L-150 | NV, July 22, 2021 (II)]

102. For the reaction

$C_2H_6 \rightarrow C_2H_4 + H_2$

the reaction enthalpy $\Delta_r H$ = ________ kJ mol^{-1}.

(Rounded off to the Nearest Integer).

[Given : Bond enthalpies in kJ mol^{-1} :

C – C : 347, C = C : 611;

C – H : 414, H – H : 436]

[NCERT : PL-150 | NV, March 18, 2021 (I)]

103. The average S-F bond energy in kJ mol^{-1} of SF_6 is ___________. (Rounded off to the nearest integer)

[Given : The values of standard enthalpy of formation of $SF_6(g)$, S(g) and F(g) are – 1100, 275 and 80 kJ mol^{-1} respectively.] **[NCERT : P L-150 | NV, Feb. 26, 2021 (II)]**

104. The ionization enthalpy of Na^+ formation from Na(g) is 495.8 kJ mol^{-1}, while the electron gain enthalpy of Br is –325.0 kJ mol^{-1}. Given the lattice enthalpy of NaBr is –728.4 kJ mol^{-1}. The energy for the formation of NaBr ionic solid is (–) ______ $\times 10^{-1}$ kJ mol^{-1}.
[NCERT : P L-151 | NV, Feb. 25, 2021 (I)]

105. Lattice enthalpy and enthalpy of solution of NaCl are 788 kJ mol^{-1} and 4 kJ mol^{-1}, respectively. The hydration enthalpy of NaCl is : **[NCERT : P L-151 | Sep. 05, 2020 (II)]**
(a) –780 kJ mol^{-1} (b) 780 kJ mol^{-1}
(c) –784 kJ mol^{-1} (d) 784 kJ mol^{-1}

106. The intermolecular potential energy for the molecules *A*, *B*, *C* and *D* given below suggests that :
[NCERT : P L-150 | Sep. 04, 2020 (I)]

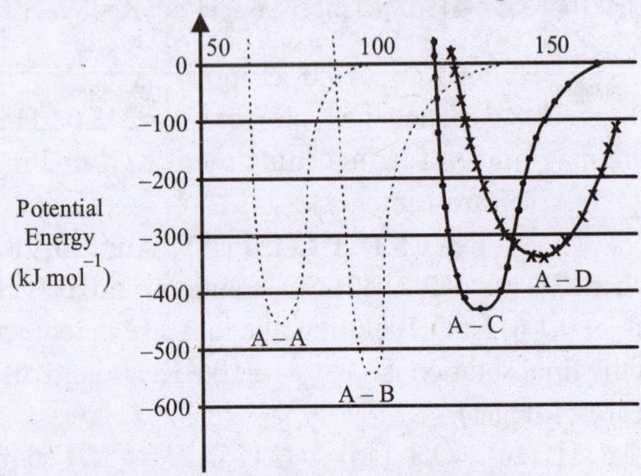

(a) A-D has the shortest bond length
(b) A-A has the largest bond enthalpy
(c) D is more electronegative than other atoms
(d) A-B has the stiffest bond

107. The internal energy change (in J) when 90 g of water undergoes complete evaporation at 100 °C is ______. (Given : ΔH_{vap} for water at 373 K = 41 kJ/ mol, R = 8.314 JK^{-1} mol^{-1}) **[NCERT : P L-150 | NV, Sep. 02, 2020 (I)]**

108. The heat of combustion of ethanol into carbon dioxide and water is – 327 kcal at constant pressure. The heat evolved (in cal) at constant volume and 27 °C (if all gases behave ideally) is (R = 2 cal mol^{-1} K^{-1}) ______.
[NCERT : P L-150 | NV, Sep. 02, 2020 (II)]

109. If enthalpy of atomisation for Br_2(l) is x kJ/mol and bond enthalpy for Br_2 is y kJ/mol, the relation between them:
[NCERT : P L-150 | Jan. 09, 2020 (I)]
(a) is $x = y$ (b) does not exist
(c) is $x > y$ (d) is $x < y$

110. The standard heat of formation ($\Delta_f H^\circ_{298}$) of ethane (in kJ/mol), if the heat of combustion of ethane, hydrogen and graphite are –1560, –393.5 and –286 kJ/mol, respectively is ______. **[NCERT : P L-150 | NV, Jan. 07, 2020 (II)]**

111. Enthalpy of sublimation of iodine is 24 cal g^{-1} at 200 °C. If specific heat of I_2(s) and I_2(vap) are 0.055 and 0.031 cal g^{-1} K^{-1} respectively, then enthalpy of sublimation of iodine at 250 °C in cal g^{-1} is :
[NCERT : P L-150 | April 12, 2019 (I)]
(a) 2.85 (b) 5.7 (c) 22.8 (d) 11.4

112. The difference between ΔH and ΔU ($\Delta H - \Delta U$), when the combustion of one mole of heptane (I) is carried out at a temperature T, is equal to :
[NCERT : P L-150 | April 10, 2019 (II)]
(a) $-4RT$ (b) $-3RT$ (c) $4RT$ (d) $3RT$

113. For silver, C_p(JK^{-1} mol^{-1}) = 23 + 0.01T. If the temperature (T) of 3 moles of silver is raised from 300 K to 1000 K at 1 atm pressure, the value of ΔH will be close to:
[NCERT : P L-150 | April 8, 2019 (I)]
(a) 62 kJ (b) 16 kJ (c) 21 kJ (d) 13 kJ

114. (i) C (graphite) + O_2(g) → CO_2(g); $\Delta_r H^0 = x$ kJ mol^{-1}

(ii) C(graphite) + $\frac{1}{2}$ O_2(g) → CO(g); $\Delta_r H^0 = y$ kJ mol^{-1}

(iii) CO(g) + $\frac{1}{2}$ O_2(g) → CO_2(g); $\Delta_r H^0 = z$ kJ mol^{-1}

Based on the above thermochemical equations, find out which one of the following algebraic relationships is correct? **[NCERT : P L-150 | Jan. 12, 2019 (II)]**
(a) $x = y + z$ (b) $z = x + y$
(c) $y = 2z - x$ (d) $x = y - z$

115. The combustion of benzene (l) gives CO_2 (g) and H_2O (l). Given that heat of combustion of benzene at constant volume is –3263.9 kJ mol^{-1} at 25 °C; heat of combustion (in kJ mol^{-1}) of benzene at constant pressure will be :
(R= 8.314 JK^{-1} mol^{-1}) **[NCERT : P L-150 | 2018]**
(a) 4152.6 (b) –452.46
(c) 3260 (d) –3267.6

116. For which of the following reactions, ΔH is equal to ΔU?
[NCERT : P L-150 | Online April 15, 2018 (I)]
(a) $N_2(g) + 3H_2(g) \rightarrow 2NH_3(g)$
(b) $2HI(g) \rightarrow H_2(g) + I_2(g)$
(c) $2SO_2(g) + O_2(g) \rightarrow 2SO_3(g)$
(d) $2NO_2(g) \rightarrow N_2O_4(g)$

117. Given
C(graphite) + O_2(g) → CO_2(g) ; $\Delta_r H^\circ$ = –393.5 kJ mol^{-1}

H_2 (g) + $\frac{1}{2}$ O_2 (g) → H_2O(l) ; $\Delta_r H^\circ$ = –285.8 kJ mol^{-1}

CO_2(g) + 2 H_2O(l) → CH_4(g) + 2 O_2(g);
$\Delta_r H^\circ$ = + 890.3 kJ mol^{-1}

Based on the above thermochemical equations, the value of $\Delta_r H^\circ$ at 298 K for the reaction,
C(graphite) + 2 H_2(g) → CH_4(g) will be :
[NCERT : P L-150 | 2017]
(a) + 74.8 kJ mol^{-1} (b) + 144.0 kJ mol^{-1}
(c) – 74.8 kJ mol^{-1} (d) – 144.0 kJ mol^{-1}

118. For a reaction, A(g) → A(l); ΔH = –3RT. The correct statement for the reaction is :
[NCERT : P L-150 | Online April 8, 2017]
(a) $\Delta H = \Delta U \neq 0$ (b) $\Delta H = \Delta U = 0$
(c) $|\Delta H| < |\Delta U|$ (d) $|\Delta H| > |\Delta U|$

119. The enthalpy change on freezing of 1 mol of water at 5 °C to ice at –5 °C is :
(Given $\Delta_{fus}H = 6$ kJ mol^{-1} at 0 °C, $C_p(H_2O, l) = 75.3$ J mol^{-1} K^{-1}, $C_p(H_2O, s) = 36.8$ J mol^{-1} K^{-1})
[NCERT : P L-150 | Online April 8, 2017]
(a) 5.44 kJ mol^{-1} (b) 5.81 kJ mol^{-1}
(c) 6.56 kJ mol^{-1} (d) 6.00 kJ mol^{-1}

120. The heats of combustion of carbon and carbon monoxide are –393.5 and –283.5 kJ mol^{-1}, respectively. The heat of formation (in kJ) of carbon monoxide per mole is:
[NCERT : P L-150 | 2016; same 2004]
(a) –676.5 (b) – 110.5
(c) 110.5 (d) 676.5

121. The heat of atomization of methane and ethane are 360 kJ/mol and 620 kJ/mol, respectively. The longest wavelength of light capable of breaking the C – C bond is (Avogadro number $= 6.02 \times 10^{23}$, $h = 6.62 \times 10^{-34}$ J s) :
[NCERT : P L-150 | Online April 10, 2015]
(a) 2.48×10^4 nm (b) 1.49×10^3 nm
(c) 2.48×10^3 nm (d) 1.49×10^4 nm

122. The standard enthalpy of formation of NH_3 is – 46.0 kJ/mol. If the enthalpy of formation of H_2 from its atoms is – 436 kJ/mol and that of N_2 is – 712 kJ/mol, the average bond enthalpy of N - H bond in NH_3 is:
[NCERT : P L-150 | Online April 9, 2014]
(a) – 1102 kJ/mol (b) – 964 kJ/mol
(c) + 352 kJ/mol (d) + 1056 kJ/mol

123. The standard enthalpy of formation ($\Delta_f H°_{298}$) for methane, CH_4 is – 74.9 kJ mol^{-1}. In order to calculate the average energy given out in the formation of a C – H bond from this, it is necessary to know which one of the following?
[NCERT : P L-150 | Online April 12, 2014; same 2006]
(a) The dissociation energy of the hydrogen molecule, H_2.
(b) The first four ionisation energies of carbon.
(c) The dissociation energy of H_2 and enthalpy of sublimation of carbon (graphite).
(d) The first four ionisation energies of carbon and electron affinity of hydrogen.

124. Given :
(I) $H_2(g) + \frac{1}{2}O_2(g) \rightarrow H_2O(l)$;
$\Delta H°_{298K} = -285.9$ kJ mol^{-1}
(II) $H_2(g) + \frac{1}{2}O_2(g) \rightarrow H_2O(g)$;
$\Delta H°_{298K} = -241.8$ kJ mol^{-1}
The molar enthalpy of vapourisation of water will be :
[NCERT : P L-150 | Online April 9, 2013]
(a) 241.8 kJ mol^{-1} (b) 22.0 kJ mol^{-1}
(c) 44.1 kJ mol^{-1} (d) 527.7 kJ mol^{-1}

125. Given

Reaction	Energy Change (in kJ)
$Li(s) \rightarrow Li(g)$	161
$Li(g) \rightarrow Li^+(g)$	520
$\frac{1}{2}F_2(g) \rightarrow F(g)$	77
$F(g) + e^- \rightarrow F^-(g)$	(Electron gain enthalpy)
$Li^+(g) + F^-(g) \rightarrow LiF(s)$	– 1047
$Li(s) + \frac{1}{2}F_2(g) \rightarrow LiF(s)$	– 617

Based on data provided, the value of electron gain enthalpy of fluorine would be :
[NCERT : P L-150 | Online April 22, 2013]
(a) – 300 kJ mol^{-1} (b) – 350 kJ mol^{-1}
(c) – 328 kJ mol^{-1} (d) – 228 kJ mol^{-1}

126. In which of the following exothermic reactions, the heat liberated per mole is the highest ?
[NCERT : P L-150 | Online April 25, 2013]
(a) $CaO + H_2O \rightarrow Ca(OH)_2$
(b) $SrO + H_2O \rightarrow Sr(OH)_2$
(c) $BaO + H_2O \rightarrow Ba(OH)_2$
(d) $MgO + H_2O \rightarrow Mg(OH)_2$

127. Given that: **[NCERT : P L-150 | Online April 25, 2013]**
(i) $\Delta_f H°$ of N_2O is 82 kJ mol^{-1}
(ii) Bond energies of $N \equiv N$, $N = N$, $O = O$ and $N = O$ are 946, 418, 498 and 607 kJ mol^{-1} respectively,
The resonance energy of N_2O is :
(a) –88kJ (b) –66kJ (c) –62kJ (d) –44kJ

128. The difference between the reaction enthalpy change ($\Delta_r H$) and reaction internal energy change ($\Delta_r U$) for the reaction:
$2C_6H_6(l) + 15O_2(g) \longrightarrow 12CO_2(g) + 6H_2O(l)$
at 300 K is ($R = 8.314$ J mol^{-1}K^{-1})
[NCERT : P L-150 | Online May 12, 2012]
(a) 0 J mol^{-1} (b) 2490 J mol^{-1}
(c) – 2490 J mol^{-1} (d) –7482 J mol^{-1}

129. The enthalpy of neutralisation of NH_4OH with HCl is –51.46 kJ mol^{-1} and the enthalpy of neutralisation of NaOH with HCl is –55.90 kJ mol^{-1}. The enthalpy of ionisation of NH_4OH is **[NCERT : P L-156 | Online May 19, 2012]**
(a) – 107.36 kJ mol^{-1} (b) –4.44 kJ mol^{-1}
(c) + 107.36 kJ mol^{-1} (d) + 4.44 kJ mol^{-1}

130. Consider the reaction :
$4NO_2(g) + O_2(g) \rightarrow 2N_2O_5(g)$, $\Delta_r H = -111$ kJ.
If $N_2O_5(s)$ is formed instead of $N_2O_5(g)$ in the above reaction, the $\Delta_r H$ value will be :
(given, ΔH of sublimation for N_2O_5 is 54 kJ mol^{-1})
[NCERT : P L-150 | 2011RS]
(a) + 54 kJ (b) + 219 kJ
(c) – 219 J (d) – 165 kJ

131. The value of enthalpy change (ΔH) for the reaction

$C_2H_5OH(l) + 3O_2(g) \rightarrow 2CO_2(g) + 3H_2O(l)$

at 27 °C is – 1366.5 k J mol^{-1}. The value of internal energy change for the above reaction at this temperature will be :

[NCERT : P L-143 | 2011RS]

(a) – 1369.0 kJ (b) – 1364.0 kJ
(c) – 1361.5 kJ (d) – 1371.5 kJ

132. The standard enthalpy of formation of NH_3 is – 46.0 kJ mol^{-1}. If the enthalpy of formation of H_2 from its atoms is – 436 kJ mol^{-1} and that of N_2 is – 712 kJ mol^{-1}, the average bond enthalpy of N – H bond in NH_3 is

[NCERT : P L-150 | 2010]

(a) – 964 kJ mol^{-1} (b) + 352 kJ mol^{-1}
(c) + 1056 kJ mol^{-1} (d) – 1102 kJ mol^{-1}

133. On the basis of the following thermochemical data : ($\Delta_f H°H^+(aq) = 0$) **[NCERT : P L-150 | 2009]**

$H_2O(l) \rightarrow H^+(aq) + OH^-(aq); \Delta H = 57.32$ kJ

$H_2(g) + \frac{1}{2}O_2(g) \longrightarrow H_2O(l); \Delta H = -286.20$ kJ

The value of enthalpy of formation of OH^- ion at 25° C is:

(a) –228.88 kJ (b) +228.88 kJ
(c) –343.52 kJ (d) –22.88 kJ

134. Oxidising power of chlorine in aqueous solution can be determined by the parameters indicated below:

$$\frac{1}{2}Cl_2(g) \xrightarrow{\frac{1}{2}\Delta_{diss}H^o} Cl(g) \xrightarrow{\Delta_{eg}H^o} Cl^-(g) \xrightarrow{\Delta_{hyd}H^o} Cl^-(aq)$$

The energy envolved in the conversion of $\frac{1}{2}Cl_2(g)$ to $Cl^{-1}(aq)$

(using the data,

$\Delta_{diss}H^o_{Cl_2} = 240$ kJ mol^{-1}, $\Delta_{eg}H^o_{Cl} = -349$ kJ mol^{-1},

$\Delta_{hyd}H^o_{Cl^-} = -381$ kJ mol^{-1}), will be

[NCERT : P L-151 | 2008]

(a) + 152 kJ mol^{-1} (b) – 610 kJ mol^{-1}
(c) – 850 kJ mol^{-1} (d) + 120 kJ mol^{-1}

135. The enthalpy changes for the following processes are listed below : **[NCERT : P L-150 | 2006]**

$Cl_2(g) = 2Cl(g)$, 242.3 kJ mol^{-1}
$I_2(g) = 2I(g)$, 151.0 kJ mol^{-1}
$ICl(g) = I(g) + Cl(g)$, 211.3 kJ mol^{-1}
$I_2(s) = I_2(g)$, 62.76 kJ mol^{-1}

Given that the standard states for iodine and chlorine are $I_2(s)$ and $Cl_2(g)$, the standard enthalpy of formation for ICl(g) is : **[NCERT : P L-150 | 2006]**

(a) +16.8 kJ mol^{-1} (b) +244.8 kJ mol^{-1}
(c) –14.6 kJ mol^{-1} (d) –16.8 kJ mol^{-1}

136. ($\Delta H - \Delta U$) for the formation of carbon monoxide (CO) from its elements at 298 K is **[NCERT : P L-150 | 2006]**

(R = 8.314 J $K^{-1} mol^{-1}$)

(a) –2477.57 J mol^{-1} (b) 2477.57 J mol^{-1}
(c) –1238.78 J mol^{-1} (d) 1238.78 J mol^{-1}

137. Consider the reaction : $N_2 + 3H_2 \rightarrow 2NH_3$ carried out at constant temperature and pressure. If ΔH and ΔU are the enthalpy and internal energy changes for the reaction, which of the following expressions is true ?

[NCERT : P L-150 | 2005]

(a) $\Delta H > \Delta U$ (b) $\Delta H < \Delta U$
(c) $\Delta H = \Delta U$ (d) $\Delta H > 0$

138. If the bond dissociation energies of XY, X_2 and Y_2 (all diatomic molecules) are in the ratio of 1 : 1 : 0.5 and ΔH_f for the formation of XY is – 200 kJ $mole^{-1}$. The bond dissociation energy of X_2 will be **[NCERT : P L-150 | 2005]**

(a) 400 kJ mol^{-1} (b) 300 kJ mol^{-1}
(c) 200 kJ mol^{-1} (d) 100 kJ mol^{-1}

139. If at 298 K, the bond energies of C — H, C — C, C = C and H — H bonds are respectively 414, 347, 615 and 435 kJ mol^{-1}, the value of enthalpy change for the reaction

[NCERT : P L-150 | 2003]

$H_2C = CH_2(g) + H_2(g) \rightarrow H_3C — CH_3(g)$ at 298 K will be

(a) – 250 kJ (b) + 125 kJ
(c) – 125 kJ (d) + 250 kJ

140. The enthalpy change for a reaction does **not** depend upon

[NCERT : P L-150 | 2003]

(a) use of different reactants for the same product
(b) the nature of intermediate reaction steps
(c) the differences in initial or final temperatures of involved substances
(d) the physical states of reactants and products

Topic 3 - Entropy and Second Law of Thermodynamics

141. Total enthalpy change for freezing of 1 mol of water at 10°C to ice at – 10°C is ______

[NCERT : P L-147 | April 7, 2025 (I)]

(Given : $\Delta_{fus}H = x$ kJ/mol

$C_p[H_2O(l)] = y$ J $mol^{-1} K^{-1}$

$C_p[H_2O(s)] = z$ J $mol^{-1} K^{-1}$

(a) $-x - 10y - 10z$ (b) $-10(100x + y + z)$
(c) $10(100x + y + z)$ (d) $x - 10y - 10z$

142. Ice at –5°C is heated to become vapor with temperature of 110°C at atmospheric pressure. The entropy change associated with this process can be obtained from :

[NCERT : P L-161 | Jan. 23, 2025 (I)]

(a) $\int_{268K}^{383K} C_p\, dT + \frac{\Delta H_{melting}}{273} + \frac{\Delta H_{boiling}}{373}$

(b) $\int_{268K}^{273K} \frac{C_{p.m}}{T} dT + \frac{\Delta H_m, \text{fusion}}{T_f} + \frac{\Delta H_{m,\ vaporisation}}{T_b} + \int_{273K}^{373K} \frac{C_{p.m} dT}{T} + \int_{373K}^{383K} \frac{C_{p.m} dT}{T}$

(c) $\int_{268K}^{383K} C_p dT + \frac{q_{rev}}{T}$

(d) $\int_{268K}^{373K} C_{p.m}dT + \frac{\Delta H_{m,\,fusion}}{T_f} + \frac{\Delta H_{m.vaporisation}}{T_b}$

$+ \int_{273K}^{373K} C_{p.m}dT + \int_{373K}^{383K} C_{p.m}dT$

143. Match List-I with List-II.

[NCERT : P L-161 | Jan. 22, 2025 (II)]

List-I (Partial Derivatives)	List-II (Thermodynamic Quantity)
(A) $\left(\frac{\partial G}{\partial T}\right)_P$	(I) C_P
(B) $\left(\frac{\partial H}{\partial T}\right)_P$	(II) –S
(C) $\left(\frac{\partial G}{\partial P}\right)_T$	(III) C_v
(D) $\left(\frac{\partial U}{\partial T}\right)_V$	(IV) V

Choose the correct answer from the option given below:

(a) (A)-(II), (B)-(I), (C)-(III), (D)-(IV)

(b) (A)-(II), (B)-(I), (C)-(IV), (D)-(III)

(c) (A)-(I), (B)-(II), (C)-(IV), (D)-(III)

(d) (A)-(II), (B)-(III), (C)-(I), (D)-(IV)

144. When $\Delta H_{vap} = 30$ kJ/mol and $\Delta S_{vap} = 75$ J mol^{-1} K^{-1}, then the temperature of vapour, at one atmosphere is ______K.

[NCERT : P L-159 | NV, April 9, 2024 (II)]

145. For the reaction at 298 K, $2A + B \rightarrow C$. $\Delta H = 400$ kJ mol^{-1} and $\Delta S = 0.2$ kJ mol^{-1} K^{-1}. The reaction will become spontaneous above______ K.

[NCERT : P L-160 | NV, April 6, 2024 (II)]

146. 30.4 kJ of heat is required to melt one mole of sodium chloride and the entropy change at the melting point is 28.4 J K^{-1} mol^{-1} at 1 atm. The melting point of sodium chloride is _______ K (Nearest Integer)

[NCERT : P L-158 | NV, April 15, 2023 (I)]

147. One mole of an ideal gas at 350K is in a 2.0 L vessel of thermally conducting walls, which are in contact with the surroundings. It undergoes isothermal reversible expansion from 2.0 L to 3.0 L against a constant pressure of 4 atm. The change in entropy of the surroundings (ΔS) is ________ J K^{-1} (Nearest integer)

Given : R = 8.314 J K^{-1}Mol^{-1}.

[NCERT : P L-158 | NV, April 12, 2023 (I)]

148. For independent process at 300 K.

[NCERT : P L-158 | NV, Jan. 24, 2023 (I)]

Process	ΔH/kJ mol^{-1}	ΔS/J K^{-1}
A	–25	–80
B	–22	40
C	25	–50
D	22	20

The number of non-spontaneous process from the following is ________.

149. For a given chemical reaction $A \rightarrow B$ at 300 K the free energy change is –49.4 kJ mol^{-1} and the enthalpy of reaction is 51.4 kJ mol^{-1}. The entropy change of the reaction is ________ J K–1 mol^{-1}.

[NCERT : P L-158 | NV, July 20, 2021 (II)]

150. For a chemical reaction $A + B \rightleftharpoons C + D$ ($\Delta_r H^o = 80$ kJ mol^{-1}) the entropy change $\Delta_r S^o$ depends on the temperature T (in K) as $\Delta_r S^o = 2T$ (J K^{-1} mol^{-1}).

Minimum temperature at which it will become spontaneous is ________ K. (Integer)

[NCERT : P L-158 | NV, Feb. 26, 2021 (I)]

151. For the reaction $A(g) \rightarrow B(g)$, the value of the equilibrium constant at 300 K and 1 atm is equal to 100.0. The value of $\Delta_r G$ for the reaction at 300 K and 1 atm in J mol^{-1} is $-xR$, where x is ________ (Rounded off to the nearest integer) ($R = 8.31$ J mol^{-1} K^{-1} and ln 10 = 2.3)

[NCERT : P L-158 | NV, Feb. 24, 2021 (I)]

152. Assuming ideal behaviour, the magnitude of log K for the following reaction at 25°C is $x \times 10^{-1}$. The value of x is ________. (Integer answer).

$3HC \equiv CH(g) \rightleftharpoons C_6H_6(l)$

[Given : $\Delta_f G^0(HC \equiv CH) = -2.04 \times 10^5$ J mol^{-1};

$\Delta_f G^0(C_6H_6) = -1.24 \times 10^5$ J mol^{-1};

$R = 8.314$ J K^{-1} mol^{-1}]

[NCERT : P L-158 | NV, Feb. 24, 2021 (II)]

153. The true statement amongst the following is :

[NCERT : P L-158 | Jan. 09, 2020 (II)]

(a) Both ΔS and S are functions of temperature.

(b) Both S and ΔS are not functions of temperature.

(c) S is not a function of temperature but ΔS is a function of temperature.

(d) S is a function of temperature but ΔS is not a function of temperature.

154. For the reaction ; $A(l) \rightarrow 2B(g)$

$\Delta U = 2.1$ kcal, $\Delta S = 20$ cal K^{-1} at 300 K.

Hence, ΔG in kcal is ________.

[NCERT : P L-158 | NV, Jan. 07, 2020 (I)]

155. Two blocks of the same metal having same mass and at temperature T_1, and T_2, respectively, are brought in contact with each other and allowed to attain thermal equilibrium at constant pressure. The change in entropy, ΔS, for this process is : **[NCERT : P L-158 | Jan. 11, 2019 (I)]**

(a) $C_P \ln\left[\frac{(T_1+T_2)^2}{4T_1T_2}\right]$ (b) $2C_P \ln\left[\frac{(T_1+T_2)^{\frac{1}{2}}}{T_1T_2}\right]$

(c) $2C_P \ln\left(\frac{T_1+T_2}{4T_1T_2}\right)$ (d) $2C_P \ln\left[\frac{T_1+T_2}{2T_1T_2}\right]$

156. The process with negative entropy change is: **[NCERT : P L-158 | Jan. 10, 2019 (II)]**

(a) Dissociation of $CaSO_4(s)$ to $CaO(s)$ and $SO_3(g)$
(b) Sublimation of dry ice
(c) Dissolution of iodine in water
(d) Synthesis of ammonia from N_2 and H_2

157. The entropy change associated with the conversion of 1 kg of ice at 273 K to water vapours at 383 K is: (Specific heat of water liquid and water vapour are 4.2 kJ K^{-1} kg^{-1} and 2.0 kJ K^{-1} kg^{-1}; heat of liquid fusion and vapourisation of water are 334 kJ kg^{-1} and 2491 kJ kg^{-1}, respectively). **[NCERT : P L-158 | Jan. 9, 2019 (II)]**
(log 273 = 2.436, log 373 = 2.572, log 383 = 2.583)

(a) 7.90 kJ kg^{-1} K^{-1} (b) 2.64 kJ kg^{-1} K^{-1}
(c) 8.49 kJ kg^{-1} K^{-1} (d) 9.26 kJ kg^{-1} K^{-1}

158. For which of the following processes, ΔS is negative? **[NCERT : P L-158 | Online April 16, 2018]**

(a) $C(\text{diamond}) \rightarrow C(\text{graphite})$
(b) $N_2(g, 1\text{ atm}) \rightarrow N_2(g, 5\text{ atm})$
(c) $N_2(g, 273\text{ K}) \rightarrow N_2(g, 300\text{ K})$
(d) $H_2(g) \rightarrow 2H(g)$

159. The molar heat capacity (C_p) of CD_2O is 10 cals at 1000 K. The change in entropy associated with cooling of 32 g of CD_2O vapour from 1000 K to 100 K at constant pressure will be: (D = deuterium, atomic mass = 2 u) **[NCERT : P L-158 | Online April 11, 2014]**

(a) 23.03 cal deg^{-1} (b) – 23.03 cal deg^{-1}
(c) 2.303 cal deg^{-1} (d) – 2.303 cal deg^{-1}

160. The (S°) of the following substances are:
CH_4 (g) 186.2 JK^{-1} mol^{-1}; O_2 (g) 205.2 JK^{-1} mol^{-1}
CO_2 (g) 213.6 JK^{-1} mol^{-1}; H_2O (l) 69.9 JK^{-1} mol^{-1}
The entropy change (ΔS°) for the reaction
$CH_4(g) + 2O_2(g) \rightarrow CO_2(g) + 2H_2O(l)$ is: **[NCERT : P L-158 | Online April 12, 2014]**

(a) – 312.5 JK^{-1} mol^{-1} (b) – 242.8 JK^{-1} mol^{-1}
(c) – 108.1 JK^{-1} mol^{-1} (d) – 37.6 JK^{-1} mol^{-1}

161. The entropy of a sample of a certain substance increases by 0.836 J K^{-1} on adding reversibly 0.3344 J of heat at constant temperature. The temperature of the sample is: **[NCERT : P L-158 | Online May 7, 2012]**

(a) 2.5 K (b) 0.3 K (c) 0.016 K (d) 0.4 K

162. One mole of an ideal gas is expanded isothermally and reversibly to half of its initial pressure. ΔS for the process in JK^{-1} mol^{-1} is [ln 2 = 0.693 and R = 8.314, J/(mol K)] **[NCERT : P L-158 | Online May 26, 2012]**

(a) 6.76 (b) 5.76 (c) 10.76 (d) 8.03

163. Standard entropy of X_2, Y_2 and XY_3 are 60, 40 and 50 J K^{-1} mol^{-1}, respectively. For the reaction,
$\frac{1}{2}X_2 + \frac{3}{2}Y_2 \rightarrow XY_3, \Delta H = -30\text{ kJ}$, to be at equilibrium, the temperature will be **[NCERT : P L-158 | 2008]**

(a) 1250 K (b) 500 K (c) 750 K (d) 1000 K

Topic 4 - Spontaneity and Gibb's Free Energy

164. Let us consider a reversible reaction at temperature, T. In this reaction, both ΔH and ΔS were observed to have positive values. If the equilibrium temperature is Te, then the reaction becomes spontaneous at : **[NCERT : P L-162 | April 4, 2025 (I)]**

(a) $T = Te$ (b) $Te > T$
(c) $T > Te$ (d) $Te = 5T$

165. Which of the following graphs correctly represents the variation of thermodynamic properties of Haber's process ? **[NCERT : P L-155 | April 2, 2025 (II)]**

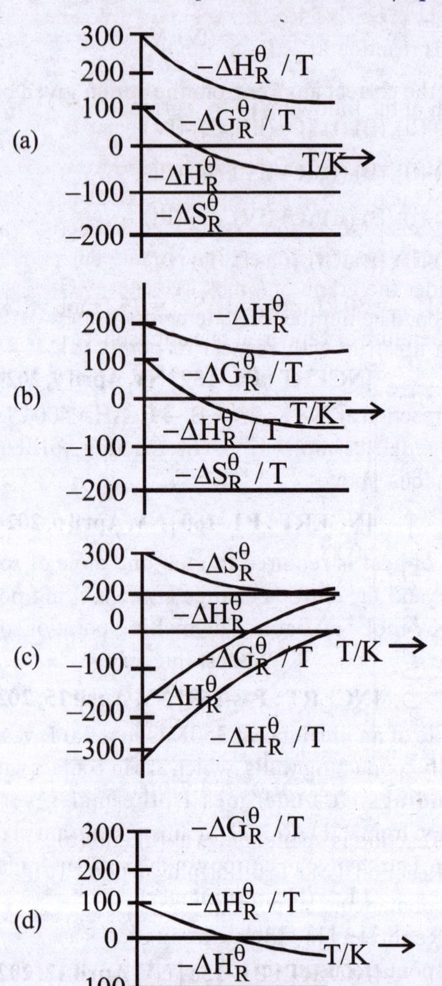

166. Let us consider an endothermic reaction which is non-spontaneous at the freezing point of water. However, the reaction is spontaneous at boiling point of water. Choose the correct option.
[NCERT : P L-162 | Jan. 24, 2025 (I)]
(a) Both ΔH and ΔS are (+ve)
(b) ΔH is (–ve) but ΔS is (+ve)
(c) ΔH is (+ve) but ΔS is (–ve)
(d) Both ΔH and ΔS are (–ve)

167. The effect of temperature on spontaneity of reactions are represented as: **[NCERT : P L-162 | Jan. 23, 2025 (II)]**

	ΔH	ΔS	Temperature	Spontaneity
(A)	+	–	any T	Non spontaneous
(B)	+	+	low T	spontaneous
(C)	–	–	low T	Non spontaneous
(D)	–	+	any T	spontaneous

The incorrect combinations are :
(a) (B) and (D) only (b) (A) and (D) only
(c) (B) and (C) only (d) (A) and (C) only

168. The standard enthalpy and standard entropy of decomposition of N_2O_4 to NO_2 are 55.0 kJ mol^{-1} and 175.0 J/K/mol respectively. The standard free energy change for this reaction at 25°C in J mol^{-1} is ______ (Nearest integer) **[NCERT : P L-162 | NV, Jan. 23, 2025 (I)]**

169. Which of the following is **not** correct?
[NCERT : P L-160 | Jan. 29, 2024 (I)]
(a) ΔG is zero for a reversible reaction
(b) ΔG is negative for a spontaneous reaction
(c) ΔG is positive for a spontaneous reaction
(d) ΔG is positive for a non-spontaneous reaction

170. Consider the graph of Gibbs free energy G vs Extent of reaction. The number of statement/s from the following which are true with respect to points (a), (b) and (c) is............ **[NCERT : P L-160 | NV, April 06, 2023 (I)]**

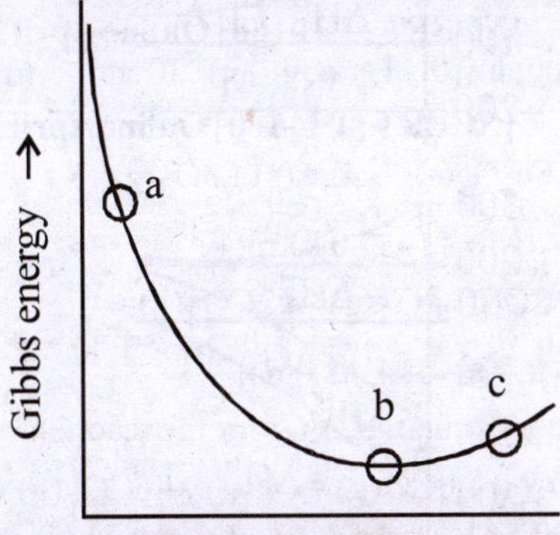

A. Reaction is spontaneous at (a) and (b)
B. Reaction is at equilibrium at point (b) and non–spontaneous at point (c)
C. Reaction is spontaneous at (a) and non–spontaneous at (c)
D. Reaction is non–spontaneous at (a) and (b)

171. Which of the following relations are correct?
(A) $\Delta U = q + p\Delta V$ (B) $\Delta G = \Delta H - T\Delta S$
(C) $\Delta S = \frac{q_{rev}}{T}$ (D) $\Delta H = \Delta U - \Delta nRT$
Choose the most appropriate answer from the options given below : **[NCERT : P L-160 | Jan. 29, 2023 (II)]**
(a) C and D only (b) B and C only
(c) A and B only (d) B and D only

172. Given below are two statements: One is labelled as **Assertion A** and the other is labelled as **Reason R**.
Assertion A: The reduction of a metal oxide is easier if the metal formed is in liquid state than solid state.
Reason R: The value of $\Delta G°$ becomes more on negative side as entropy is higher liquid state than solid state.
In the light of the above statements, choose the correct answer from the options given below:
[NCERT : P L-160 | July 28, 2022 (II)]
(a) Both A and R are correct and R is the correct explanation of A.
(b) Both A and R are correct but R is NOT the correct explanation of A.
(c) A is correct but R is not correct.
(d) A is not correct but R is correct.

173. Match List-I with List-II
[NCERT : P L-160 | June 27, 2022 (I)]

	List-I		List-II
(A)	Spontaneous process	(I)	$\Delta H < 0$
(B)	Process with $\Delta P = 0$, $\Delta T = 0$	(II)	$\Delta G_{T,P} < 0$
(C)	$\Delta H_{reaction}$	(III)	Isothermal and isobaric process
(D)	Exothermic process	(IV)	[Bond energies of molecules in reactants] - [Bond energies of product molecules]

Choose the correct answer from the options given below:
(a) (A) – (III), (B) – (II), (C) – (IV), (D) – (I)
(b) (A) – (II), (B) – (III), (C) – (IV), (D) – (I)
(c) (A) – (II), (B) – (III), (C) – (I), (D) – (IV)
(d) (A) – (II), (B) – (I), (C) – (III), (D) – (IV)

174. The standard entropy change for the reaction $4Fe(s) + 3O_2(g) \rightarrow 2Fe_2O_3(s)$ is -550 JK^{-1} at 298 K.
[Given : The standard enthalpy change for the reaction is -165 kJ mol^{-1}]. The temperature in K at which the reaction attains equilibrium is ________. (Nearest Integer) **[NCERT : P L-160 | NV, June 25, 2022 (I)]**

175. The standard free energy change ($\Delta G°$) for 50% dissociation of N_2O_4 into NO_2 at 27°C and 1 atm pressure is – x J mol^{-1}. The value of x is ________. (Nearest Integer)
[Given : R = 8.31 J K^{-1} mol^{-1}, log 1.33 = 0.1239 ln 10 = 2.3]
[NCERT : P L-160 | NV, June 25, 2022 (I)]

176. For the reaction $2NO_2(g) \rightleftharpoons N_2O_4(g)$, when $\Delta S = -176.0$ JK^{-1} and $\Delta H = -57.8$ kJ mol^{-1}, the magnitude of ΔG at 298 K for the reaction is_____ kJ mol^{-1}. (Nearest integer) **[NCERT : P L-160 | NV, Sep. 1, 2021 (II)]**

177. Consider the following cell reaction :

$$Cd_{(s)} + Hg_2SO_{4(s)} + \frac{9}{5}H_2O_{(l)} \rightleftharpoons CdSO_4.\frac{9}{5}H_2O_{(s)} + 2Hg_{(l)}$$

The value of E^0_{cell} is 4.315 V at 25°C. If $\Delta H° = -825.2$ kJ mol^{-1}, the standard entropy change $\Delta S°$ in JK^{-1} is ______. (Nearest integer)
[Given : Faraday constant = 96487 C mol^{-1}]

[NCERT : P L-162 | NV, Aug. 31, 2021 (I)]

178. Data given for the following reaction is as follows :
FeO(s) + C(graphite) → Fe(s) + CO(g)

Substance	$\Delta H°_f$ (kJ mol^{-1})	$\Delta S°$(J $mol^{-1}K^{-1}$)
FeO(s)	–266.3	57.49
C(graphite)	0	5.74
Fe(s)	0	27.28
CO(g)	–110.5	197.6

The minimum temperature in K at which the reaction becomes spontaneous is ______. (Integer answer)

[NCERT : P L-160 | NV, Aug. 27, 2021 (II)]

179. For a dimerization reaction, $2A(g) \rightarrow A_2(g)$, at 298 K, $\Delta U° = -20$ kJ mol^{-1}, $\Delta S° = -30$ J K^{-1} mol^{-1}, then the $\Delta G°$ will be ________ J.

[NCERT : P L-160 | NV, Sep. 05, 2020 (II)]

180. For the reaction ; $A(l) \rightarrow 2B(g)$
$\Delta U = 2.1$ kcal, $\Delta S = 20$ cal K^{-1} at 300 K.
Hence ΔG in kcal is ______.

[NCERT : P L-162 | NV, Jan. 07, 2020 (I)]

181. A process will be spontaneous at all temperatures if :

[NCERT : P L-160 | April 10, 2019 (I)]

(a) $\Delta H < 0$ and $\Delta S < 0$ (b) $\Delta H > 0$ and $\Delta S < 0$
(c) $\Delta H < 0$ and $\Delta S > 0$ (d) $\Delta H > 0$ and $\Delta S > 0$

182. For the chemical reaction $X \rightleftharpoons Y$, the standard reaction Gibbs energy depends on temperature T (in K) as $\Delta_r G°$ (in kJ mol^{-1}) $= 120 - \frac{3}{8}T$

The major component of the reaction mixture at T is :

[NCERT : P L-160 | Jan. 11, 2019 (I)]

(a) Y if $T = 300$ K (b) Y if $T = 280$ K
(c) X if $T = 350$ K (d) X if $T = 315$ K

183. The reaction
MgO (s) + C(s) → Mg(s) + CO (g), for which $\Delta H° = +491.1$ kJ mol^{-1} and $\Delta S° = 198.0$ $JK^{-1}mol^{-1}$, is not feasible at 298 K. Temperature above which reaction will be feasible is

[NCERT : P L-160 | Jan. 11, 2019 (II)]

(a) 2040.5 K (b) 1890.0 K
(c) 2480.3 K (d) 2380.5 K

184. The standard reaction Gibbs energy for a chemical reaction at an absolute temperature T is given by $\Delta G° = A - BT$ where A and B are non-zero constants. Which of the following is true about this reaction?

[NCERT : P L-160 | Jan. 11, 2019 (II)]

(a) Exothermic if $B < 0$
(b) Endothermic if $A > 0$
(c) Endothermic if $A < 0$ and $B > 0$
(d) Exothermic if $A > 0$ and $B < 0$

185. A process has $\Delta H = 200$ J mol^{-1} and $\Delta S = 40$ $JK^{-1}mol^{-1}$. Out of the values given below, choose the minimum temperature above which the process will be spontaneous: **[NCERT : P L-162 | Jan. 10, 2019 (I)]**

(a) 20 K (b) 12 K (c) 5 K (d) 4 K

186. Which of the following lines correctly show the temperature dependence of equilibrium constant, K, for an exothermic reaction? **[NCERT : P L-160 | 2018]**

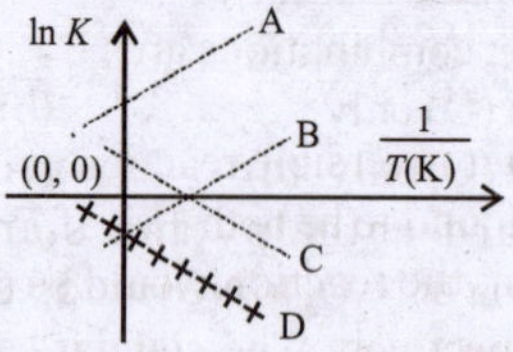

(a) A and B (b) B and C
(c) C and D (d) A and D

187. At 320 K, a gas A_2 is 20% dissociated to A(g). The standard free energy change at 320 K and 1 atm in J mol^{-1} is approximately :
($R = 8.314$ JK^{-1} mol^{-1}; ln 2 = 0.693; ln 3 = 1.098)

[NCERT : P L-160 | Online April 16, 2018]

(a) 1844 (b) 2068 (c) 4281 (d) 4763

188. $\Delta_f G°$ at 500 K for substance 'S' in liquid state and gaseous state are + 100.7 kcal mol^{-1} and + 103 kcal mol^{-1}, respectively. Vapour pressure of liquid 'S' at 500 K is approximately equal to:($R = 2$ cal $K^{-1}mol^{-1}$).

[NCERT : P L-160 | Online April 15, 2018 (II)]

(a) 100 atm (b) 1 atm (c) 10 atm (d) 0.1 atm

189. Given **[NCERT : P L-160 | Online April 15, 2018 (II)]**

(i) $2Fe_2O_3(s) \rightarrow 4Fe(s) + 3O_2(g)$;
$\Delta_r G° = +1487.0$ kJ mol^{-1}

(ii) $2CO(g) + O_2(g) \rightarrow 2CO_2(g)$;
$\Delta_r G° = -514.4$ kJ mol^{-1}

Free energy change, $\Delta_r G°$ for the reaction
$2Fe_2O_3(s) + 6CO(g) \rightarrow 4Fe(s) + 6CO_2(g)$ will be:

(a) –112.4 kJ mol^{-1} (b) –56.2 kJ mol^{-1}
(c) –208.0 kJ mol^{-1} (d) –168.2 kJ mol^{-1}

190. An ideal gas undergoes isothermal expansion at constant pressure. During the process :

[NCERT : P L-160 | Online April 9, 2017]

(a) enthalpy increases but entropy decreases.
(b) enthalpy remains constant but entropy increases.
(c) enthalpy decreases but entropy increases.
(d) Both enthalpy and entropy remain constant.

191. A reaction at 1 bar is non–spontaneous at low temperature but becomes spontaneous at high temperature. Identify the correct statement about the reaction among the following : **[NCERT : P L-160 | Online April 9, 2016]**

(a) ΔH is negative while ΔS is positive
(b) Both ΔH and ΔS are negative
(c) ΔH is positive while ΔS is negative
(d) Both ΔH and ΔS are positive.

192. The incorrect expression among the following is: **[NCERT : P L-160 | 2012]**

(a) $\dfrac{\Delta G_{system}}{\Delta S_{total}} = -T$

(b) In isothermal process, $w_{reversible} = -nRT\ell n\dfrac{V_f}{V_i}$

(c) $\ln K = \dfrac{\Delta H° - T\Delta S°}{RT}$

(d) $K = e^{-\Delta G°/RT}$

193. For a particular reversible reaction at temperature T, ΔH and ΔS were found to be both +ve. If T_e is the temperature at equilibrium, the reaction would be spontaneous when **[NCERT : P L-160 | 2010]**

(a) $T_e > T$
(b) $T > T_e$
(c) T_e is 5 times T
(d) $T = T_e$

194. In conversion of lime-stone to lime,

$CaCO_3(s) \rightarrow CaO(s) + CO_2(g)$ the values of $\Delta H°$ and $\Delta S°$ are + 179.1 kJ mol^{-1} and 160.2 J/K respectively at 298 K and 1 bar. Assuming that $\Delta H°$ and $\Delta S°$ do not change with temperature, temperature above which conversion of limestone to lime will be spontaneous is **[NCERT : P L-162 | 2007]**

(a) 1118 K
(b) 1008 K
(c) 1200 K
(d) 845 K.

195. Identify the correct statement regarding a spontaneous process: **[NCERT : P L-161 | 2007]**

(a) Lowering of energy in the process is the only criterion for spontaneity.
(b) For a spontaneous process in an isolated system, the change in entropy is positive.
(c) Endothermic processes are never spontaneous.
(d) Exothermic processes are always spontaneous.

196. In an irreversible process taking place at constant T and P and in which only pressure-volume work is being done, the change in Gibbs free energy (dG) and change in entropy (dS), satisfy the criteria **[NCERT : P L-162 | 2003]**

(a) $(dS)_{V,E} > 0, (dG)_{T,P} < 0$
(b) $(dS)_{V,E} = 0, (dG)_{T,P} = 0$
(c) $(dS)_{V,E} = 0, (dG)_{T,P} > 0$
(d) $(dS)_{V,E} < 0, (dG)_{T,P} < 0$

197. The correct relationship between free energy change in a reaction and the corresponding equilibrium constant K_c is **[NCERT : P L-162 | 2003]**

(a) $-\Delta G = RT\ln K_c$
(b) $\Delta G° = RT\ln K_c$
(c) $-\Delta G° = RT\ln K_c$
(d) $\Delta G = RT\ln K_c$

198. If an endothermic reaction is non-spontaneous at freezing point of water and becomes feasible at its boiling point, then **[NCERT : P L-162 | 2002]**

(a) ΔH is –ve, ΔS is +ve
(b) ΔH and ΔS both are +ve
(c) ΔH and ΔS both are –ve
(d) ΔH is +ve, ΔS is –ve

199. For the reactions, **[NCERT : P L-143 | 2002]**

$C + O_2 \rightarrow CO_2$; $\Delta H = -393$ J

$2\,Zn + O_2 \rightarrow 2\,ZnO$; $\Delta H = -412$ J

(a) carbon can oxidise Zn
(b) oxidation of carbon is not feasible
(c) oxidation of Zn is not feasible
(d) Zn can oxidise carbon

ANSWER KEYS

1	(d)	21	(0)	41	(a)	61	(d)	81	(278)	101	(21)	121	(b)	141	(b)	161	(d)	181	(c)
2	(304)	22	(150)	42	(c)	62	(c)	82	(499)	102	(128)	122	(c)	142	(b)	162	(b)	182	(d)
3	(d)	23	(620)	43	(b)	63	(b)	83	(610)	103	(309)	123	(c)	143	(b)	163	(c)	183	(c)
4	(b)	24	(a)	44	(c)	64	(b)	84	(173)	104	(5576)	124	(c)	144	(400)	164	(c)	184	(b)
5	(d)	25	(3)	45	(a)	65	(25)	85	(925)	105	(c)	125	(c)	145	(2000)	165	(a)	185	(c)
6	(a)	26	(2)	46	(a)	66	(d)	86	(360)	106	(d)	126	(c)	146	(1070)	166	(a)	186	(a)
7	(d)	27	(1718)	47	(c)	67	(48)	87	(54)	107	(189494)	127	(a)	147	(3)	167	(c)	187	(c)
8	(a)	28	(195)	48	(c)	68	(466)	88	(57)	108	(– 326400)	128	(d)	148	(2)	168	(2850)	188	(d)
9	(55)	29	(0)	49	(b)	69	(b)	89	(117)	109	(c)	129	(d)	149	(336)	169	(c)	189	(b)
10	(38)	30	(8630)	50	(c)	70	(200)	90	(–103.7)	110	(–192.5)	130	(c)	150	(200)	170	(2)	190	(b)
11	(274)	31	(84)	51	(d)	71	(95)	91	(2)	111	(c)	131	(b)	151	(1380)	171	(b)	191	(d)
12	(200)	32	(38)	52	(a)	72	(b)	92	(600)	112	(a)	132	(b)	152	(855)	172	(a)	192	(c)
13	(d)	33	(31)	53	(a)	73	(150)	93	(38)	113	(a)	133	(a)	153	(a)	173	(b)	193	(b)
14	(28721)	34	(50)	54	(c)	74	(6535)	94	(c)	114	(a)	134	(b)	154	(–2.70)	174	(300)	194	(a)
15	(200)	35	(2218)	55	(d)	75	(125)	95	(c)	115	(d)	135	(a)	155	(a)	175	(710)	195	(b)
16	(1200)	36	(d)	56	(c)	76	(492)	96	(82)	116	(b)	136	(d)	156	(d)	176	(5)	196	(a)
17	(a)	37	(d)	57	(c)	77	(56)	97	(3)	117	(c)	137	(b)	157	(d)	177	(25)	197	(c)
18	(4)	38	(48.00)	58	(c)	78	(800)	98	(718)	118	(d)	138	(N)	158	(b)	178	(964)	198	(b)
19	(4)	39	(6.25)	59	(d)	79	(4)	99	(82)	119	(c)	139	(c)	159	(b)	179	(– 13538)	199	(a)
20	(1200)	40	(c)	60	(b)	80	(a)	100	(101)	120	(b)	140	(b)	160	(b)	180	(–2.70)		

Equilibrium

TREND ANALYSIS
JEE MAIN (2025-2021)

	2025				2024				2023				2022				2021			
	Session II		Session I		Session II		Session I		Session II		Session I		Session II		Session I		Session II		Session I	
	MCQs	NVQs	MCQs	NVQs	MCQs	NVQs	MCQs	NVQs	MCQs	NVQs	MCQs	NVQs	MCQs	NVQs	MCQs	NVQs	MCQs	NVQs	MCQs	NVQs
Topic 1	4	2	6	1	4	–	2	2	2	3	–	4	–	2	3	3	–	10	–	3
Topic 2	–	–	1	1	1	–	–	2	–	1	2	1	–	2	–	–	–	3	–	–
Topic 3	1	1	2	–	–	–	1	–	1	1	–	4	3	–	1	2	–	1	1	–
Topic 4	–	1	–	1	1	1	–	–	–	–	–	–	2	1	–	–	–	2	1	–
Topic 5	1	1	4	–	3	–	1	3	–	2	–	3	1	2	2	2	–	3	2	4
Total	6	5	13	3	9	1	4	7	3	7	2	12	6	7	6	7	–	19	4	7

How to Code Qns for Better & Faster Revision →

Type 1: Able to solve (No code Req)

Type 2: Solved with struggle or by seeing solution **(Use code NR means Needs Revision or orange stickers)**

Type 3: Tricky/ Could not solve on my own. Need help from Teacher/ Friend. **(Use code TQ means Tricky Question or red stickers)**

You can either hand write the code in boxes or else paste coloured stickers given at end of book.

Topic 1 – Law of Mass Action, Equilibrium Constant (K_c and K_p) and its Application

1. The equilibrium constant for decomposition of $H_2O(g)$

$H_2O(g) \rightleftharpoons H_2(g) + \frac{1}{2}O_2(g) (\Delta G° = 92.34 \text{ kJ mol}^{-1})$ is 8.0×10^{-3} at 2300 K and total pressure at equilibrium is 1 bar. Under this condition, the degree of dissociation (α) of water is _____ $\times 10^{-2}$ (nearest integer value).
[Assume α is negligible with respect to 1]

[NCERT : P L-178 | NV, April 8, 2025 (II)]

2. Given below are two statements :

Statement I : When a system containing ice in equilibrium with water (liquid) is heated, heat is absorbed by the system and there is no change in the temperature of the system until whole ice gets melted.

Statement II : At melting point of ice, there is absorption of heat in order to overcome intermolecular forces of attraction within the molecules of water in ice and kinetic energy of molecules is not increased at melting point.

In the light of the above statements, choose the **correct** answer from the options given below:

[NCERT : P L-171 | April 3, 2025 (II)]

(a) Statement I is true but Statement II is false
(b) Both Statement I and Statement II are false
(c) Both Statement I and Statement II are true
(d) Statement I is false but Statement II is true

3. In the following system,
$PCl_5(g) \rightleftharpoons PCl_3(g) + Cl_2(g)$ at equilibrium, upon addition of xenon gas at constant T & P, the concentration of

[NCERT : P L-185 | April 3, 2025 (I)]

(a) PCl_5 will increase
(b) Cl_2 will decrease
(c) PCl_5, PCl_3 & Cl_2 remain constant
(d) PCl_3 will increase

4. Given below are two statements

[NCERT : P L-185 | April 3, 2025 (I)]

Statement I : A catalyst cannot alter the equilibrium constant (K_C) of the reaction, temperature remaining constant

Statement II : A homogenous catalyst can change the equilibrium composition of a system temperature remaining constant

In the light of the above statements, choose the **correct** answer from the options given below

(a) Statement I is false but Statement II is true
(b) Both Statement I and Statement II are true
(c) Both Statement I and Statement II is false
(d) Statement I is true but Statement II is false

5. Consider the following chemical equilibrium of the gas phase reaction at a constant temperature : **[NCERT : P L-178 | April 2, 2025 (II)]**

$A(g) \rightleftharpoons B(g) + C(g)$

If p being the total pressure, K_p is the pressure equilibrium constant and α is the degree of dissociation, then which of the following is true at equilibrium ?

(a) If p value is extremely high compared to K_p, $\alpha \approx 1$
(b) When p increases α decreases
(c) If k_p value is extremely high compared to p, α becomes much less than unity
(d) When p increases α increases

6. Consider the following equilibrium, **[NCERT : P L-178 | NV, April 2, 2025 (I)]**

$CO(g) + 2H_2(g) \rightleftharpoons CH_3OH(g)$

0.1 mol of CO along with a catalyst is present in a 2 dm^3 flask maintained at 500 K. Hydrogen is introduced into the flask until the pressure is 5 bar and 0.04 mol of CH_3OH is formed. The K_p^0 is ________ $\times 10^{-3}$ (nearest integer).

Given : $R = 0.08\ dm^3\ bar\ K^{-1}\ mol^{-1}$

Assume only methanol is formed as the product and the system follows ideal gas behaviour.

7. At temperature T, compound $AB_{2(g)}$ dissociates as $AB_{2(g)} \rightleftharpoons AB_{(g)} + \frac{1}{2}B_{2(g)}$ having degree of dissociation x (small compared to unity). The correct expression for x in terms of K_p and p is **[NCERT : P L-178 | Jan. 29, 2025 (I)]**

(a) $\sqrt[3]{\frac{2K_p}{p}}$ (b) $\sqrt[4]{\frac{2K_p}{p}}$

(c) $\sqrt[3]{\frac{2K_p^2}{p}}$ (d) $\sqrt{K_p}$

8. Ice and water are placed in a closed container at a pressure of 1 atm and temperature 273.15 K. If pressure of the system is increased 2 times, keeping temperature constant, then identify correct observation from following : **[NCERT : P L-181 | Jan. 28, 2025 (I)]**

(a) Volume of system increases.
(b) Liquid phase disappears completely.
(c) The amount of ice decreases.
(d) The solid phase (ice) disappears completely.

9. For the reaction, **[NCERT : P L-172 | Jan. 24, 2025 (II)]**

$H_2(g) + I_2(g) \rightleftharpoons 2HI(g)$

Attainment of equilibrium is predicted correctly by:

(a)
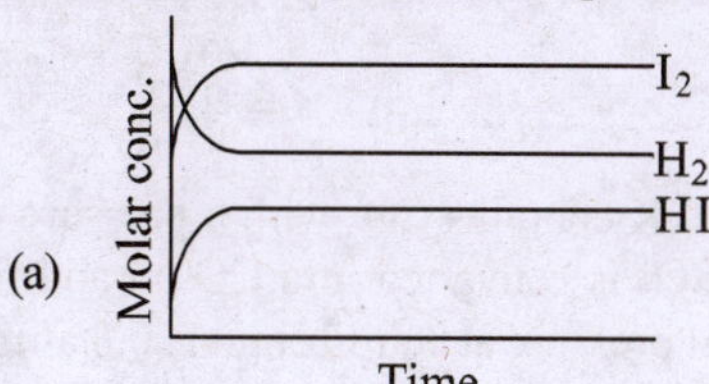

(b)
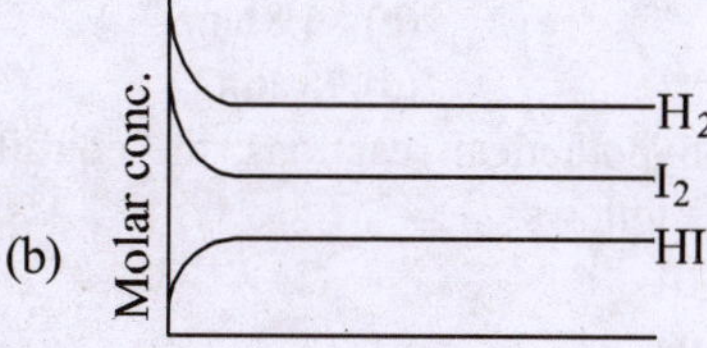

(c)
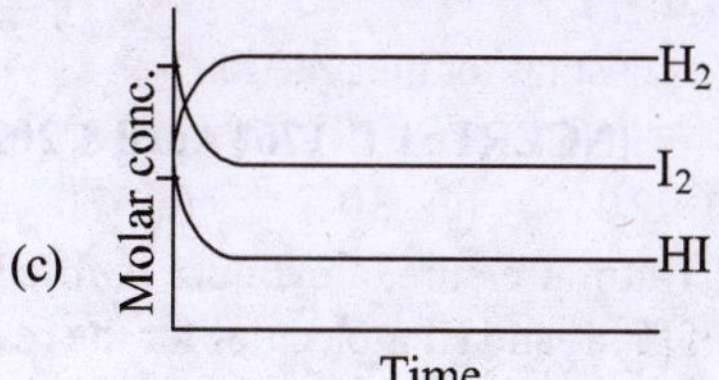

(d)
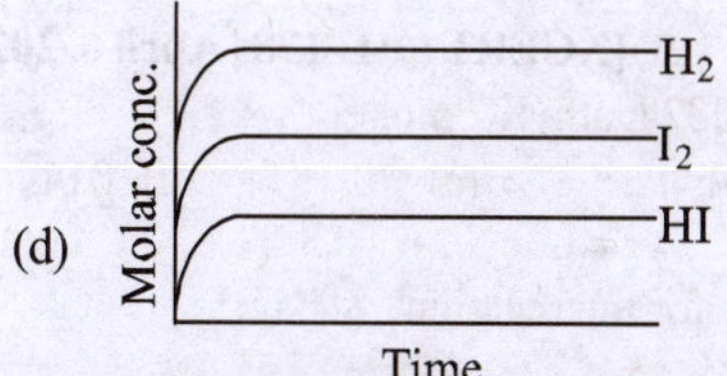

10. 37.8 g N_2O_5 was taken in a 1 L reaction vessel and allowed to undergo the following reaction at 500 K

$2N_2O_{5(g)} \rightarrow 2N_2O_{4(g)} + O_{2(g)}$

The total pressure at equilibrium was found to be 18.65 bar.

Then, K_p = ________ $\times 10^{-2}$ [nearest integer]

Assume N_2O_5 to behave ideally under these conditions

Given : $R = 0.082\ bar\ L\ mol^{-1}\ K^{-1}$. **[NCERT : P L-178 | NV, Jan. 24, 2025 (I)]**

11. For a reaction, $N_2O_{5(g)} \rightarrow 2NO_{2(g)} + \frac{1}{2}O_{2(g)}$ in a constant volume container, no products were present initially. The final pressure of the system when 50% of reaction gets completed is **[NCERT : P L-178 | Jan. 24, 2025 (I)]**

(a) 7/2 times of initial pressure
(b) 5 times of initial pressure
(c) 5/2 times of initial pressure
(d) 7/4 times of initial pressure

12. Consider the reaction **[NCERT : P L-178 | Jan. 23, 2025 (II)]**

$X_2Y(g) \rightleftharpoons X_2(g) + \frac{1}{2}Y_2(g)$

The equation representing correct relationship between the degree of dissociation (x) of $X_2Y(g)$ with its equilibrium constant Kp is________.

Assume x to be very very small.

(a) $x = \sqrt[3]{\frac{2Kp}{p}}$ (b) $x = \sqrt[3]{\frac{2Kp^2}{p}}$

(c) $x = \sqrt[3]{\frac{Kp}{2p}}$ (d) $x = \sqrt[3]{\frac{Kp}{p}}$

13. A vessel at 1000 K contains CO_2 with a pressure of 0.5 atm. Some of CO_2 is converted into CO on addition of graphite. If total pressure at equilibrium is 0.8 atm, then K_p is: **[NCERT : P L-178 | Jan. 22, 2025 (I)]**

(a) 0.18 atm (b) 1.8 atm

(c) 0.3 atm (d) 3 atm.

14. For the given hypothetical reactions, the equilibrium constants are as follows:

$X \rightleftharpoons Y; K_1 = 1.0$

$Y \rightleftharpoons Z; K_2 = 2.0$

$Z \rightleftharpoons W; K_3 = 4.0$

The equilibrium constant for the reaction

$X \rightleftharpoons W$ is **[NCERT : P L-176 | April 8, 2024 (I)]**

(a) 6.0 (b) 12.0 (c) 8.0 (d) 7.0

15. At –20 °C and 1 atm pressure, a cylinder is filled with equal number of H_2. I_2 and HI molecules for the reaction $H_2(g) + I_2(g) \rightleftharpoons 2HI(g)$, the K_P for the process is $x \times 10^{-1}$.

x = ______. **[NCERT : P L-178 | April 6, 2024 (I)]**

[Given : $R = 0.082$ L atm K^{-1} mol^{-1}]

(a) 2 (b) 1 (c) 10 (d) 0.01

16. The ratio $\frac{K_P}{K_C}$ for the reaction

$CO_{(g)} + \frac{1}{2}O_{2(g)} \rightleftharpoons CO_{2(g)}$ is:

[NCERT : P L-178 | April 6, 2024 (II)]

(a) $(RT)^{1/2}$ (b) RT (c) 1 (d) $\frac{1}{\sqrt{RT}}$

17. The equilibrium constant for the reaction

$SO_3(g) \rightleftharpoons SO_2(g) + \frac{1}{2}O_2(g)$ is $K_c = 4.9 \times 10^{-2}$. The value of K_c for the reaction given below is

$2SO_2(g) + O_2(g) \rightleftharpoons 2SO_3(g)$ is

[NCERT : P L-176 | April 4, 2024 (II)]

(a) 4.9 (b) 41.6 (c) 49 (d) 416

18. For the given reaction, choose the correct expression of K_C from the following :-

$Fe^{3+}(aq) + SCN^-(aq) \rightleftharpoons (FeSCN)^{2+}(aq)$

[NCERT : P L-175 | Jan. 31, 2024 (I)]

(a) $K_C = \frac{[FeSCN^{2+}]^2}{[Fe^{3+}][SCN^-]}$

(b) $K_C = \frac{[FeSCN^{2+}]}{[Fe^{3+}]^{-2}[SCN^-]^2}$

(c) $K_C = \frac{[Fe^{3+}][SCN^-]}{[FeSCN^{2+}]}$

(d) $K_C = \frac{[FeSCN^{2+}]}{[Fe^{3+}][SCN^-]}$

19. $A(g) \rightleftharpoons B(g) + \frac{C}{2}(g)$ The correct relationship between K_P, α and equilibrium pressure P is

[NCERT : P L-178 | Jan. 31, 2024 (II)]

(a) $K_P = \frac{\alpha^{1/2}P^{1/2}}{(2+\alpha)^{1/2}}$ (b) $K_P = \frac{\alpha^{3/2}P^{1/2}}{(2+\alpha)^{1/2}(1-\alpha)}$

(c) $K_P = \frac{\alpha^{1/2}P^{3/2}}{(2+\alpha)^{3/2}}$ (d) $K_P = \frac{\alpha^{1/2}P^{1/2}}{(2+\alpha)^{3/2}}$

20. For the reaction $N_2O_4(g) \rightleftharpoons 2NO_2(g)$, $K_p = 0.492$ atm at 300K. K_c for the reaction at same temperature is ______ $\times 10^{-2}$.

(Given : $R = 0.082$ L atm mol^{-1} K^{-1})

[NCERT : P L-178 | NV, Jan. 29, 2024 (I)]

21. The following concentrations were observed at 500K for the formation of NH_3 from N_2 and H_2. At equilibrium : $[N_2] = 2 \times 10^{-2}$ M, $[H_2] = 3 \times 10^{-2}$ M and $[NH_3] = 1.5 \times 10^{-2}$ M. Equilibrium constant for the reaction is ______.

[NCERT : P L-174 | NV, Jan. 29, 2024 (II)]

22. A mixture of 1 mole of H_2O and 1 mole of CO is taken in a 10 litre container and heated to 725 K. At equilibrium 40% of water by mass reacts with carbon monoxide according to the equation :

$CO(g) + H_2O(g) \rightleftharpoons CO_2(g) + H_2(g)$.

The equilibrium constant $K_C \times 10^2$ for the reaction is ______. (Nearest integer)

[NCERT : P L-174 | NV, April 11, 2023 (I)]

23. 4.5 moles each of hydrogen and iodine is heated in a sealed ten litre vessel. At equilibrium, 3 moles of HI were found. The equilibrium constant for $H_2(g) + I_2(g) \rightleftharpoons 2HI(g)$ is ______. **[NCERT : P L-174 | NV, April 11, 2023 (II)]**

24. The number of correct statement/s involving equilibria in physical processes from the following is

[NCERT : P L-169 | NV, April 10, 2023 (I)]

(A) Equilibrium is possible only in a closed system at a given temperature.

(B) Both the opposing processes occur at the same rate.

(C) When equilibrium is attained at a given temperature, the value of all its parameters became equal.

(D) For dissolution of solids in liquids, the solubility is constant at a given temperature.

25. For a concentrated solution of a weak electrolyte (K_{eq} = equilibrium constant) A_2B_3 of concentration 'c', the degree of dissociation 'α' is

[NCERT : P L-204 | April 06, 2023 (I)]

(a) $\left(\frac{K_{eq}}{108c^4}\right)^{\frac{1}{5}}$ (b) $\left(\frac{K_{eq}}{6c^5}\right)^{\frac{1}{5}}$

(c) $\left(\frac{K_{eq}}{5c^4}\right)^{\frac{1}{5}}$ (d) $\left(\frac{K_{eq}}{25c^2}\right)^{\frac{1}{5}}$

26. The equilibrium composition for the reaction

$PCl_3 + Cl_2 \rightleftharpoons PCl_5$ at 298 K is given below.

$[PCl_3]_{eq} = 0.2 \text{ mol L}^{-1}$ $[Cl_2]_{eq} = 0.1 \text{ mol L}^{-1}$,

$[PCl_5]_{eq} = 0.40 \text{ mol L}^{-1}$

If 0.2 mol of Cl_2 is added at the same temperature, the equilibrium concentrations of PCl_5 is ________ $\times 10^{-2} \text{ mol L}^{-1}$.

Given: K_c for the reaction at 298 K is 20

[NCERT : P L-174 | NV, April 06, 2023 (II)]

27. (i) $X(g) \rightleftharpoons Y(g) + Z(g)\ K_{p1} = 3$

(ii) $A(g) \rightleftharpoons 2B(g)\ \ K_{p2} = 1$

If the degree of dissociation and initial concentration of both the reactants X(g) and A(g) are equal, then the ratio of the total pressure at equilibrium $\left(\frac{P_1}{P_2}\right)$ is equal to x:1.

The value of x is______ (Nearest integer)

[NCERT : P L-174 | NV, Feb. 01, 2023 (I)]

28. For reaction : $SO_2(g) + \frac{1}{2}O_2(g) \rightleftharpoons SO_3(g)$

$K_P = 2 \times 10^{12}$ at 27°C and 1 atm pressure. The K_c for the same reaction is ________ $\times 10^{13}$. (Nearest integer)

(Given R = 0.082 L atm K^{-1} mol^{-1})

[NCERT : P L-178 | NV, Jan. 31, 2023 (I)]

29. Water decomposes at 2300 K

$H_2O(g) \rightarrow H_2(g) + \frac{1}{2}O_2(g)$

The percent of water decomposing at 2300 K and 1 bar is ______ (Nearest integer).

Equilibrium constant for the reaction is 2×10^{-3} at 2300 K

[NCERT : P L-174 | NV, Jan. 29, 2023 (I)]

30. At 298 K **[NCERT : P L-174 | NV, Jan. 29, 2023 (II)]**

$N_2(g) + 3H_2(g) \rightleftharpoons 2NH_3(g), K_1 = 4 \times 10^5$

$N_2(g) + O_2(g) \rightleftharpoons 2NO(g), K_2 = 1.6 \times 10^{12}$

$H_2(g) + \frac{1}{2}O_2(g) \rightleftharpoons H_2O(g), K_3 = 1.0 \times 10^{-13}$

Based on above equilibria, the equilibrium constant of the reaction,

$2NH_3(g) + \frac{5}{2}O_2(g) \rightleftharpoons 2NO(g) + 3H_2O(g)$

is ______ $\times 10^{-33}$ (Nearest integer)

31. At 600K, 2 mol of NO are mixed with 1 mol of O_2.

$$2NO(g) + O_2(g) \rightleftharpoons 2NO_2(g)$$

The reaction occurring as above comes to equilibrium under a total pressure of 1 atm. Analysis of the system shows that 0.6 mol of oxygen are present at equilibrium. The equilibrium constant for the reaction is ________. (Nearest integer)

[NCERT : P L-174 | NV, July 28, 2022 (II)]

32. At 298 K, the equilibrium constant is 2×10^{15} for the reaction:

$Cu(s) + 2Ag^+(aq) \rightleftharpoons Cu^{2+}(aq) + 2Ag(s)$

The equilibrium constant for the reaction

$\frac{1}{2}Cu^{2+}(aq) + Ag(s) \rightleftharpoons \frac{1}{2}Cu(s) + Ag^+(aq)$

is $x \times 10^{-8}$. The value of x is ________. (Nearest Integer)

[NCERT : P L-177 | NV, July 26, 2022 (I)]

33. The equilibrium constant for the reversible reaction

$2A(g) \rightleftharpoons 2B(g) + C(g)$ is K_1 and for the reaction

$\frac{3}{2}A(g) \rightleftharpoons \frac{3}{2}B(g) + \frac{3}{4}C(g)$ is K_2

K_1 and K_2 are related as:

[NCERT : P L-174, 177 | June 30, 2022 (I)]

(a) $K_1 = \sqrt{K_2}$ (b) $K_2 = \sqrt{K_1}$

(c) $K_2 = K_1^{3/4}$ (d) $K_1 = K_2^{3/4}$

34. 4.0 moles of argon and 5.0 moles of PCl_5 are introduced into an evacuated flask of 100 litre capacity at 610 K. The system is allowed to equilibrate. At equilibrium, the total pressure of mixture was found to be 6.0 atm. The K_p for the reaction is [Given : $R = 0.082$ L atm K^{-1} mol^{-1}]

[NCERT : P L-174 | June 29, 2022 (II)]

(a) 2.25 (b) 6.24 (c) 12.13 (d) 15.24

35. A box contains 0.90 g of liquid water in equilibrium with water vapour at 27 °C. The equilibrium vapour pressure of water at 27 °C 32.0 Torr. When the volume of the box is increased, some of the liquid water evaporates to maintain the equilibrium pressure. If all the liquid water evaporates, then the volume of the box must be ____ litre. [nearest integer]

(Given: $R = 0.082$ L atm K^{-1} mol^{-1})

(Ignore the volume of the liquid water and assume water vapours behave as an ideal gas.)

[NCERT : P L-174 | NV, June 29, 2022 (II)]

36. $2NOCl(g) \rightleftharpoons 2NO(g) + Cl_2(g)$

In an experiment, 2.0 moles of NOCl was placed in a one-litre flask and the concentration of NO after equilibrium established, was found to be 0.4 mol/L. The equilibrium constant at 30°C is ______ $\times 10^{-4}$.

[NCERT : P L-174 | NV, June 27, 2022 (I)]

37. For a reaction at equilibrium

$$A(g) \rightleftharpoons B(g) + \frac{1}{2}C(g)$$

the relation between dissociation constant (K), degree of dissociation (α) and equilibrium pressure (p) is given by: **[NCERT : P L-174 | June 24, 2022 (I)]**

(a) $K = \dfrac{\alpha^{\frac{1}{2}} p^{\frac{3}{2}}}{\left(1+\frac{3}{2}\alpha\right)^{\frac{1}{2}}(1-\alpha)}$

(b) $K = \dfrac{\alpha^{\frac{3}{2}} p^{\frac{1}{2}}}{(2+\alpha)^{\frac{1}{2}}(1-\alpha)}$

(c) $K = \dfrac{(\alpha p)^{\frac{3}{2}}}{\left(1+\frac{3}{2}\alpha\right)^{\frac{1}{2}}(1-\alpha)}$

(d) $K = \dfrac{(\alpha p)^{\frac{3}{2}}}{(1+\alpha)(1-\alpha)^{\frac{1}{2}}}$

38. PCl_5 dissociates as

$$PCl_5(g) \rightleftharpoons PCl_3(g) + Cl_2(g)$$

5 moles of PCl_5 are placed in a 200 litre vessel which contains 2 moles of N_2 and is maintained at 600 K. The equilibrium pressure is 2.46 atm. The equilibrium constant K_p for the dissociation of PCl_5 is_____ $\times 10^{-3}$. (nearest integer)

(Given: R = 0.082 L atm K^{-1} mol^{-1} : Assume ideal gas behaviour) **[NCERT : P L-178 | NV, June 24, 2022 (II)]**

39. The number of moles of NH_3, that must be added to 2 L of 0.80 M $AgNO_3$ in order to reduce the concentration of Ag^+ ions to 5.0×10^{-8} M ($K_{formation}$ for $[Ag(NH_3)_2]^+ = 1.0 \times 10^8$) is ________. (Nearest integer)

[Assume no volume change on adding NH_3]

[NV, Aug. 27, 2021 (I)]

40. When 5.1 g of solid NH_4HS is introduced into a two litre evacuated flask at 27 °C, 20% of the solid decomposes into gaseous ammonia and hydrogen sulphide. The K_p for the reaction at 27 °C is $x \times 10^{-2}$. The value of x is _______. (Integer answer)

[Given $R = 0.082$ L atm K^{-1} mol^{-1}]

[NCERT : P L-178 | NV, Aug. 27, 2021 (II)]

41. The equilibrium constant K_c at 298 K for the reaction $A + B \rightleftharpoons C + D$ is 100. Starting with an equimolar solution with concentrations of A, B, C and D all equal to 1M, the equilibrium concentration of D is____ $\times 10^{-2}$M.

[NCERT : P L-178 | NV, Aug. 26, 2021 (II); July 25, 2021 (I); Same 2016]

42. The reaction rate for the reaction

$[PtCl_4]^{2-} + H_2O \gg [Pt(H_2O)Cl_3]^- + Cl^-$

was measured as a function of concentrations of different species. It was observed that

$$\frac{-d\left[[PtCl_4]^{2-}\right]}{dt} = 4.8\times10^{-5}\left[[PtCl_4]^{2-}\right] - 2.4\times10^{-3}\left[[Pt(H_2O)Cl_3]^-\right]\left[Cl^-\right].$$

where square brackets are used to denote molar concentrations. The equilibrium constant K_c = _____. (Nearest integer)

[NCERT : P L-176 | NV, Aug. 26, 2021 (II)]

43. $PCl_5 \rightleftharpoons PCl_3 + Cl_2$; $K_c = 1.844$

3.0 moles of PCl_5 is introduced in a 1 L closed reaction vessel at 380 K. The number of moles of PCl_5 at equilibrium is ________ $\times 10^{-3}$.

(Round off to the Nearest Integer)

[NCERT : P L-174 | NV, July 27, 2021 (I)]

44. The equilibrium constant for the reaction

$$A(s) \rightleftharpoons M(s) + \frac{1}{2}O_2(g)$$

is $K_p = 4$. At equilibrium, the partial pressure of O_2 is _____ atm. (Round off to the nearest integer)

[NCERT : P L-178 | NV, July 27, 2021 (II)]

45. $2SO_2(g) + O_2(g) \rightarrow 2SO_3(g)$

The above reaction is carried out in a vessel starting with partial pressure

$P_{SO_2} = 250$ m bar, $P_{O_2} = 750$ m bar and $P_{SO_3} = 0$ m bar

When the reaction is complete, the total pressure in the reaction vessel is _____ m bar. (Round off of the nearest integer). **[NCERT : P L-178 | NV, July 27, 2021 (II)]**

46. For the reaction

$A + B \rightleftharpoons 2C$

the value of equilibrium constant is 100 at 298 K. If the initial concentration of all the three species is 1 M each, then the equilibrium concentration of C is $x \times 10^{-1}$ M. The value of x is ________.

[NCERT : P L-176 | NV, July 25, 2021 (I)]

(Nearest integer)

47. Value of K_p for the equilibrium reaction

$N_2O_4(g) \rightleftharpoons 2NO_2(g)$ at 288 K is 47.9. The K_C for this reaction at same temperature is _______. (Nearest integer)

[$R = 0.083$ L bar K^{-1} mol^{-1}]

[NCERT : P L-178 | NV, July 22, 2021 (II)]

48. $2SO_2(g) + O_2(g) \rightleftharpoons 2SO_3(g)$

In an equilibrium mixture, the partial pressures are

$P_{SO_3} = 43$ kPa; $P_{O_2} = 530$ Pa and $P_{SO_2} = 45$ kPa. The equilibrium constant K_P = ______ $\times 10^{-2}$. (Nearest integer)

[NCERT : P L-178 | NV, July 20, 2021 (I)]

49. Consider the reaction $N_2O_4 \rightleftharpoons 2\,NO_2$ (g). The temperature at which $K_C = 20.4$ and $K_P = 600.1$, is ________ K. (Round off to the Nearest Integer).

[Assume all gases are ideal and $R = 0.0831$ L bar K^{-1} mol^{-1}]

[NCERT : P L-178 | NV, March 17, 2021 (II)]

50. A homogeneous ideal gaseous reaction $AB_2(g) \rightleftharpoons A(g) + 2B(g)$ is carried out in a 25 litre flask at 27 °C. The initial amount of AB_2 was 1 mole and the equilibrium pressure was 1.9 atm. The value of K_p is $x \times 10^{-2}$. The value of x is ______. (Integer answer) [$R = 0.08206$ dm^3 atm K^{-1} mol^{-1}]

[NCERT : P L-178 | NV, Feb. 26, 2021 (I)]

51. At 1990 K and 1 atm pressure, there are equal number of Cl_2 molecules and Cl atoms in the reaction mixture. The value of K_P for the reaction $Cl_2\,(g) \rightleftharpoons 2Cl(g)$ under the above conditions is $x \times 10^{-1}$. The value of x is ________ (Rounded off to the nearest integer)

[NCERT : P L-178 | NV, February 24, 2021 (I)]

52. For the reaction **[NCERT : P L-178 | Sep. 06, 2020 (I)]**

$$Fe_2N(s) + \frac{3}{2}H_2(g) \rightleftharpoons 2\,Fe(s) + NH_3(g)$$

(a) $K_c = K_p(RT)$ (b) $K_c = K_p(RT)^{-\frac{1}{2}}$

(c) $K_c = K_p(RT)^{\frac{1}{2}}$ (d) $K_c = K_p(RT)^{\frac{3}{2}}$

53. The value of K_c is 64 at 800 K for the reaction $N_2(g) + 3H_2(g) \rightleftharpoons 2NH_3(g)$. The value of K_c for the following reaction is :

$$NH_3(g) \rightleftharpoons \frac{1}{2}N_2(g) + \frac{3}{2}H_2(g)$$

[NCERT : P L-177 | Sep. 06, 2020 (II)]

(a) 1/64 (b) 8 (c) 1/4 (d) 1/8

54. For a reaction $X + Y \rightleftharpoons 2Z$, 1.0 mol of X, 1.5 mol of Y and 0.5 mol of Z were taken in a 1 L vessel and allowed to react. At equilibrium, the concentration of Z was 1.0 mol L^{-1}. The equilibrium constant of the reaction is $\frac{x}{15}$. The value of x is ____________

[NCERT : P L-174 | NV, Sep. 05, 2020 (II)]

55. For the equilibrium $A \rightleftharpoons B$, the variation of the rate of the forward (a) and reverse (b) reaction with time is given by : **[NCERT : P L-174 | Sep. 04, 2020 (I)]**

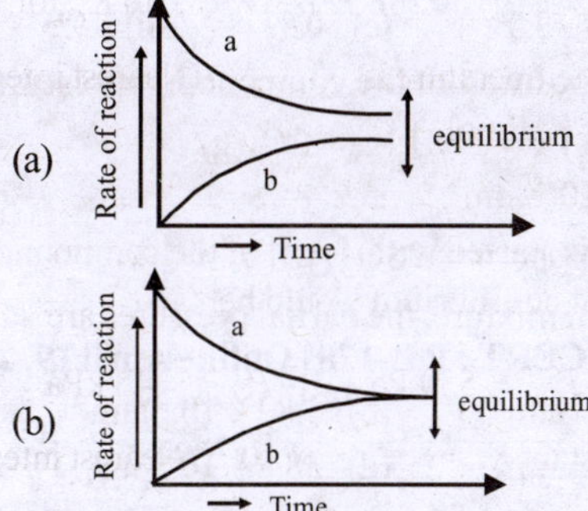

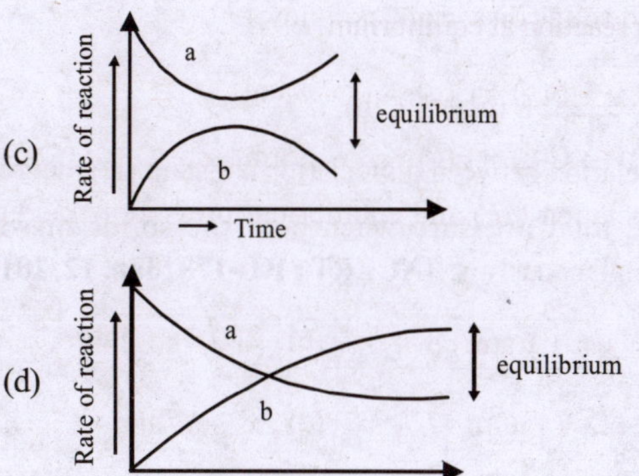

56. If the equilibrium constant for $A \rightleftharpoons B + C$ is $K_{eq}^{(1)}$ and that of $B + C \rightleftharpoons P$ is $K_{eq}^{(2)}$, the equilibrium constant for $A \rightleftharpoons P$ is: **[NCERT : P L-177 | Sep. 04, 2020 (II)]**

(a) $K_{eq}^{(1)} / K_{eq}^{(2)}$ (b) $K_{eq}^{(2)} - K_{eq}^{(1)}$

(c) $K_{eq}^{(1)} + K_{eq}^{(2)}$ (d) $K_{eq}^{(1)}\, K_{eq}^{(2)}$

57. In the figure shown below, reactant A (represented by square) is in equilibrium with product B (represented by circle). The equilibrium constant is:

[NCERT : P L-174 | Jan. 09, 2020 (II)]

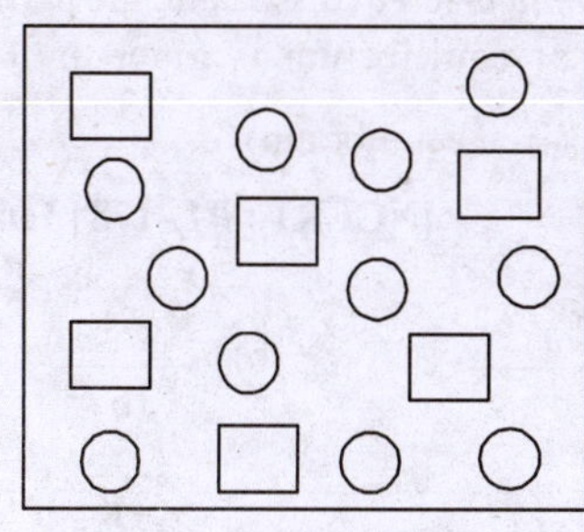

(a) 4 (b) 8

(c) 1 (d) 2

58. In which one of the following equilibria, $K_p \neq K_c$?

[NCERT : P L-178 | April 12, 2019 (II)]

(a) $2C(s) + O_2(g) \rightleftharpoons 2CO(g)$

(b) $2HI(g) \rightleftharpoons H_2(g) + I_2(g)$

(c) $NO_2(g) + SO_2(g) \rightleftharpoons NO(g) + SO_3(g)$

(d) $2NO(g) \rightleftharpoons N_2(g) + O_2(g)$

59. For the following reactions, equilibrium constants are given:

$S(s) + O_2(g) \rightleftharpoons SO_2(g);\ K_1 = 10^{52}$

$2S(s) + 3O_2(g) \rightleftharpoons 2SO_3(g);\ K_2 = 10^{129}$

The equilibrium constant for the reaction,

$2SO_2(g) + O_2(g) \rightleftharpoons 2SO_3(g)$ is :

[NCERT : P L-177 | April 8, 2019 (II)]

(a) 10^{154} (b) 10^{181}

(c) 10^{25} (d) 10^{77}

60. Two solids dissociate as follows

$A(s) \rightarrow B(g) + C(g); K_{P_1} = x \text{ atm}^2$

$D(s) \rightarrow C(g) + E(g); K_{P_2} = y \text{ atm}^2$

The total pressure when both the solids dissociate simultaneously is: **[NCERT : P L-178 | Jan. 12, 2019 (I)]**

(a) $\sqrt{x+y}$ atm (b) $2\left(\sqrt{x+y}\right)$ atm

(c) $(x+y)$ atm (d) x^2+y^2 atm

61. In a chemical reaction, $A + 2B \underset{}{\overset{K}{\rightleftharpoons}} 2C + D$, the initial concentration of B was 1.5 times of the concentration of A, but the equilibrium concentrations of A and B were found to be equal. The equilibrium constant (K) for the aforesaid chemical reaction is:

[NCERT : P L-174 | Jan. 12, 2019 (I); Same Online April 9, 2013]

(a) 4 (b) 16 (c) $\frac{1}{4}$ (d) 1

62. Consider the reaction

$N_2(g) + 3H_2(g) \rightleftharpoons 2NH_3(g)$

The equilibrium constant of the above reaction is K_P. If pure ammonia is left to dissociate, the partial pressure of ammonia at equilibrium is given by (Assume that $P_{NH_3} << P_{total}$ at equilibrium)

[NCERT : P L-178 | Jan. 11, 2019 (I)]

(a) $\frac{3^{3/2} K_P^{1/2} P^2}{16}$ (b) $\frac{K_P^{1/2} P^2}{16}$

(c) $\frac{K_P^{1/2} P^2}{4}$ (d) $\frac{3^{3/2} K_P^{1/2} P^2}{4}$

63. 5.1 g NH_4SH is introduced in 3.0 L evacuated flask at 327°C, 30% of the solid NH_4SH decomposed to NH_3 and H_2S as gases. The K_p of the reaction at 327°C is (R = 0.082 L atm $mol^{-1} K^{-1}$, molar mass of S = 32 g mol^{-1}, molar mass of N = 14 g mol^{-1}) **[NCERT : P L-178 | Jan. 10, 2019 (II)]**

(a) 0.242×10^{-4} atm^2 (b) 1×10^{-4} atm^2

(c) 4.9×10^{-3} atm^2 (d) 0.242 atm^2

64. The values of K_p/K_c for the following reactions at 300 K are, respectively: (At 300 K, $RT = 24.62$ dm^3 atm mol^{-1})

$N_2(g) + O_2(g) \rightleftharpoons 2NO(g)$

$N_2O_4(g) \rightleftharpoons 2NO_2(g)$

$N_2(g) + 3H_2(g) \rightleftharpoons 2NH_3(g)$

[NCERT : P L-178 | Jan. 10, 2019 (I)]

(a) 1, 24.62 dm^3 atm mol^{-1}, 606.0 dm^6 atm^2 mol^{-2}

(b) 1, 24.62 dm^3 atm mol^{-1}, 1.65×10^{-3} dm^{-6} atm^{-2} mol^2

(c) 1, 4.1×10^{-2} dm^{-3} atm^{-1} mol, 606 dm^6 atm^2 mol^{-2}

(d) 24.62 dm^3 atm mol^{-1}, 606.0 dm^6 atm^2 mol^{-2}, 1.65×10^{-3} dm^{-6} atm^{-2} mol^2

65. Consider the following reversible chemical reactions:

$A_2(g) + B_2(g) \underset{}{\overset{K_1}{\rightleftharpoons}} 2AB(g)......(1)$

$6AB(g) \underset{}{\overset{K_2}{\rightleftharpoons}} 3A_2(g) + 3B_2(g).....(2)$

The relation between K_1 and K_2 is:

[NCERT : P L-178 | Jan. 9, 2019 (II)]

(a) $K_1K_2 = \frac{1}{3}$ (b) $K_2 = K_1^3$

(c) $K_2 = K_1^{-3}$ (d) $K_1K_2 = 3$

66. At a certain temperature in a 5L vessel, 2 moles of carbon monoxide and 3 moles of chlorine were allowed to reach equilibrium according to the reaction, $CO + Cl_2 \rightleftharpoons COCl_2$. At equilibrium, if one mole of CO is present then equilibrium constant (K_c) for the reaction is:

[NCERT : P L-174 | Online April 15, 2018 (II)]

(a) 2.5 (b) 4 (c) 2 (d) 3

67. A solid XY kept in an evacuated sealed container undergoes decomposition to form a mixture of gases X and Y at temperature T. The equilibrium pressure is 10 bar in the vessel. K_p for this reaction is:

[NCERT : P L-179 | Online April 10, 2016]

(a) 25 (b) 100 (c) 10 (d) 5

68. Gaseous N_2O_4 dissociates into gaseous NO_2 according to the reaction

$[N_2O_4(g) \rightleftharpoons 2NO_2(g)]$

At 300 K and 1 atm pressure, the degree of dissociation of N_2O_4 is 0.2. If one mole of N_2O_4 gas is contained in a vessel, then the density of the equilibrium mixture is :

[NCERT : P L-174 | Online April 10, 2015]

(a) 1.56 g/L (b) 6.22 g/L

(c) 3.11 g/L (d) 4.56 g/L

69. For the reaction $SO_2(g) + \frac{1}{2}O_2(g) \rightleftharpoons SO_3(g)$, if $K_p = K_c(RT)^x$ where the symbols have usual meaning then the value of x is (assuming ideality):

[NCERT : P L-178 | 2014]

(a) –1 (b) $-\frac{1}{2}$ (c) $\frac{1}{2}$ (d) 1

70. At a certain temperature, only 50% HI is dissociated into H_2 and I_2 at equilibrium. The equilibrium constant is:

[NCERT : P L-174 | Online April 9, 2014]

(a) 1.0 (b) 3.0 (c) 0.5 (d) 0.25

71. For the decomposition of the compound, represented as

$NH_2COONH_4(s) \rightleftharpoons 2NH_3(g) + CO_2(g)$

the $K_p = 2.9 \times 10^{-5}$ atm^3.

If the reaction is started with 1 mol of the compound, the total pressure at equilibrium would be:

[NCERT : P L-178 | Online April 19, 2014]

(a) 1.94×10^{-2} atm (b) 5.82×10^{-2} atm

(c) 7.66×10^{-2} atm (d) 38.8×10^{-2} atm

72. The ratio $\frac{K_p}{K_c}$ for the reaction

$CO(g) + \frac{1}{2}O_2(g) \rightleftharpoons CO_2(g)$ is:

[NCERT : P L-178 | Online April 25, 2013]

(a) $\frac{1}{\sqrt{RT}}$ (b) $(RT)^{1/2}$ (c) RT (d) 1

73. (1) $N_2(g) + 3H_2(g) \rightleftharpoons 2NH_3(g), K_1$

(2) $N_2(g) + O_2(g) \rightleftharpoons 2NO(g), K_2$

(3) $H_2(g) + \frac{1}{2}O_2(g) \rightleftharpoons H_2O(g), K_3$

The equation for the equilibrium constant of the reaction

$2NH_3(g) + \frac{5}{2}O_2(g) \rightleftharpoons 2NO(g) + 3H_2O(g), (K_4)$ in terms of K_1, K_2 and K_3 is :

[NCERT : P L-174 | Online April 23, 2013]

(a) $\frac{K_1.K_2}{K_3}$ (b) $\frac{K_1.K_3^2}{K_2}$

(c) $K_1 K_2 K_3$ (d) $\frac{K_2.K_3^3}{K_1}$

74. The equilibrium constant (K_c) for the reaction $N_2(g) + O_2(g) \rightarrow 2NO(g)$ at temperature T is 4×10^{-4}. The value of K_c for the reaction

$NO(g) \rightarrow \frac{1}{2}N_2(g) + \frac{1}{2}O_2(g)$ at the same temperature is:

[NCERT : P L-178 | 2012]

(a) 0.02 (b) 2.5×10^2 (c) 4×10^{-4} (d) 50.0

75. K_1, K_2 and K_3 are the equilibrium constants of the following reactions (I), (II) and (III) respectively:

[NCERT : P L-178 | Online May 7, 2012]

(I) $N_2 + 2O_2 \rightleftharpoons 2NO_2$

(II) $2NO_2 \rightleftharpoons N_2 + 2O_2$

(III) $NO_2 \rightleftharpoons \frac{1}{2}N_2 + O_2$

The correct relation from the following is

(a) $K_1 = \frac{1}{K_2} = \frac{1}{K_3}$ (b) $K_1 = \frac{1}{K_2} = \frac{1}{(K_3)^2}$

(c) $K_1 = \sqrt{K_2} = K_3$ (d) $K_1 = \frac{1}{K_2} = K_3$

76. For the following three reactions (i), (ii) and (iii), equilibrium constants are given: **[NCERT : P L-178 | 2008]**

(i) $CO(g) + H_2O(g) \rightleftharpoons CO_2(g) + H_2(g); K_1$

(ii) $CH_4(g) + H_2O(g) \rightleftharpoons CO(g) + 3H_2(g); K_2$

(iii) $CH_4(g) + 2H_2O(g) \rightleftharpoons CO_2(g) + 4H_2(g); K_3$

Which of the following is correct?

(a) $K_1\sqrt{K_2} = K_3$ (b) $K_2K_3 = K_1$

(c) $K_3 = K_1 K_2$ (d) $K_3.K_2^3 = K_1^2$

77. 8 mol of AB_3(g) are introduced into a 1.0 dm³ vessel. If it dissociates as

$2AB_3(g) \rightleftharpoons A_2(g) + 3B_2(g)$. At equilibrium, 2 mol of A_2 are found to be present. The equilibrium constant of this reaction is **[NCERT : P L-174 | Online May 12, 2012]**

(a) 2 (b) 3 (c) 27 (d) 36

78. The value of K_p for the equilibrium reaction

$N_2O_4(g) \rightleftharpoons 2NO_2(g)$ is 2.

The percentage dissociation of $N_2O_4(g)$ at a pressure of 0.5 atm is **[NCERT : P L-178 | Online May 19, 2012]**

(a) 25 (b) 88 (c) 50 (d) 71

79. One mole of $O_2(g)$ and two moles of $SO_2(g)$ were heated in a closed vessel of one-litre capacity at 1098 K. At equilibrium 1.6 moles of $SO_3(g)$ were found. The equilibrium constant K_c of the reaction would be

NCERT : P L-174 | Online May 26, 2012]

(a) 30 (b) 40 (c) 80 (d) 60

80. The equilibrium constants K_{p_1} and K_{p_2} for the reactions $X \rightleftharpoons 2Y$ and $Z \rightleftharpoons P + Q$, respectively are in the ratio of 1 : 9. If the degree of dissociation of X and Z be equal, then the ratio of total pressures at these equilibria is

[NCERT : P L-178 | 2008]

(a) 1:36 (b) 1 : 1 (c) 1 : 3 (d) 1 : 9

81. The equilibrium constant for the reaction

$$SO_3(g) \rightleftharpoons SO_2(g) + \frac{1}{2}O_2(g)$$

is $K_c = 4.9 \times 10^{-2}$. The value of K_c for the reaction

$$2SO_2(g) + O_2(g) \rightleftharpoons 2SO_3(g)$$

will be **[NCERT : P L-178 | 2006]**

(a) 9.8×10^{-2} (b) 4.9×10^{-2}

(c) 416 (c) 2.40×10^{-3}

82. Phosphorus pentachloride dissociates as follows, in a closed reaction vessel **[NCERT : P L-174 | 2006]**

$$PCl_5(g) \rightleftharpoons PCl_3(g) + Cl_2(g)$$

If total pressure at equilibrium of the reaction mixture is P and degree of dissociation of PCl_5 is x, the partial pressure of PCl_3 will be

(a) $\left(\frac{x}{x-1}\right)P$ (b) $\left(\frac{x}{1-x}\right)P$

(c) $\left(\frac{x}{x+1}\right)P$ (d) $\left(\frac{2x}{1-x}\right)P$

83. For the reaction : **[NCERT : P L-174 | 2005]**

$2NO_2(g) \rightleftharpoons 2NO(g) + O_2(g)$,

$(K_c = 1.8 \times 10^{-6}$ at 184°C) (R = 0.0831 kJ/ (mol. K))

When K_p and K_c are compared at 184 °C, it is found that

(a) Whether K_p is greater than, less than or equal to K_c depends upon the total gas pressure

(b) $K_p = K_c$

(c) K_p is less than K_c

(d) K_p is greater than K_c

84. An amount of solid NH_4HS is placed in a flask already containing ammonia gas at a certain temperature and 0.50 atm pressure. Ammonium hydrogen sulphide decomposes to yield NH_3 and H_2S gases in the flask. When the decomposition reaction reaches equilibrium, the total pressure in the flask rises to 0.84 atm. The equilibrium constant for NH_4HS decomposition at this temperature is **[NCERT : P L-179 | 2005]**

(a) 0.11 (b) 0.17 (c) 0.18 (d) 0.30

85. For the reaction, $CO(g)+Cl_2(g) \rightleftharpoons COCl_2(g)$ the K_p/K_c is equal to **[NCERT : P L-178 | 2004]**

(a) $\sqrt{RT}$ (b) RT (c) $1/RT$ (d) 1.0

86. What is the equilibrium expression for the reaction $P_4(s)+5O_2(g) \rightleftharpoons P_4O_{10}(s)$? **[NCERT : P L-174 | 2004]**

(a) $K_c=[O_2]^5$

(b) $K_c=[P_4O_{10}]/5[P_4][O_2]$

(c) $K_c=[P_4O_{10}]/[P_4][O_2]^5$

(d) $K_c=1/[O_2]^5$

87. For the reaction equilibrium, **[NCERT : P L-174 | 2003]**

$N_2O_4(g) \rightleftharpoons 2NO_2(g)$

the concentrations of N_2O_4 and NO_2 at equilibrium are 4.8×10^{-2} and 1.2×10^{-2} mol L^{-1} respectively. The value of K_c for the reaction is

(a) 3×10^{-1} mol L^{-1} (b) 3×10^{-3} mol L^{-1}

(c) 3×10^{3} mol L^{-1} (d) 3.3×10^{2} mol L^{-1}

88. For the reaction $CO(g)+(1/2)O_2(g) \rightleftharpoons CO_2(g)$, K_p/K_c is **[NCERT : P L-178 | 2002]**

(a) RT (b) $(RT)^{-1}$

(c) $(RT)^{-1/2}$ (d) $(RT)^{1/2}$

Topic 2 - Relation between K, Q and G, and Factors Effecting Equilibrium

89. Consider the equilibrium **[NCERT : P L-184 | Jan. 29, 2025 (II)]**

$CO(g)+3H_2(g) \rightleftharpoons CH_4(g)+H_2O(g)$

If the pressure applied over the system increases by two fold at constant temperature then

(A) Concentration of reactants and products increases.

(B) Equilibrium will shift in forward direction.

(C) Equilibrium constant increases since concentration of products increases.

(D) Equilibrium constant remains unchanged as concentration of reactants and products remain same.

Choose the correct answer from the options given below :

(a) (A) and (B) only (b) (A), (B) and (D) only

(c) (B) and (C) only (d) (A), (B) and (C) only

90. Standard entropies of X_2, Y_2 and XY_5 are 70, 50 and 110 J K^{-1} mol^{-1} respectively. The temperature in Kelvin at which the reaction **[NCERT : P L-184 | NV, Jan. 24, 2025 (I)]**

$\frac{1}{2}X_2+\frac{5}{2}Y_2 \rightarrow XY_5$ $\Delta H^{\circ}=-35$ kJ mol^{-1}

Will be at equilibrium is ____ (Nearest integer)

91. The following reaction occurs in the Blast furnance where iron ore is reduced to iron metal

$Fe_2O_{3(s)}+3CO_{(g)} \rightleftharpoons Fe_{(l)}+3CO_{2(g)}$

Using the Le-chatelier's principle, predict which one of the following will not disturb the equilibrium.

[NCERT : P L-185 | April 5, 2024 (I)]

(a) Addition of Fe_2O_3 (b) Addition of CO_2

(c) Removal of CO (d) Removal of CO_2

92. For a certain reaction at 300 K, K = 10, then ΔG° for the same reaction is – ________ $\times 10^{-1}$ KJ mol^{-1}. (Given R = 8.314 JK^{-1} mol^{-1})

[NCERT : P L-184 | NV, Feb. 1, 2024 (II)]

93. Consider the following reaction at 298 K.

$\frac{3}{2}O_{2(g)} \rightleftharpoons O_{3(g)}$. $K_p=2.47\times10^{-29}$.

$\Delta_r G^{\ominus}$ for the reaction is _____ kJ. (Given R = 8.314 JK^{-1} mol^{-1}) **[NCERT : P L-184 | NV, Jan. 31, 2024 (I)]**

94. For a certain thermochemical reaction $M \rightarrow N$ at T = 400K, $\Delta H^{\ominus}$ = 77.2 kJ mol^{-1}, ΔS = 122 JK^{-1}, log equilibrium constant (log K) is – _________ $\times 10^{-1}$.

[NCERT : P L-184 | NV, Jan. 27, 2024 (II)]

95. The value of log K for the reaction $A \rightleftharpoons B$ at 298 K is – (Nearest integer)

Given : $\Delta H^{\circ}=-54.07$ kJ mol^{-1}

$\Delta S^{\circ}=10$ JK^{-1} mol^{-1}

(Take $2.303\times8.314\times298=5705$)

[NCERT : P L-184 | NV, April 06, 2023 (I)]

96. The effect of addition of helium gas to the following reaction in equilibrium state, is :

$PCl_5(g) \rightleftharpoons PCl_3(g)+Cl_2(g)$

[NCERT : P L-187 | Feb. 01, 2023 (II)]

(a) addition of helium will not affect the equilibrium.

(b) helium will deactivate PCl_5 and reaction will stop.

(c) the equilibrium will go backward due to suppression of dissociation of PCl_5.

(d) the equilibrium will shift in the forward direction and more of Cl_2 and PCl_3 gases will be produced.

97. Consider the following equation :

$2SO_2(g)+O_2(g) \rightleftharpoons 2SO_3(g)$, $\Delta H=-190$ kJ

The number of factors which will increase the yield of SO_3 at equilibrium from the following is ______

[NCERT : P L-184 | NV, Jan. 30, 2023 (II)]

A. Increasing temperature

B. Increasing pressure

C. Adding more SO_2

D. Adding more O_2

E. Addition of catalyst

98. Consider the following reaction approaching equilibrium at 27°C and 1 atm pressure

$$A + B \underset{K_r=10^2}{\overset{K_{f=10^3}}{\rightleftharpoons}} C + D$$

The standard Gibb's energy change ($\Delta_r G^\circ$) at 27°C is (–) ______ kJ mol^{-1}.
(Nearest integer).
(Given : $R = 8.3\ J\ K^{-1}\ mol^{-1}$ and ln 10 = 2.3)
[NCERT : P L-184 | NV, Jan. 29, 2023 (I)]

99. 40% of HI undergoes decomposition to H_2 and I_2 at 300 K. ΔG° for this decompostion reaction at one atmopsphere pressure is_______ J mol^{-1}. [nearest integer]
(Use $R = 8.31\ J\ K^{-1}\ mol^{-1}$; log 2 = 0.3010.
ln 10 = 2.3, log 3 = 0.477)
[NCERT : P L-184 | NV, June 26, 2022 (II)]

100. $2O_3(g) \rightleftharpoons 3O_2(g)$
At 300 K, ozone is fifty percent dissociated. The standard free energy change at this temperature and 1 atm pressure is (–) __ J mol^{-1} (Nearest integer)
[Given: ln 1.35 = 0.3 and $R = 8.3\ J\ K^{-1}\ mol^{-1}$]
[NCERT : P L-184 | NV, June 24, 2022 (I)]

101. For the reaction $A(g) \rightleftharpoons B(g)$ at 495 K, $\Delta_r G^\circ = -9.478$ kJ mol^{-1}.
If we start the reaction in a closed container at 495 K with 22 millimoles of A, the amount of B in the equilibrium mixture is _____ millimoles. (Round off to the Nearest Integer).
[$R = 8.314\ J\ mol^{-1}\ K^{-1}$; ln 10 = 2.303]
[NCERT : P L-184 | NV, March 16, 2021 (I)]

102. The gas phase reaction
$2A(g) \rightleftharpoons A_2(g)$ at 400 K has $dG^\circ = +25.2$ kJ mol^{-1}
The equilibrium constant K_c for this reaction is _____ $\times 10^{-2}$. (Round off to the nearest integer).
[Use : $R = 8.3\ J mol^{-1} K^{-1}$, ln 10 = 2.3 $\log_{10} 2 = 0.30$, 1 atm = 1 bar]
[antilog (–0.3) = 0.501]
[NCERT : P L-184 | NV, March 18, 2021 (II)]

103. For the reaction $A(g) \rightleftharpoons B(g)$ at 495 K,
$\Delta_r G^\circ = -9.478$ kJ mol^{-1}.
If we start the reaction in a closed container at 495 K with 22 millimoles of A, the amount of B in the equilibrium mixture is _____ millimoles. (Round off to the Nearest Integer).
[$R = 8.314\ J\ mol^{-1}\ K^{-1}$; ln 10 = 2.303]
[NCERT : P L-184 | NV, March 16, 2021 (I)]

104. The variation of equilibrium constant with temperature is given below: **[NCERT : P L-184 | Sep. 06, 2020 (I)]**

Temperature	**Equilibrium Constant**
$T_1 = 25\ ^\circ C$	$K_1 = 10$
$T_2 = 100\ ^\circ C$	$K_2 = 100$

The values of ΔH°, ΔG° at T_1 and ΔG° at T_2 (in kJ mol^{-1}) respectively, are close to
[use $R = 8.314\ J\ K^{-1}\ mol^{-1}$]
(a) 28.4, – 7.14 and – 5.71
(b) 0.64, – 7.14 and – 5.71
(c) 28.4, – 5.71 and – 14.29
(d) 0.64, – 5.71 and – 14.29

105. Consider the following reaction :
[NCERT : P L-185 | Sep. 05, 2020 (I)]
$N_2O_4(g) \rightleftharpoons 2NO_2(g)$; $\Delta H^0 = +58$ kJ
For each of the following cases ((i), (ii)), the direction in which the equilibrium shifts is :
(i) Temperature is decreases
(ii) Pressure is increased by adding N_2 at constant T.
(a) (i) towards product, (ii) towards product
(b) (i) towards reactant, (ii) towards product
(c) (i) towards reactant, (ii) no change
(d) (i) towards product, (ii) no change

106. The INCORRECT match in the following is :
[NCERT : P L-184 | April 12, 2019 (II)]
(a) $\Delta G^0 < 0, K > 1$
(b) $\Delta G^0 = 0, K = 1$
(c) $\Delta G^0 > 0, K < 1$
(d) $\Delta G^0 < 0, K < 1$

107. For the reaction,
$2SO_2(g) + O_2(g) \rightleftharpoons 2SO_3(g)$,
$\Delta H = -57.2$ kJ mol^{-1} and $K_c = 1.7 \times 10^{16}$
Which of the following statement is INCORRECT ?
[NCERT : P L-184 | April 10, 2019 (II)]
(a) The equilibrium constant is large suggestive of reaction going to completion and so no catalyst is required.
(b) The equilibrium will shift in forward direction as the pressure increases.
(c) The equilibrium constant decreases as the temperature increases.
(d) The addition of inert gas at constant volume will not affect the equilibrium constant.

108. For the equilibrium
$2H_2O \rightleftharpoons H_3O^+ + OH^-$; the value of ΔG° at 298 K is approximately: **[NCERT : P L-184 | Jan. 11, 2019 (II)]**
(a) 100 kJ mol^{-1}
(b) –80 kJ mol^{-1}
(c) 80 kJ mol^{-1}
(d) –100 kJ mol^{-1}

109. In which of the following reactions, an increase in the volume of the container will favour the formation of products? **[NCERT : P L-184 | Online April 15, 2018 (I)]**
(a) $4NH_3(g) + 5O_2(g) \rightleftharpoons 4NO(g) + 6H_2O(l)$
(b) $2NO_2(g) \rightleftharpoons 2NO(g) + O_2(g)$
(c) $3O_2 \rightleftharpoons 2O_3(g)$
(d) $H_2(g) + I_2(g) \rightleftharpoons 2HI(g)$

110. The gas phase reaction $2NO_2(g) \longrightarrow N_2O_4(g)$ is an exothermic reaction. The decomposition of N_2O_4, in equilibrium mixutre of $NO_2(g)$ and $N_2O_4(g)$, can be increased by : **[NCERT : P L-187 | Online April 16, 2018]**
(a) addition of an inert gas at constant pressure
(b) lowering the temperature
(c) increasing the pressure
(d) addition of an inert gas at constant volume

111. The following reaction occurs in the Blast Furnace where iron ore is reduced to iron metal :
$Fe_2O_3(s) + 3\,CO(g) \rightleftharpoons 2\,Fe(l) + 3\,CO_2(g)$
Using the Le Chatelier's principle, predict which one of the following will not disturb the equilibrium ?
[NCERT : P L-184 | Online April 9, 2017]
(a) Removal of CO (b) Removal of CO_2
(c) Addition of CO_2 (d) Addition of Fe_2O_3

112. The plot shows the variation of $-\ln K_p$ versus temperature for the two reactions. **[Online April 9, 2016]**
$M(s)+\frac{1}{2}O_2(g) \rightarrow MO(s)$ and $C(s)+\frac{1}{2}O_2(g) \rightarrow CO(s)$

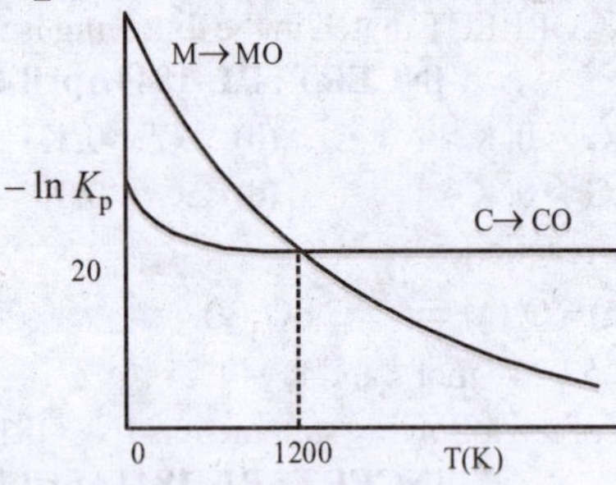

Identify the correct statement:
(a) At T < 1200 K, oxidation of carbon is unfavourable.
(b) Oxidation of carbon is favourable at all temperatures.
(c) At T < 1200 K, the reaction $MO(s) + C(s) \rightarrow M(s) + CO(g)$ is spontaneous.
(d) At T > 1200 K, carbon will reduce MO(s) to M(s).

113. For the reaction,
$A(g) + B(g) \rightarrow C(g) + D(g)$, ΔH° and ΔS° are, respectively, -29.8 kJ mol^{-1} and -0.100 kJ K^{-1} mol^{-1} at 298 K. The equilibrium constant for the reaction at 298 K is :
[NCERT : P L-184 | Online April 9, 2016]
(a) 1.0×10^{-10} (b) 10
(c) 1 (d) 1.0×10^{10}

114. The standard Gibbs energy change at 300 K for the reaction $2A \rightleftharpoons B + C$ is 2494.2 J. At a given time, the composition of the reaction mixture is $[A] = \frac{1}{2}$, $[B] = 2$ and $[C] = \frac{1}{2}$. The reaction proceeds in the : [$R = 8.314$ J/K/mol, $e = 2.718$] **[NCERT : P L-184 | 2015]**
(a) forward direction because $Q < K_c$
(b) reverse direction because $Q < K_c$
(c) forward direction because $Q > K_c$
(d) reverse direction because $Q > K_c$

115. The following reaction is performed at 298 K.
$$2NO(g) + O_2(g) \rightleftharpoons 2NO_2(g)$$
The standard free energy of formation of NO(g) is 86.6 kJ/mol at 298 K. What is the standard free energy of formation of $NO_2(g)$ at 298 K? ($K_p = 1.6 \times 10^{12}$)
[NCERT : P L-184 | 2015]
(a) $86600 - \dfrac{\ln(1.6 \times 10^{12})}{R\,(298)}$
(b) $0.5[2 \times 86{,}600 - R(298)\ln(1.6 \times 10^{12})]$
(c) $R(298)\ln(1.6 \times 10^{12}) - 86600$
(d) $86600 + R(298)\ln(1.6 \times 10^{12})$

116. The increase of pressure on ice $\rightleftharpoons$ water system at constant temperature will lead to
[NCERT : P L-184 | Online April 11, 2015]
(a) a decrease in the entropy of the system
(b) an increase in the Gibb's energy of the system
(c) no effect on the equilibrium
(d) a shift of the equilibrium in the forward direction

117. What happens when an inert gas is added to an equilibrium keeping volume unchanged?
[NCERT : P L-187 | Online April 12, 2014]
(a) More product will form
(b) Less product will form
(c) More reactant will form
(d) Equilibrium will remain unchanged

118. The exothermic formation of ClF_3 is represented by the equation :
$Cl_2(g) + 3F_2(g) \rightleftharpoons 2ClF_3(g)$; $\Delta H = -329$ kJ
Which of the following will increase the quantity of ClF_3 in an equilibrium mixture of Cl_2, F_2 and ClF_3 ?
[NCERT : P L-187 | 2005]
(a) Adding F_2
(b) Increasing the volume of the container
(c) Removing Cl_2
(d) Increasing the temperature

119. Consider the reaction equilibrium
$2\,SO_2(g) + O_2(g) \rightleftharpoons 2\,SO_3(g)$; $\Delta H^\circ = -198$ kJ
On the basis of Le Chatelier's principle, the condition favourable for the forward reaction is
[NCERT : P L-184 | 2003]
(a) increasing temperature as well as pressure
(b) lowering the temperature and increasing the pressure
(c) any value of temperature and pressure
(d) lowering of temperature as well as pressure

120. Change in volume of the system does not alter which of the following equilibria? **[NCERT : P L-184 | 2002]**
(a) $N_2(g) + O_2(g) \rightleftharpoons 2NO(g)$
(b) $PCl_5(g) \rightleftharpoons PCl_3(g) + Cl_2(g)$
(c) $N_2(g) + 3H_2(g) \rightleftharpoons 2NH_3(g)$
(d) $SO_2Cl_2(g) \rightleftharpoons SO_2(g) + Cl_2(g)$.

Topic 3 -Theories of Acids and Bases, Ionic Product of Water and pH Scale

121. An aqueous solution of HCl with pH 1.0 is diluted by adding equal volume of water (ignoring dissociation of water). The pH of HCl solution would (Given log 2 = 0.30) **[NCERT : P L-194 | April 7, 2025 (I)]**
(a) reduce to 0.5 (b) increase to 1.3
(c) remain same (d) increase to 2

122. x mg of $Mg(OH)_2$ (molar mass = 58) is required to be dissolved in 1.0 L of water to produce a pH of 10.0 at 298 K. The value of x is____ mg. (Nearest integer)
(**Given :** $Mg(OH)_2$ is assumed to dissociate completely in H_2O). **[NCERT : P L-194 | NV, April 4, 2025 (II)]**

123. A weak acid HA has degree of dissociation x. Which option gives the correct expression of $(pH - pK_a)$? **[NCERT : P L-195 | Jan. 28, 2025 (I)]**
(a) $\log(1+2x)$ (b) $\log\left(\frac{1-x}{x}\right)$
(c) 0 (d) $\log\left(\frac{x}{1-x}\right)$

124. pH of water is 7 at 25°C. If water is heated to 80°C, it's pH will: **[NCERT : P L-194 | Jan. 23, 2025 (II)]**
(a) Decrease
(b) Remains the same
(c) H^+ concentration increases, OH^- concentration decreases
(d) Increase

125. What will be the decreasing order of basic strength of the following conjugate bases ?
$^{-}OH, R\bar{O}, CH_3CO\bar{O}, C\bar{l}$
[NCERT : P L-191 | April 4, 2024 (I)]
(a) $C\bar{l} > OH > R\bar{O} > CH_3CO\bar{O}$
(b) $R\bar{O} > {}^{-}OH > CH_3CO\bar{O} > C\bar{l}$
(c) ${}^{-}OH > R\bar{O} > CH_3CO\bar{O} > C\bar{l}$
(d) $C\bar{l} > R\bar{O} > {}^{-}OH > CH_3CO\bar{O}$

126. Which of the following statement(s) is/are correct? **[NCERT : P L-194 | April 15, 2023 (I)]**
(A) The pH of 1×10^{-8} M HCl solution is 8.
(B) The conjugate base of $H_2PO_4^-$ is HPO_4^{2-}.
(C) K_w increases with increase in temperature.
(D) When a solution of weak monoprotic acid is titrated against a strong base at half neutralisation point,
$pH = \frac{1}{2}pK_a$
Choose the correct answer from the option given below.
(a) (B), (C), (D) (b) (A), (D)
(c) (A), (B), (C) (d) (B), (C)

127. An analyst wants to convert 1 L HCl of pH = 1 to a solution of HCl of pH 2. The volume of water needed to do this dilution is ________ mL. (Nearest integer) **[NCERT : P L-194 | NV, April 12, 2023 (I)]**

128. 600 mL of 0.01 M HCl is mixed with 400 mL of 0.01 M H_2SO_4. The pH of the mixture is ________ $\times 10^{-2}$. (Nearest integer)
[Given log 2 = 0.30, log 3 = 0.48
log 5 = 0.69 log 7 = 0.84
log 11 = 1.04] **[NCERT : P L-193 | NV, Jan. 30, 2023 (I)]**

129. Millimoles of calcium hydroxide required to produce 100 mL of the aqueous solution of pH 12 is x $\times 10^{-1}$. The value of x is ______ (Nearest integer).
Assume complete dissociation.
[NCERT : P L-194 | NV, Jan. 29, 2023 (I)]

130. A litre of buffer solution contains 0.1 mole of each of NH_3 and NH_4Cl. On the addition of 0.02 mole of HCl by dissolving gaseous HCl, the pH of the solution is found to be ________ $\times 10^{-3}$ (Nearest integer)
[Given : $pK_b(NH_3) = 4.745$
log 2 = 0.301 log 3 = 0.477 T = 298 K]
[NCERT : P L-202 | NV, Jan. 25, 2023 (I)]

131. The density of a monobasic strong acid (Molar mass 24.2 g/mol) is 1.21 kg/L. The volume of its solution required for the complete neutralization of 25 mL of 0.24 M NaOH is__________ $\times 10^{-2}$ mL (Nearest integer)
[NV, Jan. 25, 2023 (I)]

132. When the hydrogen ion concentration $[H^+]$ changes by a factor of 1000, the value of pH of the solution ______.
[NCERT : P L-194 | Jan. 25, 2023 (II)]
(a) increases by 1000 units (b) decreases by 3 units
(c) decreases by 2 units (d) increases by 2 units

133. 200 mL of 0.01 M HCl is mixed with 400 mL of 0.01 M H_2SO_4. The pH of the mixture is ____.
[NCERT : P L-194 | NV, July 29, 2022 (II)]
(a) 1.14 (b) 1.78 (c) 2.34 (d) 3.02

134. A compound 'X' is a weak acid and it exhibits colour change at pH close to the equivalence point during neutralization of NaOH with CH_3COOH. Compound 'X' exists in ionized form in basic medium. The compound 'X' is : **[NCERT : P L-202 | July 29, 2022 (II)]**
(a) methyl orange (b) methyl red
(c) phenolphthalein (d) erichrome Black T

135. Given below are two statements : one is labelled as **Assertion A** and the other is labelled as **Reason R**.
Assertion A : Phenolphthalein is a pH dependent indicator, remains colourless in acidic solution and gives pink colour in basic medium
Reason R : Phenolphthalein is a weak acid. It doesn't dissociate in basic medium.
[NCERT : P L-202 | July 26, 2022 (II)]
In the light of the above statements, choose the **most appropriate** answer from the options given below :
(a) Both A and R are true and R is the correct explanation of A.
(b) Both A and R are true but R is NOT the correct explanation of A.
(c) A is true but R is false.
(d) A is false but R is true.

136. pH value of 0.001 M NaOH solution is______.

[NCERT : P L-194 | NV, June 27, 2022 (II)]

137. Given below are two statements one is labelled as **Assertion A** and the other is labelled as **Reason R:**

Assertion A : The amphoteric nature of water is explained by using Lewis acid/base concept.

Reason R : Water acts as an acid with NH_3 and as a base with H_2S.

In the light of the above statements choose the correct answer from the options given below :

[NCERT : P L-192 | June 25, 2022 (II)]

(a) Both A and R are true and R is the correct explanation of A.
(b) Both A and R are true but R is NOT the correct explanation of A.
(c) A is true but R is false.
(d) A is false but R is true.

138. The pH of a solution obtained by mixing 50 mL of 1 M HCl and 30 mL of 1 M NaOH is $x \times 10^{-4}$. The value of x is ______. (Nearest integer)

[log 2.5 = 0.3979] **[NCERT : P L-194 | Aug. 31, 2021 (II)]**

139. Which of the following compound CANNOT act as a Lewis base? **[NCERT : P L-192 | March 17, 2021 (I)]**

(a) NF_3 (b) PCl_5
(c) ClF_3 (d) SF_4

140. Arrange the following solutions in the decreasing order of pOH : **[NCERT : P L-194 | Sep. 06, 2020 (I)]**

(A) 0.01 M HCl
(B) 0.01 M NaOH
(C) 0.01 M CH_3COONa
(D) 0.01 M NaCl

(a) (A) > (C) > (D) > (B)
(b) (A) > (D) > (C) > (B)
(c) (B) > (C) > (D) > (A)
(d) (B) > (D) > (C) > (A)

141. A soft drink was bottled with a partial pressure of CO_2 of 3 bar over the liquid at room temperature. The partial pressure of CO_2 over the solution approaches a value of 30 bar when 44 g of CO_2 is dissolved in 1 kg of water at room temperature. The approximate pH of the soft drink is ______ $\times 10^{-1}$. **[NV, Sep. 05, 2020 (I)]**

(First dissociation constant of $H_2CO_3 = 4.0 \times 10^{-7}$; log 2 = 0.3; density of the soft drink = 1 g mL^{-1})

142. 100 mL of 0.1 M HCl is taken in a beaker and to it 100 mL of 0.1 M NaOH is added in steps of 2 mL and the pH is continuously measured. Which of the following graphs **correctly** depicts the change in pH?

[NCERT : P L-194 | Sep. 03, 2020 (II)]

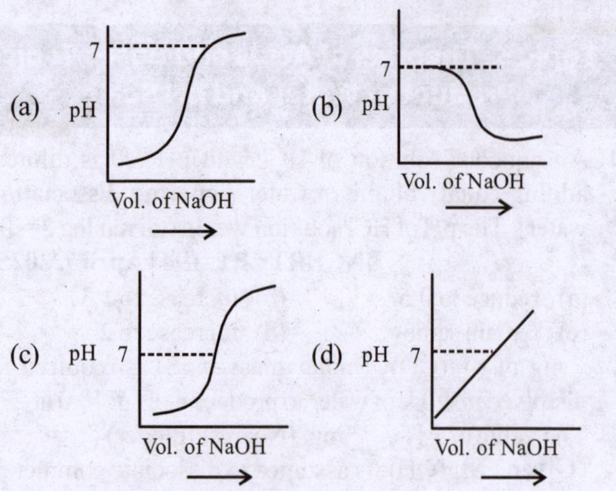

143. For the following Assertion and Reason, the correct option is: **[NCERT : P L-193 | Jan. 08, 2020 (II)]**

Assertion: The pH of water increases with increase in temperature.

Reason: The dissociation of water into H^+ and OH^- is an exothermic reaction.

(a) Both assertion and reason are true, and the reason is the correct explanation for the assertion.
(b) Both assertion and reason are false.
(c) Both assertion and reason are true, but the reason is not the correct explanation for the assertion.
(d) Assertion is not true, but reason is true.

144. Two solutions, A and B, each of 100 L was made by dissolving 4 g of NaOH and 9.8 g of H_2SO_4 in water, respectively. The pH of the resultant solution obtained from mixing 40 L of solution A and 10 L of solution B is ________. **[NV, Jan. 07, 2020 (I)]**

145. 3 g of acetic acid is added to 250 mL of 0.1 M HCl and the solution made up to 500 mL. To 20 mL of this solution, $\frac{1}{2}$ mL of 5 M NaOH is added. The pH of the solution is ________. **[NCERT : P L-194 | NV, Jan. 07, 2020 (II)]**

[Given: pK_a of acetic acid = 4.75, molar mass of acetic acid = 60 g/mol, log 3 = 0.4771]

Neglect any changes in volume.

146. Consider the following statements

[NCERT : P L-194 | April 10, 2019 (I)]

(A) The pH of a mixture containing 400 mL of 0.1 M H_2SO_4 and 400 mL of 0.1 M NaOH will be approximately 1.3.
(B) Ionic product of water is temperature dependent.
(C) A monobasic acid with $K_a = 10^{-5}$ has a pH = 5. The degree of dissociation of this acid is 50%.
(D) The Le Chatelier's principle is not applicable to common-ion effect.

The correct statements are : **[April 10, 2019 (I)]**

(a) (A), (B), and (D)
(b) (A), (B) and (C)
(c) (B) and (C)
(d) (A) and (B)

147. In an acid base titration, 0.1 M HCl solution was added to the NaOH solution of unknown strength. Which of the following correctly shows the change of pH of the titration mixture in this experiment? **[NCERT : P L-194 | April 9, 2019 (II)]**

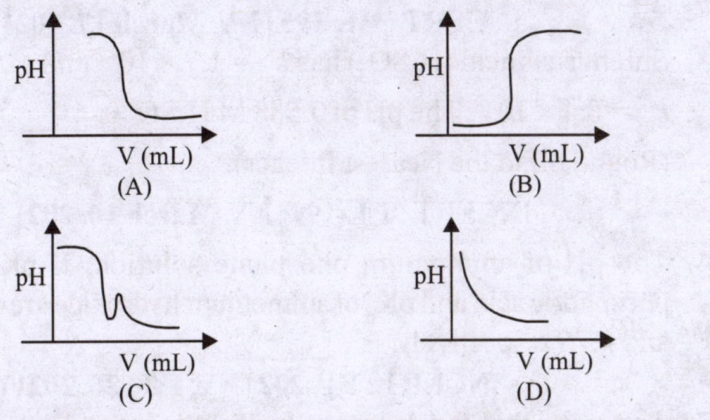

(a) (B) (b) (A) (c) (C) (d) (D)

148. Which amongst the following is the strongest acid? **[NCERT : P L-190 | Jan. 9, 2019 (I)]**

(a) $CHBr_3$ (b) CHI_3 (c) $CH(CN)_3$ (d) $CHCl_3$

149. Following four solutions are prepared by mixing different volumes of NaOH and HCl of different concentrations, pH of which one of them will be equal to 1? **[NCERT : P L-194 | Online April 15, 2018 (II)]**

(a) $55\ mL \frac{M}{10} HCl + 45\ mL \frac{M}{10} NaOH$

(b) $75\ mL \frac{M}{5} HCl + 25\ mL \frac{M}{5} NaOH$

(c) $100\ mL \frac{M}{10} HCl + 100\ mL \frac{M}{10} NaOH$

(d) $60\ mL \frac{M}{10} HCl + 40\ mL \frac{M}{10} NaOH$

150. Which of the following is a Lewis acid? **[NCERT : P L-192 | Online April 15, 2018 (I)]**

(a) PH_3 (b) NF_3 (c) NaH (d) $B(CH_3)_3$

151. 50 mL of 0.2 M ammonia solution is treated with 25 mL of 0.2 M HCl. If pK_b of ammonia solution is 4.75, the pH of the mixture will be: **[NCERT : P L-194 | Online April 9, 2017]**

(a) 3.75 (b) 4.75 (c) 8.25 (d) 9.25

152. The conjugate base of hydrazoic acid is: **[NCERT : P L-190 | Online April 12, 2014]**

(a) N^{-3} (b) N_3^- (c) N_2^- (d) HN_3^-

153. How many litres of water must be added to 1 litre an aqueous solution of HCl with a pH of 1 to create an aqueous solution with pH of 2 ? **[NCERT : P L-194 | 2013]**

(a) 0.1 L (b) 0.9 L (c) 2.0 L (d) 9.0 L

154. NaOH is a strong base. What will be pH of 5.0×10^{-2} M NaOH solution ? (log 2 = 0.3) **[NCERT : P L-194 | Online April 22, 2013]**

(a) 14.00 (b) 13.70 (c) 13.00 (d) 12.70

155. Equimolar solutions of the following compounds are prepared separately in water. Which will have the lowest pH value ? **[NCERT : P L-201 | Online April 23, 2013]**

(a) $BeCl_2$ (b) $SrCl_2$ (c) $CaCl_2$ (d) $MgCl_2$

156. What is the pH of a 10^{-4} M OH^- solution at 330*K*, if K_w at 330 K is $10^{-13.6}$? **[NCERT : P L-194 | Online April 23, 2013]**

(a) 4 (b) 9.0 (c) 10 (d) 9.6

157. Which one of the following arrangements represents the correct order of the proton affinity of the given species : **[NCERT : P L-190 | Online April 25, 2013]**

(a) $I^- < F^- < HS^- < NH_2^-$ (b) $HS^- < NH_2^- < F^- < I^-$

(c) $F^- < I^- < NH_2^- < HS^-$ (d) $NH_2^- < HS^- < I^- < F^-$

158. Four species are listed below: **[NCERT : P L-190 | 2008]**

i. HCO_3^- ii. H_3O^+ iii. HSO_4^- iv. HSO_3F

Which one of the following is the correct sequence of their acid strength?

(a) iv < ii < iii < i (b) ii < iii < i < iv
(c) i < iii < ii < iv (d) iii < i < iv < ii

159. Hydrogen ion concentration in mol/L in a solution of pH = 5.4 will be : **[NCERT : P L-194 | 2005]**

(a) 3.98×10^{-6} (b) 3.68×10^{-6}
(c) 3.88×10^{6} (d) 3.98×10^{8}

160. What is the conjugate base of OH^- ? **[NCERT : P L-190 | 2005]**

(a) O^{2-} (b) O^- (c) H_2O (d) O_2

161. The conjugate base of $H_2PO_4^-$ is **[NCERT : P L-192 | 2004]**

(a) H_3PO_4 (b) P_2O_5
(c) PO_4^{3-} (d) HPO_4^{2-}

162. Which one of the following statements is not true? **[2003]**

(a) pH + pOH = 14 for all aqueous solutions
(b) The pH of 1×10^{-8} M HCl is 8
(c) 96,500 coulombs of electricity when passed through a $CuSO_4$ solution deposits 1 gram equivalent of copper at the cathode
(d) The conjugate base of $H_2PO_4^-$ is HPO_4^{2-}

163. When rain is accompanied by a thunderstorm, the collected rain water will have a pH value **[NCERT : P L-194 | 2003]**

(a) slightly higher than that when the thunderstorm is not there
(b) uninfluenced by occurrence of thunderstorm
(c) that depends on the amount of dust in air
(d) slightly lower than that of rain water without thunderstorm.

164. Species acting as both Bronsted acid and base is **[NCERT : P L-190 | 2002]**

(a) HSO_4^- (b) Na_2CO_3 (c) NH_3 (d) OH^-

Topic 4 - Ionisation of Weak Acids and Bases and Relation between K_a and K_b

165. The pH of a 0.01 M weak acid HX ($K_a = 4 \times 10^{-10}$) is found to be 5. Now the acid solution is diluted with excess of water so that the pH of the solution changes to 6. The new concentration of the diluted weak acid is given as $x \times 10^{-4}$ M. The value of x is ________(nearest integer)

[NCERT : P L-198 | NV, April 4, 2025 (I)]

166. If 1 mM solution of ethylamine produces pH = 9, then the ionization constant (K_b) of ethylamine is 10^{-x}. The value of x is ______(nearest integer).
[The degree of ionization of ethylamine can be neglected with respect to unity.]

[NCERT : P L-197 | NV, Jan. 23, 2025 (I)]

167. K_a for CH_3COOH is 1.8×10^{-5} and K_b for NH_4OH is 1.8×10^{-5}. The pH of ammonium acetate solution will be _____. **[NCERT : P L-199 | Feb. 1, 2024 (I)]**

168. Given below are two statements :

[NCERT : P L-200 | Jan. 27, 2024 (I)]

Statement (I) : Aqueous solution of ammonium carbonate is basic.

Statement (II) : Acidic/basic nature of salt solution of a salt of weak acid and weak base depends on K_a and K_b value of acid and the base forming it.

In the light of the above statements, choose the most appropriate answer from the options given below :

(a) Both Statement I and Statement II are correct
(b) Statement I is correct but Statement II is incorrect
(c) Both Statement I and Statement II are incorrect
(d) Statement I is incorrect but Statement II is correct

169. K_a for butyric acid (C_3H_7COOH) is 2×10^{-5}. The pH of 0.2 M solution of butyric acid is ___ $\times 10^{-1}$. (Nearest integer) [Given log 2 = 0.30]

[NCERT : P L-195 | NV, July 28, 2022 (I)]

170. 20 mL of 0.1 M NH_4OH is mixed with 40 mL of 0.05 M HCl. The pH of the mixture is nearest to:
(Given : $K_b(NH_4OH) = 1 \times 10^{-5}$, log 2 = 0.30, log 3 = 0.48, log 5 = 0.69, log 7 = 0.84, log 11 = 1.04)

[NCERT : P L-201 | July 25, 2022 (I)]

(a) 3.2 (b) 4.2 (c) 5.2 (d) 6.2

171. K_{a_1}, K_{a_2} and K_{a_3} are the respective ionization constants for the following reactions (A), (B), and (C).

(A) $H_2C_2O_4 \rightleftharpoons H^+ + HC_2O_4^-$

(B) $HC_2O_4 \rightleftharpoons H^+ + HC_2O_4^{2-}$

(C) $H_2C_2O_4 \rightleftharpoons 2H^+ + C_2O_4^{2-}$

The relationship between K_{a_1}, K_{a_2} and K_{a_3} is given as

[NCERT : P L-199 | July 25, 2022 (II)]

(a) $K_{a_3} = K_{a_1} + K_{a_2}$ (b) $K_{a_3} = K_{a_1} + K_{a_2}$
(c) $K_{a_3} = K_{a_1} / K_{a_2}$ (d) $K_{a_3} = K_{a_1} \times K_{a_2}$

172. 0.01 moles of a weak acid HA ($K_a = 2.0 \times 10^{-6}$) is dissolved in 1.0 L of 0.1 M HCl solution. The degree of dissociation of HA is ______ $\times 10^{-5}$ (Round off to the Nearest Integer). [Neglect volume change on adding HA. Assume degree of dissociation <<1]

[NCERT : P L-195 | NV, March 17, 2021 (I)]

173. Sulphurous acid (H_2SO_3) has $k_{a_1} = 1.7 \times 10^{-2}$ and $k_{a_2} = 6.4 \times 10^{-8}$. The pH of 0.588 M H_2SO_3 is _______. (Round off to the Nearest Integer).

[NCERT : P L-199 | NV, March 16, 2021 (II)]

174. The pH of ammonium phosphate solution, if pK_a of phosphoric acid and pK_b of ammonium hydroxide are 5.23 and 4.75 respectively, is ________.

[NCERT : P L-202 | NV, Feb. 26, 2021 (II)]

175. Assuming that the degree of hydrolysis is small, the pH of 0.1 M solution of sodium acetate ($K_a = 1.0 \times 10^{-5}$) will be: **[NCERT : P L-202 | Online April 11, 2014]**

(a) 5.0 (b) 6.0 (c) 8.0 (d) 9.0

176. Values of dissociation constant, K_a are given as follows :

Acid	K_a
HCN	6.2×10^{-10}
HF	7.2×10^{-4}
HNO_2	4.0×10^{-4}

Correct order of increasing base strength of the base CN^-, F^- and NO_2^- will be :

[NCERT : P L-195 | Online April 22, 2013]

(a) $F^- < CN^- < NO_2^-$ (b) $NO_2^- < CN^- < F^-$
(c) $F^- < NO_2^- < CN^-$ (d) $NO_2^- < F^- < CN^-$

177. What would be the pH of a solution obtained by mixing 5 g of acetic acid and 7.5 g of sodium acetate and making the volume equal to 500 mL?
($K_a = 1.75 \times 10^{-5}$, $pK_a = 4.76$)

[NCERT : P L-202 | Online April 25, 2013]

(a) pH = 4.70
(b) pH < 4.70
(c) pH of solution will be equal to pH of acetic acid
(d) 4.76 < pH < 5.0

178. The pH of a 0.1 molar solution of the acid HQ is 3. The value of the ionization constant, K_a of the acid is :

[NCERT : P L-195 | 2012]

(a) 3×10^{-1} (b) 1×10^{-3}
(c) 1×10^{-5} (d) 1×10^{-7}

179. Given **[NCERT : P L-195 | Online May 12, 2012]**

(i) $HCN(aq) + H_2O(l) \rightleftharpoons H_3O^+(aq) + CN^-(aq)$
$K_a = 6.2 \times 10^{-10}$

(ii) $CN^-(aq) + H_2O(l) \rightleftharpoons HCN(aq) + OH^-(aq)$
$K_b = 1.6 \times 10^{-5}$

These equilibria show the following order of the relative base strength,

(a) $OH^- > H_2O > CN^-$ (b) $OH^- > CN^- > H_2O$
(c) $H_2O > CN^- > OH^-$ (d) $CN^- > H_2O > OH^-$

180. An acid HA ionises as

$$HA \rightleftharpoons H^+ + A^-$$

The pH of 1.0 M solution is 5. Its dissociation constant would be : **[NCERT : P L-195 | 2011RS]**

(a) 5 (b) 5×10^{-8}
(c) 1×10^{-5} (d) 1×10^{-10}

181. pK_a of a weak acid (HA) and pK_b of a weak base (BOH) are 3.2 and 3.4, respectively. The pH of their salt (AB) solution is **[NCERT : P L-198 | 2017]**

(a) 7.2 (b) 6.9 (c) 7.0 (d) 1.0

182. Three reactions involving $H_2PO_4^-$ are given below :

(i) $H_3PO_4 + H_2O \rightarrow H_3O^+ + H_2PO_4^-$
(ii) $H_2PO_4^- + H_2O \rightarrow HPO_4^{2-} + H_3O^+$
(iii) $H_2PO_4^- + OH^- \rightarrow H_3PO_4 + O^{2-}$

In which of the above does $H_2PO_4^-$ act as an acid ? **[NCERT : P L-199 | 2010]**

(a) (ii) only (b) (i) and (ii)
(c) (iii) only (d) (i) only

183. In aqueous solution the ionization constants for carbonic acid are

$K_1 = 4.2 \times 10^{-7}$ and $K_2 = 4.8 \times 10^{-11}$.

Select the correct statement for a saturated 0.034 M solution of the carbonic acid. **[NCERT : P L-195 | 2010]**

(a) The concentration of CO_3^{2-} is 0.034 M.
(b) The concentration of CO_3^{2-} is greater than that of HCO_3^-.
(c) The concentrations of H^+ and HCO_3^- are approximately equal.
(d) The concentration of H^+ is double that of CO_3^{2-}.

184. The pK_a of a weak acid, HA, is 4.80. The pK_b of a weak base, BOH, is 4.78. The pH of an aqueous solution of the correspondng salt, BA, will be **[NCERT : P L-202 | 2008]**

(a) 9.58 (b) 4.79 (c) 7.01 (d) 9.22

185. The first and second dissociation constants of an acid H_2A are 1.0×10^{-5} and 5.0×10^{-10} respectively. The overall dissociation constant of the acid will be **[NCERT : P L-199 | 2007]**

(a) 0.2×10^5 (b) 5.0×10^{-5}
(c) 5.0×10^{15} (d) 5.0×10^{-15}.

Topic 5 - Common Ion Effect, Salt Hydrolysis, Buffer Solutions and Solubility Product

186. Only litre buffer solution was prepared by adding 0.10 mol each of NH_3 and NH_4Cl in deionised water. The change in pH on addition of 0.05 mol of HCl to the above solution is ____ $\times 10^{-2}$, (Nearest integer)

(**Given :** pK_b of $NH_3 = 4.745$ and $\log_{10}3 = 0.477$)

[NCERT : P L-202 | NV, April 7, 2025 (II)]

187. If equal volumes of AB_2 and XY (both are salts) aqueous solutions are mixed, which of the following combination will give a precipitate of AY_2 at 300 K?

(Given K_{sp} (at 300 K) for $AY_2 = 5.2 \times 10^{-7}$)

[NCERT : P L-204 | April 2, 2025 (I)]

(a) 3.6×10^{-3} M AB_2, 5.0×10^{-4} M XY
(b) 2.0×10^{-4} M AB_2, 0.8×10^{-3} M XY
(c) 2.0×10^{-2} M AB_2, 2.0×10^{-2} M XY
(d) 1.5×10^{-4} M AB_2, 1.5×10^{-3} M XY

188. Arrange the following in increasing order of solubility product : **[NCERT : P L-204 | Jan. 28, 2025 (II)]**

$Ca(OH)_2$, AgBr, PbS, HgS

(a) PbS < HgS < $Ca(OH)_2$ < AgBr
(b) HgS < PbS < AgBr < $Ca(OH)_2$
(c) $Ca(OH)_2$ < AgBr < HgS < PbS
(d) HgS < AgBr < PbS < $Ca(OH)_2$

189. Ksp for $Cr(OH)_3$ is 1.6×10^{-30}. What is the molar solubility of this salt in water? **[NCERT : P L-204 | Jan. 24, 2025 (I)]**

(a) $\sqrt[4]{\dfrac{1.6 \times 10^{-30}}{27}}$ (b) $\dfrac{1.8 \times 10^{-30}}{27}$

(c) $\sqrt[5]{1.8 \times 10^{-30}}$ (d) $\sqrt[2]{1.6 \times 10^{-30}}$

190. Which of the following happens when NH_4OH is added gradually to the solution containing 1M A^{2+} and 1M B^{3+} ions ? **[NCERT : P L-204 | Jan. 23, 2025 (I)]**

Given: $K_{sp}[A(OH)_2] = 9 \times 10^{-10}$ and
$K_{sp}[B(OH)_3] = 27 \times 10^{-18}$ at 298 K.

(a) $B(OH)_3$ will precipitate before $A(OH)_2$
(b) $A(OH)_2$ and $B(OH)_3$ will precipitate together
(c) $A(OH)_2$ will precipitate before $B(OH)_3$
(d) Both $A(OH)_2$ and $B(OH)_3$ do not show precipitation with NH_4OH

191. The molar solubility(s) of zirconium phosphate with molecular formula $(Zr^{4+})_3\left(PO_4^{3-}\right)_4$ is given by relation:

[NCERT : P L-204 | Jan. 22, 2025 (II)]

(a) $\left(\dfrac{K_{sp}}{6912}\right)^{\frac{1}{7}}$ (b) $\left(\dfrac{K_{sp}}{5348}\right)^{\frac{1}{6}}$

(c) $\left(\dfrac{K_{sp}}{8435}\right)^{\frac{1}{7}}$ (d) $\left(\dfrac{K_{sp}}{9612}\right)^{\frac{1}{3}}$

192. For a sparingly soluble salt AB_2, the equilibrium concentrations of A^{2+} ions and B^- ions are 1.2×10^{-4} M and 0.24×10^{-3} M, respectively. The solubility product of AB_2 is : **[April 9, 2024 (II)]**

(a) 0.069×10^{-12} (b) 6.91×10^{-12}
(c) 0.276×10^{-12} (d) 27.65×10^{-12}

193. Given below are two statements:

[NCERT : P L-202 | April 8, 2024 (II)]

Statement (I): A Buffer solution is the mixture of a salt and an acid or a base mixed in any particular quantities.

Statement (II): Blood is naturally occurring buffer solution whose pH is maintained by $H_2CO_3/\ HCO_3^{\ominus}$ concentrations.

In the light of the above statements, choose the **correct** answer from the options given below:

(a) **Statement I** is false but **Statement II** is true
(b) Both **Statement I** and **Statement II** is true
(c) Both **Statement I** and **Statement II** is false
(d) **Statement I** is true but **Statement II** is false

194. Given below are two statements:

[NCERT : P L-200 | April 5, 2024 (II)]

Statement I : On passing HCl(g) through a saturated solution of $BaCl_2$, at room temperature white turbidity appears.

Statement II : When HCl gas is passed through a saturated solution of NaCl, sodium chloride is precipitated due to common ion effect.

In the light of the above statements, choose the **most appropriate** answer from the options given below:

(a) **Statement I** is correct but **Statement II** is incorrect
(b) Both **Statement I** and **Statement II** are incorrect
(c) **Statement I** is incorrect but **Statement II** is correct
(d) Both **Statement I** and **Statement II** are correct

195. Solubility of calcium phosphate (molecular mass, M) in water is Wg per 100 mL at 25° C. Its solubility product at 25°C will be approximately.

[NCERT : P L-204 | Feb. 1, 2024 (II)]

(a) $10^7\left(\frac{W}{M}\right)^3$ (b) $10^3\left(\frac{W}{M}\right)^5$

(c) $10^7\left(\frac{W}{M}\right)^5$ (d) $10^5\left(\frac{W}{M}\right)^5$

196. The pH at which Mg $(OH)_2$ $[K_{sp} = 1 \times 10^{-11}]$ begins to precipitate from a solution containing 0.10 M Mg^{2+} ions is ________ .

[NCERT : P L-204 | NV, Jan. 30, 2024 (I)]

197. The pH of an aqueous solution containing 1M benzoic acid ($pK_a = 4.20$) and 1M sodium benzoate is 4.5. The volume of benzoic acid solution in 300 mL of this buffer solution is ______ mL. (given : log 2 = 0.3)

[NCERT : P L-202 | NV, Jan. 30, 2024 (II)]

198. 20 mL of 0.1 M NaOH is added to 50 mL of 0.1 M acetic acid solution. The pH of the resulting solution is ________ $\times 10^{-2}$ (Nearest integer)

Given : pKa $(CH_3COOH) = 4.76$
log 2 = 0.30
log 3 = 0.48 **[NCERT : P L-203 | NV, April 13, 2023 (II)]**

199. 25.0 mL of 0.050 M $(Ba(NO_3)_2)$ is mixed with 25.0 mL of 0.020 M NaF. K_{sp} of BaF_2 is 0.5×10^{-6} at 298 K. The ratio of $[Ba^{2+}][F^-]^2$ and K_{sp} is ________ (Nearest integer)

[NCERT : P L-204 | NV, April 13, 2023 (I)]

200. The solubility product of $BaSO_4$ is 1×10^{-10} at 298K. The solubility of $BaSO_4$ in 0.1 M K_2SO_4(aq) solution is ______ $\times 10^{-9}$ g L^{-1} (nearest integer).

Given : Molar mass of $BaSO_4$ is 233 g mol^{-1}

[NCERT : P L-204 | NV, April 08, 2023 (II)]

201. At 298 K, the solubility of silver chloride in water is 1.434×10^{-3} g L^{-1}. The value of $-\log K_{sp}$ for silver chloride is ______.

(Given mass of Ag is 107.9 g mol^{-1} and mass of Cl is 35.5 g mol^{-1}) **[NCERT : P L-204 | NV, Jan. 31, 2023 (II)]**

202. The dissociation constant of acetic is $x \times 10^{-5}$.

When 25 mL of 0.2 M CH_3COONa solution is mixed with 25 mL of 0.02 M CH_3COOH solution, the *pH* of the resultant solution is found to be equal to 5. The value of x is ______. **[NCERT : P L-202 | NV, Jan. 24, 2023 (I)]**

203. If the pKa of lactic acid is 5, then the pH of 0.005 M calcium lactate solution at 25° C is ________ $\times 10^{-1}$ (Nearest integer) **[NCERT : P L-202 | NV, Jan. 24, 2023 (II)]**

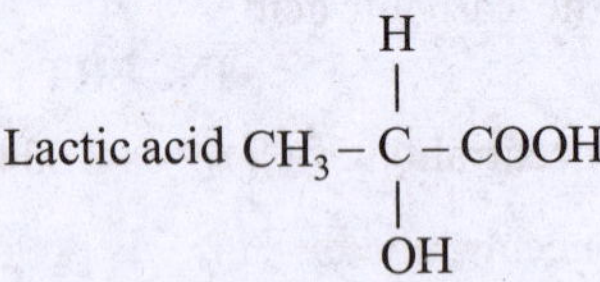

204. If the solubility product of PbS is 8×10^{-28}, then the solubility of PbS in pure water at 298 K is $x \times 10^{-16}$ mol L^{-1}. The value of x is ________. (Nearest integer)

[Given $\sqrt{2} = 1.41$]

[NCERT : P L-204 | NV, July 29, 2022 (I); Same Feb. 25, 2021 (II)]

205. At 310 K, the solubility of CaF_2 in water is 2.34×10^{-3} g / 100 mL. The solubility product of CaF_2 is _____ $\times 10^{-8}$ $(mol/L)^3$. (Given molar mass : CaF_2 = 78 g mol^{-1})

[NCERT : P L-204 | NV, July 27, 2022 (I)]

206. Class XII students were asked to prepare one litre of buffer solution of pH 8.26 by their chemistry teacher. The amount of ammonium chloride to be dissolved by the student in 0.2 M ammonia solution to make one litre of the buffer is (Given pK_b (NH_3) = 4.74; Molar mass of NH_3 = 17 g mol^{-1}; Molar mass of NH_4Cl = 53.5 g mol^{-1})

[NCERT : P L-202 | July 26, 2022 (II)]

(a) 53.5 g (b) 72.3 g
(c) 107.0 g (d) 126.0 g

207. The solubility of AgCl will be maximum in which of the following ? **[NCERT : P L-204 | June 29, 2022 (I)]**

(a) 0.01 M KCl (b) 0.01 M HCl
(c) 0.01 M $AgNO_3$ (d) Deionised water

208. A student needs to prepare a buffer solution of propanoic acid and its sodium salt with pH 4. The ratio of $\frac{[CH_3CH_2COO^-]}{[CH_3CH_2COOH]}$ required to make buffer is ....

Given : $K_a(CH_3CH_2COOH) = 1.3 \times 10^{-5}$

[NCERT : P L-202 | June 28, 2022 (II)]

(a) 0.03 (b) 0.13 (c) 0.23 (d) 0.33

209. 50 mL of 0.1 M CH_3COOH is being titrated against 0.1 M NaOH. When 25 mL of NaOH has been added, the pH of the solution will be ________ $\times 10^{-2}$. (Nearest integer)

(Given : pK_a (CH_3COOH) = 4.76)

log 2 = 0.30   log 3 = 0.48

log 5 = 0.69   log 7 = 0.84   log 11 = 1.04

[NCERT : P L-202 | NV, June 26, 2022 (I)]

210. The K_{sp} for bismuth sulphide (Bi_2S_3) is 1.08×10^{-73}. The solubility of Bi_2S_3 in mol L^{-1} at 298 K is

[NCERT : P L-204| June 25, 2022 (II)]

(a) 1.0×10^{-15} (b) 2.7×10^{-12}

(c) 3.2×10^{-10} (d) 4.2×10^{-8}

211. The molar solubility of $Zn(OH)_2$ in 0.1 M NaOH solution is $x \times 10^{-18}$ M. The value of x is _____ (Nearest integer)

(Given : The solubility product of $Zn(OH)_2$ is 2×10^{-20})

[NCERT : P L-204 | NV, Sep. 1, 2021 (II)]

212. A_3B_2 is a sparingly soluble salt of molar mass M (g mol^{-1}) and solubility x g L^{-1}. The solubility product satisfies $K_{sp} = a\left(\frac{x}{M}\right)^5$. The value of a is ________.

(Integer answer)

[NCERT : P L-204 | NV, Aug. 31, 2021 (I)]

213. The OH^- concentration in a mixture of 5.0 mL of 0.0504 M NH_4Cl and 2 mL of 0.0210 M NH_3 solution is $x \times 10^{-6}$ M. The value of x is ______. (Nearest integer)

[Given $K_w = 1 \times 10^{-14}$ and $K_b = 1.8 \times 10^{-5}$]

[NCERT : P L-202 | NV, Aug. 26, 2021 (I)]

214. In order to prepare a buffer solution of pH 5.74, sodium acetate is added to acetic acid. If the concentration of acetic acid in the buffer is 1.0 M, the concentration of sodium acetate in the buffer is ______ M.

(Rounded off to the Nearest Integer).

[Given : pK_a (acetic acid) = 4.74]

[NCERT : P L-202 | NV, March 18, 2021 (I)]

215. The solubility of $CdSO_4$ in water is 8.0×10^{-4} mol L^{-1}. Its solubility in 0.01 M H_2SO_4 solution is ________ $\times 10^{-6}$ mol L^{-1}. (Round off to the nearest integer).

(Assume that solubility is much less than 0.01 M)

[NCERT : P L-204 | NV, March 18, 2021 (II)]

216. Two salts A_2X and MX have the same value of solubility product of 4.0×10^{-12}. The ratio of their molar solubilities, i.e., $\frac{s(A_2X)}{s(MX)}$ = _____. (Round off to the Nearest Integer)

[NCERT : P L-204 | NV, March 16, 2021 (I)]

217. The solubility of $Ca(OH)_2$ in water is :

[Given : The solubility product of $Ca(OH)_2$ in water = 5.5×10^{-6}]

[NCERT : P L-204 | Feb. 25, 2021 (II)]

(a) 1.77×10^{-2} (b) 1.11×10^{-2}

(c) 1.77×10^{-6} (d) 1.11×10^{-6}

218. The solubility of AgCN in a buffer solution of pH = 3 is x. The value of x is:

[Assume: No cyano complex is formed; K_{sp} (AgCN) = 2.2×10^{-16} and K_a (HCN) = 6.2×10^{-10}]

[NCERT : P L-204 | Feb. 25, 2021 (I)]

(a) 1.9×10^{-5} (b) 1.6×10^{-6}

(c) 2.2×10^{-16} (d) 0.625×10^{-6}

219. The solubility product of PbI_2 is 8.0×10^{-9}. The solubility of lead iodide in 0.1 molar solution of lead nitrate is $x \times 10^{-6}$ mol/L. The value of x is ______. (Rounded off to the nearest integer) [Given : $\sqrt{2} = 1.41$]

[NCERT : P L-204| NV, Feb. 24, 2021 (II)]

220. If the solubility product of AB_2 is $3.20 \times 10^{-11} M^3$, then the solubility of AB_2 in pure water is ______ $\times 10^{-4}$ mol L^{-1}.

[Assuming that neither kind of ion reacts with water]

[NCERT : P L-204 | NV, Sep. 06, 2020 (II)]

221. An acidic buffer is obtained on mixing:

[Sep. 03, 2020 (I)]

(a) 100 mL of 0.1 M CH_3COOH and 100 mL of 0.1 M NaOH

(b) 100 mL of 0.1 M HCl and 200 mL of 0.1 M NaCl

(c) 100 mL of 0.1 M CH_3COOH and 200 mL of 0.1 M NaOH

(d) 100 mL of 0.1 M HCl and 200 mL of 0.1 M CH_3COONa

222. For the following **Assertion** and **Reason**, the correct option is **[NCERT : P L-204 | Sep. 02, 2020 (I)]**

Assertion (A) : When Cu (II) and sulphide ions are mixed, they react together extremely quickly to give a solid.

Reason (R) : The equilibrium constant of $Cu^{2+}(aq) + S^{2-}(aq) \rightleftharpoons CuS(s)$ is high because the solubility product is low.

(a) **(A)** is false and **(R)** is true

(b) Both **(A)** and **(R)** are false

(c) Both **(A)** and **(R)** are true but **(R)** is not the explanation for **(A)**

(d) Both **(A)** and **(R)** are true and **(R)** is the explanation for **(A)**

223. The K_{sp} for the following dissociation is 1.6×10^{-5}

$$PbCl_2(s) \rightleftharpoons Pb^{2+}(aq) + 2Cl^-(aq)$$

Which of the following choices is correct for a mixture of 300 mL 0.134 M $Pb(NO_3)_2$ and 100 mL 0.4 M NaCl?

[NCERT : P L-204 | Jan. 09, 2020 (I)]

(a) Not enough data provided

(b) $Q < K_{sp}$

(c) $Q > K_{sp}$

(d) $Q = K_{sp}$

224. The solubility product of $Cr(OH)_3$ at 298 K is 6.0×10^{-31}. The concentration of hydroxide ions in a saturated solution of $Cr(OH)_3$ will be: **[NCERT : P L-204 | Jan. 09, 2020 (II)]**

(a) $(2.22 \times 10^{-31})^{1/4}$ (b) $(18 \times 10^{-31})^{1/4}$
(c) $(18 \times 10^{-31})^{1/2}$ (d) $(4.86 \times 10^{-29})^{1/4}$

225. The stoichiometry and solubility product of a salt with the solubility curve given below is, respectively: **[NCERT : P L-204 | Jan. 08, 2020 (I)]**

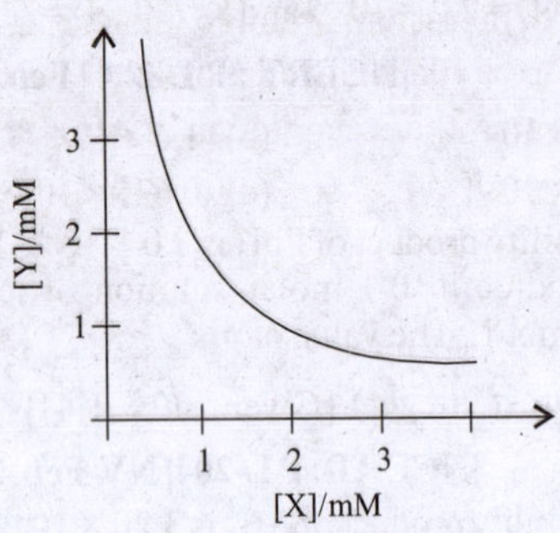

(a) $X_2Y, 2 \times 10^{-9} M^3$ (b) $XY_2, 4 \times 10^{-9} M^3$
(c) $XY_2, 1 \times 10^{-9} M^3$ (d) $XY, 2 \times 10^{-6} M^3$

226. The molar solubility of $Cd(OH)_2$ is 1.84×10^{-5} M in water. The expected solubility of $Cd(OH)_2$ in a buffer solution of pH = 12 is : **[NCERT : P L-204 | April 12, 2019 (II)]**

(a) 1.84×10^{-9} M (b) $\frac{2.49}{1.84} \times 10^{-9}$ M
(c) 6.23×10^{-11} M (d) 2.49×10^{-10} M

227. What is the molar solubility of $Al(OH)_3$ in 0.2 M NaOH solution ? Given that, solubility product of $Al(OH)_3 = 2.4 \times 10^{-24}$: **[NCERT : P L-204 | April 12, 2019 (I)]**

(a) 3×10^{-19} (b) 12×10^{-21}
(c) 3×10^{-22} (d) 12×10^{-23}

228. The pH of a 0.02 M NH_4Cl solution will be [given $K_b(NH_4OH) = 10^{-5}$ and log 2 = 0.301] **[NCERT : P L-202 | April 10, 2019 (II)]**

(a) 2.65 (b) 4.35
(c) 4.65 (d) 5.35

229. If solubility product of $Zr_3(PO_4)_4$ is denoted by K_{sp} and its molar solubility is denoted by S, then which of the following relation between S and K_{sp} is correct ? **[NCERT : P L-204 | April 8, 2019 (I)]**

(a) $S = \left(\frac{K_{sp}}{144}\right)^{1/6}$ (b) $S = \left(\frac{K_{sp}}{6912}\right)^{1/7}$
(c) $S = \left(\frac{K_{sp}}{929}\right)^{1/9}$ (d) $S = \left(\frac{K_{sp}}{216}\right)^{1/7}$

230. 20 mL of 0.1 M H_2SO_4 solution is added to 30 mL of 0.2 M NH_4OH solution. The pH of the resultant mixture is: [pK_b of $NH_4OH = 4.7$]. **[NCERT : P L-202| Jan. 9, 2019 (I)]**

(a) 5.2 (b) 9.0
(c) 5.0 (d) 9.4

231. If K_{sp} of Ag_2CO_3 is 8×10^{-12}, the molar solubility of Ag_2CO_3 in 0.1 M $AgNO_3$ is : **[NCERT : P L-204 Jan. 12, 2019 (II)]**

(a) 8×10^{-12} M (b) 8×10^{-11} M
(c) 8×10^{-10} M (d) 8×10^{-13} M

232. A mixture of 100 m mol of $Ca(OH)_2$ and 2 g of sodium sulphate was dissolved in water and the volume was made up to 100 mL. The mass of calcium sulphate formed and the concentration of OH^- in resulting solution, respectively, are : (Molar mass of $Ca(OH)_2$, Na_2SO_4 and $CaSO_4$ are 74, 143 and 136 g mol^{-1}, respectively; K_{sp} of $Ca(OH)_2$ is 5.5×10^{-6}) **[NCERT : P L-204 | Jan. 10, 2019 (I)]**

(a) 1.9 g, 0.28 mol L^{-1} (b) 13.6 g, 0.28 mol L^{-1}
(c) 1.9 g, 0.14 mol L^{-1} (d) 13.6 g, 0.14 mol L^{-1}

233. Which of the following salts is the most basic in aqueous solution? **[NCERT : P L-202 | 2018]**

(a) $Al(CN)_3$ (b) CH_3COOK
(c) $FeCl_3$ (d) $Pb(CH_3COO)_2$

234. An aqueous solution contains 0.10 M H_2S and 0.20 M HCl. If the equilibrium constants for the formation of HS^- from H_2S is 1.0×10^{-7} and that of S^{2-} from HS^- ions is 1.2×10^{-13} then the concentration of S^{2-} ions in aqueous solution is : **[2018]**

(a) 5×10^{-8} (b) 3×10^{-20}
(c) 6×10^{-21} (d) 5×10^{-19}

235. An aqueous solution contains an unknown concentration of Ba^{2+}. When 50 mL of a 1 M solution of Na_2SO_4 is added, $BaSO_4$ just begins to precipitate. The final volume is 500 mL. The solubility product of $BaSO_4$ is 1×10^{-10}. What is the original concentration of Ba^{2+}? **[NCERT : P L-204 | 2018]**

(a) 5×10^{-9} M (b) 2×10^{-9} M
(c) 1.1×10^{-9} M (d) 1.0×10^{-10} M

236. The minimum volume of water required to dissolve 0.1g lead (II) chloride to get a saturated solution (K_{sp} of $PbCl_2 = 3.2 \times 10^{-8}$; atomic mass of Pb = 207 u) is : **[NCERT : P L-204 | Online April 15, 2018 (I)]**

(a) 1.798 L (b) 0.36 L
(c) 17.95 L (d) 0.18 L

237. Addition of sodium hydroxide solution to a weak acid (HA) results in a buffer of pH 6. If ionisation constant of HA is 10^{-5}, the ratio of salt to acid concentration in the buffer solution will be : **[NCERT : P L-202 | Online April 8, 2017]**

(a) 4 : 5 (b) 1 : 10 (c) 10 : 1 (d) 5 : 4

238. In some solutions, the concentration of H_3O^+ remains constant even when small amounts of strong acid or strong base are added to them. These solutions are known as: **[NCERT : P L-202 | Online April 11, 2014]**

(a) Ideal solutions (b) Colloidal solutions
(c) True solutions (d) Buffer solutions

239. Consider the following equilibrium

$$AgCl\downarrow + 2NH_3 \rightleftharpoons \left[Ag(NH_3)_2\right]^+ + Cl^-$$

White precipitate of AgCl appears on adding which of the following? **[NCERT : P L-204 | Online April 11, 2014]**

(a) NH_3 (b) aqueous NaCl
(c) aqueous HNO_3 (d) aqueous NH_4Cl

240. Zirconium phosphate $[Zr_3(PO_4)_4]$ dissociates into three zirconium cations of charge + 4 and four phosphate anions of charge – 3. If molar solubility of zirconium phosphate is denoted by S and its solubility product by K_{sp} then which of the following relationship between S and K_{sp} is correct? **[NCERT : P L-204 | Online April 19, 2014]**

(a) $S = \{K_{sp}/(6912)^{1/7}\}$ (b) $S = \{K_{sp}/144\}^{1/7}$
(c) $S = \{K_{sp}/6912\}^{1/7}$ (d) $S = \{K_{sp}/6912\}^7$

241. Solid $Ba(NO_3)_2$ is gradually dissolved in a 1.0×10^{-4} M Na_2CO_3 solution. At which concentration of Ba^{2+}, precipitate of $BaCO_3$ begins to form ? (K_{sp} for $BaCO_3 = 5.1 \times 10^{-9}$) **[NCERT : P L-204 | 2009, Online April 9, 2013]**

(a) 5.1×10^{-5} M (b) 7.1×10^{-8} M
(c) 4.1×10^{-5} M (d) 8.1×10^{-7} M

242. Which one of the following arrangements represents the correct order of solubilities of sparingly soluble salts Hg_2Cl_2, $Cr_2(SO_4)_3$, $BaSO_4$ and $CrCl_3$ respectively ? **[NCERT : P L-204 | Online April 22, 2013]**

(a) $BaSO_4 > Hg_2Cl_2 > Cr_2(SO_4)_3 > CrCl_3$
(b) $BaSO_4 > Hg_2Cl_2 > CrCl_3 > Cr_2(SO_4)_3$
(c) $BaSO_4 > CrCl_3 > Hg_2Cl_2 > Cr_2(SO_4)_3$
(d) $Hg_2Cl_2 > BaSO_4 > CrCl_3 > Cr_2(SO_4)_3$

243. If K_{sp} of CaF_2 at 25 °C is 1.7×10^{-10}, the combination amongst the following which gives a precipitate of CaF_2 is **[NCERT : P L-204 | Online May 19, 2012]**

(a) 1×10^{-2} M Ca^{2+} and 1×10^{-3} M F^-
(b) 1×10^{-4} M Ca^{2+} and 1×10^{-4} M F^-
(c) 1×10^{-2} M Ca^{2+} and 1×10^{-5} M F^-
(d) 1×10^{-3} M Ca^{2+} and 1×10^{-5} M F^-

244. The solubility of PbI_2 at 25 °C is 0.7 g L^{-1}. The solubility product of PbI_2 at this temperature is (molar mass of PbI_2 = 461.2 g mol^{-1}) **[NCERT : P L-204 | 2003, Online May 26, 2012]**

(a) 1.40×10^{-9} (b) 0.14×10^{-9}
(c) 140×10^{-9} (d) 14.0×10^{-9}

245. The solubility (in mol L^{-1}) of AgCl ($K_{sp} = 1.0 \times 10^{-10}$) in a 0.1 M KCl solution will be **[NCERT : P L-204 | Online May 7, 2012]**

(a) 1.0×10^{-9} (b) 1.0×10^{-10}
(c) 1.0×10^{-5} (b) 1.0×10^{-11}

246. The K_{sp} for $Cr(OH)_3$ is 1.6×10^{-30}. The solubility of this compound in water is : **[NCERT : P L-204 | 2011RS]**

(a) $\sqrt[4]{1.6 \times 10^{-30}}$ (b) $\sqrt[4]{1.6 \times 10^{-30}/27}$
(c) $1.6 \times 10^{-30/27}$ (d) $\sqrt[2]{1.6 \times 10^{-30}}$

247. Solubility product of silver bromide is 5.0×10^{-13}. The quantity of potassium bromide (molar mass taken as 120 g mol^{-1}) to be added to 1 litre of 0.05 M solution of silver nitrate to start the precipitation of AgBr is **[NCERT : P L-204 | 2010]**

(a) 1.2×10^{-10} g (b) 1.2×10^{-9} g
(c) 6.2×10^{-5} g (d) 5.0×10^{-8} g

248. At 25 °C, the solubility product of $Mg(OH)_2$ is 1.0×10^{-11}. At which pH, will Mg^{2+} ions start precipitating in the form of $Mg(OH)_2$ from a solution of 0.001 M Mg^{2+} ions? **[NCERT : P L-204 | 2010]**

(a) 9 (b) 10
(c) 11 (d) 8

249. Solid $Ba(NO_3)_2$ is gradually dissolved in a 1.0×10^{-4} M Na_2CO_3 solution. At what concentration of Ba^{2+} will a precipitate begin to form? (K_{sp} for for $BaCO_3 = 5.1 \times 10^{-9}$) **[NCERT : P L-204 | 2009]**

(a) 5.1×10^{-5} M (b) 8.1×10^{-8} M
(c) 8.1×10^{-7} M (d) 4.1×10^{-5} M

250. In a saturated solution of the sparingly soluble strong electrolyte $AgIO_3$ (molecular mass = 283) the equilibrium which sets in is $AgIO_3(s) \rightleftharpoons Ag^+(aq) + IO_3^-(aq)$. If the solubility product constant K_{sp} of $AgIO_3$ at a given temperature is 1.0×10^{-8}, what is the mass of $AgIO_3$ contained in 100 mL of its saturated solution? **[NCERT : P L-204 | 2007]**

(a) 1.0×10^{-4} g (b) 28.3×10^{-2} g
(c) 2.83×10^{-3} g (d) 1.0×10^{-7} g.

251. The pK_a of a weak acid (HA) is 4.5. The pOH of an aqueous buffer solution of HA in which 50% of the acid is ionized is **[NCERT : P L-202 | 2007]**

(a) 7.0 (b) 4.5
(c) 2.5 (d) 9.5

252. The solubility product of a salt having general formula MX_2, in water, is : 4×10^{-12}. The concentration of M^{2+} ions in the aqueous solution of the salt is **[NCERT : P L-204 | 2005]**

(a) 4.0×10^{-10} M (b) 1.6×10^{-4} M
(c) 1.0×10^{-4} M (d) 2.0×10^{-6} M

253. The molar solubility (in mol L^{-1}) of a sparingly soluble salt MX_4 is 's'. The corresponding solubility product is K_{sp}. 's' is given in term of K_{sp} by the relation : **[NCERT : P L-204 | 2004]**

(a) $s = (256\,K_{sp})^{1/5}$ (b) $s = (128\,K_{sp})^{1/4}$

(c) $s = (K_{sp}/128)^{1/4}$ (d) $s = (K_{sp}/256)^{1/5}$

254. The solubility in water of a sparingly soluble salt AB_2 is 1.0×10^{-5} mol L^{-1}. Its solubility product will be **[NCERT : P L-202 | 2003]**

(a) 4×10^{-10} (b) 1×10^{-15}

(c) 1×10^{-10} (d) 4×10^{-15}

255. 1 M NaCl and 1 M HCl are present in an aqueous solution. The solution is **[NCERT : P L-202 | 2002]**

(a) not a buffer solution with pH < 7

(b) not a buffer solution with pH > 7

(c) a buffer solution with pH < 7

(d) a buffer solution with pH > 7.

256. Let the solubility of an aqueous solution of $Mg(OH)_2$ be x, then its K_{sp} is **[NCERT : P L-204 | 2002]**

(a) $4x^3$ (b) $108x^5$

(c) $27x^4$ (d) $9x$.

ANSWER KEYS

1	(5)	27	(12)	53	(d)	79	(c)	105	(c)	131	(12)	157	(a)	183	(c)	209	(476)	235	(c)
2	(c)	28	(1)	54	(16)	80	(a)	106	(d)	132	(b)	158	(c)	184	(c)	210	(a)	236	(d)
3	(d)	29	(2)	55	(b)	81	(c)	107	(a)	133	(b)	159	(a)	185	(d)	211	(2)	237	(c)
4	(c)	30	(4)	56	(d)	82	(c)	108	(c)	134	(c)	160	(a)	186	(48)	212	(108)	238	(d)
5	(b)	31	(2)	57	(d)	83	(d)	109	(b)	135	(c)	161	(d)	187	(c)	213	(3)	239	(c)
6	(74)	32	(2)	58	(a)	84	(a)	110	(a)	136	(11)	162	(b)	188	(b)	214	(10)	240	(c)
7	(c)	33	(c)	59	(c)	85	(c)	111	(d)	137	(d)	163	(d)	189	(a)	215	(64)	241	(a)
8	(d)	34	(a)	60	(b)	86	(d)	112	(c)	138	(6021)	164	(a)	190	(a)	216	(50)	242	(b)
9	(b)	35	(29)	61	(a)	87	(b)	113	(c)	139	(b)	165	(25)	191	(a)	217	(b)	243	(a)
10	(962)	36	(125)	62	(a)	88	(c)	114	(d)	140	(b)	166	(7)	192	(b)	218	(a)	244	(d)
11	(d)	37	(b)	63	(d)	89	(N)	115	(b)	141	(37)	167	(7)	193	(a)	219	(141)	245	(a)
12	(b)	38	(1107)	64	(b)	90	(700)	116	(d)	142	(c)	168	(a)	194	(d)	220	(2.0)	246	(b)
13	(b)	39	(4)	65	(c)	91	(a)	117	(d)	143	(b)	169	(27)	195	(c)	221	(d)	247	(b)
14	(c)	40	(6)	66	(a)	92	(57)	118	(a)	144	(10.60)	170	(c)	196	(9)	222	(d)	248	(b)
15	(c)	41	(182)	67	(a)	93	(163)	119	(b)	145	(5.23)	171	(d)	197	(100)	223	(c)	249	(a)
16	(d)	42	(2)	68	(c)	94	(37)	120	(a)	146	(b)	172	(2)	198	(458)	224	(b)	250	(c)
17	(d)	43	(1400)	69	(b)	95	(10)	121	(b)	147	(b)	173	(1)	199	(5)	225	(b)	251	(d)
18	(d)	44	(16)	70	(d)	96	(a,d)	122	(3)	148	(c)	174	(7.0)	200	(233)	226	(d)	252	(c)
19	(b)	45	(875)	71	(b)	97	(3)	123	(d)	149	(b)	175	(d)	201	(10)	227	(c)	253	(d)
20	(2)	46	(25)	72	(a)	98	(6)	124	(a)	150	(d)	176	(c)	202	(10)	228	(d)	254	(d)
21	(417)	47	(2)	73	(d)	99	(2735)	125	(b)	151	(d)	177	(d)	203	(85)	229	(b)	255	(a)
22	(44)	48	(172)	74	(d)	100	(747)	126	(d)	152	(b)	178	(c)	204	(282)	230	(b)	256	(a)
23	(1)	49	(354)	75	(b)	101	(20)	127	(9000)	153	(d)	179	(b)	205	(0)	231	(c)		
24	(3)	50	(72)	76	(c)	102	(166)	128	(186)	154	(d)	180	(d)	206	(c)	232	(a)		
25	(a)	51	(5)	77	(c)	103	(20)	129	(5)	155	(a)	181	(b)	207	(d)	233	(b)		
26	(48.6)	52	(c)	78	(d)	104	(c)	130	(9079)	156	(d)	182	(a)	208	(b)	234	(b)		

Redox Reactions

TREND ANALYSIS
JEE MAIN (2025-2021)

	2025				2024				2023				2022				2021			
	Session II		Session I		Session II		Session I		Session II		Session I		Session II		Session I		Session II		Session I	
	MCQs	NVQs	MCQs	NVQs	MCQs	NVQs	MCQs	NVQs	MCQs	NVQs	MCQs	NVQs	MCQs	NVQs	MCQs	NVQs	MCQs	NVQs	MCQs	NVQs
Topic 1	1	–	–	–	1	–	1	1	–	–	–	–	–	–	–	–	–	–	–	–
Topic 2	–	–	–	–	–	–	–	1	–	2	–	–	–	1	–	–	–	–	–	–
Topic 3	–	–	2	–	2	–	4	3	2	1	–	–	1	–	1	–	1	–	–	3
Topic 4	–	–	–	–	–	–	–	–	–	–	–	–	–	–	–	–	–	–	–	–
Total	1	–	2	–	3	0	5	5	2	3	–	–	1	1	1	–	1	–	–	3

How to Code Qns for Better & Faster Revision

Type 1: Able to solve (No code Req)
Type 2: Solved with struggle or by seeing solution **(Use code NR means Needs Revision or orange stickers)**
Type 3: Tricky/ Could not solve on my own. Need help from Teacher/ Friend. **(Use code TQ means Tricky Question or red stickers)**

You can either hand write the code in boxes or else paste coloured stickers given at end of book.

Topic 1 - Oxidation and Reduction Reactions

1. The group 14 elements A and B have the first ionisation enthalpy values of 708 and 715 kJ mol^{-1} respectively. The above values are lowest among their group members. The nature of their ions A^{2+} B^{4+} respectively is

[NCERT : P L-237 | April 7, 2025 (I)]

(a) both reducing (b) both oxidising
(c) reducing and oxidising (d) oxidising and reducing

2. Match **List-I** with **List-II**.

[NCERT : P L-242, 243 | April 6, 2024 (II)]

List-I (Reaction)		**List-II (Type of redox reaction)**
(A) $N_2(g) + O_2(g) \rightarrow 2NO(g)$	(I)	Decomposition
(B) $2Pb(NO_3)_2(s) \rightarrow 2PbO(s) + 4NO_2(g) + O_2(g)$	(II)	Displacement
(C) $2Na(s) + 2H_2O(l) \rightarrow 2NaOH(aq.) + H_2(g)$	(III)	Disproportionation
(D) $2NO_2(g) + 2^-OH(aq.) \rightarrow NO_2^-(aq.) + NO_3^-(aq.) + H_2O(l)$	(IV)	Combination

Choose the **correct** answer from the options given below:
(a) (A)-(I), (B)-(II), (C)-(III), (D)-(IV)
(b) (A)-(III), (B)-(II), (C)-(I), (D)-(IV)
(c) (A)-(II), (B)-(III), (C)-(IV), (D)-(I)
(d) (A)-(IV), (B)-(I), (C)-(II), (D)-(III)

3. 1 mole of PbS is oxidised by "X" moles of O_3 to get "Y" moles of O_2. X + Y = __________.

[NCERT : P L-239 | NV, Jan. 27, 2024 (II)]

4. Which of the following cannot function as an oxidising agent? **[NCERT : P L-241 | Jan. 27, 2024 (II)]**

(a) N^{3-} (b) SO_4^{2-} (c) BrO_3^- (d) MnO_4^-

5. The compound that cannot act both as oxidising and reducing agent is

[NCERT : P L-237 | NV, Jan. 09, 2020 (I)]

(a) H_3PO_4 (b) HNO_2 (c) H_2SO_3 (d) H_2O_2

6. The redox reaction among the following is: **[NCERT : P L-237 | Jan. 07, 2020 (II)]**
(a) formation of ozone from atmospheric oxygen in the presence of sunlight
(b) reaction of $[Co(H_2O)_6]Cl_3$ with $AgNO_3$
(c) reaction of H_2SO_4 with NaOH
(d) combination of dinitrogen with dioxygen at 2000 K

7. Which of the following reactions is an example of a redox reaction? **[NCERT : P L-237 | 2017, 2002 (S)]**
(a) $XeF_4 + O_2F_2 \rightarrow XeF_6 + O_2$
(b) $XeF_2 + PF_5 \rightarrow [XeF]^+ PF_6^-$
(c) $XeF_6 + H_2O \rightarrow XeOF_4 + 2\,HF$
(d) $XeF_6 + 2\,H_2O \rightarrow XeO_2F_2 + 4\,HF$

8. Copper becomes green when exposed to moist air for a long period. This is due to: **[NCERT : P L-237 | Online April 12, 2014]**
(a) the formation of a layer of cupric oxide on the surface of copper.
(b) the formation of a layer of basic carbonate of copper on the surface of copper.
(c) the formation of a layer of cupric hydroxide on the surface of copper.
(d) the formation of basic copper sulphate layer on the surface of the metal.

9. Which of the following chemical reactions depict the oxidizing behaviour of H_2SO_4? **[NCERT : P L-237 | 2006]**
(a) $NaCl + H_2SO_4 \longrightarrow NaHSO_4 + HCl$
(b) $2\,PCl_5 + H_2SO_4 \longrightarrow 2\,POCl_3 + 2\,HCl + SO_2Cl_2$
(c) $2\,HI + H_2SO_4 \longrightarrow I_2 + SO_2 + 2\,H_2O$
(d) $Ca(OH)_2 + H_2SO_4 \longrightarrow CaSO_4 + 2\,H_2O$

Topic 2 - Oxidation Number

10. From the given list, the number of compounds with + 4 oxidation state of Sulphur ________.
$SO_3, H_2SO_3, SOCl_2, SF_4, BaSO_4, H_2S_2O_7$
[NCERT : P L-239 | NV, Jan. 27, 2024 (I)]

11. In Chromyl chloride, the oxidation state of chromium is (+) ____________
[NCERT : P L-239 | NV, April 15, 2023 (I)]

12. In ammonium–phosphomolybdate, the oxidation state of Mo is ....
[NCERT : P L-239 | NV, April 06, 2023 (I)]

13. On reaction with stronger oxidizing agent like KIO_4, hydrogen peroxide with the evolution of O_2. The oxidation number of I in KIO_4 changes to ______.
[NCERT : P L-239 | NV, July 28, 2022 (II)]

14. The oxidation states of iron atoms in compounds (A), (B) and (C), respectively, are *x*, *y* and *z*. The sum of *x*, *y* and *z* is ________. **[NCERT : P L-239 | NV, Sep. 02, 2020 (I)]**

$Na_4[Fe(CN)_5(NOS)]$ (A) $\quad Na_4[FeO_4]$ (B) $\quad [Fe_2(CO)_9]$ (C)

15. The oxidation states of transition metal atoms in $K_2Cr_2O_7$, $KMnO_4$ and K_2FeO_4, respectively, are *x*, *y* and *z*. The sum of *x*, *y* and *z* is _______.
[NCERT : P L-239 | NV, Sep. 02, 2020 (II)]

16. Oxidation number of potassium in K_2O, K_2O_2 and KO_2, respectively, is: **[NCERT : P L-239 | Jan. 07, 2020 (I)]**
(a) $+2, +1$ and $+\frac{1}{2}$
(b) +1, +1 and +1
(c) +1, +4 and +2
(d) +1, +2 and +4

17. The oxidation states of Cr in $[Cr(H_2O)_6]Cl_3$, $[Cr(C_6H_6)_2]$, and $K_2[Cr(CN)_2(O)_2(O_2)(NH_3)]$ respectively are : **[NCERT : P L-239 | 2018]**
(a) +3, +4, and +6
(b) +3, +2, and +4
(c) +3, 0, and +6
(d) +3, 0, and + 4

18. In which of the following reactions, hydrogen peroxide acts as an oxidizing agent ? **[NCERT : P L-241 | Online April 8, 2017]**
(a) $HOCl + H_2O_2 \rightarrow H_3O^+ + Cl^- + O_2$
(b) $I_2 + H_2O_2 + 2OH^- \rightarrow 2I^- + 2H_2O + O_2$
(c) $2MnO_4^- + 3H_2O_2 \rightarrow 2MnO_2 + 3O_2 + 2H_2O + 2OH^-$
(d) $PbS + 4H_2O_2 \rightarrow PbSO_4 + 4H_2O$

19. Amongst the following, identify the species with an atom in + 6 oxidation state: **[NCERT : P L-239 | Online April 19, 2014]**
(a) $[MnO_4]^-$
(b) $[Cr(CN)_6]^{3-}$
(c) Cr_2O_3
(d) CrO_2Cl_2

20. Oxidation state of sulphur in anions SO_3^{2-}, $S_2O_4^{2-}$ and $S_2O_6^{2-}$ increases in the orders : **[NCERT : P L-239 | Online April 22, 2013]**
(a) $S_2O_6^{2-} < S_2O_4^{2-} < SO_3^{2-}$
(b) $SO_3^{2-} < S_2O_4^{2-} < S_2O_6^{2-}$
(c) $S_2O_4^{2-} < SO_3^{2-} < S_2O_6^{2-}$
(d) $S_2O_4^{2-} < S_2O_6^{2-} < SO_3^{2-}$

21. Consider the reaction:

$$H_2SO_3(aq) + Sn^{4+}(aq) + H_2O(l) \rightarrow Sn^{2+}(aq) + HSO_4^-(aq) + 3H^+(aq)$$

Which of the following statements is correct? **[NCERT : P L-241 | Online April 19, 2014]**
(a) Sn^{4+} is the oxidizing agent because it undergoes oxidation
(b) Sn^{4+} is the reducing agent because it undergoes oxidation
(c) H_2SO_3 is the reducing agent because it undergoes oxidation
(d) H_2SO_3 is the reducing agent because it undergoes reduction

22. Which one of the following cannot function as an oxidising agent ? **[NCERT : P L-241 | Online April 25, 2013]**
(a) I^-
(b) S (s)
(c) $NO_3^-(aq)$
(d) $Cr_2O_7^{2-}$

Topic 3 -Disproportionation and Balancing of Redox Reactions

23. Match the List-I with List-II
[NCERT : P L-242 | Jan. 28, 2025 (I)]

List-I (Redox Reaction)	List-II (Type of Redox Reaction)
(A) $CH_{4(g)} + 2O_{2(g)} \xrightarrow{\Delta} CO_{2(g)} + 2H_2O_{(l)}$	(I) Disproportionation reaction
(B) $2NaH_{(s)} \xrightarrow{\Delta} 2Na_{(s)} + H_{2(g)}$	(II) Combination reaction
(C) $V_2O_{5(s)} + 5Ca_{(s)} \xrightarrow{\Delta} 2V_{(s)} + 5CaO_{(s)}$	(III) Decomposition reaction
(D) $2H_2O_{2(aq)} \xrightarrow{\Delta} 2H_2O_{(l)} + O_{2(g)}$	(IV) Displacement reaction

Choose the correct answer from the options given below :
(a) A-II, B-III, C-IV, D-I (b) A-II, B-III, C-I, D-IV
(c) A-III, B-IV, C-I, D-II (d) A-IV, B-I, C-II, D-III

24. The species which does not undergo disproportionation reaction is : **[NCERT : P L-242 | Jan. 22, 2025 (II)]**
(a) ClO_2^- (b) ClO_4^-
(c) ClO^- (d) ClO_3^-

25. Thiosulphate reacts differently with iodine and bromine in the reaction given below : **[April 8, 2024 (I)]**

$$2S_2O_3^{2-} + I_2 \rightarrow S_4O_6^{2-} + 2I^-$$

$$S_2O_3^{2-} + 5Br_2 + 5H_2O \rightarrow 2SO_4^{2-} + 4Br^- + 10H^+$$

Which of the following statement justifies the above dual behaviour of thiosulphate?
(a) Bromine undergoes oxidation and iodine undergoes reduction by iodine in these reactions
(b) Thiosulphate undergoes oxidation by bromine and reduction by iodine in these reaction
(c) Bromine is a stronger oxidant than iodine
(d) Bromine is a weaker oxidant than iodine

26. Among the following halogens
F_2, Cl_2, Br_2 and I_2
Which can undergo disproportionation reaction?
[NCERT : P L-244 | April 8, 2024 (I)]
(a) Only I_2 (b) Cl_2, Br_2 and I_2
(c) F_2, Cl_2 and Br_2 (d) F_2 and Cl_2

27. In acidic medium, $K_2Cr_2O_7$ shows oxidising action as represented in the half reaction:

$$Cr_2O_7^{2-} + XH^+ + Ye^{\ominus} \rightarrow 2A + ZH_2O$$

X, Y, Z and A are respectively are:
[NCERT : P L-246 | Feb. 1, 2024 (I)]
(a) 8, 6, 4 and Cr_2O_3 (b) 14, 7, 6 and Cr^{3+}
(c) 8, 4, 6 and Cr_2O_3 (d) 14, 6, 7 and Cr^{3+}

28. Which of the following reactions are disproportionation reactions ? **[NCERT : P L-242 | Feb. 1, 2024 (I)]**
(A) $Cu^+ \rightarrow Cu^{2+} + Cu$
(B) $3MnO_4^{2-} + 4H^+ \longrightarrow 2MnO_4^- + MnO_2 + 2H_2O$
(C) $2KMnO_4 \longrightarrow K_2MnO_4 + MnO_2 + O_2$
(D) $2MnO_4^- + 3Mn^{2+} + 2H_2O \longrightarrow 5MnO_2 + 4H^+$
Choose the correct answer from the options given below:
(a) (A), (B) (b) (B), (C), (D)
(c) (A), (B), (C) (d) (A), (D)

29. Given below are two statements :
[NCERT : P L-242 | Jan. 31, 2024 (II)]
Statement I: S_8 solid undergoes disproportionation reaction under alkaline conditions to form S^{2-} and $S_2O_3^{2-}$
Statement II: ClO_4^- can undergo disproportionation reaction under acidic condition.
In the light of the above statements, choose the **most appropriate answer** from the options given below :
(a) Statement I is correct but statement II is incorrect.
(b) Statement I is incorrect but statement II is correct.
(c) Both statement I and statement II are incorrect.
(d) Both statement I and statement II are correct.

30. Number of moles of H^+ ions required by 1 mole of MnO_4^- to oxidise oxalate ion to CO_2 is ____.
[NCERT : P L-246 | NV, Jan. 31, 2024 (II)]

31. $2MnO_4^- + bI^- + cH_2O \rightarrow xI_2 + yMnO_2 + z\overline{O}H$
If the above equation is balanced with integer coefficients, the value of z is ________.
[NCERT : P L-246 | NV, Jan. 30, 2024 (I)]

32. Total number of species from the following which can undergo disproportionation reaction is ________.
$H_2O_2, ClO_3^-, P_4, Cl_2, Ag, Cu^{+1}, F_2, NO_2, K^+$
[NCERT : P L-242 | NV, Jan. 30, 2024 (II)]

33. Chlorine undergoes disproportionation in alkaline medium as shown below: **[NCERT : P L-246 | Jan. 29, 2024 (I)]**
$a\,Cl_2(g) + b\,OH^-(aq) \rightarrow c\,ClO^-(aq) + d\,Cl^-(aq) + \bar{e}\,H_2O(l)$
The values of a, b, c and d in a balanced redox reaction are respectively :
(a) 2, 2, 1 and 3 (b) 1, 2, 1 and 1
(c) 2, 4, 1 and 3 (d) 3, 4, 4 and 2

34. See the following chemical reaction:
[NCERT : P L-246 | NV, April 13, 2023 (II)]

$$Cr_2O_7^{2-} + XH^+ + 6Fe^{2+} \rightarrow YCr^{3+} + 6Fe^{3+} + ZH_2O$$

The sum of X. Y and Z is ________.

35. $2IO_3^- + xI^- + 12H^+ \rightarrow 6I_2 + 6H_2O$
[NCERT : P L-246 | April 08, 2023 (I)]
What is the value of x?
(a) 12 (b) 2 (c) 6 (d) 10

36. Which of the following options are correct for the reaction
$2[Au(CN)_2]^-(aq) + Zn(s) \rightarrow 2Au(s) + [Zn(CN)_4]^{-2}(aq)$
[NCERT : P L-242 | April 06, 2023 (I)]
A. Redox reaction
B. Displacement reaction
C. Decomposition reaction
D. Combination reaction
Choose the correct answer from the options given below:
(a) A and B only (b) A only
(c) C and D only (d) A and D only

37. Which of the given reactions is not an example of disproportionation reaction?
[NCERT : P L-242 | July 26, 2022 (I)]
(a) $2H_2O_2 \rightarrow 2H_2O + O_2$
(b) $2NO_2 + H_2O \rightarrow HNO_3 + HNO_2$
(c) $MnO_4^- + 4H^+ + 3e^- \rightarrow MnO_2 + 2H_2O$
(d) $3MnO_4^{2-} + 4H^+ \rightarrow 2MnO_4^- + MnO_2 + 2H_2O$

38. Which one of the following is an example of disproportionation reaction?
[NCERT : P L-242 | June 26, 2022 (II)]
(a) $3MnO_4^{2-} + 4H^+ \rightarrow 2MnO_4^- + MnO_2 + 2H_2O$
(b) $MnO_4^{2-} + 4H^+ + 4e^- \rightarrow MnO_2 + 2H_2O$
(c) $10I^- + 2MnO_4^- + 16H^+ \rightarrow 2Mn^{2+} + 8H_2O + 5I_2$
(d) $8MnO_4^- + 3S_2O_3^{2-} + H_2O \rightarrow 8MnO_2 + 6SO_4^{2-} + 2OH^-$

39. The species given below that does NOT show disproportionation reaction is :
[NCERT : P L-242 | July 20, 2021 (I)]
(a) BrO_4^- (b) BrO^- (c) BrO_2^- (d) BrO_3^-

40. $2\,MnO_4^- + b\,C_2O_4^{2-} + c\,H^+ \rightarrow x\,Mn^{2+} + y\,CO_2 + z\,H_2O$
If the above equation is balanced with integer coefficients, the value of c is_____.
[NCERT : P L-246 | NV, March 16, 2021 (I)]

41. The reaction of sulphur in alkaline medium is given below:
$S_8(s) + a\,OH^-(aq) \rightarrow b\,S^{2-}(aq) + c\,S_2O_3^{2-}(aq) + d\,H_2O(l)$
The values of 'a' is ________. (Integer answer)
[NCERT : P L-246 | NV, Feb. 24, 2021 (I)]

42. Consider the following equations :
$2\,Fe^{2+} + H_2O_2 \rightarrow x\,A + y\,B$ (in basic medium)
$2\,MnO_4^- + 6\,H^+ + 5\,H_2O_2 \rightarrow x'C + y'D + z'E$
(in acidic medium)
The sum of the stoichiometric coefficients x, y, x', y' and z' for products A, B, C, D and E, respectively, is ________.
[NCERT : P L-246 | NV, Sep. 04, 2020 (II)]

43. An example of a disproportionation reaction is:
[NCERT : P L-242 | April 12, 2019 (I)]
(a) $2\,MnO_4^- + 10\,I^- + 16\,H^+ \rightarrow 2\,Mn^{2+} + 5\,I_2 + 8\,H_2O$
(b) $2\,NaBr + Cl_2 \rightarrow 2\,NaCl + Br_2$
(c) $2\,KMnO_4 \rightarrow K_2MnO_4 + MnO_2 + O_2$
(d) $2\,CuBr \rightarrow CuBr_2 + Cu$

44. In order to oxidise a mixture of one mole of each of FeC_2O_4, $Fe_2(C_2O_4)_3$, $FeSO_4$ and $Fe_2(SO_4)_3$ in acidic medium, the number of moles of $KMnO_4$ required is:
[NCERT : P L-246 | April 8, 2019 (I)]
(a) 2 (b) 1 (c) 3 (d) 1.5

45. In the reaction of oxalate with permanganate in acidic medium, the number of electrons involved in producing one molecule of CO_2 is:
[NCERT : P L-246 | Jan. 10, 2019 (II)]
(a) 1 (b) 10 (c) 2 (d) 5

46. How many electrons are involved in the following redox reaction? **[NCERT : P L-246 | Online April 19, 2014]**
$Cr_2O_7^{2-} + Fe^{2+} + C_2O_4^{2-} \rightarrow Cr^{3+} + Fe^{3+} + CO_2$ (Unbalanced)
(a) 3 (b) 4 (c) 6 (d) 5

47. In the following balanced reaction,
$X\,MnO_4^- + Y\,C_2O_4^{2-} + Z\,H^+ \rightleftharpoons X\,Mn^{2+} + 2Y\,CO_2 + \frac{Z}{2}H_2O$
values of X, Y and Z respectively are
[NCERT : P L-246 | Online May 12, 2012; 2013]
(a) 2, 5, 16 (b) 8, 2, 5 (c) 5, 2, 16 (d) 5, 8, 4

48. Given :
$X\,Na_2HAsO_3 + Y\,NaBrO_3 + Z\,HCl \rightarrow NaBr + H_3AsO_4 + NaCl$
The values of X, Y and Z in the above redox reaction are respectively :
[NCERT : P L-246 | Online April 9, 2013, 2012 (S)]
(a) 2, 1, 2 (b) 2, 1, 3 (c) 3, 1, 6 (d) 3, 1, 4

ANSWER KEYS

1	(c)	6	(d)	11	(6)	16	(b)	21	(c)	26	(b)	31	(8)	36	(a)	41	(12)	46	(c)
2	(d)	7	(a)	12	(6)	17	(c)	22	(a)	27	(d)	32	(6)	37	(c)	42	(19)	47	(a)
3	(8)	8	(b)	13	(5)	18	(d)	23	(a)	28	(a)	33	(b)	38	(a)	43	(d)	48	(c)
4	(a)	9	(c)	14	(6)	19	(d)	24	(b)	29	(a)	34	(23)	39	(a)	44	(a)		
5	(a)	10	(3)	15	(19)	20	(c)	25	(c)	30	(8)	35	(d)	40	(16)	45	(a)		

The *p*-Block Elements (Group-13 and 14)

8

TREND ANALYSIS JEE MAIN (2025-2021)

	2025				2024				2023				2022				2021			
	Session II		Session I		Session II		Session I		Session II		Session I		Session II		Session I		Session II		Session I	
	MCQs	NVQs	MCQs	NVQs	MCQs	NVQs	MCQs	NVQs	MCQs	NVQs	MCQs	NVQs	MCQs	NVQs	MCQs	NVQs	MCQs	NVQs	MCQs	NVQs
Topic 1	5	–	–	–	6	–	3	–	2	–	2	–	3	–	1	–	1	–	2	–
Topic 2	1	–	3	–	2	–	2	–	–	–	1	–	–	–	–	–	–	–	1	–
Topic 3																				
Topic 4																				
Total	6	–	3	–	8	–	5	–	2	–	3	–	3	–	1	–	1	–	3	–

How to Code Qns for Better & Faster Revision

Type 1: Able to solve (No code Req)

Type 2: Solved with struggle or by seeing solution **(Use code NR means Needs Revision or orange stickers)**

Type 3: Tricky/ Could not solve on my own. Need help from Teacher/ Friend. **(Use code TQ means Tricky Question or red stickers)**

You can either hand write the code in boxes or else paste coloured stickers given at end of book.

Topic 1 – Boron Family

1. The correct statements from the following are :
[Old NCERT : P L-318 | April 7, 2025 (II)]

(A) Tl^{3+} is a powerful oxidising agent

(B) Al^{3+} does not get reduced easily

(C) Both Al^{3+} and Tl^{3+} are very stable in solution

(D) Tl^{+} is more stable than Tl^{3+}

(E) Al^{3+} and Tl^{+} are highly stable

Choose the **correct** answer from the options given below:

(a) (A), (B), (C), (D) and (E) (b) (A), (B), (D) and (E) only
(c) (B), (D) and (E) only (d) (A), (C) and (D) only

2. The elements of Group 13 with highest and lowest first ionisation enthalpies are respectively:
[Old NCERT : P L-318 | April 4, 2025 (II)]

(a) B & Ga (b) B & Tl (c) Tl & B (d) B & In

3. Given below are the pairs of group 13 elements showing their relation in terms of atomic radius.
(B < Al), (Al < Ga), (Ga < In) and (In < Tl)
Identify the elements present in the incorrect pair and in that pair find out the element (X) that has higher ionic radius (M^{3+}) that the other one. The atomic number of the element (X) is **[Old NCERT : P L-318 | April 4, 2025 (I)]**

(a) 31 (b) 49 (c) 13 (d) 81

4. The correct orders among the following are
[Old NCERT : P L-318 | April 3, 2025 (II)]

(A) Atomic radius : B < Al < Ga < In < Tl

(B) Electronegativity : Al < Ga < In < Tl < B

(C) Density : Tl < In < Ga < Al < B

(D) 1st Ionisation Energy : In < Al < Ga < Tl < B

Choose the **correct** answer from the options given below :

(a) B and D Only (b) A and C Only
(c) C and D Only (d) A and B Only

5. Given below are two statements :
[Old NCERT : P L-318 | April 2, 2025 (I)]

Statement (I) : The metallic radius of Al is less than that of Ga.

Statement (II) : The ionic radius of Al^{3+} is less than that of Ga^{3+}.

Note: The NCERT Page Locator (PL) note is of Old NCERT (Before rationalisation)

In the light of the above statements, choose the **most appropriate answer** from the options given below :
(a) Both **Statement I** and **Statement II** are incorrect
(b) **Statement I** is incorrect but **Statement II** is correct
(c) **Statement I** is correct but **Statement II** is incorrect
(d) Both **Statement I** and **Statement II** are correct

6. Match **List-I** with **List-II**.
[Old NCERT : P L-318 | April 9, 2024 (II)]

List-I	List-II
(A) Melting point [K]	(I) Tl > In > Ga > Al > B
(B) Ionic Radius [M^{+3}/pm]	(II) B > Tl > Al ≈ Ga > In
(C) $\Delta_{I.E}H_1$ [kJ mol^{-1}]	(III) Tl > In > Al > Ga > B
(D) Atomic Radius [pm]	(IV) B > Al > Tl > In > Ga

Choose the correct answer from the options given below:
(a) A-(III), B-(IV), C-(I), D-(II)
(b) A-(II), B-(III), C-(IV), D-(I)
(c) A-(IV), B-(I), C-(II), D-(III)
(d) A-(I), B-(II), C-(III), D-(IV)

7. Given below are two statements : One is labelled as **Assertion A** and the other is labelled as **Reason R**:
[Old NCERT : P L-318 | April 8, 2024 (I)]
Assertion A : The stability order of +1 oxidation state of Ga, In and Tl is Ga < In < Tl.
Reason R : The inert pair effect stabilizes the lower oxidation state down the group.
In the light of the above statements, choose the correct answer from the options given below :
(a) Both **A** and **R** are true and **R** is the correct explanation of **A**.
(b) **A** is true but **R** is false.
(c) Both **A** and **R** are true but **R** is NOT the correct explanation of **A**.
(d) **A** is false but **R** is true.

8. Given below are two statements :
Statement I : Gallium is used in the manufacturing of thermometers. **[Old NCERT : P L-318 | April 6, 2024 (I)]**
Statement II : A thermometer containing gallium is useful for measuring the freezing point (256 K) of brine solution.
In the light of the above statement, choose the correct answer from the options given below :
(a) Both Statement I and Statement II are false.
(b) Statement I is false but Statement II is true.
(c) Both Statement I and Statement II are true.
(d) Statement I is true but Statement II is false.

9. Given below are two statements :
Statement I: In group 13, the stability of +1 oxidation state increases down the group.
Statement II: The atomic size of gallium is greater than that of aluminium.
[Old NCERT : P L-319 | April 5, 2024 (I)]
In the light of the above statements, choose the **most appropriate** answer from the options given below:
(a) **Statement I** is incorrect but **Statement II** is correct
(b) Both **Statement I** and **Statement II** are correct
(c) Both **Statement I** and **Statement II** are incorrect
(d) **Statement I** is correct but **Statement II** is incorrect

10. The number of neutrons present in the more abundant isotope of boron is 'x'. Amorphous boron upon heating with air forms a product, in which the oxidation state of boron is 'y'. The value of x + y is ... **[April 5, 2024 (I)]**
(a) 4 (b) 6 (c) 3 (d) 9

11. The correct statements from the following are :
[Old NCERT : P L-318 | April 5, 2024 (II)]
(A) The decreasing order of atomic radii of group 13 elements is Tl > In > Ga > Al > B.
(B) Down the group 13 electronegativity decreases from top to bottom.
(C) Al dissolves in dil. HCl and liberate H_2 but conc. HNO_3 renders Al passive by forming a protective oxide layer on the surface.
(D) All elements of group 13 exhibits highly stable +1 oxidation state.
(E) Hybridisation of Al in $[Al(H_2O)_6]^{3+}$ ion is sp^3d^2.
Choose the **correct** answer from the options given below:
(a) (C) and (E) only
(b) (A), (C) and (E) only
(c) (A), (B), (C) and (E) only
(d) (A) and (C) only

12. Given below are two statements :
[Old NCERT : P L-320 | Jan. 31, 2024 (II)]
Statement (I) : Group 13 trivalent halides get easily hydrolyzed by water due to their covalent nature.
Statement (II) : $AlCl_3$ upon hydrolysis in acidified aqueous solution forms octahedral $[Al(H_2O)_6]^{3+}$ ion.
In the light of the above statements, choose the **correct answer** from the options given below :
(a) Statement I is true but statement II is false
(b) Statement I is false but statement II is true
(c) Both statement I and statement II are false
(d) Both statement I and statement II are true

13. Aluminium chloride in acidified aqueous solution forms an ion having geometry.
[Old NCERT : P L-323 | Jan. 30, 2024 (I)]
(a) Octahedral (b) Square planar
(c) Tetrahedral (d) Trigonal bipyramidal

14. Given below are two statements : one is labelled as Assertion (A) and the other is labelled as Reason (R)
[Old NCERT : P L-324 | Jan. 27, 2024 (I)]
Assertion (A) : Melting point of Boron (2453 K) is unusually high in group 13 elements.
Reason (R) : Solid Boron has very strong crystalline lattice.
In the light of the above statements, choose the most appropriate answer from the options given below :
(a) Both (A) and (R) are correct but (R) is not the correct explanation of (A)
(b) Both (A) and (R) are correct and (R) is the correct explanation of (A)
(c) (A) is true but (R) is false
(d) (A) is false but (R) is true

15. The covalency and oxidation state respectively of boron in $[BF_4]^-$, are **[Old NCERT : P L-319 | April 13, 2023 (II)]**
(a) 4 and 3 (b) 4 and 4 (c) 3 and 4 (d) 3 and 5

16. Given below are two statements :
[Old NCERT : P L-318 | April 12, 2023 (I)]
Statement – I : Boron is extremely hard indicating its high lattice energy
Statement – II : Boron has highest melting and boiling point compared to its other group members.
In the light of the above statements, choose the most appropriate answer from the options given below:
(a) Statement I is incorrect but Statement II is correct
(b) Both Statement I and Statement II is correct
(c) Statement I is correct but Statement II is incorrect
(d) Both Statement I and Statement II is incorrect

17. Which one of the following elements will remain as liquid inside pure boiling water?
[Old NCERT : P L-324 | April 06, 2023 (II)]
(a) Cs (b) Ga (c) Li (d) Br

18. The Lewis acid character of boron tri halides follows the order : **[Old NCERT : P L-320 | Jan. 31, 2023 (II)]**
(a) $BBr_3 > BI_3 > BCl_3 > BF_3$ (b) $BCl_3 > BF_3 > BBr_3 > BI_3$
(c) $BF_3 > BCl_3 > BBr_3 > BI_3$ (d) $BI_3 > BBr_3 > BCl_3 > BF_3$

19. Given below are two statements:
[Old NCERT : P L-323 | Jan. 29, 2023 (II)]
Statement I : The decrease in first ionization enthalpy from B to Al is much larger than that from Al to Ga.
Statement II : The d orbitals in Ga are completely filled.
In the light of the above statements, choose the most appropriate answer from the options given below
(a) Statement I is incorrect but statement II is correct.
(b) Both the statements I and II are correct
(c) Statement I is correct but statement II is incorrect
(d) Both the statements I and II are incorrect

20. Given below are two statements: one is labelled as **Assertion (A)** and the other is labelled as Reason (R).
Assertion (A) : Boron is unable to form BF_6^{3-}
Reason (R) : Size of B is very small.
[Old NCERT : P L-324 | July 27, 2022 (II)]
In the light of the above statements, choose the **correct** answer from the options given below:
(a) Both (A) and (R) are true and (R) is the correct explanation of (A) BF_6^{3-}.
(b) Both (A) and (R) are true but (R) is not the correct explanation of (A).
(c) (A) is true but (R) is false.
(d) (A) is false but (R) is true.

21. The metal that has very low melting point and its periodic position is closer to a metalloid is :
[Old NCERT : P L-318 | July 26, 2022 (II)]
(a) Al (b) Ga (c) Se (d) In

22. Given below are two statements.
[Old NCERT : P L-324 | July 29, 2022 (II)]
Statement I: Stannane is an example of a molecular hydride.
Statement II: Stannane is a planar molecule.
In the light of the above statement, choose the **most appropriate** answer from the options given below :
(a) Both Statement I and statement II are true.
(b) Both Statement I and statement II are false.
(c) Statement I is true but statement II is false.
(d) Statement I is false but statement II is true.

23. Match the **List-I** with **List-II**.
[Old NCERT : P L-318 | June 29, 2022 (I)]

List-I (Metal)	List-II (Application)
(A) Cs	(I) High temperature thermometer
(B) Ga	(II) Water repellent sprays
(C) B	(III) Photoelectric cells
(D) Si	(IV) Bullet proof vest

Choose the **most appropriate** answer from the option given below:
(a) (A)-(III), (B)-(I), (C)-(IV), (D)-(II)
(b) (A)-(IV), (B)-(III), (C)-(II), (D)-(I)
(c) (A)-(II), (B)-(III), (C)-(IV), (D)-(I)
(d) (A)-(I), (B)-(IV), (C)-(II), (D)-(III)

24. Given below are two statements : one is labelled as Assertion (A) and the other is labelled as Reason (R).
Assertion (A) : In $T\ell I_3$, isomorphous to CsI_3, the metal is present in +1 oxidation state.
Reason (R) : TI metal has fourteen f electrons in its electronic configuration.
In the light of the above statements, choose the most appropriate answer from the options given below :
[Old NCERT : P L-319 | Feb. 26, 2021 (II)]
(a) Both A and R are correct but R is NOT the correct explanation of A
(b) Both A and R are correct and R is the correct explanation of A
(c) A is correct but R is not correct
(d) A is not correct but R is correct

25. Which one of the following compounds of Group–14 elements is not known?
[Old NCERT : P L-324 | July 25, 2021 (I)]
(a) $[GeCl_6]^{2-}$ (b) $[Sn(OH)_6]^{2-}$
(c) $[SiCl_6]^{2-}$ (d) $[SiF_6]^{2-}$

26. Compound A is used as a strong oxidizing agent is amphoteric in nature. It is the part of lead storage batteries. Compound A is **[Old NCERT : P L-324 | Feb. 26, 2021 (I)]**
(a) Pb_3O_4 (b) $PbSO_4$ (c) PbO (d) PbO_2

27. Choose the correct stability order of group 13 elements in their +1 oxidation state. **[Old NCERT : P L-319 | June 26, 2022 (I), Same Jan. 11, 2019 (II)]**
(a) Al < Ga < In < Tl (b) Tl < In < Ga < Al
(c) Al < Ga < Tl < In (d) Al < Tl < Ga < In

28. The electronegativity of aluminium is similar to:
[Old NCERT : P L-325 | Jan. 10, 2019 (I)]
(a) Carbon (b) Beryllium (c) Boron (d) Lithium

29. Aluminium is usually found in +3 oxidation state. In contrast, thallium exists in +1 and + 3 oxidation states. This is due to: **[Old NCERT : P L-319 | Jan. 9, 2019 (I)]**
(a) inert pair effect (b) diagonal relationship
(c) lattice effect (d) lanthanoid contraction

30. The hydride that is **NOT** electron deficient is:
[Old NCERT : P L-321 | Jan. 11, 2019 (II)]
(a) SiH_4 (b) B_2H_6 (c) GaH_3 (4) AlH_3

31. Beryllium and aluminium exhibit many properties which are similar. But, the two elements differ in
[Old NCERT : P L-318 | 2004]
(a) forming covalent halides
(b) forming polymeric hydrides
(c) exhibiting maximum covalency in compounds
(d) exhibiting amphoteric nature in their oxides]

Topic 2 - Carbon Family

32. Given below are two statements :
[Old NCERT : P L-319 | April 4, 2025 (II)]
Statement (I) : The first ionisation enthalpy of group 14 elements is higher than the corresponding elements of group 13.
Statement (II) : Melting points and boiling points of group 13 elements are in general much higher than those the corresponding elements of group 14.
In the light of the above statements, choose the **most appropriate answer** from the options given below :
(a) Statement I is correct but Statement II is incorrect
(b) Statement I is incorrect but Statement II is correct
(c) Both Statement I and Statement II are incorrect
(d) Both Statement I and Statement II are correct

33. Consider the following elements In, Tl, Al, Pb, Sn and Ge. The most stable oxidation states of elements with highest and lowest first ionisation enthalpies, respectively, are
[Old NCERT : P L-323 | Jan. 28, 2025 (I)]
(a) +2 and +3 (b) +4 and +3
(c) +4 and +1 (d) +1 and +4

34. Given below are two statements :
[Old NCERT : P L-323 | Jan. 24, 2025 (II)]
Statement (I) : The first ionization energy of Pb is greater than that of Sn
Statement(II) : The first ionization energy of Ge is greater than that of Si.
In the light of the above statements, choose the correct answer from the options given below :
(a) Statement I is true but Statement II is false
(b) Both Statement I and Statement II are false
(c) Statement I is false but Statement II is true
(d) Both Statement I and Statement II are true

35. Given below are the atomic numbers of some group 14 elements. The atomic number of the element with lowest melting point is: **[Old NCERT : P L-323 | Jan. 23, 2025 (II)]**
(a) 14 (b) 6
(c) 82 (d) 50

36. Which of the following material is not a semiconductor.
[Old NCERT : P L-323 | April 6, 2024 (I)]
(a) Germanium (b) Graphite
(c) Silicon (d) Copper oxide

37. The number of ions from the following that are expected to behave as oxidising agent is:
[Old NCERT : P L-323 | April 6, 2024 (II)]
Sn^{4+}, Sn^{2+}, Pb^{2+}, Tl^{3+}, Pb^{4+}, Tl^{+}
(a) 3 (b) 4 (c) 1 (d) 2

38. Given below are two statements :
[Old NCERT : P L-324 | Feb. 1, 2024 (II)]
Statement (I): SiO_2 and GeO_2 are acidic while SnO and PbO are amphoteric in nature.
Statement (II) : Allotrophic forms of carbon are due to property of catenation and $p\pi$ - $d\pi$ bond formation.
In the light of the above statements, choose the most appropriate answer from the options given below:
(a) Statement I is false but Statement II is false
(b) Both Statement I and Statement II are false
(c) Both Statement I and Statement II are true.
(d) Statement I is true but Statement II is false.

39. Consider the oxides of group 14 elements SiO_2, GeO_2, SnO_2, PbO_2, CO and GeO. The amphoteric oxides are **[Old NCERT : P L-324 | Jan. 31, 2024 (I)]**
(a) GeO, GeO_2 (b) SiO_2, GeO_2
(c) SnO_2, CO (d) SnO_2, PbO_2

40. The correct order of bond enthalpy (kJ mol^{-1}) is:
[Old NCERT : P L-323 | Feb. 01, 2023 (II)]
(a) C – C > Si – Si > Ge – Ge > Sn – Sn
(b) C – C > Si – Si > Sn – Sn > Ge – Ge
(c) Si – Si > C – C > Sn – Sn > Ge – Ge
(d) Si – Si > C – C > Ge – Ge > Sn – Sn

41. The correct order of catenation is :
[Old NCERT : P L-325 | April 10, 2019 (I)]
(a) C > Sn > Si » Ge (b) C > Si > Ge » Sn
(c) Si > Sn > C > Ge (d) Ge > Sn > Si > C

42. The **INCORRECT** statement regarding the structure of C_{60} is **[Old NCERT : P L-324 | March 16, 2021 (II); Same April 9, 2019 (I)]**
(a) It contains 12 six-membered rings and 24 five-membered rings
(b) The six-membered rings are fused to both six and five-membered rings
(c) Each carbon atom forms three sigma bonds
(d) The five-membered rings are fused only to six-membered rings

43. The element that does NOT show catenation is :
[Old NCERT : P L-325 | Jan. 12, 2019 (II)]
(a) Ge (b) Si (c) Sn (d) Pb

44. The C – C bond length is maximum in:
[Old NCERT : P L-326 | April 12, 2019 (II)]
(a) graphite (b) C_{70}
(c) C_{60} (d) diamond

45. C_{60}, an allotrope of carbon cantains:
[Old NCERT : P L-326 | April 9, 2019 (I)]
(a) 12 hexagons and 20 pentagons.
(b) 18 hexagons and 14 pentagons.
(c) 16 hexagons and 16 pentagons.
(d) 20 hexagons and 12 pentagons.

46. The element that shows greater ability to form $p\pi - p\pi$ multiple bonds, is : **[Old NCERT : P L-326 | Jan. 12, 2019 (II)]**

(a) Sn (b) C
(c) Ge (d) Si

47. Which of the following are Lewis acids? **[Old NCERT : P L-324 | 2018]**

(a) PH_3 and BCl_3 (b) $AlCl_3$ and $SiCl_4$
(c) PH_3 and $SiCl_4$ (d) BCl_3 and $AlCl_3$

48. In graphite and diamond, the percentage of *p*-characters of the hybrid orbitals in hybridisation are respectively: **[Old NCERT : P L-325 | Online April 15, 2018 (I)]**

(a) 33 and 25 (b) 67 and 75
(c) 50 and 75 (d) 33 and 75

49. **Assertion :** Among the carbon allotropes, diamond is an insulator, whereas, graphite is a good conductor of electricity.

Reason : Hybridization of carbon in diamond and graphite are sp^3 and sp^2, respectively. **[Old NCERT : P L-323 | Online April 10, 2016]**

(a) Both assertion and reason are correct, but the reason is not the correct explanation for the assertion
(b) Both assertion and reason are correct, and the reason is the correct explanation for the assertion
(c) Both assertion and reason are incorrect
(d) Assertion is incorrect statement, but the reason is correct.

50. The catenation tendency of C, Si and Ge is in the order Ge < Si < C. The bond energies (in kJ mol^{-1}) of C – C, Si – Si and Ge – Ge bonds are respectively; **[Old NCERT : P L-325 | Online April 25, 2013]**

(a) 348, 297, 260 (b) 297, 348, 260
(c) 348, 260, 297 (d) 260, 297, 348

51. The stability of dihalides of Si, Ge, Sn and Pb increases steadily in the sequence **[Old NCERT : P L-323 | 2007]**

(a) $PbX_2 << SnX_2 << GeX_2 << SiX_2$
(b) $GeX_2 << SiX_2 << SnX_2 << PbX_2$
(c) $SiX_2 << GeX_2 << PbX_2 << SnX_2$
(d) $SiX_2 << GeX_2 << SnX_2 << PbX_2$.

52. A metal, M forms chlorides in its +2 and +4 oxidation states. Which of the following statements about these chlorides is correct? **[Old NCERT : P L-323 | 2006]**

(a) MCl_2 is more ionic than MCl_4
(b) MCl_2 is more easily hydrolysed than MCl_4
(c) MCl_2 is more volatile than MCl_4
(d) MCl_2 is more soluble in anhydrous ethanol than MCl_4

53. Graphite is a soft solid lubricant extremely difficult to melt. The reason for this anomalous behaviour is that graphite **[Old NCERT : P L-326 | 2003]**

(a) is an allotropic form of diamond
(b) has molecules of variable molecular masses like polymers
(c) has carbon atoms arranged in large plates of rings of strongly bound carbon atoms with weak interplate bonds
(d) is a non-crystalline substance

ANSWER KEYS																	
1	(b)	7	(a)	13	(a)	19	(b)	25	(c)	31	(c)	37	(d)	43	(d)	49	(b)		
2	(d)	8	(d)	14	(b)	20	(b)	26	(d)	32	(a)	38	(d)	44	(d)	50	(a)		
3	(a)	9	(d)	15	(a)	21	(b)	27	(a)	33	(b)	39	(d)	45	(d)	51	(d)		
4	(a)	10	(d)	16	(b)	22	(c)	28	(b)	34	(a)	40	(a)	46	(b)	52	(a)		
5	(b)	11	(a)	17	(b)	23	(a)	29	(a)	35	(d)	41	(b)	47	(b, d)	53	(c)		
6	(c)	12	(d)	18	(d)	24	(a)	30	(a)	36	(b)	42	(a)	48	(b)				

9 Organic Chemistry-Some Basic Principles and Techniques

TREND ANALYSIS
JEE MAIN (2025-2021)

	2025				2024				2023				2022				2021			
	Session II		Session I		Session II		Session I		Session II		Session I		Session II		Session I		Session II		Session I	
	MCQs	NVQs	MCQs	NVQs	MCQs	NVQs	MCQs	NVQs	MCQs	NVQs	MCQs	NVQs	MCQs	NVQs	MCQs	NVQs	MCQs	NVQs	MCQs	NVQs
Topic 1	4	–	3	–	4	–	7	2	1	–	–	–	2	–	2	–	1	–	1	–
Topic 2	1	1	2	1	4	–	–	–	–	–	1	–	–	–	–	1	–	1	1	–
Topic 3	8	3	7	5	10	3	16	3	11	4	5	3	1	3	2	1	1	4	1	1
Topic 4																				
Total	13	4	12	6	18	3	23	5	12	4	6	3	3	3	4	2	2	5	3	1

How to Code Qns for Better & Faster Revision →

Type 1: Able to solve (No code Req)

Type 2: Solved with struggle or by seeing solution **(Use code NR means Needs Revision or orange stickers)**

Type 3: Tricky/ Could not solve on my own. Need help from Teacher/ Friend. **(Use code TQ means Tricky Question or red stickers)**

You can either hand write the code in boxes or else paste coloured stickers given at end of book.

Topic 1 - Classification and Nomenclature of Organic Compounds

1. What is the correct IUPAC name of **[NCERT : P L-265 | April 8, 2025 (II)]**

(a) 4-Ethyl-1-hydroxycyclopent-2-ene
(b) 1-Ethyl-3-hydroxycyclopent-2-ene
(c) 1-Ethylcyclopent-2-en-3-ol
(d) 4-Ethylcyclopent-2-en-1-ol

2. Which of the following is the correct IUPAC name of given organic compound (X) ? **[NCERT : PL-265 | April 7, 2025 (I)]**

(X)

(a) 2-Bromo-2-methylbut-2-ene
(b) 3-Bromo-3-methylprop-2-ene
(c) 1-Bromo-2-methylbut-2-ene
(d) 4-Bromo-3-methylbut-2-ene

3. The IUPAC name of the following compound is – **[NCERT : P L-265 | April 4, 2025 (II)]**

$$HC \equiv C - CH_2 - \underset{}{\overset{OH}{\overset{|}{C}}}H - CH_2 - CH = CH_2$$

(a) 4-Hydroxyhept-1-en-6-yne
(b) 4-Hydroxyhept-6-en-1-yne
(c) Hept-6-en-1-yn-4-ol
(d) Hept-1-en-6-yn-4-ol

4. What is the correct IUPAC name of **[NCERT : P L-265 | April 3, 2025 (II)]**

(a) 3-Bromo-2-hydroxy-5-nitrobenzoic acid
(b) 3-Bromo-4-hydroxy-1-nitrobenzoic acid
(c) 2-Hydroxy-3-bromo-5-nitrobenzoic acid
(d) 5-Nitro-3-bromo-2-hydroxybenzoic acid

5. Match List-I with List-II. **[NCERT : P L-265 | Jan. 29, 2025 (I)]**

List-I (Structure)	List-II (IUPAC Name)
(A) $H_3CCH_2CHCH_2CHC_2H_5$ (with C_2H_5 and CH_3 branches)	(I) 4-Methylpent-1-ene
(B) $(CH_3)_2C(C_3H_7)_2$	(II) 3-Ethyl-5-methylheptane
(C) 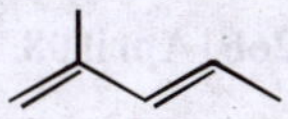 [structure]	(III) 4,4-Dimethylheptane
(D)  [structure]	(IV) 2-Methyl-1,3-pentadiene

Choose the correct answer from the options given below:
(a) (A)-(III), (B)-(II), (C)-(IV), (D)-(I)
(b) (A)-(III), (B)-(II), (C)-(I), (D)-(IV)
(c) (A)-(II), (B)-(III), (C)-(IV), (D)-(I)
(d) (A)-(II), (B)-(III), (C)-(I), (D)-(IV)

6. The alkane from below having two secondary hydrogens is : **[NCERT : P L-265 | Jan. 22, 2025 (II)]**
(a) 4-Ethyl-3, 4-dimethyloctane
(b) 2, 2, 4, 4-Tetramethylhexane
(c) 2, 2, 3, 3-Tetramethylpentane
(d) 2, 2, 4, 5-Tetramethylheptane

7. The IUPAC name of the following compound is: **[NCERT : P L-265 | Jan. 22, 2025 (I)]**

$$CH_3-\underset{|}{\overset{COOH}{CH}}-CH_2-CH_2-\overset{COOCH_3}{CH}-CH_3$$

(a) 2-Carboxy-5-methoxycarbonylhexane.
(b) Methyl-6-carboxy-2, 5-dimethylhexanoate.
(c) Methyl-5-carboxy-2-methylhexanoate.
(d) 6-Methoxycarbonyl-2,5-dimethylhexanoic acid.

8. Given below are two statements: **[NCERT : P L-269 | April 8, 2024 (I)]**

Statement I : [structure: O_2N, Cl, NO_2 substituted benzene, Compound-A]
IUPAC name of compound A is 4-chloro-1 3-dinitrobenzene:

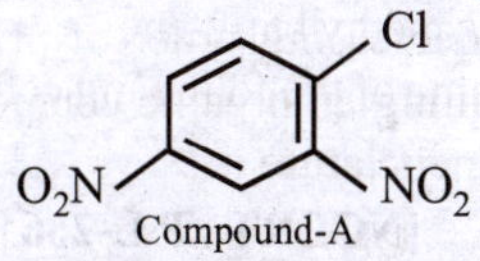

Statement II: [structure: NH_2, CH_3, C_2H_5 substituted benzene, Compound-B]
IUPAC name of Compound B is 4-ethyl-2-methylaniline.

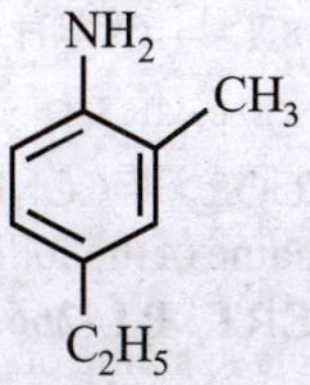

In the light of the above statements, choose the **most appropriate** answer from the options given below:
(a) Both Statement I and Statement II are correct
(b) Statement I is incorrect but Statement II is correct
(c) Statement I is correct but Statement II is incorrect
(d) Both Statement I and Statement II are incorrect

9. IUPAC name of following hydrocarbon (X) is :

$$CH_3-\underset{CH_3}{CH}-CH_2-CH_2-\underset{CH_3}{CH}-\underset{CH_3}{CH}-CH_2-CH_3 \quad (X)$$

[NCERT : P L-264 | April 8, 2024 (II)]
(a) 2-Ethyl-3,6-dimethylheptane
(b) 2-Ethyl-2,6-diethylheptane
(c) 2,5,6-Trimethyloctane
(d) 3,4,7-Trimethyloctane

10. Functional group present in sulphonic acid is : **[NCERT : P L-266 | April 6, 2024 (I)]**
(a) SO_4H (b) SO_3H
(c) $-\underset{\underset{O}{\|}}{S}-OH$ (d) $-SO_2$

11. The correct nomenclature for the following compound is: **[NCERT : P L-266 | April 5, 2024 (II)]**

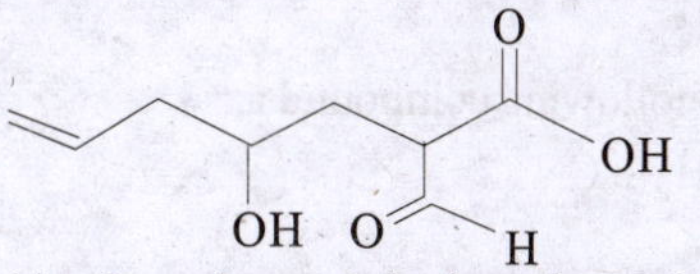

[structure with O, OH, OH, O, H labels]

(a) 2–carboxy–4–hydroxyhept–6–enal
(b) 2–carboxy–4–hydroxyhept–7–enal
(c) 2–formyl–4–hydroxyhept–6–enoic acid
(d) 2–formyl–4–hydroxyhept–7–enoic acid

12. Given below are two statements **[NCERT : P L-266 | Jan. 31, 2024 (I)]**

Statement I : IUPAC name of $HO-CH_2-(CH_2)_3-CH_2-COCH_3$ is 7-hydroxyheptan-2-one.

Statement II : 2-oxoheptan-7-ol is the correct IUPAC name for above compound.

In the light of the above statements, choose the **most appropriate answer** from the options given below:
(a) Statement I is correct but Statement II is incorrect.
(b) Statement I is incorrect but Statement II is correct.
(c) Both Statement I and Statement II are correct.
(d) Both Statement I and Statement II are incorrect.

13. Identify structure of 2,3-dibromo-1-phenylpentane. **[NCERT : P L-269 | Jan. 31, 2024 (II)]**

(a) 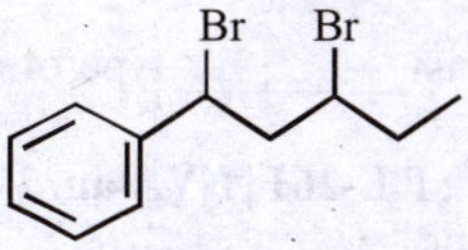 [structure: Br, Br]
(b) 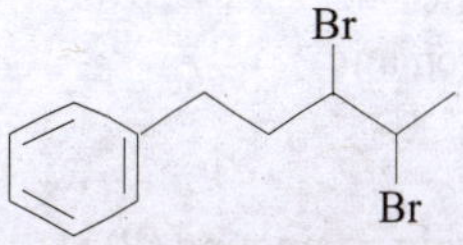 [structure: Br, Br]
(c) 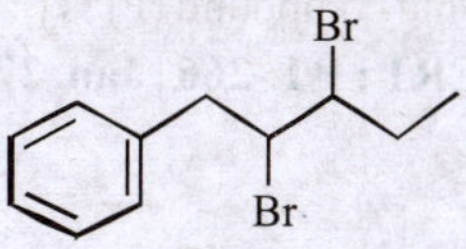 [structure: Br, Br]
(d) 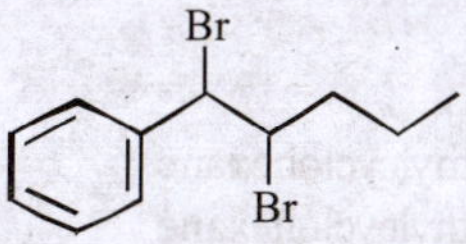 [structure: Br, Br]

14. Structure of 4- methylpent -2- enal is:

[NCERT : P L-265 | Jan. 30, 2024 (I)]

(a) $H_2C=C(H)-C(H)(CH_3)-CH_2-C(=O)-H$

(b) $CH_3-CH_2-C(CH_3)=CH-C(=O)-H$

(c) $CH_3-CH_2-CH=C(CH_3)-C(=O)-H$

(d) $CH_3-CH(CH_3)-CH=CH-C(=O)-H$

15. IUPAC name of following compound is:

$CH_3-CH(NH_2)-CH_2-CN$

[NCERT : P L-265 | Jan. 30, 2024 (II)]

(a) 2 - Aminopentanenitrile
(b) 2 - Aminobutanenitrile
(c) 3 - Aminobutanenitrile
(d) 3 - Aminopropanenitrile

16. Number of compounds among the following which contain sulphur as heteroatom is ______.

Furan, Thiophene, Pyridine, Pyrrole, Cysteine, Tyrosine

[NCERT : P L-261 | NV, Jan. 29, 2024 (I)]

17. According to IUPAC system, the compound

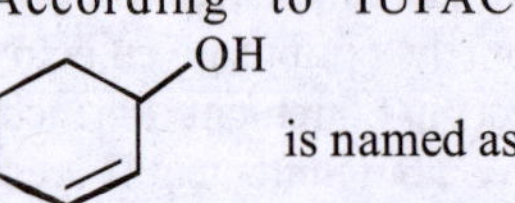

is named as

[NCERT : P L-269 | Jan. 29, 2024 (II)]

(a) Cyclohex-1-en-3-ol (b) Cyclohex-2-en-1-ol
(c) Cyclohex-1-en-2-ol (d) 1-Hydroxyhex-2-ene

18. Cyclohexene [structure] is ________ type of an organic compound. [NCERT : P L-261 | NV, Jan. 27, 2024 (I)]

(a) Benzenoid aromatic
(b) Benzenoid non-aromatic
(c) Acyclic
(d) Alicyclic

19. IUPAC name of following compound (P) is :

[NCERT : P L-266 | Jan. 27, 2024 (I)]

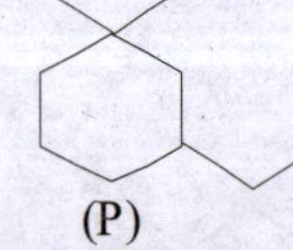

(P)

(a) l-Ethyl-5, 5-dimethylcyclohexane
(b) 3-Ethyl-1,1-dimethylcyclohexane
(c) l-Ethyl-3, 3-dimethylcyclohexane
(d) l,l-Dimethyl-3-ethylcyclohexane

20. Bond line formula of $HOCH(CN)_2$ is:

[NCERT : P L-258 | Jan. 27, 2024 (II)]

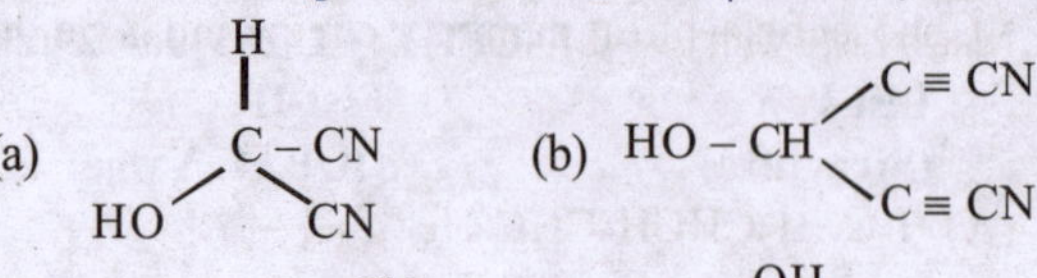

(c) HO – CH(CN)(CN) (d) OH, CN, CN

21. The correct IUPAC nomenclature for the following compound is [NCERT : P L-266 | April 08, 2023 (II)]

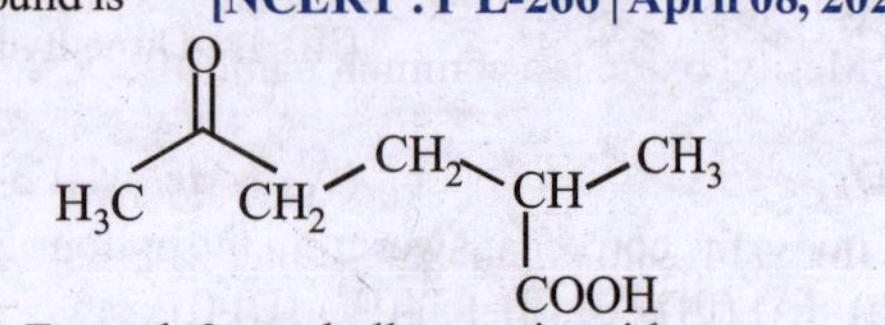

(a) 5–Formyl–2–methylhexanoic acid
(b) 2–Methyl–5–oxohexanoic acid
(c) 2–Formyl–5–methylhexan–6–oic acid
(d) 5–Methyl–2–oxohexan–6–oic acid

22. Match List-I with List-II. [July 28, 2022 (I)]

List-I	List-II
(A) [structure]	(I) Spiro compound
(B) [structure]	(II) Aromatic compound
(C)	(III) Non-planar heterocyclic compound
(D) [structure]	(IV) Bicyclo compound

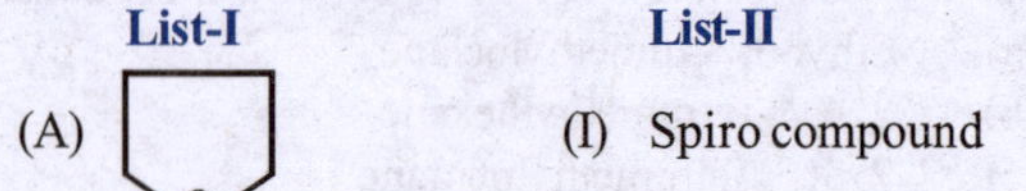

Choose the correct answer from the options given below :

(a) (A) – (II), (B) – (I), (C) – (IV), (D) – (III)
(b) (A) – (IV), (B) – (III), (C) – (I), (D) – (II)
(c) (A) – (III), (B) – (IV), (C) – (I), (D) – (II)
(d) (A) – (IV), (B) – (III), (C) – (II), (D) – (I)

23. The correct decreasing order of priority of functional groups in naming an organic compound as per IUPAC system of nomenclature is :

[NCERT : P L-266 | July 26, 2022 (II)]

(a) $-COOH > -CONH_2 > -COCl > -CHO$
(b) $-SO_3H > -COCl > -CONH_2 > -CN$
(c) $-COOH > -COCl > -NH_2 > \rangle C=O$
(d) $-COOH > -COOR > -CONH_2 > -COCl$

24. The correct IUPAC name of the following compound is :

[NCERT : P L-266 | June 28, 2022 (II)]

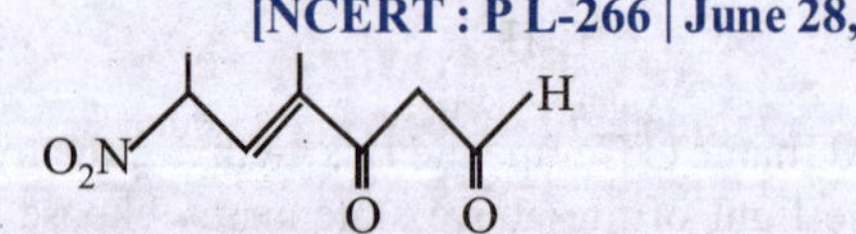

(a) 4-methyl-2-nitro-5-oxohept-3-enal
(b) 4-methyl-5-oxo-2-nitrohept-3-enal
(c) 4-methyl-6-nitro-3-oxohept-4-enal
(d) 6-formyl-4-methyl-2-nitrohex-3-enal

25. The IUPAC name of ethylidene chloride is :-
[NCERT : P L-265 | June 25, 2022 (I)]
(a) 1-Chloroethene (b) 1-Chloroethyne
(c) 1, 2-Dichloroethane (d) 1, 1-Dichloroethane

26. Choose the correct name for compound given below :
[NCERT : P L-264 | Aug. 31, 2021 (I)]

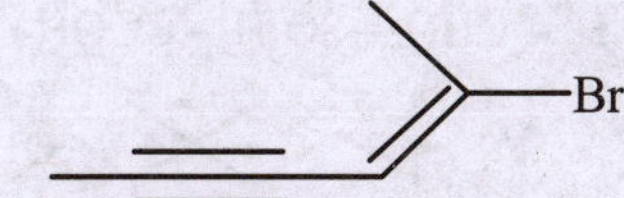

(a) (4*E*)-5-Bromohex-4-en-2-yne
(b) (2*E*)-2-Bromohex-4-yn-2-ene
(c) (2*E*)-2-Bromohex-2-en-4-yne
(d) (4*E*)-5-Bromohex-2-en-4-yne

27. Mesityl oxide is a common name of
[NCERT : P L-265 | March 17, 2021 (I)]
(a) 3-Methylcyclohexane carbaldehyde
(b) 2, 4-Dimethylpentan-3-one
(c) 2-Methylcyclohexanone
(d) 4-Methylpent-3-en-2-one

28. The IUPAC name of the following compound is:
[NCERT : P L-269 | Sep. 06, 2020 (II)]

OH, NH_2, O_2N, CHO

(a) 2-nitro-4-hydroxymethyl-5-amino benzaldehyde
(b) 3-amino-4-hydroxymethyl-5-nitrobenzaldehyde
(c) 5-amino-4-hydroxymethyl-2-nitrobenzaldehyde
(d) 4-amino-2-formyl-5-hydroxymethyl nitrobenzene

29. The IUPAC name of the following compound is :
[NCERT : P L-265 | Sep. 04, 2020 (I)]

CH_3, O, C – OH, Br

(a) 5-Bromo-3-methylcyclopentanoic acid
(b) 4-Bromo-2-methylcyclopentane carboxylic acid
(c) 3-Bromo-5-methylcyclopentanoic acid
(d) 3-Bromo-5-methylcyclopentane carboxylic acid

30. The IUPAC name for the following compound is :
[NCERT : P L-266 | Sep. 02, 2020 (I)]

CHO, CH_3, H_3C, COOH

(a) 2, 5-dimethyl-5-carboxy-hex-3-enal
(b) 2, 5-dimethyl-6-carboxy-hex-3-enal
(c) 2, 5-dimethyl-6-oxo-hex-3-enoic acid
(d) 6-formyl-2-methyl-hex-3-enoic acid

31. The IUPAC name for the following compound is :
[NCERT : P L-264 | April 12, 2019 (II)]

CH_3, H_3C, CH, CH_3, CH_2

(a) 3-methyl-4-(3-methylprop-l-enyl)-l-heptyne
(b) 3,5-dimethyl-4-propylhept-6-en-l-yne
(c) 3-methyl-4-(1-methylprop-2-ynyl)-l-heptene
(d) 3,5-dimethyl-4-propylhept-1-en-6-yne

32. The correct IUPAC name of the following compound is:
[NCERT : P L-269 | April 9, 2019 (I)]

NO_2, Cl, CH_3

(a) 5-chloro-4-methyl-1-nitrobenzene
(b) 2-chloro-1-methyl-4-nitrobenzene
(c) 3-chloro-4-methyl-1-nitrobenzene
(d) 2-methyl-5-nitro-1-chlorobenzene

33. The IUPAC name of the following compound is :

$$H_3C-\underset{}{\overset{CH_3}{\overset{|}{C}H}}-\overset{OH}{\overset{|}{C}H}-CH_2-COOH$$

[NCERT : P L-266 | April 8, 2019 (I)]
(a) 4, 4-Dimethyl-3-hydroxybutanoic acid
(b) 2-Methyl-3-hydroxypentan-5-oic acid
(c) 3-Hydroxy-4-methylpentanoic acid
(d) 4-Methyl-3-hydroxypentanoic acid

34. What is the IUPAC name of the following compound?

CH_3, CH_3, H, H, Br, CH_3

[NCERT : P L-264 | Jan. 10, 2019 (II)]
(a) 3-Bromo-1, 2-dimethylbut-1-ene
(b) 3-Bromo-3-methyl-1, 2-dimethylprop-1-ene
(c) 2-Bromo-3-methylpent-3-ene
(d) 4-Bromo-3-methylpent-2-ene

35. The IUPAC name of the following compound is :

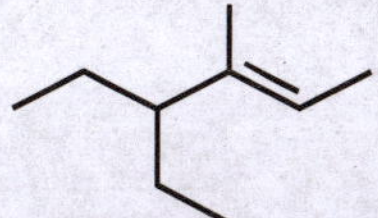

[NCERT : P L-264 | Online April 15, 2018(I); Same Online April 19, 2014]
(a) 3-ethyl-4-methylhex-4-ene
(b) 4, 4-diethyl-3-methylbut-2-ene
(c) 4-methyl-3-ethylhex-4-ene
(d) 4-ethyl-3-methylhex-2-ene

36. The IUPAC name of the following compound is :
[NCERT : P L-266 | Online April 8, 2017]

(a) 1, 1–dimethyl–2–ethylcyclohexane
(b) 2–ethyl–1, 1–dimethylcyclohexane
(c) 1–ethyl–2,2–dimethylcyclohexane
(d) 2, 2–dimethyl–1–ethylcyclohexane

37. The hydrocarbon with seven carbon atoms containing a neopentyl and a vinyl group is : **[NCERT : P L-263 | Online April 9, 2016]**
(a) 2, 2–dimethyl–4– pentene
(b) 4, 4– dimethylpentene
(c) isopropyl–2–butene
(d) 2, 2–dimethyl–3–pentene

38. In allene (C_3H_4), the type(s) of hybridization of the carbon atoms is (are): **[NCERT : P L-257 | Online April 11, 2014]**
(a) sp and sp^3 (b) sp^2 and sp
(c) only sp^2 (d) sp^2 and sp^3

39. The IUPAC name of the compound

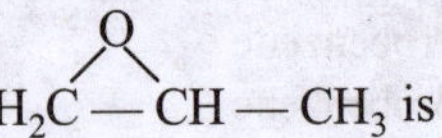

$H_2C — CH — CH_3$ is
[NCERT : P L-266 | Online May 7, 2012]
(a) 1, 2-propoxide (b) propylene oxide
(c) 1, 2-oxo propane (d) 1, 2-epoxypropane

40. The IUPAC name of the following compound is

$CH_3(H)C = C(H)C \equiv C - CH_2CH_3$

[NCERT : P L-264 | Online May 19, 2012]
(a) (*E*)-2-hepten-4-yne (b) (*Z*)-5-hepten-3-yne
(c) (*E*)-5-hepten-3-yne (d) (*Z*)-2-hepten-4-yne

41. Dipole moment is shown by **[Online May 26, 2012]**
(a) 1, 2-dichlorobenzene
(b) *trans*-2, 3-dichloro-2-butene
(c) 1, 4-dichlorobenzene
(d) *trans*-1, 2-dinitroethene

42. The IUPAC name of the following compound is **[2009]**

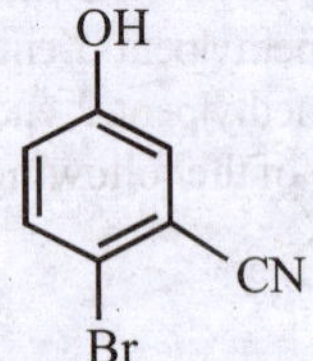

(a) 4-bromo-3-cyanophenol
(b) 2-bromo-5-hydroxybenzonitrile
(c) 2-cyano-4-hydroxybromobenzene
(d) 6-bromo-3-hydroxybenzonitrile

43. The IUPAC name of neopentane is
[NCERT : P L-263 | NCERT : P L-263 | 2009]
(a) 2, 2-dimethylpropane (b) 2-methylpropane
(c) 2, 2-dimethylbutane (d) 2- methylbutane

44. The correct decreasing order of priority for the functional groups of organic compounds in the IUPAC system of nomenclature is **[NCERT : P L-266 | 2008]**
(a) $– COOH, – SO_3H, – CONH_2, – CHO$
(b) $– SO_3H, – COOH, – CONH_2, – CHO$
(c) $– CHO, – COOH, – SO_3H, – CONH_2$
(d) $– CONH_2, – CHO, – SO_3H, – COOH$

45. The IUPAC name of [structure] is
[NCERT : P L-263 | 2007]
(a) 3-ethyl-4,4-dimethylheptane
(b) 1, 1-diethyl-2,2-dimethylpentane
(c) 4, 4-dimethyl-5,5-diethylpentane
(d) 5, 5-diethyl-4,4-dimethylpentane.

46. The IUPAC name of the compound shown below is :

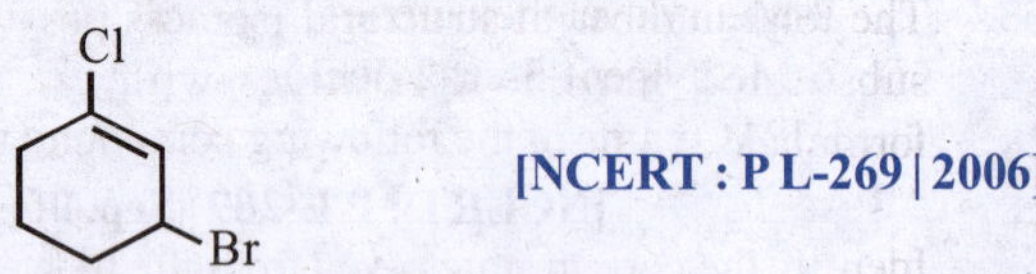

[NCERT : P L-269 | 2006]
(a) 3-bromo-1-chlorocyclohexene
(b) 1-bromo-3-chlorocyclohexene
(c) 2-bromo-6-chlorocyclohex-1-ene
(d) 6-bromo-2-chlorocyclohexene

47. The IUPAC name of C_6H_5COCl is
(a) benzene chloro ketone **[NCERT : P L-269 | 2006]**
(b) chlorobenzyl ketone
(c) chlorophenyl ketone
(d) benzene carbonyl chloride

48. The IUPAC name of the compound is
[NCERT : P L-264 | 2004]

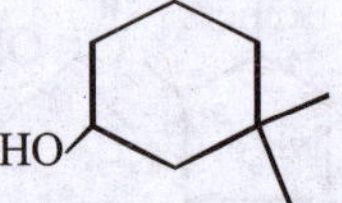

(a) 3, 3-dimethyl - 1- cyclohexanol
(b) 1, 1-dimethyl-3-hydroxy cyclohexane
(c) 3, 3-dimethyl-1-hydroxy cyclohexane
(d) 1, 1-dimethyl-3-cyclohexanol

49. Which one of the following does not have sp^2 hybridised carbon ? **[NCERT : P L-265 | 2004]**
(a) Acetonitrile (b) Acetic acid
(c) Acetone (d) Acetamide

50. The general formula $C_nH_{2n}O_2$ could be for open chain
[NCERT : P L-265 | 2003]
(a) carboxylic acids (b) diols
(c) dialdehydes (d) diketones

51. The IUPAC name of $CH_3COCH(CH_3)_2$ is
(a) 2-methyl-3-butanone **[NCERT : P L-265 | 2003]**
(b) 4-methylisopropyl ketone
(c) 3-methyl-2-butanone
(d) Isopropylmethyl ketone

52. Which of the following compounds has wrong IUPAC name? **[NCERT : P L-265 | 2002]**

(a) $CH_3–CH_2–CH_2–COO–CH_2CH_3 \rightarrow$ ethyl butanoate

(b) $CH_3 - \underset{CH_3}{\underset{|}{CH}} - CH_2 - CHO \rightarrow$ 3-methyl-butanal

(c) $CH_3 - \underset{OH}{\underset{|}{CH}} - \underset{CH_3}{\underset{|}{CH}} - CH_3 \rightarrow$ 2-methyl-3-butanol

(d) $CH_3 - \underset{CH_3}{\underset{|}{CH}} - \overset{O}{\overset{||}{C}} - CH_2 - CH_3 \rightarrow$ 2-methyl-3-pentanone

Topic 2 - Isomerism in Organic Compounds

53. The total number of structural isomers possible for the substituted benzene derivatives with the molecular formula C_9H_{12} is ______.
[NCERT : P L-270 | NV, April 3, 2025 (II)]

54. Identify the correct statements from the following
[NCERT : P L-271 | April 3, 2025 (I)]

A. (O) and (O) are metamers

B. CN and NC are functional isomers

C. OH and OH are position isomers

D. NH_2 and NH are homologous

Choose the **correct** answer from the options given below

(a) C & D only (b) B & C only
(c) A & B only (d) A, B & C only

55. Given below are two statements :
[NCERT : P L-270 | Jan. 28, 2025 (II)]

Statement (I) : and are isomeric compounds.

Statement (II) : NH_2 and NH are functional group isomers.

In the light of the above statements, choose the correct answer from the options given below :

(a) Both Statement I and Statement II are false
(b) Both Statement I and Statement II are true
(c) Statement I is true but Statement II is false
(d) Statement I is false but Statement II is true

56. The possible number of stereoisomers for 5-phenylpent-4-en-2-ol is ______.
[NCERT : P L-270 | NV, Jan. 24, 2025 (II)]

57. How many different stereoisomers are possible for the given molecule? **[NCERT : P L-270 | Jan. 22, 2025 (I)]**

$CH_3 - \underset{OH}{\underset{|}{CH}} - CH = CH - CH_3$

(a) 3 (b) 1 (c) 2 (d) 4

58. Total number of stereo isomers possible for the given structure: **[NCERT : P L-271 | April 9, 2024 (II)]**

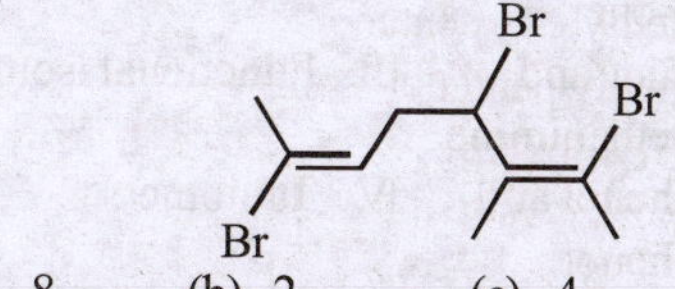

(a) 8 (b) 2 (c) 4 (d) 3

59. Which of the following is metamer of the given compound (X) ? **[NCERT : P L-271 | April 6, 2024 (I)]**

$– NH – \overset{O}{\overset{||}{C}} –$
(X)

(a) $– NH – \overset{O}{\overset{||}{C}} –$

(b) OHC – – NH –

(c) $– NH – \overset{O}{\overset{||}{C}} –$

(d) $– NH – \overset{O}{\overset{||}{C}} –$

60. The incorrect statement regarding the geometrical isomers of 2-butene is: **[April 6, 2024 (II)]**

(a) cis-2-butene and trans-2-butene are not interconvertible at room temperature.
(b) cis-2-butene has less dipole moment than trans-2-butene.
(c) trans-2-butene is more stable than cis-2-butene.
(d) cis-2-butene and trans-2-butene are stereoisomers.

61. Match **List-I** with **List-II**
[NCERT : P L-270, 271 | April 5, 2024 (II)]

	List-I (Pair of Compounds)		List-II (Isomerism)
(A)	n-propanol and Isopropanol	(I)	Metamerism
(B)	Methoxypropane and ethoxyethane	(II)	Chain Isomerism
(C)	Propanone and propanal	(III)	Position Isomerism
(D)	Neopentane and Isopentane	(IV)	Functional Isomerism

(a) (A)–(II), (B)–(I), (C)–(IV), (D)–(III)
(b) (A)–(III), (B)–(I), (C)–(II), (D)–(IV)
(c) (A)–(I), (B)–(III), (C)–(IV), (D)–(II)
(d) (A)–(III), (B)–(I), (C)–(IV), (D)–(II)

62. Match **List I** with **List II**. **[NCERT : P L-270 | Jan. 25, 2023 (II)]**

List I (Isomeric pairs)	List II (Type of isomers)
A. Propanamine and N-Methylethanamine	I. Metamers
B. Hexan-2-one and Hexan-3-one	II. Positional isomers
C. Ethanamide and Hydroxyethanimine	III. Functional isomers
D. o-nitrophenol and p-nitrophenol	IV. Tautomers

Choose the correct answer from the options given below:-
(a) A-III, B-IV, C-I, D-II (b) A-IV, B-III, C-I, D-II
(c) A-II, B-III, C-I, D-IV (d) A-III, B-I, C-IV, D-II

63. The number of acyclic structural isomers (including geometrical isomers) for pentene are ______ **[NV, July 22, 2021 (II)]**

64. Total number of possible stereoisomers of dimethyl cyclopentane is ________. **[NCERT : P L-270 | NV, June 27, 2022 (I)]**

65. Compound with molecular formula C_3H_6O can show **[NCERT : P L-270 | March 18, 2021 (I)]**
(a) Positional isomerism
(b) Both positional isomerism and metamerism
(c) Metamerism
(d) Functional group isomerism

66. The absolute configuration of

$$\begin{array}{c} CO_2H \\ H-\!\!\!-\!\!\!+\!\!\!-\!\!\!-OH \\ H-\!\!\!-\!\!\!+\!\!\!-\!\!\!-Cl \\ CH_3 \end{array}$$

is : **[2016, 2008 (S)]**
(a) (2*S*, 3*S*) (b) (2*R*, 3*R*)
(c) (2*R*, 3*S*) (d) (2*S*, 3*R*)

67. The number of structural isomers for C_6H_{14} is : **[NCERT : P L-270 | Online April 11, 2015]**
(a) 4 (b) 3 (c) 6 (d) 5

68. Which of the following pairs of compounds are positional isomers ? **[NCERT : P L-270 | Online April 11, 2015]**
(a) $CH_3-CH_2-CH_2-\underset{\underset{O}{\|}}{C}-CH_3$ and $CH_3-CH_2-\underset{\underset{O}{\|}}{C}-CH_2-CH_3$
(b) $CH_3-CH_2-CH_2-CH_2-CHO$ and $CH_3-CH_2-CH_2-\overset{\overset{O}{\|}}{C}-CH_3$
(c) $CH_3-CH_2-CH_2-\underset{\underset{O}{\|}}{C}-CH_3$ and $CH_3-\underset{\underset{CH_3}{|}}{CH}-CH_2-CHO$
(d) $CH_3-CH_2-\underset{\underset{O}{\|}}{C}-CH_2-CH_3$ and $\begin{matrix}CH_3\\CH_3\end{matrix}\!\!>CH-CH_2-CHO$

69. The optically inactive compound from the following is : **[NCERT : P L-270 |Online April 10, 2015]**
(a) 2 - chloropropanal
(b) 2 - chlorobutane
(c) 2 - chloropentane
(d) 2 - chloro - 2- methylbutane

70. Which one of the following acids does not exhibit optical isomerism? **[NCERT : P L-270 | Online April 12, 2014]**
(a) Lactic acid (b) Tartaric acid
(c) Maleic acid (d) α-amino acids

71. Maleic acid and fumaric acids are **[NCERT : P L-271 | Online May 26, 2012]**
(a) chain isomers (b) functional isomers
(c) tautomers (d) geometrical isomers

72. How many cyclic structures are possible for C_4H_6? **[Online May 7, 2012]**
(a) 3 (b) 5 (c) 6 (d) 4

73. Out of the following, the alkene that exhibits optical isomerism is **[NCERT : P L-270 | 2010]**
(a) 3-methyl-2-pentene (b) 4-methyl-1-pentene
(c) 3-methyl-1-pentene (d) 2-methyl-2-pentene

74. The alkene that exhibits geometrical isomerism is: **[NCERT : P L-271 | 2009]**
(a) 2- methyl propene (b) 2-butene
(c) 2- methyl -2- butene (d) propene

75. The number of stereoisomers possible for a compound of the molecular formula **[NCERT : P L-270 |2009]**
$CH_3-CH=CH-CH(OH)-Me$ is:
(a) 2 (b) 4 (c) 6 (d) 3

76. Which of the following molecules is expected to rotate the plane of plane-polarised light? **[NCERT : P L-271 | 2007]**

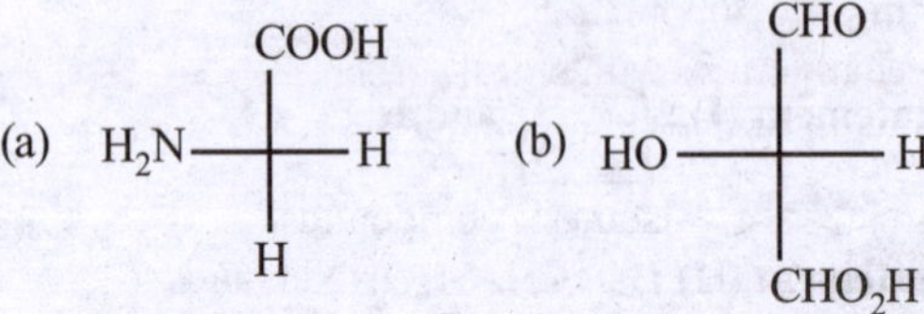

(c)
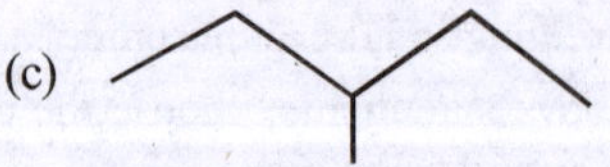

(d)
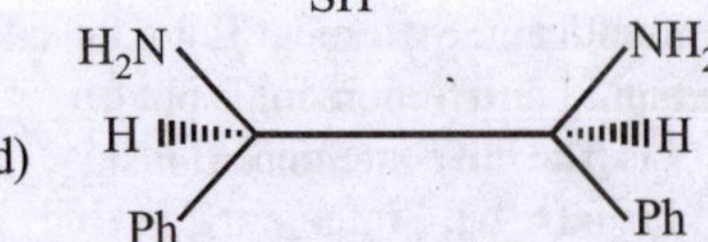

77. Which types of isomerism is shown by 2, 3-dichlorobutane? **[NCERT : P L-270 | 2005]**
(a) Structural (b) Geometric
(c) Optical (d) Diastereo

78. Which of the following will have a mesoisomer also? **[NCERT : P L-271 | 2004]**
(a) 2, 3- Dichloropentane
(b) 2, 3-Dichlorobutane
(c) 2-Chlorobutane
(d) 2-Hydroxypropanoic acid

79. Which of the following compounds is not chiral? **[NCERT : P L-270 | 2004]**
(a) 1-chloro-2-methyl pentane
(b) 2-chloropentane
(c) 1-chloropentane
(d) 3-chloro-2-methyl pentane

80. Among the following four structures I to IV, **[NCERT : P L-270 | 2003]**

$$C_2H_5-\underset{\underset{CH_3}{|}}{CH}-C_3H_7 \text{ (I)}, \quad CH_3-\overset{O}{\overset{\|}{C}}-\overset{CH_3}{\overset{|}{CH}}-C_2H_5 \text{ (II)},$$

$$H-\overset{H}{\overset{|}{\underset{\underset{H}{|}}{C^{\oplus}}}} \text{ (III)}, \quad C_2H_5-\overset{CH_3}{\overset{|}{CH}}-C_2H_5 \text{ (IV)}$$

it is true that
(a) only I and II are chiral compounds
(b) only III is a chiral compound
(c) only II and IV are chiral compounds
(d) all four are chiral compounds

81. A similarity between optical and geometrical isomerism is that **[NCERT : P L-271 | 2002]**
(a) each forms equal number of isomers for a given compound
(b) if in a compound one is present then so is the other
(c) both are included in stereoisomerism
(d) they have no similarity

82. Racemic mixture is formed by mixing two **[NCERT : P L-270 | 2002]**
(a) isomeric compounds
(b) chiral compounds
(c) meso compounds
(d) enantiomers with chiral carbon

83. Following types of compounds (as I, II) **[NCERT : P L-270 | 2002]**

$$CH_3CH=CHCH_3 \text{ (I)} \quad CH_3\underset{\underset{CH_2CH_3}{|}}{CH}OH \text{ (II)}$$

are studied in terms of isomerism in:
(a) chain isomerism (b) position isomerism
(c) conformers (d) stereoisomerism

Topic 3 -Concept of Reaction Mechanism in Organic Compounds, Purification and Quantitative Analysis

84. Match the **LIST-I** with **LIST-II** **[NCERT : P L-276 | April 8, 2025 (II)]**

LIST-I	LIST-II
A. Carbocation	I. Species that can supply a pair of electrons.
B. C-Free radical	II. Species that can receive a pair of electrons.
C. Nucleophile	III. sp^2 hybridized carbon with empty p-orbital.
D. Electrophile	IV. sp^2/sp^3 hybridized carbon with one unpaired electron.

Choose the **correct** answer from the options given below:
(a) A-IV, B-II, C-III, D-I (b) A-II, B-III, C-I, D-IV
(c) A-III, B-IV, C-II, D-I (d) A-III, B-IV, C-I, D-II

85. In Duma's method 292 mg of an organic compound released 50 mL of nitrogen gas (N_2) at 300 K temperature and 715 mm Hg pressure. The percentage composition of 'N' in the organic compound is _____ % (Nearest integer) (Aqueous tension at 300 K = 15 mm Hg) **[NCERT : P L-286 | NV, April 7, 2025 (II)]**

86. Match **List-I** with **List-II** - **[NCERT : P L-279 | April 4, 2025 (II)]**

List-I (Separation of)	List-II (Separation Technique)
(A) Aniline from aniline-water mixture	(I) Simple distillation
(B) Glycerol from spent-lye in soap industry	(II) Fractional distillation
(C) Different fractions of crude oil in petroleum industry	(III) Distillation at reduced pressure
(D) Chloroform-Aniline mixture	(IV) Steam distillation

Choose the **correct** answer from the options given below:
(a) (A)-(IV), (B)-(III), (C)-(II), (D)-(I)
(b) (A)-(I), (B)-(II), (C)-(III), (D)-(IV)
(c) (A)-(III), (B)-(IV), (C)-(I), (D)-(II)
(d) (A)-(II), (B)-(I), (C)-(IV), (D)-(III)

87. In which pairs, the first ion is more stable than the second? **[NCERT : P L-276 | April 4, 2025 (II)]**

(A) 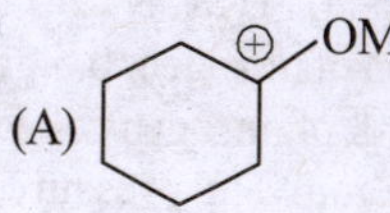

& 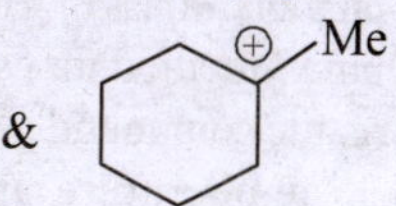

(B) 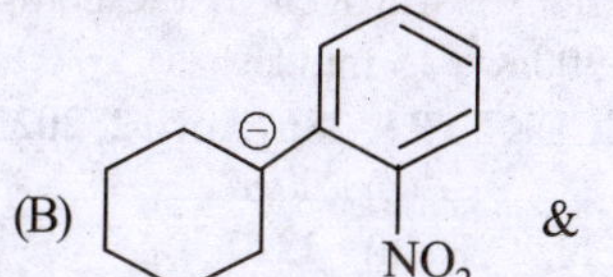

& 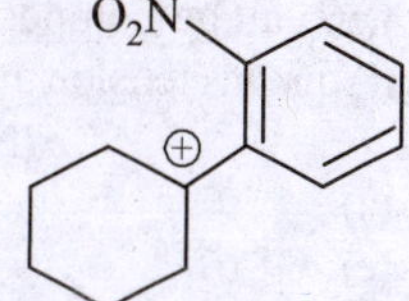

(C) [cyclohexyl–$\overset{\oplus}{C}H_2$] & [cyclohexenyl–$\overset{\oplus}{C}H_2$]

(D) $Me_2\overset{\oplus}{C}$–Me & $Me(MeO)\overset{\oplus}{C}$–Me

(a) (B) & (D) only (b) (A) & (B) only

(c) (B) & (C) only (d) (A) & (C) only

88. In Dumas' method for estimation of nitrogen 1g of an organic compound gave 150 mL of nitrogen collected at 300K temperature and 900 mm Hg pressure. The percentage composition of nitrogen in the compound is _____ % (nearest integer). **[NCERT : P L-286 | NV, April 4, 2025 (I)]** (Aqueous tension at 300 K = 15mm Hg)

89. Given below are two statements :

Statement I : Hyperconjugation is not a permanent effect.

Statement II : In general, greater the number of alkyl groups attached to a positively charged C-atom, greater is the hyperconjugation interaction and stabilization of the cation.

In the light of the above statements, choose the **correct** answer from the options given below

[NCERT : P L-277 | April 3, 2025 (II)]

(a) Statement I is true but Statement II is false

(b) Both Statement I and Statement II are false

(c) Statement I is false but Statement II is true

(d) Both Statement I and Statement II are true

90. In Dumas' method for estimation of nitrogen 0.4 g of an organic compound gave 60 mL of nitrogen collected at 300 K temperature and 715 mm Hg pressure. The percentage composition of nitrogen in the compound is (Given : Aqueous tension at 300 K = 15 mm Hg)

[NCERT : P L-286 | April 3, 2025 (II)]

(a) 15.71% (b) 20.95%

(c) 17.46% (d) 7.85%

91. During estimation of nitrogen by Dumas' method of compound X (0.42 g) :

[NCERT : P L-286 | NV, April 3, 2025 (I)]

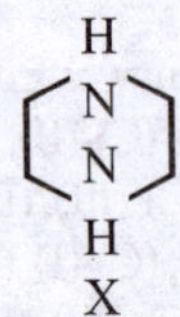

_____ mL of N_2 gas will be liberated at STP. (nearest integer)

(Given molar mass in g mol^{-1} : C : 12, H : 1, N : 14)

92. In Dumas' method for estimation of nitrogen, 0.5 gram of an organic compound gave 60 mL of nitrogen collected at 300 K temperature and 715 mm Hg pressure. The percentage composition of nitrogen in the compound (Aqueous tension at 300 K = 15 mm Hg) is

[NCERT : P L-286 | April 2, 2025 (II)]

(a) 1.257 (b) 20.87

(c) 18.67 (d) 12.57

93. Match **List-I** with **List-II**

[NCERT : P L-279 | April 2, 2025 (II)]

List-I (Purification technique)	**List-II (Mixture of organic compounds)**
(A) Distillation (simple)	(I) Diesel + Petrol
(B) Fractional distillation	(II) Aniline + Water
(C) Distillation under reduced pressure	(III) Chloroform + Aniline
(D) Steam distillation	(IV) Glycerol + Spent-lye

Choose the **correct** answer from the options given below:

(a) (A)-(II), (B)-(III), (C)-(IV), (D)-(I)

(b) (A)-(II), (B)-(IV), (C)-(I), (D)-(III)

(c) (A)-(III), (B)-(IV), (C)-(II), (D)-(I)

(d) (A)-(III), (B)-(I), (C)-(IV), (D)-(II)

94. Consider the following compound (X)

[NCERT : P L-276 | April 2, 2025 (I)]

$$\overset{I}{H-C}\equiv C-\overset{II}{CH_2}-\overset{III}{CH}(CH_3)-\overset{IV}{CH_3}$$

(X)

The most stable and least stable carbon radicals, respectively, produced by homolytic cleavage of corresponding C – H bond are :

(a) II, IV (b) III, II (c) I, IV (d) II, I

95. In the sulphur estimation 0.20 g of a pure organic compound gave 0.40 g of barium sulphate.

The percentage of sulphur in the compound is _____ $\times 10^{-1}$ %. **[NCERT : P L-289 | NV, Jan. 29, 2025 (II)]**

(Molar mass : O = 16, S = 32, Ba = 137 in g mol^{-1})

96. Given below are two statements :

Statement (I) : In partition chromatography, stationary phase is thin film of liquid present in the inert support.

Statement (II) : In paper chromatography, the material of paper acts as a stationary phase.

In the light of the above statements, choose the correct answer from the options given below :

[NCERT : P L-278 | Jan. 29, 2025 (II)]

(a) Both Statement I and Statement II are false

(b) Statement I is true but Statement II is false

(c) Both Statement I and Statement II are true

(d) Statement I is false but Statement II is true

97. Total number of nucleophiles from the following is:

[NCERT : P L-273 | Jan. 29, 2025 (I)]

NH_3, PhSH, $(H_3C)_2S$, $H_2C{=}CH_2$, $\overset{\ominus}{O}H$, $H_3O^{\oplus}$, $(CH_3)_2CO$, $\rangle{=}NCH_3$

(a) 5 (b) 4 (c) 7 (d) 6

98. The purification method based on the following physical transformation is : **[NCERT : P L-278 | Jan. 28, 2025 (II)]**

$$\underset{(X)}{\text{Solid}} \xrightarrow{\text{Heat}} \underset{(X)}{\text{Vapour}} \xrightarrow{\text{Cool}} \underset{(X)}{\text{Solid}}$$

(a) Sublimation (b) Distillation

(c) Crystallization (d) Extraction

99. Quantitative analysis of an organic compound (X) shows following % composition.
[NCERT : PL-288 | NV, Jan. 28, 2025 (I)]
C : 14.5% ; Cl : 64.46% ; H : 1.8%
(Empirical formula mass of the compound (X) is ____ $\times 10^{-1}$
(Given molar mass in g mol^{-1} of C : 12, H : 1, O : 16, Cl : 35.5)

100. The correct order of stability of following carbocations is:
[NCERT : P L-275 | Jan. 28, 2025 (I)]

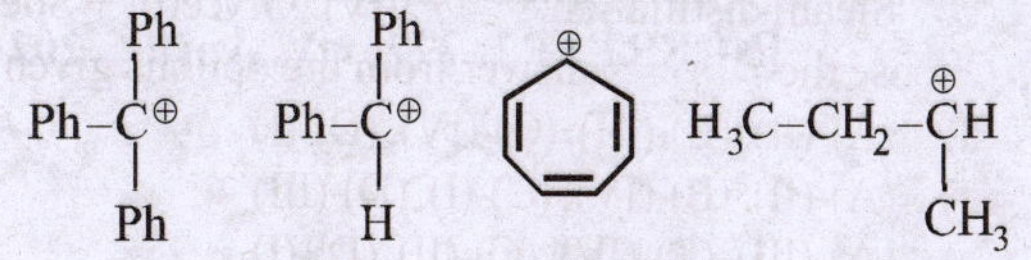

A: $Ph_3C^{\oplus}$ B: $Ph_2CH^{\oplus}$ C: cycloheptatrienyl cation D: $H_3C-CH_2-\overset{\oplus}{C}H-CH_3$

(a) A > B > C > D (b) B > C > A > D
(c) C > B > A > D (d) C > A > B > D

101. In Carius method of estimation of halogen, 0.25 g of an organic compound gave 0.15 g of silver bromide (AgBr). The percentage of Bromine in the organic compound is ________ $\times 10^{-1}$% (Nearest integer).
(Given : Molar mass of Ag is 108 and Br is 80 g mol^{-1})
[NCERT : P L-288 | NV, Jan. 24, 2025 (II)]

102. Given below are two statements I and II.
[NCERT : P L-286 | Jan. 24, 2025 (I)]
Statement I : Dumas method is used for estimation of "Nitrogen" in an organic compound.
Statement II : Dumas method involves the formation of ammonium sulphate by heating the organic compound with conc H_2SO_4.
In the light of the above statements, choose the correct answer from the options given below.
(a) Both Statement I and Statement II are true
(b) Statement I is false but Statement II is true
(c) Both Statement I and Statement II are false
(d) Statement I is true but Statement II is false

103. Which one of the carbocations from the following is most stable? **[NCERT : P L-276 | Jan. 24, 2025 (I)]**
(a) 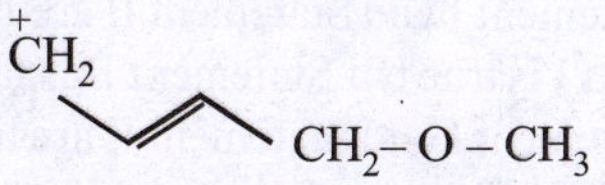 $\overset{+}{C}H_2-CH=CH-CH_2-O-CH_3$
(b) $\overset{+}{C}H_2-CH=CH-O-CH_3$
(c) $\overset{+}{C}H_2-CH=CH-O-CO-CH_3$
(d) 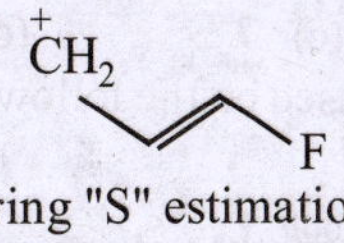 $\overset{+}{C}H_2-CH=CH-F$

104. During "S" estimation, 160 mg of an organic compound gives 466 mg of barium sulphate. The percentage of Sulphur in the given compound is ______%
(Given molar mass in g mol^{-} of Ba : 137, S : 32, O : 16)
[NCERT : P L-289 | NV, Jan. 23, 2025 (I)]

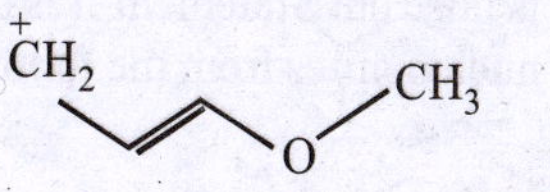

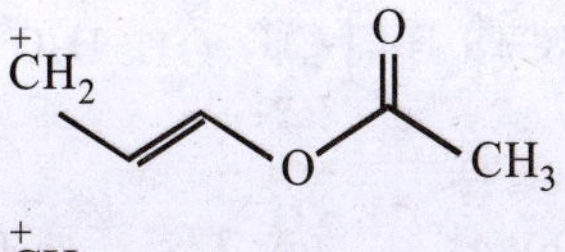

105. The most stable carbocation from the following is :
[NCERT : P L-277 | Jan. 22, 2025 (II)]
(a) 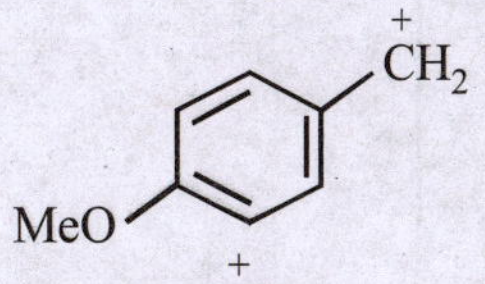 p-MeO-C_6H_4-$\overset{+}{C}H_2$
(b) m-MeO-C_6H_4-$\overset{+}{C}H_2$ (OMe)
(c) p-H_3C-C_6H_4-$\overset{+}{C}H_2$
(d) 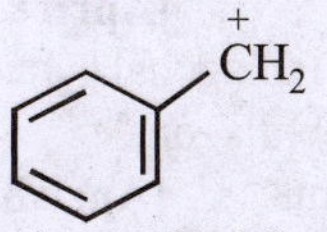 C_6H_5-$\overset{+}{C}H_2$

106. In Carius method for estimation of halogens, 180 mg of an organic compound produced 143.5 mg of AgCl. The percentage composition of chlorine in the compound is ______ %. **[NCERT : P L-288 | NV, Jan. 22, 2025 (I)]**
[Given : molar mass in g mol^{-1} of Ag : 108, Cl = 35.5]

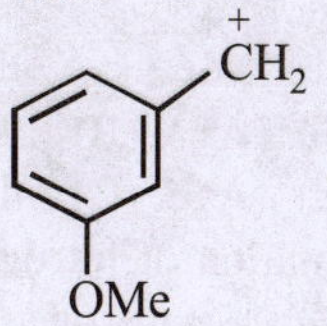

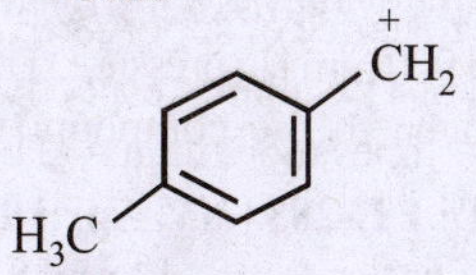

107. Methods used for purification of organic compounds are based on : **[NCERT : P L-278 | April 9, 2024 (I)]**
(a) neither on nature of compound nor on the impurity present.
(b) nature of compound only.
(c) nature of compound and presence of impurity.
(d) presence of impurity only.

108. Relative stability of the contributing structures is :

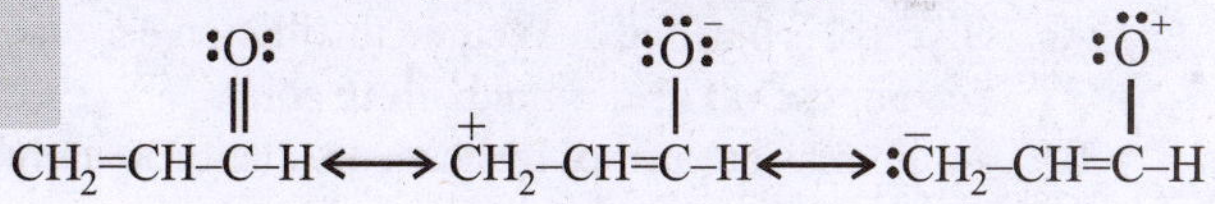

$$\underset{(I)}{CH_2=CH-\overset{\ddot{O}:}{\overset{\|}{C}}-H} \longleftrightarrow \underset{(II)}{\overset{+}{C}H_2-CH=\overset{:\ddot{O}:^-}{\overset{|}{C}}-H} \longleftrightarrow \underset{(III)}{:\overset{-}{C}H_2-CH=\overset{:\ddot{O}^+}{\overset{|}{C}}-H}$$

[NCERT : P L-276 | April 9, 2024 (I)]
(a) (I) > (III) > (II) (b) (I) > (II) > (III)
(c) (II) > (I) > (III) (d) (III) > (II) > (I)

109. How many compounds among the following compounds show inductive, mesomeric as well as hyperconjugation effects? **[NCERT : P L-274, 275, 276 | NV, April 9, 2024 (I)]**

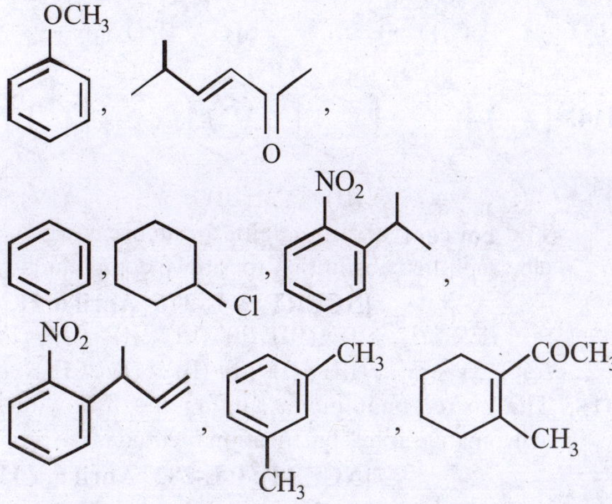

Anisole (OCH_3 on benzene), $CH_3-CH(CH_3)-CH=CH-CO-CH_3$, benzene, chlorocyclohexane (Cl), 1-isopropyl-2-nitrobenzene (NO_2), 3-(2-nitrophenyl)but-1-ene (NO_2), m-xylene (CH_3, CH_3), 1-acetyl-2-methylcyclohexene ($COCH_3$, CH_3)

110. 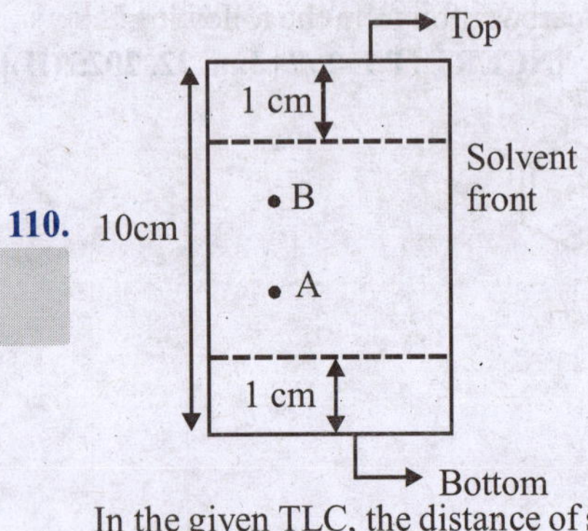

In the given TLC, the distance of spot A & B are 5 cm & 7 cm, from the bottom of TLC plate, respectively. R_f value of B is $x \times 10^{-1}$ times more than A. The value of x is_____. **[NCERT : P L-283 | NV, April 9, 2024 (II)]**

111. The shape of carbocation is : **[April 8, 2024 (II)]**

(a) trigonal planar (b) diagonal pyramidal
(c) tetrahedral (d) diagonal

112. Given below are two statements: **[April 8, 2024 (II)]**

Statement (I): Kjeldahl method is applicable to estimate nitrogen in pyridine.

Statement (II): The nitrogen present in pyridine can easily be converted into ammonium sulphate in Kjeldahl method.

In the light of the above statements, choose the **correct** answer from the options given below:

(a) Both **Statement I** and **Statement II** is false
(b) **Statement I** is false but **Statement II** is true
(c) Both **Statement I** and **Statement II** is true
(d) **Statement I** is true but **Statement II** is false

113. Which of the following statements are correct ? **[NCERT : P L-281 | April 6, 2024 (I)]**

A. Glycerol is purified by vacuum distillation because it decomposes at its normal boiling point.
B. Aniline can be purified by steam distillation as aniline is miscible in water.
C. Ethanol can be separated from ethanol water mixture by azeotropic distillation because it forms azeotrope.
D. An organic compound is pure, if mixed M.P. is remained same.

Choose the **most appropriate** answer from the options given below :

(a) A, B, C only (b) A, C, D only
(c) B, C, D only (d) A, B, D only

114. 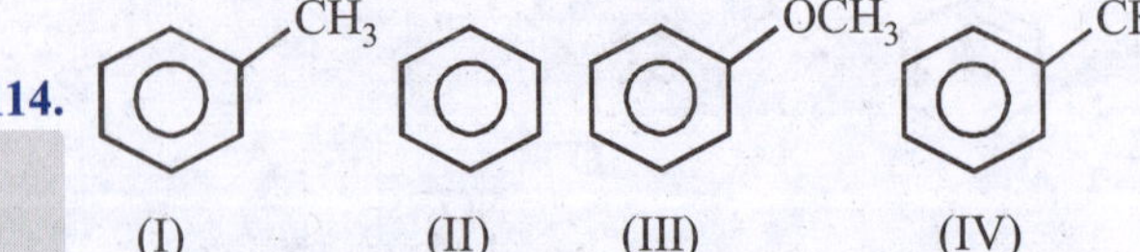

The **correct** arrangement for decreasing order of electrophilic substitution for above compounds **[NCERT : P L-276 | April 6, 2024 (II)]**

(a) (IV) > (I) > (II) > (III) (b) (III) > (I) > (II) > (IV)
(c) (II) > (IV) > (III) > (I) (d) (III) > (IV) > (II) > (I)

115. The correct statements among the following, for a "chromatography" purification method is: **[NCERT : P L-282 | April 6, 2024 (II)]**

(a) Organic compounds run faster than solvent in the thin layer chromatographic plate.
(b) Non-polar compounds are retained at top and polar compounds come down in column chromatography.
(c) R_f of a polar compound is smaller than that of a non-polar compound.
(d) R_f is an integral value.

116. Number of carbocation from the following that are **not** stabilized by hyperconjugation is……….. **[NCERT : P L-277 | NV, April 6, 2024 (II)]**

(tert-Butyl) (tert.-Butyl), $\overset{+}{C}H_3$

$\overset{+}{C}H_2-OCH_3$, $>N-\overset{+}{C}H_2$

117. Using the given figure, the ratio of R_f values of sample A and sample C is $x \times 10^{-2}$. Value of x is_____. **[NCERT : P L-283 | NV, April 5, 2024 (II)]**

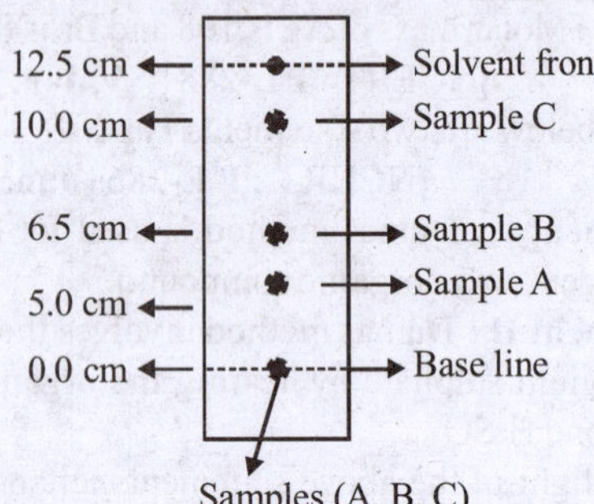

Fig : Paper chromatography of Samples

118. Match List - I with List II : **[NCERT : P L-276, 277 | April 4, 2024 (I)]**

List-I (Mechanism steps)	List-II (Effect)
(A) $\ddot{N}H_2$ ↔ $\overset{+}{N}H_2$	(I) – E effect
(B) + H^+ →	(II) – R effect
(C) 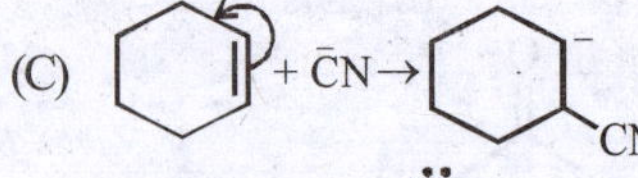 + $\bar{C}N$ → CN	(III) + E effect
(D) O ← N = O ↔ N → O	(IV) + R effect

Choose the **correct** answer from the options given below:

(a) A – (IV), B – (III), C – (I), D – (II)
(b) A – (III), B – (I), C – (II), D – (IV)
(c) A – (II), B – (IV), C) – (III), D – (I)
(d) A – (I), B – (II), C – (IV), D – (III)

119. Which among the following is **incorrect** statement? **[NCERT : P L-277 | April 4, 2024 (I)]**
(a) Electromeric effect dominates over inductive effect
(b) The electromeric effect is a temporary effect
(c) The organic compound shows electromeric effect in the presence of the reagent only
(d) Hydrogen ion (H^+) shows negative electromeric effect

120. The adsorbent used in adsorption chromatography is/are
(A) silica gel (B) alumina
(C) quick lime (D) magnesia
Choose the **most appropriate** answer from the options given below : **[NCERT : P L-278 | April 4, 2024 (II)]**
(a) B only (b) C and D only
(c) A and B only (d) A only

121. In Kjeldahl's method for estimation of nitrogen, $CuSO_4$ acts as : **[NCERT : P L-287 | Feb. 1, 2024 (I)]**
(a) reducing agent (b) catalytic agent
(c) hydrolysis agent (d) oxidising agent

122. Following Kjeldahl's method, 1g of organic compound released ammonia, that neutralised 10 mL of 2M H_2SO_4. The percentage of nitrogen in the compound is ______ %. **[NCERT : P L-287 | NV, Feb. 1, 2024 (II)]**

123. Ionic reactions with organic compounds proceed through:
(A) homolytic bond cleavage **[NCERT : P L-272 | Feb. 1, 2024 (I)]**
(B) heterolytic bond cleavage
(C) free radical formation
(D) primary free radical
(E) secondary free radical
Choose the correct answer from the options given below:
(a) (A) only (b) (C) only
(c) (B) only (d) (D) and (E) only

124. The functional group that shows negative resonance effect is: **[NCERT : P L-276 | Feb. 1, 2024 (II)]**
(a) –OH (b) –OR (c) –COOH (d) $-NH_2$

125. A species having carbon with sextet of electrons and can act as electrophile is called **[NCERT : P L-271 | Jan. 31, 2024 (I)]**
(a) pentavalent carbon (b) carbon free radical
(c) carbocation (d) carbanion

126. 'Adsorption' principle is used for which of the following purification method? **[NCERT : P L-282 | Jan. 31, 2024 (I)]**
(a) Chromatography (b) Sublimation
(c) Extraction (d) Distillation

127. Match List I with List II **[NCERT : P L-279 | Jan. 31, 2024 (I)]**

	LIST I (Technique)		LIST II (Application)
A.	Distillation	I.	Separation of glycerol from spent-lye
B.	Fractional distillation	II.	Aniline- Water mixture
C.	Steam distillation	III.	Separation of crude oil fractions
D.	Distillation under reduced pressure	IV.	Chloroform - Aniline

Choose the correct answer from the options given below:
(a) A-IV, B-I, C-II, D-III (b) A-IV, B-III, C-II, D-I
(c) A-I, B-II, C-IV, D-III (d) A-II, B-III, C-I, D-IV

128. The fragrance of flowers is due to the presence of some steam volatile organic compounds called essential oils. These are generally insoluble in water at room temperature but are miscible with water vapour in vapour phase. A suitable method for the extraction of these oils from the flowers is - **[NCERT : P L-281 | Jan. 31, 2024 (II)]**
(a) crystallisation
(b) distillation under reduced pressure
(c) distillation
(d) steam distillation

129. Which of the following molecule / species is most stable? **[NCERT : P L-275 | Jan. 30, 2024 (I)]**

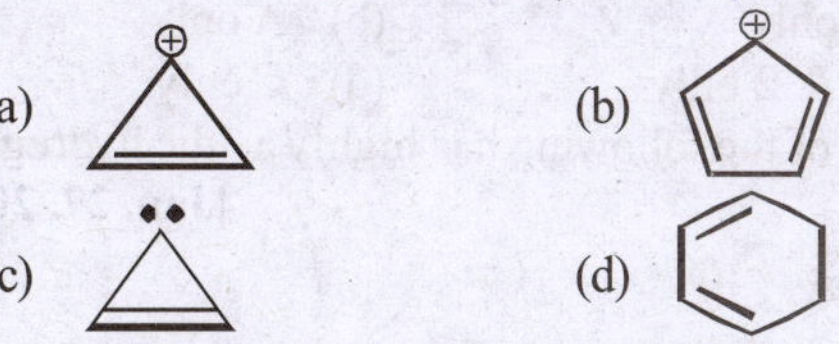

130. On a thin layer chromatographic plate, an organic compound moved by 3.5 cm, while the solvent moved by 5 cm. The retardation factor of the organic compound is ______ $\times 10^{-1}$. **[NCERT : P L-283 | NV, Jan. 30, 2024 (I)]**

131. Which among the following purification methods is based on the principle of "Solubility" in two different solvents? **[NCERT : P L-281 | Jan. 30, 2024 (II)]**
(a) Column Chromatography
(b) Sublimation
(c) Distillation
(d) Differential Extraction

132. The correct stability order of carbocations is **[NCERT : P L-271 | Jan. 30, 2024 (II)]**
(a) $(CH_3)_3C^+ > CH_3-\overset{+}{C}H_2 > (CH_3)_2\overset{+}{C}H > \overset{+}{C}H_3$
(b) $\overset{+}{C}H_3 > (CH_3)_2\overset{+}{C}H > CH_3-\overset{+}{C}H_2 > (CH_3)_3\overset{+}{C}$
(c) $(CH_3)_3\overset{+}{C} > (CH_3)_2\overset{+}{C}H > CH_3-\overset{+}{C}H_2 > \overset{+}{C}H_3$
(d) $\overset{+}{C}H_3 > CH_3-\overset{+}{C}H_2 > CH_3-\underset{CH_3}{\underset{|}{\overset{+}{C}H}} > (CH_3)_3C^+$

133. The interaction between π bond and lone pair of electrons present on an adjacent atom is responsible for **[NCERT : P L-276 | Jan. 29, 2024 (I)]**
(a) Resonance effect (b) Electrometric effect
(c) Inductive effect (d) Hyper conjugation

134. The ascending acidity order of the following H atoms is **[Jan. 29, 2024 (II)]**

$HC \equiv C-Ⓗ$ (A)

$H_2C = CH-Ⓗ$ (B)

$(H_3C)_3C-$(H) $\quad$ H_3C-CH_2-(H)

C $\quad$ D

(a) A < B < C < D $\quad$ (b) D < C < B < A
(c) A < B < D < C $\quad$ (d) C < D < B < A

135. Chromatographic technique/s based on the principle of differential adsorption is/are **[Jan. 29, 2024 (II)]**

A. Column chromatography
B. Thin layer chromatography
C. Paper chromatography

Choose the **most appropriate** answer from the options given below:

(a) B only $\quad$ (b) A only
(c) A & B only $\quad$ (d) C only

136. Which of the following has highly acidic hydrogen? **[Jan. 27, 2024 (I)]**

(a) $H_3C-CO-CO-CH_2-CH_2-CH_3$

(b) $H_3C-CO-CH_2-CH_2-CO-CH_3$

(c) $H_3C-CH_2-CH_2-CO-CH_2-CH_2-CH_3$

(d) $H_3C-CO-CH_2-CO-CH_2-CH_3$

137. Among the following, total number of meta directing functional groups is (Integer based)
$-OCH_3$, $-NO_2$, $-CN$, $-CH_3$ $-NHCOCH_3$, $-COR$, $-OH$, $-COOH$, $-Cl$ **[NV, Jan. 27, 2024 (I)]**

138. The order of relative stability of the contributing structure is: **[NCERT : P L-276 | Jan. 27, 2024 (II)]**

$CH_2=CH-C(=\ddot{O}:)-H$ (I) $\longleftrightarrow$ $\overset{\oplus}{C}H_2-CH=C(-\ddot{O}:^{\ominus})-H$ (II) $\longleftrightarrow$ $:\overset{-}{C}H_2-CH=C(-\ddot{O}:^{\oplus})-H$ (III)

Choose the correct answer from the options given below:

(a) I > II > III $\quad$ (b) II > I > III
(c) I = II = III $\quad$ (c) III > II > I

139. The technique used for purification of steam volatile water immiscible substance is: **[NCERT : P L-281 | Jan. 27, 2024 (II)]**

(a) Fractional distillation
(b) Fractional distillation under reduced pressure
(c) Distillation
(d) Steam distillation

140. Decreasing order of reactivity towards electrophilic substitution for the following compounds is : **[NCERT : P L-276 | April 15, 2023 (I)]**

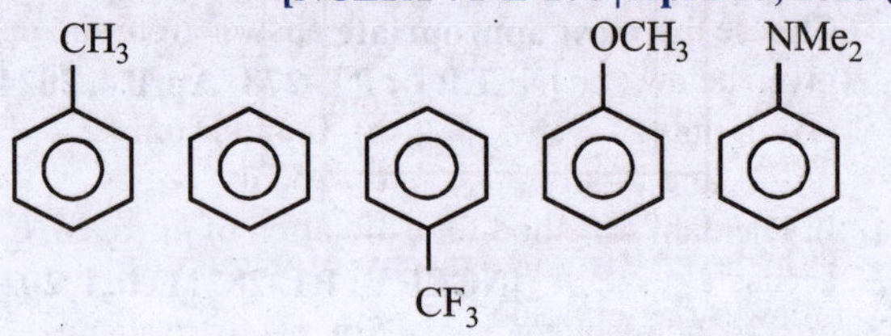

(a) c > b > a > d > e $\quad$ (b) e > d > a > b > c
(c) a > d > e > b > c $\quad$ (d) d > a > e > c > b

141. Which of the following statement is correct for paper chromatography ?**[NCERT : P L-284 | April 15, 2023 (I)]**

(a) Water present in the mobile phase gets absorbed by the paper which then forms the stationary phase.
(b) Water present in the pores of the paper forms the stationary phase.
(c) Paper sheet forms the stationary phase.
(d) Paper and water present in its pores together form the stationary phase.

142.

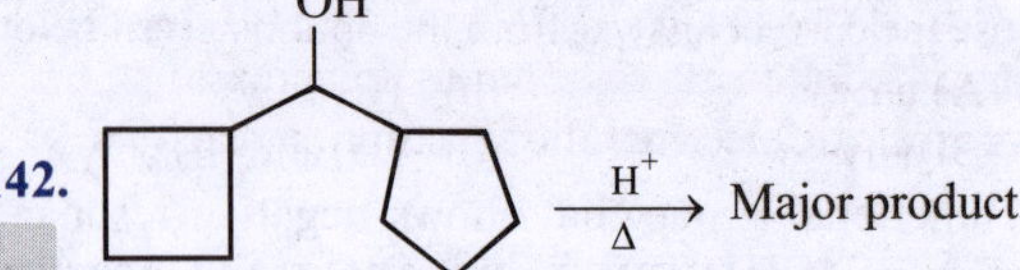

[NCERT : P L-272 | April 13, 2023 (I)]

In the above reaction. Left hand side and right hand side rings are named as 'A' and 'B' respectively. The undergo ring expansion. The correct statement for this process is:

(a) Finally both rings will become six membered each.
(b) Finally both rings will become five membered each.
(c) Only 'A' will become 6 membered.
(d) Ring expansion can go upto seven membered rings.

143. For the given reaction **[NCERT : P L-278 | NV, April 13, 2023 (I)]**

$CH_3-C(CH_3)(CH_3)-CH(OH)-CH(CH_3)-CH_3 \xrightarrow[\Delta]{H^+}$

'A'

The total number of possible products formed by tertiary carbocation of A is ______.

144. 0.400 g of an organic compound (X) gave 0.376 g of AgBr in Carius method for estimation of bromine. % of bromine in the compound (X) is ______.
(Given: Molar mass AgBr = 188 g mol^{-1} Br = 80 g mol^{-1}) **[NCERT : P L-289 | NV, April 13, 2023 (II)]**

145. Three organic compounds A, B and C were allowed to run in thin layer chromatography using hexane and gave the following result (see figure). The R_f value of the most polar compound is ______ $\times 10^{-2}$
[NCERT : P L-283 | NV, April 12, 2023 (I)]

A 6 cm B C 8 cm 4 cm 2 cm

146. Thin layer chromatography of a mixture shows the following observation :
[NCERT : P L-283 | April 11, 2023 (I)]

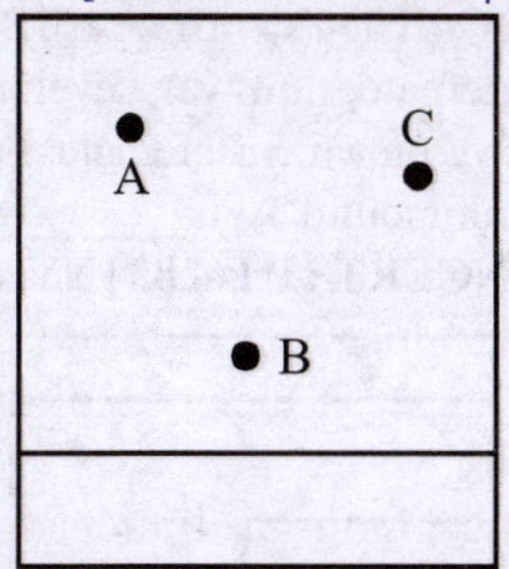

The correct order of elution in the silica gel column chromatography is
(a) A, C, B (b) B, C, A (c) C, A, B (d) B, A, C

147. Arrange the following compounds in increasing order of rate of aromatic electrophilic substitution reaction
[NCERT : P L-276 | April 11, 2023 (I)]

OH (a) OH (b) O (c) OH O (d)

(a) d, b, c, a (b) b, c, a, d
(c) c, a, b, d (d) d, b, a, c

148. Me OH $\xrightarrow{HBr}$ 'A' Major Product

The number of hyperconjugation structures involved to stabilize carbocation formed in the above reaction is ______. **[NCERT : P L-277 | NV, April 11, 2023 (I)]**

149. Compound from the following that will not produce precipitate on reaction with $AgNO_3$ is:
[NCERT : P L-273 | April 11, 2023 (II)]

(a) Br (1-bromocyclohexene)
(b) Br (benzyl bromide)
(c) Br (bromocyclopropene)
(d) C_6H_5–CH=CH–CH_2–Br

150. Using column chromatography, mixture of two compounds 'A' and 'B' was separated. 'A' eluted first, this indicates 'B' has
[NCERT : P L-283 | April 10, 2023 (I)]

(a) low R_f, weaker adsorption
(b) high R_f, stronger adsorption
(c) high R_f, weaker adsorption
(d) low R_f, stronger adsorption

151. In Carius tube, an organic compound 'X' is treated with sodium peroxide to form a mineral acid 'Y'. The solution of $BaCl_2$ is added to 'Y' to form a precipitate 'Z'. 'Z' is used for the quantitative estimation of an extra element. 'X' could be: **[NCERT : P L-289 | April 10, 2023 (II)]**

(a) Cytosine (b) Chloroxylenol
(c) A nucleotide (d) Methionine

152. The decreasing order of hydride affinity for following carbocations is: **[NCERT : P L-277 | April 10, 2023 (II)]**

A. $CH_2 = CH - \overset{+}{C}(CH_3) - CH_3$ B. $C_6H_5 - \overset{+}{C}(C_6H_5) - C_6H_5$

C. $H_3C - \overset{+}{C}(CH_3) - CH_3$ D. (tricyclopropyl carbocation, $\overset{+}{C}$)

Choose the correct answer from the options given below:
(a) A, C, B, D (b) C, A, B, D
(c) C, A, D, B (d) A, C, D, B

153. Match List I with List II
[NCERT : P L-284 | April 06, 2023 (I)]

	List I (Element detected)		**List II (Reagent used/ Product formed)**
A	Nitrogen	I.	$Na_2[Fe(CN)_5NO]$
B	Sulphur	II.	$AgNO_3$
C	Phosphorus	III.	$Fe_4[Fe(CN)_6]_3$
D	Halogen	IV.	$(NH_4)_2MoO_4$

Choose the correct answer from the options given below:
(a) A–II, B–IV, C–I, D–III (b) A–IV, B–II, C–I, D–III
(c) A–II, B–I, C–IV, D–III (d) A–III, B–I, C–IV, D-II

154. From the figure of column chromatography given below, identify incorrect statements. **[NCERT : P L-283 | April 06, 2023 (II)]**

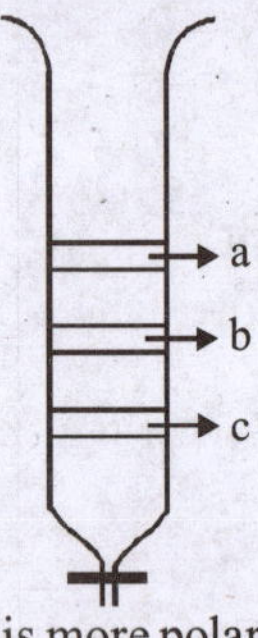

A. Compound 'c' is more polar than 'a' and 'b'
B. Compound 'a' is least polar
C. Compound 'b' comes out of the column before 'c' and after 'a'
D. Compound 'a' spends more time in the column

Choose the correct answer from the options given below:-
(a) A, B and C only (b) B, C and D only
(c) A, B and D only (d) B and D only

155. Match items of column I and II **[NCERT : P L-279, 280 | Jan. 31, 2023 (I)]**

Column I (Mixture of compounds)	**Column II (Separation technique)**
A. H_2O/CH_2Cl_2	(i) Crystallization
B. (3,4-dihydronaphthalen-1(2H)-one) / (4-nitrophenol)	(ii) Differential solvent extraction
C. Kerosene/Naphthalene	(iii) Column chromatography
D. $C_6H_{12}O_6$/NaCl	(iv) Fractional distillation

Correct match is :
(a) A-(iii), B-(iv), C-(ii), D-(i) (b) A-(i), B-(iii), C-(ii), D-(iv)
(c) A-(ii), B-(iii), C-(iv), D-(i) (d) A-(ii), B-(iv), C-(i), D-(iii)

156. In Dumas method for the estimation of N_2, the sample is heated with copper oxide and the gas evolved is passed over: **[NCERT : P L-286 | Jan. 31, 2023 (II)]**
(a) Ni (b) Copper gauze
(c) Pd (d) Copper oxide

157. What is the correct order of acidity of the protons marked A–D in the given compounds ? **[Jan. 30, 2023 (I)]**

H_D, CO_2H_C, H_A, H_B

(a) $H_C > H_D > H_B > H_A$ (b) $H_C > H_D > H_A > H_B$
(c) $H_D > H_C > H_B > H_A$ (d) $H_C > H_A > H_D > H_B$

158. The most stable carbocation for the following is: **[NCERT : P L-276 | Jan. 30, 2023 (II)]**

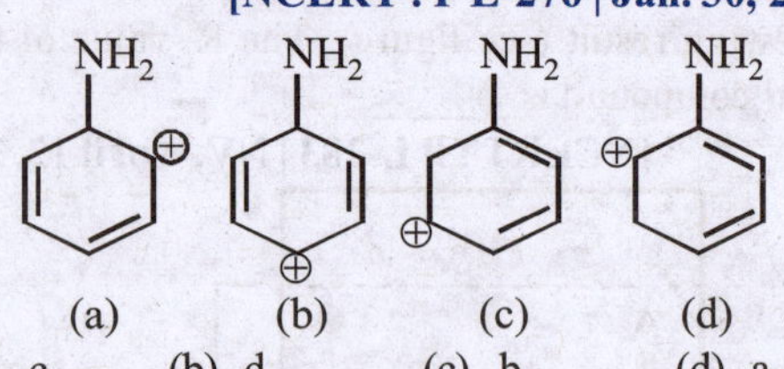

(a) c (b) d (c) b (d) a

159. Match List I with List II: **[NCERT : P L-279 | Jan. 30, 2023 (II)]**

List I (Mixture)	**List II (Separation technique)**
(A) $CHCl_3 + C_6H_5NH_2$	I Steam distillation
(B) $C_6H_{14} + C_5H_{12}$	II Differential extraction
(C) $C_6H_5NH_2 + H_2O$	III Distillation
(D) Organic compound in H_2O	IV Fractional distillation

(a) A-IV, B-I, C-III, D-II (b) A-III, B-IV, C-I, D-II
(c) A-II, B-I, C-III, D-IV (d) A-III, B-I, C-IV, D-II

160. Following chromatogram was developed by adsorption of compound 'A' on a 6 cm TLC glass plate. Retardation factor of the compound 'A' is ______ $\times 10^{-1}$. **[NCERT : P L-283 | NV, Jan. 29, 2023 (I)]**

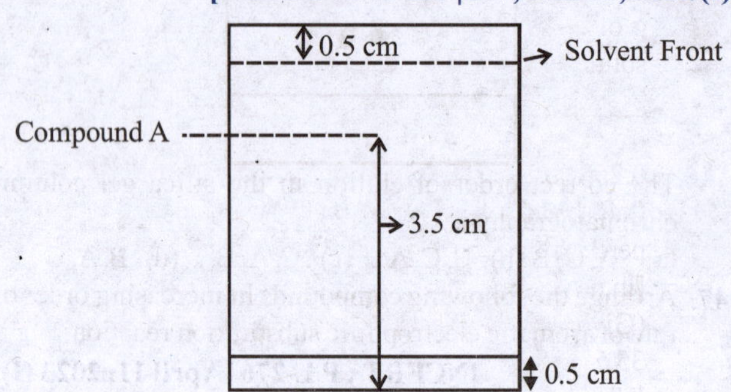

161. In sulphur estimation. 0.471 g of an organic compound gave 1.4439 g of barium sulphate.
The percentage of sulphur in the compound is ______ (Nearest Integer) **[NCERT : P L-288 | NV, Jan. 25, 2023 (I)]**
(Given: Atomic mass Ba: 137 u: S: 32 u, O: 16 u)

162. Number of moles of AgCl formed in the following reaction is ______. **[NV, Jan. 24, 2023 (I)]**

$\xrightarrow{AgNO_3}$ (A) + X AgCl↓

163. A sample of 0.125 g of an organic compound when analyzed by Duma's method yields 22.78 mL of nitrogen gas collected over KOH solution at 280 K and 759 mm Hg. The percentage of nitrogen in the given organic compound is ______. (Nearest integer)
Given : [NCERT : P L-286 | NV, July 28, 2022 (II)]
(a) The vapour pressure of water of 280 K is 14.2 mm Hg.
(b) $R = 0.082$ L atm K^{-1} mol^{-1}

164. In Carius method of estimation of halogen, 0.45 g of an organic compound gave 0.36 g of AgBr. Find out the percentage of bromine in the compound.
(Molar masses : AgBr = 188 g mol^{-1}: Br = 80 g mol^{-1})
[NCERT : P L-289 | July 27, 2022 (I); Same June 28, 2022 (I), Aug. 27, 2021 (I), July 20, 2021 (II)]
(a) 34.04% (b) 40.04% (c) 36.03% (d) 38.04%

165. In the following reaction

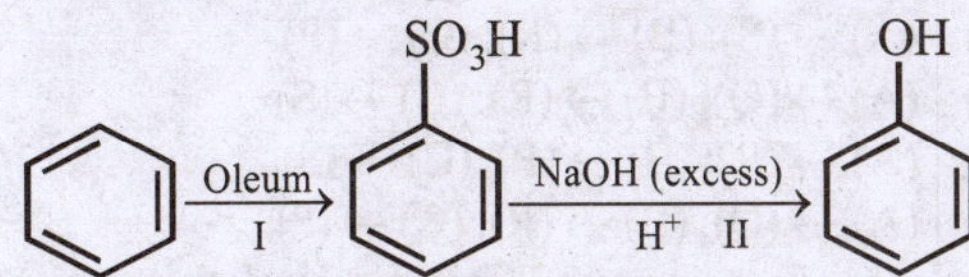

The % yield for reaction I is 60% and that of reaction II is 50%. The overall yield of the complete reaction is ___% [nearest integer] **[NV, July 27, 2022 (I)]**

166. While estimating the nitrogen present in an organic compound by Kjeldahl's method, the ammonia evolved from 0.25 g of the compound neutralized 2.5 mL of 2 M H_2SO_4. The percentage of nitrogen present in organic compound is ________ .
[NCERT : P L-287 | NV, July 25, 2022 (I)]

167. Kjeldahl's method was used for the estimation of nitrogen in an organic compound. The ammonia evolved from 0.55 g of the compound neutralised 12.5 mL of 1 M H_2SO_4 solution. The percentage of nitrogen in the compound is ________. (Nearest integer)
[NCERT : P L-287 | NV, June 29, 2022 (I)]

168. 0.25 g of an organic compound containing chlorine gave 0.40 g of silver chloride in Carius estimation. The percentage of chlorine present in the compound is ______. [in nearest integer]
(Given: Molar mass of Ag is 108 g mol^{-1} and that of Cl is 35.5 g mol^{-1}) **[NCERT : PL-289 | NV, June 27, 2022 (II)]**

169. Which one of the following techniques is not used to spot components of a mixture separated on thin layer chromatographic plate? **[June 28, 2022 (I)]**
(a) I_2 (Solid)
(b) U.V. Light
(c) Visualisation agent as a component of mobile phase
(d) Spraying of an appropriate reagent

170. Which of the following is structure of a separating funnel? **[NCERT : P L-282 | June 27, 2022 (I)]**

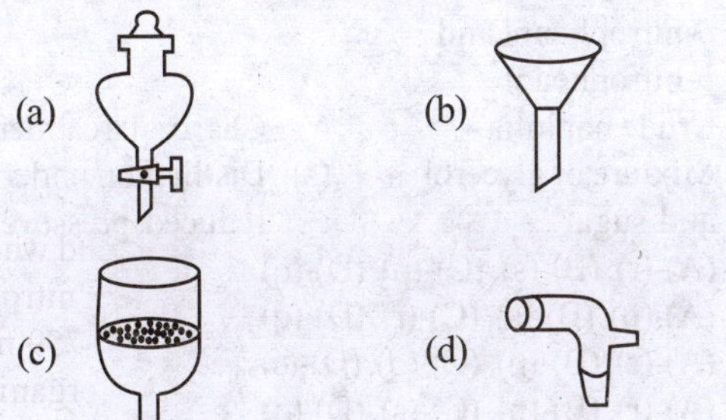

171. The transformation occurring in Duma's method is given below : **[NCERT : P L-286 | NV, Aug. 31, 2021 (II)]**

$$C_2H_7N + \left(2x + \frac{y}{2}\right)CuO \rightarrow$$
$$xCO_2 + \frac{y}{2}H_2O + \frac{z}{2}N_2 + \left(2x + \frac{y}{2}\right)Cu$$

The value of y is ________. (Integer answer)

172. The number of moles of CuO, that will be utilized in Dumas method for estimation nitrogen in a sample of 57.5g of N, N-dimethylaminopentane is ________ × 10^{-2}. (Nearest integer) **[NCERT : PL-286 | NV, Aug. 27, 2021 (I)]**

173. In the sulphur estimation, 0.471 g of an organic compound gave 1.44 g of barium sulphate. The percentage of sulphur in the compound is _____%. (Nearest integer)
(Atomic Mass of Ba = 137 u)
[NCERT : PL-288 | NV, Aug. 26, 2021 (II)]

174. An organic compound is subjected to chlorination to get compound A using 5.0 g of chlorine. When 0.5 g of compound A is reacted with $AgNO_3$ [Carius Method], the percentage of chlorine in compound A is _______ when it forms 0.3849 g of AgCl. (Round off to the nearest integer) (Atomic masses of Ag and Cl are 107.87 and 35.5 respectively) **[NCERT : PL-289 | NV, July 27, 2021 (I)]**

175. In Carius method, halogen containing organic compound is heated with fuming nitric acid in the presence of :
[NCERT : P L-289 | July 20, 2021 (II)]
(a) HNO_3 (b) $AgNO_3$ (c) $CuSO_4$ (d) $BaSO_4$

176. In Duma's method of estimation of nitrogen, 0.1840 g of an organic compound gave 30 mL of nitrogen collected at 287 K and 758 mm of Hg pressure. The percentage composition of nitrogen in the compound is________. (Round off to the nearest Integer).
[NCERT : P L-286 | NV, March 16, 2021 (II)]
[Given : Aqueous tension at 287 K = 14 mm of Hg]

177. Which of the following is a 'FALSE' statement?
[NCERT : P L-289 | Feb. 26, 2021 (I)]
(a) Carius method is used for the estimation of nitrogen in an organic compound
(b) Phosphoric acid produced on oxidation of phosphorus present in an organic compound is precipitated as $Mg_2P_2O_7$ by adding magnesia mixture
(c) Kjeldahl's method is used for the estimation of nitrogen in an organic compund
(d) Carius tube is used for the estimation of sulphur in an organic compound

178. In an estimation of bromine by Carius method, 1.6 g of an organic compound gave 1.88 g of AgBr. The mass percentage of bromine in the compound is ______.
(Atomic mass, Ag = 108, Br = 80 g mol^{-1})
[NCERT : PL-288 | NV, Sep. 06, 2020 (I)]

179. The Kjeldahl method of nitrogen estimation fails for which of the following reaction products?
[NCERT : P L-287 | Sep. 03, 2020 (I)]

(1) Nitrobenzene (NO_2) $\xrightarrow{Sn/HCl}$ (2) Benzonitrile (CN) $\xrightarrow{LiAlH_4}$

(3) 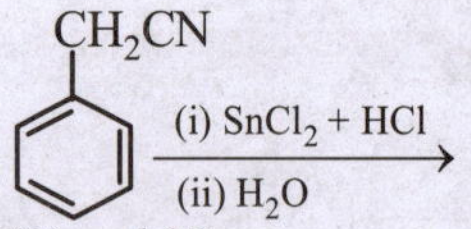 $\text{C}_6\text{H}_5\text{CH}_2\text{CN} \xrightarrow[\text{(ii) } H_2O]{\text{(i) } SnCl_2 + HCl}$ (4) 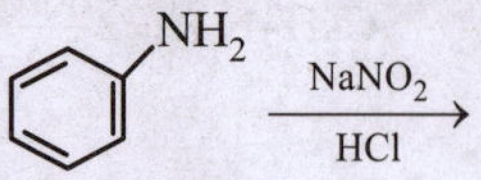 $\text{C}_6\text{H}_5\text{NH}_2 \xrightarrow[\text{HCl}]{NaNO_2}$

(a) (3) and (4) (b) (1) and (4)
(c) (1), (3) and (4) (d) (2) and (3)

180. In Carius method of estimation of halogen, 0.172 g of an organic compound showed presence of 0.08 g of bromine. Which of these is the correct structure of the compound? **[NCERT : P L-289 | Sep. 02, 2020 (I)]**

(a) H_3C-CH_2-Br (b) 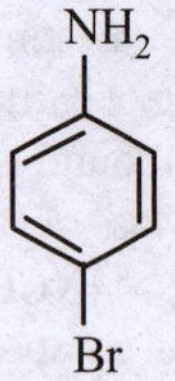 4-bromoaniline (NH_2 and Br para on benzene ring)
(c) 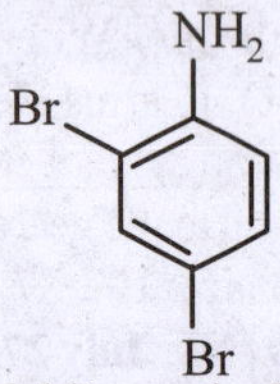 2,4-dibromoaniline (NH_2 with Br ortho and para) (d) H_3C-Br

181. Kjeldahl's method cannot be used to estimate nitrogen for which of the following compounds? **[NCERT : P L-287 | Jan. 08, 2020 (II)]**

(a) $C_6H_5NH_2$ (b) $CH_3CH_2-C\equiv N$
(c) $C_6H_5NO_2$ (d) $NH_2-\overset{O}{\overset{\|}{C}}-NH_2$

182. A chromatography column, packed with silica gel as stationary phase, was used to separate a mixture of compounds consisting of (A) benzanilide (B) aniline and (C) acetophenone. When the column is eluted with a mixture of solvents, hexane: ethyl acetate (20:80), the sequence of obtained compounds is: **[Jan. 07, 2020 (II)]**

(a) (B), (C) and (A) (b) (B), (A) and (C)
(c) (C), (A) and (B) (d) (A), (B) and (C)

183. The principle of column chromatography is : **[NCERT : P L-283 | April 10, 2019 (I)]**

(a) Gravitational force.
(b) Capillary action.
(c) Differential absorption of the substances on the solid phase.
(d) Differential adsorption of the substances on the solid phase.

184. In chromatography, which of the following statement is INCORRECT for R_f? **[NCERT : P L-283 | April 10, 2019 (II)]**

(a) R_f value depends on the type of chromatography.
(b) The value of R_f can not be more than one.
(c) Higher R_f value means higher adsorption.
(d) R_f value is dependent on the mobile phase.

185. Which of the following compounds will show the maximum 'enol' content? **[NCERT : P L-275 | April 8, 2019 (II)]**

(a) $CH_3COCH_2COOC_2H_5$ (b) $CH_3COCH_2COCH_3$
(c) CH_3COCH_3 (d) $CH_3COCH_2CONH_2$

186. The correct match between items I and II is : **[Jan. 11, 2019 (I)]**

Item - I (Mixture)	Item - II (Separation method)
(A) H_2O : Sugar	(P) Sublimation
(B) H_2O : Aniline	(Q) Recrystallization
(C) H_2O : Toluene	(R) Steam distillation
	(S) Differential extraction

(a) (A) → (S); (B) → (R); (C) → (P)
(b) (A) → (Q); (B) → (R); (C) → (S)
(c) (A) → (R); (B) → (P); (C) → (S)
(d) (A) → (Q); (B) → (R); (C) → (P)

187. An organic compound is estimated through Dumas method and was found to evolve 6 moles of CO_2, 4 moles of H_2O and 1 mole of nitrogen gas. The formula of the compound is: **[NCERT : P L-286 | Jan. 11, 2019 (I)]**

(a) $C_{12}H_8N$ (b) $C_{12}H_8N_2$
(c) $C_6H_8N_2$ (d) C_6H_8N

188. If dichloromethane (DCM) and water (H_2O) are used for differential extraction, which one of the following statements is correct? **[NCERT : P L-278 | Jan. 10, 2019 (I)]**

(a) DCM and H_2O would stay as lower and upper layer respectively in the separating funnel (S.F.)
(b) DCM and H_2O will make turbid/colloidal mixture
(c) DCM and H_2O would stay as upper and lower layer respectively in the separating funnel (S.F.)
(d) DCM and H_2O will be miscible clearly

189. The correct match between **Item- I** and **Item- II** is: **[NCERT : P L-278 | Jan. 9, 2019 (II)]**

Item -I	Item -II
(A) Benzaldehyde	(P) Mobile phase
(B) Alumina	(Q) Adsorbent
(C) Acetonitrile	(R) Adsorbate

(a) (A) → (Q) ; (B) → (P) ; (C) → (R)
(b) (A) → (R) ; (B) → (Q) ; (C) → (P)
(c) (A) → (Q) ; (B) → (R) ; (C) → (P)
(d) (A) → (P) ; (B) → (R) ; (C) → (Q)

190. The correct match between items of **List-I** and **List-II** is : **[NCERT : P L-280 | Online April 15, 2018 (I)]**

List-I	List-II
(A) Coloured impurity	(p) Steam distillation
(B) Mixture of *o*-nitrophenol and *p*-nitrophenol	(q) Fractional distillation
(C) Crude naphtha	(r) Charcoal treatment
(D) Mixture of glycerol and sugar	(s) Distillation under reduced pressure

(a) (A)-(r), (B)-(s), (C)-(p), (D)-(q)
(b) (A)-(p), (B)-(s), (C)-(r), (D)-(q)
(c) (A)-(r), (B)-(p), (C)-(q), (D)-(s)
(d) (A)-(r), (B)-(p), (C)-(s), (D)-(q)

191. Given below are two statements : One is labelled as **Assertion A** and the other is labelled as **Reason R**.

Assertion A : Thin layer chromatography is an adsorption chromatography.

Reason R : A thin layer of silica gel is spread over a glass plate of suitable size in thin layer chromatography which acts as an adsorbent.

In the light of the above statements, choose the **correct** answer from the options given below:

[NCERT : P L-283 | July 28, 2022 (II)]

(a) Both A and R are true and R is the correct explanation of A.

(b) Both A and R are true but R is NOT the correct explanation of A.

(c) A is true but R is false.

(d) A is false but R is true.

192. Match List-I with List-II

[NCERT : P L-278 | July 27, 2022 (II)]

List-I (Mixture)	List-II (Purification Process)
(A) Chloroform and Aniline	(I) Steam distillation
(B) Benzoic acid and Naphthalene	(II) Sublimation
(C) Water and Aniline	(III) Distillation
(D) Naphthalene and Sodium chloride	(IV) Crystallisation

(a) (A)-(IV), (B)-(III), (C)-(I), (D)-(II)

(b) (A)-(III), (B)-(I), (C)-(IV), (D)-(II)

(c) (A)-(III), (B)-(IV), (C)-(II), (D)-(I)

(d) (A)-(III), (B)-(IV), (C)-(I), (D)-(II)

193. Which technique, among the following, is most appropriate in separation of a mixture of 100 mg of *p*-nitrophenol and picric acid?

[NCERT : P L-278 | July 26, 2022 (I)]

(a) Steam distillation

(b) 2-5 ft long column of silica gel

(c) Sublimation

(d) Preparative TLC (Thin Layer Chromatography)

194. The number of sp^3 hybridised carbons in an acyclic neutral compound with molecular formula C_4H_5N is ________.

[NCERT : P L-261 | NV, July 25, 2022 (I)]

195. The separation of two coloured substances was done by paper chromatography. The distances travelled by solvent front, substance A and substance B from the base line are 3.25 cm, 2.08 cm and 1.05 cm respectively. The ratio of R_f values of A to B is _____.

[NCERT : P L-283 | NV, July 25, 2022 (II)]

196. TLC analysis of a mixture having 3 components (A, B, C) using silica gel as the stationary phase gave following R_f values; for A = 0.72, B = 0.48, C = 0.20.

Regarding the above observations, which one of the following statements is **not** correct for column chromatography of the mixture?

[NCERT : P L-278 | June 30, 2022 (I)]

(a) A is the most non-polar component

(b) C is the most soluble component in mobile phase

(c) A will be eluted first

(d) C is the most strongly absorbed component

197. Which of the following carbocations is most stable:

[NCERT : P L-272 | June 29, 2022 (II)]

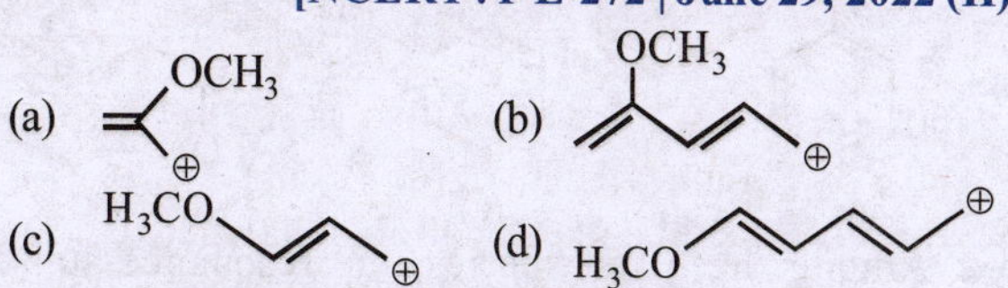

198. Which of the following is most stable?

[NCERT : P L-275 | June 27, 2022 (II)]

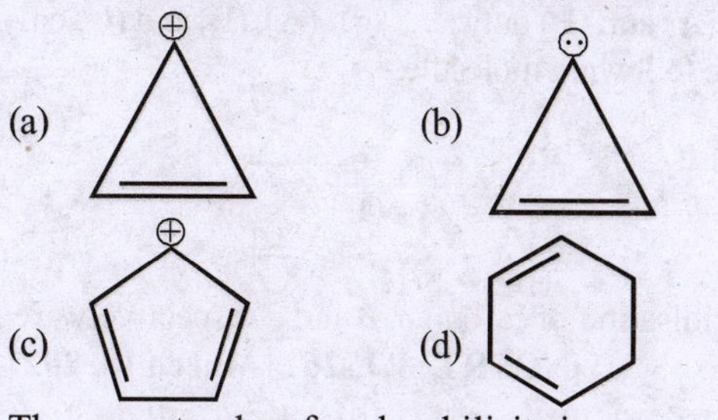

199. The correct order of nucleophilicity is

[NCERT : P L-273 | June 26, 2022 (II)]

(a) $F^- > OH^-$ (b) $H_2\ddot{O} > OH^-$

(c) $R\ddot{O}H > RO^-$ (d) $NH_2^- > NH_3$

200. Arrange the following carbocations in decreasing order of stability. **[NCERT : P L-272 | June 24, 2022 (II)]**

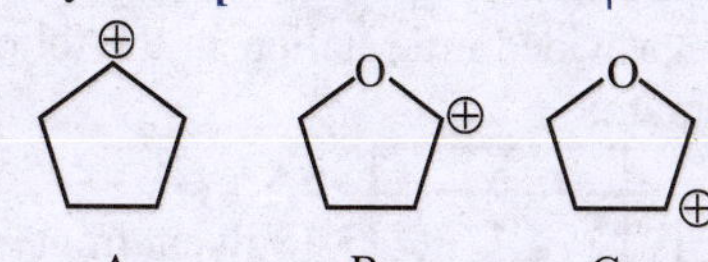

(a) A > C > B (b) A > B > C

(c) C > B > A (d) C > A > B

201. $C_6H_5\overset{\oplus}{C}H_2$ (A) $CH_2=\overset{\oplus}{C}H$ (B) $CH_3-\overset{\oplus}{C}H_2$ (C) $HC\equiv\overset{\oplus}{C}$ (D)

[NCERT : P L-271 | July 27, 2021 (I)]

The correct order of stability of given carbocation is :

(a) A > C > B > D (b) D > B > C > A

(c) D > B > A > C (d) C > A > D > B

202. Which purification technique is used for high boiling liquid compound (decomposes near its boiling point) ?

[NCERT : P L-280 | July 22, 2021 (II)]

(a) Simple distillation

(b) Steam distillation

(c) Fractional distillation

(d) Reduced pressure distillation

203. Which of the following compounds does not exhibit resonance? **[NCERT : P L-275 | July 22, 2021 (II)]**

(a) $CH_3CH_2OCH=CH_2$ (b) $C_6H_5CH_2OH$

(c) $CH_3CH_2CH_2CONH_2$ (d) $CH_3CH_2CH=CHCH_2NH_2$

204. $\overset{\oplus}{C}H_2$ (A) $\quad$ H, $\overset{\oplus}{C}H_2$, H (B) $\quad$ $\overset{\oplus}{C}H_2$ (C) $\quad$ O, $\overset{\oplus}{C}H$ (D)

Among the given species, the resonance stabilised carbocations are: **[NCERT : P L-275 | July 20, 2021 (I)]**

(a) (C) and (D) only (b) (A), (B) and (D) only

(c) (A) and (B) only (d) (A), (B) and (C) only

205. In the following molecule,

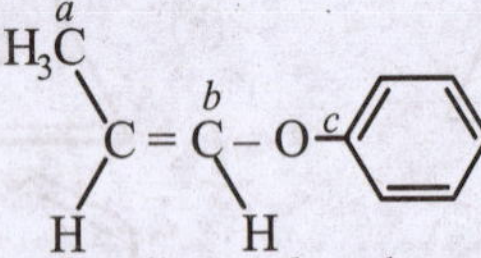

Hybridisation of carbon *a*, *b* and *c* respectively are :

[NCERT : P L-261 | March 18, 2021 (II)]

(a) sp^3, sp, sp (b) sp^3, sp^2, sp^2

(c) sp^3, sp, sp^2 (d) sp^3, sp^2, sp

206. In chromatography technique, the purification of compound is independent of

[NCERT : P L-278 | March 16, 2021 (I)]

(a) Solubility of the compound

(b) Mobility or flow of solvent system

(c) Length of the column or TLC plate

(d) Physical state of the pure compound

207. Using the provided information in the following paper chromatogram:

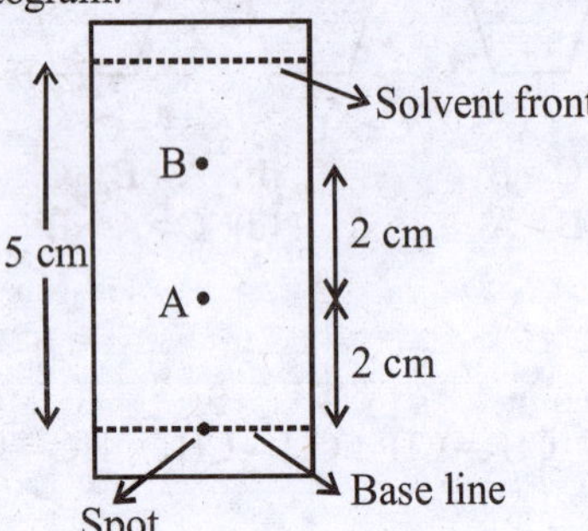

Paper chromatography for compounds A and B

The calculated R_f value of A ______ $\times 10^{-1}$.

[NCERT : P L-283 | NV, Feb. 25, 2021 (I)]

208. Which one of the following compounds possesses the most acidic hydrogen? **[NCERT : P L-274 | Sep. 03, 2020 (I)]**

(a) 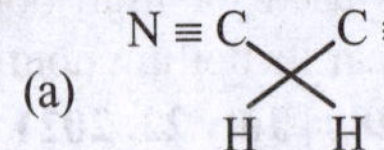

(b) $H_3C-C\equiv C-H$

(c) 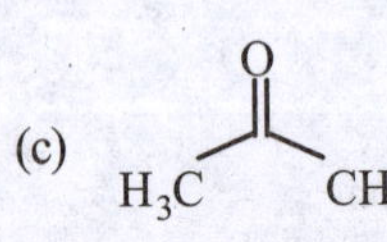

(d) 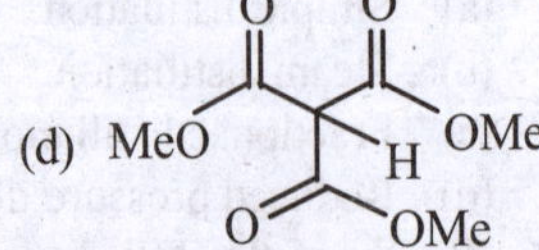

209. Glycerol is separated in soap industries by :

[NCERT : P L-280 | Sep. 03, 2020 (I); Same 2016]

(a) Fractional distillation

(b) Differential extraction

(c) Steam distillation

(d) Distillation under reduced pressure

210. The increasing order of basicity for the following intermediates is (from weak to strong)

[NCERT : P L-277 | Jan. 09, 2020 (I)]

(i) $(CH_3)_3C^{\ominus}$ $\quad$ (ii) $H_2C=CH-\overset{\ominus}{C}H_2$

(iii) $HC\equiv\overset{\ominus}{C}$ $\quad$ (iv) $\overset{\ominus}{C}H_3$ $\quad$ (v) $\overset{\ominus}{C}N$

(a) (iii) < (i) < (ii) < (iv) < (v)

(b) (v) < (i) < (iv) < (ii) < (iii)

(c) (v) < (iii) < (ii) < (iv) < (i)

(d) (iii) < (iv) < (ii) < (i) < (v)

211. A flask contains a mixture of isohexane and 3-methylpentane. One of the liquids boils at 63 °C while the other boils at 60 °C. What is the best way to separate the two liquids and which one will be distilled out first?

[NCERT : P L-278 | Jan. 08, 2020 (I)]

(a) fractional distillation, isohexane

(b) simple distillation, 3-methylpentane

(c) simple distillation, isohexane

(d) fractional distillation, 3-methylpentane

212. The correct order of stability for the following alkoxides is: **[NCERT : P L-275 | Jan. 07, 2020 (II)]**

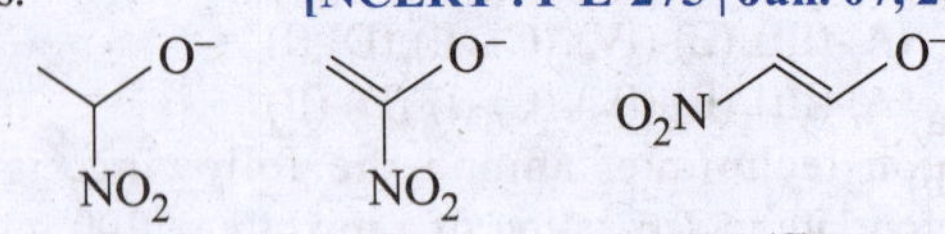

(a) (B) > (A) > (C) (b) (C) > (B) > (A)

(c) (C) > (A) > (B) (d) (B) > (C) > (A)

213. The increasing order of nucleophilicity of the following nucleophiles is : **[NCERT : P L-273 | April 10, 2019 (II)]**

(i) $CH_3CO_2^-$ (ii) H_2O

(iii) $CH_3SO_3^-$ (iv) OH^-

(a) (i) < (iv) < (iii) < (ii) (b) (ii) < (iii) < (iv) < (i)

(c) (iv) < (i) < (iii) < (ii) (d) (ii) < (iii) < (i) < (iv)

214. Which of the following compounds will be suitable for Kjeldahl's method for nitrogen estimation?

[NCERT : P L-287 | 2018]

(a) N (pyridine) (b) NH_2

(c) NO_2 (d) $N_2^+Cl^-$

215. The increasing order of basicity of the following compounds is **[NCERT : P L-275 | 2018]**

(i) NH_2 (ii) NH

(iii) NH_2, NH (iv) $NHCH_3$

(a) (i) < (ii) < (iii) < (iv) (b) (ii) < (i) < (iii) < (iv)

(c) (ii) < (i) < (iv) < (iii) (d) (iv) < (ii) < (i) < (iii)

216. The most polar compound among the following is :
[NCERT : P L-273 | Online April 16, 2018]

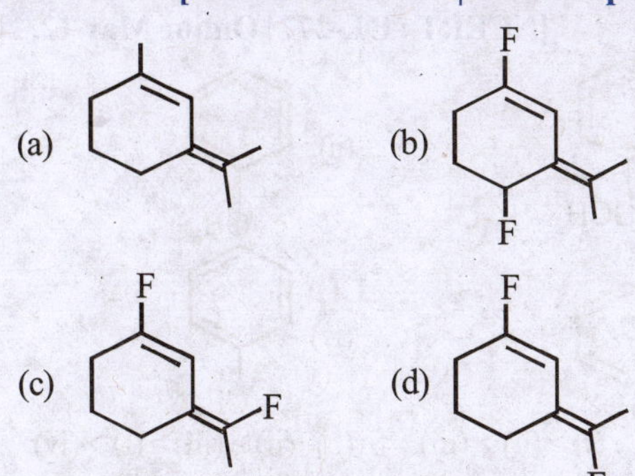

217. On the treatment of the following compound with a strong acid, the most susceptible site for bond cleavage is:

[NCERT : P L-271 | Online April 15, 2018(II)]

(a) O2 – C3 (b) O5 – C6
(c) C4 – O5 (d) C1 – O2

218. Two compounds I and II are eluted by column chromatography (adsorption of I > II). Which one of the following is a correct statement?
[NCERT : P L-283 | Online April 15, 2018 (II)]

(a) II moves slower and has higher R_f value than I
(b) II moves faster and has higher R_f value than I
(c) I moves faster and has higher R_f value than II
(d) I moves slower and has higher R_f value than II

219. Which of the following statements is not true about partition chromatography ?
[NCERT : P L-284 | Online April 8, 2017]

(a) Mobile phase can be a gas
(b) Stationary phase is a finely divided solid adsorbent
(c) Separation depends upon equilibration of solute between a mobile and a stationary phase
(d) Paper chromatography is an example of partition chromatography

220. The increasing order of the boiling points for the following compounds is :
[NCERT : P L-273 | Online April 9, 2017]

C_2H_5OH (I) C_2H_5Cl (II) $C_2H_5CH_3$ (III) $C_2H_5OCH_3$ (IV)

(a) (III) < (IV) < (II) < (I) (b) (IV) < (III) < (I) < (II)
(c) (II) < (III) < (IV) < (I) (d) (III) < (II) < (I) < (IV)

221. A mixture containing the following four compounds is extracted with 1M HCl. The compound that goes to aqueous layer is :
[NCERT : P L-278 | Online April 8, 2017]

(I) (II) (III) (IV)

(a) (I) (b) (II) (c) (III) (d) (IV)

222. Which of the following molecules is least resonance stabilized?
[NCERT : P L-275 | 2017]

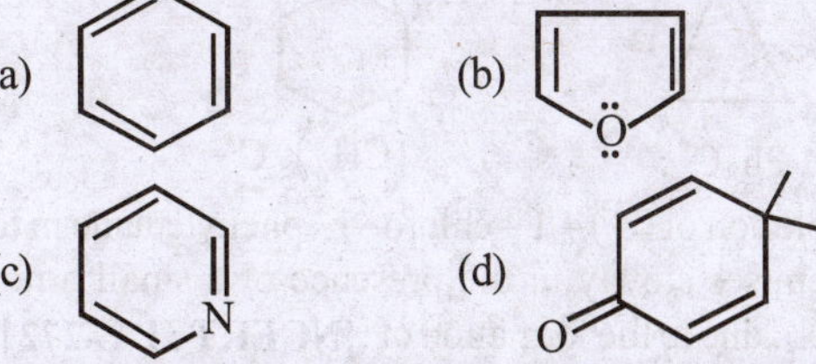

223. The increasing order of the reactivity of the following halides for the S_N1 reaction is **[NCERT : P L-272 | 2017]**

$CH_3CHCH_2CH_3$ (with Cl on C-2) (I) $CH_3CH_2CH_2Cl$ (II)

p-H_3CO–C_6H_4–CH_2Cl (III)

(a) (III) < (II) < (I) (b) (II) < (I) < (III)
(c) (I) < (III) < (II) (d) (II) < (III) < (I)

224. For the estimation of nitrogen, 1.4 g of an organic compound was digested by Kjeldahl method and the evolved ammonia was absorbed in 60 mL of $\frac{M}{10}$ sulphuric acid. The unreacted acid required 20 mL of $\frac{M}{10}$ sodium hydroxide for complete neutralization. The percentage of nitrogen in the compound is:
[NCERT : P L-287 | Online April 10, 2015]

(a) 6% (b) 10% (c) 3% (d) 5%

225. For which of the following molecule significant $\mu \neq 0$?
[NCERT : P L-274 | 2014]

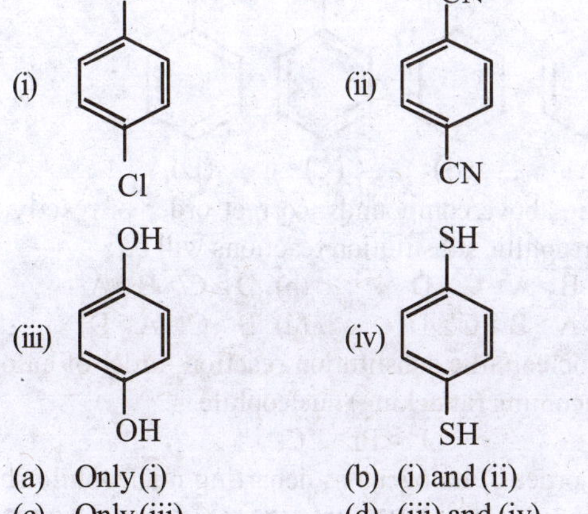

(a) Only (i) (b) (i) and (ii)
(c) Only (iii) (d) (iii) and (iv)

226. In which of the following pairs, A is more stable than B?
[NCERT : P L-272 | Online April 9, 2014]

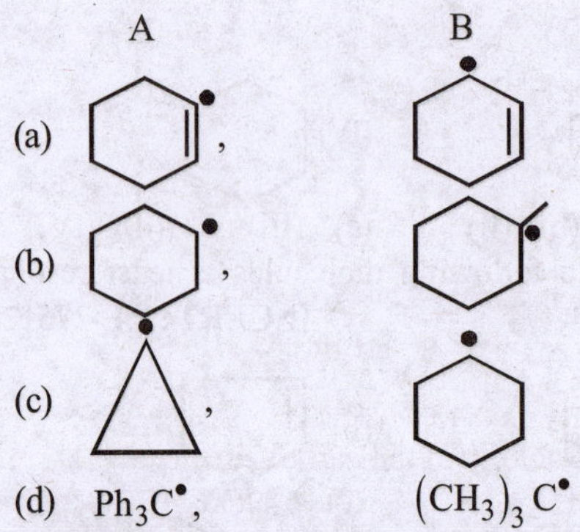

227. A solution of $(-)-1-$chloro$-1-$phenylethane in toluene racemises slowly in the presence of a small amount of $SbCl_5$, due to the formation of: **[NCERT : P L-272 | 2013]**

(a) carbanion (b) carbene
(c) carbocation (d) free radical

228. The order of stability of the following carbocations : **[NCERT : P L-271 | 2013]**

$\overset{\oplus}{CH_2}$

$CH_2 = CH - \overset{\oplus}{C}H_2$; $CH_3 - CH_2 - \overset{\oplus}{C}H_2$; [benzene ring bearing $\overset{\oplus}{CH_2}$] is :
I II III

(a) III > II > I (b) II > III > I
(c) I > II > III (d) III > I > II

229. Which one of the following is most stable ? **[NCERT : P L-271 | Online April 9, 2013]**

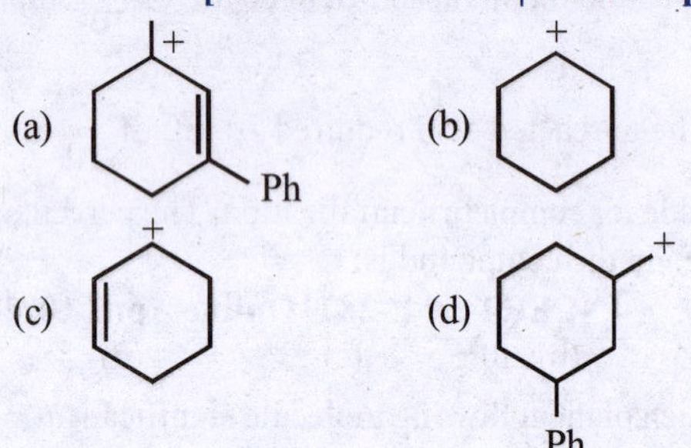

230. Given **[NCERT : P L-276 | Online April 25, 2013]**

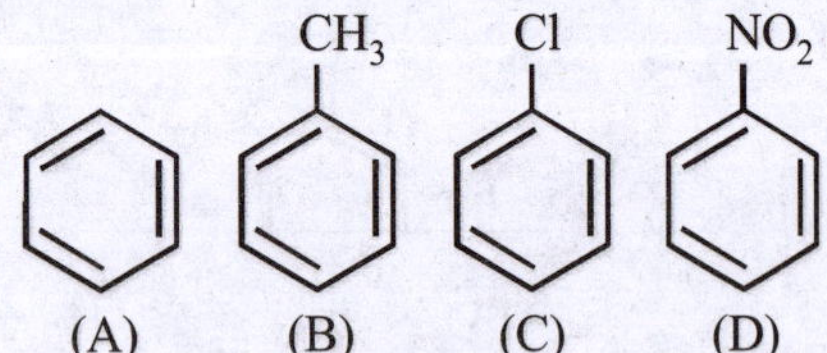

In the above compounds, correct order of reactivity in electrophilic substitution reactions will be :

(a) B > A > C > D (b) D > C > B > A
(c) A > B > C > D (d) B > C > A > D

231. In nucleophilic substitution reaction, order of halogens as incoming (attacking) nucleophile is:

$$I^- > Br^- > Cl^-$$

The order of halogens as departing nucleophile should be : **[NCERT : P L-273 | Online April 25, 2013]**

(a) $Br^- > I^- > Cl^-$ (b) $I^- > Br^- > Cl^-$
(c) $Cl^- > Br^- > I^-$ (d) $Cl^- > I^- > Br^-$

232. In the below mentioned compounds the decreasing order of reactivity towards electrophilic substitution is **[NCERT : P L-277 | Online May 12, 2012]**

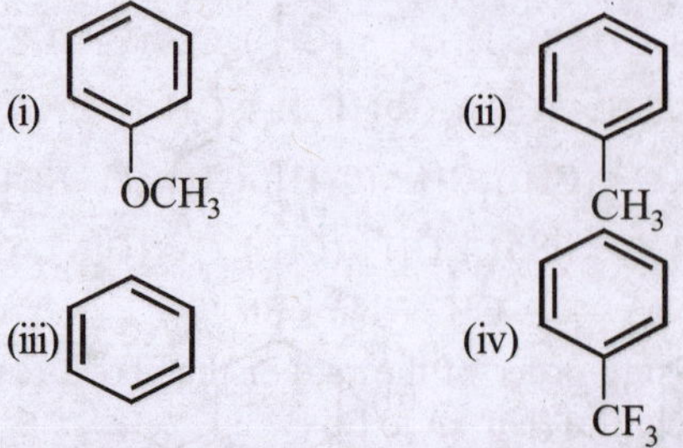

(a) (iv) > (i) > (ii) > (iii) (b) (ii) > (iii) > (i) > (iv)
(c) (iii) > (i) > (iv) > (ii) (d) (i) > (ii) > (iii) > (iv)

233. Which of the following cannot be represented by resonance structures? **[Online May 7, 2012]**

(a) Dimethyl ether (b) Nitrate anion
(c) Carboxylate anion (d) Toluene

234. The correct order of increasing basicity of the given conjugate bases ($R = CH_3$) is **[NCERT : P L-275 | 2010]**

(a) $RCO\bar{O} < HC \equiv \bar{C} < \bar{R} < \bar{N}H_2$
(b) $\bar{R} < HC \equiv \bar{C} < RCO\bar{O} < \bar{N}H_2$
(c) $RCO\bar{O} < \bar{N}H_2 < HC \equiv \bar{C} < \bar{R}$
(d) $RCO\bar{O} < HC \equiv \bar{C} < \bar{N}H_2 < \bar{R}$

235. Arrange the carbanions, **[NCERT : P L-275 | 2009]**

$(CH_3)_3\bar{C}, \bar{C}Cl_3, (CH_3)_2\bar{C}H, C_6H_5\bar{C}H_2$

in order of their decreasing stability.

(a) $(CH_3)_2\bar{C}H > \bar{C}Cl_3 > C_6H_5\bar{C}H_2 > (CH_3)_3\bar{C}$
(b) $\bar{C}Cl_3 > C_6H_5\bar{C}H_2 > (CH_3)_2\bar{C}H > (CH_3)_3\bar{C}$
(c) $(CH_3)_3\bar{C} > (CH_3)_2\bar{C}H > C_6H_5\bar{C}H_2 > \bar{C}Cl_3$
(d) $C_6H_5\bar{C}H_2 > \bar{C}Cl_3 > (CH_3)_3\bar{C} > (CH_3)_2\bar{C}H$

236. The electrophile, $E^{\oplus}$ attacks the benzene ring to generate the intermediate σ–complex. Of the following, which σ–complex has lowest energy? **[NCERT : P L-276 | 2008]**

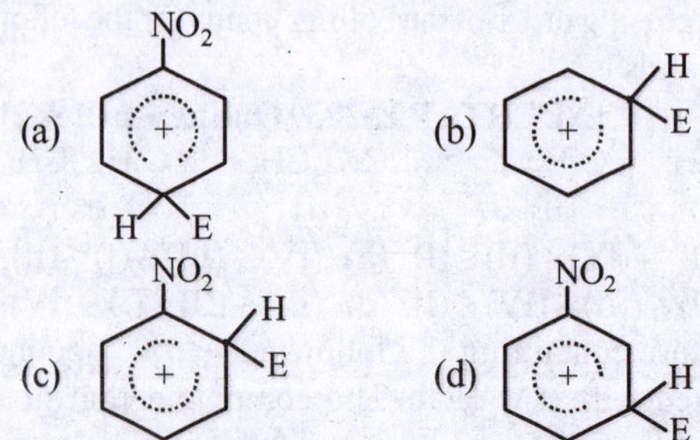

237. Presence of a nitro group in a benzene ring **[NCERT : P L-276 | 2007]**

(a) deactivates the ring towards electrophilic substitution
(b) activates the ring towards electrophilic substitution
(c) renders the ring basic
(d) deactivates the ring towards nucleophilic substitution.

238. The increasing order of stability of the following free radicals is **[NCERT : P L-272 | 2006]**

(a) $(C_6H_5)_2\dot{C}H < (C_6H_5)_3\dot{C} < (CH_3)_3\dot{C} < (CH_3)_2\dot{C}H$

(b) $(CH_3)_2\dot{C}H < (CH_3)_3\dot{C} < (C_6H_5)_2\dot{C}H < (C_6H_5)_3\dot{C}$

(c) $(CH_3)_3\dot{C} < (CH_3)_2\dot{C}H < (C_6H_5)_2\dot{C}H < (C_6H_5)_3\dot{C}$

(d) $(C_6H_5)_3\dot{C} < (C_6H_5)_2\dot{C}H < (CH_3)_3\dot{C} < (CH_3)_2\dot{C}H$

239. $CH_3Br + Nu^- \longrightarrow CH_3 - Nu + Br^-$

The decreasing order of the rate of the above reaction with nucleophiles (Nu^-) A to D is **[NCERT : P L-273 | 2006]**

[Nu^- = (A) PhO^-, (B) AcO^-, (C) HO^-, (D) CH_3O^-]

(a) A > B > C > D (b) B > D > C > A

(c) D > C > A > B (d) D > C > B > A

240. 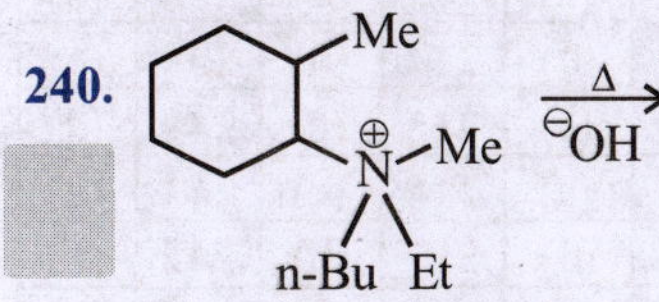

The alkene formed as a major product in the above elimination reaction is **[2006]**

(a) 1-methylcyclohexene (Me on ring double bond) (b) 3-methylcyclohexene (Me)

(c) Me–CH₂–CH=CH₂ (d) $CH_2 = CH_2$

241. Due to the presence of an unpaired electron, free radicals are: **[NCERT : P L-272 | 2005]**

(a) cations (b) anions

(c) chemically inactive (d) chemically reactive

242. The decreasing order of nucleophilicity among the nucleophiles is **[NCERT : P L-273 | 2005]**

(A) $CH_3C(=O)-O^-$

(B) CH_3O^-

(C) CN^-

(D) $H_3C-C_6H_4-S(=O)_2-O^-$ is

(a) (C) > (B) > (A) > (D) (b) (B) > (C) > (A) > (D)

(c) (D) > (C) > (B) > (A) (d) (A) > (B) > (C) > (D)

243. The reaction **[NCERT : P L-273 | 2005, 2004]**

$$R-C(=O)-X + \overset{\ominus}{Nu} \longrightarrow R-C(=O)-Nu + X^{\ominus}$$

is fastest when X is

(a) OCOR (b) OC_2H_5

(c) NH_2 (d) Cl

244. Consider the acidity of the carboxylic acids :

(1) PhCOOH (2) o-$NO_2C_6H_4COOH$

(3) p-$NO_2C_6H_4COOH$ (4) m-$NO_2C_6H_4COOH$

Which of the following order is correct ? **[NCERT : P L-275 | 2004]**

(a) 2 > 4 > 1 > 3 (b) 2 > 4 > 3 > 1

(c) 1 > 2 > 3 > 4 (d) 2 > 3 > 4 > 1

245. Which of the following is the strongest base ? **[NCERT : P L-275 | 2004]**

(a) 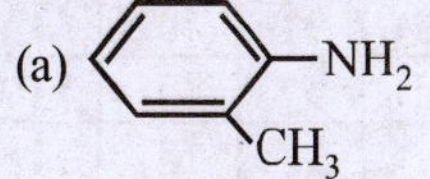

(b) 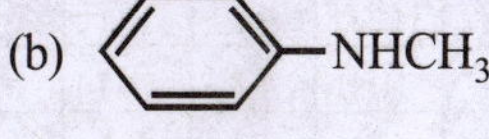

(c) $C_6H_5-NH_2$ (d) $C_6H_5-CH_2NH_2$

246. In the anion $HCOO^-$ the two carbon-oxygen bonds are found to be of equal length. What is the reason for it ? **[NCERT : P L-275 | 2003]**

(a) The C = O bond is weaker than the C — O bond

(b) The anion $HCOO^-$ has two equivalent resonating structures

(c) The anion is obtained by removal of a proton from the acid molecule

(d) Electronic orbitals of carbon atom are hybridised

247. In which of the following species is the underlined carbon having sp^3 hybridisation? **[NCERT : P L-261 | 2002]**

(a) $CH_3\underline{C}OOH$ (b) $CH_3\underline{C}H_2OH$

(c) $CH_3\underline{C}OCH_3$ (d) $CH_2 = \underline{C}H-CH_3$

248. Arrangement of $(CH_3)_3C-$, $(CH_3)_2CH-$, CH_3-CH_2- when attached to benzyl or an unsaturated group in increasing order of inductive effect is **[NCERT : P L-274 | 2002]**

(a) $(CH_3)_3C- < (CH_3)_2CH- < CH_3-CH_2-$

(b) $CH_3-CH_2- < (CH_3)_2CH- < (CH_3)_3C-$

(c) $(CH_3)_2CH- < (CH_3)_3C- < CH_3-CH_2-$

(d) $(CH_3)_3C- < CH_3CH_2- < (CH_3)_2CH-$

ANSWER KEYS

1	(d)	26	(c)	51	(c)	76	(b)	101	(255)	126	(a)	151	(d)	176	(19)	201	(a)	226	(d)
2	(c)	27	(d)	52	(c)	77	(c)	102	(d)	127	(b)	152	(b)	177	(a)	202	(d)	227	(c)
3	(d)	28	(c)	53	(8)	78	(b)	103	(b)	128	(d)	153	(d)	178	(50)	203	(d)	228	(d)
4	(a)	29	(b)	54	(c)	79	(c)	104	(40)	129	(a)	154	(a)	179	(a)	204	(c)	229	(a)
5	(c)	30	(c)	55	(b)	80	(a)	105	(a)	130	(7)	155	(c)	180	(b)	205	(b)	230	(a)
6	(c)	31	(d)	56	(4)	81	(c)	106	(20)	131	(d)	156	(b)	181	(c)	206	(d)	231	(b)
7	(d)	32	(b)	57	(d)	82	(d)	107	(c)	132	(c)	157	(b)	182	(c)	207	(4)	232	(d)
8	(b)	33	(c)	58	(a)	83	(d)	108	(b)	133	(a)	158	(a)	183	(d)	208	(d)	233	(a)
9	(c)	34	(d)	59	(d)	84	(d)	109	(4)	134	(d)	159	(b)	184	(c)	209	(d)	234	(d)
10	(b)	35	(d)	60	(b)	85	(18)	110	(15)	135	(c)	160	(6)	185	(b)	210	(c)	235	(b)
11	(c)	36	(b)	61	(d)	86	(a)	111	(a)	136	(d)	161	(42)	186	(b)	211	(a)	236	(b)
12	(a)	37	(b)	62	(d)	87	(b)	112	(a)	137	(4)	162	(2)	187	(c)	212	(b)	237	(a)
13	(c)	38	(b)	63	(6)	88	(20)	113	(b)	138	(a)	163	(22)	188	(a)	213	(d)	238	(b)
14	(d)	39	(d)	64	(6)	89	(c)	114	(b)	139	(d)	164	(a)	189	(b)	214	(b)	239	(c)
15	(c)	40	(a)	65	(d)	90	(a)	115	(c)	140	(b)	165	(30)	190	(c)	215	(c)	240	(d)
16	(2)	41	(a)	66	(d)	91	(109)	116	(5)	141	(b)	166	(56)	191	(a)	216	(c)	241	(d)
17	(b)	42	(b)	67	(d)	92	(d)	117	(50)	142	(a)	167	(64)	192	(d)	217	(b)	242	(b)
18	(d)	43	(a)	68	(a)	93	(d)	118	(a)	143	(5)	168	(40)	193	(d)	218	(b)	243	(d)
19	(b)	44	(a)	69	(d)	94	(d)	119	(d)	144	(40)	169	(c)	194	(1)	219	(b)	244	(d)
20	(d)	45	(a)	70	(c)	95	(275)	120	(c)	145	(25)	170	(a)	195	(2)	220	(a)	245	(d)
21	(b)	46	(a)	71	(d)	96	(b)	121	(b)	146	(a)	171	(7)	196	(b)	221	(b)	246	(b)
22	(c)	47	(d)	72	(b)	97	(d)	122	(56)	147	(c)	172	(1125)	197	(d)	222	(d)	247	(b)
23	(b)	48	(a)	73	(c)	98	(a)	123	(c)	148	(7)	173	(42)	198	(a)	223	(b)	248	(b)
24	(c)	49	(a)	74	(b)	99	(1655)	124	(c)	149	(a)	174	(19)	199	(d)	224	(b)		
25	(d)	50	(a)	75	(b)	100	(d)	125	(c)	150	(d)	175	(b)	200	(N)	225	(d)		

Hydrocarbons

TREND ANALYSIS JEE MAIN (2025-2021)

	2025				2024				2023				2022				2021			
	Session II		Session I		Session II		Session I		Session II		Session I		Session II		Session I		Session II		Session I	
	MCQs	NVQs	MCQs	NVQs	MCQs	NVQs	MCQs	NVQs	MCQs	NVQs	MCQs	NVQs	MCQs	NVQs	MCQs	NVQs	MCQs	NVQs	MCQs	NVQs
Topic 1	1	–	2	–	1	1	1	2	1	1	3	1	1	2	2	–	1	1	–	–
Topic 2	4	–	5	–	4	–	4	2	2	2	4	1	3	–	4	2	5	–	5	–
Topic 3	–	–	2	2	1	1	2	–	–	–	1	–	–	1	1	–	–	1	1	1
Topic 4	3	–	3	1	4	1	4	2	3	–	1	–	3	1	6	–	3	–	4	1
Total	8	–	12	3	10	3	11	6	6	3	9	2	7	4	13	2	9	2	10	2

How to Code Qns for Better & Faster Revision →

Type 1: Able to solve (No code Req)
Type 2: Solved with struggle or by seeing solution **(Use code NR means Needs Revision or orange stickers)**
Type 3: Tricky/ Could not solve on my own. Need help from Teacher/ Friend. **(Use code TQ means Tricky Question or red stickers)**

You can either hand write the code in boxes or else paste coloured stickers given at end of book.

Topic 1 - Alkanes

1. Given below are two statements : **[NCERT : P L-302 | April 2, 2025 (II)]**

Statement (I) : Neopentane forms only one monosubstituted derivative.

Statement (II) : Melting point of neopentane is higher than n-pentane

In the light of the above statements, choose the **most appropriate answer** from the options given below :

(a) **Statement I** is correct but **Statement II** is incorrect
(b) Both **Statement I** and **Statement II** are correct
(c) Both **Statement I** and **Statement II** are incorrect
(d) **Statement I** is incorrect but **Statement II** is correct

2. Propane molecule on chlorination under photochemical condition gives two di-chloro products, "x" and "y". Amongst "x" and "y", "x" is an optically active molecule. How many tri-chloro products (consider only structural isomers) will be obtained from "x" when it is further treated with chlorine under the photochemical condition? **[NCERT : P L-304 | Jan. 23, 2025 (I)]**

(a) 4 (b) 2 (c) 5 (d) 3

3. When sec-butylcyclohexane reacts with bromine in the presence of sunlight, the major product is: **[NCERT : P L-302 | Jan. 22, 2025 (II)]**

(a) Br (b) Br

(c) (d)

4. In the given compound, the number of 2° carbon atom/s is ______. **[NCERT : P L-296 | NV, April 8, 2024 (I)]**

$CH_3-\underset{|}{\overset{}{C}}(CH_3)-\underset{|}{C}H-\underset{|}{C}(CH_3)-CH_3$
H H H

(a) Three (b) One (c) Two (d) Four

5. The number of different chain isomers for C_7H_{16} is _____.
[NCERT : P L-296 | NV, April 4, 2024 (I)]

6. Number of alkanes obtained on electrolysis of a mixture of CH_3COONa and C_2H_5COONa is _____.
[NCERT : P L-301 | NV, Jan. 31, 2024 (I)]

7. Number of isomeric products formed by monochlorination of 2-methylbutane in presence of sunlight is _______.
[NCERT : P L-303 | NV, Jan. 31, 2024 (II)]

8. The incorrect statement regarding conformations of ethane is: **[NCERT : P L-305 | Jan. 27, 2024 (II)]**
(a) Ethane has infinite number of conformations
(b) The dihedral angle in staggered conformation is 60°
(c) Eclipsed conformation is the most stable conformation.
(d) The conformations of ethane are inter-convertible to one-another

9. In the following reaction 'X' is
[NCERT : P L-304 | April 13, 2023 (I)]

$CH_3(CH_2)_4CH_3 \xrightarrow[HCl.\Delta]{Anhy.AlCl_3}$ 'X' Major product

(a) $CH_3(CH_2)_4CH_2Cl$
(b) $Cl-CH_2-(CH_2)_4-CH_2-Cl$
(c) $CH_3\underset{\underset{CH_3}{|}}{C}H-(CH_2)_2CH_3$
(d) cyclohexane

10. Number of bromo derivatives obtained on treating ethane with excess of Br_2, in diffused sunlight is _____.
[NCERT : P L-302 | NV, April 06, 2023 (I)]

11. When a hydrocarbon A undergoes complete combustion it requires 11 equivalents of oxygen and produces 4 equivalents of water. What is the molecular formula of A?
[NCERT : P L-303 | Jan. 31, 2023 (II)]
(a) C_9H_8 (b) $C_{11}H_4$ (c) C_5H_8 (d) $C_{11}H_8$

12. When a hydrocarbon A undergoes combustion in the presence of air, it requires 9.5 equivalents of oxygen and produces 3 equivalents of water. What is the molecular formula of A? **[NCERT : P L-303 | Jan. 29, 2023 (II)]**
(a) C_8H_6 (b) C_9H_9 (c) C_6H_6 (d) C_9H_6

13. Which of the following conformations will be the most stable? **[NCERT : P L-305 | Jan. 25, 2023 (I)]**

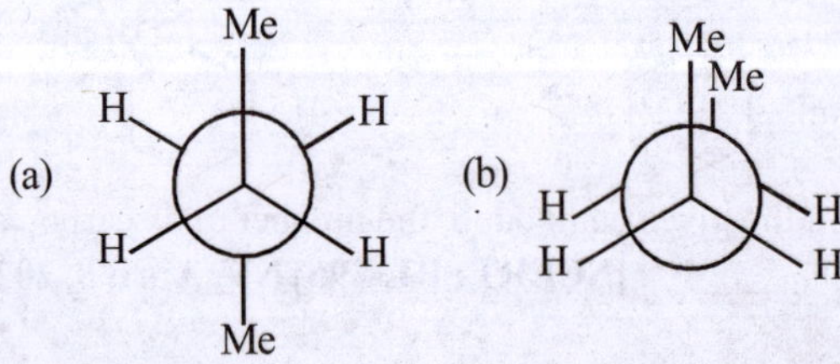

(c) Me, H, Me, H, H, H

(d) Me, H, H, Me, H, H

14. Maximum number of isomeric monochloro derivatives which can be obtained from 2,2,5,5- tetramethylhexane by chlorination is _______
[NCERT : P L-302 | NV, Jan. 24, 2023 (II)]

15. $(C_7H_5O_2)_2 \xrightarrow{hv} [X] + 2\dot{C}_6H_5 + 2CO_2$
Consider the above reaction and identify the intermediate 'X' **[NCERT : P L-301 | June 26, 2022 (I)]**
(a) $C_6H_5-\overset{\overset{O}{\|}}{C}^{\oplus}$ (b) $C_6H_5-\overset{\overset{O}{\|}}{C}-O^{\ominus}$
(c) $C_6H_5-\overset{\overset{O}{\|}}{C}-\ddot{\underset{\cdot\cdot}{O}}$ (d) $C_6H_5-\overset{\overset{O}{\|}}{C}-\ddot{\underset{\cdot\cdot}{O}}\cdot$

16. Which among the following pairs of the structures will give different products on ozonolysis?
(Consider the double bonds in the structures are rigid and not delocalized.) **[NCERT : P L-313 | July 29, 2022 (I)]**

(a) CH_3, CH_3 ⟷ CH_3, CH_3

(b) CH_3, CH_3 ⟷ CH_3, CH_3

(c) CH_3, CH_3 ⟷ CH_3, CH_3

(d) CH_3 ⟷ CH_3

17. Total number of isomers (including stereoisomers) obtained on monochlorination of methylcyclohexane is _______. **[NCERT : P L-302 | NV, July 26, 2022 (II)]**

18. The total number of monobromo derivatives formed by the alkanes with molecular formula C_5H_{12} is (excluding stereoisomers) _______. **[NV, July 25, 2022 (II)]**

19. In the following structures, which one is having staggered conformation with maximum dihedral angle?
[NCERT : P L-305 | June 25, 2022 (I)]

(a) (b)

(c) (d)

20. Arrange the following conformational isomers of *n*-butane in order of their increasing potential energy: **[NCERT : P L-206 | Aug. 31, 2021 (II)]**

I II III IV

(a) II < III < IV < I (b) I < IV < III < II
(c) II < IV < III < I (d) I < III < IV < II

21. The dihedral angle in staggered form of Newman projection of 1,1,1-Trichloroethane is .... degree. (Round off to the nearest integer) **[NCERT : P L-305 | NV, July 27, 2021 (II)]**

22. In the following skew conformation of ethane, H′ – C – C – H″ dihedral angle is : **[NCERT : P L-305 | April 12, 2019 (II)]**

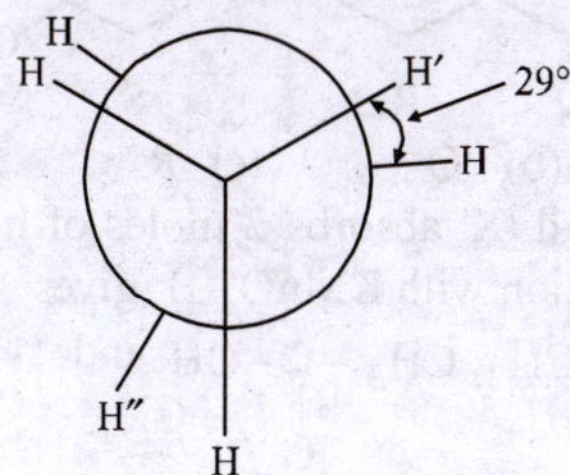

(a) 58° (b) 149° (c) 151° (d) 120°

23. 25 g of an unknown hydrocarbon upon burning produces 88 g of CO_2 and 9 g of H_2O. This unknown hydrocarbon contains: **[April 12, 2019 (II)]**
(a) 20 g of carbon and 5 g of hydrogen
(b) 22 g of carbon and 3 g of hydrogen
(c) 24 g of carbon and 1 g of hydrogen
(d) 18 g of carbon and 7 g of hydrogen

24. At 300 K and 1 atmospheric pressure, 10 mL of a hydrocarbon required 55 mL of O_2 for complete combustion, and 40 mL of CO_2 is formed. The formula of the hydrocarbon is : **[NCERT : P L-303 | April 10, 2019 (I)]**
(a) C_4H_{10} (b) C_4H_6 (c) C_4H_7Cl (d) C_4H_8

25. Which of these factors does not govern the stability of a conformation in acyclic compounds ? **[NCERT : P L-305 | April 10, 2019 (II)]**
(a) Steric interactions
(b) Torsional strain
(c) Electrostatic forces of interaction
(d) Angle strain

26. The major product obtained in the photo catalysed bromination of 2-methylbutane is: **[NCERT : PL-303 | Online April 12, 2014; Same Online May 19, 2012, 2005]**
(a) 1-bromo-2-methylbutane
(b) 1-bromo-3-methylbutane
(c) 2-bromo-3-methylbutane
(d) 2-bromo-2-methylbutane

27. Arrange in the correct order of stability (decreasing order) for the following molecules : **[NCERT : P L-305 | Online April 22, 2013]**

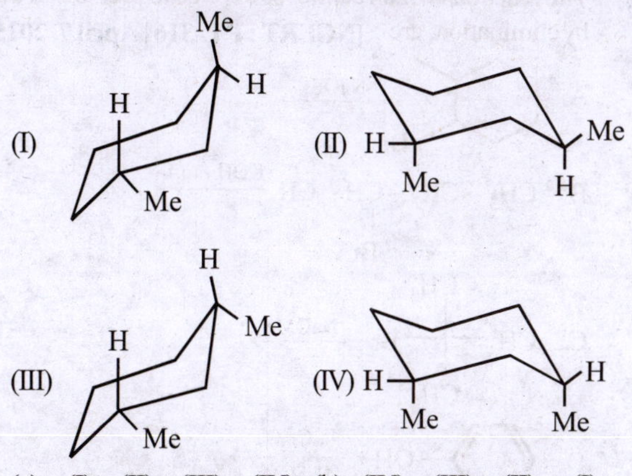

(a) (I) > (II) > (III) > (IV) (b) (IV) > (III) > (II) ≈ (I)
(c) (I) > (II) ≈ (III) > (IV) (d) (III) > (I) ≈ (II) > (IV)

28. Which branched chain isomer of the hydrocarbon with molecular mass 72u gives only one isomer of mono substituted alkyl halide ? **[2012; Same 2003]**
(a) Tertiary butyl chloride (b) Neopentane
(c) Isohexane (d) Neohexane

29. Which one of the following conformations of cyclohexane is chiral? **[NCERT : P L-305 | 2007]**
(a) Boat (b) Twist boat
(c) Rigid (d) Chair

30. Increasing order of stability among the three main conformations (i.e., Eclipse, Anti, Gauche) of 2-fluoroethanol is **[NCERT : P L-305 | 2006]**
(a) Eclipse, Anti, Gauche
(b) Anti, Gauche, Eclipse
(c) Eclipse, Gauche, Anti
(d) Gauche, Eclipse, Anti

31. Of the five isomeric hexanes, the isomer which can give two monochlorinated compounds is **[2005]**
(a) 2-methylpentane (b) 2, 2-dimethylbutane
(c) 2, 3-dimethylbutane (d) n-hexane

32. On mixing a certain alkane with chlorine and irradiating it with ultraviolet light, it forms only one monochloroalkane. This alkane could be **[NCERT : P L-302 | 2003]**
(a) pentane (b) isopentane
(c) neopentane (d) propane

Topic 2 - Alkenes

33. The number of optically active products obtained from the complete ozonolysis of the given compound is : **[NCERT : P L-313 | April 7, 2025 (II)]**

$$H_3C-CH=CH-\underset{H}{\overset{CH_3}{C}}-CH=CH-\underset{CH_3}{\overset{H}{C}}-CH=CH-CH_3$$

(a) 2 (b) 0 (c) 1 (d) 4

34. The reactions which cannot be applied to prepare an alkene by elimination, are **[NCERT : P L-316 | April 7, 2025 (I)]**

A. (1-bromo-2-methylcyclohexane) $\xrightarrow{NaOEt}$

B. $CH_3-CH_2-\underset{Br}{CH}-CH_3 \xrightarrow{KOH\ (aq.)}$

C. $CH_3-\underset{CH_3}{\overset{CH_3}{C}}-Br \xrightarrow{NaOMe}$

D. $C_6H_5-OH \xrightarrow[H_2SO_4]{Na_2Cr_2O_7}$

E. $CH_3-\underset{CH_3}{\overset{CH_3}{C}}-OH \xrightarrow[573\ K]{Cu}$

Choose the **correct** answer from the option given below :

(a) B & E Only (b) B, C & D Only

(c) A, C & D Only (d) B & D Only

35. Given below are two statements : **[NCERT : P L-313 | April 7, 2025 (I)]**

Statement I : Ozonolysis followed by treatment with Zn, H_2O of cis-2-butene gives ethanal.

Statement II : The production obtained by ozonolysis followed by treatment with Zn, H_2O of 3, 6-dimethyloct-4-ene has no chiral carbon atom. In the light of the above statements, choose the **correct** answer from the options given below

(a) Both Statement I and Statement II are true

(b) Statement I is false but Statement II are true

(c) Statement I is true but Statement II is false

(d) Both Statement I and Statement II are false

36. Which compound would give 3-methyl-6- oxoheptanal upon ozonolysis ? **[NCERT : P L-313 | April 3, 2025 (I)]**

(a) 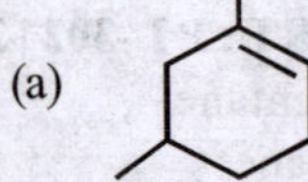 (b) 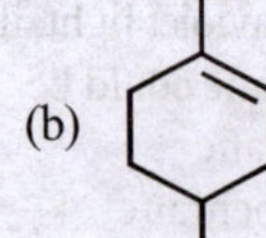

37. A molecule ("P") on treatment with acid undergoes rearrangement and gives ("Q") ("Q") on ozonolysis followed by reflux under alkaline condition gives ("R"). The structure of ("R") is given below : **[NCERT : P L-313 | Jan. 28, 2025 (I)]**

(1-acetyl-2-methylcyclopentene: O, CH_3, CH_3)

("R")

The structure of ("P") is

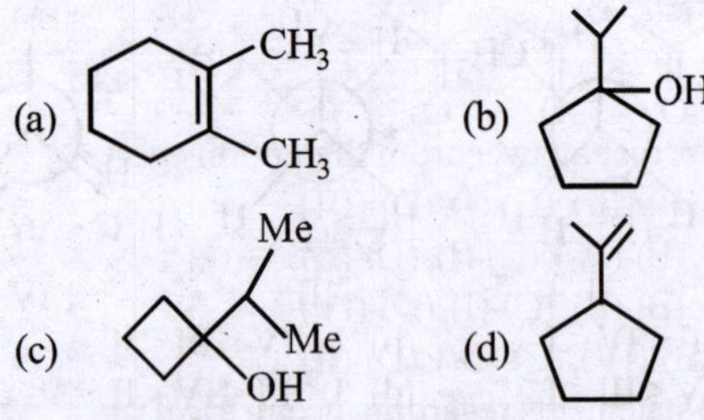

38. Following are the four molecules "P", "Q", "R" and "S". Which one among the four molecules will react with H-Br(aq) at the fastest rate? **[NCERT : P L-311 | Jan. 24, 2025 (I)]**

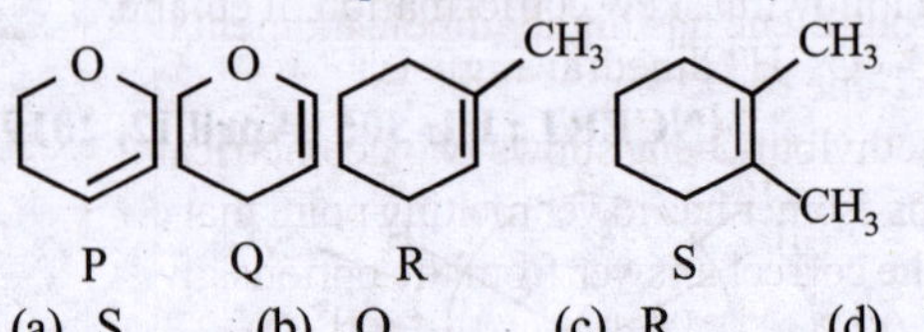

(a) S (b) Q (c) R (d) P

39. A compound 'X' absorbs 2 moles of hydrogen and 'X' upon oxidation with $KMnO_4|H^+$ gives

$$CH_3-\underset{O}{\underset{\|}{C}}-CH_3,\ CH_3-\underset{O}{\underset{\|}{C}}-OH \text{ and}$$

$$CH_3-\underset{O}{\underset{\|}{C}}-CH_2CH_2-\underset{O}{\underset{\|}{C}}-OH$$

The total number of σ bonds present in the compound 'X' is ________.

[NCERT : P L-313 | NV, Jan. 23, 2025 (II)]

40. Match List-I with List-II **[NCERT : P L-313 | Jan. 23, 2025 (II)]**

List-I
(Isomers of $C_{10}H_{14}$)

(A) 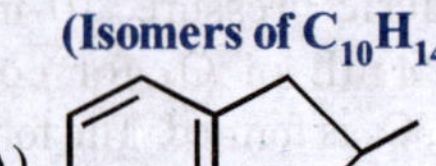 (B) 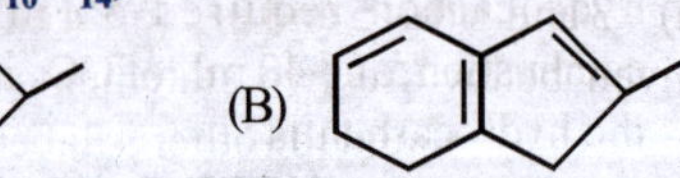

(C) 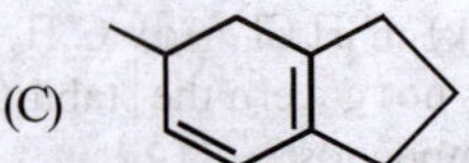 (D) 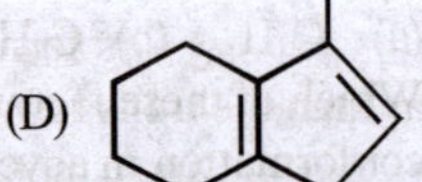

List-II
(Ozonolysis product)

(I) 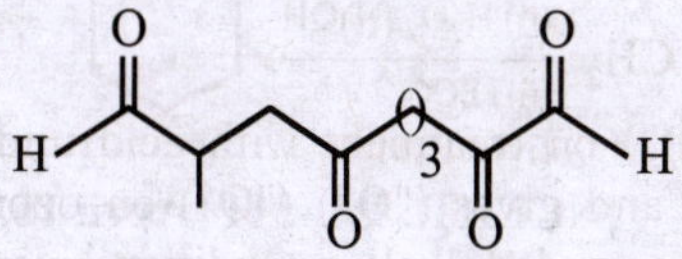 [structure]

(II) 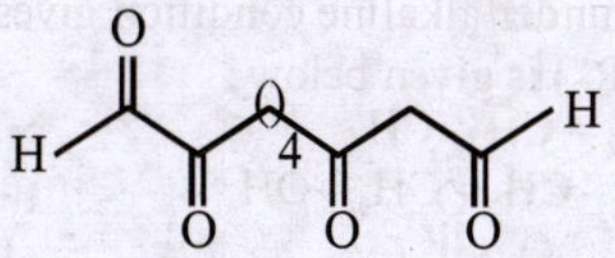 [structure]

(III) 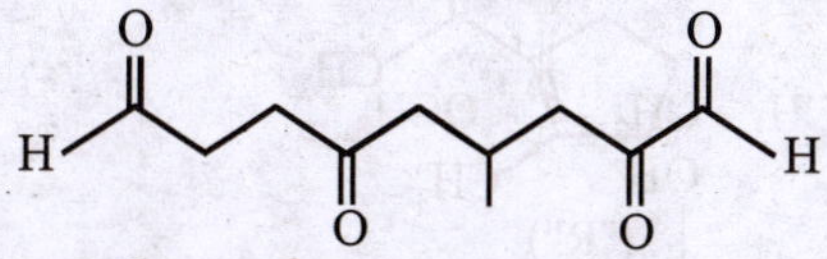 [structure]

(IV) 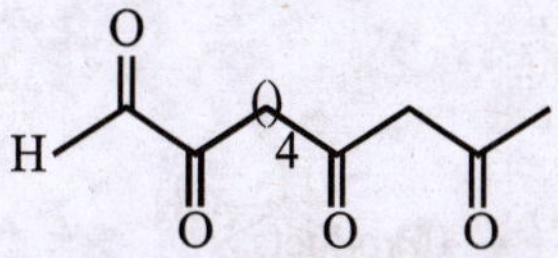 [structure]

Choose the correct answer from the options given
(a) (A)-(II), (B)-(III), (C)-(I), (D)-(IV)
(b) (A)-(III), (B)-(IV), (C)-(I), (D)-(II)
(c) (A)-(III), (B)-(II), (C)-(I), (D)-(IV)
(d) (A)-(I), (B)-(IV), (C)-(III), (D)-(II)

41. The incorrect statements regarding geometrical isomerism are: **[NCERT : P L-308 | Jan. 22, 2025 (I)]**
(A) Propene shows geometrical isomerism.
(B) Trans isomer has identical atoms/group on the opposite sides of the double bond.
(C) Cis-but-2-ene has higher dipole moment than trans-but-2-ene.
(D) 2-methylbut-2-ene shows two geometrical isomers.
(E) Trans-isomer has lower melting point that cis isomer.
Choose the correct answer from the options given below:
(a) (A), (D) and (E) only (b) (C), (D) and (E) only
(c) (B) and (C) only (d) (A) and (E) only

42. The major products from the following reaction sequence are product A and product B. **[NCERT : P L-311 | NV, April 6, 2024 (I)]**

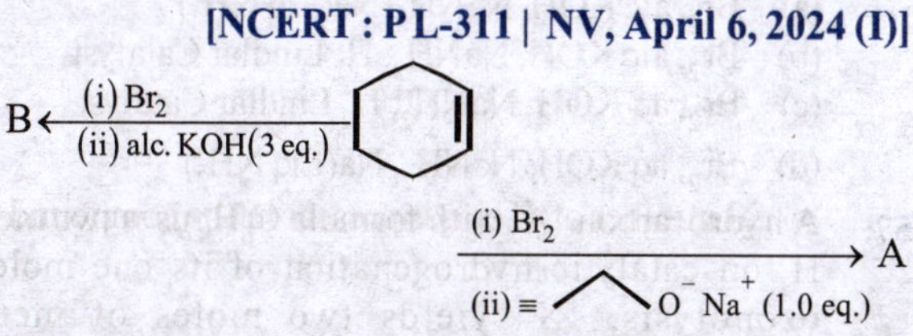

B ← (i) Br_2 (ii) alc. KOH(3 eq.) — [cyclohexene] — (i) Br_2 (ii) ≡ $\bar{O}$ $\overset{+}{Na}$ (1.0 eq.) → A

The total sum of π electrons in product A and product B are ____ (nearest integer)

43. Number of σ and π bonds present in ethylene molecule is respectively : **[NCERT : P L-306 | April 5, 2024 (I)]**
(a) 3 and 1 (b) 5 and 2 (c) 4 and 1 (d) 5 and 1

44. Given below are two statements : one is labelled as **Assertion (A)** and the other is labelled as **Reason (R).** **[NCERT : P L-308 | April 5, 2024 (I)]**
Assertion (A): Cis form of alkene is found to be more polar than the trans form
Reason (R): Dipole moment of trans isomer of 2-butene is zero.
In the light of the above statements, choose the **correct** answer from the options given below :
(a) Both **(A)** and **(R)** are true but **(R)** is **NOT** the correct explanation of **(A)**
(b) **(A)** is true but **(R)** is false
(c) Both **(A)** and **(R)** are true and **(R)** is the correct explanation of **(A)**
(d) **(A)** is false but **(R)** is true

45. [1-methylcyclohexene] —"A"→ [keto-aldehyde] —"B"→ [O^-Na^+ carboxylate aldehyde]

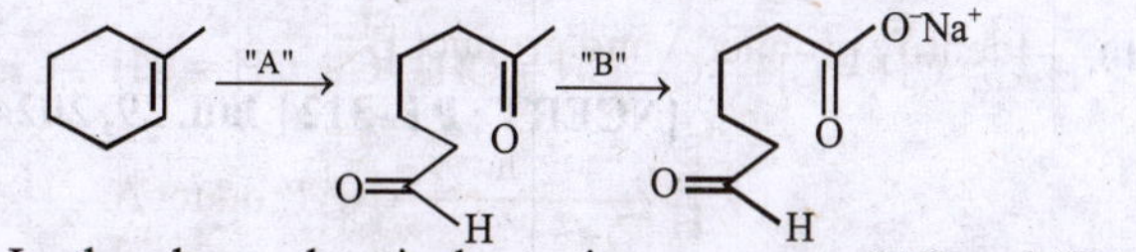

In the above chemical reaction sequence "A" and "B" respectively are : **[NCERT : P L-313 | April 4, 2024 (II)]**
(a) O_3, Zn/H_2O and NaOH(alc.) / I_2
(b) H_2O, H^+ and NaOH(alc.) / I_2
(c) H_2O, H^+ and $KMnO_4$
(d) O_3, Zn/H_2O and $KMnO_4$

46. Major product of the following reaction is – **[NCERT : P L-312 | Jan. 31, 2024 (II)]**

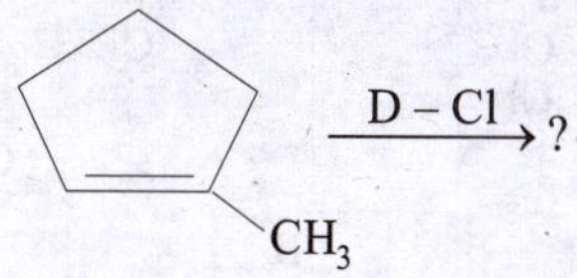

[1-methylcyclopentene, CH_3] —D – Cl→ ?

(a) 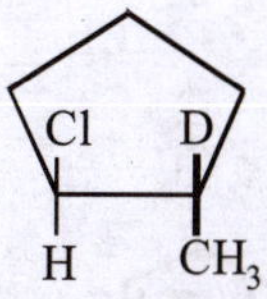 [Cl, D / H, CH_3] (b) 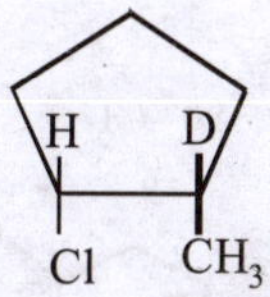 [H, D / Cl, CH_3]
(c) [D, CH_3 / H, Cl] (d) [D, Cl / H, CH_3]

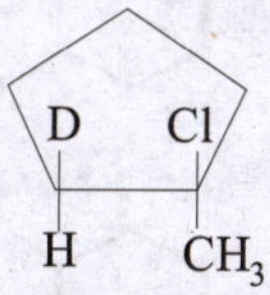

47. Products A and B formed in the following set of reactions are **[NCERT : P L-311 | Jan. 30, 2024 (II)]**

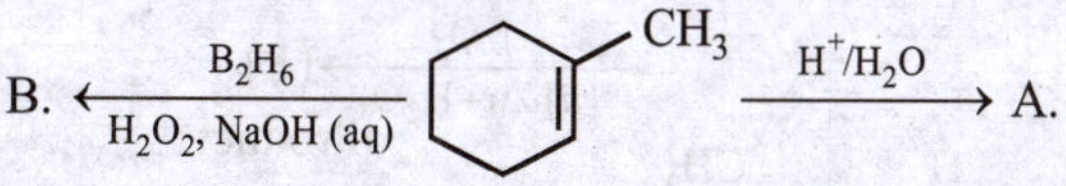

B. ← B_2H_6 / H_2O_2, NaOH (aq) — [1-methylcyclohexene, CH_3] — H^+/H_2O → A.

(a) A = [cyclohexyl CH_2OH] B = [CH_2OH, OH cyclohexane]
(b) A = [1-methylcyclohexanol, CH_3, OH] B = [2-methylcyclohexanol, CH_3, OH]
(c) A = [cyclohexyl CH_2OH] B = [2-methylcyclohexanol, CH_3, OH]
(d) A = [2-methylcyclohexanol, CH_3, OH] B = [1-methylcyclohexanol, CH_3, OH]

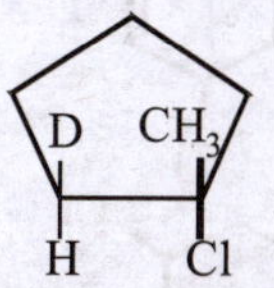

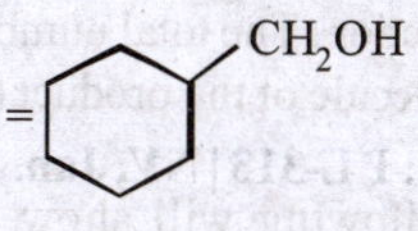

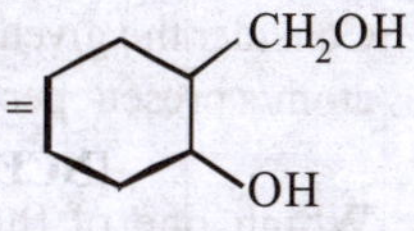

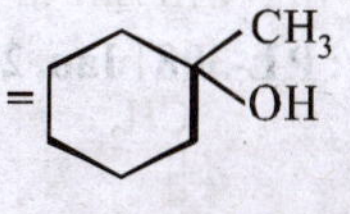

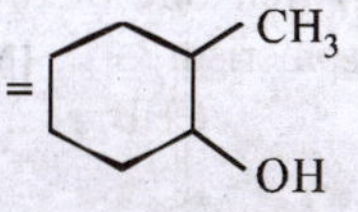

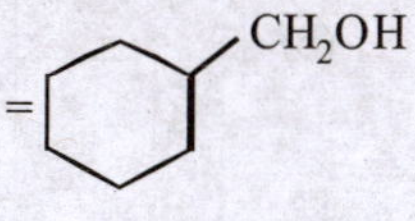

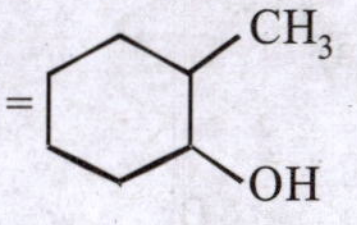

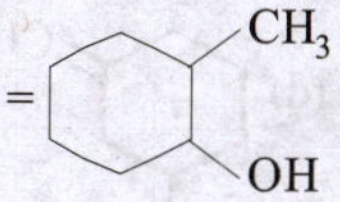

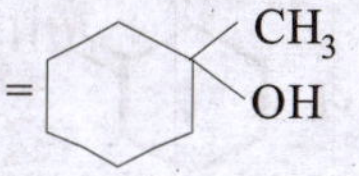

48. Number of geometrical isomers possible for the given structure is / are ________.

[NCERT : P L-308 | NV, Jan. 30, 2024 (II)]

49. Identify product A and product B :

[NCERT : P L-312 | Jan. 29, 2024 (I)]

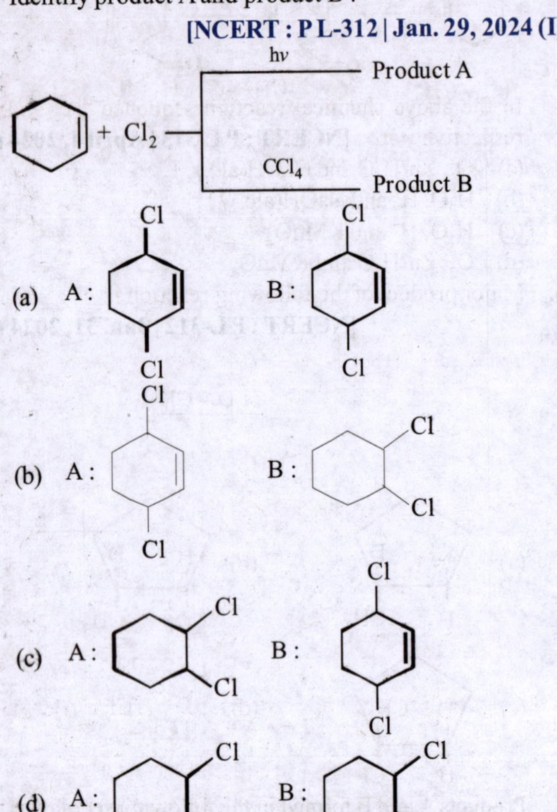

50. CH_3, H, H, CH_3 $\xrightarrow[(ii)\ Zn + H_2O]{(i)\ O_3}$ (P) Product

Consider the given reaction. The total number of oxygen atom/s present per molecule of the product (P) is ____.

[NCERT : P L-313 | NV, Jan. 29, 2024 (I)]

51. Which one of the following will show geometrical isomerism ? **[NCERT : P L-308 | Jan. 29, 2024 (II)]**

(a) CHBr, Br (b) CH_2, Br

(c) CH_2, Br (d) CHBr, Br

52. The product formed in the following multistep reaction is:

$$CH_3-CH=CH_2 \xrightarrow[(iii)\ PCC \quad (iv)\ CH_3MgBr]{(i)\ B_2H_6 \quad (ii)\ H_2O_2,\ NaOH}$$

[NCERT : P L-311 | April 15, 2023 (I)]

(a) $CH_3-CH_2-CH(OH)-CH_3$

(b) $CH_3-CH_2-CH_2-CH_2-OH$

(c) $CH_3-CH_2-C(=O)-OCH_3$

(d) $CH_3-C(OH)(CH_3)-CH_3$

53. $2-\text{Hexene} \xrightarrow[(ii)\ H_2O]{(i)\ O_3}$ Products

[NCERT : P L-313 | April 12, 2023 (I)]

The two products formed in above reaction are -

(a) Butanoic acid and acetic acid
(b) Butanal and acetic acid
(c) Butanal and acetaldehyde
(d) Butanoic acid and acetaldehyde

54. The number of possible isomeric products formed when 3-chloro-1-butene reacts with HCl through carbocation formation is ________

[NCERT : P L-311 | NV, April 11, 2023 (II)]

55. Molar mass of the hydrocarbon (X) which on ozonolysis consumes one mole of O_3 per mole of (X) and gives one mole each of ethanal and propanone is ________ g mol^{-1} (Molar mass of C : 12 g mol^{-1}, H : 1 g mol^{-1})

[NCERT : P L-313 | NV, April 08, 2023 (I)]

56. Choose the correct set of reagents for the following conversion **[NCERT : P L-315 | Jan. 31, 2023 (I)]**

trans (Ph - CH = CH − CH_3) → cis (Ph - CH = CH − CH_3)

(a) Br_2, alc KOH, $NaNH_2$, Na(Liq NH_3)
(b) Br_2, alc KOH, $NaNH_2$, H_2 Lindlar Catalyst
(c) Br_2, aq KOH, $NaNH_2$, H_2 Lindlar Catalyst
(d) Br_2, aq KOH, $NaNH_2$, Na(Liq NH_3)

57. A hydrocarbon 'X' with formula C_6H_8 uses two moles of H_2 on catalytic hydrogenation of its one mole. On ozonolysis, 'X' yields two moles of methane dicarbaldehyde. The hydrocarbon 'X' is :

[NCERT : P L-313 | Jan. 31, 2023 (II)]

(a) hexa-1, 3, 5-triene
(b) 1-methylcyclopenta-1, 4-diene
(c) cyclohexa-1, 3-diene
(d) cyclohexa-1, 4-diene

58. The major products 'A' and 'B', respectively, are

$$'A' \xleftarrow[H_2SO_4]{Cold} H_3C-C(CH_3)=CH_2 \xrightarrow[80°C]{H_2SO_4} 'B'$$

[NCERT : P L-311 | Jan. 30, 2023 (I)]

(a) $H_3C-\underset{OSO_3H}{\overset{CH_3}{C}}-CH_3$ & $CH_3-\overset{CH_3}{C}=CH-\underset{CH_3}{\overset{CH_3}{C}}-CH_3$

(b) $CH_3-\overset{CH_3}{C}=CH-\underset{CH_3}{\overset{CH_3}{C}}-CH_3$ & $CH_3-\underset{OSO_3H}{\overset{CH_3}{C}}-CH_3$

(c) $CH_3-\overset{CH_3}{CH}=CH_2CH_2-\overset{CH_3}{CH}-CH_3$ & $H_3C-\underset{OSO_3H}{\overset{CH_3}{C}}-CH_3$

(d) $H_3C-\underset{OSO_3H}{\overset{CH_3}{C}}-CH_3$ & $CH_3-\overset{CH_3}{CH}-CH_2CH_2-\overset{CH_3}{HC}-CH_3$

59. 17 mg of a hydrocarbon (M.F. $C_{10}H_{16}$) takes up 8.40 mL of the H_2 gas measured at 0ºC and 760 mm of Hg. Ozonolysis of the same hydrocarbon yields

[NCERT : P L-313 | NV, Jan. 29, 2023 (I)]

$CH_3-\underset{O}{\underset{\|}{C}}-CH_3$, $H-\underset{O}{\underset{\|}{C}}-H$, $H-\underset{O}{\underset{\|}{C}}-CH_2-CH_2-\underset{O}{\underset{\|}{C}}-\underset{O}{\underset{\|}{C}}-H$

The number of double bond/s present in the hydrocarbon is ______.

60. The one giving maximum number of isomeric alkenes on dehydrohalogenation reaction is (excluding rearrangement) **[Jan. 29, 2023 (II)]**

(a) 1-Bromo-2-methylbutane

(b) 2-Bromopropane

(c) 2-Bromopentane

(d) 2-Bromo-3,3-dimethylpentane]

61. In the following given reaction '*A*' is

[NCERT : P L-311 | Jan. 24, 2023 (I)]

(cyclobutyl)$-\overset{CH_3}{C}=CH_2$ + HBr ⟶ 'A' (major product)

(a) CH_3, Br (b) CH_3, Br

(c) Br, CH_3, CH_3 (d) CH_3, CH_3, Br

62. Choose the correct option for the following reactions.

[NCERT : P L-311, 312 | July 28, 2022 (I)]

$$B \xleftarrow[H_2O_2/OH^{\ominus}]{(BH_3)_2} H_3C-\underset{CH_3}{\overset{CH_3}{C}}-CH=CH_2 \xrightarrow[NaBH_4]{Hg(OAc)_2,\ H_2O} A$$

(a) 'A' and 'B' are both Markovnikov addition products.

(b) 'A' is Markovnikov product and 'B' is anti-Markovnikov product.

(c) 'A' and 'B' are both anti-Markovnikov products.

(d) 'B' is Markovnikov and 'A' is anti-Markovnikov product.

63. Major product 'B' of the following reaction sequence is:

[NCERT : P L-311 | July 27, 2022 (II)]

$$CH_3-\underset{CH_3}{C}=CH-CH_3 \xrightarrow[CH_3OH]{Br_2} A \xrightarrow{HI} B \text{ (major product)}$$

(a) $CH_3-\underset{CH_3}{\overset{HO}{C}}-\overset{Br}{CH}-CH_3$ (b) $CH_3-\underset{CH_3}{\overset{I}{C}}-\overset{Br}{CH}-CH_3$

(c) $CH_3-\underset{CH_3}{\overset{Br}{C}}-\overset{OH}{CH}-CH_3$ (d) $CH_3-\underset{CH_3}{\overset{Br}{C}}-\overset{I}{CH}-CH_3$

64. A compound 'A' on reaction with 'X' and 'Y' produces the same major product but different byproduct 'a' and 'b'. Oxidation of 'a' gives a substance produced by ants.

[NCERT : P L-311 | July 25, 2022 (I)]

Compound 'A': $H_2C=\overset{CH_3}{C}-CH_2-\underset{CH_3}{\overset{CH_3}{C}}-CH_3$

$\xrightarrow{X}$ a + $O=\overset{CH_3}{C}-CH_2-\underset{CH_3}{\overset{CH_3}{C}}-CH_3$

$\xrightarrow{Y}$ b + $O=\overset{CH_3}{C}-CH_2-\underset{CH_3}{\overset{CH_3}{C}}-CH_3$

'*X*' and '*Y*' respectively are

(a) $KMnO_4/H^+$ and dil. $KMnO_4$, 273 K

(b) $KMnO_4$(dilute), 273 K and $KMnO_4/H^+$

(c) $KMnO_4/H^+$ and O_3, H_2O/Zn

(d) O_3, H_2O/Zn and $KMnO_4/H^+$

65. A hydrocarbon 'X' is found to have molar mass of 80. A 10.0 mg of compound 'X' on hydrogenation consumed 8.40 mL of H_2 gas (measured at STP). Ozonolysis compound 'X' yields only formaldehyde and dialdehyde. The total number of fragments/molecules produced from the ozonolysis of compound 'X' is ______.

[NCERT : P L-313 | NV, June 30, 2022 (I)]

66. Two isomers '*A*' and '*B*' with molecular formula C_4H_8 give different products on oxidation with $KMnO_4$ in acidic medium. Isomer '*A*' on reaction with $KMnO_4/H^+$ results in effervescence of a gas and gives ketone. The compound '*A*' is **[June 29, 2022 (I)]**

(a) But-1-ene (b) *cis*-But-2-ene

(c) *trans*-But-2ene (d) 2-methyl propene

67. '**A**' and '**B**' respectively are:

$$A \xrightarrow[(2)Zn-H_2O]{(1)\ O_3} \text{Ethane}-1,2-\text{dicarbaldehyde}$$

+ Glyoxal/Oxaldehyde

$B \xrightarrow[(2) Zn-H_2O]{(1) O_3} 5-\text{Oxohexanal}$

[NCERT : P L-311 | June 27, 2022 (I)]

(a) 1-methylcyclohex-1, 3-diene & cyclopentene
(b) Cyclohex-1, 3-diene & cyclopentene
(c) 1-methylcyclohex-1,4-diene & 1-methylcyclopent-1-ene
(d) Cyclohex-1,3-diene & 1-methylcyclopent-1-ene

68. The products formed in the following reaction.

$(CH_3)_2C=CH_2 + H-C(CH_3)_3 \xrightarrow{H^{\oplus}}$? is :

[June 25, 2022 (I)]

(a) $(CH_3)_2CH-CH_2-CH_2-CH(CH_3)_2$
(b) $(CH_3)_2CH-CH_2-C(CH_3)_2-CH_3$
(c) $CH_3-CH(CH_3)-CH(CH_3)-CH(CH_3)_2$
(d) $CH_3-C(CH_3)_2-C(CH_3)_2-CH_3$

69. The major product 'A' of the following given reaction has ______ sp^2 hybridized carbon atoms.

$2,7-\text{Dimethyl}-2,6-\text{octadiene} \xrightarrow{H^+} \underset{\text{Major Product}}{A}$

[NCERT : P L-311 | NV, June 24, 2022 (I)]

70. Given below are two statements.

[NCERT : P L-306 | June 24, 2022 (II)]

Statement I : The presence of weaker π - bonds make alkenes less stable than alkanes.

Statement II : The strength of the double bond is greater than that of carbon-carbon single bond.

In the light of the above statements, choose the *correct* answer from the options given below.

(a) Both Statement I and Statement II are correct.
(b) Both Statement I and Statement II are incorrect.
(c) Statement I is correct but Statement II is incorrect.
(d) Statement I is incorrect but Statement II is correct.

71. The major product formed in the following reaction is :

2-methylbuta-1,3-diene $\xrightarrow[\text{(excess)}]{HBr}$ Major product

[NCERT : P L-311 | Aug. 26, 2021 (I)]

(a) 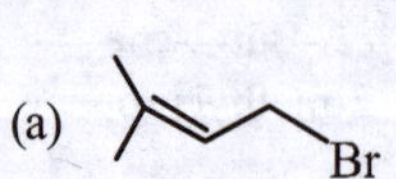
(b) 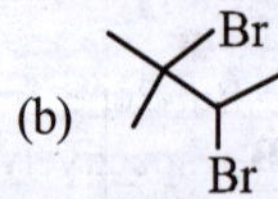
(c) (Br-substituted 3-methylbut-1-ene structure)
(d) 1,4-dibromo-2-methylbutane structure

72. An organic compound 'A' C_4H_8 on treatment with $KMnO_4/H^+$ yields compound 'B' C_3H_6O. Compound 'A' also yield compound 'B' an ozonolysis. Compound 'A' is : **[NCERT : PL-313 | July 25, 2021 (I)]**

(a) 2-Methylpropene (b) 1-Methylcyclopropane
(c) But-2-ene (d) Cyclobutane

73. cis-But-2-ene (H_3C and H_3C on same side, H and H on other) $+ Br_2 \xrightarrow{CCl_4}$ Product "P"

Consider the above chemical reaction. The total number of stereoisomers possible for product 'P' is

[NCERT : P L-311 | NV, July 25, 2021 (II)]

74. Which of the following molecules does not show stereo-isomerism? **[July 22, 2021 (II)]**

(a) 3,4-Dimethylhex-3-ene
(b) 3-Methylhex-1-ene
(c) 3-Ethylhex-3-ene
(d) 4-Methylhex-1-ene

75. Cyclohexene $\xrightarrow[H_2SO_4,\ \Delta]{KMnO_4}$ 'A' (major product)

Cyclohexene $\xrightarrow[H_2O,\ 273K]{KMnO_4}$ 'B' (major product)

For above chemical reactions, identify the correct statement from the following: **[July 20, 2021 (I)]**

(a) Both compound 'A' and compound 'B' are dicarboxylic acids
(b) Both compound 'A' and compound 'B' are diols
(c) Compound 'A' is diol and compound 'B' is dicarboxylic acid
(d) Compound 'A' is dicarboxylic acid and compound 'B' is diol

76. 1-methyl-1-vinylcyclohexane (CH_3) $\xrightarrow[CCl_4]{HBr}$ A (Major Product)

Product "A" in the above chemical reaction is

[NCERT : P L-311 | March 17, 2021 (I)]

(a) cyclohexane ring with Br and CH_3 on one carbon, CH_3 on adjacent carbon
(b) cyclohexane ring with $CH(Br)CH_3$ and adjacent CH_3
(c) cyclohexane ring with Br and $CH(CH_3)CH_3$ on same carbon (CH_3, CH_3)
(d) cyclohexane ring with CH_2CH_2Br and CH_3 on same carbon

77. Choose the **correct** statement regarding the formation of carbocation A and B given

[NCERT : P L-311 | March 17, 2021 (II)]

$CH_3-CH_2-CH=CH_2 + HBr$

$\rightarrow CH_3-CH_2-CH_2-\overset{+}{C}H_2 + Br^-$ "A"

$\rightarrow CH_3-CH_2-\overset{+}{C}H-CH_3 + Br^-$ "B"

(a) Carbocation A is more stable and formed relatively at slow rate
(b) Carbocation A is more stable and formed relatively at faster rate
(c) Carbocation B is more stable and formed relatively at slow rate
(d) Carbocation B is more stable and formed relatively at faster rate

78. 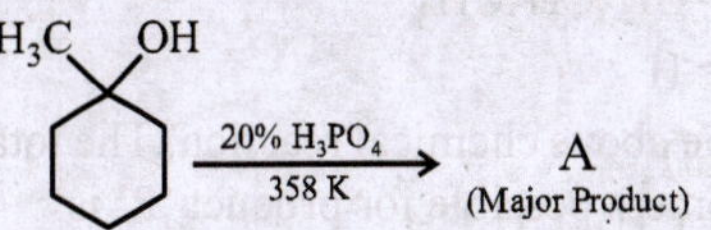

H_3C Cl (1-chloro-1-methylcyclohexane) $\xrightarrow{(CH_3)_3CO^-K^+}$ B (Major Product)

[NCERT : P L-311 | March 16, 2021 (I)]

The products "A" and "B" formed in above reactions are

(a) A— (methylenecyclohexane, CH_2) B— (1-methylcyclohexene, CH_3)
(b) A— (methylenecyclohexane, CH_2) B— (methylenecyclohexane, CH_2)
(c) A— (1-methylcyclohexene, CH_3) B— (methylenecyclohexane, CH_2)
(d) A— (1-methylcyclohexene, CH_3) B— (1-methylcyclohexene, CH_3)

79.  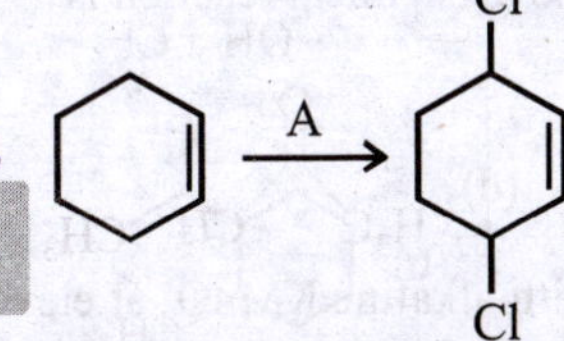

Identify the reagent(s) 'A' and condition(s) for the reaction **[NCERT : P L-312 | March 16, 2021 (II)]**

(a) $A = HCl$; Anhydrous $AlCl_3$
(b) $A = Cl_2$; UV light
(c) $A = Cl_2$; dark, Anhydrous $AlCl_3$
(d) $A = HCl, ZnCl_2$

80. In $\overset{1}{C}H_2 = \overset{2}{C} = \overset{3}{C}H - \overset{4}{C}H_3$ molecule, the hybridization of carbon 1, 2, 3 and 4 respectively, are : **[NCERT : P L-311 | Feb. 26, 2021 (II)]**

(a) sp^2, sp^2, sp^2, sp^3 (b) sp^2, sp, sp^2, sp^3
(c) sp^3, sp, sp^3, sp^3 (d) sp^2, sp^3, sp^2, sp^3

81. The major product of the following reaction is :

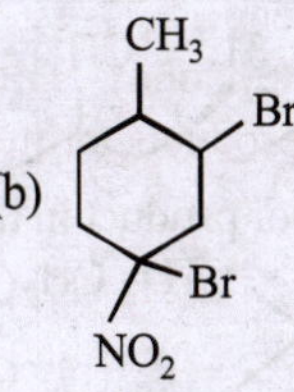

[NCERT : P L-311 | Sep. 06, 2020 (I)]

(a) 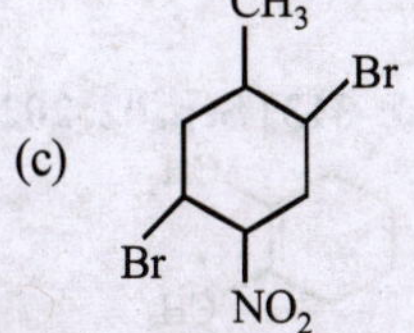 (b) 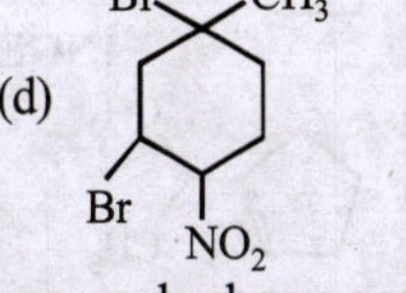

(c) (d)

82. Which of the following compounds shows geometrical isomerism? **[NCERT : P L-308 | Sep. 06, 2020 (I)]**

(a) 2-methylpent-2-ene (b) 4-methylpent-2-ene
(c) 4-methylpent-1-ene (d) 2-methylpent-1-ene

83. The increasing order of the boiling point of the major products A, B and C of the following reactions will be: **[NCERT : P L-311 | Sep. 06, 2020 (II)]**

(A) (but-1-ene) + HBr $\xrightarrow{(C_6H_5CO)_2}$ A

(B) (2-methylpropene) + HBr ⟶ B

(C) (but-2-ene) + HBr ⟶ C

(a) B < C < A (b) C < A < B
(c) A < B < C (d) A < C < B

84. The major product formed in the following reaction is :

$CH_3CH = CHCH(CH_3)_2 \xrightarrow{HBr}$

[NCERT : P L-311 | Sep. 05, 2020 (II)]

(a) $CH_3CH(Br)CH_2CH(CH_3)_2$
(b) $CH_3CH_2CH(Br)CH(CH_3)_2$
(c) $Br(CH_2)_3CH(CH_3)_2$
(d) $CH_3CH_2CH_2C(Br)(CH_3)_2$

85. Among the following compounds, geometrical isomerism is exhibited by : **[NCERT : P L-308 | Sep. 05, 2020 (II)]**

(a) (b)

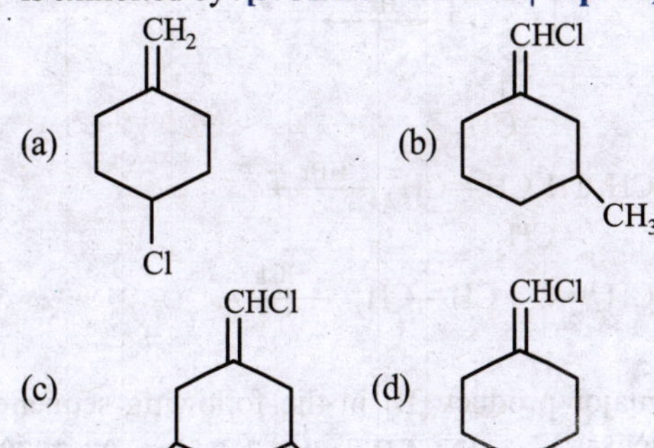

86. Which of the following compounds produces an optically inactive compound on hydrogenation? [Sep. 03, 2020 (I)]

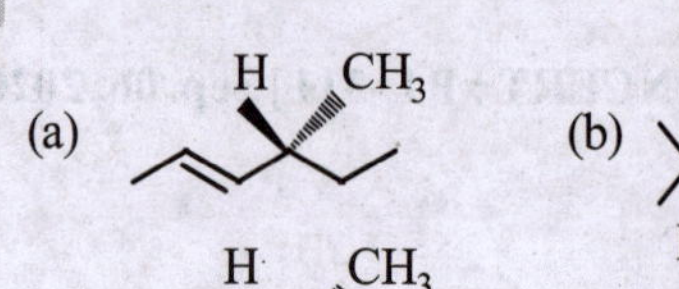

(a) H CH_3 (b) H CH_3

(c) H CH_3 (d) H CH_3

87. The major product in the following reaction is :

H_3C $CH=CH_2$ (on cyclopentane) $\xrightarrow[\text{Heat}]{H_3O^+}$

[NCERT : P L-313 | Sep. 02, 2020 (I)]

(a) CH_3, CH_3 (b)

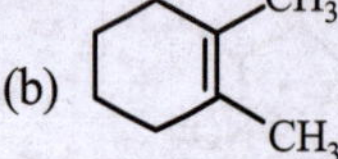

CH_3, CH_3

(c)

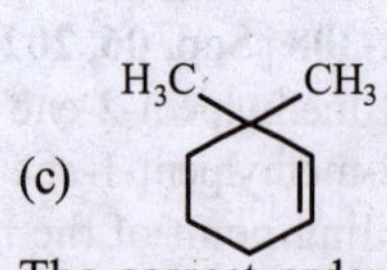

H_3C CH_3 (d) OH, H_3C, CH_3

88. The correct order of heat of combustion for following alkadienes is: [Jan. 09, 2020 (I)]

(A)

(B)

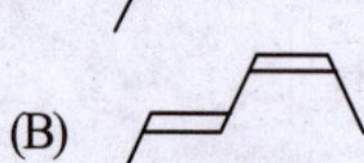

(C)

(a) (A)<(B)<(C) (b) (A)<(C)<(B)

(c) (C)<(B)<(A) (d) (B)<(C)<(A)

89. Which of the following has the shortest C–Cl bond? [Jan. 09, 2020 (II)]

(a) $Cl–CH=CH_2$ (b) $Cl–CH=CH–NO_2$

(c) $Cl–CH=CH–CH_3$ (d) $Cl–CH=CH–OCH_3$

90. Which of the following reactions will not produce a racemic product? [Jan. 09, 2020 (II)]

(a) $CH_3–\overset{O}{\overset{\|}{C}}\,CH_2\,CH_3 \xrightarrow{HCN}$

(b) H_3C (cyclohexene with CH_3) $\xrightarrow{HCl}$

(c) $CH_3CH_2CH=CH_2 \xrightarrow{HBr}$

(d) $CH_3–\underset{H}{\overset{CH_3}{\underset{|}{\overset{|}{C}}}}–CH=CH_2 \xrightarrow{HCl}$

91. The major product [B] in the following sequence of reactions is: [NCERT : P L-312 | Jan. 08, 2020 (II)]

$CH_3-\underset{CH(CH_3)_2}{\underset{|}{C}}=CH-CH_2CH_3 \xrightarrow[\text{(ii) } H_2O_2,\ OH^{\ominus}]{\text{(i) } B_2H_6} [A] \xrightarrow[\Delta]{\text{dil. } H_2SO_4} [B]$

(a) $CH_2=\underset{CH(CH_3)_2}{\underset{|}{C}}-CH_2CH_2CH_3$

(b) $CH_3-\underset{\underset{H_3C\quad CH_3}{C}}{\underset{\|}{C}}-CH_2CH_2CH_3$

(c) $CH_3-\underset{CH(CH_3)_2}{\underset{|}{C}}=CH-CH_2CH_3$

(d) $CH_3-\underset{CH(CH_3)_2}{\underset{|}{CH}}-CH=CH-CH_3$

92. The major products A and B in the following reactions are: [NCERT : P L-312 | Jan. 08, 2020 (I)]

CN $\xrightarrow[\text{Heat}]{\text{Peroxide}}$ [A]

[A] + $\longrightarrow$ B

(a) A = CN and B = CN

(b) A = CN and B = CN

(c) A = CN and B = CN

(d) A = CN and B = CN

93. The major product of the following addition reaction is

$H_3C-CH=CH_2 \xrightarrow{Cl_2/H_2O}$

[NCERT : P L-311 | April 12, 2019 (I), 2016]

(a) $H_3C-\underset{Cl}{\underset{|}{CH}}-\underset{OH}{\underset{|}{CH_2}}$ (b) $H_3C-\underset{OH}{\underset{|}{CH}}-\underset{Cl}{\underset{|}{CH_2}}$

(c) H_3C–(epoxide) (d) $H_3C-\overset{O}{\overset{\|}{C}}-CH_3$

94. But-2-ene on reaction with alkaline $KMnO_4$ at elevated temperature followed by acidification will give : [NCERT : P L-313 | April 12, 2019 (I)]

(a) $CH_3-\underset{OH}{\underset{|}{CH}}-\underset{OH}{\underset{|}{CH}}-CH_3$

(b) one molecule of CH_3CHO and one molecule of CH_3COOH

(c) 2 molecules of CH_3COOH

(d) 2 molecules of CH_3CHO

95. Which one of the following alkenes when treated with HCl yields majorly an anti Markovnikov product ? [NCERT : P L-311 | April 8, 2019 (II)]

(a) $CH_3O—CH=CH_2$ (b) $Cl—CH=CH_2$

(c) $H_2N—CH=CH_2$ (d) $F_3C—CH=CH_2$

96. The major product of the following reaction is : **[NCERT : P L-311 | Jan. 12, 2019 (II)]**

H_3C, CH_2, H $\xrightarrow{HCl}$

(a) CH_2-Cl, CH_3, H

(b) CH_3, CH_3, Cl, H

(c) CH_3, CH_2-Cl, H

(d) CH_3, CH_3, Cl, H

97. The major product of the following reaction is :

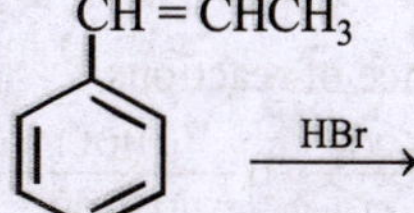

$CH=CHCH_3$ $\xrightarrow{HBr}$

[NCERT : P L-311 | Online April 16, 2018]

(a) 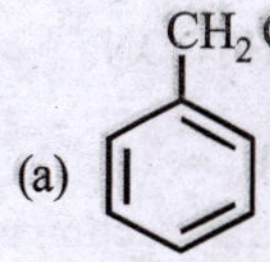

$CH_2CH_2CH_2Br$

(b) 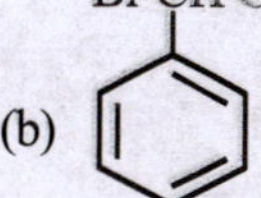

$BrCHCH_2CH_3$

(c) 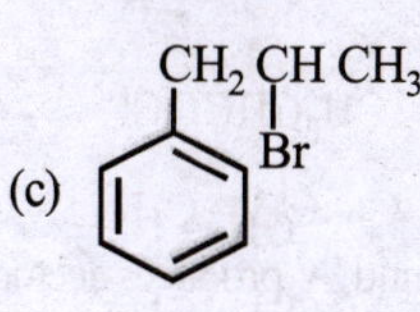

CH_2CHCH_3 with Br

(d) $CH=CHCH_3$ with Br

98. The *trans*-alkenes are formed by the reduction of alkynes with: **[2018; Same Online April 11, 2014; 2008]**

(a) H_2-Pd/C, $BaSO_4$ (b) $NaBH_4$

(c) Na/liq. NH_3 (d) Sn - HCl

99. 3-Methyl pent-2-ene on reaction with HBr in presence of peroxide forms an addition product. The number of possible stereoisomers for the product is :

[NCERT : P L-311 | 2017]

(a) Six (b) Zero (c) Two (d) Four

100. Which of the following compounds is most reactive to an aqueous solution of sodium carbonate ?

[NCERT : P L-321 | Online April 10, 2016 (S), Same Online April 9, 2017]

(a) 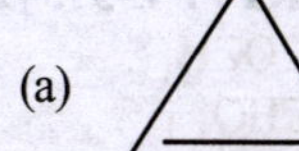

(b) 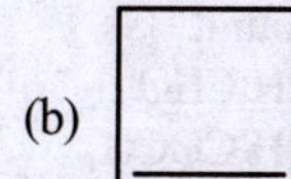

(c) (d)

101. The major product of the following reaction is : **[NCERT : P L-312 | Online April 9, 2017]**

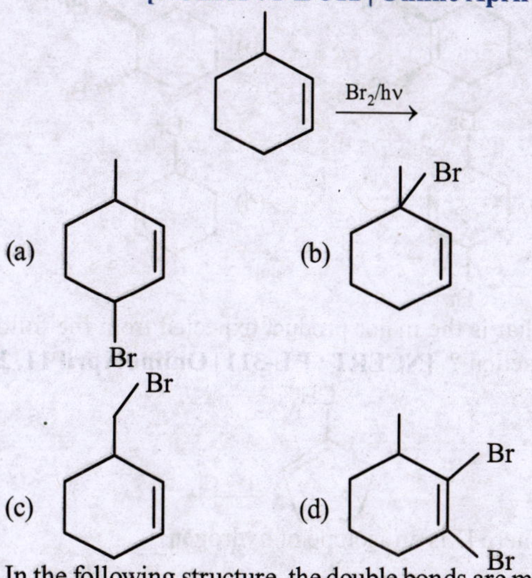

102. In the following structure, the double bonds are marked as I, II, III and IV **[NCERT : P L-308 | Online April 9, 2017]**

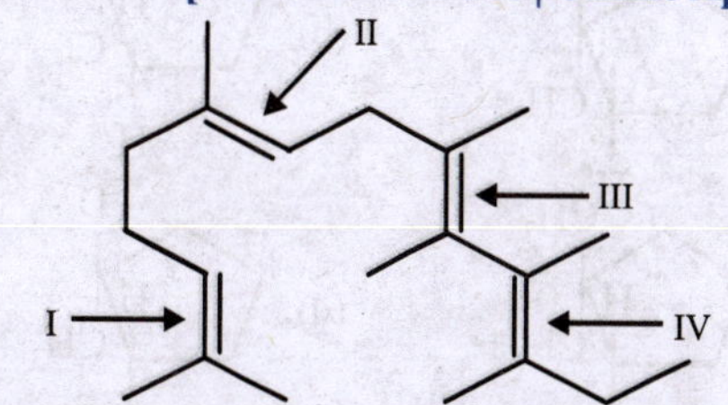

Geometrical isomerism is **not** possible at site (s) :

(a) III (b) I (c) I and III (d) III and IV

103. The product of the reaction given below is:

[NCERT : P L-312 | 2016]

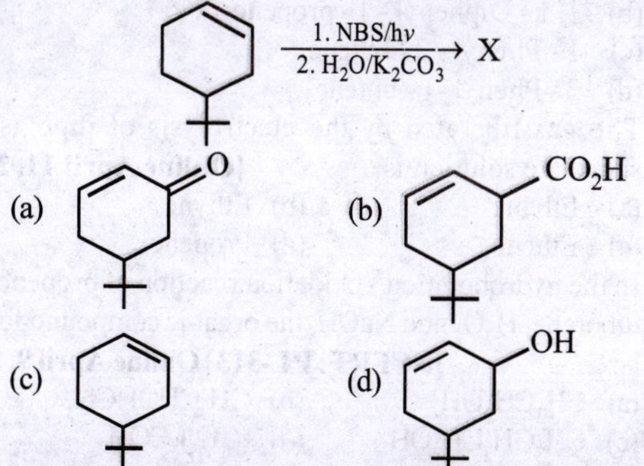

104. The reaction of propene with HOCl ($Cl_2 + H_2O$) proceeds through the intermediate: **[NCERT : P L-313 | 2016]**

(a) $CH_3-CH(OH)-CH_2^+$

(b) $CH_3-CHCl-CH_2^+$

(c) $CH_3-CH^+-CH_2-OH$

(d) $CH_3-CH^+-CH_2-Cl$

105. Bromination of cyclohexene under conditions given below yields : **[NCERT : P L-312 | Online April 10, 2016]**

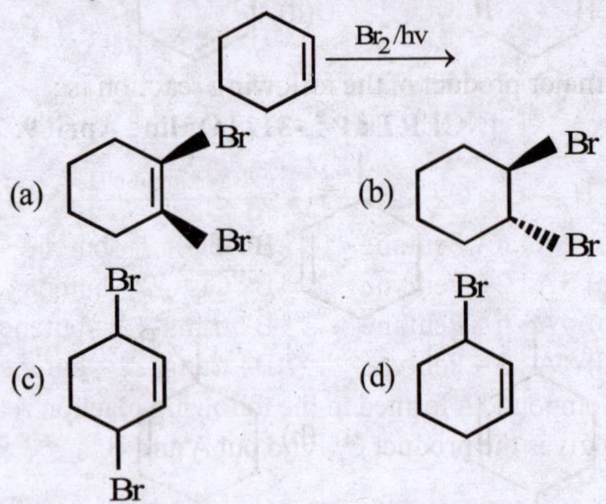

106. What is the major product expected from the following reaction ? **[NCERT : P L-311 | Online April 11, 2015]**

CH_3 (1-methylcyclopentene) $\xrightarrow{D-Cl}$

Where D is an isotope of hydrogen

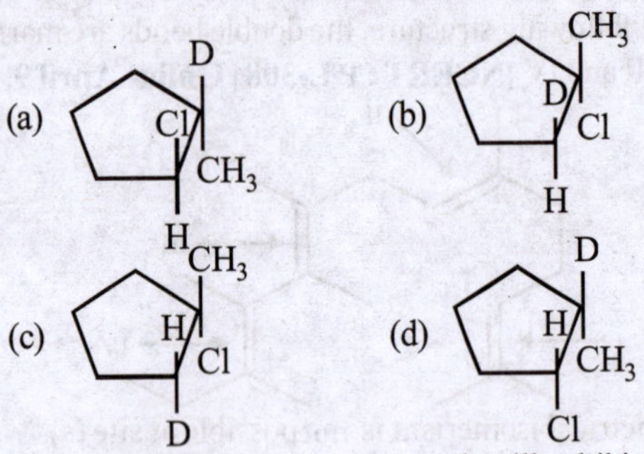

107. Which of the following compounds will exhibit geometrical isomerism ? **[2015; Same 2002]**

(a) 2 - Phenyl -1 - butene
(b) 1, 1 - Diphenyl - 1 - propene
(c) 1 - Phenyl - 2 - butene
(d) 3 - Phenyl -1 - butene

108. The gas liberated by the electrolysis of dipotassium succinate solution is: **[Online April 11, 2014]**

(a) Ethane (b) Ethyne
(c) Ethene (d) Propene

109. In the hydroboration - oxidation reaction of propene with diborane, H_2O_2 and NaOH, the organic compound formed is: **[NCERT : PL-313 | Online April 9, 2014]**

(a) CH_3CH_2OH (b) $CH_3CHOHCH_3$
(c) $CH_3CH_2CH_2OH$ (d) $(CH_3)_3COH$

110. $C_6H_5-CH_2-CH=CH_2$

On mercuration-demercuration produces the major product: **[NCERT : P L-311 | Online April 12, 2014]**

(a) $C_6H_5-CH_2-\underset{OH}{CH}-CH_3$
(b) $C_6H_5-CH_2-CH_2-CH_2-OH$
(c) $C_6H_5-CH_2-\underset{OH}{CH}-\underset{OH}{CH_3}$
(d) $C_6H_5-CH_2-COOH$

111. In the presence of peroxide, HCl and HI do not give anti-Markownikoff's addition of alkenes because: **[NCERT : P L-312 | Online April 12, 2014]**

(a) One of the steps is endothermic in HCl and HI
(b) Both HCl and HI are strong acids
(c) HCl is oxidizing and the HI is reducing
(d) All the steps are exothermic in HCl and HI

112. The addition of HI in the presence of peroxide catalyst does not follow anti-Markovnikov's rule because :

(a) HI is a strong reducing agent. **[Online April 9, 2013]**
(b) H-I bond is too strong to be broken homolytically.
(c) I atom combines with H atom to give back HI.
(d) Iodine atom is not reactive enough to add across a double bond.

113. Consider the following sequence of reactions

$$CH_3CH=CH_2 \xrightarrow[700K]{Cl_2} A \xrightarrow[420K, 12atm]{Na_2CO_3} B \xrightarrow[(ii)\ NaOH]{(i)\ HOCl} C$$

Compound 'C' is **[Online May 26, 2012]**

(a) $CH_2OH-CHOH-CH_2OH$ (b) $CH_3\underset{OH}{CH}COONa$
(c) $HOCH_2-CH=CH_2$ (d) $CH_3\underset{OH}{CH}COCl$

114. Ozonolysis of an organic compound 'A' produces acetone and propionaldehyde in equimolar mixture. Identify 'A' from the following compounds: **[2011 RS, Same 2010, 2009]**

(a) 1 – Pentene
(b) 2 – Pentene
(c) 2 – Methyl – 2 – pentene
(d) 2 – Methyl – 1 – pentene

115. One mole of a symmetrical alkene on ozonolysis gives two moles of an aldehyde having a molecular mass of 44 u. The alkene is **[NCERT : P L-313 | 2010]**

(a) propene (b) 1-butene
(c) 2-butene (d) ethene

116. In the following sequence of reactions, the alkene affords the compound 'B'

$$CH_3-CH=CH-CH_3 \xrightarrow{O_3} A \xrightarrow[Zn]{H_2O} B.$$

The compound B is **[NCERT : P L-313 | 2008]**

(a) CH_3CH_2CHO (b) CH_3COCH_3
(c) $CH_3CH_2COCH_3$ (d) CH_3CHO

117. Reaction of one molecule of HBr with one molecule of 1, 3-butadiene at 40 °C gives predominantly **[NCERT : P L-311 | 2005]**

(a) 1-bromo-2-butene under kinetically controlled conditions
(b) 3-bromobutene under thermodynamically controlled conditions
(c) 1-bromo-2-butene under thermodynamically controlled conditions
(d) 3-bromobutene under kinetically controlled conditions

118. Butene-1 may be converted to butane by reaction with **[2003]**

(a) Sn – HCl (b) Zn – Hg
(c) Pd/H_2 (d) Zn – HCl

Topic 3 -Alkynes

119. Total number of sigma (σ) ____ and pi(π) ____ bonds respectively present in hex-1-en-4-yne are : **[NCERT : P L-315 | Jan. 29, 2025 (II)]**

(a) 13 and 3 (b) 11 and 3
(c) 3 and 13 (d) 14 and 3

120. The sum of sigma (σ) and pi(π) bonds in Hex-1,3-dien-5-yne is ________. **[NCERT : P L-315 | NV, Jan. 29, 2025 (I)]**

121. Identify product [A], [B] and [C] in the following reaction sequence : **[NCERT : P L-317 | Jan. 28, 2025 (II)]**

$$CH_3-C\equiv CH \xrightarrow[H_2]{Pd/C} [A] \xrightarrow[(ii)\ Zn, H_2O]{(i)\ O_3} [B]+[C]$$

(a) [A] : CH_3–CH = CH_2, [B] : CH_3CHO, [C] : HCHO
(b) [A] : CH_2=CH_2, [B] : $H_3C-\overset{O}{\overset{\|}{C}}-CH_3$ [C] : HCHO
(c) [A] : CH_3–CH=CH_2, [B] : CH_3CHO, [C] : CH_3CH_2OH
(d) [A] : $CH_3CH_2CH_3$, [B] : CH_3CHO, [C] : HCHO

122. The compound with molecular formula C_6H_6, which gives only one monobromo derivative and takes up four moles of hydrogen per mole for complete hydrogenation has ______ π electrons. **[NCERT : P L-315 | NV, Jan. 22, 2025 (II)]**

123. The **incorrect** statement regarding ethyne is **[NCERT : P L-315 | April 9, 2024 (II)]**

(a) The C–C bonds in ethyne is shorter than that in ethene
(b) Both carbons are sp hybridised
(c) Ethyne is linear
(d) The carbon-carbon bonds in ethyne is weaker than that in ethene

124. The major product of the following reaction is P.

$$CH_3C\equiv C-CH_3 \xrightarrow[(ii)\ dil.KmnO_4\ 273K]{(i)\ Na/liq.\ NH_3} 'P'$$

Number of oxygen atoms present in product 'P' is ______ (nearest integer). **[NV, April 6, 2024 (I)]**

125. In the given reactions identify A and B **[NCERT : P L-317 | Feb. 1, 2024 (II)]**

$$H_2 + A \xrightarrow{Pd/C} \begin{matrix} CH_3 & & C_2H_5 \\ & C=C & \\ H & & H \end{matrix}$$

$$CH_3-C\equiv C-CH_3 + H_2 \xrightarrow{Na/Liquid\ NH_3} "B"$$

(a) A : n – Pentane B : Cis – 2 – butene
(b) A : 2 – Pentyne B : Cis – 2 – butene
(c) A : n – Pentane B : trans – 2 – butene
(d) A : 2 – Pentyne . B : trans – 2 – butene

126. Compound A formed in the following reaction reacts with B gives the product C. Find out A and B

$$CH_3-C\equiv CH + Na \longrightarrow A \xrightarrow{B} \underset{(C)}{CH_3-C\equiv C-CH_2-\underset{CH_3}{\underset{|}{CH_2}}} + NaBr$$

[NCERT : P L-316 | Jan. 30, 2024 (I)]

(a) $A=CH_3-C\equiv \overset{-}{C}\overset{+}{Na}$, $B=CH_3-CH_2-CH_2-Br$
(b) $A=CH_3-CH=CH_2$, $B=CH_3-CH_2-CH_2-Br$
(c) $A=CH_3-CH_2-CH_3$, $B=CH_3-C\equiv CH$
(d) $A=CH_3-C\equiv \overset{-}{C}\overset{+}{Na}$, $B=CH_3-CH_2-CH_3$

127. But-2-yne is reacted separately with one mole of hydrogen as shown below: **[NCERT : P L-315 | Feb. 01, 2023 (I)]**

$$\underline{B} \xleftarrow[liq\ NH_3]{Na} \underset{+H_2}{CH_3-C\equiv C-CH_3} \xrightarrow[\Delta]{Pd/C} \underline{A}$$

(A) A is more soluble than B.
(B) The boiling point & melting point of A are higher and lower than B respectively.
(C) A is more polar than B because dipole moment of A is zero.
(D) Br_2 adds easily to B than A.

Identify the incorrect statements from the options given below:

(a) A and B only
(b) B, C & D only
(c) A, C & D only
(d) B and C only

128. In bromination of propyne, with bromine 1, 1, 2, 2-tetrabromopropane is obtained in 27% yield. The amount of 1, 1, 2, 2-tetrabromopropane obtained from 1 g of bromine in this reaction is ______ × 10^{-1} g. (Nearest integer) **[NV, July 29, 2022 (I)]**
(Molar mass : Bromine = 80 g/mol)

129. What will be the major product of following sequence of reactions? **[NCERT : P L-315 | June 27, 2022 (II)]**

$$n\text{-Bu}-\equiv \xrightarrow[(ii)\ \text{Lindlar cat, } H_2]{(i)\ n\text{-BuLi},\ n\text{-}C_5H_{11}Cl}$$

(a) [structure: alkene with n-Bu, n-Bu and $H_{11}C_5$ substituents]

(b) [structure: alkene with n-Bu, n-Bu, C_5H_{11} substituents]

(c) [structure: $H_{11}C_5$, n-Bu cis alkene]

(d) [structure: C_5H_{11}, n-Bu trans alkene]

130. The number of sigma bonds in **[NV, July 25, 2021 (I)]**

$H_3C-\underset{|}{\overset{}{C}}=CH-C\equiv C-H$ is ______.
(H on the second carbon)

131. An unsaturated hydrocarbon X on ozonolysis gives A. Compound A when warmed with ammonical silver nitrate forms a bright silver mirror along the sides of the test tube. The unsaturated hydrocarbon X is: **[March 16, 2021 (II)]**

(a) $CH_3-\overset{CH_3}{\overset{|}{C}}=\!\!\triangleleft$ (b) $CH_3-\underset{CH_3}{\underset{|}{C}}=\underset{CH_3}{\underset{|}{C}}-CH_3$

(c) $HC\equiv C-CH_2-CH_3$ (d) $CH_3-C\equiv C-CH_3$

132. Consider the following chemical reaction.

$$CH\equiv CH \xrightarrow[\text{(2) CO, HCl, AlCl}_3]{\text{(1) Red hot Fe tube, 873K}} \text{Product}$$

The number of sp^2 hybridized carbon atom(s) present in the product is ______.

[NCERT : P L-318 | NV, Feb. 25, 2021 (I)]

133. The major product obtained from the following reaction is: **[NCERT : P L-312 | Sep. 06, 2020 (I)]**

$$O_2N-C_6H_4-C\equiv C-C_6H_4-OCH_3 \xrightarrow[H_2O]{Hg^{2+}/H^+}$$

(a) $O_2N-C_6H_4-CH_2-CO-C_6H_4-OCH_3$

(b) $O_2N-C_6H_4-CO-CH_2-C_6H_4-OH$

(c) $O_2N-C_6H_4-CH_2-CO-C_6H_4-OH$

(d) $O_2N-C_6H_4-CO-CH_2-C_6H_4-OCH_3$

134. The major product (Y) in the following reactions is: **[NCERT : P L-312 | Jan. 09, 2020 (I)]**

$$CH_3-\overset{CH_3}{\overset{|}{CH}}-C\equiv CH \xrightarrow[H_2O]{HgSO_4,\ H_2SO_4} X$$

$$\xrightarrow[\text{(ii) Conc. } H_2SO_4/\Delta]{\text{(i) } C_2H_5MgBr,\ H_2O} Y$$

(a) $CH_3-\overset{CH_3}{\overset{|}{CH}}-\underset{CH_3}{\underset{|}{C}}=CH-CH_3$

(b) $H_3C-\overset{CH_2}{\overset{\|}{C}}-\underset{C_2H_5}{\underset{|}{CH}}-CH_3$

(c) $CH_3-\overset{CH_3}{\overset{|}{CH}}-\underset{CH_2CH_3}{\underset{|}{C}}=CH_2$

(d) $CH_3-\overset{CH_3}{\overset{|}{C}}=\underset{CH_2CH_3}{\underset{|}{C}}-CH_3$

135. In the following sequence of reactions the maximum number of atoms present in molecule 'C' in one plane is ______.

$$A \xrightarrow[\text{Cu tube}]{\text{Red hot}} B \xrightarrow[\text{Anhydrous AlCl}_3]{CH_3Cl\text{ (1 eq.)}} C$$

(A is a lowest molecular weight alkyne)

[NV, Jan. 08, 2020 (II)]

136. Consider the following reaction :

$$A \xrightarrow[\Delta]{Ag_2O} \text{Ppt}$$

$$A \xrightarrow{Hg^{2+}/H^+} B \xrightarrow{NaBH_4} C \xrightarrow[\text{conc. HCl}]{ZnCl_2} \text{Turbidity within 5 minutes}$$

'A' is: **[April 12, 2019 (II)]**

(a) $CH\equiv CH$ (b) $CH_3-C\equiv C-CH_3$

(c) $CH_3-C\equiv CH$ (d) $CH_2=CH_2$

137. The increasing order of the reactivity of the following compounds towards electrophilic aromatic substitution reactions is : **[NCERT : P L-322 | April 10, 2019 (I)]**

Cl CH$_3$ COCH$_3$

(I) (II) (III)

(a) II < I < III (b) III < II < I
(c) III < I < II (d) I < III < II

138. The major product of the following reaction is:

$CH_3C\equiv CH \xrightarrow[\text{(ii) DI}]{\text{(i) DCl (1 equiv.)}}$

[NCERT : P L-317 | April 9, 2019 (I); Same 2007]

(a) $CH_3CD(I)CHD(Cl)$ (b) $CH_3CD(Cl)CHD(I)$
(c) $CH_3CD_2CH(Cl)(I)$ (d) $CH_3C(I)(Cl)CHD_2$

139. The correct order for acid strength of compounds $CH\equiv CH$, $CH_3{-}C\equiv CH$ and $CH_2{=}CH_2$ is as follows:

[NCERT : P L-316 | Jan. 12, 2019 (I)]

(a) $CH\equiv CH > CH_2{=}CH_2 > CH_3{-}C\equiv CH$
(b) $CH_3{-}C\equiv CH > CH\equiv CH > CH_2{=}CH_2$
(c) $CH_3{-}C\equiv CH > CH_2{=}CH_2 > HC\equiv CH$
(d) $HC\equiv CH > CH_3{-}C\equiv CH > CH_2{=}CH_2$

140. When 2-butyne is treated with H_2/Lindlar's catalyst, compound X is produced as the major product and when treated with Na/liq. NH_3 it produces Y as the major product. Which of the following statements is correct?

[NCERT : P L-315 | Online April 15, 2018 (II)]

(a) Y will have higher dipole moment and higher boiling point than X
(b) Y will have higher dipole moment and lower boiling point than X
(c) X will have lower dipole moment and lower boiling point than Y
(d) X will have higher dipole moment and higher boiling point than Y

141. The reagent needed for the given conversion is

$Ph-C\equiv C-Ph \longrightarrow$ (Ph)(H)C = C(H)(Ph)

[NCERT : P L-315 | Online April 11, 2014]

(a) Cat. hydrogenation (b) H_2/ Lindlar Cat.
(c) Li/NH_3 (d) $LiAlH_4$

142. Which one of the following class of compounds is obtained by polymerization of acetylene?

[NCERT : P L-318 | Online April 9, 2014]

(a) Poly-yne (b) Poly-ene
(c) Poly-ester (d) Poly-amine

143. The hydration of propyne results in formation of

[Online May 26, 2012]

(a) Acetone (b) Propanol-1
(c) Propene (d) Propanal

144. 2-Hexyne gives *trans*-2-hexene on treatment with:

[NCERT : P L-315 | 2012]

(a) Pt/H_2 (b) Li / NH_3
(c) Pd/$BaSO_4$ (d) Li AlH_4

145. The hydrocarbon which can react with sodium in liquid ammonia is **[NCERT : P L-315 | 2008]**

(a) $CH_3CH_2CH_2C\equiv CCH_2CH_2CH_3$
(b) $CH_3CH_2C\equiv CH$
(c) $CH_3CH=CHCH_3$
(d) $CH_3CH_2C\equiv CCH_2CH_3$

146. The treatment of CH_3MgX with $CH_3C\equiv C-H$ produces **[NCERT : P L-317 | 2008]**

(a) $CH_3-CH=CH_2$
(b) $CH_3C\equiv C-CH_3$
(c) $CH_3-\overset{H}{\overset{|}{C}}=\overset{H}{\overset{|}{C}}-CH_3$
(d) CH_4

147. Which of the following reactions will yield 2, 2-dibromopropane? **[NCERT : P L-317 | 2007]**

(a) $CH_3-CH=CH_2+HBr\rightarrow$
(b) $CH_3-C\equiv CH+2HBr\rightarrow$
(c) $CH_3CH=CHBr+HBr\rightarrow$
(d) $CH\equiv CH+2HBr\rightarrow$

148. Which one of the following has minimum boiling point?

[NCERT : P L-316 | 2004]

(a) 1 - Butene (b) 1 - Butyne
(c) *n*- Butane (d) Isobutane

149. What is the product when acetylene reacts with hypochlorous acid? **[NCERT : P L-316 | 2002]**

(a) CH_3COCl (b) $ClCH_2CHO$
(c) Cl_2CHCHO (d) $ClCH_2COOH$.

150. Which of these will not react with acetylene?

[NCERT : P L-316 | 2002]

(a) dil. NaOH (b) ammonical $AgNO_3$
(c) Na in liq. NH_3 (d) HCl.

Topic 4 - Aromatic Hydrocarbons

151. Choose the correct set of reagents for the following conversion. **[NCERT : P L-322 | April 8, 2025 (II)]**

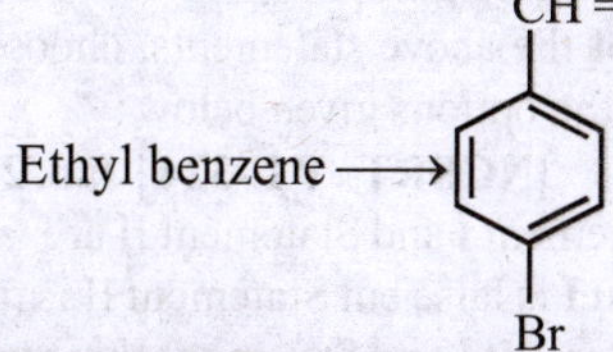

(a) Br_2/Fe ; Cl_2, Δ ; alc. KOH
(b) Cl_2/Fe ; Br_2/anhy.$AlCl_3$; aq. KOH
(c) Br_2/anhy.$AlCl_3$; Cl_2, Δ ; aq. KOH
(d) Cl_2/anhy.$AlCl_3$; Br_2/Fe ; alc. KOH

152. Benzene is treated with oleum to produce compound (*X*) which when further heated with molten sodium hydroxide followed by acidification produces compound (*Y*). The compound *Y* is treated with zinc metal to produce compound (*Z*). Identify the structure of compound (*Z*) from the following option. **[NCERT : P L-324 | April 4, 2025 (I)]**

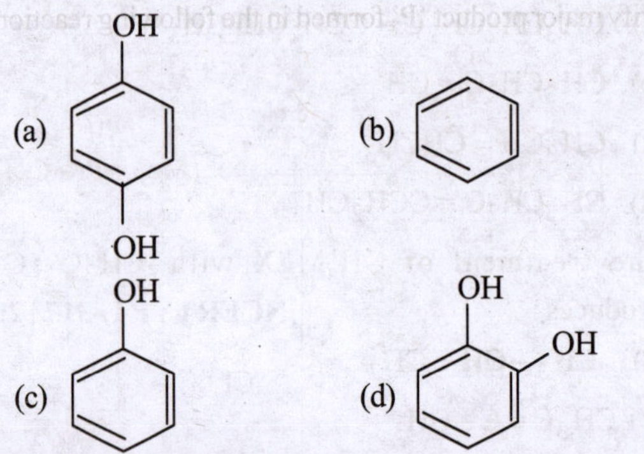

153. Designate whether each of the following compounds is aromatic or not aromatic.

[NCERT : P L-321 | April 2, 2025 (I)]

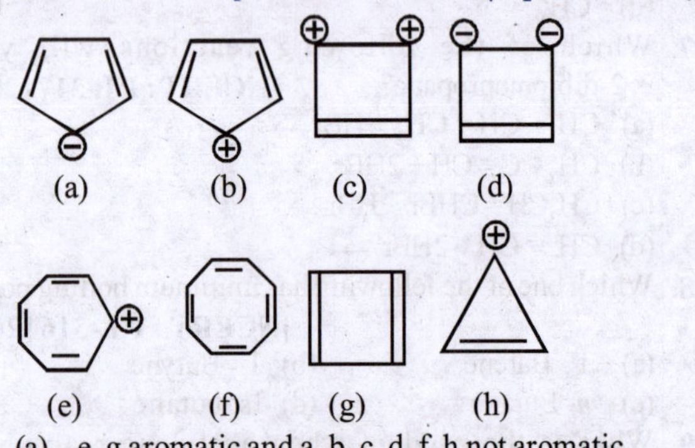

(a) e, g aromatic and a, b, c, d, f, h not aromatic
(b) b, e, f, g aromatic and a, c, d, h not aromatic
(c) a, b, c, d aromatic and e, f, g, h not aromatic
(d) a, c, d, e, h aromatic and b, f, g not aromatic

154. Isomeric hydrocarbons → negative Baeyer's test (Molecular formula C_9H_{12}).
The total number of isomers from above with four different non-aliphatic substitution sites is ______

[NCERT : P L-322 | NV, Jan. 29, 2025 (II)]

155. Given below are two statements :
Statement (I) : On nitration of *m*-xylene with HNO_3, H_2SO_4 followed by oxidation, 4-nitrobenzene-1, 3-dicarboxylic acid is obtained as the major product.
Statement (II) : CH_3 group is o/p-directing while-NO_2 group is m-directing group.
In the light of the above statements, choose the correct answer from the options given below :

[NCERT : P L-324 | Jan. 29, 2025 (II)]

(a) Both Statement I and Statement II are false
(b) Statement I is false but Statement II is true
(c) Both Statement I and Statement II are true
(d) Statement I is true but Statement II is false

156. Identify correct statement/s :

[NCERT : P L-324 | Jan. 24, 2025 (II)]

(A) —OCH_3 and —$NHCOCH_3$ are activating group
(B) —CN and —OH are meta directing group
(C) —CN and —SO_3H are meta directing group
(D) Activating groups act as ortho – and para directing groups
(E) Halides are activating groups

Choose the correct answer from the options given below :
(a) (A), (C) and (D) only (b) (A), (B) and (E) only
(c) (A) only (d) (A) and (C) only

157. The correct stability order of the following species/ molecules is: **[NCERT : P L-321 | Jan. 23, 2025 (I)]**

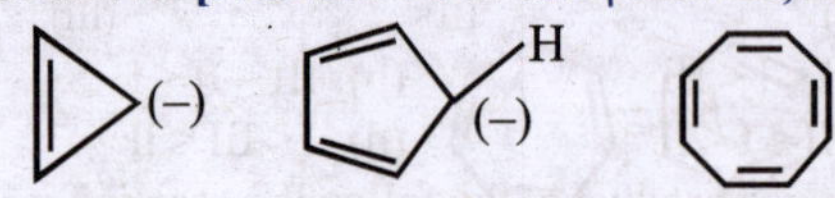

p q r

(a) q > r > p (b) r > q > p
(c) q > p > r (d) p > q > r

158. Benzene + 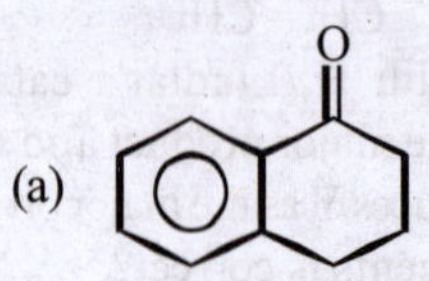 $\xrightarrow{AlCl_3}$ A $\xrightarrow[HCl]{Zn-Hg}$ B $\xrightarrow{Conc.H_2SO_4}$ C

What is the structure of C ?

[NCERT : P L-324 | April 9, 2024 (I)]

(a)

(b)

(c)

(d)

159. Which of the following are aromatic?

[NCERT : P L-321 | April 8, 2024 (I)]

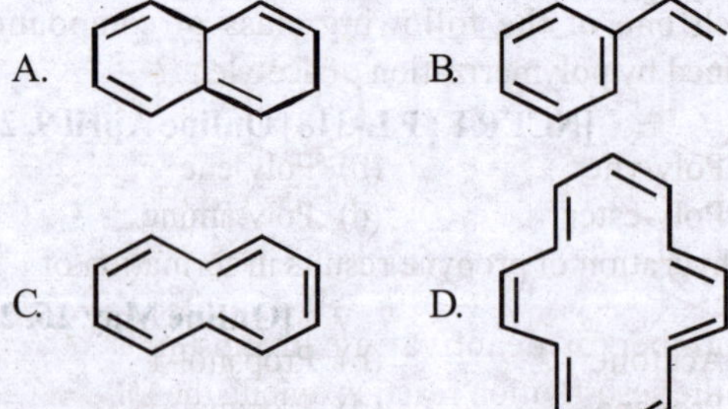

(a) B and D only (b) A and C only
(c) A and B only (d) C and D only

160. Total number of aromatic compounds among the following compounds is ______

[NCERT : P L-321 | NV, April 8, 2024 (II)]

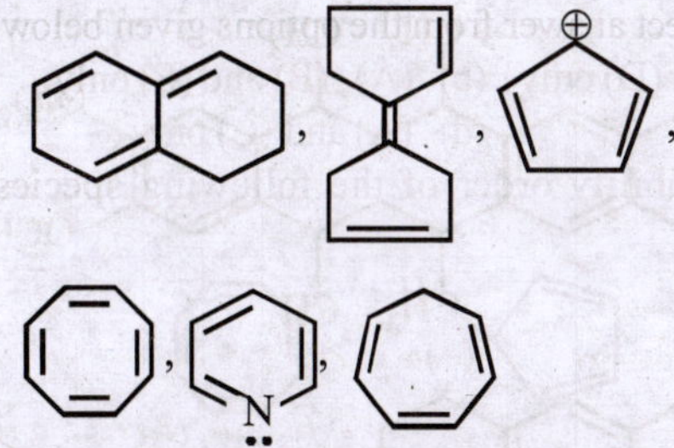

161. Given below are two statements :

[NCERT : P L-322 | April 5, 2024 (I)]

Statement I: Nitration of benzene involves the following step –

$H-\overset{H}{\overset{|}{\ddot{O}}}{}^{\ominus}-NO_2 \rightleftharpoons H_2O + \overset{\oplus}{N}O_2$

Statement II: Use of Lewis base promotes the electrophilic substitution of benzene.

In the light of the above statements, choose the **most appropriate** answer from the options given below :

(a) Both **Statement I** and **Statement II** are incorrect
(b) **Statement I** is correct but **Statement II** is incorrect
(c) Both **Statement I** and **Statement II** are correct
(d) **Statement I** is incorrect but **Statement II** is correct

162. Correct order of stability of carbanion is

[NCERT : P L-321 | April 4, 2024 (II)]

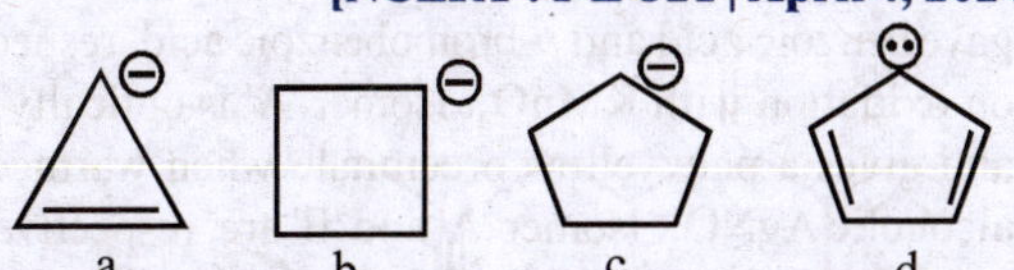

(a) C > B > D > A (b) A > B > C > D
(c) D > A > C > B (d) D > C > B > A

163. The set of meta directing functional groups from the following sets is:

[NCERT : P L-324 | Feb. 1, 2024 (II)]

(a) $-CN, -NH_2, -NHR, -OCH_3$
(b) $-CN, -CHO, -NHCOCH_3, -COOR$
(c) $-NO_2, -NH_2, -COOH, -COOR$
(d) $-NO_2, -CHO, -SO_3H, -COR$

164. Which of the following compound will most easily be attacked by an electrophile?

[NCERT : P L-325 | Feb. 1, 2024 (I)]

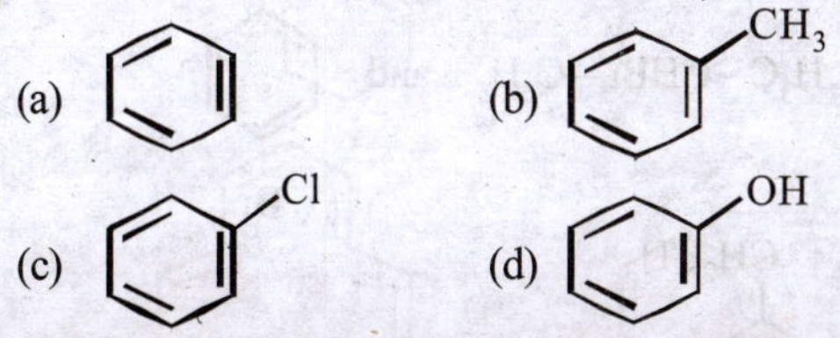

165. Total number of deactivating groups in aromatic eletrophilic susbstitution reaction among the following is _______. **[NCERT : P L-325 | NV, Feb. 1, 2024 (I)]**

$CH_3C(=O)OCH_3$, $-NH-C(=O)CH_3$, $-N(H)CH_3$, $-C\equiv N$, $-OCH_3$

166. Identify major product 'P' formed in the following reaction.

Benzene + C_6H_5COCl $\xrightarrow{\text{Anhydrous } AlCl_3}$ 'P' (Major Product)

[NCERT : P L-322 | Jan. 31, 2024 (II)]

(a) $C_6H_5-O-C_6H_4-COCl$
(b) $C_6H_5COCH_3$
(c) C_6H_5CHO
(d)

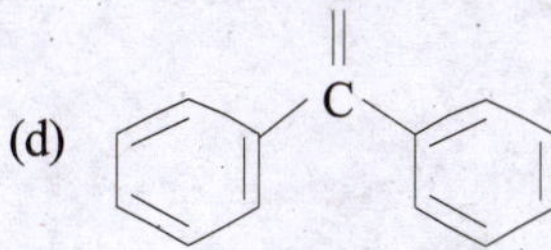

167. The correct order of reactivity in electrophilic substitution reaction of the following compounds is :

[NCERT : P L-324 | Jan. 31, 2024 (II)]

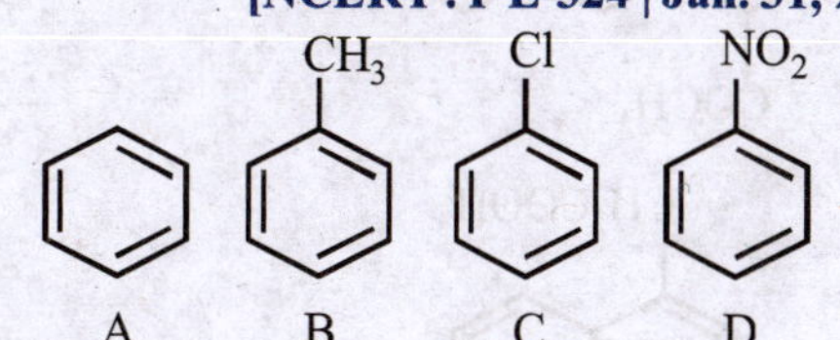

(a) B > C > A > D (b) D > C > B > A
(c) A > B > C > D (d) B > A > C > D

168. Among the given organic compounds, the total number of aromatic compounds is _____.

[NCERT : P L-321 | NV, Jan. 27, 2024 (I)]

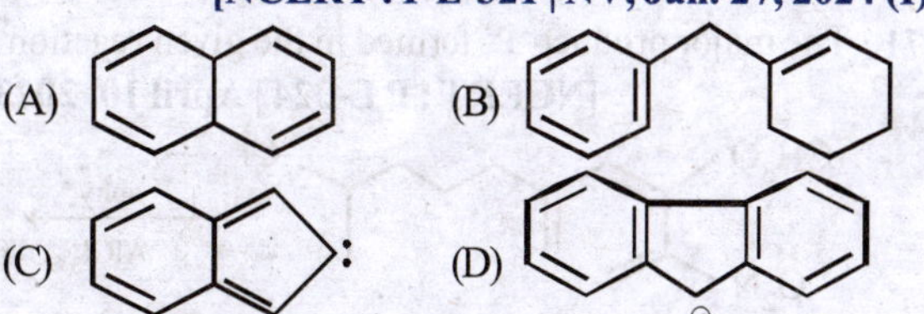

169. Given below are two statements:

[NCERT : P L-321 | April 13, 2023 (II)]

Statement-I : Tropolone is an aromatic compound and has 8π electrons.

Statement-II : π electrons of $> C = O$ group in tropolone is involved in aromaticity.

In the light of the above statements, choose the correct answer from the options given below:

(a) Both Statement I and Statement II are true
(b) Statement I is true but Statement II is false
(c) Statement I is false but Statement II is true
(d) Both Statement I and Statement II are false

170. The major product 'P' formed in the given reaction is

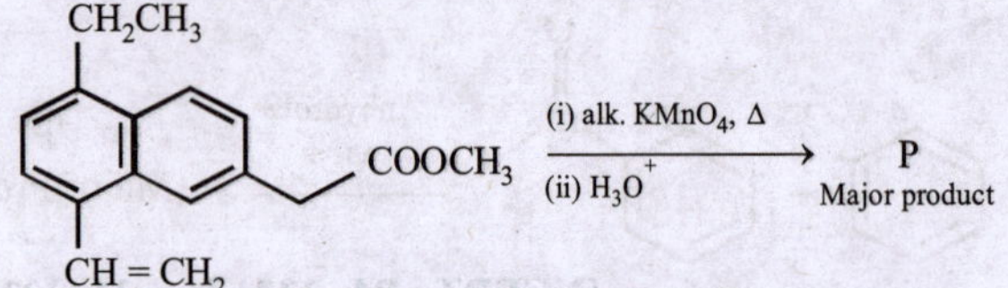

[NCERT : P L-324 | April 10, 2023 (I)]

(a) [structure]
(b) [structure]
(c) [structure]
(d) [structure]

171. The major product 'P' formed in the given reaction is:

[NCERT : P L-324 | April 10, 2023 (II)]

(a) [structure]
(b) [structure]
(c) [structure]
(d) [structure]

172. 'X' is :

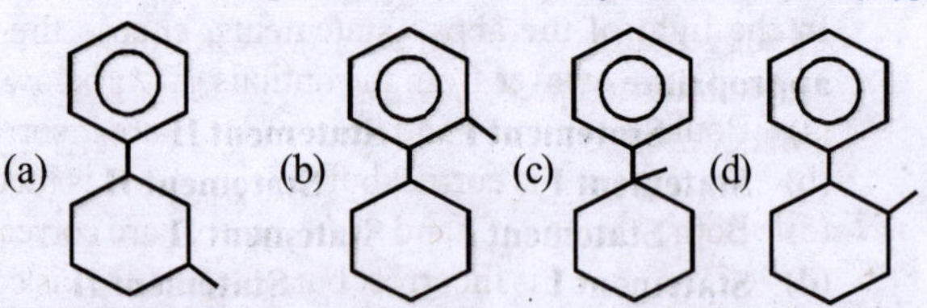

[NCERT : P L-324 | Feb. 01, 2023 (II)]

(a) (b) (c) (d)

173. Two isomers (A) and (B) with molar mass 184 g/mol and elemental composition C, 52.2%; H, 4.9% and Br 42.9% gave benzoic acid and *p*-bromobenzoic acid, respectively on oxidation with $KMnO_4$. Isomer 'A' is optically active and gives a pale yellow precipitate when warmed with alcoholic $AgNO_3$. Isomer 'A' and 'B' are, respectively :

[NCERT : P L-324 | June 29, 2022 (II)]

(a) $H_3C-CHBr-C_6H_5$ and [2-methylbenzyl bromide structure]

(b) [2-methylbenzyl bromide structure] and [4-bromoethylbenzene structure]

(c) $H_3C-CHBr-C_6H_5$ and [4-bromoethylbenzene structure]

(d) [4-bromoethylbenzene structure] and $H_3C-CHBr-C_6H_5$

174. Product 'A' of following sequence of reactions is

Ethylbenzene $\xrightarrow[\text{(b)}\,Cl_2,\,\Delta \quad \text{(c) alc. KOH}]{\text{(a)}\,Br_2.Fe}$ 'A' (Major product)

[NCERT : P L-324 | June 27, 2022 (II)]

(a) Cl, C=CH$_2$ (b) Br, C=CH$_2$

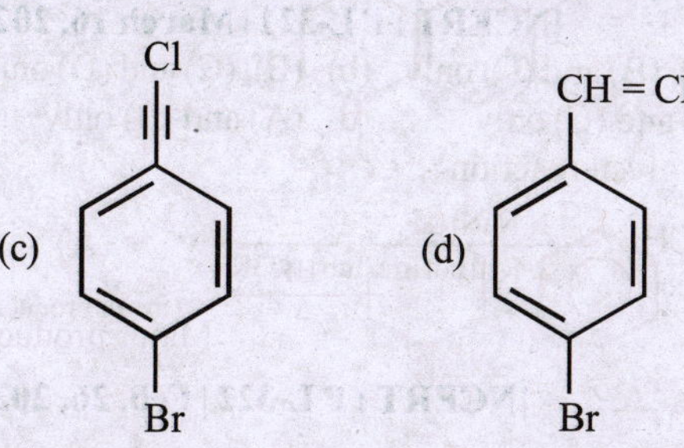

(c) Cl, Br (d) CH = CH$_2$, Br

175. Halogenation of which one of the following will yield m-substituted product with respect to methyl group as a major product? [NCERT : P L-324 | June 26, 2022 (II)]

(a) CH_3, CHO (b) CH_3, NO_2

(c) CH_3, OH (d) CH_3, NO_2

176. Given below are two statements. One is labelled as **Assertion A** and the other is labelled as **Reason R** [NCERT : P L-321 | July 27, 2022 (I)]

Assertion A : [6] Annulene. [8] Annulene and cis –[10] Annulene, are respectively aromatic, non-aromatic and aromatic.

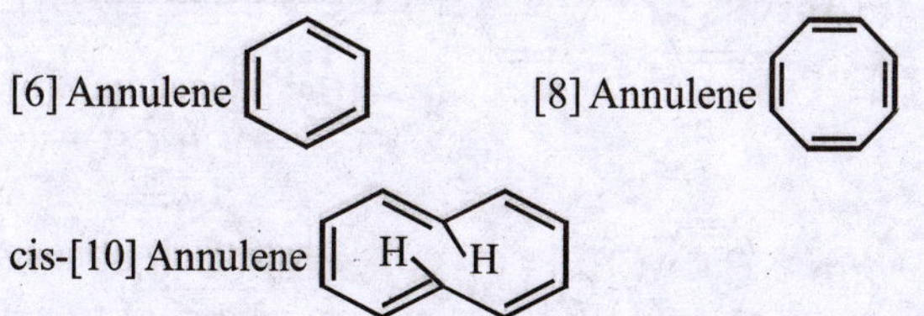

[6] Annulene [8] Annulene

cis-[10] Annulene

Reason R : Planarity is one of the requirements of aromatic systems.

In the light of the above statements, choose the **most appropriate** answer from the options given below.

(a) Both A and R are correct and R is the correct explanation of A.
(b) Both A and R are correct but R is NOT the correct explanation of A.
(c) A is correct but R is not correct.
(d) A is not correct but R is correct.

177. Which of the following compounds is **not** aromatic? [NCERT : P L-321 | July 26, 2022 (I)]

(a) (–) (b) ⊕

(c) H H (d)

178. In the presence of sunlight, benzene reacts with Cl_2 to give product, X. The number of hydrogens in X is ________. [NCERT : P L-323 | NV, July 26, 2022 (I)]

179. Which of the following is not an example of benzenoid compound ? [NCERT : P L-318 | July 26, 2022 (II)]

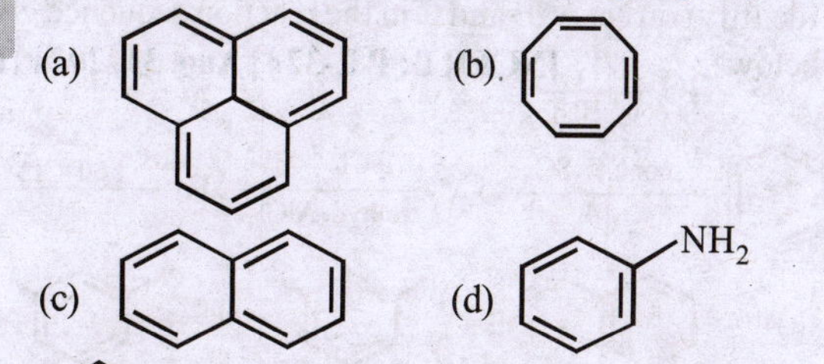

(a) (b)

(c) (d) NH_2

180. Benzene + $CH_3CH_2CH_2Cl \xrightarrow{\text{Anhydrous, } AlCl_3}$ 'A' (Major Product)

The stable carbocation formed in the above reaction is : [NCERT : P L-323 | June 29, 2022 (II)]

(a) $CH_3CH_2\overset{\oplus}{C}H_2$ (b) $CH_3\overset{\oplus}{C}H_2$

(c) $CH_3-\overset{\oplus}{C}H-CH_3$ (d) $\overset{\oplus}{C}HCH_2CH_3$ (on benzene ring)

181. Which of the following structures are aromatic in nature? [NCERT : P L-321 | June 28, 2022 (I)]

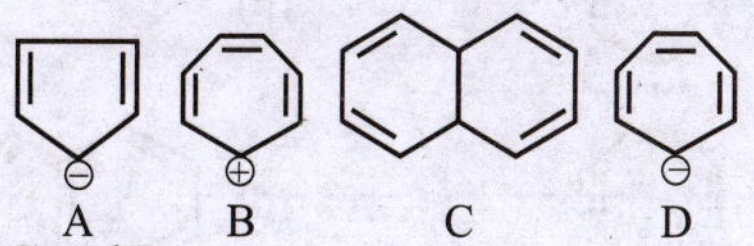

(a) A,B,C and D
(b) Only A and B
(c) Only A and C
(d) Only B, C and D

182. Given below are two statements: one is labelled as **Assertion A** and the other is labelled as **Reason R.**

Assertion A: A mixture contains benzoic acid and naphthalene. The pure benzoic acid can be separated out by the use of benzene.

Reason R: Benzoic acid is soluble in hot water.

In the light of the above statements, choose the **most appropriate** answer from the options given below. [June 25, 2022 (II)]

(a) Both A and R are true and R is the correct explanation of A.
(b) Both A and R are true but R is NOT the correct explanation of A.
(c) A is true but R is false.
(d) A is false but R is true.

183. Which one of the following compounds is aromatic in nature? **[NCERT : P L-321 | Sep. 1, 2021 (II)]**

(a) 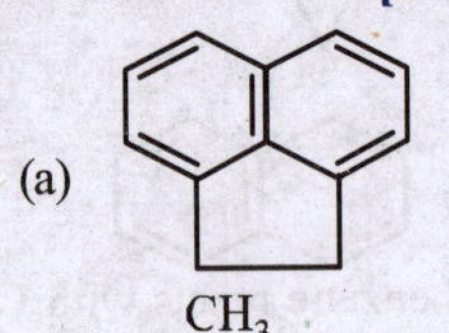 (b) 

(c) 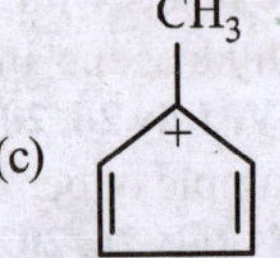 (d)

184. Identify correct A, B and C in the reaction sequence given below: **[NCERT : P L-324 | Aug. 31, 2021 (II)]**

Benzene $\xrightarrow[\Delta]{\text{conc.}HNO_3 + \text{conc.}H_2SO_4}$ A $\xrightarrow[\text{Anhyd. }AlCl_3]{Cl_2}$ B $\xrightarrow{Fe/HCl}$ C

(a) 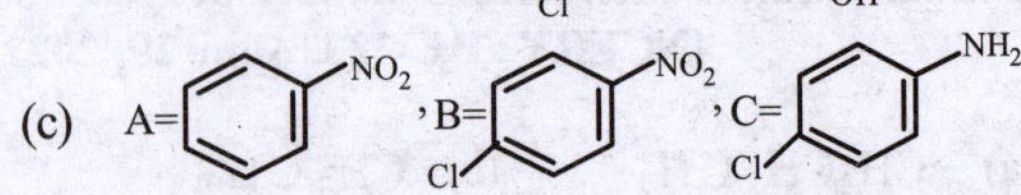

(b) 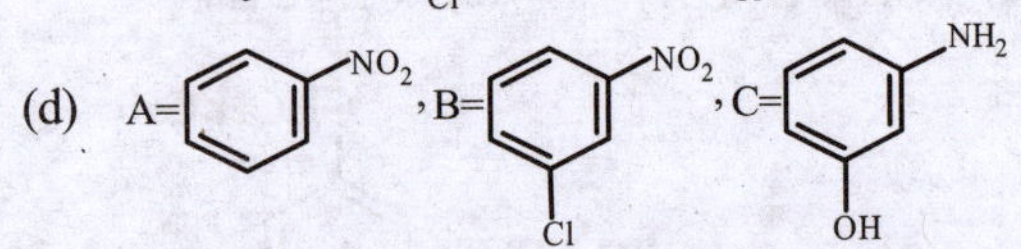

(c) (d) 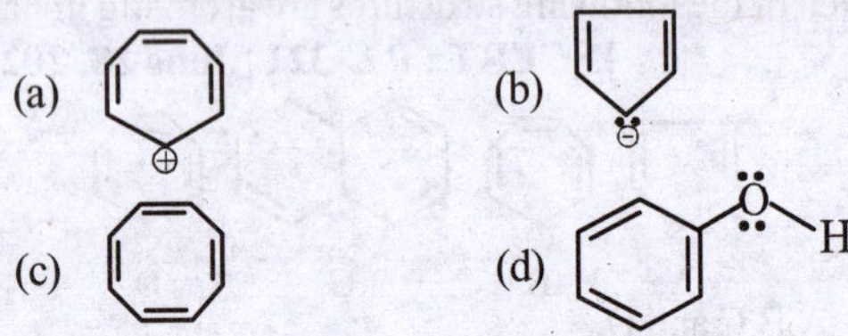

185. Which one of the following compounds is not aromatic? **[Aug. 26, 2021 (II); Same March 16, 2021 (I)]**

(a) (b)

(c) (d)

186. Benzene on nitration gives nitrobenzene in presence of HNO_3 and H_2SO_4 mixture, where: **[NCERT : P L-322 | July 20, 2021 (II)]**

(a) both H_2SO_4 and HNO_3 act as a bases
(b) HNO_3 acts as an acid and H_2SO_4 acts as a base
(c) both H_2SO_4 and HNO_3 act as an acids
(d) HNO_3 acts as a base and H_2SO_4 acts as an acid

187. Benzene $\xrightarrow[H_2SO_4]{HNO_3}$ Nitrobenzene (NO_2)

In the above reaction, 3.9 g of benzene on nitration gives 4.92 g of nitrobenzene. The percentage yield of nitrobenzene in the above reaction is ______ %. (Round off to the nearest integer)
(Given atomic mass : C : 12.0 u, H : 1.0 u, O : 16.0 u, N : 14.0 u)
[NV, March 17, 2021 (I)]

188. Among the following, the aromatic compounds are:

(A) 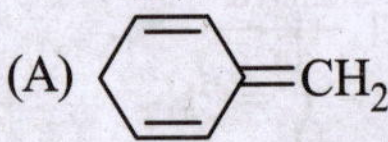 (B)

(C) (D)

Choose the correct answer from the following options: **[NCERT : P L-321 | March 16, 2021 (I)]**

(a) (A), (B) and (C) only (b) (B), (C) and (D) only
(c) (B) and (C) only (d) (A) and (B) only

189. For the given reaction:

$$\underset{CH_3}{\underset{|}{CH}}=CHBr \xrightarrow[\text{2. Red hot iron tube, 873K}]{\text{1. }NaNH_2} \underset{\text{(major product)}}{(A)}$$

What is 'A'? **[NCERT : P L-322 | Feb. 26, 2021 (I)]**

(a) (b) 1,3,5-trimethylbenzene (H_3C, CH_3, CH_3)

(c) $CH_3CH_2CH_2NH_2$ (d) $\underset{CH_3}{\underset{|}{CH}}=CH-NH_2$

190. The correct sequence of reagents used in the preparation of 4-bromo-2-nitroethyl benzene from benzene is: **[NCERT : P L-322 | Feb. 25, 2021 (II)]**

(a) $CH_3COCl/AlCl_3$, Zn-Hg/HCl, $Br_2/AlBr_3$, HNO_3/H_2SO_4
(b) HNO_3/H_2SO_4, $Br_2/AlCl_3$, $CH_3COCl/AlCl_3$, Zn-Hg/HCl
(c) $CH_3COCl/AlCl_3$, $Br_2/AlBr_3$, HNO_3/H_2SO_4, Zn/HCl
(d) $Br_2/AlBr_3$, $CH_3COCl/AlCl_3$, HNO_3/H_2SO_4, Zn/HCl

191. The number of sp^2 hybrid orbitals in a molecule of benzene is: **[NCERT : P L-320 | Jan. 09, 2020 (II)]**

(a) 24 (b) 6 (c) 18 (d) 12

192. Consider the following reactions:

(A) Benzene + Chlorobenzene $\xrightarrow{\text{anhyd. }AlCl_3}$ Biphenyl

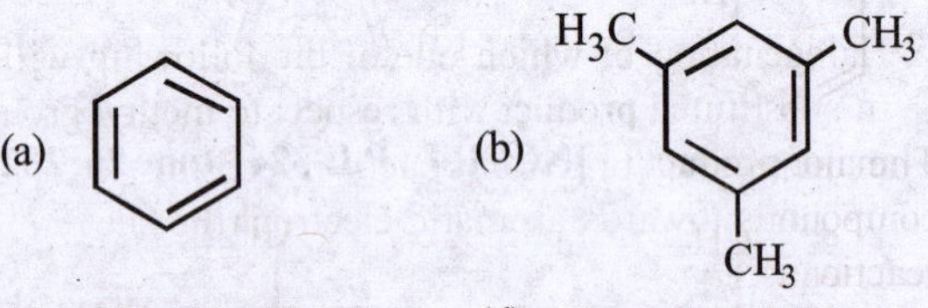

(B) Benzene + Cl_2 (excess) $\xrightarrow[\text{dark}]{\text{anhyd. }AlCl_3}$ Hexachlorobenzene (Cl, Cl, Cl, Cl, Cl, Cl)

(C) Benzene + $CH_2=CH-Cl$ $\xrightarrow[AlCl_3]{\text{anhyd.}}$ Ph$-CH=CH_2$

(D) Benzene + $CH_2 = CH - CH_2Cl \xrightarrow[AlCl_3]{anhyd.}$ $C_6H_5-CH_2-CH=CH_2$

Which of these reactions are possible ?

[NCERT : P L-322 | Jan. 07, 2020 (II)]

(a) (A) and (B) (b) (A) and (D)
(c) (B), (C) and (D) (d) (B) and (D)

193. Which of the following compounds is not aromatic?

[NCERT : P L-321 | Jan. 9, 2019 (II)]

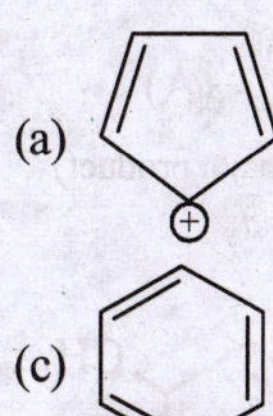

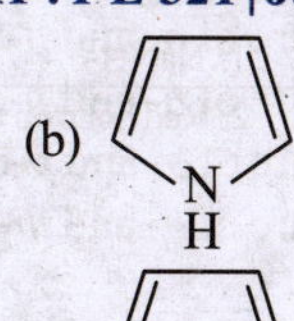

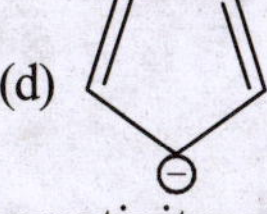

194. The increasing order of reactivity of the following compounds towards aromatic electrophilic substitution reaction is: **[NCERT : P L-324 | April 9, 2019 (I)]**

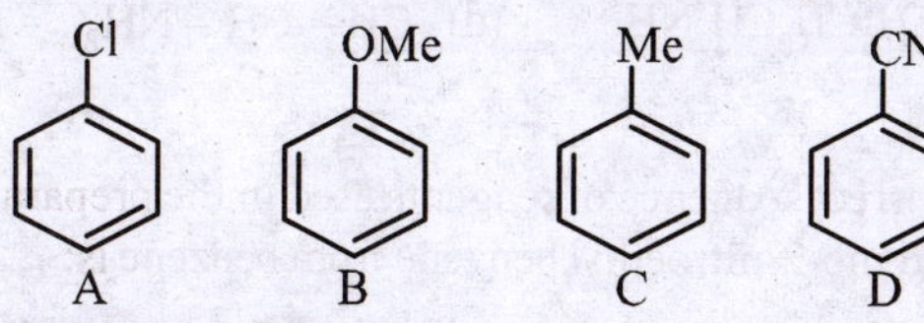

(a) D < A < C < B
(b) B < C < A < D
(c) A < B < C < D
(d) D < B < A < C

195. Which compound (s) out of the following is/are not aromatic?

[NCERT : P L-321 | Jan. 11, 2019 (I), Jan. 9, 2019 (II)]

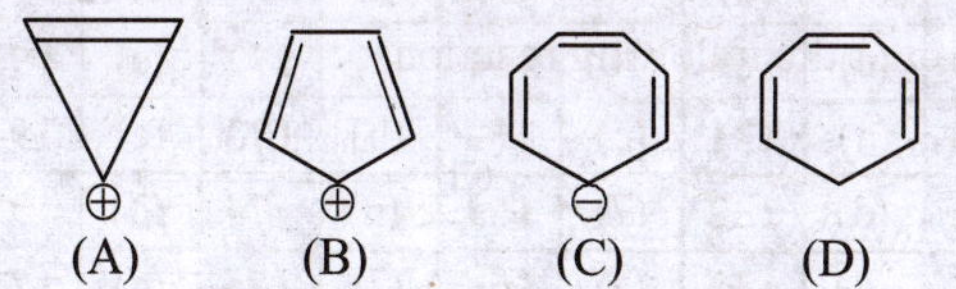

(a) (B), (C) and (D) (b) (C) and (D)
(c) (B) (d) (A) and (C)

196. $C_6H_5-CH_2-CH=CH_2$ + HCl ⟶ X,

X is : **[NCERT : P L-322 | Online April 16, 2018 Same Online April 9, 2013]**

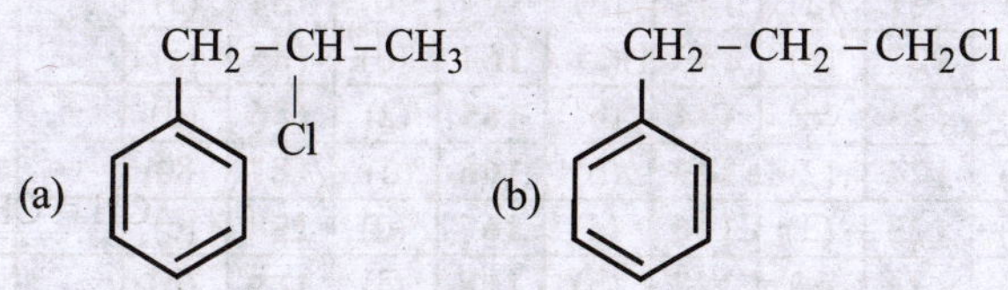

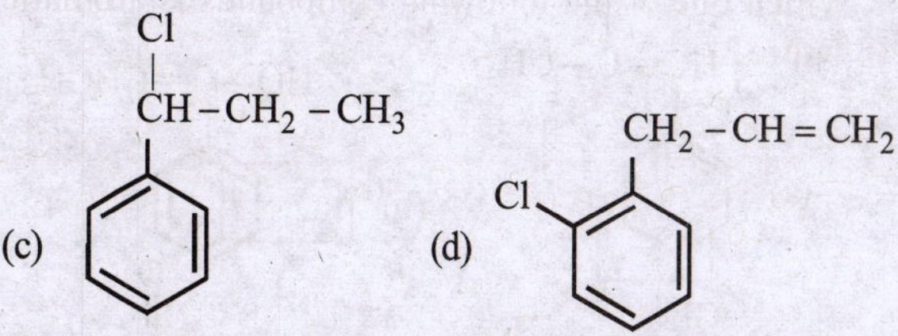

197. Which of the following compounds will show highest dipole moment ?

[NCERT : P L-321 | Online April 9, 2017]

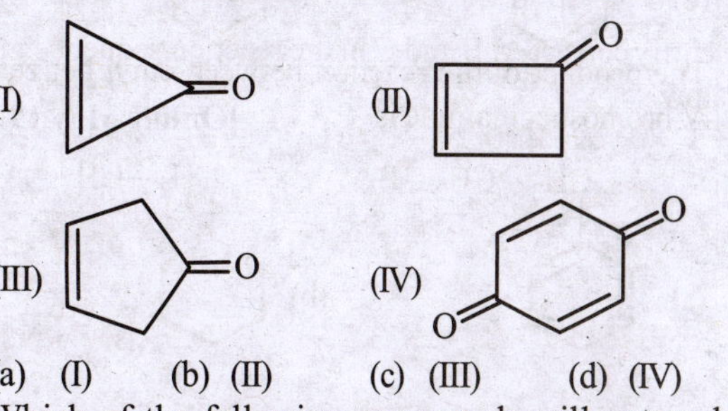

(a) (I) (b) (II) (c) (III) (d) (IV)

198. Which of the following compounds will not undergo Friedel Craft's reaction with benzene ?

[NCERT : P L-324 | Online April 8, 2017]

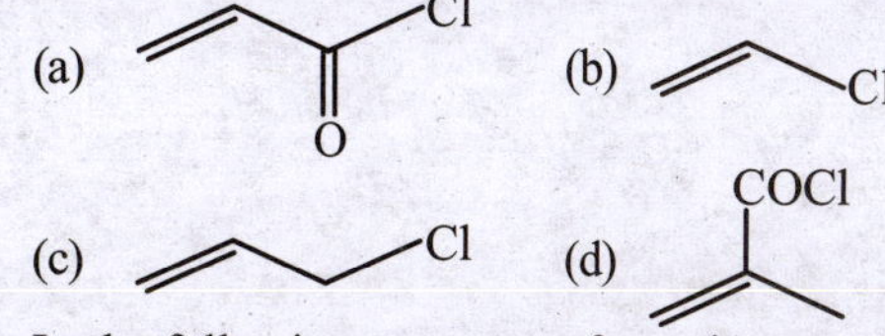

199. In the following sequence of reactions :

Toluene $\xrightarrow{KMnO_4}$ A $\xrightarrow{SOCl_2}$ B $\xrightarrow[BaSO_4]{H_2/Pd}$ C

the product C is : **[2015]**

(a) $C_6H_5CH_2OH$ (b) C_6H_5CHO
(c) C_6H_5COOH (d) $C_6H_5CH_3$

200. Which of the following would not give 2-phenylbutane as the major product in a Friedel-Crafts alkylation reaction ?

(a) 1-Butene + HF **[Online April 22, 2013]**
(b) 2-Butanol + H_2SO_4
(c) Butanoyl chloride + $AlCl_3$, then Zn, HCl
(d) Butyl chloride + $AlCl_3$

201. Which of the following compounds are antiaromatic

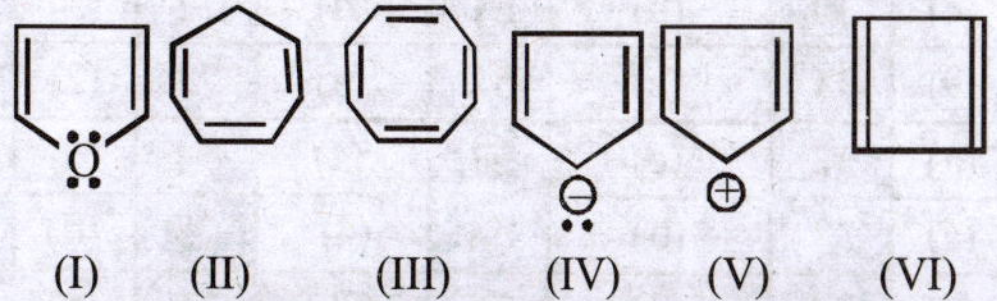

[NCERT : P L-321 | Online May 26, 2012]

(a) (I) and (V) (b) (II) and (V)
(c) (I) and (IV) (d) (III) and (VI)

202. In the given reaction, **[Online May 12, 2012]**

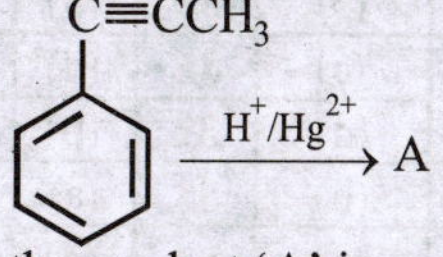

the product 'A' is

(a) $HC{=}C(OH){-}CH_3$ with C_6H_5 attached to HC

(b) $HO{-}C(C_6H_5){=}CHCH_3$

(c) $O{=}C(C_6H_5){-}CH_2CH_3$

(d) $C_6H_5CH_2{-}C({=}O){-}CH_3$

203. The product of the reaction between ethyl benzene and *N*-bromosuccinamide is **[Online May 19, 2012]**

(a) $C_6H_5{-}CH_2-CH_2-Br$

(b) CH_2-CH_3 on benzene with Br at *meta* position

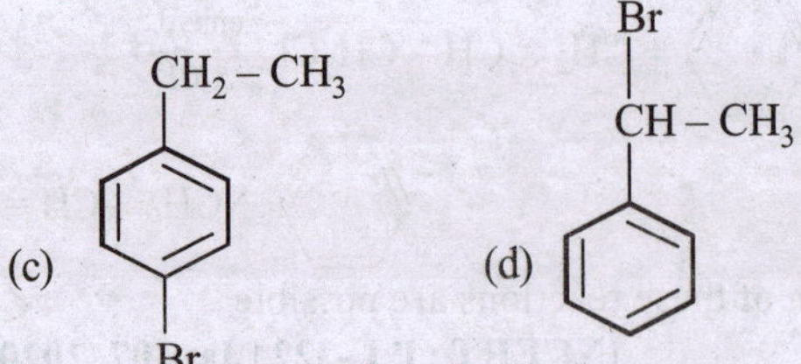

204. Toluene is nitrated and the resulting product is reduced with tin and hydrochloric acid. The product so obtained is diazotised and then heated wth cuprous bromide. The reaction mixture so formed contains **[2008]**

(a) mixture of *o*- and *p*-bromotoluenes

(b) mixture of *o*- and *p*-dibromobenzenes

(c) mixture of *o*- and *p*-bromoanilines

(d) mixture of *o*- and *m*-bromotoluenes

ANSWER KEYS

1	(b)	22	(b)	43	(d)	64	(d)	85	(b, c)	106	(b, c)	127	(b)	148	(d)	169	(b)	190	(a)
2	(d)	23	(c)	44	(c)	65	(4)	86	(d)	107	(c)	128	(3)	149	(c)	170	(b)	191	(c)
3	(d)	24	(b)	45	(a)	66	(d)	87	(b)	108	(c)	129	(c)	150	(a)	171	(d)	192	(d)
4	(b)	25	(d)	46	(c or d)	67	(d)	88	(a)	109	(c)	130	(10)	151	(a)	172	(c)	193	(a)
5	(9)	26	(d)	47	(b)	68	(b)	89	(b)	110	(a)	131	(c)	152	(b)	173	(c)	194	(a)
6	(3)	27	(d)	48	(4)	69	(2)	90	(d)	111	(d)	132	(7)	153	(d)	174	(d)	195	(a)
7	(6)	28	(b)	49	(b)	70	(a)	91	(b)	112	(d)	133	(a)	154	(2)	175	(c)	196	(c)
8	(c)	29	(b)	50	(1)	71	(a)	92	(d)	113	(a)	134	(d)	155	(c)	176	(d)	197	(a)
9	(c)	30	(a)	51	(d)	72	(a)	93	(b)	114	(c)	135	(13)	156	(a)	177	(c)	198	(b)
10	(9)	31	(c)	52	(a)	73	(2)	94	(c)	115	(c)	136	(c)	157	(a)	178	(6)	199	(b)
11	(a)	32	(c)	53	(a)	74	(c)	95	(d)	116	(d)	137	(c)	158	(a)	179	(a, b)	200	(c)
12	(a)	33	(b)	54	(4)	75	(d)	96	(d)	117	(c)	138	(d)	159	(a)	180	(c)	201	(d)
13	(a)	34	(d)	55	(70)	76	(c)	97	(b)	118	(c)	139	(d)	160	(1)	181	(b)	202	(c)
14	(3)	35	(c)	56	(b)	77	(d)	98	(c)	119	(a)	140	(d)	161	(b)	182	(d)	203	(d)
15	(d)	36	(b)	57	(d)	78	(c)	99	(d)	120	(15)	141	(c)	162	(d)	183	(d)	204	(a)
16	(c)	37	(b or d)	58	(a)	79	(b)	100	(c)	121	(a)	142	(b)	163	(d)	184	(a)		
17	(12)	38	(b)	59	(3)	80	(b)	101	(b)	122	(8)	143	(a)	164	(d)	185	(c)		
18	(8)	39	(27)	60	(c)	81	(d)	102	(b)	123	(d)	144	(b)	165	(2)	186	(d)		
19	(c)	40	(b)	61	(d)	82	(b)	103	(d)	124	(2)	145	(b)	166	(d)	187	(80)		
20	(d)	41	(a)	62	(b)	83	(a)	104	(d)	125	(d)	146	(d)	167	(d)	188	(c)		
21	(60)	42	(8)	63	(b)	84	(d)	105	(d)	126	(a)	147	(b)	168	(3)	189	(b)		

HINTS & SOLUTIONS

Some Basic Concepts of Chemistry

1. **(b)** Number of atoms $= \dfrac{\text{Mass in grams}}{\text{Molar Mas(g / mol)}} \times N_A$

Therefore for the given mass, element having the least molar mass will have the highest number of atoms.

$M_{Po} = 209, M_{Pr} = 141, M_{Pb} = 207, M_{Pt} = 195$

$\Rightarrow$ Pr has the highest number of atoms.

2. **(372)**

Molar mass = 372 g

$\therefore$ 0.1 mole $= 372 \times 10^{-1}$ g

3. **(d)**

(A) Weight is the force exerted by gravity on an object.

(B) Mass of a substance is the amount of matter present in it.

4. **(a)** $(\text{moles})_{\text{initial}} = \dfrac{x \times 10^{-3}}{44}$

$(\text{moles})_{\text{removal}} = \dfrac{10^{21}}{6.02 \times 10^{23}}$

$(\text{moles})_{\text{left}} = (\text{moles})_{\text{initial}} - (\text{moles})_{\text{removal}}$

$2.8 \times 10^{-3} = \dfrac{x \times 10^{-3}}{44} - \dfrac{10^{21}}{6.02 \times 10^{23}} \Rightarrow x = 196.2$ mg

5. **(b)** A candela is the luminous intensity of a source that emits monochromatic radiation of frequency radiation of frequency 540×10^{12} Hz and has a radiant intensity in that direction of $\dfrac{1}{683}$ w / sr. It is unit of Candela.

6. **(d)** According to dalton's atomic theory matter is made of indivisible atoms. All the atoms of a given element are same and have same mass.

7. **(a)** In Compound atoms of different elements combine in fixed ratio by mass.

8. **(11.00)** Thickness of coating = 0.05 cm

Area of coating $= 0.05\ m^2 = 0.05 \times 10^4\ cm^2$

Volume $= 0.05 \times 0.05 \times 10^4\ cm^3 = 25\ cm^3$

Mass of Ag $= 25\ cm^3 \times 7.9\ g\ cm^{-3} = 197.5$ g

Mol of Ag $= \dfrac{197.5\ g}{108\ g\ mol^{-1}} = 1.828$ mol

Atoms of Ag $= 1.828 \times 6.022 \times 10^{23} = 11.008 \times 10^{23}$

9. **(d)** Number of moles of $O_2 = \dfrac{2.8375}{22.7} = 0.125$

$\Rightarrow$ Number of molecules $= 0.125\ N_A = 7.525 \times 10^{22}$

10. **(d)** 16g CH_4 = 1 mole CH_4 contains $10 \times 6.02 \times 10^{23}$ electrons $= 60.2 \times 10^{23}$

1g H_2 = 0.5 mole H_2 gas occupy 11.35 litre volume at STP

1 mole of N_2 = 28g

0.5 mole of SO_2 = 32g

11. **(2)** $\underset{1\ mol}{X} + \underset{1\ mol}{Y} + \underset{0.05\ mol}{3Z} \rightleftharpoons XYZ_3$

Z is act as a limiting reagent

1 mol of $XYZ_3 = \dfrac{0.05}{3}$ mol of Z.

Mass of $XYZ_3 = \dfrac{0.05}{3} \times (10 + 20 + 30 \times 3) = 2$g

12. **(24)** Let the weight of Mg in the extract = w g

$\dfrac{w}{2000} \times 10^6 = 48$ [Assuming 1000 mL ≈ 1000 kg for water]

$x = 96 \times 10^{-3}$

$n_{Mg} = \dfrac{\text{weight}}{\text{molar mass}} = \dfrac{96 \times 10^{-3}}{24} = 0.004$

Number of Mg atom $= 0.004 \times 6.02 \times 10^{23}$ $= 24.08 \times 10^{20} \approx 24 \times 10^{20}$

13. **(c)** No. of Fe atoms in 3.3g of hemoglobin

$= \dfrac{0.34}{100} \times \dfrac{3.3}{56} \times 6.022 \times 10^{23} = 1.206 \times 10^{20}$

14. **(46)**

	N_2 +	$3H_2$	$\longrightarrow$ $2NH_3$
initial	56L		–
	–10L		+20L
final	46L		20L

From the stoichiometry for 2L production of ammonia 1L of N_2 gas is used.

Final volume of $N_2 = 56 - 10 = 46$L

15. **(2)** Moles of A_2B = Moles of AB_3 = 0.15

$\dfrac{w}{2a+b} = \dfrac{w}{a+3b} \Rightarrow 2a + b = a + 3b \Rightarrow a = 2b.$

16. **(c)** Let total volume = 1000 mL = 1L

total mass of solution = 1460 g

mass of HCl $= \dfrac{35}{100} \times 1460$; moles of HCl

Molarity $= \dfrac{\text{no. of moles}}{\text{Volume (1L)}} = \dfrac{35 \times 1460}{100 \times 36.5} = 14$

So molarity $= \dfrac{14\ \text{moles}}{1L} = 14$ M

17. (5418)
M.M. of $C_7H_5N_3O_6 = 84+5+42+96 = 227$

Moles of $C_7H_5N_3O_6 = \frac{681}{227} = 3$
Moles of N atoms $= 3 \times 3 = 9$
Number of N atoms $= 9 \times 6.02 \times 10^{23} = 5418 \times 10^{21}$.

18. (2) No. of atoms $= \frac{8}{23} \times 6.02 \times 10^{23} = 2.09 \times 10^{23}$
$\approx 2 \times 10^{23}$

19. (c)

	1st component	2nd component
mole	n_1	n_2
M.W.	M_1	M_2
mass	n_1M_1	n_2M_2

Mass of solution $= n_1M_1 + n_2M_2$
Mole fraction of the 2nd component

$$(x_2) = \frac{n_2}{n_1+n_2} \Rightarrow n_1 = \frac{n_2(1-x_2)}{x_2}$$

Mass of solution $= n_1M_1 + n_2M_2$

$$= \frac{n_2M_1(1-x_2)}{x_2} + n_2M_2 = \frac{n_2}{x_2}[M_2x_2 - x_2M_1 + M_1]$$

Volume of solution $= \frac{n_2[M_2x_2 - x_2M_1 + M_1]}{1000dx_2}$

$$C_2 = \frac{1000n_2dx_2}{n_2[M_2x_2 - x_2M_1 + M_1]} \Rightarrow C_2 = \frac{1000dx_2}{M_1 + x_2(M_2 - M_1)}$$

20. (14.00)
Mass percent of $HNO_3 = 63\%$
Thus, 100 g of nitric acid solution contains 63 g of nitric acid by mass.

No. of moles $= \frac{63\,g}{63\,g\,mol^{-1}} = 1$

Volume of 100 g of nitric acid solution

$$= \frac{\text{Mass}}{\text{Density}} = \frac{100\text{ g}}{1.4\text{ g/mL}} = 71.4\text{ mL}$$

$$\text{Molarity} = \frac{\text{No. of moles}}{\text{volume (mL)}} \times 1000 = \frac{1}{71.4} \times 1000 = 14\text{ M}$$

21. (c) Except (c) all postulates were given by Dalton.

22. (d) 5 mol AB_2 weighs 125 g
$\therefore$ $AB_2 = 25$ g/mol $\Rightarrow M_A + 2M_B = 25$
10 mol A_2B_2 weighs 300 g
$\therefore$ $A_2B_2 = 30$ g/mol $\Rightarrow 2M_A + 2M_B = 30$
$\therefore$ Molar mass of A$(M_A) = 5$ g or 5×10^{-3} kg
Molar mass of B $(M_B) = 10$ g or 10×10^{-3} kg

23. (d) CH_4 has one atom of carbon among 5 atoms (1C + 4H).

$\therefore$ Mole % of C $= \frac{1}{5} \times 100 = 20\%$

24. (c) No. of moles of H_2O $(n_1) = \frac{18}{18} = 1$

No. of moles of NaOH $(n_2) = \frac{8}{40} = \frac{1}{5}$

$$\text{Mole fraction of NaOH} = \frac{n_2}{n_2+n_1} = \frac{\frac{1}{5}}{\frac{1}{5}+1} = 0.167$$

$$\text{Molality} = \frac{\text{No. of moles of solute}}{\text{Mass of solvent (kg)}} = \frac{1}{5} \times \frac{1000}{18} = 11.11\text{ m}$$

25. (c) As we know,

$$\text{Molarity} = \frac{\text{No. of moles of sugar}}{\text{Volume of solution (in L)}}$$

$$\Rightarrow 0.1 = \frac{\text{No. of moles of sugar}}{2\text{L}}$$

So, no. of moles of sugar = 0.2 mole
$\therefore$ Mass of sugar = No. of moles of sugar × Molar mass of sugar
$= 0.2 \times 342 = 68.4$ g

26. (d) Given percentage of chlorine in hydrocarbon = 3.55%
i.e., 100 g of chlorohydrocarbon has 3.55 g of chlorine.
1 g of chlorohydrocarbon will have $\frac{3.55}{100} = 0.0355$ g of chlorine.
Atomic wt. of Cl = 35.5 g/mol

Number of moles of Cl $= \frac{0.0355\text{ g}}{35.5\text{ g/mol}} = 0.001$ mol

Number of atoms of Cl $= 0.001$ mol $\times 6.023 \times 10^{23}$ mol^{-1}
$= 6.023 \times 10^{20}$

27. (a) $FeCl_3$ (aq) + 3 NaOH (aq) $\longrightarrow$ $Fe(OH)_3$ (s) + 3NaCl (aq)
($FeCl_3$: limiting reagent; NaOH: Excess amount, Not behave as limiting reagent)

$$\text{Moles of } Fe(OH)_3 = \frac{\text{weight in g}}{\text{Mol. mass of } Fe(OH)_3}$$

$$= \frac{2.14\text{ g}}{107\text{ g/mol}} = 0.02\text{ mol.}$$

1.0 mole of $Fe(OH)_3$ is obtained from = 1.0 mole of $FeCl_3$
0.02 mole of $Fe(OH)_3$ will be obtained from = 0.02 mole $FeCl_3$

$$\text{Molarity} = \frac{\text{No. of moles}}{\text{Volume in L}} = \frac{0.02\text{ mole}}{0.1\text{ L}} = 0.2\text{ M}$$

28. (d) The compound undergoing combustion is an alkane. Hence, the combustion reaction can be written as

$$\underset{5L}{C_nH_{2n+2}} + \underset{25L}{\left(\frac{3n+1}{2}\right)O_2} \longrightarrow n\,CO_2 + (n+1)\,H_2O$$

Volumes are measured at constant T and P, hence according to Avogadro's law:
Volume $\propto$ mole

1 L alkane requires $= \frac{3n+1}{2}$ L of O_2

5 L alkane requires $= \left(\frac{3n+1}{2}\right)5$ L of O_2

$\Rightarrow \left(\frac{3n+1}{2}\right)5 = 25$ L (given) $\quad \therefore \; n = 3$

Hence, alkane is propane (C_3H_8).

29. (b) 2 moles of water softner require 1 mole of Ca^{2+} ion

So, 1 mole of water softner require $\frac{1}{2}$ mole of Ca^{2+} ion

Thus, $\frac{1}{2 \times 206} = \frac{1}{412}$ mol/g will be maximum uptake.

30. **(b)** From the given data,
Mass of solution = 1000 + 120 = 1120 g

$\because \quad d = \frac{\text{mass}}{V}$ or $V = \frac{\text{mass}}{d}$

Volume of solution $V = \frac{1120}{1.12} = 1000$ mL or = 1 litre

Now molarity $(M) = \frac{\text{Mass}}{\text{Mol. mass} \times V(\text{L})} = \frac{120}{60 \times 1} = 2$ M

31. **(b)** 1 mole of water contains = 16 g of oxygen
$\therefore$ 3.6 mole of water of oxygen contains
= 16 × 3.6 = 57.6 g of oxygen

32. **(c)** The relation between molarity (M) and molality (m) is

$d = M\left(\frac{1}{m} + \frac{M_2}{1000}\right)$, M_2 = Mol. mass of solute

On putting value $1.252 = 3\left(\frac{1}{m} + \frac{58.5}{1000}\right)$

On solving, $m = 2.79$

33. **(a)** 4 g of hydrogen (H_2) = 4 mole of hydrogen atoms
= 4 × 6.023 × 10^{23} atoms

71.0 g of chlorine (Cl_2) = $\frac{71.0}{71.0}$ = 1 moles of chlorine
= 2 moles of Cl atom
= 2 × 6.023 × 10^{23} atoms

127 g of iodine (I_2) = $\frac{127}{254}$ mol = $6.023 \times 10^{23} \times \frac{1}{2} \times 2$ atoms
= 6.023 × 10^{23}

48.0 g of magnesium (Mg) = $\frac{48.0}{24.0}$ mol
= 2 × 6.023 × 10^{23} atoms

$\therefore$ 4.0 g H_2 has largest number of atoms.

34. **(a)** From molarity equation
$M_1V_1 + M_2V_2 = MV$

$M = \frac{M_1V_1 + M_2V_2}{V} = \frac{2 \times 10 + 0.5 \times 200}{210}$

$M = \frac{120}{210} = 0.57$ M

35. **(a)** 95% H_2SO_4 by weight means 100 g H_2SO_4 solution contains 95 g H_2SO_4 by mass.
Molar mass of H_2SO_4 = 98 g mol^{-1}

Moles in 95 g $H_2SO_4 = \frac{95}{98} = 0.969$ mol

Volume of 100 g H_2SO_4 solution

$= \frac{\text{mass}}{\text{density}} = \frac{100\text{ g}}{1.834\text{ g cm}^{-3}} = 54.52\text{ cm}^3 = 54.52 \times 10^{-3}$ L

Molarity = $\frac{\text{Moles of solute}}{\text{Volume of solution in L}}$

$= \frac{0.969}{54.52 \times 10^{-3}} = 17.8$ M

36. **(d)**

	O_3	CO	O_2
MW:	48	28	32
given mass:	16	28	16
No. of 'O' atoms:	$\frac{16}{48} \times 3$	$\frac{28}{28} \times 1$	$\frac{16}{32} \times 2$
	= 1	= 1	= 1

Ratio of 'O' atoms = 1 : 1 : 1

37. **(a)** ppm = $\frac{\text{mass of solute (g)}}{\text{mass of solution (g)}} \times 10^6$

$= \frac{0.2}{500} \times 10^6 = 400$ ppm

38. **(c)** $\underset{78\text{g}}{C_6H_6} + HNO_3 \longrightarrow \underset{123\text{g}}{C_6H_5NO_2} + H_2O$

Now since, 78 g of benzene on nitration give = 123 g nitrobenzene
Hence 5 g of benzene on nitration give

$= \frac{123}{78} \times 5 = 7.88$ g

The nearest answer is (c), i.e., theoretical yield = 7.88 g

39. **(d)** 1 Mole of $Mg_3(PO_4)_2$ contains 8 moles of oxygen atoms
$\therefore$ 8 mole of oxygen atoms ≡ 1 mole of $Mg_3(PO_4)_2$

0.25 mole of oxygen atom ≡ $\frac{1}{8} \times 0.25$ mole of $Mg_3(PO_4)_2$
= 3.125 × 10^{-2} mole of $Mg_3(PO_4)_2$

40. **(b)** One mole is defined as exactly 6.02214076 × 10^{23} elementary entities. The definition came into force on May 2019. Hence, the value of relative amu will not change the mass of one mole.

41. **(b)** Moles of urea present in 100 mL of sol.

$= \frac{6.02 \times 10^{20}}{6.02 \times 10^{23}}$ mol

$\therefore \quad M = \frac{6.02 \times 10^{20} \times 1000}{6.02 \times 10^{23} \times 100} = 0.01\ M$

[$\because M$ = Moles of solute present in 1L of solution]

42. **(d)** $Ba(OH)_2$ HCl
$N_1V_1 = N_2V_2$
$N_1 \times 25 = 0.1 \times 35$
$N_1 = 0.14$
Since, $Ba(OH)_2$ is diacidic base

Hence, $N_1 = M_1 \times 2$ or $M_1 = \frac{N_1}{2}$; $M_1 = 0.07$ M

43. **(c)** Among all the given options, molarity changes with temperature because the term molarity involves volume which increases on increasing temperature.

44. **(a)** No. of moles of Fe = $\frac{558.5}{55.85}$ = 10 mol;
Therefore, no. of atoms = 10 N_A
No. of moles of C in 60 g of C = 60/12 = 5 mol.
$\therefore$ No of 'C' atoms = 5N_A

45. **(b)** In the combustion of organic compound, all carbon in CO_2 and all hydrogen in H_2O comes from organic compound

$$C_xH_yO_z + O_2 \longrightarrow \underset{0.307\,g}{CO_2} + \underset{0.127\,g}{H_2O}$$

Weight of Carbon in $CO_2 = \frac{12}{44} \times 0.307$
$= 0.0837$ g

All carbon in CO_2 comes from organic compound.

Weight of Hydrogen in $H_2O = \frac{2}{18} \times 0.127 = 0.0141$ g

(all hydrogen in water comes from organic compound)
% of hydrogen in the organic compound

$= \frac{0.0141}{0.21} \times 100 = 6.719\% \sim 6.72\%$

Weight of Oxygen in the organic compound
$= 0.210 - (0.0837 + 0.0141)$
$= 0.1122$

% of Oxygen in the organic compound $= \frac{0.0141}{0.21} \times 100$
$= 53.41\%$

46. **(65)**

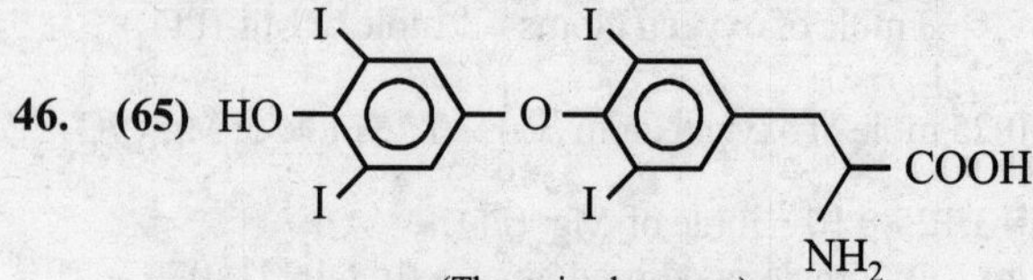

(Thyroxine hormone)

Molecular formula of Thyroxine $= C_{15}H_{11}O_4NI_4$
Molecular mass of Thyroxine $= 15 \times 12 + 11 \times 1 + 16 \times 4 + 14 \times 1 + 127 \times 4 = 777$

% of Iodine $= \frac{\text{mass of iodine}}{\text{mass of thyroxine}} \times 100 = \frac{508}{777} \times 100$
$= 65.38\% \approx 65\%$

47. **(12)**

Organic compound $\xrightarrow[\Delta]{CuO} CO_2 + H_2O$

$n_{CO_2} = \frac{220 \times 10^{-3}}{44} = 5 \times 10^{-3}$ moles

Mass of carbon $= 5 \times 10^{-3} \times 12$
Percent composition of carbon

$= \frac{5 \times 10^{-3} \times 12}{500 \times 10^{-3}} \times 100 = 12\%$

$\therefore$ Correct answer is 12

48. **(a)** Moles of carbon $= n_{CO_2} = \frac{1.46}{44} = 0.033$

Mass of carbon $= W_C = 0.033 \times 12$

Moles of Hydrogen $= 2 \times n_{H_2O} = 2 \times \frac{0.567}{18} = 0.063$

Mass of Hydrogen $W_H = 0.063$ g
Mass of Oxygen (O) $= 1 - (W_C + W_H)$
$= 1 - (0.033 \times 12 + 0.063) = 0.541$ g

Moles of Oxygen $= \frac{0.541}{16} = 0.033$
Empirical formula $= CH_2O$
Empirical formula mass $= 30$

49. **(3)** Mass of carbon $= \frac{80 \times 90}{100} = 72$ gm

Number of C-atoms $= \frac{72}{12} = 6$

Mass of hydrogen $= \frac{80 \times 10}{100} = 8$ gm

Number of H-atoms $= \frac{8}{1} = 8$

So molecular formula C_6H_8

$D.U = \frac{2C + 2 + N - H - X}{2}$

C = Carbon, N = Nitrogen
H = Hydrogen, X = Halogen

$= \frac{2 \times 6 + 2 - 8}{2} = \frac{14 - 8}{2} = \frac{6}{2} = 3$

50. **(c)**

C	:	H	:	O
$\frac{54.2}{12}$	:	$\frac{9.2}{1}$	:	$\frac{36.6}{16}$
4.516	:	9.2	:	2.287
$\frac{4.516}{2.287}$	:	$\frac{9.2}{2.287}$	:	$\frac{2.287}{2.287}$
1.97	:	4.02	:	1

$C_2H_4O \Rightarrow$ Empirical formula
E.F. mass $= 24 + 4 + 16 = 44$
and molar mass $= 132$

$n = \frac{132}{44} = 3$

Hence molecular formula $= (C_2H_4O)_3 = C_6H_{12}O_3$

51. **(d)** Let the molecular formula is $C_xH_yO_z$.
Given, $\%C = 42.1$, $\%H = 6.4$, $\%O = 51.5$

$\%C \ \frac{12x}{342} \times 100 = 42.1 \Rightarrow x = 12$

$\%H \ \frac{y}{342} \times 100 = 6.4 \Rightarrow y = 22$

$\%O = \frac{16z}{342} \times 100 = 51.5 \Rightarrow z = 11$

Hence, molecular is $C_{12}H_{22}O_{11}$

52. **(a)** 1.187 g $CaCO_3$ + 1.023 g $MgCO_3$

$$\underset{x\,g}{CaCO_3} \xrightarrow{\Delta} CaO + CO_2$$

Moles of $CaCO_3$ decomposed = moles of CaO formed

$\therefore \ \frac{x}{100}$ = moles of CaO formed

Weight of CaO formed $= \frac{x}{100} \times 56$ ...(1)

{CaO = 40 + 16 = 56}

$MgCO_3 \xrightarrow{\Delta} MgO + CO_2$

(2.21–x) gm

Moles of $MgCO_3$ decomposed = moles of MgO formed

$\frac{(2.21-x)}{84}$ = moles of MgO formed.

Weight of MgO = $\frac{2.21-x}{84} \times 40$ ...(2)

{MgO = 24 + 16 = 40}

Add equation 1 and 2.

$\therefore \frac{x}{100} \times 56 + \frac{(2.21-x)}{84} \times 40 = 1.152$

$0.56x + \frac{88.4 - 40x}{84} = 1.152$

47.04x + 88.4 – 40x = 96.768

7.04x = 8.368 ⇒ x = 1.188

$CaCO_3$ = x = 1.188 g

$MgCO_3$ = 2.21 – 1.188 = 1.022 g

53. **(56)** Moles of $CO_2 = \frac{0.22}{44} = \frac{1}{200}$

$\therefore$ Moles of carbon = (Moles of CO_2) × 1 = $\frac{1}{200}$

$\therefore$ wt. of C = $\frac{1}{200} \times 12 = 0.06$

% of C = $\frac{0.06}{W} \times 100 = 24$ (W = Wt. of Organic Compound)

⇒ W = 0.25

Moles of $H_2O = \frac{0.126}{18} = 0.007$

$\therefore$ Moles of H atom = 2 × 0.007 = 0.014

% of Hydrogen = $\frac{0.014 \times 1}{W} \times 100 = \frac{0.014 \times 1}{0.25} \times 100$

$= 5.6 = 56 \times 10^{-1}$

54. **(a)** Molecular, weight of metal chloride

$= \frac{0.57}{100} \times 22700 = 129.39$

weight of Cl = 129.39 × 0.55 = 71.1645

$\therefore$ Mole of Cl = $\frac{71.1645}{35.5} \cong 2$, Hence MCl_2

55. **(36)** Total mass of sugar in mixture of 25% of 200 and 40% of 500 g

Sugar solution = (0.25 × 200) + (0.40 × 500) = 50 + 200 = 250 g

Total mass of solution = 200 + 500 = 700 g

Mass of sugar in solution $= \frac{250}{700} \times 100 = 35.7\% \approx 36\%$

56. **(59)** Let the amount of M^{+2} in the sample be x then the amount of M^{+3} in the sample = 0.83 – x

$\therefore \quad 2x + 3(0.83 - x) = 2 \Rightarrow x = 0.49$

$\% M^{2+} = \frac{0.49}{0.83} \times 100 = 59\%$

57. **(25)** Mol. wt of uracil, $C_4N_2H_4O_2 = 112$

$\% N = \frac{28}{112} \times 100 = 25\%$

58. **(12)**

Element	Percentage	Mole	Mole ratio
C	85.8	$\frac{85.8}{12} = 7.15$	1
H	14.2	$\frac{14.2}{1} = 14.2$	2

Empirical formula $(CH_2)_n$

Empirical mass × n = Molecular Mass

14 × n = 84

n = 6

$\therefore$ Molecular formula C_6H_{12}

59. **(46)** Mole of CO_2 = Moles of C

$= \frac{\text{Weight of } CO_2}{\text{Molar mass of } CO_2} = \frac{0.793}{44}$

Weight of 'C' = moles × molar mass of carbon

$= \frac{0.793}{44} \times 12 = 0.216$ g

1 mole of H_2O = 2 moles of hydrogen.

Moles of $H_2O = \frac{\text{Weight}}{\text{Molar mass}} = \frac{0.442}{18}$

Moles of 'H' = $\frac{0.442}{18} \times 2$

Weight of 'H' = $\frac{0.442}{18} \times 2 \times 1 = 0.049$ g

$\therefore$ Weight of 'O' = 0.492 – 0.216 – 0.049 = 0.227 g

% of 'O' = $\frac{0.227}{0.492} \times 100 = 46.13\%$

60. **(d)**

C	74%	$\frac{74}{12} = 6.16$	$\frac{6.16}{1.23} = 5$
N	17.3%	$\frac{17.3}{14} = 1.23$	$\frac{1.23}{1.23} = 1$
H	8.7%	$\frac{8.7}{1} = 8.7$	$\frac{8.7}{1.23} = 7$

Emperical formula = C_5NH_7

Emperical weight = (12 × 5) + (14 × 1) + (1 × 7) = 81 amu

Given, molecular weight = 162 g/mol

Multiplying factor $= \frac{\text{Molecular weight}}{\text{Emperical weight}} = \frac{162}{81} = 2$

Molecular formula = 2 × (Emperical formula) $C_{10}N_2H_{14}$

61. **(2)**

$\% H = \frac{\text{weight of Hydrogen}}{\text{Total weight of substance}} = \frac{7.5}{116} \times 100 = 6.5$

$\% O = \frac{\text{weight of Oxygen}}{\text{Total weight of substance}} = \frac{60}{116} \times 100 = 51.7$

$\% C = \frac{\text{weight of Carbon}}{\text{Total weight of Compound}} = \frac{48.5}{116} \times 100 = 41.8$

Relative mole percentage :

H	$\frac{6.5}{1}=6.5$	$\frac{6.5}{3.25}=2$
O	$\frac{51.7}{16}=3.25$	$\frac{3.25}{3.25}=1$
C	$\frac{41.8}{12}=3.5$	$\frac{3.5}{3.25}=1.07$

Emperical formula = CH_2O

$C_2H_4O_2 = 2(CH_2O)$

$CH_2O = 1(CHO)$

62. **(18)** 44g of CO_2 contains 12 g of C

0.2 g of CO_2 contains $\left(\frac{12}{44}\times 0.2\right)$ g of C, which comes from 0.3 g of organic compound

∴ Percentage of carbon in the given compound

$=\frac{12\times 0.2}{44\times 0.3}\times 100 = 18.18\%$

63. **(d)** Given mass of organic compound = 120 g mass of $CO_2(g) = 330g$; mass of H_2O (l) = 270 g

mass of carbon = $n_{CO_2}\times 12 = \frac{330}{44}\times 12 = 90g$

% of carbon = $\frac{90}{120}\times 100 = 75\%$

mass of hydrogen = $n_{H_2O}\times 2 = \frac{270}{18}\times 2 = 30g$

% of hydrogen = $\frac{30}{120}\times 100 = 25\%$

64. **(3)** Weight of H in 210 g of $H_2O = \frac{210}{18}\times 2 = 23.333$ g

% of H = $\frac{23.333}{750}\times 100 = 3.111 \simeq 3.$

65. **(b)** 44 g of CO_2 contains 12 g of C.

So, 2.64 g of CO_2 contains $\frac{12}{44}\times 2.64 = 0.72$ g C.

1.08 g of H_2O contains $\frac{2}{18}\times 1.08 = 0.12$ g H.

∴ Mass of oxygen present = 1.80 – (0.72 + 0.12) = 0.96 g

% of O = $\frac{0.96}{1.80}\times 100 = 53.33\%$.

66. **(b)** ^{35}Cl ^{37}Cl

Molar ratio x $(1-x)$

$M_{avg.} = 35\times x + 37(1-x) = 35.5 \Rightarrow 2x = 1.5;\ x = \frac{3}{4}$

So, ratio of $^{35}Cl : {}^{37}Cl = \frac{3}{4}\Big/\frac{1}{4} = 3:1.$

67. **(5)** Mass ratio of C : H is $4:1 \Rightarrow 12:3$ and C : O is $3:4 \Rightarrow 12:16$

So,

	mass	mole	mole ratio
C	12	1	1
H	3	3	3
O	16	1	1

Empirical formula ⇒ CH_3O

As compound is saturated acyclic, so molecular formula is $C_2H_6O_2$ (CH_2OH-CH_2OH)

$$\underset{2\text{ mole}}{C_2H_6O_2} + \underset{5\text{ mole}}{\frac{5}{2}O_2(g)} \longrightarrow 2CO_2(g) + 3H_2O(l)$$

∴ Number of moles of O_2 required to oxidise 2 moles of (X) = 5.

68. **(d)**

Element	Relative mass	Relative mole	Simplest whole number ratio
C	6	$\frac{6}{12}=0.5$	1
H	1	$\frac{1}{1}=1$	2

So, $X = 1, Y = 2$

Equation for combustion of C_XH_Y

$$C_XH_Y + \left(X+\frac{Y}{4}\right)O_2 \longrightarrow XCO_2 + \frac{Y}{2}H_2O$$

Oxygen atoms required $= 2\left(X+\frac{Y}{4}\right)$

As mentioned, $Z = \frac{1}{2}\times$ oxygen atom required .

$\left(X+\frac{Y}{4}\right) = Z$

Now putting the values of X and Y

$\Rightarrow \left(1+\frac{2}{4}\right) = Z \Rightarrow Z = 1.5$

∴ Molecule ($C_XH_YO_Z$) can be written as $C_1H_2O_{3/2}$

$\Rightarrow C_2H_4O_3$

69. **(c)** Percentage (by mass) of elements given in the body of a healthy human adult is :-

Oxygen = 61.4%, Carbon = 22.9%,

Hydrogen = 10.0% and Nitrogen = 2.6%

∵ Total weight of person = 75 kg

∴ Mass due to ^1H is $= 75\times\frac{10}{100} = 7.5$ kg

If ^1H atoms are replaced by ^2H atoms.

Mass gain by person would be = 7.5 kg

70. **(c)** $\frac{V\times 45}{100} + \frac{(800-V)\times 20}{100} = \frac{800\times 29.875}{100}$

$\Rightarrow \frac{9V}{20} + 160 - \frac{V}{5} = 239 \Rightarrow V = 316$

71. **(d)** 375 ml of 20% oxygen corresponds to 375 × 20 = 75ml

The ratio of the volume of oxygen used for combustion to values of hydrocarbon $= \frac{75}{15}$

This ratio $\frac{5}{1}$ can be obtained with propane (C_3H_8)

equation: $C_3H_8 + 5O_2 \rightarrow 3CO_2 + 4H_2O$

72. **(c)** Mass of substance = 250 mg = 0.250 g

Mass of AgBr = 141 mg = 0.141 g

1 mole of AgBr ≡ 1 g atom of Br

188 g of AgBr ≡ 80 g of Br

$\therefore$ 188 g of AgBr contain bromine = 80 g

0.141 g of AgBr contain bromine = $\frac{80}{188}\times 0.141 = 0.06$ g

0.06 g of bromine is present in 0.250 g of organic compound.

$\therefore$ % of bromine = $\frac{0.06}{0.250}\times 100 = 24\%$

73. (b) $X_2O_3 \Rightarrow X^{3+}O^{2-}$

$X_2Cl_3 \Rightarrow X^{3+}Cl^{2-}$

$X_2(SO_4)_3 \Rightarrow X^{3+}SO_4{}^{2-}$

$XPO_4 \Rightarrow X^{3+}PO_4{}^{3-}$

Since Cl^{2-} does not exist. So, X_2Cl_3 is incorrect. The correct formula should be XCl_3.

74. (d) In an unknown compound containing N and H, given % of H = 12.5%

$\therefore$ % of N = 100 – 12.5 = 87.5%

Element	Percentage	Atomic ratio	Simple ratio
H	12.5	$\frac{12.5}{1} = 12.5$	$\frac{12.5}{6.25} = 2$
N	87.5	$\frac{87.5}{14} = 6.25$	$\frac{6.25}{6.25} = 1$

Empirical formula = NH_2

Mol. wt = 2 × vapour density = 16 × 2 = 32.

Molecular formula mass = n × empirical formula mass

$$n = \frac{32}{16} = 2$$

$\therefore$ Molecular formula of the compound will be = $(NH_2)_2$ = N_2H_4

75. (b) $\underset{208\text{ g}}{BaCl_2} + \underset{98\text{ g}}{H_2SO_4} \longrightarrow \underset{233\text{ g}}{BaSO_4} + \underset{73\text{ g}}{2HCl}$

Mass of $BaCl_2$ in solution $= 100\times\frac{20.8}{100} = 20.8$ g

Mass of H_2SO_4 in solution $= 50\times\frac{9.8}{100} = 4.9$ g

98 g of H_2SO_4 reacts with 208 g $BaCl_2$

4.9 g H_2SO_4 will react with $\frac{208}{98}\times 4.9 = 10.4$ g $BaCl_2$

H_2SO_4 reacts as a limiting reagent because $BaCl_2$ is given in excess.

98 g H_2SO_4 produces = 233 g $BaSO_4$

4.9 g H_2SO_4 will produce = $\frac{233}{98}\times 4.9 = 11.65$ g $BaSO_4$

76. (d) $\because$ 18 g H_2O contains = 2 g H

$\therefore$ 0.72 g H_2O contains = $\frac{2}{18}\times 0.72$ g = 0.08 g H

$\because$ 44 g CO_2 contains = 12 g C

$\therefore$ 3.08 g CO_2 contains = $\frac{12}{44}\times 3.08 = 0.84$ g C

$\therefore$ C : H = $\frac{0.84}{12} : \frac{0.08}{1}$; 0.07 : 0.08 = 7 : 8

$\therefore$ Empirical formula = C_7H_8

77. (c) Mol. wt = 2 × vapour density = 2 × 94.8 = 189.6 g

Since 74.75% is chlorine,

therefore, 189.6 g metal chloride contains

$= \frac{74.75}{100}\times 189.6 = 141.72$ g chlorine

Number of atoms of chlorine $= \frac{141.72}{35.5} = 3.99 \approx 4$

Hence, formula of metal chlorine is MCl_4.

78. (c)

Element	%	Relative no. of atoms	Simplest ratio of atoms
C	9	$\frac{9}{12} = \frac{3}{4}$	3
H	1	$\frac{1}{1} = 1$	4
N	3.5	$\frac{3.5}{14} = \frac{1}{4}$	1

Empirical formula = C_3H_4N; $(C_3H_4N)_n = 108$

$n\times(12\times 3 + 4\times 1 + 14) = 108$

$n\times(54) = 108 \Rightarrow n = \frac{108}{54} = 2$

$\therefore$ Molecular formula = $C_6H_8N_2$

79. (a)

$\underset{20\text{ ml}}{NaI\,(aq)} + \underset{\text{Excess}}{AgNO_3(aq)} \rightarrow \underset{4.74\text{ g}}{AgI(s)} + NaNO_3(aq)$

Moles of I^- in NaI = Moles of (I^-) in AgI = $\frac{4.74}{235}$

$\therefore$ Moles of NaI = $\frac{4.74}{235}$

Molarity [NaI] = $\frac{4.74}{235\times 0.02} = 1.008$ M.

80. (138) $C_4H_{10}(g) + \frac{13}{2}O_2 \rightarrow 4CO_2(g) + 5H_2O\,(l)$

Moles of butane $(C_4H_{10}) = \frac{174\times 1000}{50} = 3\times 10^3$ moles

moles of $O_2 = \frac{320\times 1000}{32} = 10\times 10^3$ moles

1 mol of C_4H_{10} reacts with $\frac{13}{2}$ mol of O_2,

3×10^3 moles of C_4H_{10} react with = $\frac{13}{2}\times 3\times 10^3$ mols of O_2

Hence O_2 is the limiting reagent.

$\therefore$ $\frac{13}{2}$ mol of O_2 form 3 mol of H_2O,

10×10^3 mol of O_2 form = $\frac{5\times 2}{13}\times 10\times 10^3$ mol of H_2O

$\therefore$ mols of H_2O formed = $\frac{10}{13}\times 10\times 10^3$

Volume of H_2O formed = $\frac{10}{13}\times 10\times 18$ L

= 138.46 L

81. **(63)** $CaCO_3 \rightarrow CaO + CO_2$

mass of $CaCO_3 = \frac{150 \times 75}{100} = 112.5$ kg
$= 112500$g

$n_{CaCO3} = \frac{112500 \text{ g}}{100 \text{ gmol}^{-1}} = 1125$

So moles of CaO = 1125

$\Rightarrow$ mass of CaO $= \frac{1125 \times 56}{1000} = 63$kg

$\therefore$ Correct answer is 63.

82. **(c)** $Mg + 2HCl \rightarrow MgCl_2 + H_2$

Volume of H_2 evolved = 220 mL

Moles of H_2 at STP $= \frac{\text{Volume}}{\text{Molar Volume at STP}}$

$= \frac{220 \times 10^{-3} \text{L}}{22.4 \text{L}}$ = moles of Mg used

$\therefore$ Mass of Mg used $= \frac{220 \times 10^{-3}}{22.4} \times 24$ g/mol
$= 235.7 \times 10^{-3}$ g
$= 235.7$ mg
≈ 236 mg

83. **(80)** Number of moles of CO_2 produced $= \frac{1.46}{44}$

Number of moles of CO_2 produced = Number of moles of carbon

$\therefore$ mass of carbon produced $= \frac{1.46}{44} \times 12 = 0.398$g

% by mass of carbon $= \frac{0.398}{0.5} \times 100 = 79.63\,\% \approx 80\%$.

84. **(c)** $CaCO_3 + 2HCl \rightarrow CaCl_2 + CO_2 + H_2O$

Moles of $CaCO_3 = \frac{1000}{100} = 10$

Moles of HCl $= 0.76 \times \frac{250}{1000} = 0.19$ (Limiting reagent)

Moles of $CaCl_2$ formed $= \frac{0.19}{2}$

Mass of $CaCl_2 = \frac{0.19}{2} \times 111 = 10.545$ g

85. **(a)** The balanced chemical reaction is

$5I^- + IO_3^- + 6H^+ \rightarrow 3I_2 + 3H_2O$

A. 200 ml of 0.1 M KI contains 0.02 mol KI

$\frac{\text{Mole of KI}}{5}$ = mole of KIO_3

$\frac{0.02}{5} = 0.004$

B. 0.02 mol of KI reacts with 0.008 mol H_2SO_4.

C. 0.5 L of KI produce 0.013 mol I_2.

D. Equivalent weight $= \frac{\text{molecular mass}}{\text{change in oxidation number}}$

86. **(c)** % w/w of $HNO_3 = 75\%$

meaning 100 gm of solution contains 75g of HNO_3

and $\left(\frac{\text{mass}}{\text{volume}}\right)_{\text{solution}} = 1.25 = \frac{100 \text{ g}}{V}$

V_{ml} of 100 gm solution $= \frac{100}{1.25}$ ml

$\because$ 75 gm of HNO_3 present in $\frac{100}{1.25}$ ml solution

$\therefore$ 30 gm of HNO_3 present in

$\frac{100}{1.25 \times 75} \times 30 = 32$ ml solution

87. **(125)** Assuming 100 gm solution contain 70 gm solute.

Volume of 100 gm solution will be $\frac{100}{1.25}$ ml.

Molarity $= \frac{70/70}{100/1.25} \times 1000 = 12.5$ or 125×10^{-1}

88. **(420)**

Moles of $Fe_3O_4 = \frac{2.32 \times 10^3 \times 10^3}{232} = 10000$ mol $= 10^4$ mol

Moles of CO $= \frac{2.8 \times 10^2 \times 10^3}{28} = 10000$ mol $= 10^4$ mol

$\underset{10^4 \text{ mol}}{Fe_3O_4} + \underset{10^4 \text{ mol}}{4CO} \rightarrow 3Fe + 4CO_2$

CO is limiting reagent

Moles of Fe $= \frac{3}{4} \times 10^4$

Mass of Fe $= \frac{3}{4} \times \frac{10^4 \times 56}{1000}$ kg $= 420$ kg

89. **(61)** $\underset{x \text{ g}}{C_6H_5COOH} + NaHCO_3 \rightarrow C_6H_5COO^- Na^+ + H_2O + \underset{11.2\text{L}}{CO_2}$

mole of C_6H_5COOH = mole of $CO_2 = \frac{11.2}{22.4} = 0.5$

mass of $C_6H_5COOH = x = 0.5 \times 122 = 61$ g

90. **(153)**

$4Al + 3O_2 \longrightarrow 2Al_2O_3$

$\frac{81}{27} = 3$ mol $\quad \frac{128}{32} = 4$ mol

Limiting reagent is Al.

$\therefore$ mole of Al_2O_3 formed $= \frac{1}{2} \times 3$ mol

$\therefore$ weight of Al_2O_3 formed $= \frac{3}{2} \times 102 = 153$ gm

91. **(100)**

Organic compound $\xrightarrow{\text{combustion}} \underset{0.9 \text{ gm}}{H_2O}$

$\therefore$ Mole of $H_2O = \frac{0.9}{18} = 0.05$ mole

$\therefore$ Mole of H in $H_2O = 0.05 \times 2 = 0.1$ mol
= mol of H in 0.01 mole Organic compound

$\therefore$ Weight of H atom in 0.01 mole compound = 0.1 × 1 = 0.1 gm

$\therefore$ Weight of H atom in one mole compound

$= \frac{0.1}{0.01} = 10$ gm

$\because$ wt. % of H $= \frac{\text{wt. of H in one mole compound}}{\text{Molar mass of compound}} \times 100$

$10 = \frac{10}{M} \times 100 \quad \therefore \quad \boxed{M = 100}$

92. (a) Phenol $\xrightarrow{3Br_2,\ water}$ 2,4,6-tribromophenol + 3HBr

Moles of phenol $= \frac{2}{94} = 0.021$

1 mole of phenol required 3 moles of bromine

$\therefore$ Moles of bromine = 0.021 × 3 = 0.064

$\therefore$ Mass of bromine = 0.064 × 160 = 10.22 g

93. (57) $M_F V_F = M_1 V_1 + M_2 V_2$

$M_F = \frac{M_1V_1 + M_2V_2}{V_1 + V_2}$

$= \frac{2 \times 20 + 0.5 \times 400}{420} = 0.571M = 57.1 \times 10^{-2} M = 57$

94. (d) 3M NaCl, $d_{sol} = 1.25$ g/mL

Molality $= \frac{C_M \times 1000}{1000d - C_M \times M_{Solute}}$

$\frac{3 \times 1000}{(1000 \times 1.25) - (3 \times 58.5)}$

$= \frac{3000}{1250 - 175.5} = 2.79$ molal

95. (45) moles of HCl $= 4 \times 10^{-3}$

moles of $Ca(OH)_2$ reacted with HCl $= 2 \times 10^{-3}$

Initial amount of $Ca(OH)_2 = 4 \times 10^{-3}$ mol

(as per the given information)

Moles of $CO_2 = 2 \times 10^{-3}$

Volume $= 2 \times 10^{-3} \times 0.0821 \times 273 = 44.8 \times 10^{-3}$ Lit $= 44.8\ cm^3$

Nearest integer = 45

96. (a) $CH_3-C\equiv CH + \underset{(excess)}{Na} \rightarrow CH_3-C\equiv \overset{-}{C}\overset{+}{Na} + \frac{1}{2}H_2\uparrow$

$CH_3-C\equiv CH + NaNH_2 \rightarrow CH_3C\equiv \overset{-}{C}\overset{+}{Na} + NH_3$

4 gm 0.1 mol

$\frac{4}{40} = 0.1$ mol = 2240 mL

Statement I is correct but statement II is incorrect

97. (b) $CH_4 + 2O_2 \rightarrow CO_2 + 2H_2O$

mol of $CO_2 = \frac{11}{44} = \frac{1}{4}$ mol

Since, stoichiometry is same for CH_4 & CO_2

Thus, mol of CH_4 required $= \frac{1}{4} = 0.25$ mol

98. (24) $\underset{72\ mmol}{3PbCl_2} + \underset{50\ mmol}{2(NH_4)_3PO_4} \longrightarrow Pb_3(PO_4)_2 + 6NH_4Cl$

3 mol $PbCl_2 \equiv$ 1 mol $Pb_3(PO_4)_2$

72×10^{-3} mol $PbCl_2 = \frac{1}{3} \times 72 \times 10^{-3}$ mol $Pb_3(PO_4)_2$

$= 24 \times 10^{-3}$ mol

99. (14)

$\underset{10\ ml}{C_xH_y} + O_2 \longrightarrow CO_2 + H_2O$

$C_xH_y + \left(x + \frac{y}{4}\right)O_2 \longrightarrow \underset{10x}{xCO_2} + \underset{5y}{\frac{y}{2}H_2O}$

$10x = 40 \Rightarrow x = 4$

$5y = 50 \Rightarrow y = 10 \quad \therefore C_4H_{10}$

Total no. of atoms = 4 + 10 = 14

100. (50) $\underset{1\ mol}{CH_4(g)} + 2O_2(g) \rightarrow \underset{44\ g}{CO_2(g)} + 2H_2O(l)$

44g CO_2 needs 1 mol CH_4

22g CO_2 needs $= \frac{1}{44} \times 22 = 50 \times 10^{-2}$ mol = 0.5 mol.

101. (4) $M_aV_a = M_bV_b = M_b$ or $M_{NaOH} = \frac{M_aV_a}{V_b}$

$= \frac{50 \times 0.5 \times 2}{25} = 2$ mM

$\Rightarrow n_{NaOH} = M_{NaOH} \times V_{NaOH} = 2 \times 50 = 100$ mmol

$= 100 \times 10^{-3}$ mol

$= 10 \times 10^{-2}$ mol

$\Rightarrow$ Mass $= n_{NaOH} \times$ Molar mass of NaOH

$= (10 \times 10^{-2}) \times 40 = 400 \times 10^{-2} = 4$ g

102. (8) $CH_4 + 2O_2 \rightarrow CO_2 + 2H_2O$

Moles of $CO_2 = \frac{22}{44} = 0.5$

So, required moles of $CH_4 = 0.5$

Mass = 0.5 × 16 = 8 gm

103. (7) $M = \frac{n_{NaOH}}{V_{sol}(\text{in L})} \Rightarrow 3 = \frac{(84/40)}{V} \Rightarrow V = 0.7$ L

$= 7 \times 10^{-1}$ L

104. (135)

$\underset{(\text{Aniline MM} = 93)}{C_6H_5NH_2} + CH_3-\overset{O}{\overset{\|}{C}}-O-\overset{O}{\overset{\|}{C}}-CH_3 \rightarrow$

$\underset{(\text{Acetanilide MM} = 135)}{C_6H_5NH-\overset{O}{\overset{\|}{C}}-CH_3} + CH_3COOH$

$n_{Acetanillide} = n_{Aniline}$

$\frac{m}{135} = \frac{9.3}{93} \Rightarrow m = 13.5\ g = 135 \times 10^{-1}$ gm

105. (100) $\underset{1\ g}{M_2CO_3} + \underset{Excess}{2HCl} \rightarrow \underset{0.02\ mole}{2MCl} + H_2O + \underset{0.01\ mole}{CO_2}$

From principle of atomic conservation of carbon atom,

Mole of $M_2CO_3 \times 1$ = Mole of $CO_2 \times 1$

$\frac{1\,g}{\text{molar mass of } M_2CO_3} = 0.01 \times 1$

$\therefore$ Molar mass of $M_2CO_3 = 100$ g/mole

106. (224) $Mg + 2HCl \rightarrow MgCl_2 + H_2\uparrow$

Number of moles $= \frac{2.4}{24} = 0.1$ mole

1 mole of gas at STP $\Rightarrow$ 22.4 L

$\therefore$ 0.1 mole of gas $= 0.1 \times 22.4 = 2.24 = 224 \times 10^{-2}$ L.

107. (11) Percentage of Carbon

$= \frac{12}{44} \times \frac{\text{mass of } CO_2 \text{ formed}}{\text{mass of compound taken}} \times 100$

$60 = \frac{12}{44} \times \frac{\text{mass of } CO_2 \text{ formed}}{0.5} \times 100$

Mass of CO_2 formed $= \frac{60 \times 44 \times 0.5}{12 \times 100}$ g

$= 1.1\text{ g} = 11 \times 10^{-1}$ g

108. (1) Hint: First, always check the limiting reagent.

$3BaCl_2 + 2Na_3PO_4 \rightarrow Ba_3(PO_4)_2 + 6NaCl$

5 moles 2 moles

Na_3PO_4 is limiting reagent.

2 mole Na_3PO_4 gives 1 mole of $Ba_3(PO_4)_2$.

109. (44) 44 g of CO_2 contain 12 g of carbon

$\therefore$ 1g of CO_2 has $\frac{12}{44}$ g of carbon

$\Rightarrow$ Weight of C in 0.792 g $CO_2 = \frac{12}{44} \times 0.792 = 0.216$

% of C in compound $= \frac{0.216}{0.492} \times 100 = 43.90\%. \approx 44$

110. (4) $Zn + 2HCl \rightarrow ZnCl_2 + H_2\uparrow$

Moles of Zn used $= \frac{11.5}{65.4}$ = Moles of H_2 evolved

Volume of $H_2 = \frac{11.5}{65.4} \times 22.7L = 3.99L \approx 4L$

111. (227) $2C(s) + O_2(g) \rightarrow 2CO(g)$

1 mol 1.5 mol

Limiting reagent is carbon. One mole carbon produces one mole CO. Hence, volume at STP is

$22.7\text{ L} = 227 \times 10^{-1}$ L.

112. (1362) Mole of $CO_2 = 0.2\text{ M} \times (300 \times 10^{-3})$L
$= 0.06$ Mole

Volume of 0.06 mole CO_2 at S.T.P

$= 0.06 \times 22.7 = 1.362\text{ L} = 1362$ mL

113. (150) $\underset{68g}{2H_2O_2} \longrightarrow 2H_2O + \underset{22.72}{O_2}$

$\therefore$ Strength in g/L $= \frac{68}{22.7} \times 50 = 149.7 \approx 150$

114. (200) Let M is the molar mass of the compound (g/mol)

mass of compound $= 0.01 \times M_{comp.}$

where $M_{comp.}$ = molar mass of compound

Mass of carbon in the compound

$= 0.01 \times M_{comp.} \times \frac{60}{100}$

moles of carbon $= \frac{0.01\, M_{comp.}}{12} \times \frac{60}{100}$

moles of CO_2 from combustion $= \frac{4.4}{44}$

It should be equal to the moles of carbon in the compound.

$\Rightarrow \frac{0.01\, M_{comp.}}{12} \times \frac{60}{100} = \frac{4.4}{44}$

$M_{comp.} = \frac{4.4}{44} \times \frac{100}{60} \times \frac{12}{0.01} = 200$ g/mol

115. (15) $\underset{1\,mol}{Na} + H_2O \rightarrow \underset{1\,mol}{NaOH} + \frac{1}{2}H_2$

0.69 g of Na forms $= \frac{0.69}{23} = 0.03$ mol of NaOH

For complete neutralization of 0.03 mol of NaOH

0.03 mol $= (0.03 \times 36.5)$g of HCl is required.

Density of HCl (d) = 73 g/L

$\therefore$ Volume of HCl required $= \frac{0.03 \times 36.5}{73}$ L

$= 1.5 \times 10^{-2}\text{ L} = 15$ mL

116. (c) Assuming, : Mass of solvent $\approx$ Mass of solution

Case I:

$0.25 = \frac{W_1}{62} \times \frac{1000}{500}$

Case II:

$0.25 = \frac{W_2}{62} \times \frac{1000}{250} \Rightarrow \frac{W_1}{W_2} = \frac{2}{1}$

117. (4) A : $Fe_{0.93}O \rightarrow Fe_2O_3$

n-factor $= \left(3 - \frac{200}{93}\right) \times 0.93$

n-factor = 0.79

B : $2x + (0.93 - x) \times 3 = 2$

$x = 0.79$

$Fe^{2+} = 0.79$, $Fe^{3+} = 0.21$

D : % $Fe^{2+} = \frac{0.79}{0.93} \times 100 = 85\%$; $Fe^{3+} = 15\%$

118. (c) $\underset{20g}{N_2(g)} + \underset{5g}{3H_2(g)} \rightleftharpoons 2NH_3(g)$

$\frac{20}{28}$ mol $\frac{5}{2}$ mol

For 3 mol of $H_2 \equiv$ 1 mol of N_2 required

For $\frac{5}{2}$ mol of $H_2 \equiv \frac{5}{6} = 0.83$ mol of N_2 req.

Moles of $N_2 = \frac{20}{28} = 0.71$; $\therefore$ N_2 is the limiting reagent.

$n(NH_3) = 2 \times n(N_2) = 2 \times \frac{20}{28} = 1.42$ moles.

119. (c) $4HNO_3(l) + 3KCl(s) \rightarrow$
$Cl_2(g) + NOCl(g) + 2H_2O(g) + 3KNO_3(g)$

$n_{KNO_3} = \frac{110}{101} = 1.09$

from stoichiometry

3 Mole KNO_3 requires = 4 mole HNO_3

1.09 mole KNO_3 requires $= \frac{4}{3} \times 1.09 = 1.45$ moles

weight of $HNO_3 = 1.45 \times 63 = 91.5g$

120. (d) $C(s) + O_2(g) \rightarrow CO_2(g) + 400\text{ kJ}$

$C(s) + \frac{1}{2}O_2(g) \rightarrow CO(g) + 100kJ$

weight of coal $= 0.6\text{ kg} = 0.6 \times 10^3\text{ g}$

weight of pure carbon $= 0.6 \times 10^3 \times \frac{60}{100}$

$$n_C = \frac{0.6 \times 10^3 \times 60}{100} \times \frac{1}{12} = \frac{6 \times 60}{12}$$

40% of C is converted into CO_2,

Energy released $= \frac{6 \times 60}{12} \times \frac{40}{100} \times 400 = 4800\text{ kJ}$

60% of C is converted into CO,

energy released $= \frac{6 \times 60}{12} \times \frac{60}{100} \times 100 = 1800\text{ kJ}$

Total energy released = 4800 + 1800 = 6600 kJ

121. (b) In 100 g solution, mass of carbon = 10.8 g

250 g solution, mass of carbon $= \frac{10.8 \times 250}{100} = 27\text{ g}$

Mass of C in one mole of glucose $(C_6H_{12}O_6) = 12 \times 6 = 72\text{ g}$

72 g of C is present in = 180 g glucose

27 g of C is present in $= \frac{180}{72} \times 27 = 67.5$ g glucose

Mass of solvent (water) = 250 – 67.5 = 182.5 g

Molality of solution $= \frac{(67.5/180)\text{ mol}}{(182.5/1000)\text{ kg}} = 2.05\text{ m}$

122. (c) 16 moles of NaOH neutralizes 16 moles of H^+

Source of $H^+ \rightarrow$ 2 moles HCl + 1 mol H_2SO_4.

$\therefore$ 1 mol SO_2Cl_2 is producing 4 moles H^+ ions

$\therefore$ No. of moles of $SO_2Cl_2 = \frac{16}{4} = 4$ moles.

123. (3810)

moles of $Cu_2O = \frac{2.86 \times 10^3}{143} = 20$

moles of $Cu_2S = \frac{4.77 \times 10^3}{159} = 30$

Cu_2O is limiting reagent

$$2Cu_2O + Cu_2S \longrightarrow 6Cu + SO_2$$

$2Cu_2O$	Cu_2S	$6Cu$	SO_2
20	30	–	–
–	20	60	10

Mass of copper = 60 × 63.5 = 3810 g

124. (c) Molecular weight of Fe_3O_4 = 236 g/mol

Moles of $Fe_3O_4 = \frac{4.640 \times 10^3}{232} = 20$

Moles of CO $= \left[\frac{\text{given mass}}{\text{molar mass}}\right] = \frac{2.52 \times 10^3}{28} = 90$

So limiting Reagent

$= Fe_3O_4$ (20 mole Fe_3O_4 required 80 mole CO)

So moles of Fe formed = 60

Weight of Fe = (moles × molar mass) = 60 × 56 = 3360 g

125. (143) CNG contains methane gas

$$CH_4(g) + 2O_2(g) \rightarrow CO_2(g) + 2H_2O\ (l)$$

$\frac{100}{16} = 6.25$ $\frac{208}{32} = 6.5$

from the stoichiometry O_2 is limiting reagent

2 moles of O_2 gives = 1 mole CO_2;

6.5 moles of O_2 gives $= \frac{6.5}{2}$ moles of CO_2

Mass of CO_2 = moles × molar mass $= \frac{6.5}{2} \times 44 = 143\text{g}$

126. (225) The balanced chemical reaction is –

$$CH_4 + 2O_2 \longrightarrow CO_2 + 2H_2O$$

Moles of water produced $(n_{H_2O}) = \frac{81}{18} = 4.5$

from the above reaction;

2 mole of H_2O requires = 1 mole of CH_4

4.5 mole of H_2O requires

$= \frac{4.5}{2} = 2.25 = 225 \times 10^{-2}$ mole of CH_4

127. (1143)

$$(C_5H_{10}) + Br_2 \longrightarrow (C_5H_{10}Br_2)$$

moles of Br_2 = moles of C_5H_{10}

$\Rightarrow \frac{w}{160} = \frac{5}{70} \Rightarrow w = \frac{5 \times 160}{70}\text{ g}$

$= 11.428\text{ g} = 1142.8 \times 10^{-2}\text{ g}$.

128. (0) Moles of H_2SO_4 in solution A = 50% of original solution = 0.02/2 = 0.01 m mol.

Moles of H_2SO_4 in final solution = 0.01 + 0.01

$= 0.02\text{ m mol} = 0.00002 \times 10^3\text{ m mol}$

Answer to the nearest integer = 0 (zero).

129. (25) There will be min. one molecule of glycine.

0.30 % glycine ≡ 75 g

1% of protein $\equiv \frac{75}{0.30}\text{ g}$

100% of protein $\equiv \frac{75}{0.30} \times 100\text{ g} = 25000\text{ g} = 25 \times 10^3\text{ g}$

130. (3) Acid (A) + Base $\longrightarrow$ Salt + H_2O

Acid: 0.1 M, 10 mL; Base $(M(OH)_2)$: 0.05 M, 30 mL

At equivalence point, equivalent of acid = equivalent of base

$0.1 \times 10 \times n = 30 \times 0.05 \times 2 \Rightarrow n = 3$.

131. (c) $C_xH_y + \left(x+\frac{y}{4}\right)O_2 \rightarrow xCO_2 + \frac{y}{2}H_2O$

$C_{15}H_{30} + \frac{45}{2}O_2 \rightarrow 15CO_2 + 15H_2O$

Mass of fuel = 0.756×1000 g

No. of moles of fuel

$= \frac{0.756\times1000}{210}$; M.W. of $C_{15}H_{30} = 210$

Wt. of oxygen $= \frac{0.756\times1000}{210}\times\frac{45}{2}\times 32 = 2592$ g

Wt. of $CO_2 = \frac{0.756\times1000}{210}\times 15\times 44 = 2376$ g

132. (525) $3Pb(NO_3)_2 + Cr_2(SO_4)_3 \rightarrow 3PbSO_4 + 2Cr(NO_3)_3$

m moles of $Pb(NO_3)_2 = 35 \times 0.15 = 5.25$ m moles

m moles of $Cr_2(SO_4)_3 = 20 \times 0.12 = 2.4$ m moles

By stoichiometry, required no. of m moles of $Pb(NO_3)_2$ $= 3 \times 2.4 = 7.2$ m. moles

Hence, $Pb(NO_3)_2$ is limiting reagent, m moles of $PbSO_4$ formed = 5.25 m moles = 525×10^{-5} moles

133. (13) $Na_2O + H_2O \rightarrow 2NaOH$

Moles of $Na_2O = \frac{20}{62}$ moles; M.W. of $Na_2O = 62$ g/mol

Moles of NaOH formed $= \frac{20}{62}\times 2 = \frac{40}{62}$

$[NaOH] = \frac{40\times1000}{62\times500} = 1.29M = 13\times10^{-1}M$

134. (19) $C_3H_8(g) + 5O_2(g) \rightarrow 3CO_2(g) + 4H_2O(l)$

	C_3H_8	O_2	CO_2	H_2O
$t=0$	$\frac{100}{44}$ mol	$\frac{1000}{32}$ mol	0	0
$t=\infty$	0	19.9 mol	6.81 mol	9.08 mol

Mole fraction of CO_2 in the final reaction mixture (heterogeneous),

$$x_{CO_2} = \frac{n_{CO_2}}{n_{O_2}+n_{CO_2}+n_{H_2O}} = \frac{6.81}{19.9+6.81+9.08}$$

$= 0.19 = 19 \times 10^{-2}$

135. (5) Consider 1 L = 1000 mL of solution

Mass of solution = Vol. × density = 1.2 × 1000 g = 1200 g

Mass of 1L solvent (water) = 1000 g $(\because \text{density} = 1\text{ g cm}^{-3})$

$\Rightarrow$ Mass of NaOH = (1200 – 1000) g = 200 g

$\Rightarrow$ Moles of NaOH $= \frac{200\text{ g}}{40\text{ g/mol}} = 5$ mol

$\Rightarrow$ Molality $= \frac{5\text{ mol}}{1\text{ kg}} = 5$ m

136. (18) milli-equivalents of Fe^{2+} = milli-equivalents of $K_2Cr_2O_7$

$\Rightarrow$ millimoles × n-factor of Fe^{2+} = millimoles × n-factor of $K_2Cr_2O_7$.

$x \times 10 \times 1 = 0.02 \times 15 \times 6$; $x = 0.18 = 18 \times 10^{-2}$ M

137. (464) $C_4H_{10} + \frac{13}{2}O_2 \longrightarrow 4CO_2 + 5H_2O$

Moles of $H_2O = \frac{72}{18} = 4$

Moles of C_4H_{10} used $= \frac{1}{5}\times 4$

Weight of C_4H_{10} used $= \frac{4}{5}\times 58 = 46.4$ g

46.4 g = 464×10^{-1} g

138. (4) Concentration of glucose in blood = 0.72g/L

$= \frac{0.72}{180} = 4\times10^{-3}$ mol / L

139. (226) Millimoles of NaOH = Volume (mL) × Molarity = 250 mL × 0.5 = 125 m mol

Millimoles of HCl = 500 mL × 1 = 500 m mol

Now reaction is

	NaOH	+ HCl	→ NaCl	+ H_2O
$t=0$	125	500	0	0
$t=t$	0	375	125	125

So, millimoles of HCl left = 375

$= 375 \times 10^{-3}$ moles

No. of HCl molecules $= 6.022 \times 10^{23} \times 375 \times 10^{-3}$

$= 225.8\times10^{21} \simeq 226\times10^{21}$

140. (1) Let moles of NaOH or Na_2CO_3 be a.

Given mixture is equimolar, thus

NaOH : a mol

Na_2CO_3 : a mol

Now, $W_{NaOH} + W_{Na_2CO_3} = 4$

$\Rightarrow 40\cdot a + 106\cdot a = 4 \Rightarrow a = \frac{4}{146}$ mol

Therefore, mass of NaOH $= \frac{4}{146}\times 40$ g $= 1.095 \approx 1$

141. (16) $(HO)CH_2{-}CH_2CHO \xrightarrow[64\%]{\Delta} C_3H_4O + H_2O$

	$(HO)CH_2{-}CH_2CHO$	C_3H_4O
M.W.	74	56
Let	x g	7.8 g
	$\frac{x}{74}$ mol	$\frac{x}{74}\times 0.64$

$\therefore \frac{x}{74}\times 0.64$ mol $= \frac{7.8}{56}$ moles $\Rightarrow x = 16.10 \approx 16$

142. (3)

$C_6H_5CH_2{-}NH_2 + 3\,CH_3{-}Br \longrightarrow C_6H_5CH_2{-}\overset{+}{N}(CH_3)_3Br^-$

Number of moles of benzyl trimethyl ammonium bromide

$= \frac{23}{230} = 0.1$

$\therefore$ No. of moles of bromomethane consumed $= 3 \times 0.1 = 3 \times 10^{-1}$

143. (18) General combustion reaction of ethane is

$$C_xH_y+\left(x+\frac{y}{4}\right)O_2 \longrightarrow xCO_2+\frac{y}{2}H_2O$$

Then for ethane is,

$$C_2H_6+\frac{7}{2}O_2 \longrightarrow 2CO_2+3H_2O$$

1 mol 3 mol

$\frac{3}{30}$ mol = 0.1 0.3 mol

$\therefore$ Number of molecules of water

$= 0.3 \times 6.023 \times 10^{23}$

$= 18.069 \times 10^{22} \approx 18 \times 10^{22}$

144. (1) Volume of Cl_2 at STP = 20 mL

No. of moles of chlorine gas $= \frac{20 \text{ mL}}{22400 \text{ mL/mol}}$

No. of molecules of Cl_2 gas $= \frac{20}{22400} \times 6.023 \times 10^{23}$

Hence, number of Cl atoms

$$= 2 \times \frac{20}{22400} \times 6.023 \times 10^{23} = 1.08 \times 10^{21} \approx 1 \times 10^{21}$$

145. (24)

$$6Fe^{2+}+Cr_2O_7^{2-}+14H^+ \rightarrow 2Cr^{3+}+6Fe^{3+}+7H_2O$$

$n=1$ $n=6$

By law of equivalence,

M_{eq} of $Fe^{2+} = M_{eq}$ of $Cr_2O_7^{2-}$

$1 \times M \times 15 = 0.03 \times 6 \times 20$

$M = 0.24 = 24 \times 10^{-2}$

146. (13) Mass of Na^+ in 50 mL = 70 × 50 mg

Millimoles of $Na^+ = \frac{70 \times 50}{23}$

Weight of $NaNO_3 = \frac{70 \times 50 \times 85 \times 10^{-3}}{23} = 12.9 \approx 13$ g

147. (2) $M = \frac{4.5/90}{250/1000} = 0.2 = 2 \times 10^{-1}$

148. (8)

$$C_xH_y(g)+\left(x+\frac{y}{4}\right)O_2(g) \rightarrow x\,CO_2(g)+\frac{y}{2}H_2O(l)$$

V $6V$ $4V$

$V_{O_2} = 6 \times V_{C_xH_y}$ (given)

$\Rightarrow V\left(x+\frac{y}{4}\right) = 6V \Rightarrow \left(x+\frac{y}{4}\right) = 6$ ...(i)

$Vx = 4V$ (given) $\Rightarrow x = 4$ ...(ii)

From relation (i) and (ii): $4+\frac{y}{4} = 6 \Rightarrow y = 8.$

149. (18) Complete combustion of hydrocarbons can be represented by the following reaction.

$$C_xH_y+\left(x+\frac{y}{4}\right)O_2 \longrightarrow xCO_2+\frac{y}{2}H_2O$$

For propane, combustion reaction is:

$$C_3H_8+\left(3+\frac{8}{4}\right)O_2 \longrightarrow 3\,CO_2+\frac{8}{2}H_2O$$

$$\therefore C_3H_8+5\,O_2 \longrightarrow 3\,CO_2+4\,H_2O$$

Similarly, for butane, Combustion reaction is

$$C_4H_{10}+\left(4+\frac{10}{4}\right)O_2 \longrightarrow 4\,CO_2+\frac{10}{2}H_2O$$

$$\therefore C_4H_{10}+\frac{13}{2}O_2 \longrightarrow 4\,CO_2+5\,H_2O$$

$\because$ For 1 mol of C_4H_{10} required $O_2 = \frac{13}{2}$ mol

$\therefore$ For 2 mol of C_4H_{10} required $O_2 = \frac{13}{2} \times 2 = 13$ mol

Number of moles of O_2 required = 5 + 13 = 18

150. (3400)

Moles of $N_2 = \frac{2800}{28} = 100$ and Moles of $H_2 = \frac{1000}{2} = 500$

	$N_2(g)$ +	$3H_2(g)$ →	$2NH_3(g)$
Initial mole	100	500	0
Final mole	0	200	200

Mass of NH_3 formed = 200 × 17 = 3400 g

151. (10)

$$\text{Normality} = \frac{\text{No. of equivalents of solute}}{\text{Volume of solution (in L)}}$$

$$0.1 = \frac{1.43}{\frac{(106+18x)}{2} \times 0.1} \Rightarrow \frac{106+18x}{2} = 143$$

$\Rightarrow 18x = 286 - 106 = 180 \Rightarrow x = 10.$

152. (47)

Let total mole of solution = 1

So, mole of glucose = 0.1

Mole of H_2O = 0.9

$$\%\,(w/w) \text{ of } H_2O = \left[\frac{0.9 \times 18}{0.9 \times 18 + 0.1 \times 180}\right] \times 100$$

$= 47.368 = 47.37.$

153. (25)

Number of mole of $x = \frac{6.022 \times 10^{22}}{6.022 \times 10^{23}} = \frac{10}{\text{Molar mass of } x}$

So molar mass of x = 100 g

Molarity $= \frac{5}{100 \times 2} = 0.025$ M.

154. (10)

Phosphinic acid is hypophosphorous acid (H_3PO_2), a monobasic acid.

$$NaOH + H_3PO_2 \longrightarrow NaH_2PO_2 + H_2O$$

For neutralisation, $(N_1V_1)_{acid} = (N_2V_2)_{base}$

$0.1 \times 10 = 0.1 \times V_{NaOH}$

$V_{NaOH} = 10$ mL

155. (10.00) Concentration $= \frac{10.3 \times 10^{-3}}{1.03 \times 1000 \text{ g}} \times 10^6 = 10$ ppm

156. (4.96) 10 ppm $= \frac{\text{Mass of Fe (in g)}}{100 \times 1000 \text{ g}} \times 10^6$ ppm

$\Rightarrow$ Mass of Fe = 1 g

Molar mass of $FeSO_4 \cdot 7H_2O = 278$

56 g of iron present in 1 mole of $FeSO_4 \cdot 7H_2O$

1 g of Fe present in $\frac{278}{56}$ g of salt = 4.96 g

157. (2130)
$NaClO_3(s) + Fe(s) \longrightarrow NaCl(s) + FeO(s) + O_2(g)$
Moles of $NaClO_3$ = Moles of O_2

Moles of $O_2 = \frac{PV}{RT} = \frac{1 \times 492}{0.082 \times 300} = 20$ moles

Mass of $NaClO_3 = 20 \times 106.5 = 2130$ g

158. (c) $NH_2CONH_2 + 2NaOH \longrightarrow Na_2CO_3 + 2NH_3$
1 mol of urea ≡ 2 mol of NH_3
60 g of urea ≡ 2 mol of NH_3

0.6 g of urea $= \frac{2}{60} \times 0.6$ mol $= 0.02$ mol of NH_3

For neutralisation: mol of NH_3 = mol of HCl
in option (c), no. of moles of HCl = 0.1 × 0.2 = 0.02

159. (a) (a) $4\,Fe + 3\,O_2 \longrightarrow 2\,Fe_2O_3$

1 g of Fe requires $= \frac{3 \times 32}{4 \times 56} = 0.43$ g of oxygen

(b) $P_4 + 5\,O_2 \longrightarrow P_4O_{10}$

1 g of P requries $= \frac{5 \times 32}{31 \times 4} = 1.3$ g of oxygen

(c) $C_3H_8 + 5\,O_2 \longrightarrow 3\,CO_2 + 4\,H_2O$

1 g of C_3H_8 requires $= \frac{5 \times 32}{44} = 3.6$ g of O_2

(d) $2\,Mg + O_2 \longrightarrow 2\,MgO$

1 g of Mg requires $= \frac{32}{2 \times 24}$ g $= 0.66$ g of O_2

160. (a) According to the stoichiometry of balanced equation, 28 g N_2 reacts with 6 g H_2.

$$\underset{\substack{1\text{ mole}\\28\text{ g}}}{N_2} + \underset{\substack{3\text{ mole}\\6\text{ g}}}{3H_2} \longrightarrow 2NH_3$$

∴ For 56 g of N_2, 12 g of H_2 is required.

161. (a) Given: 0.27 g fatty acid is present in 100 cm^3 of hexane
∴ 10 mL of aqueous solution contains only 0.027 g acid.

Volume of 0.027 g acid $= \frac{0.027}{0.9}$ mL

∴ $\pi r^2 h = \frac{0.027}{0.9}$ (given $r = 10$ cm, $\pi = 3$)

∴ $h = 10^{-4}$ cm $= 10^{-6}$ m

162. (N)

Oxalic acid		**Sodium hydroxide**
N_1V_1	=	N_2V_2
$(2 \times M_1)V_1$	=	M_2V_2
$2 \times 0.5 \times 50$	=	$M_2 \times 25$
M_2	=	2

Molarity $= \frac{\text{No. of moles}}{\text{Vol. (L)}}$

$2 = \frac{\text{No. of moles}}{50/1000} \Rightarrow$ No. of moles $= \frac{1}{10} = 0.1$

No. of moles $= \frac{W}{\text{Molar mass}}$

$W = 0.1 \times 40 = 4$ g

No option is correct.

163. (d)

$2NaHCO_3 + H_2C_2O_4 \longrightarrow Na_2C_2O_4 + 2CO_2 + 2H_2O$

Moles of CO_2 evolved $= \frac{0.25}{25 \times 10^3} = 10^{-5}$

∴ Moles of $NaHCO_3 = 10^{-5}$
∴ Mass of $NaHCO_3 = 84 \times 10^{-5}$ g
$= 0.84 \times 10^{-3}$ g $= 0.84$ mg

∴ % by weight $= 0.84 \times \frac{100}{10} = 8.4\%$

164. (a) 25 mL of HCl solution requires 30 mL of 0.1 M Na_2CO_3 solution.
∵ $N_1V_1 = N_2V_2$
∴ $25 \times N_1 = 30 \times 0.2$ (0.1 M Na_2CO_3 = 0.2 N Na_2CO_3)

$N_1 = \frac{6}{25} = 0.24$ N

Now, HCl solution is titrated with NaOH solution.
∴ $V \times 0.24 = 30 \times 0.2 \Rightarrow V = 25$ mL

165. (c) For the given reaction:
$2\,C_{57}H_{110}O_6(s) + 163\,O_2(g) \rightarrow 114\,CO_2(g) + 110\,H_2O(l)$

Moles of $C_{57}H_{110}O_6 = \frac{445}{890} = 0.5$

Now, moles of water $= \frac{110}{2} \times 0.5 = 27.5$

∴ Mass of water = 27.5 × 18 = 495 g

166. (d) No. of moles of oxygen in 0.16 g of oxygen molecule

$= \frac{0.16\text{ g}}{32\text{ g/mol}} = 0.005$ mol

$2NaClO_3 \xrightarrow{\Delta} 2\,NaCl + 3O_2$
According to the reaction,
3 moles of O_2 ≡ 2 moles of NaCl ≡ 2 moles of AgCl
Molar mass of AgCl = 143.5 g/mol

0.005 moles of O_2 will ppt. $= 0.005 \times \frac{2}{3}$ moles AgCl
$= 0.0033$ moles of AgCl

∴ Mass of AgCl (in g) obtained will be
= 143.5 g/mol × 0.0033 moles = 0.48 g.

167. (b) Given chemical equation:
$$\underset{1\text{ g}}{M_2CO_3} + 2\,HCl \rightarrow 2\,MCl + H_2O + \underset{0.01186\text{ mol}}{CO_2}$$

moles of M_2CO_3 ≡ moles of CO_2

$\frac{1}{\text{Molar mass of } M_2CO_3} = 0.01186$

∴ Molar mass of $M_2CO_3 = \frac{1}{0.01186}$

Molar mass = 84.3 g/mol

168. (d)
$$\underset{\substack{2\text{ mol}\\1\text{ mol}}}{2\,H_3AsO_4} + 5\,H_2S \xrightarrow{\text{Conc. HCl}} \underset{\substack{1\text{ mol}\\½\text{ mol}}}{As_2S_5} + 8\,H_2O$$

∵ The molar mass of H_3AsO_4 is $3 \times 1 + 79 + 4 \times 16 = 142$ g/mol

∴ Number of moles of $H_3AsO_4 = \frac{35.5}{142} = 0.25$ mol.

∴ Number of moles of $As_2S_5 = \frac{0.25}{2} = 0.125$ mol.

169. (a) Applying law of equivalence
Equivalent of acid = Equivalent of base
Equivalent of acid = Normality × volume = 0.1 × V

Another formula of equivalence = n factor × number of moles
∴ Equivalent of base = n factor of OH^- × moles of OH^-
$= 1 \times 0.04$
⇒ $0.1 \times V = 1 \times 0.04$
$V = 0.4\ L = 0.4 \times 1000 = 400\ mL$.

170. (d) Weight of hydrated $BaCl_2 = 61$ g
Weight of anhydrous $BaCl_2 = 52$ g
Loss in mass = 9 g
Assuming $BaCl_2 \cdot xH_2O$ as hydrate,
Mass of H_2O = 9 g

Moles of $H_2O = \frac{9}{18} = 0.5$ mol

Gram molecular mass of anhydrous $BaCl_2 = 208$ g/mol.

Moles of anhydrous $BaCl_2 = \frac{52}{208} = 0.25$

∵ 0.25 moles of $BaCl_2$ combines with $= 0.5$ moles of H_2O
∴ 1 mole of $BaCl_2$ combines with = 2 moles of H_2O
Therefore, $x = 2$

171. (a) $A + 2B + 3C \rightleftharpoons AB_2C_3$ .... (i)

No. of moles of A $= \frac{6.0\ g}{60\ g/mol} = 0.1$ mol

No. of moles of B $= \frac{6.00 \times 10^{23}}{6 \times 10^{23}} = 1$ mol

No. of moles of C = 0.036
Therefore, C is the limiting reagent.
The number of moles of product formed

$= \frac{0.036}{3} = 0.012$ mol

The expression for the molar mass is

$$\text{Molar mass} = \frac{\text{mass of product}}{\text{mole of product}}$$

$$60 + (2 \times x) + (3 \times 80) = \frac{4.8}{0.012}$$
$$x = 50 \text{ amu}$$

Hence, atomic mass of B is 50 amu.

172. (b) General combustion reaction for hydrocarbons is

$$C_xH_y + \left(x + \frac{y}{4}\right)O_2 \longrightarrow xCO_2 + \frac{y}{2}H_2O$$

For alkane, x = n and y = 2n

$$C_nH_{2n} + \left(n + \frac{2n}{4}\right)O_2 \longrightarrow nCO_2 + \frac{2n}{2}H_2O$$

$$C_nH_{2n} + \frac{3n}{2}O_2 \longrightarrow nCO_2 + nH_2O$$

∴ 1 mole alkene reacts with $\frac{3n}{2}$ mole of O_2
moles ∝ volume (at constant temp & spress.)
1 L alkene requires $\frac{3n}{2}$ L of O_2
6 L alkene requires 27 L of O_2

$$\frac{1}{6} = \frac{\frac{3n}{2}}{27} \Rightarrow 3n = \frac{54}{6} \Rightarrow n = 3$$

Hence alkene is propene (C_3H_6)

173. (a) Molality = Moles of solute / Mass of solvent in kg considering the density of water = 1g/mL

$$\text{Molality} = \frac{0.01/60}{0.3} = \frac{0.01}{60 \times 0.3}; = 5.55 \times 10^{-4}\ m$$

174. (d) Since molarity of solution is 3.60 M. It means 3.6 moles of H_2SO_4 is present in its 1 litre solution.
Mass of 3.6 moles of H_2SO_4 = Moles × Molecular mass
$= 3.6 \times 98\ g = 352.8$ g
∴ 1000 mL solution has 352.8 g of H_2SO_4
Given that 29 g of H_2SO_4 is present in = 100 g of solution
∴ 352.8 g of H_2SO_4 is present in

$= \frac{100}{29} \times 352.8$ g of solution = 1216 g of solution

$$\text{Density} = \frac{\text{Mass}}{\text{Volume}} = \frac{1216}{1000} = 1.216\ g/mL = 1.22\ g/mL$$

175. (a) $2Al(s) + 6HCl(aq) \rightarrow 2Al^{3+}(aq) + 6Cl^-(aq) + 3H_2(g)$
∵ 6 moles of HCl produces = 3 moles of H_2
$= 3 \times 22.4$ L of H_2 at S.T.P

∴ 1 mole of HCl produces $= \frac{3 \times 22.4}{6}$ L of H_2 at S.T.P
$= 11.2$ L of H_2 at STP

176. (a) Apply the formula $d = M\left(\frac{1}{m} + \frac{M_2}{1000}\right)$

m = molality
d = density
M_2 = molecular mass
M = molarity

$$\therefore 1.02 = 2.05\left(\frac{1}{m} + \frac{60}{1000}\right)$$

On solving we get, m = 2.288 mol/kg

177. (b) From the molarity equation
$$M_1V_1 + M_2V_2 = MV$$
Let M be the molarity of final mixture,

$$M = \frac{M_1V_1 + M_2V_2}{V} \text{ where } V = V_1 + V_2$$

$$M = \frac{480 \times 1.5 + 520 \times 1.2}{480 + 520} = 1.344\ M$$

178. (a) H_3PO_3 KOH
$N_1V_1 = N_2V_2$
Note: H_3PO_3 is dibasic,
∴ Normality = 2 × Molarity
$20 \times 0.2 = 0.1 \times V_2$
∴ $V_2 = 40$ mL

179. (a) $2BCl_3 + 3H_2 \rightarrow 2B + 6HCl$

or $BCl_3 + \frac{3}{2}H_2 \rightarrow B + 3HCl$

Now, since 10.8 g boron requires hydrogen
$= \frac{3}{2} \times 22.4$ L at S.T.P

Hence, 21.6 g boron requires hydrogen
$= \frac{3}{2} \times \frac{22.4}{10.8} \times 21.6 = 67.2$ L at S.T.P.

Structure of Atom

1. **(d)** $\lambda = \frac{c}{v} = \frac{3\times10^8}{5.16\times10^{14}} \Rightarrow \lambda = 581.39\text{ nm}$

A. λ_{Photon} is near & below yellow light therefore it can show photoelectric effect.
B. If intensity of light decreases photocurrent decreases.
C. Red light will not produce photoelectric effect.
D. $v_{Blue} > v_{yellow}$ so photoelectric current will be produced.
E. White light contain all frequencies so it will show photo electric current.

$\therefore$ Correct statements are A, B, D & E.

2. **(494)**

$E = hv = \frac{hc}{\lambda}$ For 1 atom

For 1 mol

$$E = \frac{N_A hc}{\lambda} = \frac{6.022\times10^{23}\times6.62\times10^{-34}\times3\times10^8}{242\times10^{-9}} = 494\text{kJ}$$

3. **(0)** A blackbody can emit and absorb all the wavelengths

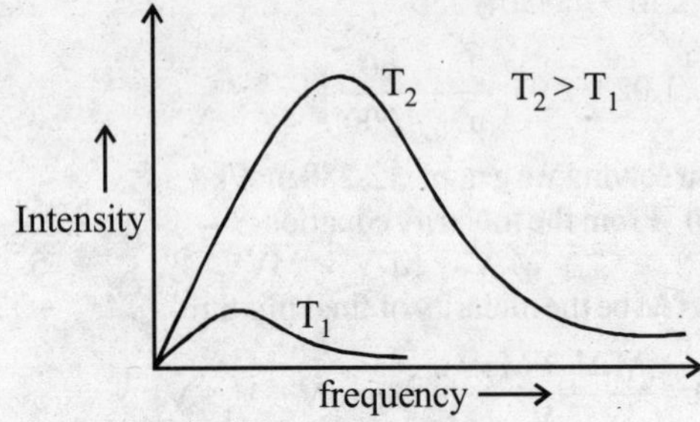

$\Rightarrow$ (A), (B), (C), (D) are correct.

4. **(2)** The spectrum of Black body radiation is explained using quantization of energy. With increase in temperature, peak of spectrum shifts to shorter wavelength or higher frequency. For above graph $\rightarrow T_1 > T_2 > T_3 > T_4$.

A black body does not have particles undergoing SHM.

$$\frac{T_1}{v_1} = \frac{T_2}{v_2} = \frac{T_3}{v_3} = \text{constant}$$

5. **(3)** $E = hv$

$$= \frac{6.62\times10^{-34}\times3\times10^8}{400\times10^{-9}} \times 6.24\times10^{18}\text{ CV}$$

$= 3.1$ eV.

$\Rightarrow$ Li, Na, K will show photoelectric effect under given condition.

6. **(a)** There is a characteristic minimum frequency, or threshold frequency, for each metal below which the photoelectric effect is not seen. The ejected electrons leave with a specific amunt of kinetic energy at a frequency $v > v_0$ with an increase in light frequency of these electron kinetic energies also rise.

7. **(d)** Both assertion and reason are correct but reason is not the correct explanation of assertion.

8. **(798)** For one photon $E = h\nu$

For one mole of photons,

$E = 6.023\times10^{23}\times6.626\times10^{-34}\times2\times10^{12}$

$= 798.16\text{ J} \approx 798\text{ J}$

9. **(c)** Energy of one mole of photons $= \frac{hc}{\lambda}\times N_A$

$$= \frac{6.63\times10^{-34}\times3\times10^8}{300\times10^{-9}}\times6.02\times10^{23}$$

$= 399.13\times10^3$ Joule/mole $= 399$ kJ/mole

10. **(2)** $\lambda = \frac{h}{\sqrt{2m(K.E.)}}$; $K.E. = \frac{h^2}{2m\lambda^2}$

$$K.E. = \frac{h^2}{2m\lambda^2} = \frac{43.9\times10^{-68}}{2\times9.1\times10^{-31}\times10.89\times10^{-20}}$$

$K.E. = 2.215\times10^{-18}$ J

$E_{abs} = E_{req} + K.E.$

Now, $E_{req} = -13.6\text{ eV} = -13.6\times1.602\times10^{-19}$ J

$$\frac{E_{abs}}{E_{req}} = 1 + \frac{K.E.}{E_{req}} = 1 + \frac{2.215\times10^{-18}}{13.6\times1.602\times10^{-19}} = 2.0166$$

11. **(300)** The minimum energy required to remove an electron from the metal surface is called work function.

$$E = hv = \frac{hc}{\lambda}$$

$$\phi = 6.63\times10^{-19}\text{ J} = \frac{hc}{\lambda} = \frac{6.63\times10^{-34}\times3\times10^8}{\lambda}$$

$\Rightarrow \lambda = 3\times10^{-7}\text{ m} = 300\text{ nm}$

12. **(c)** $W = hv = 6.6\times10^{-34}\times1.3\times10^{15} = 8.58\times10^{-19}$ J

13. **(2)** Total energy per sec. = 50 J

$$\because E = nhv = \frac{nhc}{\lambda}$$

$$50 = \frac{n\times6.63\times10^{-34}\times3\times10^8}{795\times10^{-9}};\ n \simeq 2\times10^{20}$$

14. **(50)** Power of infrared range = 1mW = 10^{-3}W or 10^{-3} Js^{-1}

$\therefore$ In 0.1 second the energy emitted = 10^{-4}J

$\because E = nh\nu = nh\frac{c}{\lambda} \Rightarrow$ then; $10^{-4} = n \times \frac{hc}{\lambda}$

$$\Rightarrow 10^{-4} = \frac{n \times 6.63 \times 10^{-34} \times 3 \times 10^8}{1000 \times 10^{-9}}$$

$\Rightarrow n = 5.02 \times 10^{14} = 50.2 \times 10^{13}$

15. **(5)** Einstein equation, $E = \phi + K.E_{max}$

$h\nu = h\nu_0 + \frac{1}{2}mv^2 \Rightarrow \frac{hc}{\lambda} = h\nu_0 + \frac{1}{2}mv^2$

$$\Rightarrow \frac{6.63 \times 10^{-34} \times 3 \times 10^8}{500 \times 10^{-9}}$$

$= 6.63 \times 10^{-34} \times 4.3 \times 10^{14} + \frac{1}{2}mv^2$

$$\Rightarrow \frac{6.63 \times 30 \times 10^{-20}}{5} = 6.63 \times 4.3 \times 10^{-20} + \frac{1}{2}mv^2$$

$\Rightarrow 11.271 \times 10^{-20} = \frac{1}{2} \times 9 \times 10^{-31} \times v^2$

$v = 5 \times 10^5$ ms^{-1}.

16. **(2)** Total energy provided by source per second

$= \frac{1000}{10} = 100$ J

Energy required to eject electron

$= \frac{hc}{\lambda} = \frac{6.626 \times 10^{-34}}{400 \times 10^{-9}} \times 3 \times 10^8$ J

Number of electrons ejected

$$= \frac{100}{\frac{6.626 \times 10^{-34} \times 3 \times 10^8}{400 \times 10^{-9}}} = \frac{400 \times 10^{-7} \times 10^{26}}{6.626 \times 3} = 2.01 \times 10^{20}$$

17. **(181)**

$E = \frac{hc}{\lambda} = \frac{6.63 \times 10^{-34} \times 3 \times 10^8}{663 \times 10^{-9}}$ J $= 3 \times 10^{-19}$ J / atom

Ionisation energy per mol

$= 3 \times 10^{-19} \times 10^{-3} \times 6.02 \times 10^{23}$ kJ mol^{-1}

$= 180.6$ kJ mol$^{-1} \approx 181$ kJ mol^{-1}

18. **(222)**

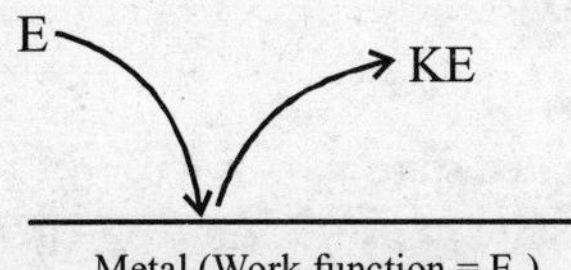

$E = E_0 + (KE)_{max}$

$\frac{hc}{\lambda} = 4.41 \times 10^{-19} + KE$

$$\frac{6.63 \times 10^{-34} \times 3 \times 10^8}{300 \times 10^{-9}} = 4.41 \times 10^{-19} + KE$$

So, $(KE)_{max} = 6.63 \times 10^{-19} - 4.41 \times 10^{-19}$

$= 2.22 \times 10^{-19}$ J $= 222 \times 10^{-21}$ J

19. **(d)** Remember the formula

$\boxed{p^2 = 2\,m\,K.E.}$ $\quad K.E. = \frac{hc}{\lambda}$

In photoelectric effect,

$\frac{hc}{\lambda} = w + KE$

Given that KE of ejected photoelectron is very high in comparison to work function w.

$\frac{hc}{\lambda} = KE$

$\frac{hc}{\lambda} = \frac{1}{2}mV^2\left(\frac{m}{m}\right) \Rightarrow \frac{hc}{\lambda} = \frac{1}{2} \cdot \frac{m^2V^2}{m}$

$\frac{hc}{\lambda} = \frac{p^2}{2m}$

New wavelength $\frac{hc}{\lambda_1} = \frac{(1.5p)^2}{2m} \Rightarrow \lambda_1 = \frac{4}{9}\lambda$

20. **(c)** $E = h\nu = \frac{hc}{\lambda}$

$$E = \frac{6.626 \times 10^{-34} \times 3 \times 10^8}{4000 \times 10^{-10}} = 4.97 \times 10^{-19} \text{ J}$$

$= \frac{4.97 \times 10^{-19}}{1.6 \times 10^{-19}} = 3.1$ eV

$KE = \frac{1}{2}mv^2 = \frac{1}{2} \times 9 \times 10^{-31}$ kg $\times (6 \times 10^5$ ms$^{-1})^2$

$= 1.62 \times 10^{-19}$ J = 1 eV

According to photoelectric effect,

$K.E. = h\nu - h\nu_0$

$h\nu_0 = h\nu - K.E.$

Work function $(W_0) = E - K.E. = 3.1 - 1 = 2.1$ eV

21. **(b, c)** After ejection of e$^-$, further increase in frequency also increases K.E. Thus, graph (b) is incorrect not all ν release e$^-$, thus (c) is incorrect.

$K.E. = h\nu - h\nu_0$

where, ν = Frequency of incident radiation

ν_0 = Threshold frequency

KE is independent of intensity but it depends on frequency of light. Intensity is directly proportional to the no. of electrons emitted.

22. **(c)** $\lambda = 250$ nm

$$E = \frac{hc}{\lambda} = \frac{6.626 \times 10^{-34} \times 3 \times 10^8}{250 \times 10^{-9}} / 1.60 \times 10^{-19} = 4.96 \text{ eV}$$

KE = stopping potential = 0.5 eV

$E = W_0 + K.E.;\ 4.96 = W_0 + 0.5$

$W_0 = 4.46 \approx 4.5$ eV

23. **(c)** $E = h\nu = 6.63 \times 10^{-34} \times 2.47 \times 10^{15}$

$= 1.640 \times 10^{-18}$ J

24. **(d)** Energy = $N_A h\nu$

$495.5 \times 10^3 = 6.023 \times 10^{23} \times 6.6 \times 10^{-34} \times \nu$

$$\nu = \frac{495.5 \times 10^3}{6.023 \times 10^{23} \times 6.6 \times 10^{-34}} = 12.4 \times 10^{14} s^{-1}$$

$= 1.24 \times 10^{15}\ s^{-1}$.

25. **(c)** The kinetic energy of the ejected electron is given by the equation

$$h\nu = h\nu_0 + \frac{1}{2}mv^2 \quad \because \nu = \frac{c}{\lambda} \text{ or } \frac{hc}{\lambda} = \frac{hc}{\lambda_0} + \frac{1}{2}mv^2$$

$$\frac{1}{2}mv^2 = \frac{hc}{\lambda} - \frac{hc}{\lambda_0} = hc\left(\frac{\lambda_0 - \lambda}{\lambda\lambda_0}\right)$$

$$\therefore \quad v^2 = \frac{2hc}{m}\left(\frac{\lambda_0 - \lambda}{\lambda\lambda_0}\right) \text{ or } v = \sqrt{\frac{2hc}{m}\left(\frac{\lambda_0 - \lambda}{\lambda\lambda_0}\right)}$$

26. **(d)** Energy required to break single Cl – Cl bond

$$= \frac{242 \times 10^3}{6.023 \times 10^{23}} = \frac{hc}{\lambda} = \frac{6.626 \times 10^{-34} \times 3 \times 10^8}{\lambda}$$

$$\therefore \quad \lambda = \frac{6.626 \times 10^{-34} \times 3 \times 10^8 \times 6.023 \times 10^{23}}{242 \times 10^3}$$

$= 0.4947 \times 10^{-6}$ m $= 494.7$ nm

27. **(54)**

E_1 of H-atom = – 13.6 eV

$$E_2 \text{ of } Be^{3+} = \frac{-13.6 \times Z^2}{n^2}$$

$$= \frac{-13.6 \times (4)^2}{(2)^2} = -54.4 \text{ eV}$$

$|E_2|$ of Be^{3+} = 54 eV

28. **(d)** The statement given in option (d) is not in agreement with quantum mechanical model of an atom.

29. **(c)** $r \propto n^2 \Rightarrow \frac{r'}{r} = \frac{n'^2}{n^2}$

Thus, $\frac{r_3}{r_1} = \frac{9}{1} \Rightarrow \frac{r_8}{r_4} = \frac{64}{16} = 4$

$\frac{r_6}{r_4} = \left(\frac{6}{4}\right)^2 = \frac{9}{4} \Rightarrow \frac{r_4}{r_2} = \left(\frac{4}{2}\right)^2 = 4$

30. **(b)** The value of energy becomes more and more negative with increase of Z. They do not follow a linear relationship. Hence (b) is the correct option.

31. **(a)** $\lambda = 900$ nm

$= 900 \times 10^{-7}$ cm $= 9 \times 10^{-5}$ cm

$R_H = 10^5$ cm^{-1}; For H-atom (Z = 1)

According to Rydberg eq.

$$\frac{1}{\lambda} = R_H Z^2 \times \left(\frac{1}{n_1^2} - \frac{1}{n_2^2}\right)$$

$$\Rightarrow \frac{1}{\lambda \times R_H} = \frac{1}{n_1^2} - \frac{1}{n_2^2}$$

$$\Rightarrow \frac{1}{9 \times 10^{-5} \text{ cm} \times 10^5 \text{ cm}^{-1}} = \left(\frac{1}{n_1^2} - \frac{1}{n_2^2}\right)$$

$$\Rightarrow \frac{1}{n_1^2} - \frac{1}{n_2^2} = \frac{1}{9}$$

It is possible when $n_1 = 3$, $n_2 = \infty$

Possible series : $\infty \rightarrow 3$

32. **(d)** $r = a_0 \frac{n^2}{Z} = a_0 . \frac{(2)^2}{2} = 2a_0$.

33. **(5)** Given : $\frac{\lambda}{4} = 1.5$ pm

$\Rightarrow \lambda = 1.5 \times 4$ pm

$= 6 \times 10^{-12}$ m

$\lambda\nu = C$

$6 \times 10^{-12} \times \nu = 3 \times 10^8$

$\Rightarrow \quad \nu = 5 \times 10^{19}$ Hz

34. **(1724)**

$$\text{Wave number} = \frac{1}{\lambda} = \frac{1}{5800 \times 10^{-8} \text{ cm}} = 17241 \text{ cm}^{-1}$$

1724.1×10 cm$^{-1} \Rightarrow x = 1724$.

35. **(34)** KE = – Total energy

$= -(-3.4)$

$= 3.4$

$= 34 \times 10^{-1}$

$x = 34$.

36. **(22)** From Bohr's theory we have

$$V_n = 2.18 \times 10^6 \times \frac{Z}{n} \text{ ms}^{-1}$$

$$= 21.8 \times 10^5 \times \frac{1}{1} \approx 22 \times 10^5 \text{ (nearest)}$$

37. **(8)** On considering wave the nature of electron.

$2\pi r_n = n\lambda_d$

$$2\pi a_0 \frac{n^2}{Z} = n\lambda_d \qquad \left\{r_n = a_0 \frac{n^2}{Z}\right\}$$

$$2\pi a_0 \frac{4^2}{1} = 4\lambda_d$$

$\lambda_d = 8\pi a_0$

38. **(10)** 5th excited state $\Rightarrow n_1 = 6$

1st excited state $\Rightarrow n_2 = 2$, $\Delta n = n_1 - n_2 = 6 - 2 = 4$

$\Rightarrow$ Maximum number of spectral lines

$$= \frac{\Delta n(\Delta n + 1)}{2} = \frac{4(4+1)}{2} = 10$$

39. **(c)** Lyman (UV) ⟶ Balmer (Visible) ⟶ Paschen (IR) ⟶ Brackett (IR) ⟶ Pfund (IR)

Thus, A – II, B – IV, C – III, D – I.

40. **(a)** According to Bohr's model the angular momentum is quantised and equal to $\frac{nh}{2\pi}$.

Heisenberg uncertainty principle explains orbital concept, which is based on probability of finding electron.

41. **(d)** $E_n = \frac{-2.18\times10^{-18} Z^2}{n^2}$

i.e. $E_n \propto \frac{1}{n^2} \Rightarrow$ Energy in the 3rd Bohr orbit is $\frac{1}{9}$ th of energy in first Bohr orbit.

42. **(1)** $\Delta E = R_H Z^2\left(\frac{1}{n_1^2}-\frac{1}{n_2^2}\right)$ $(n_1 < n_2)$

$1.47\times10^{-17} = 2.18\times10^{-18}\times9\left(\frac{1}{n^2}-\frac{1}{(n+1)^2}\right)$

$\frac{1.47}{1.96} = \frac{3}{4} = \frac{1}{n^2} - \frac{1}{(n+1)^2}$

So, $n = 1$

43. **(1)** Statement (B) is incorrect as the emission spectra of atoms in the gas phase does not show a continuous spread of wavelength.

44. **(a)** $\frac{1}{\lambda_H} = R(1)^2\left[\frac{1}{n_1^2}-\frac{1}{n_2^2}\right]$ $(n_1 < n_2)$

$\frac{1}{\lambda_{He^+}} = R(2)^2\left[\frac{1}{2^2}-\frac{1}{4^2}\right]$

Given $\lambda_H = \lambda_{He+}$

$R(1)^2\left[\frac{1}{n_1^2}-\frac{1}{n_2^2}\right] = R(4)\left[\frac{1}{2^2}-\frac{1}{4^2}\right]$

$\Rightarrow \frac{1}{n_1^2}-\frac{1}{n_2^2} = \frac{1}{1^2}-\frac{1}{2^2}$

On comparing $n_1 = 1$ and $n_2 = 2$ or $n_i = 2$, and $n_f = 1$

45. **(b)** For H : $\frac{1}{\lambda} = R_H\times1^2\left(\frac{1}{1^2}-\frac{1}{\infty^2}\right)$ ...(i)

$\frac{1}{\lambda_{He^+}} = R_H\times2^2\times\left(\frac{1}{2^2}-\frac{1}{3^2}\right)$ ...(ii)

From equation (i) & (ii) $\frac{\lambda_{He^+}}{\lambda} = \frac{9}{5}$

$\lambda_{He^+} = \lambda\times\frac{9}{5} \Rightarrow \lambda_{He^+} = \frac{9\lambda}{5}$

46. **(270)** $r \propto \frac{n^2}{Z}$ $\therefore r_{He^+} = r_H\times\frac{n^2}{Z}$

$r_{He^+} = 0.6\times\frac{(3)^2}{2} = 2.7\text{ Å} = 270\text{ pm}$

47. **(c)** $r_n = \frac{52.9n^2}{Z}$ pm

$Li^{2+}: r_2 = x = k\times\frac{2^2}{3} = \frac{4k}{2}$

$Be^{3+}: r_3 = y = k\times\frac{3^2}{4}$

$\frac{y}{x} = \frac{9}{4}\times\frac{3}{4} = \frac{27}{16} \Rightarrow y = \frac{27}{16}x$

48. **(492)** For first line of Paschen series

$\frac{1}{(\lambda_1)_p} = R_H Z^2\left(\frac{1}{9}-\frac{1}{16}\right)$

For second line of Paschen series

$\frac{1}{(\lambda_2)_p} = R_H Z^2\left(\frac{1}{9}-\frac{1}{25}\right)$

$\frac{(\lambda_2)_p}{(\lambda_1)_p} = \frac{\frac{7}{16\times9}}{\frac{16}{25\times9}} = \frac{25\times7}{16\times16}$

$(\lambda_2)_P = \frac{25\times7}{16\times16}\times720$

$(\lambda_2)_P = 492$ nm

49. **(10)** Maximum number of spectral lines

$= \frac{\Delta n(\Delta n+1)}{2} = \frac{5\times4}{2} = 10$

50. **(b)** $r = 0.529\times\frac{n^2}{Z}$ Å

$r_3 = 0.529\times\frac{3^2}{1}$; $r_4 = 0.529\times\frac{4^2}{1}$

$\frac{r_4}{r_3} = \frac{4^2}{3^2} = \frac{16}{9} \Rightarrow r_4 = \frac{16r_3}{9}$

51. **(N)** We can not calculate I.E. of lithium atom using Bohr's model because it is not a single e^- species.

52. **(3155)** $mvr = \frac{nh}{2\pi}$

K.E. $= \frac{1}{2}mv^2 \Rightarrow$ K.E. $= \frac{n^2h^2}{8\pi^2mr^2}$

For $n = 2, r = 4a_0$

$$\text{K.E.} = \frac{4h^2}{8\pi^2 m(4a_0)^2} = \left(\frac{4}{8\pi^2 \times 16}\right)\frac{h^2}{ma_0^2}$$

Comparing with given value,

$$\left(\frac{4}{8\pi^2 \times 16}\right) = \frac{1}{x} \quad \Rightarrow x = 315.507$$

$\Rightarrow 10x = 3155$ (nearest integer)

53. **(c)** According to Bohr's model :

$$v_n \propto \frac{Z}{n}$$

Z = Atomic no. (corresponds to +ve charge)
n = Principal quantum no.
This relation suggest that when the Z increases, the velocity will also increase, so statement I is wrong and as 'n' decreases, velocity will also increase. So statement II is correct.

54. **(d)** According to Bohr's theory :

(A) $\text{K.E.} = \frac{-Z^2e^4m}{8\varepsilon_0^2 n^2 h^2} \Rightarrow KE \propto \frac{Z^2}{n^2}$

(B) Velocity of electron, $v = \frac{Z}{n}\left(\frac{e^2}{2\varepsilon_0 h}\right)$

(C) $E = h\nu \Rightarrow E = -\frac{Z^2e^4m}{8\varepsilon_0^2 n^2 h^2} \quad \therefore \nu \propto \frac{Z^2}{n^2}$.

(D) $F = \frac{Ze^2}{4\pi\varepsilon_0 r^2}$ where $r = \frac{\varepsilon_0 n^2 h^2}{\pi me^2 Z}$

$$\therefore F = \frac{Z^3e^6m^2\pi}{4\varepsilon_0^3 n^4 h^4} \Rightarrow F \propto \frac{Z^3}{n^4}$$

55. **(c)** $r = 0.529\frac{n^2}{Z}$ Å

For Li^{2+},

$$(r_{Li^{2+}})_{n=4} - (r_{Li^{2+}})_{n=3} = \frac{0.529}{3}[4^2 - 3^2] = \Delta R_1$$

For He^+,

$$(r_{He^+})_{n=4} - (r_{He^+})_{n=3} = \frac{0.529}{2}[4^2 - 3^2] = \Delta R_2$$

$$\frac{\Delta R_1}{\Delta R_2} = \frac{2}{3}$$

56. **(a)** In hydrogen spectrum, maximum lines of Balmer series lies in visible region.

57. **(c)** Shortest wavelength $\rightarrow$ Max. energy ($\infty \rightarrow 1$)
For Lyman series of H atom,

$$\frac{1}{\lambda_1} = R_H(1)^2\left[\frac{1}{1} - 0\right] \Rightarrow \frac{1}{\lambda_1} = R_H \Rightarrow R_H = \frac{1}{\lambda_1}$$

For Balmer series of He^+,

$$\frac{1}{\lambda} = R_H(2)^2\left[\frac{1}{2^2} - \frac{1}{3^2}\right] \Rightarrow \frac{1}{\lambda} = R_H(4)\left(\frac{9-4}{36}\right)$$

$$\Rightarrow \frac{1}{\lambda} = \frac{5R_H}{9} \Rightarrow \lambda = \frac{9}{5R_H} = \frac{9\lambda_1}{5}$$

58. **(c)** $r = \frac{a_0 n^2}{Z}$

For Li^{2+}, $r = \frac{a_0(2)^2}{3} = \frac{4a_0}{3}$

59. **(b)** In the Balmer series of H–atom, the transition takes place from the higher orbital to $n = 2$. Therefore, the longest wavelength corresponds to $n_1 = 2$ and $n_2 = 3$. As wavelength decreases, energy increases or we can say that n_2 increases. The energy difference between two consecutive lines decreases or the lines slowly converge. Hence, statement I, II, III are the correct statements among the given options.

60. **(c)** $r = \frac{a_0 n^2}{Z}$

For Li^{2+}, $r = \frac{a_0(2)^2}{3} = \frac{4a_0}{3}$

61. **(a)** For determined shortest wavelength, n_2 = ¥

For Lyman series, $\bar{\nu}_L = \frac{1}{\lambda_L} = R\left[\frac{1}{(1)^2} - \frac{1}{\infty^2}\right]$

For Paschen series, $\bar{\nu}_P = \frac{1}{\lambda_P} = R\left[\frac{1}{(3)^2} - \frac{1}{\infty^2}\right]$

$$\frac{\bar{\nu}_L}{\bar{\nu}_P} = \frac{\lambda_P}{\lambda_L} = 9$$

62. **(b)** For H-atom

$$\bar{\nu} = R_H\left[\frac{1}{n_1^2} - \frac{1}{n_2^2}\right]$$

For Lyman series, $n_1 = 1$.

$\bar{\nu}(\max) = R_H\left(1 - \frac{1}{\infty}\right)$; $\bar{\nu}(\min) = R_H\left(1 - \frac{1}{4}\right)$ $(n_2 = 2)$

$$\therefore \bar{\nu}_{max} - \bar{\nu}_{min} = R_H\left(\frac{1}{4}\right)$$

For Balmer series, $n_1 = 2$.

$\bar{\nu}(\max) = R_H\left(\frac{1}{4} - \frac{1}{\infty}\right)$; $\bar{\nu}(\min) = R_H\left(\frac{1}{4} - \frac{1}{9}\right)$ $(n_2 = 3)$

$\therefore \quad \bar{v}_{max} - \bar{v}_{min} = R_H\left(\frac{1}{9}\right)$

So, $\frac{\Delta\bar{v}_{Lyman}}{\Delta\bar{v}_{Balmer}} = \frac{9}{4}$

63. **(d)** The total energy of the electron is minimum at a distance of a_0 from the nucleus for 1s orbital.

64. **(a)** $\frac{1}{\lambda} = R_H\left(\frac{1}{n_1^2} - \frac{1}{n_2^2}\right)$

$n_1 = 3, n_2 = \infty$

$\frac{1}{\lambda} = R_H\left(\frac{1}{9}\right) \Rightarrow \lambda = \frac{9}{R_H} = \frac{9}{10^5} = 9\times10^{-5}\,\text{cm} = 900\,\text{nm}$

65. **(c)** According to Bohr's model energy in n^{th} state

$= -13.6\times\frac{Z^2}{n^2}\,\text{eV}$

For second excited state, of He^+, $n = 3$

$\therefore \quad E_3(He^+) = -13.6\times\frac{2^2}{3^2}\,\text{eV} = -6.04\,\text{eV}$

66. **(c)** For emission spectrum,

$\bar{v} = R_H\left(\frac{1}{n_f^2} - \frac{1}{n_i^2}\right) \Rightarrow \bar{v} = R_H\left(\frac{1}{n^2} - \frac{1}{8^2}\right)$

$\Rightarrow \quad \bar{v} = \frac{R_H}{n^2} - \frac{R_H}{64}$

Comparing to $y = mx + c$, we get

$x = \frac{1}{n^2}$ and $m = +R_H$ (slope)

67. **(b)** When temperature is increased, black body emits high energy radiation from higher wavelength to lower wavelength.

68. **(b)** $r = 0.529\times\frac{n^2}{Z}\,\text{Å}$

$r = 211.6\,\text{pm} = 2.11\,\text{Å} \Rightarrow 0.529\times\frac{n^2}{1} = 2.11\,\text{Å}\ [Z = 1]$

$n = 2$ (Balmer series)

69. **(d)** For Lyman series (shortest wavelength)

$n_1 = 1, n_2 = \infty$

$\frac{1}{\lambda} = RZ^2\left(\frac{1}{n_1^2} - \frac{1}{n_2^2}\right)$

$\Rightarrow \frac{1}{A} = 1^2 R\left(\frac{1}{1} - \frac{1}{\infty}\right) \Rightarrow \frac{1}{A} = R$

Longest wavelength = 1st line

$n_1 = 3, n_2 = 4$

$\frac{1}{\lambda} = RZ^2\left(\frac{1}{3^2} - \frac{1}{4^2}\right) \Rightarrow \frac{1}{\lambda} = \frac{R7}{36}$

$R = \frac{1}{A}$

$\frac{1}{\lambda} = \frac{\frac{1}{A}\times7}{36} \Rightarrow \frac{1}{\lambda} = \frac{7}{36A} \Rightarrow \lambda = \frac{36A}{7}$

70. **(d)** Radius of nth Bohr orbit in H-atom = $0.53\,n^2$Å

Radius of II Bohr orbit = $0.53\times(2)^2 = 2.12$ Å

71. **(d)** For Lyman series (shortest wavelength)

$n_1 = 1, n_2 = \infty$

$\frac{1}{\lambda} = RZ^2\left(\frac{1}{n_1^2} - \frac{1}{n_2^2}\right)$

$\Rightarrow \frac{1}{A} = 1^2 R\left(\frac{1}{1} - \frac{1}{\infty}\right) \Rightarrow \frac{1}{A} = R$

Longest wavelength = 1st line

$n_1 = 3, n_2 = 4$

$\frac{1}{\lambda} = RZ^2\left(\frac{1}{3^2} - \frac{1}{4^2}\right) \Rightarrow \frac{1}{\lambda} = \frac{R7}{36}$

$R = \frac{1}{A}$

$\frac{1}{\lambda} = \frac{\frac{1}{A}\times7}{36} \Rightarrow \frac{1}{\lambda} = \frac{7}{36A} \Rightarrow \lambda = \frac{36A}{7}$

72. **(a)** Total energy $= -\frac{13.6}{n^2}Z^2\,\text{eV}$; where $n = 2, 3, 4 \ldots$

Putting $n = 2$

$E_T = -\frac{13.6}{4} = -3.4\,\text{eV}$

73. **(c)** For Li^{2+} ion,

$E = -13.6\times\frac{Z^2}{n^2}\,\text{eV} = -13.6\times\frac{(3)^2}{(2)^2}$

$= \frac{-13.6\times9}{4} = -30.6\,\text{eV}$

74. **(a)** $\Delta E = -2.0\times10^{-18}\times\left(\frac{1}{2^2} - \frac{1}{1^2}\right)$

$= -2.0\times10^{-18}\times\frac{-3}{4} = 1.5\times10^{-18}\,\text{J}$

$\Delta E = \frac{hc}{\lambda}$

$\lambda = \frac{hc}{\Delta E} = \frac{6.625\times10^{-34}\,\text{Js}\times3\times10^8\,\text{ms}^{-1}}{1.5\times10^{-18}\,\text{J}}$

$= 1.325\times10^{-7}\,\text{m}$

75. **(d)** Total energy of a revolving electron is the sum of its kinetic and potential energy.

Total energy = K.E. + P. E.

$$= \frac{Ke^2}{2r} + \left(-\frac{Ke^2}{r}\right) = -\frac{Ke^2}{2r}$$

76. **(a)** $\Delta E = 2.178 \times 10^{-18}\left(\frac{1}{1^2} - \frac{1}{2^2}\right) = \frac{hc}{\lambda}$

$$\Rightarrow 2.178 \times 10^{-18} \times \frac{3}{4} = \frac{hc}{\lambda}$$

$$= \frac{6.62 \times 10^{-34} \times 3 \times 10^8}{\lambda}$$

$$\lambda = \frac{6.62 \times 10^{-34} \times 3 \times 10^8 \times 4}{2.178 \times 10^{-18} \times 3}$$

$= 1.214 \times 10^{-7}$m

77. **(a)** $\bar{v} = RZ^2\left(\frac{1}{2^2} - \frac{1}{3^2}\right) = R\left(\frac{1}{4} - \frac{1}{9}\right) = \frac{5R}{36}$

78. **(a)** $\Delta E = 2.178 \times 10^{-18}\left(\frac{1}{1^2} - \frac{1}{2^2}\right) = \frac{hc}{\lambda}$

$$\Rightarrow 2.178 \times 10^{-18} \times \frac{3}{4} = \frac{hc}{\lambda}$$

$$= \frac{6.62 \times 10^{-34} \times 3 \times 10^8}{\lambda}$$

$$\lambda = \frac{6.62 \times 10^{-34} \times 3 \times 10^8 \times 4}{2.178 \times 10^{-18} \times 3}$$

$= 1.214 \times 10^{-7}$m

79. **(a)** $\nu = R\left(\frac{1}{n_1^2} - \frac{1}{n_2^2}\right)$Hz

$$\nu = 3.29 \times 10^{15}\left(\frac{1}{2^2} - \frac{1}{\infty^2}\right) = 8.22 \times 10^{14}\,s^{-1}$$

80. **(a)** For He^+,

$$v = RZ^2\left(\frac{1}{2^2} - \frac{1}{4^2}\right)\text{Hz}$$

For H,

$$v = R\left(\frac{1}{n_1^2} - \frac{1}{n_2^2}\right)\text{Hz}$$

For same frequency,

$$Z^2\left(\frac{1}{2^2} - \frac{1}{4^2}\right) = \left(\frac{1}{n_1^2} - \frac{1}{n_2^2}\right)$$

Since, $Z = 2$

$$\therefore \quad \frac{1}{n_1^2} - \frac{1}{n_2^2} = \frac{1}{1^2} - \frac{1}{2^2}$$

$\therefore \quad n_1 = 1$ & $n_2 = 2$

81. **(b)** $\text{I.E} = \frac{Z^2}{n^2} \times 13.6\,\text{eV}$ ...(i)

or $\frac{I_1}{I_2} = \frac{Z_1^2}{n_1^2} \times \frac{n_2^2}{Z_2^2}$ ...(ii)

Given $I_1 = -19.6 \times 10^{-18}$ J/atom, $Z_1 = 2$, $n_1 = 1$, $Z_2 = 3$ and $n_2 = 1$

Substituting these values in equation (ii).

$$-\frac{19.6 \times 10^{-18}}{I_2} = \frac{4}{1} \times \frac{1}{9}$$

or $I_2 = -19.6 \times 10^{-18} \times \frac{9}{4}$

$= -4.41 \times 10^{-17}$ J/atom

82. **(d)** The energy required to excite an electron in an atom of hydrogen from $n = 1$ to $n = 2$ is ΔE (difference in energy E_2 and E_1).

Values of E_2 and E_1 are,

$$E_2 = \frac{-1.312 \times 10^6 \times (1)^2}{(2)^2} = -3.28 \times 10^5\ \text{J mol}^{-1}$$

$E_1 = -1.312 \times 10^6$ J mol^{-1}

$\therefore \quad \Delta E = E_2 - E_1 = [-3.28 \times 10^5] - [-1.312 \times 10^6]$ J mol^{-1}

$= (-3.28 \times 10^5 + 1.312 \times 10^6)$ J mol^{-1}

$= 9.84 \times 10^5$ J mol^{-1}

83. **(b)** Angular momentum of an electron in n^{th} orbital is given by, $mvr = \frac{nh}{2\pi}$

For $n = 5$, we have

Angular momentum of electron $= \frac{5h}{2\pi} = \frac{2.5h}{\pi}$

84. **(c)** $\frac{1}{\lambda} = R\left(\frac{1}{n_1^2} - \frac{1}{n_2^2}\right)$

$$\frac{1}{\lambda} = 1.097 \times 10^7\left(\frac{1}{1} - \frac{1}{\infty}\right) = 1.097 \times 10^7$$

$\lambda = 91.15 \times 10^{-9}\,\text{m} \approx 91$ nm

85. **(a)** The lines falling in the visible region comprise Balmer series. Hence, the third line from red ($n_2 = 3$) would be $n_1 = 2$, $n_2 = 5$, i.e. $5 \to 2$.

86. **(a)** 2nd excited state will be the 3rd energy level

$$E_n = \frac{13.6}{n^2}\text{eV}$$

or $E_3 = \frac{13.6}{9}\text{eV} = 1.51\text{eV}.$

87. **(a)** 2nd excited state will be the 3rd energy level

$$E_n = \frac{13.6}{n^2}\text{eV}$$

or $E_3 = \frac{13.6}{9}\text{eV} = 1.51\text{eV}.$

88. (b) $\Delta x.\Delta p \geq \frac{h}{4\pi} \Rightarrow \Delta p^2 \geq \frac{h}{4\pi}$

$\Rightarrow m\Delta V \geq \sqrt{\frac{h}{4\pi}} \Rightarrow \Delta V \geq \sqrt{\frac{h}{4\pi} \times \frac{1}{m}}$

$\Rightarrow \Delta V \geq \sqrt{\frac{h}{4\pi} \times \frac{1}{m}}$

$\Rightarrow \Delta V \geq \sqrt{\frac{h}{\pi} \times \frac{1}{2m}}$

89. (b) $2\pi r_n = n\lambda$, $r_n = n^2 a_0 = (2)^2 a_0 - 4a_0$

$2\pi(4a_0) = n\lambda = \lambda = \frac{8\pi a_0}{n}$

90. (58) $m\Delta V.\Delta x = \frac{h}{4\pi} \Rightarrow \Delta V = \frac{h}{m\Delta x\, 4\pi}$

$= \frac{6.626 \times 10^{-34}}{9.1 \times 10^{-31} \times 10^{-15} \times 4 \times 3.14}$

$= 57.97 \times 10^9$ m/sec.

91. (661) $\lambda = \frac{h \times V}{mv \times V}$

$\lambda = \frac{hV}{mV^2}$

$\Rightarrow \frac{mV^2}{h} = \frac{V}{\lambda} = v$

$mV^2 = 2R_H = 2 \times 2.18 \times 10^{-18}$ J

$= 4.36 \times 10^{-18}$ J

$h = 6.6 \times 10^{-34}$ J.s (given)

$\Rightarrow v = \frac{4.36 \times 10^{-18}}{6.6 \times 10^{-34}} = 660.60 \times 10^{13}$ Hz

$\approx 661 \times 10^{13}$ Hz

92. (a) d-Broglie equation:

$\lambda = \frac{h}{p} \Rightarrow \lambda \propto \frac{1}{p}$

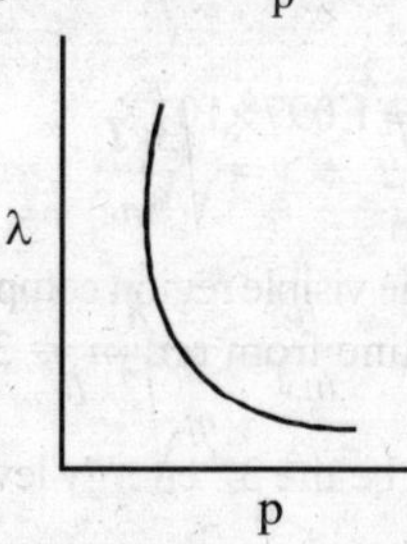

93. (7)

$\lambda_d = \frac{h}{mv} = \frac{h}{\sqrt{2mKE}} = \frac{6.6 \times 10^{-34}}{\sqrt{2 \times 9 \times 10^{-31} \times 4.5 \times 10^{-29}}}$

$= \frac{6.6 \times 10^{-34}}{\sqrt{9^2 \times 10^{-60}}}$

$= \frac{6.6 \times 10^{-34}}{9 \times 10^{-30}} = \frac{6.6}{9} \times 10^{-4} = 7.3 \times 10^{-5}$ m $\sim 7 \times 10^{-5}$ m.

94. (c) $(r_3)_H = \frac{a_0 n^2}{Z} = a_0 \times 3^2 = 9a_0$

$2\pi r = n\lambda \Rightarrow 2\pi \times 9a_0 = 3\lambda \Rightarrow \lambda = 6\pi a_0$

95. (2) de Broglie equation

$p = \frac{h}{\lambda}$ or $\lambda = \frac{h}{mv} = \frac{6 \times 10^{-34}}{9 \times 10^{-31} \times 10^3}$ $[\because v_e = 1000 \text{ m/s}]$

$= \frac{2}{3} \times 10^{-6}$ m $= 0.6667$ μm $= 666.7$ nm

The characteristic of electrons emitted do not depend on the nature of the material of electrodes.

The cathode ray starts from cathode and move towards anode.

The nature of the emitted electrons is independent on the nature of the gas present in cathode ray tube.

96. (548) $\Delta x \times \Delta p_x \geq \frac{h}{4\pi}$

$\Rightarrow 2a_0 \times m\Delta v_x = \frac{h}{4\pi}$ (for minimum uncertainty)

$\Rightarrow \Delta v_x = \frac{h}{4\pi} \times \frac{1}{2a_0} \times \frac{1}{m}$

$= \frac{6.63 \times 10^{-34}}{4 \times 3.14 \times 2 \times 52.9 \times 10^{-12} \times 9.1 \times 10^{-31}}$

$= 548273$ ms^{-1} $= 548.273$ km s^{-1}

97. (1758) $\lambda_e = \lambda_n \Rightarrow \frac{h}{p_e} = \frac{h}{p_n} \Rightarrow p_e = p_n$

$\Rightarrow m_e v_e = m_p v_n = (9.1 \times 10^{-31}) v_e = (1.6 \times 10^{-27}) v_n$

$\Rightarrow v_e = \frac{1.6 \times 10^{-27}}{9.1 \times 10^{-31}} \times v_n \Rightarrow 1758.2 \times v_n$

98. (22) $\Delta v = 2.4 \times 10^{-26}$ ms^{-1}

$\Delta x = 10^{-7}$ m

$\therefore \Delta p \cdot \Delta x = \frac{h}{4\pi} \therefore m\Delta v \cdot \Delta x = \frac{h}{4\pi}$

$\Rightarrow m \times 2.4 \times 10^{-26} \times 10^{-7} = \frac{6.626 \times 10^{-34}}{4 \times \pi}$

$\Rightarrow m = 0.02198$ kg $\Rightarrow m = 21.98$ g ≈ 22g.

99. (58) $\Delta x \cdot \Delta p = \frac{h}{4\pi}$

$\Delta x\, m\Delta v = \frac{h}{4\pi}$

$\Delta v = 5 \times 10^6 \times \frac{0.02}{100}$

$\Delta v = 10^3$ m/s

$\Delta x = \frac{h}{4\pi \times m \cdot \Delta v} = \frac{6.63 \times 10^{-34}}{4 \times 3.14 \times 9.1 \times 10^{-31} \times 10^3}$

$= 5.8 \times 10^{-8}$ m $= 58 \times 10^{-9}$ m

100. (6) de-Broglie wave-length of electron :

$\lambda_e = \frac{h}{\sqrt{2m(KE)}}$ $\left\{ \begin{array}{l} \because e^- \text{ is accelerated} \\ \text{from rest} \\ \Rightarrow KE = q \times V \end{array} \right\}$

$$\lambda = \frac{h}{\sqrt{2mqV}} = \frac{6.63\times10^{-34}}{\sqrt{2\times1.6\times10^{-19}\times9.1\times10^{-31}\times40\times10^{3}}}$$

$= 0.614\times10^{-11}\,m = 6.14\times10^{-12}\,m$

101. (9) Incident energy of photon
= Work function of metal + K.E of photoelectron

$h\nu = h\nu_0 + KE$ ...(i)

Incident energy of photon

$$= \frac{hc}{\lambda} = \frac{6.63\times10^{-34}\times3\times10^{8}}{248\times10^{-9}\times1.6\times10^{-19}}\ eV = 5\ eV$$

Now using equation (i),
$5 = 3 + KE$
$KE = 2.0\,eV$

$$\text{Now, } \lambda = \frac{h}{\sqrt{2m\ KE}} = \frac{6.63\times10^{-34}}{\sqrt{2\times9.1\times10^{-31}\times2\times1.6\times10^{-19}}}$$

$= 8.68\times10^{-10}\ m \approx 9\ Å$

102. (1) According to Heisenberg uncertainty principle,

$$\Delta x\cdot\Delta p \ge \frac{h}{4\pi}$$

$$\Delta x = \frac{h}{4\pi m\Delta v} \qquad (\because \Delta p = m\Delta v)$$

$$\Delta v = 90\times\frac{5}{100} = 4.5\ m/s$$

$$\Delta x = \frac{6.63\times10^{-34}\ Js}{4\times3.14\times10\times10^{-3}\ kg\times4.5\,m/s}\left[1J = \frac{kg.m^2}{s^2}\right]$$

$= 1.173\times10^{-33}\,m = 1\times10^{-33}\,m$

103. (2) $\lambda = \dfrac{h}{\sqrt{2mqV}}$

$$\frac{\lambda_{Li}}{\lambda_P} = \sqrt{\frac{m_P(e)V}{m_{Li}(3e)(V)}};\quad m_{Li} = 8.3m_P$$

$$\frac{\lambda_{Li}}{\lambda_P} = \sqrt{\frac{1}{8.3\times3}} = \frac{1}{5} = 0.2 = 2\times10^{-1}$$

104. (d) (a), (b) and (c) are according to quantum theory but (d) is statement of kinetic theory of gases.

105. (d) $2\pi r = n\lambda$

$$r = \frac{n^2a_0}{Z}$$

$$2\pi\times\frac{4^2}{1}a_0 = 4\lambda \Rightarrow \lambda = 2\pi\times\frac{4}{1}a_0 \Rightarrow \lambda = 8\pi a_0$$

106. (d) Given $\lambda = 1.5\pi a_0$

$n\lambda = 2\pi r$ ...(i)

Radii of stationary states (r) is expressed as:

$$r = a_0\frac{n^2}{Z} \qquad ...(ii)$$

From eqn (i) and (ii)

$$n\lambda = \frac{2\pi a_0 n^2}{Z};\ \lambda = \frac{2\pi a_0 n}{Z}$$

$$1.5\pi a_0 = 2\pi a_0\frac{n}{Z} \Rightarrow \frac{n}{Z} = \frac{1.5}{2} = 0.75$$

107. (d) According to de-Broglie wavelength equation,

$$\lambda = \frac{h}{mV} \Rightarrow \lambda \propto \frac{1}{V}$$

According to photoelectric effect,

$$h\nu - h\nu_0 = \frac{1}{2}mV^2;\ \nu-\nu_0 = \frac{1}{2}\times\frac{mV^2}{h}$$

$$\nu-\nu_0 \propto V^2 \Rightarrow V \propto (\nu-\nu_0)^{1/2}$$

$$\therefore\ \lambda \propto \frac{1}{(\nu-\nu_0)^{1/2}}$$

108. (b) First Bohr orbit of H atom has radius r = 0.529 Å
Also, the angular momentum is quantised.

$$mvr = \frac{h}{2\pi} \Rightarrow 2\pi r = \frac{h}{mv} = \lambda$$

$\therefore\ \lambda = 2\pi\times0.529\ Å$

109. (b) As electron of charge 'e' is passed through 'V' volt, kinetic energy of electron will be eV.

$$\text{Wavelength of electron wave } (\lambda) = \frac{h}{\sqrt{2m\cdot KE}}$$

$$\lambda = \frac{h}{\sqrt{2meV}}\quad \therefore\ \frac{h}{\lambda} = \sqrt{2meV}$$

110. (d) Kinetic energy of any particle $= \dfrac{3}{2}kT$

Also K.E. $= \dfrac{1}{2}mv^2$

$$\frac{1}{2}mv^2 = \frac{3}{2}kT \Rightarrow v^2 = \frac{3kT}{m} \Rightarrow v = \sqrt{\frac{3kT}{m}}$$

$$\text{de-Broglie wavelength} = \lambda = \frac{h}{mv} = \frac{h}{m\sqrt{\frac{3kT}{m}}}$$

$$\lambda = \frac{h}{\sqrt{3kTm}};\ \lambda \propto \frac{1}{\sqrt{m}}$$

Mass of electron < mass of neutron
λ (electron) > λ (neutron)

111. (a) de-Broglie wavelength $(\lambda) = \dfrac{h}{mv}$

$$= \frac{6.63\times10^{-34}\ Js}{6.63\times10^{-3}\ kg\times100\ m/s} = 10^{-33}\,m$$

112. (b) $\lambda = \frac{h}{mv}$

$h = 6.6 \times 10^{-34}$ Js

$m = 1000$ kg

$v = 36 \text{ km/hr} = \frac{36\times10^3}{60\times60}$ m/sec $= 10$ m/sec

$\therefore \lambda = \frac{6.6\times10^{-34}}{10^3\times10} = 6.6\times10^{-38}$ m

113. (b) de-Broglie wavelength is given by :

$\lambda = \frac{h}{mV}$ ... (i)

K.E. $= \frac{1}{2}mV^2 \Rightarrow V^2 = \frac{2K.E}{m} \Rightarrow V = \sqrt{\frac{2K.E}{m}}$

Substituting this in equation (i),

$\lambda = \frac{h}{m}\sqrt{\frac{m}{2K.E}} \Rightarrow \lambda = h\sqrt{\frac{1}{2m(K.E)}}$

i.e., $\boxed{\lambda \propto \frac{1}{\sqrt{K.E}}}$ → important relation

$\therefore$ When *KE* become 4 times, wavelength become 1/2.

114. (c) $r_n = a_0 n^2$

$r = a_0 \times (3)^2 = 9a_0$

$mvr = \frac{nh}{2\pi}; \; mv = \frac{nh}{2\pi r} = \frac{3h}{2\pi\times 9a_0} = \frac{h}{6\pi a_0}$

$\lambda = \frac{h}{mv} = \frac{n}{h}\times 6\pi a_0 = 6\pi a_0$

115. (a) $\lambda = \frac{h}{mv} = \frac{6.63\times10^{-34}}{1.67\times10^{-27}\times1\times10^3}$

$= 3.97\times10^{-10}$ m $= 0.397$ nm.

116. (b) According to Heisenberg uncertainty principle,

$\Delta x\cdot m\Delta v = \frac{h}{4\pi}; \; \Delta x = \frac{h}{4\pi m\Delta v}$

Here $\Delta v = \frac{600\times0.005}{100} = 0.03$ m/s

So, $\Delta x = \frac{6.6\times10^{-34}}{4\times3.14\times9.1\times10^{-31}\times0.03} = 1.92\times10^{-3}$ m

117. (a) Given $m = 9.1\times10^{-31}$ kg

$h = 6.6\times10^{-34}$ Js

$\Delta v = \frac{300\times.001}{100} = 0.003 \text{ ms}^{-1}$

From Heisenberg's uncertainity principle

$\Delta x = \frac{6.62\times10^{-34}}{4\times3.14\times0.003\times9.1\times10^{-31}}$

$= 1.92\times10^{-2}$ m

118. (d) $\lambda = \frac{h}{mv} = \frac{6.63\times10^{-34}}{60\times10^{-3}\times10}$

$= 1.105\times10^{-33} \approx 10^{-33}$ m

119. (c) $\Delta x.\Delta p = \frac{h}{4\pi}$; or $\Delta x.m.\Delta v = \frac{h}{4\pi}$

$\therefore \Delta v = \frac{6.62\times10^{-34}}{4\times3.14\times0.025\times10^{-5}} = 2.1\times10^{-28} \text{ ms}^{-1}$

120. (b) Element with atomic number 9 is Fluorine

F(9) = $1s^2$ $2s^2$ $2p^5$

⇅ ⇅ ⇅ ⇅ ↑

(A) 5 electrons can be up-spin $\left[m_s = +\frac{1}{2}\right]$ and 4 electrons can be down spin $\left[m_s = +\frac{1}{2}\right]$

(B) Unpaired electron can be in anyone of p_x, p_y or p_z orbital

(C) Last electron is in 2p subshell with $n = 2$, $l = 1$

(D) Angular node for s-orbital = 0 while for each p-orbital = 1

$\therefore$ Sum of all angular node = 3

121. (b) Extra stability of half filled is due to :

(i) Symmetrical distribution of electrons

(ii) Large exchange energy

(iii) Smaller coulombic repulsion

(iv) Smaller shielding of electrons by one another.

122. (c) Cr : $1s^2 2s^2 2p^6 3s^2 3p^6 3d^5 4s^1$

Electrons having $\ell = 1 \Rightarrow 12$

Electrons having $\ell = 2 \Rightarrow 5$

123. (d)

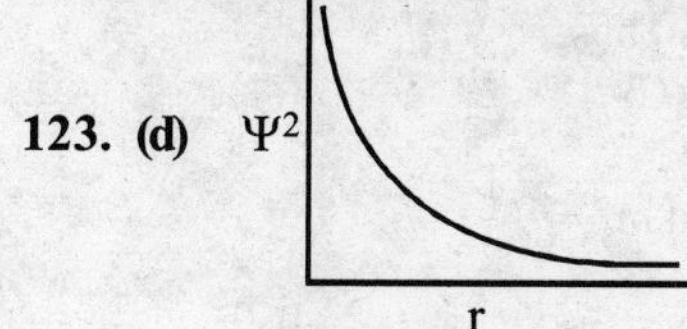

a. Probability density $(\Psi)^2$ is maximum at nucleus.

b. Electron can exist upto infinity from nucleus.

c. Statement is true.

d. Energy of electron is maximum at infinite distance from nucleus. Total energy of an electron is given by $\frac{-13.6z^2}{n^2}$. As 'n' increases, energy also increases.

124. (d) Orbital angular momentum $= \sqrt{\ell(\ell+1)}\frac{h}{2\pi}$

$\therefore$ For 2s orbital : $\ell = 0$

Orbital angular momentum = 0

$\therefore$ For 2p orbital : $\ell = 1$

Orbital angular momentum $= \sqrt{\ell(\ell+1)}\frac{h}{2\pi} = \sqrt{2}\frac{h}{2\pi}$

125. (d) Subsidiary quantum number or azimuthal quantum number ℓ defines shape of orbital.

The given surface diagram is for $2p_x$ orbital.

Mn in $[MnCl_6]^{3-}$ is in +3 oxidation state.

126. (c) In single electron system an electron's energy only depends on 'n' (principal quantum number)
So energy of 2s = 2p and energy of 3d < 4s

127. (d)

	Orbital
A : $n = 1, l = 0, m_l = 0$	1s
B : $n = 2, l = 0, m_l = 0$	2s
C : $n = 3, l = 1, m_l = 1$	3p
D : $n = 3, l = 2, m_l = 1$	3d
E : $n = 3, l = 2, m_l = 0$	3d

When there are no electric and magnetic fields, all the 3d orbitals remain degenerate.

128. (d) In hydrogen atom the orbitals in a shell are degenerate meaning the energy depends only on 'n'

$\therefore \quad E_{3p_x} = E_{3d_{x^2-y^2}} = E_{3d_{z^2}}$

129. (c) For a shell, total number of orbitals = n^2
Magnetic quantum number have values ($-l$ to $+l$) including 0.

130. (d) No spectral line will be observed for a $2p_x \to 2p_y$ transition because $2p_x$ and $2p_y$ orbitals are degenerate orbitals.

131. (d) Energy level can be determined by comparing (n + l) values
(A) $n = 4, l = 1 \Rightarrow (n + l) = 5$
(B) $n = 4, l = 2 \Rightarrow (n + l) = 6$
(C) $n = 3, l = 1 \Rightarrow (n + l) = 4$
(D) $n = 3, l = 2 \Rightarrow (n + l) = 5$
(E) $n = 4, l = 0 \Rightarrow (n + l) = 4$
For same value of (n + l), orbital having higher value of n, will have more energy. Thus,
(B) > (A) > (D) > (E) > (C)

132. (6) $n = 4; |m_l| = 1$ and $m_s = -\frac{1}{2}$

4s	4p			4d					4f						
0	−1	0	+1	−2	−1	0	+1	+2	−3	−2	−1	0	+1	+2	+3

Total orbitals with m_l = +1 or −1 = 6 orbitals

One orbital have one $-\frac{1}{2}e^-$ spin

Total e^- with $n = 4$; $|m_l| = 1$; $m_s = -\frac{1}{2} = 6$

133. (4) For $n = 4$, $\ell = 3, 2, 1, 0$

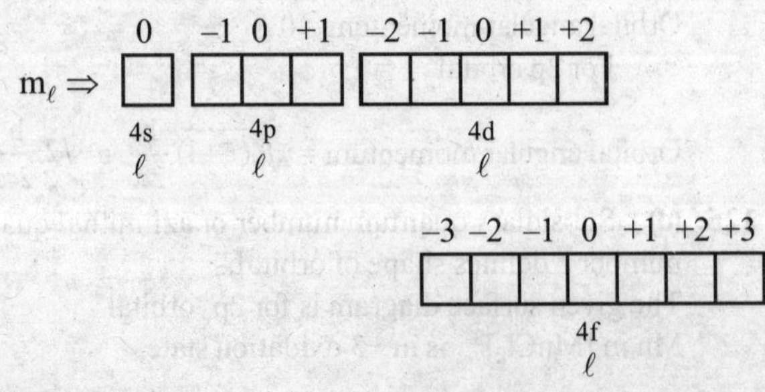

So answer is 4.

134. (c) For 3p orbital
$l = 1, n = 3$
$\therefore$ Radial node = $(n - l - 1) = (3 - 1 - 1) = 1$

135. (b) $n = 4, l = 0, m = 0, s = +½$
$_{19}K = 1s^2\ 2s^2\ 2p^6\ 3s^2\ 3p^6\ 4s^1$
Outermost orbital is 4s, containing one electron
$\therefore$ For s, $l = 0, m = 0$
single electron will have spin as + ½
$\therefore$ s = + ½

136. (a) Statement-I is true but statement-II is false. For single electron species the energy depends upon principal quantum number only.

137. (d) Rubidium (Z = 37)
Electronic configuration = [Zn] $4p^6\ 5s^1$
n = 5
For s, $l = 0$

$m_l = 2l + 1 = 1$ value i.e. 0 $\Rightarrow$ $S = +\frac{1}{2}$

The correct set of four quantum number for the given atom is 5, 0, 0, $+\frac{1}{2}$.

138. (16) n = 4 can have,

	4s	**4p**	**4d**	**4f**
Total electrons	2	6	10	14
Total electrons with $s = +\frac{1}{2}$	1	3	5	7

Thus, total electrons with n = 4 and $s = +\frac{1}{2} = 16$

139. (0) Orbital angular momentum = $\sqrt{l(l+1)}\frac{h}{2\pi}$
Value of l for s orbital = 0
$\Rightarrow$ orbital angular momentum = 0.

140. (b) 5f orbital not buried as 4f orbitals so e^- present in 5f orbital experience less nuclear attraction than e^- present in 4f orbital. Hence electrons of 5f orbital can take part in bonding to a far greater extent.

141. (3) A, D and E statements are correct.

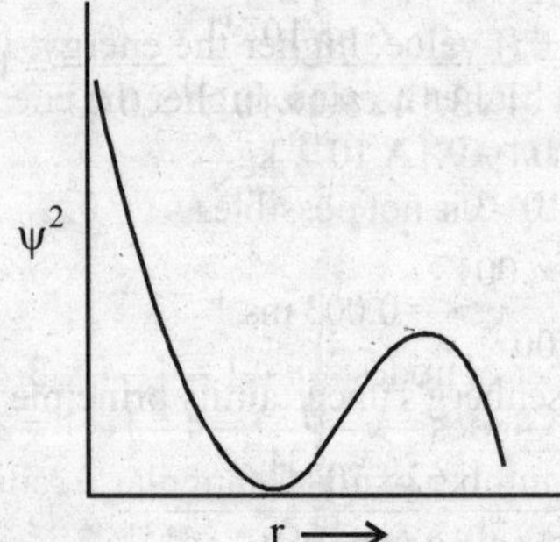

For 2s orbital, the probability density first decreases and then increases.
At any distance from nucleus the probability density of finding electron is never zero and it always have some finite value.
Number of angular nodes = ℓ
Probability density of *p*-orbital is zero at the nucleus.

142. (3) Radial node = n – ℓ – 1

Orbital	n – ℓ – 1
7s	7 – 0 – 1 = 6
7p	7 – 1 – 1 = 5
6s	6 – 0 – 1 = 5
8p	8 – 1 – 1 = 6
8d	8 – 2 – 1 = 5

So, Answer is 3.

143. (b) (A) n = 3; l = 0; n + l = 3

(B) n = 4; l = 0; n + l = 4

(C) n = 3; l = 1; n + l = 4

(D) n = 3; l = 2; n + l = 5

As per Hund's rule energy is given by (n + *l*) value. If value of (n + *l*) remains same then energy is given by n only.

Therefore the order of energy is : D > B > C > A

144. (b) The number of electrons in the orbitals of sub-shell of *n* = 4 are 4s (2), 4p (6), 4d (10), 4f (14), total 32 electrons.

145. (d) At node $\Psi_{2s} = 0$

$\therefore 2 - \frac{r_0}{a_0} = 0 \Rightarrow r_0 = 2a_0$

146. (5) p_x, p_y, p_z, d_{z^2}, and $d_{x^2-y^2}$ are axial orbitals.

147. (b) Lower the value of (n + l) for an orbital, lower is its energy. If two orbitals have same value of (n + *l*) then lower value of n will have the lower energy

A ⇒ 3d ⇒ n + *l* = 3 + 2 = 5

B ⇒ 4p ⇒ n + *l* = 4 + 1 = 5

C ⇒ 4d ⇒ n + *l* = 4 + 2 = 6

D ⇒ 3p ⇒ (n + *l*) = 3 + 1 = 4

order is D < A < B < C

148. (d) Atomic orbital is characterised by *n, l, m*.

149. (a)

150. (a)

(A) n + *l* = 3 + 0 = 3

(B) n + *l* = 4 + 0 = 4

(C) n + *l* = 3 + 1 = 4

(D) n + *l* = 3 + 2 = 5

Higher (n + *l*) value, higher the energy; if same (n + *l*) value, then higher n value, higher the energy.

Thus : D > B > C > A

151. (c) n = 3, *l* = 3 is not possible as *l* = 0, 1 ... (n – 1)

152. (a) for 4p

n = 4, *l* = 1

Total number of nodes = *n* – 1 = 4 – 1 = 3

Total radial nodes = *n* – *l* – 1 = 4 – 1 – 1 = 2

153. (d) Total number of nodes (angular + radial) = *n* – 1

154. (b) $\psi^2(r)$ ⇒ always positive

no. of radial nodes = n – ℓ –1

For 2s, n = 2, ℓ = 0

no. of radial nodes = 2 – 0 – 1 = 1

155. (c)

(A) Values of principal quantum number n = 1, 2, 3 ..... ∞

(B) Values of azimuthal quantum number ranges from *l* = 0 to (n – 1)

(C) Values of magnetic orbital quantum number $m_l = -l$ to $+l$

Total values of $m_l = 2l + 1$

(D) Values of spin quantum number $m_s = \pm\frac{1}{2}$

(E) For *l* = 5, number of orbitals = 2*l* + 1 = 2 × 5 + 1 = 11

156. (2)
- For a given value of *n*, *l* ranges from 0 to *n* – 1
- For given *l*, m_l ranges from –*l* to + *l*.

Quantum number of set (B) and (C) can be correct. (A) and (D) are wrong as *n* = *l* is not possible.

157. (a) Radial node = n – *l* – 1 = 4 – 2 – 1 = 1

Angular node (*l*) = 2.

158. (b) In option (b), both e^- belong to 3*d*-orbitals, which are degenerate.

159. (0) $Zn^+ \rightarrow 1s^2\,2s^2\,2p^6\,3s^2\,3p^6\,3d^{10}\,4s^1$.

Outermost electron is in 4s subshell, so *m* = 0

160. (0) $Ga^+ : 1s^2\,2s^2\,2p^6\,3s^2\,3p^6\,3d^{10}\,4s^2$

The azimuthal quantum number for the valence electrons (4s-subshell) of Ga^+ ion is zero(0).

161. (c)

Orbital	Angular nodes (ℓ)	Radial nodes (n–ℓ–1)
2p	ℓ = 1	2 – 1 – 1 = 0
3p	ℓ = 1	3 – 1 – 1 = 1
3s	ℓ = 0	3 – 0 – 1 = 2
2s	ℓ = 0	2 – 0 – 1 = 1

3*s* orbital has no angular nodes and two radial nodes.

162. (0) Given : *n* = 4 and $m_l = -3$

Hence, *l* value must be 3.

∴ Number of radial nodes = *n* – *l* – 1 = 4 – 3 – 1 = 0

163. (3) Possible values of *l* = 0, 1, 2, 3, 4

If *l* = 0, *m* = 0

l = 1, *m* = {–1, 0, +1}

l = 2, *m* = {–2, –1, 0, +1, +2}

l = 3, *m* = {–3, –2, –1, 0, +1, +2, +3}

l = 4, *m* = {–4, –3, –2, –1, 0, +1, +2, +3, +4}

5*d*, 5*f* and 5*g*-sub shell contains one-one orbital having (*m* = +2).

164. (c) Number of radial nodes = *n* – *l* – 1

Number of angular nodes = *l*

For 5*d*; *n* = 5, *l* = 2

5*d* orbital has two radial nodes and two angular nodes.

165. (a)

	6*s*	4*f*	5*d*	6*p*
n + *l*	6+0	4+3	5+2	6+1
	⇓	⇓	⇓	⇓
	6	7	7	7

Thus, order of orbitals filled are

6s < 4f < 5d < 6p.

166. (c) Radial node $= n - l - 1$

$\therefore$ 1s $\Rightarrow 0 (\psi^2 \neq 0)$

2s $\Rightarrow 1 (\psi^2 = 0)$

2p $\Rightarrow 0 (\psi^2 \neq 0)$

3p $\Rightarrow 1 (\psi^2 = 0)$

Probability density (ψ^2) can be zero for 3*p* orbital other than infinite distance. It has one radial node.
Thus, statement (c) is correct.

167. (c) Following the given hypothetical condition

$n = 1, l = 0, 1, 2 \Rightarrow 1s\ 1p\ 1d$
$n = 2, l = 0, 1, 2, 3 \Rightarrow 2s\ 2p\ 2d\ 2f$
$n = 3, l = 0, 1, 2, 3, 4 \Rightarrow 3s\ 3p\ 3d\ 3f\ 3g$

As per the $(n + 1)$ rule, the increasing order of energy of orbitals :

$1s < 1p < 2s < 1d < 2p < 3s < 2d < 3p ....$

$\therefore$ The electronic configurations are as follows :

Atomic No.	Electronic Configuration	Remarks
9	$1s^2\ 1p^6\ 2s^1$	2nd element of Gp.I (alkali metal)
13	$1s^2\ 1p^6\ 2s^2\ 1d^3$	*d*-orbital is not half-filled.
8	$1s^2\ 1p^6$	Outermost shell achieves 8 electrons in $n = 1$. So, it is the first noble gas.
6	$1s^2\ 1p^4$	1*p* is the valence shell, not 2*p*.

168. (b) For $n = 4$ possible values of $l = 0, 1, 2, 3$; only $l = 2$ and $l = 3$ can have $m = -2$. So possible subshells are 2 (4d and 4f).

169. (b) The possible number of orbitals in a shell in term of '*n*' is n^2. All orbitals can have $m_s = +1/2$
$\therefore n = 5 ; n^2 = 25$

170. (a) As the value of *Z* (atomic number) increases, energy of orbitals decreases (becomes more –ve value).
$\therefore$ Order of energy of 2s orbital is H > Li > Na > K.

171. (a) Probability of finding an electron will have maximum value at both '*a*' and '*c*'. There is zero probability of finding an electron at '*b*'.

172. (b) The given probability density curve is for 2s orbital due to the presence of only one radial node. 1s and 2p orbital do not have any radial node and 3s orbital has two radial nodes. Hence, option (b) is correct.

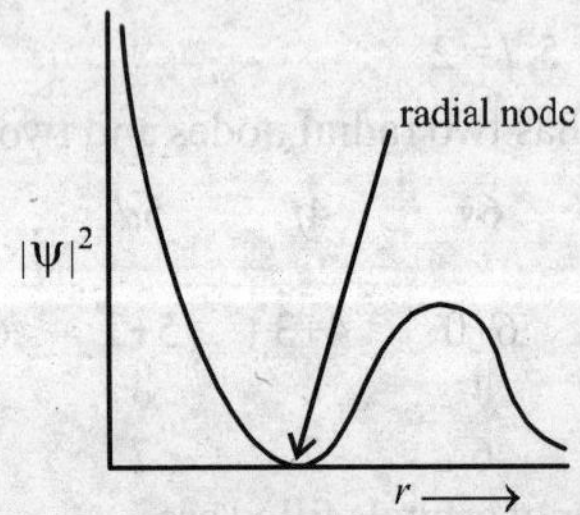

173. (d) Iso-electronic species differ in size due to different effective nuclear charge.

174. (c)

				$n + \ell$
(I)	$n = 4$	$\ell = 2$	4*d*	6
(II)	$n = 3$	$\ell = 2$	3*d*	5
(III)	$n = 4$	$\ell = 1$	4*p*	5
(IV)	$n = 3$	$\ell = 1$	3*p*	4

The energy of an atomic orbital increases with increasing $n + \ell$. For identical values of $n + \ell$, energy increases with increasing value of *n*. Therefore the correct order of energy is:

$$\underset{IV}{3p} < \underset{II}{3d} < \underset{III}{4p} < \underset{I}{4d}$$

175. (a)

(1) Angular momentum $(L) = \frac{nh}{2\pi}$
Therefore, as *n* increases, *L* also increases.

(2) $r \propto \frac{n^2}{Z}$

(3) According to wave mechanics, the orbital angular momentum, $L = \frac{h}{2\pi}\sqrt{\ell(\ell+1)}$
For, 1s orbital, $l = 0 \Rightarrow L = 0$

(4) As *l* increases, the peak of ψ vs *r* shifts towards higher '*r*' value.

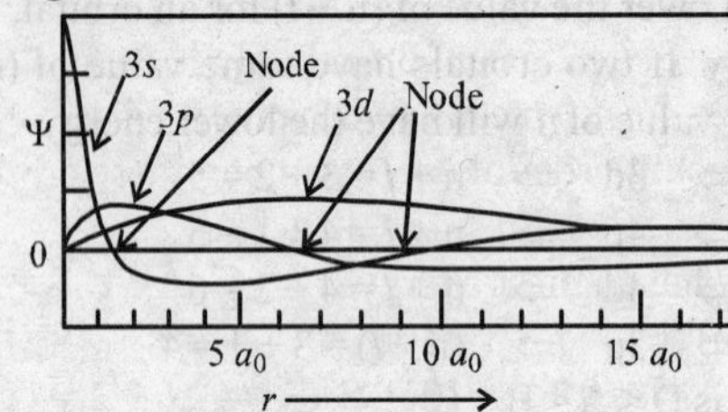

176. (b) Number of orbitals in a shell $= n^2 = (5)^2 = 25$.

177. (b) According to Aufbau principle, the sequence of filling electrons in sixth period is
$6s - 4f - 5d - 6p$ i.e., $(ns) \rightarrow (n - 2)f \rightarrow (n - 1)d \rightarrow np$

178. (a) The electronic configuration of Rubidium (Rb = 37) is

$1s^2 2s^2 2p^6 3s^2 3p^6 3d^{10} 4s^2 4p^6 5s^1$

The last electron enters in 5s orbital.

Hence, $n = 5, l = 0, m = 0, s = \pm\frac{1}{2}$

179. (c) $n = 3, l = 2$ means 3*d* orbital

+2	+1	0	–1	–2

In an atom, only one orbital can have the value $m_l = +2$.

180. (b) (i) $n = 5$ means $l = 0, 1, 2, 3, 4$
Since $m = +1$,
hence total no. of electrons will be
$= 0(\text{from } s) + 2(\text{from } p) + 2(\text{from } d) + 2(\text{from } f) + 2(\text{from } g)$
$= 0 + 2 + 2 + 2 + 2 = 8$

(ii) $n = 2, l = 1, m_l = -1, m_s = -1/2$ represent 2p orbital with one electron.

181. (c) Among isoelectronic species ionic radii increases as the negatives charge increases.
Order of ionic radii $Ca^{2+} < K^+ < Cl^- < S^{2-}$
The number of electrons remains the same but nuclear charge increases with increase in the atomic number causing decrease in size.

182. (b) (i) $4p$ (ii) $4s$
(iii) $3d$ (iv) $3p$
According to Bohr Bury's $(n+l)$ rule, increasing order of energy will be (iv) < (ii) < (iii) < (i).
Note: If the two orbitals have same value of $(n+l)$ then the orbital with lower value of n will be filled first.

183. (b) Species having same number of electrons are isoelectronic. On calculating the number of electrons in each species given here, we get.
$CN^-(6+7+1=14)$; $N_2(7+7=14)$
$O_2^{2-}(8+8+2=18)$; $C_2^{2-}(6+6+2=14)$
$O_2^-(8+8+1=17)$; $NO^+(7+8-1=14)$
$CO(6+8=14)$; $NO(7+8=15)$
From the above calculation we find that all the species listed in choice (b) have 14 electrons each so it is the correct answer.

184. (c) (a) $n=3, l=0$ means $3s$-orbital and $n+l=3$
(b) $n=3, l=1$ means $3p$-orbital $n+l=4$
(c) $n=3, l=2$ means $3d$-orbital $n+l=5$
(d) $n=4, l=0$ means $4s$-orbital $n+l=4$
Increasing order of energy among these orbitals is
$3s < 3p < 4s < 3d$
$\therefore$ $3d$ has highest energy.

185. (c) (a) $N^{3-} = 7+3 = 10e^-$, $O^{2-} \longrightarrow 8+2 = 10e^-$
$F^- = 9+1 = 10e^-$, $S^{2-} \longrightarrow 16+2 = 18e^-$
(not isoelectronic)
(b) $Li^+ = 3-1 = 2e^-$, $Na^+ = 11-1 = 10e^-$,
$Mg^{2+} = 12-2 = 10e^-$
$Ca^{2+} = 20-2 = 18e^-$ (not isoelectronic)
(c) $K^+ = 19-1 = 18e^-$, $Cl^- = 17+1 = 18e^-$,
$Ca^{2+} = 20-2 = 18e^-$, $Sc^{3+} = 21-3 = 18e^-$
(isoelectronic)
(d) $Ba^{2+} = 56-2 = 54e^-$, $Sr^{2+} = 38-2 = 36e^-$
$K^+ = 19-1 = 18e^-$, $Ca^{2+} = 20-2 = 18e^-$
(not isoelectronic)

186. (b) Calculating number of electrons

a. $BO_3^{3-} \longrightarrow 5+8\times3+3=32$
$CO_3^{2-} \longrightarrow 6+8\times3+2=32$
$NO_3^- \longrightarrow 7+8\times3+1=32$
} isoelectronic species

b. $SO_3^{2-} \longrightarrow 16+8\times3+2=42$
$CO_3^{2-} \longrightarrow 32$
$NO_3^- \longrightarrow 32$
} not isoelectronic species

c. $CN^- \longrightarrow 6+7+1=14$
$N_2 \longrightarrow 7\times2=14$
$C_2^{2-} \longrightarrow 6\times2+2=14=14$
} isoelectronic species

d. $PO_4^{3-} \longrightarrow 15+8\times4+3=50$
$SO_4^{2-} \longrightarrow 16+8+2=50$
$ClO_4^- \longrightarrow 17+8\times4+1=50$
} isoelectronic species

Hence the species in option (b) are not iso-electronic.

187. (b) (A) 4p (B) 4s (C) 3d (D) 3p
According to $(n+l)$ rule, increasing order of energy (D) < (B) < (C) < (A).

188. (a) The energy of an orbital is given by $(n+l)$ rule. $(n+l)$ value for option (E) and (D) is $(3+2)=5$ hence they will have same energy, since their n values are also same.

189. (a) The possible quantum numbers for 4f electron are
$n=4, l=3, m=-3,-2-1,0,1,2,3$ and $s=\pm\frac{1}{2}$
Of various possiblities, only option (a) is possible.

190. (c) $_{19}K^+$, $_{20}Ca^{2+}$, $_{21}Sc^{3+}$, $_{17}Cl^-$
each contains 18 electrons.

191. (b) Electronic configuration of Cr atom (Z = 24)
$= 1s^2\,2s^2\,2p^6\,3s^2\,3p^6\,3d^5\,4s^1$
when $l=1$, p-subshell,
Numbers of electrons = 12
when $l=2$, d-subshell,
Numbers of electrons = 5

192. (c) $Fe^{2+}(26-2=24) = 1s^2 2s^2 2p^6 3s^2 3p^6 4s^0 3d^6$
Hence, no. of d electrons retained is 6.
[Two 4s electron are removed.]

193. (a) For s-electron, $l=0$
$\therefore$ Orbital angular momentum $= \sqrt{0(0+1)}\frac{h}{2\pi} = 0$

194. (a) N^{3-}, F^- and Na^+ contain 10 electrons each.

Classification of Elements and Periodicity in Properties

3

1. **(b)**
   (A) Pnictogen ⇒ Mc (Moscovium), Atomic No. = 115
   (B) Chalcogen ⇒ Lv (Livermorium), Atomic No. = 116
   (C) Halogen ⇒ Ts (Tennessine), Atomic No. = 117
   (D) Noble gas ⇒ Og (Oganesson), Atomic No. = 118
2. **(d)** According to Law of octaves the elements were arranged in the increasing order of their atomic weight. Lothar Meyer plotted the physical properties such as atomic volume, melting point and boiling point against their respective atomic weight.
3. **(a)** Properties of elements are periodic function of their atomic number. Elements having similar outer electronic configuration are arranged in same group. Number of elements in a period is not equal to number of atomic orbitals available in energy level that is being filled.
   Hence, A, C & E are incorrect.
4. **(c)**

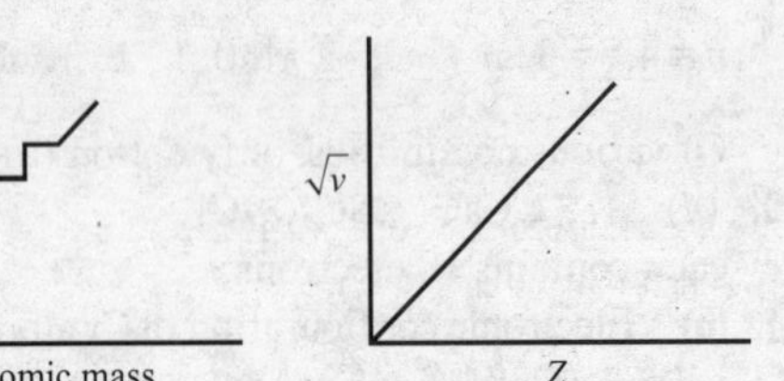

According to Henry Moseley's experiment, $\sqrt{v}$ vs z is a straight line.

5. **(a)** Palladium (Pd) ⇒ 5th period
   Iridium (Ir), Osmium (Os), Platinum (Pt) ⇒ 6th Period
6. **(b)** (A) $_7$N: [He]$2s^2 2p^3$.
   (B) $_{16}$S: [Ne]$2s^2 3p^4$
   (C) $_{35}$Br: [Ar]$3d^{10} 4s^2 4p^5$
   (D) $_{36}$Kr: [Ar]$3d^{10} 4s^2 4p^6$
7. **(11)** Unununioum refers to the element with symbol U_{44} and atomic number 111. Its IUPAC name is roentgenium. It belongs to group number 11, so n = 11.
8. **(b)** $\sqrt{v} \propto Z$
9. **(c)** According to Henry Moseley $\sqrt{v} \propto Z$

   So, $n = \dfrac{1}{2}$
10. **(d)** A - IV, B - II, C - I, D - III
11. **(b)**

Atomic Number	Electronic configuration	Remark
33	$[Ar]3d^{10} 4s^2 4p^3$	Metalloid
53	$[Kr]4d^{10} 5s^2 5p^5$	Non-metal
83	$[Xe]4f^{14} 5d^{10} 6s^2 6p^3$	Metal

12. **(d)**
13. **(b)** According to modern periodic law, the properties of the elements are repeated after certain regular intervals when these elements are arranged in order of their increasing atomic numbers.
14. **(c)**

Element	Atomic No.	Period No.	Group No.
Br	35	4	17
K	19	4	1
Ge	32	4	14
Fr	87	6	1

First ionisation energy of an element generally decreases down the group and increases from left to right along a period. Therefore, element having atomic number 87 has the lowest first ionisation energy.

15. **(b)** In a period the atomic radius decreases from left to right Al comes after Mg in period so Mg should have smaller radius.
16. **(a)**
   (A) The process of adding an electron to a neutral gaseous atom is not always exothermic it may be exothermic or endothermic.
   (C) Be ; B
   $1s^2\ 2s^2$; $1s^2 2s^2 2p^1$
   In Be 2s subshell is fully filled.
   So, high energy is required to remove electron as compared to boron.

   (D) In $CCl_4 \longrightarrow$ $\overset{\delta+}{C}$ bonded to four $Cl^{\delta-}$ atoms

   Due to partialy positive charge $z_{eff} \uparrow$, EN↑
   So, EN of C ⇒ $CCl_4 > CH_4$
   (E) Cs is most electropositive among the elements of group I.
17. **(a)** TeO_2 exhibits oxidising properties while TeH_2 act as reducing agent and acidic in nature.
18. **(d)**
   (A) $1s^2 2s^2 2p^3$ ⇒ N (B) $1s^2 2s^2 2p^4$ ⇒ O
   (C) $1s^2 2s^2 2p^5$ ⇒ F (D) $1s^2 2s^2 2p^2$ ⇒ C
   Order of electronegativity: F > O > N > C
   ∴ C > B > A > D

19. **(a)** B is smallest so it forms B_2O_3

20. **(d)** For isoelectronic species, the cations with the greater positive charge will have a smaller radius because of the greater attraction of the electrons to the nucleus. Anion with greater negative charge will have the larger radius. Therefore statement I is correct.

Electron gain enthalpy of fluorine is less negative that of chlorine due to interelectronic repulsion.

Hence the order will be Cl > F > Br > I

Therefore statement II is correct.

21. **(c)** Atomic radii is increases moving down a group in the perodic table.

Correct order of atomic radii : Be < Mg > Al > Si

22. **(b)** The IE_4 is suddenly very high therefore element must have 3 valence e^-(s) and it belongs to group 13.

23. **(a)** Statement-I: False but Statement-II is true.

On moving left to right in periodic table non-metallic character increases and we know that non-metal oxides are acidic in nature.

Non metallic character ↑ Acidic strength of oxide↑

Element on extreme left are metallic and form basic oxides.

24. **(c)** Ionic radii – $Al^{3+} < Mg^{2+} < Na^+ < F^-$

Ionisation energy – B < C < O < N

Metallic character – B < Al < Mg < K

Electronegativity – Si < P < S < Cl

25. **(a)**

	Li	Be	B	C	N	O	F
(E.N.) =	1	1.5	2	2.5	3	3.5	4.0

On pauling scale

	Na	Mg	Al	Si	P	S	Cl
(E.N.) =	0.9	1.2	1.5	1.8	2.1	2.5	3.0

Correct order Mg < Al < B < N

26. **(c)** Elements with highest electronegativity ⇒ F, O

Elements with largest atomic size ⇒ Fr, Ra

Elements which shows properties of both metal and non-metals i.e. metalloids ⇒ Ge, As

Elements with highest negative electron gain enthalpy ⇒ Cl, S.

27. **(b)** Non metalic nature ∝ Electronegativity. Across the period ionisation energy increase due to increase in effective nuclear charge.

High electronegativity difference results in ionic bond formation.

Oxides of metals are generally basic and that of non-metals are acidic in nature.

28. **(d)** For process (A), (B), (C) electron affinity value are negative.

Be = –18, N = –20.3,

$O \rightarrow O^{2-} = [141 + (-844)] = -703$

Na = +53, Al = +44

29. **(b)** Factors influencing emission wavelength

1. effective nuclear charge on the outer most electron
2. Shielding effect
3. Stability of electron in the orbitals

(A)	Li	670.8
(B)	Na	589.2
(C)	Rb	780.0
(D)	Cs	455.5

∴ Correct match of List-1 and List-II is (A)-(III), (B)-(I), (C)-(IV), (D)-(II)

30. **(a)** (A) Covalent radius of group 14 elements increases from C to Pb.

(B) Electronegatively decreases from C to Si and thereafter it remains more or less constant.

(C) Carbon can not expend octet due to unavailability of d-orbital.

(D) Heavier elements have less effective nuclear charge so unable to hold coming electrons.

(E) Carbon can show negative oxidation states when it is covalently bonded to less electronegative elements like CH_4.

31. **(d)** All species are iso electronic because each has 10 e^-. Nuclear charge depends on number of protons so all have different nuclear charge. More the number of protons high is the nuclear charge and small is the ionic radius. So order of ionic radius is :

(Ionic radius) $O^{-2} > F^- > Na^+ > Mg^{+2}$

32. **(a)** Greater the effective nuclear charge smaller will be its size, so statement I is correct but statement II is incorrect as anions are larger than parent atom.

33. **(d)** B > Tl > Ga > Al > ln

34. **(a)**

(i) Na < Li < Cl < F

	Na	Li	Cl	F
$I.E._1$ in kJ/mol	496	520	1256	1681

(ii) Na < Li < F < Cl

	Na	Li	F	Cl
$\Delta_{eg}H$ in kJ/mol	–53	–60	–328	–349

35. **(d)** Isoelectronic species have equal number of electrons but they have different number of protons in the nucleus, Nuclear charge depends on the number of protons, which affect the size of isoelectronic species, Higher the nuclear charge smaller is the size.

36. **(6)** A_2B

Lowest oxidation state of B is –2 means it can accept 2 electrons maximum to complete its octet, so it should have six valence electrons.

37. **(c)** Statement I is false but Statement II is true.

p-Block element contains metal, non-metal and metalloid but *d*-block element contain only metals.

38. **(a)**

Element	**$\Delta_{eg}H$ (kJ/mol)**
F	–333
S	–200
Br	–325
Ar	+96

39. **(b)** A, B and D only

Down the group the atomic size and ionic radii increases and along the rows the atomic radii and ionic radii decreases.

∴ Order of atomic radii : B′ < A′ < D′ < C′

$\therefore$ Order of ionic radii : $B'^{+} < A'^{+} < D'^{+} < C'^{+}$

Metallic character increases down the group and decreases along the row.

$\therefore$ Order of metallic character : $B' < A' < D' < C'$

40. **(b)** As to Bi only a small increase in covalent radius is observed, due to the presence of completely filled *d* and *f* orbitals in heavier members.

41. **(c)** In a period from left to right, the reactivity first decreases and then increases from Group 1 alkali metals to Group 17 halogens and then decreases to group 18.

Thus, Statement I is false.

Group 1 metals form basic oxide like Na_2O_2 and group 17 non-metals form acidic oxide like Cl_2O_7.

Thus, Statement II is true.

42. **(a)** Across a period, increasing nuclear charge outweights the shielding. Consequently, the outermost electrons are held more and more tightly and the first ionization enthalpy increases across a period.

43. **(b)** The electronegativity of group 14 elements from Si to Pb is approximately same. Therefore statement I is false and statement II is true.

44. **(d)** ${}_{13}Al = 1s^2\,2s^2\,2p^6\,3s^2\,3p^1$, ${}_{7}N = 1s^2\,2s^2\,2p^3$

${}_{14}Si = 1s^2\,2s^2\,2p^6\,3s^2\,3p^2$, ${}_{6}C = 1s^2\,2s^2\,2p^2$

Removal of one electron from the valence shell is most difficult in nitrogen as it disturbs the stable half-filled p-subshell configuration and therefore its first ionization energy is highest.

45. **(c)** s-block elements are highly reactive and found in combined state due to very low ionization energies. Thus, I is correct but II is false.

46. **(c)** K^+ (18), Cl^- (18), Ca^{2+}(18), Sc^{3+} (18)

47. **(5)** Isoelectronic species O^{2-}, F^-, Mg^{2+}, Na^+, Al^{3+} (all have $10e^-$)

48. **(b)** Electronegativity of an element depends on the atomic size and atomic number.

NO is a neutral oxide.

Al_2O_3 is a amphoteric oxide.

Cl has more negative electron gain enthalpy than F.

49. **(a)** According to Fajan's rule Sb^{+5} has more polarising power than Sb^{+3}. Therefore $SbCl_5$ is more covalent than $SbCl_3$.

Stability of higher oxides of halogen is primarily due to Higher oxidation state, more electronegative halogen and resonance stabilization.

50. **(c)** First I.E.

$F > N > O > C > Be > B > Li$

Generally along a period ionisation energy increases. In case of N and Be ionisation energy increases due to stable electronic configuration.

51. **(5)** $Na_2O + H_2O \rightarrow 2NaOH$

$Cl_2O_7 + H_2O \rightarrow 2HClO_4 \Rightarrow 1 + 4 = 5.$

52. **(d)** Assertion and Reason are correct and Reason is correct explanation. Hence, correct order is :

$\because$ Successive I.E. always increases.

53. **(c)** On moving from top to bottom metallic character increases while on moving from left to right metallic decreases. Hence, correct order is :

$K > Ca > Be$.

54. **(d)**

Atom	E.N.
Br	3.0
C	2.5
At	2.2
P	2.1

55. **(b)** Cl has the most negative ΔH_{eg} among all the elements and Ne has the most positive ΔH_{eg}.

56. **(c)** As electropositive character of element increases, basic character of its oxide increases.

$$\underbrace{B_2O_3}_{\text{acidic}} < \underbrace{Al_2O_3 < Ga_2O_3}_{\text{amphoteric}} < \underbrace{In_2O_3 < Tl_2O_3}_{\text{basic}}$$

57. **(d)** In isoelectronic species, size $\propto \frac{1}{Z_{eff}}$. Hence correct increasing order of ionic radii is : $Ca^{2+} < K^+ < Cl^- < S^{2-}$

58. **(b)** **For 3d series:**

	Sc	Ti	V	Cr	Mn
IE (KJ/mol)	631	656	650	653	717
	Fe	Co	Ni	Cu	Zn
IE (KJ/mol)	762	758	736	745	906

For 2nd Group:

	Be	Mg	Ca	Sr	Ba	Ra
IE (KJ/mol)	900	738	590	550	503	—

Thus, for 4th period, 3d elements have higher first ionization energies than group 2 metal that is Ca.

59. **(4)** Acidic oxides are N_2O_3, NO_2, Cl_2O_7, SO_2.

Basic oxides are CaO, Na_2O

Neutral oxides are CO, NO, N_2O

60. **(a)** Metallic character increases down the group and decreases along the period.

61. **(c)** I.E. order: $Na < Al < Mg < Si$

$\therefore$ $496 < IE\,(Al) < 737 < 786$

Only option (c), matches the condition.

62. **(c)** Rb and Cs have nearly same electron gain enthalpy $\Delta H_{eg} = -46$ kJ mol^{-1}.

Ar and Kr have same electron gain enthalpy, $\Delta H_{eg} = +96$ kJ mol^{-1}.

63. **(a)** In the periodic table; on moving down the group, the size of atom increases and tendancy to remove an electrons increases. Hence, metallic character increases and on moving left to right in a period, metallic character decreases. Therefore; the decreasing order for metallic character is $Na > Mg > Be > Si > P$.

64. **(d)** $_{31}$Ga is in group 13 with e^- configuration [Ar] $3d^{10}4s^24p^1$ whereas $_{30}$Zn has e^- configuration as [Ar] $3d^{10}4s^2$. Removal of e^- from $4p$ orbital is easier than that from $4s$ orbital. Hence, first I.E. of Ga is lower than that of Zn.

65. **(b)**
(A) $3s^2 \rightarrow$ Mg
(B) $3s^23p^1 \rightarrow$ Al
(C) $3s^23p^3 \rightarrow$ P
(D) $3s^23p^4 \rightarrow$ S
All elements have $n = 3$. Along the period I.E. increases but half-filled orbital stability of P results increase in I.E. Thus, P > S.

66. **(a)** Li ↘ Mg Diagonal relationship
Na Mg
$Li^+ \rightarrow$ Maximum hydration enthalpy in group 1 due to small size. So 'B' is Mg.

67. **(d)** For chlorine, the atomic radius and covalent radius are taken as same. Hence, statement I is incorrect.
In case of anionic species, extra electrons are added in the parent atom, so effective nuclear charge decrease. Hence, anionic radius always greater than their parent atomic radius.

68. **(d)** On moving from left to right along the period size of atom decrease, hence ionization energy increases.
Nitrogen have half-filled p-orbital and beryllium have completely filled s-orbital so their ionization energy are greater than oxygen and boron, respectively. So the order is :

$$\underset{(2p^3)}{N} > \underset{(2p^4)}{O} > \underset{(2s^2)}{Be} > \underset{(2p^1)}{B}$$

69. **(c)** Down the group, the size of atom increases, so electron gain enthalpy decreases.
In case of oxygen, due to its small size, which leads to electron-electron repulsion, results in less electron gain enthalpy than sulphur.

70. **(a)** Ionisation energy $\Rightarrow$ N > O.
$N \rightarrow 1s^22s^22p^3$ [↑|↑|↑] } half filled subshell
$O \rightarrow 1s^22s^22p^4$ [↑↓|↑|↑] } more than half filled
In oxygen atom, two of the four 2p electrons must occupy the same 2p orbital resulting in an increased electron – electron repulsion.

71. **(c)** Non-metals, metals generally form acidic oxide, basic oxides, respectively.
SO_3, $SiO_2 \rightarrow$ Acidic oxide
$CaO \rightarrow$ Basic oxide
$Al_2O_3 \rightarrow$ Amphoteric oxide

72. **(b)** Halogens generally form acidic oxides. Group 1 elements generally form basic oxides.
Cl_2O_7 – Acidic oxide; Na_2O – Basic oxide
Al_2O_3 – Amphoteric oxide; N_2O – Neutral oxide

73. **(d)** For isoelectronic species, ionic radii decrease with increase in nuclear charge. Ionic radius of O^{2-} is more than that of Mg^{2+}. Both O^{2-} and Mg^{2+} are isoelectronic with 10 electrons.

74. **(a)** For the isoelectronic species the ionic radius increase with increase of negative charged and ionic radius decrease with increase of positive charged.
$N^{3-} > O^{2-} > F^- > Na^+ > Mg^{2+}$ (Radii)

75. **(d)**

Element	Be	Mg	Ca	Sr
M.P(K)	1560	924	1124	1062

76. **(d)** Metalloids have intermediate properties between metals and non-metals.
Sc, Pb, Bi are metals Te is a metalloid.

77. **(d)** Isoelectronic species have same no. of electrons, e.g., Al^{+3}, O^{2-}, Mg^{+2} all have 10 electrons.

78. **(b)** The magnitude of e^- gain enthalpy decreases down the group but its value for Cl is more than that of F.

Element	F	Cl	Te	Po
$\Delta_{eg}H$(kJ/mol)	–328	–349	–190	–174

79. **(b)** Hg, Ga, Cs are liquid at near room temperature but Ag in solid.

80. **(c)** Al^{3+}, Mg^{2+} and Na^+ are isoelectronic ionic species. As positive charge increases, ionic size decreases.
The order of size of Na^+ and K^+ is $Na^+ < K^+$.
∴ Order of ionic radii is : $Al^{3+} < Mg^{2+} < Na^+ < K^+$.

81. **(b)** $B(OH)_3$ and H_3PO_3 are acidic.
$Al(OH)_3$ and $Be(OH)_2$ are amphoteric.
NaOH and $Ca(OH)_2$ are basic.

82. **(b)** Li ↘ Mg, Be ↘ Al, B ↘ Si
Na Mg Al Si
B–Si, Be–Al and Li–Mg exhibit diagonal relationship, whereas Li–Na belongs to same group.

83. **(c)** Ionization energy generally increases moving from left to right across the period.
So, the order is Na < Mg < ..... Cl < Ar.

84. **(c)**
CaO – Basic
SiO_2 – Acidic
B_2O_3 – Acidic
N_2O – Neutral
BaO – Basic
Generally, metal oxides are basic in nature and non-metal oxides are acidic in nature. Calcium and barium are metals, nitrogen and boron are non-metals while silicon is metalloid.

85. **(d)** Correct order of electron gain enthalpy is
S > Se > Te > O
Electron gain enthalpy of oxygen is even less than Te because of inter-electronic repulsion and effective nuclear charge in the oxygen atom.

86. **(b)** Mg and P exhibit abnormal behaviour due to extra stability of half-filled and filled electronic configuration.
First ionisation enthalpy order is :
Al < Mg < Si < S < P

87. **(b)**

Group Number						
3	4	5	6	7	...	18
$_{21}$Sc				$_{25}$Mn		$_{36}$Kr
$_{39}$Y			$_{42}$Mo			$_{54}$Xe
$_{57}$La	$_{72}$Hf					

88. **(c)** All are isoelectronic species, so more is the Z_{eff}, less will be the ionic size.

∴ Correct order of ionic radii is :

$Al^{3+} < Mg^{2+} < Na^+ < F^- < O^{2-} < N^{3-}$

Among isoelectronic species ionic radii increases as the negatives charge increases.

The number of electrons remains the same but nuclear charge increases with increase in the atomic number causing decrease in size.

89. **(b)** $Ar\ (g) + e^- \longrightarrow Ar^-\ (g)$ (endothermic)

$H(g) + e^- \longrightarrow H^-\ (g)$ (exothermic)

$Na\ (g) \longrightarrow Na^+\ (g) + e^-$ (endothermic)

$O^-\ (g) + e^- \longrightarrow O^{2-}\ (g)$ (endothermic)

- Electron gaining enthalpy (EGE) of H(g) is negative while that of Ar(g) is positive due to ns^2np^6 configuration.
- Second EGE is always positive for an atom.
- Ionization potential of an atom is positive.

90. **(b)**

	O^{2-}	F^-	Na^+	Mg^{2+}
Z	8	9	11	12
No. of e^-	10	10	10	10

In isoelectronic species greater is $Z_{effe.}$ smaller is radius so order is $O^{2-} > F^- > Na^+ > Mg^{2+}$.

91. **(a)** Charge/ radius ratio of Be and Al is same because of diagonal relationship. Remaining statements are correct.

92. **(c)** As difference in 3rd and 4th ionisation energies is high, so atom contains 3 valence electrons.

93. **(d)** On moving left to right along a period in the periodic table, atomic radius decreases while electronegativity, electron gain enthalpy and ionisation enthalpy increases.

94. **(b)** On moving left to right in a period, the acidic character of oxides increases.

3rd period element oxides.

$\underbrace{Na_2O \quad MgO}_{\text{Basic}} \quad \underbrace{Al_2O_3}_{\text{Amphoteric}} \quad \underbrace{SiO_2 \quad P_2O_5 \quad Cl_2O_7}_{\text{Acidic}}$

So, X has minimum atomic number while Z has maximum atomic number.

Thus, the correct order of the atomic number is :

X<Y<Z

95. **(c)** $_5B : 1s^2\ 2s^2\ 2p^1$

$_4Be : 1s^2\ 2s^2$

First ionisation enthalpy of B is lower than Be because Be has a stable electronic configuration. It required more energy to remove the first electron from 2s (in Be) than 2p (in B) because 2s e^- has more penetration power than 2p.

Therefore options (I), (II) and (III) are correct. Atomic radius of B is less than Be.

96. **(c)** Generally, non-metal oxides are acidic in nature and metal oxides are basic in nature, Al_2O_3 is amphoteric.

97. **(d)** A large difference between first and second ionisation enthalpies (4560–496 = 4064 kJ mol^{-1}) confirms the metal to be an alkali metal and thus is monovalent and form hydroxide of the type M(OH).

$$\underset{1\ mol}{MOH} + \underset{1\ mol}{HCl} \rightarrow MCl + H_2O$$

$$\underset{1\ mol}{2\ MOH} + \underset{1/2\ mol}{H_2SO_4} \rightarrow M_2SO_4 + H_2O$$

98. **(c)** Generally, electronegativity decreases down the group as the size increases. This can also be formulated as:

$$\text{Electronegativity} \propto \frac{1}{\text{size}}$$

99. **(a)** All given elements belong to period III and generally their ionisation energy will increase along the period but Mg will show higher ionisation potential compared to Al due to its stable configuration. Thus, correct order of ionisation energy will be : Na < Al < Mg < Si.

Ionisation energy (kJ/mol) of the given metals are

Na : 496; Al : 577; Mg : 737; Si : 786

100. **(c)** C N O F → Size decreases; F, Cl, Br, I ↓ Size increases

Correct increasing order of atomic radii is :

F < O < C < Cl < Br

101. **(c)** Fluorine has lower electron gain enthalpy value than chlorine because of its small size. The small size of the fluorine atoms results in strong interelectronic repulsion in the 2p orbital.

102. **(b)** Generally, electron affinity decreases on moving down a group. Chlorine has more electron affinity than F because of very small size of fluorine. Therefore chlorine, sulphur and Li has higher electron affinity among given groups.

103. **(d)** Nuclear charge : B > Be

$Be = 1s^2\ 2s^2$ (more stable)

$B = 1s^2\ 2s^2\ 2p^1$

∴ Ionisation energy of Be is greater than B due to ns^2 outer electronic configuration.

104. **(a)** Atomic numbers of N, O, F and Na are 7, 8, 9 and 11 respectively. Therefore, total number of electrons in each of N^{3-}, O^{2-}, F^-, and Na^+ are 10 and hence, they are isoelectronic.

105. **(c)** Generally, electronegativity decreases down the group as the size increases. This can also be formulated as:

$$\text{Electronegativity} \propto \frac{1}{\text{size}}$$

106. **(d)** Alkali metals have high difference in first and second ionisation energy as they achieve stable noble gas configuration after first ionisation.

107. (a) On going down the group, size increases while going from left to right in a period, size decreases. So order is: C < S < Al < Cs.

108. (b) Correct order of electronegativity values of the elements is
Si > Al; S > P; Se > Te; Ge > Ga.

109. (d) The second electron gain enthalpy of oxygen is positive as energy has to be added for the addition of another electron.

110. (c) All species are isoelectronic ($10e^-$).
In isoelectronic series, when negative charge increases the radius of ion increases.
$\therefore Mg^{2+} < Na^+ < F^- < O^{2-}$

111. (d) On moving from left to right across a period, the electron affinity becomes more negative. On moving from top to bottom in a group, the electron affinity becomes less negative.
Fluorine has exceptionally less negative electron affinity than chlorine, because adding an electron to fluorine (2p orbital) causes greater repulsion than adding an electron to chlorine (3p orbital) which is larger in size.
Note: The electron affinity (in kJ/mol) of Cl, F and O is –349, –328 and –141 respectively.

112. (N) All the given statements are true.

113. (a) Isoelectronic species have same no. of electrons.

Ions ⇒	O^{2-}	F^-	Na^+	Mg^{2+}
	8+2	9+1	11 – 1	12 – 2
No. of e^- ⇒	10	10	10	10

Therefore O^{2-}, F^-, Na^+, Mg^{2+} are isoelectronic.

114. (c) The smaller the atomic size, larger is the value of IP. Further, the atoms having half filled or fully filled orbitals, are comparatively more stable. Hence, more energy is required to remove the electron from such atoms.

115. (c) Due to large difference of 2nd and 3rd ionisation enthalpy, both A and B will belong to group-2. Due to low values of IE for B than that of A, B will be below A.

116. (b) Alkali metals have the lowest ionization energy in each period, on the other hand, Sc is a d - block element. Transition metals have smaller atomic radii and higher nuclear charge, leading to high ionisation energy.

117. (b) Due to extra stable half-filled p-orbital electronic configurations of Group 15 elements, larger amount of energy is required to remove electrons compared to Group 16 elements.

118. (a) For isoelectronic species, size of anion increases as negative charge increases. Thus the correct order is (a).

119. (a) For isoelectronic species, ionic radii decreases as the negative charge on ion decreases. Further on moving down in a group, ionic radii increases. Hence, the correct order is : $O^{2-} < N^{3-} < S^{2-} < P^{3-}$
Trick: Ionic radii ∝ Negative charge on ion.

120. (c) On moving down in a group atomic radii increases.

121. (b) In O_2^{2-}, there are two atoms, therefore, it will be largest in size.

122. (c) On moving down a group, size increases, hence ionisation enthalpy decreases. Hence Se < S and Ba < Ca. Further, Ar being an inert gas has maximum IE.

123. (d) On moving down in a group, atomic radii increases due to successive addition of extra, shell. Hence, order is : O < S < Se
Further, As is in group 15 having one less electron in its p orbital compared to group 16, hence As has higher atomic radii than Se.
i.e., O < S < Se < As

124. (b) The second ionization potential means removal of electron from unit charged cation.

$C^+ = 1s^2 2s^2 2p^1$; $N^+ = 1s^2 2s^2 2p^2$
$O^+ = 1s^2 2s^2 2p^3$; $F^+ = 1s^2 2s^2 2p^4$

Therefore, the correct order is :
O > F > N > C

125. (a) Due to stable $2s^2\ 2p^3$ configuration (half filled *p*-orbital). Nitrogen atom has highest energy.

126. (c) Among isoelectronic species ionic radii increases as the negatives charge increases.
Order of ionic radii $Ca^{2+} < K^+ < Cl^- < S^{2-}$
The number of electrons remains the same but nuclear charge increases with increase in the atomic number causing decrease in size.

127. (a) Due to stable $2s^2\ 2p^3$ configuration (half filled p-orbital), nitrogen atom has highest energy.

128. (a) Due to stable $2s^2\ 2p^3$ configuration (half filled *p*-orbital). Nitrogen atom has highest energy.

129. (d) Number of moles of chlorine $= \frac{1}{35.5}$ mol

Given, 1 eV = 23.06 kcal mol^{-1}
3.7 eV = 3.7 × 23.06 kcal mol^{-1}
i.e., 1 mole of chlorine releases energy = 3.7 × 23.06 kcal

$\therefore \frac{1}{35.5}$ mole of chlorine will release energy

$= \frac{1}{35.5} \times 3.7 \times 23.06 \text{ kcal} = 2.4 \text{ kcal}$

130. (b) Correct order of basic strength for group is hydride should be $NH_3 > PH_3 > AsH_3 > SbH_3$, it is because as the size of central atom increases electron density on the central atom decreases.

131. (b) As we move down in a group, electron gain enthalpy becomes less negative because the size of the atom increases and the distance of added electron from the nucleus increases. Negative electron gain enthalpy of F is less than Cl. This is due to the fact that when an electron is added to F, the added electron goes to the smaller n = 2 energy level and experiences significant repulsion from the other electrons present in this level. In Cl, the electron goes to the larger n = 3 energy level and consequently occupies a larger region of space leading to much less electron–electron repulsion. So the correct order is
Cl > F > Br > I.

132. **(d)** All the given species contains 10 e^- each i.e. isoelectronic. For isoelectronic species anion having high negative charge is largest in size and the cation having high positive charge is smallest.

133. **(b)** The correct order of ionisation enthalpies is

$$F > P > S > B$$

Note: On moving along a period ionization enthalapy increases from left to right and decreases from top to bottom in a group. But this trend breaks up in case of such element having fully or half filled stable orbitals.
In this case, P has a stable half filled electronic configuration, hence its ionisation enthalapy is greater in comparision to S. Therefore the correct order is

$$B < S < P < F.$$

134. **(b)** On moving down in group I from Li to Cs, ionisation enthalpy decreases, hence the reactivity increases. The halogens are most reactive elements due to their low bond dissociation energy, high electron affinity and high enthalpy of hydration of halide ion. Hence, their reactivity decreases with increase in atomic number.

135. **(c)** In a period the value of ionisation potential increases from left to right with breaks where the atoms have somewhat stable configuration. In the given list N has half-filled stable orbitals. Hence N has highest ionisation energy. Thus the correct order is

$$B < C < O < N$$

and not as given in option (c)

136. **(a)** CaO is basic as it forms strong base $Ca(OH)_2$ on reaction with water.

$$CaO + H_2O \longrightarrow Ca(OH)_2$$

CO_2 is acidic as it dissolves in water forming unstable carbonic acid.

$$H_2O + CO_2 \longrightarrow H_2CO_3$$

Silica (SiO_2) is insoluble in water and acts as a very weak acid.
SnO_2 is amphoteric as it reacts with both acid and base.

$$SnO_2 + 2H_2SO_4 \longrightarrow Sn(SO_4)_2 + 2H_2O$$
$$SnO_2 + 2KOH \longrightarrow K_2SnO_3 + H_2O$$

137. **(d)** As the size decreases, the basic nature of oxides changes to acidic nature, i.e., acidic nature increases.

$$\underset{\text{Acidic}}{SO_2 > P_2O_3} > \underset{\text{Weak acidic}}{SiO_2} > \underset{\text{Amphoteric}}{Al_2O_3}$$

SO_2 and P_2O_3 are acidic as their corresponding acids are H_2SO_3 and H_3PO_3.

138. **(a)** O^- ion exerts a force of repulsion on the incoming electron. Hence energy is required to overcome it.

139. **(a)** O^{2-} and F^- are isoelectronic ($10e^-$). Hence have same number of electron but different number of proton, F^- has 9 proton whereas O^{2-} has 8 proton. Therefore F^- has greater nuclear charge compared to O^{2-}. So O^{2-} will have greater ionic radius, i.e. $O^{2-} > F^-$
Further Li^+ and B^{3+} are also isoelectronic ($2e^-$) therefore, no. of proton in Li is 3 and in B is 5, So,

$$Li^+ > B^{3+}$$

Hence, the correct order of atomic size is

$$O^{2-} > F^- > Li^+ > B^{3+}$$

Therefore, O^{2-} has the highest value of ionic radius.

140. **(d)** Na_2O (basic), SO_2 and B_2O_3 (acidic) and ZnO is amphoteric

Chemical Bonding and Molecular Structure

1. **(b)** According to Fajan's rule,
Covalent character $\propto$ size of Anion
As the cation is same in all the options. Therefore covalent character will depend upon the size of anion. Covalent character : A < B < C < D; Size of anion : $F^- < Cl^- < Br^- < I^-$

2. **(c)** KCl is an ionic compound while others (PH_3, O_2, B_2H_6, and H_2SO_4) are covalent compounds.

3. **(N)**
(a) $NO^+ = 7 + 8 - 1 = 14\ e^-$ and $O_2 = 16\ e^-$
i.e., not isoelectronic.
(b) Boron forms only covalent compounds. This is due to its extremely high ionisation energy.
(c) Compounds of Tl^+ are much more stable than those of Tl^{3+}.
(d) $LiAlH_4$ is a versatile reducing agent in organic synthesis.

4. **(a)** The value of lattice energy depends on the charge present on the ions and the inter-ionic distance between them.

5. **(98)** $\Delta H_{R.E} = \Delta H_{F(exp)} - \Delta H_{F(Theo)}$
$\Delta H_{F(exp)}$ for $X_2Y_{(g)} = 80$ kJ/mole (given)

$$X_{2(g)} + \frac{1}{2}Y_{2(g)} \rightarrow X_2Y_{(g)}$$

$$\Delta H_{F(Theo)} = \left(BE_{X\equiv X} + \frac{1}{2}BE_{Y=Y}\right) - \left(BE_{X=X} + BE_{X=Y}\right)$$

$$= \left(940 + \frac{1}{2} \times 500\right) - (410 + 602) = 178 \text{ kJ/mole}$$

$\Delta H_{R.E} = 80 - 178$
$= -98$ kJ/mol
$|\Delta H_{R.E}| = 98$ k/J mol

6. **(12)** $\ddot{O} = \ddot{N} - \ddot{O}$:
12 electron

7. **(c)** (A) OH, NO_2 & (B) NH_2, NO_2
are steam volatile due to intramolecular hydrogen bonding. Compound (C) and (D) shows intermolecular hydrogen bonding.

8. **(a)** Due to resonance bond length is identical in ozone. Therefore statement I is true and statement II is false.

9. **(a)** $NO = 7e^-$, $NO_2 = 7e^-$ } incomplete octet with odd electron
$BCl_3 = 6e^-$, $AlCl_3 = 6e^-$
$H_2SO_4 = 12e^-$ } molecule with expanded octet
$PCl_5 = 10e^-$
$CCl_4 = 8e^-$, $CO_2 = 8e^-$ } molecules obeying octet rule.

10. **(6)** According to octet rule central atom should have 8 valence electrons.

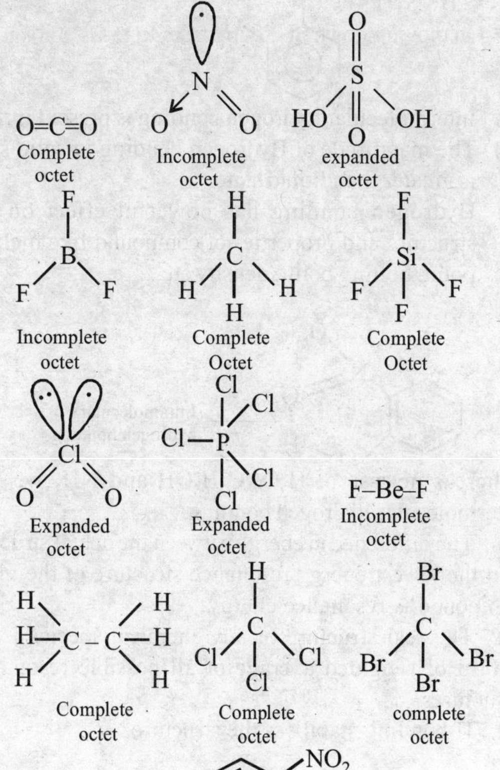

11. **(5)** CH_3OH, H_2O, (NO_2, OH) HF, NH_3
can show hydrogen bonding.

12. **(12)** $NiS + HNO_3 + HCl \longrightarrow NiCl_2 + S + NO + H_2O$
(A)

$NiCl_2 + NH_4OH + \text{Dimethylglyoxime} \longrightarrow$
(A)

$Ni(dmg)_2 + NH_4Cl_4 + H_2O$
(B)

H bond
$O^- \cdots H-O$
$CH_3-C=N \rightarrow Ni^{+2} \leftarrow N=C-CH_3$
$CH_3-C=N \rightarrow \quad \leftarrow N=C-CH_3$
$O-H \cdots O^-$
H bond

Total 12 Protons do not involve in H-bonding.

13. (8) Structure of NO_2^- is :

Number of valence electrons around N atom = 8

14. (b)
(A) Generally hydrogen bonding exists when H is covalently bonded to the highly electronegative atom like F, O, N etc,
(B) Intramolecular H bonding is present in o-nitrophenol.

(C) Intermolecular hydrogen bonding is present in HF.
(D) The magnitude of Hydrogen bonding in solid state is greater than liquid state.
(E) Hydrogen bonding has powerful effect on the structure and properties of compound like melting point, boiling point, density etc..

15. (a)

Intramolecular hydrogen bonding

Whereas in case of H_2O, C_2H_5OH and NH_3 we find intermolecular hydrogen bonding.

16. (d) The difference in energy between the actual structure and the lowest energy resonance structure of the given compound is resonance energy.

17. (c) The real structure of any chemical species is the hybrid of weighted average of all possible resonating structures.

18. (a) H-bonding stabilises the structure.

There is maximum number of H-bonding between 1, 2, 3 and 4 –OH groups

19. (3) ICl_4^-, BrF_3 and NO_2^+ do not have odd number of electron.

20. (6) Total no of lone pairs on oxygen atoms = 6

21. (b) Ice > Liquid water > Impure water
Due to impurity extent of H-Bonding decreases.

22. (N) Order should be : C < A < B < D

D is the most stable structure.

23. (b) NO → Odd electron molecule
H_2SO_4 → Expanded octet of S

$\cdot\ddot{N} = \ddot{O}:$ and

24. (c) The strength of H–bonding depends upon two factors:
(1) Electronegativity difference between the hydrogen atom and electronegative atom.
(2) Number of H-bonding interaction.
So the order of H-bonding is $CH_4 < HCN < NH_3$.

25. (c) Electron deficient species have less than 8 electrons (or two electrons for H) in their valence (incomplete octet). B_2H_6, BCl_3 have incomplete octet.

26. (b) PCl_5 forms five bonds by using the d-orbitals to "expand the octet". But NCl_5 does not exist because there are no d-orbitals in the valence shell (2nd shell). Therefore, there is no way to expand the octet.

27. (d) 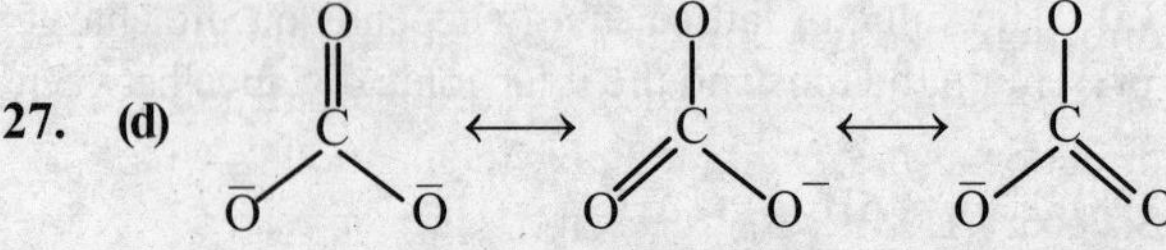

3 Canonical structures

28. (b) Compounds involved in chelation become non-polar. Consequently such compounds are soluble in non-polar solvents like ether, benzene etc. and are only sparingly soluble in water, whereas meta and para isomers are more soluble in water and less soluble in non-polar solvents.

Intra-molecular H-bonding

29. (c) Greater the difference between electronegativity of bonded atoms, stronger will be bond. Since F is most electronegative, hence F – H...... F is the strongest bond.

30. (c) In ether, there is no H-bonding while alcohols have intermolecular H-bonding.

31. (c) **Statement-I :**
$CH_3 - CH = CH - CHO$ has conjugation between the double bond and the carbonyl group, which leads to resonance stabilization and contributes to greater charge separation across the molecule. So the dipole moment is greater than Butanal.
Statement-II :

$H_3C - CH = CH - \overset{O}{\overset{\|}{C}} - H$ ↔ $H_3C - \overset{\oplus}{C}H = CH - \overset{O^{\ominus}}{\overset{|}{C}} - H$ $\equiv H_3C - \overset{\delta^+}{CH} = CH = \overset{O^{\delta^+}}{\overset{\|}{C}} - H$ ④ ③ ② ①

$C_1 - C_2$ bond has partial double bond character that means lesser bond length than $C_1 - C_2$ bond of other compound. Statement-II is false.

32. (a) Difference between ionization energy & electron gain enthalpy increases, ionic character increases.

$\therefore$ The correct order of ionic character is:

EB > EA > EC > ED

33. (a) Increasing order of Dipole moment

$NF_3 < HBr < H_2S < CHCl_3$

$\mu = 0.24D$ 0.76D 0.95 1.04D

34. (c) BF_3: trigonal planar, 120°; ClF_3: T-shaped, 87.5°; PF_3: Pyramidal, 97°

Order of bond angle is

$ClF_3 < PF_3 < BF_3$

35. (a)

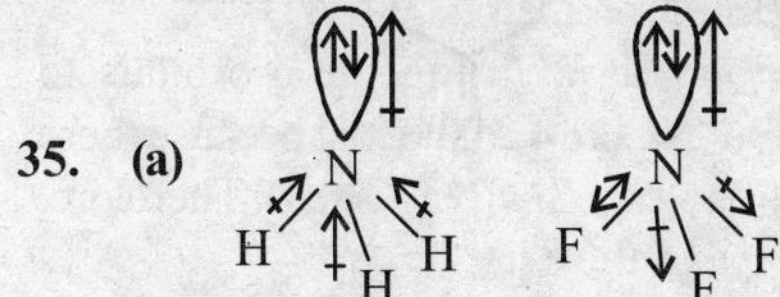

directions of dipole in such a way that net dipole moment of NH_3 is higher than NF_3 as μ of lone-pair is in same direction as the resultant dipole moment of N – H bonds. So, both (A) & (R) are true, & (R) is the correct explanation of (a).

36. (6) H—F: $\mu \neq 0$; H—H: $\mu = 0$

H_2S: $\mu \neq 0$; O=C=O: $\mu = 0$; NH_3: $\mu \neq 0$

BF_3: $\mu = 0$; CH_4: $\mu = 0$; $CHCl_3$: $\mu \neq 0$

SiF_4: $\mu = 0$; H_2O: $\mu \neq 0$; F — Be — F: $\mu = 0$

37. (c) CH_4 and PF_5, $\mu_{net} = 0$ (non-polar)

μ_{NH_3} > μ_{NF_3}

Vector addition of bond moment & lone pair moment; Vector subtraction of bond moment & lone pair moment

38. (5) Polar molecule have non-zero dipole moment.

Polar molecule NF_3, H_2O, H_2S, HBr, HCl

Non-polar molecules have zero dipole moment.

Non Polar molecule : $BeCl_2$, BCl_3, XeF_4, CCl_4, CO_2, H_2 ($\mu = 0$)

So answer is 5.

39. (c) Ionic character of a molecule increases with increase in the electronegativity difference between the atoms. The correct increasing order of ionic character is:

$N_2 < SO_2 < ClF_3 < K_2O < LiF$

40. (b) The ionic character decrease as we move from left to right in the periodic table and it increases as we move top to bottom in a group.

$\therefore$ Ionic character follows the order as

$KCl > BaCl_2 > CoCl_2 > AgCl$

$\therefore$ least ionic is AgCl.

41. (2.5) Given : $\mu = 1.2$ D

$1\ D = 10^{-18}$ esu cm $= 10^{-10}$ esu Å

d = 1Å

$\mu = q \times d \Rightarrow 1.2\ D = q \times 1Å$

1.2×10^{-10} esu Å $= q \times 1Å$

$\therefore$ $q = 1.2 \times 10^{-10}$ esu fractional acharge

$$= \frac{\text{partial charge}(q)}{\text{Total charge}(c_0)} \frac{1.2 \times 10^{-10} \text{ esu}}{4.802 \times 10^{-10} \text{ esu}} = 0.25 = 2.5 \times 10^{-1}$$

42. (b)

NF_3: Less dipole moment due to cancellation

NH_3: More dipole moment due to addition

Thus, statement I is false as NH_3 is more polar. Statement II is also false as dipoles due to lone pair and bond pairs are aligned in the same direction in NH_3 but in opposite direction in NF_3.

43. (3) Non-polar molecules (Due to symmetry):

CH_4: $\mu = 0$; BF_3: $\mu = 0$; O=C=O: $\mu = 0$

44. (d) $CHCl_3$ (Cl, C, Cl, Cl, Cl): $\mu \neq 0$

$CHCl_3$ is polar molecule and rest all molecules are non-polar due to cancellation of bond dipoles due to symmetry.

45. **(4)** The non-polar molecules are CO_2, H_2, CH_4 and BF_3

46. **(a)** Among the given compounds, the following compound has the highest dipole moment because both the +ve and –ve ends acquire aromaticity.

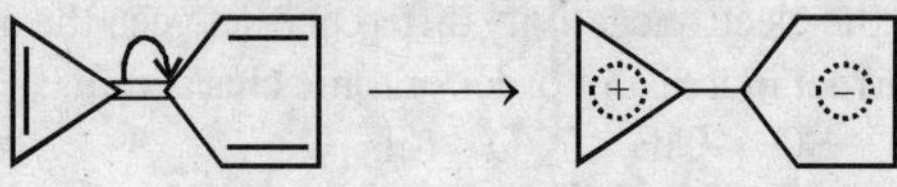

(Aromatic) (Aromatic)

47. **(a)** SO_2 and NH_3 are polar molecules. They are constituent particles of polar molecular solids. AlN and MgO are ionic compounds.

48. **(c)** **Hint:** Check for assymetry in the molecular geometry or electronegative distribution for net non-zero dipole moment.

(a) Benzene $\rightarrow$ non polar
Anisidine $\rightarrow$ polar

(b) Cl, Cl (para-dichlorobenzene) $\rightarrow$ Non polar, Cl, Cl (meta-dichlorobenzene) $\rightarrow$ polar

(c) CH_2Cl_2, $\mu_{net} \neq 0$, polar; $CHCl_3$, $\mu_{net} \neq 0$, polar

(d) 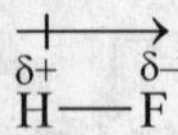 $\rightarrow$ polar

$\rightarrow$ non polar

49. **(3)** (A), (B) and (C) are factors which affect the percent covalent character of the ionic bond according to Fajan's rule

50. **(b)** **Statement I:**

$\overset{\delta+}{H}$—$\overset{\delta-}{F}$ (with crossed arrow pointing from H to F)

Tail is on the positive atom and arrow head on the negative atom.

Statement II: The crossed arrow symbolises the direction of the shift of electron density in the molecule.

51. **(c)** According to Fajan's Rule, Covalent character is higher for large size anion and small size cation.

A. KF > KI – Incorrect, LiF > KF – correct
B. KF < KI – correct, LiF > KF – correct
C. $SnCl_4 > SnCl_2$ – correct, CuCl > NaCl – correct
D. LiF > KF – correct, CuCl < NaCl – incorrect
E. KF < KI – correct, CuCl > NaCl – correct

52. **(c)** According to Fajan's rule, small size of cation and large size of anion will have high covalent character. So the order is: LiCl > NaCl > KCl > CsCl

53. **(3)** Symmetrical molecules have zero dipole moment like BF_3, BeF_2, SiF_4, CCl_4 and PF_5.
Unsymmetrical molecules have net dipole moment like – NF_3, $CHCl_3$ and H_2S

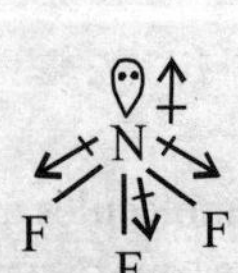

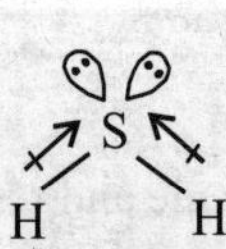

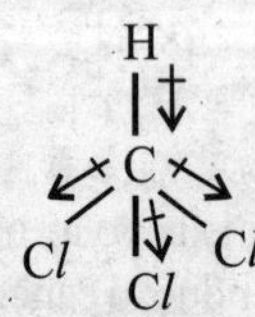

54. **(3)** BeF_2, BF_3 and $CCl_4 \Rightarrow \mu_{net} = 0$

H_2O, NH_3 and HCl $\Rightarrow \mu_{net} \neq 0$

55. **(d)** $\mu_{CCl_4} = \mu_{CH_4} = 0$ due to symmetrical structure but $\mu_{CHCl_3} \neq 0$. So dipole moment order is :

$$CHCl_3 > CH_4 = CCl_4$$

56. **(c)** Dipole moment = (Distance between opposite charges) × (charge, q)

$$\mu = q \times d$$

So, greater the distance between the opposite charges higher the dipole. Due to the resonance, the greater charge separation occurs in (c).

(Resonance structures of p-nitroaniline: $O^{\ominus}$, O, $N^{\oplus}$, NH_2 $\longleftrightarrow$ $O^{\ominus}$, $O^{\ominus}$, $N^{\oplus}$, $\overset{\oplus}{N}H_2$)

57. **(b)** Dipole moment (μ) = $q \times d$

$\Rightarrow$ 1 D $\approx 10^{-18}$ esu cm

0.38×10^{-18} esu cm = $q \times (1.617 \times 10^{-8}$cm)

$q = 2.35 \times 10^{-11}$esu

So, fractional charge $= \dfrac{\text{Partial charge}}{\text{Total charge}} = \dfrac{q}{Q}$

$$= \frac{2.35 \times 10^{-11}\text{ esu}}{4.802 \times 10^{-10}\text{ esu}} = 0.049 \approx 0.05$$

58. **(b)** According to Fajans rules, small and highly charged cation has greatest covalent character while large cation with small charge has greatest ionic character.

59. **(b)** The geometry of IF_5 is square pyramidal with an unsymmetric charge distribution, therefore this molecule is polar.

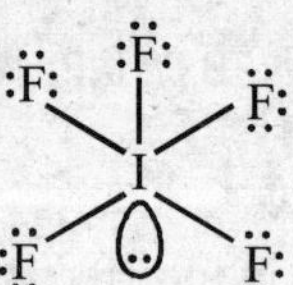

60. **(d)** Given $e = 1.60 \times 10^{-19}$ C
$d = 9.17 \times 10^{-11}$ m

From $\mu = e \times d$

$\mu = 1.60 \times 10^{-19} \times 9.17 \times 10^{-11} = 14.672 \times 10^{-30}$

$$\%\text{ ionic character} = \frac{\text{Observed dipole moment}}{\text{Dipole moment for 100\% ionic bond}}$$

$$= \frac{6.104 \times 10^{-30}}{14.672 \times 10^{-30}} \times 100 = 41.5\%$$

61. **(a)** In 1, 2-dichlorobenzene, the two dipoles are at 60° (i.e., unsymmetric). Thus, it possesses dipole moment.

62. **(d)** CCl_4 is a non-polar molecule and it has symmetrical tetrahedral structure. Although each of the C-Cl bond is polar but the resultant of all four dipole moments is zero.

63. (b) [structure: Cl and H_3C on one C, H and Cl on other C; C = C]

Very low dipole moment (μ) due to $C - CH_3$ bond.

64. (c) In nitrogen molecule, both the nitrogen atoms have same electronegativity. So it has zero polarity and hence, less tendency to break away and forms ions.

65. (d) Smaller the size and higher the charge, more will be the polarising power of the cation. Since the order of the size of cation is $K^+ > Ca^{2+} > Mg^{2+} > Be^{2+}$, so the correct order of polarising power is : $K^+ < Ca^{2+} < Mg^{2+} < Be^{2+}$

66. (b) Both NO_2 and O_3 have angular shape and hence, will have net dipole moment.

67. (a) $SO_2 \Rightarrow 2\sigma$ bond + 1 lone pair $\Rightarrow sp^2$

$NO_2^- \Rightarrow 2\sigma$ bond + 1 lone pair $\Rightarrow sp^2$

$NO_3^- \Rightarrow 2\sigma$ bond $\Rightarrow sp$

68. (b)

(A) ICl_2^- B.P : L.P = 2 : 3

(B) H_2O B.P : L.P = 2 : 2

(C) SO_2 B.P : L.P = 4 : 1

(D) XeF_4 B.P : L.P = 4 : 2

69. (b) Statement II is incorrect since lone pairs cannot occupy axial positions. So structure I is most stable.

70. (a) PF_5 : 5 bond pair + 0 $\ell p \rightarrow sp^3d$

SF_6 : 6 bond pair + 0 $\ell p \rightarrow sp^3d^2$

$Ni(CO)_4$: Ni (0)

In presence of ligand field :-

Ni(0) : [Ar] : 3d ⇅ ⇅ ⇅ ⇅ ⇅ | 4s ☐ | 4p ☐ ☐ ☐ — sp^3 hybridisation

$[PtCl_4]^{2-}$: Pt^{2+}

In presence of ligand field :-

Pt^{2+} : [Kr] : 5d ⇅ ⇅ ⇅ ⇅ ☐ | 6s ☐ | 6p ☐ ☐ ☐ — dsp^2 hybridisation

71. (c) [structure of SF_4: S with one lone pair, two axial F and two equatorial F]

$(S - F)_{axial} > (S - F)_{equitorial}$

SF_4 has 4 bond pairs of electrons and 1 lone pair of electrons.

Hybridisation of SF_4 is sp^3d

72. (a) $\overset{6}{C}H_3 - \overset{5}{C} \equiv \overset{4}{C} - \overset{3}{C}(CH_3)(CH_3) - \overset{2}{C}H = \overset{1}{C}H_2$

sp^3, sp, sp, sp^3, sp^2, sp^2; CH_3 groups: sp^3, sp^3

sp^3 C = 4, sp^2 C = 2, sp C = 2

73. (a) Given A is rarest, monoatomic, non-radioactive p-block element and form AX_4Y type of molecule.

∴ It is concluded that it is Xe.

It is given that electronegativity of A is less than X & Y

It is given that electronegativity of X & Y is highest and second highest respectively among all element.

∴ X & Y are flourine & oxygen respectively

∴ Compound is $XeOF_4$ with square pyramidal shape.

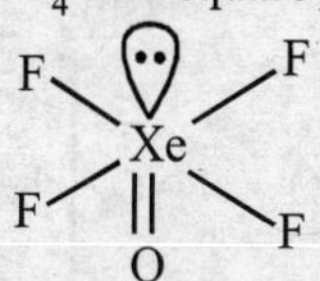

74. (d)

Molecule	Hybridisation	Dipole Moment	Lone pair on the central atom
SO_2	sp^2	Non-zero	1
NF_3	sp^3	Non-zero	1
NH_3	sp^3	Non-zero	1
XeF_2	sp^3d	zero	3
ClF_3	sp^3d	Non-zero	2
SF_4	sp^3d	Non-zero	1

75. (a)

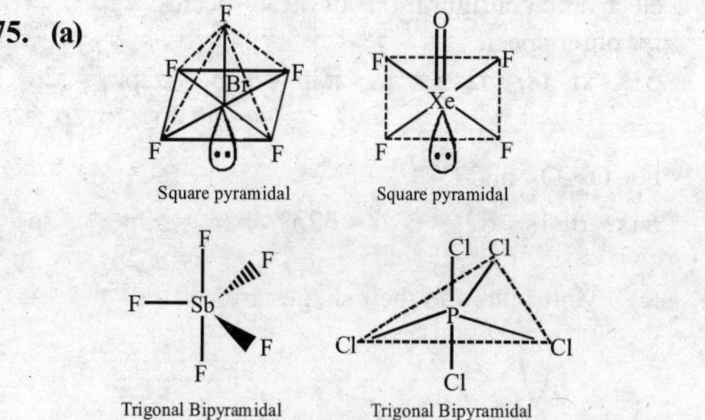

BrF_5 : Square pyramidal due to one lone pair on Br

$XeOF_4$: Square pyramidal due to one lone pair on Xe

SbF_5 : Trigonal bipyramidal

PCl_5 : Trigonal bipyramidal

76. (b)

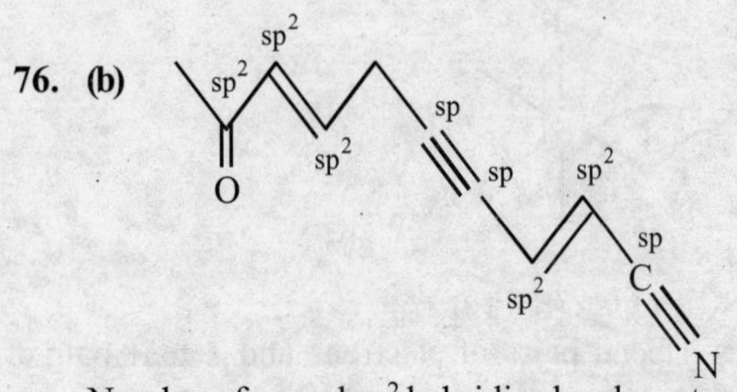

Number of sp and sp^2 hybridised carbon atom are 3 and 5.

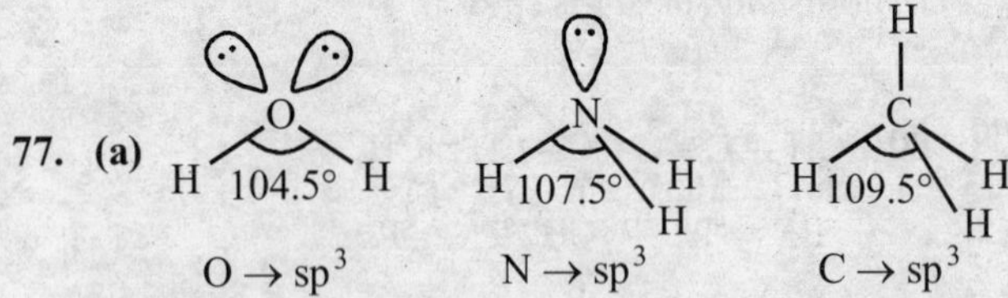

77. (a)

$O \rightarrow sp^3$ $N \rightarrow sp^3$ $C \rightarrow sp^3$

Dipole moment is due to, electronegativity difference and symmetry.

So order is $H_2O > NH_3 > CH_4$

H_2O & NH_3 are Lewis Bases.

NH_3 act as Lowry-Bronsted base.

Hence, A, B & C are correct.

78. (6) Linear species are

$Cl-Be-Cl$, $O=C=O$, $N^-=N^+=N^-$
(sp) (sp) (sp)

$F-Xe-F$, $O=\overset{+}{N}=O$, $[I-I-I]^{\ominus}$
sp^3d sp sp^3d

79. (a) $BF_3 \rightarrow sp^2(3bp)$

$NO_2^- \rightarrow sp^2.(2bp + 1lp)$

$H_2O \rightarrow sp^3(2bp + 2lp)$

$NO_2 \rightarrow sp^2(2bp + 1\ e^-$ in hybrid orbital)

$NH_2^- \rightarrow sp^3(2bp + 2lp)$

80. (4) One unpaired e^- is present in : $C_2^-; O_2^-; H_2^+; He_2^+$

N_2, O_2^{2-} and CN^- have zero while O_2 has two unpaired e^{-1}.

These calculations can be done using the molecular electronic configurations of these species.

for other species:

$$\sigma 1s < \sigma^* 1s < \sigma 2s < \sigma^* 2s < \pi 2p_x = \pi 2p_y < \sigma 2pz_1 < \pi 2p_x = \pi^* 2p_y < \sigma^* 2p_z$$

For O_2, O_2^- and O_2^- :

$$\sigma 1s < \sigma^* 1s < \sigma 2s < \sigma^* 2s < \sigma 2p_z < \pi 2p_x = \pi 2p_y < \pi^* 2p_x = \pi^* 2p_y < \sigma^* 2p_z$$

81. (c) Molecules and their shapes are:

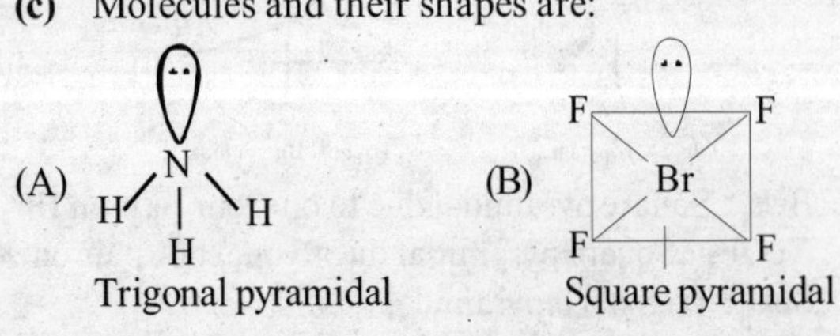

Trigonal pyramidal Square pyramidal

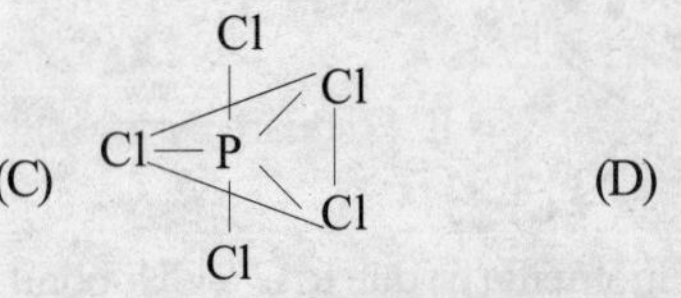

Trigonal bipyramidal Tetrahedral

82. (d) A-III, B-IV, C-I, D-II

83. (b)

	Species	Hybridisation	Shape
(A)	SF_4	sp^3d	Sec Saw
(B)	BrF_3	sp^3d	Bent T-Shape
(C)	BrO_3^-	sp^3	Pyramidal
(D)	NH_4^+	sp^3	Tetrahedral

84. (c) I $\begin{cases} PF_5 - sp^3d \\ Br\ F_5 - sp^3d^2 \end{cases}$

II $\begin{cases} SF_6 \quad sp^3d^2 \\ [Co(NH_3)_6]^{3+} \quad d^2sp^3 \end{cases}$

Hybridisation of P in PF_5 is sp^3d but the hybridisation of Br in BrF_5 is sp^3d^2. So, statement-1 is false.

Hybridisation of S in SF_6 is sp^3d^2 but the hybridisation of Co^{3+} in $[Co(NH_3)_6]^{3+}$ is d^2sp^3 and NH_3 is a strong field ligand forcing the unpaired electrons to pair up. So, statement-II false.

85. (6) SO_2, C_2H_4, BCl_3, HCHO, C_6H_6, BF_3 are sp^2 hybridised central atom

86. (c)

A ICl IV linear

B Cl–I(–Cl)–Cl I T-shape

C. 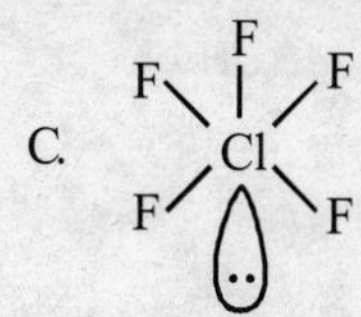 II. square pyramidal

D. (structure of IF_7) III Pentagonal bipyramidal

87. (a) $O=N(O^-)\rightarrow O$ — sp^2; BCl_3 — sp^2; ClO_2^- — sp^3; ClO_3^- — sp^3

88. (c)

SO_3^{2-} : Pyramidal; $S_2O_3^{2-}$, SO_4^{2-}, $S_2O_7^{2-}$: Tetrahedral with respect to central atom

89. (3)

PF_5 : Trigonal bipyramidal

BrF_5 : Square Pyramidal

PCl_5 : Trigonal bipyramidal

$[PtCl_4]^{2-}$: Square Planar

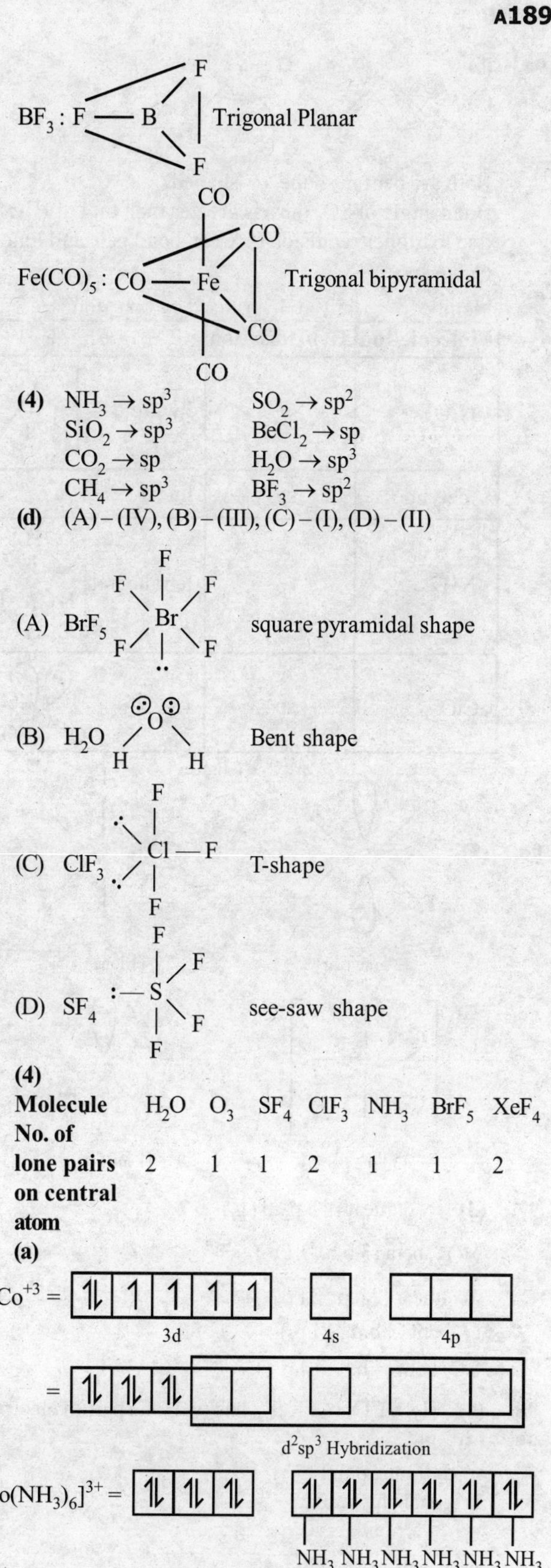

BF_3 : Trigonal Planar

$Fe(CO)_5$: Trigonal bipyramidal

90. (4) $NH_3 \rightarrow sp^3$ $SO_2 \rightarrow sp^2$
$SiO_2 \rightarrow sp^3$ $BeCl_2 \rightarrow sp$
$CO_2 \rightarrow sp$ $H_2O \rightarrow sp^3$
$CH_4 \rightarrow sp^3$ $BF_3 \rightarrow sp^2$

91. (d) (A) – (IV), (B) – (III), (C) – (I), (D) – (II)

(A) BrF_5 square pyramidal shape

(B) H_2O Bent shape

(C) ClF_3 T-shape

(D) SF_4 see-saw shape

92. (4)

Molecule	H_2O	O_3	SF_4	ClF_3	NH_3	BrF_5	XeF_4
No. of lone pairs on central atom	2	1	1	2	1	1	2

93. (a)

Co^{+3} = 3d: ⥮ 1 1 1 1 ; 4s: ☐ ; 4p: ☐☐☐

= 3d: ⥮ ⥮ ⥮ ☐☐ ; 4s: ☐ ; 4p: ☐☐☐

d^2sp^3 Hybridization

$[Co(NH_3)_6]^{3+}$ = ⥮ ⥮ ⥮ ; ⥮ ⥮ ⥮ ⥮ ⥮ ⥮ (NH_3 NH_3 NH_3 NH_3 NH_3 NH_3)

$[Co(NH_3)_6]^{+3}$ – d^2sp^3 hybridization
Similarly, BrF_5 – sp^3d^2 hybridization
$[PtCl_4]^{-2}$ – dsp^2 hybridization
SF_6 – sp^3d^2 hybridization

94. (b)

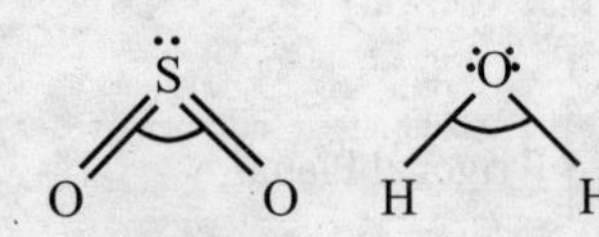

Both are bent in shape. (V-shaped)
Bond angle of SO_2 (sp^2) is greater than that of H_2O (sp^3) due to higher repulsion between bond pair and lone pair electrons in H_2O.

95. (a)

Molecule/Ion	Hybridisation	Shape
H_3O^+	sp^3	Pyramidal $[H-\ddot{O}(H)-H]^+$
Acetylide	sp	linear $\bar{C}\equiv\bar{C}$
NH_4^+	sp^3	tetrahedral $[NH_4]^+$
ClO_2^-	sp^3	Bent $O=Cl-O^-$

96. (3)

[2 lone pair]

[1 lone pair]

[3 lone pair]

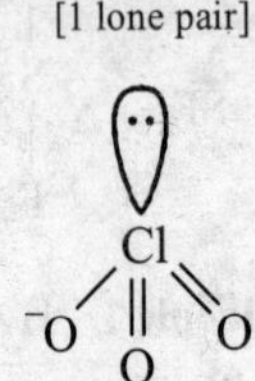

[1 lone pair]

97. (3) N_3^- linear (4 b.p., 0 l.p.)
NO_2^- bent (3 b.p., 1 l.p.)
I_3^- linear (2 b.p., 3 l.p.)
O_3 bent (3 b.p., 1 l.p.)
SO_2 bent (4 b.p., 1 l.p.)

98. (4) H_2O, CO, N_2, XeF_4, has two lone pair of electrons.

99. (3) PF_5
sp^3d (0 lone pair)
Trigonal bipyramidal

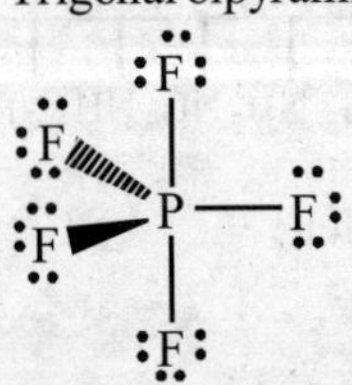

BrF_4^-
sp^3d^2 (2 lone pair)

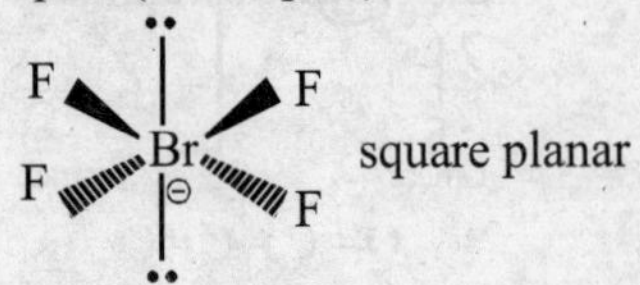

IF_5
sp^3d^2 (1 lone pair)

square pyramidal

BrF_5
sp^3d^2 (1 lone pair)

square pyramidal

$XeOF_4$
sp^3d^2 (1 lone pair)

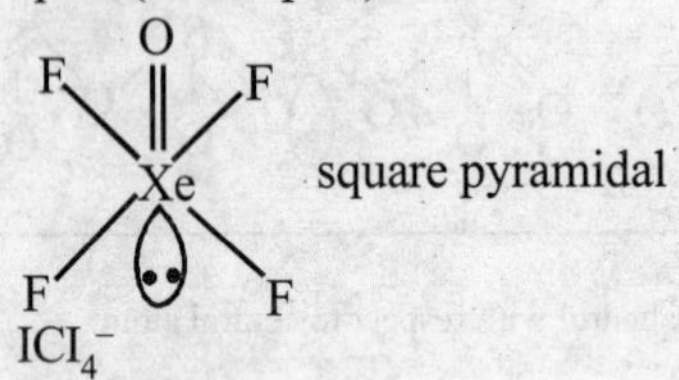

ICl_4^-
sp^3d^2 (2 lone pair)

square planar

100. (4) XeF_4, BrF_4, $[Cu(NH_3)_4]^{+2}$, $[PtCl_4]^{2-}$ has square planar shape.
$[FeCl_4]^{2-}$, SiF_4 has tetrahedral shape.
$SF_4 \rightarrow$ Trigonal bipyramidal geometry, See – Saw Shape.

101. (d)

102. (5)

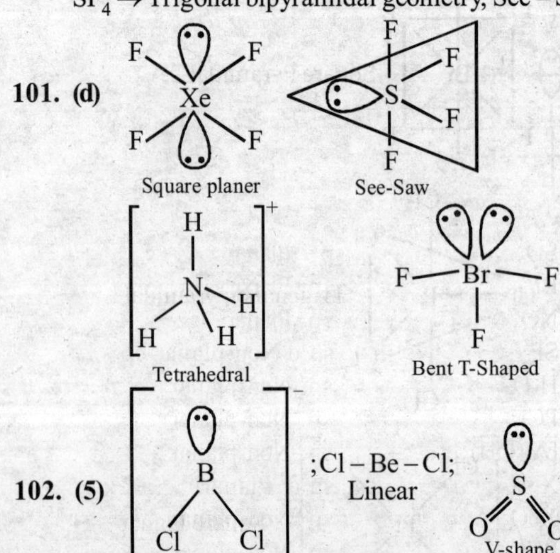

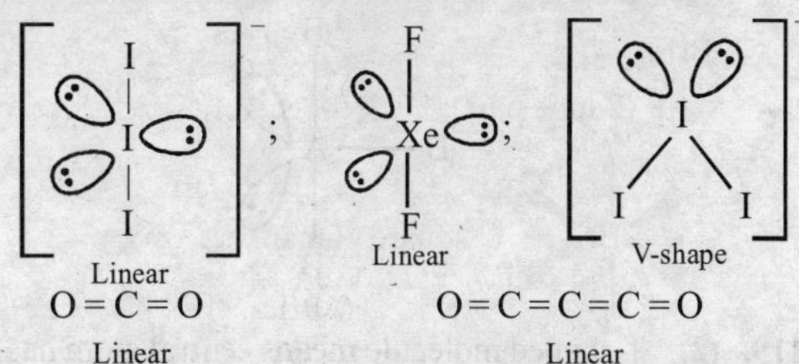

103. (b) A - IV, B - III, C - II, D - I

104. (d) $\overset{-1}{F}$ — O(+2) — $\overset{-1}{F}$

Molecule is bent 'V' shaped and oxidation state of oxygen is (+2).

105. (b)

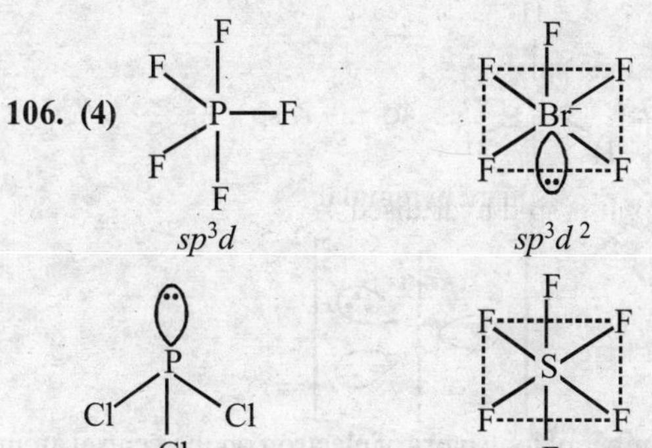

106. (4)

sp^3d sp^3d^2

sp^3 sp^3d^2

sp^3d^2 sp^3d

sp^3d^2

107. (6)

SO_3	—	sp^2 Planar
BF_3	—	sp^2 Planar
NO_3^-	—	sp^2 Planar
SF_4	—	sp^3d Non-planar
H_2O_2	—	sp^3 Non-planar
PCl_3	—	sp^3 Non-planar
$[Al(OH)_4]^-$	—	sp^3 Non-planar
XeF_4	—	sp^3d^2 Planar
XeO_3	—	sp^3 Non-planar
PH_4^+	—	sp^3 Non-planar

108. (c)

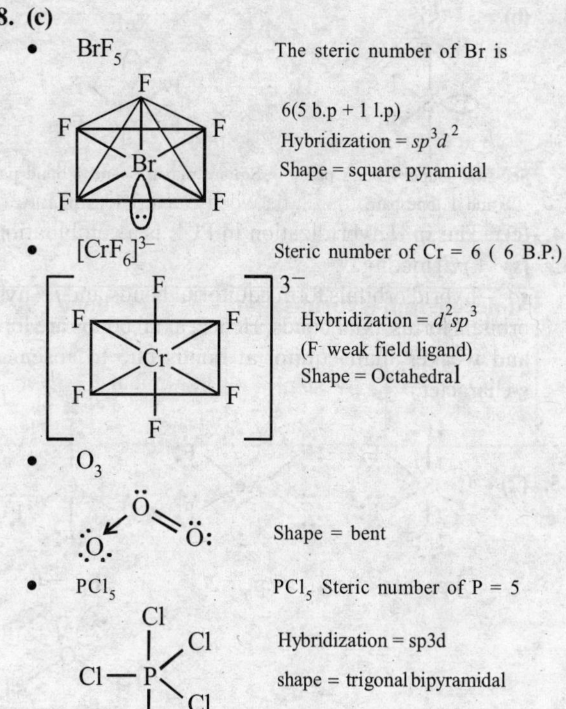

- PCl_5 — PCl_5 Steric number of P = 5

Hybridization = sp3d

shape = trigonal bipyramidal

109. (b) Isoelectronic species have same number of electrons valence

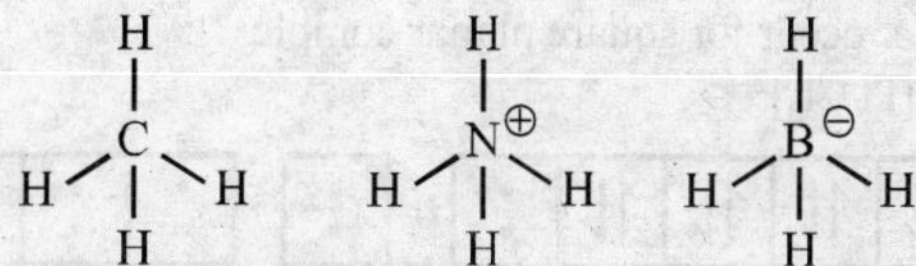

All have steric number 4.

All are tetrahedral and each have 10 electrons.

110. (c) As Fluorine is more electronegative it occupies the axial position and lone pairs occupy equitorial position

Steric no. = 5 (sp^3 d), lone pair = 2

Bent T shape.

111. (1) S.N = 5[5 b.p + 0 *l*. p]

$PF_5 \Rightarrow sp^3d$ hybridisation

112. (a) For sulphur steric number = 4 + 1 = 5

Hybridization = sp^3d, shape – see-saw

Fluorine being more electronegative preferentially occupy the axial position.

Structure is :

113. (b)

T-shaped two lone pair. Trigonal planar three bond pair.

See-saw shape 4 bond pair and 1 lone pair.

XeF_4

Square planar shape 4 bond pair and 1 lone pair.

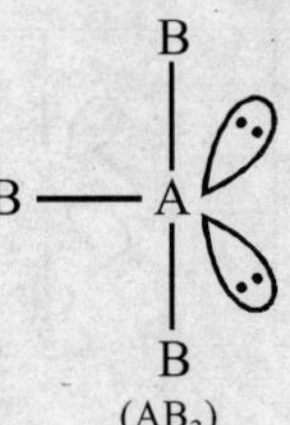

114. (c) The sp^3d hybridization in PCl_5 is a combination of $[sp^2 + pd]$ mixing.
sp^2 – hybrid orbitals form equitorial bonds and pd hybrid orbitals forms axial bonds. Hence, axial bonds are longer and weaker than equitorial bonds due to absence of s-character.

115. (2)

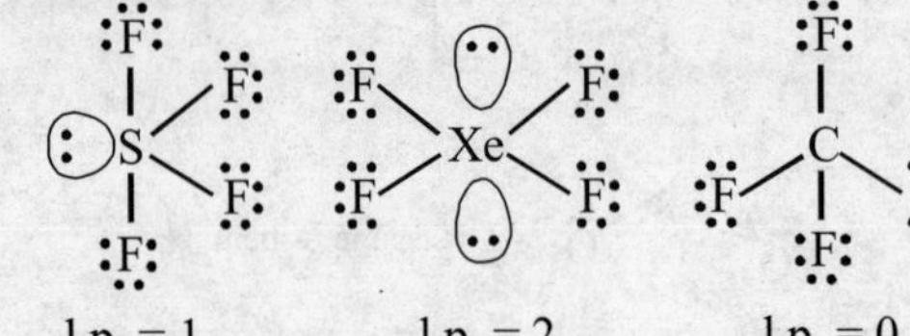

H–O–H l.p. = 2

116. (b)
(A) In case of heavier elements, i.e., $5d$ series or Pt, crystal field splitting increases such that pairing of e^- occur for square planar complex.
$[Pt\,Cl_4]^{2-}$

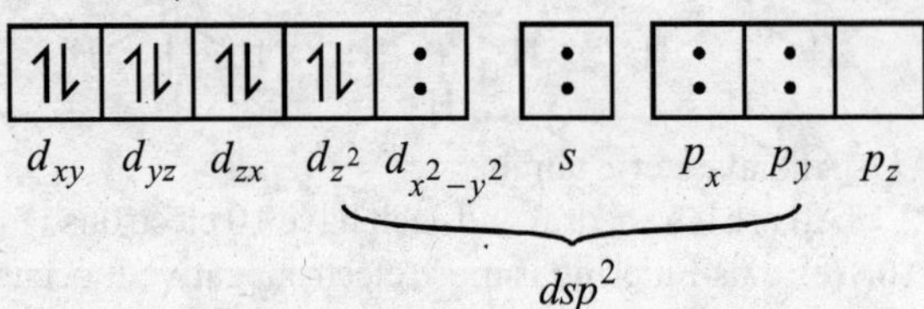

(B) $\ddot{Br}F_5$ – 5 bond pairs and 1 lone pair of e^-; Hence, sp^3d^2 hybridisation.
(C) PCl_5 – 5 bond pairs of e^-; Hence, sp^3d hybridisation.
(D) $[Co(NH_3)_6]^{3+} : Co^{3+}(d^6)$
Due to pairing of e^-:

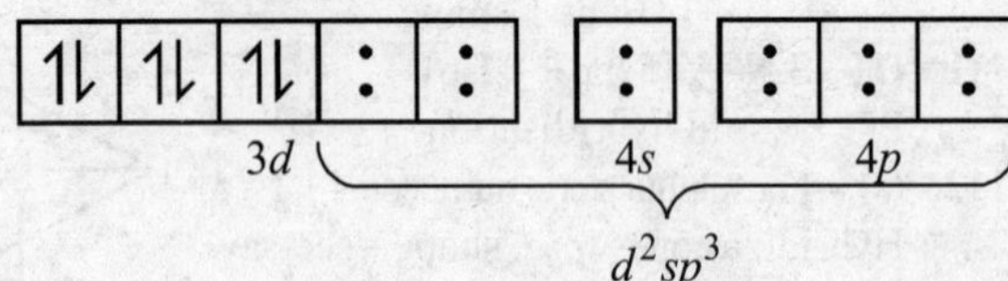

117. (3)

SO_3	NO_3^-	CO_3^{2-}	PCl_3
Trigonal planar	Trigonal planar	Trigonal planar	Pyramidal

Hence, non-pyramidal species are SO_3, NO_3^- and CO_3^{2-}.

118. (c) For 3 bond pairs and 2 lone pairs of e^-, the geometry will be trigonal bi-pyramidal and shape is T-shaped.

119. (2) T-shaped molecule means central atom has 3 sigma bond and 2 lone pairs of electron.

(AB$_3$)

120. (d) NO_2^- O=N–O$^-$ $2\sigma + 1\,lp \;\therefore\; sp^2$

$NO_2^+ \Rightarrow O = \overset{\oplus}{N} = O$ $\quad 2\sigma + 0\,lp \;\therefore\; sp$

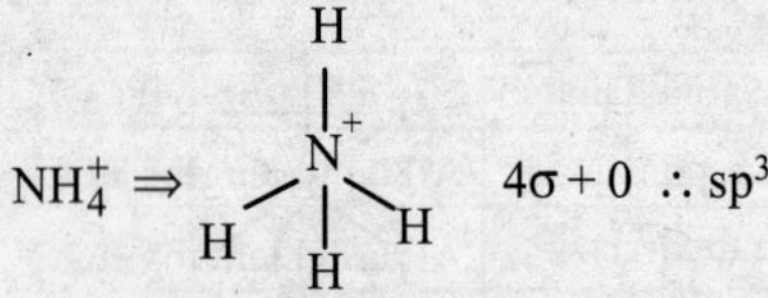

121. (3) I_3^- : It is sp^3d hydridised.

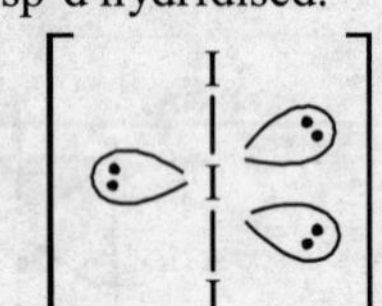

The number of lone pairs of electron on the central atom is 3.

122. (2) Short Trick:

$$\text{Lone pairs} = \frac{\text{Valence electron of central atom} + \text{charge of ion} - \text{Total bonded electron}}{2}$$

Species	Structure	Lone Pair
SF_4	F, F, S, F, F	1
BF_4^-	F, F, B, F, F	0
ClF_3	F, Cl–F, F	2
AsF_3	F–As–F, F	1

PCl_5	Cl—P(Cl)(Cl)(Cl)(Cl)	0
BrF_5	F, F, F, F, F around $\ddot{Br}$	1
XeF_4	F, F, F, F around $\ddot{Xe}$ (2 lone pairs)	2
SF_6	F, F, F, F, F, F around S	0

So, only ClF_3 and XeF_4 have two lone pairs.

123. (c)

Species	Hybridisation (shape)	Bond length
XeF_4	sp^3d^2 (Sq. planar)	All bond lengths equal
BF_4^-	sp^3 (Tetrahedral).	All bond lengths equal
SF_4	sp^3d (See-saw)	Axial B.L. > Equatorial B.L.
SiF_4	sp^3 (Tetrahedral)	All bond lengths equal

124. (d) H_2O - 104.5° (sp^3 with 2 lone pair at O)
NH_3 - 107° (sp^3 with 1 lone pair at N)
CH_4 - 109.5° (sp^3)
H_2S - 92° (sp^3 with 2 lone pair at O)
Lone pair-bond pair repulsion in H_2S will increase because 'S' has lower electronegativity than 'O'. So there will be lesser electron density on 'S' and thus H–S–H bond angle will be smaller than H_2O.

125. (a) In case of tetrahedral, rectangular planar and square planar geometries, there are only bond pair of electrons by four bonds.
Square pyramidal geometry is possible with four bond pairs and one lone pair of e^-. Hence, it will be polar due to dipole moment of these e^-.

126. (a)

(i) XeF_5^- Steric No. = Bond pair + Lone Pair
$= (5 + 2) = 7$
So, hybridisation is = sp^3d^3
and structure is pentagonal planar.

(ii) XeO_3F_2 Steric No. = 5
So, hybridisation is = sp^3d
and structure is trigonal bipyramidal.

127. (b) SF_4
Bond pair = 4
Lone pair = 1
Steric number = 5,
So, hybridisation is sp^3d.

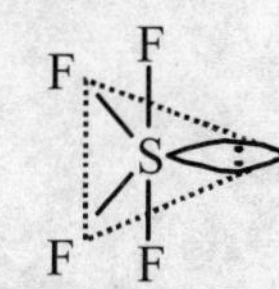

Geometry is trigonal bipyramidal but shape is "See-Saw".

128. (d) ICl_5 is sp^3d^2 hybridised
(5 *bp*, 1 *lp*)

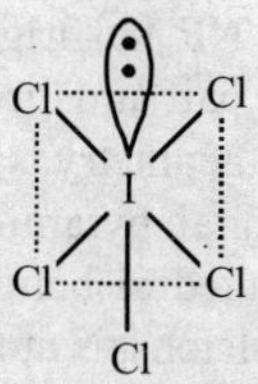

Square pyramidal

ICl_4^- is sp^3d^2 hybridised
(4 *bp*, 2 *lp*)

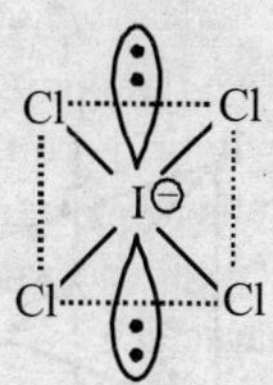

Square planar

129. (a)

Species	Hybridisation
ICl_2^-	sp^3d
ICl_4^-	sp^3d^2
BrF_2^-	sp^3d
IF_6^-	sp^3d^3

130. (a) Structure of XeO_3F_2

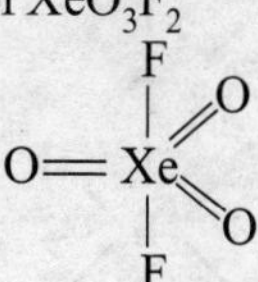

So, bond pairs = 5, π bonds = 3 and lone pairs = 0

131. (a) BF_3: B is sp^2, bond angle = 120°

$\ddot{N}H_3$: N is sp^3 with 1 *lp*, bond angle = 107°

$\ddot{P}F_3$: P is sp^3 with 1 *lp*, when central atom size↑, bond angle↓, ∴ $NH_3 > PF_3$

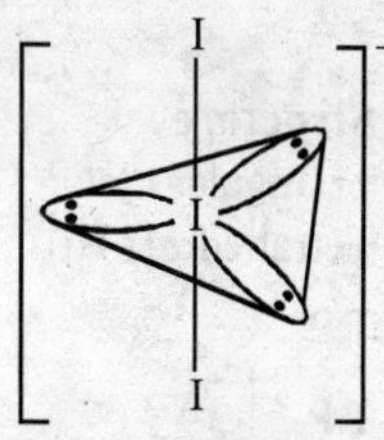

I is sp^3d (linear), bond angle = 180°

∴ Dereasing order of bond angle is $I_3^- > BF_3 > NH_3 > PF_3$

132. (c) $[:\ddot{I}:—I^{\ominus}—:\ddot{I}:]$ (three lone pairs on central I)

∴ Total number of lone pair of electrons is 9.

133. (b) $BF_3 \longrightarrow BF_4^-$

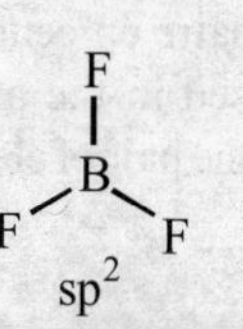

sp^2
Trigonal planar

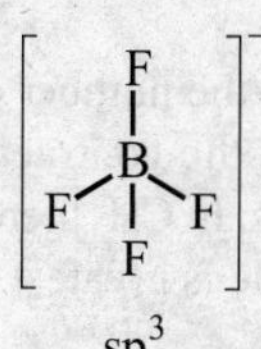

sp^3
Tetrahedral

134. (a) NF_3 has trigonal pyramidal shape. N atom has one lone pair and three bond pairs of electrons. The electron pair geometry is tetrahedral and molecular geometry is trigonal pyramidal. The bond angles are lower than tetrahedral bond angles due to lone pair - bond pair repulsions. N atom is sp^3 hybridised.

135. (d) $XeOF_2$ $XeOF_4$

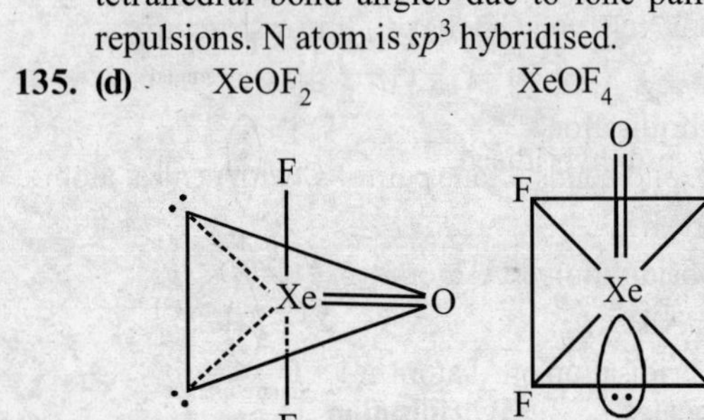

T-shape Square pyramidal

136. (d) (a) BrF_5 sp^3d^2

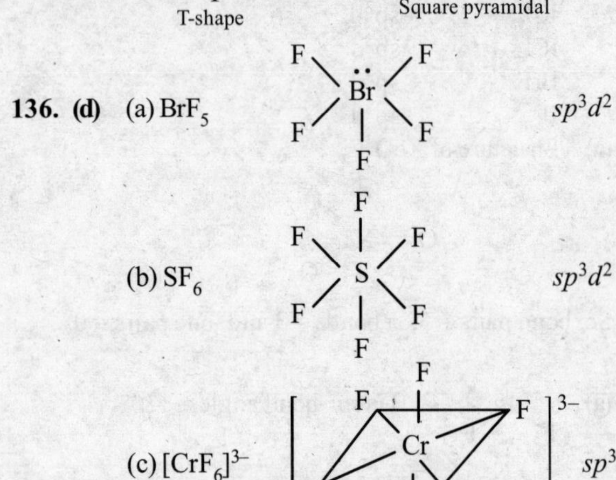

(b) SF_6 sp^3d^2

(c) $[CrF_6]^{3-}$ sp^3d^2

(d) PF_5 F—P(F)(F)(F)(F) sp^3d

137. (b)

Group	Hybridisation	Structure
(a) BF_3	sp^2	Triangular Planar (T.P.)
NF_3	sp^3	Tetrahedral (Td)
CO_3^{2-}	sp^2	T.P
(b) CO_3^{2-}	sp^2	T.P (All have same hybridisation)
NO_3^-	sp^2	T.P
SO_3	sp^2	T.P
(c) NH_3	sp^3	Td
SO_3	sp^2	T.P
CO_3^{2-}	sp^2	T.P
(d) NCl_3	sp^3	Td
BCl_3	sp^2	T.P
SO_3	sp^2	T.P

138. (b) More the number of lone pairs on central atom, the greater is the contraction caused in the angle between bond pairs. In CH_4 there is no lone pair of electrons, hence bond angle is greatest.

139. (c) NO_2^+ *sp* hybridisation
NO_2^- sp^2 hybridisation
NO_3^- sp^2 hybridisation
The Lewis structure of NO_2 shows a bent molecular geometry with trigonal planar electron pair geometry, hence the hybridization will be sp^2.

140. (d) $ClF_3 \longrightarrow$ Hybridisations sp^3d
$XeOF_2 \longrightarrow$ Hybridisation sp^3d
$XeF_3^+ \longrightarrow$ Hybridisation sp^3d
All molecules have sp^3d hybridisation and 2 lone pairs. Hence all have identical shape (T-shape).

141. (a) $XeOF_4$ (O=Xe with four F and one lone pair), sp^3d^2, no. of lone pair = 1

Square Pyramidal

142. (b) The structure of CaC_2 is $Ca^{2+}[:C\equiv C:]^{2-}$, i.e, one σ and two π bonds.

143. (d) $H \overset{\sigma}{—} C \overset{\pi,\,\sigma}{\underset{\pi}{\equiv}} N$
Therefore, HCN has 2 π and 2 σ bonds.

144. (a) The structure of IF_6^- is distorted octahedral. This is due to presence of a "weak" lone pair.

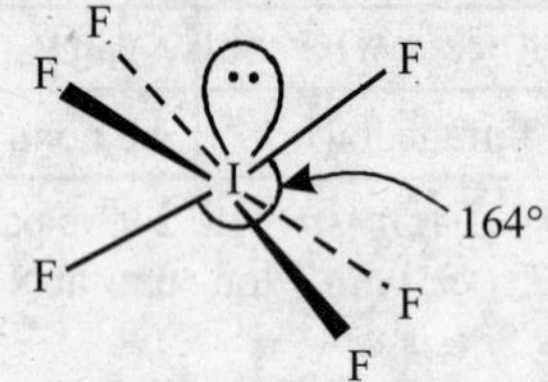

145. (d) All have tetrahedral structure.

146. (b) In BF_3, B is sp^2 hybridized with one empty p_z orbital. The empty p_z orbital of BF_3 can be filled by lone pair of molecules such as NH_3. When this occurs a tetrahedral molecule or ion is formed which is sp^3 hybridized.

147. (a) XeF_4 has square planar structure, while NH_4^+, BF_4^- and CCl_4 have tetrahedral structure.

148. (a)

Species	Hybridisation	Shape
NI_3	sp^3	Trigonal pyramidal
I_3^-	sp^3d	Linear
SO_3^{2-}	sp^3	Trigonal pyramidal
NO_3^-	sp^2	Trigonal planar

Hence, NI_3 and SO_3^{2-} have same shape.

149. (c) PF_5 is trigonal bipyramidal

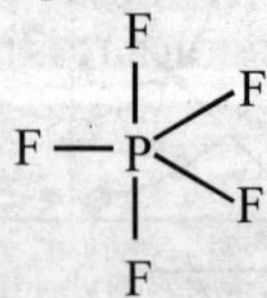

BrF_5 is square pyramidal (distorted)

F, F, Br, F, F, F

150. (b) The delocalised $p\pi - p\pi$ bonding between filled p-orbital of F and vacant p-orbital of B leads to shortening of B – F bond length which results in higher bond dissociation energy of the B – F bond.

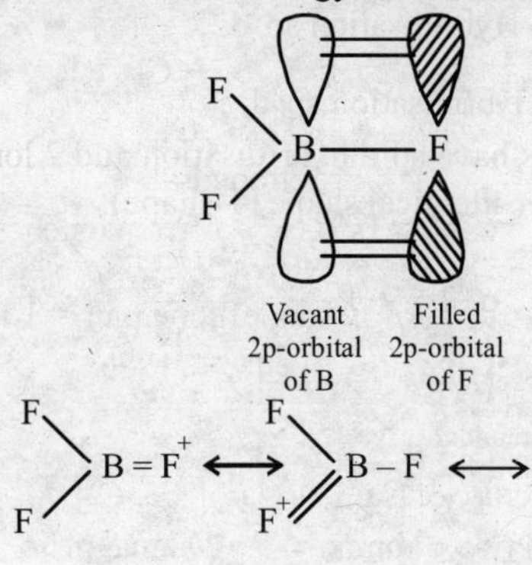

$$F_2B=\overset{+}{F} \longleftrightarrow F\overset{+}{=}B(F)-F \longleftrightarrow \ldots \equiv B\ (F^{+1/3})_3$$

151. (b) The bond angle decreases on moving down the group due to decrease in bond pair-bond pair repulsion.

NH_3	PH_3	AsH_3	SbH_3	BiH_3
107°	94°	92°	91°	90°

152. (b)

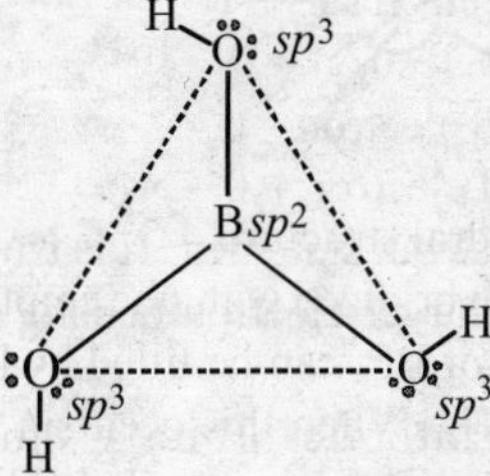

153. (d)

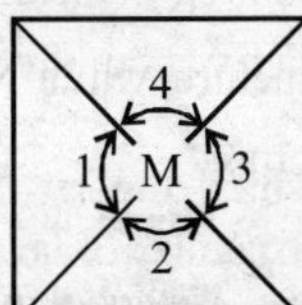

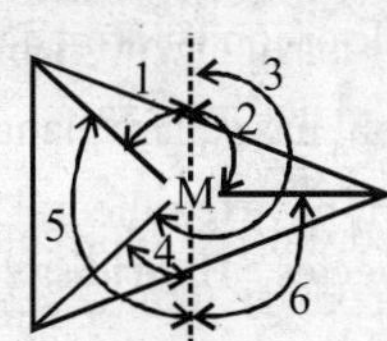

dsp^2 hybridisation Number of 90° angle between bonds = 4

sp^3d or dsp^3 hybridisation Number of 90° angle between bonds = 6

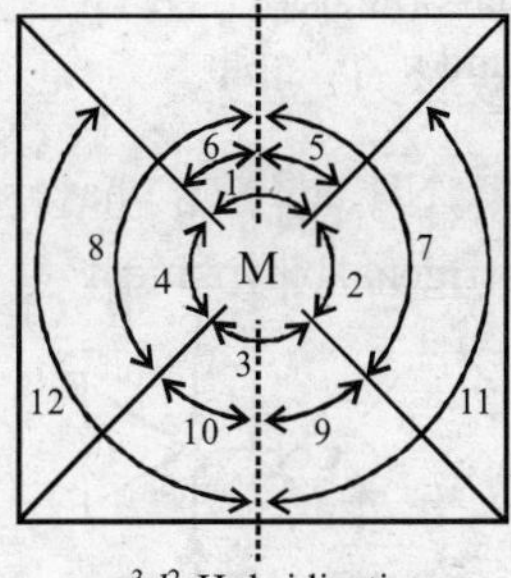

sp^3d^2 Hybridisation Number of 90° angle between bonds = 12

154. (a) $XeF_4(sp^3d^2$, square planar),

$[Ni(CN)_4]^{2-}$ (dsp^2, square planar),

BF_4^- (sp^3, tetrahedral), SF_4 (sp^3d, see saw shaped)

155. (a) Both XeF_2 and CO_2 have a linear structure.

F — Xe — F O = C = O

156. (a) Hybridisation =

No. of σ bonds + lone pair of e^- with central atom.

(a) For AlH_3,

Hybridisation of Al atom = 3 + 0 = 3 = sp^2

For AlH_4^-,

Hybridisation of Al atom = 4 + 0 = 4 = sp^3

(b) For H_2O,

Hybridisation of O atom = 2 + 2 = 4 = sp^3

For H_3O^+, Hybridisation of O atom = 3 + 1 = 4 = sp^3

(c) For NH_3, Hybridisation of N atom = 3 + 1 = 4 = sp^3

For NH_4^+, Hybridisation of N atom = 4 + 0 = 4 = sp^3

Thus, hybridisation changes only in option (a).

157. (a) In NH_3 and BF_4^-, the hybridisation is sp^3 and the bond angle is almost 109° 28'.

158. (d)

		No. of unpaired electrons
(A)	O_2	2
(B)	N_2	0
(C)	F_2	0
(D)	S_2	2
(E)	Cl_2	0

If species contain unpaired electron than it is paramagnetic. So A & D are paramagnetic. While B, C or E are diamagnetic.

159. (6) Molecules having unpaired electrons are paramagnetic which are O_2, O_2^+, O_2^-, NO, NO_2, $K_2[NiCl_4]$.

160. (N)

A: X — Z — p_z + X — Z — p_x ⇒ no molecular orbital

B: X — Z — s + X — Z — p_x ⇒ no molecular orbital

C: 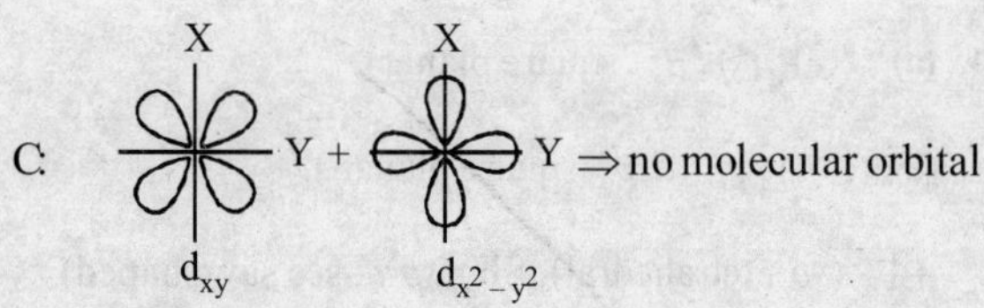

$\Rightarrow$ no molecular orbital

D: X, X; s $\oplus$ Z + p_z Z $\Rightarrow$ symmetry and energy for bidden so no molecular orbital for homonuclear diatomic molecule.

E: X, X; p_z Z + $d_{x^2-y^2}$ Y $\Rightarrow$ no molecular orbita

161. (d) Oxidation state of an element in a particular compound is defined by the charge acquired by its atom on the basis of **electronegativity** consideration from other atoms in molecule. dπ-dπ and pπ-dπ are more prevalent in 3rd period or later.

162. (6) $O_2(16e^-):(\sigma_{1s})^2\ (\sigma^*_{1s})^2(\sigma_{2s})^2\ (\sigma^*_{2s})^2$

$$(\sigma_{2p})^2\left[(\pi_{2p})^2=(\pi_{2p})^2\right],\left[(\pi^*_{2p})^1=(\pi^*_{2p})^1\right]$$

Number of e^{-1} present in (π^*) of $O_2 = 2$

Number of e^{-1} present in (π^*) of $O_2^+ = 1$

Number of e^{-1} present in (π^*) of $O_2^- = 3$

So total e^- in $(\pi^*) = 2+1+3=6$

163. (b) Antibonding molecular orbitals are formed by destructive interference of wave functions.
(ABMO) $\sigma^* = \psi_A - \psi_B$

164. (2) $C_2\ (12e^-): \sigma 1s^2, \sigma^*1s^2, \sigma 2s^2, \sigma^*2s^2\left[\pi 2p_x^2=\pi 2p_y^2\right]$

$$\text{B.O.} = \frac{8-4}{2} = 2$$

$O_2(16e^-): \sigma 1s^2, \sigma^*1s^2, \sigma 2s^2, \sigma^*2s^2, \sigma 2pz^2$

$$\left[\pi 2p_x^2 = \pi 2p_y^2\right]\left[\pi^*2p_x^1 = \pi^*2p_y^1\right]$$

$$\text{B.O.} = \frac{10-6}{2} = 2,$$

$Be_2\ (8e^-): \sigma 1s^2, \sigma^*1s^2, \sigma 2s^2, \sigma^*2s^2$

$$\text{B.O.} = \frac{4-4}{2} = 0, \text{ molecule not exist}$$

$Li_2\ (6e^-): \sigma 1s^2, \sigma^*1s^2, \sigma 2s^2$

$$\text{B.O.} = \frac{4-2}{2} = 1, \text{ One bond formed}$$

$Ne_2\ (20e^-): \sigma 1s^2, \sigma^*1s^2, \sigma 2s^2, \sigma^*2s^2, \sigma 2pz^2$

$$\left[\pi 2p_x^2 = \pi 2p_y^2\right]\left[\pi^*2p_x^2 = \pi^*2p_y^2\right]\sigma^*2p_z^2$$

$$\text{B.O.} = \frac{10-10}{2} = 0, \text{ molecule not exist}$$

$N_2\ (14e^-): \sigma 1s^2, \sigma^*1s^2, \sigma 2s^2, \sigma^*2s^2\left[\pi 2p_x^2 = \pi 2p_y^2\right]\sigma 2p_z^2$

$$\text{B.O.} = \frac{10-4}{2} = \frac{6}{2} = 3$$

$He_2\ (4e^-): \sigma 1s^2, \sigma^*1s^2$

$$\text{B.O.} = \frac{2-2}{2} = 0, \text{ Molecule not exist.}$$

165. (b)

(A) SO_2Cl_2 — sp^3 — Tetrahedral (O=S(=O)(Cl)Cl)

(B) NO(15) — $\sigma 1s^2 \sigma^*1s^2 \sigma 2s^2$ $\sigma^*2s^2\ \sigma 2p_z^2$ $\left(\pi 2p_x^2 = \pi 2p_y^2\right)$ $\left(\pi 2p_x^1 = \pi 2p_y\right)$ — Paramagnetic

(C) $NO_2(24)$ — No unpaired electron present — Diamagnetic

(D) I_3^- — sp^3d

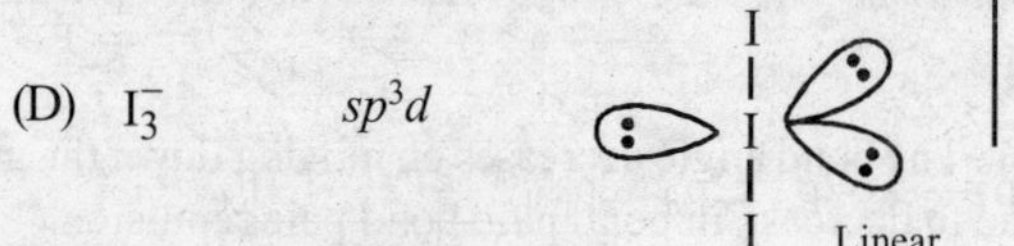

166. (2) According to M.O.T.
$O_2 \rightarrow$ no. of unpaired electrons = 2
$O_2^- \rightarrow$ no. of unpaired electron = 1
NO $\rightarrow$ no. of unpaired electron = 1
$CN^- \rightarrow$ no. of unpaired electron = 0
$O_2^{2-} \rightarrow$ no. of unpaired electron = 0

167. (d) Statement I is false but statement II is true.
A π bonding MO has larger electron density above and below the inter nuclear axis.

168. (b) (B) Combining atomic orbitals should have maximum overlap.
(D) They have same symmetry about the molecular axis.

169. (8) According to molecular orbital theory, number of molecular orbitals formed is equal to number of atomic orbitals combined, so when a diatomic molecule forms by combining of 2s and 2p orbitals 8-molecular orbitals are formed.
Two molecular orbital σ2s and σ^*2s
Six molecular orbitals $\sigma 2p_z$, σ^*2p_z
$\pi 2p_x$, $\pi 2p_y$ and π^*2p_x, π^*2p_y

170. (c)

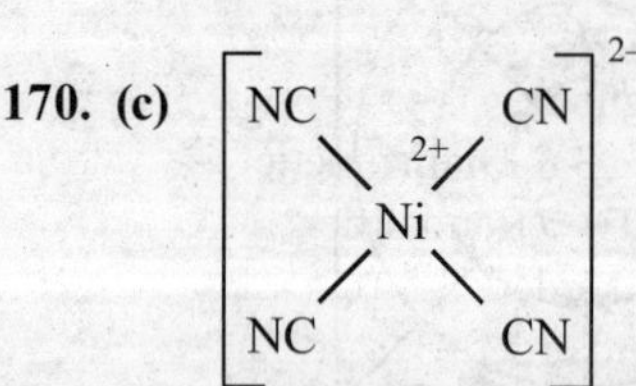

Square Planar

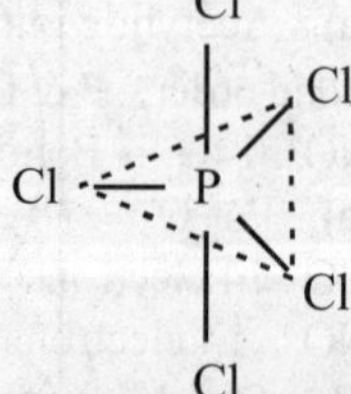

Trigonal bipyramidal

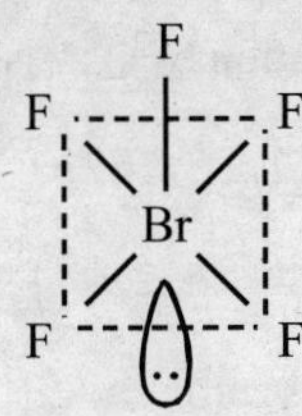

Square Pyramidal

F, F, P, F, F, F

Trigonal bipyramidal

171. (1)

Magnetic	Behaviour	Bond order
H_2 (2)	Diamagnetic	1
He_2^+ (3)	Paramagnetic	0.5
O_2^- (17)	Paramagnetic	2.5
N_2^{2-} (16)	Paramagnetic	2
O_2^{2-} (18)	Diamagnetic	1
F_2 (18)	Diamagnetic	1
Ne_2^+ (19)	Paramagnetic	0.5
B_2 (10)	Paramagnetic	1

Only B_2 has bond order 1 and paramagnetic in the given options.

172. (4) $2s \longrightarrow \sigma 2s,\ \boxed{\sigma^* 2s}$

$2p \longrightarrow \sigma 2p_z, \pi 2p_x, \pi 2p_y, \boxed{\sigma^* 2p_z, \pi^* 2p_z, \pi^* 2p_y}$

173. (22)

```
        H        H   H   O
        |        |   |   ||
    H—C—C=C—C—C—C—O—H
        |   |   |   |   |
        H   H   H   H   C—H
                        ||
                        O
```

σ-bonds = 19, π-bonds = 3, total = 19 + 3 = 22

174. (6) $CO \Rightarrow \overset{-}{C} \equiv \overset{+}{O}$: BO = 3

$NO^+ \Rightarrow N \equiv O^+$: BO = 3

175. (d)

(A) NF_3 has trigonal pyramidal shape.

(B) Bond order $\Rightarrow N_2 > O_2$

Bond length $\Rightarrow N_2 < O_2$

(C) Isoelectronic species have same bond order due to same total no. of electrons.

(D) Dipole moment $H_2O > H_2S$ due to Electronegativity difference.

176. (b) NO is paramagnetic with BO = 2.5, NO^+ is diamagnetic with BO = 3

177. (a) Acetylide ion $\rightarrow C_2^{2-}$ $(\bar{C} \equiv \bar{C})$

Bond order = 3 & Diamagnetic

NO^+, $14e^- \rightarrow$ Bond order = 3 & Diamagnetic

178. (a) For O_2^{2-}: 18 electrons $\Rightarrow$ Bond order : 1

CO : 14 electrons $\Rightarrow$ Bond order : 3

NO^+ : 14 electrons $\Rightarrow$ Bond order : 3

Short trick : for N_2 bond order is 3, no. of electrons are 14

$\therefore$ for 18 electrons bond order

$= [3 - (18 - 14) \times 0.5] = 3 - 2 = 1$

179. (a)

$N_2 - \Rightarrow 1s^2 \sigma^* 1s^2 \sigma 2s^2 \sigma^* 2s^2 \left(\pi 2p_x^2 = \pi 2p_y^2\right) \sigma 2p_z^2$ (HOMO)

$N_2^+ - \Rightarrow 1s^2 \sigma^* 1s^2 \sigma^* 2s^2 \sigma^* 2s^2 \left(\pi 2p_x^2 = \pi 2p_y^2\right) \sigma 2p_z^1$ (HOMO)

$O_2 - \Rightarrow 1s^2 \sigma^* 1s^2 \sigma 2s^2 \sigma^* 2s^2 \sigma 2p_z^2 \left(\pi 2p_x^2 = \pi 2p_y^2\right)$

$\left(\pi^* 2p_x^1 = \pi^* 2p_y^1\right)$ (HOMO)

$O_2^+ - \Rightarrow 1s^2 \sigma^* 1s^2 \sigma 2s^2 \sigma^* 2s^2 \sigma 2p_z^2 \left(\pi 2p_x^2 = \pi 2p_y^2\right)$

$\left(\pi^* 2p_x^1 = \pi^* 2p_y^0\right)$ (HOMO)

$N_2 \Rightarrow$ 0 unpaired e^- in HOMO

$N_2^+ \Rightarrow$ 1 unpaired e^- in HOMO

$O_2 \Rightarrow$ 2 unpaired e^- in HOMO

$O_2^+ \Rightarrow$ 1 unpaired e^- in HOMO

180. (4) The molecular orbital configuration of B_2 is :

$(\sigma 1s)^2 (\sigma^* 1s)^2 (\sigma 2s)^2 (\sigma^* 2s)^2 (\pi 2p_x)^1 (\pi 2p_y)^1$

The molecular orbital configuration of C_2^- is :

$(\sigma 1s)^2 (\sigma^* 1s)^2 (\sigma 2s)^2 (\sigma^* 2s)^2 (\pi 2p_x)^2 (\pi 2p_y)^2 (\sigma 2p_z)^1$

The molecular orbital configuration O_2^+ is:

$(\sigma 1s)^2 (\sigma^* 1s)^2 (\sigma 2s)^2 (\sigma^* 2s)^2 (\sigma 2p_z)^2 (\pi 2p_x)^2 (\pi 2p_y)^2 (\pi^* 2p_x)^1$

The molecular orbital configuration of He_2^+ is : $(\sigma 1s)^2 (\sigma^* 1s)^1$

Hence, B_2, C_2^-, O_2^+ and He_2^+ are paramagnetic species.

181. (a) Zero overlapping is occur due to improper orientation of orbitals and it is an out of phase overlaping.

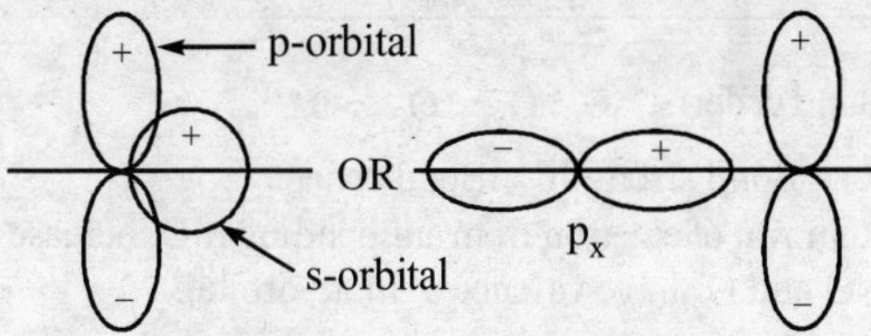

182. (3) CN^-, NO^+ and O_2^{2+} all have bond order 3.

183. (c)

(A) $\Psi_{MO} = \Psi_A - \Psi_B$ — (III) ABMO

(B) $\mu = Q \times r$ — (I) Dipole moment

(C) $\dfrac{N_b - N_a}{2}$ — (IV) Bond order

(D) $\Psi_{MO} = \Psi_A + \Psi_B$ — (II) BMO

184. (2) MO configuration :

$N_2 : \sigma_{1s}^2 \sigma_{1s}^{*2} \sigma_{2s}^2 \sigma_{2s}^{*2} \left(\pi_{2p_x}^2 = \pi_{2p_y}^2\right) \sigma_{2p_z}^2$: Diamagnetic (no unpaired e)

$N_2^+ : \sigma_{1s}^2 \sigma_{1s}^2 \sigma_{2s}^2 \sigma_{2s}^{*2} \left(\pi_{2p_x}^2 = \pi_{2p_y}^2\right) \sigma_{2p_z}^1$: Paramagnetic (1 unpaired e)

$N_2^- : \sigma_{1s}^2 \sigma_{1s}^{*2} \sigma_{2s}^2 \sigma_{2s}^{*2} \left(\pi_{2p_x}^2 = \pi_{2p_y}^2\right) \sigma_{2p_z}^2 \pi_{2p_x}^{*1}$: Paramagnetic (1 unpaired e)

$N_2^{2-}: \sigma_{1s}^2 \sigma_{1s}^{*2} \sigma_{2s}^2 \sigma_{2s}^{*2} \left(\pi_{2p_x}^2 = \pi_{2p_y}^2\right) \sigma_{2p_z}^2 \pi_{2p_x}^{*1} \pi_{2p_y}^{*1}$: Paramagnetic (2 unpaired e)

$O_2: \sigma_{1s}^2 \sigma_{1s}^{*2} \sigma_{2s}^2 \sigma_{2s}^{*2} \sigma_{2p}^2 \left(\pi_{2p_x}^4 = \pi_{2p_y}^2\right)\left(\pi_{2p_x}^{*1} = \pi_{2p_y}^{*1}\right)$: Paramagnetic (2 unpaired e)

$O_2^+: \sigma_{1s}^2 \sigma_{1s}^{*2} \sigma_{2s}^2 \sigma_{2s}^{*2} \sigma_{2p_z}^2 \left(\pi_{2p_x}^2 = \pi_{2p_y}^2\right) \pi_{2p_x}^{*1}$: Paramagnetic (1 unpaired e)

$O_2^-: \sigma_{1s}^2 \sigma_{1s}^{*2} \sigma_{2s}^2 \sigma_{2s}^{*2} \sigma_{2p_z}^2 \left(\pi_{2p_x}^2 = \pi_{2p_y}^{*2}\right)\left(\pi_{2p_x}^{*2} = \pi_{2p_y}^{*1}\right)$: Paramagnetic (1 unpaired e)

$O_2^{2-}: \sigma_{1s}^2 \sigma_{1s}^{*2} \sigma_{2s}^2 \sigma_{2s}^{*2} \sigma_{2p_z}^2 \left(\pi_{2p_x}^2 = \pi_{2p_y}^2\right)\left(\pi_{2p_x}^{*2} = \pi_{2p_y}^{*2}\right)$: Diamagnetic (no unpaired e)

185. (a)

ion / molecule	Number of e^- in BMO	Number of e^- in ABMO	Bond order
O_2^+	10	5	2.5
O_2	10	6	2
O_2^-	10	7	1.5
O_2^{2-}	10	8	1

Bond order: $O_2^{2-} < O_2^- < O_2 < O_2^+$

186. (c) Bond strength $\propto$ Bond order
Removal of electron from antibonding MO increases B.O. NO and O_2 have valence e^- in π^* orbital.

M.O. of NO^+: $\sigma_{1s}^2 \sigma_{1s}^{*2} \sigma_{2s}^2 \sigma_{2s}^{*2} \sigma_{2p_z}^2 \pi_{2p_x}^2 \pi_{2p_y}^2$

187. (b) $C_2: kk'(\sigma 2s)^2(\sigma^* 2s)^2(\pi 2p_x)^2(\pi 2p_y)^2$
$C_2^{2-}: kk'(\sigma 2s)^2(\sigma^* 2s)^2(\pi 2p_x)^2(\pi 2p_y)^2(\sigma 2p_z)^2$

$$B.O. = \frac{N_b - N_a}{2} = \frac{10-4}{2} = 3$$

$N_2: kk'(\sigma 2s)^2(\sigma^* 2s)^2(\pi 2p_x)^2(\pi 2p_y)^2(\sigma 2p_z)^2$
$N_2^{2-}: kk'(\sigma 2s)^2(\sigma^* 2s)^2(\pi 2p_x)^2(\pi 2p_y)^2(\sigma 2p_z)^2(\pi^* 2p_x)^1(\pi^* 2p_y)^1$

$$B.O. = \frac{10-6}{2} = 2$$

$O_2: kk'(\sigma 2s)^2(\sigma^* 2s)^2(\sigma 2p_z)^2(\pi 2p_x)^2(\pi 2p_y)^2(\pi^* 2p_x)^1(\pi^* 2p_y)^1$
$O_2^{2-}: kk'(\sigma 2s)^2(\sigma^* 2s)^2(\sigma 2p_z)^2(\pi 2p_x)^2(\pi 2p_y)^2(\pi^* 2p_x)^2(\pi^* 2p_y)^2$

$$B.O. = \frac{10-8}{2} = 1$$

188. (0) Molecular orbital configuration of O_2^{2-} is :
$\sigma 1s^2 \sigma^* 1s^2 \sigma 2s^2 \sigma^* 2s^2 \sigma 2p_z^2 \pi\left(2p_x^2 = \pi 2p_y^2\right)\left(\pi^* 2p_x^2 = \pi^* 2p_y^2\right)$
Total unpaired electron in O_2^{2-} is zero.

189. (a) According to the electronic configuration of O_2^- ion is
$\sigma 1s^2 \sigma^* 1s^2 \sigma 2s^2 \sigma^* 2s^2 \sigma 2p_z^2 \left(\pi 2p_x^2 = \pi 2p_y^2\right)\left(\pi^* 2p_x^2 = \pi^* 2p_y^1\right)$

$$\text{Bond order} = \frac{10-7}{2} = 1.5$$

It also exhibit paramagnetic character due to the presence of unpaired electron.

190. (0) Bond order of CO = 3
Due to isolectronic species, bond order of $NO^+ = 3$

$$\text{Difference} = 0 = \frac{x}{2} \Rightarrow x = 0$$

191. (10) M.O. Configuration of O_2^{2-} $((18\bar{e})$
$\sigma 1s^2 \sigma^* 1s^2 \sigma 2s^2 \sigma^* 2s^2 \sigma 2p_z^2 \pi 2p_x^2 = \pi 2p_y^2$
$\pi^* 2p_x^2 = \pi^* 2p_y^2$
Total B.M.O electrons = 10

192. (15) Bond order

$$= \frac{\text{No. of } e^- \text{s in BMO} - \text{No. of } e^- \text{s in ABMO}}{2} = 2.5$$

$\Rightarrow$ No. of e^-s in BMO – No. of e^-s in AMBO = 5
As AX is diatomic molecule (neutral), the only possible case is NO.
Total number of electrons = 15
Note : Total number of electrons equal to 13 will also have the 2.5 bond order. But in this case neutral diatomic molecule will not be possible.

193. (d) (A) - (III); (B) - (IV); (C) - (I); (D) - (II)

194. (b)

Species	Bond order
He_2^-	0.5
Be_2	0
He_2^+	0.5
O_2^{2-}	1

Be_2 does not exist due to zero bond order.

195. (a) (a) $[Ni(CN)_4]^{2-} = dsp^2$
(b) $BrF_5 = sp^3d^2$
(c) $XeF_4 = sp^3d^2$
(d) $[CrF_6]^{3-} = d^2sp^3$

196. (b) When two H-atoms come closer then initially due to attraction P.E. is –ve, which decreases more as atoms come closer and after reaching to a minimum value as repulsion starts dominating, P.E., increases.

197. (a) $2\,PCl_5\,(g) \longrightarrow [PCl_4]^+ \; [PCl_6]^-\,(s)$
Tetrahedral Octahedral

198. (d) Molecular orbital configuration for NO is

$$\sigma 1s^2\, \sigma^* 1s^2\, \sigma 2s^2\, \sigma^* 2s^2\, \sigma 2p_z^2\, (\pi 2p_x^2 = \pi 2p_y^2)\, (\pi^* 2p_x^1 = \pi 2p_y^{*0})$$

Species	Bond order
NO^+	3
NO^{2+}	2.5
NO^-	2
NO	2.5

Bond strength is directly proportional to the bond order, so NO^- has minimum bond strength.

199. (a) $\mu = \sqrt{n(n+2)}$ B.M.

$1.73 = \sqrt{n(n+2)}$

$n = 1$

$O_2^+ - \sigma 1s^2\, \sigma^*1s^2\, \sigma 2s^2\, \sigma^*2s^2\, \sigma 2p_z^2\, (\pi 2p_x^2 = \pi 2p_y^2)\, (\pi^*2p_x^1 = \pi^*2p_y^0)$

$O_2^- - \sigma 1s^2\, \sigma^*1s^2\, \sigma 2s^2\, \sigma^*2s^2\, \sigma 2p_z^2\, (\pi 2p_x^2 = \pi 2p_y^2)\, (\pi^*2p_x^2 = \pi^*2p_y^1)$

200. (b) Total number of electrons in $CN^- = 6 + 7 + 1 = 14$

∴ Molecular orbital distribution

$$\sigma 1s^2\, \sigma^*1s^2\, \sigma 2s^2\, \sigma^*2s^2 \begin{bmatrix} \pi 2p_x^2 \\ \pi 2p_y^2 \end{bmatrix} \sigma 2p_z^2$$

∴ Bond order $= \dfrac{10-4}{2} = 3$

CN^- is diamagnetic because all electrons are paired.

201. (c) Electronic configuration of O_2 is

$$\sigma 1s^2 \sigma^*1s^2 \sigma 2s^2 \sigma^* 2s^2 \sigma 2p_z^2 (\pi 2p_x^2 = \pi 2p_y^2)(\pi^* 2p_x^1 = \pi^* 2p_y^1)$$

When an electron is added in O_2 to form O_2^-, the incoming electron goes to π^*2p_x or π^*2p_y orbital.

202. (a) Configuration of C_2^-

$$\sigma 1s^2 \sigma^*1s^2\, \sigma 2s^2 \sigma^*2s^2\, (\pi 2p_x^2 = \pi 2p_y^2)$$

Configuration of C_2^-

$$\sigma 1s^2\, \sigma^*1s^2\, \sigma 2s^2\, \sigma^*2s^2\, (\pi 2p_x^2 = \pi 2p_y^2)\, \sigma 2p_z^1$$

$$\text{Bond order} = \frac{\text{No. of bonding } e^- - \text{No. of antibonding } e^-}{2}$$

C_2 has *s-p* mixing and the HOMO is $\pi 2p_x = \pi 2p_y$ and LUMO is $\sigma 2p_z$. So, the extra electron will occupy bonding molecular orbital and this will lead to increase in bond order.

So, C_2^- has more bond order than C_2.

203. (b) The molecules with no unpaired electrons are diamagnetic.

Molecule	No. of unpaired electrons
NO	1
CO	Zero
O_2	2
B_2	2

Since CO has no unpaired electron. Hence CO is diamagnetic.

204. (d) Bond length $\propto \dfrac{1}{\text{Bond order}}$

and diamagnetic species has no unpaired electron in their molecular orbitals.

	No. of unpaired electrons	Bond order	Magnetic character
C_2^{2-}	0	3	diamagnetic
N_2^{2-}	2	2	paramagnetic
O_2^{2-}	0	1	diamagnetic
O_2	2	2	paramagnetic

∴ C_2^{2-} has maximum bond order, least bond length and is diamagnetic.

205. (d)

$$N_2^+ = 13e^- = \sigma 1s^2 \sigma^* 1s^2 \sigma 2s^2 \sigma^* 2s^2 \left(\pi 2p_x^2 = \pi 2p_y^2\right) \sigma 2p_z^1$$

$$\text{B.O.} = \frac{\text{Bonding electrons} - \text{Antibonding electrons}}{2}$$

$$\text{B.O.} = \frac{9-4}{2} = 2.5 = 2\pi \text{ bond} + 0.5\, \sigma \text{ bond}$$

206. (c) Electronic configuratios of Li_2^+ and Li_2^-:

$Li_2^+ : \sigma 1s^2 \sigma^*1s^2 \sigma 2s^1$

$Li_2^- : \sigma 1s^2 \sigma^*1s^2 \sigma 2s^2 \sigma^*2s^1$

Now,

Bond order of $Li_2^+ = \dfrac{1}{2}(3-2) = \dfrac{1}{2}$

Bond order of $Li_2^- = \dfrac{1}{2}(4-3) = \dfrac{1}{2}$

Here, both Li_2^+ and Li_2^- have positive bond order, thus both are stable.

207. (d)

Species	No. of e^-s	Elec. conf.	Bond order
He_2^+	$(4-1=3)$	$\sigma 1s^2 \sigma^* 1s^1$	$\dfrac{2-1}{2} = 0.5$
H_2^-	$(2+1=3)$	$\sigma 1s^2 \sigma^* 1s^1$	$\dfrac{2-1}{2} = 0.5$
H_2^{2-}	$(2+2=4)$	$\sigma 1s^2 \sigma^* 1s^2$	$\dfrac{2-2}{2} = 0$
He_2^{2+}	$(4-2=2)$	$\sigma 1s^2$	$\dfrac{2-0}{2} = 1$

Molecule having zero bond order will not be a viable molecule.

208. (a) In the resonance structure of hydrogen azide, it can be seen that number of N - N bonds for bond (I) ≤ 2.

$$H-\ddot{N} \overset{(I)}{=} \overset{+(II)}{N} = \ddot{N} \longleftrightarrow H-\ddot{N} \overset{(I)}{-} \overset{+(II)}{N} \equiv N$$

Hence for bond (I), bond order will be < 2 whereas for bond (II), its bond order will be > 2.

209. (d) An antibonding π orbital best describes the given diagram of a molecular orbital. Two orbitals laterally overlap to form π bond. Out of phase combination of these two p-orbitals give π* MO.

210. (d) Total electrons in $N_2^+ = (7\times 2)-1 = 13$

$$N_2^+ \rightarrow KK\sigma(2s)^2\sigma^*(2s)^2\pi(2p_x)^2 = \pi(2p_y)^2\sigma(2p_z)^1$$

Number of electron in $\sigma 2p_z$ is 1

211. (d)

Species	Total electron
NO^+	$7+8-1=14$
CO	$6+8=14$
O_2^{2-}	$8+8+2=18$
B_2	$5+5=10$

$NO^+ \Rightarrow$

$KK\cdot\sigma(2s)^2\sigma^*(2s)^2\cdot\sigma(2p_z)^2\ (\pi 2p_x)^2 = (\pi 2p_y)^2$ diamagnetic

$CO \Rightarrow$

$KK\cdot\sigma(2s)^2\sigma^*(2s)^2\cdot(\pi 2p_x)^2 = (\pi 2p_y)^2\cdot(\sigma 2p_z)^2$ diamagnetic

$O_2^{2-} \Rightarrow KK\cdot\sigma(2s)^2\sigma^*(2s)^2\cdot\sigma(2p_z)^2(\pi 2p_x)^2$
$= (\pi 2p_y)^2\cdot\pi^*(2p_x)^2 = \pi^*(2p_y)^2$ diamagnetic

$B_2 \Rightarrow KK\cdot\sigma(2s)^2\sigma^*(2s)^2\cdot\pi(2p_x)^1 = \pi(2p_y)^1$ paramagnetic

212. (b)

(a) NO → One unpaired electron is present in π* molecular orbital, hence paramagnetic.

(b) $CO(14) \rightarrow KK\sigma 2s^2\pi 2p_x^2 = \pi 2p_y^2\sigma 2p_z^2\sigma^* 2s^2$

No unpaired electron, hence diamagnetic.

(c) $O_2(16) \rightarrow$

$KK\sigma 2s^2\sigma^* 2s^2\sigma 2p_z^2\pi 2p_x^2 = \pi 2p_y^2\pi^* 2p_x^1 = \pi^* 2p_y^1$

Two unpaired electrons, hence paramagnetic.

(d) $B_2(10) \rightarrow KK\sigma 2s^2\sigma^* 2s^2\pi 2p_x^1 = \pi 2p_y^1$

B_2 contains two unpaired electrons, hence paramagnetic.

213. (b) Assertion is correct but reason is incorrect. Bonding MO shows constructive interference of the combining electron waves.

214. (c) The bond order of N_2, O_2, and O_2^- are 3, 2 and 1.5 respectively.

Since higher bond order implies higher bond dissociation energy, hence the correct order will be

$$N_2 > O_2 > O_2^-$$

215. (a) Nitric oxide is paramagnetic in the gaseous state because of the presence of one unpaired electron in its outermost MO.

The electronic configuration of NO is

$\sigma KK\sigma 2s^2\ \sigma^* 2s^2\sigma 2p_z^2\ \pi 2p_x^2 = \pi 2p_y^2, \pi^* 2p_x^1 = \pi^* 2p_y^0$

216. (b) The molecular orbital configuration of the molecules given is

Total no. of electrons in NO = 7(N) + 8(O) = 15

Hence E.C. of NO = $KK\sigma(2s)^2\sigma^*(2s)^2\sigma 2p_z^2$
$\pi(2p_x)^2\pi(2p_y)^2\pi^*(2p_x)^1$

Due to presence of one unpaired electron NO is paramagnetic. Except NO all are diamagnetic due to absence of unpaired electrons.

217. (b) $O_2^-(17) \Rightarrow$

$KK(\sigma 2s)^2(\sigma 2s^2)(\sigma 2p_z)^2$

$(\pi 2p_x)^2 = (\pi 2p_y)^2(\pi 2p_x)^2 = (\pi 2p_y)^1$

One unpaired electron - paramagnetic.

218. (a) For NO

Total no. of electrons = 15

B.O = 2.5

Magnetic behaviour = paramagnetic

For NO^+

Total no. of electrons = 14

B.O = 3

Magnetic behaviour = diamagnetic

219. (a, b) The molecular orbital structures of C_2 and N_2 are

$N_2 = KK\sigma 2s^2\ \sigma^* 2s^2\pi 2p_x^2 = \pi 2p_y^2\ \sigma 2p_z^2$

$C_2 = KK\sigma 2s^2\ \sigma^* 2s^2\pi 2p_x^2 = \pi 2p_y^2$

Both N_2 and C_2 have paired electrons, hence they are diamagnetic.

220. (a) All options are correct,

(a) $ONCl = 8+7+17 = 32e^-$, $ONO^- = 8+7+8+1 = 24e^-$ } not isoelectronic

(b) [structure of O_3: O–Ö–O, bond lengths 1.278Å, 1.278Å, angle 116.8°] The central atom is sp^2 hybridised with one lone pair.

(c) It is a pale blue gas. At –249.7°, it forms violet black crystals.

(d) It is diamagnetic in nature due to absence of unpaired electrons.

221. (c) $H_2^{2+} = \sigma 1s^0\sigma^* 1s^0$

Bond order for $H_2^{2+} = \frac{1}{2}(0-0) = 0$

$He_2 = \sigma 1s^2\sigma^* 1s^2$

Bond order for $He_2 = \frac{1}{2}(2-2) = 0$

So both H_2^{2+} and He_2 do not exist.

222. (b) $Li_2 = \sigma 1s^2 \sigma^* 1s^2 \sigma 2s^2$; ∴ Bond order $= \frac{1}{2}(4-2) = 1$

$Li_2^+ = \sigma 1s^2 \sigma^* 1s^2 \sigma 2s^1$; B.O. $= \frac{1}{2}(3-2) = 0.5$

$Li_2^- = \sigma 1s^2 \sigma^* 1s^2 \sigma 2s^2 \sigma^* 2s^1$; B.O. $= \frac{1}{2}(4-3) = 0.5$

The bond order of Li_2^+ and Li_2^- is same but Li_2^+ is more stable than Li_2^- because Li_2^+ has 2 electrons in antibonding orbitals whereas Li_2^- has 3 electrons in antibonding orbitals.

223. (N) None of the given option is correct.
The molecular orbital configuration of the given molecules is
$H_2 = \sigma 1s^2 \sigma 1s^0$(no anti-bonding electron)
$Li_2 = \sigma 1s^2 \sigma^* 1s^2 \sigma 2s^2$(two anti-bonding electrons)
$B_2 = \sigma 1s^2 \sigma^* 1s^2 \sigma 2s^2 \sigma^* 2s^2 \left\{\pi 2p_y^1 = \pi 2p_z^1\right\}$
(4 anti-bonding electrons)
Though the bond order of all the species are same (B.O = 1) but stability is different. This is due to difference of no. of anti-bonding electrons.
Higher the no. of anti-bonding electrons, lower is the stability. Hence, the correct order is $H_2 > Li_2 > B_2$

224. (d) The bond length follows the order :

$$O_2^+ < O_2 < O_2^- < O_2^{2-}$$

According to this, the possible values are :
1.12 Å, 1.21 Å, 1.30 Å, 1.49 Å

225. (b) $NO^-(16)$ – B.O. – 2 $O_2(16)$ – B.O. – 2
$NO^+(14)$ – B.O. – 3 $NO(15)$ – B.O. – 2.5
Higher the bond order, lower is the bond length. Hence, NO^+ will have smallest bond.

226. (d) Bond order

$$= \frac{\text{No. of bonding electrons} - \text{No. of antibonding electrons}}{2}$$

Bond order in $O_2^+ = \frac{10-5}{2} = 2.5$

Bond order in $O_2^- = \frac{10-7}{2} = 1.5$

Bond order in $O_2^{2-} = \frac{10-8}{2} = 1$

Bond order in $O_2^{2+} = \frac{10-4}{2} = 3$

Since, bond order $\propto \frac{1}{\text{Bond length}}$

∴ Bond length is shortest in O_2^{2+}.

227. (a) For any species to have same bond order we can expect them to have same number of electrons. Calculating the number of electrons in various species.
O_2^- (8 + 8 + 1 = 17); CN^- (6 + 7 + 1 = 14)
NO^+(7 + 8 – 1 = 14); CN^+ (6 + 7 – 1 = 12)
We find CN^- and NO^+ both have 14 electrons, so they have same bond order.

228. (b) Diamagnetic species have no unpaired electrons whereas paramagnetic species has one or more unpaired electrons.
For electronic configuration of O_2^+, O_2 and O_2^{2-}
O_2 and O_2^+ have 2 and 1 unpaired electron respectively, while O_2^{2-} has no unpaired electron

$NO \rightarrow KK\sigma 2s^2 \sigma^* 2s^2 \sigma 2p_z^2\ \pi 2p_x^2 = \pi 2p_y^2$

$\pi^* 2p_x^1 = \pi^* 2p_y^0$

Thus no has one unpaired electron

229. (c) The distribution of electrons in MOs is as follows :

N_2^+(electrons 13) $\Rightarrow KK\sigma 2s^2 \sigma^* 2s^2 \sigma 2p_z^2 \pi 2p_x^2 = \pi 2p_y^1$
$\Rightarrow KK\ \sigma 2s^2\ \sigma^* 2s^2\ \sigma 2p_x^2 \Rightarrow O_2$ (electron 16)
$\pi 2p_x^2 = \pi 2p_y^2\ \pi^* 2p_x^1 = \pi^* 2p_y^1$
$\Rightarrow KK\ \sigma 2s^2\ \sigma^* 2s^2\ \sigma 2p_x^2 \Rightarrow O_2^{2-}$ (electron 18)
$\pi 2p_x^2 = \pi\ 2p_y^2\ \pi^* 2p_x^2 = \pi^*\ 2p_y^2$
B_2 (electrons 10) $\Rightarrow$ KK $\sigma 2s^2\ \sigma^* 2s^2\ \sigma 2p_x^2$
KK $\sigma 2s^2\ \sigma^* 2s^2\ \pi 2p_x^1 = \pi 2p_y^1$
Only O_2^{2-} does not contain any unpaired electron.

230. (c) A diamagnetic substance contains no unpaired electron. H_2 is diamagnetic as it contains all paired electrons

$H_2 = \sigma 1s^2$, $H_2^+ = \sigma 1s^1$, $H_2^- = \sigma 1s^2 \sigma 1s$
(diamagnetic) (paramagnetic) (paramagnetic)

$He_2^+ = \sigma 1s^2 \sigma^* 1s$
(paramagnetic)

231. (b) Bond order of NO^+ (3) is higher than that of NO (2.5). Thus bond length of NO^+ will be shorter.

232. (b) $O_2^+(15)$
$= KK\ \sigma 2s^2\ \sigma^* 2s^2\ \sigma 2p_z^2\ \pi 2p_x^2 = \pi 2p_y^2\ \pi^* 2p_x^1 = \pi 2p_y^0$

Bond order $= \frac{1}{2}(8-3) = \frac{5}{2} = 2.5$

$O_2(16) = KK\ \sigma 2s^2\ \sigma^* 2s^2\ \sigma 2p_z^2\ \pi 2p_x^2 = \pi 2p_y^2\ \pi^* 2p_x^1 = \pi^* 2p_y^1$

Bond order $= \frac{1}{2}(8-4) = 2$

$O_2^-(17) = KK\sigma 2s^2\ \sigma^* 2s^2\ \sigma 2p_z^2\ \pi 2p_x^2 = \pi 2p_y^2\ \pi^*\ 2p_x^2 = \pi^*\ 2p_y^1$

Bond order $= \frac{1}{2}(8-5) = 1.5$

$O_2^{2-}(18) = KK\sigma 2s^2\ \sigma^* 2s^2\ \sigma 2p_z^2\ \pi 2p_x^2 = \pi\ 2p_y^2\ \pi^* 2p_x^2 = \pi^*\ 2p_y^2$

Bond order $= \frac{1}{2}(8-6) = 1$

Note: As we know that as the bond order decreases, stability also decreases and hence the bond strength also decreases. Therefore the correct order of their increasing bond strength is

$$O_2^{2-} < O_2^- < O_2 < O_2^+$$

Thermodynamics

1. **(d)** (10L, 300K) $\xrightarrow{n=1}$ (20L, 300K)

$-q = w = -nRT \ln \frac{V_2}{V_1} = -8.3 \times 300 \times \ln\left(\frac{20}{10}\right)$

$= -1.718 \text{ kJ} \Rightarrow q = 1.718 \text{ kJ}$

$w = -1.718 \text{ kJ}$

$\Delta U = 0$ ($\therefore \Delta T = 0$; Isothermal process)

2. **(304)** The gas is monoatomic as $\bar{C}_v = \frac{3}{2}R$ and the reversible work done is calculated from the area under the curve.

$|W| = \Delta P \times \Delta U$

$= 3 \times (2000 - 1000)$

$= 3000 \text{ mL . atm} = 3 \text{ L. atm} = 3 \times 101.3 \text{ J} = 303.4 \approx 304 \text{ J}$

3. **(d)** Both solutions are having same composition. Which is 1 mole of 'x' in 1'ℓ' water, so all the intensive properties will remain same, but as total amount is greater in solution '1' compared to solution '2'. Therefore extensive properties will be different hence Gibbs free energy will be different.

4. **(b)** $|W_{\text{reversible expansion}}| > |W_{\text{irreversible expansion}}|$

$|W_{\text{reversible expansion}}| = |W_{\text{reversible Compression}}|$

$|W_{\text{irreversible expansion}}| > |W_{\text{reversible Compression}}|$

5. **(d)** It is free expansion of gas as $\Delta T = 0$ for the gas indicating vaccum in B .

$\Rightarrow P_{ext} = 0, w = 0, q = 0$ and $\Delta U = 0$

6. **(a)** **NTA Ans. (d)** $q_p = n \times C_p \times \Delta T$

$\Rightarrow \quad 500 = 0.5 \times \frac{5}{2} \times 8.3(T_f - 298)$

$\left[C_p \text{ for ideal manoatomic gas} = \frac{5}{2}R\right]$

$\Rightarrow \quad T_f \simeq 346.2 \text{ K}$

$\frac{\Delta H}{\Delta U} = \frac{C_p \Delta T}{C_v \Delta T} = \frac{C_p}{C_v} = \left(\frac{5}{3}\right)$

$\left[C_v \text{ for ideal manoatomic gas} = \frac{3}{2}R\right]$

$\Rightarrow \quad \Delta U = \frac{C_v}{C_p} \times \Delta H = \frac{3}{5} \times 500 = 300 \text{ J}$

$[q = \Delta H \text{ at constant p}]$

7. **(d)** As internal energy 'U' is a state function, its net change over a complete cycle must be zero in a cyclic process

$\therefore \quad \Delta U$ case (I) $= \Delta U$ case (II) $= \Delta U$ case (III)

8. **(a)** Thermally insulated $\Rightarrow q = 0$

from 1st law

$\Delta U = q + w$

$\Delta U = w$

$w > 0, \Delta U > 0$

9. **(55)** Work done for reversible isothermal expansion is given as:

$w = -nRT \ln\left(\frac{V_2}{V_1}\right)$

$= -1 \times 0.08206 \times 291.15 \ln\left(\frac{100}{10}\right)$

$= -55.0128 \text{ L atm}$

Work done by system $\approx$ 55 L atm

10. **(38)** $H_2O(\ell) \rightleftharpoons H_2O(g) \qquad \Delta H^o_{vap} = 40.79 \text{ kJ/mole}$

$\Delta H^o_{vap} = \Delta U^o_{vap} + \Delta n_g RT$

$40.79 = \Delta U^o_{vap} + \frac{1 \times 8.3 \times 373.15}{1000}$

$\Delta U^o_{vap} = 40.70 - 3.097$

$= 37.60$

$\Delta U^o_{vap} \simeq 38$

11. **(274)** $\Delta U = q + w$ where $(q = 0)$

$nC_v \Delta T = -P_{ext}(V_2 - V_1)$

$V_2 = 2V_1$ (given)

$\frac{nRT_2}{P_2} = \frac{2nRT_1}{P_1}$

$P_1 = 5, T_1 = 298, P_2 = \frac{5T_2}{2 \times 298}$

$n\frac{5}{2}R(T_2 - T_1) = -1\left(\frac{nRT_2}{P_2} - \frac{nRT_1}{P_1}\right)$

Put $T_1 = 298$ and $P_2 = \frac{5T_2}{2 \times 298}$

We get $T_2 = 274.16$ K

$T_2 \approx 274$ K

12. **(200)**

For isothermal irreversible process, $\Delta U = 0 \Rightarrow q = -w$

$-w = -[-P_{ext}(V_2 - V_1)]$

$q = 5(20 - 60) = -200$ atm-L.

13. **(d)** $\Delta U = q + w$

For a adiabatic process, $\boxed{q = 0}$

$\Delta U = w = -p_{ex} \Delta v$

For free expansion of ideal gas $p_{ex} = 0$

$\boxed{\Delta U = w = 0}$

As $\Delta U = C\,\Delta T = 0, \quad \boxed{\Delta T = 0}$

14. **(28721)**
In Isothermal and reversible expansion the work done is negative as the work is done by the system.

$\therefore \quad W = -2.303\ nRT\ \log\left(\frac{V_2}{V_1}\right)$

$= -2.303 \times 5 \times 8.314 \times 300 \log\left(\frac{100}{10}\right)$

$= -28720.7 \approx -28721$ J.

15. **(200)** Work done is given by area enclosed by triangle ABC.

Area of $\Delta ABC = \frac{1}{2} \times AC \times AB$

$= \frac{1}{2}(30\text{ kPa} - 10\text{kPa}) \times (30\text{dm}^3 - 10\text{dm}^3)$

$= \frac{1}{2}(20\text{ kPa})(20\text{dm}^3) = 200\text{ kPa dm}^3$

$= 200\left(\frac{1}{100}\text{bar}\right)(\text{L}) = 2\text{ L bar} \quad \left\{\begin{array}{l} 1\text{ kPa} = \frac{1}{100}\text{bar} \\ 1\text{ dm}^3 = 1\text{ L}\end{array}\right\}$

$= 200$ J $\quad \{1\text{ L bar} = 100\text{ J}\}$

16. **(1200)**
Using, first law of thermodynamics,
$\Delta U = Q + W$,
$\Delta U = 0$: Process is isothermal
$Q = -W$
$W = -P_{ext}\Delta V$ (Irreversible)
$= -80 \times 10^3 (45 - 30) \times 10^{-3} = -1200$ J

17. **(a)** Adiabatic boundary does not allow heat exchange thus heat generated in container can't escape out thereby increasing the temperature.
In case of Diathermic container, heat flow can occur to maintain the constant temperature.

18. **(4)** Extensive $\Rightarrow$ Mole, Volume, Gibbs free energy.
Intensive $\Rightarrow$ Molar mass, Molar heat capacity, Molarity, E° cell.

19. **(4)** A $\rightarrow$ Endothermic (Atomisation)
B $\rightarrow$ Endothermic (Atomisation)
C $\rightarrow$ Endothermic (Vapourisation)
D $\rightarrow$ Exothermic (Combustion)
E $\rightarrow$ Endothermic (Dissolution)

20. **(1200)** Power of heater = 60 W
Total energy emitted = $60 \times 100 = 6000$ J
Heat capacity $\times$ temp difference = 6000

Heat capacity $= \frac{6000}{5} = 1200\text{JK}^{-1}$

21. **(0)** For an, ideal gas internal energy is a function of temperature and for isothermal process temperature is constant
$\therefore$ The change in internal energy is zero.

22. **(150)** $q = 0, \Delta U = w \Rightarrow \Delta U = nCv\Delta T$
$1 \times 20 \times [T_2 - 300] = -3000$
$T_2 - 300 = -150 \Rightarrow T_2 = 150$ K.

23. **(620)** $1 \rightarrow 2 \Rightarrow$ Isobaric process
$2 \rightarrow 3 \Rightarrow$ Isochoric process
$3 \rightarrow 1 \Rightarrow$ Isothermal process

$W = W_{1\rightarrow 2} + W_{2\rightarrow 3} + W_{3\rightarrow 1}$

$= \left(-P(V_2 - V_1) + 0 + \left[-P_1V_1 \ln\left(\frac{V_2}{V_1}\right)\right]\right)$

$= \left[-1 \times (40 - 20) + 0 + \left[-1 \times 20 \ln\left(\frac{20}{40}\right)\right]\right]$

$= -20 + 20 \ln 2 = -20 + 20 \times 2.3 \times 0.3$

$= -6.2$ Lbar (1L bar = 100 J) $\Rightarrow |W| = 6.2$ L bar = 620 J

24. **(a)** H = U + PV (By definition)
$\Delta H = \Delta U + \Delta(PV)$ at constant pressure
$\Delta H = \Delta U + P\Delta V$

25. **(3)** State Variable is a variable which is independent of the path of change of state.
Internal energy, volume and enthalpy are state variable.

26. **(2)** $C_{m,p} - C_{m,v} = R$
$\Rightarrow 20.785 - C_{m,v} = 8.314 \Rightarrow C_{m,v} = 12.471\text{ JK}^{-1}\text{ mol}^{-1}$

$\Delta U = n\,C_{m,v}\,\Delta T \Rightarrow n = \frac{5000}{12.471 \times (500 - 300)} \Rightarrow n = 2$

27. **(1718)** $w = -2.303\ nRT \log\frac{V_2}{V_1}$

$P_1V_1 = P_2V_2 \quad \Rightarrow \quad \frac{V_2}{V_1} = \frac{P_1}{P_2} = \frac{6}{3} = 2$

$w = -2.303 \times 1 \times 8.3 \times 300 \times \log(2)$
$w = -1718.1$ J

28. **(195)** Moles of $N_2O = \frac{\text{mass}}{\text{molar mass}} = \frac{2.2}{44} = \frac{1}{20}$

$\Delta H = nC_p\Delta T = \frac{1}{20} \times 100\ (270 - 310)$

$= \frac{1}{20} \times 100\ (-40) = -200$ J

$\Delta U = q_p + w \quad [q_p \Rightarrow$ heat given at constant pressure]

(work done) $w = -P_{ext.}\Delta V = -1\frac{(167.75 - 217.1)}{1000} \times 101.3$J

$w = +5$J, $\quad q_p = \Delta H$
$\Delta U = -200 + 5 = -195$J

29. **(0)** For free expansion : $w = -P_{ext}dV$
$P_{ext} = 0$; $w = 0$; Isothermal expansion, $\Delta U = 0$
$\Delta U = q + w = 0$
$q = -w$ (for free expansion)
Thus, $\Delta U = q - q = 0$.

30. **(8630)** $n = 5$ mol; $T = 300$ K; $V_1 = 10$ L; $V_2 = 20$ L
Work done in isothermal process is :

$w_{rev} = -nRT \ln\frac{V_2}{V_1} = -5 \times 8.3 \times 300 \ln\frac{20}{10} = -8630.38$ J

31. **(84)** Assume the system attains a final temperature of T (such that $300 < T < 60$)

$\Rightarrow \left(\begin{array}{c}\text{Heat lost by} \\ N_2 \text{ of container} \\ \text{I}\end{array}\right) = \left(\begin{array}{c}\text{Heat gained by} \\ N_2 \text{ of container} \\ \text{II}\end{array}\right)$

$\Rightarrow n_I C_m(300-T) = n_{II} C_m(T-60)$

$\Rightarrow \left(\frac{2.8}{28}\right)(300-T) = \frac{0.2}{28}(T-60)$

$\Rightarrow T = 284$ K (final temperature)

$\Rightarrow (n_I + n_{II})_{final} = \left(\frac{3.0}{28}\right) \Rightarrow \frac{P}{RT}(V_I + V_{II}) = \frac{3.0\text{ g}}{28\text{ g/mol}}$

$P = \left(\frac{3}{28}\text{mol}\right) \times 8.31\frac{\text{J}}{\text{mol}^{-}\text{K}} \times \frac{284\text{ K}}{3\times10^{-3}} \times 10^{-5}\frac{\text{bar}}{\text{Pa}}$

$\Rightarrow 84.28 \times 10^{-2}$ bar.

32. **(38)** Given equation is

$H_2O(l) \to H_2O(g) : \Delta H = 41\text{ kJ mol}^{-1}$

$\Rightarrow$ From the relation : $\Delta H = \Delta U + \Delta n_g RT$

$\Rightarrow 41\frac{\text{kJ}}{\text{mol}} = \Delta U + (1) \times \frac{8.3}{1000} \times 373$

$\Delta U = 41 - 3.0959 = 38$ kJ/mol

33. **(31)** $H_2O_{(g)} \rightleftharpoons H_2O_{(g)}$

$\Delta H = \Delta U + \Delta n_g RT$

$\Delta H - \Delta U = \Delta n_g RT$

$\therefore \Delta n_g RT = 1\text{ mol} \times 8.31\text{ J/mol-k} \times 373\text{ K}$

$= 3099.63\text{ J} \cong 31 \times 10^2\text{ J}$

34. **(50)** $q = +50$ J

$w = -200$ J

$\Delta U = q + w$

$\Rightarrow \Delta U = 150 + (-200) = -50.$

35. **(2218)** $Fe + 2HCl \longrightarrow FeCl_2 + H_2$

No. of moles of Fe $= \frac{50}{55.85}$ moles

No. of moles of H_2 produced $= \frac{50}{55.85}$ moles

Work done $= -P_{ext} \cdot \Delta V = -\Delta n_g RT$

$= -\frac{50}{55.85} \times 8.314 \times 298 \approx -2218$ J

word done by gas = – 2218J

36. **(d)** In expansion against vacuum, $P_{ext} = 0$

$w = -P_{ext}\Delta V = 0$

37. **(d)**

(1) For ideal gas, U and H are function of temperature. $H = U + PV$ but for ideal gas, $PV = nRT$. Hence, $H = U + nRT$ or H depends only on temperature.

(2) Compressibility factor for an ideal gas is 1.

(3) $C_P - C_V = R$

(4) $\Delta U = C_v dT$ for all processes.

38. **(48.00)**

Work done is given by the area under the curve

$|w| = \frac{1}{2} \times (a+b) \times h$

$\therefore |w| = \frac{1}{2}(6+10)\frac{\text{N}}{\text{M}^2} \times 6\text{ m}^3 = 48$ J

$= 48\text{ J} = 48$ Nm

39. **(6.25)** $\Delta U = nC_v\Delta T$

$5000 = 4 \times C_v(500-300)$

$C_v = 6.25\text{ JK}^{-1}\text{ mol}^{-1}$

40. **(c)** $w = -P\Delta V = -(1\text{ bar}) \times (9\text{ L})$

$= -(10^5\text{ Pa}) \times (9 \times 10^{-3})\text{ m}^3 = -9 \times 10^2$ N.m [$\because$ 1 Pa = 1N/m^2]

$= -900\text{ J} = -0.9$ kJ

41. **(a)** We know that heat and work are not state functions but $q + w = \Delta U$ is a state function. $H - TS = G$ is also a state function.

42. **(c)** $w = +10$ kJ

$q = -2$ kJ

$\Delta U = q + w = -2 + 10 = 8$ kJ

43. **(b)** From first law of thermodynamics, $\Delta U = q + w$

For adiabatic process, $q = 0 \Rightarrow \Delta U = w$

For isothermal process, $\Delta U = 0 \Rightarrow q = -w$

For cyclic process, $\Delta U = 0 \Rightarrow q = -w$

For isochoric process, $w = 0 \Rightarrow \Delta U = q$

44. **(c)** $\Delta U = nC_v\Delta T = 5 \times 28 \times 100 = 14$ kJ

$\Delta(PV) = nR(T_2 - T_1) = 5 \times 8 \times 100 = 4$ kJ

45. **(a)** For ideal gas, C_p is independent of pressure. For polyatomic ideal gas, C_v increases slowly with increase in T.

Thus, C_p will not change with pressure.

46. **(a)** Isothermal expansion, $PV_m = K$ (graph - C)

$P = \frac{K}{V_m}$ (graph - A)

(B) P vs V_m (D) U vs V_m

Therefore, (B) and (D) are not correct representation.

47. **(c)** We know that,

$w = -P_{ext}(V_f - V_i)$

$w = -4\text{ Nm}^{-2}(1-5)\text{m}^3$

$w = 16\text{ Nm} \Rightarrow 16$ J

For isothermal compression, $\Delta U = q + w$

$\Rightarrow q = -w = -16$ J ($\because \Delta U = 0$ for isothermal process)

$q = nC_P\Delta T$

So, $16 = \frac{1 \times 24\text{J} \times \Delta T}{\text{mol K}}$

$\therefore$ Change in temperature, $\Delta T = \frac{2}{3}$ K

48. **(c)** For reversible isothermal expansion,

$w = -nRT\ln\frac{V}{V_1} \Rightarrow |w| = nRT\ln\frac{V}{V_1}$

$|w| = nRT(\ln V - \ln V_1)$

$|w| = nRT\ln V - nRT\ln V_1$

Comparing with $y = mx + c$, $m = nRT$ and $c = -nRT\ln V_1$

So, slope of curve 2 is more than curve 1 and intercept of curve 2 is more negative than curve 1.

49. **(b)** $\Delta U_{AB} = q_{AB} + W_{AB} = 2 + (-5) = -3$ kJ/mol

$\Delta U_{BC} = -5$ kJ/mol

For cyclic process, $\Delta U = 0$

$\Delta U_{AB} + \Delta U_{BC} + \Delta U_{CA} = 0$

$\Delta U_{CA} = -\Delta U_{AB} - \Delta U_{BC}$

$\Delta U_{CA} = -(-3) - (-5) = 8$ kJ/mol

$\Delta U_{CA} = q_{CA} + W_{CA}$

$8 = q_{CA} + 3 \Rightarrow q_{CA} = +5$ kJ/mol

Heat absorbed has positive sign.

50. **(c)** From 1st law of thermodynamics: $\Delta U = q + w$

For adiabatic process : $q = 0$

$\therefore \Delta U = w$

51. **(d)** $A \xrightarrow{q=+5,\ w=-8J} B$ (and back from B to A)

$\Delta U_{AB} = q + w = +5 + (-8) = -3$ J

$q = -3,\ \Delta U_{BA} = +3$

$\Delta U_{BA} = q + w \Rightarrow 3 = -3 + w \Rightarrow w = +6$

(work done on the system).

52. **(a)** $2H_2O_2(l) \rightleftharpoons 2H_2O(l) + O_2(g)$

$w = -P_{ext}(\Delta V) = -n_{O_2}RT$

$\because$ 100 mole H_2O_2 on decomposition give 50 mole O_2.

$\therefore w = -(50)(8.3)(300) = -124500\ J = -124.5$ kJ.

53. **(a)** Process is isothermal reversible expansion, hence $\Delta U = 0$, therefore $q = -w$.

Since $q = +208$ J, $w = -208$ J

54. **(c)** For isothermal reversible expansion.

$$w = -nRT \ln \frac{V_2}{V_1}$$

Trick: Either of option (b) or (c) must be not correct, depending on the work done by the system or on the system.

55. **(d)** Given $\Delta H = 41$ kJ mol^{-1} = 41000 J mol^{-1}

T = 100 °C = 273 + 100 = 373 K

$H_2O(l) \xrightarrow{\Delta} H_2O(g)$

$\Delta n_g = 1 - 0 = 1$

$\Delta U = \Delta H - \Delta nRT = 41000 - (1 \times 8.314 \times 373)$

$= 37898.88$ J mol^{-1} $\simeq 37.9$ kJmol^{-1}

56. **(c)** According to first law of thermodynamics

$\Delta U = q + w$

An isolated system is adiabatic. This means $\Delta q = 0$. The first law in this case yields

$\Delta U = w$ ..... (i)

For expansion, W is negative and hence ΔU is negative. This means T_f is less than T_i in both the cases.

For the same expansion of volume, the work done in reversible process is greater than that in ireversible one because the system has to work against friction etc. Thus,

$\Rightarrow T_{f\,irreversible} > T_{f\,reversible}$

57. **(c)** $w = -P\Delta V = -10^5[(1 \times 10^{-2}) - (1 \times 10^{-3})]$

$= -900$ J

58. **(c)** For a cyclic process, the net change in the internal energy is zero because the change in internal energy does not depend on the path.

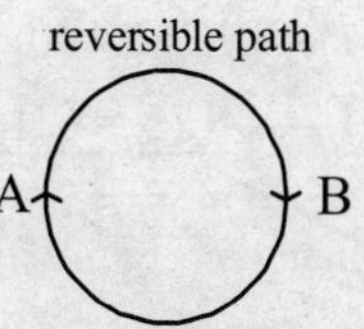

59. **(d)** According to first law of thermodynamics, energy can neither be created nor be destroyed although it can be converted from one form to another.

Hence, the given data does not violate 1st law of thermodynamics but it violates 2nd law of thermodynamics.

60. **(b)** The heat required to raise the temperature of body by 1K is called thermal capacity or heat capacity.

61. **(d)** For standard state pressure should be 1 bar and temperature is specified only.

$\Rightarrow \left(\Delta H_f^o\right)_{O_{2(g)}} = 0$

The enthalpy of formation of O_2 is zero even at high temperature.

62. **(c)** $KCl(s) + H_2O(l) \xrightarrow{\Delta H_{sol.}} K^+(aq) + Cl^{-1}(aq)$

$KCl(s) + H_2O(l) \xrightarrow{\Delta H_{sol.}} K^+(aq) + Cl^-(aq)$

L.E. (KCl(s) → $K^+(g)$ + $Cl^-(g)$); H.E ($K^+(g)$ → $K^+(aq)$); H.E ($Cl^-(g)$ → $Cl^-(aq)$)

$\Delta H_{solution} = (-\Delta H_{L.E.}) + \Delta H_{Hydration}$

$= -(-z) + (-x - y) = z - (x + y)$

63. **(b)** $H_2O(l)$ (10°C) $\longrightarrow H_2O(s)$ (– 10°C)

$H_2O(l)$ (10°C) $\xrightarrow{nC_{P(l)}\Delta T}$ $H_2O(l)$ (0°C) $\xrightarrow{-\Delta H_{fusion}}$ $H_2O(s)$ (0°C) $\xrightarrow{nC_{P(s)}\Delta T}$ $H_2O(s)$ (– 10°C)

$\Delta H = 1 \times y(0 - 10) - x \times 1000 + 1 \times z(-10° - 0°)$

$\Delta H = -10(100x + y + z)$ Joule.

64. **(b)** ΔH is negative so it means dissolution of gas HCl(g) is exothermic, hence (a) is incorrect.

$HCl(g) + 10H_2O(l) \rightarrow HCl.10H_2O$ ...(1)

$\Delta H_1 = -69.01 \frac{kJ}{mol}$

$HCl(g) + 40H_2O(l) \rightarrow HCl.40H_2O$ ...(2)

$\Delta H_2 = -72.79 \frac{kJ}{mol}$

Hence heat of solution depends the upon amount of solvent present.

$HCl.10H_2O + 30H_2O(l) \rightarrow HCl.40H_2O$

So Heat of dilution $= -72.79 - (-69.01) = -3.78 \frac{kJ}{mol}$

Hence option (c) is incorrect.

For heat of formation reactant should be in elemental form hence option (d) is incorrect

65. (25)

$$2C(s) + 2H_2(g) \longrightarrow CH_2 = CH_2(g): \left[\Delta H_F^o\right] C_2H_4(g)$$

$$\downarrow$$

$$2C(g) + 4H(g)$$

$$\left[\Delta H_F^o\right]_{C_2H_4(g)} = 2 \times \left[\Delta H_{sub}^o\right]_{C(s)} + 2 \times \Delta H_{H-H}^o - 1 \times \Delta H_{C=C}^o - 4 \times \Delta H_{C-H}^o$$

$$\Rightarrow \left[\Delta H_F^o\right]_{C_2H_4(g)} = (2 \times 710) + (2 \times 436) - 611 - 4 \times 414$$

$$\Rightarrow \left[\Delta H_F^o\right]_{C_2H_4(g)} = 25 \text{ kJ/mol}$$

66. (d) Subtracting eq. (iii) from (ii) we get eq (i)

67. (48) $\Delta H_f[CO_2(g)] = -393.5$ kJ / mol

$\Delta H_f[H_2O(l)] = -286.0$ kJ / mol

$\Delta H_c[C_6H_6] = -3267.0$ kJ / mol

$\Delta H_f(C_6H_6) = (?)$

$$C_6H_6 + \frac{15}{2}O_2(g) \longrightarrow 6CO_2(g) + 3H_2O(l)$$

$\Delta H_R = \Delta H_C = \Sigma\Delta H_f(P) - \Sigma\Delta H_f(R)$

$-3267 = [6 \times (-393.5) + 3(-286)] - [\Delta H_f(C_6H_6) + \Delta H_f(O_2)]$

$\Rightarrow \Delta H_f(C_6H_6) = 48$ kJ/mol

68. (466) $\frac{1}{2}H_2(g) \rightarrow H(g);\ \Delta_f H(H(g)) = 220$ kJ / mol

$\frac{1}{2}O_2(g) \rightarrow O(g);\ \Delta_f H(O(g)) = 250$ kJ / mol

$$H_{2(g)} + \frac{1}{2}O_{2(g)} \rightarrow H_2O_{(g)},\ \Delta_f H(H_2O) = -242 \text{ kJ/mol}$$

$2 \times 220 \downarrow$ $\quad 250 \downarrow$ $\quad -2(BE(O-H))$ (arrow upward to $H_2O_{(g)}$)

$$2H_{(g)} + O_{(g)}$$

$\Delta H_f(H_2O) = -242 = 440 + 250 - 2(B.E.(O-H))$

$\Rightarrow 2.B.E.\ (O-H) = 932$ kJ/mol

$\Rightarrow B.E.\ (O-H) = 644$ kJ/mol

69. (b) $S(g) + \frac{3}{2}O_2(g) \rightarrow SO_3(g) \quad \Delta H_1 = -2x$

$SO_2(g) + \frac{1}{2}O_2(g) \rightarrow SO_3(g) \quad \Delta H_2 = -y$

$\Delta H_f(SO_2) = (\Delta H_1) - \Delta H_2 = (-2x) - (-y) = y - 2x.$

and is shorter than a single bond but longer than a double bond as bond order is 1.5

70. (200) $M_{(s)} + \frac{1}{2}X_{2(g)} \longrightarrow M^+X^-_{(s)}$

$\downarrow \Delta H_{sub}$ $\quad \downarrow \frac{1}{2}$B.E $\quad \uparrow$ Lattice enthalpy (L.E.)

$M_{(g)} \quad X_{(g)}$

$\downarrow$ I.E. $\quad \downarrow$ E.g.

$M^+_{(g)} + X^-_{(g)}$

$\therefore \ \Delta H_f(MX) = \Delta H_{sub}(M) + I.E.(M) + \frac{1}{2}[B.E.(X-X)] + EG(X) + L.E.(MX)$

$-400 = (100) + (500) + \frac{1}{2}(B.E.) + (-300) + (-800)$ (energy released)

$\therefore \ B.E. = 200$ kJ mol^{-1}

71. (95) $2C_{(graphite)} + 3H_2(g) \rightarrow C_2H_6(g)\ \Delta H_f = ?$

$$C_2H_6(g) + \frac{7}{2}O_2(g) \rightarrow 2CO_2(g) + 3H_2O(l) \quad \Delta H_1^o = -1550$$

$$C_{(graphite)} + O_2(g) \rightarrow CO_2(g)\ \ \Delta H_2^o = -393.5$$

$$H_2(g) + \frac{1}{2}O_2(g) \rightarrow H_2O(l)\ \ \Delta H_3^o = -286$$

$$\Delta H_f^o = 2\,\Delta H_2^o + 3\,\Delta H_3^o - \Delta H_1^o = -95 \text{ kJ/mole.}$$

72. (b) Combustion of glucose takes place as :

$$C_6H_{12}O_6(s) + 6O_2(g) \rightarrow 6CO_2(g) + 6H_2O(\ell)$$

Moles of glucose in 900g of glucose $= \frac{900}{180} = 5$ mol

1 mole glucose requires 6 moles of O_2

5 moles of glucose require 30 moles of O_2

Mass of O_2 required $= 30 \times 32 = 960$ g.

73. (150) Combustion taking place as :

$$C_6H_5COOH(s) + \frac{15}{2}O_2(g) \rightarrow 7CO_2(g) + 3H_2O(\ell)$$

$\Delta n_g = 7 - \frac{15}{2} = -\frac{1}{2}$

$\Delta H = \Delta U + \Delta n_g RT$

$= -321.30 - \frac{1}{2}R \times 300$

$= (-321.30 - 150\,R)$ kJ

On comparing with $(-321.30 - xR) = x = 150.$

74. (6535) $C_6H_6(l) + \frac{15}{2}O_2(g) \rightarrow 6CO_2(g) + 3H_2O(l)$

$\Delta_f H = H_P - H_R$

$\Delta H_c = 6\Delta H_f(CO_2) + 3\Delta H_f(H_2O) - \Delta H_f(C_6H_6)$

$= 6 \times (-393.5) + 3(-286) - 48.5$

$= -3267.5$ kJ/mol

For 2 mol of benzene $= -6535$ kJ/mol

$x = 6535$

75. (125) $C_2H_4(g) + H_2(g) \rightarrow C_2H_6(g)$

$\Delta_r H = [BE(C=C) + 4BE(C-H) + BE(H-H)] - [BE(C-C) + 6BE(C-H)]$

$\Delta_r H = BE(C=C) + BE(H-H) - BE(C-C) - 2BE(C-H)$

$= 615 + 435 - 347 - 2 \times 414$

$= -125$ kJ

76. (492) Let's take:-

$$2Fe(s) + \frac{3}{2}O_2(g) \longrightarrow Fe_2O_3(s) \quad \text{... (1)}$$

$$C(s) + \frac{1}{2}O_2(g) \longrightarrow CO(g) \quad \text{... (2)}$$

$$3C(s) + Fe_2O_3(s) \longrightarrow 2Fe(s) + 3CO(g) \quad \text{... (3)}$$

Equation (3) can be obtained by subtracting (1) from (3) × (2).

Thus, $\Delta H_3^o = (3 \times (-110)) - (-822)$

$= +492 \text{ kJ mol}^{-1}$

77. **(56)** $\Delta H^0_{vap.} = 30.5 \text{ kJ mol}^{-1}$, $M_{CCl_4} = 154.82 \text{ g mol}^{-1}$

$\Rightarrow$ Heat required for 154 g CCl_4 = 30.5 kJ

$\Rightarrow$ Heat required for 284 g $CCl_4 = \frac{30.5 \times 284}{154}$

= 56.24 kJ.

78. **(800)** $A_2 + B_2 \rightarrow 2AB$; $\Delta H_f^o = -200 \text{ kJ mol}^{-1}$

$\Rightarrow \Delta H_f^o \text{ (AB)} = -200 \text{ kJ mol}^{-1}$

$\therefore \Delta H_R^o$ for reaction $A_2 + B_2 \rightarrow 2AB$ is -400 kJ mol^{-1}

Given: Bond Enthalpy of A_2, B_2 and AB is 1 : 0.5 : 1

Assuming bond enthalpy of A_2 be x kJ mol^{-1}

$\therefore$ Bond enthalpy $B_2 = 0.5 \text{ x kJ mol}^{-1}$

$\therefore$ Bond enthalpy AB = (x) kJ mol^{-1}

$A_2 + B_2 \longrightarrow 2AB$; $\Delta H_R^o = -400$ kJ/mol

x 0.5x −2x

2A + 2B

$-400 = x + 0.5x - 2x$

$-400 = -0.5x \quad \therefore x = 800$ kJ/mol

79. **(4)** $Fe_2O_3 + 2Al \longrightarrow Al_2O_3 + 2Fe$

Molar mass: 160g 27g

$$\left(\Delta H_f^o\right)_{reaction} = \left[\left(\Delta H_f^o\right)_{Al_2O_3} + 2(\Delta H_f^o)_{Fe}\right] - \left[\left(\Delta H_f^o\right)_{Fe_2O_3} + 2\left(\Delta H_f^o\right)_{Al}\right]$$

$= [-1700 + 0] - [-840 + 0] = -860$ kJ/mol

Total mass of mixture = Fe_2O_3 + Al (1 : 2 molar ratio)

= (160) + (2 × 27) = 214 g/mol

Heat evolved per gram $= \frac{860}{214} = 4 \text{ kJ / g}$

80. **(a)** $C(\text{graphite}) + \frac{1}{2}O_2(g) \rightarrow CO(g)$ .....(i)

$C(\text{graphite}) + O_2(g) \rightarrow CO_2(g)$; (ii) $\Delta H_1^o = -y$ kJ / mol

$CO_2(g) \rightarrow CO(g) + \frac{1}{2}O_2(g)$; (iii) $\Delta H_2^o = \frac{x}{2}$ kJ / mol

eq. (i) = eq. (ii) + eq. (iii) $\therefore \Delta H^o = \frac{x}{2} - y = \frac{x - 2y}{2}$

81. **(278)** $2C(s) + 3H_2(g) + \frac{1}{2}O_2(g) \rightarrow C_2H_5OH(l)$

$$(\Delta H_f)_{C_2H_5OH_{(l)}} = \sum(\Delta H_{comb})_{reactant} - \sum(\Delta H_{comb})_{product}$$

$= 2 \times (-393.5) + 3(-241.8) - (-1234.7)$

$= -277.7$ kJ/mol ≈ 278 kJ/mol

82. **(499)**

(i) $H_2O(g) \rightarrow H_2(g) + \frac{1}{2}O_2(g)$

$\Delta H_1^\circ = 242$ kJ / mol ...(1)

Multiply by 2

(ii) $H_2(g) + O_2(g) \rightarrow 2OH$, $\Delta H_2^\circ = 78 \text{ kJ mol}^{-1}$

(iii) $H_2(g) \rightarrow 2H$, $\Delta H_3^\circ = 436 \text{ kJ mol}^{-1}$

$2H_2O \rightarrow 2H + 2OH$, $\Delta H_4^\circ = 998 \text{KJmol}^{-1}$

Divide by 2

$\Rightarrow H_2O \rightarrow H + OH$

$\Delta H^\circ = \frac{998}{2} = 499 \text{ kJ mol}^{-1}$

83. **(610)** $\frac{1}{2}Cl_2(g) \rightarrow Cl(g) \rightarrow Cl^-(g) \rightarrow Cl^-(aq)$

$\Delta H_{reaction} = \frac{1}{2} \times 240 + (-350) + (-380) = -610$

84. **(173)** $\Delta_r H = \Sigma H_p - \Sigma H_R$

$CCl_4(g) + 2H_2O(g) \longrightarrow CO_2(g) + 4HCl(g)$

$= [(-394 + 4 \times (-92)] - [(-105) + (2 \times -242)] = -173$ kJ/mol

85. **(925)** Let, volume of C_2H_4 is x litre

	$C_2H_4 + 3O_2 \rightarrow 2CO_2 + 2H_2O$		
Initial	x	–	2x
Final	–		

	$CH_4 + 2O_2 \rightarrow CO_2 + 2H_2O$		
Initial	(16.8 – x)	–	
Final	–	(16.8 – x)	

Total volume of $CO_2 = 2x + 16.8 - x$

$\Rightarrow 28 = 16.8 + x$

x = 11.2 L

Volume of C_2H_4 = 4.2 L, Volume of CH_4 = (16.8 – 11.2)L = 5.6 L

$^nCH_4 = \frac{5.6}{22.4} = 0.25$ mol

$^nC_2H_4 = \frac{11.2}{22.4} = 0.5$ mol

$\therefore$ Heat evolved = 0.25 × (–900) + 0.5 (–1400) = –925 kJ

86. **(360)**

$C_6H_{12}O_6(s) + 6O_2 \rightarrow 6CO_2(g) + 6H_2O(l)$

50% energy used to convert $H_2O(l)$ into $H_2O(g)$

$= \frac{1800}{2} = 900$kJ

The energy required to evaporate 1 mole of water = 45 kJ

$\Rightarrow 900 = n_{H_2O} \times 45 \Rightarrow n_{H_2O} = \frac{900}{45} = 20$ moles

$W_{H_2O} = 20 \times 18 = 360$ g

87. **(54)** HNO_3 NaOH

600 mL × 0.2 M 400 mL × 0.1 M

= 120 mmol = 40 mmol

$HNO_3 + NaOH \rightarrow NaNO_3 + H_2O$

	HNO_3	NaOH	$NaNO_3$
Start	120	40	
End	80	0	40 mmol

$\Delta_r H = 40 \times 10^{-3}\,\text{mol} \times 57 \times 10^3 \frac{J}{mol} = 2280\,J$

$m\,S\Delta T = 2280 \Rightarrow 1000\,g \times 4.2 \times \Delta T = 2280$

[Total volume of the solution = 600 + 400 = 1000 mL]

Considering the density of water = 1g/mL.

$\Delta T = \frac{2280}{4.2} \times 10^{-3} = \frac{22800}{42} \times 10^{-3} = 542.86 \times 10^{-3}$

$= 54.286 \times 10^{-2}\,K = 54.286 \times 10^{-2}\,°C$

88. **(57)** $\Delta H = \Delta U + \Delta n_g RT$

$H_2F_2(g) \longrightarrow H_2(g) + F_2(g)$

$\Delta n_g = n_P(g) - n_R(g) = 2 - 1 = 1$

$\Delta H = -59.6 + 1 \times 8.314 \times 300 \times 10^{-3} = -57.10\,kJ\,mol^{-1}$.

89. **(117)** Molar mass of NH_3 = 17 g/mol

moles of NH_3 $\frac{85}{17} = 5$ moles

For 1 mole enthalpy change = 23.4 kJ

5 moles enthalpy change = 23.4 × 5 = 117 kJ

90. **(–103.7)**

$C_3H_8 + 5O_2 \rightarrow 3CO_2 + 4H_2O$ ...(1)

$\Delta H_1 = -2220.0\,kJ/mol$

$C + O_2 \rightarrow CO_2$ ...(2)

$\Delta H_2 = -393.5\,kJ/mol$

$H_2 + \frac{1}{2}O_2 \rightarrow H_2O$ ...(3)

$\Delta H_3 = -285.8\,kJ/mol$

Formation of C_3H_8 (1 mole):

$3C + 4H_2 \rightarrow C_3H_8$ ...(4)

Eq. (4) obtained by

3 × Eq (2) + 4 × Eq. (3) – Eq. (1)

$= 3 \times (-393.5) + 4 \times (-285.8) - (-2220.0)$

$= -1180.5 - 1143.2 + 2220.0 = -103.7\,kJ/mol$

91. **(2)** $\Delta H_{neutaralization} = -57.3\,kJ/mol$

In case of acetic acid

$\Delta H = \Delta H_{ionisation} + \Delta H_{neutralization}$

$-55.3 = \Delta H_{ionisation} - 57.3$

$\Delta H_{ioni} = 2\,kJ/mol$

92. **(600)** $Mg(s) + \frac{1}{2}O_2(g) \longrightarrow MgO(s)$

$\Delta H = \Delta U + \Delta n_g RT$

$\Delta n_g = \Sigma n_P - \Sigma n_R = 0 - \frac{1}{2} = -\frac{1}{2}$

$-601.70 \times 10^3 = \Delta U - \frac{1}{2} \times 8.3 \times 300$

$-601.70\,kJ = \Delta U - 1.245\,kJ$

$\Delta U = -600.455\,kJ \approx -600\,kJ$

93. **(38)** $H_2O\,(l) \rightarrow H_2O(g)$

$\Delta n_g = \sum n_p - \sum n_r = 1 - 0 = 1$

$\Delta H = \Delta U + \Delta n_g RT$

$\Delta U = \Delta H - \Delta n_g RT$

$= 41.1 - \frac{1 \times 8.31 \times 373}{1000}\,kJ/mol = 38\,kJ/mol$

94. **(c)** $\Delta_r H = \Sigma\Delta_c H\,(\text{Reactant}) - \Sigma\Delta_c H\,(\text{Product})$

$= 3 \times (-1300) - (-3268) = -632\,kJ\,mol^{-1}$

95. **(c)** (i) $C_2H_6(g) + \frac{7}{2}O_2(g) \longrightarrow 2CO_2(g) + 3H_2O(l)$

$\Delta H = -1560\,KJ$

(ii) $C(s) + O_2(g) \rightarrow CO_2(g)$ $\Delta H = -394\,KJ$

(iii) $H_2(g) + \frac{1}{2}O_2 \rightarrow H_2O\,(l)$ $\Delta H = -286\,KJ$.

Heat of combustion

$2C\,(s) + 2H_2\,(g) \rightarrow C_2H_6(g)$ $\Delta H = \Delta Hf\,(2H_6)$

$= \Sigma\Delta_f H_{(products)} - \Sigma\Delta_f H_{(reactants)}$

$\Delta_c H(C_2H_6, g) = 2\Delta_c H\,(C, \text{graphite}) + 3\Delta_c H\,(H_2, g) - \Delta_f H\,(O_2, g) - \Delta_f H(C_2H_6, g)$

$\Rightarrow -1560 = 2(-394) + 3\,(-286) - 0 - \Delta_f H(C_2H_6, g)$

$\Rightarrow \Delta_f H(C_2H_6, g) = -86\,KJ\,mol^{-1}$

96. **(82)**

HCl	+	$NaOH$	$\rightarrow$	$NaCl$	+	H_2O
0.04		0.03		–		–
0.07		–		0.03		0.03

Heat released (a) = 0.03 × 57.1 × 1000 = 1713 J

$q = mc\Delta T$

$1713\,J = 500\,g \times 4.18 \times \Delta T$

v = 500 mL

$\Delta T = \frac{1713}{500 \times 4.18} = 0.8196 \approx 82 \times 10^{-2}\,K$

97. **(3)** Standard electrode potential represents the overall change:

$M(s) \rightarrow M(g)$ sublimation enthalpy

$M(g) \rightarrow M^+(g) + e^-$ Ionization enthalpy

$M^+(g) + H_2O \rightarrow M^+(aq)$ hydration enthalpy

98. **(718)**

$\Delta_f H^\circ_{KCl} = \Delta_{sub} H^\circ_{(K)} + \Delta_{ionization} H^\circ_{(K)} + \frac{1}{2}\Delta_{bond}\,H^\circ_{(Cl_2)} + \Delta_{electron\ gain} H^\circ_{(Cl)} + \Delta_{lattice}\,H^\circ_{(KCl)}$

$\Rightarrow -436.7 = 89.2 + 419.1 + \frac{1}{2}(243.0) + \{-348.6\} + \Delta_{lattice}\,H^\circ_{(KCl)}$

$\Rightarrow \Delta_{lattice}\,H^\circ_{(KCl)} = -717.8\,kJ\,mol^{-1} \simeq 718\,kJ\,mol^{-1}$

99. **(82)** Moles of H^+, $(n_{H^+}) = \frac{400 \times 0.2}{1000} \times 2 = 0.16$

Moles of $\bar{O}H$, $(n_{OH^-}) = \frac{600 \times 0.1}{1000} = 0.06$

(Limiting reagent)

Total volume of solution = (400 + 600) mL

= 1000 mL or 1000 cm^3

Enthalpy of neutralization, i.e.,

heat given out by 1 mole = 57.1 kJ

∴ Heat given out by 0.06 mole = $57.1 \times 10^3 \times 0.06$ J

Now, if ΔT is increase in temperature

then, $0.06 \times 57.1 \times 10^3$

$= \left(1000\,cm^3 \times 1.0 \frac{g}{cm^3}\right) \times 4.18 \times \Delta T$

∴ $\Delta T = 0.8196\,K = 81.96 \times 10^{-2}\,K \approx 82 \times 10^{-2}\,K$

100. **(101)** $\Delta H_{sub} = \Delta H_{fus.} + \Delta H_{vap.} = 2.8 + 98.2 = 101$ kJ/mol

101. **(21)** Heat of combustion per mol (for 12 g)
$= -2.48 \times 10^2$ kJ
Heat generated for 1 g of graphite

$$= \frac{-2.48 \times 10^2}{12} = -20.66 \text{ kJ} \approx 21 \text{ kJ}$$

102. **(128)** $H_3C{-}CH_3$ (H–C–C–H with H atoms) $\longrightarrow H_2C{=}CH_2 + H{-}H$

$\Delta_r H = \Sigma B.E_{Reactant} - \Sigma B.E_{Products}$
$\Delta_r H = [(B.E)_{C-C} + 6 \times (B.E)_{C-H}] - [(B.E)_{C=C} + 4 \times (B.E)_{C-H} + (B.E)_{H-H}]$
$= [347 + 6 \times 414] - [611 + 4 \times 414 + 436] = 128 \text{ kJ mol}^{-1}$

103. **(309)** $SF_6(g) \rightarrow S(g) + 6F(g)$

$\Delta H^\circ = \Delta H_f^\circ(S) + 6\Delta H_f^\circ(F) - \Delta H_f^\circ(SF_6)$
$= 275 + 6 \times 80 - (-1100) = 1855 \text{ kJ mol}^{-1}$
Also, $\Delta H^\circ = 6\Delta H_{S-F}$

$$\therefore \quad \Delta H_{S-F} = \frac{1855}{6} = 309.17 \simeq 309 \text{ kJ mol}^{-1}$$

104. **(5576)**

$Na(g) + Br(g) \longrightarrow NaBr(s)$; IE_1 and ΔH_{eg_1} lead down to $\overset{+}{Na}(g) + \overset{-}{Br}(g)$, which goes up to $NaBr(s)$ by LE.

$\Delta H_{formation} = IE_1 + \Delta H_{eg_1} + LE$

$= 495.8 + (-325) + (-728.4)$
$= -557.6$ kJ/mol $= -5576 \times 10^{-1}$ kJ/mol.

105. **(c)** $\Delta_{sol.}H^\circ = \Delta_{lattice}H^\circ + \Delta_{Hyd.}H^\circ$
$4 = 788 + \Delta_{Hyd.}H^\circ$
$\Delta_{Hyd.}H^\circ = -784 \text{ kJ mol}^{-1}$

106. **(d)** A-B bond has highest intermolecular potential energy among the given molecules. Hence, it is strongest bond and has maximum bond enthalpy.

107. **(189494)** $\Delta H = \Delta U + \Delta n_g RT$; $n = \frac{90}{18} = 5$ mol

$H_2O(l) \rightleftharpoons H_2O(g)$; $\Delta n = 1$
$41000 = \Delta U + 1 \times 8.314 \times 373 \Rightarrow \Delta U = 37898.875$ J
For 5 moles, $\Delta U = 37898.87 \times 5 = 189494$ J

108. **(– 326400)**

$C_2H_5OH(l) + 3O_2(g) \longrightarrow 2CO_2(g) + 3H_2O(l)$
$\Delta H = \Delta U + \Delta n_g RT \Rightarrow -327 \times 10^3 = \Delta U + (-1) \times 2 \times 300$
$\Rightarrow \Delta U = -327 \times 10^3 + 600 \therefore \Delta U = -326400$ cal

109. **(c)** $\Delta H_{atomisation} = \Delta H_{vap} +$ Bond energy
Hence, $x > y$.

110. **(–192.5)** $C(s) + O_2(g) \rightarrow CO_2(g)$;
$\Delta_C H^0[C(graphite)] = -286.0$ kJ/mol ...(i)

$H_2(g) + \frac{1}{2}O_2(g) \rightarrow H_2O(g)$;
$\Delta_C H^0[H_2(g)] = -393.5$ kJ/mol ...(ii)

$C_2H_6(g) + \frac{7}{2}O_2(g) \rightarrow 2CO_2(g) + 3H_2O(g)$;
$\Delta_C H^0[C_2H_6(g)] = -1560$ kJ/mol ...(iii)

The reaction of formation of ethane is
$2C(s) + 3H_2(g) \rightarrow C_2H_6(g)$
$\therefore \Delta_f H^0$ of $C_2H_6(g) = 2 \times \Delta_C{}^0 H[C_{graphite}] + 3 \times \Delta_C H^0[H_2(g)] - \Delta_C H^0[C_2H_6(g)]$
$= 2 \times (-286.0) + 3(-393.5) - (-1560) = -192.5$ kJ/mol

111. **(c)** $I_2(s) \longrightarrow I_2(g)$
Heat of reaction depend upon temperature, i.e., it varies with temperature, as given by Kirchoff's equation,

$$\Delta H_{T_2} = \Delta H_{T_1} + \int_{T_1}^{T_2} \Delta C_p dT$$

where $\Delta C_p = C_p$ of product $- C_p$ of reactant
$\therefore \quad \Delta C_p = 0.031 - 0.055 = -0.024$ cal/g
Now, $\quad \Delta H_{T_2} - \Delta H_{T_1} = \Delta C_p(T_2 - T_1)$
$\Delta H_{(250)} - \Delta H_{(200)} = -0.024(523 - 473)$
$\Delta H_{(250)} = 24 - 50 \times 0.024 = 22.8$ cal/g

112. **(a)** $C_7H_{16}(l) + 11O_2(g) \xrightarrow{\Delta} 7CO_2(g) + 8H_2O(l)$
$\Delta H - \Delta U = \Delta n_g RT$; $\because \Delta n_g = 7 - 11 = -4$
$\therefore \Delta H - \Delta U = -4RT$

113. **(a)** **Given:** $n = 3$; $T_1 = 300$; $T_2 = 1000$
$C_p = 23 + 0.01T$
The relation between ΔH and C_p is

$$\Delta H = \int_{T_1}^{T_2} nC_p dT \quad \text{...(i)}$$

After putting all variable values in eq. (i), we get

$$\Delta H = n\int_{300}^{1000}(23 + 0.01T)dT = 3\left[23T + \frac{0.01T^2}{2}\right]_{300}^{1000}$$

$$= 3[23(1000 - 300) + \frac{0.01}{2}(1000^2 - 300^2)]$$

$= 61950$ J $= 61.95$ kJ ≈ 62 kJ

114. **(a)** Equation (i) can be obtained by adding equations (ii) and equation (iii).

$C(graphite) + \frac{1}{2}O_2(g) \rightarrow CO(g)$; $\Delta H^\circ_1 = y \text{ kJ mol}^{-1}$ ...(ii)

$CO(g) + \frac{1}{2}O_2(g) \rightarrow CO_2(g)$; $\Delta H^\circ_2 = z \text{ kJ mol}^{-1}$ ...(iii)

$C(graphite) + O_2(g) \rightarrow CO_2(g)$; $\Delta H^\circ_3 = x \text{ kJ mol}^{-1}$ ...(i)

$\because \quad \Delta H^\circ_3 = \Delta H^\circ_1 + \Delta H^\circ_2$
$x = y + z$

115. **(d)** $C_6H_6(l)+\frac{15}{2}O_2(g)\longrightarrow 6CO_2(g)+3H_2O(l)$

$\Delta n_g = 6-\frac{15}{2}=-\frac{3}{2}$

$\Delta H = \Delta U + \Delta n_g RT$

$=-3263.9+\left(-\frac{3}{2}\right)\times 8.314\times10^{-3}\times 298$

$=-3263.9+(-3.71)=-3267.6\ kJ\ mol^{-1}$

116. **(b)** $\Delta H=\Delta U+\Delta n_g RT$

$2\ HI\ (g)\rightarrow H_2(g)+I_2(g)\ ;\ \Delta n_g=(1+1)-2=0$

$\therefore\ \Delta H=\Delta U$

117. **(c)** $C(graphite)+O_2(g)\longrightarrow CO_2(g);$

$\Delta_r H^\circ_1=-393.5\ kJ/mol^{-1}$ ...(i)

$H_2(g)+\frac{1}{2}O_2(g)\longrightarrow H_2O(l);$

$\Delta_r H^\circ_2=-285.8\ kJ/mol^{-1}$ ...(ii)

$CO_2(g)+2H_2O(l)\longrightarrow CH_4(g)+2O_2(g);$

$\Delta_r H^\circ_3=+890.3\ kJ/mol^{-1}$ ...(iii)

$C(graphite)+2H_2(g)\longrightarrow CH_4(g);$

$\Delta_r H=?$ ...(iv)

[Eq. (i) + Eq. (iii)] + [2 × Eq. (ii)] = Eq. (iv)

$\therefore\ [\Delta H_1+\Delta H_3]+[2\times\Delta H_2]=\Delta H$

$[(-393.5)+(890.3)]+[2(-285.8)]=-74.8\ kJ/mol$

118. **(d)** $A(g)\longrightarrow A(l)\ ;\ \Delta H=\Delta U+\Delta n_g RT$

$\Delta H=-3RT$ and $\Delta n_g=n_p-n_r=0-1=-1$

$\Delta H=\Delta U+\Delta n_g RT\Rightarrow -3RT=\Delta U-RT\ \Rightarrow -2RT=\Delta U$

$|\Delta H|>|\Delta U|$

119. **(c)** In order to calculate the enthalpy change for H_2O at 5 °C to ice at –5 °C, we need to calculate the enthalpy change of all the transformations involved in the process.

(a) Enthalpy change of 1 mol, H_2O (l), at 5 °C $\rightarrow$ 1 mol, H_2O (l), 0 °C

(b) Enthalpy change of 1 mol, H_2O (l), at 0 °C $\rightarrow$ 1 mol, H_2O (s) (ice), 0 °C

(c) Enthalpy change of 1 mol, ice (s), at 0 °C $\rightarrow$ 1 mol, ice (s), –5 °C

Total $\Delta H=C_p[H_2O(l)]\Delta T+\Delta H\,\text{freezing}+C_p[H_2O(s)]\Delta T$

$=(75.3\ J\ mol^{-1}\ K^{-1})(0-5)K+(-6\times10^3\ J\ mol^{-1})+(36.8\ J\ mol^{-1}\ K^{-1})(-5-0)\ K$

$\Delta H=-6.56\ kJ\ mol^{-1}$ (exothermic process)

120. **(b)** Given

$C(s)+O_2(g)\rightarrow CO_2(g);\ \Delta H_1=-393.5\ kJ\ mol^{-1}$ ...(i)

$CO(g)+\frac{1}{2}O_2(g)\rightarrow CO_2(g);\ \Delta H_2=-283.5\ kJ\ mol^{-1}$ ...(ii)

$\therefore$ Heat of formation of CO $=\Delta H_1-\Delta H_2$

$=-393.5-(-283.5)=-110\ kJ$

121. **(b)** In CH_4, $4\times BE_{(C-H)}=360\ kJ/mol$

$\therefore\ BE_{(C-H)}=90\ kJ/mol$

In C_2H_6, $BE_{(C-C)}+6\times BE_{(C-H)}=620\ kJ/mol$

$\therefore\ BE_{(C-C)}=620-(6\times90)=80\ kJ/mol$

$\therefore\ BE_{(C-C)}=\frac{80\times10^3}{6.023\times10^{23}}\ J/mol$

Now, $E=\frac{hc}{\lambda}$

$\therefore\ \lambda=\frac{6.626\times10^{-34}\times3\times10^8\times6.023\times10^{23}}{80\times10^3}$

$\lambda=1.49\times10^{-6}\,m$ $(\because 1\ nm=10^{-9}\,m)$

$\therefore\ \lambda=1.49\times10^3\,nm$

122. **(c)** Given $\frac{1}{2}N_2+\frac{3}{2}H_2\rightleftharpoons NH_3;$

$\Delta_f H=-46.0\ kJ/mol$

$H+H\rightleftharpoons H_2;\ \Delta_f H=-436\ kJ/mol$

$N+N\rightleftharpoons N_2;\ \Delta_f H=-712\ kJ/mol$

$\Delta_f H(NH_3)=\frac{1}{2}\Delta H_{N-N}+\frac{3}{2}\Delta H_{H-H}-3\Delta H_{N-H}$

$-46=\frac{1}{2}(712)+\frac{3}{2}(436)-3\Delta H_{N-H}$

$\Delta H_{N-H}=352\ kJ/mol$

123. **(c)** To calculate average enthalpy of C – H bond in methane, following informations are needed:

(i) Dissociation energy of H_2

$\frac{1}{2}H_2(g)\longrightarrow H(g);\ \Delta H=x\,(\text{suppose})$

(ii) Sublimation energy of C(graphite) to C(g)

$C(graphite)\longrightarrow C(g);\ \Delta H=y\,(\text{Suppose})$

Given

$C(graphite)+2H_2(g)\longrightarrow CH_4(g);$

$\Delta H=-74.9\ kJ\ mol^{-1}$

124. **(c)** Given, $H_2(g)+\frac{1}{2}O_2(g)\longrightarrow H_2O(l);$

$\Delta H^\circ=-285.9\ kJ\ mol^{-1}$ ... (i)

$H_2(g)+\frac{1}{2}O_2(g)\longrightarrow H_2O(g);\ \Delta H^\circ=-241.8\ kJ\ mol^{-1}$ ... (ii)

We have to calculate

$H_2O(l)\longrightarrow H_2O(g);\ \Delta H^\circ=?$

On substracting eqn. (ii) from eqn. (i) we get,

$H_2O(l)\rightarrow H_2O(g);\ \Delta H^\circ=-241.8-(-285.9)=44.1\ kJ\ mol^{-1}$

125. **(c)** Applying Hess's Law,

$\Delta_f H^\circ=\Delta_{sub}H+\frac{1}{2}\Delta_{diss}H+I.E.+E.A+\Delta_{lattice}H$

$-617=161+520+77+E.A.+(-1047)$

$E.A.=-617+289=-328\ kJ\ mol^{-1}$

126. **(c)** For an alkaline earth metal M :

$MO(s)+H_2O(l)\rightarrow M(OH)_2(s)$

$\Delta_r H=\Delta_f H[M(OH)_2]-\Delta_f H[MO]-\Delta_f H[H_2O]$

$\Delta_f H[H_2O]=-285\ kJ/mol$

$\Delta_f H[MgO]=-602\ kJ/mol,\ \Delta_f H[CaO]=-635\ kJ/mol$

$\Delta_f H[SrO]=-592\ kJ/mol,\ \Delta_f H[BaO]=-548\ kJ/mol$

$\Delta_f H[Mg(OH)_2]=-925\ kJ/mol,\ \Delta_f H[Ca(OH)_2]=-985\ kJ/mol$

$\Delta_f H[Sr(OH)_2]=-959\ kJ/mol,\ \Delta_f H[Ba(OH)_2]=-945\ kJ/mol$

After calculation,
$\Delta_r H[Mg(OH)_2] = -38$ kJ/mol, $\Delta_r H[Ca(OH)_2] = -65$ kJ/mol
$\Delta_r H[Sr(OH)_2] = -82$ kJ/mol, $\Delta_r H[Ba(OH)_2] = -111$ kJ/mol
The order of liberated heat per gram is:
$Ba(OH)_2 > Sr(OH)_2 > Ca(OH)_2 > Mg(OH)_2$

127. (a) $N_2(g) + \frac{1}{2}O_2 \rightarrow N_2O(g)$

$$N \equiv N(g) + \frac{1}{2}(O=O) \rightarrow \bar{N} = \overset{+}{N} = \ddot{O}(g)$$

$\Delta_f H^o$ = [Energy required for breaking of bonds] −[Energy released for forming of bonds]

$$= (\Delta H_{N\equiv N} + \frac{1}{2}\Delta H_{O=O}) - (\Delta H_{N=N} + \Delta H_{N=O})$$

$$= (946 + \frac{1}{2} \times 498) - (418 + 607) = 170 \text{ kJ mol}^{-1}$$

Resonance energy = observed $\Delta_f H^o$ – calculated $\Delta_f H^o$
$82 - 170 = -88$ kJ mol^{-1}

128. (d) $\Delta H = \Delta U + \Delta n_g RT$
For the reaction $\Delta n_g = 12 - 15 = -3$
$\Delta H - \Delta U = -3 \times 8.314 \times 300 = -7482.6$ J mol^{-1}

129. (d) $NH_4OH(l) + HCl(l) \longrightarrow NH_4Cl(s) + H_2O(l)$
$\Delta H_1 = -51.46$ kJ/mol ... (i)
$NaOH(l) + HCl(l) \longrightarrow NaCl(s) + H_2O(l)$
$\Delta H_2 = -55.90$ kJ/mol ... (ii)
On subtracting eq. (ii) from eq. (i)
$NH_4OH(l) \longrightarrow NH_4^+(aq) + OH^-(aq)$
Enthalpy of ionisation = –51.46 – (–55.90) = 4.44 kJ/mol

130. (c) $4NO_2(g) + O_2(g) \rightarrow 2N_2O_5(g)$, $\Delta_r H = -111$ kJ

$\Delta H'$ ↓ $2N_2O_5(s)$; 2×54 kJ mol^{-1}

$\Delta_r H' = \Delta_r H - 54 = -111 - 108 = -219$ kJ

131. (b) $C_2H_5OH(l) + 3O_2(g) \longrightarrow 2CO_2(g) + 3H_2O(l)$
$\Delta n_g = 2 - 3 = -1$
$\Delta U = \Delta H - \Delta n_g RT$
$= -1366.5 - [(-1) \times 8.314 \times 300]$
$= -1366.5 + \left[(1) \times \frac{8.314}{10^3} \times 300\right]$
$= -1366.5 + 0.8314 \times 3 = -1364$ kJ

132. (b) $N_2 + 3H_2 \longrightarrow 2NH_3$ $\Delta H = 2 \times (-46.0)$ kJ mol^{-1}
Let x be the bond enthalpy of N – H bond then
[**Note :** Enthalpy of formation or bond formation enthalpy is given which is negative but the given reaction involves bond breaking hence values should be taken as positive.]
$\Delta H = \Sigma$ Bond energies of reactants
– Σ Bond energies of products
$2 \times (-46) = 712 + 3 \times (436) - 6x \Rightarrow x = +352$ kJ/mol

133. (a) $H_2O(l) \longrightarrow H^+(aq) + OH^-(aq)$;
$\Delta H = 57.32$ kJ ... (i)
$H_2(g) + \frac{1}{2}O_2(g) \longrightarrow H_2O(l)$;
$\Delta H = -286.20$ kJ ... (ii)
By adding equation (i) & (ii) we get,
$H_2(g) + \frac{1}{2}(O_2) \longrightarrow H^+(aq) + OH^-(aq)$
$\Delta H = 57.32 - 286.2 = -228.88$ kJ
Here $\Delta_f H^o$ of $H^+(aq.) = 0$ ∴ $\Delta_f H^o$ of $OH^- = -228.88$ kJ

134. (b) The energy involved in the conversion of $\frac{1}{2}Cl_2(g)$ to $Cl^-(aq)$ is given by

$$\Delta H = \frac{1}{2}\Delta_{diss}H^o_{Cl_2} + \Delta_{eg}H^o_{Cl} + \Delta_{hyd}H^o_{Cl}$$

Substituting various values from given data, we get

$$\Delta H = \left(\frac{1}{2} \times 240\right) + (-349) + (-381)$$

$$= (120 - 349 - 381) = -610 \text{ kJ mol}^{-1}$$

135. (a) $I_2(s) + Cl_2(g) \longrightarrow 2ICl(g)$
$\Delta H = [\Delta H_{I2(s) \rightarrow I2(g)} + \Delta H_{I-I} + \Delta H_{Cl-Cl}] - 2[\Delta H_{I-Cl}]$
$= 62.76 + 151.0 + 242.3 - 2 \times 211.3 = 33.46$ kJ

$$\Delta H_f^o(ICl) = \frac{33.46}{2} = 16.73 \text{ kJ/mol}$$

136. (d) For the reaction, $C(s) + \frac{1}{2}O_2(g) \longrightarrow CO(g)$
$\Delta H = \Delta U + \Delta n_g RT$ or $\Delta H - \Delta U = \Delta n_g RT$
$\Delta H - \Delta U = \frac{1}{2} \times 8.314 \times 298 = 1238.78$ J mol^{-1}

137. (b) $\Delta H = \Delta U + \Delta n_g RT$ for
$N_2 + 3H_2 \longrightarrow 2NH_3$
$\Delta n_g = 2 - 4 = -2$ ∴ $\Delta H = \Delta U - 2RT$
or $\Delta U = \Delta H + 2RT$ ∴ $\Delta U > \Delta H$

138. (N) $\frac{1}{2}X_2 + \frac{1}{2}Y_2 \longrightarrow XY$, $\Delta H = (-200)$ J
Let x be the bond dissociation energy of X_2. Then

$$\Delta H_f = -200 = \frac{1}{2}\Delta H_{X-X} + \frac{1}{2}\Delta H_{Y-Y} - \Delta H_{X-Y}$$

$$-400 = x + 0.5x - 2x \Rightarrow x = \frac{400}{0.5} = 800 \text{ kJ mol}^{-1}$$

139. (c) $CH_2 = CH_2(g) + H_2(g) \rightarrow CH_3 - CH_3(g)$
Enthalpy change = Bond energy of reactants – Bond energy of products.

$$\Delta H = 1(C=C) + 4(C-H) + 1(H-H) - 1(C-C) - 6(C-H)$$

$$= 1(C=C) + 1(H-H) - 1(C-C) - 2(C-H)$$

$$= 615 + 435 - 347 - 2 \times 414 = 1050 - 1175 = -125 \text{ kJ.}$$

140. (b) Enthalpy change for a reaction does not depend upon the nature of intermediate reaction steps.

141. (b)

$$\begin{array}{ccc} H_2O(l) & \longrightarrow & H_2O(s) \\ 10°C & & -10°C \\ \downarrow nC_{P(l)}\Delta T & & \uparrow nC_{P(s)}\Delta T \\ H_2O(l) & \xrightarrow[-\Delta H_{fusion}]{} & H_2O(s) \\ 0°C & & 0°C \end{array}$$

$\Delta H = 1 \times y(0-10) - x \times 1000 + 1 \times z(-10° - 0°)$

$\Delta H = -10(100x + y + z)$ Joule.

142. (b)

$$\underset{268K}{\text{Ice}} \xrightarrow{\Delta S_1} \underset{273K}{\text{Ice}} \xrightleftharpoons{\Delta S_2} \underset{273K}{\text{Water}} \xrightarrow{\Delta S_3} \underset{373K}{\text{Water}}$$

$$\xrightleftharpoons{\Delta S_4} \underset{373K}{\text{Water vapour}} \xrightarrow{\Delta S_5} \underset{383K}{\text{Water vapour}}$$

$\Delta S_{overall} = \Delta S_1 + \Delta S_2 + \Delta S_3 + \Delta S_4 + \Delta S_5$

$$\Delta S_1 = \int_{268}^{273} \frac{C_{p,m} dT}{T}$$

$$\Delta S_2 = \frac{\Delta H_{m\,fusion}}{273}; \; T_f = 273\,K$$

$$\Delta S_3 = \int_{273}^{373} \frac{C_{p,m} dT}{T}$$

$$\Delta S_4 = \frac{\Delta H_{m\,vaporisation}}{373}; \; T_b = 373\,K$$

$$\Delta S_5 = \int_{373}^{383} \frac{C_{p,m} dT}{T}$$

143. (b) (A) $dG = VdP - SdT$

Constant pressure

$dG = -SdT$

$$\left(\frac{\partial G}{\partial T}\right)_P = -S$$

(B) $dH = (q)_P = nCpdT$

$$\left(\frac{\partial H}{\partial T}\right)_P = C_P \text{ (for 1 mole)}$$

(C) $dG = VdP - SdT$

At constant temperature

$dG = VdP$

$$\left(\frac{\partial G}{\partial P}\right)_T = V$$

(D) $dU = nC_v dT = (q)_v$

$$\left(\frac{\partial U}{\partial T}\right)_V = C_V \text{ (for 1 mole)}$$

144. (400) At equilibrium $\Delta G_{PT} = 0$ and $\Delta G = \Delta H - T\Delta S$

$\Rightarrow \Delta H_{vap} = T\Delta S_{vap}$

$30 \times 1000 = T \times 75$

$T = 400K$

145. (2000) Applying Gibbs Helmholtz equation $\Delta G = \Delta H - T\Delta S$

For reaction to be spontaneous $\Delta G = -$ ve

For limiting case $\Delta G = 0$

$\Delta H = T\Delta S$

$$T = \frac{\Delta H}{\Delta S} = \frac{400 kJ/mol}{0.2 kJ/mol-K} = 2000\,K$$

above 2000 K reaction become spontaneous.

146. (1070) $\Delta S = \dfrac{\Delta H}{T_{mp}}$

$$28.4 = \frac{30.4 \times 1000}{T_{mp}}$$

$T_{mp} = 1070.422$ K.

147. (3) $\Delta S_{System} = nR \ln\left(\dfrac{V_2}{V_1}\right) = 1 \times 8.314 \ln\left(\dfrac{3}{2}\right)$

$\Delta S_{System} = 3.37$

$\Delta S_{Surr.} = 3.37$

148. (2) $\Delta G = \Delta H - T\Delta S$

A : $\Delta G\,(J\,mol^{-1}) = -25 \times 10^3 + 80 \times 300 \Rightarrow \Delta G = -ve$

B : $\Delta G\,(J\,mol^{-1}) = -22 \times 10^3 - 40 \times 300 \Rightarrow \Delta G = -ve$

C : $\Delta G\,(J\,mol^{-1}) = 25 \times 10^3 + 300 \times 50 \Rightarrow \Delta G = +ve$

D : $\Delta G\,(J\,mol^{-1}) = 22 \times 10^3 - 20 \times 300 \Rightarrow \Delta G = +ve$

$\Rightarrow$ Processes C and D are non-spontaneous.

149. (336) Given chemical reaction:

$A \xrightarrow[T=300K]{} B \;\; \Delta G = -49.4$ kJ/mol

$\Delta H = 51.4$ kJ/mol; $\Delta S = ?$

$\Rightarrow$ From the relation $\Delta G = \Delta H - T\Delta S$

$$\Rightarrow \Delta S = \frac{\Delta H - \Delta G}{T} = \frac{[51.4 - (-49.4)] \times 1000}{300} \frac{J}{mol\,K}$$

$\Rightarrow \Delta S = 336\,JK^{-1}\,mol^{-1}$.

150. (200) $A + B \rightleftharpoons C + D$

For a reaction to be spontaneous, $\Delta_r G° < 0$

$\Delta_r G° = \Delta_r H° - T\Delta_r S°$

$\Rightarrow \Delta_r H° - T\Delta_r S° < 0;\; 80 \times 1000 - T \times 2T < 0$

$\Rightarrow T^2 > 40000;\; T > 200 \;\therefore\; T_{min} = 200\,K$

151. (1380) $\Delta G° = RT \ln K_p$

$= -R(300)\ln(10)^2 = -R(300 \times 2 \times 2.3)$

$\Delta G° = -1380R$.

152. (855) $3HC \equiv CH(g) \rightarrow C_6H_6(l);\; \Delta G^0 = -RT\ln K$

$\Rightarrow \Delta G^0 = \Sigma(\Delta G_f^0)_P - \Sigma(\Delta G_f^0)_R$

$\Rightarrow -RT\ln K = 1 \times (-1.24 \times 10^5) - (-3 \times 2.04 \times 10^5)$

$\Rightarrow -2.303 \times R \times T \log K = 4.88 \times 10^5$

$$\Rightarrow \log K = -\frac{4.88 \times 10^5}{2.303 \times 8.314 \times 298}$$

$= -85.52 = -855 \times 10^{-1}$

$\Rightarrow$ So; magnitude of 10gk $= 855 \times 10^{-1}$

153. (a) A system at higher temperature has greater entropy (randomness). S and ΔS are related with T as:

$$S_T = \int_0^T \frac{nC \cdot dT}{T} \text{ and } \Delta S = \int \frac{dq}{T}$$

Thus, both S and ΔS are function of temperature.

154. (–2.70) $\Delta U = 2.1\text{ kcal} = 2.1\times10^3\text{ cal}$
$\Delta n_g = 2$
$\Delta H = \Delta U + \Delta n_g RT$
$= 2.1\times10^3 + 2\times2\times300 = 3300\text{ cal}$
$\Delta G = \Delta H - T\Delta S = 3300 - 300\times20$
$= -2700\text{ cals} = -2.7\text{ kcal}$

155. (a) Final temperature $= \dfrac{T_1+T_2}{2}$, let $T_2 > T_1$
(blocks have same mass)

$\therefore\ dS = \dfrac{dq}{T} = \dfrac{C_p dT}{T} \quad \therefore\ \Delta S = C_p \ln\left(\dfrac{T_f}{T_i}\right)$

$\therefore\ \Delta S_{Total} = C_p \ln\left(\dfrac{T_1+T_2}{2T_1}\right) + C_p \ln\left(\dfrac{T_1+T_2}{2T_2}\right)$

$= C_p \ln\left[\dfrac{(T_1+T_2)^2}{4T_1T_2}\right]$

156. (d) In the process of synthesis of ammonia from N_2 and H_2, number of moles decreases which implies that the change in entropy will be negative.

$N_2(g) + 3H_2(g) \rightleftharpoons 2NH_3(g)$

157. (d) $\underset{273K}{H_2O(s)} \xrightarrow{\Delta S_1} \underset{273K}{H_2O(l)} \xrightarrow{\Delta S_2} \underset{373K}{H_2O(l)}$

$\xrightarrow{\Delta S_3} \underset{373K}{H_2O(g)} \xrightarrow{\Delta S_4} \underset{383K}{H_2O(g)}$

$\Delta S_1 = \dfrac{\Delta H_{fus}}{273} = \dfrac{334}{273} = 1.22;\ \Delta S_2 = m\,C \ln\left(\dfrac{T_2}{T_1}\right)$

$= 4.2\times\ln\left(\dfrac{373}{273}\right) = 4.2\times2.303\times(\log 373 - \log 273)$

$= 4.2\times2.303\times(2.572 - 2.436) = 1.31$

$\Delta S_3 = \dfrac{\Delta H_{vap}}{373} = \dfrac{2491}{373} = 6.68$

$\Delta S_4 = m\,C\ln\left(\dfrac{T_2}{T_1}\right) = 2\times\ln\left(\dfrac{383}{373}\right)$
$= 2\times2.303\times(\log 383 - \log 373)$
$= 2\times2.303\times(2.583 - 2.572) = 0.05$

$\therefore$ Total entropy change
$\Delta S = 1.28 + 6.68 + 1.31 + 0.05 = 9.26\text{ kJ kg}^{-1}\text{ K}^{-1}$

158. (b)
(a) When diamond is converted into graphite (it is heated to 1500 °C) entropy is increased, $\Delta S > 0$
(b) When pressure increases then molecules of gas will come closer and intermolecular distance decreases, so entropy will also decrease, $\Delta S < 0$
(c) When we increase the temperature of a gas then randomness is increased as the kinetic energy is gained by molecules. So, $\Delta S > 0$
(d) H_2 molecule is converted into atoms, the no. of particles increases. Thus entropy will increase, $\Delta S > 0$

159. (b) Given, $C_p = 10$ cals at 1000 K
$T_1 = 1000\text{ K}.\ T_2 = 100\text{ K},\ m = 32\text{ g},\ \therefore\ n = \dfrac{32}{32} = 1$
$\Delta S = ?$
At constant pressure

$\Delta S = nC_p \ln\dfrac{T_2}{T_1} = 2.303\times C_p \log\dfrac{T_2}{T_1}$

$= 2.303\times10\log\dfrac{100}{1000} = -23.03\text{ cal deg}^{-1}$

160. (b) $\Delta S^\circ = S^\circ_{CO_2} + 2\times S^\circ_{H_2O} - \left(S^\circ_{CH_4} + 2\times S^\circ_{O_2}\right)$
$= (213.6 + 2\times69.9) - (186.2 + 2\times205.2)$
$= -242.8\text{ J K}^{-1}\text{ mol}^{-1}$.

161. (d) $\Delta S = \dfrac{\Delta H}{T};\ T = \dfrac{\Delta H}{\Delta S} = \dfrac{0.3344}{0.836} = 0.4\text{K}$

162. (b) For isothermal process ($\Delta T = 0$) for one mole

$\Delta S = R\ln\dfrac{P_1}{P_2} = 8.314\ln 2 = 8.314\times0.693 = 5.76$

163. (c) $\dfrac{1}{2}X_2 + \dfrac{3}{2}Y_2 \longrightarrow XY_3;\ \Delta H^\circ = -30\text{ kJ}$
Calculating ΔS° for the above reaction, we get

$\Delta S^\circ = 50 - \left[\dfrac{1}{2}\times60 + \dfrac{3}{2}\times40\right] = 50 - (30+60) = -40\text{ JK}^{-1}$

At equilibrium, $T\Delta S^\circ = \Delta H^\circ$ $\quad [\because \Delta G^\circ = 0]$
$\therefore\ T\times(-40\text{ JK}^{-1}) = -30\times1000\text{ J}$ $\quad [\because 1\text{ kJ} = 1000\text{J}]$
or $T = \dfrac{-30\times1000}{-40} = 750\text{ K}$

164. (c) For reaction to be spontaneous according to IInd law:
$\Delta G < 0$
$\Rightarrow \Delta H - T\Delta S < 0$ (ΔH or ΔS are +ve)
$\Rightarrow T > \left(\dfrac{\Delta H}{\Delta S}\right) = T_e$ ($\Delta G = -$ve at high temperature)
$\Rightarrow T > T_e$

165. (a) $N_2(g) + 3H_2(g) \rightleftharpoons 2NH_3(g)$
$\Delta S_r = -$ve, $\Delta H_r^\circ = -$ve
$\dfrac{-\Delta H_r^\circ}{T} \Rightarrow$ first decreases then constant
$\dfrac{-\Delta G_R^\circ}{T} \Rightarrow$ first decreases then constant

166. (a) Reaction is spontaneous at relatively high temperature and non-spontaneous at low temperature $\Delta G = \Delta H - T\Delta S$.
It is only possible when ΔH and ΔS both are positive as $\Delta H < T\Delta S$.

167. (c) $\because\ \Delta G = \Delta H - T\Delta S$
For spontaneity of reaction : $\Delta G = -$ve
ΔG is +ve for (B) and – ve for (C).

168. (2850)

$\Delta H^\circ_{rxn} = 55\text{ kJ/mol},\quad T = 298\text{ K}$

$\Delta S^\circ_{rxn} = 175\text{ J/k mol}$

$\Delta G^\circ_{rxn} = \Delta H^\circ_{rxn} - T\Delta S^\circ_{rxn}$

$\Rightarrow \Delta G^\circ_{rxn} = 55000\text{ J/mol} - 298\text{ K}\times175\text{ J/K 1 mol}$

$\Rightarrow \Delta G^\circ_{rxn} = 55000 - 52150$

$\Rightarrow \Delta G^\circ_{rxn} = 2850\text{ J/mol}$

169. (c) $\Delta G = 0$ for reversible reaction
$\Delta G = -ve$ for spontaneous reaction
$\Delta G = +ve$ for non-spontaneous reaction
Therefore option (c) is not correct.

170. (2) For, Spontaneous process $\Delta G < 0$
For, Equilibrium $\Delta G = 0$
For, Non-spontaneous process $\Delta G > 0$
Statement B and C are correct.

171. (b) (A) First law is given by
$\Delta U = Q + W$
If we apply reversible work at constant P.
$\Rightarrow \Delta U = Q - P\Delta V$
(D) $\Delta H = \Delta U + \Delta nRT$

172. (a) $\Delta G = \Delta H - T\Delta S$
$\because \Delta S_g > \Delta S_l > \Delta S_s$
$\therefore$ On melting the entropy increases and ΔG becomes more negative and hence it becomes easier to reduce metal.

173. (b) (A) For a spontaneous process Gibb's free energy value is negative at constant temperature and pressure i.e. $\Delta G_{T,P} < 0$.
(B) In isobaric pressure, pressure remains constant i.e., $\Delta P = 0$ while in isothermal process, temperature remains constant i.e., $\Delta T = 0$.
(C) $\Delta H_{reaction}$ = (Σ bond energies of reactants) – (Σ bond energies of products)
(D) In exothermic process, energy is released.
So, the value of enthalpy is negative i.e., $\Delta H < 0$.

174. (300) $\Delta G = \Delta H - T\Delta S = 0$ at equilibrium
$\Rightarrow -165 \times 10^3 - T \times (-505) = 0 \Rightarrow T = 300K$.

175. (710) $N_2O_4 \rightleftharpoons 2NO_2$

	N_2O_4	$2NO_2$
$t = 0$	1	mol
$t = t$	(1-0.5) mol = 0.5 mol	0.5 × 2 mol = 1 mol

Total pressure = 1 atm.

$$K_P = \frac{\left(\frac{1}{1.5} \times 1\right)^2}{\left(\frac{0.5}{1.5} \times 1\right)} = \frac{1}{0.75} = \frac{100}{75} = 1.33$$

$\Delta G^0 = -RT\ln K_P = -8.31 \times 300 \times \ln(1.33)$
$= -8.31 \times 300 \times \log_{10} 1.33 \times \log_e 10$
$= -8.31 \times 300 \times 0.1239 \times 2.3 = -710.43 \text{ J mol}^{-1}$.

176. (5) $\Delta G = \Delta H - T\Delta S$
$\Delta G = -57.8 - 298 \times (-176 \times 10^{-3}) = -5 \text{ kJ mol}^{-1}$

177. (25) $\Delta G° = -nFE° = \Delta H° - T\Delta S°$

$$\Delta S° = \frac{\Delta H° + nFE°}{T}$$

$$= \frac{(-825.2 \times 10^3) + (2 \times 96487 \times 4.315)}{298}$$

$$= \frac{-825.2 \times 10^3 + 832.682 \times 10^3}{298}$$

$$= \frac{7.483 \times 10^3}{298} = 25.11 \text{JK}^{-1}\text{mol}^{-1} \approx 25 \text{ JK}^{-1}\text{ms}^{-1}$$

178. (964) $T_{min} = \left(\frac{\Delta^o H}{\Delta^o S}\right)$

$$\Delta^o H_{rxn} = \left[\Delta_f^o H(\text{Fe}) + \Delta_f^o H(\text{CO})\right] - \left[\Delta_f^o H(\text{FeO}) + \Delta_f^o H(\text{C(graphite)})\right]$$

$= [0 - 110.5] - [-266.3 + 0] = 155.8$ kJ/mol

$\Delta^o S_{rxn} = [\Delta^o S(\text{Fe}) + \Delta^o S(\text{CO})] - [\Delta^o S(\text{FeO}) + \Delta^o S(\text{C(graphite)})]$

$= [27.28 + 197.6] - [57.49 + 5.74] = 161.65$ J/mol-K

$$T_{min} = \frac{155.8 \times 10^3 \text{ J/mol}}{161.65 \text{J/mol} - \text{K}} = 963.8 \text{ K}$$

$\simeq 964$ K (nearest integer)

179. (– 13538)
From $\Delta H° = \Delta U° + \Delta n_g RT$

$\Delta H° = -20 \times 1000 - 1 \times 8.314 \text{ J/mol.K} \times 298 \text{ K}$
$= -22477.57$ J

$\Delta G° = \Delta H° - T\Delta S° = -22477.57 - (298 \times (-30))$
$= -13538$ J

180. (–2.70)
$\Delta U = 2.1 \text{ kcal} = 2.1 \times 10^3 \text{ cal}$
$\Delta n_g = 2$
$\Delta H = \Delta U + \Delta n_g RT$
$= 2.1 \times 10^3 + 2 \times 2 \times 300$
$= 2100 + 1200$
$= 3300$ cal
$\Delta G = \Delta H - T\Delta S$
$= 3300 - 300 \times 20$
$= 3300 - 6000$
$= -2700$ cals
$= -2.7$ kcal

181. (c) A reaction is spontaneous if ΔG_{sys} is negative.
$\Delta G_{sys} = \Delta H_{sys} - T\Delta S_{sys}$
A reaction will be spontaneous at all temperatures if ΔH_{sys} is negative and ΔS_{sys} is positive.

182. (d) At 315 K; $\Delta G° = 120 - \frac{3}{8}$ T

$\Delta G° = 120 - \frac{3}{8} \times 315$ = positive

Since $\Delta G°$ is positive then $K_{eq} < 1$.

So $\frac{[Y]}{[X]} < 1$.

$\therefore [X] > [Y]$

183. (c) $MgO(s) + C(s) \rightarrow Mg(s) + CO(g)$
For a reaction to be spontaneous, $\Delta G < 0$

$\Delta H° - T\Delta S° < 0 \Rightarrow T > \frac{\Delta H°}{\Delta S°}$

$T > \frac{491.1 \times 1000}{198}$; $T > 2480.3$ K

184. **(b)** $\Delta G° = A - BT$; A and B are non-zero constants
$\therefore \Delta G° = \Delta H° - T\Delta S° = A - BT$
$\therefore$ Reaction will be endothermic if $\Delta H° > 0$.
Hence, $A > 0$

185. **(c)** $\Delta H = 200 \text{ J mol}^{-1}$
$\Delta S = 40 \text{ JK}^{-1} \text{ mol}^{-1}$
For spontaneous reaction,
$\Delta G < 0$
$\Delta H - T\Delta S < 0; \Delta H < T\Delta S$
$\frac{\Delta H}{\Delta S} < T; \frac{200}{40} < T$
$5 < T$
So, minimum temperature is 5 K

186. **(a)** From thermodynamics relation.
$\Delta G° = -RT \ln K$
$\Delta H° - T\Delta S° = RT \ln K$
$-\frac{\Delta H°}{RT} + \frac{T\Delta S°}{RT} = \ln K$ or $\ln K = -\frac{\Delta H°}{RT} + \frac{\Delta S°}{R}$
For exothermic reaction, $\Delta H° = -$ve
slope $= -\frac{\Delta H°}{R} = +$ve
So from graph, lines should be A & B.

187. **(c)**

	A_2	$\rightleftharpoons$	$2A$
initial	1		0
final	$(1 - 0.2)$		(2×0.2)

The equilibrium constant,
$K = \frac{[A]^2}{[A_2]}$; $K = \frac{[0.4]^2}{[0.8]} = 0.2$

$\Delta G° = -RT \ln K = -8.314 \text{ JK}^{-1} \text{ mol}^{-1} \times 320 \text{ K} \times \ln 0.2$
$= 4281 \text{ J/mol}$

188. **(d)** $S(\ell) \longrightarrow S(g)$; $K_p = P_{vap}$
$\Delta G^o_{rxn} = \Delta_f G°(\text{vapour}) - \Delta_f G°(\text{liquid})$
$\Delta G^o_{rxn} = 103 - 100.7 = 2.3 \text{ kcal/mol} = 2300 \text{ cal/mol}$
$\Delta G^o_{rxn} = -RT \ln K_p$
$2300 \text{ cal/mol} = -2 \text{ cal/mol/K} \times 500 \text{ K} \times \ln K_p$
$\ln K_p = -2.3$
$\log_{10} K_p = -1$
$K_p = \text{Antilog}(-1) = 0.1 \text{ atm}$
$\therefore$ Vapour pressure of liquid 'S' at 500 K is approximately equal to 0.1 atm.

189. **(b)** (i) $2Fe_2O_3(s) \rightarrow 4Fe(s) + 3O_2(g)$;
$\Delta_r G° = +1487.0 \text{ kJ mol}^{-1}$
(ii) $2CO(g) + O_2(g) \rightarrow 2CO_2(g)$;
$\Delta_r G° = -514.4 \text{ kJ mol}^{-1}$
Multiply reaction (ii) with 3, we get
(iii) $6CO(g) + 3O_2(g) \rightarrow 6CO_2(g)$;
$\Delta_r G° = 3 \times (-514.4) = -1543.2 \text{ kJmol}^{-1}$
When we add reaction (i) and reaction (iii), we get reaction (iv) :
(iv) $2Fe_2O_3(s) + 6CO(g) \rightarrow 4Fe(s) + 6CO_2(g)$
Free energy change, $\Delta_r G°$ for the reaction will be $1487.0 - 1543.2 = -56.2 \text{ kJ mol}^{-1}$

190. **(b)** $\Delta H = nC_p\Delta T = 0$ means ΔH is constant if $\Delta T = 0$
$\Delta S = nR \ln\left(\frac{V_f}{V_i}\right) \geq 0$ if $V_f > V_i$

191. **(d)** $\Delta G = \Delta H - T\Delta S$
For a spontaneous reaction $\Delta G = -$ve
which is possible when both ΔH and ΔS are positive at high temperature and hence, the reaction becomes spontaneous.

192. **(c)** $\Delta G° = \Delta H° - T\Delta S°; -RT \ln K = \Delta H° - T\Delta S°$
$\ln K = -\frac{\Delta H° - T\Delta S°}{RT}$

193. **(b)** At equilibrium $\Delta G = 0$
Hence, $\Delta G = \Delta H - T_e\Delta S = 0$
$\therefore \Delta H = T_e\Delta S$ or $T_e = \frac{\Delta H}{\Delta S}$
For a spontaneous reaction,
ΔG must be negative which is possible only if $\Delta H < T\Delta S$
or $T > \frac{\Delta H}{\Delta S}$; $T_e < T$

194. **(a)** $\Delta G° = \Delta H° - T\Delta S°$
For a spontaneous reaction $\Delta G° < 0$
or $\Delta H° - T\Delta S° < 0 \Rightarrow T > \frac{\Delta H°}{\Delta S°}$
$\Rightarrow T > \frac{179.3 \times 10^3}{160.2}$
$> 1117.9 \text{K} \approx 1118 \text{K}$

195. **(b)** Spontaneity of reaction depends on tendency to acquire minimum energy state and maximum randomness. For a spontaneous process in an isolated system, the change in entropy is positive.

196. **(a)** For spontaneous reaction, $dS > 0$ and $dG < 0$.

197. **(c)** $\Delta G° = -RT \ln K_c$ or $-\Delta G° = RT \ln K_c$

198. **(b)** For an endothermic reaction, $\Delta H > 0$
Therefore, options (a) and (c) are not possible.
At freezing point (T_f), reaction is non-spontaneous.
$\therefore \Delta G > 0 \Rightarrow \Delta H - T_f . \Delta S > 0 \Rightarrow \Delta S < \frac{\Delta H}{T_f}$
At boiling point (T_b), reaction is spontaneous.
$\therefore \Delta G < 0 \Rightarrow \Delta H - T_b . \Delta S < 0 \Rightarrow \Delta S > \frac{\Delta H}{T_b} \Rightarrow \Delta S > 0$

199. **(a)** $C + O_2 \Rightarrow CO_2$; $\Delta H_1 = -393 \text{ J}$ ..(i)
$2Zn + O_2 \Rightarrow 2ZnO$; $\Delta H_2 = -412 \text{ J}$ ...(ii)
On the basis of given values oxidation of zinc is more feasible than carbon therefore will carbon will oxidise Zn.

Equilibrium

6

1. **(5)**

$$H_2O\,(g) \rightleftharpoons H_2(g) + \frac{1}{2}\;O_2(g)$$

t = 0	1 mole	–	–
$t = t_{eq}$	$1-\alpha$	α	$\frac{\alpha}{2}$

$$n_T = 1 + \frac{\alpha}{2} = 1\,(\alpha << 1)$$

$$K_p = \frac{P_{H_2}P_{O_2}^{1/2}}{P_{H_2O}} = \frac{(\alpha \cdot P)\left(\frac{\alpha}{2}P\right)^{\frac{1}{2}}}{(1-\alpha)P} \text{ (given P = 1)}$$

$$\therefore \quad 8 \times 10^{-3} = \frac{\alpha^{3/2}}{\sqrt{2}}$$

$$\Rightarrow \alpha^{3/2} = 8\sqrt{2} \times 10^{-3} \Rightarrow \alpha^3 = 128 \times 10^{-6}$$

$$\alpha = 5.03 \times 10^{-2} \approx = 5 \times 10^{-2}$$

2. **(c)** At melting point when ice melts, supplied heat is utilised to overcome intermolecular attraction within the molecules so temperature remain constant, and kinetic energy also does not change.

3. **(d)** On addition of inert gas at constant P and T, reaction moves in the direction of greater number of moles so it will shift in forward direction, so $[PCl_5]$ decrease and $[PCl_3]$ and $[Cl_2]$ will increase.

4. **(c)** A catalyst alter the equilibrium constent (K_c) of the reaction. Therefore statement I is correct. A homogenous catalyst cannot alter the equilibrium composition of a system. Hence statement II is incorrect.

5. **(b)** According to Le-Chatelier's principle, increasing pressure favours the reaction to move in backward direction. Which results in reduction in the dissociation of A hence α decreases.

6. **(74)** $CO(g) + 2H_2(g) \rightleftharpoons CH_3OH(g)$

t = 0	0.1 mol	a mol	–
t_{eq}	0.1 – x = 0.06	a – 2x = a – 0.08	x = 0.04

$V = 2L,\ T = 500\ K,\ P_{total} = 5\ bar$

$$n_{Total} = 0.25 = \frac{1}{2}\text{ mol}$$

$$P_{total} = n_{total} \times \frac{RT}{V}$$

$$\Rightarrow \quad 5 = (0.06 + a - 0.08 + 0.04) \times \frac{0.08 \times 500}{2}$$

$$\Rightarrow \quad a = 0.23\text{ mol}$$

$$Kp = \frac{X_{CH_3OH}}{X_{CO} \times X_{H_2}^2} \times \frac{1}{(P_T)^2} = \frac{0.04}{0.06 \times (0.15)^2} \times \left[\frac{1/4}{5}\right]^2$$

$$= \frac{4}{6 \times (0.15)^2 \times 16} \times \frac{1}{25} = 0.074 = 74 \times 10^{-3}$$

7. **(c)** $AB_2(g) \rightleftharpoons AB(g) + \frac{1}{2}B_2(g)$

Initial	1	0	0
at equ.	(1–x)p	x	$\frac{x}{2}$

$$\text{Total} = 1 - x + x + \frac{x}{2}\text{ moles}$$

$$= \left(1 + \frac{x}{2}\right)\text{moles}$$

$$K_p = \frac{\left(\frac{xp}{1+\frac{x}{2}}\right) \times \left(\frac{\frac{x}{2}\cdot p}{1+\frac{x}{2}}\right)^{1/2}}{\frac{(1-x)p}{\left(1+\frac{x}{2}\right)}}$$

(p = total pressure at equilibrium)

$(p_i = x_i \times p_{total})$

$$\Rightarrow \quad x << 1 \Rightarrow \quad 1 + \frac{x}{2} \approx 1 \text{ and } 1 - x \approx 1$$

$$\therefore \quad K_p = \frac{(xp)\cdot\left(\frac{xp}{2}\right)^{1/2}}{p}$$

$$\Rightarrow K_p^2 = \frac{x^2 \cdot p^2 \cdot \frac{xp}{2}}{p^2} \Rightarrow K_p^2 = x^2 \cdot \frac{xp}{2}$$

$$\Rightarrow x^3 = \frac{2K_p^2}{p} \Rightarrow x = \sqrt[3]{\frac{2K_p^2}{p}}$$

8. **(d)**

2 atm
P
1atm
S
L
G
273.15K T

Phase diagram of H_2O

If pressure is increased by two time leads to the complete conversion of ice into water (liquid) form. Ice, being less dense than liquid water, melts under higher pressure.

9. **(b)** $H_2 + I_2 \rightleftharpoons 2HI$

Concentration of H_2 and I_2 decreases untill equilibrium condition and concentration of HI increases till equilibrium condition and after equilibrium concentration of all the reactant and products remain constant.

10. **(962)** Initial pressure of N_2O_5 $(P) = \frac{nRT}{V}$

$$= \frac{\frac{37.8}{108} \times 0.082 \times 500}{1} = 14.35 \text{ bar}$$

$$2N_2O_5 \rightleftharpoons 2N_2O_4 + O_2$$

t = 0	14.35		
t = eq	14.35–2P	2P	P

P_{total} at equilibrium = 14.35 + P = 18.65

$\Rightarrow P = 4.3$

$\Rightarrow P_{N_2O_5} = 5.75$ bar ; $P_{N_2O_4} = 8.6$ bar

$P_{O_2} = 4.3$ bar ; $k_P = \frac{(8.6)^2 \times (4.3)}{(5.75)^2} = 9.619 = x \times 10^{-2}$

$x = 961.9 \approx 962$

11. **(d)** $N_2O_5(g) \longrightarrow 2NO_2(g) + \frac{1}{2}O_2(g)$

t=0	P_0	–	–
t=eq.	$P_0 - x$	$2x$	$\frac{x}{2}$

$x = \frac{P_0}{2}$; $P_{total} = \left(P_0 - \frac{P_0}{2}\right) + P_0 + \frac{P_0}{4} = \frac{7}{4}P_0$

12. **(b)** Let's take 1 mole of X_2Y.

$$X_2Y(g) \rightleftharpoons X_2(g) + \frac{1}{2}Y_2(g)$$

Initial	1.0	0.0	0.0
Final	$1-x$ mole	x mole	$\frac{x}{2}$ mole

$\therefore P_{X_2Y} = \frac{1-x}{1+\frac{x}{2}} \times P$ (Using $P_i = x_i \times P_{total}$)

$P_{X_2} = \frac{x}{1+\frac{x}{2}} \times P$

$P_{Y_2} = \frac{x/2}{1+\frac{x}{2}} \times P$

$$\therefore K_p = \left(\frac{x}{1+\frac{x}{2}}P\right)\left(\frac{x}{2\left(1+\frac{x}{2}\right)}P\right)^{\frac{1}{2}} \Big/ \left(\frac{1-x}{1+\frac{x}{2}}\right) \times P$$

$$\therefore K_P = \left(\frac{x}{1-x}\right)\left(\frac{x}{2\left(1+\frac{x}{2}\right)}\right)^{\frac{1}{2}} \times P^{\frac{1}{2}}$$

$\because$ x is very very small

$$\therefore K_P = \frac{x^{3/2}}{(2)^{\frac{1}{2}}} \times P^{\frac{1}{2}} \quad \therefore x^{\frac{3}{2}} = \frac{K_P \times 2^{\frac{1}{2}}}{P^{\frac{1}{2}}}$$

$$\therefore x^3 = \frac{K_p^2 \times 2}{P} \quad \Rightarrow x = \left(\frac{K_p^2 \times 2}{P}\right)^{\frac{1}{3}}$$

13. **(b)** $CO_2(g) + C(s) \rightleftharpoons 2CO(g)$

0.5	–
0.5 – x	2x

$P_{total} = 0.5 + x = 0.8 = x = 0.3$

$$K_p = \frac{(0.6)^2}{0.2} = 1.8$$

14. **(c)** On adding given reactions we get desired reaction.

$X \rightleftharpoons Y$;	$K_1 = 1$ ...(1)
$Y \rightleftharpoons Z$;	$K_2 = 2$ ...(2)
$Z \rightleftharpoons W$;	$K_3 = 4$ ...(3)
$X \rightleftharpoons W$	$K_1 . K_2 . K_3$
	$K = 1 \times 2 \times 4$
	$K = 8$

Remember: When reaction (1, 2, 3) added, the rate constant will be multiply $k_1 \times k_2 \times k_3$

15. **(c)** $K_P = K_C (RT)^{\Delta n_g}$

$\Delta n_g = 0 \therefore K_P = K_c$

$$K_p = \frac{[HI]^2}{[H_2][I_2]} = 1 = 10 \times 10^{-1}$$

$\Rightarrow 1 = x \times 10^{-1}$

$\Rightarrow x = 10$

16. **(d)** $CO(g) + \frac{1}{2}O_2(g) \rightleftharpoons CO_2(g)$

K_P and K_C are related as

$K_P = K_C (RT)^{\Delta n_g}$

$\Delta n_g = n_P - n_R = 1 - 1\frac{1}{2} = -\frac{1}{2}$

Δn_g for the given reaction $= -\frac{1}{2}$

$$\therefore \frac{K_P}{K_C} = \frac{1}{\sqrt{RT}}$$

17. **(d)** $SO_3(g) \rightleftharpoons SO_2(g) + \frac{1}{2}O_2(g)$

On reversing the reaction and multiplying by 2 we get desired reaction. Thus,

$$K'_c = \left(\frac{1}{K_c}\right)^2 = \left(\frac{1}{4.9 \times 10^{-2}}\right)^2, K'_c = 416.49$$

18. **(d)** $Fe^{3+}(aq) + SCN^-(aq) \rightleftharpoons [FeSCN]^{2+}(aq)$

$$K_c = \frac{[FeSCN^{2+}]}{[Fe^{3+}][SCN^-]}$$

19. **(b)** $\therefore K_P = \frac{\alpha^{3/2} P^{1/2}}{(2+\alpha)^{1/2}(1-\alpha)}$

$$A \rightleftharpoons B + \frac{C}{2}$$

t = 0	1	–	–
t = t	$(1-\alpha)$	α	$\frac{\alpha}{2}$

$$P_t = 1-\alpha+\alpha+\frac{\alpha}{2} = \left(1+\frac{\alpha}{2}\right)$$

$$P_A = \frac{1-\alpha}{\left(1+\frac{\alpha}{2}\right)}\cdot P \qquad P_B = \frac{\alpha}{\left(1+\frac{\alpha}{2}\right)}\cdot P$$

$$P_C = \frac{\frac{\alpha}{2}}{\left(1+\frac{\alpha}{2}\right)}\cdot P$$

$$K_P = \frac{P_B\cdot (P_C)^{1/2}}{P_A}$$

$$= \frac{\frac{\alpha}{\left(1+\frac{\alpha}{2}\right)}\cdot P \times \left(\frac{\frac{\alpha}{2}\cdot P}{\left(1+\frac{\alpha}{2}\right)}\right)^{1/2}}{\frac{1-\alpha}{\left(1+\frac{\alpha}{2}\right)}\cdot P}$$

$$= \frac{\left(\frac{2\alpha}{2+\alpha}\right)\cdot P \times \left[\left(\frac{\alpha}{2+\alpha}\right)P\right]^{1/2}}{\frac{2(1-\alpha)\cdot P}{2+\alpha}}$$

$$= \frac{P^{1/2}\cdot(\alpha)\,(\alpha)^{1/2}}{(2+\alpha)^{1/2}(1-\alpha)} \Rightarrow \boxed{K_P = \frac{P^{1/2}\cdot(\alpha)^{3/2}}{(2+\alpha)^{1/2}(1-\alpha)}}$$

20. (2) $N_2O_4\,(g) \rightleftharpoons 2NO_2\,(g)$

$K_P = 0.492$ atm

$T = 300$ K

$K_c = ?$

$K_P = K_c\,(RT)^{\Delta n_g}$

$\Delta n_g = n_P - n_r = 2-1 = 1$

$\therefore K_P = K_c . RT$

$$\Rightarrow K_c = \frac{K_P}{RT} = \frac{0.492}{0.082\times 300}$$

$= 0.02 = 2\times 10^{-2}$

21. (417) $N_2\,(g) + 3H_2\,(g) \rightleftharpoons 2NH_3$

$$\Rightarrow K_c = \frac{[NH_3]^2}{[N_2][H_2]^3} = \frac{(1.5\times 10^{-2})^2}{(2\times 10^{-2})\,(3\times 10^{-2})^3}$$

$$= \frac{2.25\times 10^{-4}}{(2\times 10^{-2})\,(27\times 10^{-6})} = \frac{2.25\times 10^{-4}}{54\times 10^{-8}} = 416.67$$

22. (44) $CO(g) + H_2O(g) \rightleftharpoons CO_2(g) + H_2(g)$

	CO(g)	$H_2O(g)$	$CO_2(g)$	$H_2(g)$
t = 0;	1 mol	1 mol	0	0
at eqm.;	1–x	1–x	x	x

at equilibrium 40% by mass water reacts with CO

$x = 0.4$, $1-x = 0.6$

$$K_C = \frac{[CO_2][H_2]}{[CO][H_2O]} = \frac{0.4\times 0.4}{0.6\times 0.6} = 0.44$$

$K_C \times 10^2 = 44$

23. (1) $H_2(g) + I_2(g) \rightleftharpoons 2HI(g)$

	$H_2(g)$	$I_2(g)$	2HI(g)
t = 0 ;	4.5	4.5	–
t_{eq} ;	3	3	3

$$K_c = \frac{[HI]^2}{[H_2][I_2]} = \frac{(3)^2}{3\times 3} = \frac{9}{9} = 1$$

24. (3) (C) at equilibrium the value of parameters become constant at a given temperature and not equal. Therefore statement (C) is incorrect.

25. (a) $A_2B_3(aq.) \rightleftharpoons 2A^{3+}(aq) + 3B^{2-}(aq)$

c(1–α) 2cα 3cα

$$K_{eq} = \frac{\left[A^{3+}\right]^2\left[B^{2-}\right]^3}{\left[A_2B_3\right]} = \frac{4c^2\alpha^2\times 27c^3\alpha^3}{c(1-\alpha)}, \alpha \ll 1$$

$$\Rightarrow K_{eq} = \frac{108c^5\alpha^5}{c}, \alpha = \left(\frac{K_{eq}}{108c^4}\right)^{\frac{1}{5}}$$

26. (48.6) $PCl_3 + Cl_2 \rightleftharpoons PCl_5$

0.2M (0.1+0.2)M 0.4M

Eq^m. 0.2–x 0.3–x 0.4+x

$$\frac{(0.4+x)}{(0.2-x)(0.3-x)} = 20$$

$\Rightarrow x \simeq 0.086$

$[PCl_5]_{eq} = 0.4 + 0.086 = 0.486M = 48.6\times 10^{-2}$ M

27. (12) $X \rightleftharpoons Y(g) + Z(g);\ K_{p_1} = 3$

	X	Y(g)	Z(g)
At t = 0,	a	0	0
At equilibrium,	a – αa	αa	αa

$$\therefore K_{p_1} = \frac{p_Z\ p_Y}{p_X};\ \text{Total pressure} = P_1$$

$$\frac{\left(\frac{\alpha}{1+\alpha}\times P_1\right)^2}{\frac{1-\alpha}{1+\alpha}P_1} = 3,\ \text{or,}\ \frac{\alpha^2 P_1}{(1+\alpha)(1-\alpha)} = 3;\ \therefore\ \frac{\alpha^2P_1}{1-\alpha^2} = 3$$

$A \rightleftharpoons 2B \quad K_{p_2} = 1$

	A	2B
At t = 0,	a	0
At equilibrium,	a – αa	2αa

$$\therefore K_{p_2} = \frac{\left(\frac{2\alpha}{1+\alpha}\times P_2\right)^2}{\frac{1-\alpha}{1+\alpha}\times P_2}\ \text{or}\ \frac{4\alpha^2P_2}{1-\alpha^2} = 1$$

$$\frac{K_{p_1}}{K_{p_2}} = \frac{P_1}{4P_2}\quad \text{or}\quad \frac{P_1}{4P_2} = 3/1\ \therefore\ P_1 : P_2 = 12:1$$

28. **(1)** $SO_2(g) + \frac{1}{2}O_2(g) \rightleftharpoons SO_3(g)$

$K_P = 2 \times 10^{12}$ at 300 K

$\Delta n_g = 1 - \left(1 + \frac{1}{2}\right) = 1 - \frac{3}{2} = -\frac{1}{2}$

$K_P = K_c \times (RT)^{\Delta n_g}$

$\Rightarrow \quad 2 \times 10^{12} = K_c \times (0.082 \times 300)^{-1/2}$

$K_c = 9.92 \times 10^{12} = 0.992 \times 10^{13} \approx 1 \times 10^{13}$

29. **(2)** $H_2O(g) \rightleftharpoons H_2(g) + \frac{1}{2}O_2(g)$

At equilibrium : $P[1-\alpha]$ $\quad P\alpha \quad \frac{P\alpha}{2}$

$$K_p = \frac{(P_{H_2})(P_{O_2})^{1/2}}{P_{H_2O}}$$

$$\frac{(P\alpha)\left(\frac{P\alpha}{2}\right)^{1/2}}{P[1-\alpha]} = 2 \times 10^{-3}$$

since α is negligible w.r.t 1 so P = 1 and $1 - \alpha \approx 1$

$$\frac{\alpha\sqrt{\alpha}}{\sqrt{2}} = 2 \times 10^{-3}$$

$\alpha^{3/2} = 2^{3/2} \times 10^{-3}$

$\alpha = 2^{3/2 \times 2/3} \times 10^{-3 \times 2/3}$

$\alpha = 2 \times 10^{-2} \Rightarrow \% \alpha = 2\%$

30. **(4)** $N_2(g) + 3H_2(g) \rightleftharpoons 2NH_3(g), K_1 = 4 \times 10^5$ ...(i)

$N_2(g) + O_2(g) \rightleftharpoons 2NO(g), K_2 = 1.6 \times 10^{12}$ ...(ii)

$H_2(g) + \frac{1}{2}O_2(g) \rightleftharpoons H_2O(g), K_3 = 1.0 \times 10^{-13}$ ...(iii)

By applying (ii) + 3 × (iii) – (i), we get

$2NH_3(g) + \frac{5}{2}O_2(g) \rightleftharpoons 2NO(g) + 3H_2O(g)$

$$k_{eq} = \frac{k_2 \times k_3^3}{k_1} = \frac{1.6 \times 10^{12} \times (10^{-13})^3}{4 \times 10^5}$$

$$= \frac{1.6}{4} \times 10^{-32} = 4 \times 10^{-33}$$

31. **(2)** $2NO + O_2 \rightleftharpoons 2NO_2$

	$2NO$	O_2	$2NO_2$
Initial:	2	1	–
Change:	2–2 α	1– α	2 α
Equilibrium:	1.2	0.6	0.8

$$K_p = \frac{(x_{NO_2}P)^2}{(x_{NO}P)^2(x_{O_2}P)}$$

$$K_p = \frac{\left(\frac{0.8}{2.6}\right)^2}{\left(\frac{1.2}{2.6}\right)^2\left(\frac{0.6}{2.6}\right)} = 1.925$$

32. **(2)**

$Cu(s) + 2Ag^+(aq) \rightleftharpoons Cu^{2+}(aq) + 2Ag(s); 2 \times 10^{15} = k$

$\frac{1}{2}Cu^{2+}(aq) + Ag^+(s) \rightleftharpoons \frac{1}{2}Cu(s) + Ag^+(aq); K$

$$K = \frac{1}{\sqrt{2 \times 10^{15}}} = 2.23 \times 10^{-8}$$

33. **(c)** $2A(g) \rightleftharpoons 2B(g) + C(g) \quad K_1$ ...(i)

$\frac{3}{2}A(g) \rightleftharpoons \frac{3}{2}B(g) + \frac{3}{4}C(g) \quad K_2$ ...(ii) eq. (ii) is $\frac{3}{4}$ times of eq. (i), hence, $K_2 = (K_1)^{\frac{3}{4}}$

34. **(a)** $PCl_5 = 5$ mole; Ar = 4 mole

$$P_{Total} = \frac{9 \times 0.082 \times 610}{100} = 4.5 \text{ atm}$$

$p_y = X_y P_T$ (P_T → Total pressure; X_y → mole fraction of any species y)

$X_{PCl_5} = \frac{5}{9}; \quad X_{Ar} = \frac{4}{9}$

$p_{PCl_5} = \frac{5 \times 4.5}{9} = 2.5$ atm; $p_{Ar} = \frac{4 \times 4.5}{9} = 2$ atm

$PCl_5 \rightleftharpoons PCl_3 + Cl_2$

2.5 – P $\quad$ P $\quad$ P

At eqm., $P_{total} = 2.5 - P + P + P + P_{Ar} = 6$ atm

P = 1.5 atm

$$K_p = \frac{1.5 \times 1.5}{1} = 2.25$$

35. **(29)** moles of water $= \frac{0.90}{18}$

By ideal gas equation PV = nRT

$$V = \frac{nRT}{P} = \frac{0.90 \times 0.082 \times 300 \times 760}{18 \times 32} = 29.21 \approx 29 \text{ L}$$

36. **(125)** $2NOCl(g) \rightleftharpoons 2NO(g) + Cl_2(g)$

	$2NOCl(g)$	$2NO(g)$	$Cl_2(g)$
$t = 0$	2 M	–	–
$t = t_{eq}$	$(2 - x)$M	x M	$\frac{x}{2}$ M

$\because \; x = 0.4$M

$\therefore \; [NOCl]_{eq} = 2.0 - 0.4 = 1.6$ M; $[Cl_2]_{eq} = 0.2$ M

$$\Rightarrow K_c = \frac{[NO]^2[Cl_2]}{[NOCl]^2} = \frac{[0.4]^2[0.2]}{[1.6]^2} = 125 \times 10^{-4}.$$

37. (b)

$$A(g) \rightleftharpoons B(g) + \frac{1}{2}C(g)$$

Initial : P_i 0   0

At eq. : $P_i(1-\alpha)$ $P_i . \alpha$ $P_i \frac{\alpha}{2}$

Equilibrium pressure $(p) = p_i\left(1-\alpha+\alpha+\frac{\alpha}{2}\right)$

$$p = P_i \times \left(1+\frac{\alpha}{2}\right) \Rightarrow \therefore p_A = \left(\frac{1-\alpha}{1+\frac{\alpha}{2}}\right)p$$

$$p_B = \left(\frac{\alpha}{1+\frac{\alpha}{2}}\right)p;\ p_C = \left(\frac{\frac{\alpha}{2}}{1+\frac{\alpha}{2}}\right)p$$

$$\therefore K = \frac{p_C^{\frac{1}{2}} \times p_B}{p_A} \Rightarrow K = \frac{\alpha^{\frac{3}{2}} p^{\frac{1}{2}}}{(2+\alpha)^{\frac{1}{2}}(1-\alpha)}$$

38. (1107) Given : 2 mole of N_2 gas was present as inert gas. Equilibrium pressure = 2.46 atm

$$PCl_5(g) \rightleftharpoons PCl_3(g) + Cl_2(g)$$

$t = 0$ 5   0   0

t = Eqm. $5-x$ x x

From ideal gas equation, $PV = nRT$

$2.46 \times 200 = (5-x+x+x+2) \times 0.082 \times 600 \Rightarrow x = 3$

$$K_P = \frac{\left(\frac{3}{10}\times 2.46\right)\left(\frac{3}{10}\times 2.46\right)}{\left(\frac{2}{10}\times 2.46\right)} = 1.107 = 1107 \times 10^{-3}$$

39. (4) Let moles of NH_3 added = a

$\therefore$ concentration of $NH_3 = \frac{a}{2}$ mol/L

$$Ag^+(aq.) + 2NH_3(aq.) \rightleftharpoons Ag(NH_3)_2^+(aq).$$

$t = 0$ 0.8 $\left(\frac{a}{2}\right)$ 0

$t = \infty$ 5×10^{-8} $\left(\frac{a}{2}-1.6\right)$ 0.8

$$K_f = \frac{0.8}{\left(5\times10^{-8}\right)\left(\frac{a}{2}-1.6\right)^2} = 10^8$$

$$\Rightarrow \frac{a}{2} - 1.6 = 0.4 \Rightarrow a = 4$$

40. (6) Moles of NH_4HS initially taken $= \frac{5.1\ g}{51\ g/mol}$

$= 0.1$ mol

Volume of vessel = 2 L

$$NH_4HS(s) \rightleftharpoons NH_3(g) + H_2S(g)$$

Initial   0.1 mol   0   0

At eqm. $0.1(1-0.2)$ 0.1×0.2 0.1×0.2

$\Rightarrow$ Partial pressure of each component gaseous

$$P = \frac{nRT}{V} = \frac{0.1\times0.2\times0.082\times300}{2} = 0.246\ \text{atm}$$

$$\Rightarrow K_P = P_{NH_3} \times P_{H_2S} = (0.246)^2 = 0.060516$$

$= 6.05 \times 10^{-2}$ or 6×10^{-2}

41. (182)

$$A + B \rightleftharpoons C + D : K_{eq} = 100$$

$t=0$ 1   1   1   1

$t=t_{eq.}$ $(1-x)$ $(1-x)$ $(1-x)$ $(1-x)$

$$K_{eq} = 100 = \frac{(1+x)(1+x)}{(1-x)(1-x)} \Rightarrow 100 = \left(\frac{1+x}{1-x}\right)^2$$

$$10 = \left(\frac{1+x}{1-x}\right) \Rightarrow x = \frac{9}{11}$$

$$[D] = (1+x) = 1 + \frac{9}{11} = \frac{20}{11}$$

$= 1.8181 = 181.81 \times 10^{-2} \simeq 182 \times 10^{-2}$.

42. (2) $[PtCl_4]^{2-} + H_2O \underset{k_b}{\overset{k_f}{\rightleftharpoons}} [Pt(H_2O)Cl_3]^- + Cl^-$

$$\frac{-d[Pt\ Cl_4]^{-2}}{dt} = k_f[PtCl_4^{2+}] = k_b[Pt(H_2O)Cl_3][Cl^-]$$

$$\Rightarrow K_{eq} = \frac{k_f}{k_b} = \frac{4.8\times10^{-5}}{2.4\times10^{-3}} = 0.02 \simeq 2\times10^{-2}$$

43. (1400)

$$PCl_5(g) \rightleftharpoons PCl_3(g) + Cl_2(g); K_c = 1.844$$

$t = 0$ 3 moles   0   0

$t = \infty$ $(3-x)$ x x

$$\Rightarrow \frac{[PCl_3][Cl_2]}{[PCl_5]} = \frac{x^2}{3-x} = 1.844$$

$$\Rightarrow x^2 + 1.844x - 5.532 = 0$$

$$x = \frac{-b \pm \sqrt{D}}{2a} = \frac{-1.844 + \sqrt{(1.844)^2 + 4 \times 5.532}}{2} \cong 1.604$$

$\Rightarrow$ Moles of $PCl_5 = 3 - 1.604 \cong 1.396 \simeq 1.4 \approx 1400 \times 10^{-3}$

44. (16) $K_p = (P_{O_2})^{1/2}$

$4 = (P_{O_2})^{1/2} \Rightarrow P_{O_2} = 16$

45. (875)

	$2SO_2(g)$ +	$O_2(g)$ →	$2SO_3(g)$
Initial :	250 m bar	750 m bar	0
Final :	–250 m bar	–125 m bar	250 m bar
After completion	0	625 m bar	250 m bar

$\therefore$ Final total pressure = 625 + 250 = 875 m bar

46. (25) $A + B \rightleftharpoons 2C$

A	B	2C
1	1	1
1 – x	1 – x	1 + 2x

$$K = \frac{[C]_{eq}^2}{[A]_{eq}[B]_{eq}} = \frac{(1+2x)^2}{(1-x)(1-x)}$$

$$100 = \left(\frac{1+2x}{1-x}\right)$$

$$\left(\frac{1+2x}{1-x}\right) = 10$$

$$x = \frac{3}{4}$$

$$[C]eq. = 1 + 2x = \frac{15}{6} = 25 \times 10^{-1} M$$

47. (2) $K_P = K_C(RT)^{\Delta n_g}$

$N_2O_4(g) \rightleftharpoons 2NO_2(g)$

$\Delta n_g = 1$

$$K_C = \frac{K_P}{(RT)} = \frac{47.9}{(0.083 \times 288)} \approx 2$$

48. (172) $2SO_2(g) + O_2(g) = 2SO_3(g)$

$$K_P = \frac{\left(p_{SO_{3(g)}}\right)^2}{\left(p_{SO_{2(g)}}\right)^2 \times p_{O_{2(g)}}} = \frac{\left(43 \times 10^3\right)^2}{\left(45 \times 10^3\right)^2 \times 530}$$

$$= \frac{(43)^2}{(45)^2 \times 530 \times 10^{-3}} = 172 \times 10^{-2}$$

49. (354) $N_2O_4(g) \rightleftharpoons 2NO_2(g)$

$\Delta n_g = 2 - 1 = 1$

Relation between K_P and K_C is

$$K_P = K_C(RT)^{\Delta n_g} \Rightarrow 600.1 = 20.4(0.0831 \times T)^1$$

$$\Rightarrow T = \frac{600.1}{20.4 \times 0.0831} = 354 \text{ K}$$

50. (72) $AB_2(g) \rightleftharpoons A(g) + 2B(g)$

	AB_2	A	2B
$t = 0$	1	–	–
$t =$ eqm.	$1 - \alpha$	α	2α

$\therefore$ No. of moles at equilibrium

$$= 1 + 2\alpha = \frac{PV}{RT} = \frac{1.9 \times 25}{0.08206 \times 300} \approx 1.93$$

$\therefore \alpha = 0.465$

$$\therefore P_{AB_2} = \frac{1-\alpha}{1+2\alpha} \times P_T = \frac{1-0.465}{1+2 \times 0.465} \times 1.9 \approx 0.53 \text{ atm}$$

$$P_A = \frac{\alpha}{1+2\alpha} \times P_T = \frac{0.465}{1.93} \times 1.9 \approx 0.46 \text{ atm}$$

$$P_B = \frac{2\alpha}{1+2\alpha} \times P_T = \frac{2 \times 0.465}{1.93} \times 1.9 \approx 0.91 \text{ atm}$$

$$K_P = \frac{P_A \cdot (P_B)^2}{P_{AB_2}} = \frac{0.46 \times (0.91)^2}{0.53} \approx 0.72 \approx 72 \times 10^{-2}$$

51. (5) $Cl_2(g) \rightleftharpoons 2Cl(g)$

Let mol of both of Cl_2 and Cl be x.

$$P_{Cl} = \frac{x}{2x} \times 1 = \frac{1}{2} \text{ and } P_{Cl_2} = \frac{x}{2x} \times 1 = \frac{1}{2}$$

$$\therefore K_p = \frac{\left(\frac{1}{2}\right)^2}{\frac{1}{2}} = \frac{1}{2} = 0.5 = 5 \times 10^{-1}.$$

52. (c) $K_p = K_c(RT)^{\Delta n_g} = K_c(RT)^{1-3/2} = K_c(RT)^{-1/2}$

$\Rightarrow K_c = K_p(RT)^{1/2}$

53. (d) $N_2(g) + 3H_2(g) \rightleftharpoons 2NH_3(g)$; K_c

$2NH_3(g) \rightleftharpoons N_2(g) + 3H_2(g)$; $\frac{1}{K_c}$

$NH_3(g) \rightleftharpoons \frac{1}{2}N_2(g) + \frac{3}{2}H_2(g)$; $\frac{1}{K_c^{1/2}}$

$$\frac{1}{K_c^{1/2}} = \frac{1}{(64)^{1/2}} = \frac{1}{8}$$

54. (16)

$$X + Y \rightleftharpoons 2Z$$

$t=0$ 1 mol 1.5 mol 0.5 mol

$t=t_{eq}$ $1-a$ $1.5-a$ $0.5+2a$

$0.5+2a=1;\ a=0.25$

$$X + Y \rightleftharpoons 2Z$$

0.75 1.25 1

$$K_{eq}=\frac{[Z]^2}{[X][Y]}=\frac{1}{0.75\times[1.25]}=\frac{x}{15}$$

$$\Rightarrow x=\frac{15}{(0.75\times1.25)}=16$$

55. (b) At equilibrium, rate of forward reaction = Rate of backward reaction.

56. (d) $A \rightleftharpoons B+C;\ K_{eq}^{(1)}$ ...(i)

$B+C \rightleftharpoons P;\ K_{eq}^{(2)}$ ...(ii)

On adding equations (i) and (ii), we get

$A \rightleftharpoons P$

K_{eq} (overall) $= K_{eq}^{(1)} \cdot K_{eq}^{(2)}$

57. (d) Equilibrium constant

$$K_c=\frac{[B]}{[A]}=\frac{11}{6}\approx 2$$

58. (a) We know that, $K_p=K_c.(RT)^{\Delta n_g}$

$\therefore$ If $\Delta n_g \neq 0$ then $K_p \neq K_c$

Now, $2C(s)+O_2(g) \rightleftharpoons 2CO(g)$

$\Delta n_g = 1 \Rightarrow K_p=K_c(RT)^1$

Hence, $K_p \neq K_c$

59. (c) Given, $S+O_2 \rightleftharpoons SO_2$ ... (i); $K_1=10^{52}$

$2S+3O_2 \rightleftharpoons 2SO_3$ ... (ii); $K_2=10^{129}$

$2SO_2+O_2 \rightleftharpoons 2SO_3$ ... (iii); $K=?$

To get equation (iii) follow (ii) – 2 (i),

$2S+3O_2 \rightarrow 2SO_3$ $K_2=10^{129}$

$-(2S+2O_2 \rightarrow 2SO_2$ $K_1'=10^{104})$

$O_2 \rightarrow 2SO_3-2SO_2$ $K=10^{129-104}$

or $2SO_2+O_2 \rightarrow 2SO_3$ $K=10^{25}$

60. (b) $A(s) \longrightarrow B(g)+C(g)$

$\qquad\qquad P_1 \quad P_1+P_2$

$$K_{P_1}=P_B\times P_C$$

$$P_1(P_1+P_2)=x \quad ...(i)$$

$D(s) \longrightarrow C(g)+E(g)$

$\qquad\qquad P_1+P_2 \quad P_2$

$$K_{P_2}=P_C\times P_E$$

$$(P_1+P_2)P_2=y \quad ...(ii)$$

Adding (i) and (ii)

$\therefore\ P_1(P_1+P_2)+P_2(P_1+P_2)=x+y$

$\Rightarrow\ (P_1+P_2)(P_1+P_2)=(P_1+P_2)^2=x+y$

$\Rightarrow\ P_1+P_2=\sqrt{x+y}$

$\therefore$ Total pressure $(P_T)=P_C+P_B+P_E$

$(P_1+P_2)+P_1+P_2=2(P_1+P_2)$

$P_T=2(\sqrt{x+y})$

61. (a) $A + 2B \rightleftharpoons 2C + D$

$t=0$ 2 3 0 0

t_{eq} $2-x$ $3-2x$ $2x$ x

Given, $3-2x=2-x;\ x=1$

$\therefore\ [C]=2, [D]=1, [A]=1, [B]=1$

$$\therefore\ K_c=\left\{\frac{2^2\times1}{1^2\times1}\right\}=4$$

62. (a) $2NH_3(g) \rightleftharpoons N_2(g)+3H_2(g),\ K=\frac{1}{K_P}$

$\qquad P_{NH_3} \quad P_{N_2} \quad P_{H_2}$

$$\therefore\ K=\frac{1}{K_P}=\frac{P_{N_2}(P_{H_2})^3}{(P_{NH_3})^2} \quad ...(i)$$

$\Rightarrow P_{total}(P)=P_{N_2}+P_{H_2}+P_{NH_3}$

$\approx P_{N_2}+P_{H_2}$ ($\because P_{NH_3} << P_{total}$, remaining NH_3 at equilibrium is neglible)

Now,

Partial pressure of N_2 = mole fraction of $N_2 \times P=\frac{1}{4}P$;

Partial pressure of $H_2=\frac{3}{4}P$

$$\text{From eq (i), } \frac{1}{K_P}=\frac{\left(\frac{1}{4}P\right)\left(\frac{3}{4}P\right)^3}{(P_{NH_3})^2}$$

$$\frac{(P_{NH_3})^2}{K_P}=\frac{P}{4}\cdot\frac{P^3}{4^3}\cdot3^3;\ \frac{(P_{NH_3})^2}{K_P}=\frac{P^4}{4^4}\cdot3^3$$

$$(P_{NH_3})^2=K_P\cdot\frac{P^4}{4^4}\cdot3^3;\ P_{NH_3}=\left[K_P\cdot\frac{P^4}{4^4}\cdot3^3\right]^{1/2}$$

$$P_{NH_3}=\frac{3^{3/2}\cdot P^2\cdot K_P^{1/2}}{16}$$

63. (d) Concerned reaction:

$NH_4SH \longrightarrow NH_3(g) + H_2S(g)$

Initial moles $= \frac{5.1}{51} = 0.1$ mol

Moles at equillibrium

$NH_4SH \longrightarrow NH_3(g) + H_2S(g)$

$0.1(1 - 0.3)$ $\quad 0.1 \times 0.3$ $\quad 0.1 \times 0.3$

$\therefore K_c = [NH_3][H_2S] = \left(\frac{0.03}{3}\right)^2 = 10^{-4}$

$K_p = K_c(RT)^{\Delta n_g}$

$= 10^{-4} \times (0.082 \times 600)^2 = 0.242 \text{ atm}^2$

64. (b) $K_p = K_c\ (RT)^{\Delta n_g}$

Δn_g = No. of gaseous moles of products – No. of gaseous moles of reactants

$\frac{K_p}{K_c} = (RT)^{\Delta n_g}$

$N_2(g) + O_2(g) \rightleftharpoons 2NO(g), \Delta n_g = 0$

$\frac{K_p}{K_c} = (24.62 \text{ dm}^3 \text{ atm mol}^{-1})^0 = 1$

$N_2O_4(g) \rightleftharpoons 2NO_2(g), \Delta n_g = 1$

$\frac{K_p}{K_c} = 24.62 \text{ dm}^3 \text{ atm mol}^{-1}$

$N_2(g) + 3H_2(g) \rightleftharpoons 2NH_3(g), \Delta n_g = -2$

$\frac{K_p}{K_c} = (24.62 \text{ dm}^3 \text{ atm mol}^{-1})^{-2}$

$= 1.65 \times 10^{-3} \text{ dm}^{-6} \text{ atm}^{-2} \text{ mol}^2$

65. (c) Given: $A_2 + B_2 \overset{K_1}{\rightleftharpoons} 2AB$

$\Rightarrow\ 2AB \rightleftharpoons A_2 + B_2;\ K = \frac{1}{K_1}$

$6AB \overset{K_1}{\rightleftharpoons} 3A_2 + 3B_2;\ K_2 = \left(\frac{1}{K_1}\right)^3$

The relation between K_1 and K_2 is $K_2 = K_1^{-3}$

66. (a) Initially 2 moles of CO are present.

At equilibrium, 1 mole of CO is present

Hence, 2 – 1 = 1 moles of CO has reacted.

1 mole of CO will react with 1 mole of Cl_2 to form 1 mole of $COCl_2$.

3 – 1 = 2 moles of Cl_2 remains at equilibrium

The equilibrium constant

$$K_c = \frac{[COCl_2]}{[CO][Cl_2]} = \frac{\frac{1 \text{ mol}}{5L}}{\frac{1 \text{ mol}}{5L} \times \frac{2 \text{ mol}}{5L}} = 2.5$$

67. (a) $XY(s) \rightleftharpoons X(g) + Y(g)$ (Assume)

$\qquad\qquad\qquad\ p \qquad p$

At eqm.,

Total pressure $= 2p = 10$ bar

$\therefore\ p = 5$; Now $K_p = (p_X)(p_Y) = p^2 = 25$.

68. (c) $N_2O_4(g) \rightleftharpoons 2NO_2(g)$

	N_2O_4	NO_2
t = 0	1 mol	0
t = eqm.	1 – 0.2 = 0.8 mol	0.2 × 2 = 0.4 mol
MW :	92	46

From ideal gas equation, PV = nRT

P = 1 atm, T = 300K, $n = n_T = 0.8 + 0.4 = 1.2$

$\therefore\ V = \frac{nRT}{P} = 1.2 \times 0.082 \times 300$ L

$= 29.52$ L

Total mass inside the vessel

$= 0.8 \times (MW)_{N2O4} + 0.4\,(MW)_{NO2}$

$= (0.8 \times 92) + (0.4 \times 46) = 92$ g

$\therefore$ Density of the equilibrium mixture,

$d = \frac{m}{v} = \frac{92}{29.52} = 3.11$ g / L

69. (b) $SO_2(g) + \frac{1}{2}O_2(g) \rightleftharpoons SO_3(g)$

$K_p = K_c(RT)^x$

where $x = \Delta n_g$ = number of moles of gaseous product – number of moles of gaseous reactants

$= 1 - \left(1 + \frac{1}{2}\right) = 1 - \frac{3}{2} = -\frac{1}{2}$

70. (d) $2HI \rightleftharpoons H_2 + I_2$

At t = 0 $\quad c \quad 0 \quad 0$

At eqm. $\quad c - c\alpha \quad \frac{c\alpha}{2} \quad \frac{c\alpha}{2}$

Now, $K_c = \frac{\left(\frac{c\alpha}{2}\right)\left(\frac{c\alpha}{2}\right)}{(c - c\alpha)^2}$

$K_c = \frac{\alpha^2}{4(1-\alpha)^2};\ K_c = \frac{(0.5)^2}{4(1-0.5)^2}$

$K_c = 0.25$

71. (b) $NH_2COONH_4(s) \rightleftharpoons 2NH_3(g) + CO_2(g)$

$$K_p = \frac{(p_{NH_3})^2 \times (p_{CO_2})}{p_{NH_2COONH_4}} = (p_{NH_3})^2 \times (p_{CO_2})$$

As evident by the reaction, NH_3 and CO_2 are formed in molar ratio of 2 : 1. Thus if P is the total pressure of the system at equilibrium, then

$$p_{NH_3} = \frac{2}{3} \times P \quad p_{CO_2} = \frac{1}{3} \times P$$

$$K_p = \left(\frac{2P}{3}\right)^2 \times \frac{P}{3} = \frac{4P^3}{27}$$

Given, $K_p = 2.9 \times 10^{-5}$

$$\therefore\ 2.9 \times 10^{-5} = \frac{4P^3}{27}$$

$$P = \left(\frac{2.9 \times 10^{-5} \times 27}{4}\right)^{1/3} = 5.82 \times 10^{-2} \text{ atm}$$

72. (a) $K_p = K_c (RT)^{\Delta n_g}$

For the reaction

$$CO(g) + \frac{1}{2} O_2(g) \rightleftharpoons CO_2(g)$$

$$\Delta n_g = 1 - \left(1 + \frac{1}{2}\right) = -\frac{1}{2}$$

$$\therefore\ K_p = \frac{K_c}{\sqrt{RT}}; \quad \frac{K_p}{K_c} = \frac{1}{\sqrt{RT}}$$

73. (d) To calculate the value of K_4 in the given equation we should apply :

eqn. (2) + eqn.(3) × 3 – eqn. (1)

$$\text{Hence, } K_4 = \frac{K_2 K_3^3}{K_1}$$

74. (d) For the reaction

$N_2 + O_2 \longrightarrow 2NO \quad Kc = 4 \times 10^{-4}$

Hence for the reaction

$$NO \longrightarrow \frac{1}{2} N_2 + \frac{1}{2} O_2 \quad K'_c = \frac{1}{\sqrt{K_c}} = \frac{1}{\sqrt{4 \times 10^{-4}}} = 50$$

75. (b) (I) $N_2 + 2O_2 \overset{K_1}{\rightleftharpoons} 2NO_2$

$$K_1 = \frac{[NO_2]^2}{[N_2][O_2]^2} \quad \text{...(i)}$$

(II) $2NO_2 \overset{K_2}{\rightleftharpoons} N_2 + 2O_2$

$$K_2 = \frac{[N_2][O_2]^2}{[NO_2]^2} \quad \text{...(ii)}$$

(III) $NO_2 \overset{K_3}{\rightleftharpoons} \frac{1}{2} N_2 + O_2$

$$K_3 = \frac{[N_2]^{1/2}[O_2]}{[NO_2]}$$

$$\therefore\ (K_3)^2 = \frac{[N_2][O_2]^2}{[NO_2]^2} \quad \text{...(iii)}$$

∴ From equation (i), (ii) and (iii)

$$K_1 = \frac{1}{K_2} = \frac{1}{(K_3)^2}$$

76. (c) Reaction (iii) can be obtained by adding reactions (i) and (ii) therefore $K_3 = K_1 . K_2$

Hence (c) is the correct answer.

77. (c)

	$2AB_3(g)$	$\rightleftharpoons$	$A_2(g)$	$+ 3B_2(g)$
at t = 0	8		0	0
at eqm.	(8 – 2 × 2)		2	3 × 2
	= 4		2	6
conc.	4/V		2/V	6/V

$$\text{now } K_c = \frac{[A_2][B_2]^3}{[AB_3]^2} = \frac{2/1 \times [6/1]^3}{[4/1]^2} = 27$$

78. (d)

	$N_2O_4(g)$	$\rightleftharpoons 2NO_2(g)$
Initial moles	1	0
Moles at eqm.	(1 – α)	2α

(α = degree of dissociation)

Total no. of moles at eqm. = (1 – α) + 2α = (1 + α)

$$p_{N_2O_4} = \frac{(1-\alpha)}{(1+\alpha)} \times P$$

$$p_{NO_2} = \frac{2\alpha}{(1+\alpha)} \times P$$

$$K_p = \frac{(p_{NO_2})^2}{p_{N_2O_4}} = \frac{\left(\frac{2\alpha}{(1+\alpha)} \times P\right)^2}{\left(\frac{1-\alpha}{1+\alpha}\right) \times P}$$

$$K_p = \frac{4\alpha^2 P}{\frac{(1+\alpha)^2(1-\alpha)}{1+\alpha}} \Rightarrow K_p = \frac{4\alpha^2 P}{(1+\alpha)(1-\alpha)} = \frac{4\alpha^2 P}{1-\alpha^2}$$

Given, $K_p = 2$, $P = 0.5$ atm $\quad \therefore\ 2 = \frac{4\alpha^2 \times 0.5}{1-\alpha^2}$

$\alpha = 0.707 \approx 0.71 \Rightarrow$ 71 % dissociation

79. (c) $O_2 + 2SO_2 \rightleftharpoons 2SO_3$

At t = 0 1 mol 2 mol 0

At eqm. $(1-\alpha)$ $2(1-\alpha)$ 2α

Given at equilibrium,

$2\alpha = 1.6 \therefore \alpha = 0.8$

$$K_c = \frac{(2\alpha/v)^2}{\frac{(1-\alpha)}{v}\times\left(\frac{2-2\alpha}{v}\right)^2} = \frac{(2\alpha)^2}{(1-\alpha)(2-2\alpha)^2}$$

[Since V = 11]

$$\therefore K_c = \frac{(0.8)^2}{(1-0.8)(1-0.8)^2} = \frac{0.64}{0.002}$$

$K_c = 80$

80. (a) $\underset{1}{X} \rightleftharpoons \underset{0}{2Y}$

$1-x$ $2x \rightarrow$ Total pressure $= 1+x$

for $K_{p_1} = \frac{(2x)^2}{1-x}\frac{p_1}{1+x} = \frac{4(x^2)}{(1-x)}\frac{p_1}{1+x}$

Hence, $\underset{1-x}{Z} \rightleftharpoons \underset{x}{P} + \underset{x}{Q}$

for $K_{p_2} = \frac{(x)^2}{(1-x)}\left(\frac{p_2}{1+x}\right)$

$$\frac{K_{p_1}}{K_{p_2}} = \frac{4\times p_1}{p_2}$$

$$\frac{1}{9} = \frac{4\times p_1}{p_2}$$

$$\Rightarrow \frac{p_1}{p_2} = \frac{1}{36}$$

81. (c) $SO_3(g) \rightleftharpoons SO_2(g) + \frac{1}{2}O_2(g)$

$$K_c = \frac{[SO_2][O_2]^{1/2}}{[SO_3]} = 4.9\times10^{-2};$$

On taking the square of the above reaction

$$\frac{[SO_2]^2[O_2]}{[SO_3]^2} = 24.01\times10^{-4}$$

Now for $2SO_2(g) + O_2(g) \rightleftharpoons 2SO_3$

$$K'_c = \frac{[SO_3]^2}{[SO_2]^2[O_2]} = \frac{1}{24.01\times10^{-4}} = 416$$

82. (c) $\underset{1-x}{PCl_5(g)} \rightleftharpoons \underset{x}{PCl_3(g)} + \underset{x}{Cl_2(g)}$

Total moles after dissociation

$1-x+x+x = 1+x$

p_{PCl_3} = Mole fraction of PCl_3 × Total pressure $= \left(\frac{x}{1+x}\right)P$

83. (d) For the reaction:-

$2NO_2(g) \rightleftharpoons 2NO(g) + O_2(g)$; $\Delta n_g = 1$

Hence it is clear that, $K_p > K_c$

84. (a) $NH_4HS(s) \rightleftharpoons NH_3(s) + H_2S(g)$

At t = 0 0.5 atm 0 atm

At eqm. $0.5 + x$ atm x atm.

Then $0.5 + x + x = 2x + 0.5 = 0.84$ (given) $x = 0.17$ atm.

$p_{NH_3} = 0.5 + 0.17 = 0.67$ atm;

$p_{H_2S} = 0.17$ atm

$K_p = p_{NH_3} \times p_{H_2S} = 0.67\times0.17\ \text{atm}^2 = 0.11\ \text{atm}^2$

85. (c) $K_p = K_c(RT)^{\Delta n}$;

Here $\Delta n = 1-2 = -1$

$$\therefore \frac{K_p}{K_c} = \frac{1}{RT}$$

86. (d) For $P_4(s) + 5O_2(g) \rightleftharpoons P_4O_{10}(s)$

$K_c = \frac{1}{(O_2)^5}$. [Since solids have unit concentration]

87. (b) $K_c = \frac{[NO_2]^2}{[N_2O_4]} = \frac{[1.2\times10^{-2}]^2}{[4.8\times10^{-2}]} = 3\times10^{-3}$ mol/L

88. (c) $K_p = K_c(RT)^{\Delta n}$;

$$\Delta n = 1 - \left(1+\frac{1}{2}\right) = 1 - \frac{3}{2} = -\frac{1}{2}.$$

$$\therefore \frac{K_p}{K_c} = (RT)^{-1/2}$$

89. (N) (A) Concentration of only products increases.

(B) Equilibrium shifts in forward direction to decrease pressure.

(C) K_c remains constant at constant T.

(D) K_c is constant because of constant T and concentrations of only product change.

90. (700) $\frac{1}{2}X_2 + \frac{5}{2}Y_2 \rightleftharpoons XY_5$

$$\Delta S^0_{Rxn} = \left(\Delta S^0_{xy_5}\right) - \left(\frac{1}{2}\Delta S^0_{x_2} + \frac{5}{2}\Delta S^0_{y_2}\right)$$

$$\Delta S^0_{Rxn} = 110 - \left[\left(\frac{1}{2}\times70\right) + \left(\frac{5}{2}\times50\right)\right]$$

$$\Delta S^0_{Rxn} = 110 - 160 = -50\ \text{JK}^{-1}\ \text{mol}^{-1}$$

$\Delta G^0 = 0$ at equilibrium; $\Delta G^0 = \Delta H^0 - T\Delta S^0$

$0 = -35000 - T(-50)$; $T = 700$ K

91. **(a)** For reaction :

$Fe_2O_3(s) + 3CO(g) \rightleftharpoons Fe(\ell) + 3CO_2(g)$

Addition or removal of $Fe_2O_3(s)$ and $Fe(\ell)$ will not affect K_C and hence equilibrium

$$K_C = \frac{[CO_2]^3}{[CO]^3}$$

92. **(57)** Given: T = 300 K, K = 10

$\Delta G = -RT \ln K = -2.303\ RT \log K$

$= -2.303 \times 8.314 \times 300 \log 10$

$= -5744$ J/mol $= -5.744$ kJ/mol

$= -57 \times 10^{-1}$ kJ/mol

93. **(163)**

$\frac{3}{2}O_2(g) \rightleftharpoons O_3(g) \cdot K_P = 2.47 \times 10^{-29}$

$\Delta_r G^\ominus = -RT \ln Kp$

$= -8.314 \times 10^{-3} \times 298 \times \ln(2.47 \times 10^{-29})$

$= -8.314 \times 10^{-3} \times 298 \times (-65.87)$

$= 163.19$ kJ

94. **(37)** $\Delta G° = \Delta H° - T\Delta S°$

$= 77.2 \times 10^3 - 400 \times 122 = 28400$ J

$\Delta G° = -2.303\ RT \log K$

$\Rightarrow 28400 = -2.303 \times 8.314 \times 400 \log K$

$\Rightarrow \log K = -3.708 = -37.08 \times 10^{-1}$

95. **(10)** $\Delta G° = \Delta H° - T\Delta S°$

$\Rightarrow \Delta G° = [(-54070) - (10 \times 298)]$

Also, $\Delta G° = (-2.303\ RT \log K)$

$\Rightarrow [(-54070) - (10 \times 298)]$

$= (-2.303 \times 8.134 \times 298 \log K)$

$\Rightarrow \log K = 10$

96. **(a,d)** If the volume is kept constant and an inert gas such as He is added which does not take part in the reaction the equilibrium remain undisturbed. If P is constant, equilibrium will shift towards the forward direction.

97. **(3)** The yield of SO_3 at equilibrium will be due to :

(B) Increasing pressure

(C) Adding more SO_2

(D) Adding more O_2

98. **(6)** $\Delta G° = -RT \ln K_{eq}$

and $K_{eq} = \frac{K_f}{K_b} = \frac{10^3}{10^2} = 10$

$\therefore \Delta G = -RT \ln 10$

$= -(8.3 \times 300 \times 2.3)$

$= -5.7$ kJ mole^{-1} $\approx$ 6 kJ mol^{-1}.

99. **(2735)** $HI \rightleftharpoons \frac{1}{2}H_2 + \frac{1}{2}I_2$

	HI	H_2	I_2	
t_i	1	–	–	
t_{eq}	$1 - 0.4$	$\frac{0.4}{2}$	$\frac{0.4}{2}$	(Total no. of moles =
	= 0.6	= 0.2	= 0.2	0.6 + 0.2 + 0.2 = 1.0)

$K_c = \frac{(0.2)^{\frac{1}{2}}(0.2)^{\frac{1}{2}}}{1 - 0.4} = \frac{0.2}{0.6} = \frac{1}{3}$ [Pressure at eqm. = 1]

$\Delta G = \Delta G° + RT \ln K = 0$ [$\therefore$ at Equilibrium $\Delta G = 0$]

$\Delta G° = -RT \ln K = -8.31 \times 300 \times 2.3 \times \log\left(\frac{1}{3}\right) = 2735$ J/mol

100. **(747)** $2O_3(g) \rightleftharpoons 3O_2(g)$

	$2O_3(g)$	$3O_2(g)$
$t = 0$	2	0
t = eq.	2 – 1 = 1	3/2

Total moles at equilibrium = 1 + 3/2 = 5/2

Total pressure = 1 atm. (given)

Partial pressure = mole fracrtion × total pressure

$p_{O_3} = \frac{1}{5/2} \times 1 = \frac{2}{5}$; $p_{O_2} = \frac{3/2}{5/2} \times 1 = \frac{3}{5}$

$$K_p = \frac{(P_{O_2})^3}{(P_{O_3})^2} = \frac{(3/5)^3}{(2/5)^2} = 1.35$$

Free energy change,

$\Delta G = -RT \ln K_p = -8.3 \times 300 \times \ln 1.35 = -747$ J mol^{-1}

101. **(20)** $\Delta G° = -2.303\ RT \log K = -RT \ln K_{eq}$

$\ln K_{eq} = \frac{\Delta G°}{RT} = \frac{9.478 \times 10^3}{8.314 \times 495}$

$\ln K_{eq} = 2.303$

$\because \ln K_{eq} = \ln 10$

$\therefore K_{eq} = 10$

Now,

	A(g)	B(g)
$t = 0$	22	0
$t = t$	$22 - x$	x

$K_{eq} = \frac{[B]}{[C]}$

$10 = \frac{x}{22 - x} \Rightarrow x = 20$

102. (166) $2A(g) \rightleftharpoons A_2(g)$

For homogeneous gas phase reaction, $K_{eq} = K_p$

$\Delta G° = -RT \ln K_p$

$\Rightarrow 25.2\times10^3 = -8.3\times400\times2.3\log K_p$

$\log K_p = \dfrac{-25.2\times10^3}{8.3\times400\times2.3} \Rightarrow \log K_p = -3.3$

$K_p = \text{antilog}(-3.3)$

$K_p = 10^{-3.3}; K_p = 10^{-3}.10^{-0.3}$

$K_p = 0.501\times10^{-3} = 5\times10^{-4}$ $\because K_p = K_c(RT)^{-1}$

$\therefore K_c = 5\times10^{-4}\times8.3\times400 = 1.66 = 166\times10^{-2}$

103. (20) $\Delta G° = -RT\ln K_{eq}$

$\ln K_{eq} = \dfrac{\Delta G°}{RT} = \dfrac{9.478\times10^3}{8.314\times495} = 0.002303\times10^3$

$\ln K_{eq} = 2.303 \Rightarrow \ln K_{eq} = \ln 10 \Rightarrow K_{eq} = K_p = 10$

For the given reaction, $K_c = K_p$

	A(g) ⇌	B(g)
$t = 0$	22 m mol	0
$t = t$	$22 - x$	x

$K_c = \dfrac{[B]}{[A]}$

$10 = \dfrac{x}{22-x} \Rightarrow x = 20$

104. (c) $\Delta G° = -RT\ln K$, $T_1 = 25\,°C$, $K_1 = 10$

$\Delta G°$ at $T_1 = -8.314\times298\times2.303\times\log10 = -5.71$ kJ/mol

$\Delta G°$ at $T_2 = -8.314\times373\times2.303\times\log(100)$

$= -14.29$ kJ/mol

$\ln\left[\dfrac{K_2}{K_1}\right] = \dfrac{\Delta H}{R}\left[\dfrac{1}{T_1} - \dfrac{1}{T_2}\right]$

$\ln 10 = \dfrac{\Delta H}{8.314}\left[\dfrac{1}{298} - \dfrac{1}{373}\right]$

$\Delta H° = 28.4$ kJ/mol

105. (c) (i) As reaction is endothermic (ΔHz = +ve) so on decrease in temperature equilibrium will shift towards reactant side.

(ii) On increase in pressure by adding inert gas (N_2) at same temperature, no shifting will take place. The equilibrium changes only if the added gas is a reactant or product involved in the reaction.

106. (d) $\Delta G° = -RT\ln K$

$\therefore$ If $K > 1$ then $\Delta G° < 0$

If $K < 1$ then $\Delta G° > 0$

If $K = 1$ then $\Delta G° = 0$

107. (a) Equilibrium constant has no relation with catalyst. Catalyst only affects the rate of the reaction.

Catalyst, V_2O_5 in the given reaction, is used to speed up the reaction.

108. (c) $\Delta G = \Delta G° + RT\ln Q$

At equilibrium; $\Delta G = 0$ and $Q = K_{eq}$

$\Rightarrow \Delta G° = -2.303\ RT\log K_w$

$= -2.303\times8.314\times298\times\log10^{-14} = 79.9$ kJ/mol ≈ 80 kJ/mol

109. (b) As volume ↑ pressure ↓, reaction proceeds in that direction where the number of moles of gases increases.

110. (a) Reaction at equilibrium

$$N_2O_4 \rightleftharpoons 2NO_2$$

According to Le chatelier's principle-

(a) addition of an inert gas at constant pressure will increase volume or decrease moles per unit volume. Thus, equilibrium shifts towards more number of molecules.

111. (d)

Perturbation	Shifts reaction towards
Removal of CO	Left
Removal of CO_2	Right
Addition of CO_2	Left
Addition of Fe_2O_3	No change (This is a solid compound. Its concentration has no effect on the equilibrium.)

112. (c)

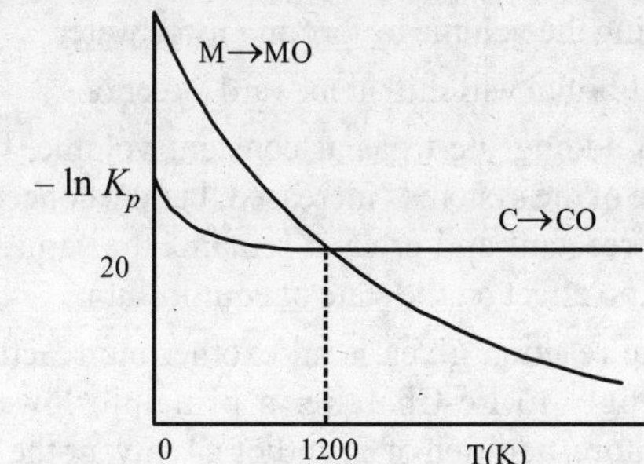

$\Delta G° = -RT\ln K_p$, So $-\ln K_p$ represents $\Delta G°$ here and thus the feasibility or spontaneity of a reaction.

At T < 1200 K, oxidation of carbon is more favourable than oxidation of M.

Hence, the following reactions will take place:

$C(s) + \frac{1}{2}O_2(g) \rightarrow CO(g)$

$MO(s) \rightarrow M(s) + \frac{1}{2}O_2(g)$

Adding these two reactions:

$C(s) + MO(s) \rightarrow M(s) + CO(g)$

113. (c) Given, $\Delta H^o = -29.8\,kJ\,mol^{-1}$
$\Delta S^o = -0.100\,kJ\,K^{-1}$
From the equation
$\Delta G^o = \Delta H^o - T\Delta S^o = -29.8 - (298 \times (-0.100))$
$= -29.8 + 29.8 = 0$
Now, $\Delta G^o = -2.303\,RT \log K_{eq}$
$0 = -2.303\,RT \log K_{eq} \quad \therefore K_{eq} = 1.$

114. (d) $\Delta G^o = 2494.2\,J$
$2A \rightleftharpoons B + C.$

$[A] = \frac{1}{2}, [B] = 2, [C] = \frac{1}{2}$

$$Q = \frac{[B][C]}{[A]^2} = \frac{2 \times 1/2}{\left(\frac{1}{2}\right)^2} = 4$$

$\Delta G^o = -2.303\,RT \log K_c$
$2494.2\,J = -2.303 \times (8.314\,J/K/mol) \times (300K) \log K_c$
$\Rightarrow \log K_c = -0.4341 \Rightarrow K_c = 0.37; \;\therefore\; Q > K_c$

115. (b) $\Delta G^\circ_r = 2\Delta G^\circ_f(NO_2) - 2\Delta G^\circ_f(NO) - \Delta G^\circ_f(O_2)$

$2\Delta G^\circ_f(NO_2) = \Delta G^\circ_r + 2\Delta G^\circ_f(NO) + \Delta G^\circ_f(O_2)$
At equilibrium, $\Delta G^\circ = -RT \ln K_p$
$\Delta G^\circ_f(O_2) = 0$ and $\Delta G^\circ_f(NO) = 86.6\,kJ/mol$

$$\therefore \Delta G^\circ_f(NO_2) = \frac{1}{2}[2 \times 86600 - R(298)\ln(1.6 \times 10^{12})]$$

116. (d) Volume of ice is greater than that of water.
On increasing pressure, the system will reduce the pressure decreasing the volume by forming more water.
So, equilibrium will shift in forward direction.

117. (d) On adding inert gas at constant volume, the total pressure of the system is increased, but the concentration of each reactant and product remains the same. Hence, there is no effect on the state of equilibrium.

118. (a) The reaction given is an exothermic reaction thus accordingly to Le-Chatalier's principle lowering of temperature, addition of F_2 and/or Cl_2 favour the forward direction or the production of ClF_3.

119. (b) Due to exothermic nature of reaction low or optimum temperature will be required. Since 3 moles are changing to 2 moles, therefore high pressure will be required.

120. (a) In reaction (a), the ratio of number of moles of reactants to products is same. Hence change in volume will not alter the number of moles at equilibrium.

121. (b) $pH = 1 \Rightarrow [H^+] = 10^{-1}$
If equal volume of water is added concentration will become half

$$[H^+]_{sol} = \frac{10^{-1}}{2} = 0.05$$

$pH = -\log 0.05 = 1.3$

122. (3) $pH = 10$
$\Rightarrow pOH = 14 - pH = 14 - 10 = 4$
$\Rightarrow [OH^-] = 10^{-4}$
Number of moles of $OH^- = 10^{-4}$

Number. of moles of $Mg(OH)_2 = \frac{10^{-4}}{2} = 5 \times 10^{-5}$

$\Rightarrow$ mass of $Mg(OH)_2 = n_{Mg(OH)_2} \times M_{mg(OH)_2}$
$= 5 \times 10^{-5} \times 58 \times 10^3\,mg = 2.9\,mg \approx 3\,mg.$

123. (d) $HA \rightleftharpoons H^{\oplus} + A^{\ominus}$

	HA	$H^\oplus$	$A^\ominus$
t = 0	a		
t = t	a(1–x)	ax	ax

$K_a = (ax)\frac{(x)}{1-x}; [H^+] = ax$

$-\log(K_a) = -\log(ax) - \log\left(\frac{x}{1-x}\right)$

$pK_a = pH - \log\left(\frac{x}{1-x}\right)$

$pH - pK_a = \log\left(\frac{x}{1-x}\right)$

124. (a) With increase in temperature, K_w of water increases
So, degree of dissociation of water increases.
$\therefore$ pH as well as pOH of water decreases.
At 80°C, pH ≈ 6.93.

125. (b) Strong acid have weak conjugate base.
Acidic strength :
$H–Cl > CH_3COOH > H_2O > R–OH$
Conjugate base strength :

$Cl^- < CH_3COO^- < \bar{O}H < RO^-$

126. (d) (A) pH of 10^{-8} M HCl is in acidic range (6.98).

(B) Conjugate base of $H_2PO_4^-$ is HPO_4^{2-}.

(C) K_w increases with increasing temperature, as the temperature increases, the dissociation of water increases.

(D) At half neutralisation point, half of the acid is present in the form of salt.

$$pH = pK_a + \log\frac{1}{1} = pK_a$$

127. (9000)
$(M_1 \times V_1) = (M_2 \times V_2)$
$10^{-1} \times 1 = 10^{-2} \times V_2$
$V_2 = 10L$
Water added = 10 – 1 = 9 Litre = 9000 mL

128. (186) Total milimoles of H^+

$= (600 \times 0.01) + (400 \times 0.01 \times 2) = 14$

$[H^+] = \frac{14}{1000} = 14 \times 10^{-3}$

$pH = -\log[H^+] \Rightarrow pH = 3 - \log 14$

$= 1.86 = 186 \times 10^{-2}$

129. (5) Considering complete dissociation of $Ca(OH)_2$

$\underset{1\,mol}{Ca(OH)_2} \rightarrow Ca^{2+} + \underset{2\,mol}{2OH^-}$

$pH = 12, \therefore pOH = 14 - 12 = 2$ (Considering T = 298 K)

$\therefore\ [OH^-] = 10^{-2}$ (M)

$[OH^-] = 10^{-2}$ (M)

$= 10^{-2}$ mol OH^- in 1000 mL

$\therefore$ In 100 mL aqueous solution has $\frac{10^{-2} \times 100}{1000}$

$= 10^{-3}$ mol of OH^-

2 mol of OH^- is produced by 1 mol $Ca(OH)_2$

10^{-3} mol OH^- is produced by $\frac{1}{2} \times 10^{-3}$ mol

$= 0.5 \times 10^{-3}$ mol $= 0.5$ m mol

$= 5 \times 10^{-1}$ m mol. of $Ca(OH)_2$

130. (9079) 0.02 mol HCl neutralize 0.02 mol of NH_3.

Rest $n_{NH_3} = 0.1 - 0.02 = 0.08$

$n_{NH_4Cl} = n_{NH_4^+} = 0.1 + 0.02 = 0.12$

$pOH = pK_b + \log\frac{[NH_4^+]}{[NH_3]} = 4.745 + \log\frac{0.12}{0.08}$

$= 4.745 + \log\frac{3}{2} = 4.745 + 0.477 - 0.301$

$pOH = 4.921; pH = 14 - pH = 9.079 = 9079 \times 10^{-3}$

131. (12) millimole of NaOH $= 0.24 \times 25$

$\therefore$ millimole of acid $= 0.24 \times 25$

$\Rightarrow$ mass of acid $= 0.24 \times 25 \times 24.2$ mg

for pure acid,

$V = \frac{w}{d}$; (d = 1.21 kg/L = 1.21 g/mL)

$\therefore V = \frac{0.24 \times 25 \times 24.2}{1.21} \times 10^{-3}$

$= 120 \times 10^{-3}$ mL $= 12 \times 10^{-2}$ mL

132. (b) $\Delta[H^+] = 1000$

$\Delta pH = -\log \Delta[H^+] = -\log 10^3$

$= -3$; (–ve) sign indicates decrease in pH.

133. (b) $[H^+] = \frac{(n_{H^+})_{HCl} + (n_{H^+})_{H_2SO_4}}{\text{Total volume}}$

$= \frac{(M_{HCl} \times n_{factor,HCl} \times V_{HCl}) + M_{H_2SO_4} \times n_{f_{H_2SO_4}} \times V_{H_2SO_4}}{600}$

$[H^+] = \frac{(0.01 \times 1 \times 200) + (0.01 \times 2 \times 400)}{600}$

$= \frac{2+8}{600} = \frac{10}{600} = \frac{1}{60}$

$pH = -\log\left[\frac{1}{60}\right] = 1.78$

134. (c) Phenolphathalein have colour change range between pH 8.3 and 10.5.

135. (c) Phenolphthalein dissociate in basic medium

$\underset{\text{(colourless)}}{HPh(aq)} \rightleftharpoons H^+ + \underset{\text{(Pink)}}{Ph^-}$

136. (11) 0.001 M NaOH

$[OH^-] = 10^{-3}$

$pOH = -\log[OH^-] \Rightarrow pOH = 3$

$pH + pOH = 14 \Rightarrow pH = 11$

137. (d) Assertion : The amphoteric nature of water can be explained by Bronsted – Lowry concept.

Reason : $\underset{Acid}{H_2S} + \underset{Base}{H_2O} \rightleftharpoons H_3O^+ + HS^-$

$\underset{Acid}{H_2O} + \underset{Base}{NH_3} \rightleftharpoons NH_4^+ + \bar{O}H$

138. (6021) $HCl(aq.) + NaOH(aq.) \rightarrow NaCl(aq.) + H_2O(\ell)$

	HCl	NaOH	NaCl	H_2O
Initial	50 m mol	30 m mol	–	–
Final	20 m mol		30	30

$[HCl] = \frac{20}{80} = \frac{1}{4} M = 2.5 \times 10^{-1} M$

$pH = -\log 2.5 \times 10^{-1} = 1 - 0.33979 = 0.6021$

$\Rightarrow\ pH = 6021 \times 10^{-4}$.

139. (b) The chemical species having capability to donate electron pair is called Lewis base. PCl_5 cannot function as a Lewis base as the central atom P does not have lone pair of electrons.

140. (b) The solution with lowest pH will have the highest value of pOH.

$[H^+]$ order : (A) > (D) > (C) > (B)

pH order : (B) > (C) > (D) > (A)

pOH order : (A) > (D) > (C) > (B)

141. (37) $CO_2 + H_2O \longrightarrow H_2CO_3$

30 bar .... 1 mol/L

3 bar .... 0.1 mol/L

$$H_2CO_3 \rightleftharpoons H^+ + HCO_3^-$$

	H_2CO_3	H^+	HCO_3^-
$t=0$	0.1	0	0
at Eqm.	$0.1(1-\alpha)$	0.1α	0.1α

$$4.0\times10^{-7} = \frac{0.1\alpha^2}{1-\alpha}$$

As $(1-\alpha)\approx 1 \Rightarrow \alpha^2 = 4\times10^{-6} \Rightarrow \alpha = 2\times10^{-3}$

$[H^+] = C\alpha = 2\times10^{-4}$ M

$pH = 4\times\log 2 = 3.7 \simeq 37\times10^{-1}$

142. (c) At equivalence point, pH is 7 and pH increases with addition of NaOH. So correct graph is (c).

143. (b) Temperature plays a significant role on pH measurements. As the temperature rises, molecular vibrations increase which results in greater ability of water to ionise and form more hydrogen ions. As a result, the pH will drop. So assertion is incorrect.

The dissociation of water molecules into ions is bond breaking and is therefore an endothermic process. So reason is also incorrect.

144. (10.60) $M_{H_2SO_4} = \frac{9.8}{98\times100} = 10^{-3}$ M

$M_{NaOH} = \frac{4}{40\times100} = 10^{-3}$ M

After neutralisation $[OH^-]$ can be calculated as

$$[OH^-] = \frac{(40\times10^{-3}) - (2\times10^{-3}\times10)}{50} \Rightarrow [OH^-] = \frac{2}{5}\times10^{-3}$$

$\Rightarrow$ $pOH = -\log[OH^-] \Rightarrow pOH = 3.397$

$pH = 14 - pOH = 14 - 3.397 = 10.603$

145. (5.23)

$3g\ CH_3COOH = \frac{3}{60} = 0.05\ mol = 50\ m\ mol$

No. of millimoles = Molarity × Volume in mL

250 mL of 0.1 M HCl = 250 × 0.1 = 25 m mol

500 mL solution = 50 m mol CH_3COOH

20 mL solution $= \frac{50}{500}\times 20 = 2$ m mol CH_3COOH

500 mL solution contains = 25 m mol HCl

20 mL solution contains $= \frac{25}{500}\times 20 = 1$ m mol HCl

$\frac{1}{2}$ mL of 5M NaOH $= \frac{1}{2}\times 5 = 2.5$ m mol NaOH

$$HCl + NaOH \longrightarrow NaCl + H_2O$$

1   2.5   1

Remaining NaOH = 2.5–1 = 1.5 m mol

$CH_3COOH + NaOH(remaining) \rightarrow CH_3COONa + Water$

CH_3COOH	NaOH	CH_3COONa	Water
2	1.5	0	0
0.5	0	1.5	–

$$pH = pK_a + \log\frac{1.5}{0.5} = 4.75 + \log 3$$
$$= 4.75 + 0.48 = 5.23$$

146. (b) (A) $H_2SO_4 + NaOH \rightarrow NaHSO_4 + H_2O$

	H_2SO_4	NaOH	$NaHSO_4$	H_2O
Initial mol	0.04	0.04	0	0
mol at eqm.	0	0	0.04	0.04

$NaHSO_4 \rightarrow Na^+ + H^+ + SO_4^{2-}$

Total volume of the solution = (400+400) mL = 800 mL = 0.8 L

$$[H^+] = \frac{0.04}{0.80} = 0.05M \Rightarrow pH = -\log[H^+] \Rightarrow pH = 1.3$$

(B) Ionic product of water increases with increase in temperature because ionisation of water is endothermic.

(C) $HA \rightleftharpoons H^+ + A^-$

	HA	H^+	A^-
Initial	C	0	0
At eqm.	C (1 – a)	Ca	Ca

Given pH = 5 $\Rightarrow -\log[H^+] = 5$

$\therefore\ [H^+] = 10^{-5}$

As we know,

$$\Rightarrow K_a = \frac{C\alpha^2}{1-\alpha} \Rightarrow 10^{-5} = \frac{C\alpha^2}{1-\alpha} = \frac{C\alpha.\alpha}{(1-\alpha)}$$

$$10^{-5} = 10^{-5}\frac{\alpha}{1-\alpha} \Rightarrow \alpha = \frac{1}{2},\ \textit{i.e.},\ 50\%$$

147. (b) Graph A and B, both represents the titration curve between strong acid and strong base, i.e., HCl and NaOH but, the pH of NaOH is more than 7 and during the titration it decreases, so graph (A) is correct.

148. (c) Due to the resonance stabilisation of the conjugate base, $CH(CN)_3$ is the strongest acid amongst the given compounds.

$$CH(CN)_3 \rightleftharpoons H^+ + \bar{C}(CN)_3$$

The conjugate bases of $CHBr_3$ and CHI_3 are stabilised by inductive effect of halogens. This is why, they are less stable. Also, the conjugate base of $CHCl_3$ involves back-bonding between 2*p* and 3*p* orbitals.

149. (b) $75\text{ mL}\frac{M}{5}HCl + 25\text{ mL}\frac{M}{5}NaOH$

$25\text{ mL}\frac{M}{5}$ NaOH will neutralise $25\text{ mL}\frac{M}{5}$ HCl

$75-25 = 50\text{ mL}\frac{M}{5}$ HCl will remain.

Total volume will be $75 + 25 = 100$ mL

When 50 mL $\frac{M}{5}$ HCl is diluted to 100 mL, molarity will reduce to half, i.e., $\frac{M}{10}$.

$[H^+] = [HCl] = \frac{1}{10}$

$pH = -\log_{10}[H^+] = -\log_{10}\frac{1}{10} = 1$

150. (d) CH_3–B(CH_3)–CH_3 (6e⁻)

Due to incomplete octet of B, it can act as a Lewis acid.

151. (d) $NH_3 + HCl \longrightarrow NH_4Cl$

Moles of HCl $= 0.2 \times 25 \times 10^{-3} = 0.005$

Moles of $NH_3 = 0.2 \times 50 \times 10^{-3} = 0.01$

Excess $NH_3 = 0.01 - 0.005 = 0.005$ moles

1 mole ammonia = 1 mole NH_4Cl

$0.005\ NH_3 \equiv 0.005\ NH_4Cl$

Total volume $= V_{HCl} + V_{NH_3} = 25 + 50 = 75$ mL

$[NH_3] = [NH_4Cl] = \frac{0.005\text{ mole}}{75\times10^{-3}\text{ L}} = 0.066\text{ M}$

$pOH = pK_b + \log\frac{[NH_4Cl]}{[NH_3]}$

$pOH = 4.75 + \log\frac{[0.066]}{[0.066]}$

$pOH = 4.75$

$pH = 14 - pOH \Rightarrow pH = 9.25$

152. (b) $\underset{\text{Hydrazoic acid}}{N_3H} \rightleftharpoons N_3^- + H^+$

153. (d) $\because pH = 1\ ;\ [H^+] = 10^{-1} = 0.1$ M

$pH = 2\ ;\ [H^+] = 10^{-2} = 0.01$ M $\therefore\ M_1 = 0.1, V_1 = 1$

$M_2 = 0.01, V_2 = ?$

From $M_1V_1 = M_2V_2$

$\Rightarrow\ 0.1 \times 1 = 0.01 \times V_2 \Rightarrow V_2 = 10$ L

$\therefore$ Volume of water added $= 10 - 1 = 9$ L

154. (d) Given $[OH^-] = 5 \times 10^{-2}$

$\therefore\ pOH = -\log 5 \times 10^{-2} = -\log 5 + 2\log 10 = 1.30$

$pH = 14 - pOH = 14 - 1.30 = 12.70$

155. (a) Metal halide on hydrolysis with water form corresponding hydroxides.

The basic strength of hydroxide increases as we move down in a group.

Hence, $Be(OH)_2$ will have lowest pH.

156. (d) Given at 330 K, $K_w = 10^{-13.6}$

$pK_w = pH + pOH \Rightarrow 13.6 = pH + pOH$

$pOH = -\log 10^{-4} \Rightarrow pOH = 4$

$\therefore\ pH = 13.6 - 4 = 9.6$

157. (a) The species with the greatest proton affinity will be the strongest base, and its conjugate acid will be the weakest acid. The weakest acid will have the smallest value of K_a. Since HI is a stronger acid than HF, which is a stronger acid than H_2S, a partial order of proton affinity is

$I^- < F^- < HS^-$

Since NH_3 is a very weak acid, NH_2^- must be a very strong base. Therefore, the correct order of proton affinity is

$I^- < F^- < HS^- < NH_2^-$

158. (c) The correct order of acidic strength of the given species is $\underset{(iv)}{HSO_3F} > \underset{(ii)}{H_3O^+} > \underset{(iii)}{HSO_4^-} > \underset{(i)}{HCO_3^-}$

or (i) < (iii) < (ii) < (iv)

159. (a) $pH = -\log[H^+] = \log\frac{1}{[H^+]}$

$5.4 = \log\frac{1}{[H^+]}$

On Solving, $[H^+] = 3.98 \times 10^{-6}$

160. (a) Conjugate acid-base pair differ by only one proton.

$OH^- \longrightarrow H^+ + O^{2-}$

161. (d) $\underset{\text{Acid}}{H_2PO_4^-} \xrightarrow{-H^+} \underset{\text{Conjugate base}}{HPO_4^{2-}}$

Conjugate acid-base differs by H^+.

162. (b) pH of an acidic solution should be less than 7. The reason is that from H_2O, $[H^+] = 10^{-7}$M which cannot be neglected in comparison to 10^{-8}M.

[**Note:** pH + pOH = 14, for water, is ture at 298 K]

163. (d) The rain water after thunderstorm contains dissolved acid and therefore, the pH is less than rain water without thunderstorm.

164. (a) $(HSO_4)^-$ can accept and donate a proton

$(HSO_4)^- + H^+ \rightarrow H_2SO_4$ (acting as base)

$(HSO_4)^- - H^+ \rightarrow SO_4^{2-}$. (acting as acid)

165. (25) After Dilution

$pH = 6 \Rightarrow [H^+] = 10^{-6}$

$[H^+] = \sqrt{k_a \cdot C}$

$10^{-6} = \sqrt{4 \times 10^{-10} \times C} \Rightarrow \frac{10^{-12}}{4 \times 10^{-10}} = C$

$0.25 \times 10^{-2} = C \Rightarrow 25 \times 10^{-4} = C$

$\therefore$ Value of x is 25.

166. (7) Given $pH = 9 \Rightarrow pOH = 14 - 9 = 5$; $[OH]^{\ominus} = 10^{-5}$ M

$$C_2H_5NH_2\,(aq) + H_2O \rightleftharpoons C_2H_2NH_3^+ + OH^{\ominus}$$

Initial	$C = 10^{-3}$ M		
Final	$C(1-\alpha)$	$C\alpha$	$C\alpha$
	$\Rightarrow C = 10^{-3}$	$= 10^{-5}$	$= 10^{-5}$

as $1 - \alpha \approx 1$

Now, $K_b = \frac{[C_2H_5NH_3^+][OH]^{\ominus}}{[C_2H_5NH_2]}$

$\Rightarrow K_b = \frac{10^{-5} \times 10^{-5}}{10^{-3}} = 10^{-7}$

167. (7) $K_a(CH_3COOH) = 1.8 \times 10^{-5}$

$pK_a = -\log(1.8 \times 10^{-5}) = 4.7$

$K_b(NH_4OH) = 1.8 \times 10^{-5}$, $pK_b = -\log(1.8 \times 10^{-5}) = 4.7$

$pH = 7 + \frac{1}{2}[pK_a - pK_b] = 7 + \frac{1}{2}[4.7 - 4.7]$

$pH = 7$

168. (a) $(NH_4)_2CO_3$ is a salt of NH_4OH and H_2CO_3. Its aqueous solution is basic as K_b of NH_4OH is greater than K_a of H_2CO_3.

pH of salt of weak acid and weak base depends on K_a and K_b value of constituent acid and the base.

169. (27) K_a of Butyric acid $= 2 \times 10^{-5}$

$pK_a = -\log K_a = -\log(2 \times 10^{-5}) = 4.7$

The pH of butyric acid is calculated as:

$pH = \frac{1}{2}pK_a - \frac{1}{2}\log c = \frac{1}{2}(4.7) - \frac{1}{2}\log(0.2) = 2.35 + 0.35$

$= 2.7$

$pH = 27 \times 10^{-1}$.

170. (c) Concentration of salt, NH_4Cl, produced by neutralization of NH_4OH and HCl.

NH_4Cl, no. of moles $= 2$m mol $= 2 \times 10^{-3}$ mol

$[NH_4Cl] = \frac{2\text{m mol}}{(20+40)\text{ml}} = \frac{1}{30}$M

$pH = 7 - \frac{1}{2}pK_b - \frac{1}{2}\log C$

$= 7 - \frac{1}{2} \times (+5) - \frac{1}{2}\log\left(\frac{1}{30}\right) = 7 - \frac{5}{2} + \frac{1.48}{2} = 5.24$

171. (d) $H_2C_2O_4 \rightleftharpoons H^+ + HC_2O_4^-$ K_{a_1} ...(i)

$H_2C_2O_4 \rightleftharpoons H^+ + C_2O_4^{2-}$ K_{a_2} ...(ii)

$H_2C_2O_4 \rightleftharpoons 2H^+ + C_2O_4^{2-}$ K_{a_3} ...(iii)

addition of eq. (i) with eq. (ii) gives eq (iii),

So $K_{a_3} = K_{a_1} \times K_{a_2}$

172. (2) $HCl \longrightarrow H^+ + Cl^-$

0.1 M 0.1 M

$HA \rightleftharpoons H^+ + A^-$ $K_a = 2.0 \times 10^{-6}$

$0.01(1-\alpha)$ $0.1 + 0.01\alpha$ 0.01α

$K_a = \frac{[H^+][A^-]}{[HA]}$

$2 \times 10^{-6} = \frac{0.1 \times 0.01\alpha}{0.01(1-\alpha)} \simeq \frac{0.1 \times 0.01\alpha}{0.01}$ [considering, $0.1 + 0.1\alpha \approx 0.1$ since α is very small]

$\alpha = 2.0 \times 10^{-5}$

173. (1) k_{a_1} of $H_2SO_3 >> k_{a_2}$ of H_2SO_3

$\therefore$ The contribution of H^+ from 2nd dissociation of H_2SO_3 can be neglected.

$$H_2SO_3 \rightleftharpoons H^+ + HSO_3^-$$

t=0	c	0	0
t=t	$c(1-\alpha)$	$c\alpha$	$c\alpha$

$\Rightarrow \frac{c\alpha^2}{1-\alpha} = 1.7 \times 10^{-2} \Rightarrow \frac{0.588\alpha^2}{1-\alpha} = 1.7 \times 10^{-2}$

$\Rightarrow 58.8\alpha^2 + 1.7\alpha - 1.7 = 0$

$\alpha = \frac{-1.7 + \sqrt{1.7^2 + 4 \times 1.7 \times 58.8}}{2 \times 58.8} = 0.156$

$[H^+] = c\alpha = 0.156 \times 0.588 = 0.092$

$pH = -\log[H^+] = 1.036 \approx 1$

174. (7.0) Since, $(NH_4)_3PO_4$ is salt of weak acid (H_3PO_4) and weak base (NH_4OH).

$pH = 7 + \frac{1}{2}(pK_a - pK_b) = 7 + \frac{1}{2}(5.23 - 4.75) = 7.24 \approx 7.0$

175. (d) Sodium acetate is a salt of strong base and weak acid.

$$\therefore\ pH = 7 + \frac{1}{2}pK_a + \frac{1}{2}\log c \quad \text{where } pK_a = -\log K_a$$

$$= 7 + \frac{5}{2} - \frac{1}{2} = 9.0$$

176. (c) Higher the value of K_a, higher will be the acidic nature. Further, since CN^-, F^- and NO_2^- are conjugate base of the acids HCN, HF and HNO_2 respectively, hence the correct order of base strength will be :

$$F^- < NO_2^- < CN^-$$

($\because$ stronger the acid, weaker will be its conjugate base)

177. (d) Conc. of CH_3COOH = 5 g in 500 mL = 10 g/L

$$[CH_3COOH] = \frac{10}{60}M = \frac{1}{6}M \ [\text{MW of } CH_3COOH = 60]$$

Concentration of CH_3COONa is computed as under.

conc. = 7.5 g in 500 mL = 15 g / L

$$[CH_3COONa] = \frac{15}{82}M \ [\text{MW of } CH_3COONa = 82]$$

$$pK_a = -\log K_a = -\log(1.75 \times 10^{-5}) = 4.76$$

$$pH = pK_a + \log\frac{[\text{salt}]}{[\text{acid}]} = 4.76 + \log\frac{15/82}{1/6} = 4.8$$

178. (c) $[H^+] = C\alpha;\ \alpha = \frac{[H^+]}{C}$ or $\alpha = \frac{10^{-3}}{0.1} = 10^{-2}$

[Since pH = 3; $[H^+] = 10^{-3}$]

$$K_a = C\alpha^2 = 0.1 \times 10^{-2} \times 10^{-2} = 10^{-5}$$

179. (b) $CN^-(aq) + H_2O(l) \rightleftharpoons HCN(aq) + OH^-(aq)$

$K_b = 1.6 \times 10^{-5}$

$$\underset{\text{base}}{OH^-(aq)} + \underset{\text{acid}}{H_2O(l)} \rightleftharpoons \underset{\text{conj.acid}}{H_2O(l)} + \underset{\text{conj.base}}{OH^-(aq)}$$

$K_b = 1$

$$\underset{\text{base}}{H_2O(l)} + \underset{\text{acid}}{H_2O(l)} \rightleftharpoons \underset{\text{conj. acid}}{H_3O^+(aq)} + \underset{\text{conj. base}}{OH^-(aq)}$$

$K_b = 1.0 \times 10^{-14}$

Hence, the relative base strength:

$OH^- > CN^- > H_2O$

180. (d) pH = 5 means

$[H^+] = 10^{-5}$

	HA	$\rightleftharpoons$	H^+	+	A^-
$t = 0$	c		0		0
t_{eq}	c (1 – α)		cα		cα

$$K_a = \frac{[H^+][A^-]}{[HA]} = \frac{(c\alpha)^2}{c(1-\alpha)} = \frac{[H^+]^2}{c - [H^+]}$$

But, $[H^+] << c$

$\therefore\ K_a = [H^+]^2 = (10^{-5})^2 = 10^{-10}$

181. (b) The salt (AB) given is a salt of weak acid and weak base. Hence the pH can be calculated by the following formula

$$\therefore\ pH = 7 + \frac{1}{2}pK_a - \frac{1}{2}pK_b$$

$$= 7 + \frac{1}{2}(3.2) - \frac{1}{2}(3.4) = 6.9$$

182. (a) (i) $\underset{\text{acid}_1}{H_3PO_4} + \underset{\text{base}_2}{H_2O} \longrightarrow \underset{\text{acid}_2}{H_3O^+} + \underset{\text{base}_1}{H_2PO_4^-}$

(ii) $\underset{\text{acid}_1}{H_2PO_4^-} + \underset{\text{base}_2}{H_2O} \longrightarrow \underset{\text{base}_1}{HPO_4^{2-}} + \underset{\text{acid}_2}{H_3O^+}$

(iii) $\underset{\text{base}_1}{H_2PO_4^-} + \underset{\text{acid}_2}{OH^-} \longrightarrow \underset{\text{acid}_1}{H_3PO_4} + \underset{\text{base}_2}{O^{2-}}$

Hence only in (ii) reaction, $H_2PO_4^-$ is acting as an acid.

183. (c) $H_2CO_3 \rightleftharpoons H^+ + HCO_3^-\ \ K_1 = 4.2 \times 10^{-7}$

$HCO_3^- \rightleftharpoons H^+ + CO_3^{2-}\ \ K_2 = 4.8 \times 10^{-11}$

Second dissociation constant (K_2) is much smaller than the first one (K_1). Just a small fraction of total HCO_3^- formed will undergo second stage of ionization. Hence in saturated solution

$[H^+] >>> [CO_3^{2-}];\ [CO_3^{2-}] \neq 0.034$ M

$[HCO_3^{2-}] >> [CO_3^{2-}]$ and $[H^+] \approx [HCO_3^-]$.

184. (c) In aqueous solution, BA(salt) hydrolyses to give

$$BA + H_2O \rightleftharpoons \underset{\text{base}}{BOH} + \underset{\text{acid}}{HA}$$

Now pH is given by $pH = \frac{1}{2}pK_w + \frac{1}{2}pK_a - \frac{1}{2}pK_b$

Substituting given values, we get

$$pH = \frac{1}{2}(14 + 4.80 - 4.78) = 7.01$$

185. (d) $H_2A \rightleftharpoons H^+ + HA^-$

$$\therefore\ K_1 = 1.0 \times 10^{-5} = \frac{[H^+][HA^-]}{[H_2A]}$$

$HA^- \rightleftharpoons H^+ + A^-$

$$\therefore\ K_2 = 5.0 \times 10^{-10} = \frac{[H^+][A^-]}{[HA^-]}$$

For the reaction, $H_2A \rightleftharpoons 2H^+ + A^{2-}$

$$K = \frac{[H^+]^2[A^{2-}]}{[H_2A]} = K_1 \times K_2$$

$$= (1.0 \times 10^{-5}) \times (5 \times 10^{-10}) = 5 \times 10^{-15}$$

186. (48) $pOH = pK_b + \log\frac{[NH_4^+]}{[NH_3]}$

$pOH = 4.745 + \log\frac{0.1}{0.1} \Rightarrow pOH = 4.745$

$\Rightarrow \quad pH = 14 - pOH = 9.255.$

on adding 0.05 mole PCl

$NH_3 + H^{\oplus} \rightarrow NH_4^{\oplus}$

0.1   0.05   0.1

0.05   0   0.15

$pOH' = = pK_b + \log\frac{0.15}{0.05}$

$= 4.745 + 0.477 = 5.222$

$pH' = 14 - pOH' = 14 - 5.222 = 8.778$

$\Delta pH = pH - pH' = 9.225 - 8.778 = 0.477$

$= 47.7 \times 10^{-2} \approx 48 \times 10^{-2}$

187. (c) When equal volumes are mixed molarity reduces to half.

For precipitation $Q_{SP} = [A^{+2}][Y^-]^2 > K_{SP}$

(a) $Q_{SP} = (1.8 \times 10^{-3})\left(\frac{5}{2} \times 10^{-4}\right)^2 < K_{SP}$

(b) $Q_{SP} = (10^{-4})(0.4 \times 10^{-3})^2 < K_{SP}$

(c) $Q_{SP} = (10^{-2})(10^{-2})^2 > K_{SP}$

(d) $Q_{SP} = \left(\frac{1.5}{2} \times 10^{-4}\right)\left(\frac{1.5}{2} \times 10^{-3}\right)^2 < K_{SP}$

188. (b) K_{sp} values determine solubility. Based on the K_{sp} values and salt analysis cation identification, the order of K_{sp} value is:

$HgS < PbS < AgBr < Ca(OH)_2$

K_{sp} values

$HgS \rightarrow 4 \times 10^{-53}$

$PbS \rightarrow 8 \times 10^{-28}$

$AgBr \rightarrow 5 \times 10^{-13}$

$Ca(OH)_2 \rightarrow 5.5 \times 10^{-6}$

Lower K_{sp} means lower solubility.

189. (a) $Cr(OH)_3(s) \rightleftharpoons Cr^{+3}(aq) + 3OH^-(aq)$

At eq: s   3s

$K_{sp} = (s).(3s)^3 = 27s^4$

$27s^4 = 1.6 \times 10^{-30}$; $s = \left(\frac{1.6}{27} \times 10^{-30}\right)^{1/4}$

190. (a) Condition for precipitation $Q_{ip} > K_{sp}$

For $[A(OH)_2]$

$[A^{2+}][OH^-]^2 > 9 \times 10^{-10}$

$[A^{+2}] = 1\,M$

$\Rightarrow [OH^-] > 3 \times 10^{-5}\,M$

For $[B(OH)_3]$

$[B^{3+}][OH^-]^3 > 27 \times 10^{-18}$

$[B^{3+}] = 1M$

$\Rightarrow [OH^-] > 3 \times 10^{-6}\,M$

So, $B(OH)_3$ will precipitate before $A(OH)_2$

191. (a) $Zr_3(PO_4)_4(s) \rightleftharpoons 3Zr^{+4}(aq) + 4PO_4^{-3}(aq)$

–   3s   4s

$K_{sp} = (3s)^3(4s)^4 = 27s^3 \times 256\,s^4 = 6912\,s^7$

$s = \left(\frac{K_{sp}}{6912}\right)^{\frac{1}{7}}$

192. (b) $AB_2(s) \rightleftharpoons A^{+2}(aq) + 2B^-(aq)$

$K_{sp} = [A^{+2}][B^-]^2$

$= 1.2 \times 10^{-4} \times (2.4 \times 10^{-4})^2$

$= 6.91 \times 10^{-12}\,M^3$

193. (a) Buffer solution prepared by mixing weak acid or base with their salt with strong base or acid.

Blood is a buffer solution of carbonic acid H_2CO_3 and bicarbonate HCO_3^-

Statement I is false but Statement II is true.

194. (d) Due to common ion effect reaction goes backward and precipitation occurs.

$Cl^{\ominus}$ + saturated solution of $BaCl_2 \rightarrow BaCl_2$ (white)

$Cl^{\ominus}$ + saturated solution of $NaCl \rightarrow NaCl$ will precipitate.

195. (c) $Ca_3(PO_4)_2 \rightleftharpoons 3Ca^{2+} + 2PO_4^{3-}$

$S = \left(\frac{W \times 1000}{M \times 100}\right) = \frac{W \times 10}{M}$

$\therefore K_{sp} = [Ca^{2+}][PO_4^{3-}]$

$= (3S)^3(2S)^2 = 108S^5$

$= 108 \times \left(\frac{W \times 10}{M}\right)^5 = 108 \times 10^5\left(\frac{W}{M}\right)$

$= 1.08 \times 10^7\left(\frac{W}{M}\right)$

$K_{sp} = 10^7\left(\frac{W}{M}\right)^5$

196. (9) $K_{sp} = 10^{-11}$

$[Mg^{2+}][OH^-]^2 = 10^{-11}$

$0.10 \times [OH^-]^2 = 10^{-11} \Rightarrow [OH^-] = 10^{-5}$

$[H^+][OH^-] = 10^{-14}$

$[H^+] = \frac{10^{-14}}{10^{-5}} = 10^{-9} \Rightarrow pH = -\log 10^{-9} = 9 \Rightarrow pH = 9$

197. (100) Using Henderson-Hasselbalch equation:

$$pH = pKa + \log\frac{[Salt]}{[Acid]}$$

$$\Rightarrow 4.5 = 4.2 + \log\frac{[V_{salt}]}{[V_{acid}]}$$

$$\Rightarrow \log\left(\frac{V_{salt}}{V_{acid}}\right) = 0.3 \Rightarrow \frac{V_{salt}}{V_{acid}} \cong 2$$

Since $V_{salt} + V_{acid} = 300$ mL, $V_{acid} = 100$ mL.

198. (458) $CH_3COOH + NaOH \rightarrow CH_3COONa + H_2O$

	CH_3COOH	$NaOH$	CH_3COONa	H_2O
Initially,	5 mmol	2mmol	0	0
after reaction,	3 mmol	0	2 mmole	2 mmole

$$pH = pK_a + \log_{10}\frac{[salt]}{[acid]}$$

$$pH = 4.76 + \log_{10}\frac{2}{3}$$

$pH = 4.58 = 458 \times 10^{-2}$

199. (5) $\left[Ba^{2+}\right] = \frac{25 \times 0.05}{50} = 0.025$ M

$$\left[F^-\right] = \frac{25 \times 0.02}{50} = 0.01 \text{ M}$$

$$\left[Ba^{+2}\right]\left[F^-\right]^2 = 25 \times 10^{-7}$$

$K_{sp} = 5 \times 10^{-7}$ (given)

$$\text{Ratio} = \frac{\left[Ba^{+2}\right]\left[F^-\right]^2}{K_{sp}} = 5$$

200. (233) $K_2SO_4 \longrightarrow 2K^+ + SO_4^{2-}$

0.1 M   0.2 M   0.1M

$BaSO_4 \rightleftharpoons Ba^{+2} + SO_4^{2-}$

a – S   S   S + 0.1 ≈ 0.1

$K_{SP} = S \times 10^{-1}$

$\Rightarrow 1 \times 10^{-10} = S \times 10^{-1}$

$\Rightarrow S = 10^{-9} \times 233$ g L^{-1}

So, Answer : 233

201. (10) $AgCl(s) \rightarrow Ag^+(aq.) + Cl^-(aq.)$

$$K_{sp} = S^2 = \left(\frac{1.43}{143.4} \times 10^{-3}\right)^2 = 10^{-10}; -\log K_{sp} = 10$$

202. (10) $pH = pK_a + \log\frac{0.2}{0.02}$

$5 = pK_a + 1$

$pK_a = 4$

$K_a = 10^{-4} = 10 \times 10^{-5}$ $\therefore x = 10$

203. (85) Concentration of calcium lactate = 0.005 M

Concentration of lactate ion = (2 × 0.005) M.

Calcium lactate is a salt of weak acid + strong base

∴ Salt hydrolysis will take place.

$$pH = 7 + \frac{1}{2}(pK_a + \log C) = 7 + \frac{1}{2}(5 + \log(2 \times 0.005))$$

$$= 7 + \frac{1}{2}[5 - 2\log 10] = 7 + \frac{1}{2} \times 3 = 8.5 = 85 \times 10^{-1}$$

204. (282) $K_{sp} = S^2$

$$S = \sqrt{K_{sp}} = \sqrt{8 \times 10^{-28}} = 2\sqrt{2} \times 10^{-14} = 2.82 \times 10^{-14}$$

$= 282 \times 10^{-16}$

205. (0) Solubility of $CaF_2 = 2.34 \times 10^{-3} \times 10$ g/L

$$s = \frac{2.34}{78} \times 10^{-2} \text{ mol / L}$$

$s = 3 \times 10^{-4}$ mol/L

$$CaF_2 \rightleftharpoons \underset{s}{Ca^{2+}} + \underset{2s}{2F^-}$$

$K_{sp} = s(2s)^2 = 4(3 \times 10^{-4})^3 = 0.0108 \times 10^{-8}$ $(mol/L)^3$

206. (c) $pH + pOH = 14$

$pOH = 14 - 8.26$

$$pOH = pK_b + \log\frac{\left[NH_4^+\right]}{[NH_3]}$$

$$\Rightarrow 5.74 = 4.74 + \log\frac{\left[NH_4^+\right]}{0.2} \Rightarrow \left[NH_4^+\right] = 2M$$

Molar mass of $NH_4Cl = 53.5$ g mol^{-1}

Amount of $NH_4Cl = 2 \times 53.5 = 107$ g.

207. (d) In deionized water no common ion effect will take place so solubility of AgCl will be maximum in it.

$KCl \longrightarrow K^+ + Cl^-$

$HCl \longrightarrow H^+ + Cl^-$

$AgNO_3 \longrightarrow Ag^+ + NO_3^-$

208. (b) According to Henderson-Hasselbalch equation

$$pH = pK_a + \log\frac{[Salt]}{[Acid]}$$

$$4 = 5 - \log 1.3 + \log\frac{[CH_3CH_2COO^-]}{[CH_3CH_2COOH]}$$

$$\log\frac{[CH_3CH_2COO^-]}{[CH_3CH_2COOH]} = \log 1.3 - 1 = \log\frac{1.3}{10}$$

$$\therefore \log A - \log(B) = \log\left(\frac{A}{B}\right)$$

$$\frac{[CH_3CH_2COO^-]}{[CH_3CH_2COOH]} = 0.13$$

209. (476) Moles of $CH_3COOH = 50 \times 0.1 = 5$ m mole

moles of $NaOH = 25 \times 0.1 = 2.5$ m mole

$NaOH + CH_3COOH \longrightarrow CH_3COONa + H_2O$

2.5 m mole   5 m mole

0   2.5 m mole   2.5 m mole

Total volume of this buffer solution = 75 mL

$$pH = pK_a + \log\left(\frac{2.5/75}{2.5/75}\right) \Rightarrow pH = 4.76 = 476 \times 10^{-2}$$

210. (a) $Bi_2S_3 \rightleftharpoons \underset{2s}{2Bi^{3+}} + \underset{3s}{3S^{2-}}$

$K_{sp} = (2s)^2 (3s)^3 = 108 (s)^5$

$$(s)^5 = \frac{1.08 \times 10^{-73}}{108} \Rightarrow s = 10^{-15}$$

211. (2) $Zn(OH)_2(s) \rightleftharpoons Zn^{2+}(aq) + 2OH^-(aq)$

$s \qquad (0.1 + 2s) \simeq 0.1$

$K_{sp} = s(0.1)^2$

$2 \times 10^{-20} = s \times 10^{-2} \Rightarrow s = 2 \times 10^{-18}$

Therefore, $x = 2$

212. (108) $A_3B_2 \rightleftharpoons \underset{3s}{3A^{2+}(aq)} + \underset{2s}{2B^{3-}(aq)}$

$K_{sp} = (3s)^3 (2s)^2 = 108s^5$

$$K_{sp} = 108\left(\frac{x}{M}\right)^5 \quad \left(\because \frac{x}{M} = s\right)$$

$$\text{(Given) } K_{sp} = a\left(\frac{x}{M}\right)^5$$

On comparing, $a = 108$

213. (3) $NH_3 + H_2O \rightleftharpoons NH_4^+ + OH^-$

$[NH_3] = 2 \times 0.0210 = 0.042$ m mol

$NH_4Cl \longrightarrow NH_4^+ + Cl^-$

$[NH_4Cl] = 5 \times 0.0504 = 0.252$ m mol

By Henderson – Hasselbalch equation,

$$pOH = pK_b + \log\frac{[Salt]}{[Base]}$$

$$= -\log(1.8 \times 10^{-5}) + \log\left(\frac{0.252}{0.042}\right) = 4.74 + 0.77$$

$pOH = 5.51$

$[OH^-] = 10^{-5.51} = 3.09 \times 10^{-6} \simeq 3 \times 10^{-6}$

214. (10) Henderson – Hasselbalch equation,

$$pH = pK_a + \log\frac{[Salt]}{[Acid]}$$

$$\Rightarrow 5.74 = 4.74 + \log\frac{[CH_3COONa]}{1}$$

$5.74 - 4.74 = \log[CH_3COONa] - \log 1$

$1 = \log[CH_3COONa] - 0$

$[CH_3COONa] = 10^1 \approx 10$ M

215. (64) In pure water, $s = 8 \times 10^{-4}$

$K_{sp} = s^2 = (8 \times 10^{-4})^2 = 64 \times 10^{-8}$

In 0.01 M H_2SO_4,

$H_2SO_4 \longrightarrow \underset{0.02}{2H^+} + \underset{0.01}{SO_4^{2-}}$

$CdSO_4 \longrightarrow \underset{s}{Cd^{2+}} + \underset{s}{SO_4^{2-}}$

Total conc. of $SO_4^{2-} = 0.01 + s$

$K_{sp} = s(s + 0.01)$

$K_{sp} = s \cdot (0.01)$ (neglecting s^2)

$$s = \frac{64 \times 10^{-8}}{0.01} = 64 \times 10^{-6}$$

216. (50) For A_2X, $A_2X \longrightarrow \underset{2s_1}{2A^+} + \underset{s_1}{X^{2-}}$

$$K_{sp} = 4s_1^3 \Rightarrow s_1 = \sqrt[3]{\frac{K_{sp}}{4}} = 10^{-4}$$

For MX, $MX \longrightarrow \underset{s_2}{M^{2+}} + \underset{s_2}{X^{2-}}$

$$K_{sp} = s_2^2 \Rightarrow s_2 = \sqrt{K_{sp}} = 2 \times 10^{-6}$$

$$\frac{s_1}{s_2} = \frac{s(A_2X)}{s(MX)} = \frac{10^{-4}}{2 \times 10^{-6}} = 50.$$

217. (b) Let s be the solubility of $Ca(OH)_2$ in water

$Ca(OH)_2 \rightleftharpoons \underset{s}{Ca^{2+}} + \underset{2s}{2OH^-}$

$K_{sp} = [Ca^{2+}][OH^-]^2 = s \times (2s)^2$

$\Rightarrow 5.5\times10^{-6} = 4s^3$

$\Rightarrow s^3 = \frac{5.5}{4}\times10^{-6}$

$s = \left[\frac{5.5}{4}\right]^{\frac{1}{3}}\times10^{-2} = 1.11\times10^{-2}$

218. (a) $AgCN \rightleftharpoons Ag^+ + CN^-$

Solubility of AgCN is x. pH = 3 ⇒ $[H^+] = 10^{-3}$

$HCN \rightleftharpoons H^+ + CN^-$

$K_a = \frac{[H^+][CN^-]}{[HCN]} \Rightarrow \frac{[10^{-3}][CN^-]}{[HCN]} = 6.2\times10^{-10}$

$\Rightarrow \frac{[CN^-]}{[HCN]} = 6.2\times10^{-7}$

Each CN– hydrolyses to give one HCN as:

$CN^- + H_2O \rightleftharpoons HCN + OH^-$

HCN is a weak acid, thus, $[HCN] >> [CN^-]$

$x = [Ag^+] = [CN^-] + [HCN]$

$\Rightarrow x = [Ag^+] \approx [HCN] \Rightarrow [CN^-] = x \times 6.2 \times 10^{-7}$

$K_{sp}(AgCN) = [Ag^+][CN^-]$

$\Rightarrow 2.2\times10^{-16} = (x)(x \times 6.2\times10^{-7})$

$\Rightarrow x^2 = 3.55\times10^{-10} \Rightarrow x = 1.88\times10^{-5}$

219. (141) Given : $[K_{sp}]_{PbI_2} = 8\times10^{-9}$

(I) $Pb(NO_3)_2 \rightarrow Pb^{2+}(aq) + 2NO_3^-(aq)$

At t = 0	0.1 M	–	–
At t = t	–	0.1 M	0.2 M

(II) $PbI_2(s) \rightleftharpoons \underset{s}{Pb^{+2}}(aq) + \underset{2s}{2I^-}(aq)$

$[Pb^{2+}] = s + 0.1 \simeq 0.1$

Now: $K_{sp} = 8\times10^{-9} = [Pb^{+2}][I^-]^2$

$\Rightarrow 8\times10^{-9} = 0.1\times(2s)^2 \Rightarrow 8\times10^{-8} = 4s^2 \Rightarrow s = \sqrt{2}\times10^{-4}$

$\Rightarrow s = 141\times10^{-6}\,M \Rightarrow x = 141.$

220. (2.0)

$AB_2 \rightleftharpoons \underset{s}{A^{2+}}(aq) + \underset{2s}{2B^-}(aq)$

$K_{sp} = 4s^3 = 3.2\times10^{-11}$

$\Rightarrow s^3 = 8\times10^{-12}$

$\Rightarrow s = 2\times10^{-4}$

221. (d)

	HCl	CH_3COONa	$\longrightarrow CH_3COOH$	$NaCl$
Millimoles (initial)	10	20	0	0
Millimoles (Final)	0	10	10	10

Buffer solution contains CH_3COONa (10 millimole) and CH_3COOH (10 millimole) which is a acidic buffer.

222. (d) CuS gets precipitated very quickly due to low K_{sp} value.

$CuS(s) \rightleftharpoons Cu^{2+} + S^{2-}; K_{sp} = [Cu^{2+}][S^{2-}]$

$Cu^{2+} + S^{2-} \rightleftharpoons CuS(s); K_{eq} = \frac{1}{[Cu^{2+}][S^{2-}]} = \frac{1}{K_{sp}}$

Due to high value of K_{eq}, CuS precipitated easily.

223. (c) $PbCl_2(s) \rightleftharpoons Pb^{2+}(aq) + 2Cl^-(aq)$

Given; $K_{sp} = 1.6\times10^{-5}$

$[Pb^{2+}] = \frac{300\times0.134}{400} = 0.1005; [Cl^-] = \frac{100\times0.4}{400} = 0.1$

$Q = [Pb^{2+}][Cl^-]^2 = 0.1005\times(0.1)^2 = 1.005\times10^{-3}$

Therefore, $Q > K_{sp}$

224. (b) $Cr(OH)_3 \rightarrow \underset{s}{Cr^{3+}} + \underset{3s}{3OH^-}$

$K_{sp} = s\cdot(3s)^3$

$\Rightarrow 6\times10^{-31} = 27\cdot s^4;\ s = \left(\frac{6}{27}\times10^{-31}\right)^{1/4}$

$[OH^-] = 3s = 3\times\left(\frac{6}{27}\times10^{-31}\right)^{1/4} = (18\times10^{-31})^{1/4}\,M$

225. (b) From the given curve,

if [X] = 1 mM then [Y] = 2 mM ∴ Salt is XY_2

$K_{sp} = [X][Y]^2 = (10^{-3})(2\times10^{-3})^2 = 4\times10^{-9}\,M^3$

226. (d) $Cd(OH)_2 \rightleftharpoons \underset{s}{Cd^{2+}} + \underset{2s}{2OH^-}$

At equilibrium, $K_{sp} = s(2s)^2 = 4s^3$

Þ $K_{sp} = 4\times(1.84\times10^{-5})^3$

Solubility in buffer solution having pH = 12

$[OH^-] = 10^{-2}$

$Cd(OH)_2 \rightleftharpoons \underset{s'}{Cd^{2+}} + \underset{2s+10^{-2}\approx10^{-2}}{2OH^-}$

\ $K_{sp} = 4\times(1.84\times10^{-5})^3 = s'(10^{-2})^2$

$\Rightarrow s' = \frac{24.9\times10^{-15}}{10^{-4}} = 2.49\times10^{-10}\,M$

227. (c) Let the solubility of $Al(OH)_3$ in 0.2M NaOH solution be s.

Then,

$Al(OH)_3 \rightleftharpoons \underset{s}{Al^{3+}} + \underset{3s}{3OH^-}$

and $\underset{0.2M}{NaOH} \rightleftharpoons \underset{0.2M}{Na^+} + \underset{0.2M}{OH^-}$

$[Al^{3+}] = s$ and $[OH^-] = 3s + 0.2 » 0.2$

$K_{sp} = 2.4 \times 10^{-24} = [Al^{3+}][OH]^3$

$2.4 \times 10^{-24} = s(0.2)^3$

$$s = \frac{2.4\times10^{-24}}{8\times10^3} = 3\times10^{-22} \text{mol/L}$$

228. (d) $pH = 7 - \frac{1}{2}pK_b - \frac{1}{2}\log C$

$$pH = 7 - \frac{5}{2} - \frac{1}{2}(\log 2\times10^{-2}) = 5.35$$

229. (b) $Zr_3(PO_4)_4 \rightleftharpoons \underset{3S}{3Zr^{4+}} + \underset{4S}{4PO_4^{3-}}$

$K_{sp} = [Zr^{4+}]^3[PO_4^{3-}]^4 = (3S)^3(4S)^4$

$K_{sp} = 6912\, S^7$

$$S = \left(\frac{K_{sp}}{6912}\right)^{1/7}$$

230. (b) m. mol of $H_2SO_4 = 20 \times 0.1 = 2$

m. mol of $NH_4OH = 30 \times 0.2 = 6$

	H_2SO_4 +	$2NH_4OH$ →	$(NH_4)_2SO_4 + 2H_2O$
Initial	2 m mol	6 m mol	0
Final	(0) = 0 m mol	(6 – 2 × 2) = 2 m mol	2 m mol

$[\overset{+}{N}H_4]$ (from $(NH_4)_2SO_4$) $= 2 \times 2 = 4$ m mol

Total volume = 30 + 20 = 50 mL

$$pOH = pK_b + \log\left[\frac{Salt}{Base}\right] = 4.7 + \log\frac{4/50}{2/50} = 4.7 + \log 2 = 5$$

pH = 14 – pOH

pH = 14 – 5 = 9

231. (c) As $AgNO_3$ dissociates completely, therefore in 0.1 M $AgNO_3$ solution, $[Ag^+] = 0.1$ M

$AgNO_3 \longrightarrow \underset{0.1}{Ag^+} + \underset{0.1}{NO_3^-}$

$Ag_2CO_3 \rightleftharpoons \underset{0.1+2s}{2Ag^+} + \underset{s}{CO_3^{2-}}$

$K_{sp} = [Ag^+]^2[CO_3^{2-}]$

$8 \times 10^{-12} = (0.1 + 2s)^2 \times s$

$0.01\, s = 8 \times 10^{-12}$; $(0.1 + 2s \times 0.1)$

$s = 8 \times 10^{-10}$ M

232. (a) $\underset{14.08 \text{ m mol}}{Na_2SO_4} + \underset{100 \text{ m mol}}{Ca(OH)_2} \rightarrow CaSO_4 + 2NaOH$

$$\text{m mol of } Na_2SO_4 = \frac{2\times1000}{142} = 14.08 \text{ m mol}$$

m mol of $CaSO_4$ formed = 14.08 m mol

Mass of $CaSO_4$ formed = $14.08 \times 10^{-3} \times 136 = 1.9$ g

m mol of NaOH ≈ 28 m mol = 0.028 mol

$Ca(OH)_2 \rightleftharpoons \underset{s}{Ca^{2+}} + \underset{2s+0.028}{2OH^-}$

Value of '*s*' will be negligible in comparison to 0.028.

$$[OH^-] = \frac{0.028 \text{ mol}}{0.1 \text{ L}} = 0.28 \text{ mol L}^{-1}$$

233. (b) CH_3COOK is a salt of weak acid (CH_3COOH) and strong base (KOH).

$FeCl_3$ is a salt of weak base $[Fe(OH)_3]$ and strong acid (HCl).

$Pb(CH_3COO)_2$ is a salt of weak base $Pb(OH)_2$ and weak acid (CH_3COOH).

$Al(CN)_3$ is a salt of weak base $[Al(OH)_3]$ and weak acid (HCN).

234. (b) $H_2S \rightleftharpoons 2H^+ + S^{2-}$, $K_{a_1} . K_{a_2} = K_{eq}$

H_2S is a very weak acid and HCl is a strong acid. Assuming that all $[H^+]$ will be from dissociation of HCl, $[H^+] = 0.2$M

$$\therefore \frac{[H^+]^2[S^{2-}]}{[H_2S]} = 1\times10^{-7} \times 1.2\times10^{-13}$$

$$\frac{[0.2]^2[S^{2-}]}{[0.1]} = 1.2\times10^{-20};\ [S^{2-}] = 3\times10^{-20}$$

235. (c) For final concentration of SO_4^{2-} in $BaSO_4$ solution,

$M_1V_1 = M_2V_2$

$1 \times 50 = M_2 \times 500$; $M_2 = \frac{1}{10}$

For just precipitation, Ionic product = K_{sp}

$[Ba^{2+}][SO_4^{2-}] = K_{sp}(BaSO_4)$

$[Ba^{2+}] \times \frac{1}{10} = 10^{-10}$

$[Ba^{2+}] = 10^{-9}$ M in 500 mL solution

For $[Ba^{2+}]$ in original solution,

$V_1 = 500 - 50 = 450$ mL $\Rightarrow M_1 \times 450 = 10^{-9} \times 500$

[where M_1 = Molarity of original solution]

$$M_1 = \frac{500}{450}\times10^{-9} = 1.11\times10^{-9} \text{ M}$$

236. (d) $(K_{sp})_{PbCl_2} = 3.2\times10^{-8} = 32\times10^{-9}$

$PbCl_2 \rightleftharpoons \underset{s}{Pb^{2+}} + \underset{2s}{2Cl^-}$; $K_{sp} = [Pb^{2+}][Cl^-]^2$

$K_{sp} = 4s^3 = 32 \times 10^{-9} \Rightarrow s = 2 \times 10^{-3}$ M

$$\frac{w}{M.W.} \times \frac{1}{V} = 2\times10^{-3}$$

$$\Rightarrow \quad \frac{0.1}{278} \times \frac{1}{V} = 2\times10^{-3}$$

$$V = \frac{0.1\times1000}{278\times2} = 0.18\ L$$

237. (c) $HA \rightleftharpoons H^+ + A^-$

(Unionized, weak acid and common ion effect)

$HA + NaOH \longrightarrow NaA + H_2O$

$NaA \longrightarrow Na^+ + A^-$ (ionized)

$$K_a = \frac{[H^+][A^-]}{[HA]}$$

Given, pH = 6, $[H^+] = 1 \times 10^{-6}$

From Henderson Hasselbalch equation

$$[H^+] = \frac{K_a\,[Acid]}{[Salt]}$$

$$\Rightarrow \quad \frac{[Salt]}{[Acid]} = \frac{K_a}{[H^+]} = \frac{10^{-5}}{10^{-6}} = 10:1$$

238. (d) Solutions which resist the change in the value of pH when small amount of acid or base is added to them are known as buffers.

239. (c) $2HNO_3(aq) + [Ag(NH_3)_2]^+(aq) + Cl^-(aq) \longrightarrow$

$AgCl(s) \downarrow + 2NH_4^+(aq) + 2NO_3^-(aq)$

When nitric acid is added to amine solution, solution is made acidic and the complex ion dissociates and liberate silver ion to recombine with chloride ion. This is the conformatory test for silver in group 1.

240. (c) $\left[Zr_3\,(PO_4)_4\right] \rightleftharpoons \underset{3s}{3Zr^{4+}} + \underset{4s}{4PO_4^{3-}}$

$K_{sp} = (3s)^3\,(4s)^4 = 6912\,s^7. \quad \therefore \quad s = \left(\frac{K_{sp}}{6912}\right)^{1/7}$

241. (a) Conc. of $Na_2CO_3 = 1.0 \times 10^{-4}$ M

$\therefore \quad [CO_3^{2-}] = 1.0 \times 10^{-4}$ M

i.e., $\quad s = 1.0 \times 10^{-4}$ M

At equilibrium, $[Ba^{2+}]\,[CO_3^{2-}] = K_{sp}$ of $BaCO_3$

$$[Ba^{2+}] = \frac{K_{sp}}{[CO_3^{2-}]} = \frac{5.1\times10^{-9}}{1.0\times10^{-4}} = 5.1 \times 10^{-5}\ M$$

242. (b) The correct order of solubility is:

$BaSO_4 > Hg_2Cl_2 > CrCl_3 > Cr_2(SO_4)_3$

243. (a) When ionic product, i.e., the product of the concentration of ions in the solution exceeds the value of solubility product, formation of precitpiate occurs.

$CaF_2 \rightleftharpoons Ca^{2+} + 2F^-$

Ionic product $= [Ca^{2+}]\,[F^-]^2 = (1\times10^{-2})\,(1\times10^{-6}) = 1\times10^{-8}$

In this case,

Ionic product $(1\times 10^{-8}) >$ solubility product (1.7×10^{-10})

244. (d) $PbI_2 \rightleftharpoons \underset{s}{Pb^{2+}} + \underset{2s}{2I^-} \qquad K_{sp} = s \times (2s)^2 = 4s^3$

$$= 4\times\left(\frac{0.7}{461.2}\right)^3 = 14.0\times10^{-9}$$

245. (a) Let solubility of AgCl = s mol/L

$AgCl \rightleftharpoons Ag^+ + Cl^-$

i.e., $K_{sp(AgCl)} = s \times s$

$KCl \longrightarrow K^+ + \underset{0.1}{Cl^-}$

$[Cl^-]$ from KCl = 0.1 M

Total $[Cl^-]$ in solution = s + 0.1

$K_{sp}(AgCl) = [Ag^+]\,[Cl^-] = s\,(s + 0.1)$

$1.0 \times 10^{-10} = s(s + 0.1)$

$1.0 \times 10^{-10} = s^2 + 0.1s$

$1.0 \times 10^{-10} = 0.1s \qquad$ (as $s^2 << 1$)

$s = 1.0 \times 10^{-9}$ mol/L

246. (b) $Cr(OH)_3(s) \rightleftharpoons \underset{s}{Cr^{3+}}(aq.) + \underset{3s}{3OH^-}(aq.)$

$27s^4 = K_{sp}$

$$s = \left(\frac{K_{sp}}{27}\right)^{1/4} = \left(\frac{1.6\times10^{-30}}{27}\right)^{1/4}$$

247. (b) $AgBr \rightleftharpoons Ag^+ + Br^-;\ K_{sp} = [Ag^+]\,[Br^-]$

For precipitation to occur : Ionic product > Solubility product

$$[Br^-] = \frac{K_{sp}}{[Ag^+]} = \frac{5\times10^{-13}}{0.05} = 10^{-11}$$

i.e., precipitation just starts when 10^{-11} moles of KBr is added to 1L $AgNO_3$ solution.

$\therefore \quad$ Number of moles of Br^- needed from KBr = 10^{-11}

$\therefore \quad$ Mass of KBr = $10^{-11} \times 120 = 1.2 \times 10^{-9}$ g

248. (b) $Mg(OH)_2 \rightleftharpoons Mg^{2+} + 2OH^-;$

$K_{sp} = [Mg^{2+}][OH^-]^2$

$1.0 \times 10^{-11} = 10^{-3} \times [OH^-]^2$

$$[OH^-] = \sqrt{\frac{10^{-11}}{10^{-3}}} = 10^{-4} \quad \therefore \quad pOH = 4$$

$\therefore$ pH + pOH = 14 $\therefore$ pH = 10

249. (a) $\underset{1\times10^{-4}M}{Na_2CO_3} \longrightarrow \underset{1\times10^{-4}M}{2Na^+} + \underset{1\times10^{-4}M}{CO_3^{2-}}$

$K_{sp(BaCO_3)} = [Ba^{2+}][CO_3^{2-}]$

$$[Ba^{2+}] = \frac{5.1\times10^{-9}}{1\times10^{-4}} = 5.1\times10^{-5}\,M$$

250. (c) $AgIO_3 \rightleftharpoons \underset{s}{Ag^+} + \underset{s}{IO_3^-}$; $K_{sp} = [Ag^+][IO_3^-] = s \times s = s^2$

Given $K_{sp} = 1\times10^{-8}$ $\therefore$ $s = \sqrt{K_{sp}} = \sqrt{1\times10^{-8}}$

$= 1.0\times10^{-4}$ mol/L $= 1.0\times10^{-4}\times283$ g/L

($\because$ Molecular mass of $AgIO_3 = 283$)

$$= \frac{1.0\times10^{-4}\times283\times100}{1000}\,g/100\,mL = 2.83\times10^{-3}\,g/100\,mL$$

251. (d) For acidic buffer, $pH = pK_a + \log\left[\frac{salt}{acid}\right]$

$$pH = 4.5 + \log\frac{[salt]}{[acid]}$$

As HA is 50% ionized so [salt] = [acid]

$\therefore$ pH = 4.5 $\because$ pH + pOH = 14

pOH = 14 – pH = 14 – 4.5 = 9.5

252. (c) $MX_2 \rightleftharpoons \underset{s}{M^{2+}} + \underset{2s}{2X^-}$

Where s is the solubility of MX_2

then $K_{sp} = 4s^3$

$4\times10^{-12} = 4s^3$

or $s = 1\times10^{-4}$

253. (d) $MX_4 \rightleftharpoons \underset{S}{M^{4+}} + \underset{4S}{4X^-}$

$K_{sp} = [s][4s]^4 = 256\,s^5$

$$\therefore s = \left(\frac{K_{sp}}{256}\right)^{1/5}$$

254. (d) $AB_2 \rightleftharpoons A^{2+} + 2B^-$

$[A] = 1.0\times10^{-5}$, $[B] = [2.0\times10^{-5}]$

$K_{sp} = [B]^2[A] = [2\times10^{-5}]^2[1.0\times10^{-5}] = 4\times10^{-15}$

255. (a) A buffer is a solution of weak acid and its salt with strong base and vice versa. HCl is strong acid and NaCl is its salt with strong base. pH is less than 7 due to HCl.

256. (a) $Mg(OH)_2 \rightarrow \underset{x}{[Mg^{2+}]} + \underset{2x}{2[OH^-]}$

$K_{sp} = [Mg][OH]^2 = [x][2x]^2 = x.4x^2 = 4x^3$.

Redox Reactions

1. **(c)** A^{2+} has the tendency to lose electrons so it is reducing in nature while B^{4+} has the tendency to gain electrons so it is oxidising in nature.

2. **(d)** (A) $\overset{0}{N_2}+\overset{0}{O_2}\longrightarrow 2NO(g)$

Combination reaction

(B) $2Pb(NO_3)_2(s)\longrightarrow 2PbO+4NO_2+O_2.$

Decomposition reaction

(C) $2Na(s)+2H_2O(l)\longrightarrow 2NaOH(aq)+H_2(g)$

Displacement reaction

(D) $2NO_2(g)+2OH^-(g)\longrightarrow$

$NO_2^-(aq)+NO_3^-(aq)+H_2O$

Nitrogen oxidises and reduces both. So it is a disproportionation reaction.

A→(IV), B→(I), C→(II), D→(III)

3. **(8)** $PbS+4O_3\rightarrow PbSO_4+4O_2$

$x=4, y=4 \therefore x+y=4+4=8$

($O_3\rightarrow O_2+O$ so we need $4O_3$ for $PbSO_4$)

4. **(a)** Oxidising agent is a species which it self undergo reduction. Out of the given option in N^{3-} ion 'N' is present in its lowest possible oxidation state hence it cannot be reduced further because of which it cannot act as an oxidising agent.

5. **(a)** In H_3PO_4 oxidation state of P is +5, which cannot be oxidised further to a higher oxidation state. Hence, it cannot act as reducing agent.

6. **(d)** $\overset{0}{N_2}+\overset{0}{O_2}\longrightarrow 2\,\overset{+2-2}{NO}$ (Redox reaction)

7. **(a)** In the reaction

$\overset{+4}{XeF_4}+\overset{+1}{O_2}F_2\longrightarrow \overset{+6}{XeF_6}+\overset{0}{O_2}$

(Oxidation: Xe +4 → +6; Reduction: O +1 → 0)

8. **(b)** Copper when exposed to moist air having CO_2, it gets superficially coated with a green layer of basic carbonate $CuCO_3\cdot Cu(OH)_2$.

9. **(c)** $2\,\overset{-1}{HI}+H_2\overset{+6}{S}O_4\longrightarrow \overset{0}{I_2}+\overset{+4}{S}O_2+2\,H_2O$

10. **(3)**

Compounds	O.S. of Sulphur
SO_3	+6
H_2SO_3	+4
$SOCl_2$	+4
SF_4	+4
$BaSO_4$	+6
$H_2S_2O_7$	+6

11. **(6)** CrO_2Cl_2

$[x-4-2]=0$

Oxidation State = +6

12. **(6)** $(NH_4)_3PO_4\cdot 12MoO_3$

Let X = oxidation state of Mo in MoO_3

$X+[(-2)\times 3]=0$

$X=+6$

13. **(5)** $\overset{+7}{IO_4^-}+H_2O_2\rightarrow \overset{+5}{IO_3^-}+O_2+H_2O$

14. **(6)** The oxidation states of iron in these compounds will be -

In A, $x+5(-1)+(-1)=-4\Rightarrow x=+2$

In B, $y+4(-2)=-4\Rightarrow y=+4$

In C, $z=0$

The sum of oxidation states will be $=4+2+0=6$.

15. **(19)**

Compound	Oxidation state of transition element
(i) $K_2Cr_2O_7$	$x=+6$
(ii) $KMnO_4$	$y=+7$
(iii) K_2FeO_4	$z=+6$

So, $(x+y+z)=6+7+6=19$.

16. **(b)** Potassium shows +1 state in all its oxides, superoxides and peroxides.

17. **(c)** $[Cr(H_2O)_6]Cl_3$

$\Rightarrow x+6\times 0+(-1)\times 3=0; x=+3$

$[Cr(C_6H_6)_2]$

$x+2\times 0=0; x=0$

$K_2[Cr(CN)_2(O)_2(O_2)(NH_3)]$

[Here $(O)_2$ is oxo form and (O_2) is peroxo form]

$2\times 1+x+2\times(-1)+2\times(-2)+(-2)+0=0$

$x=+6$

18. **(d)**

(a) $H\overset{+1}{O}Cl+H_2\overset{-1}{O_2}\longrightarrow H_3O^+ + \overset{-1}{Cl^-}+\overset{0}{O_2}$

(Oxidation / Reduction)

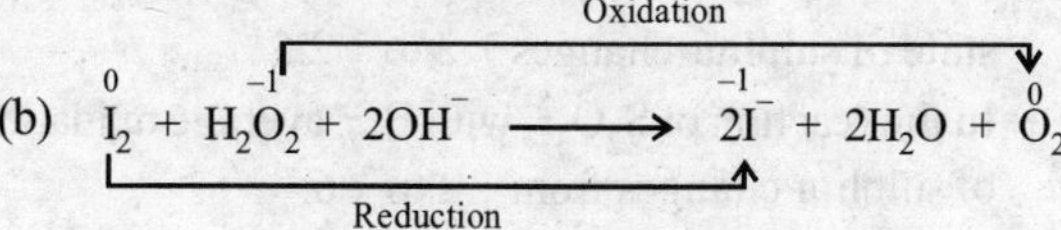

(b) $\overset{0}{I_2}+H_2\overset{-1}{O_2}+2OH^-\longrightarrow 2\overset{-1}{I^-}+2H_2O+\overset{0}{O_2}$

(Oxidation / Reduction)

(c) $\overset{+7}{2MnO_4^-} + \overset{-1}{3H_2O_2} \longrightarrow \overset{+4}{2MnO_2} + \overset{0}{3O_2} + 2H_2O + 2\,OH^-$ (Oxidation: $H_2O_2 \to O_2$; Reduction: $MnO_4^- \to MnO_2$)

(d) $\overset{-2}{PbS} + \overset{-1}{4H_2O_2} \longrightarrow \overset{+6}{PbSO_4} + \overset{-2}{4H_2O}$ (Reduction: $H_2O_2 \to H_2O$; Oxidation: $PbS \to PbSO_4$)

Notice that the oxidation state of oxygen goes from –1 on the H_2O_2 to –2 on the H_2O means H_2O_2 is being reduced. On the other hand the oxidation state of sulfur is going from –2 on the PbS to +6 on the $PbSO_4$. i.e Sulfur is being oxidised.

19. **(d)** CrO_2Cl_2

Let O. No. of Cr = x

$\therefore\ x + 2(-2) + 2(-1) = 0$

$x - 4 - 2 = 0; x = +6$

20. **(c)** In SO_3^{2-}

$x + 3(-2) = -2; x = +4$

In $S_2O_4^{2-}$

$2x + 4(-2) = -2$

$2x = 6;\quad x = +3$

In $S_2O_6^{2-}$

$2x + 6(-2) = -2$

$2x = 10;\quad x = +5$

Hence the correct order is :

$S_2O_4^{2-} < SO_3^{2-} < S_2O_6^{2-}$

21. **(c)** $\overset{+4}{H_2SO_3}(aq) + Sn^{4+}(aq) + H_2O(l) \longrightarrow$

$Sn^{2+}(aq) + \overset{+6}{HSO_4^-}(aq) + 3H^+$

Hence H_2SO_3 is the reducing agent because it undergoes oxidation.

22. **(a)** If an electronegative element is in its lowest possible oxidation state in a compound or in free state. It can function as a powerful reducing agent, e.g., I^-.

23. **(a)**

(A) Combustion of hydrocarbon → CO_2 gas release in the presence of oxygen, a type of combinatinon.

(B) Decomposition into gaseous product.

(C) Displacement of 'V' by 'Ca' atom.

(D) Disproportionation of $H_2\overset{-1}{O_2}$ into –2 and O oxidation states.

24. **(b)** $ClO_4^- \to x + -2 \times 4 = -1 \Rightarrow x = +7$

Chlorine is in its maximum oxidation state, so disproportionation not possible in ClO_4^-

25. **(c)** In the reaction of $S_2O_3^{2-}$ with I_2, average oxidation state of sulphur changes + 2 to + 2.5

In the reaction of $S_2O_3^{2-}$ with Br_2, average oxidation state of sulphur changes from + 2 to + 6.

$\therefore$ Both I_2 and Br_2 are oxidation (oxidising agent)

26. **(b)** F_2 can not disproportionate because fluorine does not show positive oxidation state however Cl_2, Br_2 and I_2 undergoes disproportionation reaction.

27. **(d)** $Cr_2O_7^{2-} + XH^+ + Ye^- \longrightarrow 2A + ZH_2O$

Balance ionic equation is :

$Cr_2O_7^{2-} + 14H^+ + 6e^- \longrightarrow 2Cr^{3+} + 7H_2O$

$X = 14, Y = 6, Z = 7, A = Cr^{3+}$

28. **(a)** In a disproportionation reaction an element in one oxidation state is simultaneously oxidised and reduced.

$Cu^+ \longrightarrow Cu^{2+} + Cu$

$\overset{+6}{3MnO_4^{2-}} + 4H^+ \longrightarrow \overset{+7}{2MnO_4^-} + \overset{+4}{MnO_2} + 2H_2O$

$\overset{+7}{2KMnO_4} \longrightarrow \overset{+6}{K_2MnO_4} + \overset{+4}{MnO_2} + O_2$ (Not a redox)

29. **(a)** Statement I is correct but statement II is incorrect.

In alkaline medium.

$S_8 + 12\,OH^{\ominus} \longrightarrow 4S^{2-} + 2S_2O_3 + 6H_2O$

$S_8^0 \longrightarrow S^{-2} + S_2^{+3}O_3$

Hence S_8 undergo disproportionation reaction.

In ClO_4^- molecule. The oxidation state of Cl is +7 hence it's in its maximum oxidation state and cannot undergo disproportionation reaction.

30. **(8)**

$2MnO_4^- + 5C_2O_4^{2-} + 16H^+ \longrightarrow 2Mn^{+2} + 10CO_2 + 8H_2O$

From the above balanced equation.

2 mole of MnO_4^- requires 16 H^+ ion to form CO_2.

$\therefore$ 1 mole of $MnO_4^- = \frac{16 \times 1}{2} H^+$ ion to form CO_2

= 8 moles of H^+ ions required

31. **(8)** $2MnO_4^- + 6I^- + 4H_2O \longrightarrow 3I_2 + 2MnO_2 + 8OH^-$

$\therefore z = 8$

32. **(6)** Species that have elements that are in intermediate oxidation states undergo disproportionation.

Such species are:

$H_2\overset{-1}{O_2}, \overset{+5}{ClO_3^-}, \overset{0}{P_4}, \overset{0}{Cl_2}, \overset{+1}{Cu^+}, \overset{+4}{NO_2}$

33. **(b)** $Cl_2 + 2OH^- \longrightarrow Cl^- + ClO^- + H_2O$

$\therefore\ a = 1, b = 2, c = 1, d = 1$

34. **(23)**

$Cr_2O_7^{2-} + 14H^+ + 6Fe^{2+} \to 2Cr^{3+} + 6Fe^{3+} + 7H_2O$

$X = 14, Y = 2, Z = 7$

Hence $(X + Y + Z) = 14 + 2 + 7 = 23$

35. (d) Number of atoms of iodine on reactant side = number of atoms of Iodine on product side

$2 + x = 6 \times 2 \therefore \Rightarrow x = 10$

$2IO_3 + 10I^- + 12H^+ \rightarrow 6I_2 + 6H_2O$

36. (a) $2\left[\overset{+1}{Au}(CN)_2\right] + \overset{0}{Zn}(s) \rightarrow 2\overset{0}{Au} + \left[\overset{+2}{Zn}(CN)_4\right]^{-2}$

Zn displaced Au^+

Reduction and Oxidation both are taking place.

37. (c) In a disproportionation reaction, an element in one oxidation state is simultaneously oxidised and reduced.

$2H_2\overset{-1}{O_2} \longrightarrow 2H_2\overset{-2}{O} + \overset{0}{O_2}$ (Disproportionation reaction)

- $2\overset{+4}{NO_2} + H_2O \longrightarrow H\overset{+5}{NO_3} + H\overset{+3}{NO_2}$ (Disproportionation reaction)
- $\overset{+7}{MnO_4^-} + 4H^+ + 3e^- \longrightarrow \overset{+4}{MnO_2} + 2H_2O$ (Reduction reaction)
- $\overset{+6}{MnO_4^{2-}} + 4H^+ \longrightarrow 2\overset{+7}{MnO_4^-} + \overset{+4}{MnO_2} + 2H_2O$ (Disproportionation reaction)

38. (a) In disproportion reaction an element in one oxidation state is simultaneously oxidised and reduced.

$$3\overset{+6}{Mn}\overset{2-}{O_4} + 4\overset{+}{H} \rightarrow 2\overset{+7}{Mn}\overset{-}{O_4} + \overset{+4}{Mn}O_2 + 2H_2O$$

Reduction (Mn +6 → +4); Oxidation (Mn +6 → +7)

39. (a) In BrO_4^-, Br is in highest oxidation state (+7). So it cannot oxidise further and hence, it cannot show disproportionation reaction.

40. (16) Writing the half reaction,

Oxidation half reaction: $MnO_4^- \longrightarrow Mn^{2+}$

Balancing oxygen: $MnO_4^- \longrightarrow Mn^{2+} + 4H_2O$

Balancing hydrogen: $MnO_4^- + 8H^+ \longrightarrow Mn^{2+} + 4H_2O$

Balancing charge:

$MnO_4^- + 8H^+ + 5e^- \longrightarrow Mn^{2+} + 4H_2O$

Reducing half reaction: $C_2O_4^{2-} \longrightarrow CO_2$

Balancing carbon and charge : $C_2O_4^{2-} \longrightarrow 2CO_2 + 2e^-$

Balancing charge in both oxidation and reduction reaction,

$MnO_4^- + 8H^+ + 5e^- \longrightarrow Mn^{2+} + 4H_2O \times 2$

$C_2O_4^{2-} \longrightarrow 2CO_2 + 2e^- \quad \times 5$

$2MnO_4^- + 5C_2O_4^{2-} + 16H^+ \longrightarrow 2Mn^{2+} + 10CO_2 + 8H_2O$

So, we get

$b = 5; c = 16; x = 2; y = 10; z = 8.$

41. (12) $16e^- + S_8 \longrightarrow 8S^{2-}$

$12H_2O + \overset{0}{S_8} \longrightarrow 4\overset{+2}{S_2}O_3^{2-} + 24H^+ + 16e^-$

$2S_8 + 12H_2O \longrightarrow 8S^{2-} + 4S_2O_3^{2-} + 24H^+$

For balancing in basic medium add OH^- equal to H^+.

$2S_8 + 12H_2O + 24OH^- \longrightarrow 8S^{2-} + 4S_2O_3^{2-} + 24H_2O$

$2S_8 + 24OH^- \longrightarrow 8S^{2-} + 4S_2O_3^{2-} + 12H_2O$

$S_8 + 12OH^- \longrightarrow 4S^{2-} + 2S_2O_3^{2-} + 6H_2O$

$\therefore a = 12.$

42. (19) $2Fe^{2+} + H_2\overset{-1}{O_2} \longrightarrow 2Fe^{3+} + 2\overset{-2}{O}H^-$

$2\overset{+7}{Mn}O_4^- + 6H^+ + 5H_2O_2 \longrightarrow 2Mn^{2+} + 8H_2O + 5O_2$

$\because x = 2,\ y = 2,\ x' = 2,\ y' = 8,\ z' = 5$

$\therefore x + y + x' + y' + z' = 19$

43. (d) $\overset{(+1)}{CuBr} \longrightarrow \overset{(0)}{Cu} + \overset{(+2)}{CuBr_2}$

It is an example of disproportionation reaction, as Cu undergoes both oxidation and reduction.

44. (a) $\overset{(+7)}{MnO_4^-} \longrightarrow Mn^{2+};\ n.f. = 5$

$\overset{(+2)}{Fe}\ \overset{(+3)}{C_2}\ O_4 \longrightarrow Fe^{3+} + \overset{(+4)}{CO_2};\ n.f. = (3-2) + 2(4-3) = 3$

$x \times 5 = 1 \times 3 \Rightarrow x = \frac{3}{5}$

$Fe_2(\overset{(+3)}{C_2}O_4)_3 \longrightarrow \overset{+4}{CO_2};\ n.f. = 6(4-3) = 6$

$y \times 5 = 1 \times 6 \Rightarrow y = \frac{6}{5}$

$\overset{(+2)}{FeSO_4} \longrightarrow Fe^{3+};\ n.f. = 1; z \times 5 = 1 \times 1 \Rightarrow z = \frac{1}{5}$

$Fe_2(SO_4)_3$ cannot be oxidised.

Hence, number of moles of $KMnO_4$ required

$$=\frac{3}{5}+\frac{6}{5}+\frac{1}{5}=2$$

45. **(a)** Reaction involved:

$$\overset{+3}{C_2O_4^{2-}} \longrightarrow 2\overset{+4}{CO_2}+2e^-$$

∴ The number of electrons involved in producing one mole of CO_2 is 1.

46. **(c)** The reaction given is

$$Cr_2O_7^{2-}+Fe^{2+}+C_2O_4^{2-} \longrightarrow Cr^{3+}+Fe^{3+}+CO_2$$

$$\overset{+6}{C}rO_7^{2-} \longrightarrow 2Cr^{3+}$$

On balancing

$$14H^+ + Cr_2O_7^{2-}+6e^- \longrightarrow 2Cr^{3+}+7H_2O \quad ......(i)$$

$$4Fe^{2+} \longrightarrow 4Fe^{3+}+4e^- \quad ......(ii)$$

$$C_2O_4^{2-} \longrightarrow 2CO_2+2e^- \quad ......(iii)$$

On adding all three equations, we get

$$Cr_2O_7^{2-}+Fe^{2+}+C_2O_4^{2-}+14H^+ \longrightarrow 2Cr^{3+}+Fe^{3+}+2CO_2+7H_2O$$

The reduction of $Cr_2O_7^{2-}$ to $2Cr^{3+}$ requires 6 electrons because on balancing the equation, we will see 2 moles of chromium ion goes from +6 to +3.

47. **(a)** $X\,MnO_4^- + Y\,C_2O_4^{2-} + Z\,H^+ \rightleftharpoons$

$$X\,Mn^{+2}+2Y\,CO_2+\frac{Z}{2}H_2O$$

First half reaction $MnO_4^- \longrightarrow Mn^{2+}$ .... (i)

On balancing

$$MnO_4^- + 8H^+ + 5e^- \longrightarrow Mn^{2+}+4H_2O \quad .... (ii)$$

Second half reaction

$$C_2O_4^{2-} \longrightarrow 2CO_2 \quad .... (iii)$$

On balancing

$$C_2O_4^{2-} \longrightarrow 2CO_2+2e^- \quad .... (iv)$$

On multiplying eqn. (ii) by 5 and (iv) by 2 and then adding we get

$$2MnO_4^- + 5C_2O_4^{2-} + 16H^+ \longrightarrow 2Mn^{2+}+10CO_2+8H_2O$$

48. **(c)** $Na_2H\overset{(+3)}{As}O_3 \longrightarrow H_3\overset{(+5)}{As}O_4$ (oxidation)

Change in O.N. = +2

$Na\overset{(+5)}{Br}O_3 \longrightarrow Na\overset{(-1)}{Br}$ (reduction)

Change in O.N. = –6

To balance the change in O.N., we will multiply first half reaction with 3 and add the second half reaction.

$$3Na_2HAsO_3+NaBrO_3 \rightarrow 3H_3AsO_4+NaBr$$

On balancing the hydrogen:

$$3Na_2HAsO_3+NaBrO_3+6H^+ \rightarrow 3H_3AsO_4+NaBr$$

On balancing Cl^- and Na^+

$$3Na_2HAsO_3+NaBrO_3+6H^++6Cl^- \rightarrow 3H_3AsO_4+NaBr+6Na^++6Cl^-$$

$$3Na_2HAsO_3+NaBrO_3+6HCl \rightarrow 3H_3AsO_4+NaBr+6NaCl$$

Hence, X = 3, Y = 1 and Z = 6

The *p*-Block Elements (Group-13 and 14)

1. **(b)**

A. Tl^+ is more stable than Tl^{3+}. So Tl^{3+} is a powerful oxidising agent and easily get reduced to Tl^+

B. Al^{3+} is highly stable. $E^o_{Al^{3+}Al} = -1.66$ V. So it is difficult to reduce Al^{3+}.

C. Tl^{3+} is unstable.

D. Tl^+ is more stable than Tl^{3+} due to inert pair effect.

E. Al^{3+} and Tl^+ are highly stable.

2. **(d)** IE order, $B > Tl > Ga > Al > In$

3. **(a)** order of atomic radius is $B < Ga < Al < In \leq Tl$

For Ionic radius incorrect pair is $Al^{3+} < Ga^{3+}$

(X) is Ga. Atomic number of Ga is 31

4. **(a)**

	B	Al	Ga	In	Tl
Atomic radius (pm)	88	143	135	167	170
Electronegativity	2	1.5	1.6	1.7	1.8
Density (g/cm^3)	2.35	2.7	5.9	7.31	11.85
Ionisation Energy (kJ/mol)	801	577	579	558	589

Atomic Radius : $Tl > In > Al > Ga > B$

Electronegativity : $B > Tl > In > Ga > Al$

Density : $Tl > In > Ga > Al > B$

First Ionisation energy : $B > Tl > Ga > Al > In$

5. **(b)** The metallic radius order of Al & Ga is

$$B < Ga < Al < In < Tl$$

This is due to poor shielding by *d*-subshell electrons in Ga.

The ionic radius order of Al^{3+} & Ga^{3+} is :

$B^{3+} < Al^{3+} < Ga^{3+} < In^{3+} < Tl^{3+}$ (No effect of poor sheilding as *d*-electrons are removed)

6. **(c)** Melting point: $B > Al > Tl > In > Ga$

Ionic radius (M^{+3}/pm): $Tl > In > Ga > Al < B$

$$(\Delta_{IE}H)_1\left(\frac{kJ}{mol}\right) : B > Tl > Al \approx Ga > In$$

Atomic radius (in pm): $Tl > In > Al > Ga > B$

7. **(a)** The relative stability of +1 oxidation state progressively increases down the group due to inert pair effect.

$\therefore$ Stability of + 1 oxidation state of group 13 is :

$Al^{+1} < Ga^{+1} < In^{+1} < Tl^{+1}$

8. **(d)** Gallium is the only metal with a low melting point (29.7°C) and high boiling point (1500 – 2000°C).

All these properties make Ga ideal for high temperature thermometers.

9. **(d)** **Statement I :** Number of d and f electrons, increases down the group and due to poor shielding of d and f electrons, s-electrons become inert and stability of lower oxidation states increases down the group

Statement II : The atomic size of aluminium is greater than that of gallium due to poor shielding of d-electrons.

10. **(d)** The most abundant isotope of boron is ^{11}B

Number of neutrons in $^{11}B = 6$

$x = 6$

$$B + O_2 \xrightarrow{\Delta} B_2O_3$$

Oxidation state of B in $B_2O_3 = +3$

So, $y = 3$

Hence $x + y = 9$

11. **(a)** (A) Group-13 size: $B < Ga < Al < In \simeq Tl$

(B) In Group-13, Al has exceptional lower EN

EN : $B > Tl > In > Ga > Al$

(C) $2Al + 6HCl \rightarrow 2AlCl_3 + 3H_2$

$Al + HNO_3 \rightarrow$ No reaction

(Due to protective oxide layer)

(D) +1 oxidation state is observed mainly for heavier Group-13 element like T1.

B and Al do not show +1 oxidation state

(E) $[Al(H_2O)_6]^{3+} \rightarrow$ Hybridisation : sp^3d^2.

12. **(d)** Both statement I and statement II are true.

In group 13 Elements trivalent halides get hydrolysed in water due to their covalent nature. The trichlorides on hydrolysis in water form tetrahedral $[M(OH)_4]^-$ species. Whereas Aluminium chloride in acidified aqueous solution forms octahedral $[Al(H_2O)]^{3+}$ ion..

13. **(a)** Aluminium chloride in acidified aqueous solution from octahedral complex with water molecules.

$$AlCl_3 \xrightarrow{\text{acidified water}} [Al(H_2O)_6]^{3+} (aq) + 3Cl^- (aq)$$

Al^{3+} utilises its vacant 3s, 3p and 3d orbitals to form sp^3d^2 hybridized octahedral complex.

14. **(b)** Solid Boron has very strong crystalline lattice so its melting point unusually high in group 13 element due to polymeric nature of the structure.

15. (a) $\left[\begin{array}{c} F \\ | \\ B \\ F \diagup | \diagdown F \\ F \end{array} \right]^-$

Number of covalent bond formed by Boron is 4

Oxidation number of fluorine is –1,

Oxidation number of B + (4 × (–1)) = –1,

Thus, Oxidation number of B = +3

16. (b) Boron is non-metallic in nature. It is extremely hard and black coloured solid. It exists in many allotropic forms. Due to very strong crystalline lattice, boron has unusually high melting point and boiling point.

	Element				
	B	Al	Ga	In	Tl
Melting point/K	2453	933	303	430	576
Boiling point/K	3923	2740	2676	2353	1730

17. (b) Li, and Cs reacts vigorously with water. Bromine changes in vapour state in boiling water (BP = 58°C). Ga do not reacts with water at 100°C.

18. (d) Down the group extent of back bonding, reduces leading to more Lewis acidic strength.

$BF_3 > BCl_3 > BBr_3 > BI_3$ (extent of back bonding)

(2p – 2p) (2p – 3p) (2p – 4p) (2p – 5p)

∴ $BF_3 < BCl_3 < BBr_3 < BI_3$ (lewis acidic character)

19. (b) The first ionization energies are as follows:

B : 801 kJ/mol; Al : 577 kJ/mol; Ga : 579 kJ/mol;

Ga(31) : $[Ar]3d^{10}4s^24p^1$

20. (b) Expansion of octet not possible for 'B' as it has outermost shell n = 2.

21. (b) **Metal** **Melting point**

Al → 933 K

Ga → 303 K

In → 430 K

Se → 490 K

Ga has very low m.pt. and its periodic position is close to metalliods.

22. (c) SnH_4 is non planar molecular hydride. It has tetraheadral shape

$$\begin{array}{c} H \\ | \\ Sn \\ H \diagup | \diagdown H \\ H \end{array}$$

23. (a) Caesium have loosely bound electrons. So it is used in devising photoelectric cells. Boron fibres are light and hard materials, so they used in making bullet-proof vest. Silicones being surrounded by non-polar alkyl groups are water repelling in nature.

Gallium alloys are less toxic and has a very high boiling point, so it is used in high temperature thermometers.

24. (a) Electronic configuration of Tl (81)

$= Xe\ 4f^{14}\ 5d^{10}\ 6s^2\ 6p^1$

Due to inert pair effect, Tl is more stable in +1 oxidation state. Hence, TlI_3 and CsI_3 are isomorphous.

25. (c) $[SiCl_6]^{2-}$ does not exist due to steric crowding of surrounding atoms and less interaction between some pair of chloride is an and Si^{++}.

26. (d) PbO_2 is strong oxidizing agent because Pb^{+4} is not stable and can be easily reduced to Pb^{+2}.

PbO_2 is used in lead storage batteries. It is also amphoteric in nature.

27. (a) Moving down the group, stability of lower oxidation state increases, so the order will be Al < Ga < In < Tl.

28. (b) Be and Al show diagonal relationship due to which these two elements have similar electronegativity.

29. (a) Due to the inert pair effect, thallium exists in more than one oxidation state. Also, for thallium + 1 oxidation state is more stable than +3 oxidation state.

30. (a) SiH_4 : Electron precise hydride

B_2H_6, GaH_3 and Al H_3 are electron deficient

31. (c) The maximum valency of beryllium is +2, while that of aluminium is +3.

32. (a) **Statement-1:** Generally left to right I.E. increases in a period.

Statement-2 : Melting point of group 14 elements is higher than group 13 elements.

33. (b) Among Al, In, Tl, Ge, Sn, Pb, the metal having highest IE_1 is of Ge 761 and lowest IE_1 is of In. 558

Most stable oxidation state of Ge is +4 and In is +3

34. (a) Order of I.E. (in kJ/mol)

C > Si > Ge > Sn < Pb

1086 786 761 708 715

35. (d) Order of M.P. of group 14 : C > Si > Ge > Pb > Sn

element	M.P. (°C)
Z = 6 = C	4373
Z = 14 = Si	1693
Z = 32 = Ge	1218
Z = 50 = Sn	505
Z = 82 = Pb	600

Sn has lowest melting point due to distorted structure and weaker bonding.

36. (b) Germanium, silicon and copper oxide are semi conductor while graphite is a conductor.

37. **(d)** Due to inert pair effect more stable ions do not lose electrons but accept electron i.e. Pb^{4+}, Tl^{+3}

38. **(d)** Statement I is true but statement II is false.

Carbon is able to show allotropic forms due to property of catenation and $p\pi - p\pi$ bond formation.

39. **(d)**

GeO - Acidic	GeO_2 - Acidic
SiO_2 - Acidic	CO - Neutral
SnO_2 - Amphoteric	PbO_2 - Amphoteric

40. **(a)** Carbon atoms have the tendency to link with one another through covalent bonds to form chains. This property is called catenation. This is because C – C bonds are very strong. Down the group, the size increases and electronegativity decreases. The order of catenation is $C >> Si > Ge \approx Sn$.

Bond	C – C	Si – Si	Ge – Ge	Sn – Sn
Bond Enthalpy (kJ/mol)	348	297	260	240

41. **(b)** The catenation property among 14th group elements is based on bond enthalpy value of bond between the same element. The decreasing order of bond enthalpy values is

Bond enthalpy $\underset{348}{C-C} > \underset{297}{Si-Si} > \underset{260}{Ge-Ge} \approx \underset{240}{Sn-Sn}$

in (kJ/mol)

$\therefore$ Decreasing order of catenation is

$C > Si > Ge \approx Sn$

42. **(a)** Buckminster fullerene is a type of fullerene with the formula C_{60}. It has a cage like fused ring structure that resembles a soccer ball, made of 20 hexagons and 12 pentagons. Each carbon atom has three bonds.

43. **(d)** Catenation power of the elements decreases as we move down in the group. Therefore, Pb does not show catenation property.

44. **(d)** Carbon-carbon bond length is maximum in diamond because diamond has all single bonds while graphite, C_{70} and C_{60} have single and double bonds.

Carbon allotrope	**C – C bond length**
Diamond	154 pm
Graphite	141.5 pm
C_{60}	138.3 pm and 143.5 pm
C_{70}	eight type of bond lengths from 137 pm to 146 pm.

45. **(d)** Fullerene (C_{60}) contains 20 hexagons (six membered) rings and 12 pentagons (five membered rings):

46. **(b)** Due to the small size of carbon atom, effective lateral overlapping between 2*p* and 2*p* occurs.

47. **(b, d)** BCl_3 and $AlCl_3$, both have vacant *p*-orbital and incomplete octet, thus they behave as Lewis acids.

$SiCl_4$ can accept lone pair of electron in *d*-orbital of silicon, hence it can act as Lewis acid.

Although the most suitable answer is (d). However, both options (b) and (d) can be considered as correct answers.

48. **(b)** % of *p*- in graphite (sp^2) $= \frac{2}{3} \times 100 = 67\%$

% of *p*- in diamond (sp^3) $= \frac{3}{4} \times 100 = 75\%$

49. **(b)** In diamond, each C-atom is covalently bonded to four other C-atoms to give a tetrahedral unit, so it shows sp^3 hybridisation. Therefore, each C-atom forms four sigma bonds with neighbouring C-atoms. In diamond, each C-atom utilizes its four unpaired electrons in bond formation. These bonding electrons are localized. Due to this reason diamond is a bad conductor of electricity.

In graphite, each C-atom is covalently bonded to three C-atoms to give trigonal geometry. Each C-atom in graphite is sp^2-hybridized. Three out of four valence electrons of each C-atom are used in bond formation while the fourth electron is free to move in the structure of graphite. Due to this reason, graphite is a good conductor of electricity.

50. **(a)** The linking of identical atoms with each other to form long chains is called catenation. However, this property decreases from carbon to lead. Decrease of this property is associated with M-M bond energy, which decreases from carbon to lead.

51. **(d)** Reluctance of valence shell electrons to participate in bonding is called inert pair effect. The stability of lower oxidation state (+2 for group 14 elements) increases on going down the group. So the correct order is:

$SiX_2 < GeX_2 < SnX_2 < PbX_2$

52. **(a)** Metal atom in the lower oxidation state forms ionic bond whereas in the higher oxidation state forms covalent bond. Higher oxidation state means small size and high polarizing power and hence, greater the covalent character. So, MCl_2 is more ionic than MCl_4.

53. **(c)** Graphite has a two dimensional sheet like structure. These various sheets are held together by van der Waals force of attraction. Due to these weak forces of attraction, one layer can slip over the other. Which makes graphite soft and a good lubricating agent. Due to strong C – C covalent bonds, graphite is extremely difficult to melt.

9 Organic Chemistry-Some Basic Principles and Techniques

1. **(d)** 4-Ethylcyclopent-2-en-1-ol

2. **(c)** 1-Bromo-2-methyl but-2-ene

3. **(d)** $\overset{7}{HC}\equiv\overset{6}{C}-\overset{5}{CH_2}-\overset{4}{CH}(OH)-\overset{3}{CH_2}-\overset{2}{CH}=\overset{1}{CH_2}$

Hept-1-en-6-yn-4-ol

4. **(a)** IUPAC name : 3-Bromo-2-hydroxy-5-nitrobenzoic acid.

5. **(c)**

(A) $\overset{1}{CH_3}-\overset{2}{CH_2}-\overset{3}{CH}(CH_2-CH_3)-\overset{4}{CH_2}-\overset{5}{CH}(CH_3)-\overset{6}{CH_2}-\overset{7}{CH_3}$

3-Ethyl-5-methylheptane

(B) $(CH_3)_2C(C_3H_7)_2$

$CH_3-\overset{4}{C}(CH_3)(\overset{5}{CH_2}-\overset{6}{CH_2}-\overset{7}{CH_3})-\overset{3}{CH_2}-\overset{2}{CH_2}-\overset{1}{CH_3}$

4, 4-Dimethylheptane

(C) 2-Methyl-1, 3-pentadiene

(D) 4-Methylpent-1-ene

6. **(c)**

Alkane	2°H
(a)	10
(b)	4
(c)	2
(d)	4

7. **(d)** $CH_3-\overset{2}{CH}(\overset{1}{CO_2H})-\overset{3}{CH_2}-\overset{4}{CH_2}-\overset{5}{CH}(\overset{6}{C}(=O)-OCH_3)-CH_3$

6-Methoxycarbonyl-2,5-dimethylhexanoic acid.

8. **(b)** Statement-I is incorrect

Correct IUPAC name is 1-chloro-2, 4-dinitrobenzene.

9. **(c)**

$\overset{1}{CH_3}-\overset{2}{CH}(CH_3)-\overset{3}{CH_2}-\overset{4}{CH_2}-\overset{5}{CH}(CH_3)-\overset{6}{CH}(CH_3)-\overset{7}{CH_2}-\overset{8}{CH_3}$

2,5,6-Trimethyloctane

10. **(b)** Group present in sulphonic acids is $-\overset{O}{\overset{\|}{\underset{\underset{O}{\|}}{S}}}-OH$.

11. **(c)** 2-formyl-4-hydroxyhept-6-enoic acid.

12. **(a)** 7-Hydroxyheptan-2-one is correct IUPAC name.

13. **(c)** 2, 3-Dibromo-1-phenylpentane.

14. **(d)** $\underset{5}{CH_3}-\underset{4}{CH}(CH_3)-\underset{3}{CH}=\underset{2}{CH}-\underset{1}{C}(=O)-H$

4-Methylpent-2-enal

15. (c) $\overset{4}{CH_3} - \overset{3}{CH}(NH_2) - \overset{2}{CH_2} - \overset{1}{CN}$

3 – Aminobutane nitrile

Priority order = $-CN > -NH_2$.

16. (2) Only 2 compounds from the given options contain sulphur as heteroatom i.e. Thiophene and Cysteine.

Thiophene

$HS - CH_2 - CH(NH_2) - COOH$

Cysteine

While tyrosine, furan has oxygen and pyridine, pyrole has N as heteroatom.

17. (b) The given compound has a six-membered ring with a double bond and a hydroxyl group.

IUPAC Name = cyclohex-2-en-1-ol.

18. (d) For Aromaticity, it must fallow Huckel's rule as per which it must have $(4n + 2)\pi$ electrons and all C atoms sp^2.

is Alicyclic (cyclic + Aliphatic)

19. (b) More branched carbon gets lower number

3-Ethyl 1, 1 -dimethylcyclohexane

20. (d) $CH(OH)(CN)_2$ is

OH, NC, CN

21. (b) H_3C–CO–CH_2–CH_2–CH(CH_3)–COOH

2–Methyl–5–oxohexanoic acid

22. (c)

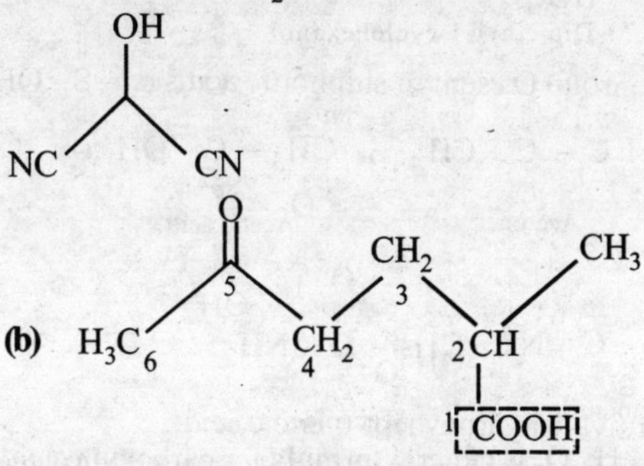

: Non-planar heterocyclic compound

: Bicyclo compound

: Spiro compound

Spiro compounds have at least two molecular rings with only one common carbon.

: Aromatic compound

23. (b) $-SO_3H > -COCl > -CONH_2 > -CN$

24. (c)
- Carbon skeleton with maximum number of carbon atoms is chosen (7).
- Carbonyl group gets priority over nitro group.

4-Methyl-6-nitro-3-oxohept-4-enal

25. (d)

1, 1-Dichloroethane is ethylidene chloride.

26. (c)

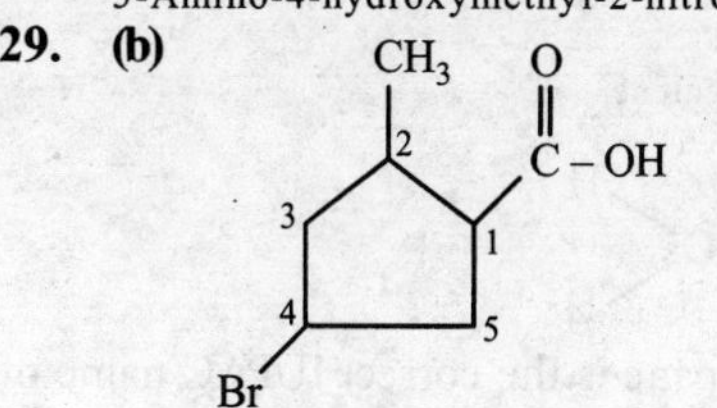

(2E)-2-Bromohex-2-en-4-yne

27. (d) Mesityl oxide (common name) is the aldol condensation product of acetone. Its structure and IUPAC name are

$CH_3 - \overset{O}{\overset{||}{C}} - CH = C(CH_3) - CH_3$

4-Methylpent-3-en-2-one

28. (c)

5-Amino-4-hydroxymethyl-2-nitrobenzaldehyde

29. (b)

4-Bromo-2-methylcyclopentane carboxylic acid

30. (c)

2,5-dimethyl-6-oxo-hex-3-enoic acid

31. (d) $\overset{1}{H_2C} = \overset{2}{CH} - \overset{3}{CH}(CH_3) - \overset{4}{CH}(CH_2CH_2CH_3) - \overset{5}{CH}(CH_3) - \overset{6}{C} \equiv \overset{7}{CH}$

3,5-dimethyl-4-propylhept-1-en-6-yne

32. (b)

2-Chloro-1-methyl-4-nitrobenzene

33. (c)

3-Hydroxy-4-methylpentanoic acid

34. (d) IUPAC name: 4-Bromo-3-methylpent-2-ene

35. (d)

4-Ethyl-3-methylhex-2-ene

36. (b)

2-Ethyl-1,1-dimethylcyclohexane

37. (b) $H_3C-C(CH_3)_2-CH_2-CH=CH_2$

4, 4 - Dimethylpentene

38. (b) $H_2C=C=CH_2$ (sp^2 sp sp^2)

39. (d) 1, 2-Epoxy propane is the correct IUPAC name of given compound.

40. (a) $\overset{1}{C}H_3-\overset{2}{C}H=\overset{3}{C}H-\overset{4}{C}\equiv\overset{5}{C}-\overset{6}{C}H_2\overset{7}{C}H_3$

(*E*) – 2 – hepten – 4 – yne

41. (a) In 1, 2-dichlorobenzene, the two dipoles are at 60° (i.e., unsymmetric). Thus, it possesses dipole moment.

42. (b)

Since – CN has the hgiest prioirty, hence the parent compound is a nitrile.

∴ IUPAC name of the compound is 2-bromo-5-hydroxybenzonitrile.

43. (a) $H_3C-C(CH_3)_2-CH_3$

Neopentane or 2,2- Dimethylpropane

44. (a) The correct order of priority for the given functional group is:

$$-COOH > -SO_3H > -\overset{O}{\overset{||}{C}}-NH_2 > -\overset{O}{\overset{||}{C}}-H$$

45. (a) $\overset{7}{C}H_3-\overset{6}{C}H_2-\overset{5}{C}H_2-\overset{4}{C}(CH_3)_2-\overset{3}{C}H(CH_2CH_3)-\overset{2}{C}H_2-\overset{1}{C}H_3$

3-Ethyl-4,4-dimethylheptane

46. (a)

3-Bromo-1-chlorocyclohexene

47. (d) C_6H_5COCl is known as benzne carbonyl (benzoyl) chloride.

48. (a)

3, 3-Dimethyl-1-cyclohexanol

49. (a) $\overset{sp^3}{H_3C}-\underset{sp^2}{\overset{O}{\overset{||}{C}}}-\overset{sp^3}{CH_3}$; $\overset{sp^3}{CH_3}-\underset{sp^2}{\overset{O}{\overset{||}{C}}}-OH$;

Acetone   Acetic acid

$\overset{sp^3}{CH_3}-\overset{sp}{C}\equiv N$; $\overset{sp^3}{CH_3}-\underset{sp^2}{\overset{O}{\overset{||}{C}}}-NH_2$

Acetonitrile   Acetamide

50. (a) $C_nH_{2n}O_2$ is general formula for carboxylic acids.

51. (c) $\overset{1}{C}H_3-\overset{2}{C}(=O)-\overset{3}{C}H(CH_3)-\overset{4}{C}H_3$

3-Methyl-2-butanone

52. (c) The correct name is 3-methylbutan-2-ol.

53. (8) Molecular formula = C_9H_{12}

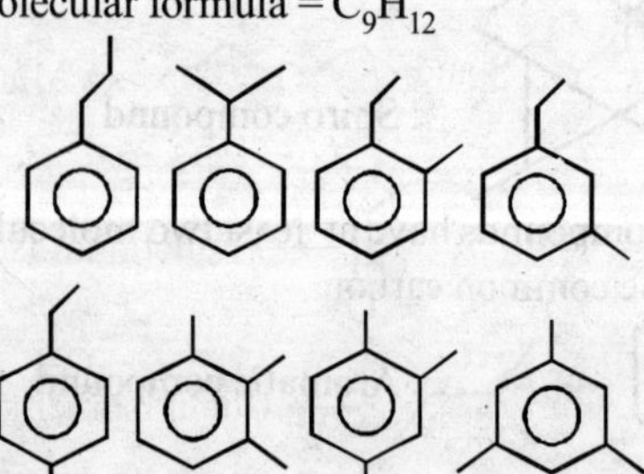

54. (c)
A. Metamers

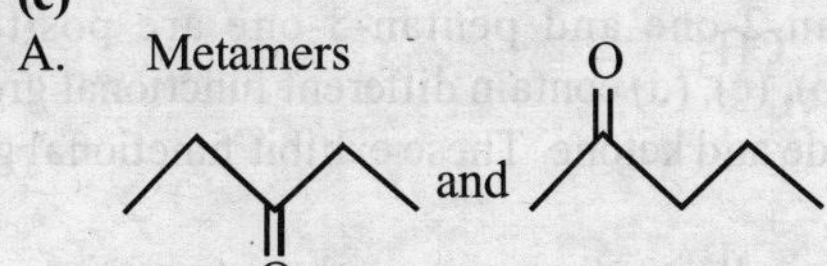

B. Functional isomers

CN and NC

55. (b) **Statement–I** is True

Both are ring chain isomers

Statement–II → True

NH_2 — NH

1° Amine 2° Amine

Both are functional group isomers. 1° Amine and 2° Amine are different functional groups.

56. (4) Ph * * OH

n(stereogenic center) = 2, $2^2 = 4$ stereoisomers are possible. (two geometrical and two optical)

57. (d) $CH_3-CH(OH)-CH=CH-CH_3$

It has 4 stereoisomers [R cis R trans; S cis S trans]

58. (a) Br Br Br

There are three stereo – centers (n = 3)

So, No. of stereoisomer = $2^n = 2^3 = 8$

59. (d) Metamer are the isomers having same molecular formula and same functional group but different alkyl or arly groups on either side of functional group.

60. (b) cis-2-butene and trans-2-butene are stereoisomers. trans-2-butene is more stable than cis-2-butene. cis-2-butene has more dipole moment than trans-2-butene. cis-2-butene and trans-2-butene are not interconvertible at room temperature due to rotational barrier.

61. (d) OH and OH → Position Isomerism

CH_3–O– and –O– → Metamerism

O and O → Functional Isomerism

and → Chain Isomerism

62. (d)
A. Propanamine N–Methylethanamine

$H_3C-CH_2-CH_2-NH_2$ $C_2H_5-NH-CH_3$ ⇒ Functional isomers

B. Hexan–2–one Hexan–3–one

$CH_3-CO-(CH_2)_3CH_3$ $CH_3-CH_2-CO-(CH_2)_2CH_3$ ⇒ Metamers

C. Ethanamide Hydroxyethanimine

$H_3C-CO-NH_2$ $CH_3-C(OH)=NH$ ⇒ Tautomers

D. *o*–Nitrophenol *p*–nitrophenol ⇒ Positional isomers

63. (6) C_5H_{10}

cis ; *trans* H CH_3

Total 6 isomers are possible.

64. (6) Dimethylcyclopentane has following isomers :

CH_3 H_3C — 1, 1-Dimethylcyclopentane no stereoisomer (i)

CH_3 CH_3 — 1, 2-Dimethylcyclopentane will show stereoisomerism (ii)

CH_3 CH_3 — 1, 3-Dimethylcyclopentane will show stereoisomerism (iii)

(ii) *cis* = *cis*

cis –1, 2 – Dimethylcyclopentane (Identical)

CH_3 CH_3 CH_3 CH_3

trans (enantiomers)

(iii) [cis-1,3-dimethylcyclopentane structures] =

cis–1, 3 – Dimethylcyclopentane
(Identical)

CH_3 CH_3

CH_3 CH_3

trans (enantiomers)

So total number of stereoisomers are 3[from (ii)] + 3[from (iii)] = 6

65. **(d)** For a compound with formula

$\Rightarrow$ $C_aH_bN_cO_dX_e$

$$\text{Degree of unsaturation} = \frac{1}{2}(2 + 2a - b + c - e)$$

For C_3H_6O, $a = 3$; $b = 6$

$$\text{Degree of unsaturation} = \frac{1}{2}(2 + 2 \times 3 - 6) = 1$$

So, given compound contains one double bond. Hence, possible isomer can be written as follows :

$CH_3 - CH_2 - \overset{O}{\overset{\|}{C}} - H$ and $CH_3 - \overset{O}{\overset{\|}{C}} - CH_3$

They are functional group isomers.

66. **(d)**

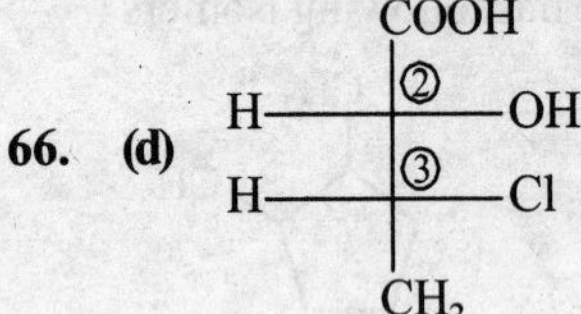

At (2),

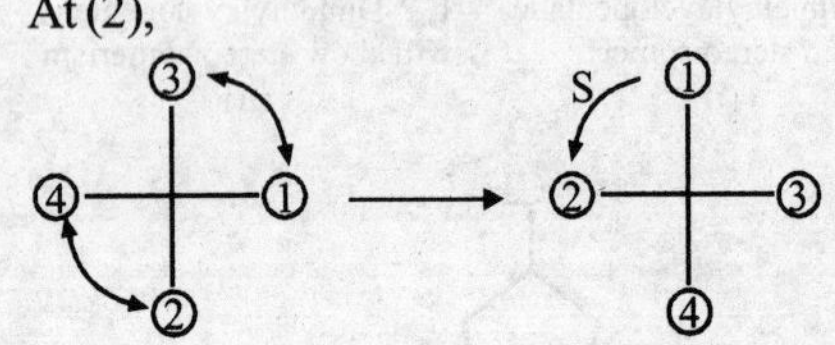

At (3),

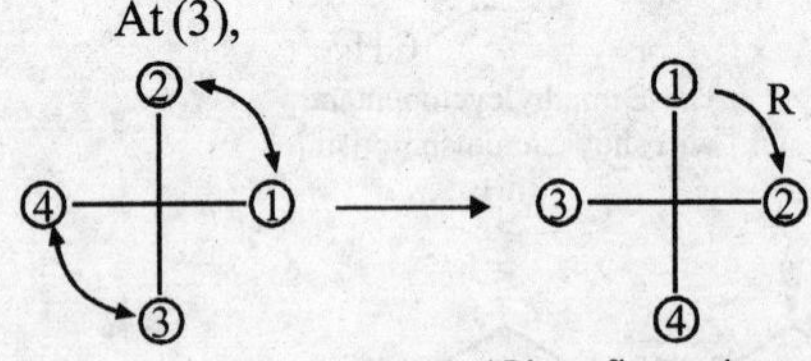

67. **(d)**

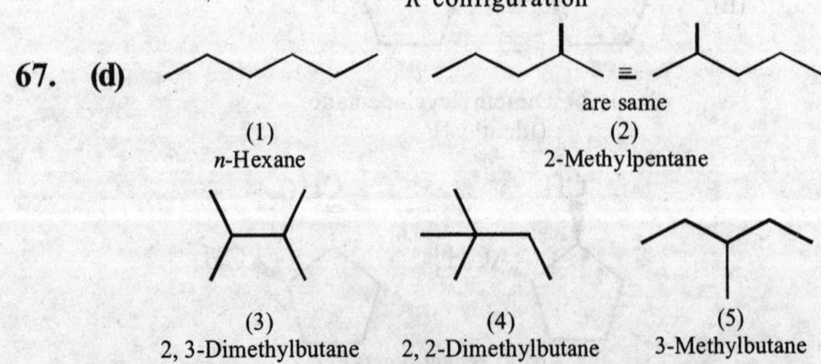

The possible structural isomers = 5

68. **(a)** Pentan-2-one and pentan-3-one are positional isomers. (b), (c), (d) contain different functional groups, *viz* aldehyde and ketone. These exhibit functional group isomerism.

69. **(d)** The optically inactive compound must contains achiral carbon atom(s). Option (d) contains achiral carbon atom

$CH_3 - \underset{H}{\overset{Cl}{C}} - CHO$ $CH_3 - CH_2 - \underset{H}{\overset{Cl}{C}} - CH_3$

2- Chloropropanol (chiral)   2-Chlorobutane (chiral)

$CH_3 - CH_2 - CH_2 - \underset{H}{\overset{Cl}{C}} - CH_3$ $CH_3 - CH_2 - \underset{Cl}{\overset{CH_3}{C}} - CH_3$

2-Chloropentane (chiral)   2-Chloro-2-methylbutane (achiral)

70. **(c)** Optically active compounds contain an asymmetric (chiral) carbon atom (a carbon atom attached to four different atoms or groups). Therefore, all acids except maleic acid exhibit optical isomerism.

71. **(d)** Maleic acid and fumaric acids are geometrical isomers.

$$\begin{matrix} H - C - COOH \\ \| \\ H - C - COOH \end{matrix} \qquad \begin{matrix} H - C - COOH \\ \| \\ HOOC - C - H \end{matrix}$$

Maleic acid   Fumaric acid

72. **(b)** Five cyclic structures are possible for C_4H_6. These are as follows.

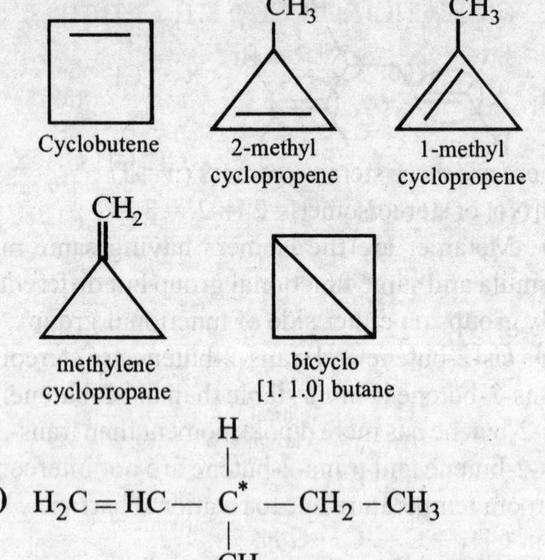

73. **(c)** $H_2C = HC - \underset{CH_3}{\overset{H}{C^*}} - CH_2 - CH_3$

3- Methyl-1-pentene

74. **(b)**

$$\begin{matrix} H_3C & & CH_3 \\ & C = C & \\ H & & H \end{matrix} \qquad \begin{matrix} H_3C & & H \\ & C = C & \\ H & & CH_3 \end{matrix}$$

cis-2-Butene   *trans*-2-Butene

75. **(b)** $CH_3 - CH = CH - \underset{OH}{\overset{*}{C}}HCH_3$

It exhibits both geometrical as well as optical isomerism.

cis - R   *cis* - S

trans - R   *trans* - S

76. **(b)** The organic compounds which have chiral carbon atom (a carbon atom attached to four different groups or atoms) and do not have plane of symmetry rotate plane polarised light.

$$HO-\overset{CHO}{\underset{CH_2OH}{C^*}}-H \quad (\text{* is asymmetric carbon})$$

77. **(c)** $CH_3-\overset{Cl}{\underset{H}{C}}-\overset{Cl}{\underset{H}{C}}-CH_3$, 2, 3-Dichlorobutane exhibits optical isomerism due to the presence of two asymmetric carbon atoms.

78. **(b)** Compounds containing two similar chiral C-atoms have plane of symmetry and can exist in *meso* form too.

CH_3; H—C—Cl; - - - - - plane of symmetry; H—C—Cl; CH_3

meso-2, 3-Dichlorobutane

$CH_3-\overset{H}{\underset{C_2H_5}{C^*}}-$(cyclopropyl)

79. **(c)** 1-chloropentane is not chiral while others are chiral in nature

(a) $ClCH_2\overset{*}{C}HCH_2CH_2CH_3$ with CH_3 on C-2 (carbons numbered 1 2 3 4 5)

(b) $H_3C\overset{*}{C}HCH_2CH_2CH_3$ with Cl on C-2 (carbons numbered 1 2 3 4 5)

(c) $ClCH_2CH_2CH_2CH_2CH_2$ (carbons numbered 1 2 3 4 5)

(d) $CH_3CH\overset{*}{C}HCH_2CH_3$ with CH_3 on C-2 and Cl on C-3 (carbons numbered 1 2 3 4 5)

80. **(a)** $C_2H_5-\overset{CH_3}{\underset{H}{C^*}}-C_3H_7$ Chiral carbon (I)

$CH_3-\overset{O}{\overset{\|}{C}}-\overset{CH_3}{\underset{H}{C^*}}-C_2H_5$ Chiral carbon (II)

$H-\overset{H}{\underset{H}{C^{\oplus}}}$ Achiral carbon (III)

$C_2H_5-\overset{CH_3}{\underset{H}{C}}-C_2H_5$ Achiral carbon (IV)

81. **(c)** Both differ in the arrangement of groups in space, therefore grouped under sterio-isomerism.

82. **(d)** A mixture of equal amount of two enantiomers is called a racemic mixture.

83. **(d)** Stereoisomerism, isomers differ in the arrangement of groups in space. The two structures show sterioisomerism. Structure I shows geometrical isomerism as it contains two different atoms or groups H and CH_3 attached to each carbon containing double bond.

$\overset{H_3C}{H}>C=C<\overset{CH_3}{H}$ *cis*-butene $\qquad \overset{H}{H_3C}>C=C<\overset{CH_3}{H}$ *trans*-butene

Structure II shows optical isomerism as it contains a chiral carbon (attached to four different groups) atom.

$H-\overset{CH_3}{\underset{CH_2CH_3}{C^*}}-OH \qquad HO-\overset{CH_3}{\underset{CH_3CH_2}{C^*}}-H$

Butyl alcohol (two enantiomers)

84. **(d)**
(A) Carbocation → sp^2 hybridised carbon with empty p-orbital
(B) Carbon free radical → sp^2/sp^3 hybridised carbon with one unpaired electron.
(C) Nuecleophile → species that can supply a pair of electron. i.e electron rich species.
(D) Electrophile → species that can receive a pair of electron i.e electron deficient species.

85. **(18)** Volume of nitrogen gas liberated (V) = 50 mL = $\frac{50}{1000}$ L

Pressure (P) = 715 mm of Hg
Temperature (T) = 300 K
Aqueous tension = 15 mm of Hg

P_{N_2} = Pressure (P) – aqueous tension

$= 715 - 15 = 700 \text{ mm of Hg} = \frac{700}{760}$ atm

$$n_{N_2} = \frac{P_{N_2}.V}{RT}$$

$$n_{N_2} = \frac{700}{760} \times \frac{50}{1000} \times \frac{1}{0.082 \times 300}$$

$$n_N = 2 \times n_{N_2}$$

Mass of N = $2 \times n_N \times 14$

$$\%N = \frac{\text{mass of N}}{\text{mass of organic compound}} \times 100$$

$$\%N = \frac{700}{760}\frac{50}{1000} \times \frac{2 \times 14}{0.0821 \times 300} \times \frac{1000}{292} \times 100$$

$$\%N = 18\%$$

86. **(a)**
(A) Aniline – H_2O : Steam Distillation (Aniline is steam volatile)
(B) Glycerol from spent-lye in soap industry: Distillation under reduced pressure
(C) Different fraction of crude oil in petroleum industry : Fractional distillation
(D) $CHCl_3$ – Aniline : Simple distillation

87. (b)

(A) [cyclohexyl cation with OMe] > [cyclohexyl cation with Me]

Stable due to resonance effect (+M)

(B) [carbanion with NO_2] > [carbocation with O_2N]

Stable due to –I and –M effect of $-NO_2$, which stabilized the carbanion

(C) [cyclohexyl $\overset{\oplus}{C}H_2$] < [cyclohexenyl $CH_2^{\oplus}$]

Stable due to resonance effect

(D) $Me\overset{\oplus}{C}H(Me)Me$ < $Me\overset{\oplus}{C}H(Me)\ddot{O}Me$

Stable due to +M effect of $-OCH_3$

88. (20) Partial pressure of $N_2 = P_{total}$ – Aqueous tension

$= (900 - 15) = 885$ mm Hg

$$\text{Mole of } N_2 = \frac{\left(\frac{885}{760} \times 0.15\right)}{(0.0821 \times 300)} = 0.0071 \text{ moles}$$

mass of nitrogen = n × molar mass of $N_2 = 0.0071 \times 28$

% of nitrogen in organic compound

$$= \frac{(0.0071 \times 28)}{1} \times 100 = 19.88\% \approx 20\%$$

89. (c) Hyperconjugation is permanent effect because external reagent is not required, so Statement-I is false. Statement-II is true because greater is the number of alkyl groups greater is the number of α-H, so more hyperconjugation which results in more stability of carbocations.

90. (a) Pressure of N_2 gas evolved = Total pressure – Aqueous tension

$$= 715 - 15 = 700 \text{ mm Hg} = \frac{700}{760} \text{ atm.}$$

∴ Moles of N_2 evolved $= \frac{PV}{RT} = \frac{700 \times 60 \times 10^{-3}}{760 \times 0.0821 \times 300}$

$= 0.0022$ mol

∴ Mass of N_2 evolved $= 0.0022 \times 28 = 0.062$ g

∴ Mass % of nitrogen in compound

$$= \frac{\text{wt. of nitrogen}}{\text{wt. of compound}} \times 100$$

$$= \frac{0.062}{0.4} \times 100$$

$= 15.5\% \cong 15.71\%$.

91. (109) moles of compound $= \frac{0.42}{86} = 0.00488$

∴ moles of $N_2 = 0.00488$

Volume of N_2 at STP $= 0.00488 \times 22.4 = 0.1093$ L $= 109.3$ mL ≈ 109 mL

92. (d) $P_{N_2} = P_{\text{moist gas}}$ – aqueous tension

$= (715 - 15)$ mm Hg

$= 700$ mm Hg

$V_{N_2} = 60$ mL $= 0.060$ L

T = 300 K

$$n_{N_2} = \frac{PV}{RT} = \frac{700 \times 0.06}{760 \times 0.0821 \times 300} = 0.022 \text{ mol}$$

Mass of $N_2 = 0.0224 \times 28 = 0.0628$

$$\% N_2 = \frac{0.0628}{0.5} \times 100 = 12.58\%$$

93. (d) Boiling point of aniline is 547 K and B.P. of $CHCl_3$ is 334 K. So, they are separated by simple distillation.

94. (d) $H - C \equiv C - \dot{C}H - C(H)(CH_3) - CH_2$ with bonds labelled I, II, III, IV

II gives most stable carbon free radical due to resonance stabilization.

I gives least stable carbon radical due to no stability factor.

95. (275) Original Compound ⟶ $BaSO_4$

0.20 g 0.40 g

$\frac{0.40}{233}$ mol ($BaSO_4$)

$\frac{0.40}{233}$ mol (Sulphur)

$\frac{0.40}{233} \times 32$ g (Sulphur)

$$\%S = \frac{\text{mass of S} \times 100}{\text{mass of organic compound}} \times 100$$

$$= \frac{\frac{0.40 \times 32}{233} \times 100}{0.20} = 27.5\% \text{ or } 275 \times 10^{-1}\%$$

96. (b) **Statement-I is true.** In partition chromatography, stationary phase is thin liquid film present in the inert support.

Statement-II is false. Because stationary phase in paper chromatography is water.

97. (d) A reagent that brings an electron pair to the reactive site is called a nucleophile.

As total six nucleophiles are there from the given options.

NH_3, PhSH, $(H_3C)_2S$, $CH_2 = CH_2$, $\overset{\ominus}{O}H$, $>= NCH_3$

98. (a) The purification process that involves a solid changing into vapour and then back into solid is called sublimation.

99. (1655)

	C:	Cl:	H:	O
%mass	14.5	64.46	1.8	19.24
Molar ratio	$\frac{14.5}{12}$ = 1.2	$\frac{64.46}{35.5}$ = 1.8	$\frac{1.8}{1}$ =1.8	$\frac{19.24}{16}$ = 1.2
Minimum integral ratio	2	3	3	2

Empiricial formula = $C_2H_3Cl_3O_2$
Mass = 165.5
Mass = 1655×10^{-1}

100. (d)

(A) $Ph-\overset{Ph}{\underset{Ph}{C}}{}^{\oplus}$ (B) $Ph-\overset{Ph}{\underset{H}{C}}{}^{\oplus}$ (C) [cycloheptatrienyl cation, ⊕]

(D) $H_3C-CH_2-\overset{\oplus}{C}H-CH_3$

Structure C is aromatic hence it is most stable.
Structure A has more resonance structures.
Structure B has less resonance structures compared to A.
Structure D have only hyperconjugation.
Preference order will be: Aromaticity > Resonance > Hyper conjugation
D < B < A < C

101. (255) $\%\text{Bromine} = \frac{\text{Molar Mass of Bromine}}{\text{Molar Mass of Silver Bromide}} \times \frac{\text{Weight of AgBr}}{\text{Weight of sample}} \times 100$

$$= \frac{80}{188} \times \frac{0.15}{0.25} \times 100 = \frac{4800}{188} = 25.53 = 255 \times 10^{-1}$$

102. (d) In Dumas method nitrogen present in organic compound is converted into N_2 gas whose volumetric analysis gives the percentage of nitrogen atom in the organic compound.

103. (b) Carbocation intermediate is stabilised by +I, +M & hyperconjugation effect. Since in option (b) carbocation is in conjugation with stronger +M group $-OCH_3$ hence it will be most stable.

$\overset{+}{C}H_2-CH=CH-\ddot{O}-CH_3 \longleftrightarrow CH_2=CH-CH=\overset{\oplus}{O}-CH_3$

104. (40) Millimoles of $BaSO_4 = \frac{466}{233} = 2$ m mol = m moles of S

$$\%S = \frac{m_1 \times 32}{m \times 233} \times 100 = \frac{n \times 32}{m} \times 100$$

$$\Rightarrow \frac{\frac{466}{233} \times 32}{160} \times 100 = 40\%$$

105. (a) [p-MeO-benzyl cation (MeÖ, +M)] > [p-H_3C-benzyl cation (+H)] > [benzyl cation] > [m-OCH_3-benzyl cation (–I)]

Due to +M effect of –OMe at para position.

[p-$CH_3\ddot{O}$-benzyl cation] will be most stable among these

106. (20) millimoles of AgCl $= \frac{143.5}{143.5} = 1$
Millimoles of Cl = 1
Mass of Cl = 35.5×10^{-3} g

$$\% \text{ by mass of Cl} = \frac{35.5}{180} \times 100 = 19.72\%$$

Nearest integer = 20%

107. (c) Organic compounds are purified based on their nature and impurities present in them.

108. (b) (1) Neutral structures are more stable than charged ones. Therefore (I) is more stable than (II) and (III).
(2) +ve charge on less electronegative atom is more stable i.e., $C^{\oplus}$ is more stable than $O^{\oplus}$
∴ Order is (I) > (II) > (III)

109. (4) [structures: conjugated enone; o-isopropylnitrobenzene (NO_2); o-nitrophenyl butene (NO_2); 2-methylcyclohexenyl methyl ketone ($COCH_3$, CH_3)]

These compounds show all these electronic effects.

110. (15)

$$R_f = \frac{\text{Distance moved by substance from base line}}{\text{Distance moved by solvent from base line}}$$

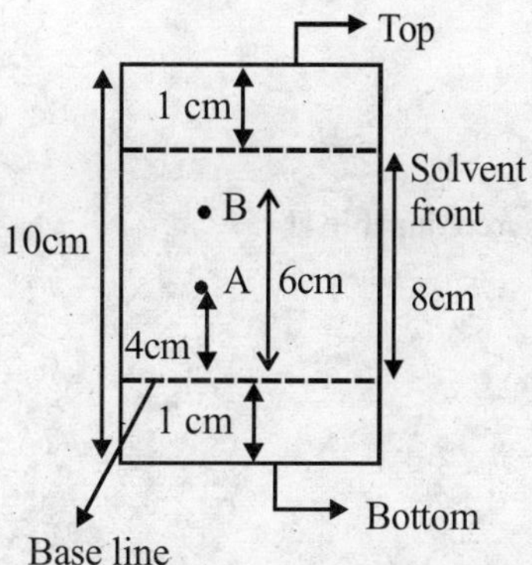

$(R_f)_A = \frac{4}{8}\ (R_f)_B = \frac{6}{8} \Rightarrow \frac{(R_f)_B}{(R_f)_A} = \frac{6}{8} \times \frac{8}{4}$

$\Rightarrow (R_f)_B = 1.5\,(R_f)_A$
$x = 15$

111. (a) Carbocation $H-\overset{\oplus}{C}(H)-H$

Trigonal planar

112. (a) Nitrogen present in pyridine can not be estimated by Kjeldahl method as the nitrogen present in pyridine can not be easily converted into ammonium sulphate.

113. (b) Aniline can be purified by steam distillation because aniline is steam volatile and immisible with water. Therefore option (b) is incorrect.

114. (b) Methoxy group strongly activates benzene ring towards electrophilic aromatic substitution due to +R effect. Methyl group weakly activates benzene ring due to its +I and +H effects. CF_3 group strongly deactivates the benzene ring due to –I and–H effects.

order as

Resonance > Hyper conjugation > Induction effect

Therefore, the correct order of the given compounds towards electrophilic aromatic substitution is

OCH_3 (III) > CH_3 (I) > (II) > CF_3 (IV)

115. (c) Organic compounds run slower than solvent in thin layer chromatography. R_f is not an integral value. It is the ratio of distance travelled by the organic compound to that of solvent. R_f value of a polar compound is smaller than that of non-polar compound as the polar compound is retained more by the adsorbent than non-polar compound.

116. (5) has 8 $\alpha - H$

no $\alpha - H$

(tert-butyl)

$\overset{+}{C}H_3$ no $\alpha - H$

, no hyperconjugable H

$\overset{\oplus}{C}H_2 - OCH_3$ no $\alpha - H$

9 α –H

$N - \overset{\oplus}{C}H_2$ no $\alpha - H$

5 carbocations are not stabilised by hyperconjugation. greater the α – H greater will be its hyperconjugating structure.

117. (50)

$$R_f = \frac{\text{Distance travelled by the substance from base line } (x)}{\text{Distance travelled by the solvent from base line } (y)}$$

$$\frac{(R_f)_A}{(R_f)_C} = \frac{\frac{5}{12.5}}{\frac{10}{12.5}} = 0.5 \Rightarrow = 50 \times 10^{-2} \Rightarrow x = 50$$

118. (a)

+R effect ($-\ddot{N}H_2$ is a electron donating group)

–R effect ($-NO_2$ is a electron withdrawing group)

$+ H^+ \longrightarrow$

+E effect (H^+ is a electron deficient species)

$+ \bar{C}N \longrightarrow$ CN

–E effect ($\bar{C}N$ is a electron rich species)

119. (d) Hydrogen ion (H^+) shows positive electromeric effect.

120. (c) The most common polar and acidic support used is adsorption chromatography is silicagel. The surface silanol groups adsorb polar compound and work particularly well for basic substances. Alumina is the example of polar and basic adsorbent that is used in adsorption chromatography.

121. (b) $CuSO_4$ acts as a catalytic agent in Kjeldahl's method for estimation of nitrogen.

122. (56)

m = 1 gm, M = 2 M, V = 10 ml

$$\% \text{ of N} = \frac{14 \times M \times 2V}{1000} \times \frac{100}{m} = \frac{14 \times 2 \times 2 \times 10}{1000} \times \frac{100}{1}$$

$= 56\%$

123. (c) Only B

Ionic reactions with organic compounds proceed through heterolytic bond cleavage.

$$A - B \xrightarrow{\text{Heterolysis}} A^{\oplus} + B^{\ominus}$$

124. (c) –COOH
Negative R effect is shown by:
$-COOH, -CHO, >C=O, -CN, -NO_2$
Positive R effect is shown by:
$-X, -OH, -OR, -OCOR, -NH_2, -NHR, -NR_2, -NHCOR$

125. (c) Carbocation has sextet of electrons. It has incomplete octet so it is a electrophile.

126. (a) Principle used in chromotography is adsorption.

127. (b) (A) Chloroform (b.p) = 61.2°C
Aniline (b.p) = 184.1°C
These have large difference in their boiling points. So these can be separated by distillation method.
(B) Petroleum products are separated by fractional distillation because they have less differences in their boiling points.
(C) Steam distillation is used to separate substances which are steam nolatile and are immiscible with water.
(D) Distillation under reduceed pressure is used to purify liquids having very high boiling pints and those, which decompose at or below their boiling points.

128. (d) Since the fragrance of flowers is due to some steam volatile organic compounds and they are insoluble in water at room temperature, therefore the suitable method for the extraction will be steam distillation.

129. (a) Species follow Huckel's rule are aromatic in nature and aromatic systems are stable.

[cyclopropenyl cation] follow $(4n + 2)\pi$ rule, it has 2π electrons in conjugation. So it is must stable.

130. (7) $R_f = \dfrac{\text{Distance moved by organic compound}}{\text{Distance moved by solvent}}$

$= \dfrac{3.5 \text{ cm}}{5 \text{ cm}} = 0.7 \Rightarrow 7 \times 10^{-1}$

131. (d) Column Chromatography → based on absorption.
Sublimation → based on volatile nature of components.
Distillation → based on relative boiling points
Differential extraction → based on relative solubilities.

132. (c) $(CH_3)_3\overset{+}{C} > (CH_3)_2\overset{+}{C}H > CH_3-\overset{+}{C}H_2 > \overset{+}{C}H_3$
tertiary secondary primary
This is due to dispersion of positive charge on more atoms available.

133. (a) The resonance effect is defined as 'the polarity produced in the molecule by the interaction of two π-bonds or between a π-bond and a lone pair of electrons present on an adjacent atom'.

134. (d) The order of acidity of the given compounds is:

$(CH_3)_3\overset{sp^3}{C}-H$ (C) (alkane with a 3° C - atom) < $CH_3-\overset{sp^3}{C}H_3$ (D) (alkane with primary C) < $CH_2=\overset{sp^2}{C}H-H$ (B) < $CH\equiv\overset{sp}{C}-H$ (A)

Due to the relative electronegativities of the carbon atoms of the corresponding hydrogens and their hybridization states, we get the order of acidic strengths as:
C < D < B < A

135. (c) Column and thin layer chromatography (TLC) are both based on the principle of differential absorption. Paper chromatography is based on the principle of differential partitioning of the components of a mixture.

136. (d) $H_3C-CO-CH_2-CO-CH_2-CH_3 \longrightarrow$ (Active methylene group between to EWG.)

$H_3C-CO-\overset{\ominus}{C}H-CO-CH_2-CH_3$

Conjugate base is more stable due to more resonance of negative charge, that results in its dispersal.

137. (4) $-NO_2, -C\equiv N, -COOH, -COR$ are meta directing.

138. (a) I > II > III, since neutral resonating structures are more stable than charged resonating structures II > III, since stability of structure (II) with –ve charge on more electronegative atom (o) is higher.

139. (d) Steam distillation is used for those liquids which are insoluble in water, containing non-volatile impurities and are steam volatile.

140. (b) Higher the electron density on benzene ring, higher is reactivity towards electrophilic substitution reaction

$C_6H_5NMe_2$ (+M) (e) > $C_6H_5OCH_3$ (+M) (d) > $C_6H_5CH_3$ (+H), +I > C_6H_6 > $C_6H_5CF_3$ (–I)

+ M in (e) is more than that of (d) due to +I effects of Me groups.

141. (b) In paper chromatography, a special quality of paper known as chromatography paper is used. Paper contains water trapped in it, which acts as the stationary phase.

142. (a)

[Reaction scheme: protonation of the alcohol (OH → $\overset{\oplus}{O}H_2$ with $H^{\oplus}$), loss of water, ring expansion, 1,2-hydride shift, further rearrangement, then loss of $H^{\oplus}$ on heating (Δ, $-H^{\oplus}$) to give the fused bicyclic alkene]

143. (5)

$$CH_3-\underset{H_3C}{\overset{CH_3}{C}}-CH-\underset{H}{\overset{CH_3}{C}}(OH)-CH_3 \xrightarrow[-H_2O]{H^+} CH_3-\underset{H_3C}{\overset{CH_3}{C}}-\overset{+}{C}H-\underset{H}{\overset{CH_3}{C}}-CH_3$$

(A) $CH_3-\underset{H_3C}{\overset{CH_3}{C}}-CH_2-\overset{+}{\overset{CH_3}{C}}-CH_3$

+

(B) $CH_3-\overset{+}{\overset{CH_3}{C}}-\overset{*}{C}H-\underset{CH_3}{\overset{CH_3}{C}H}-CH_3$

(A) →
$CH_3-\underset{CH_3}{\overset{CH_3}{C}}-CH_2-\overset{CH_3}{C}=CH_2$

$CH_3-\underset{H_3C}{\overset{CH_3}{C}}-CH=\overset{CH_3}{C}-CH_3$

(B) →
$CH_2=\overset{CH_3}{C}-\overset{*}{C}H-\underset{CH_3}{\overset{CH_3}{C}H}-CH_3$ (d + l) isomers

$CH_3-\overset{CH_3}{C}=C-\underset{CH_3}{\overset{CH_3}{C}H}-CH_3$

144. (40) mole of AgBr $= \dfrac{0.376}{188}$

mole of Br^- = mole of AgBr $= \dfrac{0.376}{188}$

mass of $Br^- = \dfrac{0.376}{188} \times 80$

$$\% \text{ of } Br^- = \frac{0.376 \times 80}{188 \times 0.4} \times 100 = 40\%$$

145. (25) More R_f, less its polarity

$$R_f = \frac{\text{Distance travelled by compound 'X'}}{\text{Distance travelled by solvent 'Y'}}$$

$$= \frac{2}{8} = 0.25 = 25 \times 10^{-2}$$

146. (a) According to the observation, A is more stable and interacts with the mobile phase more than C, and C is more drawn to the mobile phase than B. Hence, the correct order of elution in the silica gel column chromatography is – B < C < A or order of elution A, C, then B.

147. (c) Benzene becomes more reactive towards EAS when any substituent raises the electron density.

[Structures a, b, c, d]

a b

– OH has + R

c d

$-\overset{O}{\overset{\|}{C}}-$ has – R effect

–OH and – O both show +R effect

∴ Correct order
c < a < b < d

148. (7)

[Reaction scheme: protonation with H^+, Me-shift, H-shift]

including this structure there are 7 Hyperconjugation structure (Number of Hyperconjugation structure = no. α – hydrogens +1)

149. (a) [Bromocyclohexene] $\xrightarrow{AgNO_3}$ [cyclohexenyl cation] + AgBr

Unstable Carbocation

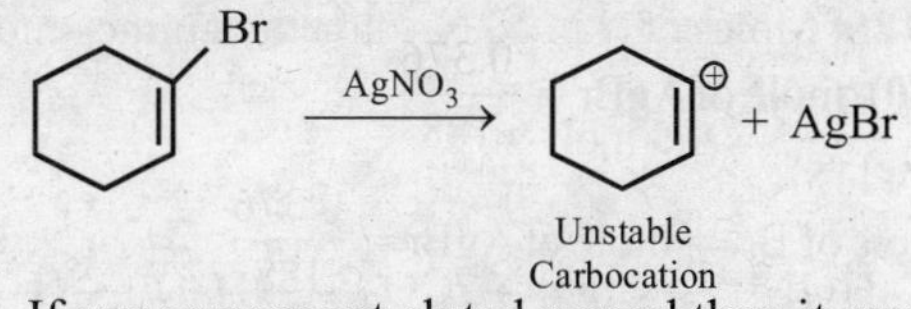

150. (d) If any component eluted second then it means that its R_f value is low and its adsorption is stronger

$$R_f = \frac{\text{distance covered by substance from base line}}{\text{total distance covered by solvent from base line}}$$

151. (d) $X \xrightarrow{Na_2O_2} Y \xrightarrow{BaCl_2} \underset{[BaSO_4]}{Z}$

Methionine: $C_5H_{11}NO_2S$ (Sulphur containing organic compound)

$H_2N-CH(COOH)-(CH_2)_2-S-CH_3$

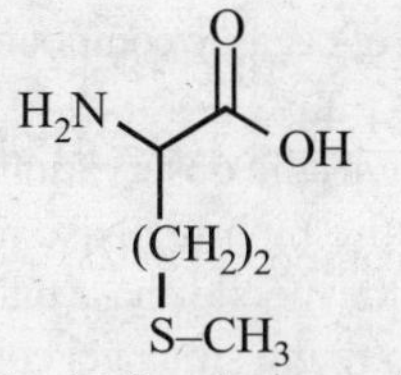

The given test checks for the quantity of sulphur.

$$\underset{(X)}{C_5H_{11}NO_2S} \xrightarrow{Na_2O_2} \underset{(Y)}{H_2SO_4} \xrightarrow{BaCl_2} \underset{(Z)}{BaSO_4}$$

152. (b) → Hydride affinity means low stability.
Thus, higher the stability of a carbocation lower will be its hydride affinity. Stability order of cations is : C < A < B < D
'A' is allylic carbocation, 'B' is resonance stabilized but – Ph is elec. withdrawing. C is stablized through hyper conjugation, D is stabilized by +I effect.
Thus the decreasing order of hydride affinity is

C > A > B > D

153. (d) $Na + C + N \rightarrow NaCN$

$6NaCN + FeSO_4 \rightarrow Na_4[Fe(CN)_6] + Na_2SO_4$

$Na_4[Fe(CN)_6] + Fe^{3+} \rightarrow \underset{\text{(Prussian blue)}}{Fe_4[Fe(CN)_6]_3}$

Sulphur detection by Sodium nitroprusside

$Na_2[Fe(CN)_5NO] + Na_2S \rightarrow \underset{\text{[Purple]}}{Na_4[Fe(CN)_5NOS]}$

Phosphorus detection by ammonium molybdate

$Na_3PO_4 + 3HNO_3 \rightarrow H_3PO_4 + 3NaNO_3$

$H_3PO_4 + 12(NH_4)_2MoO_4 + 21HNO_3 \rightarrow$
$\underset{\text{(canary yellow)}}{(NH_4)_3PO_4 \cdot 12MoO_3} + 21NH_4NO_3 + 12H_2O$

Halogens give specific coloured ppt with $AgNO_3$(aq)

$NaCl + AgNO_3(aq) \rightarrow \underset{\text{(White)}}{AgCl} + NaNO_3$

$NaBr + AgNO_3(aq) \rightarrow \underset{\text{(Pale yellow)}}{AgBr} + NaNO_3$

$NaI + AgNO_3(aq) \rightarrow \underset{\text{(Yellow)}}{AgI} + NaNO_3$

154. (a) Compound 'a' is more polar than 'b' and 'c', compound 'c' is least polar. Compound 'c' comes out first.

155. (c) Density of CH_2Cl_2 is greater than H_2O. Therefore these can be separated by differential solvent extraction.
Due to H-bonding in p-nitrophenol it can be separated from other component by column chromatography.
Due to different boiling point of kerosene and naphthalene, these can be separated by fractional distillation.
NaCl (ionic compound) and $C_6H_{12}O_6$ can be separated by crystallisation.

156. (b) The nitrogen containing organic compound, when heated with CuO in an atmosphere of CO_2, yields free N_2 in addition to CO_2 and H_2O.
Traces of nitrogen oxides formed, if any, are reduced to nitrogen by passing the gaseous mixture over heated copper gauze.

157. (b) Acidity of an acid depends upon the stability of its conjugate base

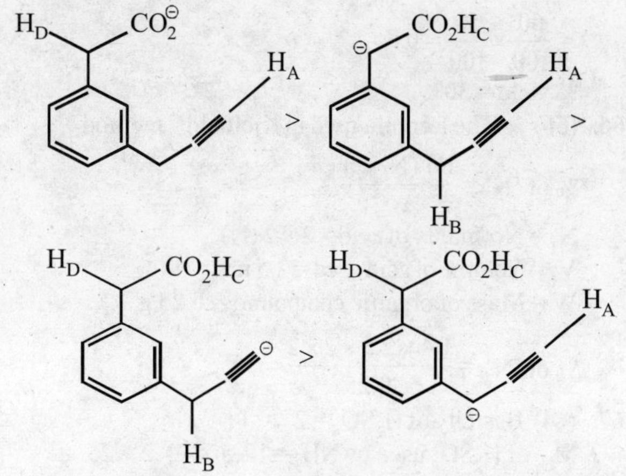

[Structures: conjugate base with H_D, $CO_2^{\ominus}$, H_A > conjugate base with carbanion, CO_2H_C, H_A, H_B > conjugate base with H_D, CO_2H_C, acetylide anion, H_B > conjugate base with H_D, CO_2H_C, H_A, carbanion]

158. (a) [Aniline-substituted carbocation: lone pair of $:NH_2$ delocalised into ring, giving $\oplus NH_2=$ cyclohexadiene]

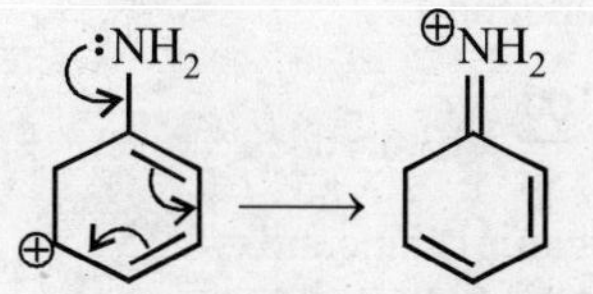

The +M effect of NH_2 is stabilizing the carbocation.

159. (b) A-(III), B-(IV), C-(I), D-(II).

160. (6)

$$R_f = \frac{\text{Distance moved by the substance from base line}}{\text{Distance moved by the solvent from base line}}$$

$$= \frac{3.0 \text{ cm}}{5.0 \text{ cm}} = 0.6 \text{ or } 6 \times 10^{-1}$$

161. (42) 233 g of $BaSO_4$ contains 32g of S

$\therefore$ 1.4439 g $BaSO_4 = \left(\frac{32}{233} \times 1.4439\right)$ g of S

$= \frac{32 \times 1.4439}{233}$ g of S Obtained from 0.471 g of organic compound.

$\therefore$ % of 'S' $= \frac{32 \times 1.4439 \times 100}{233 \times 0.471} = 42.10.$

162. (2) Benzylic and tertiary carbocations are stable.

163. (22) V = 22.78 mL, T = 280 K

$P_{total} = 759$ mmHg

$P_{N_2} = 759 - 14.2 = 744.8$ mmHg

$$n_{N_2} = \frac{744.8 \times 22.78}{760 \times 1000 \times 0.082 \times 280} = 0.00097$$

$W_{Nitrogen} = 0.0097 \times 28 = 0.02716\ (n_{N_2} \times M_{N_2})$

$$\%N = \frac{\text{Wt. of N}}{\text{Wt. of sample}} = \frac{0.02716}{0.125} \times 1000 = 21.728$$

164. (a) 188 g AgBr contains = 80 g Br

$$0.36 \text{ g AgBr contains} = \frac{80}{188} \times 0.36 = 0.1532 \text{ g}$$

$$\%\,Br = \frac{0.1532}{0.45} \times 100 = 34.04$$

165. (30) Overall yield of complete reaction

$$= \frac{60}{100} \times \frac{50}{100} = 0.3$$

% yield = 30%

166. (56) The formula used in Kjeldahl's method

$$\%\text{ of 'N'} = \frac{1.4\,(N_1V_1)}{W}$$

N_1 = Normality of acid = 2 × 2 (N)
V_1 = Volume of acid used = 2.5 mL
W = Mass of organic compound = 0.25 g

$$\%\text{ of 'N'} = \frac{1.4 \times 2.5 \times 2 \times 2}{0.25} = 56$$

167. (64) Basicity of $H_2SO_4 = 2$
M_{eq} of H_2SO_4 used by $NH_3 = 12.5 \times 1 \times 2 = 25$
% of N in the compound

$$= \frac{25 \times 10^{-3} \times 14 \times 100}{0.55} = 63.6 \approx 64\%$$

168. (40) Weight of organic compound = 0.25 g
Molar mass of AgCl = 143.5g/mol
143.5 g AgCl contains 35.5g of Cl

$$0.40 \text{ g of AgCl contains} = \frac{35.5 \times 0.4}{143.5} = 0.099 \text{ g of Cl}$$

⇒ Percentage of Cl in organic compound

$$= \frac{0.099}{0.25} \times 100 = 39.6 \approx 40\%$$

169. (c) Ninhydrin, iodine and UV rays are used for identification of spot.

170. (a) Two immiscible liquids with different densities can be separated by using this funnel.

171. (7) $C_2H_7N + \left(2x + \frac{y}{2}\right)CuO \longrightarrow$

$$xCO_2 + \frac{y}{2}H_2O + \frac{z}{2}N_2 + \left(2x + \frac{y}{2}\right)Cu \quad ...(i)$$

$C_xH_yN_z + \left(2x + \frac{y}{2}\right)CuO \longrightarrow$

$$xCO_2 + \frac{y}{2}H_2O + \frac{z}{2}N_2 + \left(2x + \frac{y}{2}\right)Cu \quad ...(ii)$$

Comparing (i) & (ii), we get : $x = 2, y = 7, z = 1$

172. (1125) Moles of N in N, N - dimethylaminopentane

$$= \left(\frac{57.5}{115}\right) = 0.5 \text{ mol}$$

$$\Rightarrow C_7H_{17}N + \frac{45}{2}CuO \rightarrow 7CO_2 + \frac{17}{2}H_2O + \frac{1}{2}N_2 + \frac{45}{2}Cu$$

For 0.5 moles of $C_7H_{17}N$ number of moles of CuO required

$$= \frac{1}{2}\left(14 + \frac{17}{2}\right) = \frac{45}{4} = 1125 \times 10^{-2}$$

173. (42) Molecular mass of $BaSO_4$ = 233 g
∵ 233 g $BaSO_4$ contains → 32 g sulphur

$$\therefore 1.44 \text{ g } BaSO_4 \text{ contains} \rightarrow \frac{32}{233} \times 1.44 \text{ g sulphur}$$

Given : 0.471 g of organic compound

$$\%\text{ of S} = \frac{32 \times 1.44}{233 \times 0.471} \times 100 = 41.98\% \approx 42\%$$

174. (19) $\underset{5.0\text{ g}}{\text{Organic compound} + Cl_2} \longrightarrow \text{Compound A}$

$$\underset{0.5\text{ g}}{\text{Compound A}} \xrightarrow{AgNO_3} \underset{0.3849\text{ g}}{AgCl}$$

$$\text{Moles of AgCl} = \text{Moles of Cl} = \frac{0.3849}{143.3}$$

$$\text{Mass of chlorine} = \frac{0.3849}{143.3} \times 35.5 = 0.095 \text{ g}$$

$$\%\text{ of chlorine in A} = \frac{0.095}{0.5} \times 100 = 19\%$$

175. (b)

176. (19) Number of moles of $N_2 = \frac{758 - 14}{760} \times \frac{30 \times 10^{-3}}{0.0821 \times 287}$

$= 1.246 \times 10^{-3}$ mol

Mass of $N_2 = 1.246 \times 10^{-3} \times 28$ g

$$\%\text{ of } N_2 = \frac{1.246 \times 10^{-3} \times 28}{0.1840} \times 100 = 18.96\% \approx 19\%$$

177. (a) Carius method is used for the estimation of halogens in organic compounds.

178. (50)
Mass of organic compound = 1.6 g
Mass of AgBr = 1.88 g

$$\text{Moles of Br} = \text{Moles of AgBr} = \frac{1.88}{188} = 0.01$$

Mass of Br = 0.01 × 80 = 0.80 g

$$\%\text{ of Br} = \frac{0.80 \times 100}{1.60} = 50\%$$

Alternate Method :

$$\%\text{ of Br} = \frac{\text{Wt. of AgBr}}{\text{Wt. of O.C.}} \times \frac{\text{Molar mass of Br}}{\text{AgBr}} \times 100$$

$$= \frac{1.88}{1.6} \times \frac{80}{188} \times 100 = 50\%$$

179. (a) 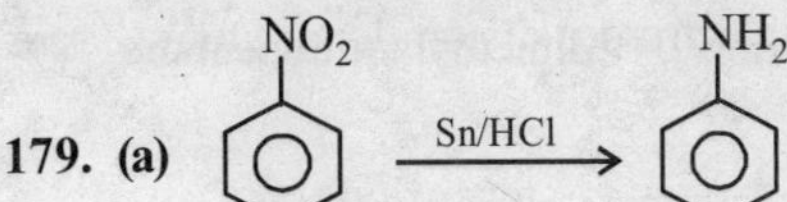

(Product will give positive Kjeldhal test due to presence of $-NH_2$ group)

(b) $C_6H_5CN \xrightarrow{LiAlH_4} C_6H_5CH_2-NH_2$

(N-present in product so it will show Kjeldhal test)

(c) $C_6H_5CH_2CN \xrightarrow[\text{(ii) } H_2O]{\text{(i) } SnCl_2 + HCl} C_6H_5CH_2CHO$ (N absent)

(d) $C_6H_5NH_2 \xrightarrow[HCl]{NaNO_2} C_6H_5N_2^+Cl^-$

($-N_2^+$ does not show Kjeldhal test)

Kjeldahl's method is not applicable to (i) compounds which do not have N-atom (ii) compounds containing nitrogen in the ring (iii) compounds containing nitrogen directly linked to oxygen atom (*e.g.*, NO_2) (iv) compounds having $-N=N-$ gp.

180. (b) Mole of bromine $= \frac{0.08}{80} = 10^{-3}$ mole

Molar mass of compound is given by the following equation,

$$\frac{0.172}{M} = 10^{-3}$$

$$\Rightarrow M = \frac{0.172}{10^{-3}} = 172 \text{ g}$$

$\because$ Molar mass of C_6H_6NBr
$= (6 \times 12) + (1 \times 6) + (1 \times 14) + (1 \times 80) = 172$ g.
172 g corresponds to the molar mass of (b).
Thus option (b) is the correct structure of compound.

181. (c) Kjeldahl's method can not be used for nitrogen determination of compounds having nitro group or azo group or nitrogen present in rings as the nitrogen of these compounds can not be converted to $(NH_4)_2SO_4$ under the condition of this method.

182. (c) Compounds which are able to form strong H-bond with the stationary phase (silica gel) will come last with the mobile phase.
H-bonding order :
Aniline > Benzanilide > Acetophenone
B > A > C
$\therefore$ Sequence of obtained compounds is (C, A, B)

183. (d) In column chromatograph; a solid adsorbent is packed on a column and a solution containing number of solute particles is allowed to flow down the column. The solute molecules get adsorbed on the surface of adsorbent and move through column at different rates based on differential adsorption of the substances on the solid phase.

184. (c) In chromatography, R_f represents retardation factor.

$$R_f = \frac{\text{Distance moved by the substance from baseline}}{\text{Distance moved by the solvent from baseline}}$$

$\therefore$ Higher R_f value means lower adsorption.

185. (b) Enolic form of acetylacetone (b) is quite stable due to H-bonding which leads to stable 6-membered ring.

186. (b) (A) Water and sugar do not react chemically, and sugar dissolves completely in water on heating. On cooling sugar solution, sugar recrystallises back **(Q)**.
(B) Aniline is insoluble in water, but it is steam volatile. Hence, it can be separated from water by steam distillation **(R)**.
(C) Toluene is insoluble as well as non-volatile in steam. Hence, it can be separated by differential extraction method **(S)**.

187. (c) $[C_xH_yN_z] \xrightarrow[\text{method}]{\text{Dumas}} 6CO_2 + 4H_2O + N_2$

Mol of $CO_2 = 6$, so mol of C is $= 6$
Mol of $H_2O = 4$, so mol of H is $= 8$
Mol of $N_2 = 1$, so mol of N is $= 2$
$\therefore$ Formula is $C_6H_8N_2$.

188. (a) Due to higher density of dichloromethane than water, DCM would be in the lower layer and water will form the upper layer in the separating funnel.

189. (b) Benzaldehyde is an adsorbate, alumina is an adsorbent (stationary phase) and acetonitrile (a solvent) is in mobile phase.

190. (c)
(A) Charcoal treatment removes coloured impurity through adsorption.
(B) Steam distillation separates the mixture of *o*-nitrophenol and *p*-nitrophenol. The *o*-nitrophenol is steam volatile (due to intramolecular hydrogen bonding), and the *para* isomer is not volatile.
(C) Fractional distillation separates crude naphtha. Naphtha is a flammable liquid hydrocarbon mixture.
(D) Distillation under reduced pressure separates mixture of glycerol and sugar. Vacuum distillation lowers the boiling point and prevents decomposition.

191. (a) Thin layer chromoatography (TLC) is type of adsorption chromatography, which is used for the separation of substance of a mixture using a thin layer of an adsorbent coated on glass plate.
A thin layer of an adsorbent (silica gel) or (alumina) is spread over a glass plate of suitable size. Hence Assertion (A) is correct and Reason (R) is correct explanation of (A)

192. (d)
(A) Chloroform + Aniline $\rightarrow$ (III) Distillation
(B) Benzoic acid + Naphthalene $\rightarrow$ (IV) Crystallisation
(C) Water + Aniline $\rightarrow$ (I) Steam distillation
(D) Naphthalene + Sodium chloride $\rightarrow$ (II) Sublimation

193. (d) Preparative thin layer chromatography is the best method of separation on a macro scale (0.05 – 1g) by using a thick layer of adsorbent (0.5 – 5) mm.

194. (1) Possible structures of acyclic neutral compound is

$$\underset{sp^3}{CH_3}-\underset{sp^2}{CH}=\underset{sp^2}{CH}-\underset{sp}{C}\equiv N \text{ or } \underset{sp^3}{CH_3}-\overset{CN}{\overset{|}{CH}}=CH_2$$

is one.

$\therefore$ No of sp^3 hybridized carbon in either of the compounds.

195. (2) $R_f = \dfrac{\text{distance travelled by the component}}{\text{distance travelled by the solvent}}$

$$\frac{R_{f_A}}{R_{f_B}} = \frac{\frac{2.08}{3.25}}{\frac{1.05}{3.25}} = \frac{2.08}{1.05} = 2$$

196. (b) The R_f (retardation factor) value is the ratio of the Solute's distance travelled to the Solvent's distance travelled.

Polar molecules will attract more toward adsorbent and cover smaller distance from the baseline.

A has maximum R_f so it will less adsorbed on the adsorbent surface and travel faster and will be eluded first.

C has minimum R_f value. So it is more strongly adsobred and less soluble in mobile phase.

197. (d) d is most stable due to maximum extended conjugation.

(a) 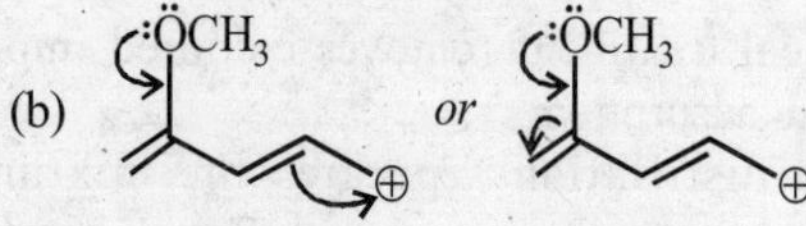

Positive charge not dispersed

(b) 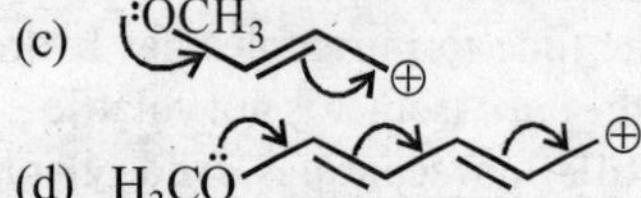

(c) [structure]

(d) [structure]

Conjugation is more extended than in (b) and (c)

198. (a) Planar molecules having $(4n+2)\pi e^-$ are aromatic in nature.

(a) is aromatic ion.

(b) is antiaromatic ion.

(c) is antiaromatic ion.

(d) is non-aromatic.

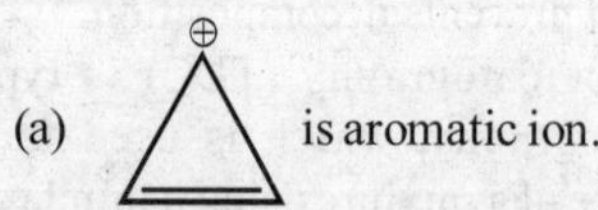

Hence (a) being aromatic, is most stable.

199. (d) Negatively charged ions of same atom are more nucleophilic than their neutral counterpart.

Nucleophilicity $\propto$ electro density on donor atom $\propto$ size of donor atom (in gas)

$$\propto \frac{1}{\text{electronegativity of atom}} \text{ (for period)}$$

200. (N) Cabocation (B) is most stable as it is stabilised by resonance. Carbocation (A) is stabilised by hyperconjugation due to 4α hydrogen. Carbocation (C) is least stable due –I effector oxygen atom.

So order will be B > A > C

201. (a)

$$C_6H_5\overset{\oplus}{C}H_2 > CH_3-\overset{\oplus}{C}H_2 > \overset{\oplus}{C}H_2=CH > CH\equiv\overset{\oplus}{C}$$

(A) Stable due to resonance; (C) +ve charge on sp^3–C; (B) +ve charge on sp^2–C; (D) +ve charge on sp–C

202. (d) Vacuum distillation or reduced pressure distillation method is used to purify compounds that can decompose before reaching their boiling point at atmospheric pressure.

203. (d)

(a) $CH_3-CH_2-\ddot{O}-CH=CH_2 \rightleftharpoons CH_3-CH_2-\overset{\oplus}{O}=CH-\overset{\ominus}{C}H_2$

(b) $C_6H_5CH_2-OH \longleftrightarrow C_6H_5CH_2OH$

(c) $CH_3-CH_2-CH_2-\overset{O}{\overset{||}{C}}-\ddot{N}H_2 \rightleftharpoons CH_3-CH_2-CH_2-\overset{\ominus O}{\overset{|}{C}}=\overset{\oplus}{N}H_2$

(d) $CH_3-CH_2-CH=CHCH_2-NH_2$

There is no conjugation of double bond with lone pair of electrons.

204. (c) (A) and (B) are in resonance.

(A) [resonance structures of benzyl cation $C_6H_5\overset{\oplus}{C}H_2$]

(B) [resonance structures of allyl cation]

The positive charge in (C) and (D) is not in conjugation with π electrons.

205. (b) H_3C [a (sp^3)], $C=C-O-C_6H_5$ with H, H; b (sp^2), c (sp^2)

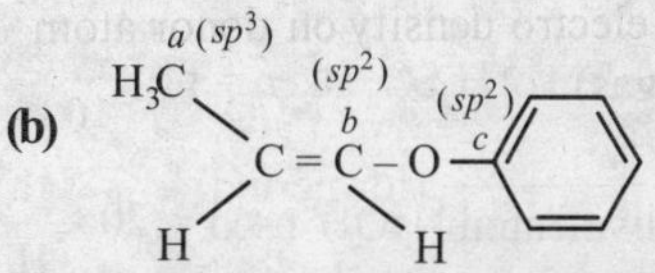

206. (d) In chromatography technique, the purification of compound is independent of physical state of the pure compound.

207. (4) Retardation factor (R_f)

$$= \frac{\text{Distance moved by the substance from base line}}{\text{Distance moved by the solvent from base line}}$$

$$= \frac{2}{5} = 4 \times 10^{-1}$$

208. (d) Acidic strength $\propto$ –I, –M effect. Due to strong – I, and –M effect of three – $COOCH_3$ groups, it has most acidic hydrogen.

209. (d) Glycerol can be separated from spent-lye (alkaline glycerol containing liquid) in soap industry by using reduced pressure distillation technique because glycerol decomposes at its boiling point.

210. (c) Since acidic order of their conjugate acid to basicity is opposite to the acidic order of their conjugate acids.

$H-C\equiv N > HC\equiv CH > H_2C=CH-CH_3 > CH_4 > CH_3-C(CH_3)_2-H$

Increasing basicity order of the given species will be

$\overset{\ominus}{C}\equiv N$ (v) $< HC\equiv\overset{\ominus}{C}$ (ii) $< H_2C=CH-\overset{\ominus}{C}H_2$ (iii) $< \overset{\ominus}{C}H_3$ (iv) $< H_3C-\overset{\ominus}{C}(CH_3)_2$ (i)

211. (a) Liquid having lower boiling point comes out first in fractional distillation. Simple distillation can't be used as boiling point difference is very small.
3-Methylpentane will show greater boiling point (63°C) comparative to isohexane due to symmetrical structure. Therefore, isohexane will distill out first.

212. (b) Electron withdrawing group like (NO_2) increases stability of alkoxide ion by dispersal of negative charge.

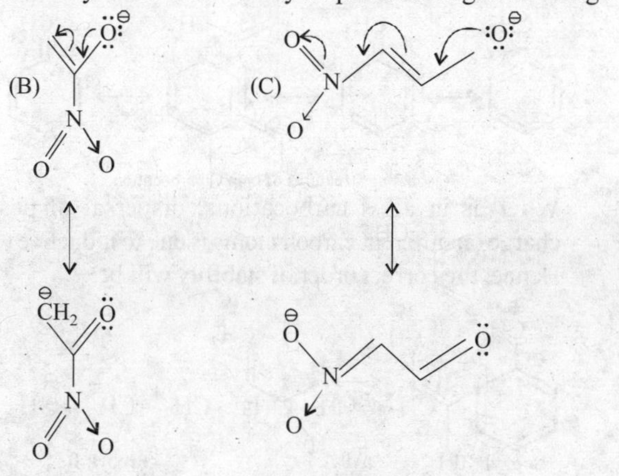

(B) conjugation and –I effect of $-NO_2$

(C) extended conjugation and –M, –I effect of $-NO_2$

Hence, (C) is more stable than (B) which is more stable than (A), (A) is least stable because negative charge is dispersed only due to – I effect.

213. (d) In all four nucleophiles, oxygen is the nucleophilic atom. Higher the basicity of a base, stronger will be its nucleophilic character. Further, a stronger base has a weaker conjugate acid. The acidic order of the conjugate acid of the given bases is:

$$H_3O^+ > CH_3SO_3H > CH_3COOH > H_2O$$

Thus, the increasing order of the basic nature or nucleophilicity of the given species is:

$$\underset{\text{(ii)}}{H_2O} < \underset{\text{(iii)}}{CH_3SO_3^-} < \underset{\text{(i)}}{CH_3COO^-} < \underset{\text{(iv)}}{OH^-}$$

214. (b) Kjeldahl's method is not applicable for compounds containing nitrogen in nitro, azo groups and nitrogen in ring, as N of these compounds does not change to ammonium sulphate under these conditions.

215. (c)

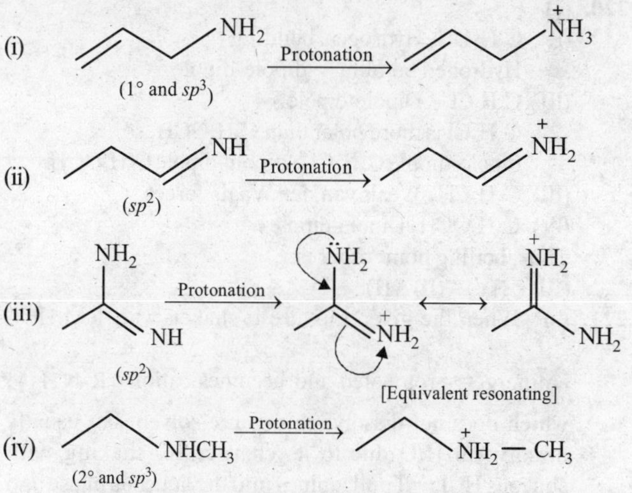

(i) NH_2 (1° and sp^3) $\xrightarrow{\text{Protonation}}$ $\overset{+}{N}H_3$

(ii) NH (sp^2) $\xrightarrow{\text{Protonation}}$ $\overset{+}{N}H_2$

(iii) NH_2, NH (sp^2) $\xrightarrow{\text{Protonation}}$ $\ddot{N}H_2$, $\overset{+}{N}H_2$ $\longleftrightarrow$ $\overset{+}{N}H_2$, NH_2 [Equivalent resonating]

(iv) $NHCH_3$ (2° and sp^3) $\xrightarrow{\text{Protonation}}$ $\overset{+}{N}H_2-CH_3$

The cation in (iii) is very stable due to equivalent resonating structures, making the parent (iii) most basic. In (ii), N is sp^2 hybridised makeup it electronegative which decreases its electron decreasing character. Consequently (i) is least basic. Among (i) and (iv), (i) being 1° is more basic than the 2° amine (iv).
Hence, correct order of basicity will be : (ii) < (i) < (iv) < (iii).

216. (c) Among the substituents attached to the given compounds, fluorine has maximum electronegativity, so it will pull electron pair towards itself. In option (b), the two F groups are attached opposite to each other, thus net dipole moment will cancel out and reduce its polarity. In option (d), the F groups are attached in slightly opposite direction. Thus, this also decreases its polarity. But in option (c), the compound has the two F groups along same direction. Thus, net dipole moment will increase in this direction and therefore, it will exhibit maximum polarity.

217. (b) The lone pair of electrons on O_2 is involved in resonance with C = C. Hence O_2 will not be protonated. The lone pair of electrons on O_5 is not involved in resonance with C = C. Hence, O_5 will be protonated. Chloride ion will then attack least substituted C atom (C_6)

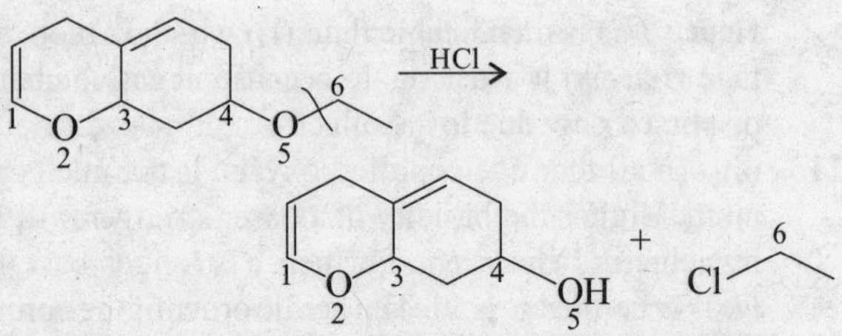

HCl →

$+ \ Cl$ (ethyl chloride) + OH product

218. (b) Since, adsorption of I > II, I is firmly attached to column (stationary phase). Hence, it moves slowly and will cover little distance. II is loosely attached to column (stationary phase). Hence, it moves faster and will cover large distance.

219. (b) Partition chromatography is the process of separation whereby the components of the mixture get distributed into two phases that may be liquid-liquid, liquid-gas but not solid-gas. Thus, it is not possible that the stationory phase is a finally devided solid adsorbent in partition chromatography.

220. (a)

(I) C_2H_5OH Hydrogen bonding
Hydrogen bonding > dipole-dipole

(II) C_2H_5Cl Dipole-dipole
C_2H_5Cl is more polar than $C_2H_5OCH_3$ so,
dipole-dipole C_2H_5Cl > dipole-dipole $C_2H_5OCH_3$

(III) $C_2H_5CH_3$ Weak van der Waals forces

(IV) $C_2H_5OCH_3$ Dipole-dipole

Thus, boiling point order is:
(III) < (IV) < (II) < (I).

221. (b) When the given mixture is shaken with 1 M HCl, amine gets protonated and becomes cation $(R_2\overset{\oplus}{N}H_2)$, which does not dissolve in organic solvent but usually dissolve in H_2O due to its charge. So, shaking with aqueous HCl will pull amines into the aqueous phase and leave all other compounds in organic layer.

222. (d) 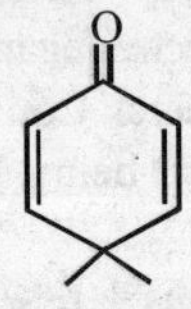 (cyclohexadienone structure) is nonaromatic and hence, least reasonance stabilized, whereas other three are aromatic.

223. (b) Since S_N1 reactions involve the formation of carbocation as intermediate in the rate determining step, more the stability of carbocation higher will be the reactivity of alkyl halides towards S_N1 route.
Since stability of carbocations follows order.

$$CH_3-CH_2-\overset{\oplus}{C}H_2 < CH_3-\overset{\oplus}{C}H-CH_2-CH_3$$
1° carbocation   2° carbocation

$$< p-H_3CO-C_6H_4-\overset{\oplus}{C}H_2$$

Max. stable due to +M effect of $-OCH_3$ group.

224. (b) Percentage of nitrogen $= \dfrac{1.4 \times NV}{W}$

$\because$ NV = milliequivalent of H_2SO_4 reacted with NH_3

Number of milliequivalent of H_2SO_4 taken $= 60 \times \frac{1}{10} \times 2 = 12$

Number of milliequivalent of NaOH' used $= 20 \times \frac{1}{10} = 2$

So, the no. of milliequivalent of H_2SO_4 reacted with NH_3 $= 12 - 2 = 10$

Percentage of nitrogen $= \dfrac{1.4 \times 10}{1.4} = 10\%$

225. (d) H–O–C_6H_4–O–H ; H–S–C_6H_4–S–H

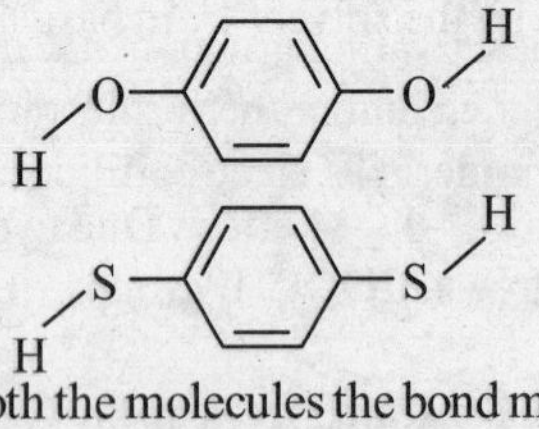

In both the molecules the bond moments are not cancelling with each other and hence, the molecules have a resultant dipole.

226. (d) $Ph_3\dot{C}$ is more stable than $(CH_3)_3\dot{C}$ because resonance stabilisation effect in $Ph_3\dot{C}$ is more pronounced as compared to hyperconjugation stabilisation effect in $(CH_3)_3\dot{C}$. Overall stability order among free radical is : Triphenylmethyl > benzyl > allyl > tertiary alkyl > secondary > primary > methyl > vinyl

227. (c) Carbocations are planar, hence can be attacked on either side to form racemic mixture.

$$Cl-\underset{\underset{(-)}{Ph}}{CH}-CH_3 \xrightarrow[\text{Toluene}]{SbCl_5} Ph-\overset{\oplus}{C}H-CH_3 + SbCl_6^- \longrightarrow$$
(carbocation)

$$Ph-\underset{Cl}{CH}-CH_3 + SbCl_5$$
$(d+\ell)$ mixture

228. (d) Higher stability of allyl and benzyl carbocations is due to dispersal of positive charge by resonance.

$$CH_2=CH-\overset{\oplus}{C}H_2 \longleftrightarrow \overset{\oplus}{C}H_2-CH=CH_2$$
Resonating structures of allyl carbocation

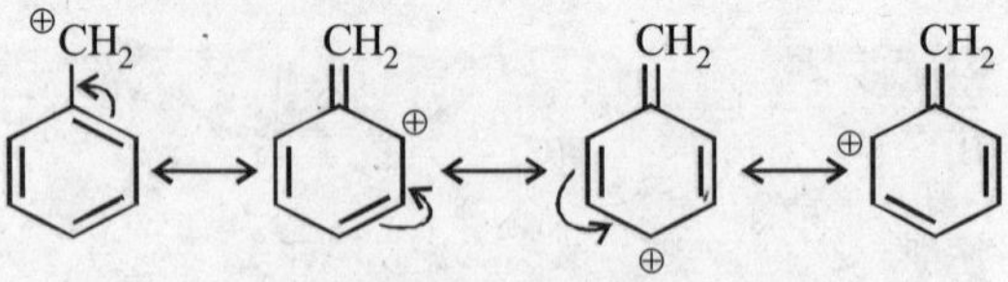

Resonating structures of benzyl carbocation

Whereas in alkyl carbocations, dispersal of positive charge on different carbon atoms is due to inductive effect. Hence, the correct order of stability will be:

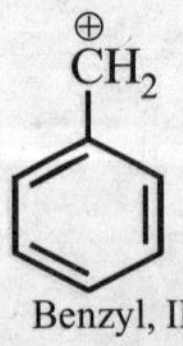

$$C_6H_5\overset{\oplus}{C}H_2 > CH_2=CH-\overset{\oplus}{C}H_2 > CH_3-CH_2-\overset{\oplus}{C}H_2$$
Benzyl, III   Allyl, I   Propyl, II

229. (a) 3° Carbocations are most stable.

230. (a) – Cl and – CH_3 groups are *o* and *p* directing. They are electron releasing due to + M and hyperconjugation effects. Further, since such groups increase electron density in the nucleus, they facilitate further electrophilic substitution and hence, known as activating group. The activating effect of these groups is in order of – CH_3 > –X but chlorine exceptionally deactive the ring due to strong – I effect. Hence, it is difficult to carry out substitution in chlorobenzene than in benzene.

Further, – NO_2 is a deactivating group, hence deactivates the benzene nucleus, i.e., hinder the further substitution. Thus, nitrobenzene undergoes electrophilic substitution with a great difficulty.

Hence, the correct order will be.

CH_3-benzene (B) > benzene (A) > Cl-benzene (C) > NO_2-benzene (D)

231. (b) Since the leaving group breaks away as a base, it is easier to displace weaker bases as compared to stronger bases. Thus less basic the substituent, the more easily it is displaced.

The basic strength of the given groups is in order:

$I^- < Br^- < Cl^-$

Thus, the order of halogen leaving groups is:

$I^- > Br^- > Cl^-$

232. (d) – OCH_3 and – CH_3 groups are activating group while – CF_3 is a deactivating group. Thus order is

OCH_3-benzene (i) > CH_3-benzene (ii) > benzene (iii) > CF_3-benzene (iv)

233. (a) Ethers, due to absence of delocalized pair of electrons do not show resonance.

234. (d) The corresponding conjugate acids are:

$$\underset{\text{Most acidic}}{CH_3COOH} > CH \equiv CH > NH_3 > \underset{\text{Least acidic}}{RH}$$

∴ the correct order of basicity is:

$$RCOO^- < CH \equiv C^- < \bar{N}H_2 < R^-$$

235. (b)

$$\overset{\ominus}{C}Cl_3 > C_6H_5\bar{C}H_2 > (CH_3)_2\bar{C}H > (CH_3)_3\bar{C}$$

–ve charge highly dispersed due to – I effect; –M effect delocalises –ve charge; +I effect of CH_3 group intensifies the –ve charge

236. (b) In option (b) the complex formed is with benzene whereas in other cases it is formed with nitrobenzene with –NO_2 group in different positions (*o*-, *m*-, *p*-).

Due to strong –I effect of –NO_2 group, it will increase the positive charge inside the ring. The complex formed with nitrobenzene in any position of –NO_2 group is less stable than the complex formed with benzene, so the most stable complex has lowest energy.

237. (a) Nitro group is electron withdrawing group, so it deactivates the ring towards electrophilic substitution.

238. (b) The order of stability of free radicals is:

$$(C_6H_5)_3\dot{C} > (C_6H_5)_2\dot{C}H > (CH_3)_3\dot{C} > (CH_3)_2\dot{C}H$$

The stabilisation of first two is due to resonance and last two is due to hyperconjugation.

239. (c)

$$\underset{e^-s\text{ are delocalised}}{CH_3COO^-} < C_6H_5O^- < OH^- < \underset{\text{Max. }e^-\text{ density on O}}{\bar{O}CH_3}$$

240. (d) Compounds bearing a quaternary nitrogen atom, when heated in presence of base like OH^- undergo E2 reaction with the formation of less substituted alkene as a major product with the loss β-hydrogen atom (**Hofmann elimeniation**, or **Hofmann rule**).

2-Me-cyclohexyl–$\overset{\oplus}{N}$(Me)(n-Bu)–CH_2–CH_2–H (H removed by OH^-) $\xrightarrow[-H_2O]{\Delta}$ 2-Me-cyclohexyl–N(Me)(n-Bu) + $CH_2=CH_2$

Note: Less sterically hindered β-hydrogen is removed.

241. (d) Free radicals are electrically neutral, unstable and very reactive on account of the presence of odd electrons.

242. (b) Weaker the acid more stable is its conjugate base and greater is the nucleophilicity of the corresponding base.

Conjugate base	Corresponding acid
$CH_3-C(=O)-O^-$ (A)	$CH_3-C(=O)-OH$
CH_3O^- (B)	CH_3OH
CN^- (C)	HCN
$H_3C-C_6H_4-SO_2-O^-$ (D)	$H_3C-C_6H_4-SO_2-OH$

Decreasing acid order of the four acids is

(D) > (A) > (C) > (B)

Hence decreasing order of nucleophilicity is

(B) > (C) > (A) > (D)

243. (d) Poor the nucleophile, better is the leaving ability of the group. Among the given species, Cl^- is the most poor neucleophile, hence it is the best leaving group among the given options.

244. (d) In carboxylic acids, presence of electron withdrawing substituent, e.g., $-NO_2$ disperses the negative charge of the anion and stabilises it and hence, increases the acidity of the parent acid.

Further, *o*-isomer will have higher acidity than corresponding *m*-and *p*-isomers due to ortho and high inductive and mesomeric effect of $-NO_2$ group. Hence, the correct order is:

COOH (benzoic acid) (1) < COOH with NO_2 at meta (–I) (4) < COOH with NO_2 at para (–M, –I) (3) < COOH with NO_2 at ortho, ortho effect (–M, –I) (2)

245. (d) Lone pair of electrons present on the nitrogen of benzyl amine is not involved in resonance.

246. (b) $HCOO^-$ exists in following resonating structures

$$H-\overset{\overset{O}{\|}}{C}-O^- \longleftrightarrow H-\overset{\overset{O^\ominus}{|}}{C}=O$$

Hence in it both the carbon oxygen bonds are found equal.

247. (b) In molecules (a), (c) and (d), the carbon atom has a multiple bond, only (b) has sp^3 hybridisation.

248. (b) $-CH_3$ group has +I effect, as number of $-CH_3$ group increases, the inductive effect increases.

Therefore, the correct order is:

$$CH_3-CH_2- < (CH_3)_2CH- < (CH_3)_3C-$$

Hydrocarbons

10

1. (b) $(CH_3)_4C$ (neopentane: central C bonded to four CH_3)

All the hydrogen in neopentane are equivalent so it will form only one monosubstituted derivative.

2. (d) $\overset{*}{C}H_3-CH(Cl)-CH_2Cl$ "x" $\xrightarrow[hv]{Cl_2}$

$CH_3-\overset{*}{C}H(Cl)-CH(Cl)-Cl + CH_3-C(Cl)_2-CH_2-Cl + Cl-CH_2-CH(Cl)-CH_2-Cl$

3. (d) $\xrightarrow[\text{Sun light}]{Br_2/hv}$ (Br attached at the tertiary carbon)

Formation of more stable tertiary free radical.

4. (b) Structure of given compound is

$\overset{1°}{C}H_3-\overset{3°}{C}H(\overset{1°}{C}H_3)-\overset{2°}{C}H_2-\overset{3°}{C}H(\overset{1°}{C}H_3)-\overset{1°}{C}H_3$

Only one 2° carbon is present in this compound.

5. (9) Different chain iosmers of C_7H_{16} are :

(i) (ii) (iii) (iv) (v) (vi) (vii) (viii) (ix)

6. (3) $CH_3COONa \xrightarrow{\text{Electrolysis}} \dot{C}H_3$

$C_2H_5COONa \xrightarrow{\text{Electrolysis}} \dot{C}_2H_5$

$2\dot{C}_2H_5 \longrightarrow \underset{(1)}{CH_3-CH_2-CH_2-CH_3}$

$2\dot{C}H_3 \longrightarrow \underset{(2)}{CH_3-CH_3}$

$\dot{C}H_3 + \dot{C}_2H_5 \longrightarrow \underset{(3)}{CH_3-CH_2-CH_3}$

7. (6) $CH_3-CH(CH_3)-CH_2-CH_3 \xrightarrow[hv]{Cl_2}$

$CH_2(Cl)-\overset{*}{C}H(CH_3)-CH_2-CH_3$ (±) $+ CH_3-C(CH_3)(Cl)-CH_2-CH_3$

$CH_3-CH(CH_3)-\overset{*}{C}H(Cl)-CH_3$ (±) $+ CH_3-CH(CH_3)-CH_2-CH_2Cl$

8. (c) Eclipsed conformation is the least stable conformation of ethane.

The electron clouds of the C–H bond come closer to each other resulting in increase in electron cloud repulsion, more energy and less stability.

9. (c) n-Alkanes on heating in presence of anhydrous $AlCl_3$ and hydrogen chloride gas isomerise to branched chain alkanes. The major product is *ter*-alkane (most stable).

10. (9) $CH_3-CH_3 + Br_2\ (\text{Excess}) \xrightarrow{hv}$

Monobromo: CH_3CH_2Br

Dibromo: $BrCH_2CH_2Br$, CH_3CHBr_2

Tribromo: $BrCH_2CHBr_2$, CH_3CBr_3

Tretrabromo: $Br_2CHCHBr_2$, $BrCH_2CBr_3$

Pentabromo: Br_2CHCBr_3

Hexabromo: Br_3CCBr_3

11. (a) $C_xH_y + \left(x+\frac{y}{4}\right)O_2 \rightarrow xCO_2 + \frac{y}{2}H_2O$

$\frac{y}{2} = 4 \Rightarrow y = 8;\ x + \frac{8}{4} = 11 \Rightarrow x = 9$

$\therefore$ Hydrocarbon will be $= C_9H_8$

12. (a) $C_xH_y + \left(x + \frac{y}{4}\right)O_2 \rightarrow xCO_2 + \frac{y}{2}H_2O$

$x + \frac{y}{4} = 9.5$ ...(i) & $\frac{y}{2} = 3$ ...(ii)

By solving equation (i) and (ii) we get.
$x = 8, y = 6$

$\therefore$ The molecular formula of A is C_8H_6.

13. (a) The staggered conformation (Newman projection: Me front top, Me back bottom, H's at other positions)

has lowest van der waal and torsional strain. Hence it must be most stable.

14. (3) $CH_3-C(CH_3)_2-CH_2-CH_2-C(CH_3)_2-CH_3 \xrightarrow{Cl_2/h\nu}$

(1) $CH_2Cl-C(CH_3)_2-CH_2-CH_2-C(CH_3)_2-CH_3$

(2) $CH_3-C(CH_3)_2-CHCl-CH_2-C(CH_3)_2-CH_3$ (±)

Total numbers of isomer = 3

15. (d) $C_6H_5-CO-O-O-CO-C_6H_5 \xrightarrow{h\nu}$ $(C_7H_5O_2)_2$

$2\,C_6H_5-CO-\dot{O}$ [X] $\longrightarrow 2\,C_6H_5^{\bullet} + 2\,CO_2$

16. (c)

o-Xylene $\xrightarrow{Ozonolysis}$ $H_3C-CO-CO-CH_3 + 2\,HC(=O)-HC=O$

o-Xylene $\xrightarrow{Ozonolysis}$ $2CH_3-CO-CO-H + CH(=O)-CH=O$

17. (12)

(Chloromethyl)cyclohexane (1); 1-chloro-1-methylcyclohexane (1); 1-chloro-2-methylcyclohexane — No. of stereoisomer $= 2^n = 2^2 = 4$ (4); 1-chloro-3-methylcyclohexane — No. of stereoisomer $= 2^n = 2^2 = 4$ (4); 1-chloro-4-methylcyclohexane (2) *cis-* and *trans*

$\Rightarrow$ Total number of isomers $= 1 + 1 + 4 + 4 + 2 = 12$

18. (8) Number of monohalogen derivatives are equal to the number of chemically different hydrogen atoms present in the compound. Possible number of structures for C_5H_{12} are-

(i) $\overset{a}{C}H_3-\overset{b}{C}H_2-\overset{c}{C}H_2-\overset{b}{C}H_2-\overset{a}{C}H_3$ *or*

3-types of hydrogen
↓
3-*Products*

(ii) $\overset{a}{C}H_3-\overset{b}{C}H(\overset{a}{C}H_3)-\overset{c}{C}H_2-\overset{d}{C}H_3$ or

4-types of hydrogen
↓
4-*Products*

(iii) $CH_3-C(\overset{a}{C}H_3)_2-\overset{a}{C}H_3$ *or*

1-type of hydrogen
↓
1-*Product*

$\therefore$ Total number of monobromo derivatives in the hydrocarbon having molecular formula $C_5H_{12} = 3 + 4 + 1 = 8$

19. (c) Conformations (*a*) and (*c*) are staggered, while (*b*) and (*d*) are eclipsed. In the anti-conformation (*c*), the dihedral angle between two methyl groups is largest (180°).

20. (d) In II and IV, both $-CH_3$ groups are at 60°. In IV, there is more repulsion between $-CH_3$ and $-H$ than that in II between $-H$ and $-H$. Potential energy $\propto \frac{1}{\text{Stability}}$

Stability order : I > III > IV > II

So, Potential energy : II > IV > III > I

21. (60) Staggered form is produced when the rearrangement of atoms or group takes place by an angle of 60°. 1, 1, 1 trichloroethane (CCl_3-CH_3) in eclipsed form on rotation by 60° gives staggered form.

Dihedral angle(ϕ) = 60°

(Newman staggered)

22. (b) A dihedral angle is the angle between two C – H bonds projected on a plane orthogonal to the C–C bond. In the given conformation, the dihedral angle H′ – C – C –H″ is the sum of $H'-C-C-H_a + H_a-C-C-H''$ i.e. $29 + 120 = 149°$

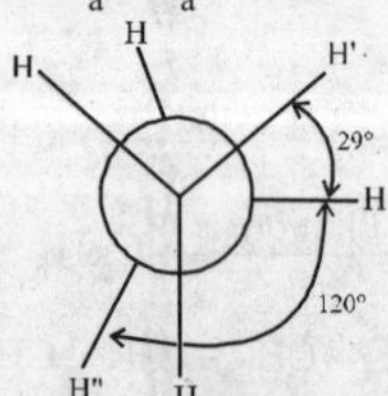

23. **(c)** Let the hydrocarbon be C_xH_y.

$$C_xH_y + \left(x+\frac{y}{4}\right)O_2 \longrightarrow xCO_2 + \frac{y}{2}H_2O$$

25g (excess) 88g 9g

2 mol $\frac{1}{2}$ mol

$x = 2y = 1$

$\because x = 2$ and $y = 1$, the hydrocarbon will be C_2H.

Hence, 2 mol carbon contains 24 g and 1 mol hydrogen contains 1g.

24. **(b)** Let the hydrocarbon be C_xH_y.

$$C_xH_y + \left(x+\frac{y}{4}\right)O_2 \longrightarrow xCO_2 + \frac{y}{2}H_2O$$

Before combustion: 10 mL 55 mL 0

After combustion: 0 $55-10\left(x+\frac{y}{4}\right)$ $10x$

Volume of CO_2, $10x = 40; x = 4$

$55-10\left(x+\frac{y}{4}\right) = 0; \quad y = 6$

$\therefore$ Hydrocarbon is C_4H_6

25. **(d)** In acyclic or open chain compounds, angle strain is absent.

26. **(d)** The order of substitution in different alkanes is :

$$3° > 2° > 1°$$

Thus, the bromination of 2-methylbutane mainly gives 2-bromo - 2 - methylbutane.

$CH_3-\overset{2°}{C}H_2-\overset{3°}{C}H(CH_3)-\overset{1°}{C}H_3 \xrightarrow{Br_2} CH_3-CH_2-C(CH_3)(Br)-CH_3$

2-Methylbutane → 2-Bromo-2-methylbutane (major)

27. **(d)** (III) Diequatorial *cis*-1,3-dimethylcyclohexane (I) *trans*-1,3-dimethylcyclohexane (one Me is equatorial and the other is axial) ≈ (II) > (IV) Diaxial *cis* -1,3-dimethyl cyclohexane.

28. **(b)** $H_3C-C(CH_3)_2-CH_3 \xrightarrow[\text{monohalogenation}]{Cl_2/hv}$ Single product

Neopentane

In neopentane (molecular mass = 72), all hydrogen atoms are equivalent.

29. **(b)** Chiral conformation will not have plane of symmetry. Since twist boat does not have plane of symmetry, it is chiral.

30. **(a)** [Newman projection: front H, H, H; back F$^{\delta-}$, H, O–H$^{\delta+}$, with H-bond between F and OH]

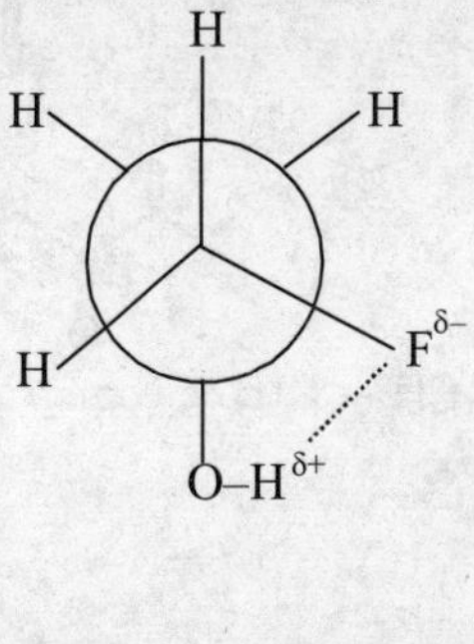

Due to hydrogen bonding between H and F, gauche conformation is most stable, hence the correct order is Eclipse, Anti, Gauche.

31. **(c)** $\overset{1°}{C}H_3-\underset{3°}{C}H(\overset{1°}{C}H_3)-\underset{3°}{C}H(\overset{1°}{C}H_3)-\overset{1°}{C}H_3$.

Since it contains only two types of H-atoms, hence it will give only two monochlorinated compounds, viz.

$Cl-CH_2-CH(CH_3)-CH(CH_3)-CH_3$ + $CH_3-C(CH_3)(Cl)-CH(CH_3)-CH_3$

1-Chloro-2, 3-dimethylbutane 2-Chloro-2,3-dimethylbutane

32. **(c)** In neopentane all H atoms are equivalent (1°).

$CH_3-C(CH_3)_2-CH_3$

33. **(b)**

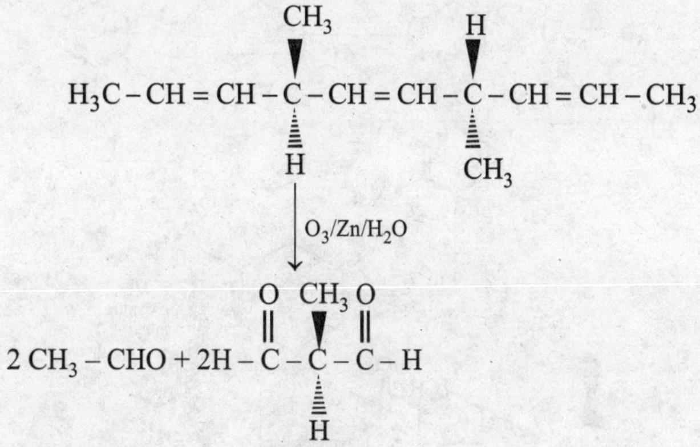

$H_3C-CH=CH-C(CH_3)(H)-CH=CH-C(H)(CH_3)-CH=CH-CH_3$

$\downarrow O_3/Zn/H_2O$

$2\,CH_3-CHO + 2H-C(=O)-C(CH_3)(H)-C(=O)-H$

Optically inactive Optically inactive

34. **(d)**

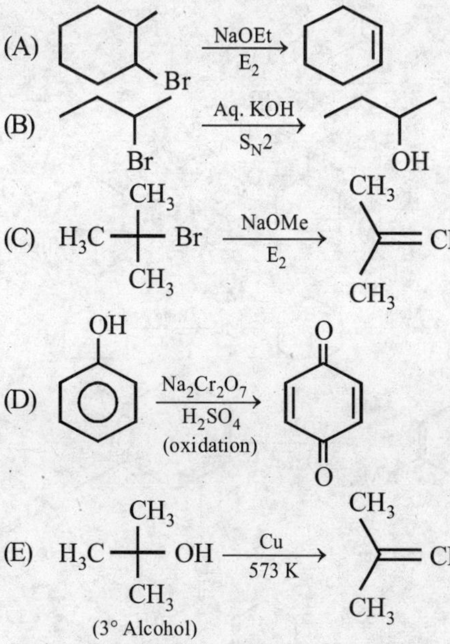

(A) 1-bromo-2-methylcyclohexane $\xrightarrow[E_2]{NaOEt}$ 1-methylcyclohexene

(B) 2-bromo-3-methylbutane (Br) $\xrightarrow[S_N2]{Aq.\ KOH}$ corresponding alcohol (OH)

(C) $H_3C-C(CH_3)_2-Br \xrightarrow[E_2]{NaOMe} (CH_3)_2C{=}CH_2$

(D) Phenol (OH) $\xrightarrow[\text{(oxidation)}]{Na_2Cr_2O_7,\ H_2SO_4}$ p-benzoquinone (O, O)

(E) $H_3C-C(CH_3)_2-OH \xrightarrow[573\ K]{Cu} (CH_3)_2C{=}CH_2$

(3° Alcohol)

Option (B) and (D) reaction are not able to form alkene as a product.

35. (c)

Statement-I :

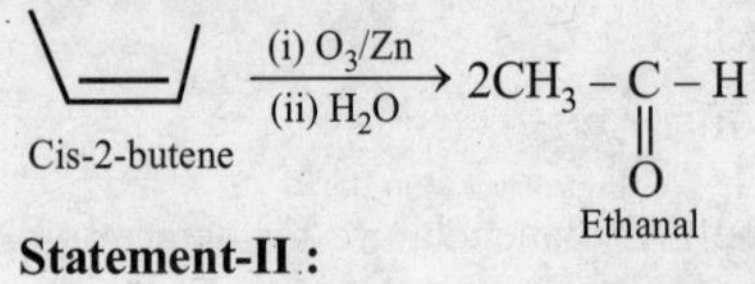

Cis-2-butene $\xrightarrow[\text{(ii) } H_2O]{\text{(i) } O_3/Zn}$ $2CH_3-\underset{\underset{O}{\|}}{C}-H$ (Ethanal)

Statement-II :

$H_3C-CH_2-\underset{CH_3}{\underset{|}{CH}}-HC=CH-\underset{CH_3}{\underset{|}{CH}}-CH_2-CH_3 \xrightarrow[\text{(ii) } H_2O]{\text{(i) } O_3/Zn} 2CH_3-CH_2-\overset{*}{\underset{CH_3}{\underset{|}{CH}}}-\underset{\underset{O}{\|}}{C}-H$

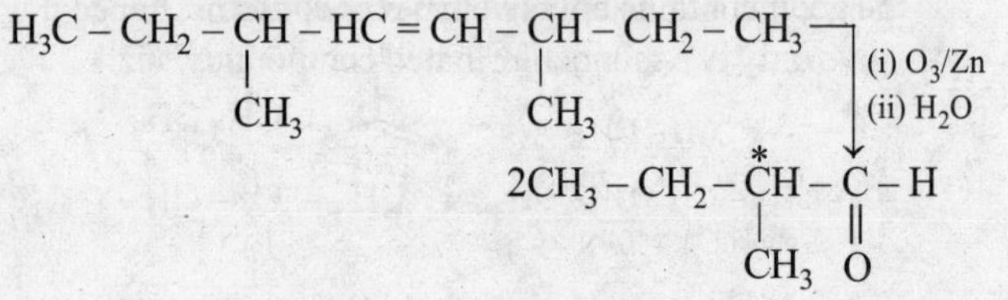

Product has chiral centre

36. (b) $\xrightarrow[Zn/H_2O]{O_3}$

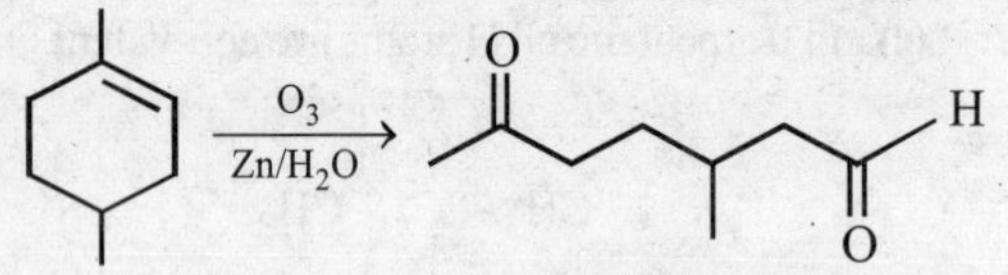

3-methyl-6-ketoheptanal

37. (b or d) $\xrightarrow[\text{From Acid}]{H^{\oplus}}$ $\xrightarrow[\text{Exp.}]{\text{Ring}}$ 1,2 Methyl shift, $-H^+$, (i) O_3 (ii) Zn, H_2O, $\overset{\ominus}{O}H$ (Alkaline med.), Intramolecular aldol rxn., $-H_2O$ Δ

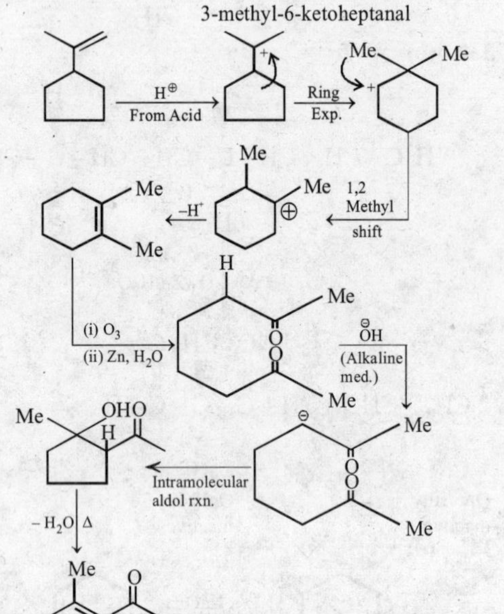

OR

$\xrightarrow[\text{From Acid}]{H^{\oplus}}$ $\xrightarrow{-H_2O}$ Hydbride shift, Ring Exp., 1,2 Methyl shift, $-H^+$, (i) O_3 (ii) Z_n/H_2O, $\overset{\ominus}{O}H/\Delta$

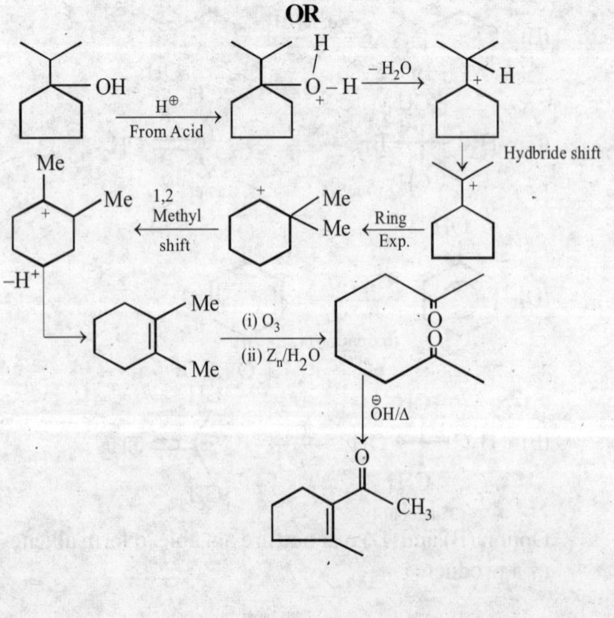

Note: In question about molecule "p" is not clear whether it is an alcohol or alkene and according to question rearrangement product is asked hence answer is either (b) or (d). Alkene also undergoes rearrangement in presence of acid and so option (b) also fulfills all conditions.

38. (b) Addition of H-Br(aq) to alkene follows electrophilic addition mechanism. In the rate determining step a carbocation intermediate is formed. Among P, Q, R & S compound Q will form most stable carbocation intermediate since it is resonance stabilized and so Q reacts fastest.

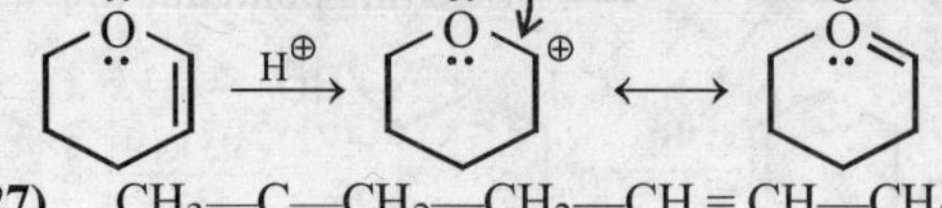

$\xrightarrow{H^{\oplus}}$ $\longleftrightarrow$

39. (27) $CH_3-\underset{\underset{\underset{CH_3\ \ CH_3}{C}}{\|}}{C}-CH_2-CH_2-CH=CH-CH_3$ (X)

$CH_3-\underset{\underset{\underset{CH_3}{|}}{CH}}{\|}{C}-CH_2-CH_2-CH=\underset{CH_3}{\underset{|}{C}}-CH_3$ (X)

$\downarrow$ $KMnO_4/H^+$ (Oxidative cleavage)

$CH_3-\underset{\underset{O}{\|}}{C}-CH_3$ + $CH_3-\underset{\underset{O}{\|}}{C}-OH$ + $CH_3-\underset{\underset{O}{\|}}{C}-CH_2CH_2-\underset{\underset{O}{\|}}{C}-OH$

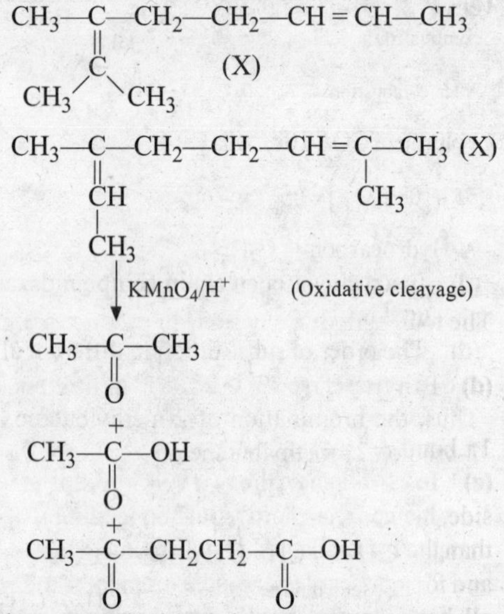

Thus, there are 27 σ-bonds in either of the 'X' molecules giving these products upon oxidative cleavage by $KMnO_4$.

40. (b) Ozonolysis product

(A) → H ... H

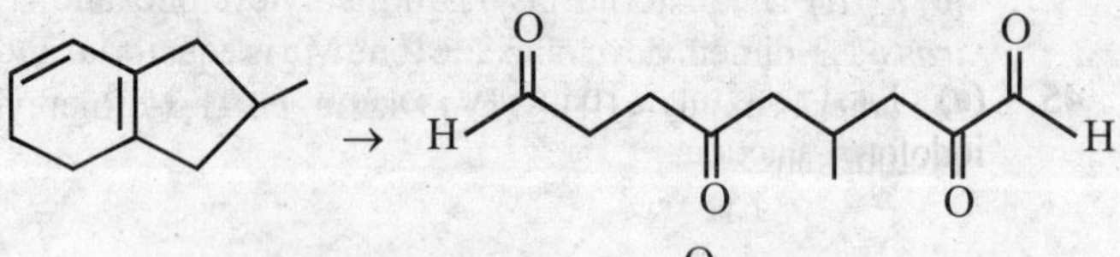

(B) → H ... 4

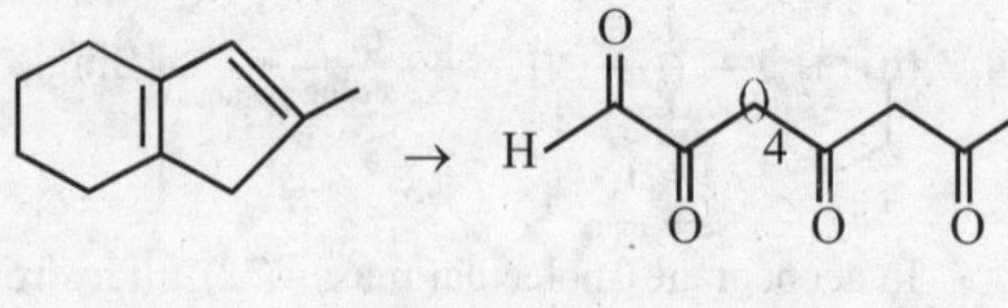

(C) → H ... 3 ... H

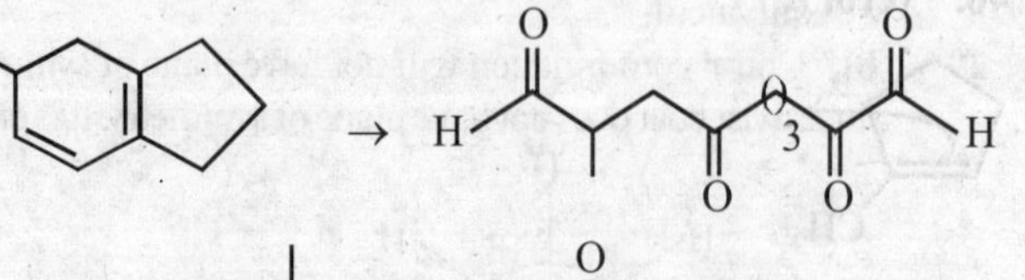

(D) → ... 4 ... H

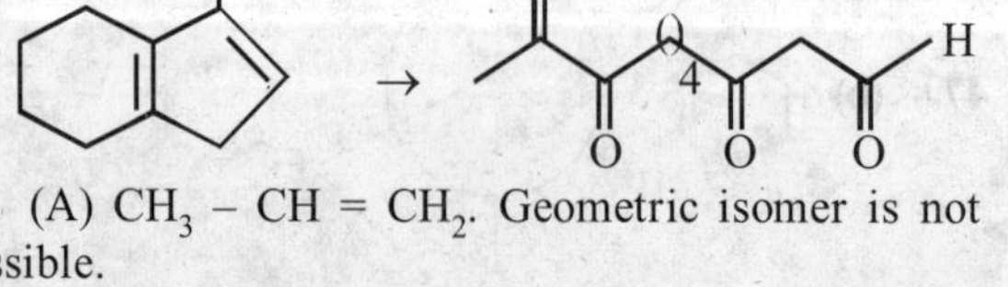

41. (a) (A) $CH_3-CH=CH_2$. Geometric isomer is not possible.

(B) Trans isomer has identical atoms/groups on the opposite side of double bond.

(C) 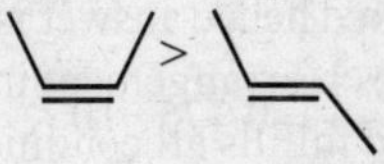 (dipole moment)

(D) $H_3C-C(CH_3)=CH-CH_3$ (does not show GI)

2 – methylbut – 2 – ene

(E) 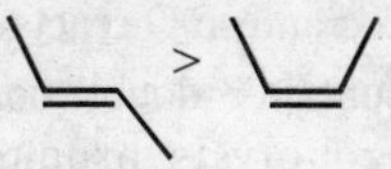 (Melting point due to symmetry)

42. **(8)** Cyclohexene $\xrightarrow{Br_2}$ 1,2-dibromocyclohexane $\xrightarrow[(3\ eq)]{alc\ KOH}$ (B)

$HC\equiv C-CH_2-O^{\ominus}Na^{+}$ → (A) 3-bromocyclohexene

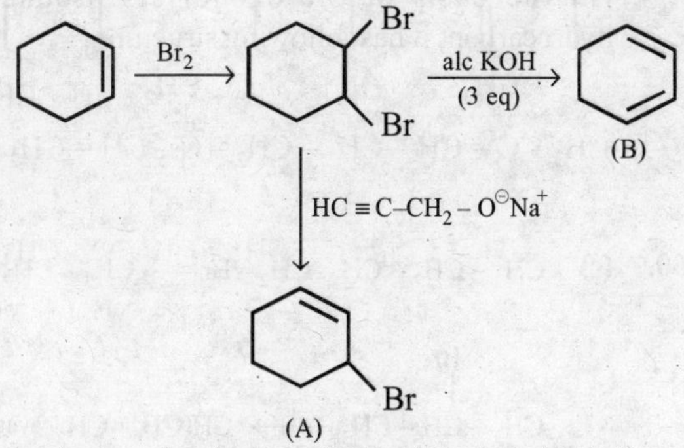

The total sum of π electrons in product A and B is 8.

43. **(d)** Ethylene is [structure] it has 5σ bonds and 1π bond.

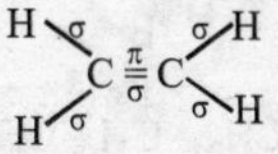

44. **(c)** In *cis*-alkenes, the two bulky groups are on the same side, hence faces more repulsion leading to its unstability than the *trans*-alkenes. Dipole moment is a vector quantity and for a molecule net dipole moment is the vector sum of all bond dipoles hence dipole moment of *Cis* form is greater than *Trans* form.

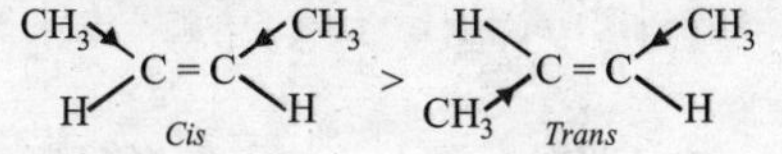

$CH_3CH=CHCH_3$ *Cis* ($\mu > 0$) > $CH_3CH=CHCH_3$ *Trans* ($\mu > 0$)

45. **(a)** First reaction is reductive ozonolysis and second is iodoform reaction.

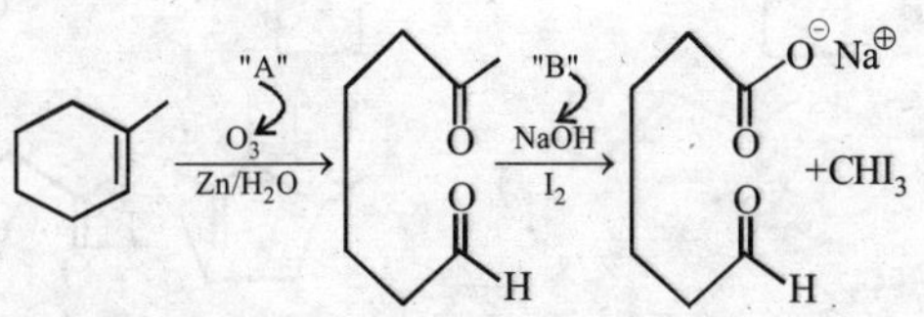

"A" O_3, Zn/H_2O; "B" NaOH, I_2; $+CHI_3$

46. **(c) or (d)**

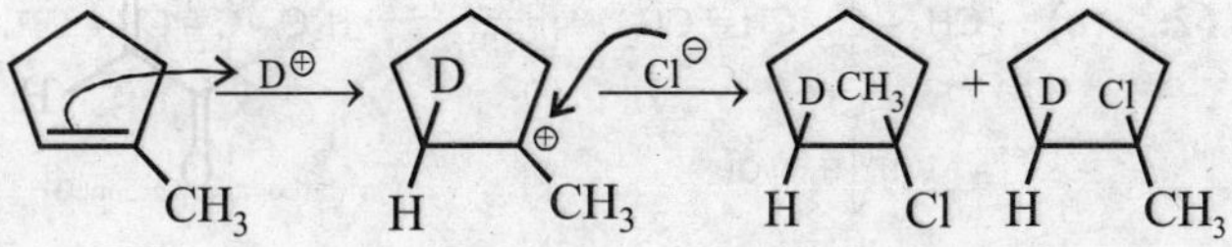

$D^{\oplus}$, $Cl^{\ominus}$

47. **(b)** $\xrightarrow[H_2O_2,\ NaOH\ (aq.)]{B_2H_6}$ (B)

(Hydro boration-oxidation, Anti-Markovnikoff)

$\xrightarrow{H^+/H_2O}$ (A)

(Hydration, follows Markovnikoff)

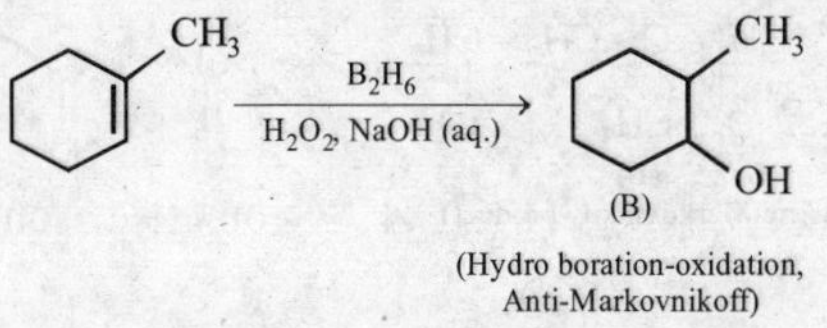

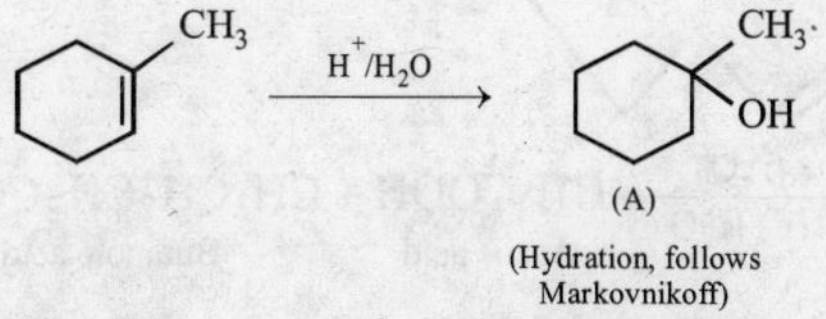

48. **(4)** We can use the E/Z nomenclature for determining the geometrical isomers of this compound.

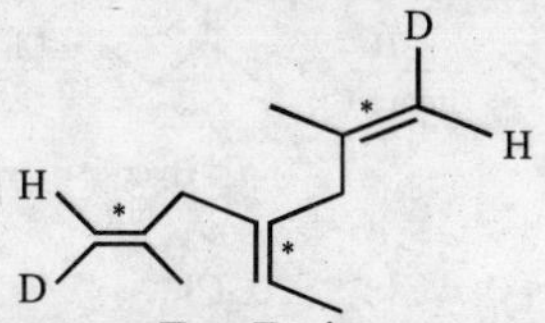

First stereocentre = E or Z – isomer.
Second stereocentre = E or Z – isomer.
Third stereocentre = We don't count it as the compound has the same two substituents and we have done it already once for one double bond.
Thus, total geometrical isomers = 4

49. **(b)** Cyclohexene $+ Cl_2$ — hv, Free radical substitution → (A); CCl_4, Electrophillic addition to double bond → (B)

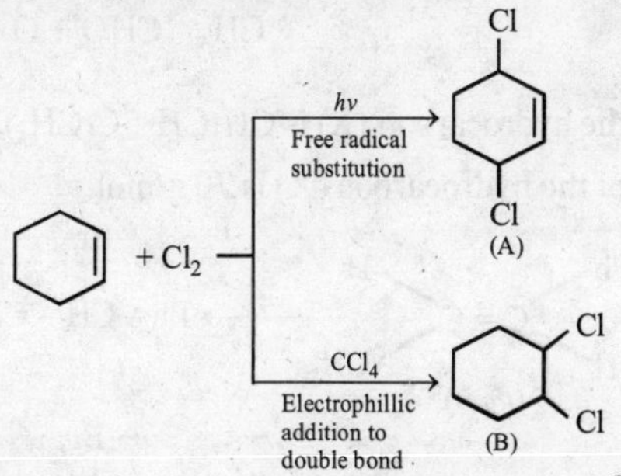

50. **(1)** $CH_3CH=CHCH_3 \xrightarrow[(ii)\ Zn + H_2O]{(i)\ O_3} 2CH_3-\overset{O}{\overset{\|}{C}}-H$

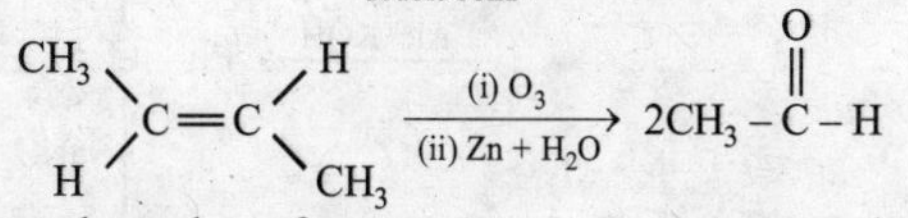

The total number of oxygen atom present per molecule of the product is 1.

51. **(d)** cis - isomer; tans - isomer

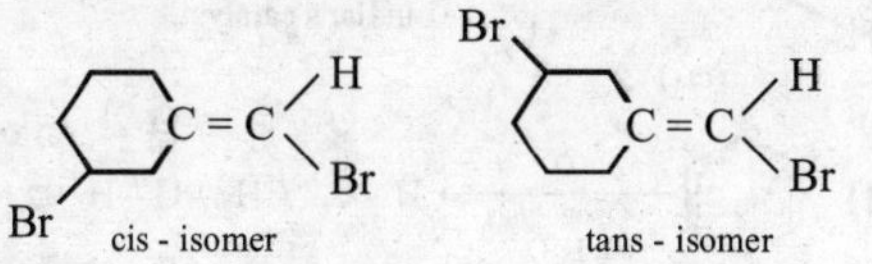

Compound in option (a) will result in the same compound when rotation along the C = C double bond is performed (formation of two such isomers, not an actual rotation along the double bond).
Same will be for compound in (d).
For (c), we cannot decide the cis- and trans- isomers as these are two H-atoms on one doubly-bonded carbon atom.

52. **(a)**

$CH_3-CH=CH_2 \xrightarrow[(ii)\ H_2O_2,\ NaOH]{(i)\ B_2H_6} CH_3-CH_2-CH_2OH$ (*anti*-Markovnikov of) hydration)

$\xrightarrow{P.C.C.} CH_3-CH_2-\overset{O}{\overset{\|}{C}}-H \xrightarrow{CH_3MgBr} CH_3-CH_2-\underset{OH}{CH}-CH_3$

53. **(a)** (structure: trans-hex-2-ene) $\xrightarrow[\text{(ii) } H_2O]{\text{(i) } O_3}$ $CH_3COOH + CH_3CH_2CH_2COOH$ (Acetic acid, Butanoic acid)

It is oxidative ozonolysis.

54. **(4)** (structure: 3-chlorobut-1-ene) $\xrightarrow[\text{(from HCl)}]{H^+}$ (secondary carbocation) $\xrightarrow{Cl^-}$ 2,3-dichlorobutane (3 Isomers)

Hydride shift → (tertiary carbocation bearing Cl) $\xrightarrow{Cl^{\ominus}}$ 2,2-dichlorobutane (1 product)

Total Possible Isomeric product = 1 + 3 = 4

55. **(70)** Hydrocarbon (X) $\xrightarrow{\text{Ozonolysis}}$

$$CH_3-CHO + O=C(CH_3)_2$$

Thus the hydrocarbon (X) is $CH_3CH=C(CH_3)_2$. Its molar mass of the hydrocarbon (X) is 70 g/mol.

56. **(b)** $Ph(H)C=C(H)CH_3$ (*trans*) $\xrightarrow{Br_2}$ $Ph-CH(Br)-CH(Br)-CH_3$ $\xrightarrow{\text{Alc. KOH}}$ $Ph-C(Br)=CH-CH_3$ $\xrightarrow{NaNH_2}$ $Ph-C\equiv C-CH_3$ $\xrightarrow[\text{Lindlar's catalyst.}]{H_2}$ $Ph(H)C=C(H)CH_3$ (*cis*)

57. **(d)** (Cyclohexa-1, 4-diene) (C_6H_8) $\xrightarrow[\text{Zn/}H_2O]{O_3 \text{ Ozonolysis}}$ $H-\overset{O}{\overset{\|}{C}}-CH_2-\overset{O}{\overset{\|}{C}}-H$ (2 mol)

58. **(a)** $H_3C-C(CH_3)=CH_2$ (1st molecule) $\xrightarrow[80°C]{H_2SO_4,\ H^{\oplus}}$ $H_3C-\overset{\oplus}{C}(CH_3)-CH_3$ + $CH_2=C(CH_3)_2$ (2nd molecule) → $(CH_3)_3C-CH_2-\overset{\oplus}{C}(CH_3)_2$ $\xrightarrow{O^-SO_3H\ (-H_2SO_4)}$ $CH_3-C(CH_3)=CH-C(CH_3)_2-CH_3$ (B)

$H_3C-C(CH_3)=CH_2 \xrightarrow[\text{cold}]{H_2SO_4} H_3C-C(CH_3)(OSO_3H)-CH_3$ (A)

$H_3C-C(CH_3)=CH_2 \xrightarrow{H^{\oplus}} H_3C-\overset{\oplus}{C}(CH_3)-CH_3 \xrightarrow{O^-SO_3H}$ (A)

59. **(3)** Moles of hydrocarbon $= \frac{17\times10^{-3}}{136} = 1.25\times10^{-4}$

Moles of H_2 gas $= \frac{8.4\times10^{-3}}{22.4} = 0.375\times10^{-3}$

$= 3.75\times10^{-4}$

Hydrogen molecule used for 1 molecule of hydrocarbon

$$= \frac{3.75\times10^{-4}}{1.25\times10^{-4}} = 3$$

Thus, hydrocarbon must contain 3 double bonds.
On the basis of the ozonolysis products of the hydrocarbon, it has following structure.

$$CH_3-C(CH_3)=CH-CH_2-CH_2-C(=CH_2)-CH=CH_2$$

60. **(c)** $CH_3-CH_2-CH(CH_3)-CH_2-Br \longrightarrow CH_3-CH_2-C(CH_3)=CH_2$ (1 alkene)

$CH_3-CH(Br)-CH_3 \longrightarrow CH_3CH=CH_2$ (1 alkene)

$CH_3-CH_2-CH_2-CH(Br)-CH_3 \longrightarrow$ $H_3C-CH_2-CH_2-CH=CH_2 + H_3C-CH_2-CH=CH-CH_3$, *cis* & *trans* (3 alkenes)

$CH_3-CH_2-C(CH_3)_2-CH(Br)-CH_3 \longrightarrow CH_3-CH_2-C(CH_3)_2-CH=CH_2$ (1 alkene)

61. **(d)** (cyclobutyl)$C(CH_3)=CH_2 \xrightarrow{H^+}$ (cyclobutyl)$\overset{\oplus}{C}(CH_3)-CH_3 \longrightarrow$ (ring expansion: 2-methylcyclopentyl cation) $\xrightarrow{Br^-}$ 1-bromo-1-methylcyclopentane

62. **(b)** $CH_3-C(CH_3)_2-CH=CH_2 \xrightarrow[NaBH_4]{Hg(OAc)_2,\ H_2O} H_3C-C(CH_3)(CH_3)-CH(OH)-CH_3$ (A) (Markovnikov product)

$CH_3-C(CH_3)_2-CH=CH_2 \xrightarrow{B_2H_6\ |\ H_2O_2/OH^-} CH_3-C(CH_3)_2-CH_2-CH_2OH$ (B) (Anti Markovnikov product)

63. (b) $Me_2C=CH-CH_3 \xrightarrow{Br_2} Me_2\overset{\oplus}{C}(-Br-)CH(CH_3)H \xrightarrow{H\ddot{O}CH_3}$

$Me_2C(Br)-C(OCH_3)(Me)H$ or $Me_2C(OCH_3)-C(Me)(Br)H$

'Br' atom attached to 3° carbon less possible due to steric hindrance

$\xrightarrow{HI \mid S_N2}$ $H_3C-C(I)(CH_3)-CH(Br)-CH_3$

Major product

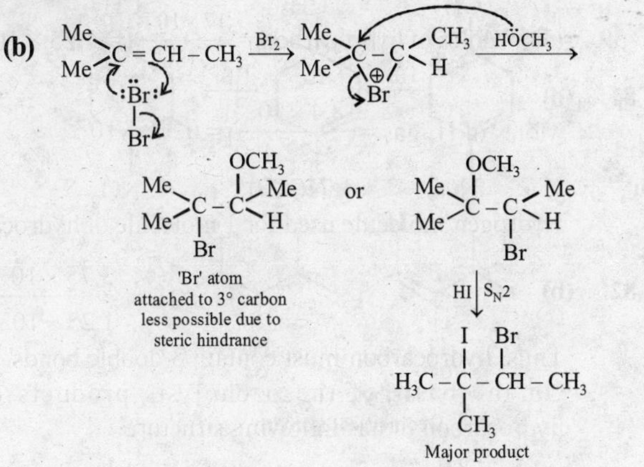

64. (d) $Me_3C-CH_2-C(CH_3)=CH_2 \xrightarrow[H^+]{KMnO_4} Me_3C-CH_2-C(CH_3)=O + H-C(=O)-OH$ (Y) (b)

Cleave

$O_3, H_2O/Zn$ [X]

$Me_3C-CH_2-C(CH_3)(O)-O-CH_2(O)$ (Cleave) $\longrightarrow Me_3C-CH_2-C(CH_3)=O + H-C(=O)-H$ (a)

$H-C(=O)-H \xrightarrow{[O]} H-C(=O)-OH$

(a) Formic acid (produced by ants)

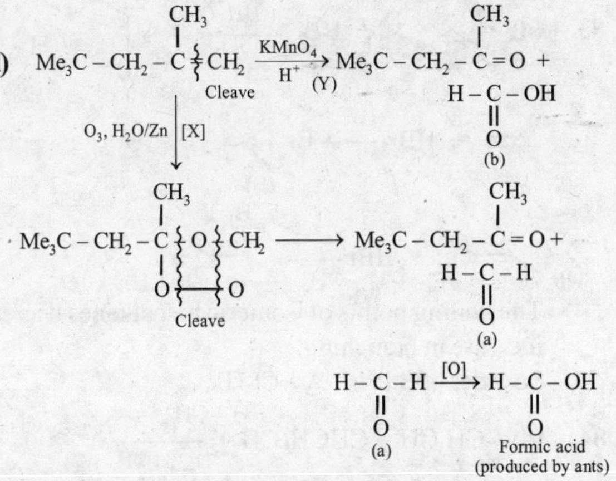

65. (4) Moles of X $= \dfrac{10 \text{ mg}}{80} = 0.125$ m mol.

Moles consumed of $H_2 = \dfrac{8.4}{22.4} = 0.375$ m mol.

$$\frac{n_{H_2}}{n_X} = \frac{0.375}{0.125} = 3$$

So, the compound X have 3 double bonds.
Ozonolysis of the compound yields only formaldehyde and dialdehyde, so the compound should be

$H_2C=CH-CH=CH-CH=CH_2$

Molecular mass $= (12 \times 6) + 1 \times 8 = 72 + 8 = 80$ amu

Ozonolysis form:

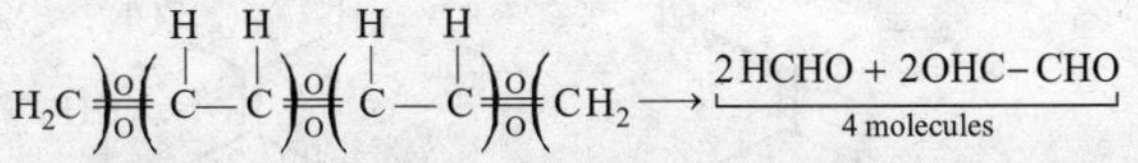

$H_2C=CH-CH=CH-CH=CH_2 \longrightarrow 2HCHO + 2OHC-CHO$ (4 molecules)

66. (d) Three isomers are possible for an oxidisable molecule having molecular formula C_4H_8.

$CH_3CH=CHCH_3$ (But-2-ene) $CH_3CH_2CH=CH_2$ (But-1-ene) $CH_3-C(CH_3)=CH_2$ (2-Methylpropene)

Oxidation with $KMnO_4$ to give a ketone along with evolution of effervescences is possibly only with 2-methylpropene

$CH_3-C(CH_3)=CH_2$ (A) $\xrightarrow{KMnO_4/H^+} CH_3-C(CH_3)=O + CO_2\uparrow$

Thus isomer A should be 2 – methylpropene.

67. (d) Cyclohex-1,3-diene (A) $\xrightarrow[2.\ Zn-H_2O]{1.\ O_3}$ Ethane-1, 2-dicarbaldehyde ($OHC-CH_2-CH_2-CHO$) + $OHC-CHO$ (Glyoxal)

1-Methylcyclopent-1-ene (B) $\xrightarrow[2.\ Zn-H_2O]{1.\ O_3}$ 5-Oxohexanal ($CH_3-CO-CH_2CH_2CH_2-CHO$)

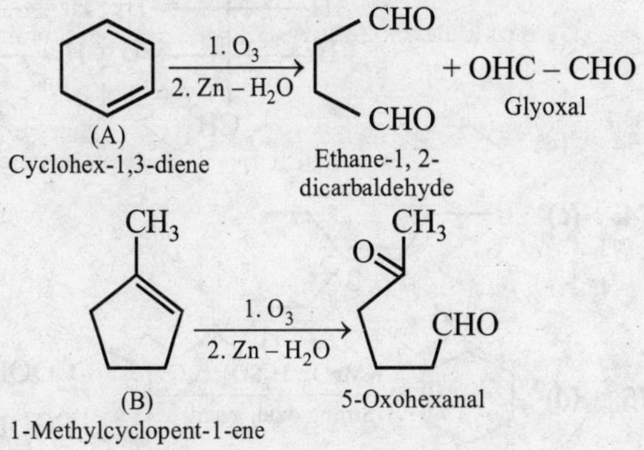

68. (b) $(CH_3)_2C=CH_2 + H^+ \longrightarrow (CH_3)_3C^+$

$(CH_3)_2C=CH_2 + (CH_3)_3C^+ \longrightarrow (CH_3)_2C^+-CH_2-C(CH_3)_3$

$(CH_3)_2C^+-CH_2-C(CH_3)_3 + H-C(CH_3)_3 \longrightarrow (CH_3)_2CH-CH_2-C(CH_3)_3 + (CH_3)_3C^+$

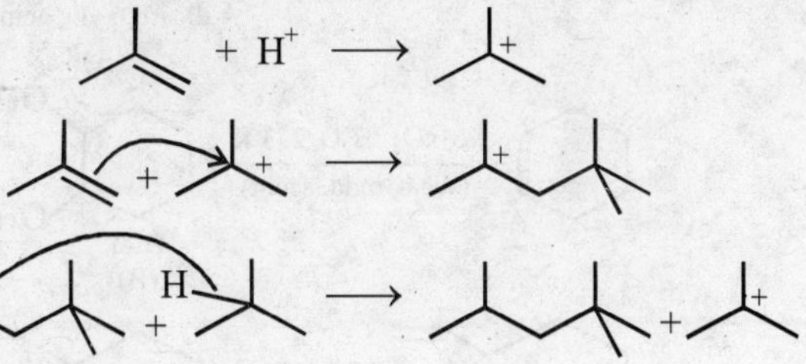

69. (2) 2, 7-Dimethyl-2, 6-octadiene $\xrightarrow{H^+}$ (carbocation at C-2) $\xrightarrow{\text{Cyclization}}$ (cyclopentane ring with carbocation at C-7) $\xrightarrow{-H^+}$ (product with C7=C6 double bond)

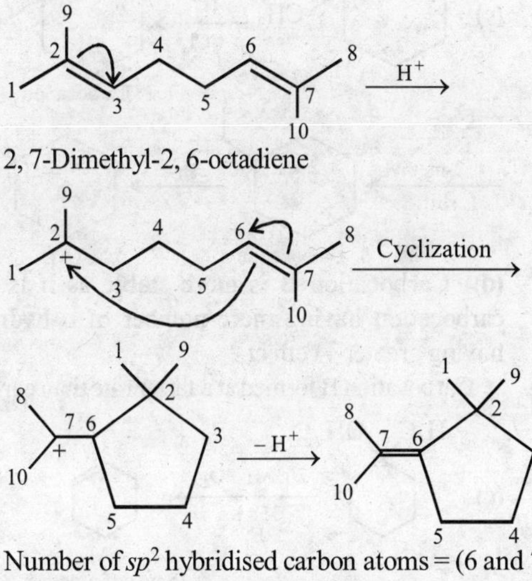

Number of sp^2 hybridised carbon atoms = (6 and 7) 2

70. (a) Both the statements are correct.

71. (a) $CH_2=C(CH_3)-CH=CH_2 \xrightarrow{HBr}$ (excess) $(CH_3)_2C=CH-CH_2Br$

1, 4 addition product (major product)

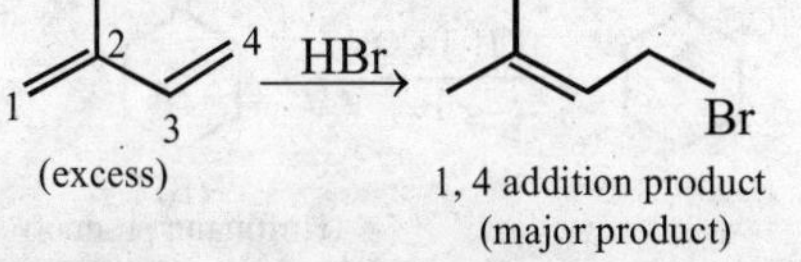

Mechanism :

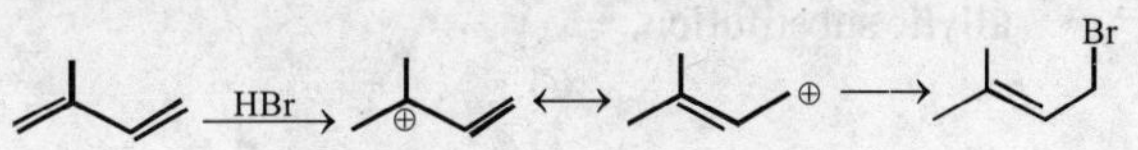

$CH_2=C(CH_3)-CH=CH_2 \xrightarrow{HBr} (CH_3)_2\overset{\oplus}{C}-CH=CH_2 \longleftrightarrow (CH_3)_2C=CH-\overset{\oplus}{C}H_2 \longrightarrow (CH_3)_2C=CH-CH_2Br$

72. (a) $H_3C-C(CH_3)=CH_2$ (2-Methylpropene) $\xrightarrow[\text{or Ozonolysis}]{KMnO_4/H^{\oplus}} (H_3C)_2C=O$ (C_3H_6O)

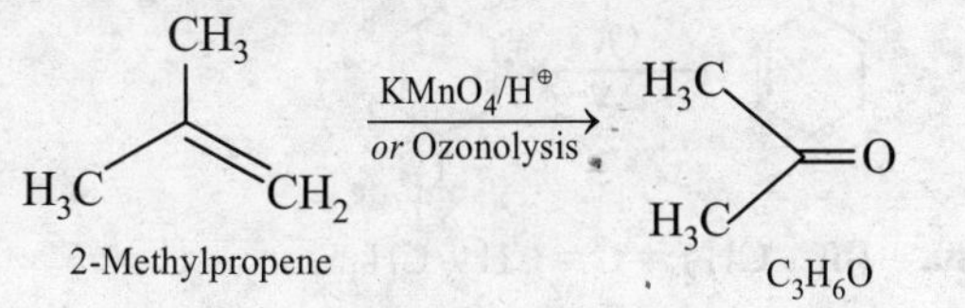

73. (2) $H_3C(H)C=C(H)CH_3$ $\xrightarrow[\text{(Anti addition)}]{Br_2/CCl_4}$

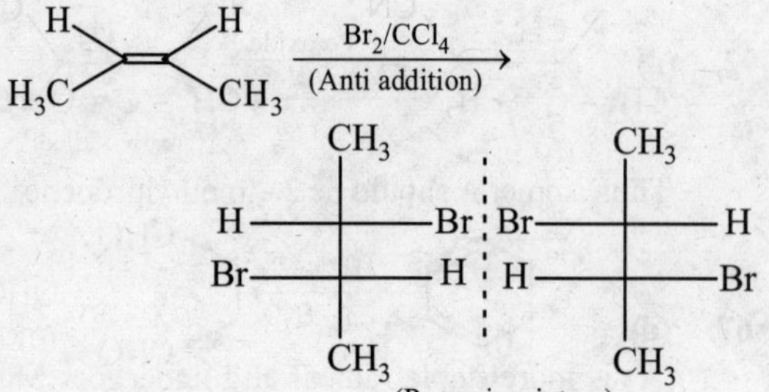

(Racemic mixture)

74. (c)

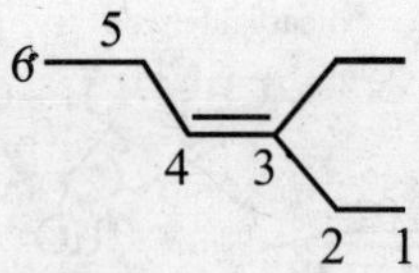

75. (d) $\xrightarrow[\text{(Strong oxid. agent)}]{KMnO_4,\ H_2SO_4,\ \Delta}$ COOH, COOH

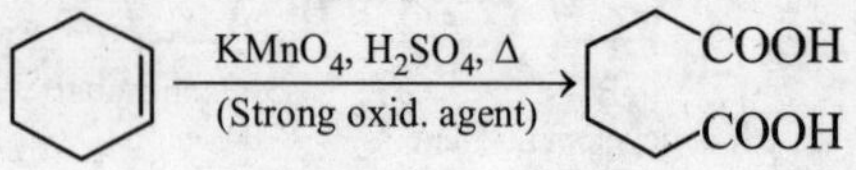

dicarboxylic acid (A)

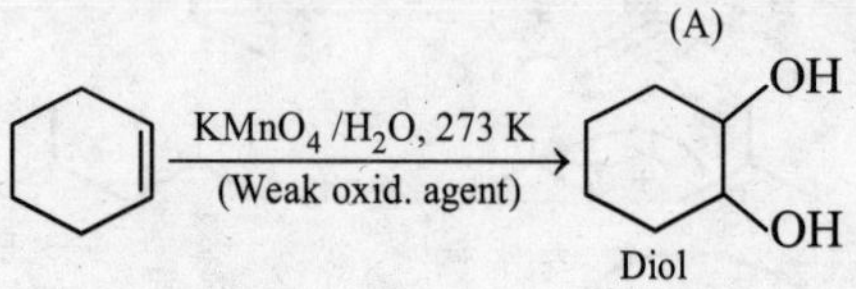

$\xrightarrow[\text{(Weak oxid. agent)}]{KMnO_4/H_2O,\ 273\ K}$ OH, OH

Diol (B)

76. (c) CH_3 $\xrightarrow[CCl_4]{H^+}$ CH_3

2° Carbocation

$\xrightarrow{\text{1, 2-methyl shift}}$ 3° Carbocation $\xrightarrow{Br^-}$ Br, CH_3, CH_3 (A)

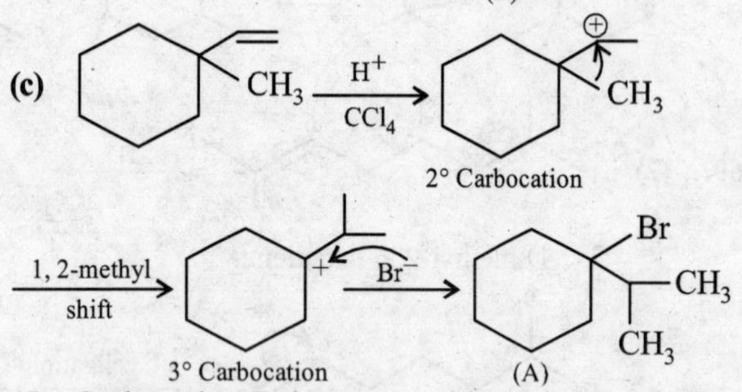

77. (d) Carbocation B is more stable as it is secondary carbocation having more number of α-hydrogens and having greater +I effect.

∴ Carbocation B formed at a faster rate than carbocation A.

78. (c) H_3C OH $\xrightarrow[358\ K,\ E_1]{20\%\ H_3PO_4}$ CH_3 (A) (Saytzeff product)

H_3C Cl $\xrightarrow[E_2]{(CH_3)_3C\overset{-}{O}\overset{+}{K}}$ CH_2 (B) (Hoffmann product)

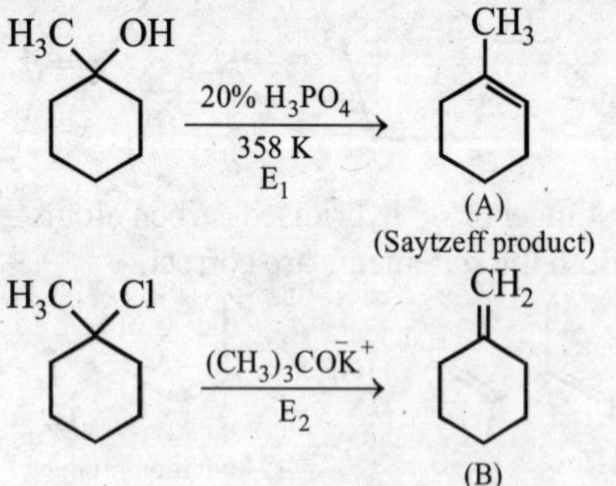

79. (b) The reagent, Cl_2; UV light is used for free radical allylic substitution.

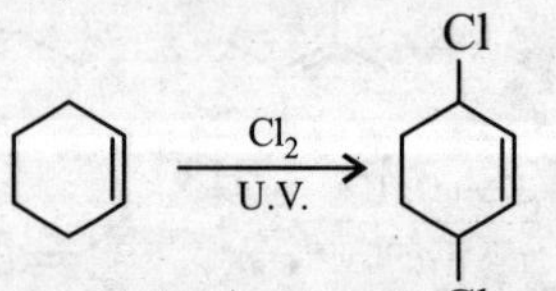

$\xrightarrow[U.V.]{Cl_2}$ Cl, Cl

80. (b) $\underset{sp^2}{CH_2} = \underset{sp}{C} = \underset{sp^2}{CH} - \underset{sp^3}{CH_3}$

81. (d) CH_3, NO_2 $\xrightarrow{HBr}$ CH_3, Br, NO_2 $\xrightarrow{HBr}$ CH_3, Br, Br, NO_2

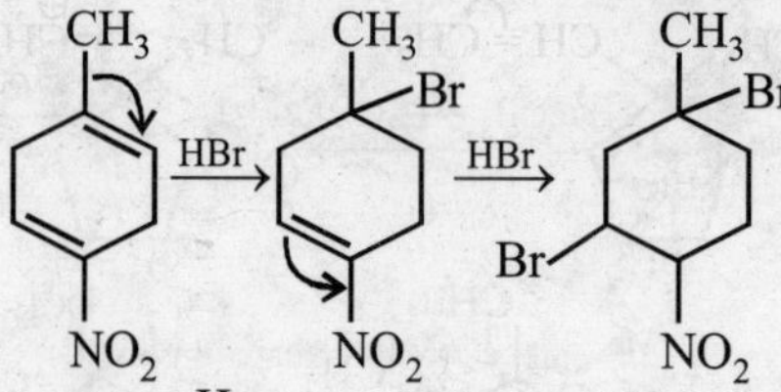

82. (b) H, H — *trans*; H, H — *cis*

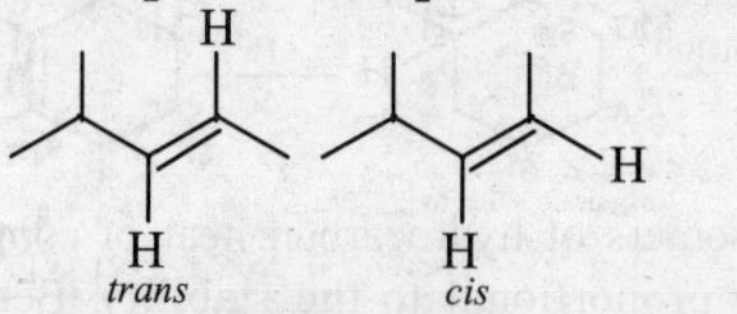

4-Methylpent-2-ene Geometrical isomers

83. (a) + HBr $\xrightarrow{(C_6H_5CO)_2}$ Br (A)

+ HBr ⟶ Br (B)

+ HBr ⟶ Br (C)

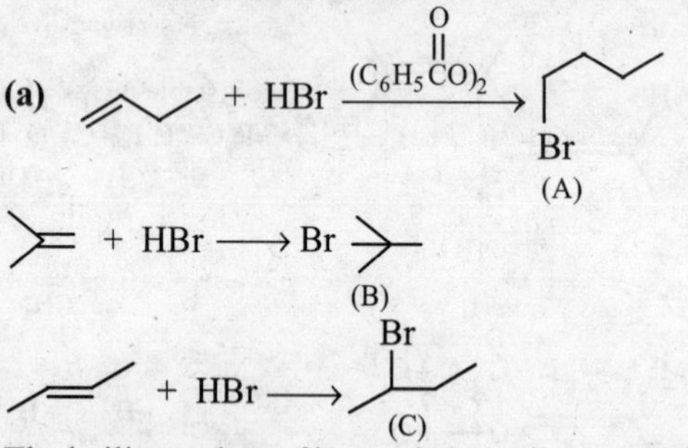

The boiling points of isomeric haloalkanes decrease with increase in branching.

So order of B.P. is : A > C > B.

84. (d) $CH_3CH = CHCH(CH_3)_2 \xrightarrow{H^+}$

$CH_3CH_2 - \overset{+}{C}H - CHCH_3(CH_3)$ $\xrightarrow{1,2\text{-}\overline{H}\ \text{shift}}$

2° carbocation

$CH_3CH_2 - CH_2 - \overset{+}{C}CH_3(CH_3)$ $\xrightarrow{Br^-}$ $CH_3CH_2 - CH_2 - C(Br)CH_3(CH_3)$

3° carbocation

85. (b, c) H, Cl, CH_3 and Cl, H, CH_3

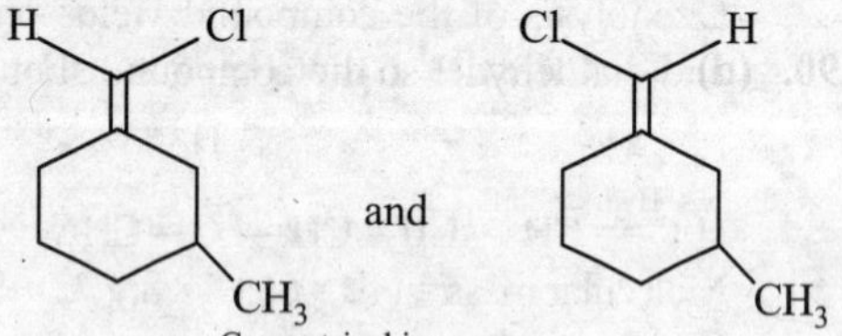

Geometrical isomers

H, Cl, H_3C, CH_3 — cis and H, Cl, H_3C, CH_3 — trans

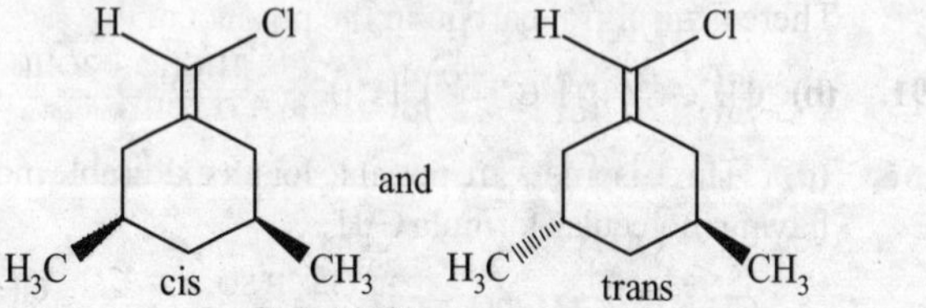

Geometrical isomers

86. (d) H, CH_3 $\xrightarrow{H_2/Ni}$ H, CH_3

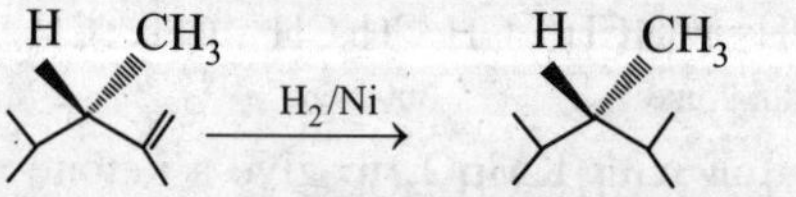

Optically inactive (No chiral centre)

87. (b) 1-methyl-1-vinylcyclopentane $\xrightarrow{H_3O^+}$ carbocation $\overset{\oplus}{C}H-CH_3$ adjacent to ring;

$\xrightarrow{\text{Ring expansion}}$ cyclohexyl cation $\xrightarrow[\Delta]{-H^+}$ 1,2-dimethylcyclohexene

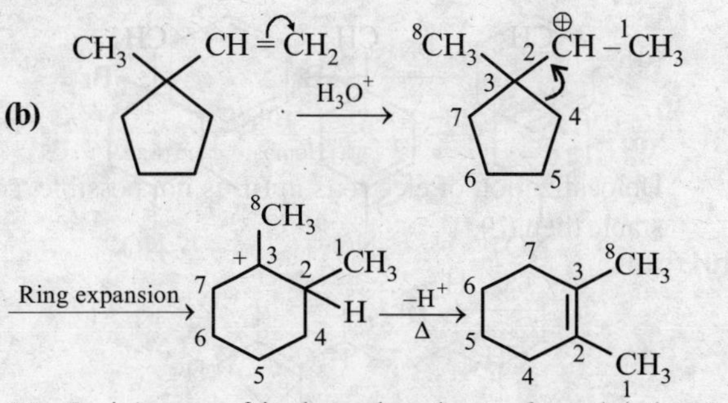

88. (a) In isomers of hydrocarbon heat of combustion is inversely proportional to the stability. Between two geometrical isomers *trans* isomer is more stable than *cis*.

(A) *trans-trans* — Most stable due to *trans-trans* arrangement

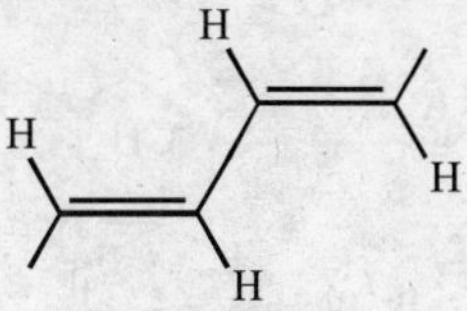

(B) *trans-cis* — Next stable due to *trans-cis* arrangement

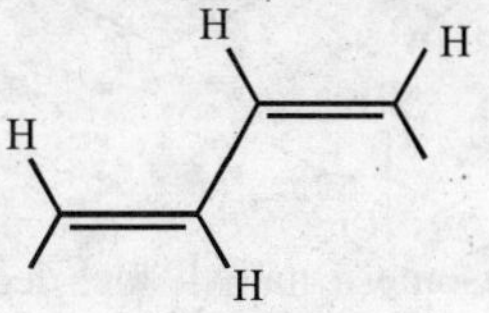

(C) *cis-cis* — Least stable due to *cis-cis* arrangement

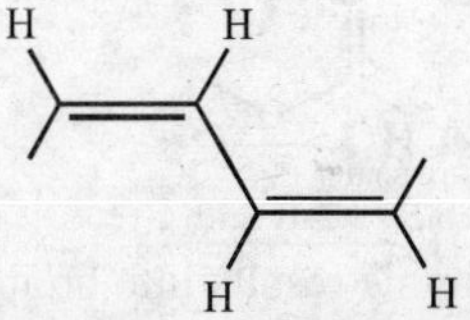

89. (b) In Cl–CH=CH–NO_2 double bond character in carbon-chlorine bond is maximum due to resonance and so the bond length is shortest.

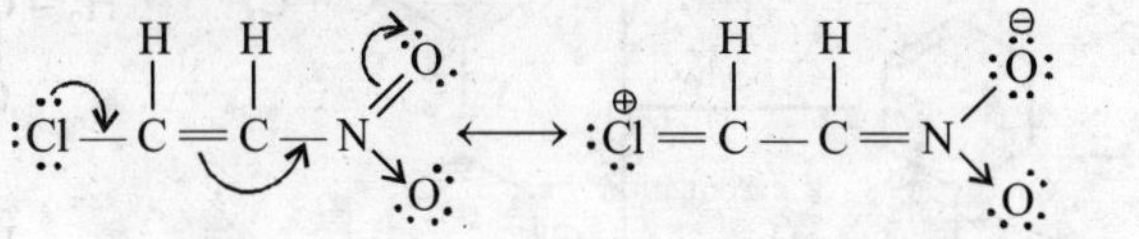

$$:\ddot{Cl}-CH=CH-NO_2 \longleftrightarrow :\overset{\oplus}{Cl}=CH-CH=N(O)O^{\ominus}$$

90. (d) $(CH_3)_2CH-CH=CH_2 \xrightarrow{HCl} (CH_3)_2CH-\overset{+}{C}H-CH_3$

$\xrightarrow{\text{Hydride shift}} CH_3-\overset{+}{C}(CH_3)-CH_2-CH_3 \rightarrow CH_3-C(Cl)(CH_3)-CH_2-CH_3$

Carbocation (planar) (±)

There is no chiral carbon in the product of this reaction.

91. (b) $CH_3-C(CH(CH_3)_2)=CH-CH_2-CH_3 \xrightarrow[\text{(ii) } H_2O_2,\ OH]{\text{(i) } B_2H_6}$

$CH_3-C(H)(CH(CH_3)_2)-CH(OH)-CH_2-CH_3$ (A) $\xrightarrow[\text{dehydration}]{\text{dil. } H_2SO_4,\Delta}$

$CH_3-C(=C(CH_3)_2)-CH_2-CH_2-CH_3$ (B)

92. (d) alkene nitrile $\xrightarrow[\Delta]{\text{Peroxide}}$ (A) [Stable radical]

(A) + pent-1-ene ⟶ (B)

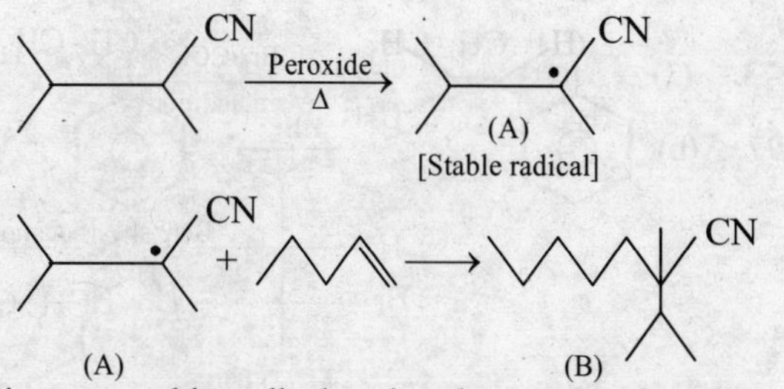

(A) is more stable radical and undergoes Markovnikov addition to form (B).

93. (b) $CH_3-CH=CH_2 \xrightarrow[H_2O]{Cl_2} CH_3-CH-CH_2$ (chloronium, $\overset{+}{Cl}$) $+ Cl^-$

$\longrightarrow CH_3-\overset{+}{C}H-CH_2Cl$ (more stable) $\xrightarrow{H_2\ddot{O}:} CH_3-CH(OH)-CH_2Cl$

94. (c) $CH_3-CH=CH-CH_3 \xrightarrow{\text{alk. } KMnO_4} [CH_3-CH(OH)-CH(OH)-CH_3]$

$\xrightarrow[\text{(ii) } H^+]{\text{(i) heated at elevated temp.}} 2CH_3COOH$

95. (d)

(a) $H_3C\ddot{O}-CH=CH_2 \xrightarrow{HCl} H_3CO-CH(Cl)-CH_3$

+ R effect — Markov. product

(b) $Cl-CH=CH_2 \xrightarrow{H^+ \text{ at } C_1} Cl-CH(H)-\overset{\oplus}{C}H_2$

$\xrightarrow{Cl^-} Cl-CH-CH_2-Cl \xrightarrow{H^+ \text{ at } C_2}$

$Cl-\overset{\oplus}{C}H-CH_3 \xrightarrow{Cl^-} Cl-CH(Cl)-CH_3$

Major

(c) $H_2\ddot{N}-CH=CH_2 \xrightarrow{HCl} H_2N-CH(Cl)-CH_3$

+ R effect — Markov. product

(d) $F_3C \leftarrow CH=CH_2 \xrightarrow{HCl} F_3C-CH_2-CH_2Cl$

Strong – I effect — *anti*-Markov. product (exclusive)

Due to lighter e^- withdrawing nature of CF_3 group it show antimarkornikoff product.

96. (d) bicyclic–CH(H)–C(CH_3)=CH_2 $\xrightarrow{\delta\text{-}\delta\ HCl}$ bicyclic–$\overset{\oplus}{C}(CH_3)_2$ $\xrightarrow{Cl^-}$ bicyclic–$C(CH_3)_2Cl$

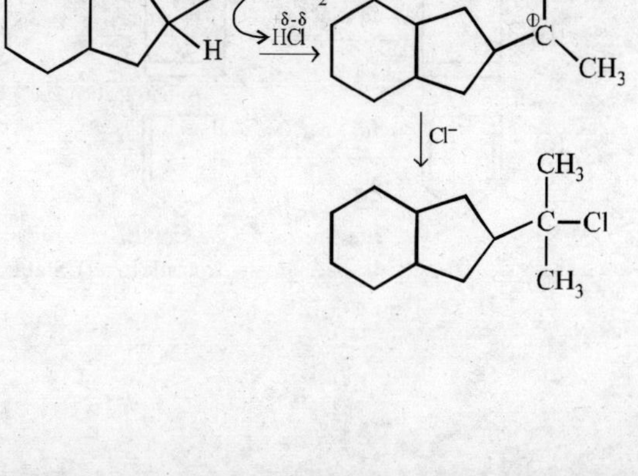

97. (b) $C_6H_5-CH=CH-CH_3 \xrightarrow{HBr} C_6H_5-\overset{+}{C}H-CH_2-CH_3$ (Benzal carbocation) $\xrightarrow{Br^-} C_6H_5-CH(Br)-CH_2-CH_3$

98. (c) Na/liq NH_3 gives trans alkene as product.

2-Butyne $\xrightarrow[\text{(Birch reduction)}]{Na/liq.NH_3}$ *trans*-2-Butene

$R-C\equiv C-R' \xrightarrow[(-Na^+)]{Na^\bullet}$ R(C=C)R' radical anion $\xrightarrow[(-NH_2)]{H-NH_2}$ *trans* - Vinylic radical (more stable than *cis*) $\xrightarrow[(-Na^+)]{Na^\bullet}$ Vinylic anion (more stable) $\xrightarrow[(-NH_2)]{H-NH_2}$ *trans* - Alkene

99. (d) Addition of HBr on 3-methylpent-2-ene in presence of peroxide takes place according to *anti*-Markownikov's rule.

$\overset{1}{C}H_3-\overset{2}{C}H=\overset{3}{C}(CH_3)-\overset{4}{C}H_2\overset{5}{C}H_3 \xrightarrow[-H_2O]{HBr}$ (3-Methylpent-2-ene)

$CH_3-\overset{*}{C}H(Br)-\overset{*}{C}H(CH_3)-CH_2-CH_3$

2-Bromo-3-methylpentane (2 chiral centres)

Since two chiral centres are present in the product, four stereoisomers ($2^n=2^2=4$) are possible.

100. (c)

(a) Cyclopropene $\xrightarrow{aq.\ Na_2CO_3}$ $\ominus Na^+$ Antiaromatic ($4\pi e^-$) **Unstable.**

(b) Cyclobutene $\xrightarrow{aq.\ Na_2CO_3}$ $\ominus Na^+$ Antiaromatic ($4\pi e^-$) **Unstable.**

(c) Cyclopentadiene $\xrightarrow{aq.\ Na_2CO_3}$ $\ominus Na^+$ Aromatic ($6\pi e^-$) **Stable.**

(d) Cyclohexa-1,3-diene $\xrightarrow{aq.\ Na_2CO_3}$ $\ominus Na^+$ Homoaromatic ($6\pi e^-$)

Delocalization of electrons in (d) is not possible, so less stable than (*c*).

101. (b)

$Br_2 \xrightarrow{h\nu} 2\dot{B}r$

1-methylcyclohexene $\xrightarrow[-\dot{H}]{\dot{B}r}$ (2° free radical); $\xrightarrow[-\dot{H}]{\dot{B}r}$ Most stable (3° free radical) $\xrightarrow{Br-Br}$ 3-bromo-3-methylcyclohexene $+\ \dot{B}r$

102. (b)

I → (structure with H_b, a, a)

Both substituents are same (G.I. not possible)

103. (d) N – bromosuccinimide results into bromination at allylic and benzylic positions.

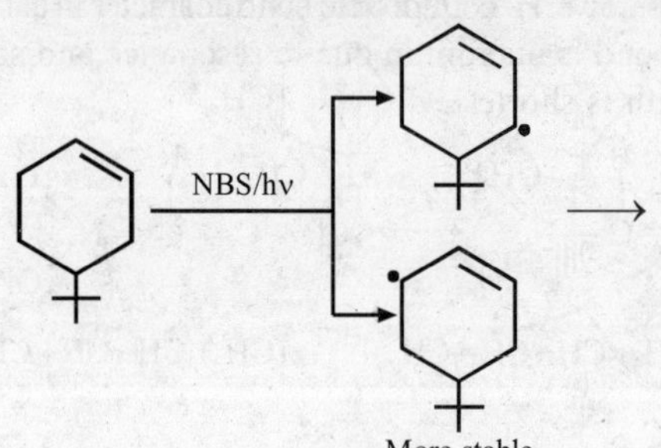

$\xrightarrow{H_2O/K_2CO_3}$ (Br → HO)

104. (d)

$CH_2=CH-CH_3+\overset{\oplus}{C}l \rightarrow CH_2(Cl)-\overset{\oplus}{C}H-CH_3$ More stable intermediate; $\rightarrow \overset{\oplus}{C}H_2-CH(Cl)-CH_3$

$CH_2(Cl)-\overset{\oplus}{C}H-CH_3 \xrightarrow{\overset{\ominus}{O}H} CH_2(Cl)-CH(OH)-CH_3$

105. (d) In presence of sunlight alkenes undergo free radical substitution.

$$\text{Cyclohexene} \xrightarrow{Br_2/hv} \text{Cyclohexenyl radical} + HBr$$

$$\text{Cyclohexenyl radical} \xrightarrow{Br_2} \text{3-Bromocyclohexene} + \dot{Br}$$

106. (b, c)

$$\text{1-Methylcyclopentene} \xrightarrow{D-Cl} \text{carbocation (}CH_3\text{, }\oplus\text{, D)}$$

Formation of above species is more favourable. The stability order of carbocations is

$$\underset{1°}{-C^{\oplus}} < \underset{2°}{-\overset{|}{C}^{\oplus}} < \underset{3°}{-\overset{|}{\underset{|}{C}}^{\oplus}}$$

Now, carbocation formed is sp^2 hybridised that is triangular planar as shown below

$Cl^{\ominus}$ ion can attack either from above or below the plane of the molecule, so there is probability of resultant product as shown below i.e., both product will form.

$$\text{(H, D, Cl, } CH_3\text{)} + \text{(H, D, } CH_3\text{, Cl)} \equiv \text{(D, H, } CH_3\text{, Cl)}$$

107. (c) $H_3C-\overset{\overset{H}{|}}{C}=CH-CH_2Ph$

108. (c) Ethene is obtained by electrolysis of dipotassium succinate as follows:

$$\underset{\text{Pot. succinate}}{\begin{matrix}CH_2COOK\\|\\CH_2COOK\end{matrix}} \xrightarrow{\text{ionization}} \begin{matrix}CH_2COO^-\\|\\CH_2COO^-\end{matrix} + 2K^+$$

$$2H_2O \xrightleftharpoons{\text{ionization}} 2OH^- + 2H^+$$

At anode :

$$\begin{matrix}CH_2COO^-\\|\\CH_2COO^-\end{matrix} \xrightarrow{-2e^-} \underset{\text{Unstable}}{\left[\begin{matrix}CH_2COO\\|\\CH_2COO\end{matrix}\right]} \longrightarrow \begin{matrix}CH_2\\||\\CH_2\end{matrix} + 2CO_2$$

At cathode :

$$2H^+ + 2e^- \longrightarrow [2H] \longrightarrow H_2$$

109. (c) $CH_3CH=CH_2 \xrightarrow[H_2O_2,NaOH]{B_2H_6} CH_3CH_2CH_2OH$

110. (a)

$$C_6H_5-CH_2-CH=CH_2 \xrightarrow[\text{(ii) }NaBH_4]{\text{(i) }Hg(OAC)_2} C_6H_5-CH_2-\underset{\underset{OH}{|}}{CH}-CH_3$$

111. (d) Anti-Markownikov addition is possible only in case of HBr and not in HCl and HI. In HBr, both the chain initiation and propagation steps are exothermic, while in HCl, first step is exothermic, and second step is endothermic and in HI, no step is exothermic. Hence HCl and HI do not undergo anti-Markownikov's addition.

112. (d) HI does not exhibit peroxide effect. HI bond although dissociates easily into iodine radicals, they being bigger in size are not much reactive but recombine together to form iodine molecule.

113. (a)

$$\underset{\text{Propene}}{\begin{matrix}CH_3\\|\\CH\\||\\CH_2\end{matrix}} \xrightarrow[500\,°C]{Cl_2} \underset{\text{Allyl chloride}}{\begin{matrix}CH_2Cl\\|\\CH\\||\\CH_2\end{matrix}} \xrightarrow{\text{aq. }Na_2CO_3} \underset{\text{Allyl alcohol}}{\begin{matrix}CH_2OH\\|\\CH\\||\\CH_2\end{matrix}}$$

$$\xrightarrow[\text{anti-Markownikov addition}]{HOCl} \underset{\text{2-Chloropro-pane-1, 3-diol}}{\begin{matrix}CH_2OH\\|\\CHCl\\|\\CH_2OH\end{matrix}} \xrightarrow{NaOH} \underset{\text{Glycerol}}{\begin{matrix}CH_2OH\\|\\CHOH\\|\\CH_2OH\end{matrix}}$$

114. (c) From the products formed it is clear that the compound has 5 carbon atoms with a double bond.

$$\underset{\text{(A)}}{\underset{\text{2–Methyl–pent–2–ene}}{CH_3-\overset{\overset{CH_3}{|}}{C}=CH-CH_2-CH_3}}$$

$$\Big\downarrow O_3/Zn,\ H_2O$$

$$\underset{\text{Acetone}}{CH_3-\overset{\overset{CH_3}{|}}{C}=O} + \underset{\text{Propionaldehyde}}{CH_3-CH_2-CHO}$$

115. (c) $CH_3CH=CHCH_3 \xrightarrow{O_3}$ (ozonide) $\xrightarrow{Zn/H_2O} 2CH_3CHO + H_2O_2$

2-Butene

116. (d) Completing the sequence of given reactions,

$CH_3-CH=CH-CH_3 \xrightarrow{O_3} H_3C-$(ozonide)$-CH_3$

'A' (ozonide)

$\xrightarrow{Zn/H_2O} 2CH_3CHO + H_2O + ZnO$

'B'

117. (c) 1, 3-Butadiene $\xrightarrow{HBr}$

$CH_2=CH-CH(Br)-CH_3$ At –80 °C the product is 1, 2-addition (kinetic product)

$CH_2(Br)-CH=CH-CH_3$ At 40 °C the product is 1, 4-addition (thermodynamic product)

118. (c) Alkenes combine with hydrogen under pressure and in presence of a catalyst (Ni, Pt or Pd) and form alkanes.

Butene - 1 $\xrightarrow{H_2/Pd}$ Butane.

119. (a) $H_2C=CH-CH_2-C\equiv C-CH_3$ (all H shown)

σ bonds = 13

π bonds = 3

120. (15) $H-CH=CH-CH=CH-C\equiv C-H$

Number of σ bond = 11

Number of π bond = 4

∴ σ + π = 11 + 4 = 15

121. (a) $CH_3-C\equiv CH \xrightarrow[H_2]{Pd/C} CH_3-CH=CH_2$ [A]

$\xrightarrow[(ii)Zn,H_2O]{(i)O_3} CH_3-CH=O + HCHO$ [B] [C]

122. (8) (diyne) $\xrightarrow[\text{derivatives}]{\text{Monobromo}}$ (bromo diyne) Br

No. of π e^- = 8

123. (d) The carbon-carbon bonds in ethyne is stronger than that in ethene.

(H–C≡C–H) Ethyne is linear and carbon atoms are sp-hybridised.

124. (2) $CH_3-C\equiv C-CH_3 \xrightarrow{Na/liq.NH_3}$ (trans) $CH_3(H)C=C(H)CH_3 \xrightarrow[273K]{dil.\ KMnO_4} CH_3-CH(OH)-CH(OH)-CH_3$ (P)

Number of oxygen atoms present in product 'P' is 2.

125. (d) A : 2 – Pentyne   B : trans-2-butene

$H_2 + CH_3-C\equiv C-C_2H_5 \xrightarrow{Pd/C}$ (cis) $CH_3(H)C=C(H)C_2H_5$

(A) 2-Pentyne   *cis*-addition

$CH_3-C\equiv C-CH_3 + H_2 \xrightarrow{Na/Liq\ NH_3}$ $H(H_3C)C=C(CH_3)H$

trans addition trans-2-butene

126. (a) Reaction takes place as:

$CH_3-C\equiv CH + Na \longrightarrow CH_3-C\equiv C^- Na^+ + H_2\uparrow$ (A)

$CH_3-C\equiv C^-Na^+ + CH_3CH_2CH_2-Br \longrightarrow$ (B)

$CH_3-C\equiv C-CH_2-CH_2-CH_3 + NaBr$ (C)

127. (b) Since *cis*-butene (A) is more polar than *trans*-butene (B) it has higher boiling point. However, *trans*-butene has better packing than *cis*-butene. So, *trans*-butene (B) has higher melting point than the *cis*-isomer (A).

trans-Butene ($\mu = 0$) $\xleftarrow[liq.NH_3]{H_2,\ Na} CH_3-C\equiv C-CH_3 \xrightarrow{H_2\ Pd/C,\ \Delta}$ *cis*-Butene ($\mu \neq 0$)

128. (3) $CH_3-C\equiv CH + 2Br_2 \rightarrow CH_3-CBr_2-CHBr_2$

(molar mass = 360)

2×160 g Br_2 produces = 360×0.27 g Compound

1 g Br_2 produces = $\frac{360\times 0.27}{2\times 160}$ g

$= 0.30375 = 3.0375 \times 10^{-1}$

129. (c) Alkyl lithium act as a base.

$$n-Bu-C\equiv CH \xrightarrow{n-BuLi} n-Bu-C\equiv C^{\ominus}Li^{\oplus}$$

$$\xrightarrow{n-C_5H_{11}Cl} n-Bu-C\equiv C-C_5H_{11} \xrightarrow[\text{(syn-addition)}]{\text{Lindlar cat., } H_2} \text{(n-Bu)HC=CH(C}_5\text{H}_{11}\text{) (cis)}$$

130. (10) $H\overset{\sigma}{-}\overset{H}{C}\overset{\sigma}{-}C\overset{\sigma}{=}C\overset{\sigma}{-}C\overset{\sigma}{\equiv}C\overset{\sigma}{-}H$ (with σ bonds to H, H, H below)

Numbers of σ bonds = 10

131. (c)

$$HC\equiv C-CH_2-CH_3 \xrightarrow[\text{(ii) Zn, } H_2O]{\text{(i) } O_3} \underset{(A)}{H-\overset{O}{\overset{\|}{C}}-\underset{O}{\underset{\|}{C}}-CH_2-CH_3}$$

Reductive ozonolysis

$$\xrightarrow{[Ag(NH_3)_2]NO_3} Ag\downarrow + {}^{-}O-\overset{O}{\overset{\|}{C}}-\underset{O}{\underset{\|}{C}}-CH_2-CH_3$$

(bright silver mirror)

132. (7) $CH\equiv CH \xrightarrow[873\text{ K}]{\text{(1) Red hot Fe tube}}$ benzene $\xrightarrow{\text{(2) CO, HCl, }AlCl_3}$ benzaldehyde (C_6H_5CHO)

In benzaldehyde, total number of sp^2 carbons are 7.

133. (a)

$$O_2N-C_6H_4-C\equiv C-C_6H_4-OCH_3 \xrightarrow[\text{(Hydration of alkyne)}]{Hg^{2+}/H^+,\ H_2O} O_2N-C_6H_4-CH=\overset{OH}{\overset{|}{C}}-C_6H_4-OCH_3$$

$$\xrightarrow{\text{Tautomerisation}} O_2N-C_6H_4-CH_2-\overset{O}{\overset{\|}{C}}-C_6H_4-OCH_3$$

134. (d)

$$CH_3-\overset{CH_3}{\overset{|}{CH}}-C\equiv CH \xrightarrow[H_2O]{HgSO_4,\ H_2SO_4} \underset{(X)}{H_3C-\overset{CH_3}{\overset{|}{CH}}-\overset{O}{\overset{\|}{C}}-CH_3}$$

$$\xrightarrow{\text{(i) EtMgBr}/H_2O} CH_3-\overset{CH_3}{\overset{|}{CH}}-\underset{CH_2-CH_3}{\underset{|}{\overset{OH}{\overset{|}{C}}}}-CH_3 \xrightarrow{\text{(ii) conc. }H_2SO_4/\Delta} \underset{(Y)}{CH_3-\overset{CH_3}{\overset{|}{C}}=\underset{CH_2-CH_3}{\underset{|}{C}}-CH_3}$$

135. (13)

$$\underset{\text{(Lowest M.W. alkyne) (A)}}{H-C\equiv C-H} \xrightarrow[\text{Cu tube}]{\text{Red hot}} \underset{(B)}{C_6H_6} \xrightarrow[\text{Anhyd. }AlCl_3]{CH_3Cl\text{ (1 eq.)}} \underset{(C)}{C_6H_5CH_3}$$

Number of atoms present in molecule (C) in one plane = 13

136. (c)

(i) Reaction of 'A' with Ag_2O to give ppt indicates $-C\equiv CH$ type of linkage.

(ii) Going backward in the second set, compound 'C' when heated with $ZnCl_2$ and conc. HCl gives turbidity within 5 minutes, it indicates that 'C' is a 2° alcohol, (C). Hence 'B' is a ketone and 'A' is prop-1-yne.

$$\underset{[A]}{CH_3-C\equiv CH} \xrightarrow{HgSO_4+H_2SO_4} CH_3-\overset{OH}{\overset{|}{C}}=CH_2 \xrightleftharpoons{\text{tautomerise}} \underset{[B]}{CH_3-\overset{O}{\overset{\|}{C}}-CH_3}$$

$$\xrightarrow{NaBH_4} \underset{[C]}{CH_3-\overset{OH}{\overset{|}{CH}}-CH_3} \xrightarrow[\text{(Lucas reagent)}]{ZnCl_2+HCl} \text{Turbidity within 5 minutes (confirmation of 2° alcohol)}$$

137. (c)

$$\underset{\text{(– M effect)}}{\underset{(III)}{C_6H_5COCH_3}} < \underset{\text{(–I effect)}}{\underset{(I)}{C_6H_5Cl}} < \underset{\text{(+ I effect)}}{\underset{(II)}{C_6H_5CH_3}}$$

138. (d)

$$CH_3-C\equiv C-H \xrightarrow[1\text{ eq}]{DCl} CH_3-\underset{Cl}{\underset{|}{C}}=\underset{D}{\underset{|}{CH}} \longrightarrow CH_3-\underset{Cl}{\underset{|}{\overset{+}{C}}}-\underset{D}{\underset{|}{CH}} \xrightarrow{\overset{+}{D}\overset{-}{I}} CH_3-\underset{Cl}{\underset{|}{\overset{I}{\overset{|}{C}}}}-\underset{D}{\underset{|}{\overset{D}{\overset{|}{CH}}}}$$

Due to combined +I effect of $-CH_3$ and +M effect of $-Cl$, positive charge is more stable at the second carbon.

139. (d) $\underset{sp}{HC\equiv CH} > H_3C\rightarrow C\equiv CH > \underset{sp^2}{CH_2=CH_2}$

140. (d) When but-2-yne is treated with H_2/Lindlar's catalyst, compound X (*cis*-but-2-ene) is produced as the major product; and when treated with Na/liq NH_3 it produces Y (*trans*-but-2-ene) as the major product. *Cis*-isomer(X) will have higher dipole moment and higher boiling point than *trans* (Y).

141. (c) With Lithium in liquid ammonia, trans-alkene is almost an exclusive product.

$$Ph-C\equiv C-Ph \xrightarrow[\text{Birch reduction}]{\text{Li in liq. } NH_3} \underset{H}{\overset{Ph}{>}}C=C\underset{Ph}{\overset{H}{<}}$$

142. (b) $\underset{\text{Acetylene}}{3\,CH\equiv CH} \longrightarrow \underset{\text{Benzene}}{C_6H_6}$ (poly-ene)

143. (a) $\underset{\text{Propyne}}{CH_3-C\equiv CH} + H_2O \xrightarrow[HgSO_4]{H_2SO_4}$

$$\left[\underset{\substack{\text{Prop-1-en-2-ol}\\ \text{(unstable)}}}{CH_3-\overset{OH}{\overset{|}{C}}=CH_2}\right] \longleftrightarrow \underset{\substack{\text{Propan-2-one}\\ \text{(Acetone)}}}{CH_3-\overset{O}{\overset{\|}{C}}-CH_3}$$

144. (b) Anti addition of hydrogen atoms to the triple bond occurs when alkynes are reduced with sodium (or lithium) metal in liq. ammonia, ethylamine, or alcohol at low temperatures. This reaction called, a dissolving metal reduction, produces an (E)- or *trans*-alkene.

Sodium in liq. NH_3 is used as a source of electrons in the reduction of an alkyne to a *trans* alkene.

$$\underset{\text{2-Hexyne}}{CH_3-CH_2-CH_2-C\equiv C-CH_3} \xrightarrow[\text{Birch reduction}]{Li/NH_3} \underset{\textit{trans}\text{-2-Hexene}}{\underset{H}{\overset{CH_3-CH_2-CH_2}{>}}C=C\underset{CH_3}{\overset{H}{<}}}$$

145. (b) Alkynes having terminal $-C\equiv H$ react with Na in liquid ammonia to yield H_2 gas

$$CH_3CH_2C\equiv CH \xrightarrow[\text{liquid } NH_3]{\text{Na in}} CH_3CH_2C\equiv C^-Na^+ + \frac{1}{2}H_2(g)$$

146. (d) $CH_3MgX + CH_3-C\equiv C-H \longrightarrow CH_3-C\equiv CMgX + CH_4(g)$

147. (b) The reaction follows Markownikoff rule which states that when unsymmetrical reagent adds across unsymmetrical double or triple bond, the negative part adds to carbon atom having lesser number of hydrogen atoms.

$$CH_3-C\equiv CH + HBr \longrightarrow$$

$$CH_3-\underset{Br}{\underset{|}{C}}=CH_2 \xrightarrow{HBr} \underset{\text{2,2-Dibromopropane}}{CH_3-\overset{Br}{\overset{|}{\underset{Br}{\underset{|}{C}}}}-CH_3}$$

148. (d) Among isomeric alkanes, the straight chain isomer has higher boiling point due to large surface area than the branched chain isomer. The greater the branching of the chain, the lower is the boiling point. Due to the presence of π electrons in alkynes, these moleculs are polar and hence, have higher boiling points than the corresponding alkanes.

Thus b.pt. follows the order :

alkynes > alkanes (straight chain) > alkenes > branched chain alkanes.

149. (c) $CH\equiv CH + HOCl \longrightarrow \begin{matrix}CHOH\\ \|\\ CHCl\end{matrix} \longrightarrow \begin{matrix}HO-\overset{\oplus}{C}-H\\ |\\ Cl-\underset{\ominus}{C}-H\end{matrix}$

$$\xrightarrow{HO^-Cl^+} \left[\begin{matrix}CH(OH)_2\\ |\\ CHCl_2\end{matrix}\right] \xrightarrow{-H_2O} \underset{\text{Dichloroacetaldehyde}}{\begin{matrix}CHO\\ |\\ CHCl_2\end{matrix}}$$

150. (a) Acetylene reacts with the other three reagents as:

$$CH\equiv CNa \xleftarrow[\text{liq. } NH_3]{Na} CH\equiv CH \xrightarrow{+HCl} \begin{matrix}CH_2\\ \|\\ CHCl\end{matrix} \xrightarrow{+HCl} \begin{matrix}CH_3\\ |\\ CHCl_2\end{matrix}$$

$$CH\equiv CH \xrightarrow{[AgNO_3+NH_4OH]} \underset{\text{white ppt.}}{AgC\equiv CAg} + NH_4NO_3$$

151. (a) $C_6H_5-CH_2-CH_3 \xrightarrow{Br_2/Fe} p\text{-}Br-C_6H_4-CH_2-CH_3 \xrightarrow{Cl_2,\,\Delta} p\text{-}Br-C_6H_4-CH(Cl)-CH_3 \xrightarrow{\text{alc. KOH}} p\text{-}Br-C_6H_4-CH=CH_2$

152. (b)

$$\underset{\text{Benzene}}{C_6H_6} \xrightarrow{\text{Oleum}} \underset{\text{Benzene Sulphonic acid}}{C_6H_5SO_3H} \xrightarrow{NaOH} \underset{\text{Sodium phenoxide}}{C_6H_5\overset{\ominus}{O}\overset{\oplus}{Na}} \xrightarrow{H^{\oplus}\text{ (acidification)}} \underset{\text{(Phenol)}}{C_6H_5OH} \xrightarrow{Zn} \underset{\text{Benzene}}{C_6H_6}$$

153. (d) Aromatic compounds :

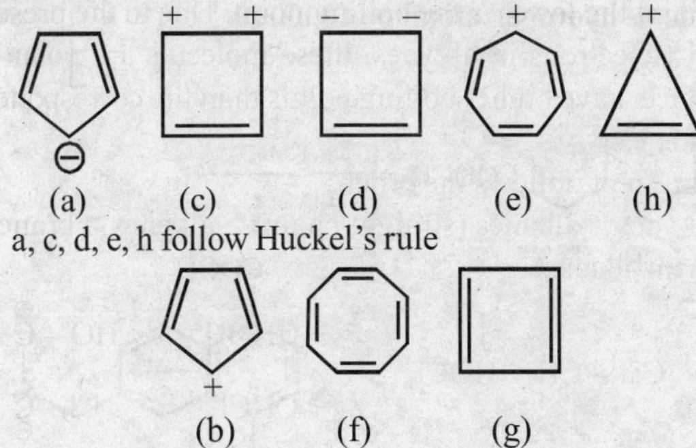

a, c, d, e, h follow Huckel's rule

(b) (f) (g)

b, f, g, are not aromatic, as these compounds do not follow Huckel's rule

154. (2) D.O.V. $= \frac{2C+2-H}{2} = 6$

Above two isomers of C_9H_{12} have four different sites for aromatic electrophilic substitution reaction.

155. (c) (m-xylene) $\xrightarrow{HNO_3/H_2SO_4}$ (nitro-m-xylene) $\xrightarrow{\text{Oxidation}}$ (4-nitroisophthalic acid)

$-CH_3$ group increase electron density at o/p position while $-NO_2$ group decrease electron density at ortho and para position.

156. (a) (B) – CN is meta directing but –OH is ortho/para directing.
(E) Halides are deactivating groups.

157. (a) q is aromatic r is non-aromatic (non-planar structure) p is antiaromatic so, order of stability is :
Aromatic > non-aromatic > Antiaromatic
(q) (r) (p)

158. (a) Benzene + succinic anhydride $\xrightarrow{AlCl_3}$ (A) $\xrightarrow[\text{(Clemmensen reduction)}]{Zn\text{-}Hg/HCl}$
Friedel-crafts Acylation

(B) $\xrightarrow[\text{Dehydration}]{Conc.H_2SO_4}$ (C)

159. (a) For a compound to be aromatic it should be planar and follow Huckel's rule of $(4n+2)\pi$ electrons.

A. → Non aromatic

B. → Aromatic

C. → Non aromatic

D. → Aromatic

160. (1)

161. (b) In nitration of benzene concentrated H_2SO_4 and HNO_3 is used as reagent which generates electrophile $\overset{\oplus}{N}O_2$ in following steps:

$$H_2SO_4 + HNO_3 \rightleftharpoons HSO_4^- + H-\overset{\oplus}{O}(H)-NO_2$$
$$\rightleftharpoons HSO_4^- + H_2O + NO_2^{\oplus}$$

Lewis acids can promote the formation of electrophiles not Lewis base

162. (d) As we know compound (4) is aromatic and the compound (1) is anti-aromatic. Hence compound (4) is most stable and compound (1) is least stable among these in compound (2) and (3) carbon atom having charge is sp^3 hybridised. On the basis of angle strain theory compound (3) is more stable than compound (2).

163. (d) $-NO_2, -CHO, -SO_3H, -COR$.
Ortho and para directing groups are:
$-NH_2, -NHR, -NHCOCH_3, -OCH_3, -CH_3, -C_2H_5$ etc.
meta directing groups are:
$-NO_2, -CN, -CHO, -COR, -COOH, -COOR, -SO_3H$ etc.

164. (d) Order of increasing reactivity towards electrophilic substitution reaction is:

Chlorobenzene < Benzene < Toluene < Phenol
(–I effect) (+I effect) (+M effect)

165. (2)

Deactivating, withdraws electron by –M effect

Weakly activating due to delocalisation of electron pair of nitrogen with carbonyl group by + M

$-\ddot{N}(H)-CH_3$: Activating due to donation of lone pair of nitrogen to ring system by +M

$-C\equiv N$: Deactivating due to –M and –I effect of CN^- group

$-\ddot{O}CH_3$: Activating due to donation of electron pair present on the oxygen to ring system by +M

166. (d) $C_6H_5COCl \xrightarrow{\text{anhydrous } AlCl_3} C_6H_5C^{\oplus}=O + AlCl_4^{\ominus}$; with benzene → (P) benzophenone

167. (d) B > A > C > D

A: benzene; B: toluene (CH_3); C: chlorobenzene (Cl); D: nitrobenzene (NO_2)

– CH_3 will show +I and +M effect.

– Cl will show –I and +M effect but inductive effect will be more influential.

– NO_2 will show –I and –M effect.

168. (3) B, C and D are aromatic.

B has one aromatic benzene ring, C has two rings fused and conjugated with all atoms sp^2-hybridised, D has three rings in conjugation.

All three have $(4n+2)\pi$ electrons.

169. (b) Tropolone is an aromatic compound and has 8π electrons ($6\pi e^-$ are endocyclic and $2\pi e^-$ are exocyclic) but 2π electrons of >C=O group in tropolone are not involved in aromaticity.

Aromatic compound ($6\ \pi e^-$)

170. (b) $KMnO_4$ oxidises benzylic carbon containing atleast one α–hydrogen atom to –COOH.

CH_2CH_3, $CH_2=CH_2$, CH_2COOCH_3 substituted naphthalene $\xrightarrow{\text{(i) alk. } KMnO_4, \Delta \text{ (ii) } H_3O^+}$ COOH, COOH, CH_2COOH substituted naphthalene

171. (d) CH_3O, O_2N, CH_3 (tetralin)

$AlCl_3$ gives rise to a carbcation here. Carbocations in the side chains of a ring undergo cyclization or ring formation.

172. (c) methylenecyclohexane $\xrightarrow{HF}$ 1-methylcyclohexyl cation $\xrightarrow[\text{electrophilic aromatic substitution}]{C_6H_6}$ 1-methyl-1-phenylcyclohexane

173. (c) (A) $C_6H_5\overset{*}{C}H(Br)-CH_3$ (Asymmetrical carbon), Optically active $\xrightarrow{KMnO_4}$ C_6H_5COOH + AgBr (Yellow ppt of AgBr) [Alc. $AgNO_3$ (Br present on side chain)]

(B) p-$BrC_6H_4CH_2-CH_3$ (Optically inactive) $\xrightarrow[\text{(Oxidation)}]{KMnO_4}$ p-BrC_6H_4COOH $\xrightarrow[\text{(Br present on ring)}]{\text{Alc. } AgNO_3}$ No ppt.

174. (d) $C_6H_5CH_2CH_3 \xrightarrow[\text{(Electrophilic substitution reaction)}]{Br_2, Fe}$ p-$BrC_6H_4CH_2CH_3$ $\xrightarrow[\text{benzylic halogenation}]{Cl_2, \Delta}$ p-$BrC_6H_4CH(Cl)-CH_3$ $\xrightarrow[\text{(Elimination reaction) (–HCl)}]{\text{alc. KOH}}$ p-$BrC_6H_4CH=CH_2$

175. (c) Electrophile will attack at ortho and para positions with respect to better electron releasing group (ERG) –OH donates through +M and $-CH_3$ donates through +H ERG : – OH > – CH_3

Para position with respect to – OH group and it will be meta position with respect to $-CH_3$ group.

176. (d) Assertion:

Annulene	Huckle's rule	Planarity	Aromaticity
[6]	$4n+2=6, n=1$	Planar	Aromatic
[8]	$4n+2\neq 8, n=1$	Non-planar	Non-aromatic
cis – [10]	$4n+2=10, n=2$	Non-Planar	Non-aromatic

[8] annulene is non-aromatic because it does not obey Huckel's rule and it is non-planar.

Reason: It is correct. Although it has $(4n+2)\pi$ electrons, i.e. 10, but it is not planar.

177. (c)

Due to the steric repulsion between the two hydrogen atoms this molecule can not achieve planarity, hence compound is not aromatic.

178. (6) Benzene $\xrightarrow[\text{sunlight}]{Cl_2}$ hexachlorocyclohexane (each carbon bearing Cl and H)

179. (a, b) Benzenoid compound contain only benzene rings. In option (a) one central carbon is sp^3 hybridised so no benzene ring is present.

180. (c) $CH_3-CH_2CH_2Cl \xrightarrow{AlCl_3} CH_3-CH_2-\overset{+}{C}H_2$

Hydride shift ↓

$CH_3-\overset{+}{C}H-CH_3$ (more stable)

181. (b)

(A) $(4n+2)\pi e^- = 6$, Planar ∴ Aromatic

(B) sp^3; $(4n+2)\pi e^- = 6$, Planar ∴ Aromatic

(C) sp^3, sp^3; $(4n+2)\pi e^- = 8$, Non-planar ∴ Non-aromatic

(D) sp^2; $(4n+2)\pi e^- = 8$, Planar ∴ Anti-aromatic

182. (d) Benzoic acid and naphthalene, both being organic compounds, are soluble in benzene, hence can't be separated by benzene. These can be effectively separated by crystallization. Benzoic acid is soluble in hot water whereas naphthalene is insoluble. Hence assertion is false but reason is true.

183. (d) In option a, saturated five membered ring therefore does not exhibit aromaticity as a whole

(a) not aromatic

(b) $\Rightarrow 4\pi e^-$ in ring conjugation $\Rightarrow$ Anti-aromatic

(c) $CH_3 \Rightarrow 4\pi e^-$ in ring conjugation $\Rightarrow$ Anti-aromatic

(d) $\Rightarrow 6\pi e^-$ in ring conjugation $\Rightarrow$ Aromatic

184. (a) Benzene $\xrightarrow[\Delta]{HNO_3 + H_2SO_4}$ Nitrobenzene [A] $\xrightarrow[\text{Anhy. } AlCl_3]{Cl_2}$ m-chloronitrobenzene [B] $\xrightarrow{Fe/HCl}$ m-chloroaniline [C]

185. (c) : Non-aromatic

All are aromatic, having 6π electrons

186. (d) Reagent for nitration of benzene

$$\underset{(Acid)}{H_2SO_4} + \underset{(Base)}{HNO_3} \rightleftharpoons HSO_4^{\ominus} + H_2\overset{\oplus}{N}O_3$$

$$H_2\overset{\oplus}{N}O_3 \rightleftharpoons H_2O + \overset{\oplus}{N}O_2$$

Benzene + $\overset{\oplus}{N}O_2$ ⟶ Nitrobenzene

Electrophilic substitution reaction

187. (80) Benzene $\xrightarrow[H_2SO_4]{HNO_3}$ Nitrobenzene (NO_2)

Moles of benzene $= \frac{3.9}{78} = 0.05$

Moles of nitrobenzene $= \frac{4.92}{123} = 0.04$

Percentage yield of nitrobenzene $= \frac{0.04}{0.05} \times 100 = 80\%$

188. (c) To be aromatic, a molecule must meet the following conditions :

(i) It must be a ring.

(ii) It must be planar.

(iii) It must have in each atom of the ring a *p*-orbital that's orthogonal to the plane of the ring. In other words, no atom in the ring can be sp^3 hybridized.

(iv) It must have a Huckel number of π electrons, following the $4n + 2$ rule. If the molecules meets the first three conditions, but only has $4n$ π electrons, the molecule is anti-aromatic.

If the molecule fails any or all of the first three conditions, then the molecule is non-aromatic.

Non-aromatic ($=CH_2$) Aromatic Aromatic ($\ominus$) Anti-aromatic ($\oplus$)

189. (b) $CH_3 - CH = CH - Br \xrightarrow[(-HBr)]{(i)\ NaNH_2} CH_3-C \equiv CH$

$\xrightarrow[\text{(Cyclization)}]{\text{(ii) Red hot Iron tube/873 K}}$ Mesitylene (CH_3, H_3C, CH_3)

190. (a) Benzene $\xrightarrow[AlCl_3]{CH_3COCl}$ Acetophenone (O, CH_3) $\xrightarrow[HCl]{Zn - Hg}$ Ethylbenzene (C_2H_5) $\xrightarrow{Br_2/AlBr_3}$ p-Bromoethylbenzene (C_2H_5, Br) $\xrightarrow{HNO_3/H_2SO_4}$ (C_2H_5, NO_2, Br)

191. (c) In benzene, each carbon atom is sp^2 hybridised. Therefore, total (3×6) 18 sp^2 hybrid orbitals are present in benzene.

192. (d) Vinyl halides and aryl halides are unreactive towards Friedel Craft's reaction.

Therefore, reactions (A) and (C) are not possible.

193. (a) Compounds (b), (c) and (d) are containing $6\pi e^-$ in complete conjugation and are aromatic.

(b) N̈H ; (c) N ; (d) $\ominus$

Compound (a) ($\oplus$) is anti-aromatic as it has $4\pi e^-$ in complete conjugation and doesn't obey Hückel rule.

194. (a) OCH_3 (+M) > CH_3 (+I and +H) > Cl (–I and +M) > CN (–M, –I)

195. (a) (cyclopropenyl cation $\oplus$) is aromatic; $(4 \times 0 + 2) = 2\ \pi e^-$ in conjugation

($\oplus$) and ($\ominus$) are anti-aromatic due to $4\ n\pi e^-$.

$4\pi e^-$ $8\pi e^-$

(cyclohexadiene) is non-aromatic due to the presence of sp^3 carbon.

196. (c) Ph–$CH_2-CH=CH_2$ $\xrightarrow{H^\oplus}$ Ph–CH(H)–$\overset{\oplus}{C}H-CH_3$ (2° Carbocation) $\xrightarrow{\text{H shift}}$ Ph–$\overset{\oplus}{C}H-CH_2-CH_3$ (Benzylic Carbocation) $\xrightarrow{Cl^-}$ Ph–CH(Cl)–CH_2-CH_3

197. (a)

A dipolar resonance structure has aromatic character in the ring and would be expected to make a major contribution to the overall structure.

Conjugation is possible only in option –(I).

198. (b) (a) $\xrightarrow{AlCl_3}$ Cl—$AlCl_3$ (stable)

(b) Cl $\xrightarrow{AlCl_3}$ Cl — $AlCl_3$ (unstable)

(c) Cl $\xrightarrow{AlCl_3}$ Cl $AlCl_3$ (stable)

(d) COCl $\xrightarrow{AlCl_3}$ O Cl — $AlCl_3$ (stable)

Hence, formation of carbocation is not possible in the case of vinyl halide.

199. (b) CH_3 $\xrightarrow{KMnO_4}$ COOH (A) $\xrightarrow{SOCl_2}$

COCl (B) $\xrightarrow[BaSO_4]{H_2/Pd}$ CHO (C)

Rosenmund's reduction

200. (c) In (*c*) Friedal-Craft's alkylation reaction will give propyl phenyl ketone which further on Clemmenson's reduction will give butylbenzene.

$$C_6H_6 + CH_3CH_2CH_2COCl \xrightarrow{AlCl_3} C_6H_5COCH_2CH_2CH_3$$

Butanoyl chloride

$$\xrightarrow{Zn\text{-}Hg/HCl} C_6H_5CH_2CH_2CH_2CH_3$$

Butylbenzene

201. (d) For a compound to be aromatic it must have $(4n + 2)\ \pi e^-$ where n is an integer. (III) have 8 and (VI) have 4 e^-. Hence are antiaromatic

202. (c) Ph–C≡C–CH$_3$ $\xrightarrow{H^+/Hg^{2+}}$ Ph–C(OH)=C(H)–CH$_3$ (Enol form) $\xrightleftharpoons{\text{tautomerisation}}$ Ph–C(=O)–CH$_2$CH$_3$ (Keto form)

203. (d) C$_6$H$_5$CH$_2$CH$_3$ + N-Bromosuccinamide ⟶ C$_6$H$_5$–CH(Br)–CH$_3$ (Br–CH–CH$_3$)

N – Bromosuccinamide

The given reagent will substitute –Br at benzylic position. (Allylic bromination).

204. (a) Toluene $\xrightarrow[\text{conc. } H_2SO_4]{\text{conc. } HNO_3}$ o-nitrotoluene (CH$_3$, NO$_2$) + p-nitrotoluene (CH$_3$, NO$_2$) $\xrightarrow{\text{Sn + HCl}}$ o-toluidine (CH$_3$, NH$_2$) + p-toluidine (CH$_3$, NH$_2$) $\xrightarrow[\text{(ii) CuBr, heat}]{\text{(i) Diazotisation}}$ p-bromotoluene (CH$_3$, Br) + o-bromotoluene (CH$_3$, Br)

CLASS XII

Solutions

TREND ANALYSIS
JEE MAIN (2025-2021)

	2025				2024				2023				2022				2021			
	Session II		Session I		Session II		Session I		Session II		Session I		Session II		Session I		Session II		Session I	
	MCQs	NVQs	MCQs	NVQs	MCQs	NVQs	MCQs	NVQs	MCQs	NVQs	MCQs	NVQs	MCQs	NVQs	MCQs	NVQs	MCQs	NVQs	MCQs	NVQs
Topic 1	3	2	–	–	3	3	2	4	2	–	–	5	–	1	1	–	–	4	–	3
Topic 2	4	–	1	–	–	–	2	–	–	2	–	2	–	3	1	2	–	3	–	2
Topic 3	4	2	9	2	1	8	3	2	1	9	3	8	3	4	1	8	–	8	–	10
Topic 4																				
Total	11	4	10	2	4	11	7	6	3	11	3	15	3	8	3	10	–	15	–	15

How to Code Qns for Better & Faster Revision →

Type 1: Able to solve (No code Req)

Type 2: Solved with struggle or by seeing solution **(Use code NR means Needs Revision or orange stickers)**

Type 3: Tricky/ Could not solve on my own. Need help from Teacher/ Friend. **(Use code TQ means Tricky Question or red stickers)**

You can either hand write the code in boxes or else paste coloured stickers given at end of book.

Topic 1 – Solubility and Concentration of Solutions

1. At the sea level, the dry air mass percentage composition is given as nitrogen gas : 70.0, oxygen gas : 27.0 and argon gas : 3.0. If total pressure is 1.15 atm, then calculate the ratio of followings respectively :

[NCERT : P L-7 | April 7, 2025 (I)]

(i) partial pressure of nitrogen gas to partial pressure of oxygen gas

(ii) partial pressure of oxygen gas to partial pressure of argon gas

(Given : Molar mass of N, O and Ar are 14, 16, and 40 g mol–1 respectively)

(a) 4.26, 19.3 (b) 2.59, 11.85

(c) 5.46, 17.8 (d) 2.96, 11.2

2. Sea water, which can be considered as a 6 molar (6 M) solution of NaCl, has a density of 2 g mL^{-1}. The concentration of dissolved oxygen (O_2) in sea water is 5.8 ppm. Then the concentration of dissolved oxygen (O_2) in sea water, is $x \times 10^{-4}$ m.

$x =$ _______. (Nearest integer)

Given: Molar mass of NaCl is 58.5 g mol^{-1}

Molar mass of O_2 is 32 g mol^{-1}.

[NCERT : P L-7 | NV, April 4, 2025 (II)]

3. Fortification of food with iron is done using $FeSO_4 \cdot 7H_2O$. The mass in grams of the $FeSO_4 \cdot 7H_2O$ required to achieve 12 ppm of iron in 150 kg of wheat is _____(Nearest integer)

[Given : Molar mass of Fe, S and O respectively are 56, 32 and 16 g mol^{-1}] **[NCERT : P L-3 | NV, April 4, 2025 (I)]**

4. 'x' g of NaCl is added to water in a beaker with a lid. The temperature of the system is raised from 1°C to 25°C. Which out of the following plots, is best suited for the change in the molarity (M) of the solution with respect to temperature ? **[NCERT : P L-4 | April 2, 2025 (II)]**

[Consider the solubility of NaCl remains unchanged over the temperature range]

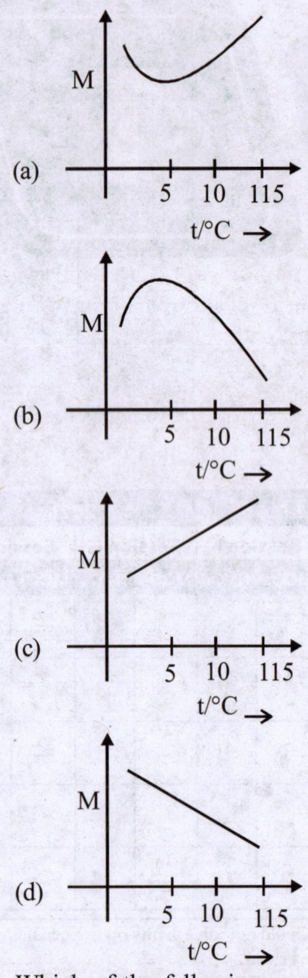

5. Which of the following graph correctly represents the plots of K_H at 1 bar gases in water versus temperature ?

[NCERT : P L-8 | April 2, 2025 (I)]

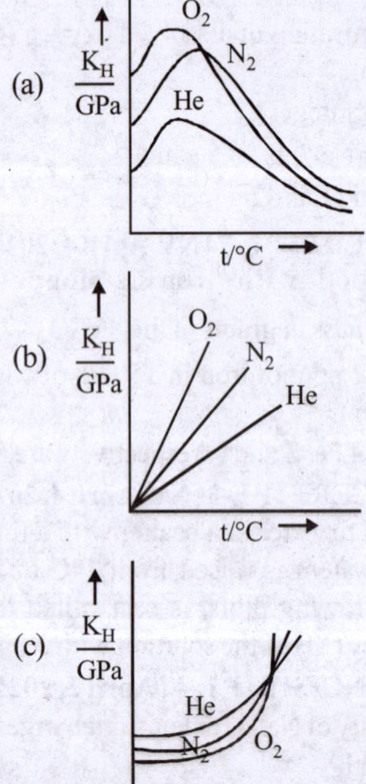

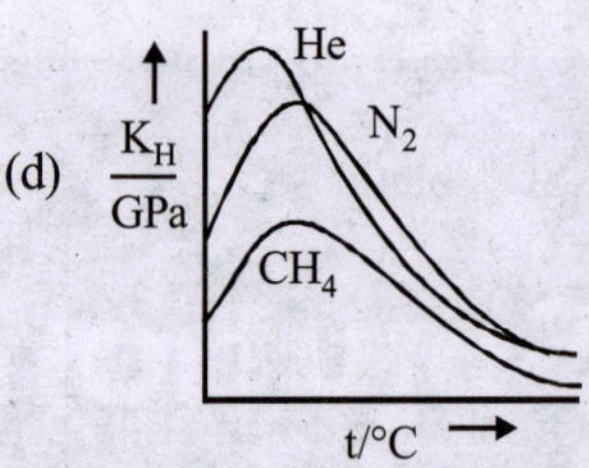

6. Molarity (M) of an aqueous solution containing x g of anhyd. $CuSO_4$ in 500 mL solution at 32 °C is 2×10^{-1} M. Its molality will be _______ $\times 10^{-3}$ m. (nearest integer).
[Given density of the solution = 1.25 g/mL.]

[NCERT : P L-2 | NV, April 9, 2024 (I)]

7. Molality of an aqueous solution of urea is 4.44 m. Mole fraction of urea in solution is $x \times 10^{-3}$.
Value of x is _______. (integer answer)

[NCERT : P L-3 | April 8, 2024 (II)]

8. A solution is prepared by adding 1 mole ethyl alcohol in 9 mole water. The mass percent of solute in the solution is _______ (Integer Answer) **[NCERT : P L-2 | NV, April 8, 2024 (II)]**
(Given : Molar mass in g mol^{-1} Ethyl alcohol : 46, water : 18)

9. The density of 'x' M solution ('x' molar) of NaOH is 1.12 g mL^{-1}. while in molality, the concentration of the solution is 3 m (3 molal). Then x is **[NCERT : P L-3 | April 6, 2024 (I)]**
(Given : Molar mass of NaOH is 40 g/mol)
(a) 3.5 (b) 3.0 (c) 3.8 (d) 2.8

10. Molality (m) of 3 M aqueous solution of NaCl is:
(Given : Density of solution = 1.25 g mL^{-1}, Molar mass in g mol^{-1} (Na-23, Cl-35.5) **[NCERT : P L-3 | April 6, 2024 (II)]**
(a) 2.90 m (b) 2.79 m (c) 1.90 m (d) 3.85 m

11. The Molarity (M) of an aqueous solution containing 5.85 g of NaCl in 500 mL water is :

[NCERT : P L-3 | April 4, 2024 (I)]

(Given : Molar Mass Na : 23 and Cl : 35.5 $gmol^{-1}$)
(a) 20 (b) 0.2 (c) 2 (d) 4

12. The molarity of 1 L orthophosphoric acid (H_3PO_4) having 70% purity by weight (specific gravity 1.54 g cm^{-3}) is _______ M.
(Molar mass of H_3PO_4 = 98 g mol^{-1})

[NCERT : P L-4 | NV, Jan. 31, 2024 (II)]

13. The mass of sodium acetate (CH_3COONa) required to prepare 250 mL of 0.35 M aqueous solution is _______ g. (Molar mass of CH_3 COONa is 82.02 g mol^{-1})

[NCERT : P L-4 | NV, Jan. 30, 2024 (I)]

14. If a substance 'A' dissolves in solution of a mixture of 'B' and 'C' with their respective number of moles as n_A, n_B and n_C, Mole fraction of C in the solution is :

[NCERT : P L-3 | Jan. 30, 2024 (II)]

(a) $\frac{n_C}{n_A \times n_B \times n_C}$ (b) $\frac{n_C}{n_A + n_B + n_C}$

(c) $\frac{n_C}{n_A - n_B - n_C}$ (d) $\frac{n_B}{n_A + n_B}$

15. A solution of H_2SO_4 is 31.4% H_2SO_4 by mass and has a density of 1.25 g/mL.
The molarity of the H_2SO_4 solution is _____ M (nearest integer)
(Given molar mass of $H_2SO_4 = 98 g mol^{-1}$]
[NCERT : P L-4 | NV, Jan. 29, 2024 (I)]

16. Molality of 0.8 M H_2SO_4 solution (density 1.06 g cm^{-3}) is ____ $\times 10^{-3}$ m. **[NCERT : P L-5 | NV, Jan. 29, 2024 (II)]**

17. The quantity which changes with temperature is:
[NCERT : P L-4 | Jan. 27, 2024 (II)]
(a) Molarity (b) Mass percentage
(c) Molality (d) Mole fraction

18. A solution is prepared by adding 2g of "X" of 1 mole of water. Mass percent of "X" in the solution is :
[NCERT : P L-2 | April 11, 2023 (II)]
(a) 20% (b) 5% (c) 2% (d) 10%

19. Given below are two statements: one is labelled as **Assertion A** and the other is labelled as **Reason R**.
Assertion A: 3.1500g of hydrated oxalic acid dissolved in water to make 250.0 mL solution will result in 0.1 M oxalic acid solution.
Reason R: Molar mass of hydrated oxalic acid is 126 g mol^{-1}.
In the light of the above statements, choose the correct answer from the options given below:
[NCERT : P L-4 | April 10, 2023 (II)]
(a) Both A and R are true but R is NOT the correct explanation of A.
(b) A is false but R is true.
(c) A is true but R is false.
(d) Both A and R are true and R is the correct explanation of A.

20. The density of 3 M solution of NaCl is 1.0 g mL^{-1}. Molality of the solution is ________ $\times 10^{-2}$m. (Nearest integer).
Given: Molar mass of Na and Cl is 23 and 35.5 g mol^{-1} respectively. **[NCERT : P L-5 | NV, Feb. 01, 2023 (I)]**

21. The molality of a 10% (v/v) solution of di-bromine solution in CCl_4 (carbon tetrachloride) is 'x'.
x = ______ $\times 10^{-2}$ M. (Nearest integer)
[Given: Molar mass of Br_2 = 160 g mol^{-1}
atomic mass of C = 12 g mol^{-1}
atomic mass of Cl = 35.5 g mol^{-1}
density of dibromine = 3.2 g cm^{-3}
density of CCl_4 = 1.6 g cm^{-3}]
[NCERT : P L-5 | NV, Feb. 01, 2023 (II)]

22. Some amount of dichloromethane (CH_2Cl_2) is added to 671.141 mL of chloroform ($CHCl_3$) to prepare 2.6×10^{-3} M solution of CH_2Cl_2 (DCM).
The concentration of DCM is ______ ppm (by mass).
Given: Atomic mass : C = 12; H : 1; Cl = 35.5 density of $CHCl_3$ 1.49 g cm^{-3}. **[NCERT : P L-4 | NV, Jan. 30, 2023 (I)]**

23. 5 g of NaOH was dissolved in deionized water to prepare a 450 mL stock solution. What volume (in *mL*) of this solution would be required to prepare 500 mL of 0.1 M solution ?
Given : Molar Mass of Na, O and H is 23, 16 and 1 g mol^{-1} respectively. **[NV, Jan. 24, 2023 (I)]**

24. The number of units, which are used to express concentration of solutions from the following is ________. **[NCERT : P L-2, 3 | NV, Jan. 24, 2023 (II)]**
Mass percent, Mole, Mole fraction, Molarity, ppm, Molality

25. When 800 mL of 0.5 M nitric acid is heated in a beaker, its volume is reduced to half and 11.5 g of nitric acid is evaporated. The molarity of the remaining nitric acid solution is $x \times 10^{-2}$ M. (Nearest Integer)
[NCERT : P L-4 | NV, July 26, 2022 (I)]
(Molar mass of nitric acid is 63 g mol^{-1})

26. Given below are two statements : one is labelled as **Assertion (A)** and the other is labelled as **Reason (R)**
Assertion (A) : At 10°C, the density of a 5M solution of KCl [atomic masses of K and Cl are 39 & 35.5 g mol^{-1}] is x g mL^{-1}. The solution is cooled to –21°C. The molality of the solution will remain unchanged.
Reason (R) : The molality of a solution does not change with temperature as mass remains unaffected with temperature.
In the light of the above statements, choose the correct answer from the options given below:
[NCERT : P L-5 | June 27, 2022 (I)]
(a) Both (A) and (R) are true and (R) is the correct explanation of (A).
(b) Both (A) and (R) are true but (R) is not the correct explanation of (A).
(c) (A) is true but (R) is false.
(d) (A) is false but (R) is true.

27. If 80 g of copper sulphate $CuSO_4 \cdot 5H_2O$ is dissolved in deionised water to make 5 L of solution. The concentration of the copper sulphate solution is $x \times 10^{-3}$ mol L^{-1}. The value of x is _________.
[NCERT : P L-4 | NV, Sep. 1, 2021 (II)]
[Atomic masses Cu: 63.54u, S: 32u, O: 16u, H: 1u]

28. The molarity of the solution prepared by dissolving 6.3 g of oxalic acid ($H_2C_2O_4.2H_2O$) in 250 mL of water in mol L^{-1} is $x \times 10^{-2}$. The value of x is ________. (Nearest integer)
[Atomic mass : H : 1.0, C : 12.0, O : 16.0]
[NCERT : P L-4 | NV, Aug. 31, 2021 (I)]

29. An aqueous KCl solution of density 1.20 g mL^{-1} has a molality of 3.30 mol kg^{-1}. The molarity of the solution in mol L^{-1} is ________ (Nearest integer)
[Molar mass of KCl = 74.5] **[NCERT : P L-4 | NV, Aug. 26, 2021(I); Same March 16, 2021 (I)]**

30. 100 mL of Na_3PO_4 solution contains 3.45 g of sodium. The molarity of the solution is _____ $\times 10^{-2}$ mol L^{-1}. (Nearest integer)
[Atomic Masses - Na : 23.0 u, O : 16.0 u, P : 31.0 u]
[NCERT : P L-4 | NV, Aug. 26, 2021 (II)]

31. The mole fraction of a solute in a 100 molal aqueous solution is ____ $\times 10^{-2}$.
(Round off to the Nearest Integer).
[Given : Atomic masses : H : 1.0 u, O : 16.0 u]
[NCERT : P L-3 | NV, March 17, 2021 (I)]

32. A 6.50 molal solution of KOH (aq.) has a density of 1.89 g cm^{-3}. The molarity of the solution is ____ mol dm^{-3}. (Round off to the Nearest Integer).
[NCERT : P L-4 | March 16, 2021 (I)]
[Atomic masses : K : 39.0 u; O : 16.0 u; H : 1.0 u]

33. The mole fraction of a solvent in aqueous solution of a solute is 0.8. The molality (in mol kg^{-1}) of the aqueous solution is : **[NCERT : P L-4 | April 12, 2019 (I)]**
(a) 13.88×10^{-2} (b) 13.88×10^{-1}
(c) 13.88 (d) 13.88×10^{-3}

34. What would be the molality of 20% (mass/mass) aqueous solution of KI? (molar mass of KI = 166 g mol^{-1})
[NCERT : P L-5 | April 9, 2019 (II)]
(a) 1.08 (b) 1.35 (c) 1.48 (d) 1.51

35. A solution of sodium sulfate contains 92 g of Na^+ ions per kilogram of water. The molality of Na^+ ions in that solution in mol kg^{-1} is: **[NCERT : P L-5 | Jan. 9, 2019 (I)]**
(a) 12 (b) 4 (c) 8 (d) 16

36. The solubility of N_2 in water at 300 K and 500 torr partial pressure is 0.01 g L^{-1}. The solubility (in g L^{-1}) at 750 torr partial pressure is : **[NCERT : P L-10 | Online April 9, 2016]**
(a) 0.0075 (b) 0.005 (c) 0.02 (d) 0.015

37. The molarity of a solution obtained by mixing 750 mL of 0.5(M) HCl with 250 mL of 2(M) HCl will be :
[NCERT : P L-4 | 2013]
(a) 0.875 M (b) 1.00 M
(c) 1.75 M (d) 0.975 M

38. The density of a solution prepared by dissolving 120 g of urea (mol. mass = 60 u) in 1000 g of water is 1.15 g/mL. The molarity of this solution is : **[NCERT : P L-4 | 2012]**
(a) 0.50 M (b) 1.78 M
(c) 1.02 M (d) 2.05 M

Topic 2 - Vapour Pressure, Laws of Solutions and Ideal, Non-ideal Solutions

39. Which of the following binary mixture does not show the behaviour of minimum boiling azeotropes ?
[NCERT : P L-14 | April 8, 2025 (II)]
(a) $H_2O + CH_3COC_2H_5$ (b) $C_6H_5OH + C_6H_5NH_2$
(c) $CS_2 + CH_3COCH_3$ (d) $CH_3OH + CHCl_3$

40. Match **List-I** with **List-II**
[NCERT : P L-14 | April 7, 2025 (II)]

List-I	List-II
(A) Solution of chloroform and acetone	(I) Minimum boiling azeotrope
(B) Solution of ethanol and water	(II) Dimerizes
(C) Solution of benzene and toluene	(III) Maximum boiling azeotrope
(D) Solution of acetic acid in benzene	(IV) $\Delta V_{mix} = 0$

Choose the **correct** answer from the options given below:
(a) (A)-(III), (B)-(I), (C)-(IV), (D)-(II)
(b) (A)-(II), (B)-(IV), (C)-(I), (D)-(III)
(c) (A)-(III), (B)-(IV), (C)-(I), (D)-(II)
(d) (A)-(II), (B)-(I), (C)-(IV), (D)-(III)

41. Liquid A and B form an ideal solution. The vapour pressure of pure liquids A and B are 350 and 750 mm Hg respectively at the same temperature. If x_A and x_B are the mole fraction of A and B in solution while y_A and y_B are the mole fraction of A and B in vapour phase then :
[NCERT : P L-12 | April 7, 2025 (II)]
(a) $\frac{x_A}{x_B} < \frac{y_A}{y_B}$ (b) $\frac{x_A}{x_B} = \frac{y_A}{y_B}$
(c) $\frac{x_A}{x_B} > \frac{y_A}{y_B}$ (d) $(x_A - y_A) < (x_B - y_B)$

42. A solution is made by mixing one mole of volatile liquid A with 3 moles of volatile liquid B. The vapour pressure of pure A is 200 mm Hg and that of the solution is 500 mm Hg. The vapour pressure of pure B and the least volatile component of the solution, respectively, are :
[NCERT : P L-12 | April 2, 2025 (I)]
(a) 1400 mm Hg, A (b) 1400 mm Hg, B
(c) 600 mm Hg, B (d) 600 mm Hg, A

43. Consider a binary solution of two volatile liquid components 1 and 2. x_1 and y_1 are the mole fractions of component 1 in liquid and vapour phase, respectively. The slope and intercept of the linear plot of $\frac{1}{x_1}$ vs $\frac{1}{y_1}$ are given respectively as:
[NCERT : P L-13 | Jan. 23, 2025 (II)]
(a) $\frac{P_1^0}{P_2^0}, \frac{P_2^0 - P_1^0}{P_2^0}$ (b) $\frac{P_2^0}{P_1^0}, \frac{P_1^0 - P_2^0}{P_2^0}$
(c) $\frac{P_1^0}{P_2^0}, \frac{P_1^0 - P_2^0}{P_2^0}$ (d) $\frac{P_2^0}{P_1^0}, \frac{P_2^0 - P_1^0}{P_2^0}$

44. Identify the mixture that shows positive deviations from Raoult's Law **[NCERT : P L-14 | Jan. 31, 2024 (I)]**
(a) $CHCl_3 + (CH_3)_2CO$ (b) $CHCl_3 + C_6H_6$
(c) $(CH_3)_2CO + C_6H_5NH_2$ (d) $(CH_3)_2CO + CS_2$

45. A solution of two miscible liquids showing negative deviation from Raoult's law will have :
[NCERT : P L-13 | Jan. 27, 2024 (I)]
(a) increased vapour pressure, increased boiling point
(b) increased vapour pressure, decreased boiling point
(c) decreased vapour pressure, decreased boiling point
(d) decreased vapour pressure, increased boiling point

46. The vapour pressure of 30% (w/v) aqueous solution of glucose is ________ mm Hg at 25°C.
[NCERT : P L-3 | NV, April 15, 2023 (I)]
[Given : The density of 30% (w/v), aqueous solution of glucose is 1.2 g cm^{-3} and vapour pressure of pure water is 24 mm Hg.]
(Molar mass of glucose is 180 g mol^{-1}.)

47. The vapour pressure vs. temperature curve for a solution solvent system is shown below. **[NCERT : P L-12 | NV, April 08, 2023 (I)]**

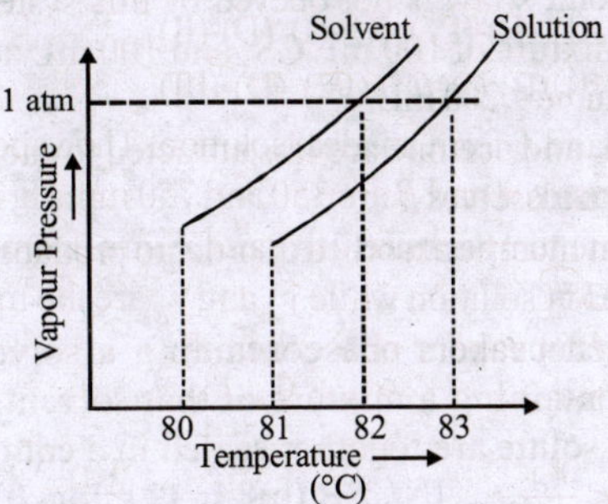

The boiling point of the solvent is ______°C

48. The total pressure of a mixture of non-reacting gases X (0.6 g) and Y (0.45 g) in a vessel is 740 mm of Hg. The partial pressure of the gas X is ______ mm of Hg. (Nearest Integer)
(Given : molar mass X = 20 and Y = 45 g mol^{-1})
[NCERT : P L-10 | NV, Jan. 31, 2023 (I)]

49. The Total pressure observed by mixing two liquid A and B is 350 mm Hg when their mole fractions are 0.7 and 0.3 respectively. **[NCERT : P L-10 | NV, Jan. 24, 2023 (II)]**
The Total pressure becomes 410 mm Hg if the mole fractions are changed to 0.2 and 0.8 respectively for A and B. The vapour pressure of pure A is ______ mm Hg. (Nearest integer)
Consider the liquids and solutions behave ideally.

50. If O_2 gas is bubbled through water at 303 K, the number of millimoles of O_2 gas that dissolve in 1 litre of water is______. (Nearest integer)
(Given : Henry's Law constant for O_2 at 303 K is 46.82 k bar and partial pressure of O_2 = 0.920 bar)
(Assume solubility of O_2 in water is too small, nearly negligible) **[NCERT : P L-7 | NV, July 29, 2022 (I)]**

51. A gaseous mixture of two substances A and B, under a total pressure of 0.8 atm is in equilibrium with an ideal liquid solution. The mole fraction of substance A is 0.5 in the vapour phase and 0.2 in the liquid phase. The vapour pressure of pure liquid A is ______ atm. (Nearest integer) **[NCERT : P L-10 | NV, July 28, 2022 (II)]**

52. A mixture of hydrogen and oxygen contains 40% hydrogen by mass when the pressure is 2.2 bar. The partial pressure of hydrogen is ______ bar. (Nearest Integer)
[NCERT : P L-10 | NV, July 26, 2022 (I)]

53. For a solution of the gases A, B, C and D in water at 298 K, the values of Henry's law constant (K_H) are 30.40, 2.34, 1.56×10^{-5} and 0.513 k bar respectively. In the given graph, the lines marked as 'p' and 's' correspond respectively to:
[NCERT : P L-7 | June 30, 2022 (I)]

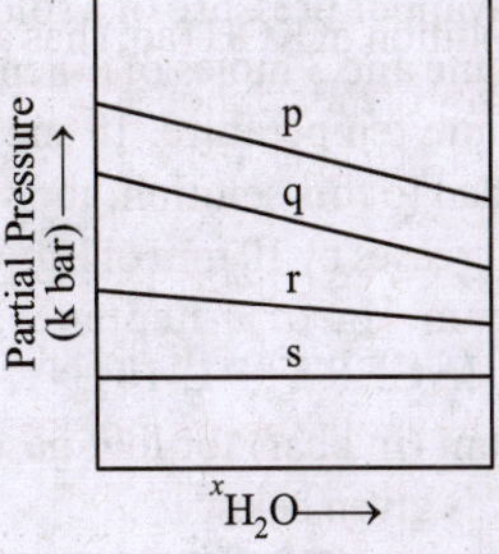

(a) A and C (b) B and A
(c) D and A (d) C and D

54. The vapour pressures of two volatile liquids A and B at 25°C are 50 Torr and 100 Torr, respectively. If the liquid mixture contains 0.3 mole fraction of A, then the mole fraction of liquid B in the vapour phase is $\frac{x}{17}$. The value of x is ______. **[NCERT : P L-10 | NV, June 28, 2022 (I)]**

55. A company dissolves 'X' amount of CO_2 at 298 K in 1 litre of water to prepare soda water X = ______ $\times 10^{-3}$g. (nearest integer)
(Given: partial pressure of CO_2 at 298 K= 0.835 bar.
Henry's law constant for CO_2 at 298 K = 1.67 kbar.
Atomic mass of H,C and O is 1, 12 and 6 g mol^{-1}, respectively) **[NCERT : P L-7 | NV, June 24, 2022 (II)]**

56. CO_2 gas is bubbled through water during a soft drink manufacturing process at 298 K. If CO_2 exerts a partial pressure of 0.835 bar then x m mol of CO_2 would dissolve in 0.9 L of water. The value of x is ______. (Nearest integer)
(Henry's law constant for CO_2 at 298 K is 1.67×10^3 bar)
[NCERT : P L-7 | NV, July 25, 2021 (I)]

57. At 20 °C, the vapour pressure of benzene is 70 torr and that of methyl benzene is 20 torr. The mole fraction of benzene in the vapour phase at 20 °C above an equimolar mixture of benzene and methyl benzene is ______ $\times 10^{-2}$. (Nearest integer) **[NCERT : P L-10 | NV, July 20, 2021 (I)]**

58. The vapour pressures of A and B at 25 °C are 90 mm Hg and 15 mm Hg respectively. If A and B are mixed such that the mole fraction of A in the mixture is 0.6, then the mole fraction of B in the vapour phase is $x \times 10^{-1}$. The value of x is ______. (Nearest integer)
[NCERT : P L-10 | NV, July 20, 2021 (II)]

59. The oxygen dissolved in water exerts a partial pressure of 20 kPa in the vapour above water. The molar solubility of oxygen in water is ______ $\times 10^{-5}$ mol dm^{-3}.
(Round off to the Nearest Integer).
[Given : Henry's law constant = $K_H = 8.0 \times 10^4$ kPa for O. Density of water with dissolved oxygen = 1.0 kg dm^{-3}]
[NCERT : P L-7 | NV, March 17, 2021 (I)]

60. At 363 K, the vapour pressure of A is 21 kPa and that of B is 18 kPa. One mole of A and 2 moles of B are mixed. Assuming that this solution is ideal, the vapour pressure of the mixture is ______ kPa.
(Round off to the Nearest Integer).
[NCERT : P L-10 | NV, March 16, 2021 (II)]

61. At 300 K, the vapour pressure of a solution containing 1 mole of *n*-hexane and 3 moles of *n*-heptane is 550 mm of Hg. At the same temperature, if one more mole of *n*-heptane is added to this solution, the vapour pressure of the solution increases by 10 mm of Hg. What is the vapour pressure in mm Hg of *n*-heptane in its pure state ______? **[NCERT : P L-10 | NV, Sep. 04, 2020 (I)]**

62. Henry's constant (in kbar) for four gases α, β, γ and δ in water at 298 K is given below : **[NCERT : P L-8 | Sep. 03, 2020 (I)]**

	α	β	γ	δ
K_H	50	2	2×10^{-5}	0.5

(density of water = 10^3 kg m^{-3} at 298 K)

This table implies that :

(a) α has the highest solubility in water at a given pressure
(b) solubility of γ at 308 K is lower than at 298 K
(c) The pressure of a 55.5 molal solution of γ is 1 bar
(d) The pressure of a 55.5 molal solution of δ is 250 bar

63. An open beaker of water in equilibrium with water vapour is in a sealed container. When a few grams of glucose are added to the beaker of water, the rate at which water molecules : **[NCERT : P L-12 | Sep. 02, 2020 (I)]**

(a) leaves the vapour increases
(b) leaves the solution increases
(c) leaves the solution decreases
(d) leaves the vapour decreases

64. A graph of vapour pressure and temperature for three different liquids *X*, *Y*, and *Z* is shown below:

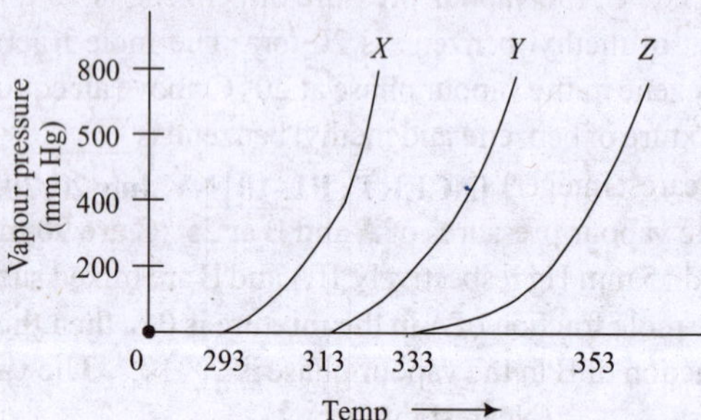

The following inferences are made: **[NCERT : P L-12 | Jan. 08, 2020 (I)]**

(A) *X* has higher intermolecular interactions compared to Y.
(B) *X* has lower intermolecular interactions compared to Y.
(C) *Z* has lower intermolecular interactions compared to Y.

The correct inference(s) is/are:

(a) (A) and (C) (b) (A)
(c) (B) (d) (C)

65. At 35 °C, the vapour pressure of CS_2 is 512 mm Hg and that of acetone is 344 mm Hg. A solution of CS_2 in acetone has a total vapour pressure of 600 mm Hg. The false statement amongst the following is: **[NCERT : P L-12 | Jan. 07, 2020 (I)]**

(a) Raoult's law is not obeyed by this system
(b) a mixture of 100 mL CS_2 and 100 mL acetone has a volume < 200 mL
(c) CS_2 and acetone are less attracted to each other than to themselves
(d) heat must be absorbed in order to produce the solution at 35 °C

66. Two open beakers one containing a solvent and the other containing a mixture of that solvent with a non volatile solute are together sealed in a container. Over time: **[NCERT : P L-12 | Jan. 07, 2020 (II)]**

(a) the volume of the solution increases and the volume of the solvent decreases
(b) the volume of the solution decreases and the volume of the solvent increases
(c) the volume of the solution and the solvent does not change
(d) the volume of the solution does not change and the volume of the solvent decreases

67. Liquid 'M' and liquid 'N' form an ideal solution. The vapour pressures of pure liquids 'M' and 'N' are 450 and 700 mmHg, respectively, at the same temperature. Then correct statement is: **[NCERT : P L-12 | April 9, 2019 (I)]**

(x_M = Mole fraction of 'M' in solution;
x_N = Mole fraction of 'N' in solution;
y_M = Mole fraction of 'M' in vapour phase;
y_N = Mole fraction of 'N' in vapour phase)

(a) $\frac{x_M}{x_N} = \frac{y_M}{y_N}$ (b) $(x_M - y_M) < (x_N - y_N)$
(c) $\frac{x_M}{x_N} < \frac{y_M}{y_N}$ (d) $\frac{x_M}{x_N} > \frac{y_M}{y_N}$

68. The vapour pressures of pure liquids A and B are 400 and 600 mmHg, respectively at 298 K. On mixing the two liquids, the sum of their initial volumes is equal to the volume of the final mixture. The mole fraction of liquid B is 0.5 in the mixture. The vapour pressure of the final solution, the mole fractions of components A and B in vapour phase, respectively are: **[NCERT : P L-12 | April 8, 2019 (I)]**

(a) 450 mmHg, 0.4, 0.6 (b) 500 mmHg, 0.5, 0.5
(c) 450 mmHg, 0.5, 0.5 (d) 500 mmHg, 0.4, 06

69. For the solution of the gases w, x, y and z in water at 298 K, the Henry's law constants (K_H) are 0.5, 2, 35 and 40 kbar, respectively. The correct plot for the given data is : **[NCERT : P L-8 | April 8, 2019 (II)]**

(a)
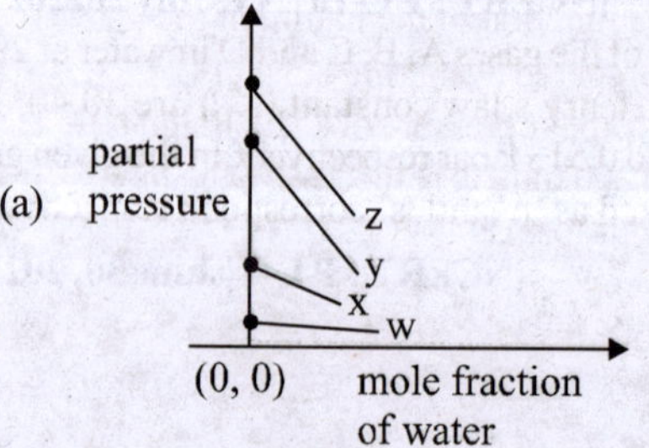

(b) 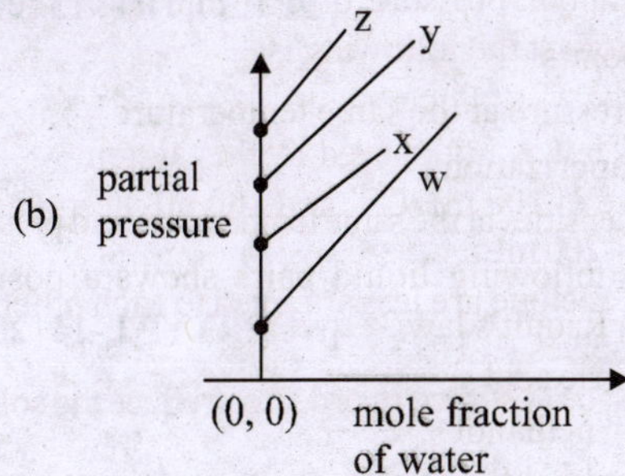

(c) 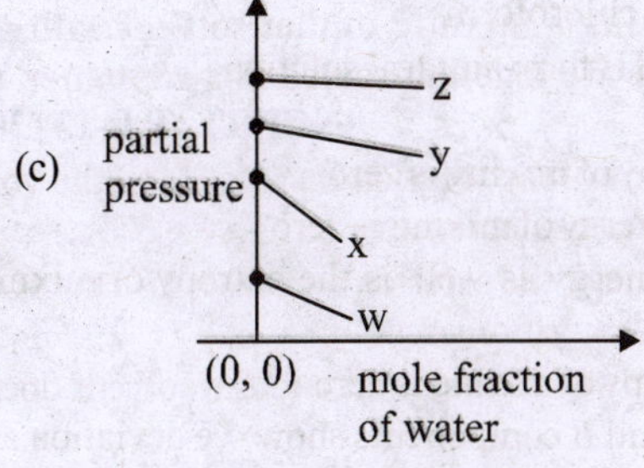

(d) 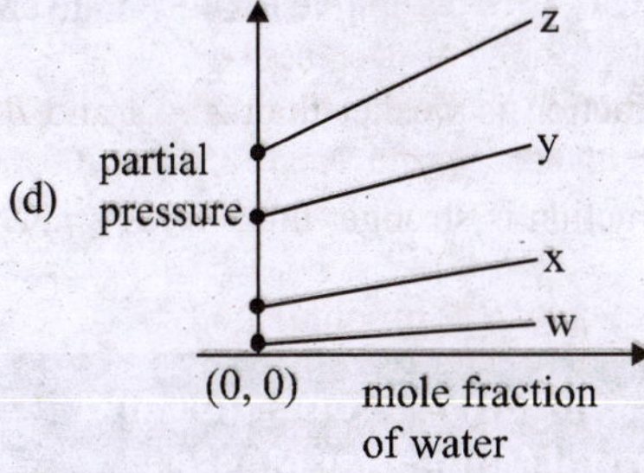

70. Liquids A and B form an ideal solution in the entire composition range. At 350 K, the vapor pressures of pure A and pure B are 7×10^3 Pa and 12×10^3 Pa, respectively. The composition of the vapour is in equilibrium with a solution containing 40 mole percent of A at this temperature is: **[NCERT : P L-10 | Jan. 10, 2019 (I)]**
(a) $x_A = 0.37; x_B = 0.63$ (b) $x_A = 0.28; x_B = 0.72$
(c) $x_A = 0.4; x_B = 0.6$ (d) $x_A = 0.76; x_B = 0.24$

71. Which one of the following statements regarding Henry's law is not correct? **[NCERT : P L-7 | Jan. 9, 2019 (I)]**
(a) Higher the value of K_H at a given pressure, higher is the solubility of the gas in liquids.
(b) Different gases have different K_H (Henry's law constant) values at the same temperature.
(c) The partial pressure of the gas in vapour phase is proportional to the mole fraction of the gas in the solution.
(d) The value of K_H increases with increase of temperature and K_H is function of the nature of the gas.

72. A solution is prepared by mixing 8.5 g of CH_2Cl_2 and 11.95 g of $CHCl_3$. If vapour pressure of CH_2Cl_2 and $CHCl_3$ at 298 K are 415 and 200 mmHg respectively, the mole fraction of $CHCl_3$ in vapour form is : **[NCERT : P L-10 | Online April 9, 2017]**
(Molar mass of Cl = 35.5 g mol^{-1})
(a) 0.162 (b) 0.675 (c) 0.325 (d) 0.486

73. The solubility of N_2 in water at 300 K and 500 torr partial pressure is 0.01 g L^{-1}. The solubility (in g L^{-1}) at 750 torr partial pressure is : **[NCERT : P L-7 | Online April 9, 2016]**
(a) 0.0075 (b) 0.005 (c) 0.02 (d) 0.015

74. A solution at 20 °C is composed of 1.5 mol of benzene and 3.5 mol of toluene. If the vapour pressure of pure benzene and pure toluene at this temperature are 74.7 torr and 22.3 torr, respectively, then the total vapour pressure of the solution and the benzene mole fraction in equilibrium with it will be, respectively : **[NCERT : P L-10 | Online April 10, 2015]**
(a) 35.8 torr and 0.280 (b) 38.0 torr and 0.589
(c) 30.5 torr and 0.389 (d) 30.5 torr and 0.480

75. Choose the correct statement with respect to the vapour pressure of a liquid among the following: **[NCERT : P L-16 | Online April 19, 2014]**
(a) Increases linearly with increasing temperature
(b) Increases non-linearly with increasing temperature
(c) Decreases linearly with increasing temperature
(d) Decreases non-linearly with increasing temperature

76. For an ideal solution of two components A and B, which of the following is true? **[NCERT : P L-13 | Online April 19]**
(a) $\Delta H_{mixing} < 0$ (zero)
(b) $\Delta H_{mixing} > 0$ (zero)
(c) A – B interaction is stronger than A – A and B – B interactions
(d) A– A, B – B and A – B interactions are identical.

77. Vapour pressure of pure benzene is 119 torr and that of toluene is 37.0 torr at the same temperature. Mole fraction of toluene in vapour phase which is in equilibrium with a solution of benzene and toluene having a mole fraction of toluene 0.50, will be : **[NCERT : P L-10 | Online April 23, 2013]**
(a) 0.137 (b) 0.237 (c) 0.435 (d) 0.205

78. Liquids A and B form an ideal solution. At 30 °C, the total vapour pressure of a solution containing 1 mol of A and 2 mol of B is 250 mm Hg. The total vapour pressure becomes 300 mm Hg when 1 more mol of A is added to the first solution. The vapour pressures of pure A and B at the same temperature are **[NCERT : P L-10 | Online May 19, 2012]**
(a) 150, 450 mm Hg (b) 125, 150 mm Hg
(c) 450, 150 mm Hg (d) 250, 300 mm Hg

79. On mixing, heptane and octane form an ideal solution. At 373 K, the vapour pressures of the two liquid components (heptane and octane) are 105 kPa and 45 kPa respectively. Vapour pressure of the solution obtained by mixing 25.0 g of heptane and 35 g of octane will be
(molar mass of heptane = 100 g mol^{-1} and of octane = 114 g mol^{-1}) **[NCERT : P L-10 | 2010]**
(a) 72.0 kPa (b) 36.1 kPa
(c) 96.2 kPa (d) 144.5 kPa

80. A binary liquid solution is prepared by mixing *n*-heptane and ethanol. Which one of the following statements is correct regarding the behaviour of the solution? **[NCERT : P L-13 | 2009]**

(a) The solution is non-ideal, showing – ve deviation from Raoult's Law.
(b) The solution is non-ideal, showing + ve deviation from Raoult's Law.
(c) n-Heptane shows + ve deviation while ethanol shows – ve deviation from Raoult's Law.
(d) The solution formed is an ideal solution.

81. Two liquids X and Y form an ideal solution. At 300 K, vapour pressure of the solution containing 1 mol of X and 3 mol of Y is 550 mmHg. At the same temperature, if 1 mol of Y is further added to this solution, vapour pressure of the solution increases by 10 mmHg. Vapour pressure (in mmHg) of X and Y in their pure states will be, respectively: **[NCERT : P L-10 | 2009]**
(a) 300 and 400 (b) 400 and 600
(c) 500 and 600 (d) 200 and 300

82. At 80 °C, the vapour pressure of pure liquid 'A' is 520 mm Hg and that of pure liquid 'B' is 1000 mm Hg. If a mixture solution of 'A' and 'B' boils at 80 °C and at 1 atm pressure, the amount of 'A' in the mixture is (1 atm = 760 mm Hg) **[NCERT : P L-10 | 2008]**
(a) 52 mol percent (b) 34 mol percent
(c) 48 mol percent (d) 50 mol percent

83. A mixture of ethyl alcohol and propyl alcohol has a vapour pressure of 290 mm at 300 K. The vapour pressure of propyl alcohol is 200 mm. If the mole fraction of ethyl alcohol is 0.6, its vapour pressure (in mm) at the same temperature will be **[NCERT : P L-10 | 2007]**
(a) 360 (b) 350 (c) 300 (d) 700

84. Equal masses of methane and oxygen are mixed in an empty container at 25 °C. The fraction of the total pressure exerted by oxygen is **[NCERT : P L-10 | 2007]**
(a) 1/2 (b) 2/3 (c) $\frac{1}{3} \times \frac{273}{298}$ (d) 1/3

85. Among the following mixtures, dipole-dipole as the major interaction, is present in **[NCERT : P L-13 | 2006]**
(a) KCl and water
(b) benzene and carbon tetrachloride
(c) benzene and ethanol
(d) acetonitrile and acetone

86. Benzene and toluene form nearly ideal solution. At 20°C, the vapour pressure of benzene is 75 torr and that of toluene is 22 torr. The partial vapour pressure of benzene at 20°C for a solution containing 78 g of benzene and 46 g of toluene in torr is **[NCERT : P L-10 | 2005]**
(a) 53.5 (b) 37.5 (c) 25 (d) 50

87. Equimolar solutions in the same solvent have **[2005]**
(a) Different boiling and different freezing points
(b) Same boiling and same freezing points
(c) Same freezing point but different boiling points
(d) Same boiling point but different freezing points

88. For which of the following parameters, the structural isomers C_2H_5OH and CH_3OCH_3 would be expected to have the same values?(Assume ideal behaviour). **[NCERT : P L-13 | 2004]**
(a) Boiling points
(b) Vapour pressure at the same temperature
(c) Heat of vaporization
(d) Gaseous densities at the same temperature and pressure

89. Which of the following liquid pairs shows a positive deviation from Raoult's law ? **[NCERT : P L-14 | 2004]**
(a) Water - nitric acid
(b) Benzene - methanol
(c) Water - hydrochloric acid
(d) Acetone - chloroform

90. If liquids A and B form an ideal solution **[NCERT : P L-13 | 2003]**
(a) the entropy of mixing is zero
(b) the free energy of mixing is zero
(c) the free energy as well as the entropy of mixing is zero
(d) the enthalpy of mixing is zero

91. In mixture A and B components show -ve deviation as **[NCERT : P L-13 | 2002]**
(a) $\Delta V_{mix} > 0$
(b) $\Delta H_{mix} < 0$
(c) $A - B$ interaction is weaker than $A - A$ and $B - B$ interaction
(d) $A - B$ interaction is stronger than $A - A$ and $B - B$ interaction.

Topic 3 –Colligative Properties and Abnormal Molecular Masses

92. $HA(aq) \rightleftharpoons H^+(aq) + A^-(aq)$
The freezing point depression of a 0.1 m aqueous solution of a monobasic weak acid HA is 0.20°C.
The dissociation constant for the acid is
Given : **[NCERT : P L-19, 24 | April 8, 2025 (II)]**
$K_f(H_2O) = 1.8$ K kg mol^{-1}, molality ≡ molarity
(a) 1.38×10^{-3} (b) 1.1×10^{-2}
(c) 1.90×10^{-3} (d) 1.89×10^{-1}

93. The percentage dissociation of a salt (MX_3) solution at given temperature (van't Hoff factor i = 2) is.................. % (Nearest integer) **[NCERT : P L-24 | April 7, 2025 (I)]**

94. Given below are two statements : **[NCERT : P L-19 | April 4, 2025 (II)]**
Statement (I) : Molal depression constant K_f is given by $\frac{M_1 R T_f}{\Delta S_{fus}}$, where symbols have their usual meaning.
Statement (II) : K_f for benzene is less than the K_f for water.
In the light of the above statements, choose the **most appropriate answer** from the options given below :
(a) Statement I is incorrect but Statement II is correct
(b) Both Statement I and Statement II are incorrect
(c) Both Statement I and Statement II are correct
(d) Statement I is correct but Statement II is incorrect

95. XY is the membrane / partition between two chambers 1 and 2 containing sugar solutions of concentration c_1 and c_2 ($c_1 > c_2$) mol L^{-1}. For the reverse osmosis to take place identify the correct condition
(Here p_1 and p_2 are pressures applied on chamber 1 and 2)
[NCERT : P L-21 | April 4, 2025 (I)]

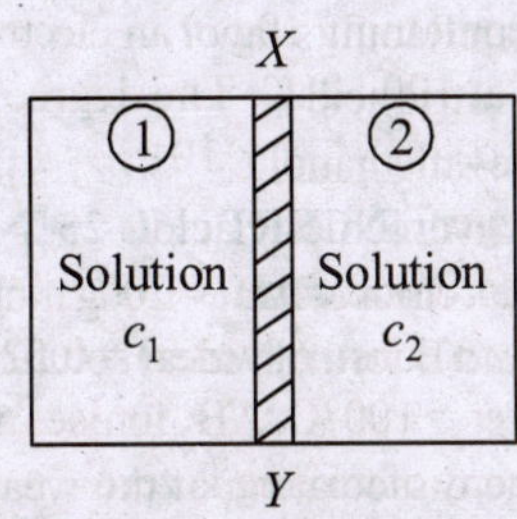

(A) Membrane/Partition ; Cellophane, $p_1 > \pi$
(B) Membrane/Partition ; Porous. $p_2 > \pi$
(C) Membrane/Partition ; Parchment paper, $p_1 > \pi$
(D) Membrane/Partition : Cellophane, $p_2 > \pi$
Choose the **correct** answer from the option given below :
(a) B and D only (b) A and D only
(c) A and C only (d) C only

96. 2 moles each of ethylene glycol and glucose are dissolved in 500 g of water. The boiling point of the resulting solution is : **[NCERT : P L-17 | April 3, 2025 (I)]**
(Given : Ebullioscopic constant of water = 0.52 K kg mol^{-1})
(a) 379.2 K (b) 377.3 K
(c) 375.3 K (d) 277.3 K

97. When 1 g each of compounds AB and AB_2 are dissolved in 15 g of water separately, they increased the boiling point of water by 2.7 K and 1.5 K respectively. The atomic mass of A (in amu) is ______ $\times 10^{-1}$(Nearest integer)
(Given : Molal boiling point elevation constant is 0.5 K kg mol^{-1}) **[NCERT : P L-17 | April 2, 2025 (II)]**

98. Given below are two statements :
[NCERT : P L-19 | Jan. 29, 2025 (II)]
Statement (I) : NaCl is added to the ice at 0°C, present in the ice cream box to prevent the melting of ice cream.
Statement (II) : On addition of NaCl to ice at 0°C, there is a depression in freezing point.
In the light of the above statements, choose the correct answer from the options given below :
(a) Statement I is false but Statement II is true
(b) Both Statement I and Statement II are true
(c) Both Statement I and Statement II are false
(d) Statement I is true but Statement II is false

99. If A_2B is 30% ionised in an aqueous solution, then the value of van't Hoff factor (i) is ______ $\times 10^{-1}$.
[NCERT : P L-24 | Jan. 29, 2025 (I)]

100. 1.24 g of AX_2 (molar mass 124 g mol^{-1}) is dissolved in 1 kg of water to form a solution with boiling point of 100.0156°C, while 25.4 g of AY_2 (molar mass 250 g mol^{-1}) in 2 kg of water constitutes a solution with a boiling point of 100.0260°C. **[NCERT : P L-24 | Jan. 29, 2025 (I)]**
$K_b(H_2O) = 0.52$ K kg mol^{-1}
Which of the following is correct ?
(a) AX_2 and AY_2 (both) are completely unionised.
(b) AX_2 and AY_2 (both) are fully ionised.
(c) AX_2 is completely unionised while AY_2 is fully ionised.
(d) AX_2 is fully ionised while AY_2 is completely unionised.

101. Assume a living cell with 0.9% $\left(\frac{\omega}{\omega}\right)$ of glucose solution (aqueous). This cell is immersed in another solution having equal mole fraction of glucose and water.
(Consider the data upto first decimal place only)
The cell will : **[NCERT : P L-21 | Jan. 28, 2025 (II)]**
(a) shrink since soluton is 0.5 % $\left(\frac{\omega}{\omega}\right)$
(b) shrink since solution is 0.45% $\left(\frac{\omega}{\omega}\right)$ as a result of association of glucose molecules (due to hydrogen bonding)
(c) swell up since solution is 1% $\left(\frac{\omega}{\omega}\right)$
(d) Show no change in volume since solution is 0.9% $\left(\frac{\omega}{\omega}\right)$

102. What is the freezing point depression constant of a solvent, 50 g of which contain 1 g non volatile solute (molar mass 256 g mol^{-1}) and the decrease in freezing point is 0.40 K?
[NCERT : P L-19 | Jan. 28, 2025 (I)]
(a) 5.12 K kg mol^{-1} (b) 4.43 K kg mol^{-1}
(c) 1.86 K kg mol^{-1} (d) 3.72 K kg mol^{-1}

103. The observed and normal masses of compound MX_2 are 65.6 and 164 respectively. The percent degree of ionisation of MX_2 is ______%. (Nearest integer)
[NCERT : P L-24 | Jan. 24, 2025 (II)]

104. Consider the given plots of vapour pressure (VP) vs temperature (T/K) Which amongst the following options is correct graphical representation showing ΔT_f, depression in the freezing point of solvent in a solution ?
[NCERT : P L-19 | Jan. 24, 2025 (I)]

(a) 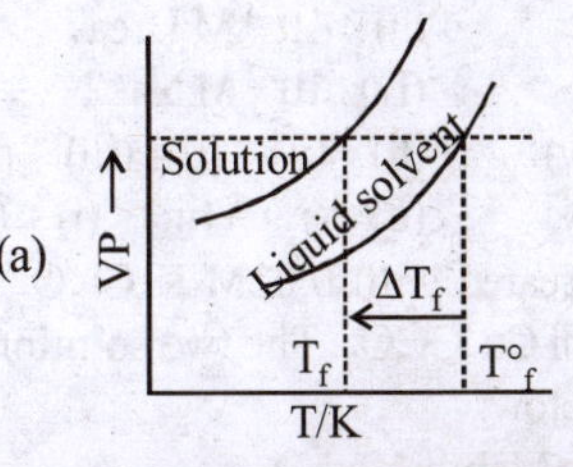

(b) 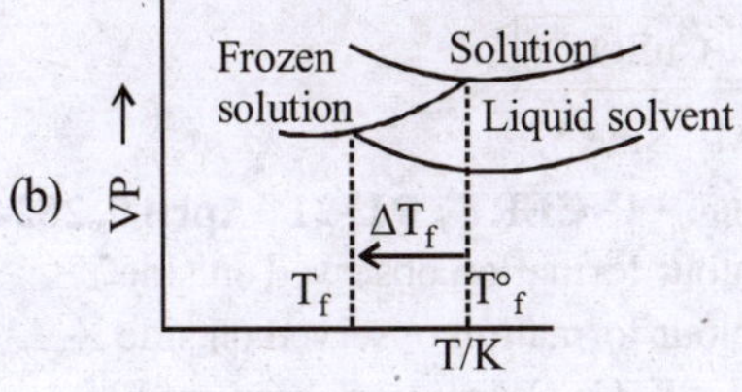

(c)

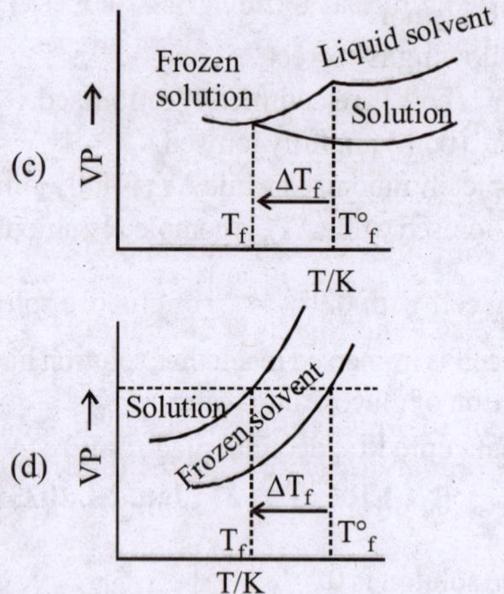

(d)

105. One mole of the octahedral complex compound $Co(NH_3)_5Cl_3$ gives 3 moles of ions on dissolution in water. One mole of the same complex reacts with excess of $AgNO_3$ solution to yield two moles of $AgCl_{(s)}$. The structure of the complex is: **[NCERT : P L-24 | Jan. 24, 2025 (I)]**

(a) $[Co(NH_3)_5Cl]Cl_2$ (b) $[Co(NH_3)_4Cl].Cl_2.NH_3$
(c) $[Co(NH_3)_4Cl_2]Cl.NH_3$ (d) $[Co(NH_3)_3Cl_3].2NH_3$

106. When a non-volatile solute is added to the solvent, the vapour pressure of the solvent decreases by 10 mm of Hg. The mole fraction of the solute in the solution is 0.2. What would be the mole fraction of the solvent if decrease in vapour pressure is 20 mm of Hg? **[NCERT : P L-16 | Jan. 23, 2025 (II)]**

(a) 0.6 (b) 0.4
(c) 0.2 (d) 0.8

107. $CrCl_3$. xNH_3 can exist as a complex. 0.1 molal aqueous solution of this complex shows a depression in freezing point of 0.558°C. Assuming 100% ionisation of this complex and coordination number of Cr is 6, the complex will be (Given K_f = 1.86 K kg mol^{-1}) **[NCERT : P L-24 | Jan. 23, 2025 (I)]**

(a) $[Cr(NH_3)_6]Cl_3$ (b) $[Cr(NH_3)_4Cl_2]Cl$
(c) $[Cr(NH_3)_5Cl]Cl_2$ (d) $[Cr(NH_3)_3Cl_3]$

108. Arrange the following solutions in order of their increasing boiling points. **[NCERT : P L-24 | Jan. 22, 2025 (I)]**

(i) 10^{-4} M NaCl (ii) 10^{-4} M Urea
(iii) 10^{-3} M NaCl (iv) 10^{-2} M NaCl
(a) (ii) < (i) < (iii) < (iv) (b) (ii) < (i) ≅ (iii) < (iv)
(c) (i) < (ii) < (iii) < (iv) (d) (iv) < (iii) < (i) < (ii)

109. 0.05M $CuSO_4$ when treated with 0.01M $K_2Cr_2O_7$ gives green colour solution of $Cu_2Cr_2O_7$. The two solutions are separated as shown below:
[SPM : Semi Permeable Membrane]

$K_2Cr_2O_7$	$CuSO_4$

Side X SPM Side Y

Due to osmosis : **[NCERT : P L-21 | April 9, 2024 (I)]**

(a) Green colour formation observed on side Y.
(b) Green colour formation observed on side X.
(c) Molarity of $K_2Cr_2O_7$ solution is lowered.
(d) Molarity of $CuSO_4$ solution is lowered.

110. The vapour pressure of pure benzene and methyl benzene at 27°C is given as 80 Torr and 24 Torr, respectively. The mole fraction of methyl benzene in vapour phase, in equilibrium with an equimolar mixture of those two liquids (ideal solution) at the same temperature is ____ × 10^{-2} (nearest integer) **[NCERT : P L-15 | NV, April 9, 2024 (II)]**

111. A solution containing 10g of an electrolyte AB_2 in 100g of water boils at 100.52°C. The degree of ionization of the electrolyte (α) is ____ × 10^{-1}.
(nearest integer) **[NCERT : P L-25 | NV, April 8, 2024 (I)]**
[Given : Molar mass of AB_2 = 200g mol^{-1}. K_b (molal boiling point elevation const. of water) = 0.52 K kg mol^{-1}, boiling point of water = 100°C; AB_2 ionises as $AB_2 \rightarrow A^{2+} + 2B^-$]

112. Consider the dissociation of the weak acid HX as given below **[NV, April 6, 2024 (I)]**
$HX(aq) \rightleftharpoons H^+(aq) + X^-(aq)$, $K_a = 1.2 \times 10^{-5}$
[K_a : dissociation constant]
The osmotic pressure of 0.03 M aqueous solution of HX at 300 K is ____ × 10^{-2} bar (nearest integer).
[Given : R = 0.083 L bar Mol| K^{-1}]

113. When 'x' × 10^{-2} mL methanol (molar mass = 32 g; density = 0.792 g/cm^3) is added to 100 mL water (density = 1 g/cm^3), the following diagram is obtained. **[NCERT : P L-18 | NV, April 6, 2024 (II)]**

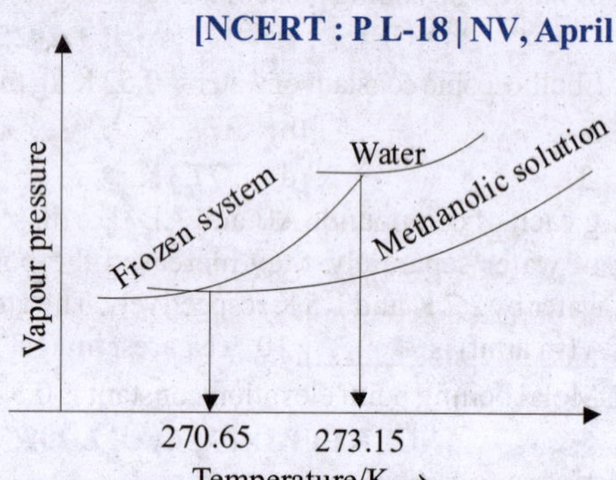

x = ...(nearest integer)
[Given: Molal freezing point depression constant of water at 273.15 K is 1.86 K kg mol^{-1}]

114. An artificial cell is made by encapsulating 0.2 M glucose solution within a semipermeable membrane. The osmotic pressure developed when the artificial cell is placed within a 0.05 M solution of NaCl at 300 K is ______ × 10^{-1} bar. (Nearest Integer) **[NCERT : P L-20 | NV, April 5, 2024 (I)]**
[Given : R = 0.083 L bar mol^{-1} K^{-1}]
Assume complete dissociation of NaCl

115. Considering acetic acid dissociates in water, its dissociation constant is 6.25 × 10^{-5}. If 5 mL of acetic acid is dissolved in 1 litre water, the solution will freeze at –x × 10^{-2} °C, provided pure water freezes at 0 °C. **[NCERT : P L-18 | NV, April 5, 2024 (II)]**
x = ________. (Nearest integer)
Given : $(K_f)_{water}$ = 1.86 K kg mol^{-1}.
density of acetic acid is 1.2 g mol^{-1}
molar mass of water = 18 g mol^{-1}.
molar mass of acetic acid = 60 g mol^{-1}.

density of water = 1 g cm^{-3}
Acetic acid dissociates as

$CH_3COOH \rightleftharpoons CH_3COO^{\ominus} + H^{\oplus}$

116. 2.5 g of a non-volatile, non-electrolyte is dissolved in 100 g of water at 25°C. The solution showed a boiling point elevation by 2°C. Assuming the solute concentration in negligible with respect to the solvent concentration, the vapour pressure of the resulting aqueous solution is _____ mm of Hg (nearest integer)

[Given : Molal boiling point elevation constant of water $(K_b) = 0.52$ K. kg mol^{-1},
1 atm pressure = 760 mm of Hg, molar mass of water = 18 g mol^{-1}] **[NCERT : P L-17 | NV, April 4, 2024 (I)]**

117. 2.7 Kg of each of water and acetic acid are mixed, The freezing point of the solution will be –x °C. Consider the acetic acid does not dimerise in water, nor dissociates in water x = _____(nearest integer)

[Given : Molar mass of water = 18 g mol^{-1}, acetic acid = 60 g mol^{-1}] **[NCERT : P L-17 | NV, April 4, 2024 (II)]**

$K_{F(H_2O)}$: 1.86 K kg mol^{-1}

$K_{F(\text{acetic acid})}$: 3.90 K kg mol^{-1}

freezing point : H_2O = 273 K, acetic acid = 290 K]

118. We have three aqueous solution of NaCl labelled as 'A' 'B' and 'C' with concentration 0.1 M, 0.01 M and 0.001 M, respectively. The value of van't Hoff factor (i) for these solution will be in the order:

[NCERT : P L-25 | Feb. 1, 2024 (I)]

(a) $i_A < i_B < i_C$ (b) $i_A < i_C < i_B$
(c) $i_A = i_B = i_C$ (d) $i_A > i_B > i_C$

119. Mass of ethylene glycol (antifreeze) to be added to 18.6 kg of water to protect the freezing point at – 24°C is _____ kg (Molar mass in g mol^{-1} for ethylene glycol 62 K_f of water = 1.86 K kg mol^{-1}).

[NCERT : P L-19 | NV, Feb. 1, 2024 (II)]

120. What happens to freezing point of benzene when small quantity of napthalene is added to benzene?

[NCERT : P L-18 | Jan. 30, 2024 (I)]

(a) Increases
(b) Remains unchanged
(c) First decreases and then increases
(d) Decreases

121. The solution from the following with highest depression in freezing point / lowest freezing point is

[NCERT : P L-19 | Jan. 30, 2024 (II)]

(a) 180 g of acetic acid dissolved in water
(b) 180 g of acetic acid dissolved in benzene
(c) 180 g of benzoic acid dissolved in benzene
(d) 180 g of glucose dissolved in water

122. The osmotic pressure of a dilute solution is 7×10^5 Pa at 273K. Osmotic pressure of the same solution at 283K is _____ $\times 10^4$ Nm^{-2}.

NCERT : P L-21 | NV, Jan. 29, 2024 (I)]

123. Solution of 12 g of non - electrolyte (A) prepared by dissolving it in 1000 mL of water exerts the same osmotic pressure as that of 0.05 M glucose solution at the same temperature. The empirical formula of A is CH_2O. The molecular mass of A is _____ g. (Nearest integer)

[NCERT : P L-21 | NV, April 13, 2023 (I)]

124. Sea water contains 29.25% NaCl and 19% $MgCl_2$ by weight of solution. The normal boiling point of the sea water is _____ °C (Nearest integer)

Assume 100% ionization for both NaCl and $MgCl_2$
Given : $K_b(H_2O) = 0.52$ K kg mol^{-1}
Molar mass of NaCl and $MgCl_2$ is 58.5 and 95 g mol^{-1} respectively. **[NCERT : P L-17 | NV, April 13, 2023 (II)]**

125. 80 mole percent of $MgCl_2$ is dissociated in aqueous solution. The vapour pressure of 1.0 molal aqueous solution of $MgCl_2$ at 38°C is _____ mm Hg. (Nearest integer)

Given : Vapour pressure of water at 38°C is 50 mm Hg

[NCERT : P L-16 | NV, April 12, 2023 (I)]

126. 0.004 M K_2SO_4 solution is isotonic with 0.01 M glucose solution. Percentage dissociation of K_2SO_4 is _____ (Nearest integer) **[NCERT : P L-21 | NV, April 11, 2023 (I)]**

127. What weight of glucose must be dissolved in 100 g of water to lower the vapour pressure by 0.20 mm Hg? (Assume dilute solution is being formed)

Given : Vapour pressure of pure water is 54.2 mm Hg at room temperature. Molar mass of glucose is 180 g mol^{-1}.

[NCERT : P L-16 | April 11, 2023 (II)]

(a) 4.69 g (b) 3.59 g
(c) 2.59 g (d) 3.69 g

128. If the degree of dissociation of aqueous solution of weak monobasic acid is determined to be 0.3, then the observed freezing point will be _____ % higher than the expected/ theoretical freezing point. (Nearest integer)

[NCERT : P L-19 | NV, April 10, 2023 (I)]

129. An aqueous solution of volume 300 cm^3 contains 0.63 g of protein. The osmotic pressure of the solution at 300 K is 1.29 mbar. The molar mass of the protein is _____ g mol^{-1}
Given : R = 0.083 L bar $K^{-1}mol^{-1}$

[NCERT : P L-21 | NV, April 10, 2023 (II)]

130. If the boiling points of two solvents X and Y (having same molecular weights) are in the ratio 2 : 1 and their enthalpy of vaporizations are in the ratio 1 : 2, then the boiling point elevation constant of X is m times the boiling point elevation constant of Y. The value of m is _____ (Nearest integer) **[NCERT : P L-17 | NV, April 08, 2023 (II)]**

131. Mass of Urea (NH_2CONH_2) required to be dissolved in 1000 g of water to reduce the vapour pressure of water by 25% is .... g. (Nearest integer)

Given: Molar mass of N, C, O and H are 14, 12, 16 and 1 g mol^{-1} respectively. **[NCERT : P L-17 | NV, April 06, 2023 (I)]**

132. Consider the following pairs of solution which will be isotonic at the same temperature. The number of pairs of solutions is/are ....

[NCERT : P L-21 | NV, April 06, 2023 (II)]

A. 1 M aq. NaCl and 2 M aq. Urea
B. 1 M aq. $CaCl_2$ and 1.5 M aq. KCl
C. 1.5 M aq. $AlCl_3$ and 2 M aq. Na_2SO_4
D. 2.5 M aq. KCl and 1 M aq. $Al_2(SO_4)_3$

133. 25 mL of an aqueous solution of KCl was found to require 20 mL of 1 M $AgNO_3$ solution when titrated using K_2CrO_4 as an indicator. What is the depression in freezing point of KCl solution of the given concentration? ____ (Nearest integer). **[NCERT : P L-19 | NV, Feb. 01, 2023 (I)]**
(Given: $K_f = 2.0$ K kg mol^{-1})
Assume 1) 100% ionization and
2) density of the aqueous solution as 1 g mL^{-1}

134. 20% of acetic acid is dissociated when its 5 g is added to 500 mL of water. The depression in freezing point of such water is ______ $\times 10^{-3}$ °C.
[NCERT : P L-19 | NV, Feb. 01, 2023 (II)]
Atomic mass of C, H and O are 12, 1 and 16 a.m.u. respectively.
[Given : Molal depression constant and density of water are 1.86 K kg mol^{-1} and 1 g cm^{-3} respectively.

135. At 27°C, a solution containing 2.5 g of solute in 250.0 mL of solution exerts an osmotic pressure of 400 Pa. The molar mass of the solute is ____ g mol^{-1} (Nearest integer)
(Given : R = 0.083 L bar K^{-1} mol^{-1})
[NCERT : P L-21 | NV, Jan. 31, 2023 (I)]

136. Evaluate the following statements for their correctness.
[NCERT : P L-17 | Jan. 31, 2023 (II)]
(A) The elevation in boiling point temperature of water will be same for 0.1 M NaCl and 0.1 M urea.
(B) Azeotropic mixtures boil without change in their composition
(C) Osmosis always takes place from hypertonic to hypotonic solution
(D) The density of 32% H_2SO_4 solution having molarity 4.09 M is approximately 1.26g mL^{-1}
(E) A negatively charged sol is obtained when KI solution is added to silver nitrate solution.
Choose the correct answer from the options given below:
(a) B, D, and E only (b) A, B, and D only
(c) A and C only (d) B and D only

137. A solution containing 2 g of a non-volatile solute in 20 g of water boils at 373.52 K. The molecular mass of the solute is ______ g mol^{-1}. (Nearest integer)
[NCERT : P L-17 | NV, Jan. 30, 2023 (I)]
Given, water boils at 373 K, K_b for water = 0.52 kg mol^{-1}.

138. Lead storage battery contains 38% by weight solution of H_2SO_4. The van't Hoff factor is 2.67 at this concentration. The temperature in Kelvin at which the solution in the battery will freeze is ____ (Nearest integer).
Given $K_f = 1.8$ K kg mol^{-1}.
[NCERT : P L-24 | NV, Jan. 30, 2023 (II)]

139. Solid Lead nitrate is dissolved in 1 litre of water. The solution was found to boil at 100.15°C. When 0.2 mol of NaCl is added to the resulting solution, it was observed that the solution froze at –0.8° C. The solubility product of $PbCl_2$ formed is ______ $\times 10^{-6}$ at 298 K. (Nearest integer)
Given : $K_b = 0.5$ K kg mol^{-1} and $K_f = 1.8$ K kg mol^{-1}.
Assume molality to be equal to molarity in all cases.
[NCERT : P L-19 | NV, Jan. 29, 2023 (I)]

140. Match List I with List II. **[NCERT : P L-24 | Jan. 29, 2023 (II)]**

List I	List II
A. van't Hoff factor, i	I. Cryoscopic constant
B. k_f	II. Isotonic solutions
C. Solutions with same osmotic pressure	III. Normal molar mass / Abnormal molar mass
D. Azeotropes	IV. Solutions with same composition of vapour above it

Choose the correct answer from the options given below:
(a) A-III, B-I, C-II, D-IV (b) A-III, B-II, C-I, D-IV
(c) A-III, B-I, C-IV, D-II (d) A-I, B-III, C-II, D-IV

141. The osmotic pressure of solutions of PVC in cyclohexanone at 300 K are plotted on the graph. The molar mass of PVC is ________ g mol^{-1} (Nearest integer) **[NCERT : P L-21 | NV, Jan. 25, 2023 (I)]**

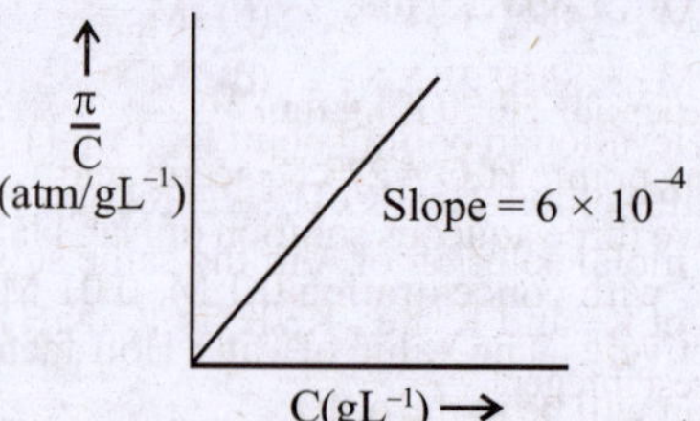

(Given : R = 0.083 L atm K^{-1} mol^{-1})

142. The number of pairs of the solution having the same value of the osmotic pressure from the following is ________.
(Assume 100% ionization)
[NCERT : P L-24 | NV, Jan. 25, 2023 (II)]
A. 0.500 M C_2H_5OH (aq) and 0.25 M KBr (aq)
B. 0.100 M $K_4[Fe(CN)_6]$ (aq) and 0.100 M $FeSO_4(NH_4)_2SO_4$ (aq)
C. 0.05 M $K_4[Fe(CN)_6]$ (aq) and 0.25 M NaCl (aq)
D. 0.15 M NaCl (aq) and 0.1 M $BaCl_2$ (aq)
E. 0.02 M KCl. $MgCl_2$. $6H_2O$ (aq) and 0.05 M KCl (aq)

143. In the depression of freezing point experiment
[NCERT : P L-19 | Jan. 24, 2023 (I)]
A. Vapour pressure of the solution is less than that of pure solvent
B. Vapour pressure of the solution is more than that of pure solvent
C. Only solute molecules solidify at the freezing point
D. Only solvent molecules solidify at the freezing point
(a) A and D only (b) B and C only
(c) A and C only (d) A only

144. 1.80 g of solute A was dissolved in 62.5 cm^3 of ethanol and freezing point of the solution was found to be 155.1 K. The molar mass of solute A is __ g mol^{-1}.
[Given: Freezing point of ethanol is 156.0 K.

Density of ethanol is 0.80 g cm^{-3}.
Freezing point depression constant of ethanol is 2.00 K kg mol^{-1}] **[NCERT : P L-19 | NV, July 29, 2022 (II)]**

145. 150 g of acetic acid was contaminated with 10.2 g ascorbic acid ($C_6H_8O_6$) to lower down its freezing point by $(x \times 10^{-1})$°C. The value of x is _______. (Nearest integer) [Given K_f = 3.9 K kg mol^{-1}; Molar mass of ascorbic acid = 176 g mol^{-1}] **[NCERT : P L-19 | NV, July 28, 2022 (I)]**

146. When a certain amount of solid A is dissolved in 100 g of water at 25°C to make a dilute solution, the vapour pressure of the solution is reduced to one-half of that of pure water. The vapour pressure of pure water is 23.76 mmHg. The number of moles of solute A added is_______. (Nearest integer) **[NCERT : P L-16 | NV, July 27, 2022 (II)]**

147. Boiling point of a 2% aqueous solution of a non-volatile solute A is equal to the boiling point of 8% aqueous solution of a non-volatile solute B. The relation between molecular weights of A and B is. **[NCERT : P L-17 | July 27, 2022 (I)]**
(a) $M_A = 4M_B$ (b) $M_B = 4M_A$
(c) $M_A = 8M_B$ (d) $M_B = 8M_A$

148. The elevation in boiling point for 1 molal solution of non-volatile solute A is 3K. The depression in freezing point for 2 molal solution of A in the same solvent is 6 K. The ratio of K_b and K_f i.e., K_b/K_f is 1 : X. The value of X is [nearest integer] **[NCERT : P L-19 | NV, July 26, 2022 (II)]**

149. The depression in freezing point observed for a formic acid solution of concentration 0.5 mL L^{-1} is 0.0405°C. Density of formic acid is 1.05 g mL^{-1}. The Van't Hoff factor of the formic acid solution is nearly:
(Given for water k_f = 1.86 K kg mol^{-1})
[NCERT : P L-24 | July 25, 2022 (I)]
(a) 0.8 (b) 1.1 (c) 1.9 (d) 2.4

150. Two solutions A and B are prepared by dissolving 1g of non-volatile solutes X and Y, respectively in 1 kg of water. The ratio of depression in freezing points for A and B is found to be 1 : 4. The ratio of molar masses of X and Y is: **[NCERT : P L-19 | July 25, 2022 (II)]**
(a) 1 : 4 (b) 1 : 0.25 (c) 1 : 0.20 (d) 1 : 5

151. 1.2 mL of acetic acid is dissolved in water to make 2.0 L of solution. The depression in freezing point observed for this strength of acid is 0.0198 °C. The percentage of dissociation of the acid is _______. (Nearest integer)
[Given : Density of acetic acid is 1.02 g mL^{-1}
Molar mass of acetic acid is 60 g mol^{-1}
$K_f(H_2O)$ = 1.85 K kg mol^{-1}]
[NCERT : P L-19 | NV, June 29, 2022 (I)]

152. Elevation in boiling point for 1.5 molal solution of glucose in water is 4K. The depression in freezing point for 4.5 molal solution of glucose in water is 4K. The ratio of molal elevation constant to molal depression constant (K_b/K_f) is____. **[NCERT : P L-24 | NV, June 29, 2022 (II)]**

153. 2.5 g of protein containing only glycine ($C_2H_5NO_2$) is dissolved in water to make 500 mL of solution. The osmotic pressure of this solution at 300 K is found to be 5.03×10^{-3} bar. The total number of glycine units present in the protein is____ **[NCERT : P L-21 | NV, June 28, 2022 (II)]**
(Given : R = 0.083 L bar K^{-1} mol^{-1})

154. 2g of a non-volatile non-electrolyte solute is dissolved in 200 g of two different solvents A and B whose ebullioscopic constants are in the ratio of 1 : 8. The elevation in boiling points of A and B are in the ratio $\frac{x}{y}$ $(x : y)$. The value of y is________ (Nearest integer)
[NCERT : P L-17 | NV, June 27, 2022 (I)]

155. A solution containing 2.5×10^{-3} kg of a solute dissolved in 75×10^{-3} kg of water boils at 373.535 K. The molar mass of the solute is ______ g mol^{-1}.
[nearest integer] (Given: $K_b(H_2O)$ = 0.52 K Kg mol^{-1}, boiling point of water = 373.15 K)
[NCERT : P L-17 | NV, June 27, 2022 (II)]

156. A 0.5 percent solution of potassium chloride was found to freeze at –0.24°C. The percentage dissociation of potassium chloride is ________. (Nearest integer)
(Molal depression constant for water is 1.80 K kg mol^{-1} and molar mass of KCl is 74.6 g mol^{-1})
[NCERT : P L-24 | NV, June 26, 2022 (I)]

157. The osmotic pressure exerted by a solution prepared by dissolving 2.0 g of protein of molar mass 60 kg mol^{-1} in 200 mL of water at 27°C is _______ Pa. [integer value]
(Use R = 0.083 L bar mol^{-1} K^{-1})
[NCERT : P L-21 | NV, June 26, 2022 (II)]

158. Solute A associates in water. When 0.7 g of solute A is dissolved in 42.0 g of water, it depresses the freezing point by 0.2°C. The percentage association of solute A in water, is **[NCERT : P L-24 | June 25, 2022 (II)]**
[Given: Molar mass of A = 93 g mol^{-1}. Molal depression constant of water is 1.86 K kg mol^{-1}]
(a) 50% (b) 60% (c) 70% (d) 80%

159. The osmotic pressure of blood is 7.47 bar at 300 K. To inject glucose to a patient intravenously, it has to be isotonic with blood. The concentration of glucose solution in gL^{-1} is_____(Molar mass of glucose = 180 g mol^{-1} R = 0.083 L bar K^{-1} mol^{-1}) (Nearest integer)
[NCERT : P L-21 | NV, June 24, 2022 (I)]

160. 1.22 g of an organic acid is separately dissolved in 100 g of benzene (K_b = 2.6 K kg mol^{-1}) and 100 g of acetone (K_b = 1.7 K kg mol^{-1}). The acid is known to dimerize in benzene but remain as a monomer in acetone. The boiling point of the solution in acetone increases by 0.17°C.
The increase in boiling point of solution in benzene in °C is $x \times 10^{-2}$. The value of x is ________.(Nearest integer)
[Atomic mass : C = 12.0, H = 1.0, O = 16.0]
[NCERT : P L-17 | NV, Aug. 31, 2021 (II)]

161. 1 kg of 0.75 molal aqueous solution of sucrose can be cooled up to –4 °C before freezing. The amount of ice (in g) that will be separated out is ________. (Nearest integer) [Given : $K_f(H_2O) = 1.86$ K kg mol^{-1}]

[NCERT : P L-19 | NV, Aug. 27, 2021 (I)]

162. 40 g of glucose (Molar mass = 180) is mixed with 200 mL of water. The freezing point of solution is _____ K. (Nearest integer)

[NV, Aug. 27, 2021 (II)]

[Given : $K_f = 1.86$ K kg mol^{-1} ; Density of water = 1.00 g cm^{-3}; Freezing point of water = 273.15 K]

163. 83 g of ethylene glycol dissolved in 625 g of water. The freezing point of the solution is _______K. (Nearest integer)

[Use : Molal Freezing point depression constant of water = 1.86 K kg mol^{-1}]

Freezing Point of water = 273 K

Atomic masses : C : 12.0 u, O : 16.0 u, H : 1.0 u]

[NCERT : P L-19 | NV, Aug. 26, 2021 (II)]

164. Of the following four aqueous solutions, total number of those solutions whose freezing point is lower than that of 0.10 M C_2H_5OH is ________(Integer answer)

[NCERT : P L-24 | NV, Aug. 26, 2021 (I)]

(i) 0.10 M $Ba_3(PO_4)_2$ (ii) 0.10 M Na_2SO_4
(iii) 0.10 M KCl (iv) 0.10 M Li_3PO_4

165. 1.46 g of a biopolymer dissolved in a 100 mL water at 300 K exerted an osmotic pressure of 2.42×10^{-3} bar. The molar mass of the biopolymer is ____ $\times 10^4$ g mol^{-1}. (Round off to the nearest integer)

[Use : $R = 0.083$ L bar mol^{-1} K^{-1}]

[NCERT : P L-21 | NV, July 27, 2021 (I)]

166. In a solvent 50% of an acid HA dimerizes and the rest dissociates. The van't Hoff factor of the acid is _____ $\times 10^{-2}$. **[NCERT : P L-24 | NV, July 27, 2021 (II)]** (Round off to the nearest integer)

167. When 3.00 g of a substance 'X' is dissolved in 100 g of CCl_4, it raises the boiling point by 0.60 K. The molar mass of the substance 'X' is _______ g mol^{-1}. (Nearest integer).

[Given K_b for CCl_4 is 5.0 K kg mol^{-1}]

[NCERT : P L-17 | NV, July 25, 2021 (II)]

168. 2 molal solution of a weak acid HA has a freezing point of –3.885 °C. The degree of dissociation of this acid is _____ $\times 10^{-3}$.

(Rounded off to the nearest integer).

[Given : Molal depression constant of water = 1.85 K kg mol^{-1}, Freezing point of pure water = 0 °C]

[NCERT : P L-19 | NV, March 18, 2021 (I)]

169. A solute A dimerizes in water. The boiling point of a 2 molal solution of A is 100.52 °C. The percentage association of A is . (Round off to the nearest integer).

[Use : K_b for water = 0.52 K kg mol^{-1}, Boiling point of water = 100 °C] **[NCERT : P L-17 | NV, March 18, 2021 (II)]**

170. A 1 molal $K_4Fe(CN)_6$ solution has a degree of dissociation of 0.4. Its boiling point is equal to that of another solution which contains 18.1 weight percent of a non-electrolytic solute A. The molar mass of A is ________ u. (Round off to the nearest integer).

[Density of water = 1.0 g cm^{-3}]

[NCERT : P L-19 | NV, March 17, 2021 (II)]

171. AB_2 is 10% dissociated in water to A^{2+} and B^-. The boiling point of a 10.0 molal aqueous solution of AB_2 is _____ °C. (Round off to the Nearest Integer)

[Given : Molal elevation constant of water $K_b = 0.5$ K kg mol^{-1} boiling point of pure water = 100°C]

[NCERT : P L-24 | NV, March 16, 2021 (I)]

172. When 12.2 g of benzoic acid is dissolved in 100 g of water, the freezing point of solution was found to be –0.93 °C ($K_f(H_2O) = 1.86$ K kg mol^{-1}). The number (n) of benzoic acid molecules associated (assuming 100% association) is________. **[NCERT : P L-19 | NV, Feb. 26, 2021 (II)]**

173. 224 mL of $SO_2(g)$ at 298 K and 1 atm is passed through 100 mL of 0.1 M NaOH solution. The non-volatile solute produced is dissolved in 36 g of water. The lowering of vapour pressure of solution (assuming the solution is dilute) $\left(P^{\circ}_{(H_2O)} = 24 \text{ mm of Hg}\right)$ is $x \times 10^{-2}$ mm of Hg, the value of x is ______. (Integer answer)

[NCERT : P L-16 | NV, Feb. 26, 2021 (I)]

174. If a compound AB dissociates to the extent of 75% in an aqueous solution, the molality of the solution which shows a 2.5 K rise in the boiling point of the solution is ____ molal. (Rounded-off to the nearest integer)

[$K_b = 0.52$ K kg mol^{-1}]

[NCERT : P L-24 | NV, Feb. 25, 2021 (II)]

175. 1 molal aqueous solution of an electrolyte A_2B_3 is 60% ionised. The boiling point of the solution at 1 atm is _______ K. (Rounded-off to the nearest integer)

[Given K_b for $(H_2O) = 0.52$ K kg mol^{-1}]

[NCERT : P L-24 | NV, Feb. 25, 2021 (I)]

176. When 9.45 g of $ClCH_2COOH$ is added to 500 mL of water, its freezing point drops by 0.5 °C. The dissociation constant of $ClCH_2COOH$ is $x \times 10^{-3}$. The value of x is ________. (Rounded off to the nearest integer)

[$K_{f(H2O)} = 1.86$ K kg mol^{-1}]

[NCERT : P L-24 | NV, Feb. 24, 2021 (I)]

177. C_6H_6 freezes at 5.5 °C. The temperature at which a solution 10 g of C_4H_{10} in 200 g of C_6H_6 freeze is ________°C. (The molal freezing point depression constant of C_6H_6 is 5.12 °C/m.) **[NCERT : P L-24 | NV, Feb. 24, 2021 (II)]**

178. A set of solutions is prepared using 180 g of water as a solvent and 10 g of different non-volatile solutes A, B and C. The relative lowering of vapour pressure in the presence of these solutes are in the order [Given, molar mass of A = 100 g mol^{-1}; B = 200 g mol^{-1}; C = 10,000 g mol^{-1}]

[NCERT : P L-16 | Sep. 06, 2020 (II)]

(a) B > C > A (b) C > B > A
(c) A > B > C (d) A > C > B

179. A set of solutions is prepared using 180g of water as a solvent and 10g of different non-volatile solutes A, B and C. The relative lowering of vapour pressure in the presence of these solutes are in the order [Given, molar mass of A = 100g mol^{-1}; B = 200g mol^{-1}; C = 10,000g mol^{-1}]

[NCERT : P L-16 | Sep. 06, 2020 (II)]

(a) B > C > A (b) C > B > A
(c) A > B > C (d) A > C > B

180. The elevation of boiling point of 0.10 m aqueous $CrCl_3.xNH_3$ solution is two times that of 0.05 m aqueous $CaCl_2$ solution. The value of x is ______.
[Assume 100% ionisation of the complex and $CaCl_2$, coordination number of Cr as 6, and that all NH_3 molecules are present inside the coordination sphere]

[NCERT : P L-24 | NV, Sep. 06, 2020 (I)]

181. The osmotic pressure of a solution of NaCl is 0.10 atm and that of a glucose solution is 0.20 atm. The osmotic pressure of a solution formed by mixing 1 L of the sodium chloride solution with 2 L of the glucose solution is $x \times 10^{-3}$ atm. x is __________. (nearest integer)

[NCERT : P L-21 | NV, Sep. 04, 2020 (II)]

182. If 250 cm^3 of an aqueous solution containing 0.73 g of a protein A is isotonic with one litre of another aqueous solution containing 1.65 g of a protein B, at 298 K, the ratio of the molecular masses of A and B is __________ $\times 10^{-2}$ (to the nearest integer).

[NCERT : P L-22 | NV, Sep. 03, 2020 (II)]

183. How much amount of NaCl should be added to 600 g of water (r = 1.00 g/mL) to decrease the freezing point of water to –0.2 °C? _______. (The freezing point depression constant for water = 2 K kg mol^{-1})

[NCERT : P L-19 | NV, Jan. 09, 2020 (I)]

184. A cylinder containing an ideal gas (0.1 mol of 1.0 dm^3) is in thermal equilibrium with a large volume of 0.5 molal aqueous solution of ethylene glycol at its freezing point. If the stoppers S_1 and S_2 (as shown in the figure) are suddenly withdrawn, the volume of the gas in litres after equilibrium is achieved will be _______.
(Given, K_f (water) = 2.0 K kg mol^{-1}, R = 0.08 dm^3 atm K^{-1} mol^{-1}) **[NCERT : P L-19 | NV, Jan. 09, 2020 (II)]**

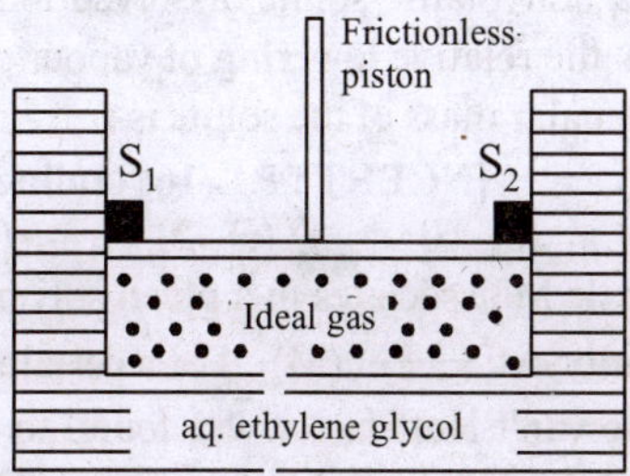

185. A solution is prepared by dissolving 0.6 g of urea (molar mass = 60 g mol^{-1}) and 1.8 g of glucose (molar mass = 180 g mol^{-1}) in 100 mL, of water at 27 °C. The osmotic pressure of the solution is : **[NCERT : P L-21 | April 12, 2019 (II)]**
(R = 0.08206 L atm K^{-1} mol^{-1})

(a) 8.2 atm (b) 2.46 atm
(c) 4.92 atm (d) 1.64 atm

186. At room temperature, a dilute solution of urea is prepared by dissolving 0.60 g of urea in 360 g of water. If the vapour pressure of pure water at this temperature is 35 mmHg, lowering of vapour pressure will be :

[NCERT : P L-16 | April 10, 2019 (I)]

(molar mass of urea = 60 g mol^{-1})

(a) 0.027 mmHg (b) 0.028 mmHg
(c) 0.017 mmHg (d) 0.031 mmHg

187. 1 g of a non-volatile non-electrolyte solute is dissolved in 100 g of two different solvents A and B whose ebullioscopic constants are in the ratio of 1 : 5. The ratio of the elevation in their boiling points, $\frac{\Delta T_b(A)}{\Delta T_b(B)}$, is :

[NCERT : P L-17 | April 10, 2019 (II)]

(a) 5 : 1 (b) 10 : 1 (c) 1 : 5 (d) 1 : 0.2

188. The osmotic pressure of a dilute solution of an ionic compound XY in water is four times that of a solution of 0.01 M $BaCl_2$ in water. Assuming complete dissociation of the given ionic compounds in water, the concentration of XY (in mol L^{-1}) in solution is:

[NCERT : P L-24 | April 9, 2019 (I)]

(a) 4×10^{-2} (b) 6×10^{-2}
(c) 4×10^{-4} (d) 16×10^{-4}

189. Molal depression constant for a solvent is 4.0 K kg mol^{-1}. The depression in the freezing point of the solvent for 0.03 mol kg^{-1} solution K_2SO_4 is:
(Assume complete dissociation of the electrolyte)

[NCERT : P L-24 | April 9, 2019 (II)]

(a) 0.18 K (b) 0.24 K (c) 0.12 K (d) 0.36 K

190. Freezing point of a 4% aqueous solution of X is equal to freezing point of 12% aqueous salution of Y. If molecular weight of X is A, then molecular weight of Y is:

[NCERT : P L-19 | Jan. 12, 2019 (I)]

(a) 3A (b) 2A (c) A (d) 4A

191. Molecules of benzoic acid (C_6H_5COOH) dimerise in benzene. 'w' g of the acid dissolved in 30 g of benzene shows a depression in freezing point equal to 2 K. If the percentage association of the acid to form dimer in the solution is 80, then w is :
(Given that K_f = 5 K kg mol^{-1}, Molar mass of benzoic acid = 122 g mol^{-1}) **[NCERT : P L-19 | Jan. 12, 2019 (II)]**

(a) 2.4 g (b) 1.0 g (c) 1.5 g (d) 1.8 g

192. The freezing point of a diluted milk sample is found to be – 0.2 °C, while it should have been – 0.5 °C for pure milk. How much water has been added to pure milk to make the diluted sample? **[NCERT : P L-19 | Jan. 11, 2019 (I)]**

(a) 1 cup of water to 2 cups of pure milk
(b) 3 cups of water to 2 cups of pure milk
(c) 1 cup of water to 3 cups of pure milk
(d) 2 cups of water to 3 cups of pure milk

193. K_2HgI_4 is 40% ionised in aqueous solution. The value of its van't Hoff factor (*i*) is :

[NCERT : P L-24 | Jan. 11, 2019 (II)]

(a) 1.6 (b) 1.8 (c) 2.0 (d) 2.2

194. Elevation in the boiling point for 1 molal solution of glucose is 2 K. The depression in the freezing point for 2 molal solution of glucose in the same solvent is 2 K. The relation between K_b and K_f is:

[NCERT : P L-19 | Jan. 10, 2019 (II)]

(a) $K_b = 1.5\,K_f$ (b) $K_b = K_f$
(c) $K_b = 0.5\,K_f$ (d) $K_b = 2\,K_f$

195. A solution containing 62 g ethylene glycol in 250 g water is cooled to – 10 °C. If K_f for water is 1.86 K kg mol^{-1}, the amount of water (in g) separated as ice is:

[NCERT : P L-19 | Jan. 9, 2019 (II)]

(a) 48 (b) 32 (c) 64 (d) 16

196. For 1 molal aqueous solution of the following compounds, which one will show the highest freezing point?

[NCERT : P L-24 | 2018]

(a) $[Co(H_2O)_6]Cl_3$ (b) $[Co(H_2O)_5Cl]Cl_2.H_2O$
(c) $[Co(H_2O)_4Cl_2]Cl.2H_2O$ (d) $[Co(H_2O)_3Cl_3].3H_2O$

197. The mass of non-volatile, non-electrolyte solute (molar mass = 50 g mol^{-1}) needed to be dissolved in 114 g octane to reduce its vapour pressure to 75%, is __________.

[NCERT : P L-24 | Online April 16, 2018]

(a) 37.5 g (b) 75 g (c) 150 g (d) 50 g

198. Two 5 molal solutions are prepared by dissolving a non-electrolyte, non-volatile solute separately in the solvents X and Y. The molecular weights of the solvents are M_X and M_Y, respectively where $M_X = \frac{3}{4}M_Y$. The relative lowering of vapour pressure of the solution in X is "*m*" times that of the solution in Y. Given that the number of moles of solute is very small in comparison to that of solvent, the value of "*m*" is:

[NCERT : P L-16 | Online April 15, 2018 (II)]

(a) $\frac{3}{4}$ (b) $\frac{1}{2}$ (c) $\frac{1}{4}$ (d) $\frac{4}{3}$

199. 5 g of Na_2SO_4 was dissolved in *x* g of H_2O. The change in freezing point was found to be 3.82°C. If Na_2SO_4 is 81.5% ionised, the value of *x*

[NCERT : P L-19 | Online April 8, 2017]

(K_f for water = 1.86°C kg mol^{-1}) is approximately :
(molar mass of S = 32 g mol^{-1} and that of Na = 23 g mol^{-1})

(a) 15 g (b) 25 g (c) 45 g (d) 65 g

200. The freezing point of benzene decreases by 0.45 °C when 0.2g of acetic acid is added to 20 g of benzene. If acetic acid associates to form a dimer in benzene, percentage association of acetic acid in benzene will be :

[NCERT : P L-19 | 2017]

(K_f for benzene = 5.12 K kg mol^{-1})

(a) 64.6% (b) 80.4% (c) 74.6% (d) 94.6%

201. An aqueous solution of a salt MX_2 at certain temperature has a van't Hoff factor of 2. The degree of dissociation for this solution of the salt is :

[NCERT : P L-24 | Online April 10, 2016]

(a) 0.50 (b) 0.33 (c) 0.67 (d) 0.80

202. 18 g glucose ($C_6H_{12}O_6$) is added to 178.2 g water. The vapour pressure of water (in torr) for this aqueous solution is:

[NCERT : P L-12 | 2016, 2006]

(a) 752.4 (b) 759.0 (c) 7.6 (d) 76.0

203. The vapour pressure of acetone at 20 °C is 185 torr. When 1.2 g of a non-volatile substance was dissolved in 100 g of acetone at 20 °C, its vapour pressure was 183 torr. The molar mass (g mol^{-1}) of the substance is :

[NCERT : P L-16 | 2015]

(a) 128 (b) 488 (c) 32 (d) 64

204. Determination of the molar mass of acetic acid in benzene using freezing point depression is affected by :

[NCERT : P L-19 | Online April 11, 2015]

(a) partial ionization (b) dissociation
(c) complex formation (d) association

205. Consider separate solutions of 0.500 M C_2H_5OH (aq), 0.100 M $Mg_3(PO_4)_2$ (aq), 0.250 M KBr (aq) and 0.125 M Na_3PO_4 (aq) at 25 °C. Which statement is **true** about these solutions, assuming all salts to be strong electrolytes?

(a) They all have the same osmotic pressure.

[NCERT : P L-24 | 2014]

(b) 0.100 M $Mg_3(PO_4)_2$(aq) has the highest osmotic pressure.
(c) 0.125 M Na_3PO_4(aq) has the highest osmotic pressure.
(d) 0.500 M C_2H_5OH(aq) has the highest osmotic pressure.

206. The observed osmotic pressure for a 0.10 M solution of $Fe(NH_4)_2(SO_4)_2$ at 25 °C is 10.8 atm. The expected and experimental (observed) values of van't Hoff factor (*i*) will be respectively: **[NCERT : P L-24 | Online April 19, 2014]**
(R = 0.082 L atm K^{-1} mol^{-1})

(a) 5 and 4.42 (b) 4 and 4.00
(c) 5 and 3.42 (d) 3 and 5.42

207. 12 g of a nonvolatile solute dissolved in 108 g of water produces the relative lowering of vapour pressure of 0.1. The molecular mass of the solute is :

[NCERT : P L-16 | Online April 9, 2013]

(a) 80 (b) 60 (c) 20 (d) 40

208. A molecule M associates in a given solvent according to the equation $M \rightleftharpoons (M)_n$. For a certain concentration of M, the van't Hoff factor was found to be 0.9 and the fraction of associated molecules was 0.2. The value of *n* is:

[NCERT : P L-24 | Online April 22, 2013]

(a) 3 (b) 5
(c) 2 (d) 4

209. How many grams of methyl alcohol should be added to 10 litre tank of water to prevent its freezing at 268 K ? (K_f for water is 1.86 K kg mol^{-1})

[NCERT : P L-19 | Online April 25, 2013]

(a) 880.07 g (b) 899.04 g
(c) 886.02 g (d) 868.06 g

210. K_f for water is 1.86 K kg mol^{-1}. If your automobile radiator holds 1.0 kg of water, how many grams of ethylene glycol ($C_2H_6O_2$) must you add to get the freezing point of the solution lowered to –2.8 °C ? **[NCERT : P L-19 | 2012]**

(a) 72 g (b) 93 g (c) 39 g (d) 27 g

211. A solution containing 0.85 g of $ZnCl_2$ in 125.0 g of water freezes at –0.23 °C. The apparent degree of dissociation of the salt is (K_f for water = 1.86 K kg mol^{-1}, atomic mass: Zn = 65.3 and Cl = 35.5)

[NCERT : P L-19 | Online May 12, 2012]

(a) 1.36% (b) 73.5%
(c) 7.35% (d) 2.47%

212. The freezing point of a 1.00 m aqueous solution of HF is found to be –1.91°C. The freezing point constant of water, K_f is 1.86 K kg mol^{-1}. The percentage dissociation of HF at this concentration is

[NCERT : P L-24 | Online May 26, 2012]

(a) 30% (b) 10% (c) 5.2% (d) 2.7%

213. A 5% solution of cane sugar (molar mass 342) is isotonic with 1% of a solution of an unknown solute. The molar mass of unknown solute in g/mol is :

[NCERT : P L-22 | 2011RS]

(a) 171.2 (b) 68.4 (c) 34.2 (d) 136.2

214. If sodium sulphate is considered to be completely dissociated into cations and anions in aqueous solution, the change in freezing point of water (ΔT_f), when 0.01 mol of sodium sulphate is dissolved in 1 kg of water, is (K_f = 1.86 K kg mol^{-1}). **[NCERT : P L-19 | 2010]**

(a) 0.372 K (b) 0.0558 K
(c) 0.0744 K (d) 0.0186 K

215. The vapour pressure of water at 20 °C is 17.5 mm Hg. If 18 g of glucose ($C_6H_{12}O_6$) is added to 178.2 g of water at 20 °C, the vapour pressure of the resulting solution will be

[NCERT : P L-16 | 2008]

(a) 17.325 mm Hg (b) 15.750 mm Hg
(c) 16.500 mm Hg (d) 17.500 mm Hg

216. A 5.25% solution of a substance is isotonic with a 1.5% solution of urea (molar mass = 60 g mol^{-1}) in the same solvent. If the densities of both the solutions are assumed to be equal to 1.0 g cm^{-3}, molar mass of the substance will be

[NCERT : P L-22 | 2007]

(a) 210.0 g mol^{-1} (b) 90.0 g mol^{-1}
(c) 115.0 g mol^{-1} (d) 105.0 g mol^{-1}.

217. Which one of the following aqueous solutions will exihibit highest boiling point ? **[NCERT : P L-24 | 2004]**

(a) 0.015 M urea (b) 0.01 M KNO_3
(c) 0.01 M Na_2SO_4 (d) 0.015 M glucose

218. In a 0.2 molal aqueous solution of a weak acid H*X*, the degree of ionization is 0.3. Taking K_f for water as 1.85, the freezing point of the solution will be nearest to

[NCERT : P L-19 | 2003]

(a) –0.360 ºC (b) –0.260 ºC
(c) +0.480 ºC (d) –0.480 ºC]

219. Which one of the following statements is **false?**

[NCERT : P L-24 | 2004]

(a) The correct order of osmotic pressure for 0.01 M aqueous solution of each compound is
$BaCl_2 > KCl > CH_3COOH >$ sucrose
(b) The osmotic pressure (π) of a solution is given by the equation π = MRT, where M is the molarity of the solution
(c) Raoult's law states that the vapour pressure of a component over a solution is proportional to its mole fraction
(d) Two sucrose solutions of same molality prepared in different solvents will have same freezing point depression

220. A pressure cooker reduces cooking time for food because

[NCERT : P L-17 | 2003]

(a) boiling point of water involved in cooking is increased
(b) the higher pressure inside the cooker crushes the food material
(c) cooking involves chemical changes helped by a rise in temperature
(d) heat is more evenly distributed in the cooking space

221. Freezing point of an aqueous solution is (–0.186) °C. Elevation of boiling point of the same solution is K_b = 0.512 °C, K_f = 1.86°C, find the increase in boiling point.

[NCERT : P L-17 | 2002]

(a) 0.186 °C (b) 0.0512 °C
(c) 0.092 °C (d) 0.2372 °C.

ANSWER KEYS

1	(d)	24	(5)	47	(82)	70	(b)	93	(33)	116	(707)	139	(13)	162	(271)	185	(c)	208	(c)
2	(2)	25	(54)	48	(555)	71	(a)	94	(d)	117	(31)	140	(a)	163	(269)	186	(c)	209	(d)
3	(9)	26	(a)	49	(314)	72	(c)	95	(c)	118	(a)	141	(41,500)	164	(4)	187	(c)	210	(b)
4	(b)	27	(64)	50	(1)	73	(d)	96	(b)	119	(15)	142	(4)	165	(15)	188	(b)	211	(b)
5	(d)	28	(20)	51	(2)	74	(b)	97	(25)	120	(d)	143	(a)	166	(125)	189	(d)	212	(d)
6	(164)	29	(3)	52	(2)	75	(b)	98	(b)	121	(a)	144	(80)	167	(250)	190	(a)	213	(b)
7	(74)	30	(50)	53	(a)	76	(d)	99	(16)	122	(72.56)	145	(15)	168	(50)	191	(a)	214	(b)
8	(22)	31	(64)	54	(14)	77	(b)	100	(d)	123	(240)	146	(6)	169	(100)	192	(b)	215	(a)
9	(b)	32	(9)	55	(1222)	78	(c)	101	(N)	124	(116)	147	(b)	170	(85)	193	(b)	216	(a)
10	(b)	33	(c)	56	(25)	79	(a)	102	(a)	125	(48)	148	(1)	171	(106)	194	(d)	217	(c)
11	(b)	34	(d)	57	(77)	80	(b)	103	(75)	126	(75)	149	(c)	172	(2)	195	(c)	218	(d)
12	(11)	35	(b)	58	(1)	81	(b)	104	(c)	127	(d)	150	(b)	173	(24)	196	(d)	219	(d)
13	(7)	36	(d)	59	(25)	82	(d)	105	(a)	128	(30)	151	(5)	174	(3)	197	(c)	220	(a)
14	(b)	37	(a)	60	(19)	83	(b)	106	(a)	129	(40535)	152	(3)	175	(375)	198	(a)	221	(b)
15	(4)	38	(d)	61	(600)	84	(d)	107	(c)	130	(8)	153	(330)	176	(35)	199	(c)		
16	(815)	39	(b)	62	(d)	85	(d)	108	(a)	131	(1111)	154	(8)	177	(1)	200	(d)		
17	(a)	40	(a)	63	(a, c)	86	(d)	109	(d)	132	(4)	155	(45)	178	(c)	201	(a)		
18	(d)	41	(c)	64	(c)	87	(b)	110	(23)	133	(3.4)	156	(99)	179	(c)	202	(a)		
19	(d)	42	(d)	65	(b)	88	(d)	111	(5)	134	(372)	157	(415)	180	(5)	203	(d)		
20	(3.64)	43	(a)	66	(a)	89	(b)	112	(76)	135	(62250)	158	(d)	181	(167)	204	(d)		
21	(139)	44	(d)	67	(d)	90	(d)	113	(543)	136	(d)	159	(54)	182	(177)	205	(a)		
22	(148)	45	(d)	68	(d)	91	(d)	114	(25)	137	(100)	160	(13)	183	(1.75)	206	(a)		
23	(180)	46	(23)	69	(a)	92	(a)	115	(19)	138	(243)	161	(518)	184	(2.18)	207	(c)		

2 Electrochemistry

TREND ANALYSIS
JEE MAIN (2025-2021)

	2025				2024				2023				2022				2021			
	Session II		Session I		Session II		Session I		Session II		Session I		Session II		Session I		Session II		Session I	
	MCQs	NVQs	MCQs	NVQs	MCQs	NVQs	MCQs	NVQs	MCQs	NVQs	MCQs	NVQs	MCQs	NVQs	MCQs	NVQs	MCQs	NVQs	MCQs	NVQs
Topic 1	2	2	1	1	5	–	1	1	–	2	1	3	2	1	–	3	1	2	–	3
Topic 2	–	1	3	1	2	–	–	5	1	1	–	–	–	1	–	2	1	1	–	1
Topic 3	1	2	6	–	5	1	–	3	3	5	2	7	–	2	3	6	–	5	–	3
Topic 4	2	–	2	–	3	–	2	–	1	–	–	–	1	–	–	–	–	–	–	–
Total	5	5	12	2	15	1	3	9	5	8	3	10	3	4	3	11	2	8	–	7

How to Code Qns for Better & Faster Revision →

Type 1: Able to solve (No code Req)
Type 2: Solved with struggle or by seeing solution **(Use code NR means Needs Revision or orange stickers)**
Type 3: Tricky/ Could not solve on my own. Need help from Teacher/ Friend. **(Use code TQ means Tricky Question or red stickers)**

You can either hand write the code in boxes or else paste coloured stickers given at end of book.

Topic 1 - Conductance and Conductivity

1. Given below are two statements :

Statement I : Mohr's salt is composed of only three types of ions-ferrous, ammonium and sulphate.

Statement II : If the molar conductance at infinite dilution of ferrous, ammonium and sulphate ions are x_1, x_2 and x_3 S cm^2 mol^{-1}, respectively then the molar conductance for Mohr's salt solution at infinite dilution would be given by $x_1 + x_2 + 2x_3$. **[NCERT : P L-45 | April 7, 2025 (I)]**

In the light of the given statements, choose the **correct** answer from the options given below :

(a) Both statements I and Statement II are false
(b) Statement I is false but Statement II is true
(c) Statement I is true but Statement II are false
(d) Both statements I and Statement II are true

2. The molar conductance of an infinitely dilute solution of ammonium chloride was found to be 185 S cm^2 mol^{-1} and the ionic conductance of hydroxyl and chloride ions are 170 and 70 S cm^2 mol^{-1}, respectively. If molar conductance of 0.02 M solution of ammonium hydroxide is 85.5 S cm^2 mol^{-1}, its degree of dissociation is given by $x \times 10^{-1}$. The value of x is _____. (Nearest integer)

[NCERT : P L-45 | NV, April 4, 2025 (II)]

3. Correct order of limiting molar conductivity for cations in water at 298 K is : **[NCERT : P L-47 | April 3, 2025 (I)]**

(a) $H^+ > Na^+ > K^+ > Ca^{2+} > Mg^{2+}$
(b) $H^+ > Ca^{2+} > Mg^{2+} > K^+ > Na^+$
(c) $Mg^{2+} > H^+ > Ca^{2+} > K^+ > Na^+$
(d) $H^+ > Na^+ > Ca^{2+} > Mg^{2+} > K^+$

4. 0.2 % (w/v) solution of NaOH is measured to have resistivity 870.0 mΩ m. The molar conductivity of the solution will be ________ $\times 10^2$ mS dm^2 mol^{-1}. (Nearest integer) **[NCERT : P L-45 | NV, April 2, 2025 (II)]**

5. The molar conductivity of a weak electrolyte when plotted against the square root of its concentration, which of the following is expected to be observed?

[NCERT : P L-47 | Jan. 29, 2025 (I)]

(a) A small decrease in molar conductivity is observed at infinite dilution.
(b) A small increase in molar conductivity is observed at infinite dilution.
(c) Molar conductivity increases sharply with increase in concentration.
(d) Molar conductivity decreases sharply with increase in concentration.

6. Given below is the plot of the molar conductivity vs $\sqrt{\text{concentration}}$ for KCl in aqueous solution.
[NCERT : P L-47 | NV, Jan. 28, 2025 (I)]

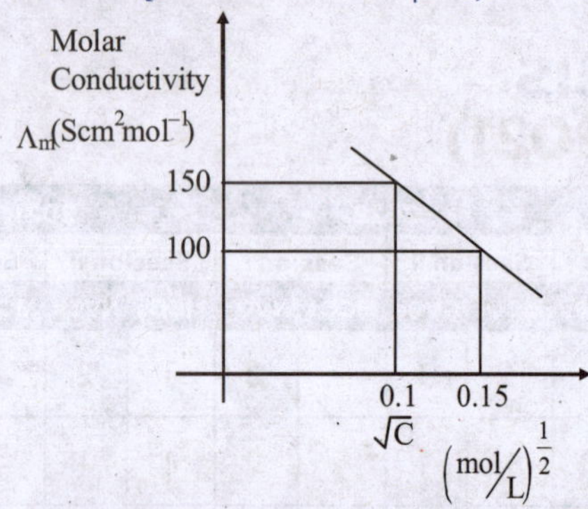

If, for the higher concentration of KCl solution, the resistance of the conductivity cell is 100Ω, then the resistance of the same cell with the dilute solution is 'x' Ω. The value of x is ________ (Nearest integer)

7. The molar conductivity for electrolytes A and B are plotted against $C^{1/2}$ as shown below. Electrolytes A and B respectively are : **[NCERT : P L-43 | April 9, 2024 (I)]**

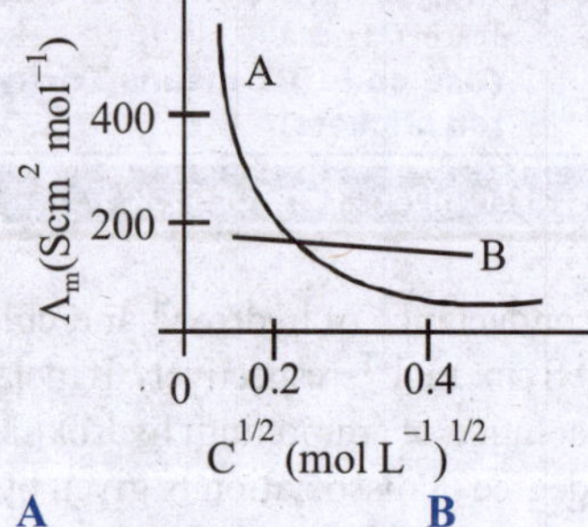

	A	B
(a)	Weak electrolyte	weak electrolyte
(b)	Strong electrolyte	strong electrolyte
(c)	Weak electrolyte	strong electrolyte
(d)	Strong electrolyte	weak electrolyte

8. Which out of the following is a correct equation to show change in molar conductivity with respect to concentration for a weak electrolyte, if the symbols carry their usual meaning :**[NCERT : P L-50 | April 9, 2024 (II)]**

(a) $\Lambda^2_m C - K_a \Lambda^{o2}_m + K_a \Lambda_m \Lambda^\circ_m = 0$

(b) $\Lambda_m - \Lambda^\circ_m + AC^{\frac{1}{2}} = 0$

(c) $\Lambda_m - \Lambda^\circ_m - AC^{\frac{1}{2}} = 0$

(d) $\Lambda^2_m C + K_a \Lambda^{o2}_m - K_a \Lambda_m \Lambda^\circ_m = 0$

9. A conductivity cell with two electrodes (dark side) are half filled with infinitely dilute aqueous solution of a weak electrolyte. If volume is doubled by adding more water at constant temperature, the molar conductivity of the cell will - **[NCERT : P L-47 | April 6, 2024 (I)]**

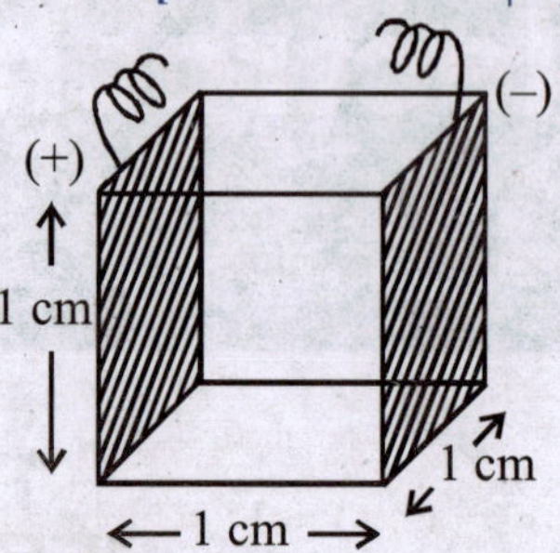

(a) increase sharply
(b) remain same or can not be measured accurately
(c) decrease sharply
(d) depend upon type of electrolyte

10. Molar ionic conductivities of divalent cation and anion are 57 $S\ cm^2\ mol^{-1}$ and 73 $S\ cm^2\ mol^{-1}$ respectively. The molar conductivity of solution of an electrolyte with the above cation and anion will be : **[NCERT : P L-49 | April 5, 2024 (I)]**

(a) $65\ S\ cm^2 mol^{-1}$ (b) $130\ S\ cm^2 mol^{-1}$
(c) $187\ S\ cm^2 mol^{-1}$ (d) $260\ S\ cm^2 mol^{-1}$

11. For a strong electrolyte, a plot of molar conductivity against $(concentration)^{1/2}$ is a straight line, with a negative slope, the correct unit for the slope is
[NCERT : P L-45 | April 4, 2024 (II)]

(a) $S\ cm^2\ mol^{-3/2}\ L^{1/2}$ (b) $S\ cm^2\ mol^{-1}\ L^{1/2}$
(c) $S\ cm^2\ mol^{-3/2}\ L$ (d) $S\ cm^2\ mol^{-3/2}\ L^{-1/2}$

12. Identify the factor from the following that **does not** affect electrolytic conductance of a solution.
[NCERT : P L-41 | Jan. 31, 2024 (I)]

(a) The nature of solvent used.
(b) Concentration of the electrolyte.
(c) The nature of the electrode used.
(d) The nature of the electrolyte added.

13. The values of conductivity of some materials at 298.15 K in Sm^{-1} are 2.1×10^3, $1.0 \times 10^{-16}, 1.2 \times 10, 3.91, 1.5 \times 10^{-2}$, $1 \times 10^{-7}, 1.0 \times 10^3$. The number of conductors among the materials is ______.**[NCERT : P L-42 |NV, Jan. 31, 2024 (II)]**

14. The number of correct statements from the following is ________ **[NCERT : P L-41 | NV, April 15, 2023 (I)]**

(A) Conductivity always decreases with decrease in concentration for both strong and weak electrolytes.
(B) The number of ions per unit volume that carry current in a solution increases on dilution.
(C) Molar conductivity increases with decrease in concentration.
(D) The variation in molar conductivity is different for strong and weak electrolytes.
(E) For weak electrolytes, the change in molar conductivity with dilution is due to decrease in degree of dissociation.

15. The specific conductance of 0.0025 M acetic acid is 5×10^{-5} S cm^{-1} at a certain temperature. The dissociation constant of acetic acid is _____ $\times 10^{-7}$. (Nearest integer) Consider limiting molar conductivity of CH_3COOH as 400 S cm^2 mol^{-1} **[NCERT : P L-41 | NV, April 10, 2023 (II)]**

16. 1×10^{-5} M $AgNO_3$ is added to 1 L of saturated solution of AgBr. The conductivity of this solution at 298 K is _____ $\times 10^{-8}$ S m^{-1}.
[Given : $K_{SP}(AgBr) = 4.9 \times 10^{-13}$ at 298 K
$\lambda^0_{Ag^+} = 6 \times 10^{-3}$ S m^{2}mol^{-1}
$\lambda^0_{Br^-} = 8 \times 10^{-3}$ S m^{2}mol^{-1}
$\lambda^0_{NO_3^-} = 7 \times 10^{-3}$ S m^{2}mol^{-1}] **[NV, Feb. 01, 2023 (II)]**

17. The resistivity of a 0.8 M solution of an electrolyte is 5×10^{-3} Ωcm. Its molar conductivity is _______ $\times 10^4$ Ω^{-1} cm^2 mol^{-1}. (Nearest integer)
[NCERT : P L-45 | NV, Jan. 31, 2023 (II)]

18. Following figure shows dependence of molar conductance of two electrolytes on concentration. Λ^0_m is the limiting molar conductivity.

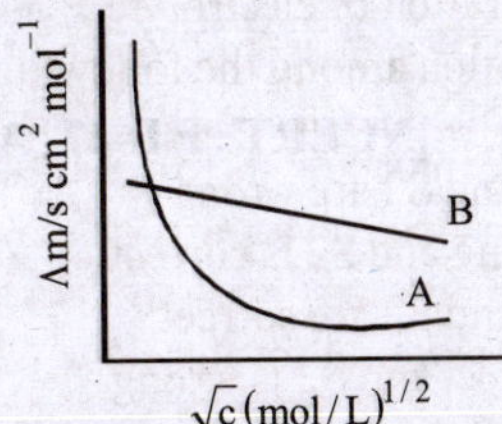

The number of incorrect statement(s) from the following is _______. **[NCERT : P L-47 | NV, Jan. 29, 2023 (I)]**
(A) Λ^0_m for electrolyte A is obtained by extrapolation
(B) For electrolyte B, Λ_m vs $\sqrt{c}$ graph is a straight line with intercept equal to Λ^0_m
(C) At infinite dilution, the value of degree of dissociation approach zero for electrolyte B.
(D) Λ^0_m for any electrolyte A or B can be calculated using λ° for individual ions.

19. Choose the correct representation of conductometric titration of benzoic acid *vs* sodium hydroxide.
[Jan. 24, 2023 (II)]

(a) Conductance vs V_{NaOH}
(b) Conductance vs V_{NaOH}
(c) Conductance vs V_{NaOH}
(d) Conductance vs V_{NaOH}

20. Resistance of a conductivity cell (cell constant 129 m^{-1}) filled with 74.5 ppm solution of KCl is 100 Ω (labelled as solution 1). When the same cell is filled with KCl solution of 149 ppm, the resistance is 50 Ω (labelled as solution 2). The ratio of molar conductivity of solution 1 and solution 2 is i.e. $\frac{\Lambda_1}{\Lambda_2} = x \times 10^{-3}$. The value of x is _____.
(Nearest integer)
Given, molar mass of KCl is 74.5 g mol^{-1}
[NCERT : P L-45 | NV, July 29, 2022 (I)]

21. Given below are two statements:
Statement I: For KI, molar conductivity increases steeply with dilution.
Statement II: For carbonic acid, molar conductivity increases slowly with dilution.
In the light of the above statements, choose the correct answer from the options given below:
[NCERT : P L-47 | July 27, 2022 (II),]
(a) Both Statement I and Statement II are true.
(b) Both Statement I and Statement II are false.
(c) Statement I is true but Statement II is false.
(d) Statement I is false but Statement II is true.

22. The molar conductivity of a conductivity cell filled with 10 moles of 20 mL NaCl solution is Λ_{m_1} and that of 20 moles another identical cell having 80 mL NaCl solution is Λ_{m_2}. The conductivities exhibited by these two cells are same. **[NCERT : P L-45 | July 25, 2022 (II)]**
The relationship between Λ_{m_2} and Λ_{m_1} is
(a) $\Lambda_{m_2} = 2\Lambda_{m_1}$ (b) $\Lambda_{m_2} = \Lambda_{m_1}/2$
(c) $\Lambda_{m_2} = \Lambda_{m_1}$ (d) $\Lambda_{m_2} = 4\Lambda_{m_1}$

23. The solubility product of a sparingly soluble salt A_2X_3 is 1.1×10^{-23}. If specific conductance of the solution is 3×10^{-5} S m^{-1}, the limiting molar conductivity of the solution is $x \times 10^{-3}$ S m^2 mol^{-1}.
The value of x is _________. **[NV, June 28, 2022 (I)]**

24. The limiting molar conductivities of NaI, $NaNO_3$ and $AgNO_3$ are 12.7, 12.0 and 13.3 mS m^2 mol^{-1}, respectively (all at 25°C). The limiting molar conductivity of AgI at this temperature is _____ mS m^2 mol^{-1}.
[NCERT : P L-49 | NV, June 27, 2022 (I)]

25. The resistance of conductivity cell containing 0.01 M KCl solution at 298 K is 1750 Ω. If the conductivity of 0.01 M KCl solution at 298 K is 0.152×10^{-3} S cm^{-1}, then the cell constant of the conductivity cell is _______ $\times 10^{-3}$ cm^{-1}.
[NCERT : P L-44 | NV, June 24, 2022 (II)]

26. Given below are two statements :
[NCERT : P L-47 | Aug. 26, 2021 (I)]
Statement I : The limiting molar conductivity of KCl (strong electrolyte) is higher compared to that of CHCOOH (weak electrolyte).
Statement II : Molar conductivity decreases with decrease in concentration of electrolyte.

In the light of the above statements, choose the **most appropriate** answer from the options given below :
(a) Statement I is true but Statement II is false.
(b) Statement I is false but Statement II is true.
(c) Both Statement I and Statement II are true.
(d) Both Statement I and Statement II are false.

27. The conductivity of a weak acid HA of concentration 0.001 mol L^{-1} is 2.0×10^{-5} S cm^{-1}. If Λ_m° (HA) = 190 S cm^2 mol^{-1}, the ionization constant (K_a) of HA is equal to ____ $\times 10^{-6}$. (Round off to the Nearest Integer) **[NV, July 27, 2021 (I)]**

28. The resistance of a conductivity cell with cell constant 1.14 cm^{-1}, containing 0.001 M KCl at 298 K is 1500 Ω. The molar conductivity of 0.001 M KCl solution at 298 K in S cm^2 mol^{-1} is_______. (Integer answer)
[NCERT : P L-45 | NV, Aug. 27, 2021 (II)]

29. The molar conductivities at infinite dilution of barium chloride, sulphuric acid and hydrochloric acid are 280, 860 and 426 S cm^2 mol^{-1} respectively. The molar conductivity at infinite dilution of barium sulphate is ______ S cm^2 mol^{-1}. (Round off to the nearest integer).
[NCERT : P L-49 | NV, March 18, 2021 (II)]

30. A KCl solution of conductivity 0.14 Sm^{-1} shows a resistance of 4.19 Ω in a conductivity cell. If the same cell is filled with an HCl solution, the resistance drops to 1.03 Ω. The conductivity of the HCl solution is ________ $\times 10^{-2}$ S m^{-1}. (Round off to the nearest integer).
[NCERT : P L-43 | NV, March 17, 2021 (II)]

31. A 5.0 m mol dm^{-3} aqueous solution of KCl has a conductance of 0.55 mS when measured in a cell of cell constant 1.3 cm^{-1}. The molar conductivity of this solution is ____________ m Sm^2 mol^{-1}.
(Round off to the nearest integer).
[NCERT : P L-45 | NV, March 16, 2021 (II)]

32. The variation of molar conductivity with concentration of an electrolyte (X) in aqueous solution is shown in the given figure. **[NCERT : P L-47 | Sep. 05, 2020 (II)]**

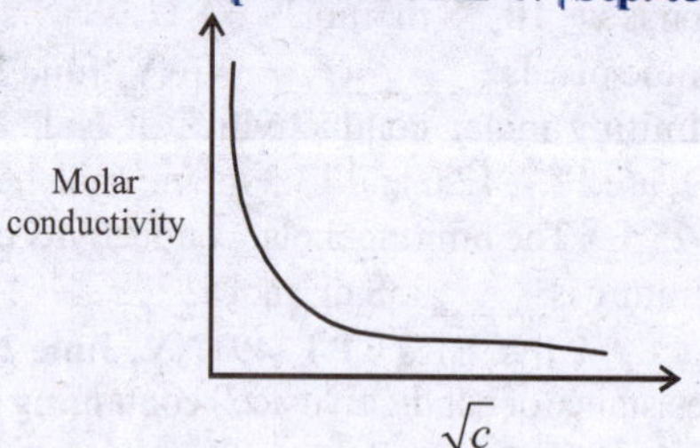

The electrolyte X is :
(a) HCl (b) NaCl (c) KNO_3 (d) CH_3COOH

33. Let C_{NaCl} and C_{BaSO_4} be the conductances (in S) measured for saturated aqueous solutions of NaCl and $BaSO_4$, respectively, at a temperature T.
Which of the following is **false**? **[Sep. 03, 2020 (I)]**
(a) Ionic mobilities of ions from both salts increase with T
(b) $C_{BaSO_4}(T_2) > C_{BaSO_4}(T_1)$ for $T_2 > T_1$
(c) $C_{NaCl}(T_2) > C_{NaCl}(T_1)$ for $T_2 > T_1$
(d) $C_{NaCl} >> C_{BaSO_4}$ at a given T

34. The equation that is incorrect is:
[NCERT : P L-49 | Jan. 07, 2020 (II)]
(a) $(\Lambda_m^0)_{NaBr} - (\Lambda_m^0)_{NaCl} = (\Lambda_m^0)_{KBr} - (\Lambda_m^0)_{KCl}$
(b) $(\Lambda_m^0)_{KCl} - (\Lambda_m^0)_{NaCl} = (\Lambda_m^0)_{KBr} - (\Lambda_m^0)_{NaBr}$
(c) $(\Lambda_m^0)_{H_2O} = (\Lambda_m^0)_{HCl} + (\Lambda_m^0)_{NaOH} - (\Lambda_m^0)_{NaCl}$
(d) $(\Lambda_m^0)_{NaBr} - (\Lambda_m^0)_{NaI} = (\Lambda_m^0)_{KBr} - (\Lambda_m^0)_{NaBr}$

35. The decreasing order of electrical conductivity of the following aqueous solutions is :
0.1 M Formic acid (A), 0.1 M Acetic acid (B),
0.1 M Benzoic acid (C).
[NCERT : P L-47 | April 12, 2019 (II)]
(a) A > C > B (b) C > B > A
(c) A > B > C (d) C > A > B

36. Consider the statements S1 and S2 :
S1 : Conductivity always increases with decrease in the concentration of electrolyte.
S2 : Molar conductivity always increases with decrease in the concentration of electrolyte.
The correct option among the following is :
[NCERT : P L-47 | April 10, 2019 (I)]
(a) Both S1 and S2 are wrong
(b) S1 is wrong and S2 is correct
(c) Both S1 and S2 are correct
(d) S1 is correct and S2 is wrong

37. Which one of the following graphs between molar conductivity (Λ_m) versus $\sqrt{C}$ is correct?
[NCERT : P L-47 | April 10, 2019 (II)]

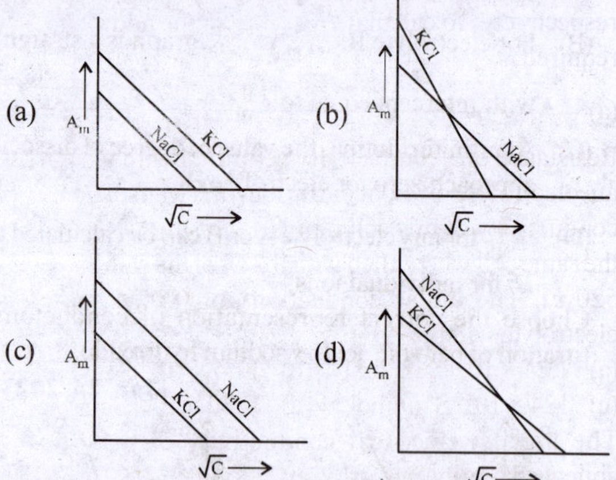

38. Λ_m° for NaCl, HCl and NaA are 126.4, 425.9 and 100.5 S $cm^2$$mol^{-1}$, respectively. If the conductivity of 0.001 M HA is 5×10^{-5} S cm^{-1}, degree of dissociation of HA is :
[NCERT : P L-49 | Jan. 12, 2019 (II)]
(a) 0.50 (b) 0.25
(c) 0.125 (d) 0.75

39. Resistance of 0.2 M solution of an electrolyte is 50 Ω. The specific conductance of the solution is 1.4 S m^{-1}. The resistance of 0.5 M solution of the same electrolyte is 280 Ω. The molar conductivity of 0.5 M solution of the electrolyte in S $m^2 mol^{-1}$ is: **[NCERT : P L-41 | 2006, Same 2011(S)]**

(a) 5×10^{-4} (b) 5×10^{-3}
(c) 5×10^{3} (d) 5×10^{2}

40. The equivalent conductance of NaCl at concentration C and at infinite dilution are λ_C and λ_∞, respectively. The correct relationship between λ_C and λ_∞ is given as: (Where the constant B is positive) **[NCERT : P L-47 | 2014]**

(a) $\lambda_C = \lambda_\infty + (B)C$ (b) $\lambda_C = \lambda_\infty - (B)C$
(c) $\lambda_C = \lambda_\infty - (B)\sqrt{C}$ (d) $\lambda_C = \lambda_\infty + (B)\sqrt{C}$

41. The equivalent conductances of two strong electrolytes at infinite dilution in H_2O (where ions move freely through a solution) at 25 °C are given below :

$\Lambda^\circ_{CH_3COONa} = 91.0\ S\ cm^2 / equiv.$

$\Lambda^\circ_{HCl} = 426.2\ S\ cm^2 / equiv.$

What additional information/ quantity one needs to calculate Λ° of an aqueous solution of acetic acid? **[NCERT : P L-49 | 2007]**

(a) Λ° of chloroacetic acid ($ClCH_2COOH$)
(b) Λ° of NaCl
(c) Λ° of CH_3COOK
(d) the limiting equivalent conductance of H^+ ($\lambda^\circ_{H^+}$).

42. The molar conductivities Λ^o_{NaOAc} and Λ^o_{HCl} at infinite dilution in water at 25°C are 91.0 and 426.2 S cm^2/mol respectively. To calculate $\Lambda^o_{CH_3COOH}$, the additional value required is **[NCERT : P L-45 | 2006]**

(a) Λ^o_{NaOH} (b) Λ^o_{NaCl} (c) $\Lambda^o_{H_2O}$ (d) Λ^o_{KCl}

43. Resistance of a conductivity cell filled with a solution of an electrolyte of concentration 0.1 M is 100 Ω. The conductivity of this solution is 1.29 S m^{-1}. Resistance of the same cell when filled with 0.2 M of the same solution is 520 Ω. The molar conductivity of 0.2 M solution of electrolyte will be **[NCERT : P L-47 | 2006]**

(a) 1.24×10^{-4} S $m^2 mol^{-1}$ (b) 12.4×10^{-4} S $m^2 mol^{-1}$
(c) 124×10^{-4} S $m^2 mol^{-1}$ (d) 1240×10^{-4} S $m^2 mol^{-1}$

44. The highest electrical conductivity of the following aqueous solutions is of **[NCERT : P L-45 | 2005]**

(a) 0.1 M difluoroacetic acid
(b) 0.1 M fluoroacetic acid
(c) 0.1 M chloroacetic acid
(d) 0.1 M acetic acid

45. The limiting molar conductivities Λ° for NaCl, KBr and KCl are 126, 152 and 150 S $cm^2 mol^{-1}$ respectively. The Λ° for NaBr is **[NCERT : P L-47 | 2004]**

(a) 278 S $cm^2 mol^{-1}$ (b) 176 S $cm^2 mol^{-1}$
(c) 128 S $cm^2 mol^{-1}$ (d) 302 S $cm^2 mol^{-1}$

46.

Electrolyte:	KCl	KNO_3	HCl	NaOAc	NaCl
$\Lambda^\infty (S\ cm^2\ mol^{-1})$:	149.9	145	426.2	91	126.5

Calculate Λ^∞_{HOAc} using appropriate molar conductances of the electrolytes listed above at infinite dilution in H_2O at 25 °C **[NCERT : P L-49 | 2004]**

(a) 217.5 (b) 390.7 (c) 552.7 (d) 517.2

47. Conductivity (unit Siemen's S) is directly proportional to area of the vessel and the concentration of the solution in it and is inversely proportional to the length of the vessel then the unit of the constant of proportionality is **[NCERT : P L-45 | 2002]**

(a) S mol^{-1} (b) S$m^2 mol^{-1}$
(c) $S^{-2} m^2$ mol (d) $S^2 m^2 mol^{-2}$.

Topic 2 - Electrolytic Cells and Types of Electrolysis

48. 1 Faraday electricity was passed through Cu_2+ (1.5 M, 1 L)/Cu and 0.1 Faraday was passed through Ag+ (0.2 M, 1 L)/Ag electrolytic cells. After this the two cells were connected as shown below to make an electrochemical cell. The emf of the cell thus formed at 298 K is- **[NCERT : P L-51 | NV, April 7, 2025 (I)]**

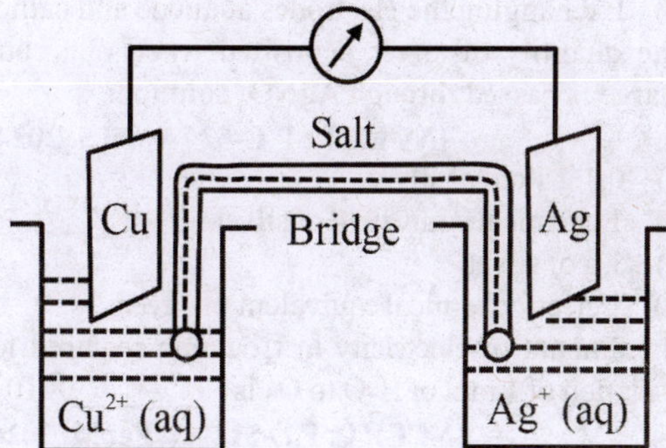

Given : $E^0_{Cu^{2+}/Cu} = 0.34\ V$

$E^0_{Ag^+/Ag} = 0.8\ V$

$\frac{2.303RT}{F} = 0.8\ V$

49. O_2 gas will be evolved as a product of electrolysis of : **[NCERT : P L-49, 50 | Jan. 29, 2025 (II)]**

(A) an aqueous solution of $AgNO_3$ using silver electrodes.
(B) an aqueous solution of $AgNO_3$ using platinum electrodes.
(C) a dilute solution of H_2SO_4 using platinum electrodes.
(D) a high concentration solution of H_2SO_4 using platinum electrodes.

Choose the correct answer from the options given below :

(a) (B) and (C) only (b) (A) and (D) only
(c) (B) and (D) only (d) (A) and (C) only

50. Electrolysis of 600 mL aqueous solution of NaCl for 5 min changes the pH of the solution to 12. The current in Amperes used for the given electrolysis is ____. (Nearest integer). **[NCERT : P L-53 | NV, Jan. 28, 2025 (II)]**

51. Which of the following electrolyte can be used to obtain $H_2S_2O_8$ by the process of electrolysis? **[NCERT : P L-53 | Jan. 22, 2025 (I)]**
(a) Dilute solution of sodium sulphate
(b) Dilute solution of sulphuric acid
(c) Concentrated solution of sulphuric acid
(d) Acidified dilute solution of sodium sulphate.

52. A solution of aluminium chloride is electrolysed for 30 minutes using a current of 2A. The amount of the aluminium deposited at the cathode is ____. [Given : molar mass of aluminium and chlorine are27 g mol^{-1} and 35.5 g mol^{-1} respectively, Faraday constant = 96500 C mol^{-1}]. **[NCERT : P L-53 | Jan. 22, 2025 (I)]**
(a) 1.660 g (b) 1.007 g
(c) 0.336 g (d) 0.441 g

53. How can an electrochemical cell be converted into an electrolytic cell ? **[NCERT : P L-51 | April 6, 2024 (II)]**
(a) Applying an external opposite potential greater than E^0_{cell}
(b) Reversing the flow of ions in salt bridge.
(c) Applying an external opposite potential lower than E^0_{cell}.
(d) Exchanging the electrodes at anode and cathode.

54. The quantity of silver deposited when one coulomb charge is passed through $AgNO_3$ solution: **[NCERT : P L-52 | April 5, 2024 (II)]**
(a) 0.1 g atom of silver
(b) 1 chemical equivalent of silver
(c) 1 g of silver
(d) 1 electrochemical equivalent of silver

55. The amount of electricity in Coulomb required for the oxidation of 1 mol of H_2O to O_2 is ________ $\times 10^5$C. **[NCERT : P L-51 | NV, Feb. 1, 2024 (II)]**

56. One Faraday of electricity liberates $x \times 10^{-1}$ gram atom of copper from copper sulphate, x is _____. **[NCERT : P L-51 | NV, Jan. 31, 2024 (I)]**

57. The mass of zinc produced by the electrolysis of zinc sulphate solution with a steady current of 0.015 A for 15 minutes is _____ $\times 10^{-4}$ g. (Atomic mass of zinc = 65.4 amu) **[NCERT : P L-51 | NV, Jan. 29, 2024 (I)]**

58. A constant current was passed through a solution of $AuCl_4^-$ ion between gold electrodes. After a period of 10.0 minutes, the increase in mass of cathode was 1.314g. The total charge passed through the solution ____ $\times 10^{-2}$ F. (Given atomic mass of Au = 197) **[NCERT : P L-51 | NV, Jan. 29, 2024 (II)]**

59. The mass of silver (Molar mass of Ag : 108 $gmol^{-1}$ displaced by a quantity of electricity which displaces 5600 mL of O_2 at S.T.P. will be ______ g. **[NCERT : P L-51 | NV, Jan. 27, 2024 (I)]**

60. A metal surface of 100 cm^2 area has to be coated with nickel layer of thickness 0.001 mm. A current of 2A was passed through a solution of $Ni(NO_3)_2$ for 'x' seconds to coat the desired layer. The value of x is ______ (Nearest integer) (ρ_{Ni} (density of Nickel) is 10 gmL^{-1}, Molar mass of Nickel is 60 $gmol^{-1}$ F = 96500 C mol^{-1}) **[NCERT : P L-52 | NV, April 13, 2023 (I)]**

61. The standard electrode potential of M^+/M in aqueous solution does not depend on **[NCERT : P L-40 | April 06, 2023 (I)]**
(a) Ionisation of a solid metal atom
(b) Sublimation of a solid metal
(c) Ionisation of a gaseous metal atom
(d) Hydration of a gaseous metal ion

62. The amount of charge in F (Faraday) required to obtain one mole of iron from Fe_3O_4 is ________. (Nearest Integer) **[NCERT : P L-51 | NV, July 26, 2022 (I)]**

63. A dilute solution of sulphuric acid is electrolysed using a current of 0.10 A for 2 hours to produce hydrogen and oxygen gas. The total volume of gases produced at STP is ________ cm^3. (Nearest integer) [Given : Faraday constant F = 96500 C mol^{-1} at STP, molar volume of an ideal gas is 22.7 L mol^{-1}] **[NCERT : P L-53 | NV, June 29, 2022 (I)]**

64. A solution of $Fe_2(SO_4)_3$ is electrolyzed for 'x' min with a current of 1.5 A to deposit 0.3482 g of Fe. The value of x is __________. [nearest integer]
Given : 1 F = 96500 C mol^{-1}
Atomic mass of Fe = 56 g mol^{-1} **[NCERT : P L-52 | NV, June 25, 2022 (II)]**

65. Consider the following reaction
$MnO_4^- + 8H^+ + 5e^- \rightarrow Mn^{2+} + 4H_2O$, E° = 1.51 V.
The quantity of electricity required in Faraday to reduce five moles of MnO_4^- is _______. **[NCERT : P L-51 | NV, Feb. 26, 2021 (I)]**

66. The product obtained from the electrolytic oxidation of acidified sulphate solutions, is : **[NCERT : P L-52 | July 27, 2021 (I)]**
(a) HSO_4^- (b) HO_3SOOSO_3H
(c) HO_2SOSO_2H (d) HO_3SOSO_3H

67. Potassium chlorate is prepared by the electrolysis of KCl in basic solution $6\,OH^- + Cl^- \rightarrow ClO_3^- + 3\,H_2O + 6\,e^-$
If only 60% of the current is utilized in the reaction, the time (rounded to the nearest hour) required to produce 10 g of $KClO_3$ using a current of 2 A is ______. (Given : F = 96,500 C mol^{-1}; molar mass of $KClO_3$ = 122 g mol^{-1}) **[NV, Sep. 06, 2020 (I) Same July 20, 2021 (II)]**

68. 250 mL of a waste solution obtained from the workshop of a goldsmith contains 0.1 M $AgNO_3$ and 0.1 M AuCl. The solution was electrolyzed at 2 V by passing a current of 1 A for 15 minutes. The metal/metals deposited will be : **[NCERT : P L-51 | Sep. 04, 2020 (II)]**
$\left(E^0_{Ag^+/Ag} = 0.80\text{ V}, E^0_{Au^+/Au} = 1.69\text{ V}\right)$

(a) only gold
(b) silver and gold in proportion to their atomic weights
(c) only silver
(d) silver and gold in equal mass proportion

69. An acidic solution of dichromate is electrolyzed for 8 minutes using 2A current. As per the following equation

$$Cr_2O_7^{2-} + 14H^+ + 6e^- \rightarrow 2Cr^{3+} + 7H_2O$$

The amount of Cr^{3+} obtained was 0.104 g. The efficiency of the process (in %) is (Take : F = 96000 C, At. mass of chromium = 52) ________.

[NCERT : P L-51 | NV, Sep. 03, 2020 (II)]

70. 108 g of silver (molar mass 108 g mol^{-1}) is deposited at cathode from $AgNO_3(aq)$ solution by a certain quantity of electricity. The volume (in L) of oxygen gas produced at 273 K and 1 bar pressure from water by the same quantity of electricity is _______.

[NCERT : P L-51 | NV, Jan. 09, 2020 (I)]

71. A solution of Ni $(NO_3)_2$ is electrolysed between platinum electrodes using 0.1 Faraday electricity. How many mole of Ni will be deposited at the cathode?

[NCERT : P L-51 | April 9, 2019 (II)]

(a) 0.05 (b) 0.20
(c) 0.15 (d) 0.10

72. The anodic half-cell of lead-acid battery is recharged using electricity of 0.05 Faraday. The amount of $PbSO_4$ electrolyzed in g during the process is : (Molar mass of $PbSO_4$ = 303 g mol^{-1}) **[NCERT : P L-55 | Jan. 9, 2019 (I)]**

(a) 22.8 (b) 15.2 (c) 7.6 (d) 11.4

73. How long (approximate) should water be electrolysed by passing through 100 amperes current so that the oxygen released can completely burn 27.66 g of diborane? (Atomic weight of B = 10.8 u) **[2018]**

(a) 6.4 hours (b) 0.8 hours
(c) 3.2 hours (d) 1.6 hours

74. When 9.65 ampere current was passed for 1.0 hour into nitrobenzene in acidic medium, the amount of *p*-amino-phenol produced is _______. **[Online April 16, 2018]**

(a) 109.0 g (b) 98.1 g
(c) 9.81 g (d) 10.9 g

75. When an electric current is passes through acidified water, 112 mL of hydrogen gas at N.T.P. was collected at the cathode in 965 seconds. The current passed, in ampere, is : **[NCERT : P L-52 | Online April 15, 2018 (I)]**

(a) 2.0 (b) 0.1 (c) 0.5 (d) 1.0

76. Two Faraday of electricity is passed through a solution of $CuSO_4$. The mass of copper deposited at the cathode is (at. mass of Cu = 63.5 amu) **[NCERT : P L-51 | 2015]**

(a) 2g (b) 127 g (c) 0 g (d) 63.5 g

77. A current of 10.0 A flows for 2.00 h through an electrolytic cell containing a molten salt of metal X. This results in the decomposition of 0.250 mol of metal X at the cathode. The oxidation state of X in the molten salt is: (F = 96,500 C)

[NCERT : P L-51 | Online April 9, 2014]

(a) 1+ (b) 2+ (c) 3+ (d) 4+

78. How many electrons would be required to deposit 6.35 g of copper at the cathode during the electrolysis of an aqueous solution of copper sulphate? (Atomic mass of copper = 63.5 u, N_A = Avogadro's constant):

[NCERT : P L-51 | Online April 12, 2014]

(a) $\frac{N_A}{20}$ (b) $\frac{N_A}{10}$ (c) $\frac{N_A}{5}$ (d) $\frac{N_A}{2}$

79. A solution of copper sulphate ($CuSO_4$) is electrolysed for 10 minutes with a current of 1.5 amperes. The mass of copper deposited at the cathode (at. mass of Cu = 63u) is:

[NCERT : P L-51 | Online April 25, 2013]

(a) 0.3892 g (b) 0.2938 g
(c) 0.2398 g (d) 0.3928 g

80. A battery is constructed of Cr and $Na_2Cr_2O_7$. The unbalanced chemical equation when such a battery discharges is following:

$$Na_2Cr_2O_7 + Cr + H^+ \rightarrow Cr^{3+} + H_2O + Na^+$$

If one Faraday of electricity is passed through the battery during the charging, the number of moles of Cr^{3+} removed from the solution is **[NCERT : P L-51 | Online May 7, 2012]**

(a) $\frac{4}{3}$ (b) $\frac{1}{3}$ (c) $\frac{3}{3}$ (d) $\frac{2}{3}$

81. The standard potentials of Ag^+/Ag, $Hg_2^{2+}/2Hg$, Cu^{2+}/Cu and Mg^{2+}/Mg electrodes are 0.80, 0.79, 0.34 and –2.37 V, respectively. An aqueous solution which contains one mole per litre of the salts of each of the four metals is electrolyzed. With increasing voltage, the correct sequence of deposition of the metals at the cathode is

[NCERT : P L-51 | Online May 19, 2012]

(a) Ag, Hg, Cu, Mg (b) Cu, Hg, Ag only
(c) Ag, Hg, Cu only (d) Mg, Cu, Hg, Ag

82. Aluminium oxide may be electrolysed at 1000 °C to furnish aluminium metal (At. Mass = 27 amu; 1 Faraday = 96,500 Coulombs). The cathode reaction is: – $Al^{3+} + 3e^- \rightarrow Al^\circ$

To prepare 5.12 kg of aluminium metal by this method we require **[NCERT : P L-51 | 2005]**

(a) 5.49×10^1 C of electricity
(b) 5.49×10^4 C of electricity
(c) 1.83×10^7 C of electricity
(d) 5.49×10^7 C of electricity

83. When during electrolysis of a solution of $AgNO_3$ 9650 coulombs of charge pass through the electroplating bath, the mass of silver deposited on the cathode will be

[NCERT : P L-53 | 2003]

(a) 10.8 g (b) 21.6 g (c) 108 g (d) 1.08 g

84. When the sample of copper with zinc impurity is to be purified by electrolysis, the appropriate electrodes are

[NCERT : P L-51 | 2002]

	Cathode	Anode
(a)	pure zinc	pure copper
(b)	impure sample	pure copper
(c)	impure zinc	impure sample
(d)	pure copper	impure sample.

Topic 3 - Cells and Electrode Potential, Nernst Equation

85. Consider the following half cell reaction

$Cr_2O_7^{2-}(aq) + 6e^- + 14H^+(aq) \rightarrow 2Cr^{3+}(aq) + 7H_2O(l)$

The reaction was conducted with the ratio of $\frac{[Cr^{3+}]^2}{[Cr_2O_7^{2-}]} = 10^{-6}$. The pH value at which the EMF of the half cell will become zero is ________.

(nearest integer value)

[Given : standard half cell reduction potential $E^o_{Cr_2O_7^{2-},H^+/Cr^{3+}} = 1.33V, \frac{2.303RT}{F} = 0.059V$]

[NCERT : P L-38 | NV, April 8, 2025 (II)]

86. Given below are two statements :

[NCERT : P L-38 | April 7, 2025 (II)]

1 M aqueous solution of each of $Cu(NO_3)_2$, $AgNO_3$, $Hg_2(NO_3)_2$; $Mg(NO_3)_2$ are electrolysed using inert electrodes,

Given : $E^\theta_{Ag^+/Ag} = 0.80$ V, $E^\theta_{Hg_2^{2+}/Hg} = 0.79$ V,

$E^\theta_{Cu^{2+}/Cu} = 0.24$ V and $E^\theta_{Mg^{2+}/Mg} = -2.37$ V

Statement (I) : With increasing voltage, the sequence of deposition of metals on the cathode will be Ag, Hg and Cu

Statement (II) : Magnesium will not be deposited at cathode instead oxygen gas will be evolved at the cathode.

In the light of the above statement, choose the **most appropriate answer** from the options given below

(a) Both statement I and statement II are incorrect
(b) Statement I is correct but statement II is incorrect
(c) Both statement I and statement II are correct
(d) Statement I is incorrect but statement II is correct

87. Consider the following electrochemical cell at standard condition.

$Au(s)|QH_2,Q|NH_4X(0.01M)||Ag^+(1M)|Ag(s)$

$E_{cell} = +0.4V$

The couple QH_2/Q represents quinhydrone electrode, the half cell reaction is given below

[NCERT : P L-38 | NV, April 2, 2025 (I)]

Q (quinone) $+ 2e^- + 2H^+ \longrightarrow$ QH_2 (hydroquinone) $\quad E^o_{Q/QH_2} = +0.7V$

$\left[\text{Given} : E^\circ_{Ag^+/Ag} = +0.8V \text{ and } \frac{2.303RT}{F} = 0.06V\right]$

The pK_b value of the ammonium halide salt (NH_4X) used here is ______. (nearest integer)

88. For a Mg | Mg^{2+} (aq) || Ag^+(aq) | Ag the correct Nernst Equation is : **[NCERT : P L-38 | Jan. 29, 2025 (I)]**

(a) $E_{cell} = E^o_{cell} - \frac{RT}{2F}\ln\frac{[Ag^+]}{[Mg^{2+}]}$

(b) $E_{cell} = E^o_{cell} + \frac{RT}{2F}\ln\frac{[Ag^+]^2}{[Mg^{2+}]}$

(c) $E_{cell} = E^o_{cell} - \frac{RT}{2F}\ln\frac{[Mg^{2+}]}{[Ag^+]}$

(d) $E_{cell} = E^o_{cell} - \frac{RT}{2F}\ln\frac{[Ag^+]^2}{[Mg^{2+}]}$

89. The standard reduction potential values of some of the p-block ions are given below. Predict the one with the strongest oxidising capacity.

[NCERT : P L-35 | Jan. 29, 2025 (I)]

(a) $E^\ominus_{Sn^{4+}/Sn^{2+}} = +1.15V$ (b) $E^\ominus_{Tl^{3+}/Tl} = +1.26V$

(c) $E^\ominus_{Al^{3+}/Al} = -1.66V$ (d) $E^\ominus_{Pb^{4+}/Pb^{2+}} = +1.67V$

90. Based on the data given below:

[NCERT : P L-35 | Jan. 24, 2025 (II)]

$E^0_{Cr_2O_7^{2-}/Cr^{3+}} = 1.33$ V $\quad E^0_{Cl_2/Cl^-} = 1.36$ V

$E^0_{MnO_4^-/Mn^{2+}} = 1.51$ V $\quad E^0_{Cr^{3+}/Cr} = -0.74$ V

the strongest reducing agent is :

(a) Mn^{2+} (b) Cr
(c) MnO_4^- (d) Cl^-

91. For the given cell **[NCERT : P L-37 | Jan. 24, 2025 (I)]**

$Fe^{2+}(aq) + Ag^+(aq) \rightarrow Fe^{3+}(aq) + Ag(s)$

The standard cell potential of the above reaction is

Given :

$Ag^+ + e^- \rightarrow Ag \qquad E^0 = xV$
$Fe^{2+} + 2e^- \rightarrow Fe \qquad E^0 = yV$
$Fe^{3+} + 3e^- \rightarrow Fe \qquad E^0 = zV$

(a) $x + y - z$ (b) $x + 2y - 3z$
(c) $y - 2x$ (d) $x + 2y$

92. Standard electrode potentials for a few half cells are mentioned below: **[NCERT : P L-37 | Jan. 23, 2025 (II)]**

$E^o_{Cu^{2+}/Cu} = 0.34V, E^o_{Zn^{2+}/Zn} = -0.76V$

$E^o_{Ag^+/Ag} = 0.80V, E^o_{Mg^{2+}/Mg} = -2.37V$

Which one of the following cells gives the most negative value of $\Delta G°$?

(a) $Zn|Zn^{2+}(1M)||Ag^+(1M)|Ag$
(b) $Zn|Zn^{2+}(1M)||Mg^{2+}(1M)|Mg$
(c) $Ag|Ag^+(1M)||Mg^{2+}(1M)|Mg$
(d) $Cu|Cu^{2+}(1M)||Ag^+(1M)|Ag$

93. $FeO_4^{2-} \xrightarrow{+2.0V} Fe^{3+} \xrightarrow{0.8V} Fe^{2+} \xrightarrow{-0.5V} Fe^0$
In the above diagram, the standard electrode potentials are given in volts (over the arrow).

The value of $E^{\ominus}_{FeO_4^{2-}/Fe^{2+}}$ is

[NCERT : P L-40 | Jan. 23, 2025 (I)]

(a) 1.7 V (b) 1.2 V (c) 2.1 V (d) 1.4 V

94. The standard reduction potentials at 298 K for the following half cells are given below :

[NCERT : P L-37 | NV, April 9, 2024 (I)]

$Cr_2O_7^{2-} + 14H^+ + 6e^- \rightarrow 2Cr^{3+} + 7H_2O, E° = 1.33V$
$Fe^{3+}(aq) + 3e^- \rightarrow Fe \quad E° = -0.04V$
$Ni^{2+}(aq) + 2e^- \rightarrow Ni \quad E° = -0.25V$
$Ag^+(aq) + e^- \rightarrow Ag \quad E° = 0.80V$
$Au^{3+}(aq) + 3e^- \rightarrow Au \quad E° = 1.40V$

Consider the given electrochemical reactions,
The number of metal(s) which will be oxidized be $Cr_2O_7^{2-}$, in aqueous solution is ________.

95. The emf of cell $Tl\left|\begin{matrix}Tl^+\\(0.001M)\end{matrix}\right|\left|\begin{matrix}Cu^{2+}\\(0.01M)\end{matrix}\right|Cu$ is 0.83 V at 298 K.

It could be increased by :

[NCERT : P L-34 | April 8, 2024 (II)]

(a) increasing concentration of Tl^+ ions
(b) increasing concentration of both Tl^+ and Cu^{2+} ions
(c) decreasing concentration of both Tl^+ and Cu^{2+} ions
(d) increasing concentration of Cu^{2+} ions

96. The reaction ; **[NCERT : P L-33 | April 8, 2024 (II)]**

$$\frac{1}{2}H_{2(g)} + AgCl_{(s)} \rightarrow H^+_{(aq)} + Cl^-_{(aq)} + Ag_{(s)}$$

Occurs in which of the following galvanic cell :
(a) $Pt|H_{2(g)}|HCl_{(soln.)}|AgCl_{(s)}|Ag$
(b) $Pt|H_{2(g)}|HCl_{(soln.)}|AgNO_{3(aq)}|Ag$
(c) $Pt|H_{2(g)}|KCl_{(soln.)}|AgCl_{(s)}|Ag$
(d) $Ag|AgCl_{(s)}|KCl_{(soln.)}|AgNO_{3(aq.)}|Ag$

97. For the electro chemical cell
$M|M^{2+}||X|X^{2-}$ **[NCERT : P L-32 | April 5, 2024 (II)]**

If $E^0_{(M^{2+}/M)} = 0.46V$ and $E^0_{(X/X^{2-})} = 0.34V$.

Which of the following is **correct** ?
(a) $E_{cell} = -0.80 V$
(b) $M + X \rightarrow M^{2+} + X^{2-}$ is a spontaneous reaction
(c) $M^{2+} + X^{2-} \rightarrow M + X$ is a spontaneous reaction
(d) $E_{cell} = 0.80 V$

98. What pressure (bar) of H_2 would be required to make emf of hydrogen electrode zero in pure water at 25°C ?

[NCERT : P L-34 | April 4, 2024 (I)]

(a) 10^{-14} (b) 10^{-7} (c) 1 (d) 0.5

99. One of the commonly used electrode is calomel electrode. Under which of the following categories calomel electrode comes ? **[NCERT : P L-34 | April 4, 2024 (I)]**

(a) Metal – Insoluble Salt – Anion electrodes
(b) Oxidation – Reduction electrodes
(c) Gas – Ion electrodes
(d) Metal ion – Metal electrodes

100. The potential for the given half cell at 298 K is (–) ______ $\times 10^{-2}$ V. **[NCERT : P L-34 | NV, Feb. 1, 2024 (I)]**

$$2H^+(aq) + 2e^- \longrightarrow H_2(g)$$

$$[H^+] = 1M, P_{H_2} = 2 \text{ atm}$$

(Given : $2.303RT/F = 0.06$ V, $\log 2 = 0.3$)

101. Consider the following redox reaction:

$$MnO_4^- + H^+ + H_2C_2O_4 \rightleftharpoons Mn^{2+} + H_2O + CO_2$$

The standard reduction potentials are given as below ($E°_{red}$):

$E°_{MnO_4^-/Mn^{2+}} = +1.51V$

$E°_{CO_2/H_2C_2O_4} = -0.49 V$

If the equilibrium constant of the above reaction is given as $K_{eq} = 10^x$, then the value of x = ________ (nearest integer) **[NCERT : P L-39 | NV, Feb. 1, 2024 (II)]**

102. The hydrogen electrode is dipped in a solution of pH = 3 at 25°C. The potential of the electrode will be – ______ $\times 10^{-2}$ V. **[NCERT : P L-34 | NV, Jan. 27, 2024 (II)]**

$$\left(\frac{2.303RT}{F} = 0.059V\right)$$

103. At 298 K, the standard reduction potential for Cu^{2+}/Cu electrode is 0.34 V.

Given : K_{sp} $Cu(OH)_2 = 1 \times 10^{-20}$

Take $\frac{2.303RT}{F} = 0.059V$

The reduction potential at pH = 14 for the above couple is (–)x $\times 10^{-2}$ V. The value of x is ______.

[NCERT : P L-38 | NV, April 13, 2023 (II)]

104. In an electrochemical reaction of lead, at standard temperature, if $E^0_{(Pb^{2+}/Pb)} = m$ Volt and $E^0_{(Pb^{4+}/Pb)} = n$ Volt, then the value of $E^0_{(Pb^{2+}/Pb^{4+})}$ is given by m – xn.

The value of x is ______.
(Nearest integer)

[NCERT : P L-34, 40 | NV, April 11, 2023 (I)]

105. The number of correct statements from the following is :

[NCERT : P L-33, 51 | NV, April 11, 2023 (II)]

A. E_{cell} is an intensive parameter.

B. A negative $E^{\ominus}$ means that the redox couple is a stronger reducing agent than the H^+/H_2 couple.

C. The amount of electricity required for oxidation or reduction depends on the stoichiometry of the electrode reaction.

D. The amount of chemical reaction which occurs at any electrode during electrolysis by a current is proportional to the quantity of electricity passed through the electrolyte.

106. The reaction **[NCERT : P L-34 | April 08, 2023 (I)]**

$\frac{1}{2}H_2(g) + AgCl(s) \rightleftharpoons H^+(aq) + Cl^-(aq) + Ag(s)$

occurs in which of the given galvanic cell.

(a) $Pt|H_2(g)|KCl(sol^n)|AgCl(s)|Ag$
(b) $Pt|H_2(g)|HCl(sol^n)|AgCl(s)|Ag$
(c) $Ag|AgCl(s)|KCl(sol^n)|AgCl(s)|Ag$
(d) $Pt|H_2(g)|HCl(sol^n)|AgNO_3(sol^n)|Ag$

107. The number of incorrect statements from the following is ______ **[NCERT : P L-40 | NV, April 08, 2023 (II)]**

(A) The electrical work that a reaction can perform at constant pressure and temperature is equal to the reaction Gibbs energy.
(B) E^0_{cell} is dependent on the pressure
(C) $\frac{dE^0 cell}{dT} = \frac{\Delta_r S^0}{nF}$
(D) A cell is operating reversibly if the cell potential is exactly balanced by an opposing source of potential difference.

108. The standard reduction potential at 298 K for the following half cells are given below:- **[NCERT : P L-37 | NV, April 06, 2023 (II)]**

$NO_3^- + 4H^+ + 3e^- \rightarrow NO(g) + 2H_2O \quad E^0 = 0.97V$

$V^{2+}(aq) + 2e^- \rightarrow V \quad E^0 = -1.19V$

$Fe^{3+}(aq) + 3e^- \rightarrow Fe \quad E^0 = -0.04V$

$Ag^+(aq) + e^- \rightarrow Ag(s) \quad E^0 = 0.80V$

$Au^{3+}(aq) + 3e^- \rightarrow Au(s) \quad E^0 = 1.40V$

The number of metal(s) which will be oxidized by NO_3^- is aqueous solution is ________.

109. At what pH, given half cell MnO_4^- (0.1 M) | Mn^{2+} (0.001 M) will have electrode potential of 1.282 V? ______ (Nearest Integer)

Given $E^o_{MnO_4^-|Mn^{2+}} = 1.54V, \frac{2.303RT}{F} = 0.059V$

[NCERT : P L-36 | NV, Feb. 01, 2023 (I)]

110. The logarithm of equilibrium constant for the reaction $Pd^{2+} + 4Cl^- \rightleftharpoons PdCl_4^{2-}$ is ______ (Nearest integer)

Given : $\frac{2.303RT}{F} = 0.06V$

$Pd^{2+}(aq) + 2e^- \rightleftharpoons Pd(s) \quad E° = 0.83V$

$PdCl_4^{2-}(aq) + 2e^- \rightleftharpoons Pd(s) + 4Cl^-(aq)$

$E° = 0.65V$ **[NCERT : P L-39 | NV, Jan. 31, 2023 (I)]**

111. Consider the cell **[NCERT : P L-36 | NV, Jan. 30, 2023 (I)]**

$Pt(s)|H_2(g, 1\,atm)|H^+(aq, 1M)||Fe^{3+}(aq), Fe^{2+}(aq)|Pt(s)$

When the potential of the cell is 0.712 V at 298 K, the ratio $[Fe^{2+}]/[Fe^{3+}]$ is ______. (Nearest integer)

Given: $Fe^{3+} + e^- = Fe^{2+}, E°_{Fe^{3+}, Fe^{2pt}} = 0.771$

$\frac{2.303\,RT}{F} = 0.06V$

112. The electrode potential of the following half cell at 298 K $X|X^{2+}(0.001\,M)||Y^{2+}(0.01\,M)|Y$ is ______ $\times 10^{-2}$ V (Nearest integer).

Given : $E^0_{X^{2+}|X} = -2.36V$; $E^0_{Y^{2+}|Y} = +0.36V$

$\frac{2.303RT}{F} = 0.06V$

[NCERT : P L-36 | NV, Jan. 30, 2023 (II)]

113. The standard electrode potential (M^{3+}/M^{2+}) for V, Cr, Mn & Co are –0.26 V, –0.41 V, + 1.57 V and +1.97 V, respectively. The metal ions which can liberate H_2 from a dilute acid are **[NCERT : P L-37 | Jan. 29, 2023 (I)]**

(a) V^{2+} and Mn^{2+}
(b) Cr^{2+} and CO^{2+}
(c) V^{2+} and Cr^{2+}
(d) Mn^{2+} and Co^{2+}

114. The equilibrium constant for the reaction $Zn(s) + Sn^{2+}(aq) \rightleftharpoons Zn^{2+}(aq) + Sn(s)$ is 1×10^{20} at 298 K. The magnitude of standard electrode potential of Sn/Sn^{2+} if $E^{\circ}_{zn^{2+}/E^{\circ}_{zn}}$ – 0.76 V is ______ $\times 10^{-2}$ V. (Nearest integer)

Given: $\frac{2.303\,RT}{F} = 0.059V$

[NCERT : P L-39 | NV, Jan. 29, 2023 (II)]

115. Consider the cell

$Pt(s)|H_2(s)(1atm)|H^+(aq, [H^+] = 1)||Fe^{3+}(aq), Fe^{2+}(q)|Pt(s)$

Given : $E^{\circ}_{Fe^{3+}/Fe^{2+}} = 0.771$ V and $E^{\circ}_{H^+/\frac{1}{2}H_2} = 0$ V, T= 298 K

If the potential of the cell is 0.712 V the ratio of concentration of Fe^{2+} to Fe^{3+} is ______ (Nearest integer)

[NCERT : P L-36 | NV, Jan. 25, 2023 (I)]

116. $Pt(s)\ H_2(g)\ (1bar)|H^+(aq)(1M)||M^{3+}(aq), M^+(aq)|Pt(s)$

The E_{cell} for the given cell is 0.1115 V at 298 K **[NCERT : P L-36 | NV, Jan. 25, 2023 (II)]**

when $\frac{[M^+(aq)]}{[M^{3+}(aq)]} = 10^a$ The value of a is ______

Given: $E^{\circ}_{M^{3+}/M^+} = 0.2$ V $\quad \frac{2.303\,RT}{F} = 0.059$ V

117. At 298 K, a 1 litre solution containing 10 mmol of $Cr_2O_7^{2-}$ and 100 mmol of Cr^{3+} shows a pH of 3.0.

Given : $Cr_2O_7^{2-} \rightarrow Cr^{3+}$; $E^0 = 1.330$ V and $\frac{2.303RT}{F} = 0.059V$

The potential for the half cell reaction is $x \times 10^{-3}$ *V*. The value of *x* is ______.

[NCERT : P L-36 | NV, Jan. 24, 2023 (I)]

118. The correct group of halide ions which can be oxidised by oxygen in acidic medium is **[NCERT : P L-37 | April 13, 2023 (II)]**
(a) Br^- only (b) Cl^-, Br^- and I^- only
(c) Br^- and I^- only (d) I^- only

119. $FeO_4^{2-} \xrightarrow{+2.2V} Fe^{3+} \xrightarrow{+0.70V} Fe^{2+} \xrightarrow{-0.45V} Fe^0$
$E^{\ominus}_{FeO_4^{2-}/Fe^{2+}}$ is $x \times 10^{-3}$ V. The value of x is ______
[NCERT : P L-37 | NV, April 10, 2023 (I)]

120. For a cell, Cu(s) $|Cu^{2+}(0.001M) \| Ag^+(0.01M)|$ Ag(s) the cell potential is found to be 0.43 V at 298 K. The magnitude of standard electrode potential for Cu^{2+}/Cu is ___ $\times 10^{-2}$ V.
$\left[\text{Given}: E^{\ominus}_{Ag^+/Ag} = 0.80 \text{ V and } \frac{2.303RT}{F} = 0.06 \text{ V}\right]$
[NV, July 29, 2022 (II)]

121. The cell potential for $Zn|Zn^{2+}(aq)\|Sn^{x+}|Sn$ is 0.801 V at 298 K. The reaction quotient for the above reaction is 10^{-2}. The number of electrons involved in the given electrochemical cell reaction is ______.
(Given: $E^\circ_{Zn^{2+}|Zn} = -0.763V$, $E^\circ_{Sn^{x+}|Sn} = +0.008$ V and $\frac{2.303RT}{F} = 0.06$ V)
[NCERT : P L-36 | NV, July 25, 2022 (I)]

122. In which of the following half cells, electrochemical reaction is pH dependent? **[June 30, 2022 (I)]**
(a) $Pt \mid Fe^{3+}, Fe^{2+}$ (b) $MnO_4^- \mid Mn^{2+}$
(c) $Ag \mid AgCl \mid Cl^-$ (d) $\frac{1}{2}F_2 \mid F^-$

123. The cell potential for the given cell at 298 K
$Pt \mid H_2(g, 1\ bar) \mid H^+(aq) \| Cu^{2+}(aq) \mid Cu(s)$ is 0.31V. The pH of the acidic solution is found to be 3, whereas the concentration of Cu^{2+} is 10^{-x} M. The value of x is ______.
(Given $E^\circ_{Cu^{2+}/Cu} = 0.34V$ and $\frac{2.303RT}{F} = 0.06V$)
[NCERT : P L-35 | NV, June 29, 2022 (II)]

124. For the given reactions
$Sn^{2+} + 2e^- \rightarrow Sn$
$Sn^{4+} + 4e^- \rightarrow Sn$
The electrode potentials are; $E^\circ_{Sn^{2+}/Sn} = -0.140$ V and $E^\circ_{Sn^{4+}/Sn} = 0.010$ V. The magnitude of standard electrode potential for Sn^{4+}/Sn^{2+} i.e. $E^\circ_{Sn^{4+}/Sn^{2+}}$ is ______ $\times 10^{-2}$ V. (Nearest integer)
[NCERT : P L-36 | NV, June 28, 2022 (II)]

125. For the reaction taking place in the cell:
$Pt(s) \mid H_2(g) \mid H^+(aq) \| Ag^+(aq) \mid Ag(s)$
$E^\circ_{Cell} = +0.5332$ V.
The value of $\Delta_f G^0$ is ________ kJ mol^{-1}.
(in nearest integer)
[NCERT : P L-40 | NV, June 27, 2022 (II)]

126. The $\left(\frac{\partial E}{\partial T}\right)_P$ of different types of half cells are as follows:

A	B	C	D
1×10^{-4}	2×10^{-4}	0.1×10^{-4}	0.2×10^{-4}

(Where E is the electromotive force)
Which of the above half cells would be preferred to be used as reference electrode ?
[NCERT : P L-40 | June 26, 2022 (I)]
(a) A (b) B (c) C (d) D

127. $Cu(s) + Sn^{2+}(0.001M) \rightarrow Cu^{2+}(0.01M) + Sn(s)$
The Gibbs free energy change for the above reaction at 298 K is $x \times 10^{-1}$ kJ mol^{-1};
The value of x is______. [nearest integer]
[Given $E^\circ_{Cu^{2+}/Cu} = 0.34V$; $E^\circ_{Sn^{2+}/Sn} = -0.14V$; $F = 96500C\ mol^{-1}$]
[NCERT : P L-40 | NV, June 26, 2022 (II)]

128. In a cell, the following reactions take place
$Fe^{2+} \rightarrow Fe^{3+} + e^-$ $E^o_{Fe^{3+}/Fe^{2+}} = 0.77$ V
$2I^- \rightarrow I_2 + 2e^-$ $E^o_{I_2/I^-} = 0.54$ V
The standard electrode potential for the spontaneous reaction in the cell is $x \times 10^{-2}$ V 298 K. The value of x is ________. (Nearest Integer)
[NCERT : P L-40 | NV, June 25, 2022 (I)]

129. The correct order of reduction potentials of the following pairs is
A. Cl_2/Cl^- B. I_2/I^- C. Ag^+/Ag
D. Na^+/Na E. Li^+/Li
Choose the correct answer from the options given below.
[NCERT : P L-37 | NV, June 25, 2022 (II)]
(a) A > C > B > D > E (b) A > B > C > D > E
(c) A > C > B > E > D (d) A > B > C > E > D

130. The cell potential for the following cell
$Pt \mid H_2(g) \mid H^+(aq) \| Cu^{2+}(0.01M) \mid Cu(s)$
is 0.576 V at 298 K. The pH of the solution is ___.
(Nearest integer) (Given: $E^\circ Cu^{2+}/Cu = 0.34V$ and $\frac{2.303RT}{F} = 0.06V$) **[NCERT : P L-35 | NV, June 24, 2022 (I)]**

131. Consider the following cell reaction :
$$Cd(s) + Hg_2SO_4(s) + \frac{9}{5}H_2O(l) \rightleftharpoons CdSO_4.\frac{9}{5}H_2O(s) + 2Hg(l)$$
The value of E^0_{cell} is 4.315 V at 25°C. If $\Delta H^\circ = -825.2$ kJ mol^{-1}, the standard entropy change ΔS° in J K^{-1} is ______. (Nearest integer) **[NV, Aug. 31, 2021 (I)]**
[Given : Faraday constant = 96487 C mol^{-1}]

132. For the galvanic cell, **[NCERT : P L-38 | NV, Aug. 26, 2021 (II)]**
$Zn(s) + Cu^{2+}(0.02\ M) \rightarrow Zn^{2+}(0.04\ M) + Cu(s)$,
E_{cell} = ________ $\times 10^{-2}$ V. (Nearest integer)
[Use: $E^0_{Cu/Cu^{2+}} = -0.34$ V, $E^0_{Zn/Zn^{2+}} = +0.76$ V, $\frac{2.303RT}{F} = 0.059$ V]

133. For the cell

$Cu(s) | Cu^{2+}(aq) (0.1M) || Ag^+(aq) (0.01M) | Ag(s)$

the cell potential $E_1 = 0.3095$ V

For the cell

$Cu(s) | Cu^{2+}(aq) (0.01 M) || Ag^+(aq) (0.001 M) | Ag(s)$

the cell potential = ______ $\times 10^{-2}$ V. (Round off the nearest integer).

[Use: $\frac{2.303\,RT}{F} = 0.059$]

[NCERT : P L-36 | NV, July 27, 2021 (II)]

134. Consider the cell at 25 °C

$Zn | Zn^{2+}(aq), (1 M) || Fe^{3+}(aq), Fe^{2+}(aq) | Pt(s)$

The fraction of total iron present as Fe^{3+} ion at the cell potential of 1.500 V is $x \times 10^{-2}$. The value of x is ______. (nearest integer)

$\left(\text{Given}: E^0_{Fe^{3+}/Fe^{2+}} = 0.77\text{ V}, E^0_{Zn^{2+}/Zn} = -0.76\text{ V}\right)$

[NCERT : P L-36 | NV, July 25, 2021 (I)]

135. Assume a cell with the following reaction

$Cu(s) + 2Ag^+(1 \times 10^{-3}\text{ M}) \rightarrow Cu^{2+}(0.250\text{ M}) + 2Ag(s)$

$E^{\circ}_{cell} = 2.97$ V

E_{cell} for the above reaction is ______ V. (Nearest integer)

[Given : log 2.5 = 0.3979, T = 298 K]

[NCERT : P L-36 | NV, July 22, 2021 (II)]

136. Emf of the following cell at 298 K in V is $x \times 10^{-2}$.

$Zn|Zn^{2+}(0.1\text{ M})||Ag^+(0.01\text{ M})|Ag$

The value of x is ______. (Rounded off to the nearest integer)

[Given : $E^{\theta}_{Zn^{2+}/Zn} = -0.76$V;

$E^{\theta}_{Ag^+/Ag} = +0.80$V; $\frac{2.303RT}{F} = 0.059$]

[NCERT : P L-34 | NV, Feb. 26, 2021 (II)]

137. Copper reduces NO_3^- into NO and NO_2 depending upon the concentration of HNO_3 in solution. (Assuming fixed $[Cu^{2+}]$ and $P_{NO} = P_{NO_2}$), the HNO_3 concentration at which the thermodynamic tendency for reduction of NO_3^- into NO and NO_2 by copper is same is 10^x M. The value of 2x is ______. (Rounded-off to the nearest integer)

[Given : $E^{\circ}_{Cu^{2+}/Cu} = 0.34$ V, $E^{\circ}_{NO_3^-/NO} = 0.96$ V,

$E^{\circ}_{NO_3^-/NO_2} = 0.79$ V and at 298 K, $\frac{RT}{F}(2 \cdot 303) = 0.059$]

[NV, Feb. 25, 2021 (II)]

138. The magnitude of the change in oxidising power of the MnO_4^- / Mn^{2+} couple is $x \times 10^{-4}$ V, if the H^+ concentration is decreased from 1 M to 10^{-4} M at 25 °C. (Assume concentration of MnO_4^- and Mn^{2+} to be same on change in H^+ concentration). The value of x is ______. (Rounded off to the nearest integer)

$\left[\text{Given}: \frac{2.303RT}{F} = 0.059\right]$ **[NV, February 24, 2021 (II)]**

139.

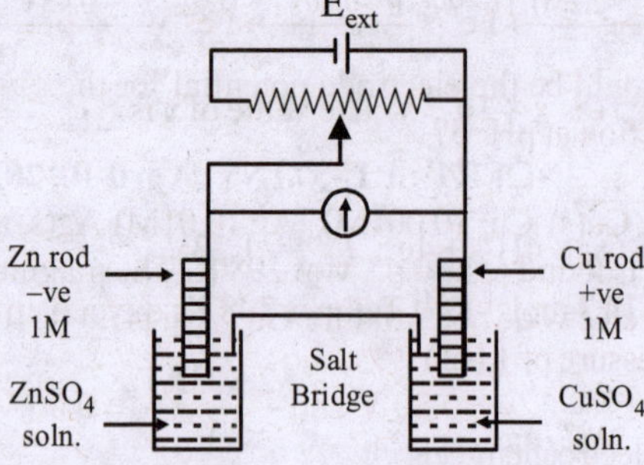

$E^o_{Cu^{2+}|Cu} = +0.34$ V; $E^o_{Zn^{2+}|Zn} = -0.76$ V

Identify the **incorrect** statement from the options below for the above cell : **[NCERT : P L-32 | Sep. 04, 2020 (I)]**

(a) If $E_{ext} > 1.1$ V, e^- flows from Cu to Zn

(b) If $E_{ext} > 1.1$ V, Zn dissolves at Zn electrode and Cu deposits at Cu electrode

(c) If $E_{ext} < 1.1$ V, Zn dissolves at anode and Cu deposits at cathode

(d) If $E_{ext} = 1.1$ V, no flow of e^- or current occurs

140. For the given cell; **[NCERT : P L-40 | Sep. 06, 2020 (II)]**

$Cu(s)|Cu^{2+}(C_1M)||Cu^{2+}(C_2M)|Cu(s)$

change in Gibbs energy (ΔG) is negative, if :

(a) $C_1 = C_2$ (b) $C_2 = \frac{C_1}{\sqrt{2}}$

(c) $C_1 = 2C_2$ (d) $C_2 = \sqrt{2}C_1$

141. An oxidation-reduction reaction in which 3 electrons are transferred has a ΔG^0 of 17.37 kJ mol^{-1} at 25°C. The value of E^0_{cell} (in V) is ______ $\times 10^{-2}$.

(1 F = 96,500 C mol^{-1})

[NCERT : P L-41 | NV, Sep. 05, 2020 (I)]

142. The photoelectric current from Na (work function, $w_0 = 2.3$ eV) is stopped by the output voltage of the cell $Pt(s)|H_2(g, 1\text{ bar})|HCl(aq.\ pH = 1)|AgCl(s)|Ag(s)$. The pH of aq. HCl required to stop the photoelectric current from K($w_0 = 2.25$ eV), all other conditions remaining the same, is ______ $\times 10^{-2}$ (to the nearest integer).

Given, **[NV, Sep. 03, 2020 (I)]**

$2.303\frac{RT}{F} = 0.06$ V; $E^0_{AgCl|Ag|Cl^-} = 0.22$ V

143. The Gibbs energy change (in J) for the given reaction at $[Cu^{2+}] = [Sn^{2+}] = 1$ M and 298 K is :

$Cu(s) + Sn^{2+}(aq) \rightarrow Cu^{2+}(aq) + Sn(s)$;

($E^0_{Sn^{2+}|Sn} = -0.16$ V, $E^0_{Cu^{2+}|Cu} = 0.34$ V,

Take F = 96500 C mol^{-1})

[NCERT : P L-40 | NV, Sep. 02, 2020 (I)]

144. For the disproportionation reaction,

$2Cu^{+}(aq) \rightleftharpoons Cu(s) + Cu^{2+}(aq)$ at 298 K, ln K (where K is the equilibrium constant) is ______ $\times 10^{-1}$.

Given **[NCERT : P L-36 | NV, Sep. 02, 2020 (II)]**

$(E^{0}_{Cu^{2+}/Cu^{+}} = 0.16\ V;\ E^{0}_{Cu^{+}/Cu} = 0.52\ V;\ \frac{RT}{F} = 0.025)$

145. What would be the electrode potential for the given half cell reaction at pH=5?

______.

$2H_2O \rightarrow O_2 + 4H^{+} + 4e^{-};\ E^{0}_{red} = 1.23V$

($R = 8.314\ J\ mol^{-1}\ K^{-1}$; Temp = 298 K; oxygen under std. atm. pressure of 1 bar)

[NCERT : P L-36 | NV, Jan. 08, 2020 (I)]

146. For an electrochemical cell,

$Sn(s)|Sn^{2+}(aq, 1M)||Pb^{2+}(aq, 1M)|Pb(s)$

the ratio $\frac{[Sn^{2+}]}{[Pb^{2+}]}$ when this cell attains equilibrium is ______. Given $E^{o}Sn^{2+}/Sn = -0.14V$, $E^{o}Pb^{2+}/Pb = 0.13V$, $\frac{2.303RT}{F} = 0.06$ **[NCERT : P L-39 | NV, Jan. 08, 2020 (II)]**

147. Given that the standard potentials (E^0) of Cu^{2+}/Cu and Cu^{+}/Cu are 0.34 V and 0.522 V respectively, the E^0 of Cu^{2+}/Cu^{+} is: **[NCERT : P L-36 | Jan. 07, 2020 (I)]**

(a) +0.182 V (b) +0.158 V
(c) −0.182 V (d) −0.158 V

148. The standard Gibbs energy for the given cell reaction in $kJ\ mol^{-1}$ at 298 K is:

$Zn(s) + Cu^{2+}(aq) \rightarrow Zn^{2+}(aq) + Cu(s)$,

$E^{\circ} = 2$ V at 298 K

(Faraday's constant, $F = 96000\ C\ mol^{-1}$)

[NCERT : P L-40 | April 9, 2019 (I)]

(a) −384 (b) 384
(c) 192 (d) −192

149. Given : **[NCERT : P L-37 | April 12, 2019 (I)]**

$Co^{3+} + e^{-} \rightarrow Co^{2+}; E^{o} = +1.81$ V

$Pb^{4+} + 2e^{-} \rightarrow Pb^{2+}; E^{o} = +1.67$ V

$Ce^{4+} + e^{-} \rightarrow Ce^{3+}; E^{o} = +1.61$ V

$Bi^{3+} + 3e^{-} \rightarrow Bi; E^{o} = +0.20$ V

Oxidizing power of the species will increase in the order:

(a) $Ce^{4+} < Pb^{4+} < Bi^{3+} < Co^{3+}$
(b) $Bi^{3+} < Ce^{4+} < Pb^{4+} < Co^{3+}$
(c) $Co^{3+} < Ce^{4+} < Bi^{3+} < Pb^{4+}$
(d) $Co^{3+} < Pb^{4+} < Ce^{4+} < Bi^{3+}$

150. Calculate the standard cell potential (in V) of the cell in which following reaction takes place :

$Fe^{2+}(aq) + Ag^{+}(aq) \rightarrow Fe^{3+}(aq) + Ag(s)$

Given that, **[NCERT : P L-36 | April 8, 2019 (II)]**

$E^{0}_{Ag^{+}/Ag} = x$ V; $E^{0}_{Fe^{2+}/Fe} = y$ V; $E^{0}_{Fe^{3+}/Fe} = z$ V

(a) $x - z$ (b) $x - y$
(c) $x + 2y - 3z$ (d) $x + y - z$

151. Given that $E^{o}_{O_2/H_2O} = +1.23$ V;

$E^{o}_{S_2O_8^{2-}/SO_4^{2-}} = 2.05$ V; $E^{o}_{Br_2/Br^{-}} = +1.09$ V

$E^{o}_{Au^{3+}/Au} = +1.4$ V

The strongest oxidising agent is :

[NCERT : P L-37 | April 8, 2019 (I)]

(a) Au^{3+} (b) O_2
(c) $S_2O_8^{2-}$ (d) Br_2

152. Consider the following reduction processes:

$Zn^{2+} + 2e^{-} \rightarrow Zn(s); E^{o} = -0.76$ V

$Ca^{2+} + 2e^{-} \rightarrow Ca(s); E^{o} = -2.87$ V

$Mg^{2+} + 2e^{-} \rightarrow Mg(s); E^{o} = -2.36$ V

$Ni^{2+} + 2e^{-} \rightarrow Ni(s); E^{o} = -0.25$ V

The reducing power of the metals increases in the order:

[NCERT : P L-37 | Jan. 10, 2019 (I)]

(a) Ca < Zn < Mg < Ni
(b) Ni < Zn < Mg < Ca
(c) Zn < Mg < Ni < Ca
(d) Ca < Mg < Zn < Ni

153. The standard electrode potential E^0 and its temperature coefficient $\left(\frac{dE^0}{dT}\right)$ for a cell are 2 V and $-5 \times 10^{-4}\ VK^{-1}$ at 300 K respectively. The cell reaction is:

$Zn(s) + Cu^{2+}(aq) \rightarrow Zn^{2+}(aq) + Cu$

The standard reaction enthalpy $(\Delta_r H^0)$ at 300 K in $kJ\ mol^{-1}$ is,

[Use $R = 8\ JK^{-1}mol^{-1}$ and $F = 96,000\ C\ mol^{-1}$]

[NCERT : P L-36 | Jan. 12, 2019 (I)]

(a) −412.8 (b) −384.0 (c) 192.0 (d) 206.4

154. For the cell $Zn(s) | Zn^{2+}(aq) || M^{x+}(aq) | M(s)$, different half cells and their standard electrode potentials are given below:

$M^{x+}(aq)/M(s)$	$Au^{3+}(aq)/Au(s)$	$Ag^{+}(aq)/Ag(s)$	$Fe^{3+}(aq)/Fe^{2+}(aq)$	$Fe^{2+}(aq)/Fe(s)$
$E^{o}_{M^{x+}/M}/(V)$	1.40	0.80	0.77	−0.44

If $E^{o}_{Zn^{2+}/Zn} = -0.76$ V, which cathode will give a maximum value of E°_{cell} per electron transferred?

[NCERT : P L-36 | Jan. 11, 2019 (I)]

(a) Ag^{+}/Ag (b) Fe^{3+}/Fe^{2+}
(c) Au^{3+}/Au (d) Fe^{2+}/Fe

155. Given the equilibrium constant K_C of the reaction :

$Cu(s) + 2Ag^{+}(aq) \rightarrow Cu^{2+}(aq) + 2Ag(s)$ is 10×10^{15}. Calculate the E^{0}_{cell} of this reaction at 298 K.

$\left[2.303\frac{RT}{F} \text{ at } 298\ K = 0.059\ V\right]$

[NCERT : P L-39 | Jan. 11, 2019 (II)]

(a) 0.04736 mV (b) 0.4736 mV
(c) 0.4736 V (d) 0.04736 V

156. In the cell,
$Pt(s) | H_2(g, 1bar)/HCl(aq)||AgCl(s)/Ag(s)|Pt(s)$, the cell potential is 0.92 V when a 10^{-6} molal HCl solution is used. The standard electrode potential of $(AgCl/Ag, Cl^-)$ electrode is: **[NCERT : P L-36 | Jan. 10, 2019 (II)]**

$$\left\{\text{Given}: \frac{2.303RT}{F} = 0.06 \text{ V at } 298 \text{ K}\right\}$$

(a) 0.94 V (b) 0.76 V (c) 0.40 V (d) 0.20 V

157. If the standard electrode potential for a cell is 2 V at 300 K, the equilibrium constant (K) for the reaction

$$Zn(s) + Cu^{2+}(aq) \rightleftharpoons Zn^{2+}(aq) + Cu(s)$$

at 300 K is approximately
$(R = 8\ JK^{-1}\ mol^{-1}, F = 96000\ C\ mol^{-1})$
[NCERT : P L-38 | Jan. 9, 2019 (II)]

(a) e^{-80} (b) e^{-160} (c) e^{320} (d) e^{160}

158. To find the standard potential of M^{3+}/M electrode, the following cell is constituted : $Pt/M|M^{3+}(0.001\ mol\ L^{-1})||Ag^+(0.01\ mol\ L^{-1})|Ag$
The emf of the cell is found to be 0.421 volt at 298 K. The standard potential of half reaction $M^{3+} + 3e^- \rightarrow M$ at 298 K will be: (Given $E^\circ_{Ag^+/Ag}$ at 298 K = 0.80 Volt)
[NCERT : P L-36 | Online April 9, 2017]

(a) 0.38 Volt (b) 0.32 Volt
(c) 1.28 Volt (d) 0.66 Volt

159. Given,

$$E^\circ_{Cl_2/Cl^-} = 1.36\ V, E^\circ_{Cr^{3+}/Cr} = -0.74\ V,$$

$$E^\circ_{Cr_2O_7^{2-}/Cr^{3+}} = 1.33\ V, E^\circ_{MnO_4^-/Mn^{2+}} = 1.51\ V.$$

Among the following, the strongest reducing agent is
[NCERT : P L-37 | 2017, Same 23, April, 2013]

(a) Cr (b) Mn^{2+} (c) Cr^{3+} (d) Cl^-

160. Consider the following standard electrode potentials (E° in volts) in aqueous solution :
[NCERT : P L-37 | Online April 8, 2017]

Element	M^{3+}/M	M^+/M
Al	–1.66	+0.55
Tl	+1.26	–0.34

Based on these data, which of the following statements is correct?

(a) Tl^+ is more stable than Al^{3+}
(b) Al^+ is more stable than Al^{3+}
(c) Tl^+ is more stable than Al^+
(d) Tl^{3+} is more stable than Al^{3+}

161. What is the standard reduction potential (E°) for $Fe^{3+} \rightarrow Fe$?
Given that : **[NCERT : P L-39 | Online April 8, 2017]**

$$Fe^{2+} + 2e^- \rightarrow Fe;\ E^\circ_{Fe^{2+}/Fe} = -0.47V$$

$$Fe^{3+} + e^- \rightarrow Fe^{2+};\ E^\circ_{Fe^{3+}/Fe^{2+}} = +0.77V$$

(a) –0.057 V (b) +0.057 V
(c) +0.30 V (d) –0.30 V

162. What will occur if a block of copper metal is dropped into a beaker containing a solution of 1M $ZnSO_4$?
[NCERT : P L-38 | Online April 9, 2016]

(a) The copper metal will dissolve with evolution of oxygen gas
(b) The copper metal will dissolve with evolution of hydrogen gas
(c) No reaction will occur
(d) The copper metal will dissolve and zinc metal will be deposited.

163. At 298 K, the standard reduction potentials are 1.51 V for $MnO_4^-|Mn^{2+}$, 1.36 V for $Cl_2|Cl^-$, 1.07 V for $Br_2|Br^-$, and 0.54 V for $I_2|I^-$. At pH = 3, permanganate is expected to oxidize : $\left(\frac{RT}{F} = 0.059\ V\right)$ **[Online April 11, 2015]**

(a) Cl^-, Br^- and I^- (b) Br^- and I^-
(c) Cl^- and Br^- (d) I^- only

164. The standard electrode potentials $\left(E^\circ_{M^+/M}\right)$ of four metals A, B, C and D are –1.2 V, 0.6 V, 0.85 V and –0.76 V, respectively. The sequence of deposition of metals on applying potential is:
[NCERT : P L-37 | Online April 9, 2014]

(a) A, C, B, D (b) B, D, C, A
(c) C, B, D, A (d) D, A, B, C

165. Given
$Fe^{3+}(aq) + e^- \rightarrow Fe^{2+}(aq);\ E^\circ = +0.77\ V$
$Al^{3+}(aq) + 3e^- \rightarrow Al(s);\ E^\circ = -1.66\ V$
$Br_2(aq) + 2e^- \rightarrow 2Br^-;\ E^\circ = +1.09\ V$
Considering the electrode potentials, which of the following represents the correct order of reducing power?
[NCERT : P L-40 | Online April 11, 2014]

(a) $Fe^{2+} < Al < Br^-$ (b) $Br^- < Fe^{2+} < Al$
(c) $Al < Br^- < Fe^{2+}$ (d) $Al < Fe^{2+} < Br^-$

166. Given below are the half-cell reactions:

$$Mn^{2+} + 2e^- \rightarrow Mn;\ E^\circ = -1.18\ V$$

$$2(Mn^{3+} + e^- \rightarrow Mn^{2+});\ E^\circ = +1.51V$$

The E° for $3Mn^{2+} \rightarrow Mn + 2Mn^{3+}$ will be:
[NCERT : P L-40 | 2014]

(a) –2.69 V; the reaction will not occur
(b) –2.69 V; the reaction will occur
(c) –0.33 V; the reaction will not occur
(d) –0.33 V; the reaction will occur

167. Given : $E^\circ_{Cr^{3+}/Cr} = -0.74\ V;\ E^\circ_{MnO_4^-/Mn^{2+}} = 1.51\ V$

$$E^\circ_{Cr_2O_7^{2-}/Cr^{3+}} = 1.33\ V;\ E^\circ_{Cl/Cl^-} = 1.36\ V$$

Based on the data given above, strongest oxidising agent will be : **[NCERT : P L-35 | 2013]**

(a) Cl (b) Cr^{3+}
(c) Mn^{2+} (d) MnO_4^-

168. The standard reduction potentials for Zn^{2+}/Zn, Ni^{2+}/Ni and Fe^{2+}/Fe are –0.76, –0.23 and –0.44 V respectively. The reaction $X + Y^{2+} \longrightarrow X^{2+} + Y$ will be spontaneous when : **[NCERT : P L-40 | 2012]**

(a) X = Ni, Y = Fe (b) X = Ni, Y = Zn
(c) X = Fe, Y = Zn (d) X = Zn, Y = Ni

169. Given

$E^{\circ}_{Cu^{2+}/Cu} = 0.34V$, $E^{\circ}_{Cu^{2+}/Cu^{+}} = 0.15V$

Standard electrode potential for the half cell Cu^+/Cu is **[NCERT : P L-33 | Online May 26, 2012]**

(a) 0.38 V (b) 0.53 V
(c) 0.19 V (d) 0.49 V

170. The Gibbs energy for the decomposition of Al_2O_3 at 500 °C is as follows :

$\frac{2}{3}Al_2O_3 \rightarrow \frac{4}{3}Al + O_2, \Delta_r G = +966\,kJ\,mol^{-1}$

The potential difference needed for electrolytic reduction of Al_2O_3 at 500 °C is at least **[NCERT : P L-40 | 2010]**

(a) 4.5 V (b) 3.0 V (c) 2.5 V (d) 5.0 V

171. Given: **[NCERT : P L-33| 2009]**

$E^{\circ}_{Fe^{3+}/Fe} = -0.036V$, $E^{\circ}_{Fe^{2+}/Fe} = -0.439\,V$

The value of standard electrode potential for the change, $Fe^{3+}(aq) + e^- \longrightarrow Fe^{2+}(aq)$ will be:

(a) 0.385 V (b) 0.770 V
(c) –0.270 V (d) –0.072 V

172. Given $E^{\circ}_{Cr^{3+}/Cr} = -0.72\,V$,

$E^{\circ}_{Fe^{2+}/Fe} = -0.42\,V$. The potential for the cell $Cr|Cr^{3+}(0.1M)||Fe^{2+}(0.01\,M)|Fe$ is

[NCERT : P L-36 | 2003 Same 2008]

(a) –0.26 V (b) 0.336 V (c) –0.339 V (d) 0.26 V

173. The cell, $Zn\,|\,Zn^{2+}(1\,M)\,||\,Cu^{2+}(1\,M)\,|\,Cu\;(E^{\circ}_{cell} = 1.10\,V)$ was allowed to be completely discharged at 298 K. The relative concentration of Zn^{2+} to Cu^{2+} $\left(\frac{[Zn^{2+}]}{[Cu^{2+}]}\right)$ is

[NCERT : P L-38 | 2007]

(a) 9.65×10^4 (b) antilog (24.08)
(c) 37.3 (d) $10^{37.3}$.

174. For a spontaneous reaction, the ΔG, equilibrium constant (K) and E°_{cell} will be respectively **[NCERT : P L-40 | 2005]**

(a) –ve, >1, –ve (b) –ve, <1, –ve
(c) +ve, >1, –ve (d) –ve, >1, +ve

175. The standard e.m.f. of a cell involving one electron change is found to be 0.591 V at 25 °C. The equilibrium constant of the reaction is ($F = 96{,}500\,C\,mol^{-1}$; $R = 8.314\,JK^{-1}\,mol^{-1}$)

[NCERT : P L-39 | 2004]

(a) 1.0×10^{10} (b) 1.0×10^5
(c) 1.0×10^1 (d) 1.0×10^{30}

176. In a cell that utilises the reaction,

$Zn(s) + 2H^+(aq) \rightarrow Zn^{2+}(aq) + H_2(g)$, addition of H_2SO_4 to cathode compartment, will

[NCERT : P L-39 | 2004]

(a) increase the E and shift equilibrium to the right
(b) lower the E and shift equilibrium to the right
(c) lower the E and shift equlibrium to the left
(d) increase the E and shift equilibrium to the left

177. Consider the following E° values

$E^{\circ}_{Fe^{3+}/Fe^{2+}} = +0.77V$; $E^{\circ}_{Sn^{2+}/Sn} = -0.14V$

Under standard conditions the potential for the reaction

$Sn(s) + 2Fe^{3+}(aq) \rightarrow 2Fe^{2+}(aq) + Sn^{2+}(aq)$ is

[NCERT : P L-36 | 2004]

(a) 0.91 V (b) 1.40 V
(c) 1.68 V (d) 0.63 V

178. The $E^{\circ}_{M^{3+}/M^{2+}}$ values for Cr, Mn, Fe and Co are – 0.41, + 1.57, + 0.77 and + 1.97V respectively. For which one of these metals the change in oxidation state from +2 to +3 is easiest? **[NCERT : P L-37 | 2004]**

(a) Fe (b) Mn (c) Cr (d) Co

179. Standard reduction electrode potentials of three metals A, B & C are respectively + 0.5 V, – 3.0 V & –1.2 V. The reducing powers of these metals are **[NCERT : P L-34 | 2003]**

(a) A > B > C (b) C > B > A
(c) A > C > B (d) B > C > A

180. For a cell reaction involving a two-electron change, the standard e.m.f. of the cell is found to be 0.295 V at 25 °C. The equilibrium constant of the reaction at 25°C will be

[NCERT : P L-39 | 2003]

(a) 29.5×10^{-2} (b) 10
(c) 1×10^{10} (d) 1×10^{-10}

181. For the redox reaction : **[NCERT : P L-39 | 2003]**

$Zn(s) + Cu^{2+}(0.1M) \rightarrow Zn^{2+}(1M) + Cu(s)$

taking place in a cell, E°_{cell} is 1.10 volt. E_{cell} for the cell will be $\left(2.303\frac{RT}{F} = 0.0591\right)$

(a) 1.80 volt (b) 1.07 volt
(c) 0.82 volt (d) 2.14 volt

182. What will be the emf for the given cell ?

$Pt\,|\,H_2(P_1)\,|\,H^+(aq)\,||\,H_2(P_2)\,|\,Pt$ **[NCERT : P L-36 | 2002]**

(a) $\frac{RT}{F}\log_e\frac{P_1}{P_2}$ (b) $\frac{RT}{2F}\log_e\frac{P_2}{P_1}$

(c) $\frac{RT}{F}\log_e\frac{P_2}{P_1}$ (d) None of these

183. Which of the following reaction is possible at anode?

[NCERT : P L-33 | 2002]

(a) $2\,Cr^{3+} + 7H_2O \rightarrow Cr_2O_7^{2-} + 14H^+$
(b) $F_2 \rightarrow 2F^-$
(c) $(1/2)\,O_2 + 2H^+ \rightarrow H_2O$
(d) none of these.

184. A variable, opposite external potential (E_{ext}) is applied to the cell $Zn|Zn^{2+}$ (1 M) || Cu^{2+} (1 M) | Cu, of potential 1.1 V. When $E_{ext} < 1.1$ V and $E_{ext} > 1.1$ V, respectively electrons flow from : **[NCERT : P L-33 | Online April 10, 2015]**
(a) anode to cathode in both cases
(b) cathode to anode and anode to cathode
(c) anode to cathode and cathode to anode
(d) cathode to anode in both cases

185. Electrode potentials (E°) are given below :
[NCERT : P L-34 | Online April 9, 2013]
$Cu^+ / Cu = +0.52$ V,
$Fe^{3+} / Fe^{2+} = +0.77$ V,
$\frac{1}{2} I_2(s) / I^- = +0.54$ V,
$Ag^+ / Ag = +0.88$V.
Based on the above potentials, strongest oxidizing agent will be :
(a) Cu^+ (b) Fe^{3+} (c) Ag^+ (d) I_2

Topic 4 – Commercial Cells and Corrosion

186. On charging the lead storage battery, the oxidation state of lead changes from x_1 to y_1 at the anode and from x_2 to y_2 at the cathode. The values of x_1, y_1, x_2, y_2 are respectively:
[NCERT : P L-55 | April 4, 2025 (I)]
(a) +4, +2, 0, +2 (b) +2, 0, +2, +4
(c) 0, +2, +4, +2 (d) +2, 0, 0, +4

187. The standard cell potential ($E^{\ominus}_{cell}$) of a fuel cell based on the oxidation of methanol in air that has been used to power television relay station is measured as 1.21 V. The standard half cell reduction potential for O_2 ($E^{\circ}_{O_2/H_2O}$) is 1.229 V. Choose the correct statement:
[NCERT : P L-56 | April 3, 2025 (II)]
(a) The standard half cell reduction potential for the reduction of CO_2 ($E^{\circ}_{CO_2/CH_3OH}$) is 19 mV
(b) Oxygen is formed at the anode.
(c) Reactants are fed at one go to each electrode.
(d) Reduction of methanol takes place at the cathode.

188. Match List-I with List-II :
[NCERT : P L-55, 56 | Jan. 29, 2025 (II)]

List-I (Applications)	List-II (Batteries/Cell)
(A) Transistors	(I) Anode - Zn/Hg ; Cathode - HgO + C
(B) Hearing aids	(II) Hydrogen fuel cell
(C) Invertors	(III) Anode – Zn; Cathode - Carbon
(D) Apollo space ship	(IV) Anode – Pb ; Cathode – Pb | PbO_2

Choose the correct answer from the options given below :
(a) (A)-(III), (B)-(I), (C)-(IV), (D)-(II)
(b) (A)-(III), (B)-(II), (C)-(IV), (D)-(I)
(c) (A)-(IV), (B)-(III), (C)-(II), (D)-(I)
(d) (A)-(II), (B)-(III), (C)-(IV), (D)-(I)

189. Given below are two statements :
[NCERT : P L-57 | Jan. 22, 2025 (II)]
Statement (I) : Corrosion is an electrochemical phenomenon in which pure metal acts as an anode and impure metal as a cathode.
Statement (II) : The rate of corrosion is more in alkaline medium than in acidic medium.
In the light of the above statements, choose the correct answer from the options given below :
(a) Both Statement I and Statement II are false
(b) Statement I is false but Statement II is true
(c) Both Statement I and Statement II are true
(d) Statement I is true but Statement II is false

190. Match List-I with List-II.
[NCERT : P L-54, 55, 56 | April 9, 2024 (II)]

List - I (Cell)	List - II (Use/Property/Reaction)
(A) Leclanche cell	(I) Converts energy of combustion into electrical energy
(B) Ni-Cd cell	(II) Does not involve any ion in solution and is used in hearing aids
(C) Fuel cell	(III) Rechargeable
(D) Mercury cell	(IV) Reaction at anode $Zn \rightarrow Zn^{2+} + 2e^-$

Choose the correct answer from the options given below:
(a) A-(I), B-(II), C-(III), D-(IV)
(b) A-(III), B-(I), C-(IV), D-(II)
(c) A-(IV), B-(III), C-(I), D-(II)
(d) A-(II), B-(III), C-(IV), D-(I)

191. The reaction at cathode in the cells commonly used in clocks involves. **[NCERT : P L-56 | April 5, 2024 (I)]**
(a) reduction of Mn from +4 to +3
(b) oxidation of Mn from +3 to +4
(c) reduction of Mn from +7 to +2
(d) oxidation of Mn from + 2 to +7

192. Fuel cell, using hydrogen and oxygen as fuels,
[NCERT : P L-56 | April 4, 2024 (II)]
(A) has been used in spaceship
(B) has as efficiency of 40% to produce electricity
(C) uses aluminium as catalysts
(D) is eco-friendly
(E) is actually a type of Galvanic cell only
(a) A, B, C only (b) A, B, D only
(c) A, B, D, E only (d) A, D, E only

193. The metals that are employed in the battery industries are
[NCERT : P L-54 | Jan. 31, 2024 (I)]
A. Fe B. Mn C. Ni D. Cr
E. Cd

Choose the correct answer from the options given below:

(a) A, B, C and D only
(b) B, C and E only
(c) B, D and E only
(d) A, B, C, D and E

194. Which of the following statements is not correct about rusting of iron? **[NCERT : P L-57 | Jan. 27, 2024 (II)]**

(a) Coating of iron surface by tin prevents rusting, even if the tin coating is peeling off.
(b) When pH lies above 9 or 10, rusting of iron does not take place.
(c) Dissolved acidic oxides SO_2, NO_2 in water act as catalyst in the process of rusting.
(d) Rusting of iron is envisaged as setting up of electrochemical cell on the surface of iron object.

195. For lead storage battery pick the correct statements **[NCERT : P L-55 | April 12, 2023 (I)]**

A. During charging of battery, $PbSO_4$ on anode is converted into PbO_2
B. During charging of battery, $PbSO_4$ on cathode is converted into PbO_2
C. Lead storage battery, consists of grid of lead packed with PbO_2 as anode
D. Lead storage battery has ~ 38% solution of sulphuric acid as an electrolyte

Choose the correct answer from the options given below:

(a) B, D only (b) B, C, D only
(c) A, B, D only (d) B, C only

196. Match List-I with List-II. **[NCERT : P L-54 | July 28, 2022 (I)]**

List-I	List-II
(A) $Cd(s) + 2Ni(OH)_3(s) \rightarrow CdO(s) + 2Ni(OH)_2(s) + H_2O(l)$	(I) Primary battery
(B) $Zn(Hg) + HgO(s) \rightarrow ZnO(s) + Hg(l)$	(II) Discharging of secondary battery
(C) $2PbSO_4(s) + 2H_2O(l) \rightarrow Pb(s) + PbO_2(s) + 2H_2SO_4(aq)$	(III) Fuel cell
(D) $2H_2(g) + O_2(g) \rightarrow 2H_2O(l)$	(IV) Charging of secondary battery

Choose the correct answer from the options given below :

(a) (A) – (I), (B) – (II), (C) – (III), (D) – (IV)
(b) (A) – (IV), (B) – (I), (C) – (II), (D) – (III)
(c) (A) – (II), (B) – (I), (C) – (IV), (D) – (III)
(d) (A) – (II), (B) – (I), (C) – (III), (D) – (IV)

197. In a fuel cell methanol is used as fuel and oxygen gas is used as an oxidizer. The reaction is

$$CH_3OH(l) + 3/2O_2(g) \longrightarrow CO_2(g) + 2H_2O(l)$$

At 298 K, standard Gibb's energies of formation for $CH_3OH(l)$, $H_2O(l)$ and and $CO_2(g)$ are –166.2, –237.2 and –394.4 kJ mol^{-1} respectively. If standard enthalpy of combustion of methanol is – 726 kJ mol^{-1}, efficiency of the fuel cell will be: **[NCERT : P L-56 | 2009]**

(a) 87% (b) 90% (c) 97% (d) 80%

198. In a hydrogen-oxygen fuel cell, combustion of hydrogen occurs to **[NCERT : P L-56 | 2004]**

(a) produce high purity water
(b) create potential difference between two electrodes
(c) generate heat
(d) remove adsorbed oxygen from elctrode surfaces

ANSWER KEYS

1	(c)	**21**	(b)	**41**	(b)	**61**	(a)	**81**	(c)	**101**	(338.9)	**121**	(4)	**141**	(– 6)	**161**	(a)	**181**	(b)
2	(3)	**22**	(a)	**42**	(b)	**62**	(3)	**82**	(d)	**102**	(18)	**122**	(b)	**142**	(142)	**162**	(c)	**182**	(d)
3	(b)	**23**	(3)	**43**	(b)	**63**	(127)	**83**	(a)	**103**	(25)	**123**	(7)	**143**	(96500)	**163**	(b)	**183**	(a)
4	(23)	**24**	(14)	**44**	(a)	**64**	(20)	**84**	(d)	**104**	(2)	**124**	(16)	**144**	(144)	**164**	(c)	**184**	(c)
5	(d)	**25**	(266)	**45**	(c)	**65**	(25)	**85**	(10)	**105**	(4)	**125**	(51)	**145**	(1.52)	**165**	(b)	**185**	(c)
6	(150)	**26**	(d)	**46**	(b)	**66**	(b)	**86**	(b)	**106**	(b)	**126**	(c)	**146**	(2.15)	**166**	(a)	**186**	(b)
7	(c)	**27**	(12)	**47**	(b)	**67**	(11)	**87**	(6)	**107**	(1)	**127**	(983)	**147**	(b)	**167**	(d)	**187**	(a)
8	(a)	**28**	(760)	**48**	(400)	**68**	(a)	**88**	(b)	**108**	(3)	**128**	(23)	**148**	(a)	**168**	(d)	**188**	(a)
9	(b)	**29**	(288)	**49**	(a)	**69**	(60)	**89**	(d)	**109**	(3)	**129**	(a)	**149**	(b)	**169**	(b)	**189**	(d)
10	(b)	**30**	(57)	**50**	(2)	**70**	(5.68)	**90**	(b)	**110**	(6)	**130**	(5)	**150**	(c)	**170**	(c)	**190**	(c)
11	(a)	**31**	(14)	**51**	(c)	**71**	(a)	**91**	(b)	**111**	(10)	**131**	(25)	**151**	(c)	**171**	(b)	**191**	(a)
12	(c)	**32**	(d)	**52**	(c)	**72**	(c)	**92**	(a)	**112**	(275)	**132**	(109)	**152**	(b)	**172**	(d)	**192**	(d)
13	(4)	**33**	(N)	**53**	(a)	**73**	(c)	**93**	(a)	**113**	(c)	**133**	(28)	**153**	(a)	**173**	(d)	**193**	(b)
14	(3)	**34**	(d)	**54**	(d)	**74**	(c)	**94**	(3)	**114**	(17)	**134**	(24)	**154**	(a)	**174**	(d)	**194**	(a)
15	(66)	**35**	(a)	**55**	(1.93)	**75**	(d)	**95**	(d)	**115**	(10)	**135**	(3)	**155**	(c)	**175**	(a)	**195**	(a)
16	(13.4)	**36**	(b)	**56**	(5)	**76**	(d)	**96**	(a)	**116**	(3)	**136**	(147)	**156**	(d)	**176**	(a)	**196**	(c)
17	(25)	**37**	(a)	**57**	(45.75)	**77**	(c)	**97**	(c)	**117**	(917)	**137**	(4)	**157**	(d)	**177**	(a)	**197**	(c)
18	(2)	**38**	(c)	**58**	(2)	**78**	(c)	**98**	(a)	**118**	(d)	**138**	(3776)	**158**	(b)	**178**	(c)	**198**	(b)
19	(b)	**39**	(a)	**59**	(108)	**79**	(b)	**99**	(a)	**119**	(1825)	**139**	(b)	**159**	(a)	**179**	(d)		
20	(1000)	**40**	(c)	**60**	(161)	**80**	(c)	**100**	(1)	**120**	(34)	**140**	(d)	**160**	(c)	**180**	(c)		

3 Chemical Kinetics

TREND ANALYSIS
JEE MAIN (2025-2021)

	2025				2024				2023				2022				2021			
	Session II		Session I		Session II		Session I		Session II		Session I		Session II		Session I		Session II		Session I	
	MCQs	NVQs	MCQs	NVQs	MCQs	NVQs	MCQs	NVQs	MCQs	NVQs	MCQs	NVQs	MCQs	NVQs	MCQs	NVQs	MCQs	NVQs	MCQs	NVQs
Topic 1	4	2	–	–	–	3	–	1	–	2	–	1	–	1	–	1	–	4	–	1
Topic 2	3	–	7	2	–	5	–	8	–	1	–	3	–	–	–	1	–	3	–	3
Topic 3	4	–	1	1	–	–	–	1	2	2	–	2	–	1	–	7	–	3	–	5
Topic 4																				
Total	11	2	8	3	–	8	–	10	2	5	–	6	–	2	–	9	–	10	–	9

How to Code Qns for Better & Faster Revision →

Type 1: Able to solve (No code Req)

Type 2: Solved with struggle or by seeing solution **(Use code NR means Needs Revision or orange stickers)**

Type 3: Tricky/ Could not solve on my own. Need help from Teacher/ Friend. **(Use code TQ means Tricky Question or red stickers)**

You can either hand write the code in boxes or else paste coloured stickers given at end of book.

Topic 1 – Rate of Reaction, Rate Laws and Rate Constant

1. In a first order decomposition reaction, the time taken for the decomposition of reactant to one fourth and one eighth of its initial concentration are t_1 and t_2 (s), respectively. The ratio $\frac{t_1}{t_2}$ will : **[NCERT : P L-73 | April 8, 2025 (II)]**

(a) $\frac{4}{3}$ (b) $\frac{3}{2}$

(c) $\frac{3}{4}$ (d) $\frac{2}{3}$

2. $A(g) \rightarrow B(g) + C(g)$ is a first order reaction. **[NCERT : P L-73 | April 7, 2025 (II)]**

Time	T	∞
P_{system}	P_t	P_∞

The reaction was started with reactant A only. Which of the following expression is correct for rate constant k ?

(a) $k = \frac{1}{t}\ln\frac{2(P_\infty - P_t)}{P_t}$ (b) $k = \frac{1}{t}\ln\frac{P_\infty}{P_t}$

(c) $k = \frac{1}{t}\ln\frac{P_\infty}{2(P_\infty - P_t)}$ (d) $k = \frac{1}{t}\ln\frac{P_\infty}{(P_\infty - P_t)}$

3. Reaction $A(g) \rightarrow 2B(g) + C(g)$ is a first order reaction. It was started with pure A **[NCERT : P L-73 | April 7, 2025 (I)]**

t/min	Pressure of system at time t/mm Hg
10	160
∞	240

Which of the following option is incorrect ?

(a) Initial pressure of A is 80 mm Hg

(b) The reaction never goes to completion

(c) Rate constant of the reaction is 1.693 min^{-1}

(d) Partial pressure of A after 10 minute is 40 mm Hg

4. Rate law for a reaction between A and B is given by $R = k[A]^n[B]^m$ **[NCERT : P L-67 | April 4, 2025 (I)]**
If concentration of A is doubled and concentration of B is halved from their initial value, the ratio of new rate of reaction to the initial rate of reaction $\left(\frac{r_2}{r_1}\right)$ is
(a) $2^{(n-m)}$ (b) $(n-m)$
(c) $(m+n)$ (d) $\frac{1}{2^{m+n}}$

5. For the reaction $A \rightarrow B$ the following graph was obtained. The time required (in seconds) for the concentration of A to reduce to 2.5 g L^{-1} (if the initial concentration of A was 50 g L^{-1}) is ______.
(Nearest integer) **[NCERT : P L-66 | NV, April 2, 2025 (II)]**
Given : log 2 = 0.3010

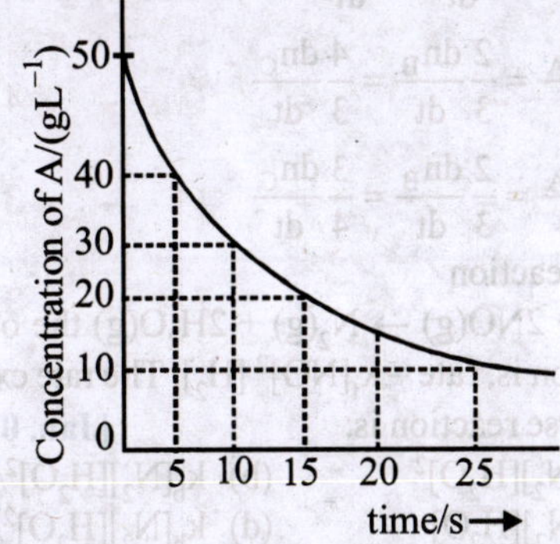

6. For the reaction $A \rightarrow$ products.

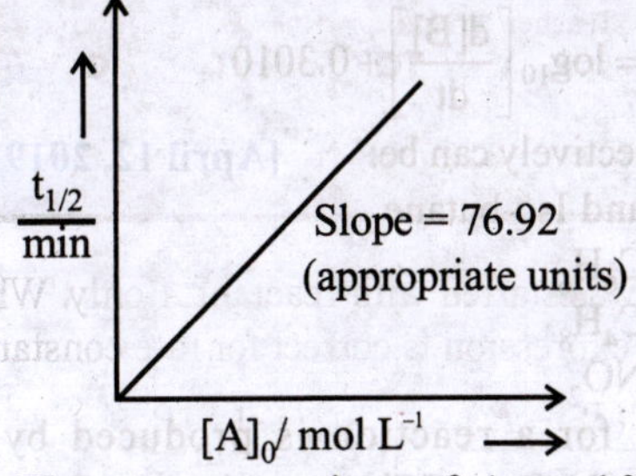

The concentration of A at 10 minutes is ____ $\times 10^{-3}$ mol L^{-1} (nearest integer).
The reaction was started with 2.5 mol L^{-1} of A.
[NCERT : P L-73 | NV, April 2, 2025 (I)]

7. Given below are two statements :
Statement I : The rate law for the reaction $A + B \rightarrow C$ is rate (r) = $k[A]^2[B]$. When the concentration of both A and B is doubled, the reaction rate is increased "x" times.
Statement II :

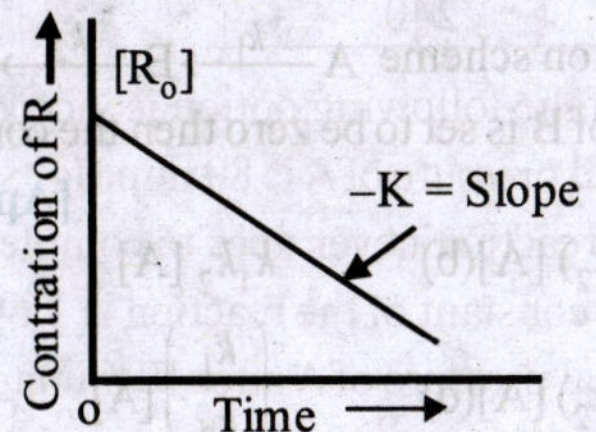

The figure is showing "the variation in concentration against time plot" for a "y" order reaction.
The value of x + y is ______
[NCERT : P L-72 | NV, April 9, 2024 (I)]

8. Consider the following first order gas phase reaction at constant temperature $A(g) \rightarrow 2B(g) + C(g)$
If the total pressure of the gases is found to be 200 torr after 23 sec. and 300 torr upon the complete decomposition of A after a very long time, then the rate constant of the given reaction is ____ $\times 10^{-2}\ s^{-1}$ (nearest integer)
[Given : $\log_{10}(2) = 0.301$]
[NCERT : P L-75 | NV, April 9, 2024 (II)]

9. Consider the following reaction
$A + B \rightarrow C$ **[NCERT : P L-72 | NV, April 8, 2024 (I)]**
The time taken for A to become 1/4th of its initial concentration is twice the time taken to become 1/2 of the same. Also, when the change of concentration of B is plotted against time, the resulting graph gives a straight line with a negative slope and a positive intercept on the concentration axis.
The overall order of the reaction is ______

10. For a reaction $A \xrightarrow{K_1} B \xrightarrow{K_2} C$
If the rate of formation of B is set to be zero then the concentration of B is given by : **[April 8, 2024 (II)]**
(a) $K_1K_2[A]$ (b) $(K_1 - K_2)[A]$
(c) $(K_1 + K_2)[A]$ (d) $(K_1/K_2)[A]$

11. NO_2 required for a reaction is produced by decomposition of N_2O_5 in CCl_4 as by equation
[NCERT : P L-66 | NV, Jan. 30, 2024 (II)]
$2N_2O_5(g) \rightarrow 4NO_2(g) + O_2(g)$
The initial concetration, of N_2O_5 is 3 mol L^{-1} and it is 2.75 mol L^{-1} after 30 minutes.
The rate of formation of NO_2 is x $\times 10^{-3}$ mol L^{-1} min^{-1}, value of x is ______. (nearest integer)

12. $A(g) \rightarrow 2B(g) + C(g)$ is a first order reaction. The initial pressure of the system was found to be 800 mm Hg which increased to 1600 mm Hg after 10 min. The total pressure of the system after 30 min will be ______ mm Hg. (Nearest integer) **[NCERT : P L-73 | NV, April 13, 2023 (II)]**

13. $KClO_3 + 6FeSO_4 + 3H_2SO_4 \rightarrow KCl + 3Fe_2(SO_4)_3 + 3H_2O$
The above reaction was studied at 300 K by monitoring the concentration of $FeSO_4$ in which initial concentration was 10 M and after half an hour became 8.8 M. The rate of production of $Fe_2(SO_4)_3$ is ______ $\times 10^{-6}$ mol L^{-1} s^{-1}.
(Nearest integer) **[NCERT : P L-66 | NV, April 11, 2023 (I)]**

14. A first order reaction has the rate constant, $k = 4.6 \times 10^{-3}$ s^{-1}. The number of correct statement/s from the following is/are ______. **[NCERT : P L-73 | NV, Jan. 25, 2023 (II)]**
Given : log 3 = 0.48

A. Reaction completes in 1000 s.

B. The reaction has a half-life of 500 s.

C. The time required for 10% completion is 25 times the time required for 90% completion.

D. The degree of dissociation is equal to $(1 - e^{-kt})$.

E. The rate and the rate constant have the same unit.

15. The reaction between X and Y is first order with respect to X and zero order with respect to Y.

[NCERT : P L-68 | NV, July 29, 2022 (I)]

Experiment	[X] mol L^{-1}	[Y] mol L^{-1}	Initial rate mol L^{-1} min^{-1}
I.	0.1	0.1	2×10^{-3}
II.	L	0.2	4×10^{-3}
III.	0.4	0.4	$M \times 10^{-3}$
IV.	0.1	0.2	2×10^{-3}

Examine the data of table and calculate ratio of numerical values of M and L. (Nearest integer)

16. For a given chemical reaction $\gamma_1 A + \gamma_2 B \rightarrow \gamma_3 C + \gamma_4 D$ Concentration of C changes from 10 m mol dm^{-3} to 20 m mol dm^{-3} in 10 seconds. Rate of appearance of D is 1.5 times the rate of disappearance of B which is twice the rate of disappearance A. The rate of appearance of D has been experimentally determined to be 9 m mol dm^{-3} s^{-1}. Therefore the rate of reaction is ________ m mol dm^{-3} s^{-1}. (Nearest Integer)

[NCERT : P L-65 | NV, June 25, 2022 (I)]

17. The reaction that occurs in a breath analyser, a device used to determine the alcohol level in a person's blood stream is **[NCERT : P L-65 | NV, Aug. 27, 2021 (I)]**

$$2K_2Cr_2O_7 + 8H_2SO_4 + 3C_2H_6O \rightarrow 2Cr_2(SO_4)_3 + 3C_2H_4O_2 + 2K_2SO_4 + 11H_2O$$

If the rate of appearance of $Cr_2(SO_4)_3$ is 2.67 mol min^{-1} at a particular time, the rate of disappearance of C_2H_6O at the same time is ________ mol min^{-1}. (Nearest integer)

18. The following data was obtained for chemical reaction given below at 975 K.

$$2NO(g) + 2H_2(g) \rightarrow N_2(g) + 2H_2O(g)$$

	[NO] mol L^{-1}	[H_2] mol L^{-1}	Rate mol $L^{-1}s^{-1}$
(A)	8×10^{-5}	8×10^{-5}	7×10^{-9}
(B)	24×10^{-5}	8×10^{-5}	2.1×10^{-8}
(C)	24×10^{-5}	32×10^{-5}	8.4×10^{-8}

The order of the reaction with respect to NO is ________. [Integer answer] **[NCERT : P L-67 | NV, Aug. 26, 2021 (I)]**

19. For a chemical reaction $A \rightarrow B$, it was found the concentration of B increased by 0.2 mol L^{-1} in 30 mn. The average rate of the reaction is ________ $\times 10^{-1}$ mol L^{-1} h^{-1}. (in nearest integer) **[NCERT : P L-63 | NV, July 25, 2021 (II)]**

20. The inactivation rate of a viral preparation is proportional to the amount of virus. In the first minute after preparation, 10% of the virus is inactivated. The rate constant for viral inactivation is ________ $\times 10^{-3}$ min^{-1}. (Nearest integer) [Use : ln 10 = 2.303 ; $\log_{10} 3 = 0.477$; property of logarithm : $\log x^y = y \log x$]

[NV, July 20, 2021 (I)]

21. The reaction $2A + B_2 \rightarrow 2AB$ is an elementary reaction. For a certain quantity of reactants. If the volume of the reaction vessel is reduced by a factor of 3, the rate of the reaction increases by a factor of ________. (Round off to the nearest integer). **[NCERT : P L-67 | NV, March 17, 2021 (II)]**

22. For the reaction $2A + 3B + \frac{3}{2}C \rightarrow 3P$, which statement is correct? **[NCERT : P L-65 | Sep. 03, 2020 (II)]**

(a) $\frac{dn_A}{dt} = \frac{3}{2}\frac{dn_B}{dt} = \frac{3}{4}\frac{dn_C}{dt}$

(b) $\frac{dn_A}{dt} = \frac{dn_B}{dt} = \frac{dn_C}{dt}$

(c) $\frac{dn_A}{dt} = \frac{2}{3}\frac{dn_B}{dt} = \frac{4}{3}\frac{dn_C}{dt}$

(d) $\frac{dn_A}{dt} = \frac{2}{3}\frac{dn_B}{dt} = \frac{3}{4}\frac{dn_C}{dt}$

23. For the reaction

$2H_2(g) + 2NO(g) \rightarrow N_2(g) + 2H_2O(g)$ the observed rate expression is, rate = $k_f[NO]^2[H_2]$. The rate expression for the reverse reaction is: **[Jan. 07, 2020 (II)]**

(a) $k_b[N_2][H_2O]^2$ (b) $k_b[N_2][H_2O]^2/[NO]$

(c) $k_b[N_2][H_2O]$ (d) $k_b[N_2][H_2O]^2/[H_2]$

24. In the following reaction : $xA \rightarrow yB$

$$\log_{10}\left[-\frac{d[A]}{dt}\right] = \log_{10}\left[\frac{d[B]}{dt}\right] + 0.3010$$

'A' and 'B' respectively can be: **[April 12, 2019 (I)]**

(a) *n*-Butane and Iso-butane

(b) C_2H_2 and C_6H_6

(c) C_2H_4 and C_4H_8

(d) N_2O_4 and NO_2

25. NO_2 required for a reaction is produced by the decomposition of N_2O_5 in CCl_4 as per the equation,

$$2N_2O_5(g) \rightarrow 4NO_2(g) + O_2(g).$$

The initial concentration of N_2O_5 is 3.00 mol L^{-1} and it is 2.75 mol L^{-1} after 30 minutes. The rate of formation of NO_2 is : **[NCERT : P L-65 | April 12, 2019 (II)]**

(a) 4.167×10^{-3} mol L^{-1} min^{-1}

(b) 1.667×10^{-2} mol L^{-1} min^{-1}

(c) 8.333×10^{-3} mol L^{-1} min^{-1}

(d) 2.083×10^{-3} mol L^{-1} min^{-1}

26. For a reaction scheme $A \xrightarrow{k_1} B \xrightarrow{k_2} C$, if the rate of formation of B is set to be zero then the concentration of B is given by : **[April 8, 2019 (II)]**

(a) $(k_1 - k_2)[A]$ (b) $k_1 k_2 [A]$

(c) $(k_1 + k_2)[A]$ (d) $\left(\frac{k_1}{k_2}\right)[A]$

27. For the reaction 2A + B → C, the values of initial rate at different reactant concentrations are given in the table below. The rate law for the reaction is :

$[A](mol L^{-1})$	$[B](mol L^{-1})$	Initial Rate$(mol L^{-1} s^{-1})$
0.05	0.05	0.045
0.10	0.05	0.090
0.20	0.10	0.72

[NCERT : P L-67 | April 8, 2019 (I)]

(a) Rate = $k[A][B]^2$ (b) Rate = $k[A]^2[B]^2$
(c) Rate = $k[A][B]$ (d) Rate = $k[A]^2[B]$

28. For an elementary chemical reaction, $A_2 \underset{k_{-1}}{\overset{k_1}{\rightleftharpoons}} 2A$, the expression for $\frac{d[A]}{dt}$ is: **[Jan. 10, 2019 (II)]**

(a) $k_1[A_2] - k_{-1}[A]^2$ (b) $2k_1[A_2] - k_{-1}[A]^2$
(c) $k_1[A_2] + k_{-1}[A]^2$ (d) $2k_1[A_2] - 2k_{-1}[A]^2$

29. The reaction of ozone with oxygen atoms in the presence of chlorine atoms can occur by a two step process shown below : **[Online April 9, 2016]**

$O_3(g) + Cl^*(g) \rightarrow O_2(g) + ClO^*(g)$ ...(i)
$[K_i = 5.2 \times 10^9 \text{ L mol}^{-1} \text{s}^{-1}]$
$ClO^*(g) + O^*(g) \rightarrow O_2(g) + Cl^*(g)$ ...(ii)
$[K_{ii} = 2.6 \times 10^{10} \text{ L mol}^{-1} \text{s}^{-1}]$

The closest rate constant for the overall reaction $O_3(g) + O^*(g) \rightarrow 2O_2(g)$ is :

(a) 1.4×10^{20} L mol^{-1} s^{-1} (b) 3.1×10^{10} L mol^{-1} s^{-1}
(c) 5.2×10^9 L mol^{-1} s^{-1} (d) 2.6×10^{10} L mol^{-1} s^{-1}

30. The rate law for the reaction below is given by the expression k [A][B]

A + B → Product

If the concentration of B is increased from 0.1 to 0.3 mole, keeping the value of A at 0.1 mole, the rate constant will be: **[Online April 10, 2016]**

(a) 3*k* (b) 9*k* (c) *k*/3 (d) *k*

31. A + 2B → C, the rate equation for this reaction is given as Rate = k[A][B]. **[Online April 11, 2015]**
If the concentration of A is kept the same but that of B is doubled what will happen to the rate itself ?

(a) halved (b) the same
(c) doubled (d) quadrupled

32. For the non - stoichiometric reaction 2A + B → C + D, the following kinetic data were obtained in three separate experiments, all at 298 K.

Initial Concentration [A]	Initial Concentration [B]	Initial rate of formation of C $(mol L^{-1} s^{-1})$
0.1 M	0.1 M	1.2×10^{-3}
0.1 M	0.2 M	1.2×10^{-3}
0.2 M	0.1 M	2.4×10^{-3}

The rate law for the formation of C is: **[2014]**

(a) $\frac{dC}{dt} = k[A][B]$ (b) $\frac{dC}{dt} = k[A]^2[B]$
(c) $\frac{dC}{dt} = k[A][B]^2$ (d) $\frac{dC}{dt} = k[A]$

33. In the reaction of formation of sulphur trioxide by contact process, $2SO_2 + O_2 \rightleftharpoons 2SO_3$, the rate of reaction was measured as:
$\frac{d[O_2]}{dt} = -2.5 \times 10^{-4} \text{ mol L}^{-1}\text{s}^{-1}$. The rate of reaction in terms of $[SO_2]$ in mol L^{-1} s^{-1} will be:
[NCERT : P L-65 | Online April 11, 2014]

(a) -1.25×10^{-4} (b) -2.50×10^{-4}
(c) -3.75×10^{-4} (d) -5.00×10^{-4}

34. For the reaction, $2N_2O_5 \rightarrow 4NO_2 + O_2$, the rate equation can be expressed in two ways $-\frac{d[N_2O_5]}{dt} = k[N_2O_5]$ and $+\frac{d[NO_2]}{dt} = k'[N_2O_5]$
k and *k'* are related as:
[NCERT : P L-65 | Online April 11, 2014]

(a) $k = k'$ (b) $2k = k'$
(c) $k = 2k'$ (d) $k = 4k'$

35. For the reaction, **[NCERT : P L-65 | Online April 19, 2014]**
$3A + 2B \rightarrow C + D$,
the differential rate law can be written as:

(a) $\frac{1}{3}\frac{d[A]}{dt} = \frac{d[C]}{dt} = k[A]^n[B]^m$
(b) $-\frac{d[A]}{dt} = \frac{d[C]}{dt} = k[A]^n[B]^m$
(c) $+\frac{1}{3}\frac{d[A]}{dt} = -\frac{d[C]}{dt} = k[A]^n[B]^m$
(d) $-\frac{1}{3}\frac{d[A]}{dt} = \frac{d[C]}{dt} = k[A]^n[B]^m$

36. The instantaneous rate of disappearance of MnO_4^- ion in the following reaction is 4.56×10^{-3} Ms^{-1}
$2MnO_4^- + 10I^- + 16H^+ \rightarrow 2Mn^{2+} + 5I_2 + 8H_2O$
The rate of appearance of I_2 is :
[NCERT : P L-65 | Online April 9, 2013]

(a) 4.56×10^{-4} Ms^{-1} (b) 1.14×10^{-2} Ms^{-1}
(c) 1.14×10^{-3} Ms^{-1} (d) 5.7×10^{-3} Ms^{-1}

37. Reaction rate between two substances A and B is expressed as following:
rate = $k[A]^n[B]^m$ **[Online May 7, 2012, Offline 2003]**
If the concentration of A is doubled and concentration of B is made half of initial concentration, the ratio of the new rate to the earlier rate will be:

(a) $m + n$ (b) $n - m$ (c) $\frac{1}{2^{(m+n)}}$ (d) $2^{(n-m)}$

38. Consider the reaction :
$Cl_2(aq) + H_2S(aq) \rightarrow S(s) + 2H^+(aq) + 2Cl^-(aq)$

The rate equation for this reaction is,

rate = $k[Cl_2][H_2S]$

Which of these mechanisms is/are consistent with this rate equation? **[2010]**

A. $Cl_2 + H_2S \rightarrow H^+ + Cl^- + Cl^+ + HS^-$ (slow)

$Cl^+ + HS^- \rightarrow H^+ + Cl^- + S$ (fast)

B. $H_2S \rightleftharpoons H^+ + HS^-$ (fast equilibrium)

$Cl_2 + HS^- \rightarrow 2Cl^- + H^+ + S$ (Slow)

(a) B only (b) Both A and B

(c) Neither A nor B (d) A only

39. For a reaction $\frac{1}{2}A \rightarrow 2B$, rate of disappearance of 'A' is related to the rate of appearance of 'B' by the expression **[NCERT : P L-65 | 2008]**

(a) $-\frac{d[A]}{dt} = \frac{1}{2}\frac{d[B]}{dt}$ (b) $-\frac{d[A]}{dt} = \frac{1}{4}\frac{d[B]}{dt}$

(c) $-\frac{d[A]}{dt} = \frac{d[B]}{dt}$ (d) $-\frac{d[A]}{dt} = 4\frac{d[B]}{dt}$

40. The differential rate law for the reaction **[NCERT : P L-65 | 2002]**

$H_2 + I_2 \rightarrow 2HI$ is

(a) $-\frac{d[H_2]}{dt} = -\frac{d[I_2]}{dt} = -\frac{d[HI]}{dt}$

(b) $\frac{d[H_2]}{dt} = \frac{d[I_2]}{dt} = \frac{1}{2}\frac{d[HI]}{dt}$

(c) $\frac{1}{2}\frac{d[H_2]}{dt} = \frac{1}{2}\frac{d[I_2]}{dt} = -\frac{d[HI]}{dt}$

(d) $-2\frac{d[H_2]}{dt} = -2\frac{d[I_2]}{dt} = \frac{d[HI]}{dt}$

41. Units of rate constant of first and zero order reactions in terms of molarity M unit are respectively **[NCERT : P L-69 | 2002]**

(a) sec^{-1}, $Msec^{-1}$ (b) sec^{-1}, M

(c) $Msec^{-1}$, sec^{-1} (d) M, sec^{-1}

Topic 2 - Order of Reaction and Half Life Period

42. A person's wound was exposed to some bacteria and then bacteria growth started to happen at the same place. The wound was later treated with some antibacterial medicine and the rate of bacterial decay (r) was found to be proportional with the square of the existing number of bacteria at any instance. Which of the following set of graphs correctly represents the 'before' and 'after' situation of the application of the medicine?

[Given : N = No. of bacteria, t = time, bacterial growth follows I[st] order kinetics.]

[NCERT : P L-77 | April 7, 2025 (I)]

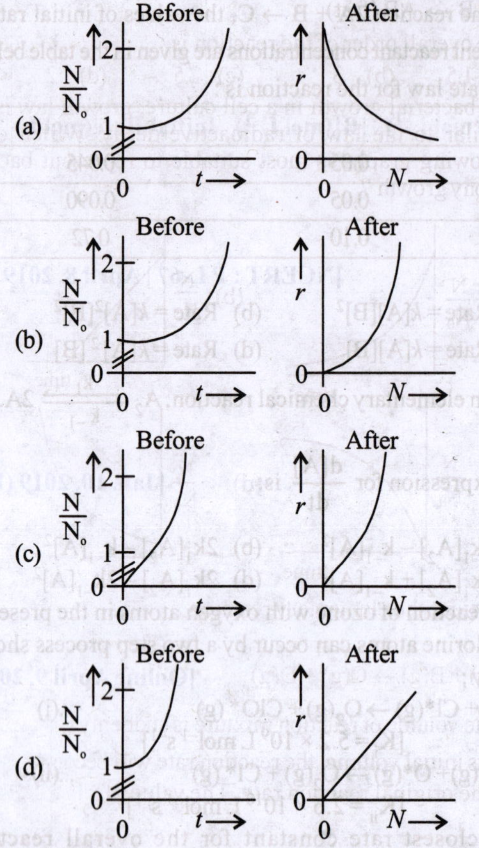

43. Half life of zero order reaction $A \rightarrow$ product is 1 hour, when initial concentration of reaction is 2.0 mol L^{-1}. The time required to decrease concentration of A from 0.50 to 0.25 mol L^{-1} is: **[NCERT : P L-77 | April 4, 2025 (II)]**

(a) 0.5 hour (b) 4 hour

(c) 15 min (d) 60 min

44. In a reaction $A + B \rightarrow C$, initial concentrations of A and B are related as $[A]_0 = 8[B]_0$. The half lives of A and B are 10 min and 40 min. respectively. If they start to disappear at the same time, both following first order kinetics, after how much time will the concentration of both the reactants be same? **[NCERT : P L-77 | April 3, 2025 (I)]**

(a) 60 min (b) 80 min

(c) 20 min (d) 40 mm

45. Drug X becomes ineffective after 50% decomposition. The original concentration of drug in a bottle was 16 mg/mL which becomes 4 mg/mL in 12 months. The expiry time of the drug in months is

Assume that the decomposition of the drug follows first order kinetics. **[NCERT : P L-73 | Jan. 29, 2025 (II)]**

(a) 12 (b) 2 (c) 3 (d) 6

46. The reaction $A_2 + B_2 \rightarrow 2\,AB$ follows the mechanism

$A_2 \underset{k_{-1}}{\overset{k_1}{\rightleftharpoons}} A + A$ (fast)

$A + B_2 \xrightarrow{k_2} AB + B$ (slow)

[NCERT : P L-70 | Jan. 29, 2025 (I)]

$A + B \rightarrow AB$ (fast)

The overall order of the reaction is :

(a) 1.5 (b) 3 (c) 2.5 (d) 2

47. For bacterial growth in a cell culture, growth law is very similar to the law of radioactive decay. Which of the following graphs is most suitable to represent bacterial colony growth ? **[NCERT : P L-73 | Jan. 28, 2025 (II)]**

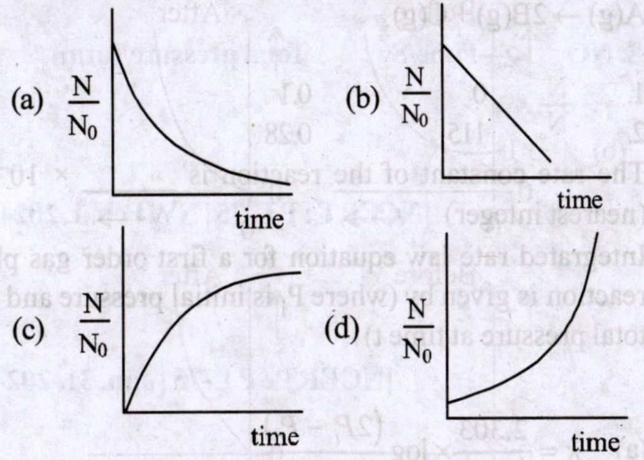

48. consider the elementary reaction

[NCERT : P L-73 | Jan. 28, 2025 (II)]

$A(g) + B(g) \rightarrow C(g) + D(g)$

If the volume of reaction mixture is suddenly reduced to $\frac{1}{3}$ of its initial volume, the reaction rate will become 'x' times of the original reaction rate. The value of x is :

(a) $\frac{1}{9}$ (b) 9 (c) $\frac{1}{3}$ (d) 3

49.

$[A]_0$ / mol L^{-1}	$t_{1/2}$ / min
0.100	200
0.025	100

For a given reaction $R \rightarrow P$, $t_{1/2}$ is related to $[A]_0$ as given in table : **[NCERT : P L-73, 77 | Jan. 28, 2025 (I)]**

Given : log 2 = 0.30

Which of the following is true ?

(A) The order of the reaction is $\frac{1}{2}$.

(B) If $[A]_0$ is 1M, then $t_{1/2}$ is $200\sqrt{10}$ min

(C) The order of the reaction changes to 1 if the concentration of reactant changes from 0.100 M to 0.500 M.

(D) $t_{1/2}$ is 800 min for $[A]_0 = 1.6$ M

Choose the correct answer from the options given below :

(a) A and C only (b) A and B only

(c) A, B and D only (d) C and D only

50. Given below are two statements :

[NCERT : P L-73 | Jan. 24, 2025 (II)]

Statement(I) :

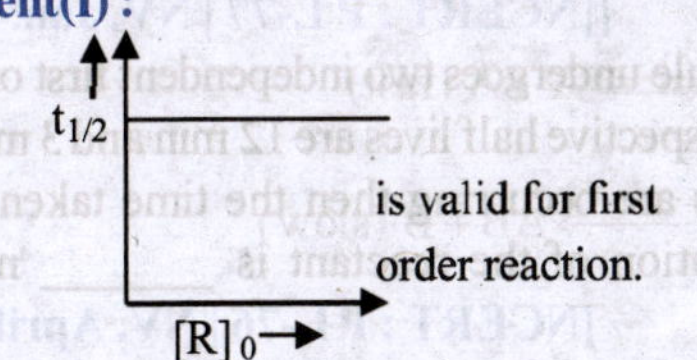

Statement(II) :

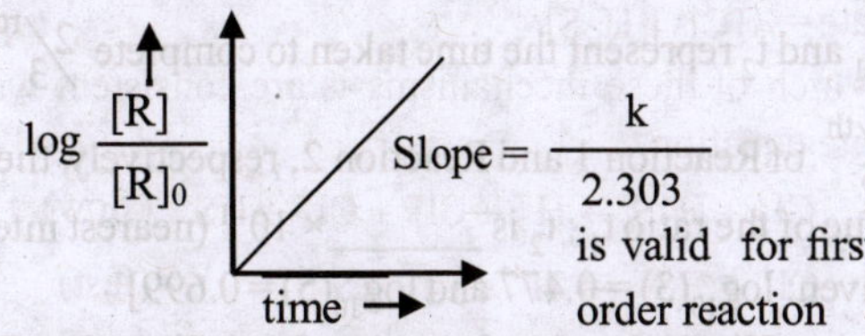

In the light of the above statements, choose the correct answer from the options given below :

(a) Both Statement I and Statement II are false

(b) Statement I is false but Statement II is true

(c) Both Statement I and Statement II are true

(d) Statement I is true but Statement II is false

51. Which of the following graphs most appropriately represents a zero order reaction ? **[NCERT : P L-72 | Jan. 23, 2025 (II)]**

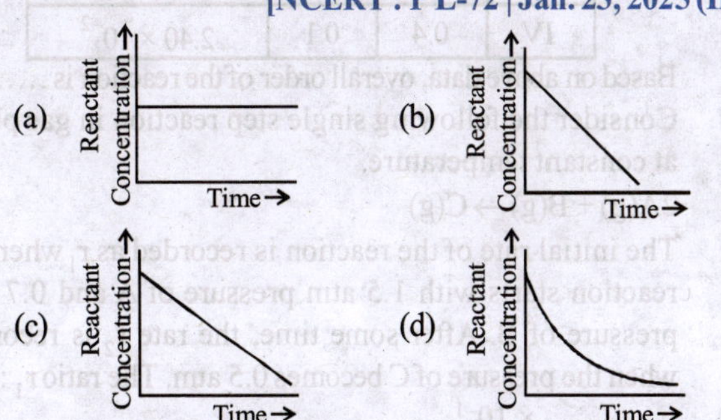

52. For the thermal decomposition of $N_2O_5(g)$ at constant volume, the following table can be formed, for the reaction mentioned below : **[NCERT : P L-72 | NV, Jan. 23, 2025 (I)]**

$2N_2O_5(g) \rightarrow 2N_2O_4(g) + O_2(g)$

S.No.	Time / s	Total pressure / (atm)
1.	0	0.6
2.	100	'x'

x = ________ $\times 10^{-3}$ atm [nearest integer]

Given : Rate constant for the reaction is 4.606×10^{-2} s^{-1}.

53. Which of the following statement is not true for radioactive decay? **[NCERT : P L-73 | Jan. 22, 2025 (I)]**

(a) Amount of radioactive substance remained after three half lives is $\frac{1}{8}$th of original amount.

(b) Decay constant does not depend upon temperature.

(c) Decay constant increases with increase in temperature.

(d) Half life is ln 2 times of $\frac{1}{\text{rate constant}}$.

54. Time required for 99.9% completion of a first order reaction is ______ times the time required for completion of 90% reaction.(nearest integer).

[NCERT : P L-77 | NV, April 6, 2024 (I)]

55. Consider the two different first order reactions given below **[NCERT : P L-77 | NV, April 6, 2024 (II)]**

$A + B \rightarrow C$ (Reaction 1)

$P \rightarrow Q$ (Reaction 2)

The ratio of the half life of Reaction 1 : Reaction 2 is 5 : 2. If t_1 and t_2 represent the time taken to complete $2/3^{rd}$ and $4/5^{th}$ of Reaction 1 and Reaction 2, respectively, then the value of the ratio $t_1 : t_2$ is ______ $\times 10^{-1}$ (nearest integer). [Given: $\log_{10}(3) = 0.477$ and $\log_{10}(5) = 0.699$]

56. During Kinetic study of reaction $2A + B \rightarrow C + D$, the following results were obtained : **[NCERT : P L-64 | NV, April 5, 2024 (I)]**

EXP	A[M]	B[M]	initial rate of formation of D
I	0.1	0.1	6.0×10^{-3}
II	0.3	0.2	7.2×10^{-2}
III	0.3	0.4	2.88×10^{-1}
IV	0.4	0.1	2.40×10^{-2}

Based on above data, overall order of the reaction is ....

57. Consider the following single step reaction in gas phase at constant temperature.

$2A(g) + B(g) \rightarrow C(g)$

The initial rate of the reaction is recorded as r_1 when the reaction starts with 1.5 atm pressure of A and 0.7 atm pressure of B. After some time, the rate r_2 is recorded when the pressure of C becomes 0.5 atm. The ratio $r_1 : r_2$ is ______ $\times 10^{-1}$.

(Nearest integer) **[NCERT : P L-75 | NV, April 5, 2024 (II)]**

58. Consider the following transformation involving first order elementary reaction in each step at constant temperature as shown below. **[NV, April 4, 2024 (I)]**

$$A + B \underset{\text{Step 3}}{\overset{\text{Step 1}}{\rightleftharpoons}} C \xrightarrow{\text{Step 2}} P$$

Some details of the above reaction are listed below.

Step energy	Rate constant (sec^{-1})	Activation ($kJ\ mol^{-1}$)
1.	k_1	300
2.	k_2	200
3.	k_3	E_{a3}

If the overall rate constant of the above transformation (k) is given as $k = \frac{k_1 k_2}{k_3}$ and the overall activation energy (E_a) is 400 $kJ\ mol^{-1}$, then the value of E_{a3} is ______ $kJ\ mol^{-1}$ (nearest integer)

59. Consider the following reaction, the rate expression of which is given below **[NCERT : P L-68 | NV, April 4, 2024 (II)]**

$A + B \rightarrow C$

rate $= k\,[A]^{1/2}\,[B]^{1/2}$

The reaction is initiated by taking 1M concentration A and B each. If the rate constant (k) is $4.6 \times 10^{-2}\ s^{-1}$, then the time taken for A to become 0.1 M is______sec. (nearest integer)

60. The ratio of $\frac{^{14}C}{^{12}C}$ in a piece of wood is $\frac{1}{8}$ part that of atmosphere. If half of ^{14}C is 5730 years, the age of wood sample is ______ years. **[NV, Feb. 1, 2024 (I)]**

61. The following data were obtained during the first order thermal decomposition of a gas A at constant volume:

$A(g) \rightarrow 2B(g) + C(g)$

S. NO	Time/S	Total pressure/(atm)
1.	0	0.1
2.	115	0.28

The rate constant of the reaction is ______ $\times 10^{-2} s^{-1}$ (nearest integer) **[NCERT : P L-75 | NV, Feb. 1, 2024 (II)]**

62. Integrated rate law equation for a first order gas phase reaction is given by (where P_i is initial pressure and P_t is total pressure at time t) **[NCERT : P L-75 | Jan. 31, 2024 (I)]**

(a) $k = \frac{2.303}{t} \times \log \frac{(2P_i - P_t)}{P_i}$

(b) $k = \frac{2.303}{t} \times \frac{P_i}{(2P_i - P_t)}$

(c) $k = \frac{2.303}{t} \times \log \frac{2P_i}{(2P_i - P_t)}$

(d) $k = \frac{2.303}{t} \times \log \frac{P_i}{(2P_i - P_t)}$

63. r = k[A] for a reaction, 50% of A is decomposed in 120 minutes. The time taken for 90% decomposition of A is ______ minutes. **[NCERT : P L-73 | NV, Jan. 31, 2024 (II)]**

64. The rate of First order reaction is 0.04 mol $L^{-1} s^{-1}$ at 10 minutes and 0.03 mol $L^{-1} s^{-1}$ at 20 minutes after initiation. Half life on the reaction is ______ minutes. (Given log2 = 0.3010, log3 = 0.4771) **[NCERT : P L-72 | NV, Jan. 30, 2024 (I)]**

65. The half-life of radioisotope bromine- 82 is 36 hours. The fraction which remains after one day is ______ $\times 10^{-2}$. (Given antilog 0.2006 = 1.587) **[NCERT : P L-76 | NV, Jan. 29, 2024 (II)]**

66. Consider the following data for the given reaction

$2HI\,(g) \rightarrow H_2(g) + I_2(g)$ **[NCERT : P L-68 | NV, Jan. 27, 2024 (I)]**

	1	2	3
HI (mol L^{-1})	0.005	0.01	0.02
Rate (mol $L^{-1}s^{-1}$)	7.5×10^{-4}	3.0×10^{-3}	1.2×10^{-2}

The order of the reaction is ______.

67. Time required for completion of 99.9% of a First order reaction is ______ times of half life ($t_{1/2}$) of the reaction. **[[NCERT : P L-77 | NV, Jan. 27, 2024 (II)]**

68. A molecule undergoes two independent first order reactions whose respective half lives are 12 min and 3 min. If both the reactions are occurring then the time taken for the 50% consumption of the reactant is ______ min. (Nearest integer) **[NCERT : P L-76 | NV, April 10, 2023 (I)]**

69. The rate constant for a first order reaction is 20 min^{-1}. The time required for the initial concentration of the reactant to reduce to its $\frac{1}{32}$ level is ______ $\times 10^{-2}$ min.
(Nearest integer) **[NCERT : PL-73 | NV, Jan. 31, 2023 (II)]**
(Given : ln10 = 2.303 log 2 = 0.3010)

70. If compound A reacts with B following first order kinetics with rate constant $2.011 \times 10^{-3}\ s^{-1}$. The time taken by A (in seconds) to reduce from 7 g to 2 g will be ______.
(Nearest Integer) **[NCERT : P L-73 | NV, Jan. 30, 2023 (I)]**
[log5 = 0.698, log 7 = 0.845, log 2 = 0.301]

71. An organic compound undergoes first order decomposition. If the time taken for the 60% decomposition is 540 s, then the time required for 90% decomposition will be is ______ s. (Nearest integer).
Given : ln 10 = 2.3; log 2 = 0.3
[NCERT : P L-73 | NV, Jan. 30, 2023 (II)]

72. A flask is filled with equal moles of A and B. The half lives of A and B are 100 s and 50 s respectively and are independent of the initial concentration. The time required for the concentration of A to be four times that of B is ______s. **[NCERT : P L-73 | NV, June 26, 2022 (I)]**
(Given : ln 2 = 0.693)

73. For a first order reaction, the ratio of the time for 75% completion of a reaction to the time for 50% completion is ______. (Integer answer)
[NCERT : P L-73 | NV, Aug. 31, 2021 (I)]

74. For a reaction of order n, the unit of the rate constant is :
[NCERT : P L-69 | July 27, 2021 (I)]
(a) $mol^{1-n}\ L^{1-n}\ s$ (b) $mol^{1-n}\ L^{2n}\ s^{-1}$
(c) $mol^{1-n}\ L^{n-1} s^{-1}$ (d) $mol^{1-n}\ L^{1-n}\ s^{-1}$

75. $N_2O_5(g) \rightarrow 2NO_2(g) + \frac{1}{2}O_2(g)$
In the above first order reaction, the initial concentration of N_2O_5 is 2.40×10^{-2} mol L^{-1} at 318 K. The concentration of N_2O_5 after 1 hour was 1.60×10^{-2} mol L^{-1}. The rate constant of the reaction at 318 K is ______ $\times 10^{-3}$ min^{-1}. (Nearest integer)
[NCERT : P L-73 | NV, July 22, 2021 (II)]

76. $PCl_5(g) \rightarrow PCl_3(g) + Cl_2(g)$
In the above first order reaction, the concentration of PCl_5 reduces from initial concentration 50 mol L^{-1} to 10 mol L^{-1} in 120 minutes at 300 K. The rate constant for the reaction at 300 K is $x \times 10^{-2}\ min^{-1}$. The value of x is ______.
[Given log 5 = 0.6989]
[NCERT : P L-73 | NV, July 20, 2021 (II)]

77. $2NO(g) + Cl_2(g) \rightleftharpoons 2NOCl(s)$
This reaction was studied at –10 °C and the following data was obtained **[NV, March 18, 2021 (I)]**

run	$[NO]_0$	$[Cl_2]_0$	r_0
1	0.10	0.10	0.18
2	0.10	0.20	0.35
3	0.20	0.20	1.40

$[NO]_0$ and $[Cl_2]_0$ are the initial concentrations and r_0 is the initial reaction rate.
The overall order of the reaction is ______.
(Rounded off to the nearest integer).

78. A reaction has a half life of 1 min. The time required for 99.9% completion of the reaction is ______ min.
(Round off to the nearest integer).
[Use : ln 2 = 0.69; ln 10 = 2.3]
[NCERT : P L-77 | NV, March 18, 2021 (II)]

79. For a certain first order reaction, 32% of the reactant is left after 570 s. The rate constant of this reaction is ______ $\times 10^{-3}\ s^{-1}$.
[NCERT : PL-73 | NV, March 17, 2021 (I)]
(Round off to the nearest integer).
[Given $log_{10}2 = 0.301$, ln10 = 2.303]

80. Gaseous cyclobutene isomerizes to butadiene in a first order process which has a '*k*' value of $3.3 \times 10^{-4} s^{-1}$ at 153 °C. The time in minutes it takes for the isomerization to proceed 40% to completion at this temperature is ______. (Rounded off to the nearest integer)
[NV, Feb. 24, 2021 (I)]

81. A bacterial infection in an internal wound grows as N′(t) = N_0exp(t), where the time t is in hours. A dose of antibiotic, taken orally, needs 1 hour to reach the wound. Once it reaches there, the bacterial population goes down as $\frac{dN}{dt} = -5N^2$. What will be the plot of $\frac{N_0}{N}$ vs. t after 1 hour?
[April 10, 2019 (I)]

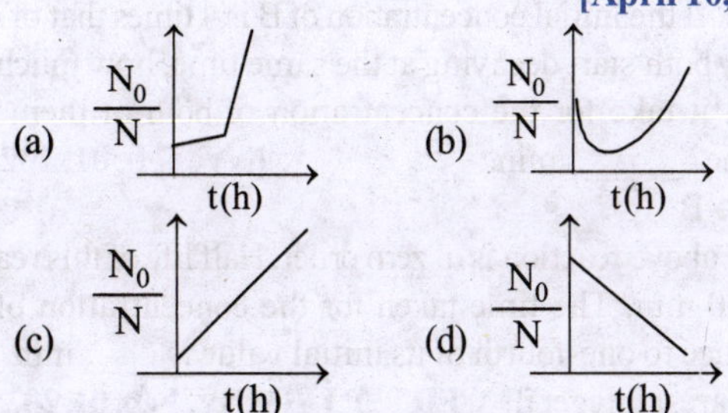

82. The given plots represents the variation of the concentration of a reactant R with time for two different reactions (i) and (ii). The respective orders of the reactions are: **[NCERT : P L-72, 74 | April 9, 2019 (I)]**

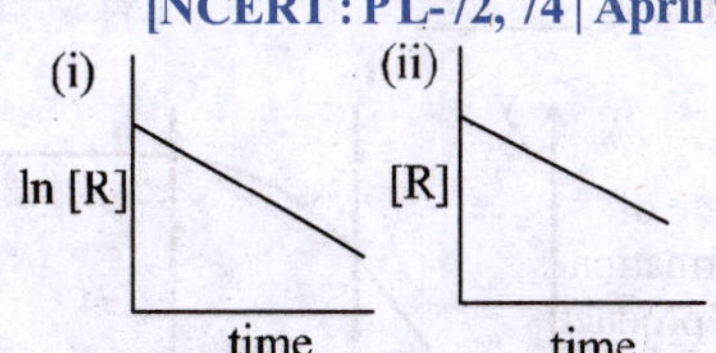

(a) 1, 0 (b) 1, 1 (c) 0, 1 (d) 0, 2

83. $t_{87.5}$ is the time required for the reaction to undergo 87.5% completion and t_{50} is the time required for the reaction to undergo 50% completion. The relation between $t_{87.5}$ and t_{50} for a first order reaction is $t_{87.5} = x \times t_{50}$
The value of x is ______. (Nearest integer)
[NV, April 13, 2023 (I)]

84. The reaction $2NO + Br_2 \rightarrow 2NOBr$
takes places through the mechanism given below :
$NO + Br_2 \Leftrightarrow NOBr_2$ (fast); $NOBr_2 + NO \rightarrow 2NOBr$ (slow)
The overall order of the reaction is ______.
[NV, April 12, 2023 (I)]

85. For a chemical reaction A + B → Product, the order is 1 with respect to A and B. **[NCERT : P L-67 | April 11, 2023 (II)]**

Rate $mol\ L^{-1}s^{-1}$	[A] $mol\ L^{-1}$	[B] $mol\ L^{-1}$
0.10	20	0.5
0.40	x	0.5
0.80	40	y

What is the value of x and y?

(a) 80 and 2 (b) 40 and 4

(c) 160 and 4 (d) 80 and 4

86. The number of incorrect statement/s from the following is **[NCERT : P L-77 | NV, April 10, 2023 (II)]**

A. The successive half lives of zero order reactions decreases with time.

B. A substance appearing as reactant in the chemical equation may not affect the rate of reaction.

C. Order and molecularity of a chemical reaction can be a fractional number.

D. The rate constant units of zero and second order reaction are $mol\ L^{-1}\ s^{-1}$ and $mol^{-1}\ Ls^{-1}$ respectively.

87. A and B are two substances undergoing radioactive decay in a container. The half life of A is 15 min and that of B is 5 min. If the initial concentration of B is 4 times that of A and they both start decaying at the same time, how much time will it take for the concentration of both of them to be same? _____ min. **[NV, Feb. 01, 2023 (I)]**

88. A → B

The above reaction is of zero order. Half life of this reaction is 50 min. The time taken for the concentration of A to reduce to one-fourth of its initial value is ____ min. (Nearest integer) **[NCERT : P L-73 | NV, Feb. 01, 2023 (II)]**

89. For certain chemical reaction X → Y, the rate of formation of product is plotted against the time as shown in the figure. The number of Correct statement/s from the following is _______ **[NV, Jan. 29, 2023 (I)]**

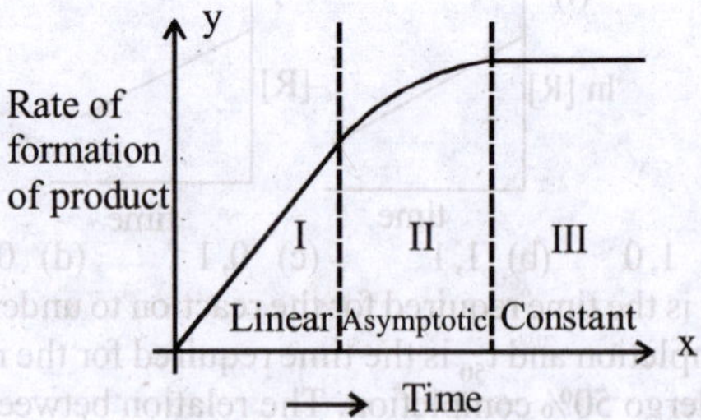

(A) Over all order of this reaction is one

(B) Order of this reaction can't be determined

(C) In region-I and III, the reaction is of first and zero order respectively

(D) In region-II, the reaction is of first order

(E) In region-II, the order of reaction is in the range of 0.1 to 0.9.

90. For conversion of compound A → B, the rate constant of the reaction was found to be $4.6 \times 10^{-5}\ L\ mol^{-1}\ s^{-1}$. The order of the reaction is _______.

[NCERT : P L-67 | NV, Jan. 29, 2023 (II)]

91. For the first order reaction A → B the half life is 30 min. The time taken for 75% completion of the reaction is ____ min. (Nearest integer)

Given : log 2 = 0.3010

log 3 = 0.4771

log 5 = 0.6989 **[NCERT : P L-77 | NV, Jan. 25, 2023 (I)]**

92. A student has studied the decomposition of a gas AB_3 at 25°C. He obtained the following data. **[Jan. 24, 2023 (II)]**

p(mm Hg)	50	100	200	400
Relative $t_{1/2}$(s)	4	2	1	0.5

The order of the reaction is

(a) 0.5 (b) 2 (c) 1 (d) 0 (zero)

93. Assuming 1μg of trace radioactive element X with a half life of 30 years is absorbed by a growing tree. The amount of X remaining in the tree after 100 years is ____ $\times 10^{-1}$μg. [Given : ln 10 = 2.303; log2 = 0.30]

[NCERT : PL-77 | NV, July 29, 2022 (II)]

94. For the given first order reaction

A → B the half life of the reaction is 0.3010 min. The ratio of the initial concentration of reactant to the concentration of reactant at time 2.0 min will be equal to _______. (Nearest integer) **[NCERT : P L-73 | NV, July 28, 2022 (I)]**

95. $2NO + 2H_2 \rightarrow N_2 + 2H_2O$

The above reaction has been studied at 800°C. The related data are given in the table below **[NV, July 27, 2022 (I)]**

Reaction serial number	Initial pressure of H_2 / kPa	Initial Pressure of NO/kPa	Initial rate $\left(\frac{-dp}{dt}\right)$ / (kPa/s)
1	65.6	40.0	0.135
2	65.6	20.1	0.033
3	38.6	65.6	0.214
4	19.2	65.6	0.106

The order of the reaction with respect to NO is _____.

96. [A] → [B]

Reactant Product

If formation of compound [B] follows the first order of kinetics and after 70 minutes the concentration of [A] was found to be half of its initial concentration. Then the rate constant of the reaction is $x \times 10^{-6} s^{-1}$. The value of x is _______. (Nearest integer)

[NCERT : P L-76 | NV, July 27, 2022 (II)]

97. For a reaction A → 2B + C the half lives are 100 s and 50 s when the concentration of reactant A is 0.5 and 1.0 mol L^{-1} respectively. The order of the reaction is _______. (Nearest Integer) **[NCERT : P L-77 | NV, July 26, 2022 (I)]**

98. At 30°C, the half life for the decomposition of AB_2 is 200 s and is independent of the initial concentration of AB_2.

The time required for 80% of the AB_2 to decompose is (Given: log 2 = 0.30; log 3 = 0.48)

[NCERT : P L-73 | July 26, 2022 (II)]

(a) 200 s (b) 323 s (c) 467 s (d) 532 s

99. The half life for the decomposition of gaseous compound A is 240 s when the gaseous pressure was 500 Torr initially. When the pressure was 250 Torr, the half life was found to be 4.0 min. The order of the reaction is ______. (Nearest integer) **[NCERT : P L-107 | NV, July 25, 2022 (I)]**

100. For the decomposition of azomethane, $CH_3N_2CH_3(g) \rightarrow CH_3CH_3(g) + N_2(g)$ a first order reaction, the variation in partial pressure with time at 600 K is given as

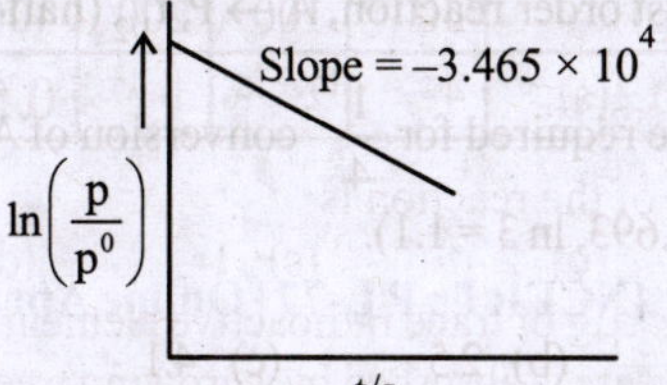

The half life of the reaction is ______ $\times 10^{-5}$s.

[Nearest integer] **[NCERT : P L-73 | NV, July 25, 2022 (II)]**

101. For a first order reaction $A \rightarrow B$, the rate constant, $k = 5.5 \times 10^{-14} s^{-1}$. The time required for 67% completion of reaction is $x \times 10^{-1}$ times the half life of reaction. The value of x is ______ (Nearest integer)
(Given : log 3 = 0.4771)

[NCERT : P L-73 | NV, June 28, 2022 (I)]

102. A radioactive element has a half life of 200 days. The percentage of original activity remaining after 83 days is ______. (Nearest integer)
(Given : antilog 0.125 = 1.333, antilog 0.693 = 4.93)

[NV, June 28, 2022 (II)]

103. At 345 K, the half life for the decomposition of a sample of a gaseous compound initially at 55.5 kPa was 340 s. When the pressure was 27.8 kPa, the half life was found to be 170 s. The order of the reaction is ______.
[integer answer] **[NV, June 25, 2022 (II)]**

104. For a first order reaction, the time required for completion of 90% reaction is 'x' times the half life of the reaction. The value of 'x' is **[June 24, 2022 (II)]**
(Given: ln 10 = 2.303 and log 2 = 0.3010)

(a) 1.12 (b) 2.43 (c) 3.32 (d) 33.31

105. For the first order reaction $A \rightarrow 2B$, 1 mole of reactant A gives 0.2 moles of B after 100 minutes. The half life of the reaction is ...... min. (Round off to the nearest integer).
[Use : ln 2 = 0.69, ln 10 = 2.3
Properties of logarithms : $\ln x^y = y \ln x$;
$\ln\left(\frac{x}{y}\right) = \ln x - \ln y$]
(Round off to the nearest integer) **[NV, July 27, 2021 (II)]**

106. A and B decompose via first order kinetics with half lives 54.0 min and 18.0 min respectively. Starting from an equimolar non reactive mixture of A and B, the time taken for the concentration of A to become 16 times that of B is ______ min. **[NV, March 16, 2021 (II)]**
(Round off to the nearest integer).

107. Sucrose hydrolyses in acid solution into glucose and fructose following first order rate law with a half-life of 3.33 h at 25 °C. After 9 h, the fraction of sucrose remaining is f. The value of $\log_{10}\left(\frac{1}{f}\right)$ is ______ $\times 10^{-2}$.
(Rounded off to the nearest integer)
[Assume : ln 10 = 2.303, ln 2 = 0.693]

[NV, Feb. 24, 2021 (II)]

108. Consider the following reactions
$A \rightarrow P1$; $B \rightarrow P2$; $C \rightarrow P3$; $D \rightarrow P4$,
The order of the above reactions are (i), (ii), (iii), and (iv), respectively. The following graph is obtained when log [rate] vs. log[conc.] are plotted : **[Sep. 06, 2020 (I)]**

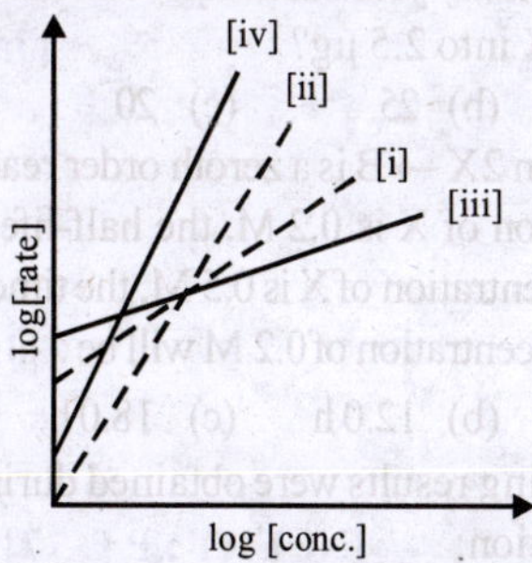

Among the following, the correct sequence for the order of the reactions is:

(a) (iv) > (i) > (ii) > (iii) (b) (i) > (ii) > (iii) > (iv)
(c) (iii) > (i) > (ii) > (iv) (d) (iv) > (ii) > (i) > (iii)

109. A flask contains a mixture of compounds A and B. Both compounds decompose by first-order kinetics. The half-lives for A and B are 300 s and 180 s, respectively. If the concentrations of A and B are equal initially, the time required for the concentration of A to be four times that of B (in s) is: (Use ln 2 = 0.693)

[NCERT : P L-77 | Sep. 05, 2020 (I)]

(a) 180 (b) 900 (c) 300 (d) 120

110. If 75% of a first order reaction was completed in 90 minutes, 60% of the same reaction would be completed in approximately (in minutes) ______.
(Take : log 2 = 0.30; log 2.5 = 0.40) **[NV, Sep. 04, 2020 (I)]**

111. It is true that : **[Sep. 03, 2020 (I)]**

(a) A second order reaction is always a multistep reaction
(b) A zero order reaction is a multistep reaction
(c) A first order reaction is always a single step reaction
(d) A zero order reaction is a single step reaction

112. The results given in the below table were obtained during kinetic studies of the following reaction :
$2A + B \rightarrow C + D$ **[Sep. 02, 2020 (II)]**

Experiment	[A]/ $molL^{-1}$	[B]/ $molL^{-1}$	Initial rate/ $molL^{-1} min^{-1}$
I	0.1	0.1	6.00×10^{-3}
II	0.1	0.2	2.40×10^{-2}
III	0.2	0.1	1.20×10^{-2}
IV	X	0.2	7.20×10^{-2}
V	0.3	Y	2.88×10^{-1}

X and Y in the given table are respectively :

(a) 0.4, 0.4 (b) 0.4, 0.3 (c) 0.3, 0.4 (d) 0.3, 0.3

113. During the nuclear explosion, one of the products is ^{90}Sr with half life of 6.93 years. If 1 μg of ^{90}Sr was absorbed in the bones of a newly born baby in place of Ca, how much time, in years, is required to reduce it by 90% if it is not lost metabolically. **[NV, Jan. 07, 2020 (I)]**

114. Decomposition of X exhibits a rate constant of 0.05 μg/year. How many years are required for the decomposition of 5 μg of X into 2.5 μg? **[Jan. 12, 2019 (I)]**

(a) 50 (b) 25 (c) 20 (d) 40

115. The reaction $2X \rightarrow B$ is a zeroth order reaction. If the initial concentration of X is 0.2 M, the half-life is 6 h. When the initial concentration of X is 0.5 M, the time required to reach its final concentration of 0.2 M will be : **[Jan. 11, 2019 (II)]**

(a) 9.0 h (b) 12.0 h (c) 18.0 h (d) 7.2 h

116. The following results were obtained during kinetic studies of the reaction; **[NCERT : P L-77 | Jan. 9, 2019 (I)]**

$2A + B \rightarrow$ Products

Experiment	[A] (in mol L^{-1})	[B] (in mol L^{-1})	Initial Rate of reaction (in mol L^{-1} min^{-1})
I	0.10	0.20	6.93×10^{-3}
II	0.10	0.25	6.93×10^{-3}
III	0.20	0.30	1.386×10^{-2}

The time (in minutes) required to consume half of A is:

(a) 5 (b) 10 (c) 1 (d) 100

117. For the reaction, 2A + B → products, when the concentrations of A and B both were doubled, the rate of the reaction increased from 0.3 mol L^{-1} s^{-1} to 2.4 mol $L^{-1}s^{-1}$. When the concentration of A alone is doubled, the rate increased from 0.3 mol $L^{-1}s^{-1}$ to 0.6 mol $L^{-1}s^{-1}$. Which one of the following statements is correct? **[NCERT : P L-73 | Jan. 9, 2019 (II)]**

(a) Total order of the reaction is 4
(b) Order of the reaction with respect to B is 2
(c) Order of the reaction with respect to B is 1
(d) Order of the reaction with respect to A is 2

118. At 518 °C, the rate of decomposition of a sample of gaseous acetaldehyde, initially at a pressure of 363 Torr, was 1.00 Torr s^{-1} when 5% had reacted and 0.5 Torr s^{-1} when 33% had reacted. The order of the reaction is : **[2018]**

(a) 2 (b) 3 (c) 1 (d) 0

119. If 50% of a reaction occurs in 100 seconds and 75% of the reaction occurs in 200 seconds, the order of this reaction is: **[NCERT : P L-73 | Online April 16, 2018]**

(a) 2 (b) 3 (c) Zero (d) 1

120. N_2O_5 decomposes to NO_2 and O_2 and follows first order kinetics. After 50 minutes, the pressure inside the vessel increases from 50 mm Hg to 87.5 mm Hg. The pressure of the gaseous mixture after 100 minutes at constant temperature will be ______. **[Online April 15, 2018 (I)]**

(a) 136.25 mm Hg (b) 106.25 mm Hg
(c) 175.0 mm Hg (d) 116.25 mm Hg

121. For a first order reaction, $A \rightarrow P$, $t_{1/2}$ (half-life) is 10 days. The time required for $\frac{1}{4}^{th}$ conversion of A (in days) is: ($\ln 2 = 0.693, \ln 3 = 1.1$). **[NCERT : P L-77 | Online April 15, 2018 (II)]**

(a) 3.2 (b) 2.5 (c) 4.1 (d) 5

122. Decomposition of H_2O_2 follows a first order reaction. In fifty minutes the concentration of H_2O_2 decreases from 0.5 to 0.125 M in one such decomposition. When the concentration of H_2O_2 reaches 0.05 M, the rate of formation of O_2 will be: **[2016]**

(a) 2.66 L min^{-1} at STP (b) 1.34×10^{-2} mol min^{-1}
(c) 6.96×10^{-2} mol min^{-1} (d) 6.93×10^{-4} mol min^{-1}

123. The reaction $2N_2O_5(g) \rightarrow 4NO_2(g) + O_2(g)$ follows first order kinetics. The pressure of a vessel containing only N_2O_5 was found to increase from 50 mm Hg to 87.5 mm Hg in 30 min. The pressure exerted by the gases after 60 min. will be (assume temperature remains constant) : **[NCERT : P L-73 | Online April 10, 2015]**

(a) 106.25 mm Hg (b) 150 mm Hg
(c) 125 mm Hg (d) 116.25 mm Hg

124. Higher order (>3) reactions are rare due to : **[NCERT : P L-69 | 2015]**

(a) shifting of equilibrium towards reactants due to elastic collisions
(b) loss of active species on collision
(c) low probability of simultaneous collision of all the reacting species
(d) increase in entropy and activation energy as more molecules are involved

125. The half-life period of a first order reaction is 15 minutes. The amount of substance left after one hour will be: **[NCERT : P L-77 | Online April 9, 2014]**

(a) $\frac{1}{4}$ of the original amount
(b) $\frac{1}{8}$ of the original amount
(c) $\frac{1}{16}$ of the original amount
(d) $\frac{1}{32}$ of the original amount

126. The rate constant of a zero order reaction is 2.0×10^{-2} mol L^{-1} s^{-1}. If the concentration of the reactant after 25 seconds is 0.5 M. What is the initial concentration ? **[NCERT : P L-72 | Online April 23, 2013]**

(a) 0.5 M (b) 1.25 M (c) 12.5 M (d) 1.0 M

127. A radioactive isotope having a half - life period of 3 days was received after 12 days. If 3 g of the isotope is left in the container, what would be the initial mass of the isotope ? **[NCERT : P L-77 | Online April 25, 2013]**

(a) 12 g (b) 36 g (c) 48 g (d) 24 g

128. For a first order reaction (A) $\rightarrow$ products, the concentration of A changes from 0.1 M to 0.025 M in 40 minutes. The rate of reaction when the concentration of A is 0.01 M is : **[NCERT : P L-73 | 2012]**

(a) 1.73×10^{-5} M/min (b) 3.47×10^{-4} M/min
(c) 3.47×10^{-5} M/min (d) 1.73×10^{-4} M/min

129. In a chemical reaction *A* is converted into *B*. The rates of reaction, starting with initial concentrations of *A* as 2×10^{-3} M and 1×10^{-3} M, are equal to 2.40×10^{-4} Ms^{-1} and 0.60×10^{-4} Ms^{-1} respectively. The order of reaction with respect to reactant *A* will be **[NCERT : P L-67 | Online May 12, 2012]**

(a) 0 (b) 1.5 (c) 1 (d) 2

130. For a reaction $A \rightarrow$ Products, a plot of log $t_{1/2}$ versus log a_0 is shown in the figure. If the initial concentration of *A* is represented by a_0, the order of the reaction is **[Online May 19, 2012]**

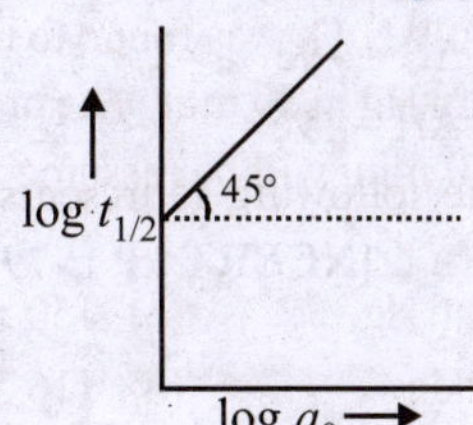

(a) one (b) zero (c) two (d) three

131. The time for half life period of a certain reaction A $\longrightarrow$ Products is 1 hour. When the initial concentration of the reactant 'A' is 2.0 mol L^{-1}, how much time does it take for its concentration to come from 0.50 to 0.25 mol L^{-1} if it is a zero order reaction ? **[NCERT : P L-77 | 2010]**

(a) 4 h (b) 0.5 h (c) 0.25 h (d) 1 h

132. The half life period of a first order chemical reaction is 6.93 minutes. The time required for the completion of 99% of the chemical reaction will be (log 2 = 0.301) **[NCERT : P L-77 | 2009]**

(a) 23.03 minutes (b) 46.06 minutes
(c) 460.6 minutes (d) 230.03 minutes

133. Consider the reaction, 2A + B $\rightarrow$ products. When concentration of B alone was doubled, the half-life did not change. When the concentration of A alone was doubled, the rate increased by two times. The unit of rate constant for this reaction is **[NCERT : P L-77 | 2007]**

(a) s^{-1} (b) L mol^{-1} s^{-1} (c) no unit (d) mol L^{-1} s^{-1}.

134. A radioactive element gets spilled over the floor of a room. Its half-life period is 30 days. If the initial velocity is ten times the permissible value, after how many days will it be safe to enter the room? **[NCERT : P L-77 | 2007]**

(a) 100 days (b) 1000 days
(c) 300 days (d) 10 days.

135. The following mechanism has been proposed for the reaction of NO with Br_2 to form NOBr :

$$NO(g) + Br_2(g) \rightleftharpoons NOBr_2(g)$$
$$NOBr_2(g) + NO(g) \longrightarrow 2NOBr(g)$$

If the second step is the rate determining step, the order of the reaction with respect to NO(g) is **[2006]**

(a) 3 (b) 2 (c) 1 (d) 0

136. A reaction was found to be second order with respect to the concentration of carbon monoxide. If the concentration of carbon monoxide is doubled, with everything else kept the same, the rate of reaction will **[NCERT : P L-68 | 2006]**

(a) increase by a factor of 4 (b) double
(c) remain unchanged (d) triple

137. A reaction involving two different reactants can never be **[NCERT : P L-69 | 2005]**

(a) bimolecular reaction (b) second order reaction
(c) first order reaction (d) unimolecular reaction

138. $t_{1/4}$ can be taken as the time taken for the concentration of a reactant to drop to 3/4 of its initial value. If the rate constant for a first order reaction is *k*, the $t_{1/4}$ can be written as **[NCERT : P L-73 | 2005]**

(a) 0.75/*k* (b) 0.69/*k*
(c) 0.29/*k* (d) 0.10/*k*

139. The rate equation for the reaction 2A + B $\rightarrow$ C is found to be: rate = *k*[A][B]. The correct statement in relation to this reaction is that the **[2004]**

(a) rate of formation of C is twice the rate of disappearance of A
(b) $t_{1/2}$ is a constant
(c) unit of *k* must be s^{-1}
(d) value of *k* is independent of the initial concentrations of A and B

140. For the reaction system : **[NCERT : P L-67 | 2003]**

$$2NO(g) + O_2(g) \rightarrow 2NO_2(g)$$

volume is suddenly reduced to half of its value by increasing the pressure on it. If the reaction is of first order with respect to O_2 and second order with respect to NO, the rate of reaction will

(a) diminish to one-eighth of its initial value
(b) increase to eight times of its initial value
(c) increase to four times of its initial value
(d) diminish to one-fourth of its initial value

141. The half-life of a radioactive isotope is three hours. If the initial mass of the isotope were 256 g, the mass of it remaining undecayed after 18 hours would be **[NCERT : P L-77 | 2003]**

(a) 8.0 g (b) 12.0 g (c) 16.0 g (d) 4.0 g

142. The integrated rate equation is **[NCERT : P L-73 | 2002]**
$Rt = \log C_0 - \log C_t$.
The straight line graph is obtained by plotting
(a) time vs $\log C_t$ (b) $\frac{1}{\text{time}}$ vs C_t
(c) time vs C_t (d) $\frac{1}{\text{time}}$ vs $\frac{1}{C_t}$

143. In a first order reaction, the concentration of the reactant, decreases from 0.8 M to 0.4 M in 15 minutes. The time taken for the concentration to change from 0.1 M to 0.025 M is **[NCERT : P L-73 | 2004]**
(a) 7.5 minutes (b) 15 minutes
(c) 30 minutes (d) 60 minutes

144. The half-life of a radioisotope is four hours. If the initial mass of the isotope was 200 g, the mass remaining after 24 hours undecayed is **[NCERT : P L-77 | 2004]**
(a) 3.125 g (b) 2.084 g
(c) 1.042 g (d) 4.167 g

145. For the reaction $A + 2B \rightarrow C$, rate is given by $R = [A][B]^2$ then the order of the reaction is **[NCERT : P L-68 | 2002]**
(a) 3 (b) 6 (c) 5 (d) 7

146. If half-life of a substance is 5 yrs, then the total amount of substance left after 15 years, when initial amount is 64 grams is **[NCERT : P L-77 | 2002]**
(a) 16 grams (b) 2 grams
(c) 32 grams (d) 8 grams

Topic 3 - Theories of Rate of Reaction

147. Consider the following plots of log of rate constant k (log k) vs $\frac{1}{T}$ for three different reactions. The correct order of activation energies of these reactions is
[NCERT : P L-80 | April 4, 2025 (II)]

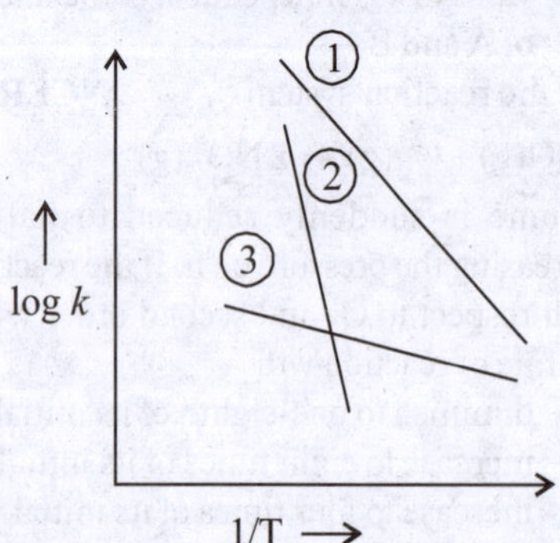

(a) $Ea_2 > Ea_1 > Ea_3$ (b) $Ea_1 > Ea_3 > Ea_2$
(c) $Ea_1 > Ea_2 > Ea_3$ (d) $Ea_3 > Ea_2 > Ea_1$

148. For $A_2 + B_2 \rightleftharpoons 2AB$
E_a for forward and backward reaction are 180 and 200 kJ mol^{-1} respectively **[NCERT : P L-79 | April 4, 2025 (I)]**
If catalyst lowers E_a for both reaction by 100 kJ mol^{-1}. Which of the following statement is correct?
(a) Catalyst does not alter the Gibbs energy change of a reaction.
(b) Catalyst can cause non-spontaneous reactions to occur.
(c) The enthalpy change for the reaction is +20 kJ mol^{-1}.
(d) The enthalpy change for the catalysed reaction is different from that of uncatalysed reaction.

149. Consider the following statements related to temperature dependence of rate constants. Identify the correct statements, **[NCERT : P L-81 | April 3, 2025 (II)]**
A. The Arrhenius equation holds true only for an elementary homogenous reaction.
B. The unit of A is same as that of k in Arrhenius equation.
C. At a given temperature, a low activation energy means a fast reaction.
D. A and E_a as used in Arrhenius equation depend on temperature.
E. When $E_a >> RT$, A and E_a become interdependent.
Choose the **correct** answer from the options given below:
(a) A, C and D Only (b) B, D and E Only
(c) B and C Only (d) A and B Only

150. Reactant A converts to product D through the given mechanism (with the net evolution of heat) :
$A \rightarrow B$ slow ; $\Delta H = +ve$
$B \rightarrow C$ fast ; $\Delta H = -ve$
$C \rightarrow D$ fast ; $\Delta H = -ve$
Which of the following represents the above reaction mechanism ? **[NCERT : P L-79 | April 2, 2025 (II)]**

(a)
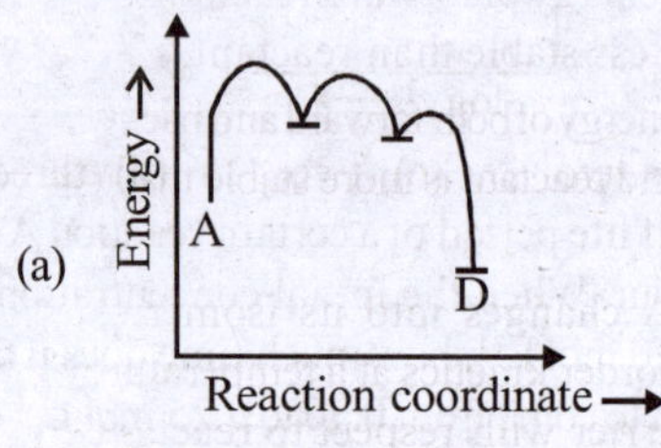

(b)
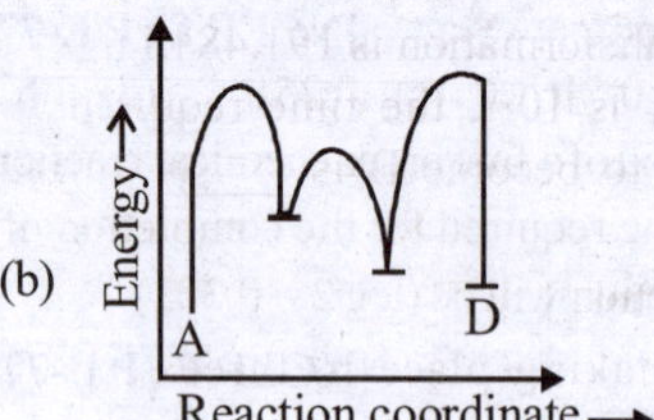

(c)
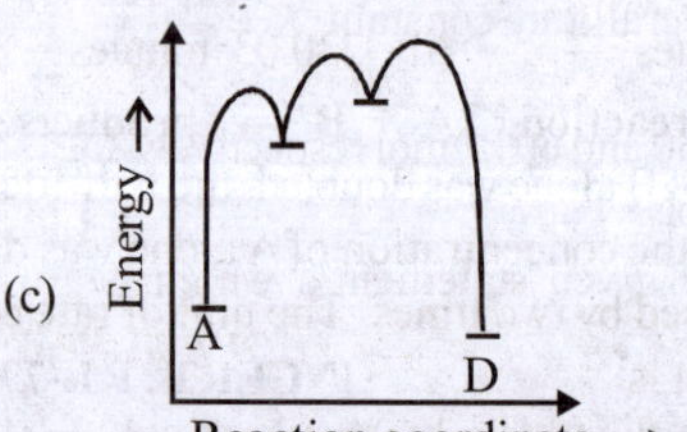

(d) 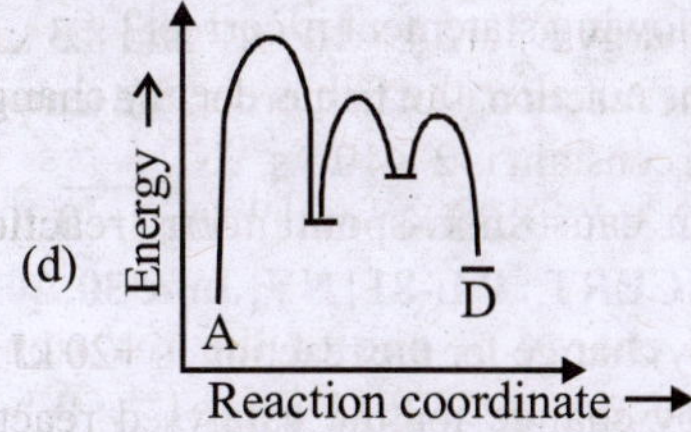

151. Consider a complex reaction taking place in three steps with rate constants k_1, k_2 and k_3 respectively. The overall rate constant k is given by the expression $k = \sqrt{\frac{k_1 k_3}{k_2}}$. If the activation energies of the three steps are 60, 30 and 10 kJ mol^{-1} respectively, then the overall energy of activation in kJ mol^{-1} is ______. (Nearest integer)
[NCERT : P L-79 | NV, Jan. 24, 2025 (II)]

152. Consider the given figure and choose the correct option:
[NCERT : P L-79 | Jan. 22, 2025 (II)]

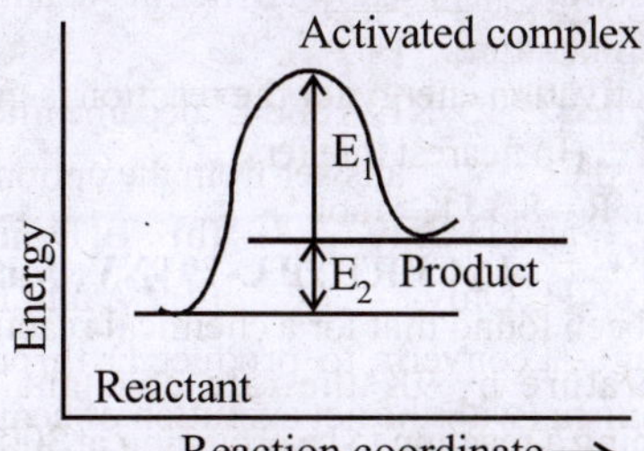

(a) Activation energy of backward reaction is E_1 and product is more stable than reactant.
(b) Activation energy of forward reaction is $E_1 + E_2$ and product is more stable than reactant.
(c) Activation energy of forward reaction is $E_1 + E_2$ and product is less stable than reactant.
(d) Activation energy of both forward and backward reaction is $E_1 + E_2$ and reactant is more stable than product.

153. $A \rightarrow B$
The molecule A changes into its isomeric form B by following a first order kinetics at a temperature of 1000 K. If the energy barrier with respect to reactant energy for such isomeric transformation is 191.48 kJ mol^{-1} and the frequency factor is 10^{20}, the time required for 50%, molecules of A to become B is ______ picoseconds (nearest integer). **[NCERT : P L-79 | NV, Jan. 22, 2025 (I)]**
[R = 8.314 J K^{-1} mol^{-1}]

154. For a reaction taking place in three steps at same temperature, overall rate constant $K = \frac{K_1 K_2}{K_3}$. If E_{a_1}, E_{a_2} and E_{a_3} are 40, 50 and 60 kJ/mol respectively, the overall E_a is ______ kJ/mol. **[NCERT : P L-81 | NV, Jan. 29, 2024 (I)]**

155. The number of given statement/s which is/are correct is____ **[NCERT : P L-81 | NV, April 08, 2023 (I)]**

(A) The stronger the temperature dependence of the rate constant, the higher is the activation energy.
(B) If a reaction has zero activation energy, its rate is independent of temperature.
(C) The stronger the temperature dependence of the rate constant, the smaller is the activation energy.
(D) If there is no correlation between the temperature and the rate constant then it means that the reaction has negative activation energy.

156. The correct reaction profile diagram for a positive catalyst reaction. **[NCERT : P L-82 | April 08, 2023 (II)]**

(a) 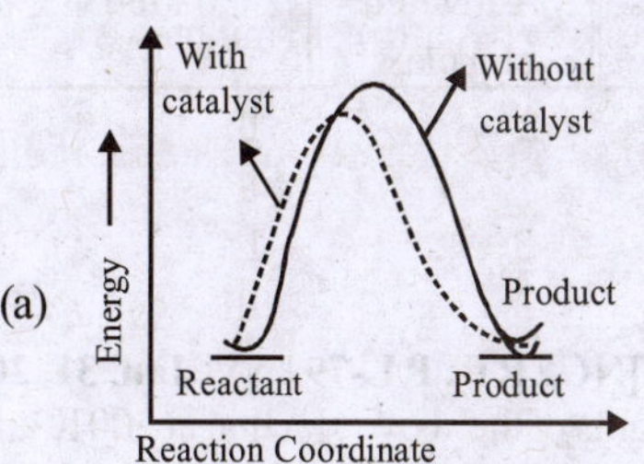

(b) 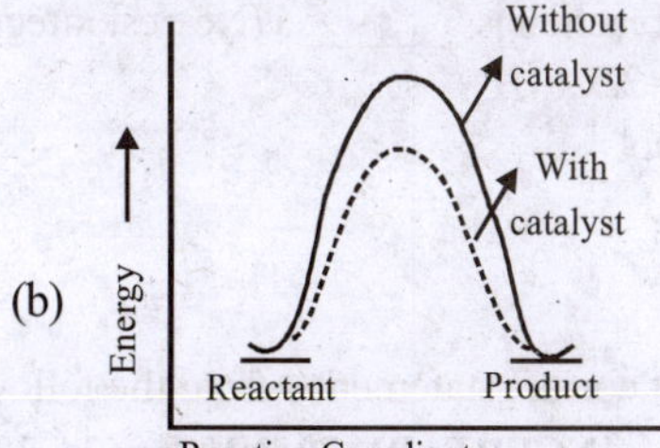

(c)
Without catalyst
With catalyst
Energy
Reactant
Product
Reaction Coordinate

(d) 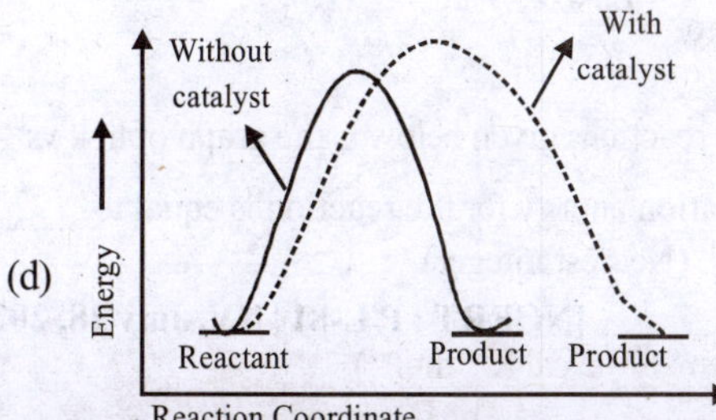

157. For the adsorption of hydrogen on platinum, the activation energy is 30 kJ mol^{-1} and for the adsorption of hydrogen on nickel, the activation energy is 41.4 kJ mol^{-1}. The logarithm of the ratio of the rates of chemisorption on equal areas of the metals at 300 K is .... (Nearest integer)
Given: ln 10 = 2.3 R = 8.3 JK^{-1} mol^{-1}
[NCERT : P L-81 | NV, April 06, 2023 (I)]

158. Consider the following reaction that goes from A to B in three steps as shown below:- **[April 06, 2023 (II)]**

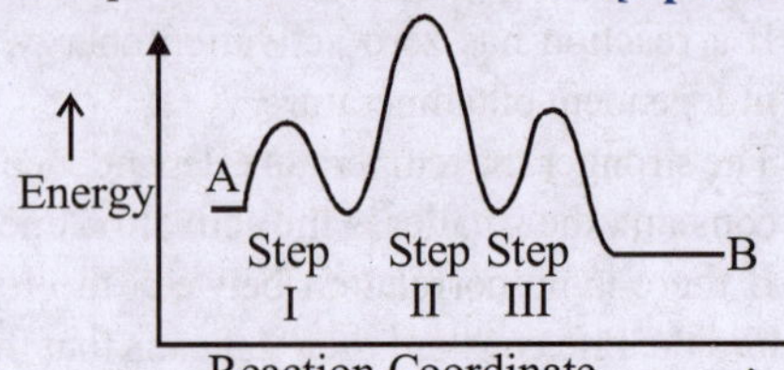

Choose the correct

Number of intermediates	Number of Activated complexes	Rate determining step
(a) 3	2	II
(b) 2	3	II
(c) 2	3	I
(d) 2	3	III

159. $A \rightarrow B$ **[NCERT : P L-79 | NV, Jan. 31, 2023 (I)]**

The rate constants of the above reaction at 200 K and 300K are 0.03 min^{-1} and 0.05 min^{-1} respectively. The activation energy for the reaction is _______ J (Nearest integer)

(Given : In 10 = 2.3

$R = 8.3 \text{ J K}^{-1} \text{ mol}^{-1}$

log5 = 0.70

log3 = 0.48

log2 = 0.30

160. The number of correct statement/s from the following is _______. **[NCERT: P L-79 | NV, Jan. 24, 2023 (I)]**

A. Larger the activation energy, smaller is the value of the rate constant.

B. The higher is the activation energy, higher is the value of the temperature coefficient.

C. At lower temperatures, increase in temperature causes more change in the value of k than at higher temperature.

D. A plot of $\ln k$ vs $\frac{1}{T}$ is a straight line with slope equal to $-\frac{Ea}{R}$

161. For a reaction, given below is the graph of ln k vs $\frac{1}{T}$. The activation energy for the reaction is equal to _______ cal mol^{-1}. (Nearest integer)

[NCERT : P L-81 | NV, July 28, 2022 (II)]

(Given : $R = 2 \text{ cal K}^{-1} \text{ mol}^{-1}$)

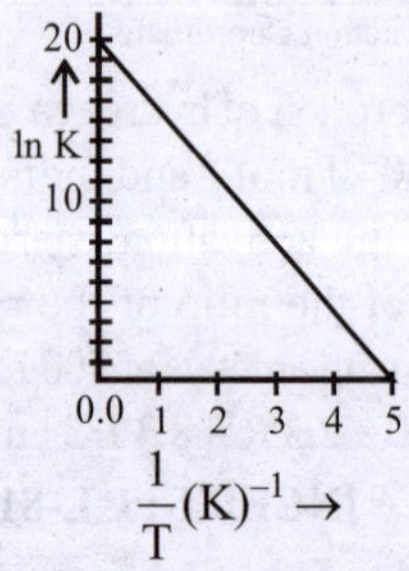

162. For the reaction $P \rightarrow B$, the values of frequency factor A and activation energy E_A are $4 \times 10^{13} \text{ s}^{-1}$ and 8.3 kJ mol^{-1} respectively. If the reaction is of first order, the temperature at which the rate constant is $2 \times 10^{-6} \text{ s}^{-1}$ is _____ $\times 10^{-1}$ K. (Given : ln 10 = 2.3, $R = 8.3 \text{ J K}^{-1} \text{ mol}^{-1}$, log 2 = 0.30)

[NCERT : P L-81 | NV, June 30, 2022 (I)]

163. The activation energy of one of the reactions in a biochemical process is 532611 J mol^{-1}. When the temperature falls from 310 K to 300 K, the change in rate constant observed is $k_{300} = x \times 10^{-3} k_{310}$. The value of x is _______. [Given: ln10 = 2.3, $R = 8.3 \text{ J K}^{-1} \text{ mol}^{-1}$]

[NCERT : P L-81 | NV, June 29, 2022 (I)]

164. The equation $k = (6.5 \times 10^{12} \text{ s}^{-1}) e^{-26000K/T}$ is followed for the decomposition of compound A. The activation energy for the reaction is _______ kJ mol^{-1}. [nearest integer] (Given: $R = 8.314 \text{ J K}^{-1} \text{ mol}^{-1}$) **[NV, June 29, 2022 (II)]**

165. The rate constant for a first order reaction is given by the following equation:

$$\ln k = 33.24 - \frac{2.0 \times 10^4 \text{ K}}{T}$$

The Activation energy for the reaction is given by _______ kJ mol^{-1}. (In nearest integer)

(Given: $R = 8.3 \text{ J K}^{-1} \text{ mol}^{-1}$)

[NCERT : P L-79 | NV, June 27, 2022 (I)]

166. It has been found that for a chemical reaction with rise in temperature by 9K the rate constant gets doubled. Assuming a reaction to be occurring at 300 K, the value of activation energy is found to be _______ kJ mol^{-1}. [nearest integer] **[NCERT : P L-81 | NV, June 27, 2022 (II)]**

(Given ln 10 = 2.3, $R = 8.3 \text{ JK}^{-1}\text{mol}^{-1}$, log 2 = 0.30)

167. Catalyst A reduces the activation energy for a reaction by 10 kJ mol^{-1} at 300 K. The ratio of rate constants, $\frac{k_t\text{, catalysed}}{k_t\text{, Uncatalysed}}$ is e^x. The value of x is _____. [nearest integer]

[Assume that the pre-exponential factor is same in both the cases. **[NCERT : P L-81 | NV, June 26, 2022 (II)]**

Given $R = 8.31 \text{ J K}^{-1} \text{ mol}^{-1}$]

168. The rate constants for decomposition of acetaldehyde have been measured over the temperature range 700 – 1000 K. The data has been analysed by plotting ℓn k vs $\frac{10^3}{T}$ graph. The value of activation energy for the reaction is ___ kJ mol^{-1}. (Nearest integer) (Given : $R = 8.31 \text{ J K}^{-1} \text{ mol}^{-1}$)

[NCERT : P L-80 | NV, June 24, 2022 (I)]

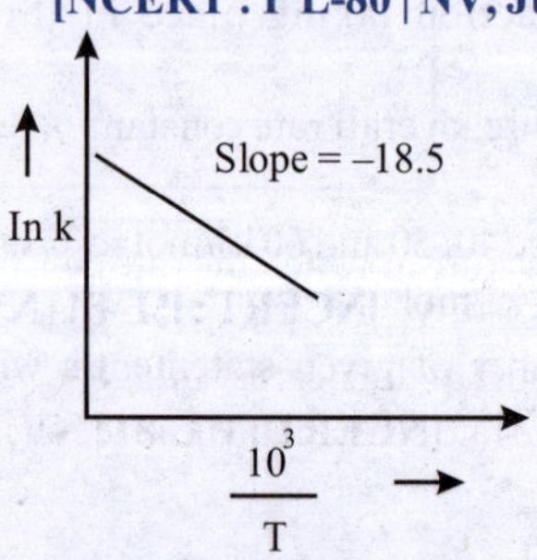

169. According to the following figure, the magnitude of the enthalpy change of the reaction $A + B \rightarrow M + N$ in kJ mol^{-1} is equal to ________. (Integer answer) [NV, Aug. 31, 2021 (I)]

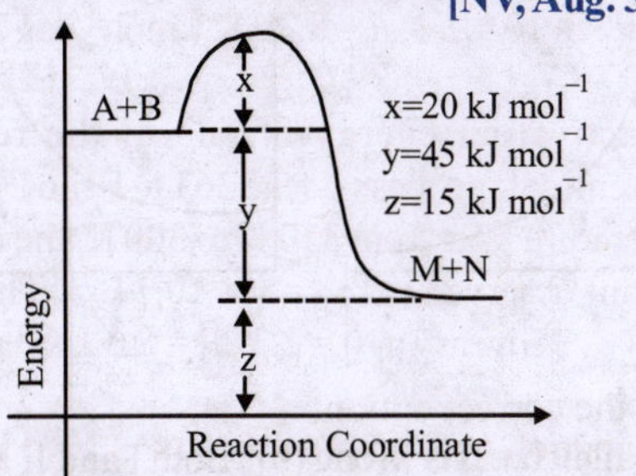

170. For the reaction $A \rightarrow B$, the rate constant k(in s^{-1}) is given by $\log_{10} k = 20.35 - \frac{(2.47 \times 10^3)}{T}$
The energy of activation in kJ mol^{-1} is ____.
(Nearest integer) [Given : $R = 8.314\ JK^{-1}\ mol^{-1}$] [NCERT : P L-79 | NV, Aug. 31, 2021 (II)]

171. The first order rate constant for the decomposition of $CaCO_3$ at 700 K is $6.36 \times 10^{-3} s^{-1}$ and activation energy is 209 kJ mol^{-1}. Its rate constant (in s^{-1}) at 600 K is $x \times 10^{-6}$. The value of x is ______. (Nearest integer)
[Given $R = 8.31\ J K^{-1} mol^{-1}$; $\log 6.36 \times 10^{-3} = -2.19$, $10^{-4.79} = 1.62 \times 10^{-5}$] [NCERT : P L-81 | NV, Aug. 27, 2021 (II)]

172. The decomposition of formic acid on gold surface follows first order kinetics. If the rate constant at 300 K is $1.0 \times 10^{-3}\ s^{-1}$ and the activation energy $E_a = 11.488$ kJ mol^{-1}, the rate constant at 200 K is ____ $\times 10^{-5}\ s^{-1}$. (Round off to the Nearest Integer) [NCERT : P L-81 | NV, March 16, 2021 (I)]

173. An exothermic reaction $X \rightarrow Y$ has an activation energy 30 kJ mol^{-1}. If energy change ΔE during the reaction is –20 kJ, then the activation energy for the reverse reaction in kJ is ______. [NV, Feb. 26, 2021 (I)]

174. If the activation energy of a reaction is 80.9 kJ mol^{-1}, the fraction of molecules at 700 K, having enough energy to react to form products is e^{-x}. The value of x is________.
[Use $R = 8.31\ J K^{-1} mol^{-1}$] [NV, Feb. 26, 2021 (II)]

175. For the reaction, $aA + bB \rightarrow cC + dD$, the plot of log k vs $\frac{1}{T}$ is given below:

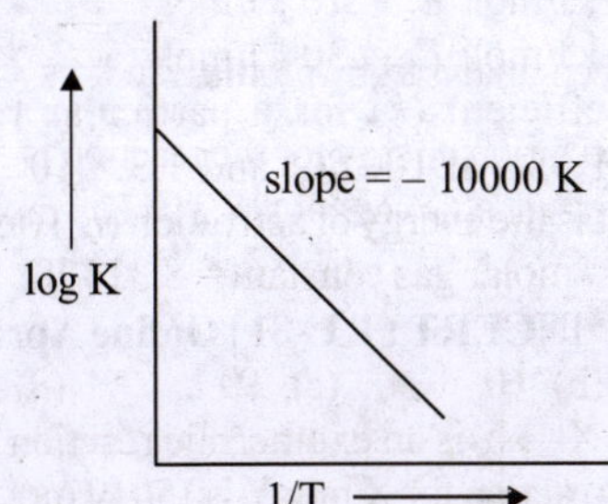

The temperature at which the rate constant of the reaction is $10^{-4} s^{-1}$ is ______ K.
(Rounded-off to the nearest integer)
[Given: The rate constant of the reaction is $10^{-5} s^{-1}$ at 500 K.] [NV, Feb. 25, 2021 (I)]

176. The rate constant of a reaction increases by five times on increase in temperature from 27 °C to 52 °C. The value of activation energy in kJ mol^{-1} is ______. (Rounded-off to the nearest integer)
[R = 8.314 J K–1 mol–1] [NCERT : P L-81 | NV, Feb. 25, 2021 (II)]

177. The rate of a reaction decreased by 3.555 times when the temperature was changed from 40ºC to 30ºC. The activation energy (in kJ mol^{-1}) of the reaction is ______. Take; $R = 8.314\ J\ mol^{-1}\ K^{-1}$ ln 3.555 = 1.268 [NCERT : P L-81 | NV, Sep. 06, 2020 (II)]

178. The rate constant (*k*) of a reaction is measured at different temperatures (*T*), and the data are plotted in the given figure. The activation energy of the reaction in kJ mol^{-1} is : (*R* is gas constant) [NCERT : P L-79 | Sep. 05, 2020 (II)]

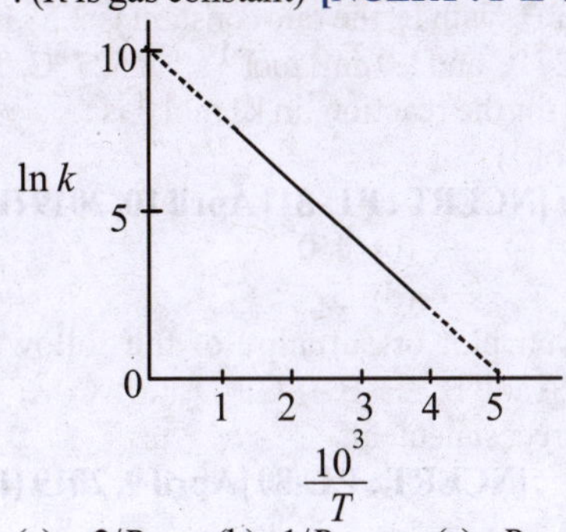

(a) 2/*R* (b) 1/*R* (c) *R* (d) 2*R*

179. The number of molecules with energy greater than the threshold energy for a reaction increases five fold by a rise of temperature from 27°C to 42°C. Its energy of activation in J/ mol is __________. (Take ln 5 = 1.6094; $R = 8.314\ J mol^{-1} K^{-1}$) [NCERT : P L-81 | NV, Sep. 04, 2020 (II)]

180. A sample of milk splits after 60 min. at 300 K and after 40 min. at 400 K when the population of *lactobacillus acidophilus* in it doubles. The activation energy (in kJ/ mol) for this process is closest to ______.
(Given, $R = 8.3\ J mol^{-1} K^{-1}$, $\ln\left(\frac{2}{3}\right) = -0.4$, $e^{-3} = 0.04$) [NV, Jan. 09, 2020 (II)]

181. For following reactions:
$A \xrightarrow{700K} Product$
$A \xrightarrow[catalyst]{500K} Product$
it was found that the E_a is decrease by 30 kJ/mol in the presence of catalyst. If the rate remains unchanged, the activation energy for catalysed reaction is (Assume pre exponential factor is same): [NCERT : P L-79 | Jan. 09, 2020 (I)]
(a) 75 kJ/mol (b) 105 kJ/mol
(c) 135 kJ/mol (d) 198 kJ/mol

182. The rate of a certain biochemical reaction at physiological temperature (*T*) occurs 10^6 times faster with enzyme than without. The change in the activation energy upon adding enzyme is: [Jan. 08, 2020 (I)]
(a) –6(2.303)RT (b) –6RT
(c) +6(2.303)RT (d) +6RT

183. Consider the following plots of rate constant versus $\frac{1}{T}$ for four different reactions. Which of the following orders is correct for the activation energies of these reactions?

[NCERT : P L-80 | Jan. 08, 2020 (II)]

(a) $E_b > E_a > E_d > E_c$ (b) $E_a > E_c > E_d > E_b$
(c) $E_c > E_a > E_d > E_b$ (d) $E_b > E_d > E_c > E_a$

184. For the reaction of H_2 with I_2, the rate constant is 2.5×10^{-4} $dm^3\,mol^{-1}\,s^{-1}$ at 327 °C and 1.0 $dm^3\,mol^{-1}\,s^{-1}$ at 527 °C. The activation energy for the reaction, in $kJ\,mol^{-1}$ is :
($R = 8.314\,J\,K^{-1}\,mol^{-1}$)

[NCERT : P L-81 | April 10, 2019 (II)]

(a) 166 (b) 150
(c) 72 (d) 59

185. Consider the given plot of enthalpy of the following reaction between A and B, $A + B \rightarrow C + D$. Identify the incorrect statement.

[NCERT : P L-80 | April 9, 2019 (II)]

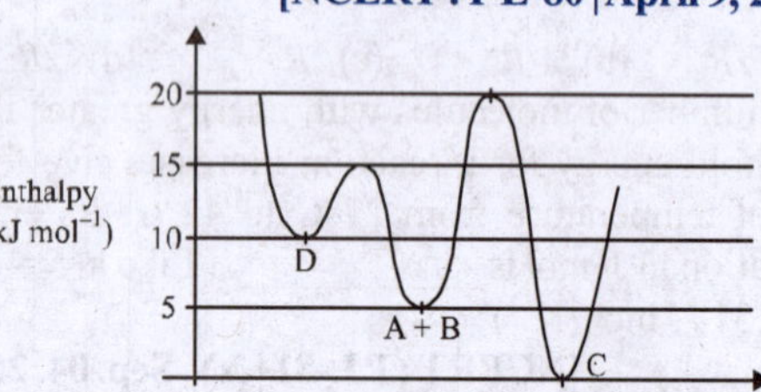

(a) Activation enthalpy to form C is 5 $kJ\,mol^{-1}$ less than that to form D.
(b) C is the thermodynamically stable product.
(c) D is kinetically stable product.
(d) Formation of A and B from C has highest enthalpy of activation.

186. For a reaction, consider the plot of ln k versus 1/T given in the figure. If the rate constant of this reaction at 400 K is $10^{-5}\,s^{-1}$, then the rate constant at 500 K is :

[NCERT : P L-80 | Jan. 12, 2019 (II)]

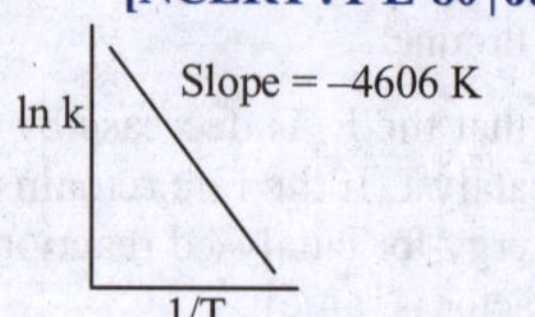

(a) $10^{-6}\,s^{-1}$ (b) $2 \times 10^{-4}\,s^{-1}$
(c) $10^{-4}\,s^{-1}$ (d) $4 \times 10^{-4}\,s^{-2}$

187. If a reaction follows the Arrhenius equation, the plot ln k vs $\frac{1}{(RT)}$ gives straight line with a gradient (– y) unit. The energy required to activate the reactant is:

[NCERT : P L-80 | Jan. 11, 2019 (I)]

(a) y/R unit (b) y unit
(c) yR unit (d) – y unit

188. Consider the given plots for a reaction obeying Arrhenius equation (0 °C < T < 300 °C) : (k and E_a are rate constant and activation energy, respectively)

[NCERT : P L-80 | Jan. 10, 2019 (I)]

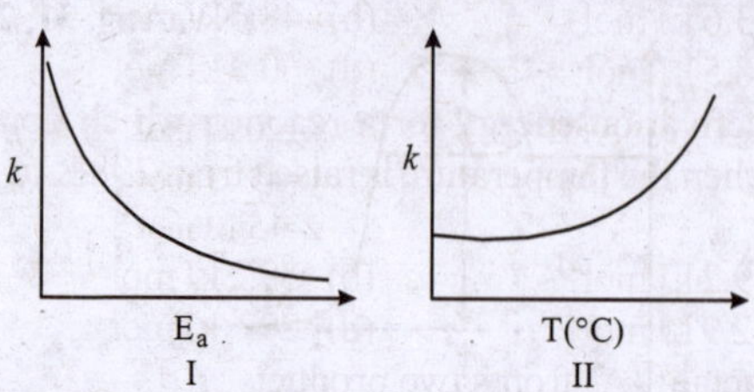

Choose the correct option:
(a) I is right but II is wrong (b) Both I and II are correct
(c) I is wrong but II is right (d) Both I and II are wrong

189. Two reactions R_1 and R_2 have identical pre-exponential factors. Activation energy of R_1 exceeds that of R_2 by 10 $kJ\,mol^{-1}$. If k_1 and k_2 are rate constants for reactions R_1 and R_2 respectively at 300 K, then $\ln(k_2/k_1)$ is equal to :
($R = 8.314\,J\,mol^{-1}K^{-1}$) **[NCERT : P L-81 | 2017]**

(a) 8 (b) 12 (c) 6 (d) 4

190. The rate of a reaction A doubles on increasing the temperature from 300 to 310 K. By how much, the temperature of reaction B should be increased from 300 K so that rate doubles if activation energy of the reaction B is twice to that of reaction A.

[NCERT : P L-81 | Online April 8, 2017]

(a) 9.84 K (b) 4.92 K (c) 2.45 K (d) 19.67 K

191. The rate of a reaction quadruples when the temperature changes from 300 to 310 K. The activation energy of this reaction is : (Assume activation energy and pre-exponential factor are independent of temperature; ln 2 = 0.693; R = 8.314 $J\,mol^{-1}\,K^{-1}$) **[NCERT : P L-81 | Online April 9, 2017]**

(a) 107.2 $kJ\,mol^{-1}$ (b) 53.6 $kJ\,mol^{-1}$
(c) 26.8 $kJ\,mol^{-1}$ (d) 214.4 $kJ\,mol^{-1}$

192. For the equilibrium, $A(g) \rightleftharpoons B(g)$, ΔH is –40 kJ/mol. If the ratio of the activation energies of the forward (E_f) and reverse (E_b) reactions is $\frac{2}{3}$ then :

[NCERT : P L-80 | Online April 11, 2015]

(a) E_f = 80 kJ/mol; E_b = 120 kJ/mol
(b) E_f = 60 kJ/mol; E_b = 100 kJ/mol
(c) E_f = 30 kJ/mol; E_b = 70 kJ/mol
(d) E_f = 70 kJ/mol; E_b = 30 kJ/mol

193. The rate coefficient (k) for a particular reactions is $1.3 \times 10^{-4}\,M^{-1}\,s^{-1}$ at 100 °C, and $1.3 \times 10^{-3}\,M^{-1}\,s^{-1}$ at 150 °C. What is the energy of activation (E_a) (in kJ) for this reaction? (R = molar gas constant = 8.314 $JK^{-1}\,mol^{-1}$)

[NCERT : P L-81 | Online April 12, 2014]

(a) 16 (b) 60 (c) 99 (d) 132

194. The reaction $X \rightarrow Y$ is an exothermic reaction. Activation energy of the reaction for X into Y is 150 $kJ\,mol^{-1}$. Enthalpy of reaction is 135 $kJ\,mol^{-1}$. The activation energy for the reverse reaction, $Y \rightarrow X$ will be :

[NCERT : P L-79 | Online April 22, 2013]

(a) 280 $kJ\,mol^{-1}$ (b) 285 $kJ\,mol^{-1}$
(c) 270 $kJ\,mol^{-1}$ (d) 15 $kJ\,mol^{-1}$

195. The rate of a reaction doubles when its temperature changes from 300 K to 310 K. Activation energy of the reaction will be : (R = 8.314 JK^{-1} mol^{-1} and log 2 = 0.301) **[NCERT : P L-81 | 2013]**

(a) 53.6 kJ mol^{-1} (b) 48.6 kJ mol^{-1}
(c) 58.5 kJ mol^{-1} (d) 60.5 kJ mol^{-1}

196. The activation energy for a reaction which doubles the rate when the temperature is raised from 298 K to 308 K is **[NCERT : P L-81 | Online May 26, 2012]**

(a) 59.2 kJ mol^{-1} (b) 39.2 kJ mol^{-1}
(c) 52.9 kJ mol^{-1} (d) 29.5 kJ mol^{-1}

197. A reactant (A) froms two products : **[NCERT : P L-79 | 2011RS]**

$A \xrightarrow{k_1} B$, Activation energy E_{a1}

$A \xrightarrow{k_2} C$, Activation energy E_{a2}

If $E_{a_2} = 2E_{a_1}$, then k_1 and k_2 are related as :

(a) $k_2 = k_1 e^{E_{a_1}/RT}$ (b) $k_2 = k_1 e^{E_{a_2}/RT}$
(c) $k_1 = Ak_2 e^{E_{a_1}/RT}$ (d) $k_1 = 2k_2 e^{E_{a_2}/RT}$

198. Rate of a reaction can be expressed by Arrhenius equation as : **[NCERT : P L-79 | 2006]**

$$k = A\, e^{-Ea/RT}$$

In this equation, E_a represents

(a) the total energy of the reacting molecules at a temperature, T
(b) the fraction of molecules with energy greater than the activation energy of the reaction
(c) the energy above which all the colliding molecules will react
(d) the energy below which all the colliding molecules will react

199. A schematic plot of $\ln K_{eq}$ versus inverse of temperature for a reaction is shown below **[NCERT : P L-79 | 2005]**

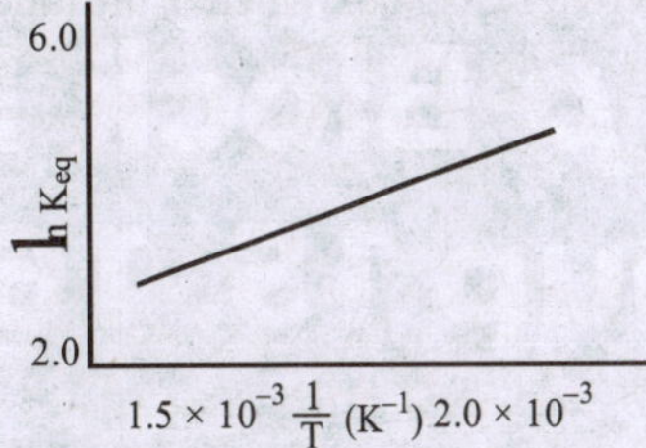

The reaction must be

(a) highly spontaneous at ordinary temperature
(b) one with negligible enthalpy change
(c) endothermic
(d) exothermic

200. Consider an endothermic reaction $X \rightarrow Y$ with the activation energies E_b and E_f for the backward and forward reactions, respectively. In general **[NCERT : P L-80 | 2005]**

(a) there is no definite relation between E_b and E_f
(b) $E_b = E_f$
(c) $E_b > E_f$
(d) $E_b < E_f$

201. In respect of the equation $k = Ae^{-E_a/RT}$ in chemical kinetics, which one of the following statements is correct ? **[NCERT : P L-79 | 2003]**

(a) A is adsorption factor
(b) E_a is energy of activation
(c) R is Rydberg's constant
(d) k is equilibrium constant

ANSWER KEYS

1	(d)	**22**	(c)	**43**	(c)	**64**	(24)	**85**	(a)	**106**	(108)	**127**	(c)	**148**	(a)	**169**	(45)	**190**	(b)
2	(c)	**23**	(d)	**44**	(d)	**65**	(63)	**86**	(1)	**107**	(81)	**128**	(b)	**149**	(c)	**170**	(47)	**191**	(a)
3	(c)	**24**	(c)	**45**	(d)	**66**	(2)	**87**	(15)	**108**	(d)	**129**	(d)	**150**	(a)	**171**	(16)	**192**	(a)
4	(a)	**25**	(b)	**46**	(a)	**67**	(10)	**88**	(75)	**109**	(b)	**130**	(b)	**151**	(20)	**172**	(10)	**193**	(b)
5	(47)	**26**	(d)	**47**	(d)	**68**	(2)	**89**	(2)	**110**	(60)	**131**	(c)	**152**	(c)	**173**	(50)	**194**	(b)
6	(2435)	**27**	(a)	**48**	(b)	**69**	(17)	**90**	(2)	**111**	(b)	**132**	(b)	**153**	(69)	**174**	(14)	**195**	(a)
7	(8)	**28**	(d)	**49**	(c)	**70**	(623)	**91**	(60)	**112**	(c)	**133**	(b)	**154**	(30)	**175**	(526)	**196**	(c)
8	(3)	**29**	(a)	**50**	(d)	**71**	(1350)	**92**	(b)	**113**	(23.03)	**134**	(a)	**155**	(2)	**176**	(52)	**197**	(c)
9	(1)	**30**	(d)	**51**	(b)	**72**	(200)	**93**	(1)	**114**	(a)	**135**	(b)	**156**	(b)	**177**	(100)	**198**	(c)
10	(d)	**31**	(c)	**52**	(897)	**73**	(2)	**94**	(100)	**115**	(c)	**136**	(a)	**157**	(2)	**178**	(d)	**199**	(d)
11	(17)	**32**	(d)	**53**	(c)	**74**	(c)	**95**	(2)	**116**	(b)	**137**	(d)	**158**	(b)	**179**	(84297.48)	**200**	(d)
12	(2200)	**33**	(d)	**54**	(3)	**75**	(7)	**96**	(165)	**117**	(b)	**138**	(c)	**159**	(2520)	**180**	(3.98)	**201**	(b)
13	(333)	**34**	(b)	**55**	(17)	**76**	(1)	**97**	(2)	**118**	(a)	**139**	(d)	**160**	(3)	**181**	(a)		
14	(2)	**35**	(d)	**56**	(3)	**77**	(3)	**98**	(c)	**119**	(d)	**140**	(b)	**161**	(8)	**182**	(a)		
15	(40)	**36**	(b)	**57**	(315)	**78**	(10)	**99**	(1)	**120**	(b)	**141**	(d)	**162**	(225)	**183**	(c)		
16	(1)	**37**	(d)	**58**	(100)	**79**	(2)	**100**	(2)	**121**	(c)	**142**	(a)	**163**	(1)	**184**	(a)		
17	(4)	**38**	(d)	**59**	(50)	**80**	(26)	**101**	(16)	**122**	(d)	**143**	(c)	**164**	(216)	**185**	(a)		
18	(1)	**39**	(b)	**60**	(17190)	**81**	(c)	**102**	(75)	**123**	(a)	**144**	(a)	**165**	(166)	**186**	(c)		
19	(4)	**40**	(d)	**61**	(2)	**82**	(a)	**103**	(0)	**124**	(c)	**145**	(a)	**166**	(59)	**187**	(b)		
20	(106)	**41**	(a)	**62**	(d)	**83**	(3)	**104**	(c)	**125**	(c)	**146**	(d)	**167**	(4)	**188**	(b)		
21	(27)	**42**	(b)	**63**	(399)	**84**	(3)	**105**	(654)	**126**	(d)	**147**	(a)	**168**	(154)	**189**	(d)		

4

The *p*-Block Elements (Group 15, 16, 17 & 18)

TREND ANALYSIS
JEE MAIN (2025-2021)

	2025				2024				2023				2022				2021			
	Session II		Session I		Session II		Session I		Session II		Session I		Session II		Session I		Session II		Session I	
	MCQs	NVQs	MCQs	NVQs	MCQs	NVQs	MCQs	NVQs	MCQs	NVQs	MCQs	NVQs	MCQs	NVQs	MCQs	NVQs	MCQs	NVQs	MCQs	NVQs
Topic 1	2	–	3	1	1	–	3	–	1	–	–	–	2	–	5	–	2	–	1	–
Topic 2	1	–	1	–	1	1	4	–	1	–	–	–	–	–	1	–	–	–	1	–
Topic 3	1	–	–	–	2	1	3	1	1	1	2	–	–	2	1	–	–	–	1	–
Topic 4	–	–	–	–	–	–	1	–	–	1	1	–	1	1	–	–	1	–	–	–
Total	4	–	4	1	4	2	11	1	3	2	3	–	3	3	7	–	3	–	3	–

How to Code Qns for Better & Faster Revision

Type 1: Able to solve (No code Req)
Type 2: Solved with struggle or by seeing solution
(Use code NR means Needs Revision or orange stickers)
Type 3: Tricky/ Could not solve on my own. Need help from Teacher/ Friend.
(Use code TQ means Tricky Question or red stickers)

You can either hand write the code in boxes or else paste coloured stickers given at end of book.

Topic 1 – Nitrogen Family

1. Given below are two statements :

Statement I : Nitrogen forms oxides with +1 to +5 oxidation states due to the formation of $p\pi$ - $p\pi$ bond with oxygen .

Statement II : Nitrogen does not form halides with +5 oxidation state due to the absence of *d*-orbital in it.

In the light of given statements, choose the **correct** answer from the options given below.

[NCERT : P L-172 | April 4, 2025 (I)]

(a) Statement I is true but Statement II is false
(b) Both Statement I and Statement II are false
(c) Statement I is false but Statement II is true
(d) Both Statement I and Statement II are true

2. Given below are two statements

[NCERT : P L-172 | April 3, 2025 (I)]

Statement I : The N–N single bond is weaker and longer than that of P–P single bond

Statement II : Compounds of group 15 elements in +3 oxidation states readily undergo disproportionation reactions.

In the light of above statements, choose the **correct** answer from the options given below

(a) Statement I is true but statement II is false
(b) Both statement I and statement II are false
(c) Statement I is false but statement II is true
(d) Both statement I and statement II are true

3. First ionisation enthalpy values of first four group 15 elements are given below. Choose the correct value for the element that is a main component of apatite family :

[NCERT : P L-173 | Jan. 29, 2025 (II)]

(a) 1012 kJ mol^{-1} (b) 1402 kJ mol^{-1}
(c) 834 kJ mol^{-1} (d) 947 kJ mol^{-1}

4. A group 15 element forms $d\pi - d\pi$ bond with transition metals. It also forms hydride, which is a strongest base among the hydrides of other group members that form $d\pi - d\pi$ bond. The atomic number of the element is _____.

[NCERT : P L-173 | NV, Jan. 28, 2025 (II)]

Note : The NCERT Page Locators (PL) here are of Old NCERT (before rationalisation)

5. The incorrect statements among the following is [NCERT : P L-173 | Jan. 23, 2025 (I)]
(a) PH_3 shows lower proton affinity than NH_3.
(b) PF_3 exists but NF_5 does not.
(c) NO_2 can dimerise easily.
(d) SO_2 can act as an oxidizing agent, but not as a reducing agent.

6. The maximum covalency of a non-metallic group 15 element 'E' with weakest E–E bond is : [NCERT : P L-173 | Jan. 22, 2025 (II)]
(a) 5 (b) 3 (c) 6 (d) 4

7. Identify the **incorrect** statements about group 15 elements: [NCERT : P L-172 | April 8, 2024 (II)]
(A) Dinitrogen is a diatomic gas which acts like an inert gas at room temperature.
(B) The common oxidation states of these elements are –3, +3 and +5.
(C) Nitrogen has unique ability to form $p\pi - p\pi$ multiple bonds.
(D) The stability of +5 oxidation states increases down the group.
(E) Nitrogen shows a maximum covalency of 6.
Choose the **correct** answer from the options given below.
(a) (A), (B), (D) only (b) (A), (C), (E) only
(c) (B), (D), (E) only (d) (D) and (E) only

8. Given below are two statements: One is labelled as **Assertion (A)** and the other is labelled as **Reason (R).**
Assertion (A) : PH_3 has lower boiling point than NH_3.
Reason (R) : In liquid state NH_3 molecules are associated through vander Waal's forces, but PH_3 molecules are associatd through hydrogen bonding.
In the light of the above statements, choose the most appropriate answer from the options given below: [NCERT : P L-73 | Feb. 1, 2024 (I)]
(a) Both (A) and (R) are correct but (R) is not the correct explanation of (A)
(b) (A) is not correct but (R) is correct
(c) Both (A) and (R) are correct and (R) is the correct explanation of (A)
(d) (A) is correct but (R) is not correct.

9. The strongest reducing agent among the following is: [NCERT : P L-73 | Feb. 1, 2024 (II)]
(a) SbH_3 (b) NH_3 (c) BiH_3 (d) PH_3

10. Choose the correct statements about the hydrides of group 15 elements. [NCERT : P L-73 | Jan. 30, 2024 (II)]
A. The stability of the hydrides decreases in the order $NH_3 > PH_3 > AsH_3 > SbH_3 > BiH_3$.
B. The reducing ability of the hydride increases in the order $NH_3 < PH_3 < AsH_3 < SbH_3 < BiH_3$.
C. Among the hydrides, NH_3 is strong reducing agent while BiH_3 is mild reducing agent.
D. The basicity of the hydrides increases in the order $NH_3 < PH_3 < AsH_3 < SbH_3 < BiH_3$.
Choose the most appropriate from the options given below:
(a) B and C only (b) C and D only
(c) A and B only (d) A and D only

11. Given below are two statements : [NCERT : P L-174 | April 11, 2023 (I)]
Statement – I : Methane and steam passed over a heated Ni catalyst produces hydrogen gas.
Statement – II : Sodium nitrite reacts with NH_4Cl to give H_2O, N_2 and NaCl.
In the light of the above statements, choose the most appropriate answer from the options given below :
(a) Both the Statements I and II are correct
(b) Both the Statements I and II are incorrect
(c) Statement I is incorrect but Statement II is correct
(d) Statement I is correct but Statement II is incorrect

12. Dinitrogen is a robust compound, but reacts at high altitude to form oxides. The oxide of nitrogen that can damage plant leaves and retard photosynthesis is : [NCERT : P L-172 | July 29, 2022 (II)]
(a) NO (b) NO_3^- (c) NO_2 (d) NO_2^-

13. Dinitrogen and dioxygen, the main constituents of air do not react with each other in atmosphere to form oxides of nitrogen because [NCERT : P L-175 | July 28, 2022 (II)]
(a) N_2 is unreactive in the condition of atmosphere.
(b) Oxides of nitrogen are unstable.
(c) Reaction between them can occur in the presence of a catalyst.
(d) The reaction is endothermic and require very high temperature.

14. $\text{A} \xrightarrow{573\text{ K}} \text{Red phosphorus} \xrightarrow[\text{under pressure}]{\text{heat; 803 K}} \text{B}$
Red phosphorus is obtained by heating "A" at 573 K, and can be converted to "B" by heating at 803 K under pressure. [NCERT : P L-181 | June 30, 2022 (I)]
A and B, respectively, are
(a) β-black phosphorus and white phosphorus.
(b) white phosphorus and β-black phosphorus.
(c) α-black phosphorus and white phosphorus.
(d) white phosphorus and α-black phosphorus.

15. Nitrogen gas is obtained by thermal decomposition of [NCERT : P L-174 | June 28, 2022 (I)]
(a) $Ba(NO_3)_2$ (b) $Ba(N_3)_2$
(c) $NaNO_2$ (d) $NaNO_3$

16. Given below are two statements : [NCERT : P L-173 | June 28, 2022 (I)]
Statement I : The pentavalent oxide of group-15 element E_2O_5 is less acidic than trivalent oxide. E_2O_3 of the same element.
Statement II : The acidic character of trivalent oxide of group 15 elements E_2O_3 decreases down the group.
In light of the above statements choose **most appropriate** answer from the options given below:
(a) Both Statement I and Statement II are true.
(b) Both Statement I and Statement II are false.
(c) Statement I true but Statement II is false.
(d) Statement I is false but Statement II is true.

17. The gas produced by treating an aqueous solution of ammonium chloride with sodium nitrite is [NCERT : P L-174 | June 27, 2022 (II)]
(a) NH_3 (b) N_2 (c) N_2O (d) Cl_2

18. The most stable trihalide of nitrogen is: [NCERT : P L-174 | June 24, 2022 (I)]
(a) NF_3 (b) NCl_3 (c) NBr_3 (d) NI_3

19. Which one of the following is formed (mainly) when red phosphorus is heated in a sealed tube at 803 K?
[NCERT : P L-181 | Aug. 27, 2021 (II)]
(a) White phosphorus (b) Yellow phosphorus
(c) β-Black phosphorus (d) α-Black phosphorus

20. Which one of the following group-15 hydride is the strongest reducing agent?
[NCERT : P L-173 | July 22, 2021 (II)]
(a) AsH_3 (b) BiH_3
(c) PH_3 (d) SbH_3

21. A group 15 element, which is a metal and forms a hydride with strongest reducing power among group 15 hydrides. The element is **[NCERT : P L-173 | March 16, 2021 (I)]**
(a) As (b) P (c) Bi (d) Sb

22. The correct statement with respect to dinitrogen is:
(a) N_2 is paramagnetic in nature.
[NCERT : P L-175 | Sep. 06, 2020 (I)]
(b) it can combine with dioxygen at 25 °C.
(c) liquid dinitrogen is not used in cryosurgery.
(d) it can be used as an inert diluent for reactive chemicals.

23. Reaction of ammonia with excess Cl_2 gives :
[Sep. 05, 2020 (II)]
(a) NH_4Cl and N_2 (b) NH_4Cl and HCl
(c) NCl_3 and NH_4Cl (d) NCl_3 and HCl

24. The number of pentagons in C_{60} and trigons (triangles) in white phosphorous, respectively, are:
[NCERT : P L-181 | April 10, 2019 (II)]
(a) 20 and 3 (b) 12 and 4
(c) 12 and 3 (d) 20 and 4

25. The compound that **does not** produce nitrogen gas by the thermal decomposition is : **[NCERT : P L-174 | 2018]**
(a) $Ba(N_3)_2$ (b) $(NH_4)_2Cr_2O_7$
(c) NH_4NO_2 (d) $(NH_4)_2SO_4$

26. For per gram of reactant, the maximum quantity of N_2 gas is produced in which of the following thermal decomposition reactions? (Given: Atomic wt. : Cr = 52 u, Ba = 137 u).
[NCERT : P L-174, 175 | Online April 15, 2018 (II)]
(a) $Ba(N_3)_2(s) \rightarrow Ba(s) + 3N_2(g)$
(b) $(NH_4)_2Cr_2O_7(s) \rightarrow N_2(g) + 4H_2O(g) + Cr_2O_3(s)$
(c) $2NH_3(g) \rightarrow N_2(g) + 3H_2(g)$
(d) $2NH_4NO_3(s) \rightarrow 2N_2(g) + 4H_2O(g) + O_2(g)$

27. A metal 'M' reacts with nitrogen gas to afford 'M_3N'. 'M_3N' on heating at high temperature gives back 'M' and on reaction with water produces a gas 'B'. Gas 'B' reacts with aqueous solution of $CuSO_4$ to form a deep blue compound. 'M' and 'B' respectively are :
[NCERT : P L-175 | Online April 8, 2017]
(a) Li and NH_3 (b) Ba and N_2
(c) Na and NH_3 (d) Al and N_2

28. The molecule having smallest bond angle is : **[2012]**
(a) NCl_3 (b) $AsCl_3$ (c) $SbCl_3$ (d) PCl_3

29. Regular use of the following fertilizers increases the acidity of soil? **[2007]**
(a) Ammonium sulphate (b) Potassium nitrate
(c) Urea (d) Superphosphate of lime.

30. Which one of the following substances has the highest proton affinity ? **[2003]**
(a) H_2S (b) NH_3 (c) PH_3 (d) H_2O

31. In case of nitrogen, NCl_3 is possible but not NCl_5 while in case of phosphorus, PCl_3 as well as PCl_5 are possible. It is due to **[NCERT : P L-173 | 2002]**
(a) availability of vacant *d* orbitals in P but not in N
(b) lower electronegativity of P than N
(c) lower tendency of H-bond formation in P than N
(d) occurrence of P in solid while N in gaseous state at room temperature.

Topic 2 - Oxygen Family

32. Given below are two statements :
[NCERT : P L-187 | April 8, 2025 (II)]
Statement I : H_2Se is more acidic than H_2Te.
Statement II : H_2Se has higher bond enthalpy for dissociation than H_2Te.
In the light of the above statements, choose the **correct** answer from the options given below.
(a) Both statement I and Statement II are false.
(b) Both statement I and Statement II are true.
(c) Statement I is true but Statement II is false.
(d) Statement I is false but Statement II is true.

33. The large difference between the melting and boiling points of oxygen and sulphur may be explained on the basis of **[NCERT : P L-188 | Jan. 24, 2025 (I)]**
(a) Atomic size (b) Atomicity
(c) Electronegativity (d) Electron gain enthalpy

34. Given below are two statements: one is labelled as **Assertion (A)** and the other is labelled as **Reason (R).**
[NCERT : P L-187 | April 9, 2024 (I)]
Assertion (A) : Both rhombic and monoclinic sulphur exist as S_8 while oxygen exists as O_2.
Reason (R) : Oxygen forms pπ-pπ multiple bonds with itself and other elements having small size and high electronegativity like C, N, which is not possible for sulphur.
In the light of the above statements, choose the **most appropriate** answer from the options given below :
(a) Both **(A)** and **(R)** are correct and **(R)** is the correct explanation of **(A).**
(b) Both **(A)** and **(R)** are correct but **(R)** is not the correct explanation of **(A).**
(c) **(A)** is correct but **(R)** is not correct.
(d) **(A)** is not correct but **(R)** is correct.

35. Number of oxygen atoms present in chemical formula of fuming sulphuric acid is ______.
[NCERT : P L-194 | NV, April 9, 2024 (II)]

36. Given below are two statements: One is labelled as Assertion A and the other is labelled as Reason R:
[NCERT : P L-188 | Jan. 30, 2024 (II)]
Assertion A : H_2Te is more acidic than H_2S.
Reason R : Bond dissociation enthalpy of H_2Te is lower than H_2S.

In the light of the above statements, choose the most appropriate from the options given below:
(a) Both A and R are true but R is NOT the correct explanation of A.
(b) Both A and R are true and R is the correct explanation of A.
(c) A is false but R is true.
(d) A is true but R is false.

37. Anomalous behavior of oxygen is due to its **[NCERT : P L-187 | Jan. 29, 2024 (II)]**
(a) small size and high electronegativity
(b) small size and low electronegativity
(c) large size and high electronegativity
(d) large size and low electronegativity

38. Choose the correct statements from the following **[NCERT : P L-187 | Jan. 31, 2024 (II)]**
A. All group 16 elements form oxides of general formula EO_2 and EO_3 where E = S, Se, Te and Po. Both the types of oxides are acidic in nature.
B. TeO_2 is an oxidising agent while SO_2 is reducing in nature.
C. The reducing property decreases from H_2S to H_2Te down the group.
D. The ozone molecule contains five lone pairs of electrons.
Choose the correct answer from the options given below:
(a) A and D only (b) B and C only
(c) C and D only (d) A and B only

39. Given below are two statements: **[NCERT : P L-186 | Jan. 27, 2024 (II)]**
Statement (I) : Oxygen being the first member of group 16 exhibits only –2 oxidation state.
Statement (II) : Down the group 16 stability of +4 oxidation state decreases and +6 oxidation state increases.
In the light of the above statements, choose the most appropriate answer from the options given below:
(a) Statement I is correct but Statement II is incorrect
(b) Both Statement I and Statement II are correct
(c) Both Statement I and Statement II are incorrect
(d) Statement I is incorrect but Statement II is correct

40. Bond dissociation energy of E–H bond of the "H_2E" hydrides of group 16 elements (given below), follows order. **[NCERT : P L-188 | Jan. 30, 2023 (II)]**
(A) O (B) S (C) Se (D) Te
(a) A > B > C > D (b) A > B > D > C
(c) B > A > C > D (d) D > C > B > A

41. The correct order of melting points of hydrides of group 16 elements is : **[NCERT : P L-188 | June 26, 2022 (I)]**
(a) $H_2S < H_2Se < H_2Te < H_2O$
(b) $H_2O < H_2S < H_2Se < H_2Te$
(c) $H_2S < H_2Te < H_2Se < H_2O$
(d) $H_2Se < H_2S < H_2Te < H_2O$

42. Among the following allotropic forms of sulphur, the number of allotropic forms, which will show paramagnetism is ________. **[NCERT : P L-192 | NV, Feb. 24, 2021 (II)]**
(A) α-sulphur
(B) β-sulphur
(C) S_2-form

43. Reaction of an inorganic sulphite X with dilute H_2SO_4 generates compound Y. Reaction of Y with NaOH gives X. Further, the reaction of X with Y and water affords compound Z. Y and Z, respectively, are: **[Sep. 06, 2020 (II)]**
(a) SO_2 and Na_2SO_3 (b) SO_3 and $NaHSO_3$
(c) SO_2 and $NaHSO_3$ (d) S and Na_2SO_3

44. If the boiling point of H_2O is 373 K, the boiling point of H_2S will be : **[NCERT : P L-188 | Sep. 03, 2020 (I)]**
(a) less than 300 K
(b) equal to 373 K
(c) more than 373 K
(d) greater than 300 K but less than 373 K

45. Identify the incorrect statement: **[NCERT : P L-193 | Online April 10, 2016]**
(a) The S–S–S bond angles in the S_8 and S_6 rings are the same.
(b) Rhombic and monoclinic sulphur have S_8 molecules.
(c) S_2 is paramagnetic like oxygen
(d) S_8 ring has a crown shape.

Topic 3 – Halogen Family

46. The property/properties that show irregularity in first four elements of group-17 is/are : **[NCERT : P L-198 | April 2, 2025 (I)]**
(A) Covalent radius (B) Electron affinity
(C) Ionic radius (D) First ionization energy
Choose the **correct** answer from the options given below:
(a) B and D only (b) A and C only
(c) B only (d) A, B, C and D

47. The F^- ions make the enamel on teeth much harder by converting hydroxyapatite (the enamel on the surface of teeth) into much harder fluoroapatite having the formula. **[NCERT : P L-181 | April 9, 2024 (I)]**
(a) $[3(Ca_3(PO_4)_2).CaF_2]$ (b) $[3(Ca_2(PO_4)_2).Ca(OH)_2]$
(c) $[3(Ca_3(PO_4)_3).CaF_2]$ (d) $[3(Ca_3(PO_4)_2).Ca(OH)_2]$

48. Number of elements from the following that CANNOT form compounds with valencies which match with their respective group valencies is ______.
B, C, N, S, O, F, P, Al, Si **[NCERT : P L-172 | April 4, 2024 (I)]**
(a) 7 (b) 5 (c) 6 (d) 3

49. The correct order of first ionization enthalpy values of the following elements is : **[NCERT : P L-171 | April 4, 2024 (I)]**
(A) O (B) N (C) Be
(D) F (E) B
Choose the correct answer from the options given below:
(a) B < D < C < E < A (b) E < C < A < B < D
(c) C < E < A < B < D (d) A < B < D < C < E

50. Among the following oxides of p - block elements, number of oxides having amphoteric nature is ______.
Cl_2O_7, CO, PbO_2, N_2O, NO, Al_2O_3, SiO_2, N_2O_5, SnO_2 **[NV, Feb. 1, 2024 (I)]**

51. Identify the incorrect pair from the following : **[Jan. 29, 2024 (I)]**
(a) Carnallite - $KCl . MgCl_2 . 6H_2O$
(b) Cryolite - Na_3AlF_6
(c) Fluoroapatite - $3 Ca_3(PO_4)_2.CaF_2$
(d) Fluorspar - BF_3

52. Given below are two statements **[NCERT : P L-199 | Jan. 29, 2024 (II)]**
Statement I : Fluorine has most negative electron gain enthalpy in its group.
Statement II : Oxygen has least negative electron gain enthalpy in its group.
In the light of the above statements, choose the **most appropriate** from the options given below:
(a) Statement I is false but Statement II is true
(b) Both Statement I and Statement II are true
(c) Both Statement I and Statement II are false
(d) Statement I is true but Statement II is false

53. Element not showing variable oxidation state is : **[NCERT : P L-202 | Jan. 27, 2024 (I)]**
(a) Bromine (b) Iodine
(c) Chlorine (d) Fluorine

54. Which of the following compounds is an example of Freon? **[NCERT : P L-200 | April 11, 2023 (II)]**
(a) $C_2Cl_2F_2$ (b) C_2HF_3 (c) $C_2H_2F_2$ (d) C_2F_4

55. The sum of lone pairs present on the central atom of the interhalogen IF_5 and IF_7 is ______ **[NCERT : P L-206 | NV, April 10, 2023 (I)]**

56. For electron gain enthalpies of the elements denoted as $\Delta_{eg}H$, the incorrect option is : **[Feb. 01, 2023 (II)]**
(a) $\Delta_{eg}H(Cl) < \Delta_{eg}H(F)$ (b) $\Delta_{eg}H(Se) < \Delta_{eg}H(S)$
(c) $\Delta_{eg}H(Te) < \Delta_{eg}H(Po)$ (d) $\Delta_{eg}H(I) < \Delta_{eg}H(At)$

57. The bond dissociation energy is highest for **[NCERT : P L-198 | Jan. 29, 2023 (I)]**
(a) Cl_2 (b) I_2 (c) Br_2 (d) F_2

58. The number of interhalogens from the following having square pyramidal structure is : **[NCERT : P L-206 | NV, July 28, 2022 (I)]**
$ClF_3, IF_7, BrF_5, BrF_3, I_2Cl_6, IF_5, ClF, ClF_5$

59. Amongst the following the number of oxide(s) which are paramagnetic in nature is
$Na_2O, KO_2, NO_2, N_2O, ClO_2, NO, SO_2, Cl_2O$ **[NV, July 27, 2022 (I)]**

60. Given below are two statements: one is labelled as **Assertion A** and the other is labelled as **Reason R**
Assertion A : Flourine forms one oxoacid.
Reason R : Flourine has smallest size amongst all halogens and is highly electronegative.
In the light of the above statements, choose the **most appropriate** answer from the options given below.0 **[NCERT : P L-200 | June 27, 2022 (II)]**
(a) Both A and R are correct and R is the correct explanation of A.
(b) Both A and R are correct but R is NOT the correct explanation of A.
(c) A is correct but R is not correct.
(d) A is not correct but R is correct.

61. The correct order of bond dissociation enthalpy of halogens is: **[NCERT : P L-198 | Feb. 25, 2021 (II)]**
(a) $Cl_2 > Br_2 > F_2 > I_2$ (b) $F_2 > Cl_2 > Br_2 > I_2$
(c) $Cl_2 > F_2 > Br_2 > I_2$ (d) $I_2 > Br_2 > Cl_2 > F_2$

62. Arrange the following bonds according to their average bond energies in descending order: C – Cl, C – Br, C – F, C – I **[NCERT : P L-198 | Jan. 08, 2020 (II)]**
(a) C – F > C – Cl > C – Br > C – I
(b) C – Br > C – I > C – Cl > C – F
(c) C – I > C – Br > C – Cl > C – F
(d) C – Cl > C – Br > C – I > C – F

63. HF has highest boiling point among hydrogen halides, because it has: **[NCERT : P L-199 | April 9, 2019 (II)]**
(a) strongest van der Waals' interactions
(b) lowest ionic character
(c) strongest hydrogen bonding
(d) lowest dissociation enthalpy

64. The non–metal that does not exhibit positive oxidation state is : **[NCERT : P L-200 | Online April 9, 2016]**
(a) Chlorine (b) Iodine
(c) Fluorine (d) Oxygen

65. Aqueous solution of which salt will not contain ions with the electronic configuration $1s^22s^22p^63s^23p^6$? **[NCERT : P L-199 | Online April 10, 2016]**
(a) NaF (b) KBr (c) NaCl (d) CaI_2

66. Which among the following is the most reactive ? **[NCERT : P L-207 | 2015]**
(a) I_2 (b) ICl (c) Cl_2 (d) Br_2

67. Shapes of certain interhalogen compounds are stated below. Which one of them is not correctly stated? **[NCERT : P L-206 | Online April 11, 2014]**
(a) IF_7 : pentagonal bipyramid
(b) BrF_5 : trigonal bipyramid
(c) BrF_3 : planar T-shaped
(d) ICl_3 : planar dimeric

68. Electron gain enthalpy with negative sign of fluorine is less than that of chlorine due to : **[NCERT : P L-198 | Online April 9, 2013]**
(a) High ionization enthalpy of fluorine
(b) Smaller size of chlorine atom
(c) Smaller size of fluorine atom
(d) Bigger size of 2*p* orbital of fluorine

69. The correct order of the thermal stability of hydrogen halides (H–X) is **[NCERT : P L-201 | 2005]**
(a) HI > HCl < HF > HBr (b) HCl < HF > HBr < HI
(c) HF > HCl > HBr > HI (d) HI < HBr > HCl < HF

70. Which among the following factors is the most important in making fluorine, the strongest oxidizing halogen? **[2004]**
(a) Hydration enthalpy (b) Ionization enthalpy
(c) Electron affinity (d) Bond dissociation energy

Topic 4 - Noble Gases

71. Given below are two statements **[NCERT : P L-209 | Jan. 31, 2024 (I)]**
Statement I : Noble gases have very high boiling points.
Statement II : Noble gases are monoatomic gases. They are held together by strong dispersion forces. Because of this they are liquefied at very low temperature. Hence, they have very high boiling points.
In the light of the above statements, choose the **correct answer** from the options given below:
(a) Statement I is false but Statement II is true
(b) Both Statement I and Statement II are true
(c) Statement I is true but Statement II is false
(d) Both Statement I and Statement II are false

72. The number of species from the following carrying a single lone pair on central atom Xenon is ____:

XeF_5^+, XeO_3, XeO_2F_2, XeF_5^-, XeO_3F_2, $XeOF_4$, XeF_4

[NCERT : P L-210 | NV, April 08, 2023 (II)]

73. Inert gases have positive electron gain enthalpy. Its correct order is **[NCERT : P L-208 | Jan. 25, 2023 (I)]**

(a) Xe < Kr < Ne < He (b) He < Ne < Kr < Xe

(c) He < Xe < Kr < Ne (d) He < Kr < Xe < Ne

74. Match List–I with List–II :

[NCERT : P L-210 | Aug. 27, 2021 (I)]

List–I (Species)	**List–II (Number of lone pairs of electrons on the central atom)**
(A) XeF_2	(i) 0
(B) XeO_2F_2	(ii) 1
(C) XeO_3F_2	(iii) 2
(D) XeF_4	(iv) 3

Choose the **most appropriate** answer from the options given below :

(a) (A)–(iv), (B)–(i), (C)–(ii), (D)–(iii)

(b) (A)–(iii), (B)–(iv), (C)–(ii), (D)–(i)

(c) (A)–(iii), (B)–(ii), (C)–(iv), (D)–(i)

(d) (A)–(iv), (B)–(ii), (C)–(i), (D)–(iii)

75. The sum of number of lone pairs of electrons present on the central atoms of

[NCERT : P L-210 | NV, July 25, 2022 (II)]

XeO_3, $XeOF_4$ and XeF_6 is ______.

76. Match List-I with List-II:

[NCERT : P L-210 | NV, July 25, 2022 (II)]

List-I (Molecule)	**List-II (hybridization; shape)**
(A) XeO_3	(I) sp^3d; linear
(B) XeF_2	(II) sp^3; pyramidal
(C) $XeOF_4$	(III) sp^3d^3;distorted octahedral
(D) XeF_6	(IV) sp^3d^2; sqaure pyramidal

Choose the correct answer from the options given below:

(a) A-II, B-I, C-IV, D-III (b) A-II, B-IV, C-III, D-I

(c) A-IV, B-II, C-III, D-I (d) A-IV, B-II, C-I, D-III

77. The noble gas that does NOT occur in the atmosphere is :

[NCERT : P L-208 | April 10, 2019 (II)]

(a) He (b) Kr (c) Ne (d) Ra

78. In XeO_3F_2, the number of bond pair(s), π-bond(s) and lone pair(s) on Xe atom respectively are:

[NCERT : P L-210 | Online April 15, 2018 (II)]

(a) 5, 3, 0 (b) 5, 2, 0

(c) 4, 2, 2 (d) 4, 4, 0

79. Which intermolecular force is most responsible in allowing xenon gas to liquefy?

[NCERT : P L-209 | Online April 9, 2016]

(a) Instantaneous dipole–induced dipole

(b) Ion–dipole

(c) Ionic

(d) Dipole – dipole

80. The geometry of $XeOF_4$ by VSEPR theory is :

[NCERT : P L-210 | Online April 10, 2015]

(a) pentagonal planar (b) octahedral

(c) square pyramidal (d) trigonal bipyramidal

81. Which one has the highest boiling point ?

[NCERT : P L-208 | 2015]

(a) Kr (b) Xe (c) He (d) Ne

82. Trigonal bipyramidal geometry is shown by :

[NCERT : P L-210 | Online April 9, 2013]

(a) XeO_2F_2 (b) XeO_3F_2

(c) $FXeOSO_2F$ (d) $[XeF_8]^{2-}$

83. Which has trigonal bipyramidal shape ?

[NCERT : P L-210 | Online April 23, 2013]

(a) $XeOF_4$ (b) XeO_3 (c) XeO_3F_2 (d) $XeOF_2$

84. The compound of xenon with zero dipole moment is

[NCERT : P L-210 | Online May 19, 2012]

(a) XeO_3 (b) XeF_4 (c) $XeOF_4$ (d) XeO_2

85. Which of the following has maximum number of lone pairs associated with Xe ? **[NCERT : P L-210 | 2011RS]**

(a) XeF_4 (b) XeF_6 (c) XeF_2 (d) XeO_3

86. In XeF_2, XeF_4, XeF_6 the number of lone pairs on Xe are respectively **[NCERT : P L-210 | 2002]**

(a) 2, 3, 1 (b) 1, 2, 3 (c) 4, 1, 2 (d) 3, 2, 1.

87. Which one of the following statement regarding helium is **incorrect** ? **[NCERT : P L-208 | 2004]**

(a) It is used to produce and sustain powerful superconducting magnets

(b) It is used as a cryogenic agent for carrying out experiments at low temperatures

(c) It is used to fill gas balloons instead of hydrogen because it is lighter and non-inflammable

(d) It is used in gas-cooled nuclear reactors

1	(d)	10	(c)	19	(d)	28	(c)	37	(a)	46	(c)	55	(1)	64	(c)	73	(c)	82	(b)
2	(b)	11	(a)	20	(b)	29	(a)	38	(d)	47	(a)	56	(b)	65	(a)	74	(d)	83	(b)
3	(a)	12	(c)	21	(c)	30	(b)	39	(c)	48	(d)	57	(a)	66	(b)	75	(3)	84	(b)
4	(15)	13	(d)	22	(d)	31	(a)	40	(a)	49	(b)	58	(3)	67	(b)	76	(a)	85	(c)
5	(d)	14	(d)	23	(d)	32	(d)	41	(a)	50	(3)	59	(4)	68	(c)	77	(d)	86	(d)
6	(d)	15	(b)	24	(b)	33	(b)	42	(1)	51	(d)	60	(a)	69	(c)	78	(a)	87	(c)
7	(d)	16	(d)	25	(d)	34	(c)	43	(c)	52	(a)	61	(a)	70	(a)	79	(a)		
8	(d)	17	(b)	26	(c)	35	(7)	44	(a)	53	(d)	62	(a)	71	(d)	80	(c)		
9	(c)	18	(a)	27	(a)	36	(b)	45	(a)	54	(a)	63	(c)	72	(4)	81	(b)		

The *d*-and *f*-Block Elements

TREND ANALYSIS
JEE MAIN (2025-2021)

	2025				2024				2023				2022				2021			
	Session II		Session I		Session II		Session I		Session II		Session I		Session II		Session I		Session II		Session I	
	MCQs	NVQs	MCQs	NVQs	MCQs	NVQs	MCQs	NVQs	MCQs	NVQs	MCQs	NVQs	MCQs	NVQs	MCQs	NVQs	MCQs	NVQs	MCQs	NVQs
Topic 1	5	2	5	1	8	6	8	–	2	–	2	–	4	3	3	1	2	1	4	1
Topic 2	–	2	4	2	2	4	9	1	4	3	6	1	4	3	1	6	1	–	1	3
Topic 3	–	–	2	–	2	1	4	1	2	–	3	–	1	–	4	–	2	2	–	–
Topic 4																				
Total	5	4	11	3	12	11	21	2	8	3	11	1	9	6	8	7	5	3	5	4

How to Code Qns for Better & Faster Revision

Type 1: Able to solve (No code Req)

Type 2: Solved with struggle or by seeing solution **(Use code NR means Needs Revision or orange stickers)**

Type 3: Tricky/ Could not solve on my own. Need help from Teacher/ Friend. **(Use code TQ means Tricky Question or red stickers)**

You can either hand write the code in boxes or else paste coloured stickers given at end of book.

Topic 1 – Characteristics of *d*-Block Elements

1. The first transition series metal 'M' has the highest enthalpy of atomisation in its series. One of its aquated ion (M^{n+}) exists in green colour. The nature of the oxide formed by the above M^{n-} ion is : **[NCERT : P L-106 | April 7, 2025 (I)]**

(a) neutral (b) acidic
(c) basic (d) amphoteric

2. The number of valence electrons present in the metal among Cr, Co, Fe and Ni which has the lowest enthalpy of atomisation is **[NCERT : P L-92 | April 7, 2025 (I)]**

(a) 8 (b) 9
(c) 6 (d) 10

3. The incorrect relationship in the following pairs in relation to ionisation enthalpies is : **[NCERT : P L-95 | April 4, 2025 (II)]**

(a) $Mn^+ < Cr^+$ (b) $Mn^+ < Mn^{2+}$
(c) $Fe^{2+} < Fe^{3+}$ (d) $Mn^{2+} < Fe^{2+}$

4. Pair of transition metal ions having the same number of unpaired electrons is : **[NCERT : P L-91 | April 4, 2025 (I)]**

(a) V^{2+}, Co^{2+} (b) Ti^{2+}, Co^{2+}
(c) Fe^{3+}, Cr^{2+} (d) Ti^{3+}, Mn^{2+}

5. Among, Sc, Mn, Co and Cu, identify the element with highest enthalpy of atomisation. The spin only magnetic moment value of that element in its +2 oxidation state is _______ BM (in nearest integer). **[NCERT : P L-92 | NV, April 3, 2025 (II)]**

6. Given below are two statements : **[NCERT : P L-96 | April 3, 2025 (II)]**

Statement I : CrO_3 is a stronger oxidizing agent than MoO_3

Statement II : Cr(VI) is more stable than Mo(VI)

In the light of the above statements, choose the **correct** answer from the options given below

(a) Statement I is false but Statement II is true
(b) Statement I is true but Statement II is false
(c) Both Statement I and Statement II are true
(d) Both Statement I and Statement II are false

7. The spin-only magnetic moment value of M^{n+} ion formed among Ni, Zn Mn and Cu that has the least enthalpy of atomisation is ______. (in nearest integer)
Here n is equal to the number of diamagnetic complexes among $K_2[NiCl_4]$, $[Zn(H_2O)_6]Cl_2$, $K_3[Mn(CN)_6]$ and $[Cu(PPh_3)_3I]$ **[NCERT : PL-92, 102 | NV, April 2, 2025 (II)]**

8. The correct option with order of melting points of the pairs (Mn, Fe), (Tc, Ru) and (Re, Os) is :
[NCERT : P L-92 | Jan. 29, 2025 (I)]
(a) Fe < Mn, Ru < Tc and Re < Os
(b) Mn < Fe, Tc < Ru and Re < Os
(c) Mn < Fe, Tc < Ru and Os < Re
(d) Fe < Mn, Ru < Tc and Os < Re

9. Consider '*n*' is the number of lone pair of electrons present in the equatorial position of the most stable structure of ClF_3. The ions from the following with '*n*' number of unpaired electrons are : **[NCERT : P L-92 | Jan. 28, 2025 (I)]**
(A) V^{3+} (B) Ti^{3+}
(C) Cu^{2+} (D) Ni^{2+} (E) Ti^{2+}
Choose the correct answer from the options given below :
(a) A and C only (b) A, D and E only
(c) B and C only (d) B and D only

10. Match List-I with List-II.
[NCERT : P L-102 | Jan. 24, 2025 (II)]

List-I (Transition metal ion)	List-II (Spin only magnetic moment (B.M.))
(A) Ti^{3+}	(I) 3.87
(B) V^{2+}	(II) 0.00
(C) Ni^{2+}	(III) 1.73
(D) Sc^{3+}	(IV) 2.84

Choose the correct answer from the options given below :
(a) (A)-(III), (B)-(I), (C)-(II), (D)-(IV)
(b) (A)-(III), (B)-(I), (C)-(IV), (D)-(II)
(c) (A)-(IV), (B)-(II), (C)-(III), (D)-(I)
(d) (A)-(II), (B)-(IV), (C)-(I), (D)-(III)

11. Match List-I with List-II.
[NCERT : P L-104 | Jan. 23, 2025 (II)]

List-I	List-II
(A) Bronze	(I) Cu, Ni
(B) Brass	(II) Fe, Cr, Ni, C
(C) UK silver coin	(III) Cu, Zn
(D) Stainless Steel	(IV) Cu, Sn

Choose the correct answer from the options given below:
(a) (A)-(IV), (B)-(II), (C)-(III), (D)-(I)
(b) (A)-(IV), (B)-(III), (C)-(I), (D)-(II)
(c) (A)-(III), (B)-(I), (C)-(IV), (D)-(II)
(d) (A)-(III), (B)-(IV), (C)-(II), (D)-(I)

12. The correct set of ions (aqueous solution) with same colour from the following is :
[NCERT : P L-103 | Jan. 23, 2025 (I)]
(a) V^{2+}, Cr^{3+}, Mn^{3+} (b) Zn^{2+}, V^{3+}, Fe^{3+}
(c) Ti^{4+}, V^{4+}, Mn^{2+} (d) $Sc^{3+}, Ti^{3+}, Cr^{2+}$

13. Niobium (Nb) and ruthenium (Ru) have "x" and "y" number of electrons in their respective 4d orbitals. The value of x + y is ____. **[NCERT : P L-91 | NV, Jan. 22, 2025 (II)]**

14. The electronic configuration of Cu(II) is $3d^9$ whereas that of Cu(I) is $3d^{10}$. Which of the following is correct ?
[NCERT : PL-90 | April 9, 2024 (I)]
(a) Cu(II) is less stable
(b) Stability of Cu(I) and Cu(II) depends on nature of copper salts
(c) Cu(II) is more stable
(d) Cu(I) and Cu(II) are equally stable

15. Given below are two statements :
[NCERT : PL-97 | April 9, 2024 (II)]
Statement I : The higher oxidation states are more stable down the group among transition elements unlike p-block elements.
Statement II : Copper can not liberate hydrogen from weak acids.
In the light of the above statements, choose the correct answer from the options given below :
(a) Both Statement I and Statement II are false
(b) Statement I is false but Statement II is true
(c) Both Statement I and Statement II are true
(d) Statement I is true but Statement II is false

16. A transition metal 'M' among Sc, Ti, V, Cr, Mn and Fe has the highest second ionisation enthalpy. The spin only magnetic moment value of M^+ ion is ______ BM (Near integer) **[NCERT : PL-102 | NV, April 9, 2024 (II)]**
(Given atomic number Sc : 21, Ti : 22, V : 23, Cr : 24, Mn : 25, Fe : 26)

17. Iron (III) catalyses the reaction between iodide and persulphate ions, in which
[NCERT : P L-104 | April 8, 2024 (I)]
A. Fe^{3+} oxidises the iodide ion
B. Fe^{3+} oxidises the persulphate ion
C. Fe^{2+} reduces the iodide ion
D. Fe^{2+} reduces the persulphate ion
Choose the **most appropriate** answer from the options given below:
(a) B and C only (b) B only
(c) A only (d) A and D only

18. The 'spin only' magnetic moment value of MO_4^{2-} is ____ BM. (Where M is a metal having least metallic radii. among Sc, Ti, V, Cr, Mn and Zn).
[NCERT : PL-102 | NV, April 8, 2024 (I)]
(Given atomic number : Sc = 21, Ti = 22, V = 23, Cr = 24, Mn = 25 and Zn = 30)

19. Arrange the following elements in the increasing order of number of unpaired electrons in it.
[NCERT : PL-90 | April 6, 2024 (II)]
(A) Sc (B) Cr (C) V
(D) Ti (E) Mn
Choose the correct answer from the options given below:
(a) (C) < (E) < (B) < (A) < (D)
(b) (B) < (C) < (D) < (E) < (A)
(c) (A) < (D) < (C) < (B) < (E)
(d) (A) < (D) < (C) < (E) < (B)

20. Among VO_2^+, MnO_4^- and $Cr_2O_7^{2-}$, the spin-only magnetic moment value of the species with least oxidising ability is .... BM (Nearest integer).

[NCERT : PL-100 | NV, April 6, 2024 (II)]

(Given atomic member V = 23, Mn = 25, Cr = 24)

21. The spin only magnetic moment value of the ion among Ti^{2+}, V^{2+}, Co^{3+} and Cr^{2+}, that acts as strong oxidising agent in aqueous solution is .... BM (Near integer).

(Given atomic numbers : Ti : 22, V : 23, Cr : 24, Co : 27)

[NCERT : PL-102 | NV, April 5, 2024 (I)]

22. The metal that shows highest and maximum number of oxidation state is: **[NCERT : PL-96 | April 5, 2024 (I)]**

(a) Fe (b) Mn
(c) Ti (d) Co

23.

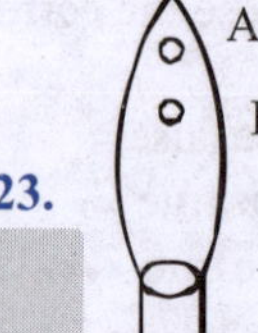

In a borax bead test under hot condition, a metal salt (one from the given) is heated at point B of the flame, resulted in green colour salt bead. The spin-only magnetic moment value of the salt is .... BM (Nearest integer)

[Given atomic number of Cu = 29, Ni = 28, Mn = 25, Fe = 26] **[NCERT : PL-102 | NV, April 5, 2024 (I)]**

24. The number of ions from the following that have the ability to liberate hydrogen from a dilute acid is ______.

Ti^{2+}, Cr^{2+} and V^{2+} **[NCERT : PL-98 | April 5, 2024 (II)]**

(a) 0 (b) 2 (c) 3 (d) 1

25. The element which shows only one oxidation state other than its elemental form is : ______

[NCERT : PL-97 | April 4, 2024 (I)]

(a) Cobalt (b) Scandium
(c) Titanium (d) Nickel

26. A first row transition metal in its +2 oxidation state has a spin-only magnetic moment value of 3.86 BM. The atomic number of the metal is

[NCERT : PL-102 | April 4, 2024 (II)]

(a) 25 (b) 26 (c) 22 (d) 23

27. A first row transition metal with highest enthalpy of atomisation, upon reaction with oxygen at high temperature forms oxides of formula M_2O_n (where n = 3,4,5). The 'spin-only' magnetic moment value of the amphoteric oxide from the above oxides is ___ BM (near integer)

[NCERT : PL-100 | NV, April 4, 2024 (II)]

(Given atomic number : Sc : 21, Ti : 22, V : 23, Cr : 24, Mn : 25, Fe : 26, Co : 27, Ni : 28 ,Cu : 29, Zn : 30)

28. The transition metal having highest 3rd ionisation enthalpy is: **[NCERT : PL-95 | Feb. 1, 2024 (II)]**

(a) Mn (b) Fe (c) Cr (d) V

29. Given below are two statements : one is labelled as Assertion (A) and the other is labelled as Reason (R).

Assertion (A) : In aqueous solutions Cr^{2+} is reducing while Mn^{3+} is oxidising in nature.

Reason (R) : Extra stability to half filled electronic configuration is observed than incompletely filled electronic configuration.

In the light of the above statements, choose the most appropriate answer from the options given below:

[NCERT : PL-91 | Feb. 1, 2024 (II)]

(a) (A) is true but (R) is false
(b) Both (A) and (R) are true and (R) is the correct explanation of (A)
(c) Both (A) and (R) are true but (R) is not the correct explanation of (A)
(d) (A) is false but (R) is true

30. Match **List - I** with **List- II**

[NCERT : PL-90 | Jan. 30, 2024 (I)]

List - I Species	List - II Electronic distribution
(A) Cr^{+2}	(I) $3d^8$
(B) Mn^+	(II) $3d^3 4s^1$
(C) Ni^{+2}	(III) $3d^4$
(D) V^+	(IV) $3d^5 4s^1$

Choose the correct answer from the options given below:

(a) (A) - (I), (B) - (II), (C) - (III), (D)- (IV)
(b) (A) - (III), (B) - (IV), (C) - (I), (D) - (II)
(c) (A) - (IV), (B) - (III), (C) - (I), (D) - (II)
(d) (A) - (II), (B) - (I), (C) - (IV), (D) - (III)

31. Match List I with List II

[NCERT : PL-114 | Jan. 29, 2024 (I)]

LIST I (Substances)	LIST II (Element Present)
A. Ziegler catalyst	I. Rhodium
B. Blood Pigment	II. Cobalt
C. Wilkinson catalyst	III. Iron
D. Vitamin B_{12}	IV. Titanium

Choose the correct answer from the options given below:

(a) A-II, B-IV, C-I, D-III (b) A-IV, B-III, C-I, D-II
(c) A-II, B-III, C-IV, D-I (d) A-III, B-II, C-IV, D-I

32. Which of the following statements are correct about Zn, Cd and Hg ? **[NCERT : PL-92 | Jan. 29, 2024 (II)]**

A. They exhibit high enthalpy of atomization as the d-subshell is full.
B. Zn and Cd do not show variable oxidation state while Hg shows +I and +II.
C. Compounds of Zn, Cd and Hg are paramagnetic in nature.
D. Zn, Cd and Hg are called soft metals.

Choose the **most appropriate** from the options given below:

(a) B, C only (b) A, D only
(c) C, D only (d) B, D only

33. Which of the following electronic configuration would be associated with the highest magnetic moment ?
[NCERT : P L-102 | Jan. 27, 2024 (I)]
(a) $[Ar]\,3d^7$ (b) $[Ar]\,3d^8$
(c) $[Ar]\,3d^3$ (d) $[Ar]\,3d^6$

34. Choose the correct option having all the elements with d^{10} electronic configuration from the following:
[NCERT : P L-90 | Jan. 27, 2024 (II)]
(a) ^{27}Co, ^{28}Ni, ^{26}Fe, ^{24}Cr
(b) ^{29}Cu, ^{30}Zn, ^{48}Cd, ^{47}Ag
(c) ^{46}Pd, ^{28}Ni, ^{26}Fe, ^{24}Cr
(d) ^{28}Ni, ^{24}Cr, ^{26}Fe, ^{29}Cu

35. Identity the incorrect pair from the following:
[NCERT : P L-114 | Jan. 27, 2024 (II)]
(a) Photography - AgBr
(b) Polythene preparation – $TiCl_4$, $Al(CH_3)_3$
(c) Haber process - Iron
(d) Wacker process – Pt Cl_2

36. Which of the following statements are correct?
[NCERT : P L-99 | April 10, 2023 (I)]
(A) The M^{3+}/M^{2+} reduction potential for iron is greater than manganese
(B) The higher oxidation states of first row d-block elements get stabilized by oxide ion.
(C) Aqueous solution of Cr^{2+} can liberate hydrogen from dilute acid.
(D) Magnetic moment of V^{2+} is observed between 4.4 – 5.2 BM
Choose the correct answer from the options given below:
(a) (B), (C) only (b) (C), (D) only
(c) (A), (B), (D) only (d) (A), (B) only

37. In chromyl chloride, the number of d-electrons present on chromium is same as in (Given at no. of Ti : 22, V : 23, Cr : 24, Mn : 25, Fe : 26) **[NCERT : P L-96 |April 08, 2023 (I)]**
(a) Ti (III) (b) Fe (III)
(c) V (IV) (d) Mn (VII)

38. Given below are two statements: one is labelled as **Assertion (A)** and the other is labelled as **Reason (R).**
Assertion (A): Cu^{2+} in water is more stable than Cu^+.
Reason (R): Enthalpy of hydration for Cu^{2+} is much less than that of Cu^+.
In the light of the above statements, choose the **correct** answer from the options given below
[NCERT : P L-101 |Feb. 01, 2023 (II)]
(a) Both **(A)** and **(R)** are correct but **(R)** is not the correct explanation of **(A)**
(b) **(A)** is not correct but **(R)** is correct
(c) Both **(A)** and **(R)** are correct and **(R)** is the correct explanation of **(A)**
(d) **(A)** is correct but **(R)** is not correct

39. The set of correct statements is:
[NCERT : P L-96 | Jan. 29, 2023 (II)]
(i) Manganese exhibits +7 oxidation state in its oxide.
(ii) Ruthenium and Osmium exhibit +8 oxidation in their oxides.
(iii) Sc shows +4 oxidation state which is oxidizing in nature.
(iv) Cr shows oxidising nature in +6 oxidation state.
(a) (ii) and (iii) (b) (i), (ii) and (iv)
(c) (i) and (iii) (d) (ii), (iii) and (iv)

40. In following pairs, the one in which both transition metal ions are colourless is : **[NCERT : P L-103 |July 29, 2022 (I)]**
(a) Sc^{3+}, Zn^{2+} (b) Ti^{4+}, Cu^{2+}
(c) V^{2+}, Ti^{2+} (d) Zn^{2+}, Mn^{2+}

41. Which of the following 3d-metal ion will give the lowest enthalpy of hydration ($\Delta_{hyd}H$) when dissolved in water ?
[NCERT : P L-92 | July 29, 2022 (II)]
(a) Cr^{2+} (b) Mn^{2+} (c) Fe^{2+} (d) Co^{2+}

42. Which of the following has least tendency to liberate H_2 from mineral acids ? **[NCERT : P L-101 | July 28, 2022 (I)]**
(a) Cu (b) Mn (c) Ni (d) Zn

43. The spin-only magnetic moment value of the compound with strongest oxidizing ability among MnF_4, MnF_3 and MnF_2 is ______ B.M. [nearest integer]
[NCERT : P L-98, 99 | NV, July 26, 2022 (II)]

44. Among Co^{3+}, Ti^{2+}, V^{2+} and Cr^{2+} ions, one if used as a reagent cannot liberate H_2 from dilute mineral acid solution, its spin-only magnetic moment in gaseous state is ______ B.M. (Nearest integer) **[NV, July 25, 2022 (I)]**

45. The spin-only magnetic moment value of M^{3+} ion (in gaseous state) from the pairs Cr^{3+}/Cr^{2+}, Mn^{3+}/Mn^2, Fe^{3+}/Fe^{2+} and Co^{3+}/Co^{2+} that has negative standard electrode potential is B.M. [Nearest integer]
[NCERT : P L-99 | NV, July 25, 2022 (II)]

46. The electronic configuration of Pt (atomic number 78) is:
[NCERT : P L-91 | June 29, 2022 (I)]
(a) $[Xe]\,4f^{14}\,5d^9\,6s^1$ (b) $[Kr]\,4f^{14}\,5d^{10}$
(c) $[Xe]\,4f^{14}\,5d^{10}$ (d) $[Xe]\,4f^{14}\,5d^8\,6s^2$

47. The reaction of zinc with excess of aqueous alkali, evolves hydrogen gas and gives :
[NCERT : P L-101 | July 29, 2022 (I)]
(a) $Zn(OH)_2$ (b) ZnO
(c) $[Zn(OH)_4]^{2-}$ (d) $[ZnO_2]^{2-}$

48. The number of statement(s) correct from the following for copper (at no. 29) is/are ______.
(A) Cu (II) complexes are always paramagnetic
(B) Cu (I) complexes are generally colourless
(C) Cu (I) is easily oxidized
(D) In Fehling solution, the active reagent has Cu (I)
[NCERT : P L-99, 101 | NV, June 27, 2022 (I)]

49. In 3*d* series, the metal having the highest M^{2+}/M standard electrode potential is
[NCERT : P L-98 | June 27, 2022 (II)]
(a) Cr (b) Fe (c) Cu (d) Zn

50. Among the following, which is the strongest oxidizing agent ? **[NCERT : P L-96 | June 25, 2022 (I)]**
(a) Mn^{3+} (b) Fe^{3+} (c) Ti^{3+} (d) Cr^{3+}

51. Identify the element for which electronic configuration in +3 oxidation state is $[Ar]3d^5$:

[NCERT : P L-91 | Sep. 1, 2021 (II)]

(a) Ru (b) Mn (c) Co (d) Fe

52. The set having ions which are coloured and paramagnetic both is - **[NCERT : P L-102, 103 | July 22, 2021 (II)]**

(a) Cu^{2+}, Cr^{3+}, Sc^{+} (b) $Cu^{2+}, Zn^{2+}, Mn^{4+}$
(c) Sc^{3+}, V^{5+}, Ti^{4+} (d) $Ni^{2+}, Mn^{7+}, Hg^{2+}$

53. Number of electrons that Vanadium (Z = 23) has in *p*-orbitals is equal to _____

[NCERT : P L-91 | NV, July 22, 2021 (II)]

54. The common positive oxidation states for an element with atomic number 24, are

[NCERT : P L-96 | March 17, 2021 (II)]

(a) +1 and +3 to +6 (b) +1 to +6
(c) +2 to +6 (d) +1 and +3

55. In the ground state of atomic Fe(Z = 26), the spin-only magnetic moment is _____ $\times 10^{-1}$ BM.
(Round off to the nearest integer).
[Given : $\sqrt{3} = 1.73, \sqrt{2} = 1.41$]

[NCERT : P L-102 | NV, March 17, 2021 (II)]

56. What is the spin-only magnetic moment value (BM) of a divalent metal ion with atomic number 25, in its aqueous solution? **[NCERT : P L-102 | March 17, 2021 (I)]**

(a) 5.92 (b) 5.26 (c) zero (d) 5.0

57. In which of the following pairs, the outer most electronic configuration will be the same?

[NCERT : P L-114 | Feb. 25, 2021 (I)]

(a) Ni^{2+} and Cu^{+} (b) Fe^{2+} and Co^{+}
(c) Cr^{+} and Mn^{2+} (d) V^{2+} and Cr^{+}

58. The major components of German Silver are:

[NCERT : P L-114 | Feb. 25, 2021 (II)]

(a) Cu, Zn and Ni (b) Zn, Ni and Ag
(c) Ge, Cu and Ag (d) Cu, Zn and Ag

59. The third ionization enthalpy is minimum for:

[NCERT : P L-95 | Jan. 08, 2020 (I)]

(a) Co (b) Fe (c) Ni (d) Mn

60. For the following Assertion and Reason, the correct option is:
Assertion: For hydrogenation reactions, the catalytic activity increases from Group 5 to Group 11 metals with maximum activity shown by Group 7-9 elements.
Reason: The reactants are most strongly adsorbed on group 7-9 elements.

[NCERT : P L-104 | Jan. 08, 2020 (II)]

(a) The assertion is true, but the reason is false.
(b) Both assertion and reason are false.
(c) Both assertion and reason are true and the reason is the correct explanation for the assertion.
(d) Both assertion and reason are true but the reason is not the correct explanation for the assertion.

61. The atomic radius of Ag is closest to:

[NCERT : P L-93 | Jan. 07, 2020 (I)]

(a) Au (b) Ni (c) Cu (d) Hg

62. The pair that has similar atomic radii is :

[NCERT : P L-94 | April 12, 2019 (II)]

(a) Mn and Re (b) Ti and Hf
(c) Sc and Ni (d) Mo and W

63. Consider the hydrated ions of Ti^{2+}, V^{2+}, Ti^{3+}, and Sc^{3+}. The correct order of their spin-only magnetic moments is:

[NCERT : P L-102 | April 10, 2019 (I)]

(a) $V^{2+} < Ti^{2+} < Ti^{3+} < Sc^{3+}$
(b) $Sc^{3+} < Ti^{3+} < Ti^{2+} < V^{2+}$
(c) $Ti^{3+} < Ti^{2+} < Sc^{3+} < V^{2+}$
(d) $Sc^{3+} < Ti^{3+} < V^{2+} < Ti^{2+}$

64. The correct order of the first ionization enthalpies is :

[NCERT : P L-95 | April 10, 2019 (II)]

(a) Ti < Mn < Zn < Ni (b) Ti < Mn < Ni < Zn
(c) Mn < Ti < Zn < Ni (d) Zn < Ni < Mn < Ti

65. Match the catalysts **(Column-I)** with products **(Column-II)**.

[NCERT : P L-113 | April 9, 2019 (I)]

Column-I (Catalyst)	Column-II (Product)
(A) V_2O_5	(i) Polyethylene
(B) $TiCl_4/Al(Me)_3$	(ii) ethanol
(C) $PdCl_2$	(iii) H_2SO_4
(D) Iron Oxide	(iv) NH_3

(a) (A)-(iii); (B)-(iv); (C)-(i); (D)-(ii)
(b) (A)-(ii); (B)-(iii); (C)-(i); (D)-(iv)
(c) (A)-(iii); (B)-(i); (C)-(ii); (D)-(iv)
(d) (A)-(iv); (B)-(iii); (C)-(ii); (D)-(i)

66. The statement that is INCORRECT about the interstitial compounds is : **[NCERT : P L-104 | April 8, 2019 (II)]**

(a) they are chemically reactive.
(b) they are very hard.
(c) they have metallic conductivity.
(d) they have high melting points.

67. The element that usually does **NOT** show variable oxidation states is: **[NCERT : P L-97 | Jan. 11, 2019 (I)]**

(a) Cu (b) Ti (c) Sc (d) V

68. The highest value of the calculated spin only magnetic moment (in BM) among all the transition metal complexes is : **[NCERT : P L-101 | Jan. 9, 2019 (I)]**

(a) 5.92 (b) 6.93 (c) 3.87 (d) 4.90

69. The transition element that has lowest enthalpy of atomisation is: **[NCERT : P L-92 | Jan. 9, 2019 (II)]**

(a) Fe (b) Cu (c) V (d) Zn

70. Which of the following ions does **not** liberate hydrogen gas on reaction with dilute acids ?

[NCERT : P L-98 | Online April 9, 2017]

(a) Ti^{2+} (b) V^{2+} (c) Cr^{2+} (d) Mn^{2+}

71. Which one of the following species is stable in aqueous solution ? [NCERT : PL-106 | Online April 9, 2016]
(a) Cr^{2+} (b) MnO_4^{2-} (c) MnO_4^{3-} (d) Cu^+

72. The reaction of zinc with dilute and concentrated nitric acid, respectively, produces: [NCERT : P L-101 | 2016]
(a) NO and N_2O (b) NO_2 and N_2O
(c) N_2O and NO_2 (d) NO_2 and NO

73. Chloro compound of Vanadium has only spin magnetic moment of 1.73 BM. This Vanadium chloride has the formula: [NCERT : PL-91 | Online April 9, 2014]
(a) VCl_2 (b) VCl_4 (c) VCl_3 (d) VCl_5

74. Which one of the following exhibits the large number of oxidation states? [NCERT : P L-96 | Online April 12, 2014]
(a) Ti (22) (b) V (23)
(c) Cr (24) (d) Mn (25)

75. Identify incorrect statement : [NCERT : PL-100 |Online April 23, 2013]
(a) Cu_2O is colourless.
(b) Copper (I) compounds are colourless except when colour results from charge transfer.
(c) Copper (I) compounds are diamagnetic.
(d) Cu_2S is black.

76. Which of the following arrangements does not represent the correct order of the property stated against it ? [NCERT : P L-93-102 | 2013]
(a) $V^{2+} < Cr^{2+} < Mn^{2+} < Fe^{2+}$: paramagnetic behaviour
(b) $Ni^{2+} < Co^{2+} < Fe^{2+} < Mn^{2+}$: ionic size
(c) $Co^{3+} < Fe^{3+} < Cr^{3+} < Sc^{3+}$: stability in aqueous solution
(d) Sc < Ti < Cr < Mn : number of oxidation states

77. Four successive members of the first row transition elements are listed below with atomic numbers. Which one of them is expected to have the highest $E^{\circ}_{M^{3+}/M^{2+}}$ value ? [NCERT : P L-99 | 2013]
(a) Cr (Z = 24) (b) Mn (Z = 25)
(c) Fe (Z = 26) (d) Co (Z = 27)

78. The element with which of the following outer electron configuration may exhibit the largest number of oxidation states in its compounds : [NCERT : P L-90 | Online April 9, 2013]
(a) $3d^5 4s^2$ (b) $3d^8 4s^2$
(c) $3d^7 4s^2$ (d) $3d^6 4s^2$

79. Iron exhibits +2 and + 3 oxidation states. Which of the following statements about iron is incorrect ? [NCERT : P L-96 | 2012]
(a) Ferrous oxide is more basic in nature than the ferric oxide.
(b) Ferrous compounds are relatively more ionic than the corresponding ferric compounds.
(c) Ferrous compounds are less volatile than the corresponding ferric compounds.
(d) Ferrous compounds are more easily hydrolysed than the corresponding ferric compounds.

80. Which pair of elements with the given atomic numbers is expected to have similar properties? [NCERT : P L-101 | Online May 19, 2012]
(a) 40, 72 (b) 20, 36
(c) 10, 28 (d) 11, 12

81. The correct order of $E^{\circ}_{M^{2+}/M}$ values with negative sign for the four successive elements Cr, Mn, Fe and Co is [NCERT : P L-98 | 2010]
(a) Mn > Cr > Fe > Co (b) Cr < Fe > Mn > Co
(c) Fe > Mn > Cr > Co (d) Cr > Mn > Fe > Co

82. Which of the following factors may be regarded as the main cause of lanthanide contraction? [NCERT : P L-93 | 2005]
(a) Greater shielding of 5*d* electrons by 4*f* electrons
(b) Poorer shielding of 5*d* electrons by 4*f* electrons
(c) Effective shielding of one of the 4*f* electrons by another in the subshell
(d) Poor shielding of one of the 4*f* electrons by another in the subshell

83. The lanthanide contraction is responsible for the fact that [NCERT : P L-93 | 2005]
(a) Zr and Zn have the same oxidation state
(b) Zr and Hf have about the same radius
(c) Zr and Nb have similar oxidation state
(d) Zr and Y have about the same radius

84. Of the following outer electronic configurations of atoms, the highest oxidation state is achieved by which one of them ? [NCERT : P L-91 |2004]
(a) $(n-1)d^3 ns^2$ (b) $(n-1)d^5 ns^1$
(c) $(n-1)d^8 ns^2$ (d) $(n-1)d^5 ns^2$

85. Which of the following ions has the maximum magnetic moment? [NCERT : P L-102 | 2002]
(a) Mn^{+2} (b) Fe^{+2}
(c) Ti^{+2} (d) Cr^{+2}.

Topic 2 - Compounds of Transition Metals

86. $KMnO_4$ acts as an oxidising agent in acidic medium. '*X*' is the difference between the oxidation states of Mn in reactant and product. '*Y*' is the number of '*d*' electrons present in the brown red precipitate formed at the end of the acetate ion test with neutral ferric chloride. The value of $X + Y$ is _____. [NCERT : P L-106 | NV, April 4, 2025 (I)]

87. Consider the following reactions

$A + NaCl + H_2SO_4 \rightarrow CrO_2Cl_2 + \text{Side Products}$

Little amount

$CrO_2Cl_{2(Vapour)} + NaOH \rightarrow B + NaCl + H_2O$

$B + H^+ \rightarrow C + H_2O$

The number of terminal 'O' present in the compound 'C' is ______. **[NCERT : P L-106 | NV, April 3, 2025 (I)]**

88. The molar mass of the water insoluble product formed from the fusion of chromite ore ($FeCr_2O_4$) with Na_2CO_3 in presence of O_2 is ______ g mol^{-1}.

[NCERT : P L-106 | NV, Jan. 29, 2025 (I)]

89. The spin only magnetic moment (μ) value (B.M.) of the compound with strongest oxidising power among Mn_2O_3, TiO and VO is ______ B.M. (Nearest integer).

[NCERT : P L-106 | NV, Jan. 28, 2025 (II)]

90. The amphoteric oxide among V_2O_3, V_2O_4 and V_2O_5 upon reaction with alkali leads to formation of an oxide anion. The oxidation state of V in the oxide anion is :

[NCERT : P L-106 | Jan. 28, 2025 (II)]

(a) +3 (b) +7 (c) +5 (d) +4

91. Which of the following oxidation reactions are carried out by both $K_2Cr_2O_7$ and $KMnO_4$ in acidic medium ?

[NCERT : P L-106 | Jan. 28, 2025 (I)]

(A) $I^- \rightarrow I_2$

(B) $S^{2-} \rightarrow S$

(C) $Fe^{2+} \rightarrow Fe^{3+}$

(D) $I^- \rightarrow IO_3^-$

(E) $S_2O_3^{2-} \rightarrow SO_4^{2-}$

Choose the correct answer from the options given below :

(a) B, C and D only (b) A, D and E only

(c) A, B and C only (d) C, D and E only

92. Preparation of potassium permanganate from MnO_2 involves two step process in which the 1st step is a reaction with KOH and KNO_3 to produce

[NCERT : P L-106 | Jan. 24, 2025 (I)]

(a) $K_4[Mn(OH)_6]$ (b) K_3MnO_4

(c) $KMnO_4$ (d) K_2MnO_4

93. Consider the following reactions

[NCERT : P L-106 | Jan. 23, 2025 (II)]

$$K_2Cr_2O_7 \xrightarrow[-H_2O]{KOH} [A] \xrightarrow[-H_2O]{H_2SO_4} [B] + K_2SO_4$$

The products [A] and [B], respectively are:

(a) $K_2Cr(OH)_6$ and Cr_2O_3 (b) K_2CrO_4 and Cr_2O_3

(c) K_2CrO_4 and $K_2Cr_2O_7$ (d) K_2CrO_4 and CrO

94. The equilibrium $Cr_2O_7^{2-} \rightleftharpoons 2CrO_4^{2-}$ is shifted to the right in: **[NCERT : P L-106 | April 8, 2024 (II)]**

(a) an acidic medium

(b) a basic medium

(c) a weakly acidic medium

(d) a neutral medium

95. Given below are two statements:

[NCERT : P L-106 | April 8, 2024 (II)]

Statement (I): Fusion of MnO_2 with KOH and an oxidising agent gives dark green K_2MnO_4.

Statement (II): Manganate ion on electrolytic oxidation in alkaline medium gives permanganate ion.

In the light of the above statements, choose the **correct** answer from the options given below:

(a) Both **Statement I** is false but **Statement II** is true

(b) Both **Statement I** and **Statement II** is false

(c) **Statement I** is true but **Statement II** is false

(d) **Statement I** and **Statement II** is true

96. The difference in the 'spin-only' magnetic moment values of $KMnO_4$ and the manganese product formed during titration of $KMnO_4$ against oxalic acid in acidic medium is ______ BM. (nearest integer)

[NCERT : P L-102 | NV, April 6, 2024 (I)]

97. Among CrO, Cr_2O_3 and CrO_3, the sum of spin-only magnetic moment values of basic and amphoteric oxides is ______ 10^{-2} BM (nearest integer).

(Given atomic number of Cr is 24)

[NCERT : P L-102 | NV, April 6, 2024 (I)]

98. The fusion of chromite ore with sodium carbonate in the presence of air leads to the formation of products A and B along with the evolution of CO_2. The sum of spin-only magnetic moment values of A and B is ______ B.M. (Nearest integer)

(Given atomic number : C : 6, Na : 11, O : 8, Fe : 26, Cr : 24)

[NCERT : P L-102 | NV, April 5, 2024 (II)]

99. Consider the following reaction

$MnO_2 + KOH + O_2 \rightarrow A + H_2O.$

Product 'A' in neutral or acidic medium disproportionate to give products 'B' and 'C' along with water. The sum of spin-only magnetic moment values of B and C is ______ BM. (nearest integer)

(Given atomic number of Mn is 25)

[NCERT : P L-102 | NV, April 4, 2024 (I)]

100. Which of the following compounds show colour due to d - d transition? **[NCERT : P L-103 |Feb. 1, 2024 (II)]**

(a) $K_2Cr_2O_7$ (b) $CuSO_4.5H_2O$

(c) $KMnO_4$ (d) K_2CrO_4

101. Identify correct statements from below:

[NCERT : P L-106 |Jan. 31, 2024 (I)]

A. The chromate ion is square planar.

B. Dichromates are generally prepared from chromates.

C. The green manganate ion is diamagnetic.

D. Dark green coloured K_2MnO_4 disproportionates in a neutral or acidic medium to give permanganate.

E. With increasing oxidation number of transition metal, ionic character of the oxides decreases.

Choose the correct answer from the options given below:

(a) B, C, D only (b) A, D, E only

(c) A, B, C only (d) B, D, E only

102. Choose the correct statements from the following **[NCERT : P L-107 |Jan. 31, 2024 (II)]**

A. Mn_2O_7 is an oil at room temperature

B. V_2O_4 reacts with acid to give VO_2^{2+}

C. CrO is a basic oxide

D. V_2O_5 does not react with acid

Choose the correct answer from the options given below:

(a) A, B and D only (b) A and C only

(c) A, B and C only (d) B and C only

103. In the reaction of potassium dichromate, potassium chloride and sulfuric acid (conc.), the oxidation state of the chromium in the product is (+)______. **[NCERT : P L-106 |Jan. 31, 2024 (II)]**

104. The orange colour of $K_2Cr_2O_7$ and purple colour of $KMnO_4$ is due to **[NCERT : P L-106 |Jan. 30, 2024 (II)]**

(a) Charge transfer transition in both.

(b) $d \rightarrow d$ transitions in $KMnO_4$ and charge transfer transitions in $K_2Cr_2O_7$.

(c) $d \rightarrow d$ transitions in $K_2Cr_2O_7$ and charge transfer transitions in $KMnO_4$.

(d) $d \rightarrow d$ transitions in both

105. Alkaline oxidative fusion of MnO_2 gives "A" which on electrolytic oxidation in alkaline solution produces B. A and B respectively are **[NCERT : P L-106 | Jan. 30, 2024 (II)]**

(a) Mn_2O_7 and MnO_4^- (b) MnO_4^{2-} and MnO_4^-

(c) Mn_2O_3 and MnO_4^{2-} (d) MnO_4^{2-} and Mn_2O_7

106. In chromyl chloride test for confirmation of Cl^- ion, a yellow solution is obtained. Acidification of the solution and addition of amyl alcohol and 10% H_2O_2 turns organic layer blue indicating formation of chromium pentoxide. The oxidation state of chromium in that is **[NCERT : P L-106 |Jan. 29, 2024 (I)]**

(a) +5 (b) +3

(c) +10 (d) +6

107. In alkaline medium, MnO_4^- oxidises I^- to **[NCERT : P L-106 |Jan. 29, 2024 (I)]**

(a) IO^- (b) IO_4^- (c) I_2 (d) IO_3^-

108. $KMnO_4$ decomposes on heating at 513K to form O_2 along with **[NCERT : P L-106 |Jan. 29, 2024 (I)]**

(a) Mn & KO_2 (b) K_2MnO_4 & MnO_2

(c) K_2MnO_4 & Mn (d) MnO_2 & K_2O_2

109. Yellow compound of lead chromate gets dissolved on treatment with hot NaOH solution. The product of lead formed is a : **[NCERT : P L-106 |Jan. 27, 2024 (I)]**

(a) Tetraanionic complex with coordination number six

(b) Neutral complex with coordination number four

(c) Dianionic complex with coordination number six

(d) Dianionic complex with coordination number four

110. Strong rducing and oxidising agents among the following, respectively, are **[NCERT: PL-109|April 06, 2023 (I)]**

(a) Ce^{4+} and Eu^{2+} (b) Ce^{4+} and Tb^{4+}

(c) Ce^{3+} and Ce^{4+} (d) Eu^{2+} and Ce^{4+}

111. The total change in the oxidation state of manganese involved in the reaction of $KMnO_4$ and potassium iodide in the acidic medium is ______ **[NCERT : PL-107 | NV, April 15, 2023 (I)]**

112. Given below are two statements **[NCERT : P L-105 | April 10, 2023 (I)]**

Statement I : Aqueous solution of $K_2Cr_2O_7$ is preferred as a primary standard in volumetric analysis over $Na_2Cr_2O_7$ aqueous solution

Statement II : $K_2Cr_2O_7$ has a higher solubility in water than $Na_2Cr_2O_7$

In the light of the above statements, choose the correct answer from the options given below:

(a) Both Statement I and Statement II are true

(b) Both Statement I and Statement II are false

(c) Statement I is true but Statement II is false

(d) Statement I is false but Statement II is true

113. In alkaline medium, the reduction of permanganate anion involves a gain of ______ electrons. **[NCERT : PL-107 | NV, April 10, 2023 (II)]**

114. Which halogen is known to cause the reaction given below: **[NCERT : P L-101 | April 08, 2023 (I)]**

$2Cu^{2+} + 4X^- \rightarrow Cu_2X_2(s) + X_2$

(a) Only Iodine (b) Only Bromine

(c) All halogens (d) Only Chlorine

115. The sum of oxidation state of the metals in $Fe(CO)_5$, VO^{2+} and WO_3 is ______ : **[NCERT : P L-98 | NV, April 08, 2023 (II)]**

116. During the reaction of permanganate with thiosulphate, the change in oxidation of manganese occurs by value of 3. Identify which of the below medium will favour the reaction. **[NCERT : PL-106 | April 06, 2023 (II)]**

(a) aqueous acidic

(b) aqueous neutral

(c) both aqueous acidic and neutral

(d) both aqueous acidic and faintly alkaline.

117. Highest oxidation state of Mn is exhibited in Mn_2O_7. The correct statements about Mn_2O_7 are. **[NCERT : P L-101 | Feb. 01, 2023 (I)]**

(A) Mn is tetrahedrally surrounded by oxygen atoms.

(B) Mn is octahedrally surrounded by oxygen atoms.

(C) Contains Mn-O-Mn bridge.

(D) Contains Mn-Mn bond.

Choose the correct answer from the options given below:

(a) A and D only (b) A and C only

(c) B and D only (d) B and C only

118. The correct order of basicity of oxides of vanadium is **[NCERT : P L-101 | Jan. 31, 2023 (I)]**

(a) $V_2O_3 > V_2O_4 > V_2O_5$ (b) $V_2O_3 > V_2O_5 > V_2O_4$

(c) $V_2O_5 > V_2O_4 > V_2O_3$ (d) $V_2O_4 > V_2O_3 > V_2O_5$

119. When Cu^{2+} ion is treated with KI, a white precipitate, X appears in solution. The solution is titrated with sodium thiosulphate, the compound Y is formed. X and Y respectively are **[NCERT : P L-101 | Jan. 31, 2023 (I)]**
(a) $X = Cu_2I_2$ $Y = Na_2S_4O_5$ (b) $X = Cu_2I_2$ $Y = Na_2S_4O_6$
(c) $X = CuI_2$ $Y = Na_2S_4O_3$ (d) $X = CuI_2$ $Y = Na_2S_4O_6$

120. The number of electrons involved in the reduction of permanganate to manganese dioxide in acidic medium is ________. **[NCERT : P L-106 | NV, Jan. 30, 2023 (I)]**

121. $KMnO_4$ oxidises I^- in acidic and neutral/faintly alkaline solution, respectively to **[NCERT : P L-106 | Jan. 30, 2023 (II)]**
(a) I_2 & IO_3^- (b) IO_3^- & I_2
(c) IO_3^- & IO_3^- (d) I_2 & I_2

122. Potassium dichromate acts as a strong oxidizing agent in acidic solution. During this process, the oxidation state changes from **[NCERT : P L-106 | Jan. 25, 2023 (II)]**
(a) +3 to +1 (b) +6 to +3
(c) +2 to +1 (d) +6 to +2

123. $K_2Cr_2O_7$ paper acidified with dilute H_2SO_4 turns green when exposed to **[NCERT : P L-106 | Jan. 24, 2023 (II)]**
(a) Carbon dioxide (b) Sulphur trioxide
(c) Hydrogen sulphide (d) Sulphur dioxide

124. In neutral or faintly alkaline medium, $KMnO_4$ being a powerful oxidant can oxidize, thiosulphate almost quantitatively, to sulphate. In this reaction overall change in oxidation state of manganese will be : **[NCERT : P L-108 | July 29, 2022 (I)]**
(a) 5 (b) 1 (c) 0 (d) 3

125. Dihydrogen reacts with CuO to give **[NCERT : P L-101 | June 28, 2022 (I)]**
(a) CuH_2 (b) Cu (c) Cu_2O (d) $Cu(OH)_2$

126. The disproportionation of MnO_4^{2-} in acidic medium resulted in the formation of two manganese compounds A and B. If the oxidation state of Mn in B is smaller than that of A, then the spin-only magnetic moment (μ) value of B in BM is ________. (Nearest integer) **[NCERT : P L-102 | NV, July 28, 2022 (I)]**

127. In neutral or alkaline solution, MnO_4^- oxidises thiosulphate to: **[NCERT : P L-106 | July 27, 2022 (II)]**
(a) $S_2O_7^{2-}$ (b) $S_2O_8^{2-}$ (c) SO_3^{2-} (d) SO_4^{2-}

128. Given below are two statements:
Statement I : Iron (III) catalyst, acidified $K_2Cr_2O_7$ and neutral $KMnO_4$ have the ability to oxidise I^- to I_2 independently.
Statement II: Manganate ion is paramagnetic in nature and involves $p\pi - p\pi$ bonding.
In the light of the above statements, choose the **correct** answer from the options given below:
[NCERT : P L-106 | July 27, 2022 (I)]
(a) Both statement I and statement II are ture
(b) Both statement I and statement II are false
(c) Statement I is true but statement II is false
(d) Statement I is false but statement II is true

129. The total number of Mn = O bonds in Mn_2O_7 is ____ **[NCERT : P L-106 | July 27, 2022 (I)]**
(a) 4 (b) 5 (c) 6 (d) 3

130. In the titration of $KMnO_4$ and oxalic acid in acidic medium, the change in oxidation number of carbon at the end point is______ **[NCERT : P L-106 | NV, July 27, 2022 (I)]**

131. The oxidation state of manganese in the product obtained in a reaction of potassium permanganate and hydrogen peroxide in basic medium is________. **[NCERT : P L-106 | NV, July 27, 2022 (II)]**

132. The number of terminal oxygen atoms present in the product *B* obtained from the following reaction is

$$FeCr_2O_4 + Na_2CO_3 + O_2 \rightarrow A + Fe_2O_3 + CO_2$$
$$A + H^+ \rightarrow B + H_2O + Na^+$$

[NCERT : P L-106 | NV, June 29, 2022 (I)]

133. An acidified manganate solution undergoes disproportionation reaction. The spin-only magnetic moment value of the product having manganese in higher oxidation state is ________ B.M. (Nearest integer) **[NCERT : P L-106 | NV, June 29, 2022 (I)]**

134. Acidified potassium permanganate solution oxidises oxalic acid. The spin-only magnetic moment of the manganese product formed from the above reaction is ______ B.M. (Nearest Integer) **[NCERT : P L-106 | NV, June 27, 2022 (I)]**

135. The difference in oxidation state of chromium in chromate and dichromate salts is ________. **[NCERT : P L-106 | NV, June 24, 2022 (I)]**

136. Manganese (VI) has ability to disproportionate in acidic solution. The difference in oxidation states of two ions it forms in acidic solution is ______ **[NCERT : P L-106 | NV, June 24, 2022 (II)]**

137. The spin–only magnetic moment value of the most basic oxide of vanadium among V_2O_3, V_2O_4 and V_2O_5 is ______ B.M. (Nearest Integer) **[NCERT : P L-102 | NV, June 26, 2022 (I)]**

138. Potassium permanganate on heating at 513 K gives a product which is : **[NCERT : P L-107 | Aug. 27, 2021 (II)]**
(a) paramagnetic and colourless
(b) diamagnetic and green
(c) diamagnetic and colourless
(d) paramagnetic and green

139. Given below are two statements :
Statement I : Potassium permanganate on heating at 573 K forms potassium manganate.
Statement II : Both potassium permanganate and potassium manganate are tetrahedral and paramagnetic in nature.

In the light of the above statements, choose the **most appropriate** answer from the options given below:

[NCERT : P L-107 | March 17, 2021 (I)]

(a) Statement I is false but statement II is true
(b) Both statement I and statement II are false
(c) Both statement I and statement II are true
(d) Statement I is true but statement II is false

140. Dichromate ion is treated with base, the oxidation number of Cr in the product formed is ____.

[NCERT : P L-106 | NV, Feb. 26, 2021 (I)]

141. In mildly alkaline medium, thiosulphate ion is oxidized by MnO_4^- to "A". The oxidation state of sulphur in "A" is ____. **[NCERT : P L-108 | NV, Feb. 26, 2021 (II)]**

142. In basic medium CrO_4^{2-} oxidises $S_2O_3^{2-}$ to form SO_4^{2-} and itself changes into $Cr(OH)_4^-$. The volume of 0.154 M CrO_4^{2-} required to react with 40 mL of 0.25 M $S_2O_3^{2-}$ is ______ mL. (Rounded-off to the nearest integer) **[NCERT : P L-106 | NV, Feb. 25, 2021 (I)]**

143. The correct electronic configuration and spin-only magnetic moment (BM) of Gd^{3+}(Z = 64), respectively, are:

[NCERT : P L-102 | Sep. 05, 2020 (I)]

(a) [Xe] $4f^7$ and 8.9 (b) [Xe] $4f^7$ and 7.9
(c) [Xe] $5f^7$ and 8.9 (d) [Xe] $5f^7$ and 7.9

144. The incorrect statement(s) among (1) - (3) is (are):

[NCERT : P L-106 | Sep. 04, 2020 (II)]

(1) W(VI) is more stable than Cr(VI).
(2) In the presence of HCl, permanganate titrations provide satisfactory results.
(3) some lanthanoid oxides can be used as phosphorus.

(a) (2) and (3) only (b) (1) and (2) only
(c) (2) only (d) (1) only

145. The **incorrect** statement is:

[NCERT : P L-106 | Sep. 03, 2020 (II)]

(a) Manganate and permanganate ions are tetrahedral
(b) In manganate and permanganate ions, the π-bonding takes place by overlap of *p*-orbitals of oxygen and *d*-orbitals of manganese
(c) Manganate and permanganate ions are paramagnetic
(d) Manganate ion is green in colour and permanganate ion is purple in colour

146. The sum of the total number of bonds between chromium and oxygen atoms in chromate and dichromate ions is ____. **[NCERT : P L-106 | NV, Jan. 09, 2020 (II)]**

147. Consider the following reactions:

$NaCl + K_2Cr_2O_7 + H_2SO_4 \text{ (Conc.)} \rightarrow (A) + \text{Side products}$

$(A) + NaOH \rightarrow (B) + \text{Side products}$

$(B) + H_2SO_4 \text{ (dilute)} + H_2O_2 \rightarrow (C) + \text{Side products}$

The sum of the total number of atoms in one molecule each of (A), (B) and (C) is ______.

[NCERT : P L-106 | NV, Jan. 07, 2020 (II)]

148. Thermal decomposition of a Mn compound (X) at 513 K results in compound Y, MnO_2 and a gaseous product. MnO_2 reacts with NaCl and concentrated H_2SO_4 to give a pungent gas Z. X, Y, and Z, respectively, are:

[NCERT : P L-107 | April 12, 2019 (II)]

(a) $KMnO_4$, K_2MnO_4 and Cl_2
(b) K_2MnO_4, $KMnO_4$ and SO_2
(c) K_3MnO_4, K_2MnO_4 and Cl_2
(d) K_2MnO_4, $KMnO_4$ and Cl_2

149. $A \xrightarrow{4\,KOH,\,O_2} \underset{\text{(Green)}}{2B} + 2H_2O$

$B \xrightarrow{4\,HCl} \underset{\text{(Purple)}}{2C} + MnO_2 + 2H_2O$

$2C \xrightarrow{H_2O,\,KI} 2A + KOH + D$

In the above sequence of reactions, A and D, respectively, are: **[NCERT : P L-106 | Jan. 11, 2019 (II)]**

(a) KI and $KMnO_4$ (b) MnO_2 and KIO_3
(c) KIO_3 and MnO_2 (d) KI and K_2MnO_4

150. When XO_2 is fused with an alkali metal hydroxide in presence of an oxidizing agent such as KNO_3, a dark green product is formed which disproportionates in acidic solution to afford a dark purple solution. X is:

[NCERT : P L-107 | Online April 16, 2018]

(a) Mn (b) Cr (c) V (d) Ti

151. Which of the following compounds is metallic and ferromagnetic? **[NCERT : P L-102 | 2016]**

(a) VO_2 (b) MnO_2 (c) TiO_2 (d) CrO_2

152. When concentrated HCl is added to an aqueous solution of $CoCl_2$, its colour changes from reddish pink to deep blue. Which complex ion gives blue colour in this reaction?

[NCERT : P L-103 | Online April 11, 2015]

(a) $[CoCl_4]^{2-}$ (b) $[CoCl_6]^{3-}$
(c) $[CoCl_6]^{4-}$ (d) $[Co(H_2O)_6]^{2+}$

153. Which of the following statements is false?

[NCERT : P L-106 | Online April 11, 2015]

(a) $Na_2Cr_2O_7$ is less soluble than $K_2Cr_2O_7$
(b) $Na_2Cr_2O_7$ is primary standard in volumetry
(c) CrO_4^{2-} is tetrahedral in shape
(d) $Cr_2O_7^{2-}$ has a Cr–O–Cr bond

154. The equation which is balanced and represents the correct product(s) is: **[NCERT : P L-104 | 2014]**

(a) $Li_2O + 2KCl \longrightarrow 2LiCl + K_2O$

(b) $[CoCl(NH_3)_5]^+ + 5H^+ \longrightarrow Co^{2+} + 5NH_4^+ + Cl^-$

(c) $[Mg(H_2O)_6]^{2+} + (EDTA)^{4-} \xrightarrow{\text{excess NaOH}} [Mg(EDTA)]^{2+} + 6H_2O$

(d) $CuSO_4 + 4KCN \longrightarrow K_2[Cu(CN)_4] + K_2SO_4$

155. Which series of reactions correctly represents chemical reactions related to iron and its compound? [NCERT : P L-101 | 2014]

(a) $Fe \xrightarrow{\text{dil. } H_2SO_4} FeSO_4 \xrightarrow{H_2SO_4, O_2} Fe_2(SO_4)_3 \xrightarrow{\text{heat}} Fe$

(b) $Fe \xrightarrow{O_2, \text{ heat}} FeO \xrightarrow{\text{dil. } H_2SO_4} FeSO_4 \xrightarrow{\text{heat}} Fe$

(c) $Fe \xrightarrow{Cl_2, \text{ heat}} FeCl_3 \xrightarrow{\text{heat, air}} FeCl_2 \xrightarrow{Zn} Fe$

(d) $Fe \xrightarrow{O_2, \text{ heat}} Fe_3O_4 \xrightarrow{CO, 600°C} FeO \xrightarrow{CO, 700°C} Fe$

156. Which of the following is not formed when H_2S reacts with acidic $K_2Cr_2O_7$ solution? [NCERT : P L-106 | Online April 9, 2014]

(a) $CrSO_4$ (b) $Cr_2(SO_4)_3$
(c) K_2SO_4 (d) S

157. Identify incorrect statement : [NCERT : P L-101 | Online April 23, 2013]

(a) Cu_2O is colourless.
(b) Copper (I) compounds are colourless except when colour results from charge transfer.
(c) Copper (I) compounds are diamagnetic.
(d) Cu_2S is black.

158. When a small amount of $KMnO_4$ is added to concentrated H_2SO_4, a green oily compound is obtained which is highly explosive in nature. Compound may be : [NCERT : P L-105 | Online April 23, 2013]

(a) $MnSO_4$ (b) Mn_2O_7
(c) MnO_2 (d) Mn_2O_3

159. The oxidation state of chromium in the final product formed by the reaction between KI and acidified potassium dichromate solution is: [NCERT : P L-106 | 2005]

(a) +3 (b) +2 (c) +6 (d) +4

160. Heating mixture of Cu_2O and Cu_2S will give [NCERT : P L-107 | 2005]

(a) Cu_2SO_3 (b) $CuO + CuS$
(c) $Cu + SO_3$ (d) $Cu + SO_2$

161. Ammonia forms the complex ion $[Cu(NH_3)_4]^{2+}$ with copper ions in alkaline solutions but not in acidic solutions. What is the reason for it ? [NCERT : P L-107 | 2003]

(a) In acidic solutions, protons coordinate with ammonia molecules forming NH_4^+ ions and thus NH_3 molecules are not available
(b) In alkaline solutions insoluble $Cu(OH)_2$ is precipitated which is soluble in excess of any alkali
(c) Copper hydroxide is an amphoteric substance
(d) In acidic solutions hydration protects copper ions

162. A red solid is insoluble in water. However it becomes soluble if some KI is added to water. Heating the red solid in a test tube results in liberation of some violet coloured fumes and droplets of a metal appear on the cooler parts of the test tube. The red solid is [NCERT : P L-107 | 2003]

(a) HgI_2 (b) HgO
(c) Pb_3O_4 (d) $(NH_4)_2Cr_2O_7$

163. Which one of the following nitrates will leave behind a metal on strong heating ? [NCERT : P L-107 | 2003]

(a) Copper nitrate (b) Manganese nitrate
(c) Silver nitrate (d) Ferric nitrate

164. What would happen when a solution of potassium chromate is treated with an excess of dilute nitric acid? [NCERT : P L-106 | 2003]

(a) $Cr_2O_7^{2-}$ and H_2O are formed
(b) CrO_4^{2-} is reduced to +3 state of Cr
(c) CrO_4^{2-} is oxidized to +7 state of Cr
(d) Cr^{3+} and $Cr_2O_7^{2-}$ are formed

165. A reduction in atomic size with increase in atomic number is a characteristic of elements of [NCERT : P L-94 | 2003]

(a) *d*-block (b) *f*-block
(c) radioactive series (d) high atomic masses

166. When $KMnO_4$ acts as an oxidising agent and ultimately forms $[MnO_4]^{-2}$, MnO_2, Mn_2O_3, Mn^{+2} then the number of electrons transferred in each case respectively is [NCERT : P L-106 | 2002]

(a) 4, 3, 1, 5 (b) 1, 5, 3, 7
(c) 1, 3, 4, 5 (d) 3, 5, 7, 1.

Topic 3 – Lanthanoids and Actinoids

167. Which of the following ions is the strongest oxidizing agent ? [NCERT : P L-111 | Jan. 24, 2025 (I)]
(Atomic Number of Ce = 58, Eu = 63, Tb = 65, Lu = 71]

(a) Lu^{3+} (b) Eu^{2+}
(c) Tb^{4+} (d) Ce^{3+}

168. Lanthanoid ions with $4f^7$ configuration are: [NCERT : P L-109 | Jan. 22, 2025 (I)]

(A) Eu^{2+} (B) Gd^{3+}
(C) Eu^{3+} (D) Tb^{3+}
(E) Sm^{2+}

Choose the correct answer from the options given below:

(a) (A) and (B) only (b) (A) and (D) only
(c) (B) and (E) only (d) (B) and (C) only

169. Number of colourless lanthanoid ions among the following is ______. [NCERT : P L-110 |NV, April 9, 2024 (I)]

Eu^{3+}, Lu^{3+}, Nd^{3+}, La^{3+}, Sm^{3+}

170. The electronic configuration of Einsteinium is :
(Given atomic number of Einsteinium = 99)
[NCERT : PL-111 | April 9, 2024 (II)]
(a) [Rn] $5f^{12}\,6d^0\,7s^2$ (b) [Rn] $5f^{11}\,6d^0\,7s^2$
(c) [Rn] $5f^{13}\,6d^0\,7s^2$ (d) [Rn] $5f^{10}\,6d^0\,7s^2$

171. The number of element from the following that do not belong to lanthanoids is :
[NCERT : PL-110 | April 6, 2024 (I)]
Eu, Cm, Er, Tb, Yb and Lu
(a) 3 (b) 4 (c) 1 (d) 5

172. Diamagnetic Lanthanoid ions are:
[NCERT : PL-110 | Jan. 30, 2024 (I)]
(a) Nd^{3+} & Eu^{3+} (b) La^{3+} & Ce^{4+}
(c) Nd^{3+} & Ce^{4+} (d) Lu^{3+} & Eu^{3+}

173. Which of the following acts as a strong reducing agent?
(Atomic number : Ce = 58, Eu = 63, Gd = 64, Lu = 71)
[NCERT : PL-110 | Jan. 29, 2024 (II)]
(a) Lu^{3+} (b) Ce^{4+} (c) Gd^{3+} (d) Eu^{2+}

174. The electronic configuration for Neodymium is :
[Atomic Number for Neodymium 60]
[NCERT : PL-110 | Jan. 27, 2024 (I)]
(a) [Xe] $4f^4\,6s^2$ (b) [Xe] $5f^4 7s^2$
(c) [Xe] $4f^6\,6s^2$ (d) [Xe] $4f^1 5d^1 6s^2$

175. Given below are two statements:
[NCERT : PL-109 | Jan. 27, 2024 (II)]
Statement (I) : In the Lanthanoids, the formation of Ce^{+4} is favoured by its noble gas configuration.
Statement (II) : Ce^{+4} is a strong oxidant reverting to the common +3 state.
In the light of the above statements, choose the most appropriate answer from the options given below:
(a) Statement I is false but Statement II is true
(b) Both Statement I and Statement II are true
(c) Statement I is true but Statement II is false
(d) Both Statement I and Statement II are false

176. Total number of ions from the following with noble gas configuration is ________.
[NCERT : PL-91 | NV, Jan. 27, 2024 (II)]
Sr^{2+} (Z = 38), Cs^+ (Z = 55), La^{2+} (Z = 57) Pb^{2+} (Z = 82), Yb^{2+} (Z = 70) and Fe^{2+} (Z = 26)

177. Strong reducing and oxidising agents among the following, respectively, are **[NCERT : PL-109 | April 06, 2023 (I)]**
(a) Ce^{4+} and Eu^{2+} (b) Ce^{4+} and Tb^{4+}
(c) Ce^{3+} and Ce^{4+} (d) Eu^{2+} and Ce^{4+}

178. The pair of lanthanides in which both elements have high third - ionization energy is:
[NCERT : PL-109 | April 13, 2023 (I)]
(a) Eu, Gd (b) Eu, Yb (c) Lu, Yb (d) Dy, Gd

179. Nd^{2+} = ________ **[NCERT : PL-110 | Jan. 31, 2023 (I)]**
(a) $4f^2 6s^2$ (b) $4f^4$ (c) $4f^3$ (d) $4f^4 6s^2$

180. Which of the following elements have half-filled f-orbitals in their ground state ?
[NCERT : PL-110 | Jan. 31, 2023 (II)]
(Given : atomic number
Sm = 62; Eu = 63; Tb = 65; Gd = 64, Pm = 61)
A. Sm B. Eu C. Tb D. Gd
E. Pm
Choose the correct answer from the options given below:
(a) B and D only (b) A and E only
(c) A and B only (d) C and D only

181. Which one amongst the following are good oxidizing agents? **[NCERT : PL-109 | Jan. 24, 2023 (II)]**
A. Sm^{2+} B. Ce^{2+} C. Ce^{4+} D. Tb^{4+}
Choose the most appropriate answer from the options given below :
(a) C only (b) D only
(c) A and B only (d) C and D only

182. Which of the following pair is not isoelectronic species?
(At. no. Sm, 62; Er, 68; Yb, 70; Lu, 71; Eu, 63; Tb, 65; Tm, 69) **[NCERT : PL-91 | July 28, 2022 (II)]**
(a) Sm^{2+} and Er^{3+} (b) Yb^{2+} and Lu^{3+}
(c) Eu^{2+} and Tb^{4+} (d) Tb^{2+} and Tm^{4+}

183. Which one of the lanthanoids given below is the most stable in divalent form? **[NCERT : PL-109 | June 28, 2022 (I)]**
(a) Ce (Atomic Number 58)
(b) Sm (Atomic Number 62)
(c) Eu (Atomic Number 63)
(d) Yb (Atomic Number 70)

184. The *f*-orbitals are half and completely filled, respectively in lanthanide ions
(Given: Atomic no. Eu, 63; Sm, 62; Tm, 69; Tb, 65; Yb, 70; Dy, 66] **[NCERT : PL-110 | June 27, 2022 (II)]**
(a) Eu^{2+} and Tm^{2+} (b) Sm^{2+} and Tm^{3+}
(c) Tb^{4+} and Yb^{2+} (d) Dy^{3+} and Yb^{3+}

185. The most common oxidation state of Lanthanoid elements is +3. Which of the following is likely to deviate easily from +3 oxidation state?
[NCERT : PL-109 | June 26, 2022 (II)]
(a) Ce (At. No. 58) (b) La (At. No. 57)
(c) Lu (At. No. 71) (d) Gd (At. No. 64)

186. Cerium (IV) has a noble gas configuration. Which of the following is correct statement about it?
[NCERT : PL-109 | June 25, 2022 (I)]
(a) It will not prefer to undergo redox reactions.
(b) It will prefer to gain electron and act as an oxidizing agent
(c) It will prefer to give away an electron and behave as reducing agent
(d) It acts as both, oxidizing and reducing agent.

187. Which one of the following lanthanides exhibits +2 oxidation state with diamagnetic nature?
(Given Z for Nd = 60, Yb = 70, La = 57, Ce = 58)
[NCERT : PL-109 | Aug. 31, 2021 (I)]
(a) Nd (b) Yb (c) La (d) Ce

188. The Eu^{2+} ion is a strong reducing agent in spite of its ground state electronic configuration
[NCERT : PL-110 | Aug. 31, 2021 (II)]
(outermost) : [Atomic number of Eu = 63]
(a) $4f^7 6s^2$ (b) $4f^6$ (c) $4f^7$ (d) $4f^6 6s^2$

189. Ge(Z = 32) in its ground state electronic configuration has x completely filled orbitals with $m_l = 0$. The value of x is ______. **[NCERT : PL-109 | NV, Aug. 31, 2021 (I)]**

190. The number of *f* electrons in the ground state electronic configuration of Np (Z = 93) is _______. (Nearest integer) **[NCERT : PL-110 | NV, Aug. 27, 2021 (I)]**

191. The number of 4*f* electrons in the ground state electronic configuration of Gd^{2+} is _______.
[Atomic number of Gd = 64] **[NCERT : P L-110 | NV, Aug. 26, 2021 (I); Same Sep. 05, 2020 (I)]**

192. Number of electrons present in 4*f* orbital of Ho^{3+} ion is __________. (Given Atomic No. of Ho = 67) **[NCERT : PL-110 | NV, July 25, 2021 (II)]**

193. Given below are two statements : **[NCERT : PL-110 | March 16, 2021 (I)]**

Statement I : The E° value for Ce^{4+}/Ce^{3+} is + 1.74 V.

Statement II : Ce is more stable in Ce^{4+} state than Ce^{3+} state.

In the light of the above statements, choose the **most appropriate** answer from the options given below.

(a) Both statement I and statement II are correct
(b) Statement I is correct but statement II is incorrect
(c) Both statement I and statement II are incorrect
(d) Statement I is incorrect but statement II is correct

194. Arrange the following metal complex/compounds in the increasing order of spin only magnetic moment. Presume all the three, high spin system. **[NCERT : PL-112 | March 16, 2021 (II)]**
(Atomic numbers Ce = 58, Gd = 64 and Eu = 63.)

(A) $(NH_4)_2[Ce(NO_3)_6]$
(B) $Gd(NO_3)_3$ and (C) $Eu(NO_3)_3$

(a) (A) < (C) < (B) (b) (A) < (B) < (C)
(c) (C) < (A) < (B) (d) (B) < (A) < (C)

195. The lanthanoid that does NOT show + 4 oxidation state is: **[NCERT : P L-109 | Sep. 06, 2020 (I)]**

(a) Dy (b) Ce (c) Eu (d) Tb

196. Mischmetal is an alloy consisting mainly of: **[NCERT : P L-111 | Sep. 06, 2020 (II)]**

(a) lanthanoid metals
(b) actinoid and transition metals
(c) lanthanoid and actinoid metals
(d) actinoid metals

197. The electronic configurations of bivalent europium and trivalent cerium are:
(atomic number : Xe = 54, Ce = 58, Eu = 63) **[NCERT : P L-110 | Jan. 09, 2020 (I)]**

(a) [Xe] $4f^2$ and [Xe] $4f^7$
(b) [Xe] $4f^7$ and [Xe] $4f^1$
(c) [Xe] $4f^7 6s^2$ and [Xe] $4f^2 6s^2$
(d) [Xe] $4f^4$ and [Xe] $4f^9$

198. The highest possible oxidation states of uranium and plutonium, respectively, are : **[NCERT : P L-112 | April 10, 2019 (II)]**

(a) 6 and 7 (b) 6 and 4 (c) 7 and 6 (d) 4 and 6

199. The maximum number of possible oxidation states of actinoides are shown by: **[NCERT : P L-112 | April 9, 2019 (II)]**

(a) Nobelium (No) and lawrencium (Lr)
(b) Actinium (Ac) and thorium (Th)
(c) Berkelium (Bk) and californium (Cf)
(d) Neptunium (Np) and plutonium (Pu)

200. The lanthanide ion that would show colour is : **[NCERT : P L-110 | April 8, 2019 (I)]**

(a) Gd^{3+} (b) Sm^{3+} (c) La^{3+} (d) Lu^{3+}

201. The correct order of atomic radii is : **[Jan. 12, 2019 (II)]**

(a) N > Ce > Eu > Ho (b) Ho > N > Eu > Ce
(c) Ce > Eu > Ho > N (d) Eu > Ce > Ho > N

202. The effect of lanthanoid contraction in the lanthanoid series of elements by and large means: **[NCERT : P L-109 | Jan. 10, 2019 (I)]**

(a) increase in both atomic and ionic radii
(b) decrease in atomic radii and increase in ionic radii
(c) decrease in both atomic and ionic radii
(d) increase in atomic radii and decrease in ionic radii

203. Which is not the correct statement? **[NCERT : P L-112 | Online May 7, 2012]**
(At. nos. Ce = 58, Lu = 71, La = 57, Yb = 70)

(a) Colour of Yb^{3+} ion is pink.
(b) La^{3+} is diamagnetic.
(c) Ce^{4+} has f^0 configuration.
(d) Lu^{3+} had f^{14} configuration.

204. The number of unpaired electrons in Gadolinium [Z = 64] is **[NCERT : P L-102 | Online May 26, 2012]**

(a) 3 (b) 8 (c) 6 (d) 2

205. Magnetic moment of Gd^{3+} ion (Z = 64) is **[Online May 12, 2012; Same May 26, 2012]**

(a) 3.62 BM (b) 9.72 BM (c) 7.9 BM (d) 10.60 BM

206. Which of the following forms stable + 4 oxidation state? **[NCERT : P L-109 | Online May 26, 2012]**

(a) La (Z = 57) (b) Eu (Z = 63)
(c) Ce (Z = 58) (d) Gd (Z = 64)

207. Knowing that the chemistry of lanthanoids (Ln) is dominated by its + 3 oxidation state, which of the following statements is incorrect? **[NCERT : P L-109 | 2009]**

(a) The ionic size of Ln (III) decreases in general with increasing atomic number
(b) Ln (III) compounds are generally colourless.
(c) Ln (III) hydroxides are mainly basic in character.
(d) Because of the large size of the Ln (III) ions, the bonding in its compounds is predominantly ionic in character.

208. Larger number of oxidation states are exhibited by the actinoids than those by the lanthanoids, the main reason being **[NCERT : P L-113 | 2008]**

(a) $4f$ orbitals more diffused than the $5f$ orbitals
(b) lesser energy difference between $5f$ and $6d$ than between $4f$ and $5d$ orbitals
(c) more energy difference between $5f$ and $6d$ than between $4f$ and $5d$ orbitals
(d) more reactive nature of the actionids than the lanthanoids

209. Identify the incorrect statement among the following: **[2007]**

(a) $4f$ and $5f$ orbitals are equally shielded.
(b) *d*-Block elements show irregular and erratic chemical properties among themselves.
(c) La and Lu have partially filled *d*-orbitals and no other partially filled orbitals.
(d) The chemistry of various lanthanoids is very similar.

210. The actinoids exhibit more number of oxidation states in general than the lanthanoids. This is because

(a) the $5f$ orbitals extend further from the nucleus than the $4f$ orbitals **[NCERT : P L-113 | 2007]**
(b) the $5f$ orbitals are more buried than the $4f$ orbitals
(c) there is a similarity between $4f$ and $5f$ orbitals in their angular part of the wave function
(d) the actinoids are more reactive than the lanthanoids.

211. Lanthanoid contraction is caused due to **[NCERT : P L-109 | 2006; Same 2005]**

(a) the same effective nuclear charge from Ce to Lu
(b) the imperfect shielding on outer electrons by $4f$ electrons from the nuclear charge
(c) the appreciable shielding on outer electrons by $4f$ electrons from the nuclear charge
(d) the appreciable shielding on outer electrons by $5d$ electrons from the nuclear charge

212. The radius of La^{3+} (Atomic number of La = 57) is 1.06Å. Which one of the following given values will be closest to the radius of Lu^{3+} (Atomic number of Lu = 71) ? **[NCERT : P L-111 | 2003]**

(a) 1.40 Å (b) 1.06 Å (c) 0.85 Å (d) 1.60 Å

213. Most common oxidation states of Ce (cerium) are **[NCERT : P L-111 | 2002]**

(a) +2, +3 (b) +2, +4 (c) +3, +4 (d) +3, +5.

214. Arrange Ce^{+3}, La^{+3}, Pm^{+3} and Yb^{+3} in increasing order of their ionic radii. **[NCERT : P L-109 | 2002]**

(a) $Yb^{+3} < Pm^{+3} < Ce^{+3} < La^{+3}$
(b) $Ce^{+3} < Yb^{+3} < Pm^{+3} < La^{+3}$
(c) $Yb^{+3} < Pm^{+3} < La^{+3} < Ce^{+3}$
(d) $Pm^{+3} < La^{+3} < Ce^{+3} < Yb^{+3}$.

215. Cerium (Z = 58) is an important member of the lanthanoids. Which of the following statements about cerium is **incorrect**? **[NCERT : P L-109 | 2004; Same 2002]**

(a) The +4 oxidation state of cerium is not known in solutions.
(b) The +3 oxidation state of cerium is more stable than the +4 oxidation state
(c) The common oxidation states of cerium are +3 and +4
(d) Cerium (IV) acts as an oxidizing agent

ANSWER KEYS																			
1	(c)	**23**	(6)	**45**	(4)	**67**	(c)	**89**	(5)	**111**	(5)	**133**	(0)	**155**	(d)	**177**	(d)	**199**	(d)
2	(c)	**24**	(c)	**46**	(a)	**68**	(a)	**90**	(c)	**112**	(c)	**134**	(6)	**156**	(a)	**178**	(b)	**200**	(b)
3	(d)	**25**	(b)	**47**	(c)	**69**	(d)	**91**	(c)	**113**	(3)	**135**	(0)	**157**	(a)	**179**	(b)	**201**	(d)
4	(a)	**26**	(d)	**48**	(3)	**70**	(d)	**92**	(d)	**114**	(a)	**136**	(3)	**158**	(b)	**180**	(a)	**202**	(c)
5	(4)	**27**	(0)	**49**	(c)	**71**	(b)	**93**	(c)	**115**	(10)	**137**	(3)	**159**	(a)	**181**	(d)	**203**	(a)
6	(b)	**28**	(a)	**50**	(a)	**72**	(c)	**94**	(b)	**116**	(b)	**138**	(d)	**160**	(d)	**182**	(a, d)	**204**	(b)
7	(0)	**29**	(b)	**51**	(d)	**73**	(b)	**95**	(d)	**117**	(b)	**139**	(d)	**161**	(a)	**183**	(c)	**205**	(c)
8	(c)	**30**	(b)	**52**	(a)	**74**	(d)	**96**	(6)	**118**	(a)	**140**	(6)	**162**	(a)	**184**	(c)	**206**	(c)
9	(b)	**31**	(d)	**53**	(12)	**75**	(a)	**97**	(877)	**119**	(b)	**141**	(6)	**163**	(c)	**185**	(a)	**207**	(b)
10	(b)	**32**	(d)	**54**	(c)	**76**	(a)	**98**	(6)	**120**	(3)	**142**	(173)	**164**	(a)	**186**	(b)	**208**	(b)
11	(b)	**33**	(d)	**55**	(49)	**77**	(d)	**99**	(4)	**121**	(a)	**143**	(b)	**165**	(b)	**187**	(b)	**209**	(a)
12	(a)	**34**	(b)	**56**	(a)	**78**	(a)	**100**	(b)	**122**	(b)	**144**	(c)	**166**	(c)	**188**	(c)	**210**	(a)
13	(11)	**35**	(d)	**57**	(c)	**79**	(d)	**101**	(d)	**123**	(d)	**145**	(c)	**167**	(c)	**189**	(7)	**211**	(b)
14	(c)	**36**	(a)	**58**	(a)	**80**	(a)	**102**	(b)	**124**	(d)	**146**	(18)	**168**	(a)	**190**	(18)	**212**	(c)
15	(c)	**37**	(d)	**59**	(b)	**81**	(a)	**103**	(6)	**125**	(b)	**147**	(18.00)	**169**	(2)	**191**	(7)	**213**	(c)
16	(6)	**38**	(a)	**60**	(a)	**82**	(d)	**104**	(a)	**126**	(4)	**148**	(a)	**170**	(b)	**192**	(10)	**214**	(a)
17	(d)	**39**	(b)	**61**	(a)	**83**	(b)	**105**	(b)	**127**	(d)	**149**	(b)	**171**	(c)	**193**	(b)	**215**	(a)
18	(0)	**40**	(a)	**62**	(d)	**84**	(d)	**106**	(d)	**128**	(c)	**150**	(a)	**172**	(b)	**194**	(a)		
19	(d)	**41**	(b)	**63**	(b)	**85**	(a)	**107**	(d)	**129**	(c)	**151**	(d)	**173**	(d)	**195**	(c)		
20	(0)	**42**	(a)	**64**	(b)	**86**	(10)	**108**	(b)	**130**	(1)	**152**	(a)	**174**	(a)	**196**	(a)		
21	(5)	**43**	(5)	**65**	(c)	**87**	(6)	**109**	(d)	**131**	(4)	**153**	(a)	**175**	(b)	**197**	(b)		
22	(b)	**44**	(5)	**66**	(a)	**88**	(160)	**110**	(d)	**132**	(6)	**154**	(b)	**176**	(2)	**198**	(a)		

Co-ordination Compounds

TREND ANALYSIS
JEE MAIN (2025-2021)

	2025				2024				2023				2022				2021			
	Session II		Session I		Session II		Session I		Session II		Session I		Session II		Session I		Session II		Session I	
	MCQs	NVQs	MCQs	NVQs	MCQs	NVQs	MCQs	NVQs	MCQs	NVQs	MCQs	NVQs	MCQs	NVQs	MCQs	NVQs	MCQs	NVQs	MCQs	NVQs
Topic 1	5	2	3	1	3	1	2	1	4	2	5	2	–	3	4	2	4	8	–	4
Topic 2	13	3	11	1	14	2	6	2	16	4	10	5	6	3	6	6	11	4	3	2
Topic 3	–	–	–	–	–	–	3	–	1	–	1	–	–	–	–	–	–	–	–	–
Topic 4																				
Total	18	5	14	2	17	3	11	3	21	6	16	7	6	6	10	8	15	12	3	6

How to Code Qns for Better & Faster Revision

Type 1: Able to solve (No code Req)
Type 2: Solved with struggle or by seeing solution
(Use code NR means Needs Revision or orange stickers)
Type 3: Tricky/ Could not solve on my own. Need help from Teacher/ Friend.
(Use code TQ means Tricky Question or red stickers)

You can either hand write the code in boxes or else paste coloured stickers given at end of book.

Topic 1 - Coordination Number, Nomenclature and Isomerism of Coordination Compounds

1. Given below are two statements :

[NCERT : PL-122 | April 8, 2025 (II)]

Statement I : A homoleptic octahedral complex, formed using monodentate ligands, will not show stereoisomerism.

Statement II : cis– and trans– platin are heteroleptic complexes of Pd.

In the light of the above statements, choose the **correct** answer from the options given below.

(a) Both statement I and Statement II are false.
(b) Statement I is false but Statement II is true.
(c) Both statement I and Statement II are true.
(d) Statement I is true but Statement II is false.

2. Match **List-I** with **List-II**

[NCERT : PL-119 | April 7, 2025 (II)]

List-I (Complex)	List-II (Primary valency and Secondary valency)	
(A) $[Co(en)_2Cl_2]Cl$	(I) 3	6
(B) $[Pt(NH_3)_2Cl(NO_2)]$	(II) 3	4
(C) $Hg[Co(SCN)_4]$	(III) 2	6
(D) $[Mg(EDTA)]^{2-}$	(IV) 2	4

Choose the **correct** answer from the options given below:

(a) (A)-(III), (B)-(I), (C)-(II), (D)-(IV)
(b) (A)-(I), (B)-(IV), (C)-(II), (D)-(III)
(c) (A)-(I), (B)-(III), (C)-(II), (D)-(IV)
(d) (A)-(II), (B)-(III), (C)-(IV), (D)-(I)

3. 'X' is the number of acidic oxides among VO_2, V_2O_3, CrO_3, V_2O_5 and Mn_2O_7. The primary valency of cobalt in $[Co(H_2NCH_2CH_2NH_2)_3]_2(SO_4)_3$ is Y. The value of $X + Y$ is:

[NCERT : PL-119 | April 7, 2025 (II)]

(a) 5 (b) 4 (c) 2 (d) 3

4. An octahedral complex having molecular composition $Co.5NH_3.Cl.SO_4$ has two isomers A and B. The solution of A gives a white precipitate with $AgNO_3$ solution and the solution of B gives white precipitate with $BaCl_2$ solution. The type of isomerism exhibited by the complex is,

[NCERT : P L-127 | April 7, 2025 (I)]

(a) Co-ordinate isomerism (b) Linkage isomerism
(c) Ionisation isomerism (d) Geometrical isomerism

5. A metal complex with a formula $MCl_4 \cdot 3NH_3$ is involved in sp^3d^2 hybridisation. It upon reaction with excess of $AgNO_3$ solution gives '*x*' moles of AgCl. Consider '*x*' is equal to the number of lone pairs of electron present in central atom of BrF_5. Then the number of geometrical isomers exhibited by the complex is________.

[NCERT : P L-125 | NV, April 4, 2025 (II)]

6. Number of stereoisomers possible for the complexes, $[CrCl_3(py)_3]$ and $[CrCl_2(ox)_2]^{3-}$ are respectively (py = pyridine, ox = oxalate)

[NCERT : P L-126 | April 4, 2025 (I)]

(a) 3 & 3 (b) 2 & 2
(c) 2 & 3 (d) 1 & 2

7. The number of optical isomers exhibited by the iron complex (A) obtained from the following reaction is ________. **[NCERT : P L-126 | NV, April 3, 2025 (I)]**

$FeCl_3 + KOH + H_2C_2O_4 \rightarrow A$

8. Identify the homoleptic complexes with odd number of d electrons in the central metal.

[NCERT : P L-122 | Jan. 29, 2025 (II)]

(A) $[FeO_4]^{2-}$ (B) $[Fe(CN)_6]^{3-}$
(C) $[Fe(CN)_5NO]^{2-}$ (D) $[CoCl_4]^{2-}$
(E) $[Co(H_2O)_3F_3]$

Choose the correct answer from the options given below :

(a) (B) and (D) only (b) (C) and (E) only
(c) (A), (B) and (D) only (d) (A), (C) and (E) only

9. The complex that shows Facial – Meridional isomerism is

[NCERT : P L-125 | Jan. 23, 2025 (I)]

(a) $[Co(NH_3)_3Cl_3]$ (b) $[Co(NH_3)_4Cl_2]^+$
(c) $[Co(en)_3]^{3+}$ (d) $[Co(en)_2Cl_2]^+$

10. The complex of Ni^{2+} ion and dimethyl glyoxime contains ______ number of Hydrogen (H) atoms.

[NCERT : P L-122 | Jan. 22, 2025 (II)]

11. Identify the homoleptic complex(es) that is/are low spin.

[NCERT : P L-122 | Jan. 22, 2025 (II)]

(A) $[Fe(CN)_5NO]^{2-}$ (B) $[CoF_6]^{3-}$
(C) $[Fe(CN)_6]^{4-}$ (D) $[Co(NH_3)_6]^{3+}$
(E) $[Cr(H_2O)_6]^{2+}$

Choose the correct answer from the options given below:

(a) (B) and (E) only (b) (A) and (C) only
(c) (C) and (D) only (d) (C) only

12. Number of ambidentate ligands among the following is ______. **[NV, April 9, 2024 (I)]**

$NO_2^-, SCN^-, C_2O_4^{2-}, NH_3, CN^-, SO_4^{2-}, H_2O.$

13. An octahedral complex with the formula $CoCl_3nNH_3$ upon reaction with excess of $AgNO_3$ solution given 2 moles of AgCl. Consider the oxidation state of Co in the complex is 'x'. The value of "x + n" is ______.

[NCERT : P L-119 | April 8, 2024 (I)]

(a) 3 (b) 6 (c) 8 (d) 5

14. The correct IUPAC name of $[PtBr_2(PMe_3)_2]$ is:

[NCERT : P L-123 | April 6, 2024 (II)]

(a) bis(trimethylphosphine)dibromoplatinum(II)
(b) bis[bromo(trimethylphosphine)]platinum(II)
(c) dibromobis(trimethylphosphine)platinum(II)
(d) dibromodi(trimethylphosphine)platinum(II)

15. The metal atom present in the complex MABXL (where A, B, X and L are unidentate ligands and M is metal) involves sp^3 hybridization. The number of geometrical isomers exhibited by the complex is:

[NCERT : P L-125 | April 5, 2024 (II)]

(a) 4 (b) 0 (c) 2 (d) 3

16. Which of the following complex is homoleptic?

[NCERT : P L-122 | Feb. 1, 2024 (I)]

(a) $[Ni(CN)_4]^{2-}$ (b) $[Ni(NH_3)_2Cl_2]$
(c) $[Fe(NH_3)_4Cl_2]^+$ (d) $[Co(NH_3)_4Cl_2]^+$

17. Number of complexes which show optical isomerism among the following is ______.

cis – $[Cr(ox)_2Cl_2]^{3-}$, $[Co(en)_3]^{3+}$, cis – $[Pt(en)_2Cl_2]^{2+}$. cis – $[Co(en)_2Cl_2]^+$, trans – $[Pt(en)_2Cl_2]^{2+}$, trans – $[Cr(ox)_2Cl_2]^{3-}$ **[NCERT : P L-126 | NV, Jan. 30, 2024 (II)]**

18. The correct IUPAC name of K_2MnO_4 is

[NCERT : P L-123 | Jan. 29, 2024 (II)]

(a) Potassium tetraoxidomanganese (VI)
(b) Potassium tetraoxidomanganate (VI)
(c) Dipotassium tetraoxidomanganate (VII)
(d) Potassium tetraoxopermanganate (VI)

19. The volume (in mL) of 0.1 M $AgNO_3$ required for complete precipitation of chloride ions present in 20 mL of 0.01 M solution of $[Cr(H_2O)_5Cl]Cl_2$ as silver chloride is ______

[NV, April 15, 2023 (I)]

20. The total number of stereoisomers for the complex $[Cr(ox)_2ClBr]^{3-}$ (where ox = oxalate) is:

[NCERT : P L-126 | April 13, 2023 (II)]

(a) 2 (b) 3 (c) 1 (d) 4

21. The set which does not have ambidentate ligand(s) is

[NCERT : P L-121 | April 11, 2023 (I)]

(a) $C_2O_4^{2-}$, ethylene diammine, H_2O
(b) $EDTA^{4-}, NCS^-, C_2O_4^{2-}$
(c) $NO_2^-, C_2O_4^{2-}, EDTA^{4-}$
(d) $C_2O_4^{2-}, NO_2^-, NCS^-$

22. Which of the following complex has a possibility to exist as meridional isomer?

[NCERT : P L-126 | April 11, 2023 (I)]

(a) $[Co(NH_3)_3(NO_2)_3]$ (b) $[Co(en)_3]$
(c) $[Co(en)_2Cl_2]$ (d) $[Pt(NH_3)_2Cl_2]$

23. The IUPAC name of $K_3[Co(C_2O_4)_3]$ is:- **[NCERT : P L-123 | April 06, 2023 (II)]**
(a) Potassium trioxalatocobaltate (III)
(b) Potassium tris(oxalato)cobalt (III)
(c) Potassium tris(oxalato) cobaltate (III)
(d) Potassium trioxalatocobalt (III)

24. Number of ambidentate ligands in a representative metal complex $[M(en)(SCN)_4]$ is [en = ethylenediamine] **[NCERT : P L-122 | NV, April 06, 2023 (I)]**

25. The complex cation which has two isomers is : **[NCERT : P L-127 | Feb. 01, 2023 (II)]**
(a) $[Co(NH_3)_5Cl]^{2+}$ (b) $[Co(NH_3)_5Cl]^{+}$
(c) $[Co(NH_3)_5NO_2]^{2+}$ (d) $[Co(H_2O)_6]^{3+}$

26. Which of the following are the example of double salt? **[NCERT : P L-120 | Feb. 01, 2023 (II)]**
(A) $FeSO_4.(NH_4)_2SO_4.6H_2O$
(B) $CuSO_4.4NH_3.H_2O$
(C) $K_2SO_4.Al_2(SO_4)_3.24H_2O$
(D) $Fe(CN)_2.4KCN$
Choose the correct answer
(a) A, B and D only (b) A and B only
(c) B and D only (d) A and C only

27. 1 L, 0.02 M solution of $[Co(NH_3)_5SO_4]Br$ is mixed with 1L, 0.02 M solution of $[Co(NH_3)_5Br]SO_4$. The resulting solution is divided into two equal parts (X) and treated with excess $AgNO_3$ solution and $BaCl_2$ solution respectively as shown below: **[Jan. 30, 2023 (II)]**
1 L Solution (X) + $AgNO_3$ solution (excess) $\rightarrow$ Y
1 L Solution (X) + $BaCl_2$ solution (excess) $\rightarrow$ Z
The number of moles of Y and Z respectively are
(a) 0.02, 0.02 (b) 0.01, 0.01
(c) 0.02, 0.01 (d) 0.01, 0.02

28. Chiral complex from the following is : **[NCERT : P L-126 | Jan. 29, 2023 (I)]**
Here en = ethylene diamine
(a) cis – $[PtCl_2(en)_2]^{2+}$ (b) trans – $[PtCl_2(en)_2]^{2+}$
(c) cis – $[PtCl_2(NH_3)_2]$ (d) trans – $[Co(NH_3)_4Cl_2]^{+}$

29. The denticity of the ligand present in the Fehling's reagent is _______ . **[NCERT : P L-122 | NV, Jan. 29, 2023 (II)]**

30. Total number of moles of AgCl precipitated on addition of excess of $AgNO_3$ to one mole each of the following complexes $[Co(NH_3)_4Cl_2]Cl$, $[Ni(H_2O)_6]Cl_2$, $[Pt(NH_3)_2Cl_2]$ and $[Pd(NH_3)_4]Cl_2$ is **[NCERT : P L-199 | NV, Jan. 25, 2023 (II)]**

31. The primary and secondary valencies of cobalt respectively in $[Co(NH_3)_5Cl]Cl_2$ are : **[NCERT : P L-118 | Jan. 24, 2023 (I)]**
(a) 3 and 5 (b) 2 and 6 (c) 2 and 8 (d) 3 and 6

32. Sum of oxidation state (magnitude) and coordination number of cobalt in $Na[Co(bpy)Cl_4]$ is____.
(Given bpy = [2,2'-bipyridine structure with two N atoms])
[NCERT : P L-121 | NV, July 29, 2022 (II)]

33. Total number of relatively more stable isomer(s) possible for octahedral complex $[Cu(en)_2(SCN)_2]$ will be _______. **[NCERT : P L-125 | NV, July 28, 2022 (I)]**

34. The conductivity of a solution of complex with formula $CoCl_3(NH_3)_4$ corresponds to 1 : 1 electrolyte, then the primary valency of central metal ion is______ **[NCERT : P L-118 | NV, July 27, 2022 (I)]**

35. In the following brown complex, the oxidation state of iron is + ____. **[NV, June 30, 2022 (I)]**
$$[Fe(H_2O)_6]^{2+} + NO \rightarrow \underset{\text{Brown complex}}{[Fe(H_2O)_5(NO)]^{2+}} + H_2O$$

36. Correct formula of the compound which gives a white precipitate with $BaCl_2$ solution, but not with $AgNO_3$ solution, is **[NCERT : P L-127 | June 30, 2022 (I)]**
(a) $[Co(NH_3)_5Br]SO_4$ (b) $[Co(NH_3)_5SO_4]Br$
(c) $[Pt(NH_3)_4Cl_2]Br_2$ (d) $[Pt(NH_3)_4Br_2]Cl_2$

37. For the reaction given below: $CoCl_3.xNH_3 + AgNO_3\,(aq) \rightarrow$
If two equivalents of AgCl precipitate out, then the value of x will be_____. **[NCERT : P L-199 | NV, June 29, 2022 (II)]**

38. Given below are two statements. **[June 29, 2022 (II)]**
Statement I : In $CuSO_4.5H_2O$, Cu–O bonds are present.
Statement II : In $CuSO_4.5H_2O$, ligands coordinating with Cu(II) ion are O-and S-based ligands.
In the light of the above statements, choose the correct answer from the options given below:
(a) Both Statement I and Statement II are correct.
(b) Both Statement I and Statement II are incorrect.
(c) Statement I is correct but Statement II is incorrect.
(d) Statement I is incorrect but Statement II is correct.

39. Number of complexes which will exist in cis trans is/are
(A) $CoCl_3{\cdot}4NH_3$ (B) $CoCl_3{\cdot}5NH_3$
(C) $CoCl_3{\cdot}6NH_3$ and (D) $CoCl(NO_3)_2{\cdot}5NH_3$
[NCERT : P L-125 | NV, June 28, 2022 (II)]

40. White precipitate of AgCl dissolves in aqueous ammonia solution due to formation of : **[June 25, 2022 (I)]**
(a) $[Ag(NH_3)_4]Cl_2$ (b) $[Ag(Cl)_2(NH_3)_2]$
(c) $[Ag(NH_3)_2]Cl$ (d) $[Ag(NH_3)Cl]Cl$

41. The sum of oxidation states of two silver ions in $[Ag(NH_3)_2][Ag(CN)_2]$ complex is_______. **[NCERT : P L-122 | NV, Sep. 1, 2021 (II)]**

42. The denticity of an organic ligand, biuret is : **[Aug. 31, 2021 (I)]**
(a) 2 (b) 4
(c) 3 (d) 6

43. The number of hydrogen bonded water molecule(s) associated with stoichiometry $CuSO_4{\cdot}5H_2O$ is_____. **[NCERT : P L-120 | NV, Aug. 31, 2021 (I)]**

44. 1 mol of an octahedral metal complex with formula $MCl_3{\cdot}2L$ on reaction with excess of $AgNO_3$ gives 1 mol of AgCl. The denticity of Ligand L is ________. (Integer answer) **[NV, Aug. 27, 2021 (I)]**

45. The number of optical isomers possible for $[Cr(C_2O_4)_3]^{3-}$ is _______. **[NCERT : P L-126 | NV, Aug. 27, 2021 (II)]**

46. The ratio of number of water molecules in Mohr's salt and potash alum is ________ $\times 10^{-1}$. (Integer answer) **[NV, Aug. 26, 2021 (I)]**

47. Indicate the complex/complex ion which did not show any geometrical isomerism : [NCERT : P L-125 | NV, Aug. 26, 2021 (II)]
(a) $[CoCl_2(en)_2]$ (b) $[Co(CN)_5(NC)]^{3-}$
(c) $[Co(NH_3)_3(NO_2)_3]$ (d) $[Co(NH_3)_4Cl_2]^+$

48. The number of geometrical isomers found in the metal complexes $[PtCl_2(NH_3)_2]$, $[Ni(CO)_4]$, $[Ru(H_2O)_3Cl_3]$ and $[CoCl_2(NH_3)_4]^+$ respectively, are : [July 27, 2021 (I)]
(a) 1, 1, 1, 1 (b) 2, 1, 2, 2
(c) 2, 0, 2, 2 (d) 2, 1, 2, 1

49. The number of geometrical isomers possible in triammine-trinitrocobalt (III) is X and in trioxalatochromate (III) is Y. Then the value of X + Y is ______. [NCERT : P L-125 | NV, July 27, 2021 (I)]

50. 3 moles of metal complex with formula $Co(en)_2 Cl_3$ gives 3 moles of silver chloride on treatment with excess of silver nitrate. The secondary valency of Co in the complex is ______. [NCERT : P L-118 | NV, July 27, 2021 (II)] (Round off to the nearest integer)

51. Three moles of AgCl get precipitated when one mole of an octahedral co-ordination compound with empirical formula $CrCl_3 \cdot 3NH_3 \cdot 3H_2O$ reacts with excess of silver nitrate. The number of chloride ions satisfying the secondary valency of the metal ion is______ [NCERT : P L-118 | NV, July 25, 2021 (I)]

52. Which one of the following metal complexes is most stable? [July 25, 2021 (II)]
(a) $[Co(en)(NH_3)_4]Cl_2$ (b) $[Co(en)_3]Cl_2$
(c) $[Co(en)_2(NH_3)_2]Cl_2$ (d) $[Co(NH_3)_6]Cl_2$

53. The total number of unpaired electrons present in the complex $K_3[Cr(oxalate)_3]$ is__________. [NCERT : P L-129 | NV, March 18, 2021 (I)]

54. On complete reaction of $FeCl_3$ with oxalic acid in aqueous solution containing KOH, resulted in the formation of product A. The secondary valency of Fe in the product A is ______. (Round off to the Nearest integer). [NV, March 17, 2021 (II)]

55. The equivalents of ethylene diamine required to replace the neutral ligands from the coordination sphere of the trans-complex of $CoCl_2 \cdot 4NH_3$ is ____ (Round off to the nearest integer). [NV, March 16, 2021 (I)]

56. The number of stereoisomers possible for $[Co(ox)_2(Br)(NH_3)]^{2-}$ is ________. [ox = oxalate] [NCERT : P L-126 | NV, Feb. 26, 2021 (II)]

57. The total number of coordination sites in ethylene-diaminetetraacetate ($EDTA^{4-}$) is ______. [NCERT : P L-121 | NV, Sep. 05, 2020 (I)]

58. Consider the complex ions, *trans*-$[Co(en)_2Cl_2]^+$ (A) and *cis*-$[Co(en)_2Cl_2]^+$ (B). The correct statement regarding them is : [NCERT : P L-126 | Sep. 05, 2020 (II)]
(a) both (A) and (B) cannot be optically active.
(b) (A) can be optically active, but (B) cannot be optically active.
(c) both (A) and (B) can be optically active.
(d) (A) cannot be optically active, but (B) can be optically active.

59. The number of isomers possible for $[Pt(en)(NO_2)_2]$ is : [NCERT : P L-127 | Sep. 04, 2020 (I)]
(a) 2 (b) 4 (c) 1 (d) 3

60. The complex that can show optical activity is : [NCERT : P L-126 | Sep. 03, 2020 (I)]
(a) *trans*-$[Cr(Cl_2)(ox)_2]^{3-}$
(b) *trans*-$[Fe(NH_3)_2(CN)_4]^-$
(c) *cis*-$[Fe(NH_3)_2(CN)_4]^-$
(d) *cis*-$[CrCl_2(ox)_2]^{3-}$ (ox = oxalate)

61. Complex A has a composition of $H_{12}O_6Cl_3Cr$. If the complex on treatment with conc. H_2SO_4 loses 13.5% of its original mass, the correct molecular formula of A is : [Given : atomic mass of Cr = 52 amu and Cl = 35 amu] [Sep. 03, 2020 (II)]
(a) $[Cr(H_2O)_6]Cl_3$
(b) $[Cr(H_2O)_3Cl_3] \cdot 3H_2O$
(c) $[Cr(H_2O)_5Cl]Cl_2 \cdot H_2O$
(d) $[Cr(H_2O)_4Cl_2]Cl \cdot 2H_2O$

62. The one that is not expected to show isomerism is : [NCERT : P L-127 | Sep. 02, 2020 (II)]
(a) $[Ni(NH_3)_4(H_2O)_2]^{2+}$ (b) $[Ni(en)_3]^{2+}$
(c) $[Ni(NH_3)_2Cl_2]$ (d) $[Pt(NH_3)_2Cl_2]$

63. Complex X of composition $Cr(H_2O)_6Cl_n$ has a spin only magnetic moment of 3.83 BM. It reacts with $AgNO_3$ and shows geometrical isomerism. The IUPAC nomenclature of X is: [NCERT : P L-122 | Jan. 09, 2020 (I)]
(a) Hexaaqua chromium (III) chloride
(b) Tetraaquadichlorido chromium (IV) chloride dihydrate
(c) Dichloridotetraaqua chromium (IV) chloride dihydrate
(d) Tetraaquadichlorido chromium (III) chloride dihydrate

64. The isomer(s) of $[Co(NH_3)_4Cl_2]$ that has/have a Cl–Co–Cl angle of 90°, is/are : [NCERT : P L-125 | Jan. 09, 2020 (II)]
(a) meridional and *trans* (b) *cis* and *trans*
(c) *trans* only (d) *cis* only

65. The complex that can show *fac*- and *mer*- isomers is: [NCERT : P L-120 | Jan. 08, 2020 (I)]
(a) $[Co(NH_3)_4Cl_2]^+$ (b) $[Pt(NH_3)_2Cl_2]$
(c) $[CoCl_2(en)_2]$ (d) $[Co(NH_3)_3(NO_2)_3]$

66. Among (1) - (4), the complexes that can display geometrical isomerism are: [NCERT : P L-125 | Jan. 08, 2020 (II)]
(1) $[Pt(NH_3)_3Cl]^+$ (2) $[Pt(NH_3)Cl_5]^-$
(3) $[Pt(NH_3)_2Cl(NO_2)]$ (4) $[Pt(NH_3)_4ClBr]^{2+}$
(a) (2) and (3) (b) (4) and (1)
(c) (3) and (4) (d) (1) and (2)

67. The IUPAC name of the complex $[Pt(NH_3)_2Cl(NH_2CH_3)]Cl$ is: [NCERT : P L-123 | Jan. 07, 2020 (I)]
(a) Diamminechlorido (methanamine) platinum (II)chloride
(b) Diammine (methanamine) chlorido platinum (II)chloride

(c) Diamminechlorido (aminomethane) platinum (II)chloride
(d) Bisammine (methanamine) chlorido platinum (II)chloride

68. The number of possible optical isomers for the complexes MA_2B_2 with sp^3 and dsp^2 hybridized metal atom, respectively, is:
Note: A and B are unidentate neutral and unidentate monoanionic ligands, respectively. **[Jan. 07, 2020 (II)]**
(a) 0 and 2 (b) 2 and 2
(c) 0 and 0 (d) 0 and 1

69. The coordination numbers of Co and Al in $[Co(Cl)(en)_2]Cl$ and $K_3[Al(C_2O_4)_3]$, respectively, are :
(en = ethane-1, 2-diamine)
[NCERT : PL-121 | April 12, 2019 (II)]
(a) 5 and 3 (b) 3 and 3 (c) 6 and 3 (d) 5 and 6

70. The species that can have a trans-isomer is :
(en = ethane – 1, 2–diamine, Ox = oxalate)
[NCERT : PL-126 | April 10, 2019 (I)]
(a) $[Zn(en)Cl_2]$ (b) $[Pt(en)Cl_2]$
(c) $[Cr(en)_2(Ox)]^+$ (d) $[Pt(en)_2Cl_2]^{2+}$

71. The maximum possible denticities of a ligand given below towards a common transition and inner-transition metal ion, respectively, are: **[NCERT : PL-122 | April 9, 2019 (II)]**

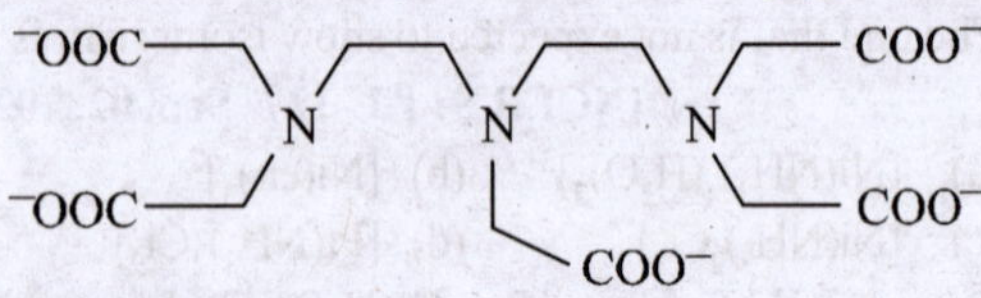

(a) 8 and 6 (b) 6 and 8 (c) 6 and 6 (d) 8 and 8

72. The one that will show optical activity is:
(en = ethane 1, 2-diamine)
[NCERT : PL-126 | April 9, 2019 (I)]

(a) 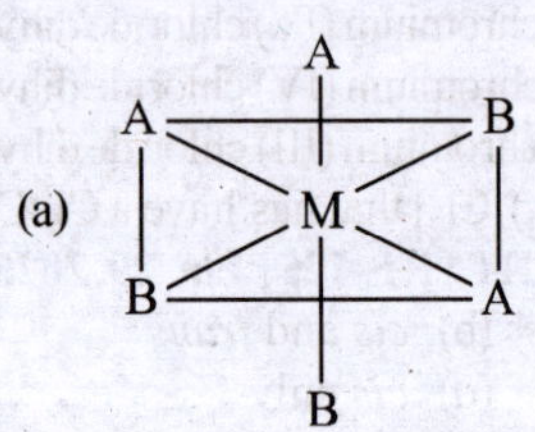

(b) 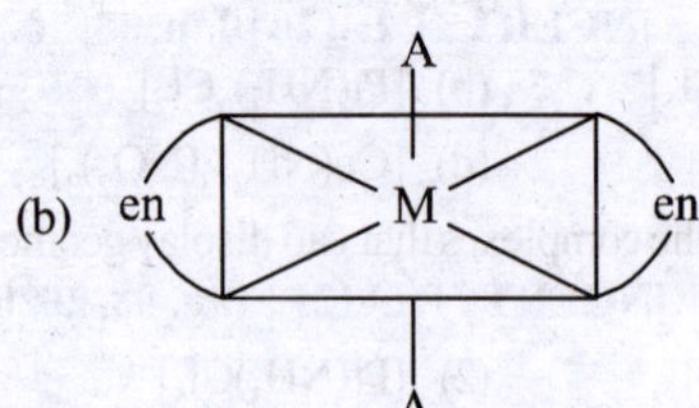

(c) 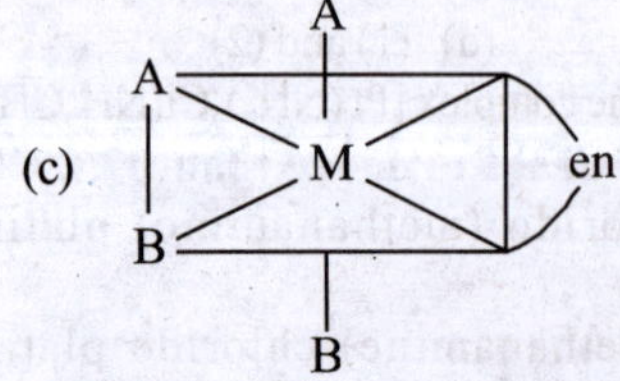

(d) 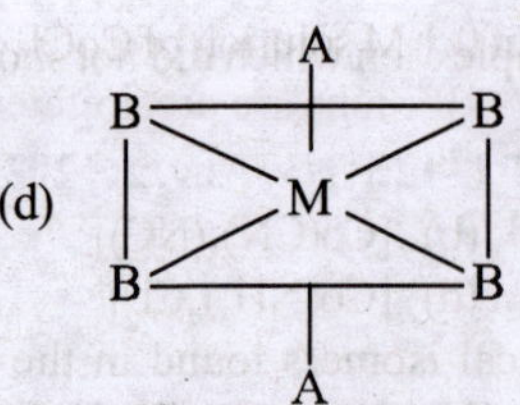

73. The following ligand is: **[April 8, 2019 (I)]**

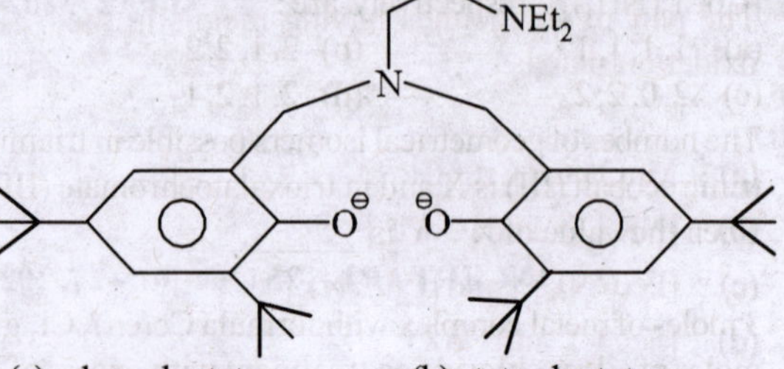

(a) hexadentate (b) tetradentate
(c) bidentate (d) tridentate

74. The coordination number of Th in $K_4[Th(C_2O_4)_4(H_2O)_2]$ is: ($C_2O_4^{2-}$ = oxalato)
[NCERT : PL-122 | Jan. 11, 2019 (II)]
(a) 14 (b) 6 (c) 8 (d) 10

75. The total number of isomers for a square planar complex $[M(F)(Cl)(SCN)(NO_2)]$ is:
[NCERT : PL-127 | Jan. 10, 2019 (I)]
(a) 16 (b) 8 (c) 4 (d) 12

76. A reaction of cobalt(III) chloride and ethylenediamine in a 1 : 2 mole ratio generates two isomeric products A (violet coloured) and B (green coloured). A can show optical activity, but, B is optically inactive. What type of isomers does A and B represent?
[NCERT : PL-125 | Jan. 10, 2019 (II)]
(a) Geometrical isomers (b) Coordination isomers
(c) Linkage isomers (d) Ionisation isomers

77. Consider the following reaction and statements: **[2018]**
$[Co(NH_3)_4Br_2]^+ + Br^- \rightarrow [Co(NH_3)_3Br_3] + NH_3$
(I) Two isomers are produced if the reactant complex ion is a *cis*-isomer.
(II) Two isomers are produced if the reactant complex ion is a *trans* -isomer.
(III) Only one isomer is produced if the reactant complex ion is a *trans* -isomer.
(IV) Only one isomer is produced if the reactant complex ion is a *cis*-isomer.
The correct statements are:
(a) (I) and (II) (b) (I) and (III)
(c) (III) and (IV) (d) (II) and (IV)

78. Which of the following complexes will show geometrical isomerism? **[NCERT : PL-125 | Online April 16, 2018]**
(a) Potassium tris(oxalato) chromate (III)
(b) Pentaaquachlorochromium (III) chloride
(c) Aquachlorobis (ethylenediamine) cobalt (II) chloride
(d) Potassium amminetrichloroplatinate (II)

79. The total number of possible isomers for square-planar $[Pt(Cl)(NO_2)(NO_3)(SCN)]^{2-}$ is:
[NCERT : PL-127 | Online April 15, 2018 (II)]
(a) 16 (b) 12 (c) 8 (d) 24

80. On treatment of 100 mL of 0.1 M solution of $CoCl_3 \cdot 6H_2O$ with excess $AgNO_3$; 1.2×10^{22} ions are precipitated. The complex is : [2017]
(a) $[Co(H_2O)_4Cl_2]Cl \cdot 2H_2O$
(b) $[Co(H_2O)_3Cl_3] \cdot 3H_2O$
(c) $[Co(H_2O)_6]Cl_3$
(d) $[Co(H_2O)_5Cl]Cl_2 \cdot H_2O$

81. The pair of compounds having metals in their highest oxidation state is :
[NCERT : P L-122 | Online April 8, 2017]
(a) MnO_2 and CrO_2Cl_2
(b) $[NiCl_4]^{2-}$ and $[CoCl_4]^{2-}$
(c) $[Fe(CN)_6]^{3-}$ and $[Cu(CN)_4]^{2-}$
(d) $[FeCl_4]^-$ and Co_2O_3

82. Which one of the following complexes will consume more equivalents of aqueous solution of $AgNO_3$?
[Online April 9, 2016]
(a) $Na_2[CrCl_5(H_2O)]$ (b) $Na_3[CrCl_6]$
(c) $[Cr(H_2O)_5Cl]Cl_2$ (d) $[Cr(H_2O)_6]Cl_3$

83. Which of the following is an example of homoleptic complex? [NCERT : P L-122 | Online April 10, 2016]
(a) $[Co(NH_3)_6]Cl_3$ (b) $[Pt(NH_3)_2Cl_2]$
(c) $[Co(NH_3)_4Cl_2]$ (d) $[Co(NH_3)_5Cl]Cl_2$

84. The number of geometric isomers that can exist for square planar complex $[Pt (Cl) (py) (NH_3) (NH_2OH)]^+$ is (py = pyridine) : [NCERT : P L-125 | 2015]
(a) 4 (b) 6 (c) 2 (d) 3

85. The correct statement on the isomerism associated with the following complex ions,
[NCERT : P L-125 | Online April 10, 2015]
(A) $[Ni(H_2O)_5NH_3]^{2+}$,
(B) $[Ni(H_2O)_4(NH_3)_2]^{2+}$ and
(C) $[Ni(H_2O)_3(NH_3)_3]^{2+}$ is :
(a) (A) and (B) show only geometrical isomerism
(b) (A) and (B) show geometrical and optical isomerism
(c) (B) and (C) show geometrical and optical isomerism
(d) (B) and (C) show only geometrical isomerism

86. Which of the following name formula combinations is not correct? [NCERT : P L-123 | Online April 11, 2014]

	Formula	Name
(a)	$K_2[Pt(CN)_4]$	Potassium tetracyanoplatinate (II)
(b)	$[Mn(CN)_5]^{2-}$	Pentacyanomangnate (II) ion
(c)	$K[Cr(NH_3)_2Cl_4]$	Potassium diammine tetrachlorochromate (III)
(d)	$[Co(NH_3)_4(H_2O)I]SO_4$	Tetraammineaquaiodido cobalt (III) sulphate.

87. An octahedral complex with molecular composition $M \cdot 5NH_3 \cdot Cl \cdot SO_4$ has two isomers, A and B. The solution of A gives a white precipitate with $AgNO_3$ solution and the solution of B gives white precipitate with $BaCl_2$ solution. The type of isomerism exhibited by the complex is: [NCERT : P L-127 | Online April 19, 2014]
(a) Linkage isomerism (b) Ionisation isomerism
(c) Coordinate isomerism (d) Geometrical isomerism

88. Which of the following complex species is not expected to exhibit optical isomerism ? [NCERT : P L-126 | 2013]
(a) $[Co(en)_3]^{3+}$ (b) $[Co(en)_2Cl_2]^+$
(c) $[Co(NH_3)_3Cl_3]$ (d) $[Co(en)(NH_3)_2Cl_2]^+$

89. Type of isomerism which exists between $[Pd(C_6H_5)_2(SCN)_2]$ and $[Pd(C_6H_5)_2(NCS)_2]$ is :
[NCERT : P L-127 | Online April 9, 2013]
(a) Linkage isomerism
(b) Coordination isomerism
(c) Ionisation isomerism
(d) Solvate isomerism

90. Which among the following will be named as dibromido bis (ethylenediamine) chromium (III) bromide?
[NCERT : P L-123 | 2012]
(a) $[Cr(en)_3]Br_3$ (b) $[Cr(en)_2Br_2]Br$
(c) $[Cr(en)Br_4]^-$ (d) $[Cr(en)Br_2]Br$

91. Which one of the following complex ions has geometrical isomers ? [NCERT : P L-125 | 2011]
(a) $[Ni(NH_3)_5Br]^+$ (b) $[Co(NH_3)_2(en)_2]^{3+}$
(c) $[Cr(NH_3)_4(en)_2]^{3+}$ (d) $[Co(en)_3]^{3+}$
(en-ethylenediamine)

92. A solution containing 2.675 g of $CoCl_3 \cdot 6NH_3$ (molar mass = 267.5 g mol^{-1}) is passed through a cation exchanger. The chloride ions obtained in solution were treated with excess of $AgNO_3$ to give 4.78 g of AgCl (molar mass = 143.5 g mol^{-1}). The formula of the complex is
(At. mass of Ag = 108 u) [2010]
(a) $[Co(NH_3)_6]Cl_3$ (b) $[CoCl_2(NH_3)_4]Cl$
(c) $[CoCl_3(NH_3)_3]$ (d) $[CoCl(NH_3)_5]Cl_2$

93. Which one of the following has an optical isomer?
[NCERT : P L-125 | 2010]
(a) $[Zn(en)(NH_3)_2]^{2+}$ (b) $[Co(en)_3]^{3+}$
(c) $[Co(H_2O)_4(en)]^{3+}$ (d) $[Zn(en)_2]^{2+}$
(en = ethylenediamine)

94. Which of the following shows optical isomerism
[NCERT : P L-126 | 2009]
(a) $[Co(en)(NH_3)_2]^{2+}$ (b) $[Co(H_2O)_4(en)]^{3+}$
(c) $[Co(en)_2(NH_3)_2]^{3+}$ (d) $[Co(NH_3)_3Cl]^+$

95. Which of the following pairs represent linkage isomers?
[NCERT : P L-127 | 2009]
(a) $[Pd(PPh_3)_2(NCS)_2]$ and $[Pd(PPh_3)_2(SCN)_2]$
(b) $[Co(NH_3)_5NO_3]SO_4$ and $[Co(NH_3)_5SO_4]NO_3$
(c) $[PtCl_2(NH_3)_4]Br_2$ and $[PtBr_2(NH_3)_4]Cl_2$
(d) $[Cu(NH_3)_4][PtCl_4]$ and $[Pt(NH_3)_4][CuCl_4]$

96. The coordination number and the oxidation state of the element 'E' in the complex $[E(en)_2(C_2O_4)]NO_2$ (where (en) is ethylene diamine) are, respectively, [NCERT : P L-121 | 2008]
(a) 6 and 2 (b) 4 and 2 (c) 4 and 3 (d) 6 and 3

97. The IUPAC name for the complex $[Co(NO_2)(NH_3)_5]Cl_2$ is : [NCERT : P L-123 | 2006]
(a) pentaamminenitrito-N-cobalt(II) chloride
(b) pentaamminenitrito-N-cobalt(III) chloride
(c) nitrito-N-pentaamminecobalt(III) chloride
(d) nitrito-N-pentaamminecobalt(II) chloride

98. The oxidation state Cr in $[Cr(NH_3)_4Cl_2]^+$ is [NCERT : P L-122 | 2005]
(a) 0 (b) +1 (c) +2 (d) +3

99. The IUPAC name of the coordination compound $K_3[Fe(CN)_6]$ is [NCERT : P L-123 | 2005]
(a) Tripotassium hexacyanoiron (II)
(b) Potassium hexacyanoiron (II)
(c) Potassium hexacyanoferrate (III)
(d) Potassium hexacyanoferrate (II)

100. The coordination number of a central metal atom in a complex is determined by [NCERT : P L-121 | 2004]
(a) the number of ligands around a metal ion bonded by sigma and pi-bonds both
(b) the number of ligands around a metal ion bonded by pi-bonds
(c) the number of ligands around a metal ion bonded by sigma bonds
(d) the number of only anionic ligands bonded to the metal ion.

101. Which one of the following has largest number of isomers? [NCERT : P L-127 | 2004]
(a) $[Ir(PR_3)_2H(CO)]^{2+}$ (b) $[Co(NH_3)_5Cl]^{2+}$
(c) $[Ru(NH_3)_4Cl_2]^+$ (d) $[Co(en)_2Cl_2]^+$
(R = alkyl group, en = ethylenediamine)

102. Among the properties (a) reducing (b) oxidising (c) complexing, the set of properties shown by CN^- ion towards metal species is [2004]
(a) c, a (b) b, c (c) a, b (d) a, b, c

103. One mole of the complex compound $Co(NH_3)_5Cl_3$, gives 3 moles of ions on dissolution in water. One mole of the same complex reacts with two moles of $AgNO_3$ solution to yield two moles of AgCl (s). The structure of the complex is [NCERT : P L-119 | 2003]
(a) $[Co(NH_3)_3Cl_3]\cdot 2NH_3$ (b) $[Co(NH_3)_4Cl_2]Cl\cdot NH_3$
(c) $[Co(NH_3)_4Cl]Cl_2\cdot NH_3$ (d) $[Co(NH_3)_5Cl]Cl_2$

104. In the coordination compound, $K_4[Ni(CN)_4]$, the oxidation state of nickel is [NCERT : P L-122 | 2003]
(a) 0 (b) +1 (c) +2 (d) –1

105. The type of isomerism present in nitropentammine chromium (III) chloride is [NCERT : P L-127 | 2002]
(a) optical (b) linkage
(c) ionization (d) polymerisation.

Topic 2 - Magnetic Moment, Valence Bond Theory and Crystal Field Theory

106. Match the **LIST-I** with **LIST-II** [NCERT : P L-128 | April 8, 2025 (II)]

LIST-I (Complex/Species)	LIST-II (Shape & magnetic moment)
A. $[Ni(CO)_4]$	I. Tetrahedral, 2.8 BM
B. $[Ni(CN)_4]^{2-}$	II. Square planar, 0 BM
C. $[NiCl_4]^{2-}$	III. Tetrahedral, 0 BM
D. $[MnBr_4]^{2-}$	IV. Tetrahedral, 5.9 BM

Choose the **correct** answer from the options given below:
(a) A-III, B-IV, C-II, D-I (b) A-I, B-II, C-III, D-IV
(c) A-III, B-II, C-I, D-IV (d) A-IV, B-I, C-III, D-II

107. The number of species from the following that are involved in sp^3d^2 hybridization is
$[Co(NH_3)_6]^{3+}$, SF_6, $[CrF_6]^{3-}$, $[CoF_6]^{3-}$, $[Mn(CN)_6]^{3-}$ and $[MnCl_6]^{3-}$ [NCERT : P L-128 | April 8, 2025 (II)]
(a) 5 (b) 6
(c) 4 (d) 3

108. The correct decreasing order of spin only magnetic moment values (BM) of Cu^+, Cu^{2+}, Cr^{2+} and Cr^{3+} ions is : [NCERT : P L-128 | April 8, 2025 (II)]
(a) $Cu^+ > Cu^{2+} > Cr^{3+} > Cr^{2+}$
(b) $Cu^{2+} > Cu^+ > Cr^{2+} > Cr^{3+}$
(c) $Cr^{2+} > Cr^{3+} > Cu^{2+} > Cu^+$
(d) $Cr^{3+} > Cr^{2+} > Cu^+ > Cu^{2+}$

109. The number of paramagnetic metal complex species among $[Co(NH_3)_6]^{3+}$, $[Co(C_2O_4)_3]^{3-}$, $[MnCl_6]^{3-}$, $[Mn(CN)_6]^{3-}$, $[CoF_6]^{3-}$, $[Fe(CN)_6]^{3-}$ and $[FeF_6]^{3-}$ with same number of unpaired electrons is ____. [NCERT : P L-128 | April 7, 2025 (II)]

110. The number of unpaired electrons responsible for the paramagnetic nature of the following complex species are respectively : [NCERT : P L-128 | April 7, 2025 (II)]
$[Fe(CN)_6]^{3-}$, $[FeF_6]^{3-}$, $[CoF_6]^{3-}$, $[Mn(CN)_6]^{3-}$
(a) 1, 5, 4, 2 (b) 1, 5, 5, 2
(c) 1, 1, 4, 2 (d) 1, 4, 4, 2

111. The number of paramagnetic complex among $[FeF_6]^{3-}$, $[Fe(CN)_6]^{3-}$, $[Mn(CN)_6]^{3-}$, $[Co(C_2O_4)_3]^{3-}$, $[MnCl_6]^{3-}$ and $[CoF_6]^{3-}$, which involved d^2sp^3 hybridization is .... [NCERT : P L-128 | NV, April 7, 2025 (I)]

112. 'X' is the number of electrons in t_{2g} orbitals of the most stable complex ion among $[Fe(NH_3)_6]^{3+}$, $[FeC_6]^{3-}$, $[Fe(C_2O_4)_3]^{3-}$ and $[Fe(H_2O)_6]^{3+}$. The nature of oxide of vanadium of the type V_2O_X is: [NCERT : P L-131 | April 4, 2025 (II)]
(a) Acidic (b) Neutral
(c) Basic (d) Amphoteric

113. The correct order of $[FeF_6]^{3-}$, $[CoF_6]^{3-}$, $[Ni(CO)_4]$ and $[Ni(CN)_4]^{2-}$ complex species based on the number of unpaired electrons present is :

[NCERT : P L-128 | April 4, 2025 (II)]

(a) $[FeF_6]^{3-} > [CoF_6]^{3-} > [Ni(CN)_4]^{2-} > [Ni(CO)_4]$
(b) $[Ni(CN)_4]^{2-} > [FeF_6]^{3-} > [CoF_6]^{3-} > [Ni(CO)_4]$
(c) $[CoF_6]^{3-} > [FeF_6]^{3-} > [Ni(CO)_4] > [Ni(CN)_4]^{2-}$
(d) $[FeF_6]^{3-} > [CoF_6]^{3-} > [Ni(CN)_4]^{2-} = [Ni(CO)_4]$

114. Which one of the following complexes will have $\Delta_0 = 0$ and $\mu = 5.96$ B.M.? **[NCERT : P L-128 | April 4, 2025 (I)]**

(a) $[Fe(CN)_6]^{4-}$ (b) $[CO(NH_3)_6]^{3+}$
(c) $[FeF_6]^{4-}$ (d) $[Mn(SCN)_6]^{4-}$

115. Identify the diamagnetic octahedral complex ions from below; **[NCERT : P L-128 | April 3, 2025 (II)]**

A. $[Mn(CN)_6]^{3-}$ B. $[Co(NH_3)_6]^{3+}$
C. $[Fe(CN)_6]^{4-}$ D. $[Co(H_2O)_3F_3]$

Choose the **correct** answer from the options given below:

(a) B and D Only (b) A and D Only
(c) A and C Only (d) B and C Only

116. The correct order of the complexes $[Co(NH_3)_5(H_2O)]^{3+}$ (A), $[Co(NH_3)_6]^{3+}$ (B), $[Co(CN)_6]^{3-}$ (C) and $[CoCl(NH_3)_5]^{2+}$ (D) in terms wavelength of light absorbed is :

[NCERT : P L-133 | April 3, 2025 (I)]

(a) D > A > B > C (b) C > B > D > A
(c) D > C > B > A (d) C > B > A > D

117. The metal ions that have the calculated spin only magnetic moment value of 4.9 B.M. are

[NCERT : P L-128 | April 3, 2025 (I)]

A. Cr^{2+} B. Fe^{2+}
C. Fe^{3+} D. Co^{2+}
E. Mn^{3+}

Choose the **correct** answer from the options given below

(a) A, C and E only (b) A, D and E only
(c) B and E only (d) A, B and E only

118. The type of hybridization and the magnetic property of $[MnCl_6]^{3-}$ are : **[NCERT : P L-128 | April 2, 2025 (II)]**

(a) d^2sp^3, paramagnetic with four unpaired electrons
(b) sp^3d^2, paramagnetic with four unpaired electrons
(c) d^2sp^3, paramagnetic with two unpaired electrons
(d) sp^3d^2, paramagnetic with two unpaired electrons

119. The d-orbital electronic configuration of the complex among $[Co(en)_3]^{3+}$, $[CoF_6]^{3-}$, $[Mn(H_2O)_6]^{2+}$ and $[Zn(H_2O)_6]^{2+}$ that has the highest CFSE is :

[NCERT : P L-131 | April 2, 2025 (II)]

(a) $t_{2g}^6e_g^0$ (b) $t_{2g}^6e_g^4$
(c) $t_{2g}^3e_g^2$ (d) $t_{2g}^4e_g^2$

120. A transition metal (M) among Mn, Cr, Co and Fe has the highest standard electrode potential (M^{3+}/M^{2+}). It forms a metal complex of the type $[M(CN)_6]^{4-}$. The number of electrons present in the e_g orbital of the complex is _____.

[NCERT : P L-131 | April 2, 2025 (I)]

121. Given below are two statements :

Statement (I) : In octahedral complexes, when $\Delta_o < P$ high spin complexes are formed. When $\Delta_o > P$ low spin complexes are formed.

Statement (II) : In tetrahedral complexes because of $\Delta_t < P$, low spin complexes are rarely formed.

In the light of the above statements, choose the **most appropriate answer** from the options given below :

[NCERT : P L-132 | April 2, 2025 (I)]

(a) **Statement I** is correct but **Statement II** is incorrect.
(b) Both **Statement I** and **Statement II** are incorrect
(c) **Statement I** is incorrect but **Statement II** is correct
(d) Both **Statement I** and **Statement II** are correct

122. Consider the following low-spin complexes

$K_3[Co(NO_2)_6]$, $K_4[Fe(CN)_6]$, $K_3[Fe(CN)_6]$, $Cu_2[Fe(CN)_6]$ and $Zn_2[Fe(CN)_6]$. **[NCERT : P L-130 | Jan. 29, 2025 (II)]**

The sum of the spin-only magnetic moment values of complexes having yellow colour is ______. B.M. (answer is nearest integer)

123. The calculated spin-only magnetic moments of $K_3[Fe(OH)_6]$ and $K_4[Fe(OH)_6]$ respectively are :

[NCERT : P L-130 | Jan. 29, 2025 (II)]

(a) 4.90 and 4.90 B.M. (b) 5.92 and 4.90 B.M.
(c) 3.87 and 4.90 B.M. (d) 4.90 and 5.92 B.M.

124. Match List-I with List-II.

[NCERT : P L-129 | Jan. 29, 2025 (I)]

List-I (Complex)	**List-II (Hybridisation & Magnetic characters)**
(A) $[MnBr_4]^{2-}$	(I) d^2sp^3 & diamagnetic
(B) $[FeF_6]^{3-}$	(II) sp^3d^2 & paramagnetic
(C) $[Co(C_2O_4)_3]^{3-}$	(III) sp^3 & diamagnetic
(D) $[Ni(CO)_4]$	(IV) sp^3 & paramagnetic

Choose the correct answer from the options given below :

(a) (A)-(III), (B)-(II), (C)-(I), (D)-(IV)
(b) (A)-(III), (B)-(I), (C)-(II), (D)-(IV)
(c) (A)-(IV), (B)-(I), (C)-(II), (D)-(III)
(d) (A)-(IV), (B)-(II), (C)-(I), (D)-(III)

125. The correct increasing order of stability of the complexes based on Δ_o value is : **[NCERT : P L-131 | Jan. 29, 2025 (I)]**

(I) $[Mn(CN)_6]^{3-}$ (II) $[Co(CN)_6]^{4-}$
(III) $[Fe(CN)_6]^{4-}$ (IV) $[Fe(CN)_6]^{3-}$

(a) II < III < I < IV (b) IV < III < II < I
(c) I < II < IV < III (d) III < II < IV < I

126. Match List-I with List-II.

[NCERT : P L-128 | Jan. 28, 2025 (II)]

List-I (Complex)	**List-II (Hybridisation of central metal ion)**
(A) $[CoF_6]^{3-}$	(I) d^2sp^3
(B) $[NiCl_4]^{2-}$	(II) sp^3
(C) $[Co(NH_3)_6]^{3+}$	(III) sp^3d^2
(D) $[Ni(CN)_4]^{2-}$	(IV) dsp^2

Choose the correct answer from the options given below :
(a) (A)-(I), (B)-(IV), (C)-(III), (D)-(II)
(b) (A)-(III), (B)-(II), (C)-(I), (D)-(IV)
(c) (A)-(I), (B)-(II), (C)-(III), (D)-(IV)
(d) (A)-(III), (B)-(IV), (C)-(I), (D)-(II)

127. The conditions and consequence that favours the t_{2g}^3, e_g^1 configuration in a metal complex are :
[NCERT : P L-131 | Jan. 24, 2025 (II)]
(a) weak field ligand, high spin complex
(b) strong field ligand, high spin complex
(c) strong field ligand, low spin complex
(d) weak field ligand, low spin complex

128. When Ethane-1,2-diamine is added progressively to an aqueous solution of Nickel (II) chloride, the sequence of colour change observed will be :
[NCERT : P L-133 | Jan. 24, 2025 (II)]
(a) Pale Blue → Blue → Green → Violet
(b) Pale Blue → Blue → Violet → Green
(c) Green → Pale Blue → Blue → Violet
(d) Violet → Blue → Pale Blue → Green

129. Identify the coordination complexes in which the central metal ion has d^4 configuration.
[NCERT : P L-122 | Jan. 23, 2025 (II)]
(A) $[FeO_4]^{2-}$
(B) $[Mn(CN)_6]^{3-}$
(C) $[Fe(CN)_6]^{3-}$
(D) $Cr_2(O-\overset{\overset{O}{\|}}{C}-Me)_4(H_2O)_2$
(E) $[NiF_6]^{2-}$
Choose the correct answer from the options given below:
(a) (C) and (E) only (b) (B), (C) and (D) only
(c) (B) and (D) only (d) (A), (B) and (E) only

130. The d-electronic configuration of an octahedral Co(II) complex having magnetic moment of 3.95 BM is:
[NCERT : P L-131 | Jan. 23, 2025 (I)]
(a) $t_{2g}^6\ e_g^1$ (b) $t_{2g}^3\ e_g^0$
(c) $t_{2g}^5\ e_g^2$ (d) $e^4 t_2^3$

131. The correct order of the following complexes in terms of their crystal field stabilization energies is:
[NCERT : P L-131 | Jan. 22, 2025 (II)]
(a) $[Co(NH_3)_4]^{2+} < [Co(NH_3)_6]^{2+} < [Co(en)_3]^{3+} < [Co(NH_3)_6]^{3+}$
(b) $[Co(NH_3)_4]^{2+} < [Co(NH_3)_6]^{2+} < [Co(NH_3)_6]^{3+} < [Co(en)_3]^{3+}$
(c) $[Co(NH_3)_6]^{2+} < [Co(NH_3)_6]^{3+} < [Co(NH_3)_4]^{2+} < [Co(en)_3]^{3+}$
(d) $[Co(en)_3]^{3+} < [Co(NH_3)_6]^{3+} < [Co(NH_3)_6]^{2+} < [Co(NH_3)_4]^{2+}$

132. From the magnetic behaviour of $[NiCl_4]^{2-}$ (paramagnetic) and $[Ni(CO)_4]$ (diamagnetic), choose the correct geometry and oxidation state: **[NCERT : P L-133 | Jan. 22, 2025 (I)]**
(a) $[NiCl_4]^{2-}$: Ni^{II}, square planar
$[Ni(CO)_4]$: Ni(0), square planar
(b) $[NiCl_4]^{2-}$: Ni^{II}, tetrahedral
$[Ni(CO)_4]$: Ni(0), tetrahedral
(c) $[NiCl_4]^{2-}$: Ni^{II}, tetrahedral
$[Ni(CO)_4]$: Ni^{II}, square planar
(d) $[NiCl_4]^{2-}$: Ni(0), tetrahedral
$[Ni(CO)_4]$: Ni(0), square planar

133. In which of the following complexes the CFSE, Δ_0 will be equal to zero? **[NCERT : P L-131 | Jan. 22, 2025 (I)]**
(a) $[Fe(NH_3)_6]Br_2$ (b) $[Fe(en)_3]Cl_3$
(c) $K_4[Fe(CN)_6]$ (d) $K_3[Fe(SCN)_6]$

134. Given below are two statements : one is labelled as **Assertion (A)** and the other is labelled as **Reason (R).**
[NCERT : P L-126 | April 9, 2024 (I)]
Assertion (A) : The total number of geometrical isomers shown by $[Co(en)_2Cl_2]^+$ complex ion is three
Reason (R) : $[Co(en)_2Cl_2]^+$ complex ion has an octahedral geometry.
In the light of the above statements, choose the **most appropriate** answer from the options given below :
(a) Both **(A)** and **(R)** are correct and **(R)** is the correct explanation of **(A).**
(b) **(A)** is correct but **(R)** is not correct
(c) **(A)** is not correct but **(R)** is correct
(d) Both **(A)** and **(R)** are correct but **(R)** is not the correct explanation of **(A)**

135. Match **List-I** with **List-II.**
[NCERT : P L-129 | April 9, 2024 (II)]

List-I	List-II
(A) $K_2[Ni(CN)_4]$	(I) sp^3
(B) $[Ni(CO)_4]$	(II) sp^3d^2
(C) $[Co(NH_3)_6]Cl_3$	(III) dsp^2
(D) $Na_3[CoF_6]$	(IV) d^2sp^3

Choose the correct answer from the options given below:
(a) A-(III), B-(I), C-(II), D-(IV)
(b) A-(III), B-(II), C-(IV), D-(I)
(c) A-(I), B-(III), C-(II), D-(IV)
(d) A-(III), B-(I), C-(IV), D-(II)

136. The coordination environment of Ca^{2+} ion in its complex with $EDTA^{4-}$ is : **[NCERT : P L-122 | April 9, 2024 (II)]**
(a) tetrahedral (b) octahedral
(c) square planar (d) trigonal prismatic

137. Given below are two statements:
[NCERT : P L-129 | April 8, 2024 (I)]
Statement I : $N(CH_3)_3$ and $P(CH_3)_3$ can act as ligands to form transition metal complexes.
Statement II: As N and P are from same group, the nature of bonding of $N(CH_3)_3$ and $P(CH_3)_3$ is always same with transition metals.
In the light of the above statements, choose the **most appropriate** answer from the options given below:

(a) Statement I is incorrect but Statement II is correct
(b) Both Statement I and Statement II are correct
(c) Statement I is correct but Statement II is incorrect
(d) Both Statement I and Statement II are incorrect

138. Number of complexes with even number of electrons in t_{2g} orbitals is - **[NCERT : P L-131 | April 8, 2024 (I)]**
$[Fe(H_2O)_6]^{2+}$, $[Co(H_2O)_6]^{2+}$, $[Co(H_2O)_6]^{3+}$, $[Cu(H_2O)_6]^{2+}$, $[Cr(H_2O)_6]^{2+}$
(a) 1 (b) 3 (c) 2 (d) 5

139. Match **List-I** with **List-II** **[NCERT : P L-130 | April 8, 2024 (II)]**

List-I (Complex ion)	**List-II (Spin only magnetic moment in B.M.)**
(A) $[Cr(NH_3)_6]^{3+}$	(I) 4.90
(B) $[NiCl_4]^{2-}$	(II) 3.87
(C) $[CoF_6]^{3-}$	(III) 0.0
(D) $[Ni(CN)_4]^{2-}$	(IV) 2.83

Choose the **correct** answer from the option given below:
(a) (A)-(I), (B)-(IV), (C)-(II), (D)-(III)
(b) (A)-(IV), (B)-(III), (C)-(I), (D)-(II)
(c) (A)-(II), (B)-(IV), (C)-(I), (D)-(III)
(d) (A)-(II), (B)-(III), (C)-(I), (D)-(IV)

140. Total number of unpaired electrons in the complex ion $[Co(NH_3)_6]^{3+}$ and $[NiCl_4]^{2-}$ is **[NCERT : P L-129 | NV, April 8, 2024 (II)]**

141. Consider the following complexes. **[NCERT : PL-132 | April 6, 2024 (I)]**
$[CoCl(NH_3)_5]^{2+}$ (A), $[Co(CN)_6]^{3-}$ (B),
$[Co(NH_3)_5(H_2O)]^{3+}$ (C), $[Cu(H_2O)_4]^{2+}$ (D)
The correct order of A, B, C and D in terms of wavenumber of light absorbed is :
(a) C < D < A < B (b) D < A < C < B
(c) A < C < B < D (d) B < C < A < D

142. Match **List-I** with **List-II** **[NCERT : P L-133 | April 6, 2024 (II)]**

List-I Tetrahedral Complex	**List-II Electronic configuration**
(A) $TiCl_4$	(I) e^2, t_2^0
(B) $[FeO_4]^{2-}$	(II) e^4, t_2^3
(C) $[FeCl_4]^-$	(III) e^0, t_2^0
(D) $[CoCl_4]^{2-}$	(IV) e^2, t_2^3

Choose the **correct** answer from the option given below:
(a) (A)-(I), (B)-(III), (C)-(IV), (D)-(II)
(b) (A)-(IV), (B)-(III), (C)-(I), (D)-(II)
(c) (A)-(III), (B)-(IV), (C)-(II), (D)-(I)
(d) (A)-(III), (B)-(I), (C)-(IV), (D)-(II)

143. Which one of the following complexes will exhibit the least paramagnetic behaviour ? **[NCERT : PL-130 | April 5, 2024 (I)]**
[Atomic number, Cr = 24, Mn = 25, Fe = 26, Co = 27]
(a) $[Co(H_2O)_6]^{2+}$ (b) $[Fe(H_2O)_6]^{2+}$
(c) $[Mn(H_2O)_6]^{2+}$ (d) $[Cr(H_2O)_6]^{2+}$

144. The correct order of ligands arranged in increasing field strength. **[NCERT : P L-132 | April 5, 2024 (I)]**
(a) $Cl^- < {}^-OH < Br^- < CN^-$
(b) $F^- < Br^- < I^- < NH_3$
(c) $Br^- < F^- < H_2O < NH_3$
(d) $H_2O < {}^-OH < CN^- < NH_3$

145. The number of complexes from the following with no electrons in the t_2 orbital is ______.
$TiCl_4$, $[MnO_4]^-$, $[FeO_4]^{2-}$, $[FeCl_4]^-$, $[CoCl_4]^{2-}$ **[NCERT : P L-133 | April 5, 2024 (II)]**
(a) 3 (b) 1 (c) 4 (d) 2

146. The correct sequence of ligands in the order of decreasing field strength is : **[NCERT : P L-132 | April 4, 2024 (I)]**
(a) $CO > H_2O > F^- > S^{2-}$
(b) ${}^-OH > F^- > NH_3 > CN^-$
(c) $NCS^- > EDTA^{4-} > CN^- > CO$
(d) $S^{2-} > {}^-OH > EDTA^{4-} > CO$

147. Number of complexes from the following with even number of unpaired "d" electrons is_____.
$[V(H_2O)_6]^{3+}$, $[Cr(H_2O)_6]^{2+}$, $[Fe(H_2O)_6]^{3+}$, $[Ni(H_2O)_6]^{3+}$, $[Cu(H_2O)_6]^{2+}$
[Given atomic numbers : V = 23, Cr = 24, Fe = 26, Ni = 28, Cu = 29] **[NCERT : P L-129 | April 4, 2024 (I)]**
(a) 2 (b) 4 (c) 5 (d) 1

148. If an iron (III) complex with the formula $\left[Fe(NH_3)_x(CN)_y\right]^-$ has no electron in its e_g orbital, then the value of x + y is **[NCERT : P L-133 | April 4, 2024 (II)]**
(a) 5 (b) 6 (c) 3 (d) 4

149. The number of unpaired d-electrons in $[Co(H_2O)_6]^{3+}$ is ______ **[NCERT : P L-133 | NV, April 4, 2024 (II)]**
(a) 4 (b) 2 (c) 0 (d) 1

150. Given below are two statements: **[NCERT : P L-134 | Feb. 1, 2024 (I)]**
Statement (I) : A solution of $[Ni(H_2O)_6]^{2+}$ is green in colour.
Statement (II) : A solution of $[Ni(CN)_4]^{2-}$ is colourless.
In the light of the above statements, choose the most appropriate answer from the options given below:
(a) Both Statement I and Statement II are incorrect
(b) Both Statement I and Statement II are correct
(c) Statement I is incorrect but Statement II is correct
(d) Statement I is correct but Statement II is incorrect.

151. $[Co(NH_3)_6]^{3+}$ and $[CoF_6]^{3-}$ are respectively known as: **[NCERT : P L-129 | Feb. 1, 2024 (II)]**
(a) Inner orbital Complex, Spin paired Complex
(b) Spin paired Complex, Spin free Complex
(c) Spin free Complex, Spin paired Complex
(d) Outer orbital Complex, Inner orbital Complex

152. The correct statements from following are:

[NCERT : P L-128 | Jan. 31, 2024 (I)]

A. The strength of anionic ligands can be explained by crystal field theory.

B. Valence bond theory does not give a quantitative interpretation of kinetic stability of coordination compounds.

C. The hybridization involved in formation of $[Ni(CN)_4]^{2-}$ complex is dsp^2.

D. The number of possible isomer(s) of cis-$[PtCl_2(en)_2]^{2+}$ is one

Choose the correct answer from the options given below:

(a) B, C only (b) A, C only

(c) A, D only (d) B, D only

153. The 'Spin only' Magnetic moment for $[Ni(NH_3)_6]^{2+}$ is ____ $\times 10^{-1}$ BM.

(given = Atomic number of Ni : 28)

[NCERT : P L-128 | NV, Jan. 31, 2024 (I)]

154. Match List I with List II

[NCERT : P L-131 | Jan. 31, 2024 (II)]

LIST - I (Complex ion)	LIST - II (Electronic Configuration)
A. $[Cr(H_2O)_6]^{3+}$	I. $t_{2g}^2 e_g^0$
B. $[Fe(H_2O)_6]^{3+}$	II. $t_{2g}^3 e_g^0$
C. $[Ni(H_2O)_6]^{2+}$	III. $t_{2g}^3 e_g^2$
D. $[V(H_2O)_6]^{3+}$	IV. $t_{2g}^6 e_g^2$

Choose the correct answer from the options given below:

(a) A-III, B-II, C-IV, D-I (b) A-IV, B-I, C-II, D-III

(c) A-IV, B-III, C-I, D-II (d) A-II, B-III, C-IV, D-I

155. Select the option with correct property -

[NCERT : P L-128 | Jan. 31, 2024 (II)]

(a) $[Ni(CO)_4]$ and $[NiCl_4]^{2-}$ both diamagnetic

(b) $[Ni(CO)_4]$ and $[NiCl_4]^{2-}$ both paramagnetic

(c) $[NiCl_4]^{2-}$ diamagnetic, $[Ni(CO)_4]$ paramagnetic

(d) $[Ni(CO)_4]$ diamagnetic, $[NiCl_4]^{2-}$ paramagnetic

156. Consider the following complex ions

[NCERT : P L-128 | Jan. 27, 2024 (I)]

$P = [FeF_6]^{3-}$

$Q = [V(H_2O)_6]^{2+}$

$R = [Fe(H_2O)_6]^{2+}$

The correct order of the complex ions, according to their spin only magnetic moment values (in B.M.) is :

(a) $R < Q < P$ (b) $R < P < Q$

(c) $Q < R < P$ (d) $Q < P < R$

157. The Spin only magnetic moment value of square planar complex $[Pt(NH_3)_2Cl(NH_2CH_3)]Cl$ is ________ B.M. (Nearest integer) **[NCERT : P L-128 | Jan. 27, 2024 (II)]**

(Given atomic number for Pt = 78)

158. The complex with highest magnitude of crystal field splitting energy (Δ_0) is

[NCERT : P L-131 | April 15, 2023 (I)]

(a) $[Cr(OH_2)_6]^{3+}$ (b) $[Ti(OH_2)_6]^{3+}$

(c) $[Fe(OH_2)_6]^{3+}$ (d) $[Mn(OH_2)_6]^{3+}$

159. The homoleptic and octahedral complex of Co^{2+} and H_2O has ______ unpaired electron(s) in the t_{2g} set of orbitals.

[NCERT : P L-132 | NV, April 15, 2023 (I)]

160. Which of the following complexes will exhibit maximum attraction to an applied magnetic field?

[NCERT : P L-128 | April 13, 2023 (II)]

(a) $[Zn(H_2O)_6]^{2+}$ (b) $[Co(H_2O)_6]^{2+}$

(c) $[Co(en)_3]^{3+}$ (d) $[Ni(H_2O)_6]^{2+}$

161. Match List I with List II

[NCERT : P L-131 | April 12, 2023 (I)]

	List I Complex		List II CFSE (Δ_0)
A.	$[Cu(NH_3)_6]^{2+}$	I.	–0.6
B.	$[Ti(N_2O)_6]^{3+}$	II.	–2.0
C.	$[Fe(CN)_6]^{3-}$	III.	–1.2
D.	$[NiF_6]^{4-}$	IV.	–0.4

Choose the correct answer from the options given below:

(a) A-I, B-IV, C-II, D-III (b) A-II, B-III, C-I, D-IV

(c) A-I, B-II, C-IV, D-III (d) A-III, B-IV, C-I, D-II

162. The complex that dissolves in water is **[April 11, 2023 (I)]**

(a) $Fe_4[Fe(CN)_6]_3$ (b) $[Fe_3(OH)_2(OAc)_6]Cl$

(c) $K_3[Co(NO_2)_6]$ (d) $(NH_4)_3[As(Mo_3O_{10})_4]$

163. The ratio of spin-only magnetic moment values $\mu_{eff}[Cr(CN)_6]^{3-} / \mu_{eff}[Cr(H_2O)_6]^{3+}$ is ______.

[NCERT : P L-128 | NV, April 11, 2023 (I)]

164. Given below are two statements, one is labelled as **Assertion A** and the other is labelled as **Reason R**.

Assertion A : $[CoCl(NH_3)_5]^{2+}$ absorbs at lower wavelength of light with respect to $[Co(NH_3)_5(H_2O)]^{3+}$

Reason R: It is because the wavelength of the light absorbed depends on the oxidation state of the metal ion.

In the light of the above statements, choose the correct answer from the options given below :

[NCERT : P L-133 | April 11, 2023 (II)]

(a) A is false but R is true.

(b) A is true but R is false.

(c) Both A and R are true and R is the correct explanation of A.

(d) Both A and R are true and R is NOT the correct explanation of A.

165. The magnetic moment is measured in Bohr Magneton (BM). **[NCERT : P L-128 | April 11, 2023 (II)]**

Spin only magnetic moment of Fe in $[Fe(H_2O)_6]^{3+}$ and $[Fe(CN)_6]^{3-}$ complexes respectively is:

(a) 6.92 B.M. in both

(b) 4.89 B.M. and 6.92 B.M.

(c) 3.87 B.M. and 1.732 B.M.

(d) 5.92 B.M. and 1.732 B.M.

166. Match List I with List II.
[NCERT : PL-133 | April 11, 2023 (II)]

	List I Complex		List II Colour
A.	$Mg(NH_4)PO_4$	I.	Brown
B.	$K_3[Co(NO_2)_6]$	II.	White
C.	$MnO(OH)_2$	III.	Yellow
D.	$Fe_4[Fe(CN)_6]_3$	IV.	Blue

Choose the correct answer from the options given below:
(a) A-II, B-III, C-I, D-IV (b) A-III, B-IV, C-II, D-I
(c) A-II, B-IV, C-I, D-III (d) A-II, B-III, C-IV, D-I

167. If Ni^{2+} is replaced by Pt^{2+} in the complex $[NiCl_2Br_2]^{2-}$, which of the following properties are expected to get changed?
[NCERT : PL-127 | April 11, 2023 (II)]
A. Geometry B. Geometrical isomerism
C. Optical isomerism D. Magnetic properties
(a) A, B and C (b) A, B and D
(c) A and D (d) B and C

168. The octahedral diamagnetic low spin complex among the following is **[NCERT : PL-132 | April 10, 2023 (I)]**
(a) $[NiCl_4]^{2-}$ (b) $[CoCl_6]^{3-}$
(c) $[CoF_6]^{3-}$ (d) $[Co(NH_3)_6]^{3+}$

169. In potassium ferrocyanide, there are ____ pairs of electrons in the t_2g set of orbitals
[NCERT : PL-133 | NV, April 10, 2023 (I)]

170. The correct order of the number of unpaired electrons in the given complexes is
[NCERT : PL-128 | April 10, 2023 (II)]
A. $[Fe(CN)_6]^{3-}$ B. $[FeF_6]^{3-}$
C. $[CoF_6]^{3-}$ D. $[Cr(oxalate)_3]^{3-}$
E. $[Ni(CO)_4]$
Choose the correct answer from the options given below:
(a) A < E < D < C < B (b) E < A < D < C < B
(c) E < A < B < D < C (d) A < E < C < B < D

171. Match List I with List II
[NCERT : PL-132 | April 10, 2023 (II)]

	List I (Complex)		List II (Crystal Field splitting energy (Δ_0))
A.	$[Ti(H_2O)_6]^{2+}$	I.	–1.2
B.	$[V(H_2O)_6]^{2+}$	II.	–0.6
C.	$[Mn(H_2O)_6]^{3+}$	III.	0
D.	$[Fe(H_2O)_6]^{3+}$	IV.	–0.8

Choose the correct answer from the options given below:
(a) A-II, B-IV, C-I, D-III (b) A-IV, B-I, C-II, D-III
(c) A-IV, B-I, C-III, D-II (d) A-II, B-IV, C-III, D-I

172. For a metal ion, the calculated magnetic moment is 4.90 BM. This metal ion has ________ number of unpaired electons. **[NCERT : PL-128 | NV, April 10, 2023 (II)]**

173. Which of the following complex is octahedral, diamagnetic and the most stable? **[April 08, 2023 (I)]**
(a) $Na_3[CoCl_6]$ (b) $[Ni(NH_3)_6]Cl_2$
(c) $K_3[Co(CN)_6]$ (d) $[Co(H_2O)_6]Cl_2$

174. The correct order of spin only magnetic moments for the following complex ions is
[NCERT : PL-130 | April 08, 2023 (I)]
(a) $[Fe(CN)_6]^{3-} < [CoF_6]^{3-} < [MnBr_4]^{2-} < [Mn(CN)_6]^{3-}$
(b) $[Fe(CN)_6]^{3-} < [Mn(CN)_6]^{3-} < [CoF_6]^{3-} < [MnBr_4]^{2-}$
(c) $[MnBr_4]^{2-} < [CoF_6]^{3-} < [Fe(CN)_6]^{3-} < [Mn(CN)_6]^{3-}$
(d) $[CoF_6]^{3-} < [MnBr_4]^{2-} < [Fe(CN)_6]^{3-} < [Mn(CN)_6]^{3-}$

175. Match List I with List II:
[NCERT : PL-128 | April 08, 2023 (II)]

	List I (Coordination complex)		List II (Number of unpaired electrons)
A.	$[Cr(CN)_6]^{3-}$	I.	0
B.	$[Fe(H_2O)_6]^{2+}$	II.	3
C.	$[Co(NH_3)_6]^{3+}$	III.	2
D.	$[Ni(NH_3)_6]^{2+}$	IV.	4

Choose the correct answer from the options given below:-
(a) A-II, B-IV, C-I, D-III (b) A-IV, B-III, C-II, D-I
(c) A-III, B-IV, C-I, D-II (d) A-II, B-I, C-IV, D-III

176. The observed magnetic moment of the complex $[Mn(\underline{N}CS)_6]^{x-}$ is 6.06 BM. The numerical value of x is ______ : **[NCERT : PL-128 | NV, April 08, 2023 (II)]**

177. Given below are two statements, one is labelled as **Assertion A** and the other is labelled as **Reason R.**
Assertion A: The spin only magnetic moment value for $[Fe(CN)_6]^{3-}$ is 1.74 BM, whereas for $[Fe(H_2O)_6]^{3+}$ is 5.92 BM.
Reason R : In both complexes, Fe is present in +3 oxidation state.
In the light of the above statements, choose the correct answer from the options given below:
[NCERT : PL-128 | April 06, 2023 (I)]
(a) Both A and R are true but R is NOT the correct explanation of A
(b) A is false but R is true
(c) A is true but R is false
(d) Both A and R are true and R is the correct explanation of A

178. Which of the following complex will show largest splitting of d-orbitals? **[NCERT : PL-131 | Feb. 01, 2023 (I)]**
(a) $[Fe(CN)_6]^{3-}$ (b) $[Fe(C_2O_4)_3]^{3-}$
(c) $[FeF_6]^{3-}$ (d) $[Fe(NH_3)_6]^{3+}$

179. The spin only magnetic moment of $[Mn(H_2O)_6]^{2+}$ complexes is ______ B.M. (Nearest integer)
(Given: Atomic no. of Mn is 25)
[NCERT : PL-128 | NV, Feb. 01, 2023 (II)]

180. Cobalt chloride when dissolved in water forms pink colored complex X which has octahedral geometry. This solution on treating with conc. HCl forms deep blue complex, Y which has a Z geometry. X, Y and Z, respectively, are
[Jan. 31, 2023 (I)]
(a) X= $[Co(H_2O)_6]^{2+}$, Y= $[CoCl_4]^{2-}$, Z = Tetrahedral
(b) X= $[Co(H_2O_6)]^{2+}$, Y= $[CoCl_6]^{3-}$, Z = Octahedral
(c) X= $[Co(H_2O)_6]^{3+}$, Y= $[CoCl_6]^{3-}$, Z = Octahedral
(d) X= $[Co(H_2O)_4Cl_2]^{+}$, Y= $[CoCl_4]^{2-}$, Z = Tetrahedral

181. If the CFSE of $\left[Ti\left(H_2O\right)_6\right]^{3+}$ is –96.0 kJ/mol, this complex will absorb maximum at wavelength ____ nm. (nearest integer)
Assume Planck's constant (h) = 6.4×10^{-34} Js Speed of light (c) = 3.0×10^8 m / s and Avogadro's constant $(N_A) = 6 \times 10^{23}$/mol.
[NCERT : PL-131 | NV, Jan. 31, 2023 (II)]

182. Which of the following is correct order of ligand field strength? **[NCERT : P L-132 | Jan. 30, 2023 (I)]**
(a) $CO < en < NH_3 < C_2O_4^{2-} < S^{2-}$
(b) $S^{2-} < C_2O_4^{2-} < NH_3 < en < CO$
(c) $NH_3 < en < CO < S^{2-} < C_2O_4^{2-}$
(d) $S^{2-} < NH_3 < en < CO < C_2O_4^{2-}$

183. Match List I with List II:
[NCERT : P L-128 | Jan. 30, 2023 (II)]

	List I (Complexes)		List II (Hybridisation)
(A)	$[Ni(CO)_4]$	I	sp^3
(B)	$[Cu(NH_3)_4]^{2+}$	II	dsp^2
(C)	$[Fe(NH_3)_6]^{2+}$	III	sp^3d^2
(D)	$[Fe(H_2O)_6]^{2+}$	IV	d^2sp^3

(a) A – II, B – I, C – III, D – IV
(b) A – I, B – II, C – III, D – IV
(c) A – II, B – I, C – IV, D – III
(d) A – I, B – II, C – IV, D – III

184. The Cl – Co – Cl bond angle values in a fac-$[Co(NH_3)_3Cl_3]$ complex is/are: **[NCERT : P L-125 | Jan. 30, 2023 (II)]**
(a) 90° & 180° (b) 90°
(c) 180° (d) 90° & 120°

185. Correct order of spin only magnetic moment of the following complex ions is:
[NCERT : P L-128 | Jan. 29, 2023 (II)]
(Given At. No. Fe: 26, Co:27)
(a) $[FeF_6]^{3-} > [CoF_6]^{3-} > \left[Co\left(C_2O_4\right)_3\right]^{3-}$
(b) $\left[Co\left(C_2O_4\right)_3\right]^{3-} > [CoF_6]^{3-} > [FeF_6]^{3-}$
(c) $[FeF_6]^{3-} > \left[Co\left(C_2O_4\right)_3\right]^{3-} > [CoF_6]^{3-}$
(d) $[CoF_6]^{3-} > [FeF_6]^{3-} > \left[Co\left(C_2O_4\right)_3\right]^{3-}$

186. How many of the following metal ions have similar value of spin only magnetic moment in gaseous state?
(Given: Atomic number : V, 23 ; Cr, 24 ; Fe, 26 ; Ni, 28)
V^{3+}, Cr^{3+}, Fe^{2+}, Ni^{3+}
[NCERT : PL-128 | NV, Jan. 25, 2023 (I)]

187. The number of paramagnetic species from the following is____. **[NCERT : PL-128 | NV, Jan. 25, 2023 (I)]**
$[Ni(CN)_4]^{2-}$, $[Ni(CO)_4]$, $[NiCl_4]^{2-}$
$[Fe(CN)_6]^{4-}$, $[Cu(NH_3)_4]^{2+}$
$[Fe(CN)_6]^{3-}$ and $[Fe(H_2O)_6]^{2+}$

188. Match List I with List II **[NCERT : P L-133 | Jan. 25, 2023 (II)]**

	List I (Coordination entity)		List II (Wavelength of light absorbed in nm)
A.	$[CoCl(NH_3)_5]^{2+}$	I.	310
B.	$[Co(NH_3)_6]^{3+}$	II.	475
C.	$[Co(CN)_6]^{3-}$	III.	535
D.	$[Cu(H_2O)_4]^{2+}$	IV.	600

Choose the correct answer from the options given below:-
(a) A-IV, B-I, C-III, D-II (b) A-III, B-II, C-I, D-IV
(c) A-III, B-I, C-II, D-IV (d) A- II, B-III, C-IV, D-I

189. The magnetic moment of a transition metal compound has been calculated to be 3.87 B.M. The metal ion is
[NCERT : P L-128 | Jan. 24, 2023 (I)]
(a) Cr^{2+} (b) Mn^{2+}
(c) V^{2+} (d) Ti^{2+}

190. The *d*-electronic configuration of $[CoCl_4]^{2-}$ in tetrahedral crystal field is $e^m t_2^n$. Sum of '*m*' and 'number of unpaired electrons is ________.
[NCERT : P L-133 | NV, Jan. 24, 2023 (I)]

191. Which of the following cannot be explained by crystal field theory? **[NCERT : P L-135 | Jan. 24, 2023 (II)]**
(a) The order of spectrochemical series
(b) Magnetic properties of transition metal complexes
(c) Colour of metal complexes
(d) Stability of metal complexes

192. The hybridization and magnetic behaviour of cobalt ion in $[Co(NH_3)_6]^{3+}$ complex, respectively is
[NCERT : P L-131 | Jan. 24, 2023 (II)]
(a) sp^3d^2 and diamagnetic
(b) d^2sp^3 and paramagnetic
(c) d^2sp^3 and diamagnetic
(d) sp^3d^2 and paramagnetic

193. $[Fe(CN)_6]^{3-}$ should be an inner orbital complex. Ignoring the pairing energy, the value of crystal field stabilization energy for this complex is (–) ________ Δ_0. (Nearest integer) **[NCERT : P L-131 | NV, July 29, 2022 (I)]**

194. Octahedral complexes of copper (II) undergo structural distortion (Jahn-Teller). Which one of the given copper (II) complexes will show the maximum structural distortion? (en–ethylenediamine; $H_2N - CH_2 - CH_2 - NH_2$)
[July 29, 2022 (II)]
(a) $[Cu(H_2O)_6]SO_4$ (b) $[Cu(en)(H_2O)_4]SO_4$
(c) *cis*-$[Cu(en)_2Cl_2]$ (d) *trans*-$[Cu(en)_2Cl_2]$

195. Match List-I with List-II

	List-I (Complex)		List-II (Hybridization)
(A)	$Ni(CO)_4$	I.	sp^3
(B)	$[Ni(CN)_4]^{2-}$	II.	sp^3d^2
(C)	$[Co(CN)_6]^{3-}$	III.	d^2sp^3
(D)	$[CoF_6]^{3-}$	IV.	dsp^2

Choose the correct answer from the options given below: **[NCERT : P L-128 | July 28, 2022 (II)]**

(a) (A) - IV, (B) - I, (C) - III, (D) - II
(b) (A) - I, (B) - IV, (C) - III, (D) - II
(c) (A) - I, (B) - IV, (C) - II, (D) - III
(d) (A) - IV, (B) - (I), (C) - II, (D) - III

196. Low oxidation state of metals in their complexes are common when ligands: **[July 27, 2022 (II)]**

(a) have good π -accepting character
(b) have good σ-donor character
(c) are having good π-donating ability
(d) are having poor σ-donating ability

197. The spin only magnetic moment of the complex present in Fehling's reagent is____ B.M. (Nearest integer). **[NCERT : P L-128 | NV, July 27, 2022 (II)]**

198. The difference between spin only magnetic moment values of $[Co(H_2O)_6]Cl_2$ and $[Cr(H_2O)_6]Cl_3$ is ______. **[NCERT : P L-128 | NV, July 26, 2022 (I)]**

199. The metal complex that is diamagnetic is (Atomic number : Fe, 26; Cu, 29) **[NCERT : P L-130 | July 26, 2022 (II)]**

(a) $K_3[Cu(CN)_4]$ (b) $K_2[Cu(CN)_4]$
(c) $K_3[Fe(CN)_4]$ (d) $K_4[FeCl_6]$

200. Consider the following metal complexes:
$[Co(NH_3)_6]^{3+}$, $[CoCl(NH_3)_5]^{2+}$
$[Co(CN)_6]^{3-}$, $[Co(NH_3)_5(H_2O)]^{3+}$
The spin-only magnetic moment value of the complex that absorbs light with shortest wavelength is_____ B.M. (Nearest integer) **[NV, July 25, 2022 (I)]**

201. The correct order of energy of absorption for the following metal complexes is **[July 25, 2022 (II)]**
A: $[Ni(en)_3]^{2+}$, B: $[Ni(NH_3)_6]^{2+}$, C: $[Ni(H_2O)_6]^{2+}$

(a) C < B < A (b) B < C < A
(c) C < A < B (d) A < C < B

202. Spin only magnetic moment (μ_s) of $K_3[Fe(CN)_6]$ is _____ B.M. (Nearest integer). **[NCERT : P L-128 | NV, June 30, 2022 (I)]**

203. Spin only magnetic moment of $[MnBr_6]^{4-}$ is____ B.M. (round off to the closest integer) **[NV, June 29, 2022 (II)]**

204. Given below are two statements : **[June 28, 2022 (I)]**
Statement I : $[Ni(CN)_4]^{2-}$ is square planar and diamagnetic complex with dsp^2 hybridization for Ni but $[Ni(CO)_4]$ is tetrahedral paramagnetic and with sp^3-hybridication for Ni.
Statement II : $[NiCl_4]^{2-}$ and $[Ni(CO)_4]$ both have same *d*-electron configuration have same geometry and are paramagnetic.
In light the above statements choose the correct answer form the options given below:

(a) Both Statement I and Statement II are true.
(b) Both Statement I and Statement II are false.
(c) Statement I is correct but Statement II is false.
(d) Statement I is incorrect but Statement II is true.

205. $[Fe(CN)_6]^{4-}$ $[Fe(CN)_6]^{3-}$ $[Ti(CN)_6]^{3-}$ $[Ni(CN)_4]^{2-}$ $[Co(CN)_6]^{3-}$
Among the given complexes, number of paramagnetic complexes is _____. **[NCERT : P L-130 | NV, June 28, 2022 (II)]**

206. Which of the following will have maximum stabilization due to crystal field? **[June 27, 2022 (I)]**

(a) $[Ti(H_2O)_6]^{3+}$ (b) $[Co(H_2O)_6]^{2+}$
(c) $[Co(CN)_6]^{3-}$ (d) $[Cu(NH_3)_4]^{2+}$

207. Arrange the following coordination compounds in the increasing order of magnetic moments. (Atomic numbers: Mn = 25; Fe = 26) **[NCERT : P L-128 | June 27, 2022 (II)]**

(A) $[FeF_6]^{3-}$ (B) $[Fe(CN)_6]^{3-}$
(C) $[MnCl_6]^{3-}$ (high spin) (D) $[Mn(CN)_6]^{3-}$

(a) A < B < D < C (b) B < D < C < A
(c) A < C < D < B (d) B < D < A < C

208. The spin – only magnetic moment value of an octahedral complex among $CoCl_3.4NH_3$, $NiCl_2.6H_2O$ and $PtCl_4.2HCl$, which upon reaction with excess of $AgNO_3$ gives 2 moles of AgCl is ________ B.M. (Nearest Integer) **[NV, June 26, 2022 (I)]**

209. Reaction of $[Co(H_2O)_6]^{2+}$ with excess ammonia and in the presence of oxygen results into a diamagnetic product. Number of electrons present in t_{2g} orbitals of the product is _______ . **[NV, June 26, 2022 (II)]**

210. If $[Cu(H_2O)_4]^{2+}$ absorbs a light of wavelength 600 nm for d–d transition, then the value of octahedral crystal field splitting energy for $[Cu(H_2O)_6]^{2+}$ will be _______ $\times 10^{-21}$ J. (Nearest Integer)
(Given : h = 6.63×10^{-34} Js and c = 3.08×10^8 ms^{-1}) **[NCERT : P L-133 | NV, June 25, 2022 (I)]**

211. The metal ion (in gaseous state) with lowest spin-only magnetic moment value is **[June 25, 2022 (II)]**

(a) V^{2+} (b) Ni^{2+} (c) Cr^{2+} (d) Fe^{2+}

212. Amongst $FeCl_3.3H_2O$, $K_3[Fe(CN)_6]$ and $[Co(NH_3)_6]Cl_3$, the spin-only magnetic moment value of the inner-orbital complex that absorbs light at shortest wavelength is _____ B.M. [nearest integer] **[NV, June 25, 2022 (II)]**

213. Transition metal complex with highest value of crystal field splitting (Δ_0) will be **[June 24, 2022 (II)]**

(a) $[Cr(H_2O)_6]^{3+}$ (b) $[Mo(H_2O)_6]^{3+}$
(c) $[Fe(H_2O)_6]^{3+}$ (d) $[Os(H_2O)_6]^{3+}$

214. The Crystal Field Stabilization Energy (CFSE) and magnetic moment (spin-only) of an octahedral aqua complex of a metal ion (M^{Z+}) are $-0.8\ \Delta_0$ and 3.87 BM, respectively. Identify (M^{Z+}) : **[NCERT : P L-131 | Sep. 1, 2021 (II)]**

(a) V^{3+} (b) Cr^{3+} (c) Mn^{4+} (d) Co^{2+}

215. The spin-only magnetic moment value of B_2^+ species is ________ $\times 10^{-2}$ BM. (Nearest integer)
[Given : $\sqrt{3} = 1.73$] **[NV, Sep. 1, 2021 (II)]**

216. Spin only magnetic moment in BM of $[Fe(CO)_4(C_2O_4)]^+$ is: **[Aug. 31, 2021 (II)]**

(a) 5.92 (b) 0 (c) 1 (d) 1.73

217. Which one of the following complexes is violet in colour? [NCERT : P L-133 | Aug. 26, 2021 (I)]
(a) $[Fe(CN)_6]^{4-}$ (b) $[Fe(SCN)_6]^{4-}$
(c) $Fe_4[Fe(CN)_6]_3 \cdot H_2O$ (d) $[Fe(CN)_5NOS]^{4-}$

218. Arrange the following Cobalt complexes in the order of increasing Crystal Field Stabilization Energy (CFSE) value.
Complexes :
$\underset{A}{[CoF_6]^{3-}}, \underset{B}{[Co(H_2O)_6]^{2+}},$
$\underset{C}{[Co(NH_3)_6]^{3+}}$ and $\underset{D}{[Co(en)_3]^{3+}}$
Choose the **correct** option : [NCERT : P L-131 | Aug. 26, 2021 (II)]
(a) A < B < C < D (b) B < A < C < D
(c) B < C < D < A (d) C < D < B < A

219. The type of hybridisation and magnetic property of the complex $[MnCl_6]^{3-}$, respectively, are : [NCERT : P L-128 | July 27, 2021 (I)]
(a) sp^3d^2 and diamagnetic
(b) d^2sp^3 and diamagnetic
(c) d^2sp^3 and paramagnetic
(d) sp^3d^2 and paramagnetic

220. Which one of the following species responds to an external magnetic field? [NCERT : P L-128 | July 25, 2021 (I)]
(a) $[Fe(H_2O)_6]^{3+}$ (b) $[Ni(CN)_4]^{2-}$
(c) $[Co(CN)_6]^{3-}$ (d) $[Ni(CO)_4]$

221. The spin only magnetic moments (in BM) for free Ti^{3+}, V^{2+} and Sc^{3+} ions respectively are
(At.No. Sc : 21, Ti : 22, V : 23) [NCERT : P L-130 | July 25, 2021 (II)]
(a) 3.87, 1.73, 0 (b) 1.73, 3.87, 0
(c) 1.73, 0, 3.87 (d) 0, 3.87, 1.73

222. The total number of unpaired electrons present in $[Co(NH_3)_6]Cl_2$ and $[Co(NH_3)_6]Cl_3$ is [NV, July 22, 2021 (II)]

223. According to the valence bond theory, the hybridization of central metal atom is dsp^2 for which one of the following compounds? [NCERT : P L-128 | July 20, 2021 (I)]
(a) $NiCl_2 \cdot 6H_2O$ (b) $K_2[Ni(CN)_4]$
(c) $[Ni(CO)_4]$ (d) $Na_2[NiCl_4]$

224. The correct order of intensity of colors of the compounds is : [NCERT : P L-133 | July 20, 2021 (I)]
(a) $[Ni(CN)_4]^{2-} > [NiCl_4]^{2-} > [Ni(H_2O)_6]^{2+}$
(b) $[Ni(H_2O)_6]^{2+} > [NiCl_4]^{2-} > [Ni(CN)_4]^{2-}$
(c) $[NiCl_4]^{2-} > [Ni(H_2O)_6]^{2+} > [Ni(CN)_4]^{2-}$
(d) $[NiCl_4]^{2-} > [Ni(CN)_4]^{2-} > [Ni(H_2O)_6]^{2+}$

225. The spin-only magnetic moment value for the complex $[Co(CN)_6]^{4-}$ is _______ BM. [NV, July 20, 2021 (I)]
[At. no. of Co = 27]

226. Spin only magnetic moment of an octahedral complex of Fe^{2+} in the presence of a strong field ligand in BM is : [July 20, 2021 (II)]
(a) 4.89 (b) 2.82 (c) 0 (d) 3.46

227. Which one of the following species doesn't have a magnetic moment of 1.73 B.M., (spin only value) ? [July 20, 2021 (II)]
(a) O_2^+ (b) CuI
(c) $[Cu(NH_3)_4]Cl_2$ (d) O_2^-

228. An aqueous solution of $NiCl_2$ was heated with excess sodium cyanide in presence of strong oxidizing agent to form $[Ni(CN)_6]^{2-}$. The total change in number of unpaired electrons on metal centre is ____. [NV, July 20, 2021 (II)]

229. $[Ti(H_2O)_6]^{3+}$ absorbs light of wavelength 498 nm during a d – d transition. The octahedral splitting energy for the above complex is _______ $\times 10^{-19}$J.
(Round off to the nearest integer).
$h = 6.626 \times 10^{-34}$ Js; $c = 3 \times 10^8$ ms^{-1} [NV, March 16, 2021 (II)]

230. The hybridization and magnetic nature of $[Mn(CN)_6]^{4-}$ and $[Fe(CN)_6]^{3-}$, respectively are [NCERT : P L-130, 129 | Feb. 25, 2021 (I)]
(a) d^2sp^3 and paramagnetic
(b) d^2sp^3 and diamagnetic
(c) sp^3d^2 and paramagnetic
(d) sp^3d^2 and diamagnetic

231. In which of the following order, the given complex ions are arranged correctly with respect to their decreasing spin only magnetic moment?
(i) $[FeF_6]^{3-}$ (ii) $[Co(NH_3)_6]^{3+}$
(iii) $[NiCl_4]^{2-}$ (iv) $[Cu(NH_3)_4]^{2+}$
[NCERT : P L-128 | Feb. 25, 2021 (II)]
(a) (iii) > (iv) > (ii) > (i) (b) (ii) > (iii) > (i) > (iv)
(c) (i) > (iii) > (iv) > (ii) (d) (ii) > (i) > (iii) > (iv)

232. The spin only magnetic moment of a divalent ion in aqueous solution (atomic number 29) is [NV, Feb. 25, 2021 (II)]

233. The calculated magnetic moments (spin only value) for species $[FeCl_4]^{2-}$, $[Co(C_2O_4)_3]^{3-}$ and MnO_4^{2-} respectively are : [NCERT : P L-128 | Feb. 24, 2021 (II)]
(a) 5.82, 0 and 0 BM (b) 4.90, 0 and 1.73 BM
(c) 5.92, 4.90 and 0 BM (d) 4.90, 0 and 2.83 BM

234. The species that has a spin-only magnetic moment of 5.9 B.M., is : (T_d = tetrahedral) [NCERT : P L-128 | Sep. 06, 2020 (I)]
(a) $[Ni(CN)_4]^{2-}$ (square planar)
(b) $[NiCl_4]^{2-}$ (T_d)
(c) $[Ni(CO)_4]$ (T_d)
(d) $[MnBr_4]^{2-}$ (T_d)

235. For a d^4 metal ion in an octahedral field, the correct electronic configuration is : [NCERT : P L-131 | Sep. 06, 2020 (II)]
(a) $t_{2g}^3 e_g^1$ when $\Delta_0 < P$ (b) $t_{2g}^3 e_g^1$ when $\Delta_0 > P$
(c) $t_{2g}^4 e_g^0$ when $\Delta_0 < P$ (d) $e_g^2 t_{2g}^2$ when $\Delta_0 < P$

236. The values of the crystal field stabilization energies for a high spin d^6 metal ion in octahedral and tetrahedral fields, respectively, are : [NCERT : P L-131 | Sep. 05, 2020 (I)]
(a) $-0.4\,\Delta_o$ and $-0.6\,\Delta_t$ (b) $-2.4\,\Delta_o$ and $-0.6\,\Delta_t$
(c) $-1.6\,\Delta_o$ and $-0.4\,\Delta_t$ (d) $-0.4\,\Delta_o$ and $-0.27\,\Delta_t$

237. Considering that $\Delta_0 > P$, the magnetic moment (in BM) of $[Ru(H_2O)_6]^{2+}$ would be ________.
[NCERT : P L-131 | NV, Sep. 05, 2020 (II)]

238. The pair in which both the species have the same magnetic moment (spin only) is: [Sep. 04, 2020 (I); Same 2016]
(a) $[Cr(H_2O)_6]^{2+}$ and $[Fe(H_2O)_6]^{2+}$
(b) $[Co(OH)_4]^{2-}$ and $[Fe(NH_3)_6]^{2+}$
(c) $[Mn(H_2O)_6]^{2+}$ and $[Cr(H_2O)_6]^{2+}$
(d) $[Cr(H_2O)_6]^{2+}$ and $[CoCl_4]^{2-}$

239. The Crystal Field Stabilization Energy (CFSE) of $[CoF_3(H_2O)_3]$ $(\Delta_0 < P)$ is :
[NCERT : P L-132 | Sep. 04, 2020 (II)]
(a) $-0.8\,\Delta_0 + 2P$ (b) $-0.4\,\Delta_0$
(c) $-0.8\,\Delta_0$ (d) $-0.4\,\Delta_0 + P$

240. The one that can exhibit highest paramagnetic behaviour among the following is : [Sep. 04, 2020 (II)]
gly = glycinato; bpy = 2, 2'-bipyridine
(a) $[Pd(gly)_2]$
(b) $[Fe(en)(bpy)(NH_3)_2]^{2+}$
(c) $[Co(ox)_2(OH)_2]^-$ $(\Delta_0 > P)$
(d) $[Ti(NH_3)_6]^{3+}$

241. The electronic spectrum of $[Ti(H_2O)_6]^{3+}$ shows a single broad peak with a maximum at 20,300 cm^{-1}. The crystal field stabilization energy (CFSE) of the complex ion, in kJ mol^{-1}, is :
(1 kJ mol^{-1} = 83.7 cm^{-1}) [Sep. 03, 2020 (I)]
(a) 145.5 (b) 242.5 (c) 83.7 (d) 97

242. The *d*-electron configuration of $[Ru(en)_3]Cl_2$ and $[Fe(H_2O)_6]Cl_2$, respectively are :
[NCERT : P L-132 | Sep. 03, 2020 (II)]
(a) $t_{2g}^6\,e_g^0$ and $t_{2g}^6\,e_g^0$ (b) $t_{2g}^4\,e_g^2$ and $t_{2g}^6\,e_g^0$
(c) $t_{2g}^6\,e_g^0$ and $t_{2g}^4\,e_g^2$ (d) $t_{2g}^4\,e_g^2$ and $t_{2g}^4\,e_g^2$

243. Consider that a d^6 metal ion (M^{2+}) forms a complex with aqua ligands, and the spin only magnetic moment of the complex is 4.90 BM. The geometry and the crystal field stabilization energy of the complex is :
[NCERT : P L-132 | Sep. 02, 2020 (I)]
(a) octahedral and $-2.4\Delta_0 + 2P$
(b) tetrahedral and $-0.6\,\Delta_t$
(c) octahedral and $-1.6\,\Delta_0$
(d) tetrahedral and $-1.6\Delta_t + 1P$

244. For octahedral Mn(II) and tetrahedral Ni(II) complexes, consider the following statements : [Sep. 02, 2020 (I)]
(I) both the complexes can be high spin.
(II) Ni(II) complex can very rarely below spin.
(III) with strong field ligands, Mn(II) complexes can be low spin.
(IV) aqueous solution of Mn(II) ions is yellow in color.
The correct statements are :
(a) (I) and (II) only (b) (I), (III) and (IV) only
(c) (I), (II) and (III) only (d) (II), (III) and (IV) only

245. Simplified absorption spectra of three complexes ((i), (ii) and (iii)) of M^{n+} ion are provided below; their λ_{max} values are marked as A, B and C respectively. The correct match between the complexes and their λ_{max} values is :
[Sep. 02, 2020 (II)]

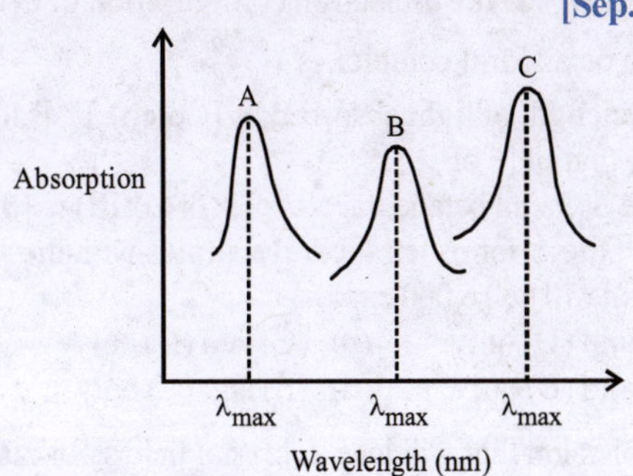

(i) $[M(NCS)_6]^{(-6+n)}$ (ii) $[MF_6]^{(-6+n)}$
(iii) $[M(NH_3)_6]^{n+}$
(a) A-(iii), B-(i), C-(ii) (b) A-(ii), B-(i), C-(iii)
(c) A-(ii), B-(iii), C-(i) (d) A-(i), B-(ii), C-(iii)

246. $[Pd(F)(Cl)(Br)(I)]^{2-}$ has *n* number of geometrical isomers. Then, the spin-only magnetic moment and crystal field stabilisation energy [CFSE] of $[Fe(CN)_6]^{n-6}$, respectively, are:
[Note : Ignore the pairing energy] [Jan. 09, 2020 (I)]
(a) 2.84 BM and $-1.6\,D_0$
(b) 5.92 BM and 0
(c) 1.73 BM and $-2.0\,D_0$
(d) 0 BM and $-2.4\,D_0$

247. The correct order of the spin-only magnetic moments of the following complexes is: [Jan. 09, 2020 (II)]
(A) $[Cr(H_2O)_6]Br_2$ (B) $Na_4[Fe(CN)_6]$
(C) $Na_3[Fe(C_2O_4)_3]$ $(\Delta_0 > P)$ (D) $(Et_4N)_2[CoCl_4]$
(a) (C) > (A) > (D) > (B) (b) (C) > (A) > (B) > (D)
(c) (A) > (D) > (C) > (B) (d) (B) ≈ (A) > (D) > (C)

248. The correct order of the calculated spin-only magnetic moments of complexes (A) to (D) is:
[NCERT : P L-131 | Jan. 08, 2020 (II)]
(A) $Ni(CO)_4$ (B) $[Ni(H_2O)_6]Cl_2$
(C) $Na_2[Ni(CN)_4]$ (D) $PdCl_2(PPh_3)_2$
(a) (A) ≈ (C) < (B) ≈ (D) (b) (C) < (D) < (B) < (A)
(c) (C) ≈ (D) < (B) < (A) (d) (A) ≈ (C) ≈ (D) < (B)

249. The volume (in mL) of 0.125 M $AgNO_3$ required to quantitatively precipitate chloride ions in 0.3 g of $[Co(NH_3)_6]Cl_3$ is ______. **[NV, Jan. 08, 2020 (I)]**

$M_{[Co(NH_3)_6]Cl_3} = 267.46$ g/mol

$M_{AgNO_3} = 169.87$ g/mol.

250. Complexes (ML_5) of metals Ni and Fe have ideal square pyramidal and trigonal bipyramidal geometries, respectively. The sum of the 90°, 120° and 180° L-M-L angles in the two complexes is ______. **[NV, Jan. 08, 2020 (II)]**

251. Among the statements (A)-(D), the incorrect ones are: **[NCERT : P L-133 | Jan. 07, 2020 (II)]**

(A) Octahedral Co(III) complexes with strong field ligands have very high magnetic moments.

(B) When $\Delta_0 < P$, the *d*-electron configuration of Co(III) in an octahedral complex is $t_{2g}^4 e_g^2$.

(C) Wavelength of light absorbed by $[Co(en)_3]^{3+}$ is lower than that of $[CoF_6]^{3-}$.

(D) If the Δ_0 for an octahedral complex of Co(III) is 18,000 cm^{-1}, the Δ_t for its tetrahedral complex with the same ligand will be 16,000 cm^{-1}.

(a) (A) and (D) only (b) (C) and (D) only
(c) (A) and (B) only (d) (B) and (C) only

252. The complex ion that will lose its crystal field stabilization energy upon oxidation of its metal to +3 state is : **[April 12, 2019 (I)]**

(Phen = [structure of 1,10-phenanthroline] and ignore pairing energy)

(a) $[Co(phen)_3]^{2+}$ (b) $[Ni(phen)_3]^{2+}$
(c) $[Zn(phen)_3]^{2+}$ (d) $[Fe(phen)_3]^{2+}$

253. Complete removal of both the axial ligands (along the *z*-axis) from an octahedral complex leads to which of the following splitting patterns ? (relative orbital energies not on scale) **[April 12, 2019 (I)]**

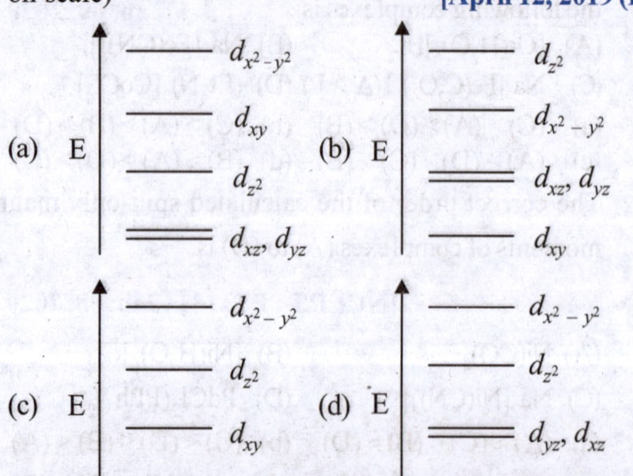

254. Three complexes, $[CoCl(NH_3)_5]^{2+}$(I), $[Co(NH_3)_5H_2O]^{3+}$(II) and $[Co(NH_3)_6]^{3+}$ (III) absorb light in the visible region. The correct order of the wavelength of light absorbed by them is : **[NCERT : P L-133 | April 10, 2019 (I)]**

(a) (III) > (I) > (II) (b) (III) > (II) > (I)
(c) (II) > (I) > (III) (d) (I) > (II) > (III)

255. The crystal field stabilization energy (CFSE) of $[Fe(H_2O)_6]Cl_2$ and $K_2[NiCl_4]$, respectively are : **[NCERT : P L-131 | April 10, 2019 (II)]**

(a) $-0.6\,\Delta_o$ and $-0.8\,\Delta_\tau$
(b) $-0.4\,\Delta_o$ and $-0.8\,\Delta_\tau$
(c) $-2.4\,\Delta_o$ and $-1.2\,\Delta_\tau$
(d) $-0.4\,\Delta_o$ and $-1.2\,\Delta_\tau$

256. The INCORRECT statement is : **[April 10, 2019 (II)]**

(a) the gemstone, ruby, has Cr^{3+} ions occupying the octahedral sites of beryl.

(b) the spin-only magnetic moment of $[Ni(NH_3)_4(H_2O)_2]^{2+}$ is 2.83 BM.

(c) the color of $[CoCl(NH_3)_5]^{2+}$ is violet as it absorbs the yellow light.

(d) the spin-only magnetic moments of $[Fe(H_2O)_6]^{2+}$ and $[Cr(H_2O)_6]^{2+}$ are nearly similar.

257. The degenerate orbitals of $[Cr(H_2O)_6]^{3+}$ are: **[NCERT : P L-128 | April 9, 2019 (I)]**

(a) d_{xz} and d_{yz} (b) d_{yz} and d_{z^2}
(c) d_{z^2} and d_{xz} (d) $d_{x^2-y^2}$ and d_{xy}

258. The correct statements among I to III are: **[NCERT : P L-131 | April 9, 2019 (II)]**

(I) Valence bond theory cannot explain the color exhibited by transition metal complexes.

(II) Valence bond theory can predict quantitatively the magnetic properties of transition metal complexes.

(III) Valence bond theory cannot distinguish ligands as week and strong field ones.

(a) (II) and (III) only (b) (I), (II) and (III)
(c) (I) and (III) only (d) (I) and (II) only

259. The calculated spin-only magnetic moments (BM) of the anionic and cationic species of $[Fe(H_2O)_6]_2[Fe(CN)_6]$, respectively, are: **[NCERT : P L-128 | April 8, 2019 (II)]**

(a) 0 and 4.9 (b) 2.84 and 5.92
(c) 4.9 and 0 (d) 0 and 5.92

260. The metal *d*-orbitals that are directly facing the ligands in $K_3[Co(CN)_6]$ are : **[NCERT : P L-128 | Jan. 12, 2019 (I)]**

(a) d_{xy} and $d_{x^2-y^2}$ (b) $d_{x^2-y^2}$ and d_{z^2}
(c) d_{xz}, d_{yz} and d_{z^2} (d) d_{xy}, d_{xz} and d_{yz}

261. The pair of metal ions that can give a spin only magnetic moment of 3.9 BM for the complex $[M(H_2O)_6]Cl_2$, is: **[NCERT : P L-128 | Jan. 12, 2019 (I)]**

(a) V^{2+} and Co^{2+} (b) V^{2+} and Fe^{2+}
(c) Co^{2+} and Fe^{2+} (d) Cr^{2+} and Mn^{2+}

262. The magnetic moment of an octahedral homoleptic Mn(II) complex is 5.9 BM. The suitable ligand for this complex is : **[Jan. 12, 2019 (II)]**

(a) Ethylenediamine (b) CN^-

(c) NCS^- (d) CO

263. Match the metals **(column I)** with the coordination compound(s)/enzyme(s) **(column II)** : **[NCERT : P L-137 | Jan. 11, 2019 (I)]**

Column I (Metals)	Column II (Coordination compound(s)/enzyme(s))
(A) Co	(i) Wilkinson catalyst
(B) Zn	(ii) Chlorophyll
(C) Rh	(iii) Vitamin B_{12}
(D) Mg	(iv) Carbonic anhydrase

(a) (A)-(iii); (B)-(iv); (C)-(i); (D)-(ii)

(b) (A)-(i); (B)-(ii); (C)-(iii); (D)-(iv)

(c) (A)-(ii); (B)-(i); (C)-(iv); (D)-(iii)

(d) (A)-(iv); (B)-(iii); (C)-(i); (D)-(ii)

264. The difference in the number of unpaired electrons of a metal ion in its high-spin and low-spin octahedral complexes is two. The metal ion is: **[Jan. 10, 2019 (II)]**

(a) Ni^{2+} (b) Fe^{2+} (c) Co^{2+} (d) Mn^{2+}

265. Two complexes $[Cr(H_2O)_6]Cl_3$ (A) and $[Cr(NH_3)_6]Cl_3$ (B) are violet and yellow coloured, respectively. The incorrect statement regarding them is: **[NCERT : P L-131 | Jan. 9, 2019 (I)]**

(a) Δ_0 values of (A) and (B) are calculated from the energies of violet and yellow light, respectively.

(b) both are paramagnetic with three unpaired electrons.

(c) both absorb energies corresponding to their complementary colors.

(d) Δ_0 value for (A) is less than that of (B).

266. The complex that has highest crystal field splitting energy (Δ) is: **[NCERT : P L-131 | Jan. 9, 2019 (II)]**

(a) $[Co(NH_3)_5(H_2O)]Cl_3$

(b) $K_2[CoCl_4]$

(c) $[Co(NH_3)_5Cl]Cl_2$

(d) $K_3[Co(CN)_6]$

267. Homoleptic octahedral complexes of a metal ion 'M^{3+}' with three monodentate ligands L_1, L_2 and L_3 absorb wavelengths in the region of green, blue and red respectively. The increasing order of the ligand strength is: **[NCERT : P L-133 | Jan. 9, 2019 (II)]**

(a) $L_3 < L_1 < L_2$ (b) $L_3 < L_2 < L_1$

(c) $L_1 < L_2 < L_3$ (d) $L_2 < L_1 < L_3$

268. In Wilkinson's catalyst, the hybridisation of central metal ion and its shape are respectively: **[NCERT : P L-128 | Online April 16, 2018]**

(a) sp^3d, trigonal bipyramidal

(b) d^2sp^3, octahedral

(c) dsp^2, square planar

(d) sp^3, tetrahedral

269. The correct combination is : **[Online April 15, 2018 (I)]**

(a) $[NiCl_4]^{2-}$-square-planar; $[Ni(CN)_4]^{2-}$-paramagnetic

(b) $[Ni(CN)_4]^{2-}$-tetrahedral; $[Ni(CO)_4]$-paramagnetic

(c) $[NiCl_4]^{2-}$-paramagnetic; $[Ni(CO)_4]$-tetrahedral

(d) $[NiCl_4]^{2-}$-dimagnetic; $[Ni(CO)_4]$-square-planar

270. The correct order of spin-only magnetic moments among the following is: (Atomic number: Mn = 25, Co = 27, Ni = 28, Zn = 30) **[Online April 15, 2018 (II)]**

(a) $[ZnCl_4]^{2-} > [NiCl_4]^{2-} > [CoCl_4]^{2-} > [MnCl_4]^{2-}$

(b) $[CoCl_4]^{2-} > [MnCl_4]^{2-} > [NiCl_4]^{2-} > [ZnCl_4]^{2-}$

(c) $[NiCl_4]^{2-} > [CoCl_4]^{2-} > [MnCl_4]^{2-} > [ZnCl_4]^{2-}$

(d) $[MnCl_4]^{2-} > [CoCl_4]^{2-} > [NiCl_4]^{2-} > [ZnCl_4]^{2-}$

271. The transition metal ions responsible for colour in ruby and emerald are, respectively : **[Online April 10, 2016]**

(a) Co^{3+} and Cr^{3+} (b) Co^{3+} and Co^{3+}

(c) Cr^{3+} and Cr^{3+} (d) Cr^{3+} and Co^{3+}

272. Identify the correct trend given below : **[NCERT : P L-131 | Online April 9, 2016]**

(Atomic No. = Ti : 22, Cr : 24 and Mo : 42)

(a) Δ_o of $[Cr(H_2O)_6]^{2+} > [Mo(H_2O)_6]^{2+}$ and Δ_o of $[Ti(H_2O)_6]^{3+} > [Ti(H_2O)_6]^{2+}$

(b) Δ_o of $[Cr(H_2O)_6]^{2+} > [Mo(H_2O)_6]^{2+}$ and Δ_o of $[Ti(H_2O)_6]^{3+} < [Ti(H_2O)_6]^{2+}$

(c) Δ_o of $[Cr(H_2O)_6]^{2+} < [Mo(H_2O)_6]^{2+}$ and Δ_o of $[Ti(H_2O)_6]^{3+} > [Ti(H_2O)_6]^{2+}$

(d) Δ_o of $[Cr(H_2O)_6]^{2+} < [Mo(H_2O)_6]^{2+}$ and Δ_o of $[Ti(H_2O)_6]^{3+} < [Ti(H_2O)_6]^{2+}$

273. The pair having the same magnetic moment is: [At. No.: Cr = 24, Mn = 25, Fe = 26, Co = 27] **[NCERT : P L-131 | 2016]**

(a) $[Mn(H_2O)_6]^{2+}$ and $[Cr(H_2O)_6]^{2+}$

(b) $[CoCl_4]^{2-}$ and $[Fe(H_2O)_6]^{2+}$

(c) $[Cr(H_2O)_6]^{2+}$ and $[CoCl_4]^{2-}$

(d) $[Cr(H_2O)_6]^{2+}$ and $[Fe(H_2O)_6]^{2+}$

274. Which of the following compounds is not colored yellow? **[2015]**

(a) $(NH_4)_3[As(Mo_3O_{10})_4]$

(b) $BaCrO_4$

(c) $Zn_2[Fe(CN)_6]$

(d) $K_3[Co(NO_2)_6]$

275. The color of $KMnO_4$ is due to : **[2015]**

(a) $L \rightarrow M$ charge transfer transition

(b) $\sigma - \sigma^*$ transition

(c) $M \rightarrow L$ charge transfer transition

(d) d – d transition

276. Which of the following complex ions has electrons that are symmetrically filled in both t_{2g} and e_g orbitals ? **[NCERT : P L-131 | Online April 11, 2015]**

(a) $[FeF_6]^{3-}$ (b) $[Mn(CN)_6]^{4-}$

(c) $[CoF_6]^{3-}$ (d) $[Co(NH_3)_6]^{2+}$

277. When concentrated HCl is added to an aqueous solution of $CoCl_2$, its colour changes from reddish pink to deep blue. Which complex ion gives blue colour in this reaction? **[Online April 11, 2015]**

(a) $[CoCl_4]^{2-}$ (b) $[CoCl_6]^{3-}$
(c) $[CoCl_6]^{4-}$ (d) $[Co(H_2O)_6]^{2+}$

278. An octahedral complex of Co^{3+} is diamagnetic. The hybridisation involved in the formation of the complex is: **[NCERT : P L-128 | Online April 9, 2014]**

(a) sp^3d^2 (b) dsp^2 (c) d^2sp^3 (d) sp^3d

279. The correct statement about the magnetic properties of $[Fe(CN)_6]^{3-}$ and $[FeF_6]^{3-}$ is (Z = 26) : **[NCERT : P L-130 | Online April 9, 2014]**

(a) both are paramagnetic.
(b) both are diamagnetic.
(c) $[Fe(CN)_6]^{3-}$ is diamagnetic, $[FeF_6]^{3-}$ is paramagnetic.
(d) $[Fe(CN)_6]^{3-}$ is paramagnetic, $[FeF_6]^{3-}$ is diamagnetic.

280. Consider the coordination compound, $[Co(NH_3)_6]Cl_3$. In the formation of this complex, the species which acts as the Lewis acid is: **[NCERT : P L-132 | Online April 11, 2014]**

(a) $[Co(NH_3)_6]^{3+}$ (b) Cl^-
(c) Co^{3+} (d) NH_3

281. Among the following species, the one which causes the highest CFSE, Δ_o as a ligand is: **[Online April 12, 2014]**

(a) CN^- (b) NH_3 (c) F^- (d) CO

282. Which one of the following complexes will most likely absorb visible light? **[Online April 12, 2014]**
(At nos. Sc = 21, Ti = 22, V = 23, Zn = 30)

(a) $[Sc(H_2O)_6]^{3+}$ (b) $[Ti(NH_3)_6]^{4+}$
(c) $[V(NH_3)_6]^{3+}$ (d) $[Zn(NH_3)_6]^{2+}$

283. Nickel (Z = 28) combines with a uninegative monodentate ligand to form a diamagnetic complex $[NiL_4]^{2-}$. The hybridisation involved and the number of unpaired electrons present in the complex are respectively: **[NCERT : P L-129 | Online April 19, 2014]**

(a) sp^3, two (b) dsp^2, zero
(c) dsp^2, one (d) sp^3, zero

284. The octahedral complex of a metal ion M^{3+} with four monodentate ligands L_1, L_2, L_3 and L_4 absorb wavelengths in the region of red, green, yellow and blue, respectively. The increasing order of ligand strength of the four ligands is: **[NCERT : P L-133 | 2014]**

(a) $L_4 < L_3 < L_2 < L_1$ (b) $L_1 < L_3 < L_2 < L_4$
(c) $L_3 < L_2 < L_4 < L_1$ (d) $L_1 < L_2 < L_4 < L_3$

285. Which of the following is diamagnetic? **[NCERT : P L-132 | Online April 22, 2013]**

(a) $[Fe(CN)_6]^{3-}$ (b) $[Co(ox)]^{3-}$
(c) $[FeF_6]^{3-}$ (d) $[CoF_6]^{3-}$

286. The magnetic moment of the complex anion $[Cr(NO)(NH_3)(CN)_4]^{2-}$ is : **[Online April 23, 2013]**

(a) 5.91 BM (b) 3.87 BM
(c) 1.73 BM (d) 2.82 BM

287. The structure of which of the following chloro species can be explained on the basis of dsp^2 hybridisation ? **[NCERT : P L-128 | Online April 25, 2013]**

(a) $PdCl_4^{2-}$ (b) $FeCl_4^{2-}$
(c) $CoCl_4^{2-}$ (d) $NiCl_4^{2-}$

288. Square-planar geometry is shown by **[NCERT : P L-128 | Online May 19, 2012]**

(a) $[PtCl_2(NH_3)_2]$ (b) $[NiCl_4]^{2-}$
(c) MnO_4^- (d) CrO_4^{2-}

289. The *d*-electron configurations of Cr^{2+}, Mn^{2+}, Fe^{2+} and Co^{2+} are d^4, d^5, d^6 and d^7 respectively. Which one of the following will exhibit the lowest paramagnetic behaviour? (Atomic no. Cr = 24, Mn = 25, Fe = 26, Co = 27). **[NCERT : P L-130 | Online May 7, 2012]**

(a) $[Co(H_2O)_6]^{2+}$ (b) $[Cr(H_2O)_6]^{2+}$
(c) $[Mn(H_2O)_6]^{2+}$ (d) $[Fe(H_2O)_6]^{2+}$

290. The correct order of ligands in the spectrochemical series is **[NCERT : P L-132 | Online May 19, 2012]**

(a) $Cl^- > en > CN^- > NCS^-$
(b) $CN^- > en > NCS^- > Cl^-$
(c) $NCS^- > CN^- > Cl^- > en$
(d) $en > CN^- > Cl^- > NCS^-$

291. Which of the following paramagnetic ions would exhibit a magnetic moment (spin only) of the order of 5 BM? **[Online May 19, 2012]**
(At. Nos. Mn = 25, Cr = 24, V = 23, Ti = 22)

(a) Mn^{2+} (b) Ti^{2+} (c) V^{2+} (d) Cr^{2+}

292. In which of the following complexes of the Co (at. no. 27), will the magnitude of Δ_0 be the highest? **[NCERT : P L-131 | 2008]**

(a) $[Co(CN)_6]^{3-}$ (b) $[Co(C_2O_4)_3]^{3-}$
(c) $[Co(H_2O)_6]^{3+}$ (d) $[Co(NH_3)_6]^{3+}$

293. Which of the following has a square planar geometry? **[NCERT : P L-128 | 2007]**

(a) $[PtCl_4]^{2-}$ (b) $[CoCl_4]^{2-}$
(c) $[FeCl_4]^{2-}$ (d) $[NiCl_4]^{2-}$
(At. nos.: Fe = 26, Co = 27, Ni = 28, Pt = 78)

294. Nickel (Z = 28) combines with a uninegative monodentate ligand X^- to form a paramagnetic complex $[NiX_4]^{2-}$. The number of unpaired electron(s) in the nickel and geometry of this complex ion are, respectively : **[NCERT : P L-128 | 2006]**

(a) one, square planar (b) two, square planar
(c) one, tetrahedral (d) two, tetrahedral

295. The "spin-only" magnetic moment [in units of Bohr magneton, (μ_B)] of Ni^{2+} in aqueous solution would be (At. No. Ni = 28) **[2006]**

(a) 6 (b) 1.73 (c) 2.82 (d) 4.90

296. The value of the 'spin only' magnetic moment for one of the following configurations is 2.84 BM. The correct one is **[NCERT : P L-128 | 2005]**

(a) d^5 (in strong ligand field)
(b) d^3 (in weak as well as in strong fields)
(c) d^4 (in weak ligand fields)
(d) d^4 (in strong ligand fields)

297. Which one of the following cyano complexes would exhibit the lowest value of paramagnetic behaviour ? **[NCERT : P L-130 | 2005]**

(a) $[Co(CN)_6]^{3-}$ (b) $[Fe(CN)_6]^{3-}$
(c) $[Mn(CN)_6]^{3-}$ (d) $[Cr(CN)_6]^{3-}$

(At. No. : Cr = 24, Mn = 25, Fe = 26, Co = 27)

298. The correct order of magnetic moments (spin only values in B.M.) among the following is **[NCERT : P L-128 | 2004]**

(a) $[Fe(CN)_6]^{4-} > [MnCl_4]^{2-} > [CoCl_4]^{2-}$
(b) $[MnCl_4]^{2-} > [Fe(CN)_6]^{4-} > [CoCl_4]^{2-}$
(c) $[MnCl_4]^{2-} > [CoCl_4]^{2-} > [Fe(CN)_6]^{4-}$
(d) $[Fe(CN)_6]^{4-} > [CoCl_4]^{2-} > [MnCl_4]^{2-}$

(Atomic nos. : Mn = 25, Fe = 26, Co = 27)

299. Which one of the following complexes is an outer orbital complex ? **[NCERT : P L-129 | 2004]**

(a) $[Co(NH_3)_6]^{3+}$ (b) $[Mn(CN)_6]^{4-}$
(c) $[Fe(CN)_6]^{4-}$ (d) $[Ni(NH_3)_6]^{2+}$

(Atomic no. : Mn = 25; Fe = 26; Co = 27, Ni = 28)

300. A square planar complex is formed by hybridisation of which atomic orbitals? **[2002]**

(a) s, p_x, p_y, d_{yz} (b) $s, p_x, p_y, d_{x^2-y^2}$
(c) s, p_x, p_y, d_{z^2} (d) s, p_y, p_z, d_{xy}

301. The most stable ion is **[2002]**

(a) $[Fe(OH)_5]^{3-}$ (b) $[Fe(Cl)_6]^{3-}$
(c) $[Fe(CN)_6]^{3-}$ (d) $[Fe(H_2O)_6]^{3+}$.

Topic 3 - Organometallic Compounds

302. Choose the correct statements from the following **[Jan. 30, 2024 (I)]**

(A) Ethane – 1, 2 - diamine is a cheating ligand.
(B) Metallic aluminium is produced by electrolysis of aluminium oxide is presence of cryolite.
(C) Cyanide ion is used as ligand for leaching of silver.
(D) Phosphine act as a ligand in Wilkinson catalyst.
(E) The stability constants of Ca^{2+} and Mg^{2+} are similar with EDTA complexes.

Choose the correct answer from the options given below:

(a) (B), (C), (E) only (b) (C), (D), (E) only
(c) (A), (B), (C) only (d) (A), (D), (E) only

303. The coordination geometry around the manganese in decacarbonyldimanganese (0) is : **[NCERT : P L-136 | Jan. 30, 2024 (II)]**

(a) Octahedral (b) Trigonal bipyramidal
(c) Square pyramidal (d) Square planar

304. In which one of the following metal carbonyls, CO forms a bridge between metal atoms ? **[NCERT : P L-136 | Jan. 29, 2024 (I)]**

(a) $[Os_3(CO)_{12}]$ (b) $[Ru_3(CO)_{12}]$
(c) $[Co_2(CO)_8]$ (d) $[Mn_2(CO)_{10}]$

305. The mismatched combinations are **[NCERT : P L-136, 137 | April 13, 2023 (I)]**

A. Chlorophyll - Co
B. Water hardness - EDTA
C. Photography - $[Ag(CN)_2]^-$
D. Wilkinson catalyst - $[(Ph_3P)_3 RhCl]$
E. Chelating ligand - D - Penicillamine

Choose the correct answer from the options given below:

(a) A and C Only (b) A and E Only
(c) D and E Only (d) A, C and E Only

306. To inhibit the growth of tumours, identify the compounds used from the following: **[NCERT : P L-137 | Jan. 30, 2023 (I)]**

(A) EDTA
(B) Coordination Compounds of Pt
(C) D – Penicillamine
(D) Cis – Platin

Choose the correct answer from the option given below:

(a) B and D Only (b) C and D Only
(c) A and B Only (d) A and C Only

307. The theory that can completely/properly explain the nature of bonding in $[Ni(CO)_4]$ is: **[Jan. 07, 2020 (I)]**

(a) Werner's theory (b) Molecular orbital theory
(c) Crystal field theory (d) Valence bond theory

308. The compound used in the treatment of lead poisoning is : **[NCERT : P L-137 | April 12, 2019 (II)]**

(a) D-penicillamine (b) desferrioxime B
(c) *Cis*-platin (d) EDTA

309. The compound that inhibits the growth of tumors is : **[NCERT : P L-137 | April 8, 2019 (II)]**

(a) *trans*-$[Pt(Cl)_2(NH_3)_2]$ (b) *cis*-$[Pd(Cl)_2(NH_3)_2]$
(c) *cis*-$[Pt(Cl)_2(NH_3)_2]$ (d) *trans*-$[Pd(Cl)_2(NH_3)_2]$

310. The number of bridging CO ligand(s) and Co-Co bond(s) in $Co_2(CO)_8$, respectively are: **[NCERT : P L-135 | Jan. 11, 2019 (II)]**

(a) 2 and 1 (b) 2 and 0 (c) 0 and 2 (d) 4 and 0

311. Wilkinson catalyst is: **[NCERT : P L-136 | Jan. 10, 2019 (I)]**

(a) $[(Ph_3P)_3IrCl]$ (b) $[(Et_3P)_3RhCl]$
(c) $[(Ph_3P)_3RhCl]$ (d) $[(Et_3P)_3IrCl]$

312. $[Co_2(CO)_8]$ displays :

[NCERT : P L-135 | Online April 9, 2017]

(a) one Co – Co bond, six terminal CO and two bridging CO
(b) one Co – Co bond, four terminal CO and four bridging CO
(c) no Co – Co bond, six terminal CO and two bridging CO
(d) no Co – Co bond, four terminal CO and four bridging CO

313. Which molecule/ion among the following cannot act as a ligand in complex compounds? Online April 10, 2015]

(a) CH_4 (b) CO (c) CN^- (d) Br^-

314. In $Fe(CO)_5$, the Fe – C bond possesses

[NCERT : P L-136 | 2006]

(a) ionic character (b) σ-character only
(c) π-character (d) both σ and π characters

315. How many EDTA (ethylenediaminetetraacetic acid) molecules are required to make an octahedral complex with a Ca^{2+} ion?

[NCERT : P L-136 | 2006]

(a) One (b) Two (c) Six (d) Three

316. Coordination compounds have great importance in biological systems. In this context which of the following statements is **incorrect**? [NCERT : P L-137 | 2004]

(a) Cyanocobalamin is B_{12} and contains cobalt
(b) Haemoglobin is the red pigment of blood and contains iron
(c) Chlorophylls are green pigments in plants and contain calcium
(d) Carboxypeptidase - A is an enzyme and contains zinc.

317. $CH_3 – Mg – Br$ is an organo metallic compound due to

[2002]

(a) Mg – Br bond (b) C – Mg bond
(c) C – Br bond (d) C – H bond

ANSWER KEYS

1	(d)	33	(3)	65	(d)	97	(b)	129	(c)	161	(a)	193	(2)	225	(2)	257	(a)	289	(a)
2	(b)	34	(3)	66	(c)	98	(d)	130	(c)	162	(b)	194	(a)	226	(c)	258	(c)	290	(b)
3	(a)	35	(1)	67	(a)	99	(c)	131	(b)	163	(1)	195	(b)	227	(b)	259	(a)	291	(d)
4	(c)	36	(a)	68	(c)	100	(c)	132	(b)	164	(a)	196	(a)	228	(2)	260	(b)	292	(a)
5	(2)	37	(5)	69	(d)	101	(d)	133	(d)	165	(d)	197	(2)	229	(4)	261	(a)	293	(a)
6	(c)	38	(c)	70	(d)	102	(a)	134	(c)	166	(a)	198	(0)	230	(a)	262	(c)	294	(d)
7	(2)	39	(1)	71	(b)	103	(d)	135	(d)	167	(N)	199	(a)	231	(c)	263	(a)	295	(c)
8	(a)	40	(c)	72	(c)	104	(a)	136	(b)	168	(d)	200	(0)	232	(2)	264	(c)	296	(d)
9	(a)	41	(2)	73	(b)	105	(b)	137	(c)	169	(3)	201	(a)	233	(b)	265	(a)	297	(b)
10	(14)	42	(a)	74	(d)	106	(c)	138	(b)	170	(b)	202	(2)	234	(d)	266	(d)	298	(c)
11	(c)	43	(1)	75	(d)	107	(d)	139	(c)	171	(b)	203	(6)	235	(a)	267	(a)	299	(d)
12	(3)	44	(2)	76	(a)	108	(c)	140	(2)	172	(4)	204	(b)	236	(a)	268	(c)	300	(b)
13	(c)	45	(2)	77	(b)	109	(2)	141	(b)	173	(c)	205	(2)	237	(0)	269	(c)	301	(c)
14	(c)	46	(5)	78	(c)	110	(a)	142	(d)	174	(b)	206	(c)	238	(a)	270	(d)	302	(c)
15	(b)	47	(b)	79	(b)	111	(2)	143	(a)	175	(a)	207	(b)	239	(b)	271	(c)	303	(a)
16	(a)	48	(c)	80	(d)	112	(d)	144	(c)	176	(4)	208	(3)	240	(c)	272	(c)	304	(c)
17	(4)	49	(2)	81	(a)	113	(d)	145	(a)	177	(a)	209	(6)	241	(d)	273	(d)	305	(a)
18	(b)	50	(6)	82	(d)	114	(d)	146	(a)	178	(a)	210	(766)	242	(c)	274	(c)	306	(a)
19	(4)	51	(0)	83	(a)	115	(d)	147	(a)	179	(6)	211	(b)	243	(b)	275	(a)	307	(b)
20	(b)	52	(b)	84	(d)	116	(a)	148	(b)	180	(a)	212	(2)	244	(c)	276	(a)	308	(d)
21	(a)	53	(3)	85	(d)	117	(d)	149	(c)	181	(480)	213	(d)	245	(a)	277	(a)	309	(c)
22	(a)	54	(6)	86	(b)	118	(b)	150	(b)	182	(b)	214	(d)	246	(c)	278	(c)	310	(a)
23	(a)	55	(2)	87	(b)	119	(a)	151	(b)	183	(d)	215	(173)	247	(c)	279	(a)	311	(c)
24	(4)	56	(3)	88	(c)	120	(1)	152	(a)	184	(b)	216	(d)	248	(d)	280	(c)	312	(a)
25	(c)	57	(6)	89	(a)	121	(d)	153	(28)	185	(a)	217	(d)	249	(26.92)	281	(d)	313	(a)
26	(d)	58	(d)	90	(b)	122	(0)	154	(d)	186	(2)	218	(a)	250	(20.0)	282	(c)	314	(d)
27	(b)	59	(d)	91	(b)	123	(b)	155	(d)	187	(4)	219	(d)	251	(a)	283	(b)	315	(a)
28	(a)	60	(d)	92	(a)	124	(d)	156	(c)	188	(b)	220	(a)	252	(d)	284	(b)	316	(c)
29	(4)	61	(d)	93	(b)	125	(c)	157	(0)	189	(c)	221	(b)	253	(a)	285	(b)	317	(b)
30	(5)	62	(c)	94	(c)	126	(b)	158	(a)	190	(7)	222	(1)	254	(d)	286	(d)		
31	(d)	63	(d)	95	(a)	127	(a)	159	(1)	191	(a)	223	(b)	255	(b)	287	(a)		
32	(9)	64	(d)	96	(d)	128	(c)	160	(b)	192	(c)	224	(c)	256	(a)	288	(a)		

Haloalkanes and Haloarenes

7

TREND ANALYSIS
JEE MAIN (2025-2021)

	2025				2024				2023				2022				2021			
	Session II		Session I		Session II		Session I		Session II		Session I		Session II		Session I		Session II		Session I	
	MCQs	NVQs	MCQs	NVQs	MCQs	NVQs	MCQs	NVQs	MCQs	NVQs	MCQs	NVQs	MCQs	NVQs	MCQs	NVQs	MCQs	NVQs	MCQs	NVQs
Topic 1	5	1	10	1	10	2	8	5	6	–	4	2	4	1	6	2	1	1	2	1
Topic 2	3	–	1	–	1	1	1	–	3	–	6	–	2	–	2	–	3	–	2	–
Topic 3	–	–	–	–	2	–	1	–	–	–	2	–	1	–	–	–	–	–	–	–
Topic 4																				
Total	8	1	11	1	13	3	10	5	9	–	12	2	7	1	8	2	4	1	4	1

How to Code Qns for Better & Faster Revision →

Type 1: Able to solve (No code Req)
Type 2: Solved with struggle or by seeing solution **(Use code NR means Needs Revision or orange stickers)**
Type 3: Tricky/ Could not solve on my own. Need help from Teacher/ Friend. **(Use code TQ means Tricky Question or red stickers)**

You can either hand write the code in boxes or else paste coloured stickers given at end of book.

Topic 1 – Preparation and Properties of Haloalkanes

1. Given below are two statements :
[NCERT : P L-163 | April 7, 2025 (II)]

Statement (I) : Cl Cl is more polar than Br Br

Statement (II) : Boiling point of Br Br is lower than Br Br but it is more polar than Br Br

In the light of the above statements, choose the **most appropriate answer** from the options given below :

(a) Statement I is correct but statement II is incorrect
(b) Statement I is incorrect but statement II is correct
(c) Both statement I and statement II are incorrect
(d) Both statement I and statement II are correct

2. Consider the following molecule (*X*).
[NCERT : P L-165 | April 4, 2025 (II)]

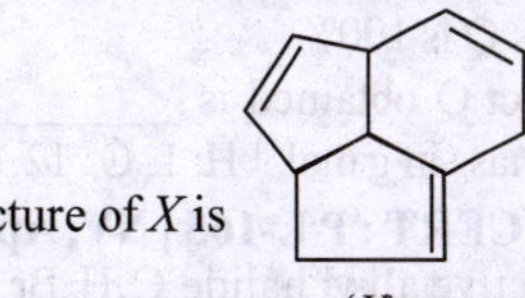

The structure of *X* is

(*X*)

The major product formed when the given molecule(X) treated with HBr (1 eq) is :

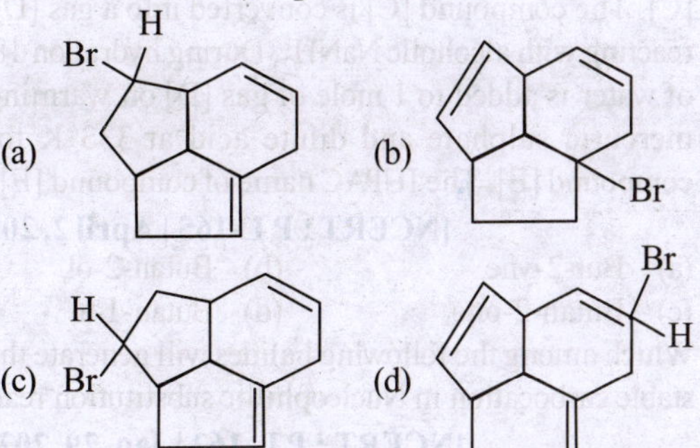

3. Given below are two statements :

[NCERT : P L-164 | April 4, 2025 (II)]

Statement (I) : Alcohols are formed when alkyl chlorides are treated with aqueous potassium hydroxide by elimination reaction.

Statement (II) : In alcoholic potassium hydroxide, alkyl chlorides form alkenes by abstracting the hydrogen from the β-carbon.

In the light of the above statements, choose the **most appropriate answer** from the options given below :

(a) Both Statement I and Statement II are incorrect
(b) Statement I is incorrect but Statement II is correct
(c) Statement I is correct but Statement II is incorrect
(d) Both Statement I and Statement II are correct

4. Predict the major product of the following reaction sequence :- **[NCERT : P L-165 | April 4, 2025 (I)]**

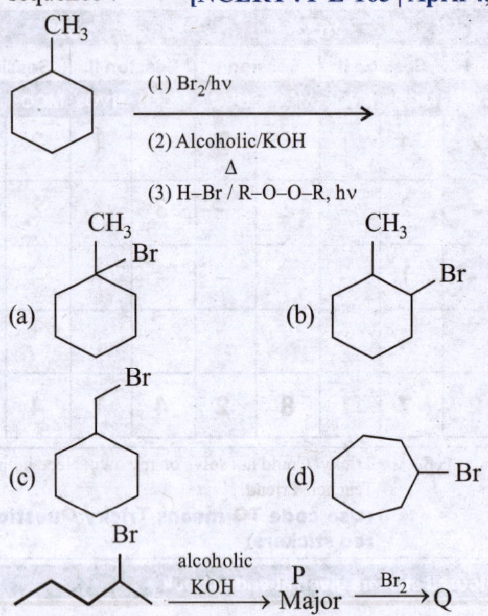

5. Br; alcoholic KOH → P Major Product $\xrightarrow{Br_2}$ Q

Consider the above sequence of reactions. 151 g of 2-bromopentane is made to react. Yield of major product P is 80% whereas Q is 100%.

Mass of product Q obtained is ________g.

(Given molar mass in g mol^{-1} H: 1, C : 12, O : 16, Br : 80)

[NCERT : P L-165 | NV, April 2, 2025 (II)]

6. An optically active alkyl halide C_4H_9Br [A] reacts with hot KOH dissolved in ethanol and forms alkene [B] as major product which reacts with bromine to give dibromide [C]. The compound [C] is converted into a gas [D] upon reacting with alcoholic $NaNH_2$. During hydration 18 gram of water is added to 1 mole of gas [D] on warming with mercuric sulphate and dilute acid at 333 K to form compound [E]. The IUPAC name of compound [E] is :

[NCERT : P L-165 | April 2, 2025 (I)]

(a) But-2-yne (b) Butan-2-ol
(c) Butan-2-one (d) Butan-1-al

7. Which among the following halides will generate the most stable carbocation in Nucleophillic substitution reaction ?

[NCERT : P L-163 | Jan. 29, 2025 (II)]

(a) Br (b) Br (c) Br (d) Ph, Ph, Ph, Br

8. The major product of the following reaction is :

[NCERT : P L-170 | Jan. 28, 2025 (II)]

Br, Br $\xrightarrow[\Delta]{KOH / EtOH(excess)}$ Major product

(a) 6-Phenylhepta-2,4-diene (b) 2-Phenylhepta-2,5-diene
(c) 6-Phenylhepta-3,5-diene (d) 2-Phenylhepta-2,4-diene

9. The product B formed in the following reaction sequence is : **[NCERT : P L-170 | Jan. 28, 2025 (II)]**

$\xrightarrow{HCl}$ (A) (Major) $\xrightarrow{AgCN}$ (B) (Major)

(a) NC (b) NC (c) CN (d) NC

10. The products A and B in the following reactions, respectively are **[NCERT : P L-170 | Jan. 28, 2025 (I)]**

$A \xleftarrow{Ag-NO_2} CH_3-CH_2-CH_2-Br \xrightarrow{AgCN} B$

(a) $CH_3–CH_2–CH_2–ONO, CH_3–CH_2–CH_2–NC$
(b) $CH_3–CH_2–CH_2–ONO, CH_3–CH_2–CH_2–CN$
(c) $CH_3–CH_2–CH_2–NO_2, CH_3–CH_2–CH_2–CN$
(d) $CH_3–CH_2–CH_2–NO_2, CH_3–CH_2–CH_2–NC$

11. Given below are two statements :

[NCERT : P L-173 | Jan. 28, 2025 (I)]

Statement I : Et, Et N–Cl will undergo alkaline hydrolysis at a faster rate than Et, Et CH–Cl

Statement II : Et, Et N–Cl intramolecular substitution takes place first by involving lone pair of electrons on nitrogen.

In the light of the above statements, choose the most appropriate answer from the options given below :

(a) Both Statement I and Statement II are incorrect
(b) Statement I is incorrect but statement II is correct
(c) Both Statement I and Statement II are correct
(d) Statement I is correct but Statement II is incorrect

12. The structure of the major product formed in the following reaction is : **[NCERT : P L-170 | Jan. 24, 2025 (II)]**

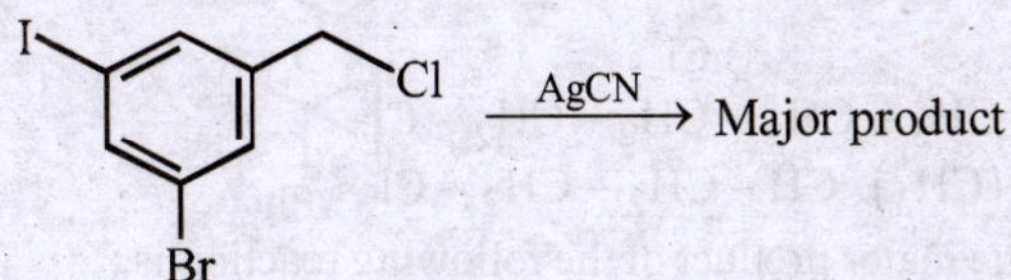

(a) [3-iodo-5-cyanobenzyl chloride: I, CH_2Cl, CN on ring] (b) [I, CH_2NC, Br on ring]

(c) [I, CH_2CN, Br on ring] (d) [CN, CH_2Cl, Br on ring]

13. Given below are two statements: **[NCERT : P L-171, 172 | Jan. 24, 2025 (I)]**

Statement-I: The conversion proceeds well in the less polar medium.

$$CH_3-CH_2-CH_2-CH_2-Cl \xrightarrow{HO^-} CH_3-CH_2-CH_2-CH_2-OH + Cl^-$$

Statement-II: The conversion proceeds well in the more polar medium.

$$CH_3-CH_2-CH_2-CH_2-Cl \xrightarrow{R_3\ddot{N}} CH_3-CH_2-CH_2-CH_2-\overset{\oplus}{N}R_3\ Cl^{\ominus}$$

In the light of the above statements, choose the correct answer from the options given below.

(a) Both statement I and statement II are true
(b) Both statement I and statement II are false
(c) Statement I is false but statement II is true
(d) Statement I is true but statement II is false

14. The ascending order of relative rate of solvolysis of following compounds is **[NCERT : P L-173 | Jan. 23, 2025 (II)]**

(A) [1-bromo-1,4,4-trimethylcyclohexane, Br] (B) [1,4-dibromotetralin, Br, Br]

(C) [bromodiphenylmethane, Br] (D) [2-bromo-3-methylpentane skeleton, Br]

(a) (D) < (A) < (B) < (C) (b) (C) < (B) < (A) < (D)
(c) (D) < (B) < (A) < (C) (d) (C) < (D) < (B) < (A)

15. Match the List-I with List-II **[NCERT : P L-165, 166, 186 | Jan. 23, 2025 (I)]**

	List-I Name reaction		List-II Product obtainable
A.	Swarts reaction	I.	Ethyl benzene
B.	Sandmeyer's reaction	II.	Ethyl iodide
C.	Wurtz Fittig reaction	III.	Cyanobenzene
D.	Finkelstein reaction	IV.	Ethyl fluoride

Choose the correct answer from the options given below:

(a) A-II, B-III, C-I, D-IV
(b) A-IV, B-I, C-III, D-II
(c) A-IV, B-III, C-I, D-II
(d) A-II, B-I, C-III, D-IV

16. $RBr \xrightarrow[\text{(ii) } H_2O]{\text{(i) Mg, dry ether}}$ [2-Methylbutane]

2-Methylbutane

The maximum number of RBr producing 2-methylbutane by above sequence of reactions is _____. (Consider the structural isomers only) **[NCERT : P L-180 | NV, Jan. 22, 2025 (II)]**

(a) 4 (b) 5 (c) 3 (d) 1

17. Given below are two statements : **[NCERT : P L-174 | Jan. 22, 2025 (I)]**

Statement I : CH_3-O-CH_2-Cl will undergo S_N1 reaction though it is a primary halide.

Statement II : $CH_3-\underset{CH_3}{\overset{CH_3}{\underset{|}{\overset{|}{C}}}}-CH_2-Cl$ will not undergo S_N2 reaction very easily though it is a primary halide.

In the light of the above statements, choose the most appropriate answer from the options given below:

(a) Statement I is incorrect but Statement II is correct.
(b) Both Statement I and Statement II are incorrect
(c) Statement I is correct but Statement II is incorrect
(d) Both Statement I and Statement II are correct.

18. Given below are two statements : one is labelled as **Assertion (A)** and the other is labelled as **Reason (R) :** **[NCERT : P L-171 | April 9, 2024 (I)]**

Assertion (A) : S_N2 reaction of $C_6H_5CH_2Br$ occurs more readily than the S_N2 reaction of CH_3CH_2Br.

Reason (R) : The partially bonded unhybridized p-orbital that develops in the trigonal bipyramidal transition state is stabilized by conjugation with the phenyl ring.

In the light of the above statements, choose the **most appropriate** answer from the options given below :

(a) **(A)** is not correct but **(R)** is correct
(b) Both **(A)** and **(R)** are correct but **(R)** is not the correct explanation of **(A)**
(c) Both **(A)** and **(R)** are correct and **(R)** is the correct explanation of **(A)**
(d) **(A)** is correct but **(R)** is not correct

19. Br, OCH_3 $\xrightarrow[\Delta]{KCN(alc)}$ Major Product 'P'

In the above reaction product 'P' is

[NCERT : P L-171 | April 9, 2024 (II)]

(a) CN, OCH_3 (b) OCH_3, CN

(c) CN, OCH_3 (d) CN, OCH_3

20. Which among the following compounds will undergo fastest S_N2 reaction. **[NCERT : P L-171 | April 8, 2024 (I)]**

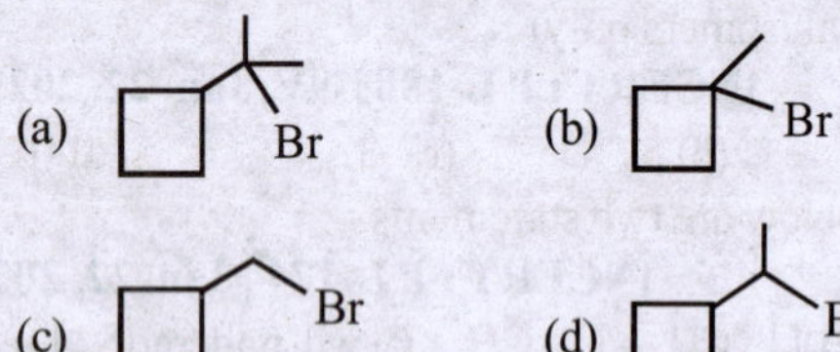

(a) Br (b) Br

(c) Br (d) Br

21. The number of optical isomers in following compound is : ______. **[NCERT : P L-175 | NV, April 8, 2024 (I)]**

Br CH_3

22. Given below are two statements :

[NCERT : P L-171 | April 8, 2024 (II)]

Statement (I) : S_N2 reactions are 'stereospecific', indicating that they result in the formation only one stereo-isomers as the product.

Statement (II) : S_N1 reactions generally result in formation of product as racemic mixtures. In the light of the above statements, choose the **correct** answer from the options given below :

(a) **Statement I** is true but **Statement II** is false

(b) **Statement I** is false but **Statement II** is true

(c) Both **Statement I** and **Statement II** is true

(d) Both **Statement I** and **Statement II** is false

23. Total number of optically active compounds from the following is ______.

[NCERT : P L-171 | NV, April 8, 2024 (II)]

$CH_3-CH(OH)-CH(OH)-CH_3$, OH OH OH,

$CH_3-CH_2-CH_2-CH_2-OH$,

$CH_3-CH_2-CH(Cl)-CH_3$

$CH_3-CH_2-CH_2-CH_2-Cl$,

$(CH_3)_2CH-CH_2-CH_2-Cl$

24. Cl, CH_3 + NaOH $\xrightarrow{H_2O}$ Major Product "A"

Consider the above chemical reaction. Product "A" is:

[NCERT : P L-173 | April 6, 2024 (II)]

(a) CH_3, OH

(b) CH_3, OH

(c) CH_3, OH

(d) OH, CH_3

25. Identify the major product in the following reaction.

Br, CH_3 $\xrightarrow[C_2H_5OH]{\bar{O}H}$ Major Product

[NCERT : P L-171 | April 5, 2024 (II)]

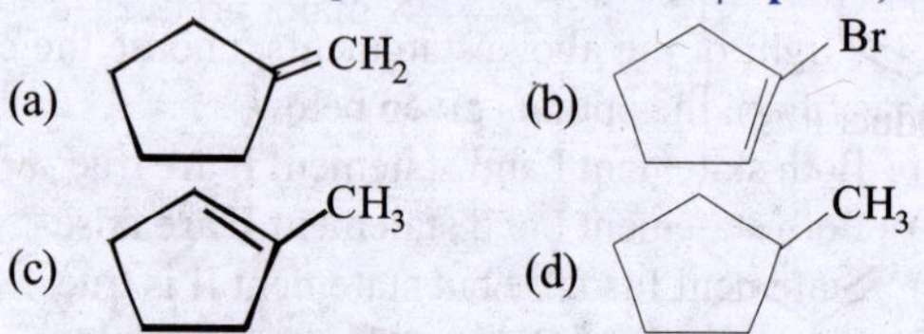

(a) $=CH_2$ (b) Br

(c) CH_3 (d) CH_3

26. Identify the correct set of reagents or reaction conditions 'X' and 'Y' in the following set of transformation.

$CH_3-CH_2-CH_2-Br \xrightarrow{'X'}$ Product $\xrightarrow{'Y'}$ $CH_3-CH(Br)-CH_3$

[NCERT : P L-171 | April 4, 2024 (I)]

(a) X = conc.alc. NaOH, 80°C, Y = $Br_2/CHCl_3$

(b) X = dil.aq. NaOH, 20°C, Y = HBr/acetic acid

(c) X = conc.alc. NaOH, 80°C, Y = HBr/acetic acid

(d) X = dil.aq. NaOH, 20°C, Y = $Br_2/CHCl_3$

27. Find out the major product formed from the following reaction. [Me: $-CH_3$]

[NCERT : P L-171 | April 4, 2024 (II)]

Br Br $\xrightarrow{Me_2NH(2equiv)}$

(a) NMe_2 NMe_2

(b) NMe_2 NMe_2

(c) NMe_2 NMe_2

(d) NMe_2 NMe_2

28. $CH_3-CH_2-CH_2-Br + NaOH \xrightarrow{C_2H_5OH}$ Product 'A'

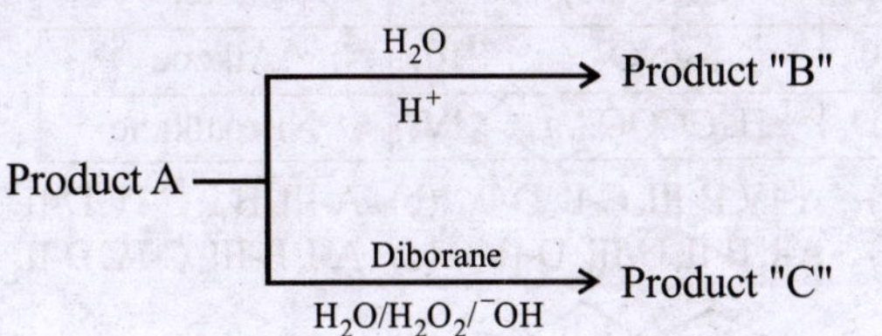

Consider the above reactions, identify product B and product C. **[April 4, 2024 (II)]**

(a) B = C = 2-Propanol
(b) B = 2-Propanol C = 1-Propanol
(c) B = 1-Propanol C = 2-Propanol
(d) B = C = 1-Propanol

29. Br $\xrightarrow[\Delta]{KOH(alc.)}$ major product "P"

Product P is

[NCERT : P L-180 | April 4, 2024 (II)]

(a) (b)

(c) (d)

30. Given below are two statements : one is labelled as **Assertion (A)** and the other is labelled as **Reason (R).**

Assertion (A) : Haloalkanes react with KCN to form alkyl cyanides as a main product while with AgCN form isocyanide as the main product.

Reason (R) : KCN and AgCN both are highly ionic compounds.

In the light of the above statements, choose the most appropriate answer from the options given below:

[NCERT : P L-170 | Feb. 1, 2024 (I)]

(a) (A) is correct but (R) is not correct.
(b) Both (A) and (R) are correct but (R) is not the correct explanation of (A)
(c) (A) is not correct but (R) is correct.
(d) Both (A) and (R) are correct and (R) is the correct explanation of (A).

31. Number of optical isomers possible for 2 - chlorobutane ______. **[NCERT : P L-175 | NV, Feb. 1, 2024 (I)]**

32. Total number of isomeric compounds (including stereoisomers) formed by monochlorination of 2- methylbutane is ______. **[NV, Feb. 1, 2024 (II)]**

33. The product (C) in the below mentioned reaction is :

$CH_3-CH_2-CH_2-Br \xrightarrow[\Delta]{KOH(alc)} A \xrightarrow{HBr}$

$B \xrightarrow[KOH(aq)]{\Delta} C$ **[NCERT : P L-179 | Jan. 31, 2024 (I)]**

(a) Propene (b) Propan-2-ol
(c) Propyne (d) Propan-1-ol

34. $CH_3CH_2Br + NaOH$ — $\xrightarrow{C_2H_5OH}$ Product A ; $\xrightarrow{H_2O}$ Product B

The total number of hydrogen atoms in product A and product B is ______.

[NCERT : P L-179 | NV, Jan. 31, 2024 (I)]

35. Example of vinylic halide is :

[NCERT : P L-161 | Jan. 30, 2024 (I)]

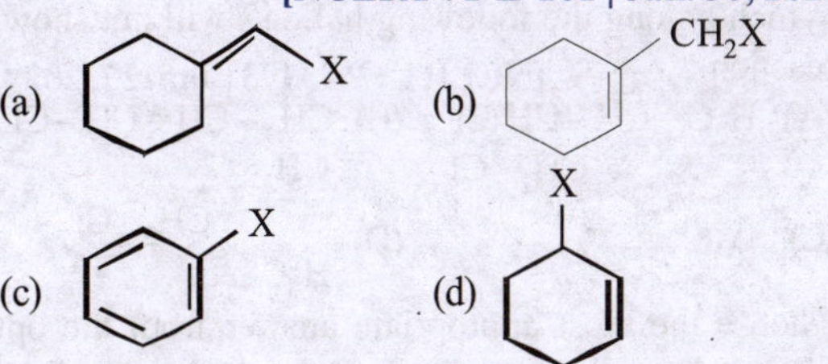

(c) X (d) X

36. Given below are two statements : One is labelled as Assertion **(A)** and the other is labelled as Reason **(R).**

[NCERT : P L-160 | Jan. 30, 2024 (I)]

Assertion (A) : $CH_2=CH-CH_2-Cl$ is an example of allyl halide.

Reason (R) : Allyl halides are the compounds in which the halogen atom is attached to sp^2 hybridised carbon atom.

In the light of the above statements, choose the most **appropriate** answer from the option given below:

(a) **(A)** is true but **(R)** is false
(b) Both **(A)** and **(R)** are true but **(R)** is not the correct explanation of **(A)**
(c) **(A)** is false but **(R)** is true
(d) Both **(A)** and **(R)** are true and **(R)** is the correct explanation of **(A)**.

37. Given below are two statements:

[NCERT : P L-170 | Jan. 30, 2024 (II)]

Statement – I: High concentration of strong nucleophilic reagent with secondary alkyl halides which do not have bulky substituents will follow S_N2 mechanism.

Statement - II: A secondary alkyl halide when treated with a large excess of ethanol follows S_N1 mechanism.
In the light of the above statements, choose the most appropriate from the options given below:
(a) Statement I is true but Statement II is false.
(b) Statement I is false but Statement II is true.
(c) Both Statement I and Statement II are false.
(d) Both Statement I and Statement II are true.

38. 2 - Chlorobutane + $Cl_2 \rightarrow C_4H_8Cl_2$ (isomers)
Total number of optically active isomers shown by $C_4H_8Cl_2$, obtained in the above reaction is ______.
[NCERT : P L-175 | NV, Jan. 30, 2024 (II)]

39. Alkyl halide is converted into alkyl isocyanide by reaction with **[NCERT : P L-170 | Jan. 29, 2024 (II)]**
(a) NaCN (b) AgCN (c) NH_4CN (d) KCN

40. The correct statement regarding nucleophilic substitution reaction in a chiral alkyl halide is :
[NCERT : P L-171, 173 | Jan. 27, 2024 (I)]
(a) Retention occurs in S_N1 reaction and inversion occurs in S_N2 reaction.
(b) Racemisation occurs in S_N1 reaction and retention occurs in S_N2 reaction.
(c) Racemisation occurs in both S_N1 and S_N2 reactions.
(d) Racemisation occurs in S_N1 reaction and inversion occurs in S_N2 reaction.

41. 3-Methylhex-2-ene on reaction with HBr in presence of peroxide forms an addition product (A). The number of possible stereoisomers for 'A' is______.
[NCERT : P L-165 | NV, Jan. 27, 2024 (I)]

42. Which among the following halide/s will not show S_N1 reaction: **[NCERT : P L-173 | Jan. 27, 2024 (II)]**
(A) $H_2C=CH-CH_2Cl$ (B) $CH_3-CH=CH-Cl$
(C) 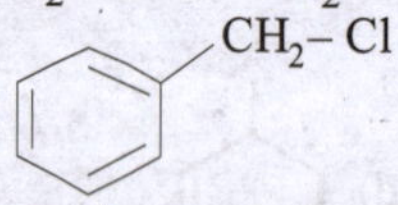 benzene ring with CH_2-Cl (D) 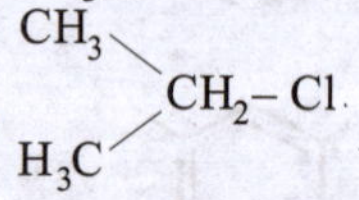 CH_3, H_3C on carbon bearing CH_2-Cl
Choose the most appropriate answer from the options given below:
(a) (A), (B) and (D) only
(b) (A) and (B) only
(c) (B) and (C) only
(d) (B) only

43. Total number of compounds with Chiral carbon atoms from following is ______.
[NCERT : P L-175 | NV, Jan. 27, 2024 (II)] 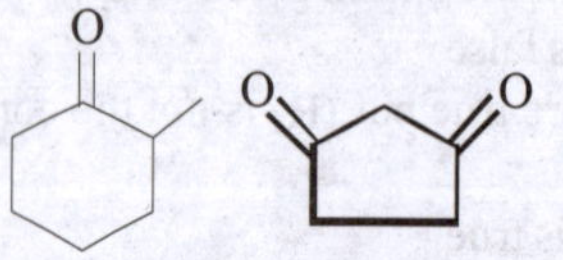
(2-methylcyclohexanone: O) (cyclopentane-1,3-dione: O, O)
$CH_3-CH_2-CH(NO_2)-COOH$
$CH_3-CH_2-CHBr-CH_2-CH_3$
$CH_3-CH(I)-CH_2-NO_2$
$CH_3-CH_2-CH(OH)-CH_2OH$
$CH_3-CH(I)-CH(I)-C_2H_5$

44. 2-Methyl propyl bromide reacts with $C_2H_5O^-$ and gives 'A' whereas on reaction with C_2H_5OH it gives 'B'. The mechanism followed in these reactions and the products 'A' and 'B' respectively are:
[NCERT : P L-171, 173 | April 13, 2023 (I)]
(a) S_N2, A = iso-butyl ethyl ether; S_N1, B = tert-butyl ethyl ether
(b) S_N1, A = tert-butyl ethyl ether; S_N1, B = 2-butyl ethyl ether
(c) S_N1, A = tert-butyl ethyl ether; S_N2, B = iso-butyl ethyl ether
(d) S_N2, A = 2-butyl ethyl ether; S_N2, B = iso-butyl ethyl ether

45. Match List I with List II **[April 13, 2023 (II)]**
1 - Bromopropane is reacted with reagents in List I to give product in List II

	List I - Reagent		List II - Product
A.	KOH (alc)	I.	Nitrile
B.	KCN (alc)	II.	Ester
C.	$AgNO_2$	III.	Alkene
D.	$H_3CCOOAg$	IV.	Nitroalkane

(a) A-IV, B-III, C-II, D-I (b) A-III, B-I, C-IV, D-II
(c) A-I, B-II, C-III, D-IV (d) A-I, B-III, C-IV, D-II

46. MeO–C_6H_4–CH_2Cl $\xrightarrow{Nu}$ OMe–C_6H_4–CH_2Nu ... (I) 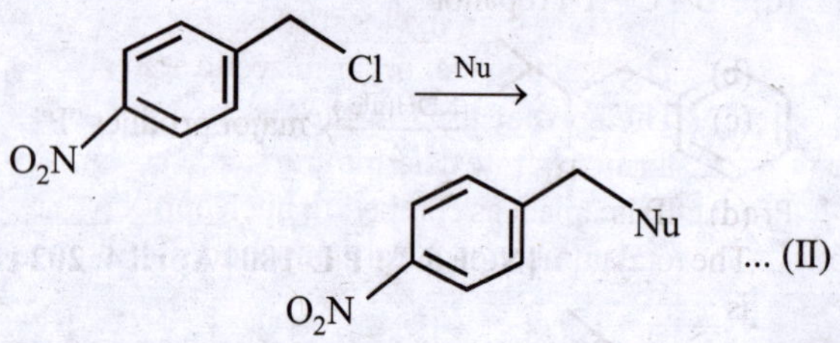
O_2N–C_6H_4–CH_2Cl $\xrightarrow{Nu}$ O_2N–C_6H_4–CH_2Nu ... (II)
Where Nu = Nucleophile
Find out the correct statement from the options given below for the above 2 reactions.
[NCERT : P L-171, 173 | April 11, 2023 (I)]
(a) Reaction (I) is of 2nd order and reaction (II) is of 1st order
(b) Reaction (I) and (II) both are of 2nd order
(c) Reaction (I) is of 1st order and reaction (II) is of 2nd order
(d) Reaction (I) and (II) both are of 1st order

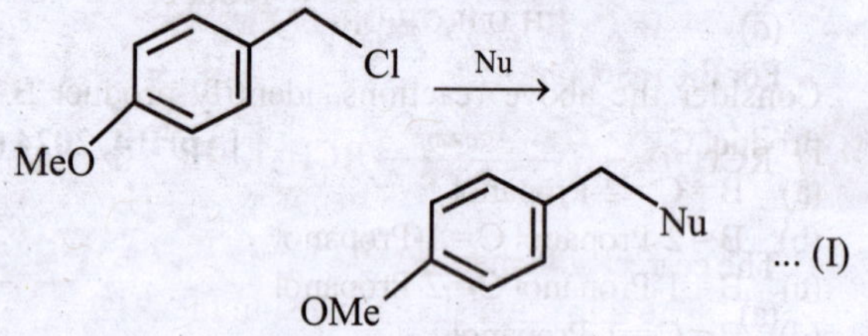

47. Given below are two statements, one is labelled as **Assertion A** and the other is labelled as **Reason R**.
Assertion A : A solution of the product obtained by heating a mole of glycine with a mole of chlorine in presence of red phosphorous generates chiral carbon atom.
Reason R: A molecule with 2 chiral carbons is always optically active.

In the light of the above statements, choose the correct answer from the options given below: **[April 11, 2023 (II)]**

(a) A is false but R is true.
(b) A is true but R is false.
(c) Both A and R are true and R is the correct explanation of A.
(d) Both A and R are true and R is NOT the correct explanation of A.

48. Choose the halogen which is most reactive towards S_N1 reaction in the given compounds (A, B, C & D)
[NCERT : P L-173 | April 08, 2023 (I)]

A. $Br_{(a)}$ $Br_{(b)}$ B. $I_{(a)}$ $I_{(b)}$

C. $Br_{(a)}$ $Br_{(b)}$ Me D. $Br_{(b)}$ $Br_{(a)}$

(a) A-$Br_{(b)}$; B-$I_{(b)}$; C-$Br_{(b)}$; D-$Br_{(b)}$
(b) A-$Br_{(a)}$; B-$I_{(a)}$; C-$Br_{(b)}$; D-$Br_{(a)}$
(c) A-$Br_{(b)}$; B-$I_{(a)}$; C-$Br_{(a)}$; D-$Br_{(a)}$
(d) A-$Br_{(b)}$; B-$I_{(a)}$; C-$Br_{(a)}$; D-$Br_{(a)}$

49. For the reaction: **[NCERT : P L-171 | April 06, 2023 (I)]**

$$RCH_2Br + I^- \xrightarrow{\text{Acetone}} \underset{\text{major}}{RCH_2I} + Br^-$$

The correct statement is:

(a) The transition state formed in the above reaction is less polar than the localised anion.
(b) The reaction can occur in acetic acid also.
(c) The solvent used in the reaction solvates the ions formed in rate determining step.
(d) Br^- can act as competing nucleophile.

50. The total number of chiral compound/s from the following is_____. **[NCERT : P L-176 | NV, Feb. 01, 2023 (I)]**

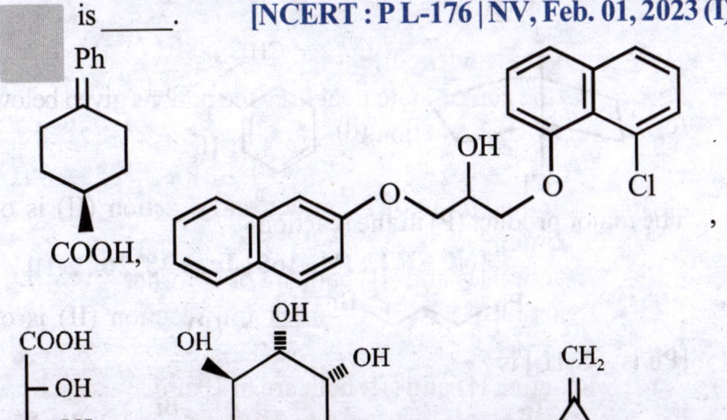

51. The graph which represents the following reaction is:

$$(C_6H_5)_3C-Cl \xrightarrow[\text{Pyridine}]{OH^-} (C_6H_5)_3C-OH$$

[NCERT : P L-171 | Feb. 01, 2023 (II)]

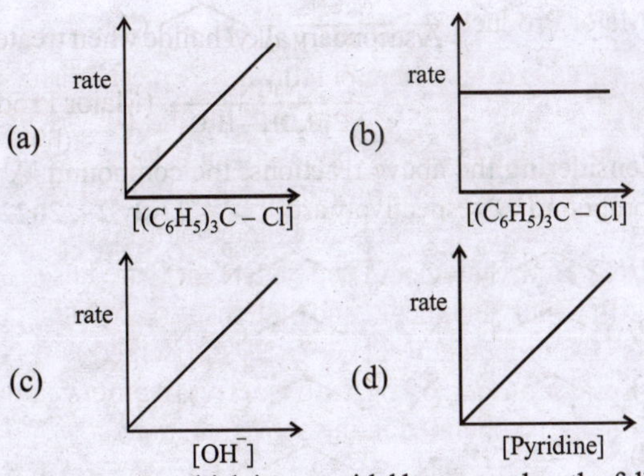

52. Testosterone, which is a steroidal hormone, has the following structure. **[NCERT : P L-175 | NV, Feb. 01, 2023 (II)]**

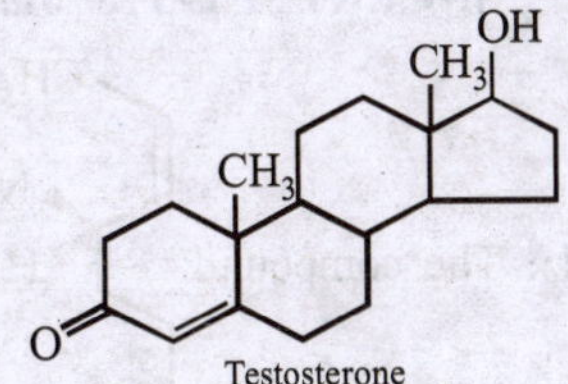

Testosterone

The total number of asymmetric carbon atom/s in testosterone is_______.

53. Identify the correct order for the given property for following compounds **[NCERT : P L-167 | Jan. 29, 2023 (I)]**

(A) Boiling Point: Cl < Cl < Cl
(B) Density Br < Cl < I
(C) Boiling Point: Br < Br Br < Br Br Br
(D) Density: I Br < Br < Cl Br
(E) Boiling Point: Cl > Cl > Cl

Choose the correct answer from the option given below :-

(a) (B), (C) and (D) only (b) (A), (C) and (E) only
(c) (A), (C) and (D) only (d) (A), (B) and (E) only

54. The isomeric deuterated bromide with molecular formula C_4H_8DBr having two chiral carbon atoms is **[NCERT : P L-175 | Jan. 25, 2023 (II)]**

(a) 2–Bromo–1–deuterobutane
(b) 2–Bromo–2–deuterobutane
(c) 2–Bromo–3–deuterobutane
(d) 2–Bromo–1–deutero–2–methylpropane

55. **Assertion A** : Hydrolysis of an alkyl chloride is a slow reaction but in the presence of NaI, the rate of the hydrolysis increases. **[Jan. 24, 2023 (I)]**

Reason R : I^- is a good nucleophile as well as a good leaving group.

In the light of the above statements, choose the correct answer from the options given below.

(a) A is false but R is true
(b) A is true but R is false
(c) Both A and R are true and R is the correct explanation of A
(d) Both A and R are true but R is NOT the correct explanation of A

56. (Major Product) (A′) $\xleftarrow[C_2H_5OH-H_2O]{AgCN}$ ⁀Cl $\xrightarrow[C_2H_5OH-H_2O]{NaCN}$ (Major Product) (B′)

Considering the above reactions, the compound 'A' and compound 'B' respectively are : **[July 29, 2022 (I)]**

(a) $\overset{\oplus}{N}\equiv\overset{\ominus}{C}$, $\overset{\oplus}{N}\equiv\overset{\ominus}{C}$

(b) $C\equiv N$, $C\equiv N$

(c) $\overset{\oplus}{N}\equiv\overset{\ominus}{C}$, $C\equiv N$

(d) $C\equiv N$, $\overset{\oplus}{N}\equiv\overset{\ominus}{C}$

57. Given below are two statements. **[NCERT : P L-175 | July 29, 2022 (II)]**

Statement I : The compound (CH_3, NO_2, H, CH_3) is optically active.

Statement II : (CH_3, O_2N, H, CH_3) is mirror image of above compound A.

In the light of the above statement, choose the **most appropriate** answer from the options given below.

(a) Both statement I and statement II are correct.
(b) Both statement I and statement II are incorrect.
(c) Statement I is correct but statement II is incorrect.
(d) Statement I is incorrect but statement II is correct.

58. Compound 'A' undergoes following sequence of reactions to give compound 'B'. The correct structure and chirality of compound 'B' is:
[where Et is $-C_2H_5$]
[NCERT : P L-180 | July 29, 2022 (II)]

Compound 'A' (Br) $\xrightarrow[(ii)\ D_2O]{(i)\ Mg,\ Et_2O}$ B

(a) (D), Achiral (b) (OD), Chiral
(c) (D), Chiral (d) (OD), Achiral

59. Optical activity of an enantiomeric mixture is +12.6° and the specific rotation of (+) isomer is +30°. The optical purity is____% **[NCERT : P L-177 | NV, July 27, 2022 (I)]**

60. Major product of the following reaction is

(O, O) $\xrightarrow{2\ HBr}$ **[July 25, 2022 (II)]**

(a) (Br, O, Br, O)
(b) (Br, Br, O, O)
(c) (Br, O, Br, O)
(d) (Br, O, Br, O)

61. Observe structures of the following compounds

(Cl) (H_2N, OH, O) (NH_2, OH, O) (OH)

The total number of structures/compounds which possess asymmetric carbon atoms is________.
[NCERT : P L-176 | NV, June 29, 2022 (I)]

62. Which one of the following compounds is inactive towards S_N1 reaction? **[NCERT : P L-173 | June 28, 2022 (I)]**

(a) $(CH_3)_3C-Cl$ (b) (Cl)

(c) 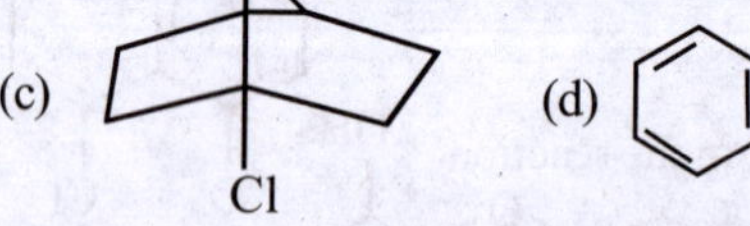 (Cl) (d) $C_6H_5-CH(CH_3)-Cl$

63. The major product (P) in the reaction **[NCERT : P L-165 | June 28, 2022 (I)]**

Ph⁀Br $\xrightarrow{HBr}$?(P)

[Ph is $-C_6H_5$] is

(a) (Ph, Br, Br) (b) (Ph, Br, Br)
(c) (Ph, Br, Br) (d) (Ph, Br, Br)

64. The major product of the following reaction contains _______ bromine atom(s). [NV, June 28, 2022 (II)]

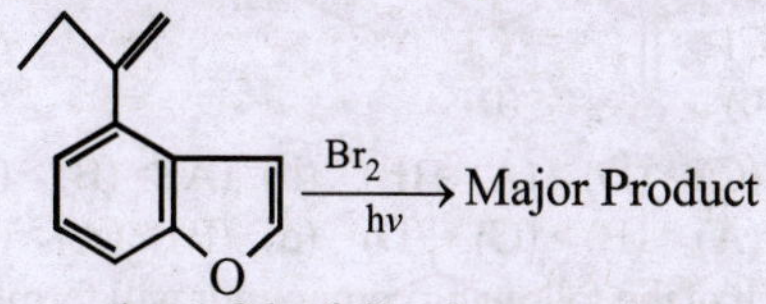

$\xrightarrow[h\nu]{Br_2}$ Major Product

65. The major product of the following reaction is:

$\xrightarrow[DMF]{Ph\overset{\ominus}{S}\overset{\oplus}{Na}}$

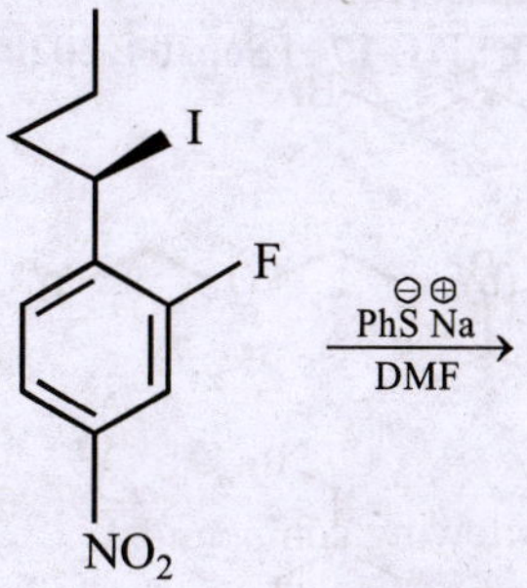

[NCERT : P L-171 | June 27, 2022 (I)]

(a) 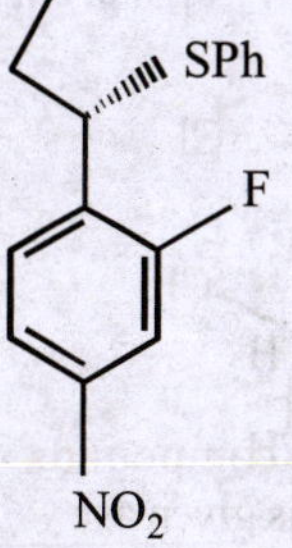 (b) 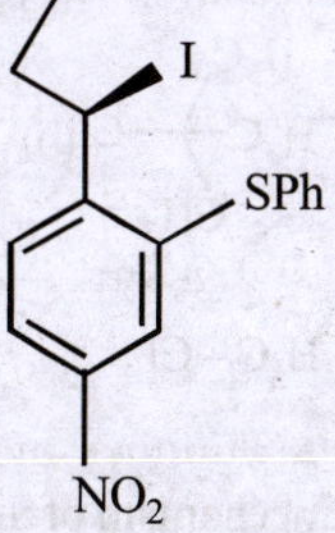 (c) 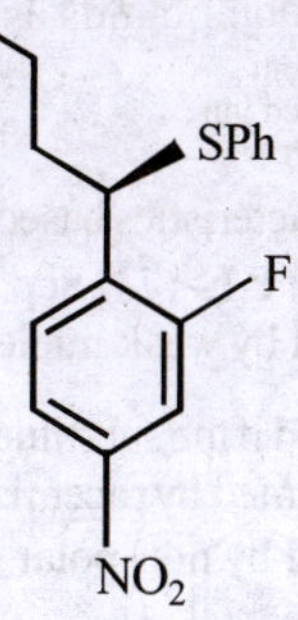 (d)

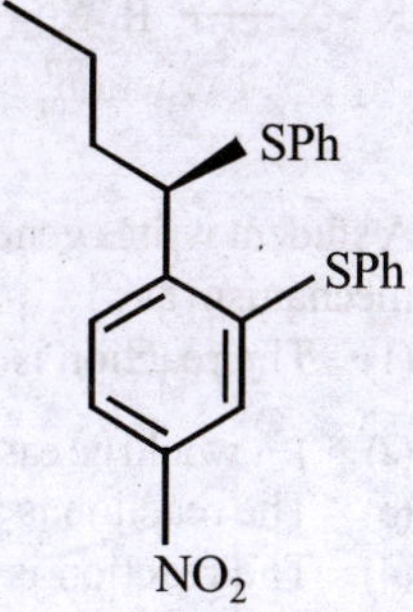

66. The major product in the reaction

$CH_3-\underset{CH_3}{\overset{CH_3}{C}}-Cl+\overset{+}{K}\overset{-}{O}-\underset{CH_3}{\overset{CH_3}{C}}-CH_3 \longrightarrow ?$ is

[NCERT : P L-173 | June 25, 2022 (I)]

(a) t-Butyl ethyl ether (b) 2, 2-Dimethyl butane
(c) 2-Methyl pent-1-ene (d) 2-Methyl prop-1-ene

67. Amongst the following, the major product of the given chemical reaction is [June 25, 2022 (II)]

$\xrightarrow[CH_3OH]{Br_2}$ Major product

(a) 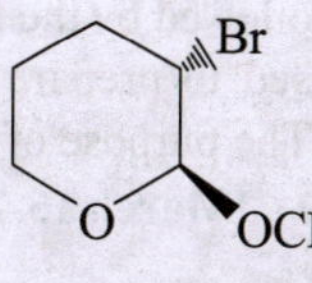 Br, OCH_3 (b) 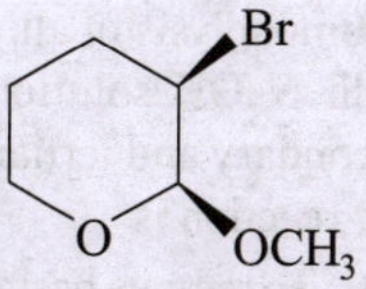 Br, OCH_3
(c) 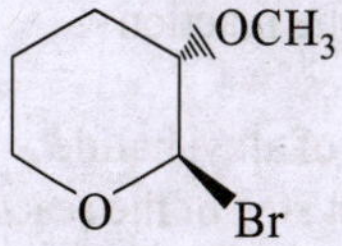 OCH_3, Br (d) 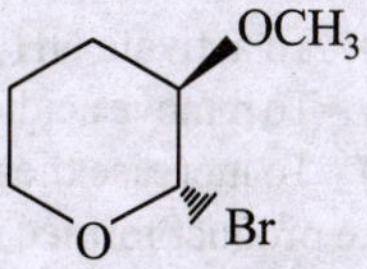 OCH_3, Br

68. In the given reactions sequence, the major product 'C' is :

$C_8H_{10} \xrightarrow[H_2SO_4]{HNO_3} A \xrightarrow[\Delta]{Br_2} B \xrightarrow[KOH]{\text{alcoholic}} C$

[June 24, 2022 (I)]

(a) 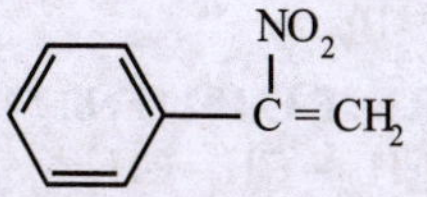 $C_6H_5-\overset{NO_2}{\overset{|}{C}}=CH_2$

(b) $O_2N-C_6H_4-CH=CH_2$

(c) $O_2N-C_6H_4-\underset{NO_2}{\underset{|}{C}}=CH_2$

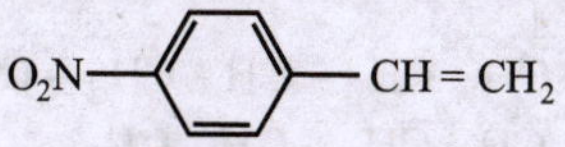

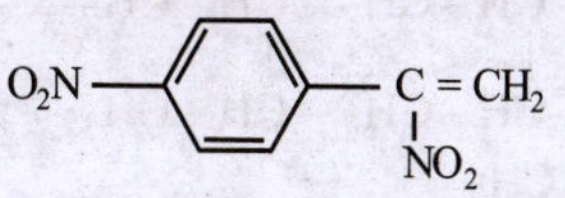

(d) 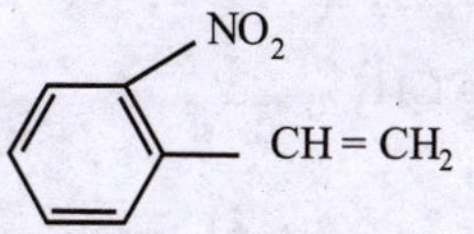 $C_6H_4(NO_2)-CH=CH_2$

69. 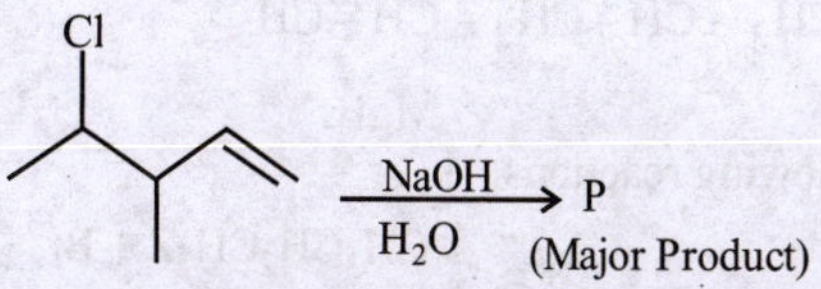 $\xrightarrow[H_2O]{NaOH}$ P (Major Product)

Consider the above reaction. The number of π electrons present in the product 'P' is____. [NV, June 24, 2022 (II)]

70. The major product of the following reaction is :

[NCERT : P L-180 | Aug. 31, 2021 (II)]

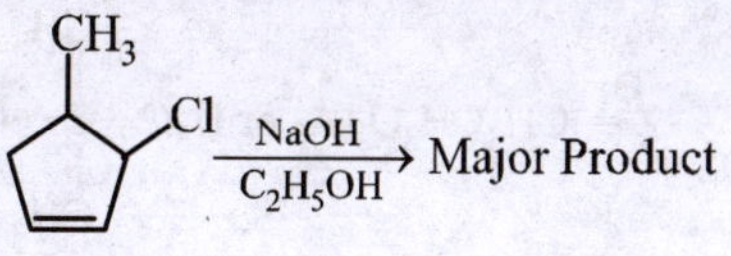

$\xrightarrow[C_2H_5OH]{NaOH}$ Major Product

(a) 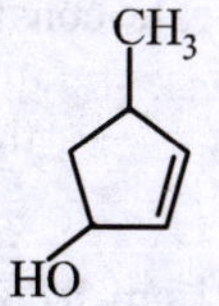 CH_3, HO (b) 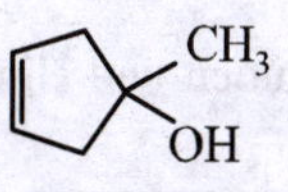 CH_3, OH
(c) 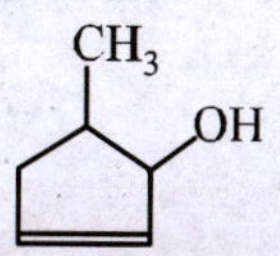 CH_3, OH (d) 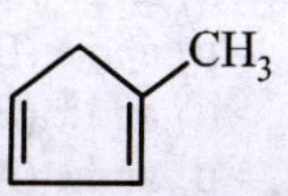 CH_3

71. A chloro compound "A".

(i) forms aldehydes on ozonolysis followed by the hydrolysis.

(ii) when vaporized completely 1.53 g of A, gives 448 mL of vapour at STP.

The number of carbon atoms in a molecule of compound A is _______. [NV, Aug. 26, 2021 (II)]

72. Ammonolysis of alkyl halides followed by the treatment with NaOH solution can be used to prepare primary, secondary and tertiary amines. The purpose of NaOH in the reaction is **[March 16, 2021 (II)]**
(a) To remove basic impurities
(b) To activate NH_3 used in the reaction
(c) To remove acidic impurities
(d) To increase the reactivity of alkyl halide

73. The product formed in the first step of the reaction of

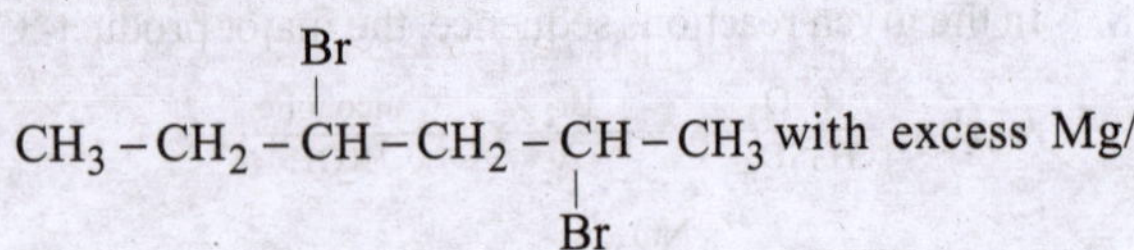

with excess Mg/ $Et_2O(Et = C_2H_5)$ is : **[NCERT : P L-181 | Feb. 24, 2021 (I)]**
(a) $CH_3-CH_2-CH-CH_2-CH-CH_3$ / $CH_3-CH-CH_2-CH-CH_2-CH_3$
(b) $CH_2-CH_2-CH-CH_2-CH-CH_2$ / $CH_3-CH_2-CH-CH_2-CH-CH_3$
(c) CH_3-CH (ring with CH_2 and $CH-CH_3$)
(d) $CH_3CH_2-CH(MgBr)-CH_2-CH(MgBr)-CH_3$

74. For the following reactions

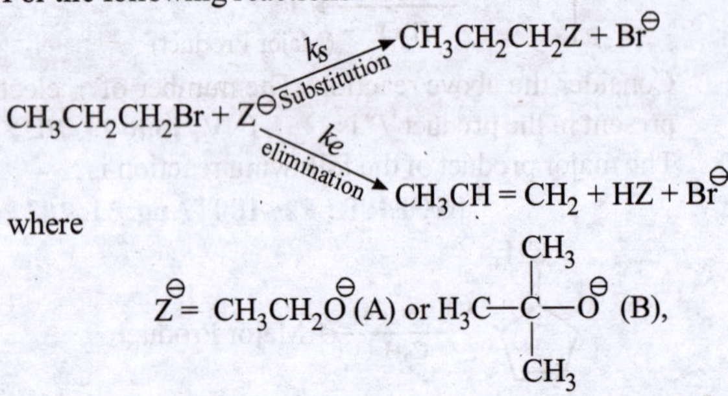

where

$Z^{\ominus} = CH_3CH_2O^{\ominus}$ (A) or $H_3C-C(CH_3)_2-O^{\ominus}$ (B),

k_s and k_e, are, respectively, the rate constants for substitution and elimination, and $\mu = \frac{k_s}{k_e}$, the correct option is ________.

[NCERT : P L-179, 181 | Jan. 07, 2020 (II)]
(a) $\mu_B > \mu_A$ and $k_e(A) > k_e(B)$
(b) $\mu_A > \mu_B$ and $k_e(B) > k_e(A)$
(c) $\mu_B > \mu_A$ and $k_e(B) > k_e(A)$
(d) $\mu_A > \mu_B$ and $k_e(A) > k_e(B)$

75. The decreasing order of reactivity of the following organic molecules towards $AgNO_3$ solution is :
[NCERT : P L-174 | Sep. 04, 2020 (I)]

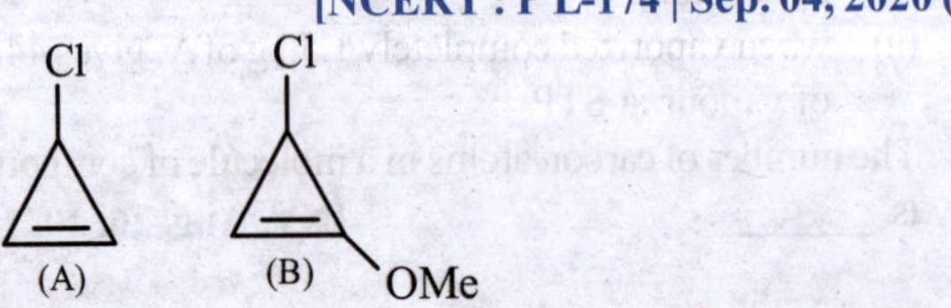

$CH_3CH(Cl)CH_3$ (C) $CH_3CH(Cl)CH_2NO_2$ (D)

(a) (C) > (D) > (A) > (B) (b) (A) > (B) > (D) > (C)
(c) (A) > (B) > (C) > (D) (d) (B) > (A) > (C) > (D)

76. Which of the following compounds will form the precipitate with aq. $AgNO_3$ solution most readily?
[NCERT : P L-174 | Sep. 04, 2020 (II)]

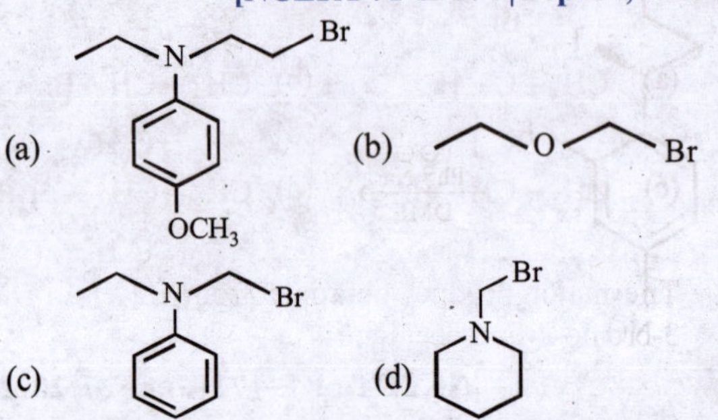

77. Among the following compounds, which one has the shortest C–Cl bond? **NCERT : P L-174 | Sep. 04, 2020 (II)]**

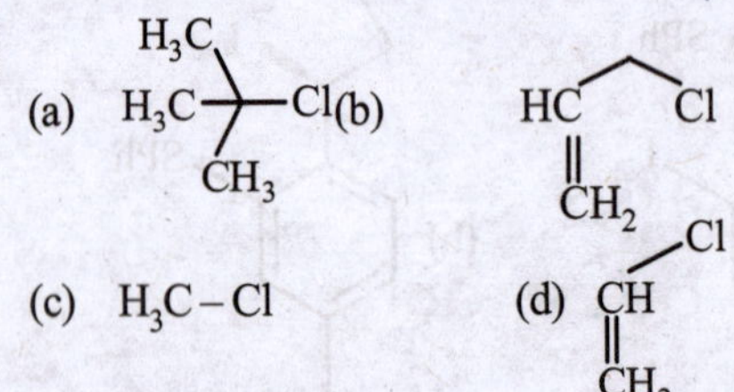

(c) H_3C-Cl (d) $CH_2=CH-Cl$

78. The mechanism of S_N1 reaction is given as :

$R-X \longrightarrow \overset{\oplus}{R}\overset{\ominus}{X} \longrightarrow \overset{\oplus}{R}\|\overset{\ominus}{X} \xrightarrow{Y^{\ominus}} R-Y+X^{\ominus}$

Ion pair; Solvent Separated ion pair

A student writes general characteristics based on the given mechanism as : **[NCERT : P L-173 | Sep. 03, 2020 (I)]**
(1) The reaction is favoured by weak nucleophiles.
(2) $\overset{\oplus}{R}$ would be easily formed if the substituents are bulky.
(3) The reaction is accompanied by racemization.
(4) The reaction is favoured by non-polar solvents.
Which observations are correct?
(a) (1) and (2) (b) (1) and (3)
(c) (1), (2) and (3) (d) (2) and (4)

79. The total number of monohalogenated organic products in the following (including stereoisomers) reaction is ________.

A (Simplest optically active alkene) $\xrightarrow[\text{(ii) } X_2/\Delta]{\text{(i) } H_2/Ni/\Delta}$ **[NV, Sep. 03, 2020 (I)]**

80. The major product in the following reaction is :

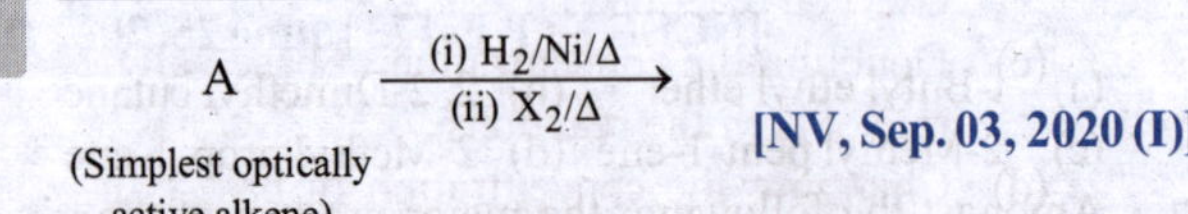

[Sep. 03, 2020 (II)]

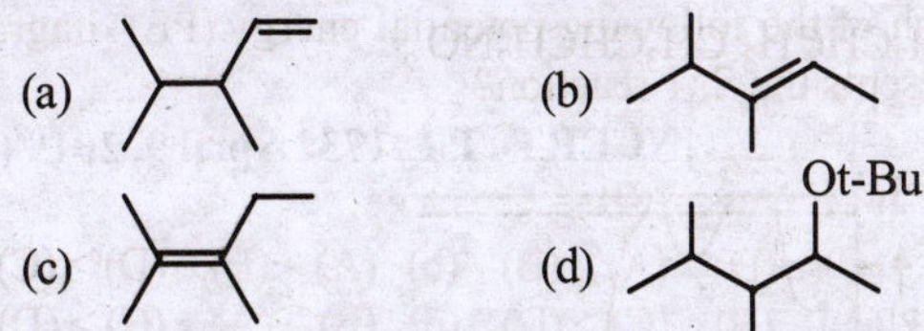

81. Which of the following compounds will show retention in configuration on nucleophilic substitution by OH^- ion ? **[NCERT : P L-173 | Sep. 02, 2020 (I)]**

(a) $CH_3-C(Br)(H)-C_6H_{13}$ (b) $CH_3-CH(C_6H_5)-Br$

(c) $CH_3-CH(CH_3)-Br$ (d) $CH_3-CH(C_2H_5)-CH_2Br$

82. The major product obtained from E2-elimination of 3-bromo-2-fluoropentane is : **[NCERT : P L-179 | Sep. 02, 2020 (II)]**

(a) $CH_3CH_2-CH(Br)-CH=CH_2$

(b) $CH_3CH_2CH=C(CH_3)-F$

(c) $CH_3-CH=CH-CH(F)-CH_3$

(d) $CH_3-CH_2-C(Br)=CH-CH_3$

83. Consider the reaction sequence given below : **[NCERT : P L-171 | Sep. 02, 2020 (II)]**

t-Bu–Br $\xrightarrow[H_2O]{OH^{\ominus}}$ t-Bu–OH + $Br^{\ominus}$ ....(1)

rate = k[t-BuBr]

t-Bu–Br $\xrightarrow[C_2H_5OH]{OH^{\ominus}}$ $H_2C=C(CH_3)_2$ + HOH + $Br^{\ominus}$ ....(2)

rate = k[t-BuBr][$OH^{\ominus}$]

Which of the following statements is true ?

(a) Changing the base from $OH^{\ominus}$ to $^{\ominus}OR$ will have no effect on reaction (2).

(b) Changing the concentration of base will have no effect on reaction (1).

(c) Doubling the concentration of base will double the rate of both the reactions.

(d) Changing the concentration of base will have no effect on reaction (2).

84. The decreasing order of reactivity towards dehydrohalogenation (E_1) reaction of the following compounds is: **[NCERT : P L-179 | Jan. 08, 2020 (I)]**

(A) Cl–CH₂CH₂CH₂CH₃ (1-chlorobutane skeleton) (B) Cl–CH₂CH₂CH=CH₂ skeleton

(C) CH₃CH(Cl)CH₂CH₃ skeleton (D) CH₃CH(Cl)CH=CH₂ skeleton

(a) D > B > C > A (b) B > D > A > C

(c) B > D > C > A (d) B > A > D > C

85. Consider the following reactions:

(1) $(CH_3)_3CCH(OH)CH_3 \xrightarrow{conc.H_2SO_4}$

(2) $(CH_3)_2CHCH(Br)CH_3 \xrightarrow{alc.KOH}$

(3) $(CH_3)_2CHCH(Br)CH_3 \xrightarrow{(CH_3)_3O^-K^{\oplus}}$

(4) $(CH_3)_2-C(OH)-CH_2-CHO \xrightarrow{\Delta}$

Which of these reaction(s) will not produce Saytzeff product? **[NCERT : P L-180 | Jan. 07, 2020 (I)]**

(a) (1), (3) and (4) (b) (4) only

(c) (3) only (d) (2) and (4)

86. In the following reaction sequence, structures of A and B, respectively will be: **[Jan. 07, 2020 (II)]**

(chroman with CH_2Br on the benzene ring) $\xrightarrow[\Delta]{HBr}$ A $\xrightarrow[Ether]{Na}$ (Intramolecular product) B

(a) Br-substituted benzene ring with $CH_2CH_2CH_2OH$ chain and CH_2Br & chroman with CH_2Br

(b) OH-substituted benzene ring with $CH_2CH_2CH_2Br$ chain and CH_2Br & OH-substituted indane with methyl

(c) OH-substituted benzene ring with $CH_2CH_2CH_2Br$ chain and CH_2Br & OH-substituted tetralin

(d) Br-substituted benzene ring with $CH_2CH_2CH_2OH$ chain and CH_2Br & Br-substituted isochroman

87. An 'Assertion' and a 'Reason' are given below. Choose the correct answer from the following options :

Assertion (A) : Vinyl halides do not undergo nucleophilic substitution easily.

Reason (R) : Even though the intermediate carbocation is stabilized by loosely held p-electrons, the cleavage is difficult because of strong bonding. **[April 12, 2019 (II)]**

(a) Both (A) and (R) are wrong statements.

(b) Both (A) and (R) are correct statements and (R) is the correct explanation of (A).

(c) Both (A) and (R) are correct statements but (R) is not the correct explanation of (A).

(d) (A) is a correct statement but (R) is a wrong statement.

88. Increasing rate of S_N1 reaction in the following compounds is : **[NCERT : P L-173 | April 10, 2019 (I)]**

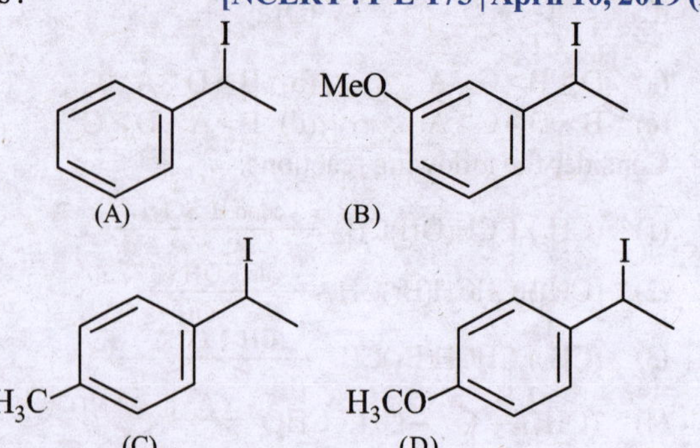

(a) (A) < (B) < (C) < (D) (b) (B) < (A) < (C) < (D)
(c) (B) < (A) < (D) < (B) (d) (A) < (B) < (D) < (C)

89. The major product of the following reaction is **[NCERT : P L-179 | April 10, 2019 (I)]**

$$CH_3-\underset{H}{\overset{CH_3}{C}}-\underset{Br}{CHCH_3} \xrightarrow{CH_3OH}$$

(a) $CH_3-\underset{H}{\overset{CH_3}{C}}-CH=CH_2$ (b) $CH_3-\overset{CH_3}{C}=CHCH_3$

(c) $CH_3-\underset{OCH_3}{\overset{CH_3}{C}}-CH_2CH_3$ (d) $CH_3-\underset{H}{\overset{CH_3}{C}}-\underset{OCH_3}{CHCH_3}$

90. The major product 'Y' in the following reaction is : **[NCERT : P L-171 | April 10, 2019 (II)]**

Cl (structure) $\xrightarrow[\text{Heat}]{\text{EtONa}}$ X $\xrightarrow{\text{HBr}}$ Y

(a) Br (structure) (b) HO (structure)
(c) Br (structure) (d) Br (structure)

91. Increasing order of reactivity of the following compounds for S_N1 substitution is: **[NCERT : P L-173 | April 9, 2019 (II)]**

(A) $(CH_3)_2CH-CH_2-Cl$ (B) H_3C (structure) Cl

(C) H_3CO (structure) Cl (D) (structure) Cl

(a) (B) < (C) < (D) < (A) (b) (B) < (C) < (A) < (D)
(c) (B) < (A) < (D) < (C) (d) (A) < (B) < (D) < (C)

92. Which of the following potential energy (PE) diagrams represents the S_N1 reaction? **[NCERT : P L-173 | April 9, 2019 (II)]**

(a)
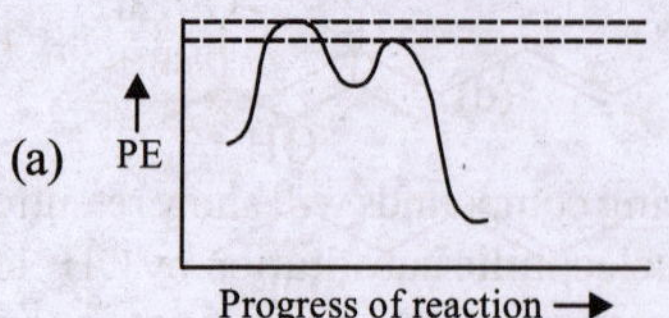

(b)
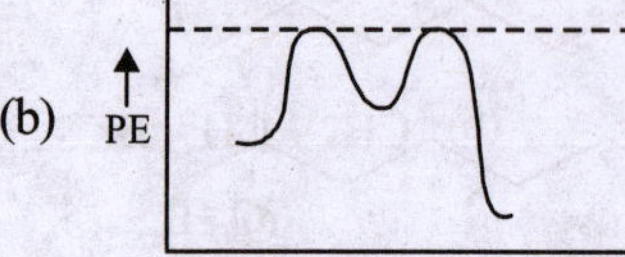

(c) PE
Progress of reaction →

(d)
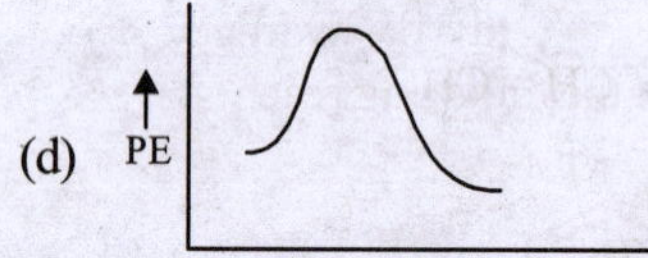

93. The major product of the following reaction is :

$$CH_3CH_2\underset{Br}{CH}-\underset{Br}{CH_2} \xrightarrow[\text{(ii) } NaNH_2 \text{ in liq. } NH_3]{\text{(i) alc. KOH}}$$

[NCERT : P L-179 | Jan. 12, 2019 (II)]

(a) $CH_3CH=C=CH_2$
(b) $CH_3CH_2\underset{NH_2}{CH}-\underset{NH_2}{CH_2}$
(c) $CH_3CH=CHCH_2NH_2$
(d) $CH_3CH_2C\equiv CH$

94. The major product of the following reaction is: **[NCERT : P L-179 | Jan. 10, 2019 (I)]**

Ph (structure) Br, Br $\xrightarrow[\Delta]{\text{KOH alc (excess)}}$

(a) Ph (structure)
(b) Ph (structure)
(c) Ph (structure)
(d) Ph (structure)

95. The major product of the following reaction is: **[NCERT : P L-168 | Jan. 10, 2019 (II)]**

CH_3N ... O ... $\xrightarrow{NaBH_4}$

(a) CH_3N ... OH

(b) CH_3N ... OH

(c) CH_3HN ... O

(d) CH_3HN ... OH

96. Which hydrogen in compound (E) is easily replaceable during bromination reaction in presence of light? **[Jan. 10, 2019 (I)]**

$$\underset{\delta}{CH_3}-\underset{\gamma}{CH_2}-\underset{\beta}{CH}=\underset{\alpha}{CH_2}$$

(a) α - hydrogen (b) γ - hydrogen

(c) δ - hydrogen (d) β - hydrogen

97. The major product of the following reaction is: **[Jan. 9, 2019 (I)]**

$\xrightarrow[\text{(ii) EtOH}]{\text{(i) } Br_2}$

(a) OEt, Br (b) OEt

(c) Br, OEt (d) OEt, OEt

98. Which of the following will most readily give the dehydrohalogenation product? **[NCERT : P L-170 | Online April 15, 2018 (I)]**

(a) Br, Ph (b) Br

(c) Br (d) Br

99. The major product of the following reaction is : **[NCERT : P L-179 | 2018]**

Br $\xrightarrow[\text{MeOH}]{\text{NaOMe}}$

(a) OMe (b)

(c) (d) OMe

100. The major product of the following reaction is: **[NCERT : P L-171 | Online April 16, 2018]**

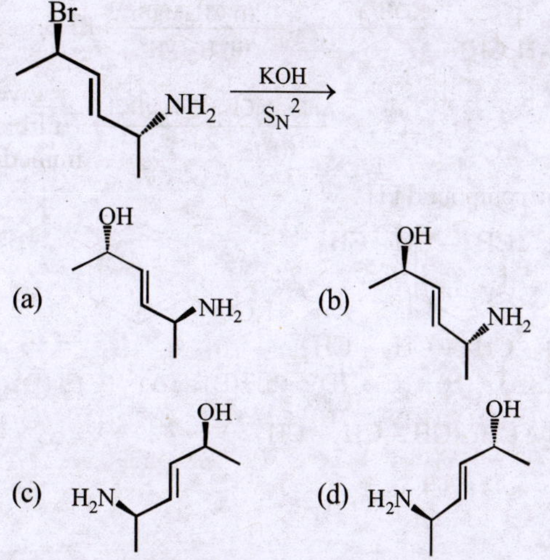

101. The major product of the following reaction is : **[NCERT : P L-180 | Online April 8, 2017]**

$$CH_3\underset{|}{\underset{Br}{C}}HCH_2\underset{|}{\underset{Br}{C}}HCH_2CH_3 \xrightarrow[\text{heat}]{KOH, CH_3OH}$$

(a) $CH_2=CHCH_2CH=CHCH_3$

(b) $CH_2=CHCH=CHCH_2CH_3$

(c) $CH_3CH=C=CHCH_2CH_3$

(d) $CH_3CH=CH-CH=CHCH_3$

102. The increasing order of the reactivity of the following halides for the S_N1 reaction is **[NCERT : P L-173 | 2017]**

$CH_3CH(Cl)CH_2CH_3$ (I) $\quad CH_3CH_2CH_2Cl$ (II)

$p\text{-}H_3CO-C_6H_4-CH_2Cl$ (III)

(a) (III) < (II) < (I) (b) (II) < (I) < (III)

(c) (I) < (III) < (II) (d) (II) < (III) < (I)

103. The major product of the following reaction is : **[NCERT : P L-179 | Online April 8, 2017]**

$$C_6H_5CH_2-\underset{Br}{\overset{CH_3}{C}}-CH_2-CH_3 \xrightarrow[C_2H_5OH]{C_2H_5ONa}$$

(a) $C_6H_5CH_2-\underset{OC_2H_5}{\overset{CH_3}{C}}-CH_2-CH_3$

(b) $C_6H_5CH=\underset{CH_3}{C}-CH_2-CH_3$

(c) $C_6H_5CH_2-\underset{CH_3}{C}=CHCH_3$

(d) $C_6H_5CH_2-\underset{CH_2CH_3}{C}=CH_2$

104. In the following reaction sequence : **[NCERT : P L-173, 180 | Online April 9, 2017]**

$$\underset{(C_3H_6Cl_2)}{I} \xrightarrow{KOH(aq)} II \xrightarrow[(ii)\ H_2O/H^+]{(i)\ CH_3MgBr} III$$

$$\xrightarrow{Anhy.ZnCl_2 + Conc.HCl} \text{gives turbidity immediately}$$

The compound I is :

(a) $CH_2-CH-CH_3$ with Cl on C_1 and C_2 ($CH_2Cl-CHCl-CH_3$)

(b) $CH_2-CH_2-CH_3$ with Cl on C_1 and C_3 ($CH_2Cl-CH_2-CH_2Cl$)

(c) $CH-CH-CH_2-CH_3$ with Cl on C_2

(d) $CH_3-CCl_2-CH_3$

105. Which one of the following reagents is not suitable for the elimination reaction ? **[NCERT : P L-170 | Online April 10, 2016]**

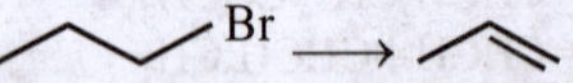

(a) NaI (b) NaOEt/ EtOH
(c) $NaOH/H_2O$ (d) $NaOH/H_2O-EtOH$

106. The synthesis of alkyl fluorides is best accomplished by : **[2015]**
(a) Finkelstein reaction
(b) Swarts reaction
(c) Free radical fluorination
(d) Sandmeyer's reaction

107. A compound A with molecular formula $C_{10}H_{13}Cl$ gives a white precipitate on adding silver nitrate solution. A on reacting with alcoholic KOH gives compound B as the main product. B on ozonolysis gives C and D. C gives Cannizzaro reaction but not aldol condensation. D gives aldol condensation but not Cannizzaro reaction. A is : **[Online April 10, 2015]**

(a) $C_6H_5-CH_2-CH_2-CH_2-CH_2-Cl$

(b) $C_6H_5-CH_2-CH_2-CH(Cl)-CH_3$

(c) $C_6H_5-CH_2-C(Cl)(CH_3)_2$

(d) benzene ring bearing $CH_2-CH_2-CH_3$ and CH_2-Cl

108. For the compounds
CH_3Cl, CH_3Br, CH_3I and CH_3F,
the correct order of increasing C-halogen bond length is: **[NCERT : P L-174 | Online April 9, 2014]**

(a) $CH_3F < CH_3Cl < CH_3Br < CH_3I$
(b) $CH_3F < CH_3Br < CH_3Cl < CH_3I$
(c) $CH_3F < CH_3I < CH_3Br < CH_3Cl$
(d) $CH_3Cl < CH_3Br < CH_3F < CH_3I$

109. In S_N2 reactions, the correct order of reactivity for the following compounds: **[NCERT : P L-171 | 2014, 2007 (s)]**
CH_3Cl, CH_3CH_2Cl, $(CH_3)_2CHCl$ and $(CH_3)_3CCl$ is:

(a) $CH_3Cl > (CH_3)_2CHCl > CH_3CH_2Cl > (CH_3)_3CCl$
(b) $CH_3Cl > CH_3CH_2Cl > (CH_3)_2CHCl > (CH_3)_3CCl$
(c) $CH_3CH_2Cl > CH_3Cl > (CH_3)_2CHCl > (CH_3)_3CCl$
(d) $(CH_3)_2CHCl > CH_3CH_2Cl > CH_3Cl > (CH_3)_3CCl$

110. The order of reactivity of the given haloalkanes towards nucleophile is : **[NCERT : P L-170 | Online April 23, 2013]**
(a) RI > RBr > RCl (b) RCl > RBr > RI
(c) RBr > RCl > RI (d) RBr > RI > RCl

111. Which of the following statements is wrong? **[NCERT : P L-170 | Online May 12, 2012]**
(a) Ethyl chloride on reduction with Zn-Cu couple and alcohol gives ethane.
(b) The reaction of methyl magnesium bromide with acetone gives butanol-2.
(c) Alkyl halides follow the following reactivity sequence on reaction with alkenes.
R–I > R–Br > R–Cl > R–I
(d) $C_2H_4Cl_2$ may exist in two isomeric forms

112. How many chiral compounds are possible on monochlorination of 2- methylbutane ? **[NCERT : P L-176 | 2012]**
(a) 8 (b) 2 (c) 4 (d) 6

113. $C_2H_5Br \xrightarrow{AgCN} X \xrightarrow[Zn-Hg/HCl]{Reduction} Y$, Here Y is **[Online May 7, 2012]**
(a) Ethyl methyl amine (b) n-propylamine
(c) Isopropylamine (d) Ethylamine

114. Out of the following, the alkene that exhibits optical isomerism is **[NCERT : P L-175 | 2010]**
(a) 3-methyl-2-pentene (b) 4-methyl-1-pentene
(c) 3-methyl-1-pentene (d) 2-methyl-2-pentene

115. Consider the following bromides :

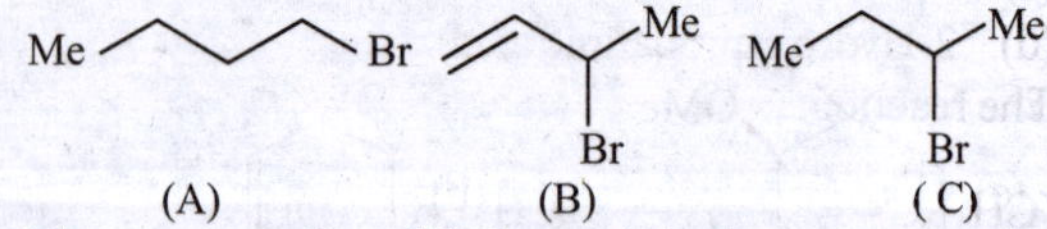

The correct order of S_N1 reactivity is **[NCERT : P L-170 | 2010]**
(a) B > C > A (b) B > A > C
(c) C > B > A (d) A > B > C

116. The number of stereoisomers possible for a compound of the molecular formula **[NCERT : P L-175 | 2009]**
$CH_3-CH=CH-CH(OH)-Me$ is:
(a) 2 (b) 4 (c) 6 (d) 3

117. The organic chloro compound, which shows complete stereochemical inversion during an S_N2 reaction, is **[NCERT : P L-178 | 2008]**
(a) $(C_2H_5)_2CHCl$ (b) $(CH_3)_3CCl$
(c) $(CH_3)_2CHCl$ (d) CH_3Cl

118. Which of the following molecules is expected to rotate the plane of plane-polarised light? **[NCERT : P L-175 | 2007]**
(a) COOH / H_2N—C—H / H
(b) CHO / HO—C—H / CH_2OH
(c) SH
(d) H_2N, NH_2, H, H, Ph, Ph

119. Tertiary alkyl halides are practically inert to substitution by S_N2 mechanism because of **[NCERT : P L-171 | 2005]**
(a) steric hindrance (b) inductive effect
(c) instability (d) insolubility

120. Alkyl halides react with dialkyl copper reagents to give **[2005]**
(a) alkenyl halides (b) alkanes
(c) alkyl copper halides (d) alkenes

121. Elimination of bromine from 2-bromobutane results in the formation of– **[NCERT : P L-180 | 2005]**
(a) Predominantly 2-butyne
(b) Predominantly 1-butene
(c) Predominantly 2-butene
(d) Equimolar mixture of 1 and 2-butene

122. Which type of isomerism is shown by 2, 3-dichlorobutane? **[2005]**
(a) Structural (b) Geometric
(c) Optical (d) Diastereo

123. Which of the following will have a mesoisomer also? **[NCERT : P L-175 | 2004]**
(a) 2, 3- Dichloropentane
(b) 2, 3-Dichlorobutane
(c) 2-Chlorobutane
(d) 2-Hydroxypropanoic acid

124. The reaction: **[NCERT : P L-170 | 2002]**
$$(CH_3)_3C-Br \xrightarrow{H_2O} (CH_3)_3-C-OH$$
(a) elimination reaction
(b) substitution reaction
(c) free radical reaction
(d) displacement reaction.

Topic 2 - Preparation and Properties of Haloarenes

125. Match **List-I** with **List-II** **[NCERT : P L-184 | April 7, 2025 (II)]**

List-I (Conversion)	List-II (Reagents, Conditions used)
(A) Chlorobenzene (Cl) → Phenol (OH)	(I) Warm, H_2O
(B) Cl, NO_2 → OH, NO_2	(II) (a) NaOH, 368 K; (b) H_3O^+
(C) Cl, NO_2, NO_2 → OH, NO_2, NO_2	(III) (a) NaOH, 443 K; (b) H_3O^+
(D) O_2N, Cl, NO_2, NO_2 → O_2N, OH, NO_2, NO_2	(IV) (a) NaOH, 623 K, 300 atm; (b) H_3O^+

Choose the **correct** answer from the options given below:
(a) (A)-(II), (B)-(III), (C)-(I), (D)-(IV)
(b) (A)-(III), (B)-(IV), (C)-(II), (D)-(I)
(c) (A)-(IV), (B)-(III), (C)-(II), (D)-(I)
(d) (A)-(IV), (B)-(III), (C)-(I), (D)-(II)

126. In the following series of reactions identify the major products A & B respectively. **[NCERT : P L-184 | April 3, 2025 (II)]**

Bromobenzene (Br) $\xrightarrow[H_2SO_4]{SO_3}$ A (Major product) $\xrightarrow{Br_2 Fe}$ B (Major product)

(a) A: Br, SO_3H B: Br, SO_3H, Br

(b) A: 4-bromobenzenesulfonic acid (SO_3H para to Br); B: 3,4-dibromobenzenesulfonic acid

(c) A: 2-bromobenzenesulfonic acid (SO_3H ortho to Br); B: 2,5-dibromobenzenesulfonic acid (Br, SO_3H, Br)

(d) A: 4-bromobenzenesulfonic acid (SO_3H para to Br); B: 2,4-dibromobenzenesulfonic acid (Br para to SO_3H, Br ortho to SO_3H)

127. Match **List-I** with **List-II**.

[NCERT : P L-165, 186 | April 2, 2025 (II)]

List-I (Reaction)	**List-II (Name of reaction)**
(A) $2\,C_6H_5{-}X + 2Na \xrightarrow[\text{Ether}]{\text{Dry}} C_6H_5{-}C_6H_5 + 2NaX$	(I) Lucas reaction
(B) $ArN_2^+X^- \xrightarrow[\text{HCl}]{\text{Cu}} ArCl + N_2\uparrow + CuX$	(II) Finkelstein reaction
(C) $C_2H_5Br + NaI \xrightarrow[\text{Acetone}]{\text{Dry}} C_2H_5I + NaBr$	(III) Fittig reaction
(D) $CH_3C(OH)(CH_3)CH_3 \xrightarrow[\text{ZnCl}_2]{\text{HCl}} CH_3C(Cl)(CH_3)CH_3$	(IV) Gatterman reaction

Choose the **correct** answer from the options given below:

(a) (A)-(III), (B)-(II), (C)-(IV), (D)-(I)
(b) (A)-(III), (B)-(IV), (C)-(II), (D)-(I)
(c) (A)-(IV), (B)-(III), (C)-(I), (D)-(II)
(d) (A)-(IV), (B)-(I), (C)-(II), (D)-(III)

128. In the following substitution reaction :

[NCERT : P L-184 | Jan. 29, 2025 (I)]

1,2-Dibromo-4-nitrobenzene $\xrightarrow[C_2H_5OH]{C_2H_5ONa}$ P Product

Product 'P' formed is

(a) 1-OC_2H_5, 2-Br, 4-NO_2 benzene
(b) 1,2-di(OC_2H_5), 4-NO_2 benzene
(c) 1-Br, 2-OC_2H_5, 4-NO_2 benzene
(d) 1-Br, 2-Br... 1-Br, 2-Br-4-OC_2H_5 benzene (Br, Br, OC_2H_5)

129. The number of halobenzenes from the following that can be prepared by Sandmeyer's reaction is ....

[NCERT : P L-166 | NV, April 5, 2024 (I)]

I: C_6H_5F II: C_6H_5Cl III: C_6H_5Br IV: C_6H_5I V: C_6H_5At

130. $p\text{-}BrC_6H_4CH_2CH_2Br$ (A) $\xrightarrow{\text{NaOH(alc)}}$ Ⓑ $\xrightarrow[\text{Ether}]{\text{HBr}}$ Ⓒ

Identify Ⓑ and Ⓒ and how are Ⓐ and Ⓒ related?

[NCERT : P L-184 | April 4, 2024 (I)]

	(B)	(C)	
(a)	$p\text{-}HOC_6H_4CH_2CH_2Br$	$p\text{-}HOC_6H_4CH_2CH_2OH$	functional group isomers
(b)	$p\text{-}HOC_6H_4CH_2CH_2Br$	$p\text{-}HOC_6H_4CH(Br)CH_3$	Derivative
(c)	$p\text{-}BrC_6H_4CH{=}CH_2$	$p\text{-}BrC_6H_4CH(Br)CH_3$	position isomers
(d)	$p\text{-}BrC_6H_4CH{=}CH_2$	$p\text{-}BrC_6H_4CH_2CH_2Br$	chain isomers

131. Identify A and B in the following reaction sequence.

[NCERT : P L-184 | Jan. 31, 2024 (II)]

$C_6H_5Br \xrightarrow{\text{Conc. HNO}_3} A \xrightarrow[\text{(ii) HCl}]{\text{(i) NaOH}} B$

(a) A = 1-bromo-2,4,6-trinitrobenzene (NO_2, Br, NO_2, NO_2), B = 2,4,6-trinitrophenol (OH, NO_2, NO_2, NO_2)

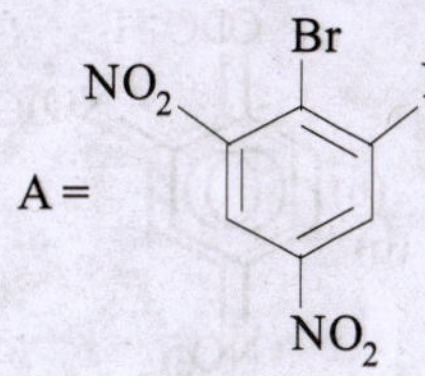

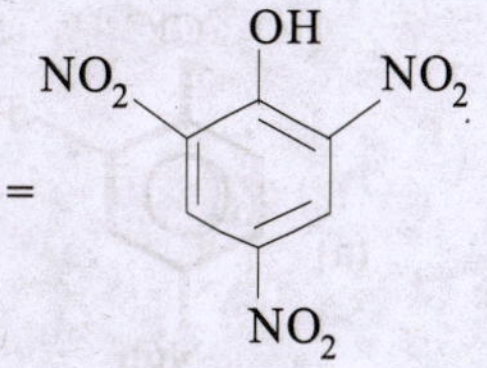

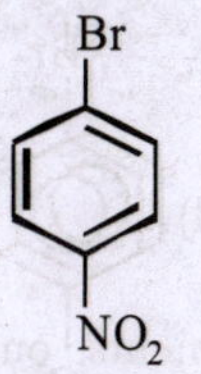

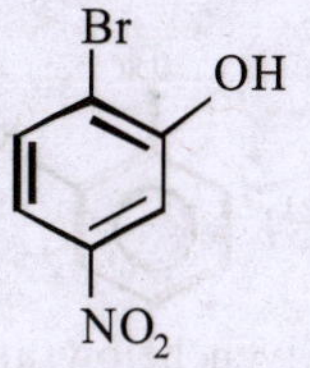

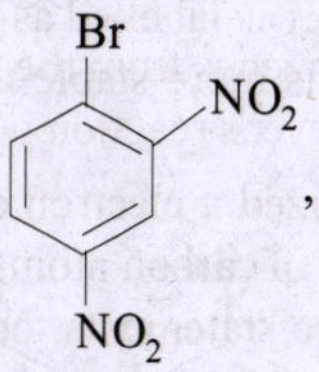

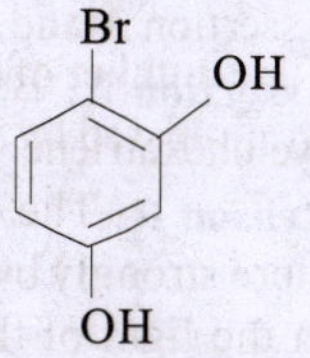

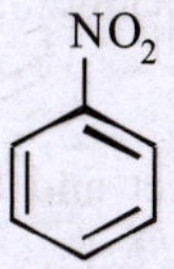

(b) A = 4-bromonitrobenzene (Br, NO_2), B = (Br, OH, NO_2)

(c) A = (Br, NO_2, NO_2), B = (Br, OH, OH)

(d) A = (NO_2), B = (NO_2, OH)

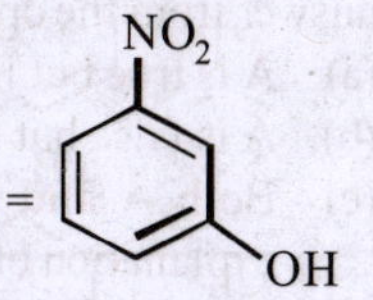

132. The major product formed in the Friedel-Craft acylation of chlorobenzene is: **[NCERT : P L-184 | April 15, 2023 (I)]**

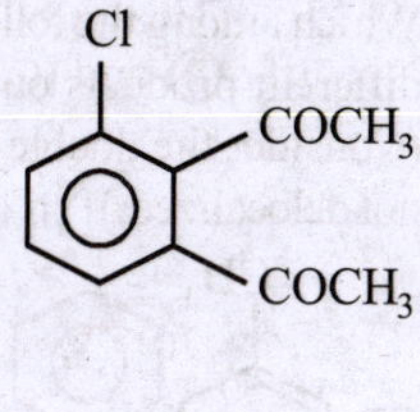

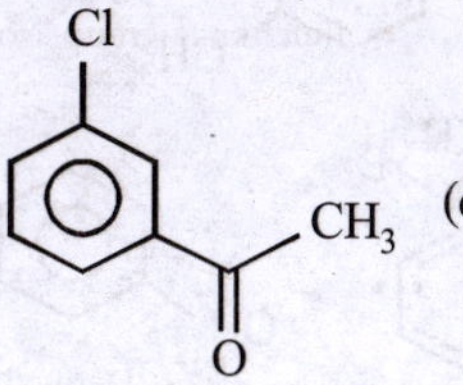

(a) 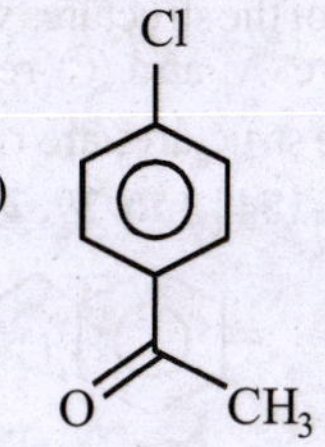 Cl, $C(=O)CH_3$ (para) (b) Cl, $COCH_3$, $COCH_3$

(c) Cl, $C(=O)CH_3$ (meta) (d) 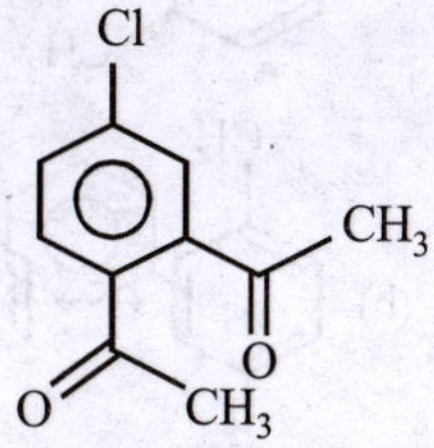 Cl, $C(=O)CH_3$, $C(=O)CH_3$

133. Identify the correct order of reactivity for the following pairs towards the respectively mechanism **[NCERT : P L-173 | April 10, 2023 (I)]**

(A) S_N2: 1-bromo-2-methylpropane (Br) > 2-bromo-2-methylpropane (Br)

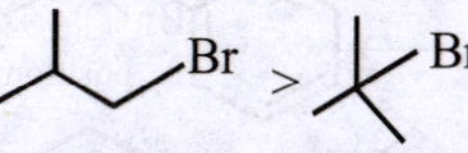

(B) S_N1: benzyl bromide (Br) > 2-phenylethyl bromide (Br)

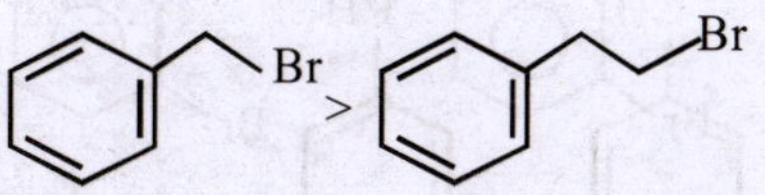

(C) Electrophilic substitution: chlorobenzene (Cl) > 1-chloro-4-nitrobenzene (Cl, NO_2)

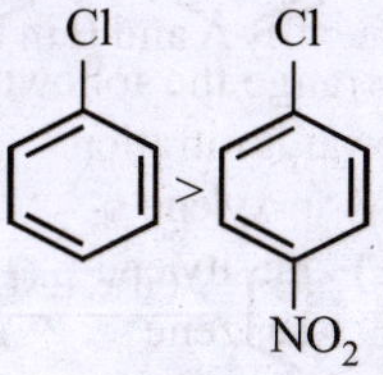

(D) Nucleophilic substitution: bromobenzene (Br) < 1-bromo-4-nitrobenzene (Br, NO_2)

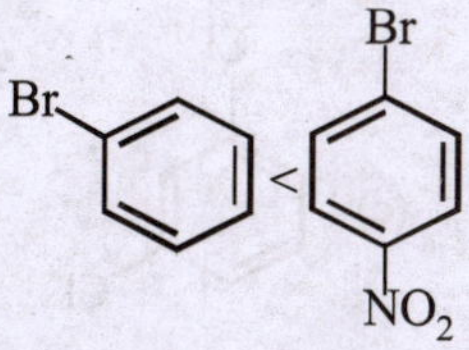

Choose the correct answer from the options given below:

(a) (A), (B) and (D) only (b) (A), (B), (C) and (D)

(c) (A), (C) and (D) only (d) (B), (C) and (D) only

134. The correct order of reactivity of following haloarenes towards nucleophilic substitution with aqueous NaOH is: **[NCERT : P L-183 | April 08, 2023 (II)]**

A. 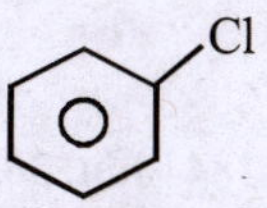 chlorobenzene (Cl) B. 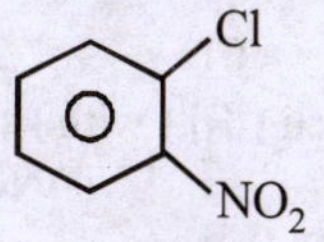 (Cl, NO_2)

C. 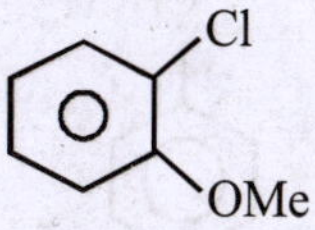 (Cl, OMe) D. 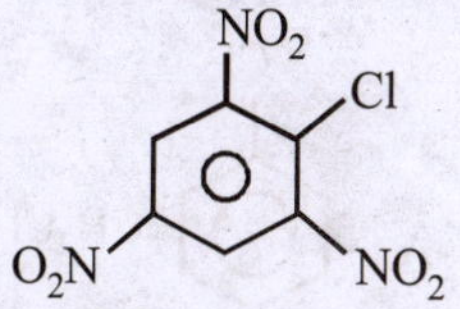 (NO_2, Cl, O_2N, NO_2)

Choose the correct answer from the options given belowe:

(a) A > B > D > C (b) C > A > D > B

(c) D > C > B > A (d) D > B > A > C

135. Identify the incorrect option from the following: **[NCERT : P L-184 | Feb. 01, 2023 (I)]**

(a) 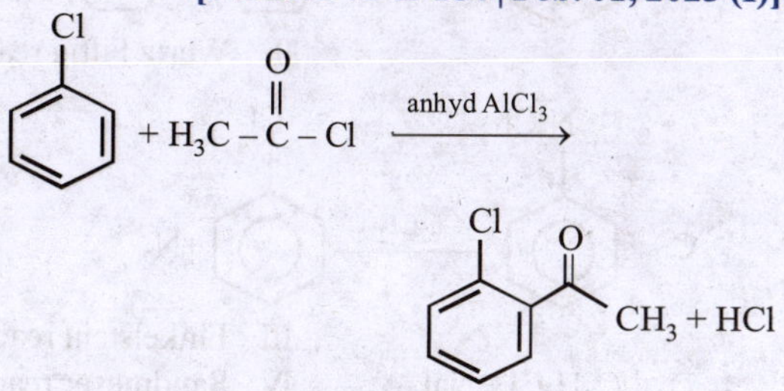 chlorobenzene (Cl) $+ H_3C-\overset{O}{\overset{\|}{C}}-Cl \xrightarrow{\text{anhyd } AlCl_3}$ (Cl, $C(=O)CH_3$) $CH_3 + HCl$

(b) 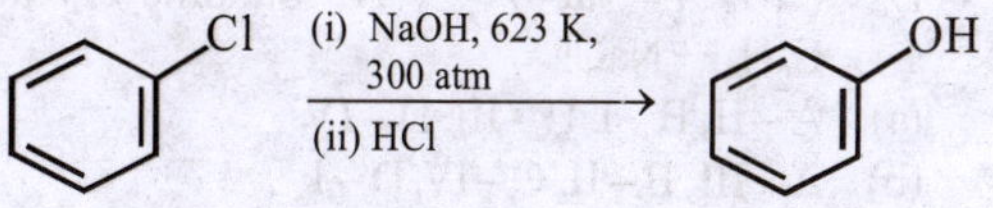 chlorobenzene (Cl) $\xrightarrow[\text{(ii) HCl}]{\text{(i) NaOH, 623 K, 300 atm}}$ phenol (OH)

(c) $\diagup\diagdown$Br + KOH (aq) ⟶ $\diagup\diagdown$OH + KBr

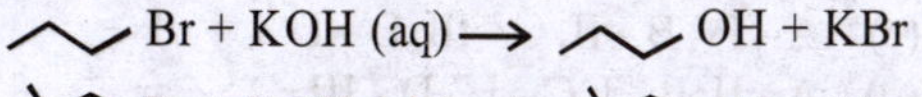

(d) 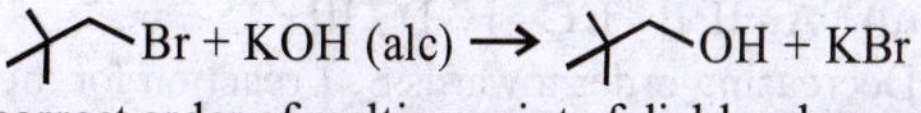 neopentyl Br + KOH (alc) ⟶ neopentyl OH + KBr

136. The correct order of melting point of dichlorobenzenes is **[NCERT : P L-167 | Jan. 31, 2023 (I)]**

(a) 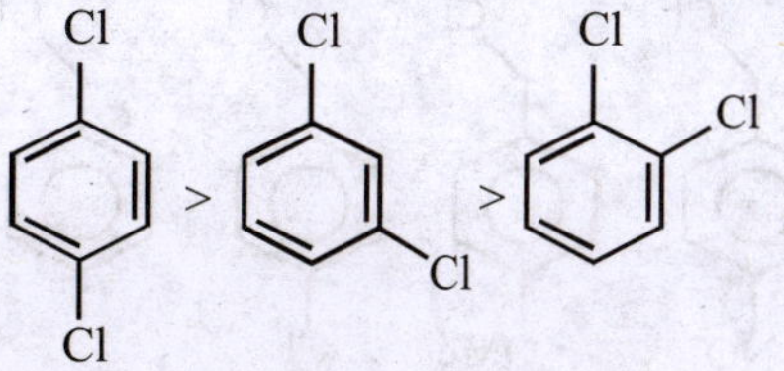 p-dichlorobenzene (Cl, Cl) > m-dichlorobenzene (Cl, Cl) > o-dichlorobenzene (Cl, Cl)

(b) 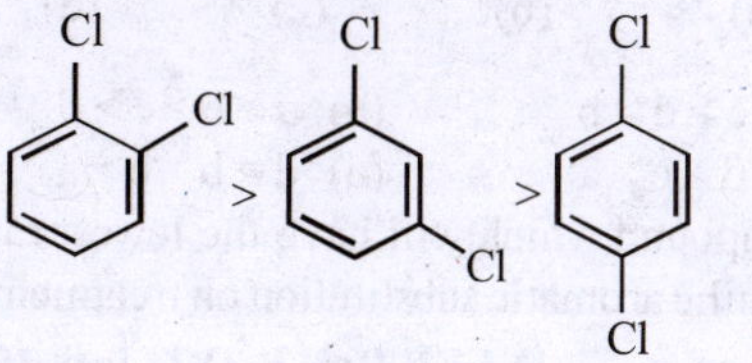 o-dichlorobenzene (Cl, Cl) > m-dichlorobenzene (Cl, Cl) > p-dichlorobenzene (Cl, Cl)

(c) m-dichlorobenzene > p-dichlorobenzene > o-dichlorobenzene

(d) p-dichlorobenzene > o-dichlorobenzene > m-dichlorobenzene

137. Match **List I** with **List II** **[NCERT : P L-184 | Jan. 30, 2023 (I)]**

List I	**List II**
A. Chlorobenzene $+ CH_3Cl \xrightarrow{Na}$ Toluene	I Fittig reaction
B. Chlorobenzene $+ 2Na \longrightarrow$ Biphenyl	II Wurtz Fittig reaction
C. $\overset{+}{N_2}\overset{-}{Cl}$ (benzenediazonium chloride) $\xrightarrow{Cu_2Cl_2}$ Chlorobenzene $+ N_2$	III Finkelstein reaction
D. $C_2H_5Cl + NaI \rightarrow C_2H_5I + NaCl$	IV Sandmeyer reaction

(a) A – II, B – I, C – III, D – IV
(b) A – III, B – II, C – IV, D – I
(c) A – IV, B – II, C – III, D – I
(d) A – II, B – I, C – IV, D – III

138. Decreasing order towards S_N1 reaction for the following compounds is: **[Jan. 30, 2023 (II)]**

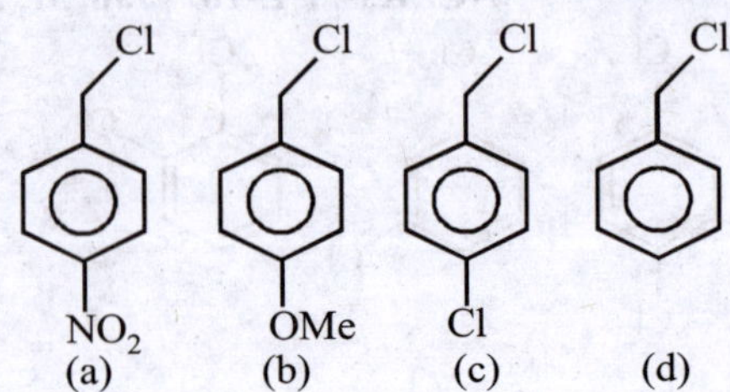

(a) a > c > d > b
(b) a > b > c > d
(c) b > d > c > a
(d) d > b > c > a

139. The compound which will have the lowest rate towards nucleophilic aromatic substitution on treatment with OH^- is **[NCERT : P L-183 | Jan. 25, 2023 (I)]**

(a) 1-chloro-4-nitrobenzene (Cl, NO_2)
(b) 1-chloro-2,4-dinitrobenzene (Cl, NO_2, NO_2)
(c) 1-chloro-2-nitrobenzene (Cl, NO_2)
(d) 1-chloro-3-nitrobenzene (Cl, NO_2)

140. Given below are two statements, one is labelled as Assertion A and the other is labelled as Reason R.

Assertion A: Benzene is more stable than hypothetical cyclohexatriene.

Reason R: The delocalized π electron cloud is attracted more strongly by nuclei of carbon atoms.

In the light of the above statements, choose the correct answer from the options given below: **[Jan. 24, 2023 (II)]**

(a) A is true but R is false.
(b) A is false but R is true.
(c) Both A and R are correct and R is the correct explanation of A.
(d) Both A and R are correct but R is NOT the correct explanation of A.

141. Which among the following pairs of the structures will give different products on ozonolysis?
(Consider the double bonds in the structures are rigid and not delocalized.) **[NCERT : P L-184 | July 29, 2022 (I)]**

(a) CH_3, CH_3 (m-xylene Kekulé structures) $\longleftrightarrow$ CH_3, CH_3
(b) CH_3, CH_3 (p-xylene Kekulé structures) $\longleftrightarrow$ CH_3, CH_3
(c) CH_3, CH_3 (o-xylene Kekulé structures) $\longleftrightarrow$ CH_3, CH_3
(d) CH_3 (toluene Kekulé structures) $\longleftrightarrow$ CH_3

142. Arrange the following in increasing order of reactivity towards nitration **[July 28, 2022 (II)]**

(A) p-xylene
(B) bromobenzene
(C) mesitylene
(D) nitrobenzene
(E) benzene

Choose the correct answer from the options given below:
(a) C<D<E<A<B (b) D<B<E<A<C
(c) D<C<E<A<B (d) C<D<E<B<A

143. Product 'A' of following sequence of reactions is

Ethylbenzene $\xrightarrow[\text{(b) } Cl_2, \Delta \text{ (c) alc. KOH}]{\text{(a) } Br_2.Fe}$ 'A' (Major product)

[NCERT : P L-184 | June 27, 2022 (II)]

(a) 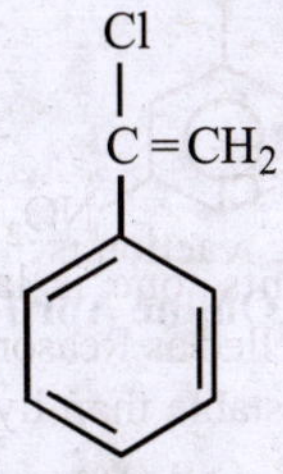 (b) 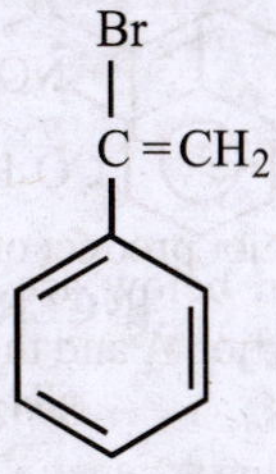 (c) 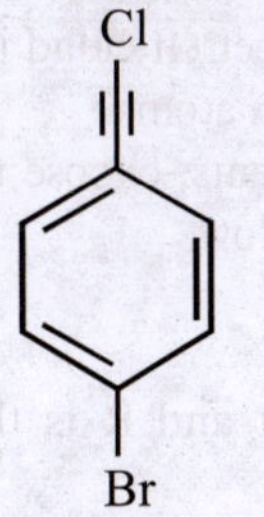 (d)

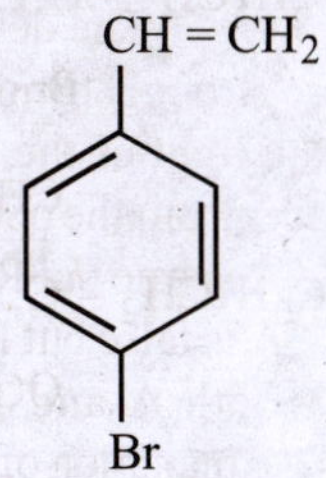

144. Halogenation of which one of the following will yield m-substituted product with respect to methyl group as a major product? **[NCERT : P L-184 | June 26, 2022 (II)]**

(a) 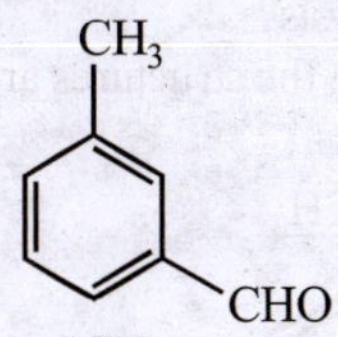 (b) 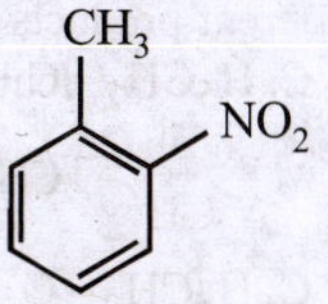 (c) 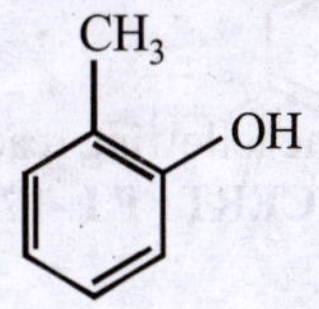 (d)

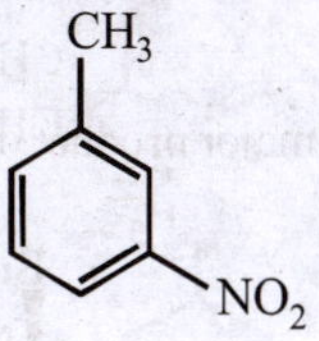

145. In the given reaction

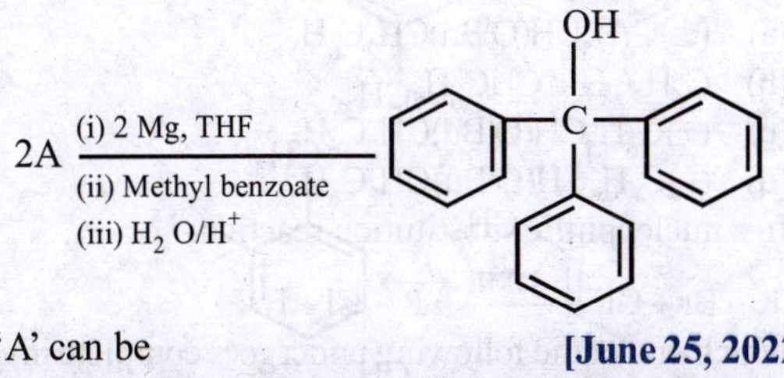

2A $\xrightarrow[\text{(ii) Methyl benzoate (iii) } H_2O/H^+]{\text{(i) 2 Mg, THF}}$ (triphenylmethanol)

'A' can be **[June 25, 2022 (II)]**
(a) benzyl bromide (b) bromobenzene
(c) cyclohexyl bromide (d) methyl bromide

146. The **correct** order of reactivity of the given chlorides with acetate in acetic acid is : **[Aug. 31, 2021 (I)]**

(a) 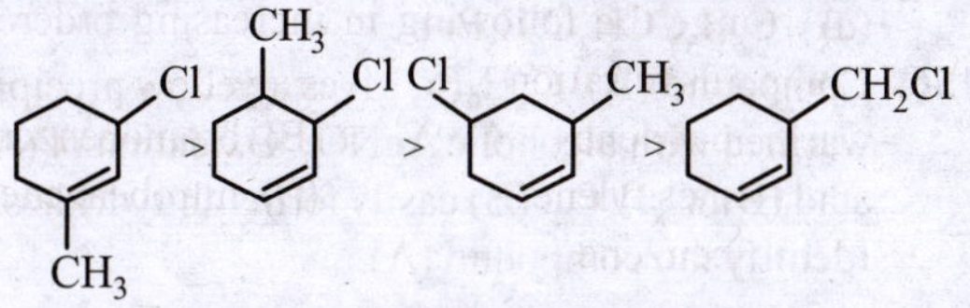 > > >
(b) 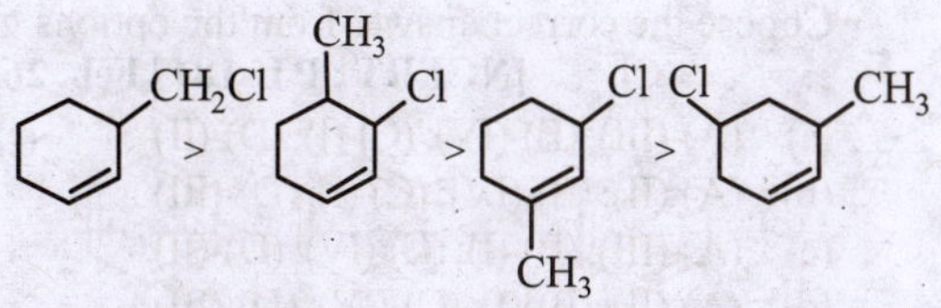 > > >
(c) 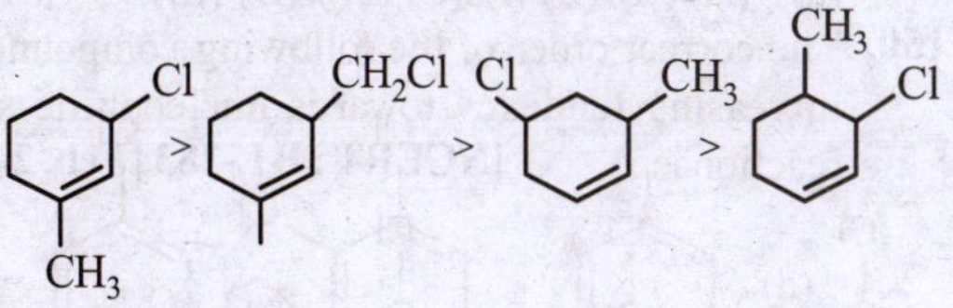 > > >
(d) 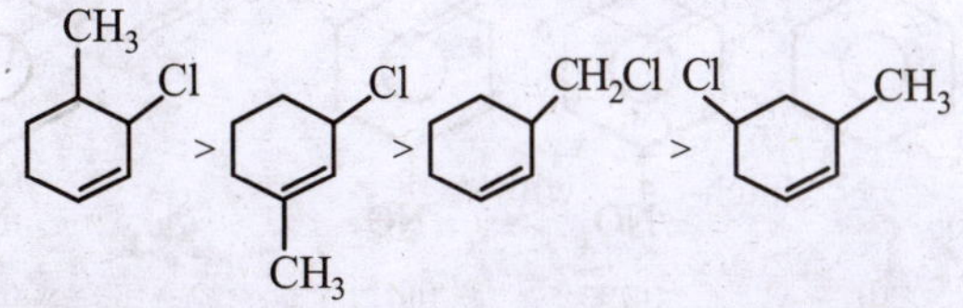 > > >

147. Among the following compounds I-IV, which one forms a yellow precipitate on reacting sequentially with (i) NaOH (ii) dil. HNO_3 (iii) $AgNO_3$? **[Aug. 26, 2021 (I)]**

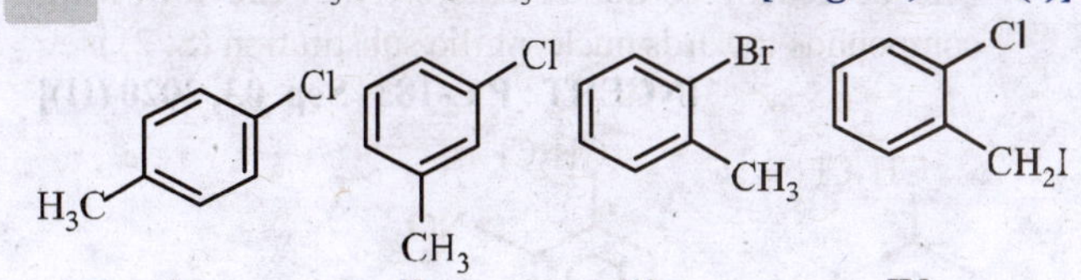

I II III IV
(a) II (b) IV (c) I (d) III

148. The correct decreasing order of densities of the following compounds is: **[NCERT : P L-167 | July 25, 2021 (II)]**

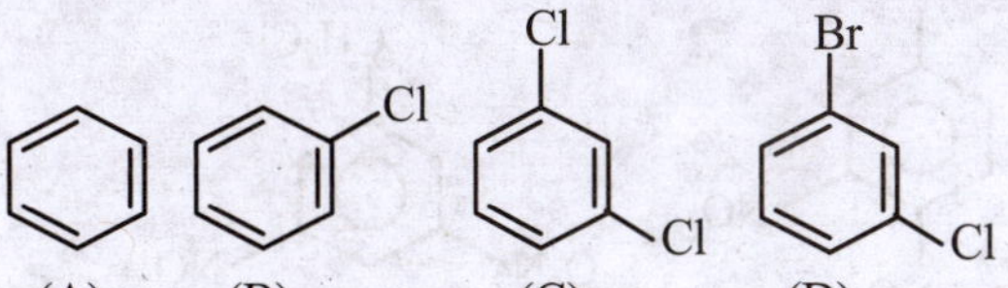

(A) (B) (C) (D)

(a) (D)>(C)>(B)>(A)
(b) (C)>(D)>(A)>(B)
(c) (C)>(B)>(A)>(D)
(d) (A)>(B)>(C)>(D)

149. Match List-I with List-II

List-I	List-II
(A) 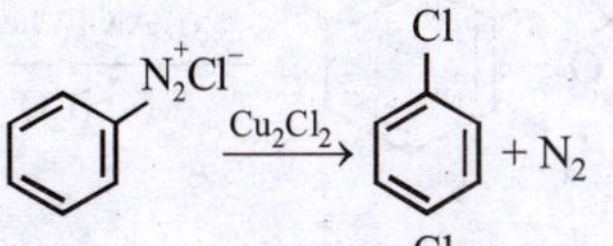 $C_6H_5N_2^+Cl^- \xrightarrow{Cu_2Cl_2} C_6H_5Cl + N_2$	(I) Wurtz reaction
(B) 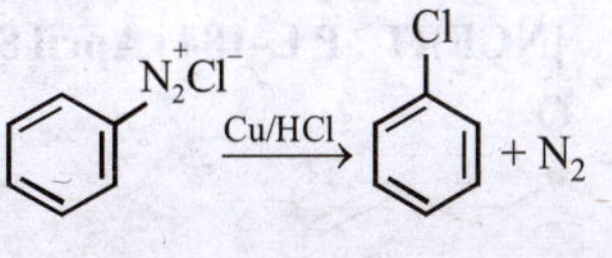 $C_6H_5N_2^+Cl^- \xrightarrow{Cu/HCl} C_6H_5Cl + N_2$	(II) Sandmeyer reaction
(C) $2CH_3CH_2Cl + 2Na \xrightarrow{Ether} C_2H_5-C_2H_5 + 2NaCl$	(III) Fittig reaction
(D) $2C_6H_5Cl + 2Na \xrightarrow{Ether} C_6H_5-C_6H_5 + 2NaCl$	(IV) Gattermann reaction

Choose the correct answer from the options given below **[NCERT : P L-181 | Feb. 26, 2021 (II)]**

(a) (A)-(III); (B)-(IV); (C)-(I); (D)-(II)
(b) (A)-(II); (B)-(IV); (C)-(I); (D)-(III)
(c) (A)-(III); (B)-(I); (C)-(IV); (D)-(II)
(d) (A)-(II); (B)-(I); (C)-(IV); (D)-(III)

150. The correct order of the following compounds showing increasing tendency towards nucleophilic substitution reaction is : **[NCERT : P L-183 | Feb. 24, 2021 (II)]**

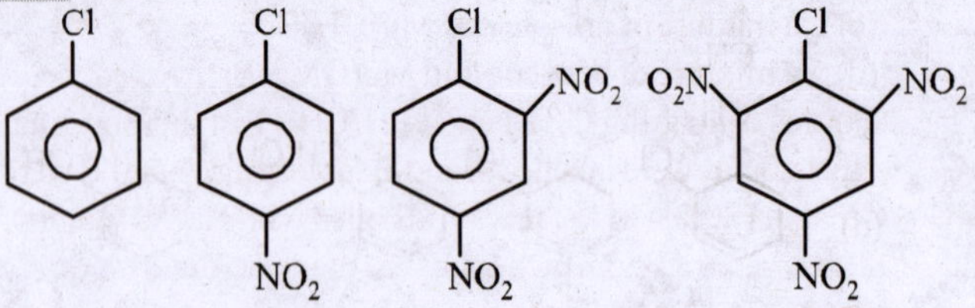

Cl, NO_2, NO_2, O_2N, NO_2, NO_2, NO_2

(i) (ii) (iii) (iv)

(a) (iv) < (iii) < (ii) < (i) (b) (iv) < (i) < (ii) < (iii)
(c) (iv) < (i) < (iii) < (ii) (d) (i) < (ii) < (iii) < (iv)

151. The decreasing order of reactivity of the following compounds towards nucleophilic substitution (S_N2) is : **[NCERT : P L-183 | Sep. 03, 2020 (II)]**

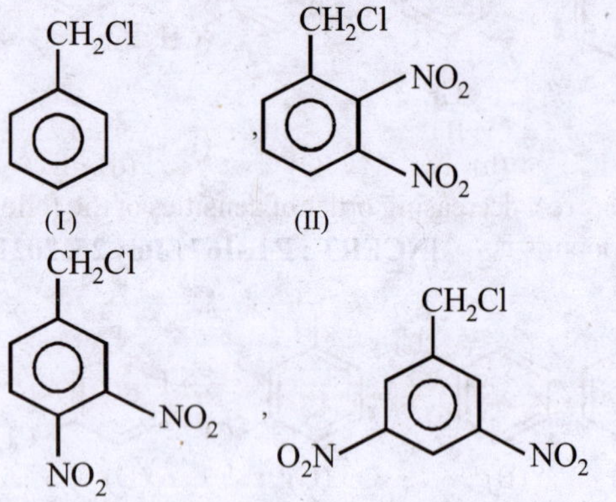

CH_2Cl, CH_2Cl, NO_2, NO_2

(I) (II)

CH_2Cl, NO_2, NO_2, CH_2Cl, O_2N, NO_2

(III) (IV)

(a) (II) > (III) > (I) > (IV) (b) (II) > (III) > (IV) > (I)
(c) (III) > (II) > (IV) > (I) (d) (IV) > (II) > (III) > (I)

152. The major product of the following reaction is :

O, O, O, Cl, + $\xrightarrow[\text{(ii) } H_2O]{\text{(i) } AlCl_3\text{, heat}}$

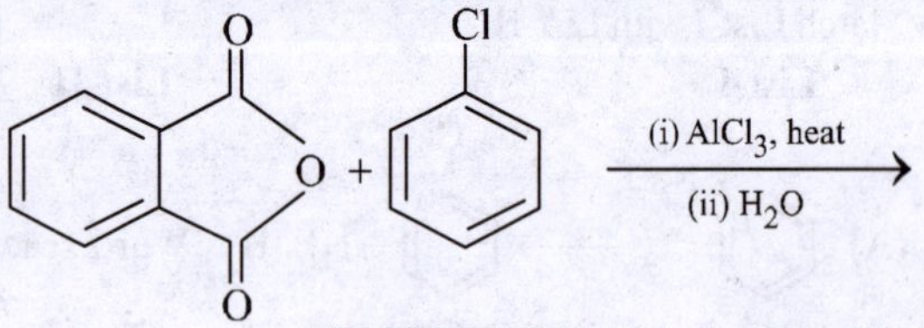

[NCERT : P L-184 | April 8, 2019 (I)]

(a) 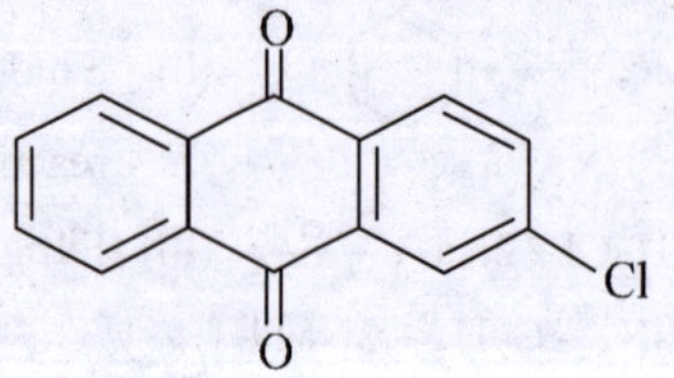 O, O, Cl

(b) 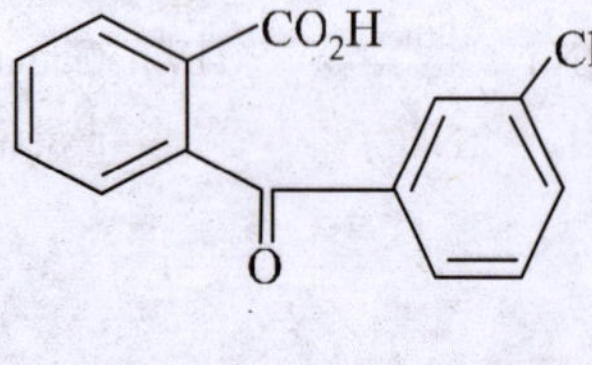 CO_2H, O, Cl

(c) 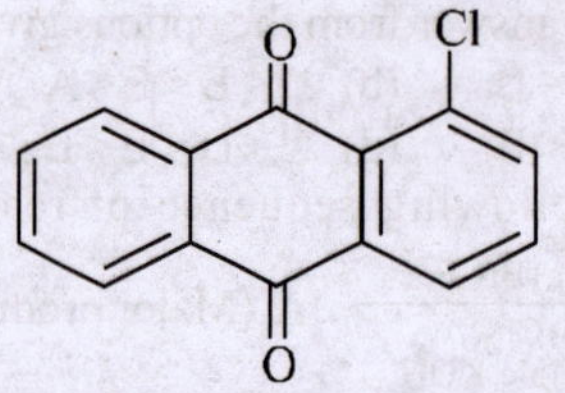 O, O, Cl

(d) 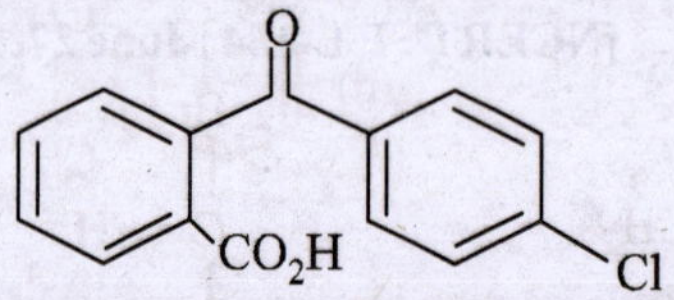 O, CO_2H, Cl

153. The major product of the following reaction is : **[NCERT : P L-180 | Online April 8, 2017]**

$$C_6H_5CH_2-\underset{Br}{\overset{CH_3}{C}}-CH_2-CH_3 \xrightarrow[C_2H_5OH]{C_2H_5ONa}$$

(a) $C_6H_5CH_2-\underset{OC_2H_5}{\overset{CH_3}{C}}-CH_2-CH_3$

(b) $C_6H_5CH=\underset{CH_3}{C}-CH_2-CH_3$

(c) $C_6H_5CH_2-\underset{CH_3}{C}=CHCH_3$

(d) $C_6H_5CH_2-\underset{CH_2CH_3}{C}=CH_2$

154. The major product obtained in the following reaction is : **[NCERT : P L-179 | 2017]**

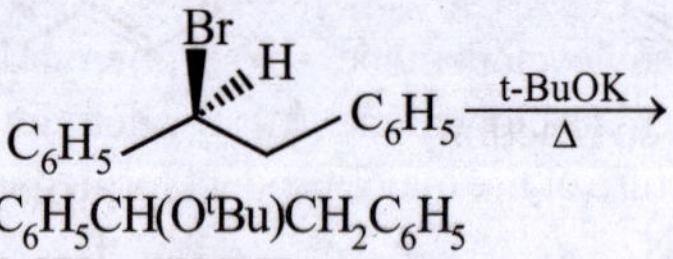

Br, H, C_6H_5, C_6H_5 $\xrightarrow[\Delta]{t\text{-BuOK}}$

(a) $(\pm)C_6H_5CH(O^tBu)CH_2C_6H_5$
(b) $C_6H_5CH=CHC_6H_5$
(c) $(+)C_6H_5CH(O^tBu)CH_2C_6H_5$
(d) $(-)C_6H_5CH(O^tBu)CH_2C_6H_5$

155. In a nucleophilic substitution reaction:

$R-Br+Cl^- \xrightarrow{DMF} R-Cl+Br^-$,

which one of the following undergoes complete inversion of configuration? **[Online April 9, 2014]**

(a) $C_6H_5CHC_6H_5Br$
(b) $C_6H_5CH_2Br$
(c) $C_6H_5CHCH_3Br$
(d) $C_6H_5CCH_3C_6H_5Br$

156. Compound (A), C_8H_9Br, gives a yellow precipitate when warmed with alcoholic $AgNO_3$. Oxidation of (A) gives an acid (B), $C_8H_6O_4$. (B) easily forms anhydride on heating. Identify the compound (A). **[2013]**

(a) CH_2Br, CH_3 (b) C_2H_5, Br

(c) CH_2Br, CH_3 (d) CH_2Br, CH_3

157. The Wurtz-Fittig reaction involves condensation of :

[NCERT : P L-186 | Online April 22, 2013]

(a) two molecules of aryl halides
(b) one molecule of each of aryl halide and alkyl halide.
(c) one molecule of each of aryl halide and phenol.
(d) two molecules of alkyl halides.

158. A major component of Borsch reagent is obtained by reacting hydrazine hydrate with which of the following ?

[Online April 22, 2013]

(a) Cl, NO_2 (b) Cl, O_2N, NO_2

(c) Cl, NO_2, NO_2 (d) Cl, O_2N, NO_2, NO_2

159. Aryl fluoride may be prepared from arene diazonium chloride using : **[NCERT : P L-166 | Online April 9, 2013]**

(a) HBF_4/Δ (b) $HBF_4/NaNO_2, Cu, \Delta$
(c) CuF/HF (d) Cu/HF

160. Reaction of *trans* 2-phenyl-1-bromocyclopentane on reaction with alcoholic KOH produces

[NCERT : P L-182 | 2006]

(a) 1-phenylcyclopentene (b) 3-phenylcyclopentene
(c) 4-phenylcyclopentene (d) 2-phenylcyclopentene

161. The structure of the major product formed in the following reaction is **[NCERT : P L-171 | 2006]**

CH_2Cl, I $\xrightarrow[DMF]{NaCN}$

(a) CH_2Cl, CN (b) CH_2CN, I

(c) CH_2CN, CN (d) CH_2Cl, CN, I

162. Fluorobenzene (C_6H_5F) can be synthesized in the laboratory **[NCERT : P L-168 | 2006]**

(a) by direct fluorination of benzene with F_2 gas
(b) by reacting bromobenzene with NaF solution
(c) by heating phenol with HF and KF
(d) from aniline by diazotisation followed by heating the diazonium salt with HBF_4

163. Phenyl magnesium bromide reacts with methanol to give

[NCERT : P L-180 | 2005]

(a) a mixture of toluene and Mg(OH)Br
(b) a mixture of phenol and Mg(Me)Br
(c) a mixture of anisole and Mg(OH)Br
(d) a mixture of benzene and Mg(OMe)Br

164. Bottles containing C_6H_5I and $C_6H_5CH_2I$ lost their original labels. They were labelled A and B for testing. A and B were separately taken in test tubes and boiled with NaOH solution. The end solution in each tube was made acidic with dilute HNO_3 and then some $AgNO_3$ solution was added. Substance B gave a yellow precipitate. Which one of the following statements is true for this experiment ? **[2003]**

(a) A is $C_6H_5CH_2I$
(b) B is C_6H_5I
(c) Addition of HNO_3 was unnecessary
(d) A is C_6H_5I

Topic 3 - Some Important Polyhalogen Compounds

165. In the following sequence of reaction, the major products B and C respectively are :

[NCERT : P L-181 | April 9, 2024 (I)]

Cl–◇–Br $\xrightarrow{Na/Et_2O}$ A $\xrightarrow[(ii)\ D_2O]{(i)\ Mg/Et_2O}$ B; A $\xrightarrow{CoF_2}$ C

(a) D–◇–◇–D and F–◇–◇–F
(b) D–◇–D and F–◇–◇–F
(c) D–◇–D and F–◇–F
(d) ◇ and F–◇–F

166. Match **List-I** with **List-II** :

[NCERT : P L-188 | April 6, 2024 (I)]

List-I (Compound)	**List-II (Uses)**
(A) Iodoform	(I) Fire extinguisher
(B) Carbon tetrachloride	(II) Insecticide
(C) CFC	(III) Antiseptic
(D) DDT	(IV) Refrigerants

Choose the **correct** answer from the options given below:

(a) A-(I), B-(II), C-(III), D-(IV)
(b) A-(III), B-(II), C-(IV), D-(I)
(c) A-(III), B-(I), C-(IV), D-(II)
(d) A-(II), B-(IV), C-(I), D-(III)

167. Match **List - I** with **List - II.**

[NCERT : P L-187 | Feb. 1, 2024 (II)]

List - I (Compound)	**List - II (Use)**
(A) Carbon tetrachloride	(I) Paint remover
(B) Methylene chloride	(II) Refrigerators and air conditioners
(C) DDT	(III) Fire extinguisher
(D) Freons	(IV) Non Biodegradable insecticide

Choose the correct answer from the options given below:

(a) (A)- (II), (B) - (III), (C) - (I), (D) - (IV)
(b) (A) - (III), (B) - (I), (C) - (IV), (D) - (II)
(c) (A) - (I), (B) - (II), (C) - (III), (D) - (IV)
(d) (A) - (IV), (B) - (III), (C) - (II), (D) - (I)

168. In the following halogenated organic compounds the one with maximum number of chlorine atoms in its structure is:

[NCERT : P L-187 | Jan. 31, 2023 (II)]

(a) Chloral (b) Gammaxene
(c) Chloropicrin (d) Freon -12

169. Which of the following is true about freons ?

[NCERT : P L-187 | Jan. 24, 2023 (I)]

(a) These are chlorofluoro carbon compounds
(b) These are chemicals causing skin cancer
(c) These are radicals of chlorine and chlorine monoxide
(d) All radicals are called freons

170. Which among the following pairs has only herbicides ?

[NCERT : P L-188 | July 29, 2022 (I)]

(a) Aldrin and Dieldrin
(b) Sodium chlorate and Aldrin
(c) Sodium arsinate and Dieldrin
(d) Sodium chlorate and sodium arsenite.

171. The major organic compound formed by the reaction of 1, 1, 1-trichloroethane with silver powder is:

[NCERT : P L-187 | 2014]

(a) Acetylene (b) Ethene
(c) 2 - Butyne (d) 2 - Butene

172. Chlorobenzne reacts with trichloroacetaldehyde in the presence of H_2SO_4.

$$2\,C_6H_5{-}Cl + H{-}\overset{O}{\overset{\|}{C}}{-}CCl_3 \xrightarrow{H_2SO_4}$$

The major product formed is:

[NCERT : P L-187 | Online April 11, 2014]

(a) $Cl{-}C_6H_4{-}C(Cl)_2{-}C_6H_4{-}Cl$
(b) $Cl{-}C_6H_4{-}C(Cl)(CH_2Cl){-}C_6H_4{-}Cl$
(c) $Cl{-}C_6H_4{-}CH(CCl_3){-}C_6H_4{-}Cl$
(d) $Cl{-}C_6H_4{-}CH(Cl){-}C_6H_4{-}Cl$

173. The major product formed when 1, 1, 1-trichloropropane is treated with aqueous potassium hydroxide, is:

[NCERT : P L-187 | Online April 19, 2014]

(a) propyne (b) 1-propanol
(c) 2-propanol (d) propionic acid

174. The compound formed on heating chlorobenzene with chloral in the presence of concentrated sulphuric acid, is

[NCERT : P L-188 | 2004]

(a) freon (b) DDT
(c) gammexene (d) hexachloroethane

ANSWER KEYS

1	(a)	19	(a)	37	(d)	55	(c)	73	(d)	91	(c)	109	(b)	127	(b)	145	(b)	163	(d)
2	(b)	20	(c)	38	(6)	56	(c)	74	(b)	92	(a)	110	(a)	128	(a)	146	(a)	164	(d)
3	(b)	21	(32)	39	(b)	57	(c)	75	(d)	93	(d)	111	(b)	129	(2)	147	(b)	165	(a)
4	(b)	22	(c)	40	(d)	58	(c)	76	(d)	94	(a)	112	(b)	130	(c)	148	(a)	166	(c)
5	(184)	23	(1)	41	(4)	59	(42)	77	(d)	95	(d)	113	(a)	131	(a)	149	(b)	167	(b)
6	(c)	24	(b)	42	(d)	60	(d)	78	(c)	96	(b)	114	(c)	132	(a)	150	(d)	168	(b)
7	(d)	25	(c)	43	(5)	61	(3)	79	(8)	97	(a)	115	(a)	133	(b)	151	(b)	169	(a)
8	(d)	26	(c)	44	(a)	62	(c)	80	(c)	98	(a)	116	(b)	134	(d)	152	(d)	170	(d)
9	(d)	27	(b)	45	(b)	63	(c)	81	(d)	99	(b)	117	(d)	135	(d)	153	(b)	171	(c)
10	(d)	28	(b)	46	(c)	64	(1)	82	(b)	100	(d)	118	(b)	136	(d)	154	(b)	172	(c)
11	(c)	29	(b)	47	(b)	65	(a)	83	(b)	101	(d)	119	(a)	137	(d)	155	(b)	173	(d)
12	(b)	30	(a)	48	(b)	66	(d)	84	(a)	102	(b)	120	(b)	138	(c)	156	(d)	174	(b)
13	(a)	31	(2)	49	(a)	67	(a)	85	(c)	103	(b)	121	(c)	139	(d)	157	(b)		
14	(a)	32	(6)	50	(2)	68	(b)	86	(c)	104	(d)	122	(c)	140	(c)	158	(c)		
15	(c)	33	(b)	51	(a)	69	(2)	87	(d)	105	(a)	123	(b)	141	(c)	159	(a)		
16	(a)	34	(10)	52	(6)	70	(d)	88	(b)	106	(b)	124	(b)	142	(b)	160	(a)		
17	(d)	35	(a)	53	(b)	71	(3)	89	(c)	107	(c)	125	(c)	143	(d)	161	(b)		
18	(c)	36	(a)	54	(c)	72	(c)	90	(c)	108	(a)	126	(b)	144	(c)	162	(d)		

8 Alcohols, Phenols and Ethers

TREND ANALYSIS
JEE MAIN (2025-2021)

	2025				2024				2023				2022				2021			
	Session II		Session I		Session II		Session I		Session II		Session I		Session II		Session I		Session II		Session I	
	MCQs	NVQs	MCQs	NVQs	MCQs	NVQs	MCQs	NVQs	MCQs	NVQs	MCQs	NVQs	MCQs	NVQs	MCQs	NVQs	MCQs	NVQs	MCQs	NVQs
Topic 1	1	–	1	–	3	1	2	–	5	–	5	–	4	3	2	–	–	1	3	–
Topic 2	–	–	1	–	7	1	9	–	3	–	5	1	4	1	8	1	3	–	2	–
Topic 3	2	–	1	–	2	1	3	–	1	–	1	–	1	–	1	–	1	–	–	–
Topic 4																				
Total	3	–	3	–	12	3	14	–	9	–	11	1	9	4	11	1	4	1	5	–

How to Code Qns for Better & Faster Revision →

Type 1: Able to solve (No code Req)
Type 2: Solved with struggle or by seeing solution **(Use code NR means Needs Revision or orange stickers)**
Type 3: Tricky/ Could not solve on my own. Need help from Teacher/ Friend. **(Use code TQ means Tricky Question or red stickers)**

You can either hand write the code in boxes or else paste coloured stickers given at end of book.

Topic 1 – Preparation and Properties of Alcohols

1. A toxic compound "*A*" when reacted with NaCN in aqueous acidic medium yields an edible cooking component and food preservative '*B*'. "*B*" is converted to "*C*" by diborane and can be used as an additive to petrol to reduce emission. "*C*" upon reaction with oleum at 140°C yields an inhalable anesthetic "*D*". Identify "*A*", "*B*", "*C*" and "*D*", respectively. **[NCERT : P L-199 | April 4, 2025 (II)]**

(a) Methanol; formaldehyde; methyl chloride; chloroform
(b) Ethanol; acetonitrile; ethylamine; ethylene
(c) Methanol; acetic acid; ethanol; diethyl ether
(d) Acetaldehyde;2- hydroxypropanoic acid; propanoic acid; dipropyl ether

2. Given below are two statements: **[NCERT : P L-203 | Jan. 23, 2025 (II)]**

Statement I: The boiling points of alcohols and phenols increase with increase in the number of C-atoms.

Statement II : The boiling points of alcohols and phenols are higher in comparison to other class of compounds such as ethers, haloalkanes.

In the light of the above statements, choose the correct answer from the options given below

(a) Both Statement I and Statement II are false
(b) Statement I is false but Statement II is true
(c) Statement I is true but Statement II is false
(d) Both Statement I and Statement II are true

3. Identify the product A and product B in the following set of reactions. **[NCERT : P L-199 | April 9, 2024 (I)]**

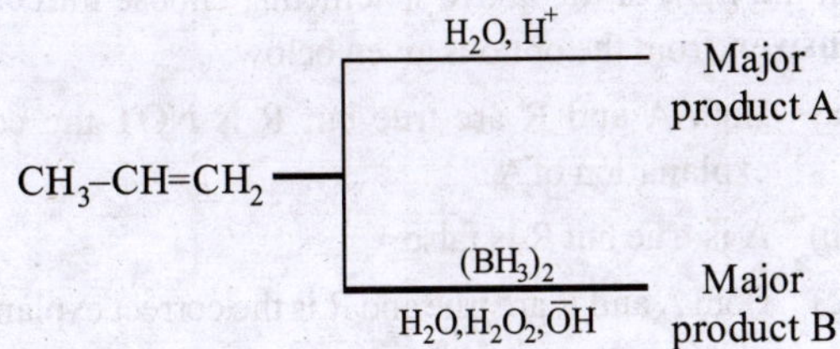

(a) A-$CH_3CH_2CH_2$-OH, B-$CH_3CH_2CH_2$-OH

(b) $A-CH_3CH_2CH_2-OH, B-CH_3CH(OH)-CH_3$

(c) $A-CH_3-CH(OH)-CH_3, B-CH_3CH_2CH_2-OH$

(d) A-$CH_3CH_2CH_3$, B-$CH_3CH_2CH_3$

4. Identify the major products A and B respectively in the following set of reactions.

[NCERT : P L-209 | April 8, 2024 (I)]

B $\xleftarrow[\text{Pyridine}]{CH_3COCl}$ 1-methylcyclohexan-1-ol (CH_3, OH) $\xrightarrow[\Delta]{\text{Conc. } H_2SO_4}$ A

(a) A = 1-methylcyclohexene (CH_3) and B = 1-methylcyclohexyl ring with CH_3, $OCOCH_3$

(b) A = 1-methylcyclohexene (CH_3) and B = ring with CH_3, OH and CH_3CO on ring

(c) A = methylenecyclohexane (CH_2) and B = ring with CH_3, $COCH_3$

(d) A = methylenecyclohexane (CH_2) and B = ring with CH_3, OH, $COCH_3$

5. Which one the following compounds will readily react with dilute NaOH? **[NCERT : P L-205 | April 8, 2024 (II)]**

(a) $C_6H_5CH_2OH$ (b) C_2H_5OH

(c) $(CH_3)_3COH$ (d) C_6H_5OH

6. Vanillin compound obtained from vanilla beans, has total sum of oxygen atoms and π electrons is _____.

[NCERT : P L-198 | NV, April 4, 2024 (II)]

7. Given below are two statements : one is labelled as **Assertion A** and the other is labelled as **Reason R :**

[NCERT : P L-204 | Jan. 31, 2024 (I)]

Assertion A : Alcohols react both as nucleophiles and electrophiles.

Reason R : Alcohols react with active metals such as sodium, potassium and aluminum to yield corresponding alkoxides and liberate hydrogen.

In the light of the above statements, choose the **correct answer** from the options given below:

(a) Both A and R are true but R is NOT the correct explanation of A

(b) A is true but R is false

(c) Both A and R are true and R is the correct explanation of A

(d) A is false but R is true

8. The final product A, formed in the following reaction sequence is: **[NCERT : P L-200 | Jan. 27, 2024 (II)]**

$$Ph-CH=CH_2 \xrightarrow[\text{(iii)HBr; (iv)Mg, ether, then } HCHO/H_3O^+]{\text{(i)}BH_3;\ \text{(ii)}H_2O_2, {}^{\ominus}OH} A$$

(a) $Ph-CH_2-CH_2-CH_3$

(b) $Ph-CH(CH_3)-CH_3$

(c) $Ph-CH(CH_2OH)-CH_3$

(d) $Ph-CH_2-CH_2-CH_2-OH$

9. Given below are two statements, one is labelled as **Assertion A** and the other is labelled as **Reason R**.

Assertion A : Order of acidic nature of the following compounds is A > B > C.

A: 2-chlorocyclohexanol (OH, Cl); B: 4-fluorocyclohexanol (OH, F); C: 3-methylcyclohexanol (OH, CH_3)

Reason R : Fluoro is a stronger electron withdrawing group than chloro group.

In the light of the above statements, choose the correct answer from the options given below:

[NCERT : P L-205 | April 13, 2023 (II)]

(a) A is false but R is true

(b) Both A and R are correct and R is the correct explanation of A

(c) Both A and R are correct but R is NOT the correct explanation of A

(d) A is true but R is false

10. $H_3C-CH_2-CH(OH)-CH_3 \xrightarrow[\text{(iii) } D_2O]{\text{(i) NaI, } H_3PO_4;\ \text{(ii) Mg, Dry ether}}$ [X] Product

Product [X] formed in the above reaction is :

[NCERT : P L-203 | April 11, 2023 (II)]

(a) $H_3C-CH_2-CH(D)-CH_3$

(b) $H_3C-CH_2-C(H)(OH)-CH_3$

(c) $H_3C-CH_2-CH=CH_2$

(d) $H_3C-CH=CH-CH_3$

11. Incorrect method of preparation for alcohols from the following is: **[NCERT : P L-203 | April 10, 2023 (II)]**
(a) Ozonolysis of alkene.
(b) Reaction of Ketone with RMgBr followed by hydrolysis.
(c) Hydroboration–oxidation of alkene.
(d) Reaction of alkyl halide with aqueous NaOH.

12. Given below are two statements: One is labelled as **Assertion A** and the other is labelled as **Reason R**.
Assertion A: Butan -1- ol has higher boiling point than ethoxyethane.
Reason R: Extensive hydrogen bonding leads to stronger association of molecules.
In the light of the above statements, choose the correct answer from the options given below:
[NCERT : P L-203 | April 08, 2023 (I)]
(a) Both **A** and **R** are true and **R** is the correct explanation of **A**.
(b) **A** is true but **R** is false.
(c) Both **A** and **R** are true but **R** is not the correct explanation of **A**.
(d) **A** is false but **R** is true.

13. In the following reaction, 'B' is
[NCERT : P L-209 | April 06, 2023 (II)]

$\xrightarrow{H_3O^+}$ 'B' major

OH

(a) (b)

(c) O (d) O

14. Compound A, $C_5H_{10}O_5$, gives a tetraacetate with Ac_2O and oxidation of A with $Br_2 - H_2O$ gives an acid, $C_5H_{10}O_6$. Reduction of A with HI gives isopentane. The possible strucutre of A is: **[NCERT : P L-203 | Jan. 31, 2023 (II)]**

(a) CHO
|
CH—OH
|
C—OH
CH_2 CH_2
| |
OH OH

(b) HOH_2C—C(=O)—CH(OH)—CH(OH)... CH_2OH

(c) CH_2OH
C=O
OH
HOH_2C CH_2OH

(d) CHO
|
$(CHOH)_3$
|
CH_2OH

15. Find out the major product for the following reaction.

CH_3 $\xrightarrow{H_3O^+}$ Major Product

OH

[NCERT : P L-209 | Jan. 29, 2023 (II)]

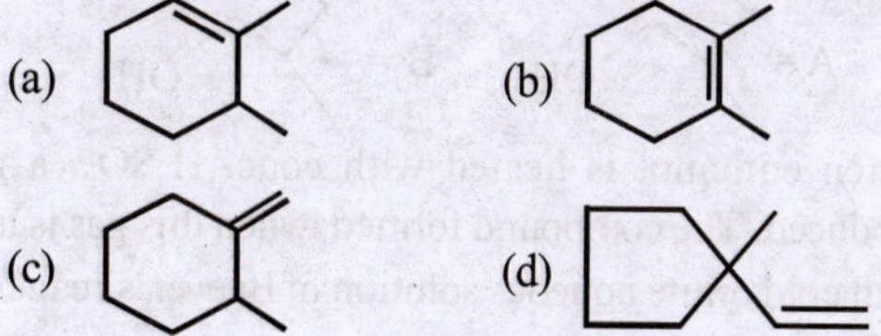

16. Find out the major product from the following reaction.

$\xrightarrow[\Delta]{H_2SO_4\text{(Concentrated)}}$

OH

[NCERT : P L-209 | Jan. 25, 2023 (II)]

(a) (b)

(c) (d)

17. Number of compounds giving (i) red colouration with ceric ammonium nitrate and also (ii) positive iodoform test from the following is ______
[NCERT : P L-203 | NV, Jan. 25, 2023 (II)]

O CH_3, OH, O H

OH , O , OH

OH , OH

18. Find out the major products from the following reactions.

B $\xleftarrow[NaBH_4]{Hg(OAc)_2, H_2O}$ (alkene) $\xrightarrow[H_2O_2/OH^-]{BH_3, THF}$ A

[NCERT : P L-200 | Jan. 24, 2023 (II)]

(a) A = (structure with OH), B = (structure with OH)

(b) A = (structure with OH), B = (structure with OH)

(c) A = (structure with OH), B = (structure with OH)

(d) A = (structure with OH), B = (structure with OH)

19. When enthanol is heated with conc. H_2SO_4, a gas is produced. The compound formed, when this gas is treated with cold dilute aqueous solution of Baeyer's reagent, is : [NCERT : P L-203 | July 29, 2022 (II)]

(a) Formaldehyde (b) Formic acid

(c) Glycol (d) Ethanoic acid

20. A 1.84 mg sample of polyhydric alcoholic compound 'X' of molar mass 92.0 g/mol gave 1.344 mL of H_2 gas at STP. The number of alcoholic hydrogens present in compound 'X' is____. [NCERT : P L-203 | NV, July 29, 2022 (II)]

21. The major product in the given reaction is

(structure)–OH $\xrightarrow[(2)\ HBr]{(1)\ H^+,\ heat}$

[NCERT : P L-209 | July 28, 2022 (II)]

(a) (structure with Br)

(b) (structure with Br)

(c) (structure with Br)

(d) (structure with Br)

22. (4-hydroxybenzyl alcohol: CH_2OH, OH) + HCl $\xrightarrow{\Delta}$ A $\xrightarrow{NaI}$ B

In the above reaction product B is:

[NCERT : P L-203 | July 27, 2022 (I)]

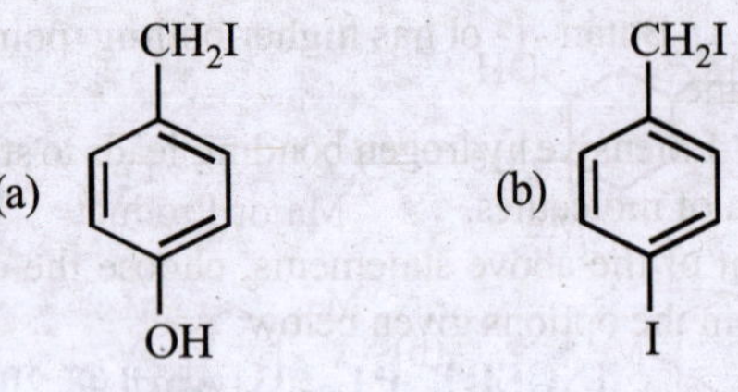

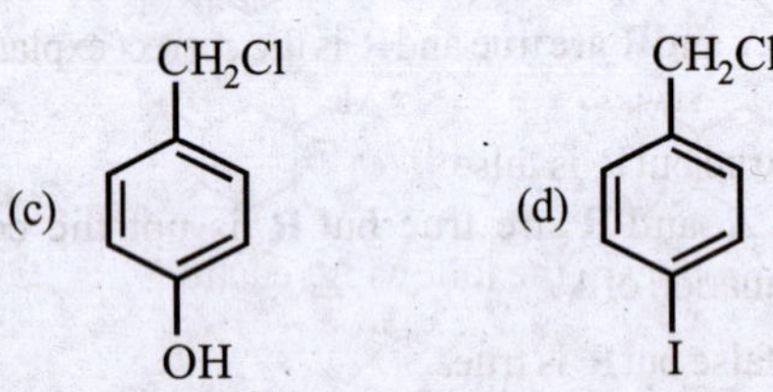

23. A 100 mL solution of CH_3CH_2MgBr on treatment with methanol produces 2.24 mL of a gas at STP. The weight of gas produced is _______ mg. [nearest integer]

[NCERT : P L-203 | NV, July 26, 2022 (II)]

24. Most stable product of the following reaction is:

[NCERT : P L-203 | July 25, 2022 (I)]

(structure with OH) $\xrightarrow[(ii)\ NaCN,\ DMF]{(i)\ H_3C-C_6H_4-SO_2Cl,\ Pyridine}$

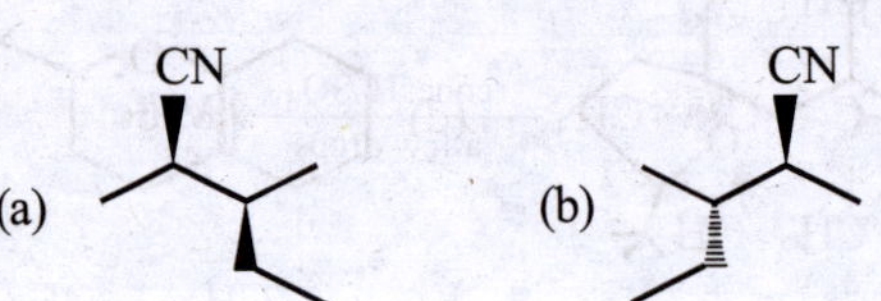

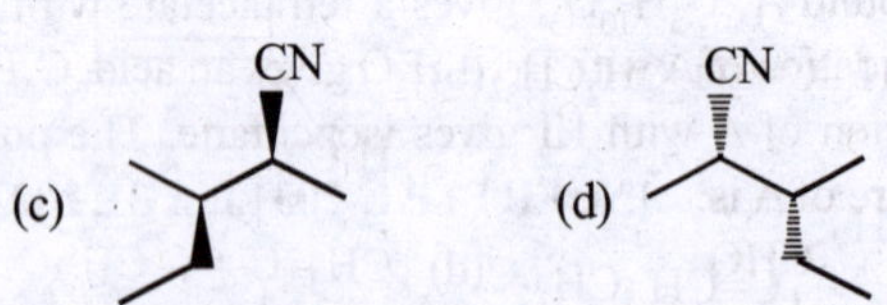

25. Given below are two statements : [July 25, 2022 (I)]

Statement I : On heating with $KHSO_4$, glycerol is dehydrated and acrolein is formed.

Statement II : Acrolein has fruity odour and can be used to test glycerol's presence.

Choose the **correct** option.

(a) Both Statement I and Statement II are correct.

(b) Both Statement I and Statement II are incorrect.

(c) Statement I is correct but Statement II is incorrect.

(d) Statement I is incorrect but Statement II is correct.

26. A sample of 4.5 mg of an unknown monohydric alcohol, R – OH was added to methylmagnesium iodide. A gas is evolved and is collected and its volume measured to be 3.1 mL. The molecular weight of the unknown alcohol is _____ g/mol. [Nearest integer] **[NV, July 25, 2022 (II)]**

27. The number of chiral alcohol(s) with molecular formula $C_4H_{10}O$ is______.
[NCERT : P L-203 | NV, June 29, 2022 (II)]

28. The major product (P) of the given reaction is (where, Me is – CH_3) **[NCERT : P L-200 | June 28, 2022 (II)]**

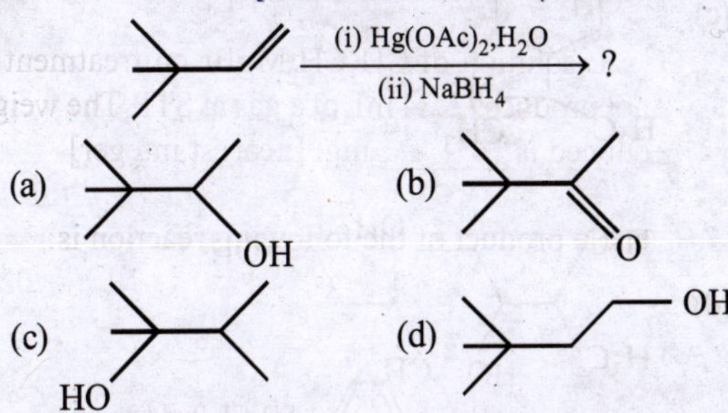

29. The major product in the following reaction
[NCERT : P L-199 | June 26, 2022 (II)]

(i) $Hg(OAc)_2, H_2O$ (ii) $NaBH_4$ → ?

30. The major product formed in the following reaction is :
[NCERT : P L-209 | Aug. 31, 2021 (I)]

$CH_3-C(CH_3)_2-CH(OH)-CH_3 \xrightarrow[\text{a few drops}]{\text{conc. } H_2SO_4}$ Major product

(a) $CH_3-C=CH-CH_2CH_3$
(b) $(CH_3)_2C=C(CH_3)_2$
(c) $(CH_3)_2C=CH-CH_3$
(d) $CH_3-C(CH_3)-CH=CH_2$

31. To synthesise 1.0 mole of 2-methylpropan-2-ol from ethylethanoate _____ equivalents of CH_3MgBr reagent will be required. (Integer value)
[NCERT : P L-200 | NV, July 20, 2021 (I)]

32. Ceric ammonium nitrate and $CHCl_3$/alc. KOH are used for the identification of functional groups present in _____ and _____ respectively.
[NCERT : P L-200 | Feb. 26, 2021 (II)]
(a) alcohol, phenol (b) amine, phenol
(c) amine, alcohol (d) alcohol, amine

33. What is 'X' in the given reaction?

CH_2OH-CH_2OH + oxalic acid $\xrightarrow{210°C}$ X (major product)

[NCERT : P L-200 | Feb. 25, 2021 (II)]

(a) CHO–CHO
(b) $CH_2=CH_2$
(c) $CH-OH \| CH_2$
(d) $CH_2OH \| CHO$

34. What is the final product (major) 'A' in the given reaction?
[NCERT : P L-209 | Feb. 24, 2021 (I)]

HCl → 'A' (major product)

(a)
(b)
(c)
(d)

35. The major product of the following reaction is :
[NCERT : P L-209 | Sep. 05, 2020 (II)]

H_2SO_4 →

(a)
(b)
(c)
(d)

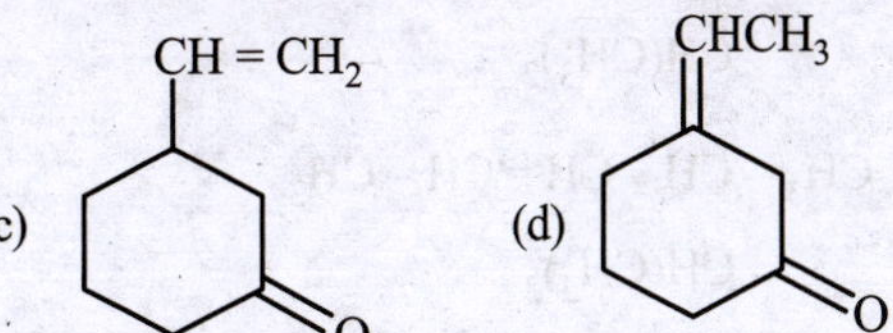

36. When neopentyl alcohol is heated with an acid, it is slowly converted into an 85 : 15 mixture of alkenes A and B, respectively. What are these alkenes?

[NCERT : P L-209 | Sep. 04, 2020 (I)]

(a) $(H_3C)_2C$–$C(=CH_2)$... H_3C, CH_2 / H_3C and H_3C, CH_3 / CH_2

(b) H_3C, CH_3 / H_3C and H_3C, CH_3 / CH_2

(c) CH_3, CH_3 / CH_2 and CH_3, CH_2 / CH_3

(d) H_3C, CH_3 / H_3C and H_3C, CH_2 / H_3C

37. The number of chiral centres present in [B] is ______.

[NCERT : P L-200 | Sep. 04, 2020 (I)]

$$\text{(3-methylcyclohexyl)}-\underset{\displaystyle CH_3}{\underset{|}{CH}}-C\equiv N \xrightarrow[\text{(ii) } H_3O^+]{\text{(i) } C_2H_5MgBr} [A] \xrightarrow[\text{(ii) } H_2O]{\text{(i) } CH_3MgBr} [B]$$

38. The major product [B] in the following sequence of reactions is: **[NCERT : P L-200 | Jan. 08, 2020 (II)]**

$$CH_3-\underset{\displaystyle CH(CH_3)_2}{\underset{|}{C}}=CH-CH_2CH_3 \xrightarrow[\text{(ii) } H_2O_2,\ OH^\ominus]{\text{(i) } B_2H_6} [A] \xrightarrow[\Delta]{\text{dil. } H_2SO_4} [B]$$

(a) $CH_2=\underset{\displaystyle CH(CH_3)_2}{\underset{|}{C}}-CH_2CH_2CH_3$

(b) $CH_3-\underset{\displaystyle C(CH_3)_2}{\underset{\|}{C}}-CH_2CH_2CH_3$

(c) $CH_3-\underset{\displaystyle CH(CH_3)_2}{\underset{|}{C}}=CH-CH_2CH_3$

(d) $CH_3-\underset{\displaystyle CH(CH_3)_2}{\underset{|}{CH}}-CH=CH-CH_3$

39. The major product of the following reaction is:

[NCERT : P L-209 | Jan. 08, 2020 (I)]

$(H_3C)_2C=CH-CH_2-CH_2-C(CH_3)=CH-CH_2OH \xrightarrow{\text{dil. } H_2SO_4}$

(a) [structure: cyclopentane ring bearing CH_3, vinyl group and $C(OH)(CH_3)$–H_3C]

(b) [structure: H_3C, H_3C, OH, OH, OH, CH_3]

(c) [structure: methylcyclohexene ring with CH_3 and $C(OH)(CH_3)$–H_3C]

(d) [structure: H_3C, H_3C, OH, OH, HO, CH_3]

40. Heating of 2-chloro-1-phenylbutane with EtOK/EtOH gives X as the major product. Reaction of X with $Hg(OAc)_2$/H_2O followed by $NaBH_4$ gives Y as the major product, Y is :

[NCERT : P L-200 | April 12, 2019 (II)]

(a) Ph–CH_2–CH(OH)–CH_2CH_3 (b) Ph–CH_2CH_2–CH(OH)–CH_3

(c) Ph–CH(OH)–$CH_2CH_2CH_3$ (d) Ph–$CH_2CH_2CH_2CH_3$

41. $CH_3CH_2-\underset{\displaystyle Ph}{\overset{\displaystyle OH}{C}}-CH_3$ cannot be prepared by:

[Jan. 12, 2019 (I)]

(a) $CH_3CH_2COCH_3 + PhMgX$
(b) $PhCOCH_2CH_3 + CH_3MgX$
(c) $PhCOCH_3 + CH_3CH_2MgX$
(d) $HCHO + PhCH(CH_3)CH_2MgX$

42. The major product 'X' formed in the following reaction is

[Structure: 2-substituted cyclohex-3-enone bearing $CH_2-\overset{O}{\overset{||}{C}}-OCH_3$] $\xrightarrow[MeOH]{NaBH_4}$ X

[NCERT : P L-200 | Jan. 10, 2019 (I)]

(a) [cyclohexenone ring] $CH_2-\overset{O}{\overset{||}{C}}-H$

(b) [cyclohexanol ring, OH] CH_2CH_2OH

(c) [cyclohexenol ring, OH] CH_2CH_2OH

(d) [cyclohexenol ring, OH] $CH_2-\overset{O}{\overset{||}{C}}-OCH_3$

43. Which is the most suitable reagent for the following transformation? [NCERT : P L-200 | Jan. 10, 2019 (II)]

$CH_3-CH=CH-CH_2-\underset{}{\overset{OH}{\overset{|}{CH}}}-CH_3 \longrightarrow$

$CH_3-CH=CH-CH_2CO_2H$

(a) Tollen's reagent (b) I_2/NaOH
(c) CrO_2Cl_2/CS_2 (d) alkaline $KMnO_4$

44. The gas evolved on heating CH_3MgBr in methanol is :
[NCERT : P L-200 | Online April 9, 2016]

(a) Methane (b) Ethane
(c) Propane (d) HBr

45. The most suitable reagent for the conversion of $R-CH_2-OH \rightarrow R-CHO$ is: [NCERT : P L-200 | 2014]

(a) $KMnO_4$
(b) $K_2Cr_2O_7$
(c) CrO_3
(d) PCC (Pyridinium chlorochromate)

46. Which one of the following statements is not correct?
[NCERT : P L-205 | Online April 11, 2014]

(a) Alcohols are weaker acids than water
(b) Acid strength of alcohols decreases in the following $RCH_2OH > R_2CHOH > R_3COH$
(c) Carbon-oxygen bond length in methanol, CH_3OH is shorter than that of C – O bond length in phenol.
(d) The bond angle C–O–H (C / O \ H) in methanol is 108.9°.

47. In the Victor-Meyer's test, the colour given by 1°, 2° and 3° alcohols are respectively:
[NCERT : P L-200 | Online April 12, 2014]

(a) Red, colourless, blue (b) Red, blue, colourless
(c) Colourless, red, blue (d) Red, blue, violet

48. In the hydroboration - oxidation reaction of propene with diborane, H_2O_2 and NaOH, the organic compound formed is: [NCERT : P L-200 | Online April 9, 2014]

(a) CH_3CH_2OH (b) $CH_3CHOHCH_3$
(c) $CH_3CH_2CH_2OH$ (d) $(CH_3)_3COH$

49. [benzene ring]$-CH_2-CH=CH_2$

On mercuration-demercuration produces the major product:
[NCERT : P L-200 | Online April 12, 2014]

(a) [benzene ring]$-CH_2-\underset{OH}{\underset{|}{CH}}-CH_3$

(b) [benzene ring]$-CH_2-CH_2-CH_2-OH$

(c) [benzene ring]$-CH_2-\underset{OH}{\underset{|}{CH}}-\underset{OH}{\underset{|}{CH_3}}$

(d) [benzene ring]$-CH_2-COOH$

50. An unknown alochol is treated with the "Lucas reagent" to determine whether the alcohol is primary, secondary or tertiary. Which alcohol reacts fastest and by what mechanism : [NCERT : P L-208 | 2013]

(a) secondary alcohol by S_N1
(b) tertiary alcohol by S_N1
(c) secondary alcohol by S_N2
(d) tertiary alcohol by S_N2

51. Rate of dehydration of alcohols follows the order:
[NCERT : P L-209 | Online April 9, 2013]

(a) $2° > 1° > CH_3OH > 3°$ (b) $3° > 2° > 1° > CH_3OH$
(c) $2° > 3° > 1° > CH_3OH$ (d) $CH_3OH > 1° > 2° > 3°$

52. Amongst the following alcohols which would react fastest with conc. HCl and $ZnCl_2$?
[NCERT : P L-208 | 2010 (s), Online April 22, 2013]

(a) pentanol (b) 2-methyl butanol
(c) 2-pentanol (d) 2-methyl butan-2-ol

53. Reagent used to convert allyl alcohol to acrolein is
[NCERT : P L-200 | Online May 19, 2012]

(a) MnO_2 (b) H_2O_2 (c) OsO_4 (d) $KMnO_4$

54. Consider thiol anion ($RS^{\ominus}$) and alkoxy anion ($RO^{\ominus}$). Which of the following statements is correct ? [2011RS]

(a) $RS^{\ominus}$ is less basic but more nucleophilic than $RO^{\ominus}$
(b) $RS^{\ominus}$ is more basic and more nucleophilic than $RO^{\ominus}$
(c) $RS^{\ominus}$ is more basic but less nucleophilic than $RO^{\ominus}$
(d) $RS^{\ominus}$ is less basic and less nucleophilic than $RO^{\ominus}$

55. The main product of the following reaction is

$C_6H_5CH_2CH(OH)CH(CH_3)_2 \xrightarrow{\text{conc.}H_2SO_4} ?$

[NCERT : P L-209 | 2010]

(a) $(H_5C_6)(H)C=C(H)(CH(CH_3)_2)$

(b) $(C_6H_5CH_2)(H)C=C(CH_3)(CH_3)$

(c) $(C_6H_5CH_2CH_2)(H_3C)C=CH_2$

(d) $(C_6H_5)(H)C=C(CH(CH_3)_2)(H)$

56. From amongst the following alcohols, the one that would react fastest with conc. HCl and anhydrous $ZnCl_2$, is [NCERT : P L-200 | 2010]

(a) 2-Butanol (b) 2-Methylpropan-2-ol

(c) 2-Methylpropanol (d) 1-Butanol

57. In the following sequence of reactions, [NCERT : P L-200 | 2007]

$CH_3CH_2OH \xrightarrow{P+I_2} A \xrightarrow[\text{ether}]{Mg} B \xrightarrow{HCHO} C \xrightarrow{H_2O} D$

the compound D is

(a) propanal (b) butanal

(c) *n*-butyl alcohol (d) *n*-propyl alcohol.

58. Among the following the one that gives positive iodoform test upon reaction with I_2 and NaOH is [NCERT : P L-200 | 2006]

(a) $CH_3-CH(CH_3)CH_2OH$

(b) $PhCHOHCH_3$

(c) $CH_3CH_2CH(OH)CH_2CH_3$

(d) $C_6H_5CH_2CH_2OH$

59. The best reagent to convert pent-3-en-2-ol into pent-3-en-2-one is [NCERT : P L-200 | 2005]

(a) Pyridinium chlorochromate

(b) Chromic anhydride in glacial acetic acid

(c) acidic dichromate

(d) Acidic permanganate

60. Among the following compounds which can be dehydrated very easily? [NCERT : P L-209 | 2004]

(a) $CH_3CH_2-C(CH_3)(OH)-CH_2CH_3$

(b) $CH_3CH_2CH_2CH(OH)CH_3$

(c) $CH_3CH_2CH_2CH_2CH_2OH$

(d) $CH_3CH_2CH(CH_3)CH_2CH_2OH$

61. During dehydration of alcohols to alkenes by heating with conc. H_2SO_4, the initiation step is [NCERT : P L-209 | 2003]

(a) formation of carbocation

(b) elimination of water

(c) formation of an ester

(d) protonation of alcohol molecule

Topic 2 - Preparation and Properties of Phenols

62. Identify the products [A] and [B], respectively in the following reaction: [NCERT : P L-205 | Jan. 23, 2025 (II)]

Chlorobenzene $\xrightarrow[\text{(ii) } H^+]{\text{(i) NaOH, 623K, 300 atm}} [A] \xrightarrow[H_2SO_4]{Na_2Cr_2O_7} [B]$

(a) [A] phenol (OH on benzene), [B] sodium phenoxide (O^-Na^+ on benzene)

(b) [A] benzene, [B] phenol (OH on benzene)

(c) [A] phenol (OH on benzene), [B] p-benzoquinone

(d) [A] benzene, [B] cyclohexadienone (C=O on ring)

63. For the given compounds, the correct order of increasing pK_a value : [NCERT : P L-207 | April 9, 2024 (I)]

(A) Phenol (C₆H₅–OH) (B) $O_2N-C_6H_4-OH$ (p-nitrophenol)

(C) $HO-C_6H_4-OCH_3$ (m-methoxyphenol, OCH_3 meta) (D) $C_6H_4(NO_2)(OH)$ (m-nitrophenol)

(E) $OH-C_6H_4-OCH_3$ (p-methoxyphenol)

(a) (E) < (D) < (C) < (B) < (A)

(b) (D) < (E) < (C) < (B) < (A)

(c) (E) < (D) < (B) < (A) < (C)

(d) (B) < (D) < (A) < (C) < (E)

64. Match **List-I** with **List-II**.

[NCERT : P L-264 | April 8, 2024 (II)]

List-I (Reactions)	**List-II (Products)**
(A) NH_2 (aniline) $\xrightarrow[\text{(ii) } H_2O\text{, warm}]{\text{(i) } NaNO_2 + HCl}$	(I) OH, CHO (salicylaldehyde)
(B) OH (phenol) $\xrightarrow[H_2SO_4]{Na_2Cr_2O_7}$	(II) OH (phenol)
(C) OH (phenol) $\xrightarrow[\text{(ii) } H^+]{\text{(i) } CHCl_3 + \text{aq NaOH}}$	(III) OH, COOH (salicylic acid)
(D) OH (phenol) $\xrightarrow{\text{(i) NaOH, (ii) } CO_2\text{, (iii) } H^+}$	(IV) O, O (benzoquinone)

Choose the **correct** answer from the options given below:

(a) (A)-(III), (B)-(II), (C)-(I), (D)-(IV)
(b) (A)-(IV), (B)-(II), (C)-(III), (D)-(I)
(c) (A)-(I), (B)-(IV), (C)-(II), (D)-(III)
(d) (A)-(II), (B)-(IV), (C)-(I), (D)-(III)

65. Given below are two statements :

[NCERT : P L-211 | April 6, 2024 (I)]

Statement I : Picric acid is 2, 4, 6-trinitrotoluene.

Statement II : Phenol-2, 4-disulphuric acid is treated with conc. HNO_3 to get picric acid.

In the light of the above statement, choose the **most appropriate** answer from the options given below :

(a) Statement I is incorrect but Statement II is correct.
(b) Both Statement I and Statement II are incorrect.
(c) Statement I is correct but Statement II is incorrect.
(d) Both Statement I and Statement II are correct.

66. In Reimer - Tiemann reaction, phenol is converted into salicylaldehyde through an intermediate. The structure of intermediate is ______.

[NCERT : P L-213 | April 6, 2024 (I)]

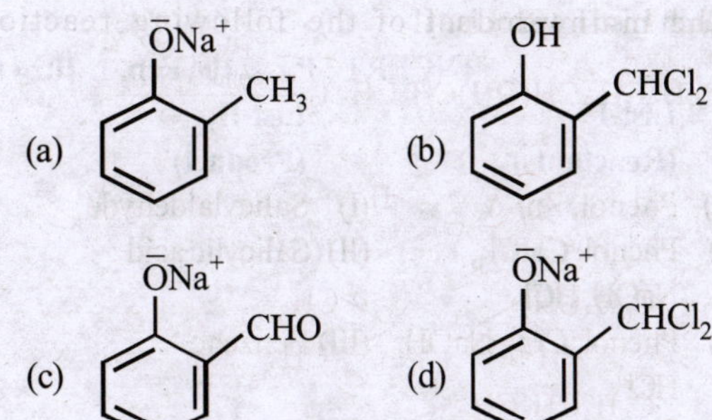

67. Identify compound (Z) in the following reaction sequence.

Cl (chlorobenzene) $+ NaOH \xrightarrow[300\text{ atm}]{623K} X \xrightarrow{HCl} Y \xrightarrow{\text{Conc. } HNO_3} Z$

[NCERT : P L-202 | April 5, 2024 (I)]

(a) OH, NO_2
(b) OH, NO_2, NO_2
(c) OH, O_2N, NO_2, NO_2
(d) OH, NO_2

68. Given below are two statement :

[NCERT : P L-212 | April 5, 2024 (I)]

Statement I : Bromination of phenol in solvent with low polarity such as $CHCl_3$ or CS_2 requires Lewis acid catalyst.

Statement II : The Lewis acid catalyst polarises the bromine to generate Br^+.

In the light of the above statements, choose the **correct** answer from the options given below :

(a) Statement I is true but Statement II is false.
(b) Both Statement I and Statement II are true
(c) Both Statement I and Statement II are false.
(d) Statement I is false but Statement II is true.

69. Consider the given chemical reaction sequence :

OH (phenol) $\xrightarrow{\text{Conc. } H_2SO_4}$ Product A $\xrightarrow{\text{Conc. } HNO_3}$ Product B

Total sum of oxygen atoms in product A and product B are ………

[NCERT : P L-205 | NV, April 5, 2024 (I)]

70. Common name of benzene-1, 2-diol is

[NCERT : P L-195 | April 4, 2024 (II)]

(a) quinol (b) resorcinol
(c) catechol (d) *o*-cresol

71. Match List-I with List-II

[NCERT : P L-210 | Feb. 1, 2024 (II)]

List-I (Reactants)	List-II (Product)
(A) Phenol, Zn/ Δ	(I) Salicylaldehyde
(B) Phenol, $CHCl_3$, NaOH, HCl	(II) Salicylic acid
(C) Phenol, CO_2, NaOH, HCl	(III) Benzene
(D) Phenol, Conc. HNO_3	(IV) Picric acid

Choose the correct answer from the options given below:

(a) (A)- (IV), (B) - (I), (C) - (II), (D) - (III)
(b) (A) - (III), (B) - (I), (C) - (II), (D) - (IV)
(c) (A) - (IV), (B) - (II), (C) - (I), (D) - (III)
(d) (A) - (III), (B) - (IV), (C) - (I), (D) - (II)

72. Given below are two statements : one is labelled as **Assertion A** and the other is labelled as **Reason R :**

[NCERT : P L-206 | Jan. 31, 2024 (I)]

Assertion A : pK_a value of phenol is 10.0 while that of ethanol is 15.9.

Reason R : Ethanol is stronger acid than phenol.

In the light of the above statements, choose the **correct answer** from the options given below:

(a) A is true but R is false
(b) Both A and R are true but R is NOT the correct explanation of A
(c) Both A and R are true and R is the correct explanation of A
(d) A is false but R is true

73. Salicylaldehyde is synthesized from phenol, when reacted with **[NCERT : P L-210 | Jan. 30, 2024 (II)]**

(a) H–C(=O)–Cl, NaOH (b) CO_2, NaOH
(c) CCl_4, NaOH (d) $HCCl_3$, NaOH

74. Given below are two statements : one is labelled as **Assertion A** and the other is labelled as **Reason R :**

[NCERT : P L-211 | Jan. 29, 2024 (I)]

Assertion A : Aryl halides cannot be prepared by replacement of hydroxyl group of phenol by halogen atom.

Reason R : Phenols react with halogen acids violently.

In the light of the above statements, choose the **most appropriate** from the options given below:

(a) **A** is false but **R** is true
(b) Both **A** and **R** are true and **R** is the correct explanation of **A**
(c) **A** is true but **R** is false
(d) Both **A** and **R** are true but **R** is NOT the correct explanation of **A**

75. Phenol treated with chloroform in presence of sodium hydroxide, which further hydrolyzed in presence of an acid results **[NCERT : P L-205 | Jan. 29, 2024 (II)]**

(a) Benzene- 1, 2-diol
(b) Benzene- 1, 3-diol
(c) 2-Hydroxybenzaldehyde
(d) Salicylic acid

76. Match List-I with List-II

[NCERT : P L-206, 207 | Jan. 29, 2024 (II)]

LIST-I (Compound)	LIST-II (pK_a value)
A. Ethanol	I. 10.0
B. Phenol	II. 15.9
C. m-Nitrophenol	III. 7.1
D. p-Nitrophenol	IV. 8.3

Choose the correct answer from the options given below:

(a) A-I, B-II, C-III, D-IV (b) A-II, B-I C-IV, D-III
(c) A-III, B-IV, C-I, D-II (d) A-IV, B-I, C-II, D-III

77. Given below are two statements :

[NCERT : P L-206 | Jan. 27, 2024 (I)]

Statement (I) : p-Nitrophenol is more acidic than m-nitrophenol and o-nitrophenol.

Statement (II) : Ethanol will give immediate turbidity with Lucas reagent.

In the light of the above statements, choose the correct answer from the options given below :

(a) Statement I is true but Statement II is false
(b) Both Statement I and Statement II are true
(c) Both Statement I and Statement II are false
(d) Statement I is false but Statement II is true

78. The ascending order of acidity of –OH group in the following compounds is : **[NCERT : P L-206 | Jan. 27, 2024 (I)]**

(A) Bu – OH

(B) O_2N–C_6H_4–OH (para)

(C) MeO–C_6H_4–OH (para)

(D) C_6H_5–OH

(E) O_2N–C_6H_3(–NO_2)–OH

Choose the correct answer from the options given below:

(a) (A) < (D) < (C) < (B) < (E)
(b) (C) < (A) < (D) < (B) < (E)
(c) (C) < (D) < (B) < (A) < (E)
(d) (A) < (C) < (D) < (B) < (E)

79. Match List-I with List-II. **[NCERT : P L-213 | Jan. 27, 2024 (II)]**

List–I (Reaction)

(A) Phenol (OH) ⟶ Salicylic acid (OH, COOH)

(B) [phenol → salicylaldehyde (2-hydroxybenzaldehyde)]

(C) [phenol → benzoquinone]

(D) [phenol → anisole]

List–II
(Reagent(s))

(I) $Na_2Cr_2O_7$, H_2SO_4
(II) (i) NaOH (ii) CH_3Cl
(III) (i) NaOH, $CHCl_3$ (ii) NaOH (iii) HCl
(IV) (i) NaOH (ii) CO_2 (iii) HCl

Choose the correct answer from the options given below:

(a) (A)-(IV), (B)-(I), (C)-(III), (D)-(II)
(b) (A)-(II), (B)-(III), (C)-(I), (D)-(IV)
(c) (A)-(II), (B)-(I), (C)-(III), (D)-(IV)
(d) (A)-(IV), (B)-(III), (C)-(I), (D)-(II)

80. Compound 'B' is **[NCERT : P L-205 | April 11, 2023 (II)]**

[thymol] $\xrightarrow[HCl]{NaNO_2}$ A $\xrightarrow[H_2O]{NH_4SH}$ B (Major)

(a) [structure] (b) [structure with H_2N, OH]
(c) [structure with SH] (d) [structure with HS, OH]

81. A compound 'X' when treated with phthalic anhydride in presence of concentrated H_2SO_4 yields 'Y'. 'Y' is used as an acid/base indicator. 'X' and 'Y' are respectively:
[NCERT : P L-210 | April 08, 2023 (II)]

(a) Carbolic acid, Phenolphthalein
(b) Anisole, Methyl orange
(c) Salicylaldehyde, Phenolphthalein
(d) Toludine, Phenolphthalein

82. The strongest acid from the following is
[NCERT : P L-206 | April 06, 2023 (II)]

(a) [OH, NO_2 (meta)] (b) [OH, Cl (meta)]
(c) [OH, CH_3 (meta)] (d) [phenol]

83. Decreasing order of dehydration of the following alcohols is **[NCERT : P L-208 | Feb. 01, 2023 (I)]**

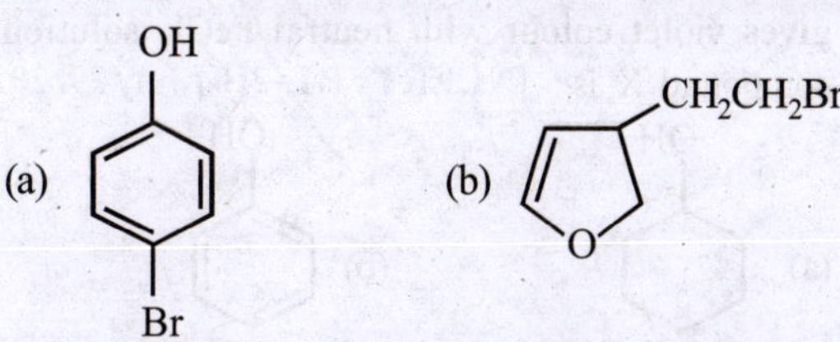

(a) D>B>C>A (b) A>D>B>C
(c) B>A>D>C (d) B>D>C>A

84. An organic compound 'A' with empirical formula C_6H_6O gives sooty flame on burning. Its reaction with bromine solution in low polarity solvent results in high yield of B. B is **[NCERT : P L-213 | Jan. 31, 2023 (I)]**

(a) [OH, Br (para)] (b) [furan with CH_2CH_2Br]
(c) [Br, Br, CH_2CH_3, O ring] (d) [OH, Br (ortho)]

85. A trisubstituted compound 'A', $C_{10}H_{12}O_2$ gives neutral $FeCl_3$ test positive. Treatment of compound 'A' with NaOH and CH_3Br gives $C_{11}H_{14}O_2$, with hydriodic acid it gives methyl iodide and with hot conc. NaOH gives a compound B, $C_{10}H_{12}O_2$. Compound 'A' also decolorises alkaline $KMnO_4$. The number of π bond/s present in the compound 'A' is ______. **[NCERT : P L-213 | NV, Jan. 30, 2023 (I)]**

86. The correct order of pK_a values for the following compounds is: **[NCERT : P L-206, 207 | Jan. 30, 2023 (II)]**

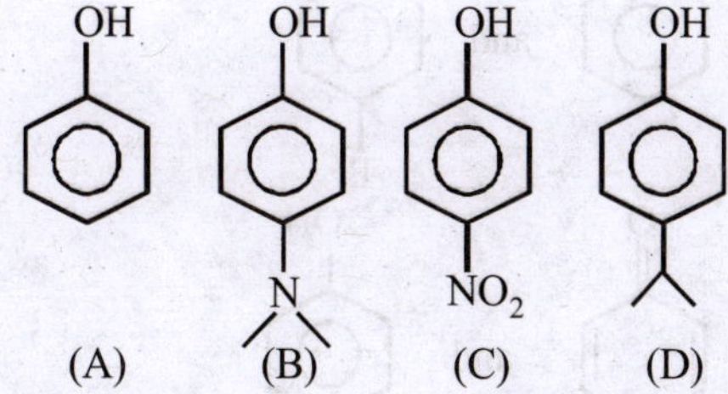

(a) C>A>D>B (b) B>D>A>C
(c) B>A>D>C (d) A>B>C>D

87. The increasing order of pK_a for the following phenols is [NCERT : P L-206, 207 | Jan. 29, 2023 (I)]

(A) 2, 4-Dinitrophenol (B) 4 - Nitrophenol
(C) 2, 4, 5- Trimethylphenol
(D) Phenol (E) 4-Chlorophenol

(a) A, E, B, D, C (b) A, B, E, D, C
(c) C, D, E, B, A (d) C, E, D, B, A

88. In the cumene to phenol preparation in presence of air, the intermediate is [NCERT : P L-202 | Jan. 25, 2023 (I)]

(a) O, CH_3 (b)

(c) (d) O–O–H

89. A compound 'X' is acidic and it is soluble in NaOH solution, but insoluble in $NaHCO_3$ solution. Compound 'X' also gives violet colour with neutral $FeCl_3$ solution. The compound 'X' is : [NCERT : P L-210 | July 29, 2022 (I)]

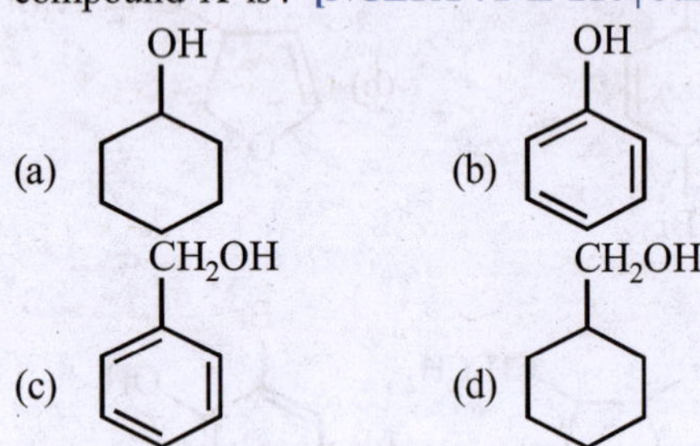

90. Identify the major product A and B for the below given reaction sequence. [NCERT : P L-210 | July 28, 2022 (I)]

(1) CH_3CHCH_3 (Cl), $AlCl_3$; (2) O_2; (3) $H^{\oplus}$, H_2O → P —$Na_2Cr_2O_7/H_2SO_4$→ A

P —Br_2 in CS_2→ (B)

(A) **(B)**

(a) OH and OH, Br

(b) O, O and OH, Br

(c) O, O and OH, Br, Br, Br

(d) OH and OH, Br, Br, Br

91. In the following reaction

Benzene —Oleum (I)→ SO_3H —NaOH (excess), H^+ (II)→ OH

The % yield for reaction I is 60% and that of reaction II is 50%. The overall yield of the complete reaction is ___% [nearest integer]

[NCERT : P L-211 | NV, July 27, 2022 (I)]

92. The difference in the reaction of phenol with bromine in chloroform and bromine in water medium is due to: [NCERT : P L-213 | July 26, 2022 (I)]

(a) Hyperconjugation in substrate
(b) Polarity of solvent
(c) Free radical formation
(d) Electromeric effect the substrate

93. Arrange the following in decreasing acidic strength. [NCERT : P L-206 | July 25, 2022 (II)]

(A) OH, NO_2 (B) OH, NO_2 (C) OH, OCH_3 (D) OH, OCH_3

(a) A > B > C > D (b) B > A > C > D
(c) D > C > A > B (d) D > C > B > A

94. A + CH_3CH_2OH ⟶ Benzene —(i) H_2SO_4 (oleum), Δ; (ii) NaOH; (iii) HCl→ B

[NCERT : P L-213 | June 30, 2022 (I)]

Consider the above reaction sequence. Identify the component A and component B:

(a) $A = C_6H_5NH_2$ B = Cl

(b) $A = C_6H_5N_2^+Cl^-$ B = OH

(c) A = chlorobenzene (benzene ring with Cl) $B = C_6H_5NH_2$

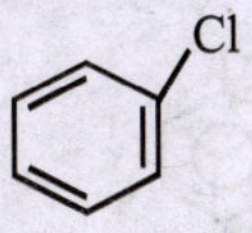

(d) A = phenol (benzene ring with OH) $B = C_6H_5N_2^+Cl^-$

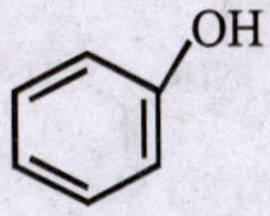

95. Given below are two statements.
[NCERT : P L-206 | June 29, 2022 (I)]

Statement I : Phenols are weakly acidic.

Statement II : Therefore they are freely soluble in NaOH solution and are weaker acids than alcohols and water.

Choose the **most appropriate** option:

(a) Both Statement I and Statement II are correct.
(b) Both Statement I and Statement II are incorrect.
(c) Statement I is correct but Statement II is incorrect.
(d) Statement I is incorrect but Statement II is correct.

96. Match List-I with List-II
[NCERT : P L-208 | June 27, 2010, 2022 (II)]

List-I	List-II
(A) 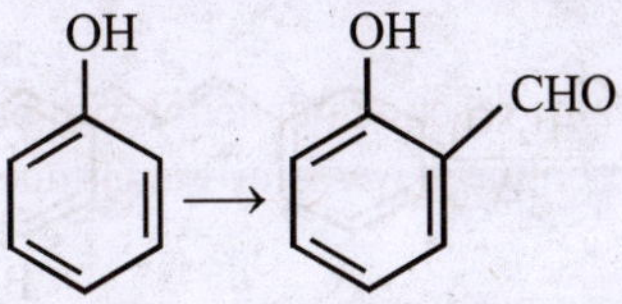 Phenol (OH) → salicylaldehyde (OH, CHO)	(I) Br_2 in CS_2
(B) 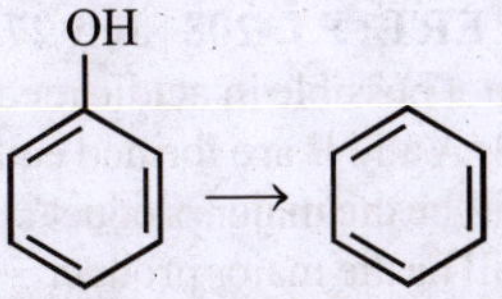 Phenol (OH) → benzene	(II) $Na_2Cr_2O_7/H_2SO_4$
(C) 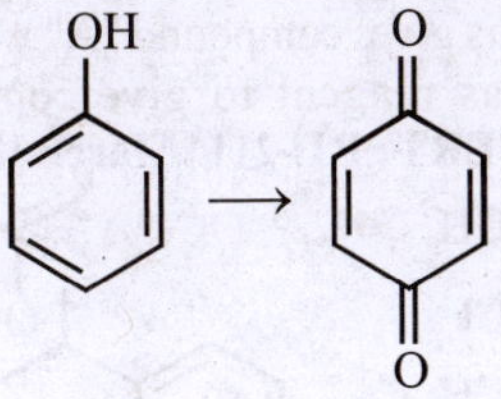 Phenol (OH) → benzoquinone (O, O)	(III) Zn
(D) 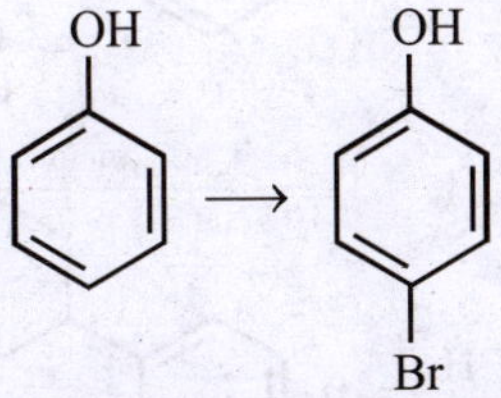 Phenol (OH) → p-bromophenol (OH, Br)	(IV) $CHCl_3/NaOH$

Choose the correct answer from the options given below:

(a) A-IV, B-III, C-II, D-I (b) A-IV, B-III, C-I, D-II
(c) A-II, B-III, C-I, D-IV (d) A-IV, B-II, C-III, D-I

97. Compound 'P' on nitration with dil. HNO_3 yields two isomers (A) and (B). These isomers can be separated by steam distillation. Isomers (A) and (B) show the intramolecular and intermolecular hydrogen bonding respectively. Compound (P) on reaction with conc. HNO_3 yields a yellow compound 'C', a strong acid. The number of oxygen atoms present in compound 'C' is ______.
[NCERT : P L-200 | NV, June 26, 2022 (I)]

98. Phenol on reaction with dilute nitric acid, gives two products. Which method will be most effective for large scale separation ? [NCERT : P L-198 | June 25, 2022 (I)]

(a) Chromatographic separation
(b) Fractional crystallisation
(c) Steam distillation
(d) Sublimation

99. The intermediate X, in the reaction

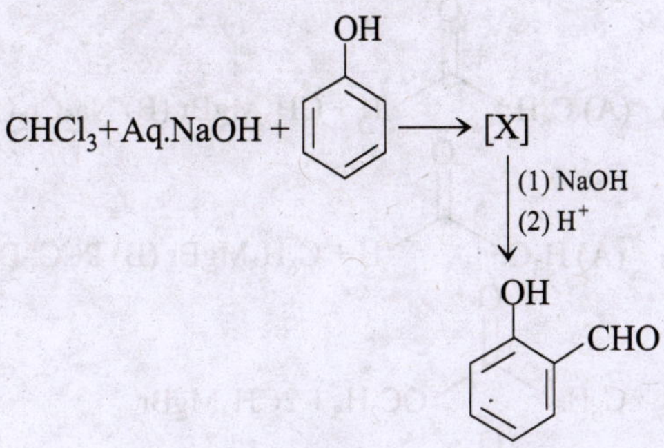

$CHCl_3$ + Aq.NaOH + phenol (OH) ⟶ [X] $\xrightarrow{(1)\ NaOH,\ (2)\ H^+}$ salicylaldehyde (OH, CHO) is :

[NCERT : P L-213 | June 25, 2022 (I)]

(a) 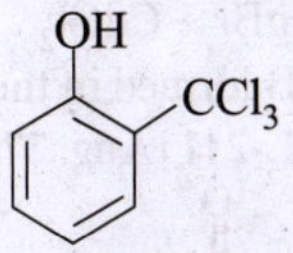 phenol with ortho CCl_3 (OH, CCl_3)
(b) 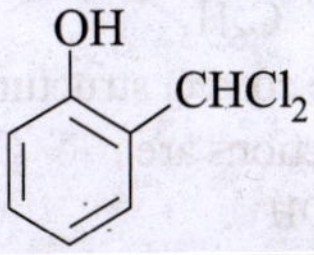 phenol with ortho $CHCl_2$ (OH, $CHCl_2$)
(c) 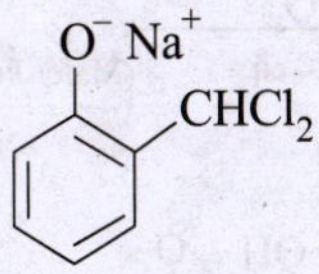 sodium phenoxide with ortho $CHCl_2$ (O^-Na^+, $CHCl_2$)
(d) 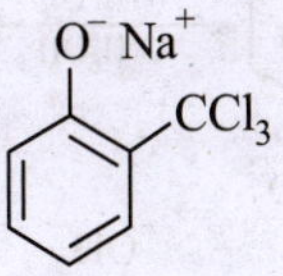 sodium phenoxide with ortho CCl_3 (O^-Na^+, CCl_3)

100. In the following reaction :

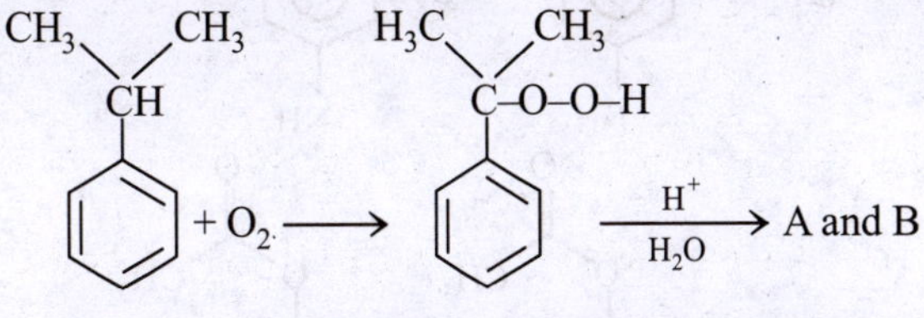

Cumene (CH_3, CH_3, CH attached to benzene) + O_2 ⟶ cumene hydroperoxide (H_3C, CH_3, C–O–O–H attached to benzene) $\xrightarrow[H_2O]{H^+}$ A and B

[NCERT : P L-202 | June 25, 2022 (I)]

The compounds A and B respectively are:

(a) 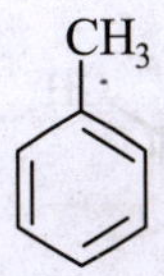 toluene (CH_3 on benzene), CH_3COOH

(b) 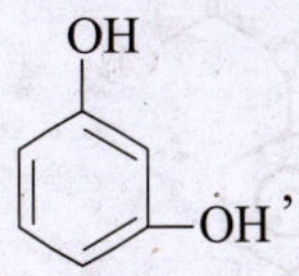 resorcinol (OH, OH on benzene), CH_3COOH

(c) [structure: phenol], CH_3COCH_3

(d) [structure: catechol, o-dihydroxybenzene], CH_3COCH_3

101. Which of the following conditions or reaction sequence will NOT give acetophenone as the major product ? **[NCERT : P L-200 | June 25, 2022 (II)]**

(a) (A) $C_6H_5CHO + CH_3MgBr$ (B) $Na_2Cr_2O_7, H^+$

(b) (A) $H_3CCHO + C_6H_5MgBr$ (B) PCC, DCM

(c) $C_6H_5COOC_2H_5 + 2CH_3MgBr$

(d) $C_6H_5COCl + CH_3MgBr + CdCl_2$

102. The correct structures of A and B formed in the following reactions are : **[NCERT : P L-211 | Aug. 27, 2021 (II)]**

p-nitrophenol $\xrightarrow[C_2H_5OH]{H_2/Pd}$ A $\xrightarrow[1.0\text{ eq.}]{(CH_3CO)_2O}$ B (Major Product)

(a) A : p-aminophenol, B : 2-acetyl-4-aminophenol

(b) A : 4-aminophenyl acetate, B : 4-acetamidophenyl acetate

(c) A : p-aminophenol, B : 4-aminophenyl acetate

(d) A : p-aminophenol, B : p-hydroxyacetanilide (4-HO–C_6H_4–NHCOCH$_3$)

103. Which one of the following phenols does not give colour when condensed with phthalic anhydride in presence of conc. H_2SO_4? **[Aug. 26, 2021 (II)]**

(a) [phenol]

(b) [p-cresol, 4-methylphenol]

(c) [resorcinol, 1,3-dihydroxybenzene]

(d) [catechol, 1,2-dihydroxybenzene]

104. $C_6H_5CH(OH)CH_2CH_3$ (1-phenylpropan-1-ol) $\xrightarrow[\Delta]{\text{Conc. } H_2SO_4}$ A ($C_6H_5CH=CHCH_3$, trans) + B ($C_6H_5CH=CHCH_3$, cis)

Consider the above reaction, and choose the correct statement : **[NCERT : P L-208 | July 27, 2021 (II)]**

(a) The reaction is not possible in acidic medium
(b) Both compounds A and B are formed equally
(c) Compound A will be the major product
(d) Compound B will be the major product

105. Reaction of Grignard reagent, C_2H_5MgBr with C_8H_8O followed by hydrolysis gives compound "A" which reacts instantly with Lucas reagent to give compound B, $C_{10}H_{13}Cl$. **[NCERT : P L-211 | March 18, 2021 (I)]**

(a) $C_6H_5C(Cl)(CH_3)CH_2CH_3$

(b) $C_6H_5CH(CH_2CH_3)CH_2Cl$

(c) $C_6H_5CH(CH_3)CH(Cl)CH_3$

(d) $C_6H_5CH(CH_3)CH_2CH_2Cl$

106. Identify A in the given reaction, **[NCERT : P L-211 | Feb. 26, 2021 (II)]**

[structure: tetrahydronaphthalene bearing aromatic OH, aromatic CH_2OH and ring HO] $\xrightarrow{SOCl_2}$ A (major product)

(a) (b) (c) (d)

107. A solution of phenol in chloroform when treated with aqueous NaOH gives compound P as a major product. The mass percentage of carbon in P is ______. (to the nearest integer) (Atomic mass: C = 12; H = 1; O = 16)

[NCERT : P L-211 | NV, Sep. 06, 2020 (II)]

108. The increasing order of boiling points of the following compounds is : **[NCERT : P L-203 | Sep. 05, 2020 (II)]**

I (CH_3), II (NO_2), III (NH_2), IV (OCH_3)

(a) I < III < IV < II (b) I < IV < III < II
(c) IV < I < II < III (d) III < I < II < IV

109. Consider the following reaction :

[NCERT : P L-213 | Sep. 03, 2020 (II)]

$\xrightarrow[\text{anhydride}]{\text{Chromic}}$ 'P'

The product 'P' gives positive ceric ammonium nitrate test. This is because of the presence of which of these –OH group(s) ?

(a) (ii) only (b) (iii) and (iv)
(c) (iv) only (d) (ii) and (iv)

110. The major product of the following reaction is :

[NCERT : P L-211 | Sep. 02, 2020 (II)]

$\xrightarrow[H_2SO_4]{\text{conc. } HNO_3 + \text{conc.}}$

(a) (b) (c) (d)

111. Arrange the following compounds in increasing order of C – OH bond length: methanol, phenol, *p*-ethoxyphenol

[NCERT : P L-198 | Jan. 08, 2020 (I)]

(a) methanol < *p*-ethoxyphenol < phenol
(b) phenol < methanol < *p*-ethoxyphenol
(c) phenol < *p*-ethoxyphenol < methanol
(d) methanol < phenol < *p*-ethoxyphenol

112. What will be the major product when *m*-cresol is reacted with propargyl bromide ($HC \equiv C — CH_2Br$) in presence of K_2CO_3 in acetone ? **[April 12, 2019 (II)]**

(a) (b) (c) (d)

113. The major product of the following reaction is :

[NCERT : P L-211 | Jan. 11, 2019 (I)]

$\xrightarrow[\text{(ii) alc.KOH}]{\text{(i) HBr}}$

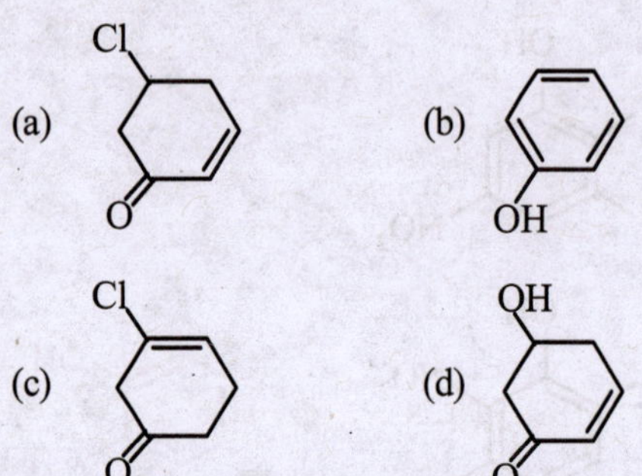

114. The major product of the following reaction is :

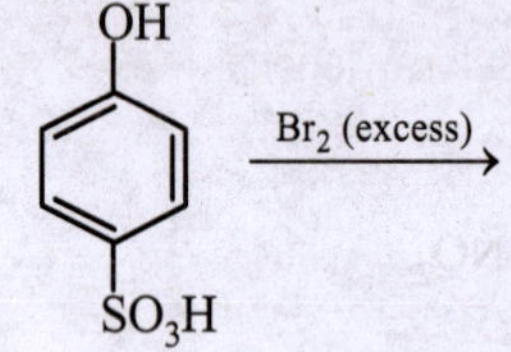

Br_2 (excess) →

[NCERT : P L-212 | Jan. 11, 2019 (I)]

(a) 2,4,6-tribromophenol (OH; Br, Br, Br)

(b) OH, Br, SO_3H

(c) OH, Br, Br, Br, Br

(d) OH, Br, Br, SO_3H

115. The major product of the following reaction is :
[NCERT : P L-213 | Jan. 11, 2019 (II)]

HO — (1) HCl, (2) $AlCl_3$(Anhyd.) →

(a) Cl

(b) Cl

(c) HO

(d) HO

116. The increasing order of the pK_a values of the following compounds is: [NCERT : P L-206 | Jan. 10, 2019 (I)]

OH	OH, NO_2	OH, NO_2	OH, OMe
A	B	C	D

(a) C < B < A < D (b) B < C < D < A

(c) D < A < C < B (d) B < C < A < D

117. The major product of the following reaction is:

CH_3, OH — (i) aq.NaOH, (ii) CH_3I →

[NCERT : P L-212 | Jan. 10, 2019 (II)]

(a) CH_3, OH, CH_3 (b) CH_3, OCH_3

(c) CH_3, OH, CH_3 (d) CH_3, OH, CH_3

118. The products formed in the reaction of cumene with O_2 followed by treatment with dil. HCl are:
[NCERT : P L-210 | Jan. 9, 2019 (II)]

(a) benzene and H_3C–CO–CH_3

(b) O, CH_3 and CH_3 – OH

(c) OH and H_3C–CO–CH_3

(d) OH and H_3C–CH(OH)–CH_3

119. Phenol on treatment with CO_2 in the presence of NaOH followed by acidification produces compound X as the major product. X on treatment with $(CH_3CO)_2O$ in the presence of catalytic amount of H_2SO_4 produces:
[NCERT : P L-208 | 2018; Same 2014]

(a) O, CH_3, CO_2H (b) O, CH_3, CO_2H

(c) 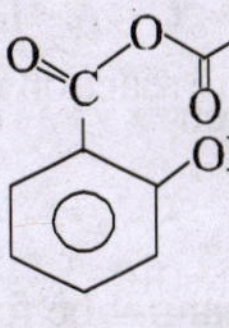 [structure] (d) 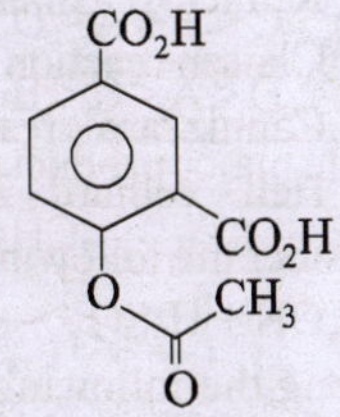 [structure]

120. Phenol reacts with methyl chloroformate in the presence of NaOH to form product A. A reacts with Br_2 to form product B. A and B are respectively : **[NCERT : P L-208 | 2018]**

(a) 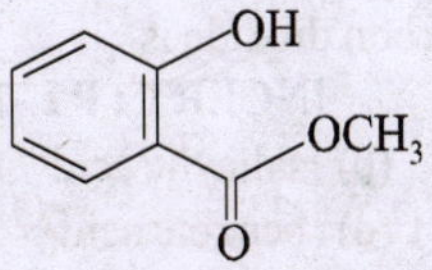 [structure] and [structure]

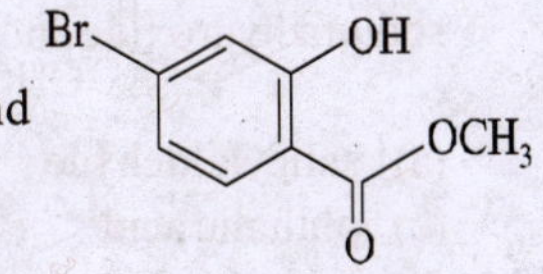

(b) 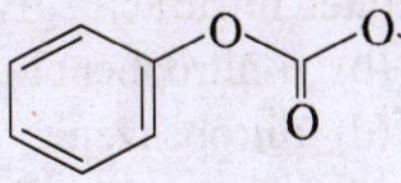 [structure] and [structure]

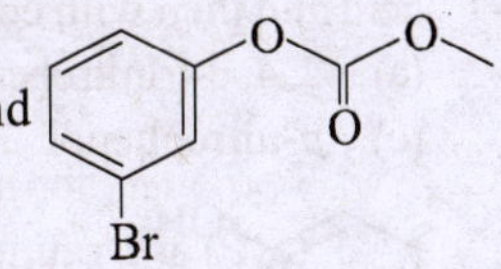

(c) 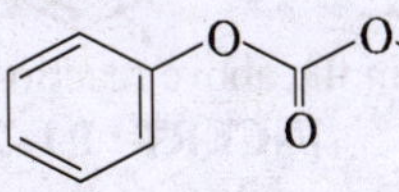 [structure] and [structure]

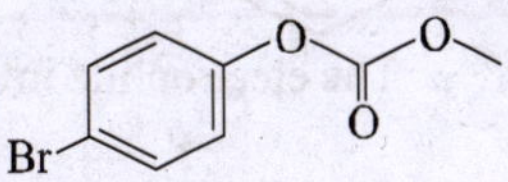

(d) 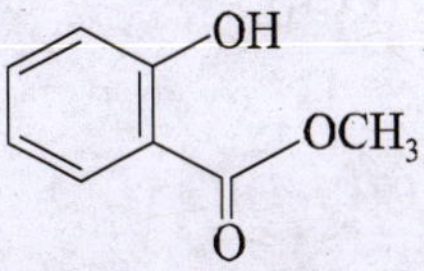 [structure] and [structure]

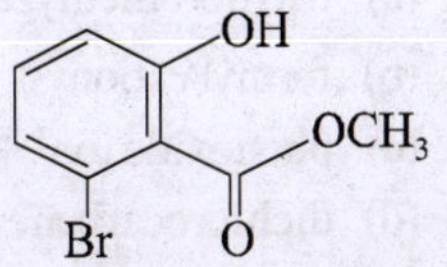

121. Which of the following, upon treatment with tert-BuONa followed by addition of bromine water, fails to decolourize the colour of bromine? **[2017]**

(a) 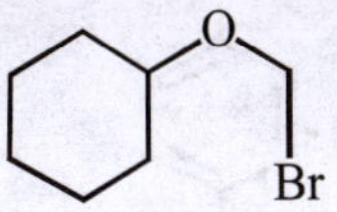 [structure] (b) 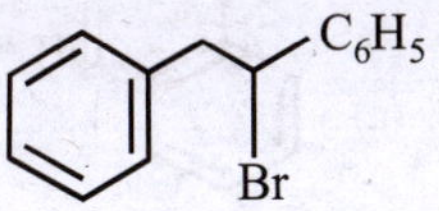 [structure]

(c) 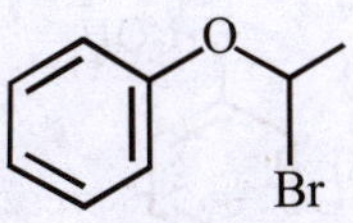 [structure] (d) 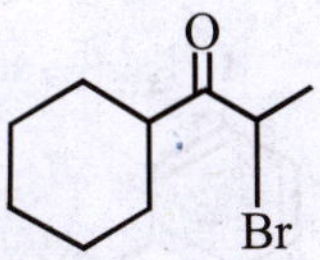 [structure]

122. The major product of the following reaction is : **[NCERT : P L-203 | Online April 8, 2017]**

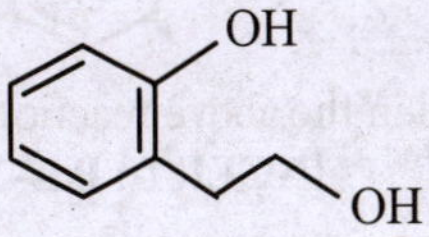

[structure] $\xrightarrow[\text{2. } CH_3I \text{ (1. eq.)}]{\text{1. } K_2CO_3}$

(a) 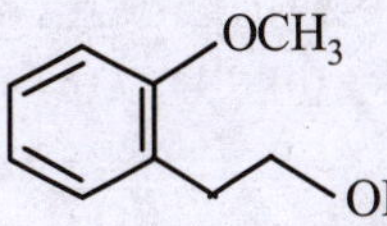 [structure] (b) 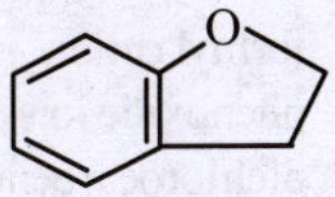 [structure]

(c) 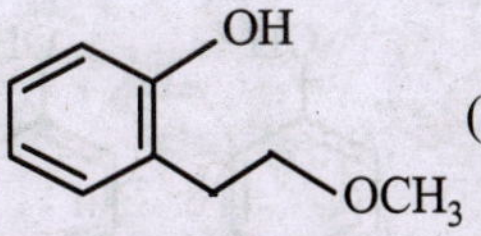 [structure] (d) 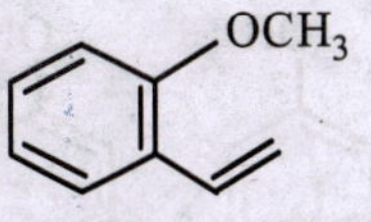 [structure]

123. The following reaction

[Phenol] + HCl + HCN $\xrightarrow[ZnCl_2]{\text{Anhyd.}}$ [o-hydroxybenzaldehyde]

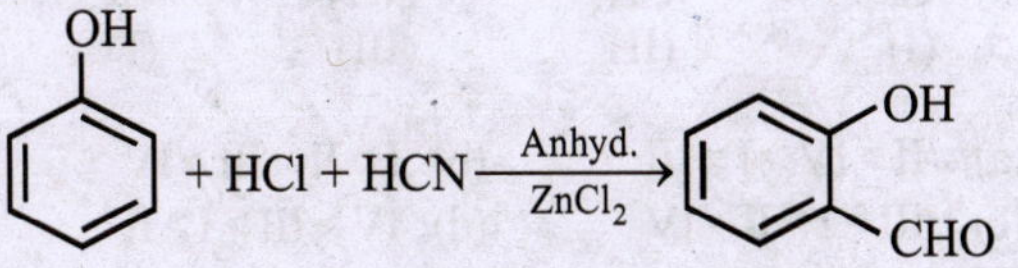

is known as: **[NCERT : P L-203 | Online April 11, 2014]**

(a) Perkin reaction
(b) Gatterman-Koch formylation
(c) Kolbe's reaction
(d) Gattermann reaction

124. Which one of the following substituents at *para*-position is most effective in stabilizing the phenoxide [structure] ion? **[NCERT : P L-207 | Online April 19, 2014]**

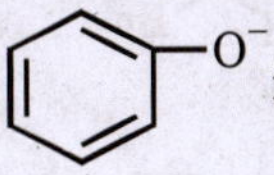

(a) $-CH_3$ (b) $-OCH_3$
(c) $-COCH_3$ (d) $-CH_2OH$

125. Which one of the following compounds will not be soluble in sodium bicarbonate? **[NCERT : P L-203 | Online April 19, 2014]**

(a) 2, 4, 6 - Trinitrophenol
(b) Benzoic acid
(c) *o* - Nitrophenol
(d) Benzene sulphonic acid

126. Sodium phenoxide when heated with CO_2 under pressure at 125 °C yields a product which on acetylation produces C **[NCERT : P L-213 | 2014]**

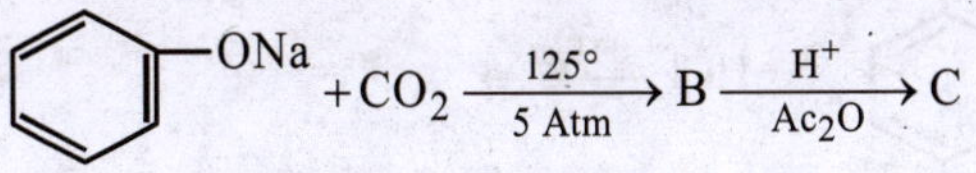

[Sodium phenoxide] $+ CO_2 \xrightarrow[\text{5 Atm}]{125°} B \xrightarrow[Ac_2O]{H^+} C$

The major product C would be

(a) 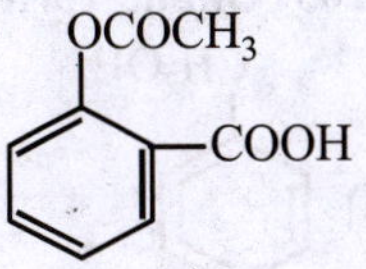 [structure] (b) 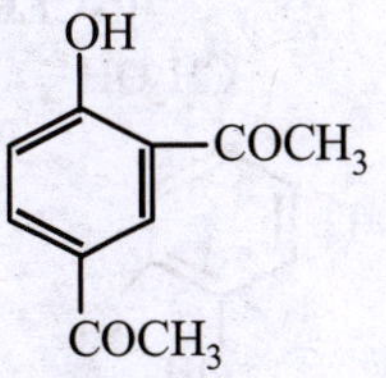 [structure]

(c) 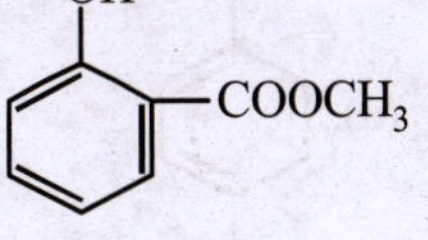 [structure] (d) 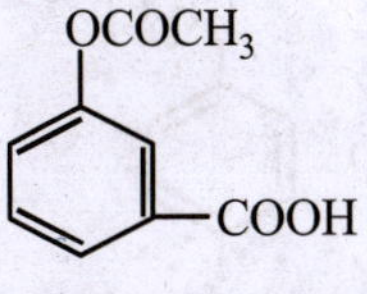 [structure]

127. Arrange the following compounds in order of decreasing acidity : **[NCERT : P L-206 | 2013; Same 2011]**

OH; OH; OH; OH;

Cl (I) CH_3 (II) NO_2 (III) OCH_3 (IV)

(a) II > IV > I > III (b) I > II > III > IV
(c) III > I > II > IV (d) IV > III > I > II

128. What is the structure of the major product when phenol is treated with bromine water ?
[NCERT : P L-212 | Online April 22, 2013]

(a) OH, Br, Br, Br (b) O, Br, Br, H, Br

(c) OH, Br (d) OH, Br

129. The reaction of phenol with benzoyl chloride to give phenyl benzoate is known as :
[NCERT : P L-203 | Online April 23, 2013]

(a) Claisen reaction
(b) Schotten-Baumann reaction
(c) Reimer-Tiemann reaction
(d) Gattermann-Koch reaction

130. The major product in the following reaction is :

CH_2OH / OH + HCl $\xrightarrow{\text{Heat}}$

[NCERT : P L-203 | Online April 25, 2013]

(a) CH_2OH, OH, Cl (b) CH_2OH, Cl

(c) CH_2Cl, Cl (d) CH_2Cl, OH

131. Phenol on heating with $CHCl_3$ and NaOH gives salicylaldehyde. The reaction is called:
[NCERT : P L-213 | Online April 25, 2013]

(a) Reimer - Tiemann reaction
(b) Claisen reaction
(c) Cannizzaro's reaction
(d) Hell - Volhard - Zelinsky reaction

132. Consider the following reaction :

$C_2H_5OH + H_2SO_4 \longrightarrow$ Product

Among the following, which one cannot be formed as a product under any conditions ?
[NCERT : P L-209 | 2011RS]

(a) Ethylene (b) Acetylene
(c) Diethyl ether (d) Ethylhydrogen sulphate

133. The major product obtained on interaction of phenol with sodium hydroxide and carbon dioxide is
[NCERT : P L-213 | 2009]

(a) salicylaldehyde (b) salicylic acid
(c) phthalic acid (d) benzoic acid

134. Phenol, when it reacts first with concentrated sulphuric acid and then with concentrated nitric acid, gives **[2008]**

(a) 2, 4, 6-trinitrobenzene (b) *o*-nitrophenol
(c) *p*-nitrophenol (d) nitrobenzene

135. OH + $CHCl_3$ + NaOH $\longrightarrow$ O^-Na^+, CHO

The electrophile involved in the above reaction is
[NCERT : P L-213 | 2006]

(a) trichloromethyl anion ($\overset{\ominus}{C}Cl_3$)
(b) formyl cation
(c) phenoxide ion
(d) dichlorocarbene

136. The structure of the compound that gives a tribromo derivative on treatment with bromine water is
[NCERT : P L-212 | 2006]

(a) CH_3, OH (b) CH_3, OH

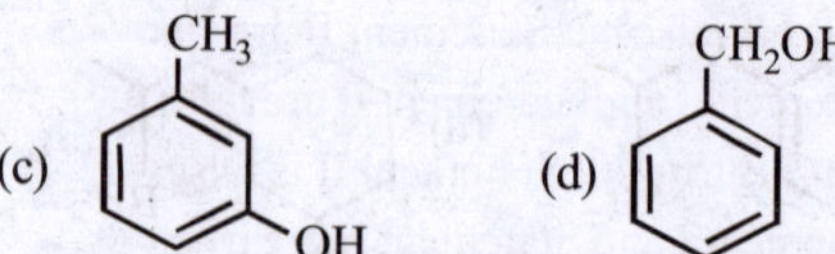

137. OH + $CHCl_3$ + NaOH $\longrightarrow$ O^-Na^+, CHO

The electrophile involved in the above reaction is
[NCERT : P L-213 | 2006]

(a) trichloromethyl anion ($\overset{\ominus}{C}Cl_3$)
(b) formyl cation
(c) phenoxide ion
(d) dichlorocarbene

138. Picric acid is: **[NCERT : P L-211 | 2002]**

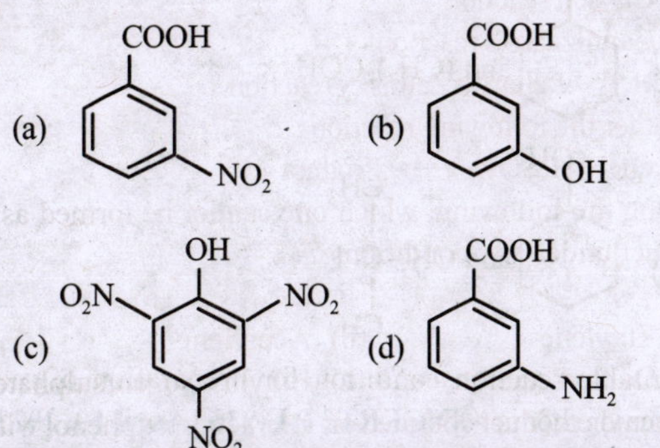

Topic 3 - Preparation and Properties of Ethers

139. Which one of the following reactions will not lead to the desired ether formation in major proportion?
(iso-Bu ⇒ isobutyl, sec-Bu ⇒ sec-butyl,
nPr ⇒ n-propyl, ${}^tBu$ ⇒ tert-butyl, Et ⇒ ethyl)
[NCERT : P L-215 | April 8, 2025 (II)]

(a) ${}^tBu\overset{\ominus}{O}\overset{\oplus}{Na} + EtBr \rightarrow {}^tBu\text{–}O\text{–}Et$

(b) Cyclohexyl–$\overset{\ominus}{O}\overset{\oplus}{Na}$ + $CH_3BR \longrightarrow$ Cyclohexyl–O–CH_3

(c) $\overset{\oplus}{Na}\overset{\ominus}{O}$–$C_6H_5$ + n–PrBr $\longrightarrow$ + n–Pr–O–C_6H_5

(d) iso-$Bu\overset{\ominus}{O}\overset{\oplus}{Na}$ + sec – BuBr → Sec–Bu–O–iso–Bu

140. Given below are two statements :
[NCERT : P L-217 | April 7, 2025 (I)]

Statement I : Dimethyl ether is completely soluble in water. However, diethyl ether is soluble in water to a very small extent.

Statement II : Sodium metal can be used to dry diethyl ether and not ethyl alcohol.

In the light of given statements, choose the *correct* answer from the options given below

(a) Statement I is false but Statement II are true
(b) Both Statement I and Statement II are false
(c) Statement I is true but Statement II is false
(d) Both Statement I and Statement II are true

141. Which one of the following, with HBr will give a phenol ?
[NCERT : P L-217 | Jan. 29, 2025 (II)]

(a) C_6H_5–CH_2–CH(OCH_3)(OCH_3)

(b) C_6H_5–O–CH_3

(c) C_6H_5–CH_2–OCH_3

(d) C_6H_5–CH_2–O–C(CH)(CH)–CH_3

142. $C_6H_5OC_2H_5 \xrightarrow{HNO_3, H_2SO_4} \underset{\text{major product}}{P} \xrightarrow{2Br_2, Fe} \underset{\text{major product}}{Q}$

The ratio of number of oxygen atoms to bromine atoms in the product Q is________ × 10^{-1}.
[NCERT : P L-219 | NV, April 6, 2024 (II)]

143. The major products formed:
[NCERT : P L-219, 220 | April 6, 2024 (II)]

$C_6H_5OCH_3 \xrightarrow{HNO_3, H_2SO_4}$ 'A' $\xrightarrow[Fe]{Br_2(\text{excess})}$ 'B'

A and B respectively are:

(a) 2-nitroanisole (OCH_3, NO_2) and 2,4-dibromo-6-nitroanisole (Br, OCH_3, NO_2, Br)

(b) 4-nitroanisole (OCH_3, NO_2) and 2,6-dibromo-4-nitroanisole (Br, OCH_3, Br, NO_2)

(c) 2-nitroanisole (OCH_3, NO_2) and 4-bromo-2-nitroanisole (OCH_3, NO_2, Br)

(d) 4-nitroanisole (OCH_3, NO_2) and 2-bromo-4-nitroanisole (OCH_3, Br, NO_2)

144. Which one of the following reactions is NOT possible?
[NCERT : P L-219, 220 | April 5, 2024 (II)]

(a) $C_6H_5OCH_3 \xrightarrow{HCl} C_6H_5OH$

(b) 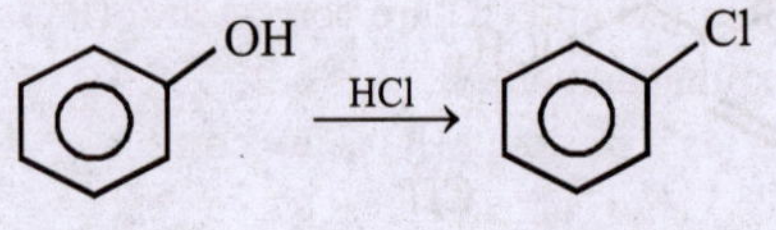 Phenol $\xrightarrow{HCl}$ Chlorobenzene

(c) 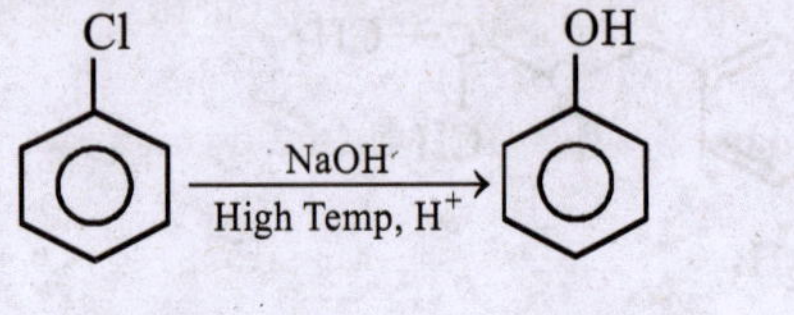 Chlorobenzene $\xrightarrow[\text{High Temp, } H^+]{NaOH}$ Phenol

(d) 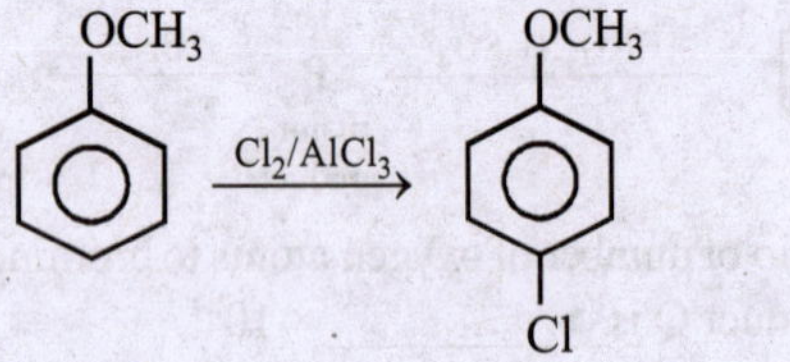 Anisole (OCH_3) $\xrightarrow{Cl_2/AlCl_3}$ p-Chloroanisole

145. Which among the following has highest boiling point? [NCERT : P L-217 | Feb. 1, 2024 (II)]

(a) $CH_3CH_2CH_2CH_2-OH$ (b) $CH_3CH_2CH_2CH_3$
(c) $CH_3CH_2CH_2CHO$ (d) $H_5C_2-O-C_2H_5$

146. The major product (P) in the following reaction is [NCERT : P L-217 | Jan. 29, 2024 (I)]

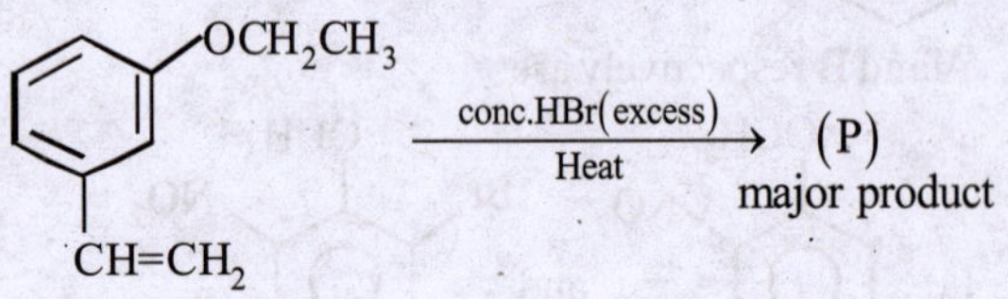

m-(OCH_2CH_3)-styrene ($CH=CH_2$) $\xrightarrow[\text{Heat}]{\text{conc.HBr(excess)}}$ (P) major product

(a) 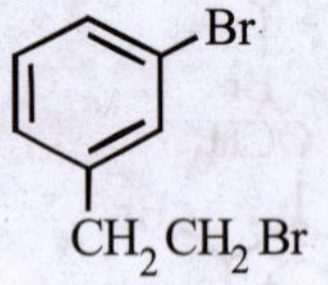 m-Br-C$_6$H$_4$-CH_2CH_2Br (b) m-OH-C$_6$H$_4$-CH_2CH_2Br
(c) m-Br-C$_6$H$_4$-$CH(Br)-CH_3$ (d) 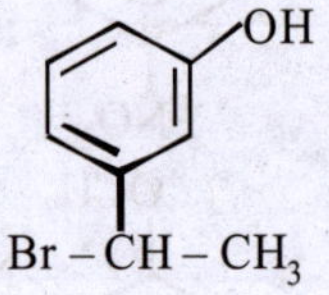 m-OH-C$_6$H$_4$-$CH(Br)-CH_3$

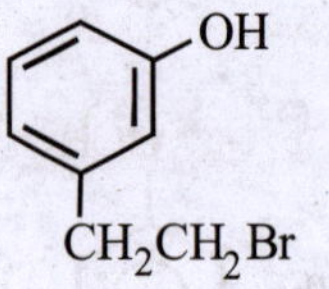

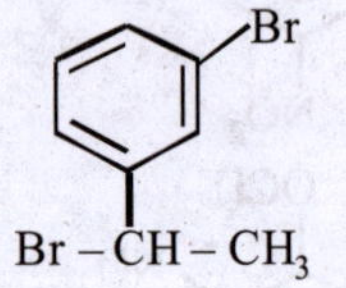

147. Major product formed in the following reaction is a mixture of: [NCERT : P L-217, 218 | Jan. 27, 2024 (II)]

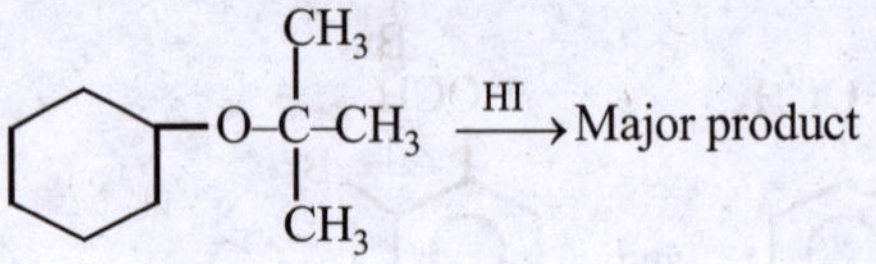

Cyclohexyl$-O-C(CH_3)_2-CH_3$ $\xrightarrow{HI}$ Major product

(a) 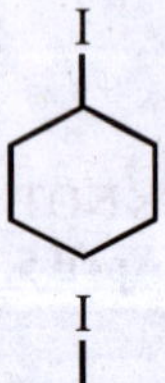 Iodocyclohexane and $(CH_3)_3Cl$
(b) 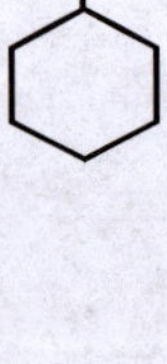 Iodocyclohexane and $(CH_3)_3COH$
(c) 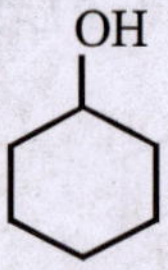 Cyclohexanol and $(CH_3)_3COH$
(d) 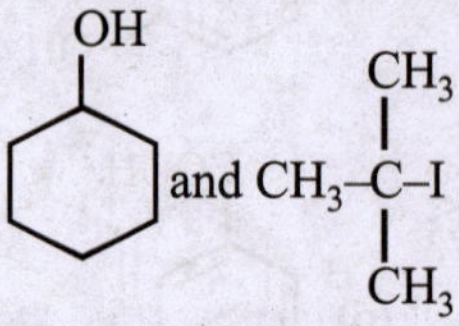 Cyclohexanol and $CH_3-C(CH_3)_2-I$

148. Suitable reaction condition for preparation of methyl phenyl ether is [NCERT : P L-215 | April 10, 2023 (I)]

(a) $Ph-Br, MeO^-Na^+$ (b) $PhO^-Na^+, MeOH$
(c) $PhO^-Na^+, MeBr$ (d) Benzene, MeBr

149. 'A' and 'B' formed in the following set of reactions are :

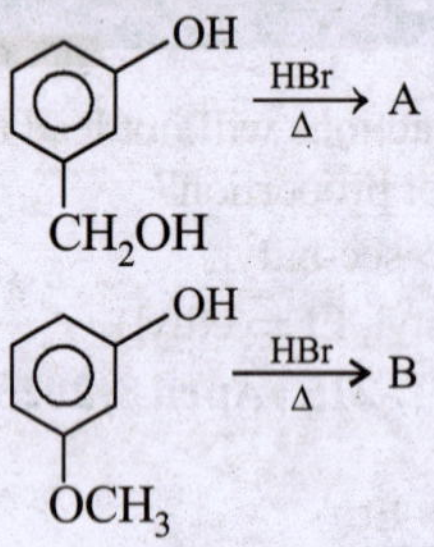

3-(CH_2OH)phenol $\xrightarrow[\Delta]{HBr}$ A

3-(OCH_3)phenol $\xrightarrow[\Delta]{HBr}$ B

[NCERT : P L-217 | Jan. 24, 2023 (I)]

(a) A = 3-Br-C$_6$H$_4$-CH_2OH, B = 1,3-dibromobenzene
(b) A = 3-(CH_2Br)phenol, B = 3-Br-C$_6$H$_4$-OCH_3
(c) A = 3-(CH_2Br)phenol, B = 3-bromophenol
(d) A = 3-(CH_2Br)phenol, B = 1,3-dihydroxybenzene

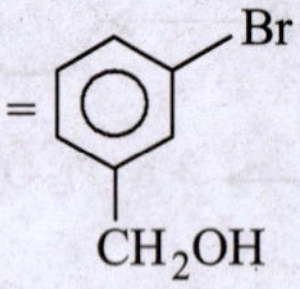

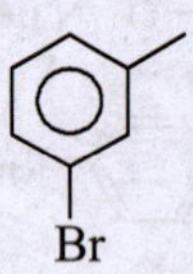

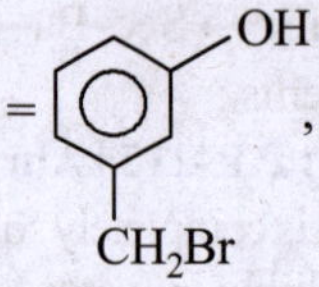

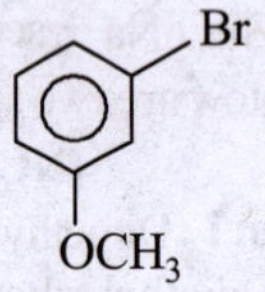

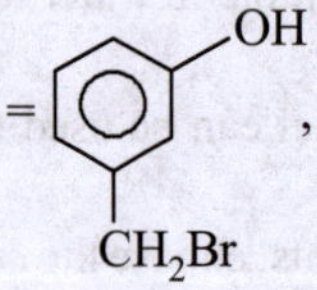

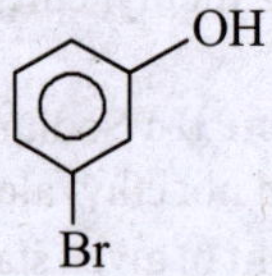

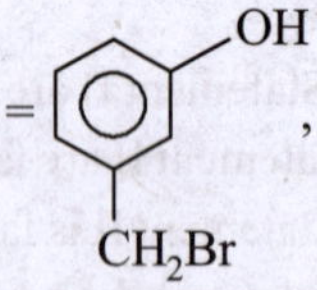

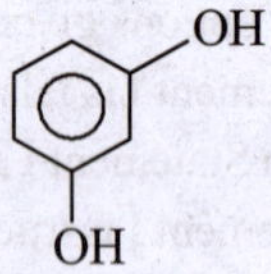

150. Compound I is heated with conc. HI to give a hydroxy compound A which is further heated with Zn dust to give compound B. Identify A and B. [NCERT : P L-217 | July 28, 2022 (II)]

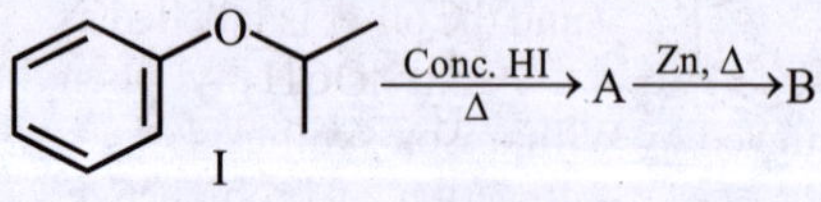

Isopropyl phenyl ether (I) $\xrightarrow[\Delta]{\text{Conc. HI}}$ A $\xrightarrow{Zn, \Delta}$ B

(a) A = propan-2-ol, B = ethane

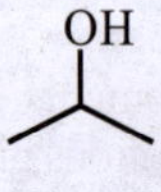

(b) A = phenol (OH on benzene ring), B = p-benzoquinone

(c) A = 2-isopropylphenol, B = isopropylbenzene

(d) A = phenol, B = benzene

151. The major product of the following reaction is [NCERT : P L-217 | July 26, 2022 (I)]

2-nitro... O_2N–C_6H_4–OCH_3 $\xrightarrow[\text{(ii) } CH_3CH_2OH]{\text{(i) Na/liq.}NH_3}$

(a) (b) (c) (d)

152. The major product formed in the following reaction, is

3-methylbutan-1-ol + 3,4-dihydro-2H-pyran $\xrightarrow{H^+}$

[NCERT : P L-217 | June 25, 2022 (II)]

(a) (b) (c) (d)

153. Given below are two statements, one is labelled as **Assertion (A)** and the other is labelled as **Reason (R)**.

Assertion (A) : Synthesis of ethyl phenyl ether may be achieved by Williamson synthesis.

Reason (R) : Reaction of bromobenzene with sodium ethoxide yields ethyl phenyl ether.

In the light of the above statements, choose the **most appropriate** answer from the options given below: [NCERT : P L-215 | Aug. 27, 2021 (I)]

(a) Both (A) and (R) are correct and (R) is the correct explanation of (A)

(b) (A) is correct but (R) is not correct

(c) (A) is not correct but (R) is correct

(d) Both (A) and (R) are correct but (R) is NOT the correct explanation of (A)

154. The major product [B] in the following reactions is : [NCERT : P L-217, 218 | Sep. 04, 2020 (II)]

$CH_3-CH_2-CH(CH_3)-CH_2-OCH_2-CH_3 \xrightarrow[\text{Heat}]{HI} [A]\text{ alcohol} \xrightarrow[\Delta]{H_2SO_4} [B]$

(a) $CH_2=CH_2$

(b) $CH_3-CH=C(CH_3)-CH_3$

(c) $CH_3-CH_2-C(CH_3)=CH_2$

(d) $CH_3-CH_2-CH=CH-CH_3$

155. The major aromatic product C in the following reaction sequence will be : [NCERT : P L-217 | Sep. 02, 2020 (I)]

2-methyl-2H-chromene $\xrightarrow[\Delta]{\text{HBr (excess)},}$ A $\xrightarrow[\text{(ii) } H^+]{\text{(i) KOH (Alc.)}}$ B $\xrightarrow[Zn/H_3O^+]{O_3}$ C

(a) 2-hydroxybenzaldehyde (OH, CHO)

(b) 2-hydroxyphenylacetic acid (OH, CH_2CO_2H)

(c) 2-bromophenylacetic acid (Br, CH_2CO_2H)

(d) 2-bromobenzaldehyde (Br, CHO)

156. An organic compound 'A' ($C_9H_{10}O$) when treated with conc. HI undergoes cleavage to yield compounds 'B' and 'C'. 'B' gives yellow precipitate with $AgNO_3$ whereas 'C' tautomerizes to 'D'. 'D' gives positive iodoform test. 'A' could be : [NCERT : P L-217 | Sep. 02, 2020 (II)]

(a) $C_6H_5-O-CH_2-CH=CH_2$

(b) $C_6H_5-O-CH=CH-CH_3$

(c) $C_6H_5-CH_2-O-CH=CH_2$

(d) $H_3C-C_6H_4-O-CH=CH_2$

157. Among the compounds A and B with molecular formula $C_9H_{18}O_3$, A is having higher boiling point than B. The possible structures of A and B are:
[NCERT : P L-217 | Jan. 08, 2020 (II)]

(a) A = HO ... OH ... HO
B = H_3CO ... OCH_3 ... OCH_3

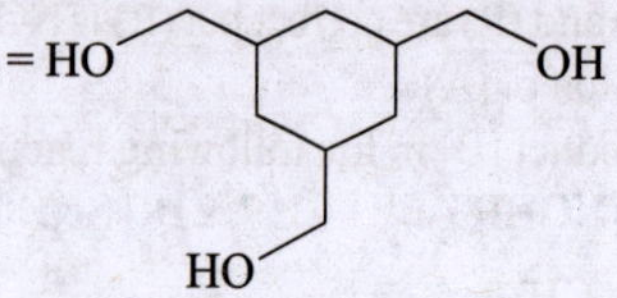

(b) A = H_3CO ... OCH_3 ... OCH_3
B = HO ... OH ... HO

(c) A = HO ... OH ... HO
B = HO ... OH ... OH

(d) A = H_3CO ... OCH_3 ... OCH_3
B = HO ... OH ... OH

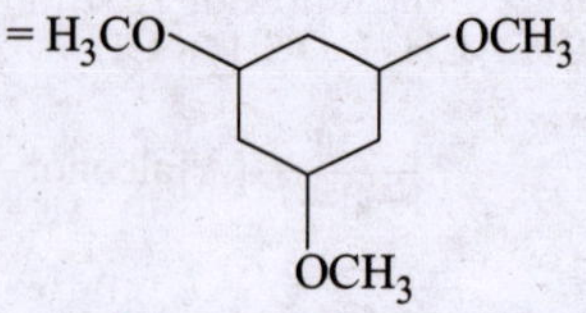

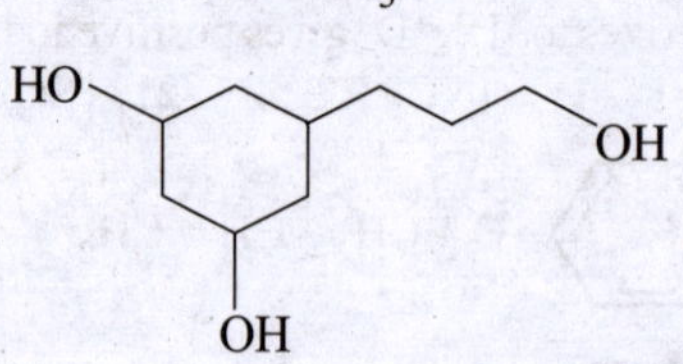

158. 1-Methylethylene oxide when treated with an excess of HBr produces: [NCERT : P L-217 | Jan. 07, 2020 (I)]

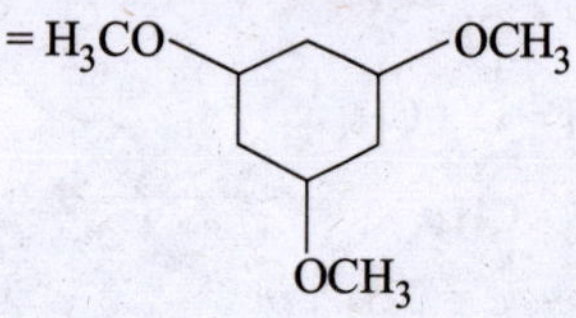

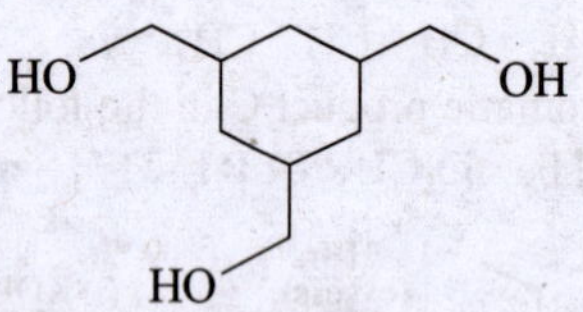

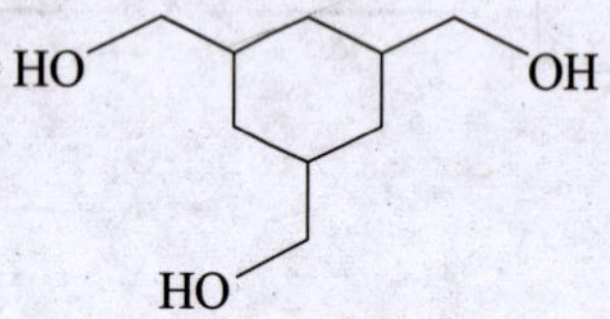

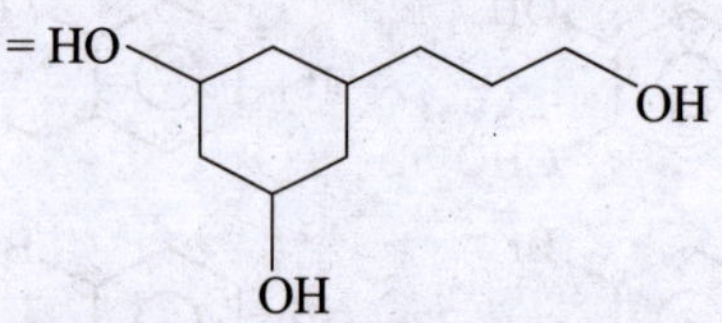

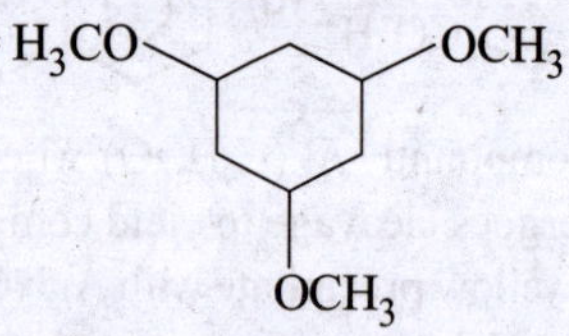

(a) 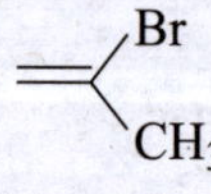 Br, CH_3
(b) 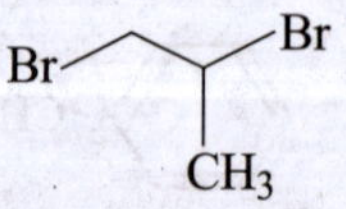 Br, Br, CH_3
(c) 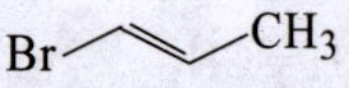 Br, CH_3
(d) 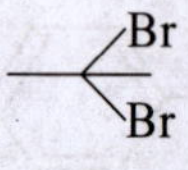 Br, Br

159. The major product of the following reaction is:
[NCERT : P L-217 | April 10, 2019 (I)]

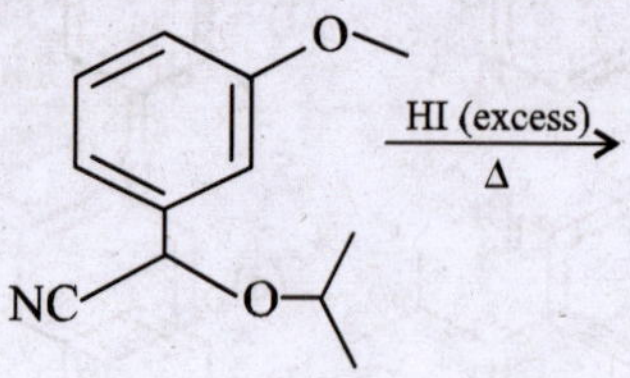

O, NC, O $\xrightarrow[\Delta]{\text{HI (excess)}}$

(a) 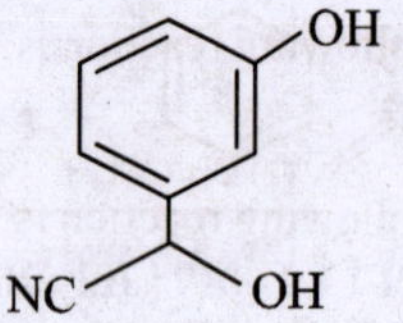 OH, NC, OH
(b) 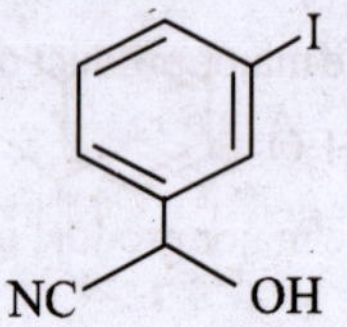 I, NC, OH
(c) 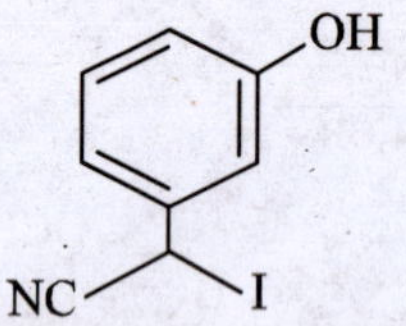 OH, NC, I
(d) 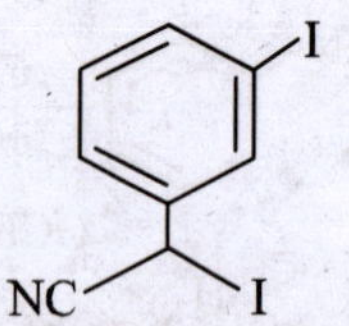 I, NC, I

160. The major product obtained in the given reaction is :
[NCERT : P L-220 | April 10, 2019 (II)]

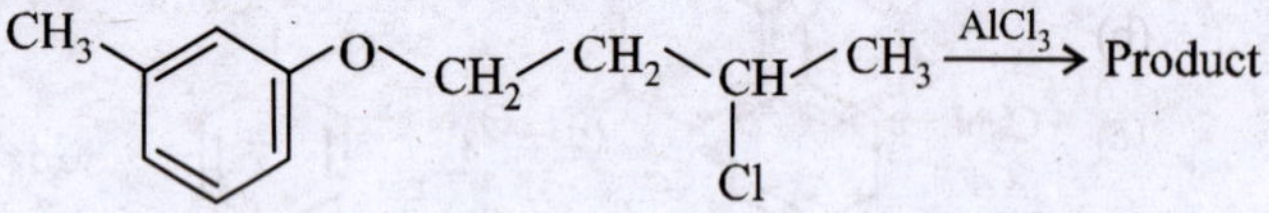

CH_3–C₆H₄–O–CH_2–CH_2–CH(Cl)–CH_3 $\xrightarrow{AlCl_3}$ Product

(a) 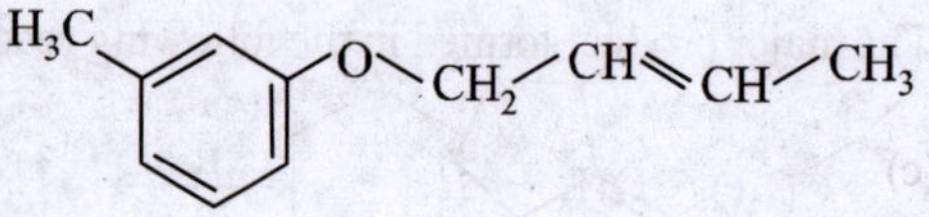 H_3C–C₆H₄–O–CH_2–CH=CH–CH_3
(b) 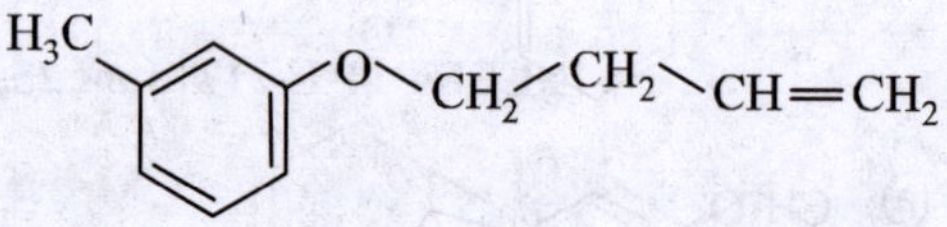 H_3C–C₆H₄–O–CH_2–CH_2–CH=CH_2
(c) 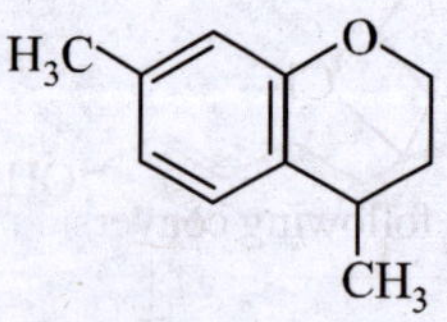 H_3C, O, CH_3
(d) 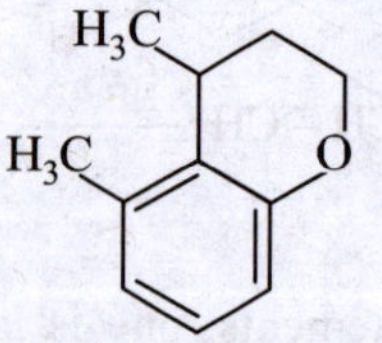 H_3C, H_3C, O

161. The major product of the following reaction is :

OCH_3, CH = CH_2 $\xrightarrow[\text{heat}]{\text{Conc HBr (excess)}}$

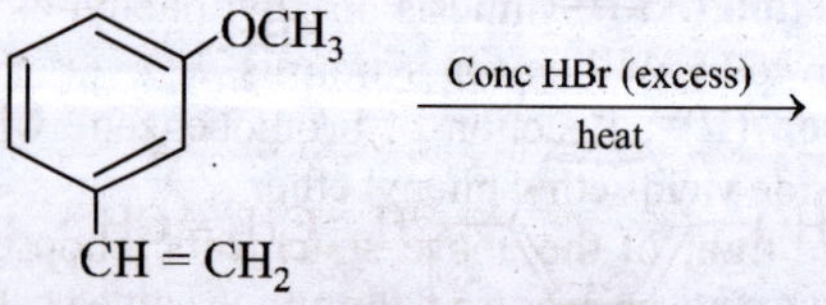

[NCERT : P L-217 | April 8, 2019 (I)]

(a) 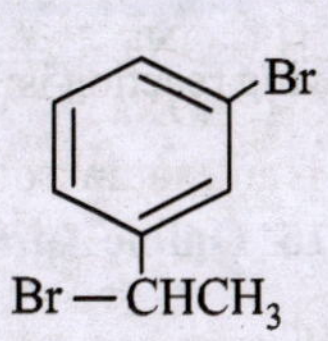 3-bromo-1-(1-bromoethyl)benzene: ring with Br; Br – CHCH$_3$

(b) 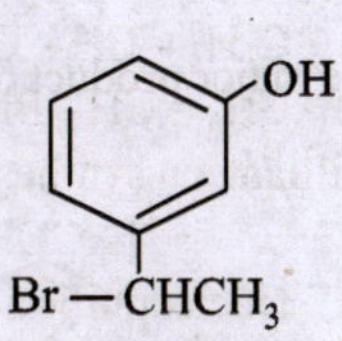 ring with OH; Br – CHCH$_3$

(c) 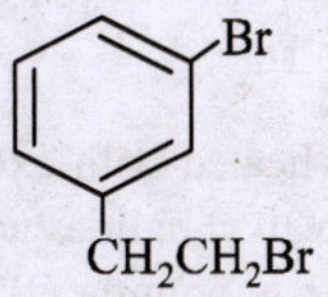 ring with Br; CH$_2$CH$_2$Br

(d) 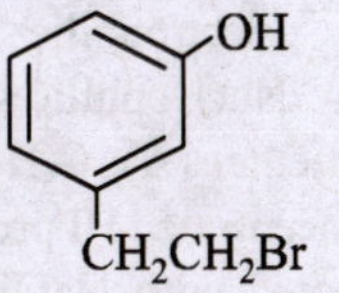 ring with OH; CH$_2$CH$_2$Br

162. The major product of the following reaction is:

CH_3O–(m-substituted benzene)–$CH_2CH_2CH=CH_2$ $\xrightarrow[(2)\ AlCl_3(anhyd.)]{(1)\ Cl_2/CCl_4}$

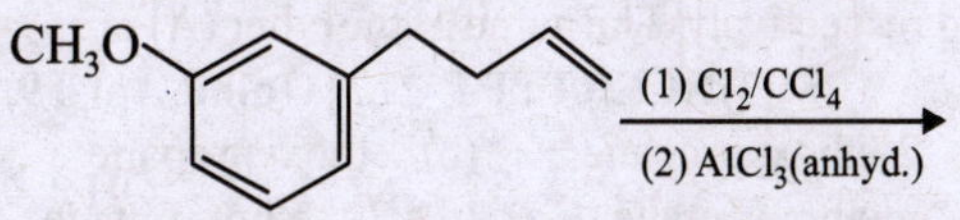

[NCERT : P L-220 | Jan. 12, 2019 (I)]

(a) 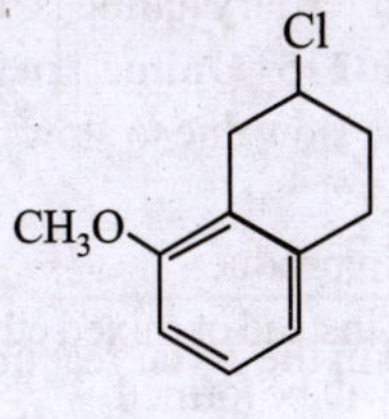 CH_3O, Cl (tetralin)

(b) 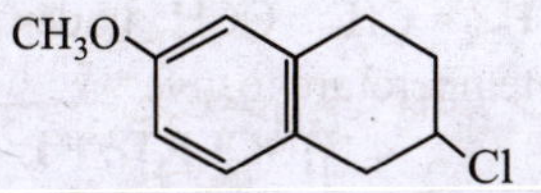 CH_3O, Cl (tetralin)

(c) 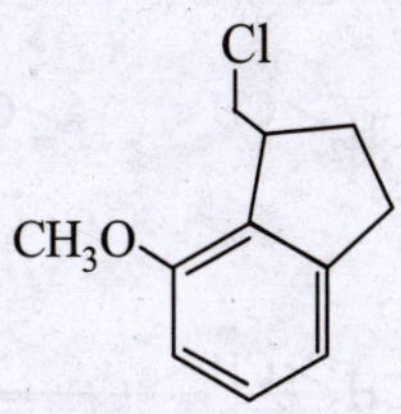 Cl, CH_3O (indane with CH_2Cl)

(d) 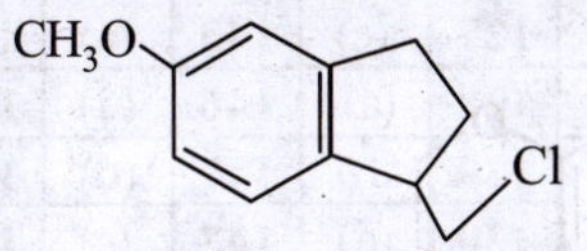 CH_3O, Cl (indane)

163. The major product in the following conversion is :

[NCERT : P L-217 | Jan. 12, 2019 (II)]

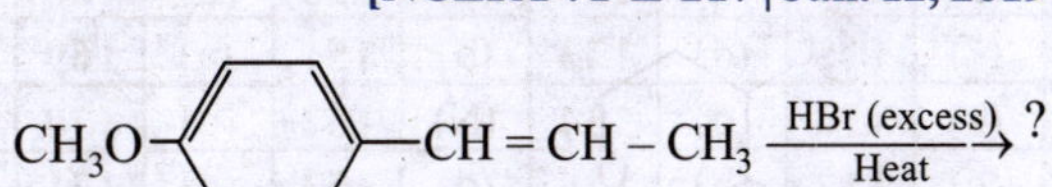

CH_3O–C_6H_4–$CH = CH – CH_3$ $\xrightarrow[Heat]{HBr\ (excess)}$?

(a) HO–C_6H_4–$CH_2 – CH – CH_3$ (Br on CH)

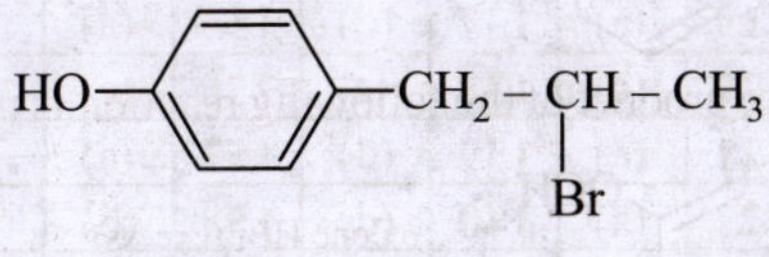

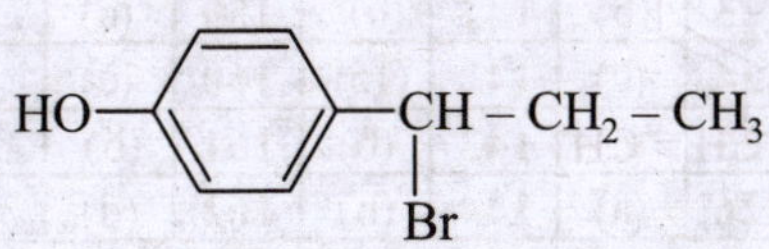

(b) HO–C_6H_4–$CH – CH_2 – CH_3$ (Br on CH)

(c) CH_3O–C_6H_4–$CH – CH_2 – CH_3$ (Br on CH)

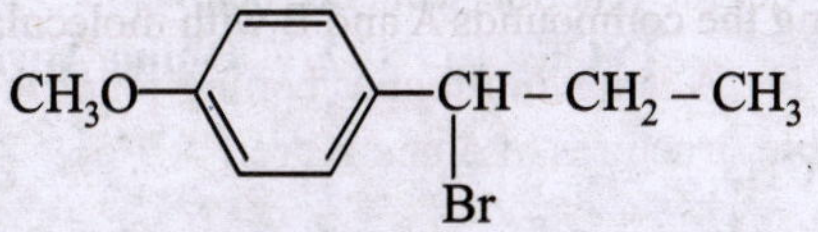

(d) 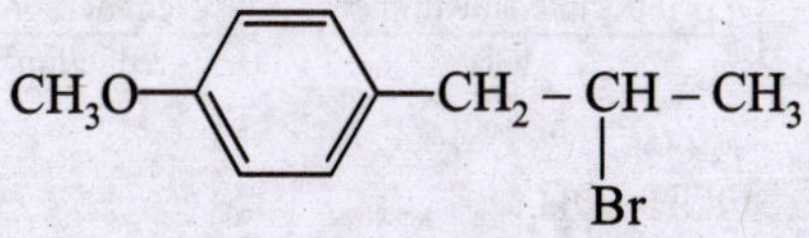 CH_3O–C_6H_4–$CH_2 – CH – CH_3$ (Br on CH)

164. The major product of the following reaction is:

[NCERT : P L-220 | Jan. 10, 2019 (I)]

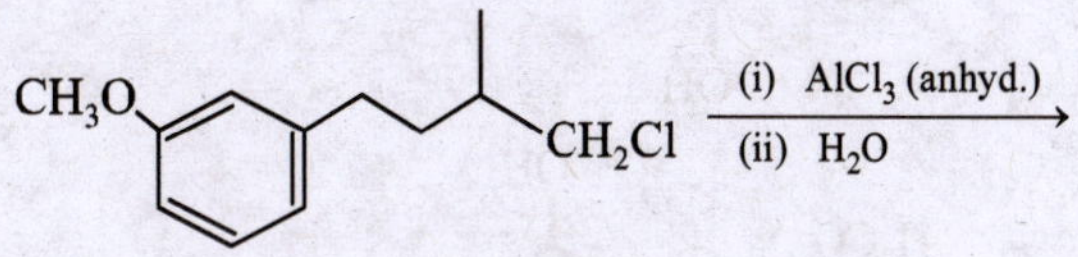

CH_3O–(m-substituted benzene)–$CH_2CH_2CH(CH_3)CH_2Cl$ $\xrightarrow[(ii)\ H_2O]{(i)\ AlCl_3\ (anhyd.)}$

(a) 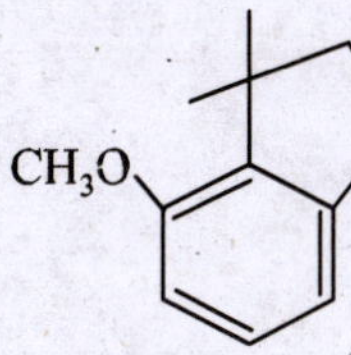 CH_3O (dimethylindane)

(b) 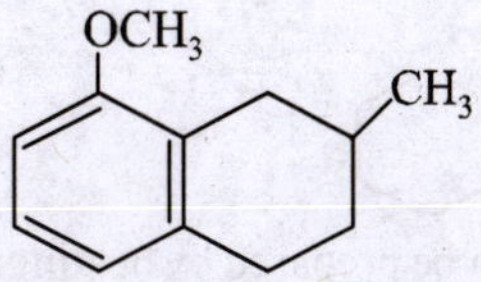 OCH_3, CH_3 (tetralin)

(c) 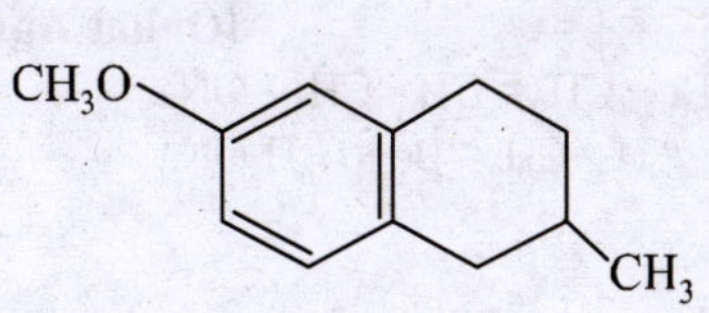 CH_3O, CH_3 (tetralin)

(d) 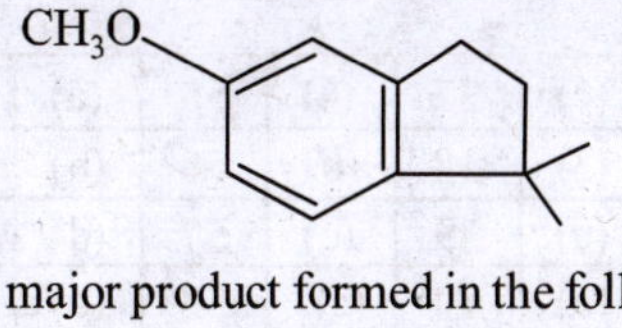 CH_3O (dimethylindane)

165. The major product formed in the following reaction is :

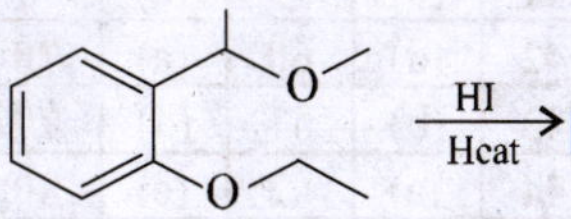

(o-substituted benzene with $CH(CH_3)OCH_3$ and OCH_2CH_3) $\xrightarrow[Heat]{HI}$ [NCERT : P L-217 | 2018]

(a) 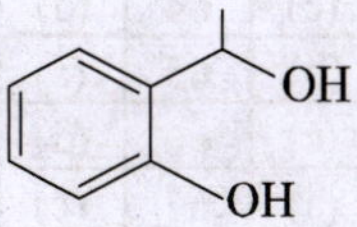 OH, OH

(b) 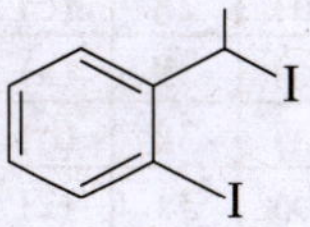 I, I

(c) 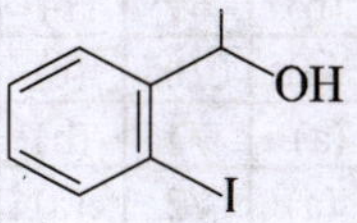 OH, I

(d) 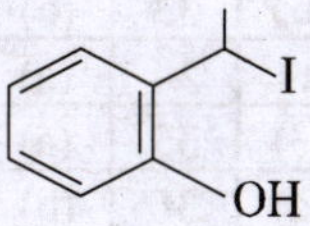 I, OH

166. Consider the reaction sequence below :

[NCERT : P L-217 | Online April 10, 2016]

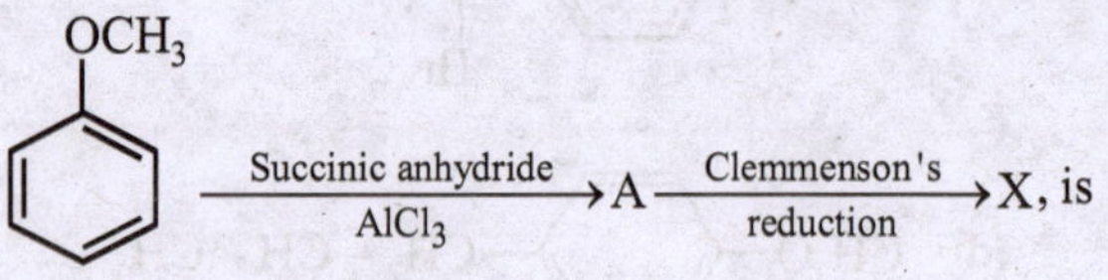

OCH_3 (benzene ring) $\xrightarrow[AlCl_3]{\text{Succinic anhydride}}$ A $\xrightarrow[\text{reduction}]{\text{Clemmenson's}}$ X, is

(a) OCH_3, OH, OH

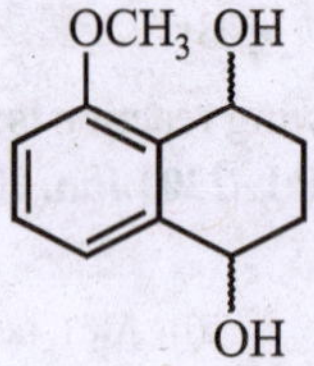

(b) 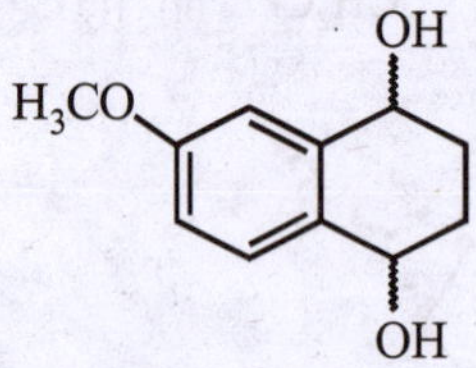 H_3CO, OH, OH

(c) 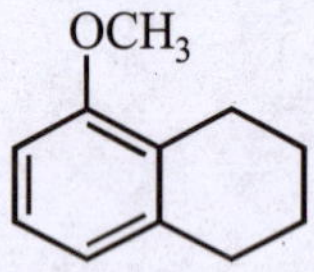 OCH_3

(d) 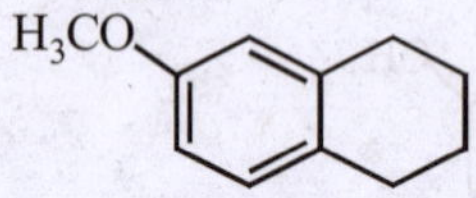 H_3CO

167. Allyl phenyl ether can be prepared by heating:

[Online April 9, 2014]

(a) $C_6H_5Br + CH_2=CH-CH_2-ONa$

(b) $CH_2=CH-CH_2-Br + C_6H_5ONa$

(c) $C_6H_5-CH=CH-Br + CH_3-ONa$

(d) $CH_2=CH-Br + C_6H_5-CH_2-ONa$

168. Williamson synthesis of ether is an example of:

[NCERT : P L-215 | Online April 19, 2014]

(a) Nucleophilic addition

(b) Electrophilic addition

(c) Electrophilic substitution

(d) Nucleophilic substitution

169. An ether (A), $C_5H_{12}O$, when heated with excess of hot concentrated HI produced two alkyl halides which when treated with NaOH yielded compounds (B) and (C). Oxidation of (B) and (C) gave propanone and ethanoic acid respectively. The name of the ether (A) is :

[NCERT : P L-217 | Online April 9, 2013]

(a) 2-ethoxypropane (b) ethoxypropane

(c) methoxybutane (d) 2-methoxybutane

170. In Williamson synthesis of mixed ether having a primary and a tertiary alkyl group, if tertiary halide is used, then :

[NCERT : P L-215 | Online April 22, 2013]

(a) Rate of reaction will be slow due to slow cleavage of carbon-halogen bond.

(b) Alkene will be the main product.

(c) Simple ether will form instead of mixed ether.

(d) Expected mixed ether will be formed.

171. HBr reacts with $CH_2 = CH - OCH_3$ under anhydrous conditions at room temperature to give

[NCERT : P L-217 | 2006]

(a) $BrCH_2-CH_2-OCH_3$

(b) $H_3C-CHBr-OCH_3$

(c) CH_3CHO and CH_3Br

(d) $BrCH_2CHO$ and CH_3OH

ANSWER KEYS

1	(c)	19	(c)	37	(4)	55	(a)	73	(d)	91	(30)	109	(a)	127	(c)	145	(a)	163	(b)
2	(d)	20	(3)	38	(b)	56	(b)	74	(c)	92	(b)	110	(c)	128	(a)	146	(d)	164	(d)
3	(c)	21	(c)	39	(c)	57	(d)	75	(c)	93	(a)	111	(c)	129	(b)	147	(d)	165	(d)
4	(a)	22	(a)	40	(c)	58	(b)	76	(b)	94	(b)	112	(a)	130	(d)	148	(c)	166	(d)
5	(d)	23	(3)	41	(d)	59	(a)	77	(a)	95	(c)	113	(b)	131	(a)	149	(d)	167	(b)
6	(11)	24	(b)	42	(d)	60	(a)	78	(d)	96	(a)	114	(a)	132	(b)	150	(d)	168	(d)
7	(a)	25	(c)	43	(b)	61	(d)	79	(d)	97	(7)	115	(d)	133	(b)	151	(a)	169	(a)
8	(d)	26	(33)	44	(a)	62	(c)	80	(b)	98	(c)	116	(d)	134	(N)	152	(d)	170	(b)
9	(c)	27	(1)	45	(d)	63	(N)	81	(a)	99	(c)	117	(b)	135	(d)	153	(b)	171	(b)
10	(a)	28	(c)	46	(c)	64	(d)	82	(a)	100	(c)	118	(c)	136	(c)	154	(b)		
11	(a)	29	(a)	47	(b)	65	(a)	83	(d)	101	(c)	119	(a)	137	(d)	155	(a)		
12	(a)	30	(b)	48	(c)	66	(d)	84	(a)	102	(d)	120	(c)	138	(c)	156	(c)		
13	(b)	31	(2)	49	(a)	67	(c)	85	(4)	103	(b)	121	(a)	139	(d)	157	(a)		
14	(a)	32	(d)	50	(b)	68	(d)	86	(b)	104	(c)	122	(a)	140	(d)	158	(b)		
15	(b)	33	(b)	51	(b)	69	(14)	87	(b)	105	(a)	123	(b)	141	(b)	159	(c)		
16	(a)	34	(a)	52	(d)	70	(c)	88	(d)	106	(d)	124	(c)	142	(15)	160	(c)		
17	(3)	35	(a)	53	(a)	71	(b)	89	(b)	107	(69)	125	(c)	143	(b)	161	(b)		
18	(a)	36	(b)	54	(a)	72	(a)	90	(b)	108	(b)	126	(a)	144	(b)	162	(d)		

9 Aldehydes, Ketones and Carboxylic Acids

TREND ANALYSIS
JEE MAIN (2025-2021)

	2025				2024				2023				2022				2021			
	Session II		Session I		Session II		Session I		Session II		Session I		Session II		Session I		Session II		Session I	
	MCQs	NVQs	MCQs	NVQs	MCQs	NVQs	MCQs	NVQs	MCQs	NVQs	MCQs	NVQs	MCQs	NVQs	MCQs	NVQs	MCQs	NVQs	MCQs	NVQs
Topic 1	–	–	2	–	2	1	4	–	–	–	–	–	3	–	2	–	1	–	1	–
Topic 2	9	1	7	2	8	1	5	2	10	4	13	3	6	1	11	4	7	1	7	2
Topic 3	5	–	1	–	6	4	3	–	4	–	1	–	2	1	6	–	2	–	2	1
Topic 4																				
Total	14	1	10	2	16	6	12	2	14	4	14	3	11	2	19	4	10	1	10	3

How to Code Qns for Better & Faster Revision →

Type 1: Able to solve (No code Req)
Type 2: Solved with struggle or by seeing solution **(Use code NR means Needs Revision or orange stickers)**
Type 3: Tricky/ Could not solve on my own. Need help from Teacher/ Friend. **(Use code TQ means Tricky Question or red stickers)**

You can either hand write the code in boxes or else paste coloured stickers given at end of book.

Topic 1 – Methods of Preparation of Carbonyl Compounds

1. Match List-I with List-II

[NCERT : P L-236 | Jan. 24, 2025 (II)]

List-I

(A) $RCN \xrightarrow[(ii)H_3O^+]{(i)SnCl_2, HCl} RCHO$

(B) $C_6H_5COCl \xrightarrow[Pd\text{-}BaSO_4]{H_2} C_6H_5CHO$

(C) $C_6H_5CH_3 \xrightarrow[(ii)\ H_3O^+]{(i)\ CrO_2Cl_2,\ CS_2} C_6H_5CHO$

(D) $C_6H_6 \xrightarrow[(ii)\ anhydrous\ AlCl_3/CuCl]{(i)\ CO,\ HCl} C_6H_5CHO$

List-II

(I) Etard reaction
(II) Gatterman – Koch reaction
(III) Rosenmund reduction
(IV) Stephen reaction

Choose the correct answer from the options given below :

(a) (A)-(IV), (B)-(III), (C)-(I), (D)-(II)
(b) (A)-(III), (B)-(IV), (C)-(II), (D)-(I)
(c) (A)-(I), (B)-(III), (C)-(II), (D)-(IV)
(d) (A)-(III), (B)-(IV), (C)-(I), (D)-(II)

2. Toluene (excess) $\xrightarrow[(ii)\ H_3O^+\ (iii)\ NaHSO_3]{(i)\ CrO_2Cl_2,\ CS_2}$ Filter $\longrightarrow$ Residue (A)

[NCERT : P L-233 | Jan. 22, 2025 (II)]

Residue (A) + HCl (dil.) $\rightarrow$ Compound (B)

Structure of residue (A) and compound (B) Formed respectively is:

(a) [A] Benzaldehyde (C_6H_5CHO) [B] Benzoic acid (C_6H_5COOH)

(b) [A] $C_6H_5CH(OCrOHCl_2)_2$ [B] C_6H_5CHO

(c) [A] C_6H_5CHO [B] C_6H_5COONa

(d) [A] $C_6H_5CH(OH)SO_3Na$ [B] $C_6H_5CH=O$

3. Identify major product "X" formed in the following reaction : [NCERT : P L-233 | April 9, 2024 (I)]

$$C_6H_6 \xrightarrow[\text{Anhydrous } AlCl_3/CuCl]{CO, HCl} \underset{\text{(Major product)}}{X}$$

(a) C_6H_5COCl (b) $C_6H_5COC_6H_5$

(c) C_6H_5CHO (d) $C_6H_5CH_2Cl$

4. Major product of the following reaction is [NCERT : P L-234 | April 9, 2024 (II)]

$$p\text{-}NC\text{-}C_6H_4\text{-}CO_2CH_3 \xrightarrow[\text{(ii) } H_3O^+]{\text{(i) } CH_3MgBr \text{ (excess)}}$$

(a) $p\text{-}NC\text{-}C_6H_4\text{-}C(OH)(CH_3)_2$ (b) $p\text{-}CH_3CO\text{-}C_6H_4\text{-}C(OH)(CH_3)_2$

(c) $p\text{-}CH_3CO\text{-}C_6H_4\text{-}CO_2CH_3$ (d) $p\text{-}NC\text{-}C_6H_4\text{-}COCH_3$

5. Two moles of benzaldehyde and one mole of acetone under alkaline conditions using aqueous NaOH after heating gives X as the major product. The number of π bonds in the product X is

[NCERT : P L-241 | NV, April 8, 2024 (II)]

6. Identify A and B in the following sequence of reaction [NCERT : P L-232, 233 | Feb. 1, 2024 (I)]

$$C_6H_5CH_3 \xrightarrow{Cl_2/h\nu} A \xrightarrow[373\text{ K}]{H_2O} B$$

(a) A = C_6H_5COCl B = C_6H_5CHO

(b) A = $C_6H_5CHCl_2$ B = C_6H_5CHO

(c) A = $C_6H_5CH_2Cl$ B = C_6H_5CHO

(d) A = $C_6H_5CHCl_2$ B = C_6H_5COOH

7. Identify the name reaction.

[NCERT : P L-232, 233 | Jan. 31, 2024 (II)]

$$C_6H_6 \xrightarrow[\text{Anhyd.}AlCl_3/CuCl]{CO, HCl} C_6H_5CHO$$

(a) Stephen reaction
(b) Etard reaction
(c) Gattermann-Koch reaction
(d) Rosenmund reduction

8. $$C_6H_5COCl \xrightarrow[Pd\text{ - }BaSO_4]{H_2} C_6H_5CHO$$

This reduction reaction is known as :

[NCERT : P L-232, 233 | Jan. 30, 2024 (I)]

(a) Rosenmund reduction
(b) Wolff - Kishner reduction
(c) Stephen reduction
(d) Etard reduction

9. In the given reactions, identify the reagent A and reagent B. [NCERT : P L-232 | Jan. 30, 2024 (I)]

Toluene: "A" + $(CH_3CO)_2O$, 273 – 283 K → [Intermediate] → H_3O^+, Δ → Benzaldehyde (CHO)

"B" + CS_2 → [Intermediate] → H_3O^+ → Benzaldehyde

(a) A - CrO_3 B - CrO_3
(b) A - CrO_3 B - CrO_2Cl_2
(c) A - CrO_2Cl_2 B - CrO_2Cl_2
(d) A - CrO_2Cl_2 B - CrO_3

10. Correct structure of γ-methylcyclohexane carbaldehyde is: [July 29, 2022 (II)]

(a) (structure) (b) (structure, CH_2CHO)
(c) (structure) (d) (structure, CH_2CHO)

11. $EtO-\overset{O}{\overset{||}{C}}-H_2C-$(cyclopentene)$-CH_2CH_2CN \xrightarrow[\text{(ii) } H_2O]{\text{(i) DiBAl-H}} ?$

Consider the above reaction and predict the major product. [NCERT : P L-236 | July 26, 2022 (II)]

(a) $OHC-H_2C-$(cyclopentene)$-CH_2CH_2CHO$
(b) $EtO-\overset{O}{\overset{||}{C}}-H_2C-$(cyclopentene)$-CH_2CH_2CHO$
(c) $EtO-\overset{O}{\overset{||}{C}}-H_2C-$(cyclopentene)$-CH_2CH_2COOH$
(d) $OHC-H_2C-$(cyclopentene)$-CH_2CH_2COOH$

12. Which one of the following reactions does **not** represent correct combination of substrate and product under the given conditions? [NCERT : P L-236 | July 25, 2022 (I)]

(a) p-methylbenzoyl chloride $\xrightarrow{H_2,\ Pd.BaSO_4}$ p-methylbenzaldehyde
(b) p-methylbenzonitrile (CN) $\xrightarrow[\text{(ii) } H_2O]{\text{(i) DIBAL-H}}$ p-methylbenzaldehyde
(c) p-methyl ethyl benzoate ($CO_2C_2H_5$) $\xrightarrow[\text{(ii) } H_2O]{\text{(i) AlH (iso Bu)}_2}$ p-methylbenzaldehyde
(d) p-methylbenzyl alcohol (OH) $\xrightarrow[\text{(ii) } H_2SO_4,\ H_2O]{\text{(i) } Na_2Cr_2O_7}$ p-methylbenzaldehyde

13. Which of the following reactions will yield benzaldehyde as a product? [NCERT : P L-236 | June 27, 2022 (I)]

(A) Benzoic acid (COOH) $\xrightarrow[\text{(ii) } H_2/Pd/BaSO_4]{\text{(i) } SOCl_2 \text{ Quinoline}}$
(B) Benzyl alcohol (CH_2OH) $\xrightarrow{CrO_3/H_2SO_4}$
(C) Methyl benzoate ($\overset{O}{\overset{||}{C}}-OCH_3$) $\xrightarrow[\text{(ii) PCC}]{\text{(i) } NaBH_4}$
(D) Toluene (CH_3) $\xrightarrow[\text{(ii) } H_3O^+,\ \Delta]{\text{(i) } CrO_3,\ (CH_3CO)_2O}$

(a) (B) and (C) (b) (C) and (D)
(c) (A) and (D) (d) (A) and (C)

14. Which of the following reagents/ reactions will convert 'A' to 'B'? [NCERT : P L-236 | June 24, 2022 (II)]

(A) 4-methyl-1-methylenecyclohexane (H_3C, $=CH_2$) ⟶ (B) 4-methylcyclohexanecarbaldehyde (H_3C, CHO)

(a) PCC oxidation
(b) Ozonolysis
(c) $BH_3, H_2O_2/^-OH$ followed by PCC oxidation
(d) HBr, hydrolysis followed by oxidation by $K_2Cr_2O_7$.

15. The major product of the following reaction is:

$$CH_3-\underset{}{\overset{CH_3}{\overset{|}{C}H}}-CH_2-CH_2-\overset{O}{\overset{||}{C}}-Cl \xrightarrow[\text{(iii) } NaNO_2,HCl \quad \text{(vi) } H_2O]{\text{(i) alcoholic } NH_3 \quad \text{(ii) NaOH, } Br_2} \text{Major product}$$

[Aug. 27, 2021 (I)]

(a) $CH_3-CH(CH_3)-CH(Br)-CH_2OH$

(b) $CH_3-CH(CH_3)-CH_2-CH_2-CH_2OH$

(c) $CH_3-CH(CH_3)-CH_2-CH_2OH$

(d) $CH_3-CH(CH_3)-CH_2-CH_2-Cl$

16. Which one of the following reactions will not form acetaldehyde? **[NCERT : P L-236 | Feb. 25, 2021 (I)]**

(a) $CH_3CH_2OH \xrightarrow{CrO_3-H_2SO_4}$

(b) $CH_2=CH_2+O_2 \xrightarrow[H_2O]{Pd(II)/Cu(II)}$

(c) $CH_3CH_2OH \xrightarrow[573K]{Cu}$

(d) $CH_3CN \xrightarrow[(ii) H_2O]{(i) DIBAL-H}$

17. The major products of the following reaction are: **[NCERT : P L-236 | Sep. 06, 2020 (I)]**

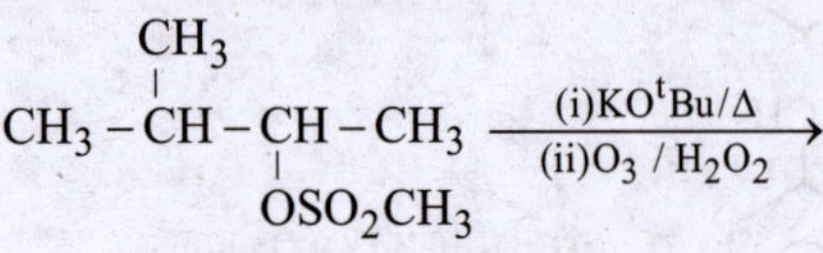

$CH_3-CH(CH_3)-CH(OSO_2CH_3)-CH_3 \xrightarrow[(ii) O_3/H_2O_2]{(i) KO^tBu/\Delta}$

(a) $(CH_3)_2C=O$ + CH_3CHO

(b) $(CH_3)_2C=O$ + CH_3COOH

(c) $(CH_3)_2CH-CHO$ + HCHO

(d) $(CH_3)_2CH-COOH$ + HCOOH

18. The correct match between Item - I (starting material) and Item - II (reagent) for the preparation of benzaldehyde is : **[NCERT : P L-236 | Sep. 06, 2020 (II)]**

Item - I	Item - II
(I) Benzene	(P) HCl and $SnCl_2$, H_3O^+
(II) Benzonitrile	(Q) H_2, Pd-$BaSO_4$, S and quinoline
(III) Benzoyl chloride	(R) CO, HCl and $AlCl_3$

(a) (I) - (Q), (II) - (R) and (III) - (P)
(b) (I) - (P), (II) - (Q) and (III) - (R)
(c) (I) - (R), (II) - (P) and (III) - (Q)
(d) (I) - (R), (II) - (Q) and (III) - (P)

19. In the following reaction A is: **[Jan. 09, 2020 (II)]**

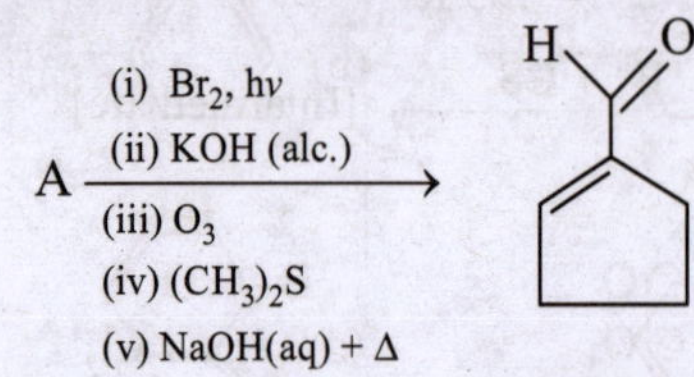

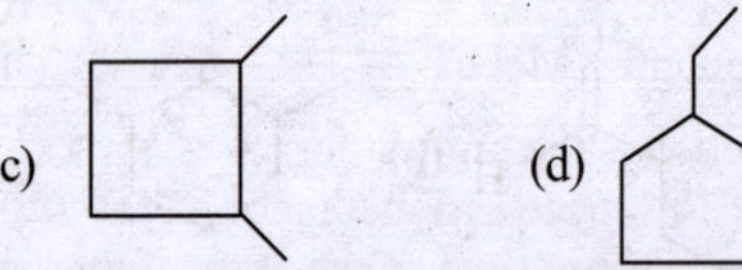

20. The major product(s) obtained in the following reaction is/are : **[NCERT : P L-237 | April 12, 2019 (I)]**

(i) KO^tBu
(ii) O_3/Me_2S

Br

(a) $OHC-CH_2-CH_2-CHO$ and OHC-CHO

(b) OHC–CH=CH–CHO

(c) $OHC-CH_2-CHO$

(d) OHC–CH₂–CH(O^tBu)–CH₂–CH₂–CHO

21. The major product of the following reaction is: **[NCERT : P L-236 | April 12, 2019 (I)]**

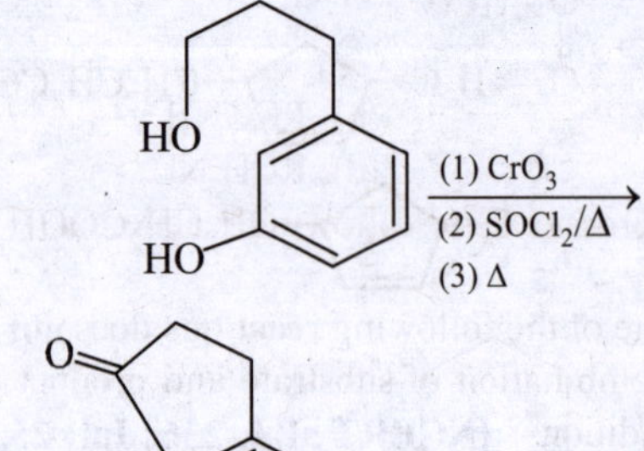

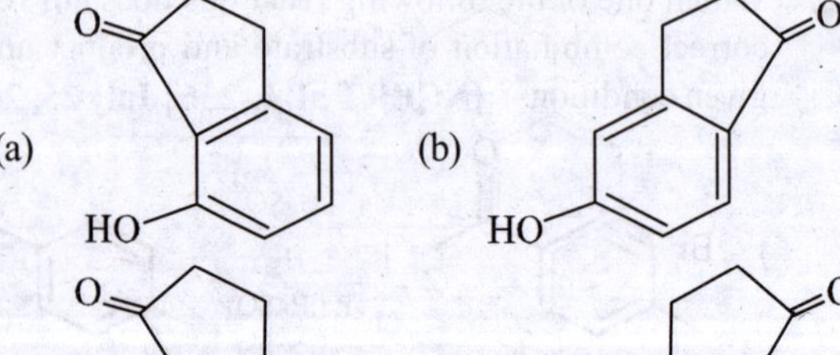

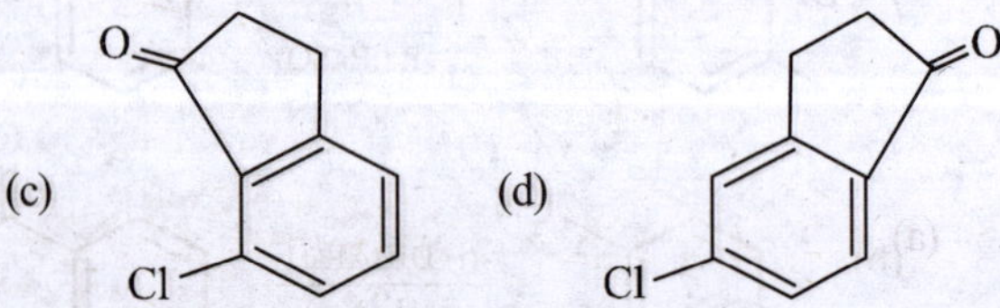

22. The major product of the following reaction is:
[NCERT : P L-236 | April 9, 2019 (I)]

1. PBr_3 2. KOH (alc.)

(a) (b) (c) (d)

23. In the following reaction
[NCERT : P L-238 | April 9, 2019 (II)]

$$\text{Carbonyl compound} + \text{MeOH} \overset{HCl}{\rightleftharpoons} \text{Acetal}$$

Rate of the reaction is the highest for:
(a) Acetone as substrate and methanol in excess.
(b) Propanal as substrate and methanol in stoichiometric amount.
(c) Propanal as substrate and methanol in excess.
(d) Acetone as substrate and methanol in stoichiometric amount.

24. In the following reaction
[NCERT : P L-238 | Jan. 12, 2019 (I)]

$$\text{Aldehyde} + \text{Alcohol} \xrightarrow{HCl} \text{Acetal}$$

Aldehyde	Alcohol
HCHO	^tBuOH
CH_3CHO	MeOH

The best combination is:
(a) CH_3CHO and ^tBuOH (b) HCHO and MeOH
(c) CH_3CHO and MeOH (d) HCHO and ^tBuOH

25. The major product of the following reaction is:

$$R - C \equiv N \xrightarrow[\text{(ii) } H_2O]{\text{(i) AlH(i–Bu)}_2}$$

[NCERT : P L-236 | Jan. 9, 2019 (I)]

(a) RCOOH (b) $RCONH_2$
(c) RCHO (d) RCH_2NH_2

26. The major product of the following reaction is:
[NCERT : P L-238 | Jan. 9, 2019 (I)]

(1) KOH (aqueous) (2) CrO_3/H^+ (3) H_2SO_4/Δ

(a) (b) (c) (d)

27. The major product of the following reaction is:

$AlCl_3, \Delta$

[Jan. 9, 2019 (II)]

(a) (b) (c) (d)

28. The total number of optically active compounds formed in the following reaction is:
[NCERT : P L-236 | Online April 15, 2018 (II)]

HBr

(a) Zero (b) Six (c) Four (d) Two

29. Which compound would give 5 - keto - 2 - methylhexanal upon ozonolysis ? **[NCERT : P L-231 | 2015]**

(a) (b)

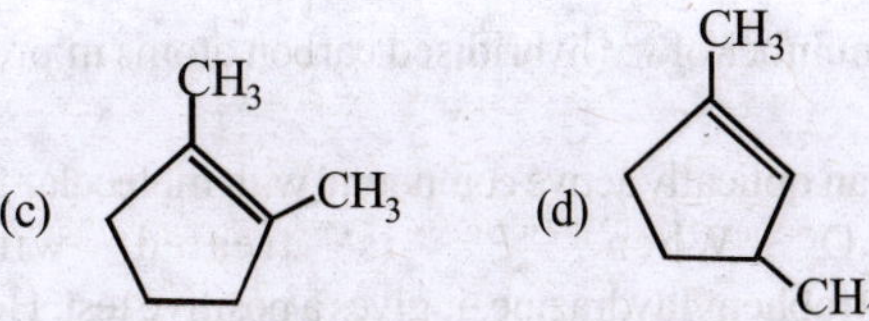

(c) (d)

30. Which of the following on heating with aqueous KOH, produces acetaldehyde? **[NCERT : P L-231 | 2009]**
(a) CH_3CH_2Cl (b) CH_2ClCH_2Cl
(c) CH_3CHCl_2 (d) CH_3COCl

Topic 2 - Properties of Carbonyl Compounds

31. When $CH_3-CO-CH_2-CH_2-CH_2-CH_2-CHO$ undergoes intramolecular aldol condensation, the major product formed is : **[NCERT : P L-241 | April 8, 2025 (II)]**

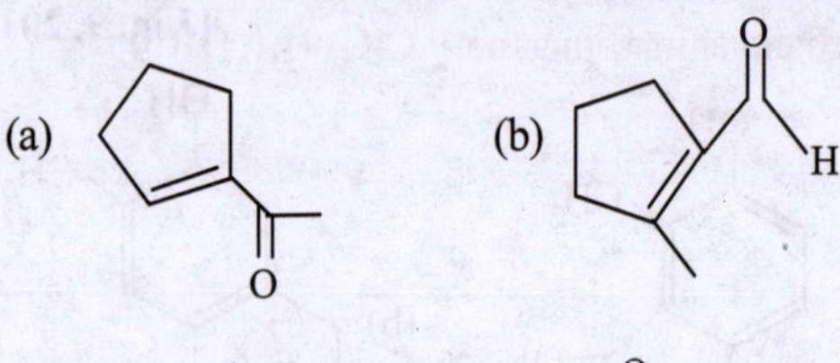

(c) (d)

32. 1,2-dibromocyclooctane $\xrightarrow[\text{(ii) NaNH}_2 \text{ (iii) Hg}^{2+}/\text{H}^+ \text{ (iv) Zn-Hg/H}^+]{\text{(i) KOH (atc.)}}$ P (Major product)

'P' is **[NCERT : P L-239 | April 8, 2025 (II)]**

(a) OH (b)

(c) (d)

33. Identify the structure of the final product (D) in the following sequence of the reactions : **[NCERT : P L-231 | NV, April 7, 2025 (II)]**

$$Ph-\overset{O}{\overset{\|}{C}}-CH_3 \xrightarrow[\Delta]{PCl_5} A \xrightarrow{3\text{ eq. } NaNH_2/NH_3} B \xrightarrow{\text{Acidify}} C \xrightarrow[2.\ H_2O_2/OH^-]{1.\ B_2H_6} D$$

Total number of sp^2 hybridised carbon atoms in product D is.

34. "P" is an optically active compound with molecular formula $C_6H_{12}O$. When "P" is treated with 2, 4-dinitrophenylhydrazine, it gives a positive test. However, in presence of Tollens reagent, "P" gives a negative test. Predict the structure of "P". **[NCERT : P L-239 | April 7, 2025 (II)]**

(a) $CH_3-\overset{O}{\overset{\|}{C}}-CH_2-CH_2-CH_2-CH_3$

(b) $CH_3-\overset{O}{\overset{\|}{C}}-\underset{CH_2-CH_3}{\underset{|}{CH}}-CH_3$

(c) $H-\overset{O}{\overset{\|}{C}}-CH_2-\underset{CH_2-CH_3}{\underset{|}{CH}}-CH_3$

(d) $CH_3-\overset{O}{\overset{\|}{C}}-CH_2-\underset{CH_3}{\underset{|}{CH}}-CH_3$

35. Which among the following compounds give yellow solid when reacted with NaOI/NaOH? **[NCERT : P L-240 | April 4, 2025 (II)]**

(A) $CH_3-\underset{}{\overset{OH}{\overset{|}{CH}}}-C_2H_5$

(B) $CH_3-CH_2-CH_2-OH$

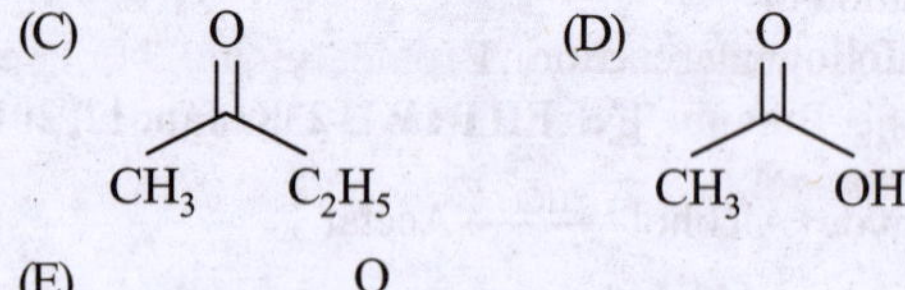

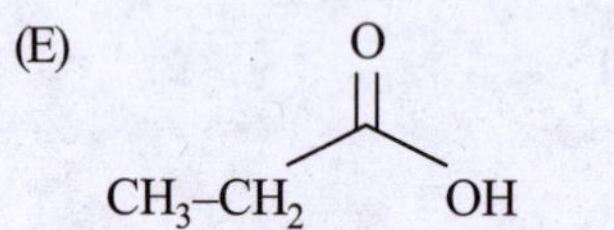

Choose the **correct** answer from the options given below :

(a) (B), (C) and (E) Only (b) (A) and (C) Only

(c) (C) and (D) Only (d) (A), (C) and (D) Only

36. An organic compound (X) with molecular formula C_3H_6O is not readily oxidised. On reduction it gives C_3H_8O (Y) which reacts with HBr to give a bromide (Z) which is converted to Grignard reagent. This Grinard reagent on reaction with (X) followed by hydrolysis give 2, 3-dimethylbutan-2-ol. **[NCERT : P L-237 | April 4, 2025 (I)]**

Compounds (X), (Y) and (Z) respectively are :

(a) $CH_3COCH_3, CH_3CH_2CH_2OH, CH_3CH(Br)CH_3$

(b) $CH_3COCH_3, CH_3CH(OH)CH_3, CH_3CH(Br)CH_3$

(c) $CH_3CH_2CHO, CH_3CH_2CH_2OH, CH_3CH_2CH_2Br$

(d) $CH_3CH_2CHO, CH_3CH=CH_2, CH_3CH(Br)CH_3$

37. Aldol condensation is a popular and classical method to prepare α, β-unsaturated carbonyl compounds. This reaction can be both intermolecular and intramolecular. Predict which one of the following is not a product of intramolecular aldol condensation ? **[NCERT : P L-241 | April 4, 2025 (I)]**

(a) [structure: bicyclic enone, octahydronaphthalenone] (b) [structure: indenone] (c) [structure: cyclohexenone] (d) [structure: 2-methylene-tetralone]

38. The major product (P) in the following reaction is :

$$PH-\underset{\underset{O}{\|}}{C}-\underset{\underset{O}{\|}}{C}-H \xrightarrow[\Delta]{KOH} \underset{\text{Major product}}{P}$$

[NCERT : P L-237 | April 3, 2025 (II)]

(a) $Ph-\underset{OH}{CH}-CH_2OH$ (b) $Ph-\underset{OH}{CH}-COO^-K^+$

(c) $Ph-\underset{\underset{O}{\|}}{C}-COO^-K^+$ (d) $Ph-\underset{\underset{O}{\|}}{C}-CH_2OH$

39. Number of molecules from below which cannot give iodoform reaction is :

Ethanol, Isopropyl alcohol, Bromoacetone, 2-Butanol, 2-Butanone, Butanal, 2-Pentanone, 3-Pentanone, Pentanal and 3-Pentanol [NCERT : P L-240 | April 3, 2025 (I)]

(a) 5 (b) 4

(c) 3 (d) 2

40. Given below are two statements : [NCERT : P L-234 | April 2, 2025 (I)]

Statement (I) : Vanillin [structure: 4-hydroxy-3-methoxybenzaldehyde, HO, OCH_3, CHO] will react with NaOH and also with Tollen's reagent.

Statement (II) : Vanillin [structure: 4-hydroxy-3-methoxybenzaldehyde, HO, OCH_3, CHO] will undergo self aldol condensation very easily. In the light of the above statements, choose the **most appropriate answer** from the options given below :

(a) **Statement I** is incorrect but **Statement II** is correct

(b) **Statement I** is correct but **Statement II** is incorrect

(c) Both **Statement I** and **Statement II** are incorrect

(d) Both **Statement I** and **Statement II** are correct

41. In the Claisen-Schmidt reaction to prepare, dibenzalacetone from 5.3 g benzaldehyde, a total of 3.51 g of product was obtained. The percentage yield in this reaction was ______ %. [NCERT : P L-241 | NV, Jan. 29, 2025 (II)]

42. [Reaction scheme: 2-(2-hydroxyethyl)cyclopentanol $\xrightarrow[\text{excess}]{CrO_3}$ P $\xrightarrow[H^+]{\text{one mole } HO\!-\!CH_2CH_2\!-\!OH}$ Q $\xrightarrow{\text{① } CH_3MgI \text{ in ether, ② } H_3O^{(+)}}$ R $\xrightarrow{\text{① } NaBH_4 \text{ reduction, ② } H_2O}$ 'S']

0.1 mole of compound 'S' will weigh ______ g. (Given molar mass in g mol^{-1} C:12, H:1, O:16)

[NCERT : P L-236 | NV, Jan. 29, 2025 (I)]

43. The product (P) formed in the following reaction is: [NCERT : P L-239 | Jan. 29, 2025 (I)]

[structure: ethyl 3-acetyl-5-formylbenzoate] $\xrightarrow[HCl]{Zn-Hg}$ Product (P)

(a) [structure: benzene with $HOCH_2CH_2-$, $-CH_2OH$ and $-CH_2OH$ groups]

(b) [structure: ethyl 3-acetyl-5-(hydroxymethyl)benzoate]

(c) [structure: ethyl 3-ethyl-5-methylbenzoate]

(d) [structure: ethyl 3-(2-hydroxyethyl)-5-(hydroxymethyl)benzoate]

44. Both acetaldehyde and acetone (individually) undergo which of the following reactions? [NCERT : P L-236 | Jan. 28, 2025 (I)]

(A) Iodoform Reaction

(B) Cannizaro Reaction

(C) Aldol condensation

(D) Tollen's Test

(E) Clemmensen Reduction

Choose the correct answer from the options given below :

(a) A, B and D only (b) A, C and E only
(c) C and E only (d) B, C and D only

45. The product (A) formed in the following reaction sequence is : **[NCERT : P L-234, 236 | Jan. 24, 2025 (I)]**

$$CH_3-C\equiv CH \xrightarrow[\text{(iii) } H_2/Ni]{\text{(i) } Hg^{2+}, H_2SO_4 \text{ (ii) HCN}} \underset{\text{Product}}{A}$$

(a) $CH_3-\overset{NH_2}{\underset{CH_3}{C}}-CH_2-OH$

(b) $CH_3-\overset{OH}{\underset{CH_3}{C}}-CH_2-NH_2$

(c) $CH_3-CH_2-\overset{NH_2}{CH}-CH_2-OH$

(d) $CH_3-CH_2-\overset{OH}{CH}-CH_2-NH_2$

46. Aman has been asked to synthesise the molecule (1-cyclopentenyl)–$\overset{O}{\overset{\|}{C}}-CH_3$ (x). He thought of preparing the molecule using an aldol condensation reaction. He found a few cyclic alkenes in his laboratory. He thought of performing ozonolysis reaction on alkene to produce a dicarbonyl compound followed by aldol reaction to prepare "x". Predict the suitable alkene that can lead to the formation of "x". **[NCERT : P L-241 | Jan. 24, 2025 (I)]**

(a) 1-methylcyclohexene (CH_3 on ring double-bond carbon)
(b) 3-methylcyclohexene (CH_3)
(c) methylenecyclohexane (CH_2)
(d) 4-methylcyclohexene (CH_3)

47. Which of the following arrangements with respect to their reactivity in nucleophilic addition reaction is correct? **[NCERT : P L-236 | Jan. 24, 2025 (I)]**

(a) benzaldehyde < acetophenone < p-nitrobenzaldehyde < p-tolualdehyde
(b) acetophenone < benzaldehyde < p-tolualdehyde < p-nitrobenzaldehyde
(c) acetophenone < p-tolualdehyde < benzaldehyde < p-nitrobenzaldehyde
(d) p-nitrobenzaldehyde < benzaldehyde < p-tolualdehyde < acetophenone

48. Given below are two statements: **[NCERT : P L-236 | Jan. 23, 2025 (II)]**

Consider the following reaction

$$R-\overset{O}{\overset{\|}{C}}-R + H_2O \underset{}{\overset{K}{\rightleftharpoons}} R-\overset{OH\ \ OH}{C}-R$$

Statement (I): In the case of formaldehyde ($H-\overset{O}{\overset{\|}{C}}-H$), K is about 2280, due to small substituents, hydration is faster.

Statement (II) : In the case of trichloro acetaldehyde ($H-\overset{O}{\overset{\|}{C}}-CCl_3$), K is about 2000 due to – I effect of – Cl.

In the light of the above statements, choose the correct answer from the options given below:

(a) Statement I true but Statement II is false
(b) Both Statement I and Statement II are true
(c) Statement I is false but Statement II is true
(d) Both Statement I and Statement II are false

49. The major product of the following reaction is : **[NCERT : P L-236 | Jan. 23, 2025 (I)]**

$$CH_3CH_2CH=O \xrightarrow[\text{reflux}]{\text{excess HCHO, alkali}} ?$$

(a) $CH_3-CH_2-CH_2-OH$

(b) $CH_3-\underset{CH_2-OH}{CH}-CH=O$

(c) $CH_3-C(CH_2-OH)_3$

(d) $CH_3-\underset{\underset{CH_2}{\|}}{C}-CH=O$

50. The correct stability order of the following resonance structures of $CH_3-CH=CH-CHO$ is **[April 9, 2024 (II)]**

$$CH_3-\overset{\ominus}{C}H-CH=\overset{:O:^{\oplus}}{C}-H \longleftrightarrow CH_3-\overset{\oplus}{C}H-CH=\overset{:O:^{\ominus}}{C}-H \longleftrightarrow :CH_3-CH=CH-\overset{:O:}{\overset{\|}{C}}-H$$

(I) (II) (III)

(a) II > III > I (b) III > II > I
(c) I > II > III (d) II > I > III

51. Which of the following compound can give positive iodoform test when treated with aqueous KOH solution followed by potassium hypoiodite.
[NCERT : P L-240 | April 9, 2024 (II)]

(a) $CH_3CH_2-\overset{O}{\overset{||}{C}}-CH_2CH_3$

(b) $CH_3CH_2-\underset{Cl}{\overset{Cl}{C}}-CH_3$

(c) $CH_3CH_2CH_2CHO$

(d) $CH_3CH_2-\overset{O}{CH-CH_2}$ (epoxide)

52. Which among the following aldehydes is most reactive towards nucleophilic addition reactions?
[NCERT : P L-236 | April 6, 2024 (I)]

(a) $H-\overset{O}{\overset{||}{C}}-H$ (b) $C_2H_5-\overset{O}{\overset{||}{C}}-H$

(c) $CH_3-\overset{O}{\overset{||}{C}}-H$ (d) $C_3H_7-\overset{O}{\overset{||}{C}}-H$

53. For the compounds: [NCERT : P L-235 | April 5, 2024 (I)]

(A) $H_3C-CH_2-O-CH_2-CH_2-CH_3$
(B) $H_3C-CH_2-CH_2-CH_2-CH_3$
(C) $CH_3-CH_2-\underset{O}{\underset{||}{C}}-CH_2-CH_3$
(D) $H_3C-\underset{OH}{\underset{|}{CH}}-CH_2-CH_2-CH_3$

The increasing order of boiling point is :
Choose the **correct** answer from the options given below:
(a) (A) < (B) < (C) < (D) (b) (B) < (A) < (C) < (D)
(c) (D) < (C) < (A) < (B) (d) (B) < (A) < (D) < (C)

54. Identify 'A' in the following reaction :
[NCERT : P L-239 | April 5, 2024 (I)]

$CH_3-\overset{O}{\overset{||}{C}}-CH_2-CH_3 \xrightarrow[\text{(ii) ethylene glycol / KOH}]{\text{(i) } N_2H_4}$ 'A'

(a) $CH_3-CH(OH)-CH_2-CH_3$ (b) $CH_3-CH_2-CH_2-CH_3$

(c) $(CH_3)(H_5C_2)C=N-NH_2$ (d) $(CH_3)_2C=N-NH_2$

55. Identify A and B in the given chemical reaction sequence : -

Benzene $\xrightarrow[\text{AlC}_3]{\text{succinic anhydride}}$ A $\xrightarrow[\text{HCl}]{\text{Zn-Hg}}$ B $\xrightarrow{H^+}$ 1-tetralone

[NCERT : P L-239 | April 5, 2024 (II)]

(a) A- $C_6H_5CH_2CH_2COOH$, B- $C_6H_5CH_2CH_2CH_2OH$

(b) A- $C_6H_5COCH_2CH_2COOH$, B- $C_6H_5CH_2CH_2CH_2COOH$

(c) A- 1,2-dioxo-tetralin (naphthalene-fused diketone), B- naphthalene

(d) A- 1,4-dioxo-tetralin, B- 1,4-dihydroxy-tetralin

56. In the Claisen-Schmidt reaction to prepare 351 g of dibenzalacetone using 87 g of acetone, the amount of benzaldehyde required is ________ g. (Nearest integer)
[NV, April 5, 2024 (II)]

57. Identify the product in the following reaction :
[NCERT : P L-239 | April 4, 2024 (I)]

3-acetylbenzaldehyde $\xrightarrow[\text{HCl}]{\text{Zn-Hg}}$ Product

(a) 3-(1-hydroxyethyl)benzyl alcohol (b) 3-methylacetophenone

(c) 3-acetylbenzyl alcohol (d) 1-ethyl-3-methylbenzene

58. Given below are two statements :
[NCERT : P L-241 | April. 4, 2024 (I)]

Statement I : Acidity of α-hydrogens of aldehydes and ketones is responsible for aldol reaction.

Statement II : Reaction between benzaldehyde and ethanal will NOT give Cross – Aldol product.

In the light of above statements, choose the most appropriate answer from the options given below.

(a) Both Statement I and Statement II are correct.
(b) Both Statement I and Statement II are incorrect.
(c) Statement I is incorrect but Statement II is correct.
(d) Statement I is correct but Statement II is incorrect.

59. The product of the following reaction is P.

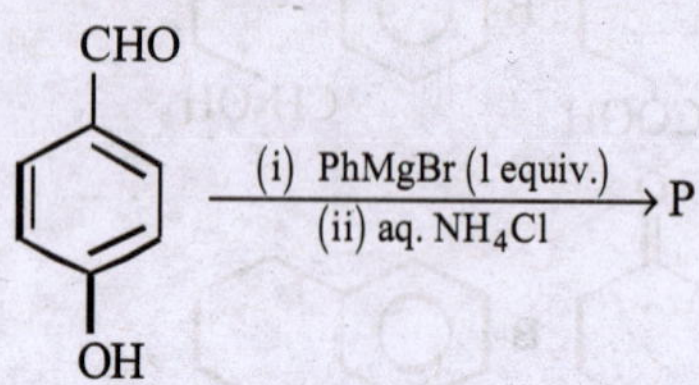

The number of hydroxyl groups present in the product P is______. **[NCERT : P L-237 | NV, Jan. 31, 2024 (I)]**

60. The compound formed by the reaction of ethanal with semicarbazide contains ______ number of nitrogen atoms. **[NV, Jan. 30, 2024 (I)]**

61. m-Chlorobenzaldehyde on treatment with 50% KOH solution yields **[NCERT : P L-242 | Jan. 30, 2024 (II)]**

(a) OH O, –CH – C–, Cl, Cl

(b) $COO^{\ominus}$, Cl + CH_2OH, Cl

(c) O, –C – H, OH + O, –C – H, OH

(d) $COO^{\ominus}$, OH + CH_2OH, OH

62. The final product A formed in the following multistep reaction sequence is **[NCERT : P L-236 | Jan. 29, 2024 (I)]**

(i) $H_2O, H^{\oplus}$
(ii) CrO_3
(ii) H_2N-NH_2, KOH
Heating
→ A

(a) OH, O

(b) N – NH_2

(c) N, NH_2

(d)

63. Identify the reagents used for the following conversion **[Jan. 29, 2024 (II)]**

OCH_3, OH, CHO, O —Ⓐ→ CHO, OH, CHO —Ⓑ→ CHO, OH —Ⓒ→

(a) A = $LiAlH_4$. B = $NaOH_{(aq)}$, C = NH_2-NH_2/KOH, ethylene glycol
(b) A = DiBAL-H, B = $NaOH_{(aq)}$, C = NH_2-NH_2/KOH, ethylene glycol
(c) A = DiBAL-H, B = $NaOH_{(alc)}$, C = Zn/HCl
(d) A = $LiAlH_4$, B = $NaOH_{(alc)}$, C = Zn/HCl

64. Highest enol content will be shown by : **[NCERT : P L-242 | Jan. 27, 2024 (I)]**

(a) O, O
(b) O, O, O
(c) O, O
(d) O, O, O

65. O, O, O —H_2O→ 'A'

[NCERT : P L-242 | April 15, 2023 (I)]

Major product 'A' formed in the above reaction is

(a) 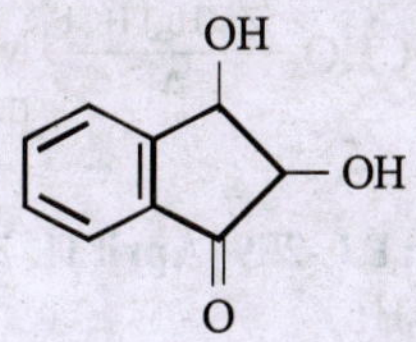 (b)

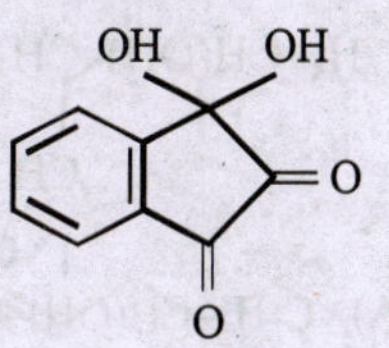

(c) 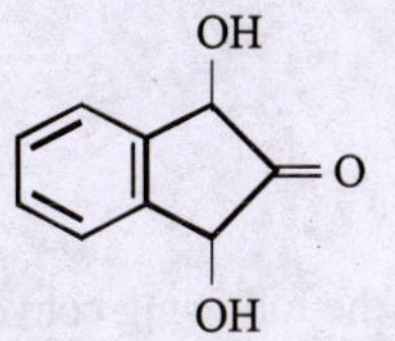 (d)

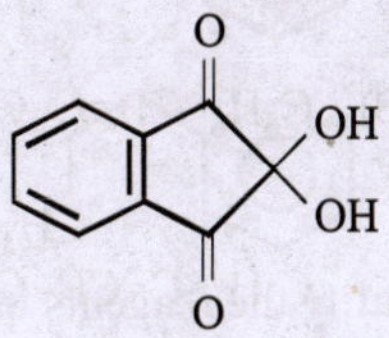

66. 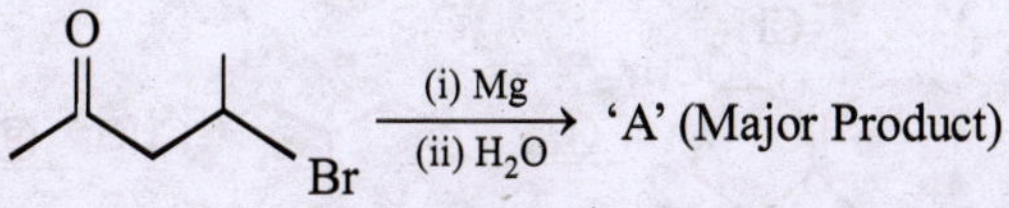 [structure: 4-bromopentan-2-one] $\xrightarrow[\text{(ii) } H_2O]{\text{(i) Mg}}$ 'A' (Major Product)

A is [NCERT : P L-237 | April 12, 2023 (I)]

(a) 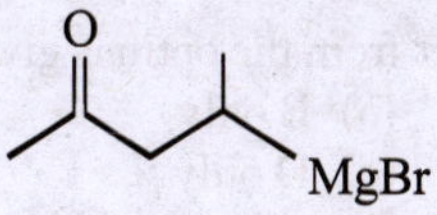 (b) 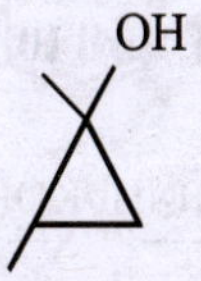 (c) 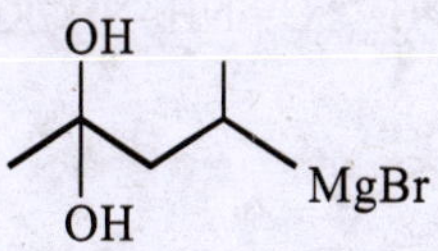 (d)

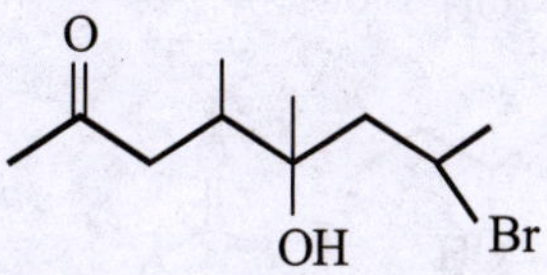

67. Correct statements for the given reaction are : [April 12, 2023 (I)]

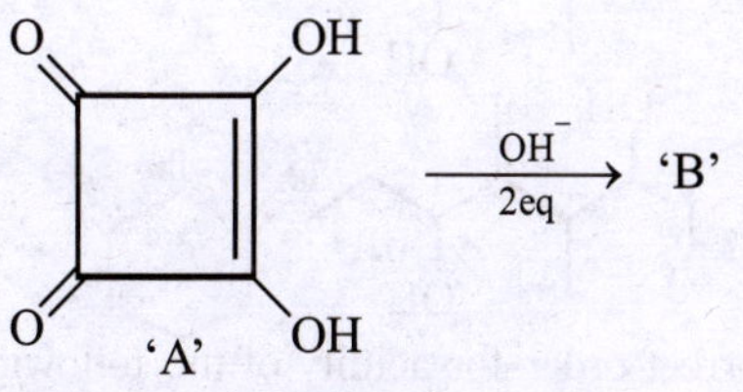

'A' $\xrightarrow[2eq]{OH^-}$ 'B'

A. Compound 'B' is aromatic
B. The completion of above reaction is very slow
C. 'A' shows tautomerism
D. The bond lengths C-C in compound B are found to be same

Choose the correct answer from the options given below:

(a) A, B and D only (b) A, B and C only
(c) B, C and D only (d) A, C and D only

68. 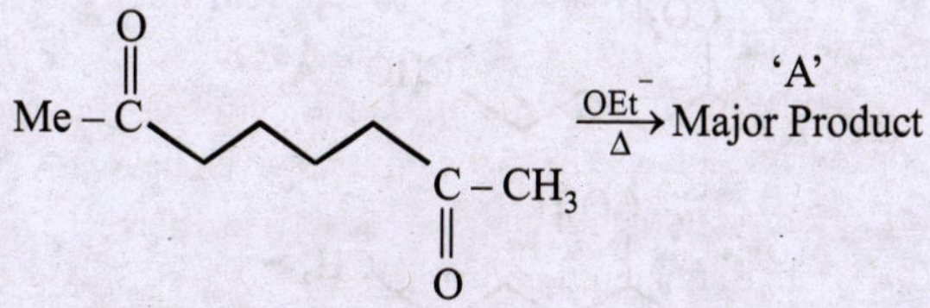 $Me-\overset{O}{\overset{\|}{C}}$–(CH₂)₄–$\underset{O}{\underset{\|}{C}}-CH_3$ $\xrightarrow[\Delta]{\overset{-}{O}Et}$ 'A' Major Product

A in the above reaction is : [NCERT : P L-237 | April 12, 2023 (I)]

(a) 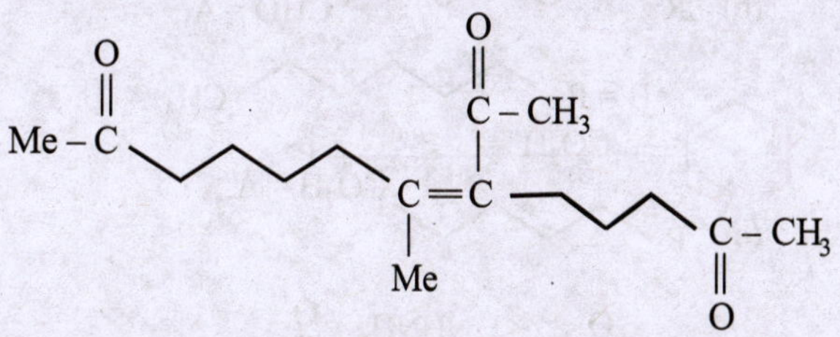 (b) 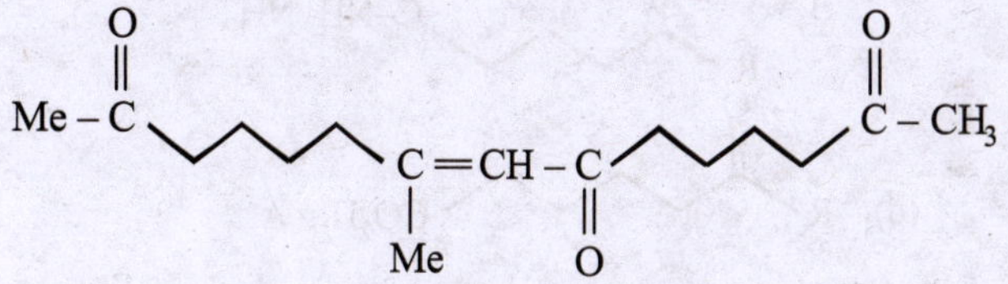 (c) 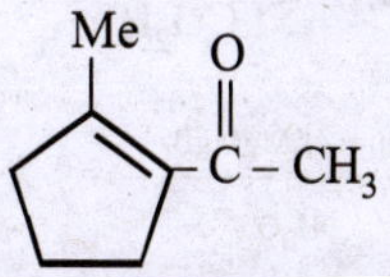 (d)

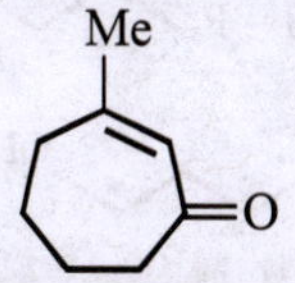

69. The mass of NH_3 produced when 131.8 kg of cyclohexanecarbaldehyde undergoes Tollen's test is ____ kg. (Nearest integer) [NCERT : P L-239 | NV, April 12, 2023 (I)]

Molar Mass of C = 12 g/mol, N = 14 g/mol, O = 16 g/mol

70. 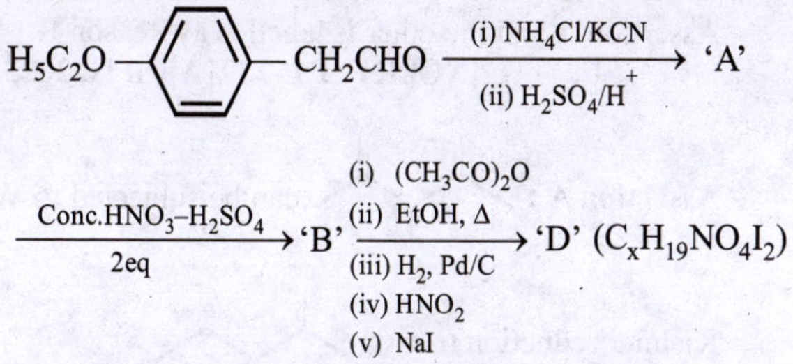 $H_5C_2O-C_6H_4-CH_2CHO \xrightarrow[\text{(ii) } H_2SO_4/H^+]{\text{(i) } NH_4Cl/KCN}$ 'A'

$\xrightarrow[2eq]{Conc.HNO_3-H_2SO_4}$ 'B' $\xrightarrow{\text{(i) }(CH_3CO)_2O\text{ (ii) EtOH, }\Delta\text{ (iii) }H_2\text{, Pd/C (iv) }HNO_2\text{ (v) NaI}}$ 'D' ($C_xH_{19}NO_4I_2$)

The value of x in compound 'D' is ____ [NCERT : P L-232 | NV, April 12, 2023 (I)]

71. 'A' and 'B' in the below reactions are : [April 11, 2023 (I)]

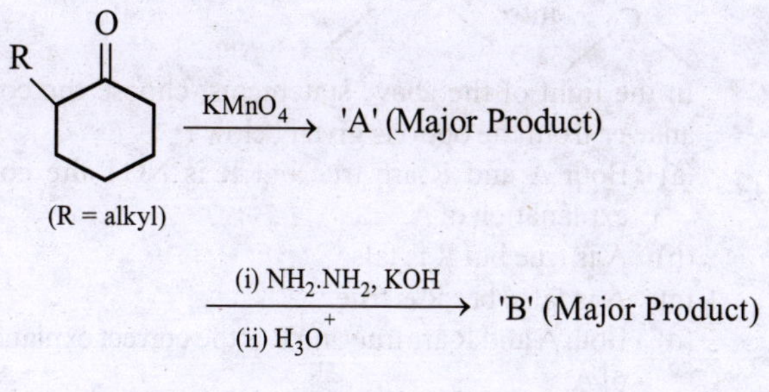

[structure: 2-alkylcyclohexanone (R = alkyl)] $\xrightarrow{KMnO_4}$ 'A' (Major Product)

$\xrightarrow[\text{(ii) } H_3O^+]{\text{(i) } NH_2.NH_2\text{, KOH}}$ 'B' (Major Product)

(a) $R-CH(CO_2H)-CH_2CH_2CH_2CH_2-CHO$ = A,

B = $R-CH(CO_2H)-CH_2CH_2CH_2CH_2-CH_3$

(b) $R-CO-CH_2CH_2CH_2CH_2CH_2-CHO$ = A,

B = $R-CH_2CH_2CH_2CH_2CH_2CH_2-CH_3$

(c) $R-CH(CO_2H)-CH_2CH_2CH_2CH_2-CO_2H$ = A,

B = $R-CH(CONHNH_2)-CH_2CH_2CH_2CH_2-C(=O)-NH-NH_2$

(d) $R-CO-CH_2CH_2CH_2CH_2CH_2-CO_2H$ = A,

B = $R-CH_2CH_2CH_2CH_2CH_2CH_2-CO_2H$

72. $H-C(=O)-CH_2CH_2CH_2-OH$ (y mole) $\xrightarrow[H_3O^+]{\text{x mol of MeMgBr}}$ $Me-CH(OH)-CH_2CH_2CH_2-OH$

The ratio x/y on completion of the above reaction is ____.

[NCERT : P L-239 | NV, April 11, 2023 (I)]

73. Given below are two statements, one is labelled as Assertion A and the other is labelled as Reason R.

[NCERT : P L-239 | April 11, 2023 (II)]

Assertion A : $(CH_3)_2CH-CH(Cl)-CO-CH_3$ can be subjected to Wolff-Kishner reduction to give $(CH_3)_2CH-CH(Cl)-CH_2CH_3$.

Reason R : Wolff-Kishner reduction is used to convert $>C=O$ into $>CH_2$.

In the light of the above statements, choose the correct answer from the options given below :

(a) Both A and R are true but R is NOT the correct explanation of A.
(b) A is true but R is false.
(c) A is false but R is true.
(d) Both A and R are true and R is the correct explanation of A.

74. The major product formed in the following reaction is:

$C_6H_5-CH(OH)-CH(CH_3)-CH_2-CHO \xrightarrow[\Delta]{Zn(Hg)/HCl}$ Major product

[NCERT : P L-239 | April 11, 2023 (II)]

(A) $C_6H_5-CH(OH)-CH(CH_3)-C_2H_5$
(B) $C_6H_5-CH=C(CH_3)-C_2H_5$
(C) $C_6H_5-C(CH_3)=CH-C_2H_5$
(D) 2-C_6H_5-3-CH_3-tetrahydrofuran (five-membered ring containing O, with C_6H_5 and CH_3 substituents)

Choose the correct answer from the options given below:

(a) A only (b) B only
(c) C only (d) D only

75. In the reaction given below:

[NCERT : P L-232 | April 10, 2023 (II)]

$H_2N-C(=O)-CH(CH_3)-C(=CH_2)-C(=O)-CH_2CH_3 \xrightarrow[(ii)\ H_3O^+]{(i)\ LiAlH_4}$ 'X'

The product 'X' is:

(a) $H_2N-CH_2-CH(CH_3)-C(=CH_2)-CH(OH)-CH_2CH_3$
(b) $H_2N-CH(OH)-CH(CH_3)-CH(CH_3)-CH(OH)-CH_2CH_3$
(c) $H_2N-CH(OH)-CH(CH_3)-CH(OH)-CH_2CH_3$
(d) $H_2N-CH(OH)-CH(CH_3)-C(=CH_2)-CH(OH)-CH_2CH_3$

76. The correct order for acidity of the following hydroxyl compounds is: [NCERT : P L-249 | April 10, 2023 (II)]

A. CH_3OH
B. $(CH_3)_3COH$
C. C_6H_5-OH
D. $MeO-C_6H_4-OH$
E. $O_2N-C_6H_4-OH$

Choose the correct answer from the options given below:

(a) E>C>D>A>B (b) D>E>C>A>B

(c) C>E>D>B>A (d) E>D>C>B>A

77. Match List I with List II

[NCERT : P L-239 | April 06, 2023 (I)]

	List I (Name of reaction)		List II (Reagent used)
A	Hell–Volhard–Zelinsky reaction	I.	$NaOH + I_2$
B	Iodoform reaction	II.	(i) CrO_2Cl_2, CS_2 (ii) H_2O
C	Etard reaction	III.	(i) Br_2/red phosphorus (ii) H_2O
D	Gatterman–Koch reaction	IV.	CO, HCl, anhyd. $AlCl_3$

Choose the correct answer from the options given below:

(a) A–III, B–II, C–I, D–IV

(b) A–III, B–I, C–IV, D–II

(c) A–I, B–II, C–III, D–IV

(d) A–III, B–I, C–II, D–IV

78. Among the following, the number of compounds which will give positive iodoform reaction is ______.

[NCERT : P L-239 | NV, April 06, 2023 (II)]

(1) 1-Phenylbutan-2-one (2) 2-Methylbutan-2-ol

(3) 3-Methylbutan-2-ol (4) 1-Phenylethanol

(5) 3,3-dimethylbutan-2-one (6) 1-Phenylpropan-2-ol

79. Find out the major product from the following reaction

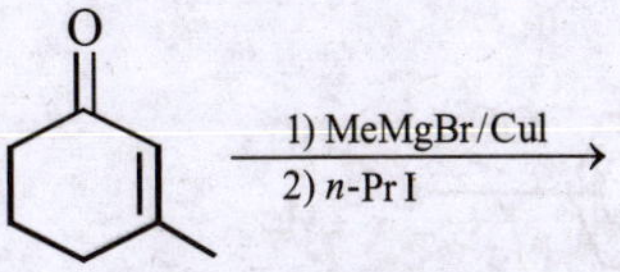

[April 06, 2023 (II)]

(3-methylcyclohex-2-en-1-one) $\xrightarrow[\text{2) } n\text{-PrI}]{\text{1) MeMgBr/CuI}}$

(a) 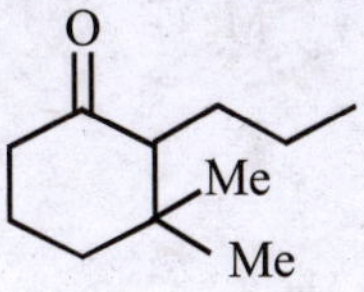 2-propyl-3,3-dimethylcyclohexanone (Me, Me)

(b) 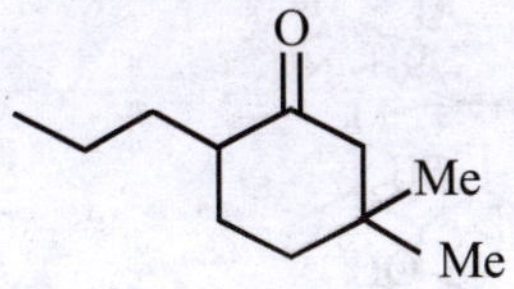 2-propyl-4,4-dimethylcyclohexanone (Me, Me)

(c) 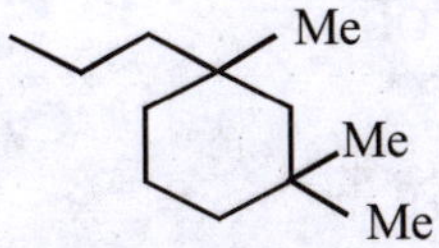 1-propyl-1,3,3-trimethylcyclohexane (Me, Me, Me)

(d) 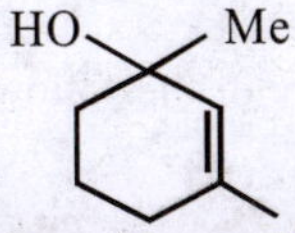 1-methyl-3-methylcyclohex-2-en-1-ol (HO, Me)

80. Number of isomeric compounds with molecular formula $C_9H_{10}O$ which (i) do not dissolve in NaOH (ii) do not dissolve in HCl. (iii) do not give orange precipitate with 2,4-DNP (iv) on hydrogenation give identical compound with molecular formula $C_9H_{12}O$ is ______.

[NCERT : P L-239 | NV, Feb. 01, 2023 (I)]

81. The structures of major products A, B and C in the following reaction sequence are. **[Feb. 01, 2023 (II)]**

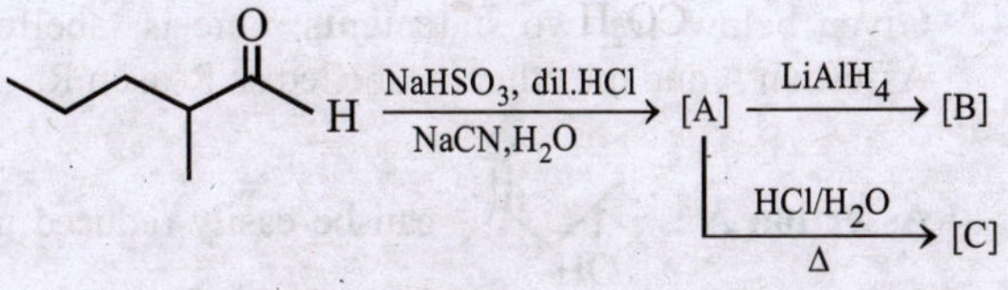

2-methylpentanal $\xrightarrow[\text{NaCN, }H_2O]{NaHSO_3\text{, dil.HCl}}$ [A] $\xrightarrow{LiAlH_4}$ [B]; [A] $\xrightarrow[\Delta]{HCl/H_2O}$ [C]

(a) A = (C with OSO_3Na, H), B = (C with OH, H), C = (C with Cl, H)

(b) A = (C with OH, CN, H), B = (C with HO, CHO, H), C = (C with HO, CO_2H, H)

(c) A = (C with HO, SO_3H, H), B = (primary alcohol, OH), C = (C with HO, SO_2Cl, H)

(d) A = (C with HO, CN, H), B = (C with HO, NH_2, H), C = (C with HO, CO_2H, H)

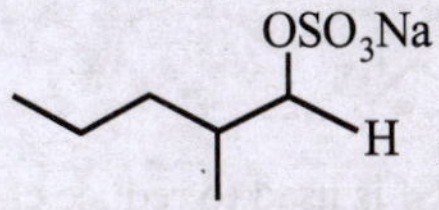

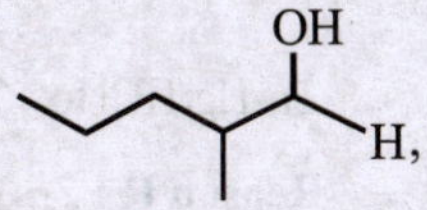

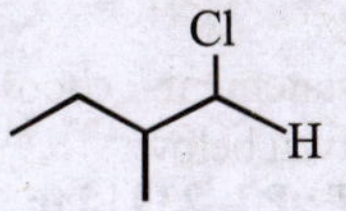

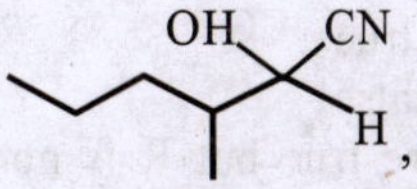

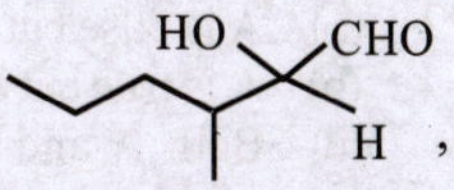

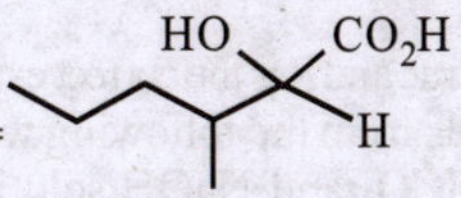

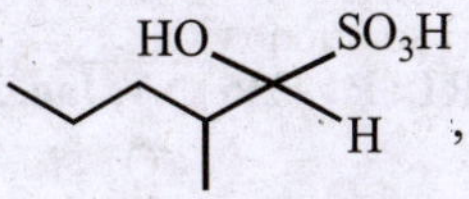

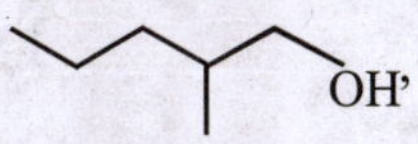

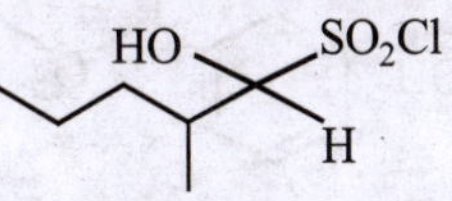

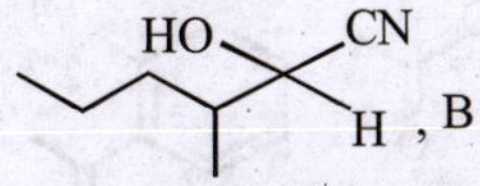

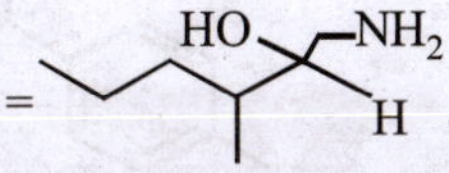

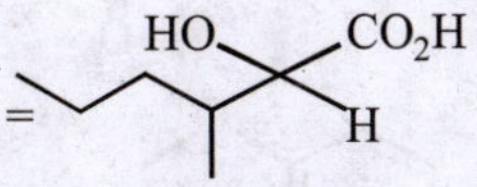

82. Consider the following reaction **[Jan. 31, 2023 (I)]**

Propanal + Methanal (X) = $\xrightarrow[\text{(ii) }\Delta\text{ (iii) NaCN (iv) }H_3O^+]{\text{(i) dil.NaOH}}$ Product B ($C_5H_8O_3$)

The correct statement for product B is. It is

(a) optically active and adds one mole of bromine

(b) racemic mixture and is neutral

(c) racemic mixture and gives a gas with saturated $NaHCO_3$ solution

(d) optically active alcohol and is neutral

83. The number of molecules which gives haloform test among the following molecules is ______.

[NCERT : P L-239 | NV, Jan. 31, 2023 (II)]

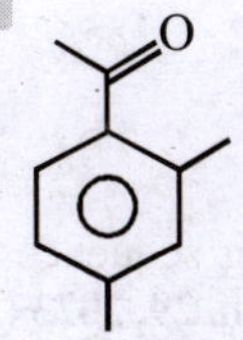

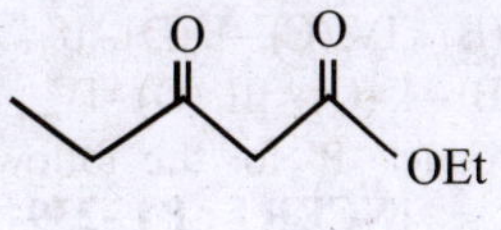

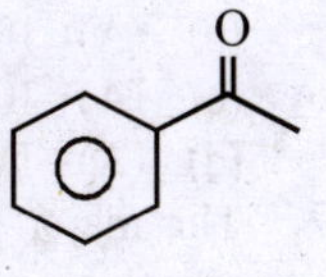

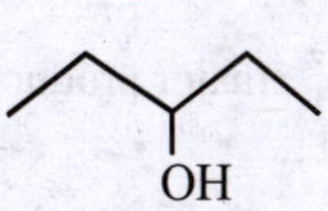

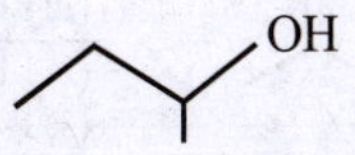

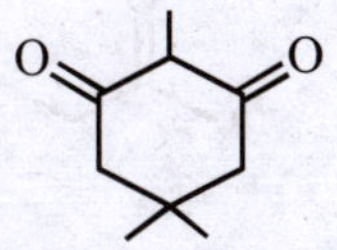

(2,4-dimethylacetophenone), (ethyl 3-oxopentanoate, OEt), (acetophenone), (pentan-3-ol, OH), (3-methylbutan-2-ol, OH), (2,5,5-trimethylcyclohexane-1,3-dione)

84. Given below are two statements: One is labelled as Assertion A and the other is labelled as Reason R.

Assertion A : [structure: 4-hydroxy-4-methylpentan-2-one] can be easily reduced using Zn-Hg/HCl to [structure: 2-methylpentan-2-ol].

Reason R : Zn-Hg/HCl is used to reduce carbonyl group to $-CH_2-$ group.

In the light of the above statements, choose the correct answer from the options given below:

[NCERT : P L-239 | Jan. 30, 2023 (II)]

(a) A is false but R is true
(b) A is true but R is false
(c) Both A and R are true but R is not the correct explanation of A
(d) Both A and R are true and R is the correct explanation of A

85. Number of compounds from the following which will not dissolve in cold $NaHCO_3$ and NaOH solutions but will dissolve in hot NaOH solution is ______.

[NCERT : P L-235 | NV, Jan. 30, 2023 (II)]

[structures: 4-methylbenzoic acid, methyl benzoate, 4-methylphenyl acetate, 2-hydroxyacetophenone, 3-hydroxy-5-methylbenzaldehyde (H, O, OH), 1-isopropyl-3-methoxybenzene (H_3C, CH_3), phenyl acetate]

86. Match List I with List II.

[NCERT : P L-239 | Jan. 29, 2023 (I)]

List-I (Reaction)	List-II (Reagents)
(A) Hoffmann degradation	(I) Conc.KOH, Δ
(B) Clemensen reduction	(II) $CHCl_3$, NaOH/H_3O^+
(C) Cannizaro reaction	(III) Br_2, NaOH
(D) Reimer-Tiemann reaction	(IV) Zn-Hg/HCl

(a) (A) – III, (B) – IV, (C) – II, (D) - I
(b) (A) – II, (B) – IV, (C) – I, (D) - III
(c) (A) – III, (B) – IV, (C) – I, (D) - II
(d) (A) – II, (B) – I, (C) – III, (D) - IV

87. The major product 'P' for the following sequence of reactions is: **[NCERT : P L-239 | Jan. 29, 2023 (I)]**

$$\text{Ph-CO-CH}_2\text{CH}_2\text{CH}_2\text{-CO-NH}_2 \xrightarrow[\text{(2) LiAlH}_4 \quad \text{(3) H}_3\text{O}^+]{\text{(1) Zn/Hg}} \text{'P' major product}$$

(a) Ph–CH(OH)–CH₂CH₂CH₂–CH₂–OH [Ph, OH, OH]
(b) Ph–CH(OH)–(CH₂)₄–NH_2 [Ph, OH, NH_2]
(c) Ph–(CH₂)₅–NH_2 [Ph, NH_2]
(d) Ph–CH(OH)–CH₂CH₂CH₂–CH(OH)–NH_2 [Ph, OH, OH, NH_2]

88. Find out the major products from the following reaction sequence. **[Jan. 29, 2023 (II)]**

$$\text{4-Cl-C}_6\text{H}_4\text{-CO-CH}_3 \xrightarrow[\text{(2) EtOH, H}_3\text{O}^\oplus]{\text{(1) NaCN}} \text{A} \xrightarrow[\text{H}_3\text{O}^\oplus]{\text{MeMgBr (excess)}} \text{B}$$

(a) A = 4-Cl-C₆H₄–C(OH)(CH₃)–CO_2H; B = Cl–C₆H₄–C(OH)(Me)–C(OH)(Me)(Me)

(b) A = 4-Cl-C₆H₄–C(OH)(CH₃)–CO_2Et; B = Cl–C₆H₄–C(OH)(Me)–C(OH)(Me)(Me)

(c) A = 4-Cl-C₆H₄–C(OH)(CH₃)–CN; B = Cl–C₆H₄–C(OH)(Me)–C(Me)(Me)(NH_2)

(d) A = 4-CN-C₆H₄–C(OH)(CH₃)–NC; B = Me–CH(OH)–C₆H₄–C(OH)(Me)–CH(OH)(Me)

89. Given below are two statements : one is labelled as Assertion A and the other is labelled as Reason R :

Assertion A : Acetal/ketal is stable in basic medium.

Reason R : The high leaving tendency of alkoxide ion gives the stability to acetal/ketal in basic medium.

In the light of the above statements, choose the correct answer from the options given below:

[NCERT : P L-236 | Jan. 25, 2023 (I)]

(a) A is true but R is false
(b) A is false but R is true
(c) Both A and R are true and R is the correct explanation of A
(d) Both A and R are true but R is NOT the correct explanation of A

90.  P $\xrightarrow{?}$ $PhCOOH + PhCH_2OH$ (Q, R)

The correct sequence of reagents for the preparation of Q and R is : **[NCERT : P L-242 | Jan. 25, 2023 (I)]**

(a) (i) Cr_2O_3, 770 K, 20 atm ; (ii) CrO_2Cl_2, H_3O^+ ; (iii) NaOH ; (iv) H_3O^+
(b) (i) CrO_2Cl_2, H_3O^+ (ii) Cr_2O_3, 770 K, 20 atm; (iii) NaOH ; (iv) H_3O^+
(c) (i) $KMnO_4, OH^-$; (ii) Mo_2O_3, A; (iii) NaOH ; (iv) H_3O^+
(d) (i) Mo_2O_3, Δ ; (ii) CrO_2Cl_2, H_3O^+; (iii) NaOH ; (iv) H_3O^+

91. 'A' in the given reaction is **[NCERT : P L-236 | Jan. 25, 2023 (II)]**

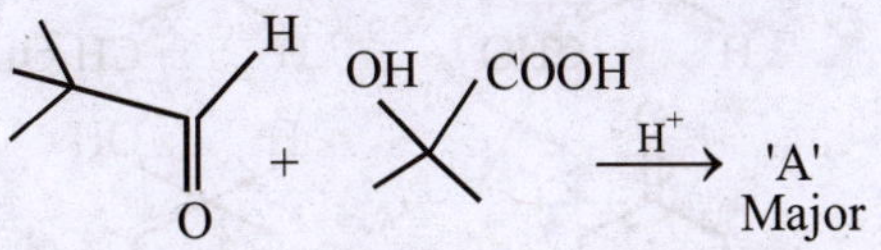

(aldehyde) + (OH, COOH) $\xrightarrow{H^+}$ 'A' Major

(a) 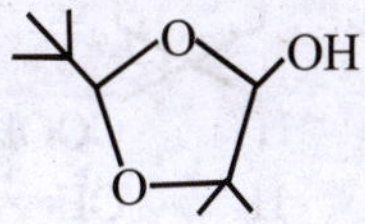 (b) 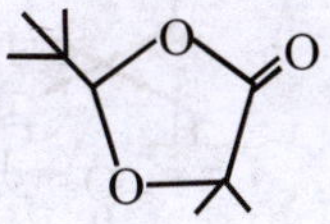 (c) 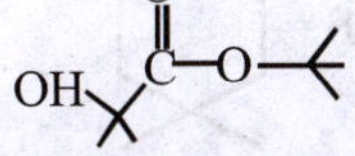 OH–C(=O)–O– (d) 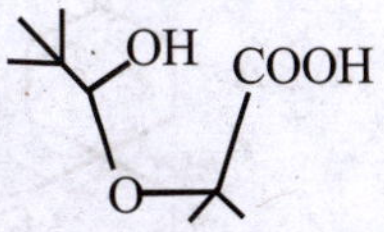 OH COOH

92. Compound (*X*) undergoes following sequence of reactions to give the lactone (*Y*), compound (*X*) is_______.

[NCERT : P L-236 | Jan. 24, 2023 (I)]

Compound (*X*) $\xrightarrow[\text{(iii) } H_3O^+]{\text{(i) HCHO, KOH; (ii) KCN (alc)}}$ 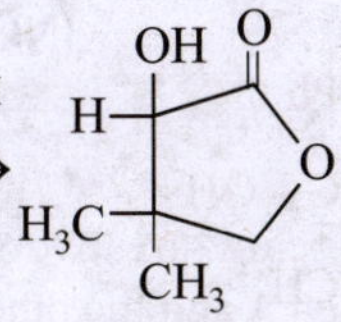 Lactone (*Y*) (OH, O, H, H_3C, CH_3)

(a) $H-C(CH_3)_2-CHO$

(b) $HOH_2C-C(CH_3)_2-CHO$

(c) $H_2C(OH)-CH_2-CH_2-CH_2-CHO$

(d) $H_2C(CH_3)-CH_2-CHO$

93. 'R' formed in the following sequence of reactions is:

[NCERT : P L-242 | Jan. 24, 2023 (I)]

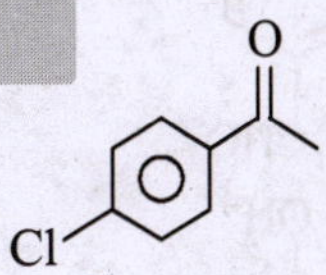

(Cl-aryl ketone) $\xrightarrow[\text{HOAc}]{\text{NaCN}}$ 'P' $\xrightarrow[H^+]{\text{EtOH}}$ 'Q' $\xrightarrow[\text{(ii) } H_3O^+]{\text{(i) 2MeMgBr}}$ 'R' major product

(a) 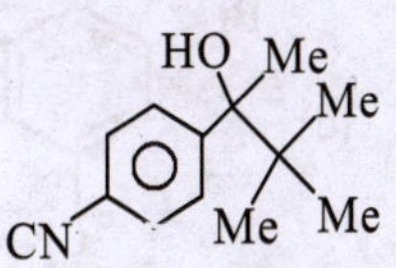 HO Me, Me, Me Me, CN (b) 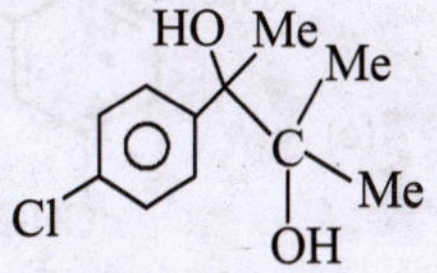 HO Me, Me, C, Me, OH, Cl (c) OH, Me Me, Cl (d) 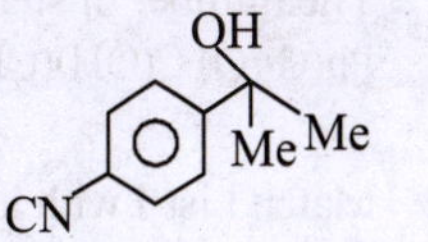 OH, Me Me, CN

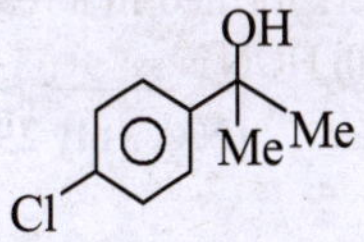

94. Given below are two statements: **[NCERT : P L-239 | Jan. 24, 2023 (II)]**

Statement I: H_2N–(O)...(O) under Clemmensen reduction conditions will give HOOC–(chain).

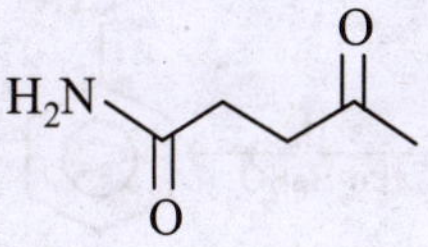

Statement II: (O, Cl) under Wolff-Kishner reduction condition will give (Cl).

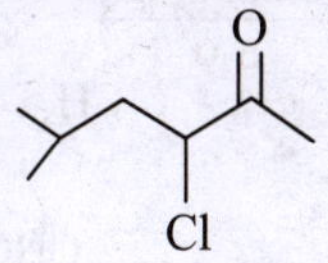

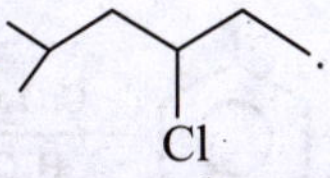

In the light of the above statements, choose the correct answer from the options given below :

(a) Statement I is false but Statement II is true.
(b) Both Statement I and Statement II are false.
(c) Statement I is true but Statement II is false.
(d) Both Statement I and Statement II are true.

95. Which will undergo deprotonation most readily in basic medium? **[NCERT : P L-242 | Jan. 24, 2023 (II)]**

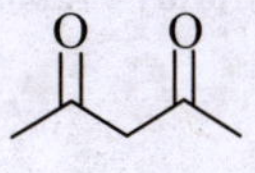

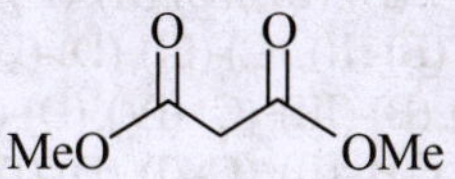

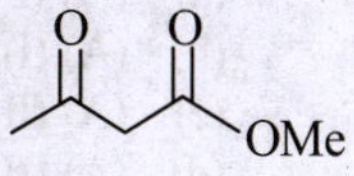

A: O O; B: MeO, OMe; C: O O, OMe

(a) A only (b) C only
(c) Both A and C (d) B only

96. p-hydroxybenzaldehyde (OH, CHO) $\xrightarrow{Br_2}$ A $\xrightarrow{NH_2OH}$ B $\xrightarrow[\Delta]{P_2O_5}$ C

Consider the above reaction sequence, the product 'C' is: **[July 29, 2022 (I)]**

(a) 3,5-dibromobenzene with CN (Br, Br, CN)
(b) OH, Br, Br, CHO
(c) Br, Br, CHO
(d) OH, Br, Br, CN

97. The number of stereoisomers formed in a reaction of (±) Ph(C=O) C(OH)(CN)Ph with HCN is______. **[NV, July 29, 2022 (II)]**

98. Match List-I with List-II.

List-I	List-II
(A) $C_6H_5COCl \xrightarrow[Pd\text{-}BaSO_4]{H_2} C_6H_5CHO$	(I) Gatterman Koch reaction
(B) $CH_3-CN \xrightarrow[H_3O^+]{SnCl_2/HCl} CH_3-CHO$	(II) Etard reaction
(C) $C_6H_5CH_3 \xrightarrow[H_3O^+]{CrO_2Cl_2} C_6H_5CHO$	(III) Stephen reaction
(D) $C_6H_6 \xrightarrow[AlCl_3\ (anhyd)]{CO,\ HCl} C_6H_5CHO$	(IV) Rosenmund reaction

Choose the **correct** answer from the options given below:

(a) (A)-(IV), (B)-(III), (C)-(II), (D)-(I) **[July 27, 2022 (II)]**
(b) (A)-(I), (B)-(II), (C)-(III), (D)-(IV)
(c) (A)-(II), (B)-(III), (C)-(IV), (D)-(I)
(d) (A)-(III), (B)-(II), (C)-(I), (D)-(IV)

99. The structure of A in the given reaction is: **[NCERT : P L-236 | July 27, 2022 (II)]**

$R-C(=O)-CH_3$ + ethyl bromide $\xrightarrow{NaOH}$ A (major product)

(a) $R-C(OH)(CH_3)-CH(CH_3)Br$
(b) $R-C(OH)(CH_3)-CH_2CH_3$
(c) $R-C(=O)-CH_2CH_2CH_3$
(d) $R-C(=O)-CH_2CH_2CH_2-OH$

100. The products formed in the following reaction, **A** and **B** are **[NCERT : P L-236 | July 26, 2022 (I)]**

4-formylcyclohexanone (H, CHO) $\xrightarrow{[Ag(NH_3)_2]^+ OH^-}$ A $\xrightarrow{NaBH_4}$ B

(a) A = cyclohexanone with H, CH_2OH; B = cyclohexane with H, OH and H, CH_2OH
(b) A = cyclohexane with H, OH and H, CHO; B = cyclohexane with H, OH and H, CH_2OH
(c) A = cyclohexanone with H, COOH; B = cyclohexane with H, OH and H, COOH
(d) A = cyclohexanone with H, COOH; B = cyclohexane with H, OH and H, CH_2OH

101. Which reactant will give the following alcohol on reaction with one mole of phenyl magnesium bromide (PhMgBr) followed by acidic hydrolysis? **[NCERT : P L-237 | July 26, 2022 (I)]**

$Ph-C(Ph)(CH_3)-OH$

(a) $CH_3-C\equiv N$
(b) $PH-C\equiv N$
(c) $CH_3-C(=O)-O-Ph$
(d) $Ph-C(=O)-CH_3$

102. What is the major product of the following reaction?

$(CH_3)_2CH-CHO \xrightarrow[H_2O]{^-OH}$

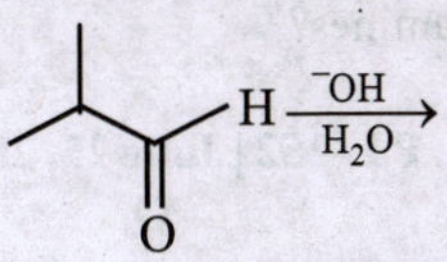

[NCERT : P L-237 | July 25, 2022 (II)]

(a) 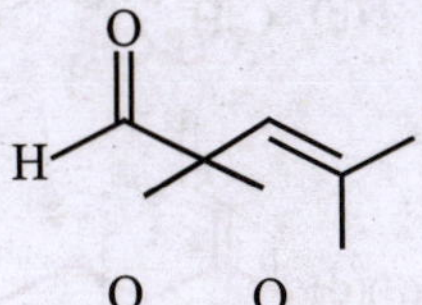 (b) 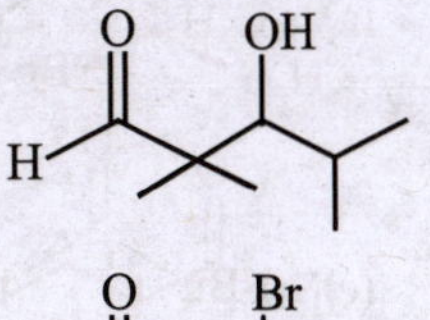 (c) 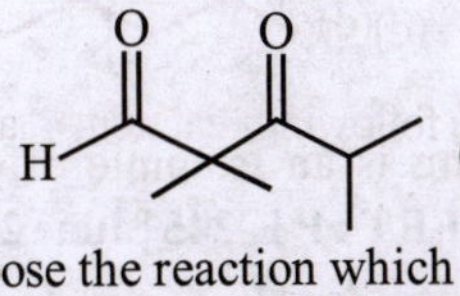 (d)

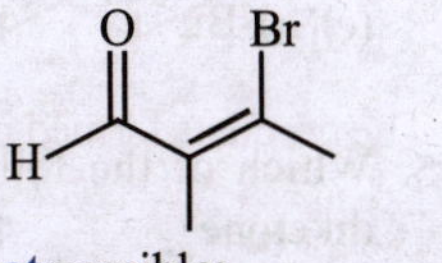

103. Choose the reaction which is **not** possible: [NCERT : P L-236 | June 30, 2022 (I)]

(a) 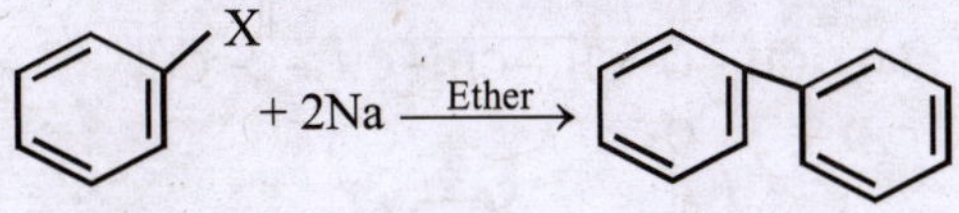 C_6H_5X + 2Na $\xrightarrow{\text{Ether}}$ biphenyl

(b) 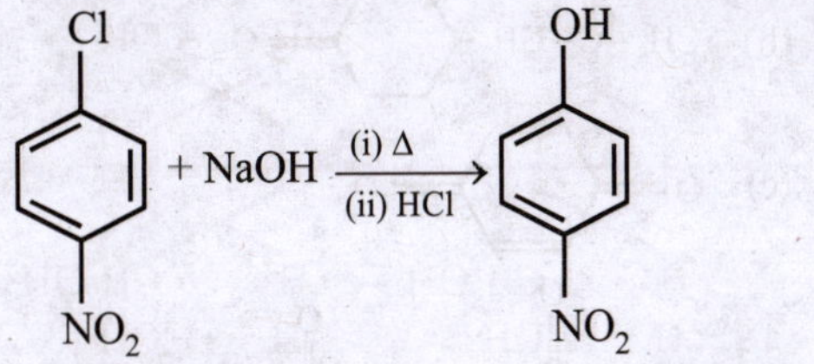 p-nitrochlorobenzene + NaOH $\xrightarrow[\text{(ii) HCl}]{\text{(i) }\Delta}$ p-nitrophenol

(c) $C_6H_5CH_2CHO + [Ag(NH_3)_2]^+ + OH^- \xrightarrow{\Delta} C_6H_5CH_2COO^-$

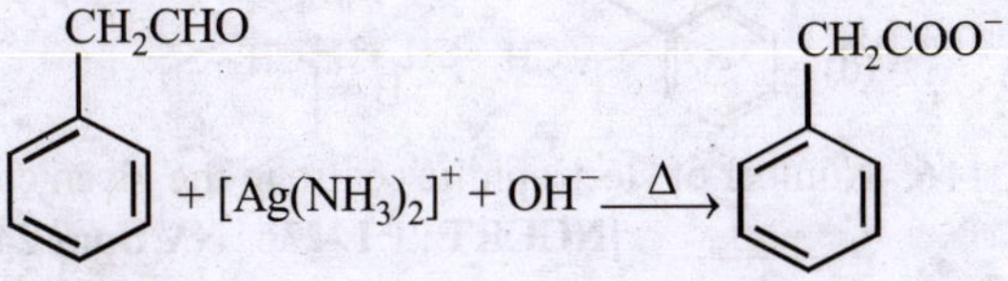

(d) 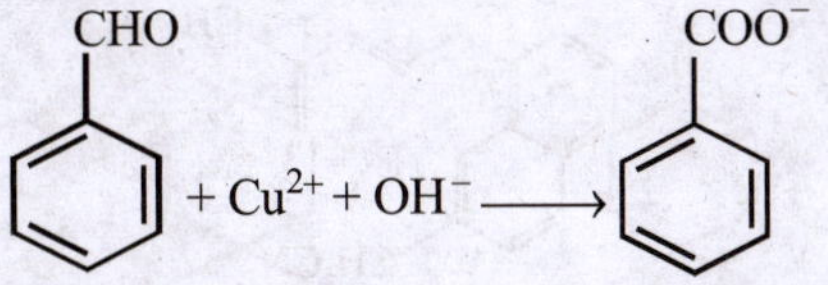 $C_6H_5CHO + Cu^{2+} + OH^- \longrightarrow C_6H_5COO^-$

104. Which among the following will be the major product of the given reaction? [June 30, 2022 (I)]

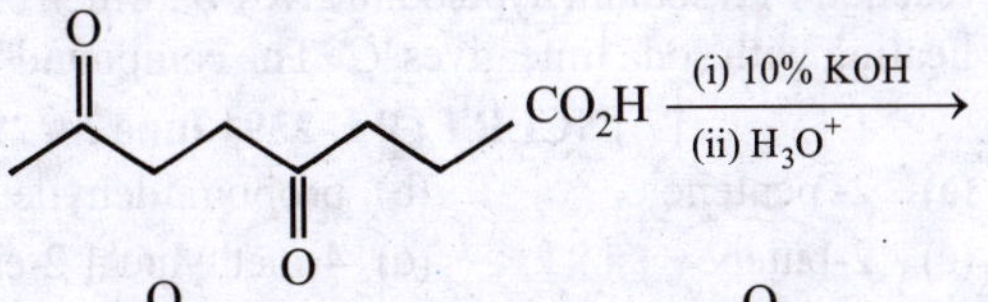

$\xrightarrow[\text{(ii) } H_3O^+]{\text{(i) 10\% KOH}}$

(a) 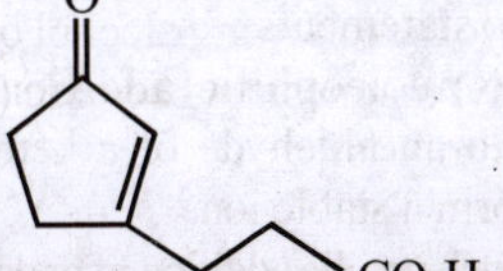 (b) 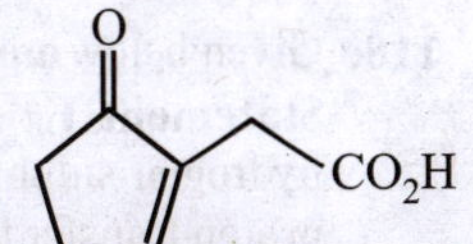 (c) 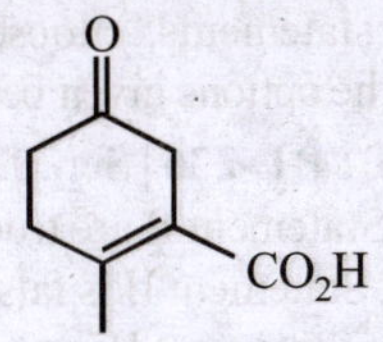 (d)

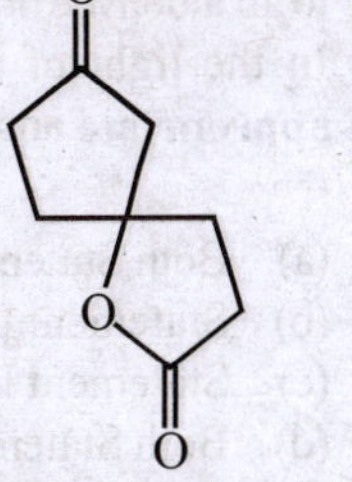

105. The correct structure of product 'A' formed in the following reaction.

$$PhCHO + PhCHO \xrightarrow[\text{in } D_2O]{NaOD} A + Ph-\overset{O}{\overset{\|}{C}}-O^-$$

(Ph is $-C_6H_5$) [NCERT : P L-242 | June 28, 2022 (I)]

(a) 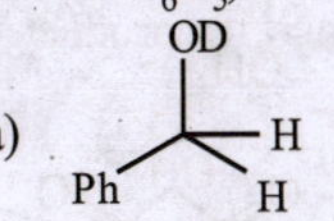 $PhCH_2OD$ (b) 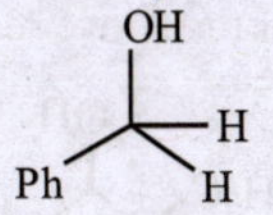 $PhCH_2OH$ (c) 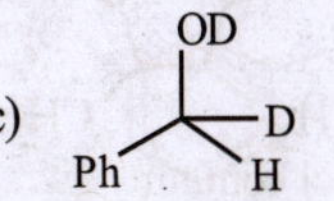 PhCHD(OD) (d) 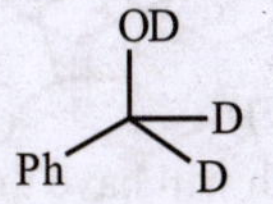 $PhCD_2OD$

106. $C_6H_{12}O_6 \xrightarrow{\text{Zymase}} A \xrightarrow[\Delta]{NaOI} B + CHI_3$

The number of carbon atoms present in the product B is ______. [NV, June 29, 2022 (I)]

107 . In the given reaciton

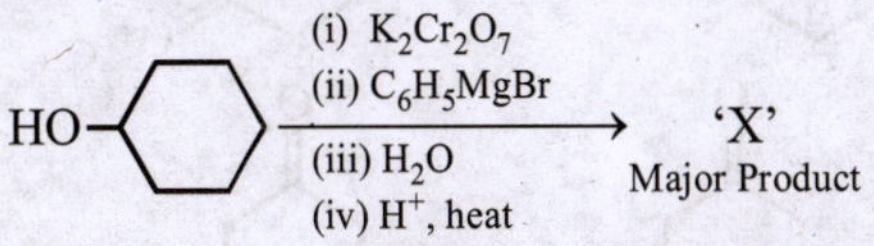

HO–cyclohexane $\xrightarrow[\text{(iv) } H^+\text{, heat}]{\text{(i) } K_2Cr_2O_7 \text{ (ii) } C_6H_5MgBr \text{ (iii) } H_2O}$ 'X' Major Product

the number of sp^2 hybridised carbon(s) in compound 'X' is ______. [NV, June 29, 2022 (II)]

108. In the given reaction

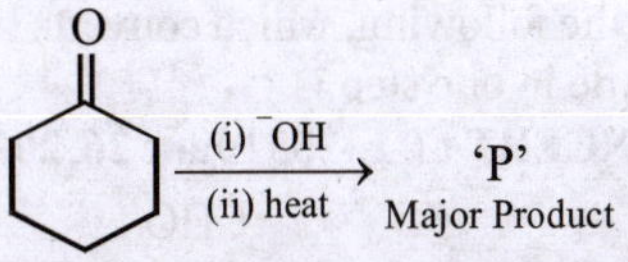

Cyclohexanone $\xrightarrow[\text{(ii) heat}]{\text{(i) } ^-OH}$ 'P' Major Product

The number of π electrons present in the product 'P' is ______. [NV, June 29, 2022 (II)]

109. Isobutyraldehyde on reaction with formaldehyde and K_2CO_3 gives compound 'A'. Compound 'A' reacts with KCN and yields compound 'B', which on hydrolysis gives a stable compound 'C'. The compound 'C' is : [June 28, 2022 (II)]

(a) 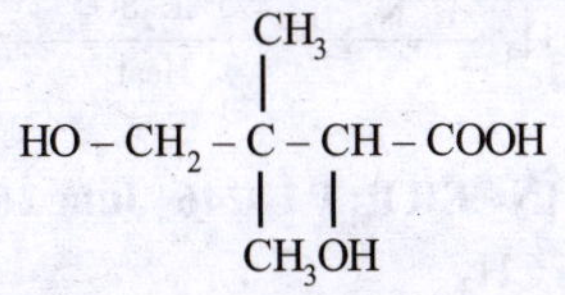 $HO-CH_2-C(CH_3)(CH_2OH)-CH-COOH$

(b) 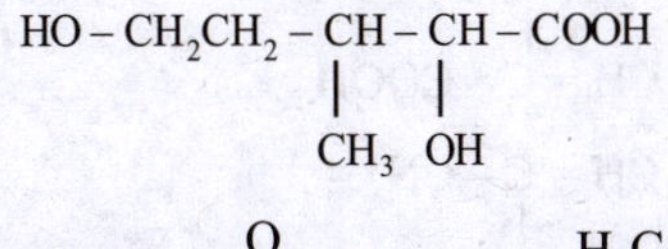 $HO-CH_2CH_2-CH(CH_3)-CH(OH)-COOH$

(c) 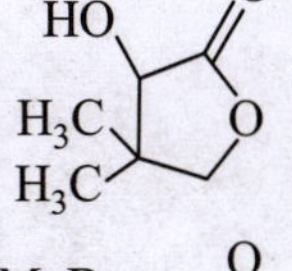 (d)

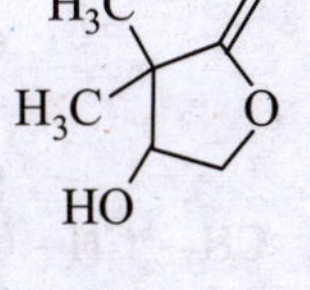

110. 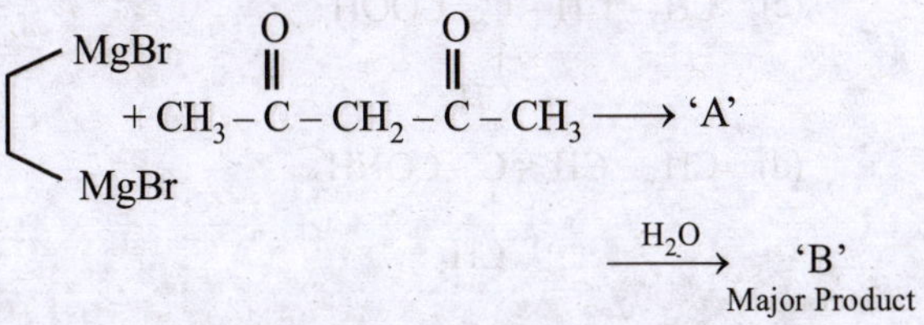 $BrMgCH_2CH_2MgBr + CH_3-\overset{O}{\overset{\|}{C}}-CH_2-\overset{O}{\overset{\|}{C}}-CH_3 \longrightarrow$ 'A' $\xrightarrow{H_2O}$ 'B' Major Product

Consider the above reaction sequence and identify the product B. **[June 26, 2022 (I)]**

(a) HO, OH, CH_3, CH_3 (substituted cyclopentane) (b) OH, OH (cyclohexane-1,4-diol)

(c) OH, H_3C, H_3C, OH (d) O, O, H_3C, CH_3

111. Which will have the highest enol content ?

[NCERT : P L-242 | June 26, 2022 (I)]

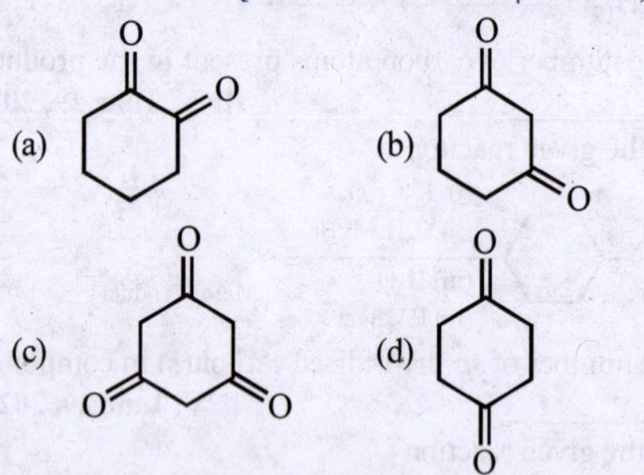

112. The reagent, from the following, which converts benzoic acid to benzaldehyde in one step is

[NCERT : P L-253 | June 26, 2022 (II)]

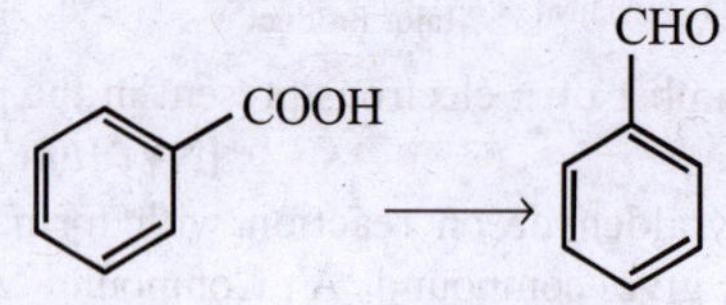

(a) $LiAlH_4$ (b) $KMnO_4$
(c) MnO (d) $NaBH_4$

113. The final product 'A' in the following reaction sequence

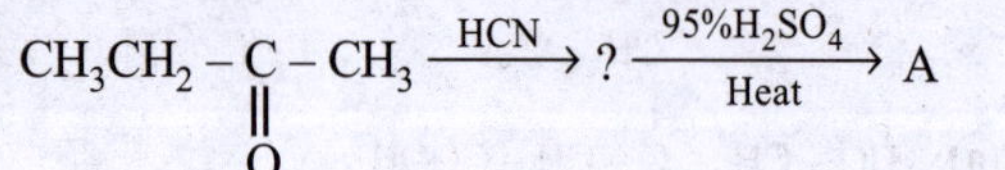

$$CH_3CH_2-\underset{\underset{O}{||}}{C}-CH_3 \xrightarrow{HCN} ? \xrightarrow[\text{Heat}]{95\%H_2SO_4} A$$

[NCERT : P L-246 | June 26, 2022 (II)]

(a) $CH_3-CH=\overset{\overset{CH_3}{|}}{C}-COOH$

(b) $CH_3-CH=\underset{\underset{CH_3}{|}}{C}-CN$

(c) $CH_3-CH-\overset{\overset{OH}{|}}{\underset{\underset{CH_3}{|}}{C}}-COOH$

(d) $CH_3-CH=\underset{\underset{CH_3}{|}}{C}-CONH_2$

114. Which of the following ketone will NOT give enamine on treatment with secondary amines?
[where t-Bu is $-C(CH_3)_3$]

[NCERT : P L-252 | June 25, 2022 (II)]

(a) $C_2H_5-\overset{\overset{O}{||}}{C}-C_2H_5$ (b) $C_2H_5-\overset{\overset{O}{||}}{C}-CH_3$

(c) $t\text{-}Bu-\overset{\overset{O}{||}}{C}-t\text{-}Bu$ (d) cyclohexanone

115. Which of the following is an example of conjugated diketone? **[NCERT : P L-245 | June 24, 2022 (I)]**

(a) $CH_3-\overset{\overset{O}{||}}{C}-CH_2-CH_2-CH_2-\overset{\overset{O}{||}}{C}-CH_3$

(b) $CH_3-\overset{\overset{O}{||}}{C}-CH_2-$(cyclohexane)$=O$

(c) $O=$(cyclohexa-2,5-diene)$=O$

(d) $C_6H_5-\overset{\overset{O}{||}}{C}-CH_2-CH_2-\overset{\overset{O}{||}}{C}-CH_3$

116. Number of electrophilic centre in the given compound is ______ **[NCERT : P L-236 | NV, June 24, 2022 (I)]**

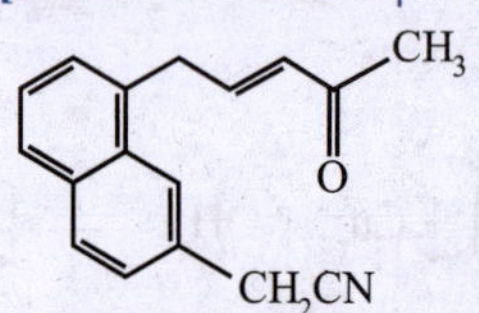

117. Hex-4-ene-2-ol on treatment with PCC gives '*A*'. '*A*' on reaction with sodium hypoiodite gives '*B*', which on further heating with soda lime gives '*C*'. The compound '*C*' is

[NCERT : P L-239 | June 24, 2022 (II)]

(a) 2- pentene (b) proponaldehyde
(c) 2-butene (d) 4-methylpent-2-ene

118. Given below are two statements :

Statement I : The nucleophilic addition of sodium hydrogen sulphite to an aldehyde or a ketone involves proton transfer to form a stable ion.

Statement II : The nucleophilic addition of hydrogen cyanide to an aldehyde or a ketone yields amine as final product.

In the light of the above statements, choose the **most appropriate** answer from the options given below :

[NCERT : P L-236 | Sep. 1, 2021 (II)]

(a) Both Statement I and Statement II are true.
(b) Statement I is true but Statement II is false.
(c) Statement I is false but Statement II is true.
(d) Both Statement I and Statement II are false.

119. For the reaction given below : [NCERT : P L-242 | Aug. 31, 2021 (II)]

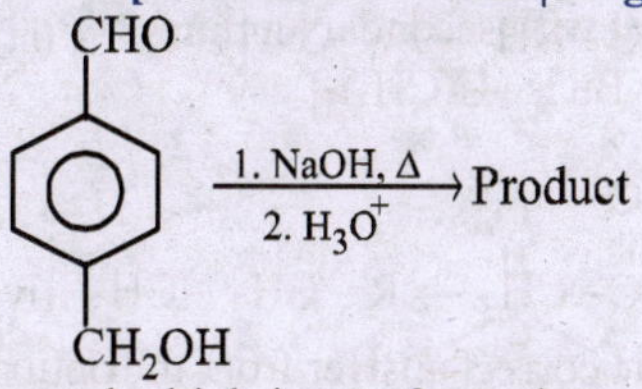

(4-(hydroxymethyl)benzaldehyde: CHO and CH_2OH para on benzene ring) $\xrightarrow[2.\ H_3O^+]{1.\ NaOH,\ \Delta}$ Product

The compound which is **not** formed as a product in the reaction is a :
(a) compound with both alcohol and acid functional groups
(b) monocarboxylic acid
(c) dicarboxylic acid
(d) diol

120. The structure of the starting compound P used in the reaction given below is :

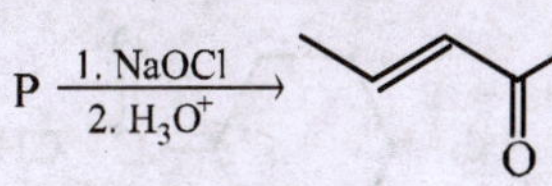

P $\xrightarrow[2.\ H_3O^+]{1.\ NaOCl}$ (but-2-enoic acid, with OH and O) [Aug. 27, 2021 (I)]

(a) 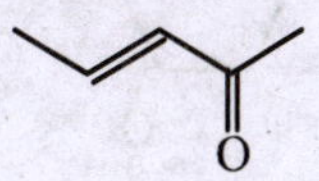 (pent-3-en-2-one structure) (b) 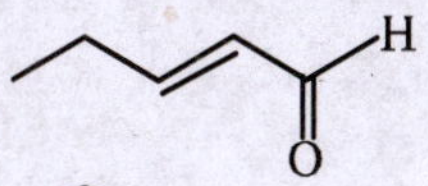 (hex-2-enal structure, with H)
(c) 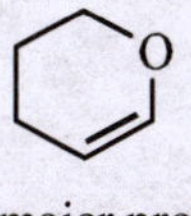 (dihydropyran ring with O) (d) 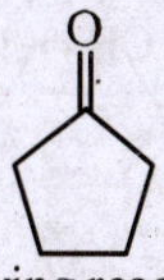 (cyclopentanone)

121. The major product of the following reaction is:

$CH_3-\underset{CH_3}{CH}-CH_2-CH_2-\overset{O}{\overset{\|}{C}}-Cl \xrightarrow[\text{(iii) } NaNO_2, HCl \text{ (iv) } H_2O]{\text{(i) alcoholic } NH_3 \text{ (ii) NaOH, } Br_2}$ Major product

[NCERT : P L-253 | Aug. 27, 2021 (I)]

(a) $CH_3-\underset{CH_3}{CH}-\overset{Br}{CH}-CH_2OH$
(b) $CH_3-\underset{CH_3}{CH}-CH_2-CH_2-CH_2OH$
(c) $CH_3-\underset{CH_3}{CH}-CH_2-CH_2OH$
(d) $CH_3-\underset{CH_3}{CH}-CH_2-CH_2-Cl$

122. Which one of the following is the major product of the given reaction? [Aug. 27, 2021 (II)]

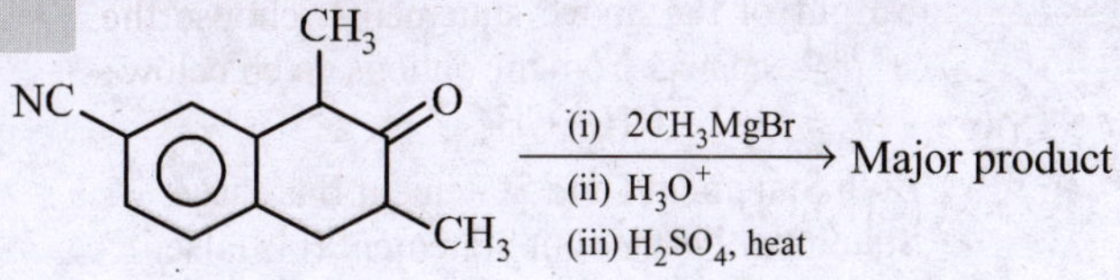

(bicyclic ketone with NC, CH_3, O and CH_3 substituents) $\xrightarrow[\text{(ii) } H_3O^+ \text{ (iii) } H_2SO_4\text{, heat}]{\text{(i) } 2CH_3MgBr}$ Major product

(a) 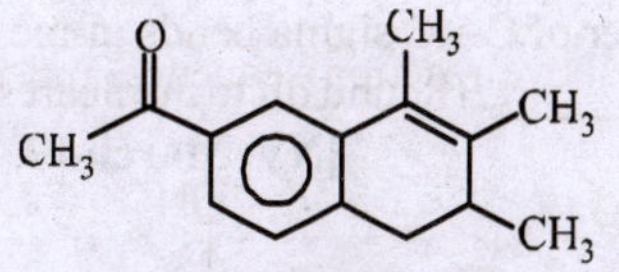 (structure with CH_3–C(=O)–, CH_3, CH_3, CH_3)
(b) (structure with NC, CH_3, CH_3, CH_3)
(c) (structure with NC, CH_3, CH_3, CH_3)
(d) 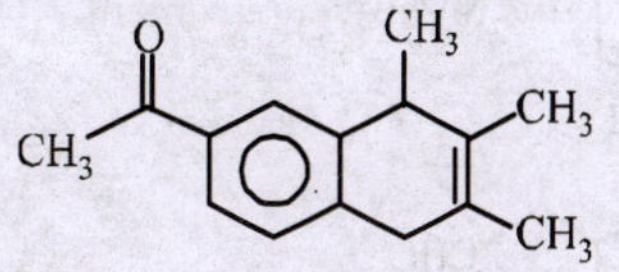 (structure with CH_3–C(=O)–, CH_3, CH_3, CH_3)

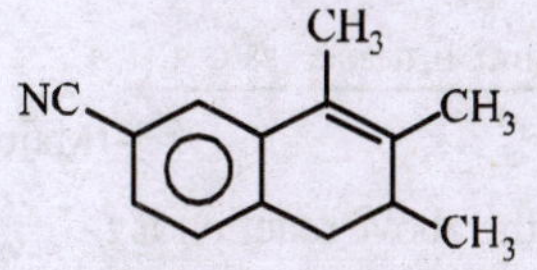

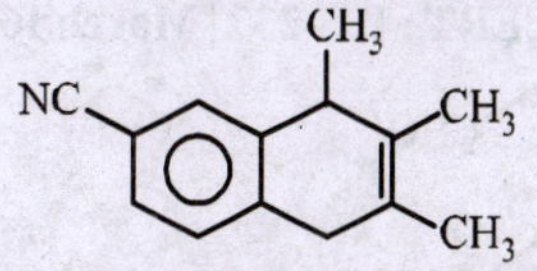

123. 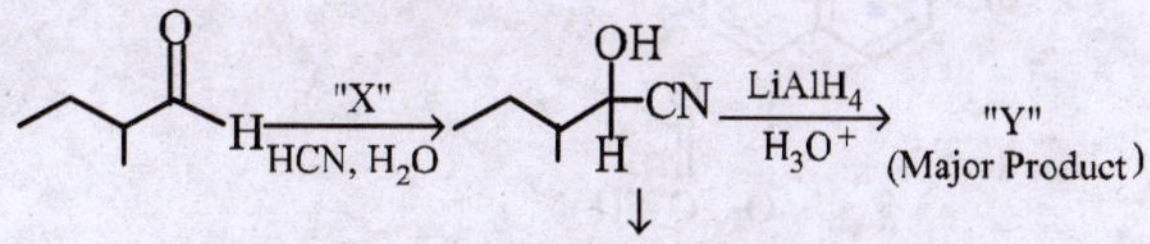 (2-methylbutanal, with O and H) $\xrightarrow[HCN,\ H_2O]{"X"}$ (cyanohydrin with OH, CN, H) $\xrightarrow[H_3O^+]{LiAlH_4}$ "Y" (Major Product)

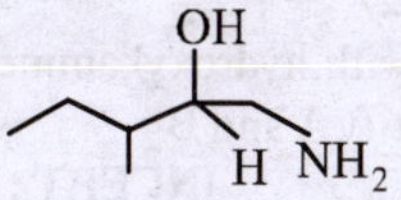

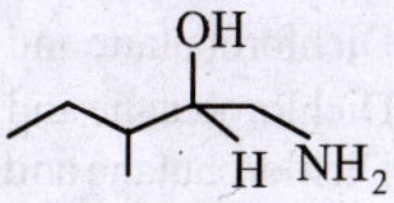

Consider the given reaction, Identify 'X' and 'Y' : [Aug. 26, 2021 (II)]

(a) X – NaOH Y – (structure with OH, NH_2, H)
(b) X – HNO_3 Y – (structure with OH, H, NH_2)
(c) X – NaOH Y – (structure with OH, H, NH_2)
(d) X – HNO_3 Y – (structure with OH, NH_2, H)

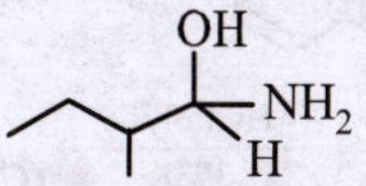

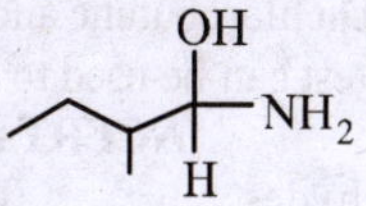

124. 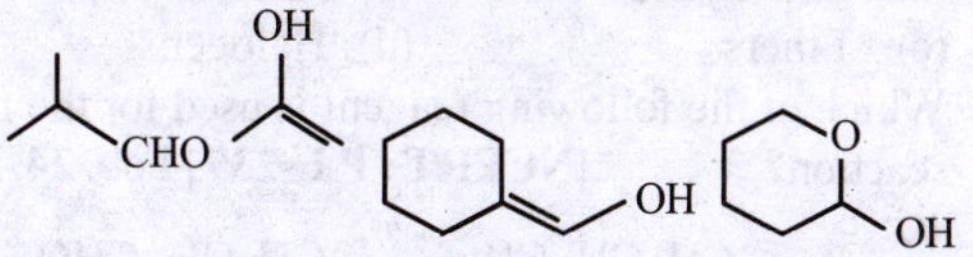 (I) (structure with CHO) (II) (structure with OH and O) (III) (structure with OH) (IV) (cyclic structure with O and OH)

Which among the above compound/s does/do not form silver mirror when treated with Tollen's reagent? [July 20, 2021 (I)]
(a) (I), (III) and (IV) only (b) Only (IV)
(c) Only (II) (d) (III) and (IV) only

125. The number of nitrogen atoms in a semicarbazone molecule of acetone is ________. [NV, July 20, 2021 (I)]

126. In Tollen's test for aldehyde, the overall number of electrons(s) transferred to the Tollen's reagent formula $[Ag(NH_3)_2]^+$ per aldehyde group to form silver mirror is ________. (Round off to the nearest integer). [NCERT : P L-239 | NV, March 18, 2021 (II)]

127. The total number of C – C sigma bonds in mesityl oxide ($C_6H_{10}O$) is ________. (Round off to the nearest integer).
[NV, March 17, 2021 (II)]

128. 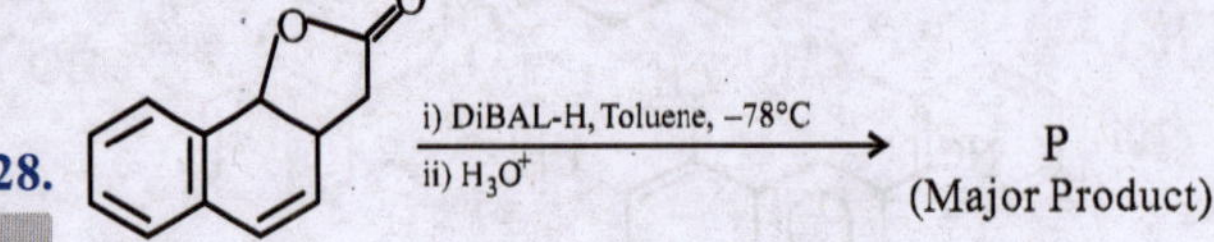

P (Major Product)

The product "P" in the above reaction is :
[NCERT : P L-237 | March 16, 2021 (I)]

(a) CHO
(b) COOH
(c) OH, CHO
(d) O – C – H (with C=O)

129. A ($C_4H_8Cl_2$) $\xrightarrow[373K]{\text{Hydrolysis}}$ B (C_4H_8O)
B reacts with hydroxyl amine but does not give Tollen's test. Identify A and B.
[NCERT : P L-239 | Feb. 26, 2021 (I)]
(a) 2, 2-Dichlorobutane and Butanal
(b) 1, 1-Dichlorobutane and Butanal
(c) 1, 1-Dichlorobutane and 2-Butanone
(d) 2, 2-Dichlorobutane and Butan-2-one

130. 2,4-DNP test can be used to identify
[NCERT : P L-239 | Feb. 26, 2021 (II)]
(a) Aldehydes (b) Amines
(c) Ethers (d) Halogens

131. Which of the following reagent is used for the following reaction? **[NCERT : P L-239 | Feb. 24, 2021 (I)]**

$$CH_3CH_2CH_3 \xrightarrow{?} CH_3CH_2CHO$$

(a) Manganese acetate
(b) Copper at high temperature and pressure
(c) Molybdenum oxide
(d) Potassium permanganate

132. Match list-I and list-II.
[NCERT : P L-239 | Feb. 24, 2021 (II)]

List-I	List-II
(A) $R-\overset{O}{\overset{\|}{C}}-Cl \rightarrow R-CHO$	(i) Br_2/NaOH
(B) $R-CH_2-COOH \rightarrow R-\underset{Cl}{\underset{\|}{CH}}-COOH$	(ii) H_2/Pd-$BaSO_4$
(C) $R-\overset{O}{\overset{\|}{C}}-NH_2 \rightarrow R-NH_2$	(iii) Zn(Hg)/ Conc. HCl
(D) $R-\overset{O}{\overset{\|}{C}}-CH_3 \rightarrow R-CH_2-CH_3$	(iv) Cl_2/Red P, H_2O

Choose the correct answer from the options given below:
(a) (A)-(ii), (B)-(i), (C)-(iv), (D)-(iii)
(b) (A)-(iii), (B)-(iv), (C)-(i), (D)-(ii)
(c) (A)-(ii), (B)-(iv), (C)-(i), (D)-(iii)
(d) (A)-(iii), (B)-(i), (C)-(iv), (D)-(ii)

133. Which one of the following carbonyl compounds cannot be prepared by addition of water on an alkyne in the presence of $HgSO_4$ and H_2SO_4?
[NCERT : P L-237 | Feb. 24, 2021 (II)]

(a) $CH_3-\overset{O}{\overset{\|}{C}}-H$ (b) cyclohexyl$-\overset{O}{\overset{\|}{C}}-CH_3$

(c) $CH_3-CH_2-\overset{O}{\overset{\|}{C}}-H$ (d) $CH_3-\overset{O}{\overset{\|}{C}}-CH_2CH_3$

134. 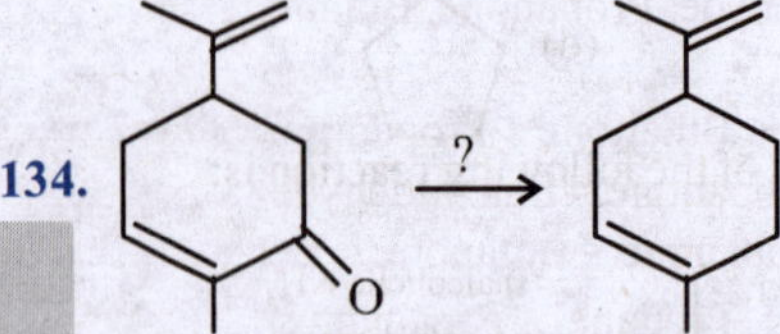

Which of the following reagent is suitable for the preparation of the product in the above reaction?
[Feb. 24, 2021 (II)]
(a) $NaBH_4$ (b) $NH_2-NH_2/C_2H_5\overset{\ominus}{O}\overset{\oplus}{Na}$
(c) Ni/H_2 (d) Red P + Cl_2

135. The major product [R] in the following sequence of reactions is : **[NCERT : P L-238 | Sep. 04, 2020 (II)]**

$$HC \equiv CH \xrightarrow[\text{(ii) } (H_3C)((CH_3)_2CH)CH-Br]{\text{(i) } LiNH_2/\text{ether}} [P] \xrightarrow[\text{(ii) } NaBH_4]{\text{(i) } HgSO_4/H_2SO_4} [Q] \xrightarrow[\Delta]{\text{Conc. } H_2SO_4} [R]$$

(a) $H_2C=C(CH(CH_3)_2)-CH_2-CH_3$

(b) $(H_3C)((CH_3)_2CH)C=CH-CH_3$

(c) $H_3C(H_3CCH_2)C=C(CH_3)_2$

(d) $H_3C((CH_3)_2CH)CH-CH=CH_2$

136. The major product [C] of the following reaction sequence will be : [NCERT : P L-239 | Sep. 04, 2020 (II)]

$$CH_2=CH-CHO \xrightarrow[\text{(ii) } SOCl_2]{\text{(i) } NaBH_4} [A] \xrightarrow[\text{Anhy. } AlCl_3]{C_6H_6} [B] \xrightarrow{DBr} [C]$$

(a) $C_6H_5-CH(Br)-CH_2-CH_2D$ (b) $C_6H_5-CH_2-CH(D)-CH_2Br$

(c) $C_6H_5-CH_2-CH(Br)-CH_2D$ (d) $C_6H_5-CH(Br)-CH(D)-CH_3$

137. The increasing order of the reactivity of the following compounds in nucleophilic addition reaction is :
Propanal, Benzaldehyde, Propanone, Butanone
[Sep. 03, 2020 (II), Online May 7, 2012 (s)]

(a) Benzaldehyde < Butanone < Propanone < Propanal
(b) Butanone < Propanone < Benzaldehyde < Propanal
(c) Propanal < Propanone < Butanone < Benzaldehyde
(d) Benzaldehyde < Propanal < Propanone < Butanone

138. The compound A in the following reactions is :
[NCERT : P L-231 | Sep. 03, 2020 (II)]

$$A \xrightarrow[\text{(ii) Conc. } H_2SO_4/\Delta]{\text{(i) } CH_3MgBr/H_2O} B \xrightarrow[\text{(ii) } Zn/H_2O]{\text{(i) } O_3} C + D$$

$$C \xrightarrow[\text{(ii) } \Delta]{\text{(i) Conc. KOH}} C_6H_5-COO^{\ominus}K^{+} + C_6H_5-CH_2OH$$

$$D \xrightarrow[\Delta]{Ba(OH)_2} H_3C-\overset{CH_3}{\overset{|}{C}}=CH-\overset{O}{\overset{||}{C}}-CH_3$$

(a) $C_6H_5-\overset{O}{\overset{||}{C}}-CH_3$

(b) $C_6H_5-\overset{O}{\overset{||}{C}}-CH(CH_3)_2$

(c) $C_6H_5-CH_2-\overset{O}{\overset{||}{C}}-CH_3$

(d) $C_6H_5-\overset{O}{\overset{||}{C}}-CH_2CH_3$

139. The increasing order of the following compounds towards HCN addition is : [NCERT : P L-237 | Sep. 02, 2020 (I)]

(i) 3-methoxybenzaldehyde (H_3CO and CHO on benzene, meta) (ii) 2-nitrobenzaldehyde (CHO, NO_2 ortho)

(iii) 2-methoxybenzaldehyde (CHO, OCH_3 ortho) (iv) 3-nitrobenzaldehyde (O_2N and CHO, meta)

(a) (i) < (iii) < (iv) < (ii) (b) (iii) < (iv) < (i) < (ii)
(c) (iii) < (i) < (iv) < (ii) (d) (iii) < (iv) < (ii) < (i)

140. Identify (A) in the following reaction sequence:
[NCERT : P L-231 | Jan. 09, 2020 (I)]

$$\underset{\text{Gives positive iodoform test}}{(A)} \xrightarrow[\text{(ii) } H^+, H_2O \text{ (ii) Conc. } H_2SO_4/\Delta]{\text{(i) } CH_3MgBr} (B) \xrightarrow{O_3/Zn, H_2O}$$ product: benzene ring bearing $-\overset{O}{\overset{||}{C}}-H$ and, ortho to it, $-C(CH_3)(H_3C)-C(=O)-CH_3$

(a) $C_6H_5-C(CH_3)=CH-C(=O)-CH_3$

(b) $C_6H_5-CH=C(CH_3)-C(=O)-CH_3$

(c) $C_6H_5-CH=CH-C(=O)-CH_3$

(d) $C_6H_5-C(CH_3)=C(CH_3)-C(=O)-CH_3$

141. Consider the following reactions

$$A \xrightarrow[\text{(ii) } H_3O^+]{\text{(i) } CH_3MgBr} B \xrightarrow[573\text{ K}]{Cu} \text{2-Methyl-2-butene}$$

The mass percentage of carbon in A is ______.
[NCERT : P L-239 | NV, Jan. 09, 2020 (II)]

142. The major product in the following reaction is:
[NCERT : P L-237 | Jan. 08, 2020 (II)]

2-methylcycloprop-2-en-1-one (C=O with ring C=C bearing CH_3) $+ H_3O^{\oplus} \longrightarrow$

(a) cyclopropanone with HO and CH_3 on the ring carbons (b) cyclopropene with C(OH)(OH) and CH_3

(c) cyclopropanone with OH and CH_3 on the same carbon (d) cyclopropenium with $\overset{\oplus}{O}H$ ... $OH^{\oplus}$ and CH_3

143. What is the product of following reaction?

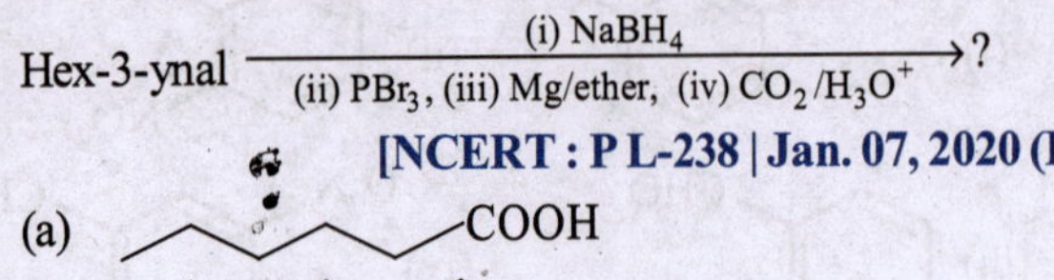

[NCERT : P L-238 | Jan. 07, 2020 (I)]

(a) COOH
(b) COOH
(c) COOH
(d) COOH

144. Major products of the following reaction are :
[NCERT : P L-242 | April 10, 2019 (I)]

CHO + HCHO $\xrightarrow[\text{(ii) } H_3O^+]{\text{(i) 50\% NaOH}}$

(a) CH_2OH and COOH
(b) CH_3OH and COOH
(c) HCOOH and CH_2OH
(d) CH_3OH and HCO_2H

145. Compound A($C_9H_{10}O$) shows positive iodoform test. Oxidation of A with $KMnO_4$/KOH gives acid B($C_8H_6O_4$). Anhydride of B is used for the preparation of phenolphthalein. Compound A is :
[NCERT : P L-248 | April 10, 2019 (II)]

(a) CH_3, O
(b) $CH_2—C(=O)—H$, CH_3
(c) CH_3, O, CH_3
(d) CH_3, CH_3, O

146. *p*-Hydroxybenzophenone upon reaction with bromine in carbon tetrachloride gives: **[April 9, 2019 (II)]**

(a) Br, O, HO
(b) Br, O, HO
(c) O, Br, HO
(d) O, Br, HO

147. The major product of the following reaction is :
[NCERT : P L-238 | April 8, 2019 (I)]

O, Br $\xrightarrow[\text{MeOH, 25 °C}]{NaBH_4}$

(a) OH, Br
(b) OH, OMe
(c) OMe
(d) O

148. Which of the following compounds will show the maximum 'enol' content? **[NCERT : P L-236 | April 8, 2019 (II)]**

(a) $CH_3COCH_2COOC_2H_5$ (b) $CH_3COCH_2COCH_3$
(c) CH_3COCH_3 (d) $CH_3COCH_2CONH_2$

149. An organic compound neither reacts with neutral ferric chloride solution nor with Fehling solution. It however, reacts with Grignard reagent and gives positive iodoform test. The compound is :
[NCERT : P L-239 | April 8, 2019 (I)]

(a) O, CH_3, OH
(b) O, CH_3, H, O
(c) O, C_2H_5, O, CH_3
(d) OH, CH_3, C_2H_5, O

150. The major product obtained in the following reaction is :
[NCERT : P L-241 | April 8, 2019 (II)]

CH_3, O, OHC $\xrightarrow[\Delta]{NaOH}$

(a) 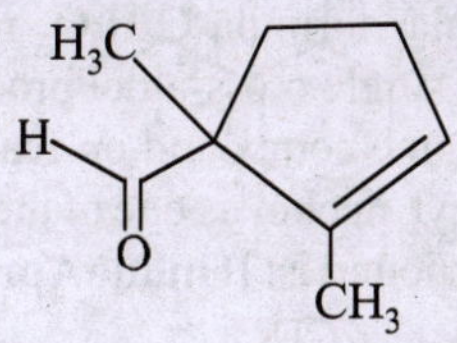 [structure: 1-methyl-2-methylcyclopent-2-ene-1-carbaldehyde; H_3C, H, O, CH_3]

(b) 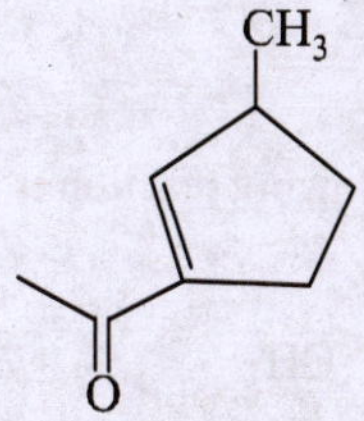 [structure: CH_3, O]

(c) 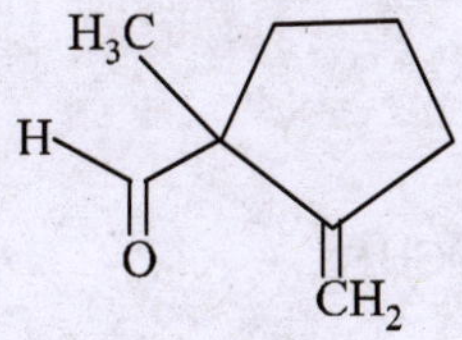 [structure: H_3C, H, O, CH_2]

(d) 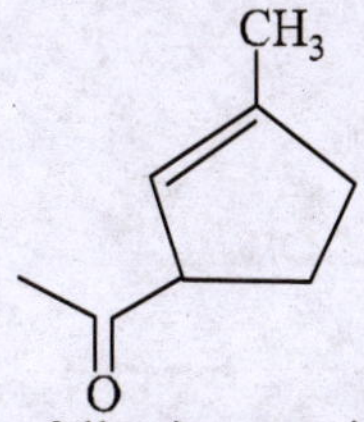 [structure: CH_3, O]

151. In the following reactions, products A and B are:

[NCERT : P L-239 | Jan. 12, 2019 (I)]

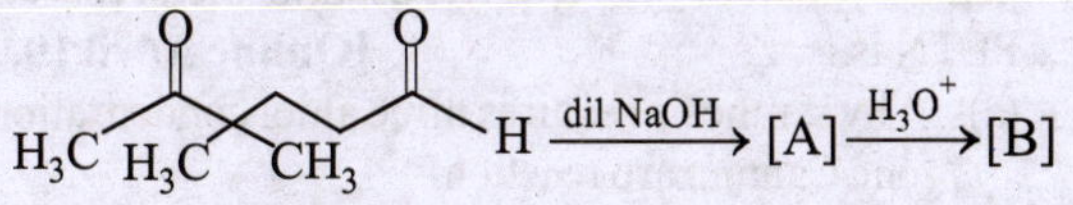

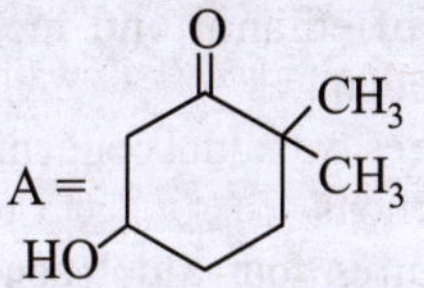

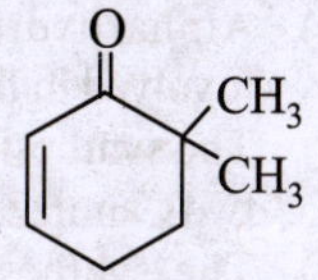

H_3C–CO–C(CH_3)(H_3C)–CH_2–CH_2–CHO $\xrightarrow{\text{dil NaOH}}$ [A] $\xrightarrow{H_3O^+}$ [B]

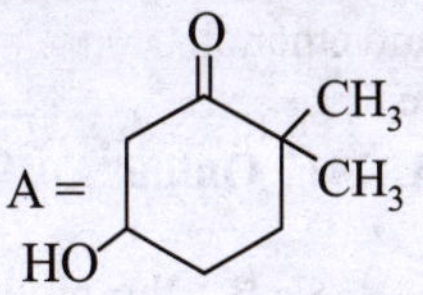

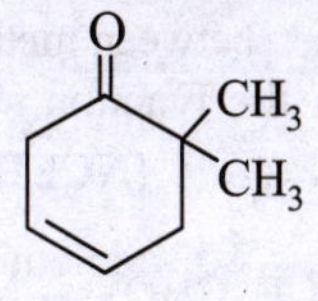

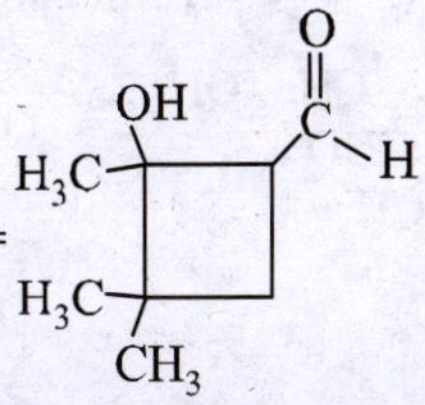

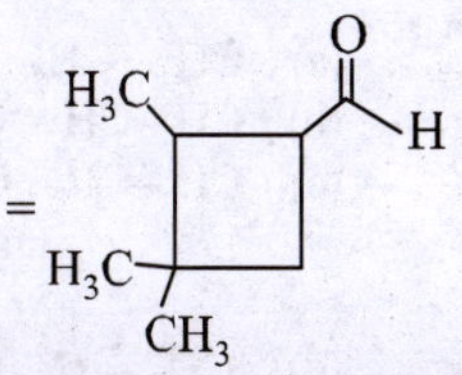

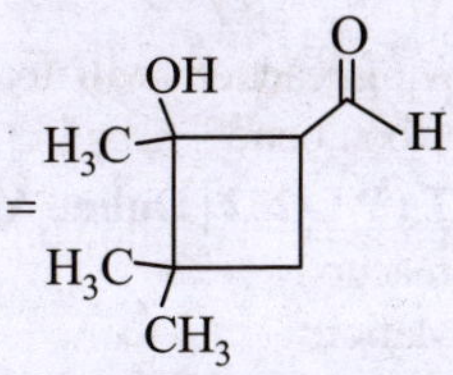

(a) A = [structure: HO, O, CH_3, CH_3] ; B = [structure: O, CH_3, CH_3]

(b) A = [structure: HO, O, CH_3, CH_3] ; B = [structure: O, CH_3, CH_3]

(c) A = [structure: OH, H_3C, H_3C, CH_3, C(O)H] ; B = [structure: H_3C, H_3C, CH_3, CHO]

(d) A = [structure: OH, H_3C, H_3C, CH_3, CHO] ; B = [structure: H_2C, H_3C, CH_3, CHO]

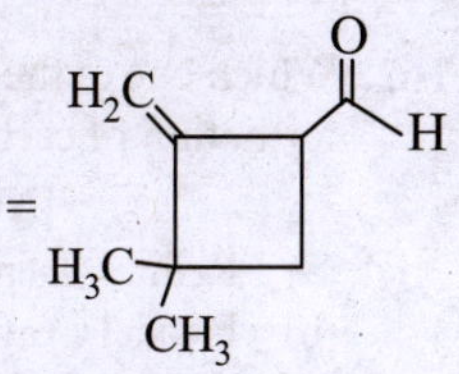

152. The major product of the following reaction is :

[NCERT : P L-238 | Jan. 12, 2019 (II)]

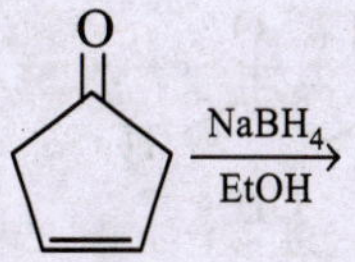

[structure: cyclopent-2-enone] $\xrightarrow[\text{EtOH}]{NaBH_4}$

(a) 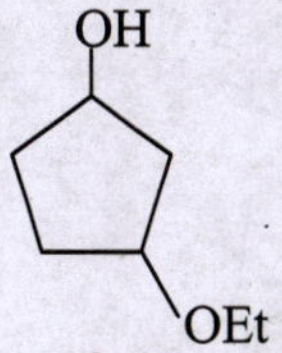 [structure: OH, OEt] (b) 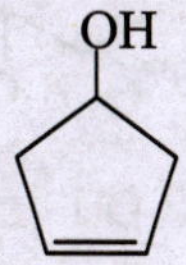 [structure: OH]

(c) 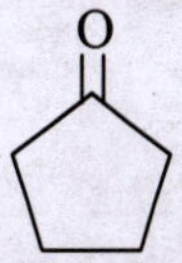 [structure: O] (d) 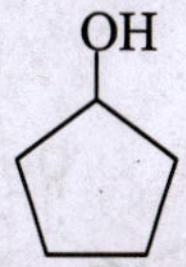 [structure: OH]

153. The aldehydes which will **not** form Grignard product with one equivalent of Grignard reagents are:

[NCERT : P L-239 | Jan. 12, 2019 (II)]

(A) 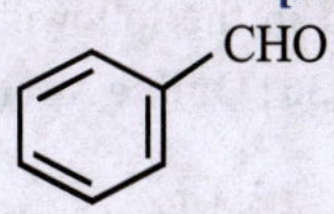 [structure: CHO]

(B) 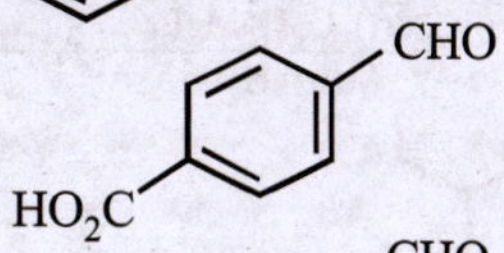 [structure: CHO, HO_2C]

(C) 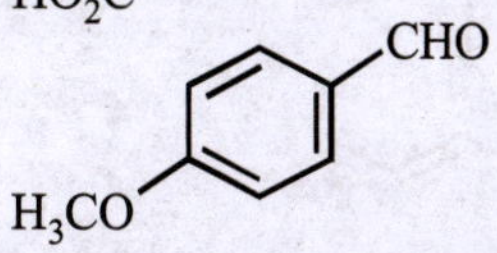 [structure: CHO, H_3CO]

(D) 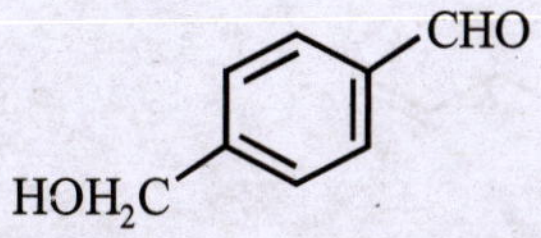 [structure: CHO, HOH_2C]

(a) (B), (D) (b) (B), (C)

(c) (B), (C), (D) (d) (C), (D)

154. The major product of the following reaction is:

[NCERT : P L-238 | Jan. 10, 2019 (II)]

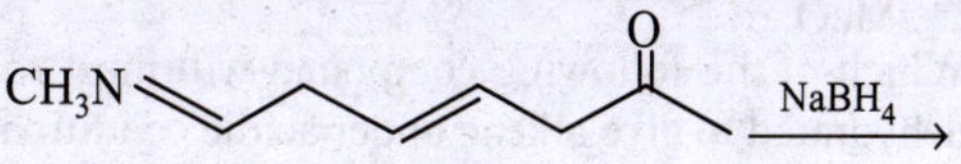

CH_3N= [structure: O] $\xrightarrow{NaBH_4}$

(a) 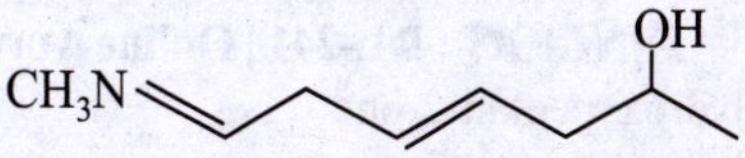 CH_3N= [structure: OH]

(b) 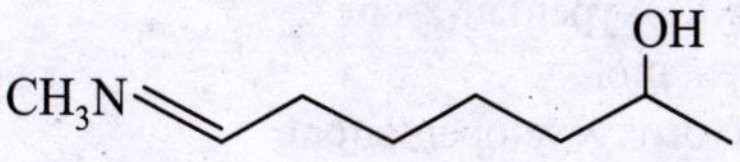 CH_3N= [structure: OH]

(c) 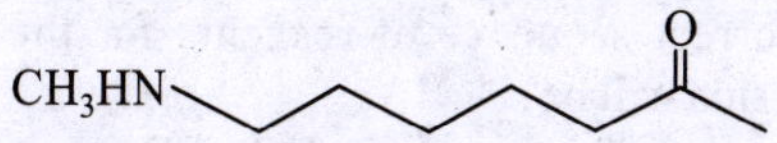 CH_3HN [structure: O]

(d) 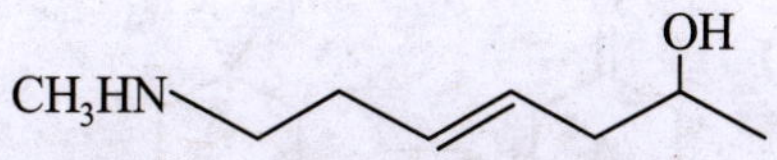 CH_3HN [structure: OH]

155. The major product formed in the following reaction is:

[NCERT : P L-241 | Jan. 9, 2019 (II)]

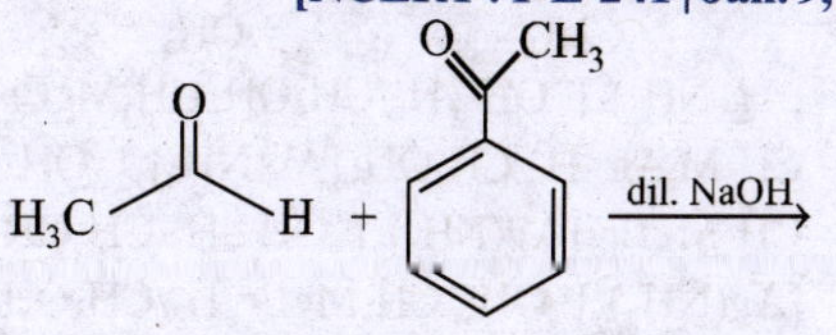

H_3C–CO–H + [structure: acetophenone, O, CH_3] $\xrightarrow{\text{dil. NaOH}}$

(a) $H_3C-CH(OH)-CH_2-CHO$

(b) $H-CO-CH_2-C(OH)(CH_3)-C_6H_5$

(c) $H_3C-CH(OH)-CH_2-CO-C_6H_5$

(d) $C_6H_5-CO-CH_2-C(OH)(CH_3)-C_6H_5$

156. The major product B formed in the following reaction sequence is: **[NCERT : P L-236 | Online April 16, 2018]**

$$MeO-C_6H_4-CHO \xrightarrow[\text{(ii) } H_2O]{\text{(i) } C_2H_5MgBr} A \xrightarrow{HCl} B$$

(a) Cl–C₆H₄–CH₂–CH(Cl)–CH₃ (para-chloro ring)

(b) MeO–C₆H₄–CH=CH–CH₃

(c) MeO–C₆H₄–CH₂–CH=CH₂

(d) MeO–C₆H₄–CH(Cl)–CH₂–CH₃

157. Which of the following compound will most readily be dehydrated to give alkene under acidic condition? **[NCERT : P L-241 | Online April 16, 2018]**

(a) 4-Hydroxypentan-2-one
(b) 3-Hydroxypentan-2-one
(c) 1-Pentanol
(d) 2-Hydroxycyclopentanone

158. The correct sequence of reagents for the following conversion will be : **[2017]**

4-formylcyclohexanone (O, CHO) $\longrightarrow$ 1,4-dimethylcyclohexane-1,4-diol (HO, CH₃ ; HO, CH₃)

(a) $[Ag(NH_3)_2]^+OH^-$, H^+/CH_3OH, CH_3MgBr
(b) CH_3MgBr, H^+/CH_3OH, $[Ag(NH_3)_2]^+OH^-$
(c) CH_3MgBr, $[Ag(NH_3)_2]^+OH^-$, H^+/CH_3OH
(d) $[Ag(NH_3)_2]^+OH^-$, CH_3MgBr, H^+/CH_3OH

159. A compound of molecular formula $C_8H_8O_2$ reacts with acetophenone to form a single cross-aldol product in the presence of base. The same compound on reaction with conc. NaOH forms benzyl alcohol as one of the products. The structure of the compound is: **[Online April 9, 2017]**

(a) $CH_3O-C_6H_4-CHO$ (para)

(b) $HO-C_6H_4-CO-CH_3$ (para)

(c) $C_6H_5-CO-OCH_3$

(d) $CH_3-C_6H_4-COOH$ (para)

160. The correct statement about the synthesis of pentaerythritol ($C(CH_2OH)_4$) used in the preparation of PETN is : **[Online April 10, 2016]**

(a) Thy synthesis requires three aldol condensations and one Cannizzaro reaction.
(b) Alpha hydrogens of ethanol and methanol are involved in this reaction.
(c) The synthesis requires two aldol condensations and two Cannizzaro reactions.
(d) The synthesis requires four aldol condensations between methanol and ethanol.

161. In the reaction sequence **[NCERT : P L-241 | Online April 11, 2015]**

$$2CH_3CHO \xrightarrow{OH^-} A \xrightarrow{\Delta} B$$; the product B is :

(a) $CH_3-CH_2-CH_2-CH_2-OH$
(b) $CH_3-CH=CH-CHO$
(c) $CH_3-CH_2-CH_2-CH_3$
(d) $CH_3-\overset{O}{\overset{\|}{C}}-CH_3$

162. Which one of the following reactions will not result in the formation of carbon-carbon bond? **[NCERT : P L-242 | Online April 9, 2014]**

(a) Reimer-Tiemann reaction
(b) Friedel Craft's acylation
(c) Wurtz reaction
(d) Cannizzaro reaction

163. Which is major product formed when acetone is heated with iodine and potassium hydroxide? **[NCERT : P L-240 | Online April 9, 2014]**

(a) Iodoacetone (b) Acetic acid
(c) Iodoform (d) Acetophenone

164. An organic compound A, C_5H_8O; reacts with H_2O, NH_3 and CH_3COOH as described below: **[Online April 11, 2014]**

$$A \xrightarrow{H_2O} \sim\sim CH-\overset{O}{\overset{||}{C}}-OH$$
$$A \xrightarrow{NH_3} \sim\sim CH-\overset{O}{\overset{||}{C}}-NH_2$$
$$A \xrightarrow{CH_3COOH} \sim\sim CH-\overset{O}{\overset{||}{C}}-O-\overset{O}{\overset{||}{C}}-CH_3$$

A is:

(a) $CH_3CH=\underset{\underset{CH_3}{|}}{C}-CHO$

(b) $CH_2=CH\,\underset{\underset{CH_3}{|}}{CH}-CHO$

(c) $CH_3-CH_2-\underset{\underset{CH_3}{|}}{C}=C=O$

(d) $CH_3-CH_2-\underset{\underset{CH_2}{||}}{C}-\underset{\underset{H}{|}}{C}=O$

165. Tischenko reaction is a modification of: **[NCERT : P L-242 | Online April 11, 2014]**

(a) Aldol condensation
(b) Claisen condensation
(c) Cannizzaro reaction
(d) Pinacol-pinacolone reaction

166. Formaldehyde can be distinguished from acetaldehyde by the use of: **[NCERT : P L-240 | Online April 9, 2013]**

(a) Schiff's reagent (b) Tollen's reagent
(c) I_2/Alkali (d) Fehling's solution

167. Clemmensen reduction of a ketone is carried out in the presence of: **[NCERT : P L-237 | Online April 22, 2013]**

(a) $LiAlH_4$ (b) Zn-Hg with HCl
(c) Glycol with KOH (d) H_2 with Pt as catalyst

168. Which of the following is the product of aldol condensation? **[NCERT : P L-241 | Online April 23, 2013]**

(a) $(CH_3)_2CH-CH_2-CO-CH_2-OH$ (b) $(CH_3)_2C(OH)-CH_2-CO-CH_3$

(c) $HO-CH_2-CH_2-CH_2-CO-CH_3$ (d) $(CH_3)_2CH-CH(OH)-CO-CH_3$

169. Cannizzaro's reaction is not given by: **[NCERT : P L-242 | Online April 25, 2013]**

(a) C_6H_5-CHO (b) 1-methylcyclohexane with CHO and CH_3 on the same carbon

(c) CH_3CHO (d) HCHO

170. Which of the following reagent(s) is/are used for the conversion? **[NCERT : P L-242 | Online April 25, 2013]**

$$CH_3-\overset{O}{\overset{||}{C}}-CH_2-CH_2-CO_2Et \longrightarrow CH_3-\overset{O}{\overset{||}{C}}-CH_2-CH_2-CH_2-OH$$

(a) glycol/$LiAlH_4$/H_3O^+ (b) glycol/NaH/H_3O^+
(c) $LiAlH_4$ (d) $NaBH_4$

171. Iodoform can be prepared from all except: **[2012]**

(a) Ethyl methyl ketone (b) Isopropyl alcohol
(c) 3-Methyl 2-butanone (d) Isobutyl alcohol

172. In the given transformation, which of the following is the most appropriate reagent? **[2012]**

$$HO-C_6H_{10}-CH=CH-COCH_3 \xrightarrow{\text{Reagent}} HO-C_6H_{10}-CH=CH-CH_2-CH_3$$

(a) $NH_2NH_2, \overset{-}{O}H$ (b) Zn – Hg/ HCl
(c) Na, Liq. NH_3 (d) $NaBH_4$

173. Among the following the order of reactivity towards nucleophilic addition is **[NCERT : P L-236 | Online May 7, 2012]**

(a) $CH_3CHO > CH_3COCH_3 > HCHO$
(b) $HCHO > CH_3CHO > CH_3COCH_3$
(c) $CH_3CHO > HCHO > CH_3COCH_3$
(d) $CH_3COCH_3 > CH_3CHO > HCHO$

174. The reaction, **[NCERT : P L-239 | Online May 12, 2012]**

$$CH_3CHO \xrightarrow{Zn(Hg)/Conc.\ HCl\ [H]} CH_3CH_3 \text{ is}$$

(a) Cannizzaro's reaction (b) Rosenmund reduction
(c) Wolf-Kishner reduction (d) Clemmenson reduction

175. Tollen's reagent and Fehling solutions are used to distinguish between **[NCERT : P L-239 | Online May 26, 2012]**

(a) acids and alcohols
(b) alkanes and alcohols
(c) ketones and aldehydes
(d) *n*-alkanes and branched alkanes

176. In Cannizzaro reaction given below

$$2PhCHO \xrightarrow{:\overset{-}{O}H} PhCH_2OH + PhC\overset{-}{O_2}$$

the slowest step is : **[NCERT : P L-239 | 2009]**

(a) the transfer of hydride to the carbonyl group
(b) the abstraction of proton from the carboxylic group
(c) the deprotonation of Ph CH_2OH
(d) the attack of : OH at the carboxyl group

177. The increasing order of the rate of HCN addition to compound A to D is **[2006]**

(A) HCHO (B) CH_3COCH_3
(C) $PhCOCH_3$ (D) PhCOPh

(a) D < C < B < A (b) C < D < B < A
(c) A < B < C < D (d) D < B < C < A

178. Reaction of cyclohexanone with dimethylamine in the presence of catalytic amount of an acid forms a compound if water during the reaction is continuously removed. The compound formed is generally known as **[2005]**

(a) an amine (b) an imine
(c) an enamine (d) a Schiff's base

179. On mixing ethyl acetate with aqueous sodium chloride, the composition of the resultant solution is **[2004]**

(a) $CH_3COCl + C_2H_5OH + NaOH$
(b) $CH_3COONa + C_2H_5OH$
(c) $CH_3COOC_2H_5 + NaCl$
(d) $CH_3Cl + C_2H_5COONa$

180. Which one of the following is reduced with zinc amalgam and hydrochloric acid to give the corresponding hydrocarbon? **[NCERT : P L-239 | 2004]**

(a) Acetamide (b) Acetic acid
(c) Ethyl acetate (d) Butan-2-one

181. Which one of the following undergoes reaction with 50% sodium hyroxide solution to give the corresponding alcohol and acid? **[NCERT : P L-242 | 2004]**

(a) Butanal (b) Benzaldehyde
(c) Phenol (d) Benzoic acid

182. Acetyl bromide reacts with excess of CH_3MgI followed by treatment with a saturated solution of NH_4Cl gives **[2004]**

(a) 2-methyl-2-propanol (b) acetamide
(c) acetone (d) acetyl iodide

183. On vigorous oxidation by permanganate solution. **[NCERT : P L-249 | 2002]**

$(CH_3)_2C = CH - CH_2 - CHO$ gives

(a) $CH_3 - C(OH)(CH_3) - CH(OH) - CH_2CHO$

(b) $(CH_3)_2C = O + HOOCCH_2COOH$

(c) $(CH_3)_2C = O + OHCCH_2COOH$

(d) $(CH_3)_2C = O + OHCCH_2CHO$

Topic 3 - Preparation and Properties of Carboxylic Acids and their Derivatives

184. Mixture of 1 g each of chlorobenzene, aniline and benzoic acid is dissolved in 50 mL ethyl acetate and placed in a separating funnel, 5 M NaOH (30 mL) was added in the same funnel. The funnel was shaken vigorously and then kept aside. The ethyl acetate layer in the funnel contains : **[NCERT : P L-245 | April 7, 2025 (II)]**

(a) benzoic acid
(b) benzoic acid and aniline
(c) benzoic acid and chlorobenzene
(d) chlorobenzene and aniline

185. Which of the following compounds is least likely to give effervescence of CO_2 in presence of aq. $NaHCO_3$? **[NCERT : P L-249 | April 7, 2025 (I)]**

(a) 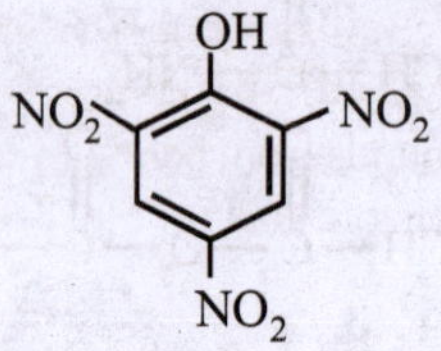 2,4,6-trinitrophenol (OH with NO_2 groups at 2, 4, 6) (b) 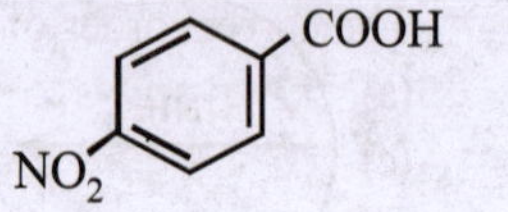 4-nitrobenzoic acid (NO_2–C_6H_4–COOH)

(c) $Ph - \overset{(+)}{N}H_3\overset{(-)}{Cl}$ (d) 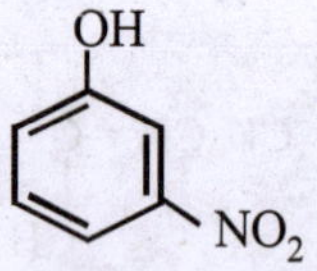 3-nitrophenol (OH and NO_2 on benzene ring)

186. The least acidic compound, among the following is: **[NCERT : P L-249 | April 3, 2025 (I)]**

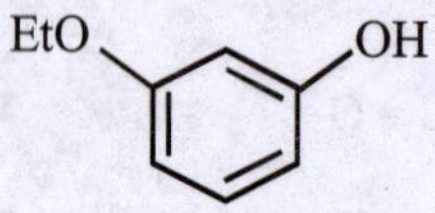

(A) 3-ethoxyphenol (EtO, OH on benzene ring)
(B) benzenesulfonic acid ($C_6H_5SO_3H$)
(C) butanoic acid (COOH)
(D) $EtO_2C-C \equiv C-H$

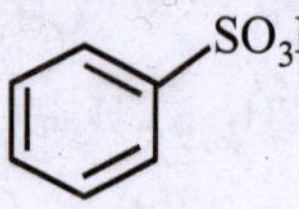

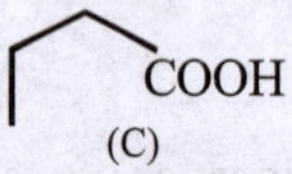

(a) D (b) A (c) B (d) C

187. Consider the following reactions. From these reactions which reaction will give carboxylic acid as a major product ? **[NCERT : P L-246 | April 2, 2025 (II)]**

(A) $R - C \equiv N \xrightarrow[\text{mild condition}]{(i) H^+ / H_2O}$

(B) $R - MgX \xrightarrow[(ii) H_3O^+]{(i) CO_2}$

(C) $R - C \equiv N \xrightarrow[(ii) H_3O^+]{(i) SnCl_2 / HCl}$

(D) $R \cdot CH_2 \cdot OH \xrightarrow{PCC}$

(E) 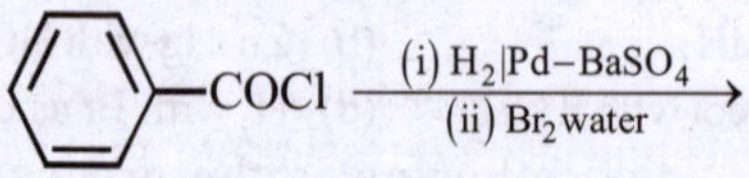 $C_6H_5-COCl \xrightarrow[(ii) Br_2 \text{ water}]{(i) H_2|Pd-BaSO_4}$

Choose the **correct** answer from the options given below :

(a) A and D only (b) A, B and E only
(c) B, C and E only (d) B and E only

188. Consider the following molecules : **[NCERT : P L-251 | April 2, 2025 (I)]**

$CH_3 - CH_2 - \overset{O}{\overset{||}{C}} - Cl$
(p)

$CH_3-CH_2-\overset{O}{\overset{||}{C}}-O-\overset{O}{\overset{||}{C}}-CH_3$ (q)

$CH_3-CH_2-\overset{O}{\overset{||}{C}}-O-CH_2-CH_3$ (r)

$CH_3-CH_2-\overset{O}{\overset{||}{C}}-NH_2$ (s)

The correct order of rate of hydrolysis is :

(a) $r > q > p > s$ (b) $q > p > r > s$
(c) $p > r > q > s$ (d) $p > q > r > s$

189. The total number of compounds from below when treated with hot $KMnO_4$ giving benzoic acid is :

[NCERT : P L-246 | Jan. 28, 2025 (II)]

(a) 3 (b) 4 (c) 6 (d) 5

190. Identify the product (P) in the following reaction:

$\text{Cyclopentane-COOH} \xrightarrow[\text{ii) } H_2O]{\text{i) } Br_2 / \text{Red P}} (P)$

[NCERT : P L-253 | April 8, 2024 (I)]

(a) 1-bromocyclopentane-1-COOH (b) cyclopentane-COBr
(c) 1-bromocyclopentane-1-CHO (d) 3-bromocyclopentane-1-COOH

191. Major product B of the following reaction has ______ π-bond. **[NCERT : P L-246 | NV, April 8, 2024 (I)]**

$C_6H_5CH_2CH_3 \xrightarrow[\Delta]{KMnO_4-KOH} (A) \xrightarrow{HNO_3/H_2SO_4} (B)$

192. The correct sequence of acidic strength of the following aliphatic acids in their decreasing order is :

[NCERT : P L-250, 251 | April 8, 2024 (II)]

$CH_3CH_2COOH, CH_3COOH, CH_3CH_2CH_2COOH, HCOOH$

(a) $HCOOH > CH_3COOH > CH_3CH_2COOH > CH_3CH_2CH_2COOH$
(b) $HCOOH > CH_3CH_2CH_2COOH > CH_3CH_2COOH > CH_3COOH$
(c) $CH_3CH_2CH_2COOH > CH_3CH_2COOH > CH_3COOH > HCOOH$
(d) $CH_3COOH > CH_3CH_2COOH > CH_3CH_2CH_2COOH > HCOOH$

193. Consider the given reaction, identify the major product P.

$CH_3-COOH \xrightarrow[\text{(iv) } H_2O/\bar{O}H,\Delta]{\text{(i) } LiAlH_4 \text{ (ii) PCC (iii) } HCN/\bar{O}H} \text{"P"}$

[NCERT : P L-253 | April 6, 2024 (II)]

(a) $CH_3-CH_2-CH_2-OH$
(b) $CH_3-CH_2-\overset{O}{\overset{||}{C}}-NH_2$
(c) $CH_3-\overset{O}{\overset{||}{C}}-CH_2CH_3$
(d) $CH_3-\overset{OH}{\overset{|}{C}H}-COOH$

194. $CH_3CH_2-OH \xrightarrow[\text{(ii) } KMnO_4 \text{ (iii) NaOH, CaO,}\Delta]{\text{(i) Jone's Reagent}}$

Consider the above reaction sequence and identify the major product P. **[NCERT : P L-245 | April 5, 2024 (II)]**

(a) Methane (b) Methanal
(c) Methoxymethane (d) Methanoic acid

195. Consider the given chemical reaction :

$\text{Cyclohexene} \xrightarrow[\text{Heat}]{KMnO_4-H_2SO_4} \text{Product "A"}$

[NCERT : P L-246 | April 5, 2024 (II)]

Product "A" is :

(a) picric acid (b) oxalic acid
(c) acetic acid (d) adipic acid

196. The product C in the following sequence of reactions has ______ π bonds. **[NCERT : P L-246 | NV, April 5, 2024 (II)]**

$C_6H_5CH_2CH_2CH_3 \xrightarrow[\Delta]{KMnO_4-KOH} (A) \xrightarrow{H_3O^+} (B) \xrightarrow[FeBr_3]{Br_2} (C)$

197. X g of ethylamine is subjected to reaction with $NaNO_2$/HCl followed by water; evolved dinitrogen gas which occupied 2.24 L volume at STP.

X is ______ $\times 10^{-1}$ g. **[NV, April 4, 2024 (I)]**

198. Given below are two statements : **[April 4, 2024 (I)]**

Statement I : Acidity of α-hydrogens of aldehydes and ketones is responsible for aldol reaction.

Statement II : Reaction between benzaldehyde and ethanal will NOT give cross – aldol product.

In the light of above statements, choose the **most appropriate** answer from the options given below.

(a) Both **Statement I** and **Statement II** are correct.
(b) Both **Statement I** and **Statement II** are incorrect.
(c) **Statement I** is incorrect but **Statement II** is correct.
(d) **Statement I** is correct but **Statement II** is incorrect.

199. The total number of 'sigma' and 'Pi' bonds in 2-oxohex-4-ynoic acid is ______. **[NV, April 4, 2024 (II)]**

200. Match List - I with List - II. **[Feb. 1, 2024 (I)]**

List-I
(Reaction)

(A) $CH_3(CH_2)_5-\underset{\underset{O}{||}}{C}-OC_2H_5 \longrightarrow CH_3(CH_2)_5CHO$

(B) $C_6H_5COC_6H_5 \rightarrow C_6H_5CH_2C_6H_5$

(C) $C_6H_5CHO \rightarrow C_6H_5CH(OH)CH_3$

(D) $CH_3COCH_2COOC_2H_5 \rightarrow CH_3\underset{\underset{H}{|}}{C}(OH)CH_2COOC_2H_5$

List-II
(Reagents)

(I) CH_3MgBr, H_2O
(II) Zn(Hg) and conc. HCl
(III) $NaBH_4, H^+$
(IV) DiBAL - H, H_2O

Choose the correct answer from the option given below:

(a) (A)- (III), (B)-(IV), (C) -(I), (D)-(II)
(b) (A)- (IV), (B)-(II), (C) -(I), (D)-(III)
(c) (A)- (IV), (B)-(II), (C) -(III), (D)-(I)
(d) (A)- (III), (B)-(IV), (C) -(II), (D)-(I)

201. 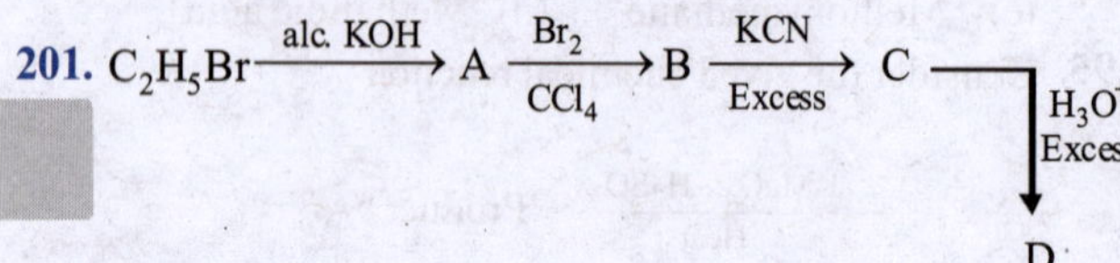

Acid D formed in above reaction is:

[NCERT : P L-245 | Feb. 1, 2024 (II)]

(a) Malonic acid (b) Oxalic acid
(c) Succinic acid (d) Gluconic acid

202. The molecular formula of second homologue in the homologous series of mono carboxylic acids is ______. **[Jan. 27, 2024 (II)]**

(a) $C_3H_6O_2$ (b) $C_2H_4O_2$
(c) CH_2O (d) $C_2H_2O_2$

203. The major product 'P' formed in the following sequence of reactions is **[NCERT : P L-249 | April 12, 2023 (I)]**

OH, O $\xrightarrow[\text{(ii) R-NH}_2 \text{ (iii) LiAlH}_4 \text{ (iv) H}_3\text{O}^+]{\text{(i) SOCl}_2}$ 'P' (Major Product)

(a) H, N – R, OH

(b) H – N – R, OH, Cl

(c) H, N – R, O

(d) H, N – R

204. The major product formed in the following reaction is:

CO_2H, CO_2Et $\xrightarrow[\text{(ii) } H_3O^+]{\text{(i) LiBH}_4 \text{ EtOH}}$ major product

[NCERT : P L-238, 254 | April 08, 2023 (I)]

(a) OH, OH
(b) OH, CO_2Et
(c) CO_2H, OH
(d) CO_2H, CO_2H

205. Major product 'P' formed in the following reaction is :-

OH, O $\xrightarrow[NaHCO_3]{Br_2}$ P Major Product

[NCERT : P L-249 | April 08, 2023 (II)]

(a) Br, Br, OH, O
(b) Br, O, O
(c) Br, OH, O
(d) Br, OH, O

206. The descending order of acidity for the following carboxylic acid is :

[NCERT : P L-251 | April 08, 2023 (II)]

A. CH_3COOH B. $F_3C–COOH$
C. $ClCH_2–COOH$ D. $FCH_2–COOH$
E. $BrCH_2–COOH$

Choose the correct answer from the options given below:

(a) D > B > A > E > C (b) E > D > B > A > C
(c) B > C > D > E > A (d) B > D > C > E > A

207. In a reaction, **[NCERT : P L-245 | Feb. 01, 2023 (II)]**

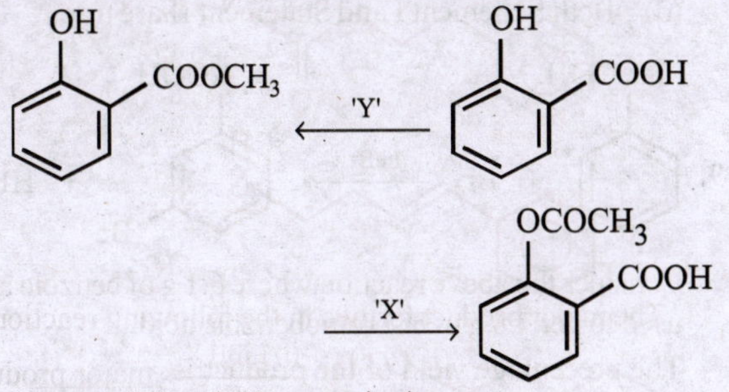

Methyl salicylate (o-HOC$_6$H$_4$COOCH$_3$) $\xleftarrow{\text{'Y'}}$ Salicylic acid (o-HOC$_6$H$_4$COOH) $\xrightarrow{\text{'X'}}$ o-$CH_3COOC_6H_4COOH$ (OCOCH$_3$, COOH)

reagents 'X' and 'Y' respectively are :

(a) $CH_3OH/H^+, \Delta$ and $(CH_3CO)_2O/H^+$
(b) $(CH_3CO)_2O/H^+$ and $CH_3OH/H^+, \Delta$
(c) $CH_3OH/H^+, \Delta$ and $CH_3OH/H^+, \Delta$
(d) $(CH_3CO)_2O/H^+$ and $(CH_3CO)_2O/H^+$

208. Among the following marked proton of which compound shows lowest pK_a value?

[NCERT : P L-251 | July 28, 2022 (I)]

(a) $H_2C(H)-COOH$
(b) $H_2C(H)-C(=O)-CH_3$
(c) 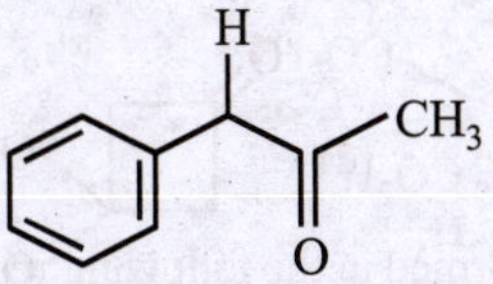 $C_6H_5-CH(H)-C(=O)-CH_3$
(d) 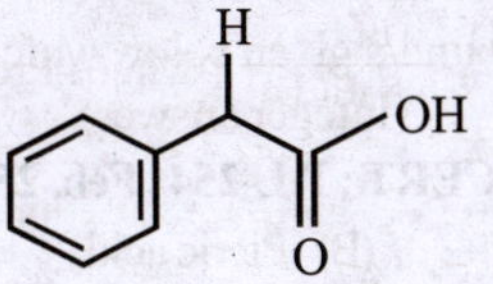 $C_6H_5-CH(H)-C(=O)-OH$

209. 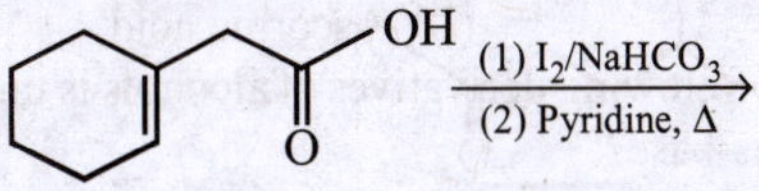 Cyclohexenylacetic acid $\xrightarrow[\text{(2) Pyridine, }\Delta]{\text{(1) } I_2/NaHCO_3}$

Find out the major product for the above reaction.

[NCERT : P L-245 | July 28, 2022 (II)]

(a) 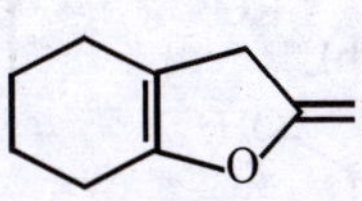 bicyclic lactone (C=O) (b) diiodo compound with CO_2H (I, I)
(c) bicyclic unsaturated lactone (C=O) (d) 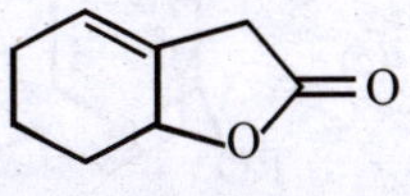 bicyclic lactone with ring double bond (C=O)

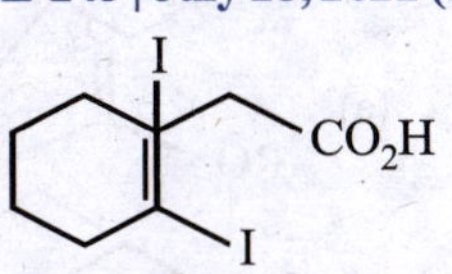

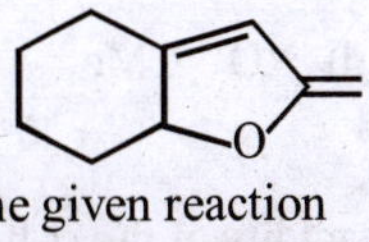

210. In the given reaction

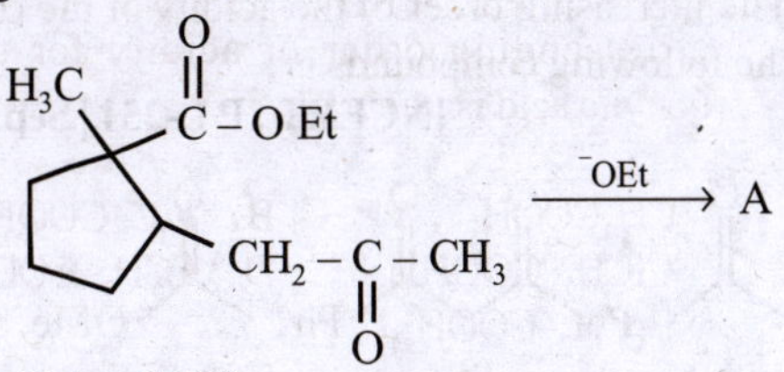

1-methyl-3-(2-oxopropyl)cyclopentane-1-carboxylate: H_3C, $C(=O)-OEt$, $CH_2-C(=O)-CH_3$ $\xrightarrow{^-OEt}$ A

(Where Et is $-C_2H_5$)

The number of chiral carbon/s in product A is ________.

[NV, July 25, 2022 (I)]

211. Consider the given chemical reaction.

[NCERT : P L-249 | June 30, 2022 (I)]

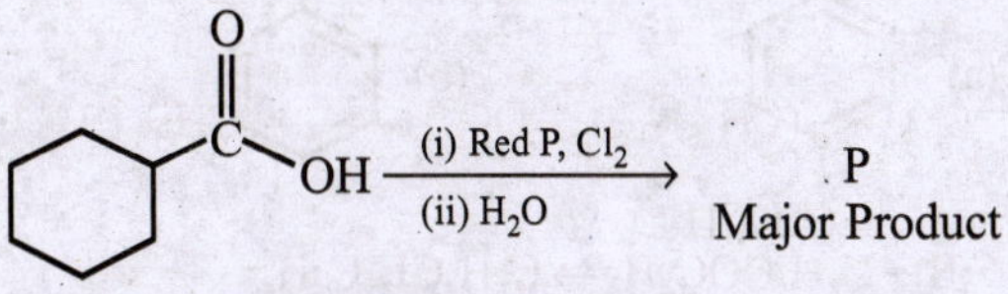

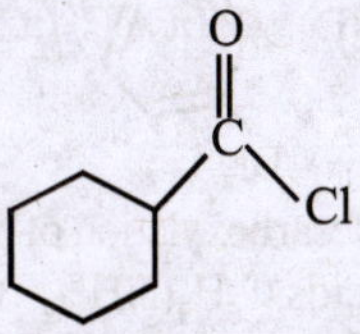

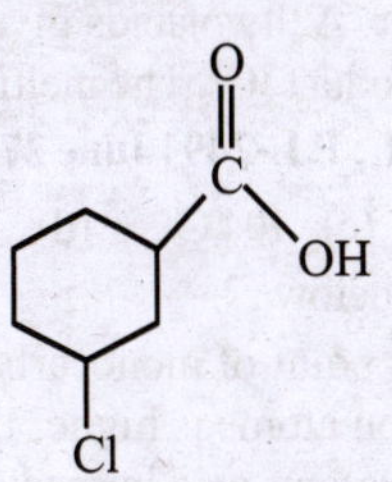

Cyclohexanecarboxylic acid ($C_6H_{11}-C(=O)-OH$) $\xrightarrow[\text{(ii) } H_2O]{\text{(i) Red P, } Cl_2}$ P (Major Product)

Identify the product P.

(a) 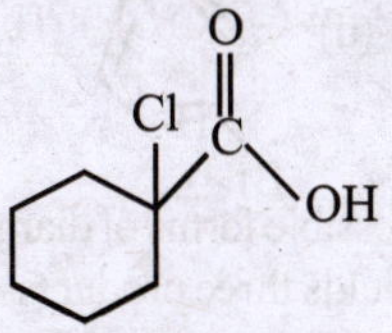 1-chlorocyclohexanecarboxylic acid (Cl, C(=O)OH on same carbon)
(b) cyclohexanecarbonyl chloride ($C_6H_{11}-C(=O)-Cl$)
(c) 3-chlorocyclohexanecarboxylic acid (Cl, C(=O)OH)
(d) 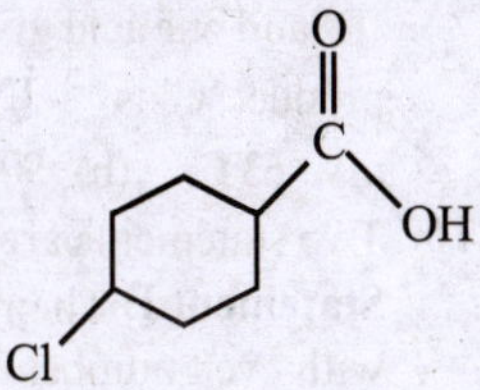 4-chlorocyclohexanecarboxylic acid (Cl, C(=O)OH)

212. 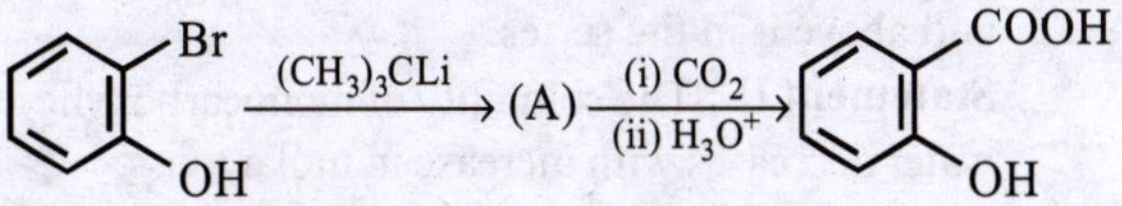 o-Bromophenol (Br, OH) $\xrightarrow{(CH_3)_3CLi}$ (A) $\xrightarrow[\text{(ii) } H_3O^+]{\text{(i) } CO_2}$ salicylic acid (COOH, OH)

In the given conversion the compound A is :

[NCERT : P L-251 | June 29, 2022 (I)]

(a) 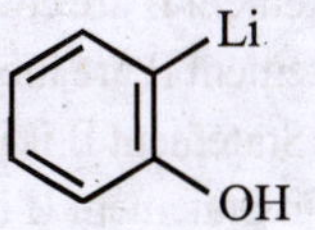 o-lithiophenol (Li, OH) (b) 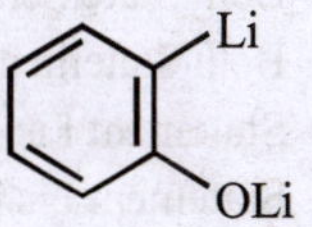 (Li, OLi)
(c) 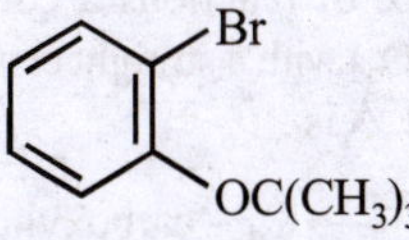 (Br, $OC(CH_3)_3$) (d) 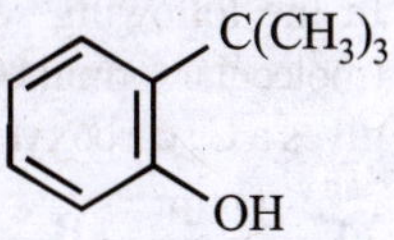 ($C(CH_3)_3$, OH)

213. Given below are two statements :

[NCERT : P L-251 | June 29, 2022 (I)]

Statement I : The esterification of carboxylic acid with an alcohol is a nucleophilic acyl substitution.

Statement II : Electron withdrawing groups in the carboxylic acid will increase the rate of esterification reaction.

Choose the **most appropriate** option :

(a) Both Statement I and Statement II are correct.
(b) Both Statement I and Statement II are incorrect.
(c) Statement I is correct but Statement II is incorrect.
(d) Statement I is incorrect but Statement II is correct.

214. A $\xrightarrow[\text{(ii) } CN^- \text{ (iii) } H_2O/H^+]{\text{(i) } Cl_2, \Delta}$ 4-Bromophenyl acetic acid.

In the above reaction 'A' is

[NCERT : P L-245 | June 28, 2022 (II)]

(a) Br, CH_2CH_3 (b) Br

(c) CH_3, Br (d) $CH=CH_2$, Br

215. Decarboxylation of all six possible forms of diaminobenzoic acids $C_6H_3(NH_2)_2COOH$ yields three products A, B and C. Three acids give a product 'A', two acids give a product 'B' and one acid gives a product 'C'. The melting point of product 'C' is **[NCERT : P L-249 | June 27, 2022 (II)]**

(a) 63°C (b) 90°C (c) 104°C (d) 142°C

216. Two statements are given below :

Statement I: The melting point of monocarboxylic acid with even number of carbon atoms is higher than that of with odd number of carbon atoms acid immediately below and above it in the series.

Statement II : The solubility of monocarboxylic acids in water decreases with increase in molar mass.

Choose the **most appropriate** option:

[NCERT : P L-249 | June 24, 2022 (I)]

(a) Both Statement I and Statement II are correct.
(b) Both Statement I and Statement II are incorrect.
(c) Statement I is correct but Statement II is incorrect.
(d) Statement I is incorrect but Statement II is correct.

217. In the following sequence of reactions a compound **A**, (molecular formula $C_6H_{12}O_2$) with a straight chain structure gives a C_4 carboxylic acid. **A** is :

$$A \xrightarrow[H_3O^+]{LiAlH_4} B \xrightarrow{\text{Oxidation}} C_4-\text{carboxylic acid}$$

[NCERT : P L-251, 245 | Sep. 1, 2021 (II)]

(a) $CH_3-CH_2-COO-CH_2-CH_2-CH_3$
(b) $CH_3-CH_2-CH(OH)-CH_2-O-CH=CH_2$
(c) $CH_3-CH_2-CH_2-COO-CH_2-CH_3$
(d) $CH_3-CH_2-CH_2-O-CH=CH-CH_2-OH$

218. Given below are two statements :

[NCERT : P L-251 | Aug. 27, 2021 (II)]

Statement I : Ethyl pent–4–yn–oate on reaction with CH_3MgBr gives a 3°–alcohol.

Statement II : In this reaction one mole of ethyl pent–4–yn–oate utilizes two moles of CH_3MgBr.

In the light of the above statements, choose the **most appropriate** answer from the options given below :

(a) Both Statement I and Statement II are false.
(b) Statement I is false but Statement II is true.
(c) Statement I is true but Statement II is false.
(d) Both Statement I and Statement II are true.

219. COOH (benzoic acid) + Br_2 $\xrightarrow{FeBr_3}$ COOH, Br (m-bromobenzoic acid) + HBr

Consider the above reaction where 6.1 g of benzoic acid is used to get 7.8 g of *m*-bromobenzoic acid. The precentage yield of the product is ______. (Round off to nearest integer).

[Given : Atomic masses : C : 120 u, H : 1.0 u, O : 16.0 u, Br : 80.0 u] **[NCERT : P L-254 | NV, March 18, 2021 (II)]**

220. $CH_3COCH_2COOC_2H_5 \xrightarrow[H^+]{\text{Ethylene glycol}}$ A (Major Product)

The product "A" in the above reaction is

[NCERT : P L-254 | March 17, 2021 (I)]

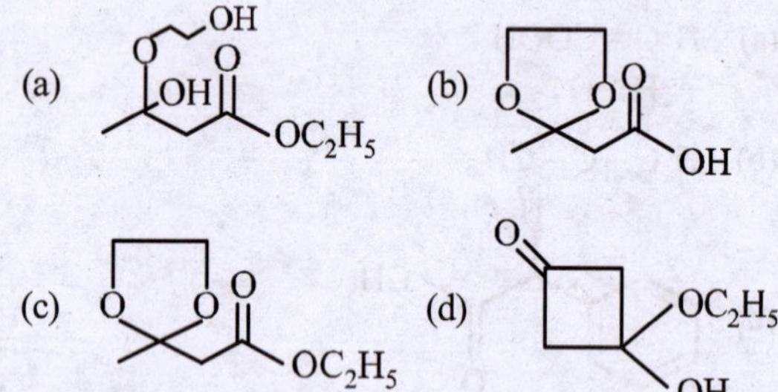

221. The number of compound/s given below which contain/s —COOH group is ____. (Integer answer)

[NCERT : P L-254 | Feb. 25, 2021 (II)]

(A) Sulphanilic acid (B) Picric acid
(C) Aspirin (D) Ascorbic acid

222. Which of the following derivatives of alcohols is unstable in an aqueous base?

[NCERT : P L-254 | Sep. 05, 2020 (I)]

(a) RO–CO–Me (b) RO–CH_2–Ph

(c) RO (tetrahydropyranyl) (d) $RO-CMe_3$

223. The increasing order of the acidity of the α-hydrogen of the following compounds is :

[NCERT : P L-251 | Sep. 05, 2020 (I)]

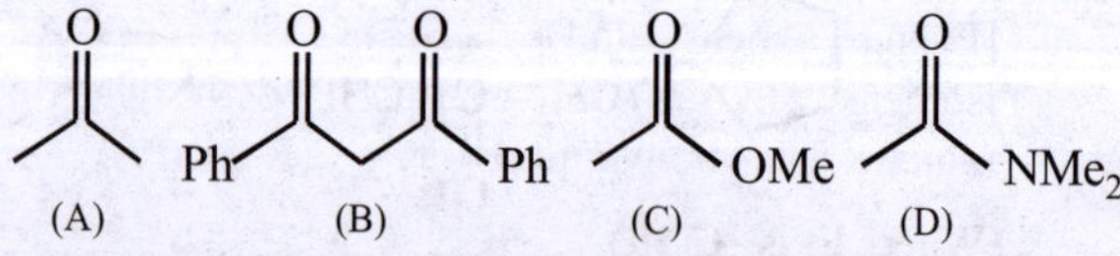

(a) (D) < (C) < (A) < (B) (b) (B) < (C) < (A) < (D)
(c) (A) < (C) < (D) < (B) (d) (C) < (A) < (B) < (D)

224. In the following reaction sequence the major products A and B are: **[NCERT : P L-249 | Sep. 05, 2020 (I)]**

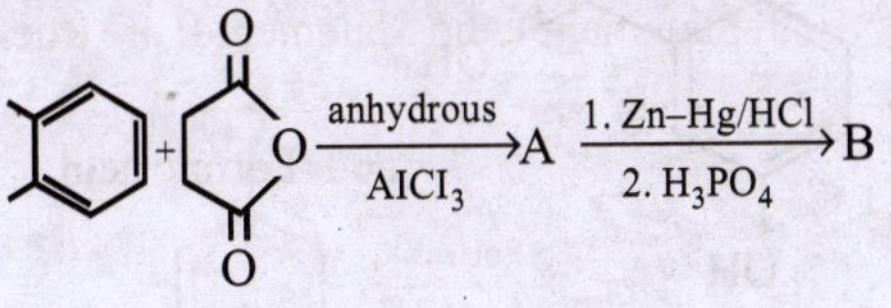

(o-xylene) + (succinic anhydride) $\xrightarrow[\text{AlCl}_3]{\text{anhydrous}}$ A $\xrightarrow[2.\ H_3PO_4]{1.\ \text{Zn–Hg/HCl}}$ B

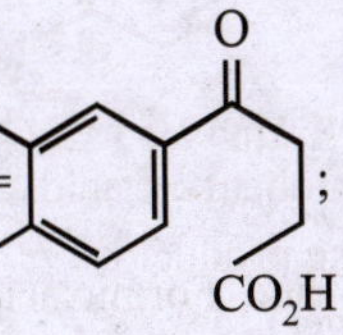

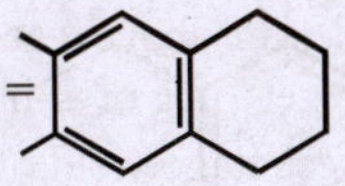

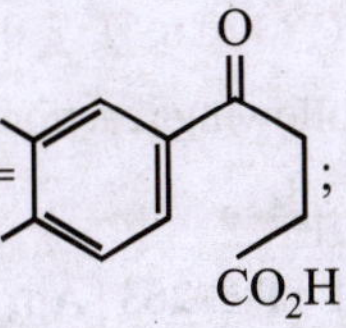

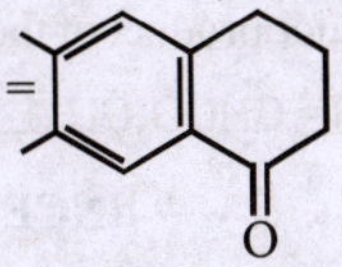

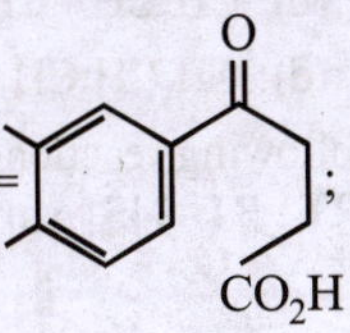

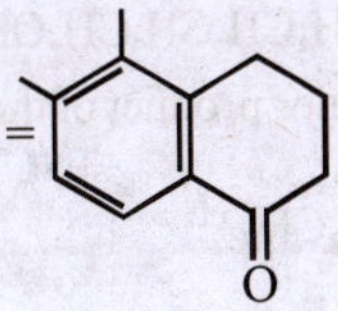

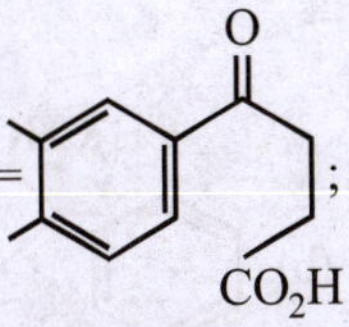

(a) A= (3,4-dimethylphenyl)–CO–CH_2CH_2–CO_2H ; B = (dimethyl tetralin)

(b) A= (3,4-dimethylphenyl)–CO–CH_2CH_2–CO_2H ; B = (dimethyl tetralone)

(c) A= (2,3-dimethylphenyl)–CO–CH_2CH_2–CO_2H ; B = (dimethyl tetralone)

(d) A= (3,4-dimethylphenyl)–CO–CH_2CH_2–CO_2H ; B = (dimethyl dihydronaphthalenone)

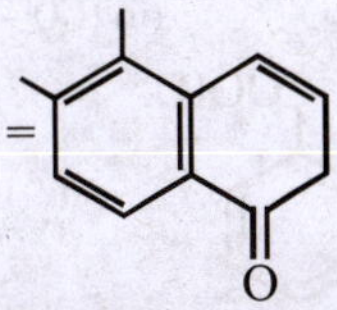

225. An organic compound (A) (molecular formula $C_6H_{12}O_2$) was hydrolysed with dil. H_2SO_4 to give a carboxylic acid (B) and an alcohol (C). 'C' gives white turbidity immediately when treated with anhydrous $ZnCl_2$ and conc. HCl. The organic compound (A) is : **[NCERT : P L-254 | Sep. 04, 2020 (I)]**

(a) 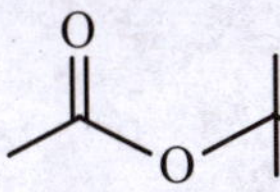 $CH_3COOC(CH_3)_3$

(b) 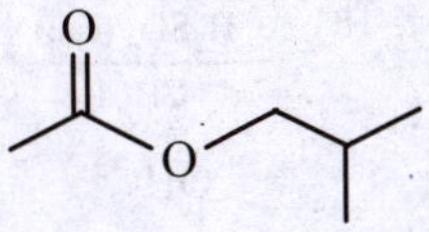 $CH_3COOCH_2CH(CH_3)_2$

(c) 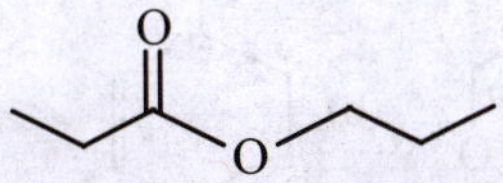 $CH_3CH_2COOCH_2CH_2CH_3$

(d) 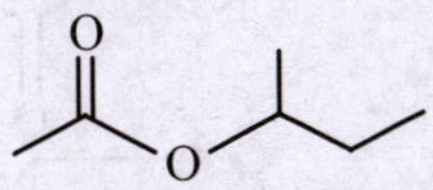 $CH_3COOCH(CH_3)CH_2CH_3$

226. [P] on treatment with $Br_2/FeBr_3$ in CCl_4 produced a single isomer $C_8H_7O_2Br$ while heating [P] with soda-lime gave toluene. The compound [P] is : **[NCERT : P L-254 | Sep. 04, 2020 (I)]**

(a) 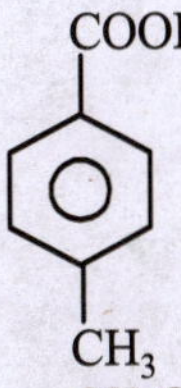 COOH on benzene ring with CH_3 para

(b) 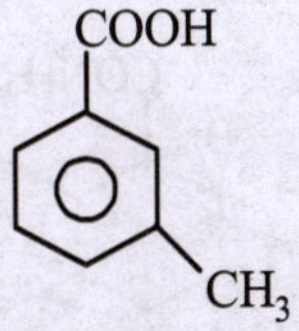 COOH on benzene ring with CH_3 meta

(c) 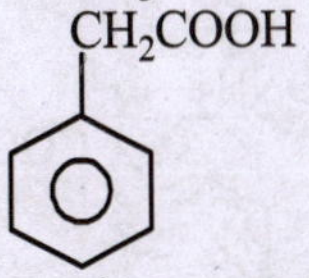 $C_6H_5CH_2COOH$

(d) 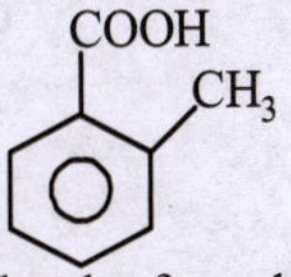 COOH on benzene ring with CH_3 ortho

227. An organic compound [A], molecular formula $C_{10}H_{20}O_2$ was hydrolyzed with dilute sulphuric acid to give a carboxylic acid [B] and an alcohol [C]. Oxidation of [C] with $CrO_3-H_2SO_4$ produced [B]. Which of the following structures are not possible for [A]? **[NCERT : P L-250 | Sep. 03, 2020 (I)]**

(a) $CH_3-CH_2-\underset{CH_3}{\underset{|}{CH}}-OCOCH_2\overset{CH_3}{\overset{|}{CH}}-CH_2CH_3$

(b) $CH_3-CH_2-\underset{CH_3}{\underset{|}{CH}}-COOCH_2-\overset{CH_3}{\overset{|}{CH}}-CH_2CH_3$

(c) $CH_3CH_2CH_2COOCH_2CH_2CH_2CH_3$

(d) $(CH_3)_3-C-COOCH_2C(CH_3)_3$

228. Consider the following molecules and statements related to them : **[NCERT : P L-251 | Sep. 03, 2020 (II)]**

(A) 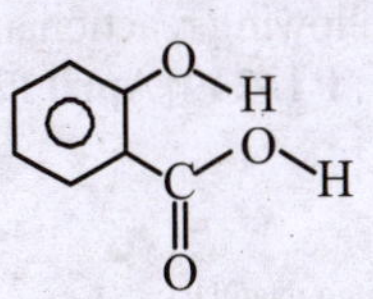 o-hydroxybenzoic acid (B) 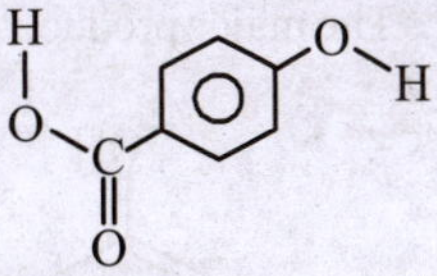 p-hydroxybenzoic acid

(1) (B) is more likely to be crystalline than (A)
(2) (B) has higher boiling point than (A)
(3) (B) dissolves more readily than (A) in water

Identify the **correct** option from below :

(a) (1) and (2) are true (b) (1) and (3) are true
(c) only (1) is true (d) (2) and (3) are true

229. Arrange the following labelled hydrogens in decreasing order of acidity : **[NCERT : P L-251 | Sep. 02, 2020 (II)]**

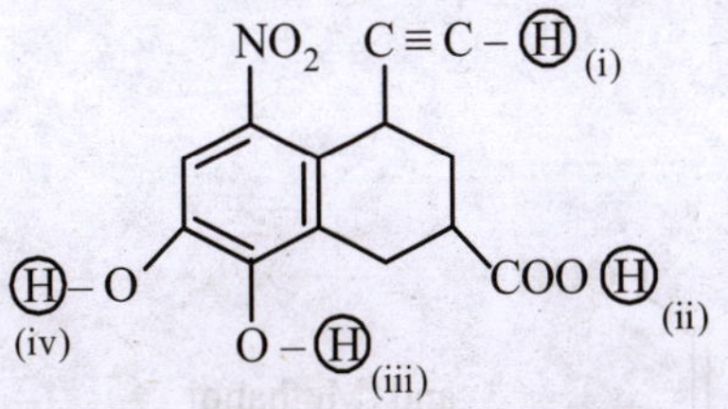

(Structure: tetralin ring bearing NO_2, C≡C–Ⓗ(i), COOⒽ(ii), O–Ⓗ(iii), Ⓗ–O(iv))

(a) (ii) > (i) > (iii) > (iv) (b) (iii) > (ii) > (iv) > (i)
(c) (ii) > (iii) > (iv) > (i) (d) (iii) > (ii) > (i) > (iv)

230. The most suitable reagent for the given conversion is: **[NCERT : P L-253 | Jan. 08, 2020 (I)]**

$\xrightarrow{?}$

(a) B_2H_6 (b) $NaBH_4$ (c) $LiAlH_4$ (d) H_2/Pd

231. An unsaturated hydrocarbon X absorbs two hydrogen molecules on catalytic hydrogenation, and also gives following reaction:

$X \xrightarrow[Zn/H_2O]{O_3} A \xrightarrow{[Ag(NH_3)_2]^+}$ B (3-oxo-hexanedicarboxylic acid)

X will be : **[NCERT : P L-231 | Jan. 08, 2020 (II)]**

(a) (b) (c) (d)

232. The major products of the following reaction are : **[NCERT : P L-249 | April 12, 2019 (I)]**

$\xrightarrow[(3)\ H_3O^+]{(1)\ CHCl_3/aq.\ NaOH,\ (2)\ HCHO,\ NaOH\ (conc.)}$

(a) and Methanol

(b) and Methanol

(c) and Formic acid

(d) and Formic acid

233. The major product of the following reaction is:

$CH_3CH=CHCO_2CH_3 \xrightarrow{LiAlH_4}$

[NCERT : P L-253 | April 9, 2019 (I)]

(a) $CH_3CH_2CH_2CO_2CH_3$ (b) $CH_3CH=CHCH_2OH$
(c) $CH_3CH_2CH_2CH_2OH$ (d) $CH_3CH_2CH_2CHO$

234. The major product of the following reaction is: **[NCERT : P L-245 | April 9, 2019 (I)]**

$\xrightarrow[(ii)\ H_3O^+]{(i)\ alkaline\ KMnO_4}$

(a) (b) (c) (d)

235. The major product of the following reaction is: **[NCERT : P L-253 | April 9, 2019 (II)]**

$\xrightarrow[CHCl_3]{H_2SO_4\ (cat.)}$

(a) (b) (c) (d)

236. The major product of the following reaction is:
[NCERT : P L-253 | April 8, 2019 (II)]

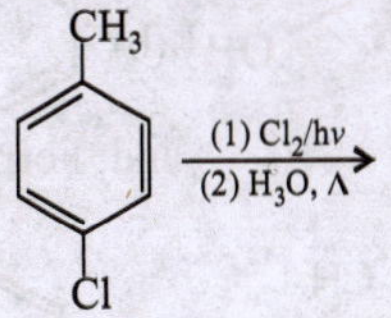

(p-chlorotoluene, CH_3 / Cl) $\xrightarrow[(2)\ H_3O,\ \Delta]{(1)\ Cl_2/h\nu}$

(a) 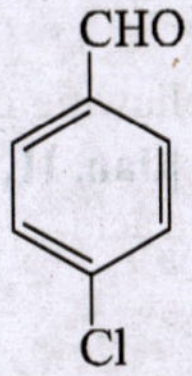 CHO / Cl
(b) 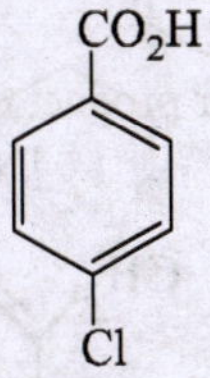 CO_2H / Cl
(c) 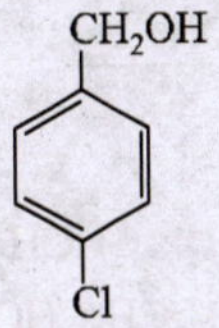 CH_2OH / Cl
(d) 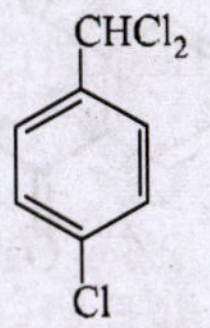 $CHCl_2$ / Cl

237. The major product of the following reaction is:
[NCERT : P L-250 | April 8, 2019 (II)]

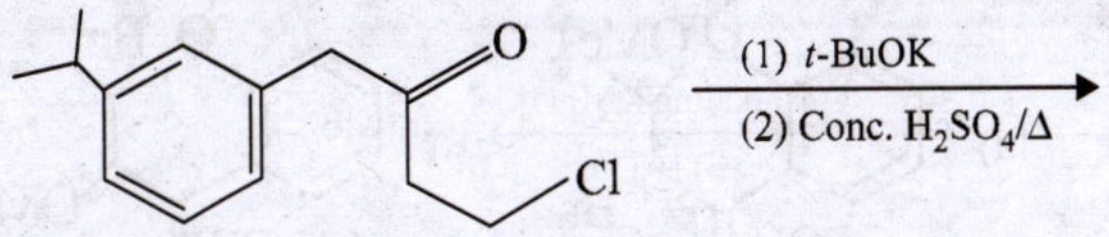

O / Cl $\xrightarrow[(2)\ \text{Conc. } H_2SO_4/\Delta]{(1)\ t\text{-BuOK}}$

(a) 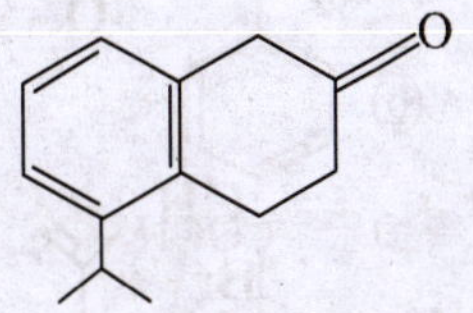 O
(b) 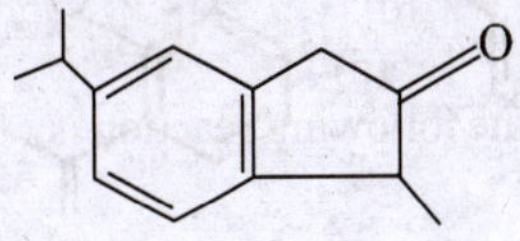 O
(c) 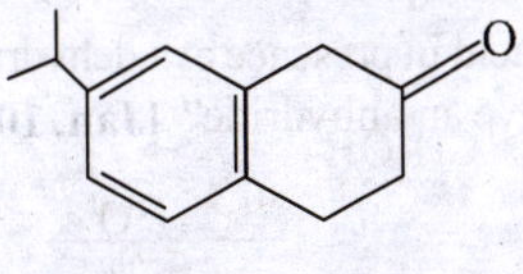 O
(d) 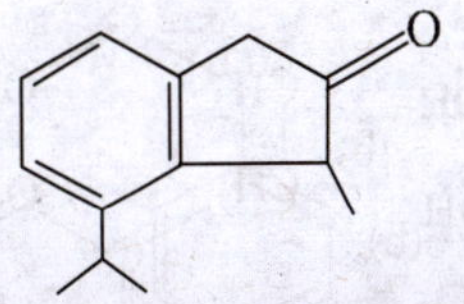 O

238. The major product of the following reaction is:
[NCERT : P L-249 | Jan. 12, 2019 (I)]

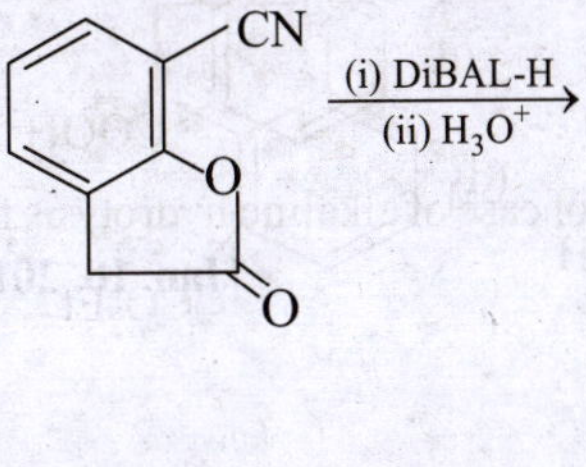

CN / O / O $\xrightarrow[(ii)\ H_3O^+]{(i)\ \text{DiBAL-H}}$

(a) 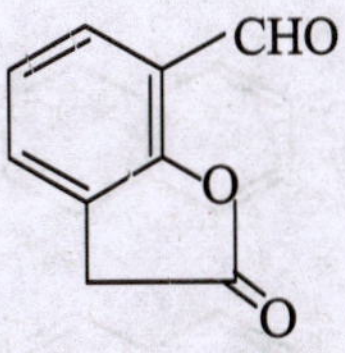 CHO / O / O
(b) 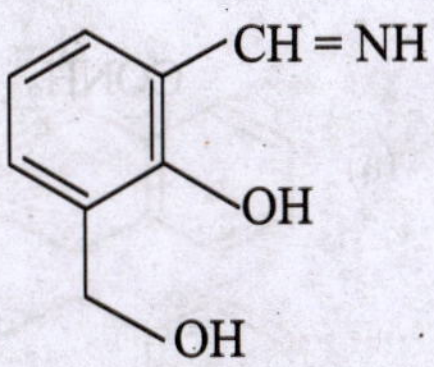 CH = NH / OH / OH
(c) 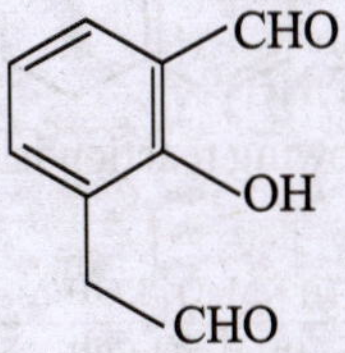 CHO / OH / CHO
(d) 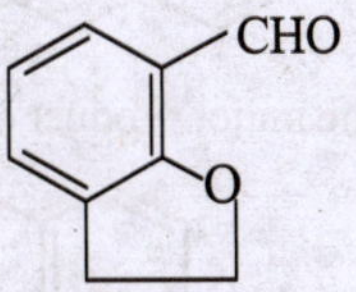 CHO / O

239. Among the following four aromatic compounds ,which one will have the lowest melting point?
[NCERT : P L-249 | Jan. 12, 2019 (I)]

(a) 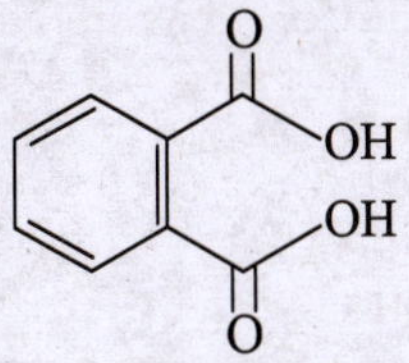 O / OH / OH / O
(b) 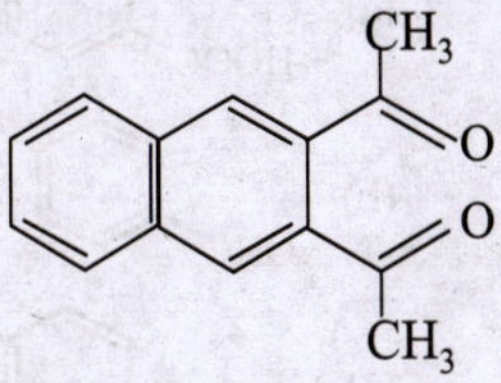 CH_3 / O / O / CH_3
(c) 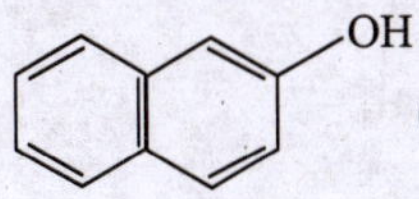 OH
(d)

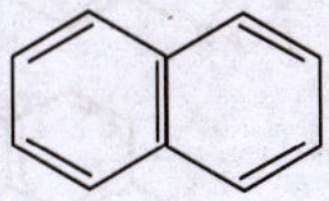

240. The major product of the following reaction is :

$$H_3C \blacktriangleright \overset{CH_2CH_3}{\underset{COOCH_2CH_3}{C}} \blacktriangleleft Cl \xrightarrow[\Delta]{NaOEt}$$

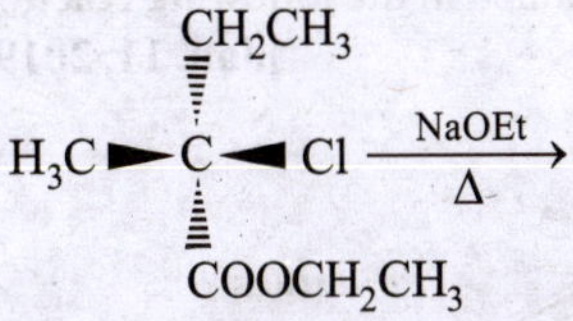

[NCERT : P L-247 | Jan. 12, 2019 (II)]

(a) $CH_3CH_2\underset{CO_2CH_2CH_3}{C} = CH_2$

(b) $CH_3 - \overset{CO_2CH_2CH_3}{C} = CH\ CH_3$

(c) 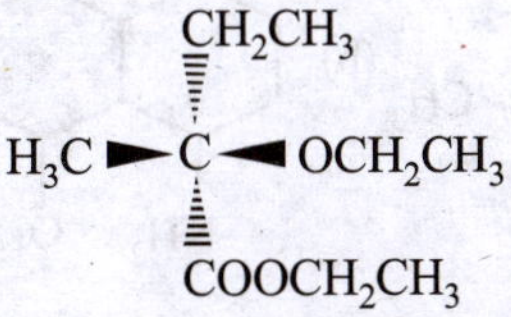 $H_3C \blacktriangleright \overset{CH_2CH_3}{\underset{COOCH_2CH_3}{C}} \blacktriangleleft OCH_2CH_3$

(d) 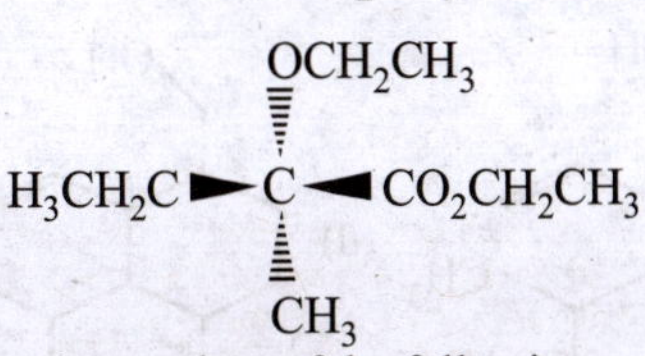 $H_3CH_2C \blacktriangleright \overset{OCH_2CH_3}{\underset{CH_3}{C}} \blacktriangleleft CO_2CH_2CH_3$

241. The major product of the following reaction is :

O / OEt / CN $\xrightarrow[(ii)\ \text{DiBAL-H}]{(i)\ Ni/H_2}$

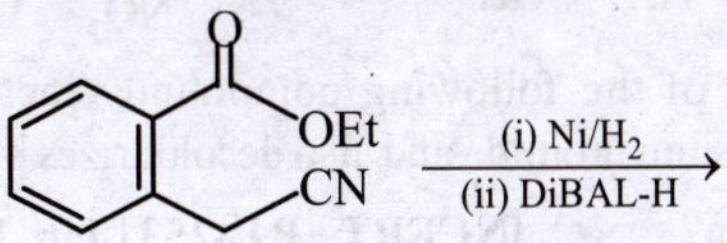

[NCERT : P L-253 | Jan. 11, 2019 (I)]

(a) H, O, CHO (b) NH

(c) OH, NH_2 (d) N

242. The major product of the following reaction is :

CH_3–C_6H_4–$COCH_3$ $\xrightarrow[\text{(ii) } H_2SO_4\text{(dil)}]{\text{(i) } KMnO_4/KOH/\Delta}$

[NCERT : P L-249 | Jan. 11, 2019 (I)]

(a) HOOC–C_6H_4–COCOOH

(b) OHC–C_6H_4–COOH

(c) HOOC–C_6H_4–COOH

(d) HOOC–C_6H_4–$COCH_3$

243. The major product obtained in the following reaction is: **[Jan. 11, 2019 (II)]**

O, OH, CH_3, NO_2, O $\xrightarrow[\text{(excess)}]{LiAlH_4}$

(a) O, OH, CH_3, NO_2, OH (b) OH, CH_3, NH_2, OH

(c) OH, CH_3, NH_2, OH (d) OH, CH_3, NO_2, OH

244. Which of the following compound reacts with ethyl magnesium bromide and also decolourizes bromine water solution? **[NCERT : P L-253 | Jan. 11, 2019 (II)]**

(a) CN, O (b) CN, $CH_2CO_2CH_3$

(c) OCH_3, CH, CH_2 (d) OH

245. The major product obtained in the following conversion is: **[NCERT : P L-253 | Jan. 11, 2019 (II)]**

CH_3, O, O, O $\xrightarrow[\text{MeOH}]{Br_2 \text{ (1 eq.)}}$

(a) CH_3, O, O, OMe, Br, O (b) CH_3, O, O, Br, OMe, O

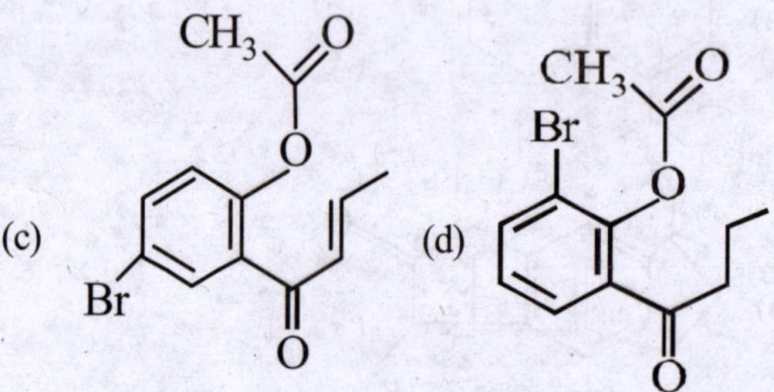

246. Which dicarboxylic acid in presence of a dehydrating agent is least reactive to give an anhydride? **[Jan. 10, 2019 (I)]**

(a) CH_2, CH_2, COOH, CH_2, COOH, CH_2 (b) O, C, OH, CH_2, CH_2, OH, C, O

(c) CO_2H, CO_2H (d) COOH, COOH

247. The decreasing order of ease of alkaline hydrolysis for the following esters is **[Jan. 10, 2019 (I)]**

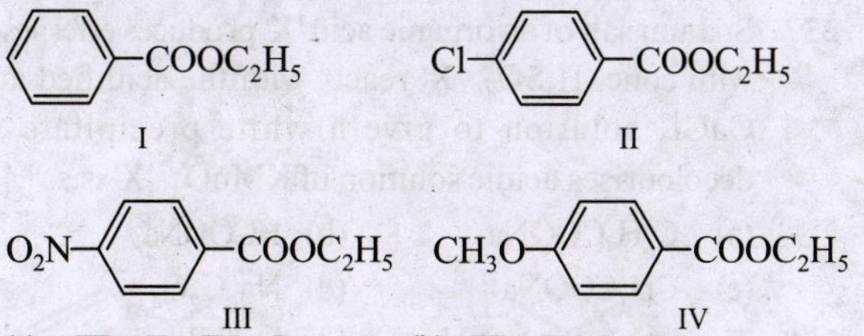

I: $C_6H_5COOC_2H_5$ II: $Cl-C_6H_4-COOC_2H_5$

III: $O_2N-C_6H_4-COOC_2H_5$ IV: $CH_3O-C_6H_4-COOC_2H_5$

(a) III > II > IV > I (b) III > II > I > IV
(c) IV > II > III > I (d) II > III > I > IV

248. An aromatic compound 'A' having molecular formula $C_7H_6O_2$ on treating with aqueous ammonia and heating forms compound 'B'. The compound 'B' on reaction with molecular bromine and potassium hydroxide provides compound 'C' having molecular formula C_6H_7N. The structure of 'A' is: **[Jan. 10, 2019 (II)]**

(a) 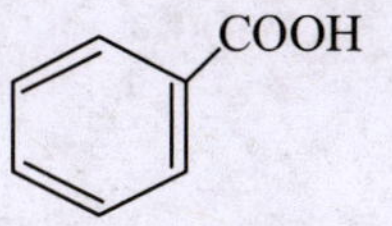 benzene ring with COOH
(b) 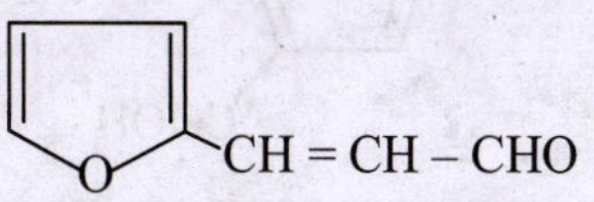 furan ring with $CH = CH - CHO$
(c) 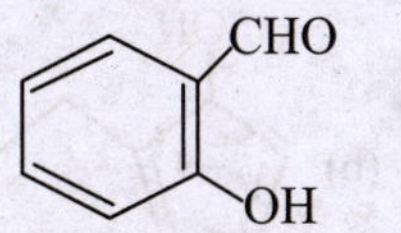 benzene ring with CHO and OH (ortho)
(d) 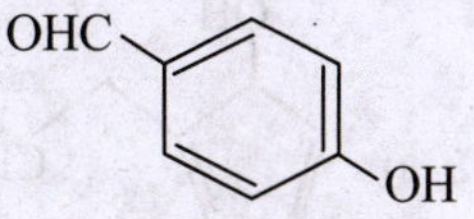 benzene ring with OHC and OH (para)

249. The major product obtained in the following reaction is:

2-oxocyclopentyl–CH_2–CO–CH_2CH_2–CO_2Et $\xrightarrow{NaOEt/\Delta}$

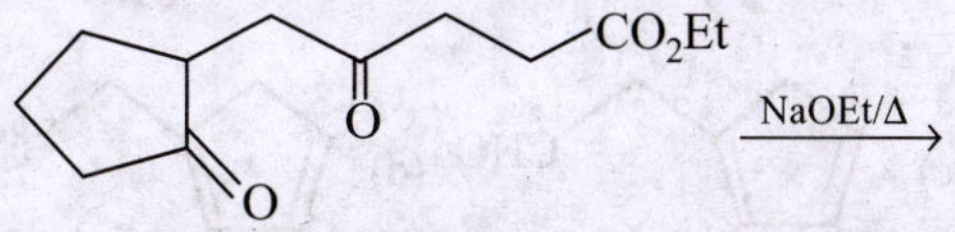

[NCERT : P L-249 | Jan. 10, 2019 (II)]

(a) 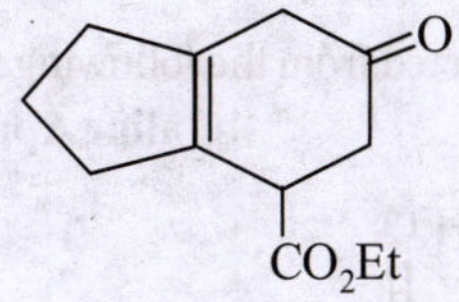 bicyclic enone (O) with CO_2Et
(b) 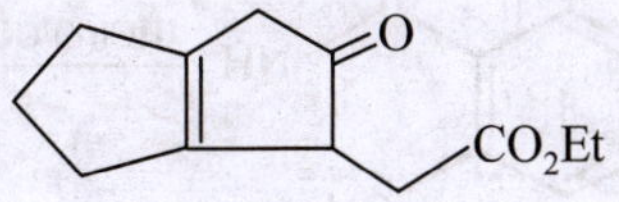 bicyclic enone (O) with CH_2CO_2Et
(c) 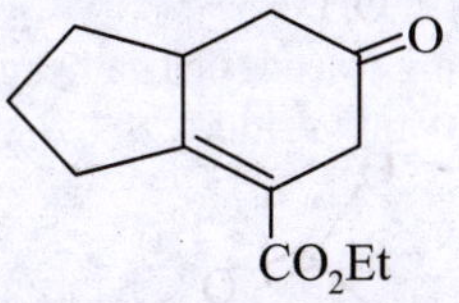 bicyclic ketone (O) with CO_2Et
(d) 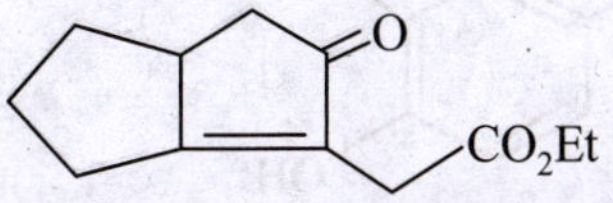 bicyclic enone (O) with CO_2Et

250. The correct decreasing order for acid strength is: **[NCERT : P L-251 | Jan. 9, 2019 (I)]**

(a) $NO_2CH_2COOH > FCH_2COOH > CNCH_2COOH > ClCH_2COOH$
(b) $FCH_2COOH > CNCH_2COOH > NO_2CH_2COOH > ClCH_2COOH$
(c) $CNCH_2COOH > NO_2CH_2COOH > FCH_2COOH > ClCH_2COOH$
(d) $NO_2CH_2COOH > CNCH_2COOH > FCH_2COOH > ClCH_2COOH$

251. The major product of the given reaction is: **[Online April 16, 2018]**

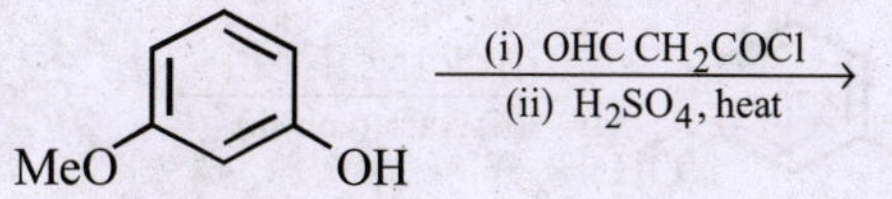

MeO–C_6H_4–OH (meta) $\xrightarrow[\text{(ii) } H_2SO_4\text{, heat}]{\text{(i) } OHC\,CH_2COCl}$

(a) chromone with OMe (O)
(b) coumarin with MeO (O)
(c) chromone with MeO (O)
(d) 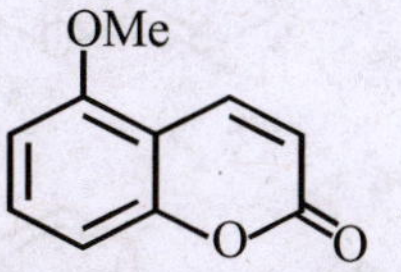 coumarin with OMe (O)

252. The major product formed in the following reaction is:

$OCOCH_3$, HO, OH substituted cyclopentene–cyclohexane $\xrightarrow[\text{CHCl}_3]{\text{PCC (Pyridinium chlorochromate)}}$

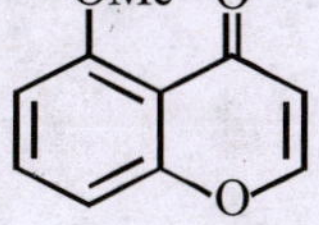

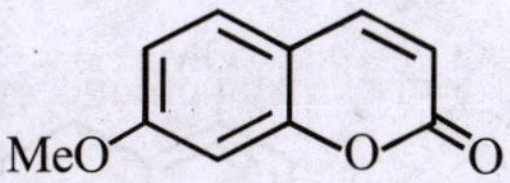

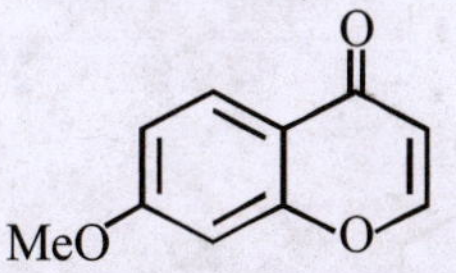

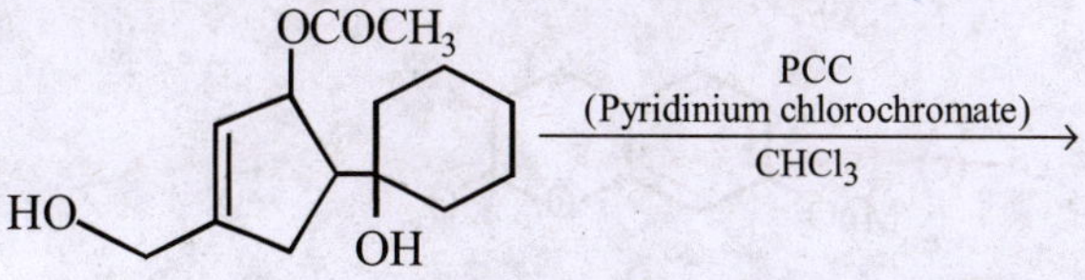

[NCERT : P L-233 | Online April 15, 2018 (II)]

(a) 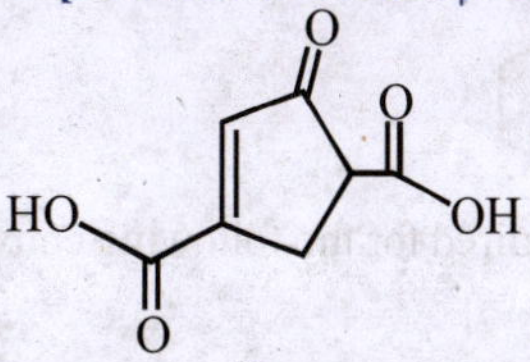 HO, OH, O (cyclopentenone dicarboxylic acid)
(b) 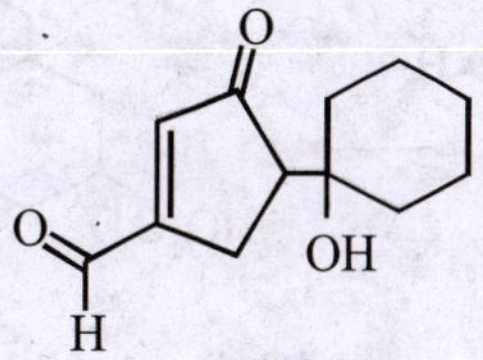 O, H, OH (aldehyde cyclopentenone with cyclohexanol)
(c) 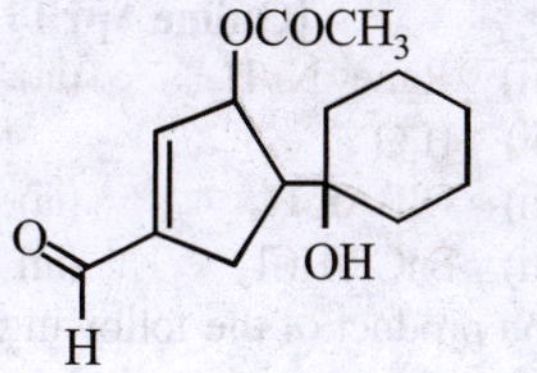 $OCOCH_3$, O, H, OH
(d) 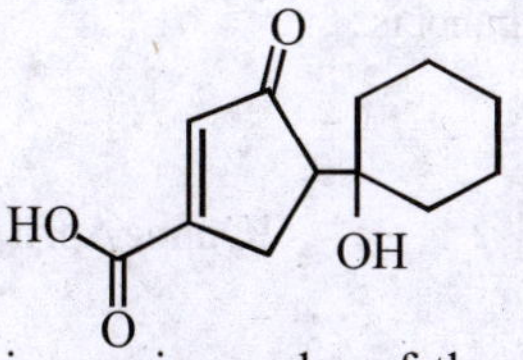 HO, O, OH (carboxylic acid cyclopentenone with cyclohexanol)

253. The increasing order of the acidity of the following carboxylic acids is: **[NCERT : P L-251 | Online April 15, 2018 (II)]**

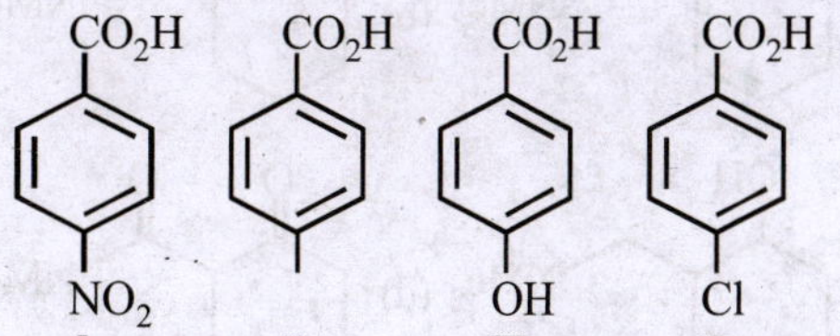

I: CO_2H–C_6H_4–NO_2 II: CO_2H–C_6H_4–CH_3 III: CO_2H–C_6H_4–OH IV: CO_2H–C_6H_4–Cl

(a) III < II < IV < I (b) I < III < II < IV
(c) IV < II < III < I (d) II < IV < III < I

254. The major product of the following reaction is

MeO–C₆H₄–OH $\xrightarrow[\text{(ii) } AlCl_3 \text{(anhyd.)}]{\text{(i) } ClCH_2CH_2\overset{O}{\overset{\|}{C}}Cl}$

[Online April 15, 2018 (I)]

(a) OMe

(b) MeO

(c) MeO

(d) OMe

255. The reagent(s) required for the following conversion are :

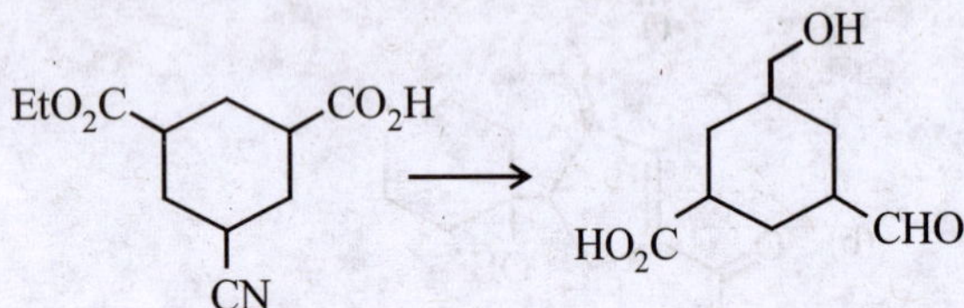

[Online April 15, 2018 (I)]

(a) (i) $NaBH_4$, (ii) Raney Ni/H_2, (iii) H_3O^+
(b) (i) $LiAlH_4$, (ii) H_3O^+
(c) (i) B_2H_6, (ii) DIBAL-H, (iii) H_3O^+
(d) (i) B_2H_6, (ii) $SnCl_2/HCl$, (iii) H_3O^+

256. The main reduction product of the following compound with $NaBH_4$ in methanol is :

NMe₂ **[Online April 15, 2018 (I)]**

(a) OH O NMe₂ (b) OH O NMe₂

(c) OH NMe₂ (d) O O NMe₂

257. Sodium salt of an organic acid 'X' produces effervescences with conc. H_2SO_4. 'X' reacts with the acidified aqueous $CaCl_2$ solution to give a white precipitate which decolourises acidic solution of $KMnO_4$. 'X' is : **[2017]**

(a) C_6H_5COONa (b) $HCOONa$
(c) CH_3COONa (d) $Na_2C_2O_4$

258. The major product obtained in the following reaction is :
[NCERT : P L-249 | 2017]

COOH $\xrightarrow{\text{DiBAL–H}}$

(a) OH, CHO, COOH (b) OH, CHO, CHO

(c) CHO, COOH (d) CHO, CHO

259. The major product expected from the following reaction is:
[Online April 8, 2017]

HO_2C, CH_2OH, O, NH_2, OH $\xrightarrow{HCl(g)/CCl_4}$

(a) HO_2C, O, O, OH

(b) CH_2OH O, HO_2C, NH_2, Cl

(c) 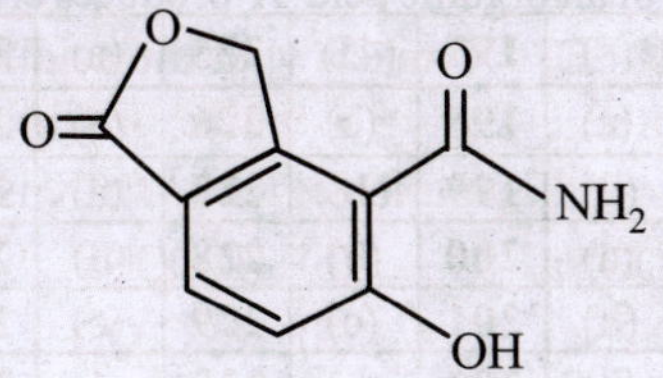 [structure: lactone ring fused to benzene bearing $CONH_2$ and OH]

(d) 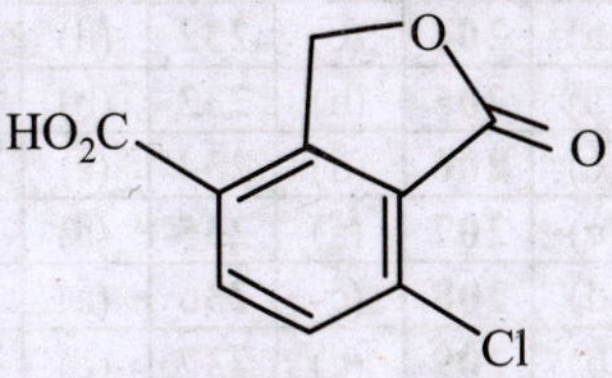 HO_2C [structure: lactone ring fused to benzene bearing Cl]

260. Bouveault–Blanc reduction reaction involves : [Online April 9, 2016]
(a) Reduction of an acyl halide with H_2/Pd
(b) Reduction of an anhydride with $LiAlH_4$
(c) Reduction of an ester with Na/C_2H_5OH
(d) Reduction of a carbonyl compound with Na/Hg and HCl.

261. 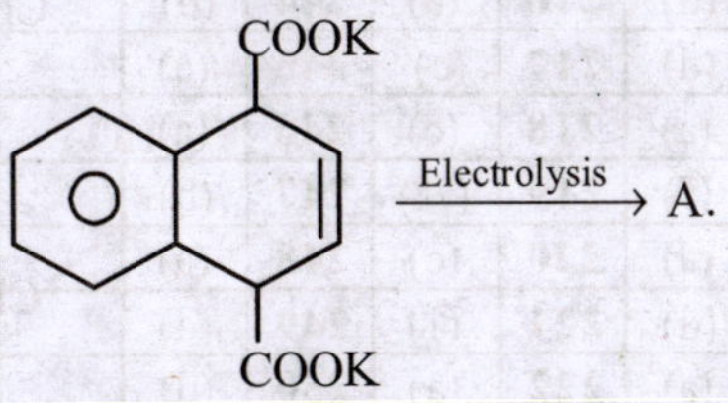 [dihydronaphthalene dicarboxylate with two COOK groups] $\xrightarrow{\text{Electrolysis}}$ A.

A is :
[NCERT : P L-253 | Online April 10, 2015]
(a) 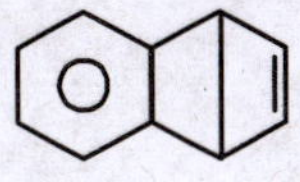 [structure] (b) 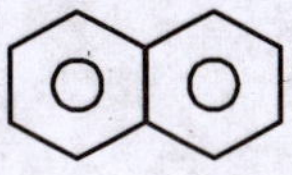 [structure: naphthalene]
(c) 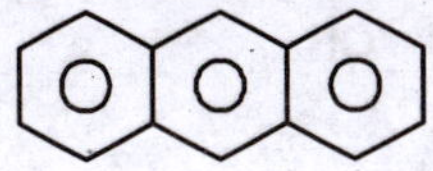 [structure: anthracene] (d) 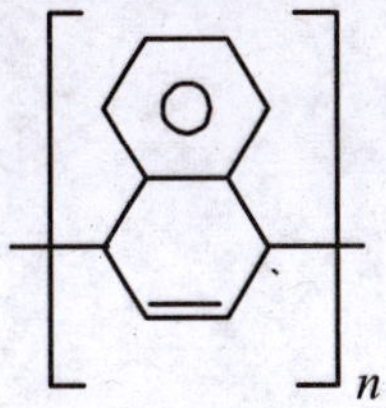 [polymer structure]$_n$

262. 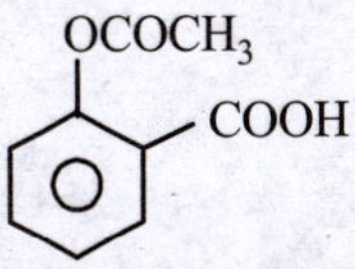 [structure: benzene with $OCOCH_3$ and COOH] is used as : [Online April 10, 2015]
(a) Analgesic (b) Insecticide
(c) Antacid (d) Antihistamine

263. In the presence of a small amount of phosphorus, aliphatic carboxylic acids react with chlorine or bromine to yield a compound in which α-hydrogen has been replaced by halogen. This reaction is known as :
[NCERT : P L-249 | Online April 10, 2015]
(a) Wolff - Kishner reaction
(b) Rosenmund reaction
(c) Etard reaction
(d) Hell - Volhard - Zelinsky reaction

264. In the reaction,

$$CH_3COOH \xrightarrow{LiAlH_4} A \xrightarrow{PCl_5} B \xrightarrow{Alc.KOH} C,$$

the product C is: [NCERT : P L-256 | 2014]
(a) Acetaldehyde (b) Acetylene
(c) Ethylene (d) Acetyl chloride

265. Phthalic acid reacts with resorcinol in the presence of concentrated H_2SO_4 to give: [Online April 12, 2014]
(a) Phenolphthalein (b) Alizarin
(c) Coumarin (d) Fluorescein

266. Monocarboxylic acids are functional isomers of:
[Online April 23, 2013]
(a) Ethers (b) Amines (c) Esters (d) Alcohols

267. Maleic acid and fumaric acids are
[NCERT : P L-244 | Online May 26, 2012]
(a) chain isomers (b) functional isomers
(c) tautomers (d) geometrical isomers

268. The correct order of increasing acid strength of the compounds [NCERT : P L-249 | 2006]
(A) CH_3CO_2H (B) $MeOCH_2CO_2H$
(C) CF_3CO_2H (D) $(Me)_2CH-CO_2H$
is
(a) D < A < B < C (b) A < D < B < C
(c) B < D < A < C (d) D < A < C < B

269. Among the following acids which has the lowest pK_a value? [NCERT : P L-249 | 2005]
(a) CH_3CH_2COOH (b) $(CH_3)_2CH-COOH$
(c) HCOOH (d) CH_3COOH

270. When $CH_2 = CH — COOH$ is reduced with $LiAlH_4$, the compound obtained will be [NCERT : P L-253 | 2003]
(a) $CH_2=CH—CH_2OH$ (b) $CH_3—CH_2—CH_2OH$
(c) $CH_3—CH_2—CHO$ (d) $CH_3—CH_2—COOH$

271. $CH_3CH_2COOH \xrightarrow[\text{red P}]{Cl_2} A \xrightarrow{\text{alc. KOH}} B$. What is *B*?
[NCERT : P L-2 | 2002]
(a) CH_3CH_2COCl (b) CH_3CH_2CHO
(c) $CH_2=CHCOOH$ (d) $ClCH_2CH_2COOH$.

ANSWER KEYS

1	(a)	29	(d)	57	(d)	85	(3)	113	(a)	141	(66.67)	169	(c)	197	(45)	225	(a)	253	(a)
2	(d)	30	(c)	58	(d)	86	(c)	114	(c)	142	(d)	170	(a)	198	(c)	226	(a)	254	(c)
3	(c)	31	(a)	59	(0)	87	(c)	115	(c)	143	(d)	171	(d)	199	(18)	227	(a)	255	(d)
4	(b)	32	(b)	60	(3)	88	(b)	116	(3)	144	(c)	172	(a)	200	(b)	228	(d)	256	(a)
5	(9)	33	(7)	61	(b)	89	(a)	117	(c)	145	(d)	173	(b)	201	(c)	229	(c)	257	(d)
6	(b)	34	(b)	62	(d)	90	(a)	118	(b)	146	(b)	174	(d)	202	(b)	230	(a)	258	(b)
7	(c)	35	(b)	63	(b)	91	(b)	119	(c)	147	(d)	175	(c)	203	(d)	231	(c)	259	(c)
8	(a)	36	(b)	64	(b)	92	(a)	120	(a)	148	(b)	176	(a)	204	(c)	232	(d)	260	(c)
9	(b)	37	(d)	65	(d)	93	(b)	121	(c)	149	(d)	177	(a)	205	(b)	233	(b)	261	(b)
10	(a)	38	(b)	66	(d)	94	(c)	122	(a)	150	(b)	178	(c)	206	(d)	234	(a)	262	(a)
11	(a)	39	(b)	67	(d)	95	(a)	123	(c)	151	(a)	179	(c)	207	(a)	235	(a)	263	(d)
12	(d)	40	(b)	68	(c)	96	(d)	124	(c)	152	(b)	180	(d)	208	(c)	236	(a)	264	(c)
13	(c)	41	(60)	69	(60)	97	(3)	125	(3)	153	(a)	181	(b)	209	(c)	237	(c)	265	(d)
14	(c)	42	(13)	70	(15)	98	(a)	126	(2)	154	(d)	182	(a)	210	(2)	238	(c)	266	(c)
15	(c)	43	(c)	71	(d)	99	(c)	127	(5)	155	(a)	183	(b)	211	(a)	239	(d)	267	(d)
16	(a)	44	(b)	72	(2)	100	(c)	128	(c)	156	(d)	184	(d)	212	(b)	240	(b)	268	(a)
17	(d)	45	(b)	73	(c)	101	(d)	129	(d)	157	(a)	185	(d)	213	(a)	241	(d)	269	(c)
18	(c)	46	(a)	74	(b)	102	(b)	130	(a)	158	(a)	186	(a)	214	(c)	242	(c)	270	(a)
19	(a)	47	(c)	75	(a)	103	(d)	131	(c)	159	(a)	187	(d)	215	(d)	243	(c)	271	(c)
20	(a)	48	(b)	76	(a)	104	(b)	132	(c)	160	(a)	188	(d)	216	(a)	244	(d)		
21	(b)	49	(c)	77	(d)	105	(a)	133	(c)	161	(b)	189	(d)	217	(c)	245	(a)		
22	(c)	50	(b)	78	(4)	106	(1)	134	(b)	162	(d)	190	(a)	218	(c)	246	(a)		
23	(c)	51	(b)	79	(a)	107	(8)	135	(c)	163	(c)	191	(5)	219	(78)	247	(b)		
24	(b)	52	(a)	80	(2)	108	(4)	136	(a)	164	(c)	192	(a)	220	(c)	248	(a)		
25	(c)	53	(b)	81	(d)	109	(c)	137	(b)	165	(c)	193	(d)	221	(1)	249	(d)		
26	(a)	54	(b)	82	(c)	110	(a)	138	(c)	166	(c)	194	(a)	222	(a)	250	(d)		
27	(c)	55	(b)	83	(3)	111	(c)	139	(c)	167	(b)	195	(d)	223	(a)	251	(b)		
28	(c)	56	(318)	84	(a)	112	(c)	140	(b)	168	(b)	196	(4)	224	(b)	252	(b)		

Amines

TREND ANALYSIS
JEE MAIN (2025-2021)

	2025				2024				2023				2022				2021			
	Session II		Session I		Session II		Session I		Session II		Session I		Session II		Session I		Session II		Session I	
	MCQs	NVQs	MCQs	NVQs	MCQs	NVQs	MCQs	NVQs	MCQs	NVQs	MCQs	NVQs	MCQs	NVQs	MCQs	NVQs	MCQs	NVQs	MCQs	NVQs
Topic 1	5	–	5	1	2	5	7	2	3	1	10	–	7	–	6	–	2	2	7	1
Topic 2	1	1	–	–	–	1	–	–	5	–	–	–	3	–	4	–	3	–	4	1
Topic 3	4	–	2	3	1	3	2	–	6	–	4	–	5	–	2	–	1	–	2	–
Topic 4																				
Total	10	1	7	4	3	9	9	2	14	1	14	–	15	–	12	–	6	2	13	2

How to Code Qns for Better & Faster Revision

Type 1: Able to solve (No code Req)
Type 2: Solved with struggle or by seeing solution **(Use code NR means Needs Revision or orange stickers)**
Type 3: Tricky/ Could not solve on my own. Need help from Teacher/ Friend. **(Use code TQ means Tricky Question or red stickers)**

You can either hand write the code in boxes or else paste coloured stickers given at end of book.

Topic 1 - Aliphatic and Aromatic Amines

1. The descending order of basicity of following amines is: **[NCERT : PL-266 | April 7, 2025 (II)]**

(A) NH_2 (aniline)

(B) NH_2, MeO (p-methoxyaniline)

(C)

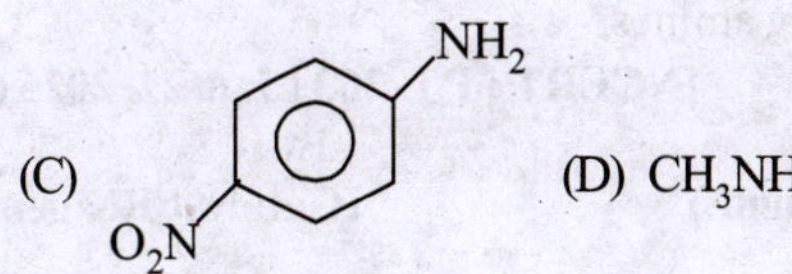

(D) CH_3NH_2

(E) $(CH_3)_2NH$

Choose the **correct** answer from the options given below:

(a) B > E > D > A > C
(b) E > D > B > A > C
(c) E > D > A > B > C
(d) E > A > D > C > B

2. Which of the following amine (s) show (s) positive carbylamines test ? **[NCERT : PL-271 | April 7, 2025 (I)]**

(A) NH_2 (aniline)
(B) $(CH_3)_2NH$
(C) CH_3NH_2
(D) $(CH_3)_3N$

(E)

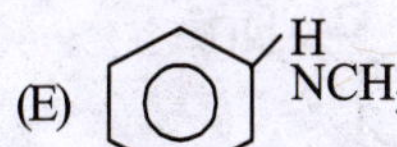

Choose the **correct** answer from the options given below :

(A) A and E Only
(B) C Only
(C) A and C Only
(D) B, C and D Only

3. The correct order of basicity for the following molecules is:

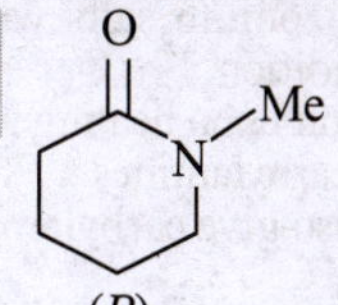

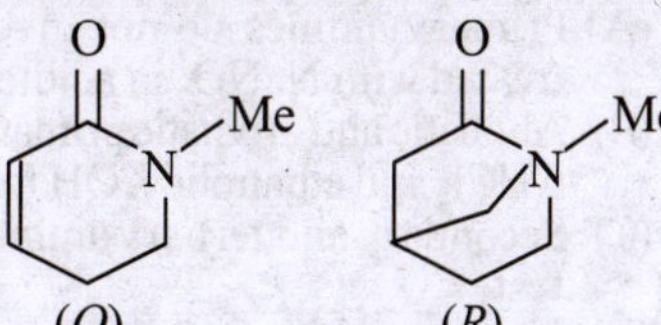

[NCERT : PL-266 | April 4, 2025 (II)]

(a) $P > Q > R$ (b) $R > P > Q$
(c) $Q > P > R$ (d) $R > Q > P$

4. The sequence from the following that would result in giving predominantly 3, 4, 5 –Tribromoaniline is :

[NCERT : P L-272, 275 | April 3, 2025 (II)]

(a) Nitrobenzene (NO_2) $\xrightarrow[\text{(ii) Sn, HCl}]{\text{(i) } Br_2\text{, acetic}}$

(b) Benzene $\xrightarrow[\text{(ii) } NH_3]{\text{(i) } Br_2\text{, } AlBr_3}$

(c) H_2N–C_6H_4–NO_2 $\xrightarrow[\text{(ii) } NaNO_2\text{, HCl, CuBr; (iii) Sn, HCl}]{\text{(i) } Br_2 \text{ (excess), acetic acid}}$

(d) Aniline (NH_2) $\xrightarrow{Br_2\text{, water}}$

5. The correct order of basic nature on aqueous solution for the bases NH_3, $H_2N–NH_2$, $CH_3CH_2NH_2$, $(CH_3CH_2)_2NH$ and $(CH_3CH_2)_3N$ is :

[NCERT : P L-266 | April 2, 2025 (I)]

(a) $NH_3 < H_2N–NH_2 < (CH_3CH_2)_3N < CH_3CH_2NH_2 < (CH_3CH_2)_2NH$
(b) $NH_3 < H_2N–NH_2 < CH_3CH_2NH_2 < (CH_3CH_2)_2NH < (CH_3CH_2)_3N$
(c) $H_2N–NH_2 < NH_3 < (CH_3CH_2)_3N < CH_3CH_2NH_2 < (CH_3CH_2)_2NH$
(d) $NH_2–NH_2 < NH_3 < CH_3CH_2NH_2 < (CH_3CH_2)_3N < (CH_3CH_2)_2NH$

6. Given below are some nitrogen containing compounds.

[NCERT : P L-269 | NV, Jan. 29, 2025 (I)]

NH_2 / NO_2 (p-nitroaniline); $CH_2–NH_2$ (benzylamine); $N – COCH_3$ / H (acetanilide); NH_2 (aniline)

Each of them is treated with HCl separately. 1.0 g of the most basic compound will consume ______ mg of HCl.
(Given molar mass in g mol^{-1} C:12, H : 1, O : 16, Cl : 35.5)

7. Identify correct statements :

[NCERT : P L-271 | Jan. 28, 2025 (II)]

(A) Primary amines do not give diazonium salts when treated with $NaNO_2$ in acidic condition.
(B) Aliphatic and aromatic primary amines on heating with $CHCl_3$ and ethanolic KOH form carbylamines.
(C) Secondary and tertiary amines also give carbylamine test.
(D) Benzenesulfonyl chloride is known as Hinsberg's reagent.
(E) Tertiary amines reacts with benzenesulfonyl chloride very easily.

Choose the correct answer from the options given below :
(a) (B) and (D) only (b) (A) and (B) only
(c) (D) and (E) only (d) (B) and (C) only

8. Consider the following sequence of reactions :

[NCERT : P L-264 | NV, Jan. 28, 2025 (I)]

Cl (Chlorobenzene) $\xrightarrow[\text{(ii) } CO_2, H_3O^+ \text{ (iii) } NH_3, \Delta]{\text{(i) Mg. dry ether}}$ A $\xrightarrow{Br_2\text{, NaOH}}$ B

Chlorobenzene

11.25 mg of chlorobenzene will produce ______ $\times 10^{-1}$ mg of product B.
(Consider the reactions result in complete conversion.)
[Given molar mass of C, H, O, N and Cl as 12, 1, 16, 14 and 35.5 g mol^{-1} respectively]

9. For reaction **[NCERT : P L-272 | Jan. 24, 2025 (II)]**

NH_2 (aniline) ⟶ NH_2, Br (Major)

The correct order of set of reagents for the above conversion is :
(a) $Br_2 | FeBr_3$, $H_2O(\Delta)$, NaOH
(b) H_2SO_4, Ac_2O, Br_2, $H_2O(\Delta)$, NaOH
(c) Ac_2O, Br_2, $H_2O(\Delta)$, NaOH
(d) Ac_2O, H_2SO_4, Br_2, NaOH

10. Which among the following react with Hinsberg's reagent? **[NCERT : P L-271 | Jan. 23, 2025 (I)]**

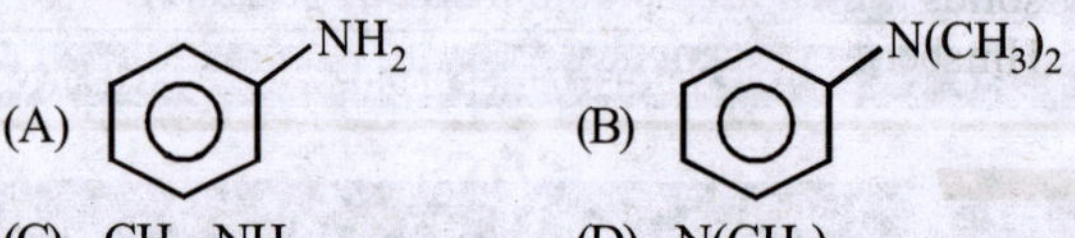

(C) $CH_3–NH_2$ (D) $N(CH_3)_3$

(E) H / N (diphenylamine)

Choose the correct answer from the options given below:
(a) B and D only (b) C and D only
(c) A, B and E only (d) A, C and E only

11. Match the Compounds (List-I) with the appropriate Catalyst/Reagents (List-II) for their reduction into corresponding amines.

[NCERT : P L-263 | Jan. 22, 2025 (II)]

List-I (Compounds)	List-II (Catalyst/Reagents)
(A) $R – \overset{O}{\overset{\|}{C}} – NH_2$	(I) NaOH (aqueous)
(B) NO_2 (nitrobenzene)	(II) H_2/Ni

(C) R–C≡N (III) $LiAlH_4, H_2O$

(D) N–R (IV) Sn, HCl

Choose the correct answer from the options given below:

(a) (A)-(III), (B)-(II), (C)-(IV), (D)-(I)
(b) (A)-(II), (B)-(IV), (C)-(III), (D)-(I)
(c) (A)-(II), (B)-(I), (C)-(III), (D)-(IV)
(d) (A)-(III), (B)-(IV), (C)-(II), (D)-(I)

12. Correct order of basic strength of Pyrrole, Pyridine and Piperidine is:

[NCERT : P L-269 | April 9, 2024 (I)]

(a) Piperidine > Pyridine > Pyrrole
(b) Pyrrole > Pyridine > Piperidine
(c) Pyridine > Piperidine > Pyrrole
(d) Pyrrole > Piperidine >Pyridine

13. Number of compounds from the following which **cannot** undergo Friedel-Crafts reactions is :____ toluene, nitrobenzene, xylene, cumene, aniline, chlorobenzene, m-nitroaniline, m-dinitrobenzene

[NCERT : P L-273 | April 9, 2024 (II)]

14. Number of amine compounds from the following giving solids which are soluble in NaOH upon reaction with Hinsberg's reagent is _____.

[NCERT : P L-271 | April 8, 2024 (I)]

NH_2, NH_2, $H_2NNH-C(=O)-NH_2$

NH–CH_3

N–H, NH_2 H_2N–C(=O)–NH_2,

NH_2, OCH_3, NH_2

N, N, NH_2

15. Given below are two statements:

[NCERT : P L-145 | April 8, 2024 (II)]

Statement (I): All the following compounds react with p-toluenesulfonyl chloride.

$C_6H_5NH_2$, $(C_6H_5)_2NH$, $(C_6H_5)_3N$

Statement (II): Their products in the above reaction are soluble in aqueous NaOH.

In the light of the above statements, choose the **correct** answer from the options given below:

(a) Both **Statement I** and **Statement II** is false
(b) **Statement I** is true but **Statement II** is false
(c) **Statement I** is false but **Statement II** is true
(d) Both **Statement I** and **Statement II** is true

16. An amine (X) is prepared by ammonolysis of benzyl chloride. On adding p-toluenesulphonyl chloride to it the solution remains clear. Molar mass of the amine (X) formed is__________ g mol^{-1}.

[NCERT : P L-262 | NV, April 6, 2024 (II)]

(Given molar mass in $gmol^{-1}$ C : 12, H : 1, O : 16, N : 14)

17. 9.3 g of pure aniline is treated with bromine water at room temperature to give a white precipitate of the product 'P'. The mass of product 'P' obtained is 26.4 g. The percentage yield is ...... %. **[NCERT : P L-272 | NV, April 5, 2024 (I)]**

18. Phthalimide is made to undergo following sequence of reactions. **[NCERT : PL-264 | NV, April 4, 2024 (II)]**

$$\text{Phthalimide} \xrightarrow[\text{(ii) Benzyl chloride}]{\text{(i) KOH}} \text{'P'}$$

Total number of π bonds present in product 'P' is/are

19. Given below are two statements:

[NCERT : P L-260 | Feb. 1, 2024 (I)]

Statements (I) : Aminobenzene and aniline are same organic compounds.

Statements (II) : Aminobenzene and aniline are different organic compounds.

In the light of the above statements, choose the most appropriate answer from the options given below:

(a) Both Statement I and Statement II are correct
(b) Statement I is correct but Statement II is incorrect
(c) Statement I is incorrect but Statement II in correct
(d) Both Statement I and Statement II are incorrect.

20. Given below are two statements:

[NCERT : P L-272 | Feb. 1, 2024 (I)]

Statement (I) : The NH_2 group in aniline is ortho and para directing and a powerful activating group.

Statement (II) : Aniline does not undergo Friedel - Craft's reaction (alkylation and acylation).

In the light of the above statements, choose the most appropriate answer from the options given below:

(a) Both Statement I and Statement II are correct
(b) Both Statement I and Statement II are incorrect.
(c) Statement I is incorrect but Statement II in correct
(d) Statement I is incorrect but Statement II in incorrect

21. Number of compounds which give reaction with Hinsberg's reagent is ________.

[NCERT : P L-271 | NV, Feb. 1, 2024 (II)]

$N_2^+Cl^-$ O NH O NH_2

N NH N NH_2

N H N NH_2 $H_2N-\overset{O}{\overset{\|}{C}}-NH_2$

22. A compound (X) with molar mass 108 g mol^{-1} undergoes acetylation to give product with molar mass 192 g mol^{-1}. The number of amino groups in the compound (X) is ______. **[NV, Jan. 31, 2024 (II)]**

23. The final product A, formed in the following multi step reaction sequence is: **[NCERT : P L-264 | Jan. 30, 2024 (I)]**

Br $\xrightarrow[\text{(ii) } NH_3,\Delta \text{ (iii) } Br_2, NaOH]{\text{(i) Mg, ether then } CO_2, H^+}$ A

(a) $C_6H_5CONH_2$ (b) $C_6H_5NH_2$

(c) C_6H_5COBr (d) C_6H_5COOH

24. The arenium ion which is not involved in the bromination of aniline is ______. **[NCERT : P L-273 | Jan. 29, 2024 (I)]**

(a) $\overset{\oplus}{N}H_2$ H Br (b) $\overset{\oplus}{N}H_2$ H Br

(c) NH_2 ⊕ H Br (d) NH_2 H Br ⊕

25. Which of the following reaction is correct? **[NCERT : P L-274 | Jan. 29, 2024 (II)]**

(a) cyclohexene $+ Br_2 \xrightarrow[\text{UV light}]{\Delta}$ 1,2-dibromocyclohexane (Br, Br)

(b) $C_2H_5CONH_2 + Br_2 + NaOH \rightarrow C_2H_5CH_2NH_2 + Na_2CO_3 + NaBr + H_2O$

(c) $CH_3CH_2CH_2NH_2 \xrightarrow[H_2O]{HNO_2, 0°C} CH_3CH_2OH + N_2 + HCl$

(d) 1-methylcyclohexene (CH_3) $+ HI \longrightarrow$ 1-iodo-1-methylcyclohexane (CH_3, I)

26. Which of the following is strongest Bronsted base? **[NCERT : P L-267, 268 | Jan. 27, 2024 (I)]**

(a) NH_2 (aniline) (b) NH (diphenylamine)

(c) H–N (pyrrole) (d) H–N (pyrrolidine)

27. Identify B formed in the reaction. **[Jan. 27, 2024 (II)]**

$Cl-(CH_2)_4-Cl \xrightarrow{\text{excess } NH_3} A \xrightarrow{NaOH} B + H_2O + NaCl$

(a) NH (pyrrolidine)

(b) $H_2N-(CH_2)_4-NH_2$

(c) $Cl\overset{+}{N}H_3-(CH_2)_4-\overset{+}{N}H_3Cl^-$

(d) N–N (bipyrrolidine)

28. In the reaction given below **[NCERT : P L-264 | April 13, 2023 (I)]**

H_2N OH $\xrightarrow[\text{(ii) KOH}]{\text{(i) HCl}}$ 'B' Major Product 'B' is:

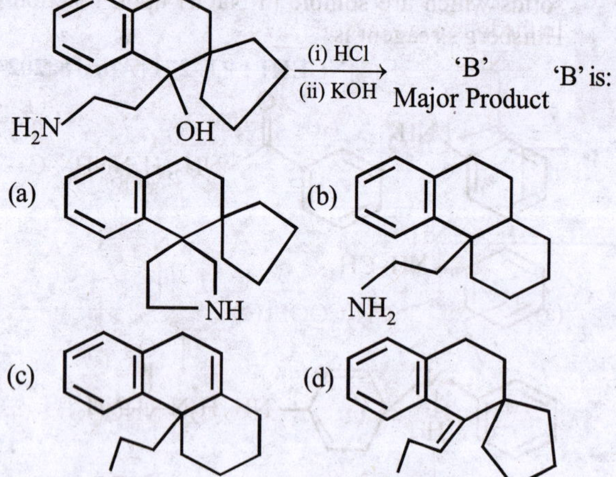

29. o-Phenylenediamine $\xrightarrow{HNO_2}$ 'X' Major Product

'X' is **[NCERT : P L-273 | April 11, 2023 (I)]**

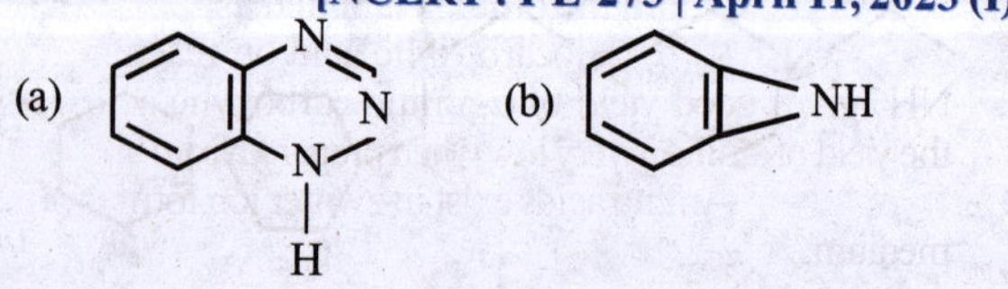

(c) [o-bis-diazonium benzene: $\overset{+}{N}\equiv N$, $\overset{+}{N_2}$] (d) [o-aminobenzenediazonium: $\overset{+}{N_2}$, NH_2]

30. Isomeric amines with molecular formula $C_8H_{11}N$ give the following tests **[NCERT : P L-264 | April 10, 2023 (I)]**
Isomer (P) ⇒ Can be prepared by Gabriel phthalimide synthesis
Isomer (Q) ⇒ Reacts with Hinsberg's reagent to give solid insoluble in NaOH
Isomer (R) ⇒ Reacts with HONO followed by β-naphthol in NaOH to give red dye.
Isomers (P), (Q) and (R) respectively are

P Q R

(a) $C_6H_5CH_2CH_2NH_2$; o-methylbenzyl–N(H)–CH_3 (with CH_3, CH_3); 4-ethylaniline (NH_2)

(b) $C_6H_5CH_2CH_2NH_2$; $C_6H_5N(CH_3)$–CH_3; 2-ethylaniline (NH_2)

(c) 2-ethylaniline (NH_2); $C_6H_5CH_2CH_2NH_2$; C_6H_5–N(H)–C_2H_5

(d) $C_6H_5CH_2NHCH_3$; $C_6H_5CH_2CH_2NH_2$; o-methylbenzyl N–H (CH_3, H)

31. Number of isomeric aromatic amines with molecular formula $C_8H_{11}N$, which can be synthesized by Gabriel phthalimide synthesis is ______
[NCERT : P L-264 | NV, April 06, 2023 (II)]

32. In the following reaction, 'A' is **[Feb. 01, 2023 (I)]**

(valinol: NH_2, CH_2OH) $\xrightarrow{EtO-C(=O)-OEt}$ 'A'
Major product

(a) NH_2 / CH_2COOEt (b) N=C(OEt)–O ring (oxazoline)

(c) NH–C(=O)–O ring (oxazolidinone) (d) NHCOOEt / CH_2OH

33. Given below are two statements: one is labelled as **Assertion (A)** and the other is labelled as **Reason (R).**
Assertion (A): α-Halocarboxylic acid on reaction with dil NH_3 gives good yield of α-amino carboxylic acid whereas the yield of amine is very low when prepared from alkyl halides.
Reason (R): Amino acids exist in zwitter ion form in aqueous medium.

In the light of the above statements, choose the **correct** answer from the options given below **[Feb. 01, 2023 (II)]**
(a) Both **(A)** and **(R)** are correct but **(R)** is not the correct explanation of **(A)**
(b) **(A)** is correct but **(R)** is not correct
(c) Both **(A)** and **(R)** are correct and **(R)** is the correct explanation of **(A)**
(d) **(A)** is not correct but **(R)** is correct

34. How many of the transformation given below would result in aromatic amines?
[NCERT : P L-262, 263, 264 | NV, Jan. 31, 2023 (I)]

(A) $C_6H_5CONH_2 + Br_2 + NaOH \longrightarrow$

(B) Potassium phthalimide (NK) + chlorobenzene (Cl) $\longrightarrow$

(C) $C_6H_5NO_2 \xrightarrow[Pd/C]{H_2}$

(D) $C_6H_5NHCOCH_3 \xrightarrow[\Delta]{dil.\ H_2SO_4}$

35. Cyclohexylamine when treated with nitrous acid yields (P). On treating (P) with PCC results in (Q). When (Q) is heated with dil. NaOH we get (R) The final product (R) is:
[Jan. 31, 2023 (II)]

(a) 2-(cyclohexen-1-yl)cyclohexanone (b) 2-cyclohexylidenecyclohexanone

(c) 2-(1-hydroxycyclohexyl)cyclohexanone (O, OH) (d) 2-(2-oxocyclohexyl)cyclohexanone

36. An organic compound [A]($C_4H_{11}N$), shows optical activity and gives N_2 gas on treatment with HNO_2. The compound [A] reacts with $PhSO_2Cl$ producing a compound which is soluble in KOH. The structure of A is:
[NCERT : P L-271 | Jan. 31, 2023 (II)]

(a) $CH_3CH_2CH_2CH_2NH_2$ (b) Et–N(Me)–Me

(c) $CH_3CH_2CH_2$–N(Me)–H (d) $CH_3CH_2CH(CH_3)NH_2$

37. Benzyl isocyanide can be obtained by :

[NCERT : P L-271 | Jan. 30, 2023 (I)]

(A) 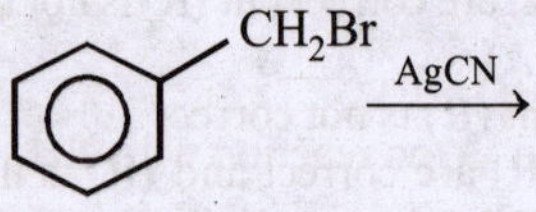 $C_6H_5CH_2Br \xrightarrow{AgCN}$

(B) $C_6H_5CH_2NH_2 \xrightarrow[Aq.\ KOH]{CHCl_3}$

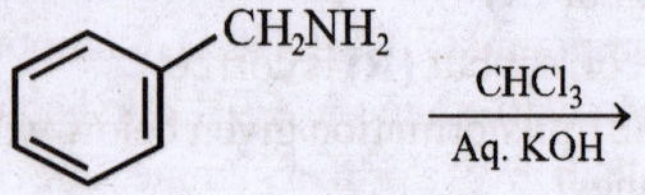

(C) $C_6H_5CH_2\text{–}NHCH_3 \xrightarrow[Aq.\ KOH]{CHCl_3}$

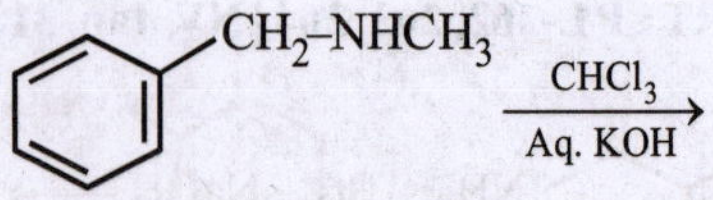

(D) 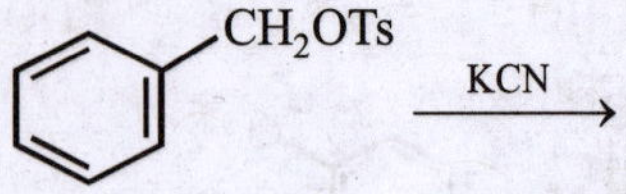 $C_6H_5CH_2OTs \xrightarrow{KCN}$

Choose the correct answer from the options given below:

(a) A and D (b) Only B

(c) A and B (d) B and C

38. Reaction of propanamide with Br_2 / KOH (aq) produces :

[NCERT : P L-264 | Jan. 29, 2023 (II)]

(a) Ethyl nitrile (b) Propyl amine

(c) Propane nitrile (d) Ethyl amine

39. The correct order in aqueous medium of basic strength in case of methyl substituted amines is :

[NCERT : P L-266, 267, 268 | Jan. 25, 2023 (I)]

(a) $Me_2NH > MeNH_2 > Me_3N > NH_3$

(b) $Me_2NH > Me_3N > MeNH_2 > NH_3$

(c) $NH_3 > Me_3N > MeNH_2 > Me_2NH$

(d) $Me_3N > Me_2NH > MeNH_2 > NH_3$

40. Match **List I** with **List II**

[NCERT : P L-266 | Jan. 25, 2023 (II)]

	List I (Amines)		List II (pK_b)
A.	Aniline	I.	3.25
B.	Ethanamine	II.	3.00
C.	N-Ethylethanamine	III.	9.38
D.	N, N-Diethylethanamine	IV.	3.29

Choose the correct answer from the options given below:-

(a) A-I, B-IV, C-II, D-III (b) A-III, B-II, C-I, D-IV

(c) A-III, B-II, C-IV, D-I (d) A-III, B-IV, C-II, D-I

41. Given below are two statements :

[NCERT : P L-265 | Jan. 24, 2023 (II)]

Statement I: Pure aniline and other aryl amines are usually colourless.

Statement II: Aryl amines get coloured on storage due to atmospheric reduction.

In the light of the above statements, choose the most appropriate answer from the options given below :

(a) Both Statement I and Statement II are incorrect.

(b) Both Statement I and Statement II are correct.

(c) Statement I is correct but Statement II is incorrect.

(d) Statement I is incorrect but Statement II is correct.

42. Which among the following is the strongest Bronsted base? [NCERT : P L-266, 267, 268 | July 29, 2022 (I)]

(a) 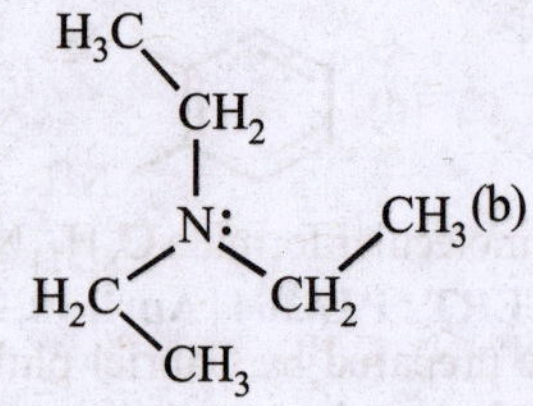 $(CH_3CH_2)_2N\text{–}CH_2CH_3$ (N: with $H_3C\text{–}CH_2$, $H_2C\text{–}CH_3$, $CH_2\text{–}CH_3$)

(b) 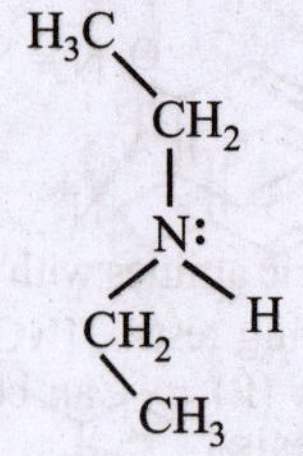 $H_3C\text{–}CH_2\text{–}\ddot{N}H\text{–}CH_2\text{–}CH_3$

(c) 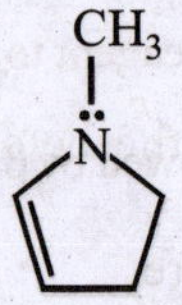 N-methyl pyrroline (ring with $\ddot{N}\text{–}CH_3$)

(d) 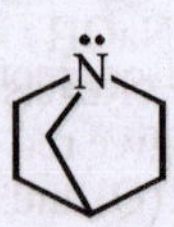 bicyclic amine with bridgehead $\ddot{N}$

43. 'A' $(C_8H_6Cl_2O) \xrightarrow{NH_3} C_8H_8ClNO \xrightarrow[NaOH]{Br_2}$ (benzene ring with CH_3, Cl and H_2N substituents)

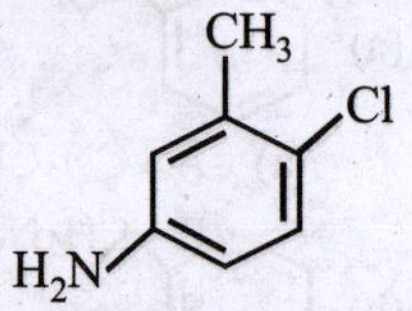

Consider the above reaction, the compound 'A' is :

[NCERT : P L-264 | July 29, 2022 (I)]

(a) 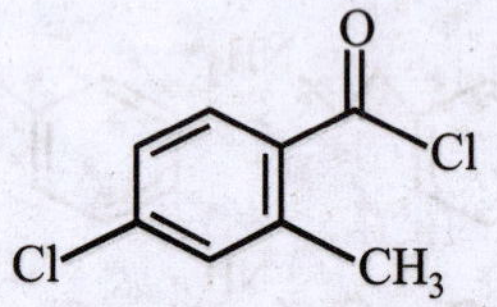 benzoyl chloride (–COCl) with Cl and CH_3 substituents on the ring

(b) 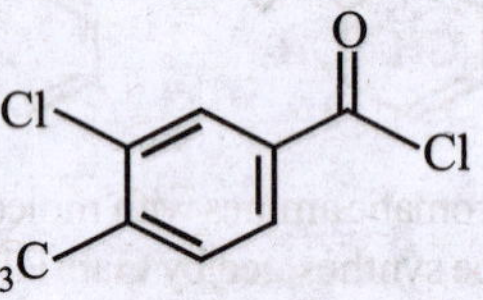 benzoyl chloride (–COCl) with Cl and H_3C substituents on the ring

(c) 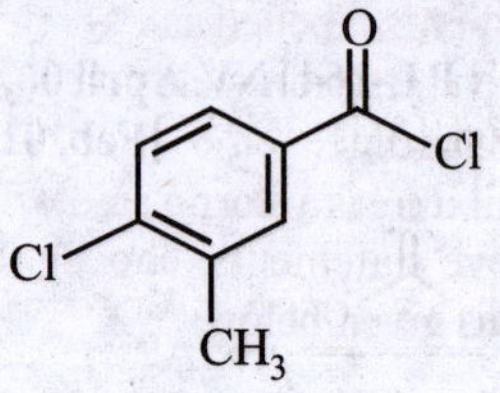 benzoyl chloride (–COCl) with Cl and CH_3 substituents on the ring

(d) 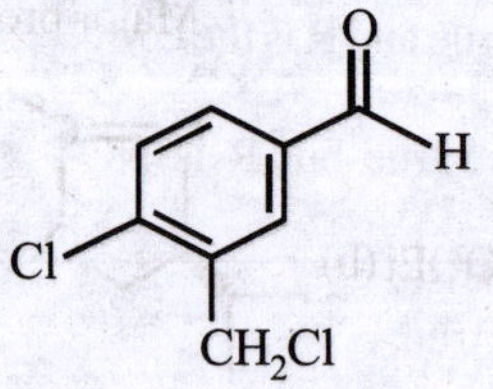 benzaldehyde (–CHO) with Cl and CH_2Cl substituents on the ring

44. The Hinsberg reagent is :

[NCERT : P L-271 | July 29, 2022 (II)]

(a) 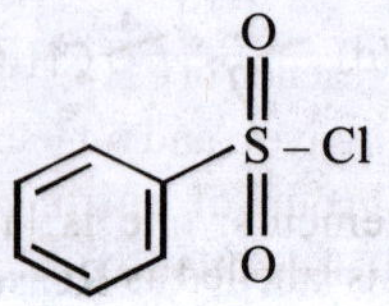 $C_6H_5\text{–}SO_2\text{–}Cl$

(b) 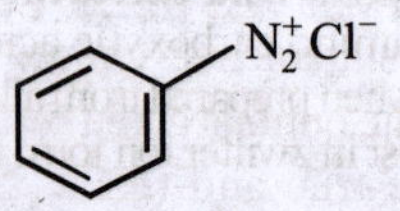 $C_6H_5\text{–}N_2^+Cl^-$

(c) 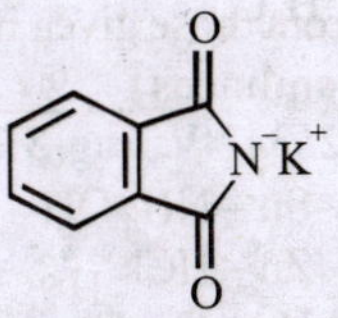 [potassium phthalimide structure: phthalimide ring with $\bar{N}K^+$]

(d) 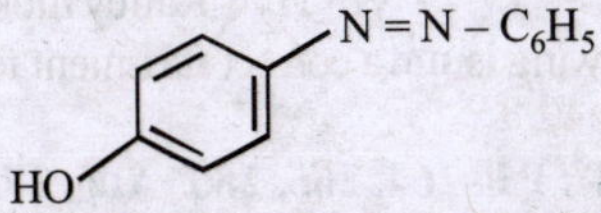 [p-hydroxyazobenzene structure: HO–C$_6$H$_4$–$N=N-C_6H_5$]

45. Identify the correct statement for the below given transformation. **[NCERT : P L-272 | July 28, 2022 (I)]**

$$CH_3-CH_2-CH_2-\underset{\overset{|}{\oplus N(CH_3)_3}}{CH}-CH_3 \xrightarrow[C_2H_5OH]{C_2H_5ONa} \underset{(Major)}{A} + \underset{(Minor)}{B}$$

(a) $A-CH_3CH_2CH=CH-CH_3$, $B-CH_3CH_2CH_2CH=CH_2$, Saytzeff products

(b) $A-CH_3CH_2CH=CH-CH_3$, $B-CH_3CH_2CH_2CH=CH_2$, = Hafmann products

(c) $A-CH_3CH_2CH_2CH=CH_2$, $B-CH_3CH_2CH=CHCH_3$, Hofmann products

(d) $A-CH_3CH_2CH_2CH=CH_2$, $B-CH_3CH_2CH=CHCH_3$, Saytzeff products

46. Given below are two statements: one is labelled as **Assertion A** and the other is labelled as **Reason R**.

Assertion A : Aniline on nitration yields ortho, meta and para nitro derivatives of aniline.

Reason R : Nitrating mixture is a storng acidic mixture.

In the light of the above statements, choose the correct answer from the options given below:

[NCERT : P L-272, 273 | July 28, 2022 (II)]

(a) Both A and R are true and R is the correct explanation of A.
(b) Both A and R are true but R is NOT the correct explanation of A.
(c) A is true but R is false.
(d) A is false but R is true.

47. Given below are two statements: one is labelled as **Assertion (A)** and the other is labelled as **Reason (R).**

[NCERT : P L-273 | July 26, 2022 (I)]

Assertion (A): Experimental reaction of CH_3Cl with aniline and anhydrous $AlCl_3$ does **not** give *o* and *p*-methylaniline.

Reason (R): The $-NH_2$ group of aniline becomes deactivating because of salt formation with anhydrous $AlCl_3$ and hence yields *m*-methyl aniline as the product.

In the light of the above statements, choose the **most appropriate** answer from the options given below:

(a) Both (A) and (R) are true and (R) is the correct explanation of (A).
(b) Both (A) and (R) are true but (R) is not the correct explanation of (A).
(c) (A) is true, but (R) is false.
(d) (A) is false, but (R) is true.

48. An organic compound 'A' on reaction with NH_3 followed by heating gives compound B. Which on further strong heating gives compound C ($C_8H_5NO_2$). Compound C on sequential reaction with ethanolic KOH, alkyl chloride and hydrolysis with alkali gives a primary amine. The compound A is: **[NCERT : P L-270 | July 25, 2022 (I)]**

(a) 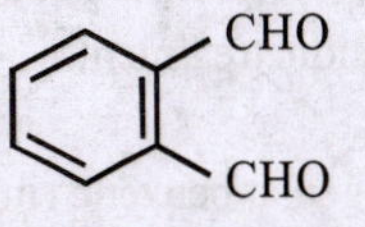 [benzene ring with ortho CHO, CHO]
(b) 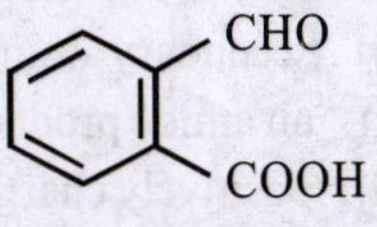 [benzene ring with ortho CHO, COOH]
(c) 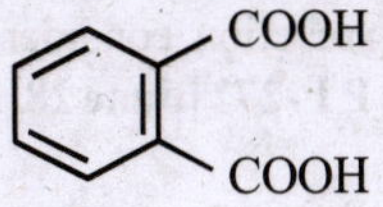 [benzene ring with ortho COOH, COOH]
(d) 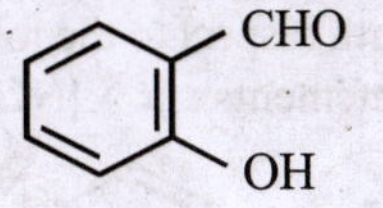 [benzene ring with ortho CHO, OH]

49. 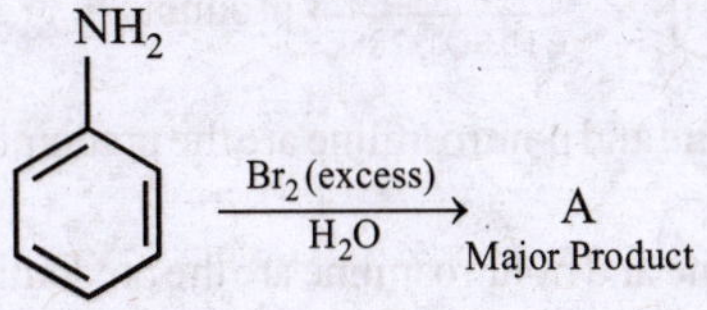 [aniline] $\xrightarrow[H_2O]{Br_2\text{ (excess)}}$ A (Major Product)

[aniline] $\xrightarrow[\text{(ii) } Br_2,\ CH_3COOH \text{ (iii) HCl}]{\text{(i) } (CH_3CO)_2O}$ B (Major Product)

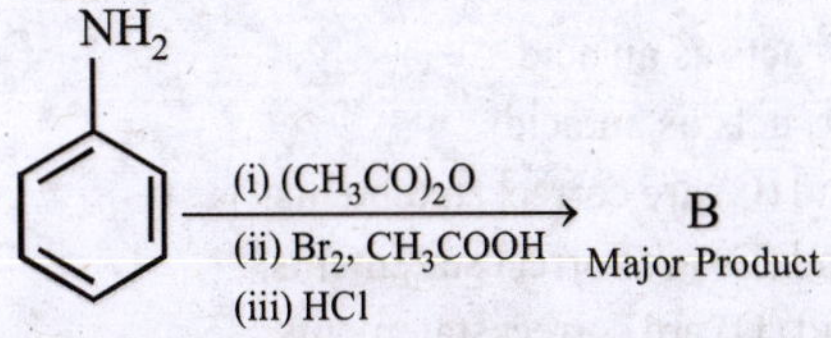

Consider the above reaction, the product A and product B respectively are **[NCERT : P L-272, 273 | June 29, 2022 (I)]**

(a) 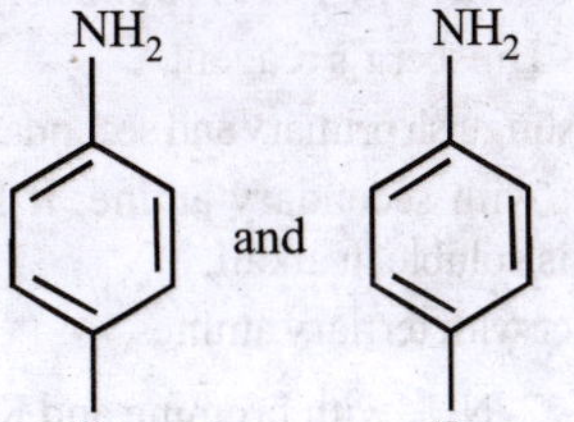 [4-bromoaniline] and [4-bromoaniline]

(b) 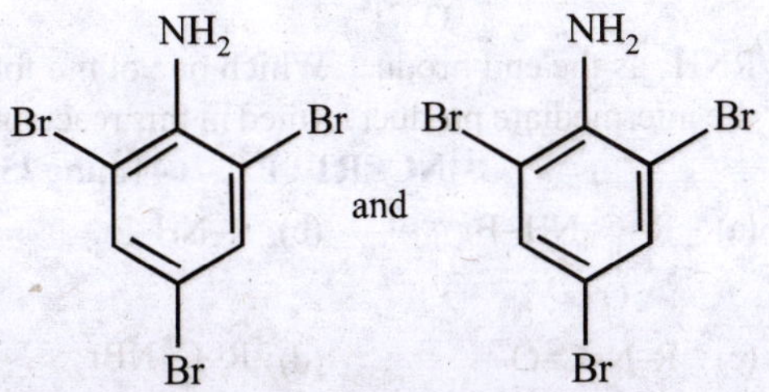 [2,4,6-tribromoaniline] and [2,4,6-tribromoaniline]

(c) 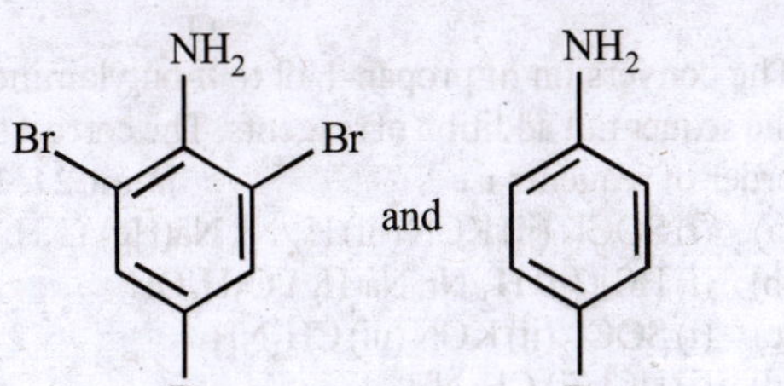 [2,4,6-tribromoaniline] and [4-bromoaniline]

(d) [p-bromoaniline structure: NH_2 at top, Br at para position] and [2,4,6-tribromoaniline structure: NH_2 at top, Br at 2, 4, 6 positions]

50. In Friedel-Crafts alkylation of aniline, one gets : **[NCERT : P L-273 | June 29, 2022 (II)]**
(a) alkylated product with ortho and para substitution.
(b) secondary amine after acidic treatment.
(c) an amide product.
(d) positively charged nitrogen at benzene ring.

51. With respect to the following reaction, consider the given statements : **[NCERT : P L-272 | June 28, 2022 (II)]**

$$C_6H_5NH_2 \xrightarrow[H_2SO_4,\ 288k]{HNO_3} \text{products}$$

(A) o-Nitroaniline and p-nitroaniline are the predominant products
(B) p-Nitroaniline and m-nitroaniline are the predominant products
(C) HNO_3 acts as an acid
(D) H_2SO_4 acts as an acid
(a) (A) and (C) are correct statements.
(b) (A) and (D) are correct statements.
(c) (B) and (D) are correct statements.
(d) (B) and (C) are correct statements.

52. Which statement is NOT correct for *p*-toluene sulphonyl chloride? **[NCERT : P L-271 | June 26, 2022 (II)]**
(a) It is known as Hinsberg's reagent.
(b) It is used to distinguish primary and secondary amines.
(c) On treatment with secondary amine, it leads to a product, that is soluble in alkali.
(d) It doesn't react with tertiary amines.

53. The reaction of $R-\overset{\overset{O}{\|}}{C}-NH_2$ with bromine and KOH gives RNH_2 as the end product. Which one of the following is the intermediate product formed in this reaction ? **[NCERT : P L-264 | June 25, 2022 (I)]**
(a) $R-\overset{\overset{O}{\|}}{C}-NH-Br$ (b) $R-NH-Br$
(c) $R-N=C=O$ (d) $R-\overset{\overset{O}{\|}}{C}-NBr_2$

54. The conversion of propan-1-ol to *n*-butylamine involves the sequential addition of reagents. The correct sequential order of reagents is. **[June 24, 2022 (II)]**
(a) (i) $SOCl_2$ (ii) KCN (iii) H_2/Ni, Na(Hg)/C_2H_5OH
(b) (i) HCl (ii) H_2/Ni, Na(Hg)/C_2H_5OH
(c) (i) $SOCl_2$ (ii) KCN (iii) CH_3NH_2
(d) (i) HCl (ii) CH_3NH_2

55. The total number of reagents from those given below, that can convert nitrobenzene into aniline is ______. (Integer answer) **[NCERT : P L-263 | NV, Aug. 31, 2021 (I)]**
I. $Sn-HCl$ II. $Sn-NH_4OH$
III. $Fe-HCl$ IV. $Zn-HCl$
V. H_2-Pd VI. H_2-Raney nickel

56. Which of the following is **not** a correct statement for primary aliphatic amines? **[NCERT : P L-264, 265, 266 | Aug. 27, 2021 (I)]**
(a) The intermolecular association in primary amines is less than the intermolecular association in secondary amines.
(b) Primary amines on treating with nitrous acid solution form corresponding alcohols except methyl amine.
(c) Primary amines are less basic than the secondary amines.
(d) Primary amines can be prepared by the Gabriel phthalimide synthesis.

57. Given below are two statements : **[NCERT : P L-267, 268 | July 27, 2021 (I)]**
Statement I : Aniline is less basic than acetamide.
Statement II : In aniline, the lone pair of electrons on nitrogen atom is delocalised over benzene ring due to resonance and hence less available to a proton.
Choose the **most appropriate** option :
(a) Statement I is true but statement II is false.
(b) Statement I is false but statement II is true.
(c) Both statement I and statement II are true.
(d) Both statement I and statement II are false.

58. In gaseous triethyl amine, the "–C–N–C–" bond angle is ______ degree. **[NCERT : P L-260 | NV, July 27, 2021 (I)]**

59. $C_6H_5NH_2 \xrightarrow[288\ K]{HNO_3,\ H_2SO_4}$ o-nitroaniline (A) + m-nitroaniline (B) + p-nitroaniline (C)

Consider the given reaction, precentage yield of : **[NCERT : P L-272, 273 | March 18, 2021 (II)]**
(a) C > B > A (b) C > A > B
(c) B > C > A (d) A > C > B

60. An organic compound "A" on treatment with benzene sulphonyl chloride gives compound B. B is soluble in dil. NaOH solution. Compound A is **[NCERT : P L-271 | March 18, 2021 (II)]**
(a) $C_6H_5-N-(CH_3)_2$ (b) $C_6H_5-CH_2NHCH_3$
(c) $C_6H_5-\underset{\underset{CH_3}{|}}{CH}-NH_2$ (d) $C_6H_5-NHCH_2CH_3$

61. Primary, secondary and tertiary amines can be separated using [March 17, 2021 (II)]
(a) para-Toluene sulphonyl chloride
(b) Acetyl amide
(c) Chloroform and KOH
(d) Benzene sulphonic acid

62. An amine on reaction with benzene sulphonyl chloride produces a compound insoluble in alkaline solution. This amine can be prepared by ammonolysis of ethyl chloride. The correct structure of amine is : [NCERT : P L-271 | Feb. 26, 2021 (I)]
(a) $CH_3CH_2CH_2-N(H)-CH_2CH_3$
(b) $CH_3CH_2NH_2$
(c) $CH_3CH_2CH_2NHCH_3$
(d) $C_6H_5-NH-CH_2CH_2CH_3$

63. A. Phenyl methanamine B. N,N-Dimethylaniline
C. N-Methyl aniline D. Benzenamine
Choose the correct order of basic nature of the above amines. [NCERT : P L-266, 267, 268 | Feb. 26, 2021 (II)]
(a) A > C > B > D (b) D > B > C > A
(c) D > C > B > A (d) A > B > C > D

64. In the following reaction, the reason why meta-nitro product also formed is [NCERT : P L-273 | Feb. 24, 2021 (I)]

Aniline $\xrightarrow[\text{Conc. } H_2SO_4,\ 288K]{\text{Conc. } HNO_3}$ p-nitroaniline [A] 51% + m-nitroaniline [B] 47% + o-nitroaniline [C] 2%

(a) low temperature
(b) $-NH_2$ group is highly meta-directive
(c) Formation of anilinium ion
(d) $-NO_2$ substitution always takes place at meta-position

65. The total number of amines among the following which can be synthesized by Gabriel synthesis is ________. [NCERT : P L-264 | NV, Feb. 24, 2021 (II)]
(A) $(CH_3)_2CH-CH_2-NH_2$
(B) $CH_3CH_2NH_2$
(C) $C_6H_5-CH_2-NH_2$
(D) $C_6H_5-NH_2$

66. Carbylamine test is used to detect the presence of primary amino group in an organic compound. Which of the following compound is formed when this test is performed with aniline? [NCERT : P L-271 | Feb. 25, 2021 (II)]

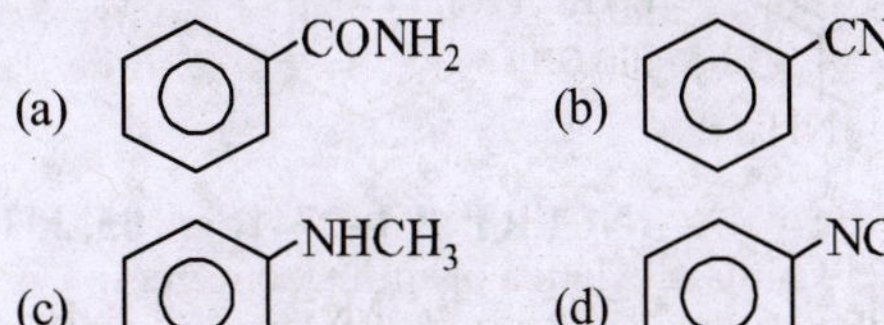

(c) $C_6H_5-NHCH_3$ (d) C_6H_5-NC

67. The increasing order of pK_b values of the following compounds is : [NCERT : P L-267 | Sep. 06, 2020 (I)]

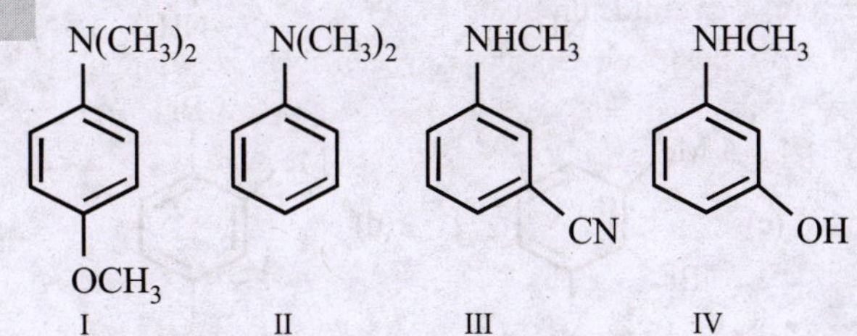

(a) II < IV < III < I (b) I < II < IV < III
(c) II < I < III < IV (d) I < II < III < IV

68. Which of the following compounds can be prepared in good yield by Gabriel phthalimide synthesis? [NCERT : P L-264 | Sep. 06, 2020 (II)]
(a) $C_6H_5CH_2NH_2$ (b) $CH_3-CH_2-NHCH_3$
(c) $C_6H_5-CH_2-C(=O)-NH_2$ (d) $C_6H_5NH_2$

69. The most appropriate reagent for conversion of C_2H_5CN into $CH_3CH_2CH_2NH_2$ is : [NCERT : P L-271 | Sep. 05, 2020 (I)]
(a) $NaBH_4$ (b) CaH_2
(c) $LiAlH_4$ (d) $Na(CN)BH_3$

70. The increasing order of basicity of the following compounds is: [Sep. 05, 2020 (I)]

(A) Pyridine (B) Pyrrole (N–H) (C) Pyrrolidine (N–H) (D) Imidazole (N–H)

(a) (A) < (B) < (C) < (D)
(b) (B) < (A) < (D) < (C)
(c) (D) < (A) < (B) < (C)
(d) (B) < (A) < (C) < (D)

71. The final major product of the following reaction is :

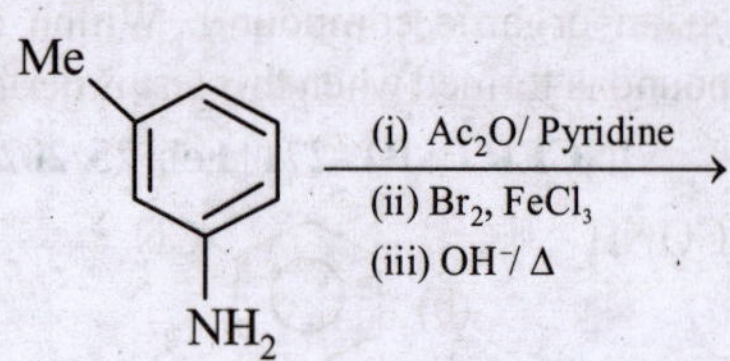

[NCERT : P L-274 | Sep. 05, 2020 (II)]

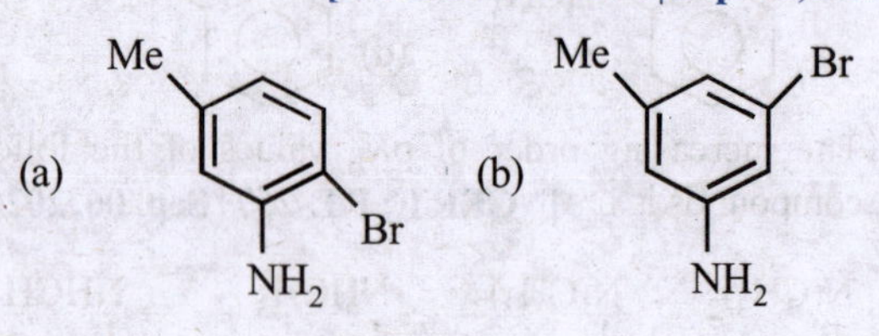

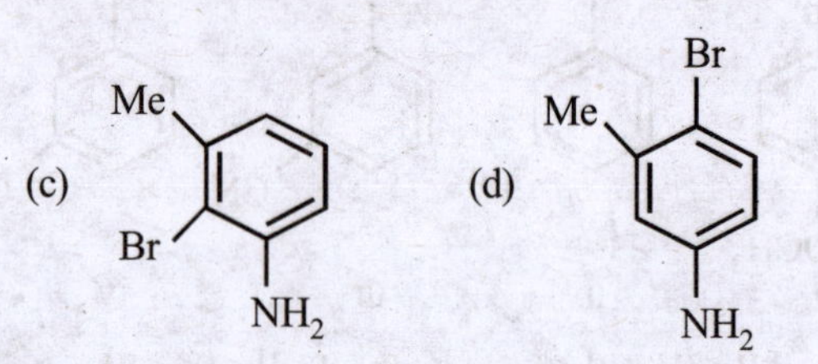

72. The decreasing order of basicity of the following amines is: [Jan. 09, 2020 (II)]

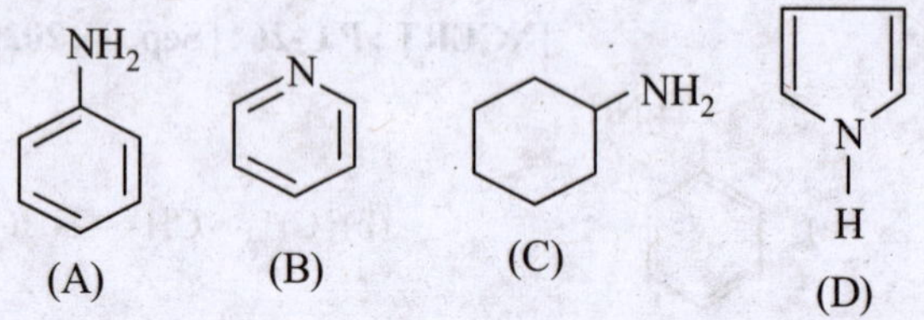

(a) (A) > (C) > (D) > (B) (b) (C) > (A) > (B) > (D)
(c) (B) > (C) > (D) > (A) (d) (C) > (B) > (A) > (D)

73. The increasing order of pK_b for the following compounds will be: [NCERT : P L-266 | Jan. 07, 2020 (I)]

(A) $NH_2-CH=NH$, (B) [cyclic guanidine: N, N, NH], (C) CH_3NHCH_3

(a) (B) < (C) < (A) (b) (A) < (B) < (C)
(c) (C) < (A) < (B) (d) (B) < (A) < (C)

74. In the following reaction sequence,

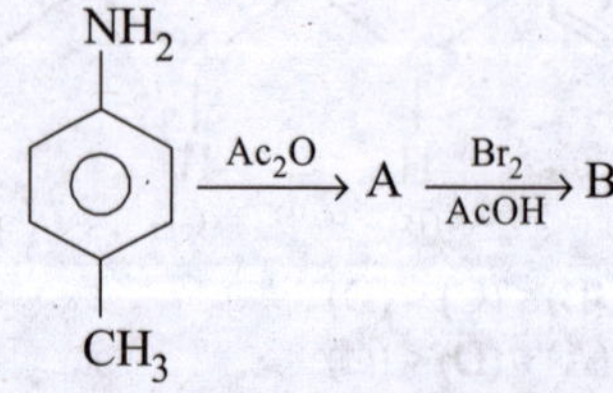

$\xrightarrow{Ac_2O}$ A $\xrightarrow[AcOH]{Br_2}$ B

the major product B is:
[NCERT : P L-272 | Jan. 07, 2020 (II)]

(a) $NHCOCH_3$, Br, CH_3 (b) $NHCOCH_3$, Br, CH_2Br
(c) $NHCOCH_3$, $COCH_3$, CH_3 (d) $NHCOCH_3$, Br, CH_3

75. The increasing order of the pK_b of the following compounds is : [NCERT : P L-267 | April 12, 2019 (I)]

(A) F–C_6H_4–NH–C(=S)–NH–C_6H_5
(B) CH_3O–C_6H_4–NH–C(=S)–NH–C_6H_5
(C) O_2N–C_6H_4–NH–C(=S)–NH–C_6H_5
(D) H_3O–C_6H_4–NH–C(=S)–NH–C_6H_5

(a) (A) < (C) < (D) < (B) (b) (C) < (A) < (D) < (B)
(c) (B) < (D) < (A) < (C) (d) (B) < (D) < (C) < (A)

76. Ethylamine ($C_2H_5NH_2$) can be obtained from N-ethylphthalimide on treatment with :
[NCERT : P L-263 | April 10, 2019 (I)]

(a) NH_2NH_2 (b) CaH_2
(c) $NaBH_4$ (d) H_2O

77. The major product of the following reaction is:
[NCERT : P L-270 | April 10, 2019 (I)]

$CH_3CH(OH)CH_2CH_2NH_2 \xrightarrow[\text{triethylamine}]{\text{ethyl formate (1 equiv.)}}$

(a) $CH_3-CH(OH)-CH=CH_2$
(b) $CH_3CH(OH)CH_2CH_2NHCHO$
(c) $CH_3CH(OCHO)CH_2CH_2NH_2$
(d) $CH_3CH=CH-CH_2NH_2$

78. Which of the following is NOT a correct method of the preparation of benzylamine from cyanobenzene ?
[NCERT : P L-263 | April 10, 2019 (II)]
(a) H_2/Ni
(b) (i) $LiAlH_4$ (ii) H_3O^+
(c) (i) $SnCl_2$ + HCl(gas) (ii) $NaBH_4$
(d) (i) HCl/H_2O (ii) $NaBH_4$

79. The major product 'Y' in the following reaction is:
[NCERT : P L-271 | April 10, 2019 (II)]

Ph–CO–CH_3 $\xrightarrow{NaOCl}$ X $\xrightarrow[\text{(ii) aniline}]{\text{(i) } SOCl_2}$ Y

(a) HN–CO–Ph on benzene ring with NH_2 para
(b) N=CH–CO–Ph attached to benzene ring
(c) NH_2 on benzene ring with CO–Ph para
(d) NH_2 on benzene ring with CO–Ph ortho

80. Hinsberg's reagent is: [April 9, 2019 (II)]
(a) C_6H_5COCl (b) $SOCl_2$
(c) $C_6H_5SO_2Cl$ (d) $(COCl)_2$

81. Polysubstitution is a major drawback in :
[NCERT : P L-270 | April 8, 2019 (II)]
(a) Friedel Craft's alkylation
(b) Reimer Tiemann reaction
(c) Acetylation of aniline
(d) Friedel Craft's acylation

82. The major product in the following reaction is:
[NCERT : P L-264 | April 8, 2019 (II)]

Adenine (NH_2, N–H) + CH_3I $\xrightarrow{base}$

(a) NH_2; N–H; $\overset{+}{N}$–CH_3
(b) NH_2; $\overset{+}{N}CH_3$; N–H
(c) $NHCH_3$; N–H
(d) NH_2; N–CH_3

83. Which of the following amines can be prepared by Gabriel phthalimide reaction ? [April 8, 2019 (I)]
(a) *n*-butylamine (b) triethylamine
(c) *t*-butylamine (d) neo-pentylamine

84. In the following compounds, the decreasing order of basic strength will be: [April 8, 2019 (I)]
(a) $C_2H_5NH_2 > NH_3 > (C_2H_5)_2NH$
(b) $(C_2H_5)_2NH > NH_3 > C_2H_5NH_2$
(c) $(C_2H_5)_2NH > C_2H_5NH_2 > NH_3$
(d) $NH_3 > C_2H_5NH_2 > (C_2H_5)_2NH$

85. The major product of the following reaction is :
[Jan. 12, 2019 (II)]

H_3C–CO–O–(benzene ring)–$CH_2CH_2CH_2$–NH_2

(i) $NaNO_2/H^+$
(ii) CrO_3/H^+
(iii) H_2SO_4 (conc.), Δ

(a) HO (indanone)
(b) H_3C–CO–O– (indanone)
(c) H_3C–CO–O– (indanone)
(d) HO (indanone)

86. In the following compound,

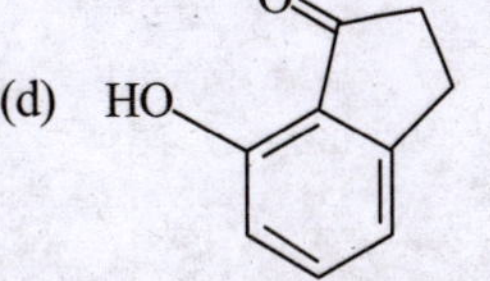

the favourable site/s for protonation is/are :
[NCERT : P L-272 | Jan. 11, 2019 (II)]
(a) (a) and (e) (b) (b), (c) and (d)
(c) (a) and (d) (d) (a)

87. The major product formed in the reaction given below will be: [Jan. 10, 2019 (I)]

(bicyclic)–CH_2–NH_2 $\xrightarrow[\text{Aq. HCl, 0– 5 °C}]{NaNO_2}$

(a) (bicyclic)–CH_2–NO_2
(b) (bicyclic)–NO_2

(c) [structure: decahydronaphthalenol, OH]

(d) [structure: hexahydroindanol, OH]

88. Arrange the following amines in the decreasing order of basicity: **[NCERT : P L-266, 267, 268 | Jan. 9, 2019 (I)]**

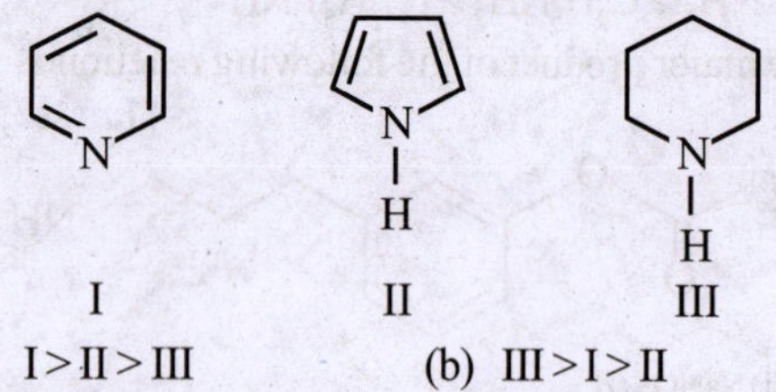

I II III

(a) I > II > III (b) III > I > II
(c) III > II > I (d) I > III > II

89. The major product obtained in the following reaction is:

$$\text{PhCH(OH)CH(NH}_2\text{)CH}_3 \xrightarrow[\text{room temp.}]{(CH_3CO)_2O/\text{pyridine (1 eq.)}}$$

[NCERT : P L-272 | Jan. 9, 2019 (II)]

(a) [Ph–C(CH₃)=C(CH₃)–NHCOCH₃ structure] $NHCOCH_3$

(b) [ring with $COCH_3$; side chain OH, NH_2]

(c) [Ph–C($OCOCH_3$)–CH(NH_2)–CH₃]

(d) [Ph–CH(OH)–CH($NHCOCH_3$)–CH₃]

90. The increasing basicity order of the following compounds is: **[NCERT : P L-266, 267, 268 | Jan. 9, 2019 (II)]**

(A) $CH_3CH_2NH_2$ (B) $CH_3CH_2NH(CH_2CH_3)$

(C) $H_3C-N(CH_3)-CH_3$ (D) $Ph-N(CH_3)-H$

(a) (D) < (C) < (B) < (A) (b) (D) < (C) < (A) < (B)
(c) (A) < (B) < (C) < (D) (d) (A) < (B) < (D) < (C)

91. Products A and B formed in the following reactions are respectively : **[Online April 16, 2018]**

$$p\text{-HO}_3S\text{–C}_6H_4\text{–}\overset{+}{N}H_2\text{–CH}_2\text{–COO}^- + HNO_2 \longrightarrow A \xrightarrow{C_6H_5NH_2} B$$

(a) p-HO_3S–C_6H_4–N=N–O–C(=O)CH_3 and p-HO_3S–C_6H_4–N=N–C_6H_4–NH_2

(b) p-HO_3S–C_6H_4–N=N–OC(=O)CH_3 and p-HO_3S–C_6H_4–N=N–NH–C_6H_5

(c) p-HO_3S–C_6H_4–N=N $COCH_3$ and p-HO_3S–C_6H_4–N=N–C_6H_5

(d) p-HO_3S–C_6H_4–N=N $COCH_3$ and HO_3S–C_6H_4–NH–C_6H_4–NH_2

92. The increasing order of diazotisation of the following compounds is: **[NCERT : P L-274 | Online April 15, 2018 (II)]**

(A) [cyclohexane with NH_2 and $COOH$ on same carbon]

(B) [aniline, NH_2]

(C) [H_3C–C(=O)–O–C₆H₄–NH_2 (meta)]

(D) [o-aminoacetophenone: $COCH_3$, NH_2]

(a) (D) < (C) < (B) < (A) (b) (A) < (D) < (B) < (C)
(c) (A) < (B) < (C) < (D) (d) (A) < (D) < (C) < (B)

93. Among the following compounds, the increasing order of their basic strength is : **[NCERT : P L-266, 267, 268 | Online April 9, 2017]**

(I) 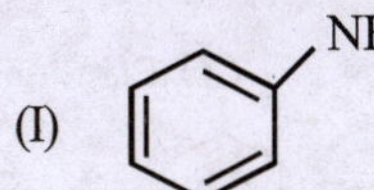

(II) pyrrole (N–H)

(III) N-methylpiperidine (CH_3 on N)

(IV) cyclohexylamine (NH_2)

(a) (I) < (II) < (IV) < (III) (b) (I) < (II) < (III) < (IV)
(c) (II) < (I) < (IV) < (III) (d) (II) < (I) < (III) < (IV)

94. Which of the following compounds will form significant amount of meta product during mono-nitration reaction ? **[2017]**

(a) OH (phenol) (b) $OCOCH_3$ (phenyl acetate)
(c) NH_2 (aniline) (d) $NHCOCH_3$ (acetanilide)

95. In the Hoffmann bromamide degradation reaction, the number of moles of NaOH and Br_2 used per mole of amine produced are : **[NCERT : P L-264 | 2016]**

(a) Two moles of NaOH and two moles of Br_2.
(b) Four moles of NaOH and one mole of Br_2.
(c) One mole of NaOH and one mole of Br_2.
(d) Four moles of NaOH and two moles of Br_2.

96. The test to distinguish primary, secondary and tertiary amines is : **[NCERT : P L-271 | Online April 9, 2016]**

(a) Sandmeyer's reaction (b) Carbylamine reaction
(c) Mustard oil test (d) $C_6H_5SO_2Cl$

97. The "N" which does not contribute to the basicity for the compound is : **[NCERT : P L-269 | Online April 10, 2016]**

(purine ring, atoms numbered 1N, 2, N 3, 4, 5, 6, N 7, 8, N 9 H)

(a) N9 (b) N3 (c) N1 (d) N7

98. Arrange the following amines in the order of increasing basicity. **[NCERT : P L-266, 267, 268 | Online April 10, 2015]**

(a) CH_3NH_2 < p-methoxyaniline (NH_2, OCH_3) < aniline (NH_2) < p-nitroaniline (NH_2, NO_2)

(b) p-methoxyaniline (NH_2, OCH_3) < p-nitroaniline (NH_2, NO_2) < aniline (NH_2) < CH_3NH_2

(c) p-nitroaniline (NH_2, NO_2) < aniline (NH_2) < p-methoxyaniline (NH_2, OCH_3) < CH_3NH_2

(d) aniline (NH_2) < p-methoxyaniline (NH_2, OCH_3) < p-nitroaniline (NH_2, NO_2) < CH_3NH_2

99. On heating an aliphatic primary amine with chloroform and ethanolic potassium hydroxide, the organic compound formed is: **[NCERT : P L-264 | 2014]**

(a) an alkanol (b) an alkanediol
(c) an alkyl cyanide (d) an alkyl isocyanide

100. Considering the basic strength of amines in aqueous solution, which one has the smallest pK_b value? **[NCERT : P L-267 | 2014; Same 2007]**

(a) $(CH_3)_2NH$ (b) CH_3NH_2
(c) $(CH_3)_3N$ (d) $C_6H_5NH_2$

101. The major product of the reaction

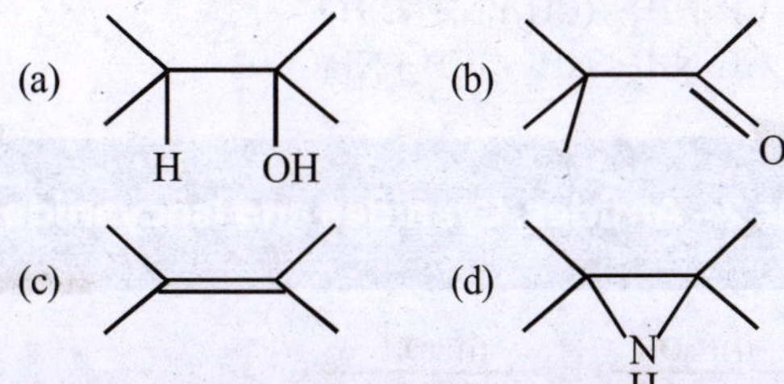

[NCERT : P L-271 | Online April 9, 2014]

(a) (H, OH) (b) (=O)
(c) (alkene) (d) (N–H aziridine)

102. The final product formed when methyl amine is treated with $NaNO_2$ and HCl is: **[NCERT : P L-274 | Online April 19, 2014]**

(a) Diazomethane (b) Methyl alcohol
(c) Methylcyanide (d) Nitromethane

103. A compound with molecular mass 180 is acylated with CH_3COCl to get a compound with molecular mass 390. The number of amino groups present per molecule of the former compound is : **[2013]**

(a) 2 (b) 5 (c) 4 (d) 6

104. The order of basicity of amines in gaseous state is :

[NCERT : P L-266, 267 | Online April 23, 2013]

(a) $1° > 2° > 3° > NH_3$ (b) $3° > 2° > NH_3 > 1°$

(c) $3° > 2° > 1° > NH_3$ (d) $NH_3 > 1° > 2° > 3°$

105. Carbylamine is formed from aliphatic or aromatic primary amine via which of the following intermediates ?

[NCERT : P L-260 | Online April 25, 2013]

(a) Carbanions (b) Carbenes

(c) Carbocations (d) Carbon radicals

106. The most basic compound among the following is

[2005, Online May 7, 2012]

(a) Acetanilide (b) Benzyl is amine

(c) *p*-Nitro aniline (d) Aniline

107. The order of basicity of the compounds is

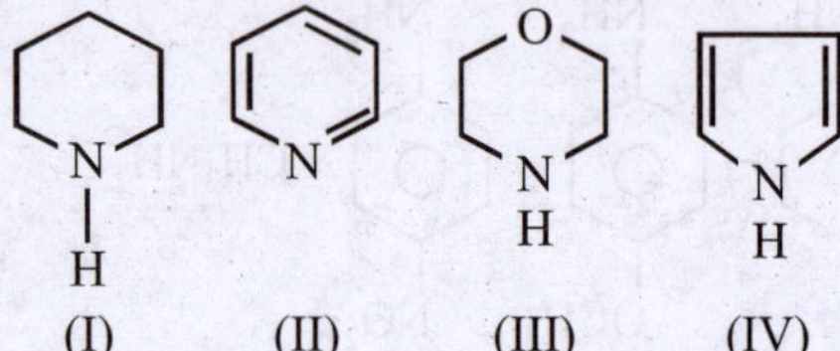

[Online May 19, 2012]

(a) IV > I > III > II (b) I > III > II > IV

(c) III > I > IV > II (d) II > I > III > IV

108. Which one of the following is the strongest base in aqueous solution ? **[2007]**

(a) Methylamine (b) Trimethylamine

(c) Aniline (d) Dimethylamine.

109. Which one of the following methods is neither meant for the synthesis nor for separation of amines? **[2005]**

(a) Curtius reaction (b) Wurtz reaction

(c) Hofmann method (d) Hinsberg method

110. The correct order of increasing basic nature for the bases NH_3, CH_3NH_2 and $(CH_3)_2NH$ is **[2003]**

(a) $(CH_3)_2NH < NH_3 < CH_3NH_2$

(b) $NH_3 < CH_3NH_2 < (CH_3)_2NH$

(c) $CH_3NH_2 < (CH_3)_2NH < NH_3$

(d) $CH_3NH_2 < NH_3 < (CH_3)_2NH$

Topic 2 - Amides, Cyanides and Isocyanides

111. $A \xrightarrow[\text{(ii)}H_3O^+]{\text{(i)NaOH}} B \xrightarrow[\text{(ii)}H_2SO_4\Delta]{\text{(i)EtOH}} C$

'A' shows positive Lassaign's test for N and its molar mass is 121.

'B' gives effervescence with aq. $NaHCO_3$.

'C' gives fruity smell.

Identify A, B and C from the following.

[NCERT : P L-271 | April 8, 2025 (II)]

(a) A = $C_6H_5CONH_2$, B = $C_6H_5CO_2H$, C = $C_6H_5CO_2Et$

(b) A = $C_6H_5CH_2NHNH_2$, B = $C_6H_5CO_2H$, C = $C_6H_5CO_2Et$

(c) A = p-$OHC-C_6H_4-NH_2$, B = $H_2N-C_6H_4-CH(OH)-CO-C_6H_4-NH_2$, C = $H_2N-C_6H_4-CH(OH)-C(OH)(OEt)-C_6H_4-NH_2$

(d) A = o-$H_2N-C_6H_4-CHO$, B = o-$H_2N-C_6H_4-CO_2H$, C = o-$EtHN-C_6H_4-CO_2Et$

112. X g of nitrobenzene on nitration gave 4.2 g of m-dinitrobenzene.

X = ________ g. (nearest integer)

[Given : molar mass (in g mol^{-1}) C : 12, H : 1, O : 16, N : 14]

[NCERT : P L-272 | NV, April 3, 2025 (II)]

113. The number of correct reaction(s) among the following is ________.

[NCERT : P L-272 | NV, April 4, 2024 (I)]

(A) $C_6H_6 + C_6H_5COCl \xrightarrow{\text{Anhyd.}AlCl_3} C_6H_5CH_2C_6H_5$

(B)

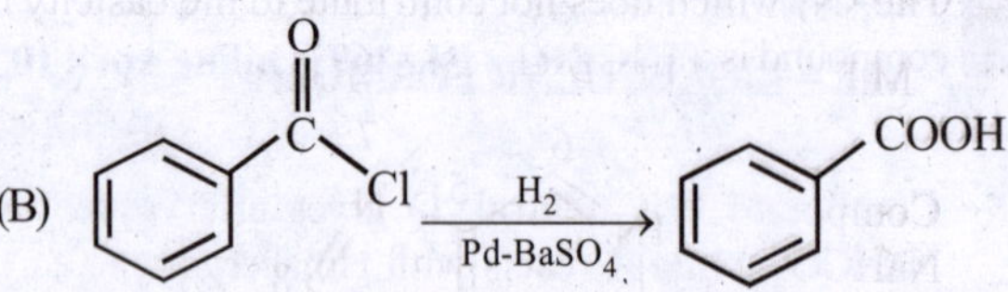

(C) $C_6H_6 \xrightarrow[\text{Anhyd. }AlCl_3/CuCl]{CO,\ HCl} C_6H_5CHO$

(D)

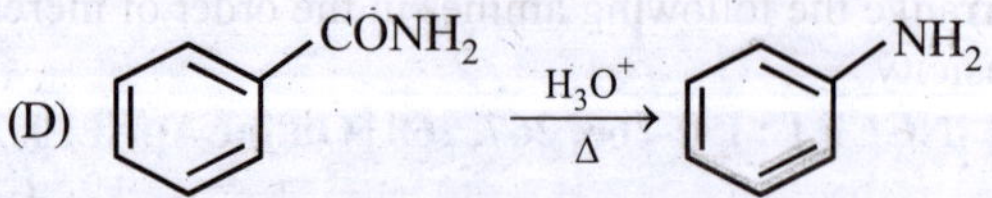

114. In the reaction given below

[NCERT : P L-264 | April 13, 2023 (I)]

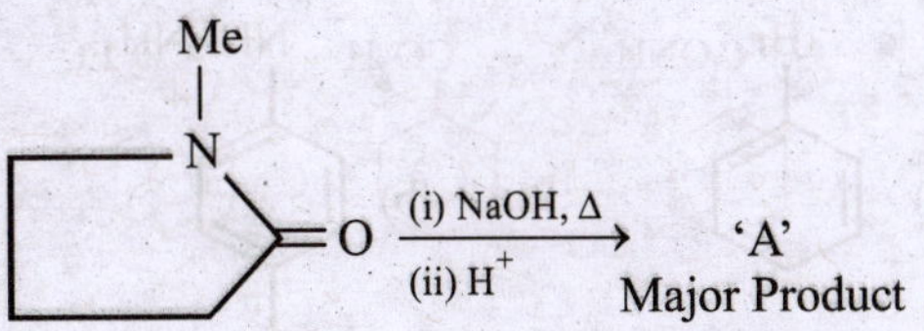

'A' is

(a)  MeHN–$CH_2CH_2CH_2$–COOH

(b)

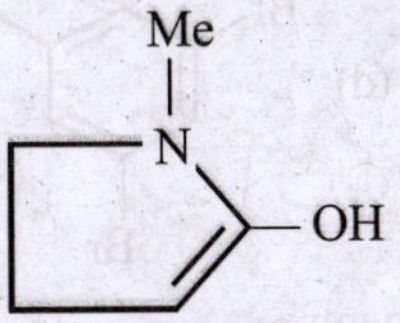

(c)

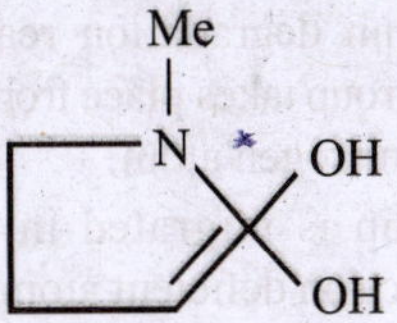

(d)  MeHN–$CH_2CH_2CH_2$–CHO

115. The major product for the following reaction is:

$CH_2=CH–CN \xrightarrow{HO\frown\frown SH}$

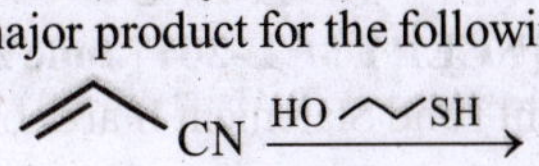

[NCERT : P L-266 | April 13, 2023 (II)]

(a) 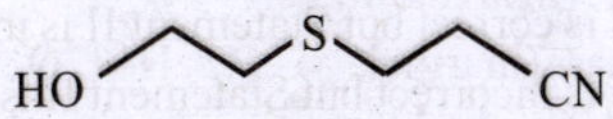 HO–S–CN

(b) 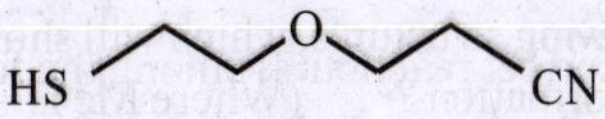 HS–O–CN

(c) 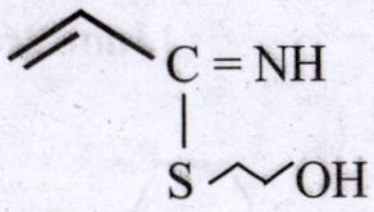 C = NH, S–OH

(d) 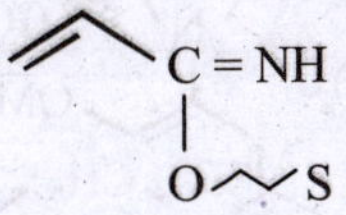 C = NH, O–S

116. Compound P (M.F. $C_{14}H_{13}ON$) $\xrightarrow{HCl, \Delta}$ Filter → Residue Q; → Filtrate $\xrightarrow{NaOH}$ Oily liquid R.

M.F = MOLECULAR FORMULA

Compound P is neutral. Q gives effervescences with $NaHCO_3$ while R reacts with Hinsbergs reagent to give solid soluble in NaOH. Compound P is

[NCERT : P L-271 | April 06, 2023 (I)]

(a) 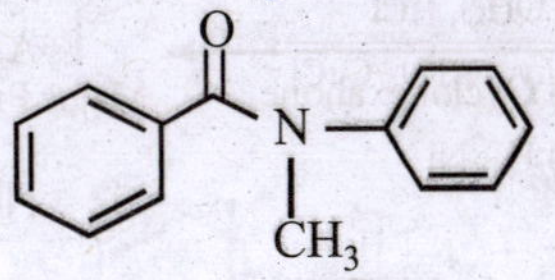 N-methyl-N-phenylbenzamide (CH_3)

(b) 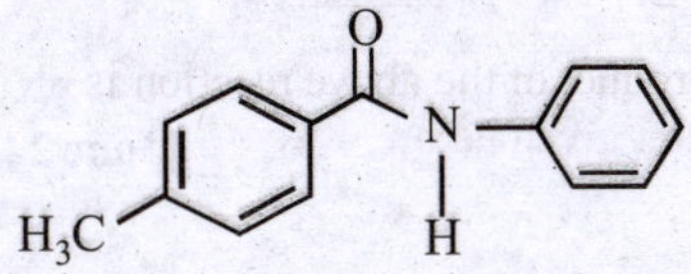 H_3C–C$_6$H$_4$–CO–NH–C$_6$H$_5$

(c) C–N–H (O, H)

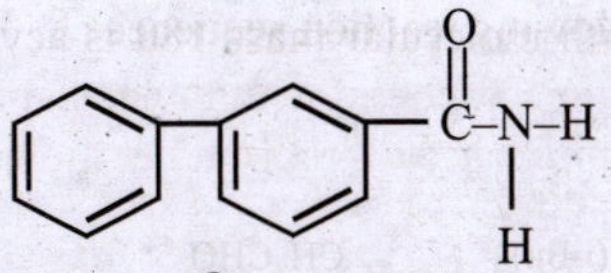

(d) 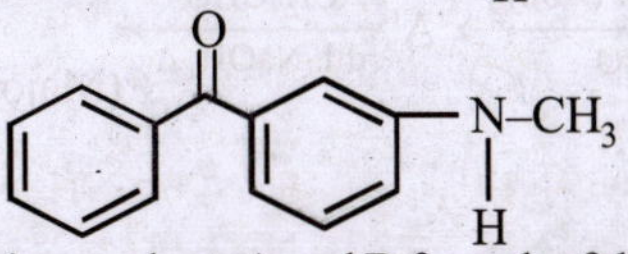 N–CH_3 (H)

117. A major products A and B from the following reactions are:

[NCERT : P L-263, 272 | April 06, 2023 (I)]

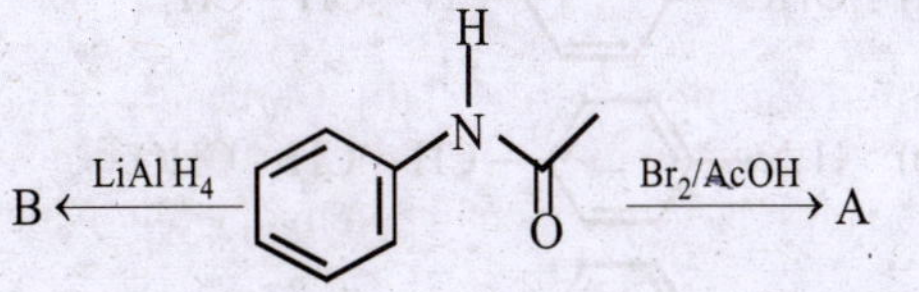

B $\xleftarrow{LiAlH_4}$ (acetanilide) $\xrightarrow{Br_2/AcOH}$ A

(a) A = 2,4,6-tribromoacetanilide, B = N–CH(OH)–CH_3 derivative (OH)

(b) A = 4-bromoacetanilide (Br), B = N–CH(OH)–CH_3 (OH)

(c) A = 2,4,6-tribromoacetanilide (Br, Br, Br), B = aniline (NH_2)

(d) A = 4-bromoacetanilide (Br), B = N-ethylaniline

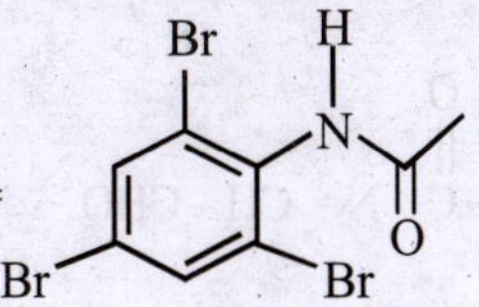

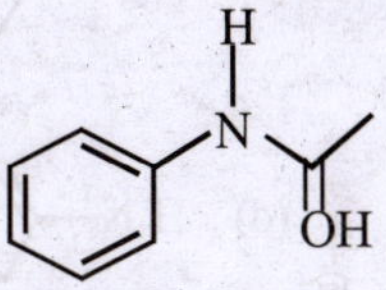

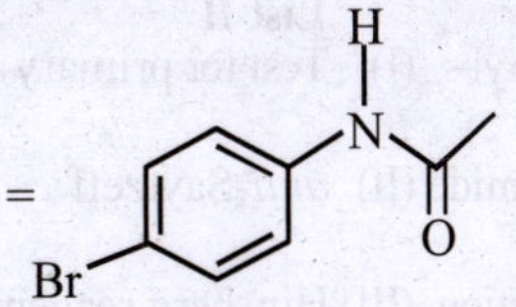

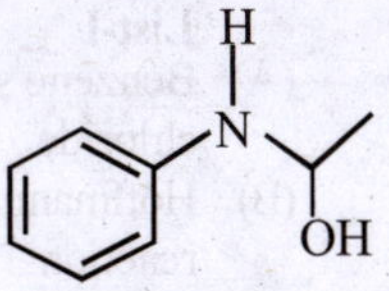

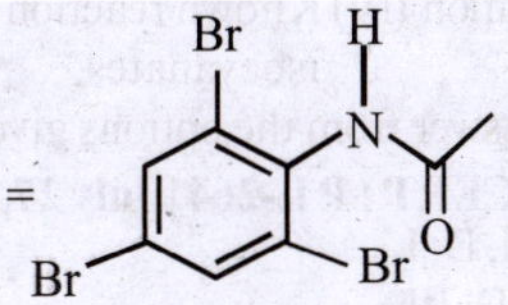

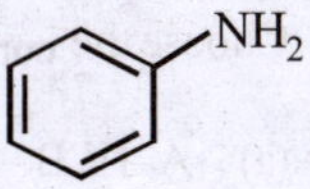

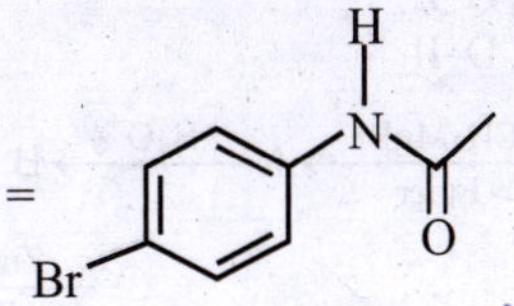

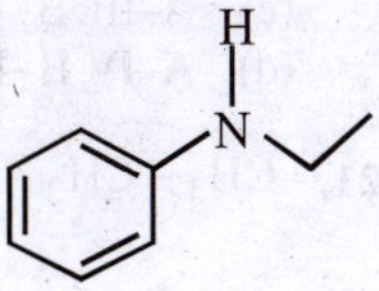

118. The major product formed in the following reaction is

$C_6H_4(CH_2CONH_2)(COOCH_3) \xrightarrow[\Delta]{Br_2/NaOH}$

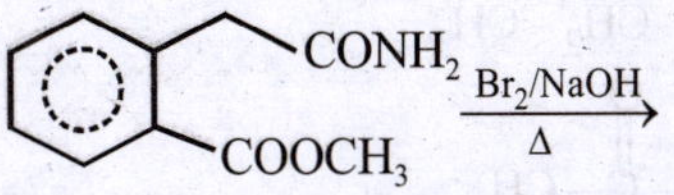

[NCERT : P L-264 | April 06, 2023 (I)]

(a) 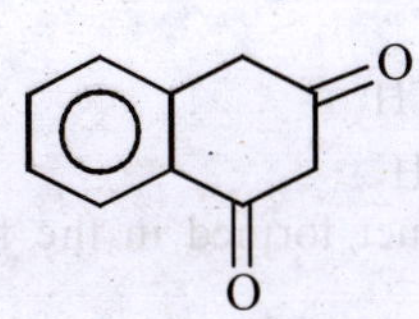 (O, O) (b) 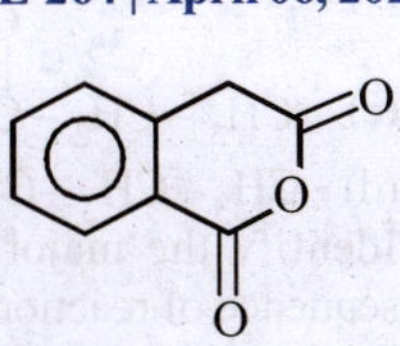 (O, O, O)

(c) 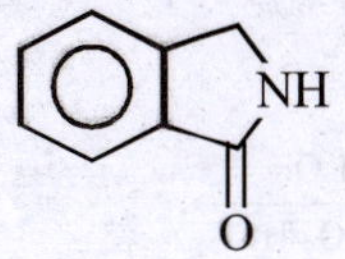 NH, O (d) 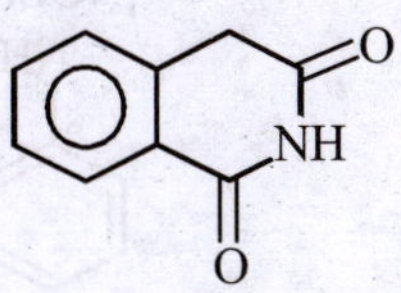 NH, O, O

119. Consider the following reaction sequence :

$$\text{NH}_2\text{-C}_6\text{H}_4\text{-CN} \xrightarrow[\text{(ii) } H_2O]{\text{(i) AlH (i-Bu)}_2} \text{'A'} \xrightarrow[\text{dil. NaOH, }\Delta]{CH_3CHO} \underset{\text{(Major Product)}}{B}$$

The product 'B' is : **[July 29, 2022 (I)]**

(a) OHC—C_6H_4—N = CH – CH_3

(b) H_2N—C_6H_4—CH = CH – CHO

(c) H_2N—C_6H_4—CH_2 – N = CH – CH_3

(d) H_2N—C_6H_4—C(=O) – N = CH – CHO

120. Match **List-I** with **List-II**

	List-I		List-II
(A)	Benzene sulphonyl chloride	(I)	Test for primary amines
(B)	Hoffmann bromamide reaction	(II)	*anti*-Saytzeff
(C)	Carbylamine reaction	(III)	Hinsberg reagent
(D)	Hoffmann orientation	(IV)	Known reaction of isocyanates.

Choose the **correct** answer from the options given below:

[NCERT : P L-264 | July 27, 2022 (I)]

(a) A–IV, B–III, C–II, D–I
(b) A–IV, B–II, C–I, D–III
(c) A–III, B–IV, C–I, D–II
(d) A–IV, B–III, C–I, D–II

121. $CH_3 - CH_2 - CN \xrightarrow[\text{Ether}]{CH_3MgBr} A \xrightarrow{H_3O^+} B \xrightarrow[HCl]{Zn-Hg} C$

The correct structure of C is **[July 25, 2022 (II)]**

(a) $CH_3 - CH_2 - CH_2 - CH_3$

(b) $CH_3 - CH_2 - \overset{O}{\overset{\|}{C}} - CH_3$

(c) $CH_3 - CH_2 - \underset{}{\overset{OH}{\overset{|}{C}H}} - CH_3$

(d) $CH_3 - CH_2 - CH = CH_2$

122. Identify the major product formed in the following sequence of reactions :

[NCERT : P L-272, 274 | June 28, 2022 (I)]

$$C_6H_5NH_2 \xrightarrow[\text{(ii) } NaNO_2/HCl \quad \text{(iii) } H_3PO_2]{\text{(i) } Br_2/H_2O}$$

(a) Bromobenzene (C_6H_5Br)

(b) $NH\cdot NH_2$ on benzene ring with Br at para position

(c) 1,3,5-tribromobenzene (Br, Br, Br)

(d) Phenol (OH) with Br, Br, Br at 2,4,6 positions

123. Given below are two statements:

Statements I : In Hofmann degradation reaction, the migration of only an alkyl group takes place from carbonyl carbon of the amide to the nitrogen atom.

Statement II : The group is migrated in Hofmann degradation reaction to electron deficient atom.

In the light of the above statement, choose the **most appropriate** answer from the options given below:

[NCERT : P L-264 | June 27, 2022 (I)]

(a) Both Statement I and Statement II are correct.
(b) Both Statement I and Statement II are incorrect.
(c) Statement I is correct but Statement II is incorrect.
(d) Statement I is incorrect but Statement II is correct.

124. Among the following structures, which will show the most stable enamine formation ? (Where Me is – CH_3)

[June 26, 2022 (I)]

(a) Piperidine with COOH (wedge); N–C(Me)=CH–OMe

(b) Pyrrolidine with COOH (wedge); N–C(Me)=CH–OMe

(c) Pyrrolidine with COOH (wedge); N–C(Me)=CH–OMe

(d) Piperidine with COOH (wedge); N–C(Me)=CH–OMe (MeO)

125. p-CH_3O–C_6H_4–CH_2Br $\xrightarrow[\text{(iii) Cyclohexanone} \quad \text{(iv) } H_2, Ni]{\text{(i) NaCN} \quad \text{(ii) } OH^-}$ $\underset{\text{Major Product}}{A}$

The major product of the above reaction is

[June 24, 2022 (I)]

(a) [structure: OCH_3-phenyl, cyclohexyl, NH_2] (b) [structure: OCH_3-phenyl, HO-cyclohexyl, NH_2]

(c) [structure: OCH_3-phenyl, cyclohexyl, $\overset{O}{\overset{\|}{C}}-NH_2$] (d) [structure: OCH_3-phenyl, HO-cyclohexyl, NH_2]

126. $R-CN \xrightarrow[(ii) H_2O]{(i) DiBAL-H} R-Y$

[NCERT : P L-271 | July 27, 2021 (II)]

Consider the above reaction and identify "Y"

(a) $-CH_2NH_2$ (b) $-CONH_2$
(c) $-CHO$ (d) $-COOH$

127. Which one of the following compounds will liberate CO_2, when treated with $NaHCO_3$? **[July 25, 2021 (I)]**

(a) $(CH_3)_3\overset{\oplus}{N}H\overset{\ominus}{Cl}$ (b) $(CH_3)_4\overset{\oplus}{N}\overset{\ominus}{O}H$

(c) $CH_3-\overset{O}{\overset{\|}{C}}-NH_2$ (d) CH_3NH_2

128. A reaction of benzonitrile with one equivalent CH_3MgBr followed by hydrolysis produces a yellow liquid "P". The compound "P" will give positive____.

[July 25, 2021 (II)]

(a) Iodoform test (b) Schiff's test
(c) Ninhydrin's test (d) Tollen's test

129. [cyclohexyl–$C\equiv N$] $\xrightarrow[H^+]{H_2O}$ "A" (Major Product) $\xrightarrow[H^+, \Delta]{H_2O}$ [cyclohexyl–COOH]

[NCERT : P L-271 | March 18, 2021 (I)]

Consider the above chemical reaction and identify product "A"

(a) cyclohexyl–CH_2NH_2 (b) cyclohexyl–$CONH_2$

(c) cyclohexyl–CH_2NO_2 (d) cyclohexyl–$\overset{H}{\overset{|}{C}}=N-OH$

130. $C_6H_5-\overset{O}{\overset{\|}{C}}-Cl + C_6H_5NHC_6H_5 \longrightarrow C_6H_5-\overset{O}{\overset{\|}{C}}-N(C_6H_5)_2$

0.140 g 0.388 g 0.210 g

Consider the above reaction. The percentage yield of amide product is ______. (Round off to the nearest integer). (Given : Atomic mass : C : 12.0 u, H : 1.0 u, N : 14. 0 u, O : 16.0 u, Cl : 35.5 u) **[NV, March 17, 2021 (II)]**

131. Which of the following is least basic?

[NCERT : P L-268 | March 16, 2021 (II)]

(a) $(CH_3CO)\ddot{N}HC_2H_5$ (b) $(CH_3CO)_2\ddot{N}H$
(c) $(C_2H_5)_2\ddot{N}H$ (d) $(C_2H_5)_3\ddot{N}$

132. For the given reaction :

[NCERT : P L-271 | Feb. 26, 2021 (I)]

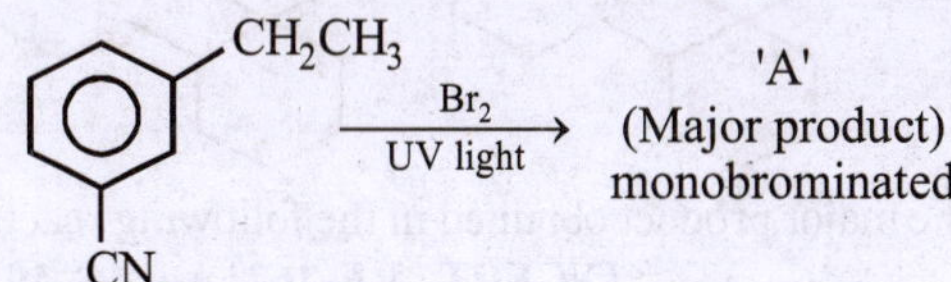

What is 'A'?

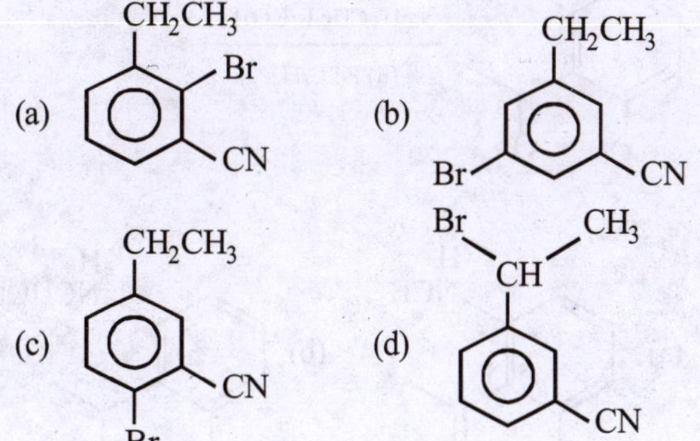

133. The major product of the following chemical reaction is:

$CH_3CH_2CN \xrightarrow[(3)\ Pd/BaSO_4,\ H_2]{(1)\ H_3O^+, \Delta \quad (2)\ SOCl_2}$

[NCERT : P L-271 | Feb. 25, 2021 (I)]

(a) $CH_3CH_2CH_3$ (b) $CH_3CH_2CH_2OH$
(c) CH_3CH_2CHO (d) $(CH_3CH_2CO)_2O$

134. Which of these will produce the highest yield in Friedel Crafts reaction? **[NCERT : P L-272 | Jan. 09, 2020 (I)]**

(a) [benzene–$CONH_2$] (b) [benzene–OH]

(c) [benzene–Cl] (d) [benzene–NH_2]

135. The major products A and B for the following reactions are, respectively: **[NCERT : P L-262, 275 | April 9, 2019 (II)]**

(2-(2-iodoethyl)cyclohexanone) $\xrightarrow[\text{DMSO}]{\text{KCN}}$ [A] $\xrightarrow{H_2/Pd}$ [B]

(a) HO, CN ... I ; HO, CH_2-NH_2 ... I

(b) O ... CN ; O ... CH_2NH_2

(c) O ... CN ; OH ... CH_2NH_2

(d) HO, CN ... I ; HO, CH_2-NH_2 ... H

136. The major product obtained in the following reaction is : **[NCERT : P L-262 | April 8, 2019 (II)]**

(NH_2, acetyl, CN substituted benzene) $\xrightarrow[\text{(ii) Pd/C/}H_2]{\text{(i) CHCl}_3\text{/KOH}}$

(a) $\overset{H}{N}CH_3$, CN, C=O

(b) $\overset{H}{N}CHCl_2$, CN, OH

(c) $\overset{H}{N}CH_3$, CN, OH

(d) $\overset{H}{N}CH_3$, H_2N, OH

137. The increasing order of reactivity of the following compound towards reaction with alkyl halides directly is: **[NCERT : P L-272 | Jan. 12, 2019 (I)]**

(A) benzamide (NH_2) (B) phthalimide (NH) (C) 2-aminobenzonitrile (CN, NH_2) (D) aniline (NH_2)

(a) (B) < (A) < (C) < (D) (b) (A) < (B) < (C) < (D)
(c) (B) < (A) < (D) < (C) (d) (A) < (C) < (D) < (B)

138. The increasing order of the reactivity of the following with $LiAlH_4$ is : **[NCERT : P L-263 | Jan. 12, 2019 (II)]**

(A) $H_5C_2CONH_2$

(B) $H_5C_2COOCH_3$

(C) H_5C_2COCl

(D) $H_5C_2COOCOC_2H_5$

(a) (B) < (A) < (C) < (D) (b) (B) < (A) < (D) < (C)
(c) (A) < (B) < (D) < (C) (d) (A) < (B) < (C) < (D)

139. A compound 'X' on treatment with Br_2/NaOH, provided C_3H_9N, which gives positive carbylamine test. Compound 'X' is: **[NCERT : P L-271 | Jan. 11, 2019 (II)]**

(a) $CH_3COCH_2NHCH_3$ (b) $CH_3CH_2COCH_2NH_2$
(c) $CH_3CH_2CH_2CONH_2$ (d) $CH_3CON(CH_3)_2$

140. What will be the major product in the following mononitration reaction? **[NCERT : P L-272 | Jan. 10, 2019 (II)]**

(N-phenylbenzamide) $\xrightarrow[\text{conc. }H_2SO_4]{HNO_3}$

(a) NO_2 para on the benzoyl ring

(b) NO_2 ortho on the benzoyl ring

(c) O_2N meta on the N-phenyl ring

(d) O_2N para on the N-phenyl ring

141. The compounds A and B in the following reaction are, respectively : **[Jan. 9, 2019 (I)]**

$$C_6H_6 \xrightarrow{HCHO + HCl} A \xrightarrow{AgCN} B$$

(a) A = Benzyl alcohol, B = Benzyl cyanide
(b) A =Benzyl chloride, B = Benzyl cyanide
(c) A = Benzyl alcohol, B = Benzyl isocyanide
(d) A = Benzyl chloride, B = Benzyl isocyanide

142. The major product of the following reaction is:

(2-ethylbenzamide, $C_6H_4(CONH_2)(CH_2CH_3)$) $\xrightarrow[\text{(ii) KOH (dil)}]{\text{(i) } Br_2/h\nu}$

[NCERT : P L-264 | Jan. 9, 2019 (II)]

(a) (benzo-fused ring with NH) (b) (benzo-fused ring with C=O and NH) (c) (benzo-fused ring with C=O, NH and CH_3) (d) (benzo-fused ring with NH and CH_3)

143. In the chemical reaction,
$CH_3CH_2NH_2 + CHCl_3 + 3KOH \rightarrow$
(A) + (B) + $3H_2O$, the compounds (A) and (B) are respectively **[2007]**
(a) C_2H_5NC and 3KCl
(b) C_2H_5CN and 3KCl
(c) $CH_3CH_2CONH_2$ and 3KCl
(d) C_2H_5NC and K_2CO_3.

144. The reaction of chloroform with alcoholic KOH and *p*-toluidine forms **[NCERT : P L-271 | 2003]**
(a) $H_3C-C_6H_4-N_2Cl$
(b) $H_3C-C_6H_4-NHCHCl_2$
(c) $H_3C-C_6H_4-NC$
(d) $H_3C-C_6H_4-CN$

145. Ethyl isocyanide on hydrolysis in acidic medium generates **[2003]**
(a) propanoic acid and ammonium salt
(b) ethanoic acid and ammonium salt
(c) methylamine salt and ethanoic acid
(d) ethylamine salt and methanoic acid

146. When a primary amine reacts with chloroform in ethanolic KOH then the product is **[2002]**
(a) an isocyanide (b) an aldehyde
(c) a cyanide (d) an alcohol.

Topic 3 - Nitrocompounds, Alkyl Nitrites and Diazonium Salts

147. The major product (*A*) formed in the following reaction sequence is **[NCERT : P L-272 | April 4, 2025 (I)]**

$C_6H_5NO_2 \xrightarrow{\text{(i) Sn, HCl; (ii) } Ac_2O\text{, Pyridine; (iii) } Br_2\text{, AcOH; (iv) NaOH (aq)}} A$

(a) NH_2 with Br at para position
(b) NH_2 with Br at meta position
(c) NH_2 with Br at both ortho positions and para position
(d) NH_2 with Br at meta and para positions

148. In the following reactions, which one is NOT correct? **[NCERT : P L-272 | April 3, 2025 (I)]**

(a) $C_6H_5\overset{+}{N}\equiv N\overset{-}{Cl} \xrightarrow{EtOH} C_6H_5OEt$

(b) $C_6H_5\overset{+}{N}\equiv N\overset{-}{Cl} \xrightarrow{H_3PO_2/H_2O} C_6H_6$

(c) $C_6H_5\overset{+}{N}\equiv N\overset{-}{Cl} \xrightarrow{\text{Potassium Iodide}} C_6H_5I$

(d) $C_6H_5\overset{+}{N}\equiv N\overset{-}{Cl} \xrightarrow{CuCN/KCN} C_6H_5CN$

149. Identify [A], [B], and [C], respectively in the following reaction sequence : **[NCERT : P L-272 | April 3, 2025 (I)]**

$$[A] \xrightarrow[273-278K]{NaNO_2,\ HCl} C_6H_5\overset{+}{N_2}\overset{-}{Cl} \xrightarrow{KI} [B] \xrightarrow[\text{Dry ether}]{2Na} [C]$$

(a) [A] aniline (NH_2) [B] chlorobenzene (Cl) [C] iodobenzene (I)

(b) [A] aniline (NH_2) [B] chlorobenzene (Cl) [C] benzene

(c) [A] aniline (NH_2) [B] iodobenzene (I) [C] biphenyl

(d) [A] nitrobenzene (NO_2) [B] iodobenzene (I) [C] biphenyl

150. When a concentrated solution of sulphanilic acid and 1-naphthylamine is treated with nitrous acid (273 K) and acidified with acetic acid, the mass (g) of 0.1 mole of product formed is : **[NCERT : P L-272 | April 2, 2025 (II)]**
(Given molar mass in g mol^{-1} H: 1, C: 12, N: 14, O: 16, S: 32)

(a) 343 (b) 330
(c) 33 (d) 66

151. Which one of the following reaction sequences will give an azo dye ? **[NCERT : P L-274 | Jan. 29, 2025 (II)]**

(a) Nitrobenzene $\xrightarrow{\text{(I) Sn/HCl; (ii) } NaNO_2/HCl\text{; (iii) } \beta\text{-naphthol, NaOH}}$

(b) Benzenesulphonic acid (SO_3H) $\xrightarrow{\text{(i) } SOCl_2\text{; (ii) } NH_3\text{; (iii) benzyl chloride } (CH_2-Cl)}$

(c) Benzonitrile (CN) $\xrightarrow{\text{(i) 70\% } H_2SO_4\text{; (ii) } PCl_5\text{; (iii) aniline } (NH_2)}$

(d) Aniline (NH_2) $\xrightarrow{\text{(i) HCl/NaNO}_2\text{; (ii) toluene } (CH_3)}$

152. Consider the following sequence of reactions.
[NCERT : P L-274 | Jan. 23, 2025 (II)]

4-Ethoxyaniline (CH_3CH_2O–C_6H_4–NH_2) $\xrightarrow[\text{0-5°C}]{\text{(i) } NaNO_2\text{, HCl}}$ (A)

(A) $\xrightarrow{\text{(i) phenol (OH), NaOH; (ii) HCl dil.}}$ (B) (Molecular formula) $C_{14}H_{14}N_2O_2$

(B) $\xrightarrow{\text{(i) NaOH; (ii) } H_3CCH_2Br}$ (C) (Molecular formula) $C_{16}H_{18}N_2O_2$

Total number of sp^3 hybridised carbon atoms in the major product C formed is ______.

153. Consider the following sequence of reactions to produce major product (A) **[NCERT : P L-274 | NV, Jan. 23, 2025 (I)]**

3-Nitrotoluene (CH_3, NO_2) $\xrightarrow{\text{i) } Br_2\text{, Fe; ii) Sn, HCl; iii) } NaNO_2\text{, HCl, 273K; iv) } H_3PO_2\text{, } H_2O}$ (A) Major Product

Molar mass of product (A) is ______ g mol^{-1}.
(Given molar mass in g mol–1 of C : 12, H : 1, O : 16, Br : 80, N : 14, P : 31)

154. Consider the following sequence of reactions :

Nitrobenzene (NO_2) $\xrightarrow{\text{(i) Sn, HCl; (ii) } NaNO_2\text{, HCl, 0°C; (iii) } Cu_2Cl_2\text{; (iv) Na, Ether}}$ A Product

Molar mass of the product formed (A) is ______ g mol^{-1}.
[NCERT : P L-275 | NV, Jan. 22, 2025 (I)]

155. The products formed in the following reaction sequence are : **[NCERT : P L-274 | Jan. 22, 2025 (I)]**

Nitrobenzene (NO_2) $\xrightarrow{\text{(i) } Br_2\text{, AcOH; (ii) Sn, HCl; (iii) } NaNO_2\text{, HCl, 273 K; (iv) } C_2H_5OH}$ A + B

(a) 3-Bromophenol (OH, Br), 3-Bromoethoxybenzene (OEt, Br)

(b) 3-Bromoethoxybenzene (OEt, Br), CH_3–COOH

(c) Bromobenzene, CH_3CHO

(d) 3-Bromophenol (OH, Br), $CH_3–CHO$

156. If 279 g of aniline is reacted with one equivalent of benzenediazonium chloride, the maximum amount of aniline yellow formed will be ____ g. (nearest integer) (consider complete conversion) **[NV, April 8, 2024 (I)]**

157. 9.3 g of pure aniline upon diazotisation followed by coupling with phenol gives an orange dye. The mass of orange dye produced (assume 100% yield/ conversion) is ______g. (nearest integer)

[NCERT : P L-275 | NV, April 6, 2024 (I)]

158. Identify the product Ⓐ in the following reaction.

[NCERT : P L-271 | April 6, 2024 (II)]

Aniline (NH_2) $\xrightarrow[\text{(iii) NaOH, 623K, 300 atm; (iv) } H^+]{\text{(i) } NaNO_2 + HCl; \text{ (ii) } Cu_2Cl_2}$ Ⓐ

(a) 4-Aminophenol (NH_2, OH) (b) Phenol (OH)

(c) 4-Chloroaniline (NH_2, Cl) (d) 4-Chlorophenol (Cl, OH)

159. X of enthanamine was subjected to reaction with $NaNO_2$/HCl followed by hydrolysis to liberate N_2 and HCl. The HCl generated was completely neutralised by 0.2 moles of NaOH. X is ____ g. **[April 5, 2024 (II)]**

160. The products A and B formed in the following reaction scheme are respectively

[NCERT : P L-274 | Jan. 30, 2024 (II)]

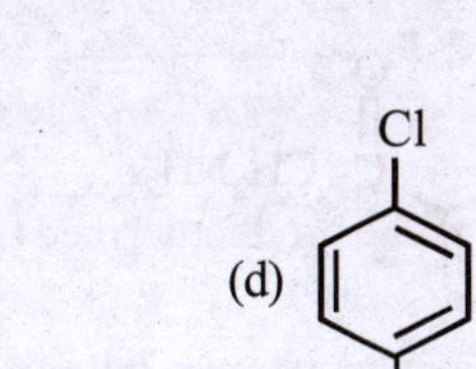

$\xrightarrow[\text{ii) Sn/HCl}]{\text{i) Con.}HNO_3/\text{Con.}H_2SO_4 \text{ 323–333K}}$ A $\xrightarrow[\text{ii)Phenol}]{\text{i)}NaNO_2\text{,HCl,273–278K}}$ B

(a) 1-Chloro-3-nitrobenzene (Cl, NO_2), HO–C_6H_4–C_6H_4–NO_2

(b) Aniline (NH_2), NH_2–C_6H_4–C_6H_4–OH

(c) Aniline (NH_2), HO–C_6H_4–N = N–C_6H_5

(d) Aniline (NH_2), 2-(N = N–C_6H_5)phenol (OH)

161. The product A formed in the following reaction is

[NCERT : P L-274, 275 | Jan. 29, 2024 (II)]

Aniline (NH_2) $\xrightarrow[\text{then } Cu_2Cl_2]{NaNO_2\text{,HCl, } 0°C}$ A

(a) 1,2-Dichlorobenzene (Cl, Cl) (b) 3-Chloroaniline (NH_2, Cl)

(c) $\overset{+}{N}H_3\bar{C}l$ (anilinium chloride) (d) Chlorobenzene (Cl)

162. Consider the following sequence of reactions:

Aniline (NH_2) $\xrightarrow[0\text{-}5°C]{NaNO_2}$ 'A' $\xrightarrow{\text{N, N-Dimethylaniline}}$ 'B'

The product 'B' is

[NCERT : P L-276 | April 15, 2023 (I)]

(a) H_2NHN–C_6H_4–C_6H_4–$N(CH_3)_2$

(b) C_6H_5–N = N–C_6H_4–$N(CH_3)_2$

(c) C_6H_5–N(H)–N(H)–C_6H_4–$N(CH_3)_2$

(d) C_6H_5–N(CH_3)–N(CH_3)–C_6H_4–NH_2

163. The incorrect statement regarding the reaction given below is [NCERT : P L-271 | April 12, 2023 (I)]

Me – N – Me (benzene ring) 'A' + $NaNO_2$ + HX ⟶ 'B'

(a) The electrophile involved in the reaction is NO^+
(b) 'B' is N-nitroso ammonium compound
(c) The reaction occurs at low temperature
(d) The product 'B' formed in the above reaction is p-nitroso compound at low temperature

164. (CH_3, NO_2 substituted benzene) ⟶ ⟶ (COOH, Br, Cl substituted benzene)

In the above conversion the correct sequence of reagents to be added is [NCERT : P L-272, 275 | April 15, 2023 (I)]
(a) (i) Fe/H^+, (ii) HONO, (iii) CuCl, (iv) $KMnO_4$, (v) Br_2
(b) (i) $KMnO_4$, (ii) Br_2/Fe, (iii) Fe/H^+, (iv) Cl_2
(c) (i) Br_2/Fe, (ii) Fe/H^+, (iii) HONO, (iv) CuCl, (v) $KMnO_4$
(d) (i) Br_2/Fe, (ii) Fe/H^+, (iii) $KMnO_4$, (iv) Cl_2

165. Compound A from the following reaction sequence is:

A. $\xrightarrow[0-5°C]{Br_2, CS_2}$ B. $\xrightarrow{NaNO_2/HCl}$ C. $\xrightarrow[\Delta]{H_3PO_2}$ (1,3,5-tribromobenzene: Br, Br, Br)

[NCERT : P L-271, 272, 273 | April 13, 2023 (II)]

(a) Benzoic Acid (b) Phenol
(c) Salicylic Acid (d) Aniline

166. Match List I with List II: [NCERT : P L-275 | April 08, 2023 (I)]

($C_6H_5N_2Cl$) is reacted with reagents in List I to form products in List II.

List-I (Reagent)	List-II (Product)
A. (aniline, NH_2)	I. (fluorobenzene, F)
B. HBF_4, Δ	II. (benzonitrile, CN)
C. Cu, HCl	III. (C_6H_5–N=N–C_6H_4–NH_2)
D. CuCN/KCN	IV. (chlorobenzene, Cl)

Choose the correct answer from the options given below:
(a) A-IV, B-III, C-II, D-I (b) A-I, B-III, C-IV, D-II
(c) A-III, B-I, C-II, D-IV (d) A-III, B-I, C-IV, D-II

167. The product (P) formed from the following multistep reaction is : [NCERT : P L-262, 271 | April 08, 2023 (II)]

(H_3C–C_6H_4–NO_2) $\xrightarrow[\text{(iii) } NaNO_2, HCl, 0°C \text{ (iv) } H_3PO_2]{\text{(i) } Br_2 \text{ (ii) } H_2/Pd}$ (P) Product

(a) (H_3C, OH, Br substituted benzene) (b) (H_3C, Br, OH substituted benzene)
(c) (H_3C, Br substituted benzene) (d) (H_3C, Br substituted benzene)

168. ($C_6H_5NO_2$) $\xrightarrow[C_2H_5OH]{H_2/Pd}$ [A] $\xrightarrow[\text{Pyridine}]{(CH_3CO)_2O}$ [B]

[NCERT : P L-272 | Jan. 31, 2023 (I)]

(a) $C_6H_5CH_2CONH_2$
(b) C_6H_5–C(=O)–CH_2NH_2
(c) $C_6H_5NHCH_2COOH$
(d) $C_6H_5NHCOCH_3$

169. (CH_3, NO_2 substituted benzene) (X) ⟶ ⟶ (CH_3, Br, Br substituted benzene) (Y)

In the above conversion of compound (X) to product (Y), the sequence of reagents to be used will be: [NCERT : P L-272 | Jan. 30, 2023 (II)]

(a) (i) Br_2, Fe (ii) Fe, H^+ (iii) $LiAlH_4$
(b) (i) $Br_2(aq)$ (ii) $LiAlH_4$ (iii) H_3O^+
(c) (i) Fe, H^+ (ii) $Br_2(aq)$ (iii) HNO_2 (iv) CuBr
(d) (i) Fe, H^+ (ii) $Br_2(aq)$ (iii) HNO_2 (iv) H_3PO_2.

170. Identify the products formed (A and E)
[NCERT : P L-275 | Jan. 25, 2023 (I)]

(4-nitrotoluene: Me, NO_2) $\xrightarrow{Br_2}$ A $\xrightarrow{Sn/HCl}$ B $\xrightarrow[273-278K]{NaNO_2/HCl}$

C $\xrightarrow{H_3PO_2/H_2O}$ D $\xrightarrow[(ii)\ H_3O^+]{(i)\ KMnO_4/KOH}$ E

(a) A = (Me, Br, Br, NO_2), E = (COOH, Br, Br)
(b) A = (Me, Br, NO_2), E = (COOH, Br)
(c) A = (Me, Br, NO_2), E = (Me, Br)
(d) A = (Me, Br, NO_2), E = (COOH, Br, OH)

171. Choose the correct colour of the product for the following reaction.
[Jan. 24, 2023 (II)]

($N=N-OOCCH_3$, SO_3H) + 1–Naphthyl amine ⟶

(a) Yellow (b) White
(c) Red (d) Blue

172. ($\overset{+}{N}\equiv NCl^-$) + 'A' $\xrightarrow{NaOH}$ Product (orange-red dye)

Which among the following represents reagent 'A'?
[NCERT : P L-276 | July 29, 2022 (I)]

(a) (OH) (b) (OH)
(c) (NH_2) (d) (NH_2)

173. An organic compound 'A' contains nitrogen and chlorine. It dissolves readily in water to give a solution that turns blue litmus red. Titration of compound 'A' with standard base indicates that the molecular weight of 'A' is 131 ± 2. When a sample of 'A' is treated with aq. NaOH, a liquid separates which contains N but not Cl. Treatment of the obtained liquid with nitrous acid followed by phenol gives orange precipitate. The compound 'A' is :
[NCERT : P L-272, 275 | July 27, 2022 (II)]

(a) ($\overset{+}{N}H_3Cl^-$) (b) (NH_2, Cl)
(c) ($CH_2\overset{+}{N}H_3Cl^-$) (d) ($\overset{+}{N}H_3Cl^-$)

174. The correct stability order of the following diazonium salt is
[NCERT : P L-275 | July 26, 2022 (I)]

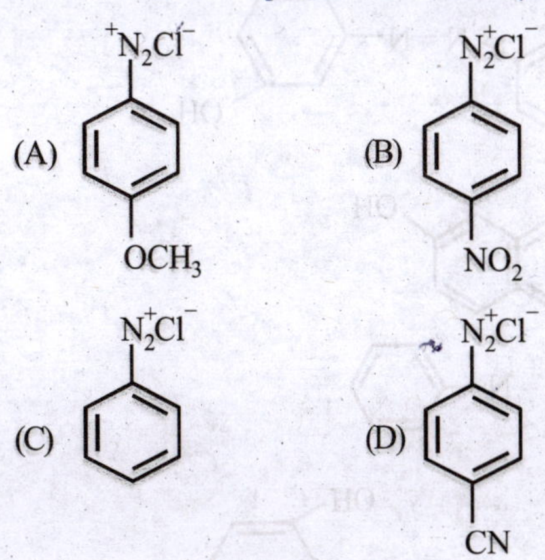

(a) (A) > (B) > (C) > (D)
(b) (A) > (C) > (D) > (B)
(c) (C) > (A) > (D) > (B)
(d) (C) > (D) > (B) > (A)

175. Hydrolysis of which compound will give carbolic acid ?
[NCERT : P L-273 | July 26, 2022 (II)]
(a) Cumene
(b) Benzene diazonium chloride
(c) Benzal chloride
(d) Ethylene glycol ketal

176. The correct sequential order of the reagents for the given reaction is : **[NCERT : P L-272 | July 26, 2022 (II)]**

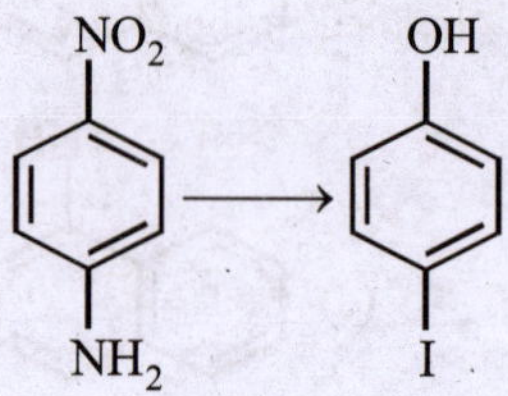

(a) HNO_2, Fe/H^+, HNO_2, KI, H_2O/H^+
(b) HNO_2, KI, Fe/H^+, HNO_2, H_2O/warm
(c) HNO_2, KI, HNO_2, Fe/H^+, H_2O/H^+
(d) HNO_2, Fe/H^+, KI, HNO_2, H_2O/warm

177. A primary aliphatic amine on reaction with nitrous acid in cold (273 K) and there after raising temperature of reaction mixture to room temperature (298 K). Gives a/an
[NCERT : P L-275 | June 28, 2022 (I)]
(a) nitrile (b) alcohol
(c) diazonium salt (d) secondary

178. The final product 'C' in the following series of reactions is
[NCERT : P L-275 | June 26, 2022 (II)]

$C_6H_5NO_2 \xrightarrow{Sn/HCl} A \xrightarrow[HCl]{NaNO_2} B \xrightarrow[NaOH]{\beta\text{-Naphthol}}$ C is

(a) Ph–N=N–(naphthyl)–OH
(b) OH–(naphthyl)–N=N–Ph
(c) Ph–N=N–(naphthyl)–OH
(d) Ph–N=N–(naphthyl)–OH

179. 'A' and 'B' in the following reactions are :
[NCERT : P L-275 | Feb. 24, 2021 (I)]

$C_6H_5NH_2 \xrightarrow[KCN]{NaNO_2/HCl} (A) \xrightarrow{SnCl_2 / HCl / H_3O^+} (B)$

(a) (A) : $C_6H_5\overset{+}{N_2}\overset{-}{Cl}$ (B) : C_6H_5CHO
(b) (A) : $C_6H_5\overset{+}{N_2}\overset{-}{Cl}$ (B) : $C_6H_5Cl_2$
(c) (A) : C_6H_5CN (B) : C_6H_5CHO
(d) (A) : C_6H_5CN (B) : C_6H_5Cl

180. Which one of the following gives the most stable diazonium salt ? **[NCERT : P L-275 | Sep. 1, 2021 (II)]**
(a) $CH_3–CH_2–CH_2–NH_2$ (b) $CH_3–C_6H_4–NH_2$
(c) $CH_3–\underset{H}{\overset{CH_3}{C}}–NH_2$ (d) $C_6H_5NHCH_3$

181. What is the correct sequence of reagents used for converting nitrobenzene into *m*-dibromobenzene?

$C_6H_5NO_2 \longrightarrow \longrightarrow \longrightarrow \longrightarrow$ m-$C_6H_4Br_2$

[NCERT : P L-272 | Feb. 24, 2021 (II)]

(a) $\xrightarrow{NaNO_2}$ / $\xrightarrow{HCl}$ / $\xrightarrow{KBr}$ / $\xrightarrow{H^+}$

(b) $\xrightarrow{Br_2/Fe}$ / $\xrightarrow{Sn/HCl}$ / $\xrightarrow{NaNO_2/HCl}$ / $\xrightarrow{CuBr/HBr}$

(c) $\xrightarrow{Sn/HCl}$ / $\xrightarrow{KBr}$ / $\xrightarrow{Br_2}$ / $\xrightarrow{H^+}$

(d) $\xrightarrow{Sn/HCl}$ / $\xrightarrow{Br_2}$ / $\xrightarrow{NaNO_2}$ / $\xrightarrow{NaBr}$

182. In the following reaction sequence, [C] is :

[NCERT : P L-274 | Sep. 04, 2020 (II)]

p-Toluidine (NH_2, CH_3) $\xrightarrow[\text{(ii) } Cu_2Cl_2 + HCl]{\text{(i) } NaNO_2 + HCl, \text{ 0-5 °C}}$ [A] $\xrightarrow[h\nu]{Cl_2}$ [B] $\xrightarrow{\text{Na+dry ether}}$ [C] (Major product)

(a) Cl–C_6H_4–CH_2–CH_2–C_6H_4–Cl

(b) Cl–C_6H_4–CH_2–C_6H_4–CH_2–Cl

(c) $ClCH_2$–C_6H_4–C_6H_4–CH_2Cl

(d) CH_3–C_6H_4–C_6H_4–CH_3

183. Three isomers A, B and C (mol. formula $C_8H_{11}N$) give the following results :

A and C $\xrightarrow{\text{Diazotization}}$ P + Q $\xrightarrow[\text{(ii) Oxidation (KMnO}_4\text{+H}^+\text{)}]{\text{(i) Hydrolysis}}$ R (product of A) + S (product of C)

R has lower boiling point than S

B $\xrightarrow{C_6H_5SO_2Cl}$ Alkali-insoluble product

A, B and C, respectively are :

[NCERT : P L-271 | Sep. 03, 2020 (II)]

(a) 2-ethylaniline (NH_2, CH_2CH_3), 3-methylbenzylamine (CH_2NH_2, CH_3), 3-ethylaniline (NH_2, CH_2CH_3)

(b) 2-ethylaniline (NH_2, CH_2CH_3), $C_6H_5CH_2NHCH_3$, 4-ethylaniline (H_2N, CH_2CH_3)

(c) 4-ethylaniline (CH_2CH_3, H_2N), $C_6H_5NHCH_2CH_3$, 2-ethylaniline (NH_2, CH_2CH_3)

(d) 3-ethylaniline (CH_2CH_3, NH_2), $C_6H_5CH_2NHCH_3$, 2-ethylaniline (NH_2, CH_2CH_3)

184. The major product Z obtained in the following reaction scheme is: **[NCERT : P L-274 | Jan. 09, 2020 (I)]**

3-Bromoaniline (NH_2, Br) $\xrightarrow[\text{273-278 K}]{NaNO_2/HCl}$ X $\xrightarrow{Cu_2Br_2}$ Y $\xrightarrow[H_2SO_4]{HNO_3}$ Z

(a) Br, O_2N, Br (2-nitro-1,3-dibromobenzene)

(b) Br, Br, NO_2 (1,3-dibromo-4-nitrobenzene)

(c) NO_2, Br, Br, NO_2

(d) NO_2, Br, Br (1,3-dibromo-5-nitrobenzene)

185. Consider the following reactions,

[P] $\xrightarrow[\text{(ii) } \beta\text{-naphthol/NaOH}]{\text{(i) } NaNO_2/HCl, \text{ 0-5 °C}}$ Coloured solid

[P] $\xrightarrow{Br_2/H_2O}$ $C_7H_6NBr_3$

The compound [P] is:

[NCERT : P L-274 | Jan. 09, 2020 (II)]

(a) NH_2, CH_3 (p-toluidine)

(b) $NHCH_3$, CH_3 (meta)

(c) NH_2, CH_3 (m-toluidine)

(d) NH_2, CH_3 (o-toluidine)

186. Benzene diazonium chloride on reaction with aniline in the presence of dilute hydrochloric acid gives :

[NCERT : P L-276 | April 12, 2019 (II)]

(a) $C_6H_5-C_6H_4-NH_2$

(b) $C_6H_5-N=N-C_6H_4$ (H_2N on ortho position)

(c) $C_6H_5-N=N-C_6H_4-NH_2$

(d) $C_6H_5-N=N-NH-C_6H_5$

187. Aniline dissolved in dilute HCl is reacted with sodium nitrite at 0 °C. This solution was added dropwise to a solution containing equimolar mixture of aniline and phenol in dil. HCl. The structure of the major product is:

[NCERT : P L-276 | April 9, 2019 (I)]

(a) $C_6H_5-N=N-C_6H_4-OH$

(b) $C_6H_5-N=N-NH-C_6H_5$

(c) $C_6H_5-N=N-C_6H_4-NH_2$

(d) $C_6H_5-N=N-O-C_6H_5$

188. Coupling of benzene diazonium chloride with 1-naphthol in alkaline medium will give :

NCERT : P L-274 | [April 8, 2019 (I)]

(a) 1-naphthol (OH) with $-N=N-C_6H_5$ at position 5

(b) 1-naphthol (HO) with $-N=N-C_6H_5$ at position 2

(c) 1-naphthol (OH) with $-N=N-C_6H_5$ at position 4

(d) 2-naphthol type structure (OH) with $-N=N-C_6H_5$

189. The major product formed in the following reaction is:

Cyclopentane bearing NO_2, Cl, CH_3 and Cl $\xrightarrow[\text{Heat}]{NaOCH_3\text{ (1 eq.)}}$

[NCERT : P L-266 | Online April 15, 2018 (II)]

(a) Cyclopentene with NO_2 on the double bond, Cl, CH_3

(b) Cyclopentene with NO_2, Cl, CH_3 on the double bond

(c) Cyclopentene with NO_2, Cl

(d) Cyclopentene with NO_2, Cl, CH_3 on the double bond

190. The increasing order of nitration of the following compounds is :

[NCERT : P L-269 | Online April 15, 2018 (I)]

(A) $C_6H_5NH_2$ (B) C_6H_5Cl (C) $C_6H_5OCH_3$ (D) $C_6H_5CH_3$

(a) (A) < (B) < (D) < (C)
(b) (A) < (B) < (C) < (D)
(c) (B) < (A) < (C) < (D)
(d) (B) < (A) < (D) < (C)

191. Fluorination of an aromatic ring is easily accomplished by treating a diazonium salt with HBF_4. Which of the following conditions is correct about this reaction?

[NCERT : P L-275 | Online April 10, 2016]

(a) NaF/Cu (b) Cu_2O/H_2O
(c) Only heat (d) $NaNO_2/Cu$

192. In the reaction

p-toluidine (4-methylaniline) $\xrightarrow[0-5\,°C]{NaNO_2/HCl}$ D $\xrightarrow[\Delta]{CuCN/KCN}$ E + N_2

the product E is : **[2015]**

(a) 4-methylcyclohexane-1-carbonitrile (CN and CH_3 on a cyclohexane ring, para)
(b) methylcyclohexane (CH_3 on a cyclohexane ring)
(c) p-toluic acid (COOH and CH_3 on a benzene ring, para)
(d) H_3C–(cyclohexane)–(cyclohexane)–CH_3

193. In a set of reactions *p*-nitrotoluene yielded a product E.

p-nitrotoluene $\xrightarrow[FeBr_3]{Br_2}$ B $\xrightarrow{Sn/HCl}$ C $\xrightarrow[HCl]{NaNO_2}$ D $\xrightarrow[HBr]{CuBr}$ E

The product E would be: **[Online April 19, 2014]**

(a) toluene with Br at 2, 4 and 6 positions (CH_3; Br, Br, Br)
(b) toluene with Br at 2 and 4 positions (CH_3; Br, Br)
(c) toluene with Br at 2 and 3 positions (CH_3; Br, Br)
(d) p-bromobenzyl bromide (CH_2Br; Br)

194. Complete reduction of benzene diazonium chloride with Zn/HCl gives: **[Online April 11, 2014]**

(a) Aniline (b) Phenylhydrazine
(c) Azobenzene (d) Hydrazobenzene

195. Conversion of benzene diazonium chloride to chlorobenzene is an example of which of the following reactions? **[NCERT : P L-275 | Online April 12, 2014]**

(a) Claisen (b) Friedel-craft
(c) Sandmeyer (d) Wurtz

196. The conversion of benzene diazonium chloride to bromobenzene can be accomplished by

[NCERT : P L-275 | Online May 12, 2012]

(a) Reimer-Tiemann reaction
(b) Friedel-Crafts reaction
(c) Gattermann reaction
(d) Azo-Coupling reaction

197. In the chemical reactions : **[2011RS]**

Aniline (NH_2) $\xrightarrow[HCl, 278K]{NaNO_2}$ A $\xrightarrow[\Delta]{CuCN}$ B,

the compounds A and B respectively are :

(a) Benzene diazonium chloride and benzonitrile
(b) Nitrobenzene and chlorobenzene
(c) Phenol and bromobenzene
(d) Fluorobenzene and phenol

198. In the chemical reactions,

Aniline (NH_2) $\xrightarrow[HCl, 278\,K]{NaNO_2}$ A $\xrightarrow[(ii)\ heat]{(i)\ HBF_4}$ B,

the compounds 'A' and 'B' respectively are

[NCERT : P L-274, 275 | 2010]

(a) nitrobenzene and fluorobenzene
(b) phenol and benzene
(c) benzene diazonium chloride and fluorobenzene
(d) nitrobenzene and chlorobenzene

ANSWER KEYS

1	(b)	21	(5)	41	(c)	61	(a)	81	(a)	101	(b)	121	(a)	141	(d)	161	(d)	181	(b)
2	(c)	22	(2)	42	(d)	62	(a)	82	(d)	102	(b)	122	(c)	142	(c)	162	(b)	182	(a)
3	(d)	23	(b)	43	(c)	63	(d)	83	(a)	103	(b)	123	(d)	143	(a)	163	(b)	183	(b)
4	(c)	24	(c)	44	(a)	64	(c)	84	(c)	104	(c)	124	(c)	144	(c)	164	(c)	184	(b)
5	(d)	25	(d)	45	(c)	65	(3)	85	(a)	105	(b)	125	(d)	145	(d)	165	(d)	185	(c)
6	(340)	26	(d)	46	(a)	66	(d)	86	(b)	106	(b)	126	(c)	146	(a)	166	(d)	186	(c)
7	(a)	27	(b)	47	(c)	67	(b)	87	(c)	107	(b)	127	(a)	147	(a)	167	(d)	187	(c)
8	(93)	28	(c)	48	(c)	68	(a)	88	(b)	108	(d)	128	(a)	148	(a)	168	(d)	188	(c)
9	(b)	29	(a)	49	(c)	69	(c)	89	(d)	109	(b)	129	(b)	149	(c)	169	(d)	189	(a)
10	(d)	30	(a)	50	(d)	70	(b)	90	(b)	110	(b)	130	(77)	150	(c)	170	(b)	190	(a)
11	(d)	31	(6)	51	(c)	71	(d)	91	(a)	111	(a)	131	(b)	151	(a)	171	(c)	191	(c)
12	(a)	32	(c)	52	(c)	72	(d)	92	(d)	112	(3)	132	(d)	152	(4)	172	(a)	192	(a)
13	(4)	33	(a)	53	(a, c)	73	(d)	93	(c)	113	(1)	133	(c)	153	(171)	173	(d)	193	(b)
14	(5)	34	(3)	54	(a)	74	(a)	94	(c)	114	(a)	134	(c)	154	(154)	174	(b)	194	(a)
15	(a)	35	(b)	55	(5)	75	(c)	95	(b)	115	(a)	135	(c)	155	(c)	175	(b)	195	(c)
16	(287)	36	(d)	56	(a)	76	(a)	96	(d)	116	(b)	136	(d)	156	(197)	176	(b)	196	(c)
17	(80)	37	(c)	57	(b)	77	(b)	97	(a)	117	(d)	137	(a)	157	(20)	177	(b)	197	(a)
18	(8)	38	(d)	58	(108°	78	(d)	98	(c)	118	(c)	138	(c)	158	(b)	178	(c)	198	(c)
19	(b)	39	(a)	59	(a)	79	(a)	99	(d)	119	(b)	139	(c)	159	(9)	179	(c)		
20	(a)	40	(d)	60	(c)	80	(c)	100	(a)	120	(c)	140	(d)	160	(c)	180	(b)		

11 Biomolecules

TREND ANALYSIS
JEE MAIN (2025-2021)

	2025				2024				2023				2022				2021			
	Session II		Session I		Session II		Session I		Session II		Session I		Session II		Session I		Session II		Session I	
	MCQs	NVQs	MCQs	NVQs	MCQs	NVQs	MCQs	NVQs	MCQs	NVQs	MCQs	NVQs	MCQs	NVQs	MCQs	NVQs	MCQs	NVQs	MCQs	NVQs
Topic 1	3	–	7	–	4	–	2	–	2	1	3	–	6	–	3	–	3	–	1	–
Topic 2	6	–	2	–	1	2	2	1	6	1	3	2	1	1	2	2	2	2	2	–
Topic 3	–	1	1	–	2	–	3	1	–	–	–	1	–	–	3	1	2	–	1	–
Topic 4	1	–	1	–	–	–	–	1	1	–	1	–	–	–	1	–	1	–	1	–
Total	10	1	11	–	7	2	7	3	9	2	7	3	7	1	9	3	8	2	5	–

How to Code Qns for Better & Faster Revision

Type 1: Able to solve (No code Req)
Type 2: Solved with struggle or by seeing solution **(Use code NR means Needs Revision or orange stickers)**
Type 3: Tricky/ Could not solve on my own. Need help from Teacher/ Friend. **(Use code TQ means Tricky Question or red stickers)**

You can either hand write the code in boxes or else paste coloured stickers given at end of book.

Topic 1 - Carbohydrates and Lipids

1. Given below are two statements :

[NCERT : P L-283 | April 7, 2025 (I)]

Statement I: D – (+) – glucose + D – (+) fructose $\xrightarrow{-H_2O}$ sucrose

sucrose $\xrightarrow{\text{Hydrolysis}}$ D – (+) – glucose + D – (+) fructose

Statement II : Invert sugar is formed during sucrose hydrolysis.

In the light of the above statements, choose the correct answer from the options given below -

(a) Both Statement I and Statement II are true.
(b) Statement I is false but Statement II are true.
(c) Statement I is true but Statement II is false.
(d) Both Statement I and Statement II are false.

2. Given below are two statements :

[NCERT : P L-289 | April 3, 2025 (II)]

Statement I : Wet cotton clothes made of cellulose based carbohydrate takes comparatively longer time to get dried than wet nylon polymer based clothes.

Statement II : Intermolecular hydrogen bonding with water molecule is more in nylon-based clothes than in the case of cotton clothes.

In the light of above statements, choose the **Correct** answer from the options given below

(a) Statement I is false but Statement II is true
(b) Statement I is true but Statement II is false
(c) Both Statement I and Statement II are true
(d) Both Statement I and Statement II are false

3. Which of the following is the correct structure of L-fructose? **[NCERT : PL-286 | April 3, 2025 (I)]**

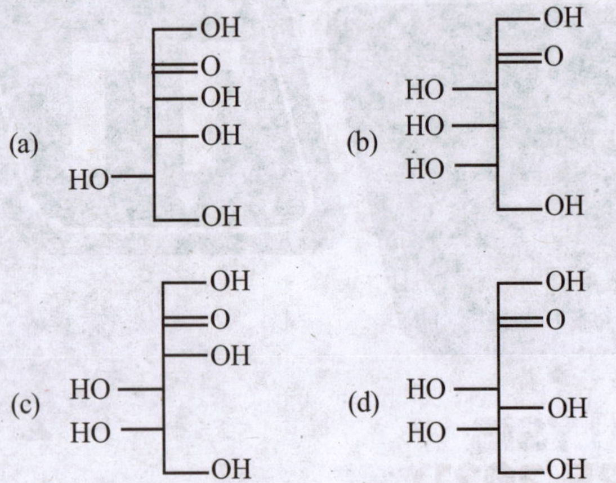

4. Match List – I with List – II.

[NCERT : P L-288 | Jan. 29, 2025 (I)]

List-I (Carbohydrate)	**List-II (Linkage Source)**
(A) Amylose	(I) β-C_1-C_4, plant
(B) Cellulose	(II) α-C_1-C_4, animal
(C) Glycogen	(III) α-C_1-C_4, α-C_1-C_6, plant
(D) Amylopectin	(IV) α-C_1-C_4, plant

Choose the correct answer form the options given below :

(a) (A)-(III), (B)-(II), (C)-(I), (D)-(IV)
(b) (A)-(IV), (B)-(I), (C)-(II), (D)-(III)
(c) (A)-(II), (B)-(III), (C)-(I), (D)-(IV)
(d) (A)-(IV), (B)-(I), (C)-(III), (D)-(II)

5. Identify correct conversion during acidic hydrolysis from the following : **[NCERT : P L-287 | Jan. 28, 2025 (II)]**

(A) starch gives galactose.
(B) cane sugar gives equal amount of glucose and fructose.
(C) milk sugar gives glucose and galactose.
(D) amylopectin gives glucose and fructose.
(E) amylose gives only glucose.

Choose the correct answer from the options given below :

(a) (C), (D) and (E) only (b) (A), (B) and (C) only
(c) (B), (C) and (E) only (d) (B), (C) and (D) only

6. Match List-I with List-II

[NCERT : P L-287 | Jan. 28, 2025 (II)]

List-I (Saccharides)	**List-II (Glycosidic-linkages found)**
(A) Sucrose	(I) $\alpha\,1-4$
(B) Maltose	(II) $\alpha\,1-4$ and $\alpha\,1-6$
(C) Lactose	(III) $\alpha\,1-\beta\,2$
(D) Amylopectin	(IV) $\beta\,1-4$

Choose the correct answer from the options given below :

(a) (A)-(III), (B)-(I), (C)-(IV), (D)-(II)
(b) (A)-(IV), (B)-(II), (C)-(I), (D)-(III)
(c) (A)-(II), (B)-(IV), (C)-(III), (D)-(I)
(d) (A)-(I), (B)-(II), (C)-(III), (D)-(IV)

7. Given below are two statements :

[NCERT : P L-283, 288 | Jan. 28, 2025 (I)]

Statement I : D-glucose pentaacetate reacts with 2, 4-dinitrophenylhydrazine.

Statement II : Starch, on heating with concentrated sulfuric acid at 100°C and 2-3 atmosphere pressure produces glucose.

In the light of the above statements, choose the correct answer from the options given below

(a) Both Statement I and Statement II are false
(b) Statement I is false but Statement II is true
(c) Statement I is true but Statement II is false
(d) Both Statement I and Statement II are true

8. The carbohydrates "Ribose" present in DNA, is

[NCERT : P L-298 | Jan. 24, 2025 (I)]

A. A pentose sugar
B. present in pyranose from
C. in "D" configuration
D. a reducing sugar, when free
E. in α-anomeric form

Choose the correct answer from the options given below:

(a) A, C and D Only (b) A, B and E Only
(c) B, D and E Only (d) A, D and E Only

9. Given below are two statements:

[NCERT : P L-286 | Jan. 23, 2025 (I)]

Statement I: Fructose does not contain an aldehydic group but still reduces Tollen's reagent

Statement II : In the presence of base, fructose undergoes rearrangement to give glucose.

In the light of the above statements, choose the correct answer from the options given below

(a) Statement I is false but Statement II is true
(b) Both Statement I and Statement II are true
(c) Both Statement I and Statement II are false
(d) Statement I is true but Statement II is false

10. Identify the number of structure/s from the following which can be correlated to D-glyceraldehyde.

[NCERT : P L-284 | Jan. 22, 2025 (II)]

(A) CHO; —OH; —OH; —OH
(B) CHO; —OH; HO—; —OH; —OH
(C) CHO; —OH; HO—; HO—; —OH
(D) HO—; HO—; —OH; HO—; CHO

(a) three (b) two (c) four (d) one

11. The **incorrect** statement about Glucose is : **[NCERT : P L-284, 285 | April 9, 2024 (II)]**
(a) Glucose is soluble in water because of having aldehyde functional group
(b) Glucose remains in multiple isomeric form in its aqueous solution
(c) Glucose is an aldohexose
(d) Glucose is one of the monomer unit in sucrose

12. 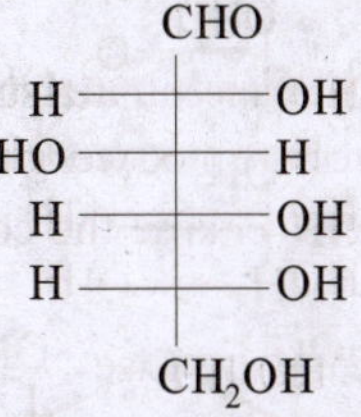

The **incorrect** statement regarding the given structure is **[NCERT : P L-283 | April 8, 2024 (I)]**
(a) can be oxidized to a dicarboxylic acid with Br_2 water
(b) despite the presence of – CHO does not give Schiff's test
(c) has 4-asymmetric carbon atoms
(d) will coexist in equilibrium with 2 other cyclic structures

13. Which of the following is the correct structure of L-Glucose ? **[NCERT : P L-285 | April 4, 2024 (I)]**

(a) CHO; HO–, –OH, HO–, HO–; CH_2OH

(b) CHO; –OH, HO–, –OH, HO–; CH_2OH

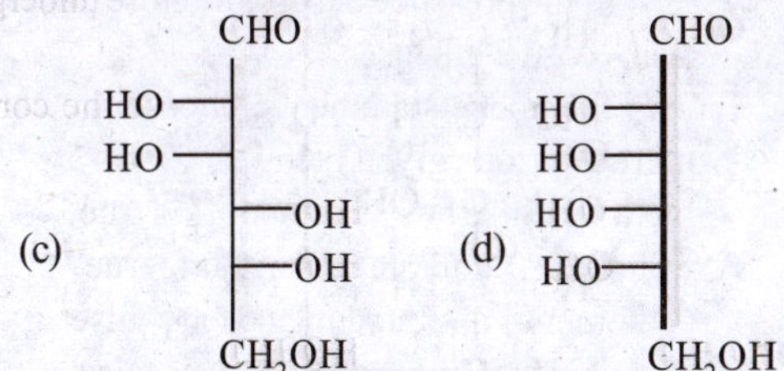

14. Match **List-I** with **List-II** **[NCERT : P L-286 | April 4, 2024 (II)]**

List-I	List-II
(A) α- Glucose and α-Galactose	I. Functional isomers
(B) α- Glucose and β-Glucose	II. Homologues
(C) α- Glucose and α-Fructose	III. Anomers
(D) α- Glucose and α-Ribose	IV. Epimers

Choose the **correct** answer from the options given below:
(a) A-(III), B-(IV), C-(II), D-(I)
(b) A-(III), B-(IV), C-(I), D-(II)
(c) A-(IV), B-(III), C-(I), D-(II)
(d) A-(IV), B-(III), C-(II), D-(I)

15. Match List I with List II **[NCERT : P L-283 | Jan. 31, 2024 (I)]**

LIST I	LIST II
A. Glucose/$NaHCO_3$/Δ	I. Gluconic acid
B Glucose/HNO_3	II. No reaction
C Glucose/HI/Δ	III. n-Hexane
D. Glucose/Bromine water	IV. Saccharic acid

Choose the correct answer from the options given below:
(a) A-IV, B-I, C-III, D-II
(b) A-I, B-IV, C-III, D-II
(c) A-II, B-IV, C-III, D-I
(d) A-III, B-II, C-I, D-IV

16. Sugar which does not give reddish brown precipitate with Fehling's reagent, is : **[Jan. 30, 2024 (I)]**
(a) Sucrose (b) Lactose (c) Glucose (d) Maltose

17. $\text{D-(+)-Glyceraldehyde} \xrightarrow[\text{(ii) } H_2O/H^+ \text{ (iii) } HNO_3]{\text{(i) HCN}}$

The products formed in the above reaction are **[NCERT : P L-283 | April 13, 2023 (I)]**
(a) Two optically active products
(b) One optically active and one meso product
(c) One optically inactive and one meso product
(d) Two optically inactive products

18. L - Isomer of tetrose X ($C_4H_8O_4$) gives positive Schiff's test and has two chiral carbons. On acetylation. 'X' yields triacetate. 'X' also undergoes following reactions

$$'A' \xleftarrow{HNO_3} 'X' \xrightarrow{NaBH_4} \underset{\text{Chiral compound}}{'B'}$$

'X' is **[April 11, 2023 (I)]**

(a) 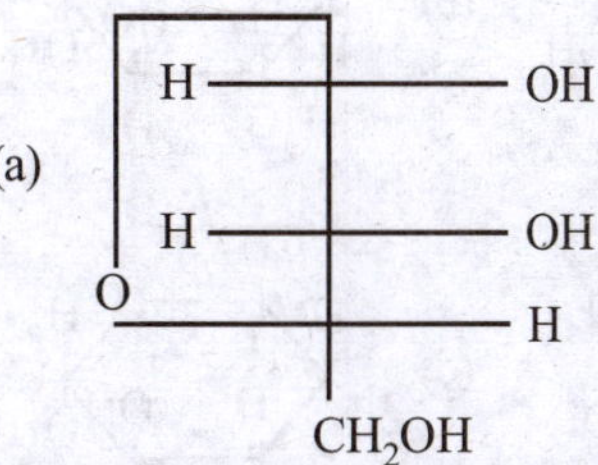

(b) 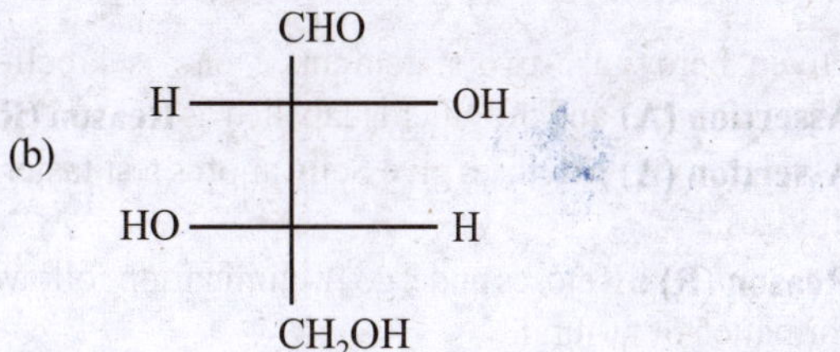

(c)

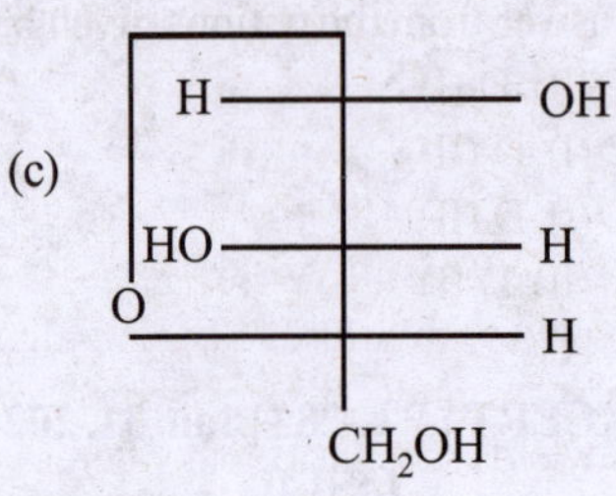

(d) CHO

HO — H

HO — H

CH_2OH

19. Number of compounds from the following which will not produce orange red precipitate with Benedict solution is ______.

Glucose, maltose, sucrose, ribose, 2-deoxyribose, amylose, lactose. **[NV, April 11, 2023 (II)]**

20. The correct representation in six membered pyranose form for the following sugar [X] is **[NCERT : P L-285 | Feb. 01, 2023 (I)]**

CHO

HO — H

HO — H

H — OH

H — OH

H_2COH

Sugar [X]

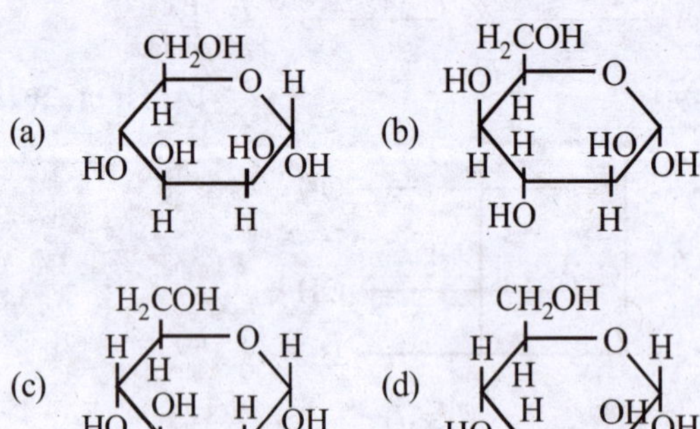

21. Given below are two statements : one is labelled as **Assertion (A)** and the other is labelled as **Reason (R)**.

Assertion (A) : Ketoses give Seliwanoff's test faster than aldoses.

Reason (R) : Ketoses undergo β-elimination followed by formation of furfural.

In the light of the above statements, choose the correct answer from the options given below : **[Jan. 30, 2023 (I)]**

(a) (A) is false but (R) is true

(b) Both (A) and (R) are true and (R) is the correct explanation of (A)

(c) (A) is true but (R) is false

(d) Both (A) and (R) are true but (R) is not the correct explanation of (A)

22. Match items of Row I with those of Row II.

Row I : **[NCERT : P L-285 | Jan. 25, 2023 (I)]**

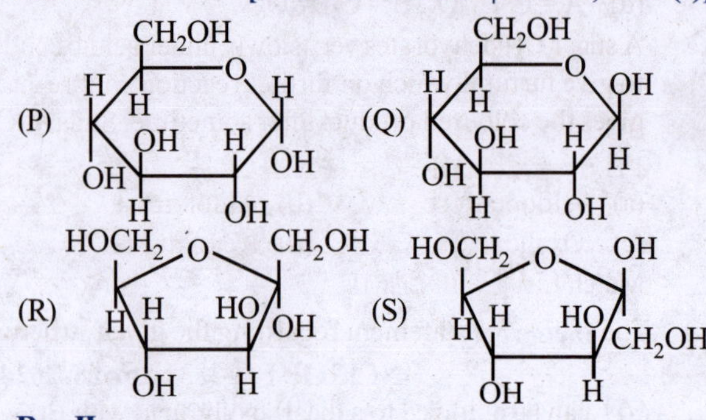

Row II :

(i) α-D-(–) Fructofuranose.

(ii) β-D-(–) Fructofuranose

(iii) α-D-(–) Glucopyranose.

(iv) β-D-(–) Glucopyranose

Correct match is

(a) P→iv, Q→iii, R→i, S→ii

(b) P→i, Q→ii, R→iii, S→iv

(c) P→iii, Q→iv, R→ii, S→i

(d) P→iii, Q→iv, R→i, S→ii

23. For the below given cyclic hemiacetal (X), the correct pyranose structure is :

[NCERT : P L-285 | July 28, 2022 (I)]

HO – C – H

HO – C – H O

H – C – OH

HO – C – H

H – C

CH_2OH

(X)

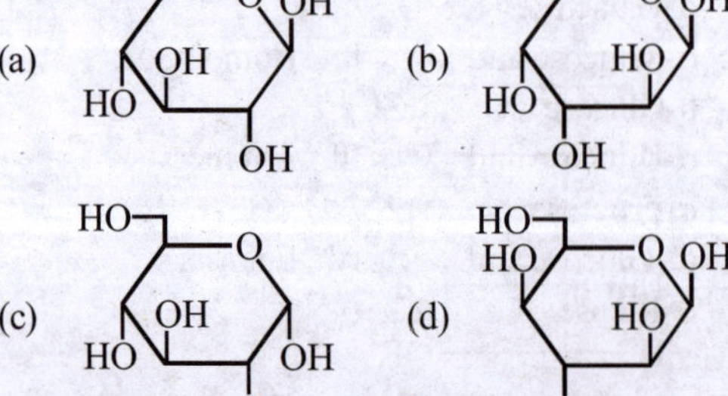

24. The formulas of A and B for the following reaction sequence

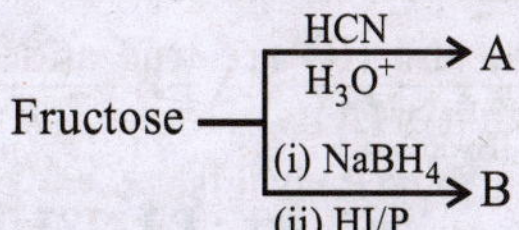

(a) $A = C_7H_{14}O_8, B = C_6H_{14}$ **[July 28, 2022 (II)]**
(b) $A = C_7H_{13}O_7, B = C_7H_{14}O$
(c) $A = C_7H_{12}O_8, B = C_6H_{14}$
(d) $A = C_7H_{14}O_8, B = C_6H_{14}O_6$

25. A sugar 'X' dehydrates very slowly under acidic condition to give furfural which on further reaction with resorcinol gives the coloured product after sometime. Sugar 'X' is **[July 27, 2022 (I)]**

(a) Aldopentose (b) Aldotetrose
(c) Oxalic acid (d) Ketotetrose

26. Match List-I with List-II **[NCERT : P L-283 | July 27, 2022 (II)]**

List-I	List-II
(A) Glucose + HI	(I) Gluconic acid
(B) Glucose + Br_2 water	(II) Glucose pentacetate
(C) Glucose + acetic anhydride	(III) Saccharic acid
(D) Glucose + HNO_3	(IV) Hexane

Choose the correct answer from the options given below:
(a) (A)-(IV), (B)-(I), (C)-(II), (D)-(III)
(b) (A)-(IV), (B)-(III), (C)-(II), (D)-(I)
(c) (A)-(III), (B)-(I), (C)-(IV), (D)-(II)
(d) (A)-(I), (B)-(III), (C)-(IV), (D)-(II)

27. Which one of the following is a reducing sugar? **[NCERT : P L-287, 288 | July 26, 2022 (I)]**

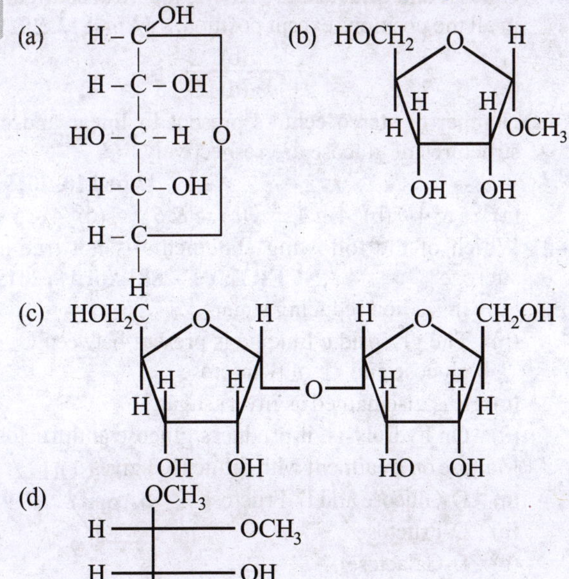

28. Glycosidic linkage between C_1 of α-glucose and C_2 of β-fructose is found in **[NCERT : P L-287, 288 | July 25, 2022 (II)]**

(a) maltose (b) sucrose
(c) lactose (d) amylose

29. When sugar 'X' is boiled with dilute H_2SO_4 in alcoholic solution, two isomers 'A' and 'B' are formed. 'A' on oxidation with HNO_3 yields saccharic acid whereas 'B' is laevorotatory. The compound 'X' is : **[NCERT : P L-283 | June 28, 2022 (II)]**

(a) Maltose (b) Sucrose
(c) Lactose (d) Starch

30. L-Isomer of a compound 'A' ($C_4H_8O_4$) gives a positive test with $[Ag(NH_3)_2]^+$. Treatment of 'A' with acetic anhydride yield triacetate derivative. Compound 'A' produces an optically active compound (B) and an optically inactive compound (C) on treatment with bromine water and HNO_3 respectively, compound (A) is: **[June 27, 2022 (I)]**

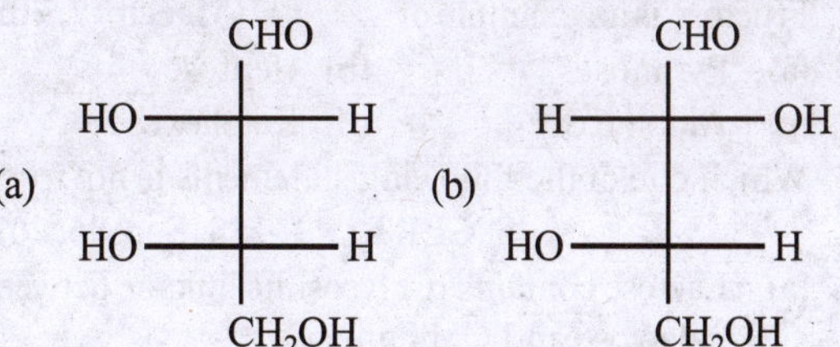

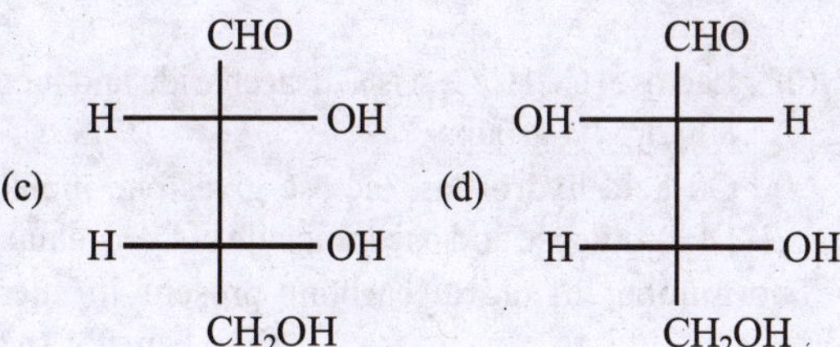

31. Given below are two statements. **[NCERT : P L-288 | June 27, 2022 (II)]**

Statement I: Maltose has two α-D-glucose units linked at C_1 and C_4 and is a reducing sugar.

Statement II: Maltose has two monosaccharides: α-D-glucose and β-D-glucose linked at C_1 and C_6 and it is a non-reducing sugar.

In the light of the above statements, choose the correct answer from the options given below.
(a) Both Statement I and Statement II are true.
(b) Both Statement I and Statement II are false.
(c) Statement I is true but Statement II is false.
(d) Statement I is false but Statement II is true.

32. Which one of the following compounds contains β-C_1-C_4 glycosidic linkage? **[NCERT : P L-288 | Aug. 31, 2021 (I)]**

(a) Lactose (b) Sucrose
(c) Maltose (d) Amylose

33. Given below are two statements, one is labelled as **Assertion (A)** and other is labelled as **Reason (R)**.
Assertion (A) : Sucrose is a disaccharide and a non-reducing sugar.
Reason (R) : Sucrose involves glycosidic linkage between C_1 of β-glucose and C_2 of β-fructose.
Choose the **most appropriate** answer from the options given below : **[NCERT : P L-288 | Aug. 26, 2021 (II)]**
(a) Both (A) and (R) are true but (R) is not the true explanation of (A)
(b) (A) is false but (R) is true.
(c) (A) is true but (R) is false
(d) Both (A) and (R) are true and (R) is the true explanation of (A)

34. Compound A gives D-galactose and D-glucose on hydrolysis. The compound A is : **[NCERT : P L-288 | July 27, 2021 (II)]**
(a) Amylose (b) Sucrose
(c) Maltose (d) Lactose

35. Fructose is an example of **[March 17, 2021 (II)]**
(a) Pyranose (b) Heptose
(c) Aldohexose (d) Ketohexose

36. Which one of the following statements is not true? **[NCERT : P L-288 | Sep. 06, 2020 (II)]**
(a) Lactose contains α-glycosidic linkage between C_1 of galactose and C_4 of glucose.
(b) Lactose is a reducing sugar and it gives Fehling's test.
(c) Lactose ($C_{11}H_{22}O_{11}$) is a disaccharide and it contains 8 hydroxyl groups.
(d) On acid hydrolysis, lactose gives one molecule of D(+)-glucose and one molecule of D(+)-galactose.

37. The number of chiral carbons present in sucrose is ________. **[NV, Sep. 05, 2020 (II)]**

38. What are the functional groups present in the structure of maltose? **[Sep. 04, 2020 (I)]**
(a) One ketal and hemiketal
(b) Two acetals
(c) One acetal and one hemiacetal
(d) One acetal and one ketal

39. Consider the following reactions : **[NCERT : P L-283 | Sep. 02, 2020 (I)]**
(i) Glucose + ROH $\xrightarrow{\text{dry HCl}}$ Acetal $\xrightarrow[(CH_3CO)_2O]{x \text{ eq. of}}$ Acetyl derivative
(ii) Glucose $\xrightarrow{Ni/H_2}$ A $\xrightarrow[(CH_3CO)_2O]{y \text{ eq. of}}$ Acetyl derivative
(iii) Glucose $\xrightarrow[(CH_3CO)_2O]{z \text{ eq. of}}$ Acetyl derivative
'x', 'y' and 'z' in these reactions are respectively.
(a) 5, 4 & 5 (b) 4, 6 & 5
(c) 4, 5 & 5 (d) 5, 6 & 5

40. The correct observation in the following reactions is :
$$\text{Sucrose} \xrightarrow[\substack{\text{cleavage} \\ \text{(Hydrolysis)}}]{\text{Glycosidic bond}} A + B \xrightarrow[\text{reagent}]{\text{Seliwanoff's}} ?$$
[NCERT : P L-283 | Sep. 02, 2020 (II)]
(a) Formation of blue colour
(b) Gives no colour
(c) Formation of red colour
(d) Formation of violet colour

41. Which of the following statement is not true for glucose? **[Jan. 08, 2020 (I)]**
(a) Glucose exists in two crystalline forms α and β
(b) Glucose gives Schiff's test for aldehyde
(c) Glucose reacts with hydroxylamine to form oxime
(d) The pentaacetate of glucose does not react with hydroxylamine to give oxime

42. Two monomers in maltose are: **[NCERT : P L-288 | Jan. 08, 2020 (II)]**
(a) α-D-glucose and β-D-glucose
(b) α-D-glucose and α-D-galactose
(c) α-D-glucose and α-D-fructose
(d) α-D-glucose and α-D-glucose

43. Which of the following statements is correct ? **[NCERT : P L-283, 284 | Jan. 07, 2020 (II)]**
(a) Gluconic acid can form cyclic (acetal/hemiacetal) structure
(b) Gluconic acid is a dicarboxylic acid
(c) Gluconic acid is a partial oxidation product of glucose
(d) Gluconic acid is obtained by oxidation of glucose with HNO_3

44. Glucose and galactose are having identical configuration in all the positions except position. **[April 12, 2019 (I)]**
(a) C – 3 (b) C – 4
(c) C – 2 (d) C – 5

45. Number of stereo centers present in linear and cyclic structures of glucose are respectively: **[April 10, 2019 (II)]**
(a) 5 & 4 (b) 4 & 4 (c) 5 & 5 (d) 4 & 5

46. Which of the following statements is not true about sucrose? **[NCERT : P L-288 | April 9, 2019 (I)]**
(a) It is a non reducing sugar
(b) The glycosidic linkage is present between C_1 of α-glucose and C_1 of β-fructose
(c) It is also named as invert sugar
(d) On hydrolysis, it produces glucose and fructose

47. Maltose on treatment with dilute HCl gives : **[April 8, 2019 (I)]**
(a) D-Glucose and D-Fructose
(b) D-Fructose
(c) D-Galactose
(d) D-Glucose

48. Fructose and glucose can be distinguished by: **[April 8, 2019 (II)]**
(a) Benedict's test (b) Fehling's test
(c) Barfoed's test (d) Seliwanoff's test

49. Glucose on prolonged heating with HI gives : **[NCERT : P L-283 | 2018]**

(a) *n*-Hexane (b) 1-Hexene
(c) Hexanoic acid (d) 6-iodohexanal

50. Among the following, the incorrect statement is: **[NCERT : P L-289 | Online April 16, 2018]**

(a) Cellulose and amylose have 1,4-glycosidic linkage
(b) Lactose contains β-D-galactose and β-D-glucose
(c) Maltose and lactose have 1,4-glycosidic linkage
(d) Sucrose and amylose have 1,2-glycosidic linkage

51. Which of the following compounds will behave as a reducing sugar in an aqueous KOH solution? **[2017]**

(a) HOH_2C, O, CH_2OH, HO, $OCOCH_3$, OH

(b) HOH_2C, O, CH_2OH, HO, OH

(c) HOH_2C, O, CH_2OH, HO, OCH_3, OH

(d) HOH_2C, O, CH_2OCH_3, OH, OH, OH

52. The **incorrect** statement among the following is : **[NCERT : P L-283, 286 | Online April 9, 2017]**

(a) α-D-glucose and β-D-glucose are anomers.
(b) α-D-glucose and β-D-glucose are enantiomers.
(c) Cellulose is a straight chain polysaccharide made up of only β-D-glucose units.
(d) The penta-acetate of glucose does not react with hydroxyl amine.

53. Accumulation of which of the following molecules in the muscles occurs as a result of vigorous exercise ? **[Online April 11, 2015]**

(a) Glycogen (b) Glucose
(c) Pyruvic acid (d) L-lactic acid

54. Which of the following will not show mutarotation? **[NCERT : P L-285 | Online April 12, 2014]**

(a) Maltose (b) Lactose
(c) Glucose (d) Sucrose

55. Glycosidic linkage is actually a (an) : **[Online April 23, 2013]**

(a) Carbonyl bond (b) Ether bond
(c) Ester bond (d) Amide bond

56. Synthesis of each molecule of glucose in photosynthesis involves : **[2013]**

(a) 18 molecules of ATP (b) 10 molecules of ATP
(c) 8 molecules of ATP (d) 6 molecules of ATP

57. Natural glucose is termed D-glucose because : **[Online April 23, 2013]**

(a) – OH on the second carbon is on the right side in Fischer projection
(b) – OH on the sixth carbon is on the right side in Fischer projection.
(c) – OH on the fifth carbon is on the right side in Fischer projection.
(d) It is dextrorotatory.

58. Which of the following statement is not correct? **[NCERT : P L-289 | Online April 25, 2013]**

(a) Amylopectin is a branched polymer of α - glucose.
(b) Cellulose is a linear polymer of β-glucose
(c) Glycogen is the food reserve of plants
(d) All proteins are polymers of α - amino acids.

59. Which of the following compounds can be detected by Molisch's test ? **[2012]**

(a) Nitro compounds (b) Sugars
(c) Amines (d) Primary alcohols

60. Which of the following is a non-reducing sugar? **[NCERT : P L-288 | Online May 19, 2012]**

(a) Lactose (b) Fructose
(c) Sucrose (d) Maltose

61. The change in the optical rotation of freshly prepared solution of glucose is known as: **[2011RS]**

(a) racemisation (b) specific rotation
(c) mutarotation (d) tautomerism

62. The two functional groups present in a typical carbohydrate are: **[NCERT : P L-282 | 2009]**

(a) – CHO and – COOH (b) $>C=O$ and – OH
(c) – OH and – CHO (d) – OH and – COOH

63. α - D-(+)-glucose and β-D-(+)-glucose are 1 **[NCERT : P L-286 | 2008]**

(a) conformers (b) epimers
(c) anomers (d) enatiomers

64. The term anomers of glucose refers to **[NCERT : P L-286 | 2006]**

(a) enantiomers of glucose
(b) isomers of glucose that differ in configuration at carbon one (C-1)
(c) isomers of glucose that differ in configurations at carbons one and four (C-1 and C-4)
(d) a mixture of (D)-glucose and (L)-glucose

Topic 2 - Amino Acids and Proteins

65.

$$\begin{array}{c} CH(CH_3)_2 \\ | \\ H_2N-CH-COOH \end{array}$$

pH = 2 ↙ ↘ pH = 10

A B

Choose the **correct** option for structures of A and B, respectively. **[NCERT : P L-292 | April 8, 2025 (II)]**

(a) $H_3\overset{+}{N}-CH(CH(CH_3)_2)-COOH$ and $H_2N-CH(CH(CH_3)_2)-CO\bar{O}H$

(b) $H_2N-CH(CH(CH_3)_2)-CO\bar{O}$ and $H_3\overset{+}{N}-CH(CH(CH_3)_2)-COOH$

(c) $H_2N-CH(CH(CH_3)_2)-CO\overset{\ominus}{O}$ and $H_3\overset{+}{N}-CH(CH(CH_3)_2)-CO\bar{O}$

(d) $H_3\overset{+}{N}-CH(CH(CH_3)_2)-CO\bar{O}$ and $H_3\overset{+}{N}-CH(CH(CH_3)_2)-COOH$

66. Given below are two statements :

[NCERT : P L-292 | April 7, 2025 (II)]

Statement (I) : On hydrolysis, oligo peptides give rise to fewer number of α-amino acids while proteins give rise to a large number of β-amino acids.

Statement (II) : Natural proteins are denatured by acids which convert the water soluble form of fibrous proteins to their water insoluble form.

In the light of the above statements, choose the **most appropriate answer** from the options given below :

(a) Both statement I and statement II are correct
(b) Statement I is incorrect but Statement II is correct
(c) Both statement I and statement II are incorrect
(d) Statement I is correct but Statement II is incorrect

67. A dipeptide, "x" on complete hydrolysis gives "y" and "z". "y" on treatment with aq. HNO_2 produces lactic acid. On the other hand "z" on heating gives the following cyclic molecule.

$$\begin{array}{ccc} & O & \\ & \| & \\ & C & \\ CH_2 & & N-H \\ | & & | \\ N & & CH_2 \\ & C & \\ H & \| & \\ & O & \end{array}$$

Based on the information given, the dipeptide x is:

[NCERT : P L-291 | April 4, 2025 (II)]

(a) valine-glycine (b) alanine-glycine
(c) valine-leucine (d) alanine-alanine

68. Identify the pair of reactants that upon reaction, with elimination of HCl will give rise to the dipeptide Gly-Ala.

[NCERT : P L-292 | April 4, 2025 (I)]

(a) NH_2-CH_2-COCl and $NH_2-CH(CH_3)-COOH$

(b) NH_2-CH_2-COCl and $\overset{\oplus}{N}H_2-CH(CH_3)-COCl$

(c) NH_2-CH_2-COOH and $NH_2-CH(CH_3)-COCl$

(d) NH_2-CH_2-COOH and $NH_2-CH(CH_3)-COOH$

69. A tetrapeptide "x" on complete hydrolysis produced glycine (Gly), alanine (Ala), valine (Val), leucine (Leu) in equimolar proportion each. The number of tetrapeptides (sequences) possible involving each of these amino acids is **[NCERT : P L-292 | April 2, 2025 (II)]**

(a) 16 (b) 32 (c) 8 (d) 24

70. Identify the correct statement among the following:

[NCERT : P L-291 | April 2, 2025 (I)]

(a) All naturally occurring amino acids except glycine contain one chiral centre.
(b) All naturally occurring amino acids are optically active.
(c) Glutamic acid is the only amino acid that contains a –COOH group at the side chain.
(d) Amino acid, cysteine easily undergo dimerization due to the presence of free SH group.

71. Identify the essential amino acids from below :

(A) Valine (B) Proline
(C) Lysine (D) Threonine
(E) Tyrosine **[NCERT : P L-291 | Jan. 29, 2025 (II)]**

Choose the correct answer from the options given below :

(a) (A), (C) and (D) only (b) (A), (C) and (E) only
(c) (B), (C) and (E) only (d) (C), (D) and (E) only

72. The α- Helix and β- Pleated sheet structures of protein are associated with its: **[NCERT : P L-293 | Jan. 23, 2025 (II)]**

(a) quaternary structure (b) primary structure
(c) secondary structure (d) tertiary structure

73. Total number of essential amino acids among the given list of amino acids is ______.

[NCERT : P L-291 | NV, April 9, 2024 (I)]

Arginine, Phenylalanine, Aspartic acid, Cysteine, Histidine, Valine, Proline.

74. The total number of carbon atoms present in tyrosine, an amino acid, is _______.
[NCERT : P L-291 | NV, April 8, 2024 (II)]

75. Coagulation of egg, on heating is because of :
[NCERT : P L-294 | April 5, 2024 (II)]
(a) Denaturation of protein occurs
(b) The secondary structure of protein remains unchanged
(c) Breaking of the peptide linkage in the primary structure of protein occurs
(d) Biological property of protein remains unchanged

76. The number of tripeptides formed by three different amino acids using each amino acid once is _________ .
[NV, Feb. 1, 2024 (II)]

77. Type of amino acids obtained by hydrolysis of proteins is:
[NCERT : P L-290 | Jan. 29, 2024 (I)]
(a) γ (b) δ (c) β (d) α

78. Which structure of protein remains intact after coagulation of egg white on boiling?
[NCERT : P L-294 | Jan. 27, 2024 (II)]
(a) Primary (b) Tertiary
(c) Secondary (d) Quaternary

79. Which is not true for arginine? **[April 15, 2023 (I)]**
(a) It is a crystalline solid.
(b) It is associated with more than one pK_a values.
(c) It has a fairly high melting point.
(d) It has high solubility in benzene.

80. The naturally occurring amino acid that contains only one basic functional group in its chemical structure is
[NCERT : P L-290 | April 13, 2023 (II)]
(a) arginine (b) lysine
(c) asparagine (d) histidine

81. In an oligopeptide named Alanylglycylphenylalanyl-isoleucine, the number of sp^2 hybridised carbons is _______.
[NV, April 12, 2023 (I)]

82. The one that does not stabilize 2° and 3° structures of proteins is **[April 10, 2023 (I)]**
(a) H-bonding (b) –S-S-linkage
(c) –O-O-linkage (d) van der Waals forces

83. Sulphur (S) containing amino acids from the following are:
[NCERT : P L-290 | April 08, 2023 (I)]
(A) isoleucine (B) cysteine
(C) lysine (D) methionine
(E) glutamic acid
(a) A, D (b) B, D (c) B, C, E (d) A, B, C

84. Match List I with List II
Choose the correct answer from the options given below:
[NCERT : P L-290 | April 08, 2023 (II)]

List I (Natural amino acid)	List II (One letter code)
A. Glutamic acid	I. Q
B. Glutamine	II. W
C. Tyrosine	III. E
D. Tryptophan	IV. Y

(a) A-II, B-I, C-IV, D-III (b) A-IV, B-III, C-I, D-II
(c) A-III, B-I, C-IV, D-II (d) A-III, B-IV, C-I, D-II

85. Match **List I** with **List II**
[NCERT : P L-290 | April 06, 2023 (II)]

List I		List II	
Natural Amino acid		**One Letter Code**	
(A)	Arginine	(I)	D
(B)	Aspartic acid	(II)	N
(C)	Asparagine	(III)	A
(D)	Alanine	(IV)	R

Choose the correct answer from the options given below:-
(a) (A)-IV, (B)-I, (C)-III, (D)-II
(b) (A)-I, (B)-III, (C)-IV, (D)-II
(c) (A)-III, (B)-I, (C)-II, (D)-IV
(d) (A)-IV, (B)-I, (C)-II, (D)-III

86. A protein 'X' with molecular weight of 70,000 u, on hydrolysis gives amino acids. One of these amino acid is
[Jan. 31, 2023 (I)]

(a) $NH_2-CH_2-\underset{\underset{CH_3}{|}}{CH}-CH_2CH_2COOH$

(b) $CH_3-\overset{\overset{CH_3}{|}}{CH}-CH_2-\underset{\underset{NH_2}{|}}{CH}-COOH$

(c) $CH_3-\overset{\overset{CH_3}{|}}{CH}-\underset{\underset{NH_2}{|}}{CH}-CH_2COOH$

(d) $CH_3-\overset{\overset{CH_3}{|}}{\underset{\underset{NH_2}{|}}{C}}-CH_2-CH_2COOH$

87. A short peptide on complete hydrolysis produces 3 moles of glycine (G), two moles of leucine (L) and two moles of valine (V) per mole of peptide. The number of peptide linkages in it are _______. **[NV, Jan. 30, 2023 (II)]**

88. Number of cyclic tripeptides formed with 2 amino acids A and B is: **[Jan. 29, 2023 (I)]**
(a) 2 (b) 3
(c) 5 (d) 4

89. Following tetrapeptide can be represented as [Jan. 29, 2023 (II)]

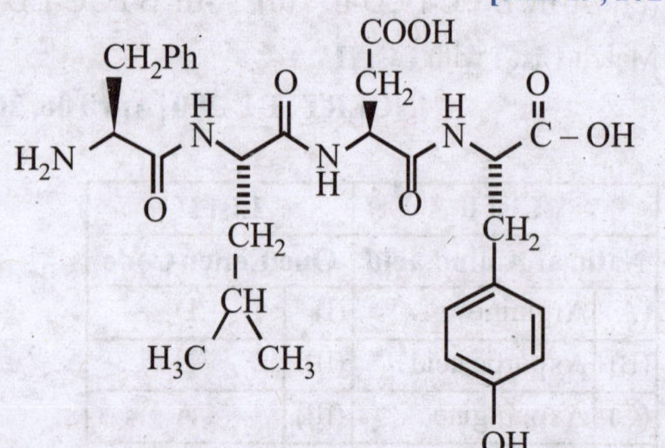

(F, L, D, Y, I, Q, P are one letter codes for amino acids)

(a) FIQY (b) FLDY (c) YQLF (d) PLDY

90. Total number of tripeptides possible by mixing of valine and proline is ________ [NV, Jan. 24, 2023 (II)]

91. Stability of α -Helix structure of proteins depends upon [NCERT : P L-57 | June 28, 2022 (I)]

(a) dipolar interaction (b) H-bonding interaction
(c) van der Waals forces (d) π -stacking interaction

92. In a linear tetrapeptide (Constituted with different amino acids), (number of amino acids) - (number of peptide bonds) is______. [NV, July 29, 2022 (I)]

93. During the denaturation of proteins, which of these structures will remain intact? [NCERT : P L-294 | July 25, 2022 (I)]

(a) Primary (b) Secondary
(c) Tertiary (d) Quaternary

94. The structure of protein that is unaffected by heating is: [NCERT : P L-294 | June 29, 2022 (II)]

(a) secondary structure (b) tertiary structure
(c) primary structure (d) quaternary structure

95. How many of the given compounds will give a positive biuret test ________? Glycine, Glycylalanine, Tripeptide, Biuret [NV, June 25, 2022 (II)]

96. In alanylglycylleucylalanylvaline, the number of peptide linkages is________. [NV, June 24, 2022 (II)]

97. A peptide synthesized by the reactions of one molecule each of Glycine, Leucine, Aspartic acid and Histidine will have ________ peptide linkages. [NV, Sep. 1, 2021 (II)]

98. Which of the following is NOT an example of fibrous protein ? [NCERT : P L-293 | Aug. 31, 2021 (II)]

(a) Keratin (b) Albumin
(c) Collagen (d) Myosin

99. The total number of negative charge in the tetrapeptide, Gly-Glu-Asp-Tyr, at pH 12.5 will be _______. (Integer answer) [NV, Aug. 26, 2021 (I)]

100. The water soluble protein is : [NCERT : P L-293 | July 25, 2021 (I)]

(a) Fibrin (b) Albumin (c) Myosin (d) Collagen

101. Seliwanoff test and xanthoproteic test are used for the identification of _______ and ______ respectively. [Feb. 26, 2021 (II)]

(a) ketoses, aldoses (b) proteins, ketoses
(c) ketoses, proteins (d) aldoses, ketoses

102. Out of the following, which type of interaction is responsible for the stabilisation of α-helix structure of proteins? [NCERT : P L-293 | February 24, 2021 (I)]

(a) Ionic bonding (b) Hydrogen bonding
(c) Covalent bonding (d) van der Waals forces

103. Which of the following is not an essential amino acid? [NCERT : P L-290 | Sep. 05, 2020 (I)]

(a) Tyrosine (b) Leucine
(c) Valine (d) Lysine

104. The number of chiral carbon(s) present in peptide, Ile-Arg-Pro, is _____. [NV, Sep. 05, 2020 (I)]

105. Which of the following will react with $CHCl_3$ + alc. KOH? [Sep. 04, 2020 (I)]

(a) Adenine and proline (b) Thymine and proline
(c) Adenine and lysine (d) Adenine and thymine

106. The number of chiral centres present in threonine is ___________. [NV, Sep. 04, 2020 (II)]

107. The number of $\rangle C = O$ groups present in a tripeptide Asp - Glu - Lys is ___________. [NV, Sep. 03, 2020 (II)]

108. The mass percentage of nitrogen in histamine is_____. [NV, Jan. 09, 2020 (I)]

109. The correct structure of product 'P' in the following reaction is: [Jan. 10, 2019 (I)]

$$\text{Asn} - \text{Ser} + \underset{\text{(excess)}}{(CH_3CO)_2O} \xrightarrow{NEt_3} P$$

(a) H_3C–CO–NH–CH(CH_2CONHCOCH_3)–CO–NH–CH(CH_2OCOCH_3)–COOH

(b) H_3C–CO–NH–CH(CH_2CONH_2)–CO–NH–CH(CH_2OCOCH_3)–COOH

(c) H_3C–CO–NH–CH(CH_2OCOCH_3)–CO–NH–CH(CH_2CONH_2)–COOH

(d) H_3C–CO–NH–CH(CH_2OCOCH_3)–CO–NH–CH(CH_2CONHCOCH_3)–COOH

110. The peptide that gives positive ceric ammonium nitrate and carbylamine tests is: **[April 9, 2019 (II)]**

(a) Ser – Lys (b) Gln – Asp
(c) Lys – Asp (d) Asp – Gln

111. Among the following compounds most basic amino acid is: **[Jan. 12, 2019 (I)]**

(a) Asparagine (b) Lysine
(c) Serine (d) Histidine

112. The correct structure of histidine in a strongly acidic solution (pH = 2) is : **[Jan. 12, 2019 (II)]**

(a) $H_3\overset{\oplus}{N}$–CH–COOH, ring: $\overset{\oplus}{N}H_2$, N
(b) $H_3\overset{\oplus}{N}$–CH–$CO\overset{\ominus}{O}$, ring: NH, $N^{\oplus}$H
(c) $H_3\overset{\oplus}{N}$–CH–COOH, ring: NH, $N^{\oplus}$H
(d) $H_3\overset{\oplus}{N}$–CH–$CO\overset{\ominus}{O}$, ring: $\overset{\oplus}{N}H_2$, N

113. The increasing order of pK_a of the following amino acids in aqueous solution is: **[Jan. 9, 2019 (I)]**

Gly Asp Lys Arg

(a) Asp < Gly < Arg < Lys
(b) Gly < Asp < Arg < Lys
(c) Asp < Gly < Lys < Arg
(d) Arg < Lys < Gly < Asp

114. The correct sequence of amino acids present in the tripeptide given below is: **[Jan. 9, 2019 (II)]**

Me, Me, H_2N, O, H, N, O, OH, Me, OH, N, H, C, OH, O

(a) Val - Ser - Thr (b) Thr - Ser - Val
(c) Leu - Ser - Val (d) Thr - Ser - Leu

115. Which of the following will not exist in zwitter ionic form at pH = 7 ? **[NCERT : P L-292 | Online April 15, 2018 (I)]**

(a) cyclohexane ring with NH_2 and COOH
(b) O, N, H, CO_2H
(c) benzene ring with NH_2 and SO_3H
(d) cyclohexane ring with NH_2 and SO_3H

116. Among the following, the essential amino acid is : **[NCERT : P L-291 | Online April 8, 2017]**

(a) Alanine (b) Valine
(c) Aspartic acid (d) Serine

117. Observation of "Ruhemann's purple" is a confirmatory test for the presence of : **[Online April 10, 2016]**

(a) Starch
(b) Reducing sugar
(c) Protein
(d) Cupric ion

118. Which one of the following statements is correct? **[NCERT : P L-290 | 2012]**

(a) All amino acids except lysine are optically active
(b) All amino acids are optically active
(c) All amino acids except glycine are optically active
(d) All amino acids except glutamic acids are optically active

119. All of the following statements apply to proteins except **[Online May 7, 2012]**

(a) Proteins generally have no definite melting point
(b) Proteins contain the grouping —CONH—
(c) Proteins have high molecular weight
(d) Proteins can only contain the elements C, H, O and N.

120. Biuret test is **not** given by **[2010]**

(a) carbohydrates (b) polypeptides
(c) urea (d) proteins

121. The secondary structure of a protein refers to **[NCERT : P L-293 | 2007]**

(a) fixed configuration of the polypeptide backbone
(b) α–helical backbone
(c) hydrophobic interactions
(d) sequence of α –amino acids.

Topic 3 - Nucleic Acid and Enzymes

122. The total number of hydrogen bonds of a DNA-double Helix strand whose one strand has the following sequence of bases is ____.

5′–G–G–C–A–A–A–T–C–G–G–C–T–A–3′.

[NCERT : P L-298 | NV, April 4, 2025 (I)]

123. Match List-I with List-II.

[NCERT : P L-298 | Jan. 24, 2025 (II)]

List-I	List-II
(A) Adenine	(I) 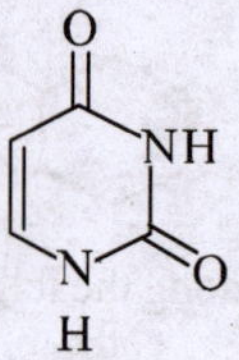
(B) Cytosine	(II) 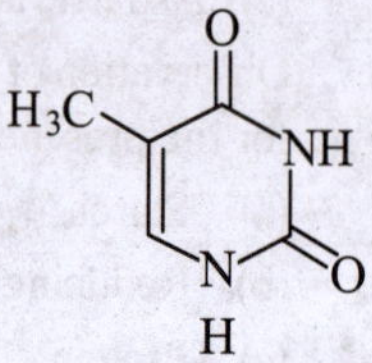
(C) Thymine	(III) 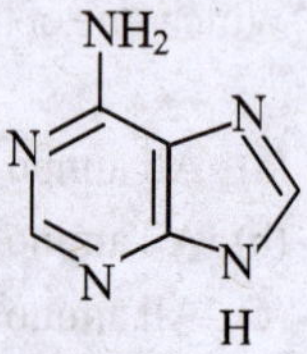
(D) Uracil	(IV) 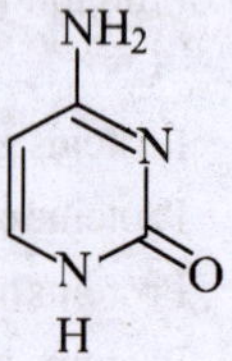

Choose the correct answer from the options given below :

(a) (A)-(III), (B)-(IV), (C)-(II), (D)-(I)
(b) (A)-(III), (B)-(I), (C)-(IV), (D)-(II)
(c) (A)-(IV), (B)-(III), (C)-(II), (D)-(I)
(d) (A)-(III), (B)-(IV), (C)-(I), (D)-(II)

124. DNA molecule contains 4 bases whoes structure are shown below. One of the structure is not correct, identify the **incorrect** base structure.

[NCERT : P L-298 | April 6, 2024 (I)]

(a)

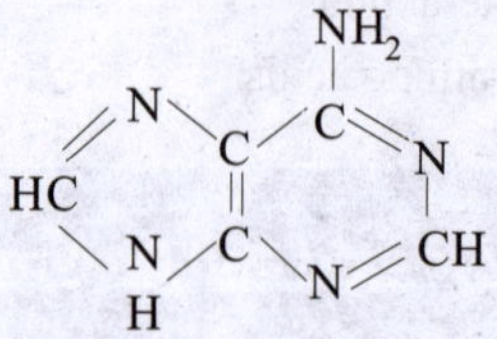

(b)

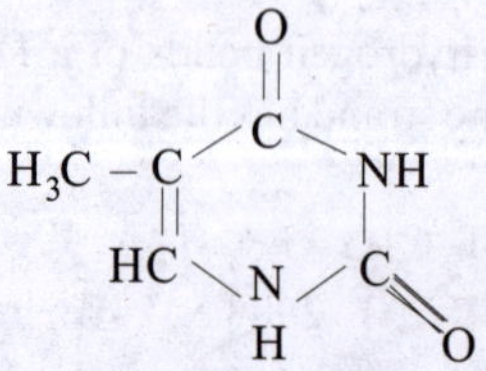

(c)

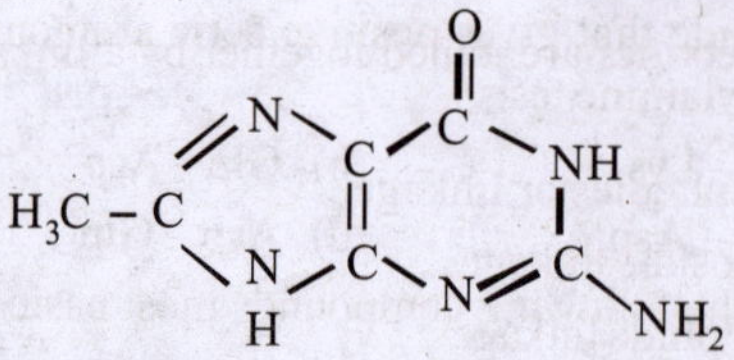

(d)

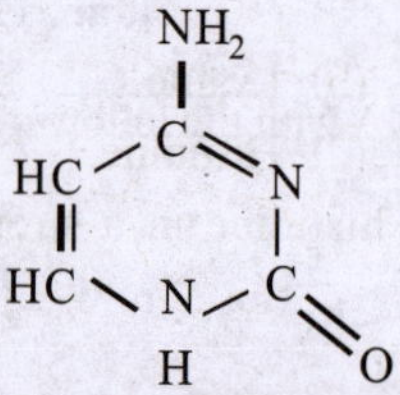

125. The incorrect statements regarding enzymes are:

[NCERT : P L-295 | April 6, 2024 (II)]

(A) Enzymes are biocatalysts.
(B) Enzymes are non-specific and can catalyse different kinds of reactions.
(C) Most Enzymes are globular proteins.
(D) Enzyme - oxidase catalyses the hydrolysis of maltose into glucose.

Choose the correct answer from the option given below:

(a) (B) and (C) (b) (B), (C) and (D)
(c) (B) and (D) (d) (A), (B) and (C)

126. If one strand of a DNA has the sequence ATGCTTCA, sequence of the bases in complementary strand is:

[NCERT : P L-298 | Feb. 1, 2024 (I)]

(a) CATTAGCT (b) TACGAAGT
(c) GTACTTAC (d) ATGCGACT

127. The total number of correct statements, regarding the nucleic acids is ______

[NCERT : P L-300 | NV, Jan. 30, 2024 (II)]

A. RNA is regarded as the reserve of genetic information
B. DNA molecule self - duplicates during cell division
C. DNA synthesizes proteins in the cell
D. The message for the synthesis of particular proteins is present in DNA
E. Identical DNA strands are transferred to daughter cells.

128. Match List I with List II

[NCERT : P L-289, 290, 298 | Jan. 29, 2024 (II)]

LIST I (Bio polymer)	LIST II (Monomer)
A. Starch	I. nucleotide
B. Cellulose	II. α-glucose
C. Nucleic acid	III. β-glucose
D. Protein	IV. α-amino acid

Choose the correct answer from the options given below:

(a) A-II, B-III, C-I, D-IV (b) A-II, B-I C-III, D-IV
(c) A-IV, B-II, C-I, D-III (d) A-I, B-III, C-IV, D-II

129. Two nucleotides are joined together by a linkage known as : **[NCERT : P L-298 | Jan. 27, 2024 (I)]**

(a) Phosphodiester linkage
(b) Glycosidic linkage
(c) Disulphide linkage
(d) Peptide linkage

130. Uracil is base present in RNA with the following structure. % of N in uracil is _______.
[NCERT : P L-298 | NV, Jan. 24, 2023 (I)]

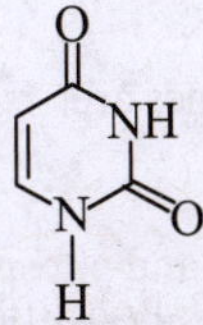

Given :
Molar mass N = 14 g mol^{-1}; O = 16 g mol^{-1};
C = 12 g mol^{-1}; H = 1 g mol^{-1}

131. The sugar produced after complete hydrolysis of DNA is **[NCERT : P L-297 | June 30, 2022 (I)]**

(a) a pentose sugar. (b) a hexose sugar.
(c) a polysaccharide. (d) a disaccharide.

132. Sugar moiety in DNA and RNA molecules respectively are **[NCERT : P L-298 | June 29, 2022 (I)]**

(a) β-D-2-deoxyribose, β-D-deoxyribose
(b) β-D-2-deoxyribose, β-D-ribose
(c) β-D-ribose, β -D-2-deoxyribose
(d) β-D-deoxyribose, β -D-2-deoxyribose

133. Stability of α -Helix structure of proteins depends upon **[NCERT : P L-293 | June 28, 2022 (I)]**

(a) dipolar interaction
(b) H-bonding interaction
(c) van der Waals forces
(d) π -stacking interaction

134. The number of oxygen present in a nucleotide formed from a base, that is present only in RNA is ________.
[NV, June 26, 2022 (I)]

135. Out of following isomeric forms of uracil, which one is present in RNA? **[NCERT : P L-297 | Aug. 27, 2021 (I)]**

(a) 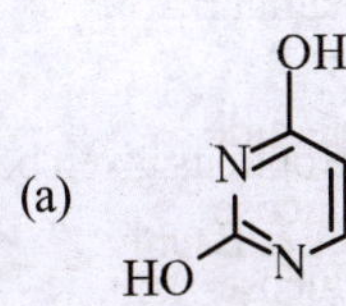
(b)
(c)
(d)

136. Which one of the following is correct structure for cytosine? **[NCERT : P L-297 | July 25, 2021 (II)]**

(a) (b)
(c) (d)

137. $C_{12}H_{22}O_{11} + H_2O \xrightarrow{\text{Enzyme B}} \underset{\text{Glucose}}{C_6H_{12}O_6} + \underset{\text{Fructose}}{C_6H_{12}O_6}$

$\underset{\text{Glucose}}{C_6H_{12}O_6} \xrightarrow{\text{Enzyme B}} 2C_2H_5OH + 2CO_2$

In the above reactions, the enzyme A and enzyme B respectively are
[NCERT : P L-295 | March 17, 2021 (II)]

(a) Amylase and Invertase (b) Invertase and Zymase
(c) Zymase and Invertase (d) Invertase and Amylase

138. Which of the following statements is not true about RNA? **[NCERT : P L-300 | April 12, 2019 (I)]**

(a) It controls the synthesis of protein.
(b) It has always double stranded helix structure.
(c) It usually does not replicate.
(d) It is present in the nucleus of the cell.

139. Among the following compounds, which one is found in RNA only? **[NCERT : P L-298 | Jan. 11, 2019 (I)]**

(a)
(b)
(c)
(d)

140. Which of the following is the correct structure of adenosine?

[NCERT : P L-298 | Online April 15, 2018 (I)]

(a) 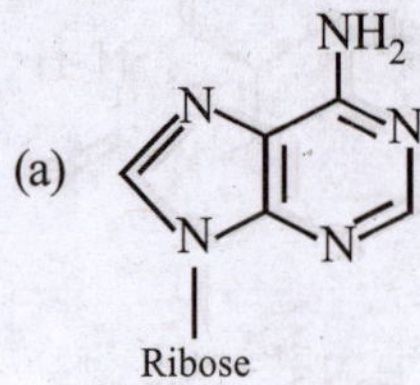

(b) 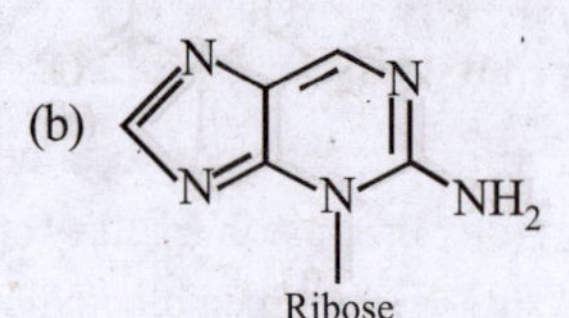

(c) 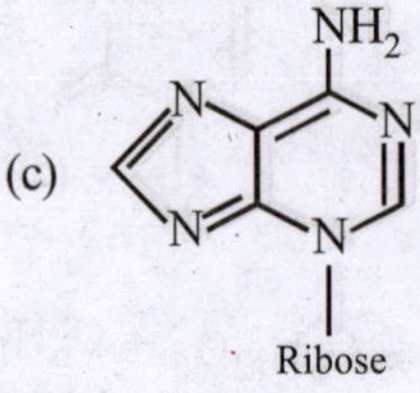

(d) 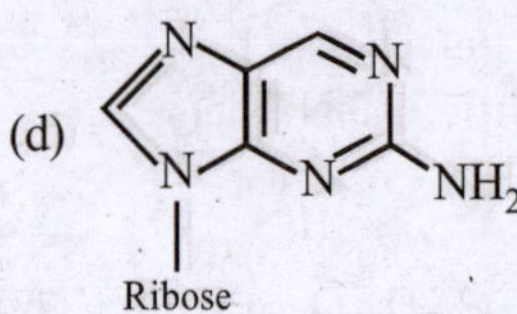

141. Thiol group is present in : [2016]

(a) Cysteine (b) Methionine

(c) Cytosine (d) Cystine

142. The reason for double helical structure of DNA is the operation of:

[NCERT : P L-299 | Online April 19, 2014; Same, 2003]

(a) Electrostatic attractions

(b) van der Waals forces

(c) Dipole - Dipole interactions

(d) Hydrogen bonding

143. Which one of the following bases is **not** present in DNA?

[NCERT : P L-298 | 2014]

(a) Quinoline (b) Adenine

(c) Cytosine (d) Thymine

144. Among the following organic acids, the acid present in rancid butter is: [Online April 19, 2014]

(a) Pyruvic acid (b) Lactic acid

(c) Butyric acid (d) Acetic acid

145. Which of the following enzyme converts starch into maltose? [NCERT : P L-295 | Online April 9, 2013]

(a) Diastase (b) Maltase

(c) Zymase (d) Invertase

146. Which of the following structures represents thymine ?

[NCERT : P L-298 | Online April 22, 2013]

(a) 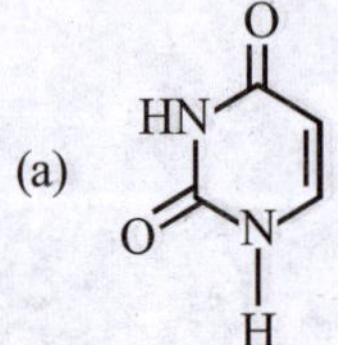

(b) 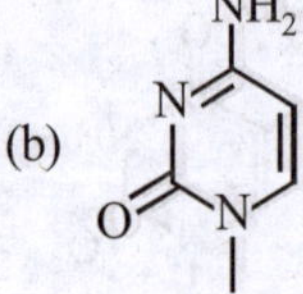

(c) 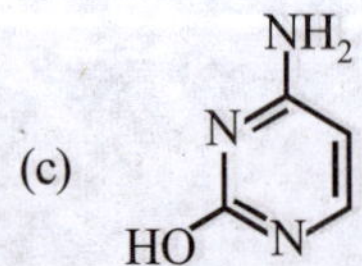

(d) 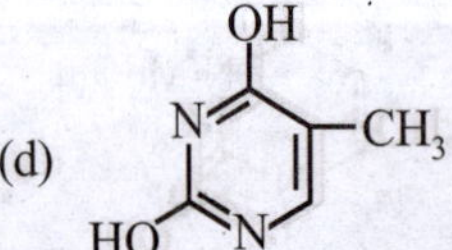

147. Which of the following statements is correct?

[NCERT : P L-300 | Online May 26, 2012]

(a) RNA controls the synthesis of proteins.

(b) The sugar present in DNA is D-(–)-ribose.

(c) RNA has double stranded α-helix structure.

(d) DNA mainly occurs in the cytoplasm of the cell.

148. The pyrimidine bases present in DNA are

[NCERT : P L-298 | 2006]

(a) cytosine and thymine

(b) cytosine and uracil

(c) cytosine and adenine

(d) cytosine and guanine

149. In both DNA and RNA, heterocyclic base and phosphate ester linkages are at – [NCERT : P L-297 | 2005]

(a) C_5' and C_1' respectively of the sugar molecule

(b) C_1' and C_5' respectively of the sugar molecule

(c) C_2' and C_5' respectively of the sugar molecule

(d) C_5' and C_2' respectively of the sugar molecule

150. Which base is present in RNA but not in DNA ?

[NCERT : P L-298 | 2004]

(a) Guanine (b) Cytosine

(c) Uracil (d) Thymine

151. RNA is different from DNA because RNA contains

[NCERT : P L-300 | 2002]

(a) ribose sugar and thymine

(b) ribose sugar and uracil

(c) deoxyribose sugar and thymine

(d) deoxyribose sugar and uracil.

Topic 4 - Vitamins and Hormones

152. Fat soluble vitamins are :

[NCERT : P L-296 | April 3, 2025 (II)]

A. Vitamin B_1 B. Vitamin C

C. Vitamin E D. Vitamin B_{12}

E. Vitamin K

Choose the **correct** answer from the options given below:

(a) C & D Only (b) A & B Only

(c) B & C Only (d) C & E Only

153. Which of the following acids is a vitamin ?

[NCERT : P L-296 | Jan. 22, 2025 (I)]

(a) Adipic acid (b) Aspartic acid

(c) Ascorbic acid (d) Saccharic acid

154. From the vitamins A, B_1, B_6, B_{12}, C, D, E and K, the number of vitamins that can be stored in our body is ______. **[NCERT : P L-296 | NV, Jan. 31, 2024 (II)]**

155. Match List I with List II **[NCERT : P L-296 | April 06, 2023 (I)]**

	List I (Vitamin)		List II (Deficiency disease)
A	Vitamin A	I.	Beri–beri
B	Thiamine	II.	Cheilosis
C	Ascorbic acid	III.	Xerophthalmia
D	Riboflavin	IV.	Scurvy

Choose the correct answer from the options given below:
(a) A–IV, B–II, C–III, D–I
(b) A–III, B–II, C–IV, D–I
(c) A–IV, B–I, C–III, D–II
(d) A–III, B–I, C–IV, D–II

156. Given below are two statements : **[NCERT : P L-300 | Jan. 24, 2023 (I)]**
Statement I : Noradrenaline is a neurotransmitter.
Statement II : Low level of noradrenaline is not the cause of depression in human.
In the light of the above statements, choose the correct answer from the options given below
(a) Statement I is correct but Statement II is incorrect
(b) Statement I is incorrect but Statement II is correct
(c) Both Statement I and Statement II are correct
(d) Both Statement I and Statement II are incorrect

157. Thiamine and pyridoxine are also known respectively as : **[NCERT : P L-296 | July 22, 2021 (II)]**
(a) Vitamin B_2 and Vitamin E
(b) Vitamin E and Vitamin B_2
(c) Vitamin B_6 and Vitamin B_2
(d) Vitamin B_1 and Vitamin B_6

158. Which one of the following is a water soluble vitamin, that is not excreted easily? **[NCERT : P L-296 | June 26, 2022 (II)]**
(a) Vitamin B_2 (b) Vitamin B_1
(c) Vitamin B_6 (d) Vitamin B_{12}

159. Deficiency of vitamin K causes : **[NCERT : P L-296 | March 18, 2021 (II)]**
(a) Cheilosis
(b) Increase in blood clotting time
(c) Increase in fragility of RBC's
(d) Decrease in blood clotting time

160. Match the following: **[NCERT : P L-296 | Jan. 07, 2020 (I)]**

(i)	Riboflavin	(a)	Beriberi
(ii)	Thiamine	(b)	Scurvy
(iii)	Pyridoxine	(c)	Cheilosis
(iv)	Ascorbic acid	(d)	Convulsions

(a) (i) – (a), (ii) – (d), (iii) – (c), (iv) – (b)
(b) (i) – (c), (ii) – (d), (iii) – (a), (iv) – (b)
(c) (i) – (c), (ii) – (a), (iii) – (d), (iv) – (b)
(d) (i) – (d), (ii) – (b), (iii) – (a), (iv) – (c)

161. Which of the vitamins given below is water soluble ? **[NCERT : P L-296 | 2015]**
(a) Vitamin E (b) Vitamin K
(c) Vitamin C (d) Vitamin D

162. Among the following vitamins the one whose deficiency causes rickets (bone deficiency) is **[NCERT : P L-296 | Online April 25, 2013]**
(a) Vitamin A (b) Vitamin B
(c) Vitamin D (d) Vitamin C

163. The change in the optical rotation of freshly prepared solution of glucose is known as: **[2011RS]**
(a) racemisation (b) specific rotation
(c) mutarotation (d) tautomerism

164. Insulin production and its action in human body are responsible for the level of diabetes. This compound belongs to which of the following categories ? **[NCERT : P L-300 | 2004]**
(a) An enzyme (b) A hormone
(c) A co-enzyme (d) An antibiotic

ANSWER KEYS																			
1	(b)	18	(b)	35	(d)	52	(b)	69	(d)	86	(b)	103	(a)	120	(a)	137	(b)	154	(5)
2	(b)	19	(2)	36	(a)	53	(d)	70	(d)	87	(6)	104	(4)	121	(b)	138	(b)	155	(d)
3	(c)	20	(a)	37	(9)	54	(d)	71	(a)	88	(d)	105	(c)	122	(33)	139	(b)	156	(a)
4	(b)	21	(b)	38	(c)	55	(b)	72	(c)	89	(b)	106	(2)	123	(a)	140	(a)	157	(d)
5	(c)	22	(d)	39	(b)	56	(a)	73	(4)	90	(8)	107	(5)	124	(c)	141	(a)	158	(d)
6	(a)	23	(d)	40	(c)	57	(c)	74	(9)	91	(b)	108	(37.84)	125	(c)	142	(d)	159	(b)
7	(b)	24	(a)	41	(b)	58	(c)	75	(a)	92	(1)	109	(b)	126	(b)	143	(a)	160	(c)
8	(a)	25	(a)	42	(d)	59	(b)	76	(6)	93	(a)	110	(a)	127	(3)	144	(c)	161	(c)
9	(b)	26	(a)	43	(c)	60	(c)	77	(d)	94	(c)	111	(b)	128	(a)	145	(a)	162	(c)
10	(a)	27	(a)	44	(b)	61	(c)	78	(a)	95	(2)	112	(c)	129	(a)	146	(d)	163	(c)
11	(a)	28	(b)	45	(d)	62	(c)	79	(d)	96	(4)	113	(c)	130	(25)	147	(a)	164	(b)
12	(a)	29	(b)	46	(b)	63	(c)	80	(c)	97	(3)	114	(a)	131	(a)	148	(a)		
13	(a)	30	(a)	47	(d)	64	(b)	81	(10)	98	(b)	115	(b)	132	(b)	149	(b)		
14	(c)	31	(c)	48	(d)	65	(a)	82	(c)	99	(4)	116	(b)	133	(b)	150	(c)		
15	(c)	32	(a)	49	(a)	66	(c)	83	(b)	100	(b)	117	(c)	134	(9)	151	(b)		
16	(a)	33	(c)	50	(d)	67	(b)	84	(c)	101	(c)	118	(c)	135	(d)	152	(d)		
17	(b)	34	(d)	51	(a)	68	(a)	85	(d)	102	(b)	119	(d)	136	(c)	153	(c)		

Principles Related to Practical Chemistry

TREND ANALYSIS
JEE MAIN (2025-2021)

	2025				2024				2023				2022				2021			
	Session II		Session I		Session II		Session I		Session II		Session I		Session II		Session I		Session II		Session I	
	MCQs	NVQs	MCQs	NVQs	MCQs	NVQs	MCQs	NVQs	MCQs	NVQs	MCQs	NVQs	MCQs	NVQs	MCQs	NVQs	MCQs	NVQs	MCQs	NVQs
Topic 1	1	–	4	–	5	1	7	1	1	–	4	–	–	1	3	4	4	2	1	1
Topic 2	3	–	4	1	8	1	8	4	6	–	8	–	2	–	5	–	2	1	2	–
Topic 3	3	1	2	–	2	3	1	–	3	3	3	–	5	3	–	3	–	2	–	2
Topic 4																				
Total	7	1	10	1	15	5	16	5	10	3	15	–	7	4	8	7	6	5	3	3

How to Code Qns for Better & Faster Revision

Type 1: Able to solve (No code Req)

Type 2: Solved with struggle or by seeing solution **(Use code NR means Needs Revision or orange stickers)**

Type 3: Tricky/ Could not solve on my own. Need help from Teacher/ Friend. **(Use code TQ means Tricky Question or red stickers)**

You can either hand write the code in boxes or else paste coloured stickers given at end of book.

Topic 1 – Detection of Extra Elements, Detection of Organic Functional Groups and Preparation of Organic Compounds

1. Match the LIST-I with LIST-II **[April 8, 2025 (II)]**

LIST-I (Reagent)	LIST-II (Functional Group detected)
A. Sodium bicarbonate solution	I. double bond/unsaturation
B. Neutral ferric chloride	II. carboxylic acid
C. Ceric ammonium nitrate	III. phenolic - OH
D. Alkaline $KMnO_4$	IV. alcoholic - OH

Choose the correct answer from the options given below:

(a) A-II, B-III, C-IV, D-I (b) A-II, B-III, C-I, D-IV

(c) A-III, B-II, C-IV, D-I (d) A-II, B-IV, C-III, D-I

2. The compounds that produce CO_2 with aqueous $NaHCO_3$ solution are : **[Jan. 28, 2025 (I)]**

(A) CO_2H (B) OH

(C) NO_2, OH, NO_2, NO_2 (D) CO_2H

(E) OH, H_3CO

Choose the correct answer from the options given below :

(a) A and C only (b) A, B and E only

(c) A, C and D only (d) A and B only

3. Given below are two statements : **[Jan. 23, 2025 (I)]**

Statement I : In Lassaigne's test, the covalent organic molecules are transformed into ionic compounds.

Statement II : The sodium fusion extract of an organic compound having N and S gives prussian blue colour with $FeSO_4$ and $Na_4[Fe(CN)_6]$

In the light of the above statements, choose the correct answer from the options given below

(a) Both Statement I and Statement II are true
(b) Both Statement I and Statement II are false
(c) Statement I is false but Statement II is true
(d) Statement I is true but Statement II is false

4. Given below are two statement: **[Jan. 22, 2025 (II)]**

Statement (I): Nitrogen, sulphur, halogen and phosphorus present in a organic compound are detected by Lassaigne's Test.

Statement (II): The elements present in the compound are converted from covalent form into ionic form by fusing the compound with Magnesium in Lassaigne's test.

In the light of the above statements, choose the correct answer from the options given below :

(a) Both Statement I and Statement II are true
(b) Both Statement I and Statement II are false
(c) Statement I is true but Statement II is false
(d) Statement I is false but Statement II is true

5. The compounds which give positive Fehling's test are : **[Jan. 22, 2025 (I)]**

(A) CHO (benzaldehyde structure)

(B) O, CH_3 (acetophenone structure)

(C) $HOCH_2–CO–(CHOH)_3–CH_2–OH$

(D) $CH_3-\overset{\overset{O}{||}}{C}-H$

(E) CHO (phenylacetaldehyde structure)

Choose the CORRECT answer from the options given below :

(a) (A),(C) and (D) Only
(b) (A),(D) and (E) Only
(c) (C), (D) and (E) Only
(d) (A), (B) and (C) Only

6. Match **List-I** with **List-II**. **[April 9, 2024 (II)]**

List-I (Test)	List-II (Observation)
(A) Br_2 water test	(I) Yellow orange or orange red precipitate formed
(B) Ceric ammonium nitrate test	(II) Reddish orange colour disappears
(C) Ferric chloride test	(III) Red colour appears
(D) 2, 4-DNP test	(IV) Blue, Green, Violet or Red colour appear

Choose the correct answer from the options given below:

(a) A-(I), B-(II), C-(III), D-(IV)
(b) A-(II), B-(III), C-(IV), D-(I)
(c) A-(III), B-(IV), C-(I), D-(II)
(d) A-(IV), B-(I), C-(II), D-(III)

7. Which of the following compounds will give silver mirror with ammoniacal silver nitrate? **[April 9, 2024 (II)]**

(A) Formic acid
(B) Formaldehyde
(C) Benzaldehyde
(D) Acetone

Choose the correct answer from the options given below:

(a) C and D only
(b) A, B and C only
(c) A only
(d) B and C only

8. Match **List-I** with **List-II**. **[April 8, 2024 (II)]**

List-I (Test)	List-II (Identification)
(A) Bayer's test	(I) Phenol
(B) Ceric ammonium nitrate test	(II) Aldehyde
(C) Phthalein dye test	(III) Alcoholic-OH group
(D) Schiff's test	(IV) Unsaturation

Choose the **correct** answer from the options given below:

(a) (A)-(III), (B)-(I), (C)-(IV), (D)-(II)
(b) (A)-(II), (B)-(III), (C)-(IV), (D)-(I)
(c) (A)-(IV), (B)-(I), (C)-(II), (D)-(III)
(d) (A)-(IV), (B)-(III), (C)-(I), (D)-(II)

9. Which of the following gives a positive test with ninhydrin ? **[April 5, 2024 (I)]**

(a) Cellulose
(b) Starch
(c) Polyvinyl chloride
(d) Egg albumin

10. Which of the following nitrogen containing compound does not give Lassaigne's test ? **[April 4, 2024 (I)]**

(a) Phenyl hydrazine
(b) Glycene
(c) Urea
(d) Hydrazine

11. From 6.55 g of aniline, the maximum amount of acetanilide that can be prepared will be _____ $\times 10^{-1}$ g. **[April 4, 2024 (II)]**

12. Lassaigne's test is used for detection of: **[Feb. 1, 2024 (II)]**

(a) Phosphorous and halogens only
(b) Nitrogen, Sulphur and Phosphorous only
(c) Nitrogen, Sulphur and phosphorous and halogens
(d) Nitrogen and Sulphur only

13. The azo-dye (Y) formed in the following reactions is

Sulphanilic acid + $NaNO_2$ + $CH_3COOH \rightarrow X$ **[Jan. 31, 2024 (II)]**

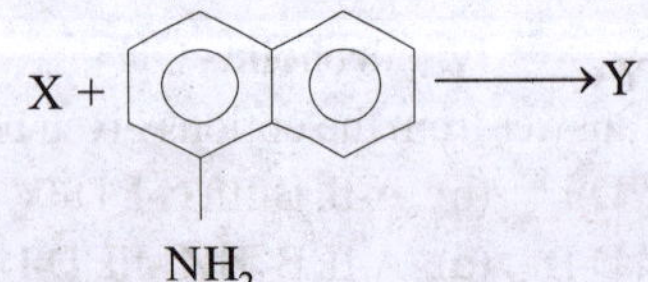

(a) HSO_3–C₆H₄–N=N–(naphthalene)–N=N–C₆H₄–SO_3H

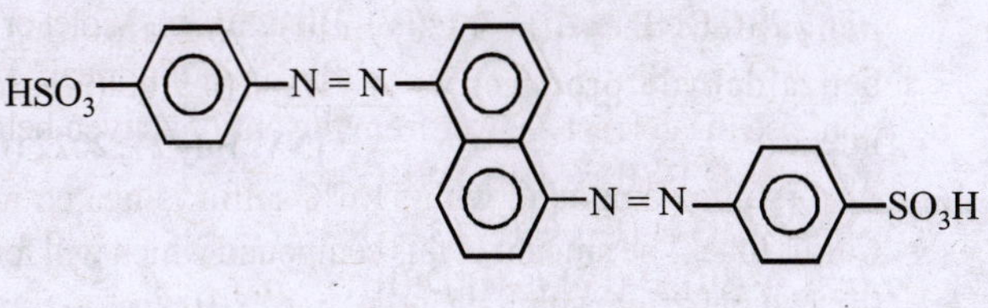

(b) HO_3S–(naphthalene)–N=N–(naphthalene)–NH_2; HO_3S–(naphthalene)–N=N–(naphthalene)

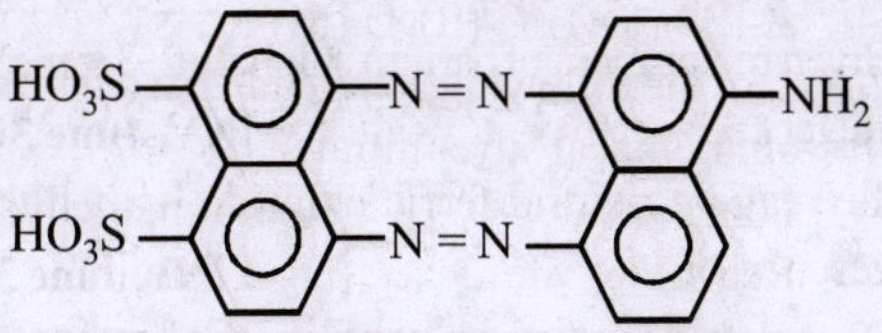

(c) HO_3S–C₆H₄–N=N–(naphthalene)–NH_2

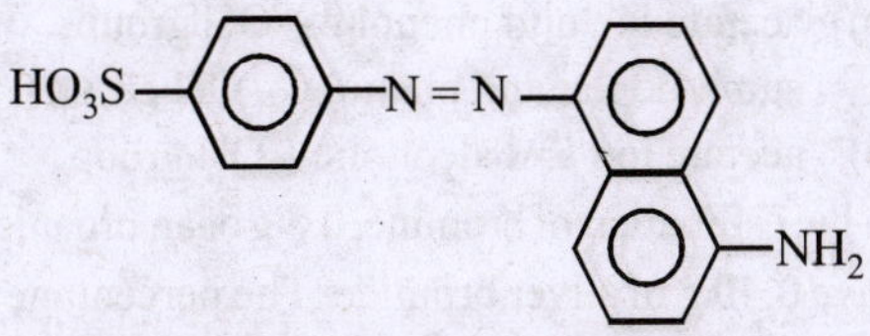

(d) HO_3S–C₆H₄–N=N–(naphthalene)–NH_2

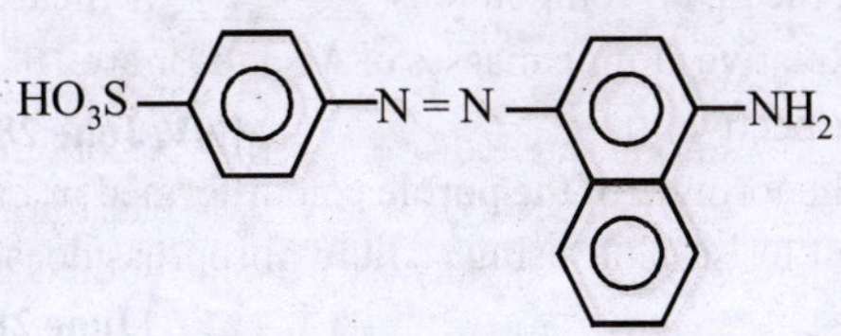

14. Given below are two statements : **[Jan. 31, 2024 (II)]**

Statement (I) : Aniline reacts with con. H_2SO_4 followed by heating at 453-473 K gives p-aminobenzene sulphonic acid, which gives blood red colour in the 'Lassaigne's test'.

Statement (II) : In Friedel - Craft's alkylation and acylation reactions, aniline forms salt with the $AlCl_3$ catalyst. Due to this, nitrogen of aniline aquires a positive charge and acts as deactivating group.

In the light of the above statements, choose the **correct answer** from the options given below :

(a) Statement I is false but Statement II is true.
(b) Both Statement I and Statement II are false.
(c) Statement I is true but Statement II is false.
(d) Both Statement I and Statement II are true.

15. Following is a confirmatory test for aromatic primary amines. Identify reagent (A) and (B). **[Jan. 30, 2024 (I)]**

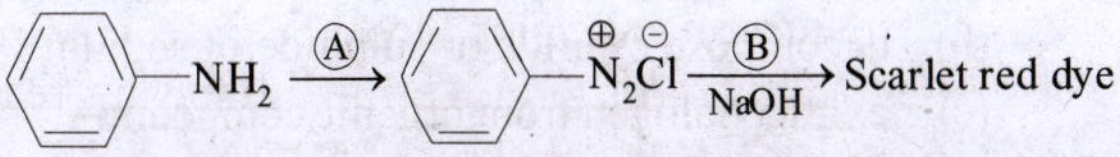

$$C_6H_5NH_2 \xrightarrow{(A)} C_6H_5\overset{\oplus}{N_2}\overset{\ominus}{Cl} \xrightarrow[NaOH]{(B)} \text{Scarlet red dye}$$

(a) $A = HNO_3/H_2SO_4$; B = phenol (C₆H₅OH)

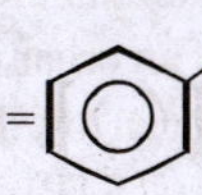

(b) $A = NaNO_2 + HCl$, 0-5°C; B = aniline (C₆H₅NH₂)

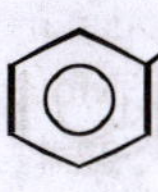

(c) $A = NaNO_2 + HCl$, 0–5°C; B = 1-naphthol (naphthalene–OH)

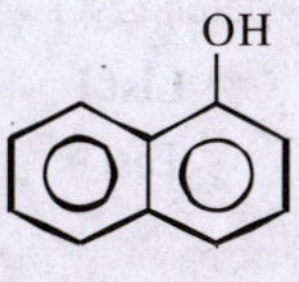

(d) $A = NaNO_2 + HCl$, 0–5°C; B = 2-naphthol (naphthalene–OH)

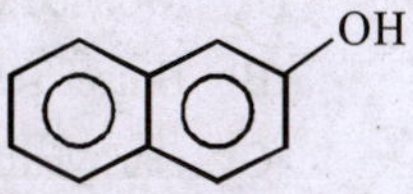

16. The Lassiagne's extract is boiled with dil HNO_3 before testing for halogens because, **[Jan. 30, 2024 (I)]**

(a) AgCN is soluble in HNO_3.
(b) Silver halides are soluble in HNO_3.
(c) Ag_2S is soluble in HNO_3.
(d) Na_2S and NaCN are decomposed by HNO_3.

17. Appearance of blood red colour, on treatment of the sodium fusion extract of an organic compound with $FeSO_4$ in presence of concentrated H_2SO_4 indicates the present of element/s **[Jan. 29, 2024 (I)]**

(a) N and S (b) N
(c) S (d) Br

18. From the compounds given below, number of compounds which give positive Fehling's test is ______.

Benzaldehyde, Acetaldehyde, Acetone, Acetophenone, Methanal, 4-nitrobenzaldehyde, cyclohexane carbaldehyde. **[NV, Jan. 29, 2024 (I)]**

19. Phenolic group can be identified by a positive: **[Jan. 27, 2024 (II)]**

(a) Phthalein dye test (b) Lucas test
(c) Tollen's test (d) Carbylamine test

20. Match List-I with List-II : **[April 08, 2023 (I)]**

List-I (Reagents used)		List-II (Compound with functional group detected)
A. Alkaline solution of copper sulphate and sodium citrate	I. 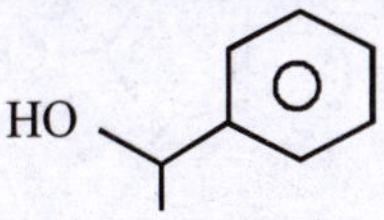	HO–CH(CH₃)–C₆H₅
B. Neutral $FeCl_3$ solution	II. 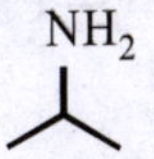	(CH₃)₂CH–NH₂
C. Alkaline chloroform solution	III. 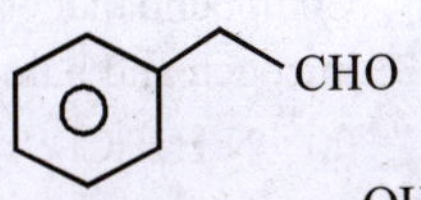	C₆H₅–CH₂–CHO
D. Potassium iodide and sodium hypochlorite	IV. 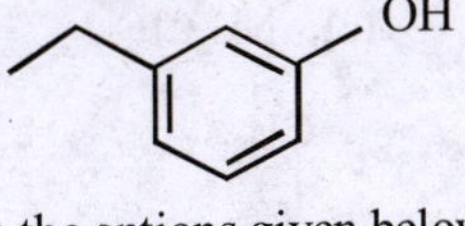	CH₃CH₂–C₆H₄–OH

Choose the correct answer from the options given below:

(a) A-II, B-IV, C-III, D-I (b) A-IV, B-I, C-II, D-III
(c) A-III, B-IV, C-I, D-II (d) A-III, B-IV, C-II, D-I

21. Match List I with List II: **[Feb. 01, 2023 (I)]**

List I Test	List II Functional group / Class of Compound
(A) Molisch's Test	(I) Peptide
(B) Biuret Test	(II) Carbohydrate
(C) Carbylamine Test	(III) Primary amine
(D) Schiff's Test	(IV) Aldehyde

Choose the **correct** answer from the options given below:

(a) (A) – (I), (B) – (II), (C) – (III), (D) – (IV)
(b) (A) – (II), (B) – (I), (C) – (III), (D) – (IV)
(c) (A) – (III), (B) – (IV), (C) – (II), (D) – (I)
(d) (A) – (III), (B) – (IV), (C) – (I), (D) – (II)

22. Given below are two statements: **[Feb. 01, 2023 (II)]**

Statement I : Sulphanilic acid gives esterification test for carboxyl group.

Statement II: Sulphanilic acid gives red colour in Lassigne's test for extra element detection.

In the light of the above statements, choose the **most appropriate** answer from the options given below.

(a) **Statement I** is incorrect but **Statement II** is correct
(b) **Statement I** is correct but **Statement II** is incorrect
(c) Both **Statement I** and **Statement II** are correct
(d) Both **Statement I** and **Statement II** are incorrect

23. Which of the following compounds would give the following set of qualitative analysis ? **[Jan. 30, 2023 (I)]**

(i) Fehling's Test : Positive

(ii) Na fusion extract upon treatment with sodium nitroprusside gives a blood red colour.

(a) [imidazole-2-CHO] (b) [oxazole-2-CHO]

(c) [thiazole-2-CHO] (d) [thiazole-2-CH_2CH_2CHO]

24. Compound that will give positive Lassaigne's test for both nitrogen and halogen is **[Jan. 29, 2023 (I)]**

(a) $N_2H_4.HCl$ (b) $CH_3NH_2.HCl$
(c) NH_4Cl (d) $NH_2OH.HCl$

25. Toluene (CH_3) 5g → Benzaldehyde (CHO) (92%)

In the above reaction, 5 g of toluene is converted into benzaldehyde with 92% yield. The amount of benzaldehyde produced is______$\times 10^{-2}$ g. (Nearest integer) **[NV, July 27, 2022 (II)]**

26. An organic compound with 51.6% sulfur is heated in a Carius tube. The amount of this compound which will form 0.752 g of barium sulphate is ______ $\times 10^{-1}$g. (Given molar mass of barium sulphate 233 g mol^{-1}) (Nearest integer). **[NV, June 30, 2022 (I)]**

27. The reagent neutral ferric chloride is used to detect the presence of ______ . **[NV, June 30, 2022 (I)]**

(a) sulphide ion and alcoholic – OH group.
(b) acetate ion and phenolic – OH group.
(c) sulphide ion and phenolic – OH group.
(d) acetate ion and alcoholic – OH group.

28. In the estimation of bromine, 0.5 g of an organic compound gave 0.40 g of silver bromide. The percentage of bromine in the given compound is ________% (nearest integer) (Relative atomic masses of Ag and Br are 108u and 80u, respectively). **[NV, June 28, 2022 (I)]**

29. The formula of the purple colour formed in Laissaigne's test for sulphur using sodium nitroprusside is **[June 28, 2022 (I)]**

(a) $NaFe[Fe(CN)_6]$ (b) $Na[Cr(NH_3)_2(NCS)_4]$
(c) $Na_2[Fe(CN)_5(NO)]$ (d) $Na_4[Fe(CN)_5(NOS)]$

30. Given below are two statements : **[June 26, 2022 (I)]**

Statement I : In 'Lassaigne's test, when both nitrogen and sulphur are present in an organic compound, sodium thiocyanate is formed.

Statement II : If both nitrogen and sulphur are present in an organic compound, then the excess of sodium used in sodium fusion will decompose the sodium thiocyanate formed to give NaCN and Na_2S.

In the light of the above statements, choose the **most appropriate** answer from the options given below :

(a) Both Statement I and Statement II are correct.
(b) Both Statement I and Statement II are incorrect.
(c) Statement I is correct but Statement II is incorrect.
(d) Statement I is incorrect but Statement II is correct.

31. During halogen test, sodium fusion extract is boiled with concentrated HNO_3 to **[June 25, 2022 (II)]**

(a) remove unreacted sodium
(b) decompose cyanide or sulphide of sodium
(c) extract halogen from organic compound
(d) maintain the pH of extract

32. 0.2 g of an organic compound was subjected to estimation of nitrogen by Dumas method in which volume of N_2 evolved (at STP) was found to be 22.400 mL. The percentage of nitrogen in the compound is____.[nearest integer] (Given: Molar mass of N_2 is 28 mol^{-1}. Molar volume of N_2 at STP : 22.4 L) **[NV, June 24, 2022 (II)]**

33. In Carius method for estimation of halogens, 0.2 g of an organic compound gave 0.188 g of AgBr. The percentage of bromine in the compound is ________. (Nearest integer)
[Atomic mass : Ag = 108, Br = 80] **[NV, Aug. 27, 2021 (I)]**

34. Which one of the following set of elements can be detected using sodium fusion extract? **[July 27, 2021 (II)]**
(a) Sulfur, Nitrogen, Phosphorus, Halogens
(b) Phosphorus, Oxygen, Nitrogen, Halogens
(c) Nitrogen, Phosphorus, Carbon, Sulfur
(d) Halogens, Nitrogen, Oxygen, Sulfur

35. 0.8 g of an organic compound was analysed by Kjeldahl's method for the estimation of nitrogen. If the percentage of nitrogen in the compound was found to be 42%, then ______ mL of 1 M H_2SO_4 would have been neutralized by the ammonia evolved during the analysis. **[NV, July 25, 2021 (II)]**

36. Methylation of 10 g of benzene gave 9.2 g of toluene. Calculate the percentage yield of toluene ______.(Nearest integer) **[NV, July 22, 2021 (II)]**

37. When 0.15 g of an organic compound was analyzed using Carius method for estimation of bromine, 0.2397 g of AgBr was obtained. The percentage of bromine in the organic compound is ______. (Nearest integer)
[Atomic mass : Silver = 108, Bromine = 80] **[NV, July 20, 2021 (II)]**

38. Which of the following compound is added to the sodium extract before addition of silver nitrate for testing of halogens? **[Feb. 25, 2021 (II)]**
(a) Hydrochloric acid
(b) Sodium hydroxide
(c) Ammonia
(d) Nitric acid

39. 1.86 g of aniline completely reacts to form acetanilide. 10% of the product is lost during purification. Amount of acetanilide obtained after purification (in g) is ____ $\times 10^{-2}$. **[NV, Feb. 24, 2021 (II)]**

40. Consider the following reactions:

'A' (C_7H_{14}) $\xrightarrow{\text{ozonolysis}}$ 'B' + 'C'

'B' $\xrightarrow[\Delta]{(I_2+NaOH)}$ yellow ppt
'B' $\xrightarrow[\Delta]{Ag_2O}$ silver mirror

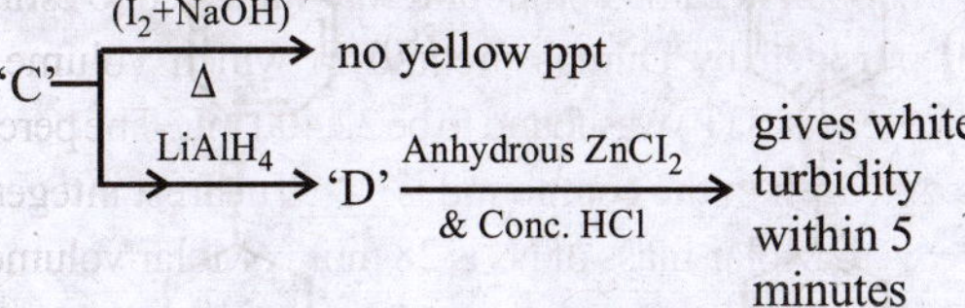

'A' is : **[Sep. 06, 2020 (I)]**

(a) (b)

(c) (d)

41. Match the following : **[Sep. 06, 2020 (II)]**

Column-I (Test / Method)	Column-II (Reagent)
(I) Lucas Test	(A) $C_6H_5SO_2Cl$ / aq. KOH
(II) Dumas method	(B) HNO_3 / $AgNO_3$
(III) Kjeldahl's method	(C) CuO / CO_2
(IV) Hinsberg test	(D) Conc. HCl and $ZnCl_2$
	(E) H_2SO_4

(a) (I)-(D), (II)-(C), (III)-(B), (IV)-(A)
(b) (I)-(B), (II)-(D), (III)-(E), (IV)-(A)
(c) (I)-(D), (II)-(C), (III)-(E), (IV)-(A)
(d) (I)-(B), (II)-(A), (III)-(C), (IV)-(D)

42. Two compounds A and B with same molecular formula (C_3H_6O) undergo Grignard's reaction with methyl-magnesium bromide to give products C and D. Products C and D show following chemical tests.

Test	C	D
Ceric ammonium nitrate test	Positive	Positive
Lucas test	Turbidity obtained after five minutes	Turbidity obtained immediately
Iodoform test	Positive	Negative

C and D respectively are : **[Sep. 02, 2020 (II)]**

(a) $C = H_3C-CH_2-\underset{}{\overset{OH}{\overset{|}{C}H}}-CH_3$;
$D = H_3C-\overset{CH_3}{\overset{|}{\underset{CH_3}{\underset{|}{C}}}}-OH$

(b) $C = H_3C-CH_2-CH_2-CH_2-OH$;
$D = H_3C-CH_2-\underset{OH}{\underset{|}{C}H}-CH_3$

(c) $C = H_3C-CH_2-CH_2-CH_2-OH$;
$D = H_3C-\overset{CH_3}{\overset{|}{\underset{CH_3}{\underset{|}{C}}}}-OH$

(d) $C = H_3C - \underset{CH_3}{\overset{CH_3}{C}} - OH;$

$D = H_3C - CH_2 - \underset{OH}{CH} - CH_3$

43. A, B and C are three biomolecules. The results of the tests performed on them are given below:

	Molisch's Test	Barfoed Test	Biuret Test
A	Positive	Negative	Negative
B	Positive	Positive	Negative
C	Negative	Negative	Positive

A, B and C are respectively: **[Jan. 09, 2020 (II)]**

(a) A = Glucose, B = Fructose, C = Albumin
(b) A = Lactose, B = Glucose, C = Albumin
(c) A = Lactose, B = Glucose, C = Alanine
(d) A = Lactose, B = Fructose, C = Alanine

44. A solution of *m*-chloroaniline, *m*-chlorophenol and *m*-chlorobenzoic acid in ethyl acetate was extracted initially with a saturated solution of $NaHCO_3$ to give fraction *A*. The left over organic phase was extracted with dilute NaOH solution to give fraction *B*. The final organic layer was labelled as fraction *C*. Fractions *A*, *B* and *C*, contain respectively: **[Jan. 07, 2020 (I)]**

(a) *m*-chlorobenzoic acid, *m*-chloroaniline and *m*-chlorophenol
(b) *m*-chlorobenzoic acid, *m*-chlorophenol and *m*-chloroaniline
(c) *m*-chlorophenol, *m*-chlorobenzoic acid and *m*-chloroaniline
(d) *m*-chloroaniline, *m*-chlorobenzoic acid and *m*-chlorophenol

45. Consider the following reaction: **[Jan. 07, 2020 (I)]**

$C_6H_5-N(CH_3)_2 + \overset{\oplus}{Na}\ \overset{\ominus}{SO_3}-C_6H_4-\overset{\oplus}{N_2}\overset{\ominus}{Cl} \xrightarrow{OH^-}$ '*X*'

The product '*X*' is used:

(a) in protein estimation as an alternative to ninhydrin
(b) in acid base titration as an indicator
(c) as food grade colourant
(d) in laboratory test for phenols

46. An organic compound 'A' is oxidized with Na_2O_2 followed by boiling with HNO_3. The resultant solution is then treated with ammonium molybdate to yield a yellow precipitate. Based on above observation, the element present in the given compound is: **[April 12, 2019 (I)]**

(a) Nitrogen (b) Phosphorus
(c) Fluorine (d) Sulphur

47. Which one of the following is likely to give a precipitate with $AgNO_3$ solution ? **[April 12, 2019 (II)]**

(a) $CH_2=CH-Cl$ (b) CCl_4
(c) $CHCl_3$ (d) $(CH_3)_3CCl$

48. The organic compound that gives following qualitative analysis is: **[April 9, 2019 (I)]**

Test	Inference
(A) Dil. HCl	Insoluble
(B) NaOH solution	Soluble
(C) Br_2/water	Decolourization

(a) [benzene ring]–OH (b) [benzene ring]–NH_2
(c) [cyclohexane ring]–NH_2 (d) [cyclohexane ring]–OH

49. An organic compound 'X' showing the following solubility profile is:

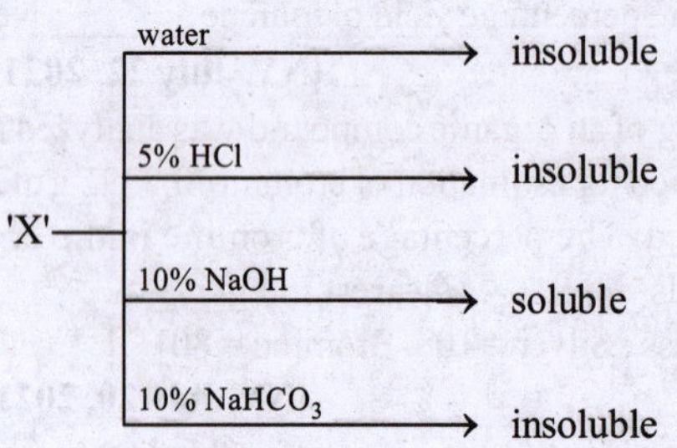

[April 8, 2019 (I)]

(a) *o*-Toluidine
(b) Oleic acid
(c) *m*-Cresol
(d) Benzamide

50. The correct match between **Item-I** and **Item-II** is: **[Jan. 11, 2019 (II)]**

Item-I	Item-II
(A) Ester test	(P) Tyr
(B) Carbylamine test	(Q) Asp
(C) Phthalein dye test	(R) Ser
	(S) Lys

(a) (A) → (Q, R); (B) → (S); (C) → (P)
(b) (A) → (R); (B) → (Q); (C) → (P)
(c) (A) → (R); (B) → (S); (C) → (Q)
(d) (A) → (Q); (B) → (S); (C) → (R)

51. Which of the following compounds will produce a precipitate with $AgNO_3$? **[Jan. 11, 2019 (II)]**

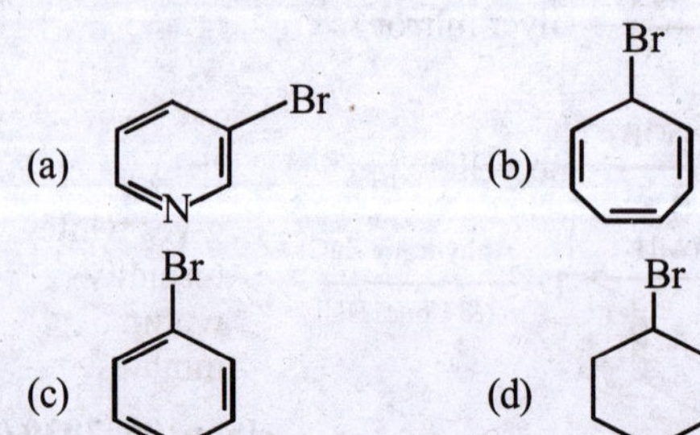

52. The correct match between **Item-I** and **Item-II** is: **[Jan. 10, 2019 (II)]**

Item-I (Compound)	Item-II (Reagent)
(A) Lysine	(P) 1-Naphthol
(B) Furfural	(Q) Ninhydrin
(C) Benzyl alcohol	(R) $KMnO_4$
(D) Styrene	(S) Ceric ammonium nitrate

(a) (A)→(Q); (B)→(P); (C)→(S); (D)→(R)
(b) (A)→(Q); (B)→(P); (C)→(R); (D)→(S)
(c) (A)→ (R); (B)→(P); (C)→(Q) : (D)→(S)
(d) (A)→(Q); (B)→(R); (C)→(S); (D)→(P)

53. Which of the following tests cannot be used for identifying amino acids? **[Jan. 10, 2019 (II)]**
(a) Biuret test (b) Barfoed test
(c) Ninhydrin test (d) Xanthoproteic test

54. The correct match between **Item-I** and **Item-II** is: **[Jan. 9, 2019 (I)]**

Item-I (Drug)	Item-II (Test)
(A) Chloroxylenol	(P) Carbylamine test
(B) Norethindrone	(Q) Sodium hydrogen carbonate test
(C) Sulphapyridine	(R) Ferric chloride test
(D) Penicillin	(S) Baeyer's test

(a) A→R ; B→P ; C→S ; D→Q
(b) A→Q ; B→S ; C→P ; D→R
(c) A→R ; B→S ; C→P ; D→Q
(d) A→Q ; B→P ; C→S ; D→R

55. The tests performed on compound X and their inferences are:

Test	Inference
(A) 2, 4 - DNP test	Coloured precipitate
(B) Iodoform test	Yellow precipitate
(C) Azodye test	No dye formation

Compound 'X' is: **[Jan. 9, 2019 (II)]**

(a) 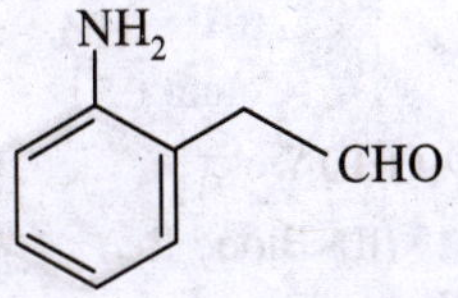 [benzene ring with NH_2 and CH_2CHO on adjacent carbons]

(b) 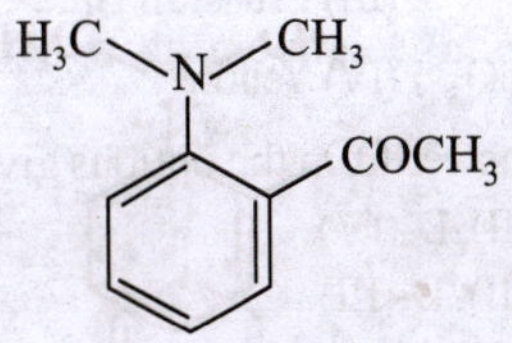 [benzene ring with $N(CH_3)_2$ (H_3C–N–CH_3) and $COCH_3$ on adjacent carbons]

(c) 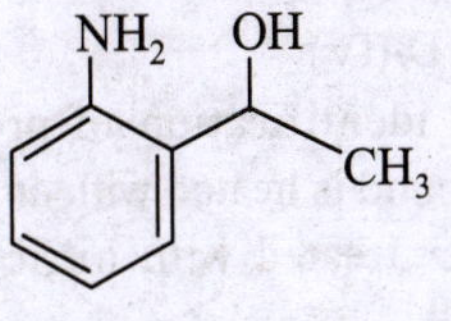 [benzene ring with NH_2 and $CH(OH)CH_3$ on adjacent carbons]

(d) 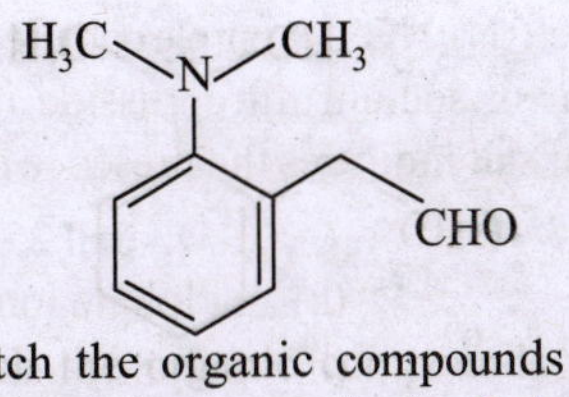 [benzene ring with $N(CH_3)_2$ (H_3C–N–CH_3) and CH_2CHO on adjacent carbons]

56. Match the organic compounds in **Column-I** with the Lassaigne's test results in **Column-II** appropriately : **[Online April 11, 2015]**

Column-I	Column-II
(A) Aniline	(i) Red colour with $FeCl_3$
(B) Benzene sulfonic acid	(ii) Violet colour with sodium nitroprusside
(C) Thiourea	(iii) Blue colour with hot and acidic solution of $FeSO_4$

(a) A-(ii); B-(iii); C-(i) (b) A-(iii); B-(i); C-(ii)
(c) A-(iii); B-(ii); C-(i) (d) A-(ii); B-(i); C-(iii)

57. Copper wire test for halogens is known as **[Online May 7, 2012]**
(a) Duma's Test (b) Beilstein's Test
(c) Liebig's Test (d) Lassigne's Test

58. Beilstein test is used for the estimation of which one of the following elements? **[Online May 19, 2012]**
(a) S (b) Cl (c) C and H (d) N

59. 29.5 mg of an organic compound containing nitrogen was digested according to Kjeldahl's method and the evolved ammonia was absorbed in 20 mL of 0.1 M HCl solution. The excess of the acid required 15 mL of 0.1 M NaOH solution for complete neutralization. The percentage of nitrogen in the compound is **[2010]**
(a) 59.0 (b) 47.4 (c) 23.7 (d) 29.5

60. An organic compound having molecular mass 60 is found to contain C = 20%, H = 6.67% and N = 46.67% while rest is oxygen. On heating it gives NH_3 along with a solid residue. The solid residue gives violet colour with alkaline copper sulphate solution. The compound is **[2005]**
(a) $CH_3CH_2CONH_2$ (b) $(NH_2)_2CO$
(c) CH_3CONH_2 (d) CH_3NCO

61. The compound formed in the positive test for nitrogen with the Lassaigne solution of an organic compound is **[2004]**
(a) $Fe_4[Fe(CN)_6]_3$ (b) $Na_3[Fe(CN)_6]$
(c) $Fe(CN)_3$ (d) $Na_4[Fe(CN)_5NOS]$

Topic 2 - Qualitative Salt Analysis and Preparation of Inorganic Compounds

62. When a salt is treated with sodium hydroxide solution it gives gas X. On passing gas X through reagent Y a brown coloured precipitate is formed. **[April 7, 2025 (I)]**
X and Y respectively, are
(a) $X = NH_3$ and $Y = HgO$
(b) $X = NH_3$ and $Y = K_2HgI_4 + KOH$
(c) $X = NH_4Cl$ and $Y = KOH$
(d) $X = HCl$ and $Y = NH_4Cl$

63. Formation of $Na_4[Fe(CN)_5NOS]$, a purple coloured complex formed by addition of sodium nitroprusside in sodium carbonate extract of salt indicates the presence of : **[April 2, 2025 (II)]**

(a) Sodium ion (b) Sulphate ion
(c) Sulphide ion (d) Sulphite ion

64. Choose the correct tests with respective observations. **[April 2, 2025 (I)]**

(A) $CuSO_4$ (acidified with acetic acid) + $K_4[Fe(CN)_6] \rightarrow$ Chocolate brown precipitate.
(B) $FeCl_3 + K_4[Fe(CN)_6] \rightarrow$ Prussian blue precipitate.
(C) $ZnCl_2 + K_4[Fe(CN)_6]$, neutralised with $NH_4OH \rightarrow$ White or bluish white precipitate.
(D) $MgCl_2 + K_4[Fe(CN)_6] \rightarrow$ Blue precipitate.
(E) $BaCl_2 + K_4[Fe(CN)_6]$, neutralised with NaOH $\rightarrow$ White precipitate.

Choose the **correct** answer from the options given below :
(a) A, D and E only (b) B, D and E only
(c) A, B and C only (d) C, D and E only

65. Identify the inorganic sulphides that are yellow in colour :
(A) $(NH_4)_2S$ (B) PbS (C) CuS
(D) As_2S_3 (E) As_2S_5 **[Jan. 28, 2025 (II)]**

Choose the correct answer from the options given below :
(a) (A) and (C) only (b) (A), (D) and (E) only
(c) (A) and (B) only (d) (D) and (E) only

66. The metal ion whose electronic configuration is not affected by the nature of the ligand and which gives a violet colour in non-luminous flame under hot condition in borax bead test is **[Jan. 28, 2025 (I)]**

(a) Ti^{3+} (b) Ni^{2+}
(c) Mn^{2+} (d) Cr^{3+}

67. Find the compound 'A' from the following reaction sequences. **[Jan. 24, 2025 (II)]**

$$A \xrightarrow{\text{aqua-regia}} B \xrightarrow[\text{(2)AcOH}]{\text{(1)}KNO_2|NH_4OH} \text{yellow ppt}$$

(a) ZnS (b) CoS
(c) MnS (d) NiS

68. Among the following cations, the number of cations which will give characteristic precipitate in their identification tests with $K_4[Fe(CN)_6]$ is :
$Cu^{2+}, Fe^{3+}, Ba^{2+}, Ca^{2+}, NH_4^+, Mg^{2+}, Zn^{2+}$. **[NV, Jan. 24, 2025 (I)]**

69. Identify A, B and C in the given below reaction sequence **[Jan. 23, 2025 (II)]**

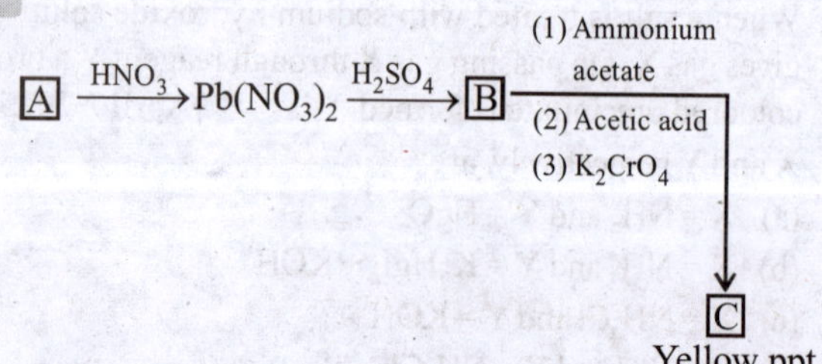

(a) $PbCl_2, PbSO_4, PbCrO_4$
(b) $PbS, PbSO_4, PbCrO_4$
(c) $PbS, PbSO_4, Pb(CH_3COO)_2$
(d) $PbCl_2, Pb(SO_4)_2, PbCrO_4$

70. On reaction of Lead Sulphide with dilute nitric acid which of the following is **not** formed ? **[April 9, 2024 (I)]**

(a) Lead nitrate (b) Sulphur
(c) Nitric oxide (d) Nitrous oxide

71. Consider the following test for a group-IV cation.

$M^{2+} + H_2S \rightarrow A$ (Black precipitate) + by product
A + aqua regia $\rightarrow B + NOCl + S + H_2O$
$B + KNO_2 + CH_3COOH \rightarrow C$ + by product

The spin only magnetic moment value of the metal complex C is ______ BM.
(Nearest integer) **[NV, April 9, 2024 (II)]**

72. Match **List-I** with **List-II** : **[April 8, 2024 (I)]**

List-I (Name of the test)		List-II (Reaction sequence involved [M is metal])
(A) Borax bead test	(I)	$MCO_3 \rightarrow MO \xrightarrow[\Delta]{Co(NO_3)_2} CoO.MO$
(B) Charcoal cavity test	(II)	$MCO_3 \rightarrow MCl_2 \rightarrow M^{2+}$
(C) Cobalt nitrate test	(III)	$MSO_4 \xrightarrow[\Delta]{Na_2B_4O_7} M(BO_2)_2 \rightarrow MBO_2 \rightarrow M$
(D) Flame test	(IV)	$MSO_4 \xrightarrow[\Delta]{Na_2CO_3} MCO_3 \rightarrow MO \rightarrow M$

Choose the **correct** answer from the options below:
(a) A-(III), B-(I), C-(IV), D-(II)
(b) A-(III), B-(II), C-(IV), D-(I)
(c) A-(III), B-(I), C-(II), D-(IV)
(d) A-(III), B-(IV), C-(I), D-(II)

73. Match **List-I** with **List-II** : **[April 8, 2024 (I)]**

List-I (Compound)	List-II (Colour)
(A) $Fe_4[Fe(CN)_6]_3.xH_2O$	(I) Violet
(B) $[Fe(CN)_5NOS]^{4-}$	(II) Blood Red
(C) $[Fe(SCN)]^{2+}$	(III) Prussian Blue
(D) $(NH_4)_3PO_4.12MoO_3$	(IV) Yellow

Choose the **correct** answer from the options given below:
(a) A-(III), B-(I), C-(II), D-(IV)
(b) A-(IV), B-(I), C-(II), D-(III)
(c) A-(II), B-(III), C-(IV), D-(I)
(d) A-(I), B-(II), C-(III), D-(IV)

74. In qualitative test for identification of presence of phosphorous, the compound is heated with an oxidising agent. Which is further treated with nitric acid and ammonium molybdate respectively. The yellow coloured precipitate obtained is : [April 8, 2024 (II)]

(a) $Na_3PO_4 \cdot 12MoO_3$
(b) $(NH_4)_3PO_4 \cdot 12(NH_4)_2MoO_4$
(c) $(NH_4)_3PO_4 \cdot 12MoO_3$
(d) $MoPO_4 \cdot 21NH_4NO_3$

75. Match **List-I** with **List-II** : [April 6, 2024 (I)]

List-I (Precipitating reagent and conditions)	List-II (Cation)
(A) $NH_4Cl + NH_4OH$	(I) Mn^{2+}
(B) $NH_4OH + Na_2CO_3$	(II) Pb^{2+}
(C) $NH_4OH + NH_4Cl + H_2S$ gas	(III) Al^{3+}
(D) dilute HCl	(IV) Sr^{2+}

Choose the **correct** answer from the options given below:

(a) A-(IV), B-(III), C-(II), D-(I)
(b) A-(IV), B-(III), C-(I), D-(II)
(c) A-(III), B-(IV), C-(I), D-(II)
(d) A-(III), B-(IV), C-(II), D-(I)

76. During the detection of acidic radical present in a salt, a student gets a pale yellow precipitate soluble with difficulty in NH_4OH solution when sodium carbonate extract was first acidified with dil. HNO_3 and then $AgNO_3$ solution was added. This indicates presence of: [April 6, 2024 (II)]

(a) Br^- (b) CO_3^{2-} (c) I^- (d) Cl^-

77. While preparing crystals of Mohr's salt, dil. H_2SO_4 is added to a mixture of ferrous sulphate and ammonium sulphate, before dissolving this mixture in water, dil. H_2SO_4 is added here to: [April 5, 2024 (II)]

(a) prevent the hydrolysis of ferrous sulphate
(b) prevent the hydrolysis of ammonium sulphate
(c) make the medium strongly acidic
(d) increase the rate of formation of crystals

78. In the precipitation of the iron group (III) in qualitative analysis, ammonium chloride is added before adding ammonium hydroxide to : [April 4, 2024 (I)]

(a) prevent interference by phosphate ions
(b) decrease concentration of ^-OH ions
(c) increase concentration of Cl^- ions
(d) increase concentration of NH_4^+ ions

79. The number of white coloured salts, among the following is ______ . [Feb. 1, 2024 (I)]

(a) $SrSO_4$ (b) $Mg(NH_4)PO_4$
(c) $BaCrO_4$ (d) $Mn(OH)_2$
(e) $PbSO_4$ (f) $PbCrO_4$
(g) $AgBr$ (h) PbI_2
(i) CaC_2O_4 (j) $[Fe(OH)_2(CH_3COO)]$

80. Given below are two statements: [Feb. 1, 2024 (II)]

Statement (I) : Dimethyl glyoxime forms a six - membered covalent chelate when treated with $NiCl_2$ solution in presence of NH_4OH.

Statement (II) : Prussian blue precipitate contains iron both in (+2) and (+3) oxidation states.

In the light of the above statements, choose the most appropriate answer from the options given below:

(a) Statement I is false but Statement II is true.
(b) Both Statement I and Statement II are true
(c) Statement I is true but Statement II is false
(d) Both Statement I and Statement II are false

81. Given below are two statements: [Feb. 1, 2024 (I)]

Statement (I) : Potassium hydrogen phthalate is a primary standard for standardisation of sodium hydroxide solution.

Statement (II) : In this titration phenophthalein can be used as indicator.

In the light of the above statements, choose the most appropriate answer from the options given below:

(a) Both Statement I and Statement II are correct
(b) Statement I is correct but Statement II is incorrect
(c) Statement I is incorrect but Statement II in correct
(d) Both Statement I and Statement II are incorrect.

82. The compound that is white in color is [Jan. 31, 2024 (I)]

(a) ammonium sulphide
(b) lead sulphate
(c) ammonium arsinomolybdate
(d) lead iodide

83. Molar mass of the salt from NaBr, $NaNO_3$, KI and CaF_2 which does not evolve coloured vapours on heating with concentrated H_2SO_4 is _____ g mol^{-1}.
(Molar mass in g mol^{-1} : Na : 23, N : 14, K : 39, O : 16, Br : 80, I : 127, F : 19, Ca : 40) [NV, Jan. 31, 2024 (I)]

84. Given below are two statements: [Jan. 30, 2024 (I)]

Statement (I) : The gas liberated on warming a salt with dil H_2SO_4, turns a piece of paper dipped in lead acetate into black, it is a confirmatory test for sulphide ion.

Statement (II) : In statement - I the colour of paper turns black because of formation of lead sulphite.

In the light of the above statements, choose the **most appropriate** answer from the options given below:

(a) Both **Statement I** and **Statement II** are false.
(b) **Statement I** is false but **Statement II** is true.
(c) **Statement I** is true but **Statement II** is false.
(d) Both **Statement I** and **Statement II** are true.

85. A and B formed in the following reactions are:

$CrO_2Cl_2 + 4NaOH \rightarrow A + 2NaCl + 2H_2O,$
$A + 2HCl + 2H_2O_2 \rightarrow B + 3H_2O$ [Jan. 30, 2024 (II)]

(a) $A = Na_2CrO_4, B = CrO_5$
(b) $A = Na_2Cr_2O_4, B = CrO_4$
(c) $A = Na_2Cr_2O_7, B = CrO_3$
(d) $A = Na_2Cr_2O_7, B = CrO_5$

86. Number of metal ions characterized by flame test among the following is _______. **[NV, Jan. 30, 2024 (II)]**
$Sr^{2+}, Ba^{2+}, Ca^{2+}, Cu^{2+}, Zn^{2+}, Co^{2+}, Fe^{2+}$

87. A reagent which gives brilliant red precipitate with Nickel ions in basic medium is **[Jan. 29, 2024 (II)]**
(a) dimethyl glyoxime (b) meta-dinitrobenzene
(c) sodium nitroprusside (d) neutral $FeCl_3$

88. On passing a gas, 'X', through Nessler's regent, a brown precipitate is obtained. The gas 'X' is **[Jan. 29, 2024 (II)]**
(a) Cl_2 (b) H_2S
(c) CO_2 (d) NH_3

89. The oxidation number of iron in the compound formed during brown ring test for NO_3^- ion is _____. **[NV, Jan. 29, 2024 (II)]**

90. NaCl reacts with conc. H_2SO_4 and $K_2Cr_2O_7$ to give reddish fumes (B), which react with NaOH to give yellow solution (C). (B) and (C) respectively are : **[Jan. 27, 2024 (I)]**
(a) CrO_2Cl_2, Na_2CrO_4 (b) Na_2CrO_4, CrO_2Cl_2
(c) $CrO_2Cl_2, KHSO_4$ (d) $CrO_2Cl_2, Na_2Cr_2O_7$

91. In the wet tests for detection of various cations by precipitation, Ba^{2+} cations are detected by obtaining precipitate of **[April 13, 2023 (II)]**
(a) Ba(ox) : Barium oxalate (b) $BaCO_3$
(c) $Ba(OAc)_2$ (d) $BaSO_4$

92. When a solution of mixture having two inorganic salts was treated with freshly prepared ferrous sulphate in acidic medium, a dark brown ring was formed whereas on treatment with neutral $FeCl_3$, it gave deep red colour which disappeared on boiling and a brown red ppt was formed. The mixture contains **[April 11, 2023 (I)]**
(a) CH_3COO^- & NO_3^- (b) $C_2O_4^{2-}$ & NO_3^-
(c) SO_3^{2-} & CH_3COO^- (d) SO_3^{2-} & $C_2O_4^{2-}$

93. 25 mL of silver nitrate solution (1 M) is added dropwise to 25 mL of potassium iodide (1.05 M) solution. The ion(s) present in very small quantity in the solution is/are **[April 11, 2023 (I)]**
(a) NO_3^- only (b) K^+ only
(c) Ag^+ and I^- both (d) I^- only

94. Prolonged heating is avoided during the preparation of ferrous ammonium sulphate to **[April 10, 2023 (I)]**
(a) prevent oxidation (b) prevent reduction
(c) prevent hydrolysis (d) prevent breaking

95. Formation of which complex, among the following, is not a confirmatory test of Pb^{2+} ions **[April 06, 2023 (II)]**
(a) lead chromate (b) lead iodide
(c) lead nitrate (d) lead sulphate

96. Element not present in Nessler's reagent is:- **[April 06, 2023 (II)]**
(a) Hg (b) I (c) K (d) N

97. Which element is not present in Nessler's reagent? **[Feb. 01, 2023 (II)]**
(a) Mercury (b) Potassium
(c) Oxygen (d) Iodine

98. A solution of $FeCl_3$ when treated with $K_4[Fe(CN)_6]$ gives a prussiun blue precipitate due to the formation of **[Feb. 01, 2023 (I)]**
(a) $Fe[Fe(CN)_6]$ (b) $Fe_3[Fe(CN)_6]_2$
(c) $Fe_4[Fe(CN)_6]_3$ (d) $K[Fe_2(CN)_6]$

99. During the qualitative analysis of SO_3^{2-} using dilute H_2SO_4, SO_2 gas is evolved which turns $K_2Cr_2O_7$ solution (acidified with dilute H_2SO_4): **[Jan. 30, 2023 (I)]**
(a) Black (b) Red (c) Green (d) Blue

100. In the wet tests for identification of various cations by precipitation, which transition element cation doesn't belong to group IV in qualitative inorganic analysis? **[Jan. 30, 2023 (I)]**
(a) Fe^{3+} (b) Zn^{2+} (c) Co^{2+} (d) Ni^{2+}

101. Formulae for Nessler's reagent is: **[Jan. 30, 2023 (II)]**
(a) KHg_2I_2 (b) $KHgI_3$ (c) K_2HgI_4 (d) HgI_2

102. During the borax bead test with $CuSO_4$, a blue green colour of the bead was observed in oxidising flame due to the formation of **[Jan. 29, 2023 (I)]**
(a) Cu_3B_2 (b) Cu
(c) $Cu(BO_2)_2$ (d) CuO

103. An ammoniacal metal salt solution gives a brilliant red precipitate on addition of dimethylglyoxime. The metal ion is : **[Jan. 24, 2023 (I)]**
(a) Cu^{2+} (b) Co^{2+} (c) Fe^{2+} (d) Ni^{2+}

104. Match the List-I with List-II : **[Jan. 25, 2023 (I)]**

Cations	**Group reaction**
$P \rightarrow Pb^{2+}, Cu^{2+}$	(i) H_2S gas in presence of dilute HCl
$Q \rightarrow Al^{3+}, Fe^{3+}$	(ii) $(NH_4)_2CO_3$ in presence of NH_4OH
$R \rightarrow Co^{2+}, Ni^{2+}$	(iii) NH_4OH in presence of NH_4Cl
$S \rightarrow Ba^{2+}, Ca^{2+}$	(iv) H_2S in presence of NH_4OH

(a) P→i, Q→iii, R→ii, S→iv
(b) P→iv, Q→ii, R→iii, S→i
(c) P→iii, Q→i, R→iv, S→ii
(d) P→i, Q→iii, R→iv, S→ii

105. When borax is heated with CoO on a platinum loop, blue coloured bead formed is largely due to : **[July 29, 2022 (II)]**
(a) B_2O_3 (b) $Co(BO_2)_2$
(c) CoB_4O_7 (d) $Co[B_4O_5(OH)_4]$

106. Fe^{3+} cation gives a prussian blue precipitate on addition of potassium ferrocyanide solution due to the formation of: **[July 27, 2022 (II)]**

(a) $[Fe(H_2O)_6]_2[Fe(CN)_6]$ (b) $Fe_2[Fe(CN)_6]_2$
(c) $Fe_3[Fe(OH)_2(CN)_4]_2$ (d) $Fe_4[Fe(CN)_6]_3$

107. A white precipitate was formed when $BaCl_2$ was added to water extract of an inorganic salt. Further, a gas '*X*' with characteristic odour was released when the formed white precipitate was dissolved in dilute HCl. The anion present in the inorganic salt is : **[June 29, 2022 (II)]**

(a) I^- (b) SO_3^{2-}
(c) S^{2-} (d) NO_2^-

108. Match List-I with List-II. **[June 27, 2022 (II)]**

List-I (Anion)		List-II (Gas evolved on reaction with dil. H_2SO_4)
(A) CO_3^{2-}	(I)	Colourless gas which turns lead acetate paper black
(B) S^{2-}	(II)	Colourless gas which turns acidified potassium dichromate solution green.
(C) SO_3^{2-}	(III)	Brown fumes which turns acidified KI solution containing starch blue.
(D) NO_2^-	(IV)	Colourless gas evolved with brisk effervescences, which turns lime water milky.

Choose the correct answer from the options given below:

(a) A-III, B-I, C-II, D-IV (b) A-II, B-I, C-IV, D-III
(c) A-IV, B-I, C-III, D-II (d) A-IV, B-I, C-II, D-III

109. Which statement is **not** true with respect to nitrate ion test? **[June 26, 2022 (I)]**

(a) A dark brown ring is formed at the junction of two solutions.
(b) Ring is formed due to nitroferrous sulphate complex.
(c) The brown complex is $[Fe(H_2O)_5(NO)]SO_4$.
(d) Heating the nitrate salt with conc. H_2SO_4, light brown fumes are evolved.

110. During the qualitative analysis of salt with cation y^{2+}, addition of a reagent (X) to alkaline solution of the salt gives a bright red precipitate. The reagent (X) and the cation (y^{2+}) present respectively are: **[June 24, 2022 (I)]**

(a) Dimethylglyoxime and Ni^{2+}
(b) Dimethylglyoxime and Co^{2+}
(c) Nessler's reagent and Hg^{2+}
(d) Nessler's reagent and Ni^{2+}

111. In the flame test of a mixture of salts, a green flame with blue centre was observed. Which one of the following cations may be present? **[June 24, 2022 (II)]**

(a) Cu^{2+} (b) Sr^{2+} (c) Ba^{2+} (d) Ca^{2+}

112. Consider the sulphides HgS, PbS, CuS, Sb_2S_3, As_2S_3 and CdS. Number of these sulphides soluble in 50% HNO_3 is ______. **[NV, Aug. 31, 2021 (I)]**

113. Match List-I with List-II :

List-I (Metal Ion)	List-II (Group in Qualitative analysis)
(A) Mn^{2+}	(i) Group - III
(B) As^{3+}	(ii) Group - IIA
(C) Cu^{2+}	(iii) Group - IV
(D) Al^{3+}	(iv) Group - IIB

Choose the most appropriate answer from the options given below : **[Aug. 31, 2021 (II)]**

(a) (A)-(i), (B)-(ii), (C)-(iii), (D)-(iv)
(b) (A)-(iii), (B)-(iv), (C)-(ii), (D)-(i)
(c) (A)-(i), (B)-(iv), (C)-(ii), (D)-(iii)
(d) (A)-(iv), (B)-(ii), (C)-(iii), (D)-(i)

114. To an aqueous solution containing ions such as Al^{3+}, Zn^{2+}, Ca^{2+}, Fe^{3+}, Ni^{2+}, Ba^{2+} and Cu^{2+} conc. HCl was added, followed by H_2S.
The total number of cations precipitated during this reaction is/are : **[July 27, 2021 (II)]**

(a) 1 (b) 3
(c) 4 (d) 2

115. Reagent, 1-naphthylamine and sulphanilic acid in acetic acid is used for the detection of **[March 18, 2021 (I)]**

(a) NO_3^- (b) N_2O
(c) NO_2^- (d) NO

116. On treating a compound with warm dil. H_2SO_4, gas X is evolved which turns $K_2Cr_2O_7$ paper acidified with dil. H_2SO_4 to a green compound Y. X and Y respectively are : **[Feb. 26, 2021 (I)]**

(a) $X = SO_3, Y = Cr_2O_3$
(b) $X = SO_3, Y = Cr_2(SO_4)_3$
(c) $X = SO_2, Y = Cr_2(SO_4)_3$
(d) $X = SO_2, Y = Cr_2O_3$

117. The incorrect statement out of the following is **[Online April 16, 2018]**

(a) Cu^{2+} ion gives chocolate coloured precipitate with potassium ferrocyanide solution
(b) Cu^{2+} and Ni^{2+} ions give black precipitate with H_2S in presence of HCl solution
(c) Ferric ion gives blood red colour with potassium thiocyanate
(d) Cu^{2+} salts give red coloured borax bead test in reducing flame.

118. A white sodium salt dissolves readily in water to give a solution which is neutral to litmus. When silver nitrate solution is added to the afore-mentioned solution, a white precipitate is obtained which does not dissolve in dil. nitric acid. The anion is : **[Online April 15, 2018 (I)]**

(a) CO_3^{2-} (b) SO_4^{2-}
(c) S^{2-} (d) Cl^-

119. A solution containing a group–IV cation gives a precipitate on passing H_2S. A solution of this precipitate in dil.HCl produces a white precipitate with NaOH solution and bluish–white precipitate with basic potassium ferrocyanide. The cation is : **[Online April 8, 2017]**
(a) Co^{2+} (b) Ni^{2+} (c) Mn^{2+} (d) Zn^{2+}

120. The hottest region of Bunsen flame shown in the figure below is : **[2016]**

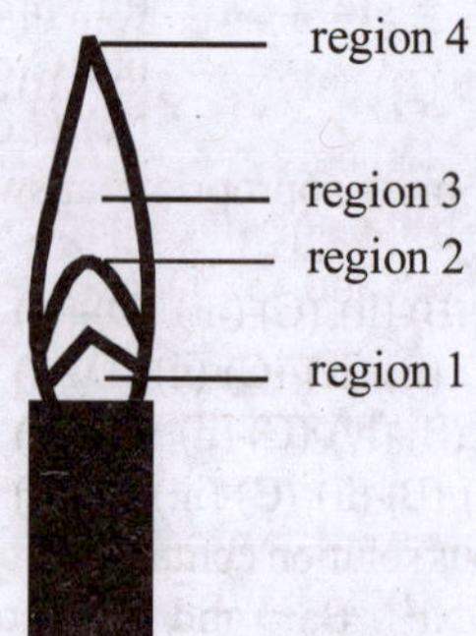

(a) region 3 (b) region 4
(c) region 1 (d) region 2

121. An aqueous solution of a salt X turns blood red on treatment with CNS^- and blue on treatment with $K_4[Fe(CN)_6]$. X also gives a positive chromyl chloride test. The salt X is : **[Online April 10, 2015]**
(a) $CuCl_2$ (b) $FeCl_3$
(c) $Cu(NO_3)_2$ (d) $Fe(NO_3)_3$

122. The cation that will not be precipitated by H_2S in the presence of dil. HCl is : **[Online April 10, 2015]**
(a) Pb^{2+} (b) Cu^{2+}
(c) Co^{2+} (d) As^{3+}

123. A pink coloured salt turns blue on heating. The presence of which cation is most likely? **[Online April 11, 2015]**
(a) Co^{2+} (b) Cu^{2+}
(c) Zn^{2+} (d) Fe^{2+}

124. Sodium carbonate cannot be used in place of $(NH_4)_2CO_3$ for the identification of Ca^{2+}, Ba^{2+} and Sr^{2+} ions (in group V) during mixture analysis because : **[Online April 9, 2013]**
(a) Mg^{2+} ions will also be precipitated.
(b) Concentration of CO_3^{2-} ions is very low.
(c) Sodium ions will react with acid radicals.
(d) Na^+ ions will interfere with the detection of Ca^{2+}, Ba^{2+}, Sr^{2+} ions.

125. Which of the following statements is incorrect ? **[Online April 22, 2013]**
(a) Fe^{2+} ion also gives blood red colour with SCN^- ion.
(b) Fe^{3+} ion also gives blood red colour with SCN^- ion.
(c) On passing H_2S into Na_2ZnO_2 solution a white ppt of ZnS is formed.
(d) Cupric ion reacts with excess of ammonia solution to give deep blue colour of $[Cu(NH_3)_4]^{2+}$ ion.

126. Potassium dichromate when heated with concentrated sulphuric acid and a soluble chloride, gives brown-red vapours of : **[Online April 9, 2013]**
(a) CrO_3 (b) $CrCl_3$ (c) CrO_2Cl_2 (d) Cr_2O_3

127. Calomel (Hg_2Cl_2) on reaction with ammonium hydroxide gives **[2005]**
(a) HgO (b) Hg_2O
(c) $NH_2-Hg-Hg-Cl$ (d) $HgNH_2Cl$

128. Which one of the following statements is correct ? **[2003]**
(a) From a mixed precipitate of AgCl and AgI, ammonia solution dissolves only AgCl
(b) Ferric ions give a deep green precipitate on adding potassium ferrocyanide solution
(c) On boiling a solution having K^+, Ca^{2+} and HCO_3^- ions, we get a precipitate of $K_2Ca(CO_3)_2$
(d) Manganese salts give a violet borax bead test in the reducing flame

129. When H_2S is passed through Hg_2S we get **[2002]**
(a) HgS (b) $HgS + Hg_2S$
(c) $Hg_2S + Hg$ (d) None of these.

130. How do we differentiate between Fe^{3+} and Cr^{3+} in group III? **[2002]**
(a) by taking excess of NH_4OH solution
(b) by increasing NH_4^+ ion concentration
(c) by decreasing OH^- ion concentration
(d) both (b) and (c)

Topic 3 – Principles Involved in Titrimetric Exercises, Calorimetry, Preparation of Colloidal Sols and Kinetic Study of Reaction Between I^- and H_2O_2

131. A sample of n-octane (1.14 g) was completely burnt in excess of oxygen in a bomb calorimeter, whose heat capacity is 5 kJ K^{-1}. As a result of combustion reaction, the temperature of the calorimeter is increased by 5 K. The magnitude of the heat of combustion of octane at constant volume is ____ kJ mol^{-1} (nearest integer). **[NV, April 3, 2025 (II)]**

132. Compounds that should not be used as primary standards in titrimetric analysis are :
A. $Na_2Cr_2O_7$ B. Oxalic acid
C. NaOH D. $FeSO_4.6H_2O$
E. Sodium tetraborate
Choose the **most appropriate** answer from the options given below: **[April 3, 2025 (II)]**
(a) B and D Only (b) D and E Only
(c) C, D and E Only (d) A, C and D Only

133. 10 mL of 2 M NaOH solution is added to 20 mL of 1 M HCl solution kept in a beaker. Now, 10 mL of this mixture is poured into a volumetric flask of 100 mL containing 2 moles of HCl and made the volume upto the mark with distilled water. The solution in this flask is : **[April 3, 2025 (II)]**

(a) 0.2 M NaCl solution (b) 20 M HCl solution
(c) 10 M HCl solution (d) Neutral solution

134. 40 mL of a mixture of CH_3COOH and HCl (aqueous solution) is titrated against 0.1 M NaOH solution conductometrically. Which of the following statement is **correct**? **[April 3, 2025 (II)]**

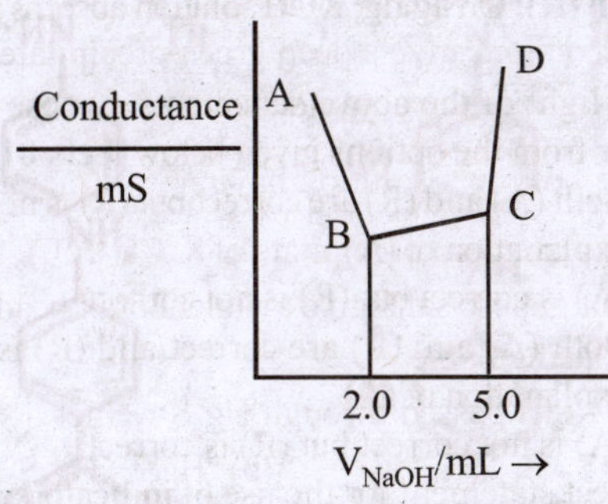

(a) The concentration of CH_3COOH in the original mixture is 0.005 M
(b) The concentration of HCl in the original mixture is 0.005 M
(c) CH_3COOH is neutralised first followed by neutralisation of HCl
(d) Point 'C' indicates the complete neutralisation HCl

135. Given below are two statements : **[Jan. 28, 2025 (I)]**

Statement I : In the oxalic acid vs $KMnO_4$ (in the presence of dil H_2SO_4) titration the solution needs to be heated initially to 60°C, but no heating is required in Ferrous ammonium sulphate (FAS) vs $KMnO_4$ titration (in the presence of dil H_2SO_4)

Statement II : In oxalic acid vs $KMnO_4$ titration, the initial formation of $MnSO_4$ takes place at high temperature, which then acts as catalyst for further reaction. In the case of FAS vs $KMnO_4$, heating oxidizes Fe^{2+} into Fe^{3+} by oxygen of air and error may be introduced in the experiment. In the light of the above statements, choose the correct answer from the options given below :

(a) Statement I is false but Statement II is true
(b) Both Statement I and Statement II are true
(c) Statement I is true but Statement II is false
(d) Both Statement I and Statement II are false

136. Which of the following mixing of 1M base and 1M acid leads to the largest increase in temperature? **[Jan. 24, 2025 (II)]**

(a) 30 mL HCl and 30 mL NaOH
(b) 30 mL CH_3COOH and 30 mL NaOH
(c) 50 mL HCl and 20 mL NaOH
(d) 45 mL CH_3COOH and 25 mL NaOH

137. Identify the **incorrect** statements regarding primary standard of titrimetric analysis **[April 9, 2024 (I)]**

(A) It should be purely available in dry form.
(B) It should not undergo chemical change in air.
(C) It should be hygroscopic and should react with another chemical instantaneously and stoichiometrically.
(D) It should be readily soluble in water.
(E) $KMnO_4$ & NaOH can be used as primary standard.

Choose the **correct** answer from the options given below:

(a) (C) and (D) only (b) (B) and (E) only
(c) (A) and (B) only (d) (C) and (E) only

138. The heat of solution of anhydrous $CuSO_4$ and $CuSO_4 \cdot 5H_2O$ are -70 kJ mol^{-1} and $+12$ kJ mol^{-1} respectively.
The heat of hydration of $CuSO_4$ to $CuSO_4 \cdot 5H_2O$ is $-x$ kJ. The value of x is_____. **[NV, April 9, 2024 (I)]**

139. When equal volume of 1M HCl and 1M H_2SO_4 are separately neutralised by excess volume of 1M NaOH solution. X and y kJ of heat is liberated respectively. The value of y/x is _______. **[NV, April 9, 2024 (I)]**

140. Given below are two statements : One is labelled as **Assertion (A)** and the other is labelled as **Reason (R)** **[April 5, 2024 (I)]**

Assertion (A): Enthalpy of neutralisation of strong monobasic acid with strong monoacidic base is always -57 kJ mol^{-1}

Reason (R): Enthalpy of neutralisation is the amount of heat liberated when one mole of H^+ ions furnished by acid combine with one mole of ^-OH ions furnished by base to form one mole of water. In the light of the above statements, choose the **correct** answer from the options given below.

(a) **(A)** is true but **(R)** is false
(b) Both **(A)** and **(R)** are true and **(R)** is the correct explanation of **(A)**
(c) **(A)** is false but **(R)** is true
(d) Both **(A)** and **(R)** are true but **(R)** is **not** the correct explanation of **(A)**

141. Only 2 mL of $KMnO_4$ solution of unknown molarity is required to reach the end point of a titration of 20 mL of oxalic acid (2 M) in acidic medium. The molarity of $KMnO_4$ solution should be _____ M. **[NV, April 4, 2024 (I)]**

142. Given below are two statements: **[Feb. 1, 2024 (I)]**

Statement (I) : Potassium hydrogen phthalate is a primary standard for standardisation of sodium hydroxide solution.

Statement (II) : In this titration phenophthalein can be used as indicator.

In the light of the above statements, choose the most appropriate answer from the options given below:

(a) Both Statement I and Statement II are correct
(b) Statement I is correct but Statement II is incorrect
(c) Statement I is incorrect but Statement II in correct
(d) Both Statement I and Statement II are incorrect.

143. 20 mL of calcium hydroxide was consumed when it was reacted with 10 mL of unknown solution of H_2SO_4. Also 20 mL standard solution of 0.5 M HCl containing 2 drops of phenolphthalein was titrated with calcium hydroxide the mixture showed pink colour when burette displaced the value of 35.5 mL whereas the burette showed 25.5 mL initially. The concentration of H_2SO_4 is _____ M (Nearest integer) **[NV, April 13, 2023 (I)]**

144. $KMnO_4$ is titrated with ferrous ammonium sulphate hexahydrate in presence of dilute H_2SO_4. Number of water molecules produced for 2 molecules of $KMnO_4$ is _______. **[NV, April 13, 2023 (I)]**

145. The titration curve of weak acid vs. strong base with phenolphthalein as indictor) is shown below. The $K_{phenolphthalein} = 4 \times 10^{-10}$. **[NV, April 08, 2023 (I)]**
Given: log2 = 0.3

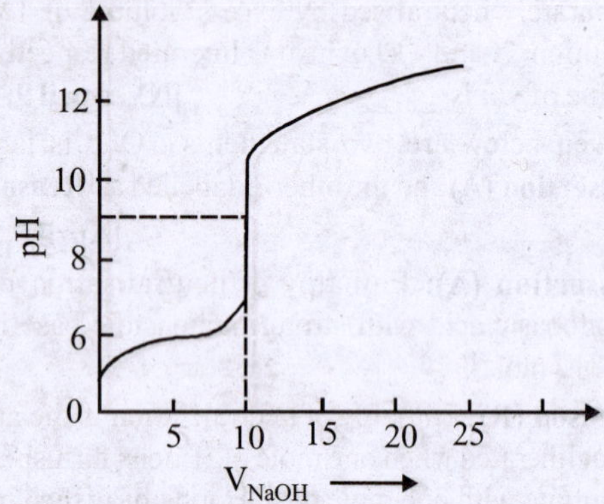

The number of following statements which is/are correct about phenolphthalein is_______

(A) It can be used as an indicator for the titration of weak acid with weak base.
(B) It begins to change colour at pH = 8.4.
(C) It is a weak organic base.
(D) It is colourless in acidic medium.

146. Given below are two statements :- **[April 08, 2023 (II)]**
Statement I : Methyl orange is a weak acid.
Statement II : The benzenoid form of methyl orange is more intense/deeply coloured than the quinonoid form.
In the light of the above statement, choose the most appropriate answer from the options given below :-
(a) Statement I is correct but Statement II is incorrect.
(b) Statement I is incorrect but statement II is correct.
(c) Both Statement I and Statement II are incorrect.
(d) Both statement I and Statement II are correct.

147. Given below are two statements :- **[April 08, 2023 (II)]**
Statement I : In redox titration, the indicators used are sensitive to change in pH of the solution.
Statement II : In acid-base titration, the indicators used are sensitive to change in oxidation potential.
In the light of the above statements, choose the most appropriate answer from the options given below
(a) Both statement I and Statement II are correct.
(b) Statement I is incorrect but statement II is correct.
(c) Statement I is correct but Statement II is incorrect.
(d) Both Statement I and Statement II are incorrect.

148. The volume of 0.02 M aqueous HBr required to neutralize 10.0 mL of 0.01 M aqueous $Ba(OH)_2$ is (Assume complete neutralization) **[April 06, 2023 (II)]**
(a) 2.5 mL (b) 5.0 mL (c) 10.0 mL (d) 7.5 mL

149. Given below are two statements: one is labelled as **Assertion (A)** and the other is labelled as **Reason (R).**
Assertion (A): An aqueous solution of KOH when used for volumetric analysis, its concentration should be checked before the use.
Reason (R): On aging, KOH solution absorbs atmospheric CO_2.
In the light of the above statements, choose the **correct** answer from the options given below **[Feb. 01, 2023 (II)]**
(a) Both (A) and (R) are correct but (R) is not the correct explanation of (A)
(b) (A) is correct but (R) is not correct
(c) Both (A) and (R) are correct and (R) is the correct explanation of (A)
(d) (A) is not correct but (R) is correct

150. Incorrect statement for the use of indicators in acid-base titration is: **[Jan. 31, 2023 (II)]**
(a) Methyl orange may be used for a weak acid vs weak base titration.
(b) Methyl orange is a suitable indicator for a strong acid vs weak base titration
(c) Phenolphthalein is a suitable indicator for a weak acid vs strong base titration
(d) Phenolphthalein may be used for a strong acid vs strong base titration.

151. An indicator 'X' is used for studying the effect of variation in concentration of iodide on the rate of reaction of iodide ion with H_2O_2 at room temp. The indicator 'X' forms blue colored complex with compound 'A' present in the solution. The indicator 'X' and compound 'A' respectively are **[Jan. 29, 2023 (II)]**
(a) Starch and iodine (b) Methyl orange and H_2O_2
(c) Starch and H_2O_2 (d) Methyl orange and iodine

152. For kinetic study of the reaction of iodide ion with H_2O_2 at room temperature : **[July 28, 2022 (I)]**
(A) Always use freshly prepared starch solution.
(B) Always keep the concentration of sodium thiosulphate solution less than that of KI solution.
(C) Record the time immediately after the appearance of blue colour.
(D) Record the time immediately before the appearance of blue colour.
(E) Always keep the concentration of sodium thiosulphate solution more than that of KI solution.
Choose the correct answer from the options given below:
(a) (A), (B), (C) only (b) (A), (D), (E) only
(c) (D), (E) only (d) (A), (B), (E) only

153. Given below are two statements: One is labelled as **Assertion A** and the other is labelled as **Reason R**.

Assertion A: Permanganate titrations are not performed in presence of hydrochloric acid.

Reason R: Chlorine is formed as a consequence of oxidation of hydrochloric acid.

In the light of the above statements, choose the correct answer from the options given below: **[July 28, 2022 (II)]**

(a) Both A and R are true and R is the correct explanation of A.

(b) Both A and R are true but R is NOT the correct explanation of A.

(c) A is true but R is false.

(d) A is false but R is true.

154. 20 mL of 0.02 M $K_2Cr_2O_7$ solution is used for the titration of 10 mL of Fe^{2+} solution in the acidic medium. The molarity of Fe^{2+} solution is ______ $\times 10^{-2}$ M. (Nearest Integer) **[NV, July 27, 2022 (I)]**

155. The plot of pH-metric titration of weak base NH_4OH vs strong acid HCl looks like: **[July 27, 2022 (II)]**

(a)
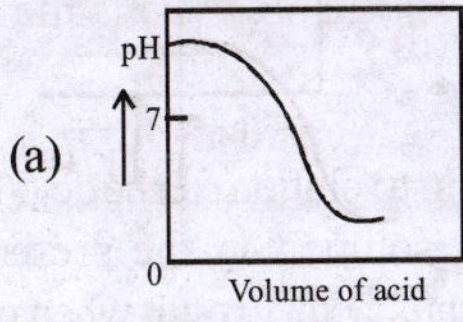

(b)
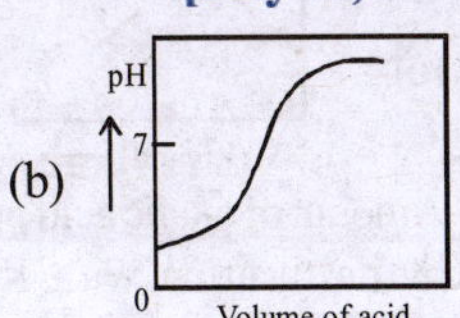

(c)
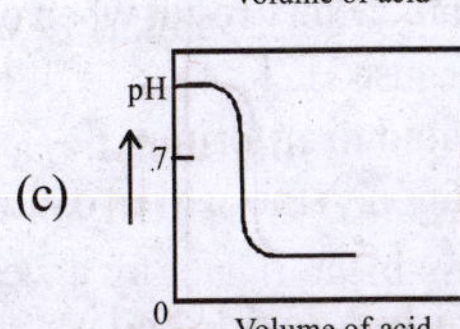

(d)
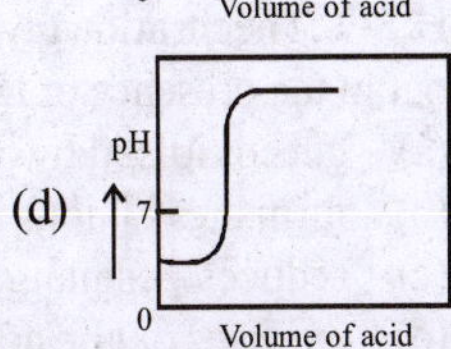

156. The normality of H_2SO_4 in the solution obtained on mixing 100 mL of 0.1 M H_2SO_4 with 50 mL of 0.1 M NaOH is____$\times 10^{-1}$ N. (Nearest integer) **[NV, July 27, 2022 (II)]**

157. 20 mL of 0.02 M hypo solution is used for the titration of 10 mL of copper sulphate solution, in the presence of excess of KI using starch as an indicator. The molarity of Cu^{2+} is found to be ______ $\times 10^{-2}$ M [nearest integer]

Given : $2Cu^{2+} + 4I^- \rightarrow Cu_2I_2 + I_2$

$I_2 + 2S_2O_3^{2-} \rightarrow 2I^- + S_4O_6^{2-}$ **[NV, July 26, 2022 (II)]**

158. The dark purple colour of $KMnO_4$ disappears in the titration with oxalic acid in acidic medium. The overall change in the oxidation number of manganese in the reaction is: **[July 26, 2022 (I)]**

(a) 5 (b) 1 (c) 7 (d) 2

159. In base vs acid titration, at the end point methyl orange is present as **[July 25, 2022 (II)]**

(a) quinonoid form (b) heterocyclic form

(c) phenolic form (d) benzenoid form

160. A 2.0 g sample containing MnO_2 is treated with HCl liberating Cl_2. The Cl_2 gas is passed into a solution of KI and 60.0 mL of 0.1 M $Na_2S_2O_3$ is required to titrate the liberated iodine. The percentage of MnO_2 in the sample is ______. (Nearest integer)

[Atomic masses (in u) Mn = 55; Cl = 35.5; O = 16, I = 127, Na = 23, K = 39, S = 32] **[NV, June 28, 2022 (I)]**

161. 0.01 M $KMnO_4$ solution was added to 20.0 mL of 0.05 M Mohr's salt solution through a burette. The initial reading of 50 mL burette is zero. The volume of $KMnO_4$ solution left in the burette after the end point is ______ mL. (nearest integer) **[NV, June 28, 2022 (II)]**

162. A 0.166 g sample of an organic compound was digested with conc. H_2SO_4 and then distilled with NaOH. The ammonia gas evolved was passed through 50.0 mL of 0.5 N H_2SO_4. The used acid required 30.0 mL of 0.25 N NaOH for complete neutralization. The mass percentage of nitrogen in the organic compound is____. **[NV, June 24, 2022 (I)]**

163. When 10 mL an aqueous solution of $KMnO_4$ was titrated in acidic medium, equal volume of 0.1 M of an aqueous solution of ferrous sulphate was required for complete discharge of colour. The strength of $KMnO_4$ in grams per litre is______ $\times 10^{-2}$. (Nearest integer)

[Atomic mass of K = 39, Mn = 55, O = 16] **[NV, Aug. 27, 2021 (I)]**

164. 10.0 mL of 0.05 M $KMnO_4$ solution was consumed in a titration with 10.0 mL of given oxalic acid dihydrate solution. The strength of given oxalic acid solution is ______ $\times 10^{-2}$ g/L. **[NV, July 27, 2021 (II)]**

165. 10 mL of Na_2CO_3 solution is titrated against 0.2 M HCl solution. The following titre values were obtained in 5 readings.

4.8 mL, 4.9 mL, 5.0 mL, 5.0 mL and 5.0 mL

Based on these readings and convention of titrimetric estimation, the concentration of Na_2CO_3 solution is ______mM

(Round off the nearest integer). **[NV, March 18, 2021 (II)]**

166. 0.4 g mixture of NaOH, Na_2CO_3 and some inert impurities was first titrated with $\frac{N}{10}$ HCl using phenolphthalein as an indicator, 17.5 mL of HCl was required at the end point. After this, methyl orange was added and titrated. 1.5 mL of same HCl was required for the next end point. The weight percentage of Na_2CO_3 in the mixture is ______. (Rounded-off to the nearest integer) **[NV, Feb. 25, 2021 (I)]**

167. Consider titration of NaOH solution versus 1.25 M oxalic acid solution. At the end point, following burette readings were obtained.

(i) 4.5 mL (ii) 4.5 mL (iii) 4.4 mL

(iv) 4.4 mL (v) 4.4 mL

If the volume of oxalic acid taken was 10.0 mL then the molarity of the NaOH solution is ____ M. (Rounded-off to the nearest integer) **[NV, Feb. 25, 2021 (II)]**

168. The volume, in mL, of 0.02 M $K_2Cr_2O_7$ solution required to react with 0.288 g of ferrous oxalate in acidic medium is ______. (Molar mass of Fe = 56 g mol^{-1}) **[NV, Sep. 05, 2020 (II)]**

169. A 20.0 mL solution containing 0.2 g impure H_2O_2 reacts completely with 0.316 g of $KMnO_4$ in acid solution. The purity of H_2O_2 (in %) is ______ (mol. wt. of H_2O_2 = 34; mol. wt. of $KMnO_4$ = 158) **[NV, Sep. 04, 2020 (I)]**

170. While titrating dilute HCl solution with aqueous NaOH, which of the following will **not** be required? **[Sep. 02, 2020 (I)]**

(a) Burette and porcelain tile
(b) Pipette and distilled water
(c) Clamp and phenophthalein
(d) Bunsen burner and measuring cylinder

171. If you spill a chemical toilet cleaning liquid on your hand, your first aid would be : **[Sep. 02, 2020 (II)]**

(a) vinegar (b) aqueous NaOH
(c) aqueous $NaHCO_3$ (d) aqueous NH_3

172. The strength of an aqueous NaOH solution is *most accurately* determined by titrating: (Note: consider that an appropriate indicator is used) **[Jan. 08, 2020 (I)]**

(a) Aq. NaOH in a pipette and aqueous oxalic acid in a burette
(b) Aq. NaOH in a burette and aqueous oxalic acid in a conical flask
(c) Aq. NaOH in a burette and concentrated H_2SO_4 in a conical flask
(d) Aq. NaOH in a volumetric flask and concentrated H_2SO_4 in a conical flask

173. An alkali is titrated against an acid with methyl orange as indicator, which of the following is a correct combination? **[2018]**

	Base	**Acid**	**End point**
(a)	Weak	Strong	Colourless to pink
(b)	Strong	Strong	Pinkish red to yellow
(c)	Weak	Strong	Yellow to Pinkish red
(d)	Strong	Strong	Pink to colourless

174. For standardizing NaOH solution, which of the following is used as a primary standard? **[Online April 16, 2018]**

(a) Sodium tetraborate
(b) Ferrous ammonium sulfate
(c) Oxalic acid
(d) dil. HCl

175. In a complexometric titration of metal ion with ligand M(Metal ion) + L(Ligand) $\longrightarrow$ C(Complex) end point is estimated spectrophotometrically (through light absorption) If 'M' and 'C' do not absorb light and only 'L' absorbs, then the titration plot between absorbed light (A) versus volume of ligand 'L' (V) would look like: **[Online April 16, 2018]**

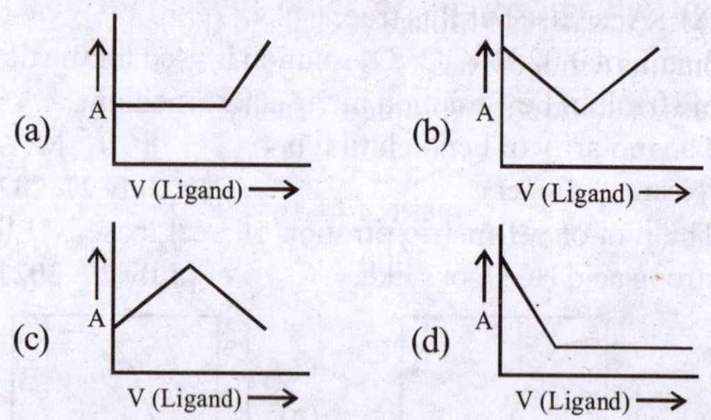

176. Amount of oxalic acid present in a solution can be determined by its titration with $KMnO_4$ solution in the presence of H_2SO_4. The titration gives unsatisfactory result when carried out in the presence of HCl, because HCl **[2008]**

(a) gets oxidised by oxalic acid to chlorine
(b) furnishes H^+ ions in addition to those from oxalic acid
(c) reduces permanganate to Mn^{2+}
(d) oxidises oxalic acid to carbon doxide and water

ANSWER KEYS

1	(a)	**19**	(a)	**37**	(68)	**55**	(b)	**73**	(a)	**91**	(b)	**109**	(b)	**127**	(d)	**145**	(2)	**163**	(316)
2	(c)	**20**	(d)	**38**	(d)	**56**	(b)	**74**	(c)	**92**	(a)	**110**	(a)	**128**	(a)	**146**	(c)	**164**	(1575)
3	(d)	**21**	(b)	**39**	(243)	**57**	(b)	**75**	(c)	**93**	(c)	**111**	(a)	**129**	(c)	**147**	(d)	**165**	(50)
4	(c)	**22**	(d)	**40**	(d)	**58**	(b)	**76**	(a)	**94**	(a)	**112**	(4)	**130**	(d)	**148**	(c)	**166**	(4)
5	(c)	**23**	(d)	**41**	(c)	**59**	(c)	**77**	(a)	**95**	(c)	**113**	(b)	**131**	(2500)	**149**	(c)	**167**	(6)
6	(b)	**24**	(b)	**42**	(b)	**60**	(b)	**78**	(b)	**96**	(d)	**114**	(a)	**132**	(d)	**150**	(a)	**168**	(50)
7	(b)	**25**	(530)	**43**	(a)	**61**	(a)	**79**	(5)	**97**	(d)	**115**	(c)	**133**	(b)	**151**	(a)	**169**	(85)
8	(d)	**26**	(2)	**44**	(b)	**62**	(b)	**80**	(a)	**98**	(c)	**116**	(b)	**134**	(b)	**152**	(a)	**170**	(d)
9	(d)	**27**	(b)	**45**	(b)	**63**	(c)	**81**	(a)	**99**	(c)	**117**	(b)	**135**	(b)	**153**	(a)	**171**	(c)
10	(d)	**28**	(34)	**46**	(b)	**64**	(c)	**82**	(b)	**100**	(a)	**118**	(d)	**136**	(a)	**154**	(24)	**172**	(b)
11	(95)	**29**	(d)	**47**	(d)	**65**	(b)	**83**	(78)	**101**	(c)	**119**	(d)	**137**	(d)	**155**	(a)	**173**	(c)
12	(c)	**30**	(a)	**48**	(a)	**66**	(b)	**84**	(c)	**102**	(c)	**120**	(d)	**138**	(82)	**156**	(1)	**174**	(c)
13	(d)	**31**	(b)	**49**	(c)	**67**	(b)	**85**	(a)	**103**	(d)	**121**	(b)	**139**	(2)	**157**	(4)	**175**	(a)
14	(d)	**32**	(14)	**50**	(a)	**68**	(4)	**86**	(4)	**104**	(d)	**122**	(c)	**140**	(b)	**158**	(a)	**176**	(c)
15	(d)	**33**	(40)	**51**	(b)	**69**	(b)	**87**	(a)	**105**	(b)	**123**	(a)	**141**	(8)	**159**	(a)		
16	(d)	**34**	(a)	**52**	(a)	**70**	(d)	**88**	(d)	**106**	(d)	**124**	(a)	**142**	(a)	**160**	(13)		
17	(a)	**35**	(12)	**53**	(b)	**71**	(0)	**89**	(1)	**107**	(b)	**125**	(a)	**143**	(1)	**161**	(30)		
18	(3)	**36**	(78)	**54**	(c)	**72**	(d)	**90**	(a)	**108**	(d)	**126**	(c)	**144**	(68)	**162**	(148)		

HINTS & SOLUTIONS

Solutions

1. **(d)** $\frac{P_{N_2}}{P_{O_2}} = \frac{X_{N_2}.P_T}{X_{O_2}.P_T} = \frac{n_{N_2}}{n_{O_2}} = \frac{70/28}{27/32} = 2.96$

$\frac{P_{O_2}}{P_{Ar}} = \frac{n_{O_2}}{n_{Ar}} = \frac{27/32}{3/40} = 11.25$

2. **(2)** Sea water is 6 Molar in NaCl, So 1000 mL of sea water contains 6 mol of NaCl.
mass of solution = Volume × density
$= 1000 \times 2 = 2000$ g

$\text{ppm} = \frac{\text{mass of } O_2}{2000\text{ g}} \times 10^6$

$\Rightarrow \text{mass of } O_2 = 5.8 \times 2000 \times \frac{1}{10^6} = 5.8 \times 2 \times 10^{-3}$

$= 1.16 \times 10^{-2}$ g

$\text{molality for } O_2 = \frac{1.16 \times 10^{-2}/32}{(2000 - 6 \times 58.5)} \times 1000$

$= \frac{1.16 \times 10}{32 \times 1649} = 0.000219 = 2.19 \times 10^{-4}$

∴ Correct answer is 2.

3. **(9)** Let mass of iron = w g

$\Rightarrow \frac{w}{150 \times 10^3} \times 10^6 = 12 \Rightarrow w = 150 \times 12 \times 10^{-3} = 1.8$ g

Let mass of $FeSO_4.7H_2O = w_1$ g
molar mass of $FeSO_4.7H_2O = 56 + 32 + 64 + 126 = 278$ g/mol.

$\Rightarrow \text{Moles of Fe} = \frac{1.8}{56}$

Moles of $FeSO_4.7H_2O = \frac{w_1}{278}$

Each mole of $FeSO_4.7H_2O$ provides 1 mole of Fe

$\Rightarrow \frac{w_1}{278} = \frac{1.8}{56} = w_1 = \frac{278 \times 1.8}{56} \Rightarrow w_1 = 8.935\text{ g} \approx 9\text{ g}.$

4. **(b)** The Volume of water is minimum at 4°C, hence molarity will be maximum at 4°C.

5. **(d)** For ideal permanent gases, as temperature increases, solubility first decreases then increases hence K_H first increases then decreases in aqueous solutions. The solubility order is $He < N_2 < CH_4$, So the K_H order will be $CH_4 < N_2 < He$.

6. **(164)** Let's take 500 mL of the solution.

$\text{Mass}_{sol^n} = V_{sol^n} \times d_{sol^n} = 500 \times 1.25 = 625\text{g}$

Mass of solute (x) = Molarity × Volume × Molar mass of $CuSO_4 = 0.2 \times 0.5 \times 159.5 = 15.95$g
$n_{solute} = 0.1$ mol.
Mass of solvent = Mass of solution − Mass of solute $= 625 - 15.95 = 609.05$g

$\text{Molility} = \frac{\text{No. of moles}}{\text{mass of solvent in kg}} = \frac{0.1}{\frac{609.05}{1000}} = 0.164$

$m = 164 \times 10^{-3}$

7. **(74)** Molality of urea is 4.44 m, that means 4.44 moles of urea present in 1000 gm of water.

$\frac{X_B}{X_A} = \frac{n_B}{n_A} \Rightarrow \frac{X_B}{1 - X_B} = \frac{4.44}{\frac{1000}{18}}\ (n_B = m)$

$X_B = 0.07992(1 - X_B) \Rightarrow X_B = 0.07992 - 0.07992\,X_B$

$1.07992X_B = 0.07992 \Rightarrow X_B = \frac{0.07992}{1.07992} = 0.0740$

$X_B = 74 \times 10^{-3}$

8. **(22)** Mass percent of Alcohol

$= \frac{\text{Mass of ethyl alcohol}}{\text{Total mass of solution}} \times 100$

$= \frac{1 \times 46}{1 \times 46 + 9 \times 18} \times 100 = \frac{4600}{208} = 22.11 \approx 22$

9. **(b)**

$\text{Molality} = \frac{1000 \times \text{Molarity}}{1000 \times \text{density} - \text{molarity} \times \text{molar mass of solute}}$

$3 = \frac{1000 \times x}{1000 \times 1.12 - (x \times 40)} \Rightarrow x = 3$

10. **(b)** $\text{Molality} = \frac{M \times 1000}{d \times 1000 - (M \times mm)}$

$= \frac{3 \times 100}{1.25 \times 1000 - (3 \times 58.5)} = 2.79\text{m}$

11. **(b)** $\text{Molarity (M)} = \frac{n_{NaCl}}{V_{sol}\text{(in L)}} \Rightarrow M = \frac{\frac{5.85}{58.5}}{0.5} = 0.2\text{M}$

12. **(11)** Given : V = 1L = 1000 mL
Specific gravity = 1.54 g/cm³
M = 98 g/mol
Mass of solution = 1.54 × 1000 = 1540 g
% purity of $H_3PO_4 = 70\%$

$\text{Weight of } H_3PO_4 = \frac{70}{100} \times 1540 = 1078\text{ g}$

No. of moles of $H_3PO_3 = \frac{1078}{98} = 11$

$\text{Molarity} = \frac{11}{1} = 11\text{M}$

13. **(7)** $\text{Molarity} = \frac{\text{Mass} \times 1000}{\text{Molar mass} \times \text{Volume (mL)}}$

$\Rightarrow \ 0.35 = \frac{\text{Mass} \times 1000}{82.02 \times 250} \Rightarrow \text{mass} = 7.176\text{ g}$

14. **(b)** Mole fraction of 'C' $= \frac{n_C}{n_A + n_B + n_C}$

$= \frac{\text{Moles of C}}{\text{Total number of moles}}$

15. **(4)** 31.4% H_2SO_4 by mass
$\Rightarrow$ 31.4 g of H_2SO_4 in 100 g of water

$\text{density} = 1.25\text{ g/mL} = \frac{100}{V} \Rightarrow V = \frac{100}{1.25}\text{ mL}$

$\text{molarity} = \frac{m}{M} \times \frac{1000}{100} \times 1.25 = \frac{31.4}{98} \times \frac{1000}{100} \times 1.25$
$= 4.0051\text{ M} \approx 4\text{ M}$

16. **(815)** Let's take 1000 mL of this solution (1000 cm^{-3})
$\Rightarrow$ Mass of this solution = density × volume
$= 1.06\text{ g cm}^{-3} \times 1000\text{ cm}^{-3} = 1.06 \times 10^3\text{ g}$
$\Rightarrow$ Mass of H_2SO_4 present = n × molar mass
$= 0.8 \times 98 = 78.4\text{ g}$
$\Rightarrow$ Mass of solvent = Mass of solution – Mass of solute
$= 1060\text{ g} - 78.4\text{ g} = 981.6\text{ g} = 0.982\text{ kg}$

$\Rightarrow \text{Molality} = \frac{n}{\text{mass of solvent in Kg}} = \frac{0.8}{0.982}$
$= 0.815\text{ mol Kg}^{-1} = 815 \times 10^{-3}\text{ mol Kg}^{-1}$

17. **(a)** $\text{Molarity} = \frac{\text{Moles of solute}}{\text{Volume of solution in 'L'}}$

Since volume depends on temperature, molarity will change upon change in temperature.

18. **(d)** Solute (X) = 2 g
Solvent (H_2O) = 1 mole = 18 g
Total mass = 2 + 18 = 20 g

$\%\text{ mass of X} = \frac{2}{20} \times 100 = 10\%$

19. **(d)** $M = \frac{3.15 \times 1000}{126 \times 250} = \frac{12.6}{126} = 0.1$

20. **(3.64)** 3M NaCl solution = 3 mol of NaCl = (3 × 58.5)g = 175.5 g of NaCl is present in 1000 ml solution = 1000 g of solution. (as density = 1g/ml)
$\therefore$ Mass of solvent = 1000g – 175.5 g = 824.5g

$\therefore$ Molality of the solution $= \frac{3 \times 1000}{824.5} = 3.64(\text{m})$

21. **(139)** 10 mL solute in 90 mL solvent
$\therefore$ mass of solute = 10 × 3.2 = 32g
mass of solvent = 90 × 1.6g

$m = \frac{32 \times 1000}{160 \times 90 \times 1.6} = 1.388$
$m = 138.8 \times 10^{-2} \approx 139 \times 10^{-2}$ molal

22. **(148)** $\text{Molarity} = \frac{\text{mole of solute}}{\text{volume of the solution (L)}}$

$2.6 \times 10^{-3} = \frac{X/85}{0.67141} \Rightarrow X = 0.148\text{ g}$

$\text{Conc. of DCM in ppm} = \frac{0.148}{1.49 \times 671.141} \times 10^6 = 148\text{ ppm.}$

23. **(180)** $M = \frac{5}{40} \times \frac{1000}{450}$

$M_1V_1 = M_2V_2$

$\left(\frac{5}{40} \times \frac{1000}{450}\right) \times V_1 = 0.1 \times 500 \Rightarrow V_1 = 180$

24. **(5)** Mass percent, mole fraction, molarity, ppm, molality are used for measuring concentration terms.

25. **(54)** Moles of nitric acid = $800 \times 10^{-3} \times 0.5 = 0.4$

Moles of nitric acid evaporated $= \frac{11.5}{63} = 0.182$

Moles left = 0.4 – 0.182 = 0.218

$\text{Molarity} = \frac{0.218 \times 10^3}{400} = 0.545\text{ M} = 54.5 \times 10^{-2}\text{ M}$

26. **(a)** Molality of solution does not depend on the temperature as it depends on the moles of solute and mass of solvent.
Hence, both assertion and reason are true.

27. **(64)** Moles of $CuSO_4.5H_2O = \frac{80}{249.54}$

$\text{Molarity} = \frac{80}{249.54} \times \frac{1}{5} = 64.117 \times 10^{-3} \simeq 64 \times 10^{-3}$

28. **(20)** $[H_2C_2O_4 \cdot 2H_2O] = \frac{\text{weight / M.W.}}{V(\text{L})}$

$\Rightarrow x \times 10^{-2} = \frac{6.3/126}{250/1000} \Rightarrow x = 20$

29. **(3)** $\text{Molality} \Rightarrow \frac{3.30\text{ mol}}{\text{kg}} = \frac{3.30\text{ mol solute}}{1000\text{ g solvent}}$

Mass of solution (solute + solvent)
$= (3.3 \times 74.5) + 1000 = 1245.85\text{ g}$

$\text{Volume of solution} = \frac{1245.85}{1.2}\text{ mL}$

$\text{So Molarity} = \frac{3.3 \times 1.2}{1245.85} \times 1000 = 3.17 \simeq 3.0\text{ M}$

30. **(50)** Mole of $Na^+ = \frac{3.45}{23} = 0.15\text{ mol}$

3 mol of Na^+ is present in 1 mole of Na_3PO_4.

0.15 mol of Na^+ is present in $\frac{0.15}{3}$ mole of Na_3PO_4.

Now,

$\text{Molarity} = \frac{0.15 \times 1000}{3 \times 100} = 0.5\text{ mol/L} = 50 \times 10^{-2}\text{ mol/L}$

31. **(64)** Molality of an aqueous solution of a solute = 100 m
Let the mole fraction of solute be x_a

$$\text{Molality} = \frac{\text{Mole fraction of solute} \times 1000}{\text{Mole fraction of solvent} \times \text{Mol. mass of solvent}}$$

Then, $\frac{x_a \times 1000}{(1-x_a)\times 18} = 100\,\text{m}$

$\Rightarrow x_a = \frac{18}{28} = 0.6428 = 64.28 \times 10^{-2} \simeq 64 \times 10^{-2}$

32. **(9)** $\text{Molality} = \frac{\text{Moles of solute}}{\text{Mass of solvent (in kg)}}$

6.50 molal solution of KOH means
6.50 moles of KOH in 1000 g of water (solvent)
So, weight of solute = 6.5 × 56 = 364 g
Weight of solution = 1000 + 364 = 1364 g

$\text{Volume of solution} = \frac{\text{Mass of solution}}{\text{Density of solution}} = \frac{1364}{1.89}$

$\text{Molarity} = \frac{\text{Moles of solute}}{\text{Volume of solution (in L)}}$

$= \frac{6.50 \times 1.89 \times 1000}{1364} = 9.00\ \text{M}$

33. **(c)** Let, 1 mole be present in the solution,

$n_{solute} = 0.2$ mol

$n_{solvent} = 0.8\ \text{mol} = 0.8 \times 18\ \text{g}$

$\text{molality} = \frac{\text{Moles of solute}}{\text{Kilogram of solvent}}$

$\therefore\ m = \frac{0.2 \times 1000}{0.8 \times 18} = \frac{1000}{4 \times 18} \approx 13.88$

34. **(d)** 20% W/W KI solution (Given)
i.e., 100 g solution contains 20 g KI
∴ Mass of solvent = 100 – 20 = 80 g

No. of moles of KI is $\frac{20}{166} = 0.121$ mol

$\therefore\ \text{Molality} = \frac{0.121 \times 1000}{80} = 1.51\ \text{m}$

35. **(b)** Number of moles in 92 g of $Na^+ = \frac{92}{23} = 4$ mol

$\text{Molality}\ (m) = \frac{\text{Number of moles}}{\text{Mass of solvent (in kg)}}$

$\therefore\ m = \frac{4}{1} = 4\ \text{mol kg}^{-1}$

36. **(d)** According to Henry's law

$\frac{P_1}{P_2} = \frac{S_1}{S_2}$

$\frac{500}{750} = \frac{0.01}{S_2}$

$\therefore S_2 = \frac{750 \times 0.01}{500} = 0.015\ \text{g/L}$

37. **(a)** From molarity equation :
$M_1V_1 + M_2V_2 = M \times V$

$M = \frac{M_1V_1 + M_2V_2}{V}$

$= \frac{750 \times 0.5 + 250 \times 2}{750 + 250} = 0.875\ \text{M}$

38. **(d)** $\text{Molarity} = \frac{\text{Moles of solute}}{\text{Volume of solution in L}}$

Mass of solution = 1000 + 120 = 1120

$V = \frac{M}{d} = \frac{1120}{1.15}\ \text{mL}$

$\text{Molarity} = \frac{120 \times 1.15}{60 \times 1120} \times 1000 = 2.05\ \text{M}$

39. **(b)** Binary mixture of C_6H_5OH and $C_6H_5NH_2$ shows negative deviation from Raoult's law. Vapour pressure of solution is less than vapour pressure of pure C_6H_5OH & $C_6H_5NH_2$. So boiling point of solution is greater than boiling point of pure C_6H_5OH & $C_6H_5NH_2$. Hence it shows maximum boiling azeotrope.

40. **(a)**
(A) Solution of chloroform and acetone shows –ve deviation, so maximum boiling azeotrope.
(B) Solution of ethanol & water shows +ve deviation. So minimum boiling azeotrope.
(C) Solution of benzene and toluene form ideal solution. $\Delta V_{mix} = 0$
(D) Acetic acid in benzene form dimer.

41. **(c)** $P_A^o = 350$ mm Hg, $P_B^o = 750$ mmHg

$\Rightarrow\ P_A^o < P_B^o \Rightarrow \frac{P_A^o}{P_B^o} < 1 \Rightarrow \frac{350}{750} < 1$

$\frac{y_A}{y_B} = \frac{P_A^o\, x_A}{P_B^o\, x_B} = \frac{350\, x_A}{750\, x_B}$

$\Rightarrow\ \frac{y_A / y_B}{x_A / x_B} < 1 \Rightarrow \frac{y_A}{y_B} < \frac{x_A}{x_B}$

42. **(d)** $P_S = \overset{\circ}{P}_A.X_A + \overset{\circ}{P}_B.X_B$

$500 = 200 \times \frac{1}{4} + \overset{\circ}{P}_B.\frac{3}{4}$

$\overset{\circ}{P}_B = 600$ mm Hg

As $\overset{\circ}{P}_A < \overset{\circ}{P}_B \Rightarrow$ A is least volatile.

43. **(a)** $\because$ For liquid solution of two liquids '1' and '2'
$P_1 = P_T y_1 = P^0_1 x_1$ (From Dalton's Law and Raoult's Law)

$$\therefore \frac{P_T}{x_1} = \frac{P_1^0}{y_1}$$

$$\therefore \frac{P_2^0 + x_1(P_1^0 - P_2^0)}{x_1} = \frac{P_1^0}{y_1}$$

$$\therefore \frac{P_2^0}{x_1} + (P_1^0 - P_2^0) = \frac{P_1^0}{y_1}$$

$$\therefore \frac{P_2^0}{x_1} = \frac{P_1^0}{y_1} - (P_1^0 - P_2^0)$$

$$\therefore \frac{P_2^0}{x_1} = \frac{P_1^0}{y_1} + (P_2^0 - P_1^0)$$

$$\therefore \frac{1}{x_1} = \left(\frac{P_1^0}{P_2^0}\right)\left(\frac{1}{y_1}\right) + \left(\frac{P_2^0 - P_1^0}{P_2^0}\right)$$

Since $y = mx + c$

$$\therefore \text{Slope} = \left(\frac{P_1^0}{P_2^0}\right) \quad \therefore \text{ Intercept} = \left(\frac{P_2^0 - P_1^0}{P_2^0}\right)$$

44. **(d)** $(CH_3)_2CO + CS_2$ exibits positive deviations from Raoult's Law.

45. **(d)** Solution with negative deviation has
$P_T < P_{A^\circ}X_A + P_{B^\circ}X_B$ (Raoult's Law) $\Rightarrow P_A < P_{A0}X_A$ and $P_B < P_{B^\circ}X_B$
If vapour pressure decreases so boiling point increases, due to stronger intermolecular forces.

46. **(23)** wt of solute = 30 g
Volume of solution = 100 mL
wt. of solution = 1.2 × 100 = 120 g
wt. of solvent = 120 – 30 = 90 g

(n_1) mole of glucose $= \frac{30}{180} = 0.1666$

(n_2) mole of water $= \frac{90}{18} = 5$

$$\frac{P^\circ - P}{P^\circ} = \frac{n_1}{n_1 + n_2} = \frac{0.1666}{5 + 0.166} = \frac{0.1666}{5.166} = 0.32$$

$$\frac{24 - P}{24} = 0.32$$

$24 - P = 0.77$
$P = 23.23$

47. **(82)** Boiling point is the temperature at which its vapour pressure is equal to the atmospheric pressure (1 atm or 1.013 bar). Therefore Boiling point of solvent is 82°C.
Boiling point of solution is 83°C

48. **(555)** $P_X = \chi_x P_T = \dfrac{\frac{0.6}{20}}{\frac{0.6}{20} + \frac{0.45}{45}} \times 740$

$P_X = 555$ mm Hg

49. **(314)** Let V.P. of pure A be $P_A{}^0$ Let V.P. of pure B be $P_B{}^0$
When $X_A = 0.7$ & $X_B = 0.3$; $P_s = 350$ mm Hg

$\Rightarrow P_A^0 \times 0.7 + P_B^0 \times 0.3 = 350$ ...(i)

When $X_A = 0.2$ & $X_B = 0.8$
$P_s = 410$ mm Hg

$\Rightarrow P_A^0 \times 0.2 + P_B^0 \times 0.8 = 410$ ...(ii)

Solving (i) and (ii)
$P^0{}_A = 314$ mm Hg; $P^0{}_B = 434$ mm Hg

50. **(1)** $p = K_H \times x$

$$0.920 \text{ bar} = 46.82 \times 10^3 \text{ bar} \times \frac{\text{mol of } O_2}{\text{mol of } H_2O}$$

$$0.920 = 46.82 \times 10^3 \times \frac{\text{mol of } O_2}{1000/18}$$

$0.920 = 46.82 \times n_{O_2} \times 18 \Rightarrow n_{O_2} = 1.09 \times 10^{-3}$

$\Rightarrow$ mmol of $O_2 = 1$

51. **(2)** mole fraction of substance A in liquid phase = 0.2
The Raoult's law equation is:

$$P_A = \left(P_A^\circ\right) \times X_A$$

$$\Rightarrow 0.4 \text{ atm} = P_A^\circ \times 0.2 \Rightarrow P_A^\circ = \frac{0.4}{0.2} = 2 \text{ atm}$$

52. **(2)** 40% Hydrogen contains 40g Hydrogen gas and 60g of Oxygen gas.

$$n_{H_2} = \frac{40}{2} = 20\,;\; n_{O_2} = \frac{60}{32} = 1.8$$

$$\chi_{H_2} = \frac{n_{H_2}}{n_{\text{Total}}} = \frac{20}{20 + 1.8} = \frac{20}{21.8} = 0.92$$

$$p_{H_2} = \chi_{H_2} \cdot P_{\text{Total}} = 0.92 \times 2.2 = 2.02 \text{ bar}$$

53. **(a)** According to Henry's law
$p = K_H \chi$
for χ constant : $p \propto K_H$; p has max partial pressures.
So it should have maximum K_H and s have minimum partial pressure, So K_H should be minimum.

54. **(14)** $P_{\text{Total}} = P_A^o x_A + P_B^o x_B$
$= (50 \times 0.3) + (100 \times 0.7) = 85$ torr
$P_B = y_B \times P_{\text{Total}} \Rightarrow 70 = y_B \times 85$

$$y_B = \frac{70}{85} = \frac{14}{17} \Rightarrow x = 14$$

55. **(1222)** $P = K_H X_{CO2}$

$$0.835 = 1.67 \times 10^3 \times \frac{n_{CO_2}}{n_{CO_2} + n_{H_2O}}$$

$$\approx 1.67 \times 10^3 \times \frac{n_{CO_2}}{n_{H_2O}}$$

$$\Rightarrow 0.5 \times 10^{-3} = \frac{n_{CO_2}}{1000/18} \Rightarrow n_{CO_2} = 27.775 \times 10^{-3}$$

$$w_{CO_2} = 44 \times 27.775 \times 10^{-3} = 1222.1 \times 10^{-3} \text{ g}$$

56. **(25)** From Henry's law,

$P_{Gas} = K_H.X_{gas}$

$$X_{gas} = \frac{n_{gas}}{n_{gas} + n_{H_2O}} \approx \frac{n_{gas}}{n_{H_2O}}$$

$$0.835 = 1.67 \times 10^3 \times \frac{n_{(CO_2)}}{\frac{0.9 \times 1000}{18}}$$

$n_{(CO_2)} = 0.025$ mole or 25 m mol

57. **(77)** $P^o_B = 70$ torr and $P^o_{MB} = 20$ torr

$P_B = P^o_B X_B$ and $P_{MB} = P^o_{MB} X_{MB}$

$P_{total} = P_B + P_{MB}$

$X_B = X_{MB} = 0.5$ [∵ mixture is equimolar]

∴ $P_{total} = (70 \times 0.5) + (20 \times 0.5) = 45$ torr

Mole fraction of benzene in the vapour phase above the equimolar mixture

$$= \frac{P^o_B X_B}{P_{total}} = \frac{70 \times 0.5}{45} = 0.77 = 77 \times 10^{-2}$$

58. **(1)** Given $P_A° = 90$ mm Hg, at 25 °C

$P_B° = 15$ mm Hg

$$\left.\begin{matrix} X_A = 0.6 \\ X_B = 0.4 \end{matrix}\right\} P_T = X_A P_A^\circ + X_B P_B^\circ$$

$= (0.6 \times 90) + (0.4 \times 15) = 54 + 6 = 60$ mm

Now mol fraction of B in the vapour phase

$$= \frac{P_B}{P_T} = \frac{X_B P_B^\circ}{60} = \frac{15 \times 0.4}{60} = 0.1 = 1 \times 10^{-1}$$

59. **(25)** P_{O_2} (over water) = 20 kPa

K_H for $O_2 = 8.0 \times 10^4$ kPa

If X_{O_2} is the mole fraction of O_2 in solution, then according to Henry's law, $P_{O_2} = K_H(X_{O_2})$

$$X_{O_2} = \frac{20}{8.0 \times 10^4} = 2.5 \times 10^{-4}$$

Mass of 1 kg of water containing O_2 = 1 L

∴ Molarity of O_2 in solution = 25×10^{-5} M

60. **(19)** Using Raoult's law,

$$P_s = \chi_A P_A^o + \chi_B P_B^o = \left(\frac{1}{3} \times 21\right) + \left(\frac{2}{3} \times 18\right) = 19 \text{ kPa}$$

61. **(600)** $P_{total} = \left(P^o_{hexane} \cdot X_{hexane}\right) + \left(P^o_{heptane} \cdot X_{heptane}\right)$

$$550 = \left([P^o_{hexane}] \times \frac{1}{4}\right) + \left([P^o_{heptane}] \times \frac{3}{4}\right)$$

$\Rightarrow 2200 = P^o_{hexane} + 3P^o_{heptane}$ ...(i)

$$560 = \left([P^o_{hexane}] \times \frac{1}{5}\right) + \left([P^o_{heptane}] \times \frac{4}{5}\right)$$

$\Rightarrow 2800 = P^o_{hexane} + 4P^o_{heptane}$ ...(ii)

From equation (i) and (ii),

$P^o_{heptane} = 600$ mm of Hg

62. **(d)** (a) From Henry's law $p = K_H(x)$

Higher the value of K_H, smaller will be the solubility of the gas, so γ is more soluble.

(b) Though solubility of gases will decrease with increase in temperature but this conclusion can not be drawn from the given table.

(c) For γ : $(p)_\gamma = (K_H)_\gamma \cdot (x)_\gamma$

$$= 2 \times 10^{-5} \left[\frac{\frac{55.5}{1000}}{55.5 + \frac{1000}{18}}\right] = 10^{-5} \text{ k bar} = 10^{-2} \text{ bar}$$

(d) For δ $\Rightarrow p_\delta = (K_H)_\delta \cdot (x)_\delta = 0.5 \left[\frac{55.5}{55.5 + \frac{1000}{18}}\right]$

$= 0.5 \times 0.5 = 0.25$ k bar = 250 bar.

63. **(a, c)** Initial : $H_2O(l) \rightleftharpoons H_2O(g)$

On addition of solute : $H_2O(l)$ + glucose $\rightleftharpoons H_2O(g)$

Hence, the rate at which water molecules leaves the vapour increases and that leaves the solution decreases.

64. **(c)** At a particular temperature, as intermolecular force of attraction increases, vapour pressure decreases. Thus, intermolecular forces are inversely proportional to vapour pressure and directly proportional to temperature. Therefore, X has lower intermolecular interactions compared to Y.

65. **(b)** (a) Mixture of carbon disulphide and acetone will show positive deviation from Raoult's Law.

(c) The dipolar interaction between solute (CS_2) and solvent (acetone) molecules in solution are weaker than that between solute-solute and solvent-solvent. So the vapour pressure of solution will be greater than the individual vapour pressure of pure components.

(b) For positive deviation, $\Delta_{mix} V > 0$ and (d) $\Delta_{mix} H > 0$.

66. **(a)** When these two beakers are sealed together, then there will be natural flow of solvent from low concentration to high concentration. Thus, solvent will flow from the beaker with pure solvent towards beaker with solution. Hence, the volume of solution increases and the volume of the solvent decreases.

67. **(d)** $P^o_M = 450$ mmHg, $P^o_N = 700$ mmHg

$$P_M = P^o_M x_M = y_M P_T \Rightarrow P°_M = \frac{y_M}{x_M}(P_T)$$

Similarly, $P^o_N = \frac{y_N}{x_N}(P_T)$; given, $P^o_M < P°_N$

$$\Rightarrow \frac{y_M}{x_M} < \frac{y_N}{x_N} \Rightarrow \frac{y_M}{y_N} < \frac{x_M}{x_N}$$

68. **(d)** $P = x_B p°_B + x_A p°_A$

$= (0.5 \times 600) + (0.5 \times 400) = 300 + 200 = 500$ mm Hg

Using the relation $p_i = y_i P_{Total}$, we can calculate the mole fractions of the components in vapour phase.

$p_B = y_B P_{total}$

$$y_B = \frac{p_B}{P_{Total}} = \frac{300}{500} = \frac{3}{5} = 0.6$$

$$y_A = \frac{p_A}{P_{Total}} = \frac{200}{500} = \frac{2}{5} = 0.4$$

69. **(a)** According to Henry's law,
$P = K_H . X_{gas}$
$\because X_{gas} + X_{H_2O} = 1 \Rightarrow X_{gas} = 1 - X_{H_2O}$
$\therefore P = K_H(1 - X_{H_2O}) \Rightarrow P = K_H - K_H . X_{H_2O}$
$y = c + mx$; $m = -K_H \Rightarrow$ slope is negative
and $c = K_H$
Therefore, the intercept on Y axis increases from w to z.

70. **(b)** $P°_A = 7 \times 10^3$; $P°_B = 12 \times 10^3$
$x'_A = 0.4$; $x'_B = 1 - 0.4 = 0.6$
$P_{total} = P°_A x'_A + P°_B x'_B$
$= 7 \times 10^3 \times 0.4 + 12 \times 10^3 \times 0.6$
$= (7 \times 0.4 + 12 \times 0.6) \times 10^3 = 10^4$

$$x_A = \frac{P_A^\circ x'_A}{P_{total}} = \frac{7 \times 10^3 \times 0.4}{10^4}$$

$\therefore x_A = 0.28, x_B = 1 - 0.28 = 0.72$

71. **(a)** The solubility of the gas in liquids decreases with the increase in value of K_H at a given pressure.

72. **(c)** Molar mass of $CHCl_3$ = 119.5 g/mol
Molar mass of CH_2Cl_2 = 85 g/mol

$$\text{Moles of } CHCl_3 = \frac{11.95}{119.5} = 0.1 \text{ mol}$$

$$\text{Moles of } CH_2Cl_2 = \frac{8.5}{85} = 0.1 \text{ mol}$$

$$\text{Mole fraction of } CHCl_3 = \frac{0.1}{0.2} = 0.5$$

$$\text{Mole fraction of } CH_2Cl_2 = \frac{0.1}{0.2} = 0.5$$

Vapour pressure of $CHCl_3$ = 200 mm Hg
Vapour pressure of CH_2Cl_2 = 415 mm Hg
$\therefore P_{(total)}$ = [Mole fraction of $CHCl_3$ × (Vapour pressure of $CHCl_3$)] + [Mole fraction of CH_2Cl_2 × (Vapour pressure of CH_2Cl_2)]
$= (0.5 \times 200) + (0.5 \times 415) = 307.5$
Mole fraction of $CHCl_3$ in vapour form

$$= \frac{\text{vapour pressure}}{P_{total}} = \frac{0.5 \times 200}{307.5} = 0.325.$$

73. **(d)** According to Henry's law,

$$\frac{P_1}{P_2} = \frac{S_1}{S_2} \Rightarrow \frac{500}{750} = \frac{0.01}{S_2}$$

$$\therefore S_2 = \frac{750 \times 0.01}{500} = 0.015 \text{ g/L}$$

74. **(b)** Total V.P. of solution = $P_A^\circ X_A + P_B^\circ X_B$

Given, $P_A^\circ = 74.7$ torr, $P_B^\circ = 22.3$ torr

$n_{benzene} = 1.5$ mol, $n_{toluene} = 3.5$ mol
$n_{solution} = 1.5 + 3.5 = 5$ mol

$$x_A = \frac{n_{benzene}}{n_{solution}} = \frac{1.5}{5} = 0.3$$

$$x_B = \frac{n_{toluene}}{n_{solution}} = \frac{3.5}{3} = 0.7$$

Total V.P. of solution = $(0.3 \times 74.7 + 0.5 \times 22.3)$ torr
= $(22.4 + 15.6)$ torr = 38 torr

$$\text{Mole fraction of benzene in vapour form} = \frac{22.4}{38} = 0.589$$

75. **(b)** If we plot a graph between the vapour pressure and the temperature, we would get a curve that rises faster as T increases, giving a curved line.

76. **(d)** Solutions in which solute - solute and solvent-solvent interactions are almost similar to solute-solvent interactions are known as ideal solution.

77. **(b)** $P_A = P_A^\circ \times X_A$ = total pressure × y_A
$P_B = P_B^\circ \times X_B$ = total pressure × y_B
where x and y represents mole fraction in liquid and vapour phase respectively.

$$\frac{P_B^\circ X_B}{P_A^\circ X_A} = \frac{y_B}{y_A}; \frac{P_B^\circ(1 - X_A)}{P_A^\circ X_A} = \frac{1 - y_A}{y_A}$$

on putting values $\frac{119(1 - 0.50)}{37 \times 0.50} = \frac{1 - y_A}{y_A}$

on solving $y_A = 0.237$

78. **(c)** Let vapour pressure of $A = P_A^\circ$
Vapour pressure of $B = P_B^\circ$
In first solution,

$$\text{Mole fraction of } A(X_A) = \frac{1}{1+2} = \frac{1}{3}$$

$$\text{Mole fraction of } B(X_B) = \frac{2}{1+2} = \frac{2}{3}$$

According to Raoult's law,
Total vapour pressure $= 250 = P_A^\circ X_A + P_B^\circ X_B$

$$250 = \frac{1}{3}P_A^\circ + \frac{2}{3}P_B^\circ \quad ...(i)$$

In second solution

$$\text{Mole fraction of } A(X_A) = \frac{2}{2+2} = \frac{2}{4} = \frac{1}{2}$$

$$\text{Mole fraction of } B(X_B) = \frac{2}{4} = \frac{1}{2}$$

$\therefore$ Total vapour pressure $= 300 = P_A^\circ X_A + P_B^\circ X_B$

$$300 = \frac{1}{2}P_A^\circ + \frac{1}{2}P_B^\circ \quad ...(ii)$$

Multiplying equation (i) by $\frac{1}{2}$ and equation (ii) by $\frac{1}{3}$

$$\frac{1}{6}P_A^\circ + \frac{2}{6}P_B^\circ = 125$$

$$\frac{1}{6}P_A^\circ + \frac{1}{6}P_B^\circ = 100$$

$$\frac{1}{6}P_B^\circ = 25$$

$P_B^\circ = 25 \times 6 = 150$ mm Hg

On substituting value of P_B° in equation (ii) we get

$$300 = P_A^\circ \times \frac{1}{2} + 150 \times \frac{1}{2}$$

$P_A^\circ = 450$ mm Hg

79. **(a)** $P_{Total} = (P^\circ{}_A X_A) + (P^\circ{}_B X_B)$

$= (P^\circ{}_{Heptane} X_{Heptane}) + (P^\circ{}_{Octane} X_{Octane})$

$$= \left(105 \times \frac{25/100}{\frac{25}{100} + \frac{35}{114}}\right) + \left(45 \times \frac{35/114}{\frac{25}{100} + \frac{35}{114}}\right)$$

$$= \left(105 \times \frac{0.25}{0.25+0.3}\right) + \left(45 \times \frac{0.3}{0.25+0.3}\right)$$

$$= \frac{105 \times 0.25}{0.55} + \frac{45 \times 0.3}{0.55} = \frac{26.25 + 13.5}{0.55} = 72\,\text{kPa}$$

80. **(b)** For this solution, intermolecular interactions between *n*-heptane and ethanol are weaker than *n*-heptane - *n*-heptane and ethanol-ethanol interactions. Hence, the solution of *n*-heptane and ethanol is non-ideal and shows positive deviation from Raoult's law.

81. **(b)** $P_{total} = P_x^\circ X_x + P_y^\circ X_y$

$$550 = P_x^\circ \times \frac{1}{4} + P_y^\circ \times \frac{3}{4}$$

$P_x^\circ + 3P_y^\circ = 550 \times 4$ ...(i)

In second case,

$$P_{total} = P_x^\circ \times \frac{1}{5} + P_y^\circ \times \frac{4}{5}$$

$P_x^\circ + 4P_y^\circ = 560 \times 5$ ...(ii)

Subtract (i) from (ii)

$\therefore P_y^\circ = 560 \times 5 - 550 \times 4 = 600$

$\because P_x^\circ = 400$

82. **(d)** At 1 atmospheric pressure, the boiling point of mixture is 80 °C.

At boiling point, the vapour pressure of mixture, $P_T = 1$ atmosphere = 760 mm Hg.

Using the relation,

$P_T = P_A^\circ X_A + P_B^\circ X_B$, we get

$\{P_B^\circ = 1000$ mm Hg, $X_A + X_B = 1\}$

or $760 = (520X_A) + (1000(1 - X_A))$

or $X_A = \frac{240}{480} = \frac{1}{2}$ or 50 mol. percent

83. **(b)** $P_A^\circ = ?$, Given $P_B^\circ = 200\,\text{mm}$, $X_A = 0.6$,

$X_B = 1 - 0.6 = 0.4$, $P = 290$

$P = P_A + P_B = (P_A^\circ X_A) + (P_B^\circ X_B)$

$290 = (P_A^\circ \times 0.6) + (200 \times 0.4)$

$\therefore P_A^\circ = 350\,\text{mm}$

84. **(d)** Let the mass of methane and oxygen = m g.

Mole fraction of O_2

$$= \frac{\text{Moles of } O_2}{\text{Moles of } O_2 + \text{Moles of } CH_4}$$

$$= \frac{m/32}{(m/32) + (m/16)} = \frac{m/32}{3m/32} = \frac{1}{3}$$

Partial pressure of O_2 = Total pressure × mole fraction of O_2

$$\Rightarrow p_{O_2} = P \times \frac{1}{3} = \frac{1}{3}P$$

85. **(d)** Acetonitrile ($\overset{\delta+}{CH_3} - C \equiv \overset{\delta-}{N}$) and acetone both are polar molecules, hence dipole-dipole interaction exists

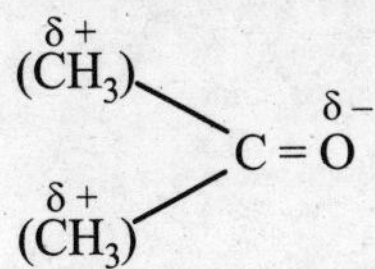

between them. Between KCl and water, ion-dipole interaction is found and in benzene-ethanol and benzene–carbon tetrachloride, dispersion force is present.

86. **(d)** Vapour pressure of benzene = 75 torr

Vapour pressure of toluene = 22 torr

$$\text{Moles of benzene} = \frac{78}{78} = 1\,\text{mol}$$

$$\text{Moles of toluene} = \frac{46}{92} = 0.5\,\text{mol}$$

Now partial pressure of benzene

$$= P^\circ{}_b \cdot X_b = 75 \times \frac{1}{1+0.5} = 75 \times \frac{1}{1.5} = 75 \times \frac{2}{3} = 50$$

87. **(b)** Equimolar solutions of normal solutes in the same solvent will have the same b. pts and same f. pts. As these are colligative properties.

88. **(d)** Gaseous densities of ethanol and dimethyl ether would be same at same temperature and pressure. The heat of vaporisation, V.P. and b.pts will differ due to H-bonding in ethanol.

89. **(b)** Positive deviations are shown by such solutions in which solvent-solvent and solute-solute interactions are stronger than the solvent-solute interactions. In such solutions, the interactions among molecules becomes weaker. Therefore, their escaping tendency increases which results in the increase in their partial vapour pressure. In a solution of methanol, there exists intermolecular H–bonding.

In this solution, benzene molecules come between methanol molecules which weaken intermolecular forces. This results in increase in vapour pressure.

90. **(d)** When A and B form an ideal solution, $\Delta H_{mix} = 0$

91. **(d)** In solution containing A and B component showing negative deviation A–A and B–B interactions are weaker than that of A–B interactions. For such solutions, $\Delta H = -ve$ and $\Delta V = -ve$

92. **(a)** $\Delta T_f = i k_f m$

$$i = \frac{\Delta T_f}{k_f m} \Rightarrow i = \frac{0.20}{1.8 \times 0.1} = 1.11$$

$$\alpha = \frac{i-1}{n-1}$$

for HA, n = 2

$$\therefore \alpha = \frac{1.11-1}{1} = 0.11$$

$$K_a = \frac{[H^+][A^-]}{[HA]}$$

$$= \frac{c\alpha^2}{1-\alpha} = \frac{0.1 \times (0.11)^2}{1-0.11} = 1.38 \times 10^{-3}$$

93. **(33)** $MX_3 \rightarrow M^{3+} + 3X^-$

$i = 1 + (n-1)\alpha$

$\Rightarrow 2 = 1 + (4-1)\alpha$

$\Rightarrow 2 - 1 = 3\alpha$

$\Rightarrow \alpha = \frac{1}{3} = 33.33\% \approx 33\%$

94. **(d)** **Statement-I:** Molar depression constant K_f

$$= \frac{M_1 R T_f^2}{\Delta H_{fus}} = \frac{M_1 R \cdot T^2}{\Delta S_{fus} \cdot T_f} = \frac{M_1 R T_f}{\Delta S_{fus}}$$

Hence statement-I is correct.

Statement-II: K_f for benzene = 5.12 K kg mol^{-1}; K_f for water = 1.86 K kg mol^{-1}.

Hence statement- II is incorrect.

95. **(c)**

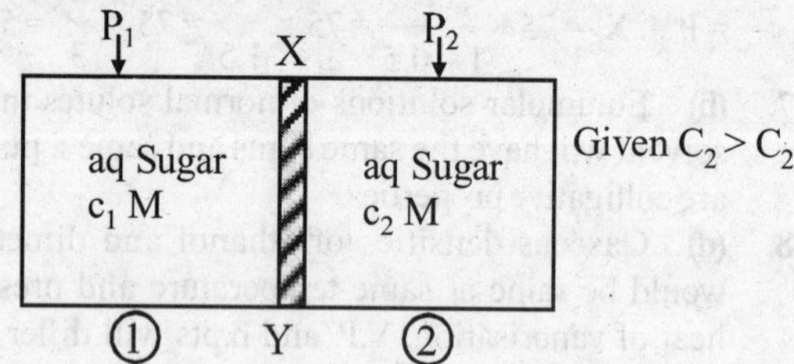

Membrane should be cellophane or parchment paper

Normal osmosis occurs from (2) to (1)

For reverse osmosis from (1) to (2)

Pressure : P_1 > Osmotic pressure (π)

$\therefore$ Correct options are A and C.

96. **(b)** $\Delta T_b = i_1 m_1 k_b + i_2 m_2 k_b$

$$= 1 \times \frac{2}{0.5} \times 0.52 + \frac{1 \times 2}{0.5} \times 0.52 = 4.16$$

$$(T_b)_{solution} = 373.16 + 4.16 = 377.3K$$

97. **(25)** For AB

$\Delta T_b = 2.7, K_b = 0.5$ K kg mol^{-1}

$$m = \frac{\Delta T_b}{K_b} = 5.4m$$

For AB_2

$$\Delta T_b = 1.5\text{ K},\ m = \frac{1.5}{0.5} = 3m$$

Moles of AB = 0.081 mol

Moles of AB_2 = 0.045 mol

$$\text{Molar mass of AB} = \frac{\text{mass}}{\text{moles}} = \frac{1}{0.081} = 12.35 \text{ g/mol}$$

$$\text{Molar mass of } AB_2 = \frac{1}{0.045} = 22.22 \text{ g/mol}$$

Let atomic mass of A = a, B = b

$\therefore$ a + b = 12.35

a + 2b = 22.22

On solving we get a = 2.48 = 24.8 × 10^{-1}

98. **(b)** Salt and ice mixture is colder so it prevents early melting due to depression in freezing point

99. **(16)** $A_2B \rightarrow 2A^+ + B^{-2}$; n = 3

$\alpha = 0.3$

$i = 1 + (n-1)\alpha$

$= 1 + (3-1)(0.3) = 1.6 = 16 \times 10^{-1}$

100. **(d)** For AX_2:- $\Delta T_b = K_b \times m \times i$

$$0.0156 = 0.52 \times \frac{0.01}{1} \times i_{AX_2}$$

$\Rightarrow i_{AX_2} = 3 \Rightarrow$ complete ionisation

For AY_2:- $\Delta T_b = K_b \times m \times i$

$$0.026 = 0.52 \times \frac{25.4}{250 \times 2} \times i_{AY_2}$$

$\Rightarrow \quad i_{AY_2} = 0.98 \simeq 1 \ \therefore$ completely unionised

101. **(N)**

NTA Ans. (d) Living cell = 0.9 gm in 100 gm of solution

% w/w = 0.9

Solution has equal moles of glucose and water = 0.5 each

Weight of solution = 0.5 × 180 + 0.5 × 18 = 99 g

$\Rightarrow$ w/w % = (90/99) × 100 = 90.9%

Concentrated solution $\Rightarrow$ Cell will shrink.

102. **(a)** $\Delta T_f = K_b . m$

$$0.4 = K_b \frac{\frac{1}{256}}{50 \times 10^{-3}}$$

K_b = 5.12 Kg/mol

103. **(75)** $MX_2 \rightarrow M^{+2} + 2X^-$

n = 3

$$i = \frac{\text{normal molar mass}}{\text{abnormal molar mass}};\ i = \frac{164}{65.6}$$

$$1 + (n-1)\alpha = \frac{164}{65.6};\ 1 + (3-1)\alpha = \frac{164}{65.6}$$

$$2\alpha = \frac{98.4}{65.6};\ \alpha = 0.75$$

Percent dissociation = 75%

104. (c) A solution will freeze when its vapour pressure equal the vapour pressure of the pure solid solvent.
On adding non-volatile solute in a solvent, the freezing point of solution decreases.

$T_f < T_f^0 \quad \therefore \quad \Delta T_f = T_f^0 - T_f$
F.P. of solution < F.P. of pure solvent
Also V.P. of solution decreases on adding non-volatile solute in a solvent.

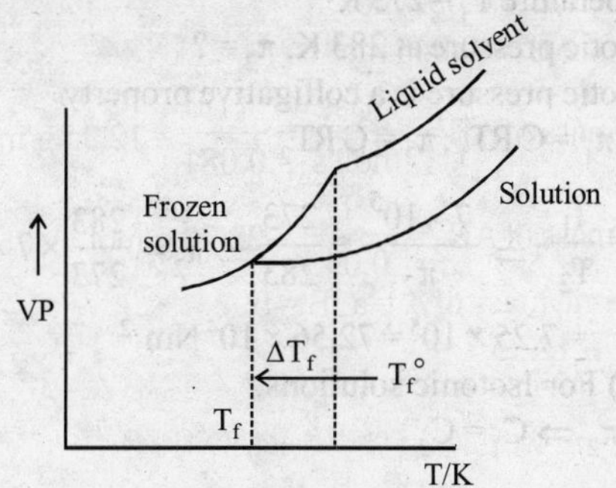

105. (a)

$$[Co(NH_3)_5Cl]Cl_2 \rightarrow \underbrace{[Co(NH_3)_5Cl]^{2+}(aq) + 2Cl^-(aq)}_{3 \text{ ions in water}}$$

$$[Co(NH_3)_5Cl]Cl_2(aq) + 2AgNO_3(aq) \rightarrow [Co(NH_3)_5Cl](NO_3)_2(aq) + 2AgCl(s)$$

106. (a) $P^\circ - P \propto x_{solute}$
$P^\circ - P = 10$ mm Hg and $x_{solute} = 0.2$
If $P^\circ - P = 20$ mm Hg, then $x_{solute} = 0.4$
$\therefore \quad x_{solvent} = 1 - x_{solute}$
$= 1 - 0.4$
$= 0.6$

107. (c) Given : $\Delta T_f = 0.558°C$, $k_f = 1.86 \dfrac{K \times kg}{mol}$,

Molality = 0.1 m
$\Delta T_f = i \times k_f \times m$
$\Rightarrow 0.558 = i \times 1.86 \times 0.1 \Rightarrow i = 3$
When 100% ionisation takes place

$$[Cr(NH_3)_5Cl]Cl_2 \xrightarrow{\text{aqueous solution}} [Cr(NH_3)_5Cl]^{2+} + 2Cl^-$$

number of ions = 3

108. (a) $\Delta T_b \propto$ i.C. Where C = concentration and i = van't Hoff factor.

Options	i.C.
(i)	2×10^{-4}
(ii)	1×10^{-4}
(iii)	2×10^{-3}
(iv)	2×10^{-2}

B.P. order:
(ii) < (i) < (iii) < (iv)

109. (d) Only solvent Molecules are allowed to pass through the SPM. Water moves from $K_2Cr_2O_7$ to $CuSO_4$ solution causing molarity of $CuSO_4$ to decrease.

110. (23) $X_{methylbenzene} = 0.5$

$$Y_{methylbenzene} = \frac{P_{methylbenzene}}{P_{total}}$$

$$= \frac{X_{methylbenzene} \times P^\circ_{methylbenzene}}{P_{total}}$$

$$= \frac{0.5 \times 24}{(0.5 \times 80) + (0.5 \times 24)} = \frac{12}{40 + 12} = 0.23 = 23 \times 10^{-2}$$

111. (5) Ionisation of AB_2 occurs as:
$AB_2 \rightarrow A^{2+} + 2B^-$
$1 - \alpha \quad \alpha \quad 2\alpha$
$i = 1 + 2\alpha$
$\Delta T_b = k_b \, im$

$$0.52 = 0.52\,(1 + 2\alpha)\frac{10}{20}$$

$$1 = (1 + 2\alpha)\frac{10}{20} \Rightarrow 2 = 1 + 2\alpha \Rightarrow \alpha = 0.5$$

Hence, $\alpha = 5 \times 10^{-1}$

112. (76) $HX \rightleftharpoons H^+ + X^-$; $K_a = 1.2 \times 10^{-5}$

	HX	H^+	X^-
Initial conc.	0.03 M		
Conc. at eq.	0.03 – x	x	x

$$K_a = 1.2 \times 10^{-5} = \frac{x^2}{0.03 - x}$$

$0.03 - x \approx 0.03$ (K_a is very small)

$$\Rightarrow \frac{x^2}{0.03} = 1.2 \times 10^{-5} \Rightarrow x = 6 \times 10^{-4}$$

Conc. of final solution $= 0.03 - x + x + x$
$= 0.03 + x = 0.03 + 6 \times 10^{-4}$
osmotic pressure, $\pi = CRT$
$\pi = (0.03 + (6 \times 10^{-4})) \times 0.083 \times 300$
$= 76.19 \times 10^{-2} \approx 76 \times 10^{-2}$ bar

113. (543) $\Delta T_f = 2.5°C$
$\Delta T_f = i \times k_f \times m$

$$2.5 = 1 \times 1.86 \times \frac{n_B}{0.1}$$

$$n_B = \frac{2.5 \times 0.1}{1.86} = 0.1344 \text{ mol}$$

mass of methanol $= 0.1344 \times 32 = 4.3$ g

$$d = \frac{m}{v} \Rightarrow v = \frac{m}{d}$$

$$v = \frac{4.3}{0.792}\text{ mL} = 5.43 \text{ mL} = 543 \times 10^{-2} \text{ mL}$$

x = 543

114. (25)

$$\underset{0.05M}{NaCl} \longrightarrow \underset{0.05M}{Na^+} + \underset{0.05M}{Cl^-}$$

Total $C_1 = 0.05 + 0.05 = 0.1$ M (NaCl)
$C_2 = 0.2$ M (glucose)
$\pi = (C_2 - C_1)RT$
$= (0.2 - 0.1) \times 0.083 \times 300 = 2.49$ bar
$= 24.9 \times 10^{-1}$ bar

115. (19) $CH_3COOH(aq) \rightleftharpoons CH_3COO^-(aq) + H^+(aq)$
$K_a = 6.25 \times 10^{-5}$

mass of acetic acid = 1.2 × 5 g

$$\text{concentration} = \frac{1.2\times5}{60\times1}\text{M}$$

$$\alpha = \sqrt{\frac{K_a}{c}} = \sqrt{\frac{6.25\times10^{-5}}{(1.2\times5)/60}}$$

$$= \sqrt{625\times10^{-6}} = 25\times10^{-3}$$

$i = \alpha n + (1-\alpha) = 1 - \alpha + n\alpha = 1 + (n-1)\alpha$

$= 1 + (2-1)\,0.025 = 1.025$

$\Delta T_f = 1.025 \times 1.86 \times 0.1 = 0.19065$

$= -19.065 \times 10^{-2} = 19$

116. (707) $\Delta T_b = k_b \times m \Rightarrow 2 = 0.52 \times m;\ m = \frac{2}{0.52}$

$$\frac{\Delta P}{P^\circ} = \frac{n_{solute}}{n_{solvent}}$$

$$\frac{\Delta P}{P^\circ} = \frac{m}{1000} \times M_{solvent}$$

$$\Delta P = P^\circ \times \frac{m}{1000} \times M_{solvent}$$

$$760 \times \frac{\frac{2}{0.52}}{1000} \times 18 = 52.615$$

$P^0 = 760 - 52.615 = 707.385$ mm of Hg

117. (31) As moles of water > moles of CH_3COOH water is solvent.

Molality of acetic acid

$$= \frac{2700}{60} \times \frac{1}{2.7}$$

$= 16.66$

$\Delta T_f = k_f \times 16.66$

$= 1.86 \times 16.66 = 31K.$

118. (a) A = 0.1 M, B = 0.01 M, C = 0.001 M

Value of i increases as solution becomes dilute so the correct order is: $i_A < i_B < i_C$

119. (15) Given: $T = (-24)°C$

$K_f = 1.86\ \text{K Kgmol}^{-1}$

$M_2 = 62\ \text{g/mol} = 0.062\ \text{Kg/mol}^{-1}$

$W_1 = 18.6$ Kg

$W_2 = ?$

$\Delta T_f = T° - T = (273 - 249)\text{ K} = 24\text{ K}$

$$\therefore\ \Delta T_f = \frac{K_f \times W_2}{M_2\ W_1} \Rightarrow W_2 = \frac{\Delta T_f \times M_2 W_1}{K_f}$$

$$= \frac{24 \times 0.062 \times 18.6}{1.86} \Rightarrow W_2 = 14.88\text{ Kg} \approx 15\text{ Kg}$$

120. (d) Depression in freezing point of benzene takes place.

121. (a)

Freezing point depression ∝ number of solute particles
(Colligative property) (Van't Hoff factor, i)

Acetic acid dissociates in water so 2i will be the van't Hoff factor. Acetic acid and benzoic acid will dimense in benzene so $\frac{i}{\alpha}$ will be the actual van't Hoff factor.

Glucose does not dissociate in water so i will be the actual van't Hoff factor.

Thus, Acetic acid in water will have the highest van't Hoff factor and thus highest depression in the freezing point.

122. (72.56) Osmotic pressure $\pi_1 = 7 \times 10^5$ Pa

Temperature $T_1 = 273$ K

Osmotic pressure at 283 K, $\pi_2 = ?$

Osmotic pressure is a colligative property.

$\therefore\ \pi_1 = C\,RT_1;\ \pi_2 = C\,RT_2$

$$\frac{\pi_1}{\pi_2} = \frac{T_1}{T_2} \Rightarrow \frac{7\times10^5}{\pi_2} = \frac{273}{283} \Rightarrow \pi_2 = \frac{283}{273} \times 7\times10^5$$

$= 7.25 \times 10^5 = 72.56 \times 10^4\ \text{Nm}^{-2}$

123. (240) For Isotonic solutions,

$\pi_1 = \pi_2 \Rightarrow C_1 = C_2$

$\frac{12}{x} = 0.05$ [x → Molar Mass of A]

$x = 240$

124. (116) Amount of solvent = 100 − (29.25 + 19) = 51.75 g

$$\Delta T_b = \left[\frac{2\times29.25\times1000}{58.5\times51.75} + \frac{3\times19\times1000}{95\times51.75}\right] \times 0.52$$

(For NaCl and $MgCl_2$)

$\Delta T_b = 16.075$

$\Delta T_b = (T_b)_{solution} - (T_b)_{solvent}$

$\Rightarrow (T_b)_{solution} = 100 + 16.07 = 116.07°C$

125. (48) $MgCl_2 \rightarrow Mg^{+2} + 2Cl^-$

$1-\alpha \quad \alpha \quad 2\alpha$

$i = 1 + 2\alpha\ (\alpha = 0.8) \Rightarrow i = 2.6$

$$\frac{\Delta p}{p^o} = \frac{i \times n_2}{n_1} \Rightarrow \Delta p = 2.34$$

$p_s = 47.66 \approx 48$

126. (75) $\pi_{K_2SO_4} = \pi_{Glucose}$ (Isotonic solutions)

$i \times 0.004 \times RT = 0.01 \times RT$

$i = 2.5$

For K_2SO_4 {for dissociation $i = 1 + (n-1)\alpha$}

$$(\alpha) = \frac{i-1}{n-1} = \frac{2.5-1}{3-1} = 0.75\ \%\text{ dissociation} = 75$$

127. (d) $\frac{P^o - P_s}{P^o} = \frac{n}{N}$ (for dilute solution)

$$\frac{0.2}{54.2} = \frac{n\times18}{100} \Rightarrow n = \frac{100}{271\times18}$$

$$w = \frac{100\times180}{271\times18} \Rightarrow w = 3.69\text{ g}$$

128. (30) $i = 1 + \alpha$ (for HA) = 1.3

$$\%\text{ increase} = \frac{(\Delta T_f)_{obs} - (\Delta T_f)_{cal}}{(\Delta T_f)_{cal}} \times 100$$

$$= \frac{(K_f \times i \times m) - (K_f \times m)}{K_f \times m} \times 100 = \frac{i-1}{1} \times 100 = 30\%$$

129. (40535) $\because \pi = CRT$

$$\pi = \frac{n}{V}\frac{RT}{M} \Rightarrow \pi = \frac{w}{V}\frac{RT}{M} \Rightarrow M = \frac{wRT}{\pi \times V}$$

$$M = \frac{0.63 \times 0.083 \times 300}{1.29 \times 10^{-3} \times 300 \times 10^{-3}}$$

M = 40535 g/moL

130. (8) $\frac{(T_B)_x}{(T_B)_y} = \frac{2}{1}; \frac{(\Delta H)_x}{(\Delta H)_y} = \frac{1}{2}$ (given)

$$\frac{(\Delta T_B)_x}{(\Delta T_B)_y} = \frac{(K_B)_x \times \text{molality}}{(K_B)_y \times \text{molality}}$$

$(K_B)_x/(K_B)_y = m$

$$K_B = \frac{R \times M_1 \times T_b^2}{1000 \times \Delta_{vap}H} \Rightarrow K_B \,\alpha\, \frac{T_b^2}{\Delta_{vap}H}$$

$$\therefore \frac{(K_B)_x}{(K_B)_y} = \frac{(T_b)_x^2}{(T_b)_y^2} \times \frac{\Delta H_y}{\Delta H_x} = m$$

$$\therefore m = \frac{2^2}{1^2} \times \frac{2}{1} = 4 \times 2 = 8$$

131. (1111) $\frac{P^\circ - P}{P^\circ} = \frac{n_{solute}}{n_{solute} + n_{solvent}}$

$$\Rightarrow \frac{100 - 75}{100} = \frac{\frac{x}{60}}{\frac{x}{60} + \frac{1000}{18}} = \frac{1}{4} = \frac{\frac{x}{60}}{\frac{x}{60} + 55.55}$$

$$\therefore \quad 4 = 1 + \frac{55.55 \times 60}{x}$$

$$\therefore \quad x = \frac{55.55 \times 60}{3}$$

x = 1111g

132. (4) $\pi = i\,c\,RT \Rightarrow \pi \propto i \times c$

$\therefore$ (A) $\pi_{NaCl} \propto 2 \times 1; \pi_{Urea} \propto 1 \times 2$

(B) $\pi_{CaCl_2} \propto 3 \times 1; \pi_{KCl} \propto 2 \times 1.5$

(C) $\pi_{AlCl_3} \propto 4 \times 1.5; \pi_{Na_2SO_4} \propto 3 \times 2$

(D) $\pi_{KCl} \propto 2 \times 2.5; \pi_{Al_2(SO_4)_3} \propto 5 \times 1$

$\therefore$ A, B. C and D are isotonic pairs.

133. (3.4) $KCl + AgNO_3 \rightarrow AgCl \downarrow + KNO_3$

1 mol   1 mol   1 mol   1 mol

At equivalence point, 1 mmol KCl = 1 mmol $AgNO_3$

$\therefore$ 20 ml of 1 M $AgNO_3$ = 20 mmol of $AgNO_3$ = 20 mmol of KCl

= $20 \times 10^{-3} \times 74.5$ g of KCl = 1.49 g of KCl

25 ml of an aqueous solution of KCl

= 25g of aqueous solution of KCl.

($\because$ density of solution = 1g/ml)

$\therefore$ Mass of solvent = 25 – 1.49 = 23·51g

$$\therefore \quad \text{Molality of the solution} = \frac{20 \times 10^{-3} \times 10^3}{23 \cdot 51} = 0.85$$

$\therefore \Delta T_f = i\,K_f m$ [i = 2 considering 100% ionization]

$= 2 \times 2 \times 0.85 = 3.4$

134. (372) 20% of acetic acid is dissociated in the solution.

$\therefore \quad i = 1 + (n-1)\alpha$

$i = 1 + 0.2(2-1) = 1.2$

$\Delta T_f = i K_f m$

$$\Delta T_f = 1.2 \times 1.86 \times \frac{5 \times 1000}{60 \times 500} = 0.372$$

$\Delta T_f = 372 \times 10^{-3}\,°C$.

135. (62250) $\pi = CRT$

$$\frac{400}{10^5} = \frac{2.5}{M_o} \times \frac{1000}{250} \times 0.83 \times 300 \Rightarrow M_o = 62250$$

136. (d) $\Delta T_b \propto i \times C$

Osmosis always takes place from hypotonic to hypertonic solution.

When KI solutions is added to $AgNO_3$ solution, positively charged solution results due to adsorption of Ag^+ ions from dispersion medium.

137. (100) $\Delta T_b = 373.52 - 373 = 0.52$ K

$\Delta T_b = K_b$ molality

$$0.52 = 0.52 \times \frac{2}{\text{molecular mass}} \times \frac{1000}{20}$$

Molecular Mass of solute = 100g/mol

138. (243) $\Delta T_f = i.K_f.m$

$$\Rightarrow \Delta T_f = 2.67 \times 1.8 \times \frac{38}{98} \times \frac{1000}{62}$$

$\Rightarrow \Delta T_f = 30.05$ K

$\therefore$ Freezing temperature = 273 – 30 = 243 K.

139. (13) Let a mole of $Pb(NO_3)_2$ be added

$$Pb(NO_3)_2 \rightarrow \underset{a}{Pb^{2+}} + \underset{2a}{2NO_3^-}$$

(a under $Pb(NO_3)_2$)

$\Delta T_b = 0.15 = 0.5\,[3a] \Rightarrow a = 0.1$

$Pb^{2+}(aq) \;+\; 2Cl^-(aq) \;\rightarrow\; PbCl_2(s)$

t = 0   0.1   0.2

t = ∞   (0.1 – x)   (0.2 – 2x)

In final solution; ions present = Pb^{2+}, Na^+, NO_3^-, Cl^-

$$\Delta T_f = 0.8 = 1.8\left[\frac{0.3 - 3x + 0.2 + 0.2}{1}\right] \Rightarrow x = \frac{2.3}{7}$$

$$K_{sp} = [Pb^{2+}][Cl^-]^2 = \left(0.1 - \frac{2.3}{27}\right)\left(0.2 - \frac{4.6}{27}\right)^2 = 13 \times 10^{-6}$$

140. (a) A-III, B-I, C-II, D-IV

141. (41,500) The relation related to the osmotic pressure:

$$\pi = \left(\frac{W/M}{V}\right)RT$$

(W = mass of PVC, M = molecular weight of PVC and V = the volume of the solution)

$$\Rightarrow \pi = \left(\frac{W}{V}\right)\left(\frac{1}{M}\right)RT = C\left(\frac{RT}{M}\right) \Rightarrow \frac{\pi}{C} = \frac{RT}{M} \neq f(C)$$

As per the given conditions:

$$\text{Slope} = \frac{RT}{M} = \frac{0.083 \times 300}{M} = 6 \times 10^{-4}$$

$$\Rightarrow M = \frac{0.083 \times 300}{6 \times 10^{-4}} = \frac{830 \times 300}{6} = 41{,}500 \text{ g/mole}$$

142. (4) $\pi = iCRT$

$\pi \propto iC$

A, B, D and E have same value of osmatic pressure.

143. (a) Vapour pressure of solvent is greater than vapour pressure of solution.

Only solvent molecules is solidify at the freezing point.

144. (80) Mass of C_2H_5OH = density × volume

$= 62.5 \times 0.8 = 50$ g

$\Delta T_f = K_f \times m$ [m = molality]

$$0.9 = 2 \times \frac{1.8 \times 1000}{M_W \times 50} \Rightarrow M_W = \frac{2 \times 1.8 \times 1000}{0.9 \times 50} = 80$$

145. (15) Moles of ascorbic acid $= \frac{10.2\text{g}}{176\text{g mol}^{-1}} = 0.058$ mol

$\Delta T_f = (x \times 10^{-1})\,°C = K_f$. molality

$$= 3.9 \times \frac{0.058}{150} \times 1000 = 1.5\,°C = 15 \times 10^{-1}\,°C$$

146. (6) $\frac{p^0 - p_s}{p^\circ} = \frac{n_{solvent}}{n_{solute}} \Rightarrow \frac{P^0 - \frac{P^0}{2}}{P^0} = \frac{n_{solute}}{n_{solvent} + n_{solute}}$

$$\frac{1}{2} = \frac{n_{solute}}{n_{solute} + n_{solvent}}$$

$$2 = 1 + \frac{5.55}{n_{solute}}$$

$n_{solute} = 5.55 \approx 6$ mol

147. (b) $\Delta_A T_b = \frac{2}{M_A \times 100} \times 1000 \times K_b$

$$\Delta_B T_b = \frac{8}{M_B \times 100} \times 1000 \times K_b$$

Given, $\Delta_A T_b = \Delta_B T_b$

$$\Rightarrow \frac{2}{M_A} = \frac{8}{M_B} \Rightarrow M_B = 4M_A$$

148. (1) $\Delta T_b = iK_b m_1; \Delta T_f = iK_f m_2$

$$\frac{\Delta T_b}{\Delta T_f} = \frac{K_b \times 1}{K_f \times 2} \Rightarrow \frac{3}{6} = \frac{1}{2} = \frac{K_b}{K_f} \times \frac{1}{2} \Rightarrow \frac{K_b}{K_f} = \frac{1}{1} \Rightarrow x = 1$$

149. (c) $\Delta T_f = 0.0405$ K

Concentration of formic acid (HCOOH) solution

= 0.5 ml/L

$= 0.5 \times 1.05$ g/L = 0.525 g ($\because d_{HCOOH} = 1.05$ g/ml)

No. of moles of HCOOH in 1 L solution

$$= \frac{0.525}{46} = 0.0114 \quad (\because Mw_{HCOOH} = 46)$$

Molality (m) of HCOOH = 0.0114 mol/kg

(considering d_{H2O} = 1kg/L)

$\Delta T_f = i K_f m$

$$\therefore i = \frac{\Delta T_f}{K_f m} = \frac{0.0405}{1.86 \times 0.0114} = 1.91 \approx 1.9$$

∴ The van't Hoff factor of the formic acid solution is nearly 1.9.

150. (b) Depression in freezing point is a colligative property and depends only on the amount of solute.

$\Delta T_f = k_f . m$

$$\frac{\Delta T_{fx}}{\Delta T_{fy}} = \frac{k_f \cdot m_x}{k_f \cdot m_y} = \frac{\frac{1}{M_x}}{\frac{1}{M_y}} = \frac{1}{1}$$

$$\Rightarrow \frac{1}{4} = \frac{M_y}{M_x} \Rightarrow M_x : M_y = 1 : 0.25$$

151. (5) $d = \frac{\text{Mass}}{\text{volume}}$

∴ Mass of acetic acid = 1.02 × 1.2 = 1.224 g

$$\text{Moles of acetic acid} = \frac{\text{Mass}}{\text{Molar mass}} = \frac{1.224}{60} = 0.0204$$

Mass of water = 2 kg [1L ≈ 1kg for water]

$$\text{Molality (m)} = \frac{0.0204}{2} = 0.0102 \text{ mol/kg}$$

Now $\Delta T_f = i \times K_f \times m$

$\Rightarrow i = \frac{\Delta T_f}{K_f \times m}$ and $i = 1 + \alpha$ for acetic acid

∴ $0.0198 = (1 + \alpha) \times 1.85 \times 0.0102$

$\alpha = 0.04928 \cong 5\%$

152. (3) Elevation in boiling point and depression in freezing point are colligative properties and their value depends only on the amount of solute present in the solution.

$\Delta T_b = iK_b m$; $\Delta T_f = iK_f m$

$$\frac{\Delta T_b}{\Delta T_f} = \frac{4}{4} = \frac{K_b 1.5}{K_f 4.5} \Rightarrow \frac{K_b}{K_f} = 3$$

153. (330) $\pi = CRT$

$5.03 \times 10^{-3} = C \times 0.083 \times 300$

$C = 0.202 \times 10^{-3}$M; $C = \frac{n}{\text{volume}}$

Moles of protein $= 0.202 \times 10^{-3} \times 0.5 = 10^{-4} \times 1.01$

$$\text{moles} = \frac{\text{given weight}}{\text{molar mass}}$$

$1.01 \times 10^{-4} = \frac{2.5}{M}$; M(molar mass of protein) = 24752

$$\therefore \text{No. of glycine units} = \frac{\text{molar mass of protein}}{\text{Normal molar mass of Glycine}}$$

$$= \frac{24752}{75} = 330.03$$

154. (8) Given: $\frac{(K_b)_A}{(K_b)_B} = \frac{1}{8}$

$$\frac{(\Delta T_b)_A}{(\Delta T_b)_B} = \frac{(K_b)_A \cdot m}{(K_b)_B \, m} = \frac{1}{8} \Rightarrow \frac{x}{y} = \frac{1}{8}$$

∴ $y = 8$ (nearest integer)

155. (45) $w_{Solute} = 2.5$ g $K_b = 0.52$ K kg mol^{-1}
$w_{solvent} = 75$g M = Mol. Wt. of solute

$T_b^\circ = 373.15$K ; $\Delta T_b = 0.385$ K

$$\Delta T_b = \frac{K_b \times 10^3\, w_{solute}}{M_{solute} \times w_{solvent}} \Rightarrow 0.385 = 0.52 \times \left(\frac{2.5}{M} \times \frac{1000}{75}\right)$$

M = 45 g mol^{-1}.

156. (99) 0.5% solution of KCl

$$\text{Molality, } m = \frac{0.5}{74.6} \times \frac{1000}{100}$$

$$\Delta T_f = i \times m \times K_f$$

$$0.24 = i \times \frac{0.5}{74.6} \times \frac{1000}{100} \times 1.8 \Rightarrow i = 1.989$$

$$\underset{1-\alpha}{KCl} \rightleftharpoons \underset{\alpha}{K^+} + \underset{\alpha}{Cl^-} \Rightarrow 1.989 = 1 + \alpha \Rightarrow \alpha = 0.989$$

$\% \alpha = 98.9\% \approx 99\%$

157. (415) $\pi = iCRT$; i = 1 here

$$\left(C = \frac{n}{v} = \frac{\text{weight}}{\text{Molar mass} \times \text{volume (1 kg = 1000 g)}}\right)$$

$$= \frac{1 \times 2}{60000 \times 0.2} \times 0.083 \times 300 = 0.00415 \text{ bar } (\because 1 \text{ bar} = 10^5 \text{Pa})$$

So, 0.00415 × 10^5 Pa = 415 Pa

158. (d) $\Delta T_f = i \times K_f \times m$

$$0.2 = i \times 1.86 \times \frac{0.7}{93} \times \frac{1000}{42} \Rightarrow i = 0.60$$

$$\begin{matrix} 2A & \rightleftharpoons & A_2 \\ 1-\alpha & & \frac{\alpha}{2} \end{matrix}$$

$$i = 1 - \alpha + \frac{\alpha}{2} \Rightarrow 1 - \frac{\alpha}{2} = 0.60 \Rightarrow \alpha = 0.80$$

⇒ % association = 80%

159. (54) $\pi = C.R.T$
$7.47 = C \times 0.083 \times 300 \Rightarrow C = 0.3$ M $= 0.3 \times 180$ gL^{-1} = 54 gL^{-1}

160. (13) K_b (benzene) = 2.6 K kg mol^{-1}
K_b (acetone) = 1.7 K kg mol^{-1}
With benzene as solvent, $\Delta T_b = iK_b m$

$$\Delta T_b = \frac{1}{2} \times 2.6 \times \frac{1.22/M_w}{100/1000} \quad ...(i)$$

With Acetone as solvent, $\Delta T_b = iK_b m$

$$0.17 = 1 \times 1.7 \times \frac{1.22/M_w}{100/1000} \quad ...(ii)$$

Dividing equation (i) by (ii)

$$\frac{\Delta T_b}{0.17} = \frac{\frac{1}{2} \times 2.6}{1 \times 1.7} \Rightarrow \Delta T_b = 13 \times 10^{-2} \Rightarrow x = 13$$

161. (518) Molality of sucrose sol. = 0.75 m, i.e., 0.75 mol/kg (solvent)
Mass of sucrose = 0.75 × 342g = 256.5 g
Mass of solution = mass of sucrose + mass of H_2O
= 1000 + 256.5 = 1256.5 g
Mass of solution in 1 kg (given) solution

$$= \frac{256.5}{1256.5} \times 1000 = 204.1 \text{ g}$$

Mass of H_2O in 1 kg solution = 1000 – 204.1 = 795.9 g
After cooling the solution to – 4°C,
$\Delta T_f = iK_f m$

$$4 = 1 \times 1.86 \times \frac{204.1 \times 1000}{342 \times w_B} \quad [\text{where } w_B = \text{Mass of } H_2O \text{ left}]$$

$\Rightarrow w_B = 277.5$ g
Mass of ice separated = 795.9 – 277.5 = 518.4 ≃ 518 g

162. (271)
$$\text{molality} = \frac{\left(\frac{40}{180}\right)\text{mol}}{0.2 \text{ kg}} = \left(\frac{10}{9}\right)\text{molal}$$

$$\Rightarrow \Delta T_f = K_f \times m = T_f - T_f' = 1.86 \times \frac{10}{9}$$

$$\Rightarrow T_f' = 273.15 - \left(1.86 \times \frac{10}{9}\right)$$

= 271.08 K ≃ 271K (nearest-integer)

163. (269) $k_f = 1.86$ k. kg/mol
Solute : $C_2H_6O_2$ (83 g) (Non dissociative solute)
$\Rightarrow \Delta T_f = k_f \times m$

$$= 1.86 \times \frac{83/62}{624/1000}$$

$$= \frac{1.86 \times 83 \times 1000}{62 \times 625} = \frac{154380}{38750}$$

$\Delta T_f = 4$K
$\because \Delta T_f = T^o - T_s$
$\therefore T_s = T^o - \Delta T_f$
= 273 – 4 = 269K

164. (4) $\Delta T_f = iK_f m$
$\Delta T_f \propto i$

Compound	i
C_2H_5OH	1
$Ba_3(PO_4)_2$	5
Na_2SO_4	3
KCl	2
Li_3PO_4	4

So, all compounds will exhibit lower freezing point as compared to C_2H_5OH.

165. (15) $\pi = CRT$

$$\pi = \frac{W_B \times R \times T}{M_B \times \text{Vol (in L)}}$$

$$\Rightarrow 2.42 \times 10^{-3} = \frac{1.46 \times 0.083 \times 300}{M_B \times 0.1}$$

$\Rightarrow M_B = 15.02 \times 10^4$ g/mol

166. (125) 50% i.e., 0.5 moles of HA dimerizes :

	2HA	→	$(HA)_2$
Initial	a		0
Final	$(1-0.5)a$		$\frac{0.5a}{2}$

Remaining 50% of HA dissociates :

	HA	→	H^+	+	A^-
Initial	0.5a		0		0
Final	0		0.5a		0.5a

van't Hoff factor (i)

$$= \frac{\text{Total no. of particles dissociated or associated}}{\text{Initial total no. of particles}}$$

$$= \frac{0.25a + 0.5a + 0.5a}{a} = 1.25 = 125 \times 10^{-2}$$

167. (250) Let molar mass of 'X' be M.

$\Delta T_b = K_b m.$

$$\Rightarrow 0.6 = 5 \times \left(\frac{\text{Wt.} \times 1000}{M \times \text{Mass of solvent (g)}} \right)$$

$$\Rightarrow 0.6 = 5 \times \frac{3 \times 1000}{M \times 100} \Rightarrow M = \frac{150}{0.6} = 250 \text{ g / mol.}$$

168. (50) Given freezing point = – 3.885 °C.

$\Delta T_f = 0 - (-3.885) = 3.885$

$\Delta T_f = iK_f \cdot m$

$\Rightarrow \quad 3.885 = i \times 1.85 \times 2 \Rightarrow \quad i = 1.05$

$$\underset{1-\alpha}{HA} \rightleftharpoons \underset{\alpha}{H^+} + \underset{\alpha}{A^-}$$

(α = degree of dissociation)

van't Hoff factor $(i) = 1 + \alpha \Rightarrow 1.05 = 1 + \alpha$

$\therefore \alpha = 0.05 = 50 \times 10^{-3}$

169. (100) $\underset{1-\alpha}{2A} \rightarrow \underset{\frac{\alpha}{2}}{A_2}$

$i = 1 - \alpha + \frac{\alpha}{2}$; Also, $\Delta T_b = iK_b m$

$\Delta T_b = T_b - T_b^0 = 100.52 - 100 = 0.52$

$\Rightarrow 0.52 = i \times 0.52 \times 2 \Rightarrow i = \frac{1}{2}$

$\Rightarrow \frac{1}{2} = 1 - \alpha + \frac{\alpha}{2} \Rightarrow 1 - \frac{\alpha}{2} = \frac{1}{2} \Rightarrow \alpha = 1$

So, % association = 100%.

170. (85) $K_4Fe(CN)_6 \rightleftharpoons 4K^+ + Fe(CN)_6^{4-}$

	$K_4Fe(CN)_6$	$4K^+$	$Fe(CN)_6^{4-}$
Initial	1	0	0
Final	$1-\alpha$	4α	α

Total moles = $1 - \alpha + 4\alpha + \alpha$

$$i = \frac{1 + 4\alpha}{1} = 1 + (4 \times 0.4) = 2.6$$

$\Delta T_{b_1} = 2.6 \times K_b \times 1$ ...(i)

For a non-electrolyte solute (A), $i = 1$

18.1 weight percent solution means 18.1 g solute is present in 100 g solution, then weight of solvent

$= 100 - 18.1 = 81.9$g

$$(\Delta T_b)_A = 1 \times \frac{18.1 \times 1000}{M_A \times 81.9} \times K_b \quad \text{...(ii)}$$

Given, $\Delta T_b = (\Delta T_b)_A$

$$2.6 \times K_b \times 1 = \frac{1 \times 18.1 \times 1000}{M_A \times 81.9} \times K_b \Rightarrow M_A = 85.$$

171. (106) $AB_2 \rightleftharpoons A^{2+} + 2B^-$

	AB_2	A^{2+}	$2B^-$
$t = 0$	1	0	0
$t = t$	$1-\alpha$	α	2α

$i = 1 + \frac{2\alpha}{1} = 1 + 2 \times 0.1 \Rightarrow i = 1.2$

$\Delta T_b = i \times k_b \times m = 1.2 \times 0.5 \times 10 = 6$

$\Delta T_b = T_s - T°$

$T_s = \Delta T_b + T° = 6 + 100 = 106°C$

172. (2) $\Delta T_f = iK_f m$

$\Delta T_f = T_f^o - T_f = 0 - (-0.93) = 0.93$

$$0.93 = i \times 1.86 \times \frac{12.2 \times 1000}{122 \times 100} \Rightarrow i = 0.5$$

n(Benzoic acid) $\longrightarrow$ $(\text{Benzoic acid})_n$

$$i = \frac{\text{Total number of particles after association}}{\text{Number of particles before association}}$$

$0.5 = \frac{1}{n} \quad \Rightarrow \quad n = 2$

173. (24) $n_{SO2} = \frac{1 \times 0.224}{0.082 \times 298} \approx 0.0092 \approx 0.01$ mol

$NaOH + SO_2 \rightarrow NaHSO_3$

$n_{NaHSO3} = 0.01$

$NaHSO_3 \rightarrow Na^+ + HSO_3^-$

Ignoring the dissociation of HSO_3^- into H^+ and SO_3^{2-}, van't Hoff factor $(i) = 2$

Relative lowering of vapour pressure,

$$\frac{P^o - P_s}{P^o} = i \times \frac{n_{NaHSO_3}}{n_{H_2O} + n_{NaHSO_3}}$$

For dilute solution, $n_{NaHSO3} << n_{H2O}$

then $\frac{P^o - P_s}{P^o} = i \times \frac{n_{NaHSO_3}}{n_{H_2O}}$

$$\Delta P = P^o \times i \times \frac{n_{NaHSO_3}}{n_{H_2O}}$$

$$= \frac{24 \times 2 \times 0.01}{36/18} = 0.24 = 24 \times 10^{-2} \text{ mmHg}$$

174. (3) $\underset{1-\alpha}{AB} \rightleftharpoons \underset{\alpha}{A^{n+}} + \underset{\alpha}{B^{n-}}$

$i = 1 + \alpha = 1 + 0.75 \, (\because \alpha = 0.75)$

$= 1.75$

$\Delta T_b = iK_b m$

$\therefore m = \dfrac{2.5}{1.75 \times 0.52} = 2.75 \text{ mol/kg} \simeq 3.$

175. (375) $\underset{1-0.6}{A_2B_3} \longrightarrow \underset{2\times0.6}{2A^{3+}} + \underset{3\times0.6}{3B^{2-}}$

$\therefore$ van't Hoff factor $= 1 - 0.6 + 1.2 + 1.8 = 3.4$

$\Delta T_b = i \cdot k_b \cdot m = 3.4 \times 0.52 \times 1 = 1.768$

$\because \Delta T_b = T_b - T_b^0 \Rightarrow \therefore T_b = 373 + 1.768 = 374.77 \approx 375$

176. (35) $ClCH_2COOH \rightleftharpoons ClCH_2COO^- + H^+$

Degree of dissociation $= \alpha$

$\Rightarrow i = 1 + \alpha$ and $\Delta T_f = iK_f m$

$\Rightarrow 0.5 = (1+\alpha)(1.86) \times \dfrac{9.45}{94.5} \times \dfrac{1000}{500}$

$\Rightarrow 0.5 = (1+\alpha)(1.86)(0.2) \Rightarrow 1 + \alpha = 1.34 \Rightarrow \alpha = 0.34$

	$ClCH_2COOH$	$\rightleftharpoons ClCH_2COO^-$	$+ H^+$
Initial	α	0	0
Final	$C - C\alpha$	$C\alpha$	$C\alpha$

$K_a = \dfrac{(C\alpha)^2}{C - C\alpha} = \dfrac{C\alpha^2}{1-\alpha}$ $[C \approx m]$

$\Rightarrow K_a = \dfrac{0.2 \times (0.34)^2}{1 - 0.34} = 0.035 \Rightarrow \therefore K_a = 35 \times 10^{-3}.$

177. (1) $\Delta T_f = K_f \times m$ (Solute is non-dissociating)

$\Rightarrow (T_f^0 - T'_f) = 5.12 \times \dfrac{10 \times 1000}{58 \times 200}$

$\Rightarrow 5.5 - T'_f = \dfrac{5.12 \times 5 \times 10}{58} \Rightarrow T'_f = 1.086\,°C \approx 1\,°C.$

178. (c) Relative lowering in vapour pressure (RLVP)

$= \dfrac{P - P_s}{P} = \dfrac{n}{n+N}$

$n \rightarrow$ moles of solute; $N \rightarrow$ moles of solvent

RLVP $\propto n$ or RLVP $\propto \dfrac{1}{\text{Mol. Mass}}$

$n_A = \dfrac{10}{100}, n_B = \dfrac{10}{200}, n_C = \dfrac{10}{10000}$

From the above relation,
RLVP(A) > RLVP(B) > RLVP(C)

179. (c) Relative lowering in vapour pressure (RLVP)

$= \dfrac{P - P_s}{P} = \dfrac{n}{n+N}$

$n \rightarrow$ moles of solute

$N \rightarrow$ moles of solvent

$n_A = \dfrac{10}{100}, n_B = \dfrac{10}{200}, n_C = \dfrac{10}{10000}$

Moles of solvent (H_2O) $= \dfrac{180}{18} = 10$ mol

$RLVP_A = \dfrac{0.1}{10.1} = \dfrac{1}{101}$

$RLVP_B = \dfrac{0.05}{10.05} = \dfrac{1}{201}$

$RLVP_C = \dfrac{10^{-3}}{10} = 10^{-4}$

From the above relation
RLVP(A) > RLVP(B) > RLVP(C)

180. (5)
Molality of $CaCl_2$ solution $= 0.05$ m

$\Delta T_b = iK_b m = 3 \times K_b \times 0.05 = 0.15 K_b$

Molality of $CrCl_3 . xNH_3 = 0.10$ m

$\Delta T_b' = iK_b \times 0.10 \Rightarrow \Delta T_b' = 2\Delta T_b$

$iK_b \times 0.10 = 2 \times 0.15 K_b \Rightarrow i = 3$

Since, co-ordination number of Cr is 6.

$\therefore$ The complex is $[Cr(NH_3)_5Cl]Cl_2$.

$\therefore x = 5.$

181. (167) For NaCl : $\pi_1 = iC_1RT \Rightarrow C_1 = \dfrac{0.10}{2RT}$

For Glucose : $\pi_2 = C_2RT \Rightarrow C_2 = \dfrac{0.20}{RT}$

When 1 L of NaCl solution and 2 L glucose solution are Mixed.

$\therefore C'_1 = \dfrac{0.1}{2RT} \times \dfrac{1}{(1+2)} = \dfrac{0.10}{6RT}$ and

$C'_2 = \dfrac{0.2}{RT} \times \dfrac{2}{(1+2)} = \dfrac{0.40}{3RT}$

$\therefore \pi_{Total} = iC'_1 RT + C'_2 RT = \left(2 \times \dfrac{0.1}{6RT} + \dfrac{0.4}{3RT}\right) RT = \dfrac{0.50}{3}$

$\pi_{Total} = 167 \times 10^{-3}$ atm

182. (177)

$\pi_A = C_A RT,\ \pi_B = C_B RT$

For isotonic solution, $\pi_A = \pi_B$

$\Rightarrow C_1 = C_2$ (For protein $i = 1$)

$\Rightarrow \dfrac{0.73 \times 1000}{M_A \times 250} = \dfrac{1.65}{M_B \times 1}$

$\therefore \dfrac{M_A}{M_B} = \dfrac{0.73 \times 4}{1.65} = 1.77 = 177 \times 10^{-2}$

183. (1.75) $\Delta T_f = iK_f m$; i(NaCl) = 2

$0.2 = 2 \times 2 \times \dfrac{w}{58.5} \times \dfrac{1000}{600}$

$w = \dfrac{0.2 \times 58.5 \times 600}{1000 \times 4} = 1.75$ g

184. (2.18)

$K_f = 2.0 \text{ K kg mol}^{-1}$; m = 0.5

$\Delta T_f = K_f m = 0.5 \times 2 = 1$

$T_{initial} = 273 - 1 = 272$ K; $n = 0.1$ mol; V = 1 dm^3

$P_{gas} = \frac{nRT}{V} = \frac{0.1 \times 0.08 \times 272}{1} = 2.176$ atm

After releasing piston, $P_1V_1 = P_2V_2$

$2.176 \times 1 = 1 \times V_2 \Rightarrow V_2 = 2.18\ dm^3$

185. (c) Osmotic pressure (π) = CRT

Since, there are two solutes i.e., urea and glucose.

$\therefore \pi = (C_1 + C_1)RT$

$= \left(\frac{0.6}{60 \times 0.1} + \frac{1.8}{180 \times 0.1}\right) \times 0.0821 \times 300$

$= 0.2 \times 0.0821 \times 300 = 4.926$ atm

186. (c) Relative lowering of vapour pressure, is given by,

$$\frac{p^\circ - p}{p^\circ} = x_A = \frac{n_A}{n_A + n_B} \simeq \frac{n_A}{n_B}$$

Given, $p^\circ = 35\,\text{mm Hg}, n_{urea} = \frac{0.60}{60}, n_{water} = \frac{360}{18}$

$$\frac{p^\circ - p}{35} = \frac{0.6 \times 18}{60 \times 360} = \frac{1}{2000} \Rightarrow \Delta p = p^\circ - p = 0.017$$

187. (c) Ebullioscopic constant (molal elevation const.) is given by,

$K_b = \frac{\Delta T_b}{m}$ and it is given that, $\frac{(K_b)_A}{(K_b)_B} = \frac{1}{5}$

$$\therefore \frac{(\Delta T_b)_A}{(\Delta T_b)_B} = \frac{(k_b)_A}{(k_b)_B} = \frac{1}{5}$$

188. (b) We know, $\pi = iCRT$; $\pi_{XY} = 4\pi_{BaCl_2}$

Since both are ionic solute, i.e.,

$XY \rightarrow X^+ + Y^- \Rightarrow i = 2$

$BaCl_2 \rightarrow Ba^{2+} + 2Cl^- \Rightarrow i = 3$

Now,

$2[XY]RT = 4 \times 3 \times [0.01]RT$

$[XY] = 0.06 = 6 \times 10^{-2}$ mol/L

189. (d) Dissociation of potassium sulphate (K_2SO_4),

$[K_2SO_4 \longrightarrow 2K^+ + SO_4^{2-}]$

i (van't Hoff factor) = 3

We know that, $\Delta T_f = iK_fm$

where, K_f is molal depression constant and m is molality.

$= 3 \times 4 \times 0.03 = 0.36$ K

190. (a) $(\Delta T_f)_X = (\Delta T_f)_Y$

$K_f . m_X = K_f . m_Y$

$m_X = m_Y$

$$\frac{\text{No. of moles of } X}{\text{wt. of solvent in kg}} = \frac{\text{No. of moles of } Y}{\text{wt. of solvent in kg}}$$

Given, 4% aqueous solution of X = 4 g of solute X present in 100 g of H_2O.

12% aqueous solution of Y = 12 g of solute Y present in 100 g of H_2O.

Now, $\frac{4\text{ g}}{m} \times \frac{1000}{100} = \frac{12\text{ g}}{m} \times \frac{1000}{100}$

$\frac{4}{A} = \frac{12}{m}$ where $[m = A]$

$m = 3A$

191. (a) $2C_6H_5COOH \longrightarrow (C_6H_5COOH)_2$

	w g	Dimer
Initial	1	0
Final	1 − α	$\frac{\alpha}{2}$

Moles at equilibrium $= 1 - \alpha + \frac{\alpha}{2} = 1 - \frac{\alpha}{2}$

$$i = \frac{\text{Total moles at equilibrium}}{\text{Initial moles}} = \frac{1 - \frac{\alpha}{2}}{1}$$

$i = 1 - \frac{0.8}{2} = 0.6 \Rightarrow \Delta T_f = iK_fm$

$$2 = 0.6 \times 5 \times \left(\frac{\frac{w}{122}}{30}\right) \times 1000 \Rightarrow w = 2.4\text{ g}$$

192. (b) Freezing point of diluted milk = −0.2 °C

$\Rightarrow \Delta T'_f = 0.2$ °C

Freezing point of pure milk = −0.5 °C

$\Rightarrow \Delta T_f = 0.5$ °C

$\frac{\Delta T_f}{\Delta T'_f} = \frac{K_f \times m}{K_f \times m'}$; $m = \frac{\text{mole of solute}}{\text{mass of solvent (kg)}}$

Moles of solute are same in both samples.

$\therefore \frac{0.5}{0.2} = \frac{W'}{W}$

$\Rightarrow \frac{W'}{W} = \frac{5}{2} \Rightarrow W' = \frac{5}{2}W$

2 cups of pure milk is mixed with 3 cups of water to make 5 cups of diluted milk.

193. (b) $K_2HgI_4 \rightleftharpoons 2K^+ + [HgI_4]^{2-}$; $n = 3$

$\because \alpha = \frac{i-1}{n-1} \Rightarrow 0.4 = \frac{i-1}{3-1} \Rightarrow i = 1.8$

194. (d) According to the question we can write,

$\Delta T_b = K_bm \Rightarrow K_b(1) = 2 \Rightarrow K_b = 2\ Km^{-1}$

$\Delta T_f = K_fm \Rightarrow K_f(2) = 2 \Rightarrow K_f = 1\ Km^{-1}$

So, $\frac{K_b}{K_f} = \frac{2}{1} \Rightarrow K_b = 2K_f$

195. (c) As we know, $\Delta T_f = K_f . m$

$\Rightarrow 10 = 1.86 \times \frac{62 \times 1000}{62 \times w} \Rightarrow w = \frac{1.86 \times 1000}{10} = 186$ g

Total amount of water = 250 g

$\therefore$ The amount of water separated as ice = 250 − 186 = 64 g

196. (d)

		Number of particles (i)
(a)	$[Co(H_2O)_6]Cl_3$	4
(b)	$[Co(H_2O)_5Cl]Cl_2 \cdot H_2O$	3
(c)	$[Co(H_2O)_4Cl_2]Cl \cdot 2H_2O$	2
(d)	$[Co(H_2O)_3Cl_3] \cdot 3H_2O$	1

Remember, greater the no. of particles, lower will be the freezing point. Compound (d) will have the highest freezing point due to least number of particles.

197. (c) Molar mass of octane = 114 g/mol
From the lowering of vapour pressure we have,

$$\frac{\Delta P}{P}=\frac{\frac{W_2}{M_2}}{\frac{W_2}{M_2}+\frac{W_1}{M_1}} \Rightarrow \frac{75}{100}=\frac{\frac{W_2}{50\text{ g/mol}}}{\frac{W_2}{50\text{ g/mol}}+\frac{114\text{ g}}{114\text{ g/mol}}}$$

$W_2 = 150$ g

198. (a) The relationship between molar masses of the two solvents is

$$M_X=\frac{3}{4}M_Y \quad \ldots(i)$$

The relative lowering of vapour pressure of the two solutions is

$$\left(\frac{\Delta P}{P}\right)_X = m\left(\frac{\Delta P}{P}\right)_Y$$

But, the relative lowering of vapour pressure of solutions is directly proportional to the mole fraction of solute.
Given 5 molal solution, means 5 moles of solute are dissolved in 1 kg (or 1000 g) of solvent.

The number of moles of solvent $=\frac{1000\text{ g}}{M}$

The mole fraction of solute $=\frac{5}{1000/M}=M\times\frac{5}{1000}$

Hence, $M_X\times\frac{5}{1000}=m\times M_Y\times\frac{5}{1000}$ ...(ii)

Substitute equation (i) in equation (ii),

$$\frac{3}{4}\times M_Y\times\frac{5}{1000}=m\times M_Y\times\frac{5}{1000} \Rightarrow m=\frac{3}{4}$$

199. (c) $Na_2SO_4 \rightleftharpoons 2Na^+ + SO_4^{2-}$

$1-\alpha \qquad 2\alpha \qquad \alpha$

$i=1-\alpha+2\alpha+\alpha=1+2\alpha$

Given, $\alpha=81.5\% \Rightarrow i=1+(2\times0.815)$

$\Rightarrow i=2.63$

$\Delta T_f=i\times K_f\times m$

$$\Rightarrow 3.82=2.63\times1.86\times\frac{5/142}{x/1000} \Rightarrow x=45.$$

200. (d) In benzene,

$2CH_3COOH \rightleftharpoons (CH_3COOH)_2$

$1-\alpha \qquad \alpha/2$

$i=1-\alpha+\alpha/2=1-\alpha/2$

Here α is degree of association

$\Delta T_f = i\times K_f\times m$

$$0.45=\left(1-\frac{\alpha}{2}\right)(5.12)\left(\frac{\frac{0.2}{60}}{\frac{20}{1000}}\right)$$

$1-\frac{\alpha}{2}=0.527$

$\alpha=0.945$

% degree of association = 94.6%

201. (a) van't Hoff factor (i)

$$=\frac{\text{No. of particles after association or dissociation}}{\text{No. of particles before association or dissociation}}$$

For $\quad MX_2 \rightleftharpoons M^{2+}+2X^-$

Initial $\quad 1 \qquad 0 \qquad 0$

Final $\quad 1-\alpha \qquad \alpha \qquad 2\alpha$

Total no. of particles $=1-\alpha+\alpha+2\alpha=1+2\alpha$

$$\therefore i=\frac{1+2\alpha}{1}=2 \Rightarrow \alpha=0.50 \text{ or } 50\%.$$

202. (a) According to Raoult's Law

$$\frac{P^\circ-P_s}{P_s^\circ}=\frac{W_B\times M_A}{M_B\times W_A} \quad \ldots(i)$$

Here P° = Vapour pressure of pure solvent,
P_s = Vapour pressure of solution
W_B = Mass of solute, W_A = Mass of solvent
M_B = Molar mass of solute, M_A = Molar Mass of solvent
Vapour pressure of pure water at 100 °C (by assumption = 760 torr)
By substituting values in equation (i) we get,

$$\frac{760-P_s}{P_s^\circ}=\frac{18\times18}{180\times178.2} \quad \ldots(ii)$$

On solving (ii) we get
$P_s = 752.4$ torr

203. (d) Using relation,

$$\frac{P^\circ-P_s}{P^\circ}=\frac{w_2M_1}{w_1M_2}$$

where w_1, M_1 = mass in g and mol. mass of solvent
w_2, M_2 = mass in g and mol. mass of solute
P° = 185 torr, P_s = 183 torr

$$\frac{185-183}{185}=\frac{1.2\times58}{100\times M_2} \quad \text{(Mol. mass of acetone = 58)}$$

$M_2 = 64.68 \approx 64$

$\therefore$ Molar mass of substance = 64

204. (d) Acetic acid contain carboxylic group – COOH which can form H-bonding, so acetic acid dimerises.

205. (a) $\pi = iCRT$

$\pi_{C_2H_5OH}=1\times0.500\times R\times T=0.5RT$

$\pi_{Mg_3(PO_4)_2}=5\times0.100\times R\times T=0.5RT$

$\pi_{KBr}=2\times0.250\times R\times T=0.5RT$

$\pi_{Na_3PO_4}=4\times0.125\times RT=0.5RT$

Hence, the osmotic pressures of all the given solutions are equal.

206. (a) Given, $\pi_{ob}=10.8$ atm
$\pi_{nor}=CRT=0.10\times0.082\times298=2.446$ atm
Now experimental value of (i)

$$=\frac{\text{Observed osmotic pressure}}{\text{Normal osmotic pressure}}=\frac{10.8}{2.446}=4.42$$

$Fe(NH_4)_2(SO_4)_2 \rightleftharpoons Fe^{2+}+2NH_4^{+}+2SO_4^{2-}$

Expected $i=5$.

207. (c) $\frac{P^\circ - P_s}{P^\circ} = \frac{W_B \times M_A}{M_B \times W_A}$

$0.1 = \frac{12}{M_B} \times \frac{18}{108} \Rightarrow M_B = \frac{12 \times 18}{0.1 \times 108} = 20$

208. (c) $nM \rightleftharpoons (M)_n$

initial 1 0

final $1-\alpha$ α/n

$i = \frac{(1-\alpha) + (\alpha/n)}{1}$

$\Rightarrow 0.9 = (1-0.2) + (0.2/n) \Rightarrow n = 2.$

209. (d) $\Delta T_f = K_f m$

where m = molality

$\Delta T_f = T_f - T_f' = K_f \times m = K_f \times \frac{w}{M \times \text{Mass or Volume}}$

$273 - 268 = 1.86 \times \frac{w}{M \times V}$

$5 = 1.86 \times \frac{w}{32 \times 10} \Rightarrow w = \frac{5 \times 32 \times 10}{1.86} = 860.2 \approx 868.06\text{ g}$

210. (b) $\Delta T_f = i \times K_f \times m$

Given $\Delta T_f = 2.8, K_f = 1.86\text{ K kg mol}^{-1}, i = 1$

(ethylene glygol is a non- electrolyte)

Wt. of solvent = 1 kg; Let of wt of solute = x

Mol. wt of ethylene glycol = 62

$2.8 = 1 \times 1.86 \times \frac{x}{62 \times 1} \Rightarrow x = \frac{2.8 \times 62}{1.86} = 93\text{ g}$

211. (b) Mol.wt.$(M_B) = \frac{K_f \times W_B \times 1000}{\Delta T_f \times W_A}$

$= \frac{1.86 \times 0.85 \times 1000}{0.23 \times 125} \approx 55\text{ g}$

Where, $W_B = 0.85$ g; $W_A = 125$ g

$\Delta T_f = 0\text{ °C} - (-23\text{ °C}) = 23\text{ °C}$

Now, $i = \frac{M_{\text{normal}}}{M_{\text{observed}}} = \frac{136.3}{55} = 2.47$

$ZnCl_2 \rightleftharpoons Zn^{2+} + 2Cl^-$

$1-\alpha$ α 2α

van't Hoff factor $(i) = \frac{1-\alpha+\alpha+2\alpha}{1} = 2.47$

$\Rightarrow \alpha = 0.735 = 73.5\%$

212. (d) $\Delta T_f = i \times K_f \times m$

$i = \frac{\Delta T_f}{K_f \times m} = \frac{1.91}{1.86 \times 1} = 1.027$

For $HF \rightleftharpoons H^+ + F^-$

$(1-\alpha)$ α α

$1 - \alpha + \alpha + \alpha = i = 1.027 \Rightarrow \alpha = 0.027$ or 2.7%

213. (b) For isotonic solutions

$\pi_1 = \pi_2$

$\frac{5/342}{0.1} = \frac{1/M}{0.1} \Rightarrow M = \frac{342}{5} = 68.4$ g/mol

214. (b) Sodium sulphate dissociates as

$Na_2SO_4(s) \longrightarrow 2Na^+ + SO_4^{2-}$

Hence van't Hoff factor, i = 3

Now $\Delta T_f = i K_f .m = 3 \times 1.86 \times 0.01 = 0.0558\text{ K}$

215. (a) Note: On addition of glucose to water, vapour pressure of water will decrease. The vapour pressure of a solution of glucose in water can be calculated using the relation

$\frac{P^\circ - P_s}{P^\circ} = \frac{\text{Moles of glucose in solution}}{\text{Moles of water in solution}}$

or $\frac{17.5 - P_s}{17.5} = \frac{18/180}{178.2/18}$ $[\because P^\circ = 17.5]$

or $17.5 - P_s = \frac{0.1 \times 17.5}{9.9}$ or $P_s = 17.325$ mm Hg.

216. (a) Osmotic pressure (π) of isotonic solutions are equal. For solution of unknown substance (π = CRT)

$C_1 = \frac{5.25/M}{V}$

For solution of urea, C_2 (concentration) $= \frac{1.5/60}{V}$

Given, $\pi_1 = \pi_2$ $\because \pi = CRT$

$\therefore C_1RT = C_2RT$ or $C_1 = C_2$ or $\frac{5.25/M}{V} = \frac{1.5/60}{V}$

$\therefore M = 210$ g/mol

217. (c) $\because \Delta T_b = T_b - T_b^\circ$

Where T_b = b.pt of solution

T_b° = b.pt of solvent or $T_b = T_b^\circ + \Delta T_b$

Note: Elevation in boiling point is a colligative property, which depends upon the no. of particles. Thus greater the number of particles, greater is it elevation and hence greater will be its boiling point.

$Na_2SO_4 \rightleftharpoons 2Na^+ + SO_4^{2-}$

Since Na_2SO_4 has maximum number of particles (3), hence has maximum boiling point.

218. (d) $\Delta T_f = i \times K_f \times m; \Delta T_f = 1.85 \times 0.2 \times 1.3 = 0.480\text{ °C}$

$(HX \rightleftharpoons H^+ + X^-, \; i = 1.3)$

$1-0.3$ 0.3 0.3

$\Delta T_f = T_f^\circ - T_f$

$\therefore T_f = 0 - 0.480\text{ °C} = -0.480\text{ °C}$

219. (d) $\Delta T_f = i \times K_f \times m$. Since K_f has different values for different solvents, even if m is same, ΔT_f will be different.

220. (a) On increasing pressure, the temperature is also increased. Thus in pressure cooker due to increase in pressure the b.p. of water increases.

221. (b) $\Delta T_b = K_b \times m \times 1000$; $\Delta T_f = K_f \times m \times 1000$;

$\frac{\Delta T_b}{\Delta T_f} = \frac{K_b}{K_f} \Rightarrow \frac{\Delta T_b}{0.186} = \frac{0.512}{1.86} \Rightarrow \Delta T_b = 0.0512\text{ °C}.$

2 Electrochemistry

1. **(c)** Mohr's salt : $FeSO_4.(NH_4)_2SO_4.6H_2O$
Using Kohlrousch law,
$\lambda_m^{\infty} = x_1 + 2x_2 + 2x_3$

2. **(3)** $\lambda_m^{\circ}(NH_4Cl) = 185\ Scm^2\ mol^{-1}$
$\Rightarrow \left(\lambda_m^{\circ}\right)_{NH_4^+} + \left(\lambda_m^{\circ}\right)_{Cl^-} = 185$
$\left(\lambda_m^{\circ}\right)_{NH_4^+} = \left(\lambda_m^{\circ}\right)_{NH_4Cl} - \left(\lambda_m^{\circ}\right)_{Cl^-}$
$= 185 - 70 = 115\ Scm^2\ mol^{-1}$
$\left(\lambda_m^{\circ}\right)_{NH_4OH} = \left(\lambda_m^{\circ}\right)_{NH_4^+} + \left(\lambda_m^{\circ}\right)_{OH^-} = 115 + 170$
$= 285\ Scm^2\ mol^{-1}$
degree of dissociation $= \dfrac{(\lambda_m)NH_4OH}{\left(\lambda_m^{\circ}\right)_{NH_4OH}}$
$= \dfrac{85.5}{285} = 0.3 = 3 \times 10^{-1}$

3. **(b)** Higher the charge of the metal cation higher is the molar conductivity. For same charge, higher is solvation lower will be the molar conductivity where $H^{\oplus}$ cation is a exceptional case.

4. **(23)** We have 0.2 g of NaOH in 100 mL solution
$M = \dfrac{0.2}{40 \times 100} \times 1000 = 0.05\ M$
$\kappa = \dfrac{1}{0.870} = 1.15 Sm^{-1} = 1.15 \times 10^{-2}\ S\ cm^{-1}$
$\Lambda_m = \dfrac{\kappa \times 1000}{m} = \dfrac{1.15 \times 10^{-2} \times 1000}{0.05} = 230\ S\ cm^2\ mol^{-1}$
$= 23 \times 10^4\ m\ S\ cm^2\ mol^{-1} = 23 \times 10^2\ m\ S\ dm^2\ mol^{-1}$

5. **(d)**

Molar conductivity (y-axis) vs $\sqrt{C}$ (x-axis)

Molar conductivity decreases sharply with the increase in concentration because the total volume V of solution containing 1 mol of electrolyte decreases

6. **(150)** $R = \rho \dfrac{l}{A}$
$\kappa = G.G^*\quad G = \dfrac{1}{R}; \kappa = \dfrac{1}{\rho}\quad G^* = \dfrac{l}{A}$
R = Resistance
ρ = Resistivity
$\dfrac{l}{A}$ = cell constant (G*)
$\dfrac{\kappa_c}{\kappa_d} = \dfrac{R_d}{R_c}; \lambda_m = \dfrac{\kappa \times 1000}{C}$
$\dfrac{\kappa_c}{\kappa_d} = \dfrac{(\lambda_m.C)_c}{(\lambda_m.C)_d} = \dfrac{R_d}{R_c}$
c = concentrated sol.
d = diluted solution
$\dfrac{100.(0.15)^2}{150.(0.1)^2} = \dfrac{R_d}{100}$ $\boxed{R_d = 150\Omega}$

7. **(c)** A $\rightarrow$ Weak electrolyte (greater change of Λ_m)
B $\rightarrow$ Strong electrolyte (less effect of C on Λ_m)

8. **(a)** $HA(aq) \rightleftharpoons H^+(aq) + A^-(aq)$
$K_a = \dfrac{\alpha^2 C}{1-\alpha}$ $\Rightarrow \alpha^2 C + K_a\alpha - K_a = 0$
$\Rightarrow \left(\dfrac{\lambda_m}{\lambda_m^{\infty}}\right)^2 C + K_a \dfrac{\lambda_m}{\lambda_m^{\infty}} - K_a = 0$
$\Rightarrow \lambda_m^2\ C + K_a\lambda_m\lambda_m^{\infty} - K_a\left(\lambda_m^{\infty}\right)^2 = 0$

9. **(b)** Solution in the conductivity cell is infinitely dilute, hence on addition of water at constant temperature, the molar conductivity of the cell will remain same.

10. **(b)** Given : $\Lambda_+ = 57\ S\ cm^2 mol^{-1}$; $\Lambda_- = 73\ S\ cm^2 mol^{-1}$
Apply Kohlrausch's law
$\Lambda_m = \lambda_+ + \lambda_- = 57 + 73 = 130\ S\ cm^2 mol^{-1}$

11. **(a)** $\Lambda_m = \Lambda_m^{o} - A\sqrt{C}$
Unit of $A\sqrt{C}$ = unit of $\Lambda_m = S\ cm^2\ mol^{-1}$
Unit of $A = S\ cm^2\ mol^{-3/2}\ L^{1/2}$.

12. **(c)** Conductivity of electrolytic cell is affected by concentration of electrolyte, nature of electrolyte and nature of solvent.

13. **(4)**

14. **(3)** (B) On dilution, the number of ions per unit volume that carry current in a solution decreases.
(E) On dilution degree of dissociation of weak electrolyte increases.
So answer is (A), (C) & (D).

15. **(66)** $\Lambda_m = \frac{k}{C} \times 1000$

Given $k = 5 \times 10^{-5}$ S cm^{-1}; C = 0.0025 M

$$\Lambda_m = \frac{5\times10^{-5}\times10^{3}}{0.0025} = \frac{5\times10^{-2}}{2.5\times10^{-3}} = 20\ \text{S cm}^2\ \text{mol}^{-1}$$

$$\alpha = \frac{20}{400} = \frac{1}{20}\ \left(\alpha = \frac{\Lambda_m}{\Lambda_m^o}\right)$$

$$K_a = \frac{C\alpha^2}{1-\alpha} = \frac{0.0025\times\frac{1}{20}\times\frac{1}{20}}{\frac{19}{20}}$$

$$= \frac{0.0025}{19\times20} = 6.6\times10^{-6} = 66\times10^{-7}$$

16. **(13.4)** $AgNO_3 \xrightarrow[\text{dissociation}]{\text{complete}} Ag^+ + NO_3^-$

10^{-5} M $\qquad 10^{-5} \qquad 10^{-5}$

$AgBr \rightleftharpoons Ag^+ + Br^-$

$\qquad S \qquad S$

$K_{sp} = [Ag^+][Br^-] = 4.9\times10^{-13}$

$\therefore\ S^2 = 4.9\times10^{-13}$

$\therefore\ S = 7\times10^{-7}$ M

$\therefore\ [Ag^+] = (7\times10^{-7} + 100\times10^{-7}) = 107\times10^{-7}$ M

$$\therefore\ [Br^-] = \frac{K_{sp}}{[Ag^+]} = \frac{4.9\times10^{-13}}{107\times10^{-7}} = 0.0458\times10^{-6}$$

$= 4.58\times10^{-8}$ M

Conductivity $= v_{Ag^+}\lambda^o_{Ag^+} + v_{Br^-}\lambda^o_{Br^-} + v_{NO_3^-}\lambda^o_{NO_3^-}$

$= (107\times10^{-7}\times6\times10^{-3}) + (4.58\times10^{-8}\times8\times10^{-3})$
$\qquad + (10^{-5}\times7\times10^{-3})$

$= 642\times10^{-10} + 36.64\times10^{-11} + 7\times10^{-8}$

$= (642 + 3.664 + 700)\times10^{-10} = 1346\times10^{-10}$ Sm^{-1}

$= 13.46\times10^{-8}$ Sm^{-1}

17. **(25)** $\Lambda_m = \frac{\kappa\times1000}{M} \Rightarrow \Lambda_m = \frac{1}{\rho}\times\frac{1000}{M}$

$$= \frac{1}{5\times10^{-3}}\times\frac{1000}{0.8} = 25\times10^4\ \Omega^{-1}\ \text{cm}^{-2}\ \text{mol}^{-1}$$

18. **(2)** Statement (A) and Statement (C) are incorrect.
(A) is weak electrocyle and (B) is strong electrolyte

19. **(b)** $\underset{(WA)}{C_6H_5COOH} + \underset{(SB)}{NaOH} \longrightarrow C_6H_5COONa + H_2O$

From (A) → (B); Free H^+ ions are replaced by $Na^{\oplus}$ which decreases conductance.
From (B) → (C); Un-dissociated benzoic acid reacts with NaOH and forms salt which increases ions & conductance increases.
From (C) → (D); After equivalence point at (3), NaOH added further increases $Na^{\oplus}$ and $OH^{\ominus}$ ions which further increases the conductance.

20. **(1000)**

$$\text{Ratio of Molarity} = \frac{M_1}{M_2}$$

$$= \frac{W_1/mV}{W_2/mV} = \frac{W_1}{W_2} = \frac{ppm_1}{ppm_2} = \frac{74.5}{149} = \frac{1}{2}$$

$$\frac{\Lambda_1}{\Lambda_2} = \frac{\kappa_1\times\frac{1000}{M_1}}{\kappa_2\times\frac{1000}{M_2}} = \frac{\kappa_1}{\kappa_2}\times\frac{M_2}{M_1} = \frac{50}{100}\times2 = 1{,}000\times10^{-3}$$

21. **(b)** **Statement I:** KI is strong electrolyte, it dissociates completely. So, λ_m increases slowly on dilution.
Statement II: In weak electrolyte λ_m increases, sharply as the dissociation increases.

22. **(a)** $\Lambda_m = \frac{\kappa}{C} \Rightarrow \kappa = \Lambda_m \times C$

from question, $\kappa_1 = \kappa_2$

$\Lambda_{m_1}\times C_1 = \Lambda_{m_2}\times C_2$

$$\Lambda_{m_1}\times\frac{10\times10^3}{20} = \Lambda_{m_2}\times\frac{20\times10^3}{80} \Rightarrow 2\Lambda_{m_1} = \Lambda_{m_2}$$

23. **(3)** $A_2X_3(s) \rightleftharpoons 2A^{+3}(aq) + 3X^{-2}(aq)$

Solubility = s M $\qquad 2s \qquad 3s$

$(2s)^2(3s)^3 = 1.1\times10^{-23}$

$108\ s^5 = 1.1\times10^{-23}$

$$s = 10^{-5}\ \text{M} = 10^{-5}\ \frac{\text{mol}}{\text{L}} = 0.01\frac{\text{mol}}{\text{m}^3}\ \therefore\ [1\text{L} = 0.001\text{m}^3]$$

$$\text{Now, } \Lambda_m \simeq \Lambda_m^{\infty} = \frac{k}{m} = \frac{k}{s}$$

$$\Rightarrow \Lambda_m^{\infty} = \frac{3\times10^{-5}}{0.01} = 3\times10^{-3}\ \text{Sm}^2/\text{mol}$$

24. **(14)**

$\lambda_m^{\infty}(AgI) = \lambda_m^{\infty}(NaI) + \lambda_m^{\infty}(AgNO_3) - \lambda_m^{\infty}(NaNO_3)$

$= 12.7 + 13.3 - 12.0 = 14.0$

25. **(266)** $k = \frac{1}{R}\times$ cell constant

$$0.152\times10^{-3} = \frac{1}{1750}\times\text{cell constant}$$

cell constant $= 266\times10^{-3}$

26. **(d)**

Ion	H^+	K^+	Cl^-	CH_3COO^-
Λ^{∞}_{m} Scm^2/mole	349.8	73.5	76.3	40.9

So $\Lambda^{\infty}_{m\,CH_3COOH} = \Lambda^{\infty}_{m\,(H^+)} + \Lambda^{\infty}_{m\,CH_3COO^-}$

$= 349.8 + 40.9$
$= 390.7$ Scm²/mole

$\Lambda^{\infty}_{m\,KCl} = \Lambda^{\infty}_{m(K^+)} + \Lambda^{\infty}_{m(Cl^-)}$

$= 73.5 + 76.3$
$= 349.3$ Scm²/mole

So statement-I is wrong.

Molar conductivity of a solution increases with decrease in concentration because of the increase in the total volume containing one mole of electrolyte. Therefore statement II is also not correct.

27. (12)

$$\Lambda_m = 1000\times\frac{\kappa}{M} = 1000\times\frac{2\times10^{-5}}{0.001} = 20\text{ S cm}^2\text{ mol}^{-1}$$

$$\Rightarrow \alpha = \frac{\Lambda_m}{\Lambda^\circ_m} = \frac{20}{190} = \left(\frac{2}{19}\right)$$

$$\underset{0.001(1-\alpha)}{HA} \rightleftharpoons \underset{0.001\alpha}{H^+} + \underset{0.001\alpha}{A^-}$$

$$K_a = 0.001\left(\frac{\alpha^2}{1-\alpha}\right) = \frac{0.001\times\left(\frac{2}{19}\right)^2}{1-\left(\frac{2}{19}\right)} = 12.3\times10^{-6}$$

28. (760) $\kappa = \frac{1}{R}\times l/A = \left(\left(\frac{1}{1500}\right)\times1.14\right)\text{S cm}^{-1}$

$$\Rightarrow \Lambda_m = \frac{1000\times\kappa}{C} = 1000\times\frac{\left(\frac{1.14}{1500}\right)}{0.001}\text{S cm}^2\text{mol}^{-1}$$

$= 760\text{ S cm}^2\text{ mol}^{-1}$

29. (288) From Kohlrausch's law

$$\Lambda^{\infty}_m BaSO_4 = \Lambda^{\infty}_m Ba^{2+} + \Lambda^{\infty}_m SO_4^{2-}$$

$$\Lambda^{\infty}_m(BaSO_4) = \Lambda^{\infty}_m(BaCl_2) + \Lambda^{\infty}_m(H_2SO_4) - 2\Lambda^{\infty}_m(HCl)$$

$= 280 + 860 - 2\times426 = 288\text{ S cm}^2\text{ mol}^{-1}$

30. (57) For KCl, $\kappa_1 = 0.14\text{ Sm}^{-1}$; $R_1 = 4.19\Omega$

For HCl, $\kappa_2 = x\text{ Sm}^{-1}$; $R_2 = 1.03\Omega$

$$\kappa = \frac{1}{R}\cdot\frac{l}{A}$$

Hence κR = Constant (cell is same)

$\therefore \kappa_1R_1 = \kappa_2R_2 \Rightarrow 0.14\times4.19 = 1.03\times\kappa_2$

$\Rightarrow \kappa_2 = 0.5695\text{ Sm}^{-1} \approx 57\times10^{-2}\text{ Sm}^{-1}$

31. (14) Conductance $= \frac{\text{Conductivity}}{\text{Cell constant}}$

$\therefore$ Conductivity $= 0.55\times10^{-3}\times1.3\text{ S cm}^{-1}$

Molar conductivity

$$= \frac{\text{Conductivity (S cm}^{-1})\times1000}{\text{Molarity (mol/L)}}(\text{cm}^3/\text{L})$$

$$= \frac{0.55\times10^{-3}\times1.3\times1000}{5\times10^{-3}} = 143\text{ S cm}^2\text{ mol}^{-1}$$

$\therefore$ $1\text{ S cm}^2\text{ mol}^{-1} = 10^{-4}\text{ S m}^2\text{ mol}^{-1}$

$143\text{ S cm}^2\text{ mol}^{-1} = 143\times10^{-4}\text{ S m}^2\text{ mol}^{-1}$

$= 143\times10^{-4}\times10^3\text{ m S m}^2\text{ mol}^{-1}$

$= 14.3\text{ mS m}^2\text{ mol}^{-1} \approx 14\text{ mS m}^2\text{ mol}^{-1}$

32. (d)

33. (N) Ionic mobility increases with increase in temperature, which increases the conductance of the solution. $BaSO_4$ is a sparingly soluble salt, hence at any T, $C_{NaCl} >> C_{BaSO_4}$.

Therefore, all options are true.

34. (d) $(\Lambda^0_m)_{NaBr} - (\Lambda^0_m)_{NaI}$

$= \Lambda^\circ_m Na^+ + \Lambda^\circ_m Br^- - (\Lambda^\circ_m Na^+ + \Lambda^\circ_m I^-) = \Lambda^\circ_m Br^- - \Lambda^\circ_m I^-$

$(\Lambda^0_m)_{KBr} - (\Lambda^0_m)_{NaBr}$

$= \Lambda^\circ_m K^+ + \Lambda^\circ_m Br^- - (\Lambda^\circ_m Na^+ + \Lambda^\circ_m Br^-)$

$= \Lambda^\circ_m K^+ - \Lambda^\circ_m Na^+$

$\therefore (\Lambda^0_m)_{NaBr} - (\Lambda^0_m)_{NaI} \neq (\Lambda^0_m)_{KBr} - (\Lambda^0_m)_{NaBr}$

35. (a) Order of acidic strength is :

$HCOOH > C_6H_5COOH > CH_3COOH$

More the acidic strength, more will be the dissociation of acid into ions and more will be the conductivity.

Thus, order of conductivity will be,

$\underset{(A)}{HCOOH} > \underset{(C)}{C_6H_5COOH} > \underset{(B)}{CH_3COOH}$

36. (b) Conductivity of an electrolyte is the conductance of 1 cm³ of the given electrolyte. It increases with the increase in concentration of electrolyte due to increase in the number of ions per unit volume.

Molar conductivity (Λ_m) is the conductance of a solution containing 1 mole of the electrolyte. It increases with the decrease of concentration due to increase in the total volume having one mole of electrolyte.

Therefore, (S_1) is wrong and (S_2) is correct.

37. (a) Since, KCl is more conducting than NaCl, therefore, graph (a) is correct.

38. (c) Given: $\Lambda^\circ_m(NaCl) = 126.4\text{ S cm}^2\text{ mol}^{-1}$, $\Lambda^\circ_m(HCl) = 425.9\text{ S cm}^2\text{ mol}^{-1}$ and $\Lambda^\circ_m(NaA) = 100.5\text{ S cm}^2\text{ mol}^{-1}$

$\Lambda^\circ_m(HA) = \lambda^\circ_{H^+} + \lambda^\circ_{A^-}$

$= \lambda^\circ_{H^+} + \lambda^\circ_{Cl^-} + \lambda^\circ_{A^-} + \lambda^\circ_{Na^+} - \lambda^\circ_{Cl^-} - \lambda^\circ_{Na^+}$

$= \Lambda^\circ_m(HCl) + \Lambda^\circ_m(NaA) - \Lambda^\circ_m(NaCl)$

$\Lambda^\circ_m(HA) = 425.9 + 100.5 - 126.4 = 400\text{ S cm}^2\text{ mol}^{-1}$

$\kappa(HA) = 5\times10^{-5}\text{ S cm}^{-1}$

$$\Lambda_m(HA) = \frac{\kappa(HA)\times1000}{\text{Molarity of HA}} = \frac{5\times10^{-5}\times1000}{0.001}$$

$= 50\text{ S cm}^2\text{ mol}^{-1}$

$$\alpha = \frac{\Lambda_m(HA)}{\Lambda^\circ_m(HA)} = \frac{50}{400} = 0.125$$

39. (a) Given for 0.2 M solution,

$$R = \rho\frac{l}{A} = \frac{1}{\kappa}\times\frac{l}{A}$$

$$\Rightarrow \frac{l}{A} = R\times\kappa = 50\times1.4\times10^{-2}\text{ cm}^{-1}$$

For 0.5 M solution,

$\frac{l}{A} = 50 \times 1.4 \times 10^{-2}\ cm^{-1} \Rightarrow R = \rho\frac{l}{A} = \frac{1}{\kappa} \times \frac{l}{A}$

$\Rightarrow \kappa = \frac{1}{280} \times 50 \times 1.4 \times 10^{-2} = \frac{1}{280} \times 70 \times 10^{-2}$

$= 2.5 \times 10^{-3}\ S\ cm^{-1}$

Now, $\Lambda_m = \frac{\kappa \times 1000}{M}$

$= \frac{2.5 \times 10^{-3}\ S\ cm^{-1} \times 1000\ cm^3/L}{0.5\ mol/L}$

$= 5\ S\ cm^2\ mol^{-1} = 5 \times 10^{-4}\ S\ m^2\ mol^{-1}$

40. **(c)** According to Debye – Huckle – Onsager equation, $\lambda_C = \lambda_\infty - B\sqrt{C}$

41. **(b)** According to Kohlrausch's law, equivalent conductivity of weak electrolyte, acetic acid (CH_3COOH) can be calculated as follows:

$\Lambda^\circ_{CH_3COOH} = \left(\Lambda^\circ_{CH_3COONa} + \Lambda^\circ_{HCl}\right) - \Lambda^\circ_{NaCl}$

$\therefore$ Value of Λ°_{NaCl} should also be known for calculating value of $\Lambda^\circ_{CH_3COOH}$.

42. **(b)** $\overset{\circ}{\Lambda}_{CH_3COOH}$ is given by the following equation

$\overset{\circ}{\Lambda}_{CH_3COOH} = (\overset{\circ}{\Lambda}_{CH_3COONa} + \overset{\circ}{\Lambda}_{HCl}) - (\overset{\circ}{\Lambda}_{NaCl})$

Hence $\overset{\circ}{\Lambda}_{NaCl}$ is required.

43. **(b)** $R = 100\,\Omega,\ \kappa = \frac{1}{R}\left(\frac{l}{a}\right)$,

$\frac{l}{a}$ (cell constant) $= 1.29 \times 100\ m^{-1}$

Given, $R = 520\Omega$,

$C = 0.2\ M = 0.2 \times 1000\ mol\ m^{-3}$

Λ_m (molar conductivity) = ?

$\Lambda_m = K \times V$ (K can be calculated as cell constant is known.)

$K = \frac{1}{R}\left(\frac{l}{a}\right)$

Hence,

$K = \frac{129 m^{-1}}{520\ \Omega} = 0.248\ Sm^{-1}$

$\Lambda_m = \frac{K}{C} = \frac{0.248\ Sm^{-1}}{0.2 \times 1000\ mol\ m^{-3}}$

$= 12.4 \times 10^{-4} Sm^2 mol^{-1}$

44. **(a)** Difluoro acetic acid being strongest acid will furnish maximum number of ions showing highest electrical conductivity. The decreasing acidic strength of the carboxylic acids given is

difluoro acetic acid > fluoro acetic acid > chloro acetic acid > acetic acid.

45. **(c)** $\Lambda^\circ_{NaCl} = \lambda^\circ_{Na^+} + \lambda_{Cl^-}$ ....(i)

$\Lambda^\circ_{KBr} = \lambda^\circ_{K^+} + \lambda^\circ_{Br^-}$ ....(ii)

$\Lambda^\circ_{KCl} = \lambda^\circ_{K^+} + \lambda_{Cl^-}$ ....(iii)

operating (i) + (ii) – (iii)

$\Lambda^\circ_{NaBr} = \lambda^\circ_{Na^+} + \lambda^\circ_{Br^-}$

$= 126 + 152 - 150 = 128\ S\ cm^2\ mol^{-1}$

46. **(b)** $\Lambda^\infty_{HCl} = 426.2$ ....(i)

$\Lambda^\infty_{AcONa} = 91.0$ ....(ii)

$\Lambda^\infty_{NaCl} = 126.5$ ....(iii)

$\Lambda^\infty_{AcOH} = (i) + (ii) - (iii)$

$= [426.2 + 91.0 - 126.5] = 390.7$

47. **(b)** Given conductivity $\propto \frac{\text{area} \times \text{conc.}}{\text{length}}$

$= \frac{Km^2 mol}{m \times m^3}$

$\therefore\ K = Sm^2\ mol^{-1}$

48. **(400)**

	Cu^{2+} +	$2e^-$ →	Cu	(1 Faraday = charge on 1 mole of electron)
t = 0	1.5	1 mole		
t = t	1	–	0.5 mole	

$[Cu^{+2}] = 1$ M after electrolysis

	Ag^+ +	e^- →	Ag
t = 0	0.2	0.1 mole	
t = t	0.1	–	

$[Ag^+] = 0.1$ M after electrolysis

Cell reaction $Cu(s) + 2Ag^+(aq) \rightarrow Cu^{+2}(aq) + 2Ag(s)$

$E = E^\circ - \frac{0.06}{n}\log\frac{[Cu^{+2}]}{[Ag^+]^2}$

$E = (0.8 - 0.34) - \frac{0.06}{2}\log\frac{1}{(0.1)^2} = 0.4V$

$\therefore$ Correct answer = 400 mV

49. **(a)** A. At cathod silver ion has low discharge potentials than hydrogen ion hence Ag deposited in preference of hydrogen

At anode silver dissolve to form silver ion

$$Ag \longrightarrow Ag^+ + e^-$$

B. At cathode silver ion have low discharge potential than hydrogen hence silver deposited

At anode hydroxide ion will decompose to give oxygen

$$4OH^-(aq) \longrightarrow 2H_2O(l) + O_2(g) + 4e^-$$

C. At cathode $2H^+ + 2e \longrightarrow H_2(g)$

At anode OH^- ion having lower discharge potential will be discharge in preference to sulphate ions.

$$OH^-(aq) \longrightarrow 2H_2O(l) + O_2(g) + 4e^-$$

D. At higher concentration of H_2SO_4 Preferred reaction is

$$2SO_4^{2-}(aq) \longrightarrow S_2O_8^{2-}(aq) + 2e^-\quad E^\circ_{cell} = 1.96\,v$$

50. (2) Electrolysic of NaCl is represented by:

$$NaCl_{(aq)} + H_2O(l) \rightarrow NaOH(aq) + \frac{1}{2}Cl_2(g) + \frac{1}{2}H_2(g)$$

Since during electrolysis, pH changes to 12
So $[OH^-] = 10^{-2}$ and $[H^+] = 10^{-12}$
According to Faraday's law:

$$\text{Mass of } OH^- \text{ formed (m)} = \frac{E}{96500} \times i \times t$$

$$\Rightarrow i = \frac{m \times 96500}{E \times t} = \frac{([OH^-] \times E \times V) \times 96500}{E \times t}$$

$$= \frac{[OH^-] \times V \times 96500}{t}$$

$$\frac{10^{-2} \times 600 \times 10^{-3} \times 96500}{5 \times 60}$$

$1.93\ A \approx 2.0\ A.$

51. (c) At anode.

$$2HSO_4^- \rightarrow H_2S_2O_8 + 2e^-$$

52. (c) $Al^{3+} + 3e^- \longrightarrow Al(s)$

3F required for 27 gm

$$W = \frac{E \times i \times t}{96500} = \frac{M \times i \times t}{n \times 96500}$$

$$= \frac{27 \times 2 \times 30 \times 60}{3 \times 96500} = 0.336 \text{ gm Al}$$

53. (a) An electrochemical can be converted into an electrolyte cell by applying an external potential greater than E°_{cell}. The entire cell reaction will be reversed. The oxidised species will be reduced and the reduced species will be oxidised.

54. (d) $Ag^+(aq) + e^- \rightarrow Ag(s)$
w = Zq; q = 1C
w = Z = 1 electrochemical equivalent

55. (1.93)
$H_2O^{-2} \longrightarrow O_2^0 + H_2; n = 2$
∴ Amount of electricity in coulomb required is = 2F
$= 2 \times 96500 = 193000 = 1.93 \times 10^5$ C

56. (5) $Cu^{2+} + 2e^- \rightarrow Cu$
2 Faraday → 1 mol Cu
1 Faraday → 0.5 mol Cu deposit
0.5 mol = 0.5 g atom = 5×10^{-1}
∴ x = 5

57. (45.75) I = 0.015 A
t = 15 min. = 15 × 60 sec.

$Zn^{2+} + 2e^- \longrightarrow Zn$

$$W = Z \times I \times t \quad \Rightarrow = \frac{65.4}{2 \times 96500} \times 0.015 \times 15 \times 60$$

$= 45.75 \times 10^{-4}$ g

58. (2) $Au^{3+} + 3e^- \longrightarrow Au(s)$

$$\Rightarrow \frac{W}{E} = \frac{Q}{1F}$$

$$\Rightarrow Q = \frac{W \times F}{E} = \frac{W}{E}F = \frac{1.314}{\frac{197}{3}}F = 0.02F = 2 \times 10^{-2}F$$

59. (108)
Eq. of Ag = Eq. of O_2 (given)
Let x gm silver be displaced,

$$\Rightarrow \frac{x \times 1}{108} = \frac{5.6}{22.7} \times 4$$

(Molar volume of gas at STP = 22.7 lit)
⇒ x = 106.57 gm
or, as per old STP data, molar volume = 22.4 lit

$$\frac{x \times 1}{108} = \frac{5.6}{22.7} \times 4, \Rightarrow x = 108 \text{ g.}$$

60. (161) W = z × i × t.

$$\text{Density} \times \text{volume} = \frac{E \times i \times t}{96500}$$

$$(10) \times (100 \times 0.0001) = \frac{\left(\frac{\text{atomic wt.}}{\text{v.f}}\right) \times 2 \times x}{96500} \quad (\text{v.f} = 2)$$

∴ x = 161 sec.

61. (a)

62. (3) Fe_3O_4 is a mixture of FeO and Fe_2O_3. The oxidation state of Fe in FeO and Fe_2O_3 are +2 and +3 respectively.

$$\overset{(+2)}{FeO} \xrightarrow{2e^-} Fe$$

$$\overset{(+3)}{Fe_2O_3} \xrightarrow{6e^-} 2Fe$$

$$Fe_3O_4 \xrightarrow{8F} 3Fe$$

3 mole Fe requires = 8F

$$1 \text{ mole Fe requires } \frac{8}{3}F = 2.6\ F \approx 3$$

63. (127) $2H_2O \rightarrow O_2(g) + 4H^+ + 4e^-$: At anode
$2H_2O + 2e^- \rightarrow H_2(g) + 2OH^-$: At cathode

$$\text{Now number of gm eq.} = \frac{i \times t}{96500}$$

$$= \frac{0.1 \times 2 \times 60 \times 60}{96500} = 0.00746$$

$$V_{O_2} = \frac{0.00746}{4} \times 22.7 = 0.0423$$

$$V_{H_2} = \frac{0.00746}{2} \times 22.7 = 0.0846$$

$V_{total} \approx 127$ mL or cc

64. (20) $Fe^{3+} + 3e^- \longrightarrow Fe$

3F ≡ 1 mole Fe is deposited
For 56 g ≡ 3 × 96500 (required charge)

$$\text{For } 0.3482 \text{ g} \equiv \frac{3 \times 96500}{56} \times 0.3482 = 1800 \text{ coulomb}$$

$Q = it \Rightarrow 1800 = 1.5\,t \Rightarrow t = 1200$ s = 20 min

65. (25) $MnO_4^- + 8H^+ + 5e^- \rightarrow Mn^{+2} + 4H_2O, E^\circ = 1.51$ V

∵ 1 mole of MnO_4^- required 5 moles of electrons or 5 F electricity.

∴ 5 moles of MnO_4^- required = 25 F electricity.

66. **(b)** Electrolysis of concentrated solution of acidified sulphate solution yields $H_2S_2O_8$.

Anode : $2H_2SO_4 \rightarrow H_2S_2O_8 + 2H^+ + 2e^-$

$2H_2O \rightarrow O_2 + 4H^+ + 4C^-$

Cathode : $H^+ + e^- \rightarrow \frac{1}{2}H_2$

67. **(11)** $6OH^- + Cl^- \longrightarrow ClO_3^- + 3H_2O + 6e^-$

For synthesis of 1 mole of ClO_3^-, 6 F of charge is required.

$\because$ Current efficiency = 60%

$\therefore$ To synthesis 1 mole of ClO_3^-, 10 F of charge is required.

To synthesis $\frac{10}{122}$ moles of $KClO_3$, charge $= \frac{10\times10}{122}$F

$Q = I \cdot t$

$t = \frac{100\times96500}{122\times2} = 39549.18\text{ s} = 10.99\text{ h} \therefore t = 11\text{ h}.$

68. **(a)** Millimoles of $Au^+ = 0.1 \times 250 = 25$

Mole of $Au^+ = \frac{25}{1000} = \frac{1}{40} = 0.025$

Similarly, moles of $Ag^+ = 0.025$

Charge passed $= I \times t = 1 \times 15 \times 60 = 900$ C

Moles of e^- passed $= \frac{900}{96500} = 0.0093$ mol.

Species with higher value of SRP will get deposited first at cathode.

$$\underset{0.025}{Au^+(aq)} + \underset{0.0093}{e^-} \longrightarrow Au(s)$$

So, only Au will get deposited.

69. **(60)** Charge $(Q) = It = 2 \times 8 \times 60 = 960$ C

$\Rightarrow \frac{960}{96000} = 0.01F$

$$Cr_2O_7^{2-} + 14H^+ + \underset{\substack{6\text{ F} \\ 0.01\text{ F}}}{6e^-} \longrightarrow \underset{\substack{2\text{ moles} \\ \frac{1}{3}\times0.01\text{ mole}}}{2Cr^{3+}} + 7H_2O$$

Theoritical mass of $Cr^{3+} = \frac{1}{3} \times \frac{960}{96000} \times 52 = 0.173$ g

So, efficiency $= \frac{W_{actual}}{W_{Theoritical}} \times 100 = \frac{0.104}{0.173} \times 100 = 60\%$

70. **(5.68)** No. of moles of silver deposited $= \frac{108}{108} = 1$ mol

$Ag^+ + e^- \rightarrow Ag$

1 F charge is required to deposit 1 mole of Ag

$H_2O \rightarrow \frac{1}{2}O_2 + 2H^+ + 2e^-$

2 F charge deposit $\rightarrow \frac{1}{2}$ moles of oxygen

1 F charge will deposit $\rightarrow \frac{1}{4}$ moles of oxygen

$V_{O_2} = \frac{nRT}{P} = \frac{1}{4} \times \frac{0.08314\text{L bar K}^{-1}\text{mol}^{-1} \times 273\text{K}}{1\text{bar}}$

$= 5.675$ L

71. **(a)** According to the Faraday's law of electrolysis, nF of current is required for the deposition of 1 mol.

According to the reaction,

$Ni(NO_3)_2 \longrightarrow Ni^{2+} + 2NO_3^-$

2 F of current deposits = 1 mol

$\therefore$ 0.1 F of current deposits $= \frac{0.1}{2} = 0.05$ mol

72. **(c)** The anodic half cell reaction when the battery is recharged :

$PbSO_4(s) + 2e^- \longrightarrow Pb(s) + SO_4^{2-}(aq)$

We require 2F for the electrolysis of 1 mol or 303 g of $PbSO_4$.

$\therefore$ Amount of $PbSO_4$ electrolysed by 0.05 F

$= \frac{303}{2} \times 0.05 = 7.575\text{ g} \approx 7.6\text{ g}$

73. **(c)** $B_2H_6 + 3O_2 \longrightarrow B_2O_3 + 3H_2O$

27.66 g of B_2H_6 (1 mole) requires 3 moles of oxygen (O_2) for complete burning.

Now the oxygen is produced by the electrolysis of H_2O.

On electrolysis : $2H_2O \xrightarrow{4F} 2H_2 + O_2$

1 mole O_2 is produced by 4F charge

$\therefore$ 3 mole O_2 will be produced by 12F charge.

$\because$ Q = It

$12 \times 96500\text{ C} = I \times t$; $12 \times 96500\text{ C} = 100 \times t$

$t = \frac{12\times96500}{100}$ second $= \frac{12\times96500}{100\times3600}$ hour $= 3.2$ hours

74. **(c)** 9.65 ampere current was passed for 1.0 hour (3600 seconds)

Number of moles of electrons passed

$= \frac{I(A)\times t(s)}{96500} = \frac{9.65A\times3600s}{96500} = 0.36$ moles

$C_6H_5NO_2 + 4e^- + 4H^+ \longrightarrow$ *p*-Aminophenol + H_2O

$\because$ 4 moles of electrons reduces 1 mole of nitrobenzene to *p*-aminophenol.

$\therefore$ 0.36 moles of electrons will reduce $\frac{0.36}{4} = 0.09$ moles of nitrobenzene to *p*-aminophenol.

p-aminophenol molar mass = 109.14 g/mol

Mass of *p*-aminophenol obtained

= 109.14 g/mol × 0.09 mol = 9.81 g

75. **(d)** Reduction at cathode:

$2e^- + 2H_2O \rightarrow H_2 + 2OH^-$ (valence factor) $H_2 = 2$

At NTP, 22400 mL of H_2 = 1 mole of H_2

112 mL of $H_2 = \frac{1}{22400} \times 112 = 0.005$ mole of H_2

Mass of H_2 produced $= \frac{I\times t}{96500} E_{H_2}$

$2 \times 0.005 = \frac{I\times965}{96500} \times \frac{2}{2}$; $\therefore I = 1.0$ A

76. **(d)** $Cu^{2+} + 2e^- \longrightarrow Cu$

2F i.e. 2×96500 C deposit Cu = 1 mol = 63.5g

77. **(c)** According to Faraday's first law of electrolysis,

$$W = \frac{E \times I \times t}{96500}$$

Where E = equivalent weight

$$= \frac{\text{mol. mass of metal}(M)}{\text{oxidation state of metal }(x)}$$

Substituting the value in the formula

$$W = \frac{M}{x} \times \frac{I \times t}{96500}$$

or $x = \frac{M}{W} \times \frac{I \times t}{96500} = \frac{10 \times 2 \times 60 \times 60}{96500 \times 0.250} = 3$

$$\left[\text{Given : no. of moles} = \frac{W}{M} = 0.250\right]$$

Hence, oxidation state of metal is (+3).

78. **(c)** $Cu \longrightarrow Cu^{2+} + 2e^-$

i.e, to deposit 1 mole of Cu at cathode from $Cu^{2+}\ SO_4^{2-}$ solution = 2 moles of electrons are required

i.e, To deposit $6.35g = \frac{6.35}{63.5} \times 2 = \frac{2}{10} = \frac{1}{5}$ mol

Thus, total no. of electrons required = $\frac{N_A}{5}$

79. **(b)** W = ZIt

where Z = Electrochemical equivalent

Eq. wt. of copper = $\frac{63}{2} = 31.5$

$$Z = \frac{31.5}{96500}$$

$$W = ZIt = \frac{31.5}{96500} \times 1.5 \times 10 \times 60 = 0.2938\text{ g}$$

80. **(c)** Reduction half reaction :

$Cr_2O_7^{2-} + 6e^- + 14H^+ \longrightarrow 2Cr^{3+} + 7H_2O$

Oxidation half reaction :

$Cr \longrightarrow Cr^{3+} + 3e^-$

Overall reaction :

$Cr_2O_7^{2-} + Cr + 14H^+ + 3e^- \longrightarrow 3Cr^{3+} + 7H_2O$

3F of electricity = 3 moles of Cr^{3+}

1F of electricity = $\frac{3}{3}$ moles of Cr^{3+}

81. **(c)** More the reduction potential, more is the deposition of metals at cathode. Cation having E° value less than – 0.83V (reduction potential of H_2O) will not deposit from aqueous solution. Hence, correct order of deposition of the metal at the cathode is:

Ag > Hg > Cu

82. **(d)** 1 mole of e^- = 1F = 96500 C

27g of Al is deposited by 3 × 96500 C

5120 g of Al will be deposited by

$$= \frac{3 \times 96500 \times 5120}{27} = 5.49 \times 10^7\text{ C}$$

83. **(a)** When 96500 coulomb of electricity is passed through the electroplating bath the amount of Ag deposited = 108g

∴ when 9650 coulomb of electricity is passed deposited Ag.

$$= \frac{108}{96500} \times 9650 = 10.8\text{ g}$$

84. **(d)** Pure metal always deposits at cathode.

85. **(10)** $Cr_2O_7^{-2}(aq) + 14H^+(aq) + 6e^- \rightarrow 2Cr^{+3}(aq) + 7H_2O(aq)$

Using Nernst equation

$$E_R = E_R^\circ - \frac{0.059}{6}\log\frac{[Cr^{+3}]^2}{[Cr_2O_7^{-2}][H^+]^{14}}$$

$$0 = 1.33 - \frac{0.059}{6}\log\frac{10^{-6}}{[H^+]^{14}}$$

$$\Rightarrow \frac{1.33 \times 6}{0.059} = \log\frac{10^{-6}}{[H]^{14}}$$

$\Rightarrow$ 135.254 = –6 – 14 log [H^+]

$\Rightarrow$ 141.254 = 14 pH

$$\Rightarrow \text{pH} = \frac{141.254}{14} = 10.08$$

86. **(b)** Ag will deposit first due to higher SRP followed by Hg and Cu. Mg will not deposit. H_2O will get reduced and H_2 gas will evolve at cathode.

87. **(6)** $QH_2 + 2Ag^+ \rightarrow 2Ag + Q + 2H^+$ (Cell reaction)

$$E = E^\circ - \frac{0.06}{2}\log[H^+]^2 \quad (\text{as } [Ag^+] = 1\text{ M})$$

$$E = E^\circ - 0.06 \times \log[H^+]$$

$$pH = -\log[H^+] = \frac{E - E^\circ}{0.06} = \frac{0.4 - 0.1}{0.06} = \frac{0.3}{0.06} = 5$$

NH_4X is electrolyte here.

$$\text{Now, pH} = 7 - \frac{1}{2}pK_b - \frac{1}{2}\log C$$

$$= 7 - \frac{1}{2}pK_b - \frac{1}{2}\log[NH_4X]$$

$$\Rightarrow 5 = 7 - \frac{1}{2} \times pK_b - \frac{1}{2}\log(10^{-2}) \Rightarrow pK_b = 6$$

88. **(b)** According to Nernst equation:-

$$E = E^\circ - \frac{RT}{nF}\ln Q$$

Cell reaction:-

$$Mg(s) + 2Ag^+(aq) \rightleftharpoons 2Ag(s) + Mg^{+2}(aq)$$

According to Nernst equation,

$$E_{cell} = E_{cell}^\circ - \frac{RT}{2F}\ln\frac{[Mg^{+2}]}{[Ag^+]^2}$$

$$\Rightarrow E_{cell} = E^\circ + \frac{RT}{2F}\ln\frac{[Ag^+]^2}{[Mg^{2+}]}$$

89. (d) Higher the standard reduction potential value (+ve) higher will be the oxidising capacity. Therefore option (d) is the corect answer.

90. (b) For strongest reducing agent
Reduction potential should be lowest
Hence Cr is the strongest reducing agent.

91. (b) $Fe^{2+}(aq) + Ag^{+}(aq) \rightarrow Fe^{3+}(aq) + Ag(s)$

$$Fe \xrightarrow{\Delta G_1^0} Fe^{+2} \xrightarrow{\Delta G_2^0} Fe^{+3}, \quad Fe \xrightarrow{\Delta G_3^0} Fe^{+3}$$

$$\Delta G_3^0 = \Delta G_1^0 + \Delta G_2^0 = -3F(-z) = -2F(-y) + \Delta G_2^0$$

$$\Delta G_2^0 = 3Fz - 2Fy\ ; \text{ Also } \Delta G_2^0 = -nFE^0_{Fe^{+2}/Fe^{+3}}$$

$$3Fz - 2Fy = -1F\left(E^0_{Fe^{+2}/Fe^{+3}}\right)$$

$E^0_{Fe^{+2}/Fe^{+3}} = 2y - 3z$; E^0_{cell} for reaction will be

$$E^0_{Ag^+/Ag} + E^0_{Fe^{+2}/Fe^{+3}} = x + 2y - 3z$$

92. (a) $\because \Delta G^\circ = -nFE^\circ$
Option (a) $E^\circ_{cell} = 0.8 + 0.76 = 1.56$ V
$\therefore\ \Delta G^\circ = -2 \times F \times 1.56 = -3.12$ V
Option (b) $E^\circ_{cell} = -2.37 + 0.76 = -1.61$ V
$\therefore\ \Delta G^\circ = -2 \times F \times (-1.61) = +3.22$ V
Option (c) $E^\circ_{cell} = -2.37 - 0.8 = -3.17$ V
$\therefore\ \Delta G^\circ = -2 \times F \times (-3.17) = +6.34$
Option (d) $E^\circ_{cell} = 0.8 - 0.34 = 0.46$ V
$\therefore\ \Delta G^\circ = -2 \times F \times 0.46 = -0.92$ V

93. (a)

$$\overset{+6}{FeO_4^{-2}} \xrightarrow[n_1=3]{E_1^0=2V} Fe^{+3} \xrightarrow[n_2=1]{E_2^0=0.8V} Fe^{+2} \xrightarrow{E_4^0=-0.5V} Fe$$

$$FeO_4^{-2} \xrightarrow[n_3=4]{E_3^0=?} Fe^{+2}$$

$$\Delta G_3^\circ = \Delta G_1^\circ + \Delta G_2^\circ$$

$$\Rightarrow -n_3FE_3^\circ = -n_1FE_1^0 - n_2FE_2^\circ$$

$$\Rightarrow +4E_3^\circ = 3\times 2 + (1\times 0.8) \qquad \Rightarrow E_3^\circ = \frac{6.8}{4} V$$

$$\Rightarrow E_3^\circ = 1.7\ V$$

94. (3) Fe, Ni, Ag will be oxidized due to lower S.R.P than that of $Cr_2O_7^{2-}$.

95. (d) Anodic Reaction at Anode : $[T\ell_{(s)} \rightarrow T\ell^+_{(aq)} + e^-]2$

Reaction at Cathode: $Cu^{+2}(aq) + 2e^- \rightarrow Cu(s)$

Overall Redox Reaction $2T\ell(s) + Cu^{+2}(aq) \rightarrow 2T\ell^+(aq) + Cu(s)$

$$E_{cell} = E^\circ_{cell} - \frac{0.0591}{2}\log\frac{\left[T\ell^+\right]^2}{\left[Cu^{+2}\right]}$$

E_{cell} increases by increasing concentration of $[Cu^{+2}]$ ions.

96. (a) At anode :
Gas - gas ion electrode

$$\frac{1}{2}H_2(g) \rightarrow H^+(aq) + e^-$$

At cathode :
Metal-metal insoluble salt anion electrode
$Ag^+(aq) + e^- \rightarrow Ag(s)$
$AgCl(s) \rightleftharpoons Ag^+(aq) + Cl^-(aq)$
$AgCl(s) + e^- \rightarrow Ag(s) + Cl^-(aq)$
Overall redox reaction

$$\frac{1}{2}H_2(g) + AgCl(s) \rightarrow H^+(aq) + Cl^-(aq) + Ag(s)$$

Cell Representation
$Pt|H_2(g)\,|HCl(Sol).|\,AgCl(s)|Ag$

97. (c) $E^\circ_{cell} = E^\circ_{cathade} - E^\circ_{anode}$
$= 0.34 - 0.46$
$= -0.12$V (negative value nonspontaneous)
Thus reverse reaction is spontaneous.

98. (a) $2H^+(aq) + 2e^- \rightarrow H_2(g)$

$$E = E^\circ - \frac{0.059}{n}\log\frac{P_{H_2}}{\left[H^+\right]^2}$$

For pure water, $[H^+] = 10^{-7}$M

$$0 = 0 - \frac{0.059}{2}\log\frac{P_{H_2}}{\left(10^{-7}\right)^2}$$

$$\log\frac{P_{H_2}}{\left(10^{-7}\right)^2} = 0 \Rightarrow \frac{P_{H_2}}{10^{-14}} = 1$$

$P_{H_2} = 10^{-14}$bar

99. (a) Calomel electrode is a metal-insoluble salt anion electrode.

100. (1) $2H^+(aq) + 2e^- \longrightarrow H_2(g)$
$[H^+] = 1$ M, $P_{H_2} = 2$ atm, n = 2

$$E_{H^+/H_2} = E^\circ_{H^+/H_2} - \frac{2.303RT}{nF}\log\frac{(P_{H_2})}{([H^+]^2)}$$

$$= 0 - \frac{2.303RT}{2F}\log\frac{(2)}{(1)^2} \qquad \{E^\circ_{H^+/H_2} = 0\}$$

$$= -\frac{0.06}{2}\log 2 = -9.03 \times 10^{-3} = -.9 \times 10^{-2}\ V$$

101. (338.9) $2MnO_4^- + 6H^+ + 5H_2C_2O_4 \rightleftharpoons 2Mn^{2+} + 8H_2O + 10CO_2$

$$E^0_{cell} = E^0_{(R.P)_C} - E^0_{(R.P)_A} = 1.51 - (-0.49) = 2.0\ V$$

At Equilibrium $E_{cell} = 0$

$$E^0_{cell} = \frac{0.059}{n}\log K \Rightarrow 2 = \frac{0.059}{10}\log K$$

$$2 = \frac{0.059}{10}\log 10^x$$

$\log 10^x = 338.9 \Rightarrow x = 338.9$

102. (18) $2H^+(aq) + 2e^- \rightarrow H_2(g)$

$pH = -\log[H^+]$

$3 = -\log[H^+] \quad \therefore [H+] = 10^{-3}$

$$E_{cell} = E^0_{cell} - \frac{0.059}{2}\log\frac{P_{H_2}}{[H^+]^2}$$

$$\therefore E_{cell} = 0 - \frac{0.059}{2}\log\frac{P_{H_2}}{[10^{-3}]^2} = 0 - \frac{0.059}{2} \times 6$$

$E_{cell} = 0 - 0.059 \times 3 = -0.177$ volts. $= -17.7 \times 10^{-2}$ V.

103. (25) $Cu(OH)_2(s) \rightleftharpoons Cu^{2+}(aq) + 2OH^-(aq)$

$K_{sp} = [Cu^{2+}][OH^-]^2$

$pH = 14; pOH = 0; [OH^-] = 1$ M

$$\therefore [Cu^{2+}] = \frac{K_{sp}}{[1]^2} = 10^{-20} \text{ M}$$

$Cu^{2+}(aq) + 2e^- \rightarrow Cu(s)$

$$E = E^o - \frac{0.059}{2}\log_{10}\frac{1}{[Cu^{2+}]}$$

$$= 0.34 - \frac{0.059}{2}\log_{10}\frac{1}{10^{-20}}$$

$= -0.25 = -25 \times 10^{-2}$

104. (2) $Pb^{2+} + 2e^- \rightarrow Pb \qquad \Delta G_1^o = -2FE_1^o$

$Pb^{4+} + 4e^- \rightarrow Pb \qquad \Delta G_2^o = -4FE_2^o$

$Pb^{2+} \rightarrow Pb^{4+} + 2e^- \qquad \Delta G_3^o = -2FE_3^o$

$\Delta G_3^o = \Delta G_1^o - \Delta G_2^o$

$\Rightarrow \quad -2FE_3^o = 2F(2n - m) \quad \Rightarrow \quad E_3^o = m - 2n = m - xn$

Hence $x = 2$

105. (4) Given statements A, B, C and D are correct. Statement 'A' and 'B' can be explained by the definition of standard electrode potential and EMF. Statement 'C' and 'D' can be explained using Faraday's Laws of electrolysis.

106. (b) Anode : $\frac{1}{2}H_2(g) \rightleftharpoons H^+(aq) + e^-$

Cathode : $AgCl(s) + e^- \rightarrow Ag(s) + Cl^-(aq)$

107. (1) Option B is incorrect, E° cell is independent from pressure. So, Answer is 1

108. (3) Metal having lower Standard Reduction potential than 0.97V will be oxidised by NO_3^-.

109. (3) MnO_4^- (0.1 M) / Mn^{2+} (0.001 M)

Reaction at cathode : $MnO^-_4 + 8H^+ + 5e^- \rightarrow Mn^{2+} + 4H_2O$

$$E = E^o - \frac{0.059}{5}\log\frac{[Mn^{2+}]}{[MnO_4^-][H^+]^8}$$

$$1.282 = 1.54 - \frac{0.059}{5}\log\frac{10^{-3}}{10^{-1} \times [H^+]^8}$$

$$\frac{0.258 \times 5}{0.059} = \log\frac{10^{-2}}{[H^+]^8}$$

$21.86 = -2 + 8\,pH$

$pH = 2.98 \approx 3$

110. (6) $\Delta G^o = -RT\ell nK$

$-nFE^o_{cell} = -RT \times 2.303(\log_{10} K)$

$$\frac{E^o_{cell}}{0.06} \times n = \log K \qquad ...(i)$$

$Pd^{+2}(aq.) + 2e^- \rightleftharpoons Pd(s), E^o_{cat, redn} = 0.83$

$Pd(s) + 4Cl^-(aq.) \rightleftharpoons PdCl_4^{2-}, (aq) + 2e^-,$

$E^o_{anode, oxidn} = 0.65$

$Pd^{2+}_{(aq)} + 4Cl^-(aq) \rightleftharpoons PdCl_4^{2-}$

Net Reaction $\rightarrow Pd^{2+}(aq.) 4Cl^-(aq.) \rightleftharpoons PdCl_4^{2-}(aq.)$

$E^o_{cell} = E^o_{cat, redn} - E^o_{anode, oxidn}$

$E^o_{cell} = 0.83 - 0.65 \Rightarrow E^o_{cell} = 0.18 \qquad ...(ii)$

Also $n = 2$ using equation (i), (ii)

$\Rightarrow \quad \log K = 6$

111. (10) $Pt(g)|H_2(g, 1\text{ atm})|H^+(aq, 1M)||Fe^{3+}(aq), Fe^{3+}(aq)|Pt(s)$

at anode $H_2 \longrightarrow 2H^+ + 2e^-$

At cathode $2Fe^{3+}(aq) + 2e^- \longrightarrow 2Fe^{2+}(aq)$

$E^o = E^o_{Fe^{3+}|Fe^{2+}} - E^o_{H_2|H^+} = 0.771$ V

$$E = E^o - \frac{0.06}{2}\log\frac{[Fe^{2+}]^2}{[Fe^{3+}]^2}[H^+]^2$$

$$0.712 = 0.771 - \frac{0.06}{2}\log\frac{[Fe^{2+}]^2}{[Fe^{3+}]^2}$$

$$\log\frac{Fe^{2+}}{Fe^{3+}} = \frac{0.059}{0.06} \approx 1 \Rightarrow \frac{Fe^{2+}}{Fe^{3+}} = 10$$

112. (275) $X + Y^{2+} \rightarrow Y + X^{2+}$

$E^0_{Cell} = 0.36 - (-2.36) = 2.72$ V

$$E_{Cell} = 2.72 - \frac{0.06}{2}\log\frac{0.001}{0.01}$$

$= 2.72 + 0.03 = 2.75 \text{ V} = 275 \times 10^{-2}$ V

113. (c) Metal cation with negative value of reduction potential (M^{+3}/M^{+2}) or with positive value of oxidation potential (M^{+2}/M^{+3}) will liberate H_2 from dilute acid.

114. (17) $Zn(s) + Sn^{2+}(aq) \rightleftharpoons Zn^{2+}(aq) + Sn(s)$

$\Delta G^o = -2.303\,RT\log_{10} K_{eq}$

$-nF(E^0_{cell}) = -2.303RT\log_{10} K_{eq},$

$$E^0_{Sn^{2+}/Sn} - E^0_{Zn^{2+}/Zn} = \frac{0.059}{2}\log_{10} K_{eq}$$

$$E^0_{Sn^{2+}/Sn} + 0.76 = \frac{0.059}{2}\log_{10} 10^{20}$$

$$E^0_{Sn^{2+}/Sn} + 0.76 = \frac{0.059 \times 20}{2}$$

$$E^0_{Sn^{2+}/Sn} = 0.59 - 0.76 = -0.17$$

$$E^0_{Sn/Sn^{2+}} = 17 \times 10^{-2}\ V$$

115. (10) Cell reaction:

$$\frac{1}{2}H_2(g) + Fe^{3+}(aq.) \longrightarrow H^+(aq) + Fe^{2+}(aq.)$$

$$E_{cell} = E^o_{cell} - \frac{0.059}{1}\log\frac{[Fe^{2+}][H^+]}{[Fe^{3+}][H_2]^{\frac{1}{2}}}$$

$$E_{cell} = E^o_{cell} - \frac{0.059}{1}\log\frac{[Fe^{2+}]}{[Fe^{3+}]};$$

(Since $[H^+] = 1$ and $H_2 = 1$ atm

$$\Rightarrow\ 0.712 = (0.771 - 0) - \frac{0.059}{1}\log\frac{[Fe^{2+}]}{[Fe^{3+}]}$$

$$\Rightarrow\ \log\frac{[Fe^{2+}]}{[Fe^{3+}]} = \frac{(0.771 - 0712)}{0.059} = 1 \Rightarrow \frac{[Fe^{2+}]}{[Fe^{3+}]} = 10$$

116. (3) At Cathode: $M^{3+} + 2e \rightarrow M^+$, $E^\circ_{M^{3+}/M^+} = 0.2V$

At Anode: $H_2 \rightarrow 2H^+ + 2e$, $E^\circ_{H_2/2H^+} = 0.0V$

Overall reaction:

$$H_2(g) + M^{3+}(aq) \rightarrow M^+(aq) + 2H^+(aq)$$

$$E_{Cell} = E^\circ_{Cathode} - E^\circ_{anode} - \frac{0.059}{2}\log\frac{[M^+]\times 1^2}{[M^{+3}]1}$$

$$0.1115 = 0.2 - \frac{0.059}{2}\log\frac{[M^+]}{[M^{+3}]}$$

$$3 = \log\frac{[M^+]}{[M^{+3}]} \Rightarrow \therefore\ a = 3$$

117. (917) $Cr_2O_7^{2-} + 14H^+ + 6e^- \rightarrow 2Cr^{3+} + 7H_2O$

$$E = 1.33 - \frac{0.059}{6}\log\frac{(0.1)^2}{(10^{-2})(10^{-3})^{14}}$$

$$E = 1.33 - \frac{0.059}{6} \times 42 = 0.917 = 917 \times 10^{-3}$$

$\therefore x = 917$

118. (d) Only I^- among halides can be oxidised to Iodine by oxygen in acidic medium due to its reduction potential.

$$4I^-(aq) + 4H^+(aq) + O_2(g) \rightarrow 2I_2(s) + 2H_2O(l)$$

119. (1825)

$$FeO_4^{2-} \xrightarrow[n=3]{2.2V} Fe^{3+} \xrightarrow[n=1]{0.7V} Fe^{2+} \xrightarrow{-0.45V} Fe$$

$E = ?$, $n = 4$ (from FeO_4^{2-} to Fe^{2+}) $[\Delta G = -nF\,E^\circ_{cell}]$

$$4 \times E^\circ = (3 \times 2.2) + (1 \times 0.7)$$

$$E^\ominus = \frac{7.3}{4} = 1.825\ V = 1825 \times 10^{-3}\ V$$

120. (34) cell reaction $\rightarrow Cu + 2Ag^+ \rightarrow Cu^{2+} + 2Ag$

According to Nernst equation

$$E_{cell} = E^\circ_{cell} - \frac{0.06}{2}\log\frac{[Cu^{2+}]}{[Ag^+]^2}$$

$$0.43 = E^\circ_{cell} - \frac{0.06}{2}\log\frac{(0.001)}{(0.01)^2};\ E^\circ_{cell} = 0.46V$$

$$E^\circ_{cell} = E^\circ_{Ag^+/Ag} - E^\circ_{Cu^{2+}/Cu} \Rightarrow 0.46 = 0.80 - E^\circ_{Cu^{2+}/Cu}$$

$$E^\circ_{Cu^{2+}/Cu} = 0.34\ \text{volt} \Rightarrow 34 \times 10^{-2}$$

121. (4) From Nernst Eq.

$$E_{cell} = E^\circ_{cell} - \frac{RT}{nF}\ln Q$$

$$\text{or, } E_{cell} = E^\circ_{cell} - \frac{2.303\,RT}{nF}\log Q$$

$$\text{or, } 0.801 = (0.008 + 0.763) - \frac{0.06 \times (-2)}{n} \quad \text{or, } n = 4$$

122. (b) Reduction of MnO_4^- is pH dependent.

In acidic medium

$$MnO_4^- + 5e^- \longrightarrow Mn^{2+}$$

In neutral medium

$$MnO_4^- + 3e^- \longrightarrow Mn^{4+}$$

In basic medium

$$MnO_4^- + e^- \longrightarrow Mn^{6+}$$

So, according to pH, the reaction and potential of cell changes.

123. (7) The cell reaction is

$$H_2(g) + Cu^{2+}(aq.) \rightarrow 2H^+(aq.) + Cu(s)$$

$n = 2$

According to Nernst equation

$$0.31 = 0.34 - \frac{0.06}{2}\log\frac{[H^+]^2}{[Cu^{2+}]}$$

$$-0.03 = -0.03\log\frac{[H^+]^2}{Cu^{2+}} \Rightarrow 10^1 = \frac{[10^{-3}]^2}{Cu^{2+}}$$

$$Cu^{2+} = 10^{-7} \Rightarrow x = 7$$

124. (16) $\Delta G = -nFE^\circ_{cell}$

$$Sn^{2+} + 2e^- \rightarrow Sn \quad \Delta G^\circ_1 = +2 \times 0.140 \times F \quad .....(i)$$

$$Sn^{4+} + 4e^- \rightarrow Sn \quad \Delta G^\circ_2 = -4 \times 0.01 \times F \quad .....(ii)$$

Subtracting (i) from (ii), we get

$$Sn^{4+} + 2e^- \rightarrow Sn^{2+} \quad \Delta G^\circ_3 = -2 \times E^\circ_{Sn^{4+}/Sn^{2+}} \times F$$

$$\Delta G^\circ_3 = \Delta G^\circ_2 - \Delta G^\circ_1$$

$$-2 \times E^\circ \times F = -(0.04 + 0.28) \times F$$

$$E^\circ = 0.16\ \text{volt} = 16 \times 10^{-2}\ V$$

125. (51) $\frac{1}{2}H_2 + Ag^+ \rightarrow H^+ + Ag$; $n = 1$

$\Delta G° = -nE° F = -1 \times 0.5332 \times 96500\ J = -51.45\ kJ/mol$

126. (c) A cell with less variation in EMF with temperature is preferred as reference electrode because it can be used for wider range of temperature without much deviation from standard value. So a cell with less $\left(\frac{\partial E}{\partial T}\right)_p$ is preferred.

127. (983) $Cu(s) + Sn^{2+}(0.001\ M) \rightarrow Cu^{2+}(0.01\ M) + Sn(s)$

$E°_{cell} = E°_{cathode} - E°_{anode} = -0.14 - (0.34) = -0.48\ V$

$$E_{cell} = E°_{cell} - \frac{0.059}{2}\log\frac{[Cu^{2+}]}{[Sn^{2+}]}, \text{ where } n = 2$$

$$= -0.48 - \frac{0.059}{2}\log\frac{0.01}{0.001} = -0.5095\ V$$

$\Delta G = -nF\ E_{cell} = -2 \times 96500 \times (-0.5095)$

$= 98333.5\ J/mol = 98.335\ kJ/mol = 983.35 \times 10^{-1}\ kJ/mol$

Nearest Integer : 983

128. (23) $Fe^{3+} + I^- \longrightarrow I_2 + Fe^{2+}$

The $E°$ value for spontaneous reaction is positive.

$E^\circ_{Cell} = E^\circ_{cathode} - E^\circ_{anode} = (0.77 - 0.54)\ V = 0.23\ V$

$= 23 \times 10^{-2}\ V.$

129. (a) According to standard electrode potential table, the order of electrode couple from top to bottom is :

$Cl_2/Cl^- > Ag^+/Ag > I_2/I^- > Na^+/Na > Li^+/Li$

130. (5) Anode : $H_2 \rightarrow 2H^+ + 2e^-$

Cathode : $Cu^{2+} + 2e^- \rightarrow Cu$

$Cu^{2+} + H_2 \rightarrow 2H^+ + Cu$

$$E_{cell} = E^0_{cell} - \frac{0.06}{2}\log\frac{[H^+]^2}{[Cu^{2+}]}$$

$$0.576 = 0.34 - \frac{0.06}{2}\log\left\{\frac{[H^+]^2}{[0.01]}\right\}$$

$\Rightarrow 7.87 = -\log[H^+]^2 + \log 10^{-2} \Rightarrow pH = 4.93 \simeq 5$

131. (25) $\Delta G° = -nFE° = \Delta H° - T\Delta S°$

$$\Delta S° = \frac{\Delta H° + nFE°}{T}$$

$$= \frac{(-825.2 \times 10^3) + (2 \times 96487 \times 4.315)}{298}$$

$= 25.11\ JK^{-1}mol^{-1} \approx 25\ JK^{-1}mol^{-1}$

132. (109) Galvanic cell:

$$Zn(s) + \underset{0.02M}{Cu^{2+}}(aq) \rightarrow \underset{0.04M}{Zn^{2+}}(aq) + Cu(s)$$

Nernst equation

$$E_{cell} = E^\circ_{cell} - \frac{0.059}{2}\log\frac{[Zn^{2+}]}{[Cu^{2+}]}$$

$$\Rightarrow E_{cell} = \left[E^\circ_{Cu^{2+}/Cu} - E^\circ_{Zn^{+2}/Zn}\right] - \frac{0.059}{2}\log\frac{0.04}{0.02}$$

$$\Rightarrow E_{cell} = [0.34 - (-0.76)] - \frac{0.059}{2}\log 2$$

$\Rightarrow E_{cell} = 1.0911 = 109.11 \times 10^{-2}$

133. (28) Cell reaction is :

$Cu(s) + 2Ag^+(aq) \longrightarrow Cu^{2+}(aq) + 2Ag(s)$

$$\text{Now, } E_{cell} = E^0_{cell} - \frac{0.059}{2}\log\frac{[Cu^{2+}]}{[Ag^+]^2} \quad ...(1)$$

For cell 1,

$$0.3095 = E^0_{cell} - \frac{0.059}{2}\log\frac{(0.1)}{(0.01)^2};\ E^0_{cell} = 0.3981$$

For cell 2,

$$E_{cell} = 0.3981 - \frac{0.059}{2}\log\frac{(0.01)}{(0.001)^2}\ \lim_{x\to\infty} = 0.28 = 28 \times 10^{-2}$$

134. (24) $Zn \longrightarrow Zn^{2+} + 2e^-$

$2Fe^{3+} + 2e^- \longrightarrow 2Fe^{2+}$

$Zn + 2Fe^{3+} \longrightarrow Zn^{2+} + 2Fe^{2+}$

$E^0_{cell} = 0.77 - (-0.76) = 1.53\ V$

$$1.50 = 1.53 - \frac{0.06}{2}\log\left(\frac{[Fe^{2+}]}{[Fe^{3+}]}\right)^2$$

$$\Rightarrow \log\left(\frac{[Fe^{2+}]}{[Fe^{3+}]}\right) = \frac{0.03}{0.06} = \frac{1}{2}$$

$$\Rightarrow \frac{[Fe^{2+}]}{[Fe^{3+}]} = 10^{\frac{1}{2}} = \sqrt{10}$$

$$\Rightarrow \frac{[Fe^{2+}] + [Fe^{3+}]}{[Fe^{3+}]} = \sqrt{10} + 1$$

$$\Rightarrow \frac{[Fe^{3+}]}{[Fe^{2+}] + [Fe^{3+}]} = \frac{1}{1 + \sqrt{10}} = \frac{1}{4.16}$$

$= 0.2402 = 24 \times 10^{-2}.$

135. (3) $Cu + \underset{10^{-3}M}{2Ag^+} \longrightarrow \underset{0.250\ M}{Cu^{2+}} + 2Ag$

$$E_{cell} = E^\circ_{cell} - \frac{0.059}{2}\log\frac{[Cu^{2+}]}{[Ag^+]^2}$$

$$= 2.97 - \frac{0.059}{2}\log\frac{(0.25)}{(10^{-3})^2} = 2.81\ V \approx 3.0V$$

136. (147) $Zn(s) \longrightarrow Zn^{2+}(aq) + 2e^-$

$2Ag^+(aq) + 2e^- \longrightarrow 2Ag(s)$

$Zn(s) + 2Ag^+(aq) \longrightarrow Zn^{2+}(aq) + 2Ag(s)$

$$E^0_{cell} = E^0_{Ag^+/Ag} - E^0_{Zn^{2+}/Zn}$$

$= 0.80 - (-0.76)$

$= 1.56V$

$$E^0_{cell} = E^0 - \frac{2.303RT}{nF}\log_{10}\frac{[Zn^{2+}]}{[Ag^+]^2}$$

$$= 1.56 - \frac{0.059}{2}\log\frac{0.1}{(0.01)^2}$$

$$= 1.56 - \frac{0.059}{2} \times 3$$

$= 1.56 - 0.0885 = 1.4715$

$= 147.15 \times 10^{-2}$

137. (4)

$$NO_3^- + 4H^+ + 3e^- \rightarrow NO + 2H_2O; E^o_{NO_3^-/NO} = 0.96V$$

$$NO_3^- + H^+ + e^- \rightarrow NO_2 + H_2O; E^o_{NO_3^-/NO_2} = 0.79V$$

Let $[HNO_3] = yM$

$[H^+] = yM, \left[NO_3^-\right] = yM$

For same thermodynamic tendency of reduction

$$E_{NO_3^-/NO} = E_{NO_3^-/NO_2}$$

$$E^o_{NO_3^-/NO^-} \frac{0.059}{3}\log\frac{p_{NO}}{y \times y^4}$$

$$= E^o_{NO_3^-/NO_2} - \frac{0.059}{1}\log\frac{p_{NO_2}}{y \times y^2}$$

$$0.96 - \frac{0.059}{3}\log\frac{p_{NO}}{y^5} = 0.79 - 0.059\log\frac{p_{NO_2}}{y^3}$$

Assume ($p_{NO} = p_{NO_2} = 1$ bar)

$$0.17 = \frac{0.0591}{1}\log\frac{p_{NO_2}}{y^3} + \frac{0.059}{3}\log\frac{p_{NO}}{y^5}$$

$$0.17 = \frac{0.059}{3}\log\frac{p^3_{NO_2}}{y^9} + \frac{0.059}{3}\log\frac{p_{NO}}{y^5}$$

$$= \frac{0.059}{3}\left[\log\frac{p_{NO}}{y^5} - \log\frac{p^3_{NO_2}}{y^9}\right]$$

$$= \frac{0.059}{3} \times \log\left[\frac{p_{NO}}{y^5} \times \frac{y^9}{p^3_{NO_2}}\right] = \frac{0.059}{3}\log y^4$$

$$\log y^4 = \frac{0.17 \times 3}{0.059} \Rightarrow y = 10^{2.16}$$

Comparing × 2.16 and 2x = 4.32

Nearest integer is 4.

138. (3776) Equation is

$$MnO_4^-(aq) + 8H^+(aq) + 5e^- \rightarrow Mn^{2+}(aq) + 4H_2O(l)$$

Nernst equation :

$$E_{cell} = E^0_{cell} - \frac{0.059}{5}\log\frac{[Mn^{2+}]}{[MnO_4^-]}\frac{1}{[H^+]^8}$$

(I) Given $[H^+] = 1M$

$$E_1 = E^0 - \frac{0.059}{5}\log\frac{[Mn^{2+}]}{[MnO_4^-]}$$

(II) Now: $[H^+] = 10^{-4}$ M

$$E_2 = E^0 - \frac{0.059}{5}\log\frac{[Mn^{2+}]}{[MnO_4^-]} \times \frac{1}{(10^{-4})^8}$$

Therefore,

$$|E_1 - E_2| = \frac{0.059}{5} \times 32 = 0.3776\text{ V} = 3776 \times 10^{-4}$$

$\therefore x = 3776.$

139. (b) $E^o_{cell} = E^o_{Cu^{2+}|Cu} - E^o_{Zn^{2+}|Zn} = 1.1V$

So, if $E_{ext.} = 1.1$ V, no electron will flow.

At, $E_{ext.} > 1.1$ V, cell act as electrolytic cell and electron will flow from Cu to Zn.

At, $E_{ext.} < 1.1$ V, cell act as electrochemical cells so Zn dissolves at anode and Cu deposits at cathode.

140. (d) For the concentration cell, $E^o_{cell} = 0$

Anode: $Cu(s) \longrightarrow Cu^{2+}(aq)$

Cathode: $Cu^{2+}(aq) \longrightarrow Cu(s)$

Overall: $\underset{(C_2M)}{Cu^{2+}(aq)} \longrightarrow \underset{(C_1M)}{Cu^{2+}(aq)}$

$\Delta G = -nFE$

If $\Delta G = -$ve, then E_{cell} is +ve.

$$E_{cell} = E^o_{cell} - \frac{RT}{2F}\ln\frac{C_1}{C_2}$$

$$E_{cell} = 0 - \frac{RT}{2F}\ln\frac{C_1}{C_2}$$

$$E_{cell} = \frac{RT}{2F}\ln\frac{C_2}{C_1}; \text{ So, } C_2 > C_1.$$

Thus, $C_2 = \sqrt{2}C_1$ relation is correct.

141. (– 6)

$$\Delta G^o = -nFE^o_{cell}$$

$$17.37 \times 10^3 = -3 \times 96500 \times E^o_{cell}$$

$$E^o_{cell} = -0.06\text{ V} \simeq -6.0 \times 10^{-2}\text{ V}$$

142. (142)
Sodium metal :
Cell reaction

Cathode : $AgCl(s) + e^- \longrightarrow Ag(s) + Cl^-(aq)$

Anode : $\frac{1}{2}H_2(g) \longrightarrow H^+(aq) + e^-$

Overall : $AgCl(s) + \frac{1}{2}H_2(g) \longrightarrow Ag(s) + H^+(aq) + Cl^-(aq)$

$E^0_{cell} = E_{AgCl/Ag/Cl^-} - E_{H^+/H_2} = 0.22 - 0 = 0.22\text{ V}$

$E_{cell} = E^0_{cell} - \frac{0.06}{1}\log[H^+][Cl^-]$

$E_{cell} = 0.22 - \frac{0.06}{1}\log[10^{-1}][10^{-1}] = 0.22 + 0.12 = 0.34\text{ V}$

$(KE)_{max} = E_{cell} = 0.34\text{ eV}$

So, $E = W_0 + (KE)_{max} = 2.3 + 0.34 = 2.64\text{ eV}$ = Energy of photon incident

For potassium metal :

$E = W_0 + (KE)_{max}$

$2.64 = 2.25 + (KE)_{max}$ or $(KE)_{max} = 0.39 = E_{cell}$

$E_{cell} = E^0_{cell} - \frac{0.06}{1}\log[H^+][Cl^-]$

$0.39 = 0.22 - 0.06\log[H^+]^2$

$0.39 = 0.22 - 0.12\log[H^+]$

$0.17 = 0.12 \times pH$

$pH = 17/12 = 1.4166 \simeq 1.42 = 142 \times 10^{-2}$

143. (96500)

$E^0_{cell} = E^0_{Sn^{2+}/Sn} - E^0_{Cu^{2+}/Cu} = -0.16 - 0.34 = -0.50\text{ V}$

$\Delta G^0 = -nFE^0_{cell} = -2 \times 96500 \times (-0.5) = 96500\text{ J}$

144. (144)

$2Cu^+(aq) \rightleftharpoons Cu(s) + Cu^{2+}(aq)$

$E^o_{cell} = E^o_{Cu^+/Cu} - E^o_{Cu^{2+}/Cu^+} = 0.52 - 0.16 = 0.36\text{ V}$

$E^o_{cell} = \frac{RT}{nF}\ln K_{eq} \Rightarrow 0.36 = \frac{0.025}{1}\ln K_{eq}$

$\Rightarrow \ln K_{eq} = 14.4 = 144 \times 10^{-1}$

145. (1.52)

$E = 1.23 - \frac{0.0591}{4}\log[H^+]^4 = 1.23 + 0.0591 \times pH$

$= 1.23 + 0.0591 \times 5 = 1.23 + 0.2955 = 1.52\text{ V}$

146. (2.15)
At equilibrium state, $E_{cell} = 0$; $E^0_{cell} = 0.01\text{ V}$

$Sn(s) + Pb^{2+}(aq) \longrightarrow Sn^{2+}(aq) + Pb(s)$

$E = E^0_{cell} - \frac{0.06}{n}\log\frac{[P]}{[R]}$

$0 = 0.01 - \frac{0.06}{2}\log\frac{[Sn^{2+}]}{[Pb^{2+}]}$

$\frac{1}{3} = \log\frac{[Sn^{2+}]}{[Pb^{2+}]}$

$\frac{[Sn^{2+}]}{[Pb^{2+}]} = 10^{1/3} = 2.15$

147. (b) $Cu^{2+} + 2e^- \longrightarrow Cu, \Delta G_1^\circ = -2F(0.34)$ ...(i)
$Cu^+ + e^- \longrightarrow Cu, \Delta G_2^\circ = -F(0.522)$ ...(ii)
Subtract (ii) from (i)
$Cu^{2+} + e^- \longrightarrow Cu^+, \quad \Delta G_3^\circ = -F(E^0)$
$\therefore \Delta G_1^\circ - \Delta G_2^\circ = \Delta G_3^\circ$
$\Rightarrow -FE^\circ = -2F(0.34) + F(0.522)$
$\Rightarrow E^\circ = 0.68 - 0.522 = 0.158\text{ V}$

148. (a) $\Delta G^\circ = -nFE^\circ_{cell}$
$= -2 \times (96000) \times 2\text{ V} = -384000\text{ J/mol} = -384\text{ kJ/mol}$

149. (b) Higher the reduction potential, higher will be oxidising power. So,
$Bi^{3+} < Ce^{4+} < Pb^{4+} < Co^{3+}$

150. (c) $Ag^+(aq) + Fe^{2+}(aq) \longrightarrow Fe^{3+}(aq) + Ag(s)$

$E^\circ_{cell} = E^\circ_{Ag^+/Ag} - E^\circ_{Fe^{3+}/Fe^{2+}}$

$E^\circ_{Fe^{3+}/Fe^{2+}} = 3z - 2y$; $E^\circ_{Ag^+/Ag} = x$ (given)

$\therefore E^\circ_{cell} = x - 3z + 2y$

151. (c) Reduction potential is maximum for $S_2O_8^{2-}$, therefore, it is the strongest oxidising agent amongst the given species.

152. (b) Higher the oxidation potential, higher will be the reducing power. So, the order of reducing behaviour is:
$Ca > Mg > Zn > Ni$

153. (a) $\Delta_r H^\circ = \Delta G_r^\circ + T\Delta S^\circ$

$\Delta_r H^\circ = -nFE^\circ + TnF\frac{dE^\circ}{dT}$

Cell reaction: $Zn(s) + Cu^{2+}(aq) \longrightarrow Zn^{2+}(aq) + Cu(s)$

$\Delta_r H^\circ = -nF\left(E^\circ - \frac{TdE^\circ}{dT}\right)$

$\Delta_r H^\circ = -2 \times 96000[2 - 300 \times (-5 \times 10^{-4})]$
$= -412.8 \times 10^3\text{ J/mol} = -412.8\text{ kJ/mol}$

154. (a) $Zn + 2Ag^+ \longrightarrow 2Ag + Zn^{2+}$ $(n = 2e^-)$
$E^\circ_{cell} = (E^\circ_{R.P.})_{cathode} - (E^\circ_{R.P.})_{anode}$
$= 0.80 - (-0.76) = 1.56\text{ V}$ for $2e^-$

$\therefore E^\circ_{cell}$ for $1e^- = \frac{1.56}{2} = 0.78\text{ V}$

155. (c) $E^0_{cell} = \frac{2.303RT}{nF}\log K_C$ or $E^0_{cell} = \frac{0.059V}{n}\log K_C$

$= \frac{0.059V}{2}\log 10^{16} = 0.4736\text{ V}$

156. (d) Given that:
$Pt(s) | H_2(g, 1\text{ bar}) / HCl(aq) \| AgCl(s) / Ag(s) | Pt(s)$
$E_{cell} = 0.92\text{ V}$

Now, $E_{cell} = E^\circ_{H_2(g)/H^+(aq)} + E^\circ_{AgCl(s)/Ag(s),Cl^-} - \frac{0.06}{n}\log Q$

Cell reaction: $\frac{1}{2}H_2(g) \longrightarrow H^+(aq) + e^-$

$AgCl(s) + e^- \longrightarrow Ag(s) + Cl^-(aq)$

Net cell reaction:

$\frac{1}{2}H_2(g) + AgCl(s) \longrightarrow Ag(s) + H^+(aq) + Cl^-(aq)$

$\therefore\ Q = \frac{[H^+][Cl^-]}{(P_{H_2})^{1/2}}$

Here, 10^{-6} molal HCl solution is used

So $Q = \frac{10^{-6} \times 10^{-6}}{1} = 10^{-12}$

(assuming molality = molarity)

Now, $0.92 = E^\circ_{AgCl(s)/Ag(s),Cl^-} - \frac{0.06}{1}\log 10^{-12}$

$E^\circ_{AgCl(s)/Ag(s),Cl^-} = 0.92 + [0.06 \times (-12)]$

$= 0.92 - 0.72 = 0.20$ V

157. (d) We know that, $\ln K = \frac{nFE^\circ_{cell}}{RT}$

After putting the given values, we get

$\ln K = \frac{2 \times 96000 \times 2}{8 \times 300} = 160 \therefore \quad K = e^{160}$

158. (b) Cell reaction :

$M(s) + 3Ag^+(aq) \longrightarrow M^{3+}(aq) + 3Ag(s)$

$E = E^0 - \frac{0.0591}{n}\log\frac{[\text{Reduced state}]}{[\text{Oxidised state}]}$

$0.421 = E^0 - \frac{0.0591}{3}\log\frac{0.001}{(0.01)^3}$; $E^0 = 0.48$ V

$E^0 = E^0_{Ag^+/Ag} - E^0_{M^{3+}/M}$

$E^0_{M^{3+}/M} = 0.8V - 0.48V = 0.32$ Volt

159. (a) $E^\circ_{Mn^{2+}/MnO_4^-} = -1.51$ V

$E^\circ_{Cl^-/Cl_2} = -1.36$ V; $E^\circ_{Cr^{3+}/C_2O_7^{2-}} = -1.33$ V

$E^\circ_{Cr/Cr^{3+}} = +0.74$ V

Since Cr is having highest oxidation potential, Cr is the best reducing agent.

160. (c) (i) $\underset{\text{Most stable}}{Al^{3+}} \xrightarrow{E^\circ = -1.66} Al \xleftarrow{E^\circ = +0.55} \underset{\text{Less stable}}{Al^+}$

(ii) $\underset{\text{Less stable}}{Tl^{3+}} \xrightarrow{E^\circ = +1.26} Tl \xleftarrow{E^\circ = -0.34} \underset{\text{More stable}}{Tl^+}$

Tl^+ has negative electrode potential ($E^\circ = -0.34$) means, it does not prefer to convert into Tl but reverse must be preferred that's why it is more stable than Tl^{3+} ($E^\circ = +1.26$). In Al, Al^{3+} is more stable ($E^\circ = -1.66$) than Al^+ ($E^\circ = +0.55$) and also from Tl^+ due to more negative value of E°. Therefore, by comparison it confirms that Tl^+ is more stable than Al^+.

161. (a) $\Delta G^\circ = -nFE^\circ$

(i) $Fe^{2+} + 2e^- \longrightarrow Fe$; $E^0 = -0.47V$;

(ii) $Fe^{3+} + e^- \longrightarrow Fe^{2+}$; $E^0 = +0.77V$;

(iii) $Fe^{3+} + 3e^- \longrightarrow Fe$

(i) $\Delta G^0 = -nFE^0 = -2(-0.47)F = 0.94F$

(ii) $\Delta G^0 = -nFE^0 = -1(+0.77)F = -0.77F$

(iii) On adding : $\Delta G^0 = +0.17$ F

$\Delta G^0 = -nFE^0$

E^0 for ($Fe^{3+} \longrightarrow Fe$) $= \frac{\Delta G^0}{-nF} = \frac{0.17F}{-3F} = -0.057$V

162. (c) No reaction will occur as the Zn is placed above Cu in electrochemical series. Hence, there will be no displacement reaction.

163. (b) $MnO_4^- + 8H^+ + 5e^- \longrightarrow Mn^{2+} + 4H_2O$

$E = E^\circ - \frac{0.059}{n}\log\frac{[Mn^{2+}]}{[MnO_4^-][H^+]^8}$

Taking Mn^{2+} and MnO_4^- in standard state, i.e., 1 M,

$E = 1.51 - \frac{0.059}{5} \times 8\log\frac{1}{[H^+]}$

$= 1.51 - \frac{0.059}{5} \times 8 \times 3 = 1.2268$ V

Hence at this pH, MnO_4^- will oxidise only Br^- and I^- as SRP of Cl_2/Cl^- is 1.36 V which is greater than that for MnO_4^-/Mn^{2+}.

164. (c) As the value of reduction potential decreases, the sequence of deposition also decreases.

C > B > D < A
(0.85) (0.6) (–0.76) (–1.2)

165. (b) The order of electrode oxidation Potential :

$Br^- < Fe^{2+} < Al$

The same order will be for their reducing power.

166. (a) $Mn^{2+} + 2e^- \rightarrow Mn$; $E^\circ = -1.18$ V; ...(i)

$Mn^{3+} + e^- \rightarrow Mn^{2+}$; $E^\circ = +1.51$V; ... (ii)

Now multiplying equation (ii) by two and subtracting from equation (i):

$3Mn^{2+} \rightarrow Mn^+ + 2Mn^{3+}$;

$E^\circ = -1.18 - (+1.51) = -2.69$ V

–ve value of EMF (i.e., ΔG = +ve) shows that the reaction is non-spontaneous.

167. (d) Higher the value of standard reduction potential, stronger is the oxidising agent, hence MnO_4^- is the strongest oxidising agent.

168. (d) For a spontaneous reaction, ΔG° must be –ve.

$\Delta G^\circ = -nFE^\circ$

Hence for ΔG° to be –ve, ΔE° has to be positive which is possible when X = Zn, Y = Ni

$Zn + Ni^{2+} \longrightarrow Zn^{2+} + Ni$

$E^\circ_{cell} = E^\circ_{Ni^{2+}/Ni} - E^\circ_{Zn^{2+}/Zn}$

$= -0.23 - (-0.76) = +0.53$ (positive)

169. (b) $Cu^{2+} + e^- \longrightarrow Cu^+$; $E_1^\circ = 0.15V; \Delta G_1^\circ = -n_1 E_1^\circ F$

$Cu^{2+} + 2e \longrightarrow Cu$; $E_2^\circ = 0.34V; \Delta G_2^\circ = -n_2 E_2^\circ F$

On subracting eq.(i) from eq. (ii) we get

$Cu^+ + e^- \longrightarrow Cu$; $\Delta G^\circ = \Delta G_2^\circ - \Delta G_1^\circ$

$-nE^\circ F = -(n_2 E_2^\circ F - n_1 E_1^\circ F)$

$$E^\circ = \frac{n_2 E_2^\circ F - n_1 E_1^\circ F}{nF}$$

$$= \frac{2 \times 0.34 - 0.15}{1} = 0.53V$$

170. (c) $\frac{2}{3} Al_2O_3 \rightarrow \frac{4}{3} Al + O_2$

The ionic reactions are,

$\frac{2}{3} \times (2Al^{3+}) + 4e^- \rightarrow \frac{4}{3} Al$

$\frac{2}{3} \times (3O^{2-}) \rightarrow O_2 + 4e^-$

Thus, no. of electron transferred $\Rightarrow$ n = 4

$\Delta G = -nFE^\circ$

or $E^\circ = \frac{\Delta G}{-nF} = \frac{966 \times 10^3}{4 \times 96500} = -2.5$ V

$\therefore$ The potential difference needed for the reduction = 2.5 V

171. (b) Given

$Fe^{3+} + 3e^- \rightarrow Fe$,

$E^\circ_{Fe^{3+}/Fe} = -0.036$ V ...(i)

$Fe^{2+} + 2e^- \rightarrow Fe$,

$E^\circ_{Fe^{2+}/Fe} = -0.439$V ...(ii)

we have to calculate

$Fe^{3+} + e^- \rightarrow Fe^{2+}$, $\Delta G^\circ = ?$

To obtain this equation subtract equ. (ii) from (i) we get

$Fe^{3+} + e^- \rightarrow Fe^{2+}$ ...(iii)

As we know that $\Delta G^\circ = -nFE^\circ$

Thus for reaction (iii)

$\Delta G_3 = \Delta G_1 - \Delta G_2$

$-nFE_3^\circ = -nFE_1^\circ - (-nFE_2^\circ)$

$-nFE_3^\circ = nFE_2^\circ - nFE_1^\circ$

$-1FE_3^\circ = 2 \times 0.439F - 3 \times 0.036F$

$-1FE_3^\circ = 0.770F$

$\therefore E_3^\circ = -0.770V$

172. (d) From the given representation of the cell, E_{cell} can be found as follows.

$2Cr + 3Fe^{2+} \longrightarrow 2Cr^{3+} + 3Fe$

$$E_{cell} = E^o_{Fe^{2+}/Fe} - E^o_{Cr^{3+}/Cr} - \frac{0.059}{6} \log \frac{[Cr^{3+}]^2}{[Fe^{2+}]^3}$$

$$= -0.42 - (-0.72) - \frac{0.059}{6} \log \frac{(0.1)^2}{(0.01)^3}$$

$$= 0.3 - \frac{0.059}{6} \log \frac{10^{-2}}{10^{-6}} = 0.3 - \frac{0.059}{6} \times 4$$

$$= 0.30 - 0.0393 = 0.26 \text{ V}$$

173. (d) $E_{cell} = 0$; when cell is completely discharged.

$$E_{cell} = E^\circ_{cell} - \frac{0.059}{2} \log \left(\frac{[Zn^{2+}]}{[Cu^{2+}]} \right)$$

$$\text{or } 0 = 1.1 - \frac{0.059}{2} \log \left(\frac{[Zn^{2+}]}{[Cu^{2+}]} \right)$$

$$\log \left(\frac{[Zn^{2+}]}{[Cu^{2+}]} \right) = \frac{2 \times 1.1}{0.059} = 37.3 \quad \therefore \left(\frac{[Zn^{2+}]}{[Cu^{2+}]} \right) = 10^{37.3}$$

174. (d) For spontaneous reaction, ΔG should be negative. Equilibrium constant should be more than one.

$\Delta G = -2.303$ RT log K_c, If $K_c = 1$ then

$\Delta G = 0$; If $K_c > 1$ then $\Delta G = -$ve.

Again, $\Delta G = -nFE^\circ_{cell}$.

E°_{cell} must be +ve to have ΔG –ve.

175. (a) $E_{cell} = E^\circ_{cell} - \frac{0.059}{n} \log K_c$

or $0 = 0.591 - \frac{0.0591}{1} \log K_c$

or $\log K_c = \frac{0.591}{0.0591} = 10$ or $K_c = 1 \times 10^{10}$

176. (a) $Zn(s) + 2H^+(aq) \rightleftharpoons Zn^{2+}(aq) + H_2(g)$

$$E_{cell} = E^\circ_{cell} - \frac{0.059}{2} \log \frac{[Zn^{2+}][H_2]}{[H^+]^2}$$

Addition of H_2SO_4 will increase $[H^+]$ and E_{cell} will also increase and thus, the equilibrium shifts towards RHS.

177. (a) $Sn(s) + 2Fe^{3+}(aq) \rightarrow 2Fe^{2+}(aq) + Sn^{2+}(aq)$

Standard potential for the given reaction

or $E^\circ_{cell} = E^\circ_{Fe^{3+}/Fe^{2+}} - E^\circ_{Sn^{2+}/Sn} = 0.77 - (-0.14) = 0.91$ V

178. (c) The given values show that Cr has maximum oxidation potental, therefore its oxidation will be easiest. (Change the sign to get the oxidation values)

179. (d)

A	B	C
+0.5V	–3.0V	–1.2V

Note: The higher the negative value of reduction potential, the more is the reducing power.

Hence B > C > A.

180. (c) The equilibrium constant is related to the standard emf of cell by the expression

$$\log K = E^{\circ}_{cell} \times \frac{n}{0.059} = 0.295 \times \frac{2}{0.059}$$

$$\log K = \frac{590}{59} = 10 \text{ or } K = 1 \times 10^{10}$$

181. (b) $E_{cell} = E^{\circ}_{cell} + \frac{0.059}{n} \log \frac{[Cu^{2+}]}{[Zn^{2+}]}$

$= 1.10 + \frac{0.059}{2} \log[0.1]$

$= 1.10 - 0.0295 = 1.07\,V$

182. (d) Oxidation half cell:-

$\underset{P_1}{H_2(g)} \longrightarrow 2H^+(1M) + 2e^-$

Reduction half cell

$2H^+(1M) + 2e^- \longrightarrow \underset{P_2}{H_2(g)}$

The net cell reaction,

$\underset{P_1}{H_2(g)} \longrightarrow \underset{P_2}{H_2(g)}$

$E^{\circ}_{cell} = 0.00\,V;\ n = 2$

$\therefore\ E_{cell} = E^{\circ}_{cell} - \frac{RT}{nF} \log_e K = 0 - \frac{RT}{nF} \log_e \frac{P_2}{P_1}$

or $E_{cell} = \frac{RT}{2F} \log_e \frac{P_1}{P_2}$

183. (a) $2Cr^{3+} + 7H_2O \rightarrow Cr_2O_7^{2-} + 14H^+$

O.S. of Cr changes from +3 to +6 by loss of electrons. At anode, oxidation takes place.

184. (c) EMF of galvanic cell = 1.1 volt

If E_{ext} < EMF then electrons flow steadily from anode to cathode, while if E_{ext} > EMF then electrons flow from cathode to anode as polarity is changed.

185. (c) Higher the value of reduction potential stronger will be the oxidising agent, hence based on the given values Ag^+ will be strongest oxidizing agent.

186. (b) During the charging of lead storage battery, the following reaction takes place.

$2PbSO_4(s) + 2H_2O(l) \rightarrow Pb(s) + PbO_2(s) + 2H_2SO_4(aq)$

At anode $PbSO_4$ reduced back to metallic Pb. At cathode $PbSO_4$ oxidised back to PbO_2.

From the balanced reacation.

$\because x_1 = +2, y_1 = 0$

$x_2 = +2, y_2 = 4$

187. (a) $CH_3OH(\ell) + \frac{3}{2}O_2(g) \rightarrow CO_2(g) + 2H_2O(\ell)$

$\because\ E^{o}_{cell} = E^{o}_{cathode} - E^{o}_{Anode} = E^{o}_{O_2/H_2O} - E^{o}_{CO_2/CH_3OH}$

$1.21 = 1.229 - E^{o}_{CO_2/CH_3OH}$

$\Rightarrow\ E^{o}_{CO_2/CH_3OH} = 0.019\,V = 19\,mV$

Fuel cell involves oxidation of methanol which will occur at anode and reduction of O_2 will occur at cathode.

188. (a)

189. (d) Corrosion is an example of electrochemical reaction due to redox reaction in which pure metal act as anode and impure metal (rusted metal) act as cathode. Corrosion is more favourable in acid medium than alkaline medium as H^+ ions S quickly accept electrons from the metal.

190. (c) Leclanche cell and Hg-cell are primary batteries while Ni – Cd is a secondary cell.

191. (a) Reaction occurs at cathode is :

$MnO_2 + NH_4^+ + e^- \longrightarrow MnO(OH) + NH_3$

In the cathode reaction manganese (Mn) is reduced from the +4 oxidation state to the +3 state.

192. (d) Fuel cell is used in spaceship and it is type of galvanic cell. It has efficiency of about 70%.

193. (b) Mn, Ni and Cd metals used in battery industries.

194. (a) As tin coating is peeled off, then iron is exposed to atmosphere and rusting of iron will take place.

195. (a) Lead storage battery consists of lead anode and a grid of lead packed with lead oxide (PbO_2) as cathode, a 38% solution of H_2SO_4 is used as an electrolyte.

On charging the battery the reaction is reversed and $PbSO_4(s)$ on anode and cathode is converted into Pb and PbO_2 respectively.

196. (c)

(a) $Cd(s) + 2Ni(OH)_3(s) \rightarrow CdO(s) + 2Ni(OH)_2(s) + H_2O(l)$
Discharge of secondary Battery

(b) $Zn(Hg) + HgO(s) \rightarrow ZnO(s) + Hg(l)$
(Primary Battery Mercury cell)

(c) $2PbSO_4(s) + 2H_2O(l) \rightarrow Pb(s) + PbO_2(s) + 2H_2SO_4(aq)$
Charging of secondary Battery

(d) $2H_2(g) + O_2(g) \rightarrow 2H_2O(l)$ – Fuel cell

197. (c) $CH_3OH\,(l) + \frac{3}{2}O_2\,(g) \rightarrow CO_2\,(g) + 2H_2O\,(l)$

$\Delta G_r = \Delta G_f(CO_2, g) + 2\Delta G_f(H_2O, l) - \Delta G_f(CH_3OH, l) - \frac{3}{2}\Delta G_f(O_2, g)$

$= -394.4 + 2(-237.2) - (-166.2) - 0$

$= -394.4 - 474.4 + 166.2 = -702.6\,kJ$

$\%\ efficiency = \frac{702.6}{726} \times 100 = 97\%$

198. (b) Cathode : $O_2(g) + 2H_2O(l) + 4e^- \longrightarrow 4OH^-(aq)$

Anode : $2H_2(g) + 4OH^-(aq) \longrightarrow 4H_2O(l) + 2e^-$

Overall : $2H_2(g) + O_2(g) \longrightarrow 2H_2O(l)$

Due to potential difference between cathode and anode, production of electricity is possible.

Chemical Kinetics

1. **(d)** According to given conditions t_1 and t_2 can be given as

$$t_1 = t_{\frac{1}{4}} = \frac{1}{k}\ln\frac{A_0}{\frac{A_0}{4}} = \frac{1}{k}\ln 4$$

$$t_2 = t_{\frac{1}{8}} = \frac{1}{k}\ln\frac{A_0}{\frac{A_0}{8}} = \frac{1}{k}\ln 8$$

$\therefore$ Ratio t_1/t_2 will be given by

$$\frac{t_1}{t_2} = \frac{\ln 4}{\ln 8} = \frac{2\ln 2}{3\ln 2} = \frac{2}{3}$$

2. **(c)**

	A(g) →	B(g) +	C(g)
t = 0	P^o	0	0
t = t	$P^o - x$	x	x
t = ∞	0	P^o	P^o

$$P_t = P^o + x \Rightarrow x = P_t - P^o = P_t - \frac{P_\infty}{2}$$

$$P_\infty = 2P^o \Rightarrow P^o = \frac{P_\infty}{2}$$

$$k = \frac{1}{t}\ell n\frac{P^o}{P^o - x}$$

$$k = \frac{1}{t}\ell n\frac{P_\infty}{2(P_\infty - P_t)}$$

3. **(c)** $A(g) \longrightarrow 2B(g) + C(g)$

	A(g)	2B(g)	C(g)
t = 0	P_0		
t → ∞	0	$2P_0$	P_0

$P_\infty = 3P_0 = 240$

$P_0 = 80$ mm of Hg

$$kt = \ln\left(\frac{P_\infty - P_0}{P_\infty - Pt}\right)$$

$$k\times 10 = \ln\left(\frac{240-80}{240-160}\right)$$

$$k = \frac{\ln 2}{10} = 0.0693\,\text{min}^{-1}$$

$\therefore$ Option (c) is incorrect

4. **(a)**

$r_1 = k[A]^n[B]^m$

Now A is doubled & B is halved in concentration

$$\Rightarrow\ r_2 = k2^n[A]^n.\frac{[B]^m}{2^m}$$

$$r_2 = k[A]^n[B]^m.2^n.\left(\frac{1}{2}\right)^m$$

Now $\frac{r_2}{r_1} = 2^n\left(\frac{1}{2}\right)^m \Rightarrow \frac{r_2}{r_1} = 2^{(n-m)}$

5. **(47)** The graph given is for 1st order reaction

$$k = \frac{2.303}{t}\log\frac{[A_0]}{[A_t]} \Rightarrow k = \frac{2.303}{25}\log\frac{[50]}{[10]}$$

$$k = \frac{2.303}{25}\log 5 = \frac{2.303\times 0.7}{25}s^{-1}$$

$$= \frac{2.303\times 0.7}{25} = \frac{2.303}{t}\log\frac{50}{2.5} \Rightarrow \frac{1.699}{25} = \frac{2.303}{t}\log 20$$

$$t = \frac{1.301\times 25\times 2.303}{0.699} = 46.46\,s \approx 47\,s$$

6. **(2435)**

$t_{1/2} = \propto [A]_0 \Rightarrow$ Order = zero

$$t_{1/2} = \frac{A_0}{2k} \Rightarrow \text{Slope} = \frac{1}{2k} = 76.92$$

$$k = \frac{1}{2\times 76.92}$$

$$[A]_{10} = -kt + A_0 = -\frac{1}{2\times 76.92}\times 10 + 2.5 = 2.435$$

$= 2435\times 10^{-3}$ mol/L

7. **(8)** $r = k[A]^2|B|$; if conc. are doubled

$r' = k[2A]^2[2B]^1 \Rightarrow r' = 8r \Rightarrow x = 8$

For a zero order reaction

$$k = \frac{[R]_0 - [R]}{t}$$

$\Rightarrow$ Zero order, y = 0

Thus, x + y = 8

8. **(3)** $A(g) \rightarrow 2B(g) + C(g)$.

$P_t = P_0 + 2x = 200$; $P_\infty = 3P_0 = 300$; $P_0 = 100$

$$K = \frac{1}{t}\ln\frac{P_\infty - P_0}{P_\infty - P_t} = \frac{2.303}{t}\log\frac{P_\infty - P_0}{P_\infty - P_t}$$

$$= \frac{2.303}{23}\log\frac{300-100}{300-200}$$

$$= \frac{2.3\times 0.301}{23} = 0.0301 = 3.01\times 10^{-2}\,\text{sec}^{-1}$$

9. **(1)** For 1st order reaction

$$A \xrightarrow{t=x} \frac{A}{2} \xrightarrow{t=x} \frac{A}{4}$$

t = 2x

75% life = 2 × 50% life

So, order with respect to A will be first order.

So, order with respect to B will be zero.

Overall order of reaction = 1 + 0 = 1

10. **(d)** Rate of formation of B is

$$\frac{d[B]}{dt} = K_1[A] - K_2[B] = 0$$

$$K_1[A] = K_2[B] \Rightarrow [B] = \frac{K_1[A]}{K_2}$$

11. **(17)** The given reaction is a first order reaction. (from its mechanism, we know that it is a first-order reaction). Rate of NO_2 formation

$$= \frac{d[NO_2]}{dt} = 2 \times \frac{d[N_2O_5]}{dt} \text{ (from stoichiometry)}$$

$$\Rightarrow \frac{d[NO_2]}{dt} = 2 \times \frac{d[N_2O_5]}{dt}$$

$$= 2 \times \frac{\Delta[N_2O_5]}{\Delta t} = 2 \times \frac{(3 - 2.75)}{30}$$

$= 1.67 \times 10^{-2} = 16.7 \times 10^{-3}$ mol L^{-1} min^{-1}

$= x \times 10^{-3}$ mol L^{-1} min^{-1} Thus, x = 16.7 or 17.

12. **(2200)** $t_{1/2} = 10$ minutes

$$(P_A)_{30\ min} = (P_A)_0 \left(\frac{1}{2}\right)^{30/10}$$

= 100 mm Hg

	A (g) →	2B (g) +	C (g)
at t = 0,	800 mm	0	0
at t = 30,	100 mm	1400 mm	700 mm

Total pressure after 30 minutes = 2200 mm Hg

13. **(333)** $KClO_3 + 6FeSO_4 + 3H_2SO_4 \longrightarrow KCl + 3Fe_2(SO_4)_3 + 3H_2O$

$$\text{Rate} = \frac{-\Delta[KClO_3]}{\Delta t} = \frac{-1}{6}\frac{\Delta[FeSO_4]}{\Delta t}$$

$$= \frac{+1}{3}\frac{\Delta[Fe_2(SO_4)_3]}{\Delta t}$$

$$\Rightarrow \frac{\Delta[Fe_2(SO_4)_3]}{\Delta t} = \frac{-1}{2}\frac{\Delta[FeSO_4]}{\Delta t} = \frac{1}{2}\frac{(10-8.8)}{30 \times 60}$$

$= 0.333 \times 10^{-3} = 333 \times 10^{-6}$ mol litre^{-1} sec^{-1}

14. **(2)** $t_{10\%} = \frac{1}{K}\ln\left(\frac{a}{a-x}\right) = \frac{1}{K}\ln\left(\frac{100}{90}\right)$

$$t_{10\%} = \frac{2.303}{K}(\log 10 - \log 9) \Rightarrow t_{10\%} = \frac{2.303}{K}(0.04)$$

$$t_{90\%} = \frac{1}{K}\ln\left(\frac{100}{10}\right) \Rightarrow t_{90\%} = \frac{2.303}{K}$$

$$\frac{t_{90\%}}{t_{10\%}} = \frac{1}{0.046} = 21.7$$

From the rate law of 1st order reaction

$$e^{kt} = \frac{a}{a-x} \Rightarrow \frac{a-x}{a} = e^{-kt}$$

$$1 - \frac{x}{a} = e^{-kt} \Rightarrow x = a(1 - e^{-kt}) \Rightarrow \alpha = \frac{x}{a} = (1 - e^{-kt})$$

15. **(40)** $r = k[X][Y]^0 = k[X]$

Using I and II:

$$\frac{4 \times 10^{-3}}{2 \times 10^{-3}} = \left(\frac{L}{0.1}\right) \Rightarrow L = 0.2$$

Using I and III:

$$\frac{M \times 10^{-3}}{2 \times 10^{-3}} = \frac{0.4}{0.1} \Rightarrow M = 8 \Rightarrow \frac{M}{L} = \frac{8}{0.2} = 40$$

16. **(1)** $\gamma_1 A + \gamma_2 B \longrightarrow \gamma_3 C + \gamma_4 D$

$$\text{Given: } +\frac{d[D]}{dt} = \frac{-3}{2}\frac{d[B]}{dt}$$

$$\Rightarrow \frac{-1}{2}\frac{d[B]}{dt} = \frac{+1}{3}\frac{d[D]}{dt}$$

$$\frac{d[B]}{dt} = -2\frac{d[A]}{dt} \Rightarrow -\frac{1}{2}\frac{d[B]}{dt} = \frac{-d[A]}{dt}$$

Given, $+\frac{d[D]}{dt} = 9$ m mol dm^{-3}s^{-1} and

$$\frac{+d[C]}{dt} = \frac{20-10}{10} = 1 \text{ m mol dm}^{-3}\text{s}^{-1}$$

$$\text{Therefore, } \frac{+d[C]}{dt} = \frac{1}{9} \times \frac{+d[D]}{dt} \Rightarrow \frac{1}{3}\frac{d[D]}{dt} = \frac{3d[C]}{dt}$$

$$\Rightarrow \frac{-d[A]}{dt} = -\frac{1}{2}\frac{d[B]}{dt} = \frac{3d[C]}{dt} = \frac{1}{3}\frac{d[D]}{dt}$$

Hence, the chemical reaction is $A + 2B \longrightarrow \frac{1}{3}C + 3D$

Assuming γ_1, γ_2, γ_3 and γ_4 are integers.

$\Rightarrow 3A + 6B \longrightarrow C + 9D$

Rate of reaction $= \frac{+d[C]}{dt} = 1$ m mol dm^{-3} s^{-1}.

17. **(4)** $\text{Rate} = -\frac{1}{3}\frac{d[C_2H_6O]}{dt} = \frac{1}{2}\frac{d[Cr_2(SO_4)_3]}{dt}$

$$-\frac{d[C_2H_6O]}{dt} = \frac{3}{2} \times 2.67 = 4 \text{ mol min}^{-1}$$

18. **(1)** Rate = k $[NO]^x [H_2]^y$

$7 \times 10^{-9} = k \times (8 \times 10^{-5})^x (8 \times 10^{-5})^y$ ...(i)

$2.1 \times 10^{-8} = k \times (24 \times 10^{-5})^x (8 \times 10^{-5})^y$ ...(ii)

Now dividing equation (i) by equation (ii):

$$\frac{7 \times 10^{-9}}{2.1 \times 10^{-8}} = \left(\frac{8 \times 10^{-5}}{24 \times 10^{-5}}\right)^x \text{ or } \frac{1}{3} = \left(\frac{1}{3}\right)^x \Rightarrow x = 1$$

First check the unit of rate constant

19. **(4)** $\text{Rate} = \frac{d[B]}{dt} = \frac{0.2}{0.5} = 0.4$

$= 4 \times 10^{-1}$

20. **(106)** As the unit of rate constant is min^{-1}, so it must be a first order reaction: $k \times t = 2.303 \log A_0/A_t$

In 1 min, 10% is inactivated so:

$A_0 = 100$, $A_t = 90$

$k \times 1 = 2.303 \times \log \frac{100}{90} = 2.303 \times (\log 10 - 2 \log 3)$

$= 2.303 \times (1 - 2 \times 0.477) = 0.10593$

$= 105.93 \times 10^{-3} \approx 106 \times 10^{-3}$

21. **(27)** $2A + B_2 \longrightarrow 2AB$

Rate of reaction is $r = k[A]^2[B_2]$

The conc. of A and B_2 will become 3 times as the volume of reaction vessel is reduced by a factor of 3 and hence, the rate becomes $3^2 \times 3 = 27$ times of initial rate.

22. **(c)** For a given reaction,

$$\text{rate} = -\frac{1}{2}\frac{dn_A}{dt} = -\frac{1}{3}\frac{dn_B}{dt} = -\frac{2}{3}\frac{dn_C}{dt}$$

$$\text{rate} = \frac{dn_A}{dt} = \frac{2}{3}\frac{dn_B}{dt} = \frac{4}{3}\frac{dn_C}{dt}$$

23. **(d)** $2H_2(g) + 2NO(g) \rightarrow N_2(g) + 2H_2O(g)$

$R_f = K_f[NO]^2[H_2]$

This means that, for forward reaction the order of reaction w.r.t. [NO] is 2 and that w.r.t. $[H_2]$ is 1.

Assuming the equilibrium expression,

$$K_{eq} = \frac{[N_2][H_2O]^2}{[H_2]^2[NO]^2} = \frac{K_f}{K_b}$$

$$\Rightarrow K_f[NO]^2[H_2] = K_b\frac{[N_2][H_2O]^2}{[H_2]}$$

At equilibrium, $R_f = R_b \Rightarrow R_b = K_b\dfrac{[N_2][H_2O]^2}{[H_2]}$

Note: In this question, the order of reaction for backward reaction is not given.

24. **(c)** xA ® yB

$$\therefore \frac{-dA}{xdt} = \frac{1}{y}\frac{dB}{dt} \Rightarrow \frac{-dA}{dt} = \frac{dB}{dt} \times \frac{x}{y}$$

$$\log\left[\frac{-dA}{dt}\right] = \log\left[\frac{dB}{dt}\right] + \log\left(\frac{x}{y}\right)$$

Comparing this equation with the equation given in question. We get,

$$\log\frac{x}{y} = 0.3010 \text{ or } \log\frac{x}{y} = \log 2 \therefore \frac{x}{y} = 2$$

∴ The reaction is of type 2A ® B.

Hence, option (c) is correct.

25. **(b)** $2N_2O_5(g) \rightarrow 4NO_2(g) + O_2(g)$

$$\text{rate of reaction} = \frac{-1}{2}\frac{d[N_2O_5]}{dt} = \frac{1}{4}\frac{d[NO_2]}{dt} = \frac{d[O_2]}{dt}$$

According to the question

$$\frac{-d[N_2O_5]}{dt} = -\frac{(2.75-3)}{30} = \frac{0.25}{30}\text{M min}^{-1}$$

$$\therefore \frac{d[NO_2]}{dt} = 2 \times \frac{-d[N_2O_5]}{dt} = 2 \times \frac{0.25}{30}$$

$= 1.67 \times 10^{-2}\text{M min}^{-1}$

26. **(d)** $A \xrightarrow{k_1} B \xrightarrow{k_2} C$ [It is a consecutive reaction]

$$\frac{d[B]}{dt} = k_1[A] - k_2[B] = 0 \Rightarrow [B] = \frac{k_1[A]}{k_2}$$

27. **(a)** $2A + B \longrightarrow C$

Rate $= k[A]^x[B]^y$

Exp-1, $0.045 = k[0.05]^x[0.05]^y$ ...(i)

Exp-2, $0.090 = k[0.1]^x[0.05]^y$ ...(ii)

Exp-3, $0.72 = k[0.2]^x[0.1]^y$ ...(iii)

Divide equation (i) by equation (ii)

$$\frac{0.045}{0.090} = \left(\frac{1}{2}\right)^x \Rightarrow x = 1$$

Divide equation (i) by equation (iii)

$$\frac{0.045}{0.72} = \left(\frac{0.05}{0.1}\right)^y\left(\frac{0.05}{0.2}\right)^1$$

$$\frac{0.045}{0.72} \times \frac{0.2}{0.05} = \left(\frac{0.05}{0.1}\right)^y$$

$$\left(\frac{1}{2}\right)^2 = \left(\frac{1}{2}\right)^y \Rightarrow y = 2$$

Rate law $= k[A]^1[B]^2$.

28. **(d)** $A_2 \underset{k_{-1}}{\overset{k_1}{\rightleftharpoons}} 2A$

$$-\frac{1}{2}\frac{d[A]}{dt} = k_1[A_2] - k_{-1}[A]^2$$

$$\Rightarrow \frac{d[A]}{dt} = 2k_1[A_2] - 2k_{-1}[A]^2$$

29. **(a)**

$$O_3(g) + Cl^{\bullet}(g) \longrightarrow O_2(g) + ClO^{\bullet}(g),\ k_i$$

$$ClO^{\bullet}(g) + O^{\bullet}(g) \longrightarrow O_2(g) + Cl^{\bullet}(g),\ k_{ii}$$

$$O_3(g) + O^{\bullet}(g) \longrightarrow 2O_2(g),\ k_{Rate} = k_i \times k_{ii}$$

$= 5.2 \times 10^9 \times 2.6 \times 10^{10} \approx 1.4 \times 10^{20}\text{ mol}^{-1}\text{ L s}^{-1}$

Note: When reaction added, the rate of the reaction are multiply

30. **(d)** Rate constant is independent of concentration.

31. **(c)** Rate $= k[A][B] = R$

$R' = k[A][2B]$

$$\frac{R}{R'} = \frac{k[A][B]}{k[A][2B]} = \frac{k[A][B]}{2k[A][B]}$$

$\Rightarrow 2R = R'$, i.e., rate becomes double.

32. **(d)** Let rate of reaction $= \dfrac{d[C]}{dt} = k[A]^x[B]^y$

k is rate constant, for formation of C.

Now from the given data,

$1.2 \times 10^{-3} = k[0.1]^x[0.1]^y$ .....(i)

$1.2 \times 10^{-3} = k[0.1]^x[0.2]^y$ .....(ii)

$2.4 \times 10^{-3} = k[0.2]^x[0.1]^y$ .....(iii)

Dividing equation (i) by (ii),

$$\frac{1.2\times10^{-3}}{1.2\times10^{-3}} = \frac{k[0.1]^x[0.1]^y}{k[0.1]^x[0.2]^y}$$

We find, $y = 0$

Now dividing equation (i) by (iii),

$$\frac{1.2\times10^{-3}}{2.4\times10^{-3}}=\frac{k[0.1]^x[0.1]^y}{k[0.2]^x[0.1]^y}$$

We find, $x=1$

Hence, $\frac{d[C]}{dt}=k[A]^1[B]^0$

Note: In this type of questions, either Divide in such away that, x on y cancel with another equation.

33. **(d)** From rate law

$$-\frac{1}{2}\frac{d[SO_2]}{dt}=-\frac{d[O_2]}{dt}=\frac{1}{2}\frac{d[SO_3]}{dt}$$

$$\therefore -\frac{d[SO_2]}{dt}=-2\times\frac{d[O_2]}{dt}=-2\times2.5\times10^{-4}$$

$=-5\times10^{-4}\ \text{mol L}^{-1}\text{s}^{-1}$

34. **(b)** Rate of disappearance of reactant = Rate of appearance of products

$$-\frac{1}{2}\frac{d[N_2O_5]}{dt}=\frac{1}{4}\frac{d[NO_2]}{dt};\ \frac{1}{2}k[N_2O_5]=\frac{1}{4}k'[N_2O_5]$$

$$\frac{k}{2}=\frac{k'}{4}\ \therefore k'=2k$$

35. **(d)** For the reaction

$3A+2B\longrightarrow C+D$

From the rate law,

$$=-\frac{1}{3}\frac{d[A]}{dt}=\frac{d[C]}{dt}=k[A]^n[B]^m$$

36. **(b)** Given $-\frac{d[MnO_4^-]}{dt}=4.56\times10^{-3}\ \text{Ms}^{-1}$

From the reaction given,

$$-\frac{1}{2}\frac{d[MnO_4^-]}{dt}=\frac{4.56\times10^{-3}}{2}\text{Ms}^{-1}$$

$$-\frac{1}{2}\frac{d[MnO_4^-]}{dt}=\frac{1}{5}\frac{dI_2}{dt}\Rightarrow-\frac{5}{2}\frac{d[MnO_4^-]}{dt}=\frac{dI_2}{dt}$$

On substituting the given value

$$\therefore \frac{dI_2}{dt}=\frac{4.56\times10^{-3}\times5}{2}=1.14\times10^{-2}\ \text{M/s}$$

37. **(d)** $\text{Rate}_1=k[A]^n[B]^m$

$$\text{Rate}_2=k[2A]^n\left[\frac{1}{2}B\right]^m$$

$$\therefore \frac{\text{Rate}_2}{\text{Rate}_1}=\frac{k[2A]^n\left[\frac{1}{2}B\right]^m}{k[A]^n[B]^m}=(2)^n\left(\frac{1}{2}\right)^m$$

$=2^n.(2)^{-m}=2^{n-m}$

38. **(d)** Since the slow step is the rate determining step, hence if we consider mechanism A. We find

Rate $=k[Cl_2][H_2S]$

Now, if we consider mechanism B. We find

Rate $=k[Cl_2][HS^-]$ ....(i)

From step 1 (fast equilibrium) of mechanism B.

$$k_1=\frac{[H^+][HS^-]}{[H_2S]}\ \text{or}\ [HS^-]=\frac{k_1[H_2S]}{[H^+]}$$

Substituting this value in equation (i) we find,

$$\text{Rate}=k[Cl_2]k_1\frac{[H_2S]}{[H^+]}=k'\frac{[Cl_2][H_2S]}{[H^+]}$$

Hence, only mechanism A is consistent with the given rate equation.

39. **(b)** The rate of reaction for the given reaction, $\frac{1}{2}A\longrightarrow 2B$ can be written either as

$-2\frac{d}{dt}[A]$ with respect to 'A' or $\frac{1}{2}\frac{d}{dt}[B]$ with respect to 'B'

From the above, we have

$$-2\frac{d}{dt}[A]=\frac{1}{2}\frac{d}{dt}[B]\ \text{or}\ -\frac{d}{dt}[A]=\frac{1}{4}\frac{d}{dt}[B]$$

40. **(d)** Rate of appearance of HI $=\frac{1}{2}\frac{d[HI]}{dt}$

Rate of disappearance of $H_2=\frac{-d[H_2]}{dt}$

Rate of disappearance of $I_2=\frac{-d[I_2]}{dt}$

hence $-\frac{d[H_2]}{dt}=-\frac{d[I_2]}{dt}=\frac{1}{2}\frac{d[HI]}{dt}$

or $-2\frac{d[H_2]}{dt}=-2\frac{d[I_2]}{dt}=\frac{d[HI]}{dt}$

41. **(a)** For a zero order reaction.

rate $=k[A]^o$ i.e. rate $=k$

hence unit of $k=\text{M.sec}^{-1}$

For a first order reaction.

rate $=k[A]$

$\therefore k=\text{M.sec}^{-1}/\text{M}=\text{sec}^{-1}$

42. **(b)** Before applying medicine

$\frac{dA}{dt}=k[A]$ (First order Rate law)

$$\frac{A}{A_0}=\frac{N}{N_0}=e^{kt}$$

After applying medicine

Active Bacteria $\rightarrow$ Inactive Bacteria
(A) (I)

$r=-\frac{dA}{dt}=k[A]^2$ (Rate law)

$y=kx^2$ Parabola

43. **(c)** For zero order reaction :

$$\text{Half life}=\frac{[A]_o}{2k}\Rightarrow 60\text{min}=\frac{2}{2k}\Rightarrow k=\frac{1}{60}\text{M/min}$$

$$\text{Now, }[A]_t=[A]_o-kt\Rightarrow t=\frac{[A]_o-[A]_t}{k}$$

$$=\frac{0.5-0.25}{1/60}=15\text{ min}$$

44. (d) Half life of A = 10 min
Half life of B = 40 min
$[A]_0 = 8[B]_0$ (given)

$$\Rightarrow [A]_0 = 8[B]_0 \xrightarrow{10\text{ min}} 4[B]_0 \xrightarrow{10\text{ min}} 2[B_0]$$

$$\xrightarrow{10\text{ min}} [B_0] \xrightarrow{10\text{ min}} \frac{[B]_0}{2}$$

$$\Rightarrow [B_0] \xrightarrow{40\text{ min}} \frac{[B]_0}{2}$$

Therefore, 40 min. is the time required for concentration to be same for both the reactants A and B.

45. (d) $\frac{t_2}{t_1} = \left(\frac{C_1}{C_2}\right)^n$ and n = 1

$C_o = 16$ mg/mL
$t_1 = ?, C_1 = 8$ mg/mL
$t_2 = 12, C_2 = 4$ mg/mL

$$\Rightarrow t_1 = \frac{t_2 X C_2}{C_1} = \frac{12 \times 4}{8} = 6$$

46. (a) $\underset{k_{-1}}{\overset{k_1}{\rightleftharpoons}} A + A$

Rate $= k_2 A_2 [A][B_2]$ .... (1) (from slow step)

$$\Rightarrow \left(\frac{k_1}{k_{-1}}\right) = K_{eq} = \left(\frac{[A]^2}{[A_2]}\right)$$

$$\Rightarrow [A] = \sqrt{\frac{k_1}{k_{-1}}} \cdot \sqrt{[A_2]}$$

Substituting in (1); we get

$$\text{Rate} = k_2 \sqrt{\frac{k_1}{k_{-1}}} \cdot [A_2]^{1/2} \cdot [B_2]$$

$$\therefore \text{ Order} = 1 + \frac{1}{2} = \left(\frac{3}{2}\right) = 1.5$$

47. (d) Because no. of bacteria initial = N_0
and No. of bacteria at any time t = N
Since bacterial growth is given as
$N = N_0 e^{Kt}$
Where K = growth constant for bacterial growth

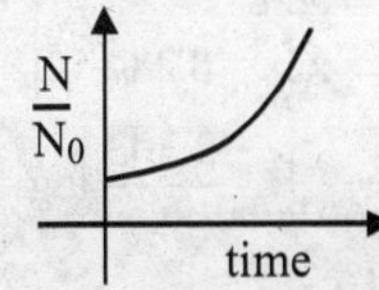

48. (b) $r_1 = k[A]^1[B]^1$

$$r_1 = k\left[\frac{n_A}{V}\right]^1 \left[\frac{n_B}{V}\right]^1$$

By Reducing the volume to 1/3 of its original volume, the concentration of rectants will becomes 3 times.

$$r_2 = k\left[\frac{3n_A}{V}\right]^1 \left[\frac{3n_B}{V}\right]^1$$

$\boxed{r_2 = 9r_1}$ Rate will become 9 times to its initial rate.

49. (c)

(A) $t_{1/2} \propto \frac{1}{A_0^{n-1}}$ $\frac{(t_{1/2})_1}{(t_{1/2})_2} = \frac{(A_0)_2^{n-1}}{(A_0)_1^{n-1}}$

$$\frac{200}{100} = \left(\frac{0.025}{0.100}\right)^{n-1} \Rightarrow 2 = \left(\frac{1}{4}\right)^{n-1}$$

$$\Rightarrow \frac{1}{4} \times 2 = \left(\frac{1}{4}\right)^n \Rightarrow 2 = 4^n \Rightarrow n = \frac{1}{2} \text{ (order)}$$

(B) $t_{1/2} \propto \sqrt{A_0}$

$$\frac{200}{t_{1/2}} = \frac{(1)^{-1/2}}{(0.1)^{-1/2}} = \frac{(0.1)^{1/2}}{(1)^{1/2}} \text{ when } A_0 = 1\text{ M}$$

$t_{1/2} = 200\sqrt{10}$ min

(C) The order would change to 0.769 if the concentration was 0.500 instead of 0.100. Thus, C is not correct.

(D) $\frac{200}{t_{1/2}} = \frac{(0.1)^{1/2}}{(1.6)^{1/2}}$ when $A_0 = 1.6$ M

$t_{1/2} = 800$ min

50. (d) For first order reaction $t_{1/2} = \frac{\ln 2}{k}$

$t_{1/2}$ is independent of $[R]_0$

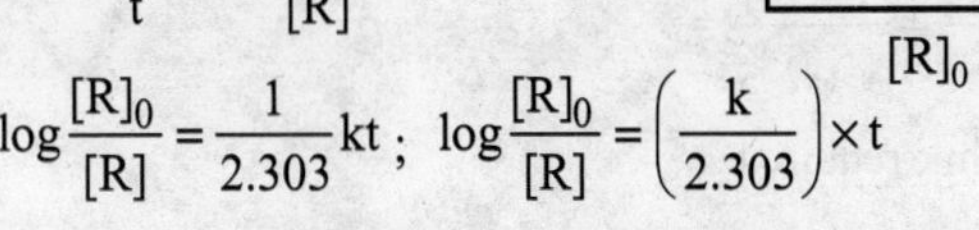

For first order reaction

$$k = \frac{2.303}{t} \log \frac{[R_0]}{[R]}$$

$$\log \frac{[R]_0}{[R]} = \frac{1}{2.303} kt \text{ ; } \log \frac{[R]_0}{[R]} = \left(\frac{k}{2.303}\right) \times t$$

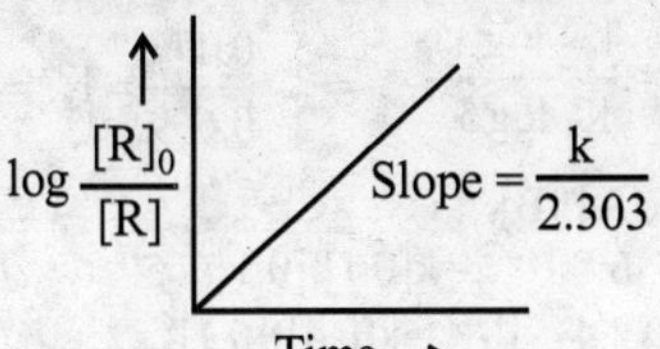

51. (b) For zero order reaction : $A \rightarrow P$

$$\text{Rate} = k \Rightarrow \frac{[A]_t - [A]_0}{t} = k$$

Thus, concentration $\propto t$ and rate is constant.

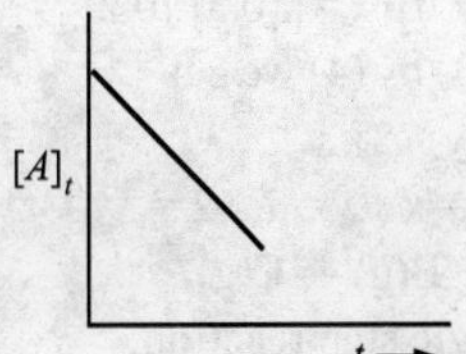

52. (897)

Given : $2N_2O_5(g) \rightarrow 2N_2O_4(g) + O_2(g)$

	$2N_2O_5$	$2N_2O_4$	O_2
t = 0	0.6	0	0
t = 100s	0.6 – y	y	y/2

$P_{Total} = (0.6 - y + y + \frac{y}{2}) = 0.6 + \frac{y}{2}$

As given in question,
$k = 4.606 \times 10^{-2} sec^{-1}$

$kt = \ln \frac{0.6}{0.6-y} = 2.303 \log \frac{0.6}{0.6-y}$

$4.606 \times 10^{-2} \times 100 = 2.303 \log \frac{0.6}{0.6-y} \Rightarrow y = 0.594$

$\Rightarrow P_{Total} = 0.6 + \frac{0.594}{2} = 0.897\ atm = 897 \times 10^{-3} atm$

Thus, $x = 897 \times 10^{-3}$ and answer is 897.

53. **(c)** Nuclear phenomenon independent of temperature. Decay constant is a nuclear property.

54. **(3)** $k = \frac{1}{t_{99.9\%}} \ell n\left(\frac{100}{0.1}\right) = \frac{1}{t_{90\%}} \ell n\left(\frac{100}{10}\right)$

$t_{99.9\%} = t_{90\%} \frac{\ell n(10^3)}{\ell n 10}$ [in $a^b = b \ln a$]

$t_{99.9\%} = t_{90\%} \times 3$

55. **(17)** $A + B \rightarrow C$ Reaction ...(1)
$P \rightarrow Q$ Reaction ...(2)

$K_1 = \frac{0.693}{(t_{1/2})_1}$; $K_2 = \frac{0.693}{(t_{1/2})_2}$

$\frac{K_2}{K_1} = \frac{(t_{1/2})_1}{(t_{1/2})_2} = \frac{5}{2}$

Time required

$\frac{(t_{2/3})_1}{(t_{4/5})} = \frac{K_2 \log 3}{K_1 \log 5} \Rightarrow = \frac{5}{2} \frac{0.477}{0.699} = 1.7 = 17 \times 10^{-1}$

56. **(3)** $r = k[A]^x[B]^y$
From (I) $6 \times 10^{-3} = k[0.1]^x[0.1]^y$ ...(1)
From (IV) $2.4 \times 10^{-2} = k[0.4]^x[0.1]^y$ ...(2)
on dividing (2) by (1) we get
$\Rightarrow 4 = (4)^x \Rightarrow x = 1$
$r = k[A]^x[B]^y$
From (III) $2.88 \times 10^{-1} = k[0.3]^x[0.4]^y$ ...(3)
From (II) $7.2 \times 10^{-2} = k[0.3]^x[0.2]^y$ ..(4)
on dividing (3) by (4) we get
$\Rightarrow 4 = 2^y \Rightarrow y = 2$
Overall order $= x + y = 1 + 2 = 3$

57. **(315)** $2A(g) + B(g) \rightarrow C(g)$
As this is single step reaction,
$r = k_f[A]^2[B]$

	2A(g)	+ B(g)	→ C(g)
t = 0	1.5	0.7	0
t = 't'	1.5 − 2x	0.7 − x	x

$\Rightarrow x = 0.5$ atm
$r_1 = k_f(0.5)^2(0.7)$; $r_2 = k_f(0.5)^2(0.2)$

$\frac{r_1}{r_2} = \frac{(1.5)^2(0.7)}{(0.5)^2(0.2)} = \frac{1.5 \times 1.5 \times 0.7}{0.5 \times 0.5 \times 0.2}$

$\frac{r_1}{r_2} = \frac{15 \times 15 \times 7}{5 \times 5 \times 2} = \frac{63}{2} = 31.5 = 315 \times 10^{-1}$

58. **(100)** $k = \frac{k_1 k_2}{k_3}$

$Ae^{\frac{-E_a}{RT}} = \frac{A_1 e^{\frac{-E_{a_1}}{RT}} A_2 e^{\frac{-E_{a_2}}{RT}}}{A_3 e^{\frac{-E_{a_3}}{RT}}}$

$Ae^{\frac{-E_a}{RT}} = \frac{A_1 A_2}{A_3} e^{\frac{-(E_{a_1} + E_{a_2} + E_{a_3})}{RT}}$

$E_a = E_{a_1} + E_{a_2} - E_{a_3} \Rightarrow 400 = 300 + 200 - Ea_3$

$E_{a_3} = 100$ kJ/ mol

Note: Direct trick: $k_1 \times k_2 \rightarrow$ Multiplication factor will act added and on division k_3 will act substracted

$E_a = E_{a_1} + E_{a_2} - E_{a_3}$

59. **(50)** $A + B \rightarrow C$
Since, $[A] = [B] \Rightarrow r = k[A]$
Apply first order rate equation.

$k = \frac{2.303}{t} \log \frac{1}{0.1} \Rightarrow 4.6 \times 10^{-2} = \frac{2.303}{t} \Rightarrow t = 50$ sec.

60. **(17190)**

$\frac{^{14}C}{^{12}C} = \frac{1}{8}$; $t_{1/2}(^{14}C) = 5730$ years

$\frac{N}{N_o} = \frac{1}{8} = \frac{1}{2^{t/t_{1/2}}} \Rightarrow \frac{N_t}{N_o} = \frac{1}{2^{t/t_{1/2}}}$

Age of wood = t = ?

$\frac{1}{8} = \frac{1}{2^{t/5730}} \Rightarrow \frac{1}{2^3} = \frac{1}{2^{t/5730}} \Rightarrow \frac{t}{5730} = 3$

t = 17190 years

61. **(2)**

	A	⟶ 2B	+ C
t = 0	0.1	–	–
t = 115 sec	0.1 − x	2x	x

$0.1 - x + 2x + x = 0.28$
$0.1 + 2x = 0.28 \Rightarrow 2x = 0.28 - 0.1 \Rightarrow x = 0.09$

$K = \frac{1}{t} \ln\left(\frac{0.1}{0.1 - 0.09}\right) = \frac{2.303}{115} \log\left(\frac{0.1}{0.01}\right) = \frac{2.303}{115} \times 1$

$K = 0.0200 = 2 \times 10^{-2} s^{-1}$

62. **(d)**

A(g)	→ B(g)	+ C(g)
P_i	0	0
$P_i - x$	x	x

$P_t = P_i + x$
$P_i - x = P_i - P_t + P_i = 2P_i - P_t$

$k = \frac{2.303}{t} \log \frac{P_i}{2P_i - P_t}$

63. **(399)**
$r = k[A]$
So this is first order reaction.
For 50% decomposition, time req is 120 min
$\therefore \quad t_{1/2} = 120\,\text{min}$
$\therefore \quad k = \frac{0.693}{t_{1/2}} = \frac{0.693}{120}$
$k = 0.00577 = 5.77 \times 10^{-3}\,\text{min}^{-1}$
For 90% decomposition
$k = \frac{2.303}{t}\log\frac{a}{(a-x)} \Rightarrow t = \frac{2.303}{0.00577}\log\left(\frac{100}{10}\right)$
$t = 399\,\text{min}$

64. **(24)**

	Rate	Time
1.	$0.04\,\text{mol L}^{-1}\text{s}^{-1}$	10 min = 600s
2.	$0.03\,\text{mol L}^{-1}\text{s}^{-1}$	20 min = 1200s

As reation follow first order kinetic,
Use, $\ln[A]_o = \ln[A]_t + kt$
$\ln[A]_o = \ln[A]_{600} + k \times 600$ ... (1)
$\ln[A]_o = \ln[A]_{1200} + k \times 1200$ ... (2)
On dividing (1) by (2) and solving we get
$\ln\frac{[A]_{600}}{[A]_{1200}} = k(1200 - 600) = 600k$
Given, $0.04 = k[A]_{600}$ ... (3)
$0.03 = k[A]_{1200}$ ... (4)
On dividing (3) by (4) we get
$\frac{[A]_{600}}{[A]_{1200}} = \frac{4}{3}$
$\ln\frac{4}{3} = 600k \Rightarrow k = 4.66 \times 10^{-4}$
$t_{1/2} = \frac{0.693}{4.66 \times 10^{-4}} = 1487\,\text{sec.} = 24\,\text{min.}$

65. **(63)** Radioactive decays are first order reactions.
Thus, $K = \frac{0.693}{t_{1/2}} = \frac{0.693}{36\,h} = 0.01925\,h^{-1}$
$\Rightarrow t = \frac{2.303}{K}\log\frac{a}{a-x}$
$\Rightarrow 24\,h = \frac{2.303}{0.01925\,h^{-1}}\log\frac{a}{a-x}$
$\Rightarrow \frac{a}{a-x} = \frac{24 \times 0.01925}{2.303} = 0.2006$
$\Rightarrow \frac{a}{a-x} = \text{antilog}\,(0.2006) = 1.587.$
$\Rightarrow \text{Fraction left} = \frac{a-x}{a} = \frac{1}{1.587} = 0.630 = 63.0 \times 10^{-2}$

66. **(2)** Let, $R = k[HI]^n$
using any two of given data,
$\frac{R_2}{R_1} = \left[\frac{[HI]_2}{[HI]_1}\right]^n \Rightarrow \frac{3 \times 10^{-3}}{7.5 \times 10^{-4}} = \left(\frac{0.01}{0.005}\right)^n$
$\Rightarrow \quad n = 2$

67. **(10)** $\frac{t_{99.9\%}}{t_{1/2}} = \frac{\frac{2.303}{k}\left(\frac{a}{a-x}\right)}{\frac{2.303}{k}\log 2}$
$= \frac{\log\left(\frac{100}{100-99.9}\right)}{\log 2} = \frac{\log 10^3}{\log 2} = \frac{3}{0.3} = 10$
$t_{99.9\%} = 10t_{1/2}$

68. **(2)** $\frac{1}{t_{1/2}} = \frac{1}{3} + \frac{1}{12} = \frac{4+1}{12} = \frac{5}{12} \Rightarrow t_{1/2} = \frac{12}{5}\,\text{min} = 2.4 \sim 2$

69. **(17)** $K = 20\,\text{min}^{-1}$
$t = \frac{2.303}{K}\log\frac{[a_0]}{a_0 \times \frac{1}{32}}$
$t = \frac{2.303}{20}\log 32 = \frac{2.303}{20} \times \log 2^5$
$= \frac{2.303}{20} \times 5 \times 0.30 = 0.1727 \approx 17 \times 10^{-2}\,\text{min}$

70. **(623)** For first order section
$t = \frac{2.303}{k}\log\frac{[A]_0}{[A]_t} = \frac{2.303}{2.011 \times 10^{-3}}\log\frac{7}{2}$
$= \frac{2.303 \times 0.544}{2.011 \times 10^{-3}} = 622.989 \approx 623$

71. **(1350)** $\frac{t_1}{t_2} = \frac{\frac{1}{K}\ln\frac{a_0}{0.4\,a_0}}{\frac{1}{K}\ln\frac{a_0}{0.1\,a_0}}$
$\frac{540}{t_2} = \frac{\ln\frac{10}{4}}{\ln 10} \Rightarrow \frac{540}{t_2} = \frac{\log 10 - \log 4}{\log 10}$
$\frac{540}{t_2} = \frac{1 - 0.6}{1} \Rightarrow \frac{540}{t_2} = 0.4 \Rightarrow t_2 = \frac{540}{0.4} = 1350\,\text{sec}$

72. **(200)** $k_A = \frac{\ln 2}{100}; k_B = \frac{\ln 2}{50}$
$A_t = A_0 \times e^{-k_A t}$
$A_t = A_0 \times e^{\left(\frac{-\ln 2}{100} \times t\right)}; \; B_t = B_0 \times e^{\left(\frac{-\ln 2}{50} \times t\right)}$
Given, $A_0 = B_0$ and $A_t = 4B_t$
$e^{\frac{-\ln 2}{100} \times t} = 4 \times e^{\frac{-\ln 2}{50} \times t} \Rightarrow e^{\frac{\ln 2}{100} \times t} = 4$
$\Rightarrow \frac{\ln 2}{100} \times t = \ln 4 = 2\ln 2 \Rightarrow t = 200\,\text{sec.}$

73. **(2)** $k = \frac{2.303}{t}\log\frac{a}{a-x}$
$\frac{2.303}{t_{50\%}}\log\frac{100}{100-50} = \frac{2.303}{t_{75\%}}\log\frac{100}{100-75}$
$t_{75\%} = 2t_{50\%}.$

74. **(c)** Rate = $k[A]^n$
comparing units

$$\frac{(mol/L)}{sec} = k\left(\frac{mol}{L}\right)^n \Rightarrow k = mol^{(1-n)}\,L^{(n-1)}\,s^{-1}$$

75. **(7)** For the first order reaction $kt = \ln\frac{[R]_0}{[R]}$

$$k\times 60 = \ln\frac{(2.4\times10^{-2})}{(1.6\times10^{-2})} = 2.303\times(\log 3 - \log 2)\frac{1}{60}$$

$$= 2.303\times(0.477-0.301)\frac{1}{60}$$

$$\Rightarrow k = 6.7\times10^{-3}\,min^{-1} \approx 7\times10^{-3}\,min^{-1}$$

76. **(1)** $PCl_5(g) \xrightarrow[300K]{1^{st}\text{ order}} PCl_3(g) + Cl_2(g)$

t = 0	50 M	0	0
t = 120 min	10 M	40 M	40 M

$$\Rightarrow k = \frac{2.303}{t}\log\frac{[A_0]}{[A_t]} \Rightarrow k = \frac{2.303}{120}\log\frac{50}{10}$$

$$\Rightarrow k = \frac{2.303}{120}\times 0.6989 = 0.013413\,min^{-1}$$

$= 1.3413\times10^{-2}\,min^{-1} \approx 1\times10^{-2}\,min^{-1}$.

77. **(3)** General form of rate law,
Rate $(r) = k[NO]^x[Cl_2]^y$
Using for all 3 exp.
$(r_1)\ 0.18 = k[0.10]^x[0.10]^y$ ... (i)
$(r_2)\ 0.35 = k[0.10]^x[0.20]^y$ ... (ii)
$(r_3)\ 1.40 = k[0.20]^x[0.20]^y$ ... (iii)
From reaction (i) and (ii),

$$\frac{0.18}{0.35} = \frac{[0.10]^x[0.10]^y}{[0.10]^x[0.20]^y}$$

$$\Rightarrow \frac{0.18}{0.35} = \left(\frac{1}{2}\right)^y \Rightarrow \left(\frac{1}{2}\right)^1 = \left(\frac{1}{2}\right)^y$$

∴ Order w.r.t. $[Cl_2] = y = 1$

From reaction (ii) and (iii), $\frac{0.35}{1.40} = \frac{[0.10]^x[0.20]^y}{[0.20]^x[0.20]^y}$

$$\Rightarrow \frac{0.35}{1.40} = \left(\frac{1}{2}\right)^x \Rightarrow \left(\frac{1}{2}\right)^2 = \left(\frac{1}{2}\right)^x$$

∴ Order w.r.t. [NO], $x = 2$
Now, $r = k[NO]^2[Cl_2]^1$
∴ Overall order of reaction = 2 + 1 = 3

78. **(10)** This question is incomplete, order is not given. If we consider, the given reaction to be first order, then

$$t_{1/2} = \frac{0.693}{k} = \frac{\ln 2}{k}$$

$$\Rightarrow k = \frac{\ln 2}{1}$$

$$\therefore t = \frac{1}{k}\ln\frac{100}{100-99.9}$$

$$\Rightarrow t = \frac{1}{\ln 2}\ln 10^3$$

$$\Rightarrow t = \frac{1\times3\times\ln 10}{\ln 2} = \frac{3\times2.3}{0.69} = 10\text{ min.}$$

79. **(2)** A → Products
Rate constant of a first order is given as,

$$k = \frac{2.303}{t}\log\frac{[A]_0}{[A]_t} = \frac{2.303}{570}\log\frac{100}{32} = 2\times10^{-3}\,s^{-1}$$

80. **(26)** $k = \frac{1}{t}\cdot\ln\frac{[A]_0}{[A]_t} \Rightarrow t = \frac{1}{k}\ln\left(\frac{100}{60}\right)$

$$\Rightarrow t = \frac{1}{3.3\times10^{-4}}(\ln 10 - \ln 6)$$

$\Rightarrow t = 1548.49$ sec = 25.81 min ≈ 26 min.

81. **(c)** From 0 to 1h, $N¢ = N_0e^t$
When drug is administered bacterial growth is given by,

$$\frac{dN}{dt} = -5N^2$$

So, at t = 1 h, $N¢ = eN_0$
Integrating the differential equation,

$$\int_{eN_0}^{N}\frac{dN}{N^2} = -5\int_1^t dt$$

$$\frac{1}{N} - \frac{1}{eN_0} = 5(t-1) \Rightarrow \frac{N_0}{N} - \frac{1}{e} = 5N_0(t-1)$$

$\frac{N_0}{N} = 5N_0t + \left(\frac{1}{e} - 5N_0\right)$ The above equation is similar to straight line equation with positive slope.

Thus $\frac{N_0}{N}$ increases linearly with t.

82. **(a)** In graph (i), ln [Reactant] vs time is linear with positive intercept and negative slope. Hence it is 1st order. In graph (ii), [Reactant] vs time is linear with positive intercept and negative slope. Hence, it is zero order.

83. **(3)** $A_t = A_0\times\frac{12.5}{100} = \frac{A_0}{8}$ [87.5% complete]

(1st order rate law)

$$A_0 \xrightarrow{t_{1/2}} \frac{A_0}{2} \xrightarrow{t_{1/2}} \frac{A_0}{4} \xrightarrow{t_{1/2}} \frac{A_0}{8}$$

$t_{87.5} = 3t_{1/2}$
∴ $x = 3$

84. **(3)** Rate determing step : $NOBr_2 + NO \rightarrow 2NOBr$
$r = K[NOBr_2][NO]$ ... (i)

$$K_{eq} = \frac{[NOBr_2]}{[NO][Br_2]} \quad ...(ii)$$

From (i) & (ii)
$r = K.K_{eq.}[NO][Br_2][NO]$

$r = K'[NO]^2[Br_2]$

Overall order = 3

85. **(a)** $r = K[A]^1[B]^1$

$0.1 = K(20)^1(0.5)^1$ ... (i)

$0.40 = K(x)^1(0.5)^1$ ... (ii)

$0.80 = K(40)^1(y)^1$ ... (iii)

From (i) and (ii)

x = 80

From (i) and (iii)

y = 2

86. **(1)** (A) For zero order $t_{1/2} = \frac{[A]_0}{2K}$ as concentration decreases half life decreases. (Correct statement)

(B) If order w.r.t. that reactant is zero then it will not affect rate of reaction. (Correct statement)

(C) Order can be fractional but molecularity can not be. (Incorrect statement)

(D) For zero order reaction unit is mol $L^{-1}s^{-1}$ and for second order reaction unit is $mol^{-1}Ls^{-1}$ (Correct statement)

87. **(15)** Suppose, at time 't', the concentration of both A and B will be same.

$[A]_t = [A]_o e^{-k_A t}$

$k_A = \frac{0.693}{t_{1/2}} = \frac{0.693}{15}$; $\therefore [A]_t = [A]_0 e^{\left(-\frac{0.693}{15}\right)t}$

and $[B]_t = [B]_o e^{-k_B t} = 4[A]_o e^{\left(-\frac{0.693}{5}\right)t}$

At time t, $\therefore$ $[A]_t = [B]_t$

or, $[A]_o e^{\left(-\frac{0.693}{15}\right)t} = 4[A]_o e^{\left(-\frac{0.693}{5}\right)t}$

or $-\frac{0.693}{15}t = \ln 4 - \frac{0.693}{5}t$

or $0.693t\left(\frac{1}{5} - \frac{1}{15}\right) = 2\ln 2$ or t = 15 min.

88. **(75)** Assume reaction starts with 1 mole A

$t_{1/2} = \frac{a}{2k}$, $k = \frac{1}{2 \times 50}$

For zero order reaction. $[R_0] - [R] = Kt$

For 75% completion

$a - \frac{a}{4} = kt \Rightarrow t = \frac{3}{4}\frac{a}{k} = \frac{3}{4} \times \frac{100}{a} = 75$

89. **(2)** (B) and (D) are correct.

90. **(2)** As unit of rate constant is $(conc.)^{1-n}\ time^{-1}$

$\Rightarrow (L\,mol^{-1}) \Rightarrow 1 - n = -1 \ \therefore n = 2$

91. **(60)** $t_{1/2} = 30$ min

Time taken for $\frac{3}{4}$th completion of the reaction

$= 2t_{1/2} = 30 \times 2 = 60$ min

92. **(b)** $t_{1/2} \propto (P_0)^{1-n}$

$\frac{(t_{1/2})_1}{(t_{1/2})_2} = \frac{(P_0)_1^{1-n}}{(P_0)_2^{1-n}} \Rightarrow \left(\frac{4}{2}\right) = \left(\frac{50}{100}\right)^{1-n}$

$\Rightarrow 2 = \left(\frac{1}{2}\right)^{1-n} \Rightarrow 2 = (2)^{n-1}$

$\Rightarrow n - 1 = 1 \Rightarrow n = 2$

93. **(1)** $t = \frac{1}{k}\ln\left(\frac{a}{a-x}\right)$

$k = \frac{\ln 2}{t_{1/2}} = \frac{\ln 2}{30} \Rightarrow 100 = \frac{30}{\ln 2}\ln\left(\frac{1}{w}\right)$

$\frac{1}{w} = 10 \Rightarrow w = 1 \times 10^{-1}\ \mu g$

94. **(100)** $A \to B$

For first order reaction;

$k = \frac{2.303}{t}\log\left[\frac{A_0}{A_t}\right] \Rightarrow \frac{0.693}{t_{1/2}} = \frac{2.303}{2}\log\left(\frac{A_0}{A_t}\right)$

$\Rightarrow \frac{0.693}{0.3010} = \frac{2.303}{2}\log\frac{A_0}{A_t}$

$\log\frac{A_0}{A_t} = 2 \ \therefore \ \frac{A_0}{A_t} = 10^2 = 100$

95. **(2)** When initial pressure of NO increases by a factor of 2 ($2 \times 20.1 \approx 40$) then initial rate increases by a factor of 4 ($4 \times 0.033 \approx 0.135$). Therefore, order of reaction w.r.t. NO = 2

96. **(165)**

$t_{1/2} = 70$ min; for Ist order reaction,

$k = \frac{0.693}{t_{1/2}} = \frac{0.693}{70 \times 60}s^{-1} = \frac{6930}{7 \times 6} \times 10^{-6}s^{-1}$

$= 165 \times 10^{-6}\,s^{-1}$

97. **(2)**

- For zero order, $t_{1/2}$ decrease with decrease in concentration of reactions.
- For first order, $t_{1/2}$ is independent of concentration of reaction
- For second order, $t_{1/2}$ increases with decrease in concentration of reactant.

$\frac{(t_{1/2})_1}{(t_{1/2})_2} = \left(\frac{[A_0]_2}{[A_0]_1}\right)^{1-n} \Rightarrow \frac{100}{50} = \left(\frac{0.5}{1.0}\right)^{1-n}$

$\Rightarrow 2 = \left(\frac{1}{2}\right)^{1-n} \Rightarrow 2 = 2^{n-1}$

$\Rightarrow n - 1 = 1 \Rightarrow n = 2$

98. **(c)** For Ist order reaction

$k = \frac{2.303}{t}\log\frac{a_o}{0.2a_o}$ also $t_{1/2} = \frac{0.693}{k}$

$k = \frac{0.693}{200} \Rightarrow \frac{0.693}{200} = \frac{2.303}{t}\log\frac{1}{0.2}$

$t = \frac{2.303}{0.693} \times 200\log\frac{1}{0.2} = 466.675 \approx 467$ sec

99. **(1)** Since the half life for the decomposition of gas is not dependent on initial concentration, the order of the reaction is one.

100. (2) For first order reaction

$$k=\frac{1}{t}\ln\left(\frac{P_\circ}{P}\right)$$

$$\ln\left(\frac{P_\circ}{P}\right)=kt \Rightarrow \ln\left(\frac{P}{P_\circ}\right)=-kt$$

On comparing with straight line equation
y = mx; k = slope = 3.465 × 10^4

$$t_{1/2}=\frac{\ln 2}{k}=\frac{0.693}{3.465\times 10^4}=2\times 10^{-5}\text{s}$$

101. (16) $t=\frac{1}{k}\ln\left(\frac{a_0}{a_0-n}\right)$ or $k=\frac{\ln(2)}{t_{1/2}}$

$$t_{67\%}=\frac{1}{k}\ln\left(\frac{1}{1-0.67}\right)=\frac{t_{1/2}}{\ln 2}\times\ln\left(\frac{1}{1-\frac{2}{3}}\right)$$

$$t_{67\%}=\frac{t_{1/2}}{\log 2}\times\log 3=\frac{t_{1/2}\times 0.4771}{0.301}$$

$\Rightarrow t_{67\%}=1.585\times t_{1/2}$ or $x\times 10^{-1}=1.585$

$\Rightarrow x=15.85\simeq 16$

102. (75) $t_{1/2}=\frac{0.693}{k}$ or $k=\frac{0.693}{t_{1/2}}$

$$t=\frac{2.303}{k}\log\frac{[A_0]}{[A]_t}=\frac{t_{1/2}}{0.3}\log\frac{[A_0]}{[A]_t}$$

$$83=\frac{200}{0.3}\log\frac{[A_0]}{[A]_t} \text{ or } 0.125=\log\frac{[A_0]}{[A]_t}$$

$$\frac{[A_0]}{[A]_t}=1.333\cong\frac{4}{3}\quad\therefore\quad\frac{[A]_t}{[A]_0}\times 100=\frac{3}{4}\times 100=75\%$$

103. (0) $t_{1/2}\propto\frac{1}{[P_0]^{n-1}}$

$$\frac{(t_{1/2})_1}{(t_{1/2})_2}=\frac{(P_2)^{n-1}}{(P_1)^{n-1}}\Rightarrow\frac{340}{170}=\left(\frac{27.8}{55.5}\right)^{n-1}$$

$$\Rightarrow 2=\frac{1}{(2)^{n-1}}\Rightarrow n=0$$

104. (c) Given $t_{0.90}=xt_{1/2}$
First order rate constant,

$$k=\frac{\ln 2}{t_{1/2}}=\frac{1}{xt_{1/2}}\ln\frac{A_0}{A_0-A_0\times\frac{90}{100}}$$

$$\frac{\ln 2}{t_{1/2}}=\frac{\ln 10}{xt_{1/2}}\Rightarrow x=\frac{\ln 10}{\ln 2}=\frac{2.303}{2.303\times 0.3010}=3.32$$

105. (654)

	A ⟶	2B
t = 0	1	0
t = 100 min	1 − x	2x
	= 0.9mol	= 0.2mol

$$2x=0.2\Rightarrow x=\frac{0.2}{2}=0.1$$

$$k=\frac{2.303}{t}\log\frac{[A_o]}{[A_t]}\Rightarrow k=\frac{2.303}{100}\log\frac{1}{0.9}\Rightarrow k=1.059\times 10^{-3}$$

$$t_{1/2}=\frac{0.693}{k}=\frac{0.693}{1.059\times 10^{-3}}=654.3$$

106. (108) Initially: $[A_0]=[B_0]=a$
After time 't' min. : $[A]=16[B]$

$$[A]=[A_0]e^{-k_At}$$

$$[B]=[B_0]e^{-k_Bt}\Rightarrow a\cdot e^{-k_At}=16a\cdot e^{-k_Bt}$$

$$\Rightarrow e^{-(k_A-k_B)t}=16\Rightarrow(k_B-k_A)t=\ln 16$$

$$\because k_A=\frac{\ln 2}{t_{1/2}}=\frac{\ln 2}{54};\therefore k_B=\frac{\ln 2}{18}$$

$$\Rightarrow\left(\frac{\ln 2}{18}-\frac{\ln 2}{54}\right)t=\ln 16\Rightarrow\ln 2\left(\frac{1}{18}-\frac{1}{54}\right)t=4\ln 2$$

$$\Rightarrow t=\frac{54\times 18\times 4}{36}=108\text{ min}$$

107. (81) Given :

$$\underset{\text{Sucrose}}{C_{12}H_{22}O_{11}}+H_2O\xrightarrow{\text{First order}}\underset{\text{Glucose}}{C_6H_{12}O_6}+\underset{\text{Fructose}}{C_6H_{12}O_6}$$

$t_{1/2}=3.33$ h
$t=0, a=[A]_0; t=9$ h., $(a-x)=[A]_t$

From first order kinetic : $\frac{k\times t}{2.303}=\log\frac{[A]_0}{[A]_t}$

$$\Rightarrow\frac{0.693\times 9}{3.33\times 2.303}=\log\left(\frac{1}{f}\right)\qquad\left(\because t_{1/2}=\frac{0.693}{k}\right)$$

$$\Rightarrow\log\left(\frac{1}{f}\right)=0.813=81.3\times 10^{-2}\approx 81\times 10^{-2}$$

108. (d) Rate = $k[A]^n$
$\log[\text{Rate}]=\log k+n\log[A]$
Slope = n [n is order of the reaction]
∴ Correct sequence for the order of the reaction is (iv) > (ii) > (i) > (iii)

109. (b) $C_t=C_0e^{-kt}; k=\frac{\ln 2}{t_{1/2}}$

$$(C_t)_A=(C_0)_Ae^{-k_At}; k_A=\frac{\ln 2}{300}$$

$$(C_t)_B=(C_0)_Be^{-k_Bt}; k_B=\frac{\ln 2}{180}$$

$$\frac{(C_t)_B}{(C_t)_A}=\frac{(C_0)_B}{(C_0)_A}\times e^{(k_B-k_A)t}\Rightarrow 4=e^{(k_B-k_A)t}$$

$$\Rightarrow 2\ln 2=\left[\frac{\ln 2}{180}-\frac{\ln 2}{300}\right]t\Rightarrow 2\ln 2=\ln 2\left[\frac{1}{180}-\frac{1}{300}\right]t$$

$$\Rightarrow 2=\left(\frac{120}{180\times 300}\right)t\Rightarrow t=\frac{2\times 180\times 300}{120}=900\text{ sec}$$

110. (60)

$$t=\frac{2.303}{k}\log\left[\frac{100}{100-x\%}\right] \quad t_{75\%}=\frac{2.303}{k}\log\left[\frac{100}{25}\right]=90$$

$$t_{60\%}=\frac{2.303}{k}\log\left[\frac{100}{40}\right]$$

$$\frac{t_{75\%}}{t_{60\%}}=\frac{2\log 2}{\log 2.5}\Rightarrow\frac{90}{t_{60\%}}=\frac{2\times 0.3}{0.4}$$

$$\Rightarrow t_{60\%}=\frac{90\times 4}{6}=60\text{ min.}$$

111. (b) For an elementary or single step reaction, order and molecularity are same. But for a reaction, molecularity $\neq 0$. Hence, the zero order reaction must be a multi-step so that overall order of reaction becomes zero.

112. (c) Rate (R) = $k[A]^a[B]^b$

Exp. I $\Rightarrow 6.0\times 10^{-3}=k[0.1]^a[0.1]^b$

Exp. II $\Rightarrow 24.0\times 10^{-3}=k[0.1]^a[0.2]^b$

Exp. III $\Rightarrow 12.0\times 10^{-3}=k[0.2]^a[0.1]^b$

Exp. IV $\Rightarrow 72\times 10^{-3}=k[X]^a[0.2]^b$

Exp. V $\Rightarrow 288\times 10^{-3}=k[0.3]^a[Y]^b$

From Exp. I and II,

$$\Rightarrow\frac{6.0\times 10^{-3}}{24\times 10^{-3}}=\frac{k[0.1]^a[0.1]^b}{k[0.1]^a[0.2]^b}$$

$$\Rightarrow\frac{1}{4}=\left(\frac{0.1}{0.2}\right)^b\Rightarrow\left(\frac{1}{2}\right)^2=\left(\frac{1}{2}\right)^b$$

$\therefore b=2$

Similarly, from exp. I and III we get, $\frac{1}{2}=\left(\frac{1}{2}\right)^a\Rightarrow a=1$

From Exp. II and IV,

$$\Rightarrow\frac{24\times 10^{-3}}{72\times 10^{-3}}=\frac{k[0.1]^a[0.2]^b}{k[X]^a[0.2]^b}$$

$$\Rightarrow\frac{1}{3}=\left(\frac{0.1}{X}\right)^a\Rightarrow\frac{1}{3}=\left(\frac{0.1}{X}\right)^1$$

$\therefore X=0.3$

From Exp. I and V,

$$\Rightarrow\frac{6.0\times 10^{-3}}{288\times 10^{-3}}=\left(\frac{0.1}{0.3}\right)^a\left(\frac{0.1}{Y}\right)^b\Rightarrow\frac{1}{48}=\left(\frac{1}{3}\right)^1\left(\frac{0.1}{Y}\right)^2$$

$$\Rightarrow\left(\frac{1}{4}\right)^2=\left(\frac{0.1}{Y}\right)^2\Rightarrow\frac{1}{4}=\frac{0.1}{Y}$$

$\therefore Y=0.4$

113. (23.03) $t_{1/2}=6.93$ years, $a=10^{-6}$ g

$$t_{1/2}=\frac{0.693}{k}\Rightarrow k=\frac{0.693}{t_{1/2}}=\frac{0.693}{6.93}=0.1$$

For 1st order reaction,

$$t=\frac{2.303}{k}\log\frac{a}{a-x}=\frac{2.303}{0.1}\log\frac{10^{-6}}{10^{-7}}=\frac{2.303}{0.1}=23.03\text{ years}$$

114. (a) Rate constant of decomposition of $X=0.05$ μg/year. Unit of rate constant confirms that the decomposition of X is a zero order reaction.

For zero order kinetics,

$[X]=[X]_0-kt$

$$kt=[X]_0-[X],\ t=\frac{[X]_0-[X]}{k}$$

$$t=\frac{5-2.5}{0.05}=\frac{2.5}{0.05}=50\text{ years}$$

115. (c) For the reaction $2X\rightarrow B$, follow zeroth order

Rate equation is

$kt=[A]_0-[A]$

For the half-life; $t=t_{1/2}$ and $[A]=0.1$

$kt_{1/2}=0.2-0.1$

$$k=\frac{0.2-0.1}{6}=\frac{0.1}{6}\text{M hr}^{-1}$$

$\therefore$ Time required to reach from 0.5 M to 0.2 M = t

$kt=[A]_0-[A]$

$$\frac{0.1}{6}\times t=(0.5-0.2)\quad\Rightarrow\quad t=18\text{ hour}$$

116. (b) From experiment I and II, it is observed that order of reaction w.r.t. (B) is zero.

From experiment II and III, x can be calculated as:

$$\frac{6.93\times 10^{-3}}{13.86\times 10^{-3}}=\left(\frac{0.1}{0.2}\right)^x\left(\frac{0.25}{0.3}\right)^y$$

$\because$ Order of reaction w.r.t. B = 0, so y = 0

$$\therefore\quad\frac{6.93\times 10^{-3}}{13.86\times 10^{-3}}=\left(\frac{0.1}{0.2}\right)^x\text{ or }\left(\frac{1}{2}\right)^1=\left(\frac{1}{2}\right)^x$$

$x=1$, Now, Rate = $k[A]^1[B]^o$ or $6.93\times 10^{-3}=k[0.1]^1[0.2]^o$

$k=6.93\times 10^{-2}$

$$\therefore\quad t_{1/2}=\frac{0.693}{k}=\frac{0.693}{6.93\times 10^{-2}}=10$$

117. (b) $\frac{r_2}{r_1}=2^x.2^y=8$ ...(i)

$$\frac{r_2}{r_1}=2^x=2\Rightarrow x=1$$

Put $x=1$ in eq(i), $2^y=4\Rightarrow y=2$

118. (a) $CH_3CHO\longrightarrow CH_4+CO$

From rate law, $r\propto(a-x)^m$

m = order of reaction, a – x = unreacted

$r_1=1$ torr s^{-1}, when 5% reacted

$r_2=0.5$ torr s^{-1}, when 33 % reacted

$(a-x_1)=0.95$(unreacted)

$(a-x_2)=0.67$(unreacted)

$$\frac{r_1}{r_2}=\left[\frac{(a-x_1)}{(a-x_2)}\right]^m\Rightarrow\frac{1}{0.5}=\left(\frac{0.95}{0.67}\right)^m$$

$$2=(1.41)^m\Rightarrow 2=(\sqrt{2})^m\Rightarrow m=2$$

119. (d) This is first order reaction as half life is constant.

120. (b) $N_2O_5 \rightarrow 2NO_2 + \frac{1}{2}O_2$

	N_2O_5	$2NO_2$	$\frac{1}{2}O_2$
At t = 0	50	0	0
At t = 50 min	$50-p_1$	$2p_1$	$\frac{p_1}{2}$

Total pressure at 50 minutes $= 50 - p_1 + 2p_1 + \frac{p_1}{2} = 87.5$

$\therefore\ p_1 = \frac{37.5\times 2}{3} = 25 = \frac{1}{2}\times 50$ mm Hg

This means, 50 min. is half-life.

For 100 minutes, i.e., 2 half lives, $50 - p_2 = 12.5$

$\therefore\ p_2 = 37.5$ mm of Hg

Total pressure at 100 minutes

$= 50 - p_2 + 2p_2 + \frac{p_2}{2} = 50 + \frac{3p_2}{2} = 50 + \frac{3}{2}\times 37.5$

$= 106.25$ mm of Hg

121. (c) The half life $t_{1/2} = 10$ days

The decay constant,

$k = \frac{0.693}{t_{1/2}} = \frac{0.693}{10 \text{ days}} = 0.0693 \text{ days}^{-1}$

The time required for one fourth conversion

$t = \frac{2.303}{k}\log_{10}\frac{a}{a-x}$

$= \frac{2.303}{0.0693\ \text{day}^{-1}}\log_{10}\frac{1}{1-(1/4)} = 4.1$ days

122. (d) $H_2O_2(aq) \rightarrow H_2O(aq) + \frac{1}{2}O_2(g)$

For a first order reaction,

$k = \frac{2.303}{t}\log\frac{a}{(a-x)}$

Given $a = 0.5$, $(a-x) = 0.125$, $t = 50$ min

$\therefore\ k = \frac{2.303}{50}\log\frac{0.5}{0.125}$

$= 2.78\times 10^{-2}\ \text{min}^{-1}$

$r = k[H_2O_2] = 2.78\times 10^{-2}\times 0.05 = 1.386\times 10^{-3}\ \text{mol min}^{-1}$

Now

$-\frac{d[H_2O_2]}{dt} = \frac{d[H_2O]}{dt} = \frac{2d[O_2]}{dt} = k[H_2O_2]$

$\therefore\ \frac{d[O_2]}{dt} = -\frac{1}{2}\times\frac{d[H_2O_2]}{dt}$

$= \frac{1.386\times 10^{-3}}{2} = 6.93\times 10^{-4}\ \text{mol min}^{-1}$

123. (a) Rate law for first order reaction $= k[N_2O_5]$

$2N_2O_5(g) \longrightarrow 4NO_2(g) + O_2(g)$

	$2N_2O_5(g)$	$4NO_2(g)$	$O_2(g)$
t = 0 min (Pressure in mm Hg)	50	0	0
t = 30 min (Pressure in mm Hg)	$50-2p$	$4p$	p

Total pressure $50 - 2p + 4p + p = 50 + 3p$

$= 87.5$ mm Hg

$\therefore\ p = 12.5$ mm Hg

$\therefore\ p_0 = 50$ & $p\ (t = 30 \text{ min}) = 25$ for N_2O_5 reactant

$\therefore\ k = \frac{2.303}{30\text{ min}}\times\log\left(\frac{50}{25}\right) = \frac{2.303}{60\text{ min}}\times\log\left(\frac{50}{x}\right)$

On solving $x = 12.5$ mm Hg $= 50 - 2p$

$\therefore\ p = 18.75$ mm Hg

$\therefore$ Total pressure $= 50 + 3p = 106.25$ mm Hg

124. (c) Reactions of higher order (>3) are very rare due to very less chances of many molecules to undergo effective collisions.

125. (c) Given $t_{1/2} = 15$ minutes

Total time $(T) = 1$ hr $= 60$ min

From $T = n\times t_{1/2} \Rightarrow n = \frac{60}{15} = 4$

Now, from the formula, $\frac{N}{N_0} = \left(\frac{1}{2}\right)^n = \left(\frac{1}{2}\right)^4 = \frac{1}{16}$

126. (d) For a zero order reaction,

$[A_t] = [A_0] - k.t$

$\Rightarrow\ 0.5 = [A_0] - 2\times 10^{-2}\times 25 \Rightarrow [A_0] = 1$ M

127. (c) Given $t_{1/2} = 3$; Total time T = 12

No. of half lives $(n) = \frac{12}{3} = 4 \Rightarrow \left(\frac{1}{2}\right)^n = \frac{N}{N_o}$

$\therefore\ \left(\frac{1}{2}\right)^4 = \frac{3}{N} \Rightarrow N = 48$ g

128. (b) For a first order reaction

$k = \frac{2.303}{t}\log\frac{a}{a-x} = \frac{2.303}{40}\log\frac{0.1}{0.025}$

$= 3.47\times 10^{-2}$

$R = k[A]^1 = 3.47\times 10^{-2}\times 0.01 = 3.47\times 10^{-4}$

129. (d) $A \longrightarrow B$

Initial concentration	Rate of reaction
2×10^{-3} M	2.40×10^{-4} Ms^{-1}
1×10^{-3} M	0.60×10^{-4} Ms^{-1}

$r = k[A]^x$

where x = order of reaction, hence,

$2.40\times 10^{-4} = k\,[2\times 10^{-3}]^x$ ......(i)

$0.60\times 10^{-4} = k\,[1\times 10^{-3}]^x$ ......(ii)

On dividing eqn.(i) from eqn. (ii) we get,

$4 = (2)^x$

$\therefore\ x = 2$, i.e., order of reaction = 2

130. (b) Plot given is for zero order reaction.

From a zero order reaction,

$[A_t] = [A_0] - k_t$

$t_{1/2} = \frac{A_0}{2k}$

$\Rightarrow\ \log t_{1/2} = -\log 2k + \log[A_0]$

131. (c) For the reaction

$A \rightarrow$ Product

Given $t_{1/2} = 1$ hour

For a zero order reaction ; $\therefore\ t_{1/2} = \frac{[A_0]}{2k}$

or $k = \frac{[A_0]}{2t_{1/2}} = \frac{2}{2\times 1} = 1\ \text{mol L}^{-1}\text{hr}^{-1}$

Further for a zero order reaction,

$$k = \frac{dx}{dt} = \frac{\text{change in concentration}}{\text{time}}$$

$$1 = \frac{0.50 - 0.25}{\text{time}} \Rightarrow \text{time} = 0.25 \text{ hr.}$$

132. (b) For first order reaction,

$$k = \frac{2.303}{t} \log \frac{100}{100-99}$$

$$\frac{0.693}{6.93} = \frac{2.303}{t} \log \frac{100}{1}$$

$$\frac{0.693}{6.93} = \frac{2.303 \times 2}{t}$$

$t = 46.06$ min

133. (b) For a first order reaction, $t_{1/2} = \frac{0.693}{K}$ i.e. for a first order reaction $t_{1/2}$ does not depend upon the concentration. From the given data, we can say that order of reaction with respect to $B = 1$ because change in concentration of B alone does not change half life.
Order of reaction with respect to $A = 1$ because rate of reaction doubles when concentration of A is doubled keeping concentration of B constant.
$\therefore$ Order of reaction = 1 + 1 = 2 and units of second order reaction are L mol^{-1} sec^{-1}.

134. (a) Suppose activity of safe working = A
Given $A_0 = 10A$

$$\lambda = \frac{0.693}{t_{1/2}} = \frac{0.693}{30}$$

$$t_{½} = \frac{2.303}{\lambda} \log \frac{A_0}{A} = \frac{2.303}{0.693/30} \log \frac{10A}{A}$$

$$= \frac{2.303 \times 30}{0.693} \times \log 10 = 100 \text{ days.}$$

135. (b) (i) $NO(g) + Br_2(g) \rightleftharpoons NOBr_2(g)$
(ii) $NOBr_2(g) + NO(g) \longrightarrow 2NOBr(g)$
Rate law equation = $k[NOBr_2][NO]$
But $NOBr_2$ is intermediate and must not appear in the rate law equation.

From Ist step, $K_c = \frac{[NOBr_2]}{[NO][Br_2]}$

$\therefore [NOBr_2] = K_c[NO][Br_2]$
$\therefore$ Rate law equation = $k.K_c[NO]^2[Br_2]$
Hence, order of reaction w.r.t. NO is 2.

136. (a) Since the reaction is 2nd order w.r.t CO. Thus, rate law is given as.
$r = k[CO]^2$
Let initial concentration of CO is a
i.e. $[CO] = a$
$\therefore \quad r_1 = k(a)^2 = ka^2$
When concentration becomes doubled, i.e. $[CO] = 2a$
$\therefore \quad r_2 = k(2a)^2 = 4ka^2 \quad \therefore r_2 = 4r_1$
So, the rate of reaction becomes 4 times.

137. (d) The molecularity of a reaction is the number of reactant molecules taking part in a single step of the reaction. Therefore, the reaction involving two different reactants can never be unimolecular.

138. (c) $t_{1/4} = \frac{2.303}{k} \log \frac{1}{3/4} = \frac{2.303}{k} \log \frac{4}{3}$

$$= \frac{2.303}{k}(2 \times 0.301 - 0.4771) = \frac{0.29}{k}$$

139. (d) The velocity constant depends on temperature only. It is independent of concentration of reactants.

140. (b) $r = k[O_2][NO]^2$. When the volume is reduced to 1/2, the conc. will double.
$\therefore$ New rate = $k[2O_2][2NO]^2 = 8k[O_2][NO]^2$
The new rate increases to eight times of its initial value.

141. (d) $t_{1/2} = 3$ h, $T = 18$ h
$\because T = n \times t_{1/2}$

$$\therefore n = \frac{18}{3} = 6$$

Initial mass $(C_0) = 256$ g

$$\therefore C_n = \frac{C_0}{2^n} = \frac{256}{(2)^6} = \frac{256}{64} = 4 \text{g}$$

142. (a) $Rt = \log C_0 - \log C_t$

$$\Rightarrow t = \frac{-\log C_t}{R} + \frac{\log C_0}{R}$$

It is clear from the equation that if we plot a graph between $\log C_t$ and time, a straight line with a slope equal to $-1/R$ and intercept equal to $\log [C_0]/R$ will be obtained.

143. (c) As the concentration of reactant decreases from 0.8 to 0.4 in 15 minutes, hence the $t_{1/2}$ is 15 minutes. To fall the concentration from 0.1 to 0.025, we need two half lives, i.e., 30 minutes.

144. (a) $N_t = N_0\left(\frac{1}{2}\right)^n$ where n is number of half life periods.

$$n = \frac{\text{Total time}}{\text{half life}} = \frac{24}{4} = 6 \therefore N_t = 200\left(\frac{1}{2}\right)^6 = 3.125 \text{ g}$$

145. (a) Order is the sum of the power of the concentration terms in rate law expression.
Hence the order of reaction is = 1 + 2 = 3

146. (d) $t_{1/2} = 5$ years, T = 15 years.

Hence total number of half life periods $= \frac{15}{5} = 3$

$$\therefore \text{Amount left} = \frac{64}{(2)^3} = 8 \text{ g}$$

147. (a) $k = Ae^{-Ea/RT}$

$$\log k = \log A - \frac{E_a}{2.303RT}$$

For graph between log k with $\frac{1}{T}$

$$|\text{Slope of curve}| = \frac{E_a}{2.303R}$$

From given graph :
Magnitude of slope ⇒ (2) > (1) > (3)
Hence $Ea_2 > Ea_1 > Ea_3$

148. (a)

$A_2 + B_2 \rightleftharpoons 2AB$
$E_f = 180 \text{ kJ mol}^{-1}$
$E_b = 200 \text{ kJ mol}^{-1}$
$\Delta H = E_f - E_b = -20 \text{ kJ mol}^{-1}$
In presence of catalyst :
$E_f = 180 - 100 = 80 \text{ kJ mol}^{-1}$
$E_b = 200 - 100 = 100 \text{ kJ mol}^{-1}$
Catalyst does not change ΔH or ΔG of a reaction.
Catalyst only reduces activation energy of the reaction.

149. (c) Arrhenius equation holds true for elementary as well as complex reactions.
Unit of A is same as unit of k. Rate of reaction is high if activation energy is low.
A and E_a are temperature independent.

150. (a) A net release of heat is observed in the process.

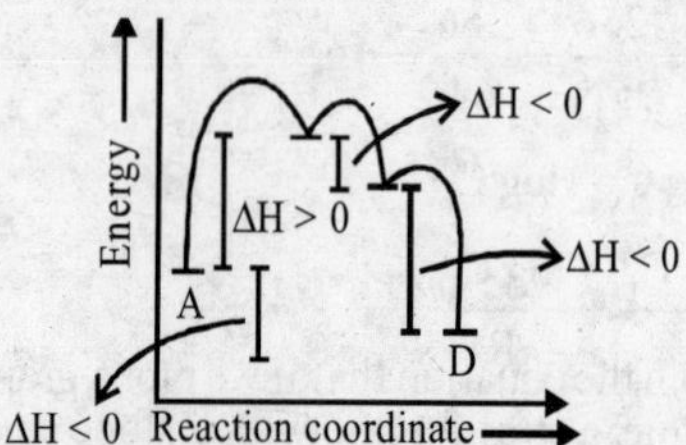

151. (20) $k = \sqrt{\dfrac{k_1 k_3}{k_2}}$

$A.e^{-Ea/RT} = \sqrt{\dfrac{A_1 e^{-Ea_1/RT} \times A_3 e^{-Ea_3/RT}}{A_2 e^{-Ea_2/RT}}}$

Taking ln and both sides and comparing exponential terms.

$\dfrac{E_a}{RT} = \dfrac{1}{2} \times \left(\dfrac{E_{a_1}}{RT} + \dfrac{E_{a_3}}{RT} - \dfrac{E_{a_2}}{RT} \right)$

$E_a = \left(E_{a_1} + E_{a_3} - E_{a_2} \right)/2$

$E_a = (60 + 10 - 30)/2 = 20 \text{ kJ/mol}^{-1}$

(**Note:** The values of A, A_1, A_2, and A_3 should also be considered here but their values are not given so we must ignore them to get overall E_a.)

152. (c) Activation energy of forward reaction = $E_1 + E_2$
Energy of product > Energy of reactant

Stability $\propto \dfrac{1}{\text{Energy}}$

153. (69) $t_{1/2} = \dfrac{0.693}{k}$; $K - Ae^{-E_a/RT}$

$= 10^{20} \times e^{-\frac{191.48 \times 10^3}{8.314 \times 1000}} = 10^{20} \times e^{-23.031} = 10^{20} \times - e^{\ln 10 \times 10}$

$= \dfrac{10^{20}}{10^{10}} = 10^{10}$ sec.

$t_{1/2} = \dfrac{0.693}{10^{10}} = 6.93 \times 10^{-11} = 69.3 \times 10^{-12}$ sec.

= 69.3 pico second

154. (30) $k = Ae^{-E_a/RT}$

Overall rate constant $K = \dfrac{k_1 k_2}{k_3}$

Trick: Used in ans 36

$E_a = E_{a_1} + E_{a_2} - E_{a_3}$

$\Rightarrow Ae^{-E_a/RT} = \dfrac{A_1 \cdot e^{-E_{a_1}/RT} \cdot A_2 \cdot e^{-E_{a_2}/RT}}{A_3 \cdot e^{-E_{a_3}/RT}}$

$= \left(\dfrac{A_1 \cdot A_2}{A_3} \right) \cdot e^{-\frac{(E_{a_1} + E_{a_2} - E_{a_3})}{RT}}$

$\therefore$ Overall $E_a = E_{a_1} + E_{a_2} - E_{a_3} = 40 + 50 - 60$
$= 90 - 60 = 30$ kJ/mol

$\left[K = A.e^{-E_a/RT} \right]$

155. (2) $k = A.e^{-E_a RT}$

$\log k = \log A - \dfrac{E_a}{2.303R} \cdot \dfrac{1}{T}$

Higher is E_a, stronger is the temperature dependence of k (i.e. steeper the slope)

$\Rightarrow \dfrac{1}{k} \dfrac{dk}{dT} = \dfrac{E_a}{R} \dfrac{1}{T^2} \Rightarrow \dfrac{dk}{dT} = A \times e^{-\frac{E_a}{R}} \cdot \dfrac{E_a}{RT^2}$

Now, if $E_a = 0 \Rightarrow \dfrac{dk}{dT} = 0$

Thus, (A) and (B) are correct.

156. (b) By using positive catalyst:
(i) ΔH does not change, (ii) Activation energy decreases

157. (2) $k = Ae^{-\frac{E_a}{RT}}$; $k_1 = Ae^{-\frac{(E_a)_1}{RT}}$

$k_2 = Ae^{-\frac{(E_a)_2}{RT}} \Rightarrow \dfrac{k_2}{k_1} = e^{\frac{(E_a)_1 - (E_a)_2}{RT}}$

$\log \dfrac{k_2}{k_1} = \dfrac{(E_a)_1 - (E_a)_2}{2.3RT} = \dfrac{(41.4 - 30) \times 1000}{2.3 \times 8.3 \times 300} = 1.99 \cong 2$

158. (b) Step II with highest activation energy is rate Determining Step.
No. of activated complex = 3

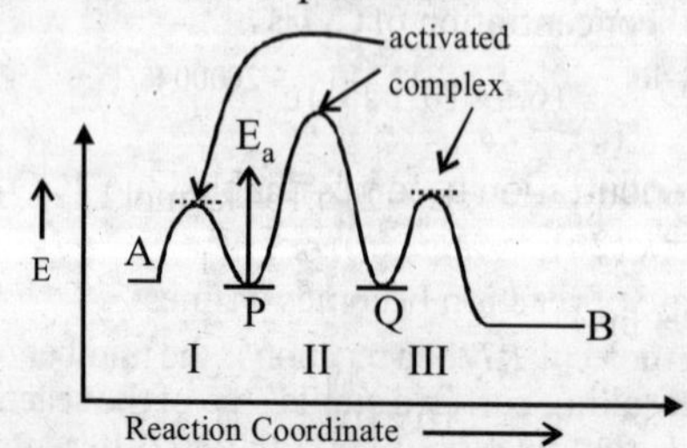

Number of intermediates = 2

159. (2520) $\log\frac{K_{300}}{K_{200}}=\frac{E_a}{2.3\times8.314}\left(\frac{1}{T_1}-\frac{1}{T_2}\right)$

$\log\frac{0.05}{0.03}=\frac{E_a}{2.3\times8.3}\times\left[\frac{1}{200}-\frac{1}{300}\right]$

$E_a=2519.88\text{ J}\Rightarrow E_a=2520\text{ J}$

160. (3) (A) $k=Ae^{\frac{-E_a}{RT}}$
As E_a increases k decreases

(B) Temperature coefficient $=\frac{k_{T+10}}{k_T}$

(C) Option (C) is wrong. Δk may be greater or lesser depending on temperature.

(D) $\ln k=\ln A-\frac{E_a}{RT}$

161. (8) $k=Ae^{-E_a/RT}$
Taking ln both sides;

$\ln k=\frac{-E_a}{RT}+\ln A$

$y=mx+c$

Slope $=\frac{E_a}{R}=\frac{20}{5}$; $E_a=4R=8$ cal/mol

162. (225) $k=Ae^{\frac{-E_a}{RT}}$ or $\ln k=\ln A-\frac{E_a}{RT}$

$\Rightarrow\ \log(2\times10^{-6})=\log(4\times10^{13})-\frac{8.3\times10^3}{8.3\times T\times2.3}$

$\Rightarrow\ \log(2)-6=2\times\log(2)+13-\frac{8.3\times10^3}{8.3\times T\times2.3}$

$\Rightarrow\ -19.3=-\frac{10^3}{T\times2.3}\Rightarrow T=225\times10^{-1}\text{ K}$

163. (1) $\ln\left(\frac{k_2}{k_1}\right)=\frac{E_a}{R}\left(\frac{1}{T_1}-\frac{1}{T_2}\right)$

$\ln\left(\frac{k_2}{k_1}\right)=\frac{532611}{8.3}\times\left(\frac{10}{300\times310}\right)$

where k_2 is at 310 K & k_1 is at 300 K

$\ln\left(\frac{k_2}{k_1}\right)=6.9$ or $=k_1=x\times10^{-3}$

$\frac{k_2}{k_1}=999.2$; $\frac{k_2}{x\times10^{-3}k_2}=999.2$

$x=1$

164. (216)

$k=Ae^{-E_a/RT}=(6.5\times10^{12}\text{ s}^{-1})e^{-26000\text{ K}/T}$

$\frac{E_a}{8.314}=26000\ \Rightarrow\ E_a=216.164$ kJ/mol.

165. (166) $\ln k=\ln A-\frac{E_a}{RT}$

Given : $\ln k=33.24-\frac{2.0\times10^4}{T}$

$\therefore$ on comparing $\frac{E_a}{R}=2.0\times10^4\Rightarrow E_a=2.0\times10^4\times8.3$ J
$=16.6\times10^4\text{ J}=166\text{ kJ}$

166. (59) $\log_{10}\frac{k_2}{k_1}=\frac{E_a}{2.303R}\left[\frac{1}{T_1}-\frac{1}{T_2}\right]$

$T_1=300\text{K}$, $T_2=309\text{K}$

$\log_{10}\frac{k_2}{k_1}=\frac{E_a}{2.303R}\left(\frac{1}{300}-\frac{1}{309}\right)$

$0.3=\frac{E_a}{2.303\times8.3}\left(\frac{9}{300\times309}\right)$

$E_a=\frac{0.3\times2.303\times8.3\times300\times309}{9}=59065.04\text{J/mol}$

$E_a=59.06$ kJ/mol

167. (4) According to the Arrhenius equation

$k=Ae^{\frac{-Ea}{RT}}$

$k_{cat}=Ae^{\frac{E_a^1}{RT}}$, $k_{uncat.}=Ae^{\frac{Ea}{RT}}$

$\frac{k_{cat}}{k_{uncat.}}=e^{\frac{Ea-E_a^1}{RT}}=e^{\frac{10\times1000}{8.3\times300}}=e^{4.009}=e^x$

$[E_a-E_a^1=10\text{ kJ}]$

$\therefore x=4$

168. (154) $\ln k=\ln A-\frac{E_a}{10^3RT}\times10^3$

$=\ln A+\frac{10^3}{T}\left[-\frac{E_a}{10^3R}\right]$

From the graph,

$\frac{-E_a}{10^3\times R}=-18.5$; $E_a=153.735=154$ kJ/mol

169. (45) ΔH = Energy of products – Energy of reactants
$=Z-(Z+Y)=15-(15+45)=-45$
$|\Delta H|=45$ kJ/mol

170. (47) Given $\log K=20.35-\frac{(2.47\times10^3)}{T}$ ...(i)

We know $\log K=\log A-\frac{E_a}{2.303\text{ RT}}$ ...(ii)

comparing equation (i) with equation (ii)

$\frac{E_a}{2.303\text{ RT}}=\frac{2.47\times10^3}{T}$

$E_a=2.47\times10^3\times2.303\times8.314$
$=47293.44$ J/mol
$=47.29=47$ kJ/mol

171. (16) $E_a=209$ kJ/mol
k_2 at 700 °C $=6.36\times10^{-3}\text{s}^{-1}$,
$k_1=x\times10^{-6}\text{ s}^{-1}$

$\log\left(\frac{k_2}{k_1}\right)=\frac{-E_a}{2.303R}\left(\frac{1}{T_2}-\frac{1}{T_1}\right)$

$$\log\left(\frac{6.36\times10^{-3}}{k_1}\right)=\frac{209\times1000}{2.303\times8.31}\left(\frac{100}{700\times600}\right)$$

$\log(6.36\times10^{-3})-\log k_1=2.6$

$\Rightarrow \log k_1=-2.19-2.6=-4.79$

$\Rightarrow k_1=10^{-4.79}=1.62\times10^{-5}=16.2\times10^{-6}\Rightarrow x=16$

172. **(10)** $\log\frac{k_2}{k_1}=\frac{E_a}{2.303R}\left[\frac{1}{T_1}-\frac{1}{T_2}\right]$

k_1 (at 200 K) = ?

k_2 (at 300 K) $=1\times10^{-3}\ s^{-1}$

$$\log\frac{1\times10^{-3}}{k_1}=\frac{11.488\times10^3}{2.303\times8.314}\left[\frac{1}{200}-\frac{1}{300}\right]$$

$$\log\frac{1\times10^{-3}}{k_1}=\frac{11.488\times10^3\times100}{2.303\times8.314\times200\times300}$$

$$\log\frac{1\times10^{-3}}{k_1}=1$$

$$\frac{1\times10^{-3}}{k_1}=10\Rightarrow k_1=10\times10^{-5}\ s^{-1}$$

173. **(50)** $X\rightarrow Y$

$\Delta E=(E_a)_f-(E_a)_b;\ -20=30-(E_a)_b\Rightarrow (E_a)_b=50$ kJ

174. **(14)** Energy of activation,

$E_a=80.9$ kJ mol^{-1} = 80900 J/mol.

Temperature of reaction, $T=700$ K

In the Arrhenius equation, $k=Ae^{-E_a/RT}$, A is the frequency factor and $e^{-E_a/RT}$ is the fraction of collision having enough energy to overcome the activation energy.

∴ Fraction of molecules having enough energy to react $=e^{-E_a/RT}=e^{-x}$ (given)

$$\therefore\ x=\frac{E_a}{RT}=\frac{80900}{8.31\times700}=13.9\simeq14$$

175. **(526)** Arrhenius equation is, $k=Ae^{-E_a/RT}$

$$\log k=\log A-\frac{E_a}{2.303R}\times\frac{1}{T}$$

Given : Slope $=-10000=\frac{-E_a}{2.303R}$

$$\because \log\frac{k_2}{k_1}=\frac{E_a}{2.303R}\left[\frac{1}{T_1}-\frac{1}{T_2}\right]$$

$$\log\frac{10^{-4}}{10^{-5}}=10000\left[\frac{1}{500}-\frac{1}{T_2}\right]$$

$$\Rightarrow 1=20-\frac{10000}{T_2}\Rightarrow T_2=\frac{10000}{19}=526.3\approx526\text{ K}$$

176. **(52)** $\log\frac{k_2}{k_1}=\frac{E_a}{2.303R}\left[\frac{1}{T_1}-\frac{1}{T_2}\right]$

$$\log5=\frac{E_a}{2.303\times8.314}\left[\frac{1}{300}-\frac{1}{325}\right]$$

$\Rightarrow E_a=52271$ J mol^{-1} = 52.271 kJ mol^{-1}

177. **(100)** The Arrhenius equation is

$$k=Ae^{\frac{E_a}{RT}}$$

Assuming A and E_a to be independent of temperature

$$\ln\frac{k_2}{k_1}=\frac{E_a}{R}\left(\frac{1}{T_1}-\frac{1}{T_2}\right)$$

$$\ln3.555=\frac{E_a}{8.314}\left(\frac{1}{303}-\frac{1}{313}\right)$$

$$\Rightarrow E_a=\frac{1.268\times8.314\times303\times313}{10}$$

$=99980.7=99.98$ kJ / mol

178. **(d)** Arrhenius equation :

$$k=Ae^{-E_a/RT}$$

$$\ln k=\ln A-\left(\frac{E_a}{R}\right)\frac{1}{T}\Rightarrow \ln k=\ln A-\left(\frac{E_a}{R\times10^3}\right)\times\frac{10^3}{T}$$

$$\text{Slope of graph }=\frac{-E_a}{R\times10^3}=\frac{-10}{5}$$

$E_a=2R\times10^3$ J $=2R$ kJ mol^{-1}

179. **(84297.48)**

$\because k=Ae^{-E_a/RT}$

$$\ln\frac{k_2}{k_1}=\frac{E_a}{R}\left(\frac{1}{T_1}-\frac{1}{T_2}\right)\Rightarrow \ln5=\frac{E_a}{R}\left(\frac{1}{300}-\frac{1}{315}\right)$$

$$\Rightarrow E_a=\frac{1.6094\times8.314\times300\times315}{15}=84297.48\text{ J/mol.}$$

180. **(3.98)**

For a first order reaction, $kt=\ln\frac{[A]}{[A_0]}$

At 300 K, $k_1\times60=\ln\frac{[A]}{[A_0]}$ ...(1)

At 400 K, $k_2\times40=\ln\frac{[A]}{[A_0]}$ ...(2)

From equation (1) and (2), $\frac{k_2}{k_1}=\frac{60}{40}$

$$\ln\frac{k_2}{k_1}=\frac{E_a}{R}\left[\frac{1}{T_1}-\frac{1}{T_2}\right];\ \ln\left(\frac{60}{40}\right)=\frac{E_a}{8.3}\times\frac{100}{400\times300}$$

$\Rightarrow E_a=0.4\times8.3\times1200=3984$ J/mol. = 3.984 kJ/mol.

181. **(a)** $A\xrightarrow{700K(k_1)}$ Product

$A\xrightarrow[\text{Catalyst}]{500K(k_2)}$ Product

Given: $k_1=k_2$

$$Ae^{-\frac{E_{a_1}}{RT_1}}=Ae^{-\frac{E_{a_2}}{RT_2}}$$

$$\frac{E_{a_1}}{T_1}=\frac{E_{a_2}}{T_2}$$

$E_{a_2} = E_{a_1} - 30$

$\Rightarrow E_{a1} = E_{a_2} + 30$

$\frac{E_{a_2} + 30}{T_1} = \frac{E_{a_2}}{T_2}$

$\frac{E_{a_2} + 30}{700} = \frac{E_{a_2}}{500}$

$\Rightarrow 150 = 2E_{a_2}$

$E_{a_2} = 75 \text{kJ / mol}$

182. (a) The rate constant of a reaction is given by,

$k = Ae^{-E_a/RT}$

The rate constant in presence of catalyst is given by,

$k' = Ae^{-E_a'/RT}$ or $\frac{k'}{k} = e^{-(E_a' - E_a)/RT}$

$10^6 = e^{-(E_a' - E_a)/RT}$; $\ln 10^6 = -\frac{(E_a' - E_a)}{RT}$

$E_a' - E_a = -6(2.303)RT$

183. (c) Arrhenius equation, $k = Ae^{-E/RT}$

$\log k = \log A - \frac{E}{2.303\ RT}$; slope $= -\frac{E}{2.303\ R}$

$\therefore$ More negative the slope, greater will be the E.

So correct order is: $E_c > E_a > E_d > E_b$

184. (a) So, $\log \frac{k_2}{k_1} = \frac{E_a}{2.303R}\left(\frac{1}{T_1} - \frac{1}{T_2}\right)$

$\log \frac{1}{2.5 \times 10^{-4}} = \frac{E_a}{8.314 \times 2.303}\left(\frac{1}{600} - \frac{1}{800}\right)$

$\Rightarrow \quad 3.6 = \frac{E_a}{8.314 \times 2.303} \times \frac{200}{600 \times 800} \Rightarrow E_a = 165.4 \text{ kJ/mol}$

185. (a) As we can see from the graph that activation enthalpy to form D from A + B is 15 – 5 = 10 kJ mol^{-1}, whereas, to form C from A + B is 20 – 5 = 15 kJ mol^{-1}. Therefore, activation enthalpy to form C is 5 kJ more than that to form D. Due to more E_a, C is thermodynamically stable and D is kinetically stable product.

186. (c) From Arrhenius equation,

$\ln K = \ln A - \frac{E_a}{RT}$

Slope $= \frac{-E_a}{R} = -4606$ K

$\log\left(\frac{K_2}{K_1}\right) = \frac{E_a}{2.303R}\left(\frac{1}{T_1} - \frac{1}{T_2}\right)$

$\log\left(\frac{K_2}{10^{-5}}\right) = \frac{1}{2.303} \times 4606\left(\frac{1}{400} - \frac{1}{500}\right)$

$= \frac{4606 \times (100)}{2.303 \times 400 \times 500} = 1$

$\Rightarrow \quad \log\left(\frac{K_2}{10^{-5}}\right) = 1; \ \frac{K_2}{10^{-5}} = \text{Antilog}(1)$

$\frac{K_2}{10^{-5}} = 10 \Rightarrow K_2 = 10^{-5} \times 10 = 10^{-4} \text{ s}^{-1}$

187. (b) From Arrhenius equation,

$k = Ae^{-E_a/RT}$; $\ln k = \ln A - \frac{E_a}{RT}$

slope $= -y$ (given) $-y = -E_a \Rightarrow E_a = y$

188. (b) From Arrhenius equation,

$k = Ae^{-E_a/RT}$

So, as E_a increases, $e^{-E_a/RT}$ decreases, k decreases

and as T increases, $\frac{E_a}{RT}$ decreases, $e^{-E_a/RT}$ increases.

189. (d) $k_1 = A \cdot e^{-E_{a_1}/RT}$ .....(i)

$k_2 = A \cdot e^{-E_{a_2}/RT}$ .....(ii)

On dividing equation (ii) by (i)

$\Rightarrow \quad \frac{k_2}{k_1} = e^{\frac{(E_{a_1} - E_{a_2})}{RT}}$

$\ln\left(\frac{k_2}{k_1}\right) = \frac{E_{a_1} - E_{a_2}}{RT} = \frac{10,000}{8.314 \times 300} = 4$

Trick: Remember the formula (direct)

$\ln\left(\frac{k_2}{k_1}\right) = \frac{E_{a_1} - E_{a_2}}{RT}$

190. (b) $\log \frac{k_2}{k_1} = \frac{E_a}{2.303R}\left[\frac{T_2 - T_1}{T_1 T_2}\right]$

For reaction *A* -

Given, $\frac{k_2}{k_1} = 2$, $T_1 = 300$ K, $T_2 = 310$ K

$\log 2 = \frac{E_a^A}{2.303R}\left[\frac{1}{300} - \frac{1}{310}\right]$ ...(i)

For reaction *B* -

Given, $\frac{k_2}{k_1} = 2$, $E_a^B = 2E_a^A$, $T_1 = 300$ K, $T_2 = ?$

$\log 2 = \frac{E_a^B}{2.303R}\left[\frac{1}{300} - \frac{1}{T_2}\right]$ ...(ii)

From equation (i) and (ii),

$\frac{2E_a^A}{2.303R}\left[\frac{1}{300} - \frac{1}{T_2}\right] = \frac{E_a^A}{2.303R}\left[\frac{1}{300} - \frac{1}{310}\right]$

$\Rightarrow 2\left[\frac{1}{300} - \frac{1}{T_2}\right] = \frac{310 - 300}{300 \times 310}$

$\Rightarrow T_2 = 304.92$ K

$T_1 = 300$ K, $T_2 = 304.92$ K

$\Delta T = T_2 - T_1 = 4.92$ K.

191. (a) $\ln\frac{k_1}{k_2}=\frac{E_a}{R}\left(\frac{1}{T_1}-\frac{1}{T_2}\right)$

$$\ln 4=\frac{E_a}{8.314}\left(\frac{310-300}{310\times300}\right)$$

$$2\ln 2=\frac{E_a}{8.314}\left(\frac{310-300}{310\times300}\right)$$

$$E_a=\frac{0.693\times2\times8.314\times300\times310}{10}=107.2\text{ kJ/mol}$$

192. (a) $A(g)\rightleftharpoons B(g)\ \Delta H=-40\text{ kJ}$

If activation energy of the reaction = x

since, $\frac{E_f}{E_b}=\frac{2}{3}$, therefore, $E_f=\frac{2x}{5}$ and $E_b=\frac{3x}{5}$

$E_b-E_f=-\Delta H=+40$

$$\Rightarrow\frac{3x}{5}-\frac{2x}{5}=40\Rightarrow x=200$$

Therefore,

$$E_b=\frac{3x}{5}=\frac{3\times200}{5}=120\text{ kJ mol}^{-1}$$

$$E_f=\frac{2x}{5}=\frac{2\times200}{5}=80\text{ kJ mol}^{-1}$$

193. (b) According to Arrhenius equation

$$\log\frac{k_2}{k_1}=\frac{E_a}{2.303R}\left(\frac{1}{T_1}-\frac{1}{T_2}\right)$$

$$\log\frac{1.3\times10^{-3}}{1.3\times10^{-4}}=\frac{E_a}{2.303\times8.314}\left[\frac{1}{373}-\frac{1}{423}\right]$$

$$1=\frac{E_a}{2.303\times8.314}\left[\frac{1}{373}-\frac{1}{423}\right]$$

$E_a=60\text{ kJ/mol}$

194. (b) $X\longrightarrow Y;\ \Delta H=-135\text{ kJ/mol}$,

$E_a=150\text{ kJ/mol}$

For an exothermic reaction,

$E_{a(F.R.)}=\Delta H+E'_{a(B.R.)}$

$\Rightarrow 150=-135+E'_{a(B.R.)}\Rightarrow E'_{a(B.R.)}=285\text{ kJ/mol}$

195. (a) Activation energy can be calculated from the equation

$$\frac{\log k_2}{\log k_1}=\frac{E_a}{2.303R}\left(\frac{1}{T_1}-\frac{1}{T_2}\right)$$

Given $\frac{k_2}{k_1}=2;\ T_2=310\text{ K};\ T_1=300\text{ K}$

$$\therefore\ \log 2=\frac{E_a}{2.303\times8.314}\left(\frac{1}{300}-\frac{1}{310}\right)$$

$E_a=53598.6\text{ J/mol}=53.6\text{ kJ/mol}$.

196. (c) Activation energy can be calculated from the equation

$$\frac{\log k_2}{\log k_1}=\frac{E_a}{2.303\,R}\left(\frac{1}{T_1}-\frac{1}{T_2}\right)$$

Given, $\frac{\log k_2}{\log k_1}=2\quad T_2=308;\quad T_1=298$

$$\therefore\ \log 2=\frac{E_a}{2.303\times8.314}\left(\frac{1}{298}-\frac{1}{308}\right)$$

$E_a=52.9\text{ kJ mol}^{-1}$

197. (c) $k_1=A_1e^{-Ea_1/RT}$ ....(i)

$k_2=A_2e^{-Ea_2/RT}$ ....(ii)

On dividing eqn. (i) by eqn. (ii),

$$\frac{k_1}{k_2}=\frac{A_1}{A_2}\exp.\{(E_{a_2}-E_{a_1})/RT\}\quad\text{....(iii)}$$

Given $E_{a_2}=2E_{a_1}$

On substituting this value in eqn. (iii),

$k_1=k_2A\times e^{Ea_1/RT}\quad[A=A_1/A_2]$

198. (c) In Arrhenius equation, $k=Ae^{-Ea/RT}$, E_a is the energy of activation, which is required by the colliding molecules to react resulting in the formation of products.

199. (d) $k_f=A_f\,e^{-E_f/RT}$ and $k_b=A_b\,e^{-E_b/RT}$

We know that $K_{eq}=\frac{k_f}{k_b}\Rightarrow K_{eq}=\frac{A_f}{A_b}e^{-(E_f-E_b)/RT}$

$$\Rightarrow\ln K_{eq}=\ln\frac{A_f}{A_b}-\frac{(E_f-E_b)}{RT}$$

$$\Rightarrow\ln K_{eq}=\frac{-\Delta H}{R}\left(\frac{1}{T}\right)+\ln\frac{A_f}{A_b}$$

Comparing this equation with $y=mx+c$

$$m=\frac{-\Delta H}{R}=+ve$$

Thus, $\Delta H<0$ or reaction is exothermic.

200. (d) Enthalpy of reaction $(\Delta H)=E_f-E_b$

For an endothermic reaction, $\Delta H=+ve$, hence, $E_b<E_f$

201. (b) In equation $k=Ae^{-E_a/RT}$;

A = Frequency factor

k = velocity constant, R = gas constant and E_a = energy of activation

The p-Block Elements (Group 15, 16, 17 & 18)

4

1. (d) Nitrogen can exhibit +5 oxidation state in its oxides because it can form pπ–pπ bond with oxygen which helps to stabilize the higher oxidation state. e.g. N_2O_5

$$O=\overset{+5}{N}(-O^-)-O-\overset{+5}{N}(=O)-O$$

Nitrogen cannot form halide in +5 oxidation state because it does not contain d-orbital. Being 2nd period element N has only s and p orbitals which cannot expand its octet. e.g. NX_5 does not exist where X = halide

2. (b) $\ddot{N}-\ddot{N}$ single bond is weaker than $\ddot{P}-\ddot{P}$ bond due to more ℓp-ℓp repulsion. Bond length of P-P bond is greater than N-N bond.

In group 15 elements only nitrogen and phosphorus show disproportionation in +3 oxidation state. As, Sb and Bi are almost inert for disproportionation in +3 oxidation state. So both statements I and II are false.

3. (a) Group of phosphate mineral are called apatite family phosphorous is main component of phosphate mineral which have ionisation enthalpy 1012 kJ mol^{-1}.

4. (15) Phosphorus belongs to 15th group and forms dπ–dπ bond with transition metal and PH_3 is strongest base among the other group members.

5. (d) SO_2 can oxidise as well as reduce as it is in + 4 state. Hence it can act as both oxidising and reducing agent.

6. (d) N – N : Single (σ) bond strength is lowest due to L.P. - L.P. repulsion. For N, the maximum possible accomodation is $8e^-$ thus covalency of nitrogen is 4.

7. (d)

(D) Due to inert pair effect lower oxidation state is more stable.

(E) Nitrogen belongs to 2nd period and cannot expand its octet due to un availability of d-orbital

8. (d) PH_3 has lower boiling point than NH_3 because NH_3 in liquid state have hydrogen bonding between molecules. PH_3 does not form hydrogen bonds.

9. (c) The stability of hydrides decreases from NH_3 to BiH_3 but their reducing character increases down the group.

10. (c) The E–H bond strength in group 15 follows the order:

$N-H > P-H > As-H > Sb-H > Bi-H$

Thus, the thermal stability will follow:

$NH_3 > PH_3 > AsH_3 > SbH_3 > BiH_3$

The reducing power will follow:

$NH_3 < PH_3 < AsH_3 < SbH_3 < BiH_3$

The basicity follows the oder:

$NH_3 > PH_3 > AsH_3 > SbH_3 > BiH_3$

This is due to N being unable to easily accommodate a lone pair due to small size while the larger atoms being able to do so.

11. (a) $$CH_4(g) + H_2O(g) \xrightarrow[\text{Steam}\quad 1270\,K]{Ni} CO(g) + 3H_2(g)$$

$$NaNO_2(aq) + NH_4Cl(aq) \longrightarrow N_2(g) + NaCl(aq) + 2H_2O(\ell)$$

12. (c) $N_2(g) + O_2(g) \rightarrow 2NO(g)$

$2NO(g) + O_2(g) \rightarrow 2NO_2(g)$

Higher concentration of NO_2 damage the leaves and retard the rate of photosynthesis.

13. (d) The reaction between dinitrogen and dioxygen is an endothermic process and feasible at high temperature.

$$N_2 + O_2 \xrightleftharpoons{(1483-2000K)} 2NO$$

14. (d) Red phosphrous is obtained by heating white phosphorous at 573 K in inert atmosphere. When red phosphorous is heated in a sealed tube at 803 K, we get α-black phosphorous.

β-Black phosphorous is prepared by heating white phosphorous at 473 K under high pressure.

15. (b) $Ba(N_3)_2 \longrightarrow Ba + 3N_2 \uparrow$

$Ba(NO_3)_2 \longrightarrow 2BaO + 4NO_2 + O_2 \uparrow$

$2NaNO_3 \longrightarrow 2NaNO_2 + O_2 \uparrow$

$2NaNO_2 \longrightarrow Na_2O + NO \uparrow + NO_2 \uparrow$

16. (d) As positive oxidation state increase, the acidic character increases.

Pentavalent oxides, E_2O_5 has E in +5 oxidation state.

Trivalent oxides; E_2O_3 has E in +3 oxidation state.

So, Acidic character $E_2O_5 > E_2O_3$

Down the group non-metallic/acidic character decreases.

17. (b) The reaction result in formation of nitrogen gas.

$$NH_4Cl(aq) + NaNO_2(aq) \rightarrow NH_4NO_2(aq) + NaCl(aq)$$
$$NH_4NO_2(aq) \rightarrow N_2\uparrow + 2H_2O(l)$$

18. (a) Due to increase in bond length, order of stability : $NF_3 > NCl_3 > NBr_3 > NI_3$

19. (d) When red phosphorus is heated in a sealed tube at 803 K, α-black phosphorus is formed.

20. (b) NH_3 is only a mild reducing agent while BiH_3 is the strongest reducing agent amongst all the hydrides due to the longest bond length.

21. (c) The reducing character of the hydrides of group 15 elements increases from NH_3 to BiH_3 because the reducing character depends upon the stability of hydrides. The greater the unstability of the hydrides, the greater is its reducing character, the BiH_3 is the least stable in the series, hence it has strongest reducing power.

22. (d)
(a) N_2 is diamagnetic in nature.
(b) $N_2 + O_2 \xrightleftharpoons{2000K} NO(g)$
(c) Liquid N_2 is used in cryosurgery.
(d) Because of its inertness, it is used where an inert atmosphere is required.

23. (d) $\underset{\text{(excess)}}{NH_3} + 3Cl_2 \longrightarrow NCl_3 + 3HCl$
If NH_3 is used in excess then NH_4Cl is formed instead of NCl_3.
$\underset{\text{(excess)}}{8NH_3} + 3Cl_2 \longrightarrow 6NH_4Cl + N_2$

24. (b) No. of pentagons in C_{60} (Buckminsterfullerene) = 12
No. of triangles in P_4 (White phosphorous) = 4

25. (d) (a) $Ba(N_3)_2 \xrightarrow{\Delta} Ba + 3N_2$
(b) $(NH_4)_2Cr_2O_7 \xrightarrow{\Delta} Cr_2O_3 + N_2 + 4H_2O$
(c) $NH_4NO_2 \xrightarrow{\Delta} N_2 + 2H_2O$
(d) $(NH_4)_2SO_4 \xrightarrow{\Delta} 2NH_3 + H_2SO_4$

26. (c) Hint: Number of moles $= \frac{\text{Weight}}{\text{Molecular mass}}$

$\underset{\substack{221g \\ 1\,g}}{Ba(N_3)_2(s)} \longrightarrow Ba(s) + \underset{\substack{3\text{ moles} \\ 3/221\text{ moles}}}{3N_2(g)}$

$\underset{1g}{(NH_4)_2Cr_2O_7(s)} \longrightarrow \underset{1/252\text{ moles}}{N_2(g)} + 4H_2O(g) + Cr_2O_3(s)$

$\underset{1g}{2\,NH_3(g)} \longrightarrow \underset{1/34\text{ moles}}{N_2(g)} + 3H_2(g)$

$\underset{1g}{2NH_4NO_3(s)} \longrightarrow \underset{1/80\text{ moles}}{2N_2(g)} + 4H_2O(g) + O_2(g)$

Hence, thermal decomposition of NH_3 will produce maximum amount of N_2.

27. (a) $6\,Li\,(s) + N_2\,(g) \underset{\text{high temperature}}{\rightleftharpoons} 2Li_3N\,(s)$

$Li_3N(s) + 3H_2O(l) \longrightarrow 3LiOH(aq) + NH_3(g)$

$CuSO_4(aq) + 4NH_3(g) + 2H_2O(l) \longrightarrow \underset{\text{deep blue compound}}{[Cu(NH_3)_4(OH_2)_2]SO_4(aq)}$

28. (c) All the members form volatile halides of the type AX_3. All halides are pyramidal in shape. The bond angle decreases on moving down the group (from NCl_3 to $SbCl_3$) due to decrease in bond pair-bond pair repulsion or increase in lone pair-bond pair repulsion.

29. (a) $(NH_4)_2SO_4 + 2H_2O \longrightarrow 2H_2SO_4 + NH_4OH$
H_2SO_4 is strong acid and increases the acidity of soil.

30. (b) Among the given compounds, $\ddot{N}H_3$ is most basic. Hence it has highest proton affinity.

31. (a) ${}_7N = 1s^2\,2s^2\,2p^3$; ${}_{15}P = 1s^2\,2s^2\,2p^6\,3s^2\,3p^3$
Note: In phosphorus the $3d$-orbitals are available. Hence phosphorus can form pentahalides but nitrogen can not form pentahalide due to absence of d-orbitals.

32. (d)
Acidic strength : $H_2Se < H_2Te$
$\Delta_{dis}H$: $\underset{[276\,kJ/mol]}{H_2Se} > \underset{[238\,kJ/mol]}{H_2Te}$

33. (b) Oxygen exists as O_2 (Atomicity = 2)
Sulphur exists as S_8 (Atomicity = 8)
Hence, Melting point & Boiling point of sulphur are significantly large compared to oxygen.

34. (c) Oxygen can from $2p\pi$-$2p\pi$ multiple bond with itself due to its small size while sulphur cannot form multiple bond with itself as $3p\pi$-$3p\pi$ bond will be unstable due to large size of sulphur, but sulphur can form multiple bond with small size atom like C and N.
eg. S=C=S
Thus, (A) is correct but (R) is incorrect

35. (7) Fuming sulphuric acid is a mixture of conc. H_2SO_4 + SO_3 or $H_2S_2O_7$
So, Number of Oxygen atoms = 7

36. (b) H_2Te is more acidic than H_2S as H – Te bond is weaker than H – S bond due to a greater size difference between H and Te atom.
Thus, both A and R are true and R explains A.

37. (a) The first member of a group exhibits anomalous behaviour and shows different physical and chemical properties than the rest of the elements due to smallest size and highest electronegativity.
E.g. Oxygen is diatomic while other members are poly atomic.

38. (d) A and B only
All group 16 element form oxides of genral formula EO_2 and EO_3, Both are acidic in nature.
The reducing property increases from H_2S to H_2Te down the group.

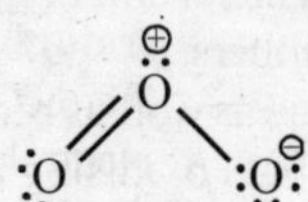

Ozone molecule contain total of six lone pair of electerons.

39. (c) **Statement (I) :** Oxygen can have oxidation state from –2 to +2
O oxidation state in elemental form –2 oxidation state in normal oxidate –1 oxidation state in peroxide +2 → oxidation state in OF_2.
So statement I is incorrect
Statement (II) : On moving down the group stability of +4 oxidation state increases whereas stability of +6 oxidation state decreases down the group, according to inert par effect.
So both statements are incorrect.

40. (a) Bond dissociation energy of E–H bond in hydrides of group 16 follows the order:
$H_2O > H_2S > H_2Se > H_2Te$

41. (a) Water molecule is significantly more polar than the other group 16 hydride molecules so the attraction between $H^{\delta+}$ and $O^{\delta-}$ between water molecules is stronger than the attraction of $H^{\delta+}$ and $E^{\delta-}$ between other group 16 hydride molecules.

42. (1) α-sulphur and β-sulphur are diamagnetic. S_2-form is paramagnetic similar to O_2.

43. (c) X = Na_2SO_3
$\underset{(X)}{SO_3^{2-}} + H_2SO_4 \rightarrow [H_2SO_3] \rightarrow H_2O + \underset{(Y)}{SO_2}$

$SO_2 + 2NaOH \rightarrow \underset{(X)}{Na_2SO_3} + H_2O$

$\underset{(X)}{Na_2SO_3} + \underset{(Y)}{SO_2} + H_2O \rightarrow \underset{(Z)}{2NaHSO_3}$

44. **(a)** At room temperature, water is liquid and has boiling point 373 K due to hydrogen bonding. Whereas H_2S is a gas and it has no hydrogen bonding. Hence boiling point of H_2S is less than 300 K (boiling point of H_2S is –60°C).

45. **(a)** The S_6 molecule has a chair-form hexagon ring with the approx same bond length as that in S_8, but with some what smaller bond angles, i.e., bond lengths are approx. same but bond angles are different.

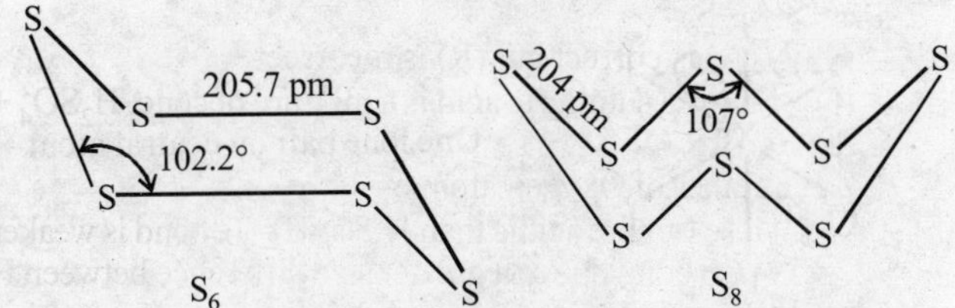

46. **(c)** F < Cl < Br < I (Covalent radius)
Cl > F > Br > I (Electron affinity) (Irregularity)
$F^- < Cl^- < Br^- < I^-$ (Ionic radius)
F > Cl > Br > I (Ist ionization energy)

47. **(a)** Fluoroapatite ⇒ $[3(Ca_3(PO_4)_2).CaF_2]$
Hydroxyapatite ⇒ $[3(Ca_3(PO_4)_2).Ca(OH)_2]$

48. **(d)** N,O, F can't extend their valencies upto their group number due to the non-availability of vacant d orbitals.

49. **(b)** Correct order of Ist IE

Li < B < Be < C < O < N < F < Ne
E < C < A < B < D

50. **(3)**

Oxides	Nature
Cl_2O_7	Acidic
CO	Neutral
PbO_2	Amphoteric
N_2O	Neutral
NO	Neutral
Al_2O_3	Amphoteric
SiO_2	Acidic
N_2O_5	Acidic
SnO_2	Amphoteric

51. **(d)** Fluorospar is CaF_2. Therefore option (d) is incorrect pair.

52. **(a)** Fluorine is the most electronegative element but due to its small size, it is not better accommodating to an electron and therefore Cl being larger in size has a higher negative value of the electron gain enthalpy.
Thus, Statement I is false.
In group 16, oxygen has the least negative electron gain enthalpy while sulphur has the most negative electron gain enthalpy.
Thus, Statement II is true.

53. **(d)** Fluorine does not show variable oxidation state as it is the most electronegative element and shows only –1 state.

54. **(a)** Freons are chlorofluorocarbons.

55. **(1)** IF_5 has 1 lone pair and IF_7 has 0 lone pair
∴ 1 + 0 = 1

56. **(b)** (a) $\Delta_{eg}H(Cl) < \Delta_{eg}H(F)$
(–349) (–333) Correct
(b) $\Delta_{eg}H(Se) < \Delta_{eg}H(S)$
(–195) (–200) Incorrect
(c) $\Delta_{eg}H(I) < \Delta_{eg}H(At)$
(–296) (–270) Correct
(d) $\Delta_{eg}H(Te) < \Delta_{eg}H(Po)$
(–190) (–174) Correct

57. **(a)** Bond energy of F_2 less than Cl_2 due to lone pair-lone pair repulsions and small size of fluorine.
∴ Bond energy order $Cl_2 > Br_2 > F_2 > I_2$

58. **(3)** AB_5 species have square pyramidal structures.
Hence, BrF_5, IF_5 and ClF_5.

59. **(4)**
(i) Due to unpaired e^- in O_2^-, KO_2 is paramagnetic.
(ii) $^{\ominus}O-\overset{+}{N}=O$ is paramagnetic due to unpaired e^- with N atom.
(iii) $O=\ddot{Cl}=O$ is paramagnetic due to unpaired e^- with Cl atom.
(iv) $\cdot\ddot{N}=O$ is paramagnetic.

60. **(a)** Due to small size and high electronegativity of fluorine it can not expand its oxidation state and show only one oxoacid i.e., HOF.

61. **(a)** Bond dissociation energy refers to the energy required to break bond between two atoms. Its value depends on the strength of the bonds. As size of atom increases, its bond strength and dissociation energy decreases. So, dissociation energy decreases down the group as size increases.
In the halogen family, size of F is very small, so it faces a strong electron repulsion between its electrons, which make its bond weak, thus, reducing its dissociation energy. Therefore, the correct order is: $Cl_2 > Br_2 > F_2 > I_2$

62. **(a)** Generally, bond energy $\propto \dfrac{1}{\text{Bond length}}$
So, bond energy order is : C–F > C–Cl > C–Br > C–I

63. **(c)** Due to strong H-bonding between HF molecules. HF has highest boiling point among the hydrogen halides.

64. **(c)** Fluorine is most electronegative element in periodic table and exhibits O.S. value of –1 only.

65. **(a)** NaF is composed of Na^+ and F^-.
$Na^+ \longrightarrow [He]2s^2 2p^6$
$F^- \longrightarrow [He]2s^2 2p^6$
Hence, configuration of Na^+ and F^- do not match with the configuration given in the question.

66. **(b)** Order of reactivity of halogens: $Cl_2 > Br_2 > I_2$
But the interhalogen compounds are generally more reactive than halogens (except F_2), since the bond between two dissimilar electronegative elements is weaker than the bond between two similar atoms, i.e, X – X.

67. **(b)** The molecular geometry of BrF_5 is square pyramidal with asymmetric charge distribution on the central atom.

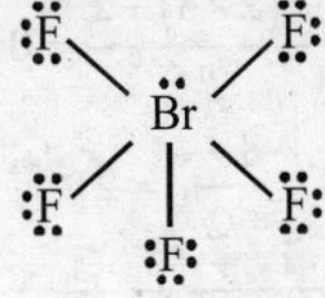

68. **(c)** The electron gain enthalpy order for halogens is
Cl > F > Br > I
Due to small size of fluorine the extra electron to be added feels more electron-electron repulsion. Therefore fluorine has less value for electron affinity than chlorine.

69. **(c)** The H–X bond strength decreases from HF to HI, i.e., HF > HCl > HBr > HI. Thus HF is most stable while HI is least stable. The decreasing stability of the hydrogen halide is also reflected in the values of dissociation energy of the H–X bond:

H – F	H – Cl	H – Br	H – I
135kcal mol^{-1}	103kcal mol^{-1}	87kcal mol^{-1}	71kcal mol^{-1}

70. **(a)** The oxidising nature of fluorine is due to its small size that result in high enthalpy of hydration

71. **(d)** Statement I and II are False.
Noble gases have low boiling points and they are held together by weak dispersion forces.

72. **(4)** XeF_5^+, ; XeO_3,

XeO_2F_2,

$XeOF_4$, ; XeF_4

So, Answer is 4

73. **(c)**

Element	$\Delta_{eg}H$[KJ / mol]
He	+48
Ne	+116
Kr	+96
Xe	+77

74. **(d)** Number of lone pair of electrons on the central atom

$$= \frac{\text{Number of valence electrons on central atom 2(Number of bivalent atoms) (Number of monovalent atoms)}}{2}$$

Species	No. of lone pair
XeF_2	$n = (8-2)/2 = 3$
XeO_2F_2	$n = \frac{8-(2\times2)-2}{2} = 1$
XeO_3F_2	$n = \frac{8-(2\times3)-2}{2} = 0$
XeF_4	$n = \frac{8-4}{2} = 2$

75. **(3)** S.N of Xe = 4 (3B.P + 1ℓ.P)

– One lone pair on central atom

S.N of Xe = 6 (5B.P + 1ℓ.P)

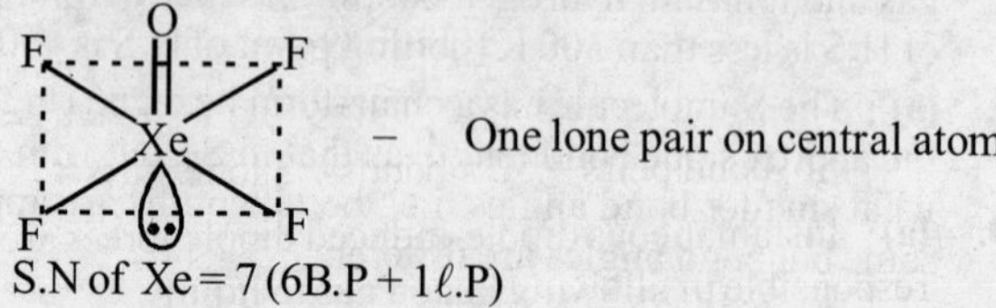

S.N of Xe = 7 (6B.P + 1ℓ.P)

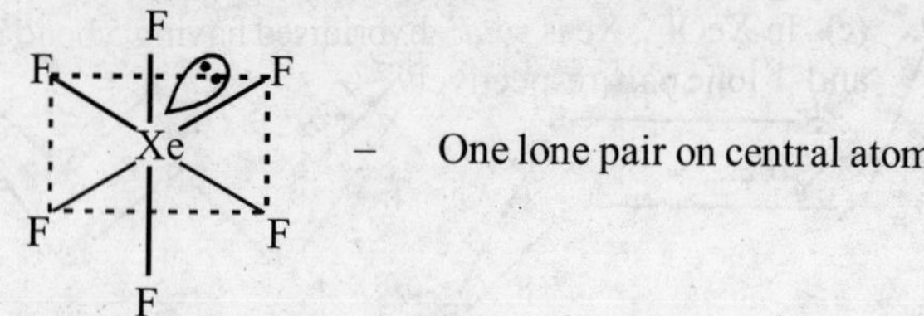

Total lone pair = 1 + 1 + 1 = 3

76. **(a)**

(A) Steric number of Xe = 4 (3B.p. + 1 l.p.)
Hybridization = sp^3 ; shape = pyramidal
geometry = tetrahedral

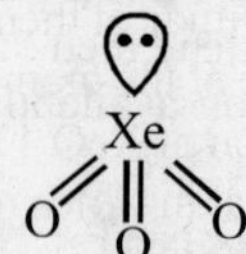

(B) Steric number of Xe = 5 (2B.p. + 3 l.p.)
Hybridization = sp^3d ; shape = linear
geometry = trigonal bipyrimidal

(C) Steric number of Xe = 6 (5B.p. + 1 l.p.)
Hybridization = sp^3d^2; shape = square pyramidal
geometry = octahedral

(D) Steric number of Xe = 7 (6B.P. + 1 l.p.)
Hybridization = sp^3d^3
Shape = distorted octahedral

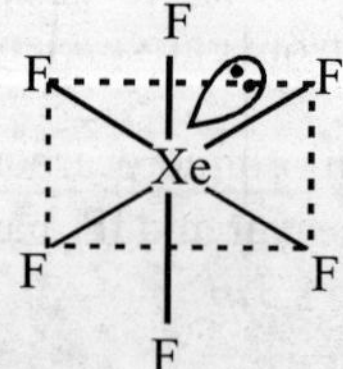

77. **(d)** Radon is radioactive element and not present in atmosphere.

78. **(a)** Structure of XeO_3F_2

So, bond pairs = 5, π bonds = 3 lone pairs = 0

79. **(a)** Instantaneous dipole-induced dipole forces are most responsible in allowing xenon gas to liquify.

80. **(c)** In $XeOF_4$, Xe is sp^3d^2, hybridised having 5 bond pairs and 1 lone pair respectively.

$XeOF_4$
Square pyramidal
(sp^3d^2)

81. **(b)** As we move down the group, the melting and boiling points show a regular increase due to corresponding increase in the magnitude of their van der Waals forces of attraction as the size of the atom increases.

82. **(b)** The hybridization of XeO_3F_2 is sp^3d and its structure is trigonal bipyramidal in which oxygen atoms are situated on the plane and the fluoride atoms are on the top and bottom.

83. **(b)** The shape of XeO_3F_2 is trigonal Pyramidal.

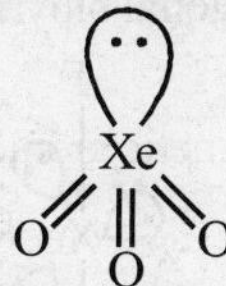

84. **(b)** XeF_4 has zero dipole moment. It has square planar structure due to which the bond moments of Xe – F cancel each other.

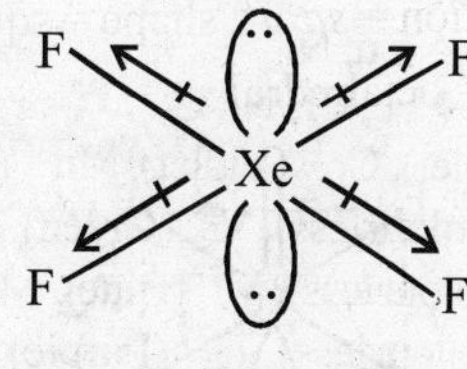

85. **(c)** XeF_2 :

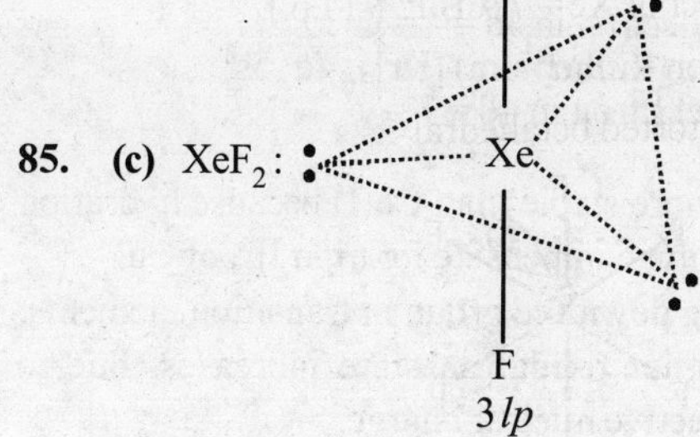

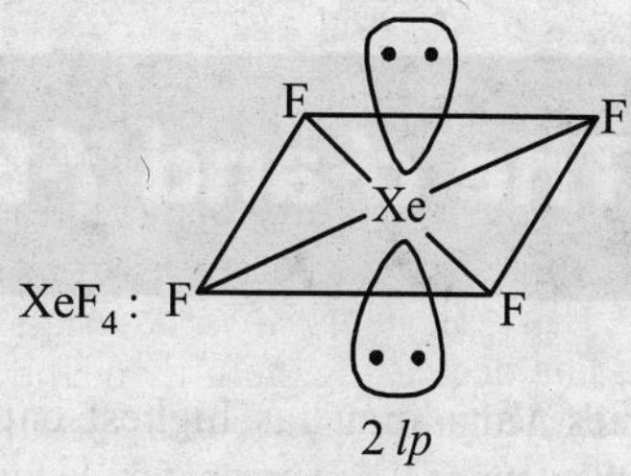

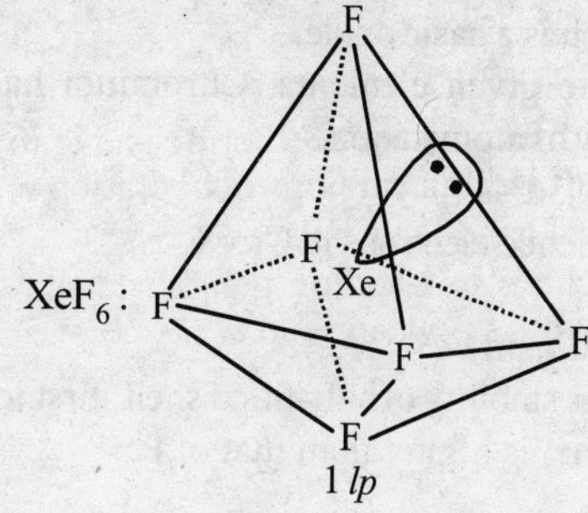

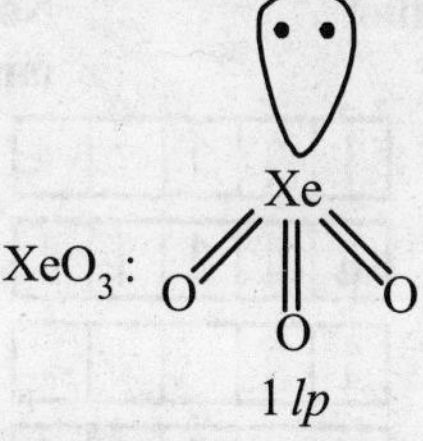

86. **(d)** In the formation of XeF_2, sp^3d hybridisation occurs which gives the molecule a trigonal bipyramidal structure.

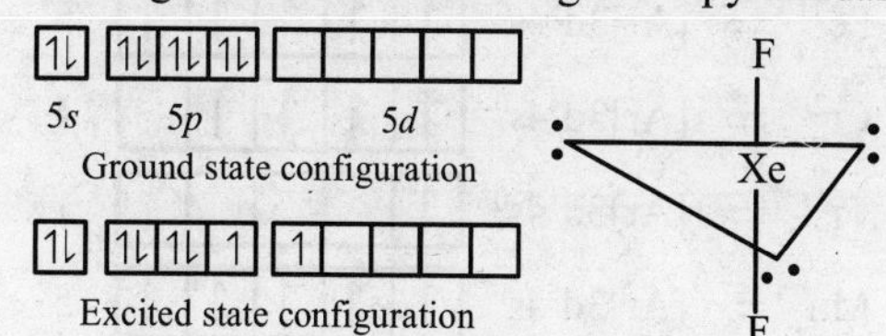

In the formation of XeF_4, sp^3d^2 hybridization occurs which gives the molecule an octahedral structure.

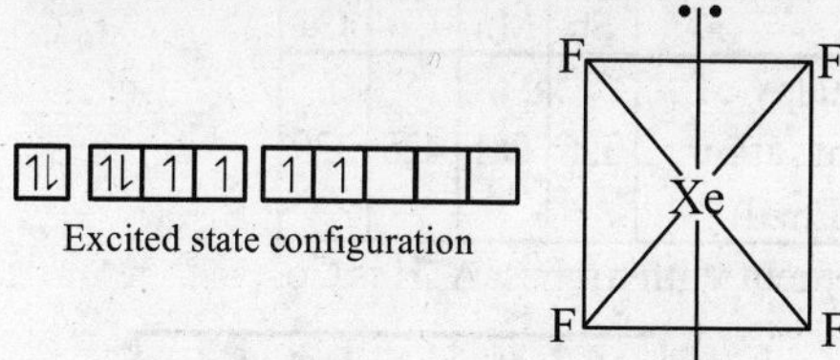

In the formation of XeF_6, sp^3d^3 hybridization occurs which gives the molecule a pentagonal bipyramidal structure.

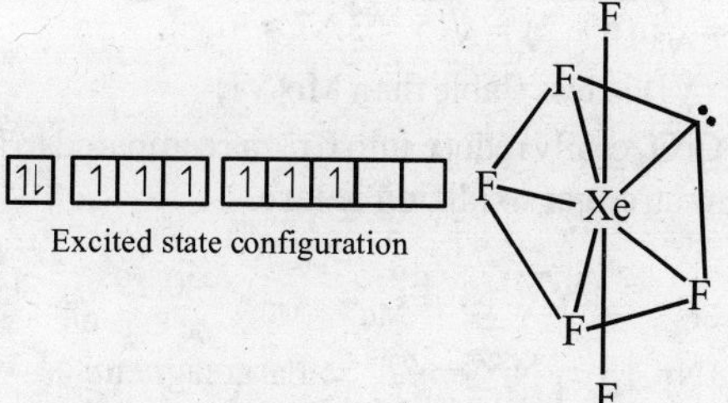

87. **(c)** Helium is heavier than hydrogen although it is non-inflammable.

The d-and f-Block Elements

1. **(c)** In 3d series Vanadium has highest enthalpy of atomization and colour of V^{3+} is green. Oxide form by V^{3+} is V_2O_3 which is a basic oxide.

2. **(c)** From the given elements, Chromium has lowest enthalpy of atomisation.
$Cr = [Ar]3d^5 4s^1$
$\therefore$ Total valence electrons in Cr = 6

3. **(d)**
$Mn^{2+} : [Ar]3d^5$ and $Fe^{2+} : [Ar]$
Due to greater stability of half-filled shell, first ionization energy of Mn^{2+} is higher than that of Fe^{2+}.

4. **(a)**

		Configuration		No. of unpaired e^-
(1) V^{3+}	$\Rightarrow$	$[Ar]3d^3 4s^0$	1 1 1 _ _	3
Co^{2+}	$\Rightarrow$	$[Ar]3d^7 4s^0$	1l 1l 1 1 1	3
(2) Ti^{2+}	$\Rightarrow$	$[Ar]3d^2 4s^0$	1 1 _ _ _	2
Co^{2+}	$\Rightarrow$	$[Ar]3d^7 4s^0$	1l 1l 1 1 1	3
(3) Fe^{3+}	$\Rightarrow$	$[Ar]3d^5 4s^0$	1 1 1 1 1	5
Cr^{2+}	$\Rightarrow$	$[Ar]3d^4 4s^0$	1 1 1 1 _	4
(4) Ti^{3+}	$\Rightarrow$	$[Ar]3d^1 4s^0$	1 _ _ _ _	1
Mn^{2+}	$\Rightarrow$	$[Ar]3d^5 4s^0$	1 1 1 1 1	5

So V^{2+} and Co^{2+} have same number of unpaired electrons.

5. **(4)**

	Sc	Mn	Co	Cu
Enthalpy of Atomisation (kJ/mol)	326	281	425	339

The element with highest $\Delta_a H$ is Co.

Co^{2+} : [Ar] $3d^7$; 1l 1l 1 1 1

$\Rightarrow n = 3$

$\mu = \sqrt{n(n+2)} = \sqrt{15} = 3.87 \approx 4$

6. **(b)** Cr (VI) is less stable than Mo(VI)
Hence, CrO_3 easily reduce into Cr^{+3} as compared to MoO_3 and show stronger oxidizing nature.

$E^o_{Cr^{6+}/Cr^{3+}} = +0.77V$; $E^o_{Mo^{6+}/Mo^{3+}} - 0.19V$

7. **(0)** $K_2[NiCl_4] \Rightarrow Ni^{2+} \Rightarrow 3d^8 \Rightarrow$ Paramagnetic
$[Zn(H_2O)_6]Cl_2 \Rightarrow Zn^{2+} \Rightarrow 3d^{10} \Rightarrow$ diamagnetic
$K_3[Mn(CN)_6] \Rightarrow Mn^{3+} \Rightarrow 3d^4 \Rightarrow$ paramagnetic
$[Cu(PPh_3)I] \Rightarrow Cu^{+1} \Rightarrow 3d^{10} \Rightarrow$ diamagnetic
Here two diamagnetic complexes
$\therefore n = 2$
Zn has least enthalpy of atomisation
$Zn^{2+} \Rightarrow \mu = 0$

8. **(c)** In any row the melting points of transition metals rise to maximum at d^5 except for anomalous values of Mn and Tc and fall regularly as the atomic number increases.
This trend is not followed by elements of 5d series due to relativistic effects and number of unpaired electrons, leading to stronger metallic bonding in Re
Therefore melting point order is:
Mn < Fe, Tc < Ru, Os < Re

9. **(b)** ClF_3

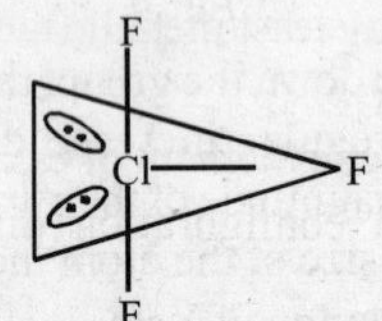

n = 2 (No of lone pair present in equatorial position)

		(Unpaired e^-)
(A)	$V^{+3} : [Ar]3d^2$	2
(B)	$Ti^{3+} : [Ar]3d^1$	1
(C)	$Cu^{+2} : [Ar]3d^9$	1
(D)	$Ni^{+2} : [Ar]3d^8$	2
(E)	$Ti^{+2} : [Ar]3d^2$	2

10. **(b)** Spin only magnetic moment $(\mu) = \sqrt{n(n+2)}$
$Sc^{+3} = 3d^0$ $\therefore$ $\mu_{spin} = 0$
$V^{+2} = 3d^3$ $\therefore$ $\mu_{spin} = 3.87$ B.M.
$Ni^{+2} = 3d^8$ $\therefore$ $\mu_{spin} = 2.84$ B.M.
$Ti^{+3} = 3d^1$ $\therefore$ $\mu_{spin} = 1.73$ B.M.

11. **(b)** Bronze $\rightarrow$ Cu, Sn
Brass $\rightarrow$ Cu, Zn
UK silver coin $\rightarrow$ Cu, Ni
Stainless steel $\rightarrow$ Fe, Cr, Ni, C

12. **(a)** (a) V^{2+} (Violet), Cr^{3+} (Violet), Mn^{3+} (Violet)
(b) Zn^{2+} (Colourless), V^{3+} (Green), Fe^{3+} (Yellow)
(c) Ti^{4+} (Colourless), V^{4+} (Blue), Mn^{2+} (Pink)
(d) Sc^{3+} (Colourless), Ti^{3+} (Purple), Cr^{2+} (Blue)

13. **(11)** $z = 41 \rightarrow$ Nb (Niobium) : $[Kr]_{36} 4d^4 5s^1$
Number of electron in 4d = 4 = x
$Z = 44 \rightarrow$ Ru (Ruthenium) $[Kr]_{36} 4d^7 5s^1$
Number of electron in 4d = 7 = y
x + y = 11

14. **(c)** Cu(II) is more stable than Cu(I) because hydration energy of Cu^{+2} ion compensate for high IE_2 of Cu.

15. **(c)** On moving down the group in transition elements, stability of higher oxidation state increases, due to increases in effective nuclear charge.

E.g:- Mo(VI), W(VI) etc.

$\Rightarrow E^{\circ}_{Cu^{+2}/Cu} = 0.34V$

$\Rightarrow E^{\circ}_{H^+/H_2} = 0$

SRP: $Cu^{2+} > H^+$. [SRP: Standard Reduction Potential]

16. **(6)** Among given metals, Cr has maximum IE_2 because second electron is removed from stable configuration $3d^5$.
Cr^+: [ar] $3d^5\,4s^0$.
$\therefore$ No of unpaired e^- in Cr^+ is 5, n = 5
So, Magnetic moment $= \sqrt{n(n+2)}$ B.M.
$= \sqrt{5(5+2)} = 5.92 \text{ BM} \approx 6$

17. **(d)** Reactions involve are:
$2Fe^{3+} + 2I^- \longrightarrow 2Fe^{2+} + I_2$
$2Fe^{2+} + S_2O_8^{2-} \longrightarrow 2Fe^{3+} + 2SO_4^{2-}$
Fe^{3+} oxidises I^- to I_2 and convert itself into Fe^{2+}. This Fe^{2+} reduces $S_2O_8^{2-}$ to SO_4^{2-} and converts itself into Fe^{3+}.

18. **(0)** Metal having least metallic radii among Sc, Ti, V, Cr, Mn and Zn is Cr.
Spin only magnetic moment of CrO_4^{2-}.
Here Cr^{+6} is in d^0 configuration (diamagnetic).
$\mu = \sqrt{n(n+2)}$ BM
$n = 0 \Rightarrow \mu = 0$

19. **(d)** Electronic configuration of the given metals
(A) $Sc \Rightarrow 3d^1\,4s^2 \Rightarrow n = 1$
(B) $Cr \Rightarrow 3d^5\,4s^1 \Rightarrow n = 6$
(C) $V \Rightarrow 3d^3\,4s^2 \Rightarrow n = 3$
(D) $Ti \Rightarrow 3d^2\,4s^2 \Rightarrow n = 2$
(E) $Mn \Rightarrow 3d^5\,4s^2 \Rightarrow n = 5$
$\therefore$ Correct order of increasing number of unpaired electrons.
(A) < (D) < (C) < (E) < (B)

20. **(0)** VO_2^+ has least oxidising ability due to more stability of their octate in penultimate shell.
$VO_2^+ \Rightarrow V^{+5} \Rightarrow 3d^0\,4s^0$
There is no unpaired electron present. So, magnetic moment = 0

21. **(5)** Among the given species
Strongest oxidising agent is Co^{3+}
No. of unpaired e^- in $Co^{3+}[3d^6] = 4$
Hence $\mu = \sqrt{n(n+2)} = \sqrt{24}$ BM = 4.89 BM
Nearest integer = 5 BM

22. **(b)** Mn shows highest oxidation state of + 7 in 3d series metals. Ti, Co and Fe shows oxidation state of +4, +4 and +6 respectively.

23. **(6)** The metaborates of transition metals have characteristic colours.
Fe^{3+} will give green coloured bead when heated at point B.
Number of unpaired electrons in $Fe^{3+} = 5$
$\mu = \sqrt{n(n+2)}$ BM $= \sqrt{5(5+2)}$ BM = 5.91 BM
Nearest integer = 6.

24. **(c)** Ions with negative SRP can liberate H_2 from a dilute acid. Ti^{2+}, Cr^{2+}, V^{2+} all can liberate H_2 gas from a dilute acid.

25. **(b)** Co, Ti, Ni can show +2, +3 and +4 oxidation state, But 'Sc' only shows +3 stable oxidation state.

26. **(d)** $M = 3.86 = \sqrt{n(n+2)}$
$\Rightarrow n = 3$ unpaired electrons.
${}_{22}Ti^{2+} \Rightarrow [Ar]3d^2$
${}_{23}V^{2+} \Rightarrow [Ar]3d^3$
${}_{25}Mn^{2+} \Rightarrow [Ar]3d^5$
${}_{26}Fe^{2+} \Rightarrow [Ar]3d^6$

27. **(0)** 'V' has highest enthalpy of atomisation (515 kJ/mol) among first row transition elements.
V_2O_5 is amphoteric oxide.
Here 'V' is in +5 oxidation state
$V^{+5} \Rightarrow 1s^2\,2s^2\,2p^6\,3s^2\,3p^6$ (no unpaired electrons)

28. **(a)** ${}_{25}Mn = [Ar]\,3d^5\,4s^2$
3rd Ionization enthalpy will be highest for Mn because when 3rd electron is removed it will loose its stability of half filled electron system.

29. **(b)** Both (A) and (R) are true and (R) is the correct explanation of (A)
Cr^{2+} is reducing as it configuration changes from d^4 to d^3 due to formation of Cr^{+3}, which has half filled t_{2g} level, on other hand the change Mn^{3+} to Mn^{2+} result half filled d^5 configuration which has extra stability.

30. **(b)** (A) – (III), (B) – (IV), (C) – (I), (D) – (II)
(A) $Cr^{2+} \to 3d^4$
(B) $Mn^+ \to 3d^5\,4s^1$
(C) $Ni^{2+} \to 3d^8$
(D) $V^+ \to 3d^3\,4s^1$

31. **(d)** $TiCl_4$ with $Al(CH_3)_3$ forms the basis of Ziegler catalysts. Blood pigment (haemoglobin) is a coordination compound of iron. Vitamin B_{12} (cyanocobalamine) is a coordination compound of cobalt. Wilkinson catalyst $[(Ph_3P)_3RhCl]$ is a complex of rhodium.

32. **(d)** Zn, Cd and Hg have low values of atomization enthalpies due to all electrons being paired which results in weaker bonding.
Zn and Cd show +2 state while Hg shows +1 and +2 states in Hg_2^{2+} and Hg^{2+} respectively.
The compounds of Zn, Cd and Hg are generally diamagnetic due to absence of unpaired electrons.
Zn, Cd and Hg are soft metals due to weak metallic bonding.
Thus, B and D are correct.

33. **(d)** $\mu = \sqrt{n(n+2)}$ B.M.

	$3d^7$	$3d^8$	$3d^3$	$3d^6$
No. of unpaired e^-	3	2	3	4
Spin only magnetic moment	$\sqrt{15}$ BM	$\sqrt{8}$ BM	$\sqrt{15}$ BM	$\sqrt{24}$ BM

34. **(b)**

35. **(d)** The catalyst used in Wacker's process is $PdCl_2$
The process of oxidation of terminal alkenes to aldehydes and ketones by using palladium and copper compounds as catalysts. Palladium chloride and copper chloride are used as catalyst.

36. **(a)** (A) The M^{3+}/M^{2+} reduction potential for manganese is greater than iron

$$E^\circ_{Fe^{+3}/Fe^{+2}} = +0.77,\ E^\circ_{Mn^{+3}/Mn^{+2}} = +1.57$$

(C) $E^\circ_{Cr^{+3}/Cr^{+2}} = -0.26$

$$\therefore Cr^{2+} + H^+ \rightarrow Cr^{3+} + \frac{1}{2}H_2$$

(D) $V^{2+} \Rightarrow 3d^3 4s^0$, unpaired electron
Magnetic Moment = 3.87 B.M

37. **(d)** In CrO_2Cl_2 oxidation state of Cr is +6
Cr (VI) = $[Ar]^{18} 3d^0$, Mn(VII) = $[Ar]^{18} 3d^0$, Fe(III) = $[Ar]^{18} 3d^5$, Ti(III) = $[Ar]^{18} 3d^1$, V(IV) = $[Ar]^{18} 3d^1$

Hence Cr (VI) and Mn (VII) have same d^0 configuration.

38. **(a)** $2Cu^+ \rightarrow Cu^{2+} + Cu$
$Cu^{2+}(aq)$ is more stable than $Cu^+(aq)$. $\Delta_{hyd}H$ of Cu^{2+} (aq) is much more negative than $Cu^+(aq)$, which compensates for the second ionisation enthalpy of Cu.

39. **(b)** Sc does not show +4 oxidation state.

40. **(a)** Sc^{3+} is with d^0 and Zn^{2+} is with d^{10} configuration.

41. **(b)**

Ion	ΔH°_{hyd}(kJ/Mole)
Cr^{2+}	– 1925
Mn^{2+}	– 1862
Fe^{2+}	– 1998
Co^{2+}	– 2079

42. **(a)** In reactivity series, Cu lies below Hydrogen and it is least electropositive among the given metals.

43. **(5)** $MnF_4 \Rightarrow +4$ state $\Rightarrow [Ar]3d^3$
$MnF_3 \Rightarrow +3$ state $\Rightarrow [Ar]3d^4$ (Strongest O.A)
$MnF_2 \Rightarrow +2$ state $\Rightarrow [Ar]3d^5$
For n = 4,

$$\therefore \quad \mu = \sqrt{4(4+2)} = \sqrt{24} = 4.89 = 5$$

44. **(5)** Co^{3+} cannot liberate H_2 from dilute mineral acid solution.

Co^{3+} (d^6) in gas phase: | ⥮ | 1 | 1 | 1 | 1 |

Unpaired electrons (n) = 4

$$M_{s.o.} = \sqrt{4(4+2)}\ B.M. = 4.89\ B.M. \approx 5\ B.M.$$

45. **(4)** In case of conversion of $Cr^{3+} \rightarrow Cr^{2+}$ the e^- is added to e_g orbital, hence electrode potential is negative.

$$E^\circ_{Cr^{+3}/Cr^{+2}} = -0.41V ; [Cr^{+3}] = 4s^0 3d^3$$

$$\mu = \sqrt{n(n+2)}B.M = \sqrt{15}\ B.M \sim 4\ B.M$$

46. **(a)** $_{78}Pt$ = [Xe] $4f^{14} 5d^9 6s^1$ (Exceptional electronic configuration)

47. **(c)** Zinc dissolves in excess of aqueous alkali

$$Zn(s) + 2OH^-(aq) + 2H_2O(l) \rightarrow \underset{\text{Tetrahydroxozincate (II) ion}}{[Zn(OH)_4]^{2-}(aq)} + H_2(g)$$

ZnO_2^{2-} is anhydrous form of $[Zn(OH)_4]^{2-}$.
So in aqueous medium best answer of this question is $[Zn(OH)_4]^{2-}$.

48. **(3)** A, B, C are correct and D is incorrect because Fehling solution has Cu(II).

49. **(c)** $Cr^{2+}/Cr \rightarrow -0.90$ V ; $Fe^{2+}/Fe \rightarrow -0.44$ V
$Cu^{2+}/Cu \rightarrow +0.34$ V ; $Zn^{2+}/Zn \rightarrow -0.76$ V
For $Cu(s) \rightarrow Cu^{2+}(aq)$
Copper has high atomization and low hydration enthalpy. So, energy required for the conversion is not balanced by hydration entropy.

50. **(a)** Strongest oxidising agent have highest reduction potential value.

$$E^\circ_{Mn^{3+}/Mn^{2+}} = 1.51\ V\ \text{(highest)}$$

51. **(d)** $Fe^{3+} = [Ar]\ 3d^5$

52. **(a)** Such ions, which have unpaired electrons exhibit colour and paramagnetic behaviour.
$Cu^{2+} = 3d^9$ (one unpaired e^-)
$Cr^{3+} = 3d^3$ (three unpaired e^-)
$Sc^+ = 3d^1 4s^1$ (two unpaired e^-)

53. **(12)** Vanadium (Z = 23); V = $1s^2 2s^2 2p^6 3s^2 3p^6 3d^3 4s^2$
Number of electrons in *p*-orbital are 12.

54. **(c)** Chromium Cr (*Z* = 24) $\Rightarrow$ [Ar] $3d^4 4s^2$
$\therefore$ Common positive oxidation state of Cr are from +2 to + 6. Where +3 and +6 are the most stable oxidation states.

55. **(49)** Fe (Z = 26) $\Rightarrow$ [Ar] $3d^6 4s^2$

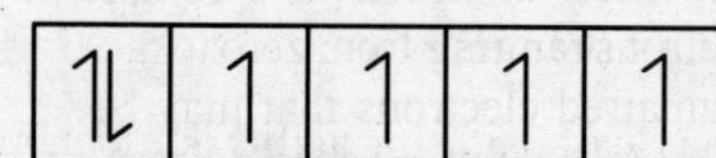

Number of unpaired electrons = 4

$$\because \mu = \sqrt{n(n+2)}\ BM$$

$$\therefore \mu = \sqrt{4(4+2)} = \sqrt{24}\ BM = 4.89 \approx 49 \times 10^{-1}\ BM$$

56. **(a)** The element having atomic number 25 is manganese.
Mn (25) = $3d^5 4s^2$; $Mn^{2+} \Rightarrow 3d^5$
In aqueous solution, it exists as $[Mn(H_2O)_6]^{2+}$ which is high spin complex due to weak field nature of H_2O ligand.
So, total no. of unpaired electrons = 5

$$\mu = \sqrt{n(n+2)}\ BM$$

$$\mu = \sqrt{5(5+2)} = \sqrt{35} = 5.92\ BM$$

57. **(c)** $Cr^+ - 4s^0 3d^5$; $Mn^{2+} - 4s^0 3d^5$
So, the outermost electronic configuration will be same in case of Cr^+ and Mn^{2+}.

58. **(a)** German silver contains Cu (50%), Zn (30%) and Ni (20%) respectively. It does not contain Ag.

59. **(b)** ${}_{26}Fe = [Ar]\,3d^6 4s^2$. Third ionisation results into stable d^5 configuration.

60. **(a)** Reactant should not be adsorbed strongly which might result into immobilisation that inhibit further adsorption on the catalyst's surface.

61. **(a)** Atomic size of elements of 4*d* and 5*d* transition series are nearly same due to Lanthanoid contraction.

62. **(d)** Mo and W belong to group-6 and period 5 (4*d* series) and 6 (5*d* series) respectively.
Due to lanthanoid contraction, radius of Mo and W are almost same, i.e., 0.140 nm and 0.141 nm respectively.

63. **(b)** Electronic configuration of the given transition metal ions are :
Sc^{3+} (Z = 21) $1s^2 2s^2 2p^6 3s^2 3p^6$
Ti^{2+} (Z = 22) $1s^2 2s^2 2p^6 3s^2 3p^6 3d^2$
Ti^{3+} (Z = 22) $1s^2 2s^2 2p^6 3s^2 3p^6 3d^1$
V^{2+} (Z = 23) $1s^2 2s^2 2p^6 3s^2 3p^6 3d^3$
Since, magnetic moment is directly proportional to the number of unpaired electrons. The correct increasing order of magnetic moment is :
$Sc^{3+} < Ti^{3+} < Ti^{2+} < V^{2+}$ because they have 0, 1, 2 and 3 unpaired electrons respectively.

64. **(b)** I.E. increases on moving left to right in a period.
$\therefore$ Ti < Mn < Ni < Zn

65. **(c)**
(A) $V_2O_5 \rightarrow$ Preparation of H_2SO_4 in contact process
(B) $TiCl_4 + Al(Me)_3 \rightarrow$ Polyethylene (Ziegler-Natta catalyst)
(C) $PdCl_2 \rightarrow$ Ethanol (Wacker's process)
(D) Iron oxide $\rightarrow NH_3$ in (Haber's process)

66. **(a)** Interstitial compounds are inert, i.e., they are chemically non-reactive.

67. **(c)** Sc shows oxidation state of +3 only.

68. **(a)** Magnetic moment, $\mu = \sqrt{n(n+2)}$ BM (where, n = no. of unpaired electrons)
As transition metal atom/ion in a complex may have unpaired electrons ranging from zero to 5. So, maximum number of unpaired electrons that may be present in a complex is 5.
$\therefore$ Maximum value of magnetic moment among all the transition metal complexes is :

$$\sqrt{5(5+2)} = \sqrt{35} = 5.92 \text{ BM}$$

69. **(d)** As zinc has no unpaired of electrons to take part in the bond, it has least enthalpy of atomisation amongst the given transition elements.

70. **(d)**

Ions	E^0 (V)
Ti^{2+}	–0.37
V^{2+}	–0.26
Cr^{2+}	–0.41
Mn^{2+}	+1.57

Negative value of E^0 means these metals librate hydrogen from dilute acid.
$(M^{2+} + 2H^+ - e^- \longrightarrow M^{3+} + H_2)$

71. **(b)** In MnO_4^{2-}, manganese is in +6 oxidation state which is having highest stability. Stability increase with increase in Solvation energy.
Solvation $\propto$ Effective nuclear charge
Since MnO_4^{2-} have +6 oxidation number and high effective nuclear charge so attracted towords H_2O and thus acquire high Solvation energy.

72. **(c)** Reaction of Zn with different concentration of HNO_3 are as follows :

$$Zn + 4HNO_3(60\%) \longrightarrow Zn(NO_3)_2 + 2NO_2\uparrow + 2H_2O$$
$$3Zn + 8HNO_3(30\%) \longrightarrow 3Zn(NO_3)_2 + 2NO\uparrow + 4H_2O$$
$$4Zn + 10HNO_3(20\%) \longrightarrow 4Zn(NO_3)_2 + N_2O\uparrow + 5H_2O$$
$$5Zn + 12HNO_3(10\%) \longrightarrow 5Zn(NO_3)_2 + N_2\uparrow + 6H_2O$$
$$4Zn + 10HNO_3(3\%) \longrightarrow 4Zn(NO_3)_2 + NH_4NO_3 + 3H_2O$$

Hence option (c) is correct.

73. **(b)** $\mu = \sqrt{n(n+2)} \Rightarrow 1.73 = \sqrt{n(n+2)}$
No. of unpaired electrons, n = 1
V(23) = $[Ar]\,3d^3 4s^2$; For one unpaired e^-, $V^{4+} = [Ar]3d^1$
$\therefore$ Its chloride has the formula is : $V^{4+}Cl^- \equiv VCl_4$

74. **(d)** Maganese exhibits the most number of oxidation states. The most common oxidation states of Mn are +2, +3, + 4, + 6 and +7.

75. **(a)** Cu_2O is yellow in colour.

76. **(a)**
(a) $V = 3d^3\,4s^2$; $V^{2+} = 3d^3 = 3$ unpaired electrons
$Cr = 3d^5\,4s^1$; $Cr^{2+} = 3d^4 = 4$ unpaired electrons
$Mn = 3d^5\,4s^2$; $Mn^{2+} = 3d^5 = 5$ unpaired electrons
$Fe = 3d^6\,4s^2$; $Fe^{2+} = 3d^6 = 4$ unpaired electrons
Hence, the correct order of paramagnetic behaviour :
$V^{2+} < Cr^{2+} = Fe^{2+} < Mn^{2+}$
(b) For the same oxidation state, the ionic radii generally decreases as the atomic number increases in a particular transition series. Hence, the order is :
$Mn^{2+} > Fe^{2+} > Co^{2+} > Ni^{2+}$
(c) Larger size and least hydrated ions are more stable in aqueous solution. As we move across the period ($Sc^{3+} \rightarrow Cr^{3+} \rightarrow Fe^{3+} \rightarrow Co^{3+}$), the ionic size usually decreases.
(d) Sc = (+ 2), (+ 3)
Ti = (+ 2), (+ 3), (+ 4)
Cr = (+ 2), (+ 3), (+ 4), (+ 5), (+ 6)
Mn = (+ 2), (+ 3), (+ 4), (+ 5), (+ 6), (+ 7)

77. **(d)** $E^\circ_{Cr^{3+}/Cr^{2+}} = -0.41$ V, $E^\circ_{Fe^{3+}/Fe^{2+}} = +0.77$ V

$E^\circ_{Mn^{3+}/Mn^{2+}} = +1.57$ V, $E^\circ_{Co^{3+}/Co^{2+}} = +1.97$ V

Cr^{3+} : $[Ar]\,3d^3$, Cr^{2+} : $[Ar]\,3d^4$;
Mn^{3+} : $[Ar]\,3d^4$, Mn^{2+} : $[Ar]\,3d^5$;
Fe^{3+} : $[Ar]\,3d^5$, Fe^{2+} : $[Ar]\,3d^6$;
Co^{3+} : $[Ar]\,3d^6$, Co^{2+} : $[Ar]\,3d^7$;

Due to highest nuclear charge, Co^{3+} can accept most easily an e^- to convert to Co^{2+}. Hence, the highest value of $E^\circ_{M^{3+}/M^{2+}}$ is for Co^{3+}.

Fe^{3+} and Mn^{2+} are more stable than Fe^{2+} and Mn^{3+} respectively. Hence, $E^\circ_{Mn^{3+}/Mn^{2+}} > E^\circ_{Fe^{3+}/Fe^{2+}}$.

Due to lowest nuclear charge in Cr^{3+} and stable half-filled t_{2g} configuration the conversion of it to Cr^{2+} is difficult. Hence, it has negative value of $E^\circ_{Cr^{3+}/Cr^{2+}}$.

78. (a) The element with outer electron configuration $3d^5 4s^2$ is Mn which exhibits oxidation states from +2 to +7.

79. (d) Fe^{3+} is easily hydrolysed than Fe^{2+} due to more positive charge.

80. (a) Zr (40), Hf (72) have similar properties as they belong to same group and have same valence shell electronic configuration.

81. (a) Across the first transition series, the negative values for standard electrode potential decrease (or the value increases) except for Mn due to the stable d^5 configuration. Thus, correct order is :

$$Mn^{2+} > Cr^{2+} > Fe^{2+} > Co^{2+}.$$

82. (d) Lanthanide contraction is associated with the intervention of the 4*f* orbitals which must be filled before the 5*d* series of elements begin. The filling of 7*f* before 5*d* orbitals results in a regular decrease in atomic radii called lanthanoid contraction which essentially compensates for the expected increase in atomic size with increasing atomic number. The factor responsible for the lanthanoid contraction is the imperfect shelding of one electron by another in the same set of orbitals.

83. (b) Note : In vertical columns of transition elements, there is an increase in size from first member to second member as expected but from second member to third member, there is very small chang in size and some times sizes are same. This is due to lanthanide contraction this is the reason for Zr and Hf to have same radius.

84. (d) $(n-1)d^5 ns^2$ attains the maximum O.S. of + 7.

85. (a) $Mn^{2+} = 5$ unpaired electrons

$Fe^{2+} = 4$ unpaired electrons

$Ti^{2+} = 2$ unpaired electrons

$Cr^{2+} = 4$ unpaired electrons

Magnetic moment $\propto$ Number of unpaired electrons

86. (10) $$\underset{\text{(O.A)}}{K\overset{+7}{Mn}O_4} \xrightarrow{\text{Acidic medium}} Mn^{2+}$$

X is difference in oxidation state. 7 – 2 = 5

So X = 5

$$6CH_3COO^{\oplus} + Fe^{3+} + H_2O \rightarrow [Fe_3(OH_2)(CH_3COO)_6]^{\oplus} + 2H^{\oplus}$$

$$[Fe_3(OH)_2(CH_3COO)_6]^{\oplus} + 4H_2O \rightarrow \underset{\text{Brown red ppt}}{[Fe(OH)_2(CH_3COO]} + CH_3COOH + H^{\oplus}$$

$Fe^{3+} \Rightarrow 3d^5 4s^0$ contains 5d electrons

So Y = 5

$\therefore$ X + Y = 5 + 5 = 10

87. (6) $$\underset{(A)}{Cr_2O_7^{2-}} + NaCl + H_2SO_4 \rightarrow CrO_2Cl_2$$

$$CrO_2Cl_2\,(\text{Vapour}) + NaOH \rightarrow \underset{(B)}{Na_2CrO_4} + NaCl + H_2O$$

$$\underset{(B)}{Na_2CrO_4} + H^{\oplus} \rightarrow \underset{(C)}{Na_2Cr_2O_7} + H_2O$$

Number of terminal Oxygen = 6

88. (160) $$4FeCr_2O_4 + 8Na_2CO_3 + 7O_2 \rightarrow 8Na_2CrO_4 + 2Fe_2O_3 + 8CO_2$$

Fe_2O_3 is water insoluble, so its molar mass

$\Rightarrow [2 \times 56 + 3 \times 16] \Rightarrow 160$ g/mol

89. (5) Strong oxidising power among the option is Mn_2O_3.

$E^\circ_{Mn^{3+}/Mn^{2+}} = +1.57$ V

$Mn^{+3} \rightarrow d^4$ configuration

$\mu = \sqrt{24}$ BM

$= 4.89$ BM $\Rightarrow 5$

90. (c) $V_2O_5 + \text{alkali} \rightarrow VO_4^{3-}$

$VO_4^{-3} = x + 4(-2) = -3 \Rightarrow x = 8 - 3 \Rightarrow x = +5$

91. (c) Both $K_2Cr_2O_7$ and $KMnO_4$ act as oxidizing agent in an acidic medium.

$$I^- \xrightarrow{H^+} I_2 \quad ; \quad I^- \xrightarrow{OH^-} IO_3^-$$

$$S^{-2} \xrightarrow{H^+} S \quad ; \quad S_2O_3^{2-} \xrightarrow{OH^-} SO_4^{2-}$$

$$Fe^{+2} \xrightarrow{H^+} Fe^{+3} \quad ; \quad S_2O_3^{2-} \xrightarrow{H^+} S\downarrow + SO_4^{2-}$$

92. (d) $$MnO_2 \xrightarrow[KNO_3,\,\Delta]{KOH} K_2MnO_4$$

93. (c)

$$K_2Cr_2O_7 \xrightarrow[-H_2O]{KOH} \underset{[A]}{K_2CrO_4} \xrightarrow[-H_2O]{H_2SO_4} K_2Cr_2O_7 + \underset{[B]}{K_2SO_4}$$

$Cr_2O_7^{2-}$ and CrO_4^{2-} exist in equilibrium in aqueous solutions and are interconvertible depending on the pH of the solution.

94. (b) Chromate and dichromate ions both are interconvertible due to pH change.

$$Cr_2O_7^{2-} \underset{H^+}{\overset{OH^-}{\rightleftharpoons}} 2CrO_4^{2-}$$

95. (d) $$MnO_2 + 4KOH + O_2 \xrightarrow{\text{fused}} \underset{\substack{\text{Potassium manganate} \\ \text{Dark green}}}{2K_2MnO_4} + 2H_2O$$

Electrolytic oxidation in alkaline medium :

At anode :

Manganate change in permanganate ion.

$$MnO_4^{2-} \rightarrow MnO_4^- + e^-$$

96. (6) Spin only magnetic moment of $\overset{+7}{Mn}$ in $KMnO_4$ is 0.
Spin only value of manganese product $(\overset{+2}{Mn})$ formed during titration of $KMnO_4$ against oxalic acid in acidic medium is 6
$\therefore$ Difference = 6 – 0 = 6

97. (877) CrO is Basic oxide
Cr_2O_3 is Amphoteric oxide
In CrO, Cr exist as Cr^{+2} and μ spin only = 4.90 BM
In Cr_2O_3, Cr exist as Cr^{+3} and μ spin only = 3.87 BM
Sum of spin only magnetic moment
$= 4.90 + 3.87 = 8.77 = 877 \times 10^{-2}$ BM.

98. (6) $4FeCr_2O_4 + 8Na_2CO_3 + 7O_2 \longrightarrow$
$8Na_2CrO_4(A) + 2Fe_2O_3(B) + 8CO_2$
No of unpaired e^- in $Na_2Cr_2O_4 = 0$
$A: \overset{+6}{Na_2CrO_4} \to Cr^{+6} \to [Ar]4s^0 3d^0 \Rightarrow \mu = 0$
No of unpaired e^- in $Fe_2O_3 = 5$
$B: \overset{+3}{Fe_2O_3} \to Fe^{+3} \to [Ar]4s^0 3d^5 \Rightarrow \mu = 5.9BM$
Sum of magnetic moments of A & B ≃ 0 + 5.9 ≃ 6 BM

99. (4) $MnO_2 + KOH + O_2 \to \underset{(A)}{K_2MnO_4} + H_2O$
$K_2MnO_4 \xrightarrow{\text{Neutral/acidic solution}} \underset{(B)}{\overset{+7}{KMnO_4}} + \underset{(C)}{\overset{+4}{MnO_2}}$
$Mn^{+4} : [Ar]3d^3$; $Mn^{+7} : [Ar]3d^0 \Rightarrow \mu = 0$
$n = 3, \mu = \sqrt{3(3+2)} = 3.87$ B.M
Nearest integer is (4)

100. (b) $CuSO_4.5H_2O$
In $K_2Cr_2O_7$ and K_2CrO_4, the Cr is in +6 oxidation state and hence contain no '*d*' electron.
In $KMnO_4$, Mn is in +7 oxidation state and this also do not contain any '*d*' electrons.
They exhibit colour due to charge transfer spectra and not due to d-d transistion.

101. (d) A. CrO_4^{2-} is tetrahedral
B. $2Na_2CrO_4 + 2H^+ \to Na_2Cr_2O_7 + 2Na^+ + H_2O$
C. Green manganate is paramagnetic with 1 unpaired electron.
D. $3MnO_4^{2-} + 4H^+ \to 2MnO_4^- + MnO_2 + 2H_2O$
$3MnO_4^{2-} + 2H_2O \to 2MnO_4^- + MnO_2 + 4OH^-$
E. As the oxidation state increases ionic character decreases.

102. (b) A and C only
V_2O_4 react with acid to give $VO.^{2+}$ V_2O_5 is amphoteric in nature and it react with acid as well as with base.

103. (6) The reaction will be as follows
$K_2Cr_2O_7 + 4KCl + \underset{\text{conc.}}{6H_2SO_4}$
$\downarrow$
$2CrO_2Cl_2 + 6KHSO_4 + 3H_2O$
Oxidation state of Cr in CrO_2Cl_2 is
$x - 4 - 2 = 0 \Rightarrow x = +6$

104. (a) Cr and Mn in $K_2Cr_2O_7$ and $KMnO_4$ are in +6 and +7 states which are highest in them and results in d^0 configurations.
Thus, their colours are due to charge transfer transitions and not d–d transitions.

105. (b) $MnO_2 \xrightarrow[\text{air / KNO}_3]{\text{Fusion with KOH,}} \underset{(A)}{K_2MnO_4}$
$\xrightarrow[\text{oxidation in alkaline medium}]{\text{electrolytic}} \underset{(B)}{KMnO_4}$

106. (d) When the yellow solution (sodium chromate) is acidified with dilute sulphuric acid and adding small amount of amyl alcohol and 10%, H_2O_2, the organic layer turns blue and confirms the formation of chromium pentoxide. The oxidation state of Cr in chromium pentoxide is +6.
$CrO_4^{2-} + 2H^+ + 2H_2O \longrightarrow \underset{\text{Chromium pentoxide}}{CrO_5} + 3H_2O$

107. (d) In faintly alkaline solution, $KMnO_4$ oxidises iodide to iodate.
$2MnO_4^- + H_2O + I^- \longrightarrow 2MnO_2 + 2OH^- + IO_3^-$

108. (b) $KMnO_4 \xrightarrow[\Delta]{513\text{ K}} K_2MnO_4 + MnO_2 + O_2$

109. (d) $PbCrO_4 + NaOH\text{ (hot excess)} \to$
$[Pb(OH)_4]^{-2} + Na_2CrO_4$
The product is dianionic complex with coordination number four.

110. (d) Ce^{4+} has empty of subshell so it gain electrons thus, strongly oxidixing.

111. (5) $\overset{(+7)}{KMnO_4} \to \overset{(+2)}{Mn^{2+}}$ (In acidic medium)
Change in oxidation state of Mn = 5

112. (c) $K_2Cr_2O_7$ is used as primary standard. The concentration of $Na_2Cr_2O_7$ changes in aq. solution. $K_2Cr_2O_7$ is less soluble than $Na_2Cr_2O_7$.

113. (3) In faintly alkaline medium,
$\overset{+7}{MnO_4^-} + 3e^- + 2H_2O \to \overset{+4}{MnO_2} + 4OH^-$
No. of electrons gained = 3

114. (a) $2Cu^{2+} + 4I^- \to Cu_2I_2(s) + I_2$

115. (10) $\overset{(0)}{Fe(CO)_5}$ $\overset{(+4)}{V}O^{2+}$ $\overset{(+6)}{W}O_3$
So, Sum of oxidation state = 0 + 4 + 6 = 10

116. (b) In neutral or weakly alkaline solution oxidation state of Mn changes by 3 unit
$\overset{+7}{MnO_4^{-1}} \to \overset{+4}{MnO_2}$

117. (b) Mn_2O_7 structure: two Mn atoms, each bonded to three O by double bonds (Mn=O), linked through a bridging O (Mn–O–Mn).

118. (a) With increase in oxygen acidic nature of oxide of an element increase and basic nature decreases.

119. (b) $Cu^{2+} + 2KI \longrightarrow \underset{\text{Unstable}}{CuI_2} \downarrow + 2K^+$
I^- is strond reducing agent it reduces Cu^{2+} to Cu^+

$$2CuI_2 \longrightarrow \underset{\text{(White)'X'}}{Cu_2I_2} \downarrow + I_2$$

$$KI + I_2 \longrightarrow K^+I_3^- \text{ (Brown solution)}$$

$$I_3^- \rightleftharpoons I_2 + I^-$$

$$KI_3 + Na_2S_2O_3 \rightarrow KI + \underset{(Y)}{Na_2S_4O_6}$$

120. (3) $\overset{+7}{Mn}O_4^- + 4H^+ + 3e^- \longrightarrow \overset{+4}{Mn}O_2 + 2H_2O$

121. (a) In acidic medium

$2MnO_4^- + 10I^- + 16H^+ \rightarrow 2MnO^{2+} + 5I_2 + 8H_2O$

In neutral/faintly alkaline solution

$2MnO_4^- + I^- + H_2O \rightarrow 2MnO_2 + 2OH^- + IO_3^-$ (ox)

122. (b) $\overset{+6}{Cr_2}O_7^{2-} + 14H^+ + 6e \longrightarrow 2Cr^{3+} + 7H_2O$

123. (d) $3SO_2 + Cr_2O_7^{2-} + 2H^+ \rightarrow 3SO_4^{2-} + \underset{\text{green}}{2Cr^{+3}} + H_2O$

124. (d) $8\overset{+7}{Mn}O_4^- + 3S_2O_3^{2-} + H_2O \rightarrow$

$$8\overset{+4}{Mn}O_2 + 6SO_4^{2-} + 2OH^-$$

Change in oxidation state of Mn is from +7 to +4 i.e., 3.

125. (b) Copper get reduced and hydrogen get oxidised

$CuO + H_2 \rightarrow Cu + H_2O$ (under hot conditions)

126. (4) $MnO_4^{2-} \xrightarrow{H^+} MnO_4^- + MnO_2$

No. of unpaired e^- $(n) = 3$

$\therefore \quad \mu = \sqrt{n(n+2)} = \sqrt{15} = 3.877 \text{B.M.}$

127. (d) $8MnO_4^- + 3S_2O_3^{2-} + H_2O \xrightarrow[\text{alk.solution}]{\text{neutral or}} 8MnO_2$

$$+ 6SO_4^{2-} + 2OH^-$$

128. (c) Statement I: Acidified $K_2Cr_2O_7$ and neutral $kMnO_4$ are strong oxidised agents and can oxidise I^- to I_2.

The standard electrode potential values are:

$E_{Fe^{3+}/Fe^{2+}} = +0.77V$ and $E_{I_2/I^-} = +0.54V$

Hence, Fe^{3+} will oxidise I^- to I_2.

Statement II: Manganate ion is MnO_4^{2-} in which Mn is in +6 O.S. Due to unpaired e^-, it is paramagnetic in nature. However, there is $p\pi - d\pi$ bonding instead of $p\pi - p\pi$ bonding.

129. (c) In manganese heptoxide, there are 6, Mn = 0 bonds.

$$O_3Mn{-}O{-}MnO_3$$

130. (1) $\overset{(+3)}{(COOH)_2} \longrightarrow \overset{(+4)}{CO_2}$ Change in O.N. = 4 − 3 = 1

131. (4) $2\overset{+7}{K}MnO_4 + 3H_2O_2 \xrightarrow{\text{basic medium}}$

$$2\overset{+4}{Mn}O_2 + 3O_2 + 2H_2O + 2KOH$$

132. (6) $4FeCr_2O_4 + 8Na_2CO_3 + 7O_2 \rightarrow \underbrace{8Na_2CrO_4}_{(A)}$

$$+ 2Fe_2O_3 + 8CO_2$$

$$2Na_2CrO_4 + 2H^+ \rightarrow \underbrace{Na_2Cr_2O_7}_{(B)} + 2Na^+ + H_2O$$

$$2Na^+ \left[O_3Cr{-}O{-}CrO_3 \right]^{2-}$$

133. (0) $3\overset{+6}{Mn}O_4^{2-} + 4H^+ \longrightarrow 2\overset{+7}{Mn}O_4^- + \overset{+4}{Mn}O_2 + 2H_2O$

In the product MnO4⁻ has highest oxidation state i.e + 7. For which number of unpaired electorns will be zero.

Hence $\mu = \sqrt{n(n+1)}$; n = number of unpaired electrons.

for $n = 0 \Rightarrow \mu = 0$ B.M.

134. (6) $2KMnO_4 + 5H_2C_2O_4 + 3H_2SO_4 \rightarrow$

$$K_2SO_4 + 2MnSO_4 + 10CO_2 + 8H_2O.$$

Mn^{2+} has 5 unpaired electrons, therefore, the magnetic moment $= \sqrt{5(5+2)} = \sqrt{35}$ BM.

135. (0) $\overset{+6}{Cr}O_4^{2-}$, $\overset{+6}{Cr_2}O_7^{2-}$

Hence, the difference is zero.

136. (3) MnO_4^{2-} disproportionates in a neutral or acidic solution to give MnO_4^- and Mn^{+4}.

$3MnO_4^{2-}(aq) + 4H^+(aq) \longrightarrow$

$$2MnO_4^-(aq) + MnO_2(s) + 2H_2O(l)$$

O.S. of Mn in MnO_4^- = +7

O.S. of Mn in MnO_2 = +4

difference = 3

137. (3) Most basic oxide is V_2O_3.

$V^{+3} \rightarrow [Ar]\,3d^2$

No. of unpaired e^-, $n = 2$

$\mu = \sqrt{n(n+2)} = \sqrt{2(2+2)} = 2.84 \text{ BM} \approx 3$

138. (d) Potassium permangnate on heating at 513 K gives potassium manganate (K_2MnO_4) and its colour is green.

$$2KMnO_4 \xrightarrow[513K]{\Delta} \underset{\text{Green}}{K_2MnO_4} + \underset{\text{Black}}{MnO_2} + O_2$$

In K_2MnO_4, manganese oxidation state is +6 and hence, it has one unpaired electron and paramagnetic.

139. (d) $KMnO_4$ on heating dissociates as

$$\underset{\text{Permanganate}}{2K\overset{+7}{Mn}O_4} \xrightarrow{\Delta} \underset{\text{Manganate}}{K_2\overset{+6}{Mn}O_4} + MnO_2 + O_2$$

$Mn^{+7} \Rightarrow 3d^0 4s^0$ (Diamagnetic)

$Mn^{+6} \Rightarrow 3d^1 4s^0$ (Paramagnetic)

Both permanganate and manganate are tetrahedral but only manganate is paramagnetic.

140. (6) $Cr_2O_7^{2-} \xrightarrow{OH^-} CrO_4^{2-}$

$\therefore$ Oxidation state of Cr in CrO_4^{2-} is +6.

141. (6) In neutral or faintly alkaline medium,

$$8MnO_4^- + 3S_2O_3^{2-} + H_2O \rightarrow 8MnO_2 + \underset{(A)}{6SO_4^{2-}} + 2OH^-$$

The oxidation state of sulphur in SO_4^{2-} is + 6.

142. (173) $\overset{+6}{Cr}O_4^{2-} + \overset{+2}{S_2}O_3^{2-} \longrightarrow \overset{+3}{Cr}(OH)_4^- + \overset{+6}{S}O_4^{2-}$

Using law of equivalence,

m. equivalents of CrO_4^{2-} used = m. equivalents of $S_2O_3^{2-}$ used

n-factor of $CrO_4^{2-} = 3$; n-factor of $S_2O_3^{2-} = 4 \times 2 = 8$

$\therefore 0.154 \times V \times 3 = 0.25 \times 40 \times 8$

$\Rightarrow V = 173.16$ mL ≈ 173 mL

143. (b) Electronic configuration of

$_{64}Gd = [Xe]\ 4f^7\ 5d^1\ 6s^2$

$_{64}Gd^{3+} = [Xe]\ 4f^7$

No. of unpaired electron = 7

$\mu = \sqrt{n(n+2)}\text{BM} = \sqrt{63} = 7.93$ BM

144. (c) (1) W(VI) is more stable than Cr(VI) due to smaller size of atoms and also due to lanthanide contraction.

(2) Permanganate titrations in presence of HCl are unsatisfactory as HCl is oxidised to Cl_2.

(3) Lanthanoid oxides are used as phosphorus.

145. (c)

Manganate :	**Permanganate :**
Paramagnetic, green in colour, Tetrahedral & contains $p\pi$-$d\pi$ bond	Diamagnetic, purple in colour, Tetrahedral & contains $p\pi$-$d\pi$ bond

Manganate ion is paramagnetic while permanganate ion is diamagnetic.

146. (18)

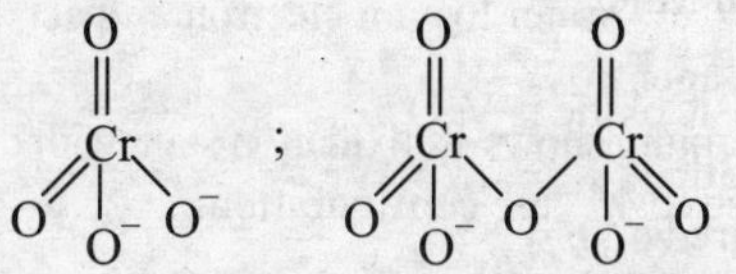

Chromate ion ; Dichromate ion

$\Rightarrow$ Total number of Cr and O bonds is 6 + 12 = 18.

147. (18.00)

$4NaCl + K_2Cr_2O_4 + 3H_2SO_4 \text{ (Conc.)} \longrightarrow 2CrO_2Cl_2 \text{ (A)} + K_2SO_4 + 2NaSO_4 + 3H_2O$

$CrO_2Cl_2 \text{ (A)} + NaOH \longrightarrow Na_2CrO_4 \text{ (B)} + H_2O + NaCl$

$Na_2CrO_4 \text{ (B)} + H_2SO_4 \text{ (dilute)} + H_2O_2 \longrightarrow CrO_5 \text{ (C)} + Na_2SO_4 + H_2O$

The sum of total no. of atoms in one molecule each of A, B and C = 5 + 7 + 6 = 18.0

148. (a) $2KMnO_4 \text{ (X)} \xrightarrow{513K} K_2MnO_4 \text{ (Y)} + MnO_2 + O_2 \text{ (gas)}$

$MnO_2 + 4NaCl + 4H_2SO_4 \text{ (Conc.)} \longrightarrow MnCl_2 + 4NaHSO_4 + Cl_2 \text{ (Z, Pungent gas)} + 2H_2O$

149. (b) $2MnO_2 \text{ (A)} \xrightarrow{KOH,\ O_2} 2K_2MnO_4 \text{ (Green) (B)} + 2H_2O$

$K_2MnO_4 \xrightarrow{4\ HCl} 2KMnO_4 \text{ (Purple) (C)} + MnO_2 + 2H_2O$

$2KMnO_4 \xrightarrow{H_2O,\ KI} 2MnO_2 \text{ (A)} + 2KOH + KIO_3 \text{ (D)}$

150. (a)

$MnO_2 + KOH + KNO_3 \longrightarrow K_2MnO_4 \text{ (dark green)} + KNO_2 + H_2O$

$MnO_4^{2-} + 4H^+ \xrightarrow[\text{tionation}]{\text{dispropor-}} 2MnO_4^- \text{ (dark purple)} + MnO_2 + 2H_2O$

151. (d) Out of all the four given metallic oxides, CrO_2 is attracted by magnetic field very strongly. The effect persists even when the magnetic field is removed. Thus CrO_2 is metallic and ferromagnetic in nature.

152. (a) Aqueous solution of $CoCl_2$ contains $[Co(H_2O)_6]^{2+}$ which is pinkish in colour so option (d) is incorrect.

Reduction potential of $Co^{3+} \longrightarrow Co^{2+}$ is high so option(b) is incorrect. Co^{2+} does not oxidises easily to Co^{3+}. It is general case that symmetrical substituted octahedral complexes are less deeper in colour than tetrahedral complexes. So $[CoCl_4]^{2-}$ is deep blue in colour.

153. (a) $Na_2Cr_2O_7$ is used as a primary standard in volumetry. Hydration energy of $Na^{\oplus}$ is greater than $K^{\oplus}$. Because of smaller size of $Na^{\oplus}$, $Na_2Cr_2O_7$ is more soluble than $K_2Cr_2O_7$.

154. (b) The equation in option (b) is correct since both charges as well as atoms are balanced. For the rest,

(a) Given reaction in unfavourable in the forward direction (K_2O is unstable, while Li_2O is stable)

(c) Given reaction is not balanced w.r.t. charge.

(d) Given reaction will give $K_3[Cu(CN)_4]$ as product instead of $K_2[Cu(CN)_4]$.

155. (d) In equation (a) $Fe_2(SO_4)_3$, and in equation (b) $FeSO_4$, on decomposition will form oxide instead of Fe. In equation (c) $FeCl_3$ cannot be reduced when heated in air. Hence equation (d) is correct.

156. (a) $K_2Cr_2O_7 + 4H_2SO_4 + 3H_2S \longrightarrow K_2SO_4 + Cr_2(SO_4)_3 + 7H_2O + 3S$

157. (a) Cu_2O is yellow in colour.

158. (b) $KMnO_4$ reacts with H_2SO_4 to form Mn_2O_7 which is highly explosive substance.

$2KMnO_4 + H_2SO_4 \longrightarrow K_2SO_4 + Mn_2O_7 + H_2O$

159. (a) $Cr_2O_7^{2-} + 6I^- + 14H^+ \longrightarrow 2Cr^{3+} + 7H_2O + 3I_2$

160. (d) $2Cu_2O + Cu_2S \longrightarrow 6Cu + SO_2$

self reduction.

161. (a) $\ddot{N}H_3 + H^+ \text{ (acid medium)} \rightleftharpoons \overset{+}{N}H_4$

162. (a)

163. (c) $AgNO_3$ on heating till red **hot** decomposes as follows:

$AgNO_3 \rightarrow Ag + NO_2 + \frac{1}{2}O_2$

164. (a) $2K_2CrO_4 + 2HNO_3 \longrightarrow K_2Cr_2O_7 + 2KNO_3 + H_2O$

165. (b) *f*-Block elements show a regular decrease in atomic size due to lanthanide/actinide contraction.

166. (c)

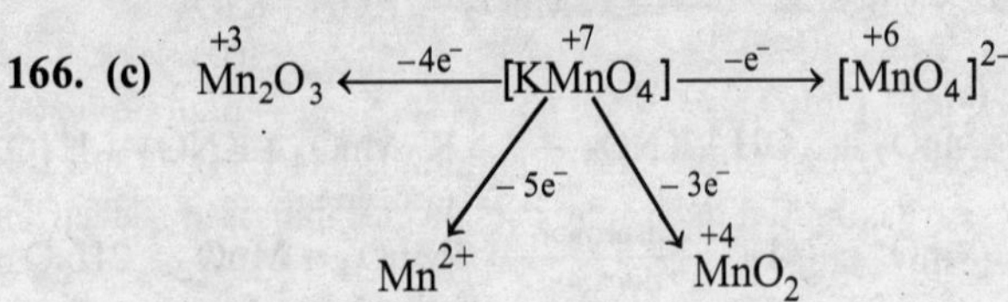

167. (c) Tb^{4+} is strongest oxidising agent due to tendency to accept electron because of its very high standard reduction potential of +3.1V.

168. (a) ${}_{63}Eu^{2+}$ – [Xe] $4f^7 6s^0$
${}_{64}Gd^{3+}$ – [Xe] $4f^7 5d^0 6s^0$
${}_{63}Eu^{3+}$ – [Xe] $4f^6 6s^0$
${}_{65}Tb^{3+}$ – [Xe] $4f^8 6s^0$
${}_{62}Sm^{2+}$ – [Xe] $4f^6 6s^0$
Eu^{2+} & Gd^{3+}

169. (2) La^{+3} and Lu^{+3} do not show any colour because no unpaired electron is present for the f–f transition.

170. (b) Einsteinium (atomic No = 99) : [Rn]$5f^{11} 6d^0 7s^2$

171. (c) Cm is Actinoid
Eu, Er, Tb, Yb, Lu are lanthanoid.

172. (b) La^{3+} and Ce^{4+} are diamagnetic

173. (d) The electronic configurations of the given species are:-
Gd^{3+} = [Xe]$4f^7$(Stable half-filled configuration)
Lu^{3+} = [Xe]$4f^{14}$ (Stable fully-filled configuration)
Ce^{4+} = [Xe]$4f^0$(Acts as oxidant, reverts back to Ce^{3+} due to high reduction potential of Ce^{4+}/Ce^{3+})
Eu^{2+} = [Xe]$4f^7$(Strong reducing agent despite stable configuration due to low reduction potential of Eu^{3+}/Eu^{2+} = –0.36V)

174. (a) Electronic configuration of Nd(Z = 60) is;
[Xe]$4f^4 6s^2$ (*f*-block)

175. (b) Statement (I) is true, Ce^{+4} has noble gas electronic configuration. as Ce^{+4} = [Xe] $4f^0 5d^0 6s^0$.
Statement (II) is also true due to high reduction potential for Ce^{4+}/Ce^{3+} (+1.74V), and stability of Ce^{3+}, Ce^{4+} acts as strong oxidizing agent.

176. (2) Noble gas configuration = $ns^2\ np^6$
$[Sr^{2+}]$ = [Kr]
$[Cs^+]$ = [Xe]
$[Yb^{2+}]$ = [Xe] $4f^{14}$
$[La^{2+}]$ = [Xe] $5d^1$
$[Pb^{2+}]$ = [Xe] $4f^{14} 5d^{10} 6s^2$
$[Fe^{2+}]$ = [Ar] $3d^6$

177. (d) Eu^{2+} can donate an electron to become Eu^{3+} and Ce^{4+} will become Ce^{3+}.

178. (b) $\left.\begin{array}{l} Eu^{+2} : [Xe]4f^7 \\ Yb^{+2} : [Xe]4f^{14} \end{array}\right\}$ High IE due to half filled & fully filled configurations

179. (b) Nd(60) = [Xe] $4f^4 5d^0 6s^2$
Nd^{2+} = [Xe] $4f^4 5d^0 6s^0$

180. (a) ${}_{62}Sm : 4f^6 6s^2$; ${}_{64}Gd : 4f^7 5d^1 6s^2$; ${}_{63}Eu : 4f^7 6s^2$; ${}_{65}Tb : 4f^9 6s^2$; ${}_{61}Pm : 4f^5 6s^2$

181. (d) Ce^{+4} and Tb^{+4} act as oxidising agent.

182. (a, d)

183. (c) $Ce^{2+} \rightarrow 4f^1$
$Sm^{2+} \rightarrow 4f^6$
$Eu^{2+} \rightarrow 4f^7 \rightarrow$ Half filled configuration
$Yb^{2+} \rightarrow 4f^{14} \rightarrow$ fully filled configuration

$$E^\circ_{M^{3+}/M^{2+}} \Rightarrow \begin{array}{cc} Eu & Yb \\ -0.35 & -1.05 \end{array}$$

Hence, due to more reduction potential in Eu as compared to Yb, it can concluded that Eu^{2+} is more stable than Yb^{2+}.

184. (c) Eu $\rightarrow$ [Xe]$4f^7 6s^2$ Tm $\rightarrow$ [Xe]$4f^{13} 6s^2$
$Eu^{2+} \rightarrow$ [Xe]$4f^7$ $Tm^{2+} \rightarrow$ [Xe]$4f^{13}$
Sm $\rightarrow$ [Xe]$4f^6 6s^2$ Tb $\rightarrow$ [Xe]$4f^9 6s^2$
$Sm^{2+} \rightarrow$ [Xe]$4f^6$ $Tb^{4+} \rightarrow$ [Xe]$4f^7$,
Yb $\rightarrow$ [Xe]$4f^{14} 6s^2$
$Yb^{2+} \rightarrow$ [Xe]$4f^{14}$

185. (a) Ce = [Xe] $4f^1 5d^1 6s^2$
Ce^{3+} = [Xe] $4f^1 5d^0$
Ce^{4+} = [Xe] $4f^0 5d^0$ (Noble gas configuration)
Ce^{3+} get easily oxidised to Ce^{4+} by acheiving noble gas configuration.

186. (b) Cerium exists in two different oxidation states +3 and +4.
$Ce^{4+} + e^- \rightarrow Ce^{3+}$ $E^0 = +1.61$ V
It shows, Ce^{4+} prefer to gain electron and acts as a strong oxidising agent.

187. (b) Ytterbium shows +2 oxidation state with diamagnetic nature due to $4f^{14}6s^2$ configuration.

188. (c) Eu $\rightarrow$ [Xe] $4f^7 6s^2$.
$Eu^{2+} \rightarrow$ [Xe] $4f^7$

189. (7)

$1s^2$ $2s^2$ $2p^6$ $3s$ $3p^6$ $4s^2$ $3d^{10}$ $4p^2$

⇅ | ⇅ | ⇅⇅⇅ | ⇅ | ⇅⇅⇅ | ⇅ | ⇅⇅⇅⇅⇅ | ↑↑_

m = 0 0 –1 0 +1 0 –1 0 +1 0 –2 –1 0 +1 +2

Completely filled orbital (*x*) with $m_\ell = 0$ are = 7

190. (18) Np [93] = [Rn] $5f^4 6d^1 7s^2$
So, no. of *f* electrons in $5f = 4$
No. of *f* electrons in $4f = 14$
Total no. of *f* electrons = 14 + 4 = 18

191. (7) Gd(64) $\Rightarrow$ [Xe] $4f^7 5d^1 6s^2 \Rightarrow Gd^{2+} \Rightarrow$ [Xe] $4f^7 5d^1 6s^0$
So, the number of 4*f* electrons in the ground state electronic configuration of Gd^{2+} is 7.

192. (10) Ho(67) = [Xe]$4f^{11} 6s^2$
Ho^{3+} = [Xe]$4f^{10}$

193. (b) $Ce^{4+} \xrightarrow{e^-} Ce^{3+}$ $E^o = +1.74$ V

Positive SRP means greater oxidising power. So, Ce^{4+} can be easily reduced to Ce^{3+}. Hence, Ce^{4+} is less stable than Ce^{3+}.

194. (a)

(A) $(NH_4)_2[Ce(NO_3)_6]$

$_{58}Ce = [Xe]\,4f^2 5d^0 6s^2$

$\Rightarrow Ce^{4+} = 4f^0$ (diamagnetic)

(B) $Gd(NO_3)_3$

$_{64}Gd = [Xe]4f^7 5d^1 6s^2 \Rightarrow Gd^{3+} = 4f^7$

$\mu = \sqrt{7(7+2)} = \sqrt{63}$ B.M.

(C) $Eu(NO_3)_3$

$_{63}Eu = [Xe]4f^7 5d^0 6s^2 \Rightarrow Eu^{3+} = 4f^6$

$\mu = \sqrt{6(6+2)} = \sqrt{48}$ B.M.

Order of spin only magnetic moment is

(B) > (C) > (A)

195. (c) Europium (Eu)
Atomic No. - 63
Electronic configuration - $[Xe]4f^7 6s^2$
It shows only +2 and +3 oxidation state.

196. (a) Mischmetal is an alloy consisting mainly of lanthanoid metals.

Lan. metal $\Rightarrow \approx 95\%$; Iron $\Rightarrow \approx 5\%$

S, C, Ca, Al $\Rightarrow$ traces

197. (b) $Eu^{2+} : [Xe]4f^7$; $Ce^{3+} : [Xe]4f^1$

198. (a) Maximum oxidation state shown by uranium is + 6 and plutonium is + 7.

199. (d) Maximum oxidation state is shown by Np and Pu.

200. (b) Sm^{3+} has unpaired electron

201. (d) Among Lanthanides, radii decrease due to lanthanoid contraction as atomic number increase.

202. (c) Due to lanthanoid contraction, size of atoms as well as ions of lanthanoid decreases.

203. (a) $Yb^{3+} : [Xe]4f^{13}$, $Ce^{3+} : [Xe]4f^1$
Unlike most of the lanthanoids, the energy gap between the *f*-orbital to the next unocupied *d*-orbital is either too low or too high for Ce^{3+} and Yb^{3+}. This is due to having only one unpaired e^- in the 4*f*-subshell.
Therefore, most of the compounds with Ce^{3+} or Yb^{3+} are colourless or very light coloured.

204. (b) Gd (64) = $[Xe]4f^7 5d^1 6s^2$

$\therefore$ No. of unpaired electrons = 8

205. (c) Gd (64) = $[Xe]\,4f^7 5d^1 6s^2$
$Gd^{3+} = [Xe]\,4f^7 5d^0 6s^0$
i.e., no. of unpaired electrons = 7

$\mu = \sqrt{n(n+2)} = \sqrt{7(7+2)} = \sqrt{63} = 7.93$ BM

206. (c) Only Ce (Z = 58) shows stable (+4) oxidation state.

207. (b) Most of the Ln^{3+} compounds except La^{3+} and Lu^{3+} are coloured due to the presence of unpaired *f*-electrons.

208. (b) There is lesser energy difference between 5*f* and 6*d* orbitals as compared to that between 4*f* and 5*d* orbitals. In case of actinoids, we can remove electrons from 5*f* as well as from 6*d* and due to this, actinoids exhibit larger number of oxidation states than lanthanoids.

209. (a) 4*f* orbital is nearer to nucleus as compared to 5*f* orbital, therefore, shielding of 4*f* is more than 5*f*.

210. (a) More the distance between nucleus and outer orbitals, lesser will be force of attraction on them. Distance between nucleus and 5*f* orbitals is more as compared to distance between 4*f* orbitals and nucleus. Therefore, there is lesser energy difference between 5*f* and 6*d* than between 4*f* and 5d orbitals.

211. (b) The imperfect shielding of *f* electrons in lanthanoids is due to the shape of *f* orbitals which is very much diffused. Thus, as the atomic number increases, the nuclear charge increases by unity at each step, while no comparable increase in the mutual shielding effect of 4*f* occurs. This causes a contraction in the size of the 4*f* subshell as a result of which atomic and ionic radii decrease gradually from La to Lu.

212. (c) Due to lanthanoid contraction, the ionic radii decreases from La^{3+} to Lu^{3+}. Thus, only possible option is (c).

213. (c) Common oxidation states of Ce(cerium) are + 3 and + 4.

214. (a) In lanthanides there is a regular decrease in the atomic radii as well as ionic radii of trivalent ions as the atomic number increases from Ce to Lu. This decrease in size of atoms and ions is known as **lanthanide contraction.** Although the atomic radii do show some irregularities but ionic radii decrease from La to Lu. Thus the correct order is.

$Yb^{3+} < Pm^{3+} < Ce^{3+} < La^{3+}$

86.8pm 97pm 102pm 103pm

215. (a) The +4 oxidation state of cerium is also known in solution due to the configuration $[Xe]4f^1 5d^1 6s^2$

Co-ordination Compounds 6

1. **(d)** Homoleptic complex of type $[Ma_6]$ cannot show geometrical as well as optical isomerism. (Where a $\Rightarrow$ monodentate ligand)
Cis-platin and *trans*-platin has formula $[Pt(NH_3)_2Cl_2]$ which is a heteroleptic complex of platinum.

NH_3, NH_3, Pt, Cl, Cl — *Cis*-platin   Cl, NH_3, Pt, NH_3, Cl — *Trans*-platin

2. **(b)** Primary valency = Oxidation state
Secondary valency = Co-ordination number

	Complex	Primary valency	Secondary
(A)	$[Co(en)_2Cl_2]Cl$	3	6
(B)	$[Pt(NH_3)_2Cl(NO_2)]$	2	4
(C)	$Hg[Co(SCN)_4]$	3	4
(D)	$[Mg(EDTA)]^{2-}$	2	6

3. **(a)** Among VO_2, V_2O_3, V_2O_5, CrO_3 and Mn_2O_7.
CrO_3 = Acidic; Mn_2O_7 = Acidic
$\therefore$ x = 2

$$\overset{III}{[Co(H_2NCH_2CH_2NH_2)_3]_2}(SO_4)_3 \rightleftharpoons 2[\overset{III}{Co}(H_2NCH_2CH_2NH_2)_3]^{3+} + 3SO_4^{2-}$$

Co have +3 charge so the primary valency = 3
$\therefore$ x + y = 5

4. **(c)**

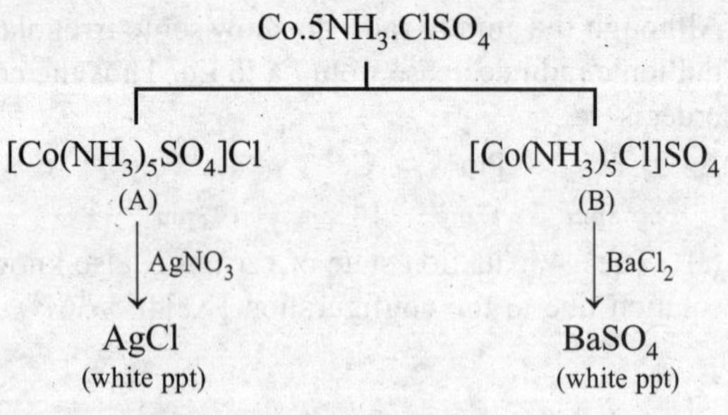

(A) complex is $[Co(NH_3)_5(SO_4)]Cl$
(B) complex is $[Co(NH_3)_5Cl]SO_4$
Both (A) and (B) are Ionisation isomers.

5. **(2)** (BrF_5 structure: Br with one lone pair and five F)

1 lone pair hence 1 mole of AgCl.
$\therefore$ Complex is $[M(NH_3)_3Cl_3]$ Cl.
It shows 2 geometrical isomers (Ma_3b_3 type); facial (fac) and meridional (Mer)

6. **(c)** $[CrCl_3(Py)_3]$ — Facial, Meridional

So total stereo isomer = 2

$[CrCl_2(ox)_2]^{3-}$ — *Cis*-isomer (Optically active), *Trans*-isomer

Geometrical isomer = 2 (1 cis + 1 trans)
Optical isomer = 3 (2 optically active + 1 optically inactive)
Stereoisomer = 3

7. **(2)** $FeCl_3 + KOH + H_2SO_4 \rightarrow \underset{(A)}{K_3[Fe(C_2O_4)_3]}$
$\Rightarrow [Fe(C_2O_4)_3]^{3-}$ is $[M(AA)_3]$ type complex. So total optical isomers = 2

8. **(a)**
A $Fe^{+6} \Rightarrow d^2$ (Homoleptic)
B $Fe^{+3} \Rightarrow d^5$ (Homoleptic)
C $Fe^{+3} \Rightarrow d^5$
D $Co^{+2} \Rightarrow d^7$ (Homoleptic)
E $Co^{+3} \Rightarrow d^6$

9. **(a)** Ma_3b_3 type complexes show Facial - Meridional isomerism
(a) $[Co(NH_3)_3Cl_3] \Rightarrow Ma_3b_3$
(b) $[Co(NH_3)_4Cl_2]^+ \Rightarrow Ma_4b_2$
(c) $[Co(en)_3]^{3+} \Rightarrow M(AA)_3$
(d) $[Co(en_2)Cl_2]^+ \Rightarrow M(AA)_2b_2$
a, *b* = NH_3, Cl^- (Monodentate ligands)
AA = en (Bidentate ligand)

10. **(14)** $[Ni(dmg)_2]$

($CH_3-C=N$, $N=C-CH_3$, Ni, O–H------O^- hydrogen-bonded structure)

Number of H-atoms = 14

11. **(c)** (A) $[Fe(CN)_5NO]^{-2} \rightarrow$ Heteroleptic, Fe^{+2}, $3d^6$, $t_{2g}^6e_g^0$, d^2sp^3, Low spin (3d series + SFL) (0) unpaired electron

(B) $[CoF_6]^{-3} \rightarrow$ Homoleptic, sp^3d^2, High spin, Co^{+3}, $3d^6$ (3d series + WFL) (4) unpaired e^-

(C) $[Fe(CN)_6]^{-4} \rightarrow$ Homoleptic Fe^{+2}, $3d^6$, d^2sp^3, $t_{2g}^6 e_g^0$ Low spin (3d series + SFL) (0) unpaired e^-

(D) $[Co(NH_3)_6]^{+3} \rightarrow$ Homoleptic, Co^{+3}, $3d^6$, d^2sp^3, $t_{2g}^6 e_g^0$, Low spin (3d series + SFL) (0) unpaired e^-

(E) $[Cr(H_2O)_6]^{+2} \rightarrow$ Homoleptic, Cr^{+2}, $3d^4$, sp^3d^2, High spin $t_{2g}^3 e_g^1$, (3d series + WFL) (4) unpaired

12. **(3)** Ligands which have two differnt donor sites but at a time connects with only one donor site to central metal are ambidentate ligands.
Ambidentate ligands are NO_2^-; SCN^-; CN^-

13. **(c)** As on reaction with excess of $AgNO_3$ solution 2 moles of AgCl formed.
So $2Cl^-$ ions should be ionisable.

$$\overset{+3}{[Co(NH_3)_nCl]}Cl_2 + \text{excess } AgNO_3 \rightarrow 2AgCl \ (2 \text{ moles})$$

oxidation state of Co:
$x + n(0) - 1 = +2$
$\Rightarrow x = +3$
Complex is octahedral so,
$n = 5$
$\therefore x + n = 8$

14. **(c)** The correct IUPAC name of $[PtBr_2(PMe_3)_2]$ is dibromobis (trimethylphosphine) platinum (II).

15. **(b)** $[MABXL] \rightarrow sp^3 \rightarrow$ Tetrahedral
Tetrahedral complexes do not show geometrical isomerism as they have same relative position.

16. **(a)** Complex having all ligands same are called homoleptic complex. $[Ni(CN)_4]^{2-}$ has only CN^- ligands.

17. **(4)** trans-$[Pt(en)_2Cl_2]^{2+}$ and trans-$[Cr(ox)_2Cl_2]^{3-}$ do not show optical isomerism as they are symmetrical in nature.

18. **(b)** Here, Mn is in +6 state and the suffix-ate is used as $[MnO_4]^{2-}$ is an anionic entity.
Thus, following the IUPAC convention, the correct IUPAC name will be Potassium tetraoxidomanganate (VI).
The term 'oxo' is a common name while 'oxido' is the IUPAC term for oxygen here.

19. **(4)** $\underset{0.01M,\ 20\ mL}{[Cr(H_2O)_5Cl]Cl_2} + \underset{0.1\ M}{2AgNO_3} \rightarrow$
$\Rightarrow$ For 0.2 milli mole $AgNO_3$ required = 0.4 milli mole
$0.4 = 0.1 \times V \Rightarrow V = 4$ mL

20. **(b)** $[Cr(ox)_2ClBr]^{-3}$
- Trans structure has plane of symmetry. So no optical isomerism will be shown.
- Cis structure does not contain plane of symmetry. So two forms d as well as l will be shown.

21. **(a)** Ambidentate ligand are the ligand which has two different donor atoms. For example NO_2^- and NCS^-.

22. **(a)** $[MA_3B_3]$ type of compound exists are facial and meridonial isomer.

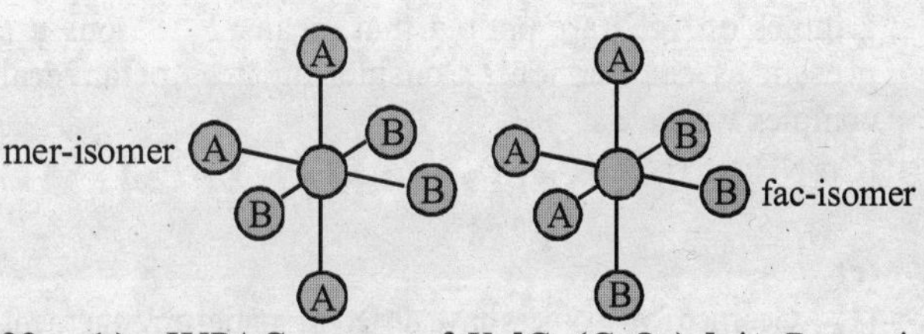

23. **(a)** IUPAC name of $K_3[Co(C_2O_4)_3]$ is Potassium trioxalatocobaltate(III)

24. **(4)** Ambidentate ligand has two donor site, so ambidentate ligand is SCN.

25. **(c)** $[Co(NH_3)_5NO_2]^{2+}$; $[Co(NH_3)_5ONO]^{2+}$
Two linkage isomers possible as NO_2 is an ambidentate ligand.

26. **(d)** The double salts dissociates completely into the constituting ions on dissolution. This is possible for $FeSO_4 \cdot (NH_4)_2SO_4 \cdot 6H_2O$ and $K_2SO_4 \cdot Al_2(SO_4)_3 \cdot 24H_2O$.

27. **(b)** $\underset{0.01\ mol}{[Co(NH_3)_5SO_4]Br} + \underset{excess}{AgNO_3} \rightarrow \underset{0.01\ Mol}{AgBr\downarrow}$
$\underset{0.01\ mol}{[Co(NH_3)_5Br]SO_4} + \underset{excess}{BaCl_2} \rightarrow \underset{0.01\ Mol}{BaSO_4\downarrow}$
Note: Concentration term is 0.02 M in 2L of solution after mining. Hence number of moles should be 0.01 for each complex.

28. **(a)** Absence of symmetry elements (plane of symmetry centre of symmetry)

29. **(4)** Denticity of the ligand is 4.

30. **(5)** $[Co(NH_3)_4Cl_2]Cl \Rightarrow$ Gives 1 mole AgCl
$[Ni(H_2O)_6]Cl_2 \Rightarrow$ Gives 2 moles AgCl
$[Pt(NH_3)_2]Cl_2] \Rightarrow$ Gives No AgCl
$[Pd(NH_3)_4]Cl_2 \Rightarrow$ Gives 2 moles AgCl
Total number of moles of AgCl = 5 mole.

31. **(d)** Oxidation number of Co is + 3 in the given complex. So primary valency is 3.
It is an octahedral complex so secondary valency is 6.

32. **(9)** Coordination no. = 6
Oxidation state is : $x - 4 = -1$
$x = +3$; Sum = 3 + 6 = 9.

33. **(3)** The isomers of $[Cu(en)_2(SCN)_2]$ are :

SCN–Cu(en)₂–SCN	NCS–Cu(en)₂–NCS	SCN–Cu(en)₂–NCS

34. **(3)** For 1 : 1 electrolyte, the complex must be $[Co(NH_3)_4Cl_2]$ Cl. Oxidation number of Co in $[Co(NH_3)_4Cl_2]^+$ is +3.

35. **(1)** $[Fe(H_2O)_6]^{2+} + NO \longrightarrow [Fe(H_2O)_5(NO)]^{2+} + H_2O$
Reaction for brown ring test
NO ligand exists as NO^+

36. **(a)** Only the counter ions are replaced easily by other reagents.
$[Co(NH_3)_5Br]SO_4 + BaCl_2 \rightarrow [Co(NH_3)_5Br]Cl_2 + BaSO_4\downarrow$ White ppt.
It does not have chloride ion as counter ion so it will not give white ppt with $AgNO_3$.

37. **(5)** $CoCl_3.xNH_3 + AgNO_3 \rightarrow \underset{2\ mol}{AgCl\downarrow}$

2 moles of AgCl are ppt out that means 2 Cl^- ions are present as counter ions. Considering it as octahedral complex we have

$[Co(NH_3)_5Cl]Cl_2 + AgNO_3 \rightarrow AgCl\downarrow$
$x = 5$ 2 mol

38. **(c)**

39. **(1)** Geometrical isomers have different spatial arrangement of atoms

(A) $CoCl_3.4NH_3 = [Co(NH_3)_4Cl_2]Cl$ can exhibit GI.

[Structures: Co with Cl (top), Cl (bottom), H_3N, H_3N, NH_3, NH_3 — *trans*-isomer; Co with Cl (top), Cl, H_3N, H_3N, NH_3, NH_3 (bottom) — *cis*-isomer]

trans-isomer *cis*-isomer

(B) $CoCl_3.5NH_3 = [Co(NH_3)_5Cl]Cl_2$ can't exhibit G.I.

(C) $CoCl_3.6NH_3 = [Co(NH_3)_6]Cl_3$ can't exhibit G.I.

(D) $CoCl(NO_3)_2.5NH_3 = [Co(NH_3)_5Cl](NO_3)_2$

OR

$= [Co(NH_3)_5(NO_3)]Cl(NO_3)$

Both can't exhibit G.I.

40. **(c)** $AgCl + 2NH_3 \rightarrow \underset{\text{soluble}}{[Ag(NH_3)_2]^+Cl^-}$

41. **(2)** $\left[\overset{+1}{Ag}(NH_3)_2\right]^+\left[\overset{+1}{Ag}(CN)_2\right]^-$

42. **(a)** $NH_2 - \overset{O}{\overset{\|}{C}} - NH - \overset{O}{\overset{\|}{C}} - NH_2$ (both NH_2 → M)

Biuret is bidenate ligand

43. **(1)** One effective hydrogen bonded H_2O molecule.

44. **(2)** $MCl_3 \cdot 2L \Rightarrow$ octahedral complex (Given)

$\underset{1\text{mole}}{MCl_3 \cdot 2L} \xrightarrow{\text{Ex. }AgNO_3} 1$ mole of AgCl (Given)

Its means that one Cl^- ion is present in ionization sphere.

$\therefore$ Formula $= [MCl_2L_2]Cl$

For octahedral complex, coordination no. is 6.

$\therefore$ L acts as bidentate ligand.

45. **(2)** The number of optical isomers for $[Cr(C_2O_4)_3]^{3-}$ is two.

46. **(5)** Mohr's salt $\Rightarrow (NH_4)_2Fe(SO_4)_2 \cdot 6H_2O$

$K_2SO_4.Al_2(SO_4) \cdot 24H_2O$ Potash alum

$\Rightarrow KAl(SO_4)_2 \cdot 12H_2O$

Required ratio $= \frac{6}{12} = 0.5 = 5\times10^{-1}, \frac{6}{24} = \frac{1}{4} = 2.5\times10^{-1}$

47. **(b)** $[Co(CN)_5(NC)]^{3-}$ cannot show geometrical isomerism.

48. **(c)** $[PtCl_2(NH_3)_2]$

[Structures: Pt with H_3N, H_3N, Cl, Cl — *cis*; and Pt with H_3N, Cl, Cl, NH_3 — *trans*]

cis *trans*

$[Ni(CO)_4]$: All ligands are same.

Hence, zero geometrical isomers.

$[Ru(H_2O)_3Cl_3]$

[Structures: Ru^{+3} with Cl, Cl, Cl, H_2O, H_2O, OH_2 — *mer–*; and Ru^{+3} with Cl, Cl, Cl, H_2O, H_2O, OH_2 — *fac–*]

mer– *fac–*

$[CoCl_2(NH_3)_4]^+$

[Structures: Co^{+3} with Cl, Cl, H_3N, H_3N, NH_3, NH_3 — *trans*; and Co^{+3} with Cl, NH_3, H_3N, H_3N, NH_3, Cl — *cis*]

trans *cis*

49. **(2)** Triamminetrinitrocobalt (III) $\rightarrow [Co(NO_2)_3(NH_3)_3]$

[Structures: Co^{+3} with NO_2, NO_2, NO_2, H_3N, H_3N, NH_3; and Co^{+3} with NO_2, NH_3, H_3N, H_3N, NO_2, NO_2]

Two geometrical isomers (X = 2)

trioxalatochromate (III) ion $\rightarrow [Cr(C_2O_4)_3]^{3-}$

Zero geometrical isomer (Y = 0)

$X + Y = 2 + 0 = 2$

50. **(6)** $3[Co(en)_2Cl_2]Cl + \underset{(\text{excess})}{3AgNO_3} \rightarrow \underset{(\text{white ppt.})}{3AgCl}$

Secondary valency (C.N.) of Co = 6

51. **(0)**

$\underset{(1\text{ mole})}{[Cr(H_2O)_3(NH_3)_3]Cl_3} \longrightarrow \underset{(1\text{ mole})}{[Cr(H_2O)_3(NH_3)_3]^{3+}} + \underset{(3\text{mole})}{3Cl^-}$

$\xrightarrow{AgNO_3\text{ (Excess)}} 3AgCl$

None of the chloride ion is directly bonded to metal ion. Hence, number of chloride ions satisfying the secondary valency of the metal ion is zero.

52. **(b)** Ethylene diamine (en) is a chelating ligand. In $[Co(en)_3]Cl_2$, three rings are formed as chelating ring by ethylene diamine, so it will be most stable.

53. **(3)** $K_3[Cr(oxalate)_3] \equiv 3K^+[Cr(C_2O_4)_3]^{3-}$

Cr (24) $\Rightarrow 3d^5 4s^1$ or $Cr^{3+} \Rightarrow 3d^3$

3*d*

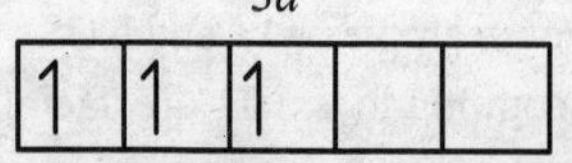

$\therefore$ No. of unpaired electrons = 3

54. (6) $Fe^{3+} + 3K^{+} + 3C_2O_4^{2-} \longrightarrow \underset{(A)}{K_3[Fe(C_2O_4)_3]}$

Secondary valency of Fe in (A) is 6.

55. (2)

$[Co(NH_3)_4Cl_2]^{+} \xrightarrow[+2\,en]{-4\,NH_3} [Co(en)_2Cl_2]^{+}$

In given complex, NH_3 is a neutral ligand. Each ethylene diamine replaces two NH_3 ligands. So, two ethylene diamine are required to replace all neutral monodentate (NH_3) ligands.

56. (3) Total number of stereoisomers possible for $[Co(ox)_2Br(NH_3)]^{2-}$ is 3.

$[Co(ox)_2Br(NH_3)]^{2-}$ (Br and NH_3 trans)

trans

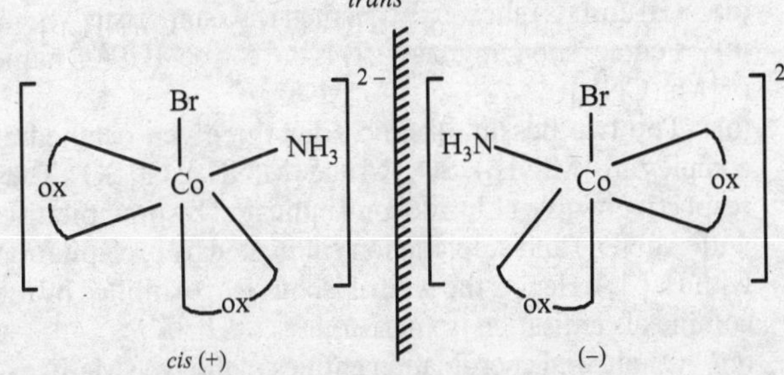

57. (6)

$$\begin{matrix} \overset{5}{^-OOC} - CH_2 \\ \overset{6}{^-OOC} - CH_2 \end{matrix} \overset{1}{\ddot{N}} - CH_2 - CH_2 - \overset{2}{\ddot{N}} \begin{matrix} CH_2 - \overset{3}{COO^-} \\ CH_2 - \overset{4}{COO^-} \end{matrix}$$

Therefore, $EDTA^{4-}$ is a hexadentate ligand.

58. (d) *trans*-$[Co(en)_2Cl_2]^{+}$ (A) contains a plane of symmetry so (A) cannot be optically active, whereas, (B) *cis* $[Co(en)_2Cl_2]^{+}$ does not contain any plane of symmetry, so (B) will be optically active.

59. (d) $[Pt(en)(NO_2)_2]$

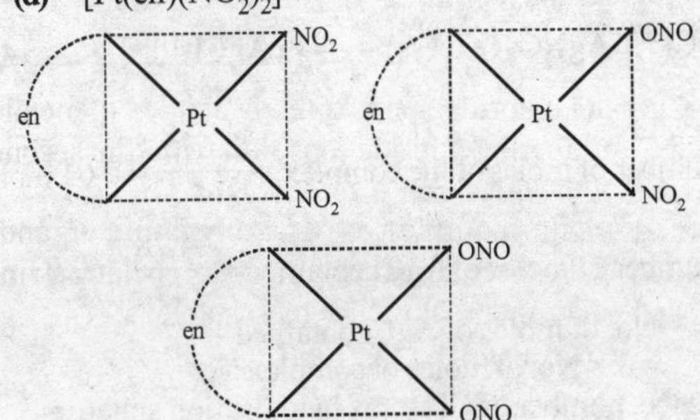

So, total possible isomers are 3.

60. (d) Only *cis*-$[CrCl_2(ox)_2]^{3-}$ shows optical isomerism while its *trans* form does not show optical isomerism due to presence of plane of symmetry.

61. (d) Conc. H_2SO_4 acts as dehydrating agent.
Molar mass of given complex = 266.5 g/mol.
On treating with conc. H_2SO_4 the mass lost by the

complex $= \dfrac{13.5}{100}(266.5) \approx 36\,g = 2$ moles of H_2O

Formula of the complex $= [Cr(H_2O)_4Cl_2]Cl \cdot 2H_2O$

62. (c)

(a) $[Ni(NH_3)_4(H_2O)_2]^{2+}$ shows geometrical isomerism.

(b) $[Ni(en)_3]^{2+}$ shows optical isomerism.

(c) $[Ni^{2+}(NH_3)_2Cl_2] \Rightarrow Ni^{2+} \Rightarrow 3d^8 4s^0$
$\Rightarrow sp^3$ hybridisation $\Rightarrow$ tetrahedral
So, $[Ni(NH_3)_2Cl_2]$ does not show isomerism.

(d) $[Pt(NH_3)_2Cl_2]$ shows geometrical isomerism.

63. (d) $\mu = \sqrt{n(n+2)}$ B.M. = 3.83 B.M.
$n = 3$ (n = No. of unpaired e^-)
Therefore, oxidation number of Cr should be +3.
Hence complex is $Cr(H_2O)_6Cl_3$.
Complex shows geometrical isomerism, therefore formula of complex is $[Cr(H_2O)_4Cl_2]Cl{\cdot}2H_2O$.
IUPAC Name:
Tetraaquadichloridochromium(III) chloride dihydrate

64. (d)

65. (d) $[Ma_3b_3]$ type complex shows facial and meridional isomerism.
So, the complex $[Co(NH_3)_3(NO_2)_3]$ will show fac and mer-isomers.

66. (c) $[Pt(NH_3)_2Cl(NO_2)]$ and $[Pt(NH_3)_4ClBr]^{2+}$ are the Ma_4bc and Ma_2bc type of complexes.
Each shows two geometrical isomers, i.e., *cis* and *trans* isomers.

67. (a) $[Pt^{(+2)}(N^{(0)}H_3)_2Cl^{(-1)}(NH_2C^{(0)}H_3)]^{+}Cl^{-}$
Diamminechlorido (methanamine) platinum (II) chloride

68. (c) Tetrahedral (sp^3) and square planar (dsp^2) complexes do not show optical isomerism. $[MA_2B_2]$ will not exhibit optical isomerism in both conditions.

69. (d) $[Co(en)_2\,Cl]\,Cl$
Cl^- - monodentate ligand
en - bidentate ligand
$\therefore$ Co-ordination Number of Co $= (2 \times 2) + 1 = 5$
$K_3[Al(C_2O_4)_3]$
$C_2O_4^{2-}$- bidentate ligand
$\therefore$ Co-ordination Number of Al $= 2 \times 3 = 6$

70. (d)

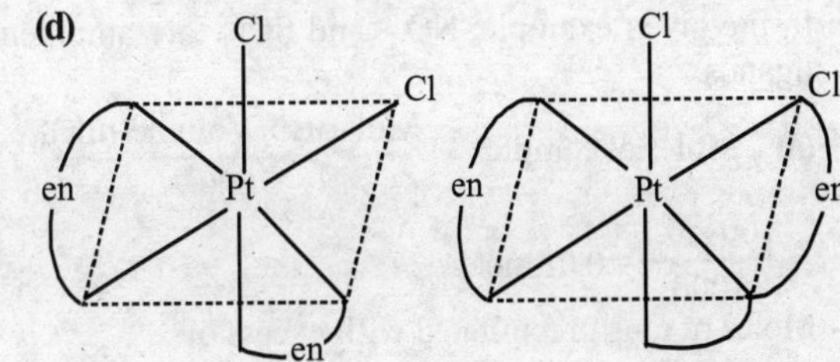

cis *trans*

cis-trans isomerism is possible with $[Pt\,(en)_2Cl_2]^{2+}$. $[Cr\,(en)_2Ox]^{+}$ shows only optical isomerism. The other two complexes, i.e., $[Pt\,(en)\,Cl_2]$ and $[Zn\,(en)Cl_2]$ do not show stereoisomerism.

71. **(b)** The maximum possible denticites of the given ligand towards transition metal ion is 6 and towards inner transition metal ion (due to greater ionic radii and more atomic orbitals) is 8.

72. **(c)**

No plane of symmetry or centre of symmetry is present. Hence, it will be optically active.

73. **(b)** It has four atoms (2 O and 2 N) containing lone pair of e^-, therefore, it will be able to donate these lone pairs and acts as a tetradentate ligand.

74. **(d)** $K_4[Th(C_2O_4)_4(H_2O)_2]$

$C_2O_4^{2-}$ (oxalato) : bidentate ligand

H_2O (aqua): monodentate

$\therefore$ Co-ordination no. of Th $= 2 \times 4 + 2 = 10$

75. **(d)**

Possible complexes	No. of isomers
$[M(F)(Cl)(SCN)(NO_2)]$	$\rightarrow$ 3 (Two *cis* and one *trans*)
$[M(F)(Cl)(SCN)(ONO)]$	$\rightarrow$ 3
$[M(F)(Cl)(NCS)(NO_2)]$	$\rightarrow$ 3
$[M(F)(Cl)(NCS)(ONO)]$	$\rightarrow$ 3
	Total = 12

76. **(a)** Reaction for the given condition can be written as:

$[CoCl_6]^{3-} + 2(en) \longrightarrow [CoCl_2(en)_2]^+$

(1:2 mole ratio) (*cis-trans*-isomer)

A = optically active (*cis*-isomer), violet

B = optically inactive (*trans*-isomer), green

77. **(b)**

78. **(c)** $[Co(H_2O)Cl(en)_2]Cl$

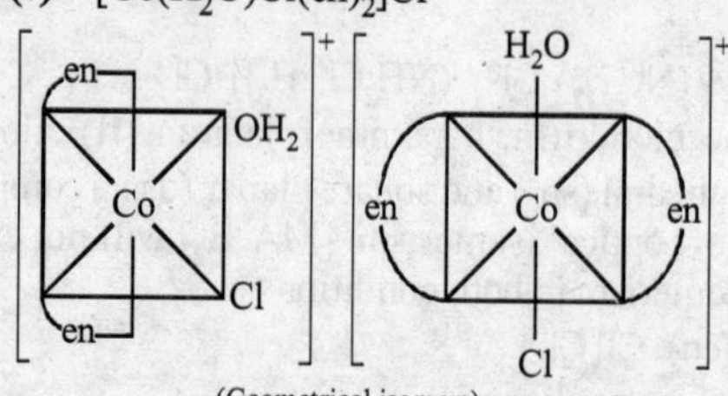

(Geometrical isomers)

79. **(b)** The square planar complex of the type $[Mabcd]^{n\pm}$, where all four ligands are different, has 3 geometrical isomers. But if one of the ligands is ambidentate, then $2 \times 3 = 6$ geometrical isomers are possible. But if two ligands are ambidentate, then $2 \times 2 \times 3 = 12$ geometrical isomers are possible.

In the given example, NO_2^- and SCN^- are ambidentate ligands.

80. **(d)** Moles of complex $= \dfrac{\text{Molarity} \times \text{Volume(mL)}}{1000}$

$= \dfrac{100 \times 0.1}{1000} = 0.01$ mole

Moles of ions precipitated with excess of

$AgNO_3 = \dfrac{1.2 \times 10^{22}}{6.02 \times 10^{23}} = 0.02$ moles

$\therefore$ Number of Cl^- present in ionisation sphere

$= \dfrac{\text{Number of moles of ions precipitated}}{\text{Number of moles of complex}} = \dfrac{0.02}{0.01} = 2$

It means $2Cl^-$ ions present in ionization sphere.

Thus, formula of the complex is, $[Co(H_2O)_5Cl]Cl_2 \cdot H_2O$

81. **(a)** $Cr\overset{2-}{O_2}\overset{-1}{Cl_2}$: Cr = +6; MnO_2 : Mn = +4

(Highest oxidation states)

(b) $[Ni\overset{-1}{Cl_4}]^{2-}$: Ni = +2; $[Co\overset{-1}{Cl_4}]^{2-}$: Co = +2

(c) $[Fe\overset{-1}{(CN)_6}]^{3-}$: Fe = +3; $[Cu\overset{-1}{(CN)_4}]^{2-}$: Cu = +2

(d) $[Fe\overset{-1}{Cl_4}]^-$: Fe = +3; $Co_2\overset{-2}{O_3}$: Co = +3

82. **(d)** More equivalents of $AgNO_3$ aqueous solution will be consumed if complex will furnish more Cl^- ions in solution. Hence, complex $[Cr(H_2O)_6]Cl_3$ will consume more equivalents of aqueous solution of $AgNO_3$.

83. **(a)** Complexes having only one type of ligands are examples of homoleptic complex.

84. **(d)** Square planar complexes of type M[ABCD] form three isomers. Their position may be obtained by fixing the position of one ligand and placing at the *trans* position any one of the remaining three ligands one by one.

85. **(d)** (B) and (C) show only geometrical isomerism.

86. **(b)** Pentacyanomangnate (III) is the correct IUPAC name of $[Mn(CN)_5]^{2-}$.

87. **(b)** The two possible isomers for the given octahedral complex are $[M(NH_3)_5SO_4]Cl$ and $[M(NH_3)_5Cl]SO_4$. They respectively give chloride ion (indicated by precipitation with $AgNO_3$) and sulphate ion (indicated by precipitation with $BaCl_2$). Hence, the type of isomerism exhibited by the complex is ionisation isomerism.

88. **(c)** Octahedral coordination entities of the type Ma_3b_3 exhibit geometrical isomerism. The compound exists as facial and meridional isomers, both contain plane of symmetry.

89. **(a)** The compound shows linkage isomerism because the ligand in the compound is an ambidenate ligand that can bond at more than one atomic site.

i.e., $:\overline{NCS}$ and $:\overline{SCN}$

90. **(b)** $[Cr(en)_2Br_2]Br$

dibromidobis(ethylenediamine) chromium (III) bromide.

91. **(b)** Shows cis and trans isomer.

92. **(a)** $\underset{2.675\ g}{CoCl_3 \cdot 6NH_3} \longrightarrow xCl^-$

$xCl^- + AgNO_3(\text{excess}) \longrightarrow \underset{4.78\ g}{xAgCl\downarrow}$

Number of moles of the complex $= \dfrac{2.675}{267.5} = 0.01$ mol

Number of moles of AgCl obtained $= \dfrac{4.78}{143.5} = 0.03$ mol

$\therefore$ No. of moles of AgCl obtained $= 3 \times$ No. of moles of complex

Hence, number of Cl atoms in ionisation sphere = 3

93. **(b)** Complexes of Zn^{2+} cannot show optical isomerism as they are tetrahedral complexes with plane of symmetry. $[Co(H_2O)_4(en)]^{3+}$ has two planes of symmetry hence it is also optically inactive.

94. **(c)**

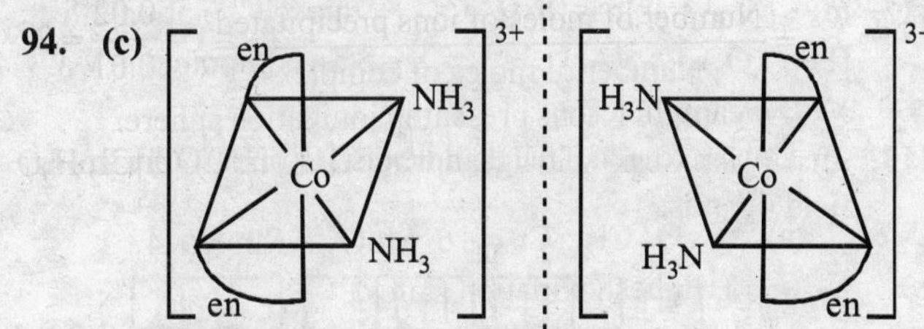

Enantiomers of *cis*-$\left[Co(en)_2(NH_3)_2\right]^{3+}$

95. **(a)** The SCN^- ion can coordinate through S or N atom giving rise to linkage isomerism

$M \leftarrow SCN$ thiocyanato; $M \leftarrow NCS$ isothiocyanato.

96. **(d)** In the given complex we have two bidentate ligands (i.e, en and C_2O_4), so coordination number of E is 6 $(2 \times 2 + 1 \times 2 = 6)$

Let the oxidation state of E in complex be x, then

$x - 2 = 1$ or $x = +3$, so its oxidation state is + 3.

97. **(b)** $[Co(NO_2)(NH_3)_5]Cl_2$

Pentaamminenitrito-N-cobalt (III) chloride

98. **(d)** Let, oxidation state of Cr in $[Cr(NH_3)_4Cl_2]^+$ be x, 1 $\times x + 4 \times 0 + 2(-1) = 1$. Therefore, $x = 3$.

99. **(c)** $K_3[Fe(CN)_6]$ is potassium hexacyanoferrate (III).

100. **(c)** The coordination number of central metal atom in a complex is equal to number of monovalent ligands, twice the number of bidentate ligands and so on, around the metal ion bonded by coordinate bonds.

Hence coordination number = No. of σ bonds formed by metals with ligands.

101. **(d)** Isomers

$[Ru(NH_3)_4Cl_2]^+$ (*cis* and *trans*), $[Co(NH_3)_5Cl]^{2+}$ (none),

$[Ir(PR_3)_2H(CO)]^{2+}$ (*cis* and *trans*), $[Co(en)_2Cl_2]^+$ (*cis* and *trans* and optical isomers)

102. **(a)** CN^- ion acts good complexing as well as reducing agent.

103. **(d)** $Co(NH_3)_5Cl_3 \rightleftharpoons [Co(NH_3)_5Cl]^{2+} + 2Cl^-$

∴ Structure is $[Co(NH_3)_5Cl]Cl_2$.

Now $[Co(NH_3)_5Cl]Cl_2 + 2AgNO_3$

$\longrightarrow [Co(NH_3)_5Cl](NO_3)_2 + 2AgCl$

104. **(a)** Let the O. N. of Ni in $K_4[Ni(CN)_4]$ be $= x$ then

$4(+1) + x + (-1) \times 4 = 0 \Rightarrow 4 + x - 4 = 0; x = 0$

105. **(b)** The chemical formula of nitropentaamminechromium (III) chloride is:

$\left[Cr(NH_3)_5(NO_2)\right]Cl_2$

It can exist in following two structures

$\left[Cr(NH_3)_5(NO_2)\right]Cl_2$ and

Pentaamminenitrochromium (III) chloride

$\left[Cr(NH_3)_5ONO\right]Cl_2$

Pentaamminenitritochromium (III) chloride

Therefore, the type of isomerism found in this compound is linkage isomerism as NO_2 group is linked through N as $-NO_2$ or through O as $-ONO$.

106. **(c)**

(A) $[Ni(CO)_4]$, Ni^0 : [Ar]$3d^8 4s^2$

Valence orbitals of Ni^0 in pre-hybridisation state :

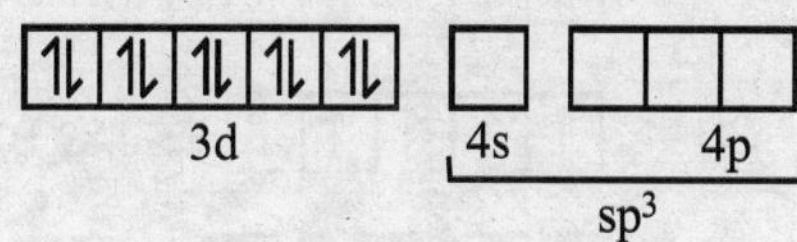

Tetrahedral, Diamagnetic, $\mu = 0$ B.M.

(B) $[Ni(CN)_4]^{2-}$, Ni^{+2} : [Ar]$3d^8 4s^0$

Valence orbitals of Ni^{+2} in pre-hybridisation state :

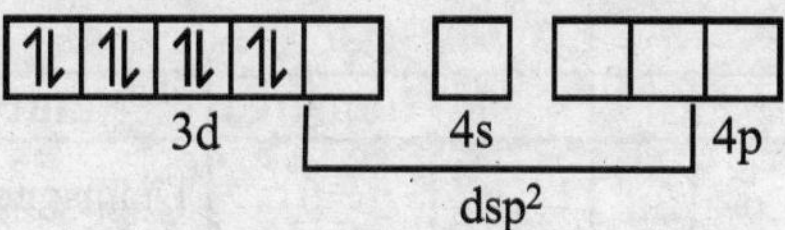

Square planar, Diamagnetic, $\mu = 0$ B.M.

(C) $[NiCl_4]^{2-}$, Ni^{+2}, [Ar]$3d^8 4s^0$

Valence orbitals of Ni^{+2} in ground state :

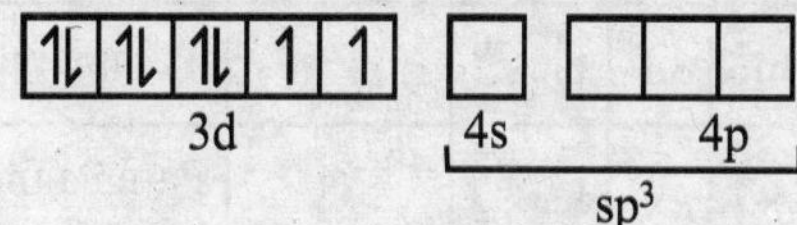

Tetrahedral, paramagnetic, $\mu = \sqrt{8} = 2.8$ B.M.

(D) $[MnBr_4]^{2-}$, Mn^{+2} : [Ar]$3d^5$

Valence orbitals of Mn^{+2} in ground state :

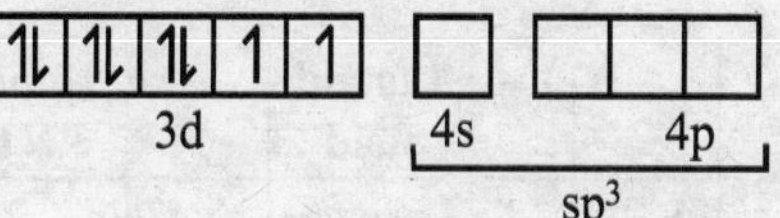

Tetrahedral, paramagnetic, $\mu = \sqrt{35} = 5.9$ B.M.

107. **(d)**

(i) $[Co(NH_3)_6]^{3+}$

$Co^{3+} \Rightarrow 3d^6$; $t_{2g}^6 e_g^0$

strong field ligand, d^2sp^3 hybridisation

(ii) SF_6

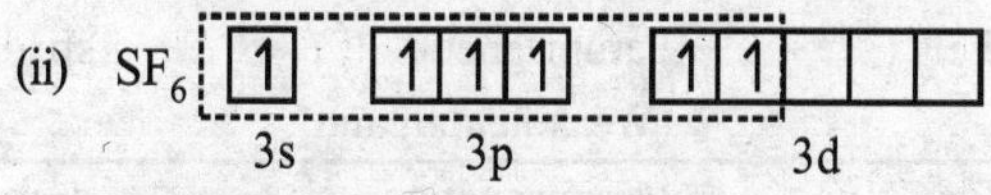

sp^3d^2 hybridisation

(iii) $[CrF_6]^{3+}$

$Cr^{3+} \Rightarrow 3d^3$; $t_{2g}^3 e_g^0$, d^2sp^3 hybridisation

(iv) $[CoF_6]^{3-}$

$Co^{3+} \Rightarrow 3d^6$; $t_{2g}^4 e_g^2$, sp^3d^2 hybridisation

(v) $[Mn(CN)_6]^{3-}$

$Mn^{3+} \Rightarrow 3d^4$; $t_{2g}^3 e_g^0$,

strong field ligand, d^2sp^3 hybridisation

(vi) $[MnCl_6]^{3-}$

$Mn^{3+} \Rightarrow 3d^4$; $t_{2g}^4 e_g^1$, sp^3d^2 hybridisation

Thus total number of sp^3d^2 hybridised molecule is 3.

108. (c) $Cu^+ \Rightarrow 3d^{10} \Rightarrow$ [⇅|⇅|⇅|⇅|⇅] $\Rightarrow 0$ BM

$Cu^{2+} \Rightarrow 3d^9 \Rightarrow$ [⇅|⇅|⇅|⇅|↑] $\Rightarrow 1.73$ BM

$Cr^{2+} \Rightarrow 3d^4 \Rightarrow$ [↑|↑|↑|↑|] $\Rightarrow 4.89$ BM

$Cr^{3+} \Rightarrow 3d^3 \Rightarrow$ [↑|↑|↑| |] $\Rightarrow 3.87$ BM

So order :

$Cr^{2+} > Cr^{3+} > Cu^{2+} > Cu^+$

109. (2)

				Unpaired	Nature
$[Co(NH_3)_6]^{3+}$	Co^{3+}	$3d^6$	$t_{2g}^6\ e_g^0$	0	Diamagnetic
$[Co(C_2O_4)_3]^{3-}$	Co^{3+}	$3d^6$	$t_{2g}^6\ e_g^0$	0	Diamagnetic
$[MnCl_6]^{3-}$	Mn^{3+}	$3d^4$	$t_{2g}^3\ e_g^1$	4	Paramagnetic
$[Mn(Cl)_6]^{3-}$	Mn^{3+}	$3d^4$	$t_{2g}^4\ e_g^0$	2	Paramagnetic
$[CoF_6]^{3-}$	Co^{3+}	$3d^6$	$t_{2g}^4\ e_g^2$	4	Paramagnetic
$[Fe(CN)_6]^{3-}$	Fe^{3+}	$3d^5$	$t_{2g}^5\ e_g^2$	1	Paramagnetic
$[FeF_6]^{3-}$	Fe^{3+}	$3d^5$	$t_{2g}^3\ e_g^2$	5	Paramagnetic

110. (a)

			Ligand field		Unpaired electrons
$[Fe(CN)_6]^{3-}$	Fe^{3+}	$3d^5$	strong	$t_{2g}^5\ e_g^0$	1
$[FeF_6]^{3-}$	Fe^{3+}	$3d^5$	weak	$t_{2g}^3\ e_g^2$	5
$[CoF_6]^{3-}$	Co^{3+}	$3d^6$	strong	$t_{2g}^4\ e_g^2$	4
$[Mn(CN)_6]^{3-}$	Mn^{3+}	$3d^4$	strong	$t_{2g}^4\ e_g^0$	2

111. (2)

$[FeF_6]^{3-}$	Paramagnetic (Weak field ligand)	sp^3d^2
$[Fe(CN)_6]^{3-}$	Paramagnetic (Strong field ligand)	d^2sp^3
$[Mn(CN)_6]^{3-}$	Paramagnetic (Strong field ligand)	d^2sp^3
$[Co(C_2O_4)_3]^{3-}$	Diamagnetic (Strong field ligand)	d^2sp^3
$[MnCl_6]^{3-}$	Paramagnetic (Weak field ligand)	sp^3d^2
$[CoF_6]^{3-}$	Paramagnetic (Weak field ligand)	sp^3d^2

Only $[Fe(CN)_6]^{3-}$ and $[Mn(CN)_6]^{3-}$ are paramagnetic and d^2sp^3 hybridised.

112. (d) Most stable is $[Fe(C_2O_4)_3]^{3-}$ due to Chelation effect. $[Fe(C_2O_4)_3]^{3-}$: [Ar] $t_{2g}^5e_g^0 \Rightarrow X = 5$ So compound is V_2O_5. V_2O_5 is amphoteric.

113. (d) F^- is a weak-field ligand while CN^- and CO are strong-field ligands.

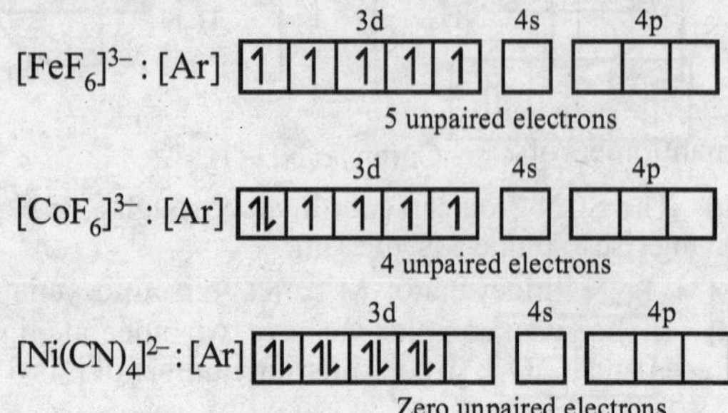

3d 4s 4p

$[Ni(CO)_4]$: [Ar] [⇅|⇅|⇅|⇅|⇅] [ ] [| |]

Zero unpaired electrons

114. (d)

(a) $[Fe(CN)_6]^{-4}$ $Fe^{2+} \Rightarrow 3d^64s^0$

$\overset{\ominus}{C}N$ is SFL — e_g [|]; t_{2g} [⇅|⇅|⇅] $\mu = 0$

CFSE $= -2.4\Delta_0$

(b) $[Co(NH_3)_6]^{3+}$ $Co^{3+} \Rightarrow 3d^64s^0$

NH_3 is strong field ligand — e_g [↑|↑]; t_{2g} [⇅|↑|↑]

$= [-0.4 \times 6 + 0.6 \times (0)]\ \Delta_0 = -2.4\ \Delta_0$

(c) $[FeF_6]^{4-}$ $Fe^{2+} \Rightarrow 3d^64s^0$

e_g [↑|↑]; t_{2g} [⇅|↑|↑]

$\mu = \sqrt{24}$ B.M. $= 4.89$B.M.

CFSE $= (-0.4 \times 4 + 0.6 \times 2)\Delta_0 = -1.2\Delta_0$

(d) $[Mn(SCN)_6]^{4-}$

$Mn^{2+} \Rightarrow 3d^5\ 4s^0$

$\overset{\ominus}{S}CN$ is weak field — [↑|↑]; [↑|↑|↑]

$\mu = \sqrt{35}$ B.M. $= 5.96$ B.M.

CFSE $= (-0.4 \times 3 + 0.6 \times 2)\ \Delta_0$

So $\Delta_0 = 0$

115. (d) (A) $[Mn(CN)_6]^{3-}$; Mn^{+3}

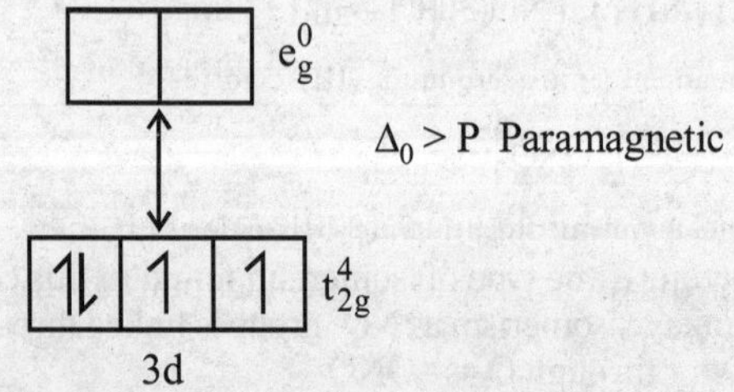

$\Delta_0 > P$ Paramagnetic

(B) $[Co(NH_3)_6]^{3+}$; Co^{+3}

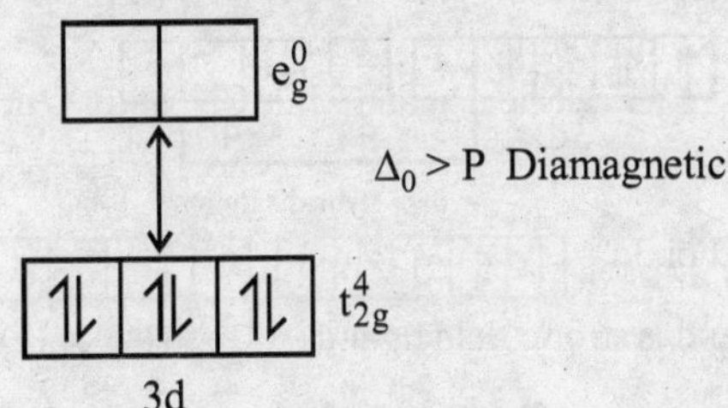

(C) $[Fe(CN)_6]^{4-}$; Fe^{+2}

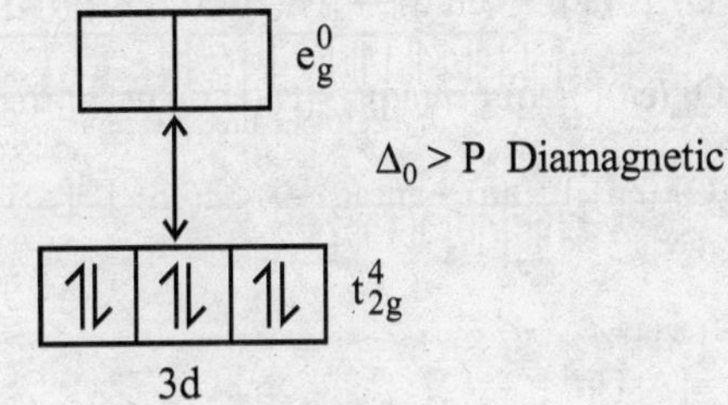

(D) $[Co(H_2O)_3F_3]$; Co^{+3}

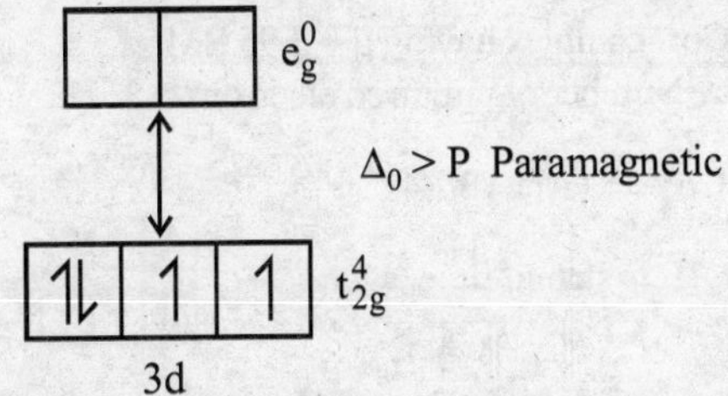

116. (a) $E = h\nu = \frac{hc}{\lambda} \Rightarrow E \propto \frac{1}{\lambda}$

Here all Co in +3 oxidation state. So, as the ligand field strength ↑, CFSE ↑

Order of field strength of ligand : $CN^- > NH_3 > H_2O > Cl^-$

∴ CFSE order : C > B > A > D

⇒ Wavelength order : D > A > B > C

117. (d) Given magnetic moment = 4.9 B.M.

Magnetic moment = $\sqrt{n(n+2)}$B.M.

Magnetic moment = 4.9 BM (given)

∴ $4.9 = \sqrt{n(n+2)} \Rightarrow n = 4$

(A) $_{24}Cr^{2+} \Rightarrow [Ar]3d^4$ (4 unpaired e^-)

(B) $_{26}Fe^{2+} \Rightarrow [Ar]3d^6$ (4 unpaired e^-)

(C) $_{26}Fe^{3+} \Rightarrow [Ar]3d^5$ (5 unpaired e^-)

(D) $_{27}Co^{2+} \Rightarrow [Ar]3d^7$ (3 unpaired e^-)

(E) $_{25}Mn^{3+} \Rightarrow [Ar]3d^4$ (4 unpaired e^-)

118. (b) $Mn^{3+} \Rightarrow 3d^4 \Rightarrow t_{2g}^3 e_g^1$

Cl^- is a weakfield ligand, so electron pairing occurs.

Hybridisation is sp^3d^2 with 4 unpaired electron and paramagnetic.

119. (a) Among the given complexes $[Co(en)_3]^{3+}$ has highest CFSE

Co^{3+} has $3d^6$ configuration.

en is a strong field ligand so the complex formed with low spin configuration in octahedral field.

$= t_{2g}^6 e_g^0$

$CFSE = (6 \times (-0.4) + 0 \times 0.6)\Delta_0 = -2.4\Delta_0$

120. (1) Co has highest standard electrode potential (M^{+3}/M^{+2}) among Mn, Cr, Co, Fe

∴ Complex is $[Co(CN)_6]^{4-}$ and its splitting is as follows.

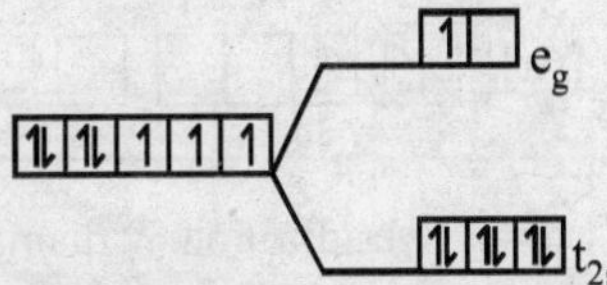

∴ electron in e_g orbital is one.

121. (d) For octahedral complex :

If $\Delta_0 < P$, then high spin complexes are formed due to weaker field strength of the ligand.

If $\Delta_0 > P$, then low spin complexes are formed. But in tetrahedral complex $\Delta_t = \frac{4}{9}\Delta_0$.

Thus, $\Delta_t < P$ and mainly high spin complexes are formed and low spin complexes are rarely formed.

122. (0)

(O)	Compound	Colour	Unpaired e^-
A	$K_3[Co(NO_2)_6]$	Yellow	0
B	$K_4[Fe(CN)_6]$	Light Yellow	0
C	$K_3[Fe(CN)_6]$	Bright red	1
D	$Cu_2[Fe(CN)_6]$	Reddish Brown	0
E	$Zn_2[Fe(CN)_6]$	Bluish – white	0

123. (b) $K_3[Fe(OH)_6]; Fe^{+3} \Rightarrow d^5 \Rightarrow t_{2g}^3 e_g^2$

⇒ 5 unpaired electron

⇒ $\mu = \sqrt{n(n+2)} = \sqrt{35} = 5.92$

$K_4[Fe(OH)_6]; Fe^{2+} \Rightarrow d^6 \Rightarrow t_{2g}^4 e_g^2$

⇒ 4 unpaired electron

$\mu = \sqrt{4(4+2)} = \sqrt{24} = 4.90$

124. (d) (A) $[MnBr_4]^{2-}$

$Mn^{+2} \Rightarrow [Ar]\ 3d^5$

In presence of weak ligand field

⇒ [Ar] 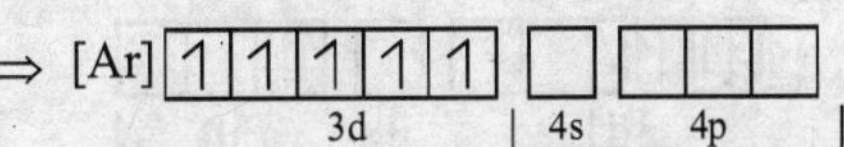

⇒ sp^3 hybridization, paramagnetic in nature

(B) $[FeF_6]^{3-}$

$Fe^{+3} \Rightarrow [Ar]\ 3d^5$

In presence of weak ligand field

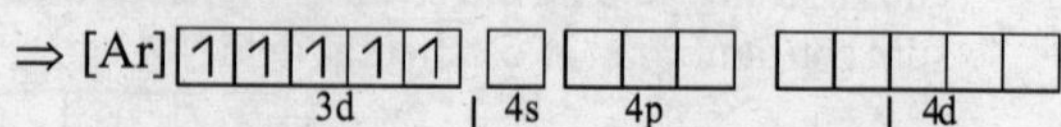

$\Rightarrow sp^3d^2$ hybridizations, & paramagnetic

(C) $[Co(C_2O_4)_3]^{3-}$

$Co^{+3} \Rightarrow [Ar]\ 3d^6$

In presence of strong ligand field

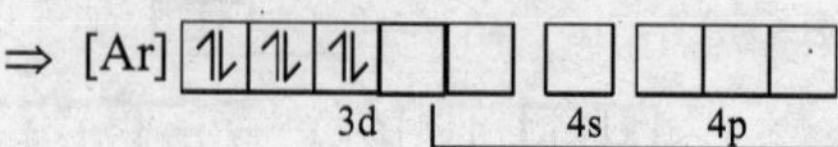

$\Rightarrow$ d^2sp^3 hybridization, & diamagnetic

(D) $[Ni(CO)_4]$

$Ni^0 \Rightarrow [Ar]\ 3d^8 4s^2$

In presence of strong ligand field

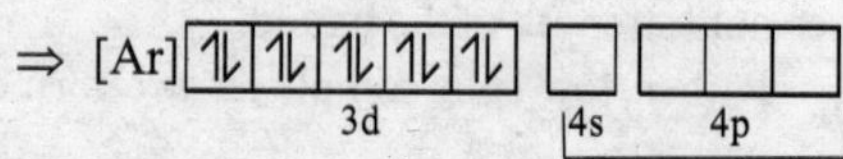

$\Rightarrow$ sp^3 hybridization, & diamagnetic

125. (c)

(I) $[Mn(CN)_6]^{3-}$; $Mn^{3+} \Rightarrow d^4 \Rightarrow t_{2g}^4 e_g^0$

$\Rightarrow (-0.4 \times 4 + 0 \times 0.6)\Delta_0 = -1.6\Delta_0$

(II) $[Co(CN)_6]^{4-}$; $Co^{+2} \Rightarrow d^7 \Rightarrow t_{2g}^6 e_g^1$

$\Rightarrow (-0.4 \times 6 + 1 \times 0.6)\Delta_0 = -1.80\Delta_0$

(III) $[Fe(CN)_6]^{4-}$; $Fe^{+2} \Rightarrow d^6 \Rightarrow t_{2g}^6 e_g^0$

$\Rightarrow (-0.4 \times 6 + 0 \times 0.6)\Delta_0 = -2.4\Delta_0$

(IV) $[Fe(CN)_6]^{3-}$; $Fe^{+3} \Rightarrow d^5 \Rightarrow t_{2g}^5 e_g^0$

$\Rightarrow (-0.4 \times 5 + 0 \times 0.6)\Delta_0 = -2.0\Delta_0$

Hence correct increasing order is:

(I) < (II) < (IV) < (III)

126. (b)

(A) $[CoF_6]^{-3}$

$Co^{3+} \to 3d^6$

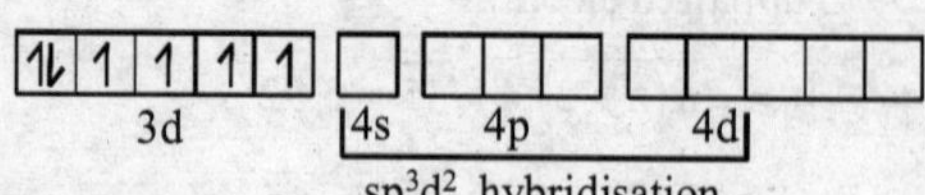

(B) $[NiCl_4]^{2-}$

$Ni^{2+} \to 3d^8$

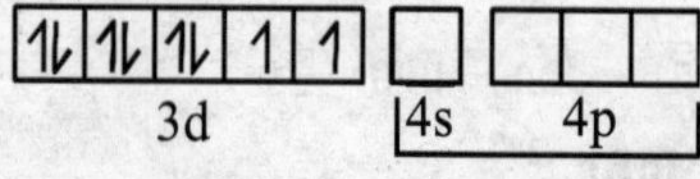

(C) $[Co(NH_3)_6]^{3+}$

$Co^{3+} \to 3d^6$

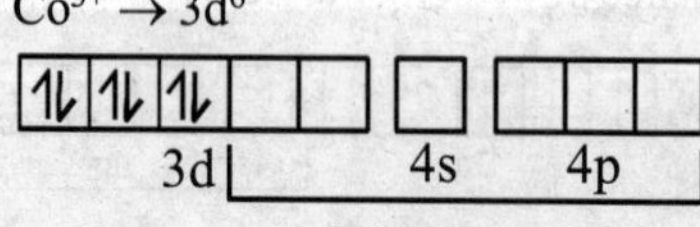

(D) $[Ni(CN)_4]^{2-}$

$Ni^{2+} \to 3d^8$

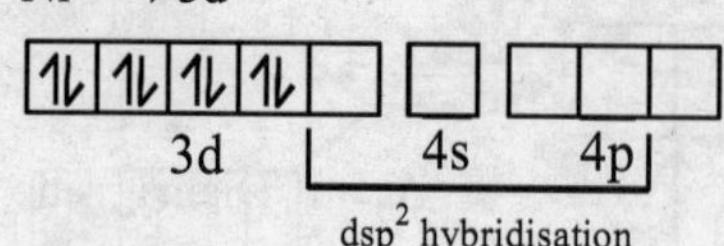

127. (a) For $3d^4$

If ligand is strong field ligand : $t_{2g}^4 e_g^0$ (low spin)

If ligand is weak field ligand : $t_{2g}^3 e_g^1$ (high spin)

128. (c)

$[Ni(H_2O)_6]^{+2}(aq) + en(aq) \to [Ni(H_2O)_4(en)]^{+2}(aq) + 2H_2O$
Green

$[Ni(H_2O)_4(en)]^{+2}(aq) + en(aq) \to [Ni(H_2O)_2(en)_2]^{+2}(aq) + 2H_2O$
Pale blue

$[Ni(H_2O)_2(en)_2]^{+2}(aq) + en(aq) \to [Ni(en)_3]^{+2}(aq) + 2H_2O$
Blue/purple Violet

129. (c)

$Fe^{+6} = [Ar]3d^2$

$Mn^{+3} = [Ar]3d^4$

$Fe^{+3} = [Ar]3d^5$

$Cr^{+2} = [Ar]3d^4$

$Ni^{+4} = [Ar]3d^6$

130. (c) Co^{2+} complex having $\mu = 3.95$ BM

$\therefore$ Number of unpaired electrons = 3

$Co^{2+} \Rightarrow (Ar)_{18}\ 3d^7 4s^0$

↿ ↿ e_g^2

⥮ ⥮ ↿ t_{2g}^5

131. (b) Crystal field splitting energy (Δ) $\propto$ charge or oxidation state of central metal atom.

Crystal field splitting energy (Δ) $\propto$ field strength of ligand (and chelation)

Crystal field stabilisation energy (CFSE) = $[-0.4\ t_{2g} + 0.6\ e_g]\ \Delta_0$ (for octahedral)

For,

$[Co(en)_3]^{3+} : Co^{3+} : t_{2g}^6 e_g^0$; CFSE = $-2.4\ (\Delta_0)_1$

$[Co(NH_3)_6]^{3+} : Co^{3+} : t_{2g}^6 e_g^0$; CFSE = $-2.4\ (\Delta_0)_2$

$[Co(NH_3)_6]^{2+} : Co^{2+} : t_{2g}^5 e_g^2$; CFSE = $-0.8\ (\Delta_0)_3$

$[Co(NH_3)_4]^{2+} : Co^{2+} : e^4 t_2^3$; CFSE = $-1.2\ \Delta_t$

$\left(\text{as} : (\Delta_t) = \frac{4}{9}(\Delta_0)_3\right)$

$\because \Delta_t < (\Delta_0)_3 < (\Delta_0)_2 < (\Delta_0)_1$

132. (b) $[NiCl_4]^{2-}$

Ni^{+2} – [Ar] $3d^8\ 4s^0 \to sp^3$, Tetrahedral

Number of unpaired electron = 2 paramagnetic

$[Ni(CO)_4]$,

Ni(0) $\to$ [Ar] $3d^{10}\ 4s^0$ (After rearrangement)

No unpaired electron sp^3, Tetrahedral, Diamagnetic

133. (d) For complex $K_3[Fe(SCN)_6]$

electron donating by sulphur in SCN^-, it is a weak field ligand.

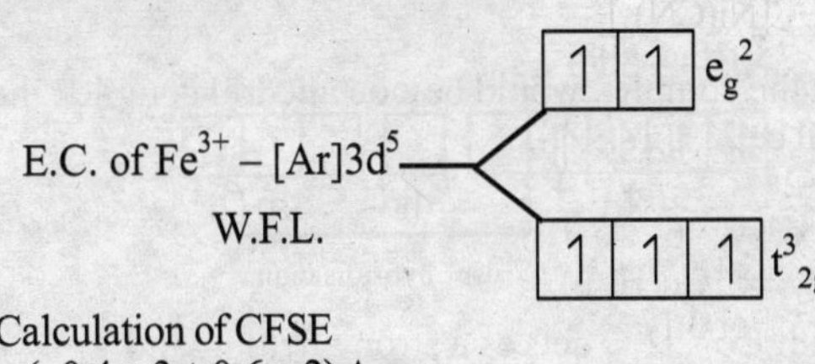

Calculation of CFSE
$= (-0.4 \times 3 + 0.6 \times 2)\ \Delta_0$
$= 0.\Delta_0$.

134. (c) $[Co(en)_2Cl_2]^+$ has octahedral geometry due to $d^2 sp^3$ hybridization with two geometrical isomers (cis and trans).

135. (d) (A) $K_2\left[\overset{+2}{Ni}(CN)_4\right]$
Ni^{2+}; $[Ar]3d^8 4s^0$, (CN^- is S.F.L)

3d 4s 4p
dsp^2

(B) $[Ni(CO)_4]$
Ni : [Ar] $3d^8 4s^2$
CO is S.F.L, so pairing occur

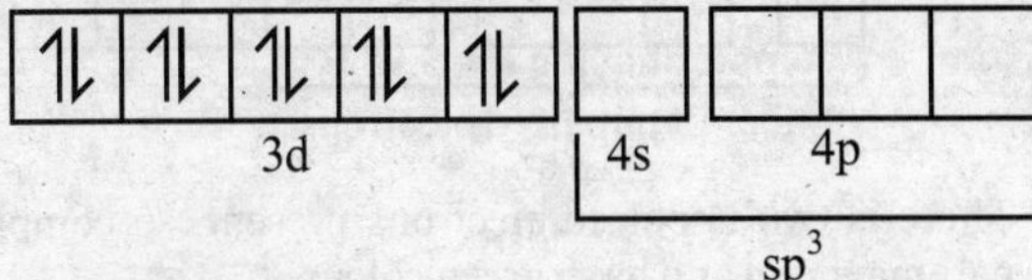

(C) $[Co(NH_3)_6]Cl_3$
Co^{+3} : $[Ar]3d^6 4s^0$
With Co^{3+}, NH_3 act as S.F.L

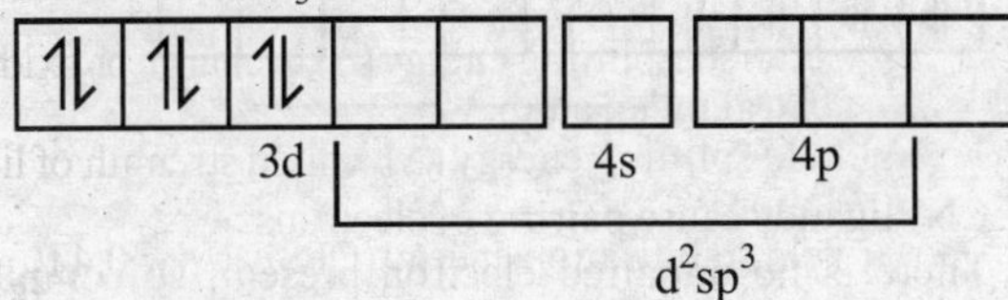

(d) $Na_3[CoF_6]$
Co^{3+} : $[Ar]3d^6$($F^{\ominus}$; W.F.L)

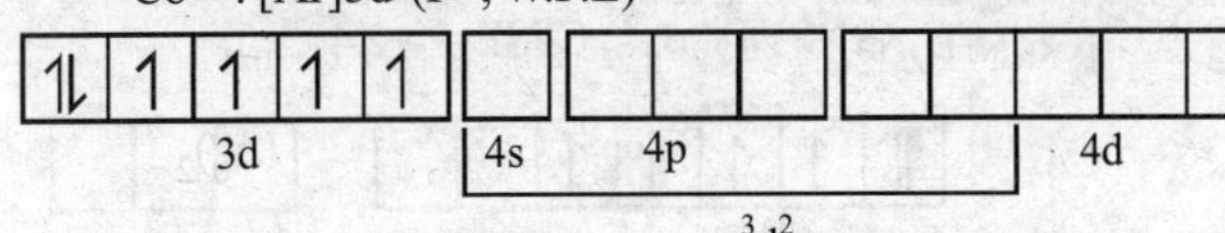

136. (b) $EDTA^{4-} \rightarrow$ Hexadentate ligand
$[Ca(EDTA)]^{2-}$
So Coordination environment is octahedral

137. (c) $N(CH_3)_3$ and $P(CH_3)_3$ both are Lewis base and act as ligand, however, $P(CH_3)_3$ has π-acceptor character due to presence of vacant d-orbitals but $N(CH_3)_3$ is not a π-acceptor due to absence of d-orbitals.

138. (b) Consider H_2O as a weak field ligand.

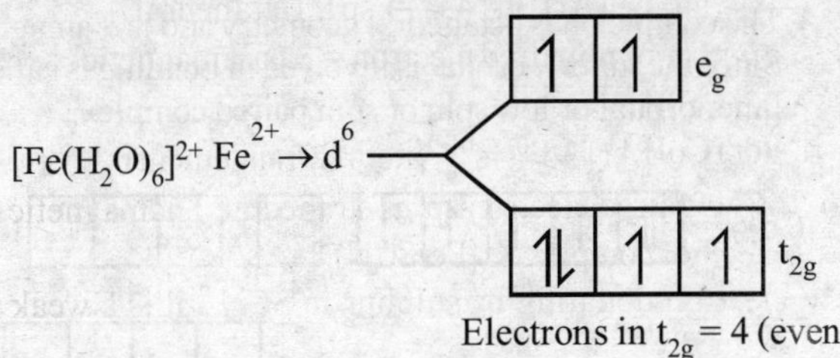

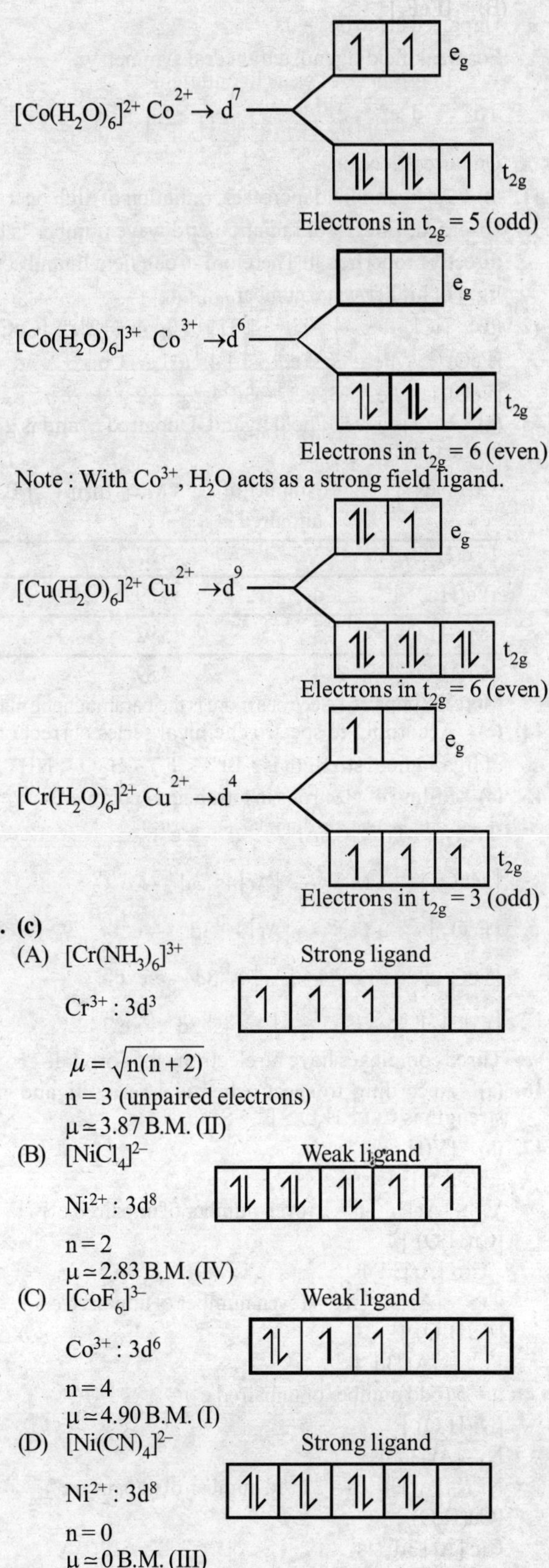

$[Co(H_2O)_6]^{2+}$ $Co^{2+} \rightarrow d^7$ (e_g, t_{2g})
Electrons in t_{2g} = 5 (odd)

$[Co(H_2O)_6]^{3+}$ $Co^{3+} \rightarrow d^6$ (e_g, t_{2g})
Electrons in t_{2g} = 6 (even)

Note : With Co^{3+}, H_2O acts as a strong field ligand.

$[Cu(H_2O)_6]^{2+}$ $Cu^{2+} \rightarrow d^9$ (e_g, t_{2g})
Electrons in t_{2g} = 6 (even)

$[Cr(H_2O)_6]^{2+}$ $Cu^{2+} \rightarrow d^4$ (e_g, t_{2g})
Electrons in t_{2g} = 3 (odd)

139. (c)
(A) $[Cr(NH_3)_6]^{3+}$ Strong ligand
Cr^{3+} : $3d^3$
$\mu = \sqrt{n(n+2)}$
n = 3 (unpaired electrons)
$\mu \simeq 3.87$ B.M. (II)

(B) $[NiCl_4]^{2-}$ Weak ligand
Ni^{2+} : $3d^8$
n = 2
$\mu \simeq 2.83$ B.M. (IV)

(C) $[CoF_6]^{3-}$ Weak ligand
Co^{3+} : $3d^6$
n = 4
$\mu \simeq 4.90$ B.M. (I)

(D) $[Ni(CN)_4]^{2-}$ Strong ligand
Ni^{2+} : $3d^8$
n = 0
$\mu \simeq 0$ B.M. (III)

140. (2) NH_3 is a strong field ligands
For octahedral symmetry
$Co^{+3} : 3d^6;\ t_{2g}^{2,2,2}\ e_g^{0,0}$

Unpaired electrons = 0

For weak field ligend, tetrahedral symmetry

$Ni^{+2}: 3d^8 e_g^{2,2} t_{2g}^{2,1,1}$

Unpaired electron = 2

141. (b) As ligand field increases, radiation of high energy is absorbed. Energy of radiation and wave number is the / directly proportional. Therefore strong field ligand absorb light of higher wave number.

142. (d) $TiCl_4 \rightarrow x - 4 = Ti = +4, Ti^{+4} = 3d^0 4s^0 (+2^0 e^0), [CoCl_4]^{2-}$ $[FeO_4]^{2-} = Fe = +6, Fe^{+6} = 3d^2 4s^0 (+2e^2)$ $Co = +2, 3d^7 4s^0$ $[FeCl_4]^- = Fe = +3, Fe^{+3} = 3d^5 4s^0 (e^2 + 2^3) (+2^3 e^4)$

143. (a) H_2O is a weak field ligand. Unpaired e^- and is given in the table as :

Complex	Number of unpaired e^-	$M = \sqrt{n(n+2)}$ B.M.
$[Co(H_2O)_6]^{2+}$	3	3.87
$[Fe(H_2O)_6]^{2+}$	4	4.89
$[Mn(H_2O)_6]^{2+}$	5	5.92
$[Cr(H_2O)_6]^{2+}$	4	4.89

More the unpaired electrons more is the paramagnetic nature.

144. (c) According to spectro chemical series correct order of ligand field strength is : $Br^- < F^- < H_2O < NH_3$

145. (a) Filling of electrons in tetrahedral complex as

$TiCl_4 \rightarrow Ti^{+4} \rightarrow [Ar]4s^0 3d^0 \rightarrow e^0 t_2^0$

$[MnO_4]^- \rightarrow Mn^{+7} \rightarrow [Ar]4s^0 3d^0 \rightarrow e^0 t_2^0$

$[FeO_4]^{2-} \rightarrow Fe^{+6} \rightarrow [Ar]4s^0 3d^2 \rightarrow e^2 t_2^0$

$[FeCl_4]^- \rightarrow Fe^{3+} \rightarrow [Ar]4s^0 3d^5 \rightarrow e^2 t_2^3$

$[CoCl_4]^{2-} \rightarrow Co^{+2} \rightarrow [Ar]4s^0 3d^7 \rightarrow e^4 t_2^3$

Three complexes have no electrons in t_2 orbital.

146. (a) According to spectrochemical series ligand field strength is $CO > H_2O > F^- > S^{2-}$

147. (a) $[V(H_2O)_6]^{3+}$

${}_{23}V$:- $[Ar]3d^3 4s^2$

V^{+3}:- $[Ar]3d^2$, n = 2 (even number of unpaired e^-)

$[Cr(H_2O)_6]^{2+}$

${}_{24}Cr$:- $[Ar]3d^5 4s^1$

Cr^{2+}:- $[Ar]3d^4$, n = 4 (even number of unpaired e^-)

$[Fe(H_2O)_6]^{3+}$

Fe^{3+} :- $[Ar]3d^5 4s^0$

n = 5 (odd number of unpaired e^-)

$[Ni(H_2O)_6]^{3+}$

Ni:- $[Ar]3d^8 4s^2$

Ni^{3+}:- $[Ar]3d^7$, n = 3 (odd number of unpaired e^-)

$[Cu(H_2O)_6]^{2+}$

Cu: $[Ar]\ 3d^{10} 4s^1$

Cu^{2+} :- $[Ar]3d^9 4s^0$

n = 1 (odd number of unpaired e^-)

148. (b) Balancing charge

$3 + x(0) + y(-1) = -1$

y = 4

As the complex would be a octahedral complex, hence x will be 2.

x = 2 y = 4

so x + y = 6

149. (c) $\Rightarrow [Co(H_2O)_6]^{3+}$

With Co^{3+} H_2O acts as a strong field ligand.

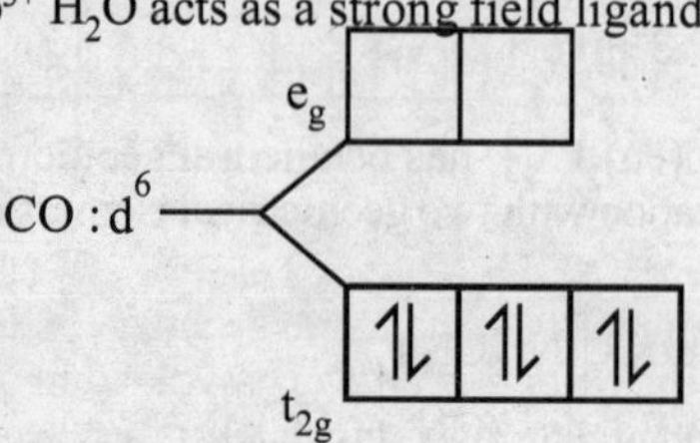

No unpaired electrons.

150. (b) H_2O is a weak field ligand so pairing of electrons does not take place

$[Ni(H_2O)_6]^{2+}$ (d^8) :

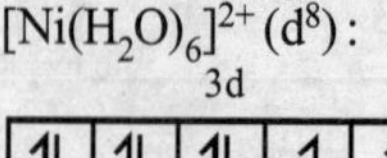

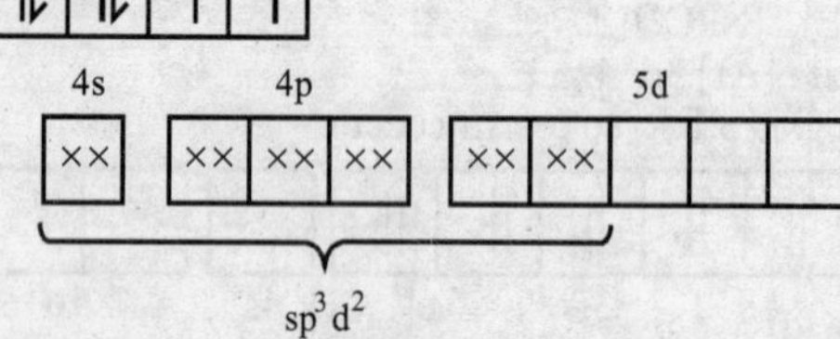

There is two unpaired electrons present, so complex is paramagnetic and have green colour.

$[Ni(CN)_4]^{2+}$ (d^8) :

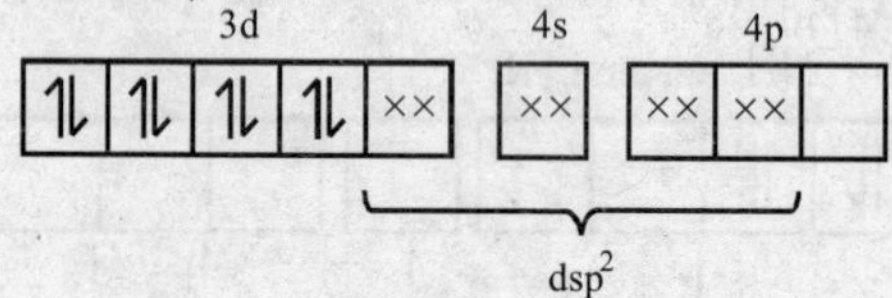

CN^- ligands cause pairing of electrons.

There is no unpaired electron present, so complex is diamagnetic and colourless.

151. (b) Spin paired complex, spin free complex $[Co(NH_3)_6]^{+3}$

$[Co(NH_3)_6]^{+3}$

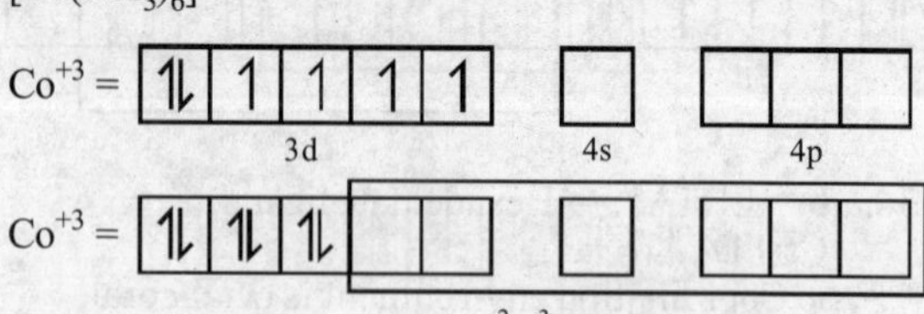

The complex has octahedral geometry and is diamagnetic. Since the inner *d* orbital is involved in bonding is called an inner orbital or low spin or spin paired complex.

For $[CoF_6]^{3-}$

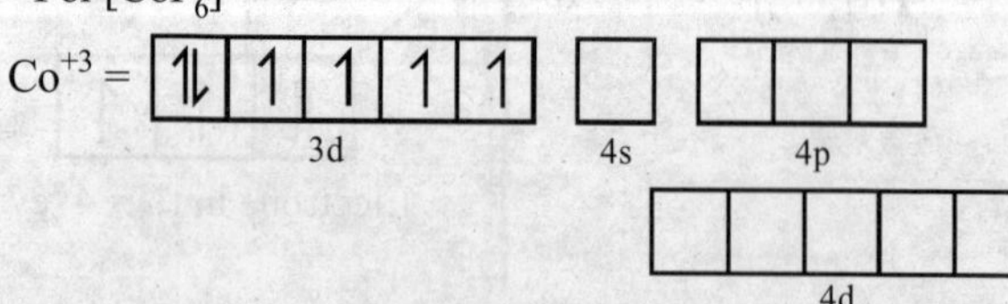

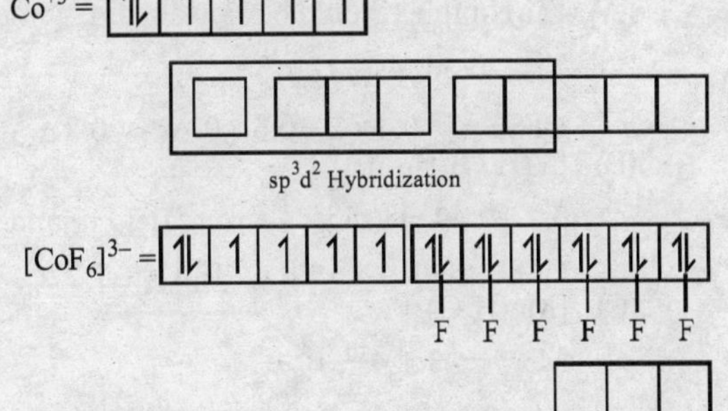

The complex has octahedral geometry and is paramagnetic. Since the outer d-orbitals is involved in bonding is called outer orbital or high spin or spin free complex.

152. (a) A. Field strength of ligands are explained by LFT and MOT.
B. VBT does not explain stability of complex.
C. Hybridisation of $[Ni(CN)_4]^{-2}$ is dsp^2.
D. Cis-$[PtCl_2(en)_2]$ also has optical isomers.

153. (28) NH_3 act as WFL with Ni^{2+}

$Ni^{2+} = 3d^8$

⥮ ⥮ ⥮ 1 1

No. of unpaired electron = n = 2

$\mu = \sqrt{n(n+2)} = \sqrt{8} = 2.82$ BM $= 28.2 \times 10^{-1}$ BM

154. (d) A-II, B-III, C-IV, D-I

$[Cr(H_2O)_6]^{3+}$ Cr^{+3} : $[Ar]3d^3 : t_{2g}^3 e_g^0$

$[Fe(H_2O)_6]^{3+}$ Fe^{+3} : $[Ar]3d^5 : t_{2g}^3 e_g^2$

$[Ni(H_2O)_6]^{2+}$ Ni^{+2} : $[Ar]3d^8 : t_{2g}^6 e_g^2$

$[V(H_2O)_6]^{3+}$ V^{+3} : $[Ar]3d^2 : t_{2g}^2 e_g^0$

155. (d) $[Ni(CO)_4]$ diamagnetic,

$[NiCl_4]^{2-}$ paramagnetic

In $[Ni(CO)_4]$, Ni = [Ar] $3d^8 4s^2$

∴ Ni = ⥮ ⥮ ⥮ 1 1 | ⥮ | □ □ □
3d 4s 4p

CO is strong field ligand so pairing will take place.

Ni = ⥮ ⥮ ⥮ ⥮ ⥮ | □ | □ □ □

sp^3 hybridization

$[Ni(CO)_4]$ = ⥮ ⥮ ⥮ ⥮ ⥮ | ⥮ | ⥮ ⥮ ⥮

CO CO CO CO

Its a diamagnetic complex.

In $[NiCl_4]^{2-}$, Ni = +2 oxidation state Cl is weak field ligand so no pairing will take place.

∴ $[NiCl_4]^{2-}$ = ⥮ ⥮ ⥮ 1 1 | ⥮ | ⥮ ⥮ ⥮

Cl Cl Cl Cl

sp^3 Hybridization

Its a paramagnetic complex.

156. (c) $[FeF_6]^{3-} : Fe^{+3} : [Ar]d^5$

F : Weak field Ligand 1 1 1 1 1

No. of unpaired electrons = 5

$\mu = \sqrt{5(5+2)} \Rightarrow \mu = \sqrt{35}$ BM

$[V(H_2O)_6]^{+2} : V^{+2} : 3d^3$ 1 1 1 □ □

No. of unpaired electrons = 3

$\mu = \sqrt{3(3+2)} \Rightarrow \mu = \sqrt{15}$ BM

$[Fe(H_2O)_6]^{+2} : Fe^{+2} : 3d^6$

H_2O : Weak field Lignad ⥮ 1 1 1 1

No. of unpaired electrons = 4

$\mu = \sqrt{4(4+2)} \Rightarrow \mu = \sqrt{24}$ BM

157. (0) Pt^{2+} (d^8)

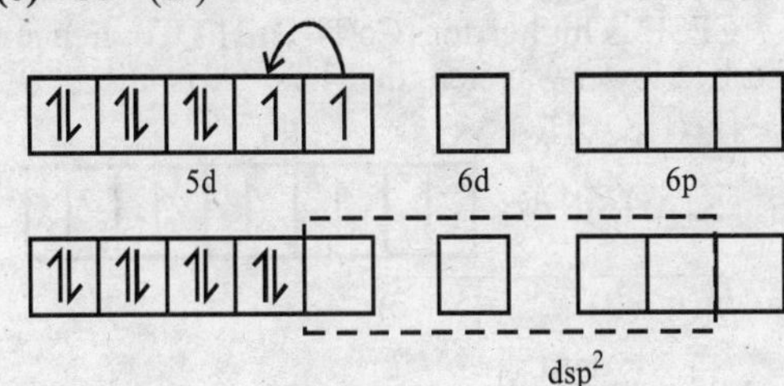

$Pt^{2+} \rightarrow dsp^2$ hybridization and have no unpaired electron.

∴ Magnetic moment = 0

158. (a) Smaller cations prefer coordinate bonding due to greater polarizing power.

Ti^{+3} = 67 pm radius, Cr^{3+} = 62 pm radius

Mn^{+3} = 65 pm radius, Fe^{+3} = 65 pm radius

So, Cr^{3+} has highest tendency to attract ligand.

159. (1) Co^{2+} : $3d^7$ configuration

$t_{2g}^{2,2,1} e_g^{1,1}$

160. (b) Complex with maximum number of unpaired electron will exhibit maximum attraction to an applied magnetic field.

$[Zn(H_2O)_6]^{2+} \rightarrow d^{10}$ system $\rightarrow t_{2g}^6 e_g^4$, 0 unpaired e^-

$[Co(H_2O)_6]^{2+} \rightarrow d^7$ system $\rightarrow t_{2g}^5 e_g^2$, 3 unpaired e^-

$[Co(en)_3]^{3+} \rightarrow d^6$ system $\rightarrow t_{2g}^6 e_g^0$, 0 unpaired e^-

$[Ni(H_2O)_6]^{2+} \rightarrow d^8$ system $\rightarrow t_{2g}^6 e_g^2$, 2 unpaired e^-

161. (a) CFSE = $(-0.4\, n t_{2g} + 0.6\, n_{eg})\Delta_o$

$n t_{2g}$ = number of electrons in t_{2g} orbital

n_{eg} = number of electrons in eg orbital

Complex	No. of at electrons	CFSE (Δ_o)
$[Cu(NH_3)_6]^{+2}$	d^9 (S.L.) $t_{2g}^{2,2,2} eg^{2,1}$	–0.6
$[Ti(H_2O)_6]^{+3}$	d^1 (W.L.) $t_{2g}^{1,0,0} eg^{0,0}$	–0.4
$[Fe(CN)_6]^{3-}$	d^5 (S.L.) $t_{2g}^{2,2,1} eg^{0,0}$	–2.0
$[NiF_6]^{4-}$	d^8 (W.L.) $t_{2g}^{2,2,2} eg^{1,1}$	–1.2

162. (b) $Fe_4[Fe(CN)_6]_3$ Prussian Blue–water insoluble $K_3[Co(NO_2)_6]$ very poorly water soluble $(NH_4)_3[As(Mo_3O_{10})_4]$ water insoluble ammonium arseno molybdate, $[Fe_3(OH)_2(OAc)_6]Cl$ is water soluble.

163. (1) Spin magnetic moment of $\left[Cr(CN)_6\right]^{3-}$ $(t_{2g}^3 e_g^0)$

$\mu_1 = \sqrt{3(3+2)} = \sqrt{15}$ BM (Considering it as an outer orbital complex)

Spin magnetic moment of $\left[Cr(H_2O)_6\right]^{3+}$ $(t_{2g}^3 e_g^0)$

$\mu_2 = \sqrt{3(3+2)} = \sqrt{15}$ BM

$$\frac{\mu_1}{\mu_2} = \frac{\sqrt{15}}{\sqrt{15}} = 1$$

164. (a) Since H_2O is strong field ligand compared to chloride and Co^{3+} ion is present.

∴ CFSE is higher for $[Co(NH_3)_5H_2O]^{+3}$, hence it will absorb at lower wavelength. Thus, 'A' is false.

165. (d) $[Fe(H_2O)_6]^{3+}$

$Fe^{3+} \Rightarrow [Ar]\ 3d^5 4s^0$, [1][1][1][1][1]

$\mu = \sqrt{n(n+2)} = \sqrt{5(5+2)}$

$\mu = \sqrt{35} = 5.92$ B.M.

$[Fe(CN)_6]^{-3}$

$Fe^{3+} \Rightarrow [Ar]\ 3d^5 4s^0$, [1][1][1][1][1]

Pairing occur due to strong field ligand CN^-

∴ Unpaired $e^- \Rightarrow 1$

$\mu = \sqrt{n(n+2)} = \sqrt{1(1+2)} = \sqrt{3} = 1.732$ B.M.

166. (a) $Mg(NH_4)PO_4 \Rightarrow$ White, $K_3[Co(NO_2)_6] \Rightarrow$ Yellow
$MnO(OH)_2 \Rightarrow$ Brown, $Fe_4[Fe(CN)_6]_3 \Rightarrow$ Blue

167. (N) $[NiBr_2Cl_2]^{2-} \rightarrow$ This complex species is tetrahedral as Br^- & Cl^- are weak field ligands.

$[PtBr_2Cl_2]^{2-} \rightarrow$ As Pt belongs to 5d series. This complex species is square planar.

Td complex are optically active and square plannar complex are optically inactive.

$[NiBr_2Cl_2]^{2-}$, being tetrahedral does not show Geometrical Isomerism.

$[PtBr_2Cl_2]^{2-}$ shows two geometrical isomers.

Ni^{2+} is paramagnetic while pt^{2+} is diamagnetic

168. (d) $[Co(NH_3)_6]^{3+}$,
$NH_3 \Rightarrow$ SFL, $Co^{+3} = [Ar]3d^6$, $t_{2g}^{\ 6} e^0_{\ g}$

[⇅][⇅][⇅][··][··] [··] [··][··][··]

∴ Diagmagnetic & Low spin complex.

169. (3) $K_4[Fe(CN)_6] \Rightarrow Fe^{+2} = [Ar]3d^6$, $CN \Rightarrow$ SFL, $t_{2g}^6\ e_g^0$

t_{2g} contain 6 electron so it become 3 pairs.

170. (b) A. $[Fe(CN)_6]^{3-}$; n = 1 (CN^- = Strong Field Ligand)
B. $[FeF_6]^{3-}$; n = 5 (F^- = Weak Field Ligand)
C. $[CoF_6]^{3-}$; n = 4 (F^- = WFL)
D. $[Cr(oxalate)_3]^{3-}$; n = 3 (oxalate = SFL)
E. $[Ni(CO)_4]$; n = 0 (CO = SFL)

171. (b) A–IV, B–I, C–II, D–III

(A) $[Ti(H_2O)_6]^{2+}$
$Ti^{2+} \Rightarrow 3d^2\ 4s^0,\ t_{2g}^2 e_g^0$
$CFSE = [-0.4 \times 2 + 0.6 \times 0]\Delta_o = -0.8\ \Delta_o$

(B) $[V(H_2O)_6]^{2+}$
$V^{2+} \Rightarrow 3d^3\ 4s^0,\ t_{2g}^3 e_g^0$
$CFSE = [-0.4 \times 3 + 0.6 \times 0]\Delta_o = -1.2\ \Delta_o$

(C) $[Mn(H_2O)_6]^{3+}$
$Mn^{3+} \Rightarrow 3d^4\ 4s^0,\ t_{2g}^3 e_g^1$
$CFSE = [-0.4 \times 3 + 0.6 \times 1]\Delta_o = -0.6\ \Delta_o$

(D) $[Fe(H_2O)_6]^{3+}$
$Fe^{3+} \Rightarrow 3d^5\ 4s^0,\ t_{2g}^3 e_g^2$
$CFSE = [-0.4 \times 3 + 0.6 \times 2]\Delta_o = 0\ \Delta_o$

172. (4) $\mu = \sqrt{n(n+2)}$BM

$4 = \sqrt{n(n+2)} \Rightarrow n = 4$

$= \sqrt{4(4+2)} = \sqrt{24} = 4.90$

173. (c) $Co^{+3} = t_{2g}^{\ 6}\ e_g^{\ 0}$
CN^-–strong field ligand
All d–electrons should be paired (μ_s = 0) Hence diamagnetic.

174. (b)

		Unpaired e⁻
$[Fe(CN)_6]^{3-}$	$Fe^{+3} \Rightarrow t_{2g}^{\ 5} e_g^{\ 0}$,	1 (Strong field ligand)
$[Mn(CN)_6]^{3-}$	$Mn^{+3} \Rightarrow t_{2g}^{\ 4} e_g^{\ 0}$,	2 (Strong field ligand)
$[CoF_6]^{3-}$	$Co^{+3} \Rightarrow t_{2g}^{\ 4} e_g^{\ 2}$,	4 (Weak field ligand)
$[MnBr_4]^{2-}$	$Mn^{+2} \Rightarrow e_g^{\ 2} t_{2g}^{\ 3}$,	5 (Weak field ligand)

Spin magnetic moment $\mu = \sqrt{n(n+2)}$ B.M

∴ Correct option is (b).

175. (a) For option (A), Cr^{+3} : $3d^3$; CN^-: SFL
⇒ No. of unpaired electrons = 3
For option (B), Fe^{+2} : $3d^6$; H_2O: WFL
⇒ No. of unpaired electrons = 4
For option (C), Co^{+3}: $3d^6$; NH_3: SFL
⇒ No. of unpaired electrons = 0
For option (D), Ni^{+2} : $3d^8$; NH_3 : SFL
⇒ No. of unpaired electrons = 2

176. (4) $[Mn(NCS)_6]^{x-}$
Number of unpaired electron = 5
So, Mn must be in +2 oxidation state (Mn^{+2})
$\Rightarrow 2 + (-6) = -x \Rightarrow -4 = -x \Rightarrow x = 4$

177. (a) $[Fe(CN)_6]^{3-}$

3d 4s 4p

Fe in ground state [⇅][1][1][1][1] [⇅] [ ][ ][ ]

Fe in +3 state [1][1][1][1][1] [ ] [ ][ ][ ]

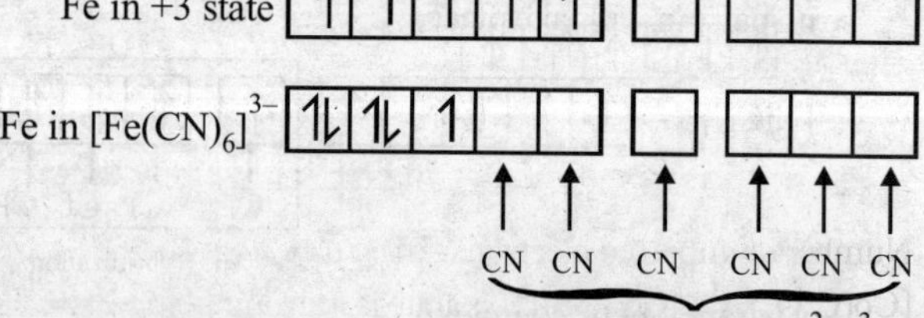

Unpaired electron = 1

$\mu = \sqrt{n(n+2)} = \sqrt{1 \times 3} = 1.74$ B.M.

$[Fe(H_2O)_6]^{3+}$ No pairing because H_2O is WFL

Number of unpaired electrons = 5, $\mu = 5.92$ B.M.

Assertion is true, Reason is true but not correct explanation.

178. (a) CN^- is the strong field ligand among the given ligands. So, the Δ_0 value of $[Fe(CN)_6]^{3-}$ will be the highest one among all. Here, the central atom is present in Fe^{3+} state in all cases.

179. (6) $[Mn(H_2O)_6]^{2+}$, $Mn^{2+} \Rightarrow 3d^5$

$\mu = \sqrt{5(5+2)} = 5.91$ BM

180. (a) $CoCl_2 + 6H_2O \longrightarrow [Co(H_2O)_6]Cl_2$

Pink(X) octahedral

$\downarrow$ +HCl(conc.)

$[CoCl_4]^{2-}$

(Y) Blue solution

(Z) Tetrahedral

181. (480) $[Ti^{+3}(H_2O)_6]^{3+}$; $Ti^{+3}: 3d^1 4s^0 + 2g^1 e_g$

C.F.S.E. $= -0.4 \times \Delta_0 + 0$

$$\Rightarrow -\frac{96\times10^3}{N_A} J = -0.4\times\Delta_0 \Rightarrow \Delta_0 = \frac{96\times10^3}{0.4\times6\times10^{23}}$$

$$\Rightarrow \frac{hc}{\lambda} = \frac{96\times10^3}{0.4\times6\times10^{23}}$$

$$\lambda = \frac{0.4\times6\times10^{23}\times6.4\times10^{-34}\times3\times10^8}{96\times10^3}$$

$= 0.48\times10^{-6} m = 480\times10^{-9} m = 480$ nm

182. (b) The increasing order of field strength of ligands (according to spectrochemical series)

$S^{2-} < C_2O_4^{2-} < NH_3 < en < CO$

183. (d) For $[Fe(NH_3)_6]^{+2}$, $\Delta_0 < P$, hence the pairing of electrons does not occur in t_{2g}. Therefore complex is outer orbital and its hybridisation is $sp^3 d^2$.

184. (b) The Cl – Co – Cl bond angle in above octahedral complex is 90°.

185. (a) $[FeF_6]^{3-}: Fe^{3+}: 3d^5$, $F^{\ominus}$ is a weak field ligand.

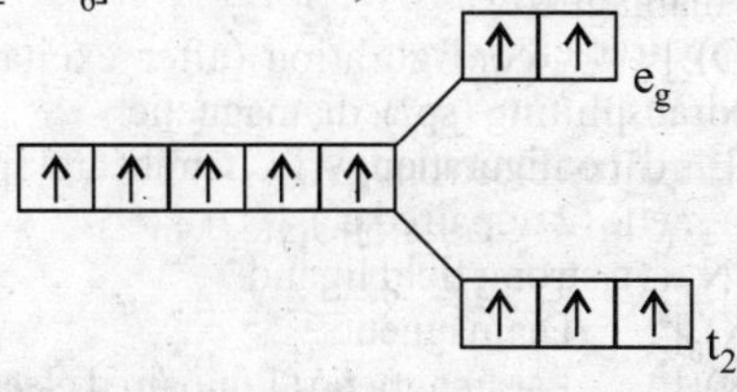

Number of unpaired electrons = 5 $\therefore \mu = \sqrt{35}$ BM

$[CoF_6]^{3-}: Co^{3+}: 3d^6$

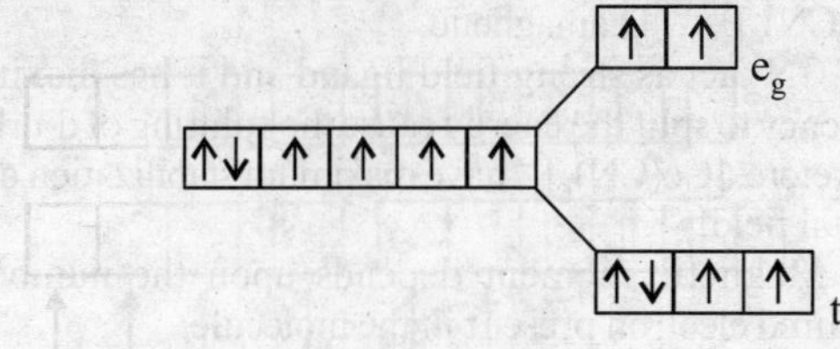

Number of unpaired electrons = 4 $\therefore \mu = \sqrt{24}$ BM

$[Co(C_2O_4)_3]^{3-}: Co^{3+}: 3d^6$ Oxalate is a strong field ligand

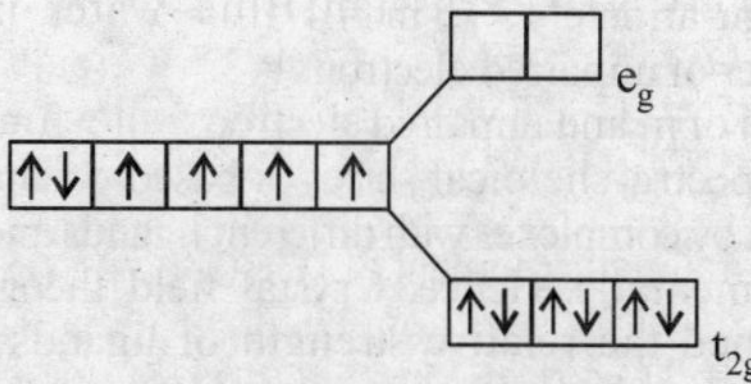

Number of unpaired electrons = 0 $\therefore \mu = 0$ BM

186. (2)

Ions:	V^{3+}	Cr^{3+}	Fe^{2+}	Ni^{3+}
Electronic Configuration:	$3d^2 4s^0$	$3d^3 4s^0$	$3d^6 4s^0$	$3d^7 4s^0$
Electrons(n):	2	3	4	3

Cr^{3+} and Ni^{3+} has the same number of unpaired electron. Hence, same magnetic moment.

187. (4) $[Ni(CN_4]^{2-}: Ni^{+2} \Rightarrow$ $3d^8$ [⇅ ⇅ ⇅ ⇅] diamagnetic

CN^- : strong field ligand

$[Ni(CO)_4]: Ni \Rightarrow$ $3d^{10}$ [⇅ ⇅ ⇅ ⇅ ⇅] diamagnetic

$[NiCl_4]^{2-}: Ni^{2+} \Rightarrow$ $3d^8$ [⇅ ⇅ ⇅ 1 1] paramagnetic

Cl^- : weak field ligand

$[Fe(CN)_6]^{4-}: Fe^{2+} \Rightarrow$ $3d^6$ [⇅ ⇅ ⇅   ] diamagnetic

CN^- : strong field ligand

$[Cu(NH_3)_4]^{+2}: Cu^{+2} \Rightarrow$ [⇅ ⇅ ⇅ ⇅ 1] paramagnetic

$[Fe(CN)_6]^{-3}: Fe^{+3} \Rightarrow$ $3d^5$ [⇅ ⇅ 1   ]

paramagnetic, CN^- : strong field ligand

$[Fe(H_2O)_6]^{2+}: Fe^{2+} \Rightarrow$ $3d^6$ [⇅ 1 1 1 1]

paramagnetic, H_2O : Weak field ligand

188. (b) Lower the wavelength absorbed = High the energy of the absorbed light $\Rightarrow$ Larger is the splitting of d – orbitals $\Rightarrow$ caused by high O.S. of central metal atom and strong field ligand.

$$E = \frac{hc}{\lambda} \Rightarrow E \propto \frac{1}{\lambda}$$

$$\Rightarrow \Delta(CFSE) \propto \frac{1}{\lambda_{absorb}} \propto \text{strength of ligand.}$$

189. (c) $Cr^{+2}: [Ar] 3d^4 4s^0; n = 4, \mu = \sqrt{4(4+2)} = 4.89$ BM

$Mn^{+2}: [Ar] 3d^5 4s^0; n = 5, \mu = \sqrt{5(5+2)} = 5.91$ BM

$V^{+2}: [Ar] 3d^3 4s^0; n = 3, \mu = \sqrt{3(3+2)} = 3.87$ BM

$Ti^{+2}: [Ar] 3d^2 4s^0; n = 2, \mu = \sqrt{2(2+2)} = 2.82$ BM

190. (7) $Co^{2+}: 3d^7 4s^0$, Cl^- is a weak field legand.

1 1 1 t_2

⇅ ⇅ e

Configuration $e^4 t_2^3 \Rightarrow m = 4$
Number of unpaired electrons = 3
$\therefore$ Sum of m and unpaired electrons = 4 + 3 = 7

191. (a) Spectro-chemical series is based on the absorption of light by complexes with different ligands and determined experimentally. Hence, Crystal field theory cannot be explained the relative strength of ligands as given in spectrochemical series.

192. (c) $[Co(NH_3)_6]^{3+}$
$Co^{+3} \Rightarrow 3d^6$
NH_3 behaves as strong field ligand in case of Co^{+3} and pairing will takes place.

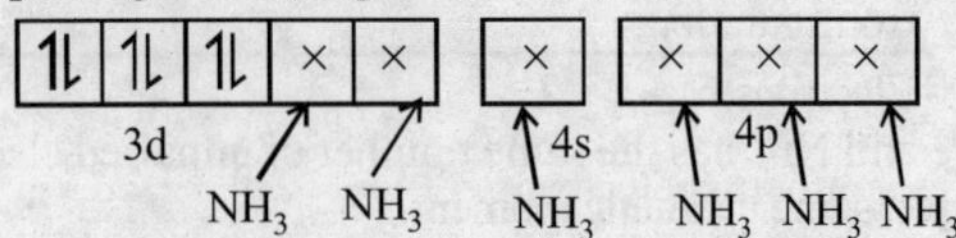

So, the hybridisation is d^2sp^3 and diamagnetic in nature.

193. (2) $[Fe(CN)_6]^{3-}$
CN^- is a strong field ligand, thus valence shell electronic configuration of Fe^{3+} will be :
$Fe^{3+} \equiv 3d^5\ (t^5_{2g} e^0_g)$
$CFSE = 5(-0.4\,\Delta_0) = -2.0\,\Delta_0$

194. (a) Strong Jahn - Teller distortion is observed in case of unsymmetrical filling of e_g orbitals.
When the ligands are same, repulsion from all side are symmetrical and maximum distortion is observed.

195. (b) CO and CN behaves as SFL i.e., they have tendency to pair up the electrons while F behaves as WFL i.e. no tendency to pair up the electrons. Therefore; A–I, B–IV, C–III, D–II.

196. (a) Metals of low oxidation state forms complexes with ligands which have π-accepting character.

197. (2) Fehling solution is a complex of Cu^{2+}
$Cu^{2+} = 3d^9$; No. of unpaired $e^- = 1$
$\mu_{S.} = \sqrt{1(1+2)} = \sqrt{3} = 1.73\,BM$

198. (0) $[Co(H_2O)_6]Cl_2$
Oxidation state of Co is :
$x + 6(0) = +2 \Rightarrow x = +2$
H_2O is a weak field ligand thus, Co^{2+} electronic configuration = $3d^7 4s^0$

energy ↑ e_g: ↿ ↿ ; t_{2g}: ⥮ ⥮ ↿

Number of unpaired $e^- = 3$
$\mu_s = \sqrt{n(n+2)} = \sqrt{3 \times 5} = \sqrt{15}$ B.M
For $[Cr(H_2O)_6]Cl_3$
Oxidation state of Cr = +3
Electronic configuration = $3d^3 4s^0$

energy ↑ e_g: – – ; t_{2g}: ↿ ↿ ↿

Number of unpaired electrons = 3
$\mu_s = \sqrt{3 \times 5} = \sqrt{15}$ B.M
Difference in spin only magnetic moment is
$= \sqrt{15} - \sqrt{15} = 0$

199. (a) $K_3[Cu(CN)_4]$
O.N. of copper is Cu^+
$Cu^+ \Rightarrow [Ar]3d^{10} \Rightarrow$ Diamagnetic

200. (0) In the given complexes, Co present in +3 O.S. and all ligands occupy inner orbitals.
Shortest wavelength is shown by complex having maximum crystal field splitting.

$[Co(CN)_6]^{2-}$ energy ↑ e_g: — — ; t_{2g}: ⥮ ⥮ ⥮

$M_s = \sqrt{0(0+2)} = 0$ B.M.

201. (a) Energy of absorption depends on the splitting of d-orbital in crystal field splitting.
A strong field ligand have high crystal filled splitting. Order of strength of ligands is
$$en > NH_3 > H_2O$$
Hence, the order of energy is
$$\underset{(A)}{[Ni(en)_3]^{2+}} > \underset{(B)}{[Ni(NH_3)_6]^{2+}} > \underset{(C)}{[Ni(H_2O)_6]^{2+}}$$

202. (2) $K_3[Fe(CN)_6]$
CN^- have –1 oxidation State.
Let oxidation state of Fe in the complex is x.
So, $x - 6 = -3 \Rightarrow x = +3$
Now, CN^- ion is a strong field ligand.

For Fe^{3+}: $-3d^5 4s^0$ Energy ↑ e_g: — — ; t_{2g}: ⥮ ⥮ ↿

Unpaired electron = 1
$\mu_B = \sqrt{n(n+2)} = \sqrt{1 \times (1+2)} = 1.7$
$\mu_B = 2$ B.M

203. (6) Oxidation state of Mn
$x - 6 = -4 \Rightarrow x = +2$
Br is weak field ligand
$Mn^{2+} \to t^3_{2g} e^2_g$; n = 5
$\mu_s = \sqrt{n(n+2)} = \sqrt{35} = 5.91 = 6$

204. (b) Ni in zero oxidation state with any ligand always form tetrahedral complex with sp^3 hybridization.
Ni^{2+} ion with strong field ligand form square planar complex with dsp^2 hybridisation.
Ni^{2+} ion with weak field ligand such as Cl^- form tetrahedral complex.
$[Ni(CN)_4]^{2-}$: d^8 configuration, SFL, sq. planar splitting (dsp^2), diamagnetic.
$[Ni(CO)_4]$: d^{10} configuration (after excitation), SFL, tetrahedral splitting (sp^3), diamagnetic.
$[NiCl_4]^{2-}$: d^8 configuration, WFL, tetrahedral splitting (sp^3), paramagnetic (2 unpaired e^-).

205. (2) CN^- is a strong field ligand.
$[Fe(CN)_6]^{4-}$ Diamagnetic
$[Fe(CN)_6]^{3-}$ Paramagnetic (1 unpaired electron)
$[Ti(CN)_6]^{3-}$ Paramagnetic (1 unpaired electron)
$[Ni(CN)_4]^{2-}$ Square planar complex diamagnetic in nature
$[Co(CN)_6]^{3-}$ Diamagnetic

206. (c) CN^- act as strong field ligand and it has maximum tendency to split the energy cause the splitting of d-orbital. Therefore, $[Co(CN)_6]^{3-}$ have maximum stabilization due to crystal field.

207. (b) Magnetic moment depends upon the number of unpaired electron present in the molecule.
(A) $[FeF_6]^{3-}$
$Fe^{3+} \to 3d^5 4s^0$; F^- is a weak field ligand.
number of unpaired electron = 5.

(B) $[Fe(CN)_6]^{3-}$
$Fe^{3+} \to 3d^5 4s^0$; (CN^-) is a strong field ligand.
$n = 1$.

(C) $[MnCl_6]^{3-}$
$Mn^{3+} \to 3d^4 4s^0$; Cl^- is a weak field ligand.
$n = 4$.

(D) $[Mn(CN)_6]^{3-}$
$Mn^{3+} \to 3d^4 4s^0$; CN^- is a strong field ligand.
$n = 2$

$\mu \Rightarrow A > C > D > B$.

208. (3) $CoCl_3 . 4NH_3 \to [Co(NH_3)_4Cl_2]Cl$
$NiCl_2 . 6H_2O \to [Ni(H_2O)_6]Cl_2$
$PtCl_4 . 2HCl \to H_2[PtCl_6]$

$$[Ni(H_2O)_6]Cl_2 \xrightarrow{2AgNO_3} 2AgCl\downarrow + [Ni(H_2O)_6](NO_3)_2$$

↑↓	↑↓	↑↓	↑	↑

3d $\quad n = 2$

$\mu = \sqrt{2(2+2)}B.M = 2.84BM \approx 3$

209. (6) $[Co(H_2O)_6]^{2+} + NH_3\,(\text{excess}) \to [Co(NH_3)_6]^{3+} + 6H_2O$
for the Co^{3+}; NH_3 ligands are strong field ligand. Hence complex will be inner orbital or low spin complex.

The configuration of Co^{3+} is $\to 3d^6 \Rightarrow t_{2g}^6 e_g^0$

Total number electrons = 6.

210. (766) $\Delta_t = \dfrac{hc}{\lambda} = \dfrac{6.63\times10^{-34} \times 3.08\times10^8}{600\times10^{-9}} = 340.34 \times 10^{-21}$ J

$\Delta_0 = \dfrac{9}{4}\Delta_t = \dfrac{9}{4} \times 340.34 \times 10^{-21}\text{ J} \approx 766 \times 10^{-21}$ J

211. (b) $V^{2+}: 1s^2 2s^2 2p^6 3s^2 3p^6 3d^3$

↑	↑	↑		

(3d) (unpaired $e^- = 3$)

$Ni^{2+}: 1s^2 2s^2 2p^6 3s^2 3p^6 3d^8$

↑↓	↑↓	↑↓	↑	↑

(3d) (unpaired $e^- = 2$)

$Cr^{2+}: 1s^2 2s^2 2p^6 3s^2 3p^6 3d^4$

↑	↑	↑	↑	

(3d) (unpaired $e^- = 4$)

$Fe^{2+}: 1s^2 2s^2 2p^6 3s^2 3p^6 3d^6$

↑↓	↑	↑	↑	↑

(3d) (unpaired $e^- = 4$)

212. (2) $[Fe(H_2O)_3Cl_3]$, $\underbrace{K_3[Fe(CN)_6], [Co(NH_3)_6]Cl_3}_{\text{inner orbital complexes}}$

$K_3[Fe(CN)_6]$ has more value of Δ_0 than that of $[Co(NH_3)_6]Cl_3$; as CN^- is a strong ligand.
More $\Delta_0 \Rightarrow$ smaller value of absorbed λ

$K_3[Fe(CN)_6]: Fe^{3+}: \quad 3d^5 \quad 4s^0 \quad 4p^0$

↑↓	↑↓	↑						

d^2sp^3

$n = 1$, $(\mu) = \sqrt{3}BM = 1.732$ BM ≈ 2 BM.

213. (d) CFSE of octahedral complexes with water is greater for 5d series metal centre ion as compared to 3d and 4d series metal centre.

214. (d) Given, $\mu = 3.87 \quad \Rightarrow \quad n = 3$
CFSE $= -0.8\Delta_0$

CFSE $= [-0.4n t_{2g} + 0.6 n_{eg}]\Delta_0$

M^{z+}	n	CFSE (Δ_0)	d-orbital configuration				
V^{3+}	2	– 0.8	↑	↑			
Cr^{3+}	3	– 1.2	↑	↑	↑		
Mn^{4+}	3	– 1.2	↑	↑	↑		
Co^{2+}	3	– 0.8	↑↓	↑↓	↑	↑	↑
			t_{2g}			e_g	

Thus, only Co^{2+} can exhibit both properties.

215. (173)

$B_2^+ \Rightarrow \sigma 1s^2 \sigma^* 1s^2 \sigma 2s^2 \sigma^* 2s^2 \pi 2p_x^1 = \pi 2p_y^0$

No. of unpaired electrons = 1

$\therefore\ \mu = \sqrt{n(n+2)}$ B.M $= \sqrt{1(1+2)}$ B.M

$= 1.73 \simeq 173 \times 10^{-2}$

216. (d) $\left[Fe(CO)_4(C_2O_4)\right]^+$
CO is strong field ligand, so pairing will take place.
$\therefore$ Spin only magnetic moment is:

$\mu = \sqrt{n(n+2)} = \sqrt{1(1+2)} = 1.73$ B.M.

217. (d) (a) $\left[Fe(CN)_6\right]^{4-} \to$ Pale yellow solution

(b) $\left[Fe(SCN)_6\right]^{4-} \to$ Blood red colour

(c) $Fe_4\left[Fe(CN_6)\right]_3 \cdot H_2O \to$ Prussian blue

(d) $\left[Fe(CN)_5 NOS\right]^{4-} \to$ Violet colour

218. (a) (A) $\left[\overset{III}{Co}F_6\right]^{3-}$; $F \to$ weak field ligand.
CFSE $= [-0.4 \times 4 + 0.6 \times 2]\Delta_0 = -0.4\Delta_0$.

(B) $\left[\overset{II}{Co}(H_2O)_6\right]^{2+}$; $H_2O \to$ weak field ligand.
CFSE $= [-0.4 \times 5 + 0.6 \times 2]\Delta_0 = -0.8\Delta_0$.

(C) $\left[\overset{III}{Co}(NH_3)_6\right]^{3+}$; $NH_3 \to$ strong field ligand.
CFSE $= [-0.4 \times 6 + 0.6 \times 0]\Delta_0$
$= -2.4\Delta_0 + 2P$ (P is pairing energy)

(D) $\left[\overset{III}{Co}(en)_3\right]^{3+}$;
CFSE $= [-0.4 \times 6 + 0.6 \times 0]\Delta_0 = -2.4\Delta_0 + 2P$
en is the strongest field ligand among the given species.
Hence, $[Co(en)_3]^{3+}$ will have highest CFSE.
Thus, order is: (A) < (B) < (C) < (D).

219. (d) $[MnCl_6]^{3-}$

$Mn^{3+} \to [Ar]$ 3d: ↑ ↑ ↑ ↑ _ ; 4s: _ ; 4p: _ _ _ ; 4d: _ _ _ _ _

sp^3d^2

Cl^- is a weak field ligand, so Mn^{3+} with weak field ligand forms high spin complex or outer orbital complex. Hence, the complex will be sp^3d^2 hybridised.

220. (a) $\left[Fe(H_2O)_6\right]^{3+}$
Fe^{3+} : $[Ar]3d^5$; Hybridisation : sp^3d^2
Magnetic nature : Paramagnetic (so this complex responds to external magnetic field)

221. (b) $Ti(22) \Rightarrow 3d^2 4s^2$; $Ti^{3+} \Rightarrow 3d^1 4s^0 \quad \Rightarrow n = 1$
$Sc(21) \Rightarrow 3d^1 4s^2$; $Sc^{3+} \Rightarrow 3d^0 4s^0 \quad \Rightarrow n = 0$
$V(23) \Rightarrow 3d^3 4s^2$; $V^{2+} \Rightarrow 3d^3 \quad \Rightarrow n = 3$
Hence, the spin only magnetic moment will increase in the order; $Sc^{3+} < Ti^{3+} < V^{2+}$

222. (1) $[Co(NH_3)_6]Cl_2$
NH_3 is strong field ligand. Hence, pairing will occur.

$\Rightarrow \quad Co^{2+} = 3d^7; t_{2g}^6\ e_g^1$

Hybridisation will be d^2sp^3 with one unpaired e^-.
$[Co(NH_3)_6]Cl_3$ – d^2sp^3 hybridised with no unpaired e^-.
$Co^{+3} \Rightarrow 3d^6\ t2_g^6 e_g o$

223. (b)
(a) $NiCl_2 \cdot 6H_2O : Ni^{+2} \rightarrow [Ar]3d^8 4s^0$
Cl and $H_2O \rightarrow$ weak field ligand

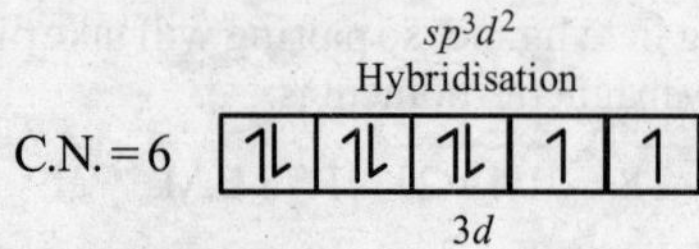

(b) $K_2[Ni(CN)_4]$
$Ni^{+2} \rightarrow [Ar]3d^8 4s^0$
C.N. = 4
$CN^- \rightarrow$ Strong field ligand
dsp^2 Hybridisation

$CN^{\ominus} \rightarrow$ pairging will takes place

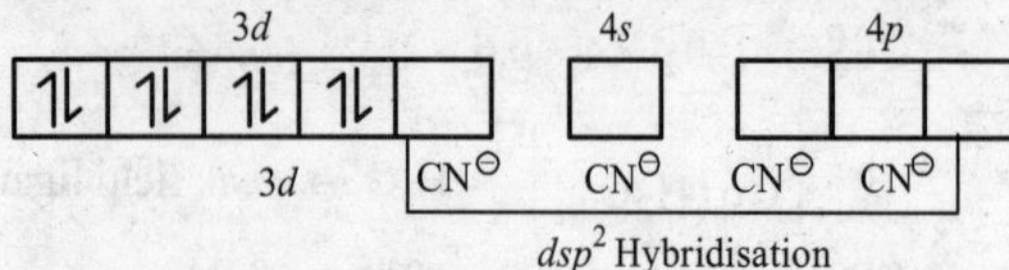

(c) $[Ni(CO)_4]$
CO - Strong field ligand
$Ni \rightarrow [Ar]\ 3d^{10}$

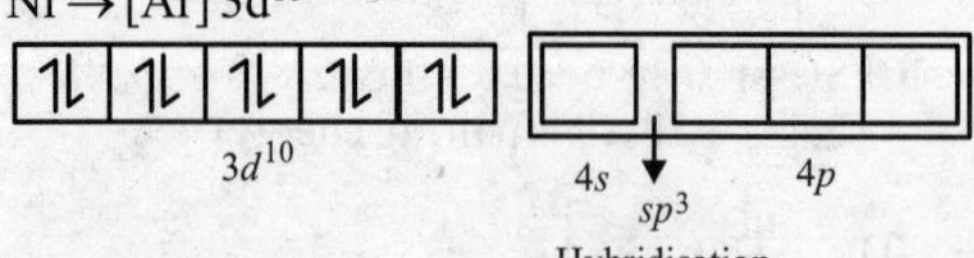

(d) $Na_2[NiCl_4] : Ni^{2+} \rightarrow [Ar]3d^8 4s^0$
sp^3 Hybridisation

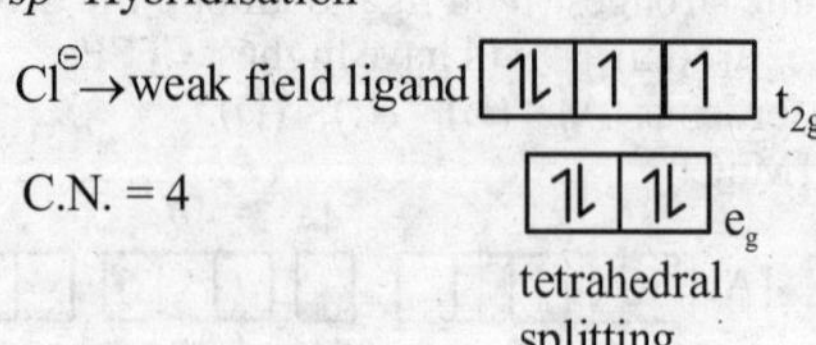

224. (c)
$\Delta_t \quad < \quad \Delta_o \quad < \quad \Delta_{sq}$
$[NiCl_4]^{2-} < [Ni(H_2O)_6]^{2+} < [Ni(CN)_4]^{2-}$
Splitting/absorbed energy order
$[NiCl_4]^{2-} > [Ni(H_2O)_6]^{2+} > [Ni(CN)_4]^{2-}$
Intensity of (complementary) colour of compound

225. (2) $[Co(CN)_6]^{4-}$
$Co^{2+} : [Ar]\ 3d^7$
and CN^- is a strong field ligand which can pair electrons of central atom. After pairing of electron, one electron will remain unpaired.
$Co^{2+} : [Ar]\ t_{2g}^{\ 6}\ e_g^{\ 1}$
It has one unpaired electron (n). So spin only magnetic moment, $\mu = \sqrt{n(n+2)}$B.M.

$\mu = \sqrt{3}$ B.M = 1.73 B.M. ≈ 2 B.M.

226. (c) $Fe^{2+} : [Ar]3d^6 4s^0$: ⇅ | ↑ | ↑ | ↑ | ↑
In the presence of strong field ligand, pairing occurs and there will be no unpaired electron.
⇅ | ⇅ | ⇅ | |

227. (b) Among given species, only CuI does not contain any unpaired electron because it furnishes Cu^+ ion. So, for Cu^+, $\mu = 0$.

228. (2) $[Ni(CN)_6]^{2-}$
$Ni^{+4} \rightarrow d^6$
↑↓ | ↑↓ | ↑↓ | |
CN^- : strong field ligand
Pairing will be there, hence zero unpaired electron.
$NiCl_2(aq) \rightarrow [Ni(H_2O)_6]Cl_2$
H_2O : Weak field ligand
↑↓ | ↑↓ | ↑↓ | ↑ | ↑ → two unpaired e^-
∴ Total change in number of unpaired electrons = 2

229. (4) Octahedral splitting energy

$$= \frac{hc}{\lambda} = \frac{6.626 \times 10^{-34} \times 3 \times 10^8}{498 \times 10^{-9}}$$

$$= 0.0399 \times 10^{-17}\,J = 3.99 \times 10^{-19}\,J \approx 4 \times 10^{-19}\,J$$

230. (a) $[Mn(CN)_6]^{4-}$

$Mn^{2+} = 3d^5 4s^0$

$[Fe(CN)_6]^{3-}$

$Fe^{3+} = 3d^5 4s^0$
CN^- is a strong field ligand, so pairing will occur in both complexes.

⇅ | ⇅ | ↑ | : | :   :   : | : | :
d^2sp^3
Paramagnetic

231. (c)

Complex ions	Unpaired e⁻ (n)	Magnetic moment
(i) $[FeF_6]^{3-}$	5	$\sqrt{35}$
(ii) $[Co(NH_3)_6]^{3+}$	0	0
(iii) $[NiCl_4]^{2-}$	2	$\sqrt{8}$
(iv) $[Cu(NH_3)_4]^{2+}$	1	$\sqrt{3}$

So, correct order of spin only magnetic moment is:
(ii) < (iv) < (iii) < (i)

232. (2) The element having atomic no. 29 is copper.
The electronic configuration of Cu^{2+} is

$Cu^{2+}: 3d^9$ 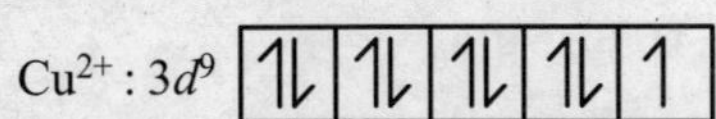

No. of unpaired electron = 1

$\mu = \sqrt{1(1+2)} = \sqrt{3} = 1.73 \text{ BM} \approx 2 \text{ BM}$

233. (b)

(i) $[FeCl_4]^{2-} \Rightarrow$

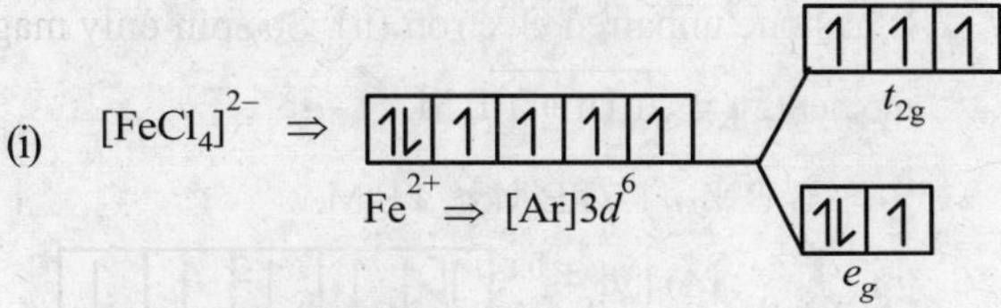

Cl^- is weak field ligand.

$\therefore \quad n = 4.$

$\mu = \sqrt{n(n+2)}$ BM

$= \sqrt{4(4+2)}$ BM $= \sqrt{24}$ BM $= 4.90$ BM

(ii) $[Co(C_2O_4)_3]^{3-}$

$C_2O_4^{2-}$ is strong field ligand.

$\therefore \quad n = 0.$

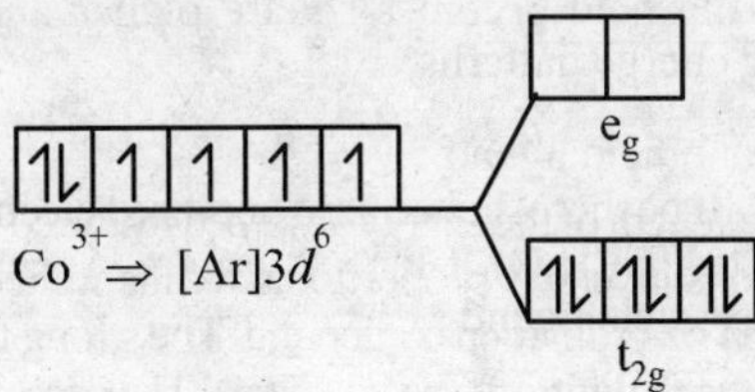

$\because \quad \mu = 0$

(iii) MnO_4^{2-}

$Mn^{+6} \Rightarrow [Ar]3d^1$

$\mu = \sqrt{1(1+2)}$ BM $= \sqrt{3}$ BM $= 1.73$ BM

234. (d) $[MnBr_4]^{2-} \Rightarrow Mn^{2+} \Rightarrow d^5$(Td)

[$\because$ Br is weak field ligand]

d^5(Td) is high spin complex.

So, $\mu = \sqrt{5(5+2)} = 5.91$ B.M.

235. (a) If $\Delta_0 > P$ then for $d^4 \Rightarrow t_{2g}^4\, e_g^0$

If $\Delta_0 < P$ then for $d^4 \Rightarrow t_{2g}^3\, e_g^1$

236. (a) For d^6 configuration, high spin complex.

(i) In case of octahedral field, $t_{2g}^4\, e_g^2$

$\text{CFSE} = [-0.4p + 0.6q]\Delta_0 + n(\text{P})$

$= [-0.4 \times 4 + 0.6 \times 2]\Delta_0 + 0 = -0.4\Delta_0$

(ii) In case of tetrahedral field, $e^3 t_2^3$

$\text{CFSE} = [-0.6p + 0.4q]\Delta_t$

$= [-0.6 \times 3 + 0.4 \times 3]\Delta_t = -0.6\Delta_t$

237. (0) $Ru^{2+} = 4d^6 = t_{2g}^6 e_g^0$ since $\Delta_0 > P$

No. of unpaired electrons = 0

$\therefore$ Magnetic moment = 0 B.M.

238. (a) $[Cr(H_2O)_6]^{2+} \Rightarrow Cr^{2+} \Rightarrow 3d^4 \Rightarrow t_{2g}^{1,1,1}, e_g^{1,0}$

No. of unpaired e^- : 4

$[Fe(H_2O)_6]^{2+} \Rightarrow Fe^{2+} \Rightarrow 3d^6 \Rightarrow t_{2g}^{2,1,1}, e_g^{1,1}$

No. of unpaired e^- : 4

$[Fe(NH_3)_6]^{2+} \Rightarrow Fe^{2+} \Rightarrow 3d^6 \Rightarrow t_{2g}^{2,1,1}, e_g^{1,1}$

No. of unpaired e^- : 4

$[Co(OH)_4]^{2-} \Rightarrow Co^{2+} \Rightarrow 3d^7 \Rightarrow e^{2,2}, t_2^{1,1,1}$

No. of unpaired e^- : 3

$[CoCl_4]^{2-} \Rightarrow Co^{2+} \Rightarrow 3d^7 \Rightarrow e^{2,2}, t_2^{1,1,1}$

No. of unpaired e^- : 3

$[Mn(H_2O)_6]^{2+} \Rightarrow Mn^{2+} \Rightarrow 3d^5 \Rightarrow t_{2g}^{1,1,1}, e_g^{1,1}$

No. of unpaired e^- : 5

So $[Cr(H_2O)_6]^{2+}$ and $[Fe(H_2O)_6]^{2+}$ have same magnetic moment (spin only).

239. (b) $\Delta_0 < P$, hence there will be no pairing of e^-.

$[CoF_3(H_2O)_3] \Rightarrow Co^{3+} \Rightarrow d^6$ or $t_{2g}^4 e_g^2$

$\text{CFSE} = -[0.4p + 0.6q]\Delta_0 + n(P)$

$= [-4 \times 0.4 + 2 \times 0.6]\Delta_0 + 0 = -0.4\,\Delta_0.$

240. (c)

Complex		No. of unpaired e^-
$[Fe(en)(bpy)(NH_3)_2]^{2+}$	:	$Fe^{2+} - 3d^6$, 0
$[Pd(gly)_2]$	:	$Pd^{2+} - 3d^8$,0
$[Co(ox)_2(OH)_2]^-$	:	$Co^{5+} - 3d^4$, 2
$[Ti(NH_3)_6]^{3+}$	:	$Ti^{3+} - 3d^1$, 1

Thus, $[Co(ox)_2(OH)_2]^-$ exhibits highest paramagnetic behaviour due to highest number of unpaired electrons.

241. (d) $[Ti(H_2O)_6]^{3+} \Rightarrow Ti^{3+} = 3d^1\, 4s^0$

$\therefore$ Electronic configuration is $t_{2g}^1 e_g^0$

$\text{CFSE} = [-0.4n_{t_{2g}} + 0.6n_{e_g}]\Delta_0$

$= [(-0.4) \times 1 + 0]\, 20300 = -8120 \text{ cm}^{-1}$

$= \dfrac{-8120}{83.7}$ kJ/mol $= -97$ kJ/mol

242. (c) $[Ru(en)_3]^{2+} \Rightarrow Ru^{2+} = 4d^6 = t_{2g}^6, e_g^0$

$[Fe(H_2O)_6]^{2+} \Rightarrow Fe^{2+} = 3d^6 = t_{2g}^4, e_g^2$

243. (b) Spin only magnetic moment $= 4 \cdot 9 = \sqrt{n(n+2)}$

$\therefore$ Number of unpaired electrons = 4

Two possible arrangements are

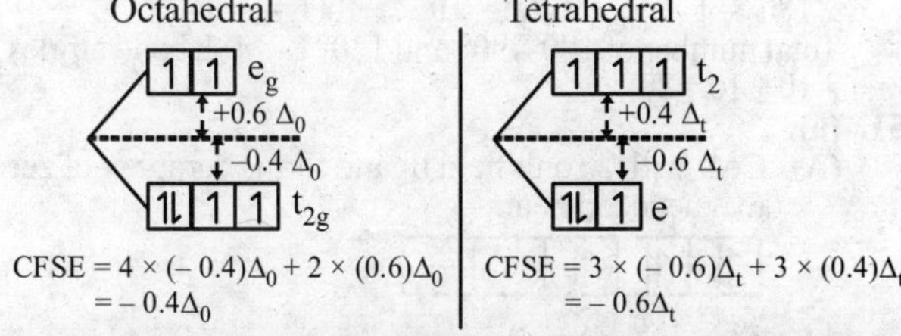

244. (c) With weak field ligands Mn(II) will be of high spin and with strong field ligands it will be of low spin. Ni(II) tetrahedral complexes will be generally of high spin due to sp^3 hybridisation. Mn(II) is of light pink colour in aqueous solution.

245. (a) Stronger the ligand, greater is splitting of d-orbitals and smaller will be wavelength of light absorbed. According to spectrochemical series, the splitting power of ligands is: $NH_3 > NCS^- > F^-$.

Thus, order of wavelength of light absorbed is:

$\lambda_{NH_3} < \lambda_{NCS^-} < \lambda_{F^-}$.

246. (c) Number of Geometrical isomers (n) in square planar $[Pd(F)(Cl)(Br)(I)]^{2-} = 3$
$[Fe(CN)_6]^{3-6} = [Fe(CN)_6]^{3-}$
$Fe^{3+} = 3d^5$

$\mu = \sqrt{n(n+2)} = 1.73$ B.M.
$CFSE = -0.4\,\Delta_0 \times n_{t2g} + 0.6\Delta_0 \times n_{eg} = -0.4\,\Delta_0 \times 5 = -2.0\,\Delta_0$

247. (c) As, $\mu_s = \sqrt{n(n+2)}$
$Na_4[Fe(CN)_6] \rightarrow Fe^{2+} \Rightarrow$ S.F.L. $\Rightarrow t_{2g}^{\ 6}\, e_g^{\ 0} \Rightarrow \mu_s = 0$
$[Cr(H_2O)_6]\,Br_2 \rightarrow Cr^{2+} \Rightarrow$ W.F.L. $\Rightarrow t_{2g}^{\ 3}\, e_g^{\ 1}$
$\Rightarrow \mu_s = \sqrt{4(4+2)} = \sqrt{24}$ B.M.
$(Et_4N)_2[CoCl_4] \equiv 2(Et_4N)^+ [CoCl_4]^{2-} \rightarrow Co^{2+}$
$\Rightarrow$ W.F.L. $\Rightarrow e^4\, t_2^{\ 3} \Rightarrow \mu_s = \sqrt{3(3+2)} = \sqrt{15}$ B.M.
$Na_3[Fe(C_2O_4)_3] \rightarrow Fe^{3+} \Rightarrow$ S.F.L. $\Rightarrow t_{2g}^{\ 5}\, e_g^{\ 0}$
$\Rightarrow \mu_s = \sqrt{1(1+2)} = \sqrt{3}$ B.M.

248. (d)

		No. of unpaired e^-		μ
(A) $Ni(CO)_4$;	$Ni = 3d^8 4s^2$(SFL)		0	0
(B) $[Ni(H_2O)_6]; Cl_2$	$Ni^{2+} = 3d^8$(WFL)		2	$\sqrt{8}$ BM
(C) $Na_2[Ni(CN)_4]$;	$Ni^{2+} = 3d^8$ (SFL)		0	0
(D) $PdCl_2(PPh_3)_2$;	$Pd^{2+} = 4d^8$		0	0

Correct order of the calculated spin only magnetic moments of complexes A to D is (A) $\approx$ (C) $\approx$ (D) < B

249. (26.92)
$[ML_6]Cl_3 + 3AgNO_3 \longrightarrow 3AgCl$
0.3g   0.125 M
V ml

Number of moles of the complex $= \dfrac{0.3}{267.46}$
Number of moles of $AgNO_3 = 0.125 \times V \times 10^{-3}$
From stoichiometry:
$$\frac{0.3}{267.46} \times 3 = 0.125 \times V \times 10^{-3}$$
$$\text{Or, } V = \frac{0.3 \times 3 \times 1000}{267.46 \times 0.125} = 26.92 \text{ mL}$$

250. (20.0)
$\angle 120° = 3; \angle 90° = 6; \quad \angle 90° = 8; \angle 180° = 2$
$\angle 180° = 1 \quad \Rightarrow$ Total = 10 $\Rightarrow$ Total = 10
Total number of 180°, 90° and 120° L–M–L bond angles $= 10 + 10 = 20$

251. (a)
(A) Co^{3+} with strong field ligand forms complex of zero magnetic moment.

1l	1l	1l		

$t_{2g}^6 \quad e_g^0$ when $\Delta_0 > p$
(B) d–e^- configuration, when $\Delta_0 < P$

1l	1	1	1	1

$t_{2g}^2 \quad e_g^2$
(C) Due to strong field ligand en, $[Co(en)_3]^{3+}$ will absorb light of high energy (frequency) or low wavelength.
(D) $\Delta_t = \left(\dfrac{4}{9}\right)\Delta_0 \Rightarrow \Delta_t = \dfrac{4}{9} \times 18000 = 8000\ cm^{-1}$

252. (d) $CFSE = [-0.4\,p + 0.6\,q]\Delta_0$
$Ni^{2+}(d^8) \longrightarrow Ni^{3+}(d^7)$
$t_{2g}^6 e_g^2 \qquad t_{2g}^6 e_g^1$
$CFSE = -1.2 \longrightarrow -1.8$
$Co^{2+}(d^7) \longrightarrow Co^{3+}(d^6)$
$t_{2g}^6 e_g^1 \qquad t_{2g}^6 e_g^0$
$CFSE = -1.8 \longrightarrow -2.4$
$Zn^{2+}(d^{10}) \longrightarrow Zn^{3+}(d^9)$
$t_{2g}^6 e_g^4 \qquad t_{2g}^6 e_g^3$
$CFSE = 0.0 \longrightarrow -0.6$
$Fe^{2+}(d^6) \longrightarrow Fe^{3+}(d^5)$
$t_{2g}^6 e_g^0 \qquad t_{2g}^5 e_g^0$
$CFSE = -2.4 \longrightarrow -2.0$
So, only Fe^{2+} will lose crystal field stabilisation energy upon oxidation to +3 state, others will gain crystal field stabilisation energy.

253. (a) After removal of both axial ligands from octahedral complex the field becomes square planar and leads to following energy patterns.
$$d_{x^2-y^2} > d_{xy} > d_{z^2} > d_{xz} = d_{yz}$$

254. (d) Wavelength of the energy absorbed by the co-ordination compound is inversely proportional to ligand field strength of the given co-ordination compound. The strong field ligand causes higher splitting of the d-orbitals. The decreasing order of ligand field strength is $NH_3 > H_2O > Cl$. Therefore, decreasing order of absorbed wavelength is:
(I) > (II) > (III).

255. (b) $[Fe(H_2O)_6]^{2+} \longrightarrow t_{2g}^{\ 4}\, e_g^2$
$CFSE_0 = (-4 \times 0.4 + 2 \times 0.6)\,\Delta_0 = -0.4\,\Delta_0$
$[NiCl_4]^{2-} \longrightarrow e_g^4 t_{2g}^4$
$CFSE_t = (-4 \times 0.6 + 4 \times 0.4)\,\Delta_t = -0.8\,\Delta_t$

256. (a) Ruby is aluminium oxide (Al_2O_3) containing about 0.5 – 1% Cr^{3+} ions, which are randomly distributed in place of Al^{3+} ions.

257. (a) Cr^{3+} has d^3 configuration and forms an octahedral inner orbitals complex, therefore the set of degenerate orbitals are (d_{xy}, d_{yz} and d_{xz}) and ($d_{x^2-y^2}$ and d_{z^2}).

258. (c) Valence bond theory cannot distinguish between weak field ligands and strong field ligands. Therefore, it cannot predict quantitatively the magnetic properties of transition metal complexes.

259. (a) $[Fe(H_2O)_6]^{2+}$: cationic species and
$[Fe(CN)_6]^{4-}$: anionic species
For the complex, $[Fe(H_2O)_6]_2\,[Fe(CN)_6]$
Now, $[Fe(H_2O)_6]^{2+} \qquad [Fe(CN)_6]^{4-}$
$n = 4 \qquad n = 0$
$\therefore\ \mu = \sqrt{4(4+2)} \qquad \mu = \sqrt{0(0+2)}$
$\mu = 4.9 \qquad \mu = 0$

260. (b) $K_3[Co(CN)_6]$ is an octahedral complex.
During splitting of d-orbitals in octahederal complexes, $d_{x^2-y^2}$ and d_{z^2} orbitals point towards the direction of ligands (i.e., they experience more repulsion and will be raised in energy by $\frac{3}{5}\Delta_0$).

261. (a) Given $\mu = 3.9$ BM

$\mu = \sqrt{n(n+2)}$ B.M. ; $3.9 = \sqrt{n(n+2)}$; $n = 3$

So, the central metal ion has 3 unpaired electrons.

$\therefore$ Configuration is either d^3 or d^7

As H_2O is a weak field ligand. V^{2+} and Co^{2+} will have 3 unpaired electrons.

V^{2+} has d^3 configuration; Co^{2+} has d^7 configuration.

262. (c) Electronic configuration of Mn^{2+} is,

$Mn^{2+} : 3d^5$ | 1 | 1 | 1 | 1 | 1 |

$3d^5$

$\mu = \sqrt{n(n+2)} = 5.9 \Rightarrow n = 5$

Presence of 5 unpaired e^- show that the complex of Mn^{2+} has only weak field ligand (NCS^-).

263. (a) Wilkinson catalyst : $[Rh(PPh_3)_3Cl]$

Chlorophyll : $C_{55}H_{72}O_5N_4Mg$

Vitamin B_{12} contains Co.

Carbonic anhydrase contains a Zn ion.

264. (c)

Metal ion	No. of unpaired elec.		Difference in the unpaired electrons
	High spin	Low spin	
Ni^{2+} $(3d^8)$	2	2	0
Mn^{2+} $(3d^5)$	5	1	4
Fe^{2+} $(3d^6)$	4	0	4
Co^{2+} $(3d^7)$	3	1	2

265. (a) E.C. of Cr^{3+} $(3d^3)$: | 1 | 1 | 1 | | |

3d

For complex A, $[Cr(H_2O)_2]^{3+}$:

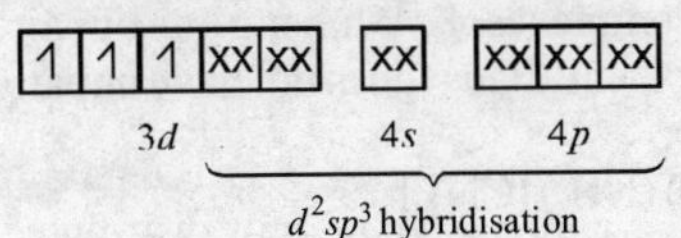

For complex B, $[Cr(NH_3)_6]^{3+}$:

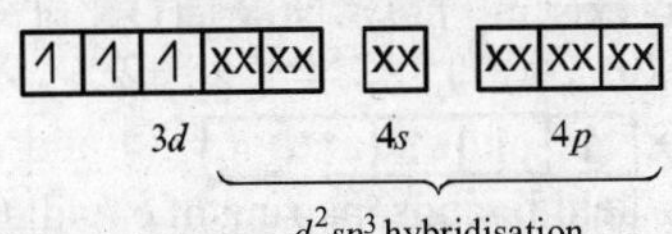

Here, both the complexes (A) and (B) are paramagnetic with 3 unpaired electrons each. H_2O is also a weak field ligand, which causes lesser splitting than NH_3, which is comparatively stronger field ligand. Hence, the (Δ_0) value of (A) and (B) are calculated from the wavelengths of light absorbed and not from the wavelengths of light emitted.

266. (d) In case of similar metal atom or ion, the value of co-ordination number and the strength of the ligands determine the value of crystal field splitting energy.

Strength of ligands : $CN^- > NH_3 > H_2O > Cl^-$

$\therefore K_3[Co(CN)_6]$ has the highest crystal field splitting energy.

267. (a) Lesser the wavelength of light absorbed (more energy) greater will be ligand strength.

Energy:	Blue	>	Green	>	Red
Ligand strength:	L_2	>	L_1	>	L_3

268. (c) Wilkinson catalyst is $[RhCl(PPh_3)_3]$.

$$[RhCl(PPh_3)_3] \equiv [\overset{(+1)}{Rh}\ \overset{(-1)}{Cl}(\overset{(0)}{PPh_3})_3]$$

$Rh^+ \rightarrow [Kr]$

| 1l | 1 | 1 | 1 | 1 | | | | | |

4d 5s 4p

PPh_3 is a strong field ligand and for square planar complex, pairing of e^- from highest energy orbital $d_{x^2-y^2}$ to e_g orbital will take place.

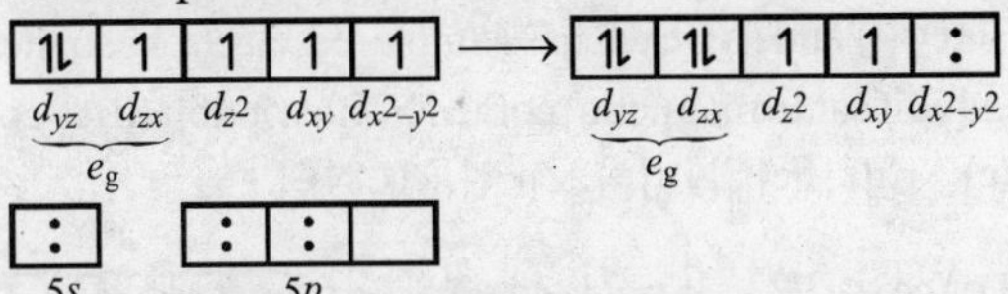

| : | | : | : | |

5s 5p

Hence, Rh will undergo dsp^2 hybridisation to form square planar complex.

269. (c) $[Ni(CN)_4]^{2-}$ is square planar, diamagnetic (0 unpaired electrons) with dsp^2 hybridisation.

$[Ni(CO)_4]$ is tetrahedral, diamagnetic (0 unpaired electrons) with sp^3 hybridisation.

$[NiCl_4]^{2-}$ is tetrahedral, paramagnetic (2 unpaired electrons) with sp^3 hybridisation.

Hence, the option (c) is the correct answer.

270. (d) The complex having higher number of unpaired electrons will have higher value of spin only magnetic moment.

In all these complexes, the central metal ion is in +2 oxidation state.

Zn^{2+} has $3d^{10}$ outer electronic configuration with 0 unpaired electron.

Ni^{2+} has $3d^8$ outer electronic configuration with 2 unpaired electrons.

Co^{2+} has $3d^7$ outer electronic configuration with 3 unpaired electrons.

Mn^{2+} has $3d^5$ outer electronic configuration with 5 unpaired electrons

Hence, the correct order of spin-only magnetic moments is: $[MnCl_4]^{2-} > [CoCl_4]^{2-} > [NiCl_4]^{2-} > [ZnCl_4]^{2-}$.

271. (c) A ruby is a crystal of alumina, i.e., aluminum oxide (Al_2O_3) containing a trace of chromium (III) ions replacing some of the aluminum ions. In ruby, each Al^{3+} ion and Cr^{3+} ion is surrounded by six oxide ions in an octahedral arrangement. The origin of the color of emeralds is similar to that of the color of rubies. However, the bulk of an emerald crystal is composed of beryl, i.e., beryllium aluminum silicate $(Be_3Al_2(SiO_3)_6)$ instead of the alumina which forms rubies. The color is produced by chromium (III) ions, which replace some of the aluminum ions in the crystal. In emeralds, the Cr^{3+} is surrounded by six silicate ions, rather than the six oxide ions in ruby. Therefore, the color (green) of emeralds is different from that of ruby.

272. (c) The splitting is affected by the oxidation state of the central metal ion. A higher oxidation state leads to larger splitting, hence,

$$\Delta_o \text{of}[Ti(H_2O)_6]^{3+} > \Delta_o \text{of}[Ti(H_2O)_6]^{2+}$$

Further Δ_o also depends of $Z_{eff.}$ and $Z_{eff.}$ of $4d$ series is more than $3d$ series.

Hence, Δ_o of $[Cr(H_2O)_6]^{2+} < \Delta_o$ of $[Mo(H_2O)_6]^{2+}$

273. (d)

	Metal ion	Unpaired electron	Magnetic moment
(i)	Cr^{2+}	4	$\sqrt{24}$ B.M.
(ii)	Fe^{2+}	4	$\sqrt{24}$ B.M.
(iii)	Co^{2+}	3	$\sqrt{15}$ B.M.
(iv)	Mn^{2+}	5	$\sqrt{35}$ B.M.

Since (i) and (ii), each has 4 unpaired electron, so they will exhibit same magnetic moment. Thus option (d) is correct.

274. (c) $Zn_2[Fe(CN)_6] \equiv 2Zn^{2+}[Fe(CN)_6]^{4-}$

$[Fe(CN)_6]^{4-} \equiv \left[Fe^{2+}\overset{-1}{(CN)_6}\right]$; Fe^{2+} ⇅ ⇅ ⇅ □ □

In presence of strong field ligand CN^-, e^- will pair and there will be no unpaired e^-. Hence the complex will not be yellow.

275. (a) $KMnO_4$ is colored because it absorbs light in the visible range of electromagnetic radiation. The permanganate ion is the source of color, as a ligand to metal (L → M) charge transfer takes place between oxygen's *p*-orbitals and the empty *d*-orbitals on the metal. This charge transfer takes place when a photon of light is absorbed which leads to the purple color of the compound.

276. (a) Symmetrically filled t_{2g} and e_g are those, which contain equal distribution of electrons.

$[FeF_6]^{3-}$ ↑ ↑ e_g; ↑ ↑ ↑ t_{2g} — equal distribution of electrons in orbitals t_{2g} and e_g

277. (a) Aqueous solution of $CoCl_2$ contains $[Co(H_2O)_6]^{2+}$ which is pinkish in colour so option (d) is incorrect.

Reduction potential of $Co^{3+} \longrightarrow Co^{2+}$ is high so option(b) is incorrect because Co^{2+} does not oxidises easily to Co^{3+}.
It is general case that symmetrical substituted octahedral complexes are less deeper in colour than tetrahedral complexes. So $[CoCl_4]^{2-}$ is deep blue in colour.

278. (c) $[Co(NH_3)_6]^{3+}$

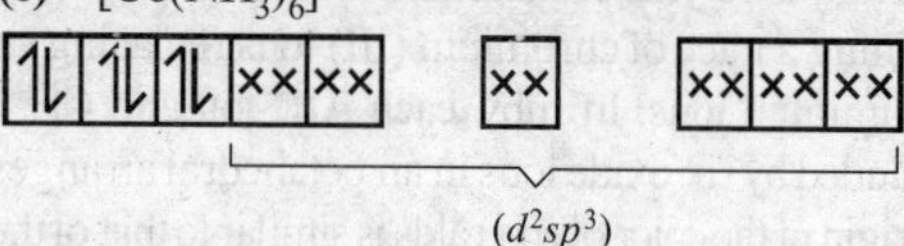

Octahedral and diamagnetic.

279. (a) Both are paramagnetic, but CN^- will result in one unpaired e^- and F^- will result in 5 unpaired electrons.

280. (c) $\underset{\text{Lewis acid}}{Co^{3+}} + \underset{\text{Lewis base}}{6NH_3} \longrightarrow \underset{\text{adduct}}{\left[Co(NH_3)_6\right]^{3+}}$

281. (d) Following is the order of increasing value of Δ_o for the ligands:
$I^- < Br^- < Cl^- < F^- < H_2O < Py = NH_3 < en < NO_2^- << CN^- < CO$

282. (c) The absorption of visible light is responsible for coloured nature of the transition metal cation due to the promotion of one or more unpaired - *d* - electron from a lower to higher level within same *d*-subshell. Hence, higher the number of unpaired electron higher will be the absorpion in visible light.
The electronic configuration of the given elements is
$Sc^{3+}(18) = 1s^2\,2s^2\,2p^6\,3s^2\,3p^6\,3d^0\,4s^0$ - no unpaired e^-.
$Ti^{4+}(18) = 1s^2\,2s^2\,2p^6\,3s^2\,3p^6\,3d^0\,4s^0$ - no unpaired e^-.
$V^{3+}(20) = 1s^2\,2s^2\,2p^6\,3s^2\,3p^6\,3d^2\,4s^0$ - Two unpaired e^-.
$Zn^{2+}(28) = 1s^2\,2s^2\,2p^6\,3s^2\,3p^6\,3d^{10}\,4s^0$ - no unpaired e.
Hence, $[V(NH_3)_6]^{3+}$ will most likely absorb visible light.

283. (b) $[NiL_4]^{2-} \Rightarrow$ Diamagnetic in nature (given)

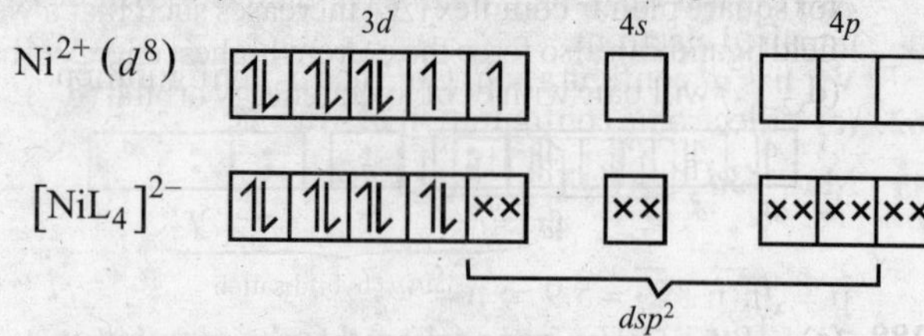

So, no. of unpaired electron = 0

284. (b)

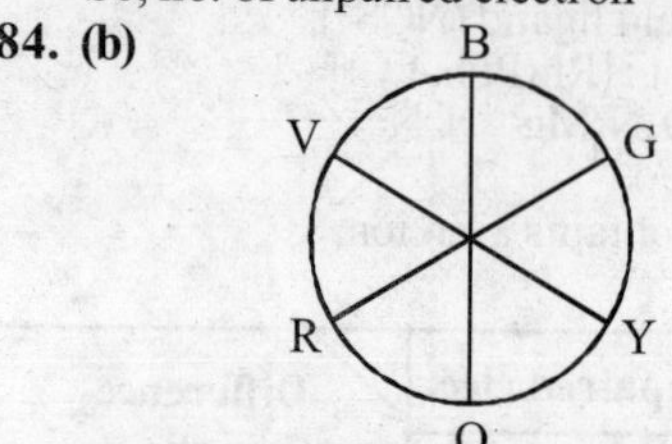

For a given metal ion, weak field ligands create a complex with smaller Δ, which will absorbs light of longer λ and thus lower frequency. Conservely, stronger field ligands create a larger Δ, absorb light of shorter λ and thus higher *v* i.e. higher energy.

Red	< Yellow	< Green	< Blue
λ = 650 nm	570 nm	490 nm	450 nm

So order of ligand strength is
$L_1 < L_3 < L_2 < L_4$

285. (b) Among four ligands, CN^- and ox^{2-} are strong field ligands. So pairing takes places in $[(FeCN)_6]^{3-}$ and $[Co(ox)_3]^{3-}$ complexes. Fe^{3+} having one unpaired electron in its complex while Co^{3+} does not have unpaired electron in its complex.

286. (d) $[Cr(NO)(NH_3)(CN)_4]^{2-}$
Let O.S. of NO be +1, if x is the O.S. of Cr.
$\therefore\ x + (+1) + (0) + 4(-1) = -2 \Rightarrow x = +1$
But Cr does not exist in +1 O.S. Now, let O.S. of NO be zero.
$\therefore\ x + (0) + (0) + 4(-1) = -2 \Rightarrow x = +2$
$Cr^{2+}(d^4) \longrightarrow$ 1 1 1 1 □
Due to strong field ligands, pairing of e^- will take place, i.e.,

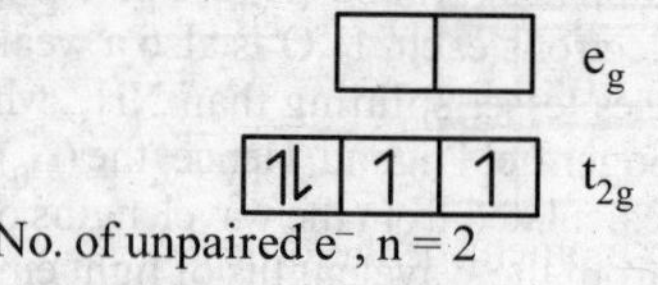

No. of unpaired e^-, n = 2
$\mu = \sqrt{n(n+2)} = \sqrt{2(2+2)} = \sqrt{8} \approx 2.82$ BM

287. (a) $[NiCl_4]^{2-}$: $Ni^{2+}(d^8)$

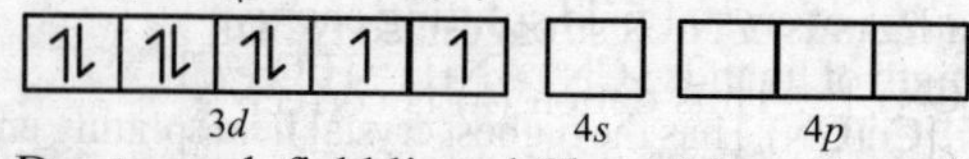

Due to weak field ligand Cl^-, pairing of e^- will not take place and the hybridisation will be sp^3.

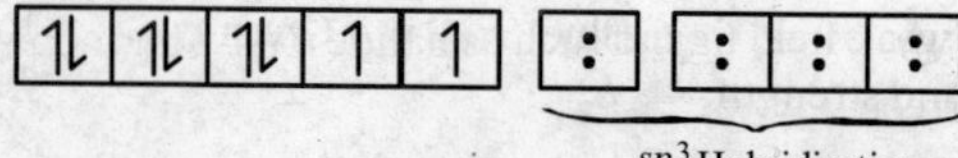

$[PdCl_4]^{2-}$: Pd^{2+} (d^8)

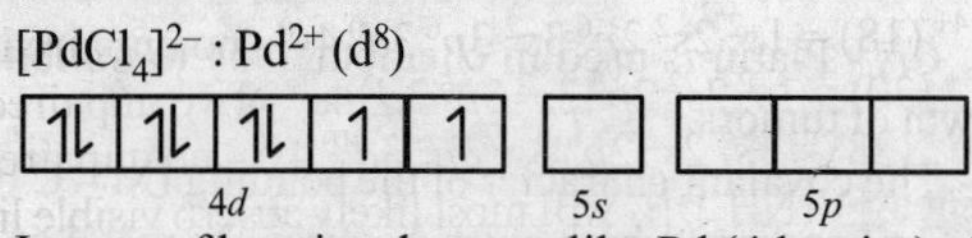

In case of heavier elements like Pd (4d series), the CFSE for square planar complex (Δ_s) increases such that a weak field ligand can also force the e^- from highest energy orbital $(d_{x^2-y^2})$ will pair with e^- of lower energy orbital (d_{xy}).

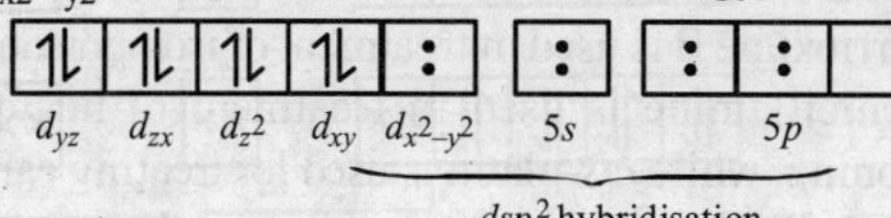

288. (a) $[PtCl_2(NH_3)_2]$ has square planar geometry.

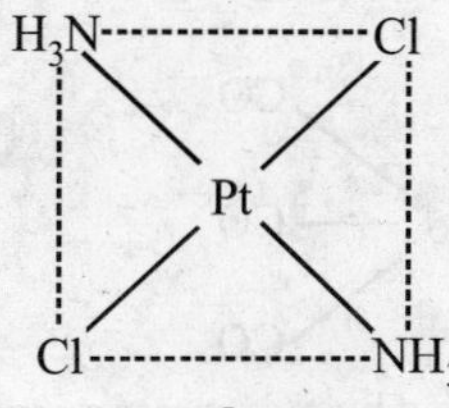

trans-Isomer

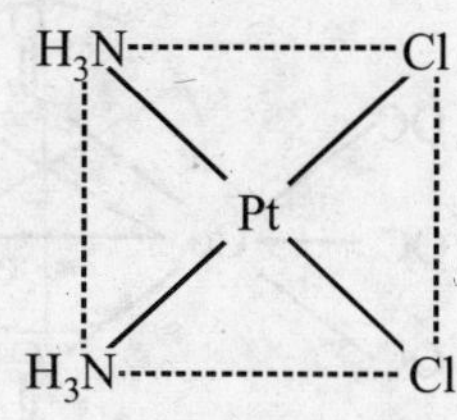

cis-isomer

289. (a)

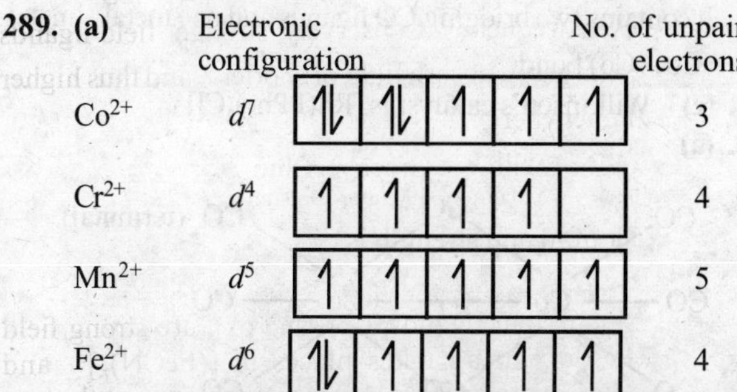

	Electronic configuration		No. of unpaired electrons
Co^{2+}	d^7	⇅ ⇅ ↿ ↿ ↿	3
Cr^{2+}	d^4	↿ ↿ ↿ ↿	4
Mn^{2+}	d^5	↿ ↿ ↿ ↿ ↿	5
Fe^{2+}	d^6	⇅ ↿ ↿ ↿ ↿	4

∴ Co^{2+} has lowest no. of unpaired electrons, hence, lowest paramagnetic behaviour is shown by $[Co(H_2O)_6]^{2+}$

290. (b) Correct order is : $CN^- > en > NCS^- > Cl^-$

291. (d)

Ion	Electronic Configuration	Magnetic moment $\left(\sqrt{n(n+2)}\right)$
Cr^{2+} d^4	1 1 1 1	$\sqrt{4(4+2)} = 4.9$
Mn^{2+} d^5	1 1 1 1 1	$\sqrt{5(5+2)} = 5.8$
Ti^{2+} d^2	1 1	$\sqrt{2(2+2)} = 2.4$
V^{2+} d^3	1 1 1	$\sqrt{3(3+2)} = 3.8$

Cr^{2+} has magnetic moment of the order of 5 B.M.

292. (a) In octahedral complex, the magnitude of Δ_0 will be highest in a complex having strongest ligand. Of the given ligands, CN^- is strongest, so Δ_0 will be highest for $[Co(CN)_6]^{3-}$. Thus option (a) is correct.

293. (a) Complexes with dsp^2 hybridisation are square planar. All the complexes of Pt^{2+} are square planar including those with weak field ligand such as halide ions. Thus option (a) is correct.

294. (d) $[Ni X_4]^{2-}$, the electronic configuration of Ni^{2+} is

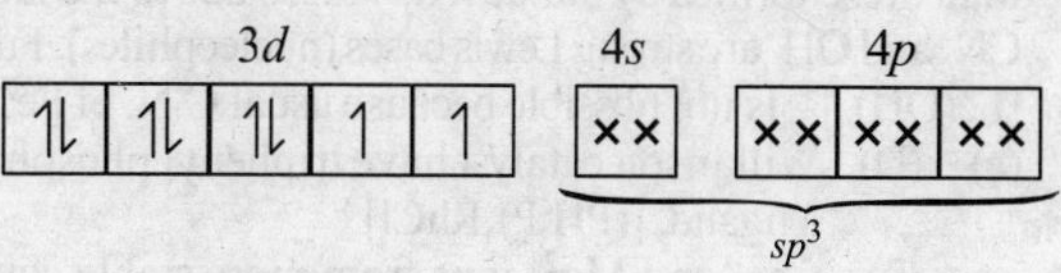

It contains two unpaired electrons and the hybridisation is sp^3 (tetrahedral).

295. (c) The number of unpaired electrons in $Ni^{2+}(aq) = 2$. Water is a weak ligand, hence no pairing will take place.

∴ spin magnetic moment

$= \sqrt{n(n+2)} = \sqrt{2(2+2)} = \sqrt{8} = 2.82$ B.M

296. (d) d^5 strong ligand field

⇅ ⇅ 1

$\mu = \sqrt{n(n+2)} = \sqrt{1(1+2)} = \sqrt{3} = 1.73$BM

d^3— in weak as well as in strong field

1 1 1

$\mu = \sqrt{3(3+2)} = \sqrt{15} = 3.87$ B.M.

d^4– in weak ligand field

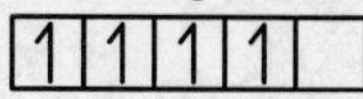

$\mu = \sqrt{4(4+2)} = \sqrt{24} = 4.89$ B.M

d^4– in strong ligand field

⇅ 1 1

$\mu = \sqrt{2(2+2)} = \sqrt{8} = 2.82$ B.M

297. (b)

Metal ion	**No. of unpaired electron**
(a) Co^{3+}	0
(b) Fe^{3+}	1
(c) Mn^{3+}	2
(d) Cr^{3+}	3

Due to strong field ligand CN^-, pairing of e^- will occur.

298. (c) $[Fe(CN)_6]^{4-} \rightarrow$ ⇅ ⇅ ⇅ : : | : | : : :

No of unpaired electron = 0

$[MnCl_4]^{2-} \rightarrow$ 1 1 1 1 1 | : | : : :

No of unpaired electrons = 5

$[CoCl_4]^{2-} \rightarrow$ ⇅ ⇅ 1 1 1 | : | : : :

No of unpaired electrons = 3

The greater the number of unpaired electrons, greater the magnitude of magnetic moment. Hence, the correct order will be :

$[MnCl_4]^{2-} > [CoCl_4]^{2-} > [Fe(CN)_6]^{4-}$

299. (d) $\underset{d^2sp^3}{[Fe(CN)_6]^{4-}}$, $\underset{d^2sp^3}{[Mn(CN)_6]^{4-}}$,

$\underset{d^2sp^3}{[Co(NH_3)_6]^{3+}}$, $\underset{sp^3d^2}{[Ni(NH_3)_6]^{2+}}$

Hence, $[Ni(NH_3)_6]^{2+}$ is outer orbital complex.

300. (b) A square planar complex is formed by hybridisation of s, p_x, p_y and $d_{x^2-y^2}$ atomic orbitals.

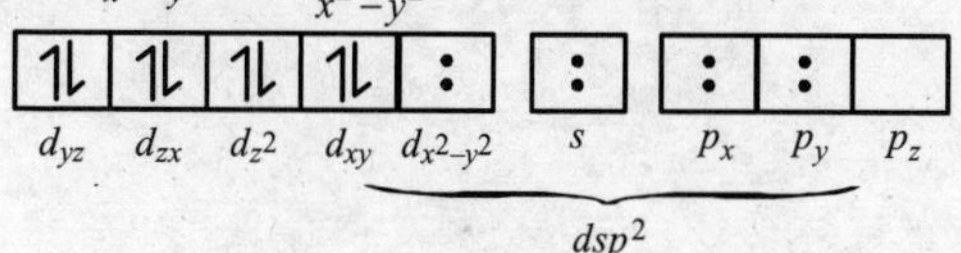

301. (c) The cyano and hydroxo complexes are far more stable than those formed by halide ion. This is due to the fact that CN^- and OH^- are strong Lewis bases (nucleophiles). Further, $[Fe(OH)_5]^{3-}$ is not possible because usual C.N. of Fe^{2+} is 6.

302. (c) (D) Wilkinson catalyst have triphenyl phosphine as a ligand, $[(PH_3P)_3RhCl]$

(E) Ca^{2+} and Mg^{2+} ions from more stable complex with EDTA.

303. (a) $Mn_2(CO)_{10}$ = decacarbonyl dimanganese (0)

octahedral geometry around Mn atom

304. (c) (a) $[Os_3(CO)_{12}]$

(b) $[Ru_3(CO)_{12}]$

(c) $[Co_2(CO)_8]$

(d) $[Mn_2(CO)_{10}]$

305. (a) Mg is present in chlorophyll and in black and white photography the developed film is fixed by washing with hypo solution which dissolves the undecomposed AgBr to form a complex ion $[Ag(S_2O_3)_2]^{3-}$

306. (a) *Cis* - Platin is used in chemotherapy to inhibits the growth of tumors.

307. (b) The covalant character of the bonding (M – C σ and M – C π bonding) which exists between the metal and the carbon atom of the CO can only be explained by the molecular orbital theory.

308. (d) EDTA is used in the treatment of lead poisoning. Deferrioxime B is used in treatment of iron poisoning and D-penicillamine is used in treatment of heavy metal poisoning, while *cis*-platin is used for treating cancer.

309. (c) *cis*-Platin is used as an anti-cancer drug.

310. (a) The structure of $Co_2(CO)_8$ is represented as

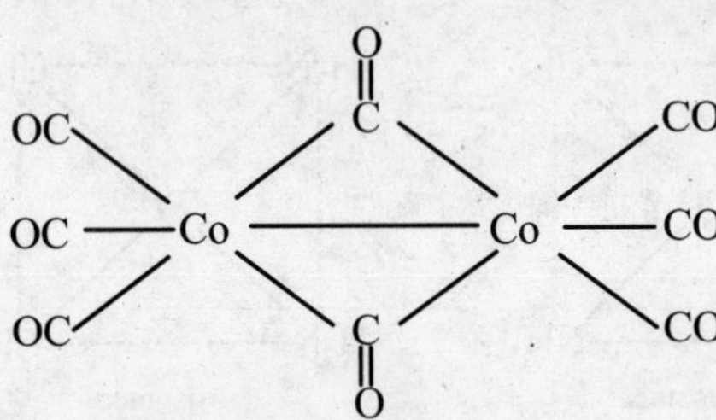

It contains two bridging CO ligands and one metal – metal (Co – Co) bond.

311. (c) Wilkinson's catalyst is $[Rh(PPh_3)_3Cl]$

312. (a)

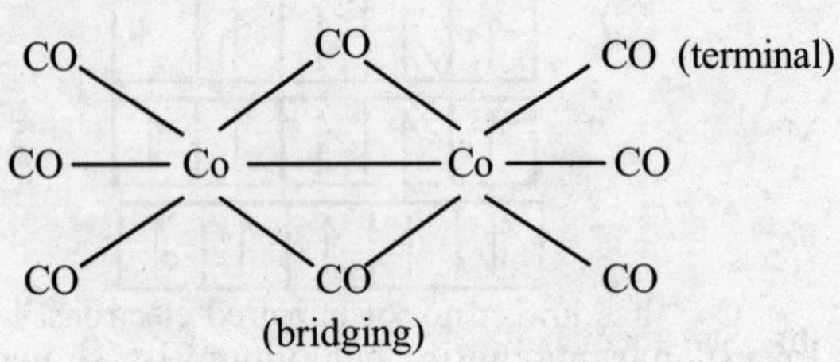

313. (a) The donor atoms, molecules or anions which donate a pair of electrons to the metal atom or ion and form a coordinate bond with it are called ligands. In methane, there are no electrons for donation to central metal atom/ion, it is stable with complete octet configuration.

314. (d) Due to some backbonding by sideways overlapping between *d*-orbitals of metal and *p*-orbital of carbon, the Fe–C bond in $Fe(CO)_5$ has both σ and π character.

315. (a) EDTA is hexadentate, four donor O atoms and 2 donor N atoms, and for the formation of octahedral complex one molecule is required.

316. (c) The chlorophyll molecule plays an important role in photosynthesis. It has porphyrin rings and the metal Mg, not Ca.

317. (b) Compounds that contain carbon-metal bond are known as organometallic compounds.

In CH_3-Mg-Br (Grignard's reagent), a bond is present between carbon and Mg (metal), hence it is an organometallic compound.

Haloalkanes and Haloarenes

7

1. (a) **Statement - I:** C—Cl bond is more polar than C—Br bond. Hence (cis-dichloroethene) is more polar than (cis-dibromoethene)

Statement - II:

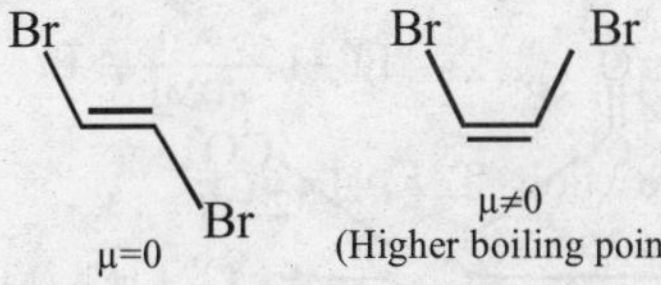

Br / Br ($\mu=0$) — Br Br ($\mu\neq0$) (Higher boiling point)

2. (b)

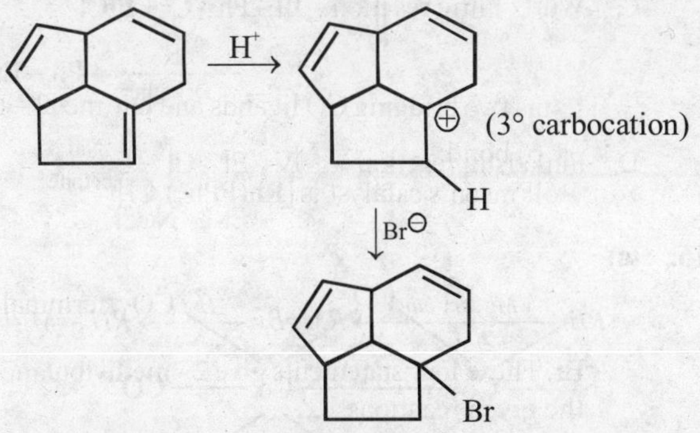

$\xrightarrow{H^+}$ (3° carbocation)

$\downarrow Br^{\ominus}$

(Major Product)

3. (b) **Statement-I:**

$$R-Cl\xrightarrow{(aq.KOH)}R-OH(S_N\ reaction)$$

Statement-II:

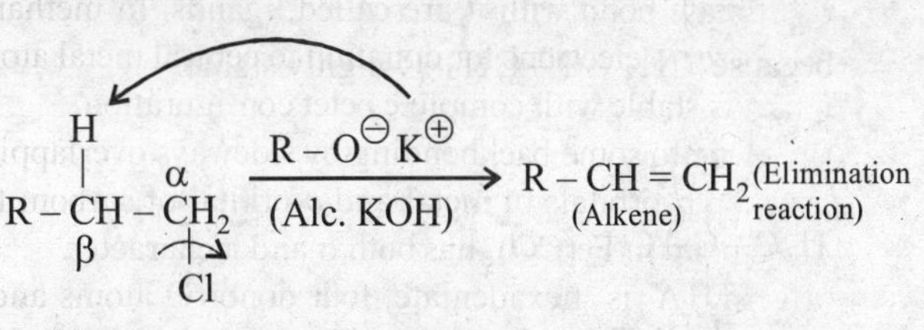

$$R-\underset{\beta}{CH}-\overset{\alpha}{CH_2}(H)(Cl)\xrightarrow[(Alc.\ KOH)]{R-O^{\ominus}K^{\oplus}}\underset{(Alkene)}{R-CH=CH_2}\ \text{(Elimination reaction)}$$

4. (b)

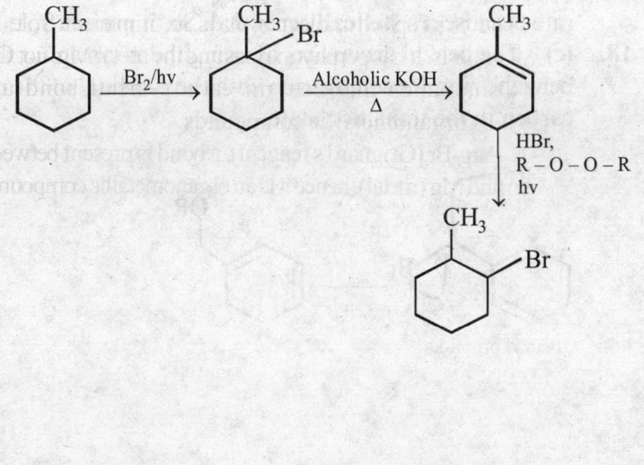

Methylcyclohexane $\xrightarrow{Br_2/hv}$ 1-bromo-1-methylcyclohexane $\xrightarrow[\Delta]{Alcoholic\ KOH}$ 1-methylcyclohexene $\xrightarrow[hv]{HBr,\ R-O-O-R}$ 1-bromo-2-methylcyclohexane

5. (184)

2-bromopentane (1 mole) $\xrightarrow{Alc.\ KOH}$ (P) (major) 0.80 mole $\xrightarrow{Br_2/CCl_4}$ (Q) (major) 0.80 mole

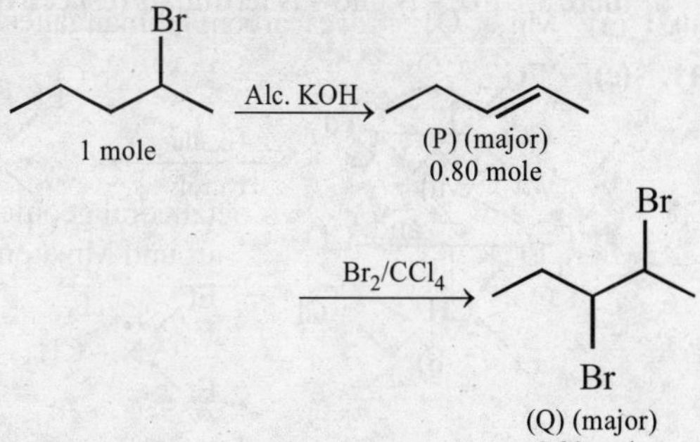

Molecular mass of Q = 230 g mol^{-1}

Mass of Q = 0.8 × 230 = 184 g

6. (c)

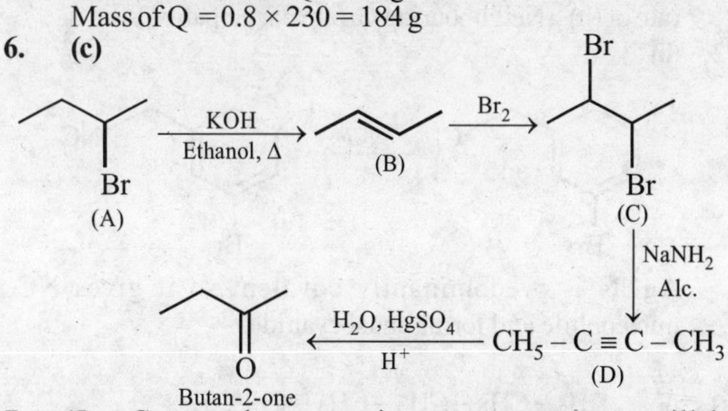

(A) $\xrightarrow[Ethanol,\ \Delta]{KOH}$ (B) $\xrightarrow{Br_2}$ (C) $\xrightarrow[Alc.]{NaNH_2}$ $CH_3-C\equiv C-CH_3$ (D) $\xrightarrow[H^+]{H_2O,\ HgSO_4}$ Butan-2-one

7. (d) Greater the resonating structure greater will be stability of carbocation.

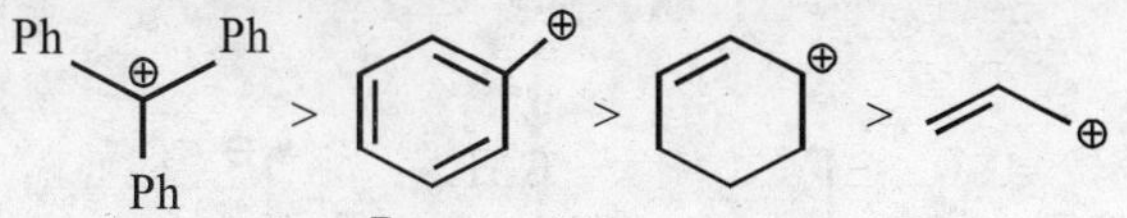

$Ph_3C^{\oplus}$ > $C_6H_5CH_2^{\oplus}$ > (cyclohexenyl cation) > $CH_2=CH-CH_2^{\oplus}$

8. (d)

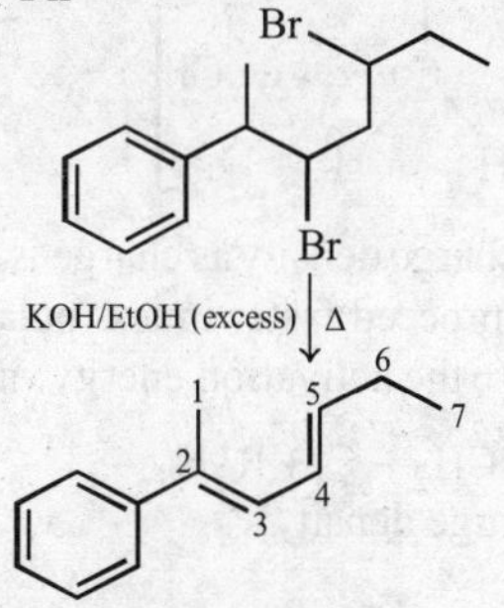

$\xrightarrow{KOH/EtOH\ (excess)\ \Delta}$ 2-Phenylhepta-2,4-diene

9. (d)

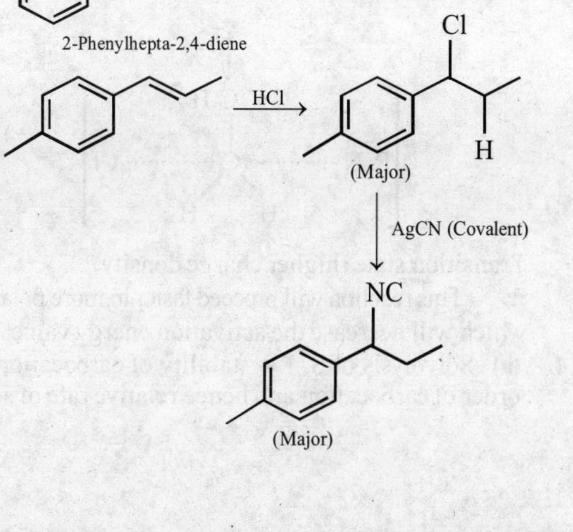

$\xrightarrow{HCl}$ (Major) $\xrightarrow{AgCN\ (Covalent)}$ (Major)

10. (d) $CH_3-CH_2-CH_2-NO_2 \xleftarrow{Ag-NO_2}$
(A)

$CH_3-CH_2-CH_2-Br \xrightarrow{AgCN} CH_3-CH_2-CH_2-NC$
(B)

$AgNO_2$ and AgCN are covalent compounds in which there are free – N and – N terminals respectively.

11. (c) $Et_2N-CH_2-CH_2-Cl$ (a) and $Et_2CH-CH_2-CH_2-Cl$ (b) $\xrightarrow{\text{Alkaline Hydrolysis}}$ $Et_2N^+(CH_2-CH_2)$ $\xrightarrow{OH^-}$ $Et_2\ddot{N}-CH_2-CH_2-OH$

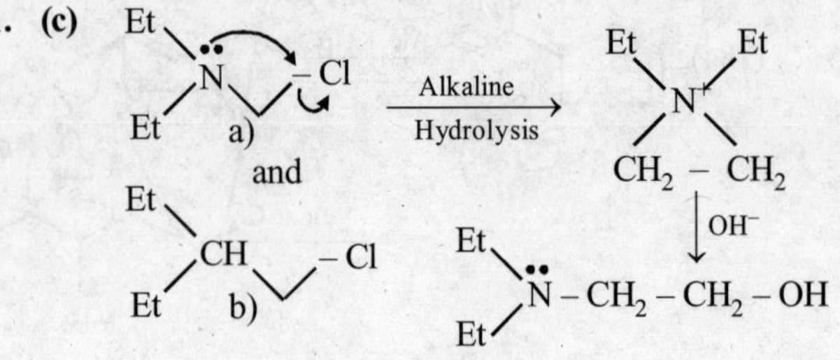

Due to intramolecular substitution rate of (0) is faster than rate of (b). (Neighbouring Group participation)

12. (b) (3-bromo-5-iodobenzyl chloride) $\xrightarrow{AgCN}$ (3-bromo-5-iodobenzyl isocyanide, $-CH_2NC$)

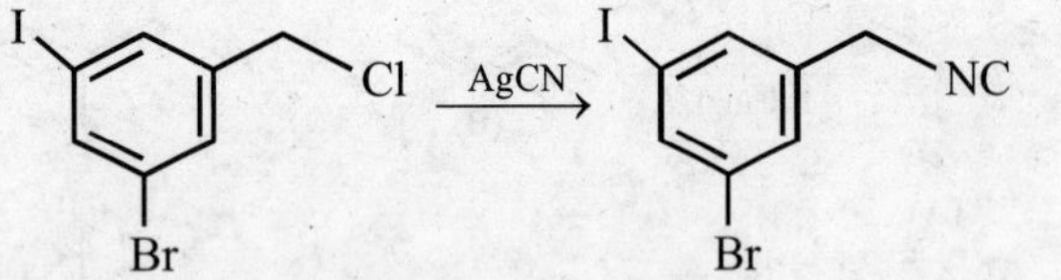

AgCN is predominantly covalent so it gives NC^- nucleophile and forms an isocyanide.

13. (a) $CH_3-CH_2-CH_2-CH_2-Cl+\overset{\ominus}{O}H$
Nucleophile (higher charge density)

$\downarrow SN^2$

$$\left[HO\cdots\cdots C(C_3H_7)(H)(H)\cdots\cdots Cl\right]^{\ominus}$$

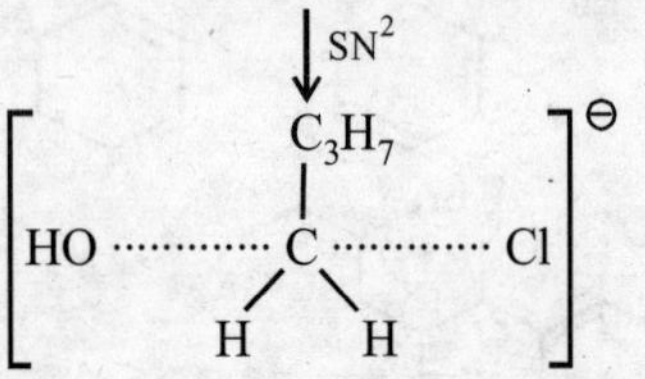

Transition state (less charge density as charge is dispersed)
⇒ This reaction will proceed faster in less polar medium which will not increase the activation energy value.

$CH_3-CH_2-CH_2-CH_2-Cl+R_3N$
Reactant (low charge density)

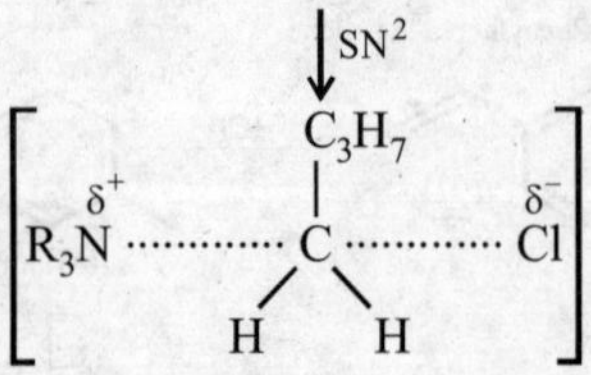

$\downarrow SN^2$

$$\left[\overset{\delta+}{R_3N}\cdots\cdots C(C_3H_7)(H)(H)\cdots\cdots \overset{\delta-}{Cl}\right]$$

Transition state (higher charge density)
⇒ This reaction will proceed faster in more polar medium which will decrease the activation energy value.

14. (a) Solvolysis of $S_N1 \propto$ stability of carbocation stability order of carbocation and hence relative rate of solvolysis.

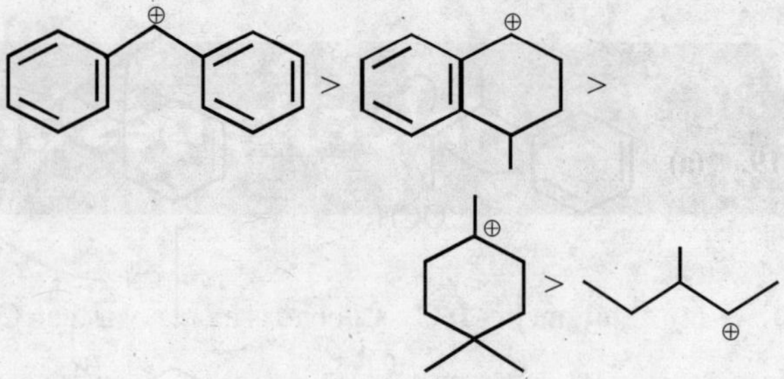

(diphenylmethyl cation) > (tetrahydronaphthyl cation) > (4,4-dimethylcyclohexyl-substituted cation) > (primary alkyl cation)

15. (c)

List-I Name reaction	List-II Product obtainabel
A. Swarts reaction	I. $Et-I \xrightarrow[DMF]{AgF} Et-F$
B. Sandmeyer's reaction	II. $PHN_2^{\oplus}Cl^- \xrightarrow{CuCN/KCN} PhCN+N_2$
C. Wurtz Fittig reaction	III. $Ph-Cl+EtCl \xrightarrow[\text{Dry ether}]{Na} Ph-Et$
D. Finkelstin reaction	IV. $Et-Cl \xrightarrow[\text{acetone}]{NaI} Et\text{-}I+NaCl$

16. (a)

$$RBr \xrightarrow{Mg,\,dry\,ether} RMgBr \xrightarrow{H_2O} RH+Mg(OH)$$

Br. These four statements give 2– methylbutane upon the given reactions.

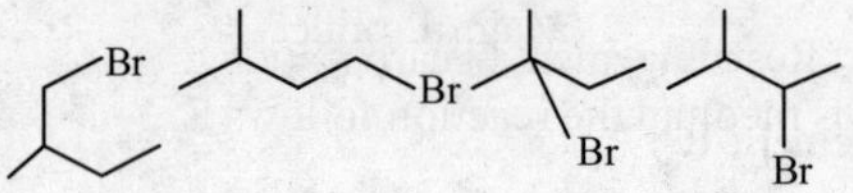

(four bromo-2-methylbutane structures)

Total number of possible structures of RBr = 4

17. (d) CH_3-O-CH_2-Cl will undergo S_N1 mechanism because $CH_3-O-\overset{+}{C}H_2$ is highly stable.

$$H_3C-\underset{CH_3}{\overset{CH_3}{C}}-CH_2-Cl$$

Neopentyl chloride will undergo S_N2 mechanism at a slow rate because it's sterically crowded. So, it prefers S_N1.

18. (c) The benzyl group acts by using the π-system of the benzene ring for conjugation with the p-orbital of the α-carbon in the transition state.

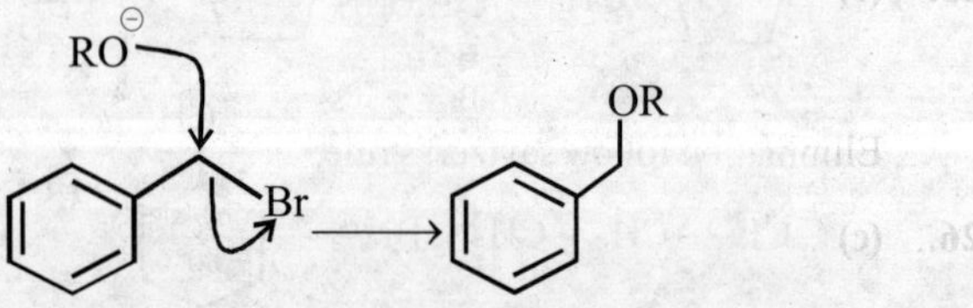

$RO^{\ominus}$ + benzyl bromide ⟶ $PhCH_2OR$

benzyl bromide

19. (a)

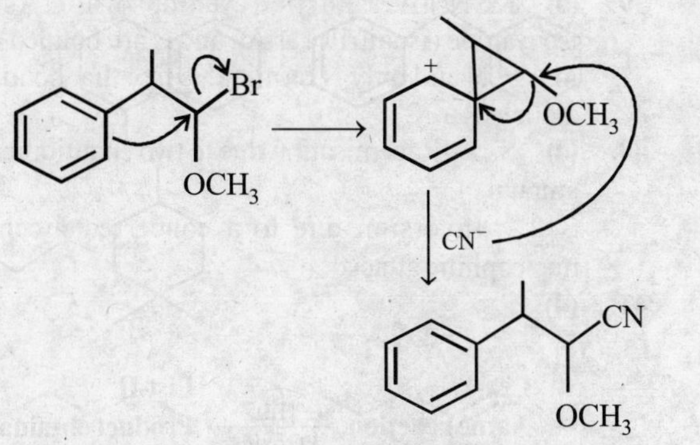

$\xrightarrow{CN^-}$

20. (c) Less the steric hindrance more will be here rate of S_N2 reaction. Order of rate of alkyl halide towards S_N2 reaction is $CH_3-X > 1° > 2° > 3°$

21. (32) Number of optical isomers is given by 2^n, n is the number of chiral centre.

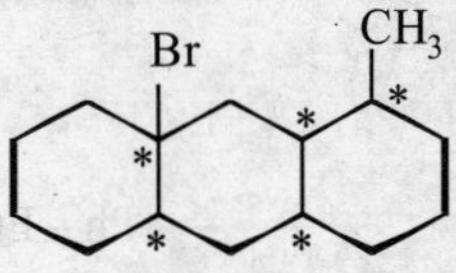

Total chiral centre = 5

No. of optical isomers = $2^n = 2^5 = 32$.

22. (c) S_N2 reactions are stereospecific and inversion of configurations observed.

S_N1 reaction generally racemisation is observed.

23. (1) Chiral compound is optically active.

$CH_3-CH_2-\overset{*}{C}H(Cl)-CH_3$

24. (b) Hint: Rearrangement of carbocation to gain stability. In aqueous medium the reaction follows S_N1 mechanism majority

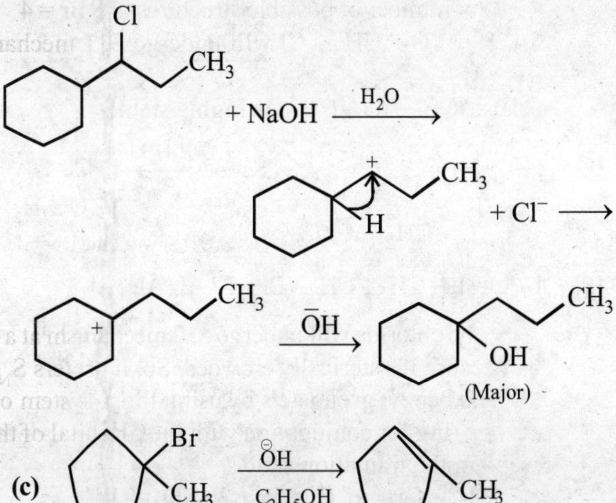

$+ NaOH \xrightarrow{H_2O}$

$+ Cl^- \longrightarrow$

$\xrightarrow{\bar{O}H}$ (Major)

25. (c) $\xrightarrow[C_2H_5OH,\ -H_2O,\ -Br^{\ominus}]{\overset{\ominus}{O}H}$

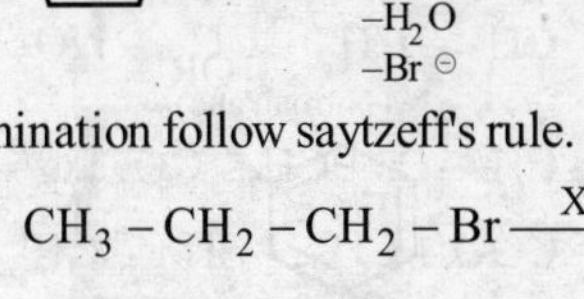

Elimination follow saytzeff's rule.

26. (c) $CH_3-CH_2-CH_2-Br \xrightarrow[80°C]{X = \text{conc. alc. NaOH}}$

$CH_3-CH=CH_2 \xrightarrow{Y = HBr/\text{Acetic acid}} CH_3-CHBr-CH_3$

27. (b) $\xrightarrow[S_N2\ -H^+]{Me_2\ddot{N}H}$

$\xleftarrow[S_N2]{Me_2\ddot{N}H}$

$\xrightarrow{-H^+}$

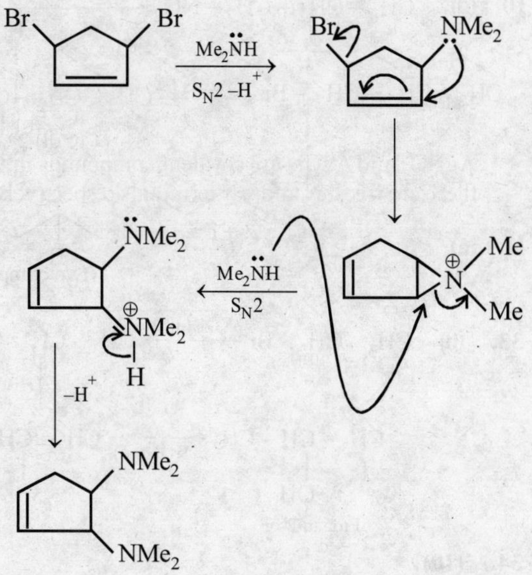

The above mechanism valid for both cis and trans isomers. So the products are same for both cis and trans isomers.

28. (b)

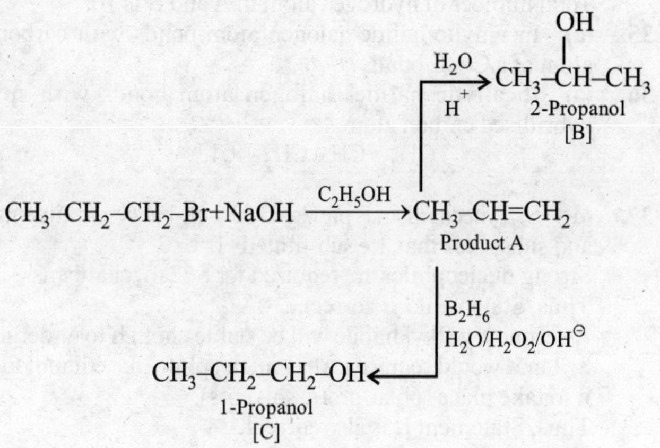

$CH_3-CH_2-CH_2-Br + NaOH \xrightarrow{C_2H_5OH} CH_3-CH=CH_2$ (Product A)

$CH_3-CH=CH_2 \xrightarrow[H^+]{H_2O} CH_3-CH(OH)-CH_3$ (2-Propanol [B])

$CH_3-CH=CH_2 \xrightarrow[H_2O/H_2O_2/OH^{\ominus}]{B_2H_6} CH_3-CH_2-CH_2-OH$ (1-Propanol [C])

29. (b) Hint: Alc. KOH abstract most acidic hydrogen

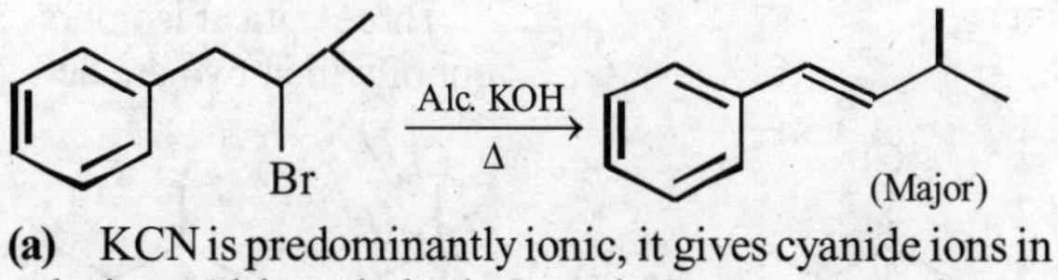

$\xrightarrow[\Delta]{\text{Alc. KOH}}$ (Major)

30. (a) KCN is predominantly ionic, it gives cyanide ions in solution. Although both C and N atoms can donate electrons, attack takes mainly through C because C – C bond formed in product is more stable than C – N bond and thus reaction mainly occurs through carbon of cyanide ion. But AgCN is mainly covalent in nature and nitrogen is free to donate electron pair foming isocyanide as the main product.

31. (2) $CH_3-\overset{*}{C}H(Cl)-CH_2-CH_3$

Optical isomer = 2^n, n = number of chiral centre

n = 1, Optical isomers = $2^1 = 2$

32. **(6)**

2-Methyl Butane $\xrightarrow{Cl_2}$ products

(± or dl) 2 compounds

(two compounds)

33. **(b)** $CH_3-CH_2-Br \xrightarrow{KOH(alc.)} CH_3-CH=CH_2 \xrightarrow{HBr} CH_3-CH(Br)-CH_3 \xrightarrow[KOH(aq)]{\Delta} CH_3-CH(OH)-CH_3$

Propan-2-ol

34. **(10)**

$CH_3CH_2Br + NaOH \xrightarrow{C_2H_5OH} CH_2=CH_2$

$CH_3CH_2Br + NaOH \xrightarrow{H_2O} CH_3-CH_2-OH$

Total number of hydrogen atom in A and B is 10.

35. **(a)** In vinylic halide halogen atom bonds with carbon atom of a C = C bond.

36. **(a)** In allylic halides halogen atom bonds with sp^3 hybridised carbon atom.

$$CH_2=CH-\underset{sp^3-C}{CH_2}-Cl$$

37. **(d)** S_N2 mechanism is preferred for non-bulky substituents and substrates that are substituted (1° or 2°).
Strong nucleophiles are required for S_N2 mechanisms.
Thus, Statement I is correct.
A secondary alkyl halide will be stable enough to undergo S_N1 and would require a weak nucleophile like ethanol for it to take place (polar protic solvents).
Thus, Statement II is also correct.

38. **(6)**

2-Chlorobutane $\xrightarrow[hv]{Cl_2}$ (A) + (B) + (C)

Three optical isomers out of which two are active

Optically inactive

Optically active isomers (each has two optical isomers or enantiomers)

Total number of optically active isomers = 2(from A) + 2 (from B) + 2 (from C) = 6.

39. **(b)** KCN gives nitrile (cyanide) while AgCN gives isocyanide (isonitrile) as Ag and C are bonded covalently in AgCN and only N terminal is free that bonds to the C-atom in R – X.

40. **(d)** S_N1 – Racemisation due to two enantiomers in equal amount.
S_N2 – Inversion due to a concerted mechanism of nucleophilic attack.

41. **(4)**

$\xrightarrow[\text{Peroxide}]{HBr}$

2 chiral centres

No. of stereoisomers $= 2^n = 2^2 = 4$

42. **(d)** sp^2 hybridized carbocation $[CH_3-CH=\overset{+}{C}H]$ is very unstable, therefore $CH_3-CH=CH-Cl$ cannot give S_N1 reaction.

43. **(5)** Chiral carbons are marked by.

, $CH_3CH_2-\overset{*}{C}H(NO_2)-COOH$,

44. **(a)** (i) $\xrightarrow[S_N2]{C_2H_5O^-}$ (A)

$C_2H_5O^-$ is strong nucleophile.

(ii) $\xrightarrow[S_N1]{C_2H_5OH}$ $\xrightarrow{\text{1,2-H shift}}$ $\xrightarrow[-H^+]{C_2H_5OH}$ (B)

C_2H_5OH is weak nucleophile.

45. **(b)** $CH_3-CH_2-CH_2-Br + KOH\,(Alc) \rightarrow CH_3-CH=CH_2$ (Alkene)

$CH_3-CH_2-CH_2-Br + KCN\,(Alc) \rightarrow CH_3-CH_2-CH_2-CN$ (Nitrile)

$CH_3-CH_2-CH_2-Br + AgNO_2 \rightarrow CH_3-CH_2-CH_2-NO_2 + AgBr\downarrow$ (Nitroalkane)

$CH_3-CH_2-CH_2-Br + CH_3-COOAg \rightarrow CH_3-COO-CH_2-CH_2-CH_3 + AgBr\downarrow$ (Ester)

46. (c) [Reaction scheme: p-methoxybenzyl chloride loses Cl^- to form a carbocation stabilised by O-Me, which is attacked by $\overset{\ominus}{Nu}$ to give p-methoxybenzyl–Nu]

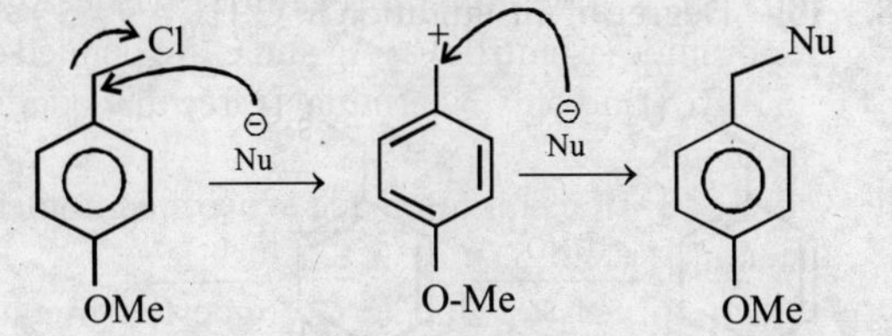

– OMe is Electron Donating group which stabilises the carbocation. The rate determining step is step one where formation of carbocation takes place.

$\Rightarrow$ S_N1 Mech., 1st order

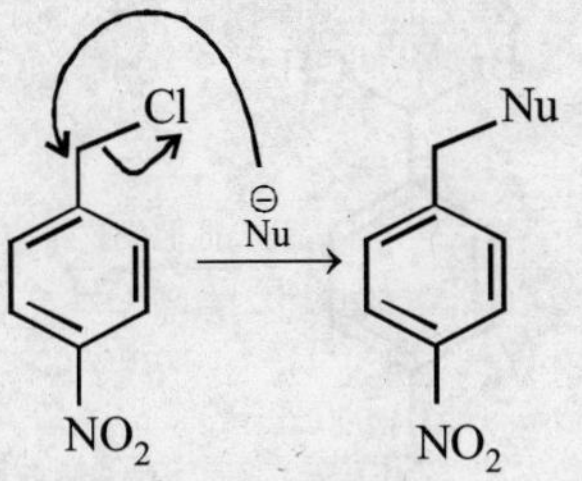

[Reaction scheme: p-nitrobenzyl chloride attacked directly by $\overset{\ominus}{Nu}$ to give p-nitrobenzyl–Nu]

$\Rightarrow$ S_N2 Mech., 2nd order

$-NO_2$ is a Electron withdrawing group

47. (b) **Assertion:**

$$\underset{\text{Glycine}}{H_2N-CH_2-COOH} \xrightarrow[\text{(HVZ)}]{\text{Red P, } Cl_2} NH_2-\overset{*}{C}H(Cl)-COOH$$

Optically active

Reason: A molecule with two chiral carbons is not always opticaly active since it may have some symmetry element present.

48. (b) Abstract the halogen where stability of carbocation moles

49. (a) This is finkelstein reaction

$$:\ddot{I}:^- \; R-\overset{\delta+}{C}H_2-\overset{\delta-}{Br} \xrightarrow{S_N2} R-CH_2-I + Br^-$$

Transition state $\left[\overset{\delta-}{I}\cdots\cdots C(R)(H)(H)\cdots\cdots\overset{\delta-}{Br}\right]$

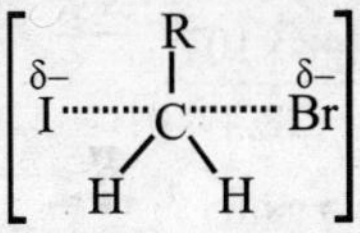

Clearly, the transition state is less polar than free anions. Br^- and I^-.

Acetic acid is protic which does not support S_N2.

Acetone does not solvate anion.

Br^- gets precipitated and hence can not compete with I^-

So only (a) is correct.

50. (2)

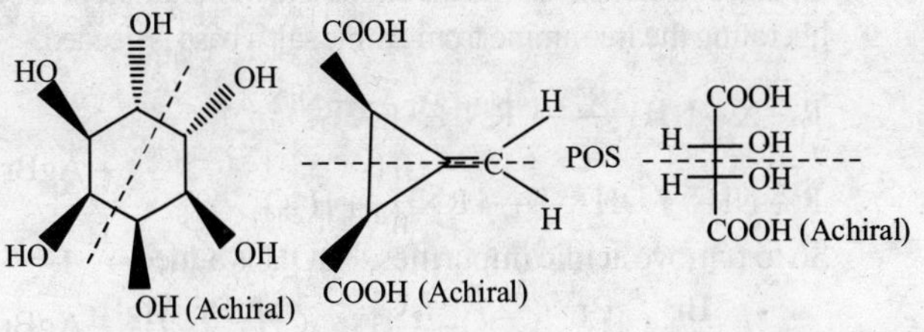

[Structures: inositol-type cyclohexane with OH groups — OH (Achiral); spiro compound with COOH groups and POS — COOH (Achiral); Fischer projection COOH, H–OH, H–OH, COOH (Achiral)]

51. (a) It is S_N1 reaction so rate of reaction depends on the concentration of alkyl halide only.

52. (6)

53. (b) Boiling point of alkyl halides increase with increase in size, mass of halogen atom and size of alkyl group. Boiling point of isomeric alkyl halides decrease with increase in branching. Density increases with increase in atomic mass of halogen atom.

54. (c) $H_3C-\overset{*}{C}(H)(Br)-\overset{*}{C}(H)(D)-CH_3$

55. (c) The rate of hydrolysis of alkyl chloride improves because of better nucleophilicity of I^-.

56. (c) NaCl is ionic and gives CN^- which attacks through carbon atom and not through N atom. AgCN is covalent in nature where nitrogen is free to donate its electron pair giving isonitrile as the main product.

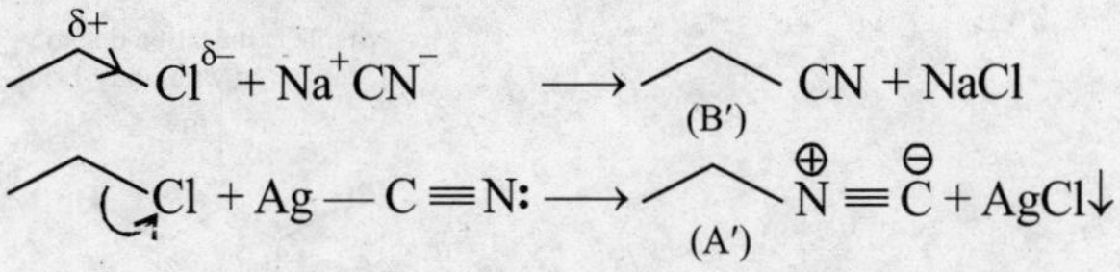

$$\overset{\delta+}{\text{CH}_3\text{CH}_2}-\overset{\delta-}{Cl} + Na^+CN^- \longrightarrow \underset{(B')}{\text{CH}_3\text{CH}_2-CN} + NaCl$$

$$\text{CH}_3\text{CH}_2-Cl + Ag-C\equiv N: \longrightarrow \underset{(A')}{\text{CH}_3\text{CH}_2-\overset{\oplus}{N}\equiv\overset{\ominus}{C}} + AgCl\downarrow$$

57. (c) Compound A has chiral carbon, hence it is optically active.

Compund B is not mirror image of compound A because both have same (*S*) configuration.

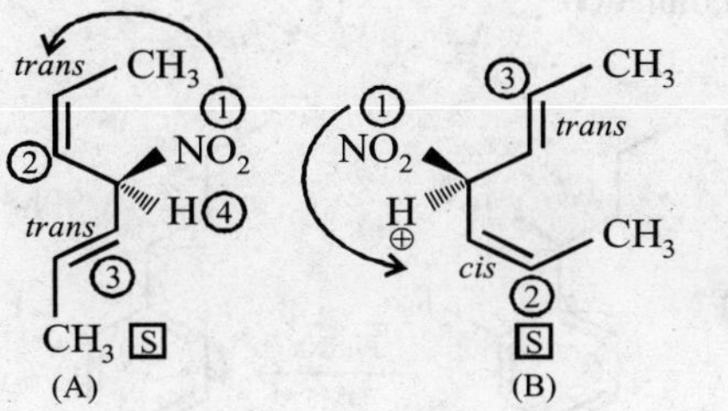

[Structures (A) and (B): chiral carbon bearing NO_2 (1), H (4), trans-propenyl and cis/trans-propenyl groups (2), (3); both labelled S]

58. (c) [Scheme: 2-bromo-2,3-dimethylbutane $\xrightarrow{\text{Mg, } Et_2O}$ Grignard (MgBr) $\xrightarrow{D_2O}$ deuterated product with chiral carbon (*) bearing D]

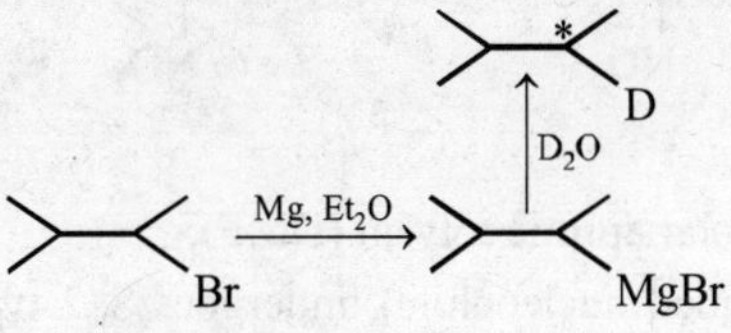

59. (42) % optical purity

$$= \frac{\text{optical rotation of mixture}}{\text{optical rotation of pure enantiomer}} \times 100$$

$$= \frac{12.6°}{30°} \times 100 = 42$$

60. (d) [Scheme: allyl-type ester of acrylic acid $\xrightarrow{2H^+}$ $\overset{\oplus}{C}H_2$–CH₂–C(=O)–O–CH₂–CH₂–$\overset{\oplus}{C}$H–CH₃ $\xrightarrow{2Br^-}$ Br–CH₂–CH₂–C(=O)–O–CH₂–CH₂–CH(Br)–CH₃]

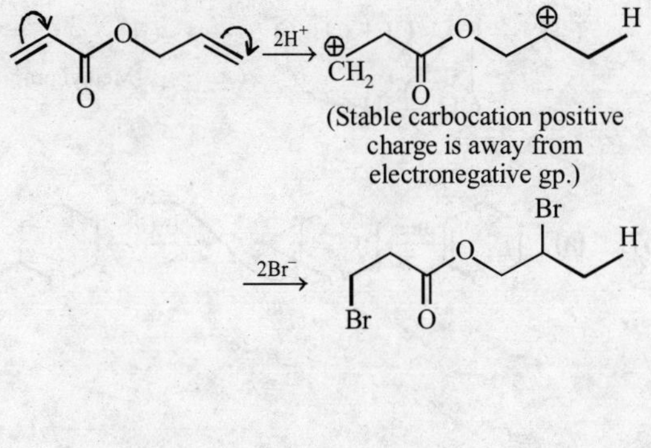

(Stable carbocation positive charge is away from electronegative gp.)

61. **(3)** Carbon attached to four different group is called asymmetric carbon.

Number of compounds containing asymmetric carbons are three.

62. **(c)** In S_N1 reaction carbocation is formed.
The carbocation formed at Bridge head in norbornene ring is very unstable due to restricted bond angle.
So it is inactive towards S_N1.

63. **(c)** Ph–CH=CH–CH₂–Br $\xrightarrow{HBr}$ Ph–$\overset{(+)}{C}$H–CH₂–CH₂–Br

Stable carbocation due to delocalization with phenyl ring.

$\downarrow Br^{\ominus}$

Ph–CH(Br)–CH₂–CH₂–Br

64. **(1)** Bromine in presence of light undergoes allylic bromination.

65. **(a)** $\xrightarrow[DMF]{Ph\bar{S}\overset{+}{N}a}$

Due to polar aprotic solvent (DMF),
PhS (a good nucleophile) undergoes S_N2 type reaction which occurs at benzylic carbon by inversion in configuration.

66. **(d)** We are given a bulky base, hence elimination will take place & not substitution.

$$CH_3-\underset{CH_2-H}{\overset{CH_3}{C}}-Cl + t\text{-}BuO^- \xrightarrow{E_2} CH_3-\overset{CH_3}{C}=CH_2$$

2-Methyl prop-1-ene

67. **(a)** $\xrightarrow{Br_2}$ (bromonium ion, Br^+) $\xrightarrow{CH_3\ddot{O}H}$ (Br, OCH_3)

68. **(b)** Degree of unsaturation in $C_8H_{10} = 9 - 5 = 4$

$C_8H_{10} \xrightarrow[H_2SO_4]{HNO_3}$ (A) (NO_2) $\xrightarrow[\Delta]{Br_2}$

(B) $\xrightarrow{alc.\ KOH}$ (C)

69. **(2)** $\xrightarrow[H_2O]{NaOH}$ (P)

Number of π electrons in $P = 2$

70. **(d)** Alcoholic-NaOH leads to elimination (E2) reaction.

$\xrightarrow[C_2H_5OH]{NaOH}$

1-Methylcyclopent-1, 3-diene

71. **(3)** $\text{Mole} = \dfrac{\text{Given mass}}{\text{Molar mass}} = \dfrac{\text{Give volume (L)}}{22.4\,(\text{L})}$

$$\frac{1.53}{\text{Molar mass}} = \frac{448 \times 10^{-3}}{22.4}$$

Molar mass = 76.5 g mol^{-1}
Possible compound is $Cl-CH_2-CH=CH_2$(A)

$$\underset{(A)}{Cl-CH_2-CH=CH_2} \xrightarrow[\text{(ii) Zn/AcOH}]{\text{(i) }O_3} ClCH_2-CHO + HCHO$$

Thus, no. of carbon atom in (A) is 3.

72. **(c)** During ammonolysis of alkyl halide, acid liberated during the reaction combines with the amine and forms amine salt. For liberating the free amine from amine salt a base is needed.

$$R-X + NH_3 \longrightarrow RNH_3^+ + X^-$$

$$R-NH_3^+ + OH^- \longrightarrow RNH_2 + H_2O$$

So to remove acidic impurities, NaOH is added.

73. **(d)** Br Br $\xrightarrow[\text{(dry ether)}]{\text{Mg (Excess)}}$ MgBr MgBr

74. **(b)** Among the given bases (A) and (B), *t*-butoxide being bulky base favours elimination reaction and ethoxide favours substitution reaction.

$\therefore$ when $Z^- = CH_3CH_2O^-$, (substitution reaction favoured)

and when $Z^- = CH_3-C(CH_3)_2-O^-$, (elimination reaction favoured)

Hence, $\mu_A > \mu_B$ and $k_e(B) > k_e(A)$

75. **(d)** Given reaction is S_N1 reaction. In S_N1 reaction

Rate of reaction $\propto$ Stability of C^+

(A) 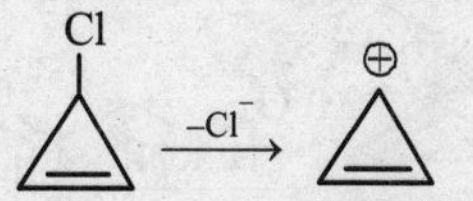 Cl-cyclopropene $\xrightarrow{-Cl^-}$ cyclopropenyl cation — Aromatic (i)

(B) 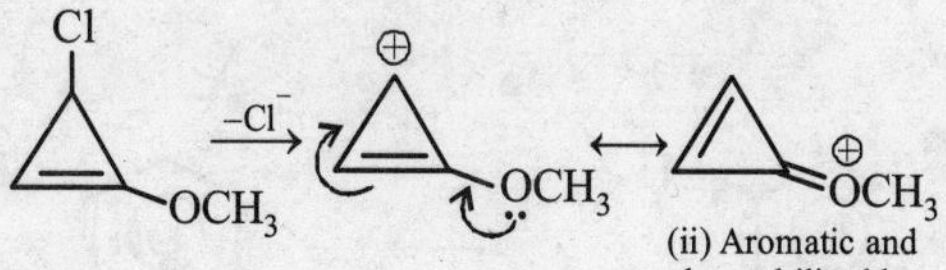 Cl, OCH_3-cyclopropene $\xrightarrow{-Cl^-}$ cation $\longleftrightarrow$ $\overset{\oplus}{O}CH_3$ — (ii) Aromatic and also stabilised by lp of –OMe

(C) $CH_3-CH(Cl)-CH_3 \xrightarrow{-Cl^-} CH_3-\overset{\oplus}{C}H-CH_3$ (2°) (iii)

(D) $CH_3-CH(Cl)-CH_2-NO_2 \xrightarrow{-Cl^-} CH_3-\overset{\oplus}{C}H-CH_2-NO_2$ (iv) (–I, –M)

Stability of C^+ : ii > i > iii > iv

Reactivity order : B > A > C > D

76. **(d)**

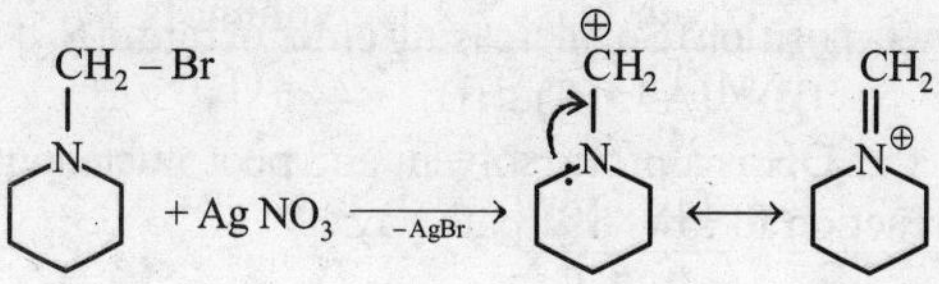

N-(bromomethyl)piperidine (CH_2-Br) + Ag NO_3 $\xrightarrow{-AgBr}$ $\overset{\oplus}{C}H_2$-piperidine $\longleftrightarrow$ $CH_2=\overset{\oplus}{N}$ piperidinium

Most stable carbocation due to +R effect of N, having a lone pair of electrons.

Ease of precipitation of AgBr depends upon the rate of formation of carbocation.

77. **(d)** Due to conjugation of lone pair of Cl with π bond, partial double bond character decreases bond length. That's why compound (d) has shortest C–Cl bond length.

78. **(c)**

(1) The reaction is favoured by weak nucleophile as strong nucleophile will favour elimination over substitution reaction.

(2) If the substituents are bulky, then it will stabilise the intermediate $R^{\oplus}$.

(3) The nucleophile $Y^{\ominus}$ can attack the planar $R^{\oplus}$ from both the sides.

Hence, the reaction is accompanied by racemization.

(4) Polar solvent helps to separate the ion pair.

79. **(8)** **Hint:** $H_2|Ni$ used for hydrogenation of alkene.

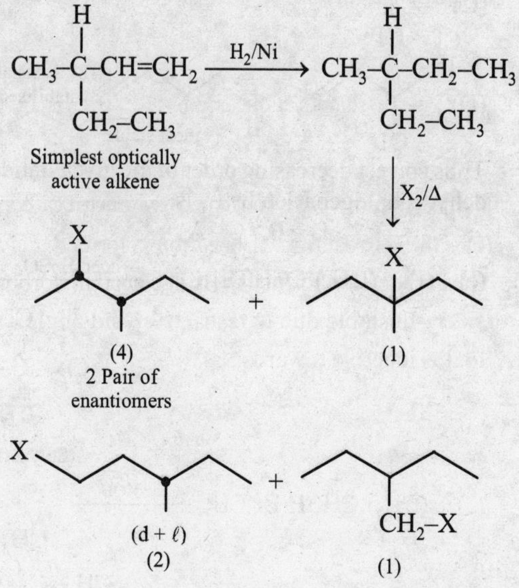

$CH_3-CH(CH_2-CH_3)-CH=CH_2 \xrightarrow{H_2/Ni} CH_3-CH(CH_2-CH_3)-CH_2-CH_3$

Simplest optically active alkene

$\downarrow X_2/\Delta$

(4) 2 Pair of enantiomers + (1)

(d + ℓ) (2) + CH_2-X (1)

80. **(c)** $\xrightarrow[\text{t-BuOH}]{\text{E1 reaction}}$ carbocation $\xrightarrow{\text{1,2-H}^- \text{ Shift}}$ $\xrightarrow{-H^+}$ alkene

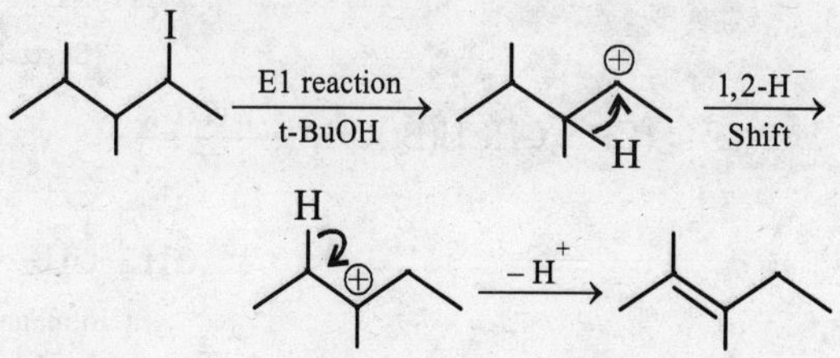

81. **(d)**

$H_3C-{}^*C(H)(C_2H_5)-C(H)(H)-Br \xrightarrow[S_N2]{^-OH} H_3C-{}^*C(H)(C_2H_5)-C(H)(H)-OH$

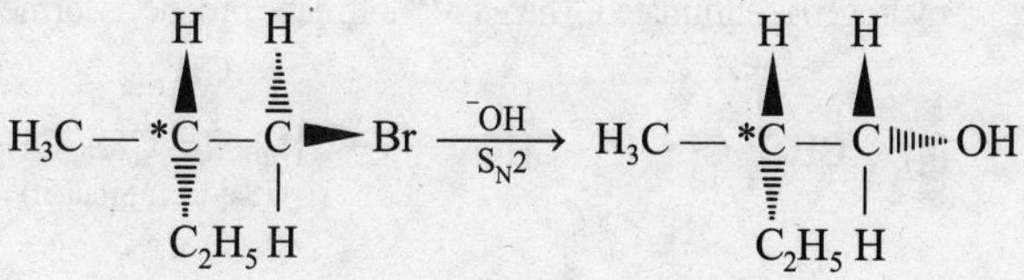

Inversion of configuration takes place when nucleophile attacks but at the stereogenic carbon.

82. **(b)** $\xrightarrow[E2]{\text{Alc. KOH}}$

More acidic (due to –I effect of F)

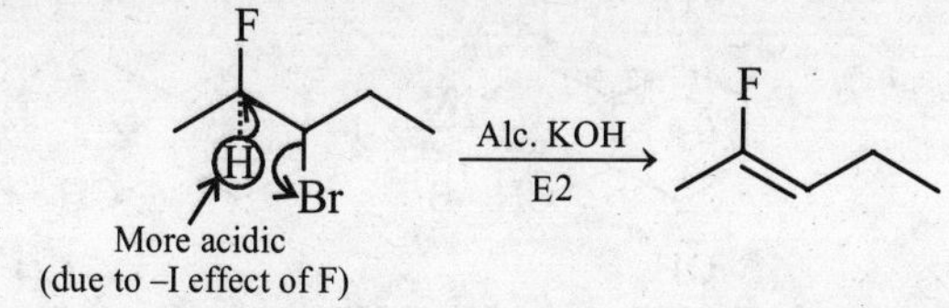

83. **(b)** First reaction is S_N1 in which rate does not depend on conc. of nucleophile but depends on reactant conc. Second reaction is E2 reaction in which rate depends on conc. of base as well as reactant conc.

Therefore, changing in the concentration of base will have no effect on rate of reaction (1).

84. **(a)** E_1 reaction proceeds via carbocation formation, therefore greater the stability of carbocation, faster will be the E_1 reaction.

(A) 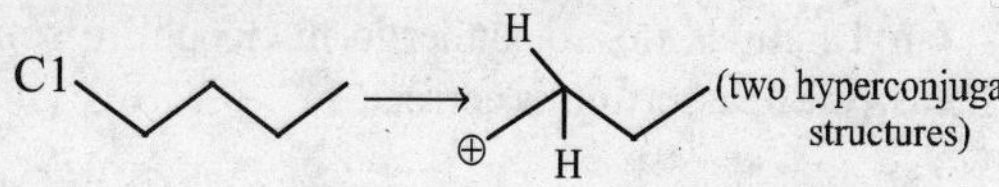 Cl ⟶ (two hyperconjugating structures)

(B) 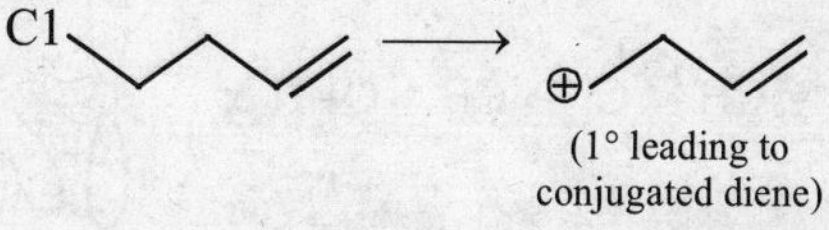 Cl ⟶ (1° leading to conjugated diene)

(C) 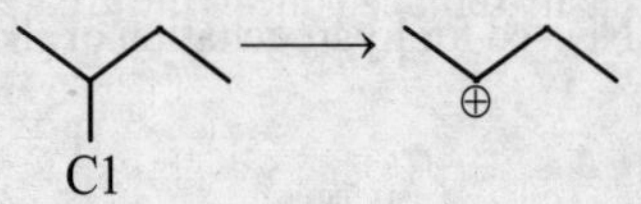 (five hyper conjugating structures)

(D) 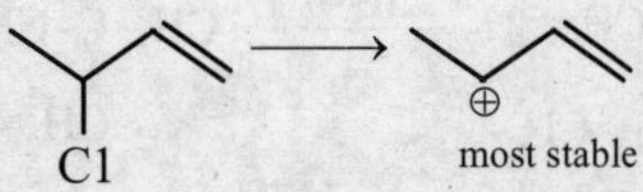 (2° Resonance stabilised) most stable

Thus correct decreasing order of the given halides towards dehydrohalogenation by E_1 is:

D > B > C > A

85. (c) (1) $(CH_3)_3CCH(OH)CH_3 \xrightarrow{conc.H_2SO_4}$ $CH_3-C(CH_3)-C(H)=CH_2$ ... $(CH_3)_2C=C(CH_3)_2$ (Saytzeff product)

(2) $(CH_3)_2CHCH(Br)CH_3 \xrightarrow{alc.KOH} CH_3-C(CH_3)=CH-CH_3$ (Saytzeff product)

(3) $(CH_3)_2CHCH(Br)CH_3 \xrightarrow{t\text{-}BuO^-K^+} CH_3-CH(CH_3)-CH=CH_2$ (Hofmann product)

Due to bulky nature of tertiary butoxide, the least hindered hydrogen is eliminated. Therefore, Hoffmann product is formed.

(4) $(CH_3)_2C(OH)-CH_2-CHO \xrightarrow{\Delta} CH_3-C(CH_3)=CH-CHO$ (Saytzeff product)

86. (c) $\xrightarrow{H^+}$ $\xrightarrow{Br^-}$ (A) $\xrightarrow[\text{Intramolecular Wurtz reaction}]{\text{Na, ether}}$ (B)

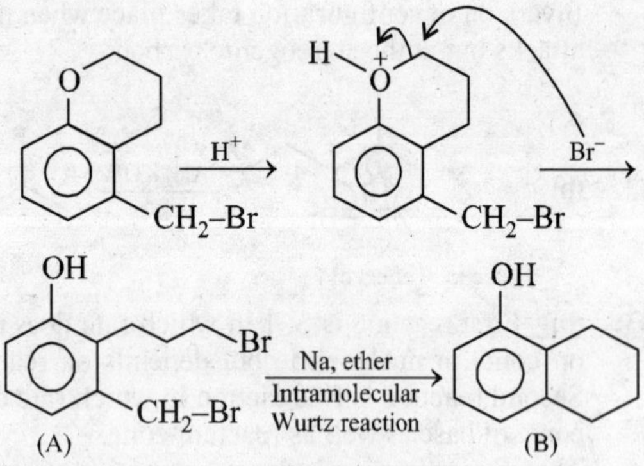

87. (d) $CH_2=CH-\ddot{Cl} \leftrightarrow \overset{\ominus}{C}H_2-CH=Cl^+$

Due to partial double bond character of C–halogen bond, halogen leaves with great difficulty, if at all it does. Hence, vinyl halides do not undergo nucleophilic substitution easily. So, assertion is correct.

$CH_2=CH-Cl \rightarrow CH_2=\overset{+}{C}-H$ or

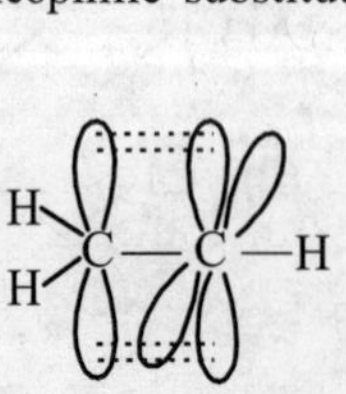

Intermediate carbocation is not stabilised by loosely held-π electrons because empty orbital, being at 90°, cannot overlap with p-orbitals of π bond. So, reason is wrong.

88. (b) The rate of S_N1 is decided by the stability of carbocation formed in the rate determining step.

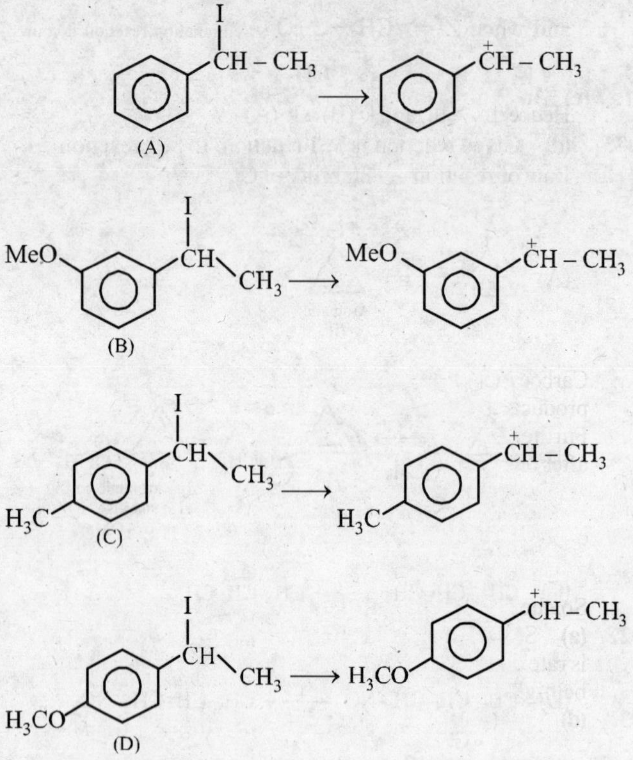

(A) $C_6H_5-CH(I)-CH_3 \longrightarrow C_6H_5-\overset{+}{C}H-CH_3$

(B) m-MeO-C_6H_4-CH(I)-$CH_3 \longrightarrow$ m-MeO-$C_6H_4-\overset{+}{C}H-CH_3$

(C) p-H_3C-C_6H_4-CH(I)-$CH_3 \longrightarrow$ p-H_3C-$C_6H_4-\overset{+}{C}H-CH_3$

(D) p-H_3CO-C_6H_4-CH(I)-$CH_3 \longrightarrow$ p-H_3CO-$C_6H_4-\overset{+}{C}H-CH_3$

Carbocation (D) is most stable due to +R effect of $-OCH_3$ group (at para position); (C) is stabilised by +I effect of the CH_3 group; (B) is least stable due to –I effect of MeO group (at meta position) So, increasing order of rate of S_N1 is:

(B) < (A) < (C) < (D)

89. (c) CH_3OH is a protic solvent and poor nucleophile, so the reaction follows S_N1 pathway.

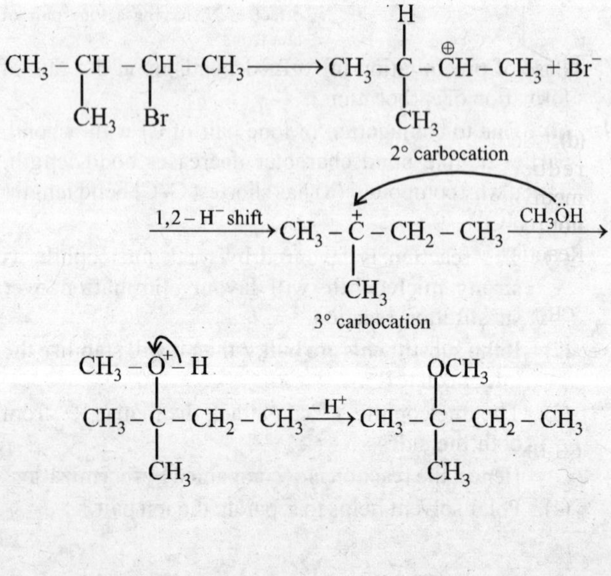

$CH_3-CH(CH_3)-CH(Br)-CH_3 \longrightarrow CH_3-C(H)(CH_3)-\overset{\oplus}{C}H-CH_3 + Br^-$ (2° carbocation)

$\xrightarrow{1,2-H^- \text{ shift}} CH_3-\overset{+}{C}(CH_3)-CH_2-CH_3$ (3° carbocation) $\xrightarrow{CH_3\ddot{O}H}$

$CH_3-C(CH_3)(\overset{+}{O}(H)CH_3)-CH_2-CH_3 \xrightarrow{-H^+} CH_3-C(CH_3)(OCH_3)-CH_2-CH_3$

90. (c) [2-chloro-3-methylbutane] $\xrightarrow[\Delta]{NaOEt}$ (X) $\xrightarrow{HBr}$ (Y) [Markovnikov product]

91. (c) In S_N1 reaction carbocation acts as an intermediate.

(C) > (D)

Carbocation produced by (C) is more stable than carbocation produced by (D) due to + M effect of $-OCH_3$ group.
Further in (A) there is formation of tertiary carbocation after rearrangement while (B) is primary carbocation.

$CH_3-\overset{+}{C}(CH_3)_2$ (A) $\quad CH_3-\overset{+}{C}H_2$ (B)

So, the correct order is (C) > (D) > (A) > (B).

92. (a) S_N1 reaction is a two step reaction, in which first step is rate determining step. Hence peak of the first step should be higher than second step.

93. (d)

$CH_3-CH_2-CH(Br)-CH_2Br \xrightarrow{(i)\ KOH\ (alc.)} CH_3CH_2C(Br)=CH_2 \xrightarrow[(liq.\ NH_3)]{(ii)\ NaNH_2} CH_3CH_2C\equiv C-H$

94. (a) Dehydrohalogenation (β–elimination) occurs as:

$\xrightarrow[\Delta]{KOH\ alc.\ (excess)}$ conjugated and more substituted alkene

95. (d) Sodium borohydride is a selective reducing agent. It reduces carbonyl group to alcoholic group, N-methylimino group (MeN = CH–) to 2° amines, but does not reduce an isolated carbon-carbon double bond. Reaction involved:

$CH_3N{=}\ldots \xrightarrow{NaBH_4} CH_3HN\ldots OH$ (major product)

96. (b) Allylic H is easily replaced due to the greater stability of allylic free radical.

$CH_3-CH_2-CH{=}CH_2 + Br_2 \xrightarrow{h\nu} CH_3-CH(Br)-CH=CH_2$

97. (a) Mechanism involved for the given reaction is:

Br–Br, EtOH

(Major product)

98. (a) Here dehydrohalogenation goes by E_1CB and most stable carbanion formation is favoured in (a).

99. (b) CH_3O^- is a strong base and strong nucleophile, so favourable condition is $S_N2/E2$.
The given alkyl halide is 2° and β carbons are 4° and 2°, so sufficiently hindered, thus E2 dominates over S_N2.

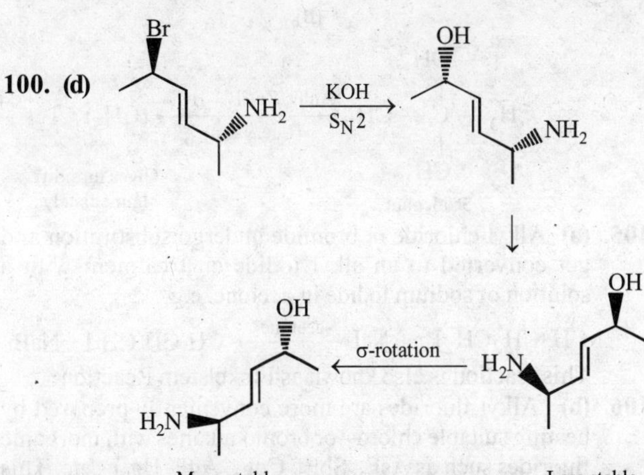

100. (d) $\xrightarrow[S_N2]{KOH}$; σ-rotation

More stable   Less stable

101. (d) $CH_3-CH(Br)-CH_2-CH(Br)-CH_2-CH_3$

$\xrightarrow[E2]{KOH,\ CH_3OH/\Delta} CH_3-CH=CH-CH=CH_2-CH_3$ (Saytzeff product)

102. (b) S_N1 reactions involve the formation of carbocation. Hence higher the stability of a carbocation more will be its reactivity towards S_N1 reactions. Hence, reactivity order is :

$CH_3-CH_2-\overset{\oplus}{C}H_2 < CH_3-\overset{\oplus}{C}H-CH_2-CH_3$
1° carbocation   2° carbocation

$< p-H_3CO-C_6H_4-\overset{\oplus}{C}H_2$

Max. stable due to +M effect of $-OCH_3$ group.

103. (b) $C_2H_5\overset{\ominus}{O}$ abstracts the more acidic H (benzylic H).

$C_6H_5-CH(H)-C(CH_3)(Br)-CH(H)-CH_3$
(–I effect)   (+I effect)

$\xrightarrow[C_2H_5OH]{C_2H_5ONa}$ (– NaBr, – C_2H_5OH)

$C_6H_5-CH=C(CH_3)-CH_2-CH_3$

104. (d) $H_3C-C(Cl)_2-CH_3 \xrightarrow[(aq.)]{KOH} CH_3-C(OH)_2-CH_3$
($C_3H_6Cl_2$) (I)   Unstable

$\xrightarrow{-H_2O} CH_3-\overset{O}{\overset{\|}{C}}-CH_3 \xrightarrow[(II)\ H_2O/H^{\oplus}]{(I)\ CH_3MgBr}$
(II)

$CH_3-C(OH)(CH_3)-CH_3 \xrightarrow[(Lucas\ test)]{Anhy.\ ZnCl_2 + HCl} (CH_3)_3Cl$
3° alcohol   Gives turbidity immediately

105. (a) Alkyl chloride or bromide undergo substitution and get converted to an alkyl iodide on treatment with a solution of sodium iodide in acetone. e.g.

$CH_3CH_2CH_2Br + NaI \xrightarrow{acetone} CH_3CH_2CH_2I + NaBr$

This reaction is also known as Finkelstein Reaction.

106. (b) Alkyl fluorides are more conveniently prepared by heating suitable chloro – or bromo-alkanes with inorganic fluorides such as AsF_3, SbF_3, CoF_2, AgF, Hg_2F_2 etc. This reaction is called **Swarts reaction.**

$CH_3Br + AgF \longrightarrow CH_3F + AgBr$

$2CH_3CH_2Cl + Hg_2F_2 \longrightarrow 2CH_3CH_2F + Hg_2Cl_2$

107. (c)

$C_{10}H_{13}Cl\ (A) \xrightarrow{alc.\ KOH} (B) \xrightarrow{Ozonolysis} C + D$

(A): White ppt with $AgNO_3$; C: Gives only Cannizaro reaction; D: Gives only aldol condensation

$C_6H_5-CH_2-C(Cl)(CH_3)_2 \xrightarrow[KOH]{alc.} C_6H_5CH=C(CH_3)_2 \xrightarrow[(ii)\ Zn/H_2O]{(i)\ O_3}$
(A)   (B)

$C_6H_5CHO + O=C(CH_3)_2$
(C)   (D)

108. (a) The correct order of increasing bond length is $CH_3F < CH_3Cl < CH_3Br < CH_3I$

109. (b) Steric hindrance around the carbon atom having Cl will slow down the S_N2 reaction, hence lesser the hindrance, faster will be the reaction. So, the order of reactivity is $CH_3Cl > (CH_3)CH_2-Cl > (CH_3)_2CH-Cl > (CH_3)_3CCl$

110. (a) For a given alkyl group, the order of reactivity is:

$R-I > R-Br > R-Cl > R-F$
→ increasing bond energy
→ decreasing halogen reactivity

This order depends on the carbon-halogen bond energy; the carbon-fluorine bond energy is maximum and thus, fluorides are least reactive while carbon-iodine bond energy is minimum, hence iodides are most reactive.

111. (b) $(CH_3)_2C=O + CH_3CH_2MgBr \longrightarrow (CH_3)_2C(OMgBr)(CH_2CH_3) \xrightarrow{H_2O} (CH_3)_2C(OH)(CH_2CH_3)$

2- Methyl butanol-2 (3° alcohol)

112. (b) $CH_2(Cl)-\overset{\star}{C}H(CH_3)-CH_2-CH_3$

$CH_3-CH(CH_3)-\overset{\star}{C}H(Cl)-CH_3$

There are only 2 chiral compounds, although four stereoisomers.

113. (a) $C_2H_5Br \xrightarrow{AgCN} C_2H_5NC + AgBr$
(X)

$\xrightarrow[Zn-Hg/HCl]{Reduction} C_2H_5NHCH_3$

(Y) Ethyl methyl amine

114. (c) $H_2C=HC-\overset{*}{C}H(CH_3)-CH_2-CH_3$

3- Methyl-1-pentene

115. (a) S_N1 reactions involve the formation of carbocations, so higher the stability of the carbocation more will be reactivity of the corresponding bromide.

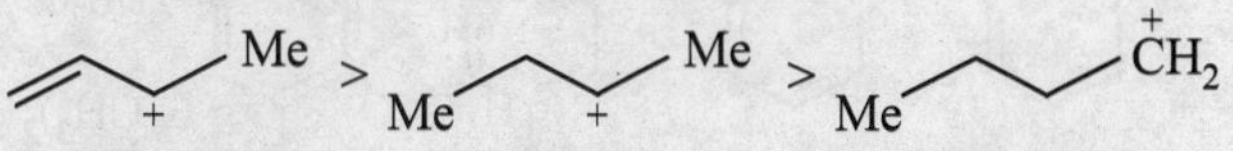

Allylic Carbocation from B > 2° Carbocation from C > 1° Carbocation from A

Hence, the reactivity order of the given bromides towards S_N1 reaction is

(B) > (C) > (A)

116. (b) $CH_3-CH=CH-\overset{*}{C}HCH_3$ (with OH on the starred carbon)

It exhibits both geometrical as well as optical isomerism.

cis - R   *cis* - S

trans - R   *trans* - S

117. (d) S_N2 reaction is favoured by small groups on the carbon atom attached to halogen.

So, the order of reactivity is:

$CH_3Cl > (CH_3)_2CHCl > (CH_3)_3CCl > (C_2H_5)_2CHCl$

118. (b) The organic compounds which have chiral carbon atom and do not have plane of symmetry hence rotate plane polarised light.

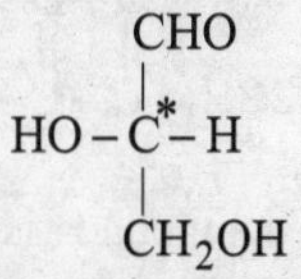

$HO-\overset{*}{C}-H$ with CHO above and CH_2OH below (* is asymmetric carbon)

119. (a) Due to steric hindrance, tertiary alkyl halides do not react by S_N2 mechanism, they react by S_N1 mechanism. S_N2 mechanism is followed in case of primary and secondary alkyl halides.

120. (b) In Corey House synthesis of alkanes, alkyl halides react with lithium dialkyl cuprate.

$R'X + LiR_2Cu \longrightarrow R'-R + RCu + LiX$

121. (c) $CH_3-\underset{Br}{CH}-CH_2-CH_3 \xrightarrow{\text{Alc. KOH}} CH_3-CH=CH-CH_3 + HBr$

The formation of 2-butene is in accordance to **Saytzeff's rule** (more substituted alkene is formed).

122. (c) 2, 3-Dichlorobutane exhibits optical isomerism due to the presence of two asymmetric carbon atoms.

123. (b) Compounds containing two similar chiral C-atoms have plane of symmetry and can exist in *meso* form too.

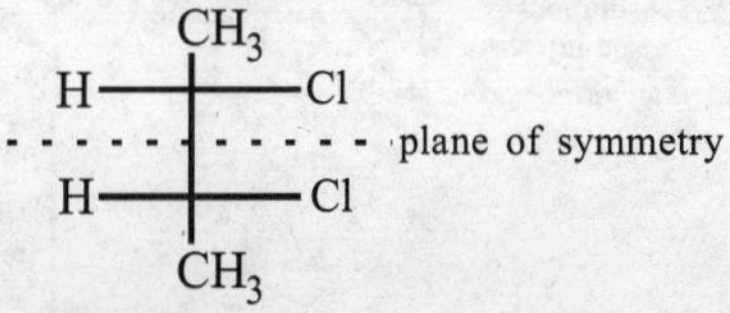

CH_3; H–C–Cl; - - - - plane of symmetry; H–C–Cl; CH_3

meso-2, 3-Dichlorobutane

124. (b) The hydrolysis of *t*-butyl bromide is an example of S_N1 reaction.

125. (c) Presence of electron withdrawing group at ortho and para position of chlorobenzene stabilize the intermediate in aromatic nucleophilic substitution and increases the rate of reaction.

126. (b)

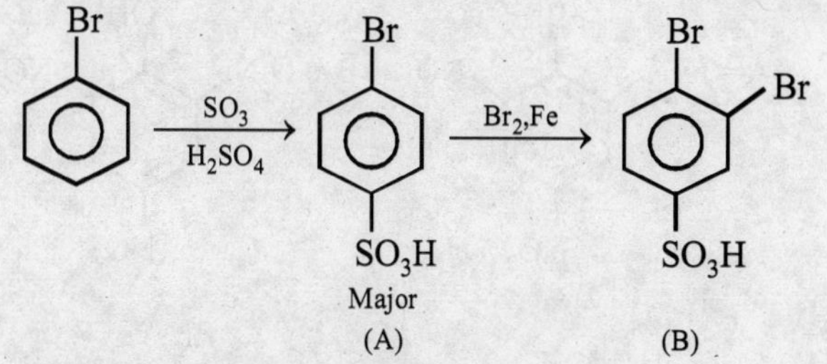

Bromobenzene $\xrightarrow[H_2SO_4]{SO_3}$ p-bromobenzenesulphonic acid (Major) (A) $\xrightarrow{Br_2, Fe}$ 3,4-dibromobenzenesulphonic acid (B)

Both reactions are electrophilic substitution reaction first is sulphonation and second is halogenation.

– Br is o- and p-directing and gives a para product (major) while – SO_3H is meta directing and so cumulative directive influence of – Br and –SO_3H give the final product (B).

127. (b)

(A) Fitting Reaction

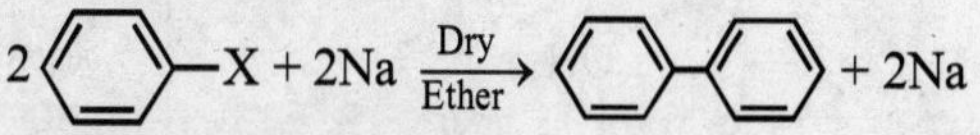

$2\,C_6H_5-X + 2Na \xrightarrow[\text{Ether}]{\text{Dry}} C_6H_5-C_6H_5 + 2Na$

(B) Crafterman Reaction

$ArN_2^+X^- \xrightarrow[HCl]{Cu} ArCl + N_2\uparrow + CuX$

(C) Finkelstein Reaction

$C_2H_5Br + NaI \xrightarrow[\text{Acetone}]{\text{Dry}} C_2H_5I + NaBr$

(D) Lucas Reaction

$CH_3C(OH)(CH_3)CH_3 \xrightarrow[ZnCl_2]{HCl} CH_3C(Cl)(CH_3)CH_3$

128. (a) 3,4-dibromonitrobenzene $\xrightarrow[C_2H_5OH]{C_2H_5ONa}$ 2-bromo-4-nitrophenetole (OC_2H_5 replaces Br para to NO_2)

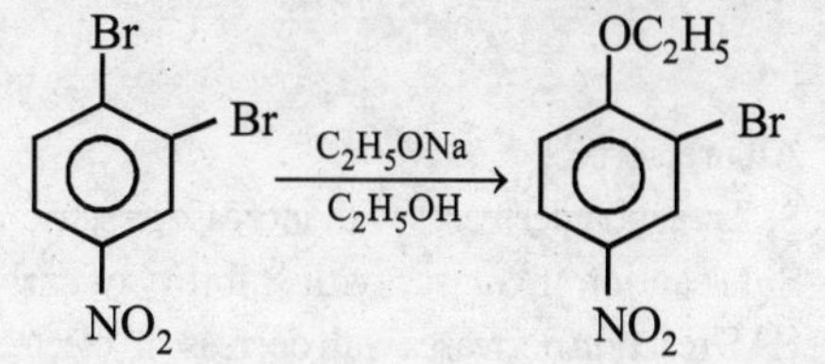

(Due to – I of – Br and – I and –M of NO_2 this Br is displaced)

129. (2) In Sandmayer reaction only bromobenzene and chlorobenzene are prepared.

130. (c)

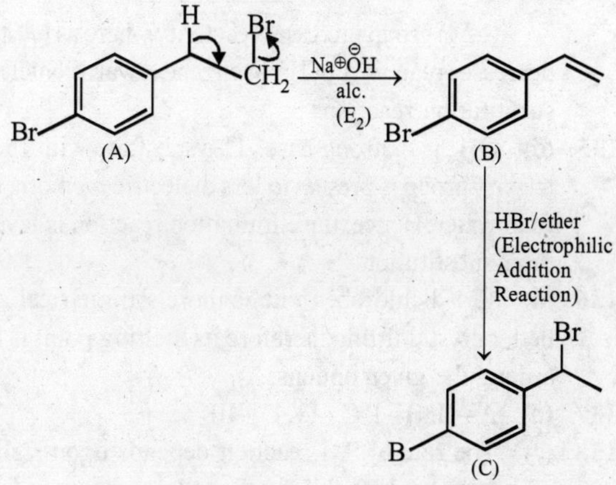

(A) p-Br-C_6H_4-CH_2CH_2Br $\xrightarrow[(E_2)]{Na^{\oplus}\overset{\ominus}{O}H \text{ alc.}}$ (B) p-Br-C_6H_4-$CH=CH_2$ $\xrightarrow{\text{HBr/ether (Electrophilic Addition Reaction)}}$ (C) p-Br-C_6H_4-$CHBrCH_3$

A and C are position isomer.

131. (a)

A = 2-bromo-1,3,5-trinitrobenzene (O_2N, Br, NO_2, NO_2); B = 2,4,6-trinitrophenol (O_2N, OH, NO_2, NO_2)

Bromobenzene $\xrightarrow{\text{Conc. } HNO_3}$ (A) (Br, O_2N, NO_2, NO_2) $\xrightarrow{\text{(i) NaOH, (ii) HCl}}$ (O_2N, OH, NO_2, NO_2)

132. (a) Chlorobenzene $\xrightarrow{CH_3-\overset{O}{\overset{||}{C}}-Cl + \text{Anhydrous } AlCl_3}$ Cl, $COCH_3$ (Major)

133. (b) All are correct

(A) S_N2 reaction decreases with increase in steric crowding.

(B) S_N1 reaction increases with stability of carbocation.

(C) EAS reaction decreases with decrease in electron density.

(D) Presence of electron withdrawing group at ortho and para–position to a halogen in haloarene increase nucleophilic aryl substitution.

134. (d) ∴ correct order is D > B > A > C

(– M) group increases reactivity where as (+M) group decreases reactivity of Halobenzene towards Nucleophilic substitution reaction.

135. (d) OH^- is a strong base. The substrate with the bulky tertiary group is present in less dielectric medium, alcohol than water. Hence, the elimination reaction is favourable than substitution.

136. (d) 1, 4-dichlorobenzene is more symmetrical and has better crystal fitting therefore its melting point is highest among the given options.

137. (d) A – II, B – I, C – IV, D – III

138. (c) The rate of S_N1 reaction depends upon stability of carbocation which follows the order.

$\overset{+}{C}H_2$ (p-OMe, +M) > $\overset{+}{C}H_2$ (benzyl) > $\overset{+}{C}H_2$ (p-Cl, –I) > $\overset{+}{C}H_2$ (p-NO_2, –I)

∴ Reactivity order : (b) > (d) > (c) > (a)

139. (d) Electron withdrawing groups are highly ineffective at meta position in nucleophilic aromatic substitution reactions. They have greater effect on ortho and para position.

140. (c) **Assertion A :** Benzene is more stable than cyclohexatriene.
Reason R : The delocalized π-electron cloud is attracted more strongly by nuclei of carbon atoms as the $\pi - e^-$ lies in bonding molecular orbital.

141. (c) o-Xylene $\xrightarrow{\text{Ozonolysis}}$ $H_3C-\overset{O}{\overset{||}{C}}-\overset{O}{\overset{||}{C}}-CH_3 + 2\,HC=O$ (HC=O)

o-Xylene $\xrightarrow{\text{Ozonolysis}}$ $2CH_3-\overset{O}{\overset{||}{C}}-\overset{O}{\overset{||}{C}}-H$

142. (b) Nitration of aromatic compounds is electrophilic substitution reaction. Sice electron donating groups (like $-CH_3$) increase electron density while electron withdrawing groups ($-NO_2 > -Br$) decrease electron density, the order of reactivity of nitration in the given compounds is

(D) NO_2-benzene < (B) Br-benzene < (E) benzene < (A) toluene < (C) m-xylene

143. (d)

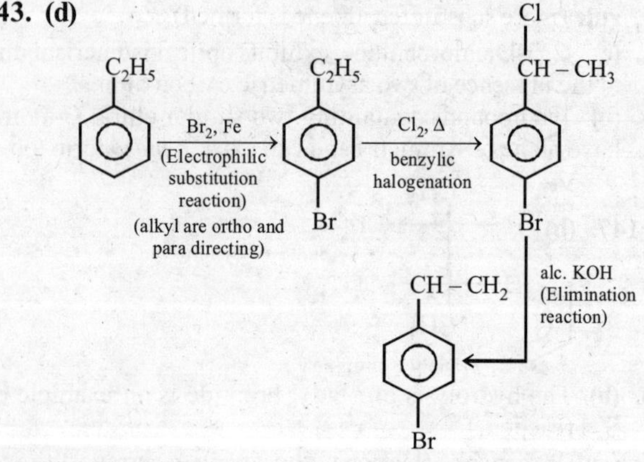

144. (c) Electrophile will attack at ortho and para position with respect to better electron releasing group (ERG)

$-OH$ donates through +M and $-CH_3$ donates through +H
ERG : $-OH > -CH_3$

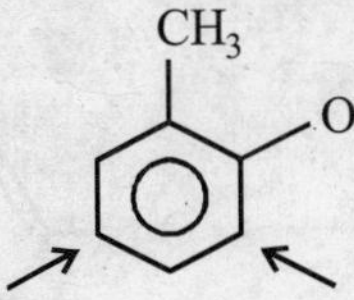

Para position with respect to $-OH$ (+R) group and it will be meta position with respect to $-CH_3$ group.

145. (b) $2\ C_6H_5Br \xrightarrow{2Mg/THF} 2\ C_6H_5MgBr$

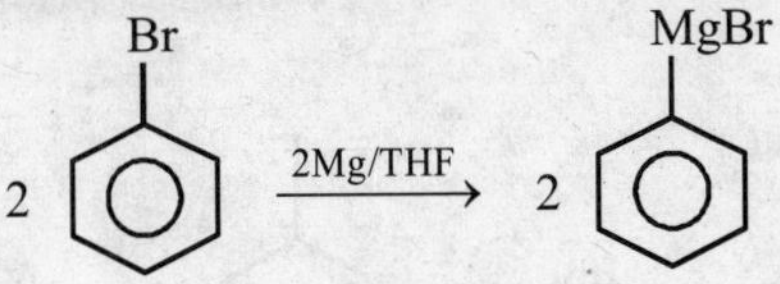

Now,

$$Ph-\overset{O}{\overset{\|}{C}}-O-CH_3 \xrightarrow{\overset{\delta-\ \delta+}{Ph\,Mg\,Br}} Ph-\overset{O^-}{\overset{|}{C}}(Ph)-O-CH_3 \longrightarrow Ph-\overset{O}{\overset{\|}{C}}-Ph$$

$$\xrightarrow{\overset{\delta-\ \delta+}{Ph\,Mg\,Br}} Ph-\overset{O^-Mg^+Br}{\overset{|}{C}}(Ph)-Ph \xrightarrow{H_2O/H^+} (C_6H_5)_3C-OH$$

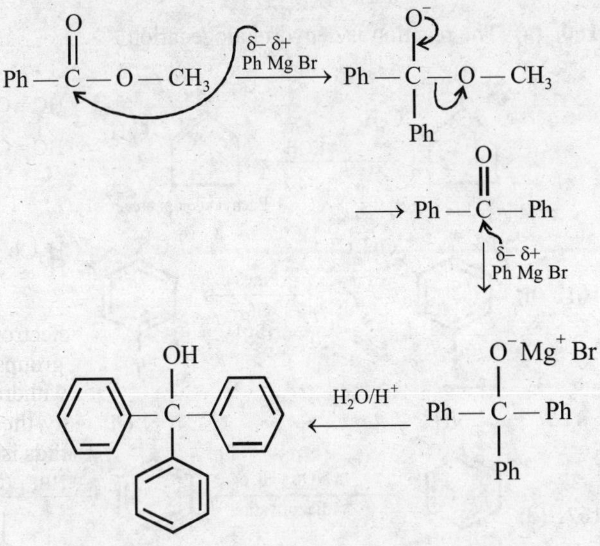

146. (a) Acetate in acetic acid with R-X gives S_N1 reaction. Allyl halide undergo faster S_N1 reaction, hence correct order is:

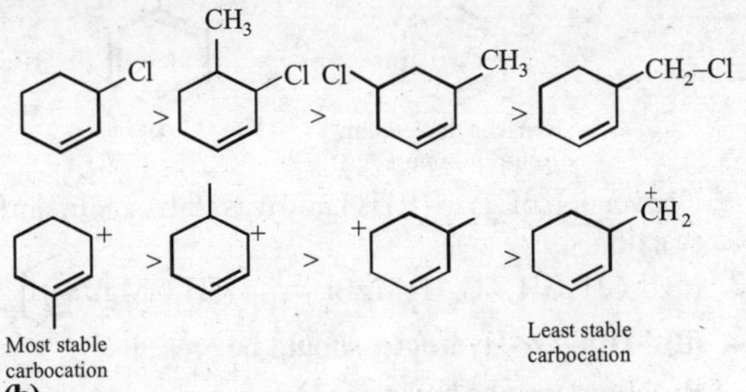

3-chlorocyclohexene > 3-chloro-2-methylcyclohexene > Cl–(cyclohexene with CH_3) > cyclohexene–CH_2–Cl

Most stable carbocation > > > Least stable carbocation ($\overset{+}{C}H_2$)

147. (b)

$$o\text{-}ClC_6H_4CH_2I \xrightarrow{(i)\ NaOH} o\text{-}ClC_6H_4CH_2OH + NaI$$

$$NaI \xrightarrow[(iii)\ AgNO_3]{(ii)\ dil.HNO_3} AgI \text{ (yellow ppt.)}$$

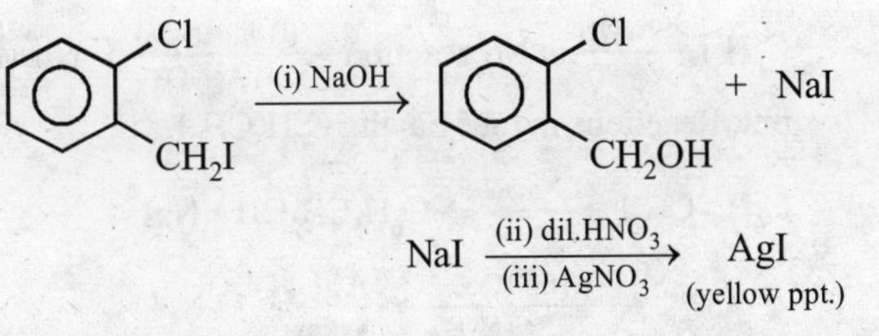

148. (a) Density is proportional to molar mass. So, correct relation is: D > C > B > A.

149. (b) (A) – Sandmeyer reaction
(B) – Gattermann reaction
(C) – Wurtz reaction
(D) – Fittig reaction
(A) - (II); (B) - (IV); (C) - (I); (D) - (III)

150. (d) For nucleophilic substitution, the correct order is:

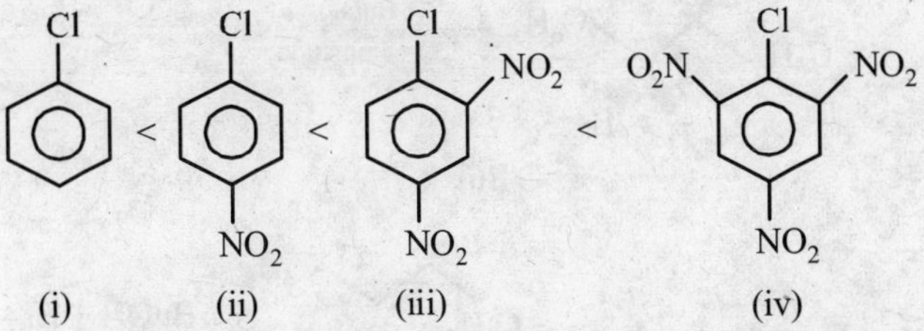

chlorobenzene (i) < 4-nitrochlorobenzene (ii) < 2,4-dinitrochlorobenzene (iii) < 2,4,6-trinitrochlorobenzene (iv)

$-NO_2$ group is a strong electron withdrawing group and it activates the aromatic ring towards nucleophillic substituion.

151. (b) Rate of S_N2 reaction $\propto \dfrac{1}{\text{Stability of carbocation}}$

Stability of carbocation $\propto \dfrac{1}{-I, -M\ \text{effect(s)}}$

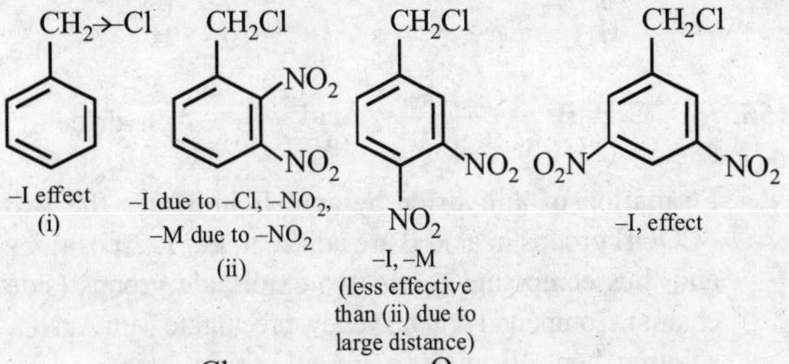

(i) $C_6H_5CH_2\rightarrow Cl$ — –I effect
(ii) CH_2Cl with NO_2 at ortho positions — –I due to –Cl, $-NO_2$, –M due to $-NO_2$
(iii) CH_2Cl with NO_2 at meta and para — –I, –M (less effective than (ii) due to large distance)
(iv) CH_2Cl with NO_2 at both meta positions — –I, effect

152. (d) Chlorobenzene + phthalic anhydride $\xrightarrow[(ii)\ H_2O]{(i)\ AlCl_3,\ Heat}$ 2-(4-chlorobenzoyl)benzoic acid

Friedel-Craft reaction

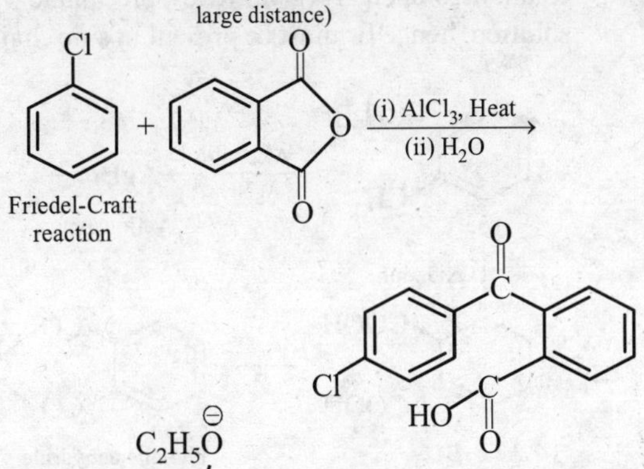

153. (b)

$$C_6H_5-\underset{\text{H}}{CH}-\overset{CH_3}{\underset{Br}{C}}-\overset{H}{CH}-CH_3$$

(–I effect) ... (+I effect); More acidic H (adjacent to C_6H_5) removed by $C_2H_5\overset{\ominus}{O}$

$$\xrightarrow[C_2H_5OH]{C_2H_5ONa} C_6H_5-CH=\overset{CH_3}{C}-CH_2-CH_3$$

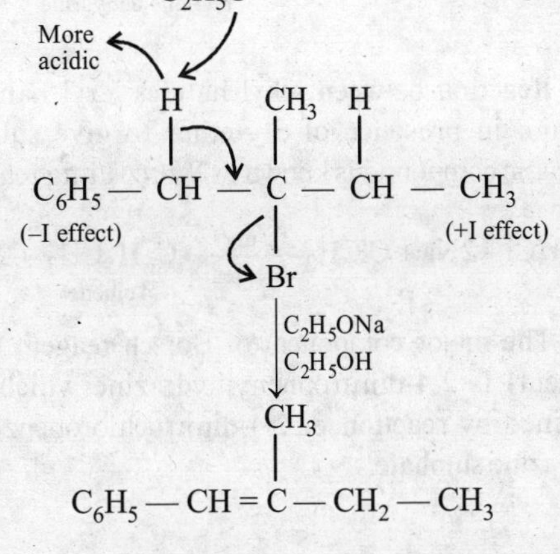

154. (b) Elimination reaction is highly favoured if
(a) Bulkier base is used
(b) Higher temperature is used
Hence, in given reaction, biomolecular elimination reaction provides major product.

C_6H_5–CH(Br)(H)–CH(H)(H)–C_6H_5 $\xrightarrow[\text{1, 2-elimination}]{\text{t-BuOK}}$ C_6H_5–CH=CH–C_6H_5 + *t*-BuOH + Br^-

155. (b) $C_6H_5CH_2Br$ being least sterically hindered alkyl bromide undergoes S_N2 nucleophilic substitution reaction most easily. Hence, it undergoes complete inversion of configuration.

$$C_6H_5-CH_2-Br + Cl^- \xrightarrow[-Br^-]{DMF} Cl-CH_2-C_6H_5$$

156. (d) C_8H_9Br (A) $\xrightarrow{\text{Oxidation}}$ Acid (B) $\xrightarrow{\Delta}$ Anhydride

Formation of anhydride by acid B indicates the two –COOH groups in acid B are adjacent, i.e. 1, 2-positions and thus compound A has two oxidisable groups (side chains). Compound A gives yellow precipitate with $AgNO_3$ solution, hence Br must be present in side chain.

(A) o-CH_2Br–C_6H_4–CH_3 $\xrightarrow[AgNO_3]{\text{Alcoholic}}$ $AgBr\downarrow$ yellow ppt.

(A) $\xrightarrow{\text{Oxidation}}$ o-$C_6H_4(COOH)_2$ Acid (B) $\xrightarrow{\Delta}$ Phthalic anhydride

157. (b) Reaction between alkyl halides, aryl halides and sodium in presence of dry ether to give substituted aromatic compound is known as Wurtz-Fittig reaction:

$$C_6H_5Cl + 2Na + ClCH_3 \xrightarrow[\text{ether}]{\text{dry}} \underset{\text{Toluene}}{C_6H_5CH_3} + 2NaCl$$

158. (c) The major component of Borsch reagent (Brady's reagent) is 2,4-dinitrophenylhydrazine which can be obtained by reaction of 2,4-dinitrochlorobenzene and hydrazine sulphate.

O_2N–$C_6H_3(NO_2)$–Cl + H – NH – $NH_2.H_2SO_4$ ⟶ O_2N–$C_6H_3(NO_2)$–NH – NH_2

2,4 -Dinitrophenylhydrazine

159. (a)

$C_6H_5N{=}NCl \xrightarrow[\text{(Balz-Schiemann's reaction)}]{HBF_4} C_6H_5F + N_2 + BF_3 + NaCl$

160. (a) The reaction is dehydrohalogenation

2-Bromo-1-phenylcyclopentane $\xrightarrow{\text{alc.KOH}}$ 1-Phenyl cyclopentene

161. (b) m-I–C_6H_4–CH_2Cl $\xrightarrow[\text{DMF}\ (S_N1)]{\text{NaCN}}$ m-I–C_6H_4–CH_2CN

162. (d) $C_6H_5NH_2 \xrightarrow[\text{0–5° diazotisation}]{Na\,NO_2+HCl} C_6H_5N_2Cl$ (I) $\xrightarrow{HBF_4}$

$C_6H_5N_2^+BF_4^-$ (Benzene diazonium tetrafluoroborate) $\xrightarrow{\Delta}$ C_6H_5F (II) $+ BF_3 + N_2$

Conversion of (I) to (II) is known as Balz-schilmann reaction.

163. (d) $CH_3OH + C_6H_5MgBr \longrightarrow CH_3OMgBr + C_6H_6$

164. (d) Hing: α-Hydrogen should be present

Reactions in tube having C_6H_5I.

$C_6H_5I \xrightarrow{NaOH}$ No reaction $\xrightarrow[\text{(ii) } AgNO_3]{\text{(i) dil·}HNO_3}$ No yellow ppt. Reactions in tube having $C_6H_5CH_2I$.

$C_6H_5–CH_2I \xrightarrow{NaOH} C_6H_5CH_2OH + \mathbf{NaI}$

$\xrightarrow[\text{(ii) } AgNO_3]{\text{(i) dil·}HNO_3} AgI\downarrow$ **yellow**

So tube having $C_6H_5CH_2I$ is B as it responds $AgNO_3$ test and the other tube having C_6H_5I is A.

165. (a)

$$Cl-\diamond-Br \xrightarrow[\text{Wurtz reaction}]{Na/Et_2O} Cl-\diamond\diamond-Cl \text{ (A)}$$

(A) $\xrightarrow[\text{Swarts reaction}]{CoF_2}$ F–◇◇–F (C)

(A) $\xrightarrow[\text{(ii) } D_2O]{\text{(i) } Mg/Et_2O}$ D–◇◇–D (B)

166. (c) A-III, B-I, C-IV, D-II

167. (b) (A) – (III), (B) – (I), (C) – (IV), (D) – (II)

Compound		Use
(A) Carbon tetrachloride	–	Fire extinguisher
(B) Methylene Chloride	–	Paint Remover
(C) DDT	–	Non Biodegradable insecticide
(D) Freons	–	Refrigerators and air conditioners

168. (b) Gammaxene ($C_6H_6Cl_6$, hexachlorocyclohexane; each ring carbon bearing one Cl)

169. (a) Freons are chlorofluoro carbon compounds.

170. (d) Sodium chlorate and Sodium arsenite are herbicides.

171. (c) $2\,Cl-\underset{Cl}{\overset{Cl}{C}}-CH_3 + 6Ag \longrightarrow CH_3C\equiv CCH_3 + 6AgCl$

1, 1, 1-Trichloroethane — But-2-yne

172. (c) Chloral on reaction with chlorobenzene in the presence of a catalytic amount of sulphuric acid forms DDT (dichloro diphenyl trichloroethane).

$$Cl_3C-\overset{H}{C}=O + 2\,H-C_6H_4-Cl \longrightarrow Cl_3C-CH(C_6H_4Cl)_2$$

Trichloro-acetaldehyde — DDT

173. (d) $Cl_3C-CH_2CH_3 + KOH \xrightarrow[(-3\,KCl)]{heat}$

1, 1, 1-Trichloropropane

$(OH)_3C-CH_2CH_3 \longrightarrow CH_3CH_2\overset{O}{\overset{\|}{C}}-OH$

Propionic acid

174. (b) DDT is prepared by heating chlorbenzene and chloral with concentrated sulphuric acid

$$CCl_3CHO + 2\,H-C_6H_4-Cl \xrightarrow[-H_2O]{H_2SO_4} CCl_3CH(C_6H_4Cl)_2$$

1,1,1-Trichloro–2,2 bis (*p*-chlorophenyl) ethane or DDT

8 Alcohols, Phenols and Ethers

1. **(c)**

$$\underset{(A)}{CH_3-OH} \xrightarrow{H_2O} CH_3-\overset{\oplus}{O}H_2 \xrightarrow[\text{(Acidic Medium)}]{Na^+CN^\ominus} CH_3-CN \xrightarrow[\text{(hydrolysis)}]{H_3O^\oplus}$$

$$\underset{(B)}{CH_3-\overset{O}{\overset{\|}{C}}-OH} \xrightarrow{B_2H_6} \underset{(C)}{CH_3-CH_2-OH} \xrightarrow[(H_2SO_4+SO_3)]{\text{Oleum}} \underset{\text{(Inhalable anesthetic)}}{\underset{(D)}{C_2H_5-O-C_2H_5}}$$

2. **(d)** B.P. $\propto$ M.W.

B.P. $\propto$ Intermolecular hydrogen bonding. Alcohols and phenols have intermolecular H-bonding so their boiling points are higher than species of comparable molecular mass.

3. **(c)** **Hydration Reaction:** (Markovnikoff)

$$CH_3-CH=CH_2+H^+ \longrightarrow \underset{\text{(More stable)}}{CH_3-\overset{+}{C}H-CH_3}$$

$$CH_3-\overset{+}{C}H-CH_3+H_2O \longrightarrow \underset{(A)}{(CH_3-\underset{OH}{\underset{|}{C}H}-CH_3)}+H^+$$

Hydroboration Oxidation Reaction:
(anti-Markovnikoff)

$$3CH_3-CH=CH_2+B_2H_6 \xrightarrow{THF} 2(CH_3CH_2CH_2)_3B$$

$$(CH_3CH_2CH_2)_3B+3H_2O_2 \xrightarrow{OH^-} \underset{(B)}{3CH_3CH_2CH_2OH}+H_3BO_3$$

4. **(a)** The major products (A) and (B) are formed as:

$$\text{1-methylcyclohexanol (CH}_3\text{, OH)} \xrightarrow{\text{Conc. } H_2SO_4} \text{(CH}_3\text{, }\overset{+}{O}H_2\text{)} \xrightarrow{\Delta} \underset{(A)}{\text{1-methylcyclohexene (CH}_3\text{)}}$$

$$\text{1-methylcyclohexanol (CH}_3\text{, OH)} \xrightarrow{\text{Pyridine}} \text{(CH}_3\text{, O}^-\text{)} \xrightarrow{CH_3-\overset{O}{\overset{\|}{C}}-Cl} \underset{(B)}{\text{(CH}_3\text{, }O-\overset{O}{\overset{\|}{C}}-CH_3\text{)}}$$

5. **(d)**

$$C_6H_5OH + NaOH \rightarrow C_6H_5O^-Na^+ + H_2O$$

Stronger acid than H_2O due to more resonating structure of phenoxide ion.

6. **(11)** Vanillin compound is an organic compound molecular formula $C_8H_8O_3$. It is a phenolic aldehyde. Its functional compounds include aldehyde, hydroxyl and ether. It is the primary component of the extract of the vanilla beans.

Total sum of oxygen atoms and π-electrons is $3+8=11$

Total number of oxygen atoms = 3

Total number of π-bonds = 4

$\therefore$ Total number of π-electrons = 8

7. **(a)**

$$\underset{(Nu)}{R-\ddot{O}-H} + \; >C^+- \longrightarrow R-\underset{+}{\overset{H}{\overset{|}{O}}}-\overset{|}{\underset{|}{C}}-$$

$$R-CH_2-OH+H^+ \longrightarrow R-CH_2-\overset{+}{O}H_2$$

$$Br^- + \underset{R}{\underset{|}{C}H_2}-OH_2^+ \;(E) \longrightarrow Br-\underset{R}{\underset{|}{C}H_2}+H_2O$$

8. **(d)**

$$PhCH=CH_2 \xrightarrow[\text{(anti-Marko. addition)}]{BH_3/H_2O_2,\,OH^-} PhCH_2CH_2OH$$

$$PhCH_2CH_2OH+HBr \longrightarrow PhCH_2CH_2Br+H_2O$$

$$PhCH_2CH_2Br \xrightarrow{\text{Mg/dry ether}} PhCH_2CH_2MgBr \xrightarrow[\text{(II) } H_3O^+]{\text{(I) } H-\overset{O}{\overset{\|}{C}}-H} PhCH_2CH_2CH_2OH$$

9. **(c)** Acidic strength $\propto -$I effect

$$\propto \frac{1}{+\text{I effect}}$$

F, Cl exerts –I effect, methyl exerts +I effect, compound C is least acidic.

Among A and B; since inductive effect is distance dependent too, extent of –I effect is higher in A followed by B even though F is stronger electron withdrawing group than Cl. Thus, A is more acidic than B.

10. (a) $H_3C{-}CH_2{-}CH(OH){-}CH_3 \xrightarrow{NaI,\ H_3PO_4} H_3C{-}CH_2{-}CH(I){-}CH_3$

Mg dry ether

$H_3C{-}CH_2{-}CH(MgI){-}CH_3 \xrightarrow{D_2O} H_3C{-}CH_2{-}CH(D){-}CH_3$

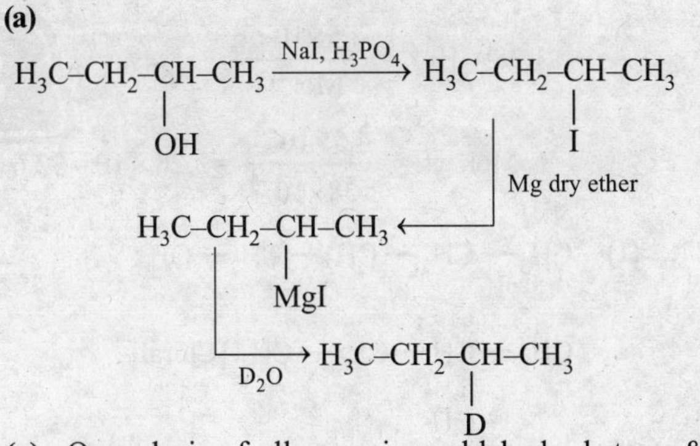

11. (a) Ozonolysis of alkene, gives aldehyde, ketone & carboxylic acid.

12. (a) Butan–1–ol ($CH_3CH_2CH_2CH_2OH$) can undergo hydrogen bonding. Ethoxyethane ($CH_3CH_2{-}O{-}CH_2CH_3$) has no hydrogen (attached with F, O, N) which can undergo hydrogen bonding. More is the extent of intermolecular H–bonding, more will be association of molecules. Thus leading to higher boiling point.
Hence both Assertion (A) and Reason(R) are true and (R) is the correct explanation of (A).

13. (b)

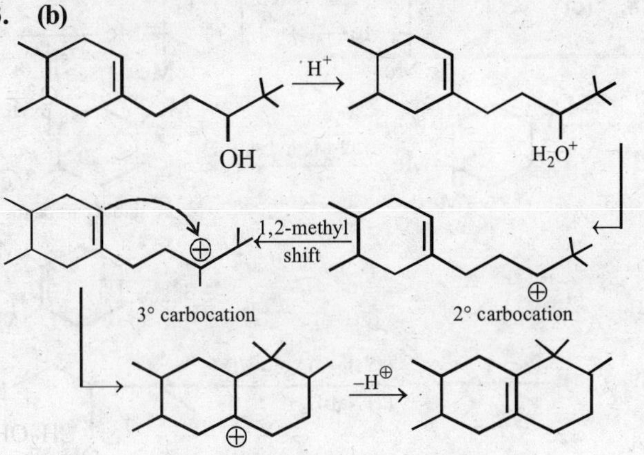

H^+, H_2O^+, 1,2-methyl shift, 3° carbocation, 2° carbocation, $-H^{\oplus}$

14. (a) A ($C_5H_{10}O_5$): CHO–CH(OH)–C(OH)(CH₂OH)(CH₂OH)

$\xrightarrow{(CH_3CO)_2O}$ CHO–CH(OAc)–C(OAc)(CH₂OAc)(CH₂OAc)

$\xrightarrow{Br_2/H_2O}$ COOH–CH(OH)–C(OH)(CH₂OH)(CH₂OH) ($C_5H_{10}O_6$)

$\xrightarrow{HI}$ $CH_3{-}CH_2{-}CH(CH_3){-}CH_3$ Isopentane

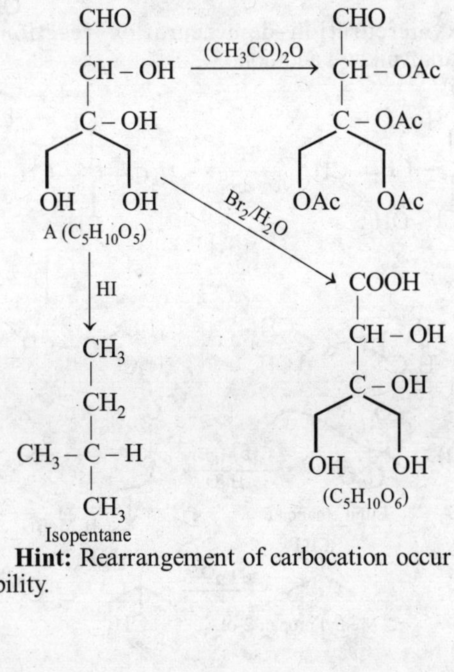

15. (b) **Hint:** Rearrangement of carbocation occur to gain stability.

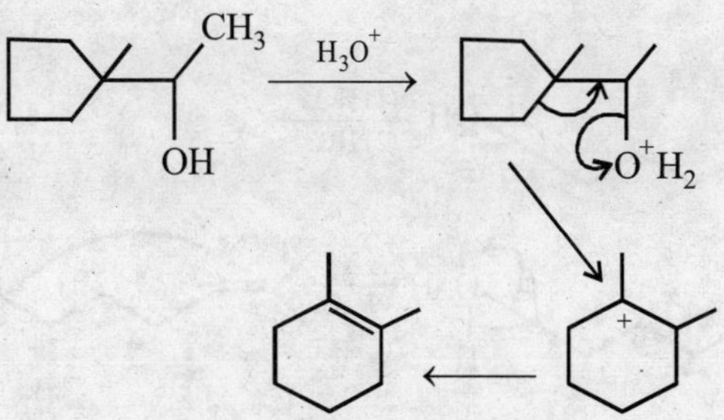

$\xrightarrow{H_3O^+}$ O^+H_2

16. (a) $\xrightarrow[(-H_2O)]{H^+}$ $\xrightarrow{-H^+}$

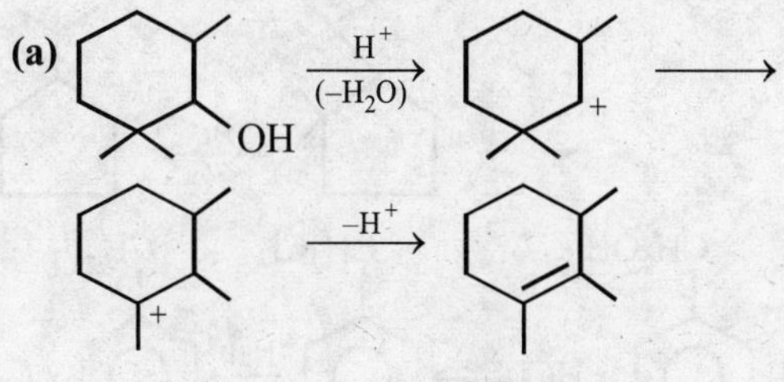

17. (3)

OH, OH, OH

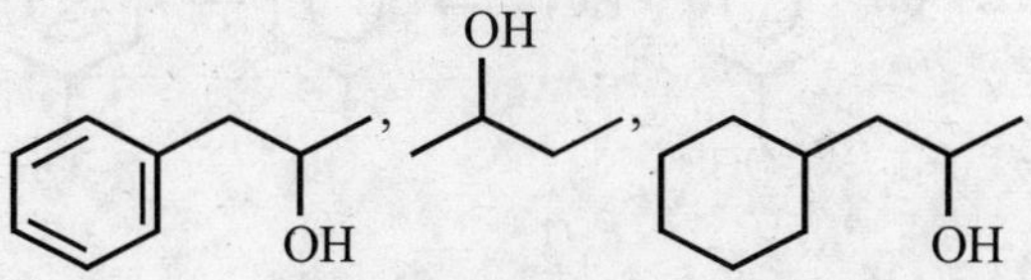

Red colouration of solution with ceric ammonium nitrate is the identification test for alcohol. Whereas the iodoform test is for the secondary alcohols having one alkyl group as $-CH_3$.

18. (a) **Hint:** In hydrocarbocation, hydrogen from BH_3 attached at stable carbocation position.

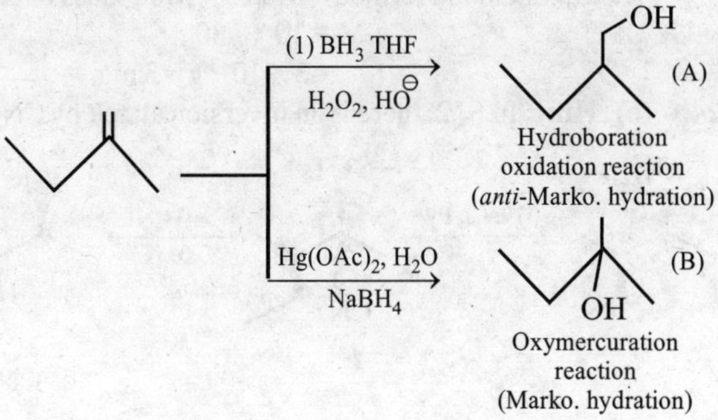

(1) BH_3 THF, H_2O_2, $HO^{\ominus}$ → (A) OH — Hydroboration oxidation reaction (*anti*-Marko. hydration)

$Hg(OAc)_2$, H_2O, $NaBH_4$ → (B) OH — Oxymercuration reaction (Marko. hydration)

19. (c) $CH_3-CH_2-OH \xrightarrow[\text{(dehydration)}]{\text{conc. } H_2SO_4,\ \Delta} CH_2=CH_2\uparrow$ Ethene

$\xrightarrow[\text{(alk-KMnO}_4)]{\text{Baeyer's reagent}}$ $CH_2(OH)-CH_2(OH)$ Glycol

20. (3) $R(OH)x \longrightarrow H_2$
Alcoholic hydrogens are involved in hydrogen gas formation. Number of alcoholic hydrogens present in x

$= \frac{1}{2}$ (numbers of hydrogen atoms in H_2 gas)

$$x\left(\frac{1.84\times10^{-3}}{92}\right)=\frac{1}{2}\left(\frac{1.344\times10^{-3}}{22.4}\right)\times 2 \Rightarrow x=3$$

21. (c)

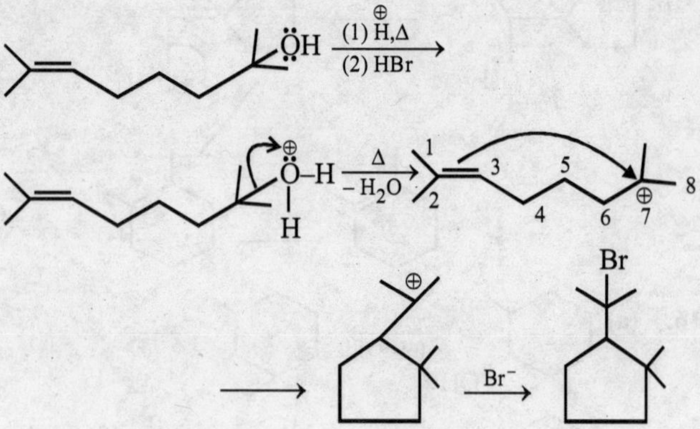

Reaction scheme: substrate with $\ddot{O}H$ treated with (1) $\overset{\oplus}{H}$, Δ (2) HBr; protonation gives $\overset{\oplus}{O}H_2$, loss of $-H_2O$ on heating (Δ) gives carbocation (carbons numbered 1–8), which cyclises; the cyclopentyl carbocation reacts with Br^- to give the bromide.

22. (a)

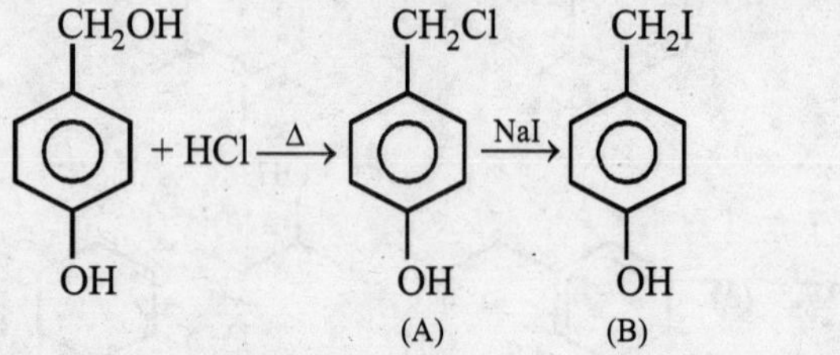

CH_2OH (p-hydroxybenzyl alcohol) $+ HCl \xrightarrow{\Delta}$ CH_2Cl (A) $\xrightarrow{NaI}$ CH_2I (B) (each with OH at para position)

23. (3)

$$CH_3CH_2MgBr + CH_3OH \longrightarrow CH_3CH_3 + Mg(Br)OCH_3$$

100 mL 2.24 mL

$$\text{Moles of ethane formed} = \frac{2.24}{22400} = 10^{-4}$$

Weight of ethane formed = Moles × Mol. mass of ethane

$= 10^{-4} \times 30$

$= 3 \times 10^{-3}\, g = 3\, mg.$

24. (b) Hint: In S_N2, there is an inversion attack by CN^-

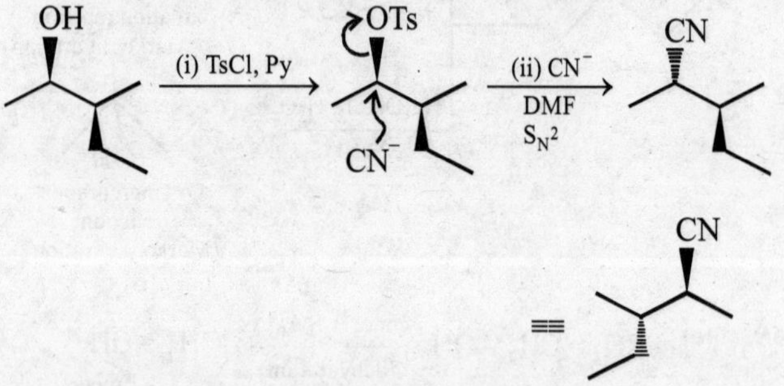

OH compound $\xrightarrow{\text{(i) TsCl, Py}}$ OTs compound (attacked by CN^-) $\xrightarrow[\text{DMF},\ S_N2]{\text{(ii) } CN^-}$ CN product ≡ CN product

25. (c) When glycerol is heated with $KHSO_5$, acrolain is formed because $KHSO_3$ act dehydrating agent and acrolain is an unsaturated aldehyde with irritating small.

Statement I is correct but Statement II is incorrect

26. (33) 1 mole of alcohol gives 1 mole of methane

$$\text{No. of moles in 3.1 mL} = \frac{1}{22400} \times 3.1$$

$$= 1.38 \times 10^{-4} \text{ moles}$$

Thus 1.38×10^{-4} moles of alcohol are used

Also, No. of moles of alcohol (n)

$$= \frac{\text{Wt. of alcohol taken}}{\text{Mol.wt. of alcohol}}$$

$$1.38 \times 10^{-4} = \frac{4.5 \times 10^{-3}\, g}{\text{Mol.wt}}$$

$$\text{Mol. wt.} = \frac{4.5 \times 10^{-3}}{1.38 \times 10^{-4}} = 3.26 \times 10 = 32.6 \approx 33.$$

27. (1) $CH_3-CH_2-CH_2-CH_2-OH$

$CH_3-\overset{*}{C}H(OH)-CH_2-CH_3$ [Chiral]

$CH_3-CH(CH_3)-CH_2-OH$

$CH_3-C(CH_3)(OH)-CH_3$

28. (c)

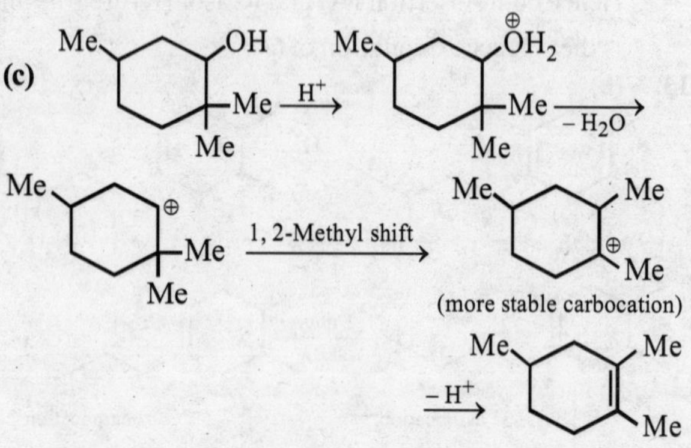

Me-substituted cyclohexanol (OH, Me, Me) $\xrightarrow{H^+}$ $\overset{\oplus}{O}H_2$ $\xrightarrow{-H_2O}$ secondary carbocation $\xrightarrow{\text{1, 2-Methyl shift}}$ tertiary carbocation (more stable carbocation) $\xrightarrow{-H^+}$ alkene (Me, Me, Me)

29. (a)

$$\text{(CH}_3)_3C-CH=CH_2 \xrightarrow[\text{(2) NaBH}_4]{\text{(1) Hg (OA}_2\text{c)}_2 \text{ HO}} \text{(CH}_3)_3C-CH(OH)-CH_2-H$$

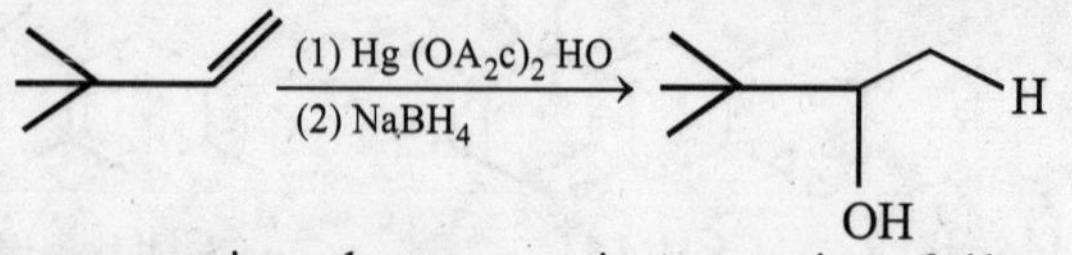

Oxymercuration–demercuration reaction follow Markovnikov's addition.

30. (b)

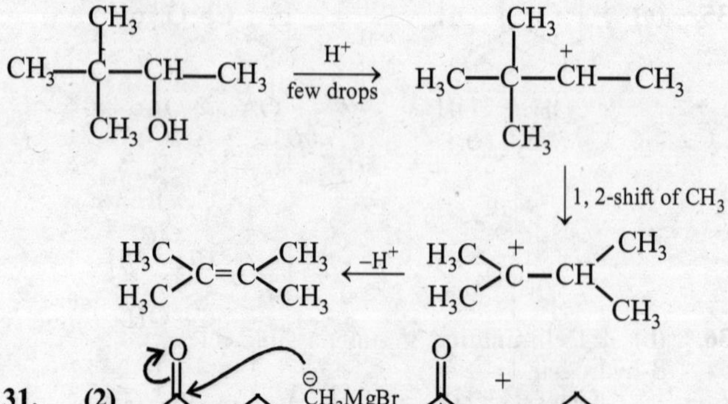

$CH_3-C(CH_3)_2-CH(OH)-CH_3 \xrightarrow[\text{few drops}]{H^+} H_3C-C(CH_3)_2-\overset{+}{C}H-CH_3$

↓ 1, 2-shift of CH_3

$(H_3C)_2C=C(CH_3)_2 \xleftarrow{-H^+} (H_3C)_2\overset{+}{C}-CH(CH_3)_2$

31. (2)

Ethyl ethanoate $\xrightarrow[H_2O]{\overset{\ominus}{C}H_3MgBr}$ CH_3COCH_3 + HO–CH_2CH_3

↓ $\overset{\ominus}{C}H_3\overset{\oplus}{MgBr}$

2–Methyl propan-2-ol $\xleftarrow{H_2O}$ $(CH_3)_3C-\overset{\ominus}{O}MgBr$

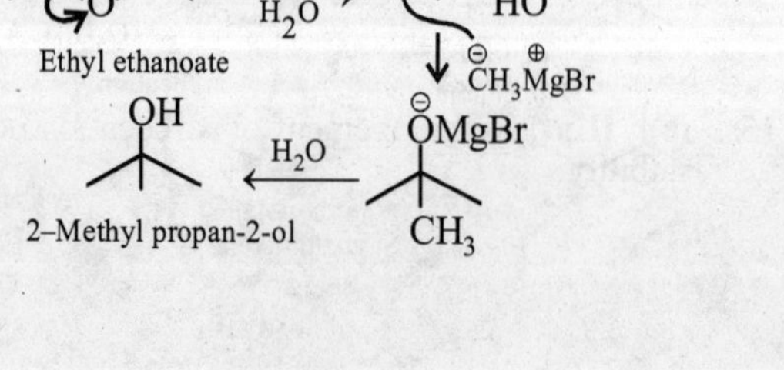

32. (d) Ceric ammonium nitrate is used for the identification of alcohol.

$$2R-OH+(NH_4)_2Ce(NO_3)_6 \longrightarrow (ROH)_2Ce(NO_3)_4+2NH_4NO_3$$

Alkoxy cerium ion (IV) Compound (Pink or Red colour)

$CHCl_3/KOH$ is used for the identification of primary amines.

$$R-NH_2+CHCl_3+3KOH(alc) \xrightarrow{warm} R-NC+3KCl+3H_2O$$

33. (b) CH_2-OH / CH_2-OH + $HO-C(=O)-C(=O)-OH$ → cyclic ester ($CH_2-O-C(=O)$ / $CH_2-O-C(=O)$) $\xrightarrow{210°C}$ $2CO_2 + CH_2=CH_2$ (major)

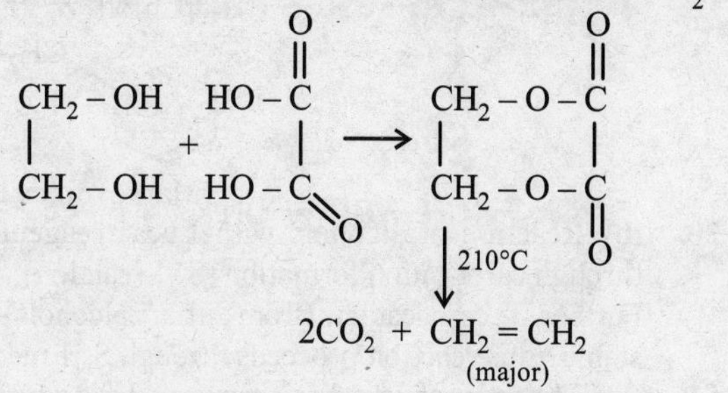

34. (a) $\xrightarrow[-Cl^-]{H-Cl}$ $\xrightarrow{-H_2O}$ 2° Carbocation $\xrightarrow{\text{1,2-hydride shift}}$ 3° Carbocation $\xrightarrow{Cl^-}$ (A)

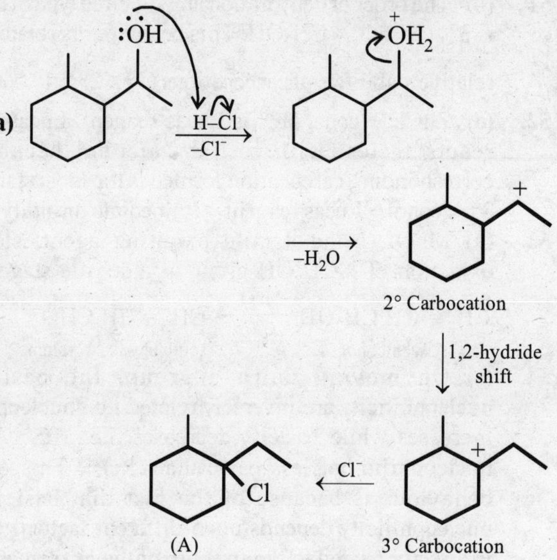

35. (a) $\xrightarrow[-H_2O]{H^+}$ $\xrightarrow{-H^+}$

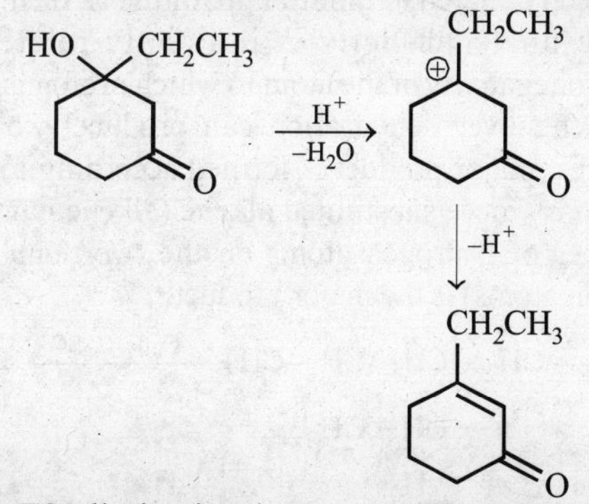

36. (b) E2 elimination is not possible due to absence of β-hydrogen.

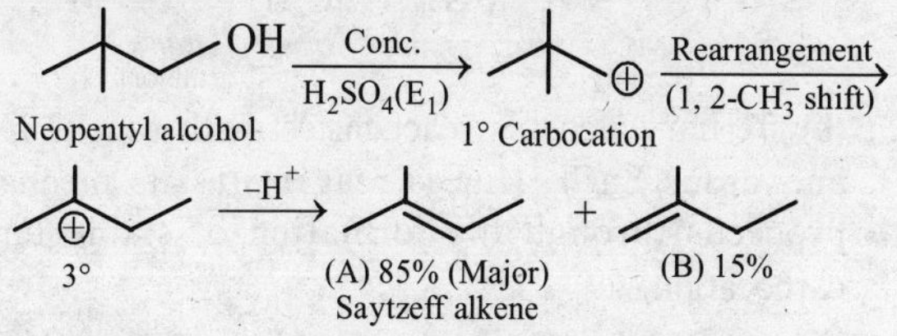

Neopentyl alcohol $\xrightarrow[H_2SO_4(E_1)]{Conc.}$ 1° Carbocation $\xrightarrow[(1,2\text{-}CH_3^-\text{ shift})]{Rearrangement}$ 3° $\xrightarrow{-H^+}$ (A) 85% (Major) Saytzeff alkene + (B) 15%

37. (4) $\xrightarrow[(ii)\ H_3O^+]{(i)\ C_2H_5MgBr}$ 'A' $\xrightarrow[(ii)\ H_2O]{(i)\ CH_3MgBr}$ 'B'

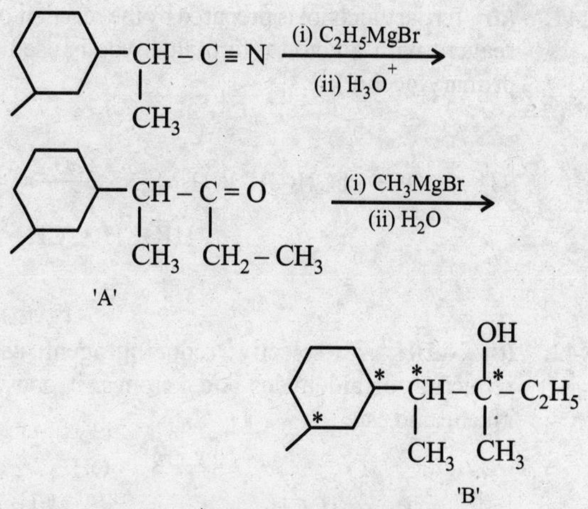

38. (b) $CH_3-C(CH(CH_3)_2)=CH-CH_2-CH_3 \xrightarrow[(ii)\ H_2O_2,\ OH^-]{(i)\ B_2H_6}$ (*anti*-Mark. hydration)

$H_3C-CH(CH(CH_3)_2)-CH(OH)-CH_2-CH_3 \xrightarrow[-H_2O]{H^+,\ E1}$

$H_3C-CH(CH(CH_3)_2)-\overset{\oplus}{C}H-CH_2-CH_3 \xrightarrow{1,2\text{-}H^-\text{ shift}}$

$H_3C-\overset{\oplus}{C}(CH(CH_3)_2)-CH_2-CH_2-CH_3 \xrightarrow[\Delta,\ -H^+]{Deprotonation}$

$H_3C-C(=C(CH_3)_2)-CH_2-CH_2-CH_3$

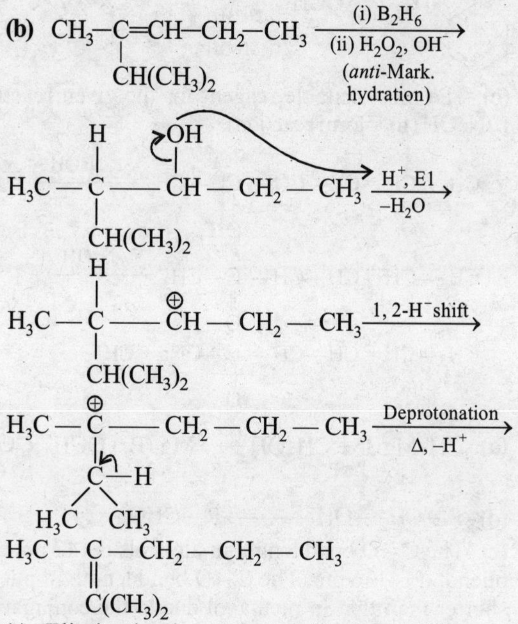

39. (c) Elimination is not favourable due to dilute acid and low temperature.

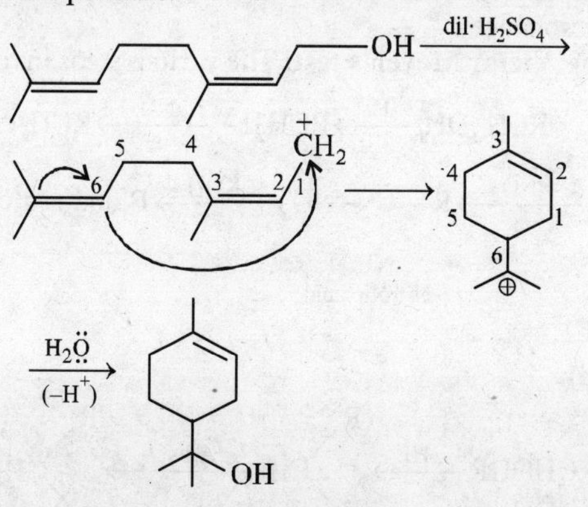

$\xrightarrow{dil\cdot H_2SO_4}$ $\longrightarrow$ $\xrightarrow[(-H^+)]{H_2\ddot{O}}$

40. (c) 2-Chloro-1-phenylbutane $\xrightarrow[(-HCl)]{EtOK/EtOH}$ (X) $\xrightarrow[(ii)\ NaBH_4]{(i)\ (CH_3COO)_2Hg,\ H_2O}$ 1-Phenylbutanol

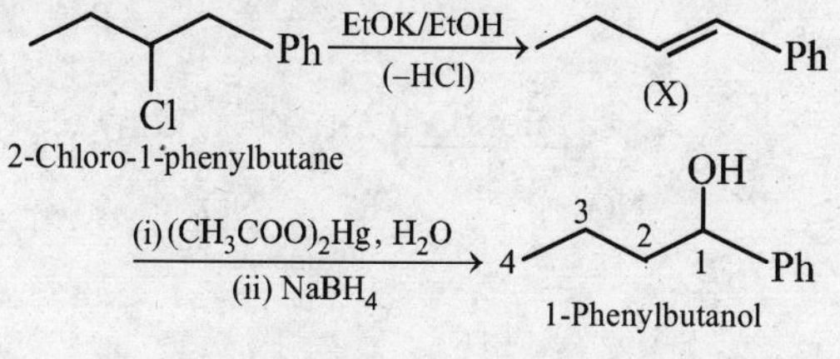

41. **(d)** Tertiary alcohol is prepared by the reaction of Grignard reagent with a ketone (formaldehyde is used to prepare primary alcohol).

$$H-\overset{O}{\overset{||}{C}}-H + PhCH(CH_3)CH_2MgX \xrightarrow{H_3O^+} \underset{1°\text{ Alcohol}}{HOCH_2-CH_2-\underset{CH_3}{\underset{|}{CH}}-Ph}$$

42. **(d)** $NaBH_4$ is a selective reducing agent, used for the reduction of aldehydes and ketones, it does not affect alkene and ester.

$$\text{(cyclohexenone with } -CH_2-\overset{O}{\overset{||}{C}}-OCH_3) \xrightarrow[MeOH]{NaBH_4} \text{(cyclohexenol with } -CH_2-\overset{O}{\overset{||}{C}}-OCH_3)$$

43. **(b)** The most suitable reagent for the given reaction is $I_2/NaOH$ (Iodoform reaction).

$$CH_3-CH=CH-CH_2-\underset{OH}{\underset{|}{CH}}-CH_3 \xrightarrow{I_2/NaOH}$$

$$CH_3-CH=CH-CH_2-\underset{O}{\underset{||}{C}}-CH_3 \xrightarrow{I_2/NaOH}$$

$$CH_3-CH=CH-CH_2-\underset{O}{\underset{||}{C}}-ONa + CHI_3$$

44. **(a)** $CH_3MgBr + CH_3OH \longrightarrow Mg(Br)OCH_3 + \underset{\text{Methane}}{CH_4\uparrow}$

45. **(d)** $R-CH_2-OH \xrightarrow{PCC} R-CHO$

46. **(c)** The C – O bond length in alcohols is 142 pm and in phenol it is 136 pm. The C – O bond length in phenol is shorter than that in methanol due to the conjugation of unshared pair of electrons on oxygen with the ring, which imparts double bond character to the C – O bond (resonance).

47. **(b) Victor Meyer's test:** The various steps involved are

(i) $RCH_2OH \xrightarrow{HI} RCH_2I \xrightarrow{AgNO_2} RCH_2NO_2$

$$\xrightarrow{HNO_2} \underset{\text{Nitrolic acid}}{R-\underset{NOH}{\underset{||}{C}}-NO_2} \xrightarrow{KOH} \text{Blood red colour}$$

(ii)

$$RCHOHR \xrightarrow{HI} R_2CHI \xrightarrow{AgNO_2} R_2CHNO_2$$

$$R_2C(H)NO_2 \xrightarrow[H_2O]{HONO} \underset{\text{Pseudonitrol}}{R_2C(NO)NO_2} \xrightarrow{KOH} \text{Blue colour}$$

(iii) $R_3COH \xrightarrow{HI} R_3CI \xrightarrow{AgNO_2} R_3CNO_2$

$$\xrightarrow{HNO_2} \text{No reaction} \xrightarrow{KOH} \text{Colourless}$$

48. **(c)** $CH_3CH=CH_2 \xrightarrow[H_2O_2, NaOH]{B_2H_6} CH_3CH_2CH_2OH$

Anti-Markovnikov hydration

49. **(a)** $C_6H_5-CH_2-CH=CH_2 \xrightarrow[\text{(ii) } H_2O, \text{ (iii) } NaBH_4/\bar{O}H]{\text{(i) } Hg(OAc)_2/THF} C_6H_5-CH_2-\underset{OH}{\underset{|}{CH}}-CH_3$

(Markovnikov hydration)

50. **(b)** Reaction of alcohols with Lucas reagent proceeds through carbocation formation (S_N1 reaction). Further, 3° carbocations (from tertiary alcohols) are highly stable, thus reaction proceeds through S_N1 mechanism.

51. **(b)** The order of dehydration among three type of alcohols is 3° > 2° > 1° > CH_3OH. This behaviour is related to the relative stabilities of carbocations (3° > 2° > 1° > $\overset{+}{C}H_3$).

52. **(d)** $ZnCl_2$ + conc. HCl is Lucas reagent. Lucas reagent reacts fastest with tertiary alcohol because the corresponding carbocation formed is the most stable (3°). 3° Alcohol + Lucas reagent = Immediate turbidity

53. **(a)** MnO_2, being a mild oxidising agent, stops the oxidation of – CH_2OH group at aldehyde stage.

$$\underset{\text{Allyl alcohol}}{CH_2=CH\cdot CH_2OH} \xrightarrow{MnO_2} \underset{\text{Acraldehyde (Acrolein)}}{CH_2=CH\cdot CHO}$$

54. **(a)** On moving down a group, the basicity & nucleophilicity are inversely related i.e., nucleophilicity increases while basicity decreases, i.e., $RS^{\ominus}$ is more nucleophilic but less basic than $RO^{\ominus}$. This opposite behaviour is because of the fact that basicity and nucleophilicity depends upon different factors. Basicity is directly related to the strength of the H–element bond, while nucleophilicity is indirectly related to the electronegativity of the atom to which proton is attached.

55. **(a)** Whenever dehydration can produce two different alkenes, major product is formed according to **Saytzeff rule,** *i.e.*, more substituted alkene (alkene having lesser number of hydrogen atoms on the two doubly bonded carbon atoms) is the major product.

$$C_6H_5-CH_2-\underset{OH}{\underset{|}{CH}}-\underset{CH_3}{\underset{|}{CH}}-CH_3 \xrightarrow{\text{Conc. } H_2SO_4}$$

$$\underset{\substack{cis\\ \text{(minor)}}}{(H)(C_6H_5)C=C(H)(CH(CH_3)_2)} + \underset{\substack{trans\\ \text{(major)}}}{(H)(C_6H_5)C=C(CH(CH_3)_2)(H)}$$

56. **(b)** Tertiary alcohols react fastest with conc. HCl and anhydrous $ZnCl_2$ (Lucas reagent) as its mechanism proceeds through the formation of stable tertiary carbocation.

Mechanism:

Step 1: $CH_3-\underset{CH_3}{\overset{CH_3}{C}}-OH \xrightarrow{HCl}$

2 Methylpropane-2-ol

$\rightleftharpoons (CH_3)_3C-\overset{+}{O}H_2+Cl^-$

Step 2: $(CH_3)_3C-\overset{+}{O}H_2 \rightleftharpoons (CH_3)_3C^+ + H_2O$

3° Carbocation

Step 3: $(CH_3)_3C^+ + Cl^- \rightleftharpoons (CH_3)_3C-Cl$

t–Butylchloride

57. **(d)** $CH_3CH_2OH \xrightarrow{P+I_2} \underset{(A)}{CH_3CH_2I} \xrightarrow[Ether]{Mg}$

$\underset{(B)}{CH_3CH_2MgI} \xrightarrow{HCHO} \underset{(C)}{H-\underset{H}{\overset{CH_2CH_3}{C}}-OMgI} \xrightarrow{H_2O}$

$\underset{(D)}{H-\underset{H}{\overset{CH_2CH_3}{C}}-OH}$

n-Propyl alcohol

58. **(b)** Ethanol and only those 2° alcohols which contain $-CHOHCH_3$ group undergo haloform reaction.

59. **(a)** $CH_3-\overset{OH}{CH}-CH=CH-CH_3 \xrightarrow{PCC}$

Pent-3-en-2-ol

$CH_3-\overset{O}{\overset{\|}{C}}-CH=CH-CH_3$

Pent-3-en-2-one

Pyridinium chlorochromate (PCC) restricts oxidation of 1° and 2° alcohols to aldehydes and ketones respectively.

60. **(a)** 3-Methylpentan-3-ol will be dehydrated most readily since it produces a very stable, tertiary carbonium ion as intermediate.

$CH_3-CH_2-\underset{OH}{\overset{CH_3}{C}}-CH_2-CH_3 \xrightarrow{H^+}$

$CH_3-CH_2-\underset{\oplus}{\overset{CH_3}{C}}-CH_2-CH_3$

ter-Carbocation (most stable)

61. **(d)** Initiation step is formation of protonated alcohol.

$CH_3-CH_2-\ddot{O}-H + H^+ \rightleftharpoons CH_3CH_2-\overset{+}{\ddot{O}}\langle^H_H$

(Protonated ethanol)

62. **(c)** A is phenol and B is para benzoquinone.
First step is nucleophilic substitution second is oxidation.

63. **(N)** Acidic strength order:- (Actual)
B > D > C > A > E
This is so because $-OCH_3$ acts as - I group at meta position while as an EDG at para position.
Thus, Correct pKa Order:
B < D < C < A < E
All options are incorrect.

64. **(d)** Diazotization followed by hydrolysis-

Aniline (NH_2) (A) $\xrightarrow[(ii)\ H_2O,\ warm]{(i)\ NaNO_2+HCl}$ Phenol (OH) (II)

→ Strong oxidising agent

Phenol (B) $\xrightarrow[H_2SO_4]{Na_2Cr_2O_7}$ p-Benzoquinone (IV)

→ Reimer Tiemann reaction

Phenol (C) $\xrightarrow[(ii)\ H^+]{(i)\ CHCl_3+aq\ NaOH}$ Salicylaldehyde (OH, CHO) (I)

→ Kolbe's Reaction

Phenol (D) $\xrightarrow{(i)\ NaOH,\ (ii)\ CO_2,\ (iii)\ H^+}$ Salicylic acid (OH, COOH) (III)

65. **(a)** Picric acid is 2, 4, 6-trinitrophenol.

Phenol $\xrightarrow{Conc.\ H_2SO_4}$ Phenol-2,4-disulphuric acid (OH, SO_3H, SO_3H) $\xrightarrow{Conc.\ HNO_3}$ Picric acid (OH, O_2N, NO_2, NO_2)

66. (d) Phenol $\xrightarrow{CHCl_3 + aq\ NaOH}$ [Intermediate: o-(O^-Na^+)$C_6H_4CHCl_2$] $\xrightarrow{NaOH}$ o-(O^-Na^+)C_6H_4CHO $\xrightarrow{H^+}$ o-HOC_6H_4CHO

Intermediate in Reimer-Tiemann reaction is substituted benzal chloride.

67. (c) Reactions taking place as :

$C_6H_5Cl + NaOH \xrightarrow[300\ atm]{623\ K} C_6H_5ONa$ (X) $\xrightarrow{HCl}$

C_6H_5OH (Y) $\xrightarrow{conc.\ HNO_3}$ 2,4,6-trinitrophenol (Z)

68. (d) Phenol is a highly activated compound due to – OH group, it can undergo bromination directly with bromine without any Lewis acid.

69. (14) Picric acid is prepared by treating phenol first with concentrated sulphuric acid which converts it to phenol-2,4-disulphonic acid and then with concentrated nitric acid to get 2, 4, 6-trinitrophenol.

Phenol $\xrightarrow[H_2SO_4]{Conc.}$ phenol-2,4-disulphonic acid (A) $\xrightarrow[HNO_3]{Conc.}$ 2,4,6-trinitrophenol (B)

Total number of oxygen atoms in A and B = 7 + 7 = 14

70. (c) o-$C_6H_4(OH)_2$

IUPAC name : Benzene- 1,2-diol; Common name : Catechol

71. (b) Phenol (A) + Zn $\xrightarrow{\Delta}$ Benzene (III) + ZnO

Phenol (B) + $CHCl_3$ + NaOH + HCl $\longrightarrow$ Salicylaldehyde (I) (Riemer Tiemann reaction)

Phenol (C) + NaOH + CO_2 + HCl $\longrightarrow$ Salicylic acid (II)

Phenol (D) + HNO_3 (Conc.) $\longrightarrow$ Picric acid (IV)

72. (a) Phenol is more acidic than ethanol because conjugate base of phenol is more stable than conjugate base of ethanol.

73. (d) Phenol $\xrightarrow[3NaOH]{CHCl_3}$ o-HOC_6H_4CHO

(Reimer-Tiemann reaction)

74. (c) Aryl halides cannot be prepared by replacement of hydroxyl group of phenol by halogen atom because the carbon-oxygen bond in phenols has a partial double bond character and is difficult to break being stronger than a single bond. Hence, phenols cannot react with halogen acids violently.

75. (c) Phenol $\xrightarrow{CHCl_3 + aq\text{-}NaOH}$ [o-(O^-Na^+)$C_6H_4CHCl_2$] $\xrightarrow{NaOH}$ o-(O^-Na^+)C_6H_4CHO $\xrightarrow{H^+/H_2O}$ Salicylaldehyde

Reimer – Tiemann reaction

76. (b) The order of acidic strength of the given compounds is:

Ethanol (Neutral) < Phenol (Acidic) < m – Nitrophenol (more acidic due to –I effect) < p – Nitrophenol (more acidic due to –I and –M effect)

Thus, the order of pK_a values will be:

D (7.1) (III) < C (8.3) (IV) < B (10.0) (I) < A (15.9) (II)

77. (a) Acidic strength

p-nitrophenol > o-nitrophenol > m-nitrophenol

[–M and I effect] [only –I effect]

Ethanol gives Lucas test after long time as it is a 1° alcohol and Lucas test follows S_N1 mechanism.

Thus, statement I is correct but II is incorrect.

78. (d)

2,4-dinitrophenol (E) –M > p-nitrophenol (B) –M > phenol (D) > p-methoxyphenol (C) +M > Bu – OH (A)

79. (d) (A) → Kolbe Schmidt Reaction (IV)

(B) → Reimer Tiemann Reaction (III)

(C) → Oxidation of phenol to p-benzoquinone (I)

(D) → $PhOH + NaOH \rightarrow H_2O + PhO^-$

$PhO^- + CH_3-Cl \longrightarrow PhOCH_3 + Cl^-$ (II)

80. (b) Hint: $NaNO_2 + HCl \rightarrow HNO_2 + NaCl$

Nitroso group will attack at para position due to steric considerations.

Thymol $\xrightarrow[HCl]{NaNO_2}$ A (para-nitroso, ON) $\xrightarrow[H_2O]{NH_4SH}$ B (H_2N, OH)

81. (a)

HO–C₆H₄–H + HO–C₆H₄–H (X) Phenol (Carbolic acid) + phthalic anhydride $\xrightarrow[-H_2O]{Conc.\ H_2SO_4}$ Phenolphthalein (Y)

82. (a) m-nitrophenol

$-NO_2$ group has more EWG than –Cl.

83. (d) The ease of dehydration of alcohols depends on the stability of the product formed during the reaction

B > D > C > A

84. (a) Aromatic compounds burns with sooty flame

(C_6H_6O) (A) $+ Br_2 \xrightarrow{CS_2}$ p-bromophenol (Major) (B) + o-bromophenol (Minor)

85. (4) The compound must be $(C_{10}H_{12}O_2)$

(A) $(C_{10}H_{12}O_2)$ ⟶ Positive $FeCl_3$ test

⟶ Decolourises alkaline $KMnO_4$ due to – C = C – linkage

$\xrightarrow{NaOH,\ CH_3Br}$ $(C_{11}H_{14}O_2)$ $\xrightarrow[NaOH]{hot\ conc.}$ OH Rest structure $C_{10}H_{12}O_2$ (B) $+ CH_3I$

$\xrightarrow{HI} CH_3I$

The number of π bonds in compound A are 4.

86. (b) Due to $-M$ effect of $-NO_2$ group, it increases acidity $+M$ effect of $N(CH_3)_2$ decreases acidity.

Hyperconjugation and +I effect of isopropyl decreases acidity.

p-nitrophenol (–M, –I) (c) > phenol (a) > p-isopropylphenol (+I, +H) (d) > p-dimethylaminophenol (+M) (b)

∴ Order of acidic strength.

(c) > (a) > (d) > (b)

Hence order of pk_a will be (b) > (d) > (a) > (c).

87. (b) Order of acidity for following phenol is

2,4-dinitrophenol (A) > p-nitrophenol (B) > p-chlorophenol (E) > phenol (D) > 2,4,6-trimethylphenol (C)

NO_2: –M, –I (A); NO_2: –M, –I (B); Cl: +M, –I (E); (D); Me: +I (C)

– M and –I effect of the substituents increase acidity
+ M and +I effect of the substituents decrease acidity.
∴ Increasing order of pKa (A) < (B) < (E) < (D) < (C).

88. (d)

Cumene $\xrightarrow[h\nu]{O_2/Air}$ Cumene hydroperoxide $\xrightarrow[H_2O]{H^+}$ (Phenol) + CH_3COCH_3

89. (b) Phenol is acidic enough to react with strong base NaOH but not with a weak base $NaHCO_3$ solution. It gives violet colour with neutral $FeCl_3$ solution.

90. (b)

Benzene $\xrightarrow[AlCl_3]{CH_3-CH(Cl)-CH_3}$ Cumene $\xrightarrow[H^+]{O_2}$ (P) Phenol

(P) $\xrightarrow[CS_2]{Br_2}$ (B) p-bromophenol, (P) $\xrightarrow[/H_2SO_4]{Na_2Cr_2O_7}$ [1,4–Benzoquinone] (A)

91. (30) Overall yield of complete reaction

$= \frac{60}{100} \times \frac{50}{100} = 0.3 \Rightarrow \%\text{ yield} = 30\%$

92. (b) Polarity of solvent

- Monosubstituted phenol is formed in case of low polarity solvent, viz. $CHCl_3$.

Phenol $\xrightarrow[CHCl_3]{Br_2}$ p-bromophenol + o-bromophenol

- In case of bromine water, water (high polarity solvent). ionises Br –Br easily.

Phenol + $3Br_2$ $\xrightarrow{(water)}$ 2,4,6-tribromophenol + 3HBr

93. (a) Electron withdrawing group increases the acidic strength and electron donating group decreases the acidic strength. So the order is

p-nitrophenol (A) > m-nitrophenol (B) > m-methoxyphenol (C) > p-methoxyphenol (D)

(A) –M, –I; (B) –I; (C) –I; (D) +M >> –I

94. (b)

(A) Diazonium salt ($N_2^+Cl^-$) $\xrightarrow{CH_3CH_2OH}$ Benzene $\xrightarrow{Oleum}$ $C_6H_5SO_3H$ $\xrightarrow[(ii)\ HCl]{(i)\ NaOH}$ (B) Phenol

95. (c) Phenol are weakly acidic. Phenol is more acidic than alcohol and H_2O because phenoxide ion is more resonance stabilized therefore readily formed from phend. Hence statement (I) is correct but (II) is incorrect.

96. (a) (A) Phenol $\xrightarrow[\text{(Reimer-Tiemann reaction)}]{CHCl_3,\ NaOH}$ Salicylaldehyde (IV)

(B) Phenol $\xrightarrow[\text{(reduction reaction)}]{Zn}$ Benzene (III)

(C) Phenol $\xrightarrow[\text{(Oxidation reaction)}]{Na_2Cr_2O_7/H_2SO_4}$ Benzoquinone (II)

(D) Phenol $\xrightarrow[\text{(Electrophilic substitution reaction)}]{Br_2 \text{ in } CS_2}$ p-Bromophenol (I)

97. (7)

Phenol (P) $\xrightarrow{\text{dil. } HNO_3}$ o-Nitrophenol (A) (Intramolecular H-bonding) + p-Nitrophenol (B) (Intermolecular H-bonding)

Phenol (P) $\xrightarrow{\text{conc. } HNO_3}$ 2,4,6-Trinitrophenol, Picric acid (C) (Strong acid, yellow compound)

Number of oxygen atoms present in compound 'C' are 7.

98. **(c)** Para product has higher boiling point than ortho as intermolecular H-bond is possible in para, whereas intramolecular H-bond is possible in ortho product. Steam distillation can separate them as ortho product is steam volatile.

99. **(c)** Mechanism explained in **Ans. 75**

100. **(c)** Given reaction is cumene-peroxide method for the preparation of phenol. Acetone is formed as by-product in this reaction.

101. **(c)**

(A) $C_6H_5-\overset{O}{\overset{\|}{C}}-H + H_3\overset{\delta-}{C}\ \overset{\delta+}{Mg}Br \rightarrow C_6H_5-\underset{CH_3}{\overset{O^-Mg^+Br}{C}}-H \xrightarrow[\text{(ii) Oxid.}]{\text{(i) } H^+} C_6H_5-\overset{O}{\overset{\|}{C}}-CH_3$ (Acetophenone)

(B) $H_3C-\overset{O}{\overset{\|}{C}}-H \xrightarrow{\overset{\delta-}{C_6H_5}\ \overset{\delta+}{MgBr}} H_3C-\underset{C_6H_5}{\overset{OH}{C}}-H \xrightarrow{\text{PCC, DCM}} H_3C-\overset{O}{\overset{\|}{C}}-Ph$ (Acetophenone)

(C) $C_6H_5-\overset{O}{\overset{\|}{C}}-O-C_2H_5 \xrightarrow{\overset{\delta-}{CH_3}\ \overset{\delta+}{MgBr}} C_6H_5-\underset{CH_3}{\overset{O^-}{C}}-O-C_2H_5 \rightarrow C_6H_5-\overset{O}{\overset{\|}{C}}-CH_3 \xrightarrow{\overset{\delta-}{CH_3}\ \overset{\delta+}{MgBr}} C_6H_5-\underset{CH_3}{\overset{OH}{C}}-CH_3$

(D) $C_6H_5-\overset{O}{\overset{\|}{C}}-Cl + CH_3MgBr + CdCl_2 \rightarrow C_6H_5-\overset{O}{\overset{\|}{C}}-CH_3$ (Acetophenone)

$CH_3MgBr + CdCl_2 \rightarrow (CH_3)_2Cd$

102. **(d)** p-Nitrophenol $\xrightarrow[C_2H_5OH]{H_2/Pd}$ p-Aminophenol (A) $\xrightarrow{\text{1 eq. } (CH_3CO)_2O}$ p-Hydroxyacetanilide ($HO-C_6H_4-NH-\overset{O}{\overset{\|}{C}}-CH_3$) (B) (Major product) + CH_3COOH

103. **(b)** It does not have free *p*-possition.

104. (c)

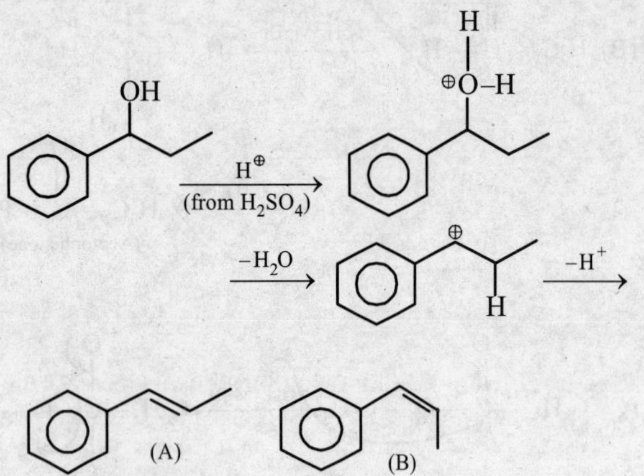

H$^{\oplus}$ (from H_2SO_4) $\longrightarrow$ $\xrightarrow{-H_2O}$ $\xrightarrow{-H^+}$

(A) [*trans*] (More stable product) (Major)

(B) [*cis*]

(Saytzeff alkenes)

105. (a) Compound (A) reacts with Lucas reagent instantly, so it should be a 3° alcohol which in turn is formed by reaction of Grignard reagent with C_8H_8O. So the latter should be a ketone.

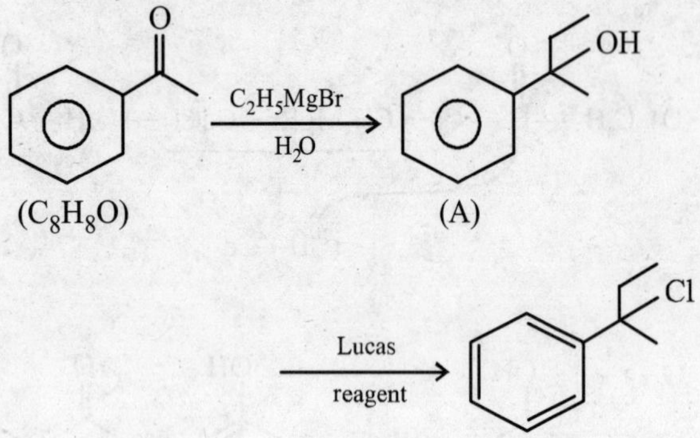

$(C_8H_8O) \xrightarrow[H_2O]{C_2H_5MgBr} (A)$

$\xrightarrow{\text{Lucas reagent}} (B)$

106. (d) $\xrightarrow{SOCl_2}$ (Major) (A)

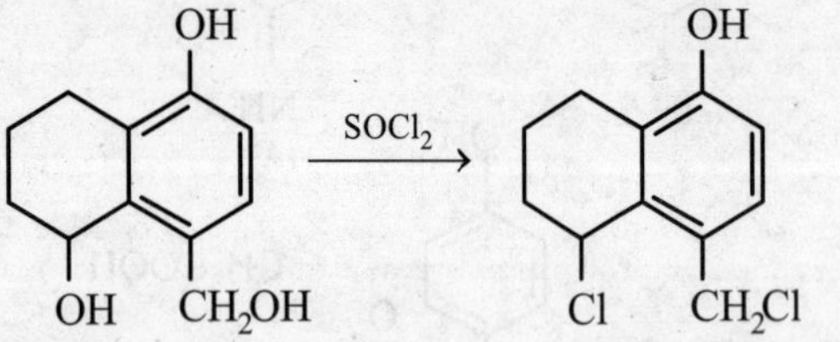

In case of –OH group present on the aromatic ring, carbon oxygen bond has partial double bond character. Hence, it is a strong bond and difficult to cleave.

107. (69) Phenol + $CHCl_3 \xrightarrow{NaOH}$ P (Salicylaldehyde)

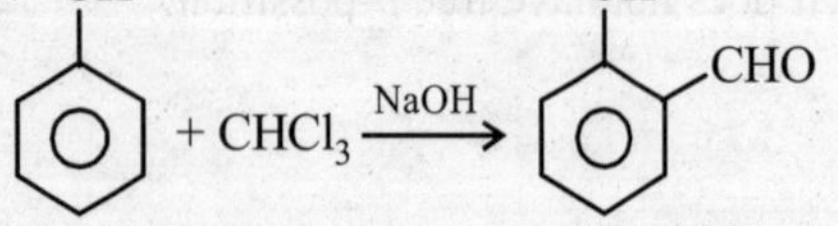

Molecular formula of product 'P' = $C_7H_6O_2$

So, mass % of C in 'P'

$$= \frac{12 \times 7}{84 + 6 + 32} \times 100 = 68.85\% \approx 69\%$$

108. (b) (II) and (III) compounds almost have same boiling point. In the given options, option (b) will be the correct answer.

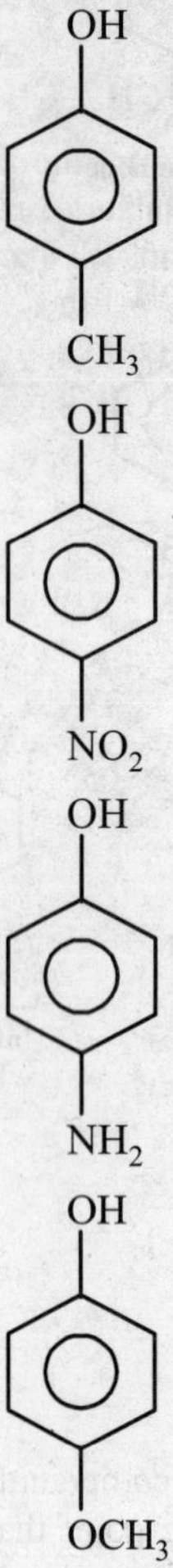

(I) B.P. → 202 °C

(II) B.P. → 279 °C

(III) B.P. → 284 °C

(IV) B.P. → 243 °C

109. (a) $\xrightarrow{\text{Chromic anhydride}}$ Product 'P'

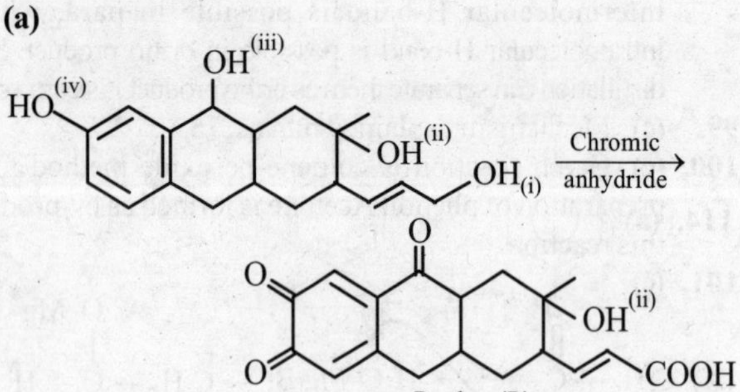

The –OH (ii), being tertiary, is not oxidised, hence it gives red colour with ceric ammonium nitrate.

110. (c) $\xrightarrow[H_2SO_4]{HNO_3}$

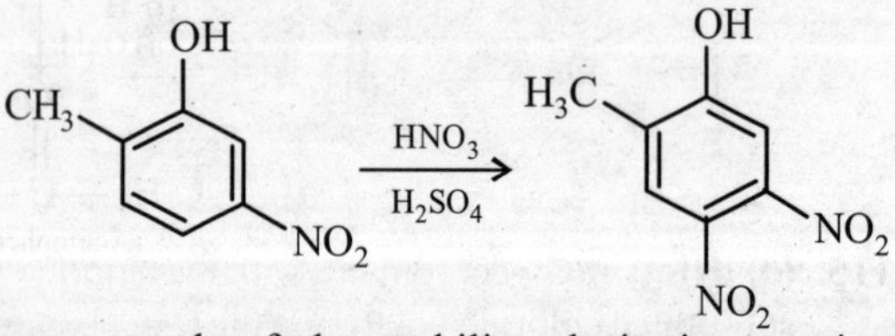

It is an example of electrophilic substitution reaction. Position of electrophile is directed by the strong ring activating group (–OH), present in the ring due to +M effect.

111. (c) Order of C–OH bond length: Phenol < p-ethoxyphenol ($O-CH_2-CH_3$) < CH_3OH

Resonance is a deciding factor to determine the order of bond length in given compounds. Phenol exhibits least C–OH bond length due to resonance, whereas methanol will show maximum bond length due to lack of resonance. *p*-Ethoxyphenol will have some intermediate value of bond length, due to conjugation with free electron pairs on O of the $-\ddot{O}CH_2CH_3$ group.

112. (a)

m-cresol (OH, H_3C) $\xrightarrow[\text{Acetone}]{K_2CO_3 \text{ in}}$ O^-K^+ (H_3C) $\xrightarrow[S_N2]{HC\equiv C-CH_2-Br}$ $OCH_2C\equiv CH$ (H_3C)

113. (b)

Cl, cyclohexenone $\xrightarrow{\text{(i) HBr}}$ Cl, Br, cyclohexanone $\xrightarrow{\text{(ii) KOH (alc.)}}$ H, dienone $\xrightarrow{\text{Tautomerization}}$ HO–phenyl (phenol)

114. (a)

OH, SO_3H $\xrightarrow{Br^+}$ $\overset{+}{O}H$, Br, $O=S(=O)-O-H$ $\xrightarrow[-H^+]{-SO_3}$ OH, Br $\xrightarrow{2Br_2}$ OH, Br, Br, Br (2,4,6-tribromophenol)

115. (d) Hint: $AlCl_3$ coordinates with the chlorine atom in the substituted chloride polarizing the C–Cl bonds and forms carbocation.

HO–phenyl–$CH_2CH_2CH=CH_2$ $\xrightarrow{\text{(1) HCl}}$ HO–phenyl–$CH_2CH_2CH(Cl)CH_3$ $\xrightarrow{\text{(2) }AlCl_3\text{(Anh.)}}$ HO–phenyl–$CH_2CH_2\overset{+}{C}HCH_3$ $+ AlCl_4^-$ ⟶ HO–(methyl indane)

(Major product)

116. (d) Electron withdrawing substituents increase the acidic strength, while electron releasing groups decrease the acidic strength.

$\therefore$ Acidic strength $\propto K_a \propto \dfrac{1}{pK_a}$

OH, NO_2 (p) > OH, NO_2 (m) > OH (phenol) > OH, OMe (p)

(B) (C) (A) (D)

pK_a : B < C < A < D

117. (b) Reaction involved:

CH_3, OH (o-cresol) $\xrightarrow{\text{(i) NaOH(aq)}}$ CH_3, O^-Na^+ $\xrightarrow{\text{(ii) }CH_3I}$ CH_3, OCH_3

118. (c) Reaction involved:

$CH_3-CH-CH_3$ (phenyl) $\xrightarrow{O_2}$ $CH_3-C(O-O-H)-CH_3$ (phenyl) $\xrightarrow{H^+}$ OH (phenyl) $+ CH_3-\overset{O}{\overset{\|}{C}}-CH_3$

Cumene — Cumene peroxide — Phenol — Acetone

119. (a)

OH (phenol) $\xrightarrow[\text{(Kolbe reaction)}]{CO_2,\ NaOH}$ OH, COOH (X) (Major) $\xrightarrow[H_2SO_4]{(CH_3CO)_2O}$ $O-\overset{O}{\overset{\|}{C}}-CH_3$, COOH

Acetyl salicylic acid (Aspirin)

120. (c) Phenol $\xrightarrow{OH^-}$ phenoxide $\xrightarrow{Cl-\overset{O}{\overset{||}{C}}-O-CH_3}$ $C_6H_5-O-\overset{O}{\overset{||}{C}}-O-CH_3$ (A) $\xrightarrow{Br_2}$ $p\text{-}Br-C_6H_4-O-\overset{O}{\overset{||}{C}}-O-CH_3$ (B)

121. (a) Cyclohexyl–O^- + BM $\xrightarrow[(-NaBr)]{\text{tert But O}^-}$ Cyclohexyl–$O-\underset{H}{\overset{H}{C}}-O-$Butyl $\xrightarrow{Br_2}$ X

Other options will lead to form a $-C=C-$ linkage which will decolorise Br_2 water color.

122. (a) 2-(2-hydroxyethyl)phenol $\xrightarrow{K_2CO_3}$

Less stable (alkoxide, $-CH_2CH_2O^{\ominus}$) + Resonance stabilized (phenoxide, $O^{\ominus}$)

Resonance stabilized phenoxide $\xrightarrow{CH_3-I}$ 2-(2-hydroxyethyl)anisole (OCH_3, CH_2CH_2OH)

123. (b) $HCN \xrightarrow[ZnCl_2]{HCl}$ $\underset{Cl}{\overset{H}{>}}C=N\overset{H}{\cdots}ZnCl_2$

Phenol + $\underset{Cl}{\overset{H}{>}}C=N\overset{H}{\cdots}ZnCl_2$ $\xrightarrow{-ZnCl_2}$ o-hydroxyphenyl–$CH=\overset{+}{N}H_2Cl^-$ $\xrightarrow[-NH_4Cl]{H_2O}$ salicylaldehyde (OH, CHO)

124. (c) Electron withdrawing group stabilises the benzene ring due to delocalisation of charge.
$-CH_3$ and $-CH_2OH$ are electron donating groups and hence, decrease the stability of benzene ring. $-OCH_3$ is a weaker electron withdrawing group than $-COCH_3$. Hence, $-COCH_3$ group at *p*-position more stabilizes the phenoxide ion.

125. (c) Due to intramolecular hydrogen bonding this will not be soluble in sodium bicarbonate.

126. (a) Sodium Phenoxide + CO_2 $\xrightarrow[5\text{ Atm}]{125\,°C}$ sodium salicylate (OH, COONa) $\xrightarrow{H^+}$ Salicyelic acid (OH, COOH)

$\xrightarrow[-CH_3COOH]{(CH_3CO)_2O}$ Aspirin (Acetyl salicyelic acid) ($OCOCH_3$, COOH)

127. (c) Electron withdrawing substituents like $-NO_2$, $-Cl$ increase the acidity of phenol while electron releasing substituents like $-CH_3$, $-OCH_3$ decrease acidity. Hence the correct order of acidity will be :

p-nitrophenol (NO_2) III (–M, –I) > p-chlorophenol (Cl) I (–I > +M) > p-cresol (CH_3) II (+I, +H) > p-methoxyphenol (OCH_3) IV (+M)

128. (a) Phenol + 3 Br_2 $\xrightarrow{\text{(water)}}$ 2, 4, 6-Tribromophenol

129. (b) Acetylation and benzoylation of phenol by acid chloride is known as Schotten-Baumann reaction.

130. (d) Substitution of –OH group attached to benzene ring will not take place due to partial double bond character.

131. (a)

132. (b) $C_2H_5-OH + H_2SO_4$

$\xrightarrow{433\text{ K}} CH_2=CH_2$ (Ethylene)

$\xrightarrow{413\text{ K}} CH_3-CH_2-O-CH_2-CH_3$ (Diethyl ether)

$\xrightarrow{383\text{ K}} CH_3CH_2HSO_4 + H_2O$ (Ethyl hydrogen sulphate)

Acetylene is not formed under any condition.

133. (b) Phenol $\xrightarrow[CO_2]{NaOH}$ salicylic acid (OH, COOH)

134. (N) Phenol on reaction with conc. H_2SO_4 gives a mixture of *o*- and *p*- products. Which on treatment with conc. HNO_3 will yield *picric acid*

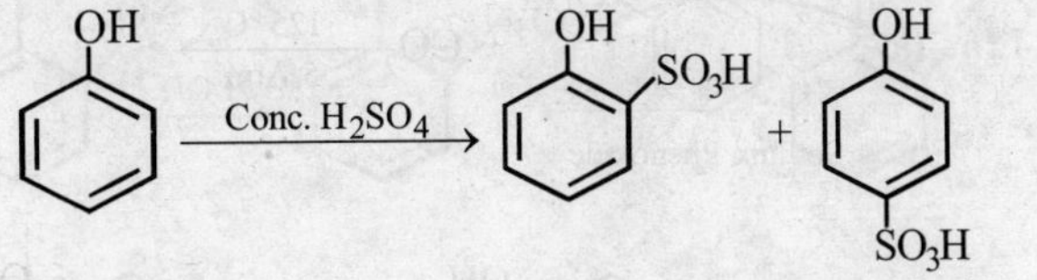

Phenol $\xrightarrow{\text{Conc. } H_2SO_4}$ o-Product (OH, SO_3H) + p-Product (OH, SO_3H)

o-Product $\xrightarrow{\text{conc. } HNO_3}$ Picric acid (OH, O_2N, NO_2, NO_2)

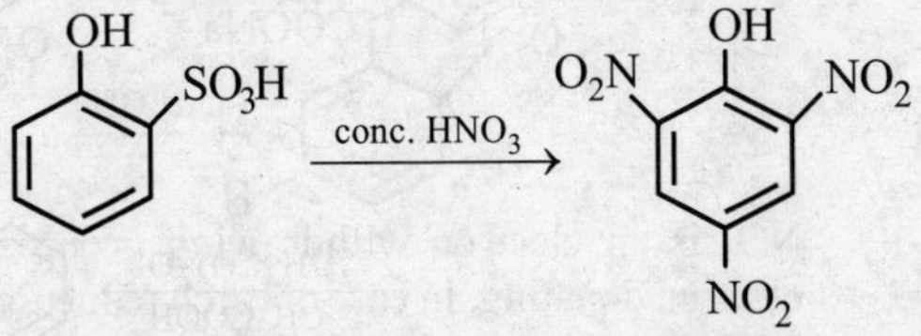

135. (d) This is Riemer-Tiemann reaction and the electrophile is dichlorocarbene.

$$H-CCl_2-Cl + NaOH \xrightarrow[(\alpha\text{-elimination})]{} Cl-\ddot{C}-Cl + NaCl + H_2O$$

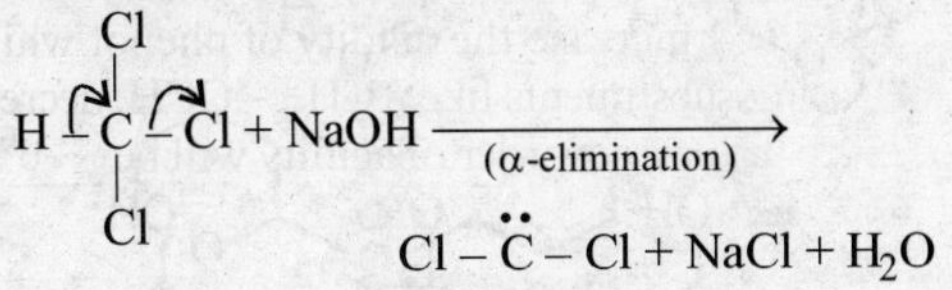

dichlorocarbene

136. (c) In (c) both groups are activating and undergo electrophilic substitution in the same positions.

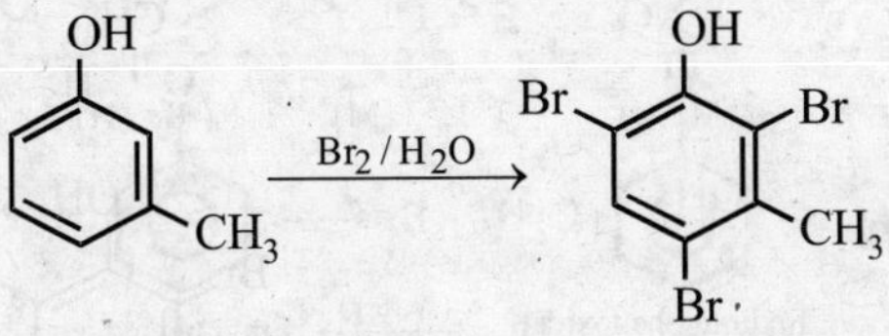

m-Cresol (OH, CH_3) $\xrightarrow{Br_2/H_2O}$ (OH, Br, Br, Br, CH_3)

137. (d) This is Riemer-Tiemann reaction and the electrophile is dichlorocarbene.

$$H-CCl_2-Cl + NaOH \xrightarrow[(\alpha\text{-elimination})]{-HCl} Cl-\ddot{C}-Cl + NaCl + H_2O$$

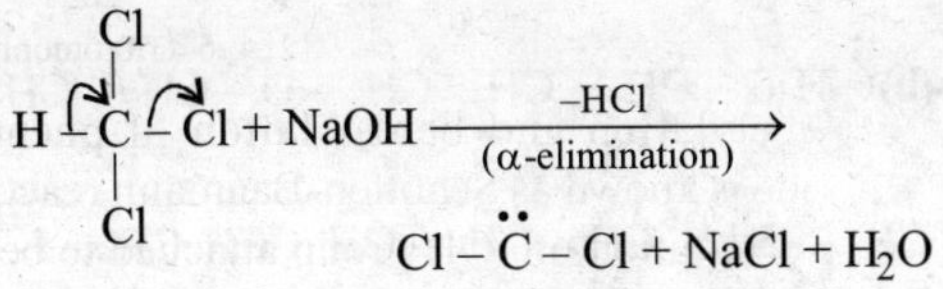

Dichlorocarbene

138. (c) 2,4,6–Trinitrophenol is also known as picric acid.

139. (d)

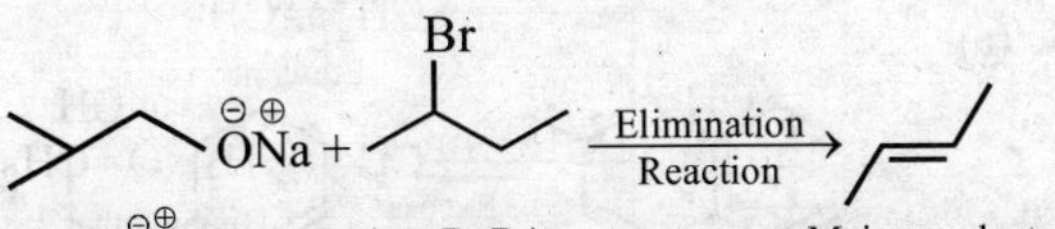

$\overset{\ominus\oplus}{ONa}$ + Br $\xrightarrow[\text{Reaction}]{\text{Elimination}}$ Major product

(iso-$Bu\overset{\ominus\oplus}{O}Na$) (sec-BuBr)

For secondary alkylhalide elimination would be more favourable than substitution.

140. (d) Statement-I: Diethylether is less soluble in water due to its larger alkylgroup.

Statement-II: Sodium metal is not used with ethyl alcohol as H_2 gas release with ethyl alcohol as below.

$$C_2H_5OH + Na \rightarrow C_2H_2ONa + \frac{1}{2}H_2$$

141. (b)

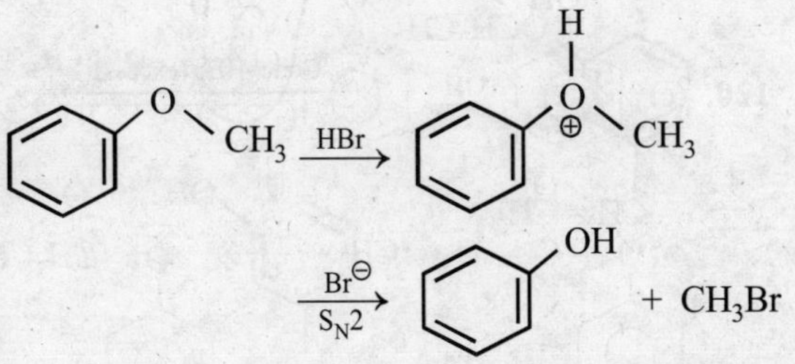

$C_6H_5-O-CH_3 \xrightarrow{HBr} C_6H_5-\overset{\oplus}{O}(H)-CH_3 \xrightarrow[S_N2]{Br^{\ominus}}$ Phenol (OH) + CH_3Br

142. (15)

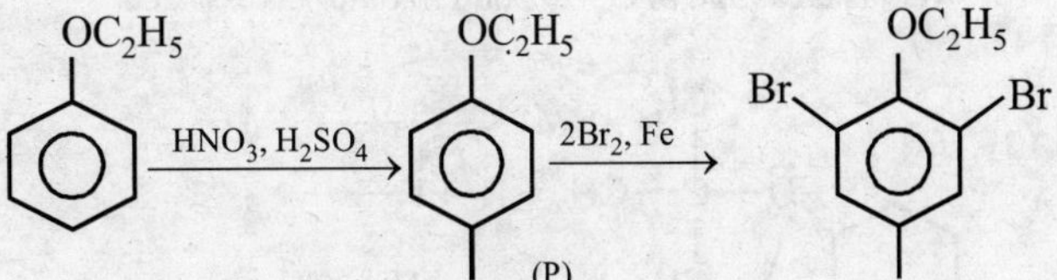

$C_6H_5OC_2H_5 \xrightarrow{HNO_3, H_2SO_4}$ (P) p-nitro (OC_2H_5, NO_2) (Major) $\xrightarrow{2Br_2, Fe}$ (Q) (OC_2H_5, Br, Br, NO_2)

Number of O-atoms in Q = 3; Number of Br-atoms = 2

$$\text{Required ratio} = \frac{3}{2} = 1.5 = 15 \times 10^{-1}.$$

143. (b)

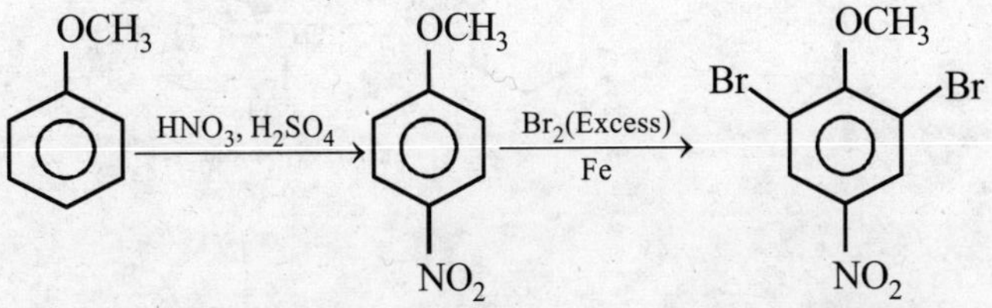

$C_6H_5OCH_3 \xrightarrow{HNO_3, H_2SO_4}$ (OCH_3, NO_2) (Major) (A) $\xrightarrow[Fe]{Br_2(\text{Excess})}$ (OCH_3, Br, Br, NO_2) (Major) (B)

Methoxy group is o, p-directing. Nitration of methoxy benzene will produce p-isomer as the major product due to steric crowding in o-isomers. Bromination of p-nitroanisole (A) with excess of Br_2 will result in disubstitution to give 2, 6-dibromo-4-nitroanisole (B) as the major product.

144. (b)

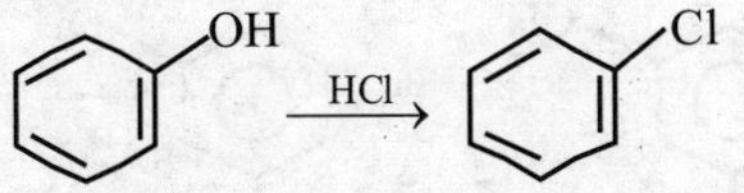

$C_6H_5OH \xrightarrow{HCl} C_6H_5Cl$

The above reaction is not possible as due to resonance C–O bond has partial double bond character. So, bond strength will increase and substitution will be less likely.

145. (a) $CH_3CH_2CH_2CH_2–OH$

Boiling point of alcohols are higher than the aldehydes having similar molecular masses due to H-bonding (intermolecular) in alcohols.

The boiling points of aldehydes are higher than hydrocarbons and ethers of comparable molecular masses.

146. (d)

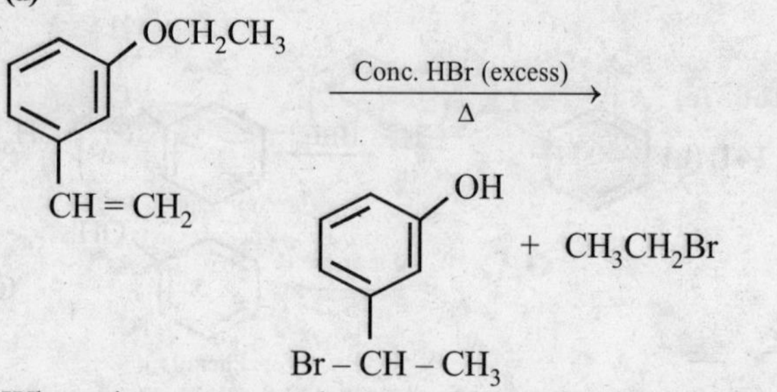

m-(OCH_2CH_3)-styrene (ring with $CH=CH_2$) $\xrightarrow[\Delta]{\text{Conc. HBr (excess)}}$ m-(HO)-C$_6$H$_4$-$CH(Br)-CH_3$ + CH_3CH_2Br

When given compound reacts with excess of conc. HBr under heat, electrophilic addition to the double bond as well as cleavage of C – O bond in ether takes place.

147. (d)

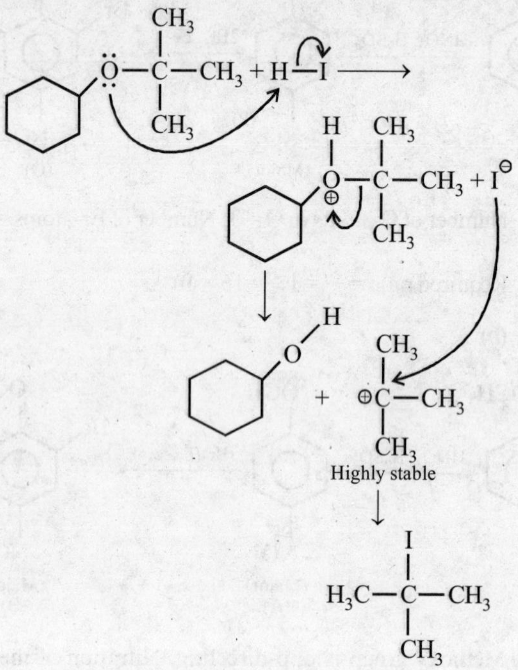

Cyclohexyl–$\ddot{O}$–$C(CH_3)_2$–CH_3 + H–I ⟶ Cyclohexyl–$\overset{\oplus}{O}(H)$–$C(CH_3)_2$–CH_3 + $I^{\ominus}$ ⟶ Cyclohexyl–O–H + $\oplus C(CH_3)_3$ (Highly stable) ⟶ $H_3C-C(I)(CH_3)-CH_3$

148. (c) $PhO^-Na^+ + Me-Br \xrightarrow{S_N2} Ph-O-Me$

Reaction (a) is not possible due to strong Ph – Br bond

Reaction (b) is not possible due to non-ionising character of the Me – OH bond.

Reaction (d) is not possible due to absence of catalyst.

149. (d)

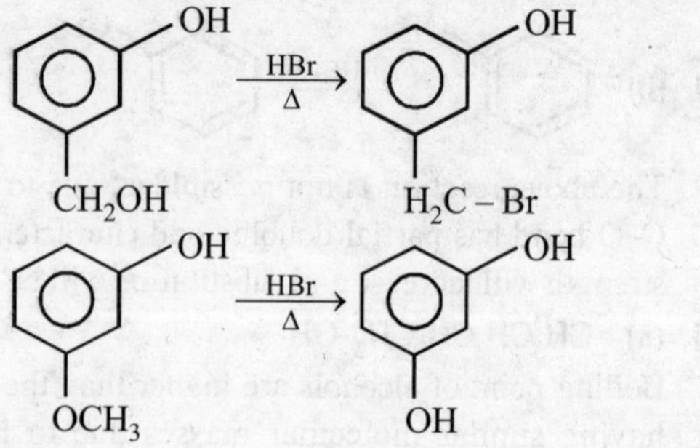

m-(CH_2OH)-phenol $\xrightarrow[\Delta]{HBr}$ m-(H_2C-Br)-phenol

m-(OCH_3)-phenol $\xrightarrow[\Delta]{HBr}$ m-(OH)-phenol

150. (d)

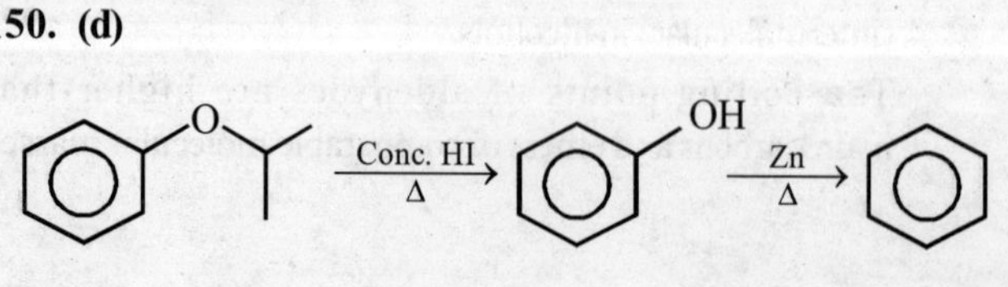

Isopropyl phenyl ether $\xrightarrow[\Delta]{\text{Conc. HI}}$ Phenol $\xrightarrow[\Delta]{Zn}$ Benzene

151. (a) **Hint:** NaI liq NH_2 is a reducing agent useful for the selective reduction of unsaturated compound.

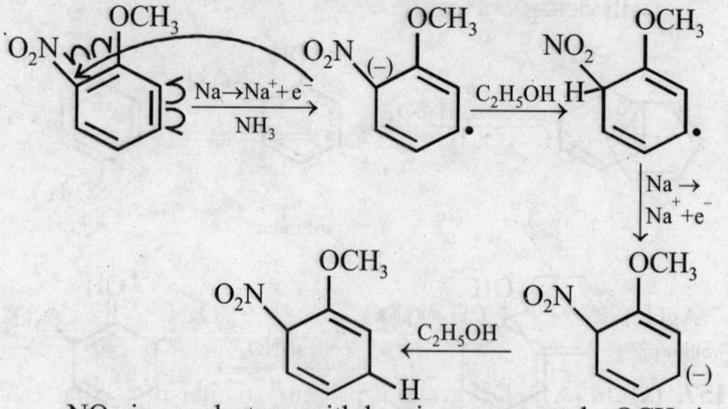

2-nitroanisole (O_2N, OCH_3) $\xrightarrow[NH_3]{Na \to Na^+ + e^-}$ radical anion $\xrightarrow{C_2H_5OH}$ radical (H added) $\xrightarrow{Na \to Na^+ + e^-}$ anion $\xrightarrow{C_2H_5OH}$ O_2N, OCH_3 cyclohexadiene (H)

NO_2 is an electron withdrawing group and $-OCH_3$ is electron donating. In case of Birch reduction, double bond is formed away from EWG and nearby EDG.

152. (d)

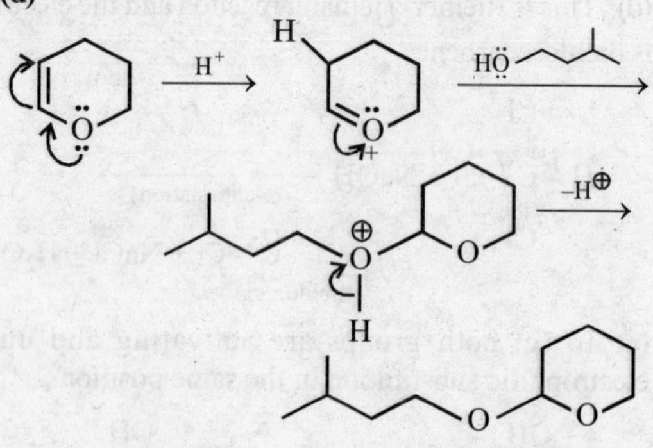

Dihydropyran $\xrightarrow{H^+}$ oxocarbenium ion $\xrightarrow{\text{HÖ (isoamyl alcohol)}}$ protonated acetal $\xrightarrow{-H^{\oplus}}$ 2-(isoamyloxy)tetrahydropyran

153. (b)

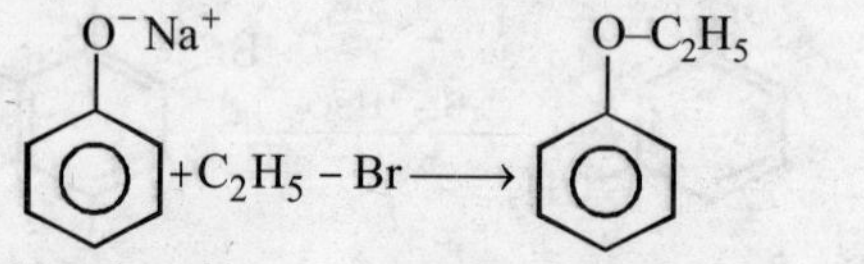

O^-Na^+ (Sodium phenoxide) $+ C_2H_5-Br \longrightarrow$ $O-C_2H_5$ (Ethyl phenyl ether)

Bromobenzene does not undergo SN reactions due to resonance which leads to partial double bond to C – Br bond.

154. (b) $H_3C-CH_2-CH(CH_3)-CH_2-O-CH_2-CH_3 \xrightarrow[Heat]{HI}$

$CH_3-CH_2-\underset{[A]}{CH(CH_3)}-CH_2-OH + CH_3CH_2-I \xrightarrow{H_2SO_4/\Delta}$ $\underset{[B]}{CH_3-CH=C(CH_3)-CH_3}$

155. (a)

2-methyl-2H-chromene $\xrightarrow[\text{Excess}]{HBr}$ (A) (phenol with Br substituents) $\xrightarrow{\text{(i) KOH (Alc.) (ii) } H^+}$ (B) (o-hydroxyphenyl diene) $\xrightarrow{O_3,\ Zn/H_3O^+}$ HCHO + CHO–CHO + (C) (salicylaldehyde, OH, CHO)

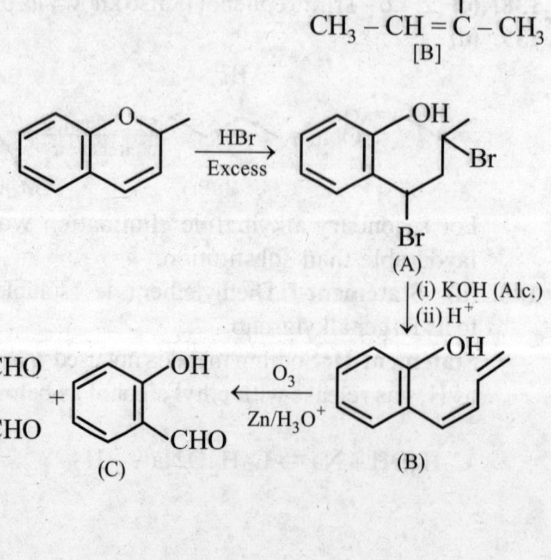

156. (c) $C_6H_5-CH_2-O-CH=CH_2 \xrightarrow{HI}$

(A) $C_9H_{10}O$

$C_6H_5-CH_2-I + CH_2=CH-OH \xrightarrow{\text{Tautomerisation}} CH_3-CH=O$

(B) (C) (D) → +ve Iodoform test

(B) $\xrightarrow{AgNO_3}$ $AgI\downarrow + C_6H_5-CH_2-ONO_2$

Yellow ppt.

157. (a) In (A), –OH group is present, so inter-molecular H-bonding is possible while in (B), due to methoxy group there is no possibility of inter-molecular H-bonding. So, A is having higher boiling point than B.

158. (b)

1-Methylethylene oxide $\xrightarrow{HBr}$ protonated epoxide, Br^- attack

$HO-CH_2-CH(CH_3)-Br \xrightarrow[SN^2]{HBr} Br-CH_2-CH(CH_3)-Br$

159. (c)

$\xrightarrow{2HI}$ (protonation of ethers, I^- attack)

$\longrightarrow$ 3-HO-C_6H_4-CH(CN)I $+ CH_3I + CH_3-CH(OH)-CH_3$

160. (c)

$Cl:\ \rightsquigarrow AlCl_3$

(Intramolecular cyclization)

$\downarrow AlCl_3$

161. (b)

3-methoxystyrene (CH=CH₂) $\xrightarrow[\text{excess}]{HBr}$ protonated ether, CH–CH₃ with Br $\xrightarrow{HBr}$ 3-(1-bromoethyl)phenol (OH, CH–CH₃, Br) $+ CH_3Br$

162. (d)

MeO-aryl-CH₂CH₂CH=CH₂ $\xrightarrow{\text{(i) } Cl_2/CCl_4}$ MeO-aryl-CH₂CH₂CHCl–CH₂Cl

$\downarrow$ (ii) $AlCl_3$ (Anhyd)

MeO-indane with CH_2Cl

163. (b) $CH_3O-C_6H_4-CH=CH-CH_3$

$\downarrow H^+$

$CH_3O-C_6H_4-\overset{+}{C}H-CH_2-CH_3$

Benzal carbocation (stable)

$\downarrow Br^-$

$H_3CO-C_6H_4-CH(Br)-CH_2-CH_3$

$\downarrow$ HBr (excess), Δ

$HO-C_6H_4-CH(Br)-CH_2-CH_3$

$+$

CH_3Br

164. (d)

MeO–C$_6$H$_4$–CH$_2$CH$_2$CH(CH$_3$)–CH$_2$–Cl $\xrightarrow{AlCl_3}$ 1° Carbocation $\xrightarrow{\text{Rearrangement}}$ 3° Carbocation ⟶ cyclized product (MeO-substituted dimethylindane)

1° Carbocation

3° Carbocation

165. (d)

$\xrightarrow{2H^+}$ $\xrightarrow{(-CH_3OH)}$

$+ H_2\overset{+}{C}CH_3 \xrightarrow{2I^-}$ $+ CH_3CH_2I$

OH OH

166. (d)

Anisole (OCH_3) + Succinic anhydride $\xrightarrow[\text{(ii) Cyclization}]{AlCl_3 \text{ (i) Friedel-Craft reaction}}$ diketone $\xrightarrow{\text{Zn – Hg, conc. HCl (Clemmensen's reduction)}}$ OCH_3-substituted tetralin

Succinic anhydride

167. (b) $C_6H_5ONa + Br-CH_2-CH=CH_2 \longrightarrow C_6H_5-O-CH_2-CH=CH_2 + NaBr$

Allyl phenyl ether

In other three options – Br is attached to sp^2 – C atom.

168. (d) This method is suitable for the preparation of a wide variety of symmetrical and unsymmetrical ethers. This nucleophilic substitution of halides with alkoxide leads to desired product.

169. (a)

$$\underset{(A)}{CH_3CH_2OCH(CH_3)_2} \xrightarrow[\text{excess}]{\text{Conc. HI}} \underset{(B)}{CH_3CH_2I} + \underset{(C)}{(CH_3)_2CHI}$$

$$(B) \xrightarrow{NaOH} CH_3CH_2OH \xrightarrow{[O]} CH_3COOH$$

$$(C) \xrightarrow{NaOH} (CH_3)_2CHOH \xrightarrow{[O]} (CH_3)_2CO$$

Hence the IUPAC name of compound (A) is

$$CH_3-CH_2-O-\overset{2}{C}H(\,^{3}CH_3)-\overset{1}{C}H_3$$

2–Ethoxypropane

170. (b) The tertiary alkyl halides undergo elimination reaction to give alkenes.

$$(CH_3)_3C-X + NaOC_2H_5 \longrightarrow CH_3-\underset{}{C}(CH_3)=CH_2$$

2-Methylpropene

171. (b) Methyl vinyl ether under anhydrous condition at room temperature undergoes addition reaction.

$$CH_2=CH-OCH_3 \xrightarrow{HBr} CH_3-\underset{Br}{\underset{|}{C}H}-O-CH_3$$

Aldehydes, Ketones and Carboxylic Acids

9

1. (a)

(A) $RCN \xrightarrow[\text{(ii) } H_3O^+]{\text{(i) } SnCl_2.HCl} RCHO$

Stephen reaction

(B) $C_6H_5COCl \xrightarrow[\text{Pd-BaSO}_4]{H_2} C_6H_5CHO$

Rosenmund reduction

(C) $C_6H_5CH_3 \xrightarrow[\text{(ii) } H_3O^+]{\text{(i) } CrO_2Cl_2,\ CS_2} C_6H_5CHO$

Etard reaction

(D) $C_6H_6 \xrightarrow[\text{(ii) anhydrous } AlCl_3/CuCl]{\text{(i) CO, HCl}} C_6H_5CHO$

Gatterman–Koch reaction

2. (d) $C_6H_5CH_3 \xrightarrow[H_3O^{\oplus}]{CrO_2Cl_2,CS_2} C_6H_5CHO \xrightarrow{NaHSO_3} C_6H_5CH(OH)SO_3Na$ (Residue) (A) $\xrightarrow{\text{dil. HCl}} C_6H_5CHO$ (B)

3. (c) This is Gattermann-Koch reaction.

$$C_6H_6 + CO + HCl \xrightarrow[CuCl]{AlCl_3} C_6H_5CHO$$

4. (b) $p\text{-}NC\text{-}C_6H_4\text{-}COOCH_3 \xrightarrow{CH_3MgBr(\text{excess})} p\text{-}[CH_3C(=\overset{-}{N}\overset{+}{Mg}Br)]\text{-}C_6H_4\text{-}C(CH_3)_2\overset{-}{O}\overset{+}{Mg}Br \xrightarrow{H_3O^+} p\text{-}[CH_3C(=O)]\text{-}C_6H_4\text{-}C(CH_3)_2OH$

5. (9) Aldol condensation reaction

$$Ph(H)C=O + CH_3-\overset{O}{\overset{\|}{C}}-CH_3 + O=C(H)Ph \xrightarrow{NaOH/\Delta} C_6H_5-CH=CH-\overset{O}{\overset{\|}{C}}-CH=CH-C_6H_5 \ (X)$$

6. (b) $C_6H_5CH_3 \xrightarrow{Cl_2/h\nu} C_6H_5CHCl_2$ (A)

$C_6H_5CHCl_2 \xrightarrow[373\text{ K}]{H_2O} C_6H_5CHO$ (B) $+ H_2O$

7. (c) Gatterman - Koch reaction.

$$C_6H_6 + CO + HCl \xrightarrow{\text{anhy. } AlCl_3/CuCl} C_6H_5CHO$$

8. (a) This reaction is known as Rosenmund reduction.

$$C_6H_5COCl \xrightarrow[\text{Pd-BaSO}_4]{H_2} C_6H_5CHO$$

9. (b)

$$C_6H_5CH_3 \xrightarrow[273-283\text{ K}]{CrO_3 + (CH_3CO)_2O} C_6H_5CH(OCOCH_3)_2 \text{ (Benzylidene diacetate)} \xrightarrow[\Delta]{H_3O^+} C_6H_5CHO$$

$$C_6H_5CH_3 \xrightarrow{CrO_2Cl_2 + CS_2} C_6H_5CH(OCrOHCl_2)_2 \text{ (Chromium complex)} \xrightarrow{H_3O^+} C_6H_5CHO$$

10. (a)

11. (a)

$$EtO-\underset{\underset{O}{\parallel}}{C}-H_2C-(\text{cyclopentene})-CH_2CH_2CN \xrightarrow{\text{(i) DiBAL–H, (ii) }H_2O} EtOH + OHC-CH_2-(\text{cyclopentene})-CH_2CH_2CHO$$

Diisobutylaluminium hydride [DiBALH or $(iBu_2)_2AlH$] reduces both cyanides as well as ester group without affecting double bond.

12. (d)

$$p\text{-}CH_3C_6H_4CH_2OH \xrightarrow[\text{[Oxidation]}]{\text{(i) }Na_2Cr_2O_7\text{, (ii) }H_2SO_4, H_2O} p\text{-}CH_3C_6H_4CHO$$

13. (c)

(A) $C_6H_5COOH \xrightarrow[\text{Quinoline}]{SOCl_2} C_6H_5COCl \xrightarrow[\text{(Rosenmund Reduction)}]{H_2/Pd/BaSO_4} C_6H_5CHO$

(B) $C_6H_5CH_2OH \xrightarrow[\text{(oxidation)}]{CrO_3/H_2SO_4} C_6H_5COOH$

(C) $C_6H_5COOCH_3 \xrightarrow{NaBH_4}$ No reduction occurs

(D) $C_6H_5CH_3 \xrightarrow[\text{Etard reaction}]{CrO_3/(CH_3CO)_2O} C_6H_5CH(OCOCH_3)_2 \xrightarrow[\Delta]{H_3O^+} C_6H_5CHO$

$NaBH_4$ is a mild reducing agent and lack, the reactivity needed to break the C = O bond in esters.

14. (c) $BH_3, H_2O_2/^-OH$ followed by PCC oxidation.

$$\text{4-methyl-1-methylenecyclohexane} \xrightarrow[\text{(Anti-Markovnikoff)}]{BH_3, H_2O_2/OH^-} \text{(4-methylcyclohexyl)}-CH_2-OH \xrightarrow{\text{PCC oxidation}} \text{(4-methylcyclohexyl)}-CHO$$

15. (c)

$$CH_3-\underset{\underset{CH_3}{|}}{CH}-CH_2-CH_2-\overset{\overset{O}{\parallel}}{C}-Cl \xrightarrow{\text{alc. }NH_3} CH_3-\underset{\underset{CH_3}{|}}{CH}-CH_2-CH_2-\overset{\overset{O}{\parallel}}{C}-NH_2$$

$$\xrightarrow{NaOH, Br_2} CH_3-\underset{\underset{CH_3}{|}}{CH}-CH_2-CH_2-NH_2 \xrightarrow{NaNO_2, HCl} CH_3-\underset{\underset{CH_3}{|}}{CH}-CH_2-CH_2-\overset{+}{N_2}\overset{-}{Cl}$$

$$\xrightarrow{H_2O} CH_3-\underset{\underset{CH_3}{|}}{CH}-CH_2-CH_2-OH \text{ (Major product)}$$

16. (a) $CH_3CH_2OH \xrightarrow[\text{Jones reagent}]{CrO_3-H_2SO_4} CH_3COOH$

17. (d) $CH_3-\underset{\underset{CH_3}{|}}{CH}-\underset{\underset{OSO_2CH_3}{|}}{CH}-CH_3 \xrightarrow{KO^tBu/\Delta} CH_3-\underset{\underset{CH_3}{|}}{CH}-CH=CH_2 \xrightarrow[\text{(oxidative ozonolysis)}]{O_3/H_2O_2} CH_3-\underset{\underset{CH_3}{|}}{CH}-COOH + HCOOH$

18. (c) (I) $C_6H_6 \xrightarrow[AlCl_3]{CO, HCl} C_6H_5CHO$ (Gattermann Koch reaction)

(II) $C_6H_5C\equiv N \xrightarrow[\text{HCl (Stephen's reduction)}]{SnCl_2 + H_3O^+} C_6H_5CH=NH \xrightarrow{H^+} C_6H_5CHO$

(III) $C_6H_5COCl \xrightarrow[\text{Quinoline, S (Rosenmund reduction)}]{H_2, Pd\text{-}BaSO_4} C_6H_5CHO$

19. (a) Cyclohexane $\xrightarrow{Br_2,\ h\nu}$ Bromocyclohexane $\xrightarrow[\Delta]{alc.\ KOH}$ Cyclohexene $\xrightarrow[Me_2S]{O_3}$ OHC–CH$_2$–CH$_2$–CH$_2$–CH$_2$–CHO $\xrightarrow[\text{(intramolecular aldol condensation)}]{OH^-,\ -H_2O}$

$\xrightarrow{cyclization}$ 2-hydroxycyclopentane-1-carbaldehyde $\xrightarrow[(-H_2O)]{\Delta}$ cyclopent-1-ene-1-carbaldehyde (CH=O)

20. (a) 4-Bromocyclohexene $\xrightarrow[(-HBr)]{ter\text{-}BuO^-}$ Conjugated diene (stable) $\xrightarrow{O_3,\ DMS}$ OHC–CH$_2$–CH$_2$–CHO + OHC–CHO

21. (b) 3-(3-hydroxyphenyl)propan-1-ol $\xrightarrow[H^+]{CrO_3}$ 3-(3-hydroxyphenyl)propanoic acid $\xrightarrow{SOCl_2}$ acid chloride (COCl) $\xrightarrow{\Delta}$ 5-hydroxyindan-1-one

22. (c) 3-Hydroxycyclohexanone $\xrightarrow{PBr_3}$ 3-Bromocyclohexanone $\xrightarrow[(-HBr)]{alc.\ OH^-}$ Cyclohex-2-enone

More stable due to conjugation

23. (c) $CH_3-\overset{O}{\overset{\|}{C}}-CH_3$ (Acetone) CH_3-CH_2-CHO (Propanal)

Generally, aldehydes are more reactive than ketones in nucleophilic addition reactions.

∴ Rate of reaction with alcohol to form acetal and ketal is

CH_3-CH_2-CHO (Propanal) > $CH_3-\overset{O}{\overset{\|}{C}}-CH_3$ (Acetone)

$$EtCHO \underset{}{\overset{H^+/MeOH}{\rightleftharpoons}} \underset{\text{Hemiacetal}}{Et(H)C(OH)(OMe)} \underset{Excess}{\overset{H^+/MeOH}{\rightleftharpoons}} \underset{\text{Acetal}}{Et(H)C(OMe)_2}$$

24. (b) Best combination is HCHO (more reactive aldehyde) and MeOH (less sterically hindered alcohol).

$$H-\overset{O}{\overset{\|}{C}}-H + CH_3OH \xrightarrow{HCl} \underset{\text{Acetal}}{H_2C(OMe)_2}$$

25. (c) $R-C\equiv N \xrightarrow[(ii)\ H_2O]{(i)\ AlH(i-Bu)_2} R-CHO$

The reduction of nitriles to aldehydes can be done using DiBAL-H[AlH(i-Bu)$_2$].

26. (a) For the given reaction condition, the major product is:

1-bromo-3-(3-bromopropyl)benzene $\xrightarrow{aq.\ KOH}$ 3-(3-bromophenyl)propan-1-ol $\xrightarrow{CrO_3/H^+}$ 3-(3-bromophenyl)propanoic acid $\xrightarrow[\Delta]{H_2SO_4}$ 5-bromoindan-1-one

27. (c) Reaction involved:

o-Cresol $\xrightarrow{AlCl_3}$ $H\overset{\oplus}{O}\overset{\ominus}{Al}Cl_3$ complex $\longrightarrow$ $H\overset{\oplus}{O}\overset{\ominus}{Al}Cl_3$ (-*m*-directing), CH_3 (*p*-directing), attack on γ-butyrolactone $\longrightarrow$ $H-\overset{\oplus}{O}\overset{\ominus}{Al}Cl_3$, CH_3, ⊕ intermediate with acyl chain ending in O^- $\longrightarrow$

$\overset{\ominus}{O}-AlCl_3$, CH_3, ⊕, H, $\ddot{O}H$ $\xrightarrow{-AlCl_3}$ $O^{\ominus}$, CH_3, $\overset{+}{O}H_2$ $\longrightarrow$ O, CH_3, H $\longrightarrow$ OH, CH_3 (tetralone product)

28. **(c)** The total number of optically active compounds formed is four. The product has two chiral C atoms. Thus, it has $2^n = 2^2 = 4$ stereoisomers.

$\xrightarrow{H^+}$ $\xrightarrow{Br^-}$

29. **(d)** $\xrightarrow[\text{(ii) Zn}-CH_3COOH]{\text{(i) }O_3-78\,°C}$ 5-Keto-2-methylhexanal

30. **(c)** $CH_3CHCl_2 \xrightarrow{\text{aq.KOH}} CH_3CH(OH)_2 \xrightarrow{-H_2O} CH_3CHO$

31. **(a)**

dil OH^-; H_2O; Δ

32. **(b)**

1, 2-dibromocyclooctane $\xrightarrow[\text{elimination}]{\text{(i) KOH (alc.)}}$ $\xrightarrow[\text{elimination}]{\text{(ii) }NaNH_2}$ $\xrightarrow{\text{(iii) }Hg^{+2}/H^{\oplus}}$ $\xrightarrow{\text{tautomerise}}$ $\xrightarrow[\text{Clemmensen Reduction}]{\text{(iv) Zn}-Hg/H^{\oplus}}$ Major Product

33. **(7)**

$Ph-\overset{O}{\overset{\|}{C}}-CH_3 \xrightarrow[\Delta]{PCl_5} Ph-CCl_2-CH_3 + POCl_3$ [A] $\xrightarrow{3eq.NaNH_2/NH_3}$

$Ph-C\equiv C^{\ominus}Na^{\oplus}$ [B] $\xrightarrow{\text{Acidify}} Ph-C\equiv C-H$ [C] $\xrightarrow[\text{2. }H_2O_2/\overset{\ominus}{O}H]{\text{1. }B_2H_6}$

$Ph-CH=CH-OH \xrightleftharpoons{\text{Tautomerisation}} Ph-CH_2-\overset{O}{\overset{\|}{C}}-H$ [D]

$\Rightarrow$ Number of sp^2 C-atoms in product D = 7

34. **(b)** P gives negative test with Tollen's reagent so it is not an aldehyde. It gives positive test with 2, 4-DNP.

$CH_3-\overset{O}{\overset{\|}{C}}-\overset{*}{C}H(CH_2-CH_3)-CH_3$ P (Chiral) $+ NH_2-NH-C_6H_3(NO_2)-NO_2$ (2, 4-DNP) $\xrightarrow{-H_2O}$ $CH_3-C(=N-NH-C_6H_3(NO_2)_2)-CH(CH_2-CH_3)-CH_3$

35. **(b)** Methyl ketones and methyl alcohols upon oxidation with NaOI/NaOH give iodoform.

36. **(b)**

$CH_3-\overset{O}{\overset{\|}{C}}-CH_3$ (X) $\xrightarrow{\text{Reduction}} CH_3-\overset{OH}{\overset{|}{C}H}-CH_3$ (Y) $\xrightarrow{HBr}$ $CH_3-\overset{Br}{\overset{|}{C}H}-CH_3$ (Z) $\xrightarrow{\text{Mg, ether}} CH_3-\overset{MgBr}{\overset{|}{C}H}-CH_3$ (Grignard reagent)

$\xrightarrow{CH_3-\overset{O}{\overset{\|}{C}}-CH_3}$ $BrMgO-C(CH_3)_2-CH(CH_3)-CH_3 \xrightarrow{\text{Hydrolysis}} CH_3-\overset{OH}{\overset{|}{C}}(CH_3)-CH(CH_3)-CH_3$

2, 3-dimethyl butan-2-ol

37. (d)

(a) (i) aq.base, (ii) Δ → (Intramolecular aldol)

(b) (i) aq.base, (ii) Δ → (Intramolecular aldol)

(c) (i) aq.base, (ii) Δ → (Intramolecular aldol)

(d) $+ H-\overset{O}{\overset{\|}{C}}-H$ (i) aq.base, (ii) Δ → (Intermolecular aldol)

38. (b) Nucleophile OH^- attacks the more reactive aldehydic group.

$Ph-\overset{O}{\overset{\|}{C}}-\overset{O}{\overset{\|}{C}}-H \xrightarrow{\overset{\oplus\ominus}{KOH}} Ph-\overset{O}{\overset{\|}{C}}-\overset{\overset{\ominus\oplus}{OK}}{\underset{OH}{C}}-H$

Hydride attack, RDS

$Ph-\underset{H}{\overset{OH}{C}}-\overset{O}{\overset{\|}{C}}-\overset{\oplus\ominus}{OK} \rightleftharpoons Ph-\underset{H}{\overset{\overset{\ominus\oplus}{OK}}{C}}-\overset{O}{\overset{\|}{C}}-O-H$

39. (b) Following compounds will not give iodoform reaction/test.

(1) Butanal $CH_3CH_2CH_2CHO$

(2) 3-Pentanone $CH_3CH_2-\overset{O}{\overset{\|}{C}}-CH_2-CH_3$

(3) Pentanal $CH_3CH_2CH_2CH_2CHO$

(4) 3-Pentanol $CH_3-CH_2-\underset{OH}{CH}-CH_2-CH_3$

40. (b) (Vanillin: $-\overset{O}{\overset{\|}{C}}-H$, HO, OCH_3)

Phenolic group soluble in NaOH

Benzaldehyde derivative reacts with Tollen's reagent.

Vanillin does not give self-aldol reaction due to lack of acidic α-H for condensation.

41. (60)

$2\,Ph-\underset{O}{\underset{\|}{C}}-H$ + (acetone) $\xrightarrow{\text{Strong base}}$ Dibenzalacetone (Product)

Benzaldehyde 5.3 g ⇓ $\frac{5.3}{106}=\frac{1}{20}$ mol

Dibenzalacetone (Product) 3.51 g ⇓ $\frac{3.51}{234}=0.015$ mol (Actual)

Theroretical yield $=\frac{1}{40}$ mol

$$\%\text{ yield}=\frac{\text{actual yield}}{\text{Theoretical yield}}\times100=\frac{0.015}{1/40}\times100=60\%$$

42. (13)

(OH, OH) $\xrightarrow[\text{(excess)}]{CrO_3}$ (diketo-aldehyde, O, O, H) $\xrightarrow[H^+]{\underset{OH}{CH_2}\ \underset{OH}{CH_2}}$ (acetal, CH) $\xrightarrow[\text{(ii) } H_3O^+]{\text{(i) } CH_3MgBr}$ (H_3C, OH, O) (R) $\xrightarrow[\text{(ii) } H_2O]{\text{(i) } NaBH_4}$ (H_3C, OH, OH) (S)

Molar mass of (S) = 130 g/mol

Mass of 0.1 mol of (S) = 0.1 × 130 = 13 g

43. (c) $\xrightarrow[\text{Clemmensen Reduction}]{\text{Zn-Hg/HCl}}$

44. (b)

S. No.	Name of Reaction	Acetaldehyde $CH_3-C(=O)-H$	Acetone $CH_3-C(=O)-CH_3$
1	Iodoform reaction	Positive	Positive
2	Cannizaro	Negative	Negative
3	Aldol reaction	Positive	Positive
4	Tollen's test	Positive	Negative
5	Clemmensen reduction	Positive	Positive

Ans. (b) A, C and E only

45. (b) $CH_3-C\equiv CH \xrightarrow{Hg^{2+}, H_2SO_4} CH_3-\overset{O}{\overset{\|}{C}}-CH_3 \xrightarrow{HCN}$

$CH_3-C(OH)(CN)-CH_3 \xrightarrow{H_2/Ni} CH_3-C(OH)(CH_3)-CH_2-NH_2$

46. (a) 1-methylcyclohexene $\xrightarrow[Zn/H_2O/\Delta]{O_3}$ 6-oxoheptanal (carbons numbered 1–7) $\xrightarrow{OH^{\ominus}/\Delta}$ 1-acetylcyclopentene (ring carbons 1–5, $-\overset{O}{\overset{\|}{C}}-CH_3$ carbons 6, 7)

47. (c) The rate of nucleophilic addition is increased by electron withdrawing group so the reactivity towards nucleophilic addition will be

(Acetophenone) < (p-tolualdehyde) < (Benzaldehyde) < (p-nitrobenzaldehyde)

48. (b) $k = 2280$ is for HCHO

$k = 2000$ is for chloral

k is much higher for both cases because HCHO and chloral are more electrophilic than R—CO—R

49. (c) Propanal undergoes aldol condensation with excess of HCHO in presence of alkali, it happens due to presence of two α-hydrogen in propanal which further react with HCHO and undergoes cannizzaro reaction.

$CH_3CH_2CH=O \xrightarrow[Alkali]{2HCHO} CH_3-C(CH_2OH)_2-CHO$

$\xrightarrow[Alkali]{HCHO} CH_3-C(CH_2OH)_2-CH_2-OH + HCOO^{\ominus}$

50. (b) $CH_3-CH=CH-\overset{O}{\overset{\|}{C}}H$ (III)

↓

Non polar R.S.

More No. of covalent bonds

$CH_3-\overset{\oplus}{C}H-CH=CH-O^{\ominus}$ (II)

↓

Having – ve charge on more electronegative atom

$CH_3-\overset{\ominus}{C}H-CH=CH-O^{\oplus}$ (I)

↓

Having – ve charge on less electronegative atom

∴ Stability order III > II > I

51. (b)

$CH_3-CH_2-C(Cl)_2-CH_3 \xrightarrow{aq.\ KOH} CH_3-CH_2-C(OH)_2-CH_3$

$\xrightarrow{-H_2O} CH_3-CH_2-\overset{O}{\overset{\|}{C}}-CH_3 \xrightarrow{KOI} CH_3-CH_2-COOK + CHI_3\downarrow$ Yellow ppt (iodoform)

52. (a) $H-\overset{O}{\overset{\|}{C}}-H$ has low steric hindrance and high partial positive charge at carbonyl carbon. Therefore it is most reactive towards nucleophilic addition reactions.

53. (b) Compounds having same number of carbon atoms follow the boiling point order as:

(Boiling point)$_{\text{Hydrogen bonding}}$ > (Boiling point)$_{\text{high polarity}}$ > (Boiling point)$_{\text{low polarity}}$ > (Boiling point)$_{\text{non polar}}$

54. (b) Reaction given is know as Wolff-Kishner reduction.

$CH_3-\overset{O}{\overset{\|}{C}}-CH_2-CH_3 \xrightarrow[\text{(ii) Ethylene glycol/ KOH}]{\text{(i) } N_2H_4} CH_3-CH_2-CH_2-CH_3$

55. (b)

Benzene + succinic anhydride $\xrightarrow{AlCl_3}$ (A) $\xrightarrow[\text{(Clemensen reduction)}]{Zn/Hg,\ HCl}$ (B) $\xrightarrow{H^+}$ α-tetralone

(ESR)

ESR: Electrophilic Substitution Reaction

56. (318) Mole of $CH_3-\overset{O}{\overset{\|}{C}}-CH$ (mol. wt = 58)

$= \frac{87}{58} = 1.5$

Mole of dibenzalacetone (mol. wt = 234)

$= \frac{351}{234} = 1.5\ \text{mol}$

$2\ C_6H_5CHO + CH_3-\overset{O}{\overset{\|}{C}}-CH_3 \xrightarrow[\Delta]{NaOH}$ $C_6H_5-CH=CH-\overset{O}{\overset{\|}{C}}-CH=CH-C_6H_5$

87g ⇒ 1.5 mol

Mol of benzaldehyde required = 1.5 × 2 = 3 mol

Mass of benzaldehyde (mol. wt. = 106) = 3 × 106 = 318 gm.

57. (d) Zn–Hg reduces both aldehyde and ketone.

3-acetylbenzaldehyde $\xrightarrow[\text{(Clemmensen reduction)}]{Zn-Hg,\ HCl}$ 1-ethyl-3-methylbenzene

58. (d) Aldehyde and ketones having acidic α-hydrogen show aldol reaction. The acidity of α-H is due to the strong electron withdrawing effect of carbonyl group.

$C_6H_5-CH=O + \overset{\alpha}{CH_3}-\overset{O}{\overset{\|}{C}}-H \xrightarrow[\Delta]{Base} C_6H_5-CH=CH-\overset{O}{\overset{\|}{C}}-H$

Benzaldehyde, Ethanal (Acidic H), Cross aldol product

So statement II is incorrect.

59. (0) Product benzene has zero hydroxyl group.

$HO-C_6H_4-CHO \xrightarrow[\text{(1 eq.)}]{PhMgBr}$ C_6H_6 + $BrMgO-C_6H_4-CHO \xrightarrow{aq.\ NH_4Cl} HO-C_6H_4-CHO$

Product (Benzenes); OMgBr; Reactant

60. (3)

$\underset{\text{Ethanal}}{CH_3(H)C=O} + \underset{\text{Semicarbazide}}{H_2N-NH-\overset{O}{\overset{\|}{C}}-NH_2} \longrightarrow$

$\underset{\text{Semicarbazone}}{CH_3(H)C=\overset{1}{N}-\overset{2}{NH}-\overset{O}{\overset{\|}{C}}-\overset{3}{NH_2}}$

61. (b)

3-chlorobenzaldehyde $\xrightarrow{50\%\ KOH}$ 3-chlorobenzoate (COO^-) + 3-chlorobenzyl alcohol (CH_2OH)

Since there are no α-hydrogens, it will undergo Cannizzaro reaction with 50% KOH and will give meta-substituted benzoic acid and benzyl alcohol.

The –Cl will not be substituted here as C – Cl bond of the ring is strong due to partial double bond character and thus will not undergo nucleophilic substitution.

62. (d)

Styrene $\xrightarrow{H_2O / H^+}$ 1-phenylethanol (OH) $\xrightarrow{CrO_3}$

Acetophenone $\xrightarrow[\text{(Wolf-Kishner reduction)}]{NH_2-NH_2}$ hydrazone ($N-NH_2$) $\xrightarrow[\Delta]{KOH}$ Ethylbenzene + N_2

63. (b)

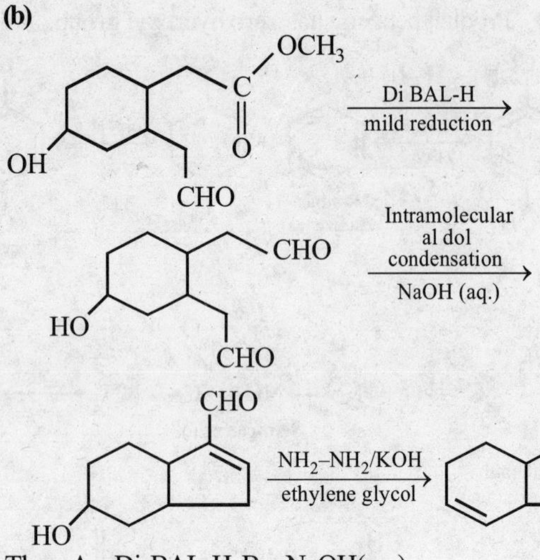

$$\xrightarrow[\text{mild reduction}]{\text{Di BAL-H}}$$

$$\xrightarrow[\text{NaOH (aq.)}]{\text{Intramolecular aldol condensation}}$$

$$\xrightarrow[\text{ethylene glycol}]{NH_2-NH_2/KOH}$$

Thus, A = Di-BAL-H, B = NaOH(aq.), C = NH_2–NH_2/KOH, ethyleneglycol.

64. (b) Keto form $\xrightleftharpoons{\text{Tautomerisation}}$ Aromatic (enol form)

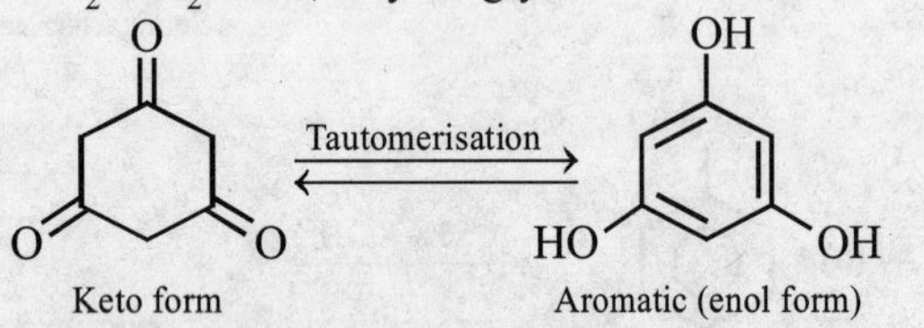

65. (d) The given reaction involves the conversion of a ketone to an acetal by hydrolysis.

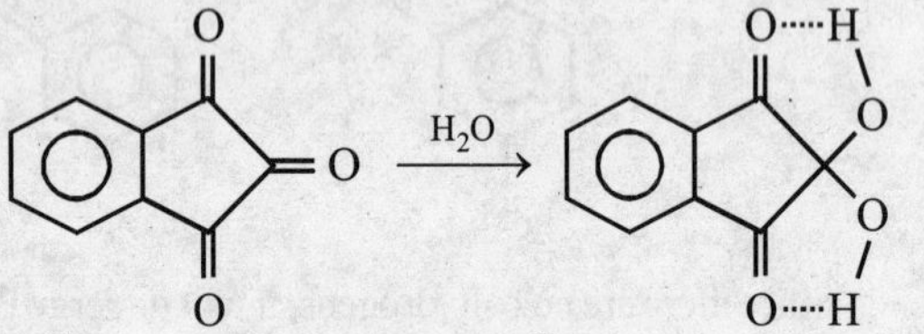

$$\xrightarrow{H_2O}$$

66. (d) $\xrightarrow{Mg}$ $\delta-$ $\delta+$ MgBr; $\xleftarrow{H_2O}$

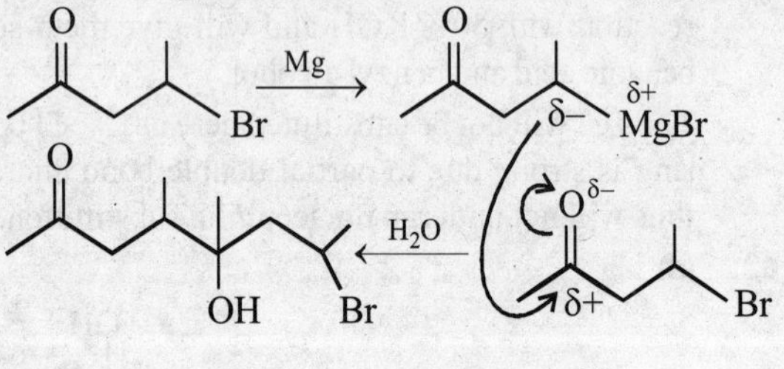

67. (d) Squaric acid $\xrightarrow[2eq.]{OH^-}$ (B) Aromatic

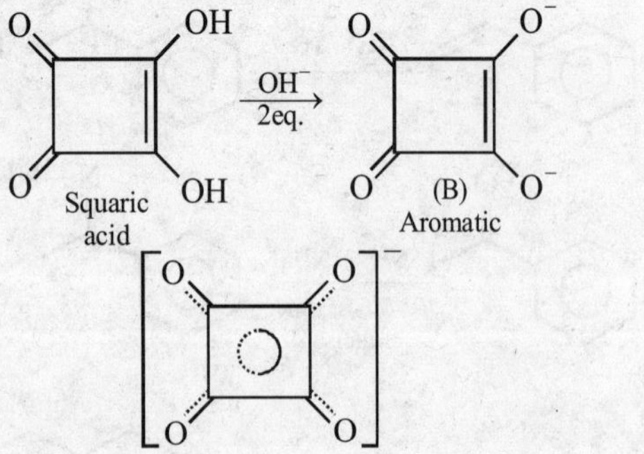

Resonance hybrid of B showing all C-C bond length same

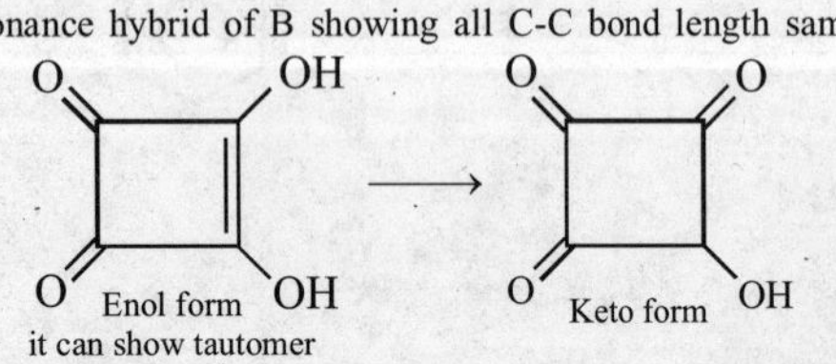

Enol form it can show tautomer $\longrightarrow$ Keto form

68. (c) $\xrightarrow{OEt^-}$

[Intramolecular alcohol condensation]

$\xleftarrow{\Delta}$

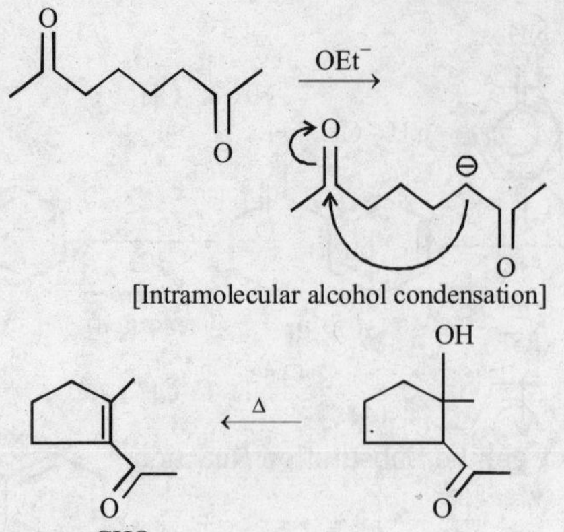

69. (60) Cyclohexanecarbaldehyde (112 kg) + $2[Ag(NH_3)_2]OH \longrightarrow$ cyclohexyl-$COONH_4$ + $3NH_3 + 2Ag + H_2O$ (51 gm)

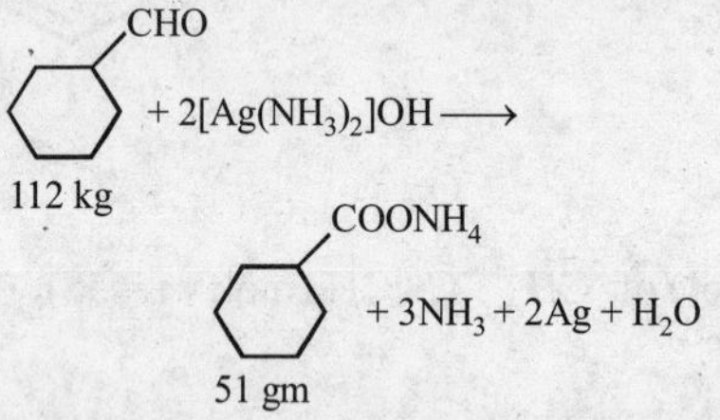

112 g cyclohexanecarbaldehyde produces 51 g NH_3

131.8 kg will produce = W_{NH_3}

$$= \frac{131.8 \times 1000}{112} \times 3 \times 17 = 60 \text{ kg}$$

70. (15) EtO–C_6H_4–CH_2–CH=O

$\downarrow NH_3/H^+ (NH_4Cl)$

EtO–C_6H_4–CH_2–CH=NH

$\downarrow$ KCN

EtO–C_6H_4–CH_2–CH($\overset{\ominus}{N}H$)(CN)

$\downarrow H_2SO_4/H^+$

EtO–C_6H_4–CH_2–CH(NH_2)(CO_2H)

$\downarrow HNO_3 + H_2SO_4$ (2eq)

EtO–$C_6H_2(NO_2)_2$–CH_2–CH(NH_2)(CO_2H)

$\downarrow Ac_2O$

EtO–$C_6H_2(NO_2)_2$–CH_2–CH(NH–CO–CH_3)(C(=O)–OH)

$\downarrow EtOH / H^+/\Delta$

EtO–$C_6H_2(NO_2)_2$–CH_2–CH(NH–CO–CH_3)(C(=O)–OEt)

$\downarrow H_2$ (Pd) / C

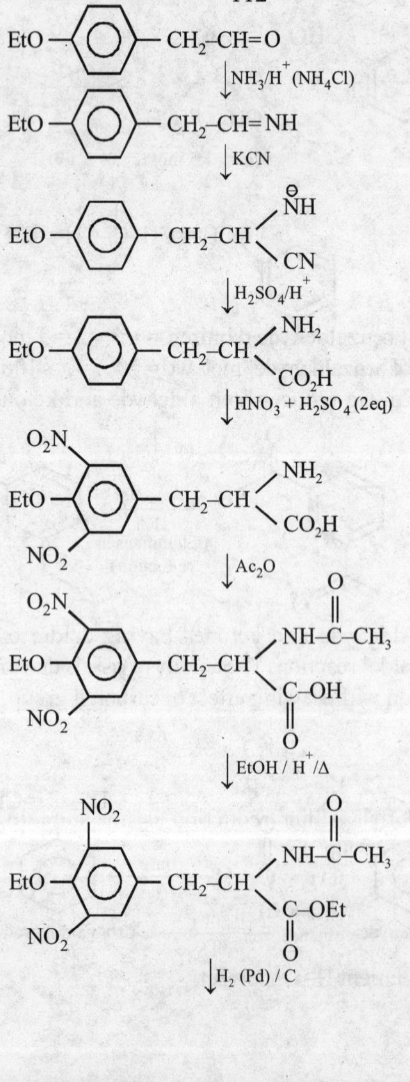

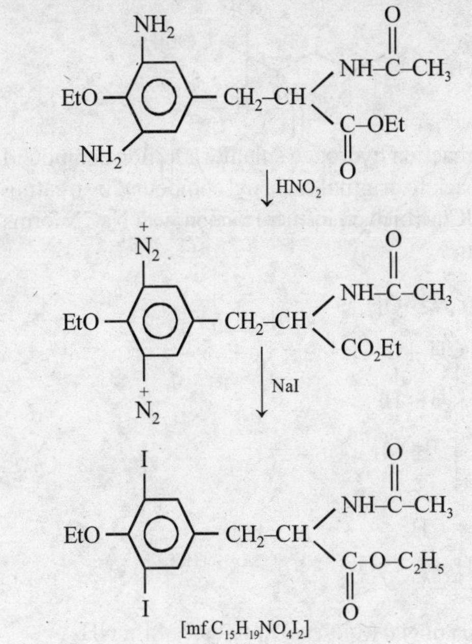

NaI

[mf $C_{15}H_{19}NO_4I_2$]

71. **(d)** KMnO$_4$, Δ oxidation

(Wolf-Kishner reduction) (i) NH_2-NH_2/OH^- (ii) H_3O^+

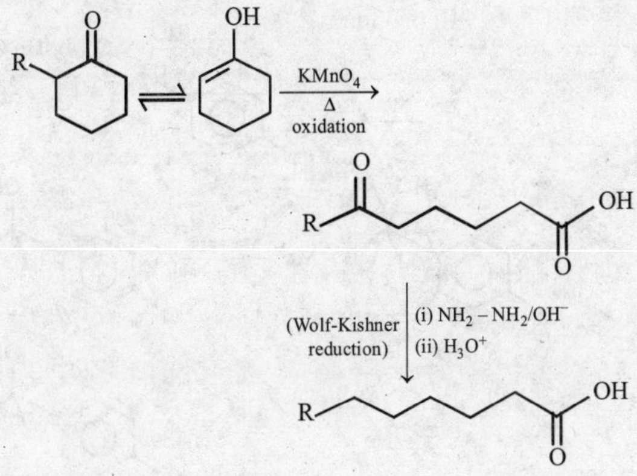

72. **(2)** y = 1 Mol

Me–Mg–Br (1 Mol)

Me–Mg–Br (1 Mol)

H_3O^+

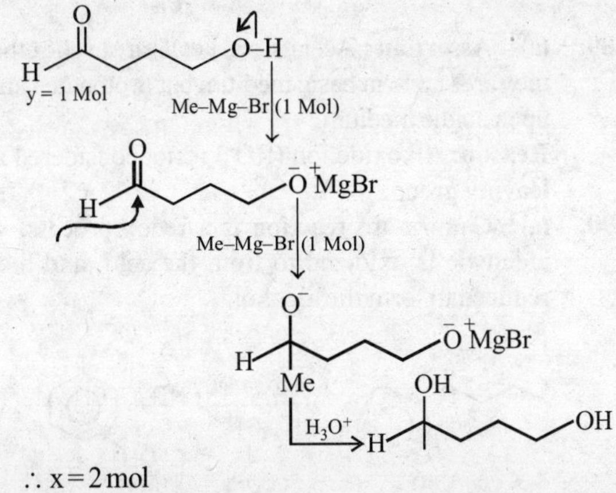

$\therefore x = 2\,mol$

$\Rightarrow \frac{x}{y} = \frac{2}{1} = 2$

73. **(c)** Wolff-Kishner reduction is not suitable for base sensitive group.

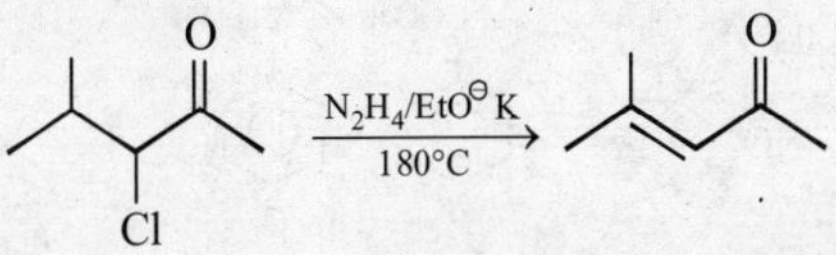

$N_2H_4/EtO^{\ominus}K$, 180°C

74. **(b)** $C_6H_5-CH(OH)-CH(CH_3)-CH_2-CHO$ $\xrightarrow{Zn(Hg)/HCl,\ \Delta}$ $C_6H_5-\overset{\oplus}{C}H-CH(CH_3)-CH_2-CH_3$ $\xrightarrow{-H^{\oplus}}$ $C_6H_5-CH=C(CH_3)-CH_2-CH_3$

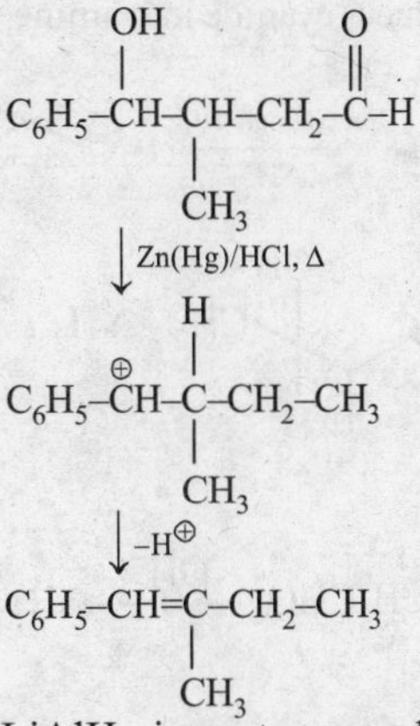

75. **(a)** $LiAlH_4$ is a strong reducing agent that reduces carbonyl and amide groups but not C = C bond.

76. **(a)** E > C > D > A > B [E is resonance stabilized with an EWG]

77. **(d)** HVZ reactions = Br_2 / red P

Iodoform reaction = NaOH + I_2

Etard reaction = (i) CrO_2Cl_2, CS_2 (ii) H_2O

Gattermann–Koch Reaction = CO, HCl, Anhydrous, $AlCl_3$

78. **(4)** (a) 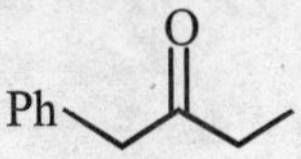 (b)

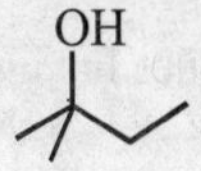

(c) 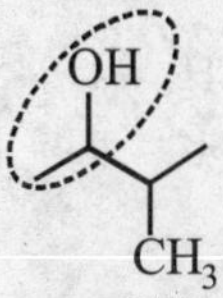 (Positive test for iodoform)

(d) 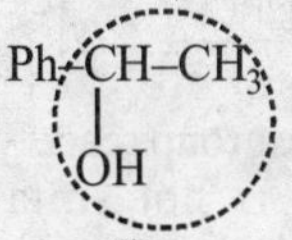 $Ph-CH(OH)-CH_3$ (Positive test for iodoform)

(e) 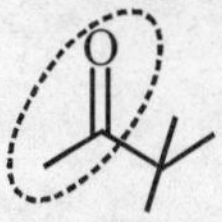 (Positive test for iodoform)

(f) 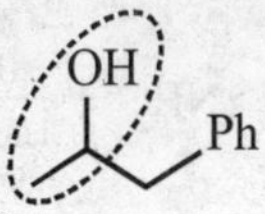 (Positive test for iodoform)

79. **(a)** $CH_3MgBr + CuI$ (1, 4-addition)

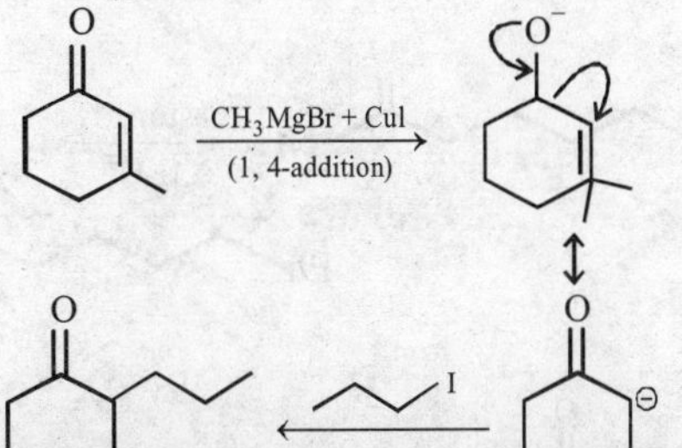

80. **(2)** As per the conditions given, the compounds do not have acidic or basic component to be dissolved in basic or acidic medium. It has at least one unsaturated bond to be hydrogenated. The molecules do not have >C = O or –CHO functional group.

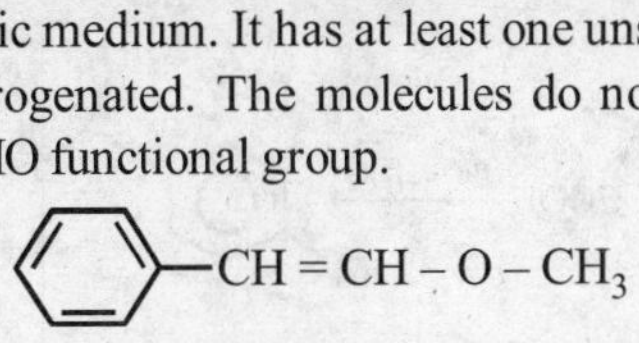

$C_6H_5-CH=CH-O-CH_3$

cis-and *trans*-isomers

81. **(d)** **Hint:** $LiAlH_4$ reduces cyanide into amine.

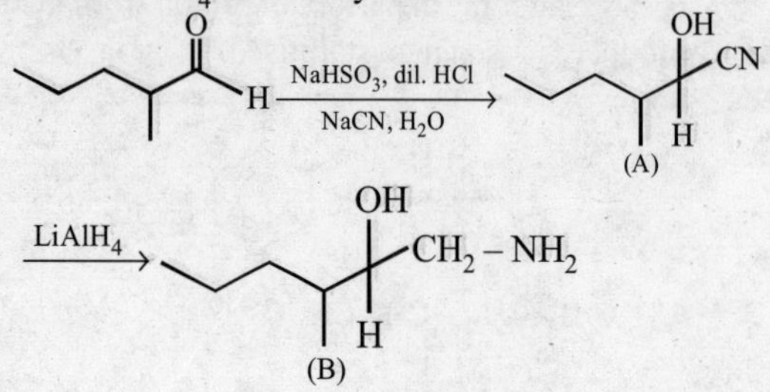

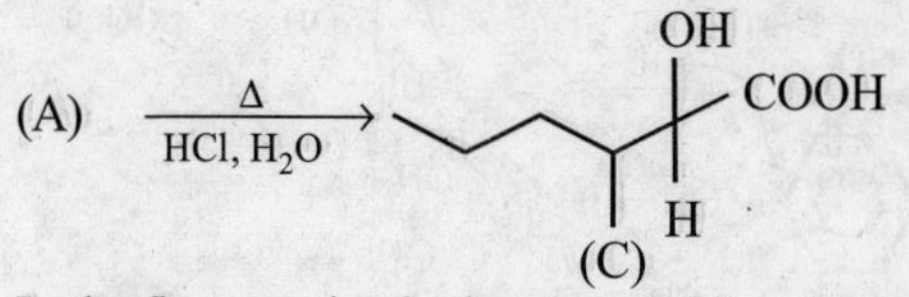

NaHSO$_3$, dil. HCl; NaCN, H_2O → (A)

(A) $\xrightarrow{LiAlH_4}$ (B) [$-CH_2-NH_2$]

(A) $\xrightarrow[HCl, H_2O]{\Delta}$ (C) [$-COOH$]

In the first reaction hydrogen sulphite addition compound converted back to original carbonyl compound by treating it with dil.HCl. which on further reaction with NaCN forms cyanohydrin.

82. **(c)** $CH_3CH_2CHO \xrightarrow{OH^-} CH_3\overset{\ominus}{C}HCHO \xrightarrow{H-C(=O)-H} H-C(=O)-CH(CH_3)-C(OH)H-H \xrightarrow{-H_2O} H-C(=O)-C(CH_3)=CH_2 \xrightarrow{NaCN} H-\overset{*}{C}(ONa)(CN)-C(CH_3)=CH_2$ (±) $\xrightarrow{H_3O^+} H-\overset{*}{C}(OH)(COOH)-C(CH_3)=CH_2$ (B) (±)

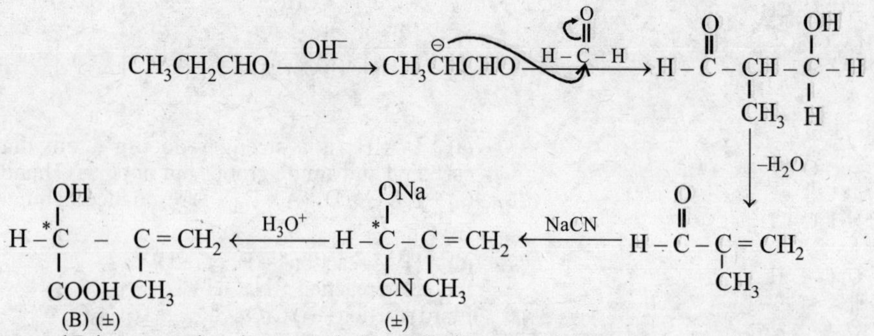

Compound B is a racemic mixture and gives off CO_2 gas with saturated solution of $NaHCO_3$.

83. **(3)** Molecules having $-\overset{O}{\overset{\|}{C}}-CH_3$ and $-\overset{OH}{\overset{|}{C}H}-CH_3$ give positive haloform test.

84. **(a)** $\xrightarrow{Zn\text{-}Hg/HCl}$

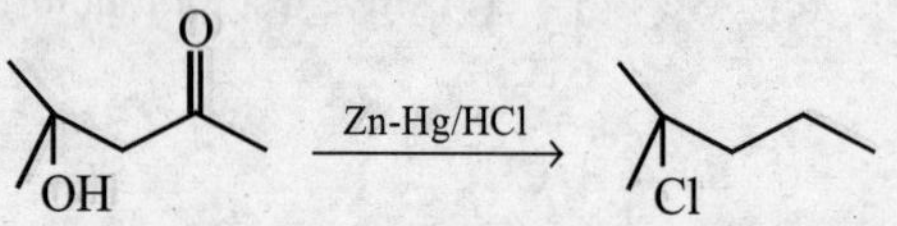

The acid sensitive alcoholic group reacts with HCl, hence Clemmensen reduction is not suitable for above conversion.

85. **(3)** Compound 2nd, 3rd, 7th.

86. **(c)** A-III, B-IV, C-I, D-II

87. **(c)** Ph–CO–(CH$_2$)$_4$–CO–NH$_2$ $\xrightarrow[\text{(Clemmensen reduction)}]{Zn/Hg, HCl}$ Ph–(CH$_2$)$_5$–CO–NH$_2$ $\xrightarrow[2.\ H_3O^+]{1.\ LiAlH_4}$ Ph–(CH$_2$)$_6$–NH$_2$

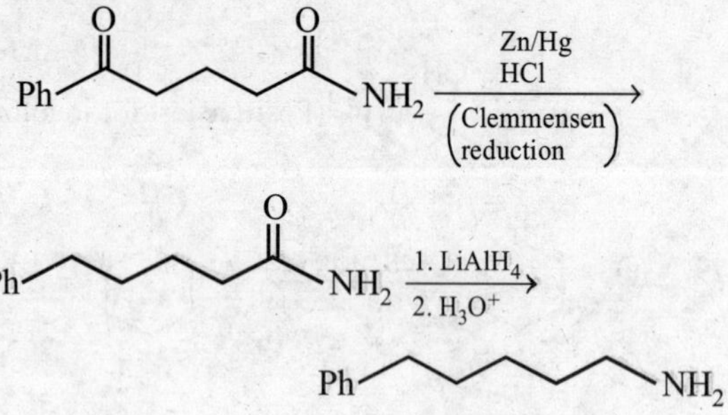

88. **(b)** $\xrightarrow{NaCN}$ ⟶ $\xrightarrow{EtOH, H_3O}$ (A) $\xrightarrow{MeMgBr}$ $\xrightarrow[-Mg(Br)OEt]{H_3O^+}$ $\xrightarrow{MeMgBr}$ $\xrightarrow{H_3O^+}$ (B)

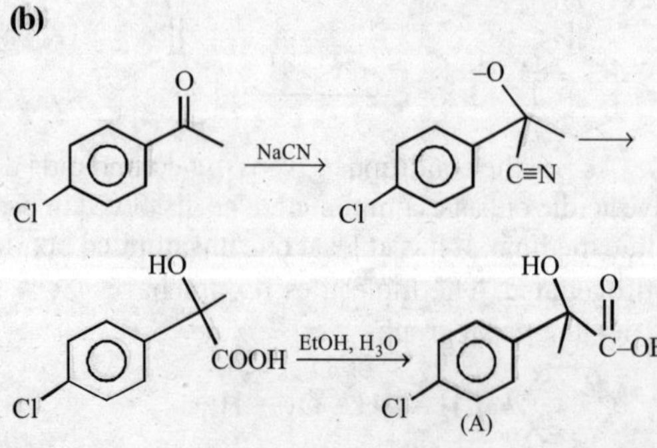

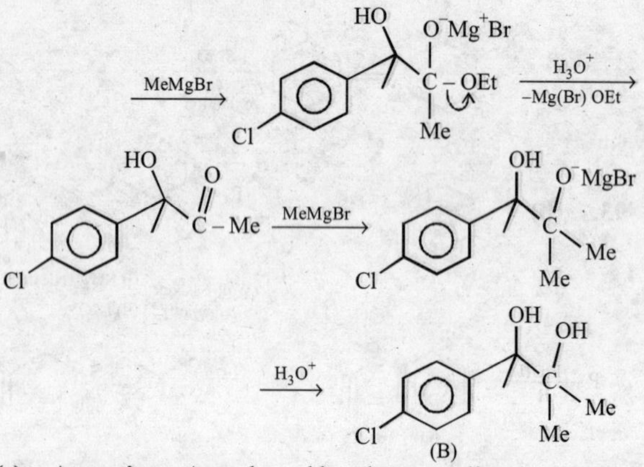

89. **(a)** **Assertion :** Acetals and ketals are cyclic ethers hence they are stable in basic medium but protonated and opened up in acidic medium.

Reason: Alkoxide ion (RO^-) is not considered as a good leaving group.

90. **(a)** Cannizzaro reaction is a redox process, where on aldehyde is oxidized to from the acid, and the other is reduced to form the alcohol.

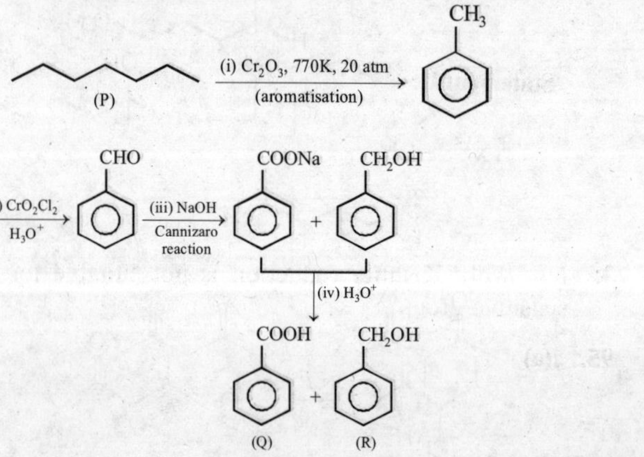

(P) $\xrightarrow[\text{(aromatisation)}]{(i)\ Cr_2O_3,\ 770K,\ 20\ atm}$ $C_6H_5CH_3$ $\xrightarrow[H_3O^+]{(ii)\ CrO_2Cl_2}$ C_6H_5CHO $\xrightarrow[\text{Cannizzaro reaction}]{(iii)\ NaOH}$ $C_6H_5COONa + C_6H_5CH_2OH$ $\xrightarrow{(iv)\ H_3O^+}$ C_6H_5COOH (Q) + $C_6H_5CH_2OH$ (R)

91. (b) Mechanism: aldehyde protonated (H^+) to give $\overset{+}{C}H(OH)$, attacked by $\ddot{O}H$ of the hydroxy acid (C(CH_3)_2(OH)COOH); then (i) $-H^+$ (ii) $-H_2O$ gives cyclic product (A).

92. (a) $CH_3-\underset{H}{\underset{|}{\overset{CH_3}{\overset{|}{C}}}}-CHO$ (X) $\xrightarrow{OH^-}$ $CH_3-\overset{CH_3}{C}^{\ominus}-CHO$ + HCHO → $CH_3-C(CH_3)(CHO)-CH_2O^-$ → $CH_3-C(CH_3)(CHO)-CH_2OH$ $\xrightarrow{KCN\ (alc.)}$ $CH_3-C(CH_3)(CH_2OH)-CH(OH)CN$ $\xrightarrow{H_3O^+}$ $CH_3-C(CH_3)(CH_2OH)-CH(OH)COOH$ $\xrightarrow{-H_2O}$ (Y)

93. (b) 4-Cl-acetophenone $\xrightarrow[HOAc]{NaCN}$ (P) (Cl-C$_6$H$_4$-C(OH)(CH$_3$)COOH)

P $\xrightarrow[H^+]{EtOH}$ (Q) Cl-C$_6$H$_4$-C(OH)(CH$_3$)-C(=O)-OC_2H_5 $\xrightarrow[(ii)\ H_3O^+]{(i)\ 2MeMgBr}$ (R)

94. (c) **Statement I:**

$H_2N-CO-CH_2CH_2-CO-CH_3$ $\xrightarrow[HCl]{Zn-Hg}$ $H_2N-CO-CH_2CH_2CH_2CH_3$ $\xrightarrow{HCl}$ HOOC-CH$_2$CH$_2$CH$_2$CH$_3$

Statement II: α-chloroketone $\xrightarrow[KOH]{H_2N-NH_2}$ chloroalkane $\xrightarrow{KOH}$ alcohol (OH) or alkene

95. (a) (A) $CH_3COCH_2COCH_3$ $\xrightarrow{B^{\ominus}}$ $CH_3CO\overset{\ominus}{C}HCOCH_3$

In structure (A), deprotonation occurs most readily because after the departure of proton from methylene, the product is resonance stabilised.

96. (d) 4-Hydroxybenzaldehyde $\xrightarrow{Br_2}$ (A) 3,5-dibromo-4-hydroxybenzaldehyde $\xrightarrow{NH_2OH}$ (B) CH=N–OH $\xrightarrow[\Delta]{P_2O_5}$ (C) CN

97. (3) $Ph-\overset{O}{\overset{||}{C}}-\underset{CN}{\underset{|}{\overset{OH}{\overset{|}{C}}}}-Ph$ (±) $\xrightarrow{HCN}$ $Ph-\overset{*}{C}(OH)(CN)-\overset{*}{C}(OH)(CN)-Ph$

Compound formed has two **similar** chiral carbon atoms, so it has 3 stereoisomers, namely RR, SS, RS.

98. (a) (A) C_6H_5COCl $\xrightarrow[Pd-BaSO_4]{H_2}$ C_6H_5CHO (Rosenmund reaction)

(B) $CH_3-C\equiv N$ $\xrightarrow[H_3O^+]{SnCl_2-HCl}$ CH_3-CHO (Stephen reaction)

(C) $C_6H_5CH_3$ $\xrightarrow[H_3O^+]{CrO_2Cl_2}$ C_6H_5CHO (Etard reaction)

(D) C_6H_6 $\xrightarrow[AlCl_3\ (anhyd)]{CO,\ HCl}$ C_6H_5CHO (Gattermann Koch reaction)

99. (c) $R-CO-CH_3$ $\xrightarrow{NaOH}$ $R-CO-\overset{\ominus}{C}H_2$ $\xrightarrow{CH_3CH_2CH_2Br}$ (A) $R-CO-CH_2CH_2CH_2CH_3$

100. (c) 4-formylcyclohexanone (H, CHO) $\xrightarrow{[Ag(NH_3)_2]^+\ OH^-}$ (A) (H, COOH) $\xrightarrow{NaBH_4}$ (It does not reduce –COOH gp) (B) 4-hydroxycyclohexanecarboxylic acid (H, OH; H, COOH)

101. (d) $Ph-\overset{O}{\overset{\|}{C}}-CH_3 \xrightarrow{\overset{\ominus}{Ph}MgBr} Ph-\overset{O-\overset{+}{Mg}Br}{\overset{|}{C}}(Ph)-CH_3 \xrightarrow{H^+, H_2O} Ph-\overset{OH}{\overset{|}{C}}(Ph)-CH_3$

102. (b) (formation of enolate ion) $\xrightarrow{OH^-}$ (aldol condensation) $\xrightarrow{H_2O}$

It does not contain α-hydrogen so no dehydration takes place.

103. (d) Fehling reagent (aq. $CuSO_4$ + alkaline sodium potassium tartarate) can not oxidise aromatic aldehydes.

104. (b) COOH $\xrightarrow{KOH}$ COOH $\xrightarrow{H_3O^+}$ COOH, OH $\xrightarrow{KOH}$ COOH

105. (a) C_6H_5CHO does not have any α-hydrogen. So it will give Cannizzaro reaction. One molecule is oxidised to acid and other is reduced to alcohol.

$$PhCH=O+PhCH=O\xrightarrow{OD/D_2O}PhCH_2OD+PhCO_2^-$$

106. (1) Zymase converts sugar molecule to alcohol and carbon dioxide.

$$C_6H_{12}O_6 \xrightarrow{Zymase} \underset{(A)}{2CH_3CH_2OH} + CO_2$$

In presense of NaOI, alcohol is oxidised to aldehyde.

$$CH_3-CH_2-OH \xrightarrow[(Oxidation)]{NaOI} \underset{(A)}{CH_3-\overset{O}{\overset{\|}{C}}-H} \longrightarrow \underset{\text{(yellow ppt.)}}{CHI_3} + \underset{(B)}{HCOONa}$$

107. (8) HO–cyclohexane $\xrightarrow{K_2Cr_2O_7}$ cyclohexanone (δ^- O=C δ^+) $\xrightarrow[H_2O]{C_6H_5MgBr}$ 1-phenylcyclohexanol (HO, C_6H_5) $\xrightarrow{H^+/Heat}$ (X)

108. (4) cyclohexanone $\xrightarrow{OH^-}$ enolate $\xrightarrow[\text{(aldol condensation)}]{\text{cyclohexanone}}$ (HO, H) $\xrightarrow{\Delta}$ (P)

109. (c) $CH_3-\overset{CH_3}{\overset{|}{C}H}-\overset{O}{\overset{\|}{C}}-H \xrightarrow[Base]{K_2CO_3} CH_3-\overset{H_3C}{\overset{|}{\underset{\ominus}{C}}}-\overset{O}{\overset{\|}{C}}-H \xrightarrow{H-\overset{O}{\overset{\|}{C}}-H}$

$\underset{(A)}{CH_3-\overset{H_3C}{\overset{|}{C}}(CH_2-OH)-\overset{O}{\overset{\|}{C}}-H} \xrightarrow{K\overset{\ominus}{C}N} \underset{(B)}{CH_3-\overset{H_3C}{\overset{|}{C}}(CH_2-OH)-\overset{O^-}{\overset{|}{C}H}-CN} \xrightarrow{H_3O^+} CH_3-\overset{CH_3}{\overset{|}{C}}(CH_2-\ddot{O}H)-\overset{OH}{\overset{|}{C}}(H)-\overset{O}{\overset{\|}{C}}-OH \longrightarrow$ Stable product (C)

110. (a) $\overset{\delta^+}{BrMg}$ δ^- ... O=C(CH_3)–CH_2–C(CH_3)=O $\longrightarrow$ $O^{\ominus}\overset{\oplus}{Mg}Br$ ... $\xrightarrow{H_2O}$ OH, OH

111. (c) (cyclohexane-1,3,5-trione) is tautomer of (1,3,5-trihydroxybenzene; HO, OH, OH) (aromatic, hence stable),

Further it has three (max.) enolic group.

112. (c)

$$C_6H_5-\overset{O}{\overset{\|}{C}}-OH+HO-\overset{O}{\overset{\|}{C}}-H \xrightarrow[\Delta]{MnO} C_6H_5-\overset{O}{\overset{\|}{C}}-H+CO_2+H_2O$$

113. (a) $CH_3CH_2-\overset{O}{\overset{\|}{C}}-CH_3 \xrightarrow{HCN} CH_3CH_2-\overset{CN}{\overset{|}{\underset{OH}{\underset{|}{C}}}}-CH_3 \xrightarrow[\text{(Hydrolysis)}]{95\%\ H_2SO_4}$

$CH_3CH_2-\overset{COOH}{\overset{|}{\underset{OH}{\underset{|}{C}}}}-CH_3 \xrightarrow[(-H_2O)]{\text{Heat}} CH_3-CH=\overset{CH_3}{\overset{|}{C}}-COOH$ (A)

114. (c) Enamine formation is an example of nucleophilic addition elimination reaction.

$(H_3C)_2(CH_3)C-\overset{O}{\overset{\|}{C}}-C(CH_3)_3$

Carbonyl group is highly sterically hindered, hence attack of nuclephile will not be possible.

115. (c) O=⟨cyclohexa-2,5-diene⟩=O is a conjugated diketone.

116. (3) [Structure: naphthalene bearing $-CH_2-CH=CH-CO-CH_3$ and CH_2CN; arrows mark 3 sites]

117. (c) $CH_3-CH=CH-CH_2-\underset{OH}{\underset{|}{CH}}-CH_3$

$\downarrow$ PCC

$CH_3-CH=CH-CH_2-\underset{O}{\underset{\|}{C}}-CH_3$ (A)

$\downarrow$ NaOI (Haloform reaction)

$CH_3-CH=CH-CH_2-COOH+CHI_3$ (B)

$\downarrow$ NaOH + CaO, $-CO_2$

$CH_3-CH=CH-CH_3$ (C)

But-2-ene

118. (b) Statement I : Correct

$>C=O \xrightarrow{NaHSO_3} >C(\overset{-}{O}Na^+)(OSO_2H) \xrightleftharpoons{\text{Proton transfer}} >C(OH)(OSO_2^-Na^+)$

Bisulphite (stable anion)

Statement II : Wrong

$>C=O \xrightarrow{HCN} >C(OH)(CN)$

Cyanohydrin

119. (c) $OHC-C_6H_4-CH_2OH \xrightarrow[\Delta]{NaOH} HOCH_2-C_6H_4-CH_2OH$ (Diol) $+ Na^+\,{}^-OOC-C_6H_4-CH_2OH \xrightarrow{H_3O^+} HOOC-C_6H_4-CH_2OH$

Monocarboxylic acid

120. (a) [P] $CH_3-CH=CH-CO-CH_3 \xrightarrow[H_3O^+]{NaOCl} CH_3-CH=CH-COOH + CHCl_3$

Haloform reaction

121. (c) $CH_3-\underset{CH_3}{\underset{|}{CH}}-CH_2-CH_2-\overset{O}{\overset{\|}{C}}-Cl \xrightarrow{\text{alc. } NH_3} CH_3-\underset{CH_3}{\underset{|}{CH}}-CH_2-CH_2-\overset{O}{\overset{\|}{C}}-NH_2$

$\xrightarrow[\text{(Hoffmann degradation)}]{NaOH,Br_2} CH_3-\underset{CH_3}{\underset{|}{CH}}-CH_2-CH_2-NH_2 \xrightarrow{NaNO_2,HCl} CH_3-\underset{CH_3}{\underset{|}{CH}}-CH_2-CH_2-\overset{+}{N_2}\overset{-}{Cl}$

$\xrightarrow{H_2O} CH_3-\underset{CH_3}{\underset{|}{CH}}-CH_2-CH_2-OH$

(Major product)

122. (a) [Tetralone with $N\equiv C-$ substituent, CH_3 groups] $\xrightarrow{2\overset{\ominus\delta}{C}H_3\overset{\oplus\delta}{Mg}Br}$ $Br\overset{\oplus}{Mg}\overset{\ominus}{N}=C(CH_3)-$[ring with CH_3, $\overset{\ominus}{O}\overset{\oplus}{Mg}Br$, CH_3, CH_3]

$\xrightarrow{H_3O^\oplus}$ CH_3CO-[ring with CH_3, OH, CH_3, CH_3] $\xrightarrow{H_2SO_4,\ \Delta(-H_2O)}$ CH_3CO-[dihydronaphthalene with CH_3, CH_3, CH_3]

123. (c) $CH_3CH_2CH(CH_3)CHO \xrightarrow[HCN/H_2O]{NaOH(X)} CH_3CH_2CH(CH_3)CH(OH)C\equiv N \xrightarrow{LiAlH_4} CH_3CH_2CH(CH_3)CH(OH)CH_2NH_2$

(Major product) (Y)

124. (c) Aldehydes give positive Tollen's test (Silver mirror test).

(I) 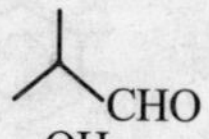 (isobutyraldehyde, CHO) gives Tollen's test.

(II)  OH (enol) $\underset{}{\overset{\text{Tauto}}{\rightleftharpoons}}$ O (Ketone) will not give Tollen's test.

(III) 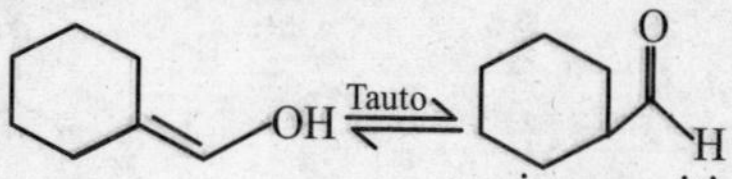 (cyclohexylidene)–OH $\overset{\text{Tauto}}{\rightleftharpoons}$ (cyclohexyl)–CHO (O, H)

gives positive Tollen's test.

(IV) 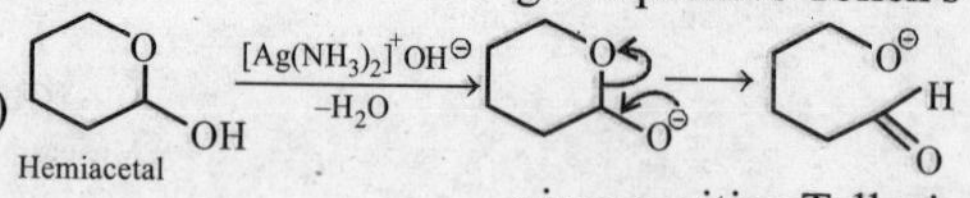 Hemiacetal (O, OH) $\xrightarrow[-H_2O]{[Ag(NH_3)_2]^+OH^\ominus}$ (O, $O^\ominus$) ⟶ ($O^\ominus$, H, O)

gives positive Tollen's test.

125. (3) 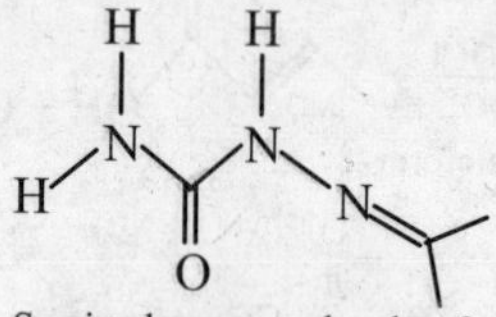 H, H, H–N, N, N, O

Semicarbazone molecule of acetone

126. (2) Two half reactions are

$$[Ag(NH_3)_2]^+ + e^- \longrightarrow Ag + 2NH_3$$

$$RCHO + 3OH^- \longrightarrow RCO_2^- + 2H_2O + 2e^-$$

Overall reaction is

$$2[Ag(NH_3)_2]^+ + RCHO + 3OH^- \longrightarrow 2Ag + RCO_2^- + 4NH_3 + 2H_2O$$

Therefore, for 1 mol of R–CHO, 2 mol of $[Ag(NH_3)_2]^+$ is reduced.

127. (5) Structure of mesityl oxide is

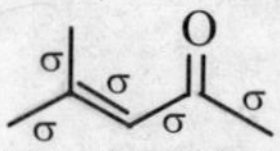

(σ σ σ σ σ, O)

Number of C – C sigma bonds = 5

128. (c) Di-isobutyl aluminium hydride (DiBAL-H) selectively reduces nitriles and esters to aldehydes and it does not reduce double bond.

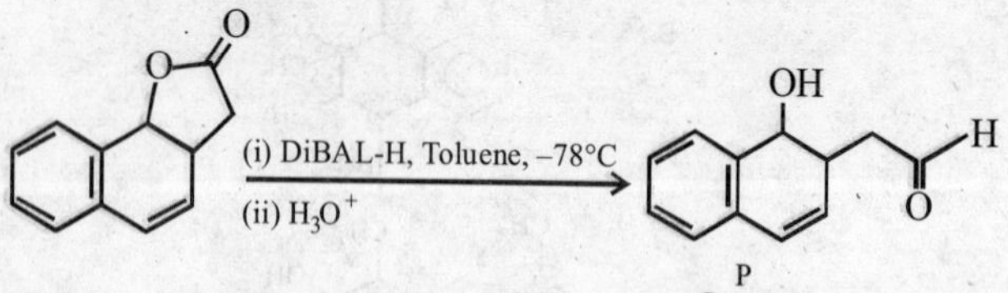

(lactone, O, O) $\xrightarrow[\text{(ii) } H_3O^+]{\text{(i) DiBAL-H, Toluene, –78°C}}$ (OH, H, O) P

129. (d) 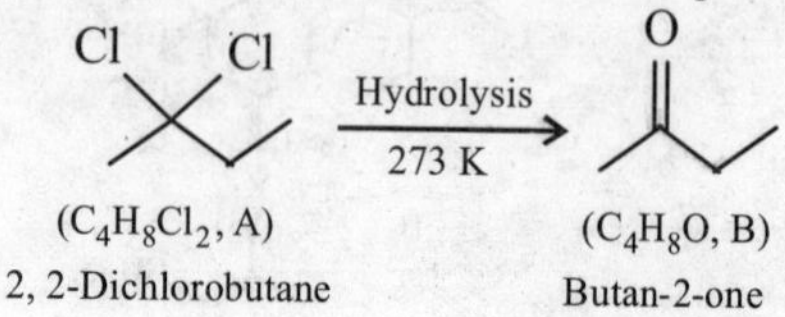 Cl Cl $\xrightarrow[\text{273 K}]{\text{Hydrolysis}}$ O

$(C_4H_8Cl_2, A)$ 2, 2-Dichlorobutane (C_4H_8O, B) Butan-2-one $\xrightarrow{Ag(NH_3)_2OH}$ No reaction

Ketones react with hydroxyl amine to give oxime.

130. (a) 2, 4-Dinitrophenylhydrazine reacts with aldehydes and ketones to form 2, 4-dinitrophenylhydrazone derivatives. It is a condensation reaction with elimination of water.

131. (c)

$$CH_3-CH_2-CH_3 \xrightarrow{Mo_2O_3} CH_3-CH_2-CHO$$

132. (c)

(A) $R-\overset{O}{\overset{\|}{C}}-Cl \xrightarrow[Pd-BaSO_4]{H_2} R-\overset{O}{\overset{\|}{C}}H$

(Rosenmund reduction)

(B) $R-CH_2-COOH \xrightarrow[H_2O]{Cl_2/P} R-\underset{Cl}{\underset{|}{CH}}-COOH$

(HVZ reaction)

(C) $R-\overset{O}{\overset{\|}{C}}-NH_2 \xrightarrow[NaOH]{Br_2} R-NH_2$

(Hoffmann bromamide reaction)

(D) $R-\overset{O}{\overset{\|}{C}}-CH_3 \xrightarrow[Conc.HCl]{Zn(Hg)} R-CH_2-CH_3$

(Clemmensen reduction)

133. (c) Reaction of $HgSO_4$/dil. H_2SO_4 with alkyne results in addition of water as per Markonikoff's rule.

(a) $HC\equiv CH \xrightarrow[H_2SO_4]{HgSO_4} CH_2=\underset{OH}{\underset{|}{CH}} \rightleftharpoons CH_3-\overset{O}{\overset{\|}{C}}H$

(b) 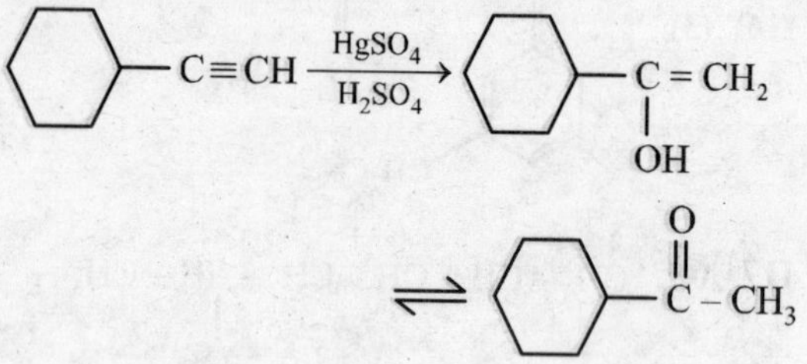 (cyclohexyl)$-C\equiv CH \xrightarrow[H_2SO_4]{HgSO_4}$ (cyclohexyl)$-\underset{OH}{\underset{|}{C}}=CH_2$

$\rightleftharpoons$ (cyclohexyl)$-\overset{O}{\overset{\|}{C}}-CH_3$

(c) $CH_3-C\equiv CH \xrightarrow[H_2SO_4]{HgSO_4} CH_3-\underset{OH}{\underset{|}{C}}=CH_2$

$\rightleftharpoons CH_3-\overset{O}{\overset{\|}{C}}-CH_3$, but not CH_3CH_2CHO

(d) $CH_3-C\equiv C-CH_3 \xrightarrow[H_2SO_4]{HgSO_4} CH_3-\underset{OH}{\underset{|}{C}}=CH-CH_3$

$\rightleftharpoons CH_3-\underset{O}{\underset{\|}{C}}-CH_2-CH_3$

134. (b) 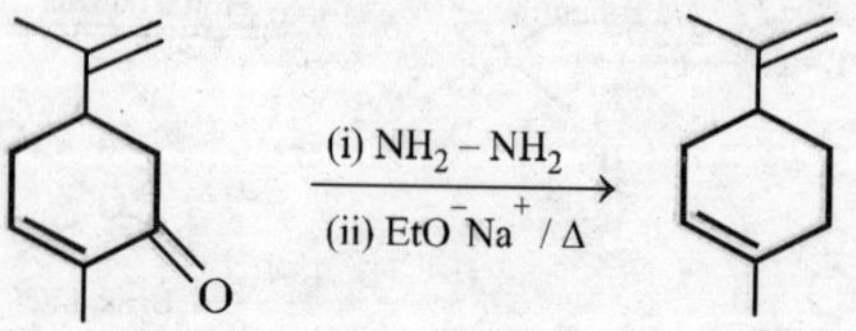 (ketone, O) $\xrightarrow[\text{(ii) } Et\overset{-}{O}\overset{+}{Na}/\Delta]{\text{(i) } NH_2-NH_2}$ (alkene)

To reduce carbonyl group into alkane ($>C=O \rightarrow >CH_2$), Wolff-Kishner reduction is used, which does not affect the double bond.

135. (c) $CH\equiv CH \xrightarrow{LiNH_2/ether} CH\equiv \overset{\ominus}{C}\overset{\oplus}{Li}$ + $CH_3-CH(Br)-CH(CH_3)_2$ → $CH_3-CH(C\equiv CH)-CH(CH_3)_2$ [P]

$CH_3-CH(C\equiv CH)-CH(CH_3)_2 \xrightarrow{HgSO_4/H_2SO_4} CH_3-CH(COCH_3)-CH(CH_3)_2$

$\xrightarrow{NaBH_4} CH_3-CH(CH(OH)CH_3)-CH(CH_3)_2$ [Q] $\xrightarrow[\Delta]{Conc.\ H_2SO_4} CH_3-CH(\overset{+}{C}H-CH_3)-CH(CH_3)_2$ (2° Carbocation)

$\xrightarrow{H^-\ shift} CH_3-\overset{+}{C}(CH_2CH_3)-CH(CH_3)_2$ (3° Carbocation) → $CH_3-C(C_2H_5)=C(CH_3)_2$ [R] (Stable)

136. (a) $CH_2=CH-CHO \xrightarrow{(i)\ NaBH_4} H_2C=CH-CH_2OH \xrightarrow{(ii)\ SOCl_2} H_2C=CH-CH_2-Cl$ [A]

$H_2C=CH-CH_2Cl$ [A] $\xrightarrow{C_6H_6,\ Anhy\ AlCl_3}$ $C_6H_5CH_2CH=CH_2$ [B] $\xrightarrow{DBr}$

$C_6H_5CH_2\overset{+}{C}H-CH_2D$ (2° Carbocation) $\xrightarrow{1,2-H^-\ shift}$ $C_6H_5\overset{+}{C}H-CH_2-CH_2D$ (2° Benzal carbocation) $\xrightarrow{Br^-}$ $C_6H_5CH(Br)-CH_2-CH_2D$ (C)

137. (b) Rate of nucleophillic addition reaction is directly proportional to the –I and –M effect of the substituents present in the substrate. Ketones are less susceptible to the nucleophillic addition, due to the presence of alkyl (R) group which has +I effect. Thus reactivity order is

CH_3CH_2CHO	CH_3COCH_3	$CH_3COCH_2CH_3$	Ph – CHO
(i)	(ii)	(iii)	(iv)
Propanal	Propanone	Butanone-2	Benzaldehyde

(i) > (iv) > (ii) > (iii)

138. (c) $C_6H_5CH_2-CO-CH_3$ (A) $\xrightarrow{(i)CH_3MgBr/H_2O}$ $C_6H_5CH_2-C(OH)(CH_3)-CH_3$ $\xrightarrow[\Delta]{(ii)\ conc.\ H_2SO_4}$

$C_6H_5CH=C(CH_3)_2$ (B) (Stable) $\xrightarrow{(i)\ O_3,\ (ii)\ Zn/H_2O}$ C_6H_5CHO (C) + $CH_3-CO-CH_3$ (D)

C_6H_5CHO (C) $\xrightarrow[\Delta]{conc.\ KOH}$ $C_6H_5CH_2OH$ + $C_6H_5\overset{-}{COO}\overset{+}{K}$ (Cannizaro reaction)

$CH_3-CO-CH_3$ (D) $\xrightarrow{Ba(OH)_2/\Delta}$ (Aldol condensation) $CH_3-C(CH_3)=CH-CO-CH_3$

139. (c) –I effect of NO_2 increases reactivity towards nucleophilic addition reaction with HCN. $-OCH_3$ group is electron donating due to resonance effect which decreases the reactivity towards nucleophilic addition.

o-nitrobenzaldehyde (ii) (–R, –I) > m-nitrobenzaldehyde (iv) (–I, –R not possible) >

m-methoxybenzaldehyde (i) (–I) > o-methoxybenzaldehyde (iii) (+R)

140. (b) (A) $\xrightarrow{(i)\ CH_3MgBr,\ (ii)\ H_3O^+}$ tertiary alcohol $\xrightarrow{Conc.\ H_2SO_4/\Delta}$ (B) $\xrightarrow{O_3,\ Zn,\ H_2O}$ 2-(CH=O)-C$_6$H$_4$-C(CH_3)$_2$-CO-CH_3

141. (66.67)

$CH_3-CO-CH_2-CH_3$ (A) $\xrightarrow{CH_3MgBr}$ $CH_3-C(O^-\overset{+}{M}gBr)(CH_3)-CH_2-CH_3$

$\xrightarrow{H_3O^+}$ $CH_3-C(OH)(CH_3)-CH_2-CH_3$ (B) $\xrightarrow{Cu,\ 573\ K}$ $CH_3-C(CH_3)=CH-CH_3$ (2-Methylbut-2-ene)

Compound A is $CH_3-CO-CH_2-CH_3$ (C_4H_8O)

Mass percentage of carbon $= \left(\frac{12\times 4}{48+16+8}\times 100\right) = 66.67$

142. (d)

Ö: + H–O⁺(H)–H → (d) protonated cyclopropenone ⟷ Stable (CH_3)

143. (d)

Hex-3-ynal (CHO) $\xrightarrow{\text{(i) NaBH}_4}$ hex-3-yn-1-ol (OH) $\xrightarrow{\text{(ii) PBr}_3}$ Br $\xrightarrow{\text{(iii) Mg/ether}}$ MgBr $\xrightarrow{\text{(iv) CO}_2}$ $C(=O)\bar{O}\overset{+}{M}gBr$ $\xrightarrow{H^+}$ COOH

144. (c) Oxidation number of carbon in – CHO group of Ph – CHO and H– CHO are + 1 and zero respectively. So, CHO is the stronger reducing agent than PhCHO. As a result, HCHO is oxidised to HCOONa (by donation of hydride, H^-) and PhCHO is reduced (by accepting H^-) to $PhCH_2OH$.

$$CH_2=O + \bar{O}H \longrightarrow H-C(O^-)(OH)-H$$

$$PhC(=O)-H + H-C(O^-)(OH)-H \xrightarrow{H^- \text{ shift}}$$

$$PhC(O^-)(H)-H + HC(=O)-OH \xrightarrow[\text{Exchange}]{\text{Proton}}$$

$$PhC(OH)(H)-H + H-C(=O)-O^- \xrightarrow{H_3O^+} PhCH_2OH + HCOOH$$

145. (d) (i) Since compound A ($C_9H_{10}O$) contains relatively very low H as compared to C, so it must contain a benzene ring.

(ii) Further, the oxidation product B ($C_8H_6O_4$) of A is a dicarboxylic acid which forms anhydride on heating, hence the acid is phthalic acid which is further confirmed by the fact that it is used in the preparation of phenolphthalein when condensed with phenol in presence of conc. H_2SO_4. So, the given compound A corresponds to option (d).

(A) $C_9H_{10}O$ (o-methylacetophenone, $C(=O)-CH_3$, CH_3) $\xrightarrow{I_2 + NaOH}$ $C(=O)-O^-Na^+$ (CH_3) + CHI_3

(A) $\xrightarrow{KMnO_4}$ B ($C_8H_6O_4$) (phthalic acid, two $C(=O)-OH$) $\xrightarrow{\Delta}$ Anhydride + H_2O

146. (b)

$Ph-C(=O)-C_6H_4-OH \xrightarrow{Br_2/CCl_4}$ (C=O: Deactivating group; OH: Activating group)

$Ph-C(=O)-C_6H_3(Br)-OH$

147. (d)

$PhC(=O)CH_2Br \xrightarrow{NaBH_4} PhCH(OH)CH_2-Br \xrightarrow[\text{(cyclisation)}]{\text{MeOH, 25 °C}}$ Ph–CH–CH_2 (epoxide, O)

148. (b) Hint: Resonance and hydrogen bonding increases enol content.

$$H_3C-\overset{O}{\overset{\|}{C}}-CH_2-\overset{O}{\overset{\|}{C}}-CH_3 \rightleftharpoons H_3C-\overset{O-H\,......\,O}{C=CH-C}-CH_3$$

Stable 6-membered ring due to H-bonding

149. (d)

$C_6H_4(CH(OH)-CH_3)(COC_2H_5) \xrightarrow{I_2/NaOH} C_6H_4(C(=O)-O^-)(COC_2H_5) + CHI_3$

Test	Observation	Reason
Reaction with Grignard reagent,	+ve	Electrophilic centre or acidic hydrogen is present
Fehling solution test	–ve	CHO group is absent
Neutral $FeCl_3$ test	–ve	phenolic group is absent
Iodoform test	+ve	$-COCH_3$ or $-CH(OH)-CH_3$ is present

150. (b)

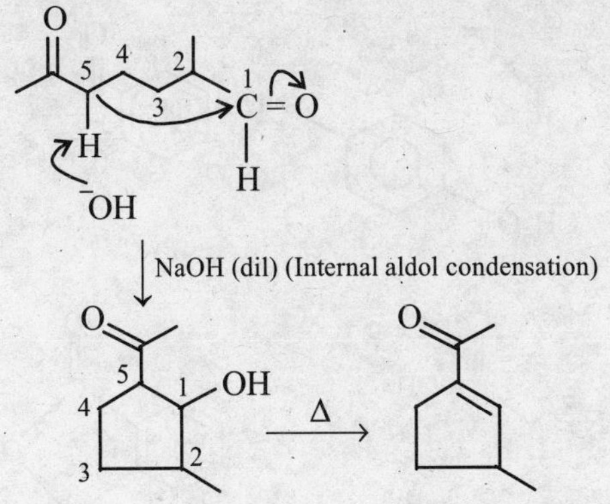

NaOH (dil) (Internal aldol condensation)

$\xrightarrow{\Delta}$

151. (a) $\xrightarrow{OH^-}$

intramolecular aldol condensation $\xrightarrow{H_2O}$ (A) $\xrightarrow{H_3O^+/\Delta}$ (B)

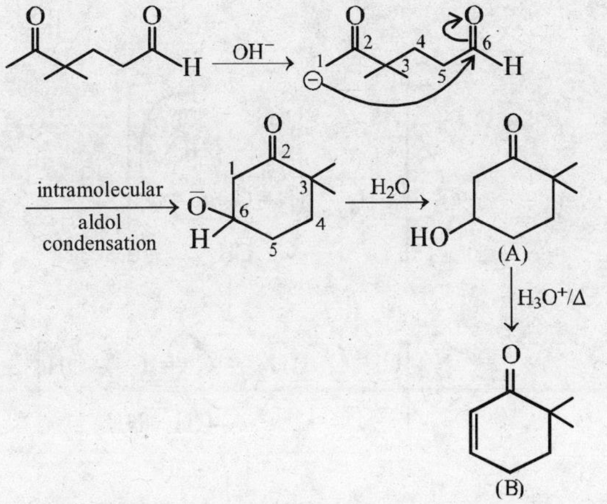

152. (b) $\xrightarrow[EtOH]{NaBH_4}$

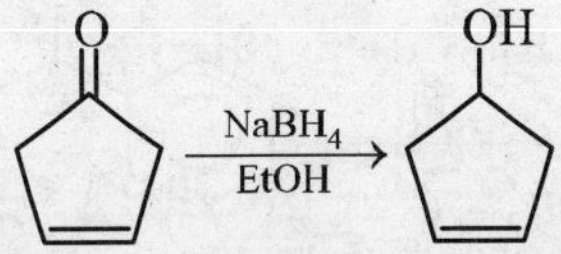

$NaBH_4$ does not reduce the double bond but can reduce keto group ($>C=O$) to —CHOH group.

153. (a) The only one equivalent of Grignard reagent will not react with aldehydes (–CHO) if it has a functional group which has acidic hydrogen. In such cases RMgX (1 equi.) will react with the active hydrogen (acid-base reaction) Thus options (B) and (D) have —COOH and —CH_2OH respectively which has acidic H-atom. Therefore, acid-base reaction occurs.

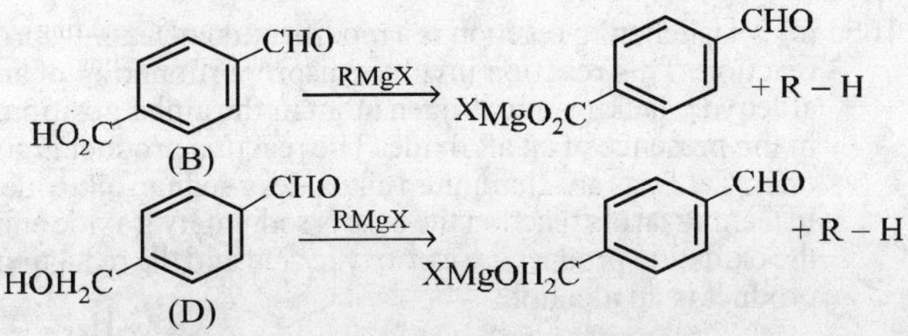

(B) $\xrightarrow{RMgX}$ $XMgO_2C$–C_6H_4–CHO + R – H

(D) $\xrightarrow{RMgX}$ $XMgOH_2C$–C_6H_4–CHO + R – H

Other options (A) and (C) will react with one equi. of RMg as usual.

154. (d) Sodium borohydride is a selective reducing agent. It reduces carbonyl group to alcoholic group, N-methylimino group (MeN= CH–) to 2° amines, but does not reduce an isolated carbon-carbon double bond.

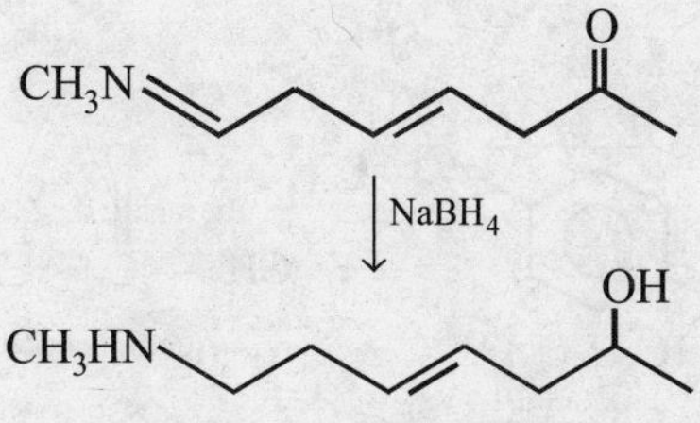

$CH_3N=$ ... $\xrightarrow{NaBH_4}$ CH_3HN ... (OH)

(major product)

155. (a) Here four products are possible.

(i) From self aldol condensation of CH_3CHO.

(ii) From self aldol condensation of $C_6H_5COCH_3$

(iii) From cross aldol condensation between CH_3CHO and $C_6H_5–COCH_3$ when CH_3CHO acts as a carbanion.

(iv) From cross aldol condensation between CH_3CHO and $C_6H_5COCH_3$ when $C_6H_5COCH_3$ acts as a carbanion.

However, CH_3CHO is more reactive than ketone so it acts as a good donor, and its carbanion is also more stable, so it also acts as a good carbanion donor. Thus the major product should be (a).

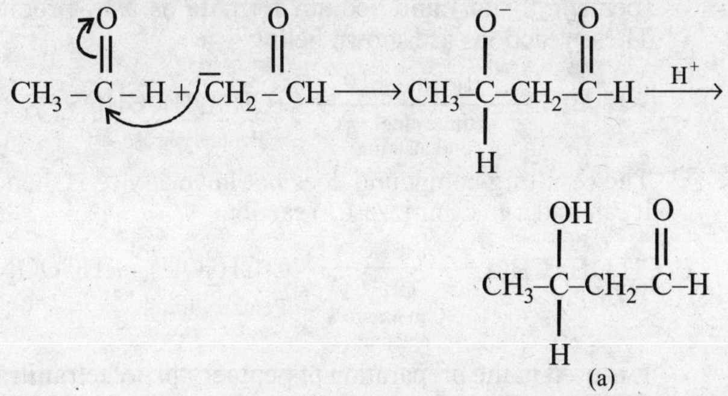

$$CH_3-\overset{O}{\overset{\|}{C}}-H + \overset{-}{C}H_2-\overset{O}{\overset{\|}{C}}H \longrightarrow CH_3-\underset{H}{\overset{O^-}{C}}-CH_2-\overset{O}{\overset{\|}{C}}-H \xrightarrow{H^+} CH_3-\underset{H}{\overset{OH}{C}}-CH_2-\overset{O}{\overset{\|}{C}}-H$$

(a)

156. (d) MeO–C_6H_4–CHO $\xrightarrow[(ii)\ H_2O]{(i)\ C_2H_5MgBr}$ (A) $\xrightarrow[S_N1]{HCl}$ (B)

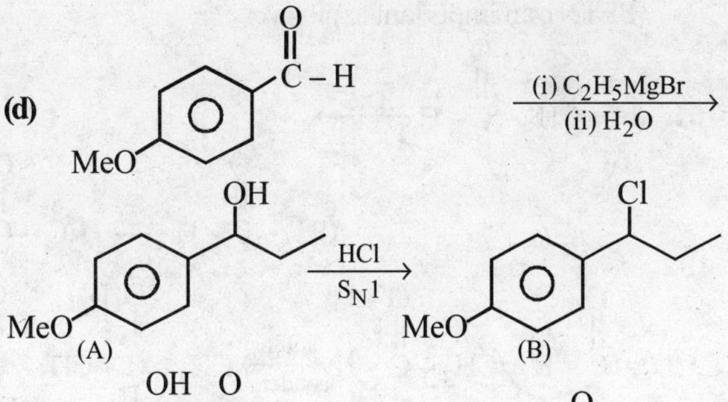

157. (a) 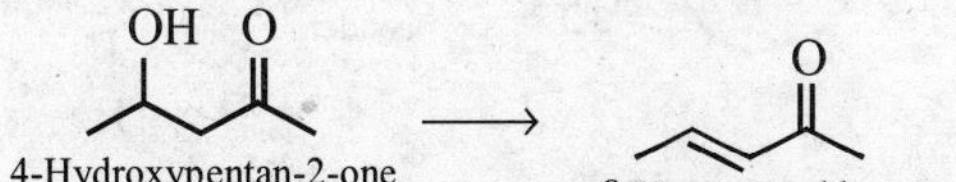 4-Hydroxypentan-2-one (a) $\longrightarrow$ α, β-Unsaturated ketone (Highly stable)

158. (a) $\xrightarrow[\text{Tollens reagent}]{[Ag(NH_3)_2]OH}$ (CHO → CO_2H) $\xrightarrow[\text{(esterification)}]{H^+/CH_3OH}$ (C(=O)OCH_3) $\xrightarrow{CH_3MgBr}$ (HO, CH_3; $H_3C-\underset{OH}{C}-CH_3$)

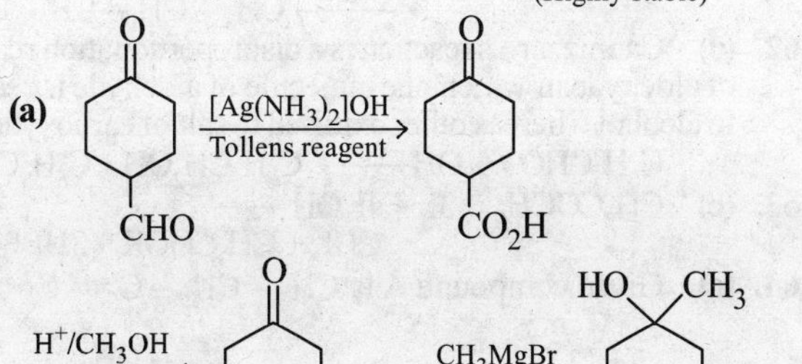

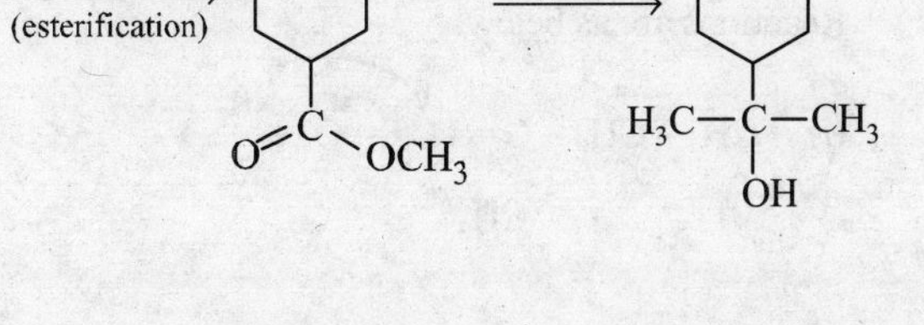

159. (a)

$$H_3CO\text{-}C_6H_4\text{-}CHO\ (C_8H_8O_2) + CH_3-\overset{O}{\overset{\|}{C}}-CH_3\ \text{(Acetophenone)} \xrightarrow[\text{Cross aldol condensation}]{\text{(base)}} CH_3O\text{-}C_6H_4\text{-}CH=CH-\overset{O}{\overset{\|}{C}}-CH_3$$

Cannizzaro reaction, Conc. NaOH:

$$H_3CO\text{-}C_6H_4\text{-}CH_2OH + H_3CO\text{-}C_6H_4\text{-}\overset{O}{\overset{\|}{C}}-\overset{\ominus}{O}\overset{\oplus}{Na}$$

160. (a) The pentaerythritol is produced via a base-catalyzed reaction of acetaldehyde with excess formaldehyde. The aldol condensation of three moles of formaldehyde with one mole of acetaldehyde followed by a crossed Cannizzaro reaction with formladehyde gives the final product (pentaerythritol) and sodium formate as a by-product. These reactions are shown below

$$CH_3CHO \xrightarrow[\text{(three aldol condensation)}]{3HCHO/OH^-} \underset{\text{Pentaerythrose}}{C(CH_2OH)_3CHO}$$

The resulting compound does not have any α–H, hence it can undergo Cannizzaro's reaction.

$$C(CH_2OH)_3CHO \xrightarrow[\substack{OH^- \\ \text{(Cannizzaro's reaction}}]{HCHO} \underset{\text{Pentaerythritol}}{C(CH_2OH)_4} + HCOONa$$

It is used in the preparation of pentaerythritol tetranitrate (PETN), an important explosive.

161. (b) $CH_3-\overset{O}{\overset{\|}{C}}-H \xrightarrow{OH^-} \left[\overset{\ominus}{C}H_2-\overset{O}{\overset{\|}{C}}-H \longleftrightarrow H_2C=\overset{O^{\ominus}}{\overset{|}{C}}-H\right]$

$$CH_3-\overset{O}{\overset{\|}{C}}-H+\overset{\ominus}{C}H_2-\overset{O}{\overset{\|}{C}}-H \xrightarrow[\text{condens.}]{\text{aldol}} CH_3-\underset{H}{\underset{|}{\overset{OH}{\overset{|}{C}}}}-CH_2-\overset{O}{\overset{\|}{C}}-H$$

$$\xrightarrow{\Delta} CH_3-CH=CH-\overset{O}{\overset{\|}{C}}-H$$

162. (d) Cannizzaro's reaction is a disproportionation reaction of aldehyde in which one molecule of aldehyde is reduced to alcohol whereas other oxidised to salt of carboxylic acid

$$C_6H_5CHO + KOH \longrightarrow C_6H_5CH_2OH + C_6H_5COOK$$

163. (c) $CH_3COCH_3 + 3I_2 + 4KOH \longrightarrow$

$$CHI_3 + CH_3COOK + 3HI + 3H_2O$$

164. (c) Given compound A is $CH_3-CH_2-\underset{CH_3}{\underset{|}{C}}=C=O$

Reactions are as below:

(*i*) $CH_3-CH_2-\underset{CH_3}{\underset{|}{C}}=C=O \xrightarrow{\ddot{N}H_3}$

$$\left[CH_3-CH_2-\underset{CH_3}{\underset{|}{C}}=\overset{\overset{+}{N}H_2}{\overset{|}{C}}-O^-\right]$$

$$\downarrow$$

$$\left[CH_3-CH_2-\underset{CH_3}{\underset{|}{C}}=\underset{NH_2}{\underset{|}{C}}-\ddot{O}H\right]$$

$$\downarrow$$

$$CH_3-CH_2-\underset{CH_3}{\underset{|}{CH}}-\overset{O}{\overset{\|}{C}}-NH_2$$

(*ii*) $CH_3-CH_2-\underset{CH_3}{\underset{|}{C}}=C=O \xrightarrow{H_2O} CH_3-CH_2-\underset{CH_3}{\underset{|}{CH}}-COOH$

(*iii*) $CH_3-CH_2-\underset{CH_3}{\underset{|}{C}}=C=O \xrightarrow{CH_3COOH} CH_3-CH_2-\underset{CH_3}{\underset{|}{CH}}-\overset{O}{\overset{\|}{C}}-O-\overset{O}{\overset{\|}{C}}-CH_3$

165. (c) Tishchenko reaction is a modification of Cannizzaro reaction. This reaction involves disproportionation of an aldehyde, lacking a hydrogen atom in the alpha position, in the presence of an alkoxide. The reaction product is an ester. Catalysts are aluminium alkoxide or sodium alkoxide. In Cannizzaro's reaction the base is sodium hydroxide and the oxidation product is a carboxylic acid and the reduction product is an alcohol.

$$R-\overset{\ddot{O}:}{\overset{\|}{C}}-H + Al(OEt)_3 \rightleftharpoons R-\overset{O-\overset{\ominus}{Al}(OEt)_3}{\overset{|}{\overset{\oplus}{C}H}}$$

$$\xrightarrow{R-CHO} R-\underset{O-\overset{\oplus}{C}HR}{\underset{|}{\overset{O-\overset{\ominus}{Al}(OEt)_3}{\overset{|}{C}}}}-H \xrightarrow{-Al(OEt)_3} R-\overset{O}{\overset{\|}{C}}-OCH_2R$$

166. (c) Only acetaldehyde and methyl ketones give iodoform test.

167. (b) $(CH_3)_2C{=}O \xrightarrow{\text{Zn-Hg/HCl}} (CH_3)_2CH_2$

Ketone Clemmensen reduction

168. (b) $CH_3-\underset{}{\overset{CH_3}{|}}C{=}O + CH_3COCH_3 \xrightarrow{OH^-} CH_3-\overset{CH_3}{\underset{OH}{C}}-CH_2-\overset{O}{\overset{||}{C}}-CH_3$

4–Hydroxy–4–methylpentan-2-one

169. (c) Only those aldehydes which do not have α-H atom undergo Cannizzaro's reaction. Hence CH_3CHO will not undergo Cannizzaro's reaction as it has 3 α H atoms.

170. (a) Ethyl 4-oxopentanoate $\xrightarrow[\text{(Protection of –CO group)}]{\text{HO–CH}_2\text{CH}_2\text{–OH}}$ ethylene ketal of ethyl 4-oxopentanoate $\xrightarrow[\text{(ii) } H_3O^+ \text{ (Release of –CO group)}]{\text{(i) } LiAlH_4}$ 5-hydroxypentan-2-one

171. (d) Iodoform test is given by methyl ketones, acetaldehyde, ethyl alcohol and methyl secondary alcohols.
Isobutyl alcohol is a primary alcohol; except ethanol, primary alcohols do not give haloform test.

$CH_3-\underset{CH_3}{\underset{|}{CH}}-CH_2-OH$

172. (a) Aldehydes and ketones can be reduced to hydrocarbons by the action (i) of amalgamated zinc and concentrated hydrochloric acid (Clemmenson reduction), or (ii) of hydrazine (NH_2NH_2) and a strong base like NaOH, KOH or potassium *tert*-butoxide in a high-boiling alcohol like ethylene glycol or triethylene glycol (Wolf-Kishner reduction). But –OH group is acid-sensitive, so Clemmenson reduction can not be used.

173. (b) Presence of alkyl group in carbonyl compounds decreases their reactivity towards nucleophilic addition. Further greater the number of such groups lesser will be the reactivity towards nucleophilic addition, thus correct order is

$HCHO > CH_3CHO > CH_3COCH_3$

174. (d) The reaction given is a Clemmenson reduction.

175. (c) Aldehydes (aliphatic and aromatic) show reaction with Tollen's reagent and aliphatic aldehydes show Fehling solution, but ketones do not show these reactions.

176. (a) $Ph-\overset{O}{\overset{||}{C}}-H + OH^- \xrightarrow{\text{Fast}} Ph-\overset{O^-}{\underset{OH}{C}}-H$

$Ph-\overset{O^-}{\underset{OH}{C}}-H \xrightarrow[\text{slow}]{Ph-\overset{O}{\overset{||}{C}}-H} Ph-\overset{O}{\overset{||}{C}}-OH + Ph-\overset{O^-}{\underset{H}{C}}-H$

$\xrightarrow[\text{fast}]{H^+ \text{ exchange}} Ph-\overset{O}{\overset{||}{C}}-O^- + Ph-\overset{OH}{\underset{H}{C}}-H$

177. (a) Addition of HCN to carbonyl compounds is nucleophilic addition reaction. The order of reactivity of carbonyl compounds is
Aldehydes (smaller to higher) > Ketones (smaller to higher). Therefore,
$HCHO > CH_3COCH_3 > PhCOCH_3 > PhCOPh$
(A) (B) (C) (D)

178. (c) Cyclohexanone $+ HN(CH_3)_2 \longrightarrow$ 1-(dimethylamino)cyclohexan-1-ol (OH, $N(CH_3)_2$) $\xrightarrow{-H_2O}$ 1-cyclohexenyl-$N(CH_3)_2$

Enamine

179. (c) There is no reaction hence the resultant mixture contains $CH_3COOC_2H_5 + NaCl$.

180. (d) It is Clemmensen's reduction

$CH_3-\overset{O}{\overset{||}{C}}-CH_2-CH_3 \xrightarrow[\text{Conc.HCl}]{\text{Zn–Hg}} CH_3CH_2-CH_2CH_3$

Butane–2–one

181. (b) This reaction is known as cannizzaro's reaction. In this reaction benzaldehyde in presence of 50%. NaOH undergoes disproportionation reaction and form one mol of benzyl alcohol (reduced product) and one mole of sod. benzoate (oxidation product)

$C_6H_5CHO \xrightarrow{50\%\ NaOH} C_6H_5CH_2OH + C_6H_5COO^-Na^+$

182. (a) $CH_3-\overset{O}{\overset{||}{C}}-Br \xrightarrow[\text{(ii) Saturated } NH_4Cl]{\text{(i) } 2CH_3MgI} CH_3-\overset{CH_3}{\underset{CH_3}{C}}-OH$

2-Methylpropan-2-ol

183. (b) Double bond is cleaved and oxidised to –COOH, –CHO is also oxidised to –COOH

$(CH_3)_2C{=}CHCH_2CHO \xrightarrow{KMnO_4} (CH_3)_2CO + HOOCCH_2COOH$

184. (d) Chlorobenzene (C_6H_5Cl) $\xrightarrow{NaOH}$ No reaction (due to the absence of an activating group)

$C_6H_5NH_2 \xrightarrow{NaOH}$ No reaction (due to basic amino group)

$C_6H_5COOH \xrightarrow{NaOH} C_6H_5COONa + H_2O$

Ethyl acetate layer in funnel are mixture of chloro benene and aniline

185. (d) Those compounds which are more acidic than H_2CO_3 can gives effervescence of CO_2 with aq. $NaHCO_3$. Compound given in option (d) is the weakest acid.

186. (a) $EtO_2C-C\equiv C-H$

Conjugate base of terminal alkyne will be sp hybridised and localised. For other compounds conjugate base will be resonance stablised.

187. (d)

(A) $R-C\equiv N \xrightarrow[\text{mild condition}]{(i)H^+/H_2O} R-\overset{O}{\overset{\|}{C}}-NH_2$

(B) $R-MgX \xrightarrow[(ii)H_3O^+]{(i)CO_2} R-COOH$

(C) $R-C\equiv N \xrightarrow[(ii)H_3O^+]{(i)SnCl_2/HCl} R-CHO$

(D) $C_6H_5-COCl \xrightarrow[(ii)Br_2\ water]{(i)H_2/Pd-BaSO_4} Ph-COOH$

188. (d) Rate of hydrolysis ∝ Leaving group ability

$\underset{(p)}{CH_3-CH_2-\overset{O}{\overset{\|}{C}}-Cl} > \underset{(q)}{Et-\overset{O}{\overset{\|}{C}}-O-\overset{O}{\overset{\|}{C}}-CH_3} > \underset{(r)}{Et-\overset{O}{\overset{\|}{C}}-OEt} > \underset{(s)}{Et-\overset{O}{\overset{\|}{C}}-NH_2}$

189. (d) Compounds with at least 1 α–H can undergo oxidation with $KMnO_4$ and resulting in the formation of benzoic acid.

Toluene, ethylbenzene, isopropylbenzene, isobutylbenzene, propylbenzene

190. (a) This reaction is an example HVZ reaction

Cyclopentanecarboxylic acid $\xrightarrow[ii)\ H_2O]{i)\ Br_2/Red\ P}$ 1-bromocyclopentanecarboxylic acid (COOH, Br)

191. (5) The group –"COOH" in benzoic acid is electron withdrawl, this ist is meta-directing.
Product (B) forms as:

$C_6H_5CH_2CH_3 \xrightarrow[\Delta]{KMnO_4-KOH} \underset{(A)}{C_6H_5-\overset{O}{\overset{\|}{C}}-OK} \xrightarrow[H_2SO_4]{HNO_3}$ $\underset{(B)}{m\text{-}O_2N-C_6H_4-\overset{O}{\overset{\|}{C}}-OH}$ (nitro group drawn as $N(=O)\rightarrow O$)

Total number of π bonds in B is 5.

192. (a) +I effect decreases the acidic nature.
Thus the correct order is:
$\underset{\text{Min. +I}}{HCOOH} > CH_3COOH > CH_3CH_2COOH > \underset{\text{Max. +I}}{CH_3CH_2CH_2COOH}$

193. (d) Reaction sequence given below

$CH_3-COOH \xrightarrow{LiAlH_4} CH_3-CH_2OH \xrightarrow{PCC} CH_3-CHO \xrightarrow{HCN/OH^-}$

$CH_3-\overset{OH}{\overset{|}{C}H}-CN \xrightarrow{H_2O/OH^-} \underset{(P)}{CH_3-\overset{OH}{\overset{|}{C}H}-COOH}$

The major product (P) formed in the given reaction is 2-hydroxy propanoic acid.

194. (a) $CH_3-CH_2-OH \xrightarrow[\text{Reagent}]{(i)\ \text{Jone's}} CH_3-CHO$

$\xrightarrow{(ii)\ KMnO_4} CH_3COOH \xrightarrow{(iii)\ NaOH,\ CaO,\ \Delta} \underset{\text{Methane}}{CH_4}$

195. (d) Cyclohexene $\xrightarrow[\text{Heat}]{KMnO_4-H_2SO_4}$ $HOOC(CH_2)_4COOH$

Adipic acid

196. (4) $C_6H_5CH_2CH_2CH_3 \xrightarrow[\Delta]{KMnO_4-KOH}$ $\underset{(A)}{C_6H_5COOK} \xrightarrow{H_3O^\oplus} \underset{(B)}{C_6H_5COOH} \xrightarrow{Br_2/FeBr_3} \underset{(C)}{m\text{-}Br-C_6H_4-COOH}$

Number of π bonds in (C) = 4

197. (45)

$$CH_3CH_2NH_2 \xrightarrow{NaNO_2 + HCl} CH_3CH_2N_2^+Cl^- \xrightarrow{H_2O} CH_3CH_2-OH + N_2$$

Mol. wt. 45g; N_2 14g

Given : Volume of N_2 evolved is 2.24 L i.e. 0.1 mole.
i.e. $CH_3CH_2NH_2$ (ethylamine) will be 0.1 mole or 4.5 g
Hence amount (X) of ethylamine = 45×10^{-1} g

198. (c) Aldehydes and ketones having acidic α-hydrogen show aldol reaction. The acidity of α-H is due to the strong electron withdrawing effect of carbonyl group.

Benzaldehyde ($C_6H_5-CH=O$) + $CH_3-C(=O)-H$ (Ethanal, Acidic H) $\xrightarrow[\Delta]{Base}$ $C_6H_5-CH=CH-C(=O)-H$ (Cross aldol product)

199. (18) $HO-C(=O)-C(=O)-CH_2-C\equiv C-CH_3$ (carbons numbered 1–6; π bonds marked)

2 – Oxohex – 4 – ynoic acid

Number of σ-bonds = 14; Number of π-bonds = 4

∴ Total = 18.

200. (b) (A) – (IV), (B) – (II), (C) – (I), (D) – (III)

(A) DiBAL-H (Di-isobutyl aluminium hydride) is a sterically hindered reducing agent, it reduces ester to aldehyde.

(B) Zn(Hg) with conc. HCl is used in Clemmensen reduction of carbonyl compound to form alkane.

(C) CH_3MgBr is a Grignard reagent, it reacts with carbonyl compound to form alcohol.

(D) $NaBH_4$ is a reducing agent, it reduces carbonyl group.

201. (c) $C_2H_5Br \xrightarrow{alc.\ KOH} H_2C=CH_2 \xrightarrow{Br_2/CCl_4} H_2C(Br)-CH_2(Br) \xrightarrow{KCN\ excess} CH_2(CN)-CH_2(CN) \xrightarrow{H_3O^+\ excess} CH_2(COOH)-CH_2(COOH)$

Succinic acid

202. (b) HCOOH, ⟶ formic acid
CH_3COOH, ⟶ Acetic acid
formula = $C_2H_4O_2$

203. (d)

$C_6H_5-CH=CH-CH_2-COOH \xrightarrow{SOCl_2} C_6H_5-CH=CH-CH_2-COCl \xrightarrow{R-NH_2} C_6H_5-CH=CH-CH_2-CO-NH-R \xrightarrow[H_3O^+]{LiAlH_4} C_6H_5-CH=CH-CH_2-CH_2-NH-R$

204. (c) o-(CH_2CH_2COOH)(CH_2COOEt)benzene $\xrightarrow[(ii)\ H_3O^+]{(i)\ LiBH_4/EtOH}$ o-(CH_2CH_2COOH)(CH_2CH_2OH)benzene

$LiBH_4$ can reduce ester selectively but not carboxylic acids. Hence correct answer is option (c).

205. (b) $CH_2=CH-CH_2-CH_2-COOH \xrightarrow{Br_2}$ bromonium intermediate $\xrightarrow{NaHCO_3}$ carboxylate ($-\overset{\ominus}{O}Na^+$) attacks C-4 ⟶ 5-(bromomethyl)-γ-lactone (ring atoms numbered 1–4)

206. (d) Acidic Strength ∝ – I effect
F > Cl > Br : – I effect order
– CH_3 gp has + I effect
So option (d) B > D > C > E > A is the correct answer.

207. (a) Salicylic acid (o-HO-C_6H_4-COOH) $\xrightarrow[(X)]{CH_3OH/H^+}$ methyl salicylate (o-HO-C_6H_4-$COOCH_3$)

Salicylic acid $\xrightarrow[(Y)]{(CH_3CO)_2O/H^+}$ o-($O-C(=O)-CH_3$)-C_6H_4-COOH

208. (c) The stronger an acid, the greater the ionization, higher the K_a value and lower will be its pK_a value.

(A) $H-CH_2-C(=O)-OH \longrightarrow \overset{\ominus}{C}H_2-C(=O)-\ddot{O}H$ (+ R effect)
Less stable (Cross resonance)

(B) $H-CH_2-C(=O)-CH_3 \longrightarrow \overset{\ominus}{C}H_2-C(=O)-CH_3$ (+ I effect)

(C) $C_6H_5-CH_2-C(=O)-CH_3 \longrightarrow C_6H_5-\overset{\ominus}{C}H-C(=O)-CH_3$ (+ I effect)
Extended conjugation (most stable)

(D) $C_6H_5-CH_2-C(=O)-OH \longrightarrow C_6H_5-\overset{\ominus}{C}H-C(=O)-\ddot{O}H$
Cross resonance

Option 'c' has lowest pK_a value as its anion is resonance stablised through benzene ring.

209. (c)

$$\text{Cyclohexenyl-}CH_2COOH \xrightarrow{I-I} \text{iodonium ion} \xrightarrow{NaHCO_3} \text{carboxylate attacks iodonium} \rightarrow \text{iodolactone} \xrightarrow{Py/\Delta} \text{unsaturated lactone } (C=O)$$

210. (2)

$$\xrightarrow{\bar{O}Et}$$

(A) 2-Chiral carbons

211. (a)

$$\text{Cyclohexane-COOH} \xrightarrow[\text{(ii) } H_2O]{\text{(i) Red P, } Cl_2} \text{1-chlorocyclohexane-COOH}$$

Halogenation takes place at α-position in presence of red phosphorus. It is known as Hell-Volhard-Zelinsky reaction.

212. (b) $(CH_3)_3CLi$ acts as a base or it can give halogen exchange reaction

$$\text{o-Bromophenol} \xrightarrow{(CH_3)_3CLi} \text{(A)} + (CH_3)_3CBr \xrightarrow{O=C=O} \text{C(=O)O}^-\text{Li, OLi} \xrightarrow{H_3O^+} \text{o-}CO_2H\text{, OH}$$

213. (a) In esterification alcohol and carboxylic acid react to form ester

$$R-OH + R-\overset{O}{\overset{\|}{C}}-OH \longrightarrow R-O-\overset{O}{\overset{\|}{C}}-R$$

alcohol acts as nucleophile; nucleophilic acyl substitution

Electron withdrawing groups on carboxylic acid will increase the rate of esterification.

214. (c)

$$\text{p-Br-}C_6H_4CH_3 \xrightarrow{Cl_2/\Delta} \text{p-Br-}C_6H_4CH_2-Cl \xrightarrow{\overset{\oplus}{CN}} \text{p-Br-}C_6H_4CH_2-CN \xrightarrow{H_3O^+} \text{p-Br-}C_6H_4CH_2-COOH$$

215. (d) Six possible isomers of $C_6H_3(NH_2)_2COOH$ are

(1) ; (2) ; (3) $\xrightarrow{-CO_2}$ (A)

(4) (5) $\xrightarrow{-CO_2}$ (B)

(6) $\xrightarrow{-CO_2}$ (C)

Product (*C*) should have higher melting point (i.e., 142°C) than products (*A*) and (*B*) due to its symmetry.

216. (a) **Statement I:** Better packing efficiency of monocarboxylic acids with even number of carbon atoms results in higher M.P. Hence statement I is correct.

Statement II: As molar mass increases, size of hydrophobic part increases, hence solubility decreases.

217. (c)

$$CH_3CH_2CH_2COOCH_2CH_3 \text{ (A) } [C_6H_{12}O_2]$$

$$\downarrow \text{(i) } LiAlH_4 \text{ (ii) } H_3O^+$$

$$CH_3-CH_2-CH_2-CH_2-OH + CH_3-CH_2-OH \text{ [B]}$$

$$\downarrow [O]$$

$$CH_3-CH_2-CH_2-\overset{O}{\underset{\|}{C}}-OH$$

[C_4 carboxylic acid]

218. (c) In this reaction one mole of ethyl pent-4-yn-oate utilizes three moles of CH_3MgBr:

$$HC \equiv C-CH_2-CH_2-\overset{O}{\overset{||}{C}}-OC_2H_5 \xrightarrow[-C_2H_5OMgBr]{CH_3MgBr(3\ mole)}$$

$$Br\overset{\oplus}{Mg}\overset{\ominus}{C} \equiv C-CH_2-CH_2-\underset{CH_3}{\overset{\overset{\ominus}{O}MgBr}{C}}-CH_3$$

$$\downarrow HOH$$

$$HC \equiv C-CH_2-CH_2-\underset{CH_3}{\overset{OH}{C}}-CH_3$$

219. (78)

$$\text{COOH-}C_6H_5 + Br_2 \xrightarrow{FeBr_3} \text{m-Br-}C_6H_4\text{-COOH} + HBr$$

Mass (in gm) 6.1 g 7.8 g

$$\text{Moles} = \frac{6.1}{122} = 0.05 \qquad \text{Moles} = \frac{7.8}{201} = 0.039$$

$$\text{Percentage yield} = \frac{0.039}{0.05} \times 100 = 77.61\% \approx 78\%.$$

220. (c) Ethylene glycol in presence of H^+ will protect ketonic group as cyclic ketal and the ester group remains intact.

$$CH_3COCH_2COOC_2H_5 \xrightarrow[H^+]{\text{Ethylene Glycol}} \text{(cyclic ketal)}CH_2COOC_2H_5$$

221. (1) Only aspirin has – COOH group.

222. (a) Esters are hydrolysed in basic medium (saponification), so it is unstable in aqueous base.

223. (a) Acidity ∝ Stability of conjugate base

Stability order of conjugate bases

$$PhCO\overset{\ominus}{\ddot{C}}HCOPh\ (B) > \overset{\ominus}{C}H_2COCH_3\ (A) > \overset{\ominus}{\ddot{C}}H_2COOCH_3\ (C) > \overset{\ominus}{C}H_2CON(Me)_2\ (D)$$

Thus increasing order of acidity is D < C < A < B.

224. (b)

$$\text{o-xylene} + \text{succinic anhydride} \xrightarrow[\text{Friedel-Craft reaction}]{\text{anhyd. } AlCl_3} \text{(A)}$$

(A) (aryl keto acid, $HOCO$) $\xrightarrow[\text{(Clemmensen reduction)}]{\text{(i) Zn-Hg/HCl}}$ (aryl butanoic acid, $HOCO$) $\xrightarrow[\text{(Cyclization)}]{\text{(ii) } H_3PO_4\ (-H_2O)}$ (B) (tetralone)

225. (a)

$$CH_3COOC(CH_3)_3\ (A) \xrightarrow{H_3O^+} CH_3COOH\ (B) + HO-C(CH_3)_3\ (C)$$

(C) 3° Alcohol (Gives Lucas test immediately)

226. (a)

$$p\text{-}CH_3C_6H_4COOH \xrightarrow{Br_2/FeBr_3} \text{3-Br-4-}CH_3C_6H_3COOH$$

Only one product

$$p\text{-}CH_3C_6H_4COOH \xrightarrow{NaOH/CaO,\Delta} C_6H_5CH_3$$

227. (a)

$$CH_3-CH_2-\underset{CH_3}{CH}-OCOCH_2-\overset{CH_3}{CH}-CH_2CH_3\ (A) \xrightarrow{H_2SO_4\ (dil.)}$$

$$CH_3-CH_2-\underset{CH_3}{CH}-OH\ (C) + HOOC-CH_2-\overset{CH_3}{CH}-CH_2-CH_3\ (B)$$

$$(C) \xrightarrow{CrO_3/H^+} CH_3-CH_2-\underset{CH_3}{C}=O \text{ which is not B.}$$

228. (d) Molecule (A) shows intramolecular H-bonding while molecule (B) shows intermolecular H-bonding. Due to presence of intermolecular H-bonding (B) has more b. pt. than molecule (A). Molecule (B) also shows intermolecule H-bonding with water which makes it more soluble than A. Since (B) is more soluble than (A) in water, so it should be less crystalline than (A).

229. (c) Acidic strength ∝ Stability of conjugate base

General order of acidic strength is

$$R\text{–}COOH > Ph\text{–}OH > R\text{–}C \equiv CH$$

In between (iii) and (iv), (iii) is more acidic due to –M effect of $-NO_2$.

Thus, decreasing order of acidity is (ii) > (iii) > (iv) > (i).

230. (a) B_2H_6 is a very selective reducing agent and usually used to reduce acid to alcohol.

231. (c)

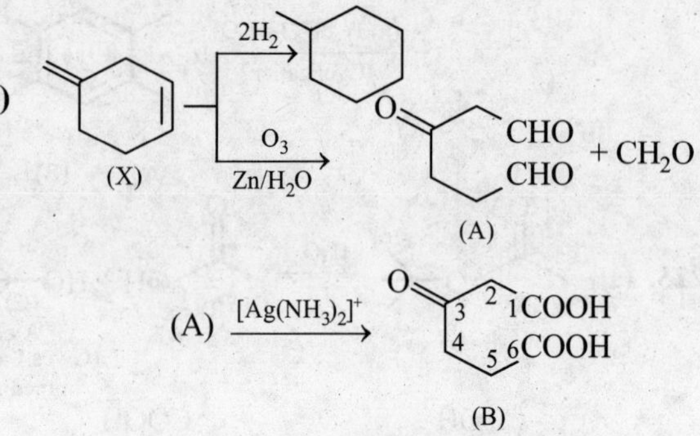

(X) $\xrightarrow{2H_2}$ methylcyclohexane

(X) $\xrightarrow{O_3,\ Zn/H_2O}$ (A) (CHO, CHO) $+ CH_2O$

(A) $\xrightarrow{[Ag(NH_3)_2]^+}$ (B) (1COOH, 6COOH)

3-Oxo-hexanedicarboxylic acid

232. (d)

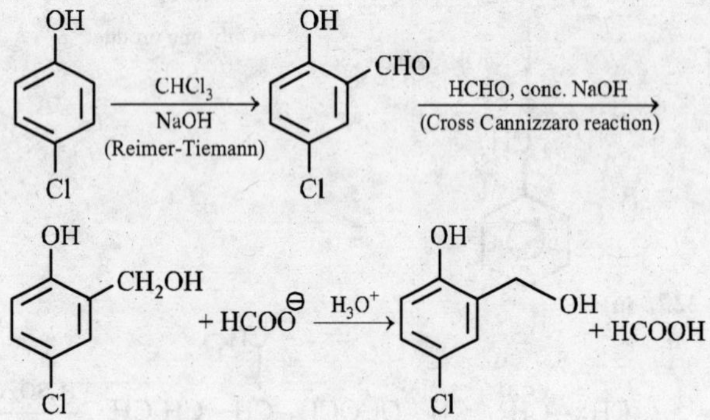

4-chlorophenol $\xrightarrow[\text{(Reimer-Tiemann)}]{CHCl_3,\ NaOH}$ 5-chloro-2-hydroxybenzaldehyde $\xrightarrow{\text{HCHO, conc. NaOH (Cross Cannizzaro reaction)}}$

CH_2OH product $+ HCOO^{\ominus} \xrightarrow{H_3O^+}$ product $+ HCOOH$

233. (b) $LiAlH_4$ reduces esters to alcohols but does not reduce C=C.

$$CH_3CH=CH-CO_2Me \xrightarrow{LiAlH_4} CH_3CH=CHCH_2OH$$

234. (a) Alkaline $KMnO_4$ converts C_6H_5-R with a benzylic hydrogen into benzoic acid.

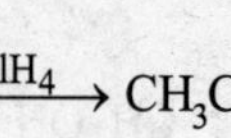

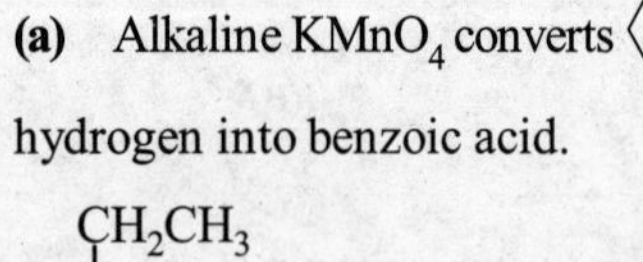

$C_6H_5CH_2CH_3 \xrightarrow[\text{(ii) } H_3O^+]{\text{(i) alkaline } KMnO_4} C_6H_5COOH$

235. (a)

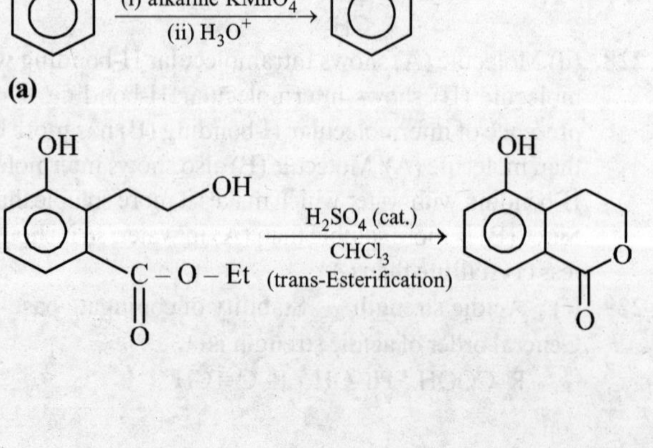

$\xrightarrow[\text{(trans-Esterification)}]{H_2SO_4\text{ (cat.)},\ CHCl_3}$

Mechanism:

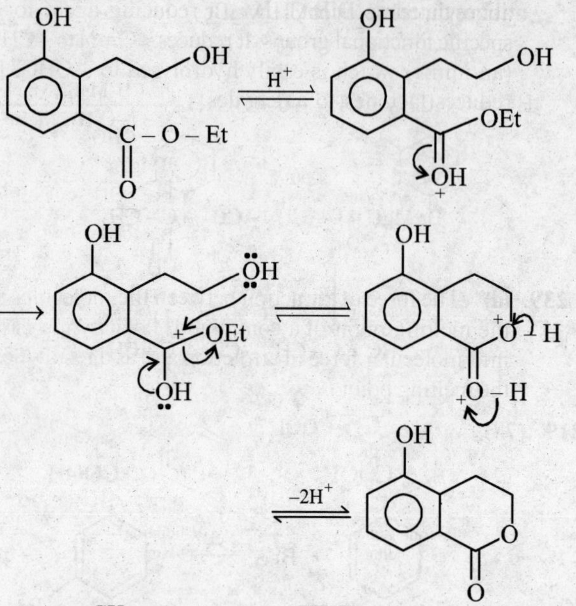

C – O – Et $\overset{H^+}{\rightleftharpoons}$ OEt, $\overset{+}{O}H$ $\longleftrightarrow$ $\overset{+}{O}Et$, OH $\rightleftharpoons$ $\overset{+}{O}$ – H, $\overset{+}{O}$ – H $\overset{-2H^+}{\rightleftharpoons}$ lactone

236. (a)

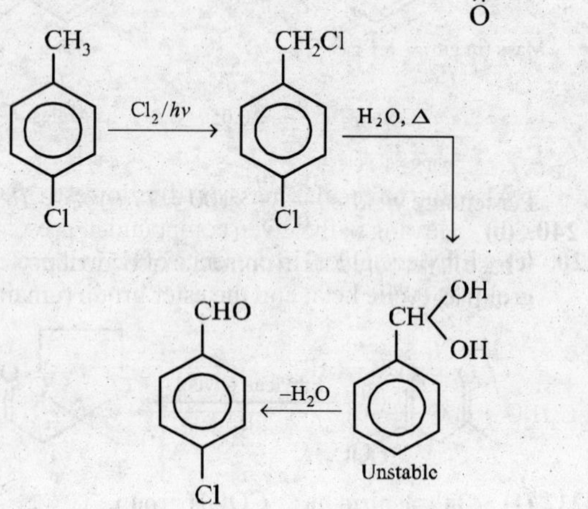

CH_3 (p-Cl) $\xrightarrow{Cl_2/hv}$ CH_2Cl (p-Cl) $\xrightarrow{H_2O,\ \Delta}$ $CH(OH)_2$ (Unstable) $\xrightarrow{-H_2O}$ CHO (p-Cl)

237. (c)

$\xrightarrow[(-\text{ HCl})]{\text{t-BuOK}}$ $\xrightarrow[\text{(Conc.} H_2SO_4)]{H^+}$

2° carbocation but less stable due to proximity of – C = O group

or

1° carbocation but more stable because – C = O group is away from positive charge

Δ

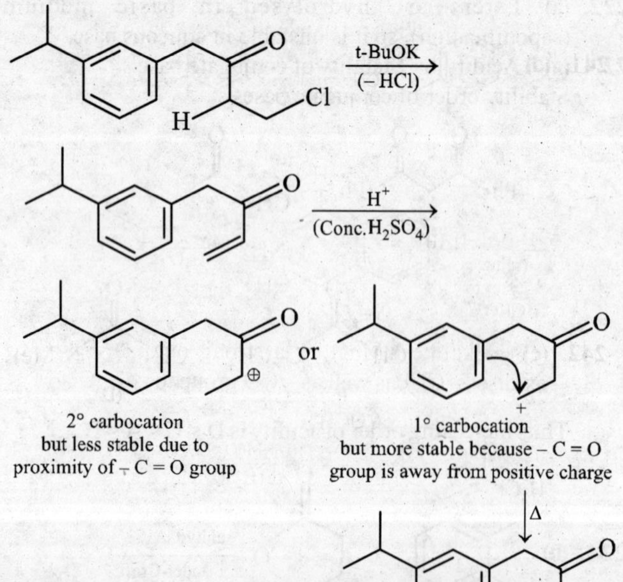

238. (c) Dissobutyl aluminium hydride, commonly abbreviated as DiBAL-H is a reducing agent for some specific functional groups. It reduces $-C\equiv N$ to $-CH=NH$ (an imine) which is easily hydrolysed to $-CHO$. It also reduces lactones to aldehydes.

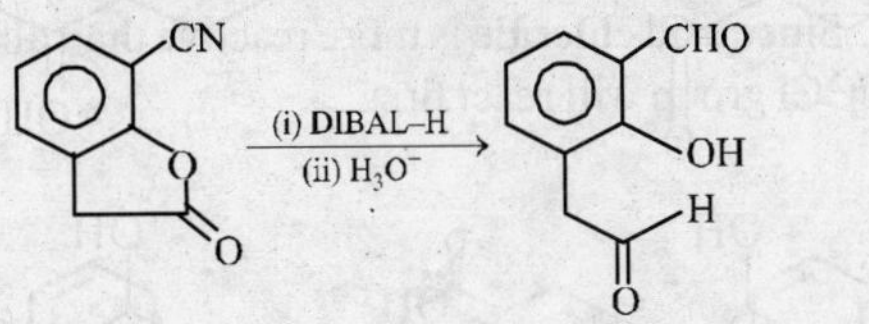

(i) DIBAL–H (ii) H_3O^-

239. (d) The force of attraction between the molecules affects the melting point of a compound. Polarity increases the intermolecular force of attraction and as a result increases the melting point.

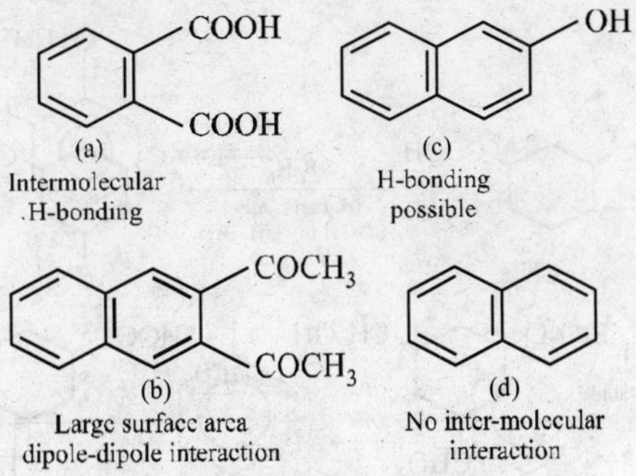

(a) Intermolecular H-bonding

(c) H-bonding possible

(b) Large surface area dipole-dipole interaction

(d) No inter-molecular interaction

Also, the molecular mass of (d) is lowest.

240. (b) Heating of the given compound in presence of strong base is favoured for elimination reaction resulting in more stable alkene.

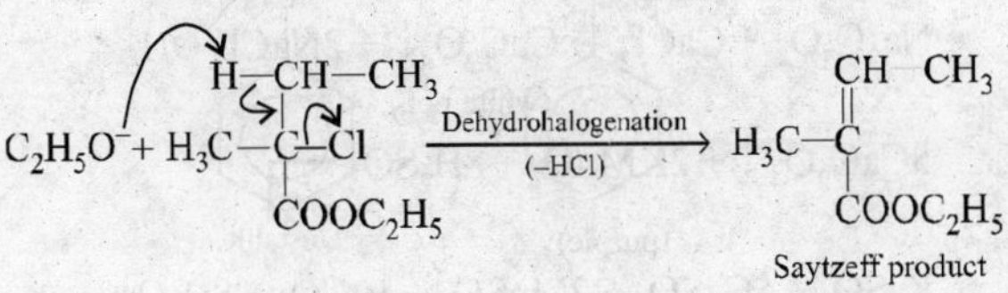

$C_2H_5O^- + H_3C-C(Cl)(COOC_2H_5)-CH_2-CH_3 \xrightarrow[(-HCl)]{\text{Dehydrohalogenation}} H_3C-C(COOC_2H_5)=CH-CH_3$

Saytzeff product

241. (d) Ni/H_2 (reduces only – CN)

DiBAL-H (reduces ester to – CHO gp.)

$-H_2O$

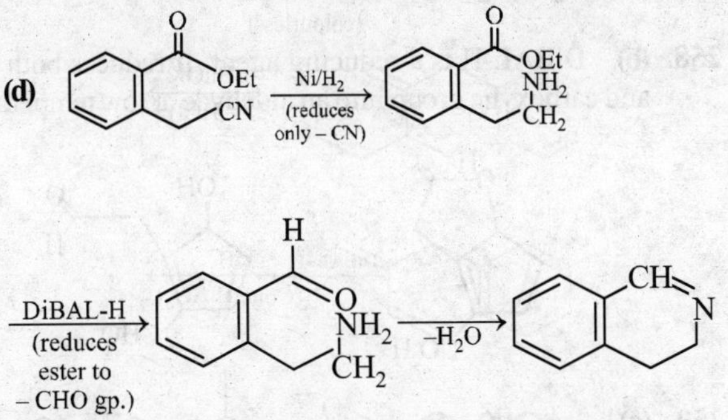

242. (c) Alkaline $KMnO_4$ is a strong oxidising agent and oxidises $-CH_3$ as well as $-CO$ group to $-COOH$.

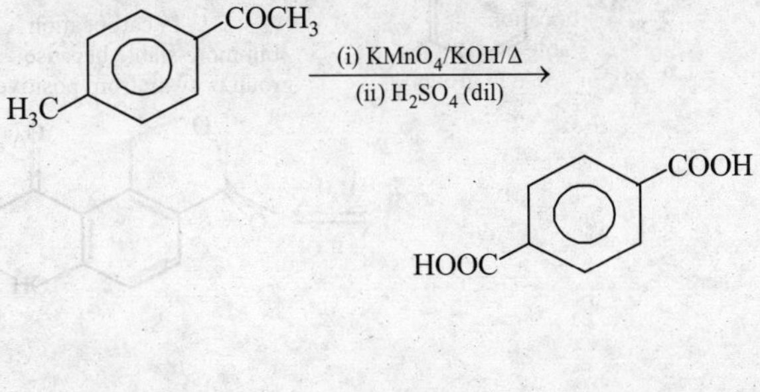

(i) $KMnO_4/KOH/\Delta$ (ii) H_2SO_4 (dil)

243. (c) $LiAlH_4$ (excess)

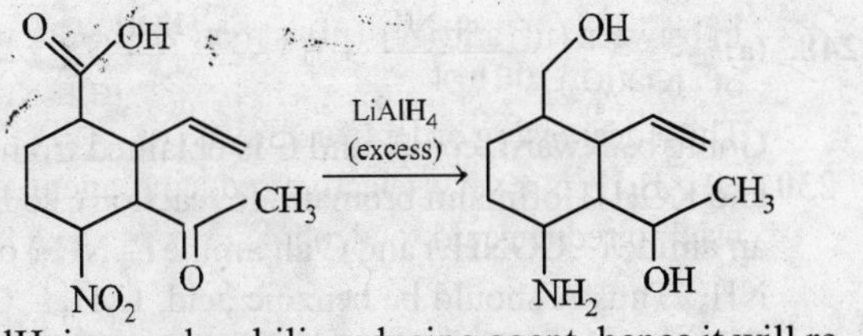

$LiAlH_4$ is a nucleophilic reducing agent, hence it will reduce $-COOH$ to CH_2OH, $-CO$ to $-CHOH$ and $-NO_2$ to NH_2 but does not reduce $C=C$ linkage.

244. (d)

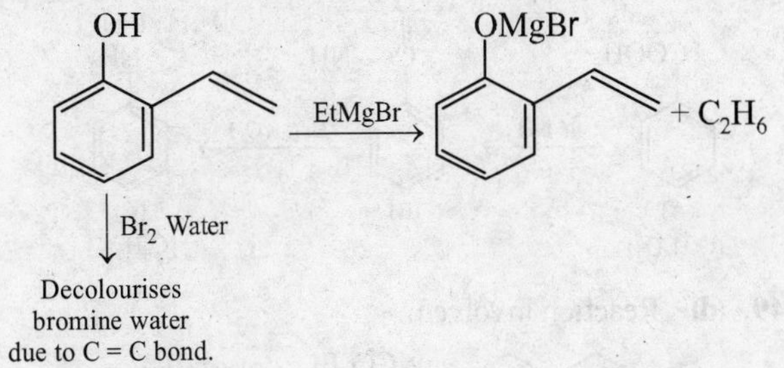

EtMgBr → + C_2H_6

Br_2 Water

Decolourises bromine water due to C = C bond.

245. (a) Br^+ (1 eqv.)

MeOH

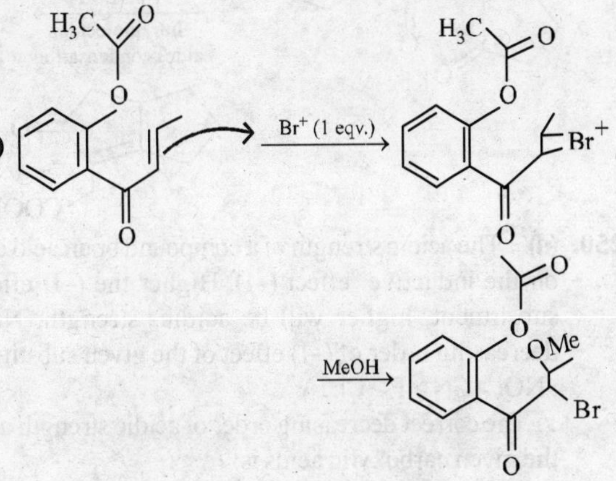

246. (a)

Adipic acid $\xrightarrow{\text{dehydrating agent}}$ Cyclopentanone $+ CO_2 + H_2O$

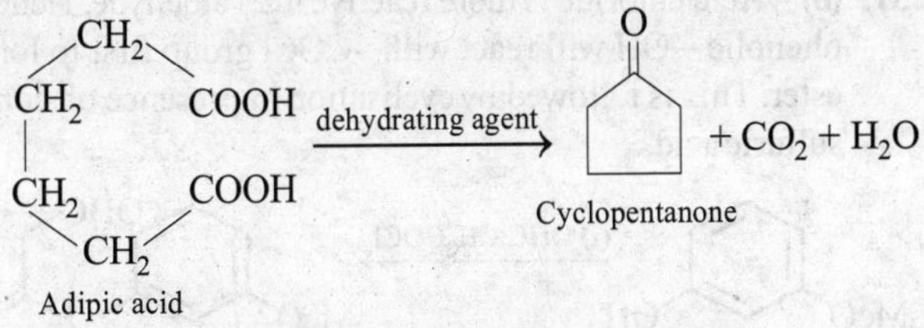

Adipic acid does not form anhydride.

247. (b) Rate of reaction ∝ Electrophilicity of carbonyl carbons, so E.W.G increases the rate, while E.R.G decreases the rate.

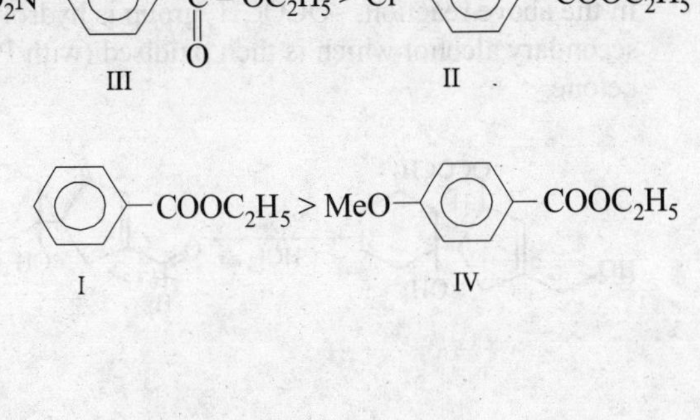

$O_2N-C_6H_4-C(=O)-OC_2H_5 > Cl-C_6H_4-COOC_2H_5$
III > II

$C_6H_5-COOC_2H_5 > MeO-C_6H_4-COOC_2H_5$
I > IV

248. (a) A $(C_7H_6O_2)$ $\xrightarrow[\text{(ii) heat}]{\text{(i) aq. } NH_3}$ B $\xrightarrow{Br_2/KOH}$ C (C_6H_7N)

Going backward, compound C is obtained from B by Br_2 and KOH (Hoffmann bromamide reaction), so B must be an amide ($-CONH_2$) and C an amine ($-NH_2$) or $C_6H_5-NH_2$. Thus A should be benzoic acid, C_6H_5-COOH or $C_7H_6O_2$.

Reaction involved:

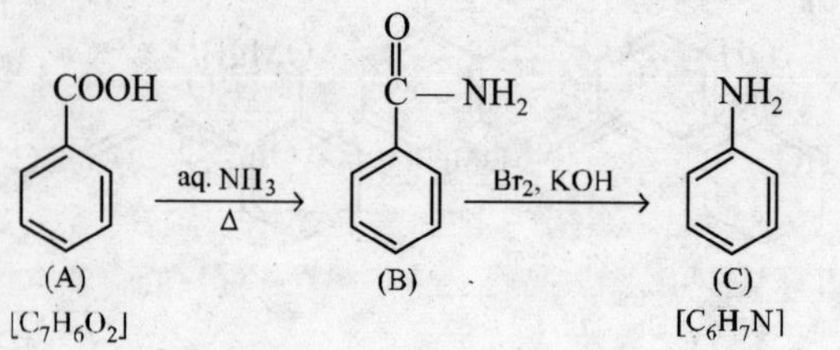

COOH (A) $[C_7H_6O_2]$ $\xrightarrow[\Delta]{\text{aq. } NH_3}$ $C(=O)-NH_2$ (B) $\xrightarrow{Br_2,\ KOH}$ NH_2 (C) $[C_6H_7N]$

249. (d) Reaction involved:

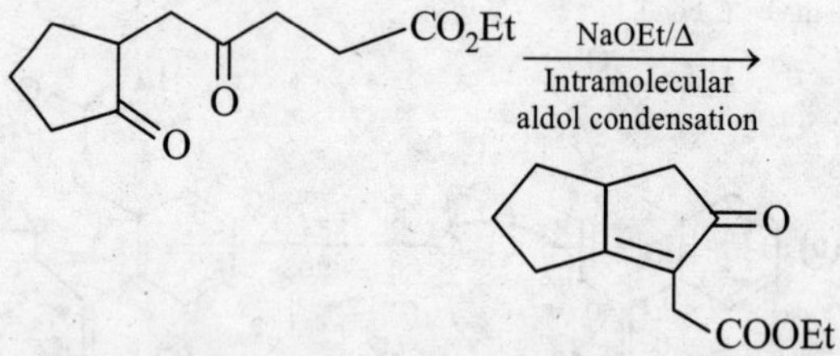

CO_2Et $\xrightarrow[\text{aldol condensation}]{\text{NaOEt}/\Delta \text{ Intramolecular}}$ COOEt

250. (d) The acidic strength of a compound or an acid depends on the inductive effect (–I). Higher the (–I) effect of a substituent, higher will be acidic strength. Now, the decreasing order of (–I) effect of the given substituents is : $NO_2 > CN > F > Cl$.

∴ The correct decreasing order of acidic strength amongst the given carboxylic acids is:

$NO_2CH_2COOH > CNCH_2COOH > FCH_2COOH > ClCH_2COOH$

251. (b) Acid chloride is more reactive than aldehyde. Hence, phenolic – OH will react with –COCl group first to form ester. This is followed by cyclisation in presence of conc. sulfuric acid.

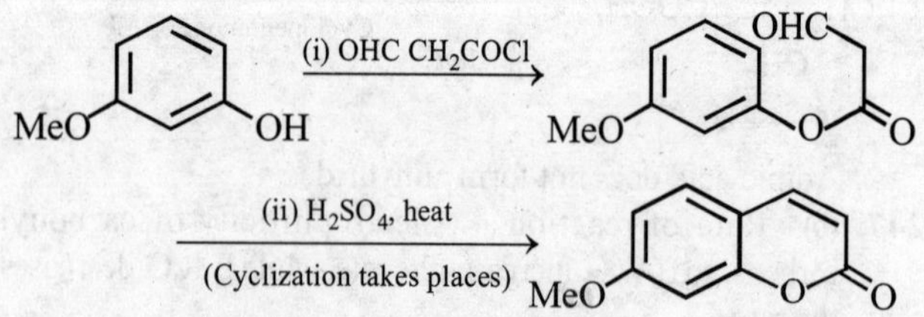

MeO, OH $\xrightarrow{\text{(i) OHC } CH_2COCl}$ MeO, OHC, O, O $\xrightarrow[\text{(Cyclization takes places)}]{\text{(ii) } H_2SO_4\text{, heat}}$ MeO, O, O

252. (b) PCC oxidizes primary alcohols to aldehydes and secondary alcohols to ketones.

In the above reaction, $-OCOCH_3$ group is hydrolyzed to secondary alcohol which is then oxidised (with PCC) to ketone.

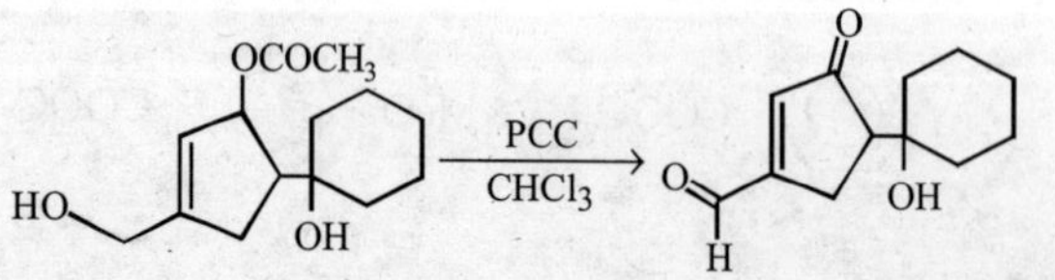

$OCOCH_3$, HO, OH $\xrightarrow[CHCl_3]{PCC}$ O, O, H, OH

253. (a) The increasing order of the acidity of the carboxylic acids is III < II < IV < I. In aromatic acids, electron withdrawing groups like –Cl, –CN, $-NO_2$ increases the acidity, whereas electron releasing groups like $-CH_3$, –OH, $-OCH_3$, $-NH_2$ decreases the acidity.

254. (c) Since acid chloride is more reactive than alkyl halide, so COCl group will react first.

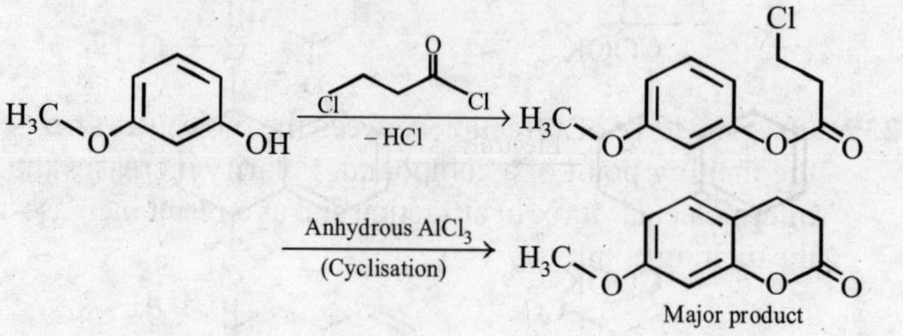

H_3C, O, OH $\xrightarrow[-HCl]{\text{Cl, O, Cl}}$ H_3C, O, O, O, Cl $\xrightarrow[\text{(Cyclisation)}]{\text{Anhydrous } AlCl_3}$ H_3C, O, O, O

Major product

255. (d)

EtO_2C, COOH, CN $\xrightarrow[\text{(reduces only } -COOH \text{ to } -CH_2OH)]{B_2H_6}$ EtO_2C, CH_2OH, CN $\xrightarrow[\text{[Reduces cyanide to aldehyde]}]{SnCl_2/HCl}$ EtO_2C, CH_2OH, CHO $\xrightarrow{\text{(iii) } H_3O^+}$ HOOC, CH_2OH, CHO

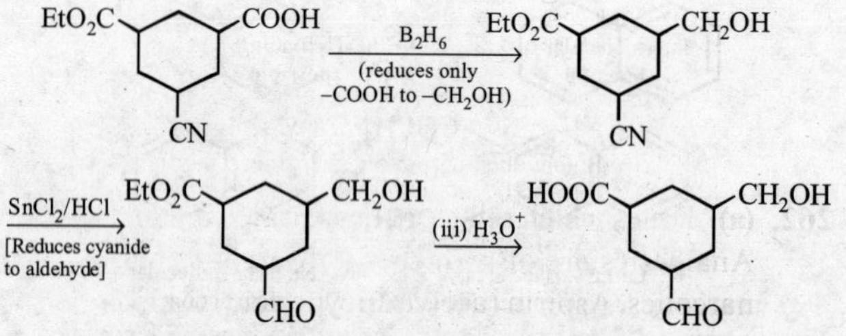

256. (a) Sodium borohydride reduces ketonic group to alcohol, but not the amide group and C = C double bond.

257. (d) $\underset{\text{'X'}}{Na_2C_2O_4} + \underset{\text{(conc.)}}{H_2SO_4} \rightarrow Na_2SO_4 + CO\uparrow + CO_2\uparrow + H_2O$

$\underset{\text{'X'}}{Na_2C_2O_4} + CaCl_2 \rightarrow \underset{\text{(white ppt.)}}{CaC_2O_4\downarrow} + 2NaCl$

$5CaC_2O_4\downarrow + \underset{\text{(purple)}}{2KMnO_4} + 8H_2SO_4 \rightarrow$

$K_2SO_4 + 5CaSO_4 + \underset{\text{(colourless)}}{2MnSO_4} + 10CO_2 + 8H_2O$

258. (b) DiBAL-H is a reducing agent. It reduces both ester and carboxylic group into an aldehyde at low temperature.

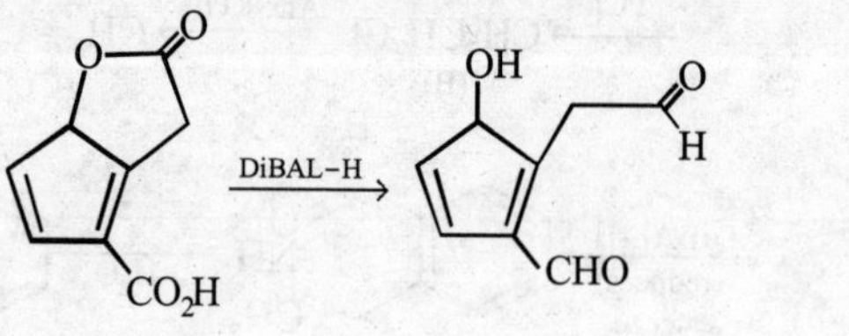

O, O, CO_2H $\xrightarrow{\text{DiBAL–H}}$ OH, O, H, CHO

259. (c) HO, HO, O, O, NH_2, OH $\underset{+H_2O}{\overset{-H_2O}{\rightleftharpoons}}$ O, O, O, NH_2, OH

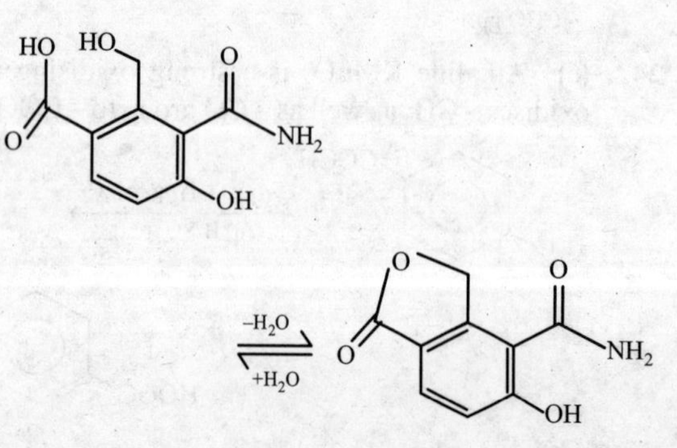

260. (c) The Bouveault reduction is a reaction in which an ester is reduced to primary alcohol using absolute ethanol and sodium.

$$R-\overset{O}{\overset{\|}{C}}-OR \xrightarrow{Na/C_2H_5OH} RCH_2OH + ROH$$

261. (b)

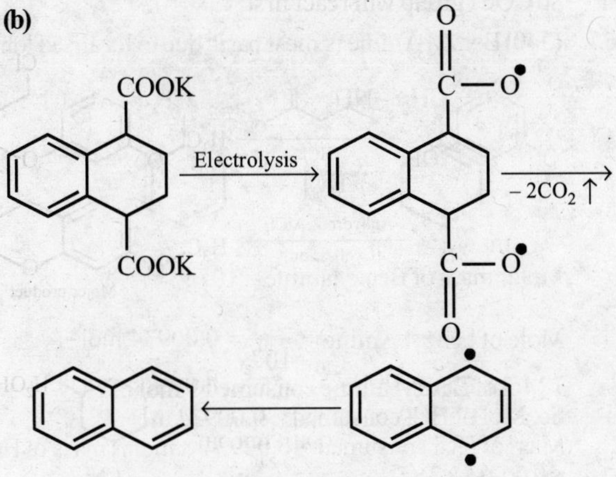

COOK, Electrolysis, $-2CO_2\uparrow$

Stable due to aromatic character

262. (a) Drugs which relieve pain are called analgesics drugs. Analgesics are of two types (i) Narcotics and (ii) Non-narcotics. Aspirin (acetylsalicylic acid) is a non-narcotic analgesic.

263. (d) α-Substitution occurs when a carboxylic acid having α-hydrogens is treated with chlorine or bromine in presence of small amount of red phosphorus. This reaction is commonly known as HVZ reaction.

$$R-CH_2COOH + X_2 \xrightarrow{P} R\underset{}{\overset{X}{\overset{|}{C}}}H-COOH + HX (X = Cl, Br)$$

264. (c) $CH_3COOH \xrightarrow{LiAlH_4} \underset{(A)}{CH_3CH_2OH} \xrightarrow{PCl_5} \underset{(B)}{CH_3CH_2Cl} \xrightarrow{Alc.\ KOH} \underset{(C)}{CH_2=CH_2}$

265. (d) Phthalic acid (COOH, COOH) $+ 2$ resorcinol (HO, OH) $\xrightarrow{Conc.\ H_2SO_4}$ Fluorescein $+ H_2O$

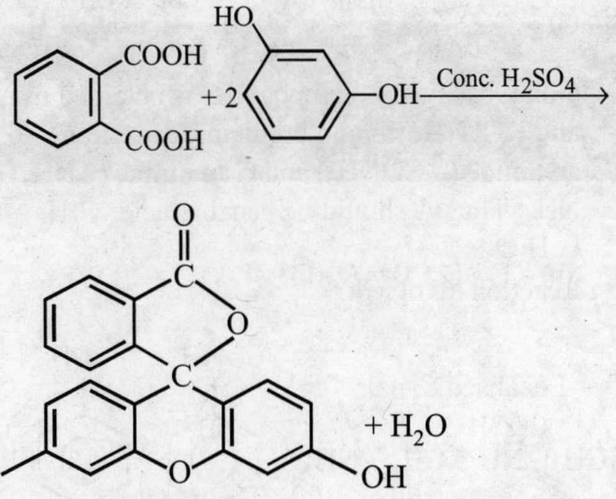

Fluorescein

266. (c) Mono-carboxylic acids are functional isomers of esters, e.g.,

$\underset{\text{Acetic acid}}{CH_3COOH}$ $\underset{\text{Methyl formate}}{HCOOCH_3}$

267. (d) Maleic acid and fumaric acids are geometrical isomers.

268. (a) The correct order of increasing acid strength is $(Me)_2CHCOOH < CH_3COOH < MeOCH_2COOH < CF_3COOH$

Note: Electron withdrawing groups increase the acid strength whereas electron donating groups decrease the acid strength.]

269. (c) $pK_a = -\log K_a$; HCOOH is the strongest acid and hence it has the highest K_a or lowest pK_a value.

270. (a) $LiAlH_4$ can reduce COOH group but not the double bond.

$$CH_2=CH-COOH \xrightarrow{LiAlH_4} CH_2=CH-CH_2OH$$

271. (c) $CH_3CH_2COOH \xrightarrow[\text{red P}]{Cl_2} CH_3CHClCOOH \xrightarrow[-HCl]{alc.KOH} \underset{\text{Acrylic acid}}{CH_2=CHCOOH}$

Amines

1. **(b)** E > D > B > A > C

Localised e^- pair

$[CH_3]_2\ddot{N}H > CH_3 - \ddot{N}H_2 >$ (p-anisidine, NH_2 / OMe) > (aniline, NH_2) > (p-nitroaniline, NH_2 / NO_2)

+I more +I less OMe +M NO_2 –I –M

delocalised e^- pair

2. **(c)** Only primary amines gives positive carbylamines test. Option (A) and (C) are primary amine.

3. **(d)**

(R) > (Q) > (P)

(R) According to Bredt's rule it is localised lone pair.

(Q) Cross-conjugation lone pair available

(P) Delocalized lone pair

4. **(c)**

p-nitroaniline $\xrightarrow[\text{acetic acid}]{\text{(i) } Br_2 \text{ (excess)}}$ 2,6-dibromo-4-nitroaniline $\xrightarrow{NaNO_2,\ HCl}$ 2,6-dibromo-4-nitrobenzenediazonium chloride ($\overset{\oplus}{N}{=}N\overset{\ominus}{Cl}$) $\xrightarrow[\text{reaction}]{\text{CuBr Sandmeyer}}$ 1,2,3-tribromo-5-nitrobenzene $\xrightarrow[\text{Reduction}]{Sn + HCl}$ 3,4,5-tribromoaniline

5. **(d)** Basic strength of amines depends on the cumulative effects of hydrogen bonding and inductive effect. For ethylamines, order is 2° > 3° > 1°. NH_3 is stronger base than $NH_2 - NH_2$ due to – I effect of $-NH_2$ groups. Thus, order will be,

$NH(Et)_2 > N(Et)_3 > NH_2Et > \ddot{N}H_3 > NH_2 - NH_2$

6. **(340)** Benzyl Amine is most basic due to localised lone pair.

$C_6H_5CH_2 - NH_2 + HCl \longrightarrow C_6H_5CH_2 - \overset{\oplus}{N}H_3\overset{\ominus}{Cl}$

(Benzyl Amine)

Molar mass of Benzylamine = 107g

Mole of benzyl Amine $= \frac{1}{107} = 0.00934$ mol

1 Mol of Benzyl amine consumed 1 mol of HCl

So, Mol of HCl consumed = 0.00934 mol

Mass of HCl consumed = 0.00934 × molar mass of HCl

= 0.00934 × 36.5

= 0.34 g = 340 mg

7. **(a)**

(A) $R - NH_2 \xrightarrow[HCl]{NaNO_2} R - N_2^{\oplus}Cl^{\ominus}$

(B) $C_6H_5NH_2 \xrightarrow[KOH(EtOH)]{CHCl_3} C_6H_5NC$

$R - NH_2 \xrightarrow[KOH(EtOH)]{CHCl_3} R - NC$

(C) Only primary amine gives carbylamine test

(D) $Ph{-}SO_2Cl \longrightarrow$ Hinsberg reagent, Benzene sulphonyl chloride

(E) Tertiary amine do not react with $Ph{-}SO_2Cl$

So correct options are (B) and (D) only

8. **(93)**

$C_6H_5Cl \xrightarrow{\text{Mg/dry ether}} C_6H_5MgCl \xrightarrow[\text{(ii) } H^{\oplus}]{\text{(i) } CO_2} C_6H_5COOH \xrightarrow{NH_3/\Delta} C_6H_5CONH_2$ (A) $\xrightarrow[\text{Hofmann's Bromamide degradation}]{Br_2 + NaOH} C_6H_5NH_2$ (B)

No. of moles of C_6H_5Cl = No. of moles of $C_6H_5NH_2$

$$\frac{11.25\times10^{-3}}{112.5} = \frac{x\times10^{-1}\times10^{-3}}{93}$$

$$0.1\times10^{-3} = \frac{x\times10^{-3}}{93} \Rightarrow x = 9.3\times10^{-3}\text{g} = 93\times10^{-1}\text{ mg}$$

9. (b) Aniline $\xrightarrow{\text{conc. } H_2SO_4}$ $\overset{+}{N}H_3 HSO_4^-$ (Anilinium hydrogen sulphate [intermediate]) $\xrightarrow{453-473\text{ K}}$

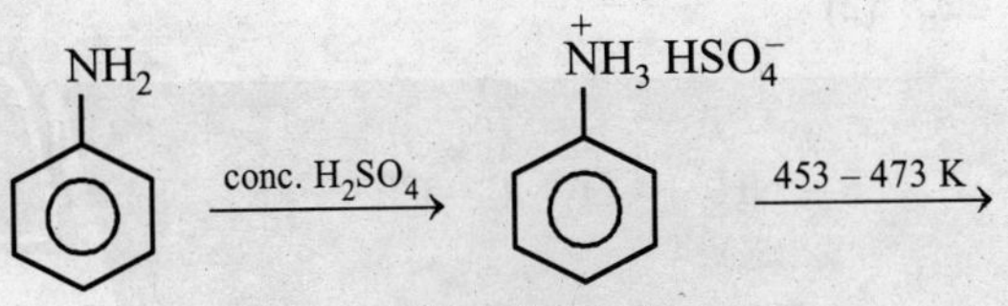

p-$NH_2C_6H_4SO_3H$ $\xrightarrow{Ac_2O}$ p-$CH_3CONHC_6H_4SO_3H$ $\xrightarrow{\text{(i) } Br_2/H_2O,\ \text{(ii) } \Delta}$ o-Br-$C_6H_4NHCOCH_3$ $\xrightarrow{NaOH}$ o-Br-$C_6H_4NH_2$

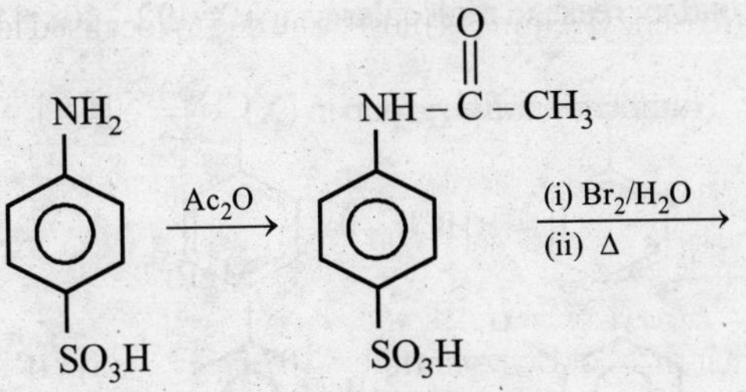

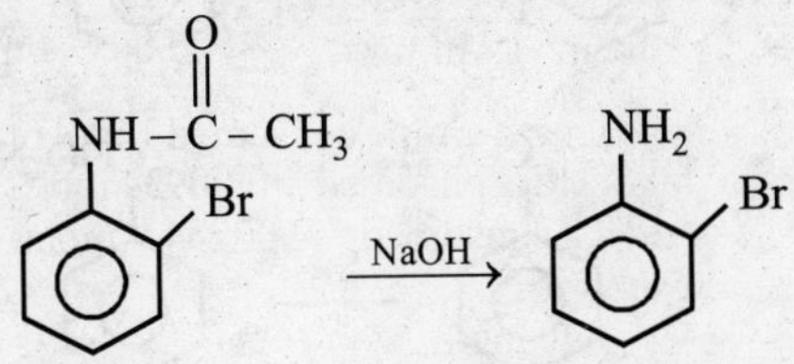

10. (d) B and D are 3° amine which does not have replaceable hydrogen on nitrogen. So they does not react. Primary and secondary amines can reacts with Hinsberg reagent.

11. (d) (A) $R-\overset{O}{\overset{\|}{C}}-NH_2 \xrightarrow[H_2O]{LiAlH_4} R-CH_2-NH_2$

(B) 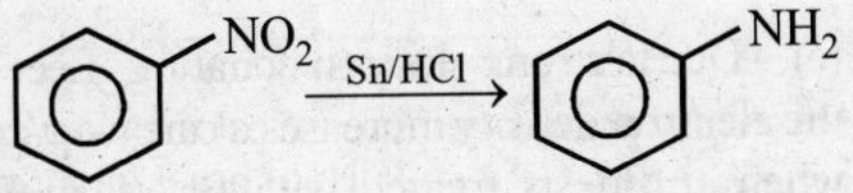 $C_6H_5NO_2 \xrightarrow{Sn/HCl} C_6H_5NH_2$

(C) $R-C\equiv N \xrightarrow{H_2/Ni} R-CH_2NH_2$

(D) 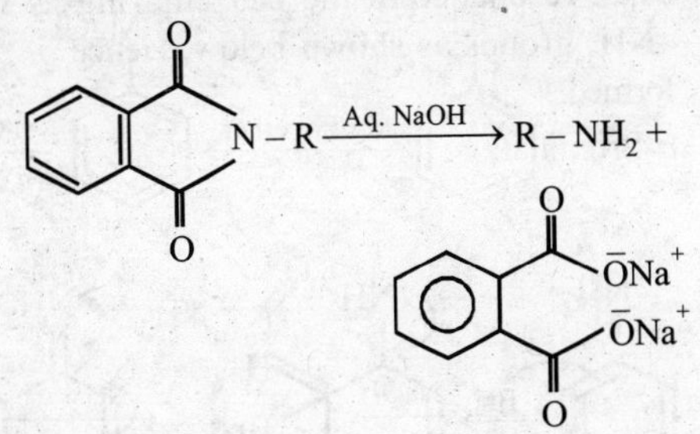 N-R phthalimide $\xrightarrow{\text{Aq. NaOH}} R-NH_2 +$ $C_6H_4(COO^-Na^+)_2$

12. (a) Stronger base has its lone pair more easily available. Thus, order of basic strength is:

N(sp^3, localized lone pair) in pipendine > N(sp^2, localized lone pair) in pyridine > N(sp^2, delocalized lone pair, aromatic) in pyrrole.

∴ Piperidine > Pyridine > Pyrrole

13. (4) Compounds which can not undergo Friedel-crafts reaction are

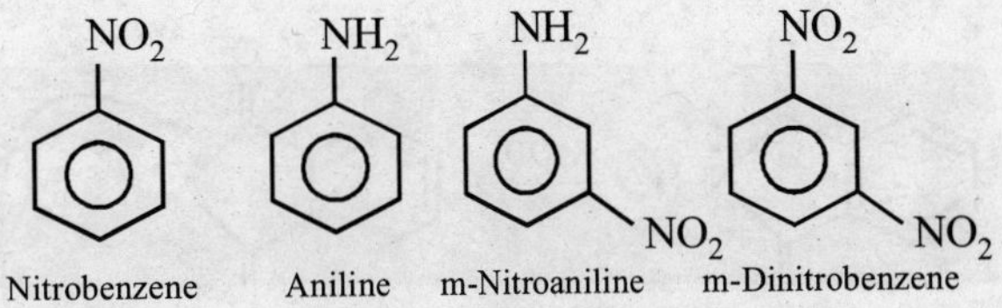

Nitrobenzene, Aniline, m-Nitroaniline, m-Dinitrobenzene

This is due to strongly deactivating groups and–NH_2 group reacts with the acid ($AlCl_3$).

14. (5) Hinsberg's reagent is benzen sulphonyl chloride it reacts with 1° amine to yield N-ethyl benzene sulphonyl amide. The hydrogen attached to nitrogen in sulphonamide is acidic. Hence it is soluble in alkali.

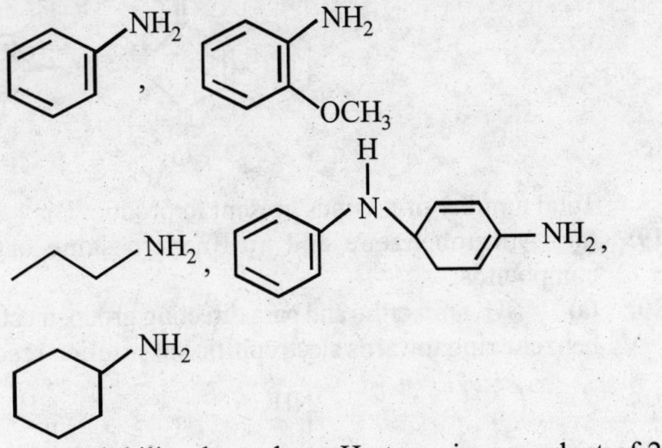

$C_6H_5NH_2$, o-$CH_3OC_6H_4NH_2$, $CH_3CH_2CH_2NH_2$, $C_6H_5-NH-C_6H_4-NH_2$, cyclohexyl-NH_2

15. (a) Solubility depends on H-atom since product of 2° amine has no H-atom thus not soluble.

1° and 2° amines give Hinsberg test but not 3° due to absence of H-atom on nitrogen.

16. (287)

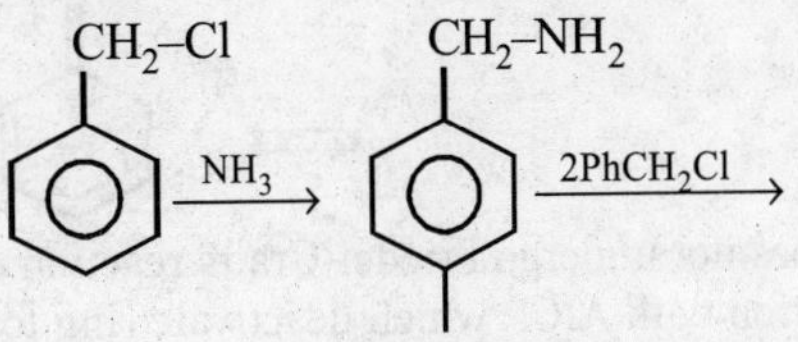

$C_6H_5CH_2-Cl \xrightarrow{NH_3} C_6H_5CH_2-NH_2$ (Benzyl amine) $\xrightarrow{2PhCH_2Cl}$

$Ph-CH_2-\underset{\underset{CH_2-Ph}{|}}{N}-CH_2-Ph$ (X)

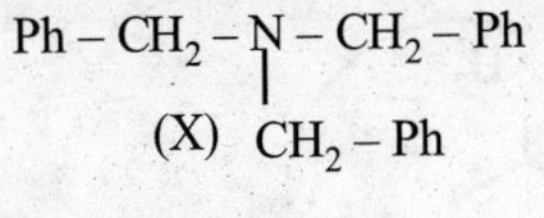

Molar mass of amine = 287 g/mol.

17. (80) Reaction is represented as :

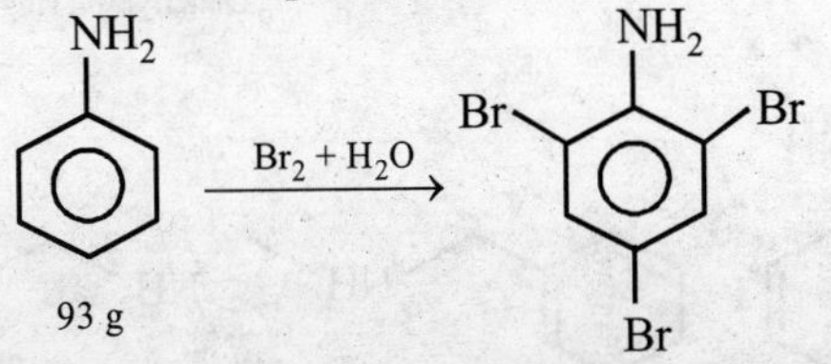

$C_6H_5NH_2$ (93 g) $\xrightarrow{Br_2+H_2O}$ 2,4,6-tribromoaniline (white ppt) 330 g

93 g of aniline produces 330 g of 2, 4, 6-tribromoaniline. Hence 9.3 g of aniline should produce 33g of 2, 4, 6-tribromoaniline.

Hence percentage yield $= \frac{26.4\times100}{33} = 80\%$

18. **(8)**

(Phthalimide) $\xrightarrow{\overset{+}{K}\ \overset{-}{OH}}$ ... $\xrightarrow{-KCl}$ ... CH_2-Cl ... (P)

Total number of π-bonds present in product P is 8.

19. **(b)** Aminobenzene and aniline are same organic compounds.

20. **(a)** $-NH_2$ is an ortho and para directing group, it activates benzene ring towards electrophilic substitution reaction.

Aniline does not undergo Friedel-Crafts reaction due to salt formation with $AlCl_3$ which deactivate ring towards electrophilic substitution reaction.

$$NH_2\text{-}C_6H_5 + AlCl_3 \longrightarrow \overset{+}{N}H_2-\overset{-}{Al}Cl_3\text{-}C_6H_5$$

Deactivated ring

21. **(5)**

NH₂ , NH , NH₂ NH₂ , NH₂ , N–H

22. **(2)**

$$\underset{(X)}{R-NH_2} + CH_3-\overset{O}{\overset{\|}{C}}-Cl \longrightarrow R-NH-\overset{O}{\overset{\|}{C}}-CH_3$$

gain in molecular weight after acetylation with one $-NH_2$ group is 42.

Total increase in molecular weight = 192 – 108 = 84

$$\therefore \text{Number of amino group in (X)} = \frac{84}{42} = 2$$

23. **(b)**

$$C_6H_5Br \xrightarrow[\text{Ether}]{Mg} C_6H_5MgBr \xrightarrow{CO_2,\ H^+} C_6H_5COOH \xrightarrow{NH_3} C_6H_5COO^-NH_4^+ \xrightarrow{\Delta} C_6H_5CONH_2 \xrightarrow[NaOH]{Br_2} C_6H_5NH_2$$

Aniline

24. **(c)** Due to strong electron donating effect of $-NH_2$ group, the electron density increases at the *o*-, *p*-positions. Thus, when aniline is treated with Br_2, then Br^+ attacks the benzene ring at *o*-, *p*-positions to form carbocation intermediates (a, b and d, but not c) which are stabilized by usual resonance of the benzene ring as well as by the $-NH_2$ groups as shown below: Hence *o, p*-products are formed.

o-Bromination:

NH₂ $\xrightarrow{Br^+}$ I ⟷ II ⟷ III ⟷ IV

(a) (d)

p-Bromination:

(b)

25. (d)

(a) Cyclohexene $+ Br_2 \xrightarrow[\text{UV light}]{\Delta}$ 3,6-dibromocyclohexene

This reaction follows free radical substitution pathway.

(b) This represents Hoffmann – Bromamide reaction which gives a primary amine with one carbon less than in the amide and therefore the given reaction has the same number of carbon atoms in the amine is incorrect.

(c) $CH_3CH_2CH_2NH_2 \xrightarrow[H_2O]{HNO_2,\ 0°C} CH_3CH_2OH + N_2 + HCl$

This is the diazotization of aliphatic amines that gives an unstable diazonium salt that decomposes to give alcohol and N_2 gas but here one C is less in alcohol which is incorrect.

(d) 1-Methylcyclohexene $+ HI \longrightarrow$ 1-iodo-1-methylcyclohexane

This is hydroiodination reaction that follows Markovnikoff's rule and therefore it is correct.

$$C_2H_5CONH_2 + Br_2 + NaOH \longrightarrow C_2H_5CH_2NH_2 + Na_2CO_3 + NaBr + H_2O$$

26. (d) Pyrrolidine N–H ← localised lone pair & sp^2 makes it easier to lose electron pair or getting protonated.

27. (b)

$$Cl-C-(CH_2)_4-Cl \xrightarrow[NH_3]{\text{excess}} Cl^- \overset{\oplus}{N}H_3-(CH_2)_4-\overset{\oplus}{N}H_3Cl^- \ (A)$$

$$\xrightarrow{NaOH} 2NaCl + 2H_2O + NH_2-(CH_2)_4-NH_2$$

28. (c)

$\xrightarrow[-H_2O]{HCl}$; $\xleftarrow[(-H_2O)]{(i)\ Cl^-,\ (ii)\ KOH}$

29. (a) o-Phenylenediamine $\xrightarrow{HNO_2}$ NH–N = O ⇌ N=N – OH $\xrightarrow{-H_2O,\ H^+}$ $\overset{+}{N}\equiv N$ $\xrightarrow{H^+}$ benzotriazole

30. (a) (P) Gabriel phthalimide synthesis is used for the preparation of aliphatic primary amines. Aromatic primary amines cannot be prepared by this method.

(Q) 2°–Amines react with Hinsberg's reagent to give solid which is insoluble in NaOH

(R) Aromatic primary amines react with nitrous acid at low temperature (273 – 298 K) to form diazonium salts, which form red dye with β–naphthol

31. (6) Aryl substituted aliphatic amines which are also considered as aromatic amines can be prepared by Gabriel-pthalimide syntheses.

$C_6H_5CH_2-CH_2-NH_2$, $C_6H_5CH_2-NH_2$

$CH_2–NH_2$ (m-methylbenzylamine), $CH_2–NH_2$ (p-methylbenzylamine)

$CH_3–\overset{*}{C}H–NH_2$ attached to phenyl (d, *l*)

32. (c) Nucleophilic attack of $\ddot{N}H_2$ of the amino alcohol on $(EtO)_2C=O$, loss of EtOH, intramolecular attack of $CH_2\ddot{O}H$ on C=O and loss of EtOH giving the cyclic carbamate (oxazolidinone) with C=O.

$\xrightarrow{-EtOH}$

33. (a) $CH_3\underset{Cl}{CH}COOH \xrightarrow{dil.NH_3} CH_3\underset{NH_2}{CH}COOH$

α-Amino carboxylic acid is stabilised by zwitter ion formation. While in case of amines obtained by the reaction of aq.NH_3 and alkyl halide, no such stabilisation is possible.

$CH_3\underset{\overset{+}{N}H_3}{C}HCOO^- \rightleftharpoons CH_3\underset{NH_2}{C}HCOOH$

34. (3) No reaction will be observed as in Gabriel phthalimide synthesis (B) because chlorobenzene is poor substrate for S_N2.

∴ Aromatic amines will be formed in A, C & D.

35. (b) Cyclohexylamine $\xrightarrow{HNO_2}$ Cyclohexanol (P) $\xrightarrow[\text{(Oxidation)}]{PCC}$

Cyclohexanone (Q) $\xrightarrow{\text{dil. NaOH}}$ enolate carbanion, which attacks another cyclohexanone $\xrightarrow{\text{(aldol condens)}}$ alkoxide → β-hydroxy ketone $\xrightarrow[-H_2O]{\Delta}$ 2-cyclohexylidenecyclohexanone (R)

36. (d) $CH_3CH_2\overset{*}{C}H(NH_2)CH_3$, A ($C_4H_{11}N$) $\xrightarrow{HNO_2}$ $CH_3CH_2CH(OH)CH_3 + N_2\uparrow$

A $\xrightarrow{PhSO_2Cl}$ $C_6H_5–SO_2–NH–CH(CH_3)CH_2CH_3$ Soluble in KOH

37. (c) $C_6H_5CH_2Br \xrightarrow{AgCN} C_6H_5CH_2–NC$ Benzyl isocyanide

$C_6H_5CH_2NH_2 \xrightarrow[\text{Aq. KOH}]{CHCl_3}$ $C_6H_5CH_2–NC$ Benzyl isocyanide (Carbylamine reaction)

$C_6H_5CH_2–NHCH_3 \xrightarrow[\text{Aq. NaOH}]{CHCl_3}$ No reaction

$C_6H_5CH_2OTs \xrightarrow{KCN} C_6H_5CH_2–CN$

38. (d) $CH_3CH_2CONH_2$ (Propanamide) $\xrightarrow{Br_2/KOH}$ $CH_3CH_2NH_2$ (Ethyl amine)

[Hoffmann-Bromamide reaction]

39. (a)

(i) In aqueous medium basic strength is dependent on donating ability of electron density on 'N'

(ii) As well as solvation of cation formed after accepting H^+. After considering all these factors overall basic strength order is

$Me_2NH > MeNH_2 > Me_3N > NH_3$

40. (d) (A) Aniline; 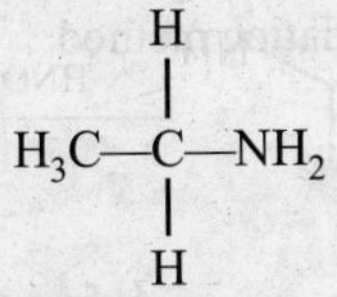 (B) Ethanamine $H_3C-CH_2-NH_2$; (C) N-Ethylethanamine $H_5C_2-NH-C_2H_5$; (D) N, N-Diethylethanamine $H_5C_2-N(C_2H_5)-C_2H_5$

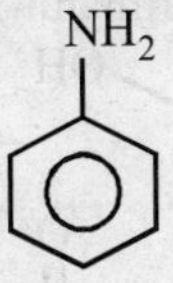

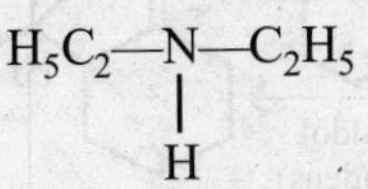

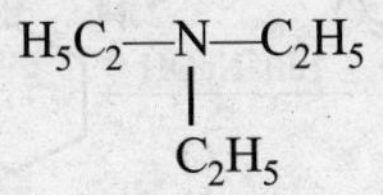

Basic Strength $\alpha \frac{1}{pK_b}$; The order of basicity: A < B < D < C

Due to steric hindrance, lone pair of electrons on N is relatively lower available in (D) than in (C)

41. (c) Pure aniline and arylamines are colourless but get coloured on storage due to atmospheric oxidation as they react readily in air and gets oxidised.

42. (d)

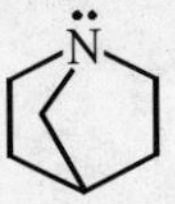

It is most basic due to no amine inversion at N atom.

43. (c) Start from the product.

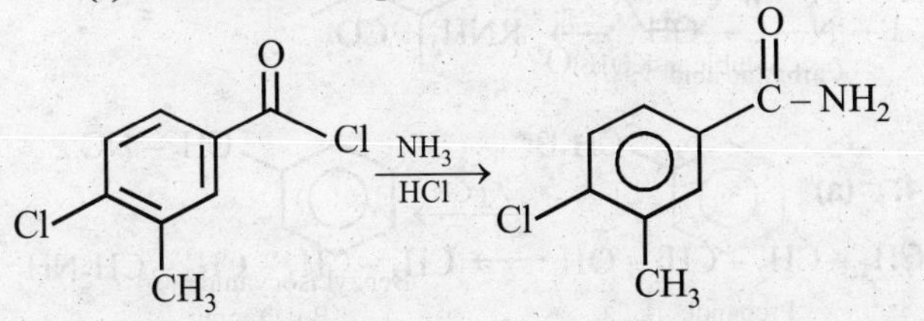

$$\text{(A) 4-Cl-3-CH}_3\text{-C}_6\text{H}_3\text{COCl} \xrightarrow[HCl]{NH_3} \text{4-Cl-3-CH}_3\text{-C}_6\text{H}_3\text{CONH}_2 \xrightarrow[NaOH]{Br_2} \text{4-Cl-3-CH}_3\text{-C}_6\text{H}_3\text{NH}_2$$

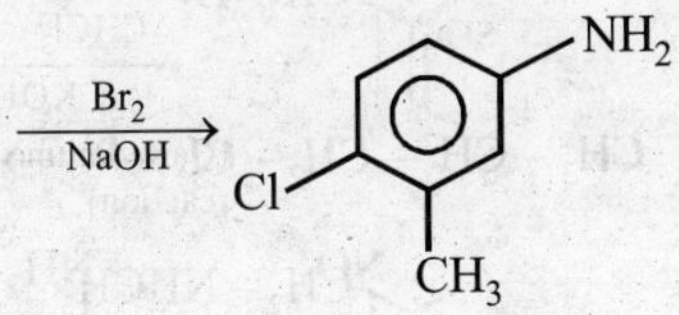

44. (a) $C_6H_5SO_2Cl$

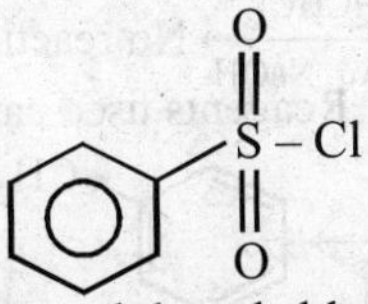

Benene sulphonyl chloride is known's Hinsberg reagent.

45. (c) In Saytzeff's product, proton is removed from the carbon atom which is less substituted.

$$CH_3CH_2CH_2CH(\overset{+}{N}(CH_3)_3)-CH_3 \xrightarrow[EtOH]{EtO^-} CH_3CH_2CH_2CH=CH_2 \text{ (major) (A) Hofmann product} + CH_3CH_2CH=CH-CH_3 \text{ (minor) (B) Saytzeff product}$$

46. (a) In acidic medium; anilinium ion is formed due to which meta product is also formed in significant amount.

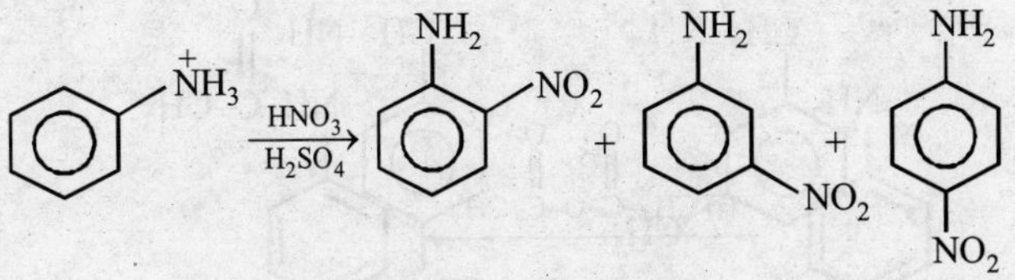

$$C_6H_5\overset{+}{N}H_3 \xrightarrow[H_2SO_4]{HNO_3} \text{o-nitroaniline (2\%)} + \text{m-nitroaniline (47\%)} + \text{p-nitroaniline (51\%)}$$

47. (c) Aniline acts as Lewis base and reacts with Lewis acid $AlCl_3$ to form anilinium hydrochloride. This results in positive charge on the nigrogen atom. Hence, it will not react further. So no *o–m–*or *p*-methyl aniline formed as a product.

48. (c)

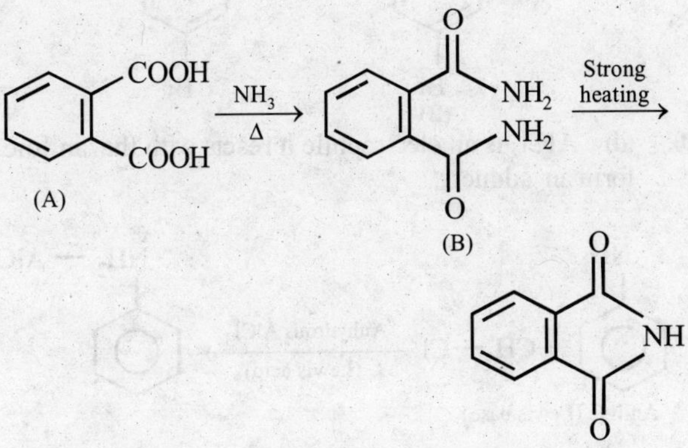

$$\text{Phthalic acid (A)} \xrightarrow[\Delta]{NH_3} \text{Phthalamide (B)} \xrightarrow{\text{Strong heating}} \text{Phthalimide (C)}$$

Gabriel phthalimide synthesis :

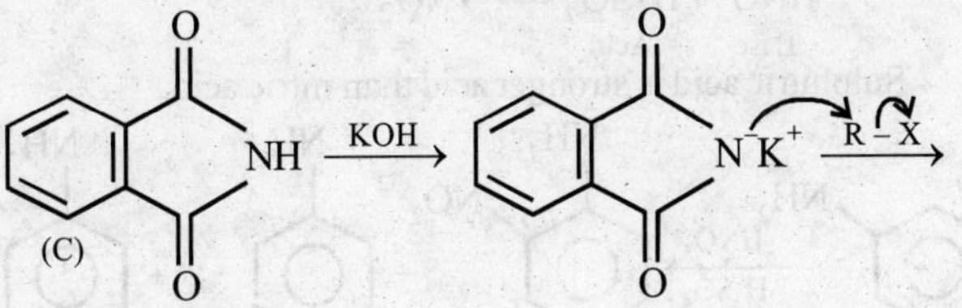

$$\text{Phthalimide (C)} \xrightarrow{KOH} \text{Potassium phthalimide } (\overset{-}{N}K^+) \xrightarrow{R-X} \text{N-R phthalimide} \xrightarrow[{[H_2O]}]{NaOH} C_6H_4(COO^-Na^+)_2 + RNH_2 \text{ (Primary amine)}$$

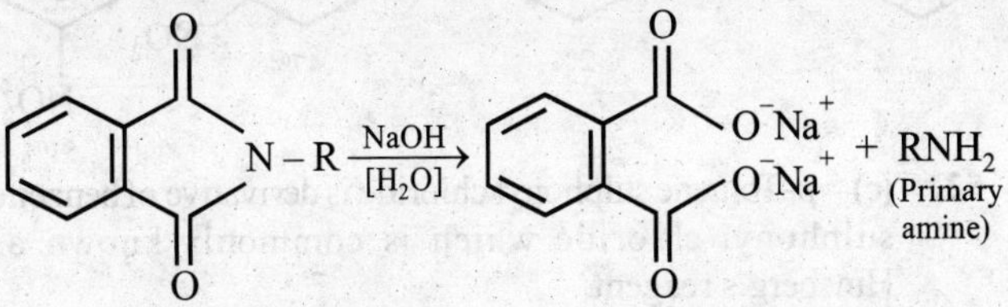

49. (c)

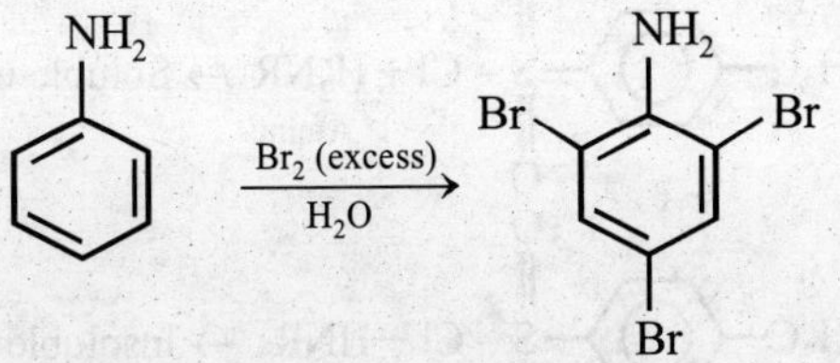

$$C_6H_5NH_2 \xrightarrow[H_2O]{Br_2\text{ (excess)}} \text{2,4,6-tribromoaniline (A)}$$

Acetic anhydride acts as protecting agent and decreases the electron donating tendency of $-NH_2$ group.

$$C_6H_5NH_2 \xrightarrow{\text{(i) } CH_3COOCOCH_3} C_6H_5NHCOCH_3 \xrightarrow{\text{(ii) } Br_2/CH_3COOH} p\text{-}BrC_6H_4NHCOCH_3 \xrightarrow{\text{(iii) HCl}} p\text{-}BrC_6H_4NH_2 \text{ (B)}$$

50. **(d)** $AlCl_3$ is an electrophile it reacts with the aniline to form an adduct

$$C_6H_5\ddot{N}H_2 + CH_3-Cl \xrightarrow[\text{(Lewis acid)}]{\text{Anhydrous } AlCl_3} C_6H_5\overset{\oplus}{N}H_2-\overset{\ominus}{Al}Cl_3$$

Aniline (Lewis base)

51. **(c)** In acidic medium the aniline gets protonated to form anilinium ion and hence substantial amount of *m*-product is also formed.

$$\underset{\text{Base}}{HNO_3} + \underset{\text{Acid}}{H_2SO_4} \longrightarrow NO_2^+$$

Sulphuric acid is stronger acid than nitric acid.

$$C_6H_5NH_2 \xrightarrow[H_2SO_4]{HNO_3} \underset{2\%}{o\text{-}NO_2C_6H_4NH_2} + \underset{47\%}{m\text{-}NO_2C_6H_4NH_2} + \underset{51\%}{p\text{-}NO_2C_6H_4NH_2}$$

52. **(c)** *p*-Toluene sulphonyl chloride is derivative of benzene sulphonyl chloride which is commonly known as Hinsberg's reagent.

$$H_3C-C_6H_4-SO_2-Cl + \underset{1^\circ \text{ Amine}}{H_2NR} \rightarrow \text{Soluble in alkali}$$

$$H_3C-C_6H_4-SO_2-Cl + \underset{2^\circ \text{ Amine}}{HNR_2} \rightarrow \text{Insoluble in alkali}$$

$$H_3C-C_6H_4-SO_2-Cl + \underset{3^\circ \text{ Amine}}{NR_3} \rightarrow \text{No reaction}$$

53. **(a, c)** The given reaction is Hoffmann-Bromamide degradation method.

$$RCONH_2 + \overset{\ominus}{O}H \rightleftharpoons RCO\bar{N}H \xrightarrow{Br-Br} RCONHBr \xrightarrow{\overset{\ominus}{O}H} RCO\ddot{\overset{\ominus}{N}}-Br \rightarrow R-\overset{\ominus}{N}-\overset{\oplus}{C}=O \rightarrow \underset{\text{Isocyanate}}{R-N=C=O} \xrightarrow{\overset{\ominus}{O}H} R-N=C(O^{\ominus})-OH$$

$$\xrightarrow{\text{Proton transfer}} \underset{\text{Carbamic acid}}{R-NH-CO-OH} \longrightarrow RNH_2 + CO_2$$

54. **(a)**

$$\underset{\text{Propanol}}{CH_3-CH_2-CH_2-OH} \xrightarrow{SOCl_2} CH_3-CH_2-CH_2-Cl \xrightarrow{KCN} CH_3-CH_2-CH_2-CN \xrightarrow{H_2/Ni,\ Na(Hg)/C_2H_5OH} \underset{n\text{-Butanamine}}{CH_3-CH_2-CH_2-CH_2NH_2}$$

55. **(5)** $C_6H_5NO_2 \longrightarrow C_6H_5NH_2$ Reagents used can be

(i) Sn + HCl (ii) Fe + HCl
(iii) Zn + HCl (iv) H_2 – Pd
(v) H_2 (Raney Ni)

56. **(a)** The intermolecular association is more prominent in case of primary amines as compared to secondary amines, due to the availability of two hydrogen atoms bounded to N-atom.

57. **(b)** Aniline is more basic than acetamide because in acetamide, lone pair of nitrogen is delocalised to more electronegative element oxygen, whereas in aniline, lone pair of nitrogen is delocalised over benzene ring.

$$C_6H_5\ddot{N}H_2, \quad CH_3-C(=O)-\ddot{N}H_2 \longleftrightarrow CH_3-C(-O^-)=\overset{+}{N}H_2$$

58. **(108°)** In gaseous triethyl amine, the "–C–N–C–" bond angle is 108 degree.

59. **(a)** Aniline $\xrightarrow[288\ K]{HNO_3,\ H_2SO_4}$ o-nitroaniline (A, 2%) + m-nitroaniline (B, 47%) + p-nitroaniline (C, 51%)

During nitration of aniline in strong acidic (HNO_3, H_2SO_4) medium, aniline changes to anilinium ion that can withdraw electron density. Its effect is felt maximum at ortho followed by meta and then para position. Consequently, very little of ortho nitrated product is formed.

60. **(c)** $C_6H_5-CH(CH_3)-NH_2 + Cl-SO_2-C_6H_5 \longrightarrow$
(A)

$C_6H_5-CH(CH_3)-NH-SO_2-C_6H_5 \xrightarrow{NaOH} C_6H_5-CH(CH_3)-\overset{\ominus}{N}(Na^{\oplus})-SO_2-C_6H_5$
(B)

Compound (B) possesses acidic hydrogen, so it is soluble in NaOH.

61. **(a)** Hinsberg reagent is benzene sulphonyl chloride. It is used for the separation of primary, secondary and tertiary amines in a given sample. Use of *p*-toluene sulphonyl-chloride is more effective than Hinsberg reagent.

62. **(a)** The amine reacts with benzene sulphonyl chloride and forms a product which is insoluble in alkali. So it should be secondary amine.

$R-NH(R) \xrightarrow{C_6H_5SO_2Cl} R-N(R)SO_2C_6H_5 \xrightarrow{alkali}$ Insoluble

Since it is prepared by ammonolysis of C_2H_5Cl, it must contain an ethyl group, therefore, the given amine is $CH_3CH_2CH_2NH-CH_2CH_3$.

$CH_3CH_2Cl \xrightarrow[(-HCl)]{NH_3} CH_3CH_2NH_2$

$\xrightarrow{ClCH_2CH_2CH_3} CH_3CH_2NHCH_2CH_2CH_3$

63. **(d)** (A) Phenyl methanamine : $C_6H_5CH_2NH_2$

(Electron pair on N not involved in resonance)
$pK_b = 4.7$

(B) N, N-Dimethylaniline : $C_6H_5N(CH_3)_2$

$pK_b = 8.92$

(C) N-Methyl aniline : $C_6H_5NH-CH_3$; $pK_b = 9.3$

(D) Benzenamine : $C_6H_5NH_2$; $pK_b = 9.38$

Electron pair on N in (*A*) is localised, while in other three (*B*, *C* and *D*) electron pair on N is delocalised in the ring. Further, alkyl groups are electron donating, hence increase the basicity of the concerned amine(s). Thus basic order of the amines is
A > B > C > D

64. **(c)** $C_6H_5NH_2 \xrightarrow[+H_2SO_4]{Conc.\ HNO_3} C_6H_5\overset{+}{N}H_3$ (Very high –I effect)
(Anilinium ion)

The positive anilinium nitrogen deactivates the benzene ring due to –I effect. –I effect is more on *-o* and *-m* position and less on *p-* position.
Therefore, *-m* product is also formed along with *-o* and *-p* products.

65. **(3)** Gabriel phthalimide synthesis is used to prepare 1° aliphatic/alicyclic amine.
Hence amine which can synthesised by Gabriel phthalimide synthesis method is :
(A) $Me_2CH-CH_2-NH_2$ (B) $CH_3CH_2NH_2$
(C) $Ph-CH_2-NH_2$

66. **(d)** $C_6H_5NH_2$ (Aniline) $+ CHCl_3 + 3KOH \longrightarrow C_6H_5NC + 3KCl + 3H_2O$

67. (b) I: $N(CH_3)_2$ with p-OCH_3; II: $N(CH_3)_2$; III: $NHCH_3$ with m-CN; IV: $NHCH_3$ with m-OH

I: +R effect and –I effect of $-OCH_3$; III: (–I effect of –CN group); IV: (–I effect of –OH group)

$-OCH_3$ group increases electron density of ring at *o*- and *p*- position making (I) most basic. (III) is least basic due to –I effect of –CN group at meta position.
Since, –I effect of –CN > –I effect of –OH group.
Hence, correct basic strength will follow the order

$$I > II > IV > III$$

$$\text{Basic strength} \propto \frac{1}{pK_b \text{ value}}$$

Order of pK_b value is: I < II < IV < III.

68. (a) Gabriel phthalimide synthesis gives 1° amine in good yield.

69. (c) $CH_3CH_2-C\equiv N \xrightarrow{LiAlH_4} CH_3CH_3-CH_2-NH_2$
$NaBH_4$ does not reduce R–CN.

70. (b) (A) pyridine — sp^2 hyb. N atom (localised L.P.)
(B) pyrrole — sp^2 hyb. N atom (delocalised L.P.)
(C) pyrrolidine — sp^3 hyb. N atom (localised L.P.)
(D) imidazole — sp^2 hyb. N atom (localised L.P. with partial negative ch.)

Thus, increasing order of basicity is

$$(B) < (A) < (D) < (C).$$

71. (d) 3-methylaniline $\xrightarrow[Py]{Ac_2O}$ $HN-C(=O)-CH_3$ (3-methylacetanilide) $\xrightarrow[FeBr_3]{Br_2}$ 4-bromo-3-methylacetanilide $\xrightarrow{OH^-/\Delta}$ 4-bromo-3-methylaniline

72. (d) Basic strength of amines depends upon availability of lone pair of electrons. Aliphatic amines are more basic than aromatic amines.

73. (d) Lower the pK_b value of a base (amine), higher will be K_b value of the base. Further basic nature of bases can best be compared by their conjugate acids. A stronger base has weak conjugate acid. The conjugate acids of the three bases are

$\ddot{N}H_2-CH=\overset{+}{N}H_2 \leftrightarrow \overset{+}{N}H_2=CH-NH_2$ (Conjugate acid of A)

Conjugate acid of B (cyclic amidinium, $\overset{+}{N}H$) — More regonance stabilized

$CH_3-\overset{+}{N}H_2-CH_3$ (Conjugate acid of C)

Thus conjugate acid of (B) is more stable than that of (A) which in turn more stable than C. Thus the basic character of the three is B > A > C and hence pK_b value follows the following order (B) < (A) < (C).

74. (a) p-toluidine (NH_2, CH_3) $\xrightarrow{Ac_2O}$ $NHCOCH_3$ (A) $\xrightarrow[AcOH]{Br_2}$ $NHCOCH_3$ with Br ortho, CH_3 para (B)

75. (c) Electron withdrawing group attached to benzene ring will reduce the basic strength and increase pK_b, while electron donating group decreases pK_b.
Therefore the correct increasing order of pK_b is

$$(B) < (D) < (A) < (C)$$

76. (a) N-Ethyl phthalimide on treatment with NH_2-NH_2 gives ethylamine (hydrazinolysis).

N-ethylphthalimide $+ NH_2-NH_2 \longrightarrow$ intermediate ($\overset{+}{N}H-NH_2$, $N-C_2H_5$) $\longrightarrow$ $C(=O)-NH-NH_2$ / $C(=O)-NH-C_2H_5$ $\longrightarrow$ cyclic intermediate (NH, $NH-H$, $\ddot{N}H-C_2H_5$, O^-) $\longrightarrow$ phthalhydrazide $+ C_2H_5-NH_2$

In place of NH_2NH_2, we can also use H_2O in presence of H^+ or OH^- as a catalyst, but the acidic or basic hydrolysis is slow, hence hydrozinolysis is preferred.

77. (b) $CH_3-CH(OH)-CH_2-CH_2-NH_2 + H-C(=O)-O-C_2H_5 \rightarrow CH_3-CH(OH)-CH_2-CH_2-\overset{+}{N}H(H)-C(H)(O^-)-O-C_2H_5 \xrightarrow{(C_2H_5)_3N} CH_3-CH(OH)-CH_2-CH_2-NH-C(=O)-H + C_2H_5OH$

78. (d)

Benzonitrile (C_6H_5CN):
- $\xrightarrow{H_2/Ni}$ $C_6H_5CH_2NH_2$
- $\xrightarrow{(i)\ LiAlH_4,\ (ii)\ H_3O^+}$ $C_6H_5CH_2NH_2$
- $\xrightarrow{(i)\ SnCl_2/HCl}$ $C_6H_5CH=NH$ $\xrightarrow{(ii)\ NaBH_4}$ $C_6H_5CH_2NH_2$
- $\xrightarrow{(i)\ HCl/H_2O}$ C_6H_5COOH $\xrightarrow{NaBH_4}$ $C_6H_5CH_2OH$

79. (a)

$C_6H_5-C(=O)-CH_3 \xrightarrow[\text{(Haloform reaction)}]{1.\ NaOCl,\ 2.\ H^+} C_6H_5-C(=O)-OH \xrightarrow{SOCl_2} C_6H_5-C(=O)-Cl \xrightarrow[\text{(aniline)}]{Ph\,NH_2} Ph-N(H)-C(=O)-Ph$

80. (c) Hinsberg's reagent is benzenesulphonyl chloride ($C_6H_5SO_2Cl$). It is used for detection of primary, secondary and tertiary amines.

81. (a) Polysubstitution is a major drawback in Friedel Craft's alkylation because the introduced alkyl group is activating and gives polyalkylated product.

82. (d)

Adenine (N–H) $\xrightarrow{Base}$ adenine anion $(-)$ $\xrightarrow{CH_3-I}$ N-methyladenine (NH_2, N–CH_3)

83. (a) Primary amines are prepared by Gabriel phthalimide synthesis

$C_6H_4(CO)_2NK + C_4H_9Cl \longrightarrow C_6H_4(CO)_2N-C_4H_9 \xrightarrow{NaOH(aq)} C_6H_4(COONa)_2 + n\text{-}C_4H_9NH_2$

Pot. phthalimide

84. (c) $(C_2H_5)_2NH > C_2H_5NH_2 > NH_3$

Two +I gp. one + I gp. No + I gp.

85. (a)

3-(3-aminopropyl)phenyl acetate $\xrightarrow{(i)\ NaNO_2/H^+}$ 3-(3-hydroxypropyl)phenyl acetate $\xrightarrow[\text{(Oxidaion)}]{(ii)\ CrO_3/H^+}$ 3-(3-acetoxyphenyl)propanoic acid $\xrightarrow[\text{hydrolysis, cyclisation}]{(iii)\ H_2SO_4\ (conc.),\ \Delta}$ hydroxy-indanone (OH, C=O)

86. (b) Higher the electronegativity ($sp^2 > sp^3$) of an atom, more will be its capacity to take proton. Thus N labelled as (b), (c) and (d) are better protonated than the N at (a) and (e).

87. (c)

Bicyclic (hydrindane)–CH_2NH_2 $\xrightarrow[\text{(ii) }-N_2]{\text{(i) }NaNO_2/\text{dil. HCl},\ 0-5°C\ \text{(Diazotisation)}}$ bicyclic–$\overset{+}{C}H_2$ (6) $\xrightarrow{\text{Ring expansion}}$ decalin carbocation (+ at C-5)

More stable ring

$\xrightarrow[(-H^+)]{H_2O}$ decalin–OH

88. **(b)** Compound, III (piperidine, N–H) is most basic as the lone pair of nitrogen is easily available for the donation.

In case of compound (I) 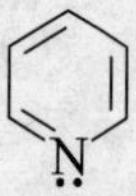 (pyridine) lone pair is not involved in resonance but nitrogen atom is sp^2 hybridsed, whereas in compound II, the lone pair of nitrogen is involved in aromaticity which makes it least basic.

89. **(d)** $-NH_2$ is a stronger base than $-OH$, hence 1 equation of acetic anhydride will acetylate only $-NH_2$ group. Reaction involved:

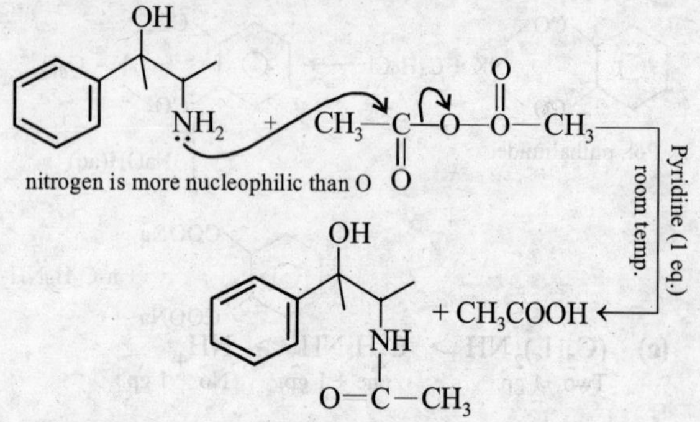

nitrogen is more nucleophilic than O

Pyridine (1 eq.) room temp.

$+ CH_3COOH$

90. **(b)**

	pK_b
(A) $EtNH_2$	3.29
(B) $(Et)_2NH$	3.00
(C) Me_3N	4.22
(D) $Ph-NH-Me$	4.7

So, order of basic strength is:

$C_2H_5 \rightarrow NH \leftarrow C_2H_5 > C_2H_5NH_2 > Me_3N > PhNH{-}Me$

2C_2H_5 groups; 1C_2H_5 group; steric factor; e^- pair delocalised over Ph - ring

(B) (A) (C) (D)

(B) > (A) > (C) > (D)

91. **(a)**

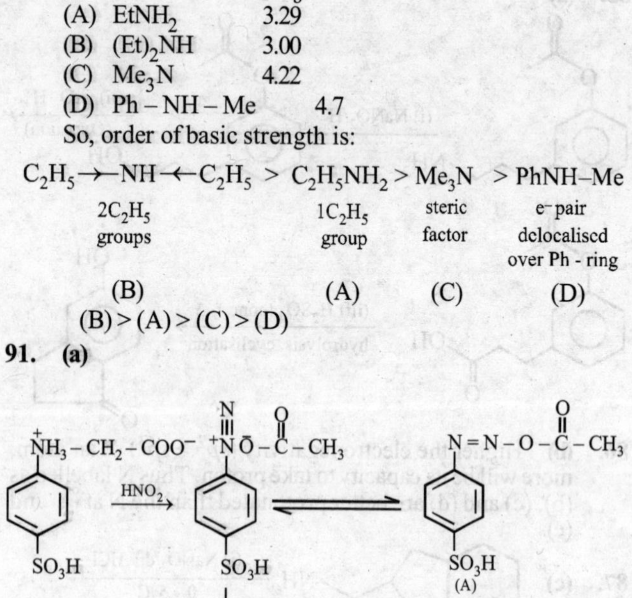

HNO_2

(A) (B)

92. **(d)** Aromatic diazonium salts are more stable than aliphatic diazonium salts. The higher stability of aryl diazonium salts is due to resonance. Electron donating substituents on benzene ring increase electron density. Hereby they increase the stability of diazonium salts. Electron withdrawing substituents decrease electron density on benzene ring. Hereby they decrease the stability of diazonium salts. $-COCH_3$ group is electron withdrawing and hence, diazonium salts from (D) is less stable than that from (B). Although $-O-COCH_3$ is electron donating substituent, but it is present in meta position. Hence, it will not have significant effect on stability.

Therefore, the increasing order of diazotisation is

(A) < (D) < (C) < (B).

93. **(c)** Basic strength of amines depends upon availability of lone pair of electrons. Aliphatic amines are more basic than armoatic amines.

(I) 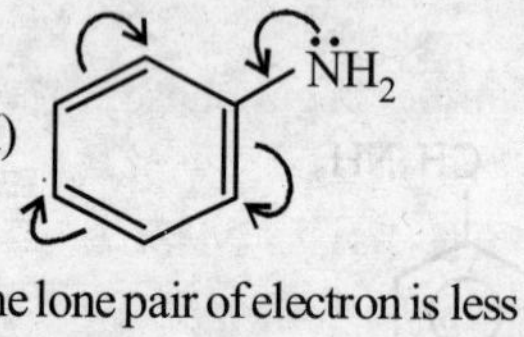 + R effect

the lone pair of electron is less easily available for protonation.

(II)  + R effect

the lone pair of electrons is part of aromatic sextet, less available for proton.

(III) 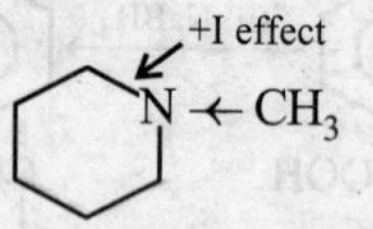 +I effect $N \leftarrow CH_3$

(IV) 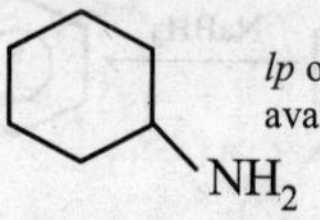 *lp* of e^- on N is most easily available for protonation

Thus, the correct order is: (II) < (I) < (IV) < (III)

94. **(c)** Nitration takes place in presence of conc. HNO_3 + conc. H_2SO_4.

In strongly acidic medium, amine is converted into anilinium ion $(-NH_3^+)$; substitution is thus controlled not by $-NH_2$ group but by $-NH_3^+$ group which, because of its positive charge, directs the entering group to the meta-position instead of ortho- and para.

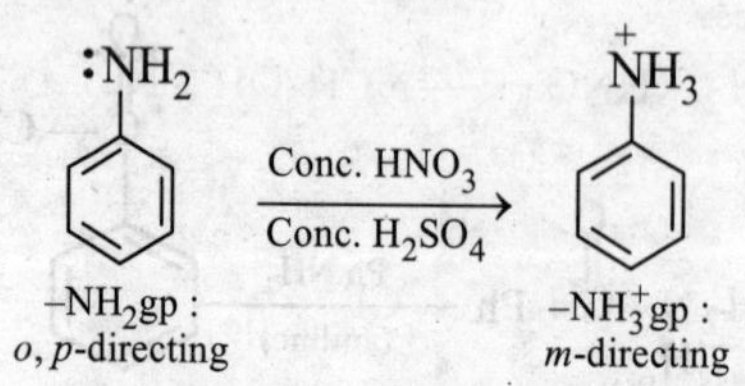

Conc. HNO_3 / Conc. H_2SO_4

$-NH_2$gp : *o, p*-directing → $-NH_3^+$gp : *m*-directing

95. **(b)** $RCONH_2 + Br_2 + 4NaOH \longrightarrow R{-}NH_2 + K_2CO_3 + 2NaBr + 2H_2O$

96. **(d)** Primary, secondary and tertiary amines are distinguished by Hinsberg reagent test. The Hinsberg reagent is an alternative name for benzene sulphonyl chloride $(C_6H_5SO_2Cl)$.

97. (a) Lone pair of electrons on N is involved in aromaticity.

98. (c) p-nitroaniline (NO_2, –I) < aniline < p-anisidine ($:OCH_3$, +M due to $-OCH_3$) < $CH_3 \rightarrow \ddot{N}H_2$ (+I)

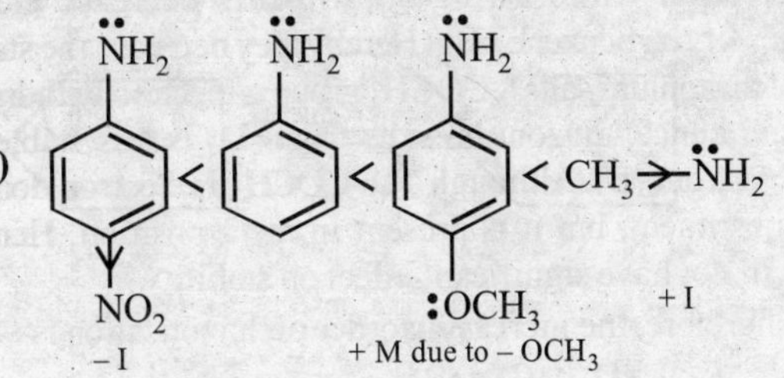

99. (d) $R-CH_2-NH_2 + CHCl_3 + 3KOH\ (alc) \xrightarrow{\text{(Carbylamine reaction)}} R-CH_2-NC + 3KCl + 3H_2O$

Alkyl isocyanide

100. (a) Relative basicities of amines in gas phase is due to +I effect of alkyl group(s).

$(CH_3)_3N > (CH_3)_2NH > CH_3NH_2 > NH_3$

However, in aqueous solution it follows the following order.

$(CH_3)_2NH > CH_3NH_2 > (CH_3)_3N > NH_3$

This is due to solvation effect. Protonated amine forms H-bonding with H_2O and stabilises the concerned conjugate acid.

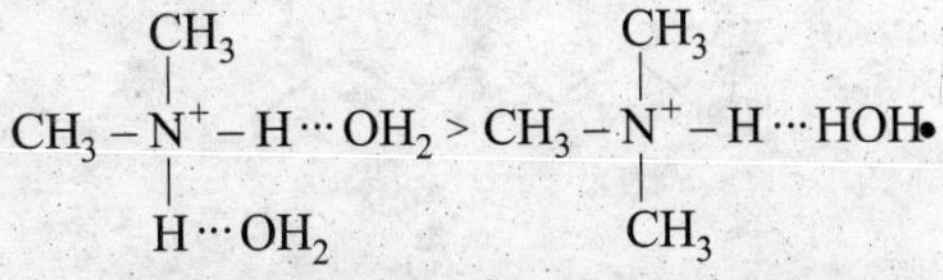

$CH_3-N^+(CH_3)(H\cdots OH_2)-H\cdots OH_2 > CH_3-N^+(CH_3)_2-H\cdots HOH$

Conjugate acid of 2° amine more stable; Conjugate acid of 3° amine forms only one H-bond

Hence the correct order is

$(CH_3)_2NH > CH_3NH_2 > (CH_3)_3N\ C_6H_5NH_2$

101. (b) Amino alcohol (NH₂, OH) $\xrightarrow[\text{(i) diazotisation (ii) } N_2]{NaNO_2/H_2SO_4}$ carbocation (OH) $\xrightarrow{1,2\ CH_3^-\ \text{shift}}$ carbocation adjacent to $\ddot{O}H$ $\xrightarrow{-H^+}$ ketone (O)

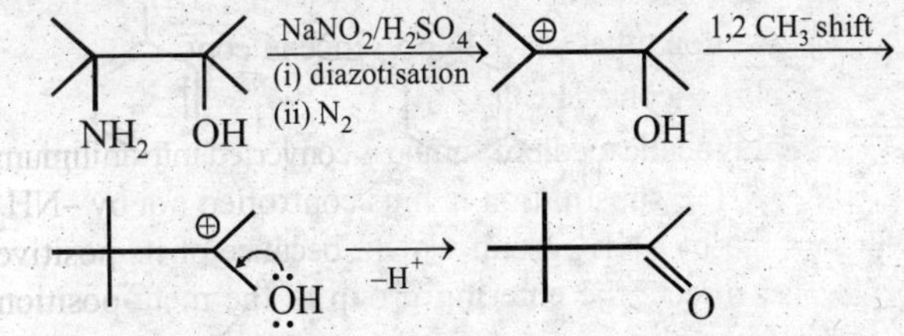

Positive charge better stabilised due to +M effect

102. (b) $CH_3NH_2 + HNO_2 \longrightarrow CH_3OH + N_2 + H_2O$

103. (b) $\underbrace{R-NH_2}_{\text{Mol.Mass}=16} + CH_3-\overset{O}{\overset{\|}{C}}-Cl \xrightarrow{-HCl} \underbrace{R-NH-\overset{O}{\overset{\|}{C}}-CH_3}_{\text{Mol.mass}=58}$

Now since the molecular mass increases by 42 unit as a result of the reaction of one mole of CH_3COCl with one-NH_2 group and the given increase in mass is (390 – 180) 210. Hence, the number of $-NH_2$ groups = 210/42 = 5.

104. (c) The correct order of relative basicity of amines in the gas phase is 3° > 2° > 1° > NH_3

The alkyl group releases electron and thus, tends to disperse the positive charge of the alkyl ammonium ion and therefore, stabilises it. Since, NH_4^+ (from NH_3) has no such alkyl group, it is not stabilised to such an extent as alkylammonium ion.

105. (b) $RNH_2 + CHCl_3 + 3KOH \longrightarrow RNC + 3KCl + 3H_2O$

Mechanism

$CHCl_3 \xrightarrow[\alpha\ \text{elimination}]{OH^-/\Delta} \ddot{C}Cl_2$ (Dichlorocarbene)

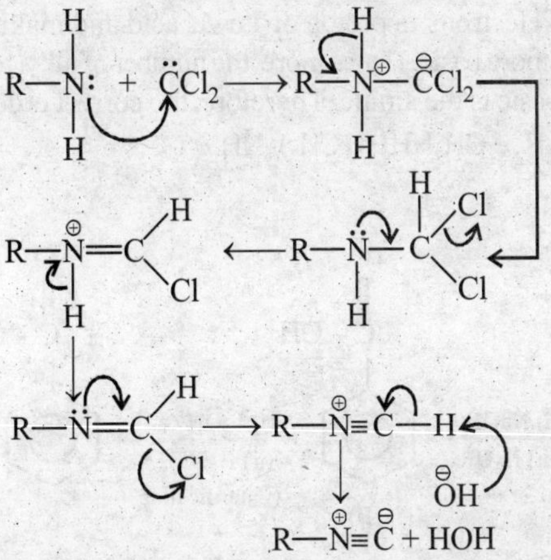

$R-\ddot{N}H_2 + \ddot{C}Cl_2 \longrightarrow R-\overset{\oplus}{N}H_2-\overset{\ominus}{C}Cl_2 \longrightarrow R-NH_2-CHCl_2 \longrightarrow R-\overset{\oplus}{N}H=CHCl \longrightarrow R-\ddot{N}=CHCl \longrightarrow R-\overset{\oplus}{N}\equiv C-H \xrightarrow{\overset{\ominus}{O}H} R-\overset{\oplus}{N}\equiv\overset{\ominus}{C} + HOH$

106. (b) 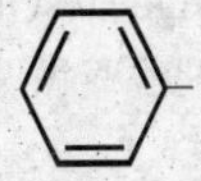 $C_6H_5-CH_2-NH_2$ is most basic due to presence of localised lone pair of electrons on nitrogen atom while in other compounds lone pair of electrons is delocalised.

107. (b) I. Piperidine: sp^3hyb. N-atom (localised *lp*)

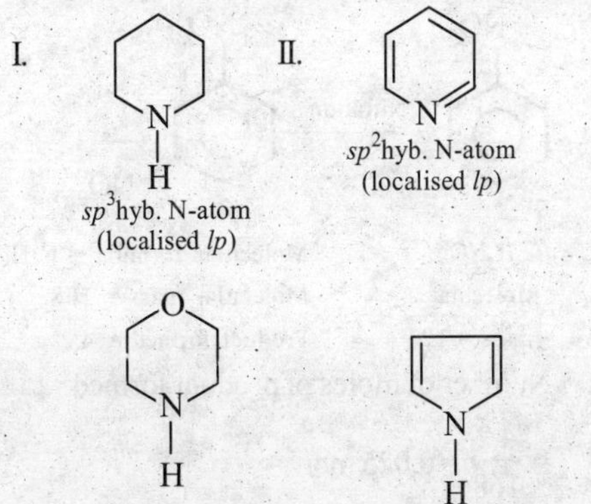

II. Pyridine: sp^2hyb. N-atom (localised *lp*)

III. Morpholine: sp^3hyb. N-atom (localised *lp*) (Due to -I effect of O it is less basic than I)

IV. Pyrrole: sp^3hyb. N-atom (localised *lp*) (*lp* on N contributes towards aromaticity)

Thus, order of basicity: I > III > II > IV

108. (d) **Note:** Aromatic amines (e.g. aniline) are less basic than aliphatic amines due to delocalisation of electrons. Among aliphatic amines the order of basicity in aq. solution is 2° > 1° > 3°. The electron density is decreased in 3° amine due to crowding of alkyl group over N atom which makes the approach and bonding by a proton relatively difficult. Therefore the basicity decreases.

∴ The correct order of basic strength is

Dimethylamine > Methyl amine > Trimethyl amine > Aniline.

109. (b) Wurtz reaction is used for the preparation of hydrocarbons from alkyl halides.

$$RX + 2Na + XR \longrightarrow R-R + 2NaX$$

110. (b) The alkyl groups are electron releasing (+ I) groups, thus increase the electron density around the nitrogen thereby increasing the availability of the lone pair of electrons to proton or Lewis acid and making the amine more basic. Hence more the number of alkyl groups, more basic is the amine. Therefore, the correct order is

$$NH_3 < CH_3NH_2 < (CH_3)_2NH$$

111. (a)

$C_6H_5CONH_2$ (A) $\xrightarrow[\text{(ii) } H_3O^{\oplus}]{\text{(i) NaOH}}$ C_6H_5COOH (B) $\xrightarrow[\text{(ii) } H_2SO_4,\ \Delta]{\text{(i) EtOH}}$ (Esterification) $C_6H_5C(O)-O-Et$ (C)

Molar mass = 121

Benzamide (A) Shows positive Lassaigne test for nitrogen and on hydrolysis benzamide gives benzoic acid (B) which gives effervescence with aq. $NaHCO_3$.

Benzoic acid with alcohol in acidic medium gives ester (C) which gives fruity smell.

112. (3) $C_6H_5NO_2 \xrightarrow[\Delta]{\text{Nitration}}$ m-$C_6H_4(NO_2)_2$

$C_6H_5NO_2$ Molecular mass = 123

Molecular formula = $C_6H_5N_2O_4$
Molecular mass = 168
Product formed = 4.2 g

∴ Number of moles of product formed

$$= \frac{4.2}{168} = 0.025 \text{ mol}$$

Required mass of nitrobenzene = 123 × 0.025 = 3.075 g

∴ Nearest integer is 3.

113. (1) Only C is correct.

114. (a)

N-Me lactam (cyclic amide) $\xrightarrow{\bar{O}H}$ tetrahedral intermediate (C(OH)O^-) → ring opened: Me–NH^- ... C(=O)OH → $MeHN\text{–}CH_2CH_2CH_2\text{–}COO^-$ $\xrightarrow{H^+}$ $MeHN\text{–}CH_2CH_2CH_2\text{–}COOH$

115. (a) $CH_2=CH-C\equiv N \xrightarrow{HO\sim SH}$ (S is better nucleophile than O)

$HO\text{–}CH_2CH_2\text{–}\overset{\oplus}{S}(H)\text{–}CH_2\text{–}CH=C=N^-$ → $HO\text{–}CH_2CH_2\text{–}S\text{–}CH_2CH_2\text{–}CN$

116. (b)

P: $H_3C\text{–}C_6H_4\text{–}CO\text{–}NH\text{–}C_6H_5 \xrightarrow[\Delta]{HCl}$ $H_3C\text{–}C_6H_4\text{–}COOH$ + $C_6H_5NH_3^+$

↓ Filter

Q: $H_3C\text{–}C_6H_4\text{–}COOH$ → CO_2 with $NaHCO_3$

Filtrate: $C_6H_5NH_3^+$ $\xrightarrow{NaOH}$ R: $C_6H_5NH_2$ $\xrightarrow{\text{Hinsberg reagent}}$ $C_6H_5NH\text{–}SO_2\text{–}C_6H_5$ $\xrightarrow{NaOH}$ Soluble

117. (d)

Acetanilide $\xrightarrow[\text{2. AcOH}]{\text{1. }Br_2}$ p-bromoacetanilide (A)

Acetanilide $\xrightarrow{LiAlH_4}$ $C_6H_5NH-CH_2CH_3$ (B)

118. (c) o-(CH_2CONH_2)$C_6H_4COOCH_3$ $\xrightarrow[Br_2/NaOH]{\text{Hoffman bromamide}}$ o-(CH_2NH_2)$C_6H_4COOCH_3$ $\xrightarrow[(-CH_3OH)]{\Delta}$ lactam (NH, C=O)

119. (b)

p-$NH_2C_6H_4CN$ $\xrightarrow[\text{(ii) }H_2O]{\text{(i) AlH (i-Bu)}_2}$ p-$NH_2C_6H_4CHO$ $\xrightarrow[\text{(Dil NaOH, }\Delta\text{ Cross aldol condensation)}]{CH_3CHO}$ p-$NH_2C_6H_4CH{=}CH{-}CHO$

120. (c) A – III, B – IV, C – I, D – II

121. (a)

$$\overset{\delta+}{CH_3CH_2-C}\equiv\overset{\delta-}{N} \xrightarrow{\overset{(-)}{CH_3}\overset{(+)}{MgBr}} CH_3CH_2-\underset{}{\overset{NMgBr}{\overset{\|}{C}}}-CH_3$$

$\downarrow H_3O^+$ (Hydrolysis)

$$CH_3CH_2CH_2CH_3 \xleftarrow[HCl]{Zn/Hg} CH_3CH_2-\overset{O}{\overset{\|}{C}}-CH_3$$

(Clemmensen reduction)

122. (c) Aniline $\xrightarrow[+H_2O]{Br_2}$ 2,4,6-tribromoaniline $\xrightarrow{NaNO_2 + HCl}$ Diazonium salt (N_2Cl^-, 2,4,6-tribromo) $\xrightarrow{H_3PO_2}$ 1,3,5-tribromobenzene

Due to high activating nature of $-NH_2$ group, aniline on reaction with bromine water results in 2, 4, 6-tribromoaniline.

123. (d) In Hofmann degradation reaction, the migration of aryl group is also possible, thus, statement-I is incorrect.

124. (c) All these enamines are interconvertible through their resonating structures. Thus, most stable form is due to least steric factor between – COOH and alkene substituent.

125. (d)

p-$CH_3OC_6H_4CH_2Br$ $\xrightarrow{NaCN}$ p-$CH_3OC_6H_4CH_2CN$ $\xrightarrow{OH^-}$ p-$CH_3OC_6H_4\bar{C}H-CN$ + cyclohexanone → alkoxide (O^-, CN) $\xrightarrow[\text{Hydrogenation}]{H_2/Ni}$ (A): 1-[2-amino-1-(4-methoxyphenyl)ethyl]cyclohexanol (NH_2, OH)

126. (c) $R-C\equiv N \xrightarrow[\text{(2) }H_2O]{\text{(1) DiBAL-H}} R-\overset{O}{\overset{\|}{C}}-H$

Here, Y is $-\overset{O}{\overset{\|}{C}}-H$ (Aldehyde).

127. (a)

$$(CH_3)_3\overset{\oplus}{N}HCl^{\ominus} + NaHCO_3 \longrightarrow H_2CO_3 + (CH_3)_3N + NaCl$$

$$H_2CO_3 \longrightarrow CO_2 + H_2O$$

128. (a)

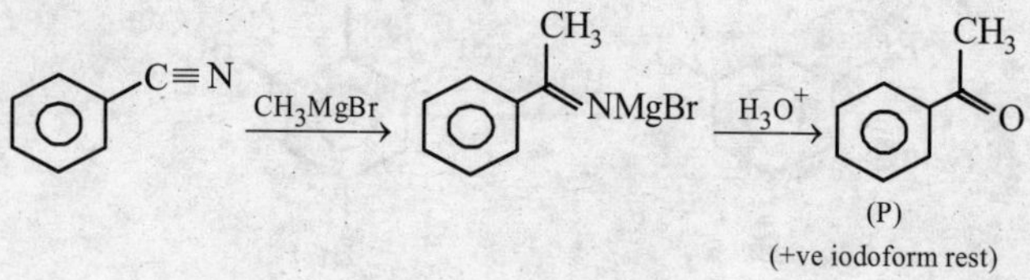

$C_6H_5C\equiv N \xrightarrow{CH_3MgBr} C_6H_5C(CH_3)=NMgBr \xrightarrow{H_3O^+} C_6H_5COCH_3$

(P)
(+ve iodoform rest)

129. (b)

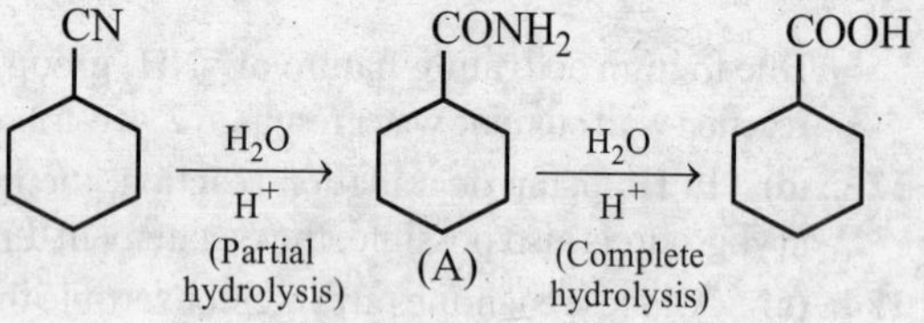

Cyclohexyl–CN $\xrightarrow[\text{(Partial hydrolysis)}]{H_2O,\ H^+}$ Cyclohexyl–$CONH_2$ (A) $\xrightarrow[\text{(Complete hydrolysis)}]{H_2O,\ H^+}$ Cyclohexyl–COOH

130. (77)

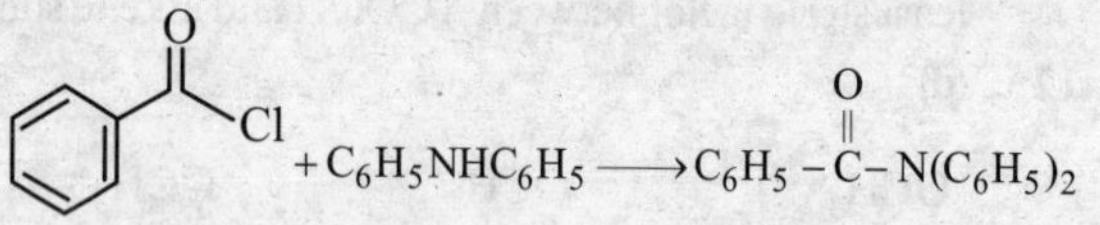

$C_6H_5COCl + C_6H_5NHC_6H_5 \longrightarrow C_6H_5-\overset{O}{\overset{||}{C}}-N(C_6H_5)_2$

	C_6H_5COCl	$C_6H_5NHC_6H_5$	$C_6H_5CON(C_6H_5)_2$
Mass (given)	0.140 g	0.388 g	0.210 g (Observed yield)
Molar mass	140.5 g/mol	169 g/mol	273 g/mol
Moles	; 10^{-3} (Limiting Reagent)	2.3×10^{-3}	Theoretical yield $= 10^{-3}\times 273 = 0.273$ g

$$\therefore \%\ \text{yield} = \frac{\text{Observed yield}}{\text{Theoretical yield}} = \frac{0.210}{0.273}\times100 \approx 77\%.$$

131. (b) Basic strength $\propto$ Availability of lone pair of e^-.

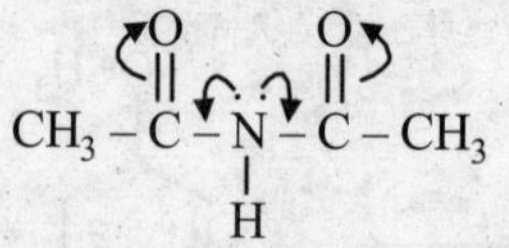

$CH_3-\overset{O}{\overset{||}{C}}-\underset{H}{\underset{|}{N}}-\overset{O}{\overset{||}{C}}-CH_3$

In this case, lone pair of electrons on nitrogen is delocalised due to conjugation with both $>C=O$ group, hence least basic.

132. (d)

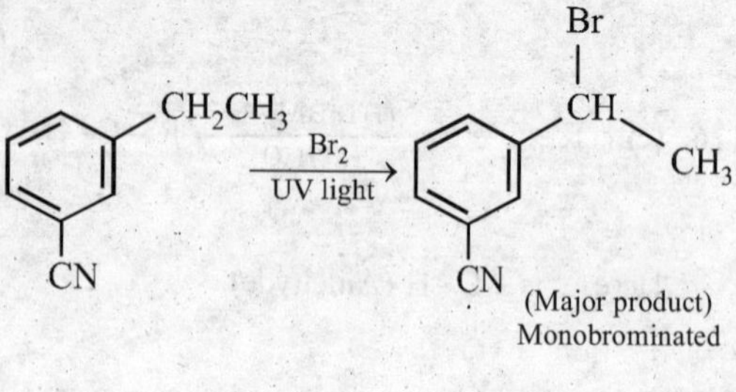

m-$NC-C_6H_4-CH_2CH_3 \xrightarrow[\text{UV light}]{Br_2}$ m-$NC-C_6H_4-CH(Br)-CH_3$

(Major product)
Monobrominated

133. (c)

$CH_3CH_2CN \xrightarrow[\Delta]{H_3O^+} CH_3CH_2COOH \xrightarrow{SOCl_2}$

$CH_3CH_2\overset{O}{\overset{||}{C}}Cl \xrightarrow[\text{(Rosenmund reduction)}]{Pd/BaSO_4,\ H_2} CH_3CH_2\overset{O}{\overset{||}{C}}-H$

134. (c) Aniline and phenol form complex with Lewis acid. Chlorobenzene produces highest yield in Friedel Craft reaction among the given options.

135. (c) Hint: H_2Pd hydrogenate both ketone and cyanide group.

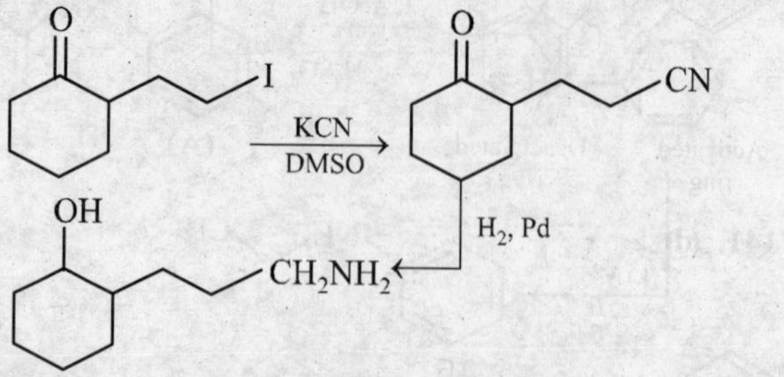

2-(3-iodopropyl)cyclohexanone $\xrightarrow[DMSO]{KCN}$ 2-(3-cyanopropyl)cyclohexanone $\xrightarrow{H_2,\ Pd}$ 2-(4-aminobutyl)cyclohexanol (OH, CH_2NH_2)

136. (d)

(Aromatic amine with NH_2, $COCH_3$, CN) $\xrightarrow[\text{(Carbylamine reaction)}]{CHCl_3,\ KOH}$ (NC, $COCH_3$, CN) $\xrightarrow{H_2/Pd-C}$ ($NH-CH_3$, $CH(OH)CH_3$, CH_2NH_2)

$NH-CH_3$, H_2N, OH

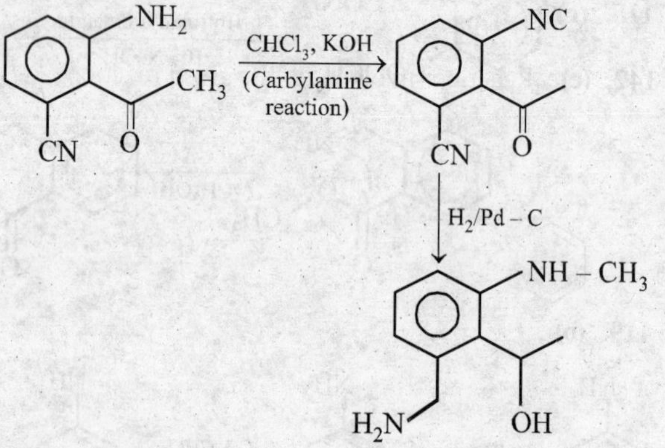

137. (a) Reactivity order of following nucleophiles.

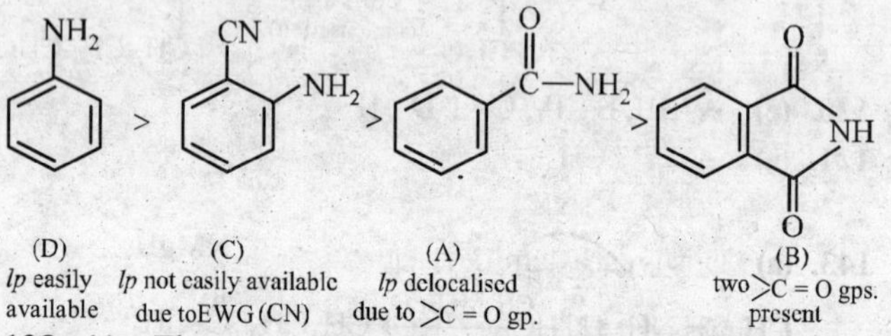

Aniline > 2-aminobenzonitrile > Benzamide ($C_6H_5-\overset{O}{\overset{||}{C}}-NH_2$) > Phthalimide

(D) *lp* easily available; (C) *lp* not easily available due to EWG (CN); (A) *lp* delocalised due to $>C=O$ gp.; (B) two $>C=O$ gps. present

138. (c) The reactivity order of carboxylic acid derivatives depends on the electrophilicity of the carbonyl carbon and leaving tendency of the leaving group.
Leaving ability of the four groups is
$-Cl > -COOC_2H_5 > -OCH_3 > -NH_2$.
Therefore, reactivity order towards $LiAlH_4$ is:
(A) < (B) < (D) < (C).

139. (c)

$\underset{(X)}{CH_3CH_2CH_2CONH_2} \xrightarrow[\text{(Hoffmann rearrangement)}]{Br_2/NaOH} \underset{(C_3H_9N)}{CH_3CH_2CH_2NH_2}$

$\xrightarrow[\text{(Carbylamine reaction)}]{CHCl_3/KOH} CH_3CH_2CH_2NC$

140. (d) In the given nitration reaction, major product will be formed as per the activating group, –NH part of $-NH-\overset{O}{\overset{||}{C}}-$ is activating, while $-\overset{O}{\overset{||}{C}}-$ part is deactivating group.

Activated ring   Deactivated ring

$\xrightarrow[\text{conc. } H_2SO_4]{\overset{+}{N}O_2}$ $O_2N-C_6H_4-NH-CO-C_6H_5$

141. (d)

$C_6H_6 + HCHO + HCl \longrightarrow C_6H_5CH_2-Cl \xrightarrow{AgCN} C_6H_5CH_2-NC$

Benzyl chloride (A)   Benzyl isocyanide (B)

142. (c) Reaction involved for the given reaction:

$C_6H_4(CONH_2)(CH_2-CH_3) \xrightarrow[hv]{Br_2} C_6H_4(CONH_2)(CH(Br)-CH_3) \xrightarrow{\text{(ii) } OH^-} C_6H_4(CO\bar{N}H)(CH(Br)-CH_3) \longrightarrow$ cyclic product (C=O, NH, CH_3)

143. (a) This is carbylamine reaction.

$$CH_3CH_2NH_2 + CHCl_3 + 3KOH \longrightarrow C_2H_5NC + 3KCl + 3H_2O$$

144. (c) $CH_3-C_6H_4-NH_2 + CHCl_3 + 3KOH \longrightarrow CH_3-C_6H_4-N\equiv C + 3KCl + 3H_2O$

145. (d) Ethyl isocyanide on hydrolysis form primary amines.

$$CH_3CH_2N\equiv C + H_2O \xrightarrow{H^+} CH_3CH_2NH_2 + HCOOH$$

146. (a) $C_2H_5NH_2 + CHCl_3 + 3KOH \longrightarrow C_2H_5N\equiv C + 3KCl + 3HCl$

(Ethyl isocyanide)

147. (a)

$C_6H_5NO_2 \xrightarrow{Sn.\ HCl} C_6H_5NH_2$ (Aniline) $\xrightarrow[Py]{CH_3-CO-O-CO-CH_3} C_6H_5NH-CO-CH_3$ (Acetanilide) $\xrightarrow{Br_2,\ CH_3COOH} p\text{-}Br-C_6H_4-NH-CO-CH_3 \xrightarrow{NaOH\ (aq.)} p\text{-}Br-C_6H_4-NH_2$ (p-Bromo aniline)

148. (a) $C_6H_5N_2Cl \xrightarrow{EtOH} C_6H_6$

149. (c) $C_6H_5NH_2$ (A) $\xrightarrow[273\text{-}278\ K]{NaNO_2.\ HCl} C_6H_5N_2Cl \xrightarrow{KI} C_6H_5I$ (B) $\xrightarrow[\text{Dry ether}]{2\ Na} C_6H_5-C_6H_5$ (C)

150. (c) $p\text{-}HO_3S-C_6H_4-\overset{+}{N}H_3CH_3COO^- + HNO_2 \longrightarrow p\text{-}HO_3S-C_6H_4-N=N-O-CO-CH_3 + 2H_2O$

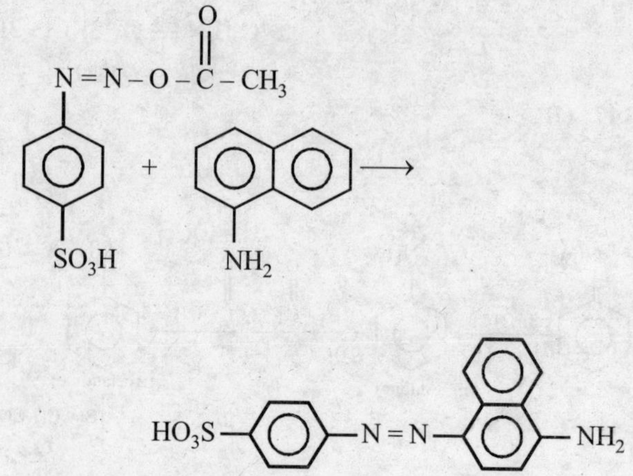

Molar mass of product formed

$=(16\times 12)+(14\times 3)+(16\times 3)+32+13=327$ g

$\therefore$ Mass of 0.1 mol product = 32.7 g ≈ 33 g

151. (a)

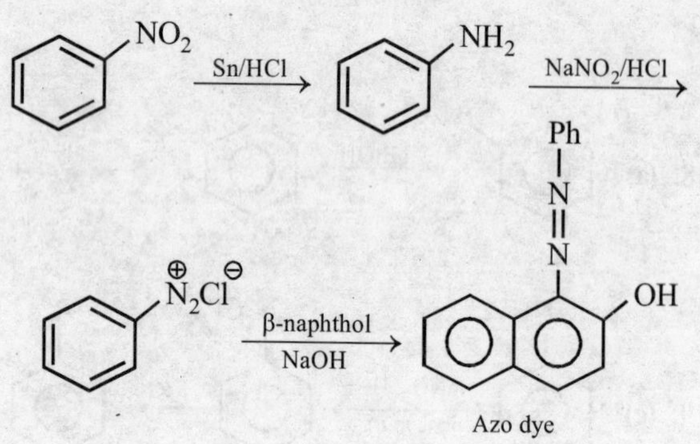

Azo dye

152. (4)

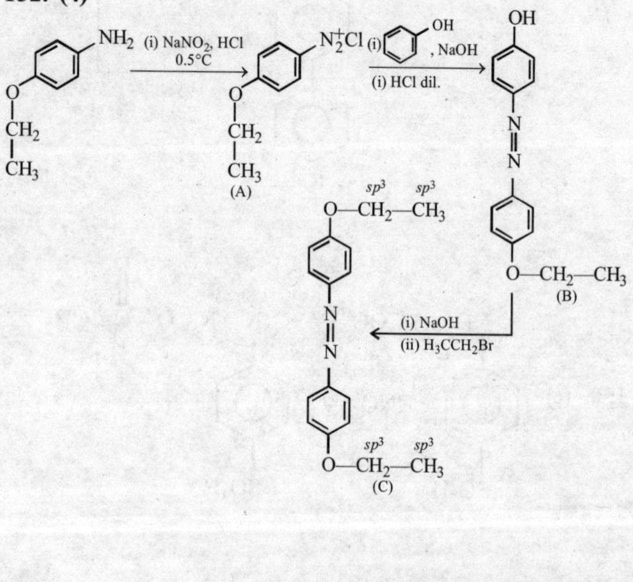

153. (171)

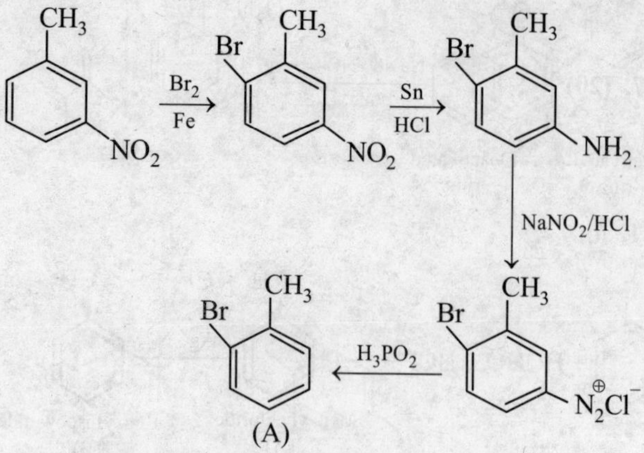

Molar mass of product (C_7H_7Br) (A)

$=12\times 7+1\times 7+80=171$ g/mol

154. (154)

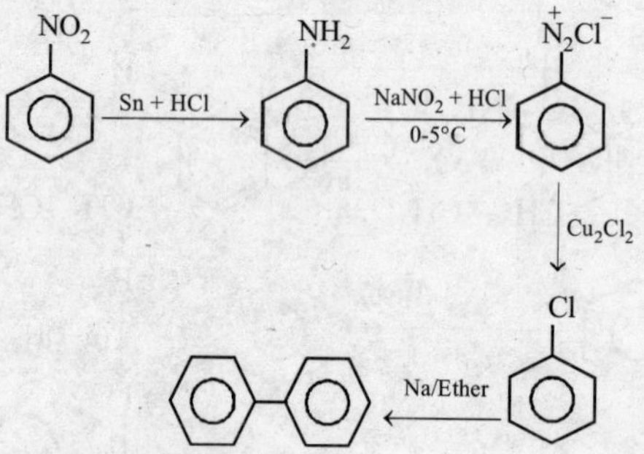

Molar mass of A = 154 gm mol^{-1}

155. (c)

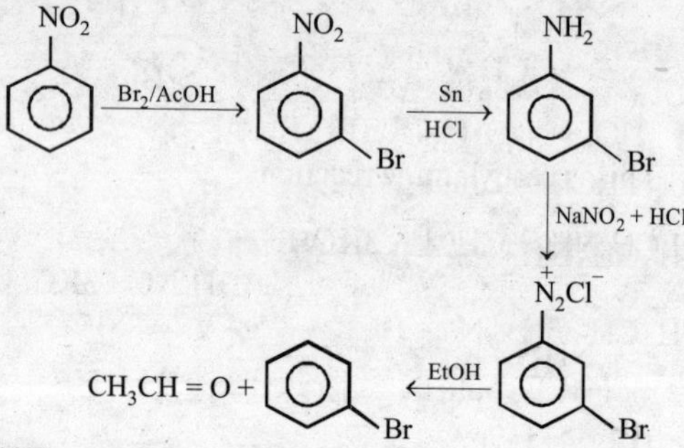

156. (197)

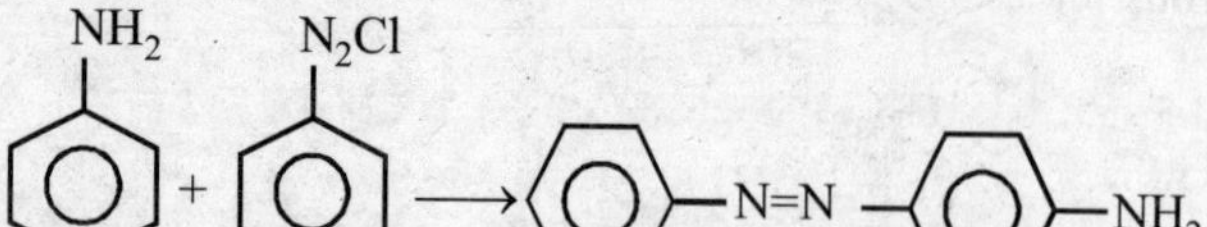

p-Amino azobenzene (yellow dye) (Aniline yellow)

Moles of aniline $=\frac{279}{93}=3$ moles

As benzenediazonium chloride (1 eq.) is limiting reagent here.

Mole of aniline yellow formed = 1 mole

Amount of aniline yellow formed = 197 g. (Molar massof aniline yellow)

157. (20)

Aniline (NH_2) $\xrightarrow[T < 5°C]{NaNO_2 + HCl}$ [$\overset{+}{N_2}Cl^{\ominus}$ (benzenediazonium chloride)] $\xrightarrow{\text{Phenol (OH)}}$ HO–C$_6$H$_4$–N=N–C$_6$H$_5$

Orange dye

(Molar mass = 199)

1 mole of aniline give 1 mole of orange dye,

So $(mol)_{aniline} = (mol)_{orange\ dye}$

$$\frac{9.3g}{93g\ mol^{-1}} = \frac{\text{mass of orange dye}}{199g\ mol^{-1}}$$

Mass of orange dye = 19.9 g ≃ 20 g.

158. (b)

Aniline (NH_2) $\xrightarrow[(0\text{-}5°C)]{NaNO_2 + HCl}$ $\overset{+}{N_2}Cl^-$ $\xrightarrow{Cu_2Cl_2}$ Chlorobenzene (Cl)

(diazotisation reaction) (Sandmeyer reaction)

$\xrightarrow[\text{623K, 300 atm (Dow's process)}]{NaOH,}$ ONa (sodium phenoxide) $\xrightarrow{H^+}$ OH (phenol)

159. (9) $HCl + NaOH \rightarrow NaCl + H_2O$

mol of HCl = 0.2 mol

$$CH_3CH_2NH_2 \xrightarrow[+HCl]{NaNO_2} CH_3CH_2N_2^+Cl \xrightarrow{H_2O}$$

$$CH_3CH_2OH + N_2 + HCl$$

0.2 mol HCl would be generated by 0.2 mol $C_2H_5NH_2$

X = 0.2 × 45 = 9g.

160. (c)

Benzene $\xrightarrow[\text{Conc. } H_2SO_4,\ 323 - 333\ K]{\text{(i) Conc. } HNO_3}$ NO_2 (nitrobenzene) $\xrightarrow[\text{(Reduction)}]{\text{(ii) Sn/HCl}}$ NH_2 (A)

(A) $\xrightarrow{\text{(i) } NaNO_2,\ HCl,\ 273 - 278\ K\ (\text{diazotization})}$ N_2Cl $\xrightarrow{\text{(ii) Phenol}}$ HO–C$_6$H$_4$–N=N–C$_6$H$_5$ (B)

(orange dyl)

161. (d)

Aniline (NH_2) $\xrightarrow[0°\ C]{NaNO_2,\ HCl}$ N_2Cl $\xrightarrow{Cu_2Cl_2}$ Cl (A) + N_2

162. (b)

Aniline (NH_2) $\xrightarrow[0\text{-}5°C]{NaNO_2}$ $\overset{\oplus}{N} \equiv N$ + $H_3C–N–CH_3$ (N,N-dimethylaniline) $\xrightarrow{\text{Diazo Coupling}}$ $H_3C–N–CH_3$–C$_6$H$_4$–N=N–C$_6$H$_5$

163. (b) $NaNO_2 + HX \rightarrow HNO_2 + NaX$

$$H - O - N = O \xrightarrow[-H_2O]{H^{\oplus}} NO^{\oplus} \text{ (Nitrosonium ion)}$$

Me–N–Me (N,N-dimethylaniline) $\xrightarrow{\overset{\oplus}{NO}}$ Me–N–Me–C$_6$H$_4$–NO

p – Nitroso product

164. (c)

CH_3–C$_6$H$_4$–NO_2 $\xrightarrow{Br_2/Fe}$ CH_3, Br, NO_2 $\xrightarrow{Fe/H^+}$ CH_3, Br, NH_2 $\xrightarrow{HONO}$ CH_3, Br, $\oplus N \equiv N$ $\xrightarrow{CuCl}$ CH_3, Br, Cl $\xrightarrow{KMnO_4}$ COOH, Br, Cl

165. (d) A = aniline ($C_6H_5NH_2$), B = 2,4,6-tribromoaniline

C = 2,4,6-tribromobenzenediazonium chloride ($N_2^+Cl^-$) $\xrightarrow[\text{(Reduction)}]{H_3PO_2}$ 1,3,5-tribromobenzene

166. (d) A-III, B-I, C-IV, D-II

167. (d)

p-Nitrotoluene + Br_2 ⟶ 2-bromo-4-nitrotoluene $\xrightarrow{H_2/Pd}$ 3-bromo-4-methylaniline $\xrightarrow{NaNO_2/HCl}$ diazonium salt ($\overset{\oplus}{N_2}\overset{\ominus}{Cl}$) $\xrightarrow{H_3PO_2}$ $N_2 + HCl +$ 2-bromotoluene

168. (d)

Nitrobenzene $\xrightarrow[C_2H_5OH]{H_2/Pd}$ aniline (NH_2) $\xrightarrow[\text{Pyridine}]{(CH_3CO)_2O}$ acetanilide ($NHCOCH_3$)

169. (d)

p-Nitrotoluene $\xrightarrow[H^\oplus]{Fe}$ p-toluidine $\xrightarrow[aq]{Br_2}$ 3,5-dibromo-4-aminotoluene $\xrightarrow{HNO_2}$ diazonium ($N_2^\oplus$) $\xrightarrow{H_3PO_2}$ 3,5-dibromotoluene

170. (b)

p-Nitrotoluene (Me, NO_2) $\xrightarrow{Br_2}$ (A) 2-bromo-4-nitrotoluene $\xrightarrow{Sn/HCl}$ (B) 3-bromo-4-methylaniline (NH_2) $\xrightarrow[\text{273-278K}]{NaNO_2\text{–}HCl}$ (C) diazonium salt ($N_2^+Cl^-$) $\xrightarrow{H_3PO_2/H_2O}$ (D) 2-bromotoluene $\xrightarrow[\text{(ii) } H_3O^+]{\text{(i) } KMnO_4/KOH}$ (E) 2-bromobenzoic acid (COOH, Br)

171. (c)

1-Naphthylamine (NH_2) + $N{=}N{-}O{-}\overset{O}{\overset{\|}{C}}{-}CH_3$ (p-SO_3H phenyl) ⟶ 4-amino-1-naphthyl–$N{=}N$–C$_6H_4$–SO_3H (Red) + CH_3COOH

172. (a)

2-Naphthol (OH) + benzenediazonium chloride ($\overset{+}{N}{\equiv}N\ Cl^-$) $\xrightarrow{^-OH}$ 1-(phenylazo)-2-naphthol ($N{=}N{-}Ph$, OH)

173. (d)

Anilinium chloride ($\overset{+}{N}H_3Cl^-$), MW = 131±2 $\xrightarrow{\text{aq. NaOH}}$ aniline (NH_2) + NaCl + H_2O

Aniline $\xrightarrow{HNO_2}$ benzenediazonium chloride ($\overset{+}{N_2}Cl^-$) $\xrightarrow{PhOH}$ $C_6H_5{-}N{=}N{-}C_6H_4{-}OH$

p-Hydroxy azobenzene
(organic dye)

174. (b) Electron donating group increases the electron density and hence stabilises the diazonium salt whereas electron withdrawing group destabilises the diazonium salt. So the order of stability is

p-methoxybenzenediazonium chloride (A) > benzenediazonium chloride (C) > p-cyanobenzenediazonium chloride (D) > p-nitrobenzenediazonium chloride (B)

$N_2^+Cl^-$ / OCH_3 (A) > $N_2^+Cl^-$ (C) > $N_2^+Cl^-$ / CN (D) > $N_2^+Cl^-$ / NO_2 (B)

175. (b) $C_6H_5N_2^+Cl^- \xrightarrow{H_2O} C_6H_5OH + N_2 + HCl$

176. (b)

p-$NO_2C_6H_4NH_2 \xrightarrow{HNO_2}$ p-$NO_2C_6H_4N_2^+Cl^- \xrightarrow{KI}$ p-$NO_2C_6H_4I \xrightarrow{Fe/H^+}$ p-$NH_2C_6H_4I \xrightarrow{HNO_2}$ p-$N_2^+Cl^-C_6H_4I \xrightarrow{H_2O/warm}$ p-HOC_6H_4I + N_2

177. (b) Diazonium salts of primary aliphatic amines on heating decompose into carbocation.

$$R-NH_2 \xrightarrow[+HCl]{NaNO_2} R-N_2^+ \rightarrow R^+ \xrightarrow[H_2O]{} \underset{\text{alcohol}}{R-OH}$$

178. (c)

$C_6H_5NO_2 \xrightarrow{Sn/HCl} C_6H_5NH_2 \xrightarrow[HCl]{NaNO_2} C_6H_5N_2Cl$ (Diazonium salt)

$C_6H_5N_2Cl$ (Diazonium salt) $\xrightarrow{\beta\text{-naphthol NaOH}}$ $C_6H_5-N=N-$(2-hydroxy-1-naphthyl)

– OH group in the β-napthol increases electron density at ortho, para position.

179. (c)

$C_6H_5NH_2 \xrightarrow[\text{(Diazotization)}]{NaNO_2 + HCl,\ 0\text{–}5\ °C} C_6H_5\overset{+}{N_2}\overset{-}{Cl} \xrightarrow{KCN} C_6H_5CN$ (A)

(A) $\xrightarrow[\text{(Stephen reduction)}]{SnCl_2 + HCl} C_6H_5CH=NH \xrightarrow{H_2O} C_6H_5CHO$ (B)

180. (b)

(a) $CH_3CH_2CH_2NH_2 \xrightarrow{NaNO_2+HCl} CH_3CH_2CH_2\overset{+}{N}\equiv N$

(b) $H_3C-C_6H_4-NH_2 \xrightarrow{NaNO_2+HCl} N\equiv\overset{\oplus}{N}-C_6H_4-CH_3$ (Most stable) (+ I effect of CH_3)

(c) $CH_3-\underset{CH_3}{\underset{|}{CH}}-NH_2 \xrightarrow{NaNO_2+HCl} H_3C-\underset{CH_3}{\underset{|}{CH}}-\overset{+}{N}\equiv N$

(d) It is not a primary aromatic amine.

$C_6H_5-NH-CH_3 \xrightarrow{NaNO_2+HCl} C_6H_5-\underset{N=O}{\underset{|}{N}}-CH_3$

N – Alkyl nitrosoamine

181. (b) $-NO_2$ group is *m*-directing.

$C_6H_5NO_2 \xrightarrow[\text{Step 1}]{Br_2/Fe}$ m-$BrC_6H_4NO_2 \xrightarrow[\text{Step 2}]{Sn + HCl}$ m-$BrC_6H_4NH_2 \xrightarrow[\text{Step 3}]{NaNO_2/HCl}$ m-$BrC_6H_4\overset{+}{N_2}\overset{-}{Cl} \xrightarrow[\text{Step 4}]{CuBr/HBr}$ m-$C_6H_4Br_2$

182. (a)

p-Toluidine (NH_2, CH_3) $\xrightarrow{\text{(i)}NaNO_2 + HCl,\ 0\text{-}5\ °C}$ $\overset{+}{N_2}Cl^-$ (CH_3) $\xrightarrow[HCl]{Cu_2Cl_2}$ Cl (CH_3) [A]

[A] $\xrightarrow[h\nu]{Cl_2}$ Cl, CH_2-Cl [B] $\xrightarrow{\text{Na + dry ether}}$ $Cl-C_6H_4-CH_2-CH_2-C_6H_4-Cl$ [C]

183. (b)

NH_2, CH_2-CH_3 (A) $\xrightarrow[0°C\text{-}5°C]{HNO_2}$ N_2Cl, CH_2-CH_3 (P)

$\xrightarrow{\text{Hydrolysis}}$ OH, CH_2-CH_3 $\xrightarrow{KMnO_4/\overset{+}{H}}$ OH, COOH (R)

$C_6H_5CH_2-NHCH_3$ (B) $\xrightarrow[\text{Hinsberg reagent}]{C_6H_5SO_2Cl}$ $Ph-CH_2-N(CH_3)-SO_2-Ph$

Insoluble in alkali
(NO 'H' on 'N')

$H_2N-C_6H_4-CH_2-CH_3$ (C) $\xrightarrow[0°C\text{-}5°C]{HNO_2}$ $ClN_2-C_6H_4-CH_2-CH_3$ (Q)

$\xrightarrow{H_2O/\overset{+}{H}}$ $HO-C_6H_4-CH_2-CH_3$ $\xrightarrow{KMnO_4/\overset{+}{H}}$ $HO-C_6H_4-COOH$ (S)

Intermolecular H–bond
(Higher b.p. than (R))

184. (b)

m-Bromoaniline (NH_2, Br) $\xrightarrow{NaNO_2 + HCl}$ $\overset{+}{N_2}\ Cl^-$, Br (X) $\xrightarrow{Cu_2Br_2}$ Br, Br (Y)

$\xrightarrow[H_2SO_4]{HNO_3}$ Br, Br, NO_2 (Z)

185. (c)

NH_2, CH_3 (P) $\xrightarrow{Br_2,\ H_2O}$ NH_2, Br, Br, Br, CH_3 ($C_7H_6NBr_3$)

P $\xrightarrow[HCl]{NaNO_2}$ $\overset{+}{N_2}Cl^-$, CH_3 $\xrightarrow{\beta\text{-Naphthol, NaOH}}$ HO, $N=N$, CH_3

Coloured compound
(Dye)

186. (c)

$C_6H_5\overset{+}{N_2}Cl^-$ + $C_6H_5NH_2$ $\xrightarrow{H^+}$ $C_6H_5-N=N-C_6H_4-NH_2$

187. (c) In acidic medium, benzene diazonium chloride reacts with aniline (a basic compound) forming a dye (c).

$H_2N-C_6H_4-N=N-C_6H_5$

(c)

It will not react with phenol in acidic medium.

188. (c) The substitution will take place at para-position.

$C_6H_5N_2^+Cl^-$ + 1-naphthol $\xrightarrow[\text{Coupling reaction}]{OH^-}$ 4-(phenylazo)-1-naphthol ($N=N-C_6H_5$)

189. (a) 4-methyl-3,5-dichloro-1-nitrocyclopentane $\xrightarrow[\text{Heat}]{NaOCH_3 (1\ eq.)}$ 1-nitro-4-methyl-5-chlorocyclopentene

$-NO_2$ gp. increases acidity of $\alpha - H$.

190. (a) Aniline (NH_2) < Chlorobenzene ($:\ddot{Cl}:$) < Toluene (CH_3) < Anisole ($:\ddot{O}CH_3$)

+ M, but becomes $-NH_3^+$ leading to strong – I (A); –I effect only (B); + I effect (D); + M effect (C)

191. (c) $C_6H_5\overset{\oplus}{N_2}\overset{\ominus}{Cl} \xrightarrow{HBF_4} C_6H_5\overset{\oplus}{N_2}\overset{\ominus}{BF_4} \xrightarrow[-N_2,\ -BF_3]{\Delta} C_6H_5F$

192. (a) p-toluidine (NH_2, CH_3) $\xrightarrow[0-5\ °C]{NaNO_2/HCl}$ $N\equiv\overset{+}{N}Cl^-$ (CH_3) (D) $\xrightarrow[\Delta]{CuCN/KCN}$ CN (CH_3) (E)

193. (b) p-nitrotoluene (CH_3, NO_2) $\xrightarrow[FeBr_3]{Br_2}$ 2-bromo-4-nitrotoluene (CH_3, Br, NO_2) $\xrightarrow{Sn/HCl}$ (CH_3, Br, NH_2) $\xrightarrow[HCl]{NaNO_2}$ (CH_3, Br, $N=\overset{+}{N}Cl^-$) $\xrightarrow[CuBr]{HBr}$ (CH_3, Br, Br)

194. (a) On reduction with Zn and HCl, $C_6H_5N_2Cl$ forms aniline as the main product.

195. (c) $C_6H_5N_2Cl$ (Benzene diazonium chloride) $\xrightarrow[\text{or Cu+HCl (Sandmeyer reaction)}]{CuCl/HCl}$ C_6H_5Cl (Chlorobenzene) $+ N_2$

196. (c) Diazonium salts react with copper powder and hydrogen bromide to form bromobenzene. This reaction is known as **Gattermann reaction.**

$C_6H_5N=\overset{+}{N}Cl^- \xrightarrow[HBr]{Cu} C_6H_5Br$

197. (a) Aniline (NH_2) $\xrightarrow[\text{HCl, 278K, Diazotization}]{NaNO_2}$ $C_6H_5N_2^+Cl^-$ Benzene diazonium chloride (A) $\xrightarrow[\text{(Sandmayer reaction)}]{CuCN}$ $C_6H_5C\equiv N$ Benzonitrile (B)

198. (c) $ArNH_2 + NaNO_2 + 2HX \xrightarrow{\text{cold}}$
1° Aromatic amine

$Ar-N\equiv N^+X^- + NaX + 2H_2O$
Arene diazonium salt

The diazonium group can be replaced by fluorine by treating the diazonium salt with fluoroboric acid (HBF_4). The precipitated diazonium fluoroborate is isolated, dried and heated until decomposition occurs to yield the aryl fluoride. This reaction is known as **Balz-Schiemann reaction.**

$$Ar-N_2^+X^- \xrightarrow[(-HX)]{HBF_4} Ar-N_2^+BF_4^-\downarrow$$

$$\xrightarrow{\text{heat}} Ar-F + BF_3 + N_2$$

Biomolecules

1. **(b)** **Statement-I :** Sucrose is formed by condensation reaction between D-(+)-glucose and D-(–)-fructose not D-(+)-fructose.
   Statement-II : Invert sugar is a 1 : 1 molar mixture of D-(+)-glucose and D-(–)-fructose.
2. **(b)** Cellulose derivative have more number of hydroxy groups, so stronger H-bonding is present with water in cellulose derivative of cotton cloths.
3. **(c)**

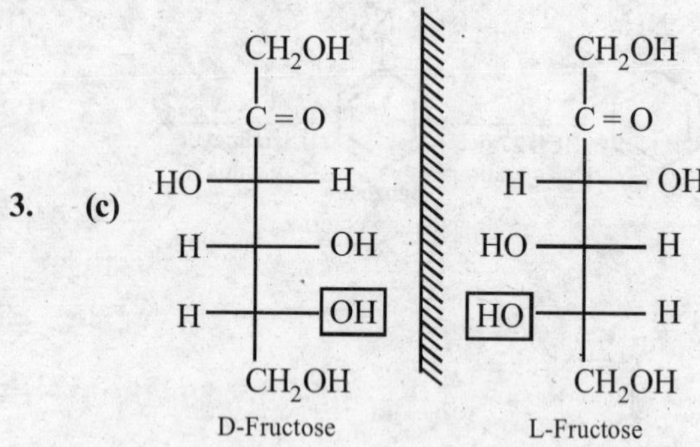

4. **(b)** Amylose $\Rightarrow \alpha - C_1$–C4 glycosidic linkage, plants
   Cellulose $\Rightarrow \beta - C_1 - C_4$ glycosidic linkage, plants
   Glycogen $\Rightarrow \alpha - C_1 - C_4$ glycosidic linkage, Animals
   Amylopectin $\Rightarrow \alpha - C_1 - C_4, \alpha - C_1 - C_6$, plants
5. **(c)**
   (A) Starch $\xrightarrow{H^+/H_2O}$ Glucose
   (B) Cane sugar (Sucrose) $\xrightarrow{H^+/H_2O}$ glucose + fructose (50% 50%)
   (C) Milk sugar (Lactose) $\xrightarrow{H^+/H_2O}$ glucose + galactose
   (D) Amylopectin $\xrightarrow{H^+/H_2O}$ Glucose
   (E) Amylose $\xrightarrow{H^+/H_2O}$ Glucose
   So, correct options are B, C and E only
6. **(a)**
   (A) Sucrose $\rightarrow \alpha_1 - \beta_2$ Glycosidic linkage
   (B) Maltose $\rightarrow \alpha$ 1– 4 Glycosidic linkage
   (C) Lactose $\rightarrow \beta$ 1– 4 Glycosidic linkage
   (D) Amylopectin $\rightarrow \alpha$ 1– 4 and α 1 – 6 Glycosidic linkage
7. **(b)** D-glucose pentaacetate does not have a free aldehyde group which is required for this reaction so does not react with 2,4-dinitrophenylhydrazine.
8. **(a)** In Ribose, carbohydrate present in DNA is β-2-Deoxy-D-Ribose whose structure is

   $HOCH_2$ O OH
   H H H H
   OH H

   which is a reducing D-sugar in β anomeric form & it is a pentose sugar.
9. **(b)** Fructose has α-hydroxy ketone group which tautomerise to aldehyde group in presence of base therefore it reduces Tollen's reagent.
   In presence of base, fructose undergo rearrangement to give glucose and mannose.

   (D-Fructose): $CH_2 - OH$; C = O; HO — H; H — OH; H — OH; CH_2OH $\underset{}{\overset{\overline{O}H}{\rightleftharpoons}}$ (Enediol): CH – OH; ‖ C – OH; HO — H; H — OH; H — OH; CH_2OH

   $\rightleftharpoons$ (D-Glucose): CH = O; H — OH; HO — H; H — OH; H — OH; CH_2OH + (D-Mannose): CH = O; HO — H; HO — H; H — OH; H — OH; CH_2OH
10. **(a)**

   (A): CHO; H — OH; H —* OH; CH_2OH
   (B): CHO; H — OH; HO — H; H —* OH; CH_2OH
   (C): CHO; H — OH; HO — H; HO —* H; CH_2OH
   (D): CHO; H — OH; HO — H; H —* OH; CH_2OH

In structure A, B, D, – OH group present in right hand (After 180° rotation) side so D configuration is assigned.

11. (a) Glucose is soluble in water due to presence of alcohol functional group and extensive hydrogen bonding.

12. (a) Glucose $\xrightarrow[\Delta]{Br_2}$ Gluconic acid

13. (a) In L isomer –OH group is on left hand side of lowest asymmetric carbon.

14. (c) α- Glucose and α- Galactose differ in configuration of only one asymmetric carbon. Thus, they are epimers. α and β-glucose differ at anomeric carbon. Thus they are anomers.
α- Glucose and α-fructose have different functional groups. Thus they are functional isomers.
α- Glucose and α-Ribose are homologues.

15. (c) Glucose $\xrightarrow[\Delta]{NaHCO_3}$ No reaction

Glucose $\xrightarrow[\Delta]{HNO_3}$ Saccharic acid

Glucose $\xrightarrow[\Delta]{HI}$ n-Hexane

Glucose $\xrightarrow[\Delta]{Br_2}$ Gluconic acid

16. (a) Reducing sugar reacts with Fehling's reagent to give reddish brown precipitate of Cu_2O. Non-reducing sugar does not reacts with Fehling's reagent. Lactose, glucose and maltose are reducing sugars but sucrose is a non-reducing sugar.

17. (b)

CHO; H–C–OH; CH_2OH (D(+) Glyceraldehyde) $\xrightarrow[\text{(ii) } H^+]{\text{(i) HCN}}$ [CN; H–C–OH; H–C–OH; CH_2OH] + [CN; HO–C–H; H–C–OH; CH_2OH]

$\downarrow H_3O^+$ each:

[COOH; H–C–OH; H–C–OH; CH_2OH] and [COOH; HO–C–H; H–C–OH; CH_2OH]

$\downarrow HNO_3$ each:

[COOH; H–C–OH; H–C–OH; COOH] Optically inactive (meso) and [COOH; HO–C–H; H–C–OH; COOH] Optically active

18. (b)

CHO; H–C*–OH; HO–C*–H; CH_2OH (X) $\xrightarrow{HNO_3}$ COOH; H–C*–OH; HO–C*–H; COOH (A)

L–Tetrose with two chiral centre

CHO; H–C*–OH; HO–C*–H; CH_2OH $\xrightarrow{NaBH_4}$ CH_2OH; H–C*–OH; HO–C*–H; CH_2OH (B) optically active

CHO; H–C*–OH; HO–C*–H; CH_2OH $\xrightarrow[Py]{CH_3COCl}$ CHO; H–C*–$OCOCH_3$; CH_3COO–C*–H; CH_2OCOCH_3

(X) gives positive Schiff's test due –CHO group

(X) is L–tetrose in the open chain structure.

19. (2) Both Amylose and Sucrose do not give Benedict's test as they are non-reducing sugars.

20. (a) The given six membered cyclic structure of sugar is represented by Haworth structure (a).

21. (b) Seliwanoff's test is a differentiating test for ketoses and aldoses. Seliwanoff's reagent is a mixture of conc. HCl and resorcinol. This test relies on the principle that the ketohexoses are more rapidly dehydrated to form 5-hydroxy-methyl furfural when heated in acidic medium which on condensation with resorcinol gives cherry red or brown red coloured complex rapidly indicating a positive test. Aldose sugars react with the Seliwanoff's reagent to give faint pink colour which intensify with time.

22. (d)

23. (d) Pyranose ring is due to hemiacetal formation between C_1 and C_5 carbon atoms. For converting Fischer structure to Haworth structure, – OH group lying on left side in Fischer structure will point upwards in Haworth structure and thus –OH group on right will point downwards. Thus for the given cyclic hemiacetal (X) the correct pyranose structure is (d).

24. (a)

$$\begin{array}{c}CH_2OH\\|\\C=O\\|\\(CHOH)_3\\|\\CH_2OH\end{array}\xrightarrow{HCN}\begin{array}{c}CH_2OH\\|\\HO-C-CN\\|\\(CHOH)_3\\|\\CH_2OH\end{array}\xrightarrow{H_3O^+}\begin{array}{c}CH_2OH\\|\\HO-C-COOH\\|\\(CHOH)_3\\|\\CH_2OH\end{array}$$

[A] $[C_7H_{14}O_8]$

$\downarrow$ 1. $NaBH_4$ 2. P/HI

$CH_3(CH_2)_4CH_3$
n-Hexane
[B] $[C_6H_{14}]$

$NaBH_4$ reduces carbonyl group to alcohol which upon further reduction with HI in presence of P is reduced to alkane.

25. (a) $HO-CH_2-CH(OH)-CH(OH)-CH(OH)-CHO$ (Aldopentose, X) $\xrightarrow[-2H_2O]{H^+}$ Furfural

Since furfural is a five membered cyclic compound having an aldedydic group, so the parent sugar (X) should be an aldopentose.

Reaction of aldose/ketose with resorcinol gives coloured compound **(Seliwanoff test)**. Ketoses respond this test immediately while aldoses after sometime.

26. (a) (A) - (IV), (B) - (I), (C) - (II), (D) - (III)

27. (a) Those carbohydrates which reduce Fehling's solution and Tollen's reagent are referred as reducing sugars. They should have a free ketonic or aldehydic group. In case of cyclic structure those have free OH at anomeric carbon, i.e., hemiacetal group $\left(\begin{array}{c}OR\ \ OH\\ \diagdown\ \diagup\\ C\\ \diagup\ \diagdown\end{array}\right)$ are reducing sugars.

28. (b)

- Maltose is composed of two glucose units.
- Sucrose has glycosidic linkage between C_1 of α-glucose and C_2 of β-fructose.
- Lactose is composed of galactose and glucose
- Amylose has glycosidic linkage between C_1 of α-glucose and C_4 of α-glucose.

29. (b) Laevorotatory rotates plane polarized light towards left.

$$\underset{\text{Sucrose}}{\underset{[\alpha]=66.6^\circ}{C_{12}H_{22}O_{11}}}+H_2O\xrightarrow{H^+}\underset{\text{(A)}}{\underset{[\alpha]=+52.7^\circ}{\underset{\text{D–Glucose}}{C_6H_{12}O_6}}}+\underset{\text{(B)}}{\underset{[\alpha]=-92.2^\circ}{\underset{\text{D–Fructose}}{C_6H_{12}O_6}}}$$

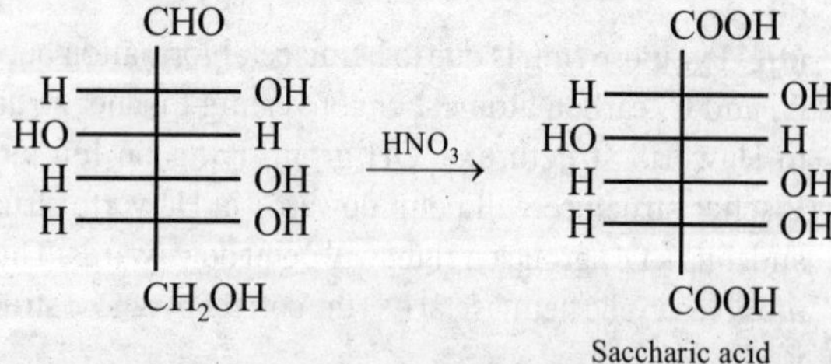

Saccharic acid

30. (a)

(i) A positive test with Tollen's reagent confirms the presence of –CHO group.

(ii) Formation of triacetate derivative confirms the presence of 3 – OH groups.

(iii) L – isomer confirms that from bottom, at C2 carbon – OH is on left and –H is on right side.

(iv) Br_2 water will convert the aldehyde group to carboxylic acid group which is optically active.

(v) HNO_3 will oxidise both aldehyde and alcohol groups to carboxylic acid group which is optically inactive due to plane of symmetry.

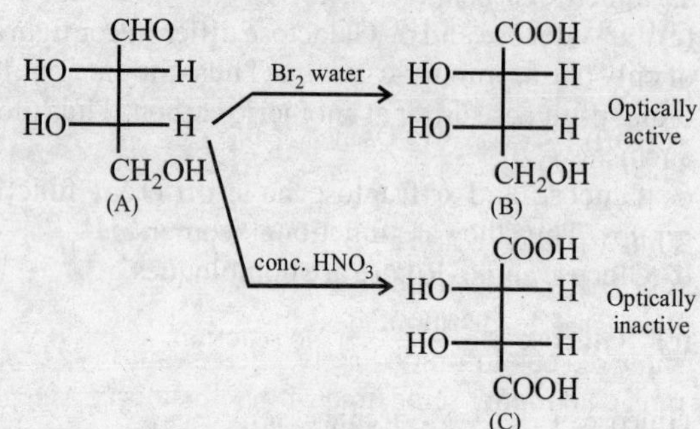

Trick : In all the options, there are 3–OH groups, 1–CHO group. Option (a) and (b) are L–isomers. But option (b) will form optically active compound in both the cases. Hence, option (a) must the correct answer.

31. (c) Maltose is a reducing sugar and glycosidic linkage is present between C_1 and C_4 carbon.

32. (a) In lactose, β Linkage is between C_1 of galactose and C_4 of glucose.

33. (c) Sucrose is an example of disaccharide and non-reducing sugar because it does not have free hemiacetal linkage. It involves glycoside linkage between C_1 of α-D-glucose C_2 of β-D-fructose.

34. (d) Lactose : It is a disaccharide of β–D–galactose and β-D-glucose with C_1 of galactose and C_4 of glucose link.

35. (d) Fructose is a six carbon monosaccharide containing ketonic group, so it is a ketohexose.

36. (a) Lactose contains β-glycosidic linkage between C_1 of galactose and C_4 of glucose.

37. (9)

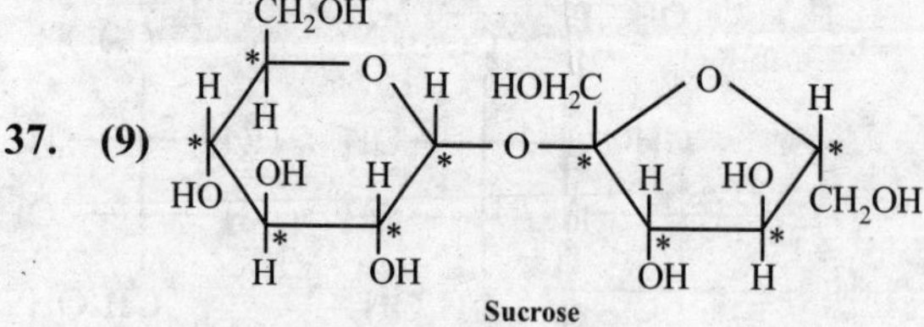

No. of chiral centres = 9.

38. (c)

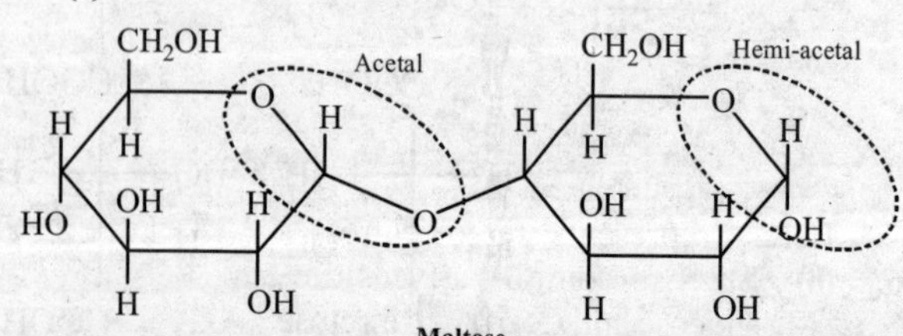

One acetal and one hemiacetal groups are present in maltose.

39. (b)

(i) 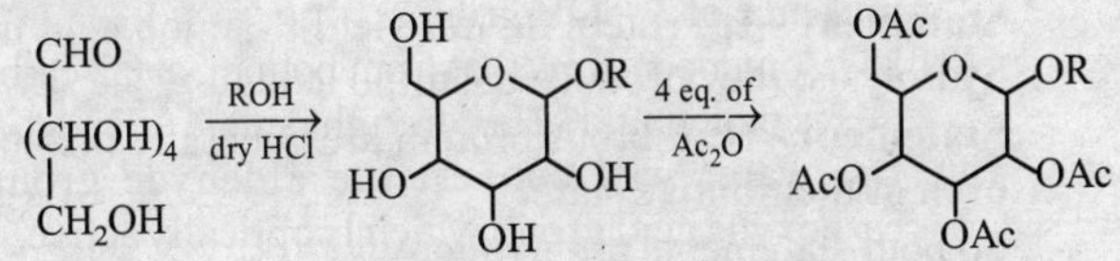 CHO–(CHOH)$_4$–CH$_2$OH $\xrightarrow[\text{dry HCl}]{\text{ROH}}$ (cyclic glycoside with OH, OR, HO, OH, OH) $\xrightarrow[\text{Ac}_2\text{O}]{\text{4 eq. of}}$ (cyclic glycoside with OAc, OR, AcO, OAc, OAc)

(ii) CHO–(CHOH)$_4$–CH$_2$OH $\xrightarrow{\text{H}_2\text{/Ni}}$ CH$_2$OH–(CHOH)$_4$–CH$_2$OH $\xrightarrow[\text{Ac}_2\text{O}]{\text{6 eq. of}}$ CH$_2$OAc–(CH–OAc)$_4$–CH$_2$OAc

(iii) CHO–(CHOH)$_4$–CH$_2$OH $\xrightarrow[\text{Ac}_2\text{O}]{\text{5 eq. of}}$ CHO–(CH–OAc)$_4$–CH$_2$OAc

Thus $x = 4$, $y = 6$ and $z = 5$

40. (c) Seliwanoff reagent → [Resorcinol + Conc. HCl] It is used to distinguish aldoses and ketoses. Ketoses show red colour immediately whereas aldoses show light pink colour after some time with Seliwanoff reagent.

41. (b) Glucose exists in cyclic form in which aldehyde group is not free, therefore it does not give Schiff's test.

42. (d) Maltose on hydrolysis gives two moles of α-D-glucose.

43. (c) Gluconic acid is obtained by partial oxidation of glucose by mild oxidising agent, e.g., Tollen's reagent, Fehling solution, Br_2 water.

$$\underset{\text{Glucose}}{\text{CHO–(CHOH)}_4\text{–CH}_2\text{OH}} \xrightarrow{\text{Br}_2\text{ water}} \underset{\text{Gluconic acid}}{\text{COOH–(CHOH)}_4\text{–CH}_2\text{OH}}$$

Gluconic acid can not form hemiacetal or acetal.

44. (b) Galactose and glucose are C_4 epimers.

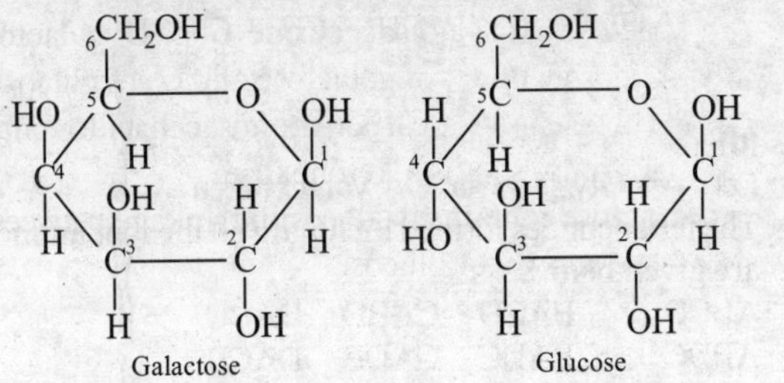

Galactose   Glucose

45. (d)

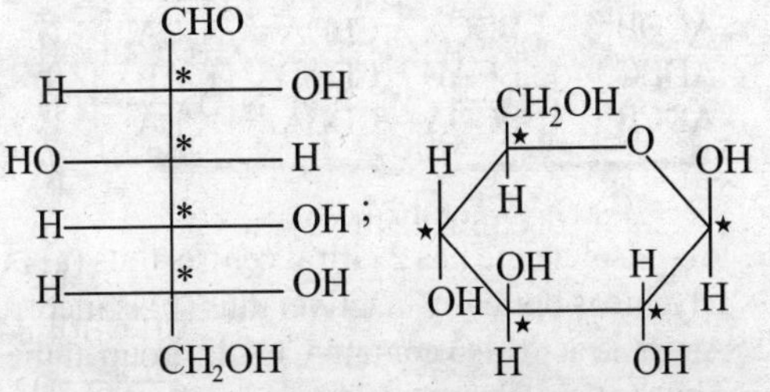

Linear structure of glucose, 4 Stereogenic centres; Cyclic structure of glucose, 5 Stereogenic centres

Here, * represents stereo centres.

46. (b) Sucrose contains glycosidic link between C_1 of α-D-glucose and C_2 of β-D-fructose.

$C_{12}H_{22}O_{11} + H_2O \longrightarrow$ Glucose + Fructose

47. (d) Hydrolysis of maltose gives glucose as maltose is composed of two α-D glucose units.

48. (d)

(a) Benedict reagent : $CuSO_4$ + Sod. citrate + Na_2CO_3 also gives red ppt. of Cu_2O with both.

(b) Fehling solution : $CuSO_4$ + Sod.–pot. tartarate (Rochelle salt) gives red ppt. of Cu_2O with glucose and fructose both.

(c) Barfoed reagent : 7%$(CH_3COO)_2Cu$ + 1% CH_3COOH + 92% H_2O gives red ppt. of Cu_2O with both.

(d) Seliwanoff's reagent : Resorcinol in conc. HCl gives red colour with both, but a ketose (fructose) reacts more quickly than an aldose (glucose).

49. (a) CHO–(CHOH)$_4$–CH$_2$OH $\xrightarrow{\text{HI, }\Delta}$ $\underset{n\text{-Hexane}}{CH_3CH_2CH_2CH_2CH_2CH_3}$

50. (d) In amylose, 1, 4-α-glycosidic linkage is present.

51. (a)

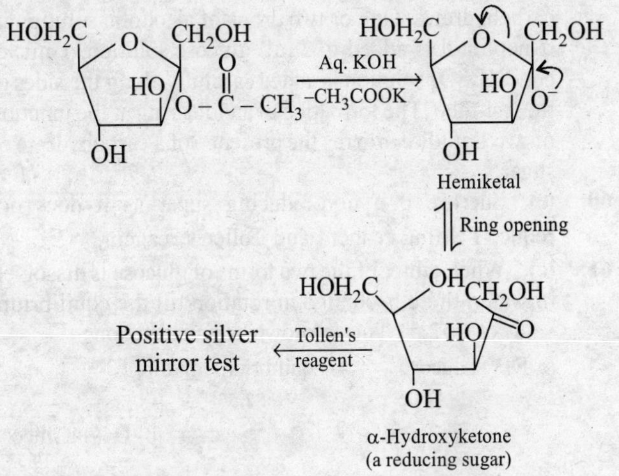

(Furanose ring with HOH_2C, CH_2OH, HO, OH and $O–\overset{O}{\overset{\|}{C}}–CH_3$) $\xrightarrow[-CH_3COOK]{\text{Aq. KOH}}$ Hemiketal $\xrightleftharpoons{\text{Ring opening}}$ α-Hydroxyketone (a reducing sugar) $\xrightarrow{\text{Tollen's reagent}}$ Positive silver mirror test

52. (b)

α-D-glucose (anomeric carbon, C-1)   β-D-glucose (anomeric carbon, C-1)

C-1 anomers

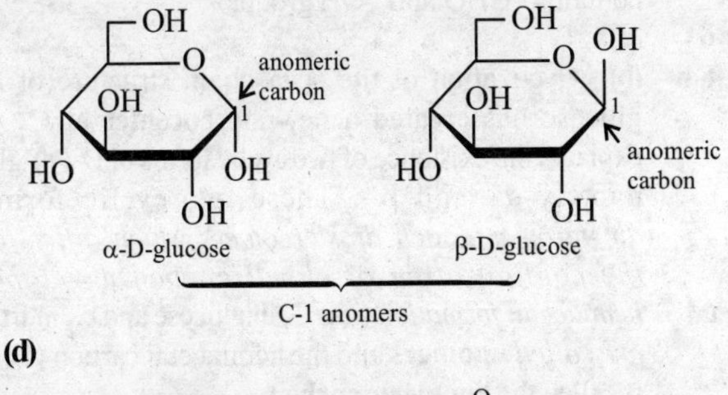

53. (d)

$$\underset{\text{Stored in the form of glycogen}}{\text{Glucose}} \xrightarrow[\text{(absence of oxygen; needs only enzymes)}]{\text{enzyme}} \text{Pyruvic acid} \xrightarrow{O_2} CO_2 + H_2O$$

Pyruvic acid ↓ Lactic acid

During vigorous exercise sufficient oxygen is not available to meet the energy demand, so energy is derived through conversion of pyruvic acid to lactic acid.

54. (d) Sucrose does not contain a free aldehydic or ketonic group, hence it does not show mutarotation.

55. (b) Glycosidic linkage is actually an ether bond as the linkage forming the rings in an oligosaccharide or polysaccharide is not just one bond, but the two bonds sharing an oxygen atom, e.g., sucrose.

56. **(a)** $6CO_2 + 12NADPH + 18ATP \longrightarrow C_6H_{12}O_6 + 12NADP + 18ADP$

57. **(c)** Fischer gave the prefix "D" to compounds whose bottom chiral has its OH to the right. So natural glucose is called D-glucose or dextrose.

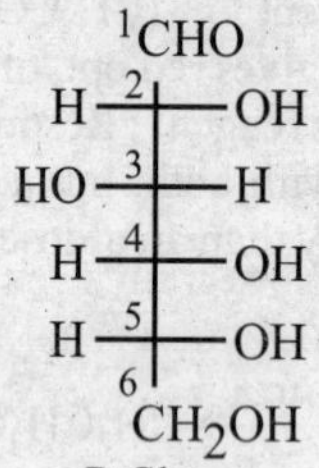

$$\begin{array}{c} {}^{1}CHO \\ H-\overset{2}{C}-OH \\ HO-\overset{3}{C}-H \\ H-\overset{4}{C}-OH \\ H-\overset{5}{C}-OH \\ {}^{6}CH_2OH \end{array}$$

D-Glucose

58. **(c)** Glycogen is called animal starch and is found in all animal cells. It constitutes the reserve food material.

59. **(b)** **Molisch's test :** This is a general test for carbohydrates. One or two drops of alcoholic solution of α-naphthol is added to 2 mL glucose solution. 1 mL of conc. H_2SO_4 solution is added carefully along the sides of the test-tube. The formation of a violet ring at the junction of two liquids confirms the presence of a carbohydrate or sugar.

60. **(c)** Sucrose is a non-reducing sugar as it does not reduce Fehling reagent and Tollen's reagent.

61. **(c)** When either of the two forms of glucose is dissolved in water, there is change in rotation till the equilibrium value of + 52.5°. This is known as mutarotation.

$$\underset{+111.5^\circ}{\alpha\text{–D(+) Glucose}} \rightleftharpoons \underset{+52.5^\circ}{\text{Equilibrium mixture}} \rightleftharpoons \underset{+19.5^\circ}{\beta\text{–D–(+)Glucose}}$$

62. **(c)** Glucose is considered as a typical carbohydrate which contains –CHO and –OH groups.

63. **(c)**

64. **(b)** Cyclization of the open chain structure of D-(+)-glucose has created a new stereocenter at C_1 which explains the existence of two cyclic forms of D-(+)-glucose, namely α– and β–. These two cyclic forms are *diasteromers, such diastereomers which differ only in the configuration of chiral carbon developed on hemiacetal formation* (it is C_1 in glucose and C_2 in fructose) *are called* **anomers** and the hemiacetal carbon (C_1 or C_2) is called the **anomeric carbon.**

65. **(a)**

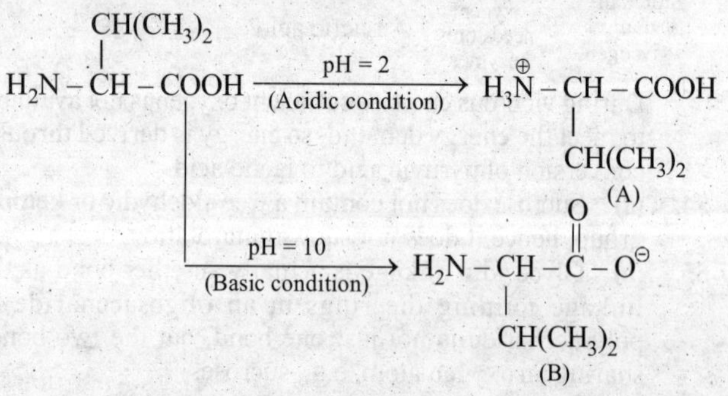

$$H_2N-\underset{|}{\overset{CH(CH_3)_2}{CH}}-COOH \xrightarrow[\text{(Acidic condition)}]{pH = 2} \underset{(A)}{H_3\overset{\oplus}{N}-\underset{CH(CH_3)_2}{CH}-COOH}$$

$$H_2N-\underset{CH(CH_3)_2}{CH}-COOH \xrightarrow[\text{(Basic condition)}]{pH = 10} \underset{(B)}{H_2N-\underset{CH(CH_3)_2}{CH}-\overset{O}{\overset{\|}{C}}-O^{\ominus}}$$

66. **(c)**

Statement - I: Protein do not yield β-amino acid upon hydrolysis, they produce α-amino acids.

Statement - II: Fibrous protein are generally insoluble, even in their native state.

So both statements are wrong.

67. **(b)**

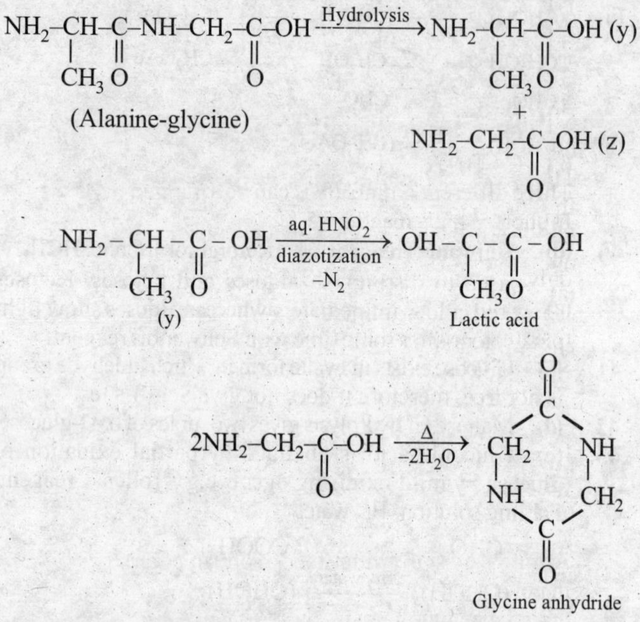

$$\underset{\text{(Alanine-glycine)}}{NH_2-\underset{CH_3}{CH}-\overset{O}{\overset{\|}{C}}-NH-CH_2-\overset{O}{\overset{\|}{C}}-OH} \xrightarrow{\text{Hydrolysis}} NH_2-\underset{CH_3}{CH}-\overset{O}{\overset{\|}{C}}-OH\ (y) + NH_2-CH_2-\overset{O}{\overset{\|}{C}}-OH\ (z)$$

$$\underset{(y)}{NH_2-\underset{CH_3}{CH}-\overset{O}{\overset{\|}{C}}-OH} \xrightarrow[\text{diazotization} \; -N_2]{aq.\ HNO_2} \underset{\text{Lactic acid}}{OH-\underset{CH_3}{C}-\overset{O}{\overset{\|}{C}}-OH}$$

$$2NH_2-CH_2-\overset{O}{\overset{\|}{C}}-OH \xrightarrow[-2H_2O]{\Delta} \text{Glycine anhydride (cyclic: } C(=O)-CH_2-NH-C(=O)-CH_2-NH\text{)}$$

Glycine anhydride

68. **(a)**

$$NH_2-CH_2-\overset{O}{\overset{\|}{C}}+\boxed{Cl + H}-\overset{H}{N}-\overset{CH_3}{CH}-COOH \xrightarrow{-HCl} \underset{\text{(Gly-Ala)}}{NH_2-CH_2-\overset{O}{\overset{\|}{C}}-NH-\overset{CH_3}{CH}-COOH}$$

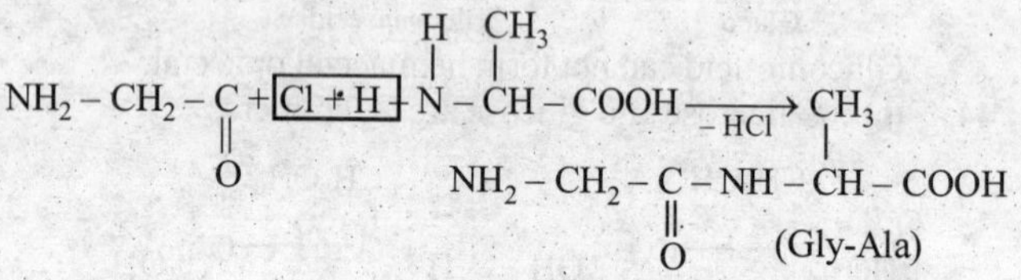

69. **(d)**

Let A = Gly, B = Ala, C = Val, D = Leu

The tetrapeptides formed including all the four amino acids are given below.

ABCD	BACD	CABD	DABC
ABDC	BADC	CADB	DACB
ACBD	BDAC	CBAD	DBAC
ACDB	BDCA	CBDA	DBCA
ADBC	BCAD	CDAB	DCAB
ADCB	BCDA	CDBA	DCBA

Total 24

70. **(d)** Isoleucine has 2 chiral centre, thus (a) is incorrect. Glycine is optically inactive, thus (b) is incorrect. Aspartic acid also contain COOH group at the side chain. Thus (c) is also incorrect. Cystine easily dimmerise due to free SH group to form cystine.

71. **(a)** Valine, Lysine and Threonine are essential amino acids. Proline and Tyrosine non-essential amino acid.

72. **(c)** α-helix and β-pleated sheet belongs to secondary structure of protein, which have hydrogen bonds.

73. **(4)** Essential amino acids are:-

Arginine, Phenylalanine, Histidine and Valine

74. **(9)** Tyrosine is a nonessential amino acid the body makes from another amino acid called phenylalanine.
Number of carbon atoms = 9

75. **(a)** On heating denaturation of protein occurs which result in coagulation of egg.

76. **(6)** A tripeptide contains 3 amino acids linked by two peptide. If A, B, C are 3 different amino acids, then 6 tripeptides bonds will be formed.
1. A – B – C
2. A – C – B
3. B – A – C
4. B – C – A
5. C – A – B
6. C – B – A

Three different amino acids can be arranged by 3!
In short, $3 \times 2 = 6$

77. **(d)** All proteins are the polymers of α-amino acids. So, only α-amino acids are obtained on hydrolysis of proteins.

78. **(a)** Boiling an egg causes denaturation of its protein resulting in loss of its quarternary, tertiary and secondary structures.

79. **(d)** $H_2N-C(=\overset{\oplus}{N}H_2)-NH-CH_2CH_2CH_2-CH(NH_2)-COO^{\ominus}$

Arginine exist is zwitterion, so solid in nature and soluble in polar solvent.

80. **(c)** Asparagine has only one basic functional group ($-NH_2$) in its chemical structure.

$H_2N-CO-CH_2-CH(NH_2)-COOH$

Others are basic amino acids with more than one basic functional group.

81. **(10)**

$$CH_3-CH(NH_2)-\overset{*}{C}(=O)-NH-CH_2-\overset{*}{C}(=O)-NH-CH(CH_2C_6H_5)-\overset{*}{C}(=O)-NH-CH(CH(CH_3)CH_2CH_3)-\overset{*}{C}(=O)-OH$$

(all six carbons of the benzene ring marked *)

82. **(c)** 2° and 3° structures of proteins are stabilized by hydrogen bonding, disulphide linkages, van der Waals force of attraction and electrostatic force of attraction.

83. **(b)** Sulphur containing amino acids

cysteine $HS-CH_2-CH(NH_2)-COOH$ and

methionine $CH_3-S-CH_2-CH_2CH(NH_2)-C(=O)-OH$.

84. **(c)** A–III, B–I, C–IV, D–II

85. **(d)** (A) – IV, (B) – I, (C) – II, (D) – III

86. **(b)** Only option (b) is an α-Amino acid all other options are not α-amino acids.

87. **(6)** Number of peptide linkage
= (No. of α-amino acids – 1) = 7 – 1 = 6

88. **(d)** Two amino acids are
$H_2N-CH(R_1)-COOH$ (A), $H_2N-CH(R_2)-COOH$ (B)

Tripeptides are formed from three amino acids. Therefore, possibilities are as follows.

A–A–A, B–B–B, A–B–A, B–A–B (middle residue A or B) : A⌣A, B⌣B, A⌣B, B⌣A

89. **(b)** Hydrolysis of the given tetrapeptide will give the following:

$Ph-CH_2-CH(NH_2)-COOH$ (F) (Phenylalanine)

$(CH_3)_2CH-CH_2-CH(NH_2)-COOH$ (L) (Leucine)

$HOOC-CH_2-CH(NH_2)-COOH$ (D) (Aspartic acid)

$HO-C_6H_4-CH_2-CH(NH_2)-COOH$ (Y) (Tyrosine)

90. **(8)** No. of possible tripeptide = x^n
where x = No. of α-acids
and n = nature of peptide
Thus, here, amino acids = 2
Hence, No. of tripeptides = $2^3 = 8$
(1) val – val – val (2) pro – pro – pro
(3) val – pro – pro (4) pro – val – pro
(5) val – val – pro (6) val – pro – val
(7) pro – pro – val (8) pro – val – val

91. **(b)** H-bonding between carbonyl oxygen and N of the amino group of every third residue in the helical turn provides stability to the α-helix structure.

92. **(1)** A tetrapeptide means, there are three peptide bonds, hence number of amino acids should be four. So
No. of amino acids – No. of peptide bonds = 4 – 3 = 1

93. **(a)** Primary structure deals with the number, nature and sequence of amino acids linked together to form polypeptide chain; secondary structure deals with the shape of the peptide chain (linear, cyclic, branched or arranged in the form of helix and forces with which the peptide chains are held together). Tertiary structure of a protein refers to its three - dimensional structure i.e. coiling (folding) of the peptide chains. After denaturation, hydrogen bonds are broken, hence secondary and tertiary structures are destroyed and only primary structure remains intact.

94. **(c)** The primary structure of protein is associated with the linear sequence of amino acids in the protein.
Primary structure of protein is unaffected by physical 'or' chemical changes.

95. **(2)** Biuret test is given by compounds having a peptide linkage. So tripeptide and glycylalanine having peptide bonds, will give biuret test.

96. **(4)** Ala-Gly-Leu-Ala-Val
Four peptide linkages to join five amino acids.

97. **(3)** No. of peptide linkage = n – 1
n = No. of reacted amino acids
Now, n = 4 (given)
No. of peptide linkage = 4 – 1 = 3.

98. **(b)** Keratin, collagen and myosin are example of fibrous protein. They contain hydrogen and disulphide linkages and are insoluble in water. Insulin and albumin are the examples of globular proteins which are water soluble.

99. **(4)** $H_2N-CH_2-C(=O)-OH$ (Glycine) + $H_2N-CH(CH_2-CH_2-COOH)-C(=O)-OH$ (Glutamic acid) +

$H_2N-CH(CH_2-COOH)-C(=O)-OH$ (Aspartic acid) + $H_2N-CH(CH_2-C_6H_4-OH)-C(=O)-OH$ (Tyrosine) $\xrightarrow{pH=12.5}$

$H_2N-CH_2-C(=O)-NH-CH(CH_2-CH_2-COO^{-(1)})-C(=O)-NH-CH(CH_2-COO^{-(2)})-C(=O)-NH-CH(CH_2-C_6H_4-O^{-(4)})-C(=O)-O^{-(3)}$

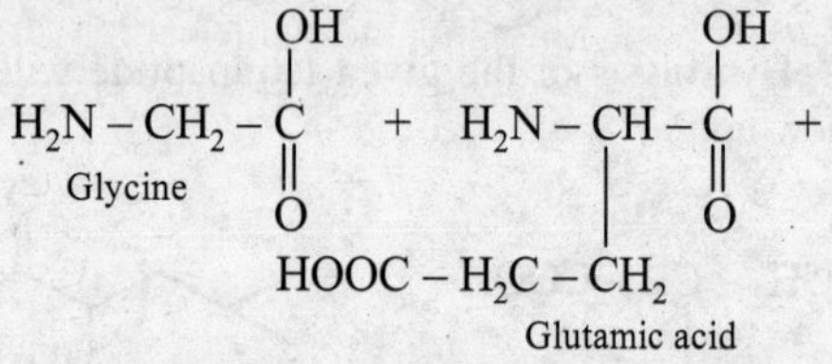

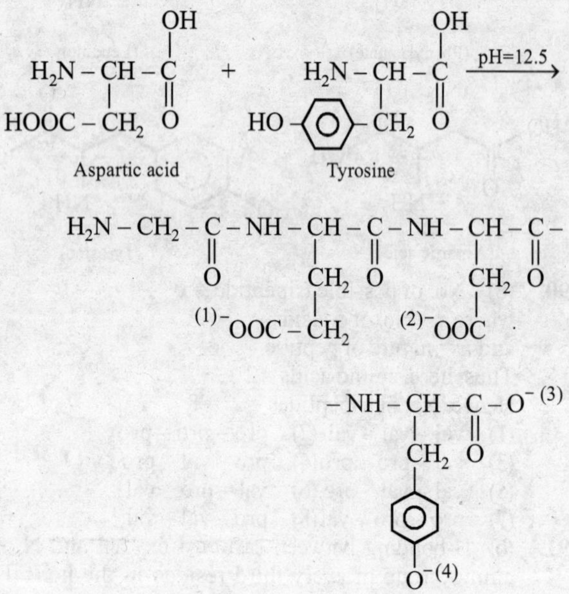

At highly basic condition, all acidic groups will lose protons. So, total number of negative charge produced = 4

100. **(b)** Albumin is a water soluble globular protein produced in liver.

101. **(c)** Seliwanoff test is used to distinguish ketoses from aldoses. The reagent of this test consists of resorcinol and conc. HCl. On treatment with a concentrated acid, ketones are dehydrated more rapidly to give furfural derivative and on condensation with resorcinol give cherry red complex.
Xanthoproteic test is used to detect amino acids containing aromatic nucleus. On heating with conc. HNO_3 it gives yellow coloured nitro derivatives.

102. **(b)** Hydrogen bonding is responsible for stabilisation of α-helix structure of proteins.

103. **(a)** Tyrosine is a non-essential amino acid.

104. **(4)**

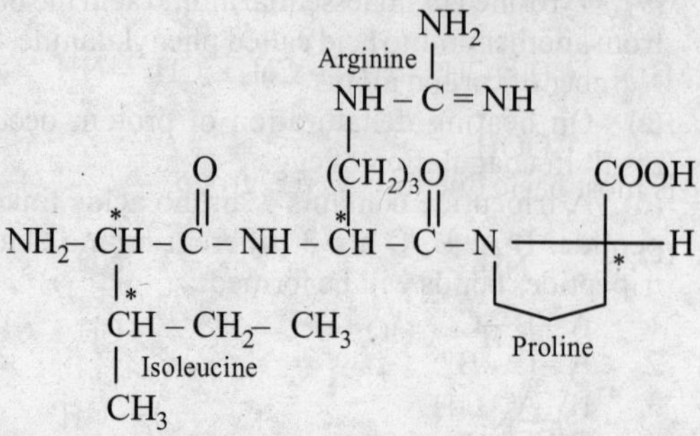

$NH_2-\overset{*}{C}H(\overset{*}{C}H(CH_3)-CH_2-CH_3)-C(=O)-NH-\overset{*}{C}H((CH_2)_3-NH-C(NH_2)=NH)-C(=O)-N$ (Proline ring, $\overset{*}{C}$(H)COOH)

Isoleucine, Arginine, Proline

* Represents chiral centre

105. **(c)** Compounds having 1° amino group give carbylamine reaction with $CHCl_3$ and alc. KOH.

106. **(2)** $HOOC-\overset{*}{C}H(NH_2)-\overset{*}{C}H(OH)-CH_3$ (Threonine)
No. of chiral centres = 2

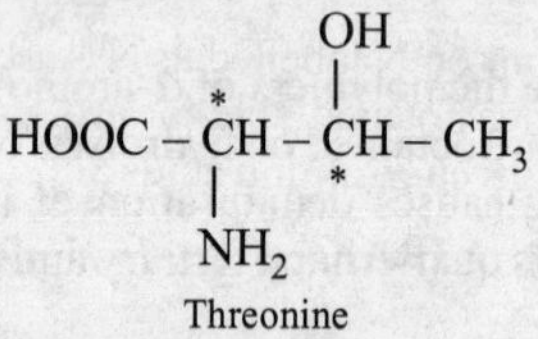

107. **(5)** Asp – Glu – Lys tripeptide is :

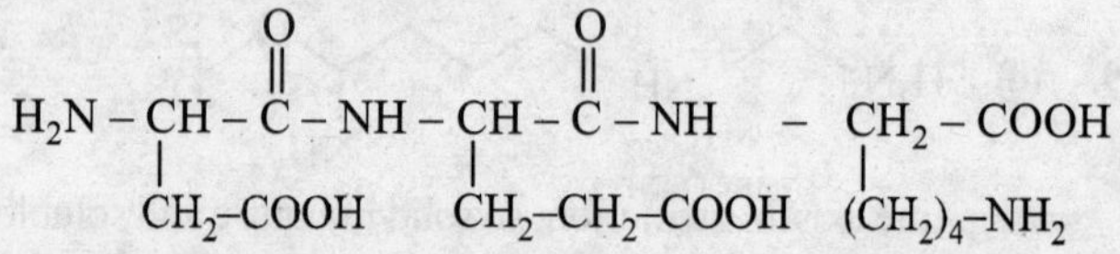

$H_2N-CH(CH_2-COOH)-C(=O)-NH-CH(CH_2-CH_2-COOH)-C(=O)-NH-CH((CH_2)_4-NH_2)-COOH$

No. of CO groups = 5

108. **(37.84)**
Molecular formula of histamine is $C_5H_9N_3$
Molecular mass of histamine = 5 × 12 + 9 × 1 + 3 × 14 = 111
Mass percentage of nitrogen in histamine

$$= \frac{42}{111} \times 100 = 37.84\%$$

109. **(b)** Asn ⇒ $NH_2-C(=O)-CH_2-C(H)(NH_2)-COOH$

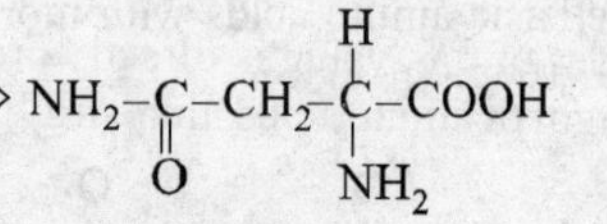

Ser ⇒ $HO-CH_2-C(H)(NH_2)-COOH$

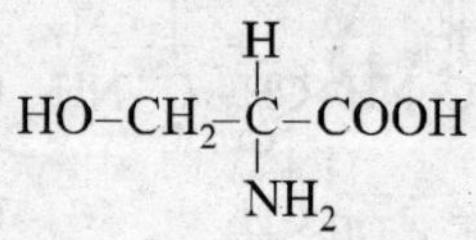

Asn - Ser

⇒ $NH_2-C(=O)-CH_2-C(H)(H_2N)-C(=O)-NH-C(H)(COOH)-CH_2-OH$

$\downarrow$ Excess $(CH_3CO)_2O$

⇒ $NH_2-C(=O)-CH_2-C(H)(CH_3CONH)-C(=O)-NH-C(H)(COOH)-CH_2OCOCH_3$

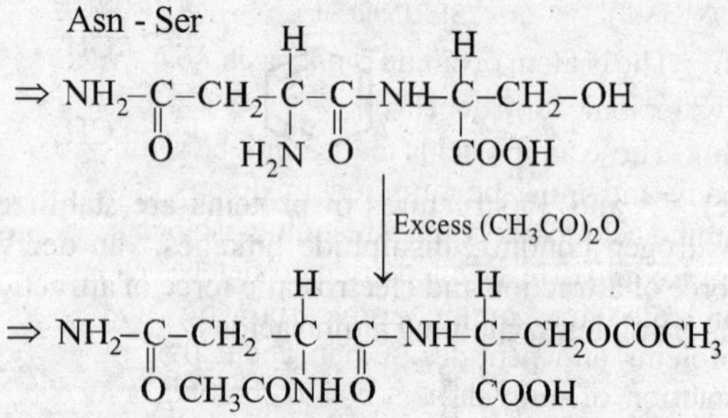

110. **(a)** Ceric ammonium nitrate test is used for detecting alcohols, while carbylamine test is for primary amines. Among the given peptides, only serine (Ser) has alcoholic group.

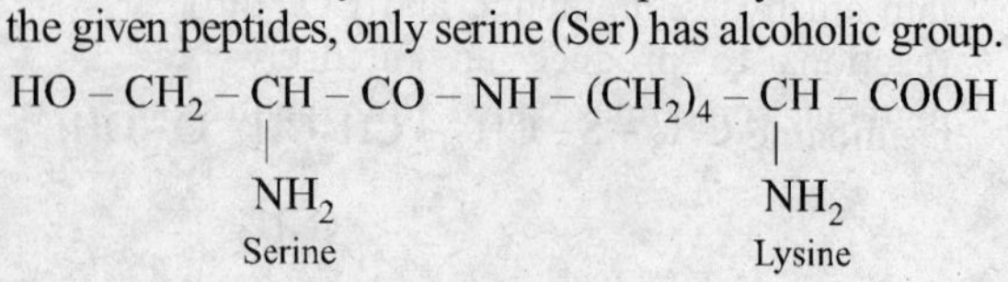

$HO-CH_2-CH(NH_2)-CO-NH-(CH_2)_4-CH(NH_2)-COOH$

Serine, Lysine

111. (b) Lysine,

$$\overset{\alpha}{H_2N-CH}-\overset{\beta}{CH_2}-\overset{\gamma}{CH_2}-\overset{\delta}{CH_2}-\overset{\in}{CH_2}-NH_2$$
(with COOH on the α-carbon)

is most basic due to $\in -NH_2$ group.

112. (c) (Zwitter ionic form of histidine) $\xrightarrow[\text{(Acidic medium)}]{\text{at pH = 2}}$ protonated form ($\overset{\oplus}{N}H_3$–CH–COO⁻ → $\overset{\oplus}{N}H_3$–CH–COOH; ring N protonated)

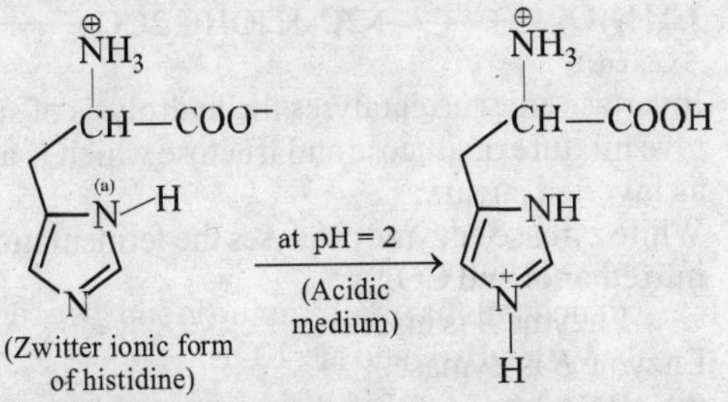

Note that *lp* of electrons on N labelled as N[(a)] are involved in delocalisation, hence not available for protonation.

113. (c) Structure of the given α-amino acids are:

$H_3N^+-CH_2-COO^-$ Glycine (Gly) **(neutral)**

$HOOC-CH_2-CH(\overset{+}{N}H_3)-COO^-$ Aspartic acid (Asp) **(acidic)**

$H_2N-(CH_2)_4-CH(\overset{+}{N}H_3)-COO^-$ Lysine (Lys) **(basic)**

$H_2N-C(=NH)-NH-(CH_2)_3-CH(\overset{+}{N}H_3)-COO^-$ Arginine (Arg) **(basic)**

Further, arginine is more basic due to stronger basic functional groups.

∴ The order of pK_a value is directly proportional to the basic strength of amino acids, thus Arg > Lys > Gly > Asp.

114. (a)

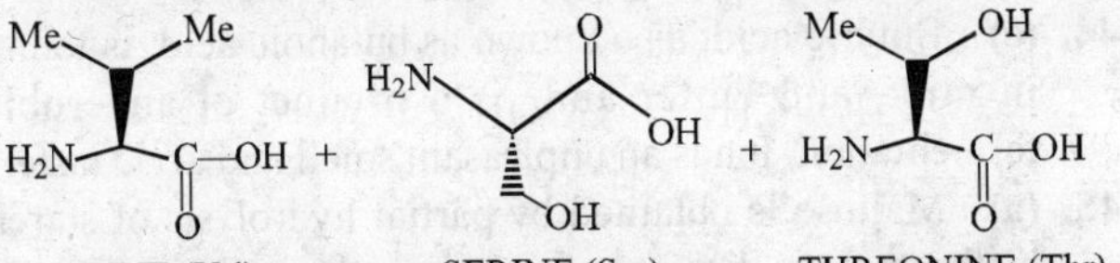

VALINE (Val) + SERINE (Ser) + THREONINE (Thr)

115. (b) The N atom of amide is not basic, so it will not exist in zwitter ionic form at pH 7.0.

116. (b) Those amino acids that cannot be synthesized in our body and must be supplied in diet is called essential amino acid for ex. valine, histidine, isoluecine etc.

117. (c) Ninhydrin is often used to detect α–amino acids and also free amino and carboxylic acid groups on proteins and peptides. When about 0.5 mL of a 0.1% solution of ninhydrin is boiled for one or two minutes with a few mL of dilute amino acid or protein solution, a blue color develops. Ninhydrin degrades amino acids into aldehydes, ammonia, and CO_2 through a series of reactions to produce an intensely blue or purple pigment, sometimes called Ruhemann's purple.

118. (c) With the exception of glycine all other amino acids have different functional groups (atom) on the central tetrahedral alpha carbon.

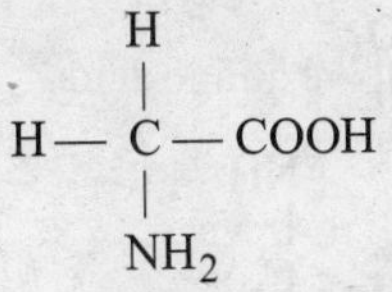

$H-C(H)(NH_2)-COOH$

Glycine

119. (d) Statement (d) is not correct. Some proteins also contain S, along with C, H, O and N.

120. (a) Biuret test produces violet colour on addition of dilute $CuSO_4$ to alkaline solution of a compound containing peptide linkage. Polypeptides, proteins and urea have $-\underset{\|}{\overset{}{C}}(O)-NH-$ (peptide) linkage, while carbohydrates have glycosidic linkages. So, test of carbohydrates should be different from that of other three.

121. (b) The secondary structure of a protein refers to the shape in which a long peptide chain can exist. There are two different conformations of the peptide linkage present in protein, these are α-helix and β-conformation. The α-helix always has a right handed arrangement.

122. (33) Adenine form two hydrogen bonds with thymine and guanine form three hydrogen bond with cytosine.

5' G–G–C–A–A–A–T–C–G–G–C–T–A–3'

In given DNA strand total seven guanine and cytosine bases are present which form total 21 H-bonds and six adenine and thymine bases are present which will form total 12 H-bonds with other DNA strand.

∴ Total no. of H bonds = 7 × 3 + 6 × 2 = 33

123. (a)

(A) Adenine (III)

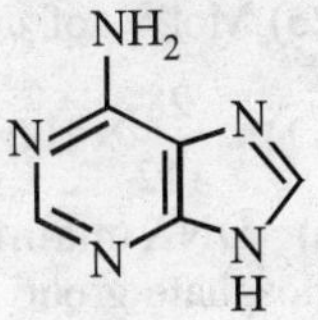

(B) Cytosine (IV)

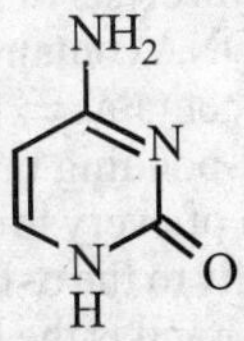

(C) Thymine (II)

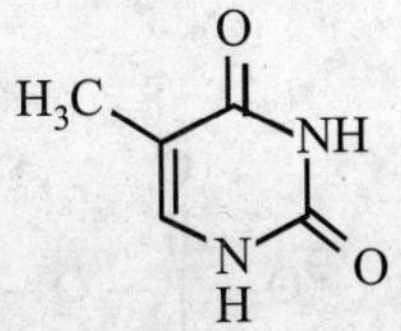

(D) Uracil (I)

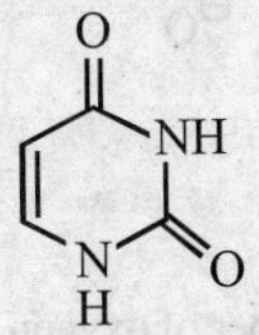

124. (c) DNA contain four bases, which are adenine, guanine, cytosine and thymine. Option (a) is adenine, option (b) is thymine option (d) is cytosine. Option (c) is not the correct structure of guanine. Correct structure of guanine is

125. (c) Enzymes are biocatalyst and are very specific for a particular reaction and for a particular substrate. Almost all enzymes are globular proteins. Enzyme maltase hydrolyses maltose into glucose.

$$\text{Maltose} \xrightarrow{\text{Maltase}} \text{Glucose}$$

126. (b) In DNA G pairs with C and A pairs with T only.

A T G C T T C A
T A C G A A G T

127. (3) A is false as DNA is regarded as the reserve of genetic information.
B is true as DNA undergoes replication during cell division.
C is false as RNA synthesizes proteins in the cell.
D is true as DNA stores the information about the protein or amino acids to be synthesized in the form of specific sequence.
E is true as DNA undergoes replication and forms strands for the daughter cells that are identical.

128. (a) (A) Starch = $(\alpha - D - \text{glucose})_n$, II
(B) Cellulose = $(\beta - D - \text{glucose})_n$, III
(C) Nucleic acid = $(\text{nucleotide})_n$, I
(D) Protein = (α – amino acid). IV

129. (a) Phosphodiester linkage joins two nucleotides.

130. (25) Mol. wt of uracil, $C_4N_2H_4O_2 = 112$

$$\% N = \frac{28}{112} \times 100 = 25\%$$

131. (a) DNA mainly consists of a nitrogeneous base, phosphate group and a pentose sugar.
So, hydrolysis of DNA produces a pentose sugar.

132. (b) DNA contains ⇒ β–D–2–deoxyribose
RNA contains ⇒ β–D–ribose

133. (b) H-bonding between carbonyl oxygen and the amine group of every third residue in the helical turn provides stability to the α-Helix structure.

134. (9) Uracil is the base which only present is RNA.

Uracil base

Number of oxygen atoms in this nucleotide = 9.

135. (d)

136. (c)

137. (b)

$$\underset{\text{Sucrose}}{C_{12}H_{22}O_{11}} + H_2O \xrightarrow{\text{Invertase}} \underset{\text{Glucose}}{C_6H_{12}O_6} + \underset{\text{Fructose}}{C_6H_{12}O_6}$$

$$\underset{\text{Glucose}}{C_6H_{12}O_6} \xrightarrow{\text{Zymase}} 2C_2H_5OH + 2CO_2$$

Invertase enzyme catalyses the hydrolysis of sucrose and give mixture of glucose and fructose which is also known as inverted sugar.
While zymase enzyme catalyses the fermentation of sugar into ethanol and CO_2.
∴ Enzyme *A* is invertase.
Enzyme *B* is Zymase.

138. (b) RNA has a single helix structure, whereas, DNA has a double helix structure.

139. (b) RNA contains adenine (A), guanine (g), cytosine (c) and uracil (u).
(a) is cytosine, found in RNA and DNA, both.
(b) is uracil, found in RNA only.
(c) is thymine, found in DNA only.
(d) is not a pyrimidine, i.e., not a constituent of RNA and DNA.

140. (a)

141. (a) Among 20 naturally occurring amino acids "cysteine" has '– SH' or thiol functional group.
⇒ General formula of amino acid

$$H_2N - \underset{\underset{R}{|}}{CH} - COOH$$

⇒ Value of R = $-CH_2-SH$ in cysteine.

142. (d) The two polynucleotide chains of DNA molecules are twisted around a common axis but run in opposite directions to form a right handed helix. The two chains are joined together by specific hydrogen bonds. Adenine-Thymine (two hydrogen bonds) and cytosine-guanine (three hydrogen bonds)

143. (a) DNA contains ATGC bases
Quinoline is not present in DNA or RNA.

144. (c) Butyric acid, also known as butanoic acid, is found in milk, and butter and is a product of anaerobic fermentation. It has an unpleasant smell and acrid taste.

145. (a) Maltose is obtained by partial hydrolysis of starch by the enzyme diastase present in malt.

$$(C_6H_{10}O_5)_n + nH_2O \xrightarrow{\text{Diastase}} nC_6H_{12}O_6 \text{ (maltose)}$$

146. (d) The correct structure of thymine is

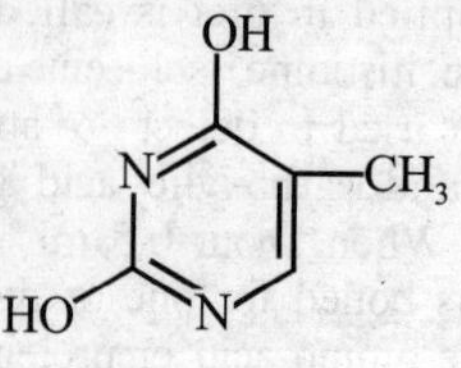

Thymine (T)

147. (a) The sugar present in DNA is deoxyribose.

148. (a) The pyrimidine bases present in DNA are cytosine and thymine.

149. (b) In DNA and RNA, heterocyclic base and phosphate ester are at C_1' and C_5' respectively of the sugar molecule.

OH
|
HO – P – O – CH_2 (5') — [sugar ring: 4', O, 1' attached to N of base; H at 4', 3', 2', 1'; OH at 3' and 2'] — purine base (N)

150. (c) RNA contains cytosine and uracil as pyrimidine bases while DNA has cytosine and thymine.
Both have the same purine bases i.e., guanine and adenine..

151. (b) In RNA, the sugar is D–ribose and base is uracil, whereas in DNA, the sugar is D-2 deoxyribose and the nitrogenous base is thymine.

152. (d) Vitamin D, E, K, A are fat soluble vitamins.

153. (c) Vitamin C is Ascorbic acid.

154. (5) Vitamins A, D, E, K and B_{12} are stored in our body.

155. (d) A – III, B – I, C – IV, D – II

156. (a) Noradrenaline is a neurotransmitter and is released by adrenal medulla. Low level of noradrenaline in the human body causes depression.

157. (d) Vitamin B_1 - Thiamin
Vitamin B_6 - Pyridoxine

158. (d) B group vitamins and vitamin C are soluble in water except vitamin B_{12} all are excreted easily.

159. (b) Vitamin K is essential for the formation of coagulation factors as well as several proteins that work together to clot the blood when injuries occur and prevent excessive bleeding. So, deficiency of vitamin K causes increase in blood clotting time.

160. (c)

	Vitamin	Deficiency	disease	
(ii)	Vitamin B_1	(thiamine)	Beriberi	(a)
(i)	Vitamin B_2	(riboflavin)	Cheilosis	(c)
(iii)	Vitamin B_6	(pyridoxine)	Convulsions	(d)
(iv)	Vitamin C	(ascorbic acid)	Scurvy	(b)

161. (c) Water soluble vitamins dissolve in water and are not stored by the body. The water soluble vitamins include the vitamin B-complex group and vitamin C.

162. (c) Deficiency of vitamin D causes rickets.

163. (c) When either of the two forms of glucose is dissolved in water, there is change in rotation till the equilibrium value of + 52.5°. This is known as mutarotation

$$\underset{+111.5°}{\alpha\text{–D(+) Glucose}} \rightleftharpoons \underset{+52.5°}{\text{Equilibrium mixture}} \rightleftharpoons \underset{+19.5°}{\beta\text{–D–(+)Glucose}}$$

164. (b) Insulin is a biochemically active peptide hormone secreted by pancreas.

Principles Related to Practical Chemistry

1. **(a)**
(A) Carboxylic acid gives effervescence with sodium bicarbonate solution.
(B) Phenolic-OH gives violet coloured complex with neutral $FeCl_3$.
(C) Alcoholic-OH gives red colour with ceric ammonium nitrate.
(D) When alkaline $KMnO_4$ reacts with an unsaturated compound (Alkene or alkyne) the purple colour of $KMnO_4$ solution disappears, indicating positive test for unsaturation.

2. **(c)** A, C, D produce CO_2 with aqueous $NaHCO_3$ solution. A, C, D acids are stronger acid than H_2CO_3 (Carbonic acid).

3. **(d)** The sodium fusion extract of organic compound having N & S gives blood red colour with $FeSO_4$ and $Na_4[Fe(CN)_6]$.
$Na + C + N + S \longrightarrow NaSCN$
$Fe^{3+}\ SCN^- \longrightarrow [Fe(SCN)]^{2+}$ (Blood red)

4. **(c)** For Lassaigne's test, sodium is used and not magnesium to convert covalent bond into ionic bond.

5. **(c)** $CH_3CH{=}O$, $PhCH_2CH{=}O$.
$HOCH_2-\underset{\underset{O}{\|}}{C}-(CHOH)_3-CH_2OH$ (Fructose)
All give positive Fehling test
Ketonic group and aromatic – CHO do not give Fehling's test.

6. **(b)** (A) Br_2 water test is test of unsaturation in which reddish orange column of bromine water disappears.
(B) Alcohols given red colour with ceric ammonium nitrate
(C) Phenol gives violet column with natural ferric chlorid
(D) Aldehyde and ketone give yellow/orange red colour with 2, 4–DNP.

7. **(b)** Apart from aldehyde, Formic acid ($H-\overset{\overset{O}{\|}}{C}-OH$) also gives silver mirror test with ammonical silver nitrate.

8. **(d)**

9. **(d)** Ninhydrin test is a chemical test useful to identify ammonia, primary and secondary amine, or amino acids. Egg albumin contains protein which is a natural polymer of amino acids which will show positive ninhydrin test.

10. **(d)** Hydrazine ($NH_2–NH_2$) have no carbon so does not show Lassaigne's test.

11. **(95)** $C_6H_5NH_2 + CH_3-\overset{\overset{O}{\|}}{C}-Cl \longrightarrow C_6H_5NH-\overset{\overset{O}{\|}}{C}-CH_3$

93 g aniline gives = 135 g acetanlide

So, 6.55 g anilne will give $= \frac{135}{93} \times 6.55 = 9.5$ g acetanlide
$= 95 \times 10^{-1}$ acetanlide.

12. **(c)** Nitrogen, sulphur, phosphorous and halogens
Lassaigne's test is used for the detection of nitrogen, sulphur, phosphorous and halogen present in an organic compound.

13. **(d)** $HSO_3-C_6H_4-N{=}N-C_{10}H_6-NH_2$

$NH_2-C_6H_4-SO_3H + NaNO_2 + CH_3COOH \longrightarrow CH_3-\overset{\overset{O}{\|}}{C}-O-N{=}N-C_6H_4-SO_3H$ (X)

$HO_3S-C_6H_4-N-N-O-\overset{\overset{O}{\|}}{C}-CH_3 + H-C_{10}H_6-NH_2$
$\downarrow$
$HO_3S-C_6H_4-N{=}N-C_{10}H_6-NH_2$

14. **(d)** Both statement I and statement II are true.

15. **(d)**
$C_6H_5NH_2 \xrightarrow[(A)]{NaNO_2 + HCl} C_6H_5\overset{+}{N}{\equiv}N\overset{-}{Cl}$
Benzene diazonium chloride

$C_6H_5\overset{+}{N}{\equiv}N\overset{-}{Cl}$ + β-Napthol (B) $\xrightarrow{NaOH}$ $C_6H_5-N{=}N-C_{10}H_6-OH$

β-napthol azo-dye (scarlet red)

16. **(d)** If nitrogen or sulphur is also present in the compound, the sodium fusion extract is boiled with conc. HNO_3 to decompose cyanide or sulphide of sodium formed, these ions would otherwise interfere with silver nitrate test for halogens.

17. **(a)** When sodium fusion extract of an organic compound treated with $FeSO_4$ in presence of conc. H_2SO_4, appearence of blood red colour indicates the presence of both N and S in the organic compound. This happens because there is no free cyanide ions.

$$Na + C + N + S \longrightarrow NaSCN$$

$$Fe^{3+} + SCN^- \longrightarrow \underset{\text{Blood red}}{[Fe(SCN)]^{2+}}$$

18. **(3)** Acetaldehyde, methanal and cyclohexane carbaldehyde gives positive Fehling solution test among the given options. Aldehydes gives positive Fehling solution test. Aromatic aldehydes do not respond to this test. Therefore benzaldehyde and substituted benzaldehyde gives negative Fehling solution test.

19. **(a)** Carbylamine Test-Identification of primary amines
Lucas Test - Differentiation between 1°, 2° and 3° alcohols
Tollen's Test - Identification of Aldehydes
Phthalein Dye Test - Identification of phenols

20. **(d)** A. Alkaline solution of copper sulphate and sodium citrate is known as Benedict's solution and it is used to test aliphatic aldehydes. Hence it can be used to test compound (III)
B. Neutral $FeCl_3$ solution is used to test phenolic compound (IV)
C. Alkaline chloroform solution is used to test primary amines (II)
D. $2KI + NaOCl + H_2O \rightarrow NaCl + I_2 + 2KOH$
Potassium iodide and sodium hypochlorite gives (I_2 + KOH) which is used to test those compounds which have $CH_3-\overset{\overset{O}{||}}{C}-$ or $CH_3-\overset{\overset{OH}{|}}{CH}-$ group (iodoform test). Hence the compound is (I)

21. **(b)** (A) – (II), (B) – (I), (C) – (III), (D) – (IV)

22. **(d)** It does not show esterification test. Presence of both sulphur and nitrogen give red colour in Lassigne's test.

23. **(d)** Aromatic aldehydes do not give Fehling's test. Both nitrogen and sulfur must be present to obtain blood red colour with sodium nitroprusside.

24. **(b)** $CH_3NH_2 . HCl \xrightarrow[\text{fusion}]{Na} NaCN$ and NaCl
NaCN gives + ve test for nitrogen and NaCl gives +ve test for halogen.

25. **(530)** Toluene ($C_6H_5CH_3$) $\xrightarrow{\text{92\% yield}}$ Benzaldehyde (C_6H_5CHO)

MW of toluene = 92 ; MW benzaldehyde = 106.

Number of moles toluene $= \frac{5}{92}$

Number of moles of C_6H_5CHO produced

$$= \frac{5}{92} \times \frac{92}{100} = 5 \times 10^{-2}$$

$\therefore$ Mass of C_6H_5CHO

$$= 106 \times 5 \times 10^{-2} = 5.3\text{ g} = 530 \times 10^{-2}\text{ g}$$

26. **(2)** 233g $BaSO_4$ contains = 32 g sulphur

$$0.752\text{g } BaSO_4 \text{ contains} = \frac{32}{233} \times 0.752 = 0.1005\text{g}$$

$$\text{Sulfur\%} = \frac{\text{Weight of Sulphur}}{\text{Weight of Compound}} \times 100$$

$$51.6 = \frac{0.1005}{\text{Weight of compound}} \times 100$$

$$\Rightarrow \text{Weight of compound} = \frac{0.1005}{51.6} \times 100 = 1.9 \times 10^{-1}\text{g}$$

$$\simeq 2 \times 10^{-1}\text{ g}$$

27. **(b)** Acetate ions gives deep red colour ppt on reaction with netural ferric chloride solution.

$$6CH_3COO^- + 3Fe^{3+} + 2H_2O \longrightarrow [Fe_3(OH)_2(CH_3COO)_6]^+ + 2H^+$$

$$[Fe_3(OH)_2(CH_3COO)_6]^+ 4H_2O \longrightarrow \underset{\text{Brown red ppt}}{3[Fe(OH)_2(CH_3COO)]} + 3CH_3COOH + H^+$$

Phenol reacts with freshly prepared ferric chloride solution and gives violet colour complex.

$$6C_6H_5OH + FeCl_3 \longrightarrow \underset{\text{Violet colour}}{[Fe(C_6H_5O)_6]^{3-}} + 3HCl + 3H^+$$

28. **(34)** $\underset{0.5\text{g}}{O.C} \longrightarrow \underset{0.4\text{g}}{AgBr}$

Molar mass at AgBr = 108 + 88 = 188 u

$$\text{mol of Br} = \text{mol of AgBr} = \frac{\text{weight of AgBr}}{\text{molar mass of AgBr}} = \frac{0.4}{188}$$

$$\%\text{ Br} = \frac{\text{weight of Br}}{\text{weight of organic compound}} \times 100$$

$$= \frac{\frac{0.4}{188} \times 80}{0.5} \times 100 = 34.04\%$$

29. **(d)** $Na_2S + \underset{\text{Sodium nitroprusside}}{Na_2[Fe(CN)_5NO]} \rightarrow \underset{\text{Sodium thionitroprusside (Violet colour)}}{Na_4[Fe(CN)_5NOS]}$

30. **(a)** Both statement I and statement II are correct.

31. **(b)** Sodium fusion extract is boiled with concentrated HNO_3 to remove sodium cyanide and sodium sulphide.

32. **(14)** Weight of organic compound = 0.2 g

$$\text{Mass of } N_2\text{(g) evolved} = \frac{22.4 \times 10^{-3}}{22.4} \times 28 = 28 \times 10^{-3}\text{ g}$$

$$\%\text{ of N} = \frac{28 \times 10^{-3}}{0.2} \times 100 = 14$$

33. **(40)** Mass of organic compound = 0.2g
Mass of AgBr = 0.188g
Molar mass of AgBr = 108 + 80 = 188

Mass of Br $= \frac{80}{188} \times 0.188 = 0.08g$

$Br\% = \frac{0.08}{0.2} \times 100 = 40$

34. **(a)** By sodium fusion extract we can detect sulphur, nitrogen, phosphorus and halogens, because they are converted into their ionic form with sodium metal.

35. **(12)** $\% N = \frac{1.4\ N \times V}{W} \Rightarrow 42 = \frac{1.4 \times (1 \times 2) \times V}{0.8}$

$\Rightarrow V = 12\,mL$

36. **(78)** 10 g of $C_6H_6 = \frac{10}{78}$ moles of C_6H_6

Moles of methyl benzene obtained from

$= \frac{10}{78}$ mole of $C_6H_6 = \frac{10}{78} \times 92$ g

$\%$ yield $= \frac{9.2}{10 \times 92} \times 78 \times 100 = 78\%$

37. **(68)** % of $Br = \frac{\text{Atomic mass} \times Br \times m_1}{\text{Molecular mass of AgBr} \times m} \times 100$

m_1 = mass of AgBr obtained

m = mass of organic compound taken

Now,

$\%$ of Br $= \frac{80 \times 0.2397}{188 \times 0.15} \times 100 = 68$

38. **(d)** For the test of halogens, it is necessary to remove NaCN and Na_2S from the sodium extract if nitrogen and sulphur are present. This is done by boiling the sodium extract with conc. HNO_3. If NaCN and Na_2S are not decomposed, white or black ppt of AgCN and Ag_2S respectively are formed with $AgNO_3$ solution.

39. **(243)**

$$\underset{\substack{\text{(Aniline)} \\ \text{(M.M=93 g)}}}{C_6H_5NH_2} \xrightarrow{\text{90\% efficiency}} \underset{\substack{\text{(Acetanilide)} \\ \text{(M.M.=135 g)}}}{C_6H_5-NH-\overset{\overset{O}{\|}}{C}-CH_3}$$

Given : The amount of aniline reacts to form acetanilide = 1.86 g

Since, 1 mol $C_6H_5NH_2$ gives 1 mol $C_6H_5NH\overset{\overset{O}{\|}}{C}CH_3$.

$\therefore$ Moles of $C_6H_5NH_2$ = mole of $C_6H_5NH\overset{\overset{O}{\|}}{C}CH_3$

$\Rightarrow \frac{1.86}{93} = \frac{W_{acetanilide}}{135}$

$\Rightarrow W_{acetanilide} = \frac{1.86 \times 135}{93} g = 2.70\,g$

But efficiency of reaction is 90% only.

$\therefore$ Mass of acetanilide produced $= 2.70 \times \frac{90}{100} g$

$= 2.43\,g = 243 \times 10^{-2}\,g.$

40. **(d)** (A) $\xrightarrow{\text{Ozonolysis}}$ (B) H–CHO (acetaldehyde skeleton) + (C) $C_2H_5-\overset{\overset{O}{\|}}{C}-C_2H_5$

(B) $\xrightarrow{I_2 + NaOH}$ $HCOO^-Na^+ + CHI_3$ (Yellow ppt)

(B) $\xrightarrow{Ag_2O}$ $CH_3COO^- + Ag\downarrow$ (Silver mirror)

(C) $\xrightarrow{I_2 + NaOH}$ No yellow ppt

(C) $\xrightarrow{LiAlH_4}$ $\underset{(D)}{C_2H_5-CH(OH)-C_2H_5} \xrightarrow[\text{HCl}]{ZnCl_2}$ (Lucas test) $C_2H_5-CH(Cl)-C_2H_5$ (Turbidity in 5 min.)

41. **(c)** (I) Lucas reagent - Conc. $HCl/ZnCl_2$

(II) Dumas method - CuO/CO_2

(III) Kjeldahl's method - H_2SO_4

(IV) Hinsberg test - $C_6H_5SO_2Cl$/aq. KOH

42. **(a)** C_3H_6O : $\underset{(B)}{CH_3COCH_3}$ and $\underset{(A)}{CH_3-CH_2CHO}$

(B) $\xrightarrow{CH_3MgBr}$ $\underset{(D)}{CH_3-\overset{\overset{CH_3}{|}}{\underset{\underset{OH}{|}}{C}}-CH_3}$

(A) $\xrightarrow{CH_3MgBr}$ $\underset{(C)}{CH_3-CH_2-\underset{\underset{OH}{|}}{CH}-CH_3}$

Test	D	C	Conformation
Iodoform test	–ve	+ve	Presence of $R-\underset{}{\overset{\overset{OH}{|}}{CH}}-CH_3$ group
Lucas test	Immediate	after 5-10 Min.	Presence of 3° & 2°, –OH groups respectively
Ceric ammonium nitrate test	+ve	+ve	Presence of –OH group

43. **(b)** Molisch's test is used to check the presence of carbohydrates while Barfoed test is used for detecting the monosaccharides. Biuret test is used to detect the peptide bonds. Therefore, A and B (monosaccharide) are carbohydrates while C is a protein.

44. **(b)** m-chlorobenzoic acid (COOH, Cl) ; m-chlorophenol (OH, Cl) ; m-chloroaniline (NH_2, Cl)

$\xrightarrow{\text{aq } NaHCO_3}$ Soluble: (A) m-chlorobenzoate (COO^-Na^+, Cl); Insoluble: [m-chlorophenol (OH, Cl) + m-chloroaniline (NH_2, Cl)]

$\xrightarrow{\text{dil NaOH}}$ Insoluble: (C) m-chloroaniline (NH_2, Cl); Soluble: (B) sodium m-chlorophenoxide (O^-Na^+, Cl)

45. (b) H_3C, H_3C N–C_6H_4 + $Cl^- N_2^+$–C_6H_4–$SO_3^- Na^+$ $\xrightarrow{OH^-}$ $(H_3C)_2N$–C_6H_4–N=N–C_6H_4–$SO_3^- Na^+$

Methyl orange

Methyl orange is used as an indicator in acid base titrations.

46. (b) Phosphorus is detected in the form of yellow ppt of ammonium phospho molybdate on reaction with ammonium molybdate.

$$\underset{(A)}{2P} + 3Na_2O_2 + O_2 \xrightarrow{\text{fusion}} 2Na_3PO_4$$

$$Na_3PO_4 + 3HNO_3 \longrightarrow H_3PO_4 + 3NaNO_3$$

$$H_3PO_4 + \underset{\text{(Ammonium molybdate)}}{12(NH_4)_2MoO_4} + 21HNO_3 \longrightarrow$$

$$\underset{\substack{\text{(Ammonium phosphomolybdate)} \\ \text{(canary yellow ppt.)}}}{(NH_4)_3PO_4 \cdot 12MoO_3 \downarrow} + 21NH_4NO_3 + 12H_2O$$

47. (d) $(CH_3)_3C^+$ from $(CH_3)_3CCl$ is quite stable and hence, easily formed and thus, Cl^- ion forms white precipitate with $AgNO_3$.

48. (a) Phenol $\xrightarrow{HCl}$ No reaction

Phenol $\xrightarrow{NaOH}$ $C_6H_5O^- Na^+$

Phenol $\xrightarrow[H_2O]{Br_2}$ 2,4,6-tribromophenol (OH; Br, Br, Br)

49. (c) Phenols (e.g. *m*-cresol), being weak acid, are soluble in dil NaOH, but insoluble in $NaHCO_3$.

m-cresol (CH_3, OH) $\xrightarrow{NaOH(10\%)}$ (CH_3, ONa)

50. (a)

	Name	Abrev-ations	Functional group	Test
(P)	Tyrosine	Tyr	($-CH_2-C_6H_4-OH$)	Phthalein dye test (C)
(Q)	Aspartic acid	Asp	($-CH_2COOH$)	Ester test (A)
(R)	Serine	Ser	($-CH_2OH$)	Ester test (A)
(S)	Lysine	Lys	($-(CH_2)_4-NH_2$)	Carbylamine test (B)

51. (b) Only those halides give precipitate with $AgNO_3$ which easily ionises to give halide ion and a stable carbocation. Corresponding carbocation from the four bromides are

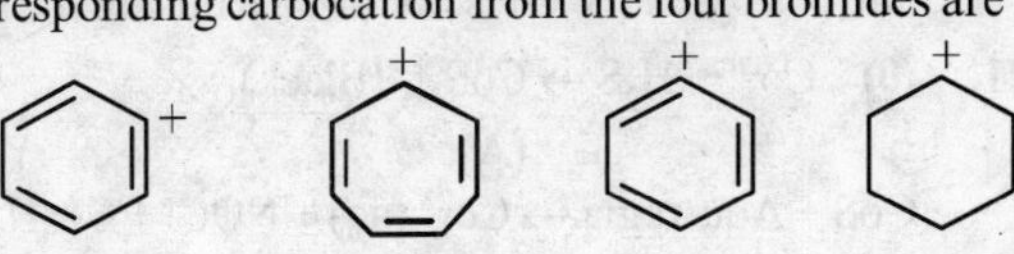

Unstable, (a) Arylcarbocations | Stable, (b) (Aromatic, 6π *es*) | Unstable, (c) | Unstable, (d)

52. (a) Lysine being an amino acid reacts with ninhydrin to give a coloured product (blue purple).

Furfural test is used to distinguish between glucose and fructose. In this test, dilute sugar solution is added to 1-naphthol (in alcohol) and conc. HCl.

Alcohols are oxidised to aldehydes using ceric ammonium nitrate.

Styrene is converted to C_6H_5–CH(OH)–CH_2–OH using $KMnO_4$.

53. (b) Barfoed test is used to test the reducing nature of sugar (carbohydrate).

54. (c) As chloroxylenol contains phenolic group, so it gives positive ferric chloride test.

Norethindrone has double bond, thus it will give Baeyer's test.

Sulphapyridine contains a primary aromatic $-NH_2$ group, so it will gives carbylamine test.

Penicillin contains $-COOH$ group, so it will give sodium hydrogen carbonate ($NaHCO_3$) test.

55. (b) Reaction involved in the tests performed is as follows:

$(H_3C)_2N$–C_6H_4–$COCH_3$ $\xrightarrow{\text{2, 4-DNP test}}$ $(H_3C)_2N$–C_6H_4–C(CH_3)=N–NH–$C_6H_3(NO_2)_2$ (Coloured ppt.)

$(H_3C)_2N$–C_6H_4–$COCH_3$ $\xrightarrow[\text{Iodoform test}]{I_2/NaOH}$ $(H_3C)_2N$–C_6H_4–COO^-Na^+ + CHI_3 (Yellow ppt.)

Thus, $N(CH_3)_2$–C_6H_4–$COCH_3$ will not give positive azo-dye test due to the presence of an electron-withdrawing group.

56. (b) In Lassaigne's test, fusion with sodium take place and following species formed respectively.

(a) Aniline $\longrightarrow CN^-$

(b) Benzene sulfonic acid $\longrightarrow$ $C_6H_5SO_3H$

(c) Thiourea $\longrightarrow S^{2-}$

Reaction of CN^- with hot and acidic solution of $FeSO_4$ leads to formation of $Fe_4[Fe(CN)_6]_3$ which is blue in colour. It contains iron in both II and III oxidation state.

Reaction of S^{2-} with sodium nitroprusside

$$Na_2S + Na_2[Fe(CN)_5NO] \rightarrow \underset{\text{(violet in colour)}}{Na_4[Fe(CN)_5NOS]}$$

Phenoxide ion on reacting with $FeCl_3$ gives red colour with $FeCl_3$.

57. **(b)** Beilstein's test : Organic compounds containing halogens when heated over Cu wire loop, give blue or green colour flame due to formation of volatile copper halides.

58. **(b)** Beilstein's test is used for halogens.

59. **(c)** Moles of HCl taken $= 20 \times 0.1 \times 10^{-3} = 2 \times 10^{-3}$

Moles of HCl neutralised by NaOH solution $= 15 \times 0.1 \times 10^{-3} = 1.5 \times 10^{-3}$

Moles of HCl neutralised by ammonia $= 2 \times 10^{-3} - 1.5 \times 10^{-3} = 0.5 \times 10^{-3}$

Moles of NH_3 absorbed $= 0.5 \times 10^{-3}$

Moles of N in $NH_3 = 0.5 \times 10^{-3}$

Mass of N in organic compound $= 14 \times 0.5 \times 10^{-3}$ g

$$\% \text{ N in organic compound} = \frac{14 \times 0.5 \times 10^{-3}}{29.5 \times 10^{-3}} \times 100 = 23.7\%$$

60. **(b)**

Element	%	Relative number of atoms	Simple ratio of atoms
C	20	20/12 = 1.66	1
H	6.67	6.67 / 1 = 6.67	4.16
N	46.7	46.67/14 = 3.33	2.02
O	26.6	26.64 / 16 = 1.66	1

$\therefore$ The empirical formula is CH_4N_2O

Empirical weight = 60; Mol. wt. = 60; $\Rightarrow \therefore n = \frac{60}{60} = 1$

$\therefore$ Molecular formula $= CH_4N_2O$; $NH_2 - \overset{O}{\overset{||}{C}} - NH_2$ (urea)

On heating, urea loses ammonia to give biuret.

$$2NH_2CONH_2 \longrightarrow \underset{\text{Solid residue (Biuret)}}{H_2NCONHCONH_2} + NH_3$$

Biuret with alkaline $CuSO_4$ gives violet colour, i.e., test for –CONH– group.

61. **(a)** Prussian blue $Fe_4[Fe(CN)_6]_3$ is formed in Lassaigne test for nitrogen.

$$\underbrace{Na + C + N}_{\text{(From org. comp.)}} \xrightarrow{\text{Fuse}} NaCN$$

$$FeSO_4 + 2NaOH \longrightarrow \underset{\text{Green ppt.}}{Fe(OH)_2\downarrow} + Na_2SO_4$$

$$Fe(OH)_2 + 2NaCN \longrightarrow Fe(CN)_2 + 2NaOH$$

$$Fe(CN)_2 + 4NaCN \longrightarrow \underset{\text{Sod. ferrocyanide}}{Na_4[Fe(CN)_6]}$$

$$3Na_4[Fe(CN)_6] + 4FeCl_3 \longrightarrow \underset{\text{Ferricferrocyanide (Prussian blue)}}{Fe_4[Fe(CN)_6]_3} + 12NaCl$$

62. **(b)** $NH_4^+ + NaOH \longrightarrow H_2O + \underset{(X)}{NH_3\uparrow}$

$$\underset{(Y)}{2[HgI_4]^{2-}} + \underset{(Y)}{3OH^-} + \underset{(X)}{NH_3} \longrightarrow$$

$$\underset{\text{Brown ppt}}{HgO.Hg\langle^{NH_2}_{I}} + 7I^- + 2H_2O$$

63. **(c)** Test for sulphide ion

$$S^{2-} + Na_2[Fe(CN)_5NO] \longrightarrow \underset{\text{Deep violet compound}}{[Fe(CN)_5NOS]^{4+}}$$

The formation of this violet complex confirms the presence of sulphide ion in the salt.

64. **(c)** $2CuSO_4 + K_4[Fe(CN)_6] \xrightarrow{CH_3COOH} \underset{\text{(Chocolate brown ppt.)}}{Cu_2[Fe(CN)_6]} + 2K_2SO_4$

$$4FeCl_3 + 3K_4[Fe(CN)_6] \longrightarrow \underset{\text{(Prussian Blue ppt.)}}{Fe_4[Fe(CN)_6]_3} + 12KCl$$

$$3ZnCl_2 + 2K_4[Fe(CN)_6] \xrightarrow{NH_4OH} \underset{\text{(White or Bluish white ppt.)}}{K_2Zn_3[Fe(CN)_6]_2} + 6KCl$$

Mg^{2+} and Ba^{2+} do not give ferrocyanide tests.

65. **(b)** As_2S_3 and As_2S_5 are yellow colour sulphides, $(NH_4)_2S$ is colourless to yellow. PbS is black, CuS is black in colour

66. **(b)** Ni^{+2} has d^8 configuration which does not depend on nature of ligand present in octahedral complex. No significant high spin or low spin effect occurs.

$Ni^{+2} : t_{2g}^6 e_g^2$

Ni^{+2} gives violet coloured bead in non-luminous oxidizing flame under hot conditions.

67. **(b)** $CoS \xrightarrow{\text{Aqua regia}} CoCl_2 \xrightarrow[\text{2.AcOH}]{\text{1. } KNO_2/NH_4OH} \underset{\text{yellow}}{K_3[Co(NO_2)_6]\downarrow}$

68. **(4)**

NTA Ans. (3)

Zinc ferrocyanide; $Zn_2[Fe(CN)_6]$ = White precipitate

Potassium ferricferrocyanide; $KFe[Fe(CN)_6]$ = Deep blue precipitate (prussian blue)

Cupric ferrocyanide; $Cu_2[Fe(CN)_6]$ = Chocolate brown precipitate

Calcium ferrocyanide; $Ca_2[Fe(CN)_6]$ = White precipitate

69. **(b)**

$$\underset{\boxed{A}}{\boxed{PbS}} \xrightarrow{HNO_3} Pb(NO_3)_2 \xrightarrow{H_2SO_4} \underset{\boxed{B}}{\boxed{PbSO_4}} \xrightarrow{\text{(1) Ammonium acetate (2) Acetic acid (3) } K_2CrO_4} \underset{\text{Yellow ppt}}{\underset{\boxed{C}}{\boxed{PbCrO_4}}}$$

70. **(d)** $3PbS + 8HNO_3 \rightarrow 3Pb(NO_3)_2 + 2NO + 3S + 4H_2O$

Nitrous oxide (N_2O) is not formed during the reaction.

71. **(0)** $Co^{2+} + H_2S \rightarrow \underset{(A)}{CoS\downarrow}$ (Black)

$$\underset{(A)}{CoS} + \text{Aqu-regia} \rightarrow \underset{(B)}{Co^{2+}(aq)} + NOCl + S + H_2O$$

$$Co^{2+}(aq) + KNO_2 + CH_3COOH$$

$$\downarrow$$

$K_3[Co(NO_2)_6] + NO + S + H_2O$

In $K_3[Co(NO_2)_6]$, $Co^{+3}: 3d^6 4s^0$

Co^{3+}: d^2sp^3 Hybridisation

Number of unpaired $e^- = 0$ (NO_2 = strong Field Ligand)

$\Rightarrow$ Magnetic moment $= \sqrt{n(n+2)} = 0$ B.M

72. (d)

(A) Borax bead test

$$MSO_4 \xrightarrow[\Delta]{Na_2B_4O_7} M(BO_2)_2 \longrightarrow MBO_2 \longrightarrow M$$

(B) Charcoal cavity test

$$MSO_4 \xrightarrow[\Delta]{Na_2CO_3} MCO_3 \longrightarrow MO \longrightarrow M$$

(C) Cobalt nitrate test

$$MCO_3 \longrightarrow MO \xrightarrow[\Delta]{CO(NO_3)_2} COO.MO$$

(D) Flame test

$$MCO_3 \longrightarrow MCl_2 \longrightarrow M^{2+}$$

So, correct option is (d).

73. (a) Complexes with their colours are:

$Fe_4[Fe(CN)_6]_3.xH_2O \Rightarrow$ Prussian Blue

$[Fe(CN)_5NOS]^{4-} \Rightarrow$ Violet

$[Fe(SCN)]^{2+} \Rightarrow$ Blood Red

$(NH_4)_3PO_4.12MoO_3 \Rightarrow$ Yellow

74. (c)

$$PO_4^{3-} + (NH_4)_2MoO_4 \xrightarrow{H^+} (NH_4)_3PO_4.12MoO_3 \downarrow$$

Or HPO_4^- — Ammonium Molybdate — Canary yellow ppt. (Ammonium phopho molybdate)

75. (c)

Precipitating reagent	Group
$NH_4Cl + NH_4OH$	III ($Fe^{3+}, Al^{3+}, Cr^{3+}$)
$NH_4OH + Na_2CO_3$	V ($Ba^{2+}, sr^{2+}, ca^{2+}$)
$NH_4OH + NH_4Cl + H_2S$ gas	IV ($Ni^{2+}, Mn^{2+}, Co^{2+}, Zn^{2+}$)
dil. HCl	I (Pb^{2+}, Ag^+)

76. (a) The Br^- ion present in the salt gives pale yellow ppt. of AgBr with $AgNO_3$, insoluble in dil HNO_3 but sparingly soluble in aq. NH_4OH.

$$Br^- + AgNO_3 \rightarrow \underset{\text{Pale yellow}}{AgBr} \downarrow + NO_3^-$$

$$AgBr + 2NH_4OH \rightarrow \underset{\text{Sparingly Soluble}}{[Ag(NH_3)_2]Br} + 2H_2O$$

77. (a) While preparing crystal of Mohr's salt, dil. H_2SO_4 is added to prevent hydrolysis of ferrous sulphate.

78. (b) $NH_4OH \rightleftharpoons NH_4^+ + OH^-$

$NH_4Cl \rightarrow NH_4^+ + Cl^-$

Due to common ion effect of NH_4^+,

$[OH^-]$ decreases in such extent that only group-III cations can be precipitated, due to their very low K_{sp} in the range of 10^{-38}.

79. (5) (a) $SrSO_4$ = White

(b) $Mg(NH_4)PO_4$ = White

(c) $BaCrO_4$ = Yellow

(d) $Mn(OH)_2$ = White

(e) $PbSO_4$ = White

(f) $PbCrO_4$ = Yellow

(g) AgBr = Brown

(h) PbI_2 = Yellow

(i) CaC_2O_4 = White

(j) $Fe(OH)_2(CH_3COO)$ = Reddish brown

80. (a) Statement I is false but statement II is true Dimethyl glyoxime forms a four membered chelate with $NiCl_2$ solution in presence of NH_4OH

81. (a) Potassium hydrogen phthalate is a primary standard for standardisation of sodium hydroxide solution as it is economical and its concentration does not changes with time. Phenolphthalein can acts as indicator in acid-base titration as it shows colour in pH range 8.3 to 10.1.

82. (b) Lead sulphate-white

Ammonium sulphide-yellow

Lead iodide-Bright yellow

Ammonium arsinomolybdate-yellow

83. (78) CaF_2 does not evolve any gas with concentrated H_2SO_4.

NaBr gives brown coloured gas with cone. H_2SO_4.

$NaNO_3$ gives brown coloured gas with cone. H_2SO_4.

KI gives violet coloured I_2 gas with cone. H_2SO_4.

Molar mass of CaF_2 is 78 mol^{-1}.

84. (c) Salts having S^{2-} ion give H_2S on warming with dil. H_2SO_4.

$$H_2S + (CH_3COO)_2Pb \longrightarrow \underset{\substack{\text{Lead Sulphide}\\ \text{Black ppt.}}}{PbS} + 2CH_3COOH$$

85. (a) $CrO_2Cl_2 + 4NaOH \longrightarrow \underset{(A)}{Na_2CrO_4} + 2NaCl + 2H_2O$

$$\underset{(A)}{Na_2CrO_4} + 2H_2O_2 + 2HCl \longrightarrow \underset{(B)}{CrO_5} + 3H_2O + 2NaCl$$

86. (4) Flame test is applicable to metals with low ionization energy. Such as metals are:

$Sr^{2+}, Ba^{2+}, Ca^{2+}, Cu^{2+}$

$Zn^{2+}, Co^{2+}, Fe^{2+}$ do not give flame test.

87. (a) $Ni^{2+} + 2dmg^{2-} \xrightarrow{NH_3} \underset{\text{red ppt.}}{[Ni(dmg)_2]} \downarrow$

dmg = dimethyl glyoxime.

88. (d) $\underset{\text{Nessler's reagent}}{2K_2HgI_4} + 3KOH + \underset{X}{NH_3} \longrightarrow$

$$\underset{\text{brown}}{[OHg_2.NH_2]I} \downarrow + 7KI + H_2O$$

NH_3 = Zero group basic radical.

89. (1) $\underset{\text{brown ring}}{[Fe(H_2O)_5NO]^{2+}}$

Oxidation state of Fe = +1. (NO exists as NO^+ in the given complex)

90. (a) $4NaCl + 6H_2SO_4 + K_2Cr_2O_7 \rightarrow$
$2CrO_2Cl_2\uparrow + 2KHSO4 + 4NaHSO4 + 3H_2O$
(B)
Reddish brown

$CrO_2Cl_2 + 4NaOH \rightarrow Na_2CrO_4 + 2NaCl + 2H_2O$
(C)
Yellow colour

91. (b) In wet testing, $(NH_4)_2CO_3$ is used as group reagent for 5th group cations ($Ba^{2+}, Ca^{2+}, Sr^{2+}$)

$Ba^{+2} + (NH_4)_2CO_3 \rightarrow BaCO_3\downarrow + NH_4^{\oplus}$
(white precipitate)

92. (a) $CH_3COO^- + FeCl_3 \rightarrow Fe(CH_3COO)_3$ or $[Fe_3(OH)_2(CH_3COO)_6]$
Blood red colour
$\downarrow \Delta$
$Fe(OH)_2(CH_3COO)\downarrow$
Red–brown precipitate

$2NO_3^- + 4H_2SO_4 + 6Fe^{2+} \longrightarrow$
$6Fe^{3+} + 2NO\uparrow + 4SO_4^{2-} + 4H_2O$

$[Fe(H_2O)_6]^{2+} + NO \rightarrow [Fe(H_2O)_5(NO)]^{2+} + H_2O$
Brown

93. (c) $AgNO_3 + KI \rightarrow AgI\downarrow + KNO_3$

$AgI \rightarrow \underset{S}{Ag^+} + \underset{S+0.625}{I^-}$

AgI is a insoluble salt so concentration Ag^+ and I^- will be negligible.

94. (a) Prolonged heating will cause oxidation of Fe^{+2} to Fe^{+3}.

95. (c) $Pb(NO_3)_2$ is a soluble colourless compound so it cannot be used in confirmatory test of Pb^{+2} ion.

96. (d) Nessler's reagent is – $K_2[HgI_4]$
Nitrogen is not present in Nessler's reagent

97. (d) Nessler's Reagent $\rightarrow K_2[HgI_4]$

98. (c) $Fe_4[Fe(CN)_6]_3 \Rightarrow$ prussiun blue precipitate

99. (c) $Cr_2O_7^{2-} + SO_3^{2-} \xrightarrow{H^+} \underset{Green}{Cr^{3+}} + SO_4^{2-}$

100. (a) $Zn^{2+}, Co^{2+}, Ni^{2+}$ = IVth Group, Fe^{3+} = IIIrd Group

101. (c) Nessler's reagent is K_2HgI_4.

102. (c) Blue green colour is due to formation of $Cu(BO_2)_2$

$CuSO_4 \xrightarrow{\Delta} CuO + SO_3$

$CuO + B_2O_3 \longrightarrow Cu(BO_2)_2$

103. (d) $Ni^{+2} + 2DMG \xrightarrow{NH_3\,(aq)} [Ni(DMG)_2]$
Rosy Red complex

104. (d)

Cations	Group No.	Group reagent
Pb^{+2}, Cu^{+2}	II	$H_2S(g)$ in presence of dil HCl
Al^{+3}, Fe^{+3}	III	NH_4OH in presence of NH_4Cl
Co^{2+}, Ni^{+2}	IV	H_2S in presence of NH_4OH
Ba^{+2}, Ca^{+2}	V	$(NH_4)_2CO_3$ in presence of NH_4OH

105. (b) $Na_2B_4O_7.10H_2O \xrightarrow{\Delta} Na_2B_4O_7$
$\downarrow \Delta$
$2NaBO_2 + B_2O_3$

$B_2O_3 + CoO \longrightarrow Co(BO_2)_2$
Blue coloured

106. (d) $4\overset{III}{Fe^{3+}} + 3[\overset{II}{Fe}(CN)_6]^{-4} \rightarrow \overset{III}{Fe_4}[\overset{II}{Fe}(CN)_6]_3$
Prussian Blue

107. (b) Salt is dissolved in water to form a solution. Barium chloride test is performed for the confirmation of sulphite ions. If the ppt formed after addition of $BaCl_2$ are dissolved in dil. HCl then sulphite ions are present and if the ppt remain undissolved then the ion is sulphate.

$BaCl_2 + SO_3^{2-} \rightarrow \underset{white}{BaSO_3\downarrow} \xrightarrow{dil.HCl} SO_2\uparrow$
burning sulphur like smell

108. (d) Each anions have particular confirmatory test.

- CO_3^{2-} ion will evolve $CO_2(g)$ which will turns lime water milky.
- S^{2-} ion will give $H_2S(g)$, that turns lead acetate paper black.
- SO_3^{2-} ion will give $SO_2(g)$, which turns acidified potassium dichromate solution green.
- NO_2^- ion will give brown fumes of $NO_2(g)$, which turns KI solution blue.

109. (b) Ring is formed due to formation of nitrosoferrous sulphate.

110. (a) $Ni^{2+} + 2DMG \rightarrow [Ni(DMG)_2]\downarrow$
(Bright red precipitate)

111. (a)

	Ion	Colour of the flame
(A)	Cu^{+2}	green flame with blue centre
(B)	Sr^{2+}	Crimson Red
(C)	Ba^{2+}	Apple green
(D)	Ca^{2+}	Brick red

112. (4) PbS, CuS, As_2S_3, CdS are soluble in 50% HNO_3. HgS, Sb_2S_3 are insoluble in 50% HNO_3.

113. (b) (A)-(iii), (B)-(iv), (C)-(ii), (D)-(i)

114. (a) Al^{3+} and Fe^{3+} sulphides hydrolyse in water Ni^{2+} and Zn^{2+} require basic medium with H_2S to form ppt.; Ca^{2+} and Ba^{2+} sulphides are soluble, hence only Cu^{2+} ions will form the ppt of CuS.

115. (c) 1-Naphthyl amine and sulphanilic acid in acetic acid are used for the detection of NO_2^-

$NO_2^- + CH_3COOH \longrightarrow HNO_2 + CH_3COO^-$

$$\underset{\text{(p-position)}}{\overset{+}{NH_3}\cdot CH_3COO^- \text{-}C_6H_4\text{-}SO_3H} + HNO_2 \longrightarrow N=N-OCOCH_3\text{-}C_6H_4\text{-}SO_3H + 2H_2O$$

$$\xrightarrow{\text{1-naphthylamine } (C_{10}H_7NH_2)} HO_3S-C_6H_4-N=N-C_{10}H_6-NH_2 + CH_3COOH$$

(Red azo-dye)

116. (b) $SO_2 + \text{dil. } H_2SO_4 \rightarrow \underset{(X)}{SO_3(g)}$

$$SO_3 + K_2Cr_2O_7 \xrightarrow[H_2SO_4]{\text{dil.}} \underset{\substack{\text{(Green)}\\(Y)}}{Cr_2(SO_4)_3}$$

117. (b) Due to common ion effect, insufficient concentration of sulphide ion (S^{2-}) is produced. Thus the ionic product of Ni^{2+} and S^{2-} ions is less than the K_{sp} of NiS, so NiS will not precipitate out.

118. (d) $Cl^- + Na^+ \longrightarrow \underset{\text{(Neutral)}}{NaCl} \xrightarrow{AgNO_3} AgCl$ (white ppt)

(Insoluble in dilute HNO_3)

119. (d) In presence of HCl

$$\left.\begin{matrix} ZnS \\ MnS \end{matrix}\right\} \xrightarrow{HCl} \begin{matrix} ZnCl_2 \\ MnCl_2 \end{matrix}$$

$$\left.\begin{matrix} CoS \\ NiS \end{matrix}\right\} \xrightarrow{HCl} \text{No reaction}$$

$$ZnS \xrightarrow{HCl} ZnCl_2 \xrightarrow{NaOH} \underset{\text{White ppt}}{Zn(OH)_2}$$

$$ZnCl_2 \xrightarrow{K_4[Fe(CN)_6]} \underset{\text{Bluish white ppt}}{Zn_2[Fe(CN)_6]}$$

$$MnS \xrightarrow{HCl} MnCl_2 \xrightarrow{NaOH} Mn(OH)_2 \xrightarrow{[O]} \underset{\text{(Brown colour)}}{MnO(OH)_2}$$

$$MnCl_2 \xrightarrow{K_4[Fe(CN)_6]} \text{No reaction}$$

The above reaction path is confirming the presence of Zn^{2+} ion.

120. (d) Region 2 (blue flame) will be the hottest region of Bunsen flame as shown in given figure.

121. (b) Fe^{3+}, radical gives blood red with CNS^-

$Fe^{3+} + 3CNS^- \rightarrow Fe(CNS)_3$ Ferric sulphocyanide (Blood red colouration)

Fe^{3+} gives blue colour on treatment with $K_4[Fe(CN)_6]$

$$4Fe^{3+} + 3K_4[Fe(CN)_6] \rightarrow 12K^+ + \underset{\substack{\text{Ferricferrocyanide}\\ \text{(Prussian blue)}}}{Fe_4[Fe(CN)_6]_3}$$

Cl^- radical gives positive chromyl chloride test.

122. (c) Co^{2+} ion is precipitated by H_2S in presence of NH_4OH which is a group reagent of group IV in cationic analysis. Other three cations are palced in group II and are precipitated as sulphides by H_2S in the presence of dil. HCl.

123. (a) Zn^{2+} salts are white. Usually Fe^{2+} salts are rarely pink. Cu^{2+} salts are usually blue in hydrated form. Co^{2+} is pink in aqueous solution.

124. (a) The reason for this is the high ionization of Na_2CO_3 as compared to $(NH_4)_2CO_3$, in water into Na^+ and CO_3^{2-}. Now the higher concentration of CO_3^{2-} is available which exceeds the solubility product of group V radicals as well as that of magnesium radicals. Leading to the precipitation of group V radicals, as well as Mg^{2+}.

125. (a) Only Fe^{3+} ions give blood red colouration with SCN^- ions.

$$Fe^{3+} + 3SCN^- \longrightarrow Fe(SCN)_3$$

126. (c) Solid potassium dichromate when heated with concentrated sulphuric acid and a soluble chloride gives orange red vapours of a volatile oily liquid CrO_2Cl_2.

$$K_2Cr_2O_7 + 4NaCl + 6H_2SO_4 \longrightarrow 2KHSO_4 + 4NaHSO_4 + \underset{\text{chromyl chloride}}{2CrO_2Cl_2}$$

127. (d) $Hg_2Cl_2 + 2NH_4OH \longrightarrow Hg + HgNH_2Cl + NH_4Cl + 2H_2O$

128. (a) Between AgCl and AgI, AgI is less soluble, hence ammonia can dissolve ppt. of AgCl only due to formation of complex as given below:

$$AgCl + 2NH_3 \longrightarrow [Ag(NH_3)_2]Cl$$

(b) $Fe^{3+} + \overset{(+2)}{[Fe(CN)_6]^{4-}} \longrightarrow \underset{\text{Prussian blue}}{\overset{(+3)\ (+2)}{Fe_4[Fe(CN)_6]_3}}$

(c) Both $KHCO_3$ and $Ca(HCO_3)_2$ are soluble in water and on boiling we get a precipitate of $CaCO_3$.

(d) Mn salts give a colourless borax bead test in the reducing flame.

129. (c) When H_2S is passed through Hg_2S we get a mixture of mercurous sulphide and mercury ($Hg_2S + Hg$).

130. (d) When we add NH_4Cl, it suppresses the ionisation of NH_4OH and prevents the precipitation of higher group hydroxide in gp(III) due to common ion effect.

Note: Further, ferric chloride and chromium chloride form different colour precipitates with NH_4OH.

$$FeCl_3 + 3NH_4OH \longrightarrow \underset{\text{Reddish brown}}{Fe(OH)_3\downarrow} + 3NH_4Cl$$

$$CrCl_3 + 3NH_4OH \longrightarrow \underset{\text{Bluish green}}{Cr(OH)_3\downarrow} + 3NH_4Cl$$

Decreasing the concentration of OH^- ions will also do the same thing by shifting the equilibrium towards the supression of ionization of NH_4OH.

Thus, both (b) and (c) are correct, and therefore (d) is correct option.

131. (2500)

$$\text{Mols of octane} = \frac{1.14}{114} = 0.01 \text{ mol}$$

Heat evolved = $C \times \Delta T = 5 \times 5$ kJ = 25 kJ

$\therefore$ Magnitude of Heat of combusion

$$= \frac{\text{Heat}}{\text{Number of moles}} = \frac{25}{0.01} = 2500 \text{ kJ/mol}$$

132. (d) The primary standard is a highly pure and stable compound with a known exact composition that can be accurately weighed and dissolved to creat a solution of known concentration.

NaOH is hygroscopic and can't be used.

$FeSO_4 \cdot 6H_2O$ is unstable and can be easily oxidised.

$Na_2Cr_2O_7$ is hygroscopic and can't be used.

133. (b) When 10 mL, 2M NaOH solution is added to 20 mL of 1M HCl solution :

$$NaOH + HCl \rightarrow NaCl + H_2O$$

Initial : MV = 2 × 0.01 = 0.02 mol; MV = 1 × 0.02 = 0.02 mol

Final : 0; 0

$\therefore$ Resulting solution becomes neutral.

Now, 10 mL of above neutral solution is poured into a flask containing 2 mol HCl and made solution 100 mL with distilled water.

$$\Rightarrow \text{Molarity of HCl} = \frac{2}{100} \times 1000 = 20M$$

134. (b) HCl is strong acid and will be neutralised first. From the given graph 2 mL NaOH solution is used for neutralisation of HCl and 3 mL NaOH solution is used for neutralisation of CH_3COOH.

This replaces H^+ with Na^+ and conductance decreases first.

$\therefore$ Mol of HCl = Mol of NaOH used

$M \times 40 = 0.1 \times 2$

$M = 0.005$

Mol of CH_3COOH = Mol of NaOH used

$M \times 40 = 0.1 \times 3$

$M = 0.0075.$

135. (b) For the titration: Oxalic acid v/s $KMnO_4$

$$2MnO_4^- + 5(COO)_2^{2-} + 16H^+ \rightarrow 10CO_2 + 2Mn^{2+} + 8H_2O$$

At room temperature this reaction proceeds at slow rate but upon heating around 60° C, the reaction becomes fast. Manganese(II) ions catalyse the reaction; thus, the reaction is autocatalytic; once manganese(II) ions are formed, it becomes progressively faster.

No heating is required for Mohr's salt as the reaction is already fast. Also heating may rsult into oxidatin of Fe^{2+} into Fe^{3+}.

136. (a) Higher the number of milli moles of acid or base reacted higher will be temperature rise.

Option (a) n_{acid} or n_{base} reacted = 30 m mol

Option (b) n_{acid} or n_{base} reacted = 30 m mol

but less energy will be released by neutralisation reaction of weak acid hence option (b) can not be correct.

Option (c) $\Rightarrow$ 20 m mol

Option (d) $\Rightarrow$ 25 m mol

137. (d) $KMnO_4$ & NaOH $\rightarrow$ Secondary standard.

Primary standard should not be Hygroscopic.

138. (82) (1) $CuSO_4(s) + 5H_2O \xrightarrow{-x} CuSO_4.5H_2O$

(2) $CuSO_4.5H_2O + H_2O \xrightarrow{+12} CuSO_4(aq)$

$$CuSO_4 + H_2O \xrightarrow{-70} CuSO_4(aq)$$

from (1) & (2)

$-70 = -x + 12$

$\Rightarrow x = 82.$

139. (2) $H^+ + OH^- \rightarrow H_2O \Rightarrow x$ (for neutralization of HCl)

$2H^+ + 2OH^- \rightarrow 2H_2O \Rightarrow y = 2x$ (for H_2SO_4) $\Rightarrow y/x = 2.$

140. (b) Enthalpy of neutralization of strong acid with strong base is always –57 kJ / mol because strong monoacidic base gives one mole of OH^- ion and strong monobasic acid gives one mole of H^+ ion which form one mole of water.

141. (8) Reaction taking place is :

$$2MnO_4^- + 16H^+ + 5C_2O_4^{2-} \longrightarrow 2Mn^{2+} + 10CO_2 + 8H_2O$$

eq.($KMnO_4$) = eq.($H_2C_2O_4$)

$M \times 2 \times 5 = 2 \times 20 \times 2$

$M = 8M$

142. (a) Potassium hydrogen phthalate is a primary standard for standardisation of sodium hydroxide solution as it is economical and its concentration does not changes with time. Phenolphthalein can acts as indicator in acid-base titration as it shows colour in pH range 8.3 to 10.1.

143. (1) Reaction with HCl

$Ca(OH)_2 + 2HCl \rightarrow CaCl_2 + 2H_2O$

Volume of $Ca(OH)_2$ = 10 ml, Volume of HCl = 20 ml

Concentration of HCl = 0.5 M

No. of milli moles of HCl = 10, No. of milli moles of $Ca(OH)_2$ = 5

$$\text{i.e. } M_{Ca(OH)_2} = \frac{\text{no. of milli moles}}{V\text{ (ml)}} = \frac{5}{10} = 0.5\text{ M}$$

Reaction with H_2SO_4

$Ca(OH)_2 + H_2SO_4 \rightarrow CaSO_4 + 2H_2O$

No. of milli moles of $Ca(OH)_2 = 20 \times 0.5 = 10$

i.e. no. of milli moles of $H_2SO_4 = 10$

$$\Rightarrow M_{H_2SO_4} = \frac{\text{no. of milli moles}}{V\text{ (ml)}} = \frac{10}{10} = 1\text{ M}$$

144. (68) By balancing the redox reaction we get

$$10[FeSO_4.(NH_4)_2SO_4.6H_2O] + 2KMnO_4 + 8H_2SO_4 \downarrow$$

$$5Fe_2(SO_4)_3 + 2MnSO_4 + 10(NH_4)_2SO_4 + K_2SO_4 + 68H_2O$$

145. (2) $pk_{In} = -\log(4 \times 10^{-10}) = 9.4$

Indicator range

$\Rightarrow pk_{In} \pm 1$, i.e. 8.4 to 10.4

$\therefore$ Statement (B) is correct.

In acidic medium, phenolphthalein is in unionized form and is colourless. Hence statement (D) is correct.

It is used as an indicator for the titration of strong acid with strong base and for weak acid with strong base. Phenolphthalein is a weak acid.

146. (c) Methyl orange is weak base.

Benzenoid structure $\rightleftharpoons$ Quinonoid structure

(yellow coloured) (Red coloured) (more intense)

147. (d) In redox titration, indicators are sensitive to change in oxidation potential and in acid base titration, indicators are sensitive to change in pH of solution. Both statements are false.

148. (c) $N_1V_1 = N_2V_2$

$\Rightarrow 0.02V_1 = 0.02 \times 10 \Rightarrow V_1 = 10\,mL$

149. (c) KOH absorb excess CO_2 from the atmosphere and form $KHCO_3$ which remains soluble in the solution. Therefore concentration of KOH should be checked before use.

150. (a)

Indicator	pH range
Methyl orange	3.2 – 4.5
Phenolpthalein	8.3 – 10.5

Methyl orange may be used for a strong acid vs strong base and strong acid vs weak base titration. Phenolphthalein may be used for a strong acid vs strong base and weak acid vs strong base titration.

151. (a) $I^- + H_2O_2 \longrightarrow \underset{(A)}{I_2} + H_2O$

$I_2 + \underset{(Indicator)}{Starch} \longrightarrow Blue$

152. (a) For kinetic study of reaction of iodide ion with H_2O.

- The starch solution should be used freshly prepared.
- The conc. of sodium thiosulphate solution should be less than KI solution.
- The time should be recorded after the appearance of blue colour.

153. (a) $2KMnO_4 + 16HCl \rightarrow 2MnCl_2 + 2KCl + 8H_2O + Cl_2$

HCl gets oxidised by $KMnO_4$ into Cl_2

154. (24) $\overset{(+6)}{Cr_2} O_7^{2-} \rightarrow Cr^{3+}$; $Fe^{2+} \rightarrow Fe^{3+}$

Equivalents of $K_2Cr_2O_7 \equiv$ Eq. of Fe^{2+}

(Molarity × n.f. × volume) of $K_2Cr_2O_7 \equiv$ (Molarity × n.f. × volume) of Fe^{2+}

$\Rightarrow 0.02 \times 6 \times 20 = M \times 1 \times 10 \Rightarrow M = 0.24 = 24 \times 10^{-2}$

155. (a) Titration curve of NH_4OH vs HCl

156. (1) No. of equivalents of $H_2SO_4 = 100 \times 0.1 \times 2 = 20$

No. of equivalents of NaOH = $50 \times 0.1 = 5$

No. of equivalents of H_2SO_4 left = $20 - 5 = 15$

Total volume = $(100 + 50) = 150$ mL $\Rightarrow 150 \times x = 15$

$x = \frac{1}{10} = 0.1N = 1 \times 10^{-1}N$

157. (4) $n_{eq.}$ of $I_2 = n_{eq}$ of $Na_2S_2O_3 = 20 \times 0.002 \times 1$

$2 \times n_{mol}$ of $I_2 = 0.4$

n_{mol} of $I_2 = 0.2$ m mol

n_{mol} of $Cu^{2+} = 0.2 \times 2 \times 10^{-3}$

$[Cu^{2+}] = \frac{0.4 \times 10^{-3}}{10 \times 10^{-3}} = 0.04 = 4 \times 10^{-2}$

158. (a) In acidic medium the reaction for $KMnO_4$ with oxalic acid is

$\overset{+7}{MnO_4^-} + 8H^+ + 5e^- \longrightarrow \overset{+2}{Mn^{2+}} + 4H_2O$

Change in oxidation state is = 7 – 2 = 5.

159. (a) $(CH_3)_2\ddot{N}-C_6H_4-N=N-C_6H_4-SO_3^-Na^+$

Methyl (orange) (yellow)

$\downarrow H^+$

$Me_2N^+=C_6H_4=N-NH-C_6H_4-SO_3^-Na^+$

Quinonid form (Red colour)

160. (13) $\overset{+4}{MnO_2} + HCl \longrightarrow Cl_2 + Mn^{2+}$ ($n_f = 2$)

$Cl_2 + KI^{(-)} \longrightarrow Cl^{(-)} + I_2^0$

$I_2 + Na_2S_2O_3 \longrightarrow I^{(-)} + Na_2S_4O_6$

n_{factor} for $Na_2S_2O_3 = 1$

Moles of $Na_2S_2O_3$ = Molarity/Volume $= 0.1 \times 60\,mL = 6$ m mol

From the above equation,

$(n_{eq})_{MnO_2} = (n_{eq})_{Cl_2} = (n_{eq})_{I_2} = (n_{eq})_{Na_2SO_3}$

$(n_{eq})_{MnO_2} = 6$ m mol

$\left(\underset{moles}{n} \times n_{factor}\right)_{MnO_2} = 6 \times 10^{-3}$

$n = \frac{6}{2} \times 10^{-3}$ mol; $\frac{Weight}{Molar\ mass} = 3 \times 10^{-3}$,

Weight of $MnO_2 = 3 \times 10^{-3} \times 87$ g

$\% MnO_2 = \frac{weight}{sample\ weight} \times 100 = \frac{3 \times 10^{-3} \times 87}{2} \times 100$

$= 13.05\% \simeq 13\%$

161. (30) The reaction is

$2KMnO_4 + 10FeSO_4(NH_4)SO_4 \cdot 6H_2O + 8H_2SO_4 \longrightarrow K_2SO_4 + 2MnSO_4 + 5Fe_2(SO_4)_3 + 10(NH_4)_2SO_4 + 68H_2O$

Fe in Mohr's salt is oxidised as :

$Fe^{2+} \longrightarrow Fe^{3+} + 1e^-$

$n_{factor} = 1$

Mn in $KMnO_4$ is reduced as :

$MnO_4^- + 8H^+ + 5e^- \longrightarrow Mn^{2+} + H_2O$

$n_{factor} = 5$

at end point; $N_1V_1 = N_2V_2$

$\Rightarrow M_1 \times n_{f_1} \times V_1 = M_2 \times n_{f_2} \times V_2$

$0.01 \times 5 \times V_1 = 0.005 \times 1 \times 20$

$V_1 = 20$ mL used

$\therefore$ Volume left = 50 – 20 = 30 mL

162. (148) Milli eq. of NaOH used = 30×0.25

Milli eq. of H_2SO_4 taken = 50×0.5

$\therefore$ Milli eq. of H_2SO_4 used = $(50 \times 0.5) - (30 \times 0.25)$ = 17.5 m mol of NH_3

$\therefore \% N = \frac{17.5 \times 10^{-3} \times 14}{0.166} \times 100 = 147.59\% \approx 148\%$

Which is more than 100%

163. (316) Let molarity of $KMnO_4 = x$

$KMnO_4 + FeSO_4 \rightarrow Fe_2(SO_4)_3 + Mn^{2+}$

$n = 5$ $n = 1$ (n-factor)

Equivalents of $KMnO_4$ = Equivalents of $FeSO_4$

$(5 \times x \times 10\text{ mL}) = 1 \times 0.1 \times 10\text{ mL} \Rightarrow x = 0.02\text{ M}$

Molar mass of $KMnO_4 = 158$ g/mol

Strength = Molarity × Molar mass = $0.02 \times 158 = 3.16$ g/L

164. (1575) According to law of chemical equivalence, equivalent of $KMnO_4$ = equivalent of oxalic acid

$N_1V_1 = N_2V_2$

$M_1 \times nf \times V_1 = M_2 \times nf \times V_2$ $(N = M \times nf)$

$0.05 \times 5 \times 10 = M_2 \times 2 \times 10$

$$M_2 = \frac{0.05 \times 5}{2} = 0.125$$

Strength of oxalic acid = Molarity × Molar mass $= 0.125 \times 126 = 15.75 = 1575 \times 10^{-2}$ g/L

165. (50) $Na_2CO_3 + 2HCl \longrightarrow 2NaCl + CO_2 + H_2O$

Most precise value of HCl is 5 mL.

At equivalent point,

$M_{eq.}$ of $Na_2CO_3 = M_{eq.}$ of HCl $\Rightarrow 2 \times M \times 10 = 1 \times 0.2 \times 5$

$$\Rightarrow M = \frac{0.2 \times 5}{20} = 0.05\text{ M} = 50 \times 10^{-3}\text{ M} = 50\text{ mM}$$

166. (4) For first end point:

$$\underset{(a)}{NaOH} + \underset{(b)}{Na_2CO_3} + \underset{(c)}{HCl} \longrightarrow \underset{(b)}{NaHCO_3} + NaCl + H_2O$$

Both NaOH and Na_2CO_3 are of basic character, thus they will not react with each other. But both of them will react with HCl.

By law of equivalence:

Equivalents of NaOH(a) + Equivalents of $Na_2CO_3(b)$ = Equivalents of HCl (c)

and Equivalents of $Na_2CO_3(b)$ = Equivalents of $NaHCO_3(b)$

For second end point:

$$\underset{(b)}{NaHCO_3} + \underset{(d)}{HCl} \longrightarrow H_2CO_3 + NaCl$$

Again, by law of equivalence:

Equivalents of $NaHCO_3$ (b) = Equivalents of HCl (d)

$$\Rightarrow b = \frac{1}{10} \times 1.5 = 0.15$$

From first end point, equivalents of $Na_2CO_3 = 0.15$

$\Rightarrow$ Mass of Na_2CO_3 = No. of eq. × Eq. mass

$$= 0.15 \times \frac{106}{1} \times 10^{-3}\text{ g} = 0.016\text{ g}$$

$$\text{Weight \% of } Na_2CO_3 = \frac{0.016}{0.4} \times 100 = 4\%.$$

167. (6) Average volume of NaOH solution used at end point

$$= \frac{4.5 + 4.5 + 4.4 + 4.4 + 4.4}{5} = 4.44\text{ mL}$$

At the end point, equivalents of NaOH = Equivalents of oxalic acid

$N_1V_1 = N_2V_2 \Rightarrow N_1 \times 4.44 = (1.25 \times 2) \times 10$

$$\Rightarrow N_1 = \frac{1.25 \times 2 \times 10}{4.44} = 5.63 \simeq 6$$

Molarity of NaOH = N × acidity of NaOH = $6 \times 1 = 6$ M.

168. (50) M. eq. of $K_2Cr_2O_7$ = M. eq. of FeC_2O_4

$$\overset{(+2)}{Fe}\ \overset{(+3)}{C_2O_4} + \overset{(+6)}{Cr_2O_7^{2-}} \longrightarrow Fe^{3+} + \overset{(+4)}{CO_2} + Cr^{3+}$$

n-factor of $K_2Cr_2O_7 = 2(6-3) = 6$

n-factor of $FeC_2O_4 = 1(3-2) + 2(4-3) = 3$

By law of equivalence,

n-factor of $K_2Cr_2O_7$ × no. of moles = n-factor of FeC_2O_4 × no. of moles

$$\Rightarrow \frac{V}{1000} \times 0.02 \times 6 = \frac{0.288}{144} \times 3 \Rightarrow V = 50\text{ mL}$$

169. (85) $\overset{-1}{H_2O_2} + \overset{(+7)}{MnO_4^-} + H^+ \longrightarrow Mn^{2+} + \overset{0}{O_2} + H_2O$

$$\text{Moles of } KMnO_4 = \frac{0.316}{158} = 2 \times 10^{-3}$$

Equivalents of H_2O_2 = Equivalents of $KMnO_4$

Equivalents of $KMnO_4 = 2 \times 10^{-3} \times 5 = 0.01$

$$\text{Moles of } H_2O_2 = \frac{\text{Equivalents of } H_2O_2}{n\text{-factor of } H_2O_2} = \frac{0.01}{2} = 0.005$$

Mass of pure $H_2O_2 = 0.005 \times 34 = 0.170$ g

$$\text{Percentage of purity} = \frac{0.17}{0.2} \times 100 = 85\%.$$

170. (d) In this acid base titration, bunsen burner and measuring cylinder are of no use while other laboratory equipments will be required, i.e., phenol phthalein, burette and pipette.

171. (c) In toilet cleaning liquid, the main constituent is HCl, which can cause skin burn, so it should be treated with $NaHCO_3$, which can easily neutralise the acid.

172. (b) Oxalic acid is a primary standard solution whereas H_2SO_4 is a secondary standard solution. Fresh oxalic acid is prepared in a conical flask and NaOH is mixed drop by drop from a burette.

173. (c) pH range for methyl orange is

$\longleftarrow$ 3.9 – 4.5 $\longrightarrow$ (Pinkish red ← | → Yellow)

Generally, weak bases have pH greater than 7. When methyl orange is added to a weak basic solution, solution becomes yellow. This solution is then titrated against a strong acid, at the end point pH will be less than 3.1.

$\therefore$ Solution becomes pinkish red.

174. (c) Oxalic acid is used as a primary standard for NaOH standardizing.

175. (a) The titration plot of absorbed right (A) vs volume of ligand (L) (V) will initially remain constant and then increased sharply often the end point reached.

176. (c) The titration of oxalic acid with $KMnO_4$ in presence of HCl gives unsatisfactory result because of the fact that $KMnO_4$ can also oxidise HCl along with oxalic acid. HCl on oxidation gives Cl_2 and HCl reduces $KMnO_4$ to Mn^{2+}, thus the correct answer is (c).

STICKERS TO CODE QUESTIONS

NR NR NR NR NR NR NR NR NR NR NR NR NR NR NR NR NR NR NR NR NR NR NR NR NR NR NR
NR NR NR NR NR NR NR NR NR NR NR NR NR NR NR NR NR NR NR NR NR NR NR NR NR NR NR
NR NR NR NR NR NR NR NR NR NR NR NR NR NR NR NR NR NR NR NR NR NR NR NR NR NR NR
NR NR NR NR NR NR NR NR NR NR NR NR NR NR NR NR NR NR NR NR NR NR NR NR NR NR NR
NR NR NR NR NR NR NR NR NR NR NR NR NR NR NR NR NR NR NR NR NR NR NR NR NR NR NR
NR NR NR NR NR NR NR NR NR NR NR NR NR NR NR NR NR NR NR NR NR NR NR NR NR NR NR
NR NR NR NR NR NR NR NR NR NR NR NR NR NR NR NR NR NR NR NR NR NR NR NR NR NR NR
NR NR NR NR NR NR NR NR NR NR NR NR NR NR NR NR NR NR NR NR NR NR NR NR NR NR NR
NR NR NR NR NR NR NR NR NR NR NR NR NR NR NR NR NR NR NR NR NR NR NR NR NR NR NR
NR NR NR NR NR NR NR NR NR NR NR NR NR NR NR NR NR NR NR NR NR NR NR NR NR NR NR
NR NR NR NR NR NR NR NR NR NR NR NR NR NR NR NR NR NR NR NR NR NR NR NR NR NR NR
NR NR NR NR NR NR NR NR NR NR NR NR NR NR NR NR NR NR NR NR NR NR NR NR NR NR NR
NR NR NR NR NR NR NR NR NR NR NR NR NR NR NR NR NR NR NR NR NR NR NR NR NR NR NR
NR NR NR NR NR NR NR NR NR NR NR NR NR NR NR NR NR NR NR NR NR NR NR NR NR NR NR
NR NR NR NR NR NR NR NR NR NR NR NR NR NR NR NR NR NR NR NR NR NR NR NR NR NR NR
NR NR NR NR NR NR NR NR NR NR NR NR NR NR NR NR NR NR NR NR NR NR NR NR NR NR NR
NR NR NR NR NR NR NR NR NR NR NR NR NR NR NR NR NR NR NR NR NR NR NR NR NR NR NR
NR NR NR NR NR NR NR NR NR NR NR NR NR NR NR NR NR NR NR NR NR NR NR NR NR NR NR
NR NR NR NR NR NR NR NR NR NR NR NR NR NR NR NR NR NR NR NR NR NR NR NR NR NR NR
NR NR NR NR NR NR NR NR NR NR NR NR NR NR NR NR NR NR NR NR NR NR NR NR NR NR NR
NR NR NR NR NR NR NR NR NR NR NR NR NR NR NR NR NR NR NR NR NR NR NR NR NR NR NR
TQ TQ TQ TQ TQ TQ TQ TQ TQ TQ TQ TQ TQ TQ TQ TQ TQ TQ TQ TQ TQ TQ TQ TQ TQ TQ TQ
TQ TQ TQ TQ TQ TQ TQ TQ TQ TQ TQ TQ TQ TQ TQ TQ TQ TQ TQ TQ TQ TQ TQ TQ TQ TQ TQ
TQ TQ TQ TQ TQ TQ TQ TQ TQ TQ TQ TQ TQ TQ TQ TQ TQ TQ TQ TQ TQ TQ TQ TQ TQ TQ TQ
TQ TQ TQ TQ TQ TQ TQ TQ TQ TQ TQ TQ TQ TQ TQ TQ TQ TQ TQ TQ TQ TQ TQ TQ TQ TQ TQ
TQ TQ TQ TQ TQ TQ TQ TQ TQ TQ TQ TQ TQ TQ TQ TQ TQ TQ TQ TQ TQ TQ TQ TQ TQ TQ TQ
TQ TQ TQ TQ TQ TQ TQ TQ TQ TQ TQ TQ TQ TQ TQ TQ TQ TQ TQ TQ TQ TQ TQ TQ TQ TQ TQ
TQ TQ TQ TQ TQ TQ TQ TQ TQ TQ TQ TQ TQ TQ TQ TQ TQ TQ TQ TQ TQ TQ TQ TQ TQ TQ TQ